2021 ACI COLLECTION

OF CONCRETE CODES, SPECIFICATIONS, AND PRACTICES

PART 6

**Part 6 contains current committee reports and standards from:
ACI 408R-03(12) to ACI 447.1R-18**

New editions of each part of the *ACI Collection* are issued annually and include the latest ACI standards and committee reports.

American Concrete Institute
Always advancing

american concrete institute
38800 Country Club Dr.
Farmington Hills, MI 48331

ACI Certification Programs

The final quality of a concrete structure depends on qualified people to construct it. ACI certification programs identify craftsmen, technicians, and inspectors who have demonstrated their qualifications. The following programs are administered by ACI to fulfill the growing demand in the industry for certified workers. More information about these programs is available at **www.acicertification.org.**

Adhesive Anchor Installer
Post-Installed Concrete Anchor Installation Inspector
Tilt-Up Supervisor & Technician
Shotcrete Inspector
Shotcrete Nozzleman (Dry-Mix Process)
Shotcrete Nozzleman (Wet-Mix Process)
Concrete Flatwork Finisher, Advanced Finisher, and Associate
Specialty Commercial/Industrial Concrete Flatwork Finisher
Concrete Field Testing Technician—Grade I
Concrete Strength Testing Technician
Concrete Laboratory Testing Technician—Levels 1 & 2
Aggregate Base Testing Technician
Aggregate Testing Technician—Levels 1 & 2
Concrete Construction Special Inspector
Concrete Transportation Construction Inspector
CSA-Based Concrete Field Testing Technician—Grade I
Concrete Quality Technical Manager
Masonry Lab & Field Testing Technician
Residential Concrete Foundation Technician
Cement Physical Tester
Decorative Concrete Flatwork Finisher & Associate
Self-Consolidating Concrete Testing Technician
EN-Standards Concrete Field Testing Technician
Nondestructive Testing Specialist—Concrete Strength

This document may already contain reference to these ACI certification programs, which can be incorporated into project specifications or quality control procedures. If not, suggested guide specifications are available on request from the ACI Certification Department.

Enhancement of ACI Documents

ACI documents are subject to periodic review, and suggestions for their improvement are invited. All comments regarding ACI documents should be addressed to the Managing Director of Engineering at ACI.

AMERICAN CONCRETE INSTITUTE
www.concrete.org

The reports and standards herein were the latest approved version at the time this edition was published. The content of each report or standard is subject to periodic review and to revision whenever the developments in concrete design and construction warrant a change. Since this is a continuing process, some reports or standards in this volume may have been superseded in the interim since publication. Inquiries concerning revisions or additional material in a subject area are welcome and should be directed to Institute headquarters.

ACI committee documents are intended for the use of individuals who are competent to evaluate the significance and limitations of its content and recommendations and who will accept responsibility for the application of the material it contains. Individuals who use this publication in any way assume all risk and accept total responsibility for the application and use of this information.

All information in this publication is provided "as is" without warranty of any kind, either expressed or implied, including but not limited to, the implied warranties of merchantability, fitness for a particular purpose, or non-infringement.

ACI and its members disclaim liability for damages of any kind, including any special, indirect, incidental, or consequential damages, including, without limitation, lost revenues or lost profits, which may result from the use of this publication.

It is the responsibility of the user of this document to establish health and safety practices appropriate to the specific circumstances involved with its use. ACI does not make any representations with regard to health and safety issues and the use of this document. The user must determine the applicability of all regulatory limitations before applying the document and must comply with all applicable laws and regulations, including but not limited to, United States Occupational Safety and Health Administration (OSHA) health and safety standards.

Participation by governmental representatives in the work of the American Concrete Institute and in the development of Institute standards does not constitute governmental endorsement of ACI or the standards that it develops.

Printed in the United States of America
ISSN 0065-7875
ISBN 978-1-64195-132-6

Kelli R. Slayden, Editor

ACI's guides, reports, specifications, and codes are based upon the most current information and research available, and every effort is made to publish error-free documents. Occasionally, however, mistakes are uncovered after publication of a document. When this happens, a list of corrections, called an errata, is issued. All errata for ACI's technical committee documents and SPs (symposium publications or special publications) are now available online. Go to www.concrete.org, click on "Publications" on the top menu bar, then on "Errata."

Contents—Part 6

ACI 408R-03(12) to ACI 447.1R-18

*Metric version available for this document at **www.concrete.org**.

ABOUT THE AMERICAN CONCRETE INSTITUTE

Founded in 1904 with a headquarters in Farmington Hills, MI, USA; a certification and training resource center in San Bernardino, CA, USA; and a regional office in Dubai, UAE, the American Concrete Institute is a leading authority and resource worldwide for the development, distribution, and adoption of consensus-based standards and technical resources.

ACI publishes thousands of codes, specifications, reports, guides, and articles dedicated to improving concrete materials, design, and construction. This information is authored by diverse committees, global experts, and industry practitioners. ACI also publishes a monthly magazine, *Concrete International*, and two research journals, the *ACI Materials Journal* and *ACI Structural Journal*.

The American Concrete Institute offers 30 certification programs for personnel employed within the concrete construction industry; face-to-face events, seminars, and training programs; and on-demand courses through the online ACI University. The ACI Career Center is specifically targeted to connect great candidates with great job opportunities in the concrete industry.

A membership-based professional organization for individuals and organizations who share a commitment to pursuing the best use of concrete, ACI has 100 chapters, 215 student chapters, and 30,000 members spanning over 120 countries. Learn more and join today at **www.concrete.org**.

ACI 408R-03(12)
to
ACI 447.1R-18

Bond and Development of Straight Reinforcing Bars in Tension

Reported by ACI Committee 408

David Darwin[*]
Chair

Adolfo B. Matamoros[*]
Secretary

John H. Allen	Rolf Eligehausen	Roberto T. Leon[*]	Stavroula J. Pantazopoulou
Atorod Azizinamini	Fernando E. Fagundo	LeRoy A. Lutz[†]	Max L. Porter
Gyorgy L. Balazs	Anthony L. Felder[†]	Steven L. McCabe	Julio A. Ramirez
JoAnn Browning[*†]	Robert J. Frosch[†]	John F. McDermott	John F. Silva
James V. Cox	Bilal S. Hamad[*]	Denis Mitchell	Jun Zuo[*]
Richard A. DeVries[*]	Neil M. Hawkins		

[*]Members of the subcommittee who prepared this report.
[†]Members of the editorial subcommittee.

The performance of reinforced concrete structures depends on adequate bond strength between concrete and reinforcing steel. This report describes bond and development of straight reinforcing bars under tensile load. Bond behavior and the factors affecting bond are discussed, including concrete cover and bar spacing, bar size, transverse reinforcement, bar geometry, concrete properties, steel stress and yield strength, bar surface condition, bar casting position, development and splice length, distance between spliced bars, and concrete consolidation. Descriptive equations and design provisions for development and splice strength are presented and compared using a large database of test results. The contents of the database are summarized, and a protocol for bond tests is presented.

Test data and reliability analyses demonstrate that, for compressive strengths up to at least 16,000 psi (110 MPa), the contribution of concrete strength to bond is best represented by the compressive strength to the 1/4 power, while the contribution of concrete to the added bond strength provided by transverse reinforcement is best represented by compressive strength to a power between 3/4 and 1.0. The lower value is used in proposed design equations. These values are in contrast with the square root of compressive strength, which normally is used in both descriptive and design expressions. Provisions for bond in ACI 318-02 are shown to be unconservative in some instances; specifically, the 0.8 bar size factor for smaller bars should not be used and a ϕ-factor for bond is needed to provide a consistent level of reliability against bond failure. Descriptive equations and design procedures developed by Committee 408 that provide improved levels of reliability, safety, and economy are presented. The ACI Committee 408 design procedures do not require the use of the 1.3 factor for Class B splices that is required by ACI 318.

Keywords: anchorage; **bond**; concrete; **deformed reinforcement**; **development length**; reinforced concrete; reinforcement; relative rib area; **splice**; stirrup; tie.

CONTENTS

ACI 408R-03 became effective September 24, 2003.

PREFACE

The bond between reinforcing bars and concrete has been acknowledged as a key to the proper performance of reinforced concrete structures for well over 100 years (Hyatt 1877). Much research has been performed during the intervening years, providing an ever-improving understanding of this aspect of reinforced concrete behavior. ACI Committee 408 issued its first report on the subject in 1966. The report emphasized key aspects of bond that are now well understood by the design community but that, at the time, represented conceptually new ways of looking at bond strength. The report emphasized the importance of splitting cracks in governing bond strength and the fact that bond forces did not vary monotonically and could even change direction in regions subjected to constant or smoothly varying moment. Committee 408 followed up in 1979 with suggested provisions for development, splice, and hook design (ACI 408.1R-79), in 1992 with a state-of-the-art report on bond under cyclic loads (ACI 408.2R-92), and in 2001 with design provisions for splice and development design for high relative rib area bars (bars with improved bond characteristics) (ACI 408.3-01). This report represents the next in that line, emphasizing bond behavior and design of straight reinforcing bars that are placed in tension.

For many years, bond strength was represented in terms of the shear stress at the interface between the reinforcing bar and the concrete, effectively treating bond as a material property. It is now clear that bond, anchorage, development, and splice strength are structural properties, dependent not only on the materials but also on the geometry of the reinforcing bar and the structural member itself. The knowledge base on bond remains primarily empirical, as do the descriptive equations and design provisions. An understanding of the empirical behavior, however, is critical to the eventual development of rational analysis and design techniques.

Test results for bond specimens invariably exhibit large scatter. This scatter increases as the test results from different laboratories are compared. Research since 1990 indicates that much of the scatter is the result of differences in concrete material properties, such as fracture energy and reinforcing bar geometry, factors not normally considered in design. This report provides a summary of the current state of knowledge of the factors affecting the tensile bond strength of straight reinforcing bars, as well as realistic descriptions of development and splice strength as a function of these factors. The report covers bond under the loading conditions that are addressed in Chapter 12 of ACI 318; dynamic, blast, and seismic loading are not covered.

Chapter 1 provides an overview of bond behavior, including bond forces, test specimens, and details of bond response. Chapter 2 covers the factors that affect bond, discussing the impact of structural characteristics as well as bar and concrete properties. The chapter provides insight not only into aspects that are normally considered in structural design, but into a broad range of factors that control anchorage, development, and splice strength in reinforced concrete members. Chapter 3 presents a number of widely cited descriptive equations for development and splice strength, including expressions recently developed by ACI Committee 408. The expressions are compared for accuracy using the test results in the ACI Committee 408 database. Chapter 4 summarizes the design provisions in ACI 318, ACI 408.3, the 1990 CEB-FIP Model Code, as well as design procedures recently developed by Committee 408. The design procedures are compared for accuracy, reliability, safety, and economy using the ACI Committee 408 database. The observations presented in Chapters 3 and 4 demonstrate that $f_c'^{1/4}$ provides a realistic representation of the contribution of concrete strength to bond for values up to at least 16,000 psi (110 MPa), while $f_c'^{3/4}$ does the same for the effect of concrete strength on the increase in bond strength provided by transverse reinforcement. This is in contrast to $\sqrt{f_c'}$, which is used in most design provisions. The comparisons in Chapter 4 also demonstrate the need to modify the design provisions in ACI 318 by removing the bar size γ factor of 0.8 for small bars and addressing the negative impact on bond reliability of changing the load factors while maintaining the strength reduction factor for tension in the transition from ACI 318-99 to ACI 318-02. Design procedures recommended by ACI Committee 408 that provide both additional safety and economy are presented. Chapter 5 describes the ACI Committee 408 database, while Chapter 6 presents a recommended protocol for bond tests. The expressions within the body of the report are presented in inch-pound units. Expressions in SI units are presented in Appendix A.

A few words are appropriate with respect to terminology. The term *bond force* represents the force that tends to move a reinforcing bar parallel to its length with respect to the surrounding concrete. *Bond strength* represents the maximum bond force that may be sustained by a bar. The terms *development strength* and *splice strength* are, respectively, the bond strengths of bars that are not spliced with other bars and of bars that are spliced. The terms *anchored length*, *bonded length*, and *embedded length* are used interchangeably to represent the length of a bar over which bond force acts; in most cases, this is the distance between the point of maximum force in the bar and the end of the bar. Bonded length may refer to the length of a lap splice. *Developed length* and *development length* are used interchangeably to represent the bonded length of a bar that is not spliced with another bar, while *spliced length* and *splice length* are used to represent the bonded length of bars that are lapped spliced. When used in design, development length and splice length are understood to mean the "length of embedded reinforcement required to develop the design strength of reinforcement at a critical section," as defined in ACI 318.

CHAPTER 1—BOND BEHAVIOR

In reinforced concrete construction, efficient and reliable force transfer between reinforcement and concrete is required for optimal design. The transfer of forces from the reinforcement to the surrounding concrete occurs for a deformed bar (Fig. 1.1) by:

- Chemical adhesion between the bar and the concrete;
- Frictional forces arising from the roughness of the interface, forces transverse to the bar surface, and relative slip between the bar and the surrounding concrete; and
- Mechanical anchorage or bearing of the ribs against the concrete surface.

After initial slip of the bar, most of the force is transferred by bearing. Friction, however, especially between the concrete and the bar deformations (ribs) plays a significant role in force transfer, as demonstrated by epoxy coatings, which lower the coefficient of friction and result in lower bond capacities. Friction also plays an important role for plain bars (that is, with no deformations), with slip-induced friction resulting from transverse stresses at the bar surface caused by small variations in bar shape and minor, though significant, surface roughness. Plain bars with suitably low allowable bond stresses were used for many years for reinforced concrete in North America and are still used in some regions of the world.

When a deformed bar moves with respect to the surrounding concrete, surface adhesion is lost, while bearing

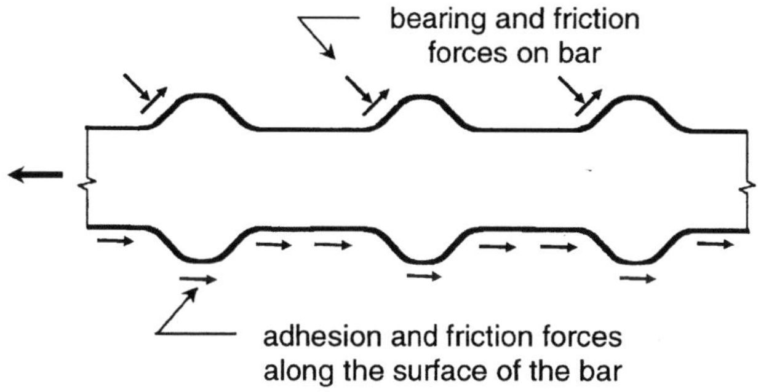

Fig. 1.1—Bond force transfer mechanisms.

forces on the ribs and friction forces on the ribs and barrel of the bar are mobilized. The compressive bearing forces on the ribs increase the value of the friction forces. As slip increases, friction on the barrel of the reinforcing bar is reduced, leaving the forces at the contact faces between the ribs and the surrounding concrete as the principal mechanism of force transfer. The forces on the bar surface are balanced by compressive and shear stresses on the concrete contact surfaces, which are resolved into tensile stresses that can result in cracking in planes that are both perpendicular and parallel to the reinforcement, as shown in Fig. 1.2(a) and 1.2(b). The cracks shown in Fig. 1.2(a), known as Goto (1971) cracks, can result in the formation of a conical failure surface for bars that project from concrete and are placed in tension. They otherwise play only a minor role in the anchorage and development of reinforcement. The transverse cracks shown in Fig. 1.2(b) form if the concrete cover or the spacing between bars is sufficiently small, leading to splitting cracks, as shown in Fig. 1.2(c). If the concrete cover, bar spacing, or transverse reinforcement is sufficient to prevent or delay a splitting failure, the system will fail by shearing along a surface at the top of the ribs around the bars, resulting in a "pullout" failure, as shown in Fig. 1.2(d). It is common, for both splitting and pullout failures, to observe crushed concrete in a region adjacent to the bearing surfaces of some of the deformations. If anchorage to the concrete is adequate, the stress in the reinforcement may become high enough to yield and even strain harden the bar. Tests have demonstrated that bond failures can occur at bar stresses up to the tensile strength of the steel.

From these simple qualitative descriptions, it is possible to say that bond resistance is governed by:

- The mechanical properties of the concrete (associated with tensile and bearing strength);
- The volume of the concrete around the bars (related to concrete cover and bar spacing parameters);
- The presence of confinement in the form of transverse reinforcement, which can delay and control crack propagation;
- The surface condition of the bar; and
- The geometry of the bar (deformation height, spacing, width, and face angle).

A useful parameter describing bar geometry is the so-called relative rib area R_r, illustrated in Fig. 1.3, which is the ratio of the bearing area of the bar deformations to the

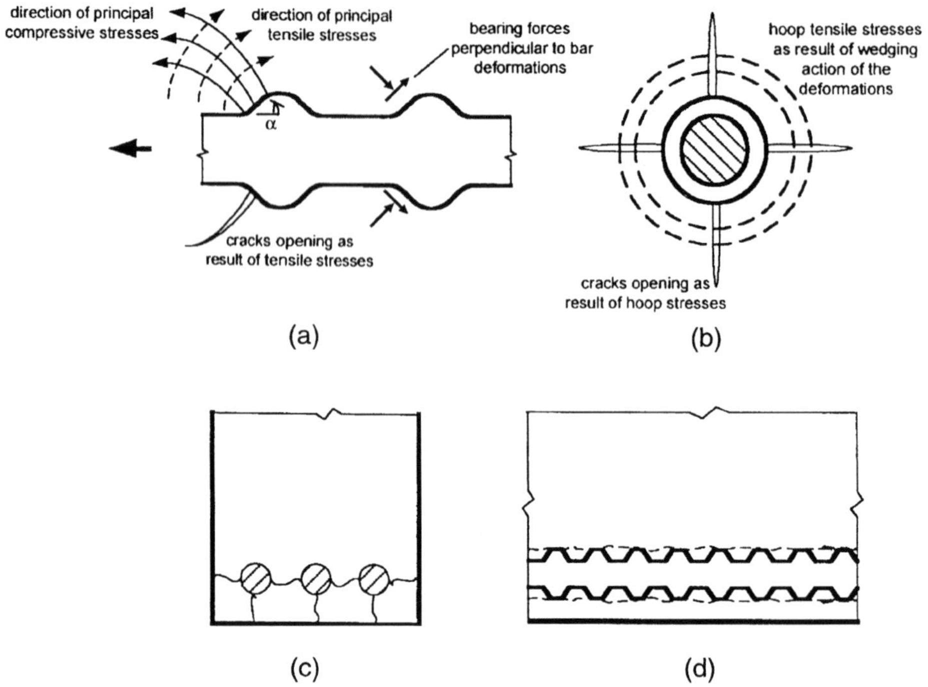

Fig. 1.2—Cracking and damage mechanisms in bond: (a) side view of a deformed bar with deformation face angle α showing formation of Goto (1971) cracks; (b) end view showing formation of splitting cracks parallel to the bar; (c) end view of a member showing splitting cracks between bars and through the concrete cover; and (d) side view of member showing shear crack and/or local concrete crushing due to bar pullout.

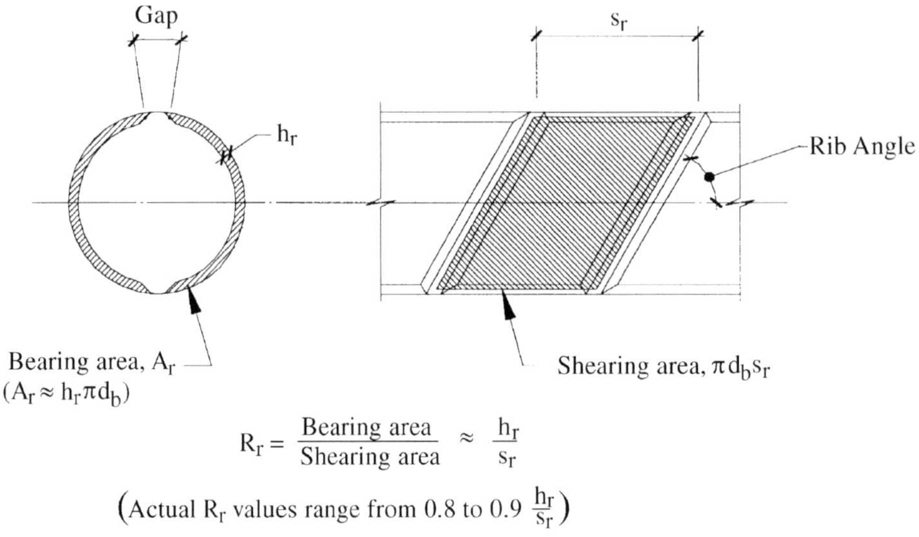

Fig. 1.3—Definition of R_r *(ACI 408.3).*

shearing area between the deformations (in U.S. practice, this is taken as the ratio of the bearing area of the ribs to the product of the nominal bar perimeter and the average spacing of the ribs). Relative rib area is discussed at greater length in Section 2.2.2.

1.1—Bond forces—background

To understand the design procedures used for selecting development and splice lengths of reinforcement, it is instructive to review the nature of bond forces and stresses in a reinforced concrete flexural member. Historically, the difference in tensile force ΔT between two sections located at flexural cracks along a member (Fig. 1.4) was calculated as

$$\Delta T = T_1 - T_2 = \frac{M_1}{jd_1} - \frac{M_2}{jd_2} \tag{1-1}$$

where T_i ($T_2 > T_1$), M_i ($M_2 > M_1$), and jd_i are the tensile force, moment, and internal moment arm at section i (i = 1, 2). For

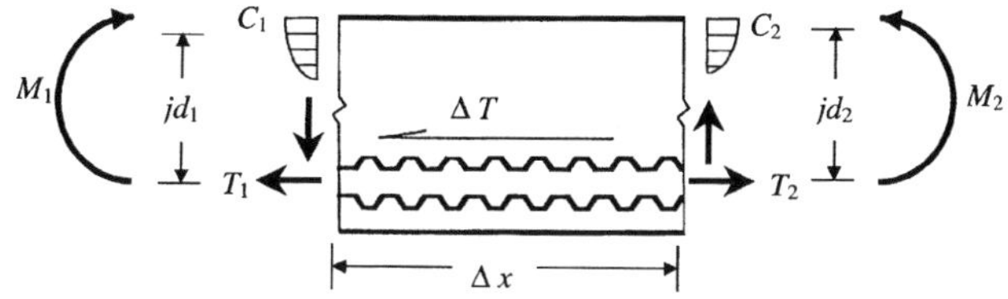

Fig. 1.4—Variation in bar force due to changes in moment in a beam.

an infinitesimally small distance between Sections 1 and 2, Eq. (1-1) becomes

$$dT = \frac{dM}{jd} \quad (1\text{-}2)$$

If the bond force per unit length U is defined as the change in tensile force per unit length, then

$$U = \frac{dT}{dl} = \frac{1}{jd}\frac{dM}{dl} \quad (1\text{-}3a)$$

$$U = \frac{1}{jd}V \quad (1\text{-}3b)$$

where V is the shear on the section.

Equation (1-3b) indicates that, away from concentrated loads, bond forces vary as a function of the applied shear along the length of reinforced concrete flexural members, and for many years, the bond force used in design U was based on this expression. Over time, however, it became apparent that the change in force in reinforcing bars dT does not vary strictly with the change in moment per unit length, as suggested in Eq. (1-3a), but simply with the force in the bar T, which varies from a relatively high value at cracks to a low value between cracks, where the concrete shares the tensile force with the reinforcing steel. Using the definition $U = dT/dl$, bond forces vary significantly along the length of a member, even varying in direction, as shown in Fig. 1.5. The real distribution of bond forces along the length of a bar, therefore, cannot be predicted because it depends on the locations of the flexural cracks and the amount of tensile load carried by the concrete—neither of which can be calculated. Given these facts and because a principal goal of design is to ensure that the bar is adequately anchored so that failure will manifest itself in some way other than in bond, it is both convenient and realistic for design purposes to treat bond forces as if they were uniform over the anchored, developed, or spliced length of the reinforcement.

Until adoption of the 1971 ACI Building Code (ACI 318-71), bond design was based on bond stress u, which is equal to bond force per unit length U divided by the sum of the perimeters of the bars developed at a section Σ_o.

$$u = \frac{U}{\Sigma_o} = \frac{\Delta T}{\Delta l \Sigma_o} = \frac{\Delta f_s A_b}{\Delta l \Sigma_o} = \frac{\Delta f_s d_b}{4\Delta l} \quad (1\text{-}4)$$

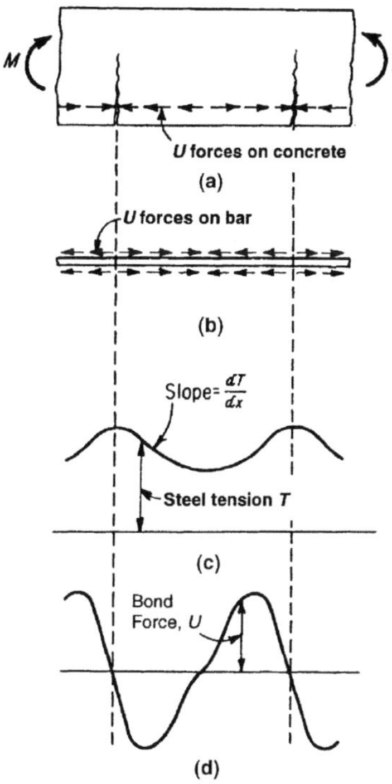

Fig. 1.5—Variation of steel and bond forces in reinforced concrete member subjected to pure bending: (a) cracked concrete segment; (b) bond stresses acting on reinforcing bar; (c) variation of tensile force in steel; and (d) variation of bond force along bar (adapted from Nilson et al. [2004]).

where A_b = area of bars; d_b = diameter of bars; and Δf_s = change in steel stress over length Δl.

For design purposes, the change in stress Δf_s equals the yield stress of the steel f_y and Δl equals the development length l_d. In ACI 318-63, the maximum bond stress was set at*

$$u = 9.5\frac{\sqrt{f_c'}}{d_b} \leq 800 \text{ psi} \quad (1\text{-}5)$$

Substituting Eq. (1-5) into Eq. (1-4), solving for $\Delta l = l_d$, and multiplying the resultant value by 1.2 to account for the reduced bond strength of closely spaced bars (due to the interaction of splitting cracks) gives the development length

$$l_d = 0.04A_b\frac{f_y}{\sqrt{f_c'}} \quad (1\text{-}6)$$

Equation (1-6) was used for design, beginning with ACI 318-71, until a design approach that more closely matched observed behavior was adopted in ACI 318-95.

While convenient, equations for development length [like Eq. (1-6) and some of those presented in Chapter 4] have led many designers to believe that the real force that must be developed is equal to the product of the area and yield strength of a bar. In fact, the basis for the expressions for

*SI conversions of equations that contain terms that depend on units of measure are presented in Appendix A.

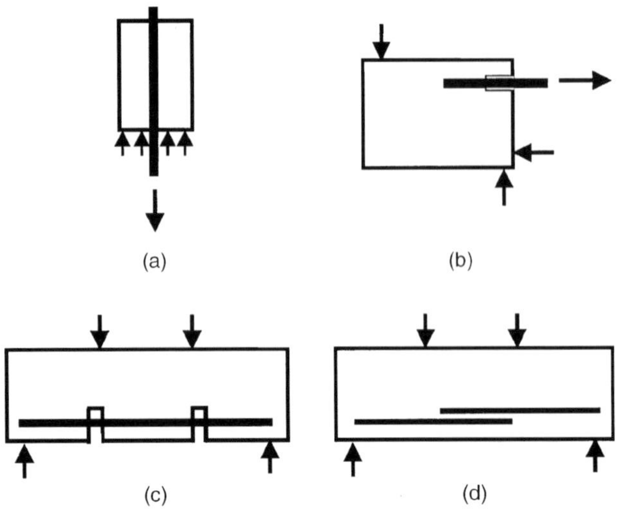

Fig. 1.6—Schematic of: (a) pullout specimen; (b) beam-end specimen; (c) beam anchorage specimen; and (d) splice specimen.

development length l_d lies in Eq. (1-3) and (1-4), which are based on the change in bar force ΔT, the result of externally applied load. The force in the bar $A_b f_y$ used in the Eq. (1-6) is the designer-selected value for ΔT. If, for example, a bar in a flexural member has a higher yield strength than specified (the usual case), a longer development length will be needed to ensure that a ductile bending failure will occur before a brittle bond failure.

1.2—Test specimens

A variety of test specimen configurations have been used to study bond between reinforcing bars and concrete. The four most common configurations are shown in Fig. 1.6. The details of the specimen affect not only the measured bond strength, but also the nature of the bond response.

The pullout specimen (Fig. 1.6(a)) is widely used because of its ease of fabrication and the simplicity of the test. The specimen often incorporates transverse reinforcement to limit splitting when the bar is placed in tension. This specimen is the least realistic of the four shown in Fig. 1.6 because the stress fields within the specimen match few cases in actual construction. As the bar is placed in tension, the concrete is placed in compression. Further, compressive struts form between the support points for the concrete and the surface of the reinforcing bar, placing the bar surface in compression. This stress state differs markedly from most reinforced concrete members, in which both the bar and the surrounding concrete are in tension, and the bearing surfaces of the bar ribs are subjected to a compressive force due to relative movement of the bar with respect to the concrete, not due to the basic load application. In cases where bar surface properties (such as epoxy coatings) or bar surface strength (such as for fiber-reinforced polymer reinforcement) are important, the compression developed at the bar surface in the pullout test reduces the applicability of the test results to structural design. Thus, the use of pullout test results as the sole basis for determining development length is inappropriate and not recommended by Committee 408.

The specimens shown in Fig. 1.6(b) through (d) provide more realistic measures of bond strength in actual structures. The modified cantilever beam, or beam-end specimen, shown in Fig. 1.6(b), provides a relatively simple test that generally duplicates the stress state obtained in reinforced concrete members; the reinforcing steel and the surrounding concrete are simultaneously placed in tension. To achieve the desired stress state, the compressive force must be located away from the reinforcing bar by a distance approximately equal to the embedded or bonded length of the bar within the concrete. To prevent a conical failure surface from forming, a small length of bar near the surface is usually unbonded. A specimen like that shown in Fig. 1.6(b), proportioned to satisfy the spacing requirements between the bar and the compressive force and reinforced to ensure bond rather than flexural or shear failure, is specified in ASTM A 944. The shear reinforcement is placed in the specimen so as not to intercept longitudinal splitting cracks that occur at bond failure. Transverse reinforcement can be added in cases where its effect is of interest (Darwin and Graham 1993a,b). The bond strengths obtained with the test specimen closely match those obtained in other specimens designed to represent full-scale reinforced concrete members.

Beam anchorage and splice specimens shown in Fig. 1.6(c) and (d), respectively, represent larger-scale specimens designed to directly measure development and splice strengths in full-size members. The anchorage specimen simulates a member with a flexural crack and a known bonded length. Based on concern that increased normal stresses at the bar surface, caused by the reactions, may increase bond strength, some anchorage specimens have been designed so that the reactions are displaced laterally from the centerline of the beam. The splice specimen, normally fabricated with the splice in a constant moment region, is easier to fabricate and produces similar bond strengths to those obtained with the anchorage specimen. Because of both its relative simplicity of fabrication and realistic stress-state in the vicinity of the bars, the splice specimen has provided the bulk of the data used to establish the current design provisions for development length, as well as splice length, in ACI 318 (starting with ACI 318-95).

Other specimens have also been used to study bond strength. These include the wall specimen, to determine the "top-bar" effect, and the tension specimen, consisting of a bar surrounded by concrete, with tension applied to both ends of the bar, which project from the concrete. Variations on the beam-end specimen have also been used in which the compressive force is placed relatively close to the bar, resulting in higher bond strengths due to the compressive strut reaching the bar surface. These specimens generally lack the realism obtained with the specimens shown in Fig. 1.6(b) through (d).

1.3—Details of bond response

Bond force-slip and bond stress-slip curves can be used to better understand the nature of bond response. In their simplest and most widely used form, the curves are based on known bar forces, such as obtained in the beam-end and beam anchorage specimens (Fig. 1.6(b) and (c)). Bar forces are compared with the external slip of the reinforcing bar, measured with respect to the concrete at either the loaded or unloaded end of the bar. Examples of bar force-loaded end and unloaded end slip curves are shown in Fig. 1.7(a) and (b), respectively. The loaded end bond force-slip curve shows a lower initial stiffness than the unloaded end curve. The difference represents the lengthening of the reinforcing bar between the two points of slip measurement.

More detailed information, at a smaller scale along the length of a bar, can be obtained by placing strain gages on the bar as a method to determine changes in bar force ΔT, which can be converted to bond force per unit length, $U \approx \Delta T/\Delta l$, and bar stress, $u = U/\Sigma_o$. In the most detailed studies, the strain gages are installed by splitting the bar in half, forming a small channel along the centerline, installing the strain gages and wires in the channel, and welding the bar back together. As an acceptable alternative, strain gages and wires are placed in longitudinal grooves cut at the location of longitudinal ribs. These types of installation minimize disturbances at the interface between the bar and the concrete.

A bond stress versus slip curve for a bar loaded monotonically and failing by pullout is shown in Fig. 1.8 (Eligehausen, Popov, and Bertero 1983). Bond force and bond stress-slip curves, like bond strength, are structural properties that depend on both the geometry of the bar and the details of the concrete member, including the cover, transverse reinforcement, and state of stress in the concrete surrounding the reinforcement. As shown in Fig. 1.7 and 1.8, bond force and bond stress-slip curves are initially very steep because of adhesion.

Because of concrete shrinkage, which is restrained by the reinforcing bar, small internal cracks exist immediately adjacent to the reinforcing bar. These cracks can act as stress raisers and points of crack initiation at the bar ribs at relatively low loads. Because cracks tend to form in front of the ribs, small splitting cracks may begin to propagate from the ribs, as shown in Fig. 1.2(a). If the reinforcing bar is placed in tension from a free surface, such as a beam-end specimen, it is possible for the crack to propagate to the surface, separating a roughly conical region of concrete from the rest of the specimen.

As loading continues, longitudinal splitting cracks form, as shown in Fig. 1.2(b) and (c), leading to a softening in the bond force-slip curve. In regions where these splitting cracks open and where confinement by transverse reinforcement is limited, the bar may slip with little local damage to the concrete at the contact surface with the bar ribs. In regions of greater confinement, higher rates of loading, or both, the concrete in front of the reinforcing bar ribs may crush as the bar moves, forming effective ribs with a reduced face angle (less than α in Fig. 1.2(a)). For higher values of confinement, all of the concrete between the ribs may crush or a shear crack may form along the periphery of the bar, or both, resulting in a pullout failure (Fig. 1.2(d)). Depending on the details of the member and the loading, bond failure may entail a combination of the failure modes. Transverse reinforcement, however, rarely yields during a bond failure

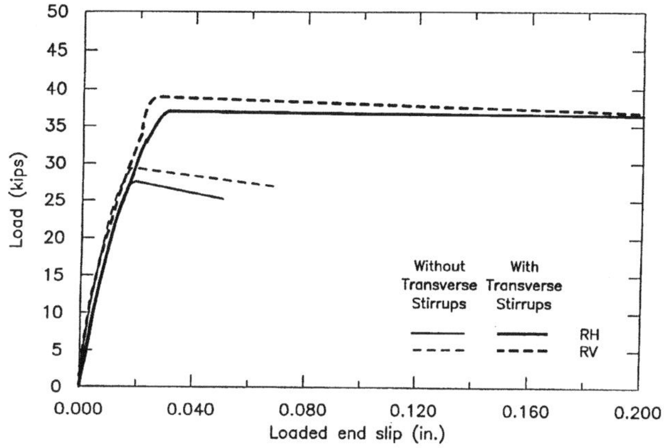

Fig. 1.7(a)—Average load-loaded end slip curves for No. 8 (No. 25) ASTM A 615 reinforcing bars in an ASTM A 944 beam-end specimen. RH and RV represent bars with longitudinal ribs oriented horizontally and vertically, respectively (Darwin and Graham 1993a). (Note: 1 in. = 25.4 mm; 1 kip = 4.45 kN)

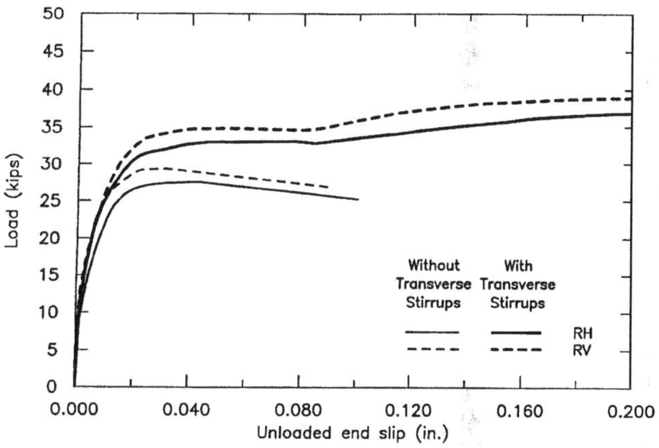

Fig. 1.7(b)—Average load-unloaded end slip curves for No. 8 (No. 25) ASTM A 615 reinforcing bars in an ASTM A 944 beam-end specimen. RH and RV represent bars with longitudinal ribs oriented horizontally and vertically, respectively (Darwin and Graham 1993a). (Note: 1 in. = 25.4 mm; 1 kip = 4.45 kN)

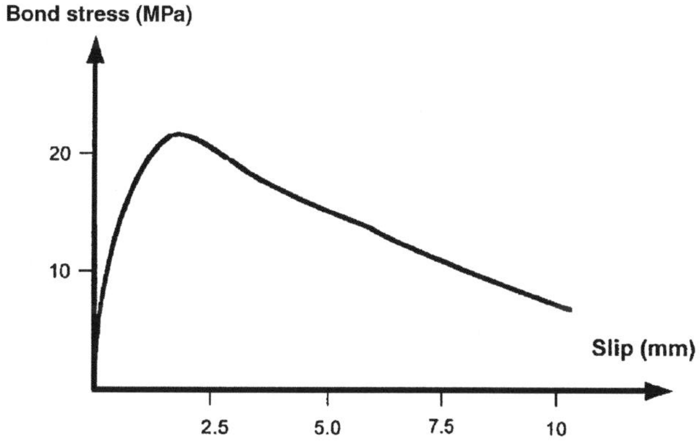

Fig. 1.8—Bond stress-slip curve for bar loaded monotonically and failing by pullout (Eligehausen, Popov, and Bertero 1983). (Note: 1 MPa = 145 psi; 1 mm = 0.0394 in.)

(Maeda, Otani, and Aoyama 1991; Sakurada, Morohashi, and Tanaka 1993; Azizinamini, Chisala, and Ghosh 1995).

Pullout failure (Fig. 1.2(d)) occurs primarily in cases of high confinement and low bonded lengths. In most structural applications, however, splitting failure (Fig. 1.2(b) and (c)) is more common (Clark 1950; Menzel 1952; Chinn, Ferguson, and Thompson 1955; Ferguson and Thompson 1962; Losberg and Olsson 1979; Soretz and Holzenbein 1979; Johnston and Zia 1982; Treece and Jirsa 1989; Choi et al. 1991). For this reason, data used for design are normally limited to members with a minimum development (or splice) length to bar diameter ratio of 15 or 16 and specified maximum values of confinement provided by the concrete and the transverse reinforcement (the latter will be discussed at greater length in Chapter 3). When extended to higher values of confinement, design expressions based on splitting give unrealistically high bond strengths (ACI 318; Orangun, Jirsa, and Breen 1977; Darwin et al. 1996a).

The role played by splitting cracks in bond failure emphasizes the importance of the tensile properties of concrete in controlling bond strength. As will be discussed in Section 2.3.3, the tensile properties involve more than strength, and include fracture energy; that is, the capacity of the concrete to dissipate energy as a crack opens.

1.4—Notation

A_b = area of bar being developed or spliced
= area of largest bar being developed or spliced (CEB-FIP 1990)

A_{tr} = area of each stirrup or tie crossing the potential plane of splitting adjacent to the reinforcement being developed, spliced, or anchored

b_w = width of the web of a beam

c = spacing or cover dimension
= $c_{min} + d_b/2$

c_b = bottom concrete cover for reinforcing bar being developed or spliced

c_{max} = maximum (c_b, c_s)

c_{med} = median $(c_{so}, c_b, c_{si} + d_b/2)$ [that is, middle value], (Esfahani and Rangan 1998a,b)

c_{min} = minimum cover used in expressions for the bond strength of bars not confined by transverse reinforcement
= minimum $(c_{so}, c_b, c_{si} + d_b/2)$ (Esfahani and Rangan 1998a,b)

c_{min} = smaller of minimum concrete cover or 1/2 of the clear spacing between bars
= minimum (c_b, c_s)

c_s = minimum $[c_{so}, c_{si} + 0.25$ in. (6.35 mm)]

c_{si} = 1/2 of the bar clear spacing

c_{so} = side concrete cover for reinforcing bar

C_R = $44 + 330(R_r - 0.10)$(ACI 408.3)

d_b = diameter of bar

E_c = modulus of elasticity of concrete

f_c = stress in concrete

f'_c = concrete compressive strength based on 6 x 12 in. (150 x 300 mm) cylinders
= specified compressive strength of concrete

f_{ct} = average splitting tensile strength of concrete, based on split cylinder test

f_s = stress in reinforcing bar

f_y = yield strength of steel being developed or spliced

f_{yt} = yield strength of transverse reinforcement

G_f = fracture energy

h_r = average height of deformations on reinforcing bar

jd_i = internal moment arm at section i

K = constant used in CEB-FIP design expression for development length

K_1 = constant used to calculate T_s

K_{tr} = transverse reinforcement index
= $\dfrac{A_{tr}f_{yt}}{1500sn}, \left(\dfrac{A_{tr}f_{yt}}{10.34sn}\right)^*$ (ACI 318)
= $35.3\, t_r t_d A_{tr}/sn$ (ACI T2-98)
= $C_R(0.72d_b + 0.28)\dfrac{A_{tr}}{sn}, \left(C_R(0.0283d_b + 0.28)\dfrac{A_{tr}}{sn}\right)^*$ (ACI 408.3)
= $(0.52t_r t_d A_{tr}/sn)f_c'^{\,1/2}$, $[(6.26t_d A_{tr}/sn)f_c'^{\,1/2}]^*$ (see Section 4.3)

l_d = development or splice length

$l_{d,min}$ = minimum development length

$l_{s,min}$ = minimum splice length

M = constant used in expressions for the bond strength of bars not confined by transverse reinforcement
= $\cosh\left(0.00092 l_d \sqrt{rf'_c/d_b}\right)$, $\left[\cosh\left(0.0022 l_d \sqrt{rf'_c/d_b}\right)\right]^*$ (Esfahani and Rangan 1998a,b)

M = ratio of the average yield strength to the design yield strength of the developed bar (CEB-FIP)

M_i = internal moment at section i

n = number of bars being developed or spliced

N = the number of transverse stirrups, or ties, within the development or splice length

p = nominal perimeter of bar

p = power of f'_c

r = constant used in expressions for the bond strength of bars not confined by transverse reinforcement; a function of R_r
= 3 for conventional reinforcement ($R_r \approx 0.07$) (Esfahani and Rangan 1998a,b)

R_r = relative rib area of the reinforcement

s = spacing of transverse reinforcement

s_r = average spacing of deformations on reinforcing bar

t_d = term representing the effect of bar size on T_s
= $0.72d_b + 0.28$, $(0.028d_b + 0.28)^*$ (Darwin et al. 1996a,b)
= $0.78d_b + 0.22$, $(0.03d_b + 0.22)^*$ (Zuo and Darwin 1998, 2000) (see Section 4.3)

t_r = term representing the effect of relative rib area on T_s
= $9.6R_r + 0.28$ (Darwin et al. 1996a,b; Zuo and Darwin 1998, 2000) (see Section 4.3)

T_b = total bond force of a developed or spliced bar
= $T_c + T_s$

---

*SI expressions.

T_c = concrete contribution to total bond force, the bond force that would be developed without transverse reinforcement
T_i = tensile force at section i
T_s = steel contribution to total bond force, the additional bond strength provided by the transverse steel
u = bond stress
u_b = bond strength of a bar confined by transverse reinforcement
= $u_c + u_s$
u_c = average bond strength at failure of a bar not confined by transverse reinforcement
u_s = bond strength of a bar attributed to the confinement provided by the transverse reinforcement
U = bond force per unit length
V = shear on a section
α = rib face angle
α = reinforcement location factor
α_b = factor used to increase the development length of a bar for lap splices (CEB-FIP 1990)
= coating factor
β = angle between transverse rib and longitudinal axis of the bar
β = reliability index
Δf_s = change in steel stress over length Δl
ΔT = change in the bar force as a result of an externally applied load
ε_c = strain in concrete in uniaxial compression
ε_o = concrete strain at maximum concrete stress under uniaxial compression
γ = reinforcement size factor
λ = lightweight aggregate concrete factor
ϕ = capacity-reduction factor
ϕ_b = capacity-reduction factor for bond
ϕ_d = effective ϕ-factor for bond
= $\phi_b/\phi_{tension}$
$\phi_{tension}$ = capacity-reduction factor for section in tension
ΣA_r = total area of ribs around bar perimeter measured on the longitudinal section of each rib using the trapezoidal method to approximate the area under the curve
ΣA_{tr} = area of transverse reinforcement along l_d
Σgaps = sum of the gaps between ends of transverse deformations on reinforcing bar
Σ_o = perimeters of the bars anchored at a section
ω = $0.1\frac{c_{max}}{c_{min}} + 0.9 \leq 1.25$ (ACI 408.3) (see Section 4.3)

CHAPTER 2—FACTORS AFFECTING BOND

Many factors affect the bond between reinforcing bars and concrete. The major factors are discussed in this chapter. Background research, bond behavior, and relationships between bond strength and geometric and material properties are presented under three major subject headings: structural characteristics, bar properties, and concrete properties. The structural characteristics addressed include concrete cover and bar spacing, the bonded length of the bar, the degree of transverse reinforcement, the bar casting position, and the use of noncontact lap splices. The bar properties covered include bar size and geometry, steel stress and yield strength, and bar surface condition, while the concrete properties include compressive strength, tensile strength and fracture energy, aggregate type and quantity, concrete slump and workability, and the effects of admixtures, fiber reinforcement, and degree of consolidation.

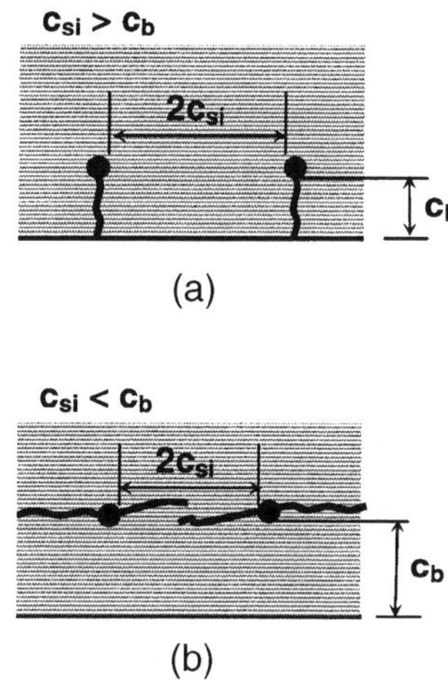

Fig. 2.1—Bond cracks: (a) $c_{si} > c_b$*; and (b)* $c_{si} < c_b$.

2.1—Structural characteristics

2.1.1 *Concrete cover and bar spacing*—Bond force-slip curves become steeper and bond strength increases as cover and bar spacing increase. The mode of failure also depends on cover and bar spacing (Untrauer 1965; Tepfers 1973; Orangun, Jirsa, and Breen 1977; Eligehausen 1979; Darwin et al. 1996a). For large cover and bar spacing, it is possible to obtain a pullout failure, such as shown in Fig. 1.2(d). For smaller cover and bar spacing, a splitting tensile failure occurs (Fig. 1.2(c)), resulting in lower bond strength. The latter failure mode is the type expected to govern for most structural members. Splitting failures can occur between the bars, between the bars and the free surface, or both. Pullout-like failures can occur with some splitting if the member has significant transverse reinforcement to confine the anchored steel.

With bond failures involving splitting of the concrete for bars that are not confined by transverse reinforcement, the peak load is governed by the tensile response of the concrete, which depends on both tensile capacity and energy dissipation capacity, normally described as fracture energy G_f. As described in Section 2.3.1, concrete exhibiting higher fracture energies provide improved bond capacities, even if the concrete has similar tensile strengths.

When splitting failures occur, the nature of the splitting failure depends, in general, on whether the concrete cover c_b is smaller than either the concrete side cover c_{so} or 1/2 of the bar clear spacing c_{si}. (In this case, the symbol for bottom cover c_b is used, but the discussion applies equally to bottom, top, and side cover.) When c_b is smaller than c_{so} and c_{si}, the splitting crack occurs through the cover to the free surface (Fig. 2.1(a)). When c_{so} or c_{si} is smaller than c_b, the splitting crack forms through the side cover or between the reinforcing bars (Fig. 2.1(b)), respectively.

In ACI 318, the effective value of c_{si} for development length calculations is equal to the actual value of c_{si}. In the Canadian requirements for reinforced concrete design (CSA Standard A23.3-94), however, a greater value [2/3 of the center-to-center spacing of the bars being developed minus 1/2 of the bar diameter = $(2/3)(2c_{si} + d_b) - (1/2)d_b = (4/3)c_{si} + (1/6)d_b$] is used. In an analysis of bars not confined by transverse reinforcement, Zuo and Darwin (1998) found that the best match with tests is obtained: 1) using $1.6c_{si}$ as the effective value of c_{si} when using a multiple of c_{si}; and 2) using c_{si} + 0.25 in. (c_{si} + 6.4 mm) as the effective value of c_{si} when a constant value is added to c_{si}. Of the two procedures, $1.6c_{si}$ provides the better predictions. The fact that the effective value of c_{si} is greater than the actual value is most likely "due to the longer effective crack lengths that occur when concrete splits between bars" (Darwin et al. 1996a) (Fig. 2.1(b)), which make the bars behave as if they have an increased separation.

Although $1.6c_{si}$ gives the best match for development and splice strengths for bars not confined by transverse reinforcement, the value tends to overestimate the effective crack length between bars confined by transverse reinforcement (Zuo and Darwin 1998). In the latter case, c_{si} + 0.25 in. (c_{si} + 6.4 mm), provides a better match with tests. This observation suggests that there is a small but significant difference in the effect of cracks between bars on bond strength in the presence of confining reinforcement. In the presence of transverse reinforcement, the effective crack length between bars is still greater than the clear distance between the bars, but not as great as for similar members without confining transverse reinforcement.

Orangun, Jirsa, and Breen (1975) and Darwin et al. (1992, 1996a) observed that, although the minimum value of c_b, c_{so}, and c_{si} has the principal effect on bond strength, the relative value of c_{so} or c_{si} to c_b is also important. For bars not confined by transverse reinforcement, Darwin et al. (1996a) found that, compared to cases in which the minimum value of c_{so} or c_{si} equals c_b, the bond strength of bars for which the minimum value c_{so} or c_{si} does not equal c_b increases by the ratio

$$\left(0.1\frac{c_{max}}{c_{min}} + 0.9\right) \leq 1.25 \qquad (2\text{-}1)$$

where

c_{max} = maximum (c_b, c_s);
c_{min} = minimum (c_b, c_s);
c_b = bottom cover;
c_s = minimum [c_{so}, c_{si} + 0.25 in. (6.4 mm)];
c_{so} = side cover; and
c_{si} = 1/2 of the bar clear spacing.

In addition to cover thickness, the nature of the cover plays a role in bond strength. With emphasis on methods of construction for reinforced concrete bridge decks, Donahey and Darwin (1983, 1985) evaluated the bond strength of bars with 3 in. (76 mm) of monolithic cover and bars with laminar cover, consisting of 3/4 in. (19 mm) monolithic cover topped with a 2-1/4 in. (57 mm) high-density concrete overlay. The bars with the 3 in. (76 mm) laminar cover achieved about the same bond strengths (average = 97%) as achieved by bars with the same thickness of monolithic cover, even though greater bond strengths would have been expected based on the compressive strength of the overlay concrete, which ranged between 110 and 155% of the compressive strength of the base concrete.

2.1.2 *Development and splice length*—Increasing the development or splice length of a reinforcing bar will increase its bond capacity. The nature of bond failure, however, results in an increase in strength that is not proportional to the increase in bonded length. The explanation starts with the observations that bond forces are not uniform (Fig. 1.5) and that bond failures tend to be incremental, starting in the region of the highest bond force per unit length. In the case of anchored bars, longitudinal splitting of the concrete initiates at a free surface or transverse flexural crack where the bar is most highly stressed. For spliced bars, splitting starts at the ends of the splice, moving towards the center. For normal-strength concrete, splitting may also be accompanied by crushing of the concrete in front of the ribs as the bar moves (or slips) with respect to the concrete. For higher-strength concrete and for normal-strength concrete in which the bars are epoxy coated, the degree of crushing in front of the ribs is significantly decreased. For splice specimens studied after failure, it is common to see no crushed concrete at ribs near the tensioned end of a spliced bar, with the crushed concrete located at the end of the bar, indicating that failure occurred by a slow wedging action followed by rapid final slip of the bar at failure. Because of the mode of bond failure, the nonloaded end of a developed or spliced bar is less effective than the loaded end in transferring bond forces, explaining the nonproportional relationship between development or splice length and bond strength.

Although the relationship between the bond force and the bonded length is not proportional, it is nearly linear, as illustrated in Fig. 2.2 for No. 4 to 14 (No. 13 to 43) bars. Figure 2.2 indicates that bars will have measurable bond strengths even at low embedded lengths. This occurs because, in the tests, there is always at least one set of ribs that forces the concrete to split before failure. When failure occurs, a significant crack area is opened in the member due to splitting (Brown, Darwin, and McCabe 1993; Darwin et al. 1994; Tholen and Darwin 1996). As the bonded length of the bar increases, the crack surface at failure also increases in a linear but not proportional manner with respect to the bonded length. Thus, the total energy needed to form the crack and, in turn, the total bond force required to fail the member, increase at a rate that is less than the increase in bonded length. Therefore, the common design practice (ACI 318) of establishing a proportional relationship between bond force and development or splice length is conservative for short bonded lengths, but becomes progressively less conservative, and eventually unconservative, as the bonded length and stress in the developed or spliced bar increase.

2.1.3 *Transverse reinforcement*—Transverse reinforcement confines developed and spliced bars by limiting the progression of splitting cracks and, thus, increasing the bond force required to cause failure (Tepfers 1973; Orangun, Jirsa and

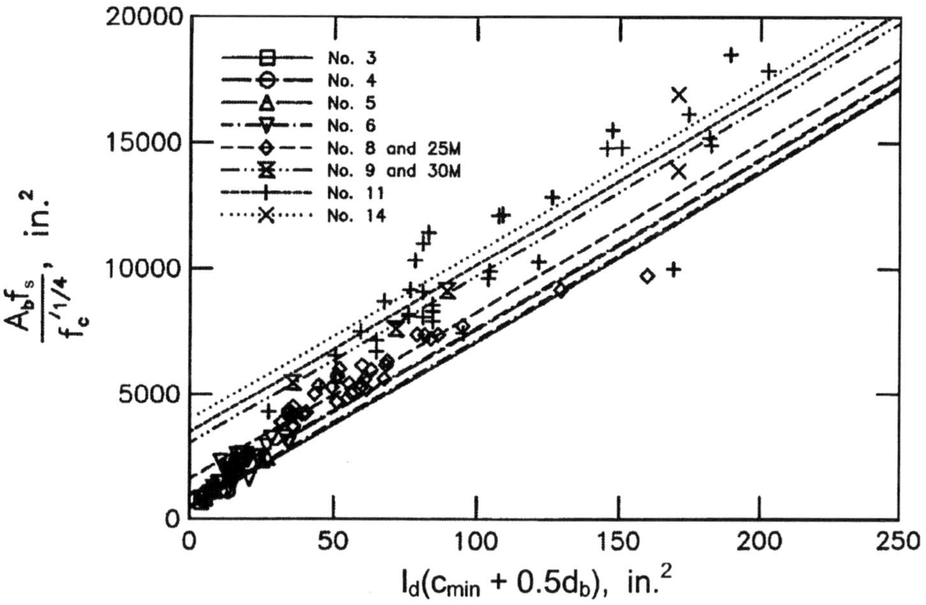

Fig. 2.2—Bond strength $A_b f_s$ normalized with respect to $f_c'^{1/4}$ versus the product of the development or splice length l_d and the smaller of the minimum concrete cover to the center of the bar or 1/2 of the center-to-center bar spacing (c_{min} + $0.5d_b$) (Darwin et al. 1996b). (Note: 1 in.^{2} = 645 mm^2)

Breen 1977; Darwin and Graham 1993a,b). An increase in transverse reinforcement results in an increase in bond force, eventually converting a splitting failure to a pullout failure. Additional transverse reinforcement, above that needed to cause the transition from a splitting to a pullout failure, becomes progressively less effective, eventually providing no increase in bond strength (Orangun, Jirsa, and Breen 1977).

The total bond force of a developed or spliced bar T_b can be represented as the sum of a concrete contribution T_c, representing the bond force that would be developed without the transverse reinforcement, plus a steel contribution T_s, representing the additional bond strength provided by the transverse steel

$$T_b = T_c + T_s \quad (2\text{-}2)$$

The value of the concrete contribution is affected somewhat by the presence of the transverse steel, as discussed in Section 2.1.1, because the effective crack length between bars is reduced as bar slip continues in the process of mobilizing the additional bond strength provided by the transverse reinforcement (Zuo and Darwin 1998, 2000). The effect of transverse reinforcement on T_c is small but measurable.

The value of the steel contribution T_s is a function of the area of reinforcing steel that crosses potential crack planes, the strength of the concrete, and both the size and deformation properties of the developed or spliced reinforcement. It can be represented in the form (Zuo and Darwin 1998, 2000)

$$T_s = K_1 t_r t_d \frac{N A_{tr}}{n} f_c'^{\,p} \quad (2\text{-}3)$$

where

K_1 = a constant;

t_r = a factor that depends on the relative rib area R_r of the reinforcement;

t_d = a factor that depends on the diameter d_b of the developed or spliced bar;

N = the number of transverse stirrups, or ties, within the development or splice length;

A_{tr} = area of each stirrup or tie crossing the potential plane of splitting adjacent to the reinforcement being developed or spliced;

n = number of bars being developed or spliced along the plane of splitting;

f_c' = concrete compressive strength based on 6 x 12 in. (150 x 300 mm) cylinders; and

p = power of f_c' between 0.75 and 1.00 (see Section 3.3 for values of K_1 and p).

In the case shown in Fig. 2.3, illustrating a two-leg stirrup confining three spliced reinforcing bars in the same plane, A_{tr} = two times the area of the stirrup A_t and $n = 3$ if c_{so} or the effective value of c_{si} is less than c_b. $A_{tr} = A_t$ and $n = 1$ if c_{so} and the effective value of c_{si} are greater than c_b.

The values of t_r and t_d can be represented as linear functions of R_r and d_b, respectively (Zuo and Darwin 1998, 2000).

$$t_r = 9.6R_r + 0.28 \quad (2\text{-}4)$$

$$t_d = 0.78d_b + 0.22 \quad (2\text{-}5)$$

with d_b in inches and $R_r \leq 0.14$.

The relationships given in Eq. (2-4) and (2-5) suggest that an increase in the wedging action of the bars, resulting from

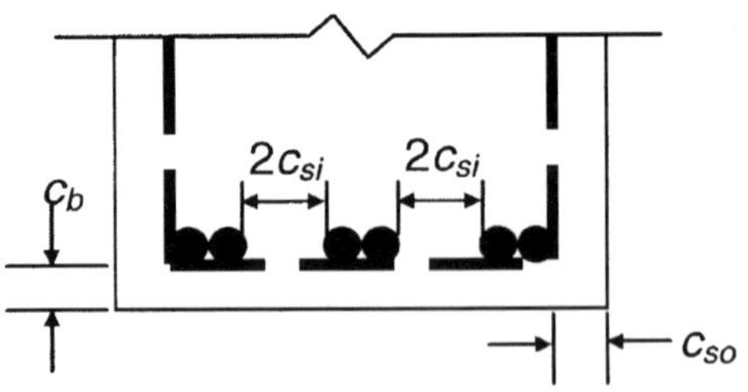

Fig. 2.3—Beam cross sections showing definition of c_b, c_{si}, *and* c_{so}.

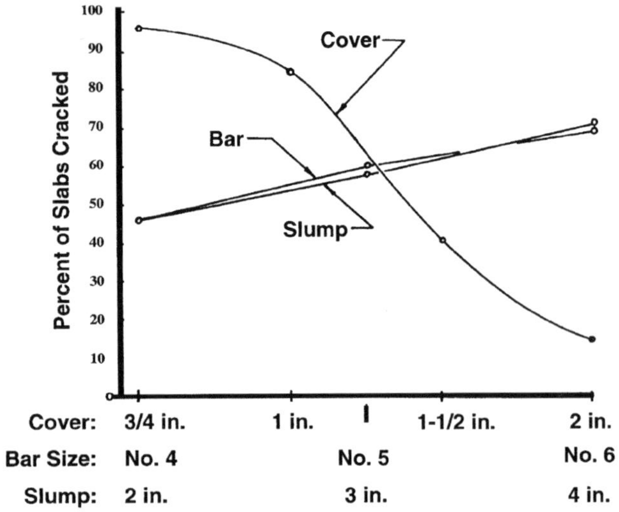

Fig. 2.4—Settlement cracking as a function of bar size, slump, and cover (Dakhil et al. 1975). (Note: 1 in. = 25.4 mm)

both an increase in R_r (a relative measure of rib size and spacing) and an increase in bar size d_b (an absolute measure of rib size), will increase the stress in the stirrups, resulting in an increase in confining force. The relationship between confinement and the degree of wedging action is in concert with the observation that stirrups rarely yield (Maeda, Otani, and Aoyama 1991; Sakurada, Morohashi, and Tanaka 1993; Azizinamini, Chisala, and Ghosh 1995), allowing an increase in lateral displacement caused by wedging to be translated into an increase in confining force. As a result, the yield strength of the transverse reinforcement f_{yt} does not play a role in the steel contribution to bond force, T_s. The effect of concrete strength on T_s is discussed further in Section 2.3.1.

2.1.4 *Bar casting position*—As early as 1913, Abrams observed that bar position during concrete placement plays an important role in the bond strength between concrete and reinforcing steel. Top-cast bars have lower bond strengths than bars cast lower in a member (Clark 1946, 1949; Collier 1947; Larnach 1952; Menzel 1952; Menzel and Woods 1952; Ferguson and Thompson 1962, 1965; CUR 1963; Untrauer 1965; Welch and Patten 1965; Untrauer and Warren 1977; Thompson et al. 1975; Jirsa and Breen 1981; Luke et al. 1981; Zekany et al. 1981; Donahey and Darwin 1983, 1985; Altowaiji, Darwin, and Donahey 1984, 1986; Brettmann, Darwin, and Donahey 1984, 1986; Jeanty,

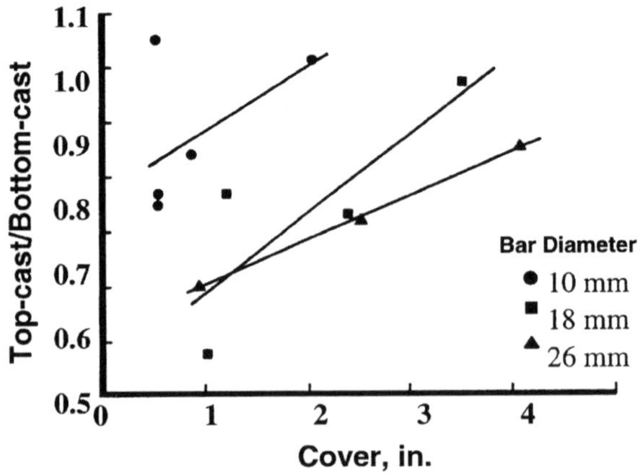

Fig. 2.5—Ratio of top-cast to bottom-cast bar bond strength versus cover (CUR 1963). (Note: 1 in. = 25.4 mm)

Mitchell, and Mirza 1988). This behavior is recognized in ACI 318 and the AASHTO Bridge Specifications (1996). Top-reinforcement, horizontal reinforcement with more than 12 in. (300 mm) of fresh concrete cast in the member below the development length or splice, requires a 30% increase in development length (ACI 318). Most research, however, indicates that while an increased depth of concrete below a bar reduces bond strength, the effect of shallow top cover is of greater significance. The impact of shallow top cover on the top-cast bar effect is emphasized by the fact that the strength reduction becomes progressively greater as cover is decreased.

The lower bond strength of top-cast bars may be explained as follows: the greater the depth of concrete below a bar, the greater will be the settlement and accumulation of bleed water at the bar, because there is more concrete beneath the bar to settle and bleed. The effects of settlement and bleeding on the concrete around a bar are aggravated by increased concrete slump and decreased cover above the bar. Dakhil, Cady, and Carrier (1975) found that longitudinal settlement cracking increased above top-cast bars with increased slump and bar size and especially with decreased top cover (Fig. 2.4).

Menzel (1952) observed settlement cracks over 1 in. (25 mm) diameter top-cast bars with a 2 in. (50 mm) cover for specimens placed with 6 in. (150 mm) slump concrete that was consolidated by hand. Cracks were not observed in lower slump concrete that was vibrated internally. Menzel (1952) also observed that the bond strength of top-cast reinforcement decreases as specimen depth increases. He observed the greatest reduction for high-slump concrete consolidated by hand rodding and the least for low-slump concrete consolidated with vibration.

The importance of cover on the reduction in bond strength for top-cast bars is demonstrated by tests in the Netherlands (CUR 1963). As shown in Fig. 2.5, the ratio of top-cast bar to bottom-cast bar bond strength decreased significantly as cover decreased.

Ferguson and Thompson (1965) conducted beam tests to compare the ratio of top-cast to bottom-cast bar bond strength. The bond strength of top-cast bars decreased with increasing slump and decreasing top cover.

Zekany et al. (1981) conducted splice tests in 16 in. (406 mm) deep beams with both top- and bottom-cast bars. Splice strength decreased for both top- and bottom-cast bars with increasing slump, but the decrease was consistently greater for the top-cast bars.

In a study of casting position, Luke et al. (1981) clearly demonstrated that as the depth of concrete below a bar increases, the bond strength decreases (Fig. 2.6). They observed the greatest incremental decrease in strength for top-cast bars. As will be discussed in more detail in Section 2.3.5, they also observed that bond strength decreased with increasing slump, but that the decrease was most pronounced for top-cast bars. For bars cast below the specimen mid-depth, slump appeared to have little effect.

Brettmann, Darwin, and Donahey (1984, 1986) also found that as the amount of concrete below a bar increases, bond strength decreases. The measured decrease was least for low-slump concrete and greatest for high-slump concrete obtained without the use of a high-range, water-reducing admixture. Bond strength decreased for bars with as little as 8 in. (200 mm) of concrete below the bar, bars that would not be defined as top reinforcement under the provisions of ACI 318. In similar tests run by Zilveti et al. (1985), a "top-bar effect" was obtained for top-cast bars with as little 6 in. (150 mm) of concrete below the bars. Brettmann, Darwin, and Donahey (1984, 1986) found that bond strength was similar for bars placed 8 and 15 in. (200 and 380 mm) above the bottom of the forms. Bond strength was lowest for bars placed 36 in. (915 mm) above the bottom of the forms, with the largest reductions obtained for high slump, nonvibrated specimens, as shown in Fig. 2.7. In all cases (even for bars with only 8 in. [200 mm] of concrete below the bars), the decrease in bond strength between bottom-cast and top-cast bars was greater than between top-cast bars with 8 and 36 in. (203 and 914 mm) of concrete below the bars. These results indicate that the choice of 12 in. (300 mm) of concrete below a bar for the 30% increase in development length for top reinforcement is arbitrary. Based on concrete depth alone, there seems to be a gradual decrease in bond strength with no sharp drop-off.

2.1.5 *Noncontact lap splices*—A noncontact lap splice (also called a spaced splice) provides continuity of reinforcement by overlapping the ends of the steel bars without the bars touching each other. According to ACI 318, "bars spliced by noncontact lap splices in flexural members shall not be spaced transversely farther apart than one-fifth the required lap splice length, nor 6 in. (150 mm)." This provision was first incorporated in ACI 318-71. The commentary to ACI 318 argues that if individual bars in noncontact lap splices are too widely spaced, an unreinforced section is created, and that forcing a potential crack to follow a zigzag line (5 to 1 slope) is considered to be a minimum precaution. The commentary points out that the 6 in. (150 mm) maximum spacing is added because most research available on lap splices of deformed bars was conducted with reinforcement within this spacing.

In earlier codes, a minimum clear spacing was actually required. ACI 318-47 specified that the minimum clear

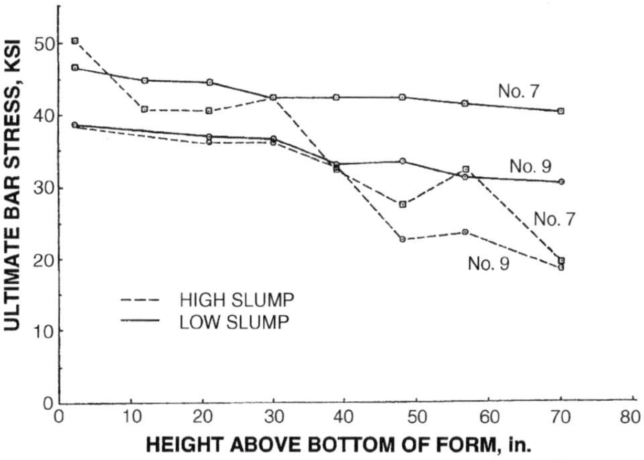

Fig. 2.6—Bond strength as a function of bar location within a wall specimen. High slump = 8-1/2 in. (215 mm). Low slump = 3 in. (75 mm) (Luke et al. 1981). (Note: 1 ksi = 6.89 MPa; 1 in. = 25.4 mm)

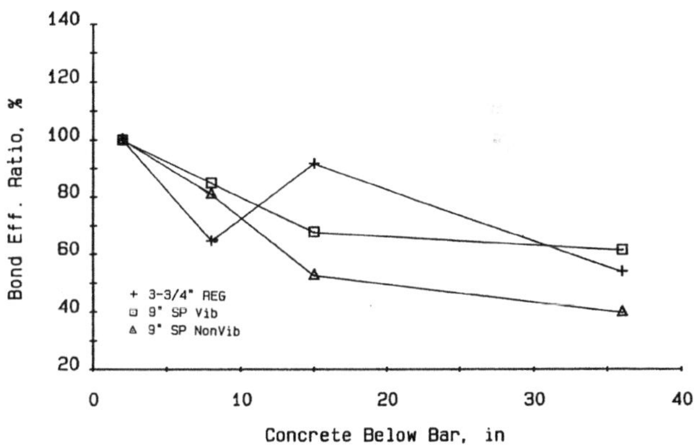

Fig. 2.7—Bond efficiency ratio versus concrete depth below the bar for lower temperature concrete (53 °F [12 °C]). REG = concrete without a superplasticizer; and SP = superplasticized concrete (Brettman, Darwin, and Donahey 1986) (Note: 1 in. = 25.4 mm)

spacing between spliced bars must not be less than $1.5d_b$ for round bars or 1-1/3 times the maximum size of aggregate, and at least 1 in. (25 mm) in any case. ACI 318-51 retained these requirements, except that the $1.5d_b$ requirement was changed to $1.0d_b$. Engineering practice before 1950 usually required that an allowance be made for a reduction in bonded area for tied lap splices to account for the fact that concrete does not completely surround spliced bars that are in contact. This was provided by lengthening the splice. It was only in 1963 that the ACI Building Code (ACI 318-63) allowed both spaced and contact lap splices.

In a lap splice, the force in one bar is transferred to the concrete which, in turn, transfers it to the adjacent bar. This transfer of forces from one bar to another in a noncontact splice can be seen from the crack pattern, as shown in Fig. 2.8 (MacGregor 1997).

Walker (1951) reported on a series of pullout and beam tests to compare the performance of spaced and tied lap splices. Two levels of concrete strength and three types of deformed reinforcing bars were studied. In all tests where the

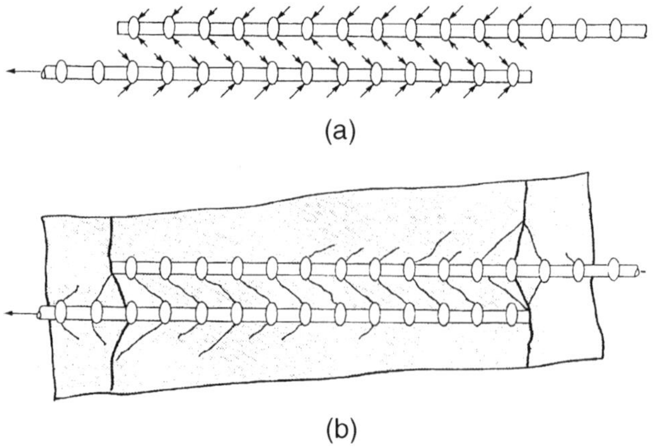

Fig. 2.8—Noncontact tension lapped splices: (a) forces on bars at splice; and (b) internal cracks at splice (MacGregor 1997).

spliced bars were spaced, the clear spacing was $1.5d_b$. Beam tests showed no significant difference between the two splicing methods (zero spacing and $1.5d_b$ spacing), but at loads close to ultimate, there was some indication that spaced spliced bars might be slightly preferable, showing less center beam deflection and end slip at a given load level. In the pullout tests, there was no weakening of bond at the tied splice. Considering all of the data, however, Walker concluded that within the scope of his study there was no significant loss of bond when deformed bars were tied together at the splice.

Chamberlin (1952) investigated the effect of spacing of spliced bars in tension pull-out specimens. The tests were designed to provide data on the effect of spacing of lapped bars on bond and also on the effect of length of overlap in relation to effectiveness of stress transfer from one bar to another at a splice. Confinement was provided by spiral wire to prevent splitting. Based on test results, Chamberlin concluded deformed bars developed better average bond stress in adjacent tied splices (zero spacing) than in spaced splices, but that the differences in bond strength for clear spacings of one-bar diameter and three-bar diameters were not significant.

Chinn, Ferguson, and Thompson (1955) reported on research investigating the effects of many variables on the bond capacity of spliced reinforcement including the clear spacing between spliced bars, with values of 0 (contact splice), 0.75, 1.0, 1.25, and 1.88 in. (20, 25, 32, and 48 mm). Forty beam specimens were tested. Each beam contained either one or two splices, placed in a constant moment region at the center of the beam. They concluded that their tests confirmed the earlier tests by Walker (1951) and Chamberlin (1952), showing little difference in strength between contact and spaced lap splices.

Chamberlin (1958) studied the effect of spacing of lapped bars on bond strength and the effect of length of lap on the load-carrying capacity of small beams. Twenty-one beams were tested with no restraint against splitting. All beams were 6 x 6 in. (150 x 150 mm) in cross section and 36 in. (915 mm) in length, simply supported under symmetrical two-point loading. The clear spacing between lapped bars was either 1/2 or 1 in. (12 or 25 mm). Chamberlin concluded that there was little difference in strength between adjacent and spaced splices.

Sagan, Gergely, and White (1991) studied the behavior of noncontact lap splices subjected to monotonic and repeated inelastic loading. Forty-seven full-scale flat-plate specimens were tested. Each specimen contained two splices. Variables included splice-bar spacing, concrete compressive strength, bar size, the amount and distribution of transverse reinforcement, and the lap length. The specimens were loaded slowly in direct tension. If a specimen survived the first loading to yield, it was unloaded slowly and then reloaded to yield; this process was repeated until failure. Sagan, Gergely, and White observed that:

1. A noncontact lap splice can be modeled as a plane truss, with load transferred between the two spliced bars through compressive struts in the concrete; the tension elements are provided by the transverse reinforcement and surrounding concrete;

2. An in-plane splitting crack forms between the bars of spaced bar splices. The crack results from bond-induced bursting stresses and Poisson strains generated by the compression stress field. Diagonal surface cracking of the concrete between the spliced bars becomes more prominent as the bar spacing increases. Even the large cracks that formed at the ends of the splice were diagonal;

3. The splice strength of the monotonically loaded specimens increased when transverse reinforcement was provided. Also, the number of inelastic load cycles sustained by a tension splice depends on the amount of confinement provided by transverse reinforcement; and

4. The strength of a splice is independent of the splice-bar spacing up to at least six times the bar diameter for monotonic loading. Under repeated loading up to the yield strength of the splice bars, the maximum load (equal to the yield load) is also independent of bar spacing, up to eight times the bar diameter for both No. 6 and No. 8 (No. 19 and No. 25 mm) bars.

To check the validity of the ACI 318 provisions for noncontact splices, Hamad and Mansour (1996) tested 17 slabs in positive bending. Reinforcement on the tension side consisted of three reinforcing bars spliced at the center of the span. No transverse reinforcement was provided in the splice region. The clear spacing between lapped bars varied between 0 and 50% of the splice length for slabs reinforced with 0.55 and 0.63 in. (14 and 16 mm*) bars, and between 0 and 40% of the splice length for 0.80 in. (20 mm) bars. For eight specimens, the clear transverse spacing between the spliced bars was at least 20% of the splice length specified by ACI 318. The splice lengths were limited to 11.8 in. (300 mm) for slabs reinforced with 0.55 and 0.63 in. (14 and 16 mm) bars, and 13.8 in. (350 mm) for 0.8 in. (20 mm) bars. They observed that the noncontact splices developed greater bond strength than the contact splices, up to an optimum clear spacing of about $5d_b$. They concluded that the ACI limit on the transverse spacing of noncontact tensile lap

*The bars used in these tests were sized in millimeters.

splices to 20% of the lap length is conservative and that a limit of 30% could be used.

2.2—Bar properties

2.2.1 *Bar size*—The relationship between bar size and bond strength is not always appreciated. The reason is that, while (1) a longer development or splice length is required as bar size increases, and (2) for a given development or splice length, larger bars achieve higher total bond forces than smaller bars for the same degree of confinement.

Addressing the second point first, for a given bonded length, larger bars require larger forces to cause either a splitting or pullout failure, as illustrated in Fig. 2.2, for bars not confined by transverse reinforcement. The result is that the total force developed at bond failure is not only an increasing function of concrete cover, bar spacing, and bonded length, but also of bar area (Orangun, Jirsa, and Breen 1977; Darwin et al. 1992, 1996a). The bond force at failure, however, increases more slowly than the bar area, which means that a longer embedment length is needed for a larger bar to fully develop a given bar stress (the first point). When evaluated in terms of bond stress (Eq. (1-4) and (1-5)), smaller bars appear to have even a greater advantage; thus, conventional wisdom suggests that it is desirable to use a larger number of small bars rather than a smaller number of large bars; this is true until bar spacings are reduced to the point that bond strength is decreased (Ferguson 1977; Rehm and Eligehausen 1979).

The size of a developed bar also plays an important role in the contribution of confining transverse reinforcement to bond strength. As larger bars slip, higher strains and, thus, higher stresses, are mobilized in the transverse reinforcement, providing better confinement. As a result, the added bond strength provided by transverse reinforcement increases as the size of the developed or spliced bars increases, as shown in Fig. 2.9, which compares the relative effect of transverse reinforcement on bond strength M, normalized to the effect of the relative rib area t_r, versus bar diameter, for No. 5, No. 8, and No. 11 (No, 16, No. 25, and No. 36) bars with a wide range of relative rib area R_r. The term M is the ratio of the additional bond force provided by the transverse steel T_s, normalized with respect to $f_c'^{3/4}$ (see Section 2.3.2), to the area of transverse steel confining the bar.

2.2.2 *Bar geometry*—Historically, there have been widely conflicting views of the effect of bar geometry on bond strength. Some studies indicate that deformation patterns have a strong influence on bond strength. Other studies show that deformation pattern has little influence, and it is not uncommon for bars with different patterns to produce nearly identical development and splice strengths. Over time, however, the effects of bar geometry on bond behavior have become increasingly clear, as will be described in this section.

The earliest study on bond resistance of plain and deformed reinforcing bars was done by Abrams (1913) using pullout and beam specimens. The test results showed that deformed bars produced higher bond resistance than plain (smooth) bars. Abrams found that in pullout tests of plain bars, bond resistance reached its maximum value at a loaded

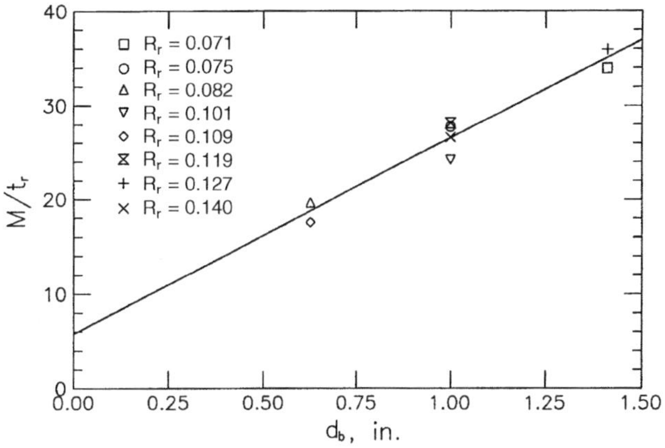

Fig. 2.9—Relative contribution of transverse reinforcement M*, normalized with respect to the relative rib area* (t_r = 9.6R_r + 0.28) *versus bar diameter (Zuo and Darwin 1998). (Note: 1 in. = 25.4 mm)*

end slip of about 0.01 in. (0.25 mm). For deformed bars, the load-slip performance was the same as for plain bars up to the slip corresponding to the maximum bond resistance of the plain bars. As slip continued, the ribs on deformed bars increased the bond resistance by bearing directly on the adjacent concrete. Abrams observed that the ratio of the bearing area of the projections (projected area measured perpendicular to the bar axis) to the entire surface area of the bar in the same length could be used as a criterion for evaluating the bond resistance of deformed bars. To improve bond resistance, he recommended that this ratio not be less than 0.2, resulting in closer spacings of the projections than were used in commercial deformed bars at the time.

Over 30 years later, Clark (1946, 1949) investigated 17 commercial deformation patterns using pullout and beam tests. The bond performance for each pattern was evaluated based on the bond stress developed at predetermined values of slip. Based on Clark's investigations, standard deformation requirements were introduced for the first time in the Tentative Specification ASTM A 305-47T, later appearing as ASTM A 305-49. The requirements included a maximum average spacing of deformations equal to 70% of the nominal diameter of the bar and a minimum height of deformations equal to 4% of the nominal diameter for bars with a nominal diameter of 1/2 in. (13 mm) or smaller, 4.5% of the nominal diameter for bars with a nominal diameter of 5/8 in. (16 mm), and 5% for larger bars. These requirements remain unchanged in the current ASTM specifications for reinforcing bars (ASTM A 615; ASTM A 706; ASTM A 767; ASTM A 955; ASTM A 996).

In his study, Clark (1946, 1949) found that bond performance improved for bars with lower ratios of shearing area (bar perimeter times center-to-center distance between ribs) to bearing area (projected rib area normal to the bar axis) and recommended that the ratio of shearing area to bearing area be limited to a maximum of 10 and, if possible, 5 or 6. In current practice, this criterion is described in terms of the inverse ratio, that is, the ratio of the bearing area to the shearing area, which is known alternately as the rib area, related rib area, or relative rib area (DIN 488; Soretz and Holzenbein 1979; ACI 408.3). Relative rib area R_r is the term used in U.S. practice (ASTM A 775, ASTM A 934, ASTM A 944, ACI 408.3).

$$R_r = \frac{\text{projected rib area normal to bar axis}}{\text{nominal bar perimeter} \times \text{center-to-center rib spacing}} \quad (2\text{-}6)$$

Clark's recommendations then become a minimum value of R_r equal to 0.1, with desirable values of 0.2 or 0.17, which are not so different from Abrams' (1913) recommendations. These later recommendations, however, were not incorporated in the ASTM requirements, so that the typical values of relative rib area for bars currently used in the U.S. range between 0.057 and 0.087 (Choi et al. 1990).

Rehm (1961) reported that one of two failure modes, splitting or pullout, can occur when a reinforcing bar slips with respect to the concrete. If the ratio of rib spacing to rib height was greater than 10 and the rib face angle (the angle between the face of the rib and the longitudinal axis of the bar, *a* in Fig. 1.2(a)) is greater than 40 degrees, he observed that the concrete in front of the rib crushes, forming wedges and then inducing tensile stress perpendicular to the bar axis. This results in transverse cracking and splitting of surrounding concrete. If the ribs had a spacing to height ratio less than 7, with a rib face angle greater than 40 degrees, he observed that the concrete in front of ribs gradually crushes, causing a pullout failure.

Lutz, Gergely, and Winter (1966), and Lutz and Gergely (1967) found that for a deformed bar with a rib face angle greater than 40 degrees, slip occurs by progressively crushing concrete in front of the ribs, producing a region of crushed concrete with a face angle of 30 to 40 degrees, which acts as a wedge. Lutz, Gergely, and Winter also showed that no crushing of concrete occurs if the rib face angle is less than 30 degrees. These observations were supported by Skorobogatov and Edwards (1979). Based on tests using bars with face angles of 48.5 and 57.8 degrees, Skorobogatov and Edwards showed that these differences in face angle do not affect bond strength because the high face angle is flattened by crushed concrete in front of the ribs.

Losberg and Olsson (1979) tested three commercial deformation patterns used in Sweden and some machined bars with different values of rib spacing and rib height. They found that the bond forces produced by the three patterns were obviously different in pullout tests in which a pullout failure governed. If splitting failure governed, however, as in beam-end and ring pullout tests, there was little difference in the bond forces obtained using the three patterns. Losberg and Olsson concluded that pullout tests are not suitable to study bond performance because the state of stress in a pullout test resulting from the additional confinement provided to the concrete does not represent the state of stress in actual structures where splitting failure typically governs. Their test results also showed that the splitting force is not sensitive to rib spacing and that ribs oriented perpendicular to the longitudinal axis of the bar give slightly higher splitting force than inclined ribs.

Soretz and Holzenbein (1979) studied the effect of rib height and spacing, rib inclination, and the cross-sectional shape of ribs on bond. Three bars were machined with different rib heights and spacings but with the same rib bearing area per unit length. In pullout tests, Soretz and Holzenbein found that, for the three patterns, the bond forces showed no significant differences up to 0.04 in. (1 mm) of slip. Once the slip was greater than 0.04 in. (1 mm), however, the bond force for the bar with the lowest rib height was about 20% smaller than that of the other two patterns. They recommended a combination of a minimum rib height of 0.03 bar diameters and a rib spacing of 0.3 bar diameters as the optimum geometry for deformed bars to limit splitting and to increase bond strength.

Darwin and Graham (1993a,b) studied the effect of deformation pattern on bond strength using beam-end specimens. The principal parameters in the study were rib height, rib spacing, relative rib area, and degree of confinement from concrete cover and transverse reinforcement. Both specially machined and conventional 1 in. (25 mm) diameter bars were used in the study. The machined bars had three different rib heights, 0.050, 0.075, and 0.100 in. (1.27, 1.91, and 2.54 mm), with center-to-center rib spacings ranging from 0.263 to 2.2 in. (6.68 to 55.9 mm), producing relative rib areas of 0.20, 0.10, and 0.05. Darwin and Graham concluded that bond strength is independent of deformation pattern if the bar is under relatively low confinement (small concrete cover and no transverse reinforcement) and bond strength is governed by a splitting failure in the concrete. If additional confinement is provided by transverse reinforcement, however, bond strength increases with an increase in relative rib area. They found that the bond force-slip response of bars is related to the relative rib area of the bars, but independent of the specific combination of rib height and spacing. The initial stiffness of the load-slip curve increases with an increase in relative rib area. Darwin and Graham also observed that, when tested in beam-end specimens, bars with the longitudinal ribs oriented in a plane parallel to the splitting cracks provide higher bond strength than bars with the longitudinal ribs oriented in a plane perpendicular to the splitting cracks.

Cairns and Jones (1995) investigated 14 different bar geometries using lapped bar test specimens. The lapped bars were confined by stirrups. The relative rib area R_r of the tested bars ranged from 0.031 to 0.090. The inclination of the transverse ribs with respect to the longitudinal axis of the bar varied from 40 to 90 degrees and the rib face angle varied from 28 to 51 degrees. Bars were placed in two ways, either alignment A0 (with the plane of two longitudinal ribs parallel to the concrete splitting face) or alignment A90 (with the plane of longitudinal ribs perpendicular to the concrete splitting face). Cairns and Jones reported that there were no significant effects of rib inclination and rib face angle on bond strength, but that, as observed by Darwin and Graham (1993a,b), the alignment of longitudinal ribs influenced bond strength: the bond force for alignment A0 was higher than for alignment A90. They also found that relative rib area plays an important role in bond strength. The test results indicated that doubling relative rib area could reduce development and splice length by 20%.

Darwin et al. (1996b), Tan et al. (1996), and Zuo and Darwin (1998, 2000) used splice and beam-end specimens to study the effect of relative rib area on bond strength. The tests in these studies involved commercially produced high R_r reinforcing bars with relative rib areas ranging from 0.101 to 0.141 and conventional bars with relative rib areas ranging from 0.068 to 0.087. The tests included top and bottom-cast bars plus specimens to study the effect of relative rib area on the splice strength of epoxy-coated bars. The test results indicated that the splice strength of uncoated bars is not affected by the deformation pattern if the bars are not confined by transverse reinforcement. For bars confined by transverse reinforcement, splice strength increases with increasing bar diameter and relative rib area. As shown in Fig. 2.9 and Eq. (2-5) for d_b and Eq. (2-4) for R_r, the contribution of transverse reinforcement to bond strength T_s increases linearly with increases in d_b and R_r.

For epoxy-coated bars under all conditions of confinement, bond strength increases with relative rib area (Darwin et al. 1996b; Tan et al. 1996; Zuo and Darwin 1998). For bars with $R_r \geq 0.10$ and concrete with $f_c' \leq 10{,}000$ psi (69 MPa), development or splice length should be increased by 20%, in contrast to the 50% increase in length needed for conventional reinforcement. For $f_c' > 10{,}000$ psi (69 MPa), a 50% increase in development or splice length appears warranted, even for high R_r bars (Zuo and Darwin 1998).

2.2.3 *Steel stress and yield strength*—For a number of years, concern existed that bars that yielded before bond failure produced average bond stresses significantly lower than higher strength steel in similar test specimens that did not yield (Orangun, Jirsa, and Breen 1975). As a result, test specimens were often deliberately configured to ensure that the bars did not yield prior to bond failure.

As it turns out, the bond strengths of bars that yield average only about 2% less when not confined by transverse reinforcement and about 10% greater when confined by transverse reinforcement than similar bars with the same bonded lengths made of higher strength steel that does not yield (Darwin et al. 1996a; Zuo and Darwin 1998, 2000).

2.2.4 *Bar surface condition*—The surface of a reinforcing bar plays an important role in bond because of its effect on the friction between reinforcing steel and concrete and the ability of ribs to transfer force between the two materials. Bar surface condition involves the cleanliness of reinforcement, the presence of rust on the bar surface, and the application of epoxy coatings to protect the reinforcement from corrosion.

2.2.4.1 *Bar cleanliness*—To prevent a reduction in the bond strength, ACI 318 requires that reinforcement must be free of mud, oil, and other nonmetallic coatings that decrease bond strength. Based on the work of Kemp, Brezny, and Unterspan (1968), steel reinforcement with rust, mill scale, or a combination of the two is considered satisfactory provided that the minimum dimensions, including the height of the deformations, and the weight of a hand-wire-brushed test specimen, comply with the applicable ASTM specifications.

2.2.4.2 *Epoxy-coated bars*—Epoxy coatings are used to improve the corrosion resistance of reinforcing bars. Under the provisions of ASTM A 775 and ASTM A 934, at least 90% of coating thickness measurements must be between 7 and 12 mils (175 and 300 μm). Bars are rejected if more than 5% of the coating thickness measurements are below 5 mils (125 μm). Epoxy coatings tend to reduce bond strength.

In the earliest study on the bond of epoxy-coated bars, Mathey and Clifton (1976) investigated the effect of coating thickness on bond strength using pullout tests. For bars with epoxy coatings between 1 to 11 mils (25 to 280 μm) in thickness, bond strength was about 6% lower than for uncoated bars. For bars with a coating thickness of 25 mils (635 μm), however, the peak bond force was between 34 and 60% of the strength of uncoated bars.

Johnston and Zia (1982) studied the effect of epoxy coating on bond strength using slab and beam-end specimens. Coating thickness was between 6.7 and 11.1 mils (170 to 282 μm). The bars were confined by transverse reinforcement. The slab specimens with coated bars had slightly larger deflections and wider cracks than those with uncoated bars. Compared with uncoated bars, the bond strength of coated bars was about 4% lower for the slab specimens and 15% lower for the beam-end specimens. Johnston and Zia recommended an increase of 15% in the development length when coated bars are used in place of uncoated bars.

Treece and Jirsa (1989) tested 21 beam-splice specimens without transverse reinforcement in the splice region. Twelve of the specimens contained epoxy-coated bars with coating thicknesses between 4.5 and 14 mils (114 to 356 μm). Seventeen specimens contained top-cast bars; four contained bottom-cast bars. Concrete strength ranged from 3860 to 12,600 psi (27 to 87 MPa). An average reduction in bond strength of 34% compared to uncoated bars was observed in the tests. The deformation patterns used in the study were discontinued shortly after the tests due to difficulties in obtaining a uniform coating thickness. The work by Treece and Jirsa is the basis of the development length modification factors for epoxy-coated bars in ACI 318 and the AASHTO bridge specifications. In ACI 318, development length is multiplied by a factor of 1.5 for epoxy-coated bars with a cover of less than $3d_b$ or clear spacing between bars less than $6d_b$ and a factor of 1.2 for other cases, with a maximum of 1.7 for the product of top-bar factor and epoxy-coating factors. In the AASHTO bridge specifications (1996), the three factors are 1.5, 1.15, and 1.7, respectively. The 1.2 (ACI) and 1.15 (AASHTO) factors were selected based on the work of Johnston and Zia (1982).

DeVries, Moehle, and Hester (1991) and Hadje-Ghaffari et al. (1992, 1994) found the maximum of 1.7 for the product of the top factor and epoxy-coating factor to be too conservative and recommended a value of 1.5.

Choi et al. (1990, 1991) studied the roles of coating thickness, bar size, and deformation pattern on the bond strength of epoxy-coated bars. They observed that coating thickness has little effect on the reduction in bond strength due to epoxy coating for No. 6 (No. 19) and larger bars. For No. 5 (No. 16) and smaller bars, however, the bond strength ratio of coated to uncoated bars (the C/U ratio) decreases with increasing coating thickness. Their tests also showed that, in general, the C/U ratio decreases as bar size increases, and epoxy coating is less detrimental to the bond strength of bars with higher relative rib areas. The average bond strength ratio for epoxy-coated bars to uncoated bars, C/U, was observed to be 0.82 for 15 splice specimens. Cleary and Ramirez (1993) obtained similar results for bars in slabs.

In a small study (12 splice specimens), Hamad and Jirsa (1993) observed that an increase in confinement provided by transverse reinforcement reduced the negative impact of epoxy coating on bond strength. Subsequent studies representing more than 140 splice specimens (Hester et al. 1993; Darwin et al. 1996b; Tan et al. 1996; Zuo and Darwin 1998), however, demonstrated no measurable effect of transverse reinforcement on the C/U ratio.

Idun and Darwin (1999) found that epoxy coating is less detrimental to bond strength for high relative rib area bars, matching the results of Choi et al. (1991). Idun and Darwin also measured the coefficient of friction of both uncoated and coated reinforcing steel, obtaining values of 0.56 and 0.49, respectively. These values are similar to values of 0.53 and 0.49 for uncoated and coated steel plates, respectively (Cairns and Abdullah 1994). Using the results of the coefficient of friction tests and a theoretical relation between C/U ratio and rib face angle developed by Hadje-Ghaffari, Darwin, and McCabe (1991), Idun and Darwin observed that epoxy coating should cause the least reduction (theoretically no reduction) in bond strength for rib face angles greater than 43 degrees, a finding supported by their experimental results. Rib face angles steeper than 40 degrees, however, are hard to produce in practice.

Tan et al. (1996) and Zuo and Darwin (1998) found, for concrete with $f_c' \leq 10{,}000$ psi (69 MPa), that an increase in the relative rib area improves the splice strength of epoxy-coated bars relative to uncoated bars, whether or not the splices are confined by transverse reinforcement. The presence of transverse reinforcement does not affect the relative splice strength. The relative splice strengths of coated high R_r bars in concrete with $f_c' > 10{,}000$ psi (69 MPa) were increased less than for the same bars in lower strength concrete. For normalweight concrete, they recommended the use of a development length modification factor of 1.2 for epoxy-coated high relative rib area bars ($R_r \geq 0.10$) in concrete with $f_c' \leq 10{,}000$ psi (69 MPa) and 1.5 for conventional bars for all concrete strengths and for high R_r bars in concrete with $f_c' > 10{,}000$ psi (69 MPa). These factors apply for all values of cover and bar spacings.

2.3—Concrete properties

A number of concrete properties affect bond strength. While compressive strength and the use of lightweight concrete are normally considered in design, other properties, such as tensile strength and fracture energy, aggregate type and quantity, the use of admixtures, concrete slump, and fiber reinforcement, also play a role. Each of these will be discussed in this section.

2.3.1 *Compressive strength*—Traditionally, in most descriptive (Tepfers 1973; Orangun, Jirsa, and Breen 1977; Darwin et al. 1992; Esfahani and Rangan 1998a,b) and design (ACI 318; AASHTO; CEB-FIP) expressions, the effect of concrete properties on bond strength is represented using the square root of the compressive strength $\sqrt{f_c'}$. This representation has proven to be adequate as long as concrete strengths remain below about 8000 psi (55 MPa). For higher-strength concrete, however, the average bond strength at failure, normalized with respect to $\sqrt{f_c'}$, decreases with an increase in compressive strength (Azizinamini et al. 1993; Azizinamini, Chisala, and Ghosh 1995; and Zuo and Darwin 1998, 2000; Hamad and Itani 1998). Azizinamini et al. (1993) and Azizinamini, Chisala, and Ghosh (19995) observed that the rate of decrease becomes more pronounced as splice length increases. They noted that the bearing capacity of concrete (related to f_c') increases more rapidly than tensile strength (related to $\sqrt{f_c'}$) as compressive strength increases. For high-strength concrete, the higher bearing capacity prevents crushing of the concrete in front of the bar ribs (as occurs for normal-strength concrete), which reduces local slip. They concluded that because of the reduced slip, fewer ribs transfer load between the steel and the concrete, which increases the local tensile stresses and initiates a splitting failure in the concrete before achieving a uniform distribution of the bond force. Esfahani and Rangan (1996) observed that, when no confining transverse reinforcement is used, as concrete strength increases, the degree of crushing decreases, with no concrete crushing observed for $f_c' \geq 11{,}000$ psi (75 MPa). In contrast to the other studies, they found that the average bond stress at failure, normalized with respect to $\sqrt{f_c'}$, is higher for higher-strength concrete than for normal-strength concrete.

The use of $\sqrt{f_c'}$ has not been universal. Zsutty (1985) found that $f_c'^{1/3}$ provided an improved match with data, compared to $\sqrt{f_c'}$. Darwin et al. (1996a) combined their own test results with a large international database and observed that a best fit with existing data was obtained using $f_c'^{1/4}$ to represent the effect of concrete compressive strength on development and splice strength. That work was continued by Zuo and Darwin (1998, 2000), who added significantly to the number of tests with high-strength concrete and incorporated additional tests into the database. Zuo and Darwin also observed that $f_c'^{1/4}$ provides the best representation for the effect of compressive strength on the concrete contribution to bond strength T_c. The ability of $f_c'^{1/4}$ to represent the effect of concrete strength on the concrete contribution T_c is demonstrated in Fig. 2.10, which is based on comparisons with 171 test specimens with bottom-cast bars not confined by transverse reinforcement. Two best-fit lines are shown, comparing test-prediction ratios versus f_c' based on two optimized descriptive expressions for bond strength, one using $f_c'^{1/2}$ and the other using $f_c'^{1/4}$. The best-fit line based on $f_c'^{1/2}$ has a negative slope, decreasing as f_c' increases, while the best-fit line based on $f_c'^{1/4}$ has nearly a horizontal slope, indicating that the 1/4 power provides an unbiased representation of the effect of concrete strength on bond strength. As will be demonstrated in Section 3.6, the advantage of the 1/4 power over the 1/2 power does not depend on the specific expressions used for this comparison.

For bars confined by transverse reinforcement, Zuo and Darwin (1998, 2000) found that $f_c'^{1/4}$ significantly underestimates the effect of concrete strength on the additional bond strength provided by transverse reinforcement T_s. They observed that $f_c'^{3/4}$ provides a good representation of the influence of compressive strength on bond strength. Figure 2.11 shows best-fit lines of test-prediction ratios based on $f_c'^p$, with p = 1/4, 1/2, 3/4, and 1, versus f_c'. Of the four values, $f_c'^{3/4}$ provides a nearly horizontal best-fit line. Using f_c' (p = 1) overestimates the effect of concrete strength on T_s, while using $f_c'^{1/2}$ underpredicts the effect of concrete strength on T_s. The small positive slope for the $f_c'^{3/4}$ line indicates that a power slightly greater than 3/4 would provide a slightly better match with the data.

The observation that $f_c'^{1/2}$ does not accurately represent the effect of concrete strength on bond means that many earlier interpretations of the effects of parameters other than compressive strength on bond strength need to be re-examined. This reexamination is necessary because test results have often been normalized with respect to $f_c'^{1/2}$ to compare results for different concrete strengths. For example, changes in concrete properties, such as caused by the addition of a high-range water-reducing admixture or the use of silica fume as a cement replacement, often result in changes in compressive strength. When bond strengths are normalized with respect to $f_c'^{1/2}$, the effect of concrete strength is exaggerated, resulting in an overestimation of bond strength for higher strength concretes. A reexamination of earlier test results often indicates much less of an effect and, in some cases, no effect on bond strength due to changes in mixture proportions.

2.3.2 *Aggregate type and quantity*—For bars not confined by transverse reinforcement, Zuo and Darwin (1998, 2000) observed that a higher-strength coarse aggregate (basalt) increased T_c by up to 13% compared with a weaker coarse aggregate (limestone). This observation was explained based on studies using the same materials (Kozul and Darwin 1997; Barham and Darwin 1999) that showed that concrete containing the basalt had only slightly higher flexural strengths, but significantly higher fracture energies (more than two times higher) than concrete of similar compressive strength containing limestone for compressive strengths between 2900 and 14,000 psi (20 and 96 MPa). The higher fracture energy provided by the basalt resulted in increased resistance to crack propagation, which delays splitting failure and increases bond strength. Zuo and Darwin observed no effect of coarse aggregate quantity on T_c.

For bars confined by transverse reinforcement, increases in both the strength and the quantity of coarse aggregate have been observed to increase the contribution of transverse reinforcement to bond strength (Darwin et al. 1996b; Zuo and Darwin 1998), with differences in T_s as high as 45%. The effects of aggregate strength and quantity on T_s explain some of the wide scatter observed for test results obtained in

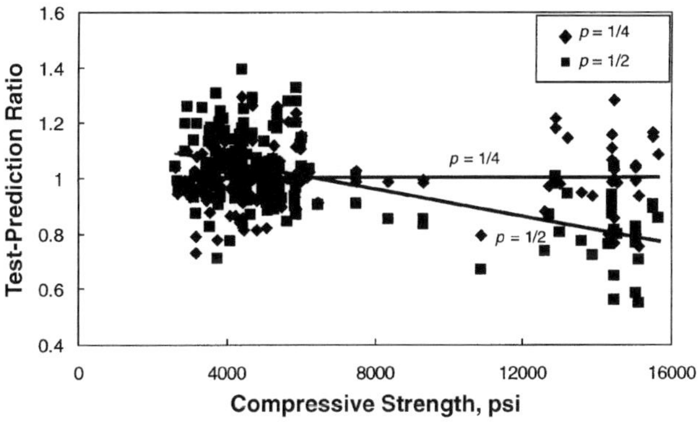

Fig. 2.10—Variation of test-prediction ratio versus compressive strength for developed/spliced bars not confined by transverse reinforcement. The contribution of concrete to bond strength is characterized as $f_c'^p$*, with* p = *1/4 and 1/2. (Note: 1 psi = 0.00689 MPa)*

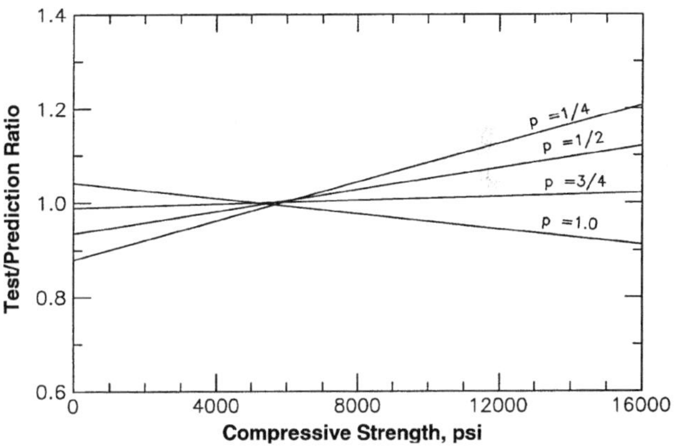

Fig. 2.11—Best-fit lines for test-prediction ratios versus compressive strength for developed/spliced bars confined by transverse reinforcement. $f_c'^p$ *is used to represent the influence of compressive strength on the additional bond strength provided by transverse reinforcement* T_s. *(Note: 1 psi = 0.00689 MPa)*

different studies, where the scatter in values of T_s far exceeds the scatter observed for T_c.

2.3.3 *Tensile strength and fracture energy*—The observed effects of aggregate strength and quantity and of concrete compressive strength on bond strength strongly indicate that the tensile properties of concrete play a significant role in determining bond strength. The concrete contribution T_c increases approximately with $f_c'^{1/4}$. This contrasts with the relationship between compressive strength and tensile strength, where it is generally agreed that tensile strength increases approximately with $f_c'^{1/2}$. [In some studies dealing with high-strength concrete, a power higher than 1/2 has been observed to relate f_c' to tensile strength (Ahmad and Shah 1985; Kozul and Darwin 1997)].

If tensile strength alone were the key governing factor in bond strength, $f_c'^{1/2}$ should provide a good representation of the relationship between compressive strength and bond strength, and aggregate strength should have little effect on T_c. The actual relationships appear to be directly related to the fracture energy of concrete. As observed earlier, higher-strength aggregates produce concrete with both higher fracture energy and higher bond strengths. For both high and low-strength aggregates, however, fracture energy increases very little, and, in fact, may decrease as compressive strength increases (Niwa and Tangtermsirikul 1997; Kozul and Darwin 1997; Barham and Darwin 1999; Darwin et al. 2001). Overall, as concrete compressive strength increases, bond strength increases at a progressively slower rate, while the failure mode becomes more brittle. Higher fracture energy, such as may be provided by high-strength fibers, should increase the bond strength of reinforcement (see Section 2.3.7).

2.3.4 *Lightweight concrete*—Due to the lower strength of the aggregate, lightweight concrete should be expected to have lower tensile strength, fracture energy, and local bearing capacity than normalweight concrete with the same compressive strength. As a result, the bond strength of bars cast in lightweight concrete, with or without transverse reinforcement, is lower than that of bars cast in normalweight concrete.

Previous reports by Committee 408 (1966, 1970) have emphasized the paucity of experimental data on the bond strength of reinforced concrete elements made with lightweight aggregate concrete. ACI 318 includes a factor for development length of 1.3 to reflect the lower tensile strength of lightweight-aggregate concrete, when compared with normalweight concrete with the same compressive strength, and allows that factor to be taken as $6.7\sqrt{f_c'}/f_{ct} \geq 1.0$ if the average splitting strength f_{ct} of the lightweight-aggregate concrete is specified. Although design provisions, in general, require longer development lengths for lightweight-aggregate concrete (CEB-FIP 1999), test results from previous research are contradictory, in part, because of the different characteristics associated with different aggregates and mixture designs.

The majority of published experimental results found in the literature are from different configurations of pullout tests. Early research by Lyse (1934), Petersen (1948), and Shideler (1957) concluded that the bond strength of reinforcing steel in lightweight-aggregate concrete was comparable to that of normalweight concrete. Lyse conducted pullout tests of 3/4 in. (19 mm) bars embedded in 6 x 12 in. (150 x 300 mm) cylinders. The mixture designs used by Lyse included natural sand for fine aggregate and gravel or slag for coarse aggregate. Among his conclusions, Lyse stated that "the compressive, bond, flexural, and diagonal tension strengths of the concrete were very nearly the same for slag and for gravel aggregates." Petersen tested beams made with expanded shale and expanded slag and concluded that the bond strength of reinforcement in lightweight-aggregate concrete was comparable to that of reinforcement in normalweight concrete. The tests by Petersen used No. 8 (25 mm) bars with embedment lengths of 10, 20, and 30 in. (250, 510, and 760 mm). Shideler (1957) conducted pullout tests on 9 in. (230 mm) cube specimens with six different types of aggregates. No. 6 (19 mm) bars were embedded in specimens with compressive strengths of 3000 and 4500 psi (21 and 31 MPa), and No. 9 (29 mm) bars were used in 9000 psi (62 MPa) specimens. Although the bond strength of normalweight concrete specimens was slightly higher than that of lightweight concrete specimens, Shideler stated that the difference was not significant.

Similar behavior has been observed in more recent studies. Based on a series of pullout tests, Martin (1982) concluded that there was no significant difference between the bond strength in normalweight and lightweight-aggregate concrete. Berge (1981) obtained similar results from pullout tests; although in a limited testing program involving beams, he observed lower bond strengths in specimens made with lightweight-aggregate concrete. The observed difference in bond strength was approximately 10%.

Clarke and Birjandi (1993) used a specimen developed by the British Cement Association (Chana 1990) and tested four lightweight aggregates with various densities available in the United Kingdom: Lytag (sintered pulverized fuel ash), Leca (expanded clay), Pellite (pelletized expanded blast furnace slag), and Fibo (expanded clay). In addition to the type of aggregate, the study investigated the effect of casting position. The fine aggregate in all mixtures was natural sand. Test results indicated that, with the exception of the lightest aggregate (Fibo), all specimens had higher bond strengths than those of specimens made with normalweight aggregate. This behavior was partially attributed by the authors to the fact that natural sand, as opposed to lightweight aggregate, was used as fine aggregate.

In contrast to the studies just described, there are several studies that indicate significant differences between bond strengths in lightweight- and normalweight-aggregate concrete. In pullout tests, Baldwin (1965) obtained bond strengths for lightweight concrete that were only 65% of those obtained for normalweight concrete. These results contradicted the prevailing assumption at the time that bond strength in lightweight-aggregate concrete was similar to that of normalweight concrete (ACI Committee 408 1966).

Robins and Standish (1982) conducted a series of pullout tests to investigate the effect of lateral stresses on the bond strength of plain and deformed bars in specimens made with lightweight-aggregate (Lytag) concrete. As the lateral pressure applied to the specimens increased, the mode of failure changed from splitting to pullout. Bond strength increased with confining pressure for both normalweight and lightweight concrete. For specimens that failed by splitting, bond strength was 10 to 15% higher for normalweight concrete than for lightweight concrete. When the lateral pressure was large enough to prevent a splitting failure, however, the difference in bond strength was much higher, on the order of 45%.

Mor (1992) tested No. 6 (19 mm) bars embedded in 3 x 3 x 20 in. (76 x 76 x 508 mm) pullout specimens to investigate the effect of condensed silica fume on the bond strength of normalweight and lightweight-aggregate concrete. His specimens had compressive strengths of about 10,000 psi (70 MPa). In specimens without silica fume, the maximum bond stress for specimens made with lightweight concrete was 88% of that of specimens made with normalweight concrete. For concrete with 13 to 15% condensed silica fume, the ratio was 82%. The specimens made with lightweight concrete developed splitting failures at 75 to 80% of the slip of specimens made with normalweight concrete. The use of silica fume had little effect on bond strength, with an increase of 2% for normalweight concrete and a decrease of 5% for lightweight concrete.

Overall, the data indicate that the use of lightweight concrete can result in bond strengths that range from nearly equal to 65% of the values obtained with normalweight concrete.

2.3.5 *Concrete slump and workability admixtures*—The workability of concrete, generally measured by slump, affects the bond strength between concrete and reinforcing steel (Darwin 1987). After concrete is cast, it continues to settle and bleed. Settlement leaves a void below rigidly held bars. Bleed water collects below bars, whether rigidly held in place or not. The higher the concrete slump, the greater the tendency to settle and bleed. Water reducers and high-range water-reducing admixtures extend the time during which settlement and bleeding occur.

Properly consolidated, low-slump concrete usually provides the best bond with reinforcing steel. For normal-strength concrete, high slump, used primarily where it is desirable to use little or no consolidation effort, results in decreased bond. The bond strength of top-cast bars (bars near the upper surface of a concrete placement) appears to be especially sensitive to slump. Top-cast bars may or may not be top reinforcement, defined as "horizontal reinforcement so placed that more than 12 in. (300 mm) of fresh concrete is cast in the member below the reinforcement" (ACI 318).

Menzel (1952) studied the effect of slump on bond strength for top-cast bars. He observed a marked reduction in bond strength as the height of top-cast bars was increased from 2-1/8 to 33-1/8 in. (54 to 841 mm) when using 5 to 6 in. (127 to 152 mm) slump, hand-rodded concrete. The rate of decrease in bond strength with the increasing height of the top bars was greatly reduced by decreasing the concrete slump to a range of 2 to 3 in. (51 to 76 mm).

Zekany et al. (1981) studied the effect of concrete slump on top-cast and bottom-cast splices. They found that the bond strength of both top-cast and bottom-cast bars decreased with increasing slump. The effect was most pronounced for the top-cast bars. Luke et al. (1981) studied the influence of casting position on development and splice strength using 72 in. (1.83 m) deep wall specimens (Fig. 2.6). As shown in Fig. 2.6, the bond of top-cast reinforcement was reduced significantly when high (8-1/2 in. [215 mm]) slump concrete was used, as compared to when low (3 in. [75 mm]) slump concrete was used. The bond strength of these bars decreased 40 to 50% due to the increase in slump alone.

Donahey and Darwin (1983, 1985) studied the bond strength of top-cast bars in bridge decks. The bars had different amounts and types of cover. These included two monolithic covers, 3/4 and 3 in. (19 and 75 mm), and a laminar cover consisting of a 3/4 in. (19 mm) monolithic concrete topped with a 2-1/4 in. (57 mm) high-density concrete overlay. Eight inches (203 mm) of concrete was used below the reinforcement. Increasing slump resulted in decreased bond strength.

High-slump concrete can be obtained in a number of ways. It can be obtained by the addition of water, in which case the strength of the concrete is reduced. It can be obtained by the addition of water and cement, in which case the strength of the concrete remains approximately constant. Or, it can be obtained by the addition of a high-range water-reducer or superplasticizer, in which case the strength is usually increased.

Brettmann, Darwin, and Donahey (1984, 1986) used beam-end specimens, with three depths and concretes varying in slump from 1-3/4 to 9 in. (44 to 229 mm). The key variables were degree of consolidation, concrete slump with and without high-range water-reducing admixture, concrete temperature, and bar position. Based on their work, Brettmann, Darwin, and Donahey (1986) concluded that, if cast at a relatively high temperature (resulting in a short setting time), properly vibrated (ACI 309R) high-slump superplasticized concrete and its low-slump nonsuperplasticized base concrete (the concrete before the superplasticizer is added) provide approximately the same bond strength. The equal bond strength is due largely to the increased concrete strength obtained with the high-range water-reducing admixture. Brettmann, Darwin, and Donahey, however, observed that for concrete with the same strength, high-slump concrete made with a high-range water-reducer has lower bond strength than low-slump concrete; the observed differences varied widely, but averaged about 10%. Brettmann, Darwin, and Donahey also observed that if high-slump, superplasticized concrete is cast at a low temperature (resulting in a longer setting time), or if high-slump, high cement content, nonsuperplasticized concrete is used, bond strength decreases regardless of concrete strength.

Musser et al. (1985) and Zilveti et al. (1985) also studied the effect of high-slump, superplasticized concrete on bond strength. Specimens were consolidated using an internal vibrator. Musser et al. considered the anchorage of deformed bars in wall specimens. Bond strength was measured using a straight pull-out procedure. Zilveti et al. used beam-end specimens. Musser et al. found little effect of superplasticizers on bond strength. Zilveti et al. concluded that the addition of a high-range water-reducing admixture simply to improve workability has no effect on bond strength. When corrected for increased concrete strength, however, the bond strength obtained with high-slump, superplasticized concrete was lower than the bond strength obtained with low-slump concrete. Also, like Brettmann, Darwin, and Donahey, Zilveti et al. observed that high slump is more detrimental to bond strength when the temperature of the concrete is initially low. The lower temperature provides a longer period during which the concrete remains plastic and, thus, a longer period during which settlement and bleeding occur.

Hamad and Itani (1998) studied the effects of silica fume, high-range water-reducing admixture dosage, and bar position on splice strength in high-strength concrete. Due to the range of mixture proportion variables in the study, the splice strength of individual specimens varied significantly. When normalized with respect to the $f_c'^{1/4}$, doubling the high-range water-reducing admixture dosage from 0.5 to 1 gal./yd^3 (2 to 4 L/m^3) resulted in a 32% decrease in splice strength for one pair of bottom-cast bars as the slump increased from 1-1/2 to

8-1/4 in. (40 to 210 mm). For two other pairs of bottom-cast bars, splice strength dropped by 10% as concrete slump increased from 5-1/2 to 7 in. (140 to 180 mm) and by 3% as slump increased from 4-1/2 to 8-1/2 in. (115 to 215 mm) for the same increase in high-range water-reducing admixture dosage. For three pairs of top-cast bars, splice strength decreased by 3% or less for increases in slump (6-1/4 to 7 in. [160 to 180 mm], 1-1/2 to 8-1/2 in. [40 to 210 mm], and 5 to 8-1/2 in. [125 to 215 mm]) with an increase in high-range water-reducing admixture dosage.

Overall, an increase in slump and the use of workability-enhancing admixtures tends to have a negative effect on bond strength. The longer that concrete has time to settle and bleed, the lower the bond strength. This effect is especially important for top bars.

2.3.6 *Mineral admixtures*—Most studies of the effect of mineral admixtures on bond strength have been limited to the effects of silica fume, the principal mineral admixture used in high-strength concrete. Because of significantly increased compressive strength for many of the concretes containing silica fume, comparisons have usually been based on bond strengths that are normalized with respect to $f_c'^{1/2}$. Because this value overestimates the effect of compressive strength, the conclusion has often been made that silica fume has a negative effect on bond strength. If the test results are normalized with respect to $f_c'^{1/4}$, however, the apparent negative impact of silica fume on bond is significantly decreased. The comparisons that follow are based on results normalized with respect to $f_c'^{1/4}$.

Gjorv, Monteiro, and Mehta (1990) observed that silica fume increases the bond strength between concrete and reinforcing steel, as measured using the ASTM C 234* pullout test (ASTM 1991). For concrete strengths between 3000 and 12,000 psi (21 and 83 MPa) and silica fume replacements of 0, 8, and 16% by weight (mass) of cement, they concluded that (as expected) bond strength increases with compressive strength and that top-cast bars develop lower bond strength than bottom-cast bars due to the accumulation of bleed water and air below the bars. Increasing the silica fume content resulted in an increase in pullout strength. They felt that the major effects of silica fume involve reduction in bleed water and strengthening of the cement paste in the transition zone adjacent to the reinforcing bars.

Hwang, Lee, and Lee (1994) evaluated the effect of silica fume using eight splice specimens. The specimens were tested in pairs in which one specimen included a 10% replacement of portland cement by an equal weight (mass) of silica fume. In this case, bond strength decreased by an average of 7-1/2% with the addition of silica fume. Test results obtained by DeVries, Moehle, and Hester (1991) and Olsen (1990a,b) show little or no negative effects of silica fume on bond strength. In fact, Olsen's tests show a 10% increase in splice strength for concrete containing silica fume and a 17% increase for concrete containing fly ash.

*This test method has been withdrawn by ASTM International.

Hamad and Itani (1998) carried out 16 splice tests that included both top- and bottom-cast bars. The test strengths showed reductions in bond strength averaging 5% for silica fume replacements of portland cement between 5 and 20%.

2.3.7 *Fiber reinforcement*—For centuries, fibers have been added to concrete and mortar to improve the inherently low tensile strength of these materials. Fibers such as straw or horsehair were used originally. Asbestos was used starting at the beginning of the twentieth century, and today, ACI 544.1R lists many different types of fiber that can be used to enhance the toughness of concrete and mortar.

ACI Committee 544 divides modern fiber-reinforced concrete (FRC) into four categories based on the type of fiber used: steel fiber-reinforced concrete (SFRC), glass fiber-reinforced concrete (GFRC), synthetic fiber-reinforced concrete (SNFRC), and natural fiber-reinforced concrete (NFRC).

Within each category, many types and lengths of fiber are available to achieve different material properties. For example: steel fibers may or may not have hooked ends; many chemical compositions of glass fibers exist; synthetic fibers may be manufactured with carbon, acrylic, polyester or other materials; and natural fibers can be obtained from many sources such as coconut or jute. In addition, the use of a new category of FRC using fibers recycled from industrial wastes from the manufacture of carpet, plastics, and other processes is gaining acceptance.

One reason to add fibers is to increase the tensile strength of concrete. The increase, however, is small. For example, Wafa and Ashour (1992) report a 10 to 20% increase in modulus of rupture for FRC over plain concrete for concrete with nominal compressive strength of 14,000 psi (100 MPa). Even with these increases, the tensile strength remains significantly less than the compressive strength.

A more important goal of adding fibers is to increase the post-cracking resistance of the concrete, which allows FRC to be used in applications where crack control is important. Fibers bridge across cracks and allow some tensile stress to be transferred. At failure, the fibers pull out of the concrete, increasing the energy required to open and propagate the cracks.

The provisions in ACI 318 for the development of deformed bars are based on bond strengths that are governed by a splitting failure of the concrete around the bar. Factors that affect the splitting resistance of the concrete, such as concrete tensile strength, transverse reinforcement, and amount of cover, are considered in design. Theoretically, the use of FRC should improve resistance to splitting cracks and reduce required development lengths.

A study of the test results described in the balance of this section indicates that fibers, especially steel fibers, behave as both transverse and longitudinal reinforcement, the former having the major effect on bond strength. The fiber volume fractions used in these tests are generally high compared with values used for conventional transverse reinforcing bars.

Davies (1981) explored the bond strength of steel reinforcing bars in polypropylene fiber-reinforced concrete. In his study, pullout specimens with No. 4 or No. 6 (No. 13 or No. 19) Grade 60 (420 MPa) reinforcement and nominal concrete compressive strengths between 4000 and 6000 psi

(28 and 41 MPa) were used. In addition, the percent of fiber by volume and length of individual fibers were variables. In general, the addition of longer fibers (3-1/2 in. [90 mm]) resulted in higher average bond strengths. The increase in bond strength, however, was not greater than the increase in $\sqrt{f'_c}$ resulting from the increase in fiber volume. Increases in the volume of shorter fibers (2-1/4 in. [57 mm]) reduced average bond strengths in some cases.

Ezeldin and Balaguru (1989) studied the effects of steel fibers on the bond strength of deformed bars in normal and high-strength concrete. Fiber content ranged up to 0.75% by volume. Using pullout specimens in which the concrete was placed in tension under load, they observed that using a fiber content of 0.25% resulted in a decrease in bond strength compared with concrete without fibers. The addition of fiber contents of 0.5 and 0.75%, however, resulted in increases in bond strength of up to 18%. Improvements in bond strength were greater for No. 5 and No. 6 bars than for No. 3 bars. Increases in fiber content and fiber length resulted in improved ductility following the peak load.

Soroushian, Mirza, and Alhozaimy (1994) tested the bond strength of a $4d_b$ embedded length of bar centered in a $15d_b$ long pullout specimen. They observed that local bond strength increased by about one-third for a 0.5% volume addition of steel fibers. Further increases in fiber volume, up to 1.5%, provided only an additional 5% increase in bond strength. The fibers reduced slip at the peak bond stress, while fiber aspect ratio and type had little effect on bond.

Harajli, Hout, and Jalkh (1995) measured the local bond stress-slip behavior for bars embedded in pullout specimens. They evaluated the effect of bar diameter (No. 6 and No. 8 [No. 19 and No. 25] bars), mode of failure (pullout, splitting), and type, volume fraction, and aspect ratio of the fibers. They observed that fiber contents up to 2% by volume resulted in an increase in bond strengths of about 20%. As in other studies, fibers significantly increased the bond force after the peak load. Polypropylene fibers provided about 1/3 of the increase in bond resistance provided by hooked steel fibers after the peak load had been reached. As would be expected, fibers had little effect on the strength or behavior of specimens that failed by pullout rather than splitting.

In a study of the bond strength of deformed bars in slurry infiltrated fiber-reinforced concrete (SIFCON), Hamza and Naaman (1996) obtained a 150% increase in bond strength for a 5% volume content of steel fibers. Bond forces equal to 50% of the peak bond force were maintained at slips equal to 10 times the maximum slip obtained for plain concrete specimens. The initial bond stiffness obtained for SIFCON was 2.5 times the value measured for plain concrete.

Harajli and Salloukh (1997) evaluated the effects of fibers on the splice strength of reinforcing bars using 15 beams, each with a lap splice at midspan. The beams were loaded in positive bending with a constant moment in the splice region. The results of the tests demonstrated that steel fibers (up to 2% by volume) increased member strength by up to 55%. The increase included the effects of the fibers both on bond and on the contribution of the concrete to the local flexural capacity of the section. The presence of fibers increased the number of cracks formed around the splices, delayed splitting cracks, and improved the ductility of members undergoing a bond failure. Polypropylene fibers improved performance in the post splitting range, but had less effect on bond strength. At 0.6% by volume, polypropylene fibers provided between 0 and a 25% increase in the failure load.

Hota and Naaman (1997) compared the bond strengths of deformed bars in plain concrete, FRC, and SIFCON using pullout specimens. The bonded length was 4 in. (100 mm). Concrete containing a 2% volume fraction of steel fibers produced a peak bond strength about twice that produced by plain concrete, while SIFCON, containing a 9% steel fiber volume fraction, produced a bond strength equal to three times that obtained with the plain concrete specimens.

Because most of the bond tests involving FRC have been pullout tests, rather than splice or development tests, and because fibers affect the failure load for splice or development specimens by altering the contribution of the concrete to the flexural strength of a section, as well as to bond, significantly more work is needed before data is available to properly judge the effect of fibers on bond strength.

2.3.8 *Consolidation*—Adequate consolidation is a key factor in quality concrete construction. The role of consolidation is usually described in terms of removing voids that may have been entrapped during handling and placement. In terms of bond, adequate consolidation, usually obtained with high frequency internal vibration, plays the additional role of reducing the effects of settlement and bleeding, which result in the accumulation of bleed water and low density, weak concrete just below horizontal reinforcement. By disturbing the concrete, vibration helps restore local uniformity. Both the removal of entrapped air and the restoration of local uniformity play an important role in improving bond strength. The balance of this section discusses the effects of initial vibration, delayed vibration, construction-induced vibration, and revibration.

2.3.8.1 *Vibration*—Davis, Brown, and Kelly (1938) studied the effects of delayed vibration and sustained jigging* on bond strength. Using 4 in. (100 mm) slump concrete, vibration was applied from 0 to 9 h after concrete placement in pullout specimens consisting of deformed bars held vertically in cylindrical specimens. Delayed vibration, up to 9 h, improved bond strength compared with initial vibration. Increases in bond strength of up to 62% were recorded. This improved bond strength has been used as evidence of the positive effect of revibration (Tuthill and Davis 1938; Tuthill 1977). Because the specimens were not initially vibrated, however, the improved bond strength must be attributed to delayed vibration not revibration. Using similar specimens for sustained jigging tests, Davis, Brown, and Kelly obtained increases in bond strength as the period of jigging* increased from 1/2 to 2 h, after which there was no significant change up to a maximum of 6 h.

Robin, Olsen, and Kinnane (1942) compared bond strengths of horizontally cast bars consolidated with external

*Jigging involves raising opposite sides of a form and allowing the form to drop so as to sharply strike the floor.

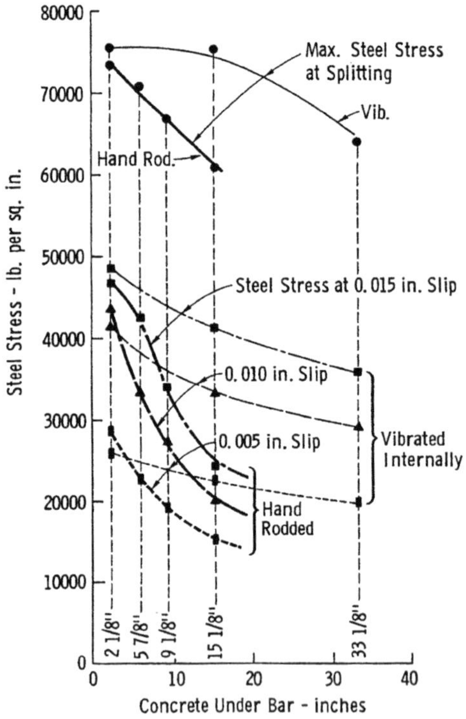

Fig. 2.12—Steel stress in beam-end specimens for vibrated and hand-rodded concrete (Menzel 1952). (Note: 1 in. = 25.4 mm; 1 psi = 0.00689 MPa)

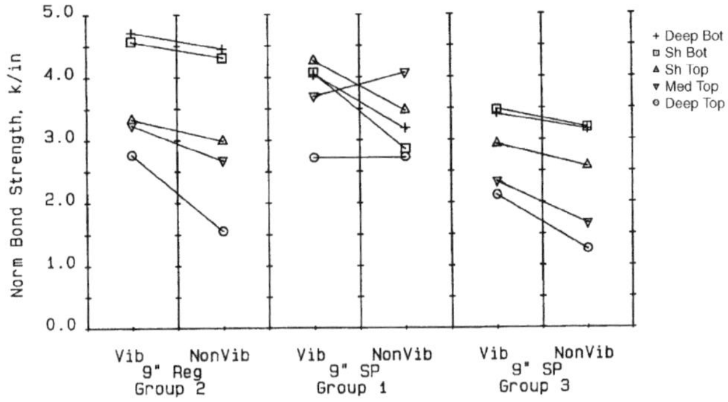

Fig. 2.13—Comparison of average normalized bond strengths for vibrated and nonvibrated high-slump concretes in beam-end specimens. REG = concrete without a high-range water-reducing admixture; and SP = superplasticized concrete. The bond strengths normalized by multiplying by $(4000/f_c')^{1/2}$ (Brettmann, Darwin, and Donahey 1986). (Note: 1 in. = 25.4 mm; 1 kip/in. = 175 kN/m)

vibration and hand rodding. External vibration of 3 to 4 in. (75 to 100 mm) slump concrete produced lower strengths than hand rodding for bars 1-1/2 or 3 in. (38 or 75 mm) from the bottom of the forms, but higher bond strengths than hand rodding for bars 6 or 7-1/2 in. (150 or 190 mm) from the bottom of the forms.

Using 2 in. (50 mm) slump concrete and both top-cast and bottom-cast deformed bars, Menzel (1952) found that internal vibration significantly increased bond strength compared with hand rodding (Fig. 2.12). The relative improvement in bond strength with vibration increased as the distance of the bar above the base of the specimen increased.

Donahey and Darwin (1983, 1985) considered the effects of the density of vibration on bond strength. They compared procedures using high-density vibration (in which the vibrator radii of influence overlapped) and low-density vibration (in which the radii of influence did not overlap). They found that high-density internal vibration improves bond strength. Brettmann, Darwin, and Donahey (1984, 1986) observed that vibration is especially important when high-slump concrete is used, whether the high slump is produced with the addition of water and cement or with the addition of a high-range water-reducing admixture. Brettmann, Darwin, and Donahey used internal vibration and followed standard practices for vibration (ACI 309R). For 9 in. (230 mm) slump concrete obtained without a high-range water-reducing admixture (Group 2 in Fig. 2.13), bond strengths averaged 14% lower for nonvibrated specimens than for vibrated specimens. For bottom-cast bars, the average decrease was only 6% for nonvibrated specimens, largely due to the consolidation provided by the concrete above the bars. For top-cast bars, bond strength decreased 23% when the concrete was not vibrated. For 9 in. (230 mm) slump (Groups 1 and 3 in Fig. 2.13) superplasticized concrete, nonvibrated specimens also exhibited lower bond strength. There were two exceptions involving top-cast bars in high-temperature (84 °F [29 °C]) (Group 1) concrete, which received additional consolidation due to finishing operations. For the high-temperature concrete (Group 1), nonvibrated bottom-cast bars had 25% lower bond strength than vibrated bars. For lower-temperature (54 °F [12 °C]) concrete (Group 3), the lack of vibration caused a decrease in bond strength that ranged from 8% for bottom-cast bars to 41% for top-cast bars in deep specimens.

2.3.8.2 *Construction-related vibrations*—Both field and laboratory studies have considered the effects of external construction-related vibrations. The concern has been that these vibrations, which occur while the concrete is setting, damage the bond between the reinforcement and the concrete.

Furr and Fouad (1981) studied the effects of maintaining traffic on existing lanes of a bridge while the bridge was being widened. In both field and laboratory investigations, they found that it was possible to have some reduction in bond strength. This reduction, however, was limited to areas close to longitudinal construction joints, where relative movement between the concrete and the steel was greatest.

Hulshizer and Desai (1984) evaluated the effects of "simulated blast loading" on bond strength using pullout specimens subjected to vibrations ranging from 50 to 150 Hz while the concrete set. They found that pullout strength increased slightly due to the vibrations.

Harsh and Darwin (1984, 1986) studied the effects of simulated traffic-induced vibrations on bond strength in bridge deck repairs. They found that, if low-slump concrete is used, bond strength actually increases. For slumps in excess of 4 in. (100 mm), however, traffic-induced vibrations result in a reduction in bond strength, in some cases by over 10%. Overall, traffic-induced vibrations do not appear to be detrimental to bond strength in bridge deck repairs if low-slump concrete is used.

2.3.8.3 *Revibration*—When properly executed, internal vibration improves the quality of bond between concrete and steel. Much less is known about the effects of revibration, the process in which a vibrator is reapplied to concrete at some time after initial vibration.

Vollick (1958) found that internal revibration increases concrete compressive strength from 7 to 19%, depending on the concrete mixture. He did not study the effect of revibration on bond strength. Larnach (1952) studied the effects of external revibration on both bond and compressive strength, finding that external revibration reduced the bond strength of horizontally cast plain bars by 6 to 33%, depending on the time after initial consolidation. He obtained reductions in compressive strength of about 15%.

Menzel (1952) studied the effects of internal revibration on the bond strength of deformed bars with 2 to 3 in. (50 to 75 mm) slump concrete. He found that revibration after 1 h had no adverse effect on bottom-cast bars, but reduced the bond strength of top-cast bars by 28 to 60%.

Altowaiji, Darwin, and Donahey (1986) studied the effects of internal revibration on bond strength for concretes with initial slumps between 2-3/4 and 7-1/2 in. (70 and 190 mm). Forty-five and 90 minute intervals were used between initial vibration and revibration. They found that revibration improves the bond strength of top-cast bars and bars placed in high-slump concrete by 5 to 20%. Revibration, however, can reduce the bond strength of bars cast in well-consolidated, low-slump concrete by 10 to 20%. Revibration is almost universally detrimental to the bond strength of bottom-cast bars.

The bars benefiting most from revibration, top-cast bars placed in high-slump concrete, are the bars that are affected most by settlement and bleeding. Revibration helps to reconsolidate the concrete adjacent to the bars, reducing voids caused by settlement and bleeding and, thus, improving bond. From a practical standpoint, structures in which revibration has its greatest advantage (structures placed with high-slump concrete) are the least likely to receive proper consolidation at any stage.

The detrimental effects of revibration to bottom-cast bars are probably due to the fact that settlement and bleeding help to consolidate the concrete around these bars, while revibration only serves to disrupt the concrete. Thus, full-depth revibration appears to be a poor construction practice. When used, revibration should be limited to the upper portions of a placement, no deeper than about 6 in. (150 mm) below the concrete surface. This limitation allows the effects of settlement and bleeding to be counteracted around top-cast reinforcement without damaging the bond of deeper bars. The use of a vibrator to tie together two lifts of concrete is a common application of revibration under these guidelines.

2.4—Summary

Many factors affect the bond of reinforcing bars to concrete. As was discussed in this chapter, bond force increases with increasing concrete cover and bar spacing, development and splice length, and the use of confining transverse reinforcement. Top-cast bars have a lower bond strength than bottom-cast bars, while for realistic bar spacing, noncontact lap splices provide a higher strength than contact lap splices. For a given length of bar, the bond force mobilized by both concrete and transverse reinforcement increases as the bar diameter increases. The bond strength of bars confined by transverse reinforcement increases with an increase in the relative rib area. The bond strength of developed and spliced bars is little affected if the stress in the bar exceeds the yield strength. Epoxy coatings reduce the bond strength of bars, but for all conditions of confinement, the higher the relative rib area of the bar, the lower the reduction. Bond strength increases with increasing concrete compressive strength for bars not confined by transverse reinforcement approximately with the 1/4 power (rather than the 1/2 power, as often assumed) of the compressive strength. The additional bond strength provided by transverse reinforcement increases approximately with the 3/4 power of the compressive strength. An increase in aggregate strength and quantity results in an increase in bond strength. This increase appears to be tied to the effect of aggregate properties on tensile strength and, even more importantly, the fracture energy of the concrete. Lightweight aggregate concrete develops lower bond strength than normalweight concrete with the same compressive strength. An increase in concrete slump and the use of workability-enhancing admixtures, in general, have a negative impact on bond strength. The longer the period that concrete has to settle and bleed, the lower the bond strength. This effect is especially important for top-cast bars. The use of silica fume has been observed to increase bond strength in some cases and decrease bond strength in others for a given compressive strength. Fiber reinforcement, especially steel fibers, tends to act as transverse reinforcement, providing increased bond strength to reinforcing steel. Improved consolidation of concrete results in improved bond strength.

CHAPTER 3—DESCRIPTIVE EQUATIONS

An accurate theory-based analysis procedure for development and splice strength has yet to be formulated. As a result, expressions for bond strength have been developed based on comparisons with test results. The expressions presented in this chapter capture the key aspects of bond behavior and produce reasonably accurate predictions of bond strength. The chapter concludes with a comparison of the expressions using the test results for bottom-cast bars in ACI Committee 408 Database 10-2001 (refer to Chapter 5), a comparison that illustrates the relative merits of each and demonstrates the advisability of representing the effect of concrete strength on bond by a factor other than $\sqrt{f_c'}$.

3.1—Orangun, Jirsa, and Breen

Using statistical techniques, Orangun, Jirsa, and Breen (1975, 1977) developed expressions to describe the bond strength of bars without and with confining transverse reinforcement. For bars not confined by transverse reinforcement, a regression analysis based on 62 beams, including four with side-cast bars, one with top-cast bars, and 57 with bottom-cast bars, produced an expression for the average bond stress at failure.

$$\frac{u_c}{\sqrt{f'_c}} = 1.22 + 3.23\frac{c_{min}}{d_b} + 53\frac{d_b}{l_d} \quad (3\text{-}1)$$

where c_{min} = smaller of minimum concrete cover or 1/2 of the clear spacing between bars; l_d = development or splice length; and d_b = bar diameter. In their analysis, the coefficients in Eq. (3-1) were rounded as follows

$$\frac{u_c}{\sqrt{f'_c}} = 1.2 + 3\frac{c_{min}}{d_b} + 50\frac{d_b}{l_d} \quad (3\text{-}2)$$

The bond strength of a bar confined by transverse reinforcement was represented by

$$\frac{u_b}{\sqrt{f'_c}} = \frac{u_c + u_s}{\sqrt{f'_c}} = 1.2 + 3\frac{c_{min}}{d_b} + \frac{50d_b}{l_d} + \frac{A_{tr}f_{yt}}{500snd_b} \quad (3\text{-}3)$$

where A_{tr} = area of transverse reinforcement normal to the plane of splitting through the anchored bars; f_{yt} = yield strength of transverse reinforcement; s = spacing of transverse reinforcement; and n = number of bars developed or spliced at the same location.

In terms of total bond force, Eq. (3-3) may be written as

$$\frac{T_b}{\sqrt{f'_c}} = \frac{T_c + T_s}{\sqrt{f'_c}} = \frac{A_b f_s}{\sqrt{f'_c}} = \quad (3\text{-}4)$$

$$3\pi l_d(c_{min} + 0.4d_b) + 200A_b + \frac{\pi l_d A_{tr}}{500sn}f_{yt}$$

where A_b = area of developed or spliced bar.

Equations (3-3) and (3-4) are limited to cases in which splitting failure, rather than pullout, governs. This leads to a restriction in applying the expressions that

$$\frac{1}{d_b}\left(c_{min} + 0.4d_b + \frac{A_{tr}f_{yt}}{1500sn}\right) \le 2.5 \quad (3\text{-}5)$$

As shown in Section 4.1, Eq. (3-3) and (3-4) serve as the basis for the design expression for development length that first appeared in ACI 318-95.

3.2—Darwin et al.

Darwin et al. (1992) reanalyzed the data used by Orangun, Jirsa, and Breen (1975, 1977) to establish Eq. (3-1) and (3-2) for bars not confined by transverse reinforcement. Darwin et al. (1992) incorporated the effect of the relative value of c_{max} and c_{min} (refer to Section 2.1) to obtain

$$\frac{T_c}{\sqrt{f'_c}} = \frac{A_b f_s}{\sqrt{f'_c}} = \quad (3\text{-}6)$$

$$6.67l_d(c_{min} + 0.5d_b)\left(0.08\frac{c_{max}}{c_{min}} + 0.92\right) + 300A_b$$

Darwin et al. (1996a) used a larger database, consisting of 133 splice and development specimens in which the bars were not confined by transverse reinforcement and 166 specimens in which the bars were confined by transverse reinforcement. All specimens contained bottom-cast bars. They observed that $f'^{1/4}_c$ provided a better representation of concrete strength on development and splice strength than the more traditional $f'^{1/2}_c$. They also incorporated the effect of relative rib area R_r, which they observed to have a significant effect on the bond strength of bars confined by transverse reinforcement. Based on their studies, the best-fit equation for the bond strength of bars not confined by transverse reinforcement was

$$\frac{T_c}{f'^{1/4}_c} = \frac{A_b f_s}{f'^{1/4}_c} = \quad (3\text{-}7)$$

$$[63l_d(c_{min} + 0.5d_b) + 2130A_b]\left(0.1\frac{c_{max}}{c_{min}} + 0.9\right)$$

where $\left(0.1\frac{c_{max}}{c_{min}} + 0.9\right) \le 1.25$.

The best-fit equation for bars confined by transverse reinforcement was

$$\frac{T_b}{f'^{1/4}_c} = \frac{T_c + T_s}{f'^{1/4}_c} = \frac{A_b f_s}{f'^{1/4}_c} = \quad (3\text{-}8)$$

$$[63l_d(c_{min} + 0.5d_b) + 2130A_b]\left(0.1\frac{c_{max}}{c_{min}} + 0.9\right)$$

$$+ 2226t_r t_d\frac{NA_{tr}}{n} + 66$$

where c_{max} and c_{min} are defined following Eq. (2-1), N = number of transverse bars in the development or splice length, $t_r = 9.6R_r + 0.28$, and $t_d = 0.72d_b + 0.28$.

Like the expressions developed in Section 3.1, Eq. (3-7) and (3-8) are applicable to cases in which a splitting failure, rather than pullout, governs. The following restriction applies to the expressions.

$$\frac{1}{d_b}\left[(c_{min} + 0.5d_b)\left(0.1\frac{c_{max}}{c_{min}} + 0.90\right) + \left(\frac{35.3t_r t_d A_{tr}}{sn}\right)\right] \le 4.0 \quad (3\text{-}9)$$

3.3—Zuo and Darwin

Zuo and Darwin (1998, 2000) expanded the work of Darwin et al. (1996a) by increasing the database and adding substantially to the percentage of test specimens containing high-strength concrete ($f_c' > 8000$ psi [55 MPa]). The database included 171 specimens containing bars not confined by transverse reinforcement and 196 specimens containing bars confined by transverse reinforcement. All bars were bottom cast. Their analysis supported the earlier observations that $f_c'^{1/4}$ realistically represents the contribution of concrete strength to bond strength for bars not confined by transverse reinforcement. As discussed in Section 2.3.1, however, they observed that $f_c'^p$, with p between 3/4 and 1.0, best represents the effect of concrete strength on T_s, the contribution of confining transverse reinforcement to bond strength. They selected $p = 3/4$ for their descriptive equations. For bars not confined by transverse reinforcement, the best-fit equation describing development and splice strength is

$$\frac{T_c}{f_c'^{1/4}} = \frac{A_b f_s}{f_c'^{1/4}} = \qquad (3\text{-}10)$$

$$[59.8l_d(c_{min} + 0.5d_b) + 2350A_b]\left(0.1\frac{c_{max}}{c_{min}} + 0.90\right)$$

For bars confined by transverse reinforcement, the descriptive equation is

$$\frac{T_b}{f_c'^{1/4}} = \frac{T_c + T_s}{f_c'^{1/4}} = \frac{A_b f_s}{f_c'^{1/4}} = \qquad (3\text{-}11)$$

$$[59.8l_d(c_{min} + 0.5d_b) + 2350A_b]\left(0.1\frac{c_{max}}{c_{min}} + 0.90\right)$$

$$+\left(31.14t_r t_d\frac{NA_{tr}}{n} + 4\right)f_c'^{1/2}$$

where $t_r = 9.6R_r + 0.28$ and $t_d = 0.78d_b + 0.22$.

To limit applicability of Eq. (3-10) and (3-11) to cases in which a splitting failure governs

$$\frac{1}{d_b}\left[(c_{min} + 0.5d_b)\left(0.1\frac{c_{max}}{c_{min}} + 0.90\right)\right. \qquad (3\text{-}12)$$

$$\left. + \left(\frac{0.52t_r t_d A_{tr}}{sn}\right)f_c'^{1/2}\right] \leq 4.0$$

3.4—Esfahani and Rangan

Extending the local bond stress theory of Tepfers (1973) to splices, Esfahani and Rangan (1998a,b) developed expressions for the bond strength of bars not confined by transverse reinforcement.

For $f_c' < 7250$ psi (50 MPa),

$$\frac{T_c}{\sqrt{f_c'}} = \frac{A_b f_s}{\sqrt{f_c'}} = \qquad (3\text{-}13)$$

$$32.45\pi l_d \frac{(c_{min} + 0.5d_b)\left(1 + \frac{1}{M}\right)}{\left(\frac{c_{min}}{d_b} + 3.6\right)(1.85 + 0.024\sqrt{M})}\left(0.12\frac{c_{med}}{c_{min}} + 0.88\right)$$

For $f_c' \geq 7250$ psi (50 MPa),

$$\frac{T_c}{\sqrt{f_c'}} = \frac{A_b f_s}{\sqrt{f_c'}} = \qquad (3\text{-}14)$$

$$56.97\pi l_d \frac{(c_{min} + 0.5d_b)\left(1 + \frac{1}{M}\right)}{\left(\frac{c_{min}}{d_b} + 5.5\right)(1.85 + 0.024\sqrt{M})}\left(0.12\frac{c_{med}}{c_{min}} + 0.88\right)$$

where c_{min} = minimum (c_{so}, c_b, $c_{si} + d_b/2$), c_{med} = median (c_{so}, c_b, $c_{si} + d_b/2$) [that is, middle value], and M = cosh $(0.00092l_d\sqrt{rf_c'/d_b})$. For conventional reinforcement ($R_r \approx 0.07$), $r = 3$. The hyperbolic cosine enters the relationship based on the assumed variation in bond stress along the developed or spliced length of a bar.

3.5—ACI Committee 408

Using ACI 408 Database 10-2001 (Section 5.2), the committee has updated Eq. (3-10) and (3-11) with only minor changes.

$$\frac{T_c}{f_c'^{1/4}} = \frac{A_b f_s}{f_c'^{1/4}} = \qquad (3\text{-}15)$$

$$[59.9l_d(c_{min} + 0.5d_b) + 2400A_b]\left(0.1\frac{c_{max}}{c_{min}} + 0.90\right)$$

$$\frac{T_b}{f_c'^{1/4}} = \frac{T_c + T_s}{f_c'^{1/4}} = \frac{A_b f_s}{f_c'^{1/4}} = \qquad (3\text{-}16)$$

$$[59.9l_d(c_{min} + 0.5d_b) + 2400A_b]\left(0.1\frac{c_{max}}{c_{min}} + 0.90\right)$$

$$+\left(30.88t_r t_d\frac{NA_{tr}}{n} + 3\right)f_c'^{1/2}$$

The restriction in Eq. (3-12) applies to Eq. (3-15) and (3-16).

3.6—Comparisons

The six descriptive equations for bars not confined by transverse reinforcement [Eq. (3-4), (3-6), (3-7), (3-10), ((3-13) or (3-14)), and (3-15)] and the four equations for bars confined by transverse reinforcement [Eq. (3-4), (3-8), (3-11), and (3-16)] are compared with the test results from ACI 408 Database 10-2001 in Table 3.1 and 3.2 and Fig. 3.1 and 3.2. The tables compare the test-prediction ratios and include the maximum, minimum, and average values. The tables also

Table 3.1—Test-prediction ratios for bars not confined by transverse reinforcement

	OJB* (Eq. (3-4))	DMIS† (Eq. (3-6))	DZTI‡ (Eq. (3-7))	Z & D§ (Eq. (3-10))	E & R‖ (Eq. (3-13) Eq. (3-14))	ACI 408R (Eq. (3-15))
Maximum	1.55	1.383	1.342	1.304	2.790	1.288
Minimum	0.505	0.528	0.719	0.729	0.545	0.724
Average	1.030	1.014	1.020	1.010	0.938	1.00
Standard deviation	0.208	0.189	0.118	0.113	0.190	0.111
Coefficient of variation	0.202	0.187	0.116	0.111	0.202	0.111

*Orangun, Jirsa, and Breen (1977).
†Darwin et al. (1992).
‡Darwin et al. (1996b).
§Zuo and Darwin (2000).
‖Esfahani and Rangan (1998a,b).

Table 3.2—Test-prediction ratios for bars confined by transverse reinforcement

	OJB* (Eq. (3-4))	DZTI† (Eq. (3-8))	Z & D‡ (Eq. (3-11))	ACI 408R (Eq. (3-16))
Maximum	1.902	1.479	1.309	1.333
Minimum	0.595	0.776	0.739	0.755
Average	1.074	1.052	0.989	1.002
Standard deviation	0.255	0.132	0.119	0.121
Coefficient of variation	0.238	0.125	0.121	0.120

*Orangun, Jirsa, and Breen (1977).
†Darwin et al. (1996b).
‡Zuo and Darwin (2000).

include the standard deviations and coefficients of variation of the ratios. The figures compare test-prediction ratios as a function of compressive strength—test-prediction ratio should not be highly sensitive to compressive strength.

Table 3.1 shows that for bars not confined by transverse reinforcement the average test-prediction ratio is within 3% of 1.0, with the exception of the predictions based on Esfahani and Rangan (1998a,b), which produce an average test-prediction ratio of 0.94. The comparisons for Esfahani and Rangan also exhibit the highest extremes between maximum and minimum test-prediction values. The lowest scatter is exhibited by the expressions obtained by Zuo and Darwin (2000) and Committee ACI 408, both of which produce a coefficient of variation of 0.111. The highest coefficients of variation, 0.202 and 0.207, are obtained using the expressions of Orangun, Jirsa, and Breen (1977) and Esfahani and Rangan, respectively. The Esfahani and Rangan expressions were not developed using tests with large values of l_d and appear to be more appropriate for predicting bond strength for shorter development and splice lengths. For example, if M [defined following Eq. (3-14)] is limited to 100 by excluding specimens with combinations of l_d and f_c' that produce higher values, the mean changes only slightly (0.94), but the coefficient of variation drops to 0.14.

Figure 3.1 demonstrates why average test-prediction ratios and coefficients of variation do not always provide an adequate evaluation of descriptive expressions. The three

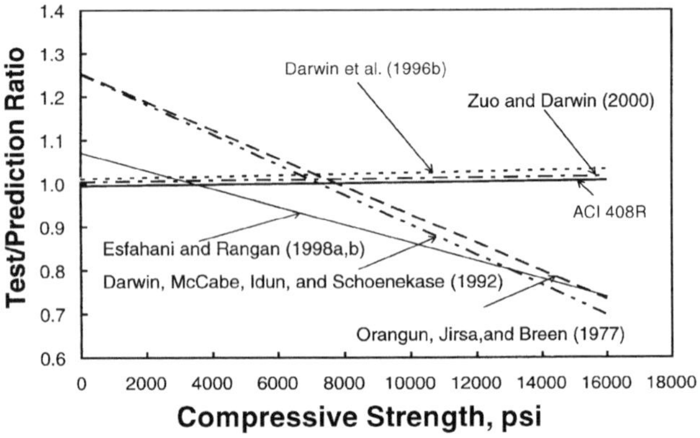

Fig. 3.1—Test-prediction ratios for six descriptive equations for bars not confined with transverse reinforcement. (Note: 1 psi = 0.00689 MPa)

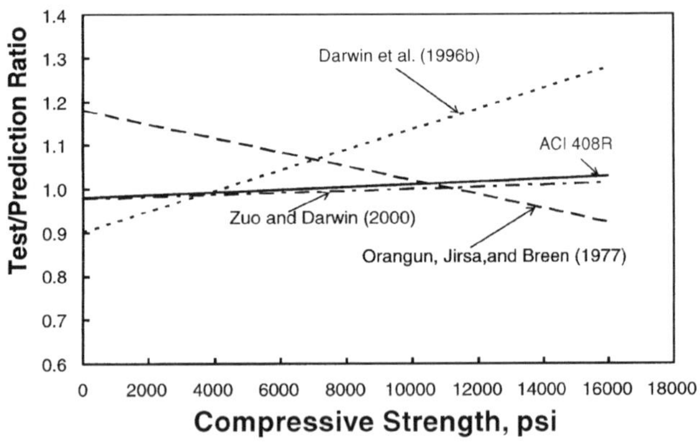

Fig. 3.2—Test-prediction ratios for descriptive equations with confinement provided by transverse reinforcement. (Note: 1 psi = 0.00689 MPa)

expressions in which concrete strength is characterized based on the $\sqrt{f_c'}$, Orangun, Jirsa, and Breen (1977), Darwin et al. (1992), and Esfahani and Rangan (1998a,b) exhibit a significant decrease in test-prediction ratio with increasing compressive strength. The Orangun, Jirsa, and Breen (1977) and Darwin et al. (1992) expressions were not developed using tests with the wide range of compressive strengths now available. In contrast, the three expressions based on $f_c'^{1/4}$ exhibit much less variation in test-prediction ratio and have only small increases in test-prediction ratio as compressive strength increases from 3000 to 16,000 psi (21 to 110 MPa). The figures reemphasize the observations made earlier about Fig. 2.10.

The test-prediction ratios for the four expressions for bars confined by transverse reinforcement are shown in Table 3.2. In this case, the average test-prediction ratios range from a high of 1.07 for the Orangun, Jirsa, and Breen (1977) to a low of 0.99 for Zuo and Darwin (2000). The best overall match is obtained by the ACI 408R expression, with an average test-prediction ratio of 1.00 and a coefficient of variation of 0.12. As shown in Fig. 3.2, the expressions developed by Zuo and Darwin and Committee 408 are largely unbiased with respect to compressive strength, whereas the other two expressions give biased results that depend upon compressive strength. Equation (3-4) by Orangun, Jirsa, and Breen (1977), which treats the contributions of both the concrete and the transverse reinforcement as functions of $\sqrt{f'_c}$, exhibits high test-prediction ratios for lower-strength concretes and low test-prediction ratios for high-strength concretes. In contrast, the expression developed by Darwin et al. (1996a) [Eq. (3-8)], in which the contributions of both the concrete and transverse reinforcement are functions of $f_c'^{1/4}$, provides low test-prediction ratios for lower-strength concrete and progressively higher test-prediction ratios as concrete strength increases. Both Eq. (3-11) (Zuo and Darwin 2000) and Eq. (3-16) (ACI Committee 408), which characterize the contribution of the concrete as a function of $f_c'^{1/4}$ and the contribution of transverse reinforcement as a function of $f_c'^{3/4}$, provide test-prediction ratios that are essentially independent of concrete compressive strength. The small increase in test-prediction ratio with increasing compressive strength shows that a power of f'_c between 3/4 and 1.0 for the transverse steel contribution would give a slightly better representation (Zuo and Darwin 2000).

CHAPTER 4—DESIGN PROVISIONS

Design provisions should be understandable, simple to apply, and result in an acceptably low probability of failure. This chapter presents the tension development and splice provisions adopted by ACI Committee 318, this committee, and the 1990 CEB-FIP Model Code. The provisions are compared for the relative accuracy, safety, and economy using ACI Committee 408 Database 10-2001. Specific aspects of the provisions of ACI 318, such as the bar size factor γ, are evaluated along with the reliability provided by the ACI 318 and ACI 408R design procedures. Special attention is given to the impact of the new load factors and ϕ-factors adopted in ACI 318-02. Comparisons are then made of the development and splice lengths obtained using the design procedures of the two committees. The procedures developed by ACI Committee 408 provide improved reliability and economy compared with those in ACI 318-02, especially for members with higher compressive strengths or transverse reinforcement and for all spliced bars.

4.1—ACI 318

The design provisions in ACI 318 for development and splice length of straight reinforcement in tension are based on the expressions developed by Orangun, Jirsa, and Breen (1975, 1977) described in Section 3.1. Solving Eq. (3-4) for the ratio of the development length l_d to the bar diameter d_b and replacing $(c_{min} + 0.4\, d_b)$ with $c = (c_{min} + 0.5\, d_b)$ gives

$$\frac{l_d}{d_b} = \frac{\frac{f_s}{\sqrt{f'_c}} - 200}{12\left(\frac{c + K_{tr}}{d_b}\right)} \tag{4-1}$$

where $K_{tr} = \dfrac{A_{tr} f_{yt}}{1500 sn}$

In Eq. (4-1), l_d is the length of bar required to develop f_s, the stress in the reinforcement resulting from the loads applied to the structure. At nominal capacity, $f_s = f_y$, and Eq. (4-1) is simplified to give the expression used in ACI 318, by removing the 200 from the numerator and changing the constant multiplying the expression from 1/12 (= 0.0833...) to 3/40 (= 0.075).

$$\frac{l_d}{d_b} = \frac{3}{40} \frac{f_y}{\sqrt{f'_c}\left(\frac{c + K_{tr}}{d_b}\right)} \tag{4-2}$$

To limit the probability of a pullout failure, ACI 318 requires that

$$\frac{c + K_{tr}}{d_b} \leq 2.5 \tag{4-3}$$

Due largely to a lack of data on concretes with compressive strengths in excess of 10,000 psi (69 MPa) at the time that Eq. (3-4) was developed, the value of $\sqrt{f'_c}$ is limited to a maximum value of 100 psi (8.3 MPa).

The effects of bar location, epoxy coating, and lightweight concrete on bond are included by multiplying l_d by the factors α, β, and λ, where

α = reinforcement location factor (1.3 for reinforcement placed so that more than 12 in. (300 mm) of fresh concrete is cast below the development length or splice; 1.0 for other reinforcement);

β = coating factor (1.5 for epoxy-coated reinforcement with cover less than $3d_b$ or clear spacing less than $6d_b$; 1.2 for other epoxy-coated reinforcement; 1.0 for uncoated reinforcement); with $\alpha\beta \leq 1.7$; and

λ = lightweight-concrete factor (1.3 for lightweight concrete; 1.0 for normalweight concrete; $6.7\sqrt{f'_c}/f_{ct} \geq 1.0$ for lightweight concrete with splitting tensile strength f_{ct} specified).

Based on limited comparisons for bars with very short development or splice lengths, l_d is also multiplied by a factor for bar size γ.

Table 4.1—Development length requirements in ACI 318 that may be used instead of Eq. (4-2)*

	No. 6 (No. 19) and smaller bars and deformed wires	No. 7 (No. 22) and larger bars
Clear spacing of bars being developed or spliced not less than d_b, clear cover not less than d_b, and stirrups or ties throughout l_d not less than the code minimum	$\frac{l_d}{d_b} = \frac{f_y \alpha\beta\lambda}{25\sqrt{f'_c}}$	$\frac{l_d}{d_b} = \frac{f_y \alpha\beta\lambda}{20\sqrt{f'_c}}$
or		
Clear spacing of bars being developed or spliced not less than $2d_b$ and clear cover not less than d_b	$\left[\frac{l_d}{d_b} = \frac{12 f_y \alpha\beta\lambda}{25\sqrt{f'_c}}\right]$	$\left[\frac{l_d}{d_b} = \frac{3 f_y \alpha\beta\lambda}{5\sqrt{f'_c}}\right]$
Other cases	$\frac{l_d}{d_b} = \frac{3 f_y \alpha\beta\lambda}{50\sqrt{f'_c}}$ $\left[\frac{l_d}{d_b} = \frac{18 f_y \alpha\beta\lambda}{25\sqrt{f'_c}}\right]$	$\frac{l_d}{d_b} = \frac{3 f_y \alpha\beta\lambda}{40\sqrt{f'_c}}$ $\left[\frac{l_d}{d_b} = \frac{9 f_y \alpha\beta\lambda}{10\sqrt{f'_c}}\right]$

*SI equations in brackets.

γ = reinforcement size factor (0.8 for No. 6 [No. 19] and smaller bars; 1.0 for No. 7 [No. 22] and larger bars).

As will be discussed in Section 4.5.2, the data does not support the use of $\gamma = 0.8$, and Committee 408 does not recommend its use in design.

In lieu of Eq. (4-2), ACI 318 allows the use of a simplified table defining l_d/d_b as a function of bar size and confinement provided by transverse reinforcement and concrete cover (Table 4.1). The center column in Table 4.1 is for No. 6 (No. 19) and smaller bars, corresponding to $\gamma = 0.8$, while the right-hand column is for No. 7 (No. 22) and larger bars, corresponding to $\gamma = 1.0$.

The simplifications shown in Table 4.1 involve the selection of combinations of minimum transverse reinforcement, concrete cover (described in the left-hand column), or both, that allow the term $(c + K_{tr})/d_b$ in Eq. (4-2) to be replaced by either 1.0 or 1.5. Replacement by 1.0 ("Other cases" in Table 4.1) is consistent with Eq. (4-2), if the structure meets the other requirements of ACI 318 for cover and bar spacing, as is replacement by 1.5 when the bars have a clear spacing of not less than $2d_b$ and a cover of not less than d_b. For bars with a clear spacing of less than $2d_b$, however, stirrups or ties that meet the code minimum do not always provide a K_{tr} value that is high enough so that $(c + K_{tr})/d_b$ is at least 1.5 (Nilson, Darwin, and Dolan 2004). The reason is that code minimum transverse reinforcement is typically based on requirements for shear or bar size and spacing criteria to provide confinement for longitudinal column reinforcement. These criteria do not take into account the number of bars being developed or spliced, n, as reflected in the term $K_{tr} = A_{tr}/1500sn$. As a result, the expressions in Table 4.1 may provide a value of l_d that is less than the value calculated using Eq. (4-2).

ACI 318 allows l_d to be reduced by the ratio (A_s required)/(A_s provided) when the reinforcement in a flexural member exceeds that required by analysis, except in cases where anchorage or development for the yield strength is specifically required or the reinforcement is designed for certain seismic applications. The basis for allowing a reduction in l_d when using a greater quantity of reinforcement than required is that an increase in steel area will reduce the stress in the steel to a value below f_y at the load corresponding to the nominal strength of the section. A lower value of f_s (refer to Eq. (4-1)) leads to a lower development length, l_d. In all cases, the minimum development length is 12 in. (300 mm).

Lapped splices are classified as Class A or Class B. Class A splices are those in which the ratio (A_s provided)/(A_s required) is equal to or greater than 2, and 50% or less of the steel is spliced within the lap length. All other splices are designated as Class B. Splice lengths of $1.0l_d$ and $1.3l_d$ are used for Class A and Class B splices, respectively. Splice lengths may not be reduced by the ratio of (A_s provided)/(A_s required) and must be at least 12 in. (300 mm). The extra length required for Class B splices is not based on strength criteria but rather is used as an incentive for designers to stagger splice locations. As will be discussed in Section 4.5, however, the 1.3 factor for Class B splices provides strength that helps make up for some unconservative aspects of the ACI 318 bond provisions. The lack of a ϕ-factor in the development of Eq. (4-2) is also discussed in Section 4.5.

4.2—ACI 408.3

Based on the work by Darwin et al. (1996a,b) (Eq. (3-7) and (3-8)), Innovation Task Group 2 (ITG 2) of the ACI TAC Technology Transfer Committee developed design provisions (ACI T2-98) for spliced and developed high relative rib area bars, defined as bars with $0.10 \le R_r \le 0.14$. The provisions were subsequently adopted as ACI 408.3.

The provisions in ACI 408.3 where obtained by incorporating a strength-reduction factor ϕ into Eq. (3-8) [because Eq. (3-8) is based on average strength] by multiplying the right side of the equation by ϕ, replacing f_s by f_y, and then solving the expression for the ratio of development length l_d to bar diameter d_b

$$\frac{l_d}{d_b} = \frac{\dfrac{f_y}{\phi f_c'^{1/4}} - 2130\left(0.1\dfrac{c_{max}}{c_{min}} + 0.9\right)}{80.2\left(\dfrac{c + K_{tr}}{d_b}\right)} = \frac{\dfrac{f_y}{f_c'^{1/4}} - \phi 2130\left(0.1\dfrac{c_{max}}{c_{min}} + 0.9\right)}{\phi 80.2\left(\dfrac{c + K_{tr}}{d_b}\right)} \tag{4-4}$$

where $K_{tr} = 35.3\, t_r t_d A_{tr}/sn$, $c = c_{min}[0.1(c_{max}/c_{min}) + 0.9] \le 1.25c_{min}$, and c_{max} and c_{min} are defined following Eq. (2-1).

Using a reliability-based factor ϕ of 0.9 (Darwin et al. 1998), Eq. (4-4) becomes

$$\frac{l_d}{d_b} = \frac{\dfrac{f_y}{0.9f_c'^{1/4}} - 2130\left(0.1\dfrac{c_{max}}{c_{min}} + 0.90\right)}{80.2\left(\dfrac{c + K_{tr}}{d_b}\right)} \quad (4\text{-}5a)$$

$$\frac{l_d}{d_b} = \frac{\dfrac{f_y}{f_c'^{1/4}} - 1900\left(0.1\dfrac{c_{max}}{c_{min}} + 0.90\right)}{72\left(\dfrac{c + K_{tr}}{d_b}\right)} \quad (4\text{-}5b)$$

A comparison between development lengths obtained with Eq. (4-4) and (4-5) shows that, for Grade 60 (414 MPa) bars, l_d increases by about 16% due to the incorporation of ϕ = 0.9 rather than 11%, the value that would be expected if l_d were proportional to bar stress (refer to Section 2.1.2). The application of ϕ-factor for bond is discussed at greater length in Sections 4.3 and 4.5.

ITG 2 modified Eq. (4-5) to

$$\frac{l_d}{d_b} = \frac{(f_y/f_c'^{1/4} - 1900\omega)\alpha\beta\lambda}{72\left(\dfrac{c\omega + K_{tr}}{d_b}\right)} \quad (4\text{-}6)$$

where c is redefined as $c_{min} + 0.5d_b$ (4-7)

$$\omega = 0.1\frac{c_{max}}{c_{min}} + 0.9 \le 1.25 \quad (4\text{-}8)$$

$$K_{tr} = C_R(0.72d_b + 0.28)\frac{A_{tr}}{sn} \quad (4\text{-}9)$$

$$C_R = 44 + 330(R_r - 0.10) \quad (4\text{-}10)$$

c_{max} and c_{min} are defined following Eq. (2-1), and $(c\omega + K_{tr})/d_b \le 4.0$.

ACI 408.3 limits $f_c'^{1/4} \le 11.0$, and $f_y \le 80$ ksi. A value of 1.0 may be used for the variable ω in place of the value calculated in Eq. (4-8).

The value of l_d is multiplied by the same factors α and λ as used in ACI 318 (Section 4.1). The factor for excess reinforcement is also the same as used in ACI 318. The coating factor β is 1.2 for all epoxy-coated bars and 1.0 for uncoated reinforcement; $l_d \ge 12$ in. (300 mm), and $\ge 16d_b$.

Lapped splice criteria are similar to those in ACI 318, except that when the splice length is confined with transverse reinforcement at two or more locations with a spacing not greater than 10 in. (250 mm), providing a value of K_{tr}/d_b of at least 0.5, the provisions for a Class A splice govern. The provisions for high R_r reinforcement may only be applied to No. 11 (No. 36) and smaller bars.

4.3—Recommendations by ACI Committee 408

The splice and development length criteria recommended by this committee are based on the work by Zuo and Darwin (1998, 2000), as extended in Section 3.5. They apply to both conventional and high relative rib area reinforcement. The design expressions given in this section are based on Eq. (3-15) and (3-16).

Converting empirical, best-fit equations to design expressions involves the incorporation of strength-reduction ϕ-factors to ensure a realistically low probability of failure. As demonstrated by Darwin et al. (1998), the ϕ-factor for bond depends upon the ϕ-factor used for tension, as well as the load factors used in design. ACI 318-02 contains two separate sets of load factors and ϕ-factors, necessitating the use of two different ϕ-factors for bond. Based on a Monte Carlo analysis (described in Section 4.5.3)

ϕ = 0.82 is obtained for a dead load factor of 1.2, live load factor of 1.6, and $\phi_{tension} = 0.9$ (corresponding to Chapter 9 of ACI 318-02); and

ϕ = 0.92 is obtained for a dead load factor of 1.4, live load factor of 1.7, and $\phi_{tension} = 0.9$ (corresponding to Appendix C of ACI 318-02 and Chapter 9 of ACI 318-99), and a dead load factor = 1.2, live load factor = 1.6, and $\phi_{tension} = 0.8$ (corresponding to Appendix C of ACI 318-99).

The derivation that follows, based on the use of $\phi = 0.92$ (corresponding to ACI 318-99 and Appendix C of 318-02), is presented first. Expressions based on $\phi = 0.82$ (Chapter 9 of 318-02) are presented in Section 4.3.2.

4.3.1 l_d *for ACI 318-99 and Appendix C of ACI 318-02*—Solving Eq. (3-16) for l_d/d_b, substituting f_y for f_s, and incorporating ϕ and the factors for bar location, epoxy coating, and lightweight aggregate concrete gives

$$\frac{l_d}{d_b} = \frac{\left(\dfrac{f_y}{\phi f_c'^{1/4}} - 2400\omega\right)\alpha\beta\lambda}{76.3\left(\dfrac{c\omega + K_{tr}}{d_b}\right)} = \frac{\left(\dfrac{f_y}{f_c'^{1/4}} - \phi 2400\omega\right)\alpha\beta\lambda}{\phi 76.3\left(\dfrac{c\omega + K_{tr}}{d_b}\right)} \quad (4\text{-}11a)$$

For $\phi = 0.92$,

$$\frac{l_d}{d_b} = \frac{\left(\dfrac{f_y}{f_c'^{1/4}} - 2210\omega\right)\alpha\beta\lambda}{70.2\left(\dfrac{c\omega + K_{tr}}{d_b}\right)} \quad (4\text{-}11b)$$

where c and ω are defined in Eq. (4-7) and (4-8), respectively

$$K_{tr} = (0.52t_r t_d A_{tr}/sn)f_c'^{1/2} \quad (4\text{-}12)$$

$$t_r = 9.6R_r + 0.28 \le 1.72 \quad (4\text{-}13)$$

$$t_d = 0.78d_b + 0.22 \quad (4\text{-}14)$$

and $(c\omega + K_{tr})/d_b \le 4.0$.

For conventional reinforcement, $K_{tr} = (0.5t_d A_{tr}/sn)f_c'^{1/2}$, corresponding to an average R_r value of 0.0727.

Equation (4-11b) may be simplified in cases in which: 1) the clear spacing of the bars being developed or spliced is not less than d_b, the cover is not less than d_b, and the stirrups or ties throughout l_d provide a value $K_{tr}/d_b \ge 0.5$; or 2) the clear spacing of the bars being developed or spliced is not less than $2d_b$ and the cover is not less than d_b. In this case, $(c\omega + K_{tr})/d_b \ge 1.5$. Setting $\omega = 1.0$, Eq. (4-11b) becomes

$$\frac{l_d}{d_b} = \left(\frac{f_y}{105.3f_c'^{1/4}} - 20.99\right)\alpha\beta\lambda \quad (4\text{-}15)$$

Rounding the values in Eq. (4-15) gives

$$\frac{l_d}{d_b} = \left(\frac{f_y}{105f_c'^{1/4}} - 21\right)\alpha\beta\lambda \quad (4\text{-}16)$$

For cases not meeting the spacing, cover and confinement criteria, Eq. (4-11b) may be simplified to

$$\frac{l_d}{d_b} = \left(\frac{f_y}{70f_c'^{1/4}} - 31\right)\alpha\beta\lambda \quad (4\text{-}17)$$

To match the rounded values in Eq. (4-16) and (4-17), Eq. (4-11b) becomes

$$\frac{l_b}{d_b} = \frac{\left(\dfrac{f_y}{f_c'^{1/4}} - 2200\omega\right)\alpha\beta\lambda}{70\left(\dfrac{c\omega + K_{tr}}{d_b}\right)} \quad (4\text{-}18)$$

Equation (4-18) may be further simplified by setting $\omega = 1.0$ and dropping the 0.25 in. term in the definition of the effective value c_{si} [refer to Eq. (2-1)]. The values of α, β, and λ and the term for excess reinforcement are as defined in Section 4.2, except that $\beta = 1.5$ for epoxy-coated bars for concrete with $f_c' > 10{,}000$ psi (70 MPa), based on observations by Zuo and Darwin (1998). The values of l_d in Eq. (4-16) to (4-18) are used for both development and lapped splice lengths. In addition, $l_d \ge 12$ in. (300 mm), $\ge 16d_b$.

4.3.2 l_d *based on Chapter 9 of ACI 318-02*—Using $\phi = 0.82$ for bond, Eq. (4-16), (4-17), and (4-18) are as follows:

For cases in which the clear spacing of the bars being developed or spliced is not less than d_b, the cover is not less than d_b, and the stirrups or ties throughout l_d provide a value of $K_{tr}/d_b \ge 0.5$; or the clear spacing of the bars being developed or spliced is not less $2d_b$ and the cover is not less than d_b

$$\frac{l_d}{d_b} = \left(\frac{f_y}{93f_c'^{1/4}} - 21\right)\alpha\beta\lambda \quad (4\text{-}19)$$

For cases not meeting the spacing, cover, and confinement criteria

$$\frac{l_d}{d_b} = \left(\frac{f_y}{62f_c'^{1/4}} - 31\right)\alpha\beta\lambda \quad (4\text{-}20)$$

Alternatively, development length may be calculated using Eq. (4-11a) with $\phi = 0.82$.

$$\frac{l_d}{d_b} = \frac{\left(\dfrac{f_y}{f_c'^{1/4}} - 1970\omega\right)\alpha\beta\lambda}{62\left(\dfrac{c\omega + K_{tr}}{d_b}\right)} \quad (4\text{-}21)$$

All other criteria presented in Section 4.3.1 remain unchanged.

4.4—CEB-FIP Model Code

The factors included in the 1990 CEB-FIP Model Code for calculating development and splice lengths of straight reinforcing bars are similar to those included in the ACI procedures discussed in Sections 4.1 through 4.3, with some additional considerations. The CEB-FIP provisions are presented here using notation that is compatible with that used in the earlier sections.

The CEB-FIP design expression for development length is

$$\frac{l_d}{d_b} = \quad (4\text{-}22)$$

$$\frac{1}{950}\left(1.15 - 0.15\frac{c_{min}}{d_b}\right)\left(1 - K\frac{\Sigma A_{tr} - \Sigma A_{tr,\min}}{A_b}\right)\frac{Mf_y}{\left(\dfrac{f_c' - 400}{1450}\right)^{2/3}}$$

where the first two terms in parentheses are between 0.7 and 1.0; $K = 0.10$ for a bar confined at a corner bend of a stirrup or tie, $K = 0.05$ for a bar confined by a single leg of a stirrup or tie, and = 0 for a bar that is not confined; ΣA_{tr} = area of transverse reinforcement along l_d; $\Sigma A_{tr,\min} = 0.25A_b$ for beams and 0 for slabs; A_b = area of the largest bar being developed or spliced; and M = ratio of the average yield strength to the design yield strength of the developed bar (values close to 1.15 are typical in U.S. practice).

The term containing f_c' is based on a characteristic strength, which is 1160 psi (8.0 MPa) below the average (not the specified) strength. For simplicity, the average strength is assumed (in this report) to be 760 psi (5.2 MPa) greater than the specified compressive strength f_c', resulting in a characteristic strength of $f_c' - 400$ psi (2.75 MPa). The value of l_d may be multiplied by $0.7 \leq (1 - 0.00028p) \leq 1.0$, where p = the transverse pressure in psi at the ultimate limit state along l_d, perpendicular to the splitting plane. For p in MPa, the term is $(1 - 0.04p)$.

The effect of bar location is included by dividing l_d by 0.7 for bars that are both more than 10 in. (250 mm) from the bottom and less than 12 in. (300 mm) from the top of a concrete layer during placement.

As for the other design methods, the development length may be reduced by the ratio (A_s required)/(A_s provided).

l_d is increased for lapped splices by multiplying by a factor $\alpha_b = 1.2$ when ≤ 20% of the bars are spliced at one location. α_b increases to 1.4 for 25%, 1.6 for 33%, 1.8 for 50%, and 2.0 for ≥ 50% of the bars spliced at one location.

The minimum development length in tension is

$$l_{d,min} = \max\left[\frac{0.3}{950}\frac{Mf_y}{\left(\frac{f_c' - 400}{1450}\right)^{2/3}}, 10d_b, 4 \text{ in.}\right] \quad (4\text{-}23)$$

The minimum splice length is

$$l_{s,min} = \max\left[\frac{0.3\alpha_b}{950}\frac{Mf_y}{\left(\frac{f_c' - 400}{1450}\right)^{2/3}}, 15d_b, 8 \text{ in.}\right] \quad (4\text{-}24)$$

4.5—Structural reliability and comparison of design expressions

The principles of structural reliability help ensure that structures possess an acceptably small probability of failure. In this section, ACI Committee 408 Database 10-2001 is used to compare the accuracy and relative safety of the four design procedures. Because high relative rib area bars are not in general production, most comparisons are based on conventional reinforcement, although tests of high relative rib area bars not confined by transverse reinforcement are used in the analysis because they exhibit the same bond strength as conventional bars. Only tests with development or splice lengths of at least 12 in. (300 mm) and $l_d/d_b \geq 16$ are included to ensure that the comparisons only involve realistic development and splice lengths. The database used for the comparisons consists of 157 tests in which the developed or spliced bars are not confined by transverse reinforcement and 163 tests in which they are.

4.5.1 *Comparison with data*—Comparisons of the design expressions (the predictions) with the test data are summarized in Fig. 4.1 and 4.2 and Table 4.2. Figure 4.1 and 4.2 show the distribution in test-prediction ratios for ACI 318 (Eq. (4-2)), ACI 408.3 (Eq. (4-6)), ACI 408R (Eq. (4-18)) and CEB-FIP [Eq. (4-22)]. Table 4.2 gives the maximum, minimum, and average test-prediction ratios, along with the standard deviation and coefficient of variation, for the expressions shown in Fig. 4.1 and 4.2 plus the second ACI 408R equation, Eq. (4-21). The table presents the results for both the full range of compressive strengths and for specimens with compressive strengths below 10,000 psi (69 MPa).

The comparisons demonstrate that the test-prediction ratios for ACI 318 and CEB-FIP exhibit more scatter than those for ACI 408.3 and 408R. The greater scatter is especially apparent for CEB-FIP, which exhibits a range in test-prediction ratios from 0.32 to 2.48. As shown in Table 4.2, many of the lower test-prediction ratios result from the inaccuracy of the CEB-FIP expression for bond in high-strength concrete. When test results for f_c' > 10,000 psi (69 MPa) are removed, the minimum test-prediction ratio for Eq. (4-22) increases to 0.79, in line with the values produced by the other expressions.

Overall, the ACI 318 and CEB-FIP provisions provide higher average test-prediction ratios and higher coefficients of variation than do those of ACI 408.3 and ACI 408R. The result is that, on the average, ACI 318 and the CEB-FIP Model Code require longer development and splice lengths

Table 4.2—Test-prediction ratios for design provisions

Bars not confined by transverse reinforcement					
	ACI 318 (Eq. (4-2))	ACI 408.3 (Eq. (4-6))	ACI 408R (Eq. (4-18))	ACI 408R (Eq. (4-21))	CEB-FIP (Eq. (4-22))
For all f_c' (157 tests)					
Maximum	2.369	1.500	1.404	1.575	2.011
Minimum	0.624	0.768	0.749	0.843	0.322
Average	1.229	1.135	1.092	1.227	1.095
Standard deviation	0.281	0.138	0.125	0.140	0.325
Coefficient of variation	0.228	0.121	0.114	0.114	0.296
For f_c' ≤ 10,000 psi (69 MPa) (114 tests)					
Maximum	1.908	1.472	1.401	1.574	2.011
Minimum	0.756	0.768	0.749	0.843	0.791
Average	1.229	1.128	1.089	1.224	1.230
Standard deviation	0.229	0.118	0.108	0.121	0.232
Coefficient of variation	0.186	0.104	0.099	0.099	0.188
Bars confined by transverse reinforcement					
	ACI 318 (Eq. (4-2))	ACI 408.3 (Eq. (4-6))	ACI 408R (Eq. (4-18))	ACI 408R (Eq. (4-21))	CEB-FIP (Eq. (4-22))
For all f_c' (163 tests)					
Maximum	2.192	1.668	1.525	1.715	2.482
Minimum	0.704	0.850	0.802	0.903	0.457
Average	1.232	1.185	1.132	1.273	1.392
Standard deviation	0.299	0.172	0.163	0.183	0.372
Coefficient of variation	0.242	0.145	0.144	0.144	0.267
For f_c' ≤ 10,000 psi (69 MPa) (128 tests)					
Maximum	2.192	1.524	1.525	1.715	2.482
Minimum	0.704	0.850	0.845	0.951	0.894
Average	1.237	1.176	1.150	1.294	1.503
Standard deviation	0.311	0.162	0.155	0.175	0.306
Coefficient of variation	0.251	0.137	0.135	0.135	0.204

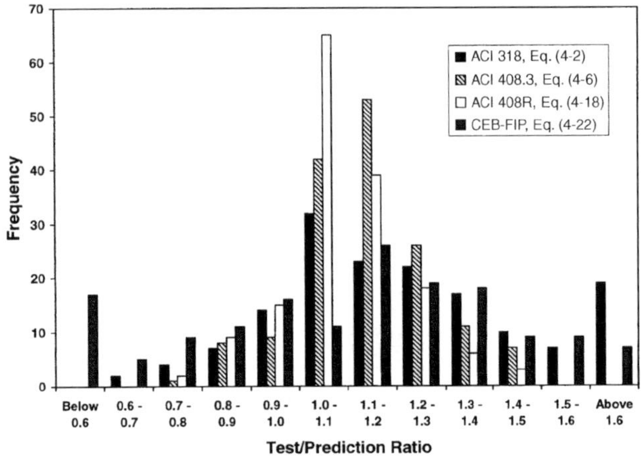

Fig. 4.1—Test-prediction ratios for design provisions for bars not confined by transverse reinforcement.

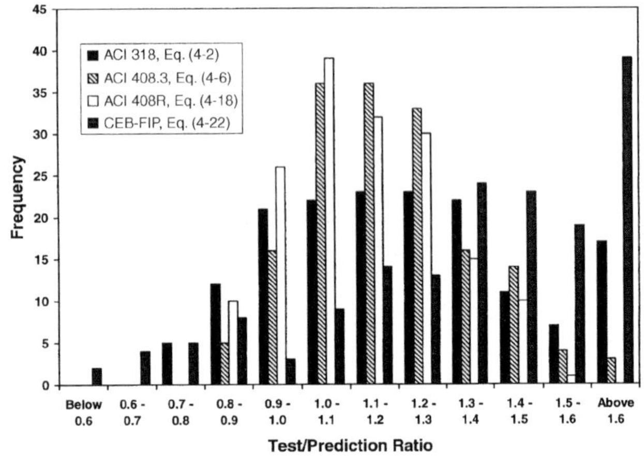

Fig. 4.2—Test-prediction ratios for design provisions for bars confined by transverse reinforcement.

than ACI 408.3 or ACI 408R, while simultaneously having a greater probability of low strength.

Equation (4-21), the ACI 408R equation for use with the load factors and strength reduction factors in Chapter 9 of ACI 318-02, provides average test-prediction ratios that are approximately the same or slightly higher than those provided by ACI 318. The range in test-prediction ratios obtained with Eq. (4-21), however, is considerably narrower. The test-prediction ratios obtained with ACI 318 have both higher maximums (2.37 and 2.19 for bars not confined and confined by transverse reinforcement, respectively) than Eq. (4-21) (1.58 and 1.72, respectively) and lower minimums (0.62 and 0.70 for bars not confined and confined by transverse reinforcement, respectively) than Eq. (4-21) (0.84 and 0.90, respectively).

Fortunately, the higher probability of low bond strength provided by ACI 318 and the CEB-FIP Model Code has not translated into structural failures because of the safety margin provided by load factors and by the ϕ-factors for bending and axial load, as well as other design provisions, such as those that limit the location of splices. The overall effect is that ACI 318 and the CEB-FIP Model Code provide lower levels of reliability in bond, when compared to those in bending and axial load, than would be achieved if the development length provisions had been formulated based on a desired level of reliability, as done with the ACI 408.3 and ACI 408R design procedures.

4.5.2 *γ-factor in ACI 318*[*]—As described in Section 4.1, the development length provisions in ACI 318 include a factor for bar size γ, with $\gamma = 0.8$ for No. 6 (No. 19) and smaller bars and 1.0 for No. 7 (No. 22) and larger bars. The commentary in ACI 318 justifies the use of $\gamma = 0.8$ based on a "comparison with past provisions and a check of a database of experimental results maintained by ACI Committee 408" which "indicated that for No. 6 [No. 19] deformed bars and smaller, as well as deformed wire, development length could be reduced 20% using $\gamma = 0.8$." At the time these comparisons were made, the ACI 408 database only included tests of small bars with short development and splice lengths, most of which were less than 12 in. (300 mm).

As discussed in Section 2.2, expressions such as Eq. (4-2), which are based on a proportional relationship between bond force and bonded length, conservatively predict the bond strength of bars with low development and splice lengths. Equation (4-2) becomes progressively less conservative, however, as more realistic splice lengths are used.

In the years since the original comparison was made, the number of tests in the ACI 408 database for No. 6 (No. 19) and smaller bars with development and splice lengths in excess of 12 in. (300 mm) has increased to 63—22 for bars without confining transverse reinforcement, and 41 for bars with confining transverse reinforcement. Figure 4.3 and 4.4 compare test-prediction ratios for No. 6 (No. 19) and smaller bars for the ACI 318 provisions using $\gamma = 0.8$ and 1.0 for bars without and with confining transverse reinforcement, respectively. The Class B splice factor of 1.3 is not used in the comparisons. For $\gamma = 0.8$, seven tests, or 32% of the bars without confining transverse reinforcement, have test-prediction ratios of less than 1.0 (Fig. 4.3). For bars with confining transverse reinforcement, 23 tests, or 56% of the data, have test-prediction ratios of less than 1.0 (Fig. 4.4). Changing γ to 1.0 substantially decreases the number of low test-prediction ratios to 1, or 4.5%, for bars without confining transverse reinforcement and to 4, or 10%, for bars with confining transverse reinforcement. Thus, based on bars with realistic splice and development lengths, it is the position of ACI Committee 408 that there is no justification for using γ less than 1.0.

4.5.3 *Reliability—ACI 318 and ACI 408R*—This section summarizes the development of strength-reduction factors ϕ for bond for use with the provisions in ACI 318 (Eq. (4-2)) and for deriving the ACI 408R expressions (Eq. (4-18) and (4-21)) based on Eq. (3-16). The analyses follow the procedures described by Ellingwood et al. (1980), Mirza and MacGregor (1986), Lundberg (1993), and Darwin et al. (1998). Separate Monte Carlo simulations are carried out for bars without and with confining transverse reinforcement.

[*]Section 4.5.2 draws heavily on the discussion by Darwin and Zuo (2002) of the changes proposed for ACI 318-02.

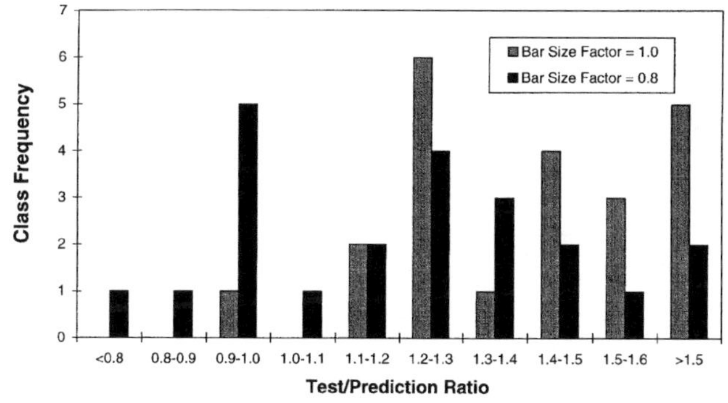

Fig. 4.3—Test-prediction ratios for spliced and developed No. 6 (No. 19) and smaller bars without confining transverse reinforcement for ACI 318; bar size factors γ = 0.8 and 1.0.

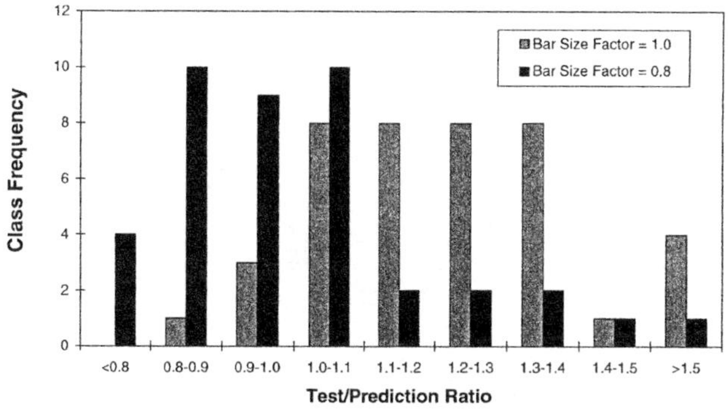

Fig. 4.4—Test-prediction ratios for spliced and developed No. 6 (No. 19) and smaller bars with confining transverse reinforcement for ACI 318; bar size factors γ = 0.8 and 1.0.

The analyses use dead load, live load, and tension ϕ-factor combinations of:

(a) 1.4, 1.7, and 0.9 (ACI 318-99, Chapter 9, and ACI 318-02, Appendix C);

(b) 1.2, 1.6, and 0.8 (ACI 318-99, Appendix C) and

(c) 1.2, 1.6, and 0.9 (ACI 318-02, Chapter 9).

A reliability index $\beta = 3.5$ is used, providing a probability of failure in bond equal to about 1/5 of the probability of failure in bending (β = 3.0). Comparisons are made for bottom-cast bars with development and splice lengths ≥ 12 in. (300 mm) and $16d_b$. Live-to-dead load ratios of 0.5, 1.0, and 1.5 are used. All analyses use a bar size factor $\gamma = 1.0$ for small as well as large bars for the reasons discussed in Section 4.5.2.

For each combination of variables, two ϕ-factors are calculated: one, a basic bond ϕ-factor, ϕ_b, represents the capacity-reduction factor that would be obtained based strictly on development and splice strength as a structural property. An effective ϕ-factor, ϕ_d, is then calculated by dividing ϕ_b by the ϕ-factor used for bars in tension $\phi_{tension}$. The higher effective ϕ-factor for bond and development ϕ_d recognizes that maximum bond force is independent of the flexural capacity and that the area of the bars has been increased (and the bar stress decreased) based on the $\phi_{tension}$.

The results of the analyses are summarized for Eq. (4-2) and Eq. (3-16) in Table 4.3 and 4.4, respectively. As shown in Table 4.3, the results justify the application of a ϕ-factor $\phi_d \approx 0.85$ to the development and splice length criteria in ACI 318-99 and Appendix C of ACI 318-02. ϕ_d drops to values of about 0.75 for Chapter 9 of ACI 318-02.

Darwin and Zuo (2002) observed that the provisions of ACI 318-99 provide a reliability index β = 2.8 to 3.0 for bond, somewhat lower than the desired reliability index for flexure. β drops to 2.4 to 2.6 for Chapter 9 of ACI 318-02. The conclusion is that, under the provisions of ACI 318-02, there is a greater probability of a bond failure than a flexural failure. This situation is mitigated to a large extent for Class B splices because of the 1.3 factor. The 1.3 factor is, however, not applied for developed bars and Class A splices.

The ϕ-factors used with Eq. (3-16) to produce the two ACI 408R expressions, Eq. (4-18) and (4-21), are presented

Table 4.3—Strength-reduction (ϕ) factors for bond using ACI 318-99 and ACI 318-02; β = 3.5

$\bar{r}$	1.23			1.23		
V_r	0.23			0.24		
	Without confining transverse reinforcement			With confining transverse reinforcement		
	Dead load factor = 1.4; live load factor = 1.7; $\phi_{tension}$ = 0.9					
$(Q_L/Q_D)_n$	0.50	1.00	1.50	0.50	1.00	1.50
$\bar{q}$	0.67	0.65	0.63	0.67	0.65	0.63
$V_{\phi q}$	0.10	0.13	0.15	0.10	0.13	0.15
ϕ_b	0.76	0.76	0.75	0.73	0.73	0.72
ϕ_d	0.84	0.84	0.83	0.81	0.81	0.80
	Dead load factor = 1.2; live load factor = 1.6; $\phi_{tension}$ = 0.8					
$(Q_L/Q_D)_n$	0.50	1.00	1.50	0.50	1.00	1.50
$\bar{q}$	0.76	0.72	0.69	0.76	0.72	0.69
$V_{\phi q}$	0.10	0.13	0.15	0.10	0.13	0.15
ϕ_b	0.68	0.68	0.68	0.65	0.66	0.65
ϕ_d	0.84	0.86	0.85	0.81	0.82	0.82
	Dead load factor = 1.2; live load factor = 1.6; $\phi_{tension}$ = 0.9					
$(Q_L/Q_D)_n$	0.50	1.00	1.50	0.50	1.00	1.50
$\bar{q}$	0.76	0.72	0.69	0.76	0.72	0.69
$V_{\phi q}$	0.10	0.13	0.15	0.10	0.13	0.15
ϕ_b	0.68	0.68	0.68	0.65	0.66	0.65
ϕ_d	0.75	0.76	0.76	0.72	0.73	0.73

Notes:
$\bar{r}$ = mean test-prediction ratio comparing Eq. (4-2) with results from Database 10-2001.
V_r = coefficient of variation of resistance random variable r.
$(Q_L/Q_D)_n$ = nominal ratio of live load to dead load.
$\bar{q}$ = mean value of random loading variable.
$V_{\phi q}$ = coefficient of variation of loading random variable q.
$\phi_b = \frac{\bar{r}}{\bar{q}} e^{-(V_r^2 + V_{\phi q}^2)^{1/2}\beta}$.
$\phi_d = \phi_b/\phi_{tension}$.

in Table 4.4. The table includes separate results for high relative rib area bars (average R_r = 0.1275) confined by transverse reinforcement and for bars with conventional deformation patterns. As discussed in Section 4.3, values of ϕ_d = 0.92 and 0.82 are selected for application with ACI 318-99 and Appendix C of ACI 318-02 (Eq. (4-18)) and Chapter 9 of ACI 318-02 (Eq. (4-21)), respectively.

The lower values of ϕ_d for use with Chapter 9 of ACI 318-02, 0.75 for ACI 318 and 0.82 for ACI 408, reflect the

Table 4.4—Strength-reduction (ϕ) factors for bond using Eq. (3-16) to develop Eq. (4-18) and (4-21); β = 3.5

$\bar{r}$	1.00			1.00					
V_r	0.11			0.12					
	Without confining transverse reinforcement			With confining transverse reinforcement					
				R_r = 0.727			R_r = 0.1275		
	Dead load factor = 1.4; live load factor = 1.7; $\phi_{tension}$ = 0.9								
$(Q_L/Q_D)_n$	0.50	1.00	1.50	0.50	1.00	1.50	0.50	1.00	1.50
$\bar{q}$	0.67	0.65	0.63	0.67	0.65	0.63	0.67	0.65	0.63
$V_{\phi q}$	0.10	0.13	0.15	0.10	0.13	0.15	0.10	0.13	0.15
ϕ_b	0.87	0.85	0.82	0.86	0.83	0.81	0.85	0.83	0.80
ϕ_d	0.97	0.94	0.91	0.95	0.93	0.90	0.95	0.92	0.89
	Dead load factor = 1.2; live load factor = 1.6; $\phi_{tension}$ = 0.8								
$(Q_L/Q_D)_n$	0.50	1.00	1.50	0.50	1.00	1.50	0.50	1.00	1.50
$\bar{q}$	0.76	0.72	0.69	0.76	0.72	0.69	0.76	0.72	0.69
$V_{\phi q}$	0.10	0.13	0.15	0.10	0.13	0.15	0.10	0.13	0.15
ϕ_b	0.78	0.77	0.75	0.76	0.75	0.74	0.76	0.75	0.73
ϕ_d	0.97	0.96	0.93	0.95	0.94	0.92	0.95	0.94	0.92
	Dead load factor = 1.2; live load factor = 1.6; $\phi_{tension}$ = 0.9								
$(Q_L/Q_D)_n$	0.50	1.00	1.50	0.50	1.00	1.50	0.50	1.00	1.50
$\bar{q}$	0.76	0.72	0.69	0.76	0.72	0.69	0.76	0.72	0.69
$V_{\phi q}$	0.10	0.13	0.15	0.10	0.13	0.15	0.10	0.13	0.15
ϕ_b	0.78	0.77	0.75	0.76	0.75	0.74	0.76	0.75	0.73
ϕ_d	0.86	0.85	0.83	0.84	0.84	0.82	0.84	0.83	0.82

Notes:
$\bar{r}$ = mean test-prediction ratio comparing Eq. (4-2) with results from Database 10-2001.
V_r = coefficient of variation of resistance random variable r.
$(Q_L/Q_D)_n$ = nominal ratio of live load to dead load.
$\bar{q}$ = mean value of random loading variable.
$V_{\phi q}$ = coefficient of variation of loading random variable q.

$$\phi_b = \frac{\bar{r}}{\bar{q}} e^{-(V_r^2 + V_{\phi q}^2)^{1/2}\beta}.$$

$\phi_d = \phi_b/\phi_{tension}$.

combined effects of the higher value of $\phi_{tension}$ and the lower load factors.

4.5.4 *Comparisons of development and splice lengths ACI 318, ACI 408.3, and ACI 408*—Development and splice lengths based on the design provisions of ACI 318 (Eq. (4-2)), ACI 408.3 (Eq. (4-6)), and ACI 408R (Eq. (4-18)) and (4-21)) are expressed in terms of l_d/d_b in Table 4.5 for concrete strengths of 3000 to 8000, 10,000, 12,000, and 15,000 psi (21 to 55, 69, 83, and 103 MPa) and Grade 60 (414 MPa) reinforcing bars. The comparisons are for conventional reinforcing bars, and although ACI 408.3 is limited (by its provisions) to high relative rib area bars ($R_r \geq 0.10$), it is applied in this case to conventional reinforcement (average R_r = 0.0727). For simplicity, the comparisons are made for No. 8 (No. 25) bars (γ= 1.0 for ACI 318 and t_d = 1.0 for ACI 408.3 and ACI 408) and ω= 1.0 for ACI 408.3 and ACI 408. Three cases are covered:

1. Minimum confinement, $(c + K_{tr})/d_b = 1.0$;
2. c/d_b = 1.0 and A_{tr}/sn = 0.0125, which corresponds to $(c + K_{tr})/d_b$ = 1.5 for ACI 318, values less than 1.5 for ACI 408.3, and values less than or greater than 1.5 for ACI 408, depending on f_c'; and
3. $(c + K_{tr})/d_b$ = maximum allowed by the provisions, = 2.5 for ACI 318 and = 4 for ACI 408.3 and ACI 408.

The comparisons of development and Class A splice lengths in Table 4.5 show that the values obtained with ACI 408.3 and ACI 408R are greater than those required by ACI 318 when $(c + K_{tr})/d_b$ = 1.0 for the range of concrete strengths evaluated. In this case, the development lengths required by ACI 408.3 exceed those required by ACI 318 by values that range from 5% for f_c' = 3000 psi (21 MPa) to 27% for f_c' = 10,000 psi (69 MPa), dropping to 9% for f_c' = 15,000 psi (103 MPa). The values of l_d provided by the ACI 408R design procedures for use with ACI 318-99 and Appendix C of ACI 318-02 (Eq. (4-18)) range from 3% greater than required by ACI 318 for f_c' = 3000 psi (21 MPa) to 21% greater at 10,000 psi (69 MPa), dropping to 2% greater at f_c' = 15,000 psi (103 MPa). The development and Class A splice lengths based on the ACI 408R design procedures for Chapter 9 of ACI 318-02 (Eq. (4-21)) range from 20% greater at 3000 psi (21 MPa) to 44% greater at 10,000 psi (69 MPa), dropping to 24% greater at 15,000 psi (103 MPa). The longer development lengths required by ACI 408.3 and ACI 408R provide a higher margin of safety compared with those calculated using ACI 318 and are a principal reason for the lower number of low test-prediction ratios obtained with these provisions (Section 4.5.1). For cases in which $1.0 < \omega \leq 1.25$, the development lengths l_d based on ACI 408.3 and

Table 4.5—Comparison of development and splice lengths for ACI 318, ACI 408.3, and ACI 408R; l_d/d_b for No. 8 (No. 25) Grade 60 (414 MPa) reinforcing bars; $\omega = 1$

	Development and Class A splice length, in.				Class B splice length, in.			
f_c', psi	ACI 318 Eq. (4-2)	ACI 408.3* Eq. (4-6)	ACI 408R Eq. (4-18)	ACI 408R Eq. (4-21)	ACI 318 Eq. (4-2)	ACI 408.3* Eq. (4-6)	ACI 408R Eq. (4-18)	ACI 408R Eq. (4-21)
$(c + K_{tr})/d_b = 1$								
3000	82.2	86.2	84.4	99.0	106.8	86.2	84.4	99.0
4000	71.2	78.4	76.4	89.9	92.5	78.4	76.4	89.9
5000	63.6	72.7	70.5	83.3	82.7	72.7	70.5	83.3
6000	58.1	68.3	66.0	78.2	75.5	68.3	66.0	78.2
7000	53.8	64.7	62.3	74.0	69.9	64.7	62.3	74.0
8000	50.3	61.7	59.2	70.6	65.4	61.7	59.2	70.6
10,000	45.0	56.9	54.3	65.0	58.5	56.9	54.3	65.0
12,000	45.0	53.2	50.5	60.7	58.5	53.2	50.5	60.7
15,000	45.0	48.9	46.0	55.7	58.5	48.9	46.0	55.7
$c/d_b = 1$ and $A_{tr}/sn = 0.0125$—corresponds to $(c + K_{tr})/d_b = 1.5$ for ACI 318								
3000	54.8	60.0	62.9	73.7	71.2	60.0	62.9	73.7
4000	47.4	54.5	54.7	64.4	61.7	54.5	54.7	64.4
5000	42.4	50.6	48.9	57.8	55.2	50.6	48.9	57.8
6000	38.7	47.5	44.4	52.7	50.3	47.5	44.4	52.7
7000	35.9	45.0	40.9	48.6	46.6	45.0	40.9	48.6
8000	33.5	42.9	38.0	45.3	43.6	42.9	38.0	45.3
10,000	30.0	39.6	33.4	40.0	39.0	39.6	33.4	40.0
12,000	30.0	37.0	30.0	36.0	39.0	37.0	30.0	36.0
15,000	30.0	34.0	26.1	31.5	39.0	34.0	26.1	31.5
$(c + K_{tr})/d_b$ = maximum, = 2.5 for ACI 318, = 4 for ACI 408.3 and ACI 408R								
3000	32.9	21.6	21.1	24.7	42.7	21.6	21.1	24.7
4000	28.5	19.6	19.1	22.5	37.0	19.6	19.1	22.5
5000	25.5	18.2	17.6	20.8	33.1	18.2	17.6	20.8
6000	23.2	17.1	16.5	19.5	30.2	17.1	16.5	19.5
7000	21.5	16.2	16.0	18.5	28.0	16.2	16.0	18.5
8000	20.1	16.0	16.0	17.6	26.2	16.0	16.0	17.6
10,000	18.0	16.0	16.0	16.3	23.4	16.0	16.0	16.3
12,000	18.0	16.0	16.0	16.0	23.4	16.0	16.0	16.0
15,000	18.0	16.0	16.0	16.0	23.4	16.0	16.0	16.0

*Applied to conventional reinforcing bars.
Note: 1 psi = 6.895×10^{-3} MPa; 1 in. = 25.4 mm.

ACI 408R could be reduced by as much as 35% from the values shown in Table 4.5.

For $c/d_b = 1$ and $A_{tr}/sn = 0.0125$, ACI 408.3 and ACI 408R again require longer development lengths than ACI 318 by values that range up to 32% for ACI 408.3 and up to 15% and 36%, respectively, for ACI 408R Eq. (4-18) and (4-21). In this case, the ratios are greater than for $(c + K_{tr})/d_b = 1.0$ until f_c' exceeds 10,000 psi (69 MPa), because ACI 318 gives greater credit to transverse reinforcement (a higher value of K_{tr}) than do the ACI 408.3 and ACI 408R expressions at lower values of f_c'. This is another reason that ACI 318 may overpredict bond strength, as illustrated in Section 4.5.1.

Because K_{tr} for ACI 408R is a function of f_c', it exceeds K_{tr} for ACI 318 for $f_c' > 6400$ psi (44 MPa). The values of l_d for ACI 318 and ACI 408R Eq. (4-18) are equal at f_c' = 12,000 psi (83 MPa), with the value for Eq. (4-18) equal to 87% of that for ACI 318 at f_c' = 15,000 psi (103 MPa).

For cases in which $(c + K_{tr})/d_b$ equals the maximum allowed by each of the provisions (= 2.5 for ACI 318, = 4 for ACI 408.3 and ACI 408), the values of l_d for the ACI 408.3 and ACI 408R provisions are consistently below the value for ACI 318 by from 11 to 36%. As f_c' increases, l_d for ACI 408.3 and ACI 408R is constrained by the minimum l_d/d_b ratio = 16. This occurs at 8000, 7000, and 12,000 psi (55, 48, and 83 MPa) for ACI 408.3 and ACI 408R Eq. (4-18) and (4-21), respectively.

The Class B splice lengths based on the ACI 408.3 and ACI 408R design procedures are, in most cases, below those required by ACI 318 because the 1.3 factor is not required. The ACI 408.3 provisions (Eq. (4-6)) provide splice lengths that range between 50 and 102% of those required by ACI 318. The ACI 408R procedures (Eq. (4-18)) provide splice lengths that range between 49 and 93% of those required by ACI 318, while those calculated using Eq. (4-21) range from 58 to 111% of the splice lengths required by ACI 318. The lowest relative values are, of course, obtained when $(c + K_{tr})/d_b$ is equal to the maximum allowed.

An additional relative decrease in both development and splice length occurs for the ACI 408.3 and ACI 408R design procedures once $f_c' > 10{,}000$ psi (69 MPa) because of the limitation of $\sqrt{f_c'}$ to 100 psi (8.3 MPa) within the provisions of ACI 318. As demonstrated in Fig. 3.1 and 3.2, such a limitation is appropriate, since Eq. (4-2) is based on Eq. (3-4), which shows a decreasing test-prediction ratio with increasing compressive strength.

Overall, the development and Class A splice lengths required by the ACI 408.3 and ACI 408R design procedures are greater than those required by ACI 318 for conditions of low cover and confinement and lower concrete strengths. The values of l_d obtained with the ACI 408.3 and ACI 408R design procedures decrease with respect to those obtained with ACI 318 as cover, confinement, and compressive strength increase and when $\omega > 1.0$. The ACI 408.3 and ACI 408R design procedures include reliability-based ϕ-factors that are embedded in the expressions.

The shorter Class B splice lengths required under the ACI 408.3 and ACI 408R design procedures result from the fact that a 1.3 factor is not used, because the values of l_d are based principally on splice tests in which 100% of the bars are spliced, and thus, do not require an increase for the conditions applying to Class B splices. Although the ACI 318 Class B splice factor is not based on strength, it is clear from the analyses in this section that it does provide improved reliability and should not be set to 1.0 if the other development and splice provisions in ACI 318 remain unchanged.

CHAPTER 5—DATABASE

Databases maintained by ACI Committee 408 have played an important role in the development of design expressions adopted in ACI 318. The databases, however, have often been maintained on an ad hoc basis, and it has only been since 1997 that a formal database has been in existence. This chapter describes the procedures used to determine bar stresses at the time of failure in development and splice tests and provides a general description of the database used for the analyses presented in Chapters 3 and 4. As will be described in Section 5.2, the database is available in a form that can be readily used by researchers.

5.1—Bar stresses

The force in a developed or spliced bar at bond failure is a structural property of the same type as flexural or shear strength. Because most development and splice tests involve bending, determination of the force in the bar at failure involves a calculation that depends on both the bending moment corresponding with failure and the properties of the nonlinear, inelastic materials involved. Historically, three approaches have been used to calculate the force in reinforcement at bond failure: the working stress method, the strength method, and the moment-curvature method. In the working stress method, the stress in the concrete and steel can be determined based on static equilibrium and compatibility of strain using a transformed section in which the steel area is replaced by an equivalent area of concrete based on the ratio of the moduli of elasticity. In the strength method, an average concrete compressive stress of 0.85 is assumed to be uniformly distributed over a stress block, the depth of which is equal to a portion of the depth to the neutral axis. The bar stress at failure can be calculated based on this assumption and equilibrium. In the moment-curvature method, stress-strain relationships are assumed for the concrete and the reinforcing steel. In all three methods, strains in the materials are assumed to vary linearly over the depth of the member.

Zuo and Darwin (1998) compared the three methods and concluded that the moment-curvature method provided the most realistic results. They observed that in cases where the bar stress is below the yield strength of the steel, the working stress method overestimates bar stresses for high-strength concrete and underestimates bar stresses for normal-strength concrete, compared to the moment-curvature method, especially for f_c' less than 3000 psi (21 MPa). The strength method, however, underestimates bar stresses in almost all cases.

For members in which the calculated bar stress exceeds the yield strength of the steel, the working stress method consistently overestimates bar stresses for high-strength concrete ($f_c' \geq 10{,}000$ psi [70 MPa]) and in slightly more than half the cases for beams made with normal-strength concrete compared to the moment-curvature method. In contrast, the strength method underestimates the bar stresses for high-strength concrete and in about 40% of the tests for beams made with normal-strength concrete. The evaluation included a total of 439 development or splice test specimens.

Based on this analysis, the bar stresses at failure in the ACI Committee 408 database are based on moment-curvature calculations. In about 3% of the tests, the bending capacity at failure in a splice or development test exceeds the value calculated based on the moment-curvature analysis. In these cases, the stress assigned to the steel at failure is based on the working stress method if the calculated stress is below the yield stress of the steel and the strength method if it is equal to or greater than the yield strength of the steel.

For calculations with the database, the expression in Section 8.5 of ACI 318 for normalweight concrete is used to determine the modulus of elasticity of concrete (all tests in the current database involve normalweight concrete).

$$E_c = 57{,}000\sqrt{f_c'} \tag{5-1}$$

A modulus of elasticity E_s of 29,000 ksi (200,000 MPa) is used for steel bars. Concrete is treated as a material with no tensile strength and the area of the steel A_s is taken as the area of the bars assuming that they are continuous rather than spliced.

A parabolic equation (Hognestad 1951) is used for the relationship between concrete stress f_c and strain ε_c

$$f_c = f_c'\left[\frac{2\varepsilon_c}{\varepsilon_o} - \left(\frac{\varepsilon_c}{\varepsilon_o}\right)^2\right] \tag{5-2}$$

where ε_o is the concrete strain at maximum concrete stress. The value of ε_o is a function of f_c' and is obtained from the experimental curves shown by Nilson (1997) for f_c' from 3000 to 12,000 psi (20.7 to 82.7 MPa); ε_o is nearly a linear function

Table 5.1—Values of concrete strain at maximum concrete stress ε_o for values of compressive strength $f_c' \geq 3000$ psi (20.7 MPa)

f_c'		ε_o
psi	MPa	με
3000	20.7	1919
4000	27.8	2065
5000	34.5	2200
6000	41.4	2330
7000	48.3	2440
8000	55.2	2550
9000	62.1	2650
10,000	69.0	2750
11,000	75.9	2850
12,000	82.8	2950
13,000	89.7	3050
14,000	96.6	3015
15,000	103.4	3025
16,000	110.3	3035

of f_c' for high-strength concrete ($f_c' \geq 8000$ psi [55.2 MPa]). Based on this observation, values of ε_o for $f_c' > 12{,}000$ psi (82.7 MPa) are obtained by extrapolation of the best-fit line for f_c' of 8000, 10,000, and 12,000 psi (55.2, 69.0, and 82.7 MPa). For concrete with compressive strength less than 3000 psi (20.7 MPa), the values of ε_o are determined using (Bashur and Darwin 1976, 1978).

$$\varepsilon_o = \frac{f_c'}{363{,}000 + 400f_c'} \quad (5\text{-}3)$$

For values of f_c' greater than 3000 psi (20.7 MPa), the values of ε_o are given in Table 5.1. Intermediate values are obtained by interpolation.

Representative stress-strain curves for reinforcing steel (Nilson, Darwin, and Dolan 2004) are used to establish the stress-strain curves for moment curvature calculations. The steel strain at the initiation of strain hardening ε_{sh} is equal to 0.0086 for Grade 60 (414 MPa) steel and 0.0035 for Grade 75 (517 MPa) and above. The value of ε_{sh} varies linearly between the two grades. There is no yield plateau in the stress-strain curve for $f_y \geq 101.5$ ksi (700 MPa). The modulus of elasticity for strain hardening E_h is 614 ksi (4244 MPa), (0.021 E_s) for f_y = 60 ksi (414 MPa), 713 ksi (4916 MPa) (0.025 E_s) for f_y = 75 ksi (517 MPa), and 1212 ksi (8357 MPa) (0.042 E_s) for $f_y \geq 90$ ksi (620 MPa). The values of E_h for f_y between 60 and 75 ksi (412 and 517 MPa) and between 75 and 90 ksi (517 and 620 MPa) are obtained using linear interpolation. The idealized stress-strain curves used are shown in Fig. 5.1.

5.2—Database

At this writing, the ACI Committee 408 database includes the results of 635 development and splice tests of uncoated reinforcing bars. The data are currently limited to normalweight concrete specimens for which compressive strength was obtained using cylinders that were tested in accordance with ASTM C 39. Tests for which compressive strength was measured solely on the basis of cubes are not included because of the highly variable relationship between cube and cylinder strength. Tests in the database are categorized based on bar placement, with individual databases maintained for bottom-cast (478 tests), top-cast (111 tests), and side-cast bars (46 tests).

The database is maintained by Subcommittee D of ACI Committee 408 and available in Microsoft® Excel™ from ACI headquarters or Committee 408. The reference list for the database is maintained in Microsoft® Word™.

Because the database is actively maintained and updated regularly, the database used for comparisons should be identified based on the month and year. Database 10-2001 was used in the analyses described in Chapters 3 and 4. The studies used to establish Database 10-2001 are listed in Table 5.2. As the database is upgraded, earlier versions of the database are maintained.

Table 5.2—References in Database 10-2001 for development and splice tests of reinforcing bars in tension*

Azizinamini et al. (1993)
Azizinamini, Chisala, and Ghosh (1995)
Azizinamini et al. (1999)
Chinn, Ferguson, and Thompson (1955)
Chamberlin (1956)
Chamberlin (1958)
Choi et al. (1990)
Choi et al. (1991)
Darwin et al. (1996a)
DeVries, Moehle, and Hester (1991)
Ferguson and Breen (1965)
Ferguson and Briceno (1969)
Ferguson and Krishnaswamy (1971)
Ferguson and Thompson (1962)
Ferguson and Thompson (1965)
Hamad and Jirsa (1990)
Hamad and Itani (1998)
Hasan, Cleary, and Ramirez (1996)
Hester et al. (1991)
Hester et al. (1993)
Kadoriku (1994)
Mathey and Watstein (1961)
Rezansoff, Konkankar, and Fu (1991)
Rezansoff, Akanni, and Sparling (1993)
Thompson et al. (1975)
Treece and Jirsa (1989)
Zekany, Neumann, Jirsa, and Breen (1981)
Zuo and Darwin (1998)
Zuo and Darwin (2000)

*Bars cast in normalweight concrete. Compressive strength based on cylinders. Includes bottom-, side-, and top-cast bars.

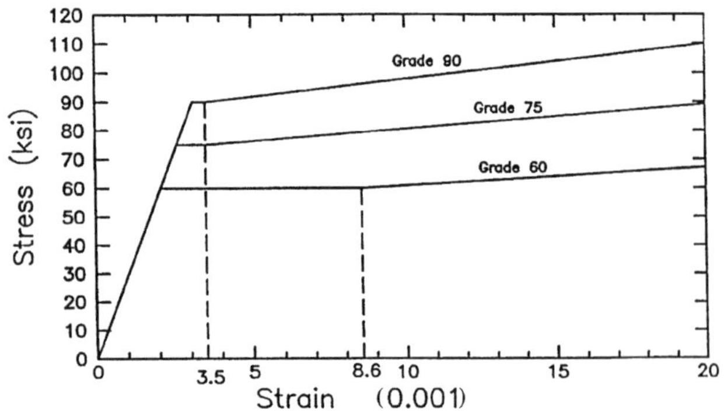

Fig. 5.1—Idealized stress-strain curves for reinforcing bars used to determine stress in development and splice tests (Zuo and Darwin 1998). (Note: 1 psi = 0.00689 MPa)

CHAPTER 6—TEST PROTOCOL

A single test procedure has not been established for determining the bond strength between reinforcing steel and concrete. As discussed in Section 1.2, however, a limited number of test specimen types are used in most studies. To be of the greatest use in understanding the bond behavior of reinforced concrete members, the results of the tests should be reported with at least a minimum level of detail. This chapter presents recommended testing and reporting criteria to allow for the equitable comparison and clear exchange of bond test data.

6.1—Reported properties of reinforcement

The properties of the reinforcing steel are required for basic identification and, in most cases, are needed to fully characterize the steel used in the tests. The following information should be provided for each heat or production run of reinforcing steel: the standard (ASTM, DIN, ISO, and so on) under which the bars were manufactured, the nominal diameter, bar designation, yield strength, tensile strength, proof strength (if applicable), elongation at failure, Poisson's ratio (if nonferrous), weight (mass) per unit length, rib spacing and rib height (according to the standard under which the bar was manufactured as well as the average value), relative rib area (refer to Section 6.6), rib angle β (the included angle between the rib and the bar axis), rib-face angle α, and type of coating and coating thickness (if applicable).

6.2—Concrete properties

The properties of concrete play a critical role in the bond between reinforcement and concrete. Therefore, certain key properties must be known to properly characterize or model the concrete. The following information should be provided: the source of the concrete, the mixture proportions (including identification of the components: cement type, mineral admixtures, chemical admixtures, fine and coarse aggregates, and their properties, such as specific gravity (saturated surface dry) and absorption of aggregates, specific gravity and percent solids of chemical admixtures), the concrete compressive strength as obtained from a standard concrete cylinder (cylinder should be cured side-by-side with and in the same manner as the bond test specimens), the size of the compressive strength specimens, the type and thickness of the cylinder caps, if used, and the age of testing. In addition, if possible, the flexural strength and fracture energy of the concrete should be recorded along with the specimen and test method used to establish each of these properties.

6.3—Specimen properties

A description of the test specimen should include: exterior dimensions, location of reinforcement (including the effective depth), the bottom or top cover, the side cover, the clear spacing between bars, the length of the specimen, the length of the developed or spliced bars, the number of developed or spliced bars, the quantity and nature of the transverse reinforcement used in the region of the developed or spliced bars, the average spacing of transverse reinforcement, the yield and tensile strengths of the transverse reinforcement, and the load (tensile or bending) on the specimen at the time of failure, including the specimen self-weight and the weight of the test system.

Specimen dimensions should be measured after casting. Cover and bar spacing should be measured before casting. If possible, cover should also be measured after casting, testing, or both. The most reliable values for cover and bar spacing should be reported.

6.4—Details of test

Based on what was considered to be the current understanding of bond behavior, investigators have often left out important aspects of the test in their description of the research. As the degree of understanding improves or changes with time, more details are often needed to fully understand earlier test results. With this in mind, the following information should be recorded for each test: a description of the test system; the weight of the loading system; the rate of loading, the time of test, or both; the presence or absence of strain gages on the developed or spliced reinforcing bars, the full load-deflection curves for the tests, and details of system calibration (calibration should be referenced to an accepted standard).

6.5—Analysis method

Results should be presented in terms of the average bar force at the time of failure. For flexural specimens, the calculated bar force should be determined based on moment-curvature calculations, such as described in Section 5.1, using the nominal area of the continuous steel.

Realistic stress-strain curves for the concrete and reinforcing steel should be used in the moment-curvature calculation. The relations used should be reported. For cases in which the analysis indicates that bond failure occurred at a load that exceeds the flexural capacity of the member, average bar force should be determined using a stress block (strength) approach based on the net area of the reinforcement and the strength of the concrete (this will provide a value for the bar force at failure). The basis for the calculations should be included in the report.

6.6—Relative rib area

The relative rib area R_r is the ratio of the projected rib area normal to the bar axis to the product of the nominal bar perimeter and the average center-to-center rib spacing (refer also to Fig. 1.3, Section 2.2.2, and Eq. (2-6)).

Following the procedure described in ACI 408.3, R_r may be calculated as

$$R_r = \frac{h_r}{s_r}\left(1 - \frac{\Sigma \text{gaps}}{p}\right) \quad (6\text{-}1)$$

where h_r = average height of deformations (measured according to Section 6.6.1); s_r = average spacing of deformations; Σgaps = sum of the gaps between ends of transverse deformations, plus the width of any continuous longitudinal lines used to represent the grade of the bar multiplied by the ratio of the height of the line to h_r; and p = nominal perimeter of bar. When transverse deformations tie into longitudinal ribs, the gap may be measured across the longitudinal rib at the midheight of the transverse deformation.

In European standards, R_r is calculated (for non-twisted bars) as

$$R_r = \frac{\Sigma A_r \sin\beta}{s_r p} \quad (6\text{-}2)$$

where ΣA_r = total area of ribs around bar perimeter measured on the longitudinal section *of each rib* using the trapezoidal method for approximating the area under a curve, and β = angle between transverse rib and longitudinal axis of the bar.

Methods that are more accurate than Eq. (6-1) or (6-2) may also be used (Tholen and Darwin 1996).

6.6.1 *Measuring deformation height*—The average height of deformations or ribs, h_r, should be based on measurements made on not less than two typical deformations on each side of the test bar. Determinations should be based on five measurements per deformation, one at the center of the overall length, two at the ends of the overall length, and two located halfway between the center and the ends. The measurements at the ends of the overall length are averaged to obtain a single value, and that value is combined with the other three measurements to obtain the average rib height h_r. Deformation measurements should be made using a depth gage with a knife-edge support that spans not more than two adjacent ribs. A knife edge is required to allow measurements to be made at the ends of the overall length of deformations, usually adjacent to a longitudinal rib. The calculation of h_r is based on a knife edge that spans only two ribs because measurements made with a longer knife edge result in higher average rib heights and, thus, an overestimate of the relative rib area of some bars.

CHAPTER 7—REFERENCES

7.1—Referenced standards and reports

The standards and reports listed below were the latest editions at the time this document was prepared. Because these documents are revised frequently, the reader is advised to contact the proper sponsoring group if it is desired to refer to the latest version.

AASHTO
Standard Specifications for Highway Bridges

American Concrete Institute

309R	Guide for Consolidation of Concrete
318	Building Code Requirements for Structural Concrete
408.3	Splice and Development Length of High Relative Rib Area Reinforcing Bars in Tension
544.1R	State-of-the-Art Report on Fiber Reinforced Concrete

ASTM International

A 615/A 615M	Standard Specification for Deformed and Plain Billet-Steel Bars for Concrete Reinforcement
A 706/A 706M	Standard Specification for Low-Alloy Steel Deformed and Plain Bars for Concrete Reinforcement
A 767/A 767M	Standard Specification for Zinc-Coated (Galvanized) Steel Bars for Concrete Reinforcement
A 775	Standard Specification for Epoxy-Coated Steel Reinforcing Bar
A 934	Standard Specification for Epoxy-Coated Prefabricated Steel Reinforcing Bars
A 944	Standard Test Method for Comparing Bond Strength of Steel Reinforcing Bars to Concrete Using Beam-End Specimens
A 955/A 955M	Standard Specification for Deformed and Plain Stainless Steel Bars for Concrete Reinforcement
A 996/A 996M	Standard Specification for Rail-Steel and Axle-Steel Deformed Bars for Concrete Reinforcement
C 39	Standard Test Method for Compressive Strength of Cylindrical Concrete Specimens

These publications may be obtained from these organizations:

American Association of State Highway and Transportation Officials
P.O Box 96716
Washington, DC 20090

American Concrete Institute
P.O. Box 9094
Farmington Hills, MI 48333-9094

ASTM International
100 Barr Harbor Dr.
West Conshohocken, PA 19428

7.2—Cited references

Abrams, D. A., 1913, "Tests of Bond between Concrete and Steel," *Bulletin* No. 71, Engineering Experiment Station, University of Illinois, Urbana, Ill., 105 pp.

ACI Committee 318, 1947, "Building Code Requirements for Reinforced Concrete (ACI 318-47)," American Concrete Institute, Farmington Hills, Mich., 64 pp.

ACI Committee 318, 1951, "Building Code Requirements for Reinforced Concrete (ACI 318-51)," American Concrete Institute, Farmington Hills, Mich., 63 pp.

ACI Committee 318, 1963, "Building Code Requirements for Reinforced Concrete (ACI 318-63)," American Concrete Institute, Farmington Hills, Mich., 144 pp.

ACI Committee 318, 1971, "Building Code Requirements for Reinforced Concrete (ACI 318-71)," American Concrete Institute, Farmington Hills, Mich., 78 pp.

ACI Committee 318, 1995, "Building Code Requirements for Reinforced Concrete (ACI 318-95) and Commentary (318R-95)," American Concrete Institute, Farmington Hills, Mich., 369 pp.

ACI Committee 318, 1999, "Building Code Requirements for Structural Concrete (318-99) and Commentary (318R-99)," American Concrete Institute, Farmington Hills, Mich., 391 pp.

ACI Committee 408, 1966, "Bond Stress—The State of the Art," ACI JOURNAL, *Proceedings* V. 63, No. 11, Nov., pp. 1161-1188.

ACI Committee 408, 1970, "Opportunities in Bond Research," ACI JOURNAL, *Proceedings* V. 67, No. 11, Nov., pp. 857-867.

ACI Committee 408, 1979, "Suggested Development, Splice, and Standard Hook Provisions for Deformed Bars in Tension (ACI 408.1R-79)," *Concrete International*, V. 1, No. 7, July, pp. 44-46.

ACI Committee 408, 1992, "State-of-the-Art Report: Bond under Cyclic Loads (ACI 408.2R-92 (Reapproved 1999))," American Concrete Institute, Farmington Hills, Mich., 32 pp.

ACI Committee 544, 1999, "Fiber Reinforced Concrete (ACI 544.1R-96)," American Concrete Institute, Farmington Hills, Mich., 66 pp.

ACI Innovation Task Group 2, 1998, "Splice and Development Length of High Relative Rib Area Reinforcing Bars in Tension (ACI ITG-2-98)," American Concrete Institute, Farmington Hills, Mich., 5 pp.

Ahmad, S. H., and Shah, S. P., 1985, "Standard Properties of High Strength Concrete and Its Implications for Precast Prestressed Concrete," *PCI Journal*, V. 30, No. 6, Nov.-Dec., pp. 92-119.

Altowaiji, W. A. K.; Darwin, D.; and Donahey, R C., 1984, "Preliminary Study of the Effect of Revibration on Concrete-Steel Bond Strength," *SL Report* No. 84-2, University of Kansas Center for Research, Lawrence, Kans., Nov., 29 pp.

Altowaiji, W. A. K.; Darwin, D.; and Donahey, R. C., 1986, "Bond of Reinforcement to Revibrated Concrete," ACI JOURNAL, *Proceedings* V. 83, No. 6, Nov.-Dec., pp. 1035-1042.

ASTM A 305T, 1947, "Tentative Specifications for Minimum Requirements for Deformations of Deformed Steel Bars for Concrete Reinforcement, A 305-47T," ASTM International, West Conshohocken, Pa.

ASTM A 305, 1949, "Specifications for Minimum Requirements for Deformations of Deformed Steel Bars for Concrete Reinforcement, A 305-49," ASTM International, West Conshohocken, Pa.

ASTM C 234-91a, 1991, "Standard Test Method for Comparing Concretes on the Basis of the Bond Developed with Reinforcing Steel, C 234-91a," ASTM International, West Conshohocken, Pa., withdrawn Feb. 2000.

Azizinamini, A.; Chisala, M.; and Ghosh, S. K., 1995, "Tension Development Length of Reinforcing Bars Embedded in High-Strength Concrete," *Engineering Structures*, V. 17, No. 7, pp. 512-522.

Azizinamini, A.; Pavel, R.; Hatfield, E.; and Ghosh, S. K., 1999, "Behavior of Spliced Reinforcing Bars Embedded in High Strength Concrete," *ACI Structural Journal*, V. 96, No. 5, Sept.-Oct., pp. 826-835.

Azizinamini, A.; Stark, M.; Roller, J. J.; and Ghosh, S. K., 1993, "Bond Performance of Reinforcing Bars Embedded in High-Strength Concrete," *ACI Structural Journal*, V. 90, No. 5, Sept.-Oct., pp. 554-561.

Baldwin, J. W., 1965, "Bond of Reinforcement in Lightweight Aggregate Concrete," *Preliminary Report*, University of Missouri, Mar., 10 pp.

Barham, S., and Darwin, D., 1999, "Effects of Aggregate Type, Water-to-Cementitious Material Ratio, and Age on Mechanical and Fracture Properties of Concrete," *SM Report* No. 56, University of Kansas Center for Research, Lawrence, Kans., 95 pp.

Bashur, F. K., and Darwin, D., 1976, "Nonlinear Model for Reinforced Concrete Slabs," *CRINC Report* SL-76-03, University of Kansas Center for Research, Lawrence, Kans., Dec., 128 pp.

Bashur, F. K., and Darwin, D., 1978, "Nonlinear Model for Reinforced Concrete Slabs," *Journal of the Structural Division*, ASCE, V. 104, No. ST1, Jan., pp. 157-170.

Berge, O., 1981, "Reinforced Structures in Lightweight Aggregate Concrete," *Publication* 81.3, Chalmers University of Technology, Goteborg.

Brettmann, B. B.; Darwin, D.; and Donahey, R. C., 1984, "Effect of Superplasticizers on Concrete-Steel Bond Strength," *SL Report* No. 84-1, University of Kansas Center for Research, Lawrence, Kans., Apr., 32 pp.

Brettmann, B. B.; Darwin, D.; and Donahey, R C., 1986, "Bond of Reinforcement to Superplasticized Concrete," ACI JOURNAL, *Proceedings* V. 83, No. 1, Jan.-Feb., pp. 98-107.

Brown, C. J.; Darwin, D.; and McCabe, S. L., 1993, "Finite Element Fracture Analysis of Steel-Concrete Bond," *SM Report* No. 36, University of Kansas Center for Research, Lawrence, Kans., Nov., 100 pp.

Cairns, J., and Abdullah, R., 1994, "Fundamental Tests on the Effect of an Epoxy Coating on Bond Strength," *ACI Materials Journal,* V. 91, No. 4, July-Aug., pp. 331-338.

Cairns, J., and Jones, K., 1995, "Influence of Rib Geometry on Strength of Lapped Joints: An Experimental and Analytical Study," *Magazine of Concrete Research*, V. 47, No. 172, Sept., pp. 253-262.

CEB-FIP, 1990, "Model Code for Concrete Structures," Comité Euro-International du Béton, c/o Thomas Telford, London.

CEB-FIP, 1999, "Lightweight Aggregate Concrete, Codes and Standards," State-of-Art Report prepared by Task Group 8.1, Fédération Internationale du Béton, Lausanne, Switzerland.

Chamberlin, S. J., 1952, "Spacing of Spliced Bars in Tension Pull-Out Specimens," ACI JOURNAL, *Proceedings* V. 49, No. 3, Nov., pp. 261-274.

Chamberlin, S. J., 1956, "Spacing of Reinforcement in Beams," ACI JOURNAL, *Proceedings* V. 53, No. 1, July, pp. 113-134.

Chamberlin, S. J., 1958, "Spacing of Spliced Bars in Beams," ACI JOURNAL, *Proceedings* V. 54, No. 8, Feb., pp. 689-698.

Chana, P. S., 1990, "A Test Method To Establish Realistic Bond Stresses," *Magazine of Concrete Research*, V. 42, No. 151, pp. 83-90.

Chinn, J.; Ferguson, P. M.; and Thompson, J. N., 1955, "Lapped Splices in Reinforced Concrete Beams," ACI JOURNAL, *Proceedings* V. 52, No. 2, Oct., pp. 201-213.

Choi, O. C.; Hadje-Ghaffari, H.; Darwin, D.; and McCabe, S. L., 1990, "Bond of Epoxy-Coated Reinforcement to Concrete: Bar Parameters," *SM Report* No. 25, University of Kansas Center for Research, Lawrence, Kans., July, 217 pp.

Choi, O. C.; Hadje-Ghaffari, H.; Darwin, D.; and McCabe, S. L., 1991, "Bond of Epoxy-Coated Reinforcement: Bar Parameters," *ACI Materials Journal*, V. 88, No. 2, Mar.-Apr., pp. 207-217.

Clark, A. P., 1946, "Comparative Bond Efficiency of Deformed Concrete Reinforcing Bars," ACI JOURNAL, *Proceedings* V. 43, No. 4, Dec., pp. 381-400.

Clark, A. P., 1950, "Bond of Concrete Reinforcing Bars," ACI JOURNAL, *Proceedings* V. 46, No. 3, Nov., pp. 161-184.

Clarke, J. L., and Birjandi, F. K., 1993, "Bond Strength Tests For Ribbed Bars in Lightweight Aggregate Concrete," *Magazine of Concrete Research*, V. 45, No. 163, pp. 79-87.

Cleary, D. B., and Ramirez, J. A., 1993, "Epoxy-Coated Reinforcement under Repeated Loading," *ACI Structural Journal*, V. 90, No. 4, July-Aug., pp. 451-458.

Collier, S. T., 1947, "Bond Characteristics of Commercial and Prepared Reinforcing Bars," ACI JOURNAL, *Proceedings* V. 43, No. 10, June, pp. 1125-1133.

CSA Standard A23.3-94, 1994, "Design of Concrete Structures," Canadian Standards Association, Ontario, Canada.

CUR, 1963, Commissie voor Uitvoering van Research Ingesteld door de Betonvereniging, "Onderzoek naar de samenwerking van geprofileerd staal met beton," *Report* No. 23, The Netherlands (Translation No. 112, 1964, Cement and Concrete Association, London, "An Investigation of the Bond of Deformed Steel Bars with Concrete"), 28 pp.

Dakhil, F. H.; Cady, P. D.; and Carrier, R. E., 1975, "Cracking of Fresh Concrete as Related to Reinforcement," ACI JOURNAL, *Proceedings* V. 72, No. 8, Aug., pp. 421-428.

Darwin, D., 1987, "Effects of Construction Practice on Concrete-Steel Bond," *Lewis H. Tuthill International Symposium on Concrete and Concrete Construction*, SP-104, G. T. Halvorsen, ed., American Concrete Institute, Farmington Hills, Mich., pp. 27-56.

Darwin, D.; Barham, S.; Kozul, R.; and Luan, S., 2001, "Fracture Energy of High-Strength Concrete," *ACI Materials Journal*, V. 98, No. 5, Sept.-Oct., pp. 410-417.

Darwin, D., and Graham, E. K., 1993a, "Effect of Deformation Height and Spacing on Bond Strength of Reinforcing Bars," *ACI Structural Journal*, V. 90, No. 6, Nov.-Dec., pp. 646-657.

Darwin, D., and Graham, E. K., 1993b, "Effect of Deformation Height and Spacing on Bond Strength of Reinforcing Bars," *SL Report* 93-1, University of Kansas Center for Research, Lawrence, Kans., Jan., 68 pp.

Darwin, D.; Idun, E. K.; Zuo, J.; and Tholen, M. L., 1998, "Reliability-Based Strength Reduction Factor for Bond," *ACI Structural Journal*, V. 95, No. 4, July-Aug., pp. 434-443.

Darwin, D.; McCabe, S. L.; Brown, C. J.; and Tholen, M. L., 1994, "Fracture Analysis of Steel-Concrete Bond," *Fracture and Damage in Quasibrittle Structures: Experiment, Modelling, and Computer Analysis*, Z. P. Bazant, Z. Bittnar, M. Jirasek, and J. Mazars, eds., E&FN Spon, London, pp. 549-556.

Darwin, D.; McCabe, S. L.; Idun, E. K.; and Schoenekase, S. P., 1992, "Development Length Criteria: Bars Not Confined by Transverse Reinforcement," *ACI Structural Journal*, V. 89, No. 6, Nov.-Dec., pp. 709-720.

Darwin, D.; Tholen, M. L.; Idun, E. K.; and Zuo, J., 1996a, "Splice Strength of High Relative Rib Area Reinforcing Bars," *ACI Structural Journal*, V. 93, No. 1, Jan.-Feb., pp. 95-107.

Darwin, D.; Zuo, J.; Tholen, M. L.; and Idun, E. K., 1996b, "Development Length Criteria for Conventional and High Relative Rib Area Reinforcing Bars," *ACI Structural Journal*, V. 93, No. 3, May-June, pp. 347-359.

Darwin, D., and Zuo, J., 2002, Discussion of proposed changes to ACI 318 in "ACI 318-02 Discussion and Closure," *Concrete International*, V. 24, No. 1, Jan., pp. 91, 93, 97-101.

Davies, R., Jr., 1981, "Bond Strength of Mild Steel in Polypropylene Fiber Reinforced Concrete (PFRC)," master's thesis, Marquette University, Milwaukee, Wis., Aug.

Davis, R. E.; Brown, E. H.; and Kelly, J. W., 1938, "Some Factors Influencing the Bond Between Concrete and Reinforcing Steel," *Proceedings of the Forty-First Annual Meeting of the American Society for Testing and Materials*, V. 38, Part II, Philadelphia, Pa., pp. 394-406.

DeVries, R. A.; Moehle, J. P.; and Hester, W., 1991, "Lap Splice of Plain and Epoxy-Coated Reinforcements: An Experimental Study Considering Concrete Strength, Casting Position, and Anti-Bleeding Additives," *Report* No. UCB/SEMM-91/02 Structural Engineering Mechanics and Materials, University of California, Berkeley, Calif., Jan., 86 pp.

DIN 488, 1984, 1986, *Reinforcing Steel*, Parts 1-7, Deutsches Institut für Normung, Berlin.

Donahey, R. C., and Darwin, D., 1983, "Effects of Construction Procedures on Bond in Bridge Decks," *SM Report* No. 7, University of Kansas Center for Research, Lawrence, Kans., Jan., 125 pp.

Donahey, R. C., and Darwin, D., 1985, "Bond of Top-Cast Bars in Bridge Decks," ACI JOURNAL, *Proceedings* V. 82, No. 1, Jan.-Feb., pp. 57-66.

Eligehausen, R., 1979, "Bond in Tensile Lapped Splices of Ribbed Bars with Straight Anchorages," *Publication* 301, German Institute for Reinforced Concrete, Berlin, 118 pp. (in German)

Eligehausen, R.; Popov, E. P.; and Bertero, V. V., 1983, "Local Bond Stress-Slip Relationships of Deformed Bars under Generalized Excitations," *Report* No. UCB/EERC-82/23, Earthquake Engineering Research Center, University of California at Berkeley, Calif., 169 pp.

Ellingwood, B.; Galambos, T. V.; MacGregor, J. G.; and Cornell, C. A., 1980, "Development of a Probability Based Criterion for American National Standard A58," *NBS Special Publication* 577, U.S. Dept. of Commerce, Washington, D.C., June, 222 pp.

Esfahani, M. R., and Vijaya Rangan, B., 1996, "Studies on Bond between Concrete and Reinforcing Bars," School of Civil Engineering, Curtin University of Technology, Perth, Western Australia, 315 pp.

Esfahani, M. R., and Vijaya Rangan, B. V., 1998a, "Local Bond Strength of Reinforcing Bars in Normal Strength and High-Strength Concrete (HSC)," *ACI Structural Journal*, V. 95, No. 2, Mar.-Apr., pp. 96-106.

Esfahani, M. R., and Vijaya Rangan, B. V., 1998b, "Bond between Normal Strength and High-Strength Concrete (HSC) and Reinforcing Bars in Splices in Beams," *ACI Structural Journal*, V. 95, No. 3, May-June, pp. 272-280.

Ezeldin, A. S., and Balaguru, P. N., 1989, "Bond Behavior of Normal and High-Strength Fiber Concrete," *ACI Materials Journal*, V. 86, No. 5, Sept.-Oct., pp. 515-524.

Ferguson, P. M., 1977, "Small Bar Spacing or Cover—A Bond Problem for the Designer," ACI JOURNAL, *Proceedings* V. 74, No. 9, Sept., pp. 435-439.

Ferguson, P. M., and Breen, J. E., 1965, "Lapped Splices for High-Strength Reinforcing Bars," ACI JOURNAL, *Proceedings* V. 62, No. 9, Sept., pp. 1063-1078.

Ferguson, P. M., and Briceno, A., 1969, "Tensile Lap Splices—Part 1: Retaining Wall Type, Varying Moment Zone," *Research Report* No. 113-2, Center for Highway Research, The University of Texas at Austin, July.

Ferguson, P. M., and Krishnaswamy, C. N., 1971, "Tensile Lap Splices—Part 2: Design Recommendation for Retaining Wall Splices and Large Bar Splices," *Research Report* No. 113-2, Center for Highway Research, The University of Texas at Austin, Tex., Apr., 60 pp.

Ferguson, P. M., and Thompson, J. N., 1962, "Development Length for Large High Strength Reinforcing Bars in Bond," ACI JOURNAL, *Proceedings* V. 59, No. 7, July, pp. 887-922.

Ferguson, P. M., and Thompson, J. N., 1965, "Development Length for Large High Strength Reinforcing Bars," ACI JOURNAL, *Proceedings* V. 62, No. 1, Jan., pp. 71-94.

Furr, H. L., and Fouad, F. H., 1981, "Bridge Slab Concrete Placed Adjacent to Moving Live Load," *Research Report* No. 226-lF, Texas Dept. of Highways and Public Transportation, Jan., pp. 131.

Gjorv, O. E.; Monteiro, P. J. M.; and Mehta, P. K.,1990, "Effect of Condensed Silica Fume on the Steel-Concrete Bond," *ACI Materials Journal*, V. 87, No. 6, Nov.-Dec., pp. 573-580.

Goto, Y., 1971, "Cracks Formed in Concrete Around Deformed Tension Bars," ACI JOURNAL, *Proceedings* V. 68, No. 4, Apr., pp. 244-251.

Hadje-Ghaffari, H.; Darwin, D.; and McCabe, S. L., 1991, "Effects of Epoxy-Coating on the Bond of Reinforcing Steel to Concrete," *SM Report* No. 28, University of Kansas Center for Research, Inc., Lawrence, Kans., July, 288 pp.

Hadje-Ghaffari, H.; Choi, O. C.; Darwin, D.; and McCabe, S. L., 1992, "Bond of Epoxy-Coated Reinforcement to Concrete: Cover, Casting Position, Slump, and Consolidation," *SL Report* 92-3, University of Kansas Center for Research, Inc., Lawrence, Kans., June, 42 pp.

Hadje-Ghaffari, H.; Choi, O. C.; Darwin, D.; and McCabe, S. L., 1994, "Bond of Epoxy-Coated Reinforcement: Cover, Casting Position, Slump, and Consolidation," *ACI Structural Journal*, V. 91, No. 1, Jan.-Feb., pp. 59-68.

Hamad, B. S., and Itani, M. S., 1998, "Bond Strength of Reinforcement in High-Performance Concrete: The Role of Silica Fume, Casing Position, and Superplasticizer Dosage," *ACI Materials Journal*, V. 95, No. 5, Sept.-Oct., pp. 499-511.

Hamad, B. S., and Jirsa, J. O., 1990, "Influence of Epoxy Coating on Stress Transfer from Steel to Concrete," *Proceedings*, First Materials Engineering Congress, ASCE, New York, V. 2, pp. 125-134.

Hamad, B. S., and Jirsa, J. O., 1993, "Strength of Epoxy-Coated Reinforcing Bar Splices Confined with Transverse Reinforcement," *ACI Structural Journal*, V. 90, No.1, Jan.-Feb., pp. 77-88.

Hamad, B. S., and Mansour, M. Y., 1996, "Bond Strength of Noncontact Tension Lap Splices," *ACI Structural Journal*, V. 93, No. 3, May-June, pp. 316-326.

Hamza, A. M., and Naaman, A. E., 1996, "Bond Characteristics of Deformed Reinforcing Steel Bars Embedded in SIFCON," *ACI Materials Journal*, V. 93, No. 6, Nov.-Dec., pp. 578-588.

Harajli, M. H.; Hout, M.; and Jalkh, W., 1995, "Local Bond Stress-Slip Behavior of Reinforcing Bars Embedded in Plain and Fiber Concrete," *ACI Materials Journal*, V. 92, No. 4, July-Aug., pp. 343-354.

Harajli, M. H., and Salloukh, K. A., 1997, "Effect of Fibers on Development/Splice Strength of Reinforcing Bars in Tension," *ACI Materials Journal*, V. 94, No. 4, July-Aug., pp. 317-324.

Harsh, S., and Darwin, D., 1984, "Effects of Traffic-Induced Vibrations on Bridge Deck Repairs," *SM Report* No. 9, University of Kansas Center for Research, Lawrence, Kans., Jan., 60 pp.

Harsh, S., and Darwin, D., 1986, "Traffic-Induced Vibrations and Bridge Deck Repairs," *Concrete International*, V. 8, No. 5, May, pp. 36-42.

Hasan, H. O.; Cleary, D. B.; and Ramirez, J. A., 1996, "Performance of Concrete Bridge Decks and Slabs Reinforced with Epoxy-Coated Steel under Repeated Loading," *ACI Structural Journal*, V. 93, No. 4, July-Aug., pp. 397-403.

Hester, C. J.; Salamizavaregh, S.; Darwin, D.; and McCabe, S. L., 1991, "Bond of Epoxy-Coated Reinforcement to Concrete: Splices," *SL Report* 91- 1, University of Kansas Center for Research, Lawrence, Kans., May, 66 pp.

Hester, C. J.; Salamizavaregh, S.; Darwin, D.; and McCabe, S. L., 1993, "Bond of Epoxy-Coated Reinforcement: Splices," *ACI Structural Journal*, V. 90, No. 1, Jan.-Feb., pp. 89-102.

Hognestad, E., 1951, "A Study of Combined Bending and Axial Load in Reinforced Concrete Members," *Bulletin Series* No. 399, University of Illinois Engineering Experiment Station, Urbana, Ill.

Hota, S., and Naaman, A. E., 1997, "Bond Stress-Slip Response of Reinforcing Bars Embedded in FRC Matrices Under Monotonic and Cyclic Loading," *ACI Structural Journal*, V. 94, No. 5, Sept.-Oct., pp. 525-537.

Hulshizer, A. J., and Desai, A. J., 1984, "Shock Vibration Effects on Freshly Placed Concrete," *Journal of Construction Engineering and Management*, V. 110, No. 2, June, pp. 266-285.

Hwang, S.-J.; Lee, Y.-Y.; and Lee, C.-S., 1994, Effect of Silica Fume on the Splice Strength of Deformed Bars of High-Performance Concrete," *ACI Structural Journal*, V. 91, No. 3, May-June, pp. 294-302.

Hyatt, T., 1877, *An Account of Some Experiments with Portland-Cement-Concrete Combined with Iron, as a Building Material*, Chiswick Press, London, 47 pp.

Idun, E. K., and Darwin, D., 1999, "Bond of Epoxy-Coated Reinforcement: Coefficient of Friction and Rib Face Angle," *ACI Structural Journal*, V. 96, No. 4, July-Aug., pp. 609-615.

Jeanty, P. R.; Mitchell, D.; and Mirza, M. S., 1988, "Investigation of 'Top Bar' Effects in Beams," *ACI Structural Journal*, V. 85, No. 3., May-June, pp. 251-257.

Jirsa, J. O., and Breen, J. E., 1981,"Influence of Casting Position and Shear on Development and Splice Length—Design Recommendation," *Research Report* No. 242-3F, Center for Transportation Research, The University of Texas at Austin, Tex.

Johnston, D. W., and Zia, P., 1982,"Bond Characteristics of Epoxy Coated Reinforcing Bars," *Report* No. FHWA-NC-82-002, Federal Highway Administration, Washington, D.C., 163 pp.

Kadoriku, J., 1994, "Study on Behavior of Lap Splices in High-Strength Reinforced Concrete Members," doctorate thesis, Kobe University, Mar., Japan, 201 pp.

Kemp, E. L.; Brezny, F. S.; and Unterspan, J. A., 1968, "Effect of Rust and Scale on the Bond Characteristics of Deformed Reinforcing Bars," ACI JOURNAL, *Proceedings* V. 65, No. 9, Sept., pp. 743-756.

Kozul, R., and Darwin, D., 1997, "Effects of Aggregate Type, Size, and Content on Concrete Strength and Fracture Energy," *SM Report* No. 43, University of Kansas Center for Research, Inc., Lawrence, Kans.

Larnach, W. J., 1952, "Changes in Bond Strength Caused by Revibration of Concrete and the Vibration of Reinforcement," *Magazine of Concrete Research*, London, No. 10, July, pp. 17-21.

Losberg, A., and Olsson, P.-A., 1979, "Bond Failure of Deformed Reinforcing Bars Based on the Longitudinal Splitting Effect of the Bars," ACI JOURNAL, *Proceedings* V. 76, No. 1, Jan., pp. 5-18.

Luke, J. J.; Hamad, B. S.; Jirsa, J. O.; and Breen, J. E., 1981, "The Influence of Casting Position on Development and Splice Length of Reinforcing Bars," *Research Report* No. 242-1, Center for Transportation Research, Bureau of Engineering Research, University of Texas at Austin, Tex., June, 153 pp.

Lundberg, J. E., 1993, "The Reliability of Composite Columns and Beam Columns," *Structural Engineering Report* No. 93-2, University of Minnesota, Minneapolis, Minn., June, 233 pp.

Lutz, L. A., and Gergely, P., 1967, "Mechanics of Bond and Slip of Deformed Bars in Concrete," ACI JOURNAL, *Proceedings* V. 64, No. 11, Nov., pp. 711-721.

Lutz, L. A.; Gergely, P.; and Winter, G., 1966, "The Mechanics of Bond and Slip of Deformed Reinforcing Bars in Concrete," *Report* No. 324, Department of Structural Engineering, Cornell University, Ithaca, N.Y., Nov., pp. 711-721.

Lyse, I., 1934, "Lightweight Slag Concrete," ACI JOURNAL, *Proceedings* V. 31, No. 1, pp. 1-20.

MacGregor, J. G., 1997, *Reinforced Concrete, Mechanics & Design*, Prentice Hall, 3rd Edition, Upper Saddle River, N.J., 939 pp.

Maeda, M.; Otani, S.; and Aoyama, H., 1991, "Bond Splitting Strength in Reinforced Concrete Members," *Transactions of the Japan Concrete Institute*, V. 13, pp. 581-588.

Martin, H., 1982, "Bond Performance of Ribbed Bars," *Bond in Concrete—Proceedings of the International Conference on Bond in Concrete*, Paisley, Applied Science Publishers, London, pp. 289-299.

Mathey, R., and Watstein, D., 1961, "Investigation of Bond in Beam and Pull-Out Specimens with High-Yield-Strength Deformed Bars," ACI JOURNAL, *Proceedings* V. 58, No. 9, Mar., pp. 1071-1090.

Mathey, R., and Clifton, J. R., 1976, "Bond of Coated Reinforcing Bars in Concrete," *Journal of the Structural Division*, ASCE, V. 102, No. 1, Jan., pp. 215-229.

Menzel, C. A., 1952, "Effect of Settlement of Concrete on Results of Pullout Tests," *Research Department Bulletin* 41, Research and Development Laboratories of the Portland Cement Association, Nov., 49 pp.

Menzel, C. A., and Woods, W. M., 1952, "An Investigation of Bond, Anchorage and Related Factors in Reinforced Concrete Beams," *Research Department Bulletin* 42, Portland Cement Association, Nov., 114 pp.

Mirza, S. A., and MacGregor, J. G., 1986, "Strength Variability of Bond of Reinforcing Bars in Concrete Beams," *Civil Engineering Report Series* No. CE-86-1, Lakehead University, Thunder Bay, Ontario, Jan., 35 pp.

Mor, A., 1992, "Steel-Concrete Bond in High-Strength Lightweight Concrete," *ACI Materials Journal*, V. 89, No. 1, Jan.-Feb., pp. 76-82.

Musser, P. L.; Carrasquillo, R. L.; Jirsa, J. O.; and Klingner, R. E., 1985, "Anchorage and Development of Reinforcement in Concrete Made Using Superplasticizers,"

Research Report 383-1, Center for Transportation Research, Bureau of Engineering Research, University of Texas at Austin, Tex., Sept., 189 pp.

Nilson, A. H., 1997, *Design of Concrete Structures*, 12th Edition, with contributions by D. Darwin, McGraw-Hill, New York, 780 pp.

Nilson, A. H.; Darwin, D.,; and Dolan, C. W., 2004, *Design of Concrete Structures*, 13th Edition, McGraw-Hill, New York, 779 pp.

Niwa, J., and Tangtermsirikul, 1997, "Fracture Properties of High-Strength and Self-Compacting High Performance Concretes," *Transactions of the Japan Concrete Institute*, V. 19, pp. 73-80.

Olsen, N. H., 1990a, "Strength of Lapped Splices in High-Strength Concrete," *High-Strength Concrete*, Proceedings of the Second International Symposium, SP-121, W. T. Hester, ed., American Concrete, Institute, Farmington Hills, Mich., pp. 179-193.

Olsen, N. H., 1990b, "Strength of Overlapped Deformed Tensile Reinforcement Splices in High-Strength Concrete," *Report Series R*, No. 234, Department of Structural Engineering, Technical University of Denmark, 160 pp.

Orangun, C. O.; Jirsa, J. O.; and Breen, J. E., 1975, "The Strength of Anchored Bars: A Reevaluation of Test Data on Development Length and Splices," *Research Report* No. 154-3F, Center for Highway Research, The University of Texas at Austin, Tex., Jan., 78 pp.

Orangun, C. O.; Jirsa, J. O.; and Breen, J. E., 1977, "Reevaluation of Test Data on Development Length and Splices," ACI JOURNAL, *Proceedings* V. 74, No. 3, Mar., pp. 114-122.

Petersen, P. H., 1948, "Properties of Some Lightweight Aggregate Concretes with and without an Air-Entraining Admixture," *Building Materials and Structures Report* BMS112, U.S. Department of Commerce, National Bureau of Standards, Aug., 7 pp.

Rehm, G., 1961, "Uber die Grunlagen des Verbundes Zwischen Stahl und Beton," *Deutscher Ausschuss fur Stahlbeton*, No. 1381, 59 pp. (C & CA Library Translation No. 134, 1968, "The Basic Principle of the Bond between Steel and Concrete.")

Rehm, G., and Eligehausen R., 1979, "Bond of Ribbed Bars Under High Cycle Repeated Loads," ACI JOURNAL, *Proceedings* V. 76, No. 2, Feb., pp. 297-309.

Rezansoff, T.; Akanni, A.; and Sparling, B., 1993, "Tensile Lap Splices under Static Loading: A Review of the Proposed ACI 318 Code Provisions," *ACI Structural Journal*, V. 90, No. 4, July-Aug., pp. 374-384.

Rezansoff, T.; Konkankar, U. S.; and Fu, Y. C., 1991, "Confinement Limits for Tension Lap Splices Under Static Loading," *Report* No. S7N 0W0, University of Saskatchewan, Aug.

Robin, R. C.; Olsen, P. E.; and Kinnane, R. F., 1942, "Bond Strength of Reinforcing Bars Embedded Horizontally in Concrete," *Journal of the Institute of Engineers*, London, V. 14, Sept., pp. 201-217.

Robins, P. J., and Standish, I. G., 1982, "Effect of Lateral Pressure on Bond of Reinforcing Bars in Concrete," *Bond in Concrete—Proceedings of the International Conference on Bond in Concrete*, Paisley, Applied Science Publishers, London, pp. 262-272.

Sagan, V. E.; Gergely, P.; and White, R. N., 1991, "Behavior and Design of Noncontact Lap Splices Subjected to Repeated Inelastic Tensile Loading," *ACI Structural Journal*, V. 88, No. 4, July-Aug., pp. 420-431.

Sakurada, T.; Morohashi, N.; and Tanaka, R., 1993, "Effect of Transverse Reinforcement on Bond Splitting Strength of Lap Splices," *Transactions of the Japan Concrete Institute*, V. 15, pp. 573-580.

Shideler, J. J., 1957, "Lightweight-Aggregate Concrete for Structural Use," ACI JOURNAL, *Proceedings* V. 54, No. 4, Oct., pp. 299-328.

Skorobogatov, S. M., and Edwards, A. D., 1979, "The Influence of the Geometry of Deformed Steel Bars on Their Bond Strength in Concrete," *Proceedings*, The Institution of Civil Engineers, V. 67, Part 2, June, pp. 327-339.

Soretz, S., and Holzenbein, H., 1979, "Influence of Rib Dimensions of Reinforcing Bars on Bond and Bendability," ACI JOURNAL, *Proceedings* V. 76, No. 1, Jan., pp. 111-127.

Soroushian, P.; Mirza, F.; and Alhozaimy, A., 1994, "Bonding of Confined Steel Fiber Reinforced Concrete to Deformed Bars," *ACI Materials Journal*, V. 91, No. 2, Mar.-Apr., pp. 141-149.

Tan, C.; Darwin, D.; Tholen, M. L.; and Zuo, J., 1996, "Splice Strength of Epoxy-Coated High Relative Rib Area Bars," *SL Report* 96-2, University of Kansas Center for Research, Lawrence, Kans., May, 69 pp.

Tepfers, R., 1973, "A Theory of Bond Applied to Overlapping Tensile Reinforcement Splices for Deformed Bars," *Publication* 73:2, Division of Concrete Structures, Chalmers University of Technology, Goteborg, Sweden, 328 pp.

Tholen, M. L., and Darwin, D., 1996, "Effects of Deformation Properties on the Bond of Reinforcing Bars," *SM Report* No. 42, University of Kansas Center for Research, Inc., Lawrence, Kans.

Thompson, M. A.; Jirsa, J. O.; Breen, J. E.; and Meinheit, D. F., 1975, "The Behavior of Multiple Lap Splices in Wide Sections," *Research Report* No. 154-1, Center for Highway Research, University of Texas at Austin, Tex., Feb., 75 pp.

Treece, R. A., and Jirsa, J. O., 1989, "Bond Strength of Epoxy-Coated Reinforcing Bars," *ACI Materials Journal*, V. 86, No. 2, Mar.-Apr., pp. 167-174.

Tuthill, L. H., and Davis, H. E., 1938, "Overvibration and Revibration of Concrete," ACI JOURNAL, *Proceedings* V. 35, No. 4, Sept., pp. 41-47.

Tuthill, L. H., 1977, "Revibration Reexamined," *Concrete Construction*, V. 22, No. 10, Oct., pp. 537-539.

Untrauer, R. E., 1965, Discussion of "Development Length for Large High Strength Reinforcing Bars," ACI Journal, *Proceedings* V. 62, No. 9, Sept., pp. 1153-1154.

Untrauer, R. E., and Warren, G. E., 1977, "Stress Development of Tension Steel in Beams," ACI JOURNAL, *Proceedings* V. 74, No. 8, Aug., pp. 368-372.

Vollick, C. A., 1958, "Effects of Revibrating Concrete," ACI Journal, *Proceedings* V. 54, No. 9, Mar., pp. 721-732.

Wafa, F. F., and Ashour, S. A., 1992, "Mechanical Properties of High-Strength Fiber Reinforced Concrete," *ACI Materials Journal*, V. 89, No. 5, Sept.-Oct., pp. 449-455.

Walker, W. T., 1951, "Laboratory Tests of Spaced and Tied Reinforcing Bars," ACI JOURNAL, *Proceedings* V. 47, No. 5, Jan., pp. 365-372.

Welch, G. B., and Patten, B. J. F., 1965, "Bond Strength of Reinforcement Affected by Concrete Sedimentation," ACI JOURNAL, *Proceedings* V. 62, No. 2, Feb., pp. 251-263.

Zekany, A. J.; Neumann, S.; Jirsa, J. O.; and Breen, J. E., 1981, "The Influence of Shear on Lapped Splices in Reinforced Concrete," *Research Report* 242-2, Center for Transportation Research, Bureau of Engineering Research, University of Texas at Austin, Tex., July, 88 pp.

Zilveti, A.; Sooi, T. K.; Klingner; R. E.; Carrasquillo, R. L.; and Jirsa, J. O., 1985, "Effect of Superplasticizers on the Bond Behavior of Reinforcing Steel in Concrete Members," *Research Report* 383-2F, Center for Transportation Research, Bureau of Engineering Research, University of Texas at Austin, Tex., Nov., 104 pp.

Zsutty, T., 1985, "Empirical Study of Bar Development Behavior," *Journal of Structural Engineering*, ASCE, V. 111, No. 1, Jan., pp. 205-219.

Zuo, J., and Darwin, D., 1998, "Bond Strength of High Relative Rib Area Reinforcing Bars," *SM Report* No. 46, University of Kansas Center for Research, Lawrence, Kans., 350 pp.

Zuo, J., and Darwin, D., 2000, "Splice Strength of Conventional and High Relative Rib Area Bars in Normal and High-Strength Concrete," *ACI Structural Journal*, V. 97, No. 4, July-Aug., pp. 630-641.

APPENDIX A—SI EQUATIONS

The equations presented in this appendix are (primarily soft) SI conversions of equations in the report that contain terms that depend on the units of measure.

$$u = 20\frac{\sqrt{f'_c}}{d_b} \leq 5.52 \text{ MPa} \tag{1-5}$$

$$l_d = 0.019A_b\frac{f_y}{\sqrt{f'_c}} \tag{1-6}$$

$$t_d = 0.03d_b + 0.22 \tag{2-5}$$

$$\frac{u_c}{\sqrt{f'_c}} = 0.101 + 0.268\frac{c_{min}}{d_b} + 4.4\frac{d_b}{l_d} \tag{3-1}$$

$$\frac{u_c}{\sqrt{f'_c}} = 0.10 + 0.25\frac{c_{min}}{d_b} + 4.15\frac{d_b}{l_d} \tag{3-2}$$

$$\frac{u_b}{\sqrt{f'_c}} = \frac{u_c + u_s}{\sqrt{f'_c}} = \tag{3-3}$$

$$0.10 + 0.25\frac{c_{min}}{d_b} + \frac{4.15d_b}{l_d} + \frac{A_{tr}f_{yt}}{41.5snd_b}$$

$$\frac{T_b}{\sqrt{f'_c}} = \frac{T_c + T_s}{\sqrt{f'_c}} = \frac{A_bf_s}{\sqrt{f'_c}} = \tag{3-4}$$

$$0.25\pi l_d(c_{min} + 0.4d_b) + 16.6A_b + \frac{\pi l_d A_{tr}}{41.5sn}f_{yt}$$

$$\frac{1}{d_b}\left(c_{min} + 0.4d_b + \frac{A_{tr}f_{yt}}{10.34sn}\right) \leq 2.5 \tag{3-5}$$

$$\frac{T_c}{\sqrt{f'_c}} = \frac{A_bf_s}{\sqrt{f'_c}} = \tag{3-6}$$

$$0.554l_d(c_{min} + 0.5d_b)\left(0.08\frac{c_{max}}{c_{min}} + 0.92\right) + 24.9A_b$$

$$\frac{T_c}{f_c'^{1/4}} = \frac{A_bf_s}{f_c'^{1/4}} = \tag{3-7}$$

$$[1.5l_d(c_{min} + 0.5d_b) + 51A_b]\left(0.1\frac{c_{max}}{c_{min}} + 0.90\right)$$

$$\frac{T_b}{f_c'^{1/4}} = \frac{T_c + T_s}{f_c'^{1/4}} = \frac{A_bf_s}{f_c'^{1/4}} = \tag{3-8}$$

$$[1.5l_d(c_{min} + 0.5d_b) + 51A_b]\left(0.1\frac{c_{max}}{c_{min}} + 0.90\right)$$

$$+ 53.3t_rt_d\frac{NA_{tr}}{n} + 1019$$

where $t_d = 0.028d_b + 0.28$.

$$\frac{T_c}{f_c'^{1/4}} = \frac{A_bf_s}{f_c'^{1/4}} = \tag{3-10}$$

$$[1.43l_d(c_{min} + 0.5d_b) + 56.2A_b]\left(0.1\frac{c_{max}}{c_{min}} + 0.90\right)$$

$$\frac{T_b}{f_c'^{1/4}} = \frac{T_c + T_s}{f_c'^{1/4}} = \frac{A_bf_s}{f_c'^{1/4}} = \tag{3-11}$$

$$[1.43l_d(c_{min} + 0.5d_b) + 56.2A_b]\left(0.1\frac{c_{max}}{c_{min}} + 0.90\right)$$

$$+\left(9t_r t_d \frac{NA_{tr}}{n} + 744\right) f_c'^{\,1/2}$$

where $t_d = 0.03d_b + 0.22$.

$$\frac{1}{d_b}\left[(c_{min} + 0.5d_b)\left(0.1\frac{c_{max}}{c_{min}} + 0.90\right)\right. \qquad (3\text{-}12)$$

$$\left. + \left(\frac{6.26t_r t_d A_{tr}}{sn}\right) f_c'^{\,1/2}\right] \le 4.0$$

$$\frac{T_c}{\sqrt{f_c'}} = \frac{A_b f_s}{\sqrt{f_c'}} = \qquad (3\text{-}13)$$

$$2.7\pi l_d \frac{(c_{min} + 0.5d_b)\left(1 + \frac{1}{M}\right)}{\left(\frac{c_{min}}{d_b} + 3.6\right)(1.85 + 0.024\sqrt{M})}\left(0.12\frac{c_{med}}{c_{min}} + 0.88\right)$$

$$\frac{T_c}{\sqrt{f_c'}} = \frac{A_b f_s}{\sqrt{f_c'}} = \qquad (3\text{-}14)$$

$$M = \cosh(0.0022 l_d \sqrt{r f_c' / d_b})$$

$$4.73\pi l_d \frac{(c_{min} + 0.5d_b)\left(1 + \frac{1}{M}\right)}{\left(\frac{c_{min}}{d_b} + 5.5\right)(1.85 + 0.024\sqrt{M})}\left(0.12\frac{c_{med}}{c_{min}} + 0.88\right)$$

$$\frac{T_c}{f_c'^{\,1/4}} = \frac{A_b f_s}{f_c'^{\,1/4}} = \qquad (3\text{-}15)$$

$$[1.43 l_d (c_{min} + 0.5d_b) + 57.4A_b]\left(0.1\frac{c_{max}}{c_{min}} + 0.90\right)$$

$$\frac{T_b}{f_c'^{\,1/4}} = \frac{T_c + T_s}{f_c'^{\,1/4}} = \frac{A_b f_s}{f_c'^{\,1/4}} = \qquad (3\text{-}16)$$

$$[1.43 l_d (c_{min} + 0.5d_b) + 57.4A_b]\left(0.1\frac{c_{max}}{c_{min}} + 0.90\right)$$

$$+\left(8.9t_r t_d \frac{NA_{tr}}{n} + 558\right) f_c'^{\,1/2}$$

$$\frac{l_d}{d_b} = \frac{\frac{f_s}{\sqrt{f_c'}} - 16.6}{\left(\frac{c + K_{tr}}{d_b}\right)} \qquad (4\text{-}1)$$

where $K_{tr} = \dfrac{A_{tr} f_{yt}}{10.34sn}$.

$$\frac{l_d}{d_b} = 0.9\frac{f_y}{\sqrt{f_c'}\left(\frac{c + K_{tr}}{d_b}\right)} \qquad (4\text{-}2)$$

The value of $\sqrt{f_c'}$ is limited to a maximum value of 8.3 MPa.

$$\frac{l_d}{d_b} = \frac{\frac{f_y}{\phi f_c'^{\,1/4}} - 51\left(0.1\frac{c_{max}}{c_{min}} + 0.90\right)}{1.92\left(\frac{c + K_{tr}}{d_b}\right)} = \qquad (4\text{-}4)$$

$$\frac{\frac{f_y}{f_c'^{\,1/4}} - \phi 51\left(0.1\frac{c_{max}}{c_{min}} + 0.90\right)}{\phi 1.92\left(\frac{c + K_{tr}}{d_b}\right)}$$

where $K_{tr} = 35.3t_r t_d A_{tr}/sn$.

$$\frac{l_d}{d_b} = \frac{\frac{f_y}{0.9f_c'^{\,1/4}} - 51\left(0.1\frac{c_{max}}{c_{min}} + 0.90\right)}{1.92\left(\frac{c + K_{tr}}{d_b}\right)} \qquad (4\text{-}5a)$$

$$\frac{l_d}{d_b} = \frac{\frac{f_y}{f_c'^{\,1/4}} - 45.5\left(0.1\frac{c_{max}}{c_{min}} + 0.90\right)}{1.72\left(\frac{c + K_{tr}}{d_b}\right)} \qquad (4\text{-}5b)$$

$$\frac{l_d}{d_b} = \frac{(f_y/f_c'^{\,1/4} - 45.5\omega)\alpha\beta\lambda}{1.72\left(\frac{c\omega + K_{tr}}{d_b}\right)} \qquad (4\text{-}6)$$

$$K_{tr} = C_R(0.0283d_b + 0.28)\frac{A_{tr}}{sn} \qquad (4\text{-}9)$$

ACI 408.3 limits $f_c'^{\,1/4} \cdot 3.2$, and $f_y \cdot 552$ MPa.

$$\frac{l_d}{d_b} = \frac{\left(\frac{f_y}{\phi f_c'^{\,1/4}} - 57.4\omega\right)\alpha\beta\lambda}{1.83\left(\frac{c\omega + K_{tr}}{d_b}\right)} = \qquad (4\text{-}11a)$$

$$\frac{\left(\frac{f_y}{f_c'^{\,1/4}} - \phi 57.4\omega\right)\alpha\beta\lambda}{\phi 1.83\left(\frac{c\omega + K_{tr}}{d_b}\right)}$$

$$\frac{l_d}{d_b} = \frac{\left(\frac{f_y}{f_c'^{1/4}} - 52.9\omega\right)\alpha\beta\lambda}{1.68\left(\frac{c\omega + K_{tr}}{d_b}\right)} \quad (4\text{-}11b)$$

$$K_{tr} = (6.26 t_r t_d A_{tr}/sn) f_c'^{1/2} \quad (4\text{-}12)$$

$$t_d = 0.03 d_b + 0.22 \quad (4\text{-}14)$$

For conventional reinforcement, $K_{tr} = (6 t_d A_{tr}/sn) f_c'^{1/2}$.

$$\frac{l_d}{d_b} = \left(\frac{f_y}{2.51 f_c'^{1/4}} - 20.99\right)\alpha\beta\lambda \quad (4\text{-}15)$$

$$\frac{l_d}{d_b} = \left(\frac{f_y}{2.5 f_c'^{1/4}} - 21\right)\alpha\beta\lambda \quad (4\text{-}16)$$

$$\frac{l_d}{d_b} = \left(\frac{f_y}{1.67 f_c'^{1/4}} - 31\right)\alpha\beta\lambda \quad (4\text{-}17)$$

$$\frac{l_b}{d_b} = \frac{\left(\frac{f_y}{f_c'^{1/4}} - 52.6\omega\right)\alpha\beta\lambda}{1.67\left(\frac{c\omega + K_{tr}}{d_b}\right)} \quad (4\text{-}18)$$

$$\frac{l_d}{d_b} = \left(\frac{f_y}{2.2 f_c'^{1/4}} - 21\right)\alpha\beta\lambda \quad (4\text{-}19)$$

$$\frac{l_d}{d_b} = \left(\frac{f_y}{1.48 f_c'^{1/4}} - 31\right)\alpha\beta\lambda \quad (4\text{-}20)$$

$$\frac{l_d}{d_b} = \frac{\left(\frac{f_y}{f_c'^{1/4}} - 47.1\omega\right)\alpha\beta\lambda}{1.48\left(\frac{c\omega + K_{tr}}{d_b}\right)} \quad (4\text{-}21)$$

$$\frac{l_d}{d_b} = \frac{1}{6.55}\left(1.15 - 0.15\frac{c_{\min}}{d_b}\right)\left(1 - K\frac{\Sigma A_{tr} - \Sigma A_{tr,\min}}{A_b}\right) \quad (4\text{-}22)$$

$$\frac{Mf_y}{\left(\frac{f_c' - 2.75}{10}\right)^{2/3}}$$

$$l_{d,\min} = \max\left[\frac{0.3}{6.55}\frac{Mf_y}{\left(\frac{f_c' - 2.75}{10}\right)^{2/3}}, 10d_b, 100\text{ mm}\right] \quad (4\text{-}23)$$

$$l_{s,\min} = \max\left[\frac{0.3\alpha_b}{6.55}\frac{Mf_y}{\left(\frac{f_c' - 2.75}{10}\right)^{2/3}}, 15d_b, 200\text{ mm}\right] \quad (4\text{-}24)$$

$$E_c = 4734\sqrt{f_c'} \quad (5\text{-}1)$$

$$\varepsilon_o = \frac{f_c'}{2500 + 400 f_c'} \quad (5\text{-}3)$$

ACI 408.2R-12

Report on Bond of Steel Reinforcing Bars Under Cyclic Loads

Reported by Joint ACI-ASCE Committee 408

Adolfo B. Matamoros
Chair

Michael Keith Thompson
Secretary

Theresa M. Ahlborn
Robert W. Barnes*
Jean-Jacques Braun
JoAnn P. Browning†
James L. Caldwell
Douglas B. Cleary*
Louis J. Colarusso
James V. Cox
Christian Dahl
David Darwin
Richard A. Devries

Rolf Eligehausen
Alvin C. Ericson
Anthony L. Felder
Scott K. Graham
Bilal S. Hamad
Neil M. Hawkins
Steven E. Holdsworth
James M. LaFave
Roberto T. Leon
LeRoy A. Lutz
Stavroula J. Pantazopoulou

Max L. Porter
Julio A. Ramirez
John F. Silva
Robert G. Smith
Ahmet Koray Tureyen*
William H. Zehrt Jr.
Jun Zuo

Consulting members
Gyorgy L. Balazs
William C. Gallenz
Brian C. Gerber
Allen J. Hulshizer
Kenneth A. Luttrell
Denis Mitchell
Mikael P. J. Olsen
Conrad Paulson
Melvyn Precious
Richard A. Ramsey

*Member, Subcommittee on Repeated Load Effects
†Chair and Editor, Subcommittee on Repeated Load Effects
Note: The contributions of Brett Baker are gratefully acknowledged.

This report summarizes research on bond strength and behavior of steel reinforcing bars under cyclic loads. The report provides a background to bond problems, discusses the main variables affecting bond performance, and describes bond behavior under cyclic loads. Two general types of cyclic loads are addressed: high-cycle (fatigue) and low-cycle (earthquake and similar). The anchorage behaviors of straight bars, hooked bars, and lap splices are included. Analytical bond models are described, design recommendations are provided for both high- and low-cycle fatigue, and suggestions for further research are given.

Keywords: anchorage; bar slip; bond; bond models; cyclic loads; design recommendations; development length; fatigue; hooks; earthquake loads; splices.

CONTENTS

ACI 408.2R-12 supersedes ACI 408.2R-92 and was adopted and published September 2012.

CHAPTER 1—INTRODUCTION AND SCOPE

1.1—Introduction

The transfer of forces across the interface between concrete and steel by bond stresses is of fundamental importance to many aspects of reinforced concrete behavior. Satisfactory bond performance is an essential goal in detailing reinforcement in structural components. Many detailing provisions in ACI 318 are aimed at preventing bond failures.

Bond stresses in reinforced concrete members arise from two distinct situations. The first is anchorage or development where bars are terminated. The second is flexural bond or the change of force along a bar due to a change in bending moment along the member. Bond performance under static monotonically increasing deformations—referred to as monotonic loading—has been summarized in ACI 352R, ACI 408R, and ACI Committee 408 (1966, 1970, 1979). Bond behavior under cyclic loads received little attention until design for earthquake and wave loads became important design topics (ACI Committee 408 1979). Investigations over the past 40 years have clarified some of the important parameters influencing bond behavior under cyclic loads. However, the influence of many of these parameters is understood only qualitatively.

In this report, "bar" means "reinforcing bar" and "ribs" refer to the deformations on deformed reinforcing bars. Longitudinal deformations on reinforcing bars are not classified as ribs. "Bond stress" refers to the stresses along the bar-concrete interface. The steel stresses along the length of the reinforcing bar are modified by transfer of force between the bar and the surrounding concrete along the interface (refer to Fig. 1.1).

The change in bar tensile force ΔF between two cracked sections along a flexural member ΔF is given by

$$\Delta F = T_1 - T_2 = \left(\frac{M_1}{jd_1}\right) - \left(\frac{M_2}{jd_2}\right) \quad (1.1a)$$

The average bond stress u_b is usually expressed as

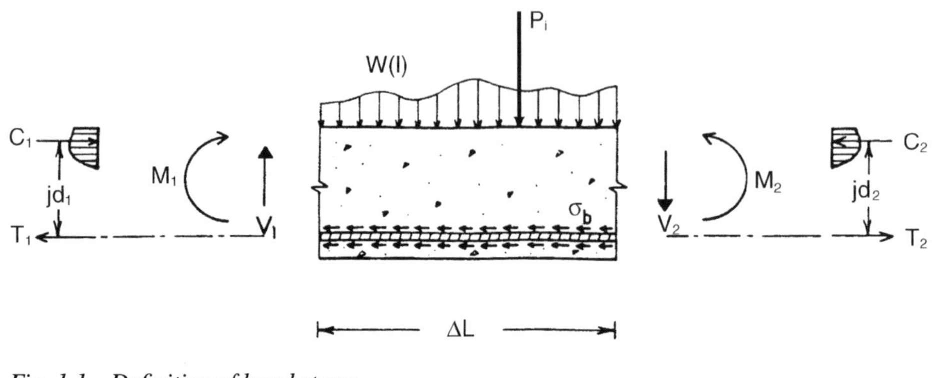

Fig. 1.1—Definition of bond stress.

$$u_b = \frac{q}{\Sigma_o} = \frac{\Delta F}{\Delta L \Sigma_o} = \frac{\Delta f_s\left(\pi d_b^2/4\right)}{\Delta L\left(\pi d_b\right)} = \frac{d_b \Delta f_s}{4\Delta L} \quad (1.1b)$$

Cyclic loadings are divided into two general categories. The first is designated low-cycle or low-cycle, high-stress loading, or a load history containing less than 100 cycles and having bond stress ranges (u_r) greater than 600 psi (4 MPa). Low-cycle loadings commonly result from earthquake and high-wind loadings. The second category is designated high-cycle or fatigue loading with a load history containing thousands or millions of cycles, but at a low bond stress range (u_r) typically less than 300 psi (2 MPa). Bridge members, offshore structures, and members supporting vibrating machinery are often subjected to high-cycle or fatigue loading. High-cycle loadings can be a problem at service load levels whereas low-cycle loadings can produce problems at the ultimate limit state.

Bond behavior under cyclic loading can further be subdivided according to the type of stress applied. The first is repeated or unidirectional loading, which implies that the bar stress does not reverse from tension to compression during a load cycle, the usual situation for fatigue loading. The second is stress reversal, where the bar is subjected alternatively to tension and compression. Earthquake loading typically causes stress reversals.

1.2—Scope

This report reviews bond and anchorage of steel reinforcing bars in normal and lightweight concrete, with emphasis on bond under cyclic loading. Although the amount of information about the bond behavior of epoxy-coated reinforcing bars subjected to cyclic loading is limited, available references on this topic were reviewed and are presented in the document. Bond and anchorage of prestressing steel and headed reinforcing bars are not addressed in this report. This report serves both designers and researchers, and is organized accordingly. Chapters 3 and 4 present background information on the issue of bond under cyclic loading. Chapters 5, 6, and 7 deal with results of research and development of analytical bond models. Chapters 8 and 9 review international design guidelines dealing with bond under cyclic loads and should be of interest to designers. This report also introduces designers to the basic mechanisms involved in bond, the variables that affect those mechanisms, and differences in bond behavior under cyclic and noncyclic loads.

CHAPTER 2—NOTATION AND DEFINITIONS

2.1—Notation

A_b = cross-sectional area of reinforcing bar, in.^{2} (mm^2)
A_c = area of concrete cross section, in.^{2} (mm^2)
A_{tr} = area of transverse reinforcement, in.^{2} (mm^2)
BI = representation of the severity of bond stresses relative to the bond strength
C = concrete cover, in. (mm)
c_b = smaller of: (a) the distance from center of a bar or wire to nearest concrete surface; and (b) one-half the center-to-center spacing of bars or wires being developed, in. (mm)
D = development length according to AIJ (1990), in. (mm)
DCH = ductility class high, based on the maximum behavior factor
DCM = ductility class medium, based on the maximum behavior factor
d = depth from extreme compression fiber to centroid of tensile reinforcement, in. (mm)
d_b = diameter of bar, or diameter of bar being developed, in. (mm)
d_{bL} = nominal diameter of longitudinal bars, in. (mm)
f_b = static bond strength as used for offshore structures, psi (MPa)
f_{br} = bond strength at 2,000,000 cycles as used for offshore structures, psi (MPa)
f_c' = specified concrete compressive strength, psi (MPa)
f_{cd} = design concrete compressive strength, psi (MPa)
f_{cr} = stress range in concrete, psi (MPa)
f_{ctm} = mean value of the tensile strength of concrete, psi (MPa)
f_{min} = minimum stress level, ksi (MPa)
f_r = stress range due to live loads and impact recommended by AASHTO, ksi (MPa)
f_{rup} = static modulus of rupture, psi (MPa)
f_s = existing reinforcement strength based on available development length, psi (MPa)
f_{sos} = yield strength divided by a factor of safety (typically = 1.15), offshore structures, psi (MPa)
f_y = yield strength of bar being developed, psi (MPa)
f_{yd} = design value of the yield strength of bars, psi (MPa)
f_{yt} = yield strength of transverse reinforcement, psi (MPa)
h = depth of member, in. (mm)
h_c = width of the column parallel to the bars, in. (mm)
h_j = joint dimension parallel to the bar, in. (mm)
jd_i = internal moment arm at section i, in. (mm)
K_{tr} = transverse reinforcement index
k_D = factor reflecting the ductility class equal to 1 for DCH and to 2/3 for DCM
ℓ = total length of the member, in. (mm)
ℓ_b = provided length of straight development, lap splice, or standard hook, in. (mm)
ℓ_d = development length, in. (mm)
ℓ_d' = equivalent straight bar anchorage length corresponding to a hooked bar anchorage, in. (mm)
ℓ_{dh} = lead embedment length for a hook, in. (mm)
ℓ_e = length of embedment of reinforcement, in. (mm)
M_i = moment at section i, lb-in. (N-m)
N = number of cycles to failure
N_{ed} = design axial force, lbf (N)
q = change of bar force per unit length of bar, lb/in. (N/mm)
r/h = ratio of base radius to height of rolled transverse rib
S = applied stress range, psi (MPa)
S_{bmax} = maximum bond strength as used for offshore structures, psi (MPa)
S_{br} = stress range as used for offshore structures, psi (MPa)

S_{max} = maximum flexural tensile stress, psi (MPa)
S_{min} = minimum flexural tensile stress, psi (MPa)
S_r = stress range in straight deformed bars as used in offshore structures, psi (MPa)
s = spacing of transverse reinforcement, in. (mm)
s_{max} = peak slip, in. (mm)
T_i = tensile force at section i, lb (N)
u = bond stress range, psi (MPa)
u_b = shear force per unit area of nominal bar surface between two sections along the bar, lb/in. (N/mm)
u_{bb} = maximum bond stress of the beam bars over the column width, psi (MPa)
u_r = bond stress range—the difference between bond stresses at the maximum and minimum load, considering the loading direction, psi (MPa)
α = stress multiplier for longitudinal bars at the joint or member interface for Type 2 joints
Δf_s = change of steel stress over length ΔL, psi (MPa)
ΔL = length of bar over which bond stress is computed, in. (mm)
γ_{Rd} = model uncertainty factor on the design value of resistances taken as being equal 1.2 or 1.0, respectively, for DCH or DCM (due to overstrength owing to strain-hardening of the longitudinal steel in the beam)
μ = constant to determine development length, with suggested values ranging from 10 to 12.5
ν_d = normalized design axial force in the column, taken with its minimum value for earthquake design situation ($\nu_d = N_{ed}/f_{cd}A_c$)
ψ_e = factor accounting for presence of epoxy coating on bar being developed
ψ_t = factor accounting for casting position of bar being developed
ψ_s = factor accounting for size of bar being developed
ρ' = compression reinforcement ratio of the beam bars passing through the joint
ρ_{max} = maximum allowed tension reinforcement ratio
σ_{bf} = frictional bond resistance during cycling, psi (MPa)
$\sigma_{b\,max}$ = monotonic bond strength, psi (MPa)
Σ_o = nominal perimeter of the bar, in. (mm)
τ = average bond stress at failure measured under cyclic loading, psi (MPa)
τ_m = safety factor = 1.25 as used for offshore structures
τ_{max} = bond strength achieved under monotonic loading, psi (MPa)

2.2—Definitions

ACI provides a comprehensive list of definitions through an online resource, "ACI Concrete Terminology," http://terminology.concrete.org.

CHAPTER 3—BOND AND ANCHORAGE

3.1—Average bond stresses

Average bond stress u refers to the bond stress averaged over a length of bar embedded in concrete for at least 18 bar diameters, and not to the local stress at an individual bar deformation or along the interface between bar and concrete. The limit of 18 bar diameters is arbitrary and constitutes a lower bound to typical anchorage lengths. However, this limit is important in discussing bond stress because most experimental data available refer to local bond stresses measured over much shorter distances (one to three bar diameters). These data indicate that local bond stresses can be four to five times higher than the average bond stress determined over longer lengths. Various techniques are used to obtain bond measurements, and the results from different research studies may not be directly comparable.

Values for maximum bond stress near ultimate load conditions that have been measured over short distances vary from approximately 1500 to 3000 psi (10 to 21 MPa). Under service load conditions, bond stresses are limited to much lower levels to avoid excessive deflections and localized damage, such as splitting cracks, and other serviceability problems. ACI 318 does not require specific checks for bond stresses. Required development lengths are based on an assumed uniformly distributed bond stress over the entire length of the bars being developed or lap spliced. The lower and upper bounds of corresponding average bond stresses at the ultimate limit state for required development lengths, using the ACI 318 approach for deformed bars in tension, are approximately 70 and 1050 psi (0.5 and 7.2 MPa), respectively. Equation (3.1a) provides the formula for development length in ACI 318.

$$\ell_d = \left(\frac{3}{40} \frac{f_y}{\lambda \sqrt{f_c'}} \frac{\psi_t \psi_e \psi_s}{\left(\frac{c_b + K_{tr}}{d_b} \right)} \right) d_b \quad \text{(psi)}$$

(3.1a)

$$\ell_d = \left(\frac{9}{10} \frac{f_y}{\lambda \sqrt{f_c'}} \frac{\psi_t \psi_e \psi_s}{\left(\frac{c_b + K_{tr}}{d_b} \right)} \right) d_b \quad \text{(MPa)}$$

Equation (3.1a) was derived by rearranging and simplifying the expression for average bond stress (Eq. (3.1b)) proposed by Orangun et al. (1977).

$$u = \left[1.2 + \frac{3C}{d_b} + \frac{50 d_b}{\ell_d} + \frac{A_{tr} f_{yt}}{500 s d_b} \right] \sqrt{f_c'} \quad \text{(psi)}$$

(3.1b)

$$u = \left[0.10 + \frac{C}{4 d_b} + \frac{4.2 d_b}{\ell_d} + \frac{A_{tr} f_{yt}}{42 s d_b} \right] \sqrt{f_c'} \quad \text{(MPa)}$$

Equation (3.1b) was developed from a statistical analysis of available data and accounts for the bar stress to be developed, tensile strength of concrete, concrete cover over the bars, bar spacing and size, and amount of transverse reinforcement crossing potential splitting planes along the bars being developed.

Equation (3.1a) retains all factors identified by Orangun et al. (1977) as relevant for determining average bond strength in a simplified and more user-friendly format. Furthermore, Eq. (3.1a) accounts for casting position of the bars, effect of epoxy coating, and whether the concrete is normalweight or lightweight. Equation (3.1a) includes an implicit relationship between development lengths and average bond stresses, and thereby provides flexibility for design with bars of different yield strengths and in cases where development of the yield strength of the bar is not required.

3.2—Components of bond resistance

Although the concept of average bond stress is convenient, the total bond force is a combination of resistance due to chemical adhesion (V_a), mechanical anchorage due to bearing of the ribs (V_b), and frictional resistance (V_f), as indicated in Fig. 3.2. Chemical adhesion between the steel and the concrete is lost at low levels of stress in the reinforcing bar. Frictional resistance is associated with micro-irregularities along the surface of the steel, wedging of granular material between the bar and the concrete, and bearing forces oriented perpendicular to the rib face that arise as the bar is loaded. Transverse components of the bearing forces can lead to a bond failure characterized by splitting of the concrete cover.

Typical adhesion values reported in the literature for test specimens subjected to monotonic loading range from 70 to 150 psi (0.5 to 1.0 MPa), whereas reported friction values range from 60 to 1450 psi (0.4 to 10.0 MPa) (Chinn et al. 1955; Eligehausen et al. 1983). It is commonly assumed that bearing forces alone are the primary load-transfer mechanism at load levels resulting in bond failure because contact between the bar and concrete necessary for friction and chemical adhesion may be lost at such load levels. Contrary to this assumption, data comparing the performance of plain and epoxy-coated reinforcing bars under monotonic loads showed that adhesion and frictional resistance may still play a significant role in anchorage failures governed by splitting of the concrete cover (Treece and Jirsa 1987).

Under cyclic loads, most of the bond stress is transferred mechanically by bearing of the bar ribs against surrounding concrete. Tensile and compressive strengths of the concrete, geometry and spacing of bar ribs, concrete cover and bar spacing, and amount of transverse reinforcement are important in controlling bond behavior for this loading case.

Although limited efforts have been made to categorize the components of bond resistance through direct measurements of bond degradation under cyclic loads, nondestructive methods (acoustic emissions and electrical resistance measurement) may further the understanding of bond properties. Acoustic emission monitoring of bond resistance

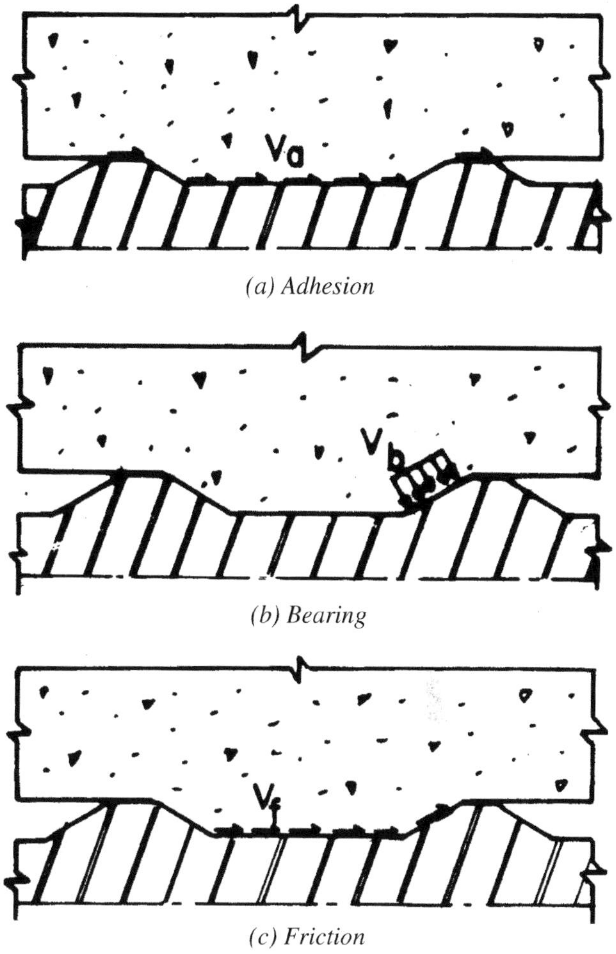

Fig. 3.2—Idealized force transfer mechanisms.

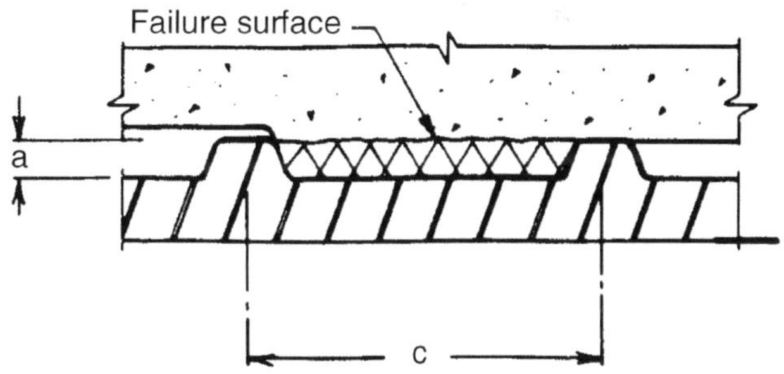

Fig. 3.3a—Pullout failure.

may permit quantification of internal damage at the steel-to-concrete interface, as damage is caused by fatigue under low- and high-cycle fatigue (Balazs 1998). Bond degradation during high-cycle (fatigue) loading has been investigated (Cao and Chung 2001) through measurement of electrical

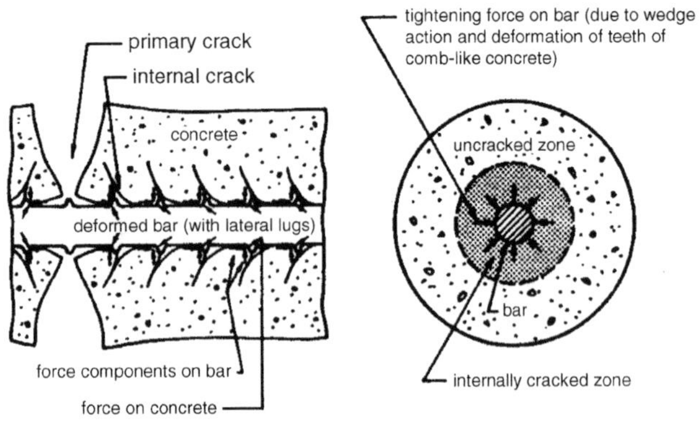

Fig. 3.3b—Splitting failure (Eligehausen et al. 1983).

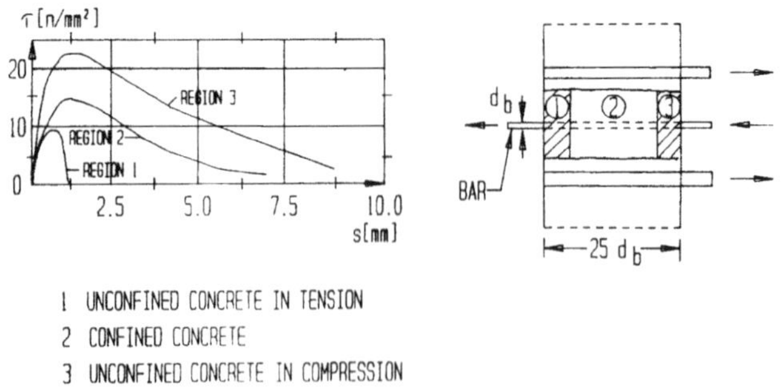

Fig. 3.4a—Monotonic bond stress-versus-slip curve for pullout failure (Eligehausen et al. 1983).

resistance along the bond interface—a method that may also have application in low-cycle fatigue investigations.

3.3—Failure modes

Under monotonic loading, two types of bond failures are typical. The first is a pullout failure, as depicted in Fig 3.3a, due to the shearing of the concrete between the ribs. This occurs when sufficient concrete cover prevents splitting or where sufficient transverse reinforcement keeps splitting cracks small. Pullout resistance depends primarily on concrete strength, pattern and geometry of bar ribs, depth of concrete cover, and the size and amount of transverse reinforcement.

The second type of bond failure is splitting of the concrete cover. This occurs when cover or confinement is insufficient to attain full pullout strength. As depicted in Fig. 3.3b, splitting failure is due primarily to tensile radial stresses caused by deformation bearing forces. In this case, splitting propagates to the edges of the member, resulting in loss of concrete cover and bond (Brown 1966; Ferguson 1979; Fujii and Morita 1981; Guiriani 1981; Hawkins et al. 1982; Lutz 1970; Lutz and Gergely 1967; Morita and Kaku 1979; Robins and Standish 1982; Robinson 1965; Untrauer and Henry 1965; Watstein and Bresler 1974).

Failure modes under low-cycle loading are similar to those under monotonic loading. Under high-cycle loading, similar failure modes can occur, but fatigue failures of both the bar and concrete need to be considered.

3.4—Bond mechanisms

Bond efficiency can be quantified creating bond stress versus bar slippage curves, which depict the change in local bond stress on the bar (τ) versus total movement of the bar relative to the surrounding concrete. Because there is no direct way to experimentally measure bond stress, most researchers measure the difference in strain between closely spaced strain gauges to compute bond stress. Similarly, slip is measured as the bar movement relative to some fixed reference frame. Bar slip includes three possible components: elastic deformation of the bar, inelastic deformation of the bar, and rigid body motion of the bar relative to the concrete. Only relative bar movement, as opposed to elongation, constitutes slip, but experimentally it is difficult to separate the three components. Slip can be measured either at the end of the bar, or near the application of load. For experimental measurements, gauges should be located and applied so that movement of the bar under load is not affected.

Bond stress-versus-slip curves for a bar loaded monotonically and failing by pullout are shown in Fig. 3.4a (Eligehausen et al. 1983). The relationship shown depends on the degree of confinement and the state of concrete stress surrounding the bar, whether tension or compression. Initially, the curve is steep because of adhesion. Because of shrinkage, small internal cracks always exist immediately adjacent to the reinforcing bar. These cracks act as stress concentrations and initiate crack propagation from the bar ribs at relatively low loads. Small internal bond cracks will begin to form at bond stresses ranging from 300 to 450 psi (2.1 to 3.1 MPa) (Chinn et al. 1955).

Because stress concentrations occur at the front of the ribs, a splitting crack will propagate from the top of the ribs at approximately one-third of the maximum pullout bond stress (Fig. 3.3b). Splitting has occurred between 300 and 1000 psi (2.1 to 6.9 MPa) (Chinn et al. 1955). As loading continues, crack growth leads to a softening of the curve. The rate of crack growth is affected significantly by confining reinforcement. As the maximum bond stress is reached, shear cracks form in the concrete between ribs until the concrete is sheared and bar slip occurs. Due to friction and interface shear, some resistance is maintained even at large values of bar slip. The magnitude of the slip corresponding to maximum bond stress is highly variable and depends primarily on initial bond stiffness (defined as the slope of the bond stress/displacement curve upon initial loading). This, in turn, depends on deformation geometry and casting position of the bar.

The failure process described previously refers to bars that fail by pullout, but bond failures are often governed by splitting of the concrete rather than by pullout. Splitting failure is initiated by the wedging action of the ribs as the bar moves with respect to the concrete. Splitting failures generally occur for bars with three diameters of concrete cover or less, but may occur for bars with more cover. Orientation of the splitting cracks depends on the number and configuration of bars within the member. For layers where bars have a concrete cover less than half the bar spacing, splitting cracks occur parallel to the bar and perpendicular to the surface. For bars in which the concrete cover is greater than twice the bar spacing, splitting cracks usually form in the plane of the

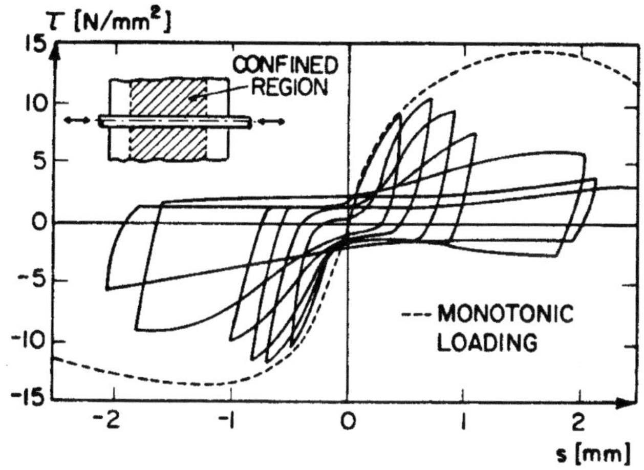

Fig. 3.4b—Cyclic bond stress-versus-slip curve with pullout failure (Eligehausen et al. 1983).

bars. In many cases, however, both types of splitting cracks occur in the same member.

The local bond stress-slip response for the pullout failure of a bar loaded cyclically is shown in Fig. 3.4b (Eligehausen et al. 1983). The initial part of the curve follows the monotonic envelope. When the load is reversed, after the bond stress exceeds approximately half of its maximum value, a significant permanent slip remains when the bar is unloaded. When loading in the opposite direction occurs, the bar must experience some rigid-body motion before bearing in the opposite direction. As cycling progresses, concrete in front of the ribs is crushed and sheared. When the load is reversed, large slip occurs before the bar ribs bear against the concrete and then bond stresses increase. The flat portion of the curve has been the source of much concern in the design of earthquake-resistant structures, especially where large bond stresses must be mobilized. In moment frame structures, where most of the joints have been subjected to inelastic load reversals, bond loss in beam-column joints can lead to large drifts. The maximum sustained bond stress under reversed cyclic loading follows a reduced monotonic bond stress-slip envelope (Balazs 1991). This reduction is a function of the load history, the number of reversed load cycles, and the maximum previous slip.

The slip increase due to constant amplitude cyclic loading tends to decrease after initial loading cycles. The Palmgren-Miner Hypothesis states that damage accumulates linearly with the number of cycles applied at a particular load level. In the context of this document, unless otherwise noted, the term "damage" refers to the degradation of bond due to the effect of cyclic loading. In the case of deformed bars, the degradation of bond is caused by crushing of the concrete in the area of the ribs (Rehm and Eligehausen 1979) or splitting of the concrete. As indicated in Fig. 3.4c, Balazs (1991, 1998) found that damage accumulates more rapidly in the initial and final phases of pullout failure. Between initial and final phases, constant-amplitude cyclic loading results in a constant rate of slip until the slip value corresponding to monotonic bond strength is reached. Under further cyclic loading, slip increases rapidly until pullout failure

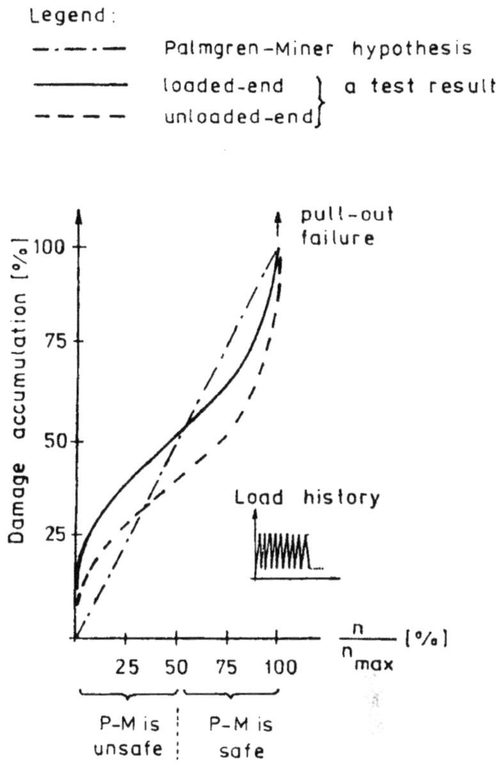

Fig. 3.4c—Bond damage accumulation with respect to the Palmgren-Miner Hypothesis (Balazs 1991).

occurs. Slip progression for reversed cyclic loading is more pronounced than for constant-amplitude cycles.

For variable or random cyclic loading, only load levels approaching or exceeding the previous highest level induce considerable slip increase, and the slip contribution of decreasing load levels is minimal (Balazs 1998).

3.5—Factors affecting bond strength under cyclic loads

The main factors influencing bond strength under cyclic loads are as follows.

3.5.1 *Concrete compressive strength*—The tensile and shear strength of concrete are seldom measured directly but have a strong correlation to the square root of the compressive strength (f_c') for concretes having f_c' less than 10,000 psi (70 MPa) (Carino and Lew 1982; ACI 363R). Because force is transferred by bearing and bond, and failure can occur by tensile splitting and shearing of the concrete, compressive strength is a key parameter in bond behavior (Orangun et al. 1977).

Bond strength increases with an increase in compressive strength (Alavi-Fard and Marzouk 2002). Additionally, they report that for concrete with compressive strengths of 7000 to 14,000 psi (50 to 95 MPa), the reinforcement bond is proportional to the cubic root of the compressive strength, and the square root of the compressive strength is not the ideal representation.

3.5.2 *Concrete cover*—The type of bond failure observed under monotonic loading in laboratory tests is strongly dependent on the concrete cover and bar spacing (Eligehausen 1979; Orangun et al. 1977; Tepfers 1973). When concrete cover is small, a splitting tensile failure occurs at a shorter anchorage length than that required for a pullout, or bond shear, failure. This effect is probably more important for cyclic than for monotonic loading because more severe cracking will probably occur and load repetitions will increase the energy available for crack propagation.

3.5.3 *Bar size*—To develop the same force in two bars with different sizes and the same development length, the large bar develops a lower bond stress than the small bar. The opposite is true if the aim is to develop the same bond stress in bars of two different sizes within a given development length. For improved bond performance, it is more effective to use many small bars rather than to use few large bars when reasonable clear distances between bars can be maintained. Although evidence shows that the apparent maximum bond strength decreases as bar size increases (Rehm and Eligehausen 1979), analysis of data in the ACI 408 Database indicates that accounting for an increased bond strength for smaller-diameter bars is not justified (Zuo and Darwin 2000). Because the tests in this database were performed under monotonic loading, the effect of bar size on bond performance of reinforcing bars subjected to reversed cyclic loads needs further study.

3.5.4 *Anchorage length*—Under monotonic loading, increases in bond force at failure are not proportional to increases in anchorage length. The maximum bond stress decreases with increasing anchorage length. Under cyclic loading, however, the number of cycles to a pullout failure should increase as anchorage length increases. There is no conclusive data available to develop guidelines in this area.

3.5.5 *Deformation (transverse rib) geometry*—Relative rib area, or ratio of transverse rib bearing area (area of ribs perpendicular to the bar axis) to transverse rib shearing area (perimeter of the ribs times rib spacing), has had a significant effect on initial bond stiffness and, thus, on bond performance. Bond strength and stiffness increase as the relative rib area increases (Balazs 1986; Eligehausen 1979; Rehm and Eligehausen 1977, 1979; Soretz and Holzenbein 1979). Typical values of this ratio vary from 0.05 to 0.08. However, high-relative rib area bars, defined as bars with relative rib area ratios between 0.1 and 0.14, inclusive, are produced and used more today than in the past. Reverse cyclic testing indicates that increasing relative rib area from 0.085 to 0.119 reduces unloaded end slip by 50 to 70 percent and loaded end slip by 30 to 40 percent (Zuo and Darwin 2000).

In addition, bar fatigue performance (high-cycle loading) is related to the radii at the bases of transverse reinforcement ribs due to stress concentrations at those radii. Transverse ribs normal to the bar axis reduce fatigue life in comparison with bars where ribs are angled to the bar axis (Zheng and Abel 1998).

3.5.6 *Steel yield strength*—Because bond stress is directly related to the force in the bar, using lower-strength steel may reduce the bond stress demand under monotonic loading. Most bars used today have nominal yield strengths from 60 to 70 ksi (410 to 480 MPa). Under monotonic loading, bond stress demand can be reduced by using bars with 40 ksi (280 MPa) yield strength. This is especially true under low-cycle loading where several reversed yielding cycles may be experienced. However, using lower-yield-strength bars may not improve bond strength because using lower-yield-strength steel requires using more or larger bars to achieve the required tensile force for a section. Using more bars results in closer spacing and reduces developed bond stress. For structures detailed for large load reversals, severe reinforcement congestion may cause difficulties in casting and achieving development. Strain hardening of the reinforcement under high-stress cycles should be considered, as it may result in larger bond stress demands than anticipated in the original design.

3.5.7 *Amount and distribution of transverse steel*—Splitting tensile failures are strongly dependent on amount and distribution of transverse reinforcement (Eligehausen 1979; Orangun et al. 1977; Tepfers 1973). Properly detailed transverse reinforcement confines the concrete after cracking and significantly increases the resistance to splitting failure, especially under cyclic loads. The upper bound to this improvement is the pullout failure of the bar.

3.5.8 *Casting position, vibration, and revibration*—Casting position and the effects of vibration and revibration can significantly influence bond strength under monotonic loading (Abrams 1913; Altowaiji et al. 1986). Proper consolidation of the concrete around the reinforcing bar is required for optimum bond strength. Consolidation is difficult to ensure in the field, particularly when bars are placed near the top of a deep member. In this case, voids and water pockets will form under the bar as the concrete flows during placement, and bond strength will decrease. In addition, tensile strength of the concrete near the top of a placement lift may be reduced if bleedwater significantly increases the local water-cement ratio (w/c).

3.5.9 *Strain (or stress) range*—Under cyclic loading, the amount of cumulative damage and the number of cycles at pullout are influenced primarily by the stress range (Bertero and Bresler 1968; Holmen 1982). For bond, both the stress range and strain (or slip) range are important (RILEM 1986), and evidence shows that the minimum stress level influences fatigue failure of bars.

3.5.10 *Type and rate of loading*—Research data indicate that the type of cyclic loading, whether fully reversed or one-directional, may have a large influence on bond performance (Rehm and Eligehausen 1977). Test data also show that a fast loading rate can result in significantly higher bond strengths and less damage in anchorages than a slow loading rate (Shah and Chung 1986). Concrete compressive, tensile, and shear strengths increase with the rate of loading. Loading frequency also affects the cumulative concrete damage under cyclic loading.

3.5.11 *Temperature*—Low temperatures can significantly increase bond strength. The strength gain and deterioration mechanisms at cryogenic temperatures can be explained by fracture mechanics of concrete (Shih et al. 1986, 1988). When a bar is subjected to multiple temperature cycle reversals, the capacity to store and dissipate energy through bond is reduced under cyclic loading (Shih et al. 1986, 1988).

This is reflected in the shape of the bond stress-slip relationships. Conversely, a loss of bond strength can be observed at elevated temperatures or after the concrete member has been exposed to fire (Diederichs and Schneider 1981; Morley and Royles 1983).

3.5.12 *Surface condition and coatings*—Evidence indicates that the bond strength of bars decreases when bars are coated with epoxy or similar substances. Monotonic tests show that pullout strength decreases approximately 15 percent, and splitting strength decreases by as much as 35 percent (Johnson and Jirsa 1981; Johnston and Zia 1982; Treece and Jirsa 1987). Tests indicate that epoxy coatings on bars reduce adhesion and friction, and lead to large splitting forces.

High-cycle testing of splices indicates that epoxy-coated bars have a smaller average bond stress at failure than the average bond stress of uncoated bars. This reduction ranges between 0.67 and 0.88, depending on concrete cover, bar size, deformation pattern, coating thickness, and confinement (Cleary and Ramirez 1993). A cover-to-bar-diameter ratio over 3.0 requires no additional precautions beyond those suggested by ACI 215R and ACI 408R in designing members with epoxy-coated bars for high-cycle fatigue or repeated loading in the service load range usually encountered in bridge decks (Cleary and Ramirez 1993). Accordingly, the larger cover-to-bar-diameter ratios recommended in harsh environments should not be reduced with the expectation that the epoxy coating will protect the reinforcing bars (Cleary and Ramirez 1993).

Epoxy-coated reinforcing bars under repeated loading does not result in significantly increased deflection despite the increased bar slip that is indicated by larger flexural crack widths (Cleary and Ramirez 1993; Hasan et al. 1996). Fewer cracks negate increased rotation and, thus, deflection, due to additional initial slip. The loss of deformation area due to extra thick epoxy coating can be critical during repeated load application if stresses are close to the upper bond limit (Cleary and Ramirez 1993).

3.5.13 *Lightweight aggregate concrete*—Research indicates that the fatigue strength of lightweight-aggregate concrete is similar to or better than that of normalweight-aggregate concrete of similar strength properties (Mor et al. 1992). Adding silica fume to lightweight concrete improved bond strength by 100 percent and improved fatigue life by over 60 percent. No significant improvement was observed when silica fume was added to normalweight concrete.

Research indicates that bond strength for high-strength lightweight concrete is not significantly affected by cyclic loading, provided the maximum cyclic displacement is less than the peak load slip in monotonic tests (Mitchell and Marzouk 2007). Once displacement exceeds this peak load slip, however, bond strength rapidly deteriorates.

3.5.14 *Corrosion*—Research by Fang (2006) indicates that the maximum corrosion level, in terms of mass loss, that improves cyclic bond is approximately 5 to 7 percent. Significant corrosion provides the most bond improvement in the first five cycles. Subsequent bond deterioration occurs as load cycles increase.

3.5.15 *Alkali-silica reaction*—The fatigue life of tension reinforcement within a lap splice is reduced when alkali-silica reactions are present due to a bond reduction resulting from expansive forces within the concrete (Ahmed et al. 1999).

3.5.16 *Fiber-reinforced concrete*—Experimental evidence suggests that introducing steel or polymer fiber reinforcement into concrete improves both maximum and residual bond strength (Krstulovic-Opara et al. 1994; Hota and Naaman 1997; Parra-Montesinos et al. 2005; Harajli 2005). Recent research on evaluating the bond of reinforcing bars embedded in fiber-reinforced cement composites (FRCC) showed that FRCC, compared with concrete with conventional transverse reinforcement, significantly enhances the bond performance in terms of bond strength, stiffness, and ductility under cyclic loading. This is particularly true for FRCC, which exhibits strain-hardening behavior in tension (Chao 2005). In concrete members confined by spiral or rectangular hoops, a minimum lateral expansion, which leads to cracking, is required in the cementitious matrix before transverse reinforcement effectively provides confinement. These initial cracks may reduce the bond contribution from friction and mechanical interlocking, especially under reversed cyclic loading. Alternatively, the confinement and bridging effects provided by fibers in FRCC after cracking can increase the tensile stress and strain capacity of the concrete matrix and limit crack width, which leads to increased bond strength of reinforcing bars embedded in such composites.

Parra-Montesinos et al. (2005) tested large-scale beam-column subassemblies constructed with strain-hardening FRCC in the connection and adjacent beam regions. No confinement (transverse) reinforcement was used in the connection region. The researchers reported that no signs of deterioration of bond between longitudinal beam bars and surrounding strain-hardening FRCC material were observed in the test specimens, even though the column depth represented 18.7 beam bar diameters and the beam bars were subjected to large inelastic strain reversals. A peak average bond stress of approximately 1500 psi (10 MPa) was developed in beam longitudinal bars with no noticeable reinforcement slip.

3.5.17 *Bond in grouted ducts used in precast structures*—Raynor et al. (2002) addressed the bond behavior of bars grouted in ducts used to connect precast concrete frames subjected to cyclic loading. They proposed a relationship between local bond stress and slip, and this relationship was used to develop a design equation for the equivalent debonded region used in the connection of precast concrete frame elements.

3.6—Summary

A single, simple model cannot represent the complex problem of bond strength under cyclic loading. Many factors affect bond strength under cyclic loading, and their influence on bond strength and failure mechanisms is understood only qualitatively in many cases. The following chapters summarize the research available for bond strength under high- and low-cycle fatigue and present models for bond behavior under cyclic loading. This report describes the research

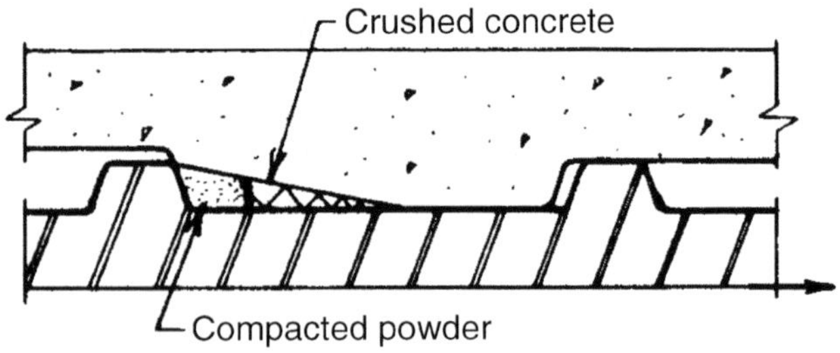

Fig. 4.1a—Crushing of concrete in front of ribs.

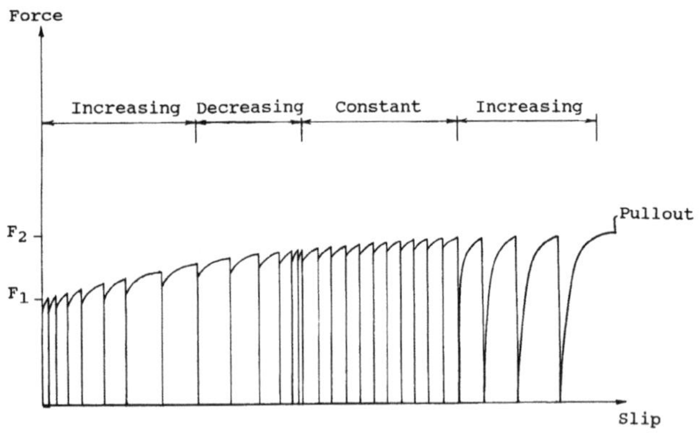

Fig. 4.1b—Typical bond fatigue failure (Balazs 1986).

efforts that are the most complete and accessible. Existing design recommendations and possible additional guidelines for design are discussed.

CHAPTER 4—BOND BEHAVIOR UNDER CYCLIC LOADS

Repeated loads can reduce bond strength at failure and accelerate the rate of bond deterioration as the stress range or the number of cycles increase. The following sections discuss the mechanisms that influence bond deterioration under cyclic loads.

4.1—High-cycle fatigue

Fatigue is a progressive permanent internal structural weakening of a material subjected to repetitive stresses (ACI 215R) and can lead to failure of a material at a stress lower than the maximum stress under monotonic loading. Fatigue loading on a specimen or member of a structure may be thought of as a load sine wave that varies periodically, such as a superimposed on a static load. The number of cycles to failure is a function of both the static load and the varying superimposed load. The most severe fatigue loading occurs when the mean static load is zero and the varying load causes alternating tension and compression. When the magnitude of reversed loading exceeds the yield strength of the material, then alternating plasticity produces failures after a few load cycles.

Analysis of systems under fatigue loading is complicated by the changes in material properties that occur over time. Conditions in a concrete member change over time due to the interactive static and dynamic influences of creep, shrinkage, temperature, and the applied stress range.

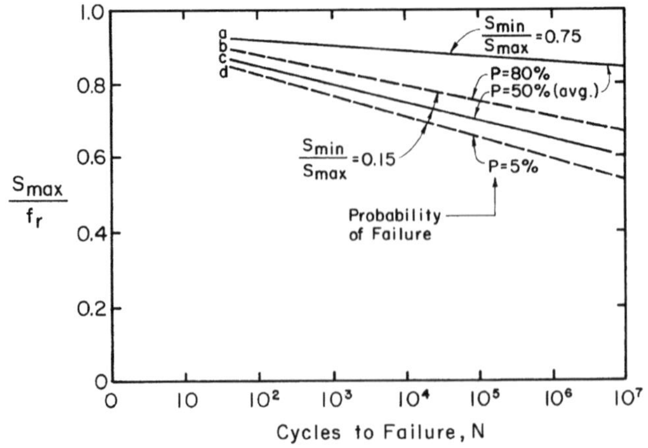

*Fig. 4.1c—*S-N *curve for plain concrete (ACI 215R).*

Therefore, comprehensive design guidelines for fatigue loading are difficult to derive. Recent data indicate that the maximum bond strength under fatigue loading relates directly to internal concrete damage. Thus, repeated loads have a similar influence on the bond strength and slip as on deformation and failure of unreinforced concrete (Rehm and Eligehausen 1979). The main mechanism of bond deterioration seems to be progressive crushing of concrete in front of the ribs, which is depicted in Fig. 4.1a. Figure 4.1b indicates four behavioral stages apparent in most bond fatigue tests.

1. A fast increase in slip due to initial crushing of the concrete
2. A rapid reduction in the slip rate due to the stabilization of the process
3. A long portion with a constant slip rate
4. A fast increase in the slip rate as the failure approaches.

Such a response is typical of a pullout failure. A splitting failure would result in a sudden drop in load-carrying capacity.

Fatigue test data are commonly analyzed using stress-versus-fatigue life diagrams, commonly called *S-N* diagrams. In these diagrams, applied stress or stress range (S) is plotted against the number of cycles to failure (N); N is usually plotted on a logarithmic scale. Figure 4.1c indicates the fatigue resistance of plain concrete beams subjected to flexural loading. The ordinate is the ratio of maximum flexural tensile stress (S_{max}) to the static modulus of rupture (f_{rup}). The abscissa represents N plotted on a logarithmic scale. Solid lines represent the fatigue life of the concrete when subjected to two different stress ranges relative to S_{max}. For a given number of cycles, a smaller stress range, indicated by a larger value of S_{min}/S_{max} in the figure, corresponds to increased fatigue strength. Each solid line represents a 50 percent probability of failure for the given stress range. Dashed lines represent different probabilities of failure for $S_{min}/S_{max} = 0.15$. For reinforcing steel, the *S-N* curve in Fig. 4.1d indicates that, regardless of the number of cycles, fatigue does not appear to be a problem below a certain stress range. This is the ideal working stress range for high-cycle fatigue applications, and design guidelines can be developed for this area of the *S-N* curves.

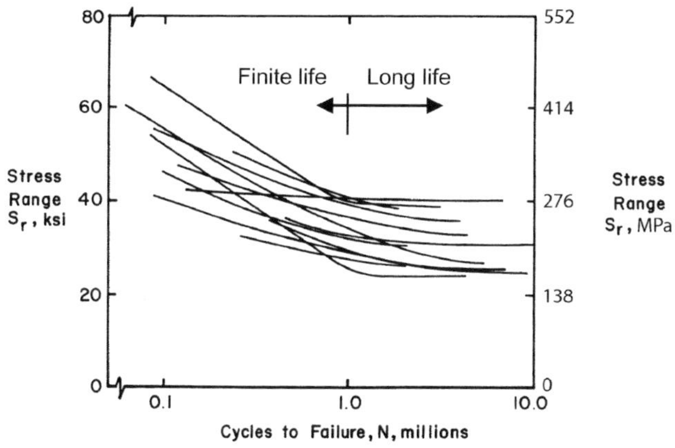

*Fig. 4.1d—*S-N *curve for reinforcing bars (ACI 215R).*

Failure of concrete members under high-cycle fatigue can be triggered by fatigue of reinforced concrete in compression, the reinforcing bars in tension, or bond failure. The first two aspects are detailed in ACI 215R and are not discussed herein. Fatigue life of reinforcing bars, shown in Fig. 4.1d, can be considered as the lower limit for bond failure in fatigue, as most tests show that reinforcing bars fracture before bond strength is exhausted. An upper-bound bond failure in fatigue is given by a line connecting 40 ksi (280 MPa) at zero load cycles with 20 ksi (140 MPa) for one million cycles (Fig. 4.1d). This range is the finite-life region. Below the 20 ksi (14 MPa) stress range, bars are insensitive to fatigue and assumed to be in the long-life region. The large scatter of results in both figures indicates the need for more systematic research and conservatism in design.

When evaluating available data for bond under fatigue loading, determining realistic stress ranges for reinforced concrete structures is essential. Reinforced concrete structures are typically not subjected to completely reversed loading under high-cycle fatigue. More commonly, a relatively small live load, which provides the stress range, is superimposed on a substantial dead load, which can be considered static. Because the stress ratio in this case is usually less than $0.4f_y$ and anchorage lengths are long, pullout failures under high-cycle fatigue are rare. If a fatigue failure of bond does not occur, then previous repeated loadings do not negatively affect the local bond stress-slip relationship near ultimate load compared with the static loading relationship, as indicated in Fig. 4.1e (Rehm and Eligehausen 1977, 1979).

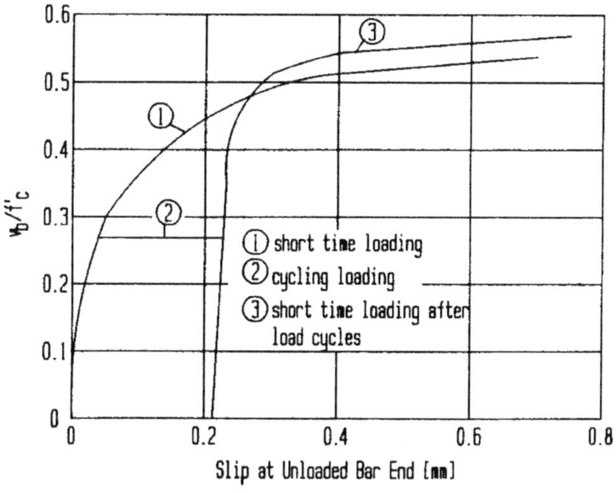

Fig. 4.1e—Effect of cyclic load on static bond strength (Rehm and Eligehausen 1979).

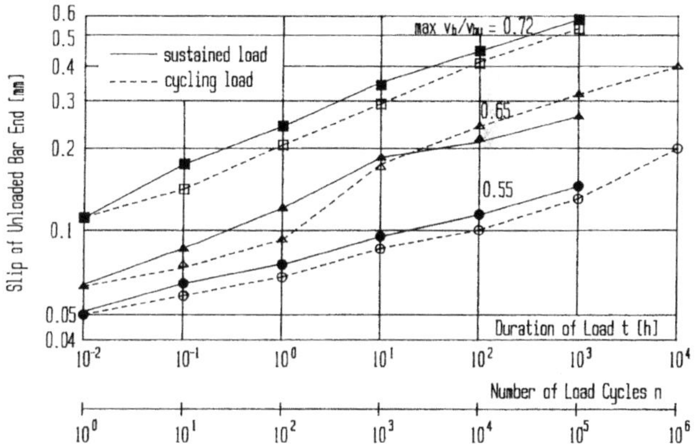

Fig. 4.2a—Comparison between sustained and cyclic loading (Rehm and Eligehausen 1979).

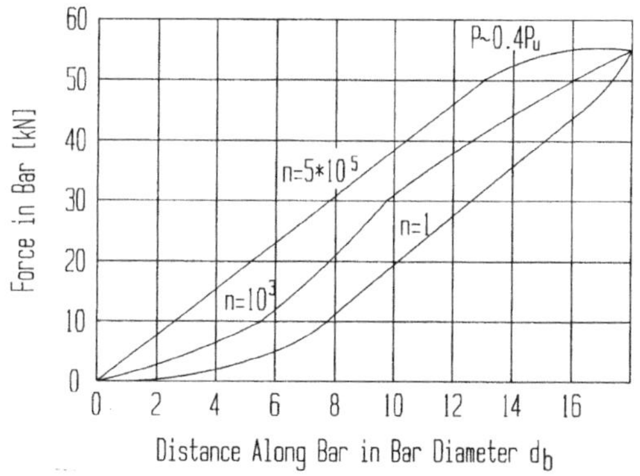

Fig. 4.2b—Redistribution of bond stresses along embedment length (Rehm and Eligehausen 1979).

4.2—Correlation between fatigue and sustained loading

A comprehensive quantitative model describing the mechanisms of bond deterioration under high-cycle fatigue is not available. Given the similarity between the deterioration of bond strength and compressive strength of plain concrete under cyclic loads, some compressive strength models may be applicable (Rehm and Eligehausen 1977, 1979). Rehm and Eligehausen (1979) compared bond specimens subjected to sustained loads of 0.55, 0.65, and 0.72 times the static failure load to bond specimens subjected to cyclic loading. Figure 4.2a shows the mean slip of the three test series as a function of the loading time in hours. In the same figure, results of comparable test specimens subjected to cyclic loads with a maximum load corresponding to the sustained load and a minimum load corresponding to 0.10 times the pullout load are shown as a function of the number of load cycles.

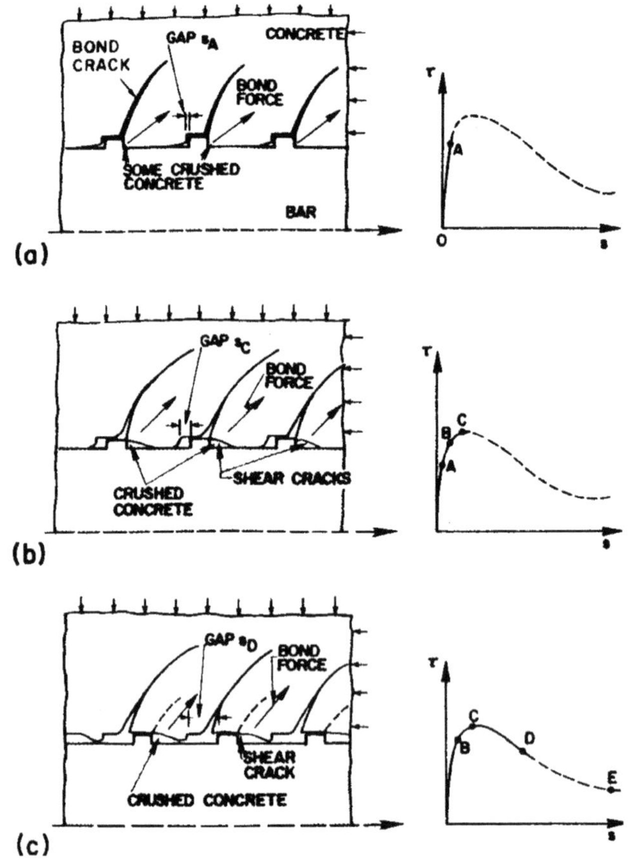

Fig. 4.3a—Monotonic envelope for bond stress (Eligehausen et al. 1983).

As far as the slip is concerned, the results in Fig. 4.2a show that cyclic loading can be a time accelerator, and increase the rate of bond deterioration, compared with a sustained load. Similar performance was found for creep tests with unreinforced concrete specimens (Holmen 1982). However, as indicated in Fig. 4.2b, repeated loadings under this loading condition influence the bond behavior under service load by redistributing bond stresses along the embedment length (Rehm and Eligehausen 1979). Expect similar redistribution under sustained load of the same magnitude.

4.3—Low-cycle loading

In contrast to high-cycle fatigue loadings, where the stress range is generally within the static elastic range, low-cycle loading has a few large deviations into the inelastic range. High-cycle fatigue failures occur under service loads. Low-cycle loading failures occur at loads approaching or exceeding the unfactored design strength of the structure. In low-cycle loading situations, the objective is not to prevent but to limit the amount of slip and damage to the concrete surrounding the bar.

In low-cyclic loading, the type of loading, whether load or strain cycling, and the rate of loading, are important for bond strength (RILEM 1986; Shah and Chung 1986). Earthquake loading is the most common case of low-cycle loading. It represents an intermediate step between load cycling and strain cycling and has wide frequency content. It is difficult to perform accurate analyses for bond strength under earthquake loads, so a conservative approach is required when developing design recommendations.

Many models have been proposed to qualitatively and quantitatively describe bond deterioration under low-cycle loading (refer to Chapter 7). The model proposed by Eligehausen et al. (1983) for pullout failures in well-confined concrete is discussed in the following. Comparable models are not available for bars with small concrete cover where failure is due to splitting.

When loading for the first time, the assumed bond stress-slip relationship follows the monotonic envelope, which is valid for monotonically increasing slip (path OABCD in Fig. 4.3a). Initial slope (with all other variables being equal) depends on the related deformation area. At low bond stresses, inclined cracks propagate from the top of the ribs. The confining pressure provided by transverse reinforcement contains the growth and size of these cracks. Transfer of forces between the concrete and the bar will be mostly by bearing of the ribs, with a shallow angle of inclination (α = 30 degrees). As loading passes Point A, inclined cracks begin to form at the top of the ribs. When maximum bond stress is attained, the concrete key is sheared off, forming a cone with a length of approximately four times the deformation height. At this point, the line of action of the force is at an angle of approximately 45 degrees. With increasing slip, the bond stress drops slowly to 85 percent of the maximum at three times the slip at the maximum bond stress. As the bond shear cracks reach the bottom of the adjacent deformation (Point D), bond stress continues to drop, and by the time the slip is equal to the deformation spacing, only the frictional component remains (Point E).

Under load reversals, the initial loading follows the monotonic curve, but the cyclic load behavior is sensitive to the slip level at which the reversal occurs (Fig. 4.3b). Three possible qualitative models have been proposed, depending on whether or not inclined cracks have formed. In the first case (Fig. 4.3b(a)), imposing a slip reversal at a slip value below the level at which inclined cracking occurs results in a stiff unloading branch (Path AF) because only a small part of the slip is caused by inelastic concrete deformation. As soon as slip in the opposite direction is imposed, the friction branch is reached (Path FH). The slope of this portion of the curve is small because the concrete surface surrounding the bar is smooth. As soon as the cracks close, the stiffness is similar to that of the monotonic envelope (Path HI).

Unloading from Point I, where the slip in the two directions is approximately equal, the curve (Path IKL) is similar to that of the initial unloading curve (AFH). Because of previous cracking and crushing of the concrete in front of the ribs, the point at which the bond stresses begin to pick up load again (Point L) is shifted right of the origin. The deformation will not bear fully until Point M is reached. Further loading follows the bond-slip curve up to the monotonic envelope.

In the second case, unloading occurs after inclined cracks have formed (Fig. 4.3b(b)). Therefore, near the slip at which maximum bond stress has been attained, the unloading path is similar to that of the first case up to Point F. Because there is more damage to the concrete, higher frictional resistance is mobilized (Point G). When loading is reversed, the deformation presses against a key that has a lower resistance as the result of inclined cracks over part of its length induced by the first half-cycle. Splitting cracks created in the first half-cycle close at a higher load than those of the first case (Point H) and lead to earlier formation of splitting cracks in the opposite direction. Splitting cracks combined with existing inclined cracks along the bar result in a reduced envelope (Path HI) and reduced bond strength in the second direction (Point I). Unloading from this peak (Path IKLMN) and reversing the load results in reduced stiffness and strength because only the remaining uncrushed concrete between the ribs must be sheared. The bond strength at Point N is substantially lower than that of Point C, and lower than that of Point I.

When unloading occurs after the slip has reached a value much larger than the slip at maximum strength (Point C), bond strength will decrease (Fig. 4.3b(c)). Because more damage has occurred, the frictional resistance (Point G) will be larger than for either the first or the second case. Because the concrete between the ribs is completely sheared, however, little force can be transmitted by bond when the direction of loading is reversed (Path HIJ). Unloading and reloading in the opposite direction (Path IKLMN) results in little additional bond strength beyond that provided by friction because most of the mechanical anchorage has been lost.

Melek and Wallace (2004) performed experimental investigations of lap splices in cantilever column specimens subject to constant axial load and reversed cyclic lateral displacements at the top of the column. Their studies focused on identifying effective rehabilitation measures for columns with inadequate lap splice lengths. Investigation of moment-rotation responses of the specimens indicated that rotation caused by longitudinal bar bond slippage accounted for a significant portion of total rotation. Melek and Wallace (2004) concluded that substantial slip rotation occurs before observed strength degradation, and that bond slip dominates rotational response after initiation of bond deterioration.

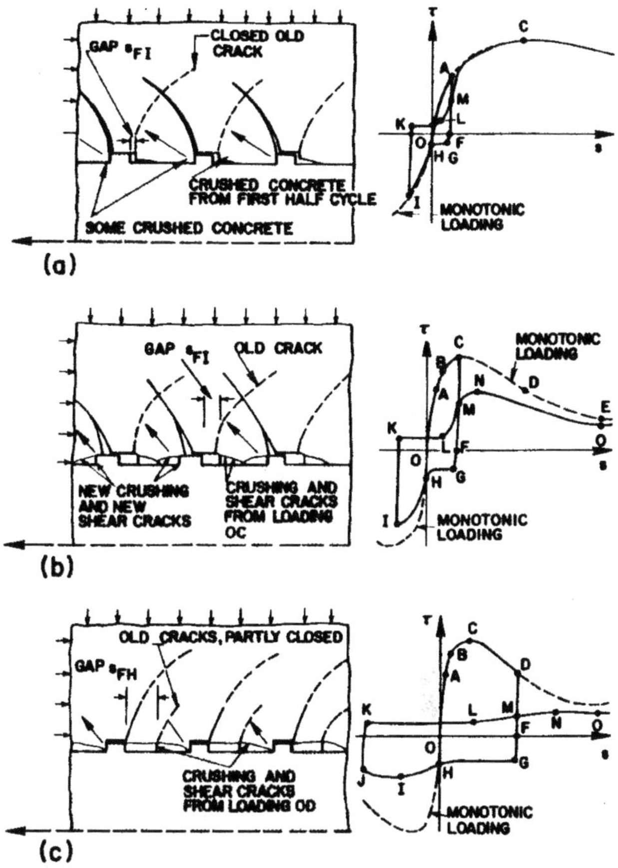

Fig. 4.3b—Cyclic envelope for bond stress (Eligehausen et al. 1983).

4.4—Influence of cracking on bond strength

Concrete contains various internal flaws and cracks that form during the hydration process and may extend over time. These cracks are often found at the aggregate-paste interface, known as the interfacial transition zone (Maso 1980). When concrete is loaded, stresses at these preexisting crack tips can exceed the tensile strength of the hydrated cement paste, even though the tensile stress in the surrounding concrete is low. Under repeated cycles of compressive loading, these cracks grow, resulting in hysteretic response (Fig. 4.4). Part of the area within each loop represents the unrecoverable strain energy associated with crack propagation. Hysteretic response varies depending on load magnitude. For cycling to compressive stress levels less than the static compressive nominal bond strength, Fig. 4.4(a) indicates that the energy lost in each cycle first decreases and stabilizes. Some of the energy lost in early cycles is likely attributable to viscous flow and the growth of microcracks in the interfacial transition zone. The energy supplied by repeated loadings, however, eventually increases damage at the crack tips until the cracks propagate and cause bond failure. As the fatigue limit is approached, the hysteretic loops increase in size, indicating energy lost as the cracks coalesce and extend. For cycling to slips larger than that corresponding to maximum bond stress, the area in each cycle tends to decrease as damage accumulates, similar to behavior indicated in Fig. 4.4(b). A comprehensive model to explain this phenomenon is not available, but a large portion of bond deterioration under cyclic loads can be attributed to formation and propagation of cracks.

4.5—Debonding to increase shear strength at connections

Shear strength of reinforced concrete connections under low-cyclic loading may be increased by debonding of the longitudinal reinforcement within the shear span zone (zone subjected to uniform shear force). This can alter the failure mechanism from a shear failure condition to a more ductile flexure condition. Debonding the longitudinal reinforcement may increase the shear strength of reinforced concrete connections even with minimal shear reinforcement. Furthermore, a debonded region can improve the ductility of reinforced concrete members by a factor of 3. To achieve such improvements, the entire shear span zone should be debonded so that flexural cracks will not form

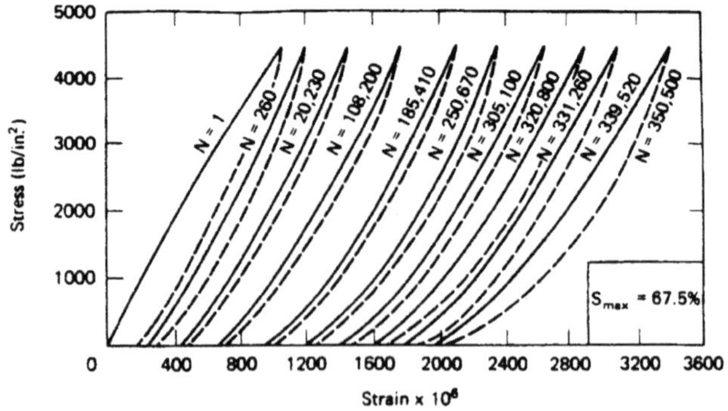

(a) Cycling below strain at nominal bond stress

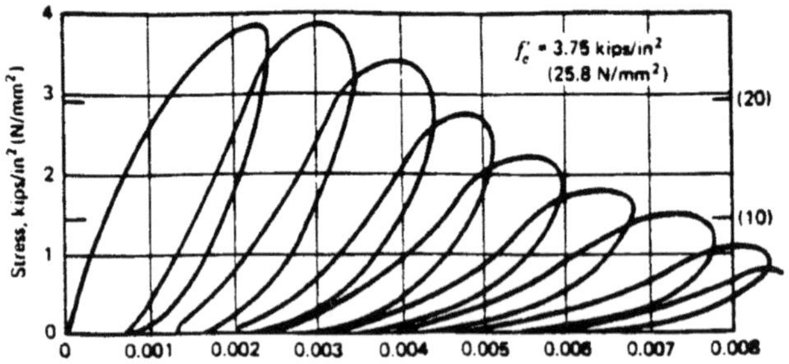

(b) Cycling above strain at nominal bond stress

Fig. 4.4—Effect of cyclic loads on plain concrete (Morita and Kaku 1973).

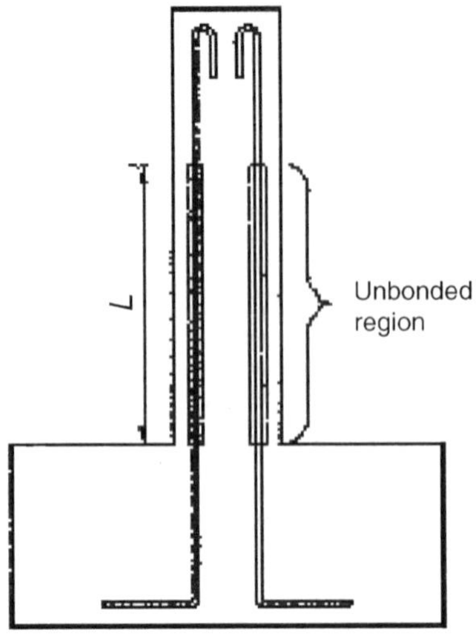

Fig. 4.5—Illustration of debonded zone (Pandey and Mutsuyoshi 2005).

in the debonded region and lead to certain shear failure. In addition, hysteretic energy dissipation will decrease in reinforced concrete connections using debonded regions. Figure 4.5 illustrates the debonded zone used in investigations by Pandey and Mutsuyoshi (2005).

Similar debonding of longitudinal beam bars in wide beam-column connections has been used to reduce the torsion generated on side faces of the column under reversed cyclic loading (Stehle et al. 2001).

CHAPTER 5—ANCHORAGE UNDER HIGH-CYCLE FATIGUE

Most research discussed in this chapter concerns high-cycle fatigue for reinforcing bars under unidirectional repeated loading. This load regimen is less detrimental to reinforced concrete than fully reversed cyclic loads.

5.1—Straight bar anchorages

Research studies on high-cycle fatigue under unidirectional repeated loading were conducted in the 1940s (Muhlenbruch 1945), but the most useful recent data come from the work of Verna and Stelson (1963), Perry and Jundi (1969), Corley et al. (1978), and Rehm and Eligehausen (1979).

Verna and Stelson (1963) reported the results of 16 small reinforced concrete beams tested to destruction under both monotonic and high-cycle loading. These beams contained reinforcing bars with f_y = 46.5 ksi (314 MPa) and failed either in diagonal tension, compression, bond, or steel fatigue (Fig. 5.1a shows the results of the 11 specimens that were tested using high-cycle loading). Beams that exhibited bond failure under monotonic loading also failed in bond when subjected to repeated loading. In these cases, bond failure was typified by splitting tensile cracking and followed by a rapid increase in deflection and a decrease of load-carrying capacity. The bond failure load under high cyclic loading conditions was lower than the strength under monotonic loading (labeled as the ultimate static load in Fig. 5.1a).

Perry and Jundi (1969) investigated bond stress distribution along a No. 6 (19M) bar embedded 9 in. (230 mm) into an eccentric pullout specimen subjected to repeated loading. A control specimen was tested monotonically to failure and 56 percent of the ultimate load sustained by this control specimen was applied to the cyclic loading specimens. These specimens were loaded at 0.0833 Hz with either a slowly varying or an impulsive load. Figure 5.1b shows two typical sets of bond stress distribution curves. Besides the peak bond stresses near the loaded and unloaded ends, of interest is the increasing bond stress near the unloaded end and decreasing bond stress near the loaded end with increasing loading cycles. This indicates that the bar is slipping as the number of cycles increase. Average maximum bond stress ranged from 274 to 735 psi (1.89 to 5.07 MPa). Peak bond stress at failure ranged from 590 to 1400 psi (4.07 to 9.65 MPa).

Corley et al. (1978) and Helgason et al. (1976) tested over 350 concrete beams reinforced with a single bar. The main variables were the bar stress range, bar minimum stress, bar diameter, bar yield strength, deformation geometry, and beam effective depth. Concrete cover to the bar centerline was 2 in. (50 mm), and a crack was preformed at midspan of the beam. Beams were designed to test bar fatigue resistance, and bond failures were not reported in any test. All failures were due to bar fracture or to run out of the test after 5 million cycles. Data generated in this report do not deal directly with bond fatigue, but design recommendations from these studies, as discussed in Chapter 8, help reduce the possibility of high-cycle fatigue bond failures.

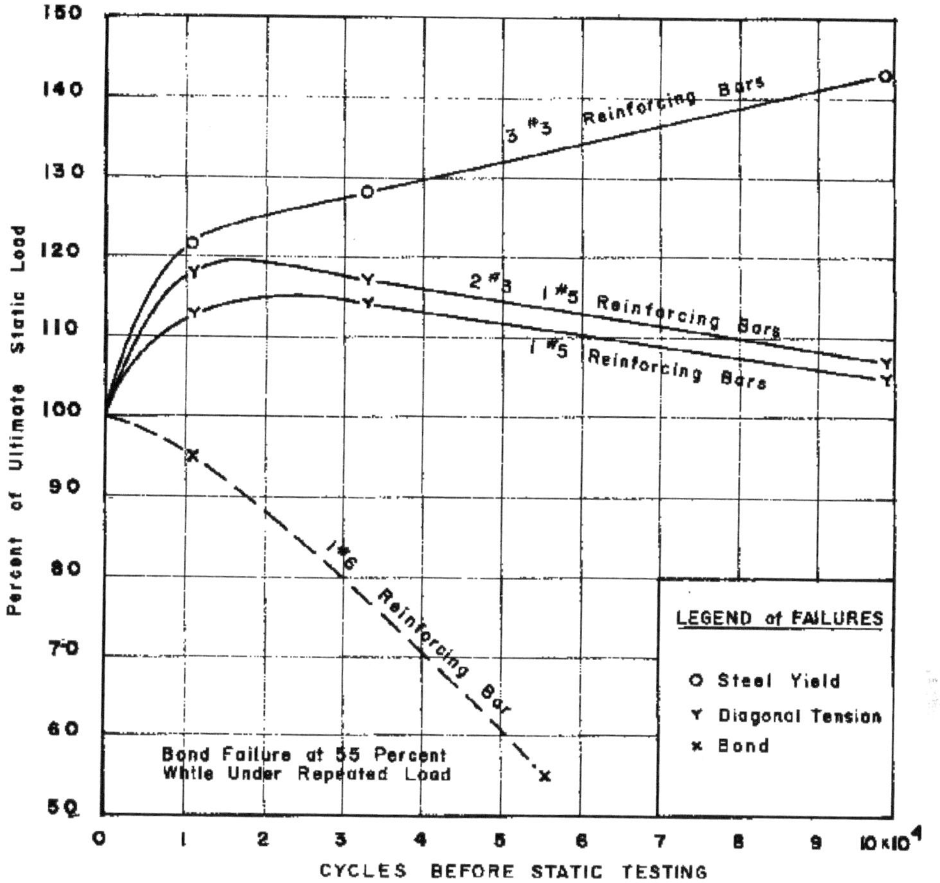

Fig. 5.1a—Effects of cyclic loading on nominal strength from Verna and Stelson (1963).

Rehm and Eligehausen (1977, 1979) performed a detailed study of bond fatigue behavior using pullout specimens. They observed that the number of load reversals required to cause fatigue failure increased with decreasing peak load when the static load remained constant. This can be seen in Fig. 5.1c, where the ratio of the average bond stress at failure measured under cyclic loading (τ) normalized by the bond strength achieved under monotonic loading (τ_{max}) is plotted versus the number of cycles to failure. The tests were performed with concrete strengths of 3400 and 7000 psi (23.5 and 48.0 MPa), and embedment lengths varied from 8 to 28 bar diameters. If no bond failure occurred, the experimental point in the diagram is marked with an arrow. A number beside the symbol corresponds to the number of similar results. A straight line in semi-logarithmic diagram can approximate test results. The scatter of the results corresponds approximately to those expected from fatigue tests. A clear influence of the concrete strength and the bar diameter on bond fatigue strength cannot be observed.

The effect of repeated loading on slip and bond strength of deformed bars is similar to the effects of repeated loading on the rib and the resulting failure behavior of unreinforced concrete loaded in compression. Bond strength decreases with an increasing number of peak load cycles between constant bond stresses due to static load. Slip under peak load and residual slip increase as the number of cycles increases. If no fatigue failure of bond occurs during cycling and the load is increased afterward, the monotonic envelope is reached again and followed thereafter (Fig. 4.1e). Therefore, when the peak load is smaller than the load corresponding to the fatigue strength in bond, a bar that has previously undergone cyclic loading will have modified bond strength under service load but similar bond strength near failure.

Increasing the static load or reducing the peak load under otherwise constant conditions increases service life. Therefore, mean stress and stress range can be used as reference values for describing the influence of repeated loading on bond.

5.2—Hooked bar anchorages

The effect of high-cycle fatigue on hooked bar anchorages has not been extensively investigated. A study by Graf and Brenner (1939) on plain bars with and without hooks found that the maximum load leading to bond failure was on the order of 10 to 15 percent higher for bars with hooks than for straight bars under fatigue loading with a stress ratio ($\sigma_{min}/\sigma_{max}$) between 0.05 and 0.28. Graf and Brenner (1939) concluded that the influence of fatigue increased with the strength of the concrete, being more significant for straight bars that for those with hooks. Because hooks are formed by cold bending of reinforcing bars, it is likely that cracks and other zones of weakness are built into hooks by the fabrication process. Hooks are probably more sensitive to

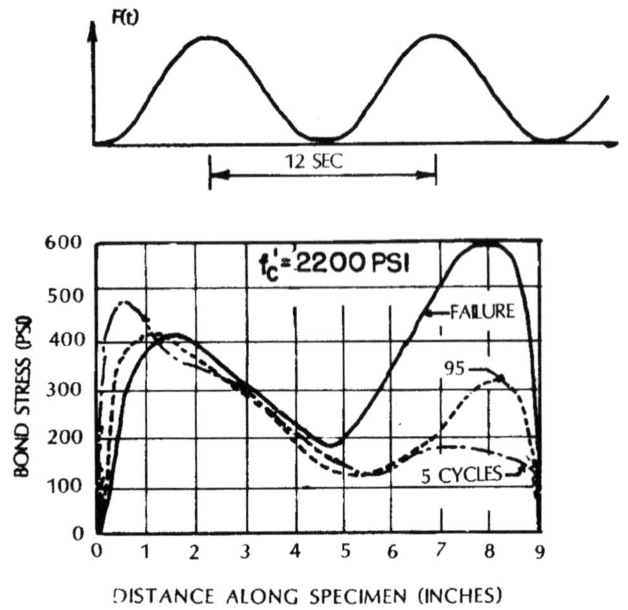

(a) Static loading

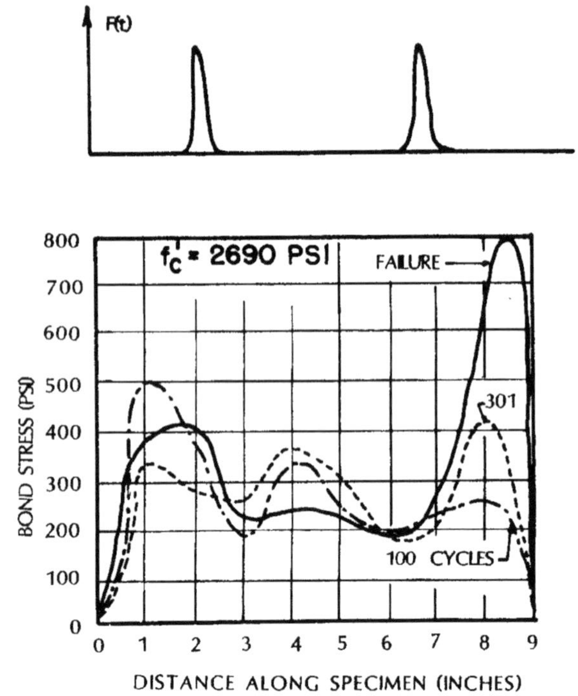

(b) Dynamic loading

Fig. 5.1b—Changes in bond stress due to static and dynamic loads (Perry and Thompson 1966).

steel fatigue rather than to bond failures. Moreover, stress distribution in a hook is different from that of a straight bar anchorage, leading to faster degradation in zones of high bond stress.

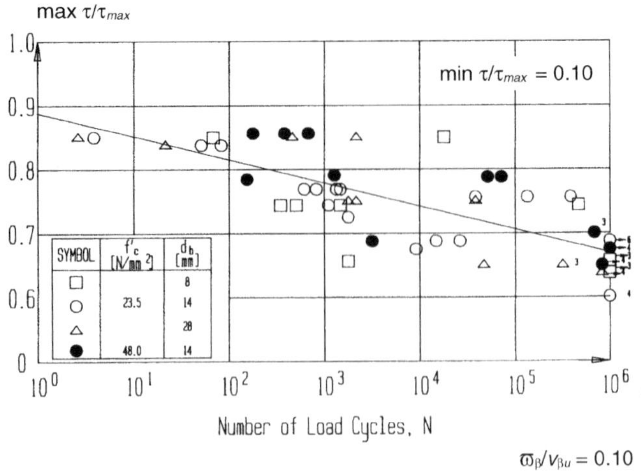

Fig. 5.1c—Influence of ratio of bond stress (Rehm and Eligehausen 1979).

5.3—Lap splices

Several investigations have applied high-cyclic loading to lap splices. Cleary and Ramirez (1993) and Hasan et al. (1996) studied lap splices of epoxy-coated reinforcing bars under repeated loading. Two types of lap splice or bond failures can occur: direct pullout of the bar when ample confinement is provided, or splitting along the bar when confinement is insufficient. Small internal cracks, however, initiate at bar ribs at relatively low loads. For fully reversed loading, small inclined cracks at the bars extend in both directions and can lead to significant concrete deterioration around the bar or splice, especially for large-diameter bars.

Tepfers (1973) tested small beams containing No. 5 (16 mm) bars, with lap splice lengths equal to 29 in. (732 mm), subjecting them to repeated flexural loading between 25 and 75 percent of the capacity under monotonic loading. Most of the beams tested by Tepfers under monotonic loading experienced splice failure, although there were beams for which the reported mode of failure was shear. The load range was defined by Tepfers based on the maximum load recorded at failure under monotonic loading. Bond failures during fatigue loading occurred in all but one case. The splices were not confined by stirrups. Rezansoff et al. (1993), working with 1.2 in. (30 mm) diameter bars and 35 to 38 in. (900 to 975 mm) lap lengths under comparable stress ranges, found that stirrup confinement improved lap splice fatigue life. With heavy confinement (stirrups spaced at $d/4$ and 35 in. [900 mm] lap length), the failure mode was failure of reinforcing bars in fatigue. When confinement was reduced (stirrups at $d/2$ and 38 in. [975 mm] lap length), the failure mode was bond fatigue. Pacholka et al. (1999) found that better confinement distribution (stirrups with three or four legs) improved bond strength under repeated loading. Rezansoff et al. (1997) did not observe this trend in a previous study. Cyclic stresses were higher than a typical service load range in each study cited previously.

Bennett (1982) compared the behavior of reinforced concrete beams containing lap splices and mechanical splices. Tests on $40d_b$ lap splices were performed at a stress range of 18.9 ksi (130 MPa). When lapped bars were offset bent, the beam failed early due to fatigue fracture of the bar after 100,000 cycles. When the lapped bars were straight, the beam withstood 4 million cycles, but crack widths at the lap splice ends were greater than the ACI 318 serviceability limits in place at that time. When bars were mechanically spliced by cold-swaged sleeves and subjected to the same cyclic stress range, the concrete beam withstood more than 4 million cycles without failure and deflection and crack widths stayed within acceptable limits. When an obsolete mechanical splice called a screw coupler was tested, fatigue fracture occurred after 750,000 cycles. When the stress range was reduced to 11.6 ksi (80.0 MPa), concrete beams with the screw coupler withstood more than 4 million cycles without failure.

Repeated loading has little effect on bond strength in well-confined lap splices when the force is less than 60 percent of the bond strength under monotonic loading. The most significant effect of high-level repeated loads is to reduce the bond at failure, whereas reversed cyclic stresses tend to deteriorate bond at a higher rate and precipitate failure at a smaller number of cycles or at lower loads.

Hasan et al. (1996) and Cleary and Ramirez (1993) considered the effects of repeated loading on beams reinforced with epoxy-coated bars. Repeated loading tended to equalize or reduce the difference in deflections and crack widths that were initially observed to be higher in beams with epoxy-coated bars. Specimens tested were cycled for 1 million cycles or until a fatigue failure and then tested statically to failure. The peak stresses used were 40, 50, and 60 percent of f_y and the stress range was 15 ksi (100 MPa) or lower. Lap splice lengths were approximately $20d_b$. Failures during cycling occurred only when the epoxy coating was unusually thick and a diamond rib pattern was present or, in two instances, with a larger peak stress or a larger stress range. Although the sample size of monotonic tests was small, specimens tested to failure following repeated loading showed no apparent strength reduction when compared with specimens loaded monotonically to failure.

CHAPTER 6—ANCHORAGE UNDER LOW-CYCLE LOADING

Research evaluating headed reinforcing bars in beam-column joints has demonstrated that the performance of straight and hooked bar anchorages under low-cycle loading can be similar or even superior to hooked bar performance (Chun et al. 2007), but a full review of these types of end anchorages is outside the scope of this report.

6.1—Straight bar anchorages

Bond behavior under unidirectional loads has been investigated by Rehm and Eligehausen (1977, 1979), Bertero and Bresler (1968), Tassios (1979), Edwards and Yannopoulos (1978), and Eligehausen et al. (1983). Extensive studies have been conducted by Viwathanatepa et al. (1979) for bars under reversed loading. Seventeen specimens of single bars embedded in column stubs with large amounts of transverse reinforcement were tested under reversed or unidirectional loadings. Bar sizes No. 6, 8, and 10 (19, 25, and 32) and column widths 15 to 25 in. (380 to 635 mm) were used. The study indicated that:

a) Cyclic load causes stiffness degradation characterized by pinched hysteresis loops, even at the service load range. Greater degradation of stiffness occurs with larger amplitudes and increased cycles of reversed loading

b) Gradual incremental cyclic loading decreases pullout strength. Maximum strength under cyclic loading ranges from 71 to 91 percent of the monotonic strength

c) Severe load reversals reduce ductility. No cases were observed where monotonically loaded specimens pulled out at displacements less than that of cyclic loading

Tests with reversed loading were also performed by Brown and Jirsa (1971), Ismail and Jirsa (1972a,b), Morita and Kaku (1973, 1979), Hassan and Hawkins (1977a,b), Gosain and Jirsa (1977), Plaines et al. (1982), and Eligehausen et al. (1983). Most tests were performed under displacement control, only a few cycles were applied, and bond failure was caused by pullout. As reported by CEB (1982) and by Eligehausen et al. (1983), the following general observations can be made:

a) Reversed loading cycles produce more severe degradation of bond strength and bond stiffness than for the same number of cycles with unidirectional loading. Degradation primarily depends on the maximum value of peak slip reached previously in either direction (Fig. 6.1a). Two important parameters are the number of cycles and the difference in consecutive peak values of slip. Under otherwise constant conditions, the largest deterioration occurs for full reversals of slip.

b) If peak bond stress during cycling does not exceed 70 to 80 percent of the monotonic bond strength σ_{bmax}, the ensuing bond stress-slip relationship (at first loading in the reverse direction and at slip values larger than the one at which the specimen was cycled) is not significantly affected by up to 10 repeated cycles (Fig. 6.1a(a)). Bond strength at peak slip deteriorates moderately with an increasing number of cycles.

c) Loading to slip values inducing a σ_b larger than 80 percent of the monotonically obtained σ_{bmax} in either direction leads to degradation in the bond stress-slip behavior in the reverse direction (Fig. 6.1a(b) and (c)). As peak slip s_{max} increases, deterioration of bond resistance increases. Deterioration increases as the number of cycles increases, and is larger for full reversals of slip than for half cycles. Furthermore, full reversals produce pronounced bond deterioration at slip values smaller than or equal to the peak slip value. A quantitative comparison is not possible because of different test conditions.

d) Cycling a specimen at different increasing values of slip has a cumulative effect on the deterioration of bond stiffness and bond resistance. Figure 6.1b compares the monotonic envelope and envelopes obtained after cycling at different slip values (labeled "a" through "e") either one or 10 times. Alternately, additional cycles between smaller slip values than the peak value in the previous cycle do not significantly influence the bond behavior at the larger peak value.

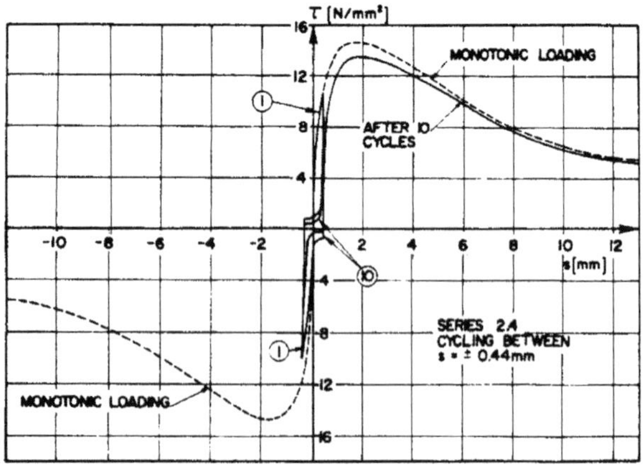

(a) Series 2.4 cycling between s = ±0.44 mm

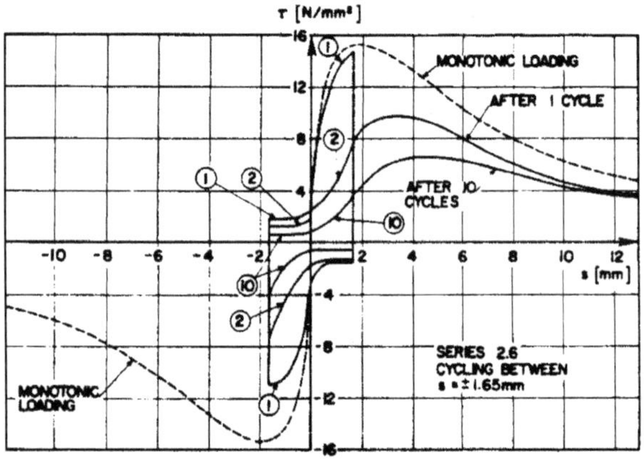

(b) Series 2.6 cycling between s = ±1.65 mm

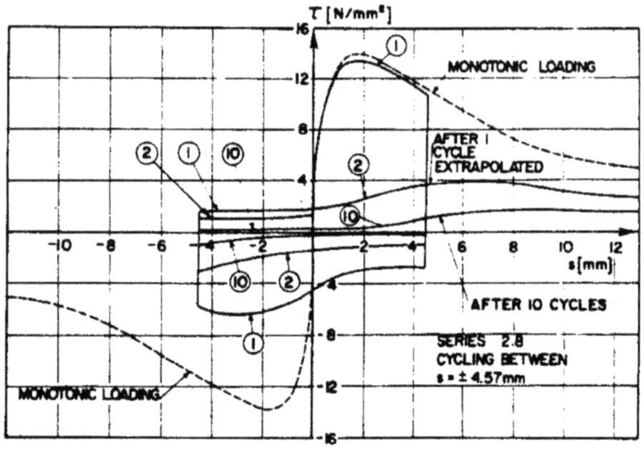

(c) Series 2.8 cycling between s = ±4.57 mm

Fig. 6.1a—Bond stress-slip curves for cycling at different s_{max} (Eligehausen et al. 1983).

e) Frictional bond resistance σ_{bf} during cycling depends on the value of peak slip s_{max} and the number of cycles. With repeated cycles, σ_{bf} deteriorates rapidly.

To identify the cause of the differences between monotonic and cyclic loading, typical bond stress and local displacements at different locations along a bar can be compared at stress levels before yielding, at yielding, and at pullout (Fig. 6.1c). For the preyielding and yielding range, local displacement along the bar in cyclic loading is larger than local displacement under monotonic loading. Larger local movement in cyclic loading implies more damage to surrounding concrete. There is no significant difference in bond stress distribution at preyielding and yielding stress levels. Smaller pullout strength under cyclic loading compared to monotonic loading is reflected in Fig. 6.1c(c), where maximum bond stress along the bar was achieved. Except for the region near the pulled end, cyclic bond stress for the remainder of the bar is below that obtained for monotonic loading.

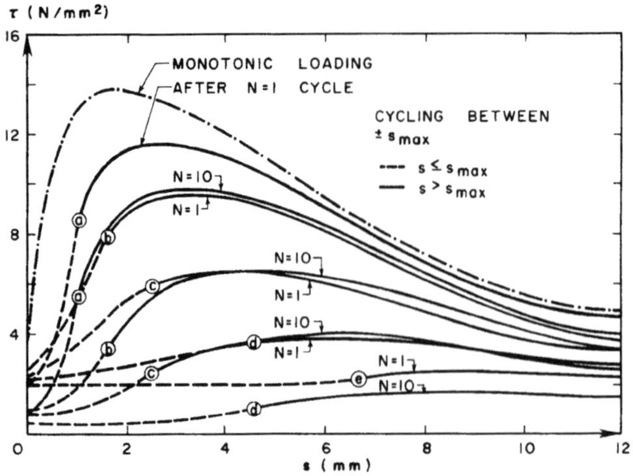

Fig. 6.1b—Effect of number of cycles on bond stress-slip curves (Eligehausen et al. 1983).

6.2—Hooked bar anchorages

Most work on hooked bar anchorages under monotonic load has been conducted by Johnson and Jirsa (1981), Marques and Jirsa (1975), and Minor and Jirsa (1975). Eligehausen et al. (1982a) presented data for cyclic loads. The main characteristics of monotonic behavior are summarized as follows:

a) Failure of a standard hooked bar anchorage of embedment $7d_b$ in a wall is governed by a loss of concrete cover in front of the hook, rather than by pulling out or side splitting, as in beam-column joints.

b) Cone-type failure is similar to that observed in tensile tests of headed studs or anchor bolts.

c) Hooked bar anchorage strength is directly proportional to $\sqrt{f_c'}$ for the range of concrete strength tested (2.5 to 5.5 ksi [17 to 38 MPa]).

d) Beam depth is the principal factor affecting the strength of short hooked bars in beam-wall joints. Anchorage strength is inversely proportional to beam or slab depth for depths of 8 to 18 in. (200 to 460 mm) because the compressive component of the force couple transferred to the wall provides more confinement of the anchor as depth decreases.

e) The strength and stiffness of hooked bar anchorages increase as lead embedment length (ℓ_{dh}, as controlled by the wall thickness) increases.

f) For a constant wall thickness, small-diameter hooked bar anchorages are most efficient in developing strength.

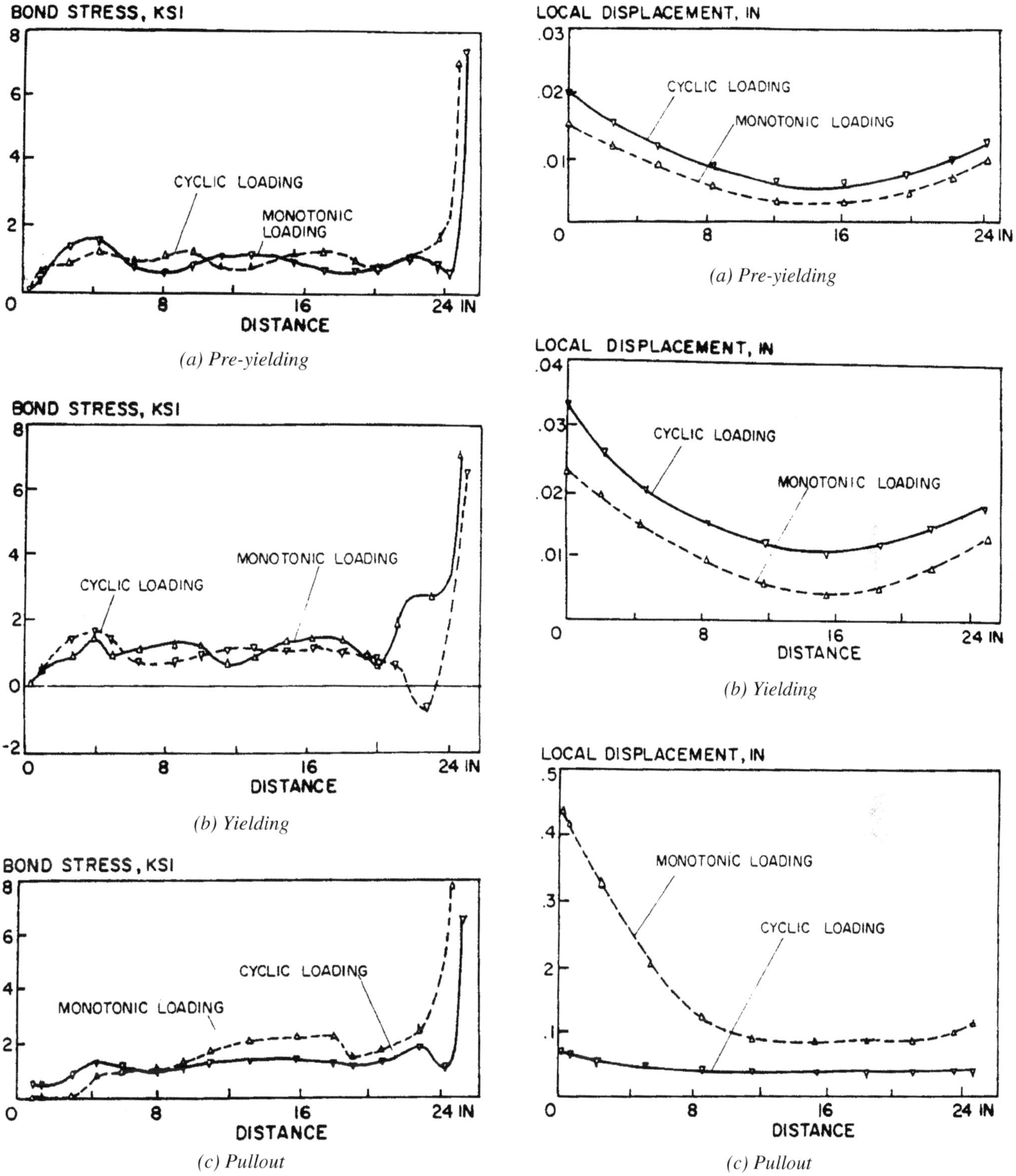

Fig. 6.1c—Distribution of bond stress and local displacement (Eligehausen et al. 1983).

g) A transverse bar placed in front of the hook has little, if any, influence on the maximum anchorage strength. The anchorage, however, appears to have more deformation capacity before failure.

h) Although the lead length affected the maximum strength of the connection, in 8 in. (203 mm) walls with the same lead length, the strength was not affected by whether the lead length was bonded or unbonded.

i) Hooked bars at spacings in which the anchorage failure zones show a decrease in maximum strength. The width of the failure zone is approximately equal to the effective depth of the beam or slab.

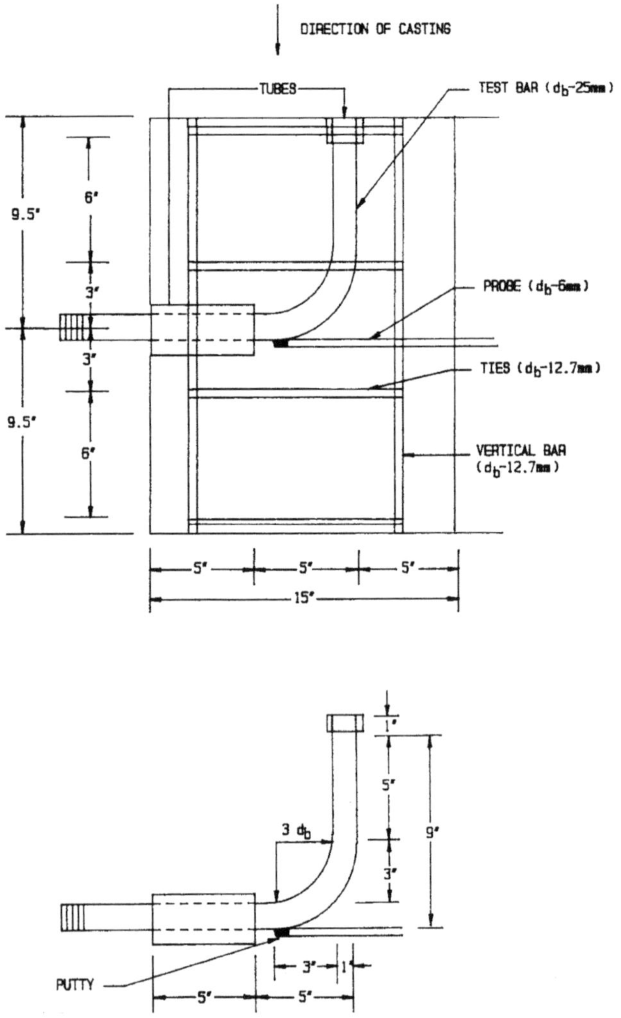

Fig 6.2a—Test specimens for hooked bar anchorages (Eligehausen et al. 1983).

Typical hooked bar specimens tested by Eligehausen et al. (1982a) are shown in Fig. 6.2a, and the results are shown in Fig. 6.2b. Test results are summarized as follows:

a) In well-confined situations, hooked bar anchorages subjected to the compression portion of the cyclic loading behave slightly better than hooked bar anchorages subjected to the tension portion.

b) Maximum resistance of a hooked bar anchorage subjected to monotonic loading is approximately 60 to 70 percent larger than that of a straight bar with an anchorage length corresponding to the equivalent development length (ℓ_d'). This is due to the larger bonded hook length, which was more than $2\ell_d'$.

c) Cyclic loading produces significant deterioration of stiffness and strength of anchored hooks at slip values smaller than the peak slip values. At peak slip, resistance deteriorates at a rate similar to that of straight bar bond resistance. However, the force-slip relationship, at slip values larger than peak slip during previous cycles, follows the monotonic loading curve closely. This behavior is dependent on the peak slip value. By contrast, straight bars cycled between large slip values did not reach the monotonic envelope again.

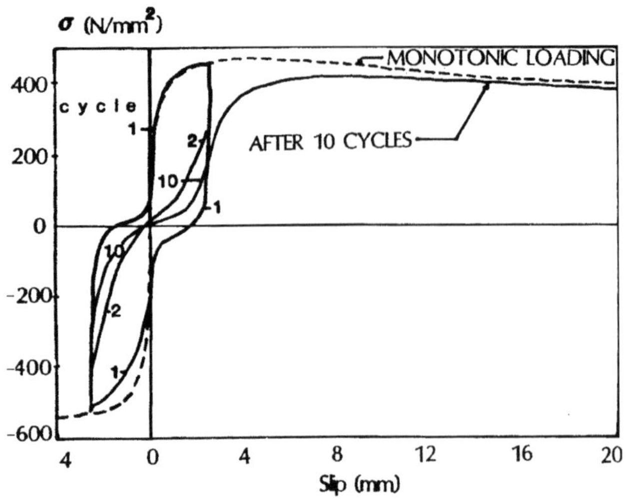

Fig. 6.2b—Bond stress-slip relationship for hooked bar anchorages (Eligehausen et al. 1983).

d) The frictional bond resistance of hooked bar anchorages during cyclic loading is significantly smaller than for straight bars.

e) Hooks bent in the direction of the concrete casting reach maximum strength at larger slips than hooks bent against the direction of casting because bleedwater and air voids tend to lodge near the bend. Otherwise, the influence of the hook position during casting on the load-slip behavior under monotonic and cyclic loading is negligible.

6.3—Lap splices

Much research has been carried out on lap splice behavior (Chinn et al. 1955; Gergely and White 1980). Following early studies by Muhlenbruch (1945, 1948), some codes required a splice length for cyclic loading that was much greater than that required for static loading. Recent evidence indicates that this requirement for extra length may not be needed. The amount of transverse reinforcement above which there is no further benefit, however, is larger under cyclic loading than under static loading. Unlike static loading, where stirrups near the ends of a lap splice are more effective than those near the center, uniform stirrup spacing provides the most effective confinement for cyclic loading.

For tension lap splices, tests showed that Class C splices conforming to the requirements of The Uniform Building Code (International Conference of Building Officials 1976) (the splice length is greater than or equal to $1.7\ell_d$) performed well under both monotonic and reversed cyclic loads. Specimen yield strength varied from 92 to 94 percent of the average tensile strength of the longitudinal bars. Ultimate loads ranged from 169 to 174 percent of the design yield strength (Aristizabal-Ochoa et al. 1977). Load history did not significantly affect specimen strength and all specimens experienced large post-yield elongations. Using offset bars at the end of the lap, however, caused severe local distress, particularly for large-diameter bars. These tests also showed

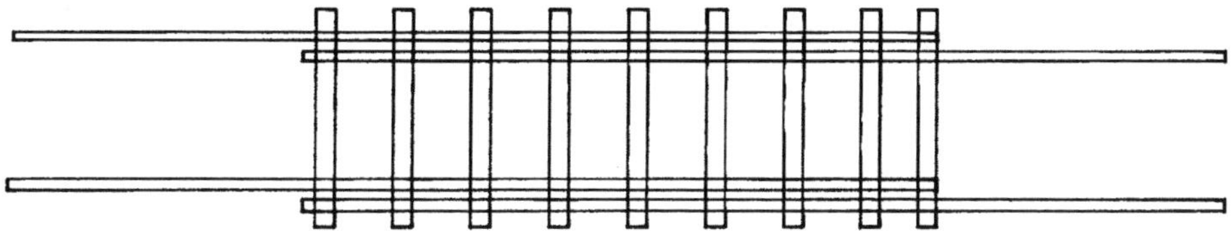

(a) Transverse reinforcement distribution from Cornell studies (Muhlenbruch 1948)

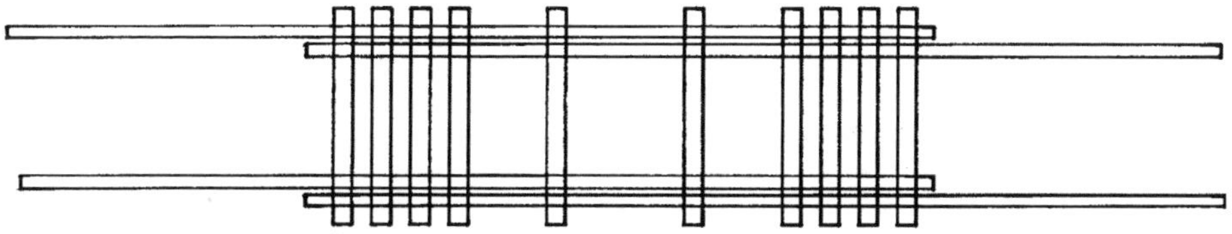

(b) Transverse reinforcement distribution from PCA studies (Aristizabal-Ochoa et al. 1977)

Fig. 6.3—Transverse reinforcement distributions from different studies.

that transverse reinforcement effectively controlled longitudinal splitting and bar slip as well as yield penetration along the spliced bars. The amount and distribution of transverse reinforcement was critical on lap splice behavior. In contrast to the work by Muhlenbruch (1948), they found that hoops at the end of a splice were more effective than hoops nearer the middle of the splice in resisting the splitting and bursting of concrete. For the hoops to be effective, moreover, they had to be in contact with the longitudinal bars.

Comprehensive research on splice behavior at Cornell University showed that lapped splices of at least 30 bar diameters can be designed to sustain repeated loading to at least twice the yield deflection (Fagundo 1979; Gergely et al. 1979). For equal side and bottom concrete cover, bottom splitting occurred first. Bottom splitting created vertical cantilevers between the splitting cracks and the sides of the beam. Due to the bursting effect, these cantilevers bent outward and led to sudden side splitting.

Excellent behavior was noted when the splices were confined by closely spaced stirrups. As stresses at the ends of the spliced region approached yield, bursting forces generated by spliced bars tended to be uniformly distributed along the splice length. As yield penetrated along the bars, the bursting forces over the middle elastic portion of the spliced region exceeded those at first yield. It was concluded that for best results, stirrups should be uniformly spaced over the splice length, matching Muhlenbruch (1948) rather than Aristizabal-Ochoa et al. (1977) (refer to Fig. 6.3).

Researchers at Cornell (Fagundo 1979; Gergely et al. 1979) found the ACI Committee 408 (1979) recommendations were adequate for monotonic loads up to yield and for repeated loads below 80 percent of the monotonic failure load. Unless the maximum amount of stirrups recommended by ACI Committee 408 is used, however, spliced regions will probably fail during the first 100 cycles at or above 80 percent of the monotonic bond failure load.

Tocci (1981) found that the earthquake provisions in most codes are unnecessarily restrictive in prohibiting the use of lap splices in regions where flexural yielding is anticipated. According to this research, understanding dowel forces that result after development of transverse shear cracking is key to understanding shear and bond interaction. Dowel action significantly influences splice strength because dowel forces approaching the dowel strength of a section rapidly reduce reinforcement anchorage.

Reversed cyclic loads are more detrimental to lap splice performance than repeated loads. This is particularly true for large-diameter bars because cyclic, post-yield loading induces progressive deterioration of the force-transfer mechanism, yield penetration along the splice length, and, for members with typical amounts of transverse (confining) reinforcement, progressive longitudinal splitting. Closer stirrup spacing along the splice length makes concrete cover less important at influencing splice strength. For monotonic loading design provisions, ACI 408R recommends adding the contribution of concrete to the contribution of transverse reinforcement to obtain total splice confinement. However, cumulative concrete cover damage makes the contribution of concrete cover at ultimate load unreliable for cyclic, post-yield loading. Finite-element fracture analyses results indicated that lap-spliced bars required greater confinement for equivalent bond lengths or, inversely, that splice lengths need to be longer than corresponding development lengths for equal amounts of confinement.

Rezansoff and Sparling (1986) and Rezansoff et al. (1988) verified the Cornell research results. Their investigations indicated that the usual overstrength of steel reinforcing bars should be accounted for in design by increasing lap length, transverse confinement, or both. Failure of a given lap splice configuration under reversed loading generally occurred when critical total deformation was achieved. Failure did not correlate with the number of cycles imposed or the hysteretic energy absorbed. Minor changes in the peak load levels during cycling produced major differences in the number of load cycles needed to reach critical deformations that caused splice failure.

Lukose et al. (1982) and Sivakumar et al. (1982, 1983) indicate that lapped splices for column-type specimens can be designed to sustain inelastic reversed cyclic loading within specified limits of ductility. The number and distribution of stirrups over the lap splice length and outside the end of the splice corresponding to the higher applied moment are crucial in ensuring ductility. Lap splices in columns subjected to combined bending moment and shear can fail either by a longitudinal concrete cover splitting mechanism along the splice length, or by localized shear-dowel failure at the high moment end. Specimens with closely spaced stirrups beyond the high moment region exhibit significant ductility even for shear-dowel-type failures. For shear levels of approximately 120 psi (0.82 MPa) or less, transverse reinforcement over the splice resists shear forces and radial bursting stresses. Very closely spaced stirrups inhibit longitudinal concrete cover splitting. This leads to shear-dowel failure just beyond the splice end corresponding to the higher applied moment. Closely spaced stirrups at this critical location control the extent of localized shear damage. Stirrups over the splice length should be uniformly spaced rather than concentrated at the ends.

As with lap splices in flexural members, reversed cycling at and above yield results in cumulative concrete deterioration and continuous changes in cyclic energy absorption characteristics and in the load-displacement relationship.

The depth of cast concrete affects bond strength, particularly for the more workable concrete mixtures. Less dense top layers in a horizontally cast beam or column specimen are less resistant to longitudinal concrete cover splitting than compacted bottom layers. In addition, higher-strength concrete resists larger compressive forces and improves compression lap splice behavior, but results in lower energy absorption in comparison with lower-strength concretes.

Results of these studies, and those by Park and Paulay (1975), Paulay et al. (1978), and Cairns and Arthur (1979), indicate that it is possible to design lap splices in columns to resist forces from large lateral loads.

CHAPTER 7—ANALYTICAL BOND MODELS

Much of the research defining analytical bond models has focused on behavior under low-cycle loading, detailed in Sections 7.1 through 7.4. Section 7.5 introduces an approach to modeling behavior under high-cycle loading.

7.1—Early models of local bond behavior

Early research in bond modeling includes a number of analytical investigations on the bond-slip behavior of single reinforcing bars under cyclic excitation. In general, these studies can be divided into three categories:

Model Category 1—A finite difference or finite element scheme is used to solve the differential equations of bond (Ciampi et al. 1982; Edwards and Yannopoulos 1979; Filippou 1985, 1986; Filippou et al. 1983, 1986; Hawkins et al. 1982; Hawkins 1974; Plaines et al. 1982; Tassios and Koroneos 1984; Viwathanatepa et al. (1979); Vos and Reinhart 1982).

Model Category 2—This approach assumes a bond stress distribution function along the bar anchorage length. This scheme has been used to model the monotonic pullout of reinforcing bars by using a stress distribution dependent on the

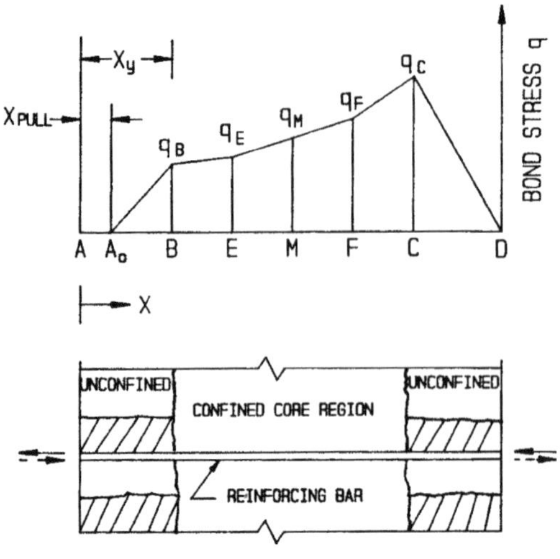

Fig. 7.1—Analytical bond model (Filippou 1986).

magnitude of bar pullout (Bertero and Bresler 1968; Bertero and Popov 1977; Bertero et al. 1978). Guiriani (1981) used an analogous approach to study the tension stiffening effect around cracks. Shah and Somayaji (1981) assumed an exponential bond stress distribution in predicting the cracking response of tension members.

Model Category 3—Schemes based on fracture mechanics have been developed (Gerstle et al. 1982; Gylltoft et al. 1982).

The number and complexity of the models proposed do not permit an individual discussion of each model in this report, and detailed comparisons between these models are not available. A sophisticated bond model was proposed by Eligehausen et al. (1983). Other models have been proposed by Ciampi et al. (1982). Filippou (1985, 1986) presented a simple Category 2 model describing the hysteretic behavior of a single reinforcing bar anchored in an interior beam-column joint. An extension of the model to exterior joints is straightforward. The model focuses on stress transfer between reinforcing bars and surrounding concrete through bond. The portion of single reinforcing bar between the beam-column interfaces of an interior joint is depicted in Fig. 7.1. The model proposed is as follows:

a) The bar is subdivided by Points A, B, C, D, and M into five basic segments. Points A and D are located at the ends of the bar. Point M is at the bar midpoint, and Points B and C mark the transition between the confined and unconfined concrete region at the end of the pullout cone. Points E and F can be inserted to increase numerical accuracy of integrations.

b) Bond stress is always zero at the bar ends. As the magnitude of bar pullout increases, bond is gradually destroyed at the bar ends. This region of zero bond stress spreads into the joint with increasing magnitude of bar pullout. At the bar end that is pulled out, a linear bond stress distribution is assumed between Point A and Point B located at the transition between the confined and the unconfined concrete region at the end of the pullout cone. At the bar end that is pushed in, a linear bond stress distribution is assumed

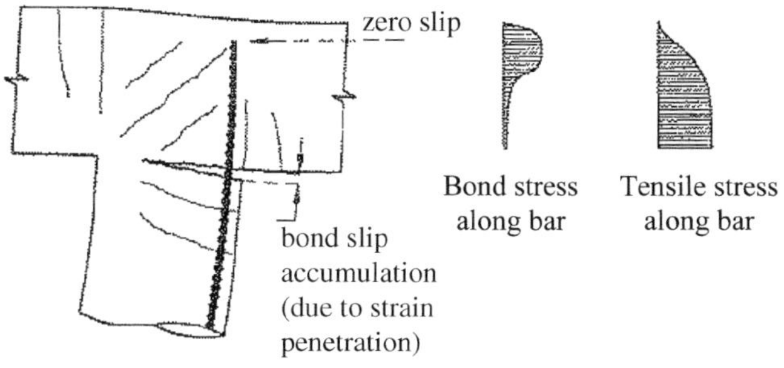

Fig. 7.2(a)—Adequate anchorage and limited slip due to strain penetration (Sritharan et al. 2000).

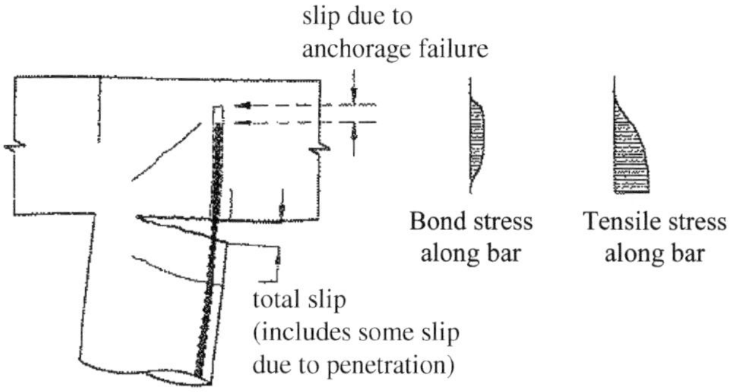

Fig. 7.2(b)—Bond slip due to anchorage failure (Sritharan et al. 2000).

between the end of the bar, Point D, and Point C, which is located at the transition between the confined and the unconfined concrete region.

c) Bond stress distribution in the confined concrete region between Points B and C is assumed piecewise linear and is established iteratively by satisfying the equilibrium and compatibility conditions.

Most models in all three categories, including the simplified Filippou model, are intended for research rather than design purposes. For most cases, a single bar is analyzed, and the effects of spacing and concrete cover are neglected. The effect of transverse reinforcement is generally accounted for in the concrete properties but not directly in the analysis. Most of these models cannot presently be applied to design of congested areas, such as beam-column joints. A comprehensive discussion of bond modeling concepts is presented by the Bond Models Task Group of *fib* (2000) primarily for static loading, but cyclic bond issues are addressed.

7.2—Models for local bond-slip effects

Several bond models have been developed using constitutive relationships for bond; for example, establishing guidelines for the bond-stress verses bond-slip relationship under cyclic loading. Typically, the constitutive relationship is based on a multi-linear or piecewise linear bond stress-slip curve. Yannopoulos and Tassios (1991) and Yankelevsky et al. (1992) developed constitutive models for single, isolated bars.

Saatcioglu et al. (1992) presented a constitutive model for reinforced concrete members based on the moment-anchorage slip-rotation relationship for bidirectional loading under axial compression. This model accounts for the inelastic strain distribution within the reinforcement and is only valid for unidirectional loading once inelastic cycles have occurred. Reinforced concrete column connections subject to axial compression and inelastic lateral deformation develop rotations due to bond-slip that are as significant as those due to flexure (Saatcioglu et al. 1992). Accounting for these rotations involves the inclusion of inelastic reinforcement strain penetration within the model even when full reinforcement slippage does not occur. Sritharan et al. (2000) made similar conclusions regarding the importance of accounting for the strain incompatibility between reinforcement and concrete for fully anchored bars within finite element models for reinforced concrete connections, as illustrated in Fig. 7.2.

Monti et al. (1997a,b) verified the performance of a nonlinear finite element model for the analysis of the hysteretic behavior of anchored reinforcing bars. This model is based on the interpolation of the steel stress distribution and includes the effect of bond damage. Advances in this model include a departure from using displacement interpolation functions for relative slip, and instead uses force interpolation functions. Rationale for this approach is based on research by Filippou (1986) and Zulfiqar and Filippou (1990) that identified it is numerically advantageous to approximate bond or steel stress distribution rather than the relative slip along the anchored reinforcing bar. Zhao and Sritharan (2007) presented a hysteretic model for reinforcing bars fully anchored in concrete with adequate confinement. The relationship between stresses applied to a bar and the bar slip can be integrated into fiber-based analysis of concrete structures using a zero-length section element (Monti and Spacone 2000). The envelope curve was based on experimental data. Loading-unloading rules were established based on observations of column tests in the literature. Zhao and Sritharan (2007) advocate that a zero-length section element placed at the end of a beam-column element accounts for strain penetration effects without full bar slippage and the resulting member end rotation.

7.3—Contribution of bond-slip for modeling member and structural system behavior

Developments in finite-element modeling of reinforced concrete structural systems under cyclic loads have acknowledged the significance of nonlinear behavior of the steel reinforcement-concrete bond relationship and the necessity to include bond-slip within the model development. The bond-slip relationship has also been included in models for evaluating beam-column connections and footing-column connections where the importance of including bond-slip within the model development has been confirmed by comparing model results with and without bond slip. To reduce the computational demand during modeling of these reinforced concrete systems, fiber-based models have evolved. Concrete and reinforcement fibers are used to construct the members within the model.

Monti and Spacone (2000) merged the bond-slip proposed by Monti et al. (1997a,b) and the force-based fiber element proposed by Spacone et al. (1996), resulting in a model that accounts for components from both bar strain and bond slip. The combined model traces the response of each bar within a section and is suitable for cross sections of various shapes, including circular, and for sections under biaxial loading.

A simplified displacement-based fiber model for reinforced concrete members that accounts for bond slip has also been presented by Spacone and Limkatanyu (2000). This model represents the reinforcement bond slip in columns and beams and between columns and foundations. This model addresses the behavior of externally bonded plates, such as external bonding of fiber-reinforced polymer to increase flexural strength. Behavior of beam-column joints was not represented by the model.

Kwak and Kim (2001) presented a hysteric moment-curvature relationship to simulate the behavior of reinforced concrete beams under cyclic loading that accounts for stiffness degradation and bond-slip effects. This approach properly assesses the energy-absorbing capacity of the beams at large deformations. This model does not consider strength degradation under cyclic loading beyond yield strength, which may be a requirement for beams with brittle behavior. Kwak and Kim (2001) make the following recommendations:

a) Inclusion of the pinching effect is important in members dominantly affected by shear

b) Accurately predicting structural behavior of the beam-to-column subassemblage where the nonlinear response is concentrated requires modification of the moment-curvature relationship to consider fixed-end rotation.

Lowes and Altoontash (2003) presented a simple two-dimensional joint model comprised of a four-node shear panel and bond-slip spring components. The model uses a constitutive relationship to define the load-deformation behavior of the shear panel and bar-slip components as a function of material properties, joint geometry, and joint reinforcement layout. The model is not appropriate for simulating response under purely monotonic loading.

Similar to the fiber-based models, general three-dimensional finite-element models confirm the contribution of bond-slip in the improved response of reinforced concrete structural systems. Limkatanyu et al. (2004) evaluated analytical models for reinforced concrete frames with and without bond slip. The models that accounted for bond-slip better represented the structural frequency, general waveform of the system response, and maximum base shear. Girard and Bastien (2002) proposed a nonlinear finite-element model using a displacement method of analysis to investigate the response of reinforced concrete columns subject to cyclic loads. They confirmed that using a finite element with a degree of freedom for slip appropriately represents the general column behavior due to deterioration of the steel-concrete bond at the column base under cyclic loading.

7.4—Bond modeling used in evaluation of existing structures with short lap splices or discontinuous bottom bars through joints

Recent analytical modeling has included the analysis of the beam-column connections with short lap splices, similar to those constructed prior to the 1980s. These efforts have progressed with the anticipation that a practical model can be developed to analyze the behavior of columns with short lap splices, which tend to exhibit a loss of lateral resistance and low ductility under reversed cyclic loading. An understanding of column behavior with short lap splices improves rehabilitation requirements.

Cho and Pincheira (2006) presented a two-dimensional, nonlinear model based on the local bond stress-slip relationship developed by Harajli and Mabsout (2002) to estimate bar stress and deformation at splice failure. The model also includes degradation of stiffness with increasing reversed cyclic loading. Cho and Pincheira (2006) identify that bond-slip in a splice region with lateral displacement can be substantial and should not be ignored in the analysis of older reinforced concrete structures.

Discontinuous bottom bars through reinforced concrete joints are often found in frames that were detailed primarily to resist gravity loads. Celik and Ellingwood (2008) observed that using rigid joints to model such frames did not adequately represent the response to strong ground motion. They developed a softened beam-column joint model that was calibrated with joint shear stresses and strains measurements from beam-column joint specimens subjected to cyclic loads. A quad-linear curve was used to define the envelope of the panel shear stress-strain relationship. Bond-slip was included by reducing the envelope of joint shear stress-strain behavior.

7.5—Modeling of high-cycle (fatigue) loading

A finite element two-dimensional damage constitutive model of reinforced concrete members under fatigue loading has been presented by Teng and Wang (2001). In this model, the bar and the surrounding concrete have been simulated as perfectly bonded, and so bond failures were excluded. Concrete deformational behavior with load repetitions has been accounted for by the concept of damage accumulation. Once the stress state reaches the fatigue tensile strength, a crack is assumed to form. Likewise, the strength degradation of the steel reinforcement with load repetition has also been included in the model. When the fatigue strength of the steel reinforcement is reached, a brittle failure is assumed.

CHAPTER 8—DESIGN AND ANALYSIS APPROACHES

8.1—High-cycle fatigue

Because tests have shown that concrete or reinforcement will fail in fatigue before the bond fatigue limit is reached, most design equations refer to the stress range in the concrete or steel rather than to any bond stress limit. It is assumed in these recommendations that the loading does not consist of stress reversals.

8.1.1 *Buildings and other structures (ACI 215R)*—ACI 215R recommends that flexural stresses within reinforced concrete beams using hot-rolled deformed bars (ASTM A615/A615M) be limited as follows (with flexural stresses in the concrete and steel calculated in accordance with the provisions of ACI 318).

a) The stress range f_{cr} in concrete should not exceed $0.40f_c'$ when the minimum stress is zero; the stress range reduces linearly as the minimum stress increases so that the permitted stress range is zero when the minimum stress is $0.75f_c'$.

b) The stress range in straight deformed bars should not exceed the value computed from the following expression

$$S_r = 23.4 - 0.33S_{min} \quad \text{(ksi)} \qquad (8.1.1)$$

$$S_r = 161 - 0.33S_{min} \quad \text{(MPa)}$$

The value of S_r, which is the stress range in straight deformed bars that can be used without inducing fatigue distress, need not be taken less than 20 ksi (138 MPa). For bent bars or bars to which auxiliary reinforcement has been tack welded, the stress range should be reduced 50 percent.

Equation (8.1.1) does not incorporate the number of cycles as a variable, but is assumed valid in the infinite life region of the *S-N* curve, where *N* is greater than 1 million cycles. The expression is based on an equation derived from statistical analysis at 95 percent probability that the specimens will not fail. For straight deformed bars, the stress range should not exceed 21 ksi (145 MPa) (ACI 215R).

8.1.2 *Bridges (AASHTO HB-17, Helgason et al. (1976))*—Helgason et al. (1976) recommend that the stress range due to live loads and impact in ASTM A615/A615M bars should not exceed (Fig. 8.1.2a demonstrates *r* and *h* values)

$$f_r = 21 - 0.33f_{min} + 8(r/h) \quad \text{(ksi)} \qquad (8.1.2)$$

$$f_r = 144.8 - 0.33f_{min} + 55.2(r/h) \quad \text{(MPa)}$$

AASHTO HB-17 has adopted Eq. (8.1.2) for use in bridge design. When the *r*/*h* ratio is unknown—the most common case in design—it should be taken as 0.3. Thus, the maximum allowed stress range is approximately 23.4 ksi (161 MPa), and the recommendation becomes identical to that of ACI 215R (Fig. 8.1.2b). Helgason et al. (1976) also recommend avoiding bends in primary longitudinal reinforcement in regions of high strength range. The limits proposed are for the fatigue limit of the bars and are not based on bond failures. There does not appear to be data indicating that bond will be a problem under fatigue loading when bar stress limits are satisfied.

8.1.3 *Offshore structures* (Det Norske Veritas 1977)—Design standards from the 1970s stated that anchorage lengths of straight bars to resist cyclic loads should be double those used for static loading, provided that the number of load repetitions exceeds 10,000 and the stress range S_{br} at

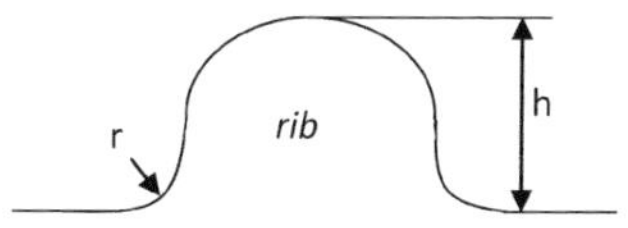

Fig. 8.1.2a—Base radius r *and transverse deformation height* h.

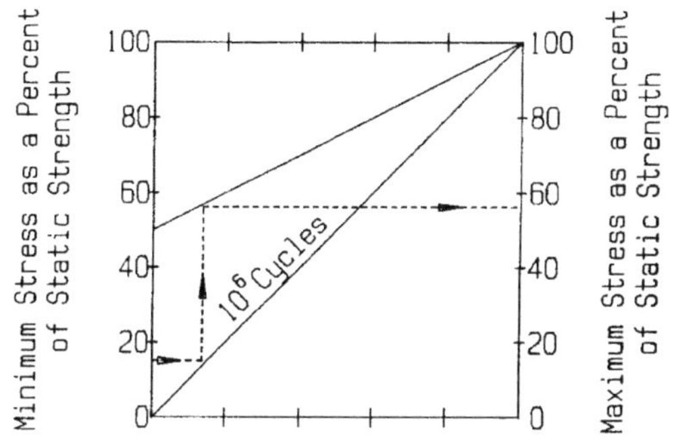

Fig. 8.1.2b—Recommended stress ranges for concrete (ACI 215R).

10,000 cycles exceeds the bond strength f_{br} at 2 million cycles. Bond strength f_{br} may be taken as

$$f_{br} = \left[1 - \frac{S_{bmax}}{(f_b/\tau_m)}\right]\left[\frac{f_b}{\tau_m}\right] \le (1/2)\left[\frac{f_b}{\tau_m}\right] \qquad (8.1.3a)$$

Using Eq. (8.1.3a) would result in requirements for straight bar anchorage lengths from that time being doubled to account for bond fatigue. Det Norske Veritas (1977) provisions suggest that fatigue limit for fracture of the reinforcing bars should be taken as

$$\log_{10}N = 6.5 - 2.3[S_r/f_{sos}] - 1.4 \times 10^{-5}S_{min} \quad \text{(psi)} \qquad (8.1.3b)$$

$$\log_{10}N = 6.5 - 2.3[S_r/f_{sos}] - 0.002S_{min} \quad \text{(MPa)}$$

Newer provisions for the fatigue life of reinforcement subjected to cyclic loading can be found in M200 of DNV-OS-C502. In the newer provisions, the logarithm of the fatigue life, *N*, is a function of the stress range and two coefficients C_3 and C_4 that depend on the type of reinforcement, the bend radius, and the corrosiveness of the environment.

A comparison of this limitation with those of some oil companies for use in concrete structures in the North Sea are given in Fig. 8.1.3 (Lenschow 1984).

8.2—Low-cycle loading (earthquake loads)

The following recommendations apply to bar anchorage in members of structures subjected to cyclic loads resulting from earthquakes. In all cases, provisions are intended for well-confined concrete sections.

8.2.1 *ACI 318-11, Chapter 21*—Chapter 21 of ACI 318-11 prohibits lap splices in special moment frames within joints within a distance twice the member depth from the joint

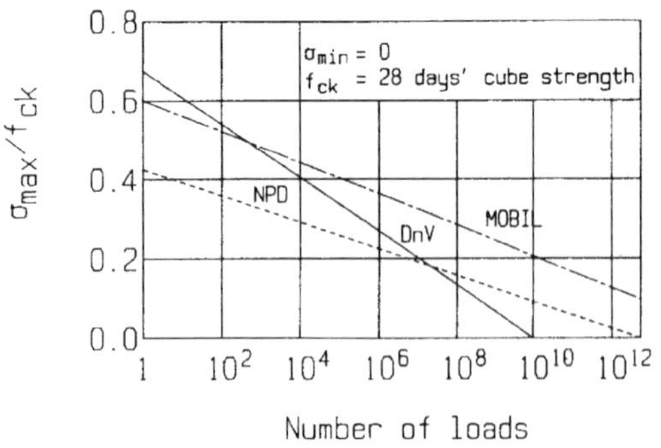

Fig. 8.1.3—European offshore fatigue recommendations for concrete (Lenschow 1984).

face, and in regions where yielding is expected. Lap splices are designed as tension lap splices enclosed with additional transverse reinforcement.

For the development length of No. 3 through No. 11 bars (10 through 36) in tension in exterior joints of special moment frames, Chapter 21 requires a standard 90-degree hook not less than the largest of $8d_b$, 6 in. (150 mm), and

$$\ell_{dh} = \frac{f_y d_b}{65\sqrt{f_c'}} \quad \text{(psi)}$$

$$\ell_{dh} = \frac{f_y d_b}{5.4\sqrt{f_c'}} \quad \text{(MPa)} \qquad (8.2.1)$$

For bars cast with lightweight aggregate, Chapter 21 requires a standard ℓ_{dh} not less than the largest of $10d_b$, 7.5 in. (190 mm), and 1.25 times the value defined by Eq. (8.2.1).

For straight bars with no more than 12 in. (300 mm) of concrete cast below the bar, the aforementioned requirements are multiplied by 2.5. For bars with more than 12 in. (300 mm) of concrete cast below the bar, the value for standard hook development length is multiplied by 3.25. The factors in ACI 318-11, Chapter 12, for epoxy-coated bars under static loading also apply.

8.2.2 *Joint ACI-ASCE Committee 352*—ACI 352R has issued recommendations for the anchorage requirements in beam-column joints subjected to large load reversals, called Type 2 Joints, as follows:

a) For hooked bar anchorages terminating in exterior joints

$$\ell_{dh} = \frac{\alpha \cdot f_y \cdot d_b}{75\sqrt{f_c'}} \quad \text{(in. and psi)}$$

$$\ell_{dh} = \frac{\alpha \cdot f_y \cdot d_b}{6.2\sqrt{f_c'}} \quad \text{(mm and MPa)} \qquad (8.2.2a)$$

Development length ℓ_{dh} is measured from the critical section as defined in Fig. 8.2.2, and α is the stress multiplier for longitudinal bars at the joint or member interface for Type 2 joints.

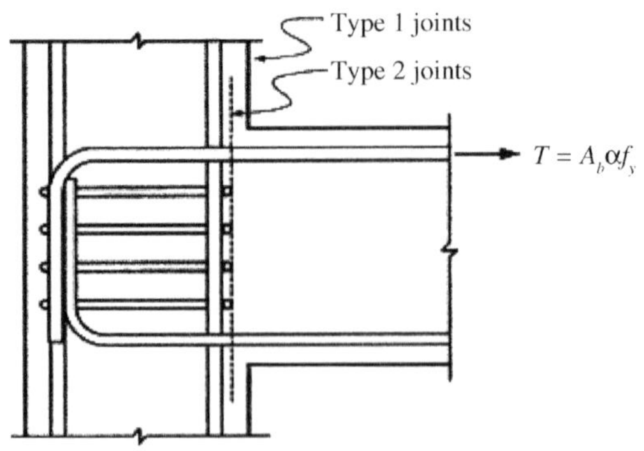

Fig. 8.2.2—Critical section for development of beam longitudinal bars terminating in the joint (ACI 352R).

b) Straight bar anchorages terminating within an exterior joint are not recommended for Type 2 joints, which require sustained strength under deformation reversals into the inelastic range.

c) For straight bar anchorages in interior Type 2 joints, the following recommendations are made.

i. Where columns are wider than beams, all straight beam and column bars passing through the joint should be selected such that

$$\frac{h_{(column)}}{d_{b(beam\ bars)}} \geq 20\frac{f_y}{60{,}000} \geq 20 \quad \text{(psi)}$$

$$\frac{h_{(column)}}{d_{b(beam\ bars)}} \geq 20\frac{f_y}{420} \geq 20 \quad \text{(MPa)} \qquad (8.2.2b)$$

and

$$\frac{h_{(beam)}}{d_{b(column\ bars)}} \geq 20\frac{f_y}{60{,}000} \geq 20 \quad \text{(psi)}$$

$$\frac{h_{(beam)}}{d_{b(column\ bars)}} \geq 20\frac{f_y}{420} \geq 20 \quad \text{(MPa)} \qquad (8.2.2c)$$

ii. For wide-beam construction, beam bars passing outside the joint core should be selected such that

$$\frac{h_{(column)}}{d_{b(beam\ bars)}} \geq 24\frac{f_y}{60{,}000} \geq 24 \quad \text{(psi)}$$

$$\frac{h_{(column)}}{d_{b(beam\ bars)}} \geq 24\frac{f_y}{420} \geq 24 \quad \text{(MPa)} \qquad (8.2.2d)$$

8.2.3 *Joint ACI-ASCE Committee 352*—ACI 352.1R has issued recommendations for anchorage requirements in slab-column joints subjected to large load reversals for Type 2 joints as follows.

a. For hooked bar anchorages terminating within exterior joints

$$\ell_{dh} = \frac{(f_y \cdot d_b)}{\left(50\sqrt{f_c'}\right)} \geq 8d_b \text{ or 6 in. (psi)}$$

$$\ell_{dh} = \frac{(f_y \cdot d_b)}{\left(4.2\sqrt{f_c'}\right)} \geq 8d_b \text{ or 150 mm (MPa)} \quad (8.2.3a)$$

All bars terminating in the joint should end with a 90-degree hook within the transverse reinforcement of the joint. When transverse reinforcement in the joint is provided at a spacing less than or equal to three times the diameter of the bar being developed, ℓ_{dh} may be reduced by 20 percent within the joint. Where significant strain-hardening of reinforcement is anticipated due to inelastic deformation, $1.25f_y$ should be substituted for f_y in Eq. (8.2.2d).

b) Straight bar anchorages terminating within exterior joints are not recommended for Type 2 joints.

c) Straight bars that pass through interior Type 2 joints should be selected such that

$$\frac{h_j}{d_b} \geq 15 \quad (8.2.3b)$$

where h_j is the joint dimension parallel to the bar. Straight bars should not be terminated within the connection region of the slab.

8.2.4 *New Zealand Standards (NZS 3101:2006)*—For members in frames subjected to earthquake loading, the New Zealand Standards governing the design of beam-column joints, lap splices, and hooked bar anchorages are generally more conservative than those proposed by ACI 352R and Section 8.3.2, and are modeled closely to those of Paulay and Priestley (1992).

Where beam longitudinal bars pass through a column at an interior beam-column joint, the allowable ratio of bar diameter to column depth is affected by the following variables:

a) Reinforcement yield strength and overstrength factor

b) Square root of concrete compressive strength (an indirect measure of the tensile strength of concrete)

c) Whether the joint is part of a one- or two-way frame

d) Design ductility of the frame

e) Depth of fresh concrete under beam bars at casting

f) Axial compression stress in the column

g) Ratio of positive to negative beam longitudinal reinforcement passing through the column

The range of variables makes it difficult to give example values of anchorage requirements. Considering reinforcing bars with yield strengths of either 44 or 73 ksi (300 or 500 MPa) and a range of concrete strengths from 4.4 to 7.3 ksi (30 to 50 MPa), the range of required straight bar anchorage lengths would be from $9d_b$ to $53d_b$, although values at the extremes of this range represent unlikely designs. Typical anchorage requirements would range from $15d_b$ to $40d_b$.

Because columns are designed to remain elastic under large cyclic loads, anchorage requirements for column longitudinal bars passing through a beam-column joint are less stringent. The ratio of bar diameter to beam depth is affected only by reinforcement and concrete strength, and it ranges between $10d_b$ for f_y = 44 ksi (300 MPa) and f_c' = 7.3 ksi (50 MPa), and $28d_b$ for f_y = 73 ksi (500 MPa) and f_c' = 4.4 ksi (30 MPa).

8.2.5 *fib Model Code*—For structures expected to develop stable mechanisms with large energy dissipation capacity, known by CEB (1987) as Ductility Level 111, at least 39 bar diameters for anchorage of flexural reinforcement and 25 bar diameters for the development of longitudinal column reinforcement are required.

8.2.6 *BS EN 1998-1:2004 (Eurocode 8)*—European standards address the bond of longitudinal bars passing through beam-column joints based on the following criteria.

To prevent bond failure for interior beam-column joints

$$\frac{d_{bL}}{h_c} = \frac{7.5 f_{ctm}}{\gamma_{Rd} f_{yd}} \frac{1+0.8\nu_d}{1+0.75 k_D \rho' / \rho_{max}} \quad (8.2.6a)$$

To prevent bond failure for exterior beam-column joints

$$\frac{d_{bL}}{h_c} = \frac{7.5 f_{ctm}}{\gamma_{Rd} f_{yd}} (1+0.8\nu_d) \quad (8.2.6b)$$

8.2.7 *Japanese recommendations* (AIJ 1990)—Current Japanese recommendations suggest that

$$D/d_b < \left[f_y/\left(\mu\sqrt{f_c'}\right)\right] \quad (8.2.7)$$

where D is the recommended development length. The value of the constant μ is not given in the recommendations, and values ranging from 10 to 12.5 have been suggested by Japanese researchers for stress values in Eq. (8.2.7) measured in kgf (ranging between 2.65 and 3.31 for stress in units of psi, and ranging between 32 and 50 when the stress is in units of MPa).

8.2.8 *AASHTO HB-17*—AASHTO HB-17 provides requirements for bar splices and anchorages as follows.

8.2.8.1 *Columns*— For a Class A splice, the development length l_d for tension lap splices should be computed as follows (modification factors are not presented herein)

No. 11 (36) bars and smaller

$$\frac{0.04A_b f_y}{\sqrt{f_c'}} \text{ (in. and psi)}$$
$$\frac{0.019A_b f_y}{\sqrt{f_c'}} \text{ (mm and MPa)} \quad (8.2.8.1a)$$

but not less than

$$0.0004d_b f_y \text{ (in. and psi)}$$
$$0.06d_b f_y \text{ (mm and MPa)} \quad (8.2.8.1b)$$

No. 14 (43) bars

$$\frac{0.085 f_y}{\sqrt{f_c'}} \text{ (in. and psi)}$$
$$\frac{26 f_y}{\sqrt{f_c'}} \text{ (mm and MPa)} \quad (8.2.8.1c)$$

No. 18 (57) bars

$$\frac{0.11 f_y}{\sqrt{f_c'}} \text{ (in. and psi)}$$
$$\frac{34 f_y}{\sqrt{f_c'}} \text{ (mm and MPa)} \quad (8.2.8.1d)$$

In all cases, the development length l_d shall not be less than 12 in. (305 mm).

Under Seismic Design Categories C and D, the lap splice length should not be less than

a) 16 in. (406 mm)

b) 60 bar diameters

Lap splices are only permitted within the center half of the column.

8.2.8.2 *Column connections with vertical extension of the column into adjoining member)*—Development length ℓ_d for all longitudinal bars is that required for a steel stress of $1.25f_y$.

8.2.9 *FEMA 310*—Guidelines for seismic evaluation of buildings recommend the following for concrete moment frames for proper development and confinement of reinforcing bars subjected to cyclic loads:

a) Lap splice lengths of bars in columns should be greater than $35d_b$ for life safety and $50d_b$ for immediate occupancy

b) Frame columns should have tie spacing at or less than $d/4$ throughout their length and at or less than $8d_b$ at potential plastic hinge locations

c) Lap splices for longitudinal reinforcing bars should be located within $\ell_b/4$ of the joint and should not be located in the vicinity of potential plastic hinge locations

d) Beams should have stirrups spaced at or less than $d/2$ throughout their length. At potential plastic hinge locations, stirrups should be spaced at or less than the minimum of $8d_b$ or $d/4$

8.2.10 *FEMA 356*—Development lengths within yielding regions should be calculated according to ACI 318 Chapter 21. FEMA guidelines for the rehabilitation of buildings recommend that, for reinforced concrete members with lap-spliced bars not meeting the development requirements of ACI 318, existing reinforcement strength should be calculated as follows:

$$f_s = \frac{\ell_b}{\ell_d} \cdot f_y \quad (8.2.10a)$$

where ℓ_d is the development length calculated according to Eq. (12-1) of ACI 318.

Strength of deformed straight, discontinuous bars embedded in concrete sections or beam-column joints with at least $3d_b$ clear concrete cover should be calculated as

$$f_s = \frac{2500}{d_b}\ell_e \le f_y \quad (8.2.10b)$$

Cho and Pincheira (2006) contend that recommendations by FEMA 356 (Eq. (8.2.10a)) underestimate the strength of lap splices and propose an alternate nonlinear relationship

$$f_s = \left(\frac{\ell_b}{0.8\ell_d}\right)^{2/3} \cdot f_y \quad (8\text{-}17)$$

where ℓ_d is the development length calculated according to Eq. (12-1) of ACI 318.

Cho and Pincheira (2006) observed that the conservatism in FEMA 356 might result in a column behavior where shear failure governs over splice failure. Similar conclusions were reached by Melek and Wallace (2004). Their research indicated that interior columns exhibit substantially less lateral strength degradation after bond deterioration than implied in FEMA 356.

8.2.11 *Other recommendations*—Bonacci and Pantazopoulou (1993) compared research from the United States, New Zealand, and Japan for reinforced concrete beam-column connections subject to cyclic loading. This research improved the understanding of joint resistance in reinforced concrete framed structures. They recommended minimizing connection flexibility by dimensioning the column with a bond index less than 20 psi (0.13 MPa). The bond index *BI* represents the severity of bond stresses relative to the bond strength (Kitayama et al. 1991). For beam bars

$$BI = \frac{u_{bb}}{\sqrt{f_c'}} = \frac{f_y\left(\frac{d_b}{h_c}\right)}{2\sqrt{f_c'}} \quad (8.2.11a)$$

Bond indexes for column bars are computed similarly.

As the bar strength and concrete strength increase, the corresponding increase in bond strength is not proportional. As a result, a reduced ductility under earthquake loading may manifest in the splitting bond failure of continuous bars through ductile columns or beams with hinge regions at both ends, even if longitudinal bar ends are well anchored. Ichinose (1995) addressed this condition and recommended that main bars that pass through ductile beams and columns with hinge regions at the two ends should conform to the following development length

$$\ell_d < \frac{1}{2} - 0.7d \quad (8.2.11b)$$

The coefficient 0.7 should be larger if the member is required to have a large ductility or the bars have a small yield plateau.

8.3—Recommendations

8.3.1 *High-cycle fatigue*—Live load is likely to be a small fraction of the dead load in situations of high-cycle fatigue discussed in this section. For realistic ratios of dead-to-live load and where the limits for materials given by ACI 215R are not exceeded, bond fatigue failures should not occur when ACI 318 anchorage provisions are followed. Designers should check only Eq. (8.3.1a), with r/h equal to 0.3, with an upper limit of 21 ksi (145 MPa)

$$f_r = 23.4 - 0.33f_{min} < 21 \quad \text{(ksi)}$$

$$f_r = 161 - 2.3f_{min} < 145 \quad \text{(MPa)} \quad (8.3.1a)$$

An allowable bond stress range can be derived using ACI 318 Section 12.2.2, and assuming a uniform bond stress exists over the development length ℓ_d. For bottom bars that are uncoated and cast in normalweight concrete, an allowable bond stress range is:

For No. 6 (19) bars and smaller

$$u = 1.6\sqrt{f_c'} < 300 \text{ psi}$$

$$u = 0.13\sqrt{f_c'} < 2.1 \text{ MPa} \quad (8.3.1b)$$

For No. 7 (22) bars and larger

$$u = 1.3\sqrt{f_c'} < 300 \text{ psi}$$

$$u = 0.11\sqrt{f_c'} < 2.1 \text{ MPa} \quad (8.3.1c)$$

The bond stress range given in Eq. (8.3.1b) and (8.3.1c) and the upper limit of 300 psi (2.1 MPa) is based on the fact that few failures have been observed when the average bond stress range is less than 60 percent of nominal bond stress (800 psi [5.5 MPa]), and on the customary 1.6 load factor for live loads.

Although cyclic load data for hooked bar anchorages are limited, the stress ranges in the steel should not exceed 15 ksi (103 MPa) for standard hooks.

8.3.2 *Low-cycle fatigue*—The following limits for anchorages under low-cycle loads are recommended

a) For hooked bar anchorages, straight bar anchorages, and lap splices, the provisions provided by ACI 318 for beams and joints in special moment frames, Chapter 21, are adequate.

b) For beam-column Type 2 joints in structural frames, the provisions of ACI 352R should be applied.

CHAPTER 9—CONCLUSIONS AND RECOMMENDATIONS

9.1—Monotonic loading

Under static monotonically increasing loads, the most important factors affecting bond strength are concrete strength, construction quality, yield strength of flexural reinforcement, bar size, concrete cover, transverse reinforcement, coatings, and bar spacing.

9.2—Cyclic loading

Most parameters that are important under monotonic loading are also important under cyclic loading. In addition, bond stress range, type of loading, and maximum imposed bond stress are important under cyclic loads. The following conclusions can be made from the data currently available.

9.2.1 *High-cycle fatigue*—The most significant effect of high-level repeated loading is bond reduction at failure. Stress ranges exceeding 40 percent of the yield strength of the reinforcement in anchorages consistent with ACI 318 appear to reduce bond strength. These losses can be as high as 50 percent of the monotonic pullout bond strength.

Reversed cyclic loads tend to deteriorate bond at a higher rate and precipitate failure, even at fewer cycles or at lower loads than monotonic failure load level. An important factor in high-cycle fatigue is the fatigue strength of the concrete itself. Internal damage through the propagation of microcracks with repeated loading is the most important parameter affecting bond strength.

The mechanism governing failure is a progressive crushing of the concrete in front of the ribs. Test data indicate a similar behavior under both fatigue and sustained loading. Models developed for sustained loading likely can be extended to model damage due to fatigue.

Bond failures under fatigue loading are unlikely if current ACI 318 provisions for anchorage lengths under monotonic loading and the ACI 215R limits for concrete and steel fatigue are followed.

9.2.2 *Low-cycle loading*—Low-cycle loading gives rise to the problem of bond deterioration, particularly at the interior joints of the moment-resisting frames. Similarly, cyclic loading places severe demands on the strength and ductility of lap splice regions. Various observations about the low-cycle loading can be summarized as follows.

a) The higher the load amplitude, the larger the additional slip, especially after the first cycle. Some permanent damage seems to occur when 60 to 70 percent of the monotonic bond strength is achieved. For design, a damage threshold can be suggested to equal 50 percent of the monotonic bond strength (400 psi [2.8 MPa]).

b) Upon completion of a load cycle of a bar to an arbitrary bond stress or slip value below the damage threshold and unloading to zero, the monotonic stress-slip relationship can be attained again during reloading. This behavior also occurs for a large number of loadings, provided no bond failure occurs during cyclic loadings.

c) Loading a bar to a bond stress higher than 80 percent of ultimate results in significant permanent slip. Loading beyond the slip corresponding to the nominal bond stress results in large losses of stiffness and bond strength.

d) Bond deterioration under large amplitude cyclic loads cannot be prevented, except by using very long anchorage lengths and substantial transverse reinforcement. Even in this case, bond damage near the most highly stressed areas cannot be eliminated.

9.3—Recommendations for future research

Extensive coordinated research should be conducted to investigate the effect of cyclic loading on bond. Some of the areas where research is most urgently needed are as follows.

a) *Bond strength in high-strength concretes*—As with shear and tensile strength and bond strength under monotonic loads, bond strength under cyclic loads may not scale with $\sqrt{f_c'}$ as the compressive strength exceeds 10,000 psi (69 MPa). Mitchell and Marzouk (2007) indicate the bond strength of high-strength concrete is more proportional to the cube root rather than $\sqrt{f_c'}$.

b) *Development of standard test methods for behavior of reinforcing bars under high-cycle and low-cycle loading*—Bond test specimens and test methods vary widely for determining the bond behavior under repeated loads. An evaluation of these methods and the effects on reported results is needed.

c) *Behavior of hooked bar anchorages under high-cycle fatigue*—Normally, hooked bar anchorages will not be subjected to large fatigue loads, but some data in this area will be helpful for special design circumstances.

d) *Behavior of large bar lap splices under high-cycle loads*—Limitations in using No. 14 and No. 18 (43 and 57) bar lap splices have not been satisfactorily explained in terms of bond stress behavior.

e) *Effect of admixtures on cyclic bond strength*—No long-term data are available on the effect of admixtures on bond strengths under cyclic loads.

f) *Influence of casting position on cyclic bond strength*—While this effect has been shown to be important for monotonic loads, no corresponding data are available for cyclic loads.

g) *Modeling strain rate effects at material level*—Although research continues to develop relationships for detailed analysis of reinforced concrete structures, comprehensive yet simple constitutive models for concrete, reinforcing bars, and bond are not yet available. The data available indicate that a simple model may not be feasible.

h) *Effect of bond deterioration on performance of redundant structures*—The main concern for low-cycle fatigue is the loss of stiffness in joints of ductile moment-resisting frames. Although significant analytical studies have been completed, most physical tests have involved determinate, single-joint specimens. Tests on full floor subassemblages incorporating more than one joint do not show the same amount of deterioration for equivalent story drifts.

i) *Bond-controlled reinforcement*—Debonding longitudinal bars within the shear span zone for reinforced concrete members has the potential to improve earthquake performance. Additional investigation is necessary to establish potential applications within design practices.

j) *High-strength lightweight concrete*—The necessity to increase the development length required for lightweight concrete with a concrete density factor as stipulated in ACI 318 may require further investigation.

k) *FRP concrete composite systems and fiber-reinforced cement composites*—The contribution to bond performance under cyclic loading by the addition of fiber-reinforced polymer composites and fiber reinforcement warrants additional investigation as using these materials showed promising performance in bond, particularly for low-cycle fatigue applications (Porter and Harries 2007).

CHAPTER 10—REFERENCES

Committee documents are listed first by document number and year of publication followed by authored documents listed alphabetically.

American Association of State Highway and Transportation Officials (AASHTO)

HB-17-02—Standard Specifications for Highway Bridges

American Concrete Institute

215R-92 Considerations for Design of Concrete Structures Subjected to Fatigue Loading (Reapproved 1997)

318-11—Building Code Requirements for Reinforced Concrete

352R-02—Recommendations for Design of Beam-Column Connections in Monolithic Reinforced Concrete Structures (Reapproved 2010)

352.1R-11—Guide for Design of Slab-Column Connections in Monolithic Concrete Structures

363R-10—Report on High-Strength Concrete

408R-03—Bond and Development of Straight Reinforcing Bars in Tension (Reapproved 2012)

ASTM International

A615/A615M-12—Standard Specification for Deformed and Plain Carbon-Steel Bars for Concrete Reinforcement

British Standards Institution

BS EN 1998-1:2004 Eurocode 8—Design of Structures for Earthquake Resistance, Part 1: General Rules, Seismic Actions and Rules for Buildings

Det Norske Veritas

DNV-OS-C502-10—Offshore Concrete Structures

Federal Emergency Management Agency (FEMA)

FEMA 356-00—Prestandard and Commentary for the Seismic Rehabilitation of Buildings

FEMA 310-98—Handbook for the Seismic Evaluation of Buildings

Standards Association of New Zealand

NZS 3101:2006 Concrete Structures Standard

Abrams, D. A., 1913, "Tests of Bond between Concrete and Steel," *Bulletin* 71, Engineering Experiment Station, University of Illinois, Urbana, IL.

ACI Committee 408, 1966, "Bond Stress—The State of the Art," ACI Journal, *Proceedings* V. 63, No. 11, pp. 1161-1189.

ACI Committee 408, 1970, "Opportunities in Bond Research," ACI Journal, *Proceedings* V. 67, No. 11, pp. 857-867.

ACI Committee 408, 1979, "Suggested Development, Splice, and Standard Hook Provisions for Deformed Bars in Tension," *Concrete International*, V. 1, No. 7, July, pp. 44-46.

Ahmed, T.; Burley, E.; and Rigden, S., 1999, "Effect of Alkali-Silica Reaction on Tensile Bond Strength of Reinforcement in Concrete Tested under Static and Fatigue Loading," *ACI Materials Journal*, V. 96, No. 4, July-Aug., pp. 419-428.

Alavi-Fard, M., and Marzouk, H., 2002, "Bond Behavior of High Strength Concrete under Reversed Pull-out Cyclic Loading," *Canadian Journal of Civil Engineering*, V. 29, pp. 191-200.

Altowaiji, W. A. K.; Darwin, D.; and Donahey, R. C., 1986, "Bond of Reinforcement to Revibrated Concrete," ACI Journal, *Proceedings* V. 83, No. 6, Nov.-Dec., pp. 1035-1042.

Architectural Institute of Japan, 1990, "Design Guidelines for Earthquake Resistant Reinforced Concrete Buildings based on Ultimate Strength Concept," Oct., 337 pp. (in Japanese)

Aristizabal-Ochoa, S. D.; Fiorato, A. E.; and Corley, W. G., 1977, "Tension Lap Splices Under Severe Load Reversals," *Research and Development Bulletin* RD077.01D, Portland Cement Association, Skokie, IL.

Balazs, G. L., 1986, "Bond Behavior under Repeated Loads," Studi e Ricerche—Corso Flli, Pesenti, Politecnico di Milano, V. 8, pp. 395-430.

Balazs, G. L., 1991, "Fatigue of Bond," *ACI Materials Journal*, V. 88, No. 6, Nov.-Dec., pp. 620-629.

Balazs, G. L., 1998, "Bond Under Repeated Loading," *Bond and Development of Reinforcement—A Tribute to Dr. Peter Gergely*, SP-180, R. Leon, ed., American Concrete Institute, Farmington Hills, MI, pp. 125-144.

Bennett, E. W., 1982, "Fatigue Tests of Spliced Reinforcement in Concrete Beams," SP-75, *Fatigue of Concrete Structures*, S. P. Shah, ed., American Concrete Institute, Farmington Hills, MI, pp. 177-194.

Bertero, V. V., and Bresler, B., 1968, "Behavior of Reinforced Concrete Under Repeated Load," *Journal of the Structural Division*, V. 94, No. ST6, June, pp. 1567-1590.

Bertero, V. V., and Popov, E. P., 1977, "Seismic Behavior of Ductile Moment-Resisting Reinforced Concrete Frames," *Reinforced Concrete Structures in Seismic Zones*, SP-53, American Concrete Institute, Farmington Hills, MI, pp. 247-292.

Bertero, V. V.; Popov, E. P.; and Viwathanatepa, S., 1978, "Bond of Reinforcing Steel: Experiments and a Mechanical Model," *IASS Symposium on Non-Linear Behavior of Reinforced Concrete Spatial Structures*, Darmstadt, Germany, V. 2, July, pp. 3-17.

Bonacci, J., and Pantazopoulou, S., 1993, "Parametric Investigation of Joint Mechanics," *ACI Structural Journal*, V. 90, No. 1, Jan.-Feb., pp. 61-71.

Brown, C. B., 1966, "Bond Failure Between Steel and Concrete," *Journal of the Franklin Institute*, V. 282, No. 5, Nov., pp. 271-290.

Brown, R. H., and Jirsa, J. O., 1971, "Reinforced Concrete Beams Under Load Reversals," ACI Journal, *Proceedings* V. 68, No. 5, May, pp. 380-390.

Cairns, J., and Arthur, P. D., 1979, "Strength of Lapped Splices in Reinforced Concrete Columns," ACI Journal, *Proceedings* V. 76, No. 2, Feb., pp. 277-296.

Cao, J., and Chung, D. D. L., 2001, "Degradation of the Bond Between Concrete and Steel Under Cyclic Shear Loading, Monitored by Contact Electrical Resistance Measurement," *Cement and Concrete Research*, V. 31, No. 4, pp. 669-671.

Carino, N. J., and Lew, H. S., 1982, "Re-Examination of the Relation between Splitting Tensile and Compressive Strength of Normal Weight Concrete," ACI Journal, *Proceedings* V. 79, No. 3, May-June, pp. 214-219.

CEB, 1982, "State of the Art Report: Bond Action and Bond Behavior of Reinforcement," *Bulletin No. 151*, Comite Euro-International Du Beton, Paris, Dec.

CEB, 1987, "Seismic Design of Concrete Structures," Comite Euro-International du Beton, Gower Technical Press, Hants, England, 298 pp.

Celik, A. C., and Ellingwood, B. R., 2008, "Modeling Beam-Column Joints in Fragility Assessment of Gravity Load Designed Reinforced Concrete Frames," *Journal of Earthquake Engineering*, V. 12, No. 3, pp. 357-381.

Chao, S.-H., 2005, "Bond Characterization of Reinforcing Bars and Prestressing Strands in High Performance Fiber Reinforced Cementitious Composites under Monotonic and Cyclic Loading," PhD dissertation, University of Michigan, Ann Arbor, MI, 475 pp.

Chinn, J.; Ferguson, P. M.; and Thompson, J. N., 1955, "Lapped Splices in R.C. Beams," ACI Journal, *Proceedings* V. 52, No. 2, Oct., pp. 201-214.

Cho, J. Y., and Pincheira, J. A., 2006, "Inelastic Analysis of Reinforced Concrete Columns with Short Lap Splices Subjected to Reversed Cyclic Loads," *ACI Materials Journal*, V. 103, No. 2, Mar.-Apr., pp. 280-291.

Chun, S. C.; Lee, S. J.; Kang, T. H.-K.; Oh, B.; and Wallace, J. W., 2007, "Mechanical Anchorage in Exterior Beam-Column Joints Subjected to Cyclic Loading," *ACI Structural Journal*, V. 104, No. 1, Jan.-Feb., pp. 102-112.

Ciampi, V.; Eligehausen, R.; Bertero, V. V.; and Popov, E. P., 1982, "Analytical Model for Concrete Anchorages of Reinforcing Bars Under Generalized Excitations," *Report No. EERC 82-83*, Earthquake Engineering Research Center, University of California, Berkeley, CA, Dec.

Cleary, D. B., and Ramirez, J. A., 1993, "Epoxy-Coated Reinforcement Under Repeated Loading," *ACI Structural Journal*, V. 90, No. 4, July-Aug., pp. 451-458.

Corley, W. G.; Hanson, J. M.; and Helgason, T., 1978, "Design of Reinforced Concrete for Fatigue," *Journal of the Structural Division*, ASCE, V. 104, No. ST6, June, pp. 921-932.

Det Norske Veritas, 1977, *Rules for the Design, Construction, and Inspection of Offshore Structures*, Oslo, Norway, 67 pp.

Diederichs, U., and Schneider, U., 1981, "Bond Strength at High Temperatures," *Magazine of Concrete Research*, V. 33, No. 115, pp. 75-84.

Edwards, A. D., and Yannopoulos, P. J., 1978, "Local Bond Stress-Slip Relationship Under Repeated Loading," *Magazine of Concrete Research*, V. 30, No. 103, June, pp. 62-72.

Edwards, A. D., and Yannopoulos, P. J., 1979, "Local Bond-Stress to Slip Relationships for Hot Rolled Deformed Bars and Mild Steel Plain Bars," ACI Journal, *Proceedings* V. 76, No. 3, May-June, pp. 405-420.

Eligehausen, R., 1979, "Bond in Tensile Lapped Splices of Ribbed Bars with Straight Anchorages," *Publication 301*, German Institute for Reinforced Concrete, Berlin, 118 pp. (in German)

Eligehausen, R.; Bertero, V. V.; and Popov, E. P., 1982a, "Hysteretic Behavior of Reinforcing Deformed Hooked Bars in R/C Joints," *Proceedings of the Seventh European Conference on Earthquake Engineering*, V. 4, Athens, GA, Sept. pp. 171-178.

Eligehausen, R.; Popov, E. P.; and Bertero, V. V., 1982b, "Local Bond Stress-Slip Relationships of Deformed Bars Under Generalized Excitations," *Proceedings of the Seventh European Conference on Earthquake Engineering*, V. 4, Athens, GA, Sept. pp. 69-80.

Eligehausen, R.; Popov, E. P.; and Bertero, V. V., 1983, "Local Bond Stress-Slip Relationships of Deformed Bars Under Generalized Excitations," *Report No. UCBEERC 83-23*, Earthquake Engineering Research Center, University of California, Berkeley, CA, Oct.

Fagundo, F., 1979, "Behavior of Lapped Splices in R.C. Beams Subjected to Cyclic Loads," PhD thesis, Cornell University, Ithaca, NY, Jan.

Fang, C. Q., 2006, "Bond Strength of Corroded Reinforcement Under Cyclic Loading," *Magazine of Concrete Research*, V. 58, No. 7, Sept., pp. 437-446.

Ferguson, P. M., 1979, "Small Bar Spacing or Cover—A Problem for the Designer," ACI Journal, *Proceedings* V. 74, No. 9, Sept., pp. 435-439.

fib, 2000, "Bond of Reinforcement in Concrete: State-of-Art Report," *Bulletin* No. 10, Federal Institute of Technology, Lausanne, Switzerland.

Filippou, F. C., 1985, "A Simple Model for Reinforcing Bar Anchorages Under Cyclic Excitations," *Report No. UCBEERC 85-05*, Earthquake Engineering Research Center, University of California, Berkeley, CA, Mar.

Filippou, F. C., 1986, "A Simple Model For Reinforcing Bar Anchorages Under Cyclic Excitations," *Journal of Structural Engineering*, V. 112, No. 7, pp. 1639-1659.

Filippou, F. C.; Popov, E. P.; and Bertero, V. V., 1983, "Modeling of R/C Joints Under Cyclic Excitations," *Journal of Structural Engineering*, V. 109, No. 11, Nov., pp. 2666-2684.

Filippou, F. C.; Popov, E. P.; and Bertero, V. V., 1986, "Hysteretic Behavior of R/C Joints," *Journal of Structural Engineering*, V. 112, No. 7, pp. 1605-1622

Fujii, S., and Morita, S., 1981, "Effect of Transverse Reinforcement on Splitting Bond Strength," *Transactions of the Japan Concrete Institute*, V. 3, pp. 237-244.

Gergely, P.; Fagundo, F.; and White, R. N., 1979, "Bond and Splices in Reinforced Concrete for Seismic Loading," *CEB Bulletin d'Information No. 132*, AICAP-CEB Symposium, Rome, May.

Gergely, P., and White, R. N., 1980, "Seismic Design of Lapped Splices in Reinforced Concrete," *Proceedings of the 7th WCEE*, IAEE, V. 4, Sept., pp. 281-288.

Gerstle, W.; Ingraffea, A. R.; and Gergely, P., 1982, "Tension Stiffening: A Fracture Mechanics Approach," *Bond in Concrete*, P. Bartos, ed., Applied Science Publishers, London, pp. 97-106.

Girard, C., and Bastien, J., 2002, "Finite-Element Bond-Slip Model for Concrete Columns under Cyclic Loads," *Journal of Structural Engineering*, V. 128, No. 12, Dec., pp. 1502-1510.

Gosain, N. K., and Jirsa, J. O., 1977, "Bond Deterioration in Reinforced Concrete Members Under Cyclic Loads," *Proceedings of the Sixth World Conference on Earthquake Engineering*, V. 3, New Delhi.

Graf, O., and Brenner, E., 1939, "Versuche zur Ermittlung des Gleitwiderstandes von Eiseneinlagen im Beton bei stetig steigender und oftmals wiederholter Belastung," *Deutscher Ausschuss fur Eisenbeton*, Heft 93, Berlin, p. 28. (in German)

Guiriani, E., 1981, "Experimental Investigation on the Bond-Slip Law of Deformed Bars in Concrete," *IABSE Colloquium Delft*, IABSE, Zurich, pp. 121-142.

Gylltoft, K.; Cederwall, K.; Elfgren, L.; and Nilsson, G., 1982, "Bond Failure in Reinforced Concrete under Monotonic and Cyclic Loading: A Fracture Mechanics Approach," *Fatigue of Concrete Structures*, SP-75, S. P. Shah, ed., American Concrete Institute, Farmington Hills, MI, pp. 269-288.

Harajli, M. H., 2005, "Bond Strengthening of Steel Bars Using External FRP Confinement: Implications on the Static and Cyclic Response of R/C Members," *7th International Symposium on Fiber-Reinforced (FRP) Polymer Reinforcement for Concrete Structures*, SP-230, American Concrete Institute, Farmington Hills, MI, pp. 579-596.

Harajli, M. H., and Mabsout, M. E., 2002, "Evaluation of Bond Strength of Steel Reinforcing Bars in Plain and Fiber-Reinforced Concrete," *ACI Structural Journal*, V. 99, No. 4, July-Aug., pp. 509-517.

Hasan, H. O.; Cleary, D. B.; and Ramirez, J. A., 1996, "Performance of Concrete Bridge Decks and Slabs Reinforced with Epoxy-Coated Steel under Repeated Loading," *ACI Structural Journal*, V. 93, No. 4, July-Aug., pp. 397-403.

Hassan, F. M., and Hawkins, N. M., 1977a, "Anchorage of Reinforcing Bars for Seismic Forces," *Reinforced Concrete Structures in Seismic Zones*, SP-53, American Concrete Institute, Farmington Hills, MI, pp. 387-416.

Hassan, F. M., and Hawkins, N. M., 1977b, "Prediction of the Seismic Loading Anchorage Characteristics of Reinforcing Bars," *Reinforced Concrete Structures in Seismic Zones*, SP-53, American Concrete Institute, Farmington Hills, MI, pp. 417-438.

Hawkins, N. M., 1974, "Fatigue Characteristics in Bond and Shear of Reinforced Concrete Beams," *Fatigue of Concrete*, SP-41, American Concrete Institute, Farmington Hills, MI, pp. 203-236.

Hawkins, N. M.; Lin, I. J.; and Jeang, F. L., 1982, "Local Bond Strength of Concrete for Cyclic Reversed Loadings," *Bond in Concrete*, P. Bartos, ed., Applied Science Publishers Ltd., pp. 151-161.

Helgason, T.; Hanson, J. M.; Somes, N. F.; Corley, G. W.; and Hognestad, E., 1976, "Fatigue Strength of High-Yield Reinforcing Bars," *NCHRP Program Report 164*, Transportation Research Board, National Research Council, Washington, DC.

Holmen, J. O., 1982, "Fatigue of Concrete by Constant and Variable Amplitude Loading," *Fatigue of Concrete Structures*, SP-75, S. P. Shah, ed., American Concrete Institute, Farmington Hills, MI, pp. 71-110.

Hota, S., and Naaman, A. E., 1997, "Bond Stress-Slip Response of Reinforcing Bars Embedded in FRC Matrices under Monotonic and Cyclic Loading," *ACI Structural Journal*, V. 94, No. 5, Sept.-Oct., pp. 525-537.

Ichinose, T., 1995, "Splitting Bond Failure of Columns Under Seismic Action," *ACI Structural Journal*, V. 92, No. 6, Nov.-Dec., pp. 535-542.

International Conference of Building Officials, 1976, "Uniform Building Code," Whittier, CA.

Ismail, M. A., and Jirsa, J. O., 1972a, "Behavior of Anchored Bars Under Low Cyclic Overloads Producing Inelastic Strains," ACI Journal, *Proceedings* V. 69, No. 7, July, pp. 433-438.

Ismail, M. A., and Jirsa, J. O., 1972b, "Bond Deterioration in Reinforced Concrete Subjected to Low-cycle Loads," ACI Journal, *Proceedings* V. 69, No. 6, June, pp. 334-343.

Johnson, L. A., and Jirsa, J. O., 1981, "The Influence of Short Embedment and Close Spacing on the Strength of Hooked Bar Anchorages," *PMFSEL Report No. 81-2*, The University of Texas at Austin, Austin, TX, Apr.

Johnston, D. W., and Zia, P., 1982, "Bond Characteristics of Epoxy-Coated Reinforcing Bars," *ERSD-110-79-2*, Center for Transportation Engineering Studies, Department of Civil Engineering, North Carolina State University, Raleigh, NC, Aug.

Kitayama, K.; Otani, S.; and Aoyama, H., 1991, "Development of Design Criteria for RC Interior Beam-Column Joints," *Design of Beam-Column Joints for Seismic Resistance*, SP-123, American Concrete Institute, Farmington Hills, MI, pp. 97-109.

Krstulovic-Opara, N.; Watson, K. A.; and LaFave, J. M., 1994, "Effect of Increased Tensile Strength and Toughness on Reinforcing Bar Bond Behavior," *Cement and Concrete Composites*, V. 16, No. 2, pp. 129-141.

Kwak, H. G., and Kim, S. P., 2001, "Nonlinear Analysis of RC Beam Subject to Cyclic Loading," *Journal of Structural Engineering*, V. 127, No. 12, Dec., pp. 1436-1444.

Lenschow, R., 1984, "Long Term Random Dynamic Loading of Concrete Structures," *Materiaux et Constructions*, V. 13, No. 75, pp. 274-278.

Limkatanyu, S.; Samakrattakit, A.; and Spacone, E., 2004, "Nonlinear Dynamic Analysis of Reinforced Concrete Frames including Bond-Slip Effects," Asia Conference on Earthquake Engineering (ACEE-2004), Mar., Manila, Philippines.

Lowes, L., and Altoontash, A., 2003, "Modeling Reinforced Concrete Beam-Column Joints subjected to Cyclic Loading," *Journal of Structural Engineering*, V. 129, No. 12, Dec., pp. 1686-1697.

Lukose, K.; Gergely, P.; and White, R. N., 1982, "Behavior of Reinforced Concrete Lapped Splices Under Inelastic Cyclic Loading," ACI Journal, *Proceedings* V. 79, No. 5, Sept.-Oct., pp. 355-365.

Lutz, L. A., 1970, "Analysis of Stresses in Concrete Near a Reinforcing Bar Due to Bond and Transverse Cracking," ACI Journal, *Proceedings* V. 67, No. 10, Oct., pp. 778-787.

Lutz, L. A., and Gergely, P., 1967, "Mechanics of Bond and Slip of Deformed Bars in Concrete," ACI Journal, *Proceedings* V. 64, No. 11, Nov., pp. 711-721.

Marques, J. L. G., and Jirsa, J. O., 1975, "A Study of Hooked Bar Anchorages in Beam-Column Joints," ACI Journal, *Proceedings* V. 72, No. 5, May, pp. 198-209.

Maso, J. C., 1980. "The Bond between Aggregates and Hydrated Cement Pastes," *Proceedings of the Seventh International Congress on the Chemistry of Cements*, Paris, pp. 3-15.

Melek, M., and Wallace, J. W., 2004, "Cyclic Behavior of Columns with Short Lap Splices," *ACI Structural Journal*, V. 101, No. 6, Nov.-Dec., pp. 802-811.

Minor, J., and Jirsa, J. O., 1975, "Behavior of Bent Bar Anchorages," ACI Journal, *Proceedings* V. 72, No. 4, Apr., pp. 141-149.

Mitchell, D. W., and Marzouk, H., 2007, "Bond Characteristics of High-Strength Lightweight Concrete," *ACI Structural Journal*, V. 104, No. 1, Jan.-Feb., pp. 22-29.

Monti, G.; Filippou, F. C.; and Spacone, E., 1997a, "Analysis of Hysteretic Behavior of Anchored Reinforcing Bars," *ACI Structural Journal*, V. 94, No. 3, May-June, pp. 248-261.

Monti, G.; Filippou, F. C.; and Spacone, E., 1997b, "Finite Element for Anchored Bars under Cyclic Load Reversals," *Journal of Structural Engineering*, V. 123, No. 5, May, pp. 614-623.

Monti, G., and Spacone, E., 2000, "Reinforced Concrete Fiber Beam Element with Bond-Slip," *Journal of Structural Engineering*, V. 126, No. 6, June, pp. 654-661.

Mor, A.; Hester, W. T.; and Gerwick, B. C., 1992, "Fatigue of High-Strength Reinforced Concrete," *ACI Materials Journal*, V. 89, No. 2, Mar.-Apr., pp. 197-207.

Morita, S., and Kaku, T., 1973, "Local Bond Stress-Slip Relationship and Repeated Loading," *Resistance and Ultimate Deformability of Structures Acted on by Well Defined Repeated Loads*, IABSE Symposium, Lisbon, pp. 221-227.

Morita, S., and Kaku, T., 1979, "Splitting Bond Failures of Large Deformed Reinforcing Bars," ACI Journal, *Proceedings* V. 76, No. 1, Jan., pp. 93-110.

Morley, P. D., and Royles, R., 1983, "Response of Bond in Reinforced Concrete to High Temperatures," *Magazine of Concrete Research*, V. 35, No. 123, pp. 67-74.

Muhlenbruch, C. W., 1945, "The Effect of Repeated Loading on the Bond Strength of Concrete," *ASTM Proceedings*, V. 45, pp. 824-845.

Muhlenbruch, C. W., 1948, "The Effect of Repeated Loading on the Bond Strength of Testing Materials," *ASTM Proceedings*, V. 48, pp. 977-985.

Orangun, C. O.; Jirsa, J. O.; and Breen, J. E., 1977, "Re-evaluation of Test Data on Development Length and Splices," ACI Journal, *Proceedings* V. 74, No. 3, Mar., pp. 114-122.

Pacholka, K.; Rezansoff, T.; and Sparling, B. F., 1999, "Stirrup Distribution Across the Beam Width in Tension Lap Splices," *Canadian Journal of Civil Engineering*, V. 26, pp. 83-95.

Pandey, G. R., and Mutsuyoshi, H., 2005, "Seismic Performance of Reinforced Concrete Piers with Bond-Controlled Reinforcements," *ACI Structural Journal*, V. 102, No. 2, Mar.-Apr., pp. 295-304.

Park, R., and Paulay, T., 1975, *Reinforced Concrete Structures*, John Wiley & Sons, New York.

Parra-Montesinos, G. J.; Peterfreund, S. W.; and Chao, S.-H., 2005, "Highly Damage-Tolerant Beam-Column Joint Through Use of High-Performance Fiber-Reinforced Cement Composites," *ACI Structural Journal*, V. 102, No. 3, May-June, pp. 487-495.

Paulay, T.; Park, R.; and Priestley, M. J. N., 1978, "Reinforced Concrete Beam Column Joints Under Seismic Actions," ACI Journal, *Proceedings* V. 75, No. 11, Nov., pp. 585-593.

Paulay, T., and Priestley, M. J. N., 1992, *Seismic Design of Reinforced Concrete and Masonry Buildings*, John Wiley & Sons, New York.

Perry, E. S., and Jundi, N., 1969, "Pullout Bond Stress Distribution Under Static and Dynamic Repeated Loadings," ACI Journal, *Proceedings* V. 66, No. 5, May, pp. 377-380.

Perry, E. S., and Thompson, J. N., 1966, "Bond Stress Distribution on Reinforcing Steel in Beams and Pullout Specimens," ACI Journal, *Proceedings* V. 63, No. 8, Aug., pp. 866-875.

Plaines, P.; Tassios, T.; and Vintzeleou, E., 1982, "Bond Relaxation and Bond-Slip Creep Under Monotonic and Cyclic Actions," *Bond in Concrete*, P. Bartos, ed., Applied Science Publishers, London, pp. 193-204.

Porter, M., and Harries, K., 2007, "Future Directions for Research in FRP Composites in Concrete Construction," *Composites Construction Journal*, American Society of Civil Engineers, Special Forum Paper, V. 11, No. 3, May-June.

Raynor, D. J.; Lehman, D. E.; and Stanton, J. F., 2002, "Bond-Slip Response of Reinforcing Bars Grouted in Ducts," *ACI Structural Journal*, V. 99, No. 5, Sept.-Oct., pp. 568-576.

Rehm, G., and Eligehausen, R., 1977, "Bond Ribbed Bars under Repeated Loads," *Report 291*, German Institute for Reinforced Concrete, Berlin. (in German)

Rehm, G., and Eligehausen, R., 1979, "Bond Ribbed Bars Under High-Cycle Repeated Loads," ACI Journal, *Proceedings* V. 76, No. 2, Feb., pp. 297-310.

Rezansoff, T., and Sparling, B. F., 1986, "The Effect of Confinement on Lap Splices in Reversed Cyclic Loading," *Canadian Journal of Civil Engineering*, V. 13, No. 4, Dec., pp. 681-692

Rezansoff, T.; Zacaruk, J.; and Afseth, J., 1993, "High Cycle (Fatigue) Resistance of Reinforced Concrete Beams with Lap Splices," *Canadian Journal of Civil Engineering*, V. 20, pp. 642-649.

Rezansoff, T.; Zacaruk, J.; and Topping, R., 1988, "Tensile Lap Splices in Reinforced Concrete Beams under Inelastic Cyclic Loading," ACI Structural Journal, V. 85, No. 1, pp. 46-52.

Rezansoff, T.; Zhang, S.; and Sparling, B. F., 1997, "Influence of Different Stirrup Configurations on Lap Splices in Beams," *Canadian Journal of Civil Engineering*, V. 24, No. 1, Jan., pp. 106-114.

RILEM, 1986, "Dynamic Behavior of Concrete Structures," *Report of the RILEM 65 MDB Committee*, G. P. Tilly, ed., Elsevier, Amsterdam.

Robins, P. J., and Standish, I. G., 1982, "Effect of Lateral Pressure on Bond of Reinforcing Bars in Concrete," *Bond in Concrete*, P. Bartos, ed., Applied Science Publishers, London.

Robinson, J. R., 1965, "Influence of Transverse Reinforcement on Shear and Bond Strength," ACI Journal, *Proceedings* V. 62, No. 3, Mar., pp. 345-362.

Saatcioglu, M.; Alsiwat, J.; and Ozcebe, G., 1992, "Hysteretic Behavior of Anchorage Slip in R/C Members," *Journal of Structural Engineering*, ASCE, V. 118, No. 9, pp. 2439-2458.

Shah, S. P., and Chung, L., 1986, "Effect of Cyclic Loading Rate on Response of Model Beam-Column Joints and Anchorage Bond," *Proceedings of the 3rd U.S. National*

Conference on Earthquake Engineering, EERI, San Francisco, CA.

Shah, S. P., and Somayaji, S., 1981, "Bond Stress versus Slip Relationships and Cracking Response of Tension Members" ACI Journal, *Proceedings* V. 78, No. 3, Mar., pp. 217-225.

Shih, T. S.; Lee, G. C.; and Chang, K. C., 1986, "Local Concrete-Steel Bond Behavior at Low Temperatures," *Journal of Structural Engineering*, V. 113, No. 11, pp. 2278-2289.

Shih, T. S.; Lee, G. C.; and Chang, K. C., 1988, "Effect of Freezing Cycles on Bond Strength of Concrete," *Journal of Structural Engineering*, V. 114, No. 3, Mar., pp. 717-726.

Sivakumar, B.; Gergely, P.; and White, R. N., 1983, "Suggestions for the Design of Lapped Splices for Seismic Loading," *Concrete International*, V. 5, No. 2, Feb, pp. 46-50.

Sivakumar, B.; White, R. N.; and Gergely, P., 1982, "Behavior and Design of Reinforced Concrete Column-Type Lapped Splices Subjected to High-Intensity Cyclic Loading," *Report 82-11*, Department of Structural Engineering, Cornell University, Ithaca, NY, Oct., 201 pp.

Soretz, S., and Holzenbein, H., 1979, "Influence of Rib Dimensions of Reinforcing Bars on Bond and Bendability," ACI Journal, *Proceedings* V. 76, No. 1, Jan., pp. 111-127.

Spacone, E.; Filippou, F. C.; and Taucer, F., 1996, "Fiber Beam-Column Model for Nonlinear Analysis of R/C Frames. Part I: Formulation," *Earthquake Engineering & Structural Dynamics*, V. 25, pp. 711-725.

Spacone, E., and Limkatanyu, S., 2000, "Responses of Reinforced Concrete Members Including Bond-Slip Effects," *ACI Structural Journal*, V. 97, No. 6, Nov.-Dec., pp. 831-839.

Sritharan, S.; Priestley, N.; and Seible, F., 2000, "Nonlinear Finite Element Analysis of Concrete Bridge Joint Systems Subjected to Seismic Actions," *Finite Elements in Analysis and Design*, V. 36, pp. 215-233.

Stehle, J. S.; Goldsworthy, H.; and Mendis, P., 2001, "Reinforced Concrete Interior Wide-Band Beam-Column Connections Subjected to Lateral Earthquake Loading," *ACI Structural Journal*, V. 98, No. 3, May-June, pp. 270-279.

Tassios, T. P., 1979, "Properties of Bond Between Concrete and Steel Under Load Cycles Idealizing Seismic Actions," *Structural Concrete Under Seismic Actions, Comité Euro-International Du Béton, Bulletin D'Information*, No. 131, V. 1, AICAP-CEB Symposium, Rome, Italy.

Tassios, T. P., and Koroneos, E. G., 1984, "Local Bond-Slip Relationships by Means of the Moire Method," ACI Journal, *Proceedings* V. 81, No. 1, Jan-Feb., pp. 27-34.

Teng, S., and Wang, F., 2001, "Finite Element Analysis of Reinforced Concrete Deep Beams under Fatigue Loading," *ACI Structural Journal*, V. 98, No. 3, May-June, pp. 315-323.

Tepfers, R., 1973, "A Theory of Bond, Applied to Overlapped Tensile Reinforcement Splices for Deformed Bars," *Publication 73.2*, Division of Concrete Structures, Chalmers University, Göteborg.

Tocci, A. D., 1981, "The Behavior and Strength of Lapped Splices in Reinforced Concrete Beams Subjected to Cyclic Loading," PhD thesis, Cornell University, Ithaca, NY, May.

Treece, R. A., and Jirsa, J. O., 1987, "Bond Strength of Epoxy-Coated Reinforcing Bars," *PMFSEL Report 87-1*, Ferguson Structural Engineering Laboratory, Department of Civil Engineering, The University of Texas at Austin, Austin, TX, Jan.

Untrauer, R. E., and Henry, R. L., 1965, "Influence of Normal Pressure on Bond Strength," ACI Journal, *Proceedings* V. 62, No. 5, May, pp. 577-585.

Verna, J. R., and Stelson, T. E., 1963, "Repeated Loading Effect on Ultimate Static Strength of Concrete Beams," ACI Journal, *Proceedings* V. 60, No. 6, June, pp. 743-749.

Viwathanatepa, S.; Popov, E. P.; and Bertero, V. V., 1979, "Effects of Generalized Loadings on Bond of Reinforcing Bars Embedded in Confined Concrete Blocks," Earthquake Engineering Research Center, *Report No. UCB/EERC-79/22*, Aug.

Vos, E., and Reinhart, H. W., 1982, "Bond-Stress Behavior of Deformed Bars, Plain Bars and Strand Under Impact Loading," *Bond in Concrete*, P. Bartos, ed., Applied Science Publishers, London, pp. 173-182.

Watstein, D., and Bresler, B., 1974, "Bond and Cracking in Reinforced Concrete," Chapter 4, *Reinforced Concrete Engineering*, John Wiley International.

Yankelevsky, D. Z.; Adin, M. A.; and Farhey, D. N., 1992, "Mathematical Model for Bond-Slip Behavior Under Cyclic Loading," *ACI Structural Journal*, V. 89, No. 6, Nov.-Dec., pp. 692-698.

Yannopoulos, P. J., and Tassios, T. P., 1991, "Reinforced Concrete Axial Elements Analyzed Under Monotonic and Cyclic Actions," *ACI Structural Journal*, V. 88, No. 1, Jan.-Feb., pp. 3-11.

Zhao, J., and Sritharan, S., 2007, "Modeling of Strain Penetration Effects in Fiber-Based Analysis of Reinforced Concrete Structures," *ACI Structural Journal*, V. 104, No. 2, Mar.-Apr., pp. 133-141.

Zheng, H., and Abel, A., 1998, "Stress Concentration and Fatigue of Profiled Reinforcing Steels," *International Journal of Fatigue*, V. 20, No. 10, pp. 767-773.

Zulfiqar, N., and Filippou, F. C., 1990, "Model of Critical Regions in Reinforced Concrete Frames under Earthquake Excitations," *EERC Report 90-06*, Earthquake Engineering Research Center, University of California, Berkeley, CA.

Zuo, J., and Darwin, D., 2000, "Bond Slip of High Relative Rib Area Bars Under Cyclic Loading," *ACI Structural Journal*, V. 97, No. 2, Mar.-Apr., pp. 331-334.

Guide for Lap Splice and Development Length of High Relative Rib Area Reinforcing Bars in Tension and Commentary

Reported by Joint ACI-ASCE Committee 408

Adolfo B. Matamoros
Chair

Theresa M. Ahlborn	David Darwin	Neil M. Hawkins	Max L. Porter
Gyorgy L. Balazs	Richard A. Devries	Roberto T. Leon	Julio A. Ramirez
Robert W. Barnes	Rolf Elighausen	LeRoy A. Lutz	John F. Silva
JoAnn P. Browning	Anthony L. Felder	Denis Mitchell	Ahmet Koray Tureyen
Douglas B. Cleary	Robert J. Frosch	Stravroula J. Pantazopoulou	Jun Zuo
James V. Cox	Bilal S. Hamad		

This guide provides recommended requirements to determine tension development and splice lengths for high relative rib area bar reinforcement, which are not in commercial production in the U.S. at the time of publication. When such requirements are applicable for a project, the user may incorporate the information in this guide into a legal document; however, this guide is not an ACI standard and shall not be adopted by reference into legal documents.

To assist the user to incorporate the information into a legal document, the relevant portion of the document is written in mandatory language. A commentary to the mandatory language portion is provided for additional information. Commentary provisions begin with an "R," such as "R1.1.1," and are shown in italics.

Keywords: bar reinforcement; bar ribs; bond; development length; high relative rib area; lap splice length.

CONTENTS

ACI 408.3R-09 supersedes ACI 408.3-01 and was adopted and published October 2009.

CHAPTER 1—INTRODUCTION

This guide is provided to help designers take advantage of higher bond strength exhibited by high relative rib area bars (Darwin and Graham 1993; Darwin et al. 1996a; Zuo and Darwin 1998) in the calculation of tension lap splice length and development length of bar reinforcement. It includes expressions for development and lap splice length applicable only for high relative rib area bars. This document has been updated to reflect improvements to the provisions of ACI 408.3-01 (ACI Committee 408 2001) proposed by Darwin et al. (2005). Because experimental data are available only to evaluate the development and lap splice lengths of straight high relative rib area bars in tension, the integrity of high relative rib area bars in compression or as hooked bars in tension should be evaluated using appropriate Chapter 12 provisions of ACI 318-08 (ACI Committee 318 2008).

To use this guide, ASTM A615/A615M for carbon-steel reinforcing bars should have the supplementary requirements imposed by the "Recommended Supplement to ASTM A615/A615M for High Relative Area Bars" that is appended to this document. With modifications to the section reference numbers, this supplement can be adapted for use with ASTM A706/A706M for low-alloy steel reinforcing bars.

CHAPTER 2—SCOPE

2.1

Evaluation of lap splice and development lengths of coated and uncoated bar reinforcement in tension shall be limited to bars having a high relative rib area, conforming to Sections 2.1.1 through 2.1.5:

2.1.1 The relative rib area is at least 0.10, but no larger than 0.14;

2.1.2 The ribs shall be in accordance with (a), (b), and (c).

(a) Ribs shall be at an angle of 45 to 65 degrees inclusive with respect to the axis of the bar;

(b) Ribs shall not cross; and

(c) Use of X patterns and diamond patterns for ribs is not permitted.

2.1.3 The rib spacing is at least 0.44 of the nominal diameter d_b of the bar reinforcement.

2.1.4 The average rib width is not greater than one-third of the average rib spacing.

2.1.5 The bar size does not exceed No. 11 (No. 36).

R2.1

A high relative rib area bar is defined as a bar reinforcement with R_r greater than or equal to 0.10, as conventionally deformed reinforcement has relative rib areas of 0.06 to 0.085. Based on available experimental results, the use of these provisions is limited to bar reinforcement with a maximum R_r = 0.14. Furthermore, consistent with the smallest spacing used in tests, the rib spacing s_r should not be less than 44% of the nominal bar diameter, as indicated in Section 2.1.3. A lower limit on width of the concrete between ribs is indirectly prescribed in Section 2.1.4 to avoid having a reduction in bond capacity due to a local shear failure of the concrete between the ribs. The variables in Section 2.1.4 are illustrated in Fig. R2.1. For calculating the average rib width, the width at 0.75 of the rib height, as illustrated in Fig. R2.1, was chosen in the recommended supplement to ASTM A615/A615M due to possible presence of rounded corners on the ribs. The bar reinforcements used in the experimental studies (Darwin and Graham 1993; Darwin et al. 1996a; Zuo and Darwin 1998) leading up to this guide were either machined or special rolled; both bar types had ribs with flat outer faces. The supporting research results, therefore, are based on the actual width of the outer face.

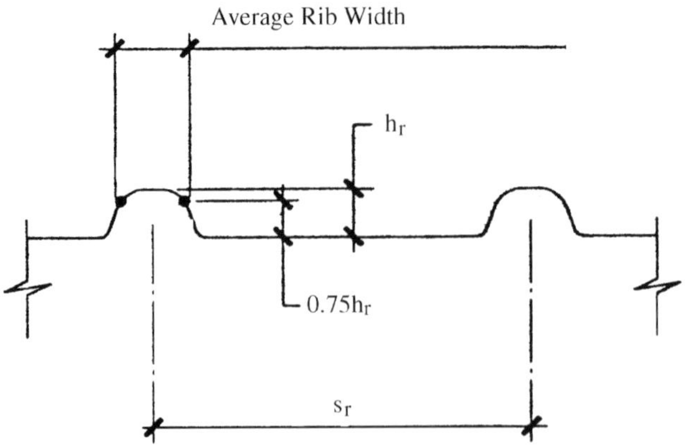

Fig. R2.1—Definition of average rib width.

Bar with X and diamond deformation patterns are excluded from this guide because their bond properties are markedly lower than bars with parallel ribs. Earlier bond strength tests on X pattern No. 6 and 11 (No. 19 and 36) epoxy-coated bars (Treece and Jirsa 1989) gave the lowest bond strengths reported in the literature, even though the bars had relative rib areas of 0.099 and 0.110, respectively. These bond values were significantly lower than values measured for epoxy-coated bars with parallel ribs and lower relative rib areas. Also, X pattern bars are not allowed in the Canadian Code (CSA 1992) because Canadian bond pullout tests on X pattern bars gave significantly lower strengths than did parallel rib bars. In addition to the bond strength issue, a National Cooperative Highway Research Program (NCHRP) study (Helgason et al. 1976) indicated that X pattern bars have lower fatigue life than bars with other types of deformation patterns.

The bamboo pattern for ribs (ribs oriented at 90 degrees to the bar axis) are also excluded by the angle restrictions adopted in this guide because of problems associated with the bending of conventionally deformed bars with this rib orientation.

No. 11 (No. 36) bars were the largest high relative rib area bars used in the experimental program that formed the basis for these provisions. Thus, these provisions are not intended to be used for No. 14 and 18 (No. 43 and 57) bars.

2.2

The design provisions in this guide shall be used only to calculate the development length or lap splice length of high relative rib area bars in reinforced concrete members made with normalweight concrete.

R2.2

The design provisions presented in this guide are intended for use in elements with normalweight concrete. The reason for this limitation in scope is that there is no research to support the validity of these design provisions in elements made with other types of concrete. According to the definition adopted in ACI 318-08, normalweight concrete typically has a density (unit weight) between 135 and 160 lb/ft^3(2155 and 2555 kg/m^3) and is normally taken as 145 to 150 lb/ft^3 (2315 to 2400 kg/m^3).

CHAPTER 3—NOTATION AND DEFINITIONS

3.1—Notation

A_r = projected rib area normal to bar reinforcement axis, in.^{2} (mm^2)

A_s = area of nonprestressed longitudinal tension reinforcement, in.^{2} (mm^2)

A_{tr} = total cross-sectional area of all transverse reinforcement that is within the spacing s that crosses the potential plane of splitting through the reinforcement being developed, in.^{2} (mm^2)

C_R = relative rib factor as defined by Eq. (4-4)

c_{bb} = clear cover of reinforcement being developed, measured to tension face of member, in. (mm)

c_{max} = maximum value of c_s or c_{bb}, in. (mm)

c_{min} = minimum value of c_s or c_{bb}, in. (mm)

c_s = minimum value of c_{si} + 0.25 in. (6 mm) or c_{so}, in. (mm); c_{si} may be used instead of c_{si} + 0.25 in. to compute c_s

c_{si} = one-half of average clear spacing between bars or lap splices in a single layer, in. (mm)

c_{so} = clear cover of reinforcement being developed or lap spliced, measured to side face of member, in. (mm)

d_b = nominal bar diameter, in. (mm)

f_c' = specified compressive strength of concrete, psi (MPa)

$f_c'^{\,1/4}$ = fourth root of f_c', expressed in psi (MPa) units

f_{ct} = average splitting tensile strength of lightweight aggregate concrete, psi (MPa)

f_y = specified yield strength of reinforcement being lap spliced or developed, psi (MPa)

K_{tr}' = transverse reinforcement index for high relative rib area bars as defined by Eq. (4-3)

ℓ_d = development length in tension of deformed bar, in. (mm)

ℓ_s = lap splice length in tension of deformed bar, in. (mm)

n = number of bars being developed or lap spliced along plane of splitting

R_r = relative rib area; ratio of projected rib area normal to bar axis to product of nominal bar perimeter and average center-to-center rib spacing

s = maximum center-to-center spacing of transverse reinforcement within ℓ_d or ℓ_s, in. (mm)

ψ_e = factor used to modify development length based on reinforcement coating; see Section 4.3.

ψ_t = factor used to modify development length based on reinforcement location; see Section 4.3.

ω_c = factor used to modify development length based on cover and/or spacing perpendicular to controlling cover and/or spacing; see Section 4.2.

R3.1—Notation

A_b = area of an individual bar, in.^{2}(mm^2)

A_r = projected rib area normal to reinforcing bar axis, in.^{2}(mm^2)

h_r = average height of deformations, in. (mm)

s_r = average center-to-center rib spacing, in. (mm)

In the notation section it is stated that, for the purpose of calculating the value of c_s, it is permitted to use c_{si} instead of c_{si} + 0.25. This is provided for simplification. Use of c_{si} instead of c_{si} + 0.25 results in a value of c_s that is less, and consequently produces a more conservative development or lap splice length.

3.2—Definitions

high relative rib area bars—deformed bar reinforcement with a relative rib area equal to 0.10 or larger.

normalweight concrete—concrete containing only aggregate that conforms to ASTM C33/C33M.

relative rib area—ratio of projected rib area normal to bar axis to product of nominal bar perimeter and average center-to-center rib spacing.

rib—transverse protrusion or deformation on a deformed bar. The term "rib" is used uniformly throughout this document.

R3.2—Definitions

The definition of relative rib area presented in Section 3.2 of this guide is based on measurements of A_r, d_b, and s_r (illustrated in Fig. R3.1). Figure R3.1 includes expressions for the approximate values of A_r and R_r.

An alternative approach to calculate the relative rib area is presented in Appendix A, "Recommended supplement to ASTM A615/A615M for high relative area bars." Equation (A-1) of Appendix A is a simplification of Eq. (4-6) of this guide which provides a close approximation of the relative rib area based on more commonly available measurements of bar deformations.

CHAPTER 4—DEVELOPMENT OF HIGH RELATIVE RIB AREA BAR REINFORCEMENT IN TENSION

4.1

Development length ℓ_d in tension shall be determined from 4.2, but ℓ_d shall not be less than the larger of 16d_b or 12 in. (300 mm).

R4.1

In bars with short development or lap splice lengths, the controlling mode of failure is pullout instead of splitting (ACI Committee 408 2003). For these cases, calculating bond strength using equations calibrated on the basis of tests with splitting failures will lead to overestimating bond strength. The minimum development length of 16d_b in Section 4.1 was established to reflect the limiting bond strength from bar pullout for No. 7 (No. 22) or larger bars. This limiting criteria was applied by Zuo and Darwin (2000) and by Darwin et al. (2005) in selecting the data set used to

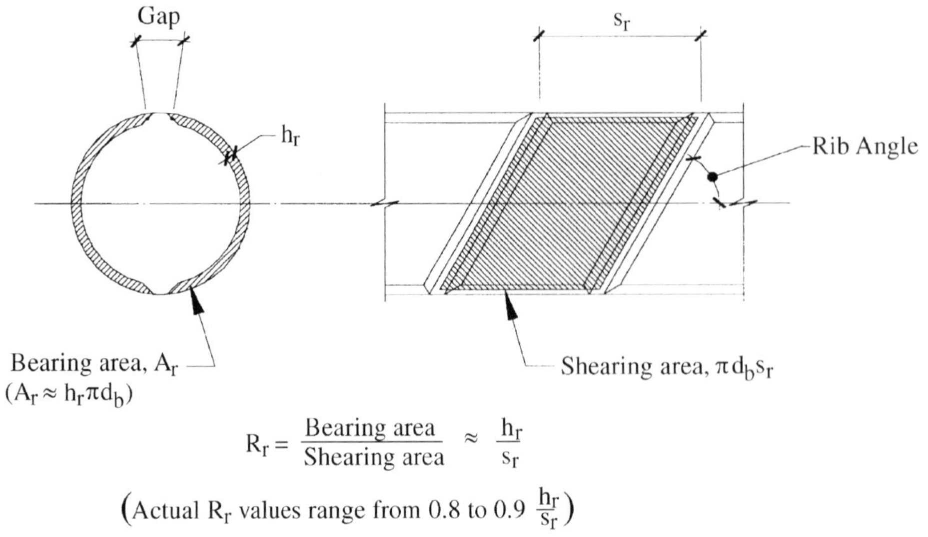

Fig. R3.1—Definition of R_r

calibrate Eq. (4-1) to (4-5). In the case of lap splices the minimum development length specified in Section 4.1 shall be disregarded in calculating the length of the lap splice and the minimum lap splice length specified in Section 5.1 shall apply.

4.2

The development length of high relative rib area bars in tension ℓ_d divided by the bar diameter d_b shall be taken as

$$\frac{\ell_d}{d_b} = \frac{(f_y/f_c'^{1/4} - 2000\omega_c)\psi_t\psi_e}{62\left(\frac{c_b\omega_c + K_{tr}'}{d_b}\right)} \quad (4\text{-}1)$$

in which the term $(c_b\omega_c + K_{tr}')/d_b$ shall not be taken greater than 4.

For calculation purposes, the value of f_c' used in these provisions shall not exceed 16,000 psi (110 MPa) for uncoated bars in normalweight concrete. For epoxy-coated bars, the value of f_c' used in these provisions shall not exceed 10,000 psi (70 MPa). The value of f_y shall not exceed 80,000 psi (550 MPa).

The variable ω_c shall be taken as 1.0 or evaluated as

$$\omega_c = 0.1\frac{c_{max}}{c_{min}} + 0.9 \le 1.25 \quad (4\text{-}2)$$

The variable K_{tr}' shall be evaluated as

$$K_{tr}' = C_R(0.72d_b + 0.28)\frac{A_{tr}\sqrt{f_c'}}{2sn} \quad (4\text{-}3)$$

where

$$C_R = 0.3 + 9.2R_r \le 1.6 \quad (4\text{-}4)$$

Table R4.1—Database size for bottom-cast uncoated bars

Bar pattern	No. of specimens: f_c' < 6000 psi (40 MPa)	No. of specimens: $f_c' \ge$ 6000 psi (40 MPa)
Conventional	207	54
High R_r	68	25

with

$$0.10 \le R_r \le 0.14 \quad (4\text{-}5)$$

$$R_r = \frac{A_r}{\pi d_b s_r} \quad (4\text{-}6)$$

Alternatively, it shall be permitted to take $K_{tr}' = 0$.

R4.2

Equation (4-1) represents the beneficial effect of the high relative rib bar reinforcement as well as the influence of other pertinent variables (Darwin and Graham 1993; Darwin et al. 1996a,b). Equation (4-1) was derived by statistical analysis of experimental data and does not represent a mechanistic model for bond behavior. Thus, it should not be extended to cases other than those explicitly covered in these provisions. The relative rib area R_r *would be fixed at a specified value for the reinforcement being used.*

Table R4.1 indicates the size of the database used in developing the provisions in Section 4.2 for high relative rib area bars in normal- and high-strength concrete. For comparison purposes, the current size of the database for conventionally deformed bars is also provided.

Figure R4.1 summarizes the development length test results for the uncoated high R_r *bars using a histogram of test/prediction ratio. A similar histogram of the test results for coated high* R_r *bars is given in Fig. R4.2.*

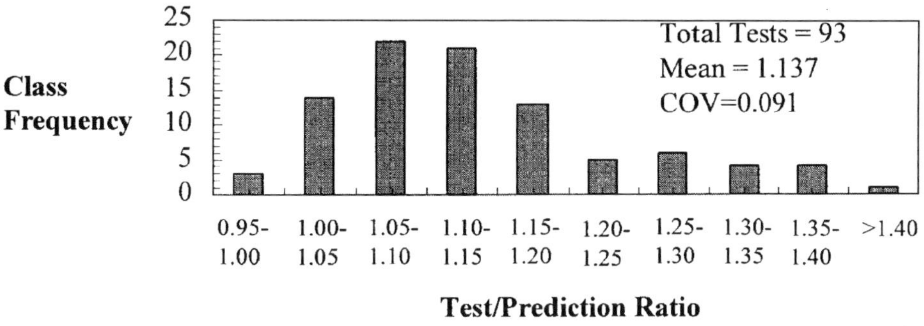

Fig. R4.1—Histogram of test/prediction ratio for all uncoated high R_r *bars (No. 5, 8, and 11 [No. 16, 25, and 36] bars).*

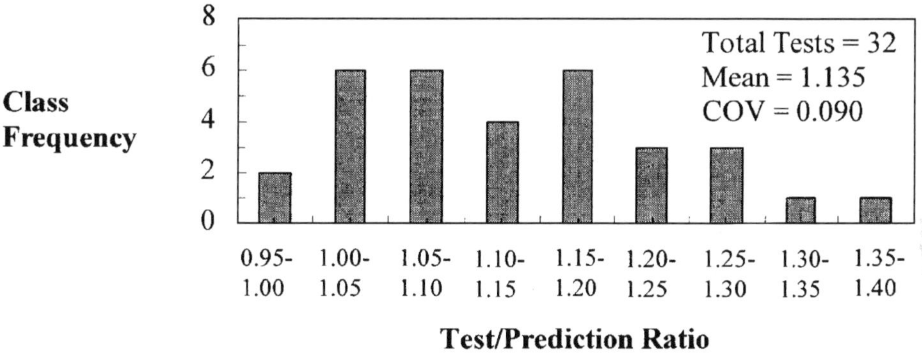

Fig. R4.2—Histogram of test/prediction ratio for all coated high R_r *bars (No. 5, 8, and 11 [No. 16, 25, and 36] bars).*

Limiting values for concrete compressive strength used in Eq. (4-2) were established due to a paucity of experimental results from specimens with concrete compressive strengths above the specified limits. The yield strength of the reinforcement is limited to 80,000 psi (550 MPa) because the maximum bar stress in tests used to calibrate the design equations was 81,000 psi (560 MPa).

The upper limit on the confinement parameter $(c_b\omega_c + K'_{tr})/d_b$ *of 4 is specified because higher values of the parameter correspond to pullout failures. No specific limit is placed on the concrete cover or the transverse reinforcement terms in the parameter.*

The K'_{tr} *parameter includes the influence of the high relative rib properties as well as the amount of transverse reinforcement confining the developing bar. The yield strength of the transverse reinforcement is not present in the* K'_{tr} *parameter because research indicates that the transverse reinforcement seldom reaches the yield value when confining the developing bar (Darwin et al. 1996b; Zuo and Darwin 2000).*

The parameter ω_c *typically reflects the benefit of wide spacing when the cover to the tension face,* c_{bb}*, dictates the value of* c_b*. It can, however, also reflect the benefit of large cover when close spacing dictates the value of* c_b*.*

Evaluations of crack width and crack spacing outside the lap splice region have indicated no measurable difference between conventional and high relative rib area bars. Compared with conventionally deformed bars, epoxy-coated high relative rib area bars typically produce fewer cracks with larger crack widths than uncoated bars.

Studies at PCA (Helgason et al. 1976) established that the principal geometric variable in the fatigue life of bar reinforcement is the ratio of the radius at the base of a deformation r *to its height* h*. The absolute values of rib height and rib spacing were not found to be critical parameters. The results of the work by Helgason et al. (1976) are incorporated in the American Association of State Highway and Transportation Officials (AASHTO) Bridge Specifications (AASHTO Subcommittee on Bridges and Structures 1996) in an expression that establishes the maximum allowable stress range as a function of* r/h *and the minimum stress. As with conventional bars, that expression should be applied when fatigue is of concern.*

High relative rib area bars have thus far exhibited no problems when subjected to standard bend tests at the producing mills or in fabrication tests in the research used to develop this guide. No modifier factors are included in this guide for bundled bars in tension. Although no testing has been conducted for high relative rib area bars in bundles, it is reasonable to assume that the length modifiers for bundled bars in Section 12.4 of ACI 318-08 are appropriate for use with high relative rib area bars. The justification for this assumption is that the factors in Section 12.4 of ACI 318-08 reflect the percentage of the area of the bars in a bundle that is not in contact with the concrete, and that percentage is independent of the relative rib area of the bars.

4.3

The factors used in the expressions for development of high relative rib area bars in tension are as follows:

ψ_t = bar location factor
Horizontal bar placed so that more than 12 in. (300 mm) of fresh concrete is cast in the member below the development length or lap splice 1.3
Other bars 1.0

ψ_e = coating factor
Epoxy-coated bars 1.2
Uncoated bars 1.0

R4.3

The presence of the higher ribs, and specifically ribs with a larger relative rib area, and the elimination of rib patterns with poor bond properties produces a beneficial effect on the bond of epoxy-coated bars (Choi et al. 1991). The epoxy coating thickness has less impact on the bearing area with high rib bars. The resulting reduced bearing stress decreases the difference between the behavior of uncoated and coated bars, which leads to using the 1.2 factor for all situations.

The use of high ribs has little effect on the ratio of the embedment length for top-cast bars to the embedment length for bottom-cast bars. With this information, it was felt there was no basis for changing the 1.3 reinforcement location factor.

These provisions differ with respect to those in ACI 318-08 in that a bar size factor ψ_s is not used to reduce the calculated development length of No. 6 (No. 19) and smaller bars. Reference ACI 408R-03 for further discussion of this issue.

4.4—Excess reinforcement

Reduction in ℓ_d by the ratio (A_s required)/(A_s provided) shall be permitted where reinforcement in a flexural member is in excess of that required by analysis, except where anchorage or development for f_y is specifically required or the reinforcement is designed under provisions of Section 21.1.1.6 of ACI 318-08.

CHAPTER 5—LAP SPLICES OF HIGH RELATIVE RIB AREA BAR REINFORCEMENT IN TENSION

5.1

Minimum length of lap for tension lap splices ℓ_s shall be as required for Class A or B splices, but not less than the larger of $16d_b$ or 12 in. (300 mm), where:

Class A splice .... $1.0\ell_d$
Class B splice...... $1.0\ \ell_d$, where ℓ_d is calculated with $\omega_c = 1.0$

where ℓ_d is the tensile development length for the specified yield strength f_y in accordance with Section 4.2 without the modification factor of Section 4.4.

R5.1

A minimum length of lap of $16d_b$ or 12 in. (300 mm) is specified in Section 5.1. This requirement is similar to that in Section 4.1 of this guide and the reasons for adopting this limit are the same as those stated in R4.1, that is, to reflect the limiting bond strength from bar pullout for No. 7 (No. 22) or larger bars.

Analyses of test data (Orangun et al. 1977; Darwin et al. 1996b; Zuo and Darwin 2000; ACI Committee 408 2003) have concluded that the lap splice length and the development length can be calculated using the same expression when conditions are the same. Consequently, Eq. (4-1) produces lap splice and development lengths that have a similar level of reliability against bond failure. For this reason, the use of a splice factor to modify the lap splice length calculated using Eq. (4-1) would have the effect of producing lap splices that have a different level of reliability against bond failure than developed bars. More specifically, the use of a splice factor greater than unity in the provisions of this guide would result in lap splice lengths significantly larger than those calculated with the provisions in the ACI 318-08 Code, in which a factor of 1.3 is used for Class B splices to encourage less congested reinforcement details and to improve reliability. Development lengths required by the provisions of this guide are, in most cases, longer than those required by the provisions in ACI 318-08. Setting ω_c = 1.0 in Eq. (4-1) produces lap splice lengths that are approximately equal to Class B lap splice lengths calculated using the provisions of ACI 318-08 (Darwin et al. 2005). For this reason, lap splice length and development length can remain the same under these provisions without any loss in reliability.

5.2

Lap splices shall be classified as Class B except as provided for in Section 5.3.

R5.2

According to the provisions in ACI 318-08, lap splices must be classified as Class B if "A_s provided" is less than twice "A_s required," or when more than 50% of the bars are spliced at one location. For this reason, Class B lap splices are used in the majority of cases under the provisions of ACI 318-08. A similar approach is adopted in this provision. Lap splices are classified as Class B when K'_{tr}/d_b is less than 1.0 or more than half of the reinforcement is spliced at one location. A classification similar to that used in ACI 318-08 was adopted in this provision to maintain a level of reliability similar to that associated with the provisions in ACI 318-08. A value of ω_c = 1.0 is used in calculating ℓ_d for Class B lap splices for the same reasons stated in Section R5.1.

5.3

Class A splices shall be allowed when the criteria of Sections 5.3.1, 5.3.2, or 5.3.3 are met.

5.3.1 Where the lap splice length is confined with transverse reinforcement at two or more locations with a spacing not greater than 12 in. (300 mm), providing a K'_{tr}/d_b of at least 1.0.

5.3.2 Where no more than one-half of the total reinforcement is spliced within the required lap length.

5.3.3 Where horizontal bar reinforcement in walls are not being used as in-plane flexural or tensile members.

R5.3

The development length computed using Eq. (4-1) provides the proper reliability provided that the bar location factor of 1.3 is employed, even if all bars are spliced at the same location (Darwin et al. 2005). If a wall is used as a structural member with prescribed flexural or tensile loads, the lap splices should be proportioned as Class B lap splices if the provisions of Section 5.3.2 are not met.

5.4

Tension lap splices shall not be used for high relative rib area bars larger than No. 11 (No. 36).

R5.4

This restriction is identical to Section 12.14.2.1 in ACI 318-08 for tension lap splices of conventionally deformed bar reinforcement.

CHAPTER 6—REFERENCES

6.1—Referenced standards and reports

The standards and reports listed below were the latest editions at the time this document was prepared. Because these documents are revised frequently, the reader is advised to contact the proper sponsoring group if it is desired to refer to the latest version.

ASTM International

A615/A615M	Standard Specification for Deformed and Plain Carbon-Steel Bars for Concrete Reinforcement
A706/A706M	Standard Specification for Low-Alloy Steel Deformed and Plain Bars for Concrete Reinforcement
C33/C33M	Standard Specification for Concrete Aggregates

The previous publications can be obtained from the following organization:

ASTM International
100 Barr Harbor Dr.
West Conshohocken, PA 19428
www.astm.org

6.2—Cited references

AASHTO Subcommittee on Bridges and Structures, 1996, “Standard Specifications for Highway Bridges,” sixteenth edition, American Association of State Highway and Transportation Officials, Washington, DC, 676 pp.

ACI Committee 318, 2008, “Building Code Requirements for Structural Concrete (ACI 318-08) and Commentary,” American Concrete Institute, Farmington Hills, MI, 473 pp.

ACI Committee 408, 2001, “Splice and Development Length of High Relative Rib Area Reinforcing Bars in Tension and Commentary (408.3-01/408.3R-01),” American Concrete Institute, Farmington Hills, MI, 6 pp.

ACI Committee 408, 2003, “Bond and Development of Straight Reinforcing Bars in Tension (ACI 408R-03),” American Concrete Institute, Farmington Hills, MI, 49 pp.

Choi, O. C.; Hadje-Ghaffari, H.; Darwin, D.; and McCabe, S., 1991, “Bond of Epoxy-Coated Reinforcement: Bar Parameters,” *ACI Materials Journal*, V. 88, No. 2, Mar.-Apr., pp. 207-217.

CSA, 1992, “Billet-Steel Bars for Concrete Reinforcement (CAN/CSA-G30.18-M92),” Canadian Standards Association, Rexdale (Toronto), ON, Canada, 18 pp.

Darwin, D., and Graham, E. K., 1993, “Effect of Deformation Height and Spacing on Bond Strength of Reinforcing Bars,” *ACI Structural Journal*, V. 90, No. 6, Nov.-Dec., pp. 646-657.

Darwin, D.; Lutz, L. A.; and Zuo, J., 2005, “Recommended Provisions and Commentary on Development and Lap Splice Lengths for Deformed Reinforcing Bars in Tension,” *ACI Structural Journal,* V. 102, No. 6, Nov.-Dec., pp. 892-900.

Darwin, D.; Tholen, M. L.; Idun, E. K.; and Zuo, J., 1996a, “Splice Strength of High Relative Rib Area Reinforcing Bars,” *ACI Structural Journal*, V. 93, No. 1, Jan.-Feb., pp. 95-107.

Darwin, D.; Zuo, J.; Tholen, M. L.; and Idun, E. K., 1996b, “Development Length Criteria for Conventional and High Relative Area Reinforcing Bars,” *ACI Structural Journal*, V. 93, No. 3, May-June, pp. 347-359.

Helgason, T.; Hanson, J. M.; Somes, N. F.; Corely, W. G.; and Hognestad, E., 1976, “Fatigue Strength of High Yield Reinforcing Bars,” *NCHRP Report* No. 164, Transportation Research Board, Washington, DC, 90 pp.

Orangun, C. O.; Jirsa, J. O.; and Breen, J. E., 1977, “Re-Evaluation of Test Data on Development Length and Splices,” ACI JOURNAL, *Proceedings* V. 74, No. 3, Mar., pp. 114-122.

Treece, R. A., and Jirsa, J. O., 1989, “Bond Strength of Epoxy-Coated Reinforcing Bars,” *ACI Materials Journal*, V. 86, No. 2. Mar.-Apr., pp. 167-184.

Zuo, J., and Darwin, D., 1998, “Bond Strength of High Relative Rib Area Reinforcing Bars,” *SM Report* No. 46, University of Kansas Center for Research, Inc., Lawrence, KS, Jan., 350 pp.

Zuo, J., and Darwin, D., 2000, “Splice Strength of Conventional and High Relative Rib Area Bars in Normal- and High-Strength Concrete,” *ACI Structural Journal*, V. 97, No. 4, July-Aug., pp. 630-641.

APPENDIX A—RECOMMENDED SUPPLEMENT TO ASTM A615/A615M FOR HIGH RELATIVE RIB AREA BARS

This appendix is intended for use in the production of high relative rib area bars according to ASTM standard A615/A615M. The appendix provides requirements additional to those in ASTM A615/A615M that are specific to high relative rib area bars.

The following supplementary requirements shall apply only when specified in the purchase order or contract.

A.1—Requirements for deformations

A.1.1 The deformations on high relative rib area bars shall meet all requirements in Section 7 of ASTM A615/A615M.

A.1.2 In addition, the relative rib area (as defined in Section A.2.1) shall meet the requirements and limitations of Section 2.1 of ACI 408.3R.

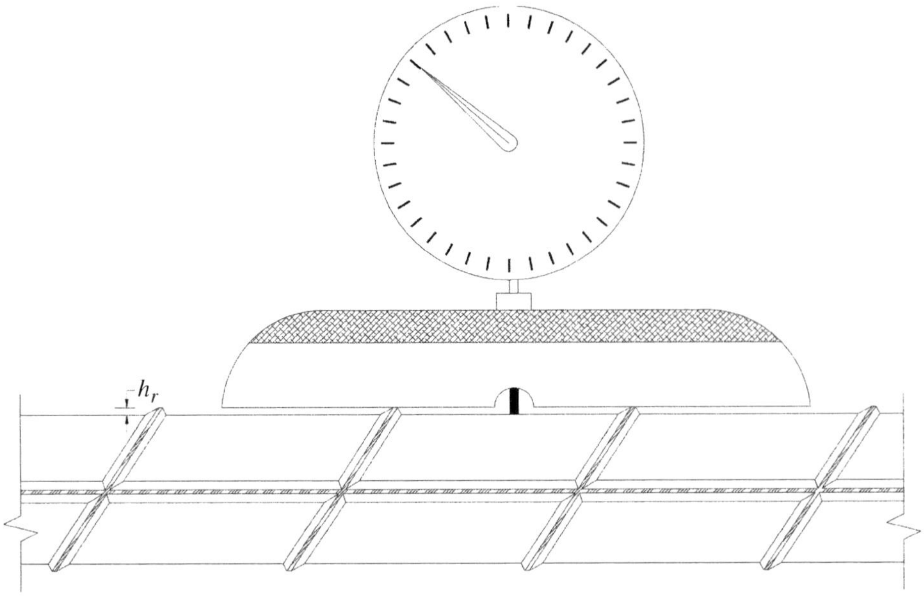

Fig. A.1—Measurement of h_r *with knife-edge depth gauge spanning two ribs.*

A.2—Relative rib area

A.2.1 The relative rib area R_r is defined in Section 3.1 of ACI 408.3R. The value of R_r should be specified by the purchaser.

A.2.2 For bars that meet the requirements of Section A.1, it shall be permitted to calculate R_r using Eq. (A-1).

$$R_r = \frac{h_r}{s_r}\left(1 - \frac{\Sigma gaps}{p_{bar}}\right) \quad \text{(A-1)}$$

where

h_r = average height of deformations (measured according to Section A.2.3), in. (mm)

s_r = average spacing of deformations, in. (mm)

$\Sigma gaps$= sum of the gaps between ends of deformations as defined in Section 7.4, plus the width of any continuous longitudinal lines used to represent the grade of the bar, multiplied by the ratio of the height of the line to hr, in. (mm)

p_{bar} = nominal perimeter to the bar (Table 1), in. (mm)

A.2.3 The average height of deformations shall be determined from measurements made on not less than two typical deformations on each side of the bar. Determinations shall be based on five measurements per deformation: one at the center of the overall length, two at the ends of the overall length, and two located halfway between the center and the ends. The measurements at the ends of the overall length shall be averaged to obtain a single value, and that value shall be combined with the other three measurements to obtain the average rib height h_r. Deformation height shall be measured using a depth gauge with a knife-edge support that spans not more than two adjacent ribs. Alternatively, it shall be permitted to use a knife edge that spans more than two adjacent ribs, in which case the average rib height shall be multiplied by 0.95 before use in Eq. (A-1).

A.2.4 The average rib width shall be determined from measurements made on not less than two typical deformations on each side of the bar. Determinations shall be based on three measurements per deformation: one at the center and one at each end. The measurements shall be taken at three-quarters of the rib height at each location. The average of the measurements at the ends shall be averaged with the center measurement to obtain a value for the one side of the deformation.

Note A.2—A knife edge is required to allow measurements to be made at the ends of the overall length of deformations, usually adjacent to a longitudinal rib. The calculation of h_r *is based on a knife edge that spans only two ribs (Fig A.1) because measurements made with a longer knife edge result in unrealistically high average rib heights and an overestimate of the relative rib area for some bars. When a longer knife edge is used,* h_r *shall be reduced by 5%.*

A.3—Type of steel

A.3.1 All bars produced to these supplementary requirements shall be identified by the letter *H*, in place of the letter *S* specified in 19.3.3, indicating that the bar was produced to meet both the specification and these supplementary requirements.

A.4—Referenced documents

A4.1 *Other standard*—"Splice and Development Length of High Relative Rib Area Reinforcing Bars in Tension and Commentary (ACI 408.3R)." (available from American Concrete Institute, 38800 Country Club Drive, Farmington Hills, MI 48333)

Guide for Shear Reinforcement for Slabs

Reported by Joint ACI-ASCE Committee 421

Mustafa A. Mahamid, Chair

Amy M. Reineke Trygestad, Secretary

Simon J. Brown
Pinaki R. Chakrabarti
William L. Gamble
Ramez Botros Gayed
Amin Ghali
Hershell Gill
Neil L. Hammill
Mahmoud E. Kamara
Theodor Krauthammer
Faris A. Malhas
Marvin D. Max
Sami Hanna Megally
Miguel C. Mota
Edward G. Nawy
Daniel Reider
Aly Said
Eugenio M. Santiago
Myoungsu Shin
Matthew Smith
Ying Tian
Stanley C. Woodson

Consulting Members

Eugene Paul Holland
J. Leroy Hulsey
James S. Lai
Sidney H. Simmonds

Special acknowledgments to Simon J. Brown, Ramez Botros Gayed, Amin Ghali, Neil L. Hammill, Mahmoud E. Kamara, James S. Lai, Mustafa A. Mahamid, and Amy M. Reineke Trygestad for their contributions to this guide.

Tests have established that punching shear in slabs can be effectively resisted by reinforcement consisting of vertical rods mechanically anchored at top and bottom of slabs. ACI 318 sets out the principles of design for slab shear reinforcement and makes specific reference to stirrups and headed studs. This guide reviews available types and makes recommendations for their design. The application of these recommendations is illustrated through numerical examples.

Keywords: column-slab connection; concrete flat plate; headed shear studs; moment transfer; prestressed concrete; punching shear; shear heads; shear stresses; slabs; two-way slabs.

CONTENTS

ACI 421.1R-20 supersedes ACI 421.1R-08 and was adopted and published January 2020.

CHAPTER 1—INTRODUCTION AND SCOPE

1.1—Introduction

In flat-plate floors, slab-column connections are subjected to high shear stresses produced by the transfer of the internal forces between the columns and the slabs. ACI 318-14 Section 8.7.6.1 allows the use of shear reinforcement for slabs and footings in the form of bars, as in the vertical legs of stirrups. ACI 318 emphasizes the importance of anchorage details and accurate placement of the shear reinforcement, especially in thin slabs. ACI 318-14 Section 8.7.7.1 permits headed shear stud reinforcement conforming to ASTM A1044/A1044M. A general procedure for evaluation of the punching shear strength of slab-column connections is given in ACI 318-14 Section 8.5.3.

Shear reinforcement consisting of vertical rods (studs) or the equivalent, mechanically anchored at each end, can be used. In this guide, all types of mechanically anchored shear reinforcement are referred to as shear stud or stud. To be fully effective, the anchorage should be capable of developing the specified yield strength of the studs. The mechanical anchorage can be obtained by heads or strips connected to the studs by welding. The heads can also be formed by forging the stud ends.

1.2—Scope

Recommendations in this guide are for the design of shear reinforcement in slabs. The design is in accordance with ACI 318. Numerical design examples are included.

1.3—Evolution of the practice

Extensive tests (Dilger and Ghali 1981; Andrä 1981a,b; Van der Voet et al. 1982; Mokhtar et al. 1985; Elgabry and Ghali 1987; Mortin and Ghali 1991; Dilger and Shatila

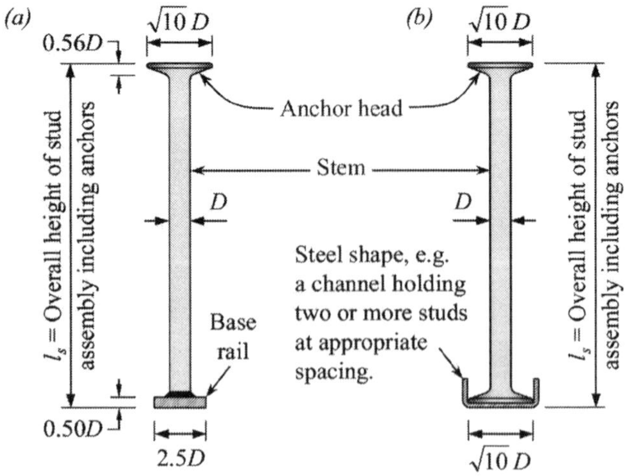

Fig. 1.3a—Stud assemblies conforming to ASTM A1044/A1044M: (a) single-headed studs welded to a base rail; and (b) double-headed studs crimped into a steel channel.

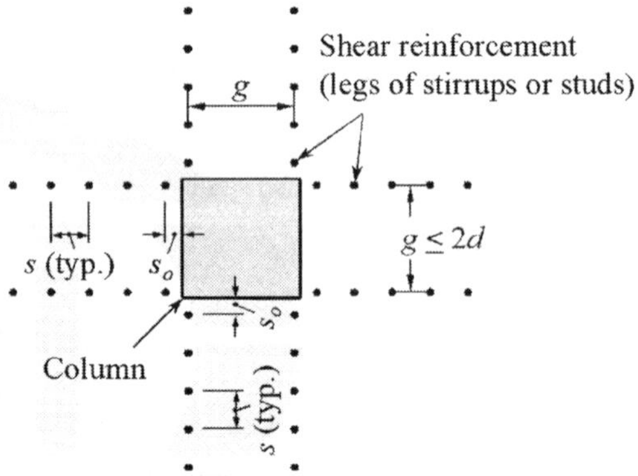

Fig. 1.3b—Top view of flat plate showing arrangement of shear reinforcement in vicinity of interior column.

1989; Cao 1993; Brown and Dilger 1994; Megally 1998; Birkle 2004; Ritchie and Ghali 2005; Gayed and Ghali 2006) have confirmed the effectiveness of mechanically-anchored shear reinforcement, such as shown in Fig. 1.3a, in increasing the strength and ductility of slab-column connections subjected to concentric punching or punching combined with moment. Stud assemblies consisting of either a single-head stud attached to a steel base rail by welding (Fig. 1.3a(a)) or double-headed studs mechanically crimped into a nonstructural steel channel (Fig. 1.3a(b)) are specified in ASTM A1044/A1044M. Figure 1.3b is a top view of a slab that shows a typical arrangement of shear reinforcement (stirrup legs or studs) in the vicinity of an interior column. ACI 318 requires that the spacing g between adjacent stirrup legs or studs, measured on the first peripheral line of shear reinforcement, be equal to or less than $2d$. Recommendations for the distances s_o and s are given in Chapter 4.

CHAPTER 2—NOTATION AND DEFINITIONS

2.1—Notation

A_c = area of concrete of assumed critical section, in.^{2} (mm^2)

A_v = cross-sectional area of shear reinforcement on one peripheral line parallel to perimeter of column section, in.^{2} (mm^2)
b_o = length of perimeter of critical section, in. (mm)
c_b, c_t = clear concrete cover of reinforcement to bottom and top slab surfaces, respectively, in. (mm)
c_x, c_y = size of rectangular column measured in two orthogonal span directions, x and y, respectively, in. (mm)
D = stud diameter, in. (mm)
d = effective depth of slab; average of distances from extreme compression fiber to centroids of tension reinforcements running in two orthogonal directions, in. (mm)
d_b = nominal diameter of flexural reinforcing bars, in. (mm)
f_c' = specified compressive strength of concrete, psi (MPa)
f_{ct} = average splitting tensile strength of lightweight aggregate concrete, psi (MPa)
f_{pc} = average value of compressive stress in concrete in two directions (after allowance for all prestress losses) at centroid of cross section, psi (MPa)
f_{yt} = specified yield strength of shear reinforcement, psi (MPa)
g = distance between adjacent stirrup legs or studs, measured in a direction parallel to a column face, in. (mm)
h = overall thickness of slab, in. (mm)
J_c = property of assumed critical section, defined by ACI 318 as analogous to polar moment of inertia, in.^{4} (mm^4)
J_{cx}, J_{cy} = property of assumed critical section of any shape, equal to d multiplied by second moment of perimeter about x or y axis, respectively (Appendix B), in.^{4} (mm^4)
$J_{c\overline{xy}}$ = d times product of inertia of assumed shear critical section about nonprincipal axes $\overline{x}$ and $\overline{y}$, in.^{4} (mm^4)
L = length of segment of assumed critical section, in. (mm)
l_s = overall specified height of headed stud assembly including anchors, in. (mm)
l_x, l_y = projections of assumed critical section on principal axes x and y, in. (mm)
l_{x1}, l_{y1} = length of sides in x- and y-directions of critical section at d/2 from column face, in. (mm)
l_{x2}, l_{y2} = length of sides in x- and y-directions of critical section at d/2 outside outermost legs of shear reinforcement, in. (mm)
$M_{sc\text{-}x}$, $M_{sc\text{-}y}$ = factored slab moments resisted by supporting column about centroidal principal axes x and y of assumed critical section, lb-in. (N-m)
$M_{sc-\overline{x}}$, $M_{sc-\overline{y}}$ = factored slab moments resisted by column about the centroidal nonprincipal $\overline{x}$ or $\overline{y}$ axis, lb-in. (N-m)
M_{scOx}, M_{scOy} = factored slab moment resisted by column about x or y axis through the column's centroid O, lb-in. (N-m)
n = number of studs or stirrup legs per line running in x- or y-direction
s = spacing between peripheral lines of shear reinforcement, in. (mm)
s_o = spacing between first peripheral line of shear reinforcement and column face, in. (mm)
V_p = vertical component of all effective prestress forces crossing the critical section, lb (N)
V_u = factored shear force, lb (N)
v_c = nominal shear strength provided by concrete in presence of shear reinforcement, psi (MPa)
v_n = nominal shear strength at critical section, psi (MPa)
v_s = nominal shear strength provided by shear reinforcement, psi (MPa)
v_u = maximum shear stress due to factored forces, psi (MPa)
w_{sd} = super-imposed dead load per unit area, lb/ft^2 (kN/m^2)
x, y = coordinates of point on perimeter of shear critical section with respect to centroidal principal axes x and y, in. (mm)
$\overline{x}$, $\overline{y}$ = coordinates of point on perimeter of shear critical section with respect to centroidal nonprincipal axes $\overline{x}$ and $\overline{y}$, in. (mm)
α_s = dimensionless coefficient equal to 40, 30, and 20 for interior, edge, and corner columns, respectively
β = ratio of long side to short side of column cross section
γ_{conc} = density of concrete, lb/ft^3 (kN/m^3)
γ_{vx}, γ_{vy} = factor used to determine moment about the axes x and y between slab and column that is transferred by shear stress at assumed critical section
λ = modification factor reflecting the reduced mechanical properties of lightweight concrete, all relative to normalweight concrete of the same compressive strength
ϕ = strength reduction factor = 0.75
κ = fraction of service dead load to be balanced by effective prestressing

2.2—Definitions

Please refer to the latest version of "ACI Concrete Terminology" for a comprehensive list of definitions. Definitions provided herein complement that resource.

headed shear stud reinforcement—reinforcement conforming to ASTM A1044/A1044M and composed of vertical rods anchored mechanically near the bottom and top surfaces of the slab.

shear cap—thickened portion of the slab around the column with plan dimensions not conforming with the ACI 318 requirements for drop panels.

shear critical section—cross section, having depth d and perpendicular to the plane of the slab, where shear stresses should be evaluated; two shear critical sections should be considered: (1) at d/2 from column periphery, and (2) at d/2 from the outermost peripheral line of shear reinforcement (if provided).

slab moment resisted by supporting column—sum of moments at the ends of the column above and below a slab-column joint.

CHAPTER 3—ROLE OF SHEAR REINFORCEMENT

Shear reinforcement is required to intercept shear cracks and prevent them from widening. The intersection of shear reinforcement and cracks can be anywhere over the height of the shear reinforcement. The strain in the shear reinforcement is highest at that intersection.

Effective anchorage is essential, and its location should be as close as possible to the structural member's outer surfaces. This means that the vertical part of the shear reinforcement should be as tall as possible to avoid the possibility of cracks passing above or below it. When the shear reinforcement is not as tall as possible, it may not intercept all inclined shear cracks. Anchorage of shear reinforcement in slabs is achieved by mechanical ends (heads), bends, and hooks.

CHAPTER 4—PUNCHING SHEAR DESIGN EQUATIONS

4.1—Strength requirement

This chapter presents the design procedure of ACI 318 when stirrups or headed studs are required in the slab in the vicinity of a column transferring moment and shear. The equations of 4.3.2 and 4.3.3 apply when stirrups and headed studs are used, respectively.

Design of critical slab sections perpendicular to the plane of a slab should be based upon

$$v_u \leq \phi v_n \quad (4.1)$$

in which v_u (in psi or MPa) is the shear stress in the critical section caused by the transfer, between the slab and the column, of factored shearing force or factored shearing force combined with moment; v_n is the nominal shear strength (psi or MPa); and ϕ is the strength reduction factor.

Equation (4.1) should be satisfied at a critical section perpendicular to the plane of the slab at a distance $d/2$ from the column perimeter and located so that its perimeter b_o is minimum (Fig. 4.1(a)). It should also be satisfied at a critical section at $d/2$ from the outermost peripheral line of the shear reinforcement (Fig. 4.1(b)), where d is the average of distances from extreme compression fiber to the centroids of the tension reinforcements running in two orthogonal directions. Figure 4.1(a) indicates the positive directions of the internal force V_u and moments $M_{sc\text{-}x}$, and $M_{sc\text{-}y}$ that the column exerts on the slab.

4.2—Calculation of factored shear stress V_u

ACI 318 requires that the shear stress resulting from moment transfer by eccentricity of shear be assumed to vary linearly about the centroid of the shear critical section; the code's equation was developed considering uniaxial moment transfer between the slab and its supporting columns. However, this guide recommends the use of Eq. (4.2a) that accounts for the general case of biaxial moment transfer. The shear stress distribution, calculated by Eq. (4.2a), satisfies the linear variation requirement of the code. The maximum factored shear stress v_u at a critical section produced by the

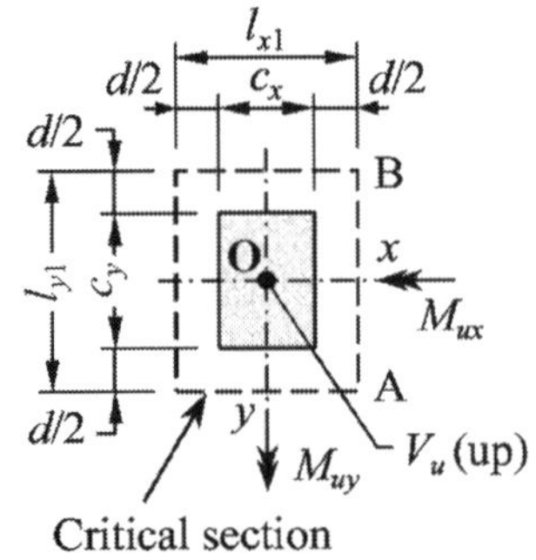

***(a)** At d/2 from column face*

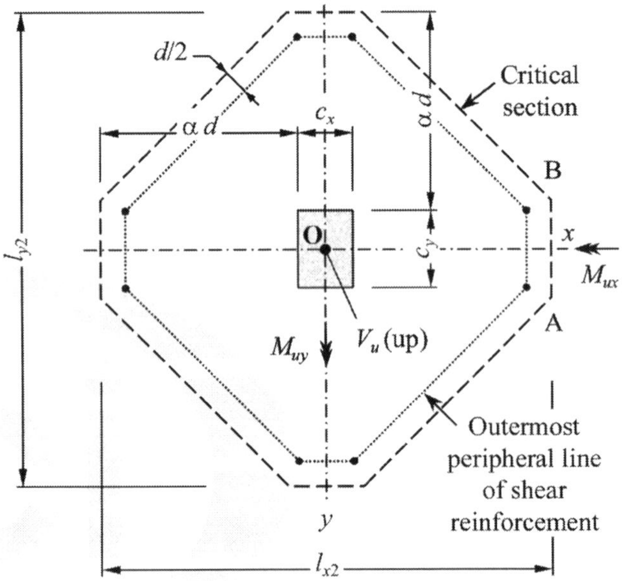

***(b)** At d/2 from outermost peripheral line of shear reinforcement*

Fig. 4.1—Critical sections for shear in slab in vicinity of interior column. Positive directions of V_u, M_{ux}, *and* M_{uy} *are indicated in (a).*

combination of factored shear force V_u and moments $M_{sc\text{-}x}$ and $M_{sc\text{-}y}$ is

$$v_u = \frac{V_u}{A_c} + \frac{\gamma_{vy} M_{sc-y}}{J_{cy}} x + \frac{\gamma_{vx} M_{sc-x}}{J_{cx}} y \quad (4.2a)$$

The coefficients γ_{vx} and γ_{vy} are given by

$$\left.\begin{aligned} \gamma_{vx} &= 1 - \frac{1}{1 + \frac{2}{3}\sqrt{l_{y1}/l_{x1}}} \\ \gamma_{vy} &= 1 - \frac{1}{1 + \frac{2}{3}\sqrt{l_{x1}/l_{y1}}} \end{aligned}\right\} \quad (4.2b)$$

where l_{x1} and l_{y1} are lengths of the sides in the x- and y-directions of a rectangular critical section at $d/2$ from the column face (Fig. 4.1(a)). For a shear critical section outside the shear-

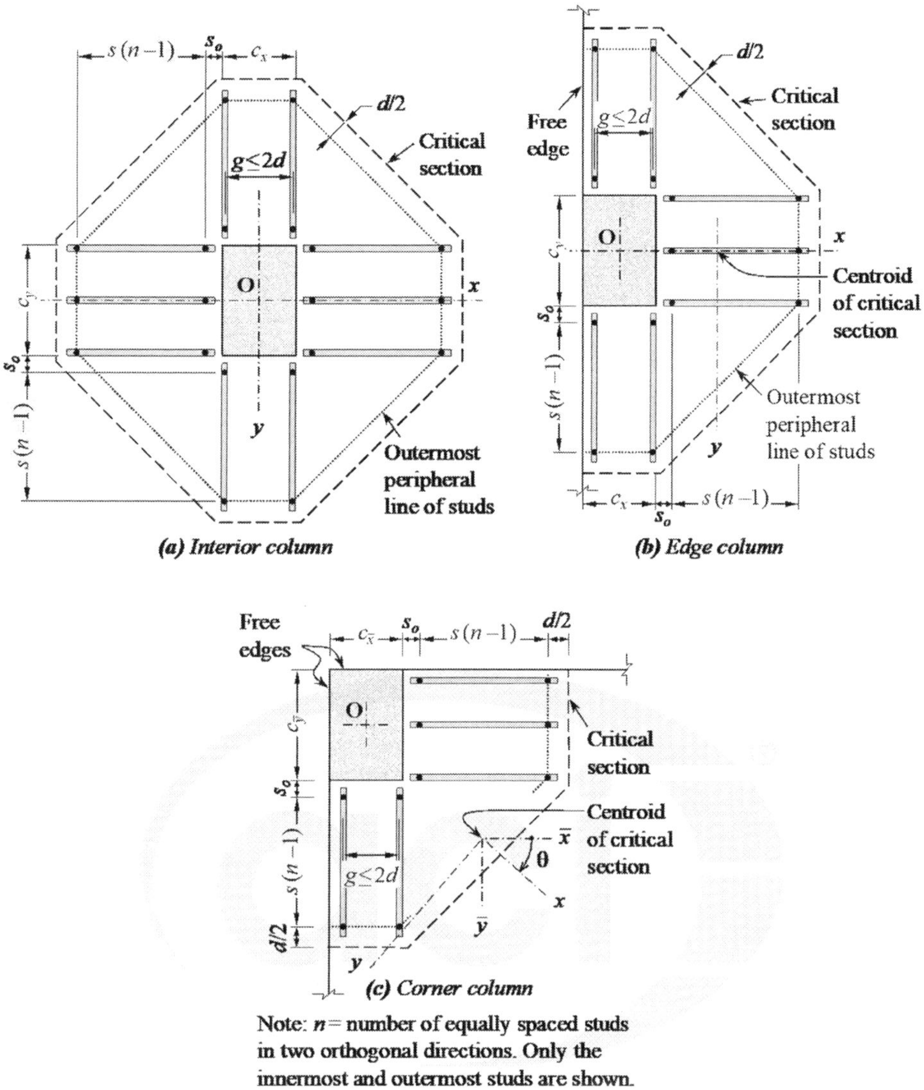

Fig. 4.2—Typical arrangement of shear studs and critical sections outside shear-reinforced zone.

reinforced zone of an interior column (Fig. 4.1(b)), Eq. (4.2b) can be used with l_{x1} and l_{y1} replaced by l_{x2} and l_{y2}, respectively. Appendix B gives equations for J_{cx}, J_{cy}, γ_{vx}, and γ_{vy} for a shear critical section of any shape. For a shear critical section in the shape of a closed rectangle, the shear stress due to V_u combined with $M_{sc\text{-}y}$, ACI 318 gives Eq. (4.2a) with $M_{sc\text{-}x} = 0$ and J_{cy} replaced by J_c, which is defined as property of assumed critical section analogous to polar moment of inertia. For the closed rectangle in Fig. 4.1(a), ACI 318 gives

$$J_c = d\left[\frac{l_{x1}^{\ 3}}{6}+\frac{l_{y1}l_{x1}^{\ 2}}{2}\right]+\frac{l_{x1}d^3}{6} \tag{4.2c}$$

The first term on the right-hand side of this equation is equal to J_{cy}; the ratio of the second term to the first is commonly less than 3 percent. The value of v_u obtained by the use of J_{cy} in Eq. (4.2a) differs on the safe side from the value obtained with J_c.

When the centroid of the shear critical section does not coincide with O, the centroid of the column (Fig. 4.2(b) and (c)), the slab moment $M_{sc\text{-}x}$ or $M_{sc\text{-}y}$ about x- or y-axis through the centroid of shear critical section is related to the slab moment $M_{sc\text{-}Ox}$ or $M_{sc\text{-}Oy}$ about the x- or y-axis through O by

$$M_{sc\text{-}x} = M_{sc\text{-}Ox} + V_u y_O;\ M_{sc\text{-}y} = M_{sc\text{-}Oy} + V_u x_O \tag{4.2d}$$

where (x_O, y_O) are the coordinates of O with respect to the centroid of the shear critical section along the centroidal principal x- and y-axes.

For the shear critical section in Fig. 4.2(c), the moments about the centroidal nonprincipal axes $\bar{x}$ and $\bar{y}$ ($M_{sc-\bar{x}}$ and $M_{sc-\bar{y}}$) are equivalent to the moments about the x and y axes ($M_{sc\text{-}x}$ and $M_{sc\text{-}y}$) that are given by Eq. (4.2e).

$$\begin{aligned} M_{sc-x} &= M_{sc-\bar{x}}\cos\theta - M_{sc-\bar{y}}\sin\theta; \\ M_{sc-y} &= M_{sc-\bar{x}}\sin\theta + M_{sc-\bar{y}}\cos\theta \end{aligned} \tag{4.2e}$$

where θ is the angle of rotation of the axes $\bar{x}$ and $\bar{y}$ to coincide with the principal axes.

4.3—Calculation of shear strength v_n

Whenever the specified compressive strength of concrete f_c' is used in Eq. (4.3.1a), (4.3.1b), (4.3.1c), (4.3.1d), and (4.3.2b), its value must be in pounds per square inch; when f_c' is in MPa, the SI unit equations are used (ACI 318-14 Section 22.6). For prestressed slabs, refer to Chapter 5.

4.3.1 *Shear strength without shear reinforcement*—For nonprestressed slabs, the shear strength of concrete at a critical section at $d/2$ from column face, where shear reinforcement is not provided, should be the smallest of

$$v_n = \left(2+\frac{4}{\beta}\right)\lambda\sqrt{f_c'} \quad \text{(in.-lb units)}$$
$$v_n = 0.17\left(1+\frac{2}{\beta}\right)\lambda\sqrt{f_c'} \quad \text{(SI units)} \qquad (4.3.1a)$$

where β is the ratio of long side to short side of the column cross section; λ is a reduction factor for light-weight concretes as defined in ACI 318-14 Section 19.2.4.2.

$$v_n = \left(\frac{\alpha_s d}{b_o}+2\right)\lambda\sqrt{f_c'} \quad \text{(in.-lb units)}$$
$$v_n = 0.083\left(2+\frac{\alpha_s d}{b_o}\right)\lambda\sqrt{f_c'} \quad \text{(SI units)} \qquad (4.3.1b)$$

where α_s is 40 for interior columns, 30 for edge columns, 20 for corner columns, and

$$v_n = 4\lambda\sqrt{f_c'} \quad \text{(in.-lb units)}$$
$$v_n = 0.33\lambda\sqrt{f_c'} \quad \text{(SI units)} \qquad (4.3.1c)$$

At a critical section outside the shear-reinforced zone,

$$v_n = 2\lambda\sqrt{f_c'} \quad \text{(in.-lb units)}$$
$$v_n = 0.17\lambda\sqrt{f_c'} \quad \text{(SI units)} \qquad (4.3.1d)$$

Equation (4.1) should be checked first at a critical section at $d/2$ from the column face (Fig. 4.1(a)). If Eq. (4.1) is not satisfied, shear reinforcement or some other change is needed.

4.3.2 *Shear strength with stirrups*—ACI 318-14 Section 22.6.7.1 permits the use of stirrups as shear reinforcement when the slab effective depth $d \geq 6$ in. (152 mm) but not less than 16 times the diameter of the stirrups. When stirrup shear reinforcement is used, ACI 318 requires that the maximum factored shear stress at $d/2$ from column face satisfy: $v_u \leq 6\sqrt{f_c'}$ (in.-lb units) ($0.5\sqrt{f_c'}$ [SI units]). The shear strength at a critical section within the shear-reinforced zone should be computed by

$$v_n = v_c + v_s \qquad (4.3.2a)$$

in which

$$v_c = 2\lambda\sqrt{f_c'} \quad \text{(in.-lb units)}$$
$$v_c = 0.17\lambda\sqrt{f_c'} \quad \text{(SI units)} \qquad (4.3.2b)$$

and

$$v_s = \frac{A_v f_{yt}}{b_o s} \qquad (4.3.2c)$$

where A_v is the cross-sectional area of the shear reinforcement legs on one peripheral line parallel to the perimeter of the column section, and s is the spacing between peripheral lines of shear reinforcement.

The upper limits, permitted by ACI 318, of s_o and the spacing s between the peripheral lines are

$$s_o \leq 0.5d \qquad (4.3.2d)$$

$$s \leq 0.5d \qquad (4.3.2e)$$

where s_o is the distance between the first peripheral line of shear reinforcement and the column face (Fig. 4.2). The upper limit of s_o is intended to eliminate the possibility of shear failure between the column face and the innermost peripheral line of shear reinforcement. Similarly, the upper limit of s is to avoid failure between consecutive peripheral lines of stirrups. A line of stirrups too close to the column can be ineffective in intercepting shear cracks; thus, s_o should not be smaller than $0.35d$.

The shear reinforcement should extend away from the column face so that the shear stress v_u at a critical section at $d/2$ from outermost peripheral line of shear reinforcement (Fig. 4.1(b) and 4.2) does not exceed ϕv_n, where v_n is calculated using Eq. (4.3.1d).

4.3.3 *Shear strength with studs*—ACI 318-14 Section 25.7.1.1 requires that:

> Stirrups shall extend as close to the compression and tension surfaces of the member as cover requirements and proximity of other reinforcement permits and shall be anchored at both ends. Where used as shear reinforcement, stirrups shall extend a distance ***d*** from extreme compression fiber.

Test results (Dilger and Ghali 1981; Andrä 1981a,b; Van der Voet et al. 1982; Mokhtar et al. 1985; Elgabry and Ghali 1987; Mortin and Ghali 1991; Dilger and Shatila 1989; Cao 1993; Brown and Dilger 1994; Megally 1998; Birkle 2004; Ritchie and Ghali 2005; Gayed and Ghali 2006) showed that studs, with anchor heads of area equal to 10 times the cross-sectional area of the shank, clearly satisfied the anchorage requirement. Furthermore, the use of shear studs, shown in Fig. 1.3a, demonstrated a higher shear capacity than that obtained with stirrups. Other researchers (Gayed and Ghali 2004; McLean et al. 1990; Muller et al. 1984; Ghali et al. 1974) successfully applied other configurations. Based on these results, ACI 318 permits the values given as follows when the shear reinforcement is composed of headed studs with mechanical anchorage capable of developing the yield strength of the rod.

The nominal shear strength provided by the concrete in the presence of headed shear studs, using Eq. (4.3.2a), is taken as

$$v_c = 3\lambda\sqrt{f_c'} \quad \text{(in.-lb units)} \qquad v_c = 0.25\lambda\sqrt{f_c'} \quad \text{(SI units)} \qquad (4.3.3a)$$

instead of $2\lambda\sqrt{f_c'}$ (in.-lb units) ($0.17\lambda\sqrt{f_c'}$ [SI units]). Discussion on the design value of v_c is given in Appendix C. The nominal shear strength v_n (psi or MPa) resisted by concrete and steel in Eq. (4.3.2a) can be taken as high as $8\sqrt{f_c'}$ (in.-lb units) ($0.66\sqrt{f_c'}$ [SI units]) instead of $6\sqrt{f_c'}$ (in.-lb units) ($0.5\sqrt{f_c'}$ [SI units]). This enables the use of thinner slabs. Experimental data showing that the higher value of v_n can be reached are included in Appendix C.

ACI 318 permits upper limits for s based on the value of v_u at the critical section at $d/2$ from column face

$$s \leq 0.75d \text{ when } \frac{v_u}{\phi} \leq 6\sqrt{f_c'} \text{ (in.-lb units)} \quad (0.5\sqrt{f_c'} \text{ [SI units]}) \qquad (4.3.3b)$$

$$s \leq 0.5d \text{ when } \frac{v_u}{\phi} > 6\sqrt{f_c'} \text{ (in.-lb units)} \quad (0.5\sqrt{f_c'} \text{ [SI units]}) \qquad (4.3.3c)$$

When stirrups are used, ACI 318 limits s to $d/2$. The higher limit for s given by Eq. (4.3.3b) for headed shear stud spacing is again justified by tests (Seible et al. 1980; Andrä 1981a,b; Van der Voet et al. 1982; Mokhtar et al. 1985; Elgabry and Ghali 1987; Institut für Werkstoffe im Bauwesen 1996; Regan 1996a,b; Sherif 1996).

Section 7.3 will show that a vertical branch of a stirrup is less effective than a stud in controlling shear cracks for two reasons: 1) the shank of the headed stud is straight over its full length, whereas the ends of the stirrup leg are curved; and 2) the anchor heads at the top and the bottom of the stud ensure that the specified yield strength is provided at all sections of the shank. In a stirrup, the specified yield strength can be developed only over the middle portion of the vertical legs when they are sufficiently long.

ACI 318-14 Section 20.2.2.4 limits the design yield strength for stirrups as shear reinforcement to 60,000 psi (414 MPa). Research (Otto-Graf-Institut 1996; Regan 1996a; Institut für Werkstoffe im Bauwesen 1996) has indicated that the performance of higher-strength studs as shear reinforcement in slabs is satisfactory. In this experimental work, the stud shear reinforcement in slab-column connections reached a yield stress higher than 72,000 psi (500 MPa), without excessive reduction of shear resistance of concrete. Thus, when studs are used, f_{yt} can be as high as 72,000 psi (500 MPa). In ASTM A1044/A1044M, the minimum specified yield strength of headed shear studs is 51,000 psi (350 MPa) based on what has been commercially available in 2005; higher yield strengths are expected in future versions of ASTM A1044/A1044M. ACI 318 requires conformance with ASTM A1044/A1044M; thus, it limits f_{yt} to 51,000 psi (350 MPa).

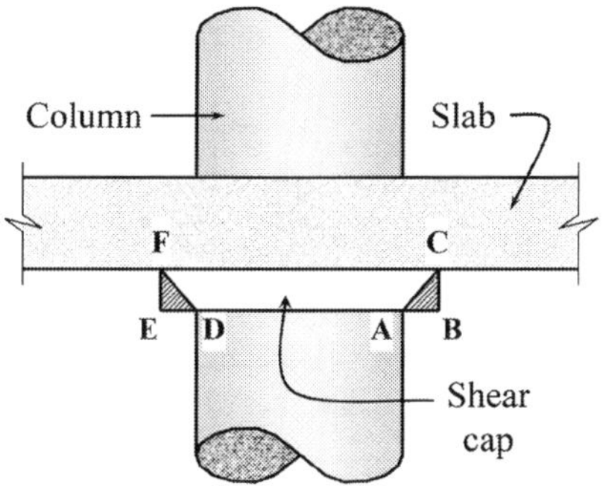

***(a)** Non-recommended design.*

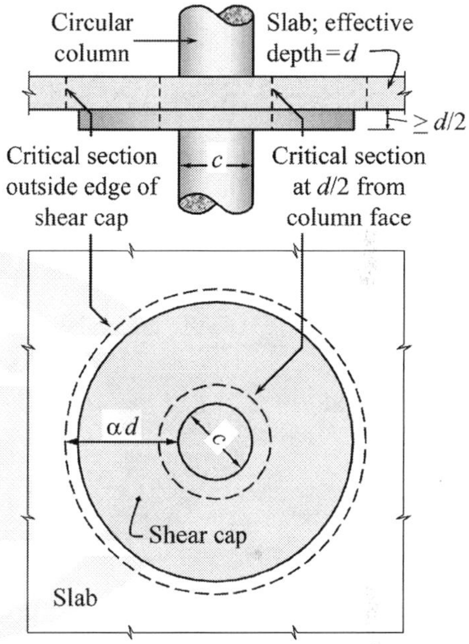

***(b)** Example of design recommended in Section 4.3.4.*

Fig. 4.3.4a—Shear cap design.

4.3.4 *Shear caps*—Figure 4.3.4a(a) shows, for simplicity of presentation, a circular shear cap whose purpose is to increase the shear capacity without using shear reinforcement. The plan dimensions of the shear cap are governed by assuming that the shear strength at $d/2$ from the edges of the cap is governed by Eq. (4.3.1a) to (4.3.1c). This type of shear cap rarely contains reinforcement other than the vertical bars of the column because its plan dimensions are small; with or without reinforcement, this practice is not recommended. Experiments (Megally and Ghali 2002) show that the failure of the shear cap is accompanied by a sudden separation of wedges ABC and DEF from the shear cap and brittle failure of the connection; such failure may occur in service. The

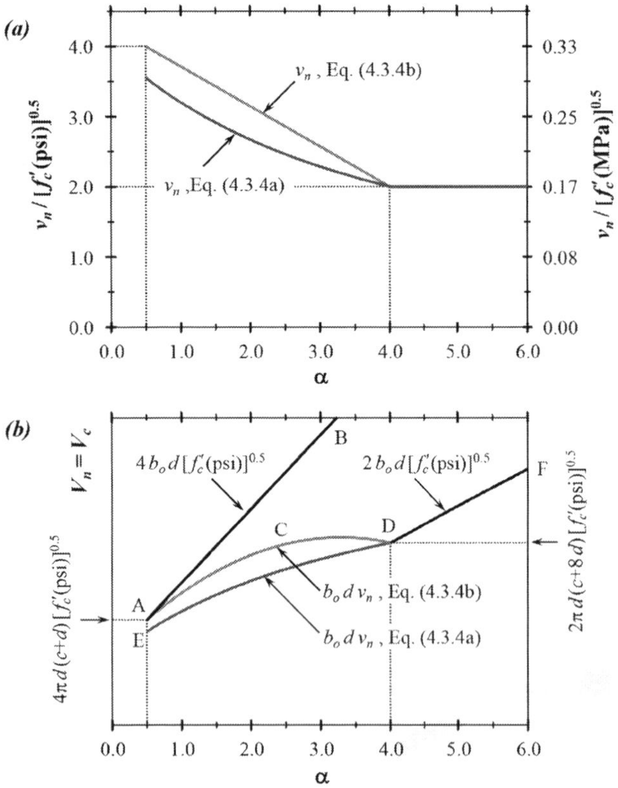

Fig. 4.3.4b—Variation of: (a) v_n*; and (b)* V_n *with the distance between the shear critical section and the column face (=* α*d).*

volume of concrete within the wedges ABC and DEF is too small to offer significant anchorage of the reinforcement that may be provided in the shear cap to prevent the separation of the wedges. Analyses and finite-element studies indicate that when moment, caused by wind or earthquake, is great enough to produce net downward shear stress, an inverted punching failure occurs; in such case, the additional depth provided by the shear cap is not effective to augment the punching shear capacity (Wey and Durrani 1992; Megally and Ghali 2002).

The plan dimensions of the shear cap should be sufficiently large such that the maximum shear stresses at two critical sections (Fig. 4.3.4a(b)) satisfy Eq. (4.1). The critical sections are at $d/2$ from the column face (where d is the total depth within the cap) within the shear cap, and at $d/2$ outside the edges of the shear cap (where d is the slab depth without cap). At $d/2$ from the column, v_n is calculated by Eq. (4.3.1a) to (4.3.1c) in the absence of shear reinforcement. At $d/2$ outside the edges of the shear cap, v_n is calculated by Eq. (4.3.1d). The extent of the shear cap should be the same as the extent of the shear reinforcement when it is used in place of the shear cap.

Based on experimental data, Eligehausen (1996) and Dilger and Ghali (1981) proposed Eq. (4.3.4a) and (4.3.4b), respectively, for the shear strength at critical sections at αd from the column faces.

$$v_n(\alpha) = \left(\frac{1}{1+0.25\alpha}\right)4\sqrt{f_c'};\ 0.5 < \alpha < 4.0 \text{ (in.-lb units)}$$

$$v_n(\alpha) = \left(\frac{1}{1+0.25\alpha}\right)\frac{\sqrt{f_c'}}{3};\ 0.5 < \alpha < 4.0 \text{ (SI units)} \tag{4.3.4a}$$

$$v_n(\alpha) = \left(\frac{7.5-\alpha}{7}\right)4\sqrt{f_c'};\ 0.5 < \alpha < 4.0 \text{ (in.-lb units)}$$

$$v_n(\alpha) = \left(\frac{7.5-\alpha}{7}\right)\frac{\sqrt{f_c'}}{3};\ 0.5 < \alpha < 4.0 \text{ (SI units)} \tag{4.3.4b}$$

At α > 4, the one-way shear strength (Eq. (4.3.1d)) is assumed. Accordingly, as α is increased, the shear strength (psi or MPa) drops (Fig. 4.3.4b(a)), while the area of the shear critical section increases. Figure 4.3.4b(b) shows the variation of the shear strength, $V_n = v_n(\alpha)b_od$ for a circular column of diameter c, transferring shearing force without moment. Line AB represents V_n when v_n (psi) = $4\sqrt{f_c'}$ (where v_n is independent of α); this greatly overestimates V_n compared with line ACDF or EDF calculated by Eq. (4.3.4a) or (4.3.4b), respectively. Line DF represents V_n with v_n (psi) = $2\sqrt{f_c'}$. Because within Zone A to D, the variation of v_n is not established, and the increase in V_n with α is not substantial, it is herein recommended to extend the shear cap to the zone where v_n is known to be not less than the one-way shear strength.

As a design example, consider a circular column of diameter c, transferring a shearing force, V_u (lb) = $6\sqrt{f_c'}\,b_od$; where $b_o = \pi(c + d)$ = the perimeter of the critical section at $d/2$ from the column face in absence of shear cap. The shear cap that satisfies the recommended design should have an approximate effective depth ≥ 1.5d, extending such that α = 1.5(c/d) + 2. It can be verified that this design will satisfy Eq. (4.1) at the critical sections at $d/2$ from the column face and at $d/2$ outside the edge of the shear cap.

For further justification of the recommendations in this section, consider the slab-column connection in Fig. 4.3.4c(a), having a 10 in. (250 mm) square column supporting a 7 in. (175 mm) slab having d = 6 in. (150 mm). Based on the potential crack AB (Fig. 4.3.4c(a)), ACI 318 permits

$$\phi V_n = \phi b_o d 4\sqrt{f_c'};$$
$$\phi V_n^{(a)} = (384 \text{ in.}^2)\phi 4\sqrt{f_c'} \qquad \text{(in.-lb units)}$$

$$\phi V_n = 0.33\phi b_o d\sqrt{f_c'};$$
$$\phi V_n^{(a)} = (240\times 10^3 \text{ mm}^2)0.33\phi\sqrt{f_c'} \quad \text{(SI units)}$$

The design in Fig. 4.3.4c(b) is not recommended as means of increasing strength. If the ϕV_n equation is applied to the potential crack CD (Fig. 4.3.4c(b)), the predicted strength would be

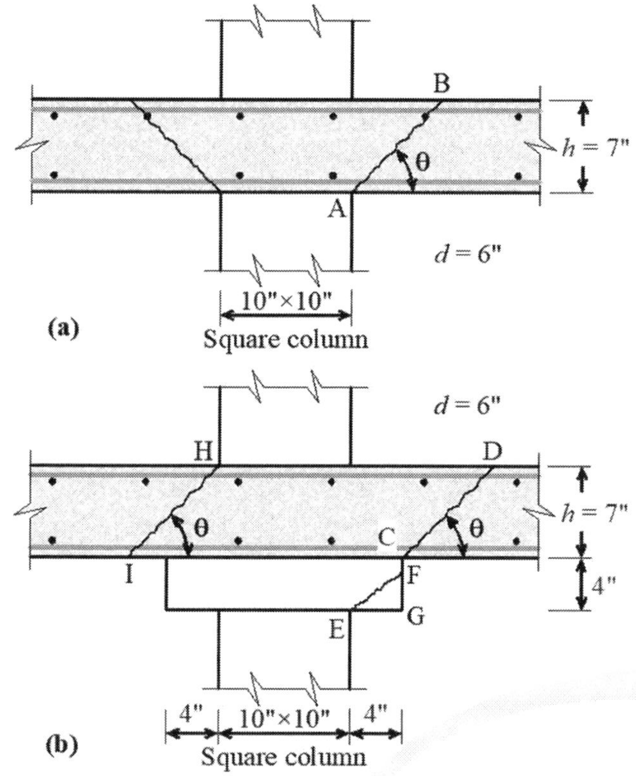

Fig. 4.3.4c—Potential shear cracks. Example of connections: (a) without shear cap; and (b) with shear cap.

$$\phi V_n^{(b)} = (576 \text{ in.}^2)\phi 4\sqrt{f_c'} \quad \text{(in.-lb units)}$$

$$\phi V_n^{(b)} = (270\times10^3 \text{ mm}^2)0.33\phi\sqrt{f_c'} \quad \text{(SI units)}$$

This guide considers the potential failure at EF, whose slope is any angle less than or equal to 45 degrees. It is obvious that the probability of failure at EF is far greater than at CD in a design that considers the shear strength, $\phi V_n = \phi V_n^{(b)} = (576 \text{ in.}^2)\phi 4\sqrt{f_c'}$. This is because EF is shorter than CD and CD crosses top and bottom flexural reinforcements whose amounts are specified by ACI 318 whereas EF may not cross any reinforcement. Although that separation of the wedge EFG (at a shearing force < $\phi V_n^{(b)}$) may not produce collapse, it should not be an acceptable failure. For further justification of recommending against the design in Fig. 4.3.4c(b), consider the potential crack at HI that does not intercept the shear cap. This crack can occur due to high slab moment in a direction that produces compressive stress in the column in the vicinity of H. This guide consistently recommends a shear-reinforced zone of the same size by the provision of shear reinforcement or by drop panels (a shear cap with sufficient plan extension (Fig. 4.3.4a(b)).

4.3.5 *Limitation on shear strength* v_n—Tests (Guandalini et al. 2009; Bayrak and Jirsa 2009) on slabs having low flexural reinforcement ratios in the vicinity of their supports show that v_n may not be reached. In these tests, failure occurred by a local yield-line mechanism, followed typically by a secondary punching failure (Peiris and Ghali 2012). Thus, the nominal strength is to be limited to the value of v_n at which the yield line mechanism occurs.

4.4—Design procedure

The values of f_c', f_{yt}, M_{ux}, M_{uy}, V_u, h, and d are given. The design of shear reinforcement can be performed by the following steps (refer to design examples in Appendix D).

1. At a critical section at $d/2$ from column face, calculate v_u and v_n by Eq. (4.2a) and (4.3.1a) to (4.3.1c). If $(v_u/\phi) \leq v_n$, no shear reinforcement or further check is required. If $(v_u/\phi) > 8\sqrt{f_c'}$ (in.-lb units) ($0.66\sqrt{f_c'}$ [SI units]), the slab thickness or column size is not sufficient; when $(v_u/\phi) \leq 8\sqrt{f_c'}$ (in.-lb units) ($0.66\sqrt{f_c'}$ [SI units]), go to Step 2.

2. When $(v_u/\phi) \leq 6\sqrt{f_c'}$ (in.-lb units) ($0.5\sqrt{f_c'}$ [SI units]), ACI 318 permits stirrups or headed studs. When $(v_u/\phi) > 6\sqrt{f_c'}$ (in.-lb units) ($0.5\sqrt{f_c'}$ [SI units]), ACI 318 permits only headed studs.

Calculate the contribution of concrete v_c to the shear strength (Eq. (4.3.2b) or (4.3.3a)) at the critical section at $d/2$ from column face. The difference $[(v_u/\phi) - v_c]$ gives the shear stress v_s to be resisted by stirrups or headed studs.

3. Select s_o and s within the limitations of Eq. (4.3.2d), (4.3.2e), (4.3.3b), and (4.3.3c), and calculate the required shear reinforcement area for one peripheral line A_v, by solution of Eq. (4.3.2c). Find the minimum number of headed studs or legs of stirrups per peripheral line.

4. Repeat Step 1 at a trial critical section at αd from column face to find the section where $(v_u/\phi) \leq 2\lambda\sqrt{f_c'}$ (in.-lb units) ($0.17\lambda\sqrt{f_c'}$ [SI units]). No other section needs to be checked, and s is to be maintained constant. Select the distance between the column face and the outermost peripheral line of shear reinforcement to be $\geq [\alpha d - (d/2)]$.

The position of the critical section can be determined by selection of the number of headed studs or stirrup legs per line, n running in x- or y-direction (Fig. 4.2). For example, the distance in the x- or y-direction between the column face and the critical section is equal to $s_o + (n - 1)s + d/2$. The number n should be ≥ 2.

5. Arrange studs to satisfy the detailing requirements described in Appendix A.

The trial calculations involved in the aforementioned steps are suitable for computer use (Decon 1996).

CHAPTER 5—PRESTRESSED SLABS

5.1—Nominal shear strength

When a slab is prestressed in two directions, the shear strength of concrete at a critical section at $d/2$ from the column face where shear reinforcement is not provided, is given by ACI 318-14 Section 22.6.5

$$v_c = \text{least of } \{3.5;\ [1.5 + (\alpha_s d/b_o)]\}\lambda\sqrt{f_c'} + 0.3 f_{pc} + \frac{V_p}{b_o d}$$

(in.-lb units)

$$v_c = \text{least of } \{0.29;\ 0.083[1.5 + (\alpha_s d/b_o)]\}\lambda\sqrt{f_c'} + 0.3 f_{pc} + \frac{V_p}{b_o d}$$

(SI units)

(5.1)

where f_{pc} is the average value of compressive stress in the two directions (after allowance for all prestress losses) at centroid of cross section; and V_p is the vertical component of all effective prestress forces crossing the critical section. Equation (5.1) is applicable only if the following are satisfied:

a) No portion of the column cross section is closer to a discontinuous edge than four times the slab thickness h

b) $\sqrt{f_c'}$ in Eq. (5.1) is not taken greater than 70 psi (5.8 MPa)

c) f_{pc} in each direction is not less than 125 psi (0.9 MPa), nor taken greater than 500 psi (3.5 MPa).

If any of the aforementioned conditions are not satisfied or if the slab is prestressed in one direction, the slab should be treated as nonprestressed, and Eq. (4.3.1a) to (4.3.1c) apply. Within the shear-reinforced zone, v_n is to be calculated by Eq. (4.3.2a); the equations and the design procedure in 4.3.2, 4.3.3, and 4.4 apply.

In thin slabs, V_p is small with practical tendon profiles and the slope of the tendon is hard to control. Special care should be exercised in computing V_p in Eq. (5.1) due to the sensitivity of its value to the as-built tendon profile. When it is uncertain that the actual construction will match the design assumption, a reduced or zero value for V_p should be used in Eq. (5.1). Section D.4 is an example of the design of the shear reinforcement in a prestressed slab.

CHAPTER 6—TOLERANCES

Shear reinforcement, in the form of stirrups or studs, can be ineffective if the specified distances s_o and s are not controlled accurately. Tolerances for these dimensions should not exceed ±0.5 in. (±13 mm). If this requirement is not met, a punching shear crack could traverse the slab thickness without intersecting the shear reinforcing elements. Tolerance for the distance between column face and outermost peripheral line of studs should not exceed ±1.5 in. (±38 mm).

Tests (Dilger and Ghali 1981; Andrä 1981a,b; Van der Voet et al. 1982; Mokhtar et al. 1985; Elgabry and Ghali 1987; Mortin and Ghali 1991; Dilger and Shatila 1989; Cao 1993; Brown and Dilger 1994; Megally 1998; Birkle 2004; Ritchie and Ghali 2005; Gayed and Ghali 2006) show that headed studs, anchored as close as possible to the top and bottom of slabs, are effective in resisting punching shear. The designer should specify the overall height of the stud assemblies having the most efficiency

$$l_s = h - c_t - c_b \tag{6}$$

where h is the thickness of the member and c_t and c_b are the specified concrete covers at top and bottom, respectively. ACI 318 permits a manufacturing tolerance: the actual overall height can be shorter than l_s by no more than $d_b/2$, where d_b is the diameter of the tensile flexural reinforcement (Fig. 6; ACI 318-14 Section 20.6.1.3.5). In slabs in the vicinity of columns, the tensile flexural reinforcement is commonly at the top; in footings, the tensile flexural reinforcement is commonly at the bottom.

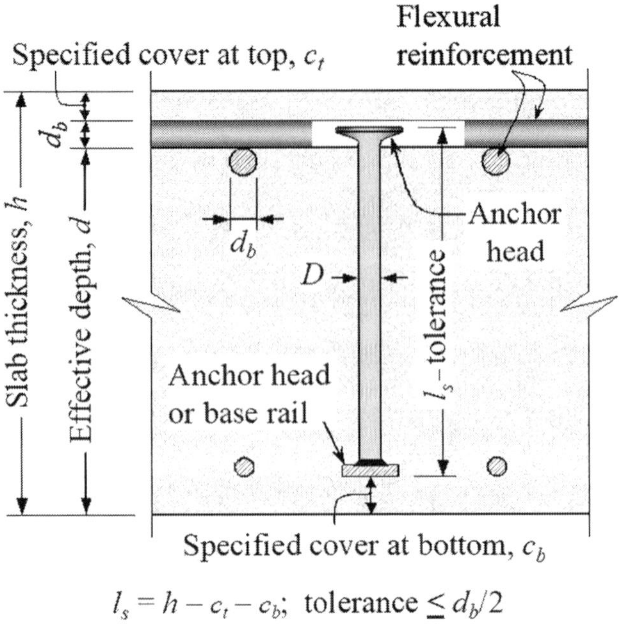

Fig. 6—Section in slab perpendicular to shear stud line.

CHAPTER 7—MOMENT TRANSFER AT SLAB-COLUMN CONNECTION

7.1—Shear stress due to moment transfer M_{sc}

ACI 318-14 Section R8.4.4.2.3 calculates the shear stress distribution in a rectangular shear critical section by the equation

$$v_{u,AB} = v_{ug} + \frac{\gamma_v M_{sc}}{J_c} c_{AB};\ v_{u,CD} = v_{ug} - \frac{\gamma_v M_{sc}}{J_c} c_{CD} \tag{7.1}$$

The code defines v_{ug} as the factored shear stress on the slab critical section for two-way action due to gravity loads without moment transfer. The second terms in Eq. (7.1) represent the shear stress due to the transfer of moment M_{sc} about a symmetry axis; where M_{sc} is the factored slab moment that is resisted by the supporting column; c_{AB} and c_{CD} are dimensions of the shear critical section. The commentary equation for J_c (Eq. (4.2c)) satisfies equilibrium when the last term is omitted. Equation (4.2a), combined with Eq. (B.8) and (B.9), gives the distribution of shear stress whose resultants are equal to the transferred force V_u and moment $\gamma_v M_{sc}$; these equations apply to shear critical sections of any shape where the transfer of the shearing force and the moment to the supporting column is caused by all factored loads (not only gravity loads) with M_{sc} in any direction. Equilibrium is satisfied by using the components of M_{sc} about the centroidal principal x- and y-axes (respectively, $M_{sc\text{-}x}$ and $M_{sc\text{-}y}$) combined with (J_{cx}, J_{cy}) and (γ_{vx}, γ_{vy}), which refer to the same axes (Ghali et al. 2015).

7.1.1 *Calculation of shear stress*—ACI 318 requires satisfying principles of equilibrium and compatibility. Equation (4.2a) satisfies these principles. It is not always safe to use empirical equations that cannot be justified with these principles. Cheng et al. (2010) applied moment about a principal axis that was substituted by two of its components about orthogonal axes; the shear stress was calculated using one of the two components. As a result, the correct maximum shear stress is higher than the value used by the researcher in interpretation of the test results (Ghali et al. 2015). Analyzing a vector in two components and then ignoring one of the two cannot be justified.

7.2—Reduction of γ_v

The parameter γ_v specifies the fraction ($\gamma_v M_{sc}$) to be transferred by eccentricity of shear in design (Eq. (4.2a)). At any point on the periphery of the shear critical section, the resultant of the shear stress due to ($\gamma_v M_{sc}$) is equal to the shearing force per unit length; thus, specifying the value of γ_v gives the magnitude of a shearing force to be used in design. The value of γ_v has been determined by linear analyses (Mast 1970a,b; Elgabry and Ghali 1996; Nazief et al. 2010) and by nonlinear numerical computations (Megally and Ghali 2000; Gayed and Ghali 2008). These analyses do not show that γ_v can be reduced. No experiment has measured the value of γ_v. For design, ACI 318 gives an equation of γ_v to be used in place of performing the necessary analysis. ACI 318-14 Section 8.4.2.3.4 permits, as an option, a reduced range of γ_v for edge and corner columns with specified limitations, including $\gamma_v = 0$; thus, the section assigns a greater portion of M_{sc} to be transferred by flexure. However, the shearing force is equal to the derivative of the bending moment with respect to the distance from the support. This fundamental equilibrium principle applies for beams as well as slabs (in linear and nonlinear stages). The reduction of γ_v, as optionally permitted, ignores equilibrium.

Removal of the optional reduction of γ_v from ACI 318 has been recommended by ACI Committee 421 and by the public during document review; the justification provided for the reduction was a description of nonlinear behavior of slabs in tests. Critical sections have to be designed for internal forces determined by analysis satisfying equilibrium; this is a requirement in all codes. No description of experimental behavior can be adequate to determine the magnitude of the shearing force that must be used in the design of a shear critical section.

Experimental data of four edge and corner column-slab connections subjected to V_u combined with M_{sc} showed that the optional reduction of γ_v can be less conservative compared to mandatory code provisions (Ritchie et al. 2006).

7.3—Anchorage of shear reinforcement

Shear reinforcement typically consists of bars traversing the thickness of the slab to control widening of inclined shear cracks that induce punching. The design considers that the shear stress v_u induced by factored load is resisted by v_c and v_s, provided by concrete and by shear reinforcement, respectively. The shear reinforcing bars should be securely anchored as close as possible to the exterior faces of the slab; the values of v_c and v_s—that can be used in the design—depend upon the anchorage of the bars to develop the tensile force in the bars and to provide the necessary confinement enabling the cracked concrete to resist shear stress. Without effective anchorage, the values of v_c and v_s permitted in design should not be used.

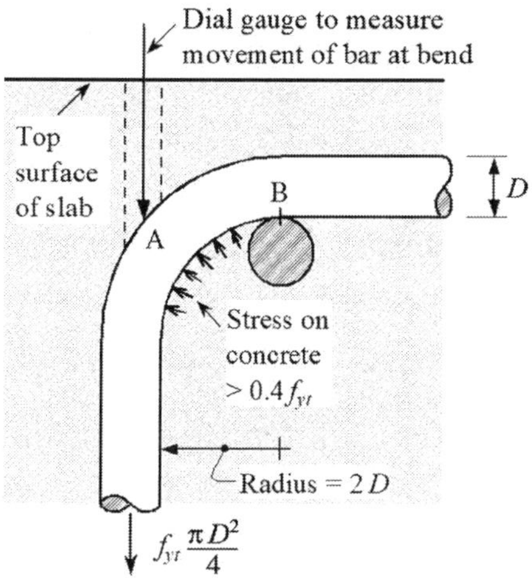

Fig. 7.3—Geometrical and stress conditions at bend of shear reinforcing bar.

ACI 318's commentary recognizes that the anchorage of shear reinforcement can be critical in slabs (because they are typically thinner than beams). Multi-leg stirrups, having the mechanically enhanced anchorage, are permitted by the Code; a bar running in a direction, perpendicular to the plane of the stirrups, has to be engaged inside each bent to provide the necessary mechanical anchorage. A multi-leg stirrup is a continuous bar bent to have multiple vertical legs connected by horizontal segments of equal lengths. The radius of the bends is controlled to avoid crushing of the concrete when the design tensile force is developed in the vertical legs. Tests (Marti 1990) have shown that movement occurs at the bends of shear reinforcement, at Point A of Fig. 7.3, before the yield strength can be reached in the shear reinforcement, causing a loss of stiffness. Furthermore, the concrete within the bend in the stirrups is subjected to stresses that could potentially exceed 0.4 times the stirrup's yield strength f_{yt}, causing concrete crushing. If f_{yt} is 60 ksi (414 MPa), the average compressive stress on the concrete under the bend has to reach $0.4f_{yt}$ for equilibrium. However, because this high stress can crush the concrete, slip occurs before the development of the full f_{yt} in the leg of the stirrup at its connection with the bend. These difficulties, including the consequences of improper stirrup details, have also been discussed (Marti 1990; Joint ASCE-ACI Committee 426 1974; Hawkins 1974; Hawkins et al. 1975). The movement at the end of the vertical leg of a stirrup can be reduced by attachment to a flexural reinforcement bar, as shown at Point B of Fig. 7.3. The flexural reinforcing bar, however, cannot be placed any closer to the vertical leg of the stirrup without reducing the effective slab depth d. Flexural reinforcing bars can provide such improvement to shear reinforcement anchorage only if attachment and direct contact exists at the intersection of the bars (Point B of Fig. 7.3). Under normal construction, however, it is very difficult to ensure such conditions for all stirrups. Thus, such support is normally not fully effective, and the end of the vertical leg of the stirrup can move. The amount of movement is the same for a short or long shear-reinforcing bar. Therefore, the loss in tension is important and the stress is unlikely to reach yield in short shear reinforcement (in thin slabs). These problems are largely avoided if shear reinforcement is provided with mechanical anchorage.

A significant force is developed in a leg of a stirrup only when it is intersected by a punching shear crack; the intersection can occur at any section in the leg. Thus, mechanical anchorage caused by deformations on the surface of the leg or other means cannot be relied upon in the design force carried by the leg. Similarly, ACI 318-14 Fig. R20.5.1 illustrates the geometry of the heads of headed stud shear reinforcement (Fig. 1.3a) and excludes the type of head that is deemed not adequate.

7.3.1 *Stirrup material permitted by ACI 318*—ACI 318 specifies the steel that can be used for stirrups and other shear reinforcements. Also, ACI 318 specifies the geometry of the bends for different bar diameters. ASTM A505 applies to thin steel strips with no surface deformations for applications other than stirrups.

7.4—Layout of shear reinforcement

At an interior column, ACI 318 commentary shows a layout of stirrup legs or headed studs in the shape of a cross. The spacing s between adjacent peripheral lines and the distance between column face and outermost peripheral line of shear reinforcement are determined by rules specified in ACI 318 based on extensive experimental and analytical research. The layout has been widely used in practice.

The radial layout proposed by researchers is difficult to install in practice. Potential interference of the shear reinforcement with two orthogonal top and bottom layers of flexural reinforcement is difficult to avoid. With stirrups, the radial layout requires additional (undesirable) top and bottom layers of flexural bars running in diagonal directions; these are the bars engaged in the hooks and the bends and needed to satisfy anchorage requirement of stirrups (Fig. 7.3).

The use of assemblies of headed studs as punching shear reinforcement originated by research at the University of Calgary in 1980s and transferred worldwide. Subsequent international research developed the crucifix layout and its detailing; these were adopted in ACI 318-08. Satisfying the provisions of ACI 318 and adhering to the crucifix layout as detailed in ACI 318 commentary provides adequate strength and ductility. Limited and nonconclusive research recommending the radial layout of headed stud assemblies is reported in Dam et al. (2017).

7.5—Minimum shear reinforcement

ACI 352.1R recommends a minimum amount of shear reinforcement that provides a nominal strength given by

$$V_s \geq 3.5\sqrt{f_c'}b_o d \quad \text{(in.-lb units)}$$
$$V_s \geq 0.29\sqrt{f_c'}b_o d \quad \text{(SI units)} \qquad (7.5)$$

This minimum is recommended for all slab-column connections; the distance between the face of the support and the outermost shear reinforcement should be $\geq 4h$ or $3h$ for nonprestressed or post-tensioned slabs, respectively. ACI 318 and ACI 421.2R discuss a minimum amount of shear reinforcement only in certain cases of seismic design. The recommendation for minimum shear reinforcement has been justified by researches; the minimum extension of the shear-reinforced zone should account for ductility at the limit state.

CHAPTER 8—REFERENCES

American Concrete Institute

ACI 318-14—Building Code Requirements for Structural Concrete

ACI 352.1R-11—Guide for Design of Slab-Column Connections in Monolithic Concrete structures

ACI 421.2R-10—Guide to Seismic Design of Punching Shear Reinforcement in Flat Plates

ASTM International

ASTM A505-16—Standard Specification for Steel, Sheet and Strip, Alloy, Hot-Rolled and Cold-Rolled, General Requirements for

ASTM A1044/A1044M-16—Standard Specification for Steel Stud Assemblies for Shear Reinforcement of Concrete

Canadian Standards Association

CSA A23.3-14—Design of Concrete Structures for Buildings

Authored documents

Andrä, H. P., 1981a, "Strength of Flat Slabs Reinforced with Stud Rails in the Vicinity of the Supports (Zum Tragverhalten von Flachdecken mit Dübelleisten-Bewehrung im Auflagerbereich)," *Beton und Stahlbetonbau*, Berlin, V. 76, No. 3, Mar., pp. 53-57.

Andrä, H. P., 1981b, "Strength of Flat Slabs Reinforced with Stud Rails in the Vicinity of the Supports (Zum Tragverhalten von Flachdecken mit Dübelleisten-Bewehrung im Auflagerbereich)," *Beton und Stahlbetonbau*, Berlin, V. 76, No. 4, Apr., pp. 100-104.

Bayrak, W. O., and Jirsa, J. O., 2009, "Two-Way Shear Strength of Slab-Column Connection: Reexamination of ACI 318 Provisions," *ACI Structural Journal*, V. 106, No. 2, Mar.-Apr., pp. 160-170.

Birkle, G., 2004, "Punching of Slabs: Thickness and Stud Layout," PhD dissertation, University of Calgary, Calgary, AB, Canada, 152 pp.

Brown, S., and Dilger, W. H., 1994, "Seismic Response of Flat-Plate Column Connections," *Proceedings*, Canadian Society for Civil Engineering Conference, V. 2, Winnipeg, MB, Canada, pp. 388-397.

Cao, H., 1993, "Seismic Design of Slab-Column Connections," MSc thesis, University of Calgary, Calgary, AB, Canada, 188 pp.

Cheng, M.-Y.; Parra-Montesinos, G. J.; and Shield, C. K., 2010, "Shear Strength and Drift Capacity of Fiber-Reinforced Concrete Slab-Column Connections Subjected to

Biaxial Displacements," *Journal of Structural Engineering*, V. 136, No. 9, Sept., pp. 1078-1088. doi: 10.1061/(ASCE) ST.1943-541X.0000213

Dam, T. X.; Wight, J. K.; and Parra-Montesinos, G. J., 2017, "Behavior of Monotonically Loaded Slab-Column Connections Reinforced with Shear Studs," *ACI Structural Journal*, V. 114, No. 1, Jan.-Feb., pp. 221-232. (with discussions in V. 114, No. 6, Nov.-Dec. 2017, pp. 1660-1664)

Decon, 1996, "STDESIGN," Computer Program for Design of Shear Reinforcement for Slabs, DECON, Brampton, ON, Canada.

Dilger, W. H., 2014, "Inclined Stirrups and Inclined Stud Shear Reinforcement in Zones of High Shear," *Proceedings*, 9th International Conference on Short and Medium Span Bridges (SMSB), Calgary, AB, Canada, 12 pp.

Dilger, W. H., and Ghali, A., 1981, "Shear Reinforcement for Concrete Slabs," *Proceedings*, ASCE, V. 107, No. ST12, Dec., pp. 2403-2420.

Dilger, W. H., and Shatila, M., 1989, "Shear Strength of Prestressed Concrete Edge Slab-Column Connections with and without Stud Shear Reinforcement," *Canadian Journal of Civil Engineering*, V. 16, No. 6, pp. 807-819. doi: 10.1139/ l89-124

Dyken, T., and Kepp, B., 1988, "Properties of T-Headed Reinforcing Bars in High-Strength Concrete," *Publication No.* 7, Nordic Concrete Research, Norske Betongforening, Oslo, Norway, Dec.

Elgabry, A. A., and Ghali, A., 1987, "Tests on Concrete Slab-Column Connections with Stud Shear Reinforcement Subjected to Shear-Moment Transfer," *ACI Structural Journal*, V. 84, No. 5, Sept.-Oct., pp. 433-442.

Elgabry, A. A., and Ghali, A., 1996, "Moment Transfer by Shear in Slab-Column Connections," *ACI Structural Journal*, V. 93, No. 2, Mar.-Apr., pp. 187-196.

Eligehausen, R., 1996, "Bericht über Zugversuche mit Deha Kopfbolzen, (Report on Pull Tests on Deha Anchor Bolts)," Institut für Werkstoffe im Bauwesen, University of Stuttgart, *Report* No. DE003/01-96/32, Sept.

Gayed, R. B., and Ghali, A., 2004, "Double-Head Studs as Shear Reinforcement in Concrete I Beams," *ACI Structural Journal*, V. 101, No. 4, July-Aug., pp. 549-557.

Gayed, R. B., and Ghali, A., 2006, "Seismic-Resistant Joints of Interior Columns with Prestressed Slabs," *ACI Structural Journal*, V. 103, No. 5, Sept.-Oct., pp. 710-719.

Gayed, R. B., and Ghali, A., 2008, "Unbalanced Moment Resistance in Slab-Column Joints: Analytical Assessment," *Journal of Structural Engineering*, V. 134, No. 5, May, pp. 859-864. doi: 10.1061/(ASCE)0733-9445(2008)134:5(859)

Ghali, A.; Gayed, R. B.; and Dilger, W. H., 2015, "Design of Concrete Slabs for Punching Shear: Controversial Concepts," *ACI Structural Journal*, V. 112, No. 4, July-Aug., pp. 505-514. doi: 10.14359/51687707

Ghali, A.; Sargious, M. A.; and Huizer, A., 1974, "Vertical Prestressing of Flat Plates around Columns," *Shear in Reinforced Concrete*, SP-42, Farmington Hills, MI, pp. 905-920.

Guandalini, S.; Burdet, O. L.; and Muttoni, A., 2009, "Punching of Slabs with Low Reinforcement Ratios," *ACI Structural Journal*, V. 106, No. 1, Jan.-Feb., pp. 87-95.

Hawkins, N. M., 1974, "Shear Strength of Slabs with Shear Reinforcement," *Shear in Reinforced Concrete*, SP-42, American Concrete Institute, Farmington Hills, MI, pp. 785-815.

Hawkins, N. M.; Mitchell, D.; and Hanna, S. H., 1975, "The Effects of Shear Reinforcement on Reversed Cyclic Loading Behavior of Flat Plate Structures," *Canadian Journal of Civil Engineering*, V. 2, No. 4, Dec., pp. 572-582. doi: 10.1139/l75-052

Hoff, G. C., 1990, "High-Strength Lightweight Aggregate Concrete—Current Status and Future Needs," *Proceedings of the 2nd International Symposium on Utilization of High-Strength Concrete*, Berkeley, CA, pp. 20-23.

Institut für Werkstoffe im Bauwesen, 1996, "Bericht über Versuche an punktgestützten Platten bewehrt mit DEHA Doppelkopfbolzen und mit Dübelleisten (Test Report on Point Supported Slabs Reinforced with DEHA Double Head Studs and Studrails)," *Universität—Stuttgart, Report* No. AF 96/6 – 402/1, Germany, DEHA, 81 pp.

Joint ASCE-ACI Committee 426, 1974, "The Shear Strength of Reinforced Concrete Members-Slabs," *Journal of the Structural Division*, V. 100, Aug., pp. 1543-1591.

Leonhardt, F., and Walter, R., 1965, "Welded Wire Mesh as Stirrup Reinforcement: Shear on T-Beams and Anchorage Tests," *Bautechnik*, V. 42, p. 1-10. (in German)

Marti, P., 1990, "Design of Concrete Slabs for Transverse Shear," *ACI Structural Journal*, V. 87, No. 2, Mar.-Apr., pp. 180-190.

Marti, P.; Parlong, J.; and Thurlimann, B., 1977, "Schubversuche and Stahlbeton-Platten," Institut fur Baustatik aund Konstruktion, ETH Zurich, *Bericht* Nr. 7305-2, Birkhauser Verlag, Basel und Stuttgart, Germany.

Mast, P. E., 1970a, "Stresses in Flat Plates near Columns," *ACI Journal Proceedings*, V. 67, No. 10, Oct., pp. 761-768.

Mast, P. E., 1970b, "Plate Stresses at Columns near the Free Edge," *ACI Journal Proceedings*, V. 67, No. 11, Nov., pp. 898-902.

McLean, D.; Phan, L. T.; Lew, H. S.; and White, R. N., 1990, "Punching Shear Behavior of Lightweight Concrete Slabs and Shells," *ACI Structural Journal*, V. 87, No. 4, July-Aug., pp. 386-392.

Megally, S. H., 1998, "Punching Shear Resistance of Concrete Slabs to Gravity and Earthquake Forces," PhD dissertation, University of Calgary, Calgary, AB, Canada, 468 pp.

Megally, S. H., and Ghali, A., 1996, "Nonlinear Analysis of Moment Transfer between Columns and Slabs," *Proceedings*, V. IIa, Canadian Society for Civil Engineering Conference, Edmonton, AB, Canada, pp. 321-332.

Megally, S. H., and Ghali, A., 2000, "Punching of Concrete Slabs Due to Column Moment Transfer," *Journal of Structural Engineering*, V. 126, No. 2, Feb., pp. 180-189. doi: 10.1061/(ASCE)0733-9445(2000)126:2(180)

Megally, S. H., and Ghali, A., 2002, "Cautionary Note on Shear Capitals," *Concrete International*, V. 24, No. 3, Mar., pp. 75-83.

Mokhtar, A. S.; Ghali, A.; and Dilger, W. H., 1985, "Stud Shear Reinforcement for Flat Concrete Plates," *ACI Journal Proceedings*, V. 82, No. 5, Sept.-Oct., pp. 676-683.

Mortin, J., and Ghali, A., 1991, "Connection of Flat Plates to Edge Columns," *ACI Structural Journal*, V. 88, No. 2, Mar.-Apr., pp. 191-198.

Muller, F. X.; Muttoni, A.; and Thurlimann, B., 1984, "Durchstanz Versuche an Flachdecken mit Aussparungen (Punching Tests on Slabs with Openings)," *Research Report* No. 7305-5, ETH Zurich, Birkhauser Verlag, Basel and Stuttgart, Germany.

Nazief, M. A.; Rashed, Y. F.; and El-Degwy, W. M., 2010, "Boundary Element Method Calculation of Moment Transfer Parameters in Slab-Column Connections," *ACI Structural Journal*, V. 107, No. 2, Mar.-Apr., pp. 164-169.

Otto-Graf-Institut, 1996, "Durchstanzversuche an Stahlbetonplatten mit Rippendübeln und Vorgefertigten Grossflächentafeln (Punching Shear Tests on Concrete Slabs with Deformed Studs and Large Precast Slabs)," *Report* No. 21-21634, University of Stuttgart, Germany, July.

Peiris, C., and Ghali, A., 2012, "Flexural Reinforcement Essential for Punching Shear Resistance of Slabs," *Recent Development in Reinforced Concrete Slab Analysis, Design and Serviceability*, SP-287, American Concrete Institute, Farmington Hills, MI, pp. 83-98.

Regan, P. E., 1996a, "Double Headed Studs as Shear Reinforcement—Tests of Slabs and Anchorages," University of Westminster, London, UK, Aug.

Regan, P. E., 1996b, "Punching Test of Slabs with Shear Reinforcement," University of Westminster, London, UK, Nov.

Ritchie, M., and Ghali, A., 2005, "Seismic-Resistant Connections of Edge Columns with Prestressed Slabs," *ACI Structural Journal*, V. 102, No. 2, Mar.-Apr., pp. 314-323.

Ritchie, M.; Ghali, A.; Dilger, W. H.; and Gayed, R. B., 2006, "Unbalanced Moment Resistance by Shear in Slab-Column Connections: Experimental Assessment," *ACI Structural Journal*, V. 103, No. 1, Jan.-Feb., pp. 74-82.

Seible, F.; Ghali, A.; and Dilger, W. H., 1980, "Preassembled Shear Reinforcing Units for Flat Plates," *ACI Journal Proceedings*, V. 77, No. 1, Jan.-Feb., pp. 28-35.

Sherif, A., 1996, "Behavior of R.C. Flat Slabs," PhD dissertation, University of Calgary, Calgary, AB, Canada, 397 pp.

Van der Voet, F.; Dilger, W. H.; and Ghali, A., 1982, "Concrete Flat Plates with Well-Anchored Shear Reinforcement Elements," *Canadian Journal of Civil Engineering*, V. 9, No. 1, pp. 107-114. doi: 10.1139/l82-011

Wey, E. H., and Durrani, A. J., 1992, "Seismic Response of Interior Slab-Column Connections with Shear Capitals," *ACI Structural Journal*, V. 89, No. 6, Nov.-Dec., pp. 682-691.

APPENDIX A—DETAILS OF SHEAR REINFORCEMENT

A.1—Geometry of shear reinforcement

Shear reinforcement in slabs consists of vertical or inclined bars to traverse cracks to prevent punching failure. The bars are mechanically anchored at their ends to develop specified yield strength f_{yt}. Several types and configurations of shear studs have been reported in the literature. Shear studs mounted on a continuous steel strip, as discussed in the main text of this guide, have been developed and investigated (Dilger and Ghali 1981; Andrä 1981a,b; Van der Voet et al. 1982; Mokhtar et al. 1985; Elgabry and Ghali 1987; Mortin and Ghali 1991; Dilger and Shatila 1989; Cao 1993; Brown and Dilger 1994; Megally 1998; Birkle 2004; Ritchie and Ghali 2005; Gayed and Ghali 2006). Headed reinforcing bars were developed and applied in Norway (Dyken and Kepp 1988) for high-strength concrete structures, and it was reported that such applications improved the structural performance significantly (Gayed and Ghali 2004; Hoff 1990). Another type of headed shear reinforcement was implemented for increasing the punching shear strength of lightweight concrete slabs and shells (McLean et al. 1990). Several other approaches for mechanical anchorage in shear reinforcement can be used (Marti 1990; Muller et al. 1984; Marti et al. 1977; Ghali et al. 1974). Several types are depicted in Fig. A.1. ACI 318 permits stirrups in slabs with $d \geq 6$ in. (152 mm), but not less than 16 times the diameter of the stirrups. In the stirrup details, shown in Fig. A.1(a) (from ACI 318), a bar has to be lodged in each bend to provide the mechanical anchorage necessary for the development of f_{yt} in the vertical legs. Matching this detail and the design spacing, s_o and s in actual construction ensure the effectiveness of stirrups as assumed in design.

The anchor plates shown in Fig. A.1(c) should be circular or rectangular, and their area should be sufficient to develop the specified yield strength of studs f_{yt}. ASTM A1044/A1044M specifies an anchor head area equal to 10 times the cross-sectional area of the stud. It is recommended that the performance of the shear stud reinforcement be verified before their use.

ACI 318 permits shear reinforcement placed perpendicular to the plane of the slab. Inclined shear reinforcement (Fig. A.1(d)) anchored as close as possible to the top and bottom of the slab is also permitted; the inclination of the bars enhances the effectiveness of shear reinforcement (Dilger 2014).

A.2—Stud arrangements

Shear studs in the vicinity of rectangular columns should be arranged on peripheral lines. The term "peripheral line" is used in this report to mean a line running parallel to and at constant distance from the sides of the column cross section. Figure 4.2 shows a typical arrangement of stud shear reinforcement in the vicinity of a rectangular interior, edge, and corner columns. Tests (Dilger and Ghali 1981) showed that studs are most effective near column corners. For this reason, shear studs in Fig. 4.2 are aligned with column faces. In the direction parallel to a column face, the distance g between lines of shear studs should not exceed $2d$, where d is the effective depth of the slab. When stirrups are used, the same limit for g should be observed (Fig. A.1(a)).

The stud arrangement for circular columns is shown in Fig. A.2. The minimum number of peripheral lines of shear

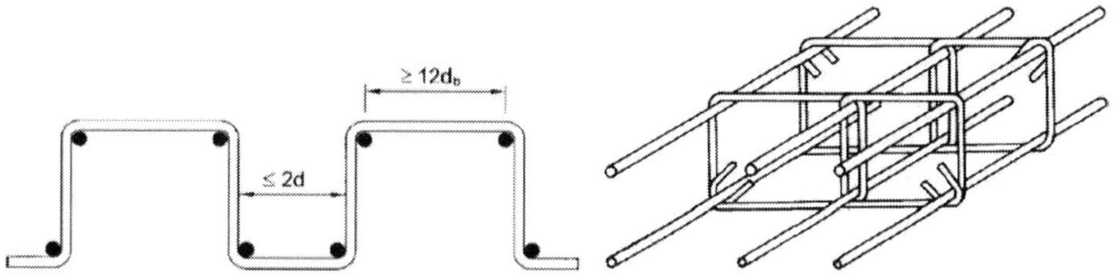

***(a)** Slab stirrups*

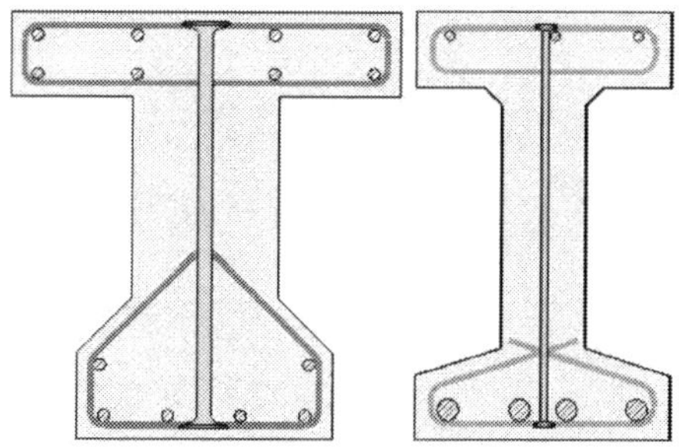

***(b)** Beam tests*

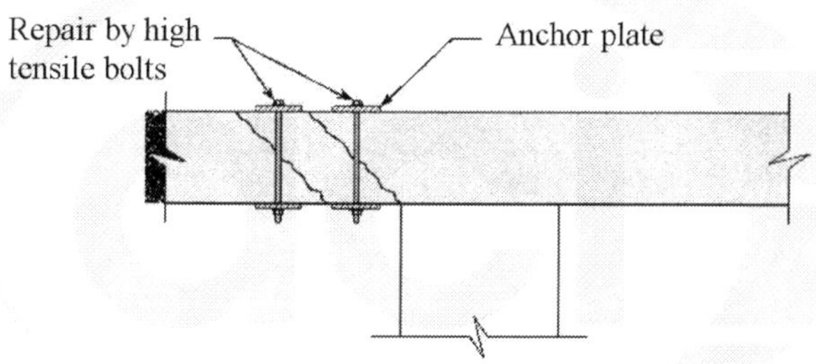

***(c)** High tensile bolts used as shear reinforcement*

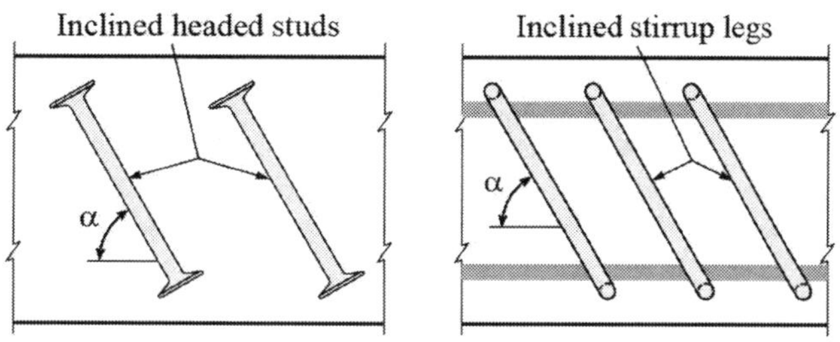

***(d)** Inclined shear reinforcement*

Fig. A.1—Shear reinforcement type: (a) is copied from ACI 318. Types (b) to (f) are from Gayed and Ghali (2004), McLean et al. (1990), Muller et al. (1984), Ghali et al. (1974), and Dilger (2014).

studs, in the vicinity of rectangular and circular columns, is two.

A.3—Stud length

The studs are most effective when their anchors are as close as possible to the top and bottom surfaces of the slab. Unless otherwise protected, the minimum concrete cover of the anchors should be as required by ACI 318. The cover of the anchors should not exceed the minimum cover plus one-half bar diameter of flexural reinforcement (Fig. 6). The mechanical anchors should be placed in the forms above reinforcement supports, which ensures the specified concrete cover.

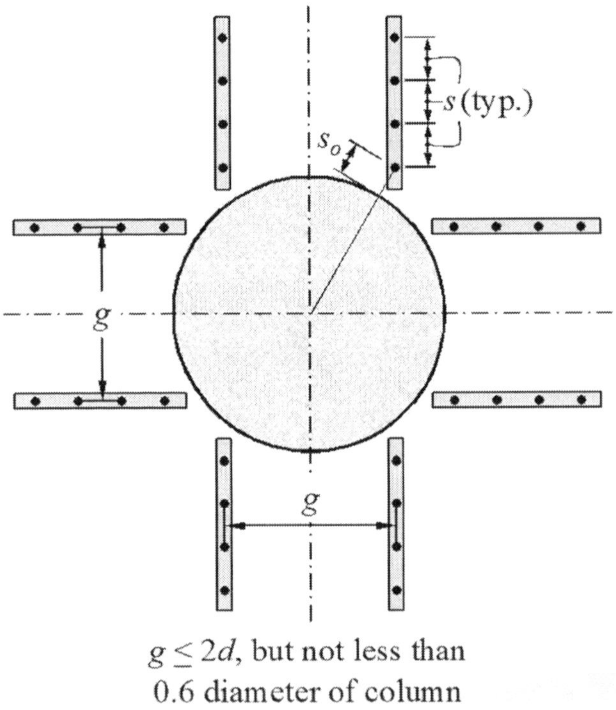

Fig. A.2—Shear headed stud reinforcement arrangement for circular columns.

APPENDIX B—PROPERTIES OF CRITICAL SECTIONS OF GENERAL SHAPE

Figure B shows the top view of critical sections for shear in slabs. The centroidal principal x and y axes of the critical sections, V_u, M_{ux}, and M_{uy} are shown in their positive directions. The shear force V_u is acting at the column centroid; V_u, M_{ux}, and M_{uy} represent the effects of the column on the slab; and l_x and l_y are projections of the shear critical sections on directions of principal x- and y-axes.

The coefficients γ_{vx} and γ_{vy} are given by Eq. (B.1) to (B.6). ACI 318 gives Eq. (B.1) and (B.2); Eq. (B.3) to (B.6) are based on finite-element studies (Elgabry and Ghali 1996; Megally and Ghali 1996).

Interior column-slab connections (Fig. B(a))

$$\gamma_{vx} = 1 - \frac{1}{1 + \frac{2}{3}\sqrt{l_y / l_x}} \tag{B.1}$$

$$\gamma_{vy} = 1 - \frac{1}{1 + \frac{2}{3}\sqrt{l_x / l_y}} \tag{B.2}$$

Edge column-slab connections (Fig. B(b))

$$\gamma_{vx} = 1 - \frac{1}{1 + \frac{2}{3}\sqrt{l_y / l_x}} \tag{B.3}$$

$$\gamma_{vy} = 1 - \frac{1}{1 + \frac{2}{3}\sqrt{\frac{l_x}{l_y} - 0.2}} \quad \text{but } \gamma_{vy} = 0 \text{ when } \frac{l_x}{l_y} < 0.2 \tag{B.4}$$

Corner column-slab connections (Fig. B(c))

$$\gamma_{vx} = 0.4 \tag{B.5}$$

$$\gamma_{vy} = 1 - \frac{1}{1 + \frac{2}{3}\sqrt{\frac{l_x}{l_y} - 0.2}} \quad \text{but } \gamma_{vy} = 0 \text{ when } \frac{l_x}{l_y} < 0.2 \tag{B.6}$$

Equations (B.7) to (B.9) give the values of A_c, J_x, and J_y that determine by Eq. (4.2a) the distribution of shear stress v_u, whose resultant components are exactly V_u, $\gamma_{vx}M_{ux}$, and $\gamma_{vy}M_{uy}$. Generally, the critical section perimeter can be considered as composed of straight segments. The values of A_c, J_{cx}, and J_{cy} can be determined by summation of the contribution of the segments

$$A_c = d\sum l \tag{B.7}$$

$$J_{cx} = d\sum\left[\frac{l}{3}(y_i^2 + y_i y_j + y_j^2)\right] \tag{B.8}$$

$$J_{cy} = d\sum\left[\frac{l}{3}(x_i^2 + x_i x_j + x_j^2)\right] \tag{B.9}$$

where x_i, y_i, x_j, and y_j are coordinates of points i and j at the extremities of a typical segment whose length is l. For a circular shear critical section, $A_c = 2\pi d$ (radius) and $J_{cx} = J_{cy} = \pi d$ (radius)^3. The values of A_c, J_{cx}, and J_{cy} are used in Eq. (4.2a) to calculate the distribution of shear stress v_u, whose resultant components are V_u, $\gamma_{vx}M_{sc\text{-}x}$, and $\gamma_{vx}M_{sc\text{-}y}$.

When the critical section has no axis of symmetry, such as in Fig. 4.2(c), the centroidal principal axes can be determined by the rotation of the centroidal nonprincipal $\bar{x}$- and $\bar{y}$-axes an angle θ, given by

$$\tan 2\theta = \frac{-2J_{c\bar{x}\bar{y}}}{J_{c\bar{x}} - J_{c\bar{y}}} \tag{B.10}$$

The absolute value of θ is less than $\pi/2$; when the value is positive, θ is measured in the clockwise direction. $J_{c\bar{x}}$ and $J_{c\bar{y}}$ can be calculated by Eq. (B.8) and (B.9), substituting $\bar{x}$ and $\bar{y}$ for x and y. $J_{c\bar{x}\bar{y}}$ is equal to d times the product of inertia of the perimeter of the critical section about the centroidal non-principal $\bar{x}$- and $\bar{y}$-axes.

$$J_{c\bar{x}\bar{y}} = d\sum\left[\frac{l}{6}(2\bar{x}_i\bar{y}_i + \bar{x}_i\bar{y}_j + \bar{x}_j\bar{y}_i + 2\bar{x}_j\bar{y}_j)\right] \tag{B.11}$$

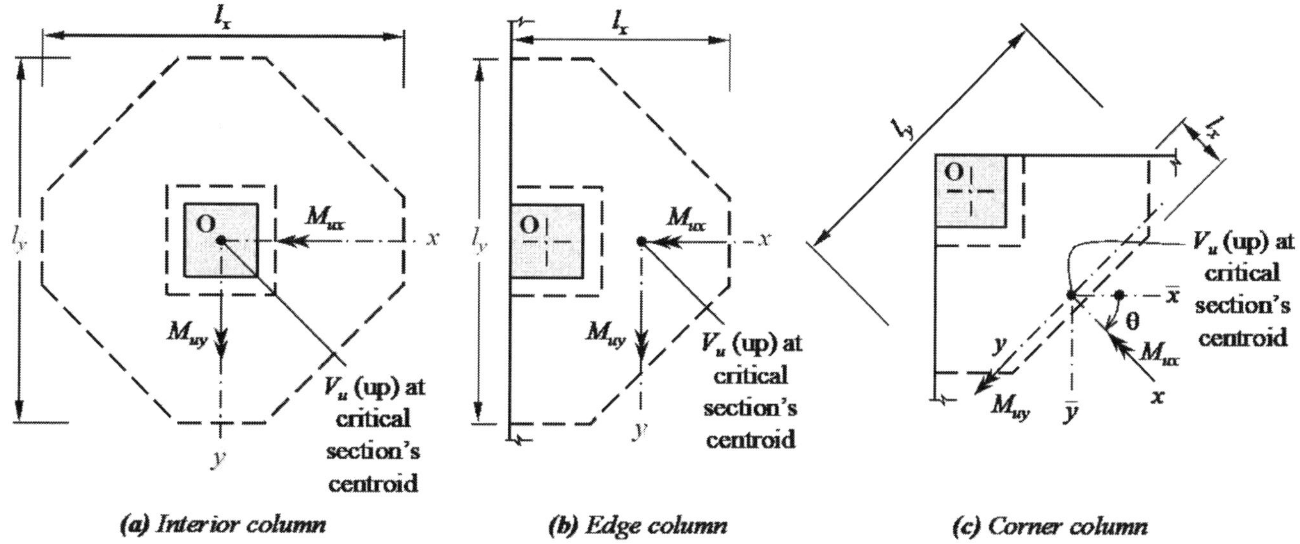

Fig. B—Shear critical sections outside shear-reinforced zones and sign convention of factored internal forces transferred from columns to slabs.

The coordinates of any point on the perimeter of the critical section with respect to the centroidal principal axes can be calculated by Eq. (B.12) and (B.13).

$$x = \bar{x}\cos\theta + \bar{y}\sin\theta \quad \text{(B.12)}$$

$$y = -\bar{x}\sin\theta + \bar{y}\cos\theta \quad \text{(B.13)}$$

The x and y coordinates, determined by Eq. (B.12) and (B.13), can now be substituted in Eq. (B.8) and (B.9) to give the values of J_{cx} and J_{cy}.

APPENDIX C—VALUES OF v_c WITHIN SHEAR-REINFORCED ZONE

This design procedure of the shear reinforcement requires calculation of $v_n = v_c + v_s$ at the critical section at $d/2$ from the column face. The value allowed for v_c is $2\sqrt{f_c'}$ (in.-lb units) ($0.17\sqrt{f_c'}$ [SI units]) when stirrups are used, and $3\sqrt{f_c'}$ (in.-lb units) ($0.25\sqrt{f_c'}$ [SI units]) when headed shear studs are used. The reason for the higher value of v_c for slabs with headed shear stud reinforcement is the almost slip-free anchorage of the studs. In structural elements reinforced with conventional stirrups, the anchorage by hooks or 90-degree bends is subject to slip, which can be as high as 0.04 in. (1 mm) when the stress in the stirrup leg approaches its yield strength (Leonhardt and Walter 1965). This slip is detrimental to the effectiveness of stirrups in slabs because of their relative small depth compared with beams. The influence of the slip is manifold:

a) Increase in width of the shear crack

b) Extension of the shear crack into the compression zone

c) Reduction of the shear resistance of the compression zone

d) Reduction of the shear friction across the crack

These effects reduce the shear capacity of the concrete in slabs with stirrups. To reflect the stirrup slip in the shear resistance equations, refinement of the shear failure model is required. The empirical equation $v_n = v_c + v_s$, adopted in almost all codes, is not the ideal approach to solve the shear design problem. A mechanics-based model that is acceptable for codes is not presently available. There is, however, enough experimental evidence that use of the empirical equation $v_n = v_c + v_s$ with $v_c = 3\sqrt{f_c'}$ (in.-lb units) ($0.25\sqrt{f_c'}$ [SI units]) gives a safe design for slabs with shear headed stud reinforcement. This approach is adopted in CSA A23.3.

Numerous test slab-column connections reinforced with headed studs are reported in the literature (Table C.1). In most of these tests, the failure is at sections outside the shear-reinforced zone. Table C.2 lists only the tests in which the failure occurred within the shear-reinforced zone. Column 12 of Table C.2 gives the ratio v_{test}/v_{code}; where v_{code} is the value allowed by ACI 318, with $v_c = 3\sqrt{f_c'}$ (in.-lb units) ($0.25\sqrt{f_c'}$ [SI units]) (Eq. (4.3.3a)). The values of v_{test}/v_{code} being greater than 1.0 indicate there is safety of design with $v_c = 3\sqrt{f_c'}$ (in.-lb units) ($0.25\sqrt{f_c'}$ [SI units]).

Table C.3 summarizes experimental data of numerous slabs in which the maximum shear stress v_u obtained in test, at the critical section at $d/2$ from column face, reaches or exceeds $8\sqrt{f_c'}$ (in.-lb units) ($0.66\sqrt{f_c'}$ [SI units]). Table C.3 indicates that v_n can be safely taken equal to $8\sqrt{f_c'}$ (in.-lb units) ($0.66\sqrt{f_c'}$ [SI units]) (4.3.3).

Table C.4 gives the experimental results of slabs having stud shear reinforcement with the spacing between headed studs greater or close to the upper limit given by Eq. (4.3.3b). In Table C.4, v_{code} is the nominal shear stress calculated by ACI 318, with the provisions given in 4.3.3. The value v_{code} is calculated at $d/2$ from column face when failure is within the shear-reinforced zone, or at a section at $d/2$ from the outermost studs when failure occurs outside the shear-reinforced zone. The ratio v_{test}/v_{code} being greater than 1.0 indicates that it is safe to use headed studs spaced at the upper limit set by Eq. (4.3.3b) and to calculate the strength with the provisions in 4.3.3.

Table C.1—List of references on slab-column connections tests using stud shear reinforcement

Experiment no.	Reference no.	Experiment no.	Reference no.	Experiment no.	Reference no.
1 to 5	Andrä 1981a,b	16 to 18	Regan 1996a	26 to 29	Elgabry and Ghali 1987
6, 7	*	19, 20	Regan 1996b	30 to 36	Mokhtar et al. 1985
8, 9	Otto-Graf-Institut 1996	21 to 24, 37	Sherif 1996	42, 43	Seible et al. 1980
10 to 15	Institut für Werkstoffe im Bauwesen 1996	25, 38 to 41	Van der Voet et al. 1982	—	—

*"Grenzzustände der Tragfäkigheit für Durchstanzen von Platten mit Dübelleistein nach EC2 (Ultimate Limit States of Punching of Slabs with Studrails According to EC2)," Stuttgart, Germany, 1996, 15 pp.

Table C.2—Slabs with stud shear reinforcement failing within shear-reinforced zone

Experiment	Square column size, in. (mm)	f_c', psi (MPa)	d, in. (mm)	s/d	Tested capacities: V_u, kip (kN)	Tested capacities: M_u, kip-in. (kN-m)	M at critical section centroid, kip-in. (kN-m)	Maximum shear stress v_u, psi (MPa)	f_{yt}, ksi (MPa)	A_v, in.^{2} (mm^2)	v_{test}/v_{code}†	Remarks
(1)	(2)	(3)	(4)	(5)	(6)	(7)	(8)	(9)	(10)	(11)	(12)	(13)
20	7.87 (200)	5660 (39.0)	6.30 (160)	0.75	214 (952)	0 (0)	0 (0)	599 (4.13)	64.1 (442)	1.402 (905)	1.00	Interior column
21	9.84 (250)	4100 (28.3)	4.49 (114)	0.70	47.4 (211)	651 (73.6)	491 (55.5)	528 (3.64)	55.1 (380)	0.66 (426)	1.14	Edge column
22	9.84 (250)	4030 (27.8)	4.49 (114)	0.70	52.8 (235)	730 (82.5)	552 (62.4)	590 (4.07)	55.1 (380)	0.66 (426)	1.28	Edge column
23	9.84 (250)	4080 (28.1)	4.49 (114)	0.70	26.0 (116)	798 (90.2)	708 (80.0)	641 (4.42)	55.1 (380)	0.66 (426)	1.39	Edge column
24	9.84 (250)	4470 (30.8)	4.49 (114)	0.70	27.2 (121)	847 (95.7)	755 (85.3)	693 (4.78)	55.1 (380)	0.66 (426)	1.48	Edge column
26	9.84 (250)	4890 (33.7)	4.49 (114)	0.75	34.0 (151)	1434 (162.0)	1434 (162.0)	570 (3.93)	66.7 (460)	1.570 (1013)	1.02	Interior column
27	9.84 (250)	5660 (39.0)	4.49 (114)	0.75	67.0 (298)	1257 (142.0)	1257 (142.0)	641 (4.42)	66.7 (460)	1.570 (1013)	1.06	Interior column
28	9.84 (250)	5920 (40.8)	4.49 (114)	0.5 and 0.95	67.0 (298)	1328 (150.1)	1328 (150.1)	665 (4.59)	66.7 (460)	0.880 (568)	1.08	Interior column
29	9.84 (250)	6610 (45.6)	4.49 (114)	0.5 and 0.97	101 (449)	929 (105)	929 (105)	673 (4.64)	66.7 (460)	0.880 (568)	1.03	Interior column
30*	9.84 (250)	5470 (37.7)	4.49 (114)	0.75	117 (520)	0 (0)	0 (0)	454 (3.13)	40.3 (278)	1.320 (852)	1.02	Interior column
39	9.84 (250)	4210 (29.0)	4.49 (114)	0.88	113 (507)	0 (0)	0 (0)	444 (3.06)	47.1 (325)	0.460 (297)	1.52	Interior column
Mean			1.18									
Coefficient of variation			0.17									

*Semi-lightweight concrete; $\sqrt{f_c'}$ is replaced in calculation by $f_{ct}/6.7$; f_{ct} is average splitting tensile strength of lightweight aggregate concrete; f_{ct} used here = 377 psi (2.60 MPa), determined experimentally.

†v_{code} is smaller of $8\sqrt{f_c'}$, psi ($0.66\sqrt{f_c'}$, MPa) and ($3\sqrt{f_c'} + v_s$, psi) ($0.25\sqrt{f_c'} + v_s$, MPa); where $v_s = A_v f_{yt}/(b_o s)$.

Table C.3—Tests with maximum v_u at critical section of $d/2$ from column face exceeding $8\sqrt{f_c'}$, psi ($2\sqrt{f_c'}/3$, MPa) (slabs with stud shear reinforcement)

Experiment	Column size, in. (mm)*	f_c', psi (MPa)	$8\sqrt{f_c'}$, psi ($0.66\sqrt{f_c'}$, MPa)	Tested capacities		d, in. (mm)	M at critical section centroid, kip-in. (kN-m)	Maximum shear stress v_u, psi (MPa)	$\frac{v_u}{8\sqrt{f_c'}}\left(\frac{v_u}{0.66\sqrt{f_c'}}\right)$
				V, kip (kN)	M, kip-in. (kN-m)				
(1)	(2)	(3)	(4)	(5)	(6)	(7)	(8)	(9)	(10)
1	11.81 sq. (300) sq.	6020 (41.5)	621 (4.28)	476 (2120)	0 (0)	9.06 (230)	0 (0)	629 (4.24)	1.07
2	11.81 sq. (300) sq.	5550 (38.3)	589 (4.06)	428 (1900)	0 (0)	8.86 (225)	0 (0)	585 (4.03)	1.00
3	11.81 sq. (300) sq.	3250 (22.4)	456 (3.14)	346 (1540)	0 (0)	8.66 (220)	0 (0)	488 (3.37)	1.07
4	19.68 cr. (500) cr.	5550 (38.3)	589 (4.06)	665 (2960)	0 (0)	10.51 (267)	0 (0)	667 (4.60)	1.13
5	14.57 sq. (370) sq.	6620 (45.7)	651 (4.49)	790 (3510)	0 (0)	11.22 (285)	0 (0)	682 (4.70)	1.05
6	12.60 cr. (320) cr.	5870 (40.5)	613 (4.23)	600 (2670)	0 (0)	9.33 (237)	0 (0)	934 (6.44)	1.52
7	12.60 cr. (320) cr.	6020 (41.5)	621 (4.28)	620 (2760)	0 (0)	9.33 (237)	0 (0)	965 (6.66)	1.55
8	10.23 sq. (260) sq.	3120 (21.5)	447 (3.08)	271 (1200)	0 (0)	8.07 (205)	0 (0)	459 (3.17)	1.03
9	10.23 sq. (260) sq.	3270 (22.6)	457 (3.15)	343 (1530)	0 (0)	8.07 (205)	0 (0)	582 (4.01)	1.27
10	7.48 cr. (190) cr.	3310 (22.8)	460 (3.17)	142 (632)	0 (0)	5.83 (148)	0 (0)	582 (4.01)	1.26
11	7.48 cr. (190) cr.	3260 (22.5)	456 (3.14)	350 (1560)	0 (0)	9.60 (244)	0 (0)	679 (4.68)	1.48
12	7.48 cr. (190) cr.	4610 (31.8)	543 (3.74)	159 (707)	0 (0)	6.02 (153)	0 (0)	623 (4.30)	1.14
13	7.48 cr. (190) cr.	3050 (21.0)	441 (3.04)	128 (569)	0 (0)	5.91 (150)	0 (0)	516 (3.56)	1.17
14	7.48 cr. (190) cr.	3340 (23.0)	462 (3.19)	278 (1240)	0 (0)	9.72 (247)	0 (0)	530 (3.66)	1.14
15	7.48 cr. (190) cr.	3160 (21.8)	449 (3.10)	255 (1130)	0 (0)	9.76 (248)	0 (0)	482 (3.32)	1.07
16	9.25 cr. (235) cr.	4630 (31.9)	544 (3.75)	207 (921)	0 (0)	5.94 (151)	0 (0)	728 (5.02)	1.34
17	9.25 cr. (235) cr.	5250 (36.2)	580 (4.00)	216 (961)	0 (0)	6.14 (156)	0 (0)	725 (5.00)	1.25
18	9.25 cr. (235) cr.	5290 (36.5)	582 (4.01)	234 (1040)	0 (0)	6.50 (165)	0 (0)	725 (5.00)	1.24
19	7.87 sq. (200) sq.	5060 (34.9)	569 (3.92)	236 (1050)	0 (0)	6.30 (160)	0 (0)	661 (4.56)	1.16
20	7.87 sq. (200) sq.	5660 (39.0)	601 (4.14)	214 (952)	0 (0)	6.30 (160)	0 (0)	599 (4.13)	1.00
21†	9.84 sq. (250) sq.	4100 (28.3)	513 (3.54)	47.4 (211)	651 (73.6)	4.49 (114)	491 (55.5)	528 (3.64)	1.03
22†	9.84 sq. (250) sq.	4030 (27.8)	508 (3.50)	52.8 (235)	730 (82.5)	4.49 (114)	552 (62.4)	590 (4.07)	1.16
23†	9.84 sq. (250) sq.	4080 (28.1)	511 (3.52)	26.9 (120)	798 (90.2)	4.49 (114)	708 (80.0)	641 (4.42)	1.25
24†	9.84 sq. (250) sq.	4470 (30.8)	535 (3.69)	27.2 (121)	847 (95.7)	4.49 (114)	755 (85.3)	693 (4.78)	1.29

Table C.3(cont.)—Tests with maximum v_u at critical section of $d/2$ from column face exceeding $8\sqrt{f_c'}$, psi ($2\sqrt{f_c'}/3$, MPa) (slabs with stud shear reinforcement)

25	9.84 sq. (250) sq.	4280 (29.5)	523 (3.61)	135 (600)	0 (0)	4.45 (113)	0 (0)	532 (3.67)	1.02
26	9.84 sq. (250) sq.	4890 (33.7)	559 (3.86)	33.7 (150)	1434 (162.0)	4.49 (114)	1434 (162.0)	570 (3.93)	1.02
27	9.84 sq. (250) sq.	5660 (39.0)	602 (4.15)	67.4 (300)	1257 (142.0)	4.49 (114)	1257 (142.0)	641 (4.42)	1.06
28	9.84 sq. (250) sq.	5920 (40.8)	615 (4.24)	67.4 (300)	1328 (150.1)	4.49 (114)	1328 (150.1)	665 (4.59)	1.08
29	9.84 sq. (250) sq.	6610 (45.6)	651 (4.49)	101 (449)	929 (105)	4.49 (114)	924 (104)	673 (4.64)	1.03
Mean		1.17							
Coefficient of variation		0.13							

*Column 2 gives side dimension of square (sq.) columns or diameter of circular (cr.) columns.

†Edge slab-column connections. Other experiments are on interior slab-column connections.

Table C.4—Slabs with stud shear reinforcement having s approximately equal to or greater than $0.75d$

Experiment	Column size,* in. (mm)	f_c',† psi (MPa)	d, in. (mm)	s/d	Tested capacities: V, kip (kN)	Tested capacities: M, kip-in. (kN-m)	M at critical section centroid, kip-in. (kN-m)	Maximum shear stress v_u,‡ psi (MPa)	f_{yt}, ksi (MPa)	A_v, in.^{2} (mm^2)	$(v_u)_{outside}$,§ psi (MPa)	V_{test}/V_{code} ‖
(1)	(2)	(3)	(4)	(5)	(6)	(7)	(8)	(9)	(10)	(11)	(12)	(13)
3	11.81 sq. (300) sq.	3250 (22.4)	8.66 (220)	0.55 and 0.73	346 (1540)	0 (0)	0 (0)	488 (3.37)	47.9 (330)	?	214 (1.48)	1.77
12	7.48 cr. (190) cr.	4610 (31.8)	6.02 (153)	0.75	159 (707)	0 (0)	0 (0)	623 (4.30)	67.6 (466)	1.09 (703)	195 (1.34)	1.42
13	7.48 cr. (190) cr.	3050 (21.0)	5.91 (150)	0.77	128 (569)	0 (0)	0 (0)	517 (3.66)	57.6 (397)	1.09 (703)	160 (1.10)	1.43
16	9.25 cr. (235) cr.	4630 (31.9)	5.94 (151)	0.66	207 (921)	0 (0)	0 (0)	728 (5.02)	72.5 (500)	1.46 (942)	182 (1.26)	1.34
17	9.25 cr. (235) cr.	5250 (36.2)	6.14 (156)	0.65	216 (961)	0 (0)	0 (0)	725 (5.00)	72.5 (500)	1.46 (942)	180 (1.24)	1.26
18	9.25 cr. (235) cr.	5290 (36.5)	6.50 (165)	0.61	234 (1040)	0 (0)	0 (0)	725 (5.00)	42.5 (293)	1.46 (942)	181 (1.25)	1.26
19	7.87 sq. (200) sq.	5060 (34.9)	6.30 (160)	0.75	236 (1050)	0 (0)	0 (0)	661 (4.56)	54.1 (373)	1.40 (903)	165 (1.14)	1.08
21	9.84 sq. (250) sq.	4100 (28.3)	4.49 (114)	0.70	47.4 (211)	651 (73.6)	491 (55.5)	528 (3.64)	55.1 (380)	0.66 (426)	—	1.07
22	9.84 sq. (250) sq.	4030 (27.8)	4.49 (114)	0.70	52.8 (235)	730 (82.5)	552 (62.4)	590 (4.07)	55.1 (380)	0.66 (426)	—	1.20
23	9.84 sq. (250) sq.	4080 (28.1)	4.49 (114)	0.70	26.9 (120)	798 (90.2)	708 (80.0)	641 (4.42)	55.1 (380)	0.66 (426)	—	1.30
24	9.84 sq. (250) sq.	4470 (30.8)	4.49 (114)	0.70	27.2 (121)	847 (95.7)	755 (85.3)	693 (4.78)	55.1 (380)	0.66 (426)	—	1.38
26	9.84 sq. (250) sq.	4890 (33.7)	4.49 (114)	0.75	33.7 (150)	1434 (162.0)	1434 (162.0)	570 (3.93)	66.7 (460)	1.57 (1010)	—	1.02
27	9.84 sq. (250) sq.	5660 (39.0)	4.49 (114)	0.75	67.4 (300)	1257 (142.0)	1257 (142.0)	641 (4.42)	66.7 (460)	1.57 (1010)	—	1.06
30#	9.84 sq. (250) sq.	5470 (37.7)	4.49 (114)	0.75	117 (520)	0 (0)	0 (0)	454 (3.13)	40.3 (278)	1.32 (852)	—	1.02
31	9.84 sq. (250) sq.	3340 (23.0)	4.49 (114)	0.75	123 (547)	0 (0)	0 (0)	476 (3.28)	40.3 (278)	1.32 (852)	136 (0.94)	1.18
32	9.84 sq. (250) sq.	5950 (41.0)	4.49 (114)	0.75	131 (583)	0 (0)	0 (0)	509 (3.51)	70.9 (489)	1.32 (852)	145 (1.00)	0.94

Table C.4(cont.)—Slabs with stud shear reinforcement having s approximately equal to or greater than 0.75d

33	9.84 sq. (250) sq.	5800 (40.0)	4.49 (114)	0.75	131 (583)	0 (0)	0 (0)	509 (3.51)	40.3 (278)	1.32 (852)	145 (1.00)	0.95
34	9.84 sq. (250) sq.	4210 (29.0)	4.49 (114)	0.75	122 (543)	0 (0)	0 (0)	473 (3.26)	70.9 (489)	1.32 (852)	166 (1.14)	1.28
35	9.84 sq. (250) sq.	5080 (35.0)	4.49 (114)	0.75 and 1.50	129 (574)	0 (0)	0 (0)	500 (3.45)	40.3 (278)	1.32 (852)	143 (0.99)	1.00
36	9.84 sq. (250) sq.	4350 (30.0)	4.49 (114)	0.75	114 (507)	0 (0)	0 (0)	444 (3.06)	70.9 (489)	1.32 (852)	178 (1.23)	1.35
38	9.84 sq. (250) sq.	4790 (33.0)	4.49 (114)	0.70	48 (214)	637 (72.0)	476 (53.8)	522 (3.60)	55.1 (380)	0.66 (426)	—	1.03
39	9.84 sq. (250) sq.	4210 (29.0)	4.45 (113)	0.88	113 (503)	0 (0)	0 (0)	444 (3.06)	47.1 (325)	0.46 (297)	—	1.52
40	9.84 sq. (250) sq.	4240 (29.2)	4.45 (113)	1.00	125 (556)	0 (0)	0 (0)	492 (3.39)	52.3 (361)	1.74 (1120)	253 (1.74)	1.94
41	9.84 sq. (250) sq.	5300 (36.6)	4.45 (113)	0.88	133 (592)	0 (0)	0 (0)	523 (3.61)	49.2 (339)	0.99 (639)	221 (1.52)	1.52
42	9.84 sq. (250) sq.	5380 (37.1)	4.45 (113)	0.88	133 (592)	0 (0)	0 (0)	523 (3.61)	49.2 (339)	1.48 (955)	273 (1.88)	1.86
43	12.0 sq. (305) sq.	4880 (33.7)	4.76 (121)	1.00	134 (596)	0 (0)	0 (0)	419 (2.89)	73.0 (503)	1.54 (994)	270 (1.86)	1.93
Mean				1.31								
Coefficient of variation				0.23								

*Column 2 gives side dimension of square (sq.) columns, or diameter of circular (cr.) columns.

†For cube strengths, concrete cylinder strength in Column 3 calculated using $f_c' = 0.83 f_{cube}'$.

‡Column 9 is maximum shear stress at failure in critical section at $d/2$ from column face.

§$(v_u)_{outside}$ in Column 12 is maximum shear stress at failure in critical section at $d/2$ outside outermost studs; $(v_u)_{outside}$ not given for slabs that failed within stud zone.

‖v_{code} is value allowed by ACI 318-14, 4.3.3. v_{code} calculated at $d/2$ from column face when failure is within stud zone and at section at $d/2$ from outermost studs when failure is outside shear-reinforced zone.

#Slab 30 is semi-lightweight concrete. $\sqrt{f_c'}$ replaced in calculations by $f_{ct}/6.7$; f_{ct} is average splitting tensile strength of lightweight aggregate concrete; f_{ct} used herein = 377 psi (2.60 MPa), determined experimentally.

APPENDIX D—DESIGN EXAMPLES

The design procedure, presented in Chapter 4, is illustrated by numerical examples for connections of nonprestressed slabs with interior, edge and corner columns. Section D.4 is a design example of shear reinforcement for a connection of an interior column with a prestressed slab.

D.1—Interior column-slab connection

The design of headed studs, conforming to ASTM A1044/A1044M, is required at an interior column (Fig. D.1) based on the following data:

column size c_x x c_y = 12 x 20 in.^{2} (305 x 508 mm^2)
slab thickness h = 7 in. (178 mm)
concrete cover = 0.75 in. (19 mm)
f_c' = 4000 psi (27.6 MPa)
yield strength of studs f_{yt} = 51 ksi (350 MPa)
flexural reinforcement nominal diameter = 5/8 in. (16 mm)

The factored forces transferred from the column to the slab are: V_u = 110 kip (489 kN) and M_{uy} = 600 kip-in. (67.8 kN-m). The five steps of design outlined in 4.4 are followed:

Step 1—The effective depth of slab

$$d = 7 - 0.75 - (5/8) = 5.62 \text{ in. (143 mm)}$$

Properties of a critical section at $d/2$ from column face shown in Fig. 4.1(a): b_o = 86.5 in. (2197 mm); A_c = 486 in.^{2} (314 × 10^3 mm^2); J_y = 28.0 × 10^3 in.^{4} (11.7 × 10^9 mm^4); l_{x1} = 17.62 in. (448 mm); and l_{y1} = 25.62 in. (651 mm).

The fraction of moment transferred by shear (Eq. (4.2b))

$$\gamma_{vy} = 1 - \frac{1}{1 + \frac{2}{3}\sqrt{\frac{17.62}{25.62}}} = 0.36$$

The maximum shear stress occurs at x = 17.62/2 = 8.81 in. (224 mm), and its value is (Eq. (4.2a))

$$v_u = \frac{110 \times 10^3}{486} + \frac{0.36(600 \times 10^3)8.81}{28.0 \times 10^3} = 294 \text{ psi (2.03 MPa)}$$

$$\frac{v_u}{\phi} = \frac{294}{0.75} = 392 \text{ psi} = 6.2\sqrt{f_c'} \text{ (2.70 MPa} = 0.52\sqrt{f_c'})$$

The nominal shear stress that can be resisted without shear reinforcement at the critical section considered (Eq. (4.3.1a) to (4.3.1c))

$$v_n = \left(2 + \frac{4}{1.67}\right)\sqrt{f_c'} = 4.4\sqrt{f_c'} \text{ (or } 0.37\sqrt{f_c'})$$

$$v_n = \left(\frac{40(5.62)}{86.5} + 2\right)\sqrt{f_c'} = 4.6\sqrt{f_c'} \text{ (or } 0.38\sqrt{f_c'}\text{)}$$

$$v_n = 4\sqrt{f_c'} \text{ (or } 0.33\sqrt{f_c'}\text{)}$$

Use the smallest value: $v_n = 4\sqrt{f_c'} = 253$ psi ($0.33\sqrt{f_c'} = 1.74$ MPa).

Step 2—The quantity (v_u/ϕ) is greater than v_n, indicating that shear reinforcement is required; the same quantity is less than the upper limit $v_n = 8\sqrt{f_c'}$ (in.-lb units) ($0.66\sqrt{f_c'}$ [SI units]), which means that the slab thickness is adequate. Stirrups are not permitted by ACI 318 because (v_u/ϕ) is greater than $6\sqrt{f_c'}$ (in.-lb units) ($0.5\sqrt{f_c'}$ [SI units]).

The shear stress resisted by concrete in the presence of headed studs at the critical section at $d/2$ from column face

$$v_c = 3\sqrt{f_c'} = 190 \text{ psi (1.31 MPa)}$$

Use of Eq. (4.1), (4.3.2a), and (4.3.2c) gives

$$v_s \geq \frac{v_u}{\phi} - v_c = 392 - 190 = 202 \text{ psi (1.39 MPa)}$$

$$\frac{A_v}{s} \geq \frac{v_s b_o}{f_{yt}} = \frac{202(86.5)}{51{,}000} = 0.34 \text{ in. (8.7 mm)}$$

Step 3—Studs' arrangement are governed by

$$s_o \leq 0.5d = 2.8 \text{ in. (71 mm)}; s \leq 0.5d = 2.8 \text{ in. (71 mm)}$$

This example has been provided for one specific type of headed shear stud reinforcement, but the approach can be adapted and used also for other types mentioned in Appendix A.

Using 3/8 in. (9.5 mm) diameter studs welded to a bottom anchor strip 3/16 x 1 in.^{2} (5 x 25 mm^2), along with a cover of 3/4 in. (19 mm) at top and bottom, results in a specified overall height of headed stud assembly (having most efficiency) (Eq. (6))

$$l_s = 7 - 2(3/4) = 5.5 \text{ in. (140 mm)}$$

The actual overall height (considering manufacturing tolerance) should not be less than l_s – 1/2 the diameter of flexural reinforcement bars (5/8 in.) = 5-3/16 in. (132 mm)

With 10 studs per peripheral line, choose the spacing between peripheral lines, s = 2.75 in. (70 mm), and the spacing between column face and first peripheral line, s_o = 2.25 in. (57 mm) (Fig. D.1)

$$\frac{A_v}{s} = \frac{10(0.11)}{2.75} = 0.40 \text{ in. (10.1 mm)}$$

This value is greater than 0.34 in. (8.7 mm), indicating that the choice of studs and their spacing are adequate.

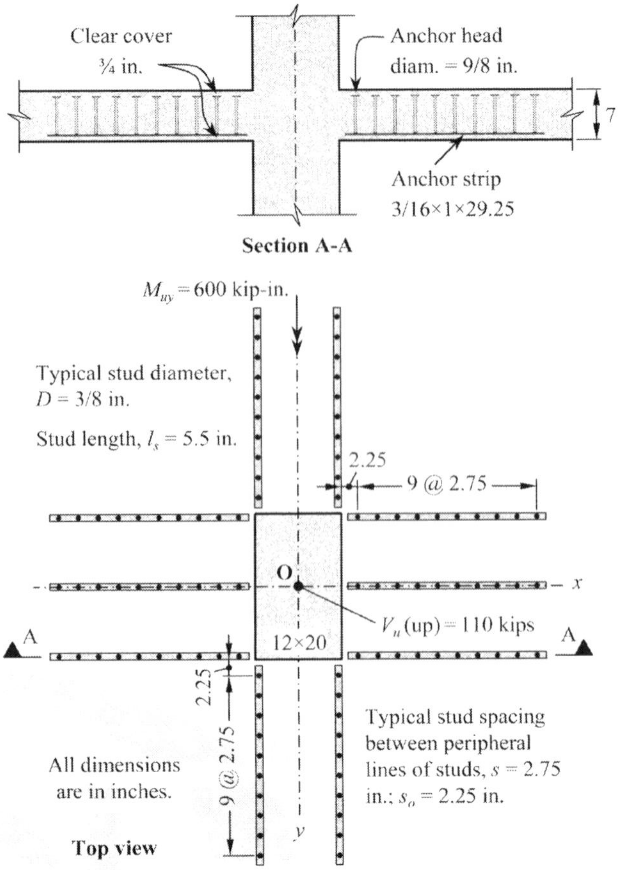

Fig. D.1—Example of interior column-slab connection: stud arrangement. (Note: 1 in. = 25.4 mm; 1 kip = 4.448 kN.)

Step 4—Selection of the outer shear critical section is an iterative process. As a guess, assume a critical section at 4.5d from column face (Fig. 4.1(b)):

$\alpha = 4.5$; $\alpha d = 4.5(5.62) = 25.3$ in. (643 mm)
$l_{x2} = 62.6$ in. (1590 mm); $l_{y2} = 70.6$ in. (1793 mm); $\gamma_{vy} = 0.39$ (Eq. (B.2))
$l = 34.1$ in. (866 mm); $b_o = 209.8$ in. (5329 mm)
$A_c = 1179$ in.^{2} (760.6×10^3 mm^2); $J_y = 547.3 \times 10^3$ in.^{4} (227.8×10^9 mm^4)

The maximum shear stress in the critical section occurs on line AB at: $x = 62.6/2 = 31.3$ in. (795 mm); Eq. (4.2a) gives

$$v_u = \frac{110 \times 10^3}{1179} + \frac{0.39(600 \times 10^3)31.3}{547.3 \times 10^3} = 107 \text{ psi (0.74 MPa)}$$

$$\frac{v_u}{\phi} = \frac{107}{0.75} = 142 \text{ psi (0.98 MPa)}$$

$$v_n = 2\sqrt{f_c'} = 126 \text{ psi} \quad (0.17\sqrt{f_c'} = 0.87 \text{ MPa})$$

The value (v_u/ϕ) = 142 psi (0.98 MPa) is greater than v_n = 126 psi (0.87 MPa), which indicates that shear stress should be checked at $\alpha > 4.5$. Try 10 peripheral lines of studs; the distance between column face and outermost peripheral line of studs is

$$s_o + 9s = 2.25 + 9(2.75) = 27 \text{ in. (686 mm)}$$

Check shear stress at a critical section at a distance from column face

$$\alpha d = 27 + d/2 = 27 + 5.62/2 = 29.8 \text{ in. (757 mm)}$$

$$\alpha = \frac{29.8}{d} = \frac{29.8}{5.62} = 5.3$$

$$\frac{v_u}{\phi} = 125 \text{ psi (0.86 MPa)}$$

$$v_n = 2\sqrt{f_c'} = 126 \text{ psi (0.87 MPa)}$$

Step 5—The value of (v_u/ϕ) is less than v_n, which indicates that the extent of the shear-reinforced zone, shown in Fig. D.1, is adequate.

The value of V_u used to calculate the maximum shear stress could have been reduced by the counteracting factored load on the slab area enclosed by the critical section; this reduction is ignored in D.2 to D.4.

D.2—Edge column-slab connection

Design the studs required at the edge column-slab connection in Fig. D.2(a), based on the following data: column cross section, $c_x \times c_y = 18 \times 18$ in.^{2} (457 x 457 mm^2); the values of h, c_t, d, f_c', f_{yt}, D, and d_b, in Section D.1 apply herein. The connection is designed for gravity loads combined with wind load in positive or negative x-direction. Cases I and II are considered, which produce extreme stresses at Points B and A of the shear critical section at $d/2$ from the column or at D and C of the shear critical section at $d/2$ from the outermost peripheral line of studs (Fig. D.2(a) and (b)). The factored forces, due to gravity load combined with wind load, are given.

Case I—Wind load in negative *x*-direction

$V_u = 36$ kip (160 kN);
$M_{uOy} = 1720$ kip-in. (194 kN-m);
$M_{uOx} = 0$

For the shear critical section at $d/2$ from column face, $x_O = -5.17$ in., and Eq. (4.2d) gives

$$M_{uy} = 1720 + 36\,(-5.17) = 1530 \text{ kip-in. (173 kN-m)};$$
$$M_{ux} = 0$$

Case II—Wind load in positive *x*-direction

$V_u = 10$ kip (44 kN);
$M_{uOy} = -900$ kip-in. (–102 kN-m)
$M_{uy} = -900 + 10\,(-5.17) = -952$ kip-in. (–107 kN-m)

The five steps of design outlined in 4.4 are followed.

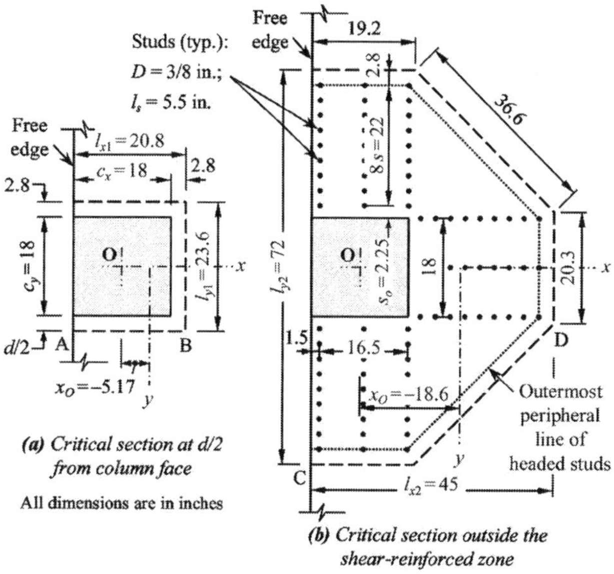

Fig. D.2—Example of edge column-slab connection: shear critical sections and stud arrangement. (Note: 1 in. = 25.4 mm.)

Table D.2—Shear stresses* (psi) due to factored loads; edge column-slab connection (Fig. D.2)

Shear critical section	Case I		Case II	
At $d/2$ from column face	$(v_u)_A$	$(v_u)_B$	$(v_u)_A$	$(v_u)_B$
	–338	302	299	–100
At $d/2$ from outermost peripheral line of studs	$(v_u)_C$	$(v_u)_D$	$(v_u)_C$	$(v_u)_D$
	–13	87	77	–27

*v_u represents stress exerted by column on slab, with a positive sign indicating upward stress.

Note: 1 MPa = 145 psi.

Step 1—Properties of the shear critical section at $d/2$ from column face shown in Fig. D.2(a) are:
$b_o = 65.25$ in. (1581 mm)
$A_c = 367$ in.^{2} (237×10^3 mm^2)
$J_y = 17.63 \times 10^3$ in.^{4} (7.338×10^6 mm^4)
$l_{x1} = 20.81$ in. (529 mm)
$l_{y1} = 23.82$ in. (600 mm)

The fraction of moment transferred by shear (Eq. (B.4))

$$\gamma_{vy} = 1 - \frac{1}{1 + \frac{2}{3}\sqrt{\frac{20.81}{23.62} - 0.2}} = 0.36$$

The shear stress at Points A and B, calculated by Eq. (4.2a) with $x_A = -14.17$ in. or $x_B = 6.64$ in., are given in Table D.2.

The maximum shear stress, in absolute value, occurs at Point A (Case I) and $|(v_u/\phi)_A| = 338/0.75 = 451$ psi $= 7.1\sqrt{f_c'}$ (3.13 MPa $= 0.59\sqrt{f_c'}$).

The nominal shear stress that can be resisted without shear reinforcement at the shear critical section, $v_n = 4\sqrt{f_c'} = 253$ psi ($0.33\sqrt{f_c'} = 1.74$ MPa).

Step 2—Because the value (v_u/ϕ) exceeds v_n, shear reinforcement is required; the same quantity is less than the upper limit $v_n = 8\sqrt{f_c'}$ psi ($0.66\sqrt{f_c'}$ MPa), indicating that the slab thickness is adequate.

The shear stress resisted by concrete in presence of headed studs at the shear critical section at $d/2$ from the column face is

$$v_c = 3\sqrt{f_c'} = 190 \text{ psi } (0.25\sqrt{f_c'} = 131 \text{ MPa})$$

Use of Eq. (4.1), (4.3.2a), and (4.3.2c) gives

$$v_s \geq \frac{v_u}{\phi} - v_c = 451 - 190 = 261 \text{ psi } (1.80 \text{ MPa})$$

$$\frac{A_v}{s} \geq \frac{v_s b_o}{f_{yt}} = \frac{261(65.25)}{51{,}000} = 0.33 \text{ in. } (8.5 \text{ mm})$$

Step 3—Studs' arrangement are governed by

$$s_o \leq 0.5d = 2.8 \text{ in. } (71 \text{ mm}); s \leq 0.5d = 2.8 \text{ in. } (71 \text{ mm})$$

Using 3/8 in. (9.5 mm) diameter studs, arranged as shown in Fig. D.2(b), with $s_o = 2.25$ in. (57 mm) and $s = 2.75$ in. (70 mm) gives: $(A_v/s) = 9(0.11)/2.75 = 0.36$ in. (9.1 mm). This value is greater than 0.33 in. (8.5 mm), indicating that the choice of studs and their spacing are adequate.

Step 4—Try nine peripheral lines of studs; the properties of the shear critical section at $d/2$ from the outermost peripheral line of studs are:

$b_o = 132$ in. (3353 mm)
$A_c = 742$ in.^{2} (479×10^3 mm^2)
$J_y = 142.9 \times 10^3$ in.^{4} (59.48×10^9 mm^4)
$l_{x2} = 45$ in. (1143 mm)
$l_{y2} = 72$ in. (1829 mm)
$\gamma_{vy} = 0.30$ (Eq. (B.4))
$x_C = -27.6$ in. (–701 mm)
$x_D = 17.4$ in. (445 mm)
$x_O = -18.6$ in. (–472 mm)

The factored shearing force and slab moment at an axis, passing through the centroid of the shear critical section outside the shear-reinforced zone, are (Eq. (4.2d)):

Case I:
$V_u = 36$ kip (160 kN);
$M_{uy} = 1720 + 36(-18.6) = 1050$ kip-in. (118 kN-m)

Case II:
$V_u = 10$ kip (44 kN);
$M_{uy} = -900 + 10(-18.6) = -1090$ kip-in. (–123 kN-m)

Equation (4.2a) gives the shear stresses at Points C and D, listed in Table D.1 for Cases I and II.

The maximum shear stress, in absolute value, occurs at Point D (Case I) and $|(v_u/\phi)_D| = 87/0.75 = 116$ psi $= 1.8\sqrt{f_c'}$ (0.80 MPa $= 0.15\sqrt{f_c'}$). The nominal shear strength outside the shear-reinforced zone (psi or MPa), $v_n = 2\sqrt{f_c'} = 126$ psi ($0.17\sqrt{f_c'} = 0.87$ MPa).

Step 5—The value of (v_u/ϕ) is less than v_n, indicating that the extent of the shear-reinforced zone, as shown in Fig. D.2(b), is adequate.

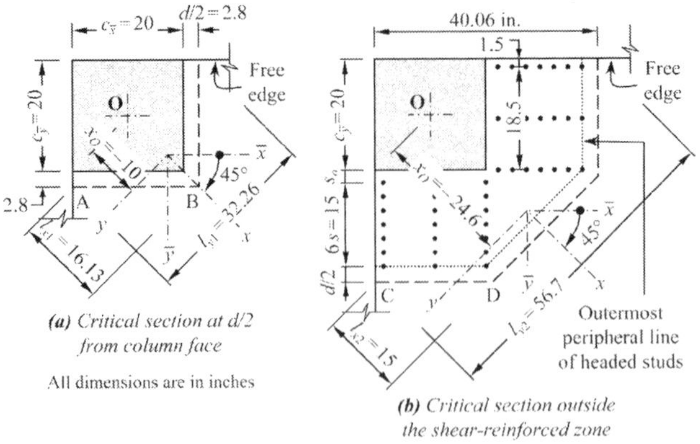

Fig. D.3—Example of corner column-slab connection: shear critical sections and stud arrangement. (Note: 1 in. = 25.4 mm.)

D.3—Corner column-slab connection

The corner column-slab connection in Fig. D.3(a) is designed for gravity loads combined with wind load in positive or negative $\bar{x}$-direction. The cross-sectional dimensions of the column are $c_{\bar{x}} = c_{\bar{y}} = 20$ in. (508 mm) (Fig. D.3(a)). The same values of: h, c_t, d, f_c', f_{yt}, D, and d_b, in D.1 apply in this example. Two cases (I and II) are considered, producing extreme shear stresses at Points A and B of the shear critical section at $d/2$ from the column or at C and D of the shear critical section at $d/2$ from the outermost peripheral line of studs (Fig. D.3(a) and (b)). The factored forces, due to gravity loads combined with wind load are given.

Case I—Wind load in positive $\bar{x}$-direction
$V_u = 6$ kip (27 kN);
$M_{uO\bar{y}} = -338$ kip-in. (–38 kN-m);
$M_{uO\bar{x}} = 238$ kip-in. (27 kN-m)

For the shear critical section at $d/2$ from column face, $\bar{x}_O = \bar{y}_O = -7.11$ in. (–181 mm) and $\theta = 45$ degrees; thus, Eq. (4.2d) and (4.2e) give

$$M_{u\bar{y}} = -338 + 6(-7.11) = -381 \text{ kip-in.};$$
$$M_{u\bar{x}} = 238 + 6(-7.11) = 195 \text{ kip-in.}$$

$$M_{uy} = -132 \text{ kip-in. } (-15 \text{ kN-m});$$
$$M_{ux} = 407 \text{ kip-in. } (46 \text{ kN-m})$$

The aforementioned moment values are calculated by substitution of $M_{uO\bar{x}}$, $M_{uO\bar{y}}$, $\bar{x}_O$, $\bar{y}_O$, and θ in Eq. (4.2d) and (4.2e).

Case II—Wind load in negative $\bar{x}$-direction
$V_u = 22$ kip (97 kN);
$M_{uO\bar{y}} = 953$ kip-in. (108 kN-m);
$M_{uO\bar{x}} = 377$ kip-in. (43 kN-m)

$$M_{u\bar{y}} = 953 + 22(-7.11) = 797 \text{ kip-in.};$$
$$M_{u\bar{x}} = 377 + 22(-7.11) = 221 \text{ kip-in.}$$

$M_{uy} = 720$ kip-in. (81 kN-m);
$M_{ux} = -407$ kip-in. (–46 kN-m)

The five steps of design, outlined in 4.4, are followed.

Step 1—Properties of the shear critical section in Fig. D.3(a) are:
$b_o = 45.63$ in. (1159 mm)
$A_c = 257$ in.^{2} (166×10^3 mm^2)
$J_x = 22.26 \times 10^3$ in.^{4} (9.27×10^9 mm^4)
$J_y = 5.57 \times 10^3$ in.^{4} (2.32×10^9 mm^4)

The projections of the critical section on the x and y-axes are: $l_{x1} = 16.13$ in. (410 mm) and $l_{y1} = 32.26$ in. (820 mm). The fractions of slab moments transferred by shear are (Eq. (B.5) and (B.6))

$$\gamma_{vy} = 1 - \frac{1}{1+(2/3)\sqrt{(l_{x1}/l_{y1}) - 0.2}} = 0.267; \ \gamma_{vx} = 0.4$$

The factored shear stress at Point A (–8.07, 16.13 in.) in Case I is (Eq. (4.2a))

$$(v_u)_A = \frac{6\times10^3}{257} + \frac{0.4(407\times10^3)16.13}{22.26\times10^3} + \frac{0.267(-132\times10^3)(-8.07)}{5.57\times10^3} = 192 \text{ psi (1.33 MPa)}$$

Similar calculations give the values of v_u at Points A and B (8.07, 0 in.) for Cases I and II, which are listed in Table D.3.

The maximum shear stress, in absolute value, occurs at Point B (Case II) and $|(v_u/\phi)_B| = 364/0.75 = 485$ psi $= 7.7\sqrt{f_c'}$ (3.35 MPa $= 0.64\sqrt{f_c'}$). The nominal shear stress that can be resisted without shear reinforcement at the shear critical section, $v_c = 4\sqrt{f_c'} = 253$ psi ($\sqrt{f_c'}/3 = 1.74$ MPa) (Eq. (4.3.1a) to (4.3.1c)).

Step 2—Because the value (v_u/ϕ) exceeds v_n, shear reinforcement is required; the same quantity is less than the upper limit, $v_n = 8\sqrt{f_c'}$, psi ($0.66\sqrt{f_c'}$, MPa), indicating that the slab thickness is adequate.

The shear stress resisted by concrete in presence of headed studs at the shear critical section at $d/2$ from the column face is

$$v_c = 3\sqrt{f_c'} = 190 \text{ psi } (0.25\sqrt{f_c'} = 1.31 \text{ MPa})$$

Use of Eq. (4.1), (4.3.2a), and (4.3.2c) gives

$$v_s \geq \frac{v_u}{\phi} - v_c = 485 - 190 = 295 \text{ psi (2.03 MPa)}$$

$$\frac{A_v}{s} \geq \frac{v_s b_o}{f_{yt}} = \frac{295(45.63)}{51{,}000} = 0.26 \text{ in. (6.7 mm)}$$

Step 3—Studs' arrangement is governed by

$$s_o \leq 0.5d = 2.8 \text{ in. (71 mm)}; \ s \leq 0.5d = 2.8 \text{ in. (71 mm)}$$

Using 3/8 in. (9.5 mm) diameter studs, arranged as shown in Fig. D.3(b), with $s_o = 2.25$ in. (57 mm) and $s = 2.5$ in. (64 mm) gives: $(A_v/s) = 6(0.11)/2.5 = 0.26$ in. (6.7 mm). This value is the same as that calculated in Step 2, indicating that the choice of studs and their spacing are adequate.

Table D.3—Shear stresses* (psi) due to factored loads; corner column-slab connection (Fig. D.3)

Shear critical section	Case I		Case II	
At $d/2$ from column face	$(v_u)_A$	$(v_u)_B$	$(v_u)_A$	$(v_u)_B$
	192	–28	–312	364
At $d/2$ from outermost peripheral line of studs	$(v_u)_C$	$(v_u)_D$	$(v_u)_C$	$(v_u)_D$
	89	19	–46	65

*v_u represents stress exerted by column on slab, with a positive sign indicating an upward stress.

Note: 1 MPa = 145 psi.

Step 4—As a guess, try seven peripheral lines of studs; the properties of the shear critical section at $d/2$ from the outermost peripheral line of studs (Fig. D.3(b)) are:
$\bar{x}_O = \bar{y}_O = -17.37$ in. (–441 mm);
$\theta = 45$ degrees
$b_o = 69$ in. (1754 mm);
$A_c = 388$ in.^{2} (251×10^3 mm^2)
$J_x = 116.9 \times 10^3$ in.^{4} (48.64×10^9 mm^4);
$J_y = 9.60 \times 10^3$ in.^{4} (4.00×10^9 mm^4)

As an example of application of Eq. B.8, calculate the J_x-value for the outer shear critical section in Fig. D.3(b). The coordinates of the points C to F are:

Point	x (in.)	y (in.)
C	–10.38	28.35
D	4.59	13.36
E	4.59	–13.36
F	–10.38	–28.35

Substituting these values in Eq. (B.8) gives the contributions of CD, DE, and EF to $J_x = 54.0 \times 10^3$, 8.9×10^3, and 54.0×10^3 in.^{4}, respectively. The sum gives: $J_x = 116.9 \times 10^3$ in.^{4}.

$l_{x2} = 15.0$ in. (380 mm)
$l_{y2} = 56.7$ in. (1439 mm)
$\gamma_{vx} = 0.40$ (Eq. (B.5))
$\gamma_{vy} = 0.14$ (Eq. (B.6))

The factored shearing force and slab moment about the centroidal principal axes of the shear critical section outside the shear-reinforced zone (Eq. (4.2d) and (4.2e)), are

Case I:
$V_u = 6$ kip (27 kN);
$M_{ux} = 407$ kip-in. (46 kN-m);
$M_{uy} = -218$ kip-in. (–25 kN-m)

Case II:
$V_u = 22$ kip (97 kN);
$M_{ux} = -407$ kip-in. (–46 kN-m);
$M_{uy} = 402$ kip-in. (45 kN-m)

Use of Eq. (4.2a) gives the values of v_u at Points C (–10.38, 28.33 in.) and D (4.59, 13.36 in.) for Cases I and II, listed in Table D.3.

The maximum shear stress, in absolute value, occurs at Point C (Case I) and $|(v_u/\phi)_C| = 89/0.75 = 119$ psi $= 1.88\sqrt{f_c'}$ (0.82 MPa $= 0.16\sqrt{f_c'}$). The nominal shear stress outside the shear-reinforced zone, $v_n = 2\sqrt{f_c'} = 126$ psi ($0.17\sqrt{f_c'} = 0.87$ MPa).

Step 5—The value of $|(v_u/\phi)$ is less than v_n, indicating that the extent of the shear-reinforced zone, as shown in Fig. D.3(b), is sufficient.

D.4—Prestressed slab-column connection

Design the shear reinforcement required for an interior column, transferring $V_u = 110$ kip (490 kN) combined with slab moment $M_{sc\text{-}y} = 550$ kip-in. (62 kN-m) to a post-tensioned flat plate of thickness, $h = 7$ in. (178 mm). The slab has equal spans 280 x 280 in.^{2} (7.1 x 7.1 m^2). The column size is 16 x 16 in.^{2} (406 × 406 mm^2). The values of c_t, f_c', f_{yt}, and D, in D.1 apply herein. Tendon profiles are commonly composed of parabolic segments, for which the average effective prestress f_{pc}, required to balance a fraction κ of the self-weight, $(h\gamma_{conc})$ per unit area, plus the superimposed dead load of intensity w_{sd} can be calculated as (Gayed and Ghali 2006) (Fig. D.4(a))

$$f_{pc} = \frac{\kappa(1-2\alpha)(\gamma_{conc}h + w_{sd})L^2}{8hh_c} \quad \text{(D.4)}$$

where L is the panel length, and geometrical parameters: α and h_c are defined in Fig. D.4(a). Choose the values:

$\kappa = 0.85$
$\alpha = 0.1$
$\gamma_{conc} = 153$ lb/ft^3 (24 kN/m^3)
$w_{sd} = 27$ lb/ft^2 (1.3 kPa)
$L = 280$ in. (7.1 m)
$h = 7$ in. (178 mm)
$h_c = 3.8$ in. (97 mm)

Equation (D.4) gives $f_{pc} = 202$ psi (1.39 MPa). This level of prestressing is closely acquired by ten 0.6 in. seven-wire post-tensioned non-bonded strands per panel. The cross-sectional area per strand = 0.217 in.^{2} (140 mm^2); the average value of the effective compressive stress provided by 10 tendons in each of two directions is

$$f_{pc} = \frac{10(38\times10^3)}{280(7)} = 194 \text{ psi (1.34 MPa)}$$

Prestressing tendons are typically placed in bands over support lines in one direction and uniformly distributed in the perpendicular direction. In the current example, the prestressing tendons are banded in the x-direction and uniformly distributed in the y-direction (Fig. D.4(b)). ACI 318 requires that at least two tendons should pass through the column cage in each direction; the arrangement of the tendons as shown in Fig. D.4(b) satisfies this requirement. ACI 318 requires a minimum amount of bonded top flexural reinforcing bars in the vicinity of the column; choose eight

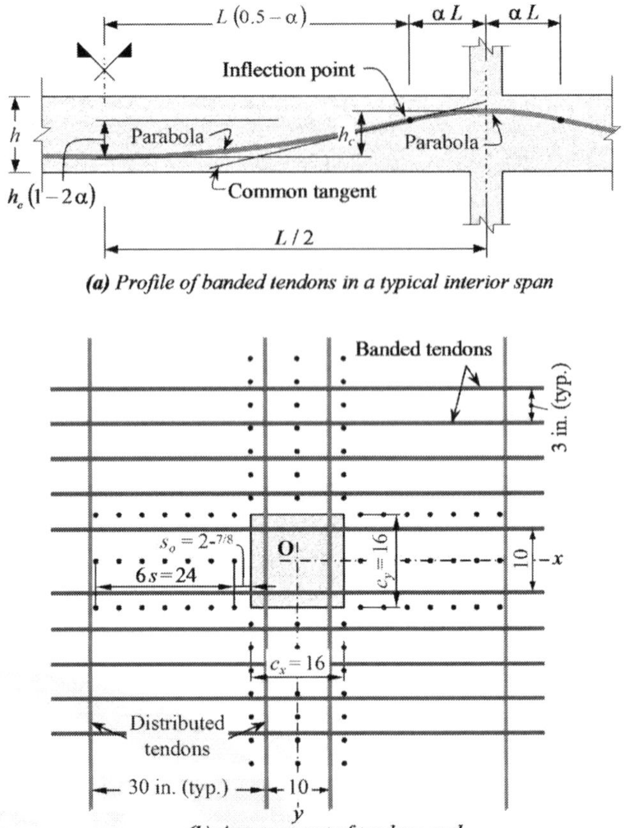

(a) Profile of banded tendons in a typical interior span

(b) Arrangement of tendons and headed stud shear reinforcement

Fig. D.4—Example connection of interior column-prestressed slab. (Note: 1 in. = 25.4 mm.)

bars of diameter $d_b = 1/2$ in.; for clarity, the bonded bars are not shown in Fig. D.4. A check that the cross-sectional areas of the bonded and nonbonded reinforcements satisfy the ultimate flexural strength required is necessary but is beyond the scope of the present report.

Punching shear design: $V_u = 110$ kip (490 kN); $M_{uy} = 550$ kip-in. (62 kN-m).

The five steps of design, outlined in 4.4, are followed.

Step 1—Properties of the shear critical section at $d/2$ from the column are:

$d = h - c_t - d_b = 7 - 3/4 - 1/2 = 5.75$ in. (146 mm)
$b_o = 87$ in. (2210 mm)
$A_c = 500$ in.^{2} (323 × 10^3 mm^2)
$J_y = 39.4 \times 10^3$ in.^{4} (16.4 × 10^9 mm^4)
$l_{x1} = l_{y1} = 21.75$ in. (552 mm)
$\gamma_{vy} = 0.4$ (Eq. (B.2))

The maximum shear stress occurs at $x = 21.75/2 = 10.88$ in. (276 mm), and its value is (Eq. (4.2a))

$$v_u = \frac{110\times10^3}{500} + \frac{0.40(550\times10^3)10.88}{39.4\times10^3} = 281 \text{ psi (1.94 MPa)}$$

$$\frac{v_u}{\phi} = \frac{281}{0.75} = 375 \text{ psi} = 5.9\sqrt{f_c'} \;\; (2.59 \text{ MPa} = 0.49\sqrt{f_c'})$$

The three conditions, warranting the use of Eq. (5.1a), are satisfied at the considered connection. Two tendons from each direction intercept the critical section at $d/2$ from the column; the sum of the vertical components of these tendons at the location of the shear critical section, V_p = 6 kip (26 kN). It is uncertain that the actual cable profiles, in the x- and y-directions, will have slopes matching those used in calculating V_p (≈ 0.02); thus, for safety, assume that $V_p = 0$. Substituting the values of f_{pc} and V_p in Eq. (5.1a) gives

$$v_n \equiv v_c = 3.5\sqrt{4000} + 0.3(194) + 0 = 280 \text{ psi (1.93 MPa)}$$

Step 2—The quantity (v_u/ϕ) is greater than v_n, indicating that shear reinforcement is required; the same quantity is less than the upper limit $v_n = 8\sqrt{f_c'}$ (in.-lb units) ($0.66\sqrt{f_c'}$ [SI units]), which means that the slab thickness is adequate.

The shear stress resisted by concrete in the presence of headed studs at the critical section at $d/2$ from column face

$$v_c = 3\sqrt{f_c'} = 190 \text{ psi (1.31 MPa)}$$

Use of Eq. (4.1), (4.3.2a), and (4.3.2c) gives

$$v_s \geq \frac{v_u}{\phi} - v_c = 375 - 190 = 185 \text{ psi (1.28 MPa)}$$

$$\frac{A_v}{s} \geq \frac{v_s b_o}{f_{yt}} = \frac{185(87)}{51{,}000} = 0.32 \text{ in. (8.0 mm)}$$

Step 3—(v_u/ϕ) < $6\sqrt{f_c'}$ (psi); thus, stirrups or headed studs can be used. For ease of installation of the prestressing tendons, use studs with $s \leq 0.75d$. Because the column width is large with respect to d, eight studs per peripheral line will not satisfy the requirement $g \leq 2d$ (Fig. 1.3b); choose 12 studs per peripheral line.

$$s_o \leq 0.5d = 2\text{-}7/8 \text{ in. (71 mm)}; s \leq 0.75d = 4\text{-}5/8 \text{ in. (117 mm)}$$

With twelve 3/8 in. studs per peripheral line and spacing, s = 4 in. (102 mm)

$$\frac{A_v}{s} = \frac{12(0.11)}{4} = 0.33 \text{ in. (8.4 mm)}$$

This value is greater than 0.32 in. (8 mm), indicating that the choice of studs and their spacing are adequate.

Step 4—As a guess, assume seven peripheral lines of studs. Properties of critical section at $d/2$ from the outermost peripheral line of studs (Fig. D.4(b)) are:

$l_{x1} = l_{y1}$ = 18.3 in.
$l_{x2} = l_{y2}$ = 75.5 in.
γ_{vy} = 0.4 (Eq. (B.2))
l = 40.4 in.; b_o = 235 in.
A_c = 1351 in.^{2}
$J_y = 848.2 \times 10^3$ in.^{4}

The maximum shear stress in the critical section occurs at: x = 75.5/2 = 37.8 in. (959 mm); Eq. (4.2a) gives

$$v_u = \frac{110 \times 10^3}{1351} + \frac{0.40(550 \times 10^3)37.8}{848.2 \times 10^3} = 91 \text{ psi (0.63 MPa)}$$

$$v_n = 2\sqrt{f_c'} = 126 \text{ psi } (v_n = 0.17\sqrt{f_c'} = 0.87 \text{ MPa})$$

Step 5—The value of (v_u/ϕ) = 91/0.75 = 121 < 126 psi, indicating that the extent of the shear-reinforced zone, as shown in Fig. D.4(b), is adequate.

Guide to Seismic Design of Punching Shear Reinforcement in Flat Plates

Reported by Joint ACI-ASCE Committee 421

Mahmoud E. Kamara
Chair

Simon J. Brown	Amin Ghali*	James S. Lai	Edward G. Nawy
Pinaki R. Chakrabarti	Hershell Gill	Mark D. Marvin	Eugenio M. Santiago
William L. Gamble*	Neil L. Hammill	Sami Hanna Megally*	Thomas C. Schaeffer
Ramez Botros Gayed*	Theodor Krauthammer	Michael C. Mota	Stanley C. Woodson

*Member of the subcommittee that prepared this guide.
The committee would like to thank Frieder Seible for his contribution to this guide.

During an earthquake, the unbalanced moments transferred at flat plate-column connections can produce significant shear stresses that increase the vulnerability of these connections to brittle punching shear failure. This guide provides recommendations for designing flat plate-column connections with sufficient ductility to withstand lateral drift without punching shear failure or loss of moment transfer capacity. This guide treats reinforced concrete flat plates with or without post-tensioning.

Keywords: ductility; flat plate; post-tensioning; punching shear; seismic design; shear reinforcement; stud shear reinforcement.

CONTENTS

ACI 421.2R-10 supersedes ACI 421.2R-07 and was adopted and published April 2010.

CHAPTER 1—INTRODUCTION

1.1—General

Brittle punching failure can occur due to the transfer of shear forces combined with unbalanced moments between slabs and columns. During an earthquake, significant horizontal displacement of a flat plate-column connection may occur, resulting in unbalanced moments that induce additional slab shear stresses. As a result, some flat plate structures have collapsed by punching shear in past earthquakes (Berg and Stratta 1964; Yanev et al. 1991; Mitchell et al. 1990, 1995). During the 1985 Mexico earthquake (Yanev et al. 1991), 91 waffle-slab and solid-slab buildings collapsed, and another 44 buildings suffered severe damage. Hueste and Wight (1999) studied a building with a post-tensioned flat plate that experienced punching shear failures during the 1994 Northridge, CA, earthquake. Their study provided a relationship between the level of gravity load and the maximum story drift ratio that a flat plate-column connection can undergo without punching shear failure. The displacement-induced unbalanced moments and resulting shear forces at flat plate-column connections, although unintended, should be designed to prevent brittle punching shear failure. Even when an independent lateral-force-resisting system is provided, flat plate-column connections should be designed to accommodate the moments and shear forces associated with the displacements during earthquakes.

1.2—Scope

In seismic design, the displacement-induced unbalanced moment and the accompanying shear forces at flat plate-column connections should be accounted for. This demand may be effectively addressed by changes in dimensions of certain members, or their material strengths (for example, shear walls and column sizes), or provision of shear reinforcement or a combination thereof. This guide does not address changes in dimensions and materials of such members, but focuses solely on the punching shear design of flat plates with or without shear reinforcement.

This guide, supplemental to ACI 421.1R, focuses on the design of flat plate-column connections with or without shear reinforcement that are subject to earthquake-induced displacement; reinforced concrete flat plates with or without post-tensioning are treated in the guide. Slab shear reinforcement can be structural steel sections, known as shearheads, or vertical rods. Although permitted in ACI 318, shearheads are not commonly used in flat plates. Stirrups and shear stud reinforcement (SSR), satisfying ASTM A1044/A1044M, are the most common types of shear reinforcement for flat plates. Shear stud reinforcement is composed of vertical rods anchored mechanically near the bottom and top surfaces of the slab. Forged heads or welded plates can be used as the anchorage of SSR; the area of the head or the plate is sufficient to develop the yield strength of the stud, with negligible slip at the anchorage. The design procedure recommended in this guide was developed based on numerical studies (finite element method) and experimental research on reinforced concrete slabs subjected to cyclic drift reversals that simulate seismic effects. The finite element analyses, supplemental to the experimental research, used software, constitutive relations, and models that were subject to extensive verifications by comparing the results with the behavior observed in tests (Megally and Ghali 2000b).

Structural integrity reinforcement near the bottom of the slab extending through the columns should be provided as required by ACI 318. This document supplements ACI 352.1R and ACI 421.1R, which, respectively, include recommendations such as extending a minimum amount of bottom integrity reinforcement through the column core and provide details of design for shear reinforcement in flat plates. ACI 352.1R also provides recommendations for the design of flat plate-column connections without slab shear reinforcement subjected to moment transfer in the inelastic-response range. The equations of this guide predict punching shear strength and drift capacity, assuming that adequate flexural reinforcement is provided at the flat plate-column connections; the present guide does not address the required flexural reinforcement.

1.3—Objective

The objective is to provide a design recommendation for flat plate-column connections with sufficient ductility to accommodate the displacement of the selected lateral-force-resisting system without punching shear failure or loss of moment transfer capacity. The objective covers reinforced concrete slabs with or without post-tensioning.

1.4—Remarks

This guide gives recommendations for the design of shear reinforcement, considering ductility, that supplement the provisions of ACI 318 for punching shear design. The term "ductility," used throughout this guide, is the ratio of displacement at ultimate strength to the displacement at which yielding of the flexural reinforcement occurs. For flat plate-column connections, there is no unique definition for these two displacements. Pan and Moehle (1989) define the ultimate and yield displacements by a graphical bilinear idealization of the experimental load-displacement response, considering the displacement at a specified load level beyond the peak load.

ACI 318 allows the analysis of flat plate-column frames as equivalent plane frames. When the frame is not designated as part of the lateral-force-resisting system and is subjected to horizontal displacements, the width of slab strip to be included in the frame model and how to account for cracking (ACI 318, Section R13.5.1.2) are modeling parameters that significantly affect the resulting computed values of the moments transferred between slabs and columns. This guide contains a procedure that determines an upper limit moment that can be transferred between the slab and column when the connection is subjected to an earthquake.

Chapter 3 defines the story drift that should be considered in design. Chapter 4 recommends a minimum amount of shear reinforcement for certain cases. Chapter 5 describes means of increasing the shear strength of a flat plate-column connection and compares the associated ductilities. Chapter 6 presents a method to calculate the unbalanced moment required for design. Chapter 7 and Appendix D discuss relevant provisions of ACI 318 and provide the design procedure and examples for interior, edge, and corner flat plate-column connections. Chapter 8, expanded in this edition, provides recommendations relevant to post-tensioned flat plates.

CHAPTER 2—NOTATION AND DEFINITIONS

2.1—Notation

A_s = area of flexural reinforcing bars, in.^{2} (mm^2)

A_v = cross-sectional area of shear reinforcement on one peripheral line, in.^{2} (mm^2)

b_o = length of perimeter of shear-critical section, in. (mm)

C_d = displacement amplification factor (ASCE/SEI 7)

c_1 to c_6 = dimensions used in Fig. 8.3, defining a post-tensioning tendon profile

c_x, c_y = column dimensions in the x- and y-directions, respectively, in. (mm)

DR_u = ultimate story drift ratio at peak strength (in experiments), or design story drift ratio of a flat plate-column connection

d = average of distances from extreme compression fiber to the centroid of the tension reinforcement positioned in two orthogonal directions, in. (mm)

E_c = elastic modulus of concrete, psi (MPa)

e = tendon eccentricity, measured from the midsurface of the slab to the centroid of the post-tensioned tendon; e is positive for a tendon situated below midsurface, in. (mm)

f'_c = specified concrete strength, psi (MPa)

f_{pc} = average in-plane compressive stress produced by effective post-tension forces in two orthogonal directions, psi (MPa)

f_{ps} = stress in post-tensioned reinforcement at nominal flexural strength of slab, psi (MPa)

f_{py} = specified yield strength of post-tensioned reinforcement, psi (MPa)

f_{se} = effective stress in a post-tensioned tendon after accounting for all post-tension losses, psi (MPa)

f_y = specified yield strength of flexural reinforcement, psi (MPa)

f_{yt} = specified yield strength of shear reinforcement, psi (MPa)

h = slab thickness, in. (mm)

I_c = moment of inertia of gross section of column, in.^{4} (mm^4)

I_E = occupancy importance factor (ASCE/SEI 7)

I_{ec} = equivalent moment of inertia of columns accounting for torsional members in accordance with ACI 318, Section 13.7.5 of (refer to plane frame idealization [Fig. 6.1])

I_s = second moment of area of slab accounting for cracking (refer to plane frame idealization [Fig. 6.1])

J_c = property of assumed critical section analogous to polar moment of inertia, defined by Eq. (C-1), taken from ACI 318

J_x, J_y = property of assumed critical section of any shape, equal to d multiplied by the second moment of perimeter about principal x- or y-axis, respectively

$J_{\overline{xy}}$ = d times product of inertia of assumed shear-critical section about nonprincipal axes $\bar{x}$ and $\bar{y}$
K_c = end rotational stiffness of column, moment per unit rotation
K_{ec} = end rotational stiffness of equivalent column, moment per unit rotation
ℓ = span length, in. (mm)
ℓ_c = story height, in. (mm)
ℓ_x, ℓ_y = projections of shear-critical section on its principal axes x and y, respectively, in. (mm)
M = unbalanced moment transferred between the slab and the column, in.-lb (N·mm)
M_{Ox}, M_{Oy} = unbalanced moment about an axis parallel to the principal x- or y-axis and passing through O, the column's centroid; positive M_{Ox} or M_{Oy} is in the same direction as positive M_{ux} or M_{uy} (Fig. 3.2), in.-lb (N·m)
M_p' = primary bending moment due to post-tensioning, in.-lb (N·mm)
M_p'' = secondary (indeterminate) post-tensioned bending moment, in.-lb (N·mm)
M_{pr} = probable flexural strength, in.-lb (N·mm) (refer to Section 6.3)
$M_{prestress}$ = bending moment produced by post-tensioning forces, in.-lb (N·mm)
M_u = ultimate unbalanced moment, at peak strength, transferred between the slab and the column at shear-critical section centroid, in.-lb (N·mm); this definition applies where capacity is considered. Where demand is considered, M_u is factored unbalanced moment in design
M_{ux}, M_{uy} = components of the unbalanced moment M_u transferred between the slab and the column; positive directions of M_{ux} and M_{uy} are defined in Fig. 3.2
P_e = absolute value of effective post-tensioning force per unit slab width, in.-lb (N·mm)
s = spacing between peripheral lines of shear reinforcement, in. (mm)
s_o = spacing between the first peripheral line of shear reinforcement and column face, in. (mm)
V = shear force transferred between the column and the slab, lb (N)
V_c = nominal punching shear capacity of a flat plate-column connection with no shear reinforcement, lb (N)
V_p = shear force produced by post-tensioning forces or vertical component of effective post-tensioning force crossing the shear-critical section at d/2 from column face, lb (N)
V_p' = primary shear force due to post-tensioning, lb (N)
V_p'' = secondary (indeterminate) post-tensioning shear force, lb (N)
V_u = ultimate shear force transferred between the slab and the column, lb (N)
($V_u/\phi V_c$) = in presence of shear reinforcement (Eq. (7-1)), v_c is nominal shear strength (expressed in stress units) provided by concrete, psi (MPa). In the absence of shear reinforcement, v_c is nominal shear capacity expressed in stress units = $V_c/(b_o d)$ (V_c is given by Eq. (3-1) to (3-3))
v_n = nominal shear strength (expressed in stress units), psi (MPa)
v_s = nominal shear strength (expressed in stress units) provided by shear reinforcement, psi (MPa)
v_u = maximum shear stress at critical section, psi (MPa)
w_c = weight of concrete per unit volume
w_D = service dead load per unit area
w_L = service live load per unit area
w_{sd} = superimposed dead load per unit area
x, y = coordinates of point of maximum shear stress on the critical section with respect to centroidal principal axes x and y, respectively. Also as subscripts, x and y refer to the same principal axes
$\bar{x}$, $\bar{y}$ = axes parallel to slab edges at the centroid of the shear-critical section at a corner column
α_m = factor used in calculation of the design moment for earthquake-resistant flat plate-column connections
α_s = factor that adjusts v_c for support type
β = ratio of long side to short side of concentrated load or reaction area
β_r = aspect ratio of the shear-critical section at d/2 from column face
γ_v = fraction of unbalanced moment transferred by vertical shear stresses at flat plate-column connections
δ_e = story drift used in elastic frame analysis, in. (mm)
δ_u = design story drift, including inelastic deformations, in. (mm)
θ = clockwise rotation angle of $\bar{x}$-axis to x-axis, or $\bar{y}$-axis to y-axis
θ_p = angle between tangent to the tendon and the centroidal axis
λ = modification factor reflecting the reduced mechanical properties of lightweight concrete, relative to normalweight concrete of the same compressive strength
ϕ = strength reduction factor according to ACI 318
κ = fraction of service dead load to be balanced by effective post-tensioning
ρ = slab reinforcement ratio
ρ_p = ratio of post-tensioning reinforcement

2.2—Definitions

ACI provides a comprehensive list of definitions through an online resource, "ACI Concrete Terminology," http://terminology.concrete.org. Definitions provided herein complement that resource.

design displacement—total lateral displacement expected for the design-basis earthquake as required by the governing code for earthquake-resistant design.

design story drift—design displacement of one level, or floor, relative to the level above or below.

design story drift ratio—design story drift divided by the story height.

ductility—ratio of displacement at ultimate strength to the displacement at which yield of the flexural reinforcement occurs.

flat plate—flat slab without column capitals or drop panels.

lateral-force-resisting system—portion of the structure composed of members designed to resist forces related to earthquake effects.

shear stud reinforcement (SSR)—reinforcement composed of vertical rods anchored mechanically near the bottom and top surfaces of the slab (ASTM A1044/A1044M).

CHAPTER 3—LATERAL STORY DRIFT

3.1—Lateral-force-resisting systems

Flat plate-column frames without beams and without a lateral-force-resisting system consisting of more rigid elements that limit the lateral displacements are not permitted by ACI 318 in Seismic Design Categories (SDC) D, E, and F. In SDC D, E, and F, flat plate buildings need to rely on a lateral-force-resisting system that limits lateral displacement. Flat plates will experience the same lateral displacements as the lateral-force-resisting system, and should be designed to do so without losing their capability to support gravity loads during or after an earthquake.

The lateral-force-resisting system should have sufficient stiffness to control the lateral displacement of the structure and limit the maximum lateral story drift to the values prescribed by the general building code, and as discussed in Section 3.2. The story drift is defined as the lateral displacement of one level, or floor, relative to the level above or below. The story drift ratio is defined as the story drift divided by the story height. The story height is defined as the distance between the midsurfaces of the consecutive flat plates at top and bottom of the story of interest.

Unbalanced moments due to design displacements are transferred between slabs and columns and are a direct result of the lateral displacement experienced by the entire building. The transfer of shear forces and unbalanced moments increases the risk of a punching shear failure during an earthquake.

ACI 352.1R provides connection design recommendations for flat plate-column frames and considers that such frames may be adequate as the lateral-force-resisting system in regions of low and moderate seismic risk. This guide excludes the selection or the design of the lateral-force-resisting systems; it includes the punching shear design of flat plates, with and without shear reinforcement, subjected to gravity loads combined with unbalanced moments induced by the lateral drift of the flat plate-column connections.

3.2—Limits on story drift ratio

It has been frequently recommended that flat plate structures should have the capability to withstand a design story drift ratio of at least 0.015, including inelastic deformations (Sozen 1980; ACI 352.1R; Pan and Moehle 1989).

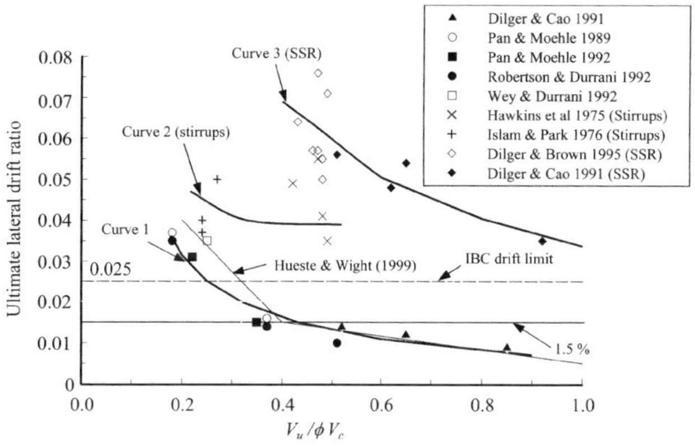

Fig. 3.1—Effect of gravity loads on lateral drift capacity of interior flat plate-column connections (Megally and Ghali 1994, 2000d; Hueste and Wight 1999).

In the force-based design approach, a static elastic analysis of the lateral-force-resisting system is performed to determine the elastic story drift δ_e (due to forces specified in the International Building Code (IBC 2006), or ASCE/SEI 7). This value is then multiplied by factors, given in IBC 2006 or ASCE/SEI 7, to obtain the design story drift δ_u.

IBC 2006 requires that the calculated δ_u (depending on the lateral-force-resisting system and the seismic design category), including inelastic deformation, not exceed 0.007 to 0.025 of the story height. The code also requires that the flat plate-column connection be adequate to carry the vertical load and the induced moments and shears resulting from the calculated design story drift.

The design story drift ratio may reach the upper limit of IBC 2006 when the flat plate-column connections are provided with slab shear reinforcement or when the gravity load produces low punching shear stress as discussed in Chapter 4.

3.3—Effects of gravity loads on story drift capacity

Figure 3.1 (Megally and Ghali 1994, 2000d; Hueste and Wight 1999) shows the variation of the ultimate story drift ratio DR_u for interior flat plate-column connections transferring gravity shear forces V_u with the ratio ($V_u/\phi V_c$) when subjected to reversals of cyclic drift. The experimental values of DR_u are compared with the design story drift δ at peak strength, divided by the story height. V_u is the ultimate shear force transferred between the slab and the column at failure; V_c is the nominal punching shear strength of the flat plate-column connection without shear reinforcement in the absence of moment transfer. Each data point represents the combination of ($V_u/\phi V_c$) and the corresponding ultimate story drift ratio DR_u at which a test specimen failed in punching shear.

Curves 1, 2, and 3 shown in Fig. 3.1 fit the results of experiments reported in the literature on interior flat plate-column connections without shear reinforcement, with conventional stirrups, and with SSR, respectively (Pan and Moehle 1989, 1992; Robertson and Durrani 1992; Wey and Durrani 1992; Hawkins et al. 1975; Islam and Park 1976;

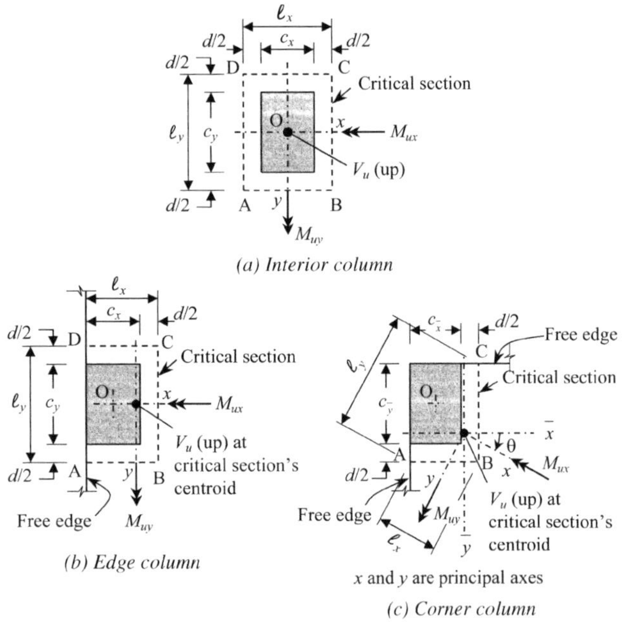

Fig. 3.2—Critical sections for punching shear at d/2 from column face. The arrows for V_u *and* M_u *represent the forces exerted by the column on the slab in their positive directions.* x *and* y *are centroidal principal axes of the critical section.*

Dilger and Brown 1995; Brown 2003; Dilger and Cao 1991; Cao 1993). The strength reduction factor ϕ is taken equal to 1.0 in this figure because the strength of materials is known for the experiments. In preparing Fig. 3.1, V_c is used as the smallest of Eq. (3-1) through (3-3)

Inch-pound units

$$V_c = \left(2 + \frac{4}{\beta}\right) b_o d \lambda \sqrt{f_c'} \tag{3-1}$$

$$V_c = \left(\frac{\alpha_s d}{b_o} + 2\right) b_o d \lambda \sqrt{f_c'} \tag{3-2}$$

$$V_c = 4 b_o d \lambda \sqrt{f_c'} \tag{3-3}$$

SI units

$$V_c = \left(1 + \frac{2}{\beta}\right) \frac{b_o d \lambda \sqrt{f_c'}}{6} \tag{3-1}$$

$$V_c = \left(\frac{\alpha_s d}{b_o} + 2\right) \frac{b_o d \lambda \sqrt{f_c'}}{12} \tag{3-2}$$

$$V_c = \frac{1}{3} b_o d \lambda \sqrt{f_c'} \tag{3-3}$$

where b_o is the perimeter of the critical section for shear at (d/2) from the column face (Fig. 3.2); d is the average of the distances from extreme compression fiber to the centroid of tension reinforcement in two orthogonal directions; f_c' is the

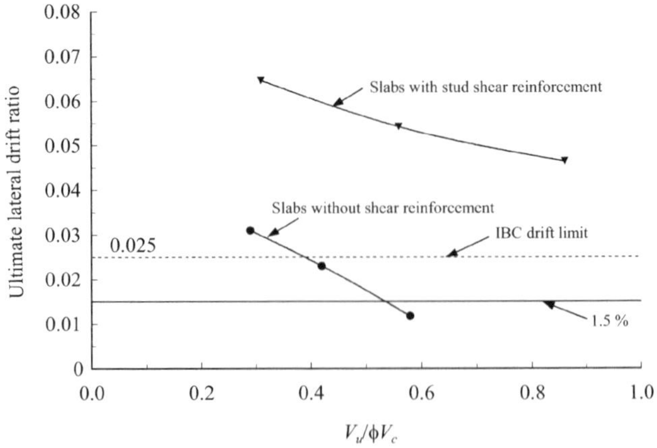

Fig. 3.3—Effect of value of V_u *on lateral drift capacity of edge flat plate-column connections (Megally and Ghali 2000a).*

specified concrete compressive strength; β is the ratio of the long side to the short side of the column; and α_s = 40, 30, and 20 for interior, edge, and corner flat plate-column connections, respectively. The arrows for V_u and M_u shown in Fig. 3.2(a) through (c) represent the forces exerted by the column on the slab in their positive directions.

Figure 3.1 indicates that the flat plate-column connection capability to experience story drift without failure decreases with increasing magnitude of the applied gravity load. The solid horizontal line shown in Fig. 3.1 represents the design story drift ratio of 0.015, which is frequently adopted as a minimum drift capacity. This horizontal line intersects Curve 1 at $(V_u/\phi V_c) \cong 0.40$, indicating that slabs without shear reinforcement can satisfy the required 0.015 design story drift ratio only if V_u does not exceed $0.40\phi V_c$. The same conclusion can be seen from a similar graph containing more data (Kang and Wallace 2005). The horizontal dashed line plotted in Fig. 3.1, corresponding to the 0.025 design story drift ratio limit specified by IBC 2006 (for structures satisfying specified conditions), intersects Curve 1 at approximately $(V_u/\phi V_c) \cong 0.25$. This suggests that for $DR_u = 0.025$, the slab should be designed with shear reinforcement when V_u exceeds approximately $0.25\phi V_c$. Section 3.4 conservatively recommends provision of shear reinforcement when DR_u = 0.025 and $(V_u/\phi V_c) > 0.20$. This complies with ACI 318, Section 21.13.6.

Figure 3.3 is similar to Fig. 3.1, but the two curves shown in Fig. 3.3 represent the results of experiments of edge flat plate-column connections (Megally and Ghali 2000a). The graphs indicate that for $DR_u = 0.015$, the flat plate should be designed with shear reinforcement when $V_u > 0.40\phi V_c$. A comparison of Fig. 3.1 and 3.3 indicates that a somewhat higher limit on $(V_u/\phi V_c)$ can be allowed for edge connections than for interior connections for the same design story drift ratio. For simplicity, a single limit on V_u is recommended in Section 3.4. In addition to the limitation on V_u, the absence of shear reinforcement should be allowed only when the maximum shear stress is less than that given in Chapter 4.

Figures 3.1 and 3.3 show that the curves representing experiments of flat plate-column connections with shear reinforcement fall well above the horizontal lines. Curves 2 and 3 of Fig. 3.1 are not perfect fits because of the scatter of experimental data. The uncertainty of the curves, however, does not change the conclusion that no limit on $(V_u/\phi V_c)$ is needed for slabs with shear reinforcement exceeding the minimum amount given in Chapter 4 to sustain the 0.025 design story drift ratio required by IBC 2006. Figure 3.1 also shows that the drift capacity for slabs with SSR is higher than slabs with stirrups.

3.4—Design recommendations for flat plates with and without shear reinforcement

As discussed in Section 3.1, the lateral-force-resisting structural system should have sufficient stiffness to control the story drift. Slab shear reinforcement is required when the maximum shear stress at $d/2$ from the column face exceeds ϕv_c, where v_c is the value given by Eq. (3-1) to (3-3) divided by $b_o d$. In addition, flat plate-column connections should have shear reinforcement equal to or exceeding the minimum amount given in Chapter 4, except when the value of V_u is less than $0.20\phi V_c$, and $\phi = 0.75$ (ACI 318, Section 9.3.2.3). This requirement ensures that the connections can sustain the design story drift ratio $DR_u = 0.025$. If it can be shown by analysis that when the maximum story drift ratio DR_u, including the inelastic deformations, is between 0.015 and 0.025, shear reinforcement is required when V_u exceeds $\phi V_c(0.70 - 20DR_u)$. This is the same as required by ACI 318, Section 21.13.6(b), that shear reinforcement be provided when DR_u exceeds $[0.035 - 0.05V_u/(\phi V_c)]$.

Curve 1 in Fig. 3.1 shows that for slabs without shear reinforcement, DR_u can reach 0.015 only when $V_u \leq 0.40\phi V_c$. The same curve also shows conservatively that DR_u can reach 0.025 without shear reinforcement when $V_u \leq 0.20\phi V_c$. The same two conclusions can be reached if Curve 1 is replaced by the bilinear graph of Hueste and Wight (1999), plotted in the same figure. The bilinear graph represents an approximation of the same test data used in Fig. 3.1 for slabs without shear reinforcement combined with data from seven more references.

CHAPTER 4—MINIMUM SHEAR AND INTEGRITY REINFORCEMENTS IN FLAT PLATES

Seismic design of non-prestressed flat plates can be without shear reinforcement only when $(V_u/\phi V_c)$ is limited, as discussed in Section 3.4. In addition, the maximum shear stress due to V_u, combined with that due to the unbalanced moment M_u, computed according to the linear stress distribution assumption of ACI 318, should not exceed ϕv_c, expressed in stress units (the smallest of Eq. (3-1) through (3-3) divided by $b_o d$), at the critical section at $d/2$ from the column face. For ductility, connections that do not satisfy both conditions should have slab shear reinforcement that satisfies Eq. (4-1).

Inch-pound units

$$v_s = \frac{A_v f_{yt}}{b_o s} \geq 3\sqrt{f_c'} \tag{4-1}$$

SI units

$$v_s = \frac{A_v f_{yt}}{b_o s} \geq \frac{1}{4}\sqrt{f_c'} \tag{4-1}$$

where A_v is the area of shear reinforcement in each peripheral line parallel to the column faces (typical peripheral lines are shown in Fig. 4.1(a) to (c)); s is the spacing between peripheral lines of shear reinforcement($s \leq 0.5d$ for stirrups; $s \leq 0.75d$ for SSR); b_o is the perimeter of the critical section at $d/2$ from the column face (Fig. 3.2); and f_{yt} is the specified yield strength of the shear reinforcement. The distance between the column face and the outermost peripheral line of shear reinforcement (Fig. 4.1) should not be less than $3.5d$. The case that requires the minimum reinforcement for ductility is further clarified in Section 7.2.

ACI 318 and ACI 421.1R provide equations to calculate the maximum shear stress at the shear-critical section $d/2$ from the column face; when the maximum shear stress exceeds $\phi V_c/(b_o d)$, shear reinforcement is required for strength. Also for strength, the extent of the shear-reinforced zone is determined by limiting the maximum shear stress at the critical section at $d/2$ from the outermost peripheral line of shear reinforcement $2\phi\sqrt{f_c'}$ psi $(0.17\phi\sqrt{f_c'}$ MPa) (ACI 318, Section 11.11.7.2). When shear reinforcement is required for strength, its extent should exceed the minimum $3.5d$.

The minimum shear reinforcement recommended previously was based on a review of experiments by Islam and Park (1976); Hawkins et al. (1975); Dilger and Brown (1995); Dilger and Cao (1991); Megally and Ghali (2000a); and Megally (1998). In these experiments, flat plate-column connections with different amounts of stirrups or SSR sustained drift ratios higher than 0.025 without a punching shear failure. Appendix A summarizes the results of these experiments.

The minimum shear reinforcement recommended increases the shear strength and ensures that the flat plate-column connections can support factored gravity loads after experiencing inelastic deformations due to cyclic drift during an earthquake. Experiments (Megally 1998) have shown that this can be achieved by the provision of shear reinforcement, as recommended in the present guide. The design of shear reinforcement, other than the minimum amount, is discussed in Chapter 7.

The structural integrity reinforcement defined in Section 5.3 of ACI 352.1R should also be provided to increase the resistance of the structural system to progressive collapse. This reinforcement can be more than that required by the detailing rules in Chapter 13 of ACI 318.

CHAPTER 5—ASSESSMENT OF DUCTILITY

The methods for increasing punching shear strength of flat plate-column connections include increasing the flat plate thickness, improving the material properties, providing shear reinforcement, or a combination of these methods. The effects of these different methods for enhancing shear strength on ductility are substantially different.

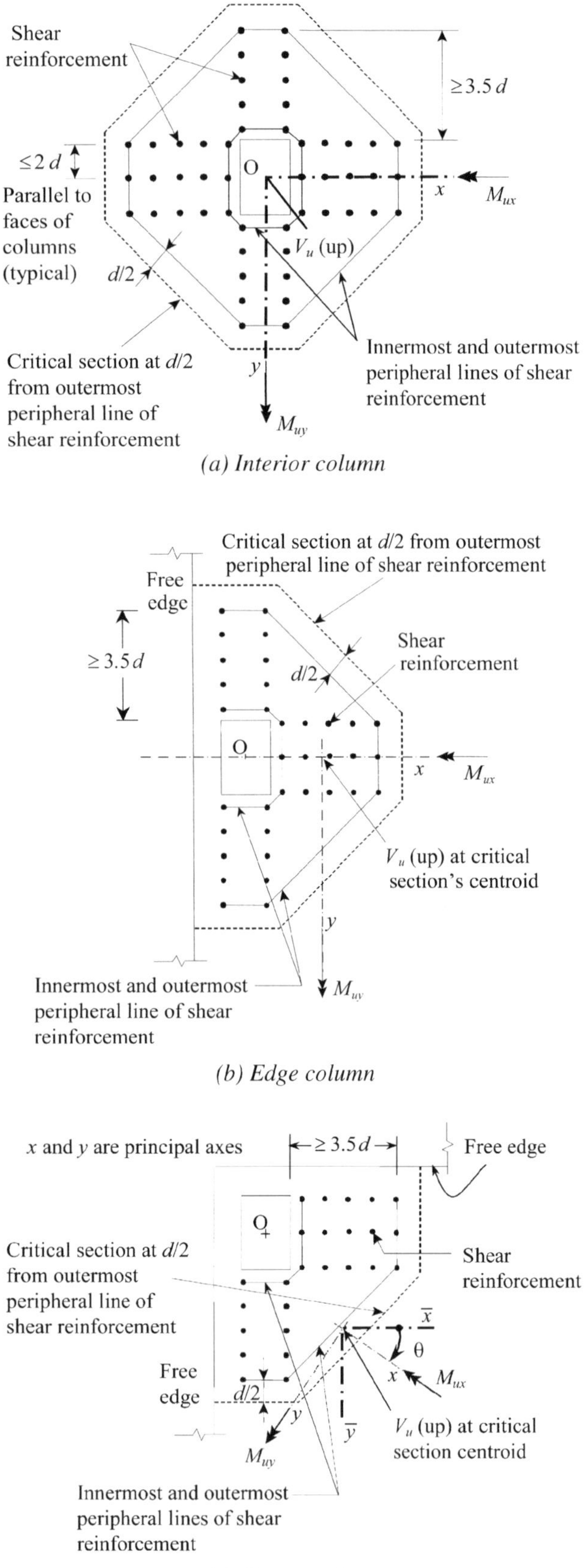

Fig. 4.1—Critical sections for punching shear outside the shear-reinforced zone.

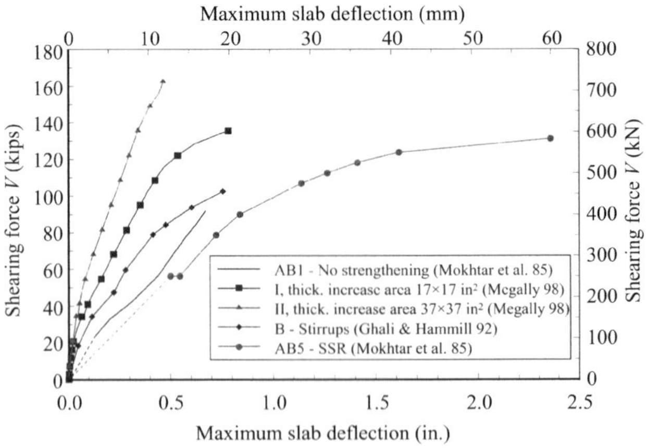

Fig. 5.1—Load-deflection curves of slabs with different means of increasing punching shear strength. Slabs subjected only to shear force V *with no unbalanced moment. The deflection is measured in the direction of* V *(Megally and Ghali 2000a).*

The most common types of shear reinforcement in flat plates are the vertical legs of stirrups and SSR. Stirrups may not be as effective as SSR in flat plates because of inefficient anchorage of the stirrups, unless they are closed and detailed in accordance with ACI 318, Section R11.11.3, which states: "Anchorage of stirrups according to the requirements of 12.13 is difficult in slabs thinner than 10 in. (250 mm)." With SSR, the anchorage is provided mechanically by forged heads or by a forged head at one end and a steel strip (called a rail) welded at the other end.

Figure 5.1 (Megally and Ghali 2000a) shows load-deflection responses for five slabs with a thickness of 6 in. (150 mm). The tested specimens, with plan dimensions of 75 x 75 in.^{2} (1.9 x 1.9 m^2), represent full-size connections of slabs with interior columns of size 10 x 10 in.^{2} (250 x 250 mm^2) and square panels of spans equal to 16 ft (4.8 m). Axial load is monotonically applied on the column, transferring a shear force, but no unbalanced moment, to the slab, which is simply supported on four edges. The slabs have the same flexural reinforcement layout and properties; the flexural reinforcement ratio is equal to 0.014 in each of the two orthogonal directions. The concrete compressive strengths f_c' ranged from 4200 to 5800 psi (29 to 40 MPa). The slabs differ only in the method used to increase punching shear strength. Control Slab AB1 (Mokhtar et al. 1985) has no means for increasing punching shear strength. Figure 5.1, taken from Megally and Ghali (2000a), includes Slabs I and II having drop caps. Slabs I and II (Megally 1998) have thickness increases from 6 to 9 in. (150 to 225 mm) over areas, 17 x 17 in.^{2} (430 x 430 mm^2) and 37 x 37 in.^{2} (950 x 950 mm^2), respectively, with the column at their centers. Slab B (Ghali and Hammill 1992) has closed stirrups, detailed according to ACI 318, Fig. R11.11.3(c). (It is noted that since 2002, ACI 318 has permitted stirrups only in slabs with $d \geq 6$ in. [150 mm]). Slab AB5 (Mokhtar et al. 1985) has SSR. The nominal shear strengths, expressed in stress units, provided by the shear reinforcement (Eq. (4-1)) in Specimens B and AB5, are: 440 psi (3.0 MPa) for Specimen B, based on measured f_{yt}, and 270 psi (1.9 MPa) for Specimen AB5, based on nominal f_{yt}. The corresponding nominal shear strengths provided by shear reinforcement, $V_s = v_s b_o d = A_v f_{yt} d/s$ are 110 and 70 kips (480 and 320 kN). Punching shear failure occurred at a section within the shear-reinforced zone in Specimen B and at a section outside the shear-reinforced zone in Specimen AB5. The extent of the shear reinforcement zones from column faces was 3.8d and 5.7d in Specimens B and AB5, respectively. The initial portions of the experimental curves shown in Fig. 5.1 for Slabs AB1 and AB5 are unavailable. The conclusions drawn from these curves, however, are unaffected.

Figure 5.1 indicates that the punching shear strength is increased slightly with the presence of stirrups; however, increases in slab thickness and SSR provided a more significant increase in punching shear strength. The specimens with increased slab thickness, Slabs I and II, experienced a brittle failure with small slab deflections compared with the slab with SSR, for which the failure may be considered ductile. The stirrups provided an insignificant increase in ductility for such a thin slab. Stirrups are assumed more efficient when used in slabs thicker than 10 in. (250 mm), as suggested in ACI 318, Section R11.11.3; however, no experimental data on slabs with this thickness could be found. Figure 5.1 and results of experiments conducted on flat plate-column connections subjected to cyclic moment reversals (Dilger and Brown 1995; Dilger and Cao 1991; Megally and Ghali 2000a,b) indicate that SSR effectively increases punching shear strength for earthquake-resistant flat plate-column connections even when the slab is relatively thin. Note that Fig. 5.1, taken from Megally and Ghali (2000a), includes Slabs I and II having drop caps; however, this guide is limited to flat plates.

CHAPTER 6—UNBALANCED DESIGN MOMENT

Moments are transferred between flat plates and their supporting columns as the flat plate-column connections experience the lateral displacement induced on the lateral-force-resisting system. Thus, flat plate-column connections should have a punching shear strength able to resist the factored shear force V_u and factored unbalanced moment M_u due to gravity loads combined with the design story drift during an earthquake.

The methods in Sections 6.1 and 6.2 are recommended for the calculation of the value of M_u. An upper limit for M_u is given in Section 6.3 and Appendix B.

6.1—Frame analysis

A flat plate-column frame, including its lateral-force-resisting system, can be analyzed for gravity and lateral seismic loads using linear frame analysis techniques (Ghali et al. 2003). Unless the layout of columns is highly irregular, the moments of inertia of the slab and its supporting columns are determined according to the equivalent frame method of ACI 318. A static analysis can be performed using the seismic lateral forces specified by IBC 2006 or ASCE/SEI 7; alternatively, a dynamic analysis may be performed according to the two references.

To account for the effect of cracking in non-prestressed flat plates, Vanderbilt and Corley (1983) recommend considering the moment of inertia of the slab as equal to one-third the value of the uncracked slab strip—from panel centerline to panel centerline—to obtain a conservative estimate of the story drift. ACI 318, Section R13.5.1.2, recommends the use of 25 to 50% of the uncracked moment of inertia of the slab to compute the elastic story drift δ_e. According to ASCE/SEI 7, δ_e is related to the maximum design story drift δ_u, including inelastic deformations by

$$\delta_u = \delta_e\left(\frac{C_d}{I_E}\right) \quad (6\text{-}1)$$

where C_d (= 1.25 to 6.5) and I_E (= 1.0 to 1.5) are dimensionless factors specified by ASCE/SEI 7, depending on the inherent inelastic deformability of the lateral-force-resisting system and the occupancy importance of the structure, respectively. To include the P-Δ effect, the factor inside the brackets in Eq. (6-1) should be multiplied by an appropriate coefficient specified in ASCE/SEI 7.

Flat plate-column frames not designated as part of the lateral-force-resisting system experience the same lateral displacements as those of the lateral-force-resisting system. Thus, flat plate-column connections should be designed to transfer shears and moments associated with the δ_u values obtained from Eq. (6-1). Accurate determination of shears and moments associated with δ_u is not possible, but for design purposes, the simplified approach presented in Section 6.2 is deemed adequate. The design moments and shears for flat plate-column connections should be determined from the aforementioned elastic frame analysis according to the IBC 2006 or ASCE/SEI 7 static lateral force analysis method. The unbalanced moment caused by the factored vertical forces that exist during an earthquake should be added to the moments obtained by the analysis discussed in this section. The value of M_u is the smaller of the upper limit given in Section 6.3 and the total factored unbalanced moment determined as described previously.

To avoid underestimation of the unbalanced moments transferred between the slab and the columns for a given value of δ_e, the unbalanced moments should be determined by an elastic analysis, but with the moment of inertia of the slab equal to 50% of the value of the uncracked slab and using the values of the moments of inertia of the uncracked columns (ACI 318, Section R13.5.1.2). There is no consensus on the 50% value in this recommendation; a one-third value has been also recommended.

6.2—Simplified elastic analysis

Instead of the analysis suggested in Section 6.1, the unbalanced moments transferred between the slab and the columns under the elastic story drift δ_e can be determined by a linear analysis of simplified equivalent frames as shown in Fig. 6.1(a) and (b) for interior and exterior columns (edge or corner), respectively. The slab is assumed to be simply supported at locations of contraflexure lines assumed at midspan. The column is assumed to have hinged supports at contraflexure points assumed approximately at midheight of the story. For the frame models shown in Fig. 6.1, it is assumed that the lengths of spans adjacent to the column are equal ℓ and the story heights above and below the considered level are equal ℓ_c. The moments of inertia of the column and the slab are calculated according to ACI 318, Sections 13.5.1.2 and 13.7.5 (considering torsional members connecting the column to the beam). Figure 6.1(c) indicates the second moment of areas of the frame members.

The horizontal displacement δ_e is introduced at the upper ends of the columns as shown in the figure. The value of δ_e, representing the elastic story drift, can be estimated by any rational analysis; for example, it can be calculated using an elastic analysis of the lateral-force-resisting system subjected to the lateral forces as specified by IBC 2006 or ASCE/SEI 7. The unbalanced moment due to the horizontal seismic forces transferred between the column and the connected slab is equal to the sum of the end moments at the column ends above and below the flat plate-column connection. Additional unbalanced moments caused by factored vertical forces that can exist during an earthquake should also be considered. The smaller of the total unbalanced moment calculated using this procedure and the upper limit given in Section 6.3 should be used.

6.3—Upper limit for M_u

The value of the unbalanced moment corresponding to the displacement δ_e can be higher than the value that produces ductile flexural failure. Provision of shear reinforcement in this case would not increase the value of the unbalanced moment strength. For slabs having low flexural reinforcement ratios, Eq. (6-2) may limit the value of M_u. Based on finite element analyses (Megally and Ghali 2000b) and experiments (Megally and Ghali 2000a) on flat plate-column connections transferring shear combined with moment reversals, an upper limit can be set for the value of δ_e when computing the shear strength of earthquake-resistant flat plate-column connections

$$M_u \leq \frac{M_{pr}}{\alpha_m} \tag{6-2}$$

where M_{pr} is the sum of the absolute values of the probable flexural strengths of opposite critical section sides of width $(c_x + d)$ or $(c_y + d)$ when the transferred moment is about the x- or y-axis, respectively. The variables c_x and c_y are the column dimensions in the x- and y-directions, respectively (Fig. 3.2). The probable flexural strength should be based on the assumption that the force in the flexural tensile reinforcement is $1.25(A_s f_y)$ (ACI 318, Chapter 21), where f_y is the specified yield strength of the reinforcement, and A_s is the cross-sectional area of the reinforcing bars normal to the two opposite sides of the shear-critical section (Fig. 3.2). The top bars on one side plus the bottom bars on the opposite side within the aforementioned width should be included, with careful consideration of the anchorages of bars. The right-hand side

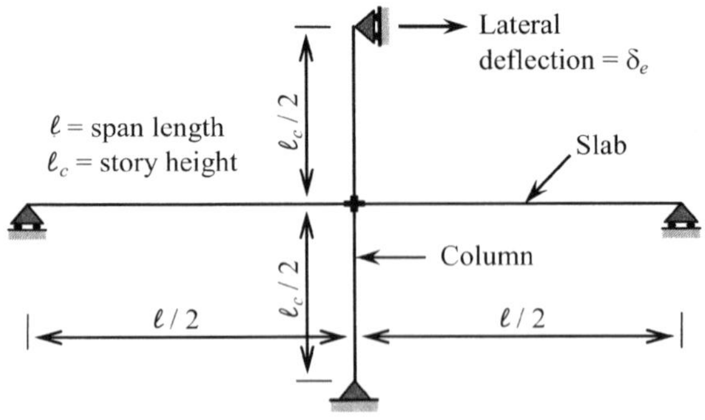

(a) Interior slab-column connection.

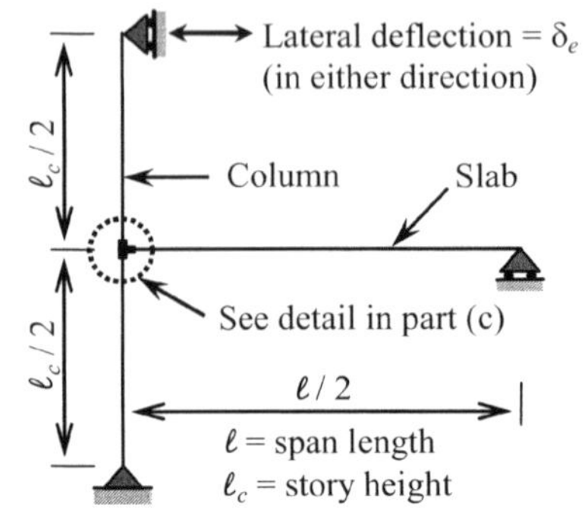

(b) Edge or corner slab-column connection.

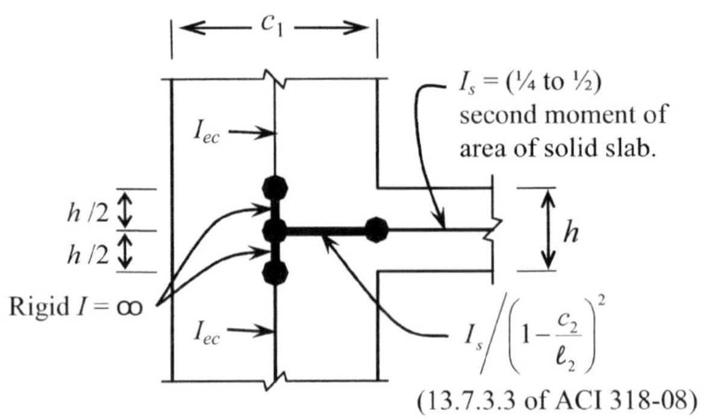

I_{ec} = equivalent second moment of area of uncracked column with a torsional member (13.7.5 of ACI 318-08).

c_2 and ℓ_2 are column and span length in direction perpendicular to the plane of the frame.

(c) Detail at slab-column connection in the equivalent plane frame, according to ACI 318.

Fig. 6.1—Plane frame idealization of flat plate-column connections.

of Eq. (6-2) represents the magnitude of the unbalanced moment that will develop the yield strength of the flexural reinforcement. When this occurs without a punching shear failure, the flat plate-column connection will experience substantial drift and not lose the ability to transfer gravity loads, thus avoiding collapse. Based on the finite element results, the empirical coefficient α_m (Eq. (6-2)) is expressed by Eq. (6-3) or (6-4)

$$\alpha_m = 0.85 - \gamma_v - \left(\frac{\beta_r}{20}\right) \quad (6\text{-}3)$$

$$\alpha_m = 0.55 - \gamma_v - \left(\frac{\beta_r}{40}\right) + 10\rho \quad (6\text{-}4)$$

where γ_v is the fraction of moment transferred by vertical shear stresses in the slab. Equations for γ_v, depending on the shape of the critical section, are given in Appendix C. Equation (6-3) applies for interior connections transferring M_{ux} or M_{uy} (Fig. 3.2(a)) and also for edge and corner connections transferring M_{ux} (with $M_{uy} = 0$ [Fig. 3.2(b) and (c)]). Equation (6-4) applies for exterior connections transferring M_{uy} (with $M_{ux} = 0$). In Eq. (6-3) and (6-4), β_r is equal to (ℓ_y/ℓ_x) or (ℓ_x/ℓ_y) when the transferred moment is about the x- or y-axis, respectively, where ℓ_x and ℓ_y are projections of the critical section at $d/2$ from the column face on its principal axes x and y, respectively (Fig. 3.2). In Eq. (6-4), ρ is the reinforcement ratio of tensile flexural reinforcing bars passing through the projection of the shear-critical section (Fig. 3.2) in the direction in which moments are transferred. In Fig. 3.2(b), for example, ρ is the reinforcement ratio of the tensile flexural reinforcing bars passing through Side BC of the critical section for the edge flat plate-column connection (when $M_{ux} = 0$). For the corner flat plate-column connection shown in Fig. 3.2(c), α_m is calculated by Eq. (6-4) with ρ given by

$$\rho = \rho_{\bar{x}}\cos^2\theta + \rho_{\bar{y}}\sin^2\theta \quad (6\text{-}5)$$

where $\rho_{\bar{x}}$ is the reinforcement ratio for $\bar{x}$-direction bars crossing critical section Side BC; $\rho_{\bar{y}}$ is the reinforcement ratio for $\bar{y}$-direction bars crossing critical section Side AB; and θ is the angle between the nonprincipal and principal axes. The value of M_u, given by Eq. (6-2), can be substantially greater than M_{pr}, meaning that the transfer of unbalanced moment can mobilize the flexural strength of the slab over a width considerably greater than $(c_x + d)$ or $(c_y + d)$.

When Eq. (6-2) is used to determine the upper limit for M_{uy} at an interior connection (with $M_{ux} = 0$ [Fig. 3.2(a)]), M_{pr} is the sum of the absolute values of the positive and negative probable flexural strength of the two opposite Sections AD and BC, respectively (Fig. 3.2(a)). For an edge flat plate-column connection (with $M_{ux} = 0$ [Fig. 3.2(b)]), M_{pr} should be calculated for the negative or the positive probable flexural strength of the critical section side parallel to the free edge (Side BC). The connection should be designed to resist V_u combined with each of the two moments. For an edge flat plate-column connection transferring M_{ux} (with $M_{uy} = 0$ [Fig. 3.2(b)]), M_{pr} is the sum of the absolute values of the positive and negative probable flexural strength of the two opposite Sections CD and AB, respectively.

For a corner flat plate-column connection, M_{pr} is calculated for the negative or positive probable flexural strength of slab strips with widths equal to projections of the critical section on its principal axes. The connection should be designed for punching shear considering the positive or negative probable flexural strength when M_{pr} is transferred about the principal y-axis (Fig. 3.2(c)). The connection should be checked considering the sum of the absolute values of the positive and negative probable flexural strength when M_{pr} is transferred about the principal x-axis (Fig. 3.2(c)). When the unbalanced moment transfer has each of its components (M_{ux} or $M_{uy} \neq 0$), the upper limit for each can be conservatively considered separately. This means that for the calculation of the upper limit of M_{ux}, M_{uy} can be ignored. Similarly the calculation of the upper limit of M_{uy} can be done assuming $M_{ux} = 0$ (using Eq. (6-2) and (6-4)).

In a flat plate-column connection (Fig. 3.2(a) or (b)) transferring constant V_u combined with M_{uy} of increasing magnitude, yielding is reached in the flexural reinforcing bars passing through critical section Sides AD and BC in Fig. 3.2(a) or Side BC in Fig. 3.2(b). The sum of the absolute values of the positive probable flexural strength of Side AD and negative probable flexural strength of Side BC (Fig. 3.2(a)) represents the product $\alpha_m M_{uy}$. Similarly, $\alpha_m M_{uy}$ represents the negative probable flexural strength of Side BC in Fig. 3.2(b). The quantity $\alpha_m M_{uy}$ can be determined from the results of nonlinear finite element analyses. The results of such analyses, which have helped in developing the empirical Eq. (6-3) and (6-4), are presented in Appendix B. The validity of the finite element software (ANATECH Consulting Engineers 1995) and the finite element models were verified by physical experiments (Megally and Ghali 2000b).

CHAPTER 7—DESIGN OF SHEAR REINFORCEMENT

7.1—Strength design

The computation of punching shear strength should follow the recommendations found in ACI 421.1R. When shear reinforcement is provided, the nominal shear strength (expressed in stress units) is given by

$$v_n = v_c + v_s \quad (7\text{-}1)$$

where v_c and v_s are the nominal shear strengths (expressed in stress units) provided by the concrete and shear reinforcement, respectively. ACI 421.1R limitsv_n to $8\sqrt{f_c'}$ or $6\sqrt{f_c'}$ psi ($0.67\sqrt{f_c'}$ or $0.5\sqrt{f_c'}$ MPa), respectively, when the shear reinforcement is SSR or stirrups. These two limits should be increased by 25% in seismic design when the shear stress due to V_u/ϕ alone does not exceed $4\sqrt{f_c'}$ psi ($0.33\sqrt{f_c'}$ MPa). This is because the maximum shear stress is caused mainly by M_u, rather than by V_u, occurring at only a point or a side of the critical section. The limit $v_n = 10\sqrt{f_c'}$ psi ($0.83\sqrt{f_c'}$ MPa) has been exceeded in tests (Ritchie and Ghali 2005). The value of v_c in Eq. (7-1) is limited to $1.5\lambda\sqrt{f_c'}$ psi ($0.125\lambda\sqrt{f_c'}$ MPa) in seismic design (ACI 421.1R).

With shear reinforcement and absence of unbalanced moment, the shear reinforcement can increase V_u/ϕ from V_c to the upper limit $2V_c$ or $1.5V_c$ with SSR or stirrups, respectively. Equation (7-1) recommends increasing the allowable shear stress due to the combination of V_u and M_u

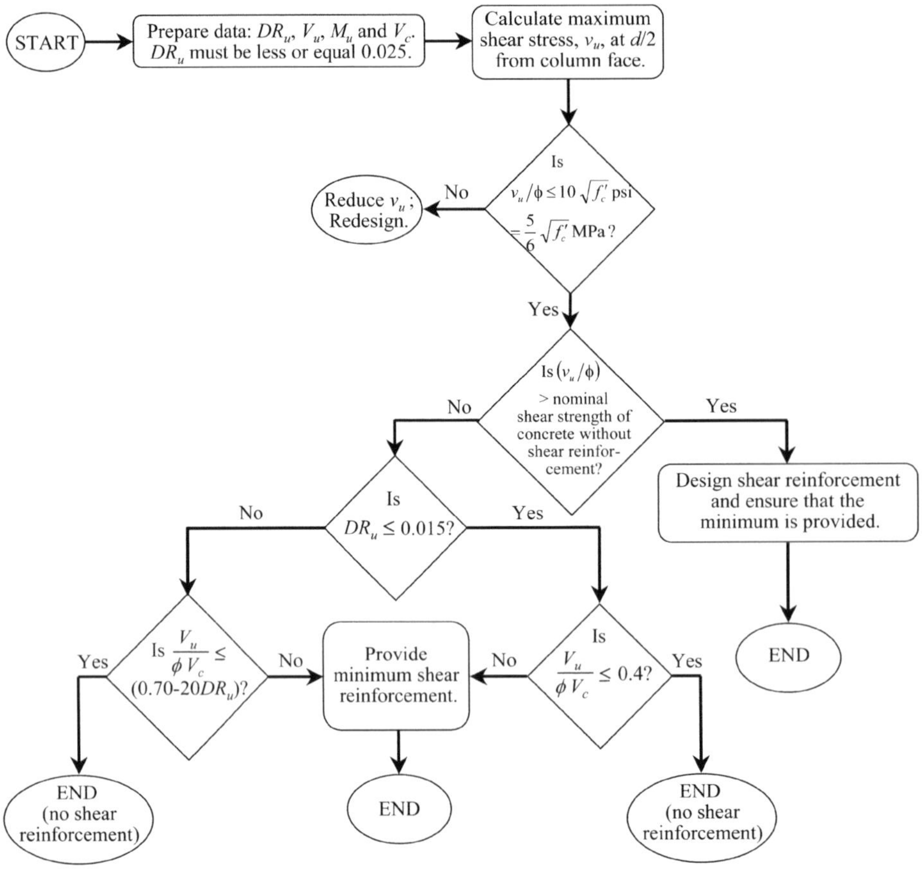

Fig. 7.1—Steps for punching shear design of earthquake-resistant flat plate-column connections.

only when the shear stress due to V_u/ϕ alone is relatively small compared with its upper limit.

7.2—Summary of design steps

The design story drift, including inelastic deformations, is controlled by the lateral-force-resisting structural system, as described previously. The steps explained previously for computing the punching shear strength of flat plate-column connections are illustrated by the flowchart shown in Fig. 7.1. In the penultimate step in the figure, M_u should be calculated taking into account the upper limit specified in Section 6.3. Then the equations found in ACI 421.1R should be used to calculate the maximum shear stress and the amount of shear reinforcement, when required, and to verify that the minimum amount of reinforcement is provided. Design examples of slab shear reinforcement required in flat plate-column connections are presented in Appendix D.

7.3—ACI 318 provisions

ACI 318-08, Section 21.13.6, is relevant to shear reinforcement at flat plate-column connections not designated as part of the lateral-force-resisting system. ACI 421.1R recommendations satisfy the requirements of ACI 318, including Section 21.13.6. Shear reinforcement greater than the requirements of ACI 318 is recommended when $V_u/(\phi V_c) > 0.4$, according to Sections 21.13.6 and R21.13.6.

Section 21.13.6 requires that the shear-reinforced zone extend at least four times the slab thickness from the face of the column, and that v_s shall be not less than $3.5\sqrt{f_c'}$ psi ([7/24]$\sqrt{f_c'}$ MPa). ACI 318 waives these requirements when the design of shear reinforcement satisfies Section 11.12.6.2.

Section R21.13.6 comments that calculation of the unbalanced moment due to the design displacement is not needed when, in Fig. 7.2, the point $\{V_u/(\phi V_c), DR_u\}$ falls below the bilinear boundary (A-B-C). When the point falls within Zone 1 (Fig. 7.2), the minimum shear reinforcement according to Chapter 4 of the present guide is recommended. When the point falls in Zone 2, ACI 318-08 requires shear reinforcement.

A point in Zones 1 and 2 represents the case when $V_u > 0.4\phi V_c$. With such a high value of V_u combined with unbalanced moment reversals, ductility can be ensured only with shear reinforcement. For interior column-flat plate connections, refer to Cao (1993) and for edge column-flat plate connections, refer to Megally and Ghali (2000a,c). This guide is consistent with ACI 318-08 in recommending shear reinforcement in Zone 3 (zone of relatively high DR_u). In addition, for ductility purposes, the minimum shear reinforcement, according to Chapter 4, should be provided.

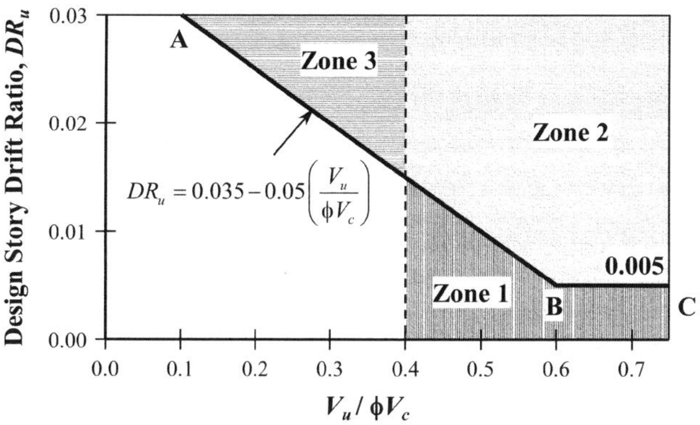

Fig. 7.2—Requirement for shear reinforcement criterion (ACI 318).

CHAPTER 8—POST-TENSIONED FLAT PLATES

8.1—General

The design recommendations of Chapter 7 are applied in this chapter to post-tensioned flat plate-column connections. Section 8.4 discusses additional internal forces—induced by post-tensioning tendons in statically indeterminate structures—that need to be considered in the design for punching shear. The compressive stress, produced by effective post-tensioned forces, is required in the design equations. A simplified equation that calculates the effective compressive stress is presented in Section 8.4. Design equations applicable for post-tensioned connections are summarized in Section 8.5. Findings of research conducted on behavior of post-tensioned flat plate-column connections subjected to earthquake-simulated effects are discussed in Section 8.6. Design examples of the connection of a post-tensioned flat plate with an interior and an edge column are presented in Sections D.7 through D.9.

Figure 8.1 shows a scheme for placing the post-tensioning tendons in two-way flat plates that is recommended by ACI 423.3R. In this scheme, tendons are closely spaced in narrow bands over support lines in one direction—commonly the longer one—and evenly distributed in the perpendicular direction. In addition to the constructibility advantage of this scheme, both directions can be designed for the maximum possible tendon drape. The banded and distributed tendons do not touch each other, except over the supports.

8.2—Sign convention

The positive sign convention for bending moment $M_{prestress}$ and shear force V_p produced by post-tensioning is indicated in Fig. 8.2(a) and (b). The variables V_p and $M_{prestress}$ depend mainly on the effective post-tensioning force and the profile of the tendons, expressed by eccentricity e, where e is measured from the midsurface of the slab to the centroid of the post-tensioning tendon; e is positive for a tendon situated below midsurface (Fig. 8.2(c)).

8.3—Post-tensioning effects

In a statically determinate structure, post-tensioning produces zero reactions. In statically indeterminate structures, post-tensioning produces reactions and internal forces due to the

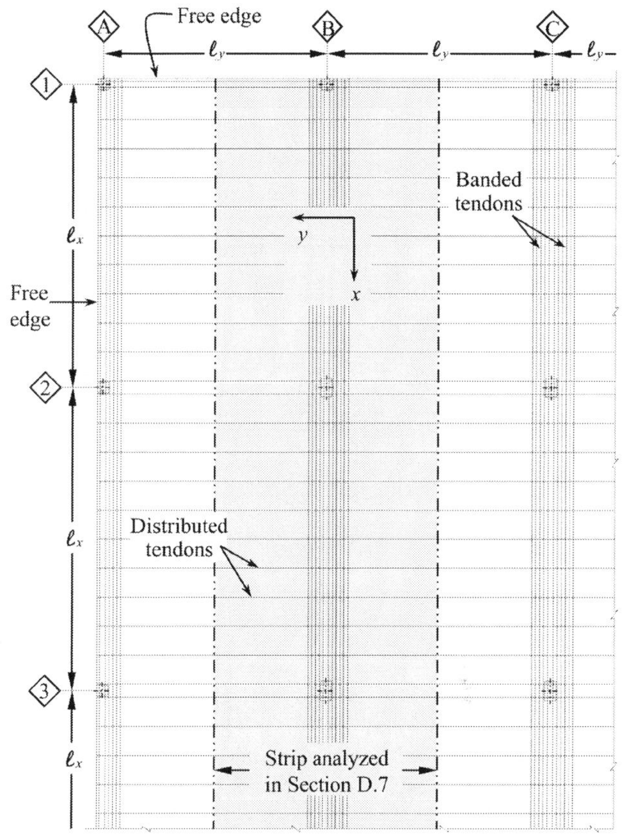

Fig. 8.1—Banded-distributed tendons.

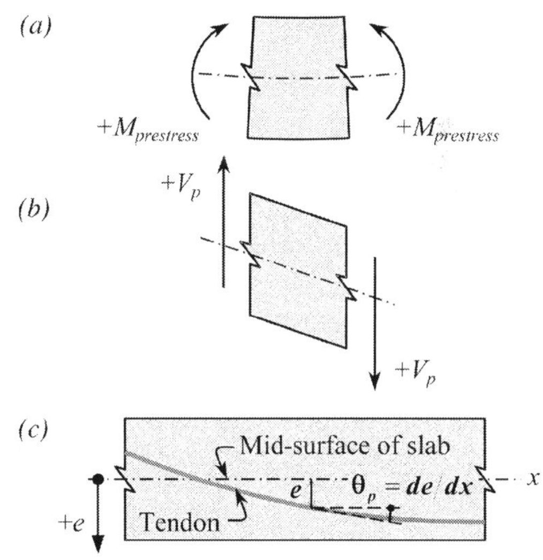

Fig. 8.2—Sign convention of: (a) $M_{prestress}$*; (b)* V_p*; and (c) tendon eccentricity and slope.*

restraints imposed at the supports. The reactions and the corresponding internal forces are called "secondary." For checking the design strength, the secondary shear force V_p'' and the secondary bending moment M_p'' are required by ACI 318.

Tendon profiles are commonly composed of parabolic segments. Figure 8.3(a) depicts a typical profile of post-tensioning tendons in adjacent spans of a slab. Figure 8.3(b) represents the forces exerted by tendons on the concrete.

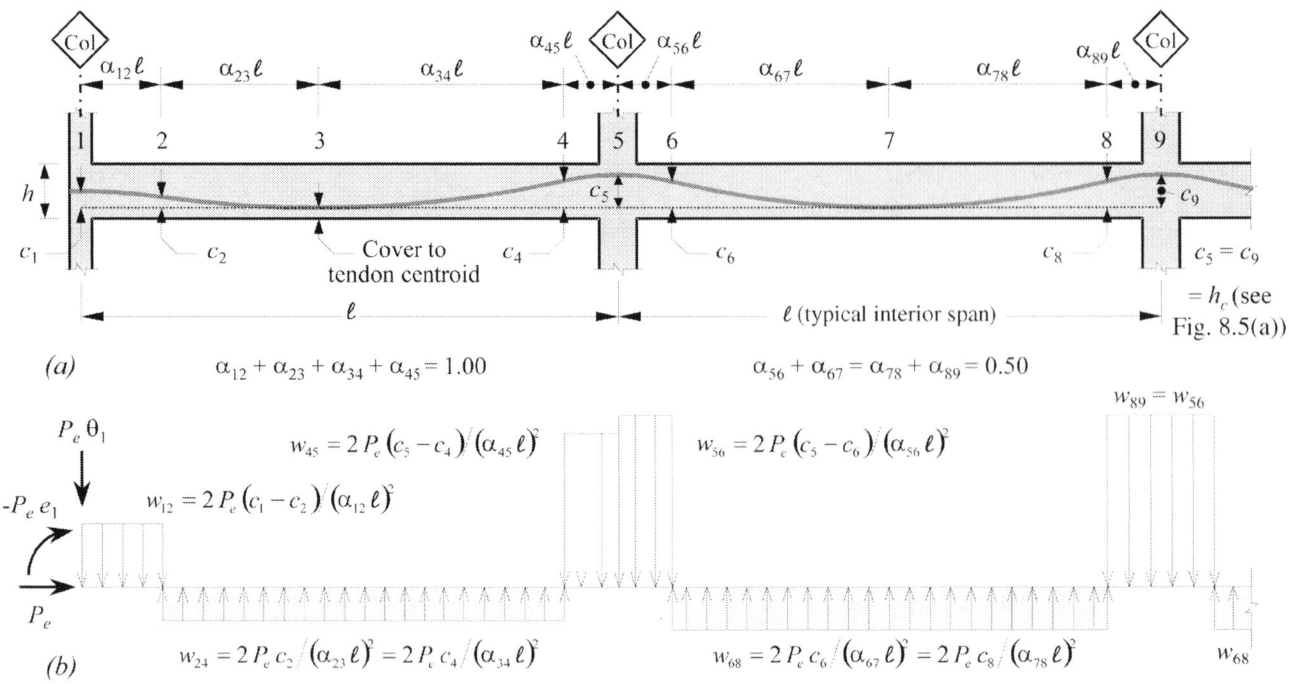

Fig. 8.3—Cross section in a unit width of a post-tensioned flat plate: (a) tendon profile; and (b) post-tensioned equivalent forces.

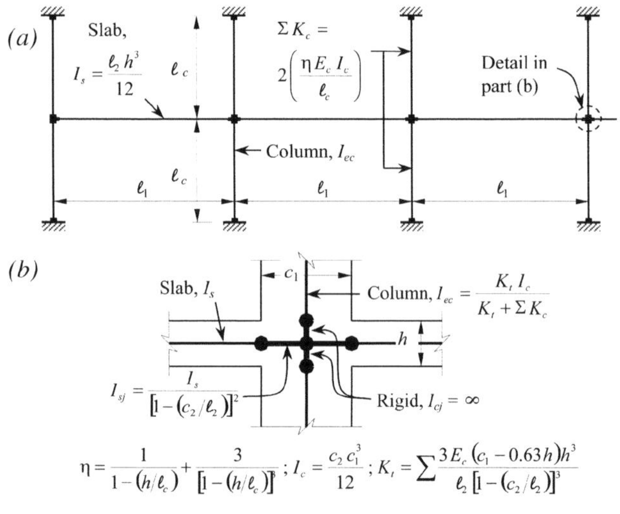

Fig. 8.4—Nonprismatic modeling of flat plate-column frames for effect of gravity loads and post-tensioning effects.

Before using this system of forces in the analysis, it should be verified that it is self-equilibrating; that is, it produces no reactions if applied on a statically determinate structure. The symbol P_e in Fig. 8.3(b) is the absolute value of the effective post-tensioning force per unit slab width, assumed constant over the length of the tendon. $\theta_p = (de/dx)$ is the angle between the tangent to the tendon and the centroidal axis (Fig. 8.2(c)), whereas e is the eccentricity of the tendon. The effective prestressing force is equal to the jacking force less the losses due to anchor setting, friction, shrinkage and creep of concrete, and relaxation of post-tensioning steel.

In defining the geometry of the tendon profile (Fig. 8.3(a)), the α-values should be selected so that the tendon has the same slope on both sides of the inflection points: 2, 4, 6, and 8. The equations in Fig. 8.3(b) are based on the assumption that the tendon has zero slope at the end ($\theta_{p1} = 0$) and each of the segments 2-3-4 and 6-7-8 is a continuous parabola. The length of the inverted segments over internal supports is typically between 20 to 30% of the span of the panel, ℓ (for example, $\alpha_{45} + \alpha_{56} \approx 0.20$ to 0.30). Section 8.6 gives an example of a tendon profile. For ease of construction and for maximum load balancing by post-tensioning, tendons should be placed in the banded direction so that their highest and lowest points are at the same level as the innermost bars of the top and bottom flexural reinforcement mesh. The distributed tendons at their highest points over the supports should be below and touching the banded tendons. At anchorages, the tendons are commonly at the midsurface of the slab.

The effect of post-tensioning is used in the analysis of the equivalent frame in Fig. 8.4. Where justified by construction sequence, V_p'' and M_p'' can be calculated by applying the post-tensioning forces (Fig. 8.3(b)) on the equivalent frame of Fig. 8.4. Reactions will form a set of forces in equilibrium, representing the secondary (statically indeterminate) reactions. The shear force V_p and the bending moment $M_{prestress}$ resulting from the analysis are the internal forces including the secondary forces. Thus, the secondary shear force and bending moment at any section are

$$V_p'' = V_p - V_p' \quad (8\text{-}1)$$

$$M_p'' = M_{prestress} - M_p' \quad (8\text{-}2)$$

where V_p' and M_p' are the primary shear force and bending moment due to post-tensioning. Their values at any section are

$$V_p' = -P_e\theta_p = -P_e(de/dx) \quad (8\text{-}3)$$

$$M_p' = -P_e e \quad (8\text{-}4)$$

The secondary shear force and bending moment diagrams are composed of straight lines whose ordinates can be calculated by considering only the secondary reactions on the structure as a free body.

Commonly the value of P_e is chosen to produce upward load (that is, w_{24} and w_{68} [Fig. 8.3(b)]) that balances 75 to 85% of the dead load. Section 8.5 gives an equation for the post-tensioning level necessary to balance a fraction κ of the dead load in an interior span.

To maintain integrity, ACI 318 requires, as minimum, either: 1) two tendons pass through the column cage in each direction, or 2) two bottom non-prestressed bars pass in each direction within the column core and be anchored at exterior supports. ACI 318 also specifies arrangement of the non-prestressed reinforcing bars required to supplement post-tensioned tendons to meet the flexural strength demand and a minimum amount of non-prestressed reinforcement to control cracking in the vicinity of the column.

8.4—Effective compressive stress f_{pc}

The compressive stress f_{pc} (= P_e/h) is the average in-plane stress produced by effective post-tension forces in one of two orthogonal directions. Consider the tendon profile in a typical interior panel of a flat plate (Fig. 8.5(a)). The force P_e that balances a fraction κ of the service dead load per unit area of the slab's surface, w_D, is

$$P_e = \kappa w_D \frac{(1-2\alpha)^2\ell^2}{8(1-2\alpha)h_c} \quad (8\text{-}5)$$

The tendon is assumed to have a common tangent at the point of inflection in Fig. 8.5(a). The geometric symbols α, h, ℓ, and h_c are defined in Fig. 8.5(a); w_c is the weight of concrete per unit volume; w_{sd} is the superimposed dead load per unit area of the floor. Consistent units for force and length should be used. For example, the basic units lb and ft (or N and m) can be used for all the symbols in Eq. (8-5). Figure 8.5(b) shows the variation of f_{pc} with ℓ and α, for which the cover of post-tensioning tendon at top or bottom is taken as $(h - h_c)/2$ = 1.57 in. (40.0 mm). Figure 8.5(b) shows that for the range of ℓ = 20 to 39 ft (6 to 12 m), f_{pc} varies between 175 and 230 psi (1.21 and 1.59 MPa). The same ordinates of the graph in Fig. 8.5(a) can be used for any value of w_{sd} by noting that f_{pc} is proportional to $(w_c h + w_{sd})$.

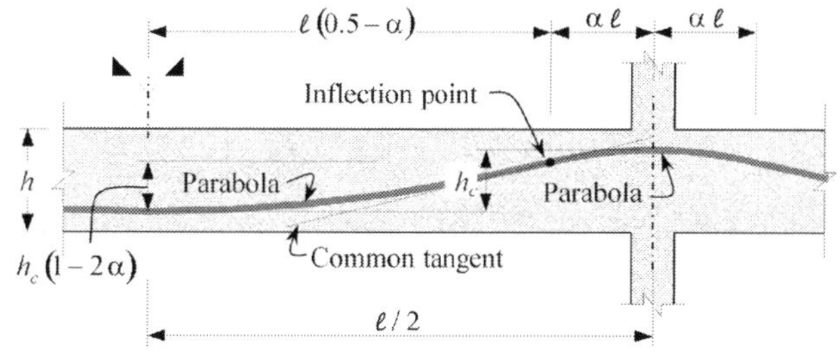

(a)

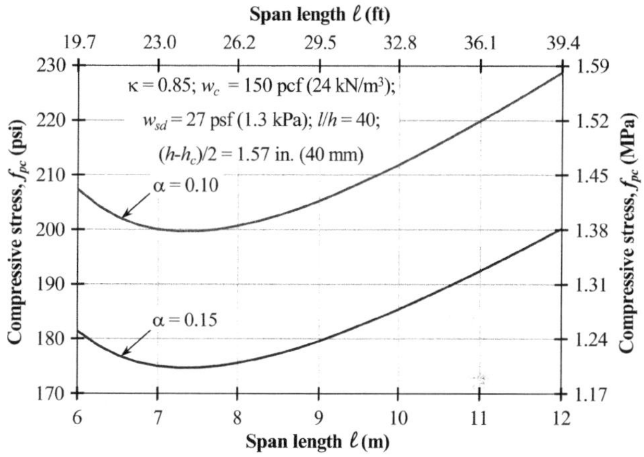

(b)

Fig. 8.5—(a) Tendon profile in typical interior span; and (b) variation of f_{pc} *with span length* ℓ*, and the geometric parameter* α*.*

8.5—Extension of punching shear design procedure to post-tensioned flat plates

The shear strength of concrete (in stress units) at a critical section at $d/2$ from the column face where shear reinforcement is not provided, is given by ACI 318 as

Inch-pound units

$$v_c = \beta_p\lambda\sqrt{f_c'} + 0.3f_{pc} + \frac{V_p}{b_o d} \quad (8\text{-}6)$$

SI units

$$v_c = \beta_p\lambda\frac{\sqrt{f_c'}}{12} + 0.3f_{pc} + \frac{V_p}{b_o d} \quad (8\text{-}6)$$

where β_p is the smaller of 3.5 and $[(\alpha_s d/b_o) + 1.5]$; λ is a modification factor related to unit weight of concrete; f_{pc} is the average value of the compressive stresses at centroid of cross section in two directions (after allowance for all post-tension losses); V_p is the vertical component of the effective post-tension forces crossing the shear-critical section. Equation (8-6) is applicable only if:

a. No portion of the column cross section is closer to a discontinuous edge than $4h$;

b. $\sqrt{f_c'}$ in Eq. (8-6) is not taken greater than 70 psi (5.8 MPa); and

c. f_{pc} in each direction is not less than 125 psi (0.86 MPa), nor taken greater than 500 psi (3.45 MPa).

If any of the above conditions is not satisfied, the shear strength is taken the same as for non-post-tensioned flat plates (Eq. (3-1), (3-2), and (3-3)). When shear reinforcement is required for post-tensioned flat plate-column connections, its design is the same as for non-post-tensioned flat plates (Eq. (4-1)).

Special care should be exercised in computing V_p in Eq. (8-6), due to the sensitivity of its value to the in-place tendon profile. When it is uncertain that the actual construction will match design assumptions, a reduced or zero value for V_p should be used in Eq. (8-6). It is noted that post-tensioning tendons in the banded direction commonly produce much higher confining stress in the vicinity of the column than the average stress, f_{pc} used by the ACI 318 Code in Eq. (8-6).

In calculating the upper limit of M_u for post-tensioned flat plates (Section 6.3, Eq. (6-2)), the probable flexural strength M_{pr} should be determined using a tensile stress $1.25f_y$ in the non-post-tensioned steel and f_{pc} in the post-tensioned tendons; with f_{ps} being the stress in unbonded tendons at the nominal strength (ACI 318), calculated by Eq. (8-7) or (8-8) depending on the span-depth ratio of the slab.

Inch-pound units

For $\ell/h \leq 35$

$$f_{ps} = f_{se} + 10{,}000 + \frac{f_c'}{100\rho_p} \leq \text{the least of: } \begin{cases} f_{py} \\ f_{se} + 60{,}000 \text{ psi} \end{cases} \quad (8\text{-}7)$$

For $\ell/h > 35$

$$f_{ps} = f_{se} + 10{,}000 + \frac{f_c'}{300\rho_p} \leq \text{the least of: } \begin{cases} f_{py} \\ f_{se} + 30{,}000 \text{ psi} \end{cases} \quad (8\text{-}8)$$

SI units

For $\ell/h \leq 35$

$$f_{ps} = f_{se} + 70 + \frac{f_c'}{100\rho_p} \leq \text{the least of: } \begin{cases} f_{py} \\ f_{se} + 414 \text{ MPa} \end{cases} \quad (8\text{-}7)$$

For $\ell/h > 35$

$$f_{ps} = f_{se} + 70 + \frac{f_c'}{300\rho_p} \leq \text{the least of: } \begin{cases} f_{py} \\ f_{se} + 207 \text{ MPa} \end{cases} \quad (8\text{-}8)$$

In Eq. (8-7) and (8-8), ρ_p is the ratio of the post-tensioning reinforcement; f_{se} is the effective stress in tendons after accounting for all post-tension losses; and f_{py} is the yield strength of the post-tensioning steel.

8.6—Research on post-tensioned flat plates

Hawkins (1981) tested a number of unbonded post-tensioned flat plate-column connections subjected to a combination of shear force and unidirectional moment reversals of increasing magnitudes until failure. A non-prestressed reinforced concrete specimen with stirrup shear reinforcement was included in this test series. The purpose of this control specimen was to compare its behavior, in terms of ductility and energy absorption, with the post-tensioned specimens. A comparison of the response of the post-tensioned flat plate-column connections with the control specimen showed that, after significant reversed cyclic loading, the post-tensioned specimens exhibited more ductility than the non-prestressed specimen. Performance of post-tensioned connections in residual shear capacity tests was better than that of the control specimen.

Martinez-Cruzado et al. (1994) tested a number of post-tensioned flat plate-column connections under simulated biaxial earthquake loading. The test specimens were subjected to different levels of gravity load to investigate the influence of gravity load on the response to biaxial lateral load of increasing magnitude. The tendons were banded in one direction and distributed in the perpendicular direction, and non-prestressed reinforcement was provided in the column vicinity to satisfy the ACI 318 requirements. No reinforcement was provided near the bottom slab surface. The test results showed that the level of gravity load applied on the slab was an important factor in the performance of the connection. Similar to non-prestressed concrete slabs (Fig. 3.1 and 3.3), the strength and the deformation of post-tensioned flat plate-column connections decreased as the gravity load increased. In addition, the results of the tests implicitly showed that providing post-tensioned tendons in accordance with ACI 318, Section 18.12.4 was effective in preventing complete collapse of the flat plate.

Experimental and analytical (finite-element) research (Kang and Wallace 2005; Ritchie and Ghali 2005; Gayed and Ghali 2006; and Gayed 2005) considered the seismic design procedure for punching shear reinforcement (stud shear reinforcement) in post-tensioned flat plates. Two test series, representing full-scale edge and interior column-slab connections, were conducted by Ritchie and Ghali (2005) and Gayed and Ghali (2006), respectively. The connections in both series were reinforced with post-tensioned tendons and non-prestressed steel bars. The amount of the two reinforcement types was varied in each series, without changing the sum of their contribution to flexural strength of the slab. The effective compressive stress f_{pc} varied between 0 and 160 psi (0.0 and 1.1 MPa). All connections in both series were provided with headed SSR, designed as outlined in Chapters 6 and 7. The test specimens, subjected to shearing force combined with drift reversals to simulate earthquake effects, exhibited ductile behavior (DR_u > 0.025) without much difference due to the variation of f_{pc}. The trend of the results as well as nonlinear finite-element analyses (Gayed 2005) indicated that increasing f_{pc} up to 200 psi (1.4 MPa) had no adverse effect on either the strength or the ductility that could be achieved without punching failure. When Eq. (8-6) or Eq. (3-1), (3-2), and (3-3), combined with the condition $v_u < \phi v_c$ is not satisfied, shear reinforcement is required and its design, as recommended in Chapters 6 and 7, applies for post-tensioned flat plates with f_{pc} = 60 to 200 psi (0.4 to 1.4 MPa). It is noted that Gayed (2005) limited f_{pc} to 200 psi (1.4 MPa), although higher values may occur in practice.

Kang and Wallace (2005) conducted two shake-table tests of one-third scale, two-story, two-bay flat plate-column frames. The slabs in the two test specimens were post-tensioned in one and non-post-tensioned in the other. The two frames were subjected to gravity loads combined with uniaxial base accelerations of increasing intensity. Both tests were provided with headed shear stud reinforcement, designed in accordance with ACI 421.1R. Drift ratio DR_u = 0.04 and 0.03 were reached, without loss of strength, in the post-tensioned and the non-post-tensioned connections, respectively.

CHAPTER 9—REFERENCES

9.1—Referenced standards and reports

The standards and reports listed below were the latest edition at the time this document was prepared. Because these documents are revised frequently, the reader is advised to contact the proper sponsoring group if it is desired to refer to the latest version.

American Concrete Institute

318 Building Code Requirements for Structural Concrete

352.1R Recommendations for Design of Slab-Column Connections in Monolithic Reinforced Concrete Structures

421.1R Guide to Shear Reinforcement for Slabs

423.3R Recommendations for Concrete Members Prestressed with Unbonded Tendons

American Society of Civil Engineers

ASCE/SEI 7 Minimum Design Loads for Buildings and Other Structures

ASTM International

A1044/1044M Standard Specification for Steel Stud Assemblies for Shear Reinforcement of Concrete

International Code Council

IBC 2006 International Building Code

The above publications may be obtained from the following organizations:

American Concrete Institute
38800 Country Club Drive
Farmington Hills, MI 48331
www.concrete.org

American Society of Civil Engineers
1801 Alexander Bell Drive
Reston, VA 20191
www.asce.org

ASTM International
100 Barr Harbor Drive
West Conshohocken, PA 19428
www.astm.org

International Code Council
500 New Jersey Avenue, NW
6th Floor
Washington, DC 20001
www.iccsafe.org

9.2—Cited references

ANATECH Consulting Engineers, 1995, "ANATECH Concrete Analysis Program ANACAP," Version 2.1, User's Guide, San Diego, CA.

Berg, G. V., and Stratta, J. L., 1964, "Anchorage and the Alaska Earthquake of March 27, 1964," American Iron and Steel Institute, New York, 63 pp.

Brown, S. J., 2003, "Seismic Response of Slab-Column Connections," PhD thesis, Department of Civil Engineering, University of Calgary, Calgary, AB, Canada.

Cao, H., 1993, "Seismic Design of Slab-Column Connections," MSc thesis, Department of Civil Engineering, University of Calgary, Calgary, AB, Canada.

Dilger, W. H., and Brown, S. J., 1995, "Earthquake Resistance of Slab-Column Connections," Institut für Baustatik und Konstruktion, ETH Zürich, Sept., pp. 22-27.

Dilger, W. H., and Cao, H., 1991, "Behaviour of Slab-Column Connections Under Reversed Cyclic Loading," *Proceedings of the 2nd International Conference of High-Rise Buildings*, China.

Gayed, R. B., 2005, "Design for Punching Shear at Interior Columns in Prestressed Concrete Slabs: Resistance to Earthquakes," PhD dissertation, University of Calgary, Calgary, AB, Canada, 318 pp.

Gayed, R. B., and Ghali, A., 2006, "Seismic-Resistant Joints of Interior Columns with Prestressed Slabs," *ACI Structural Journal*, V. 103, No. 5, Sept.-Oct., pp. 710-719.

Ghali, A., and Hammill, N., 1992, "Effectiveness of Shear Reinforcement in Slabs," *Concrete International*, V. 14, No. 1, Jan., pp. 60-65.

Ghali, A.; Neville, A. M.; and Brown, T. G., 2003, *Structural Analysis: A Unified Classical and Matrix Approach*, fifth edition, Spon Press, 844 pp.

Hawkins, N. M., 1981, "Lateral Load Resistance of Unbonded Post-Tensioned Flat Plate Construction," *PCI Journal*, V. 26, No. 1, Jan., pp. 94-117.

Hawkins, N. M.; Mitchell, D.; and Hanna, S. N., 1975, "Effects of Shear Reinforcement on the Reversed Cyclic Loading Behaviour of Flat Plate Structures," *Canadian Journal of Civil Engineering*, V. 2, pp. 572-582.

Hueste, M. B. D., and Wight, J. K., 1999, "Nonlinear Punching Shear Failure Model for Interior Slab-Column Connections," *Journal of Structural Engineering*, ASCE, V. 125, No. 9, Sept., pp. 997-1008.

Islam, S., and Park, R., 1976, "Tests on Slab-Column Connections with Shear and Unbalanced Flexure," *Journal of Structural Division*, ASCE, V. 102, No. ST3, Mar., pp. 549-568.

Kang, T. H. K., and Wallace, J. W., 2005, "Dynamic Responses of Flat Plate Systems with Shear Reinforcement," *ACI Structural Journal*, V. 102, No. 5, Sept.-Oct., pp. 763-773.

Martinez-Cruzado, J. A.; Qaisrani, A. N.; and Moehle, J. P., 1994, "Post-Tensioned Flat Plate Slab-Column Connections

Subject to Earthquake Loading," *5th U.S. National Conference on Earthquake Engineering*, V. 2, Chicago, IL, pp. 139-148.

Megally, S., 1998, "Punching Shear Resistance of Concrete Slabs to Gravity and Earthquake Forces," PhD dissertation, Department of Civil Engineering, University of Calgary, Calgary, AB, Canada, 468 pp.

Megally, S., and Ghali, A., 1994, "Design Considerations for Slab-Column Connections in Seismic Zones," *ACI Structural Journal*, V. 91, No. 3, May-June, pp. 303-314.

Megally, S., and Ghali, A., 2000a, "Seismic Behavior of Slab-Column Connections," *Canadian Journal of Civil Engineering*, V. 27, No. 1, Feb., pp. 84-100.

Megally, S., and Ghali, A., 2000b, "Punching of Concrete Slabs Due to Column's Moment Transfer," *Journal of Structural Engineering*, ASCE, V. 126, No. 2, Feb., pp. 180-189.

Megally, S., and Ghali, A., 2000c, "Seismic Behavior of Edge Column-Slab Connections with Stud Shear Reinforcement," *ACI Structural Journal*, V. 97, No. 1, Jan.-Feb., pp. 53-60.

Megally, S., and Ghali, A., 2000d, "Punching Shear Design of Seismic-Resistant Slab-Column Connections," *ACI Structural Journal*, V. 97, No. 5, Sept.-Oct., pp. 720-730.

Mitchell, D.; DeVall, R. H.; Saatcioglu, M.; Simpson, R.; Tinawi, R.; and Tremblay, R., 1995, "Damage to Concrete Structures Due to the 1994 Northridge Earthquake," *Canadian Journal Civil Engineering*, V. 22, No. 4, Apr., pp. 361-377.

Mitchell, D.; Tinawi, R.; and Redwood, R. G., 1990, "Damage to Buildings Due to the 1989 Loma Prieta Earthquake—A Canadian Code Perspective," *Canadian Journal of Civil Engineering*, V. 17, No. 10, Oct., pp. 813-834.

Mokhtar, A. S.; Ghali, A.; and Dilger, W. H., 1985, "Stud Shear Reinforcement for Flat Concrete Plates," ACI JOURNAL, *Proceedings* V. 82, No. 5, Sept.-Oct., pp. 676-683.

Pan, A. D., and Moehle, J. P., 1989, "Lateral Displacement Ductility of Reinforced Concrete Flat Plates," *ACI Structural Journal*, V. 86, No. 3, May-June, pp. 250-258.

Pan, A. D., and Moehle, J. P., 1992, "Experimental Study of Slab-Column Connections," *ACI Structural Journal*, V. 89, No. 6, Nov.-Dec., pp. 626-638.

Ritchie, M., and Ghali, A., 2005, "Seismic-Resistant Connections of Edge Columns with Prestressed Slabs," *ACI Structural Journal*, V. 102, No. 2, Mar.-Apr., pp. 314-323

Robertson, I. N., and Durrani, A. J., 1992, "Gravity Load Effect on Seismic Behavior of Interior Slab-Column Connections," *ACI Structural Journal*, V. 89, No. 1, Jan.-Feb., pp. 37-45.

Sozen, M. A., 1980, "Review of Earthquake Response of Reinforced Concrete Buildings With a View to Drift Control, State-of-the-Art in Earthquake Engineering," *7th World Conference on Earthquake Engineering,* Istanbul, Turkey, pp. 119-174.

Vanderbilt, M. D., and Corely, W. G., 1983, "Frame Analysis of Concrete Buildings," *Concrete International*, V. 5, No. 12, Dec., pp. 33-43.

Wey, E. H., and Durrani, A. J., 1992, "Seismic Response of Interior Slab-Column Connections with Shear Capitals," *ACI Structural Journal*, V. 89, No. 6, Nov.-Dec., pp. 682-691

Yanev, P. I.; Gillengerten, J. D.; and Hamburger, R. O., 1991, "The Performance of Steel Buildings in Past Earthquakes," American Iron and Steel Institute and EQE Engineering, Inc., 97 pp.

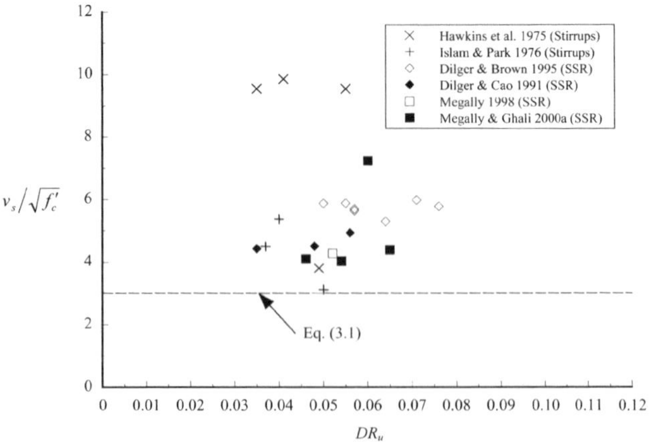

Fig. A.1—Shear strength provided by slab shear reinforcement.

APPENDIX A—VERIFICATION OF PROPOSED MINIMUM AMOUNT OF SHEAR REINFORCEMENT FOR EARTHQUAKE-RESISTANT FLAT PLATE-COLUMN CONNECTIONS

As discussed in Chapter 4, the recommended minimum amount of shear reinforcement is based on experiments (Islam and Park 1976; Hawkins et al. 1975; Dilger and Brown 1995; Dilger and Cao 1991; Megally and Ghali 2000a; Megally 1998). In these experiments, flat plate-column connections with different amounts of stirrups or SSR experienced substantial drifts without punching shear failure. The slab thickness in the tests was 6 in. (150 mm) except in Islam and Park's tests, where the thickness was 3.5 in. (89 mm). Table A.1 and Fig. A.1 summarize results of these experiments; the figure is a plot of the drift capacity versus ($v_s/\sqrt{f'_c}$). The figure shows that all of the tested flat plate-column connections have a ratio ($v_s/\sqrt{f'_c}$) exceeding 3. The table indicates that even when the SSR is spaced at $s = 0.75d$, flat plate-column connections can experience story drift ratio DR_u much higher than 0.025 without punching shear failure or significant loss of moment transfer capacity. This holds true for cases even when ($V_u/\phi V_c$) is relatively high. For this reason, the minimum amount of shear reinforcement specified by Eq. (4-1) is somewhat smaller than what has been used in the experiments.

APPENDIX B—VERIFICATION OF UPPER LIMIT TO UNBALANCED MOMENT TO BE USED IN PUNCHING SHEAR DESIGN

As mentioned in Section 6.3, the quantity $\alpha_m M_u$ can be determined from the results of a nonlinear finite element analysis (Megally and Ghali 2000b). Results of such an analysis, which have helped in selecting the empirical Eq. (6-3) and (6-4), are presented in Tables B.1 and B.2 for interior and edge flat plate-column connections, respectively (Fig. B.1(a) and (b)). The slab is simply supported on four or three edges (Fig. B.1(a) or (b), respectively). The magnitude of the constant force V_u, applied through the axis of the column, is one of the variables in the tables. The bottom reinforcement ratio in all slabs is 0.65%, whereas the top reinforcement ratio ρ is variable. The yield strength of the flexural reinforcing bars is 60 ksi (414 MPa), except as noted otherwise.

Table A.1—Ultimate story drift ratios of flat plate-column connections transferring cyclic moment reversals

(1)	(2)	(3)	(4)	(5)	(6)	(7)	(8)	(9)
Slab	Reference	Type of shear reinforcement	f_c', psi (MPa)	$V_u/\phi V_c$	s/d	f_{yt}, ksi (MPa)	$v_s/\sqrt{f_c'}$	$100DR_u$
6CS	Islam and Park (1976)	Stirrups	4090 (28.2)	0.24	0.55	42.1 (290)	5.37	4.0
7CS			4310 (29.7)	0.24	0.55	44.1 (304)	4.50	3.7
8CS			3210 (22.1)	0.27	0.55	42.5 (293)	3.11	5.0
SS1	Hawkins et al. (1975)	Stirrups	4000 (27.6)	0.49	0.33	66.6 (459)	9.54	3.5
SS3			3750 (25.9)	0.48	0.33	66.0 (455)	9.85	4.1
SS4			4000 (27.6)	0.47	0.33	66.0 (455)	9.54	5.5
SS5			4670 (32.2)	0.42	0.33	67.1 (463)	3.80	4.9
SJB-1	Dilger and Brown (1995)†	SSR	4669 (32.2)	0.48	0.57	66.7 (460)	5.88	5.5
SJB-2			4974 (34.3)	0.47	0.57	66.7 (460)	5.70	5.7
SJB-3			4698 (32.4)	0.48	0.57	66.7 (460)	5.87	5.0
SJB-4			5757 (39.7)	0.43	0.57	66.7 (460)	5.30	6.4
SJB-5			4843 (33.4)	0.47	0.57	66.7 (460)	5.78	7.6
SJB-8*			5075 (35.0)	0.46	0.57	66.7 (460)	5.64	5.7
SJB-9*			4524 (31.2)	0.49	0.57	66.7 (460)	5.98	7.1
CD3	Dilger and Cao (1991)	SSR	5162 (35.6)	0.92	0.79	53.8 (371)	4.43	3.5
CD4			4974 (34.3)	0.62	0.79	53.8 (371)	4.51	4.8
CD6			4553 (31.4)	0.65	0.39	53.8 (371)	12.3	5.4
CD7			4147 (28.6)	0.51	0.79	53.8 (371)	4.94	5.6
MG-3	Megally and Ghali (2000a)	SSR	4872 (33.6)	0.56	0.75	54.7 (377)	4.03	5.4
MG-4			4684 (32.3)	0.86	0.75	54.7 (377)	4.11	4.6
MG-5			4104 (28.3)	0.31	0.75	54.7 (377)	4.39	6.5
MG-6			4365 (30.1)	0.59	0.44	54.7 (377)	7.23	6.0
MG-10	Megally (1998)	SSR	4307 (29.7)	0.60	0.75	54.7 (377)	4.28	5.2

*Flat plate-column connections transferring biaxial moments $M_{ux} = M_{uy}$ (Fig. 3.2(a)).

†Test variable in this series was concentration of slab flexural reinforcement in column vicinity.

Table B.1—Nonlinear finite element results of fraction of unbalanced moment transferred by flexure in interior flat plate-column connections

(1)	(2)	(3)	(4)	(5)	(6)	(7)
Slab	ℓ_x/ℓ_y	ρ, %	V_u, kip (kN)	$\alpha_{m\text{-FE}}$	α_m	$\alpha_{m\text{-FE}}/\alpha_m$
I1	1.00	0.61	0 (0)	0.40	0.40	1.00
I2	1.00	0.61	16.9 (75.2)	0.39	0.40	0.98
I3	1.00	0.61	33.7 (149.9)	0.37	0.40	0.93
I4	1.00	0.61	67.4 (299.8)	0.43	0.40	1.08
I5	1.00	1.22	0 (0)	0.39	0.40	0.98
I6	1.00	1.22	16.9 (75.2)	0.39	0.40	0.98
I7	1.00	1.22	33.7 (149.9)	0.37	0.40	0.93
I8	1.00	1.22	67.4 (299.8)	0.44	0.40	1.10
I9*	1.00	1.22	33.7 (149.9)	0.37	0.40	0.93
I10†	1.00	1.22	33.7 (149.9)	0.39	0.40	0.98
I11	1.00	2.44	0 (0)	0.45	0.40	1.13
I12	1.00	2.44	16.9 (75.2)	0.42	0.40	1.05
I13	1.00	2.44	33.7 (149.9)	0.44	0.40	1.10
I14	1.00	2.44	67.4 (299.8)	0.52	0.40	1.30
I15	1.00	2.44	101.2 (450.2)	0.57	0.40	1.43
I16‡	1.00	2.44	0 (0)	0.44	0.40	1.10
I17§	1.00	2.44	33.7 (149.9)	0.47	0.40	1.18
I18‖	1.00	2.44	33.7 (149.9)	0.44	0.40	1.10
I19‡	1.00	2.44	33.7 (149.9)	0.43	0.40	1.08
I20#	1.00	2.44	33.7 (149.9)	0.45	0.40	1.13
I21	1.00	2.44	101.2 (450.2)	0.55	0.40	1.38
I22	0.39	1.22	0 (0)	0.64	0.54	1.19
I23	0.39	1.22	22.5 (100.1)	0.64	0.54	1.19
I24	0.39	1.22	33.7 (149.9)	0.62	0.54	1.15
I25	0.39	1.22	45.0 (200.2)	0.60	0.54	1.11
I26	0.39	1.22	67.4 (299.8)	0.64	0.54	1.19
I27	0.66	1.22	0 (0)	0.53	0.47	1.15
I28	0.66	1.22	33.7 (149.9)	0.47	0.47	1.00
I29	0.66	1.22	67.4 (299.8)	0.55	0.47	1.17
I30	1.52	1.22	0 (0)	0.34	0.32	1.06
I31	1.52	1.22	33.7 (149.9)	0.30	0.32	0.94
I32	1.52	1.22	67.4 (299.8)	0.37	0.32	1.16
I33	2.57	2.44	0 (0)	0.19	0.20	0.95
I34	2.57	2.44	16.9 (75.2)	0.19	0.20	0.95
I35	2.57	2.44	33.7 (149.9)	0.20	0.20	1.00
					Mean	1.09
					Standard deviation	0.12
					Coefficient of variation	0.11

*f_y = 44 ksi (303 MPa).

†f_y = 73 ksi (503 MPa).

‡Slab with shear reinforcement of 3/8 in. (9.5 mm) diameter (12 shear reinforcing bars in each peripheral line).

§Slab with shear reinforcement of 1/8 in. (3.2 mm) diameter (12 shear reinforcing bars in each peripheral line).

‖Slab with shear reinforcement of 1/4 in. (6.4 mm) diameter (12 shear reinforcing bars in each peripheral line).

#Slab with shear reinforcement of 1/2 in. (12.7 mm) diameter (12 shear reinforcing bars in each peripheral line).

Table B.2—Nonlinear finite element results of fraction of unbalanced moment transferred by flexure in edge flat plate-column connections

(1)	(2)	(3)	(4)	(5)	(6)	(7)
Slab	ℓ_x/ℓ_y	ρ, %	V_u, kip (kN)	$\alpha_{m\text{-FE}}$	α_m	$\alpha_{m\text{-FE}}/\alpha_m$
E1	0.84	0.61	0	0.24	0.24	1.00
E2	0.84	0.61	16.9 (75.2)	0.22	0.24	0.92
E3	0.84	0.61	22.5 (100.1)	0.21	0.24	0.88
E4	0.84	0.61	34.1 (151.7)	0.25	0.24	1.04
E5	0.84	1.22	0 (0)	0.34	0.30	1.13
E6	0.84	1.22	16.9 (75.2)	0.29	0.30	0.97
E7	0.84	1.22	22.5 (100.1)	0.31	0.30	1.03
E8	0.84	1.22	28.1 (125.0)	0.32	0.30	1.07
E9	0.84	1.22	33.7 (149.9)	0.32	0.30	1.07
E10	0.84	1.22	38.1 (169.5)	0.37	0.30	1.23
E11	0.84	1.22	16.9 (75.2)	0.36	0.30	1.20
E12	0.30	1.22	16.9 (75.2)	0.70	0.49	1.43
E13	0.50	1.22	16.9 (75.2)	0.49	0.39	1.26
E14	1.28	1.22	16.9 (75.2)	0.25	0.23	1.09
E15	1.91	1.22	16.9 (75.2)	0.14	0.16	0.88
E16	2.33	1.22	16.9 (75.2)	0.13	0.12	1.08
E17	0.92	1.22	16.9 (75.2)	0.30	0.29	1.03
I18*	0.84	1.22	16.9 (75.2)	0.26	0.30	0.87
E19†	0.84	1.22	16.9 (75.2)	0.36	0.30	1.20
E20‡	0.84	1.22	46.1 (205.1)	0.38	0.30	1.27
E21	0.84	1.83	16.9 (75.2)	0.34	0.36	0.94
E22	0.84	2.44	0 (0)	0.51	0.43	1.19
E23	0.84	2.44	16.9 (75.2)	0.45	0.43	1.05
E24	0.84	2.44	22.5 (100.1)	0.45	0.43	1.05
E25	0.84	2.44	33.7 (149.9)	0.46	0.43	1.07
E26	0.84	2.44	45.0 (200.2)	0.48	0.43	1.12
E27	0.84	2.44	56.9 (253.1)	0.52	0.43	1.21
E28‡	0.84	2.44	0 (0)	0.40	0.43	0.93
E29‡	0.84	2.44	16.9 (75.2)	0.40	0.43	0.93
E30‡	0.84	2.44	33.7 (149.9)	0.43	0.43	1.00
E31‡	0.84	2.44	67.7 (301.1)	0.54	0.43	1.26
					Mean	1.08
					Standard deviation	0.14
					Coefficient of variation	0.13

*f_y = 43.5 ksi (300 MPa).

†Slab with twice the number of reinforcing bars but with ρ = 1.22%.

‡Slab with shear reinforcement of 3/8 in. (9.5 mm) diameter (seven shear reinforcing bars in each peripheral line).

The ratio $\alpha_{m\text{-FE}}$ (finite element) to α_m (Eq. (6-3) or (6-4)) in Column 7 of Tables B.1 or B.2 has an average value of 1.09 or 1.08, respectively, with the coefficient of variation equal to 0.11 or 0.13, respectively. These results substantiate the proposed Eq. (6-2) through (6-4).

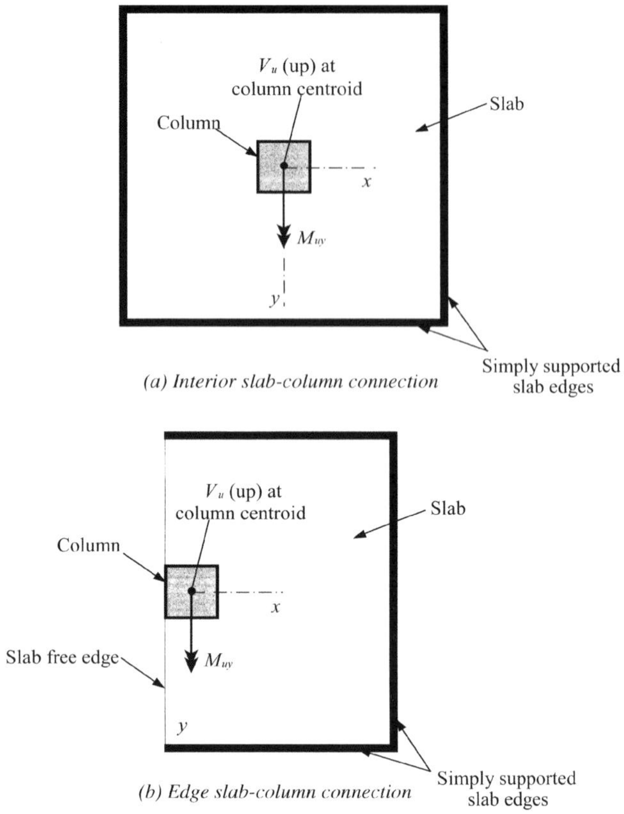

Fig. B.1—Plan views of flat plate-column connections analyzed by the finite element method.

APPENDIX C—NOTES ON PROPERTIES OF SHEAR-CRITICAL SECTION

The equations in Section C.1 and C.2 are taken from ACI 421.1R. They are given as follows because they are used in the present document.

C.1—Second moments of area

In this report, the shear stresses caused by unbalanced moments are computed using parameters J_x and J_y rather than the J_c property that has been used for several decades. The value of J_y is smaller than J_c; in most cases, the difference is only a few percent. The use of J_x or J_y instead of J_c for a critical section of general shape is recommended in ACI 421.1R; the relevant material is summarized in the present subsection.

General expressions for J_x and J_y are given, but first the expressions for J_c and J_y for the case of interior rectangular column are compared (Fig. 3.2(a)). ACI 318 defines J_c as a property of the assumed critical section analogous to polar moment of inertia and gives Eq. (C-1), which is compared with Eq. (C-2) for J_y for the same critical section:

$$J_c = \frac{d(c_x+d)^3}{6} + \frac{d(c_y+d)(c_x+d)^2}{2} + \frac{d^3(c_x+d)}{6} \quad \text{(C-1)}$$

$$J_y = \frac{d(c_x+d)^3}{6} + \frac{d(c_y+d)(c_x+d)^2}{2} \quad \text{(C-2)}$$

The expression for J_y is the same as the expression for J_c, with the last term dropped. The third term on the right-hand side of Eq. (C-1) is normally much smaller than the other two terms.

General expressions for b_o, J_x, and J_y of a shear-critical section composed of straight segments are

$$b_o = \sum \ell_{ij} \quad \text{(C-3)}$$

$$J_x = \frac{d}{3}\sum[\ell_{ij}(y_i^2 + y_i y_j + y_j^2)] \quad \text{(C-4)}$$

$$J_y = \frac{d}{3}\sum[\ell_{ij}(x_i^2 + x_i x_j + x_j^2)] \quad \text{(C-5)}$$

where b_o is the length of the perimeter of the shear-critical section; J_x and J_y = d multiplied by the second moments of the perimeter of the shear-critical section about its centroidal principal axes x and y, respectively; x_i and y_i are coordinates of i; x_j and y_j are coordinates of j, with i and j being the extremities of a typical straight segment of the perimeter; and ℓ_{ij} is the length of the segment ij.

ACI 318 requires that the shear stress, v_u due to V_u, $(\gamma_{vx}M_{ux})$ and $(\gamma_{vy}M_{uy})$ vary linearly over the shear-critical section. The shear stress distribution that satisfies this requirement is expressed as

$$v_u = \frac{V_u}{b_o d} + \gamma_{vx}\frac{M_{ux}}{J_x}y + \gamma_{vy}\frac{M_{uy}}{J_y}x \quad \text{(C-6)}$$

where x and y are coordinates, with respect to the centroidal principal axes, of any point on the perimeter of the shear-critical section. The centroidal principal axes, the only orthogonal axes about which the product of inertia is nil, must be used to obtain a stress v_u whose resultants are: V_u, $(\gamma_{vx}M_{ux})$ and $(\gamma_{vy}M_{uy})$. Figure 3.2(c) shows an example of a shear-critical section whose centroidal principal axes x and y do not coincide with the centroidal axes $\bar{x}$ and $\bar{y}$, that are parallel to the sides of the column. The principal axis x or y makes an angle θ (measured in clockwise direction) with the axis $\bar{x}$ or $\bar{y}$, given by

$$\tan 2\theta = \frac{-2J_{\bar{x}\bar{y}}}{J_{\bar{x}} - J_{\bar{y}}} \quad \text{(C-7)}$$

where $J_{\bar{x}}$ and $J_{\bar{y}}$ = d multiplied by the second moments of the perimeter of the shear-critical section about the axes $\bar{x}$ and $\bar{y}$, respectively (Eq. (C-4) and (C-5), substituting $\bar{x}$ and $\bar{y}$ for x and y). For a shear-critical section whose perimeter is composed of straight segments, the product of inertia is given by

$$J_{\bar{x}\bar{y}} = \frac{d}{6}\sum[\ell_{ij}(2\bar{x}_i\bar{y}_i + \bar{x}_i\bar{y}_j + \bar{x}_j\bar{y}_i + 2\bar{x}_j\bar{y}_j)] \quad \text{(C-8)}$$

where $\bar{x}_i$ and $\bar{y}_i$ are coordinates of i, and $\bar{x}_j$ and $\bar{y}_j$ are coordinates of j, with i and j being the extremities of a typical straight segment, whose length is ℓ_{ij}. The coordinates (x,y) of any point are related to the coordinates $(\bar{x},\bar{y})$ by

$$x = \bar{x}\cos\theta + \bar{y}\sin\theta \quad \text{(C-9)}$$

$$y = -\bar{x}\sin\theta + \bar{y}\cos\theta \quad \text{(C-10)}$$

C.2—Equations for γ_v

For interior columns, the fraction γ_v of the unbalanced moment that it transmitted by shear is defined in ACI 318-05 and in ACI 421.1R by Eq. (C-11) and (C-12). The latter source, however, also gives γ_v for edge and corner columns. Considering unbalanced moments M_{ux} or M_{uy} (Fig. 3.2), the equations for γ_{vx} and γ_{vy} are

Interior columns

$$\gamma_{vx} = 1 - \frac{1}{1 + \frac{2}{3}\sqrt{\frac{\ell_y}{\ell_x}}} \quad \text{(C-11)}$$

$$\gamma_{vy} = 1 - \frac{1}{1 + \frac{2}{3}\sqrt{\frac{\ell_x}{\ell_y}}} \quad \text{(C-12)}$$

Edge columns

$$\gamma_{vx} = 1 - \frac{1}{1 + \frac{2}{3}\sqrt{\frac{\ell_y}{\ell_x}}} \quad \text{(C-13)}$$

$$\gamma_{vy} = 1 - \frac{1}{1 + \frac{2}{3}\sqrt{\frac{\ell_x}{\ell_y} - 0.2}} \quad (\gamma_{vy} = 0 \text{ when } \ell_x/\ell_y \le 0.2) \quad \text{(C-14)}$$

Corner columns

$$\gamma_{vx} = 0.4 \quad \text{(C-15)}$$

$$\gamma_{vy} = \text{same as for edge column} \quad \text{(C-16)}$$

APPENDIX D—DESIGN EXAMPLES

D.1—General

The examples are worked out only in inch-pound units to make it easier for the reader; conversion factors are given in Appendix E; Example 1 is repeated in Example 5 using SI units. The examples consider a flat plate for a building, with equal span lengths ℓ = 20 ft (6.1 m) in two orthogonal directions, story height ℓ_c = 12 ft (3.66 m), and the slab thickness h = 8 in. (203 mm). The lateral-force-resisting structural system is composed of shear walls that limit the story drift ratio, including inelastic deformations, to DR_u = 0.02. The following are design examples of the shear reinforcement required for interior, edge, and corner flat plate-column connections, for V_u combined with earthquake-factored

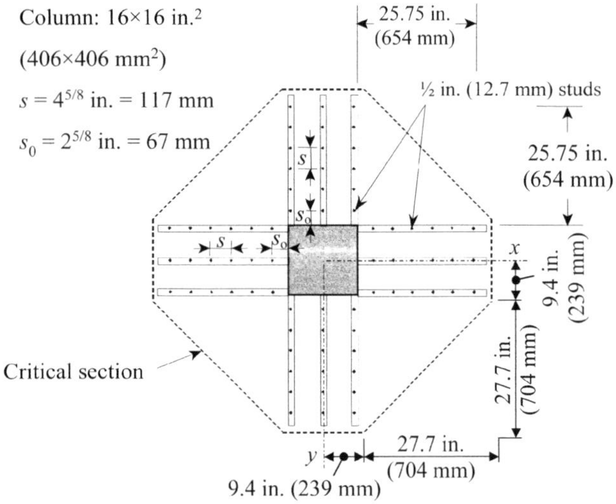

Fig. D.1—Arrangement of SSR in interior flat plate-column connection of Examples 1 and 5.

moment M_{uy} (Fig. 3.2). Other data are concrete cover = 0.75 in. (19 mm); $f_c' = 4000$ psi (27.6 MPa); $E_c = 3600$ ksi (24.8 GPa); $f_y = 60$ ksi (414 MPa); flexural reinforcement nominal diameter = 0.625 in. (15.9 mm); diameter of shear reinforcement = 0.50 in. (12.7 mm); $f_{yt} = 50$ or 60 ksi (345 or 414 MPa) for SSR or stirrups, respectively; and $d = 8 - 0.75 - (0.625) = 6.625$ in. (168 mm). The slab has been designed to support gravity loads using the Direct Design Method of ACI 318, and the flexural reinforcement ratios have been determined. The strength reduction factor ϕ according to ACI 318-08, Section 9.3.2.3, is 0.75. The value of the dimensionless coefficient (C_d/I_E) is 3.85. The dead load D is 120 lb/ft^2 (5.75 kN/m^2) and the live load L is 50 lb/ft^2 (2.39 kN/m^2). Load factors as given in ACI 318-08, Section 9.2.1 are used.

D.2—Example 1: Interior flat plate-column connection

The value of $(V_u)_{D+L+E}$, given as follows, corresponds to the load combination that governs the design. The load factors used to calculate $(V_u)_{D+L+E}$ are in accordance with ACI 318. The values of $(M_y)_D$ and $(M_y)_L$ due to nonfactored dead and live loads, respectively, are given. The given values are $(V_u)_{D+L+E} = 78$ kips, $(M_y)_D = 0.0$, and $(M_y)_L = 0.0$. The cross-sectional dimensions of the column are $c_x = c_y = 16$ in. (Fig. D.1); the reinforcement ratios of the top and bottom flexural bars are 1.0 and 0.5%, respectively.

The critical section shown in Fig. 3.2(a) has $b_o = 90.5$ in.; V_c is governed by Eq. (3-3): $V_c = 4(90.5)(6.625)\sqrt{4000}/1000 = 151.7$ kips; and $(V_u/\phi V_c) = 78/(0.75 \times 151.7) = 0.69$.

The unbalanced moment caused by seismic forces is obtained from an elastic analysis of the frame shown in Fig. 6.1(a). The prescribed elastic story drift δ_e is given by Eq. (6-1)

$$\delta_e = \frac{DR_u\ell_c}{(C_d/I_E)} = \frac{0.02 \times 12 \times 12}{3.85} = 0.75 \text{ in.}$$

where $(C_d/I_E) = 3.85$ according to the ASCE/SEI 7 (assuming that the lateral-force-resisting system consists of shear walls). For the frame shown in Fig. 6.1(a), the moment of inertia of the slab $I_s = 5120$ in.^{4} This value is equal to one-half the gross moment of inertia of an 8 in. slab that is 20 ft wide. The reduction is made to reflect some of the effects of cracking on the force distribution. The equivalent moment of inertia of the column $I_{ec} = 1320$ in.^{4} This value is obtained according to the equivalent frame method of ACI 318 by assuming that the flexibility of the equivalent column equals the sum of the flexibility of the actual column and the torsional strip of the slab. In the determination of the equivalent moment of inertia, the column lengths were taken as 72 in. (Fig. 6.1(a)) with 4 in. long rigid parts at the joints and with the other ends simply supported, as shown in Fig. 6.1(c). The equivalent moment of inertia is computed as $I_{ec} = I_c(K_{ec}/\Sigma K_c)$, where K_{ec} and K_c are end rotational stiffnesses (moment per unit rotation) of equivalent column and the actual column, respectively.

The unbalanced moment due to δ_e, obtained from elastic analysis of the frame in Fig. 6.1(a), is $(M_y)_E = 1670$ in.-kip According to ACI 318-08, Section 9.2, the factored unbalanced moment is

$$M_{uy} = 1.2(M_y)_D + 1.0(M_y)_L + 1.0(M_y)_E \quad \text{(D-1)}$$

$$M_{uy} = 1.2(0) + 1.0(0) + 1.0(1670) = 1670 \text{ in.-kip}$$

The flexural strengths provided by the top and bottom flexural reinforcement, based on a stress of $1.25f_y$, of critical section Sides BC and AD (Fig. 3.2(a)), are 660 and 350 in.-kip, respectively. Thus, $M_{pr} = 660 + 350 = 1010$ in.-kip. For this interior connection, $\gamma_{vy} = 0.4$; $\beta_r = 1.0$; and $\alpha_m = 0.4$ (Eq. (6-3)). It follows that the upper limit of M_{uy} is (Eq. (6-2))

$$(M_{uy})_{\text{upper limit}} = \frac{660 + 350}{0.4} = 2530 \text{ in.-kip}$$

Because the upper limit is higher than $M_{uy} = 1670$ in.-kip, the flat plate-column connection should be designed to resist $V_u = 78$ kips and $M_{uy} = 1670$ in.-kip. In this structure, the drift due to seismic forces does not change V_u for this interior column.

The maximum factored shear stress due to V_u and M_{uy} (computed according to the eccentric shear stress model of ACI 318) at the critical section at $d/2$ from the column face (Fig. 3.2(a)) is $v_u = 278$ psi and $(v_u/\phi) = 5.9\sqrt{f_c'}$ psi. This value is less than $6\sqrt{f_c'}$ psi; therefore, stirrups or SSR can be used. The SSR shown in Fig. D.1 is designed according to ACI 421.1R. For the chosen shear reinforcement, $A_v = 2.36$ in.^{2} (12 studs of diameter 1/2 in.); $s = 4.625$ in., the nominal shear strength (expressed in stress units) provided by shear reinforcement (Eq. (4-1)) is $v_s = [2.36(50{,}000)]/[90.5(4.625)] = 282$ psi, which is greater than the minimum: $3\sqrt{f_c'} = 190$ psi. The distance between the column face and the outermost peripheral line of SSR is 25.75 in., which is greater than ($3.5d = 23.2$ in.). The details of the stud arrangement shown in Fig. D.1 satisfy the requirement for the minimum amount of shear reinforcement. ACI 421.1R recommends that $s_o \leq 0.4d$; the value used in this example and Examples 2, 3, and

4 is s_o = 2.625 in. The nominal shear strength v_n multiplied by ϕ should exceed v_u = 278 psi. In this example

$$v_n = v_s + v_c = 282 + 1.5\sqrt{4000} = 377 \text{ psi}$$

Thus, ϕv_n = 0.75(377) = 283 psi, which is greater than 278 psi. The maximum shear stress at the critical section at $d/2$ from the outermost peripheral line of studs is

$$v_u = 80 \text{ psi} < 2\phi\sqrt{f_c'} = 95 \text{ psi}$$

Thus, the stud shear reinforcement and its extension (Fig. D.1) are adequate.

D.3—Example 2: Edge flat plate-column connection

The edge flat plate-column connection in Fig. 3.2(b) is designed for DR_u = 0.02 in the x-direction (same drift ratio as in Example 1). The cross-sectional dimensions of the column are $c_x = c_y$ = 16 in. (Fig. D.2). Ratios of top and bottom flexural reinforcements, running perpendicular to the slab edge, are 0.6 and 0.4%, respectively.

Consider Cases I and II, which will produce extreme stresses at Side BC and at the outer extremities (Points A and D) of the shear-critical section (Fig. 3.2(b)). For this example, the outer column face is at the slab edge. The unbalanced moments M_y and M_{Oy} about axes through the centroid of the shear-critical section at $d/2$ from the column face and at column's centroid are related by

$$M_y = M_{Oy} + Vx_O \quad \text{(D-2)}$$

where x_O = –5.22 in. = the x-coordinate of the column's centroid, O (Fig. 3.2(b)).

The values of V_D, V_L, $(M_y)_D$, and $(M_y)_L$ without load factors are: V_D = 22.6 kips; V_L = 9.4 kips; $(M_y)_D$ = 296 in.-kip; and $(M_y)_L$ = 123 in.-kip These moments cause tension on top of the slab.

The unbalanced moment caused by seismic forces is obtained from an elastic analysis of the frame shown in Fig. 6.1(b) subjected to lateral displacement in each of two directions. The prescribed elastic story drift δ_e is ±0.75 in. as for the interior column in Example 1. For the frame shown in Fig. 6.1(b), I_s = 5120 in.^{4} and I_{ec} = 1320 in.^{4} The unbalanced moment and shearing force obtained from the frame analysis, considering their statical equivalents at the centroid of the shear-critical section (Eq. (D-2)), are

$$(M_y)_E = \pm 1212 \text{ in.-kip}; (V)_E = \pm 10.6 \text{ kips}$$

Using ACI 318 load factors, the factored unbalanced moment for Case I (1.2D +1.0L + 1.0E) is

$$(M_{uy})_{\text{Case I}} = 1.2(296) + 1.0(123) + 1.0(1212) = 1691 \text{ in.-kip}$$

The factored unbalanced moment for Case II (0.9D + 1.0E) is

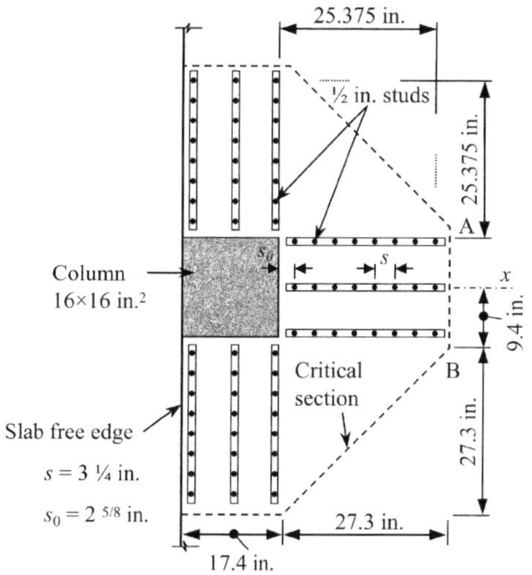

Fig. D.2—Arrangement of SSR in the edge flat plate-column connection of Example 2.

$$(M_{uy})_{\text{Case II}} = 0.90(296) + 1.0(-1212) = -946 \text{ in.-kip}$$

The probable flexural strengths provided by the top and bottom flexural reinforcement of critical section Side BC (Fig. 3.2(b)) are 417 and 285 in.-kip, respectively. For this edge connection, γ_{vy} = 0.35 (Eq. (C-14)); β_r = 0.85; and α_m = 0.24 (Eq. (6-4)) for negative moment of resistance, and α_m = 0.21 (Eq. (6-4)) for positive moment of resistance. Therefore, the upper limits of $|M_{uy}|$ are given by (Eq. (6-2))

Upper limit $|M_{uy}|$ = 417/0.24 = 1740 in.-kip > 1691 in.-kip

Upper limit $|M_{uy}|$ = 285/0.21 = 1360 in.-kip > 946 in.-kip

Thus, the connection should be designed to resist:

(a) Case I: V_u = 1.2(22.6) + 1.0(9.4) + 1.0(10.6) = 47.1 kips, combined with M_{uy} = 1691 in.-kip; and

(b) Case II: V_u = 0.9(22.6) + 1.0(–10.6) = 9.7 kips, combined with M_{uy} = –946 in.-kip.

The perimeter of the critical section b_o = 61.25 in. The parameter J_y of the critical section about the centroidal principal y-axis is J_y = 16,770 in.^{4} The fraction of the unbalanced moment transferred by vertical shear stresses is (ACI 421.1R)

$$\gamma_{vy} = 1 - \frac{1}{1 + \frac{2}{3}\sqrt{\frac{\ell_x}{\ell_y} - 0.2}} \quad \text{(D-3)}$$

The projections of the critical section on the principal axes are ℓ_x = 19.3 in. and ℓ_y = 22.6 in., and Eq. (D-3) gives γ_{vy} = 0.35. The shear stress at Point A or B (Fig. 3.2(b)) is determined by

$$(v_u)_{A \text{ or } B} = \frac{V_u}{b_o d} + \gamma_{vy}\frac{M_{uy}}{J_y}(x_A \text{ or } x_B) \quad \text{(D-4)}$$

Substitution of the values V_u combined with M_{uy} in Cases I and II, with $d = 6.625$ in., and $x_A = -13.22$ in. or $x_B = 6.09$ in., gives the maximum shear stresses

Case I: $(v_u)_B = 0.331$ ksi = 331 psi; $(v_u)_A = 0.351$ ksi = 351 psi

Case II: $(v_u)_A = 0.285$ ksi = 285 psi

The maximum factored shear stress at the critical section at $d/2$ from the column face is $v_u = 351$ psi at A or D (Fig. 3.2(b), Case I); $(v_u/\phi) = 7.4\sqrt{f_c'}$ psi ($0.62\sqrt{f_c'}$ MPa). The ratio v_u/ϕ is higher than $6\sqrt{f_c'}$ psi ($0.5\sqrt{f_c'}$ MPa); thus, stirrups cannot be used, but SSR can be used. The SSR shown in Fig. D.2 is designed in accordance with ACI 421.1R. For the chosen shear reinforcement, $A_v = 1.77$ in.^{2} (nine studs); $f_{yt} = 50$ ksi; and $s = 3.25$ in. The nominal shear strength (expressed in stress units) provided by shear reinforcement (Eq. (4-1)) is v_s = (177 × 50,000)/(61.25 × 3.25) = 444 psi. This value of v_s is greater than the minimum of $3\sqrt{f_c'} = 190$ psi. The distance between the column face and the outermost peripheral line of SSR = 25.375 in., which is greater than $3.5d = 23.2$ in. Thus, details of stud arrangement shown in Fig. D.2 satisfy the requirement of minimum amount of shear reinforcement.

The nominal shear strength v_n multiplied by ϕ must exceed $v_u = 351$ psi. In this example

$$v_n = v_s + v_c = 444 + 1.5\sqrt{4000} = 539 \text{ psi}$$

Thus, $\phi v_n = 0.75(539) = 404$ psi, which is greater than 351 psi. The maximum shear stress at the critical section at $d/2$ from the outermost peripheral line of studs (Fig. D.2) occurs at its Side AB in Case I

$$v_u = 70 \text{ psi} < 2\phi\sqrt{f_c'} = 95 \text{ psi}$$

Thus, the stud shear reinforcement and its extension (Fig. D.2) are adequate.

D.4—Example 3: Corner flat plate-column connection

The corner column-flat plate connection in Fig. 3.2(c) is designed assuming $DR_u = 0.02$ in the $\bar{x}$-direction (same DR_u value as in Examples 1 and 2). The outer faces of the column are at the slab edges. The cross-sectional dimensions of the column are $c_{\bar{x}} = c_{\bar{y}} = 20$ in. (Fig. D.3). The ratios of top and bottom flexural reinforcements, in the directions of the edges of the slab, are 0.6 and 0.4%, respectively. It is assumed that the geometry and the loading are the same in the $\bar{x}$- and $\bar{y}$-directions; the design is for DR_u in the $\bar{x}$-direction only. Consider Cases I and II producing extreme shear stresses at Points B and A, respectively (Fig. 3.2(c)).

The perimeter of the critical section is $b_o = 46.625$ in. The principal axes x and y, at the critical section's centroid, are at an angle $\theta = 45$ degrees (Fig. 3.2(c)). The coordinates of the column's centroid are $x_O = -10.59$ in. and $y_O = 0$. The parameters J_x and J_y of the critical section at $d/2$ from column faces, determined by Eq. (C-4) and (C-5), are J_x =

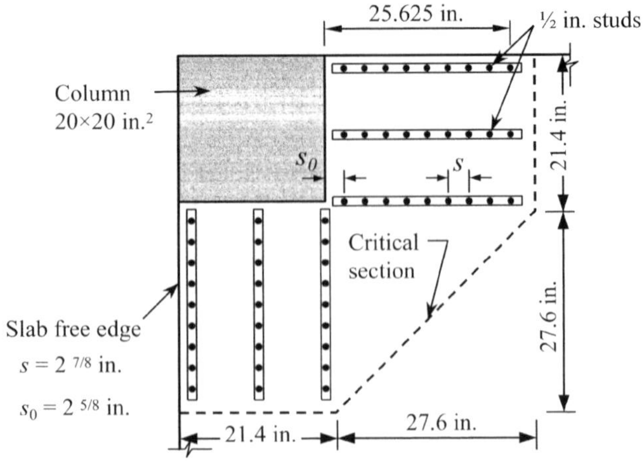

Fig. D.3—Arrangement of SSR in the corner flat plate-column connection of Example 3.

27,960 in.^{4} and $J_y = 6990$ in.^{4} The moments $M_{O\bar{y}}$ and $M_{\bar{y}}$ about axes through O and through the centroid of the critical section are related by

$$M_{\bar{y}} = M_{O\bar{y}} + V_u\bar{x}_O$$

where $\bar{x}_O = -7.48$ in. = $\bar{x}$ coordinate of O, the centroid of the column.

The values of V_D, V_L, $(M_{\bar{y}})_D$, and $(M_{\bar{y}})_L$ are $V_D = 12.7$ kips; $V_L = 5.3$ kips; $(M_{\bar{y}})_D = (M_{\bar{x}})_D = 193$ in.-kip; and $(M_{\bar{y}})_L = (M_{\bar{x}})_L = 80$ in.-kip

The unbalanced moment caused by seismic forces is obtained from an elastic analysis of the frame shown in Fig. 6.1(b) subjected to lateral displacement δ_e. The prescribed elastic story drift δ_e is ±0.75 in. as for the interior column in Example 1. For the frame shown in Fig. 6.1(b), I_s = 2770 in.^{4} and $I_c = 1150$ in.^{4} The moment of inertia of the slab is one-half that of a gross slab strip with a width = 10 ft + $c/2$ = 130 in. The moment of inertia of the column is calculated according to the equivalent frame method of ACI 318 by considering flexibility of an equivalent column combined with a torsional strip.

Elastic analysis for the frame in Fig. 6.1(b) for $\delta_e = \pm 0.75$ in. gives

$$(M_{\bar{y}})_E = \pm 783 \text{ in.-kip}; (M_{\bar{x}})_E = \mp 52 \text{ in.-kip}; (V_u)_E = \pm 7.0 \text{ kips}$$

Using the load factors of ACI 318, consider:

Case I: $1.2D + 1.0L + 1.0E$;

$$\begin{cases} (M_{u\bar{y}})_{\text{Case I}} = 1.2(193) + 1.0(80) + 1.0(783) = 1095 \text{ in.-kip} \\ (M_{u\bar{x}})_{\text{Case I}} = 1.2(193) + 1.0(80) + 1.0(-52) = 260 \text{ in.-kip} \end{cases}$$

Case II: $0.9D + 1.0E$;

$$\begin{cases} (M_{u\bar{y}})_{\text{Case II}} = 0.9(193) + 1.0(-783) = -609 \text{ in.-kip} \\ (M_{u\bar{x}})_{\text{Case II}} = 0.9(193) + 1.0(52) = 226 \text{ in.-kip} \end{cases}$$

In the directions of the principal axes (Fig. 3.2(c)), the statical equivalents to $M_{\bar{x}}$ combined with $M_{\bar{y}}$ are given by

$$M_x = M_{\bar{x}}\cos\theta - M_{\bar{y}}\sin\theta;\ M_y = M_{\bar{x}}\sin\theta + M_{\bar{y}}\cos\theta$$

With θ = 45 degrees:

Case I: M_{ux} = –590 in.-kip; M_{uy} = 958 in.-kip

Case II: M_{ux} = 590 in.-kip; M_{uy} = –271 in.-kip

The probable flexural strengths provided by the top and bottom flexural reinforcements crossing the projection of the shear-critical section on the principal y-axis (Fig. 3.2(c)) are 608 and 415 in.-kip, respectively. For this corner connection, $\gamma_{vy} = 0.27$; $\beta_r = 0.5$; and α_m = 0.31 and 0.30 for negative and positive moment resistances, respectively (Eq. (6-4)). It follows that the upper limits of $|M_{uy}|$ are (Eq. (6-2))

$$\text{Upper limit}\quad |M_{uy}|_{\text{Case I}} = \frac{608}{0.31} = 1960 \text{ in.-kip} > 958 \text{ in.-kip}$$

$$\text{Upper limit}\quad |M_{uy}|_{\text{Case II}} = \frac{415}{0.30} = 1380 \text{ in.-kip} > 271 \text{ in.-kip}$$

Similar calculation for upper limits for M_{ux} indicates that they do not govern the design. Therefore, the connection should be designed to resist:

(a) Case I: V_u = 1.2(12.7) + 1.0(5.3) + 1.0(7.0) = 27.5 kips, combined with M_{uy} = 958 in.-kip and M_{ux} = –590 in.-kip; and

(b) Case II: V_u = 0.9(12.7) + 1.0(–7.0) = 4.4 kips, combined with M_{uy} = –271 in.-kip and M_{ux} = 590 in.-kip

The shear stress at any point (x,y) on the shear-critical section is

$$v_u = \frac{V_u}{b_o d} + \gamma_{vx}\frac{M_{ux}}{J_x}y + \gamma_{vy}\frac{M_{uy}}{J_y}x \qquad \text{(D-5)}$$

The projections of the critical section on the x- and y-axes are ℓ_x = 16.48 in. and ℓ_y = 32.97 in. The fractions of unbalanced moments transferred by shear are (Eq. (C-15) and (C-16))

$$\gamma_{vy} = 1 - \frac{1}{1 + \frac{2}{3}\sqrt{\frac{\ell_x}{\ell_y} - 0.2}} = 0.267;\ \gamma_{vx} = 0.4$$

The maximum shear stress at Point A (–8.24 in., 16.48 in. [Fig. 3.2(c)]) occurs in Case I. Its value is

$$(v_u)_A = \frac{27.5}{46.625(6.625)} + \frac{0.4(-590)}{27{,}960}(16.48) + \frac{0.267(958)}{6990}(-8.24)$$

$$= -0.352 \text{ ksi}$$

The maximum shear stress at Point B (8.24 in., 0 in.) occurs also in Case I. Its value is

$$(v_u)_B = \frac{27.5}{46.625(6.625)} + \frac{0.4(-590)}{27{,}960}(0) + \frac{0.267(958)}{6990}(8.24) = 0.391 \text{ ksi}$$

The maximum factored shear stress at the critical section at $d/2$ from the column face is v_u = 391 psi (at Point B, Fig. 3.2(c), Case I); $(v_u/\phi) = 8.2\sqrt{f_c'}$ psi ($0.69\sqrt{f_c'}$ MPa). The SSR, shown in Fig. D.3, is designed in accordance with ACI 421.1R. For the chosen shear reinforcement, A_v = 1.18 in.^{2} (six studs); f_{yt} = 50 ksi; and s = 2.875 in. The nominal shear strength (expressed in stress units) provided by shear reinforcement (Eq. (4-1)) is v_s = (1.18 × 50,000)/(46.625 × 2.875) = 441 psi, which is greater than the minimum of $3\sqrt{f_c'}$ = 190 psi. The distance between the column face and the outermost peripheral line of SSR = 25.625 in., which is greater than $3.5d$ = 23.2 in. Thus, details of stud arrangement, shown in Fig. D.3, satisfy the requirement for the minimum amount of shear reinforcement.

The nominal shear strength v_n multiplied by ϕ must exceed v_u = 391 psi. In this example

$$v_n = v_s + v_c = 441 + 1.5\sqrt{4000} = 536 \text{ psi}$$

Thus, ϕv_n = 0.75(536) = 402 psi, which is greater than v_u = 391 psi.

In each of Examples 1, 2, and 3, the shear reinforcement has capacity equal to or exceeding the minimum requirement, the excess is not large enough to allow the use of fewer studs. Using the properties of the shear-critical section at $d/2$ from the outermost studs (Fig. D.1 to D.3), it can be verified by Eq. (D-5) that the maximum shear stress does not exceed $2\phi\sqrt{f_c'}$ = 95 psi.

D.5—Example 4: Use of stirrups—Interior flat plate-column connection

Repeat the design of the interior flat plate-column connection of Example 1, with DR_u = 0.005. Assume that $c_x = c_y$ = 12 in. and V_u = 78 kips. The remaining data are as in Example 1. An elastic analysis of the frame in Fig. 6.1(a) for $\delta_e = 0.005[\ell_c/(C_d/I_E)]$ = 0.19 in. gives $(M_y)_E$ = 243 in.-kip, using I_s = 5120 in.^{4} and I_{ec} = 650 in.^{4} (equivalent second moment of area calculated by combining the column with a torsional strip, in accordance with ACI 318-08, Section 13.7.5).

Assuming that the diameter of the stirrups is 3/8 in. and maintaining the 3/4 in. cover above the stirrups, the effective depth of the flexural reinforcement (having a diameter of 5/8 in.) is: d = 8 – 3/4 – 5/8 – 3/8 = 6.25 in. The properties of the shear-critical section at $d/2$ from column face are: b_o = 73 in.; J_y = 25,330 in.^{4}; and γ_{vy} = 0.4.

$$\phi V_c = \phi(4\lambda\sqrt{f_c'}\, b_o d)$$

$$\phi V_c = 0.75[4(1.0)\sqrt{4000}(73)(6.25)]/1000 = 87 \text{ kips}$$

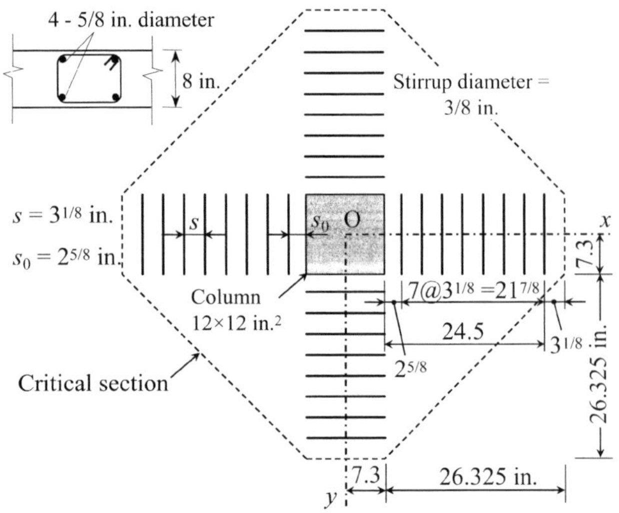

Fig. D.4—Arrangement of stirrups in interior flat plate-column connection of Example 4.

Because V_u is greater than $0.4\phi V_c = 87 \times 0.4 = 35$ kips, shear reinforcement is required. The maximum shear stress is

$$v_u = \frac{V_u}{b_o d} + \frac{\gamma_{vy} M_{uy} x}{J_y}$$

$$v_u = \frac{78}{73 \times 6.25} + \frac{0.4 \times 243 \times (18.25/2)}{25{,}330}$$

$$= 0.206 \text{ ksi} = 206 \text{ psi}$$

$$4\sqrt{f'_c} < (v_u/\phi) = 4.3\sqrt{f'_c} \text{ psi} < 6\sqrt{f'_c}$$

Shear reinforcement is required and stirrups or SSR can be used; choose stirrups.

With the choice and the arrangement of stirrups shown in Fig. D.4, $A_v = 0.88$ in.^{2}. The nominal shear strength provided by the stirrups is

$$v_s = \frac{A_v f_{yt}}{b_o s}$$

$$v_s = \frac{(0.88)(60{,}000)}{73(3.125)} = 231 \text{ psi}$$

$$v_n = v_s + 1.5\lambda\sqrt{f'_c} = 326 \text{ psi}$$

The distance between the column face and the outermost peripheral line of stirrups is 24.5 in. > $3.5d$ = 3.5(6.25) = 21.9 in.

D.6—Example 5: Interior flat plate-column connection of Example 1, repeated using SI units

The value of $(V_u)_{D+L+E}$, given as follows, corresponds to the load combination that governs the design. The load factors used to calculate $(V_u)_{D+L+E}$ are in accordance with ACI 318. Also given are the values of $(M_y)_D$ and $(M_y)_L$ due to nonfactored dead and live loads. The given values are $(V_u)_{D+L+E}$ = 347 kN, $(M_y)_D$ = 0.0, and $(M_y)_L$ = 0.0. The cross-sectional dimensions of the column are (Fig. D.1) $c_x = c_y = 406$ mm; the reinforcement ratios of the top and bottom flexural bars are 1.0 and 0.5%, respectively.

The critical section shown in Fig. 3.2(a) has $b_o = 2.299$ m and V_c (Eq. (3-3)) = $0.33(2.299)(0.168)\sqrt{27.6}\,(10^3) = 675$ kN; and $V_u/\phi V_c = 347/(0.75 \times 675) = 0.69$.

The unbalanced moment caused by seismic forces is obtained from an elastic analysis of the frame, shown in Fig. 6.1(a). The prescribed elastic story drift δ_e is given by Eq. (6-1)

$$\delta_e = \frac{DR_u \ell_c}{(C_d/I_E)} = \frac{0.02 \times 3.66}{3.85} = 19 \text{ mm}$$

where (C_d/I_E) = 3.85 according to the ASCE/SEI 7 (assuming that the lateral-force-resisting system consists of shear walls). For the frame shown in Fig. 6.1(a), the moment of inertia of the slab $I_s = 2.13 \times 10^{-3}$ m^4. This value is equal to one-half the gross moment of inertia of a 203 mm slab that is 6.1 m wide. The reduction is made to reflect some of the effects of cracking on the force distribution. The equivalent moment of inertia of the column is $I_{ec} = 0.549 \times 10^{-3}$ m^4. This value is obtained according to the equivalent frame method of ACI 318 by assuming that the flexibility of the equivalent column equals the sum of the flexibility of the actual column and the torsional strip of the slab. In the determination of the equivalent moment of inertia, the column lengths were taken as 1.83 m (Fig. 6.1(a)) with 102 mm long rigid parts at the joints and with the other ends simply supported, as shown in Fig. 6.1(c). The equivalent moment of inertia is computed as $I_{ec} = I_c(K_{ec}/\Sigma K_c)$, where K_{ec} and K_c are end rotational stiffnesses (moment per unit rotation) of equivalent column and the actual column, respectively.

The unbalanced moment due to δ_e, obtained from elastic analysis of the frame in Fig. 6.1(a), is $(M_y)_E$ = 189 kN·m. According to ACI 318-08, Section 9.2, the factored unbalanced moment is

$$M_{uy} = 1.2(M_y)_D + 1.0(M_y)_L + 1.0(M_y)_E \qquad \text{(D-1)}$$

$$M_{uy} = 1.2(0) + 1.0(0) + 1.0(189) = 189 \text{ kN·m}$$

The flexural strengths provided by the top and bottom flexural reinforcement, based on a stress of $1.25f_y$, of critical section Sides BC and AD (Fig. 3.2(a)) are 74.6 and 39.5 kN·m, respectively. Thus, M_{pr} = 74.6 + 39.5 = 114.1 kN·m. For this interior connection, $\gamma_{vy} = 0.4$; $\beta_r = 1.0$; and $\alpha_m = 0.4$ (Eq. (6-3)). It follows that the upper limit of M_{uy} is (Eq. (6-2))

$$(M_{uy})_{\text{upper limit}} = \frac{74.6 + 39.5}{0.4} = 285.3 \text{ kN·m}$$

Because the upper limit is higher than M_{uy} = 189 kN·m, the flat plate-column connection should be designed to resist V_u = 347 kN and M_{uy} = 189 kN·m. In this structure, the drift due to seismic forces does not change V_u for this interior column.

The maximum factored shear stress due to V_u and M_{uy} (computed according to the eccentric shear stress model of ACI 318) at the critical section at $d/2$ from the column face (Fig. 3.2(a)) is v_u = 1.92 MPa and $(v_u/\phi) = 0.49\sqrt{f_c'}$ MPa. This value is less than $0.5\sqrt{f_c'}$ MPa; therefore, stirrups or SSR can be used. The SSR shown in Fig. D.1 is designed according to ACI 421.1R. For the chosen shear reinforcement, A_v = 1520 mm^2 (12 studs of diameter 12.7 mm); and s = 117 mm, the nominal shear strength (expressed in stress units) provided by shear reinforcement (Eq. (4-1)) is v_s = [1520(345)]/[2299(117)] = 1.94 MPa, which is greater than the minimum $0.25\sqrt{f_c'}$ = 1.31 MPa. The distance between the column face and the outermost peripheral line of SSR is 654 mm, which is greater than $3.5d$ = 589 mm. Therefore, the details of the stud arrangement shown in Fig. D.1 satisfy the requirement for the minimum amount of shear reinforcement. ACI 421.1R recommends that $s_o \leq 0.4d$; the value used in this example and Examples 2, 3, and 4 is s_o = 67 mm. The nominal shear strength v_n multiplied by ϕ should exceed v_u = 1.92 MPa. In this example

$$v_n = v_s + v_c = 1.94 + 0.125\sqrt{27.6} = 2.60 \text{ MPa}$$

Thus, ϕv_n = 0.75(2.60) = 1.95 MPa, which is greater than 1.92 MPa. The maximum shear stress at the critical section at $d/2$ from the outermost peripheral line of studs is

$$v_u = 0.55 \text{ MPa} < 0.167\phi\sqrt{f_c'} = 0.66 \text{ MPa}$$

Thus, the stud shear reinforcement and its extension (Fig. D.1) are adequate.

D.7—Post-tensioned flat plate structure

The design of earthquake-resistant post-tensioned flat plate-column connections for punching shear, according to the recommendations of this guide, is illustrated by the following numerical examples. Consider a post-tensioned floor following the generic geometric arrangement in Fig. 8.1, having: $\ell_x \times \ell_y$ = 26.25 × 26.25 ft^2 (8.000 × 8.000 m^2) slab thickness; h = 7.87 in. (200 mm) interior columns' size, c_x = 19.7 in. (500 mm), c_y = 11.8 in. (300 mm); square edge columns with side dimension = 11.8 in. (300 mm); and story height ℓ_c = 11.48 ft (3.500 m). The earthquake-excited motion is in the x-direction, where the structure has five bays, with many bays in the y-direction. The structure has a lateral-force-resisting system that limits the interstory drift ratio, including inelastic deformations, to DR_u = 0.015; assume that the ratio is $C_d/I_E = DR_u\ell_c/\delta_e$ = 4.0. The entire structure is assumed to be of normal-density concrete, w_c = 153 lb/ft^3 (24 kN/m^3), with a specified compressive strength of f_c' = 4350 psi (30 MPa) (E_c = 3580 ksi [24.7 GPa]); and concrete cover = 0.790 in. (20.0 mm). Non-prestressed No. 5 reinforcement is of nominal diameter = 0.625 in. (16 mm) and yield strength f_y = 60 ksi (414 MPa); the reinforcement ratios of the top and bottom non-prestressed flexural bars are 0.7 and 0.5%, respectively. Post-tensioned reinforcement is 0.6 in., seven-wire strands having the following properties: area of tendon A_p = 0.217 in.^{2} (140 mm^2); ultimate strength f_{pu} = 270 ksi (1861 MPa); yield strength f_{py} = 243 ksi (1675 MPa); effective post-tensioning force per tendon = 35.1 kips (156 kN); post-tensioning is required to balance no less than 85% of the total dead load (that is, $\kappa \geq 0.85$, Section 8.4). The floor is designed for a superimposed dead load, w_{sd} = 27 lb/ft^2 (1.3 kPa) and a service live load, w_L = 50 lb/ft^2 (2.4 kPa). If shear reinforcement is required, use headed SSR satisfying ASTM A1044/A1044M-05 with nominal diameter = 1/2 in. and f_{yt} = 50 ksi (345 MPa). The average of distances from extreme compression fiber to the centroid of the tension non-prestressed reinforcement positioned in two orthogonal directions is d = 7.87 – 0.79 – 0.625 = 6.46 in. (164 mm). It is required to design for punching shear at the first interior column (B-2) and at the edge column (B-1) (Fig. 8.1; Sections D.8 and D.9, respectively). In calculating d, consider only the non-prestressed reinforcement. At the shear-critical sections, the tendons can be far away from the slab surfaces.

Elastic analysis of the equivalent frame in Fig. 8.4 subjected to service dead and live loads gives shearing forces and unbalanced moments

At the first interior column (B-2)

V_D = 97 kips; V_L = 38 kips; $(M_y)_D$ = –265 in.-kip; $(M_y)_L$ = –104 in.-kip.

At the edge column (B-1)

V_D = 36 kips; V_L = 14 kips; $(M_{Oy})_D$ = 523 in.-kip; $(M_{Oy})_L$ = 206 in.-kip.

where M_{Oy} is unbalanced moment about an axis through O, the column's centroid, parallel to the y-axis in Fig. 3.2(b); positive M_{Oy} is in the same direction as positive M_{uy}. From an elastic analysis of the equivalent frames in Fig. 6.1(a) and (b), subjected to an elastic imposed lateral displacement $\delta_e = DR_u\ell_c/(C_d/I_E)$ = 0.015(11.48 × 12)/4 = 0.512 in. (13.0 mm), the shearing force and unbalanced moment are

At the first interior column (B-2)

V_E = ±0.382 kips and $(M_y)_E$ = ±1117 in.-kip

At the edge column (B-1)

V_E = ±3.272 kips and $(M_{Oy})_E$ = ±492 in.-kip

For the banded tendons, substitution of κ = 0.85; α = 0.1; $w_D = w_c h + w_{sd}$ = 127 lb/ft^2 (6.1 kPa); ℓ = 26.25 ft (8.00 m); h = 7.87 in. (200 mm); and h_c = 4.33 in. (110 mm) in Eq. (8-5) gives $(f_{pc})_{required} = (P_e)_{required}/h$ = 219 psi (1.51 MPa). Use sixteen 0.6 in. seven-wire tendons per panel to provide effective post-tensioning force = 16(35.1 × 10³)/(26.25 × 12) = 1782 lb/in. The average compressive stress provided = 1782/7.87 = 227 psi (1.56 MPa). Similar calculations with h_c = 3.62 in. (92.0 mm) will show that 19 distributed tendons, providing a confining stress of 269 psi (1.85 MPa), are sufficient to balance approximately the same load. In the banded direction, $h_c = h$ minus the sum of the cover at top and bottom, one bar diameter of the outermost top and bottom flexural reinforcing bars and one diameter of the post-tensioning sheathing (assumed 0.71 in. [18 mm]), h_c = 7.87 – 2 × 0.79 – 2 × 0.625 – 0.71 = 4.33 in. (110 mm). The distance from the nearest slab surface to the tendons' centroid at Points 3, 5, and 7 (Fig. 8.3(a)) = 1.77 in. (45 mm). In the direction of the distributed tendons, h_c is smaller by one diameter of the post-tensioning sheathing (h_c = 4.33 – 0.71 = 3.62 in. [92.0 mm]). The average in-plane compression produced by the effective post-tension forces in the two orthogonal directions, f_{pc} = 247 psi (1.71 MPa). The selected post-tensioning, (16 × 35.1 × 10³)/(26.25 × 12) = 1782 lb/in. (312 kN/m), produces in the end span a smaller balancing load given by (Fig. 8.3(b))

$$w_{23} = \frac{2P_e c_2}{(\alpha_{23}\ell)^2} = \frac{2(1782)(1.43)}{(0.291 \times 26.52)^2} = 85.6 \text{ lb/ft}^2 \text{ (4.1 kPa)}$$

The post-tensioning tendons are banded in the x-direction and uniformly distributed in the y-direction (Fig. 8.1). Table D.1 gives values of the parameters defining the banded tendons' profile, and the forces exerted by the post-tensioning tendons on the floor (Fig. 8.3(a) and (b)). Application of the forces in Table D.1 and Fig. 8.3(b) to the equivalent frame in Fig. 8.4(a) results in a set of reactions, at the supports, that is in self-equilibrium. The internal forces of the equivalent frame (Fig. 8.4 subjected to the forces in Table D.1 and Fig. 8.3(b)) minus the primary internal forces give the secondary internal forces, V_p'' and M_p''.

At the first interior column (B-2)

V_p'' = –0.171 kips and M_p'' = ±4.831 in.-kip

At the edge column (B-1)

V_p'' = 0.140 kips and M_{Op}'' = –217 in.-kip

In this example, it is seen that V_E, V_p'', and M_p'' are of insignificant magnitudes.

D.8—Example 6: Post-tensioned flat plate connection with interior column

The design steps, outlined in Section 7.2 (Chapter 7), are followed here for the interior column-flat plate connection B-2 (Fig. 8.1). The load factors used to calculate V_u and M_{uy} are in accordance with Sections 9.2.1 and 18.10.3 of ACI 318-08. Consider two cases varying the sign of V_E and M_E

Table D.1—Parameters defining profile of the banded tendons and the corresponding set of self-equilibrating forces exerted on a unit width of floor (Fig. 8.3)

Units	c_1	c_2	c_4	$c_5 = c_9$	$c_6 = c_8$
in. (mm)	2.17 (55.0)	1.43 (36.3)	3.56 (90.3)	4.33 (110)	3.47 (88.0)
Units	α_{12}	α_{23}	α_{34}	$\alpha_{45} = \alpha_{56}$	$\alpha_{67} = \alpha_{78}$
–	0.150	0.291	0.459	0.100	0.400
Units	w_{12}	w_{24}	w_{45}	$w_{56} = w_{89}$	w_{68}
lb/ft² (kPa)	–169.3 (–8.107)	87.27 (4.179)	–400.6 (–19.185)	–447.9 (–21.450)	112.0 (5.363)

Case I: (1.2D + 0.5L + 1.0E + 1.0 Secondary effect of post-tensioning)

$$V_u = 1.2V_D + 0.5V_L + V_E + V_p'' = 1.2(97) + 0.5(38) + (-0.382) + (-0.171)$$
$$= 134 \text{ kips (598 kN)}$$

$$M_{uy} = 1.2(M_y)_D + 0.5(M_y)_L + (M_y)_E + M_p'' = 1.2(-265) + 0.5(-104) + (-1117) + 4.83$$
$$= -1482 \text{ in.-kip } (-167 \text{ kN·m})$$

Case II: (0.9D + 1.0E + 1.0 Secondary effect of post-tensioning)

$$V_u = 0.9V_D + 0.5V_E + V_p'' = 0.9(97) + 0.382 - 0.171$$
$$= 87 \text{ kips (388 kN)}$$

$$M_{uy} = 0.9(M_y)_D + (M_y)_E + M_p'' = 0.9(-265) + 1117 + 4.83$$
$$= 833 \text{ in.-kip (100 kN·m)}$$

The shear-critical section at d/2 from the column face has the following properties: b_o = 88.8 in. (2260 mm); ℓ_x = 26.1 in. (664 mm); ℓ_y = 18.3 in. (464 mm); γ_{vy} = 0.444 (Eq. (C-12)); J_y = 59.5 × 10³ in.⁴ (24.8 × 10⁹ mm⁴) (Eq. (C-5)). The three conditions warranting the use of Eq. (8-6) are satisfied. Two tendons in each of the x-direction and the y-direction intercept the shear-critical section at d/2 from the column (Fig. D.5(a)); the sum of the vertical components of these tendons at the location of the shear-critical section V_p = 4.7 kips (21 kN). Ignoring V_p (refer to Section 8.5), Eq. (8-6) gives: v_c = 306 psi (2.11 MPa) and V_c = 175 kips (780 kN). The values of $V_u/(\phi V_c)$ for Cases I and II are: 1.02 and 0.66, respectively. Case II with $(M_y)_E$ = –1117 in.-kip would result in M_{uy} = –1351 in.-kip, which is smaller than the value of Case I.

The upper limit for M_u (Section 8.5) is greater than M_{uy} in Case I and II. Thus, the punching shear design will proceed with the values of V_u and M_{uy} determined above.

The maximum factored shear stress v_u at d/2 from the column face, corresponding to Case I and II, is (Eq. (C-6))

Case I: $v_u = 379 \text{ psi} = 7.7\phi\sqrt{f_c'}$ $(2.61 \text{ MPa} = 0.64\phi\sqrt{f_c'})$

Case II: $v_u = 238 \text{ psi} = 4.8\phi\sqrt{f_c'}$ $(1.64 \text{ MPa} = 0.40\phi\sqrt{f_c'})$

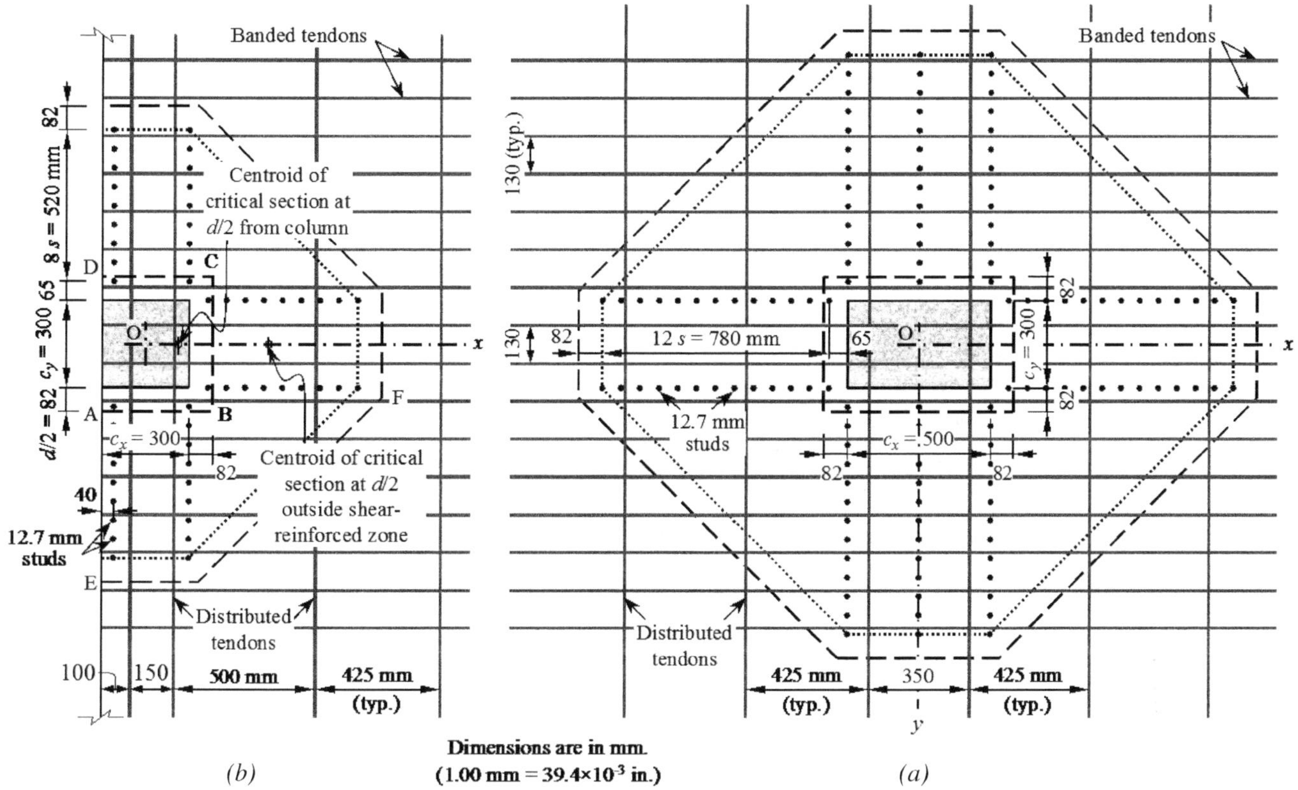

Fig. D.5—Arrangement of tendons and headed SSR at: (a) interior column-flat plate connection B-2 (Fig. 8.1; Example 6); and (b) edge column-flat plate connection B-1 (Fig. 8.1; Example 7).

(v_u/ϕ) in Case I is greater than v_c = 304 psi (2.10 MPa); thus, shear reinforcement is required; the same value is greater than $6\sqrt{f'_c}$ psi ($0.5\sqrt{f'_c}$ MPa), thus, stirrups cannot be used.

The shear reinforcement in Fig. D.5(a) provides nominal shear strength (Eq. (7-1), Eq. (4-1), and Section 7.1)

$$\phi v_n = 0.75\left[1.5\sqrt{4350} + \frac{10(0.196)(50{,}000)}{88.8(2.56)}\right] = 399 \text{ psi (2.75 MPa)}$$

This value exceeds v_u = 379 psi (2.61 MPa), indicating that the choice of studs and their spacing is adequate. Outside the shear-reinforced zone, the shear-critical section has the properties (Fig. D.5(a)): b_o = 272.6 in. (6.923 m); $J_y = 1.608 \times 10^6$ in.^{4} (669.3×10^{-3} m^4); γ_{vy} = 0.411; x_{max} = 46.34 in. (1177 mm); the maximum shear stresses are

Case I: $v_u = 94 \text{ psi} = 1.90\phi\sqrt{f'_c}$ $(0.648 \text{ MPa} = 0.158\phi\sqrt{f'_c})$

Case II: $v_u = 60 \text{ psi} = 1.21\phi\sqrt{f'_c}$ $(0.414 \text{ MPa} = 0.101\phi\sqrt{f'_c})$

The values of (v_u/ϕ) are less than than $2\sqrt{f'_c}$ psi ($0.17\sqrt{f'_c}$ MPa), indicating that the extent of the shear-reinforced zone is adequate. The studs are placed on three lines running perpendicular to the long side of the column; with the spacing between the lines <2d, which is a requirement of ACI 318-08, Section 11.12.

D.9—Example 7: Post-tensioned flat plate connection with edge column

Two load combinations are considered

Case I: (1.2D + 0.5L + 1.0E + 1.0 Secondary effect of post-tensioning)

$$V_u = 1.2V_D + 0.5V_L + V_E + V''_p = 1.2(36) + 0.5(14) + 3.272 + 0.140$$
$$= 54.0 \text{ kips (240 kN)}$$

$$M_{uOy} = 1.2(M_{Oy})_D + 0.5(M_{Oy})_L + (M_{Oy})_E + M''_{Op}$$
$$= 1.2(523) + 0.5(206) + 492 + (-217) = 1006 \text{ in.-kip (113.6 kN·m)}$$

Case II: (0.9D + 1.0E + 1.0 Secondary effect of post-tensioning)

$$V_u = 0.9V_D + V_E + V''_p = 0.9(36) - 3.272 + 0.140 = 29.3 \text{ kips (131 kN)}$$

$$M_{uOy} = 0.9(M_{Oy})_D + (M_{Oy})_E + M''_{Op} = 0.9(523) - 492 - 217$$
$$= -238.3 \text{ in.-kip } (-26.97 \text{ kN·m})$$

The shear-critical section at d/2 from the column face has the following properties: b_o = 48.3 in. (1230 mm); ℓ_x = 15.0 in. (382 mm); ℓ_y = 18.3 in. (464 mm); x_O = –4.45 in. (–113 mm); x_A = –10.4 in. (–263 mm); x_B = 4.69 in. (119 mm); γ_{vy} = 0.345 (Eq. (C-14)); $J_y = 7.81 \times 10^3$ in.^{4} (3.25×10^9 mm^4) (Eq. (C-5)). The first of the three conditions warranting the use of Eq. (8-6) is not satisfied. Thus, the shear strength of the connection will be calculated as for reinforced concrete connections without prestressing; v_c is governed by Eq. (3-3): $v_c = 4\sqrt{4350} = 265$ psi (1.83 MPa). The unbalanced moments at the centroid of the shear-critical section at $d/2$ from the column, corresponding to the above two cases are (Eq. (D-2))

Case I: $M_{uy} = M_{uOy} + V_u x_O = 1006 + 54(-4.45) = 765$ in.-kip (86.4 kN·m)

Case II: $M_{uy} = M_{uOy} + V_u x_O = -238.3 + 29.3(-4.45) = -370$ in.-kip (-41.8 kN·m)

The absolute values of the probable flexural strengths provided by the top and bottom flexural reinforcement of critical section side BC (Fig. 3.2(b)), based on stresses $1.25f_y$ and f_{ps}, respectively, for the non-post-tensioned and post-tensioned reinforcement, are: 681 and 613 in.-kip (77.0 and 69.3 kN·m), respectively. For this edge connection, γ_{vy} = 0.345, β_r = 0.82, and α_m = 0.254 for negative moment resistance and α_m = 0.235 for positive moment resistance (Eq. (6-4)). The upper limits of $|M_{uy}|$ are (Eq. (6-2))

Upper limit $|M_{uy}| = 681/0.254 = 2681$ in.-kip (303.0 kN·m) (> 765 in.-kip [86.4 kN·m])

Upper limit $|M_{uy}| = 613/0.235 = 2609$ in.-kip (294.8 kN·m) (> 370 in.-kip [41.8 kN·m])

The upper limits are greater than M_{uy}, calculated for Case I and II. Thus, the punching shear design will proceed with V_u and M_{uy} determined above for the two cases.

The maximum factored shear stress v_u at $d/2$ from the column face, corresponding to Case I and II, is (Eq. (C-6) and Fig. D.5(b))

Case I: $(v_u)_{BC} = 331$ psi (2.28 MPa)

Case II: $(v_u)_{A \text{ and } D} = 264$ psi (1.82 MPa)

The value of $(v_u/\phi)_{BC}$ in Case I is greater than $v_c = 265$ psi (1.83 MPa); thus, shear reinforcement is required. The same value is greater than $6\sqrt{f'_c}$ psi ($0.5\sqrt{f'_c}$ MPa); thus, stirrups cannot be used.

Using the arrangement of the headed SSR in Fig. D.5(b), it can be verified that the nominal shear strength $v_n = 576$ psi (3.97 MPa); $\phi v_n = 432$ psi (2.98 MPa), exceeds $(v_u)_{BC}$ = 331 psi (2.28 MPa), indicating that the choice of studs and their spacing are adequate.

Outside the shear-reinforced zone, the shear-critical section has the properties (Fig. D.5(b)): b_o = 111.3 in. (2.826 m); $J_y = 96.11 \times 10^3$ in.^{4} (40.00×10^{-3} m^4); γ_{vy} = 0.294; $x_E = -22.72$ in. (-577 mm); $x_F = 15.35$ in. (390 mm); x_O = -16.81 in. (-427 mm). The unbalanced moments M_{uy} (by Eq. (D-2)), combined with the corresponding V_u and the maximum shear stresses, are

Case I: $V_u = 54$ kips; $M_{uy} = 1006 + 54(-16.81)$ = 98.26 in.-kip

$$(v_u)_F = 80 \text{ psi} = 1.61\phi\sqrt{f'_c} \quad (0.55 \text{ MPa} = 0.134\phi\sqrt{f'_c})$$

Case II: $V_u = 29.3$ kips; $M_{uy} = -238.3 + 29.3(-16.81)$ = -730.8 in.-kip

$$(v_u)_E = 92 \text{ psi} = 1.85\phi\sqrt{f'_c} \quad (0.64 \text{ MPa} = 0.153\phi\sqrt{f'_c})$$

The values of (v_u/ϕ) are less than $2\sqrt{f'_c}$ psi ($0.17\sqrt{f'_c}$ MPa), indicating that the extent of the shear-reinforced zone is adequate.

APPENDIX E—CONVERSION FACTORS

1 in.	=	25.4 mm
1 ft	=	0.3048 m
1 kip	=	4.448 kN
1 in.-kip	=	0.113 kN·m
1 psi = 6.89 kPa = 6.89×10^{-3} MPa		
$\sqrt{f'_c}$, psi	=	$\sqrt{f'_c}/12$, MPa

Guide to Design of Reinforced Two-Way Slab Systems

Reported by Joint ACI-ASCE Committee 421

Mustafa A. Mahamid,* Chair

Simon J. Brown
Pinaki R. Chakrabarti
William L. Gamble
Ramez Botros Gayed
Amin Ghali
Hershell Gill
Neil L. Hammill
Mahmoud E. Kamara*

Theodor Krauthammer
James S. Lai*
Faris A. Malhas
Mark D. Marvin
Sami Hanna Megally
Michael C. Mota
Edward G. Nawy*
Daniel Reider

Aly Said
Eugenio M. Santiago
Myoungsu Shin*
Matthew Smith
Ying Tian
Amy M. Reineke Trygestad
Stanley C. Woodson

*Authors and editorial team.

Consulting Members
Eugene Paul Holland
J. Leroy Hulsey
Sidney H. Simmonds

This guide presents analysis methods, design procedures, slab reinforcement and detailing practices, and strength and serviceability considerations, as well as information for the resistance to lateral forces for slab-column frames. It also covers the design for flexure and shear and torsion, as well as the effect of openings. Both two-way nonprestressed slabs and post-tensioned slabs are included.

Keywords: analysis method; deflection; direct design; flat plates; flat slabs; post-tension; reinforcement; shear; shearhead; slab-column frame; two-way slabs.

CONTENTS

ACI 421.3R-15 was adopted and published October 2015.

CHAPTER 1—INTRODUCTION AND SCOPE

1.1—Introduction: history of two-way slab system

Two-way flat slab construction in the United States evolved, and was invented and patented, in the early 1900s (Cohen and Heun 1979). Early two-way flat slab construction was built and subjected to load tests in place and scaled models were later tested in laboratories. While the amount of reinforcement in slab construction varied dramatically, flat slab systems were found to be economical for heavy live load occupancy. As the number of flat slab projects increased steadily worldwide, design rules were established and formalized (Sozen and Seiss 1963).

Prior to the 1950s, two-way waffle slabs and two-way flat slabs were designed and constructed with column capitals and some with drop panels. The hollow tile and concrete slab is a type of waffle slab that dates back to at least 1918 (Gamble et al. 1964). Column capitals were used to increase slab shear strength and drop panels to reduce the flexural reinforcement over columns, which allowed for thinner slabs. In the post-1970s era, field labor to construct formwork for column capitals and drop panels became costly; the introduction of reusable forms led to construction of flat plates, which are two-way flat slabs without column capitals or drop panels.

The lift-slab system for multistory construction was popular in the 1960s and 70s, but is no longer commonly used. The slabs were cast in a stack at ground level, post-tensioned, and then lifted to their final elevations using jacks lifting on steel collars embedded in the slabs.

Draped post-tensioning can be designed to balance part of the gravity loads. Combining unbonded post-tensioned tendons and nonprestressed reinforcement results in reduced slab thickness. In addition, the use of nonprestressed reinforcement supplements prestressed tendons to meet the required nominal strength and control slab cracking.

1.2—Scope

The performance record of various two-way slab systems is well established based on results of extensive tests and practical construction improvements in the twentieth century. The ACI Building Code permits design of slab systems, both nonprestressed and post-tensioned, based directly on fundamental principles of structural mechanics that satisfy equilibrium and compatibility. This guide provides classic solutions based on linearly elastic continuum, as well as prescriptive procedures used in common practice for analysis and design of slab systems. The fundamental principles in this guide are applicable to all planar structural slab systems subjected to gravity loads and, in certain conditions, those combined with lateral forces.

This guide addresses recommended practice in the selection and distribution of flexural reinforcement, and guidelines to transmit loads from slabs to columns by flexure, torsion, and shear. Detailing practices for post-tensioned two-way slabs are found in ACI 423.3R-05. This guide also discusses aspects and parameters where two-way slabs without beams are incorporated in ordinary or intermediate moment frames with ductile detailing and toughness.

While two-way slab systems have more than 100 years of service history, various practical refinements and research programs continue to develop new materials and technologies that support sustainable construction of two-way slabs.

CHAPTER 2—NOTATION AND DEFINITIONS

2.1—Notation

A_{cf} = larger gross cross-sectional area of the slab-column strips in the two orthogonal equivalent frames intersecting at a column in a two-way slab, ft^2 (m^2)

A_{sb} = area of reinforcement through the column core used as integrity reinforcement

b_1 = dimension of the critical section b_o measured in the direction of the span for which moments are determined, in. (mm)

b_2 = dimension of the critical section b_o measured in the direction perpendicular to b_1, in. (mm)

b_e = effective slab width, in. (mm)

b_o = perimeter of critical section at $d/2$ from face of support, in. (mm)

C = cross-sectional constant to define torsional properties of slab and beam

c_1 = dimension of rectangular or equivalent rectangular column, capital, or bracket measured in the direction of the span for which moments are being determined, in. (mm)

c_2 = dimension of rectangular or equivalent rectangular column, capital, or bracket measured in the direction perpendicular to c_1, in. (mm)

c_t = distance from the interior face of the column to slab edge measured parallel to c_1, but not exceeding c_1, in. (mm)

d = distance from extreme compression fiber to the centroid of tension reinforcement, in. (mm)

E_{cb} = modulus of elasticity of beam concrete, psi (MPa)

E_c = modulus of elasticity of slab concrete, psi (MPa)

f_c' = specified compressive strength of concrete, psi (MPa)

f_{pc} = average compressive stress in the two directions at centroid of concrete cross section after allowing for all prestress losses, psi (MPa)

f_y = specified yield stress of reinforcement, psi (MPa)

g = distance between adjacent stirrup legs or studs, measured in a parallel direction to a column face

h = slab thickness, in. (mm)

I_b = moment of inertia of gross section of beam about centroidal axis, in.^{4} (mm^4)

I_s = moment of inertia of gross section of slab about centroidal axis defined for calculating α_f and β_t, in.^{4} (mm^4)

K_c = stiffness of columns based on moment of inertia at any cross section outside the joint

K_{ec} = stiffness of equivalent column

K_{FP} = modification factor accounting for reduction in joint confinement at exterior connections

K_t = torsional stiffness

ℓ_1 = length of span in direction that moments are being determined, measured center-to-center of supports, in. (mm)

ℓ_2 = length of span in direction perpendicular to ℓ_1, measured center-to-center of supports, in. (mm)

ℓ_3 = distance measured from the column centerline to the edge of the slab, in. (mm)

ℓ_n = length of clear span measured face-to-face of supports, in. (mm)

ℓ_t = span of member under load test, taken as the shorter span for two-way slab systems, in. (mm); span is the smaller of: a) distance between centers of supports; and b) clear distance between supports plus thickness h of member.

M_o = total factored static moment, in.-lb (kNm)

M_{sc} = portion of slab factored moment balanced by support moment, in.-lb (kNm)

N_c = resultant tensile force acting on the portion of the concrete cross section that is subjected to tensile stresses due to the combined effects of service loads and effective prestress, lb (N)

q_{Du} = factored dead load per unit area, lb/ft^2 (kPa)

q_{Lu} = factored live load per unit area, lb/ft^2 (kPa)

q_u = factored load per unit area, lb/ft^2 (kPa)

x = shorter overall dimension of rectangular part of cross section, in. (mm)

y = longer overall dimension of rectangular part of cross section, in. (mm)

V_c = nominal shear strength provided by concrete, lb (N)

V_p = vertical component of all effective prestress forces crossing the critical section, lb (N)

V_{se} = unfactored shear force, but not less than twice the unfactored dead load shear, lb (N)

V_{ug} = factored shear force on the slab critical section for two-way action due to gravity loads, lb (N)

α_f = ratio of flexural stiffness of beam section to flexural stiffness of a width of slab bounded laterally by centerlines of adjacent panels (if any) on each side of the beam

α_{fm} = average value of α_f for all beams on edges of a panel

α_{f1} = α_f in direction of ℓ_1

α_{f2} = α_f in direction of ℓ_2

α_s = constant used to compute V_c in slabs

β = ratio of long side-to-short side of the column, concentrated load, or reaction area

β_p = factor used to compute V_c in prestressed slabs

β_t = ratio of torsional stiffness of edge beam section to flexural stiffness of a width of slab equal to span length of beam, center-to-center of supports

γ_f = factor used to determine the unbalanced moment transferred by flexure at slab-column connections

γ_v = factor used to determine the unbalanced moment transferred by eccentricity of shear at slab-column connections

ρ_b = ratio of A_s to b_d at balanced condition when concrete and reinforcement both reach their respective yield strain

$\sum A_{sb}$ = total area of reinforcing steel passing through the column core, summed on all four sides of the column of an interior column

χ = ℓ_2/ℓ_1

2.2—Definitions

ACI provides a comprehensive list of definitions through an online resource, "ACI Concrete Terminology," https://www.concrete.org/store/productdetail.aspx?ItemID=CT13. Definitions provided herein complement that source.

design story drift—design displacement of one level, or floor, relative to the level above or below.

design story drift ratio—design story drift divided by the story height.

headed shear stud reinforcement—individual headed studs or groups of studs with anchorage provided by a head at each end or a common base rail consisting of a steel plate or shape.

lateral-force-resisting system—portion of the structure composed of members designed to resist forces related to wind or earthquake effects.

podium—a thick slab that supports a light frame building above parking or commercial levels.

shear cap—a projection below the slab used to increase the slab shear strength.

slab-on-girder system—a grid of girders in both plan directions with a solid slab spanning between girders; may include beams between girders in both directions.

slab panel—a slab bounded by column, beam, or wall centerlines on all sides.

two-way slab—slab with or without beams that meets a particular aspect ratio.

two-way wide-band system—slab with paneled ceiling with shallow and wide beams spanning between columns in each direction.

waffle slab—two-way concrete joist construction using prefabricated pan (dome) forms.

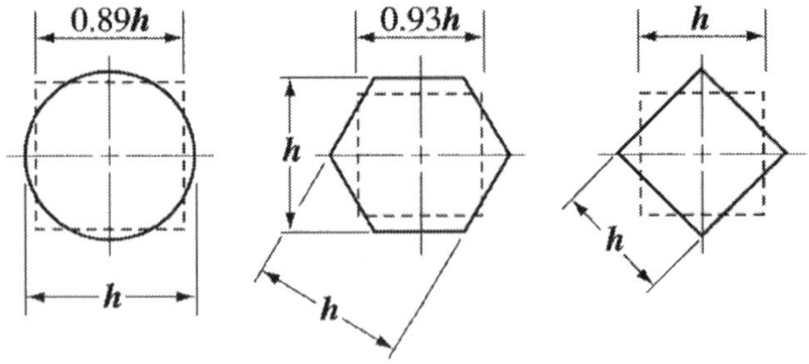

Fig. 3.1.3—Example of equivalent square section for supporting members.

CHAPTER 3—ANALYSIS METHODS

3.1—General

3.1.1 *Fundamental principles of structural mechanics*—Analysis of two-way slab can be based directly on fundamental principles of structural mechanics, provided it can be demonstrated explicitly that all strength and serviceability criteria in Chapters 4 and 6 of this guide are satisfied. Based on theory of elasticity, slab design should satisfy force equilibrium and strain compatibility. Slab analysis can be achieved through the use of classic solutions based on linearly elastic continuum theories; numerical solutions based on discrete elements; yield-line analyses; or strip method analyses, including in all cases evaluation of the stress conditions around the supports in relation to shear and torsion, as well as flexure. Regardless of analysis method, deviations in physical dimensions of the slab from common practice should be justified on the basis of knowledge of the expected loads, reliability of the calculated stresses, and deformations of the structure.

Analysis of a two-way slab system should consider the aspect ratio of each slab panel and the relative stiffness of the slab panels, supporting beams (if any), and supporting columns or walls. Analysis and design of two-way slab systems using the Direct Design Method (DDM) and Equivalent Frame Method (EFM) are discussed in 3.2.

3.1.2 *Slab stiffness*—During the life of a structure, construction loads, ordinary occupancy loads, anticipated overloads, and volume changes can cause cracking of slabs. Excessive cracking exposes concrete to moisture infiltration, which can cause corrosion of reinforcement and deterioration of structural elements. Under sustained gravity loads, cracking can lead to large vertical deflection resulting in damage to nonstructural elements.

Cracking reduces stiffness of the slabs, and increases lateral displacement when lateral forces act on the structure. Cracking of slabs should be considered in stiffness assumptions so drift caused by wind or earthquake is not grossly underestimated. Conservatively, analysis of slab-column connections can use a stiffness reduction factor that provides higher reduction in stiffness to ensure lateral displacement and that the design forces are not underestimated. Refer to 7.2.2 for further discussions.

3.1.3 *Total factored static moment*—Total factored static moment M_o for a span of a rectilinear interior panel may be determined in a strip bounded laterally by the centerline of a panel on each side of the centerline of supports. The absolute sum of positive and average negative factored moments in each direction should not be less than

$$M_o = \frac{q_u \ell_2 \ell_n^2}{8} \quad (3.1.3)$$

In the case of an edge panel, use $0.5\ell_2$, where ℓ_2 *is* the transverse span.

Equation (3.1.3) follows directly from Nichols (1914) with the simplifying assumption that the reactions are concentrated along the faces of the support perpendicular to the span considered. In general, the designer will find it expedient to calculate the static moment for two adjacent half panels, which includes a column strip with a half middle strip along each side. The clear span ℓ_n is taken as the distance between columns, capitals, brackets, or walls. The value of ℓ_n used in Eq. (3.1.3) should not be less than $0.65\ell_1$. Circular or regular polygon-shaped supports should be treated as square supports with the same area (Fig. 3.1.3).

3.2—Analysis methods

Both the DDM and EFM are based on analysis of the results of an extensive series of tests (Jirsa et al. 1963) and the well-established performance record of various slab systems. This guide covers these two methods in depth.

3.2.1 *Direct Design Method (DDM)*—The DDM consists of a set of rules for distributing moments to slab and beam sections to satisfy strength requirements. The DDM can be used for slab systems with regular column layouts subjected to gravity loads only. Three fundamental steps are involved:

1) Determination of M_o

2) Distribution of M_o along each span to supports and midspan for negative factored moments and positive factored moment, respectively

3) Distribution of the negative and positive factored moments transversely to the column and middle strips and to the beams, if any. (The same distribution of moments to column and middle strips is also used in EFM, which is discussed in 3.2.2.6.)

3.2.1.1 *Limitations*—The DDM was developed from considerations of theoretical procedures for the determination of moments in slabs with and without beams, requirements for simple analysis and construction procedures, and precedents supplied by performance of slab systems. Consequently, based on ACI 318-14, Section 8.10.1.1, the analysis of slab systems using the DDM should conform to the following limitations:

a) Slab system has a minimum of three continuous spans in each direction;

b) Panels are rectangular, with a ratio of longer-to-shorter panel span taken from center-to-center of each panel not greater than 2;

c) Successive span lengths center-to-center of supports in each direction do not differ by more than one-third the longer span;

d) Offset of columns by a maximum of 10 percent of the span (in direction of offset) from either axis between centerlines of successive columns is permitted;

e) All loads are due to gravity only and uniformly distributed over an entire panel and unfactored live load does not exceed two times unfactored dead load;

f) For a panel with beams between supports on all sides, Eq. (3.2.1.1a) should be considered for beams in the two perpendicular directions:

$$0.2 \leq \frac{\alpha_{f1}\ell_2^2}{\alpha_{f2}\ell_1^2} \leq 5.0 \quad (3.2.1.1a)$$

where α_{f1} and α_{f2} are calculated in accordance with Eq. (3.2.1.1b)

$$\alpha_f = \frac{E_{cb}I_b}{E_{cs}I_s} \quad (3.2.1.1b)$$

g) Moment redistribution as permitted by ACI 318-14, Section 6.6.5, is not applicable for slab systems analyzed by the DDM;

h) Variations from these limitations are permitted if demonstrated by analysis from 3.1.1 of this guide.

Rules given for the DDM assume that the slab system at the first interior support is neither fixed against rotation nor discontinuous. The primary reason for limitation a), which requires a minimum of three spans, is that the magnitude of the negative moments at the interior support in a structure with only two continuous spans is higher by approximately 10 percent. Under limitation b), when the ratio of the two spans (long span/short span) of a panel exceeds 2, the slab resists the moment in the shorter span essentially as a one-way slab.

Limitation c) relates to the possibility of developing negative moments beyond the point where negative moment reinforcement is terminated, as discussed in a subsequent section. Limitation d) permits columns to be offset within specified limits from a regular rectangular array. A cumulative total offset of 20 percent of the span is established as the upper limit when using the DDM.

Limitation e) is included because the DDM is based on tests (Jirsa et al. 1969) for uniform gravity loads and resulting column reactions determined by statics. Lateral loads such as wind or seismic require a frame analysis. Inverted foundation mats analyzed as two-way slabs involve application of known column loads. Therefore, even where the soil reaction is assumed to be uniform, a frame analysis should be performed. The limit of applicability of the DDM for ratios of live-to-dead load was reduced in ACI 318-95 from 3 to 2. Because in most slab systems the live-to-dead load ratio is less than 2, it is unnecessary to check the effects of pattern loading. For a slab system supporting a nonmovable load (such as a water reservoir in which the load on all panels is expected to be the same), the designer need not consider the live load imitation e) because there is no need to consider pattern loading.

The elastic distribution of moments will deviate significantly from those assumed in the DDM unless the stiffness under limitation f) is met. Limitation g) forbids the use of moment redistribution as permitted by ACI 318-14, Section 6.6.5, which is not intended for use where approximate values for bending moments are used. For the DDM, 10 percent modification is allowed as described in 3.2.2.5 of this guide.

Limitation h) permits a designer to use the DDM even when the structure does not fit the limitations in this section, provided it can be shown by analysis that the particular limitation does not apply to that structure.

Table 3.2.1.2—Distribution coefficient of end span total factored static moment, M_o

	(1)	(2)	(3)	(4)	(5)
	Exterior edge unrestrained	Slab with beams between all supports	Slab without beams between interior supports		Exterior edge fully restrained
Interior negative factored moment	0.75	0.70	0.70	0.70	0.65
Positive factored moment	0.63	0.57	0.52	0.50	0.35
Exterior negative factored moment	0	0.16	0.26	0.30	0.65

3.2.1.2 *Distribution of total factored moments*—As stated in 3.1.3, the clear span is based on distance between faces of supports. If a supporting member does not have a rectangular cross section or if the sides of the rectangle are not parallel to the spans, it should be treated as a square support having the same area, as illustrated in Fig. 3.1.3.

In an interior span, for M_o, calculated based on Eq. (3.1.3) should be distributed as follows:

a) Negative factored moment: 0.65

b) Positive factored moment: 0.35

The M_o, in an end span, should be distributed to slab midspan and supports as shown in Table 3.2.1.2.

The moment coefficients for an end span are based on the equivalent column stiffness expressions (Corley et al. 1961; Jirsa et al. 1963; Corley and Jirsa 1970). The coefficients for

Table 3.2.1.3a—Distribution in percent of interior negative factored moment to column strips

ℓ_2/ℓ_1	0.5	1.0	2.0
$(\alpha_{f1}\ell_2/\ell_1) = 0$	75	75	75
$(\alpha_{f1}\ell_2/\ell_1) \geq 1.0$	90	75	45

Note: Linear interpolations can be made between values shown.

an unrestrained edge would be used, for example, if the slab is simply supported on a masonry or concrete wall. Those for a fully restrained edge would apply if the slab is constructed integrally with a concrete wall having a flexural stiffness so large compared to that of the slab that little rotation occurs at the slab-to-wall connection. Note that Table 3.2.1.2 is a major simplification of values that is computed on the basis of relative stiffnesses (ACI 318-77) and that, in some cases, these moments could deviate substantially larger from theoretic values. This is especially true for the exterior negative moment case because there are no limits on column stiffness other than pinned or fixed.

For other than unrestrained or fully restrained edges, coefficients in Table 3.2.1.2 were selected to be near the upper bound of the range for positive moments and interior negative moments. As a result, exterior negative moments were usually closer to a lower bound. The exterior negative moment strength for most slab systems is governed by spacing of minimum reinforcement to control cracking. The final coefficients in the table have been adjusted so that the absolute sum of the positive and average negative moments equal M_o.

For two-way slab systems with beams between supports on all sides (two-way slabs), moment coefficients of Column (2) of Table 3.2.1.2 apply. For slab systems without beams between interior supports (flat plates and flat slabs), the moment coefficients of Columns (3) or (4) apply, without or with an edge (spandrel) beam, respectively. Design should be based on the larger of the two interior negative factored moments on either side of a common column or other type of support. If an analysis is made to distribute unbalanced moments, flexural stiffness could be obtained on the basis of the gross concrete section of the members involved.

The gravity load moment transferred by eccentricity within the critical section for shear between slab and edge column should be $0.3M_o$. Proportion edge beams or edges of slab to resist exterior negative factored moments. Torsional stresses caused by the moment assigned to the slab should be investigated.

3.2.1.3 *Factored moments in column strips*—The rules for assigning moments to the column strips, beams, and middle strips are based on studies (Gamble 1972) of moments in linearly elastic slabs with different beam stiffness tempered by the moment coefficients that have been used successfully. Column strips should be proportioned to resist the portions in percent tabulated in Table 3.2.1.3a for interior negative factored moments, and in Table 3.2.1.3b for exterior negative factored moments.

In Table 3.2.1.3b, linear interpolations can be made between values shown, where β_t is calculated in Eq. (3.2.1.3a) and C is calculated in Eq. (3.2.1.3b).

Table 3.2.1.3b—Distribution in percent of exterior negative factored moment to column strips

ℓ_2/ℓ_1		0.5	1.0	2.0
$(\alpha_{f1}\ell_2/\ell_1) = 0$	$\beta_t = 0$	100	100	100
	$\beta_t \geq 2.5$	75	75	75
$(\alpha_{f1}\ell_2/\ell_1) \geq 1.0$	$\beta_t = 0$	100	100	100
	$\beta_t \geq 2.5$	90	75	45

Table 3.2.1.3c—Distribution in percent of positive factored moment to column strips

ℓ_2/ℓ_1	0.5	1.0	2.0
$(\alpha_{f1}\ell_2/\ell_1) = 0$	60	60	60
$(\alpha_{f1}\ell_2/\ell_1) \geq 1.0$	90	75	45

Note: Linear interpolations can be made between values shown.

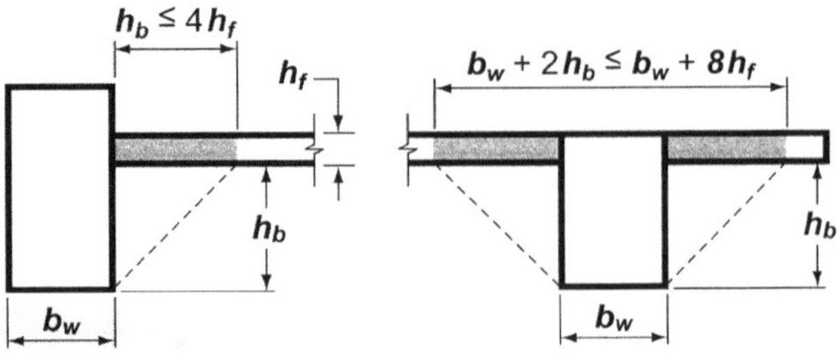

Fig. 3.2.1.3—Example of portion of slab to be included with beam.

$$\beta_t = \frac{E_{cb}C}{2E_{cs}I_s} \quad (3.2.1.3a)$$

$$C = \Sigma\left(1 - 0.63\frac{x}{y}\right)\frac{x^3 y}{3} \quad (3.2.1.3b)$$

The constant C for T- or L-sections can be evaluated by dividing the section into separate rectangular parts, as defined in Fig. 3.2.1.3, and summing the values of C for each part.

The effect of the torsional stiffness parameter β_t is to assign all of the exterior negative factored moment to the column strip, and none to the middle strip, unless the beam torsional stiffness is high relative to the flexural stiffness of the supported slab. In the definition of β_t, the shear modulus has been taken as $E_{cb}/2$. Where walls are used as supports along column lines, they can be regarded as very stiff beams with a $\alpha_{f1}\ell_2/\ell_1$ value greater than 1. Where the exterior support consists of a wall perpendicular to the direction in which moments are being determined, β_t may be taken as zero if the wall is of masonry without torsional resistance, and β_t may be taken as 2.5 for a concrete wall with large torsional resistance that is monolithic with the slab.

For the purpose of establishing moments in the half column strip adjacent to an edge supported by a wall, ℓ_n in Eq. (3.1.3) may be assumed equal to ℓ_n of the parallel adjacent column-to-column span, and the wall may be considered as a beam having a moment of inertia, I_b ,equal to infinity. Where supports consist of columns or walls extending for a distance equal to or greater than $(3/4)\ell_2$ used to compute

Table 3.2.1.6—Distribution in percent of factored moment in beams

$(\alpha_{f1}\ell_2/\ell_1) \geq 1.0$	85
$(\alpha_{f1}\ell_2/\ell_1) = 0$	0

Note: Linear interpolations can be made between values shown.

M_o, negative moments should be considered to be uniformly distributed across ℓ_2.

Column strips should be proportioned to resist the portions in percent of positive factored moments shown in Table 3.2.1.3c.

3.2.1.4 *Factored moments in middle strips*—That portion of the negative and positive factored moments not resisted by column strips should be proportionately assigned to corresponding half middle strips. Each middle strip should be proportioned to resist the sum of the moments assigned to its two half middle strips. A middle strip adjacent to and parallel with a wall-supported edge should be proportioned to resist twice the moment assigned to the half middle strip corresponding to the first row of interior supports.

3.2.1.5 *Modification of factored moments*—A modification of 10 percent in negative and positive factored moments, calculated in accordance with 3.2.2.2, is permitted provided the total static moment for a panel in the direction considered is not less than M_o required by Eq. (3.1.3). For example, a 10 percent reduction of negative factored moment should follow with a corresponding increase of positive factored moment. This is intended to recognize a limited amount of inelastic behavior and moment redistribution can occur in slabs that were analyzed with the DDM.

3.2.1.6 *Factored moments in beams*—Factored moments in beams between supports are proportioned in percent as shown in Table 3.2.1.6.

Loads assigned directly to beams are in addition to the uniform dead load of the slab; uniform superimposed dead loads, such as the ceiling, floor finish, or assumed equivalent partition loads; and uniform live loads. All of these loads are normally included with q_u in Eq. (3.1.3). Linear loads applied directly to beams include partition walls over or along beam centerlines and additional dead load of the projecting beam stem. Concentrated loads include posts above or hangers below the beams or induced force from seismic overturning. For the purpose of assigning directly applied loads, only loads located within the width of the beam stem should be considered as directly applied to the beams. The effective width of a beam as defined in Fig. 3.2.1.3 is solely for strength and relative stiffness calculations. Line loads and concentrated loads located on the slab away from the beam stem require special consideration to determine their apportionment to slab and beams. For slabs with beams between supports, the slab portion of column strips should be proportioned to resist that portion of the column strip moments not resisted by beams.

3.2.1.7 *Factored shear in slab systems with beams*—Resistance to total shear occurring on a panel based on computation of slab shear strength on the assumption that loads are distributed to supporting beams is permitted. Shear strength for beams should satisfy ACI 318-14, Sections 9.5.3 and 22.5.

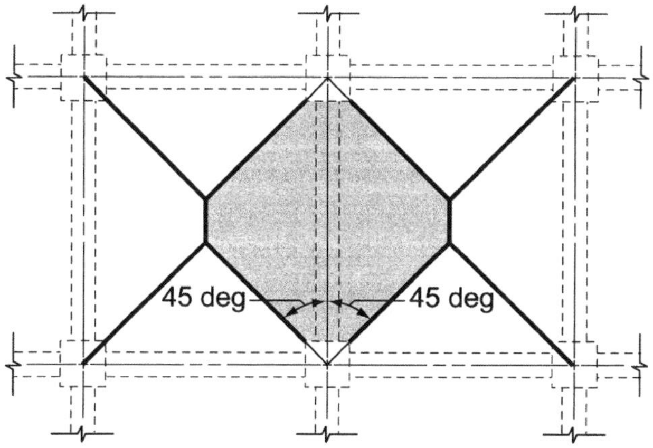

Fig. 3.2.1.7—Tributary area for shear on an interior beam.

Beams with $\alpha_{f1}\ell_2/\ell_1$ equal to or greater than 1.0 are proportioned to resist shear caused by factored loads on tributary areas that are bounded by 45-degree lines drawn from the corners of the panels and the centerlines of the adjacent panels parallel to the long sides. The tributary area for computing shear on an interior beam is shown shaded in Fig. 3.2.1.7.

If the stiffness for the beam, $\alpha_{f1}\ell_2/\ell_1$, is less than 1.0, the shear on the beam can be obtained by linear interpolation, assuming beams resist no load at $\alpha_{f1} = 0$. In such cases, the beams framing into the column will not account for all of the shear force applied on the column. The remaining shear force will produce shear stresses in the slab around the column, which should be checked in the same manner as for flat slabs.

In addition to shears calculated from slab tributary areas, beams should be proportioned to resist shear caused by factored loads applied directly on beams. Note that the proportional shear in slab systems with beams described in this section do not apply to calculation of torsional moments on the beams. Torsional moments should be based on the calculated flexural moments acting on the sides of the beam.

3.2.1.8 *Factored moments in columns and walls*—Columns and walls built integrally with a slab system should resist moments caused by factored loads on the slab system. At an interior support, columns or walls above and below the slab should resist the factored moment specified by Eq. (3.2.1.8) in direct proportion to their stiffnesses unless a general analysis is made.

$$M_u = 0.07[(q_{Du} + 0.5q_{Lu})\ell_2\ell_n^2 - q'_{Du}\ell_2'(\ell_n')^2 \quad (3.2.1.8)$$

where q'_{Du}, ℓ_2', and ℓ_n' refer to the shorter span.

Equation (3.2.1.8) refers to two adjoining spans, with one span longer than the other, and with full dead load plus one-half live load applied on the longer span with only dead load applied on the shorter span. Design and detailing of the reinforcement transferring the moment from the slab to edge column is critical to both performance and safety of flat

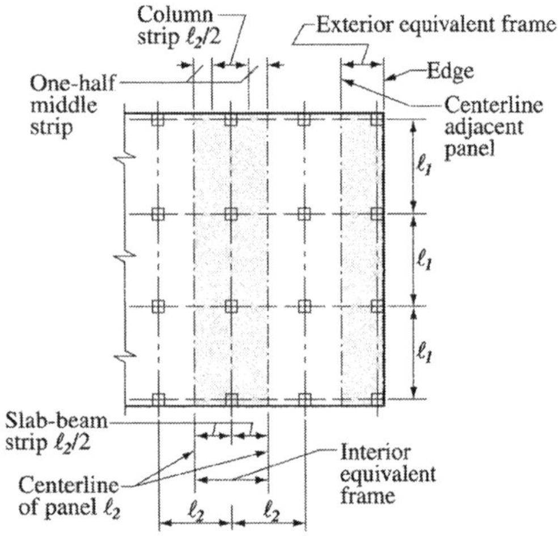

Fig. 3.2.2—Definitions of equivalent frame.

slabs or flat plates without edge beams, or cantilever slabs. Complete details are shown on construction documents, such as concentration of reinforcement over the column by closer spacing or additional reinforcement.

3.2.2 *Equivalent Frame Method (EFM)*—The EFM involves the representation of a three-dimensional slab system by a series of two-dimensional frames that are then analyzed for loads acting in plane of frames. The negative and positive moments so determined at the critical design sections of the frame are distributed to the slab sections in accordance with 3.2.1.3 (column strips), 3.2.1.4 (middle strips), and 3.2.1.6 (beams). The EFM is based on studies by Corley et al. (1961), Jirsa et al. (1963), and Corley and Jirsa (1970). Unlike the DDM, the EFM can be used for slab systems with irregular column layouts and pattern loading with or without lateral forces.

The EFM was developed when the primary structural analysis tool was the moment-distribution method. Specialized computer programs have been developed for this analysis, and give a way of using ordinary plane-frame structural analysis programs for this task (Schaeffer 1999).

Application of the equivalent frame to a regular structure is illustrated in Fig. 3.2.2. The three-dimensional building is divided into a series of two-dimensional frame bents (equivalent frames) oriented along column or support centerlines, with each frame extending the full height of the building. The width of each equivalent frame is bounded by the centerlines of the adjacent panels. The complete analysis of a slab system for a building consists of analyzing a series of equivalent (interior and exterior) frames spanning longitudinally and transversely through the building.

The equivalent frame comprises three parts:

1) The horizontal slab strip, including any beams spanning in the direction of the frame

2) The columns or other vertical supporting members, extending above and below the slab

3) The elements of the structure that provide moment transfer between the horizontal and vertical members.

Where metal column capitals are used, it is permitted to consider their contributions to stiffness and resistance to moment and shear. Change in column and slab length due to direct stress and deflections due to shear can be neglected.

3.2.2.1 *Analysis and design considerations*—All sections of slabs and supporting members analyzed by EFM should be proportioned for moments and shears obtained:

a) The structure is considered to be made up of equivalent frames on column lines taken longitudinally and transversely through the building

b) Each frame consists of a row of columns or supports and slab-beam strips, bounded laterally by the centerline of panel on each side of the centerline of columns or supports

c) Columns or supports are assumed to be attached to slab-beam strips by torsional members (3.2.2.4) transverse to the direction of the span for which moments are being determined and extending to bounding lateral panel centerlines on each side of a column

d) Frames adjacent and parallel to an edge are bounded by that edge and the centerlines of adjacent panels

e) Analysis of each equivalent frame in its entirety is permitted. Alternatively, for gravity loading, a separate analysis of each floor or roof with far ends of columns considered fixed is also permitted

f) Where slab-beams are analyzed separately, determination of the moment at a given support, assuming that the slab-beam is fixed at any support two panels distant therefrom, is permitted, provided the slab continues beyond the points of fixity.

The simplifications permitted under e) and f) were concessions to manual calculation procedures such as moment distribution.

3.2.2.2 *Slab-beams*—In a slab-beam system, a support is defined as a column, capital, bracket, or wall; a beam is not considered to be a support member for the equivalent frame. Determining the moment of inertia of slab-beams at any cross section outside of joints or column capitals using the gross area of concrete is permitted. Consider variation in moment of inertia along the axis of slab-beams. The moment of inertia of slab-beams from the center column to the face of the column, bracket, or capital is assumed equal to the moment of inertia of the slab-beam at the face of the column, bracket, or capital divided by the quantity $(1 - c_2/\ell_2)^2$, where c_2 and ℓ_2 are measured transverse to the direction of the span for which moments are being determined.

3.2.2.3 *Columns*—Column stiffness is based on column length from middepth of the slab above to middepth of the slab below. The column moment of inertia is computed on the basis of its cross section, taking into account the increase in stiffness provided by the capital, if any. Consider variation in moment of inertia along the axis of columns. Moment of inertia of columns from top to bottom of the slab-beam at a joint can be assumed to be infinite.

When slab-beams are analyzed separately for gravity loads, the concept of an equivalent column, combining the stiffness of the slab-beam and torsional member into a composite element, is used. The column flexibility is modified to account for the torsional flexibility of the slab-

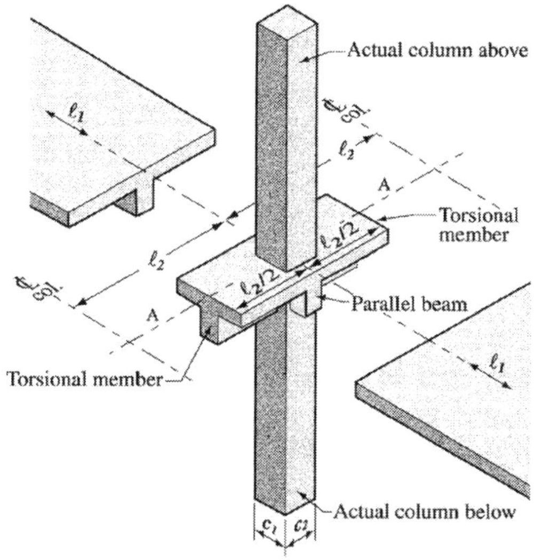

Fig. 3.2.2.3—Definitions of equivalent frame.

to-column connection that reduces its efficiency for transmission of moments. The equivalent column consists of the actual columns above and below the slab-beam, plus attached torsional members on each side of the columns extending to the centerline of the adjacent panels, as shown in Fig. 3.2.2.3.

3.2.2.4 *Torsional members*—Torsional members (3.2.2.1) should be assumed to have a constant cross section throughout their length consisting of the larger of a), b), and c):

a) A portion of slab having a width equal to that of the column, bracket, or capital in the direction of the span for which moments are being determined

b) For monolithic or fully composite construction, the portion of slab specified in a), plus that part of the transverse beam above and below the slab

c) The transverse beam as defined in Fig. 3.2.1.3.

Computation of the stiffness of the torsional member requires several simplifying assumptions. If no transverse-beam frames into the column, a portion of the slab equal to the width of the column or capital is assumed to be the torsional member. If a beam frames into the column, T-beam or L-beam action is assumed, with the flanges extending on each side of the beam a distance equal to the projection of the beam above or below the slab but not greater than four times the thickness of the slab. Additionally, it is assumed that no torsional rotation occurs in the beam over the width of the support.

The member sections used for calculating the torsional stiffness have been defined previously. An approximate expression for the stiffness of the torsional member, based on the results of three-dimensional analyses of various slab configurations (Corley et al. 1961; Jirsa et al. 1963; Corley and Jirsa 1970) is given in Eq. (3.2.2.4a).

$$K_t = \Sigma \frac{9E_{cs}C}{\ell_2\left(1-\frac{c_2}{\ell_2}\right)^3} \qquad (3.2.2.4a)$$

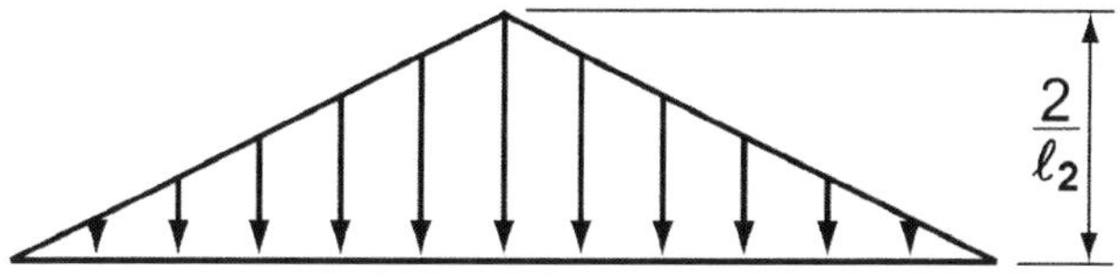

Fig. 3.2.2.4—Distribution of unit twisting moment along column centerline.

where $$C = \Sigma\left(1-0.63\frac{x}{y}\right)\frac{x^3y}{3} \qquad (3.2.2.4b)$$

Studies of three-dimensional analyses of various slab configurations suggest that a reasonable value of the torsional stiffness can be obtained by assuming a moment distribution along the torsional member that varies linearly from a maximum at the center of the column to zero at the middle of the panel. The assumed distribution of unit twisting (or torsional) moment along the column centerline is shown in Fig. 3.2.2.4, in which the unit for $2/\ell_2$ is angle per unit length; and because the length is 12, the resultant is unity.

Where beams frame into columns in the direction of the span for which moments are being determined, the torsional stiffness should be multiplied by the ratio of the moment of inertia of the slab with such a beam to the moment of inertia of the slab without such a beam.

Stiffness of the equivalent frame based on the flexibility of the column and torsional member can be expressed by Eq. (3.2.2.4c) (Schaeffer 1999)

$$\frac{1}{K_{ec}} = \frac{1}{\Sigma K_c} + \frac{1}{K_t} \qquad (3.2.2.4c)$$

3.2.2.5 *Arrangement of live load*—When the loading pattern is known, the equivalent frame is analyzed for that load. Except for the loading condition described in the next paragraph, maximum positive factored moment near midspan of a panel can be assumed to occur with three-fourths of the full factored live load on the panel and on alternate panels. The maximum negative factored moment in the slab at a support can be assumed to occur with three-fourths of the full factored live load on adjacent panels only. Factored moments should be taken not less than those occurring with full factored live load on all panels.

When the unfactored live load is variable but does not exceed three-fourths of the unfactored dead load, or the nature of live load is such that all panels will be loaded simultaneously, it is acceptable to assume that maximum factored moments occur at all sections with full factored live load on the entire slab system.

The use of only three-fourths of the full factored live load for maximum moment loading patterns is based on the fact that maximum negative and maximum positive live load moments cannot occur simultaneously and that redistribution of maximum moments is, therefore, possible before failure occurs. This procedure, in effect, permits some local overstress under the full factored live load when distributed in the prescribed manner, but still ensures that the ultimate strength of the slab system after redistribution of moment is not less than that required to resist the full factored dead and live loads on all panels.

3.2.2.6 *Factored moments*—At interior supports, the critical section for negative factored moment in both column and middle strips is at the face of rectilinear supports, but not farther away than $0.175\ell_1$ from the center of a column. At exterior supports with brackets or capitals, the critical section for negative factored moment in the span perpendicular to an edge is at a distance from the face of supporting element not greater than one-half the projection of bracket or capital beyond the face of the supporting element. Circular or regular polygon-shaped supports are treated as square supports with the same area for location of the critical section for negative moment, as illustrated in Fig. 3.1.3.

Using the centerline moment and shear from equivalent frame analysis, adjust the negative factored moments to the face of the supports. At interior supports, the adjustment is commonly taken as $V_u c_1/3$. At an exterior support, the adjustment is commonly taken as $V_u c_1/4$ to limit reductions in the exterior negative moment.

Where slab systems within limitations of 3.2.1.1 are analyzed by EFM, it is permitted to reduce the resulting computed moments in such proportion that the absolute sum of the positive and average negative moments need not exceed the value obtained from Eq. (3.1.3). This relaxation is based on long satisfactory experience with analyses when applicable limitations are met. Distribution of moments at critical sections across the slab-beam strip of each frame to column strips, middle strips and beam, and (as described in 3.2.1.3, 3.2.1.4, and 3.2.1.6) may be considered in conjunction with limitation f) of 3.2.1.1.

3.3—Finite element analysis

The finite element method is a tool for analyzing reinforced concrete slabs, particularly for complex slab systems. This method is widely used by practicing engineers and researchers. There are many commercial structural analysis software packages that have the finite element analysis option. Because this method is an approximate numerical method, it is essential to understand the relationship between the actual behavior of the structure and the numerical simulation. Results of the finite element model should be verified before being used for design applications. As a minimum, verification should include a check of summation of vertical and horizontal reactions and applied forces satisfying force equilibrium.

The finite element method is based on several assumptions, including:

1) Most commercial finite element packages use linear analysis, whereas reinforced concrete is a nonlinear material

2) The finite element method is based on isotropic and homogenous material, whereas reinforced concrete is a composite material

3) Derivations of the various plate elements used for slab modeling are based on small displacement and rotation assumptions. The term "small displacement" means, in general, small relative to the thickness of the slab; this assumption allows the use of linear analysis that ignores geometric nonlinearity.

In finite element modeling, selecting element type, mesh size, material properties, and boundary conditions are critical to perform sufficiently accurate simulation of reinforced concrete slabs. The mesh size or number of elements used in the model can significantly affect reliability of the results. In general, the finer the mesh or the higher the number of elements, the more reliable the results.

The type of finite element used in the analysis can significantly affect the results because the various elements used for slab analysis are derived from different assumptions. For example, Kirchoff plate bending is derived for thin plate applications and accounts for flexural deformation only, whereas Mindlin plate elements are derived for moderately thick plate applications and accounts for both flexural and shear deformations (Logan 2002). Thin plate refers to a plate with thickness that is much smaller than its in-plane dimensions, and thick plate refers to a plate with thickness that is greater than one-tenth the span of the plate. Finite element formulation is based on shape functions that interpolate the solution between the nodes. Elements with higher-order shape functions are more accurate than elements with lower-order shape functions, and elements with mid-nodes are more accurate than elements with edge nodes only.

Another important factor that may affect the finite element results is selection of boundary conditions. The types of boundary conditions used to model slabs pertain to support boundary conditions and to slab-to-column connections. There are two approaches to model boundary conditions: in the first approach, the columns supporting the slab and the columns above the slab are modeled using frame elements or three-dimensional solid elements. This approach accounts for the flexural effects of the columns as well as the columns above the slab. In the second approach, however, the column supports are modeled as pin or as fixed supports, or instead as springs with finite elastic stiffness to improve the behavior of the model. Gentry (1986) investigated these two approaches and concluded that using frame elements or three-dimensional elements are preferable to using pin or fixed supports to simulate columns. For modeling slab-supporting beams, the two approaches mentioned previously may be used. Using frame elements of three-dimensional elements is preferable to using a series of pin or fixed supports along the beam lines. Note that using pin or fixed supports does not accurately simulate the actual stiffness of the connection between the slab and the supporting beam or column. Also, another consideration in modeling the boundary conditions is the physical size of connection versus what is used in the model. When the supporting member is connected to a node, the connection occurs over an infinitesimal area, where the physical connection occurs over a defined area. This will result in an unrealistic stress concentration at the support, an inaccurate clear span that results in greater moment in the slab, and in an inaccurate moment at the face of the column.

Once the finite element analysis is performed and the solution converged, verify the results before using them

Fig. 3.4a—Rectangular slab panel (courtesy of Nawy [2011]).

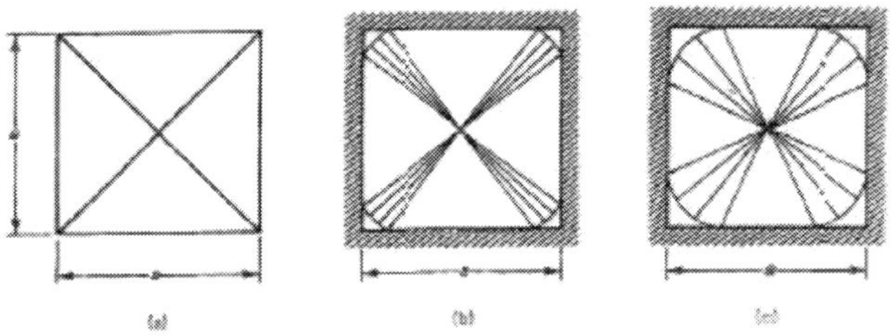

Fig. 3.4b—Idealized failure mechanism of a square panel with different boundary conditions (courtesy of Nawy [2011]).

for design purposes. Check the equilibrium of the structure to verify the sum of the reactions versus the applied load. Global displacement of the structure should be verified and confirmed as small in comparison to the structure's geometry. The engineer should critically investigate the resulting output from the model. This would include the deflected shape, and stress and displacement contour plots.

3.4—Yield-line theory

The yield-line theory is an analysis method that selects the applicable upper-bound model that gives the peak ultimate load capacity of structural slab or plate in a flexural mode of failure and perfect plasticity. Such a load level is occasionally termed as an "upper-bound" solution. A succession of hinge bands that develop at such a load level are idealized by lines, hence, the name "yield-line theory" (Johansen 1962). The theory assumes rigid-plastic behavior; that is, each concrete plate or slab segment stays planar up to collapse, producing rigid planar segments (Fig. 3.4(a) and 3.4(b)). Consequently, deflection is not considered in theory, nor are the compressive or tensile membrane forces that may act in the plane of the slab or plate being analyzed. The plates or slabs are considered significantly under-reinforced in this procedure and, therefore, require ductility. This is achieved by limiting the reinforcement ratio to 1.0 percent of the section needed for the controlling flexural yield-line moment and steel strain within a range of 0.010. This higher reinforcement percentage is necessitated by the dominance of shear-flexure failure mechanism at the column boundaries in the flat plates.

Slab thickness obtained by a yield-line analysis is often thinner than one from other lower-bound methods, such as DDM, EFM, or the strip method. Consequently, it is important to rigorously apply the serviceability requirements for deflection and crack control as recommended in ACI 224R and ACI 435R when determining the thickness of two-way slab or plate when this theory is used in analysis and design.

One distinct advantage of yield-line theory is the ability to provide rapid solutions for any shape of plate; most of the other approaches discussed are applicable only to rectangular shapes. Extending them to other shapes requires more rigorous computations for boundary effects. This theory enables an engineer to obtain the load capacity for a triangular, trapezoidal, rectangular, circular, or any other shape subjected to distributed or concentrated loads, provided that the failure mechanism is known or predictable. Because most failure patterns are presently identifiable, solutions are readily obtained (Nawy 2011; Park and Gamble 1999).

3.5—Strip method analysis

Other methods of analysis, such as the strip method, that conform to the fundamental principles of structural behavior of two-way action slabs and plates are also available, as discussed in 3.1.1. The strip method, in contrast to the yield-line method, is a lower-bound solution to the collapse load, where twisting moments are considered absent from the analysis as the plate is segmented into beam strips. The solution is, therefore, based on beam action only through satisfying equilibrium at all points in the slab. Consequently, it is applicable for proportioning rectangular two-way slabs and plates as compared to the yield-line method's versatility in its application to any conceivable shape, as discussed in 3.4. The strip method allows some freedom to the designer for choosing how the slab should resist load, as well as for analyzing and designing the slab to meet the loading conditions (Schaeffer 1999).

CHAPTER 4—DESIGN PROCEDURES

4.1—General

As stated in 3.1.1, two-way slab design is based directly on the fundamental principles of structural mechanics, provided it can be demonstrated explicitly that all safety and serviceability criteria are satisfied. For a slab system supported by columns or walls, the effective support area defined by the intersection of the bottom surface of the slab or drop panel (if there is one), is based on the dimensions c_1 and c_2 and the clear span ℓ_n. It includes the largest right circular cone, right pyramid, or tapered wedge with surfaces located within the column and capital or bracket, and oriented no greater than 45 degrees to the axis of the column. Circular or regular polygon-shaped supports should be treated as square supports with the same area as shown in Fig. 3.1.3. The design of two-way slabs considers the following general provisions:

a) A column strip is a design strip with a width on each side of a column centerline equal to $0.25\ell_1$ or $0.25\ell_2$, whichever is less

b) The column strip includes beams, if any

c) The middle strip is a design strip bounded by two column strips

d) A panel is bounded by column, beam, or wall centerlines on all sides

e) For monolithic or fully composite construction, a beam includes that portion of slab on each side of the beam web extending a distance equal to the projection of the beam above or below the slab, whichever is greater, but not greater than four times the slab thickness.

4.2—Gravity loading

For gravity loads, analysis of a slab system should consider the aspect ratio of each slab panel and the relative stiffness of the slab panels, supporting beams (if any), and supporting columns or walls. Gravity dead load includes self-weight plus long-term superimposed loads, such as flooring, ceiling, suspended nonstructural elements, or floor-mounted equipment. Live load is based on building occupancy and is prescribed in the governing building code. Permanent partitions can be included as either dead load or additional live load based on the type of occupancy, or as prescribed in the governing building code.

For light-frame construction supported on a podium flat plate or flat slab, engineering judgment should be exercised in specifying superimposed gravity dead and live loads resulting from bearing walls. Line load perpendicular to the direction of span should be included as a concentrated load in design. Line load parallel to the direction of the span should be included as a distributed load over an effective slab width not to exceed four times the thickness of slab.

4.3—Flexural design

As stated in 3.1.1, the slab design can be achieved through the combined use of classic solutions based on a linearly elastic continuum and evaluation of the stress conditions around the supports in relation to shear and torsion as well as flexure. The designer should recognize the limitations on the applicability of simplified design assumptions. Selection of physical dimensions of the slab should be compared to common office practice on the basis of knowledge of the expected loads and the reliability of the calculated stresses and deformations of the structure. Apply appropriate load factors from ACI 318-14, Section 5.2, unless other requirements are given in the governing building code.

Engineers note that current specified load and ϕ-factors lead to tensile reinforcement flexure stresses approximately 10 percent higher than codes prior to 2002. The increase reinforcement would lead to a decrease of the crack widths, but could simultaneously increase the number of cracks of narrower width, which is the preferable end result. This is especially of concern at an interior column of a flat plate, where the local reinforcement stresses were already high due to stress concentrations of flexural, shear, and torsion that have traditionally been ignored. Flexural reinforcement in column strips should be proportioned on the basis of the full negative moment value without reduced redistribution of negative moments permitted in ACI 318-14, Section 6.6.5.1, as an adequate safety margin.

4.4—Two-way action slab shear

The design of two-way slab shear is described in subsequent sections of this guide. For slabs of uniform thickness, it is sufficient to check two-way slab shear at one critical section. When shear controls at the support, provide stirrups or stud reinforcement. For slabs with changes in thickness, such as the edge of drop panels or shear caps, check shear at several critical sections. For edge columns at points where the slab cantilevers beyond the column, the critical perimeter will either be three- or four-sided.

4.5—Critical section

The critical section for shear in slabs subjected to bending in two directions follows the perimeter close to the edge of the effective support area. The shear stress acting on this section at factored loads is a function of $\sqrt{f_c'}$ and the ratio of the side dimension of the column to the effective slab depth. A much simpler design equation is derived by assuming a pseudo-critical section located at a distance $d/2$ from the periphery of the effective support area. When this is done, the shear strength is almost independent of the ratio of column size to slab depth. For rectangular columns, this critical section is defined by straight lines drawn parallel to and at a distance $d/2$ from the column edges. ACI 318-14, Section 22.6.4.1.1, allows the use of a rectangular critical section.

4.6—Openings in slab systems

In general, openings of any size are permitted in slab systems if it can be shown by analysis that the design strength is at least equal to the required strength set forth in ACI 318-14, Sections 8.4 and 8.5, and that all serviceability conditions, including the limits on deflections, are met. The locations of the effective portions of the critical section near typical openings and free edges are discussed further in 4.6.2 of this guide (Joint ACI-ASCE Committee 426 1974).

4.6.1 *Permissible openings without analysis*—As an alternate to analysis, openings are permitted in slab systems without beams if in accordance with the following conditions:

a) Openings of any size are permitted in the area common to the intersecting middle strips, provided the total amount of reinforcement required for the panel without the opening is maintained;

b) In the area common to the intersecting column strips, not more than one-eighth the width of column strip in either span should be interrupted by openings. An amount of reinforcement equivalent to that interrupted by an opening should be added on the sides of the opening;

c) In the area common to one column strip and one middle strip, not more than one-fourth of the reinforcement in either strip may be interrupted by openings. An amount of reinforcement equivalent to that interrupted by an opening should be added on the sides of the opening;

d) Shear guidelines at a critical section described in 4.5 reduced by the effects of openings should be followed.

4.6.2 *Permissible openings with analysis*—When openings in slabs are located at a distance less than 10 times the slab thickness from a concentrated load or reaction area, or when openings in flat slabs are located within column strips, the critical slab sections for shear should be modified as follows:

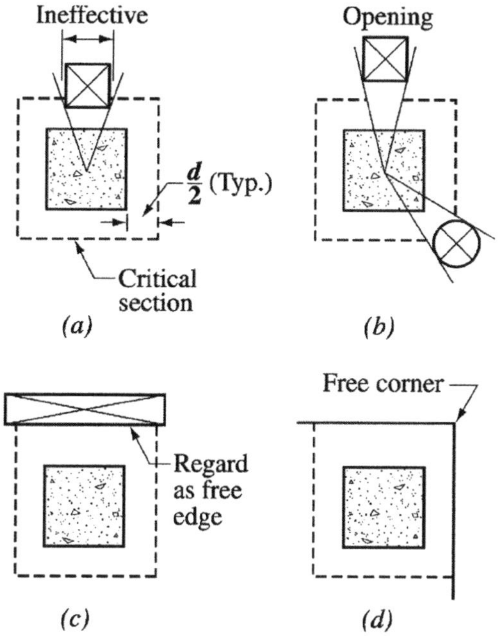

Fig. 4.6.2—Effect of openings and free edges. (Note: effective perimeter shown with dashed lines.)

a) For slabs without shearheads, that part of the perimeter of the critical section that is enclosed by straight lines projecting from the centroid of the column, concentrated load, or reaction area and tangent to the boundaries of the openings should be considered ineffective;

b) For slabs with shearheads, the ineffective portion of the perimeter should be one-half of that defined in 4.6.1, condition a).

The location of the effective portions of the critical section near typical openings and free edges is shown by the dashed line in Fig. 4.6.2.

4.7—Unbalanced moments

For slab systems without beams, tests (Hanson and Hanson 1968) and experience have shown that measures should be taken to resist the torsional and shear stresses. The fraction of unbalanced moment transferred by eccentricity of shear at slab-column connections is defined as γ_v and is equal to $(1 - \gamma_f)$. The calculated shear stresses in the slab critical section should conform to the requirements of ACI 318-14, Section 8.4.4.2 (refer to design examples in ACI 421.1R).

A fraction of the unbalanced moment given by $\gamma_f M_{sc}$ should be considered to be transferred by flexure, where M_{sc} is the moment to be transferred from slab to column. All reinforcement resisting that part of the moment to be transferred to column by flexure should be placed within an effective slab width between lines that are one and one-half slab or drop panel thicknesses ($1.5h$) outside opposite faces of the column or capital, except $1.5h$ of the slab thickness should be used where shear cap is used. For a shear critical section in the shape of a closed rectangle of sides b_1 and b_2, ACI 318-14, Section 8.4.2.3.2 gives

$$\gamma_f = \frac{1}{1+(2/3)\sqrt{b_1/b_2}} \tag{4.7a}$$

and requires that the remainder of the unbalanced moment, given by $\gamma_v M_u$, should be considered transferred by eccentricity of shear about the centroid of the shear critical section, where

$$\gamma_v = 1 - \gamma_f \tag{4.7b}$$

ACI 421.1R gives equations for γ_v for shear critical sections of any shape and equations for planar shear stress distribution satisfying ACI 318 for a nonrectangular critical section of general shape. Some flexibility in distribution of unbalanced moments transferred by shear and flexure at both exterior and interior supports is possible. Interior, exterior, and corner supports refer to slab-column connections for which the critical perimeter for rectangular columns has four, three, or two sides, respectively. The shear critical section at $d/2$ from the outer peripheral lines of shear reinforcement generally follows a nonrectangular shape.

At exterior supports, for unbalanced moments about an axis parallel to the edge, the portion of moment transferred by eccentricity of shear, $\gamma_v M_u$, may be reduced, provided that the factored shear at the support (excluding the shear produced by moment transfer) does not exceed 75 percent of the design shear strength ϕV_c as defined in ACI 318-14, Section 22.6.5.1, for edge columns or 50 percent for corner columns. This reduction of γ_v and the corresponding increase in γ_f (Eq. (4.7b)) is permitted by ACI 318, but is not recommended in this guide; the analyses and the experimental data in these references do not justify the reduction in γ_v (Gayed and Ghali 2008; Ritchie et al. 2006).

In slab-column connections, a large degree of ductility is required because the interaction between shear and unbalanced moment is critical. When the factored shear is large, the column-slab joint cannot always mobilize all of the reinforcement provided in the effective width. The reduction of γ_v and the increase of γ_f for edge, corner, or interior slab-column connections are permitted only if the strain in the flexural reinforcement within the effective width develops a minimum net tensile strain of 0.010.

When a reversal of moments occurs at opposite faces of an interior support, both top and bottom reinforcement should be concentrated within the effective width. A ratio of top to bottom reinforcement of approximately two is considered a good practice.

4.8—Shear strength

For nonprestressed slabs, V_c is calculated by the smallest of the following

$$V_c = \left(2+\frac{4}{\beta}\right)\lambda\sqrt{f_c'}b_o d \quad \text{(in.-lb)}$$
$$V_c = \left(2+\frac{4}{\beta}\right)\lambda\sqrt{f_c'}b_o d/12 \quad \text{(SI)} \tag{4.8a}$$

$$V_c = \left(\frac{\alpha_s d}{b_o}+2\right)\lambda\sqrt{f_c'}b_o d \quad \text{(in.-lb)}$$
$$V_c = \left(\frac{\alpha_s d}{b_o}+2\right)\lambda\sqrt{f_c'}b_o d/12 \quad \text{(SI)} \tag{4.8b}$$

$$V_c = 4\lambda\sqrt{f_c'}b_o d \quad \text{(in.-lb)}$$
$$V_c = \lambda\sqrt{f_c'}b_o d/3 \quad \text{(SI)} \tag{4.8c}$$

where α_s is 40 for interior columns, 30 for edge columns, and 20 for corner columns (interior, edge, and corner columns refer to critical sections with four, three, and two sides, respectively).

For square columns, the shear stress due to factored loads in slabs subjected to bending in two directions is limited to value given in Eq. (4.8c). However, tests (Joint ACI-ASCE Committee 426 1974) have indicated that the value of $4\lambda\sqrt{f_c'}$ is not conservative when the ratio β of the lengths of the long and short sides of a rectangular column or loaded area is larger than 2.0. In such cases, the actual shear stress on the critical section at punching shear failure varies from a maximum of approximately $4\lambda\sqrt{f_c'}$ around the corners of the column or loaded area, down to $2\lambda\sqrt{f_c'}$ or less along the long sides between the two end sections. Other tests (Vanderbilt 1972) indicated that V_c decreases as the ratio b_o/d increases. Equations (4.8a) and (4.8b) were developed to account for these two effects.

For shapes other than rectangular, β is taken to be the ratio of the longest overall dimension of the effective loaded area to the largest overall perpendicular dimension of the effective loaded area, as illustrated for an L-shaped reaction area in Fig. 4.8. The effective loaded area is that area totally enclosing the actual loaded area, for which the perimeter is a minimum.

4.9—Post-tensioned slabs

Typical nonprestressed flat-plates, flat-slabs, or flat-plates with shear cap can be designed using ACI 318-14, Section 8.2.5. However, for the design of post-tensioned flat plates or flat slabs with unbonded tendons, there are additional items to consider. Most two-way post-tensioned slabs in the United States use unbonded tendons. Average effective prestress compressive stress levels are limited to a minimum 125 psi (0.86 MPa), per ACI 318-14, Section 8.6.2.1.

The moment due to unbalanced load, usually as a result of live load or unequal spans, can be distributed to each of the column-strip and middle-strip locations according to ACI 318-14, Section 8.4. The average compressive stress and the unbalanced moment should be checked at each critical

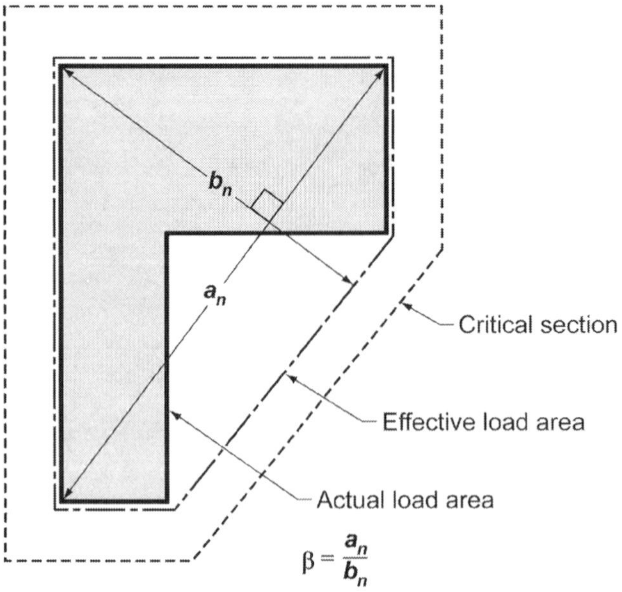

Fig. 4.8—Value of β for nonrectangular loaded area.

section to ensure they are less than the prescribed stresses in ACI 318-14, Section 22.4.2.3. Provide the minimum temperature reinforcement or nonprestressed reinforcement required to resist tensile stress as required in ACI 318-14, Section 8.6.2. Use nonprestressed reinforcement and unbonded post-tensioning to find the nominal flexural strength of the critical sections of the column strip and middle strip. Required flexural strength shall be the sum of factored load moments and secondary moments using a load factor of 1.

At columns of two-way prestressed, shear strength is given by

$$V_c = (\beta_p\lambda\sqrt{f_c'} + 0.3f_{pc})b_o d + V_p \quad \text{(in.-lb)}$$
$$V_c = (\beta_p\lambda\sqrt{f_c'}/12 + 0.3f_{pc})b_o d + V_p \quad \text{(SI)} \tag{4.9}$$

where β_p is the smaller of 3.5 and $(\alpha_s d/b_o + 1.5)$. The term V_c may be computed by Eq. (4.9) if the following conditions are met; otherwise, 4.8 applies:

a) No portion of the column cross section should be closer to a discontinuous edge than four times the slab thickness

b) The value of $\sqrt{f_c'}$ used in Eq. (4.9) should not be taken greater than 70 psi (5.8 MPa)

c) In each direction, f_{pc} should not be less than 125 psi (0.86 MPa) nor be taken greater than 500 psi (3.45 MPa).

CHAPTER 5—SLAB REINFORCEMENT AND DETAILING

5.1—General

The area of reinforcement in each direction for two-way slab systems is determined from factored moments at critical sections. The requirements for flexural reinforcement are intended to provide nominal moment strength for slab

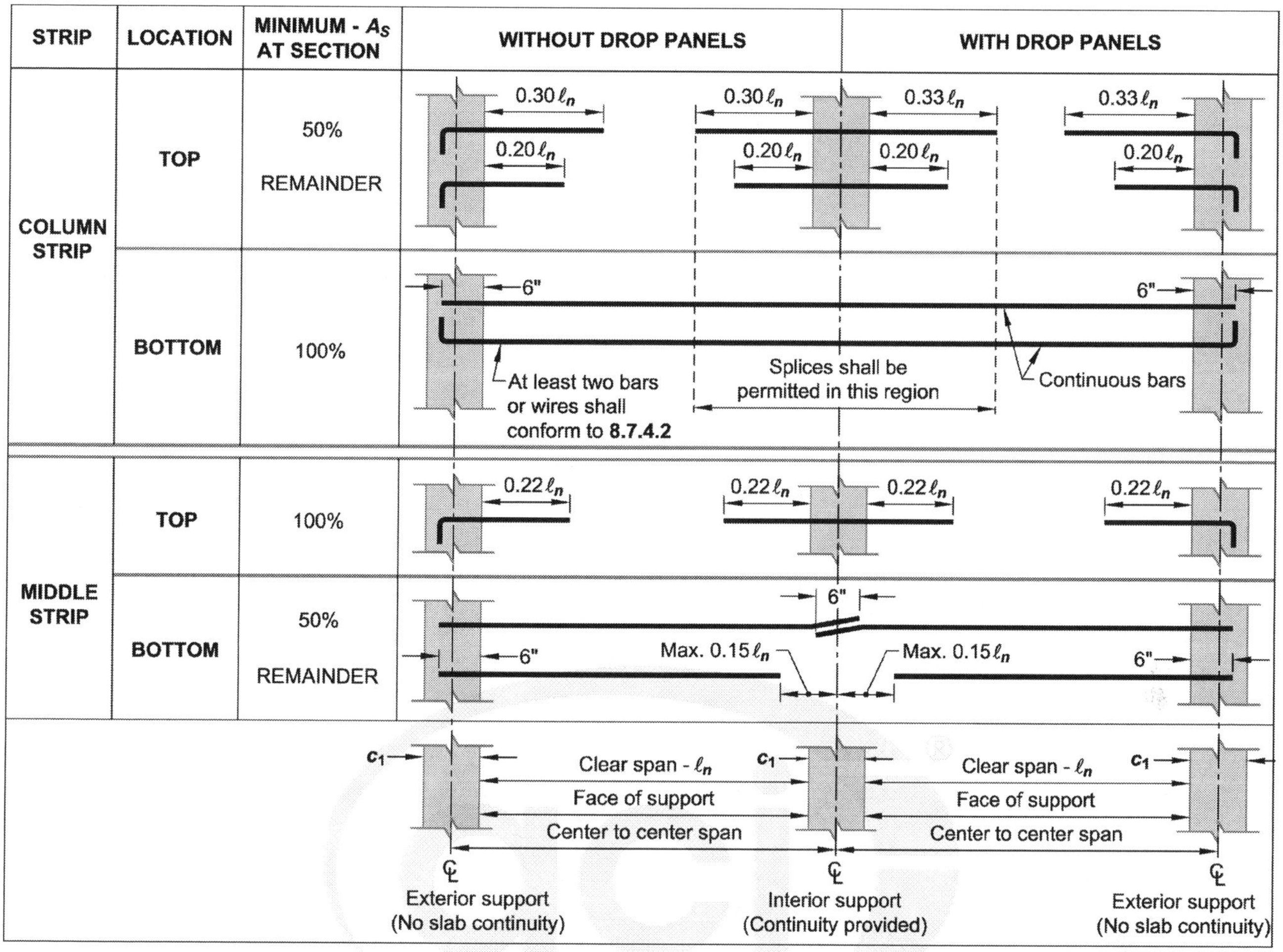

Fig. 5.2—Minimum extensions for deformed reinforcement in two-way slabs without beams (Fig. 8.7.4.1.3a, ACI 318-14).

action, reduce cracking, and provide for the possibility of loads concentrated on small areas of the slab.

5.1.1 *Minimum ratio of flexural reinforcement*—The minimum ratio of flexural reinforcement should be no less than the minimum shrinkage and temperature reinforcement and provided and distributed at the tension side of slabs in each direction. The ratio of area of reinforcement to gross area, b_h, of slab should not be less than the following:

(a) Where Grade 40 (280 MPa) or 50 (340 MPa) deformed bars are used: 0.0020

(b) Where Grade 60 (420 MPa) deformed bars or welded wire reinforcement are used: 0.0018

(c) Where reinforcement with f_y > 60,000 psi (420 MPa) (at 0.35 percent strain) are used: 0.0018 × 60,000/f_y (0.0018 × 420/f_y).

5.1.2 *Maximum reinforcement spacing*—Spacing of reinforcement required for flexure in solid slabs should not exceed 2*h*. Cracks start to generate at approximately 33 percent of the factored load and are generally wide, contributing to the reduction of the E_cI_g stiffness of the two-way slab or plate, and thereby leading to excessive deflections both short and long term. Serviceability control of cracking in two-way slabs and plates is best achieved when cracks follow the spacings of the orthogonal reinforcement rather than wide flexural crack patterns at early loading stages. A limit of 12 in. (300 mm) is recommended for maximum spacing of reinforcement in both orthogonal directions, as demonstrated in more than 100 large-scale two-way slab tests to failure (ACI 224R; Nawy 2001). The principle of using smaller-diameter bars at closer spacing rather than large-diameter bars at greater spacing tend to optimize the use of flexural reinforcement without the need for exceeding the total area of reinforcement required in design. At the same time, this helps to control cracking.

5.2—Slabs without beams

Minimum extension of reinforcement in slabs without beams is shown in Fig. 5.2. Where adjacent spans are unequal, extensions of negative moment reinforcement beyond the face of support should be based on requirements of the longer span.

5.3—Corner reinforcement

Unrestrained corners of two-way slabs tend to lift when loaded. If this lifting tendency is restrained by edge walls or beams, bending moments result in the slab. Provide reinforcement at the top and bottom exterior corners when α_f is greater than 1.0 to resist these moments and control cracking.

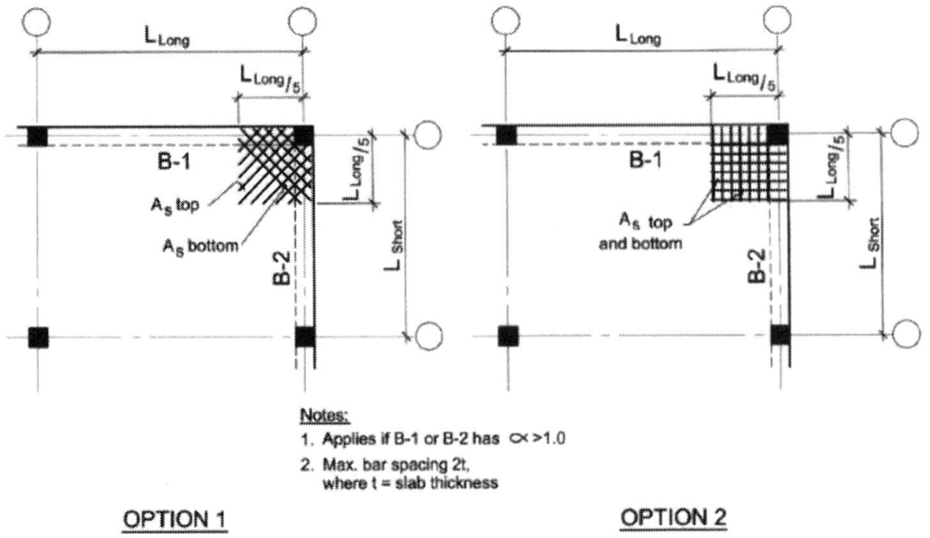

Fig. 5.3—Detail of corner reinforcement.

This reinforcement should be sufficient to resist a moment-per-unit width equal to the maximum positive moment-per-unit width in the slab at the exterior corners. The moment in the top of the slab may be assumed to act about an axis perpendicular to the diagonal from the corner. The moment in the bottom of the slab may be assumed to act about an axis parallel to the diagonal from the corner.

The corner reinforcement should be provided for a distance in each direction from the corner equal to one-fifth the longer span. The special reinforcement may be placed in a band parallel to the diagonal in the top of the slab and a band perpendicular to the diagonal in the bottom of the slab. Alternatively, reinforcement provided for flexure in the primary directions can be used if the special reinforcement is placed in two layers parallel to the sides of the slab in both the top and bottom of the slab (Fig. 5.3).

5.4—Slab with drop panel

When a drop panel is used to reduce the amount of negative moment reinforcement over the column of a flat slab, the thickness of the drop panel below the slab cannot be assumed to be greater than one-fourth the distance from the edge of the drop panel to the face of the column or column capital in computing the required slab reinforcement. When the drop panel extends less than one-sixth the span length from center-to-center of supports in each direction, the projection may be used as a shear cap to increase the shear strength of the slab but not the flexural strength. ACI 421.1R-08, Section 4.3.4, warns against specifying a small increase in slab thickness over a small area surrounding the column as a means to increase the critical perimeter and to reduce the shear stress under the required factored shear loads. For such conditions, potential shear cracks away from the shear cap and unbalanced moment should be investigated in detail.

5.5—Column strip reinforcement

All bottom bars or wires within the column strip in each direction should be continuous or spliced with Class B tension splices or with mechanical or welded splices satisfying ACI 318-14, Section 8.7.4.2.1. In general, splices should be located as shown in Fig. 5.2. At least two of the column strip bottom bars or tendons in each direction are required to pass within the column core and be anchored at exterior supports. The two continuous column strip bottom bars or tendons through the column are called integrity steel, and are provided to give the slab some residual strength following a single punching shear failure (Mitchell and Cook 1984).

Although ACI 318-14, Section 8.7.4.2 does not include an equation for the design of such integrity steel, reference may be made to CSA A23.3-04. The provision requires calculation of the area of steel passing through the column core and is given by Eq. (5.5)

$$\sum A_{sb} = 2V_{se}/f_y \quad (5.5)$$

When there is insufficient clearance for bottom bars to pass under the shearhead and through the column, such as in lift slab construction, bottom bars should pass through holes in the shearhead arms or within the perimeter of the lift collar.

5.6—Middle strip reinforcement

Middle strip bottom reinforcement may be stopped at a distance of $0.15\ell_n$ from the centerline of the interior support. Unless otherwise noted on construction document, middle strip bottom reinforcement should have a minimum extension of 6 in. (150 mm) from the face of the exterior support.

5.7—Bent bars

Bent bars were common in the earlier days of flat slabs to serve as surrogate shear reinforcement. In 1989, the provision for bent bars was removed from ACI 318-14, Section R8.7.4.1.3; however, bent bars are permitted when the depth-span ratio permits the use of bends of 45 degrees or

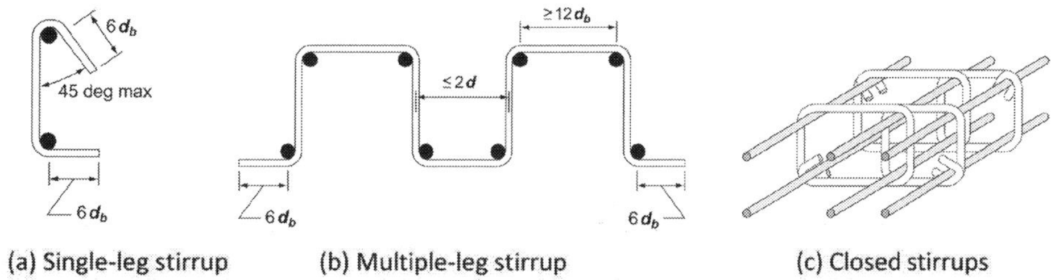

Fig. 5.8.1a—Single, multi-leg, or closed stirrup slab shear reinforcement.

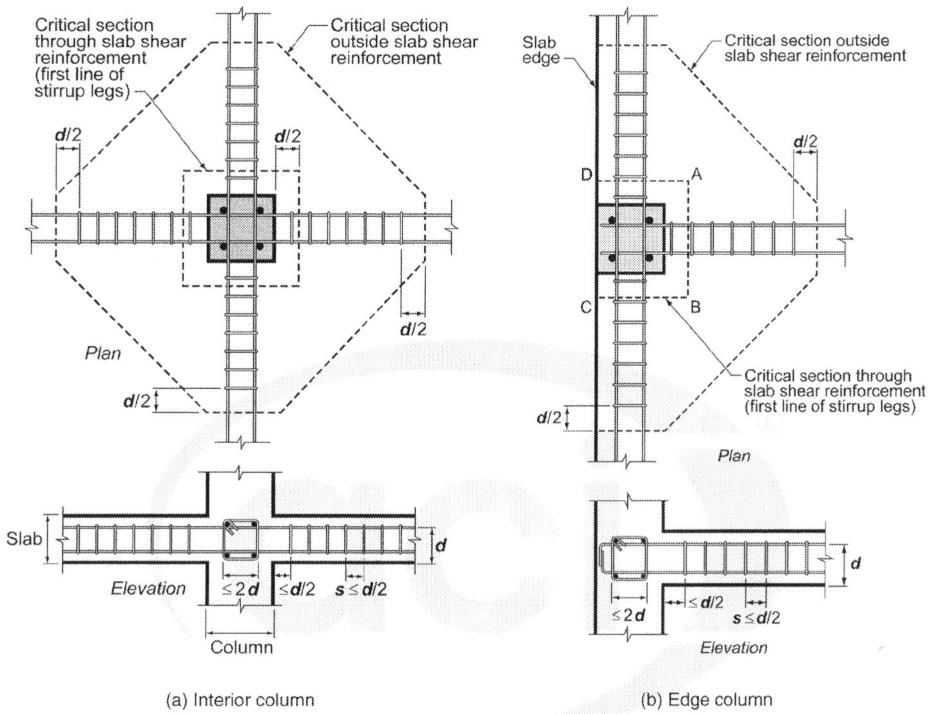

Fig. 5.8.1b—Arrangement of stirrup shear reinforcement.

less. In North America, this form of shear reinforcement in two-way slabs is rarely used.

5.8—Slab shear reinforcement

5.8.1 *Single or multiple leg stirrup type*—ACI 318-14, Section R8.7.6 sets out the principles of design for slab shear reinforcement and makes specific reference to stirrups, headed studs, and shearheads. Shear reinforcement consisting of properly anchored bars, wires and single- or multiple-leg stirrups, or closed stirrups can increase the punching shear resistance of slabs (Hawkins 1974). It is essential that shear reinforcement engage longitudinal reinforcement at both the top and bottom of the slab, as shown for typical details in Fig. 5.8.1a(a), (b), and (c). The minimum slab effective design depth d should not be less than 6 in. (300 mm) or 16 times the shear reinforcement diameter when such shear reinforcement is used. Anchorage of shear reinforcement, according to the requirements of ACI 318-14, Section 8.7.6.2, is difficult in slabs thinner than 10 in. (250 mm).

Shear reinforcement should be symmetrical about the centroid of the critical section when the unbalanced moment transfer is negligible for a slab-column connection. In accordance with ACI 318-14, Section 8.7.6.3, spacing limits for interior columns are illustrated in Fig. 5.8.1b(a). At edge columns or for interior connections where moment transfer is significant, closed stirrups are recommended in a pattern as symmetrical as possible. Although the average shear stresses on faces *AD* and *BC* of the exterior column in Fig. 5.8.1b(b) are lower than on face *AB*, the closed stirrups extending from faces *AD* and *BC* provide some torsional strength along the edge of the slab. The spacing limits shown correspond to slab shear reinforcement details that have been shown effective.

5.8.2 *Headed shear stud reinforcement*—Tests have established that punching shear in slabs can be effectively resisted by reinforcement consisting of vertical rods mechanically anchored at the top and bottom of slabs (ACI 421.1R). All types of mechanically anchored shear reinforcement are referred to as shear studs or studs. To be fully effective, the anchorage should be capable of developing the specified yield strength of the studs. The mechanical anchorage can be obtained by heads or strips connected to the studs

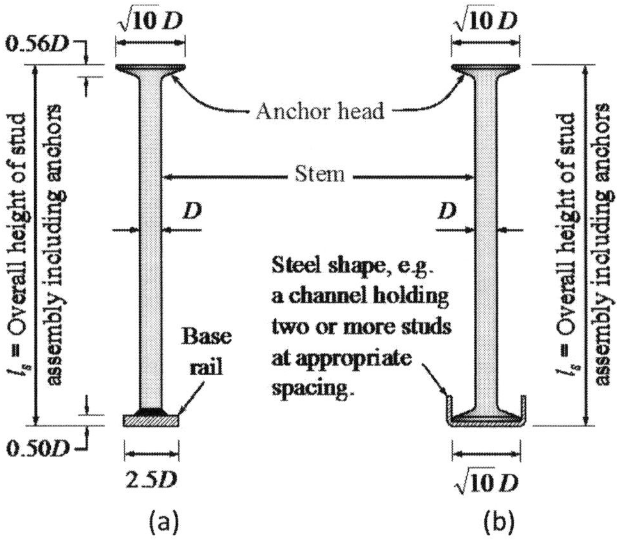

Fig. 5.8.2a—Stud assemblies: (a) single-headed studs welded to a base rail; and (b) double-headed studs crimped into a steel channel.

by welding. The heads can also be formed by forging the stud ends. Stud assemblies consisting of either a single-head stud attached to a steel base rail by welding is shown in Fig. 5.8.2a(a); a double-headed stud mechanically crimped into a nonstructural steel channel is shown in Fig. 5.8.2a(b). Mechanical properties of studs are specified in ASTM A1044/A1044M. Figure 5.8.2b is a top view of a slab that shows a typical arrangement of shear reinforcement (stirrup legs or studs) in the vicinity of an interior column. The shear reinforcement should be arranged on straight lines perpendicular to the column faces at the column corners. ACI 318 requires that the spacing g measured parallel to the column faces between corner lines shall not exceed $2d$. For optimum effectiveness, the overall height of the studs should be as great as possible. Ideally, the heads or the rail should have the minimum cover required for protection.

For full effectiveness, the head area of the studs has to be at least 10 times the area of the stem. This permits the design of headed stud shear reinforcement to be based on the specified yield strength f_{yt}. For headed stud shear reinforcement, ACI 318-14, Section 22.6.6.1 requires:

$$V_c = 3\lambda\sqrt{f_c'}b_o d \text{ and } V_n \leq 8\lambda\sqrt{f_c'}b_o d \quad \text{(in.-lb)}$$

$$V_c = \lambda\sqrt{f_c'}b_o d/4 \text{ and } V_n \leq 2\lambda\sqrt{f_c'}b_o d/3 \quad \text{(SI)}$$

When $v_u > 6\phi\sqrt{f_c'}$, $s \leq d/2$ ($v_u > 0.5\phi\sqrt{f_c'}$)
When $v_u \leq 6\phi\sqrt{f_c'}$, $s \leq 0.75d$ ($v_u \leq 0.5\phi\sqrt{f_c'}$)
For stirrups, ACI 318-14, Section 22.6.6.1 requires:

$$V_c = 2\lambda\sqrt{f_c'}b_o d \text{ and } V_n \leq 6\lambda\sqrt{f_c'}b_o d \quad \text{(in.-lb)}$$

$$V_c = 0.17\lambda\sqrt{f_c'}b_o d \text{ and } V_n \leq 0.5\lambda\sqrt{f_c'}b_o d \quad \text{(SI)}$$

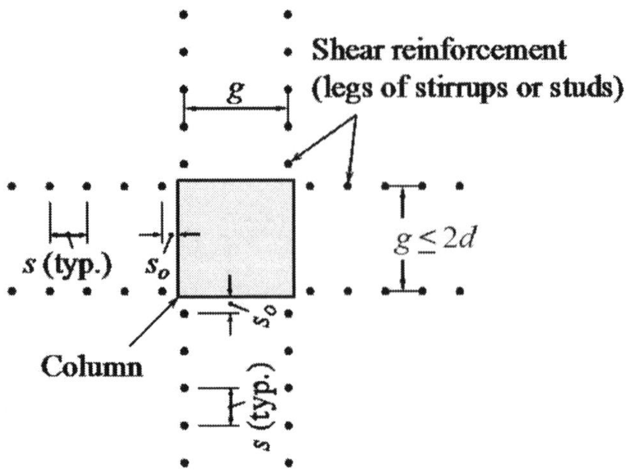

Fig. 5.8.2b—Top view of flat plate showing shear reinforcement arrangement in vicinity of interior column.

$s \leq d/2$.

Refer to ACI 421.1R for detailed information on the application of headed shear studs.

5.9—Post-tensioned slabs

Until the mid-1970s, the placement of unbonded post-tensioning strands in flat plates and flat slabs was done using the patterns of the placement of reinforcement in reinforced concrete flat plates and flat slabs. This technique was labor intensive. Banded placement of strands developed as a result, which was much more cost-effective. In one direction, almost all the strands are banded within a small width (approximately $L/6$ distance) and are placed with a given drape. In the opposite direction, the strands are placed with uniform spacing, usually placed above the banded strands. The entire placement of strands looks more like a one-way slab in one direction and a wide beam, which are the banded strands, in the other direction. This placement results in equivalent strength as the previous placement. Banded placement of strands is the predominant practice in the United States today. The use of banded slabs also enhances the use of slab-column frames, as described in Chapter 7. It is important that at least two tendons pass through the column cores in the uniformly distributed tendon direction, and that these are anchored as near the slab edge as possible. These two tendons passing the column cores also serve as integrity reinforcement to meet the requirements of ACI 318-14, Section 8.7.5.6.

5.10—Bonded reinforcement in post-tensioned slabs

For two-way post-tensioned flat slab systems, the requirements for minimum area and distribution of bonded reinforcement in the positive and negative moment areas over supporting columns are described as follows.

In the positive moment areas where the extreme fiber stress in tension in the precompressed tensile zone at service loads does not exceed $2\sqrt{f_c'}$ after allowance for all prestressed losses, bonded reinforcement is not required.

Table 6.1.1—Minimum thickness of slabs

	Without drop panels			With drop panels		
	Exterior panels			Exterior panels		
f_y, psi (MPa)	Without edge beams	With edge beams	Interior panels	Without edge beams	With edge beams	Interior panels
40,000 (280)	$\ell_n/33$	$\ell_n/36$	$\ell_n/36$	$\ell_n/36$	$\ell_n/40$	$\ell_n/40$
60,000 (420)	$\ell_n/30$	$\ell_n/33$	$\ell_n/33$	$\ell_n/33$	$\ell_n/36$	$\ell_n/36$
75,000 (520)	$\ell_n/28$	$\ell_n/31$	$\ell_n/31$	$\ell_n/31$	$\ell_n/34$	$\ell_n/34$

Note: For f_y between the values given, minimum thickness should be determined by linear interpolation. For slabs with beams along exterior edges, the value of α_f for the edge beam should not be less than 0.8.

When the computed tensile stress in concrete at service load exceeds $2\sqrt{f_c'}$, the minimum area of bonded reinforcement that should be provided is

$$A_s = \frac{N_c}{0.5f_y} \quad (5.10a)$$

where the value of f_y should not exceed 60,000 psi (420 MPa). Uniformly distribute bonded reinforcement across the precompressed tensile zone. Length of bonded reinforcement should be one-third the clear span length ℓ_n and centered in the positive moment area.

In the negative moment area over column supports, the minimum area of bonded reinforcement, A_s, required in the top of the slab in each direction is

$$A_s = 0.00075A_{cf} \quad (5.10b)$$

Distribute bonded reinforcement between lines that are 1.5h outside opposite faces of the column support. Provide at least four bars or wires in each direction. Reinforcement spacing should not exceed 12 in. (300 mm). Bonded reinforcement should extend one-sixth of the clear span ℓ_n on each side of support.

CHAPTER 6—SERVICEABILITY CONSIDERATIONS

6.1—Minimum slab thickness

Serviceability issues should be considered in the design of two-way flat slabs. These include determination of minimum slab thickness and immediate and long-term deflections. Long-to-short span ratio for two-way slabs is limited to 2-to-1. For slabs using nonprestressed reinforcement, the minimum slab thickness should conform to the prescriptive guideline based on 6.1.1 and 6.1.2. For deviation from the prescriptive guideline, slab deflection should be investigated based on 6.2.

6.1.1 *Slabs without interior beams spanning between supports*—For slabs without interior beams spanning between supports, the recommended minimum slab thicknesses are given in Table 6.1.1 as a function of ℓ_n, which is the length of clear span in the long direction measured face-to-face of supports in slabs without beams, and face-to-face of beams or other supports in other cases. For slabs without drop panels, minimum thickness should be at least 5 in. (125 mm). For slabs with drop panels, minimum thickness should be at least 4 in. (100 mm).

The minimum thicknesses described above have been developed through engineering and observational experience. Slabs conforming to these limits have not resulted in systematic problems related to stiffness for short- and long-term loads. Consider range of loads, material types and properties, boundary conditions, and environmental effects in determining the slab thickness through computational analysis as prescribed in ACI 435R-95. Slab deflection due to sustained gravity loads for heavily loaded slabs and for light frame construction supported on a podium flat plate or flat slab should be analyzed in accordance with 6.2.

6.1.2 *Slabs with beams spanning between supports on all sides*—For slabs with beams spanning between the supports on all sides, the minimum thickness h should meet the following requirements:

(a) For an average value of beam flexural stiffness α_{fm} equal to or less than 0.2, the provisions of 6.1.1 should apply.

(b) For an average value of beam flexural stiffness α_{fm} greater than 0.2 but not greater than 2.0, h should not be less than

$$h = \frac{\ell_n\left(0.8 + \frac{f_y}{200{,}000}\right)}{36 + 5\beta(\alpha_{fm} - 0.2)} \quad \text{(in.-lb)}$$

$$h = \frac{\ell_n\left(0.8 + \frac{f_y}{1400}\right)}{36 + 5\beta(\alpha_{fm} - 0.2)} \quad \text{(SI)} \quad (6.1.2a)$$

and not less than 5 in. (125 mm).

(c) For an average value of beam flexural stiffness α_{fm} greater than 2.0, h should not be less than

$$h = \frac{\ell_n\left(0.8 + \frac{f_y}{200{,}000}\right)}{36 + 9\beta} \quad \text{(in.-lb)}$$

$$h = \frac{\ell_n\left(0.8 + \frac{f_y}{1400}\right)}{36 + 9\beta} \quad \text{(SI)} \quad (6.1.2b)$$

and not less than 3.5 in. (90 mm).

(d) At discontinuous edges, provide edge beam with a stiffness ratio α_f not less than 0.80, or increase the minimum thickness required by Eq. (6.1.2a) or (6.1.2b) by at least 10 percent in the panel with a discontinuous edge.

The value of ℓ_n in (b) and (c) is the length of the clear span in the long direction measured face-to-face of the beams. The term β in (b) and (c) is the ratio of clear spans in the long-to-short direction of the slab.

For panels having a ratio of long-to-short span greater than 2, the use of Eq. (6.1.2a) and (6.1.2b), which express the minimum thickness as a fraction of the long span, may give unreasonable results. Use the rules for one-way construction for these.

6.2—Deflection analysis

Slab thickness less than the minimum required by 6.1.1 and 6.1.2 is permitted where computed deflections do not exceed the maximum permissible computed deflections limits of ACI 318-14, Section 8.3.2.1 and Table 24.2.2. When computing deflections, consider the size and shape of the panel, conditions of support, nature of restraints at the panel edges, and state of cracking.

The calculation of deflections for slabs is complicated even when linear elastic behavior can be assumed. For immediate deflections of nonprestressed slabs, the values of E_cI and I_e specified in ACI 318-14, Section 24.2.3 can be used. For prestress slabs, I_g may be used when the calculated tensile stress is less than the modulus of rupture. However, other procedures and values of the stiffness E_cI may be used if they result in predictions of deflection in reasonable agreement with the results of comprehensive tests. For lightly reinforced sections, which are common in slab positive moment regions, the I_e calculated based on ACI 318-14, 24.2.3.5 for beams overestimates the slab stiffness. For a slab with minimum reinforcement, the fully cracked section moment of inertia, I_{cr}, is approximately one-tenth the gross section moment of inertia, I_g. For typical beams, the ratio of I_{cr} is one-fourth to one-third the effective moment of inertia, I_e. Tests conducted in Canada showed that the traditional Branson's equation, ACI 318-14, Eq. (24.2.3.5a) for the calculation of I_e, underestimates the deflection of slabs (Bischoff and Scanlon 2008).

ACI 435R presents three approaches in calculating the immediate deflection: 1) classical solution; 2) simplified crossing beam analogies; and 3) finite elements. Under the crossing beam analogy, deflection of two-way slabs can be approximated by considering the column and middle strips in each of the orthogonal directions. The total static moment given in Eq. (3.1.3) is divided among column strips and middle strips using methods described under 3.2.1 or 3.2.2. Deflection can then be calculated in accordance with ACI 318-14, Sections 24.2.3.5 and 24.2.3.6.

6.2.1 *Long-term deflection*—Because available data on long-term deflections of slabs are too limited to justify more elaborate procedures, the additional long-term deflection for two-way construction can be computed using the multipliers given in ACI 318-14, Section 24.2.4.1.1. Engineers are cautioned that the long-term multipliers, which work reasonably well for beams, can be much too small for lightly reinforced sections or when the positive reinforcement ratio approaches minimum reinforcement ratio. A major factor in the deflection performance of slabs and plates as compared to beams is due to the impact of their relative stiffness effect on their cracking development. Uncracked sections subjected to initial loads produce tensile stresses approaching the modulus of rupture. Thereafter, a section is partially cracked and becomes fully cracked under long-term sustained loads with increased comparative deflection behavior in the slabs. Note also that the residual deflection of concrete slabs depends on the sequence of construction, shoring and reshoring, and proper curing techniques during construction. To preserve the integrity of the structural slab and plate, serviceability requirements through control of cracking and deflection in the applicable provisions of ACI 435R and ACI 224R should be used.

Limited data exist on long-term deflection of slabs and plates so as to justify elaborate computational procedures, including accounting for factors affecting time-dependent deflections. An alternative method is the use of ACI 318-14, Section 24.2.4.1.1. However, understand that the long-term multipliers, while working reasonably well for beams, can be too small for lightly-reinforced sections or when the positive reinforcement ratio approaches the minimum reinforcement ratio.

6.2.2 *Post-tensioned slabs*—Unlike nonprestressed concrete flat plates or flat slabs, many variations are possible in unbonded post-tensioned flat plates/slabs. Various strand placement patterns are possible, although most slabs are built with banded tendon arrangements. Various strand profiles are used, defined by the positive and negative eccentricities and points of contraflexure of the tendons. Treating discontinuous ends and strand terminations at slab openings requires careful consideration. The minimum average effective compressive stress is 125 psi (0.9 MPa) according to ACI 318-14, Section 8.6.2.1, but higher values may be required for specific cases. Although deflection is not a dominant problem, deflection calculations are required by ACI 318, Sections 24.2.3.8 and 24.2.3.9, for post-tensioned slabs. Slabs are generally thinner and spans are longer than nonprestressed concrete slabs; slab thicknesses of $L/45$ are common. Because thinner slabs are more flexible, they may be more susceptible to issues of floor vibration.

6.3—Crack control in reinforced two-way action structural slabs and plates

Crack control in concrete slabs at service load levels is as important as controlling deflection. It is closely connected to reducing sectional stiffness, upholding the integrity of the structure, retaining reinforcement corrosion, and preventing other detrimental effects that lead to ultimate loss of serviceability. Microcracking in concrete starts at an early load level of approximately 10 percent of the service load, which eventually leads to visible macrocracks.

Design equations for beams underestimate the crack width developed in two-way slabs and plates, and do not educate

Table 6.3—Guide to tolerable flexural crack widths for reinforced concrete under service loads*

Exposure condition	Crack width, in. (mm)
Dry air or protective membrane	0.016 (0.41)
Humidity, moist air, soil	0.012 (0.30)
Deicing chemicals	0.007 (0.18)
Seawater and seawater spray, wetting and drying	0.006 (0.15)
Water-retaining structures†	0.004 (0.10)

*Nawy (1968) and ACI 224R-01, Table 4.1.

†Excludes nonpressure pipes.

Note: Expect a portion of the cracks in the structure to exceed these values. With time, a significant portion could exceed these values. These are general guidelines for design to be used in conjunction with sound engineering judgment.

the design engineer on how to space reinforcement. Extensive research and tests to failure of more than 100 two-way plates have shown that cracks in such structural members are controlled primarily by the steel stress level and the spacing of reinforcement in the two perpendicular directions. In addition, the clear concrete cover in two-way slabs and plates is nearly constant (20 mm [3/4 in.] for most interior slabs), whereas it is a major variable in crack control equations for beams (Nawy and Blair 1971; Nawy 2011; ACI 224R).

The yield-line cracks described in 3.4 are generally wide and almost fully developed at approximately 30 percent of the service load level. To delay their development until the load reaches the expected nominal flexural strength, design closely-spaced reinforcement in the two orthogonal directions so the crack pattern in the two-way slab or plate at service and low overload levels is an image of the orthogonal reinforcement. The crack widths in such designs will, therefore, be controlled within the tolerable limits for various environmental conditions, as shown in Table 6.3.

When selecting reinforcement spacing for a two-way floor system, specify flexural reinforcement in the N-S and E-W directions for the same volume of steel using smaller-diameter bars at closer spacing. It is significantly more effective in reducing the crack width to an accepted level through the proper selection of the reinforcement grid and by placing closest to the tensile face of the slab. The crack width expression in terms of grid spacing, reinforcement percentage ratio, bar diameter, and magnitude of concrete cover as presented in Section 4.3 of ACI 224R (Eq. (4-15) and (4-16)) should facilitate the choice of proper proportioning of the structural slab with the choice of the appropriate size and spacing of the bars in the orthogonal directions for effective crack control. A 12 in. (300 mm) maximum spacing of the steel bar reinforcement grid is recommended to prevent the formation of detrimental wide yield-line cracks until the ultimate load is reached.

CHAPTER 7—DESIGN OF SLAB-COLUMN FRAMES UNDER LATERAL FORCES

7.1—General

This section addresses two-way slab systems without beams only, or flat plates, and recommends analysis, design, and reinforcement detailing methods for slab-column frames under combined effects of lateral and gravity loads. Slab-column frames could be used as part of an ordinary or intermediate seismic-force-resisting system, although they are usually only used to resist wind loads in combination with gravity loads. A slab-column structure acting as the seismic-force-resisting system is likely to be far too flexible for higher seismic design categories (SDCs) and would not deliver performance consistent with other requirements.

For earthquake ground motions, slab-column framing systems designed according to Chapters 1 through 6 can be used in ordinary moment frames, and are appropriate for SDC A or B. If a slab-column frame meets additional requirements described in this chapter, it may be considered as an intermediate moment frame and is allowed for structures assigned to SDC C (ASCE 7-10, Section 12.2). Additional requirements are related to the distribution of slab moments, arrangement of slab reinforcement, and punching-shear-related issues.

For structures assigned to SDC D, E, or F, slab-column frames without beams are generally not permitted as part of seismic-force-resisting systems (ACI 318-14, Section R18.2), with limited exceptions under ASCE 7-10, Section 12.2. For example, one exception would be intermediate moment frames in a dual system with special reinforced concrete shear walls less than 100 ft (30.5 m) in height. The reasoning behind this is the slab-column frames cannot be detailed for the level of energy dissipation and ductility demanded for special moment frames. In the case of ductility demand, this chapter describes punching-shear-related recommendations for slab-column connections, along with the design story drift estimated for the seismic-force-resisting system.

Under wind loads, all structural members are typically expected to behave essentially within their elastic range of response. Therefore, while the design forces induced by wind in slab-column frames may be determined based on similar analysis procedures that are used for the earthquake-resistant design (7.2), other design and detailing recommendations intended for ductility and redundancy of frames under earthquake ground motions do not apply for wind design (7.3 through 7.5).

7.2—Analysis of slab-column frames under lateral forces

For lateral forces, analysis of frames should consider effects of cracking and reinforcement on stiffness of frame members (7.2.2).

7.2.1 *Effective slab width model*—An equivalent slab-beam is a flexural member having a rectangular section with its width and depth dimensions equal to the effective slab width and slab thickness. Previous research (Vanderbilt and Corley 1983; Grossman 1997; Hwang and Moehle 2000; Dovich and Wight 2005) proposed various methods for determining the effective slab width in consideration of slab-column connection geometry and expected drift ratio under serviceability and ultimate limit states.

In the modified equation for b_e by Grossman (1997), clear span lengths ℓ_n are used in place of center-to-center span lengths ℓ_1, with the assumption that slab-column joints are rigid. The effective slab width can be expressed as

$$b_e = \left[0.3\ell_n + c_1 x + \frac{(c_2 - c_1)}{2}\right]\left(\frac{d}{0.9h}\right)K_{FP} \quad (7.2.1)$$

where χ—the ratio of span ℓ_2/ℓ_1—should not be taken greater than 1.0, and K_{FP}, which is modification factor accounting for joint confinement, is equal to 0.8 and 0.6 for edge and corner connections, respectively. Here, ℓ_1 is taken as the average of the lengths of the two spans in front and back of the column, and ℓ_2 is the average of the lengths of the two transverse spans at the sides of the column, where both ℓ_1 and ℓ_2 are measured center-to-center of supports parallel and perpendicular to lateral loading, respectively. For exterior or corner connections with the slab edge parallel to the direction of lateral loading, the effective slab width calculated by Eq. (7.2.1) is adjusted by multiplying by $(\ell_3 + \ell_2/2)/\ell_2$, where ℓ_3 is the distance from column centerline to edge of slab. The width of an equivalent slab-beam supported by two adjacent columns is then taken equal to the average of the two values determined by Eq. (7.2.1) at the supports. More detailed model descriptions are provided in the paper by Grossman (1997). Different stiffness degradation levels of flat plates at various drifts were also proposed based on the tests by Hwang and Moehle (2000).

Equation (7.2.1) is intended for slab-column frames subjected to service wind or earthquake-induced forces, reduced by the response reduction factor R, which are expected to cause drift levels of approximately 0.25 percent. Therefore, using this model for service level analysis is considered appropriate. For strength-based analysis, flexural stiffness of equivalent slab-beams was considered to be reduced by 30 percent on average due to more cracking in the slab (Grossman 1997). Also, the effective slab width estimated using Eq. (7.2.1) should be modified when slab openings exist near supports.

7.2.2 *Slab stiffness reduction*—The analysis of slab-column frames under lateral forces should consider effects of cracking and reinforcement on the stiffness of frame members. Cracking reduces the stiffness of slab members and increases lateral flexibility when lateral forces act on the structure. The selection of appropriate effective stiffness values for reinforced concrete frame members has dual purposes to: 1) provide realistic estimates of lateral deflections; and 2) determine distribution of forces and moments on the frame members. A detailed nonlinear analysis of the structure would adequately capture these two effects. An approximate method to estimate an equivalent nonlinear lateral deflection using linear analysis is to reduce the modeled stiffness of the concrete members in the structure. One reasonable option that considers the reduced stiffness of the elements is to calculate the secant stiffness value to the point of yielding of reinforcement for the member, or the secant value to a point before yielding of the reinforcement if analysis demonstrates yielding is not expected for the given loading condition. When the analysis is used to determine design drifts or moment magnifications caused by wind or earthquake-induced forces, lower-bound slab stiffnesses should be assumed. When the analysis is used to study interactions of slabs with other framing elements, such as structural walls, it is appropriate to consider a range of slab stiffnesses so that the relative importance of slabs on those interactions can be assessed. For nonprestressed slabs, it is normally appropriate to reduce slab bending stiffness to between one-half and one-fourth of uncracked stiffness values based on gross section properties or based on limits prescribed in ACI 318-14, Section 6.6.3.1.1. For prestressed construction, stiffnesses greater than those of cracked, nonprestressed slabs may be appropriate.

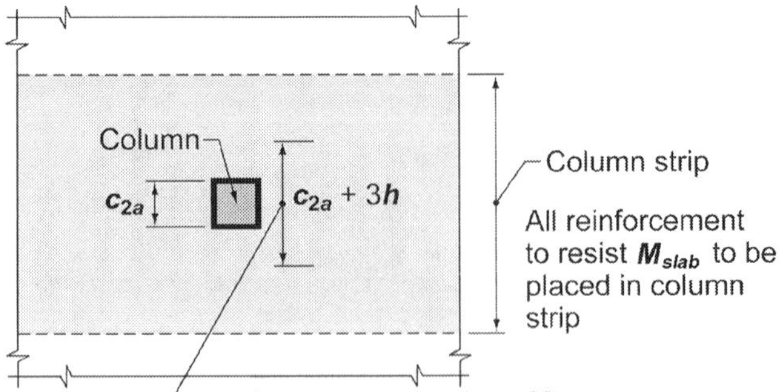

Fig. 7.2.1—Location of reinforcement in slabs.

7.3—Arrangement of reinforcement in slabs for intermediate moment frames

This section details the arrangement of slab reinforcement for two-way slab-column framing systems subjected to earthquake-induced forces and used as intermediate moment frames. Application of the provisions for two-way slabs without beam is illustrated in Fig. 7.2.1, 7.3a, and 7.3b.

7.3.1 *Slab reinforcement at support*—The factored slab moments at supports, including earthquake effects, E, are determined for load combinations $1.2D + 1.0E + 1.0L + 0.2S$ and $0.9D + 1.0E$. Using these load combinations may result in moments requiring top and bottom reinforcement at the supports. Reinforcement provided to resist M_{sc} should be placed within the column strip. The moment M_{sc} refers (for a given design load combination with E acting in one horizontal direction) to that portion of the factored slab moment that is balanced by the supporting members at a joint. It is not necessarily equal to the total design moment at a support for a load combination including earthquake effect.

As described in 4.7, only a fraction of the moment M_{sc} is assigned to the slab effective width. Reinforcement placed within the effective width should resist $\gamma_f M_{sc}$. The effective slab width for exterior and corner connections should not extend beyond the column face a distance greater than c_t measured perpendicular to the slab span. For edge and

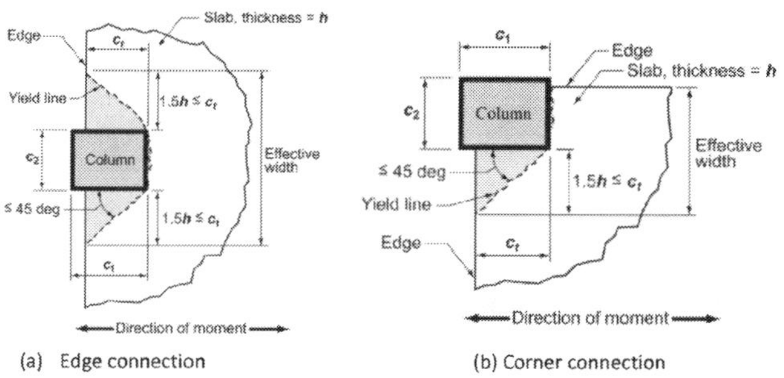

Fig. 7.3a—Effective width for reinforcement placement in edge and corner connections.

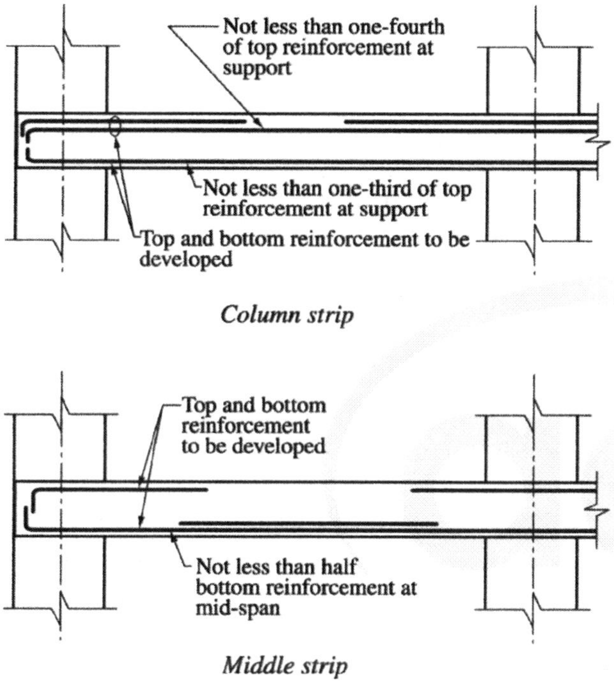

Fig. 7.3b—Arrangement of reinforcement in slabs.

corner connections, flexural reinforcement perpendicular to the edge is not considered fully effective unless it is placed within the effective slab width. At least one-half of the reinforcement in the column strip at the support should be placed within the effective slab width given in 4.7.

7.3.2 *Reinforcement continuity*—At least one-fourth of the top reinforcement at the support in the column strip should be continuous throughout the span. Also, continuous bottom reinforcement in the column strip should be at least one-third of the top reinforcement at the support in the column strip. At least one-half of all bottom middle strip reinforcement and all bottom column strip reinforcement at midspan needs to be continuous and should develop f_y at face of support. At discontinuous edges of the slab, all top and bottom reinforcement at support should be developed at the face of the support (Fig. 7.3b).

7.3.3 *Limit for factored gravity shear at support*—At the critical sections for columns defined in 4.5, two-way shear caused by factored gravity loads should not exceed $0.4\phi V_c$, where V_c should be calculated as defined in 4.8 for nonprestressed slabs and in 4.9 for prestressed slabs. This can be waived if the slab design follows 7.4.

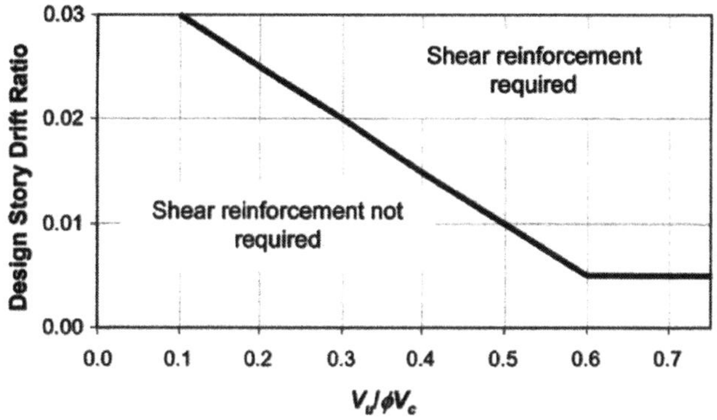

Fig. 7.4—Illustration of slab-column criterion.

7.4—Slab-column frames not designated as part of the seismic-force-resisting system

This section applies only to structures assigned to seismic design category (SDC) D, E, or F. According to ASCE 7-10, Section 12.12.5, all structural members not designated as a part of the seismic-force-resisting system should be designed to support gravity loads while subjected to design displacement. The principle behind this provision is to allow flexural yielding of slabs under design displacement, and to provide sufficient shear strength in slabs that yield so that the slabs continue to support gravity loads.

For slab-column connections of two-way slabs without beams in structures assigned to SDC D, E, or F, slab shear reinforcement (5.8.1 and 5.8.2), providing V_s not less than $3.5\sqrt{f_c'}\,b_o d$ (psi) ($0.29\sqrt{f_c'}\,b_o d$ [MPa]), should extend at least four times the slab thickness from the face of the support, unless:

a) The information in 4.7 using the design shear V_{ug} and the induced moment transferred between the slab and column under the design displacement is used

b) The design story drift ratio does not exceed the larger of 0.005 and $(0.035 - 0.05(V_{ug}/\phi V_c))$ (Fig. 7.4).

The design story drift ratio should be taken as the larger of the design story drift ratios of the adjacent stories above and below the slab-column connection. The value of V_{ug} is calculated for the load combination $1.2D + 1.0L + 0.2S$. The load factor on the live load, L, needs to be reduced to 0.5 except for garages, areas occupied as places of public assembly, and all areas where L is greater than 100 lb/ft^2 (5.0 kPa).

The induced moment is calculated to occur at the slab-column connection when subjected to design displacement. Effects of cracking and reinforcement on stiffness of frame members should be considered in analysis of slab-column frames in resisting seismic forces.

Condition b) does not require calculation of induced moments, and is based on research that identifies the likelihood of punching shear failure considering the story drift ratio and shear due to gravity loads. Figure 7.4 illustrates condition b). This can be accomplished by adding slab shear reinforcement, increasing slab thickness, changing the design to reduce the design story drift ratio, or a combination of these. If column capitals, drop panels, shear caps, or other changes in slab thickness are used, the guidelines in this section need to be evaluated at all potential critical sections, as described in 4.4.

The optional design permitted by ACI 318-14, Section 18.14.5.1 is concerned with the ductility (design story drift); it does not calculate shear forces or unbalanced moments associated with design story drift and, therefore, can permit connections without verifying that they possess the strength required in other sections of the Code. For this reason, the design option for condition a) is recommended because it verifies adequate strength. It is suggested that minimum shear reinforcement should be specified, including the amount of reinforcement and the zone where it should be placed, to ensure adequate ductility is provided.

7.5—Transfer of moments to column

Brittle punching failure can occur due to the transfer of shear forces combined with unbalanced moments between slabs and columns. During an earthquake, significant horizontal displacement of a flat plate-column connection may occur, resulting in unbalanced moments that induce additional slab shear. The displacement-induced unbalanced moments and resulting shear forces at flat plate-column connections should be considered in design to prevent brittle punching shear failure. Even when an independent lateral-force-resisting system is provided, flat plate-column connections should be designed to accommodate the moments and shear forces associated with the displacements during earthquakes (ACI 421.2R). Consideration of detailing ensures ductile behavior of building system.

CHAPTER 8—REFERENCES

ACI Committee documents and documents published by other organizations are listed first by document number, full title, and year of publication followed by authored documents listed alphabetically.

American Concrete Institute

224R-01(08)—Control of Cracking in Concrete Structures

318-11—Building Code Requirements for Structural Concrete and Commentary

318-14—Building Code Requirements for Structural Concrete and Commentary

318-77—Building Code Requirements for Reinforced Concrete and Commentary

318-95—Building Code Requirements for Structural Concrete and Commentary

421.1R-08—Guide to Shear Reinforcement for Slabs

421.2R-10—Guide to Seismic Design of Punching Shear Reinforcement in Flat Plates

423.3R-05—Recommendations for Concrete Members Prestressed with Unbonded Tendons

435R-95(00)—Control of Deflection in Concrete Structures (Appendix B added 2003)

American Society of Civil Engineers

ASCE 7-10—Minimum Design Loads for Buildings and Other Structures

ASTM International

A1044/A1044M-05(2015)—Standard Specification for Steel Stud Assemblies for Shear Reinforcement of Concrete

Canadian Standards Association

CSA A23.3-04(2010)—Design of Concrete Structures

Authored documents

Bischoff, P., and Scanlon, A., 2008, "Shrinkage Restraint and Loading History Effects on Deflections of Flexural Members," *ACI Structural Journal*, V. 105, No. 4, July, pp. 498-506.

Cohen, E., and Heun, R. D., 1979, "100 Years of Concrete Building Construction in the United States," *Concrete International*, V. 1, No. 3, Mar., pp. 38-46.

Corley, W. G., and Jirsa, J., 1970, "Equivalent Frame Analysis for Slab Design," *ACI Journal Proceedings*, V. 67, No. 11, Dec., pp. 875-884.

Corley, W. G.; Sozen, M. A.; and Siess, C. P., 1961, "The Equivalent Frame Analysis for Reinforced Concrete Slabs," *Structural Research Series* No. 218, Civil Engineering Studies, University of Illinois, Urbana, IL, June, 166 pp.

Dovich, L. M., and Wight, J. K., 2005, "Effective Slab Width Model for Seismic Analysis of Flat Slab Frames," *ACI Structural Journal*, V. 102, No. 6, Nov.-Dec., pp. 868-875.

Gamble, W. L., 1972, "Moments in Beam Supported Slabs," *ACI Journal Proceedings*, V. 69, No. 3, Mar., pp. 149-157.

Gamble, W. L.; Hatcher, D. S.; and Wheeler, W. H., 1964, discussion of "Investigation of Multi-Panel Reinforced Concrete Floor Slabs: Design Methods—Their Evolution and Comparison," by M. A. Sozen and C. P. Siess, *ACI Journal Proceedings*, V. 61, Mar, pp. 1965-1978.

Gayed, R. B., and Ghali, A., 2008, "Unbalanced Moment Resistance in Slab-Column Joints: Analytical Assessment," *Journal of Structural Engineering*, V. 134, No. 5, May, pp. 859-864. doi: 10.1061/(ASCE)0733-9445(2008)134:5(859)

Gentry, T., R., 1986, "The Use of Elastic Finite Elements in the Design of Reinforced Concrete Flat Plates," Master Thesis, Georgia Institute of Technology, Aug.

Grossman, J. S., 1997, "Verification of Proposed Design Methodologies for Effective Width of Slabs in Slab-Column Frames," *ACI Structural Journal*, V. 94, No. 2, Mar.-Apr., pp. 181-196.

Hanson, N. W., and Hanson, J. M., 1968, "Shear and Moment Transfer between Concrete Slabs and Columns," *Journal*, PCA Research and Development Laboratories, V. 10, No. 1, Jan., 16 pp.

Hawkins, N., 1974, "Shear Strength of Slabs with Shear Reinforcement," *Shear in Reinforced Concrete*, SP-42, American Concrete Institute, Farmington Hills, MI, 949 pp.

Hwang, S. J., and Moehle, J. P., 2000, "Models for Laterally Loaded Slab-Column Frames," *ACI Structural Journal*, V. 97, No. 2, Mar.-Apr., pp. 345-353.

Jirsa, J. O.; Sozen, M. A.; and Seiss, C. P., 1963, "Effects of Pattern Loadings on Reinforced Concrete Floor Slabs," *Structural Research Series*, No. 269, Civil Engineering Studies, University of Illinois, Urbana, IL, July, 166 pp.

Jirsa, J. O.; Sozen, M. A.; and Seiss, C. P., 1969, "Pattern Loadings on Reinforced Concrete Floor Slabs," *Proceedings*, V. 95, No. ST6, pp. 1117-1137.

Johansen, K. W., 1962, *Yield-Line Theory*, Cement and Concrete Association, London, UK, 181 pp.

Joint ACI-ASCE Committee 426, 1974, "The Shear Strength of Reinforced Concrete Slabs," *Proceedings*, V. 100, No. ST8, Aug., pp. 1543-1592.

Logan, D., L., 2002, "A First Course in the Finite Element Method," Pacific Grove, CA: Brooks/Cole, third edition.

Mitchell, D., and Cook, W., 1984, "Preventing Progressive Collapse of Slab Structures," *Journal of Structural Engineering*, V. 110, No. 7, July, pp. 1513-1532. doi: 10.1061/(ASCE)0733-9445(1984)110:7(1513)

Nawy, E. G., 1968, "Crack Control in Reinforced Condrete Structures," *ACI Journal Proceedings*, V. V65, No. 10, Oct., pp. 825-836.

Nawy, E. G., 2001, "Design for Crack Control in Reinforced and Prestressed Concrete Beams, Two-way Slabs and Circular Tanks—A State of the Art," *Design and Construction Practices to Mitigate Cracking*, SP-204, American Concrete Institute, Farmington Hills, MI, 280 pp.

Nawy, E. G., 2011, *Reinforced Concrete—A Fundamental Approach*, sixth edition, third printing, Prentice Hall, Upper Saddle River, NJ, 915 pp.

Nawy, E. G., and Blair, K. W., 1971, "Further Studies on Flexural Crack Control in Structural Slab Systems," *Cracking, Deflection and Ultimate Load of Concrete Slab Systems*, SP-30, American Concrete Institute, Farmington Hills, MI, pp. 1-41.

Nichols, J., 1914, "Statistical Limitations upon the Steel Requirement in Reinforced Concrete Flat Slab Floors," *Transactions of the American Society of Civil Engineers*, V. 77, pp. 1670-1736.

Park, R., and Gamble, W. L., 1999, *Reinforced Concrete Slabs*, second edition, John Wiley and Sons, New York, 736 pp.

Ritchie, M. H.; Ghali, A.; Dilger, W. H.; and Gayed, R. B., 2006, "Unbalanced Moment Resistance by Shear in Slab-Column Connections: Experimental Assessment," *ACI Structural Journal*, V. 103, No. 1, Jan.-Feb., pp. 74-82.

Schaeffer, T. C., ed., 1999, *The Design of Two Way Slabs*, SP-183, American Concrete Institute, Farmington Hills, MI, 183 pp.

Sozen, M., and Seiss, C., 1963, "Investigation of Multiple Panel Reinforced Concrete Slabs: Design Methods—Their Evolution and Comparison," *ACI Proceedings*, V. 60, No. 8, Aug., 999-1027 pp.

Vanderbilt, D., 1972, "Shear Strength of Continuous Plates," *Journal of the Structural Division*, V. 98, Apr., pp. 961-973.

Vanderbilt, M. D., and Corley, W. G., 1983, "Frame Analysis of Concrete Buildings," *Concrete International*, V. 5, No. 12, Dec., pp. 33-43.

APPENDIX A—EXCERPT BUILDING CODE PROVISIONS

The following text are excerpts from ACI 318-14, included in this guide to enhance the use of this guide as a source document in design of two- way slabs. Future revisions of this guide will merge the current commentary into the guide.

A.1—Direct design method (ACI 318-14, Section 8.10)

8.10—Direct design method

8.10.1 *General*

8.10.1.1 Two-way slabs satisfying the limits in 8.10.2 shall be permitted to be designed in accordance with this section.

8.10.1.2 Variations from the limitations in 8.10.2 shall be permitted if demonstrated by analysis that equilibrium and geometric compatibility are satisfied, the design strength at every section is at least equal to the required strength, and serviceability conditions, including limits on deflection, are met.

8.10.1.3 Circular or regular polygon-shaped supports shall be treated as square supports with the same area.

8.10.2 *Limitations for use of direct design method*

8.10.2.1 There shall be at least three continuous spans in each direction.

8.10.2.2 Successive span lengths measured center-to-center of supports in each direction shall not differ by more than one-third the longer span.

8.10.2.3 Panels shall be rectangular, with the ratio of longer to shorter panel dimensions, measured center-to-center of supports, not to exceed 2.

8.10.2.4 Column offset shall not exceed 10 percent of the span in direction of offset from either axis between centerlines of successive columns.

8.10.2.5 All loads shall be due to gravity only and uniformly distributed over an entire panel.

8.10.2.6 Unfactored live load shall not exceed two times the unfactored dead load.

8.10.2.7 For a panel with beams between supports on all sides, Eq. (8.10.2.7a) shall be satisfied for beams in the two perpendicular directions.

$$0.2 \le \frac{\alpha_{f1}\ell_2^2}{\alpha_{f2}\ell_1^2} \le 5.0 \qquad (8.10.2.7a)$$

where α_{f1} and α_{f2} are calculated by:

$$\alpha_f = \frac{E_{cb}I_b}{E_{cs}I_s} \qquad (8.10.2.7b)$$

8.10.3 *Total factored static moment for a span*

8.10.3.1 Total factored static moment M_o for a span shall be calculated for a strip bounded laterally by the panel centerline on each side of the centerline of supports.

8.10.3.2 The absolute sum of positive and average negative M_u in each direction shall be at least:

Table 8.10.4.2—Distribution coefficients for end spans

	Exterior edge unrestrained	Slab with beams between all supports	Slab without beams between interior supports		Exterior edge fully restrained
			Without edge beam	With edge beam	
Interior negative	0.75	0.70	0.70	0.70	0.65
Positive	0.63	0.57	0.52	0.50	0.35
Exterior negative	0	0.16	0.26	0.30	0.65

$$M_o = \frac{q_u \ell_2 \ell_n^2}{8} \quad (8.10.3.2)$$

8.10.3.2.1 In Eq. (8.10.3.2), ℓ_n is the clear span length in the direction that moments are considered, shall extend from face to face of columns, capitals, brackets, or walls, and shall be at least $0.65\ell_1$.

8.10.3.2.2 In Eq. (8.10.3.2), if the transverse span of panels on either side of the centerline of supports varies, ℓ_2 shall be taken as the average of adjacent transverse spans.

8.10.3.2.3 In Eq. (8.10.3.2), if the span adjacent and parallel to a slab edge is being considered, the distance from edge to panel centerline shall be substituted for ℓ_2.

8.10.4 *Distribution of total factored static moment*

8.10.4.1 In an interior span, M_o shall be distributed as follows: $0.65M_o$ to negative moment and $0.35M_o$ to positive moment.

8.10.4.2 In an end span, M_o shall be distributed in accordance with Table 8.10.4.2.

8.10.4.3 Modification of negative and positive factored moments by up to 10 percent shall be permitted if the total factored static moment for a panel, M_o, in the direction considered is at least that calculated by Eq. (8.10.3.2). Moment redistribution in accordance with 6.6.5 is not permitted.

8.10.4.4 Critical section for negative M_u shall be at the face of rectangular supports.

8.10.4.5 Negative M_u shall be the greater of the two interior negative M_u calculated for spans framing into a common support unless an analysis is made to distribute the unbalanced moment in accordance with stiffnesses of adjoining elements.

8.10.4.6 Edge beams or edges of slabs shall be designed to resist in torsion their share of exterior negative M_u.

8.10.5 *Factored moments in column strips*

8.10.5.1 The column strip shall resist the portion of interior negative M_u in accordance with Table 8.10.5.1.

8.10.5.2 The column strip shall resist the portion of exterior negative M_u in accordance with Table 8.10.5.2.

$$\beta_t = \frac{E_{cb}C}{2E_{cs}I_s} \quad (8.10.5.2a)$$

Table 8.10.5.1—Portion of interior negative M_u in column strip

	ℓ_2/ℓ_1		
$\alpha_{f1}\ell_2/\ell_1$	0.5	1.0	2.0
0	0.75	0.75	0.75
≥1.0	0.90	0.75	0.45

Note: Linear interpolations shall be made between values shown.

Table 8.10.5.2—Portion of exterior negative M_u in column strip

		ℓ_2/ℓ_1		
$\alpha_{f1}\ell_2/\ell_1$	β_t	0.5	1.0	2.0
0	0	1.0	1.0	1.0
	≥2.5	0.75	0.75	0.75
≥1.0	0	1.0	1.0	1.0
	≥2.5	0.90	0.75	0.45

Note: Linear interpolations shall be made between values shown. β_t is calculated using Eq. (8.10.5.2a), where C is calculated using Eq. (8.10.5.2b).

Table 8.10.5.5—Portion of positive M_u in column strip

	ℓ_2/ℓ_1		
$\alpha_{f1}\ell_2/\ell_1$	0.5	1.0	2.0
0	0.60	0.60	0.60
≥1.0	0.90	0.75	0.45

Note: Linear interpolations shall be made between values shown.

$$C = \Sigma\left(1 - 0.63\frac{x}{y}\right)\frac{x^3 y}{3} \quad (8.10.5.2b)$$

8.10.5.3 For T- or L-sections, it shall be permitted to calculate the constant C in Eq. (8.10.5.2b) by dividing the section, as given in 8.4.1.8, into separate rectangular parts and summing the values of C for each part.

8.10.5.4 If the width of the column or wall is at least $(3/4)\ell_2$, negative M_u shall be uniformly distributed across ℓ_2.

8.10.5.5 The column strip shall resist the portion of positive M_u in accordance with Table 8.10.5.5.

8.10.5.6 For slabs with beams between supports, the slab portion of column strips shall resist column strip moments not resisted by beams.

8.10.5.7 *Factored moments in beams*

8.10.5.7.1 Beams between supports shall resist the portion of column strip M_u in accordance with Table 8.10.5.7.1.

8.10.5.7.2 In addition to moments calculated according to 8.10.5.7.1, beams shall resist moments caused by factored loads applied directly to the beams, including the weight of the beam stem above and below the slab.

8.10.6 *Factored moments in middle strips*

8.10.6.1 That portion of negative and positive factored moments not resisted by column strips shall be proportionately assigned to corresponding half middle strips.

8.10.6.2 Each middle strip shall resist the sum of the moments assigned to its two half middle strips.

8.10.6.3 A middle strip adjacent and parallel to a wall-supported edge shall resist twice the moment assigned to the half middle strip corresponding to the first row of interior supports.

Table 8.10.5.7.1—Portion of column strip M_u in beams

$\alpha_{f1}\ell_2/\ell_1$	Distribution coefficient
0	0
≥1.0	0.85

Note: Linear interpolation shall be made between values shown.

8.10.7 *Factored moments in columns and walls*

8.10.7.1 Columns and walls built integrally with a slab system shall resist moments caused by factored loads on the slab system.

8.10.7.2 At an interior support, columns or walls above and below the slab shall resist the factored moment calculated by Eq. (8.10.7.2) in direct proportion to their stiffnesses unless a general analysis is made.

$$M_{sc} = 0.07[(q_{Du} + 0.5q_{Lu})\ell_2\ell_n^2 - q_{Du}'\ell_2'(\ell_n')^2] \quad (8.10.7.2)$$

where q_{Du}', ℓ_2', and ℓ_n' refer to the shorter span.

8.10.7.3 The gravity load moment to be transferred between slab and edge column in accordance with 8.4.2.3 shall not be less than $\mathbf{0.3}M_o$.

8.10.8 *Factored shear in slab systems with beams*

8.10.8.1 Beams between supports shall resist the portion of shear in accordance with Table 8.10.8.1 caused by factored loads on tributary areas in accordance with Fig. 8.10.8.1.

8.10.8.2 In addition to shears calculated according to 8.10.8.1, beams shall resist shears caused by factored loads applied directly to the beams, including the weight of the beam stem above and below the slab.

8.10.8.3 Calculation of required slab shear strength based on the assumption that load is distributed to supporting beams in accordance with 8.10.8.1 shall be permitted. Shear resistance to total V_u occurring on a panel shall be provided.

A.2—Equivalent frame method (ACI 318-14, Section 8.11)

8.11—Equivalent frame method

8.11.1 *General*

8.11.1.1 All sections of slabs and supporting members in two-way slab systems designed by the equivalent frame method shall resist moments and shears obtained from an analysis in accordance with 8.11.2 through 8.11.6.

8.11.1.2 Live load shall be arranged in accordance with 6.4.3.

8.11.1.3 It shall be permitted to account for the contribution of metal column capitals to stiffness, resistance to moment, and resistance to shear.

8.11.1.4 It shall be permitted to neglect the change in length of columns and slabs due to direct stress, and deflections due to shear.

8.11.2 *Equivalent frames*

8.11.2.1 The structure shall be modeled by equivalent frames on column lines taken longitudinally and transversely through the building.

8.11.2.2 Each equivalent frame shall consist of a row of columns or supports and slab-beam strips bounded laterally by the panel centerline on each side of the centerline of columns or supports.

Table 8.10.8.1—Portion of shear resisted by beam

$\alpha_{f1}\ell_2/\ell_1$	Distribution coefficient
0	0
≥1.0	1.0

Note: Linear interpolation shall be made between values shown.

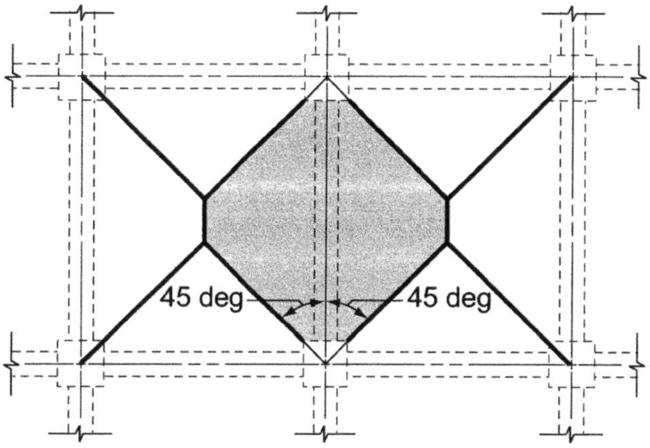

Fig. 8.10.8.1—Tributary area for shear on an interior beam.

8.11.2.3 Frames adjacent and parallel to an edge shall be bounded by that edge and the centerline of the adjacent panel.

8.11.2.4 Columns or supports shall be assumed to be attached to slab-beam strips by torsional members transverse to the direction of the span for which moments are being calculated and extending to the panel centerlines on each side of a column.

8.11.2.5 Analysis of each equivalent frame in its entirety shall be permitted. Alternatively, for gravity loading, a separate analysis of each floor or roof with the far ends of columns considered fixed is permitted.

8.11.2.6 If slab-beams are analyzed separately, it shall be permitted to calculate the moment at a given support by assuming that the slab-beam is fixed at supports two or more panels away, provided the slab continues beyond the assumed fixed supports.

8.11.3 *Slab-beams*

8.11.3.1 The moment of inertia of slab-beams from the center of the column to the face of the column, bracket, or capital shall be assumed equal to the moment of inertia of the slab-beam at the face of the column, bracket, or capital divided by the quantity $(1 - c_2/\ell_2)^2$, where c_2 and ℓ_2 are measured transverse to the direction of the span for which moments are being determined.

8.11.3.2 Variation in moment of inertia along the axis of slab-beams shall be taken into account.

8.11.3.3 It shall be permitted to use the gross cross-sectional area of concrete to determine the moment of inertia of slab-beams at any cross section outside of joints or column capitals.

8.11.4 *Columns*

8.11.4.1 The moment of inertia of columns from top to bottom of the slab-beam at a joint shall be assumed to be infinite.

8.11.4.2 Variation in moment of inertia along the axis of columns shall be taken into account.

8.11.4.3 It shall be permitted to use the gross cross-sectional area of concrete to determine the moment of inertia of columns at any cross section outside of joints or column capitals.

8.11.5 *Torsional members*

8.11.5.1 Torsional members shall be assumed to have a constant cross section throughout their length consisting of the greatest of (a) through (c):

(a) A portion of slab having a width equal to that of the column, bracket, or capital in the direction of the span for which moments are being determined.

(b) For monolithic or fully composite construction, the portion of slab specified in (a) plus that part of the transverse beam above and below the slab.

(c) The transverse beam in accordance with 8.4.1.8.

8.11.5.2 Where beams frame into columns in the direction of the span for which moments are being calculated, the torsional stiffness shall be multiplied by the ratio of the moment of inertia of the slab with such a beam to the moment of inertia of the slab without such a beam.

8.11.6 *Factored moments*

8.11.6.1 At interior supports, the critical section for negative M_u in both column and middle strips shall be taken at the face of rectilinear supports, but not farther away than $0.175\ell_1$ from the center of a column.

8.11.6.2 At exterior supports without brackets or capitals, the critical section for negative M_u in the span perpendicular to an edge shall be taken at the face of the supporting element.

8.11.6.3 At exterior supports with brackets or capitals, the critical section for negative M_u in the span perpendicular to an edge shall be taken at a distance from the face of the supporting element not exceeding one-half the projection of the bracket or capital beyond the face of the supporting element.

8.11.6.4 Circular or regular polygon-shaped supports shall be assumed to be square supports with the same area for location of critical section for negative design moment.

8.11.6.5 Where slab systems within limitations of 8.10.2 are analyzed by the equivalent frame method, it shall be permitted to reduce the calculated moments in such proportion that the absolute sum of the positive and average negative design moments need not exceed the value obtained from Eq. (8.10.3.2).

8.11.6.6 It shall be permitted to distribute moments at critical sections to column strips, beams, and middle strips in accordance with the direct design method in 8.10, provided that Eq. (8.10.2.7a) is satisfied.

Recommendations for Concrete Members Prestressed with Single-Strand Unbonded Tendons

Reported by Joint ACI-ASCE Committee 423

Carin L. Roberts-Wollmann, Chair

Amy M. Reineke Trygestad, Secretary

Theresa M. Ahlborn
Rashid Ahmed*
Robert W. Barnes
Florian G. Barth
Asit N. Baxi*
Roger J. Becker
Charles W. Dolan*
James P. Donnelly
Pierre Esselinck
Martin J. Fradua
William L. Gamble
Harry A. Gleich
Pawan R. Gupta
William M. Hale
H. R. Trey Hamilton III
Carol Hayek*
Mohammad Iqbal
Donald P. Kline*
Larry B. Krauser
Jason J. Krohn
Theodore L. Neff
Sami H. Rizkalla
James Rogers
Brandon Ross
Bruce W. Russell
Thomas C. Schaeffer*
Richard W. Stone
Miroslav Vejvoda
Jeffrey S. Volz
H. Carl Walker
Zuming Xia
Paul Zia

Consulting Members

Kenneth B. Bondy*
Robert N. Bruce Jr.
Ned H. Burns
Chunsheng "Steve" Cai
Steven R. Close
Henry J. Cronin Jr.
Ward N. Marianos Jr.
Hani Melhem
Antoine E. Naaman
Thomas E. Nehil
Andrea J. Schokker

*Subcommittee members involved in updating this report.

This report provides information for the design of flexural concrete members in buildings post-tensioned with single-strand unbonded tendons. The report is intended to complement the commentary in ACI 318 and to provide suggestions for revisions and additions to ACI 318. Consideration is given to design for gravity and lateral loads, determination of fire endurance, design for seismic forces, and design for catastrophic loadings. Recommendations concerning details and properties of tendons, protection against corrosion, and construction procedures are presented.

Keywords: concrete slabs; cracking; fire resistance; joints; punching shear; unbonded post-tensioning.

CONTENTS

ACI 423.3R-17 supersedes ACI 423.3R-05 and was adopted and published July 2017.

CHAPTER 1—INTRODUCTION

1.1—General

Since the publication of previous ACI 423.3R reports, many of their recommendations have been incorporated into ACI 318. All references to ACI 318 refer to ACI 318-14 unless specifically noted. The recommendations in this report have been prepared to provide a comprehensive guide for design, materials, and construction for concrete members prestressed with single-strand unbonded tendons. Suggested revisions and additions to ACI 318 are also included in this report. Mandatory language used in this report reflects requirements in ACI 318.

1.2—Objective

This report presents recommendations for design, materials, and construction for concrete structures prestressed with unbonded tendons that are commensurate with the strength, serviceability, and safety requirements of ACI 318. Practitioners should use their judgment when applying the recommendations of this report. This report is not intended for reference in a specification or a code.

1.3—Scope

Recommendations pertinent to design with single-strand unbonded tendons considered in this report include the design of slabs, beams, and continuous members; details and properties of tendons and anchors; and protection from corrosion during construction and throughout the life of the structure.

The recommendations in this report are not intended for unbonded construction stages of elements using bonded tendons; for multistrand unbonded tendons used as external tendons; for members subject to direct tension, such as tiebacks, cable stays, arch ties, or circumferential tendons for containment structures; or for ground-supported post-tensioned slabs for light residential construction.

CHAPTER 2—NOTATION AND DEFINITIONS

2.1—Notation

A_b = net bearing plate area. in.^{2} (mm^2)
A_b' = maximum area of the portion of the concrete anchorage surface that is geometrically similar to and concentric with the area of the anchorage, in.^{2} (mm^2)
A_c' = cross-sectional area of the slab, perpendicular to the slab edge, between the center of the exterior span and the slab edge, in.^{2} (mm^2)
A_{cf} = larger of gross cross-sectional areas of the slab-beam strips of the two orthogonal equivalent frames intersecting at the column, in.^{2} (mm^2)
A_{ps} = area of prestressed longitudinal tension reinforcement, in.^{2} (mm^2)
$A_{s,min}$ = minimum bonded reinforcement in negative moment areas of two-way systems
b = width of compression face of member, in. (mm)
b_f = total flange width
b_n = effective overhang flange width for normal forces, in. (mm)
b_o = perimeter of critical section for two-way shear in slabs, in. (mm)
b_w = effective flange width for normal forces at post-tension anchor, in. (mm)
d_p = distance from extreme compression fiber to centroid of prestressing reinforcement, in. (mm)
E_s = modulus of elasticity of prestressing reinforcement, psi (MPa)
f_{ci}' = strength of concrete at time of initial prestress, psi (MPa)
f_{cp} = permissible concrete compressive stress, psi (MPa)
f_{pc} = compressive stress in concrete, after allowance for all prestress losses, at centroid of cross section resisting externally applied loads or at junction of web and ñange where the centroid lies within the ñange, psi (MPa); in a composite member, f_{pc} is the resultant compressive stress at centroid of composite section, or at junction of web and ñange where the centroid lies within the ñange, due to both prestress and moments resisted by precast member acting alone, psi (MPa)

f_{ps} = stress in prestressing reinforcement at nominal flexural strength, psi (MPa)
$f_{ps\theta}$ = stress in post-tensioned tendons at nominal strength at high temperatures, psi (MPa)
f_{pu} = specified tensile strength of prestressing reinforcement, psi (MPa)
$f_{pu\theta}$ = tensile strength of tendons at high temperatures, psi (MPa)
f_{se} = effective stress in prestressing reinforcement, after allowance for all prestress losses, psi (MPa)
f_y = specified yield stress of reinforcement, psi (MPa)
h = slab thickness, in. (mm)
L = span length of beam or one-way slab; clear projection of cantilever, in. (mm)
ℓ = length of tendon, in. (mm)
M = total applied static moment (unfactored), lb-in. (kN-m)
M_{tq}^+ = retained midspan moment, lb-in. (kN-m)
$M_{t1\theta}^-$ = retained negative moment capacity at Column 1, lb-in. (N-m)
$M_{t2\theta}^-$ = retained negative moment capacity at Column 2, lb-in. (N-m)
N_c = resultant tensile force acting on the portion of the concrete cross section that is subjected to tensile stresses due to the combined effects of service loads and effective prestress, lb (N)
P = average force in the prestressing reinforcement, lb (N)
V_p = vertical component of effective prestress forces crossing the critical section
v_c = stress corresponding to nominal two-way shear strength provided by concrete, psi (MPa)
v_{cw} = nominal shear stress provided by concrete where diagonal cracking results from high principal tensile stress in web, lb (N)
w = uniformly distributed load, plf (kN/m)
α_s = constant used to calculate V_c in slabs, lb (MPa)
Δ_ℓ = tendon elongation, in. (mm)
ϕ = strength reduction factor
λ = modification factor to reflect the reduced mechanical properties of lightweight concrete
ρ_p = ratio of A_{ps} to bd_p

2.2—Definitions

ACI provides a comprehensive list of definitions through an online resource, "ACI Concrete Terminology," https://www.concrete.org/store/productdetail.aspx?ItemID=CT16. Definitions provided herein complement that source.

encapsulated tendon—a tendon that is completely enclosed in a watertight covering from end to end, including anchorages, sheathing, post-tensioning coating, sleeves, and an encapsulation cap over the strand tail at each end.

post-tensioning coating—material between prestressing reinforcement and sheathing used to protect against corrosion and reduce friction.

prestressing reinforcement—strand, wire, or bar tensioned to impart forces to the concrete.

sheathing—material encasing prestressing reinforcement to prevent bonding of the prestressing reinforcement with the surrounding concrete, to provide corrosion protection, and to contain the corrosion-inhibiting coating.

CHAPTER 3—MEMBER DESIGN

3.1—General

The design provisions of ACI 318 apply to the contents of this chapter, but some recommendations are offered that complement those of ACI 318. Significant changes were made in ACI 318-02, ACI 318-08, ACI 318-11, and ACI 318-14 that impact the design of prestressed concrete members and unify the design approach for prestressed and nonprestressed members. ACI 318 limits for flexural reinforcement and moment redistribution are specified in terms of the net tensile strain in the extreme tension reinforcement ε_t. Prestressed members are grouped into three design categories: uncracked (U), transition (T), and cracked (C). Each category is defined according to the maximum calculated flexural tensile stress under service load conditions. For C members, there is no upper limit on calculated flexural tensile stresses under service loads, and any combination of stressed and unstressed reinforcement is permitted. Serviceability requirements related to C and T members are given in ACI 318.

3.2—One-way systems

One-way slabs are addressed in ACI 318-14 Chapter 7. The chapter organization provides design limits, required strength, required design strength, reinforcement limits, and reinforcing detailing.

3.2.1 *Minimum bonded reinforcement*—The minimum bonded reinforcement required in ACI 318-14, 7.6.2, is considered adequate to limit crack widths due to dead load and live load by providing adequate crack distribution (Burns et al. 1978; Yamazaki et al. 1969; Burns and Pierce 1967). This amount of reinforcement also provides an independent load-carrying system in the event of a catastrophic failure of the tendons or abnormal loading in one span of a continuous one-way post-tensioned element with unbonded tendons. For this reason, ACI 318-14, 7.7.4.2, requires that bonded reinforcement used as part of the design moment strength be detailed in accordance with the provisions of ACI 318-14, 7.7.3. Such bonded reinforcement provides an independent load path in one-way systems.

All prestressed concrete slabs (Classes U, T, and C) in ACI 318-11 were exempt from the reinforcement spacing limits in ACI 318-11, 7.6.5 (deformed flexural reinforcement spacing to be less than $3h$ and 18 in. [610 mm]). However, this requirement was changed in 7.7.2 of ACI 318-14 for Class C prestressed concrete slabs to make it similar to nonprestressed. Based on field experience with the performance of Class U and Class T unbonded post-tensioned slabs, the maximum spacing of the deformed reinforcement should be limited to six times the slab thickness or 36 in. (910 mm), whichever is less. Extension requirements for reinforcement that is not part of the moment strength are given in ACI 318-14, 7.7.4.4.

In one-way slabs, the use of service load design tensile stresses in the range of $9\sqrt{f_c'}$ to $12\sqrt{f_c'}$ psi ($0.75\sqrt{f_c'}$ to $1.0\sqrt{f_c'}$ MPa) usually results in bonded reinforcement strength requirements reasonably similar to those required by ACI 318-14, 7.6.2.3. Tests of slabs with this level of design tensile stress in conjunction with the bonded reinforcement requirements of ACI 318-14, 7.6.2.3, indicate that such slabs have performed satisfactorily (Burns et al. 1978). Prestressed concrete members categorized as Class C in ACI 318 have no upper limit on flexural tensile stresses, and the use of any combination of prestressed and nonprestressed reinforcement is permitted. The use of lower design tensile stresses, however, is preferable from the durability standpoint for applications in aggressive environments, such as parking structure decks in severe climates (Walker 1980).

One-way post-tensioned beams and slabs with unbonded tendons are waived from the $1.2M_{cr}$ requirement of ACI 318-14, 7.6.2.1, because the undesirable behavior due to sudden transfer of tensile force from concrete to prestressing reinforcement at first cracking in lightly reinforced sections does not occur in members with unbonded tendons.

3.2.2 *Tendon spacing*—The minimum bonded reinforcement requirements for one-way slabs under ACI 318-14, 7.6.2.3, typically result in the use of No. 4 (No. 13) bars at 21 in. (500 mm) centers for both positive and negative moments for a 4-1/2 in. (115 mm) thick slab. For an 8 in. (200 mm) thick one-way slab, No. 4 (No. 13) bars are required at approximately 12 in. (300 mm) centers; larger bars can be used at somewhat wider spacing. In consideration of this amount and spacing of bonded reinforcement, a maximum tendon spacing of eight times the slab thickness but not more than 5 ft (1.5 m) is recommended for one-way slabs with normal live loads and uniformly distributed loads (refer also to 3.3.5). Tendon spacing may need to be adjusted for slabs with concentrated loads, based on the size, configuration, and magnitude of the load concentration. ACI 318 contains no requirement for maximum tendon spacing or minimum axial compressive stress in one-way prestressed slabs. It is common practice to use an average minimum effective axial precompression stress of 125 psi (0.86 MPa) at every cross section tributary to the tendon, which can be further reduced to 100 psi (0.69 MPa) for short spans with little or no tension stresses.

In certain cases, such as external tendon retrofits, tendon spacing greater than eight times the slab thickness or 5 ft (1.5 m) may be applicable. In such cases, these limits may be exceeded, provided it can be shown by analysis that the slab system can adequately carry the design loads.

3.2.3 *Minimum shear reinforcement*—A minimum amount of shear reinforcement is necessary in all post-tensioned joists to provide a means of supporting the tendon design profile in the beam cross section. Where tendons are not adequately supported by stirrups, local deviations of the tendons from the assumed smooth parabolic curvature can result during placement of the concrete. When such tendons are stressed, the deviations from the intended curvature tend to straighten out, and this process can impose large tensile stresses in webs of post-tensioned beams or joists. Severe cracking has been observed in several instances where no stirrups were provided. Unintended curvature of the tendons can be avoided by securely tying tendons to stirrups that are rigidly held in place by other elements of the reinforcing cage.

3.2.4 *Prestressed shrinkage and temperature reinforcement*—ACI 318-14, 24.4.4, is the result of an extensive revision in ACI 318-11. The gross concrete area is now defined as the total beam web area, including the slab thickness and the area of the slab tributary to the beam. Beam tendons may be included in providing the required 100 psi (0.7 MPa) minimum average compression stress on this gross concrete area. The spacing of shrinkage and temperature tendons should not exceed 6 ft (1.8 m), measured between adjacent tendons, between faces of beams, or between walls and the closest tendon. When tendon spacing exceeds 4.5 ft (1.4 m), nonprestressed reinforcement, parallel to the tendons, is added at the slab edge in accordance with ACI 318-14, 7.7.6. At least one tendon is required in the slab between the faces of adjacent beams or walls.

Tendons used for shrinkage and temperature reinforcement should be positioned parallel to the beams and vertically as close as practicable to the center of the slab. Where shrinkage and temperature tendons support or conflict with the principal slab tendons, variation from the slab centroid is permissible, but the resultant eccentricity of the total shrinkage and temperature tendons within the slab should not extend outside the kern limits of the slab cross section. Figure 3.2.4 illustrates details for the use of unbonded tendons as shrinkage and temperature reinforcement in one-way beam and slab construction.

3.2.5 *Structural integrity*—ACI 318 contains no specific recommendations for structural integrity of one-way prestressed slabs with unbonded tendons.

3.3—Two-way systems

Two-way slabs are addressed in ACI 318-14 Chapter 8. The chapter organization provides design limits, design strength, required resistance, reinforcement limits, and reinforcing detailing.

3.3.1 *Analysis*—Slab systems prestressed in more than one direction for flexure should be analyzed in accordance with the provisions of equivalent frame or by more precise methods, including finite element techniques or classical elastic theory. The equivalent frame method of analysis has been shown by tests of large structural models to satisfactorily predict factored moments and shears in prestressed slab systems (Post-Tensioning Institute 1977; Burns and Hemakom 1977; Scordelis et al. 1959; Cleland et al. 1979). The referenced research also shows that the yield-line theory accurately predicts the flexural strength of two-way post-tensioned slabs. Analysis using prismatic sections or other approximations of stiffness, which differ substantially from the equivalent frame method, may provide erroneous and unsafe results. Moment redistribution for prestressed slabs is covered in ACI 318-14, 6.6.5. ACI 318-14, 8.11.6.5,

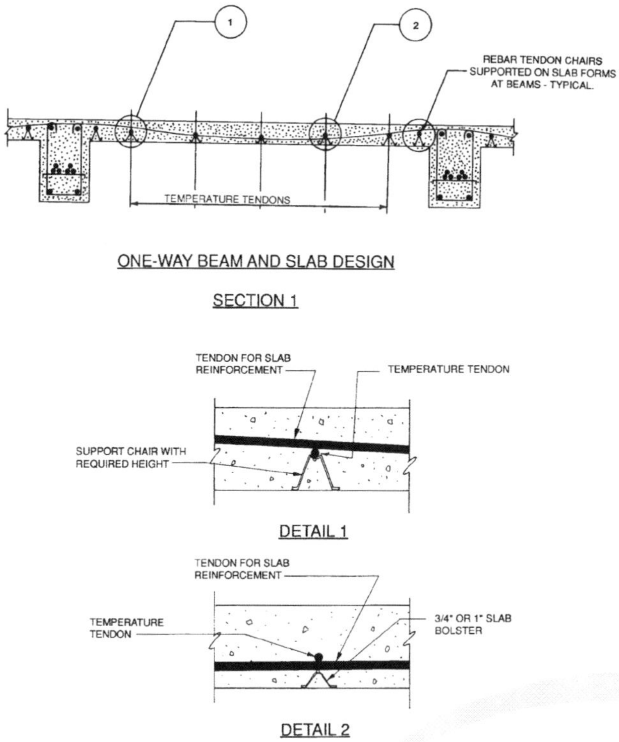

Fig. 3.2.4—Details for use of unbonded tendons as shrinkage and temperature reinforcement in one-way beam and slab construction. (Note: 1 in. = 25.4 mm.)

does not apply to two-way prestressed slabs. ACI 318-14, 8.11.6.6, is excluded because the distribution of moments between column strips and middle strips is based on tests of reinforced concrete slabs and the concept of column strips and middle strips is virtually never used in prestressed slabs, almost all of which use a banded tendon distribution (Fig. 3.3.1). Further, simplified methods of analysis using average coefficients do not apply to prestressed concrete slab systems. All other provisions of ACI 318-14, 8.11, specifically including the arrangement of live loads specified in ACI 318-14, 6.4.3, are applicable for the analysis of post-tensioned slabs.

ACI 318 requires that two-way prestressed slab systems be designed as Class U, with a maximum permissible concrete flexural tensile stress of $6\sqrt{f_c'}$ psi ($0.5\sqrt{f_c'}$ MPa) under service loads. The lateral load stiffness should be assessed considering the effects of slab cracking (ACI 318-14, 8.11; Kang and Wallace 2008b; Han et al. 2006). For lateral load analysis, the cracked section bending stiffness should be used and the torsional stiffness taken as 1/10 of the torsional member when using the equivalent frame method of analysis and where there is the potential for extensive cracking; where slabs have highly irregular shapes; where there are long pour lengths between construction, expansion joints, or both; or where slabs have unfavorable locations of walls. The cracked section bending stiffness should always be used for the computation of drift under seismic loads. Strength under lateral loads should be evaluated using the load factor combinations given in ACI 318-14, 5.3, in conjunction with the provisions of ACI 318-14, 8.4.1. Evaluation of strength

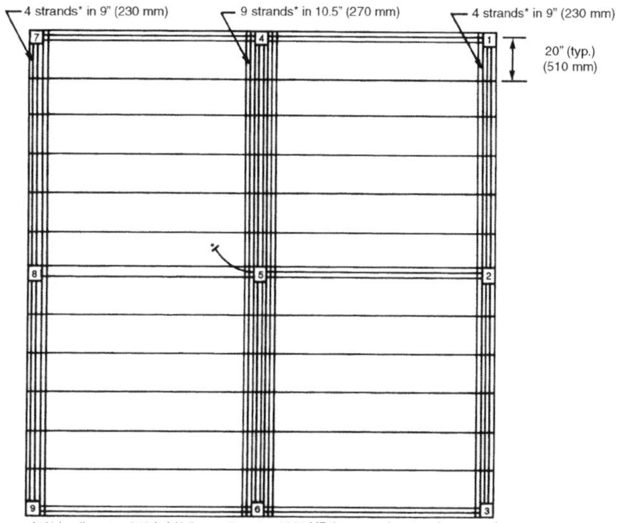

*1/4 in. diameter 240 ksi (6.5 mm diameter 1660 MPa) seven-wire strand was used to simulate strand tendons in model. Partial length tendon over interior column used for experimental purposes.

Fig. 3.3.1—Banded tension distribution (Kosut et al. 1985).

requirements under lateral loads may disclose the need for reinforcement to resist moment reversals (Han et al. 2006). Such reinforcement should be located within a distance of $1.5h$ outside opposite faces of the column, where h is the slab thickness.

3.3.2 *Limits for reinforcement*—Two-way post-tensioned slabs with unbonded tendons are waived from the $1.2M_{cr}$ requirement of ACI 318-14, 8.6.2.2, because the undesirable behavior due to the sudden transfer of tensile force from concrete to prestressing reinforcement at first cracking in lightly reinforced slabs does not occur in members with unbonded tendons.

Only at load levels beyond the factored design load is the additional cracking extensive enough to cause an abrupt change in the load-deflection pattern. Tests have also shown that it is not possible to rupture, or even yield, unbonded post-tensioning tendons in two-way slabs before a punching shear failure (Post-Tensioning Institute 1977; Burns and Hemakom 1977; Hemakom 1975; Kosut et al. 1985; Scordelis et al. 1959; Gerber and Burns 1971; Smith and Burns 1974; Kosut 1977). The use of unbonded tendons in combination with the minimum bonded reinforcement requirements of ACI 318-14, 8.6.2.3 and 8.7.5, has been shown to ensure post-cracking ductility and that a brittle failure mode will not develop at first cracking.

3.3.3 *Minimum bonded reinforcement*—Significant changes were made in ACI 318-99 to the requirement for minimum bonded reinforcement in negative moment regions of two-way systems. Prior to ACI 318-99, the amount of reinforcement was based on the span parallel to the direction of the reinforcement. In rectangular panels, this resulted in more reinforcement parallel to the longer span; however, it could also result in less reinforcement than tests have shown is necessary parallel to the shorter span (Burns and Hemakom 1977; Hemakom 1975; Kosut et al. 1985; Smith and Burns 1974; Kosut 1977; Foutch et al. 1990). It also

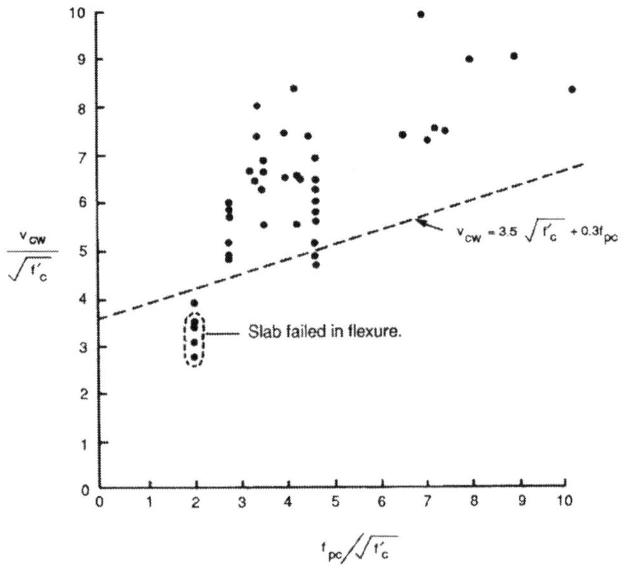

Fig. 3.3.4—Two-way post-tensioned flat plate shear test data versus Eq. (25.8.3.2) of ACI 318-14.

required substantially more reinforcement than tests have shown necessary in square panels parallel to the slab edge at exterior columns. To correct this, minimum bonded reinforcement in negative moment areas of two-way systems, $A_{s,min}$, is given by Eq. (3.3.3a)

$$A_{s,min} = 0.00075A_{cf} \quad (3.3.3a)$$

where A_{cf} is the larger of the gross cross-sectional areas of the slab-beam strips of the two orthogonal equivalent frames intersecting at the column. This addresses the problem of reinforcement parallel to the short span in rectangular panels, and results in an overall increase in reinforcement when compared with ACI 318 Codes prior to 1999.

The aforementioned change does not, however, address the edge panel condition and will still result in roughly twice the amount of minimum reinforcement required parallel to the slab edge at exterior columns. This amount of bonded reinforcement is required within a slab width between lines that are 1.5*h* outside opposite faces of the column support, where *h* is the slab thickness. Tests (Burns and Hemakom 1977; Hemakom 1975; Kosut et al. 1985; Smith and Burns 1974; Kosut 1977; Foutch et al. 1990) show that it is appropriate to provide bonded top reinforcement parallel to the slab edge at exterior columns on the basis of $0.00075A_c'$, where A_c' is the cross-sectional area of the slab, perpendicular to the slab edge, between the center of the exterior span and the slab edge. At exterior columns of flat plates with square panels and no projection of the slab beyond the exterior column face, the minimum bonded reinforcement parallel to the slab edge should logically be 50 percent of the bonded reinforcement perpendicular to the slab edge. It is recommended that ACI Committee 318 reconsider the minimum reinforcement requirement for this condition and modify the requirement to be more consistent with test results.

It is emphasized that the minimum bonded reinforcement addressed in this section is for the minimum amount required for crack distribution and to ensure sufficient punching shear strength. Flexural strength considerations may require more reinforcement than these minimum amounts.

Where the computed tensile stress in the concrete at service load exceeds $2\sqrt{f_c'}$ psi ($0.17\sqrt{f_c'}$ MPa), required bonded reinforcement in positive moment areas of two-way flat slabs is given by Eq. (3.3.3b) or ACI 318-14, 8.6.2.3.

$$A_s = \frac{N_c}{0.5f_y} \quad (3.3.3b)$$

where the value of f_y of bonded reinforcement is limited to 60,000 psi (420 MPa), and N_c is the tensile force, including the tensile component of the bending moment, in concrete due to unfactored dead load plus live load, plus effective prestress. Placement details for this reinforcement are included in 3.3.5. Slab reinforcement spacing requirements provided in ACI 318-14, 8.7.2.2, do not apply to bonded reinforcement in unbonded post-tensioned slabs.

3.3.4 *Shear and moment transfer*—ACI 318 revised the shear calculations in two-way members to be based on shear stress rather than force. The revision clarified the torsional stresses generated by the slab rotation at the column interface.

Figure 3.3.4 shows the results of single column-slab specimen punching shear tests and results of multi-panel slabs tested in shear (Kosut et al. 1985). Concrete shear capacity (Eq. (3.3.4)) expressed in terms of the perimeter of critical section for slabs b_o is

$$v_c = \beta_p \lambda \sqrt{f_c'} + 0.3f_{pc} + V_p/b_o d \quad (3.3.4)$$

where β_p is the smaller of 3.5 (0.29) or $(\alpha_s d/b_o + 1.5)$ $[(\alpha_s d/b_o + 1.5)/12]$; $\alpha_s = 40$ for interior columns, 30 for edge columns, and 20 for corner columns; b_o is perimeter of critical section defined in ACI 318-14, 8.5.3.1.2 and 22.6.4; f_{pc} is average value of f_{pc} for the two directions, with a minimum of 125 psi (0.9 MPa) and a maximum of 500 psi (3.5 MPa); V_p is the vertical component of all effective prestress forces crossing the critical section; and λ is a modification factor to reflect the reduced mechanical properties of lightweight concrete relative to normalweight concrete of the same compressive strength.

It is assumed that no portion of the column cross section is closer to a discontinuous edge than four times the slab thickness, and f_c' does not exceed 5000 psi (35 MPa). This effectively limits the use of Eq. (3.3.4) to interior columns. Edge and corner columns are required to be evaluated using ACI 318-14, 22.6.5.1.

For values of f_{pc} less than 125 psi (0.9 MPa), V_c is limited to the value obtained using ACI 318-14, 22.6.5.1, for nonprestressed construction. In practice, V_p is commonly and conservatively ignored in shear calculations.

Moment transfer from prestressed concrete slabs to interior columns can be evaluated using the procedures of ACI 318-14, 22.6 and 8.4.2.3 (Parme 1972). In this case, the factored shear stress v_u should not exceed the value of ϕv_c calculated from ACI 318-14, 22.6.5.5. All reinforcement within a distance of 1.5 times the slab thickness on either side of the column is effective for transferring the portion of the moment not transferred by shear. Effective tendon stress f_{se} should be used in moment transfer capacity calculations. Tendons bundled through the column or over the lifting collar in lift slabs are an effective means of increasing the moment transfer strength of lift-slab connections. The moment transfer strength of lift-slab connections is also controlled by details of the lift slab connection. ACI 318-14, 22.6 and 8.4.2.3, are also applicable to calculations of the moment transfer from prestressed concrete slabs to exterior column connections for moments normal to a discontinuous edge. Bonded reinforcement, detailed as closed ties or hooked so that it can act as torsional reinforcement, should be provided when the calculated upward factored shear stress v_u at the discontinuous edge exceeds $2\sqrt{f_c'}$ psi ($0.17\sqrt{f_c'}$ MPa). Until further research data become available, the maximum calculated shear stress at such exterior columns should be limited to those specified in ACI 318-14 Table 22.6.5.2. Tests of four edge column specimens of a post-tensioned flat plate with banded tendon details support the use of ACI 318-14 Eq. (22.6.5.5a) for shear design (Foutch et al. 1990). Code provisions for the use of headed shear stud reinforcement for two-way shear strength in slabs were introduced in ACI 318-08 and are provided in ACI 318-14, 22.6.8. Recommendations for the use of shearbands for two-way shear strength were introduced in ACI 352.1R. The limited test data available (Kosut 1977; Pillai et al. 1982) do not show beneficial effects on shear strength resulting from the use of nonprestressed shear reinforcement with anchorage device details such as conventional headed studs in post-tensioned flat plates. The use of headed shear studs or stirrup reinforcement with modified anchorage details has been shown to increase shear strength (Dilger and Ghali 1981; Mokhtar et al. 1985; Muller et al. 1984; ACI 421.1R; Pilakoutas and Li 2003; Kang and Wallace 2008a; Kang and Park 2012).

3.3.5 *Tendon and bonded reinforcement distribution and spacing*—Within the limits of tendon distributions that have been tested, research indicates that the moment and shear strength of two-way prestressed slabs is controlled by total tendon strength in a given bay and by the amount and location of nonprestressed reinforcement rather than by horizontal tendon distribution (Ritz et al. 1975; Burns and Hemakom 1977; Hemakom 1975; Kosut et al. 1985; Parme 1972; Cleland et al. 1979). While it is important that some tendons are located within the shear perimeter over columns, distribution elsewhere is not critical, and any rational method that satisfies static equilibrium may be used. For uniform loading, ACI 318-14, 8.7.2.3, specifies that the maximum spacing of single tendons or groups of tendons in one direction shall not exceed eight times the slab thickness, with a maximum spacing of 8h or 5 ft (1.5 m). In addition, tendons are spaced to provide a minimum average prestress of 125 psi (0.9 MPa) on the local slab section tributary to the tendon or tendon group, the section being one-half of the spacing on either side of the center of the tendon or tendon group. The spacing of individual single-strand tendons is usually governed by the minimum average prestress requirements. For groups of two or more tendons, the 8h criterion usually controls maximum tendon spacing. Tendon spacing may need to be adjusted to accommodate concentrated loads.

Horizontal curvature of bundled single-strand tendons should be avoided. If this is not possible, additional transverse reinforcement, such as hairpins, and other accessories may be required at points of horizontal curvature to maintain the horizontal plane of tendon bundles during stressing. Bundles should be limited to a maximum of four tendons where significant horizontal curvature is required.

Transverse reinforcement may also be required to control horizontal splitting cracking that can occur due to in-plane forces from horizontally-curved banded tendons, particularly those curved to avoid openings.

3.3.5.1 *Recommended details of reinforcement for banded tendon distribution*—The predominant and recommended method of placing tendons in two-way slab systems is the banded distribution illustrated in Fig. 3.3.1. The use of a banded tendon distribution greatly simplifies the process of placing tendons, and therefore provides a significant reduction in field labor cost. The number of tendons required in the design strip—that is, the center-to-center of adjacent panels—may be banded close to the column line in one direction and distributed in the other direction. At least two tendons are required by ACI 318-14, 8.7.5.6, to be through the region bounded by the longitudinal reinforcement of the columns in each direction.

For lift-slab construction, the same general details of tendon distribution apply and provisions should be made for tendons to pass through or over the lifting heads. Even though no tendons are provided in one direction between bands, the maximum spacing of tendons normal to bands ensures one-way reinforcement for this part of the slab. Except for small triangular sections adjacent to the slab edge and closure strips, the area between bands is also prestressed in both directions. For widely spaced tendons, consideration should be given to placing reinforcement similar to ACI 318 (Fig. 3.3.5.1) in these triangular regions.

Recommended details for nonprestressed reinforcement are:

a) Minimum bonded slab reinforcement $A_{s,min}$ at columns is $A_{s,min} = 0.00075A_{cf}$, given in Table 8.6.2.3 of ACI 318-14, where A_{cf} is the larger of cross-sectional areas of the slab-beam strips of the two orthogonal equivalent frames intersecting at the column (refer to 3.3.3). At least four No. 5 (No. 16) bars should be provided in each direction in negative moment areas at columns. The minimum area of bonded reinforcement parallel to the slab edge at exterior columns should be based on $0.00075A_c'$, where A_c' is the cross-sectional area of the slab, perpendicular to the slab edge, between the center of the exterior span and the slab edge.

b) Bonded reinforcement should be placed within a slab width between lines that are 1.5h outside opposite faces of columns (ACI 318-14, 8.7.5.6.3). Maximum spacing of these bars is 12 in. (300 mm).

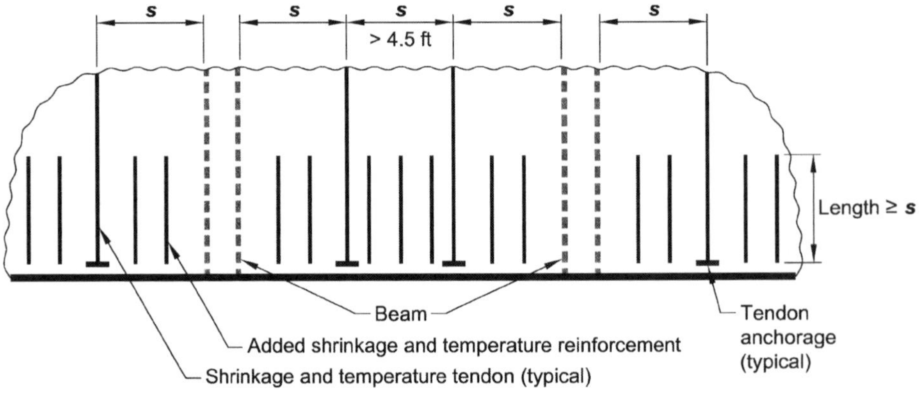

Fig. 3.3.5.1 Plan view at slab edge showing added shrinkage and temperature reinforcement (ACI 318-14, Fig. R7.7.6.3.2)

c) The minimum length for negative moment bars is one-sixth the clear span on each side of support. Refer to Item (f) for development considerations.

d) Where service load positive moment produces stress exceeding $2\sqrt{f_c'}$ psi ($0.17\sqrt{f_c'}$ MPa), minimum bonded reinforcement is given by Eq. (3.3.5.1) of this document or ACI 318-14 Table 8.6.2.3b

$$A_{s,min} = N_c/0.5f_y \tag{3.3.5.1}$$

as discussed in 3.3.3.

e) The minimum reinforcement for positive moment, when required, should have a length at least one-third the clear span, with the bars centered in the positive moment area. Refer to Item (f) for development considerations.

f) Where bonded reinforcement is required along with unbonded tendons to satisfy strength requirements, in addition to minimum $A_{s,min}$, this reinforcement is to be developed in conformance with ACI 318-14 Chapter 25.

g) Some licensed design professionals provide a continuous bottom mat of nonprestressed reinforcement in two-way slabs regardless of the bottom tensile stress to mitigate cracking. This reinforcement can be used to satisfy serviceability and strength requirements and for tendon supports in one direction.

3.3.6 *Minimum shear reinforcement*—A minimum amount of shear reinforcement is necessary in all post-tensioned two-way joists. The shear reinforcement additionally provides a means of supporting the tendon design profile in the beam cross section. Where tendons are not adequately supported by stirrups, local deviations of the tendons from the assumed parabolic curvature can result during placement of the concrete. When such tendons are stressed, the deviations from the intended curvature tend to straighten out, and this process can impose large tensile stresses that can result in cracking.

3.3.7 *Structural integrity*—ACI 318-14, 8.7.5.6.1, requires that at least two tendons with 1/2 in. (13 mm) diameter or larger strand should be placed in each direction at columns. Tendons are placed in the region bounded by the longitudinal reinforcement of the column and should be anchored within the region of the column bounded by the longitudinal reinforcement of the column, and the anchorage shall, according to ACI 318, be located beyond the column centroid and away from the anchored span. In case this provision cannot be satisfied, alternative structural integrity bonded reinforcement is provided by means of bottom reinforcement satisfying ACI 318-14, 8.7.5.6.3. This alternative provision first appeared in ACI 318-08. ACI 318-14, R8.7.5.6.1, notes that the placement of tendons to satisfy structural integrity provisions additionally prevents bursting through the top of the slab.

3.4—Beams

Beams are addressed in ACI 318-14 Chapter 9. The chapter organization provides design limits, design strength, required resistance, reinforcement limits, and reinforcing detailing.

3.4.1 *Beam tendon layout*—To maintain the tendon profile, the tendon should be supported along its length. Bundles of two to four single-strand tendons should be tied to minimum No. 3 (No. 10) stirrups or support bars at a recommended maximum spacing of 2 ft 6 in. (760 mm) on centers. Bundles of five or six single-strand tendons should be tied to minimum No. 4 (No. 15) stirrups or support bars at a recommended maximum spacing of 3 ft 6 in. (1.1 m) on center. Bundles of more than six tendons are not recommended. Bundles should also be limited to two layers. More than two layers or bundles of more than five tendons increase the possibility of tendons riding on top of, below, or between the adjacent tendons when stressed due to the initial space taken by the sheathing and coating. This size and spacing of stirrups is recommended even when the magnitude of the shear stress is such that no stirrups are required under the provisions of ACI 318-14, 9.7.6.2. In most cases, closer stirrup spacing will be required to satisfy the shear reinforcement requirements of ACI 318.

3.4.2 *T-beam flange width*—The effective flange width b_f of post-tensioned T-beams in bending may be taken in accordance with ACI 318-14, 9.2.4.2, or may be based on elastic analysis procedures. Flange widths exceeding those specified for nonprestressed reinforced concrete T-beams in ACI 318-14, 9.2.4.2, have been used with satisfactory results. The effective flange width for axial forces near post-tensioning

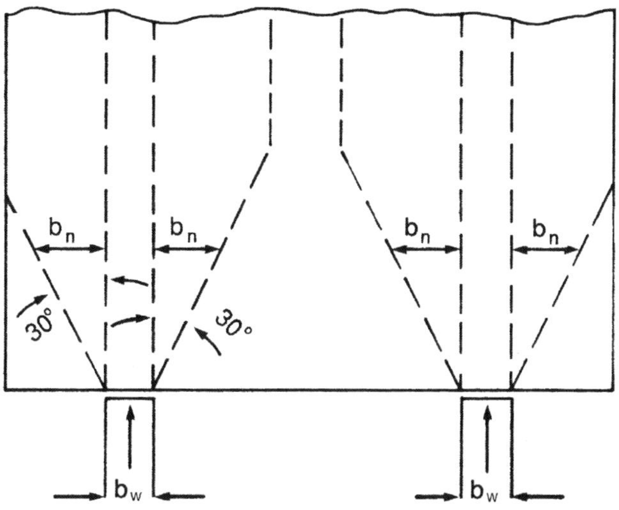

Fig. 3.4.2—Effective flange widths for normal forces.

anchorages may be assumed in accordance with Fig. 3.4.2 as $2b_n + b_w$. The transitional width $2b_n + b_w$ increases, as seen in Fig. 3.4.2, until it reaches the effective flange width b_f or the edge of the member.

Once selected, the design flange width should remain consistent throughout the design for gravity loads, including section properties for flexural concrete stress calculations and compression widths for strength design. Different flange widths may be appropriate in design for effective member stiffness in the lateral load analysis.

3.5—Effects of supporting walls and columns on prestress forces

When columns and walls have significant stiffness in the direction of prestress, consideration should be given to the effects of the mutual restraining actions of the slab, columns, and walls (Aalami and Barth 1989). These restraining actions may result in cracks in either the slab, the supporting elements, or both. This effect can be quite serious for long slabs with high shrinkage and creep (Branson and Christiason 1971). Likewise, the effects of the prestress forces on stiff supporting elements should be investigated. Design and construction options are available to reduce the effects of the shortening on both the slab and the supporting elements. The moments or stresses that occur over time due to creep and shrinkage shortening are reduced approximately 50 percent compared to the results of elastic analysis due to creep (Ghali et al. 1969). Dimensional changes due to temperature variation occur over a relatively short period, and their effect would not be reduced by creep of the concrete. The restraining effects due to dimensional changes can be accommodated in the following ways:

a) Design or locate supporting elements to minimize restraint. Relatively long, flexible columns may reduce restraint forces to the point where the forces can be accommodated by the column. This sometimes can be achieved by lowering the column footing. Lateral-load-resisting elements can often be located near the center of movement,

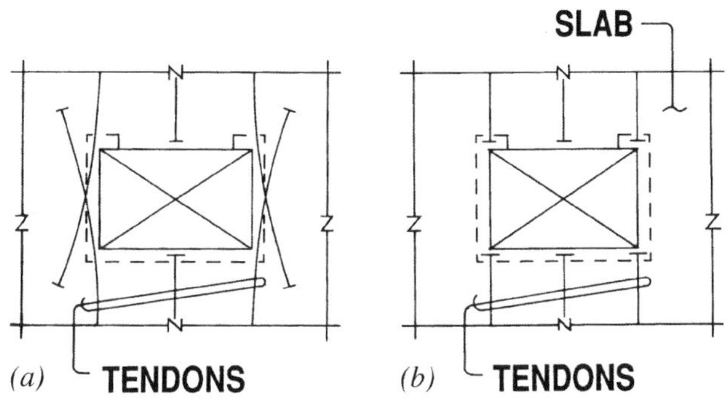

Fig. 3.5—Plan view showing tendon layouts: (a) crack-inhibiting layout; and (b) crack-promoting layout.

the point toward which all other points move, to mitigate restraint-to-shortening effects. For irregular layouts where a small slab area cannot deform with the deformations of the adjoining slab, it is advisable to provide a complete structural separation between the two slab areas. Small areas may be designed as nonprestressed reinforced concrete when it is determined that post-tensioning cannot be used effectively.

b) Where slab lengths between construction joints exceed approximately 150 ft (45 m), consideration should be given to the effects of slab shortening on restraining walls and columns. In such cases, the structure may be segmented with closure strips or temporary joints to minimize the movement and restraint developed during post-tensioning and due to early volume changes. Reinforcement, either prestressed or nonprestressed, should be provided to achieve continuity when the closure strip is closed with concrete. These strips should preferably be left open for a sufficient length of time to help minimize the effects of slab shortening. Allowing the slab to creep and shrink for 30 to 60 days is sometimes adequate. The design of reinforcement should be based on the amount of reinforcement required to achieve continuity, taking into consideration the deflection or camber that is expected to occur before casting the closure strip. Temporary shoring can be used to ensure full continuity for both dead and live loads.

c) Detail the connection between the flexural elements and columns to permit movement. Use this method with care to ensure that it is effective and easily constructed, because such details often cause more problems than they solve.

d) Improve the layout of bonded nonprestressed reinforcement or add bonded nonprestressed reinforcement. Bonded reinforcement placed parallel to restraining walls is effective in distributing potential restraint cracks. A reinforcement ratio of 0.15 percent with bars placed half at the top and half at the bottom of the slab over a width of approximately one-third of the span normal to the wall can be considered adequate for this purpose. Potential diagonal cracks at slab corners, reentrant slab corners, and corners of walls can be similarly distributed by providing either diagonal or orthogonal bonded reinforcement. Overlapping of tendons is recommended around openings to counteract potential diagonal cracks at the corners of the openings in accordance with Fig. 3.5.

e) Sand pocket column bases have also been used in alleviating volume change movement. A sand pocket column base is a temporary hinge created at the bottom of selected columns that are most susceptible to volume change effects—that is, those furthest from the center of rigidity, exceptionally stiff columns, or both. This temporary hinge is created by blocking out the bottom of the column form by filling a prescribed height, typically 1 to 2 ft (300 to 600 mm), with sand that is removed after the forms are removed, leaving a length of column supported only by the vertical reinforcement. Similar to closure strips in the slab, the pocket is left open for an allotted period to allow elastic shortening and early-age shrinkage and creep in the floor to dissipate before the pocket is cast. The capacity of the vertical reinforcement to support the anticipated column loads expected prior to casting the pocket and restoring full column capacity should be verified. A sand pocket is accounted for in design by assuming that all elastic shortening and early-age shrinkage and creep has been dissipated through the temporary hinge. The column is then only subjected to volume change that occurs after the pocket is cast.

f) The effects of elastic shortening, shrinkage, and creep of the floor system cause a sustained shortening load in the structure; however, the shortening from shrinkage and creep takes place gradually over time. For typical columns, the effect of shortening on column design is lessened due to creep and micro-cracking of columns. Thus, for cast-in-place post-tensioned structures with typical columns, the computed strain values for elastic shortening, shrinkage, and creep can be reduced by dividing by a factor of 2 for analysis and design, as suggested by Fintel (1985). Further consideration of the effects of floor system shortening may be necessary for structures with more rigid elements such as walls and large columns.

In two-way flat plates, the average prestress is often on the order of 125 psi (0.86 MPa). Prestress of this magnitude does not usually produce large dimensional changes due to elastic shortening or concrete creep. Even at this prestress level, however, care should be exercised when the building dimensions or the dimensions between joints become large, or when the flexural elements are supported by rigid elements that could produce substantial restraint forces if not properly detailed.

CHAPTER 4—ANCHORAGE ZONES

4.1—Anchorage zone reinforcement

Anchorage zone reinforcement should be designed in accordance with ACI 318-14, 25.9. This section considers the anchorage zone to be divided into two zones, as seen in Fig. 4.1a.

a) The local zone is the rectangular prism, or equivalent rectangular prism for circular or oval anchorages, of concrete immediately surrounding the anchorage device and any associated confining reinforcement.

b) The general zone in post-tensioned members is the portion of the member through which the concentrated prestress force is transferred to the concrete and distributed more uniformly across the section. For intermediate anchorage devices, the anchorage zone includes the disturbed regions ahead of and behind the anchorage devices. The general zone includes the local zone.

For the local zone, bearing plates are defined as either basic or special anchorage devices. Basic anchorage devices are typically large, stiff, bearing plates that develop bearing stresses less than $0.4f_c'$ and can function adequately with no confining reinforcement. The requirements for basic devices are specified in 5.10.9.7.2 of AASHTO (2014). Special anchorage devices are typically smaller bearing plates with more complex geometries, multiple bearing planes, integral wedge plates, and stiffening ribs, which develop high bearing stresses and require confining reinforcement to function properly. Special devices are required to pass acceptance tests outlined in 10.3.2 of AASHTO (2015).

For single-strand anchors, ACI 318 allows anchorage devices that meet the bearing strength requirements of ACI 423.7. Anchorage dimensions should be such that average bearing stresses on concrete should not exceed permissible concrete compressive stress at transfer, given as

$$f_{cp} = 0.75 f_{ci}' \sqrt{\frac{A_b'}{A_b}} \leq 1.25 f_{ci}' \quad (4.1a)$$

and at service

$$f_{cp} = 0.6 f_c' \sqrt{\frac{A_b'}{A_b}} \leq f_c' \quad (4.1b)$$

where f_{cp} is permissible concrete compressive stress; f_{ci}' is strength of concrete at time of initial prestress; f_c' is specified concrete compressive strength; A_b' is maximum area of the portion of the concrete anchorage surface that is geometrically similar to and concentric with the area of the anchorage; and A_b is net bearing plate area.

The average bearing stress P/A_b in the concrete is calculated by dividing the force P of the prestressing steel by the net bearing area A_b between concrete and bearing plate or other structural element of the anchorage that has the function of transferring force into the concrete.

Tests have shown that to function properly at this stress level, with an acceptable factor of safety against failure, backup bars are required as described in ACI 318-14, 25.9.4.4 (Sanders et al. 1987; Falconer and Wilson 1988). An acceptable detail is illustrated in Fig. 4.1b. The two No. 4 (No. 13) horizontal bars are usually adequate for single-strand anchors.

For more complex general zones, ACI 318 allows the following methods for analysis and design:

a) Equilibrium-based plasticity models; for example, strut-and-tie models

b) Linear stress analysis including finite element analysis or equivalent

c) Simplified equations where applicable

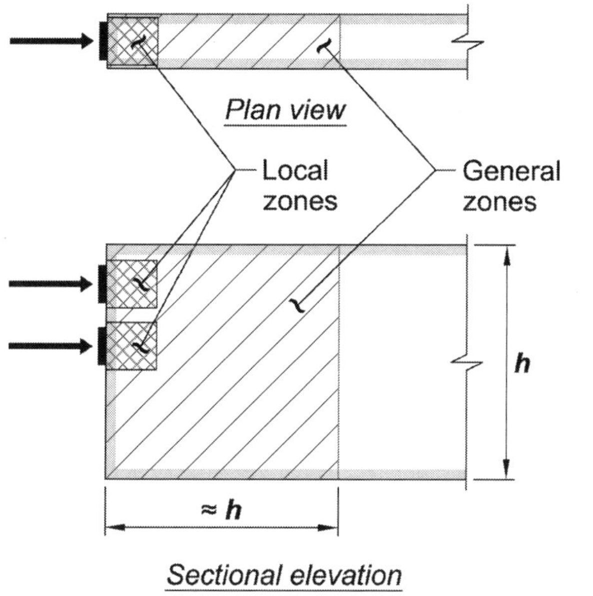

Fig. 4.1a—Local and general zones (ACI 318-14, R25.9.1.1a.)

For design, ACI 318-14, 25.9.2.1, stipulates a load factor of 1.2 be used on the maximum tendon jacking force. The strength reduction factor for post-tensioning anchorage zones for normalweight concrete is 0.85 (ACI 318-14 Table 21.2.1). Additional guidance for anchorage zone design can be found in Sanders et al. (1987), Breen et al. (1994), and Post-Tensioning Institute (2000a).

Anchorage zone reinforcement can be spirals, stirrups, orthogonal reinforcement, headed shear stud reinforcement, or combinations of these. Groups of anchorages should be restrained with reinforcement in a direction perpendicular to tendons and extending through the entire group of anchorages. All orthogonal reinforcement should be mechanically anchored around reinforcement running parallel to the tendons. Spirals, stirrups, or orthogonal reinforcement should be fully developed. The clear distance between bars or pitch of spirals used as anchorage zone reinforcement should be not less than the maximum size of the aggregate plus 1/2 in. (13 mm) or 1-1/2 in. (40 mm).

CHAPTER 5—SEISMIC DESIGN

5.1—Earthquake loading

5.1.1 *Code requirements*—Most concrete structures located in areas subject to seismic activity that include post-tensioned elements in the gravity load-carrying structural system are provided with shearwalls, braced frames, or nonprestressed special moment frames for resisting lateral forces due to wind and earthquakes. The International Building Code (International Code Council 2015) contains minimum seismic design criteria based on the requirements and commentary published by the Structural Engineers Association of California (1999), NEHRP (1991), or both.

While the International Building Code (International Code Council 2015) permits the use of unbonded post-tensioning tendons in structural elements carrying gravity or vertical

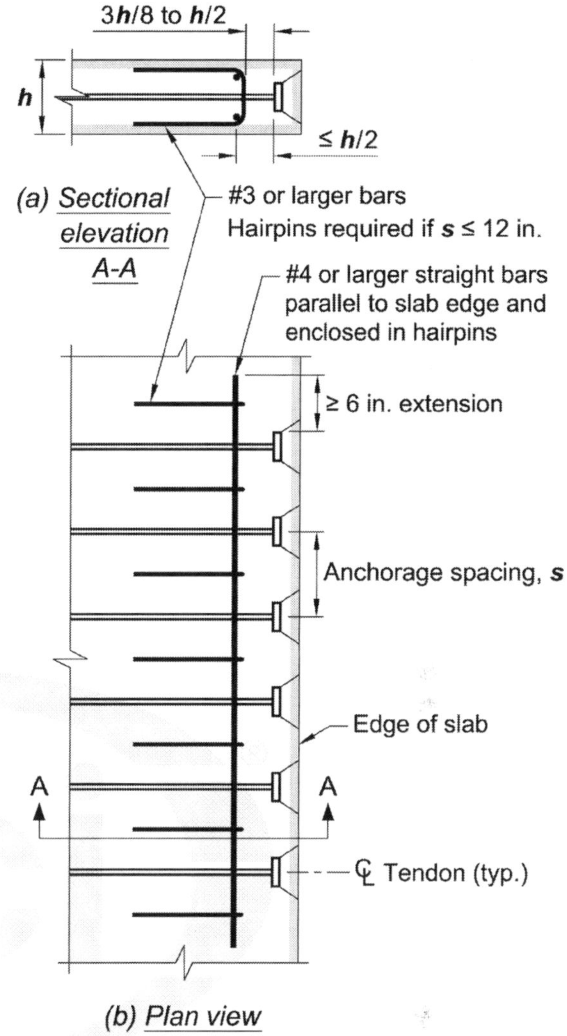

Fig. 4.1b—Anchorage zone reinforcement for groups of 1/2 in. (13 mm) diameter 270k (1860 MPa) single-strand tendon anchorages in slabs (ACI 318-14, R25.9.4.4.6). (Note: 1 in. = 25.4 mm.)

loads only, and in horizontal diaphragms connecting lateral-load-resisting elements under earthquake loading, it limits the amount of prestress force in frame members resisting earthquake forces.

ACI 318-14, 18.6.3.5, states

> Prestressing tendons shall be permitted in flexural members of frames provided the average prestress f_{pc}, calculated for an area equal to the member's shortest cross-sectional dimension multiplied by the perpendicular dimension, must be the lesser of 500 psi (3.45 MPa) or $f_c'/10$ at locations of nonlinear action where prestressing tendons are used in members of frames.

Figure 5.1.1 illustrates the applicable cross-sectional area.

ACI 318-14, 18.6.3.5c and d, require that for members in which prestressing tendons are used together with mild reinforcement to resist earthquake-induced forces, prestressing

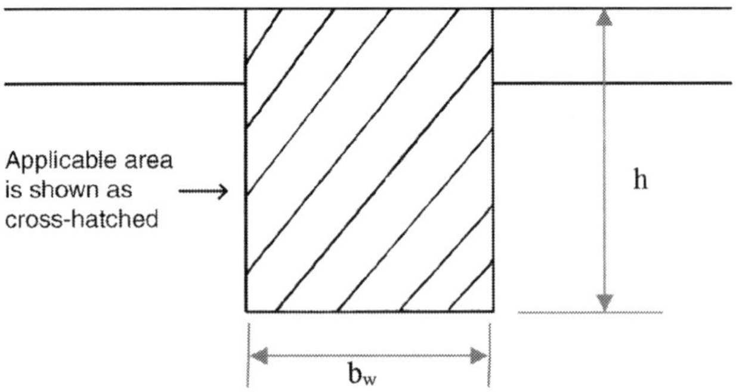

Fig. 5.1.1—Area of determining maximum prestress force in seismic frame T-sections.

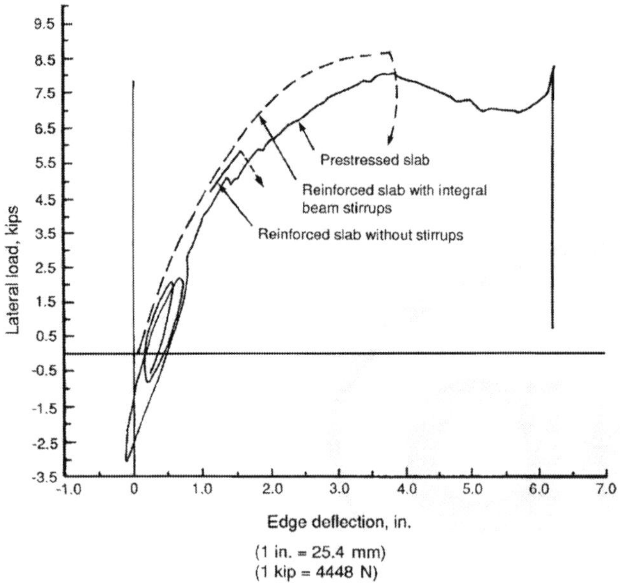

Fig. 5.1.2—Comparison of lateral load-edge deflection relationships for reinforced and prestressed concrete slab-interior column specimen (Hawkins 1981).

tendons shall not provide more than one-fourth of the strength for both positive moments and negative moments at the joint face and shall extend through exterior joints and be anchored at the exterior face of the joint or beyond. Anchorages for tendons should pass dynamic and fatigue testing requirements. Anchorage assemblies are to withstand, without failure, a minimum of 50 cycles of loading ranging between 40 and 85 percent of the minimum specified strength of the tendon.

The International Building Code (International Code Council 2015) and ACI 318 contain provisions that all framing elements not required by design to be part of the lateral-force-resisting system should be capable of resisting moments induced by the displacements of the structure resulting from lateral forces in addition to the moments caused by vertical loads; this applies to prestressed concrete elements as well as to those constructed of other materials.

Many tall concrete buildings, some with more than 40 stories, have been built in recent years in California with unbonded post-tensioned flat-plate floor systems and external precast concrete moment frames using unbonded post-tensioning tendons as part of the primary beam reinforcement. Code requirements were satisfied with performance criteria. One such building, the Paramount Apartments at 3rd and Mission Streets in San Francisco, was, at the time it was built, the tallest concrete building in Seismic Design Categories D, E, and F (Englekirk 2002).

ACI 550.3 and ACI ITG-5.2 both address specific performance and qualification requirements for specific applications of unbonded post-tensioning. ACI 550.3 provides design guidance for precast post-tensioned frame structures, whereas ACI ITG-5.2 reports design methodology for precast post-tensioned shear wall structures.

5.1.2 *Behavior*—Under-reinforced prestressed concrete elements—that is, tension controlled in accordance with ACI 318-14, 21.2.2—can generally satisfy ductility requirements of the International Building Code (International Code Council 2015), Section 1910.5.2 (Parme 1972). Figure 5.1.2 (Hawkins 1981) shows that after low-intensity reversed cyclic loading of interior column-slab specimens, conventionally reinforced slabs required the addition of closely-spaced stirrup reinforcement to attain ductility comparable to that of a post-tensioned slab. Because strains in an unbonded tendon are distributed over the tendon length, the tendons would not be expected to be stressed beyond the elastic range, even in a severe earthquake. As a result, under large member deformation, unbonded tendons absorb, store, and release substantial amounts of energy elastically; however, they do not dissipate much energy in the form of heat or member degradation. This elastic response results in a substantial decrease in associated damage to the member. ASCE/SEI 7 requires that slabs and diaphragm collector elements and the connections to the vertical seismic-force-resisting system use of an overstrength factor Ω_0 to promote elastic slab/diaphragm response and to limit the inelastic behavior to the elements specifically designed and detailed for ductile behavior. Some designers have extrapolated this requirement to the entire diaphragm design, where possible, and unbonded post-tensioning often provides for this without any added cost to the structure.

Laboratory tests of post-tensioned structural elements have indicated that energy dissipation characteristics under seismic loadings conforming to accepted standards can be achieved by appropriate combinations of prestressed and nonprestressed bonded reinforcement (Chung 1973; Miyamoto 1973; Inomata 1971; Blakeley and Park 1971; Park and Thompson 1977; Priestley and Tao 1993; Priestley and MacRae 1994). In addition to these laboratory tests, which deal with members having both bonded and unbonded tendons, several mid-rise and high-rise structures incorporating unbonded tendons in earthquake-resisting frame members resisted high lateral forces with no structural distress during the 1971 San Fernando, the 1989 Loma Prieta, and the 1994 Northridge earthquakes (Watry 1994). In the design of these structures, the contribution of the tendons as tensile reinforcement under seismic loading was neglected, but the moments induced in the frame by tendon action were considered. Grade 60 (Grade 420) reinforcing bars were provided for moment capacity and to supply energy dissipation. Because the tendons were not stressed beyond the elastic range, they reduced the deterioration of shear capacity by providing a nearly constant shear friction force at beam-column joints.

Post-tensioned beams are commonly proportioned to be more slender than conventionally-reinforced members. The resulting reduction in beam stiffness is offset by the increase in stiffness resulting from the reduction in flexural cracking and reduced inelastic hinge lengths, such that the overall performance of the frame compares favorably with conventional ductile frames.

Results of high-intensity reversed cyclic loading tests (Hawkins and Ishizukra 1988) of specimens representing concrete ductile moment-resistant frames with unbonded post-tensioned beams indicated similar results to specimens containing only nonprestressed reinforcement. This report recommends that the average prestress, based on the rectangular cross-sectional area of the beam, should be limited to approximately 350 psi (2.4 MPa). The stiffness after seismic loading of the post-tensioned frame specimens was larger than the stiffness of the non-post-tensioned specimen. Post-tensioning also improved the behavior of the beam-column joint.

Specifications for anchorage systems for unbonded tendons (Post-Tensioning Institute 2000a; ACI 423.7) contain static and fatigue test requirements that are more severe than would be anticipated in an earthquake of high intensity.

The construction failure of a flat plate lift-slab structure demonstrated the integrity of unbonded tendon anchorages even after collapse of the structure, despite the tensile failure of the strand and shattering of the concrete immediately ahead of the anchorages (Suarez and Schupack 1988).

CHAPTER 6—DESIGN CONSIDERATIONS

6.1—General

Strength and serviceability limitations, including stresses, are to conform to the provisions of ACI 318, but some recommendations are offered that differ from ACI 318 to provide equivalent strength and serviceability, or relate to areas not covered by ACI 318.

6.2—Design strength

The strength of prestressed members are to be at least equal to the required strength provisions contained in the member chapters of ACI 318.

6.3—Tendon stress at factored load

Equation (6.3a) or Table 20.3.2.4.1 of ACI 318-14 was developed primarily from test results of beams and is limited to members with span-depth ratios of 35 or less (Mattock et al. 1971)

$$f_{ps}\ (\text{psi}) = f_{se} + 10{,}000 + \frac{f'_c}{100\rho_p} \leq f_{se} + 60{,}000$$

$$f_{ps}\ (\text{MPa}) = f_{se} + 70 + \frac{f'_c}{100\rho_p} \leq f_{se} + 420 \qquad (6.3a)$$

where f_{ps} is the stress in prestressed reinforcement at nominal strength; f_{se} is the effective stress in prestressed reinforcement after allowance for all prestress losses; ρ_p is the ratio of prestressed reinforcement A_{ps}/bd_p; and f_{py} is the specified yield strength of prestressing reinforcement, psi (MPa). In all cases, f_{ps} should not exceed f_{py}.

Tests have shown that Eq. (6.3a) overestimates the amount of stress increase in unbonded tendons in one-way slabs, two-way flat plates, and flat slabs with higher span-depth ratios (Mojtahedi and Gamble 1978). For members with span-depth ratios greater than 35, f_{ps} is calculated using Eq. (6.3b) or ACI 318-14 Table 20.3.2.4.1 for design stress in unbonded tendons

$$f_{ps}\ (\text{psi}) = f_{se} + 10{,}000 + \frac{f'_c}{300\rho_p} \leq f_{se} + 30{,}000$$

$$f_{ps}\ (\text{MPa}) = f_{se} + 70 + \frac{f'_c}{300\rho_p} \leq f_{se} + 210 \qquad (6.3b)$$

Research (Burns et al. 1991) indicates that redistribution over time, or equalizing of unbonded tendon stresses f_{se}, is not fully achieved. This research recommends that Eq. (6.3a) and (6.3b) be used to calculate f_{ps} at each individual design section along the member, known as the variable force method, rather than averaging the values between positive and negative moment sections in each span.

Variable force along the tendon is a function of the type of tendon, total wobble and curvature, and construction care. Bondy (1992) indicates that for most typical configurations, designs using variable force will not vary significantly from designs using the constant force method, where the design at all points along the member is based on the average force required to produce the initial tendon elongation, less long-term losses. Given the Bondy (1992) work and the overwhelming experience with design of buildings using this method, the average force method is generally accepted for most building configurations (Kang and Wallace 2008b). In some cases, such as unusually long tendons, tendons with more angular changes in the profile, or tendons with greater friction, the engineer may opt for a variable force design and specify the force requirements accordingly.

ACI 318 eliminated the requirement that the tendon stress after transfer be less than $0.82f_{py}$ or less than $0.74f_{pu}$. Whereas this condition could theoretically exist in the interior of a beam away from anchor locations, it was not observed to result in any field performance issues.

Additional studies have been conducted to refine the aforementioned equations that appear in ACI 318. MacGregor (1989) constructed and tested a one-fourth-scale, three-span continuous segmental box-girder bridge, post-tensioned with external tendons. His measurements and observations influenced the development of the AASHTO (2003, 2014) equation for the stress increase in unbonded tendons at ultimate. The derivation of the equation is based on the assumption that, at ultimate, a simple span beam or slab prestressed with unbonded tendons will form a mechanism comprising two rigid members connected at a hinge. All tendon elongation is assumed to be equal to the length of the opening hinge.

Naaman et al. (2002) published a document on this subject. The methodology and prediction equation proposed in this study is an attempt to provide a rational solution for both steel- and fiber-reinforced polymer (FRP) tendons. The proposed equation accounts for numerous parameters that are not entirely covered in other codes, and accommodates the combined presence of reinforcing steel and prestressing reinforcement, the span-depth ratio, nonmetallic reinforcements, and differences in load patterns. It also provides a theoretical framework in cases where unusual conditions may be encountered. The document was reviewed but not accepted by Joint ACI-ASCE Committee 423 due to its complexity.

6.4—Prestress losses

Prestress losses consider friction, elastic shortening, wedge seating, creep, shrinkage, and relaxation, and should be calculated by the design engineer and stated on the design drawings. Zia et al. (1979) made it possible to calculate reasonably accurate values for the various code-defined sources of loss without excessive effort. For preliminary design, the values of prestress loss given in Table 6.4 may be used instead of more detailed loss calculations (Zia et al. 1979). The loss values in Table 6.4 are based on the use of normalweight concrete and on average values of concrete strength, prestress level, and exposure conditions. Significant deviations in loss values have little effect on strength or performance, because even large variations in losses up to 100 percent are relatively small percentages of tendon effectiveness and strength stress levels. A complete discussion of prestress losses and their interaction with strength design can be found in ACI 423.10R.

Prestress losses may vary significantly above or below the values in Table 6.4 in cases where the concrete is stressed at low strengths, where concrete is highly prestressed, or in very dry or very wet exposure conditions. The loss values in Table 6.4 do not include losses due to friction or wedge seating. Some portion of the friction loss can usually be offset by use of temporary initial tendon stresses exceeding $0.70f_{pu}$. Special consideration, such as detailed friction loss calculations, more precise wobble and curvature coefficients, and resultant reductions in effective tendon forces, should be given to friction losses whenever tendons in excess of approximately 100 ft (30 m) long are stressed from only one end.

For calculation of friction losses for unbonded strand tendons with post-tensioning coating in plastic sheathing using the formulas in ACI 423.10R, the friction factor μ, usually ranging from 0.05 to 0.15, and the wobble factor K, usually ranging from 5 to 15 × 10^{-4}/ft (15 to 50 × 10^{-4}/m), can be used for design calculations. It may be necessary to obtain more precise values for the friction-factor and wobble-factor coefficients to calculate tendon elongations during stressing for comparing tendon force as measured by gauge pressure and tendon elongation.

6.5—Average prestress

6.5.1 *Minimum average prestress*—The average prestress is defined as the total prestress force, after losses, divided by the total concrete cross section, as defined in ACI 318-14, 8.6.2.1. There has been satisfactory experience in one- and two-way slabs with an average prestress of at least 125 psi (0.9 MPa). Lower values have been used successfully for short-span applications. These short-span applications can be characterized as having flexural stresses substantially below $6\sqrt{f_c'}$ psi ($0.5\sqrt{f_c'}$ MPa), minimal volume-change effects, and vertical element stiffness such that restraint to slab shortening is minimized. It should be noted that the 125 psi (0.9 MPa) minimum average compressive stress for two-way slabs in ACI 318-14, 8.6.2.1, was based on punching shear criteria (Burns and Hemakom 1977) and not flexural behavior.

Table 6.4—Approximate prestress loss values (Post-Tensioning Institute 1990)

Post-tensioning material	Prestress loss, psi (MPa)	
	Slabs	Beams and joists
Stress-relieved 270k strand and stress-relieved 240k wire	30,000 (207)	35,000 (240)
Bar	20,000 (138)	25,000 (172)
Low-relaxation 270k strand	15,000 (103)	20,000 (138)

ACI 318-14, 8.6.2.1, clarified that the 125 psi (0.9 MPa) minimum prestress must apply at all slab sections along the tendon group. The 318-14 Commentary states

> If the slab thickness varies along the span of a slab or perpendicular to the span of a slab, resulting in a varying slab cross section, the 125 psi minimum effective prestress and the maximum tendon spacing is required at every cross section tributary to the tendon or group of tendons along the span, considering both the thinner and the thicker slab sections. This may result in higher than the minimum f_{pc} in thinner cross sections, and tendons spaced at less than the maximum in thicker cross sections along a span with varying thickness, due to the practical aspects of tendon placement in the field.

This provision requires the minimum level of prestress in changes of slab thickness, drop panels, and shear panels.

In view of the amount and distribution of bonded reinforcement required in one-way slabs as discussed in 3.2.1 and 3.2.2, minimum average prestress is considered less significant for one-way slabs than for two-way slabs, which usually do not have bonded positive moment reinforcement in interior panels. For applications such as parking structures, where control of cracking is significant from the standpoint of improving durability against application of deicing chemicals, average prestress levels of the order 200 psi (1.4 MPa) in each direction are recommended. In this case, the designer is cautioned to implement available measures to mitigate cracking that can result from restraint to axial shortening, which could be aggravated if providing 200 psi (1.4 MPa) results in an increase in prestress force (refer to 3.5).

6.5.2 *Maximum average prestress*—A high value of average prestress may induce excessive shortening due to elastic deformation and creep. Values of average prestress in slabs higher than 300 psi (2.1 MPa) should be used with caution and, in such cases, available means should be employed to mitigate the effects of cracking due to the restraint of shortening. Detailing, as discussed in 3.5, to ensure that restraint of immediate and long-term shortening does not interfere with the imposition of the calculated average prestress in the concrete, is of increasing importance as the average prestress is increased.

6.6—Serviceability requirements

Design for performance at service loads should consider the factors included in ACI 318-14 Chapter 24. Specified limits on calculated service level deflections are provided in ACI 318-14 Table 24.2.2. The deflection limits in each member chapter of ACI 318 are based on calculated deflections only and not measurements made on the actual structure. Deflections can be influenced by many construction factors that are beyond the control of the designer. Consequently, field surveys may be unable to isolate deflections caused by applied loads from deflections resulting from construction practices.

ACI 318-14, 24.2.3, requires that the immediate deflections of two-way prestressed slabs be calculated using methods or formulas for elastic deflections. ACI 318-14, R.24.2.23, states that the calculations of deflections for two-way slabs is challenging, even if linear elastic behavior can be assumed. Simplified approaches to deflection calculations are available (Scanlon 1999; Chao and Naaman 2006).

6.7—Continuous members

6.7.1 *One-way members*—For slabs or beams continuous over two or more spans with one-way prestressing only, a loading condition or fire exposure that causes failure of all the tendons in one span will lead to a loss of prestress and much of the load-carrying capacity in the other spans. Consideration should be given to the consequence of such a catastrophic failure in any specific span to the overall stability of the structural system. ACI 318 has responded to this concern and to other considerations, such as crack width limitation, in ACI 318-14, 9.6.2. This section specifies minimum bonded reinforcement equal to 0.40 percent of the area of that part of the cross section between the flexural tension face and the center of gravity of the gross section. Grade 60 (Grade 420) reinforcement should be used for this purpose. This amount of bonded reinforcement is approximately equal to the minimum flexural reinforcement requirement for nonprestressed slabs (ACI 318-14, 7.6.1).

One-way slabs can also incorporate unbonded partial length tendons, lapped tendons, or tendons with intermediate anchorages that would serve to limit the extent of the loss of load-carrying capacity. Reinforcement required by ACI 318-14, 7.6.2, is not required to satisfy Chapter 25 of ACI 318-14; the top and bottom reinforcement is not even required to overlap. In ACI 318-14, 7.7.4.4, positive moment bars need only be $L/3$ in length, centered in the span, and negative moment bars need only extend one-sixth the clear span on each side of the support. Restraint-to-shortening cracking has been observed in the space between the ends of the top and bottom bars detailed in accordance with ACI 318-14, 7.7.4.4. Some licensed design professionals extend at least one-third of the bottom reinforcement in single-span slabs or one-fourth of the reinforcement in continuous span slabs to the full span length as an added element for structural integrity.

In accordance with ACI 318-14, 7.7.4.2 or 7.7.4.3, bonded reinforcement required for strength in both beams and slabs is required to be detailed in accordance with the provisions of ACI 318-14 Chapter 25, with sufficient development length for the positive and negative moment bars, to ensure that the bonded reinforcement will function as an independent load-carrying system.

6.7.2 *Two-way members*—In the case of two-way slabs of the usual proportions, catastrophic loading beyond design capacity in one bay is generally not as critical to other spans as in one-way systems. For two-way slabs, the load-carrying capacity of the tendons in each direction should be considered. Tests (Post-Tensioning Institute 1977; Ritz et al. 1975; Burns and Hemakom 1977; Hemakom 1975) have demonstrated that two-way flexural behavior of slabs with significant loss of prestress force still has adequate strength. This favorable behavior is recognized in ACI 318 and the International Building Code (International Code Council 2015), and some local building codes by a reduction of the amount of bonded reinforcement required in comparison with one-way systems. Additional information regarding crack control in continuous post-tensioned slabs is presented in Aalami and Barth (1989).

6.8—Corrosion protection

Unbonded prestressing tendons are required to be protected against corrosion during fabrication, transit, storage, construction, and after installation. ACI 318-14, 20.6.3, indicates minimum requirements for corrosion protection and requires encapsulated tendons. Recommendations for encapsulated tendons in prior editions of ACI 318 varied, as did the description of extreme environments. More detailed requirements for encapsulated tendons are found in ACI 423.7. ACI 423.4R provides a history of the development of corrosion protection methods and a description of problems that can arise due to inadequate corrosion protection systems.

ACI 423.7 requires that all anchorage assemblies be encapsulated for tendons used in applications governed by either ACI 318 or ACI 350 to protect against corrosion. The encapsulation includes a watertight connection of the sheathing to the anchorage, and a watertight closure of the wedge cavity and prestressing steel. For tendons used in other applications, it is permissible to use non-encapsulated anchorages unless otherwise specified by the licensed design professional. Once fabricated, tendons are to be protected from exposure to moisture, deicing salts, and other corrosive elements during handling, shipping, and storage. Further guidance for the protection of tendons during storage, transit, and installation can be found in Post-Tensioning Institute (2000a). Concrete cover for unbonded tendons should be detailed considering the factors discussed in 7.4.

6.9—Fire resistance

Fire resistance ratings can be determined in accordance with the heat transmission and dimensional provisions of PCI MNL-124-11, 6.9.1, or by the rational design procedures for determining fire endurance discussed in PCI MNL-124-11, 6.9.2. Additional references are found in Post-Tensioning Institute (2000b), ACI 216.1, and ASTM E119. ASTM E119 includes a guide for classifying construction as restrained or unrestrained. A restrained member may develop considerable axial forces due to the lateral stiffness of the supporting elements. This allows the member to arch and may improve its ability to support load. Similarly, rotational end restraint may assist load-carrying capacity. If the supporting members are too flexible and cannot restrain the expansion of the member, or are incapable of resisting the axial forces caused by the expansion of the member, the member should be treated as unrestrained.

6.9.1 *Minimum dimensions for various fire resistive classifications*

6.9.1.1 *Slabs*—To meet minimum heat-transmission requirements—that is, a temperature rise of 250°F (140°C) of the unexposed surface—the thickness requirements for concrete slabs should be the same whether the concrete is plain, reinforced, or prestressed. Table 6.9.1.1a gives the slab thickness recommended for this purpose. The required thickness for post-tensioning tendons in unrestrained slabs is determined by the elapsed time during a fire test until the tendons reach a critical temperature. For cold-drawn prestressing steel, that temperature is 800°F (430°C). For restrained slabs, there are no steel temperature limitations, but the heat transmission end-point temperature limitation (250°F [140°C]) is the same as for unrestrained slabs. Fire tests of axially restrained slabs indicate that slabs with unbonded post-tensioned reinforcement behave approximately the same during a fire as reinforced concrete slabs of the same dimensions. Accordingly, the required cover for post-tensioning tendons in axially restrained slabs could be essentially the same as the cover specified by the International Building Code (International Code Council 2015) for mild reinforcement in slabs. Applying these criteria to post-tensioned slabs, the suggested slab thickness is provided in Table 6.9.1.1a and cover is as recommended in Table 6.9.1.1b (ACI 216.1; Bailey and Ellobody 2009).

Table 6.9.1.1a—Suggested thickness requirements for concrete slabs for fire endurance (From ACI 216.1-14 Table 4.2)

Aggregate type	1 hour, in. (mm)	1-1/2 hours, in. (mm)	2 hours, in. (mm)	3 hours, in. (mm)	4 hours, in. (mm)
Carbonate	3.2 (80)	4.0 (101)	4.6 (116)	5.7 (145)	6.6 (167)
Siliceous	3.5 (81)	4.3 (109)	5.0 (127)	6.2 (157)	7.0 (177)
Semi-lightweight	2.7 (69)	3.3 (84)	3.8 (97)	4.6 (119)	5.4 (137)
Lightweight	2.5 (63)	3.1 (79)	3.6 (91)	4.4 (112)	5.1 (130)

6.9.1.2 *Beams*—Minimum dimensions for beams with post-tensioned reinforcement for various fire endurances are functions of the types of steel, concrete, beam width, and cover. For restrained beams spaced more than 4 ft (1.2 m) on center, the temperature of 800°F (430°C) for cold-drawn prestressing reinforcement should not be exceeded to achieve a fire endurance classification of 1 hour or less. For classifications longer than 1 hour, this temperature should not be exceeded for the first half of the classification period or 1 hour, whichever is longer. The recommended cover in Table 6.9.1.2 is based on these criteria. For post-tensioned beams or joists less than 8 in. (200 mm) wide using strand tendons, ACI 216.1 can be used. Beams or joists that are narrower than 8 in. (200 mm) with post-tensioned high-strength alloy steel bars should have the same cover as reinforced concrete joists of the same size and fire endurance.

6.9.1.3 *Anchorage device protection*—The cover to the prestressing reinforcement at the anchorage should be at least 1/4 in. (6 mm) greater than that required in Tables 6.9.1.1b and 6.9.1.2. Minimum cover to the reinforcement bearing plate or anchor casting should be at least 1 in. (25 mm) in beams and 3/4 in. (20 mm) in slabs.

6.9.2 *Rational design for fire endurance*—Rational analytical procedures for the determination of the fire endurance of post-tensioned prestressed concrete structures have been developed from evaluation of results of fire tests conducted in accordance with the criteria for standard fire tests, ASTM E119. Basic data on the strength-temperature relationships for steel and concrete are used together with information on temperatures within concrete beams and slabs during standard fire tests. Rational design procedures for concrete

Table 6.9.1.1b—Suggested minimum cover for slabs prestressed with post-tensioned reinforcement (Post-Tensioning Institute 2006)

Restrained or unrestrained	Aggregate type	Cover thickness, in. (mm)				
		1 hour	1-1/2 hours	2 hours	3 hours	4 hours
Unrestrained	Carbonate	3/4 (20)	1-1/16 (30)	1-3/8 (35)	1-7/8 (50)	—
Unrestrained	Siliceous	3/4 (20)	1-1/4 (35)	1-1/2 (40)	2-1/8 (55)	—
Unrestrained	Lightweight	3/4 (20)	1 (25)	1-1/4 (35)	1-5/8 (40)	—
Restrained	Carbonate	3/4 (20)	3/4 (20)	3/4 (20)	1 (25)	1-1/4 (35)
Restrained	Siliceous	3/4 (20)	3/4 (20)	3/4 (20)	1 (25)	1-1/4 (35)
Restrained	Lightweight	3/4 (20)	3/4 (20)	3/4 (20)	3/4 (20)	1 (25)

Table 6.9.1.2—Suggested cover thickness for beams prestressed with post-tensioned reinforcement (Post-Tensioning Institute 2006)

Restrained or unrestrained	Steel type	Concrete type	Beam width, in. (mm)*	Cover thickness, in. (mm), for the fire endurance of:				
				1 hour	1-1/2 hours	2 hours	3 hours	4 hours
Unrestrained	Cold-drawn	Normalweight	8 (200)	1-3/4 (45)	2 (50)	2-1/2 (65)	4-1/2 (115)	—
Unrestrained	Cold-drawn	Lightweight	8 (200)	1-1/2 (40)	1-3/4 (45)	2 (50)	3-1/4 (85)	—
Unrestrained	HSA† bars	Normalweight	8 (200)	1-1/2 (40)	1-1/2 (40)	1-1/2 (40)	2-1/2 (65)	—
Unrestrained	HSA bars	Lightweight	8 (200)	1-1/2 (40)	1-1/2 (40)	1-1/2 (40)	2-1/4 (60)	—
Restrained	Cold-drawn	Normalweight	8 (200)	1-1/2 (40)	1-1/2 (40)	1-3/4 (45)	2 (50)	2-1/2 (65)
Restrained	Cold-drawn	Lightweight	8 (200)	1-1/2 (40)	1-1/2 (40)	1-1/2 (40)	1-3/4 (45)	2 (50)
Restrained	HSA bars	Normalweight	8 (200)	1-1/2 (40)	1-1/2 (40)	1-1/2 (40)	1-1/2 (40)	1-1/2 (40)
Restrained	HSA bars	Lightweight	8 (200)	1-1/2 (40)	1-1/2 (40)	1-1/2 (40)	1-1/2 (40)	1-1/2 (40)
Unrestrained	Cold-drawn	Normalweight	>12 (300)	1-1/2 (40)	1-3/4 (45)	2 (50)	2-1/2 (65)	3 (75)
Unrestrained	Cold-drawn	Lightweight	>12 (300)	1-1/2 (40)	1-1/2 (40)	1-3/4 (45)	2 (50)	2-1/2 (65)
Unrestrained	HSA bars	Normalweight	>12 (300)	1-1/2 (40)	1-1/2 (40)	1-1/2 (40)	1-1/2 (40)	2 (50)
Unrestrained	HSA bars	Lightweight	>12 (300)	1-1/2 (40)	1-1/2 (40)	1-1/2 (40)	1-1/2 (40)	2 (50)
Restrained	Cold-drawn	Normalweight	>12 (300)	1-1/2 (40)	1-1/2 (40)	1-1/2 (40)	1-3/4 (45)	2 (50)
Restrained	Cold-drawn	Lightweight	>12 (300)	1-1/2 (40)	1-1/2 (40)	1-1/2 (40)	1-1/2 (40)	1-3/4 (45)
Restrained	HSA bars	Normalweight	>12 (300)	1-1/2 (40)	1-1/2 (40)	1-1/2 (40)	1-1/2 (40)	1-1/2 (40)
Restrained	HSA bars	Lightweight	>12 (300)	1-1/2 (40)	1-1/2 (40)	1-1/2 (40)	1-1/2 (40)	1-1/2 (40)

*For beams with widths between 8 and 12 in. (200 and 300 mm), cover thickness can be determined by interpolation.

†HSA = high-strength alloy.

beams and slabs that are post-tensioned with unbonded tendons are essentially the same as those for pretensioned prestressed concrete elements (PCI MNL-124-11). Curved tendons, rather than straight or harped tendons, introduce only minor differences that do not change the design procedures. Tests of post-tensioned elements (Post-Tensioning Institute 1990) indicate that the temperatures of the tendons in positive moment regions at the end of a fire test can be considered essentially the same, regardless of whether the tendons are bonded or unbonded. Further, these tests indicate that the prestressing reinforcement stress $f_{ps\theta}$ at failure during fire tests can be estimated as a function of the ultimate steel strength at temperature θ by the relationship

$$\frac{f_{ps\theta}}{f_{pu\theta}} = \frac{f_{ps}}{f_{pu}} \quad (6.9.2a)$$

where f_{ps} is the stress in post-tensioning tendons at nominal strength, psi (MPa), which may be calculated for unbonded tendons by ACI 318-14 Table 20.3.2.4.1 (refer to 6.3); f_{pu} is the specified tensile strength of prestressing reinforcement, psi (MPa); $f_{ps\theta}$ is the stress in post-tensioned tendons at nominal strength at high temperatures, psi (MPa); and $f_{pu\theta}$ is the tensile strength of tendons at high temperatures, psi (MPa).

ASCE/SEI 7 provides a minimum required load combination for extraordinary events such as fire. All the current fire rating and testing (under ASTM E119 rules) have been performed based on the load combination of 1.0D + 1.0L, and this is the basis for the current building codes.

For continuous beams or slabs using continuous draped unbonded tendons exposed to fire from below, the value of $f_{ps\theta}$ in the negative moment regions should be taken the same as those in the positive moment region. The capacity at any point along the length of an unbonded tendon is limited by the capacity at the point where the steel temperature is highest.

On this basis, it is possible to determine the retained theoretical moment strength at a specified period of fire endurance (for example, 2 hours) in the positive moment region and in both negative moment regions of a given panel in a building. The maximum slab moment capacity at exterior columns cannot exceed that which can be transmitted to and resisted by the column. To evaluate the retained theoretical moment strength, it may be assumed that if a fire exposes the slab soffit to high temperatures, a redistribution of moments will occur, yielding the negative moment bonded reinforcement. If the applied midspan moment is less than the retained moment capacity after redistribution, the fire endurance will be adequate. This moment, based on uniformly applied loads, is given as

$$M = M_{t\theta}^{+} + (1/2)(M_{t1\theta}^{-} + M_{t2\theta}^{-}) \quad (6.9.2b)$$

where M is the total applied static moment (unfactored) = $wL^2/8$; $M_{t\theta}^{+}$ is the retained midspan moment capacity; $M_{t1\theta}^{-}$ is the retained negative moment capacity at Column 1; and $M_{t2\theta}^{-}$ is the retained negative moment capacity at Column 2.

If, however, the applied static moment is greater than the total retained moment capacity, changes should be made to the design. Several options for improving the fire endurance are available, including:

a) Increasing the concrete cover in the positive moment region

b) Increasing the number of prestressing tendons

c) Adding bonded positive moment reinforcement

d) Adding bonded negative moment reinforcement

Other solutions include the use of a thicker slab, the use of lightweight aggregate concrete, the addition of a fire-resistant ceiling, or the application of an appropriate insulating material to the slab soffit.

CHAPTER 7—MATERIALS

7.1—Tendons

ACI 423.7 specifies the testing protocols for tendons and anchorages. Key elements of the specification are summarized in the following sections.

7.1.1 *Prestressing reinforcement*—Prestressing reinforcement for unbonded post-tensioning tendons should meet the requirements of ACI 318-14, 20.3. The total elongation under ultimate load of the tendon and anchorage assembly should not be less than 2 percent measured in a minimum gauge length of 3 ft (915 mm) between two points at least 3 in. (75 mm) from each anchorage. Bondy (2008) points out that the 2 percent strain is overly conservative and is often not possible to achieve at the minimum required tendon strength of $0.95f_{pu}$ (7.1.2).

7.1.2 *Anchorage devices*—ACI 423.7 and ACI 318-14, 25.8.1, require the anchorage devices for unbonded tendons to develop 95 percent of the specified strength f_{pu}. The actual breaking strength of a length of prestressing reinforcement is an unknown quantity, known after the fact, only for the sample tested. For a discussion of actual-versus-specified strength in this context, refer to Bondy (2008). In either case, this is much higher than the maximum permissible design stress for unbonded tendons, f_{ps}, discussed in 6.3.

7.1.3 *Tests of tendons and anchorage device fittings*

7.1.3.1 *Static tests*—The test assembly should consist of standard production-quality components, and the tendons should be at least 3.5 ft (1.1 m) long between anchorages. The test assembly should be tested in a manner to allow accurate determination of the yield stress, tensile strength, and percent elongation of the complete tendon. The specimen used for the static test need not be one that has been subjected to fatigue loading.

7.1.3.2 *Fatigue tests*—A fatigue test should be performed on tendon specimens with standard production quality components and with a minimum length of 3 ft (900 mm) between anchorages. The tendon should withstand without failure 500,000 cycles from 60 to 66 percent of its minimum specified tensile strength, and 50 cycles from 40 to 80 percent of its minimum specified tensile strength. Each cycle involves the change from the lower stress level to the upper stress level and back to the lower. The specimen used for the fatigue test need not be one that has been subjected to static testing. The specimen used for the second fatigue test (50 cycles) need not be the same used for the first fatigue test (500,000 cycles).

7.1.4 *Couplers*—Unbonded tendons should be coupled only at locations specifically indicated or approved by the licensed design professional. Couplers should not be used at points where tendon radius of curvature is less than 480 strand diameters. All couplers should develop at least 95 percent of the specified tensile strength of the prestressing reinforcement. The couplers should not reduce the elongation at rupture below the requirements of the tendon itself.

Couplers should meet the fatigue test recommendations in 7.1.3.2. When used, couplers, components, or both, should be enclosed in housings long enough to permit the necessary movement of the tendon. All coupler components should be completely encapsulated before final encasement in concrete. Anchorages and couplers in aggressive environments should be encapsulated.

7.2—Protection materials

7.2.1 *Post-tensioning coating*—Unbonded single-strand tendons should use a corrosion-inhibiting coating in conformance with ACI 423.7.

7.2.2 *Sheathing*—Sheathing for unbonded single-strand tendons should conform to the requirements of ACI 423.7.

7.3—Protection of anchorage zones

The anchorages of unbonded tendons should be protected adequately from corrosion and fire. Except in special cases, anchorages should preferably be encased in concrete complying with ACI 301.

Where concrete or grout encasement cannot be used, the tendon anchorage should be completely coated with a corrosion-resistant elastomeric coating equivalent to that applied to the tendons. A suitable enclosure should be placed where necessary to prevent the penetration of moisture or the deterioration or removal of this coating. The anchorage encasement should provide fire resistance at least equal to that required for the structure.

7.4—Concrete cover

Specification of concrete cover for unbonded tendons should consider the placement tolerances specified in ACI 318-14, 26.6.2.1a, and the exposure conditions. The use of high-quality concrete, adequate cover, good construction practices, and a limit on the amount of water-soluble chloride ions in the concrete (ACI 318-14, 19.3) are all necessary to ensure long-term durability, particularly in aggressive environments. At the minimum, the use of at least the additional cover specified in ACI 318-14, 20.6.1, and consideration of a somewhat higher average prestress level, is recommended for applications exposed to deicing chemicals or for locations in the immediate vicinity of coastal regions. What is considered as the immediate vicinity is left to the judgment of the licensed design professional. Some guidance is given in ACI 362.1R. Extra cover cannot be a substitute for good quality concrete. At angled slab edges, minimum concrete covers should be maintained to all parts of the anchorage. Angled pocket formers or angled block-outs should be such so that the anchors and anchorage cavity areas have adequate cover for protection against corrosion and fire.

CHAPTER 8—CONSTRUCTION

8.1—Construction joints

Construction joints may be used to divide the floor system into segments of suitable size for placement of concrete. Intermediate stressing anchors can be used at construction joints, or the tendons may run through the joint without anchors. The distance between joints should not exceed approximately 150 ft (46 m). If it does, then consideration should be given to the effects of slab shortening as described in 3.5(b). Care should be taken to ensure a watertight joint to prevent leaks and subsequent corrosion of tendon components and reinforcement in the joint vicinity.

8.2—Closure strips

Closure strips, also known as pour strips or delay strips, are open strips that temporarily separate adjacent slabs during construction, as discussed in 3.5. Temporary shoring may be required for one or more levels to ensure full continuity of the slab for both live and dead loads. The location and details of closure strips should be clearly indicated on the construction documents.

8.3—Placement of tendons

8.3.1 *Tendon profile*—The placement of tendons should closely follow the specified profile within the tolerances recommended in 8.3.2. Prestressing tendons should be firmly supported at intervals not exceeding 4 ft (1.2 m). Any inadvertent local reversed curvature should be corrected before concreting. Tendon profiles are maintained by tying to reinforcement, chairs, or other supports with wire ties. Ties should be installed so that they do not visibly imprint or dent the extruded sheathing. Recommendations for spacing of ties to stirrups for bundles of unbonded tendons in beams are presented in 3.4.1. In beams, the tendons should be placed in bundles in accordance with the spacing limitations given in ACI 318-14, 25.6.2.1 and 26.4.4.1. The placement of beam tendons in wide, flat bundles or layers should be avoided where it prohibits the proper placement, consolidation, and continuity of the concrete in the member.

8.3.2 *Tolerances*—Vertical deviations in tendon location should be kept to ±1/4 in. (6 mm) for slab thickness dimensions less than 8 in. (200 mm), ±3/8 in. (10 mm) in concrete with dimensions between 8 in. and 2 ft (200 and 600 mm), and ±1/2 in. (13 mm) in concrete with dimensions over 2 ft (600 mm). These tolerances should be considered in establishing minimum specified tendon cover dimensions, particularly in applications exposed to deicing chemicals or saltwater environments where use of additional cover is recommended to compensate for placing tolerances. Slab behavior is relatively insensitive to horizontal location of tendons.

8.3.3 *Openings*—Deviations of tendons in the horizontal plane may be necessary to avoid interferences such as openings, ducts, chases, and inserts. Horizontal deviations around openings may result in concrete cracking due to the lateral force component of the deviated tendon. Appropriate means to avoid or limit cracking include an adequately-large radius of curvature, a minimum of 480 strand diameters, or 20 ft (6 m) for 1/2 in. (12.7 mm) diameter strands, sufficient clearance of the tendons from the edge of an opening, a straight tendon extension beyond the opening corners, and hairpin reinforcement to transfer the lateral forces to the surrounding concrete (Falconer and Wilson 1988).

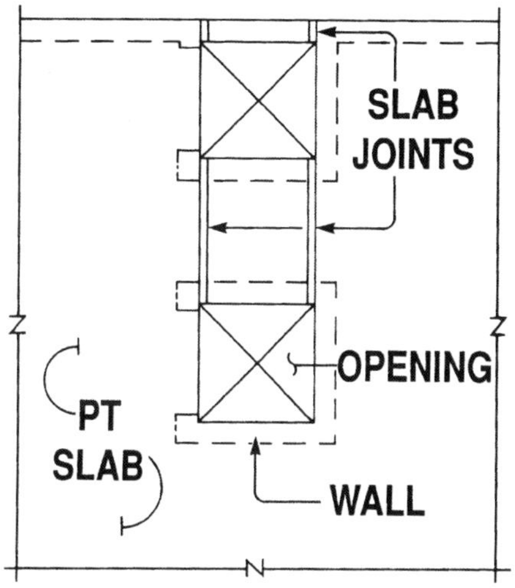

Fig. 8.3.3—Plan view showing isolated slab areas.

For larger openings, when it is necessary to terminate some tendons at the opening, the crack-inhibiting layout of tendons shown in Fig. 3.5(a) (Aalami and Barth 1989) is recommended rather than the crack-promoting layout in Fig. 3.5(b). In some cases, it may be preferable to isolate small slab sections adjacent to openings with slab joints, as shown in Fig. 8.3.3 (Aalami and Barth 1989). The isolated slab sections should be reinforced as required with nonprestressed bonded reinforcement.

For larger openings, reinforce the top and bottom of the slab at openings with diagonal bars to control cracking initiated at the corners of the opening. In some cases, additional perimeter reinforcement may be necessary to distribute loads applied at the slab opening. Loads at openings can normally be accommodated by use of tendons and additional bonded reinforcement around the perimeter. Beams, however, may be required to carry the loads at perimeters of openings, and a structural analysis should be made to determine whether these loads can be carried by use of additional tendons and additional bonded reinforcement, or whether beams are required. It is generally preferable to locate openings in the midspan areas of one- and two-way slabs to minimize the effect of the opening on the shear capacity of the slab at walls or columns. When openings are located where they reduce shear capacity, a more exact analysis of the capacity of the actual slab configuration is essential.

8.4—Concrete placement and curing

Concrete should be placed in such a manner that tendon alignment and reinforcement positions remain unchanged. Attention should be given to concrete vibration at tendon anchorages to ensure uniform consolidation at these locations. Voids behind the bearing plate, or insufficient concrete strength, will cause concrete failure or blowout. Careful vibration and proper curing will eliminate most of these difficulties. Voids behind the anchorage should be repaired before the stressing operation.

Curing in accordance with the recommendations in ACI 308.1 should be followed to avoid various types of shrinkage-related cracking and to ensure proper quality concrete.

Calcium chloride or additives containing calcium chloride or other chlorides should not be used in prestressed concrete construction or in the material used to protect end anchorages. Set accelerators that do not contain calcium chloride are commercially available and may be used when required.

8.5—Stressing operations

The stressing operation may begin when test cylinders cured under jobsite conditions and representative of the concrete strength in the immediate vicinity of the anchorages indicate that the concrete has attained the strength specified for stressing. Post-tensioning anchor assembly suppliers often specify a minimum compressive strength of 3000 psi (21 MPa) or, if no recommendation is given, 50 to 80 percent of the 28-day strength may be used. Nondestructive testing, maturity methods, or other monitoring methods may be used to verify the approximate in-place concrete strength. ACI 318 does not currently include nondestructive methods in its criteria for acceptance of concrete. Such methods, however, may be satisfactory for evaluating the concrete in the immediate vicinity of the anchorages to ensure meeting anchorage bearing requirements.

Stressing of tendons should be monitored in two ways. First, the gauge reading on the pump should be translated into force in the tendon at the anchorage. This information is generally provided in a tendon stressing data table or curve supplied as part of the shop drawings. Second, the elongation of the tendon Δ_ℓ, before losses, may be estimated using Eq. (8.5) and then compared to field measured elongation

$$\Delta_\ell = \frac{P\ell}{A_{ps}E_s} \quad (8.5)$$

where Δ_ℓ is the elongation, in. (mm); P is the average prestress force (considering friction effects along the length of the tendon), lb (N); ℓ is the length of tendon, in. (mm); A_{ps} is the area of prestressing reinforcement, in.^{2} (mm^2); and E_s is the modulus of elasticity of prestressing reinforcement, psi (MPa).

The moduli of elasticity of various post-tensioning tendon materials can be assumed as follows (Post-Tensioning Institute 1990):

Seven-wire strand: E_s = 28,000,000 psi (193 GPa)

Wire: E_s = 29,000,000 psi (200 GPa)

Bars: E_s = 30,000,000 psi (207 GPa)

A table of elongation values for various tendons on a project, a graphical presentation, or both, of expected elongations should be provided as part of the shop drawings for a project.

It is a requirement of ACI 318 that the tendon force measured by gauge pressure agrees within 7 percent of the tendon force calculated by elongation measurements. Price (2007) states that for tendons less than 25 ft (7.6 m) long, a fixed tolerance value of ±1/4 in. (6 mm) is more consistent with what is observed on site. The cause of variation in force determination in excess of 7 percent is to be ascertained and corrected. The modulus of elasticity of seven-wire strand varies somewhat from the 28,000,000 psi (193 GPa) average value suggested. Because a variation of 1,000,000 psi (6.9 GPa) in the reinforcement modulus of elasticity represents a difference of approximately 4 percent in elongation, ACI 318 requires the use of the actual modulus of elasticity of the strand used on the project when comparing tendon elongation and gauge pressure in the field.

The tendon elongation is affected by the variation in force throughout the tendon length due to friction losses. For this reason, friction losses should be considered in translating tendon elongation measurements into tendon forces. The elongation measurement provides a measure of the average force throughout the length of the tendon, whereas the gauge pressure gives the force in the tendon at the anchorage. Methods for calculating the effects of friction along the length of the tendon are presented in ACI 423.10R.

ACI 318-14, 26.10.2, requires that the gauges used to measure jacking forces be calibrated. This calibration should be done by an approved laboratory within 6 months of use.

Stressing equipment for post-tensioning tendons incorporates reasonable factors of safety. Occasionally, flaws in material are undetected or the equipment is misused. For this reason, established safety protocols should be exercised at all times, as stressing is carried out at extremely high pressure and stored strain energy. The primary safety rule is to keep personnel from being directly ahead, behind, over, or under stressing equipment.

Occasionally it is necessary to detension a fully stressed tendon. A detensioning operation that applies additional force to the tendon to free the grippers can apply additional force on concrete that has not previously been tested during the original stressing operation—that is, the concrete between the anchor surface and the bearing side of the concrete. Detensioning should only be done when absolutely necessary and with trained personnel using appropriate equipment and safety procedures. The use of detensioning as a quality-control measure is not recommended. The procedure known as a lift-off, where an anchored strand is reengaged and pulled until the wedges become loose, can cause excessive notching of the strand. This may lead to failure of the strand or individual wires.

Failure during the stressing, or detensioning, operation may cause serious injury to any personnel in back of or in the immediate vicinity of the stressing equipment. If a concrete compression failure occurs during stressing, all hydraulic pressure on the stressing equipment should be released at once, the faulty concrete should be removed, and the void should be patched with suitable material. The patching material should attain the required strength before the tendon is restressed. Calcium chloride or admixtures that contain chloride ions should not be used in the patching operation, as required by ACI 318.

According to ACI 318-14, 26.10.2(g), loss of prestress force due to unreplaced broken prestressed reinforcement shall not exceed 2 percent of the total prestress force in a prestressed concrete member. For cast-in-place post-tensioned slab systems, a member is that portion considered as an element in the design, such as the T-beam/joist or effective slab width in one-way beam and slab systems, or the slab width of an equivalent frame in two-way slab systems. The slab strip width used for effective prestress force evaluations should consider the effect of concentrated loads and slab openings.

8.6—Form removal and reshoring

Shoring should be left in place until the stressing operation is completed. Edge or pocket forms and bulkheads should be removed ahead of the stressing operation. Beam or side forms may be removed before stressing with permission from the licensed design professional.

Removal of shoring and forms may follow immediately after the stressing operation, subject to an appropriate strength gain in the concrete and the approval of the licensed design professional. After stressing, reshoring may be required to prevent overloading during additional construction. Reshoring should not be wedged beyond a snug fit against prestressed members.

8.7—Welding and burning

When welding or burning near tendons, care should be exercised to prevent the prestressing reinforcement from overheating, to keep electric arc jumps from occurring, and to keep molten slag from coming in contact with the prestressing reinforcement. Grounding of welding equipment to the prestressing reinforcement should not be allowed.

When using oxyacetylene burning of the tendon tails, the oxyacetylene flame should not be directed toward the wedges. Besides rapid oxyacetylene burning, other methods to remove tendon tails are hydraulic shears, plasma cutting, and abrasive wheel. Plasma cutting and abrasive wheel cutting may require modified anchor pockets. Construction practice requires encapsulation caps be placed immediately after the strand tail is cut off. If the tail is burned, the strands should be allowed to cool sufficiently to prevent the cap from melting.

8.8—Protection of tendons

Guidance for the protection of tendons during storage, transit, and tendon finishing can be found in ACI 423.7 and ACI 301.

CHAPTER 9—REFERENCES

Committee documents are listed first by document number and year of publication followed by authored documents listed alphabetically.

American Concrete Institute

ACI 216.1-14—Code Requirements for Determining Fire Resistance of Concrete and Masonry Construction Assemblies

ACI 301-16—Specifications for Structural Concrete

ACI 308.1-11—Specification for Curing Concrete

ACI 318-99—Building Code Requirements for Structural Concrete and Commentary

ACI 318-02—Building Code Requirements for Structural Concrete and Commentary

ACI 318-08—Building Code Requirements for Structural Concrete and Commentary

ACI 318-11—Building Code Requirements for Structural Concrete and Commentary

ACI 318-14—Building Code Requirements for Structural Concrete and Commentary

ACI 350-06—Code Requirements for Environmental Engineering Concrete Structures

ACI 352.1R-11—Guide for Design of Slab-Column Connections in Monolithic Concrete Structures

ACI 362.1R-12—Guide for the Design and Construction of Durable Concrete Parking Structures

ACI 421.1R-08—Guide to Shear Reinforcement for Slabs

ACI 423.4R-14—Report on Corrosion and Repair of Unbonded Single-Strand Tendons

ACI 423.7-14—Specification for Unbonded Single-Strand Tendon Materials

ACI 423.10R-16—Guide to Estimating Prestress Losses

ACI 550.3-13—Design Specification for Unbonded Post-Tensioned Precast Concrete Special Moment Frames Satisfying ACI 374.1

ACI ITG-5.2-09—Requirements for Design of a Special Unbonded Post-Tensioned Precast Shear Wall Satisfying ACI ITG-5.1 and Commentary

American Society of Civil Engineers

ASCE/SEI 7-10—Minimum Design Loads for Buildings and Other Structures

ASTM International

ASTM E119-16—Standard Test Methods of Fire Tests of Building Construction and Materials

Precast/Prestressed Concrete Institute

PCI MNL-124-11—Design for Fire Resistance of Precast/Prestressed Concrete

Authored documents

Aalami, B., and Barth, F., 1989, "Restraint Cracks and Their Mitigation in Unbonded Post-Tensioned Building Structures," *Cracking in Prestressed Concrete Structures*, SP-113, G. T. Halvorsen and N. H. Burns, eds., American Concrete Institute, Farmington Hills, MI, pp. 157-202.

AASHTO, 2003, "Guide Specifications for Design and Construction of Segmental Concrete Bridges," second edition, American Association of State Highway and Transportation Officials, Washington, DC, 8 pp.

AASHTO, 2014, "LRFD Bridge Design Specifications," seventh edition, American Association of State Highway and Transportation Officials, Washington, DC, 2160 pp.

AASHTO, 2015, "LRFD Bridge Construction Specifications," third edition, American Association of State Highway and Transportation Officials, Washington, DC, 3 pp.

Bailey, C. G., and Ellobody, E., 2009, "Fire Tests on Unbonded Post-Tensioned One-Way Concrete Slabs," *Magazine of Concrete Research*, V. 61, No. 1, pp. 67-76. doi: 10.1680/macr.2008.00005

Blakeley, R. W. G., and Park, R., 1971, "Seismic Resistance of Prestressed Concrete Beam-Column Assemblies," *ACI Journal Proceedings*, V. 68, No. 9, Sept., pp. 677-692.

Bondy, K. B., 1992, "Variable Prestress Force in Unbonded Post-Tensioned Members," *Concrete International*, V. 14, No. 1, Jan., pp. 27-33.

Bondy, K. B., 2008, "Requirements for Strength and Ductility of Unbonded Post-Tensioning Tendons," *PTI Journal*, Feb., pp. 61-62

Branson, D. E., and Christiason, M. L., 1971, "Time Dependent Concrete Properties Related to Design—Strength and Elastic Properties, Creep, and Shrinkage," *Designing for Effects of Creep, Shrinkage, Temperature in Concrete Structures*, SP-27, American Concrete Institute, Farmington Hills, MI, pp. 257-277.

Breen, J. E.; Burdet, O. L.; Roberts, C. L.; Sanders, D. H.; and Wollmann, G., 1994, *Anchorage Zone Reinforcement for Post-Tensioned Concrete Girders*, NCHRP Report No. 356, 205 pp.

Burns, N. H., and Hemakom, R., 1977, "Test of Scale Model Post-Tensioned Flat Plate," *Proceedings, ASCE*, V. 103, No. ST6, June, pp. 1237-1255.

Burns, N. H., and Pierce, D. M., 1967, "Strength and Behavior of Prestressed Concrete Members with Unbonded Tendons," *PCI Journal*, V. 12, No. 5, pp. 15-29. doi: 10.15554/pcij.10011967.15.29

Burns, N. H.; Charney, F. A.; and Vines, W. R., 1978, "Tests of One-Way Post-Tensioned Slabs with Unbonded Tendons," *PCI Journal*, V. 23, No. 5, pp. 66-83. doi: 10.15554/pcij.09011978.66.83

Burns, N.; Helwig, T.; and Tsujimoto, T., 1991, "Effective Prestress Force in Continuous Post-Tensioned Beams with Unbonded Tendons," *ACI Structural Journal*, V. 88, No. 1, Jan.-Feb., pp. 84-90.

Chao, S.-H., and Naaman, A. E., 2006, "Simplified Calculation of Short-Term Deflection in Prestressed Two-Way Flat Slabs," *ACI Structural Journal*, V. 103, No. 6, Nov.-Dec., pp. 850-856.

Chung, H. W., 1973, "Unbonded Tendons in Post-Tensioned Concrete Beams under Repeated Loading," *ACI Journal Proceedings*, V. 70, No. 12, Dec., pp. 814-816.

Cleland, D. J.; Franklin, Y.; and Long, A. E., 1979, "The Punching Strength of Unbonded Post-Tensioned Slabs at the Columns," International Conference on Concrete Slabs, University of Dundee, 14 pp.

Dilger, W. H., and Ghali, A., 1981, "Shear Reinforcement for Concrete Slabs," *Proceedings*, ASCE, V. 107, No. ST12, Dec., pp. 2403-2420.

Englekirk, R., 2002, "Design-Construction of the Paramount—A 39-Story Precast Prestressed Concrete Apartment Building," *PCI Journal*, July-Aug., pp. 56-71.

Falconer, D. W., and Wilson, P. W., 1988, "Inspection of Unbonded Tendons," *Concrete International*, V. 10, No. 2, Feb., pp. 53-56.

Fintel, M., 1985, *Handbook of Concrete Engineering*, second edition, Van Nostrand Reinholt, New York, 738 pg.

Foutch, D. A.; Gamble, W. L.; and Sunidja, H., 1990, "Tests of Post-Tensioned Concrete Slab-Edge Column Connections," *ACI Structural Journal*, V. 87, No. 2, Mar.-Apr., pp. 167-179.

Gerber, L. L., and Burns, N. H., 1971, "Ultimate Strength Tests of Post-Tensioned Flat Plates," *PCI Journal*, V. 16, No. 6, pp. 40-58. doi: 10.15554/pcij.11011971.40.58

Ghali, A.; Dilger, W.; and Neville, A. M., 1969, "Time-Dependent Forces Induced by Settlement of Supports in Continuous Reinforced Concrete Beams," *ACI Journal Proceedings*, V. 66, No. 11, Nov., pp. 907-915.

Han, S. W.; Kee, S. H.; Kang, T. K.; Ha, S. S.; Wallace, J. W.; and Lee, L. H., 2006, "Cyclic Behaviour of Interior Post-Tensioned Flat Plate Connections," *Magazine of Concrete Research*, V. 58, No. 10, pp. 699-711. doi: 10.1680/macr.2006.58.10.699

Hawkins, N. M., 1981, "Lateral Load Resistance of Unbonded Post-Tensioned Flat Plate Construction," *PCI Journal*, V. 26, No. 1, pp. 94-117. doi: 10.15554/pcij.01011981.94.117

Hawkins, N. M., and Ishizukra, T., 1988, "Concrete Ductile Moment Resistant Frames," *Proceedings*, Ninth World Conference on Earthquake Engineering, Tokyo, Japan.

Hemakom, R., 1975, "Strength and Behavior of Post-Tensioned Flat Plates with Unbonded Tendons," PhD dissertation, The University of Texas at Austin, Austin, TX, 272 pp.

Inomata, S., 1971, "Comparative Study on Behavior of Prestressed and Reinforced Concrete Beams Subject to Loading Reversals," *PCI Journal*, V. 16, No. 1, pp. 21-37. doi: 10.15554/pcij.01011971.21.37

International Code Council, 2015, "International Building Code," ICC, Washington, DC.

Kang, T. H.-K., and Park, H.-G., 2012, "Performance of Shearbands in Concrete Slab-Column Connections," SP-287, *Recent Development in Reinforced Concrete Slab Analysis, Design, and Serviceability*, American Concrete Institute, Farmington Hills, MI (CD-ROM).

Kang, T. H.-K., and Wallace, J. W., 2008a, "Seismic Performance of Reinforced Concrete Slab-Column Connections with Thin Plate Stirrups," *ACI Structural Journal*, V. 105, No. 5, Sept.-Oct., pp. 617-625.

Kang, T. H.-K., and Wallace, J. W., 2008b, "Stresses in Unbonded Tendons of Post-Tensioned Flat Plate Systems under Dynamic Excitation," *PTI Journal*, V. 6, No. 1, Feb., pp. 31-44.

Kosut, G. M., 1977, "Shear Strength of a Post-Tensioned Flat Plate at the Column Connections," MS thesis, The University of Texas at Austin, Austin, TX, 104 pp.

Kosut, G. M.; Burns, N. H.; and Winter, C. V., 1985, "Test of Four-Panel Post-Tensioned Flat Plate," *Proceedings*, ASCE, V. 111, No. 9, Sept., pp. 1916-1929.

MacGregor, R. J. G., 1989, "Strength and Ductility of Externally Post-Tensioned Segmental Box Girders," PhD dissertation, The University of Texas at Austin, Austin, TX.

Mattock, A. H.; Yamazaki, J.; and Kattula, B. T., 1971, "Comparative Study of Prestressed Concrete Beams, with and without Bond," *ACI Structural Journal*, V. 68, No. 2, Feb., pp. 116-125.

Miyamoto, Y., 1973, "Bending Fatigue Strength of Unbonded Prestressed Concrete Beams," *Railway Technical Research Report*, No. 844, Japan National Railway. (in Japanese)

Mojtahedi, S., and Gamble, W. L., 1978, "Ultimate Steel Stresses in Unbonded Prestressed Concrete," *Proceedings, ASCE*, V. 104, pp. 1159-1165.

Mokhtar, A.-S.; Ghali, A.; and Dilger, W. H., 1985, "Stud Shear Reinforcement for Flat Concrete Plates," *ACI Journal Proceedings*, V. 82, No. 5, Sept.-Oct., pp. 676-683.

Muller, F.-X.; Muttoni, A.; and Thurlimann, B., 1984, "Durchstanzversuche an Flachdeken mit Aussparungen," Institut für Baustatik und Konstrucktion ETH, Zürich, Dec.

Naaman, A. E.; Burns, N.; French, C.; Gamble, W. L.; and Mattock, A. H., 2002, "Stresses in Unbonded Prestressing Tendons at Ultimate: Recommendation," *ACI Structural Journal*, V. 99, No. 4, July-Aug., pp. 518-529.

NEHRP, 1991, "Recommended Provisions for the Development of Seismic Regulations for New Buildings," Federal Emergency Management Agency, Washington, D.C.

Park, R., and Thompson, K. J., 1977, "Cyclic Load Tests on Prestressed and Partially Prestressed Beam-Column Joints," *PCI Journal*, V. 22, No. 5, Sept.-Oct., pp. 84-110. doi: 10.15554/pcij.09011977.84.110

Parme, A. L., 1972, "American Practice in Seismic Design," *PCI Journal*, V. 17, No. 4, pp. 31-44. doi: 10.15554/pcij.07011972.31.44

Pilakoutas, K., and Li, X., 2003, "Alternative Shear Reinforcement for Reinforced Concrete Flat Slabs," *Journal of Structural Engineering*, V. 129, No. 9, Sept., pp. 1164-1172. doi: 10.1061/(ASCE)0733-9445(2003)129:9(1164)

Pillai, S. U.; Kirk, W.; and Scavuzzo, L., 1982, "Shear Reinforcement at Slab-Column Connections in a Reinforced Concrete Flat Plate Structure," *ACI Journal Proceedings*, V. 79, No. 1, Jan.-Feb., pp. 36-42.

Priestley, M. J. N., and MacRae, G. A., 1994, "Precast Seismic Resisting Frames Using Unbonded Prestressing Tendons," FIP XII Congress, *Proceedings*, pp. E10-E15.

Priestley, M. J. N., and Tao, J., 1993, "Seismic Response of Precast Prestressed Concrete Frames with Partially Debonded Tendons," *PCI Journal*, V. 38, No. 1, pp. 58-69. doi: 10.15554/pcij.01011993.58.69

Post-Tensioning Institute, 1977, "Design of Post-Tensioned Slabs," Post-Tensioning Institute, Farmington Hills, MI, 52 pp.

Post-Tensioning Institute, 1990, *Post Tensioning Manual*, fifth edition, Post Tensioning Institute, Farmington Hills, MI.

Post-Tensioning Institute, 2006, *Post Tensioning Manual*, sixth edition, Post Tensioning Institute, Farmington Hills, MI.

Post-Tensioning Institute, 2000a, "Specification for Unbonded Single Strand Tendons," second edition, Post-Tensioning Institute, Farmington Hills, MI, 16 pp.

Post-Tensioning Institute, 2000b, "Field Procedures Manual for Unbonded Single Strand Tendons," third edition, Post-Tensioning Institute, Farmington Hills, MI, 62 pp.

Price, R., 2007, "Field Elongation Measurements," PTI FAQ, Post-Tensioning Institute, Farmington Hills, MI.

Ritz, P.; Marti, P.; and Thurlimann, B., 1975, "Experiments on Flexure of Unbonded Prestressed Plates (Versuche über das Biegeverhalten von vorgespanntgen Platten Ohne Verbund)," Institut für Baustatik und Konstruktion, Zürich, 114 pp.

Sanders, D. H.; Breen, J. E.; and Duncan, R. R., III, 1987, "Strength and Behavior of Closely Spaced Post-Tensioned Monostrand Anchorages," Post-Tensioning Institute, Farmington Hills, MI, Dec.

Scanlon, A., 1999, "Design and Construction of Two-way Slabs for Deflection Control," *The Design of Two-Way Slabs*, T. C. Schaffer, ed., SP-183, American Concrete Institute, Farmington Hills, MI, pp. 145-160.

Scordelis, A. C.; Lin, T. Y.; and Itaya, R., 1959, "Behavior of a Continuous Slab Prestressed in Two Directions," *ACI Journal Proceedings*, V. 56, No. 12, pp. 441-460.

Smith, S. W., and Burns, N. H., 1974, "Post-Tensioned Flat Plate to Column Connection Behavior," *PCI Journal*, V. 19, No. 3, pp. 74-91. doi: 10.15554/pcij.05011974.74.91

Structural Engineers Association of California, 1999, "Recommended Lateral Force Requirements and Commentary," seventh edition, Seismology Committee, Structural Engineers Association of California, Sacramento, CA, 440 pp.

Suarez, M. G., and Schupack, M., 1988, "Evaluation of the Performance of Single Strand Unbonded Tendons in the Collapse of the L'Ambiance Plaza Lift-Slab Building," *Project Report*, Post-Tensioning Institute, Farmington Hills, MI, Apr., 4 pp.

Walker, H. C., 1980, "Durability of Parking Structure Floors," *Concrete International*, V. 2, No. 3, Mar., pp. 47-50.

Watry, C. N., 1994, "The Performance of Reinforced Concrete Structures during the Northridge Earthquake," ACI Northern California and Western Nevada Chapter, Mar., 18 pp.

Yamazaki, J.; Kattula, B. T.; and Mattock, A. H., 1969, "Comparison of the Behavior of Post-Tensioned Prestressed Concrete Beams With and Without Bond," *Structures and Mechanics Report* No. SM69-3, University of Washington, Seattle, WA, 92 pp.

Zia, P.; Preston, H. K.; Scott, N. L.; and Workman, E. B., 1979, "Estimating Prestress Losses," *Concrete International*, V. 1, No. 6, June, pp. 32-38.

Report on Corrosion and Repair of Unbonded Single-Strand Tendons

Reported by Joint ACI-ASCE Committee 423

H. R. Trey Hamilton III, Chair

Carin L. Roberts-Wollman, Secretary

Theresa M. Ahlborn
Robert W. Barnes
Florian G. Barth
Roger J. Becker
Kenneth B. Bondy*
Charles W. Dolan
James P. Donnelly*
Martin J. Fradua*
William L. Gamble
Harry A. Gleich
Shawn P. Gross
Pawan R. Gupta
William M. Hale

Mohammad Iqbal
L. S. Paul Johal
Donald P. Kline
Larry B. Krauser
Jason J. Krohn
Mark E. Moore†
Theodore L. Neff
Sami H. Rizkalla
James Rogers
Bruce W. Russell
David H. Sanders
Thomas C. Schaeffer
Andrea J. Schokker

Morris Schupack*
Richard W. Stone
Amy M. R. Trygestad
Miroslav F. Vejvoda
Jeffery S. Volz
H. Carl Walker
Zuming Xia
Paul Zia

Consulting Members
Robert N. Bruce Jr.
Ned H. Burns
Chunsheng "Steve" Cai

Steven R. Close
Henry J. Cronin Jr.
Ward N. Marianos Jr.
Hani Melhem
Antoine E. Naaman
Thomas E. Nehil

†Chair of subcommittee responsible for preparation of report.
*Member of subcommittee responsible for preparation of report.

This report gives general information regarding evaluating corrosion damage in structures reinforced with unbonded single-strand post-tensioning tendons. Historical development of those parts of the building code dealing with durability and corrosion protection is explained. Evolution of the types and components of unbonded tendons is described. Specific aspects of corrosion in unbonded single-strand tendons are described, and common problems in structures reinforced with these tendons are discussed. Methods are presented for repairing, replacing, and supplementing tendons.

Keywords: allowable stresses; anchorage; carbonation; concrete construction; corrosion; corrosion protection; cover; durability; embrittlement; external post-tensioning; post-tensioned concrete; prestressed concrete; sheathing; single-strand tendons; unbonded post-tensioning.

CONTENTS

ACI 423.4R-14 supersedes ACI 423.4R-98 and was adopted and published November 2014.

CHAPTER 1—INTRODUCTION

1.1—General

This report provides general information on the evaluation of known or suspected corrosion problems in unbonded single-strand tendons, historical information on the types and components of unbonded tendons and on the durability and corrosion protection provisions in the building code, and describes typical repair methods in use today. Expertise in design, construction, evaluation, and repair of structures using single-strand unbonded tendons is strongly recommended for a team undertaking evaluation and repair of corrosion problems.

Historically, there have been corrosion problems with other types of pre- and post-tensioning systems (Nehil 1991); however, certain aspects of corrosion of unbonded single-strand tendons are unique. The causes and effects of corrosion of unbonded single-strand tendons are, in several respects, different from those of bonded conventional reinforcing or other post-tensioning (PT) systems, so the methods for evaluating and repairing corrosion of single-strand tendons are also different. For example, because the tendons are largely isolated from the surrounding concrete, they may not be affected by deleterious materials such as chlorides and moisture in the concrete. The surrounding concrete does not stop corrosion alone because, if water gains access to the inside of the sheathing, PT coating is rendered ineffective. Measures taken to repair and protect the surrounding concrete may not repair or reduce deterioration of the prestressing steel where corrosion has been initiated. The tendons usually require separate evaluation and repair.

1.2—Background

Unbonded PT systems were introduced to North America in the 1950s. At that time, there were no accepted standards for design or material specifications for prestressing steels. Guidance came in the form of tentative recommendations from Joint ACI-ASCE Committee 323 (1958), Bureau of Public Roads (1954), or Structural Engineers Association of Northern California (SEA/NC) (1959). Unbonded tendons in the early systems used bundles of wires or strands, sometimes inaccurately called cables, of various diameters, and were assumed to be protected by PT coating and paper sheathing that were sometimes applied by hand (Nehil 1991; Schupack 1991a).

The use of unbonded tendons became more common during the late 1950s and early 1960s, as progress was made in establishing design and materials standards. Acceptance of the concept was regional at first and was largely the result of sales efforts and design tutoring by tendon suppliers. The use of post-tensioning increased rapidly during the late 1960s and 1970s as advantages of the system were demonstrated. For many types of structures, these advantages included shorter construction time, reduced structural depth, increased stiffness, and savings in overall cost. In addition to their use in enclosed buildings, unbonded PT systems were used in parking structures and slabs-on-ground, and bonded post-tensioning was used on water tanks, bridges, dams, and soil tie-back systems. Unbonded multi-wire and multi-strand tendons have been used extensively in nuclear power structures.

Incidents of corrosion of unbonded single-strand tendons began to surface during the 1970s. It was believed that corrosion protection would be provided by the PT coating during shipping, handling, and installation, and by the concrete thereafter. The early PT coatings, however, often did not provide the corrosion-inhibiting characteristics that are presently required in ACI 423.7 or PTI M10.2. In the early 1980s, the Post-Tensioning Institute (PTI) recognized the structural implications of corrosion and began implementing measures to increase the durability of unbonded PT systems (Post-Tensioning Institute 1985). Relying on the experience and practice of the nuclear industry's use of corrosion-inhibiting hydrophobic grease, similar performance standards for grease were adopted. In 2000, PTI published the second edition of its "Specification for Unbonded Single Strand Tendons (PTI M10.2-00)." In ACI 318-89, measures were incorporated that related the required protection of the tendons and the quality of the concrete to the environmental conditions that could promote corrosion of the post-tensioning.

In 2007, ACI published ACI 423.7, which provides detailed specifications for PT coating. With the adoption of this specification, the term "grease" was replaced with the more generic term "PT coating" to address other materials used as protection systems inside the sheathing. These improved coatings were introduced by PTI M10.2 long before the publication of ACI 423.7. Structures built before the adoption of these standards, especially those in aggressive environments, are more likely to experience corrosion of the PT system than those designed and built since then.

Tendons that are broken or damaged by corrosion can be repaired or supplemented by several methods. The more difficult task is determining the extent of corrosion damage and degree to which tendon repairs are needed. This report provides guidance on the evaluation of suspected or known corrosion problems and describes repair methods currently in use.

1.3—Scope

This report reviews the following:

a) Codes and code changes affecting unbonded PT tendons

b) Past and present corrosion protection systems and how they have changed to enhance corrosion protection

c) Types of corrosion damage found in prestressing steel

d) Methods for evaluating structures that are suspected of, or known to have, corrosion damage in the PT system

e) Basic repair options

1.4—Limitations

This report presents a summary of typical problems experienced with unbonded PT systems and includes general guidelines for evaluating and repairing single-strand tendons. While the methods presented are general in nature, they are not universally applicable. Standard specifications and details are not included because each structure is unique and should be analyzed accordingly.

This report is not intended to be included as a part of specifications for investigations and repairs. Presently, there is no practical method to determine the total extent of damage to a PT system. The unpredictable nature of tendon failures exhibited by inadequately protected, corroding strand makes estimating tendon life uncertain.

A wide variation exists in the durability and rate of deterioration of older PT systems. This is due, in part, to the composition of the parts of the tendon: strand, anchors, PT coating, and sheathing, and in part to the quality of the surrounding concrete, environmental exposure, and type of maintenance performed on the structure. The investigator should rely on available information on the historical performance of similar structures and should be experienced in interpreting external evidence that may give an indication of latent internal problems.

CHAPTER 2—DEFINITIONS

2.1—Definitions

ACI provides a comprehensive list of definitions through an online resource, "ACI Concrete Terminology," http://www.concrete.org/Tools/ConcreteTerminology.aspx. Definitions provided herein complement that resource.

post-tensioning (PT) coating—material used to protect the prestressing steel against corrosion and reduce friction between prestressing steel and sheathing; this term has been adopted to refer to grease or other materials, or a combination of both, that may be used inside the sheathing to protect the strand.

CHAPTER 3—REVIEW OF CODE REQUIREMENTS AND CHANGES

3.1—General

When evaluating corrosion damage in post-tensioned structures with unbonded tendons, the investigator should consider the age of the structure and the standards of practices available to the licensed design professional and contractor at the time of construction.

Although ACI published building regulations for reinforced concrete as early as 1920, ACI 318-47 was the first to acknowledge the significance of environmental exposure. ACI 318-47, ACI 318-51, and ACI 318-56 also recognized the importance of clear cover and concrete quality in providing adequate corrosion protection to the nonprestressed bonded reinforcement.

Joint ACI-ASCE Committee 323 (1958) addressed the protection of prestressing steel in three areas of recommended practice: concrete cover; allowable tensile stresses; and, for unbonded systems, protection of the strand or wire with PT coating and a sheathing material. Since 1958, provisions for prestressed concrete have included requirements for corrosion protection. The PT coating and sheathing were viewed, by most, primarily as a lubricant and bond breaker, and secondarily as a corrosion deterrent during shipping, handling, and placing. Long-term corrosion protection was viewed by some as being provided by the uncracked concrete cover.

Prestressed concrete was first included in ACI 318-63 with provisions for concrete cover, allowable tensile stresses, and strand protection. These items were modified occasionally, but the substantive change came in ACI 318-89 when durability was emphasized.

3.2—Cover requirements for unbonded tendons

ACI 318-63 required the following:

a) The following minimum thickness of concrete cover shall be provided for prestressing steel, ducts, and nonprestressed steel:

	Cover, in. (mm)
Concrete surface in contact with ground	2 (50)
Beams and girders:	
Prestressing steel and main reinforcing bars	1.5 (40)
Stirrups and ties	1 (25)
Slabs and joists exposed to weather	1 (25)
Slabs and joists not exposed to weather	.75 (20)

b) In extremely corrosive atmosphere or other severe exposures, the amount of protection shall be suitably increased.

Eight years later, in ACI 318-71, the cover requirements for prestressed members cast against and permanently exposed to earth increased from 2 in. (50 mm) to 3 in. (75 mm). In addition, a requirement for concrete protection for reinforcement was introduced that increased cover requirements 50 percent in members with allowable concrete tensile stresses above $6\sqrt{f_c'}$ psi ($0.5\sqrt{f_c'}$ MPa). The provisions for concrete cover in ACI 318-63 were also revised to require that density and nonporosity, in addition to cover of the concrete, be considered when increasing concrete protection. The provisions for cover of unbonded prestressing strand did not change in ACI 318-77, 318-83, or 318-89. In ACI 318-83 Commentary, a discussion was added as follows:

> **R7.7.5—Corrosive environments**
> When concrete will be exposed to external sources of chlorides in service, such as deicing salts, brackish water, seawater, or spray from these sources, concrete must be proportioned to satisfy the special exposure requirements of code Section 4.5. These include minimum air content, maximum water-cement ratio (or minimum strength of lightweight concrete), maximum chloride-ion in concrete, and cement type. Additionally, for corrosion protection, a minimum concrete cover for reinforcement of 2 in. (50 mm) for walls and slabs and 2-1/2 in. (60 mm) for other members is recommended. For precast concrete manufactured under plant control conditions, a minimum cover of 1-1/2 and 2 in. (40 and 50 mm), respectively, is recommended.

This discussion calls the designer's attention to the importance of air content, maximum water-cementitious material ratio (*w/cm*), maximum chloride ion content, and cement type when designing for corrosion protection.

While not specific to prestressed concrete, R7.7.5 of ACI 318-83 makes reference to increasing concrete cover when concrete is exposed to external sources of chlorides in service. Section 7.7.6.1 of ACI 318-08 requires increased cover for Class T and C prestressed concrete members in corrosive environments, and that two-way floor systems must satisfy Class U requirements of 18.3.3.

Finally, a distinction between plant-cast members and other prestressed concrete members was made in ACI 318-71 with the addition of distinct cover requirements for plant-cast members.

3.3—Allowable tensile stresses in concrete

In aggressive environments, designing to minimize cracking was used to improve durability by reducing ingress of corrosive elements. Though a properly coated tendon in an intact sheathing may not be affected at first by a crack in the surrounding concrete, corrosion of nearby conventional reinforcement can cause spalling that may expose the tendon to physical damage, which may then lead to corrosion of the strand. For consideration of long-term durability and corrosion protection, the maximum allowable tensile stresses in the concrete at service loads, after allowance for all prestress losses, are of most interest. The initial recommendations (Joint ACI-ASCE Committee 323 1958) limited allowable tensile stresses in pretensioned members to $6\sqrt{f_c'}$ psi ($0.5\sqrt{f_c'}$ MPa) where not exposed to weather or corrosive environments. In post-tensioned members not exposed to weather or corrosive environments, tensile stresses were limited to $3\sqrt{f_c'}$ psi ($0.25\sqrt{f_c'}$ MPa). No mention was made of allowable stresses in unbonded systems.

In ACI 318-63, allowable tensile stresses in concrete were limited to $6\sqrt{f_c'}$ psi ($0.5\sqrt{f_c'}$ MPa) for members not exposed to freezing temperatures or to a corrosive environment, provided that the members contained bonded reinforcement—prestressed or nonprestressed—to control cracking. For all other members, no tensile stresses were allowed at service load levels.

ACI 318-71 required that tensile stresses in concrete be limited to $6\sqrt{f_c'}$ psi ($0.5\sqrt{f_c'}$ MPa) for most members, but allowed tensile stresses up to $12\sqrt{f_c'}$ psi ($1.0\sqrt{f_c'}$ MPa) in the concrete, provided that computations were made using the cracked transformed section and a bilinear moment-deflection relationship to confirm that long-term deflection of the member satisfied the Code. In addition, the allowable tension limits could be exceeded, provided that experimental or analytical work could show that performance would not be impaired.

In ACI 318-77, for an allowable tensile stress up to $12\sqrt{f_c'}$ psi ($1.0\sqrt{f_c'}$ MPa), a provision was added requiring that the concrete cover for prestressed and nonprestressed steel be increased for prestressed members exposed to earth, weather, or corrosive environments. Provisions of ACI 318-77 required a 50 percent increase in cover for members exposed to weather, earth, or corrosive environments and with a tensile stress greater than $6\sqrt{f_c'}$ psi ($0.5\sqrt{f_c'}$ MPa).

In ACI 318-02, the allowable stress provisions were modified to classify prestressed concrete elements as Class U, T, or C members, depending on the calculated tensile stress at the extreme fiber due to service loading. Class U members (uncracked) would have a tensile stress less than or equal to $7.5\sqrt{f_c'}$ psi ($0.625\sqrt{f_c'}$ MPa), whereas Class C members (cracked) would have a tensile stress exceeding $12\sqrt{f_c'}$ psi ($1.0\sqrt{f_c'}$ MPa). Class T members (transition) would be in the transition zone between cracked and uncracked elements, and would have an extreme fiber tensile stress of between $7.5\sqrt{f_c'}$ and $12\sqrt{f_c'}$ psi ($0.625\sqrt{f_c'}$ and $1.0\sqrt{f_c'}$ MPa). These revisions allow service-load tensile stresses to exceed the cracking stress in prestressed flexural members, but those members are then required to be analyzed as cracked members. Furthermore, Class C members must have sufficient nonprestressed reinforcement to control cracking. Prestressed two-way slab systems, however, are still required to be designed as Class U members.

ACI 318-05 adjusted the allowable stress for prestressed two-way slab system back to $6\sqrt{f_c'}$ psi ($0.5\sqrt{f_c'}$ MPa), and this change applied to both bonded and unbonded prestressed flexural members requiring their design as Class U members. This requirement remains valid in ACI 318-11.

3.4—Protection of unbonded tendons

The protection of unbonded prestressing strands was initially described by Joint ACI-ASCE Committee 323

(1958) to consist of a grease- or asphalt-impregnated material enclosed in a sheathing. Specification and approval of the method of protection was left to the licensed design professional. ACI 318-63 required that "Unbonded steel shall be permanently protected from corrosion." However, no specific information was provided relative to the sheathing material and the required corrosion protection. As noted previously, concrete cover was sometimes interpreted as providing the permanent protection from corrosion.

ACI 318-71 stipulated that "Unbonded tendons shall be completely coated with suitable material to ensure corrosion protection." It also required that wrapping should be continuous over the entire unbonded zone of tendons to prevent bonding with surrounding concrete and loss of the coating material during concrete placement.

In 1972, the Post-Tensioning Institute (PTI) published its first edition of the *Post-Tensioning Manual* (Post-Tensioning Institute 1972), which included a specification for PT materials. Subsequently, this specification was revised and published as a guide specification in the second (Post-Tensioning Institute 1976) and third (Post-Tensioning Institute 1981) editions. In 1985, PTI published "Specification for Unbonded Single Strand Tendons," the first detailed specification to be based on experience (Post-Tensioning Institute 1985).

Significant provisions of the "Specification for Unbonded Single Strand Tendons" (Post-Tensioning Institute 1985) are discussed in the following.

(a) *Definition of exposure conditions*: The specification addressed tendons in normal (noncorrosive) environments, and tendons in aggressive (corrosive) environments. Normal environments were defined as those present in nearly all enclosed buildings with dry interiors, and exposed structures in areas with very little or no snow. Aggressive environments were defined as those that would expose the structure to direct or indirect applications of deicer chemicals, seawater, brackish water, or spray from these sources.

(b) *Anchorages and couplings*: Anchorages were required to have watertight connection to the sheathing and a watertight closing of the wedge cavity. Couplings were required to be protectively coated the same as the strand because they become part of the strand.

(c) *Sheathing*: Sheathing material thickness was specified according to exposure, and was given as 0.025 in. (0.6 mm) for normal environment and 0.040 in. (0.8 mm) for aggressive environment. The inside diameter of the finished sheathing was required to be at least 0.010 in. (0.3 mm) greater than the diameter of the strand.

(d) *Corrosion-preventive coating*: The minimum weight of PT coating coverage of the strand was given, as was a list of test requirements for the PT coating itself. The corrosion protection performance was based on ASTM B117 for the time until Rust Grade 7 was developed and was set as 720 hours minimum for normal environment and 1000 hours for aggressive environment.

PTI reissued this specification (Post-Tensioning Institute 1993) after substantial revisions were made to all chapters. Major changes are summarized in the following.

(a) *Definition of exposure conditions*: The terms "normal (noncorrosive)" and "aggressive (corrosive)" used in previous specifications were changed to "normal" and "aggressive." The definitions of normal and aggressive remained similar to the 1985 specification except that exposed structures in areas with very little or no snow were no longer mentioned as being in a normal environment. The designer was advised to "...evaluate the conditions carefully to determine if the environment in which a structure is located is considered aggressive in any way."

(b) *Definition of exposure conditions*: A stipulation was added that stressing pockets not maintained in a dry condition after construction should be considered exposed to an aggressive environment.

(c) *Prestressing steel*: Protection requirements were added for packaging and identification. A criterion was added that limited surface rust to pits no more than 0.002 in. (0.05 mm) diameter or length. (This type of rust can be removed with fine steel wool and might not be felt with the fingernail.)

(d) *Anchorages and couplers (formerly couplings)*: Static test criteria were clarified and linked to ACI 318, and dynamic test requirements were added. Design criteria were added for bearing stresses on concrete. A stipulation was added that required anchorages intended for use in aggressive (corrosive) environments have watertight connections between all components. Encapsulation of the anchorage, the connection of the sheathing to the anchorage encapsulation, and the seal of the wedge cavity were required to sustain a hydrostatic water pressure of 1.25 psi (0.0086 MPa) for 24 hours.

(e) *Sheathing*: The thinner sheathing previously allowed for tendons to be exposed to normal environments was removed; the minimum thickness of sheathing was specified as 0.040 in. (1.0 mm) for both environments. The size of the annular space between the outside of the strand and the inside of the sheathing was increased from 0.010 to 0.040 in. (0.3 to 0.8 mm). Complete encapsulation of tendons to be used in aggressive environments was specified with the same watertightness requirements given for anchorages. A statement was added that calls for the designer to specify the amount of unsheathed strand permitted at the anchorages for tendons exposed to normal environments.

(f) *Corrosion-inhibiting coating (formerly corrosion-preventive coating, commonly referred to as grease, and now referred to as PT coating)*: All tendons are to use PT coating that meets the ASTM B117 Rust Grade 7 criteria after 1000 hours. (The shorter test time previously permitted for tendons to be used in normal environments was dropped.)

(g) *Installation requirements*: The minimum cover requirement for anchorages was added and was specified to be at least 1.50 in. (40 mm) for normal environments and 2 in. (50 mm) for aggressive environments.

(h) *Tendon finishing*—Permissible length of strand projection from the face of the wedges was reduced. The previous projection allowed was 0.75 in. (20 mm) minimum and 1.25 in. (30 mm) maximum; the revised projection limits are 0.50 in. (12 mm) minimum and 0.75 in. (20 mm) maximum. The PT coating is now required by ACI 423.7 to:

a) Provide lubrication between the strand and the sheathing

a) Multiple wire button head anchor (Post Tensioning Institute 2006)

b) Five-wire button head anchor. Courtesy of Wiss, Janney, Elstner Associates, Inc.

Fig. 4.1a—Button head anchor system.

b) Resist flow caused by gravity within anticipated temperature range of exposure

c) Provide continuous nonbrittle coating at lowest anticipated temperature of exposure

d) Be chemically stable and nonreactive with prestressing steel, reinforcing steel, sheathing material, and concrete

CHAPTER 4—UNBONDED TENDONS

4.1—Historical development of unbonded tendons

The first building structures to use unbonded tendons in North America were lift slabs built during the mid-1950s (Peterson and Braunfield 1957). The early, unbonded tendons were coated and helically wrapped with paper. They used high-strength wires, generally 0.25 in. (6 mm) in diameter, with an ultimate strength of 240 ksi (1650 MPa) with button-head type anchorages, as shown by Fig. 4.1a. This system had large cumbersome anchorages, large trumpet transitions, and a fixed distance between buttons at each end of the tendon. Systems also evolved during this period that used single and multiple strands.

Fig. 4.1b—Typical strand anchor consisting of casting and two-piece wedges.

Fig. 4.1c—Push-through plastic tube for sheathing (Schupack 1991a).

In the early 1960s, the convenience of placing unbonded single-strand tendons was realized and the number of suppliers increased. The marketplace found the competitiveness of this system favorable, and the use of unbonded single-strand tendons increased significantly. Anchorage hardware varied considerably and included high-strength spirals, barrels, or castings with wedges and fittings that were swedged (mechanically attached) to the prestressing steel. The type of anchor that has become the most common is a casting that contained a recess to house a two-piece wedge for use with a single strand (Fig. 4.1b). This is currently the predominant system for unbonded single-strand tendons.

By the late 1960s, plastic began replacing paper sheathing. Three different processes were used (Fig. 4.1c and 4.1d). In one system, the strand was covered by a preformed push-through plastic tube. In a second system, the strand was wrapped longitudinally with a heat-sealed strip of sheathing. Currently, the coated strand is covered by molten plastic that is continuously extruded around it (Schupack 1991b). Initially, the thickness and composition of the sheathing material were left to the supplier and were not uniform in the industry. The Post-Tensioning Institute (1985) regulated the type of sheathing material, the types of coatings used, and the dimensional tolerances of the system as it is applied to the strand. ACI 423.7 summarizes the more comprehensive specifications for both the sheathing and the coating.

Initially, there were no corrosion protection standards for the PT coating (except in the nuclear industry [ACI 349]), so the tendon manufacturers used the PT coating of their choice. Corrosion-resisting properties of the PT coating were not specified, nor were there standards for the uniformity and

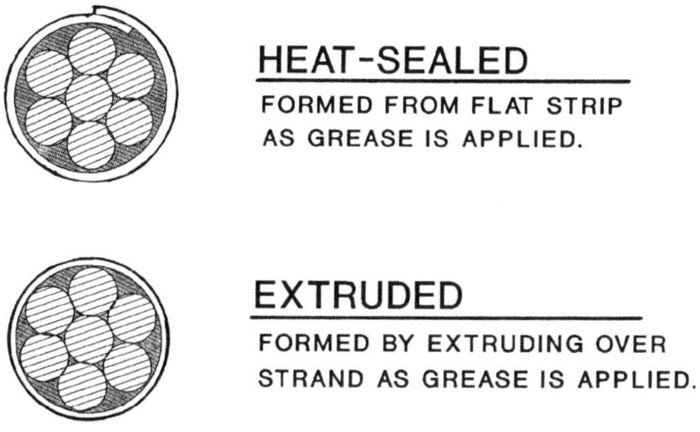

Fig. 4.1d—Heat sealed and extruded sheathing system (Schupack 1991a).

amount of PT coating to be applied to the prestressing steel. Deterioration of the PT coating was not expected, but as problems became known, they were addressed by the Post-Tensioning Institute (1993).

4.2—Sheathing problems

Advances in the sheathing systems were generally implemented in direct response to short-comings of the systems, difficulties in manufacturing, or improvements in the manufacturing systems. One of the earliest forms of sheathing was the paper-wrapped systems. These systems used a paraffin- or wax-impregnated paper that was typically helically wound around the tendon with little to no attention given to the start or termination of the wrap.

As a result, the paper wrapping would not necessarily be sealed at the overlaps in the wrapping, thereby allowing any moisture or chlorides that entered the concrete a path to reach the tendon wires. Furthermore, at cracks that crossed the tendon, water could infiltrate and erode the paper-wrapping, leaving gaps in the sheathing at locations of moisture infiltration. In addition, the paper-wrapping could be easily damaged during shipping or installation. To address these issues and provide a more durable sheathing material, the industry changed to plastic-based sheathing systems such as the push-through, heat-sealed, and extruded systems.

The push-through system required that the sheathing be sufficiently oversized to allow the coated strand to be inserted without too much difficulty. This system often resulted in larger-than-desirable gaps between the strand and the sheathing, and the coating systems could not be relied on to completely occupy this interstitial space. This resulted in a tendon with many air voids inside the sheathing and allowed infiltration of water during storage, shipping, installation, and in service.

With the heat-sealed system, the sheathing was provided in rolls of flat plastic tape that was usually 20 to 40 mils (0.5 to 1.0 mm) thick. During tendon fabrication, the strand was taken from the pack and passed through an extrusion head that applied PT coating to the strand. The tape was then folded over the coated strand and the lapped seam welded shut with a flame. While this system generally provided sheathing more closely wrapped to the strand, this method

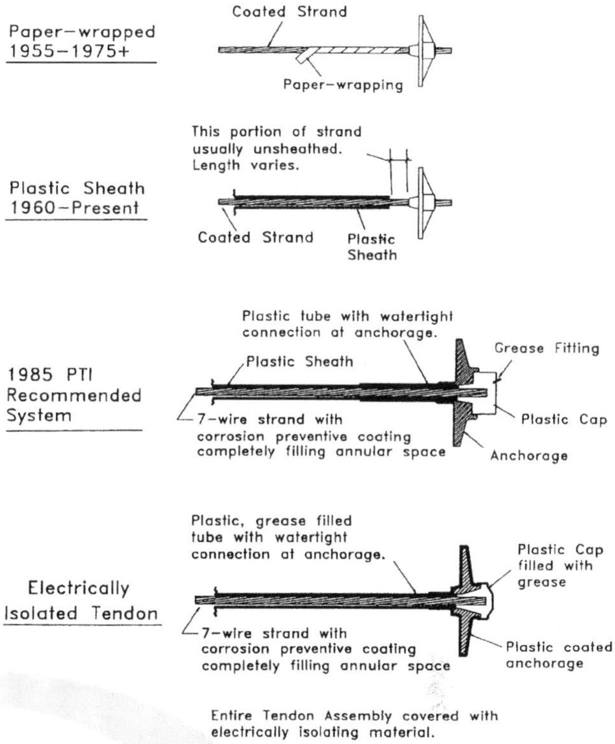

Fig. 4.3a—Comparison of typical sheathing treatment at anchorage. Note past practice of stopping sheathing short of anchorages (Schupack 1991a).

still resulted in a slightly oversized sheathing that could have air voids. Additionally, there were occasional issues with the integrity of the heat-sealed strip. The seam weld was interrupted each time there was a pause in the process, and sometimes the sealed seam pulled apart during handling or installation. This system is frequently found to have gaps from one cause or another that expose the coated strand to contact with the concrete.

Seamless extruded sheathing first appeared in the early 1970s. The extruded sheathing minimized gaps between the strand and sheathing by providing a snug, seamless sheathing around the coated strand, thereby limiting the problems experienced with the push-through and heat-sealed systems.

4.3—Detailing practices

Certain details that were initially considered acceptable have been shown to be insufficient to provide the degree of durability required by recent versions of ACI 318. It is now recognized that concrete cannot provide reliable protection to the strand, even if the strand is coated.

It is common practice in both detailing and installation to allow the sheathing to stop short of the anchorages, as shown by Fig. 4.3a. During stressing operations, the strand at the stressing end moves approximately 8 in. per 100 ft (200 mm per 30 m) of length. Direct contact between the unsheathed strand and the hardened concrete is disturbed at the stressing end during this elongation, thus providing a path for water to find its way to the strand and into the sheathing (Schupack 1991a). ACI 423.7 limits the length of exposed strand to 1

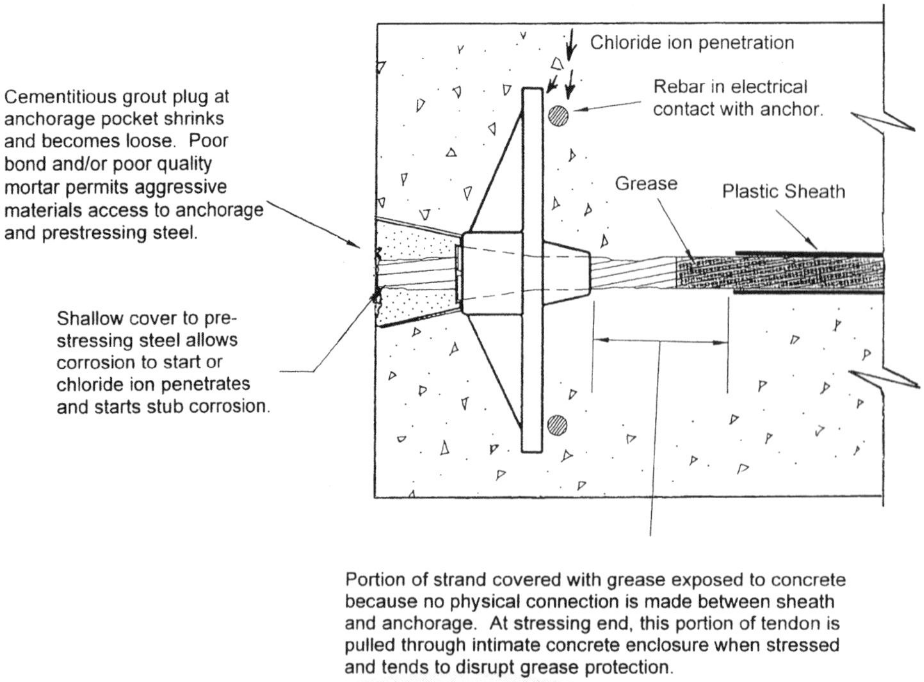

Fig. 4.3b—Typical defects in anchor region.

in. (25 mm) on the stressing end and 12 in. (300 mm) on the fixed end.

Because the fixed-end anchor does not move during stressing, it is considered acceptable for the strand near the fixed end to be exposed to the concrete. However, this also allowed the end of the tendon to be exposed to dirt and water during storage, handling, and placement before placing the protective concrete. The end of the sheathing was, for practical purposes, open and the voids inside the tendon sheathing provided a means for any available water to gain access to the tendon. Even if the storage time at the site was relatively brief (which was not always the case), there was ample opportunity for water to get into the tendons due to any snow and rain that might occur while the tendons were stored on the ground or on the formwork. For those portions of the strand not adequately protected by PT coating, water and oxygen could cause corrosion and then could be further exacerbated by the presence of chloride ions. Some typical defects contributing to corrosion in the end anchor region are shown in Fig. 4.3b.

The stressing side of the anchorage was protected by filling the stressing pocket in which the anchorage was recessed with a protective cementitious grout (Fig. 4.3c). The casting, wedges, and strand tail within the stressing pocket were typically not coated. In some cases, the strand tail was not cut off short enough for it to be fully embedded in the protective grout. Often, shrinkage of the grout plug in the stressing pocket caused a space between the side of the stressing pocket and the grout that allowed water to gain access to the tendon anchorage. In some cases, a nondurable grout was used, or grout was inadvertently omitted, leaving the end anchor, wedges, and strand tail directly exposed. In instances where moisture reached the anchorage, corrosion

Fig. 4.3c—Cementitious material used to fill stressing pocket.

of the bearing plate casting was often found, but not failure of the casting itself. The primary problem has been either corrosion of the wedges or the migration of water down the tendon in the void between the wires or around the outside edge of the strand. Corrosion failures of the strand have occurred in the unsheathed length in front of the anchor, at low points in the profile where water collects, and at points where sheathing was damaged and permitted direct access of corrosive materials.

At intermediate anchorages, features of both fixed-end and stressing-end anchorages are found (Fig. 4.3d). The sheathing may not be adequately sealed to the anchorage on the first pour side. The strand and anchorages may be in direct contact with concrete on the second pour side of

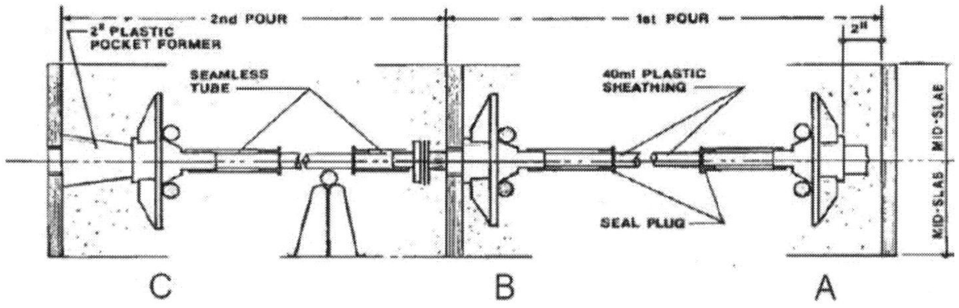

Fig. 4.3d—Intermediate stressing anchor, at location B.

the anchorage. Water can gain access to the tendon during shipping, storage, and installation, or after construction has been completed. Problems have developed at the intermediate anchorages when the construction joint over the anchorages is not sealed to prevent leaking of water (sometimes containing chlorides) through the construction joint in elements exposed to the weather, such as parking structures and balconies. The result is significant chloride contamination of the concrete. Corrosion of the upper surface of the anchorage castings and of backup bars can cause corrosion-induced spalling at an early age. Such spalls then act as small reservoirs for water, further accelerating the deterioration process. Corrosion of the backup bars, bearing plates or anchorages, and on occasion, the strand, is promoted by this frequent supply of water and oxygen adjacent to the joint. With the earlier bearing plate and intermediate barrel anchorage hardware, the bearing plate was occasionally located outside the bulkhead of the slab, positioning the construction joint directly over the plate, which allowed water to come in contact with strand end anchorage.

The choice of locations for anchorages sometimes led to corrosion problems. Locating anchorages in gutter lines, expansion joints of parking structures, or at exposed edges of slabs in buildings (such as in balconies) has resulted in water infiltration into stressing pockets and corrosion of anchorage components (Fig. 4.3e). Without adequate waterproofing, construction joints at beam-column connections have, on occasion, allowed water to enter tendon sheathing through the anchorages in joints in columns. Although these joints are seldom sealed, they are frequently in the drainage flow path and can be exposed to rain and melting snow. Water running down the face of a column can gain access to the tendon anchorage if consolidation of the concrete at the construction joint is poor. Honeycombed, permeable concrete, ponding areas, leaking construction joints, misplaced conduits, and other construction defects can all contribute to water access to tendons.

Fig. 4.3e—Tube-up detail in roof structure.

4.4—Storage, handling, and construction problems

Though the tendon sheathing is durable, the wear and tear to which it is sometimes subjected during shipping, storage, handling, and concrete placement can be severe. The result is damage to the corrosion protection provided by the PT coating and sheathing. Typical problems during the storage, handling, and installation of tendons are summarized in the following.

4.4.1 *Storage of tendons*—Between manufacture and installation in the structure, tendons should be properly stored to protect them from possible damage that could affect their long-term serviceability. To help limit possible corrosion issues, it is particularly important to protect the tendons from the infiltration of moisture, deicing salts, or other corrosive agents. This could occur if left in direct contact with the ground, if exposed to rain or snow, or if covered in a manner that allows condensation to form. While stored at a construction site, damage by construction traffic or other construction activities can occur. In addition, tendons should not be left exposed to direct sunlight if stored for more than 1 month.

4.4.2 *Handling of tendons*—Damage to the tendons that occurs during shipping and handling is often associated with a lack of care. If unpadded slings or forklift booms are used to lift tendon bundles when moving them from the truck to the storage area, and from the storage area to the formwork where they are to be placed, tearing or cutting of the sheathing can result. Using wire or metal bands without padding to hold the tendons in coils can also cause tearing, cutting, or both, of the sheathing. Metallic slings, chokers, or chains can result in similar damage.

4.4.3 *Construction issues during installation*—Significant damage to the tendons can also occur during their installation. Rough handling during placement of tendons in formwork and coiling of strand bundles on formwork at construc-

Fig. 4.4.3a—Encapsulated end anchor with oversized sleeve. (Courtesy of Wiss, Janney, Elstner Associates, Inc.)

Fig. 4.4.3b—Fit-up problems with early encapsulation systems.

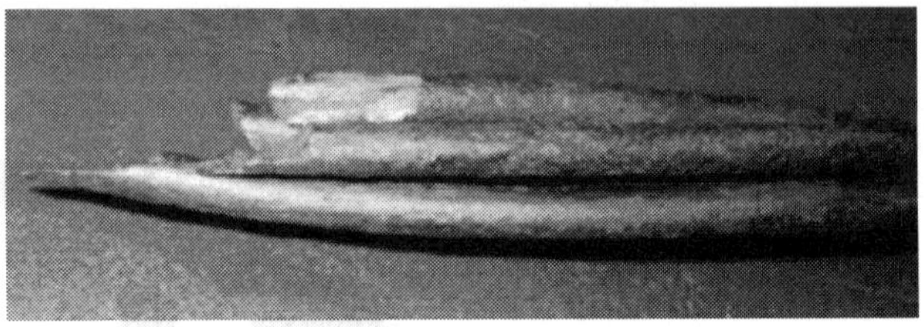

Fig. 4.5.1—Brittle fractures in failed strand.

tion joints can cause splits and tears in the sheathing, similar to that described in 4.4.2. Damage to the tendon sheathing should be repaired before concrete placement in accordance with project specifications, and preferably as soon as it is noticed, to prevent moisture infiltration. Other obvious sources of moisture infiltration should also be minimized, such as unsealed tendon ends exposed to the weather, either before concrete placement or after stressing.

Inadequate cover over the tendon at high and low points of the tendon drape, at the anchorages, and at cut-off strand tail at stressing-end pockets can allow for moisture infiltration and possible corrosion over the long term. Improper grouting or missing grout at the stressing pockets can have a similar effect, and the use of a grout that contains chlorides or other chemicals can accelerate corrosion of the strand. If moisture can enter the tendon anchor and sheathing, voids inside sleeves or trumpets can allow water to collect.

Some encapsulation systems used in aggressive environments incorporate oversized sleeves or trumpets to assist in sealing the tendon at the transition from the sheathing to the anchorage (Fig. 4.4.3a). These oversized sleeves and trumpets must be filled with PT coating to eliminate voids where water might collect. Some early systems relied on friction rather than on a mechanical connection between the anchor and sleeves. ACI 423.7 requires a mechanical connection between the sleeve and the anchor for encapsulation systems. Friction-fit sleeves have been observed to work loose and pull away from the anchorage before or during concrete placement (Fig. 4.4.3b). Final inspection with reattachment as necessary has been required to achieve the intent of PTI M10.2 for protection of unbonded tendons.

Damage to the PT coating and sheathing or exposure to moisture during periods of construction could adversely affect future performance of the post-tensioning, but neither responsibility for protection nor specific measures for achieving protection were defined in the earlier industry-wide specifications and procedures. These issues were partially addressed in PTI M10.3.

4.5—Deterioration mechanisms

4.5.1 *General deterioration mechanisms*—In most cases, the corrosion mechanism occurs when water and oxygen are present. If a PT coating completely covers the strand and the PT coating is not affected by water, corrosion generally will not occur. There is always the possibility that the PT coating will be discontinuous, so corrosion can begin if water and oxygen are available. Corrosion can be accelerated if chlorides are present in the water. As the wire(s) corrodes and the cross-sectional area decreases, the stress in the remaining section rises past its ultimate tensile strength, and the wire(s) fails.

Failure by embrittlement of the strand wires can also occur if other aggressive materials, such as nitrates and sulfides, are present (Schupack 1991a,b; Schupack and Suarez 1982). Embrittlement failures can occur without significant loss of cross section in the strand. It is not uncommon for the failure surfaces of one or two wires at a strand break to be jagged, which is typical of embrittlement failures, whereas the remaining wires have the cup-cone configurations typical of ductile failures. It is not necessary, or likely, that all of the wires at a given cross section will be similarly affected at the same time (Fig. 4.5.1). Detailed information regarding the corrosion of prestressing strand is found in ACI 222.2R.

4.5.2 *PT coating*—PT coating has corrosion-preventive properties that should help protect the strand. PT coating, however, can be broken down and emulsified by the exposure to moisture, thereby eliminating this protection. Moisture can then more easily wash the PT coating out of the sheathing, if a path is available.

Fig. 4.5.3—Common strand failure mode.

4.5.3 *Strand corrosion*—As mentioned in 4.5.1, corrosion of the strand results in a loss of cross-sectional area and, with continued corrosion, the eventual failure of the wires. In the early stages of tendon corrosion, the corrosion is mainly surficial, and the cross-sectional area of the wires is not significantly affected. As corrosion advances, however, section loss will occur and the corrosion may form pits in the wires. Because of the prestressing force in the strand, any pitting that does occur during corrosion results in localized high stresses in the wire(s) around the pits. This can cause a corroded wire to fail in a brittle fashion before reaching its ultimate stress across the entire wire cross section. This is similar to what is called stress corrosion cracking in ACI 222.2R, although stress corrosion cracking can occur with little or no corrosion-related section loss. ACI 222.2R indicates that almost all cases of strand failure in post-tensioned concrete reported in the literature are associated with general and pitting corrosion, and not hydrogen embrittlement or stress corrosion cracking.

Corrosion may not necessarily affect all wires of a seven-wire strand to the same degree, so it is common to observe that one or two of the wires have failed where corrosion has occurred, whereas the other wires remain intact. The remaining wires must carry the additional load that is transferred from the broken wires. The ability of the remaining wires to carry the load depends on their condition and the amount of overload transferred to the wires. As multiple wires fail, there is increasing likelihood that the remaining wires in the strand will fail (Fig. 4.5.3). Likewise, when an entire strand is lost, the design load must be carried by the remaining intact strands in the member.

4.5.4 *Strand failure*—Typically, a strand failure is contained entirely within the concrete, so multiple strands may have failed in a structure with a tendon corrosion problem without any direct visual evidence. However, a strand can burst from the concrete when it fails if the cover is small, usually less than approximately 3/4 in. (20 mm) (Fig. 4.5.4a and 4.5.4b). Whether this happens depends on the drape of the tendon, the type of sheathing, and the presence of perpendicular reinforcement between the broken tendon and the surface of the concrete. Occasionally the strand is released by the anchor and projected past the edge of the structure (Schupack 1991a; Aalami and Barth 1989).

4.5.5 *Vehicular impact*—In parking structures, vehicles can damage tendons if the concrete cover spalls off or is abraded away. This most commonly happens where the concrete cover is less than 3/4 in. (20 mm) and can be the result of misplacement of the tendon or poor screeding and finishing of the concrete. The sheathing is not intended to resist direct contact of in-service traffic and is easily breached under these circumstances. Thereafter, the PT coating can quickly be washed away, and contaminated water is free to enter the sheathing and cause corrosion.

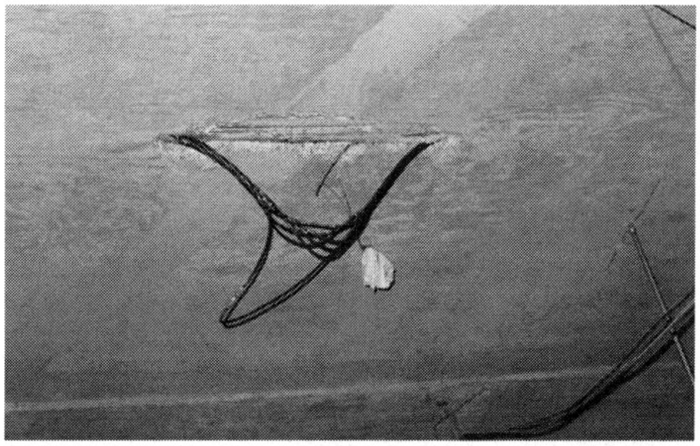

Fig. 4.5.4a—Strand failures in structure. Strand rupture occurred away from the breakout. (Courtesy of Wiss, Janney, Elstner Associates, Inc.)

Fig. 4.5.4b—Strand failure in structure. Failed strands projecting from structure.

4.6—Performance record

In general, tendons with extruded-type plastic sheathing provide superior corrosion protection when compared with tendons from other unbonded systems and other types of sheathing. Improvements in the quality and application methods of the PT coating that occurred at approximately the same time have also contributed to improved performance. The extruded-type sheathing now predominates in the industry, probably as a result of Post-Tensioning Institute (1985). Experience has shown that aggressive materials, including water, entering through the sheathing or anchorage can cause corrosion.

There are no recorded incidents of sudden collapse of structures using unbonded tendons while the structure is in service. Demolition of these structures has shown that they possess greater reserve of strength than is shown by structures that are not post-tensioned. One characteristic of structures that use unbonded tendons is their propensity to develop significant catenary action even while the main parts of the slabs and beams are being pulverized or broken away (Freyermuth 1989; Suarez and Schupack 1988). While there is no guarantee against a sudden partial collapse, it is likely that a post-tensioned structure will continue to perform in a ductile manner even when a significant number of the tendons have failed.

Strand breakage can occur during original construction, over time due to corrosion or other deterioration, or during the re-stressing of a tendon being repaired. During initial stressing or re-stressing, this could be an indication of damage to the tendon at an unknown location along its length, overstressing, or a problem with the seating or gripping of the wedges at an end anchor. If more than one wire breaks, the remaining member strength may need to be evaluated. Although many structures have sufficient excess design strength and would not be affected by a loss of one or more strands, such strand failures should be brought to the attention of the licensed design professional to determine if the effect on the calculated strength of the structure is significant and therefore warrants repair.

CHAPTER 5—CONDITION ASSESSMENT AND TESTING

5.1—General

Almost any structure can have its useful life extended by the use of appropriate maintenance procedures and suitable repairs, whether the structure is of post-tensioned concrete or some other material. The cost of repairs can usually be estimated reasonably accurately and a judgment made as to the feasibility of that investment. For concrete structures that do not use unbonded tendons, damage caused by corrosion is estimated by examination of rust-stained or spalled areas. It is assumed, based on spot-checking and on past performance of similar conditions, that the reinforcement is serviceable in areas where the concrete is not stained, spalled, or delaminated. Copper-copper sulfate half-cell potential testing or other corrosion activity survey techniques can also be used to identify areas of likely corrosion in portions of the structure not already exhibiting delaminations or spalling of the concrete. That same logic applies to evaluation of deterioration of bonded reinforcing bars in structures using unbonded tendons, but it is of less help in evaluating the tendons themselves.

An appropriate evaluation of the structure's condition should be performed to minimize the risk of overlooking something significant. Ultimately, decisions about the types of repairs and the need to augment the reinforcing system will have to be made based on an engineering evaluation of the data collected. These decisions should be made by an experienced investigator who understands that all latent deficiencies may not have been identified. As in any repair, the objective is to fix the obvious defects, eliminate the causes of deterioration where practical, slow continued deterioration, and determine requirements for monitoring and future maintenance.

5.2—Condition surveys of concrete

Evaluations should include a careful and well-documented visual inspection to identify deterioration and distress and their causes. If available, the original design and shop drawings should be reviewed to identify problems that might be attributed to the design, detailing, or material selection. Drawings should be compared with the findings of the condition survey to assess the in-service performance and determine whether suspected problems are local or widespread. It is normal for the as-built conditions to differ from the drawings to some degree. Refer to ACI 364.1R and ACI 437R for additional guidelines.

5.2.1 *Visual survey*—A visual survey of the structure can help identify the nature of any structural distress and the locations of particular types of deterioration, and provide insight to the possible causes. The visual survey should identify the location and nature of any observed cracking. Cracks can be caused by structural distress, by insufficient cover to reinforcing bars or tendons, by incipient delaminations due to corrosion of reinforcement or embedded conduit, or by restrained shrinkage. Notes should be made to document the width and length of the cracks. The visual survey should also note the presence of leakage, efflorescence, and rust stains.

5.2.2 *Delamination survey*—A survey should be performed to locate delaminations in slabs and other structural members using methods that are appropriate to the conditions. The delamination survey can be used to estimate the extent and distribution of corrosion damage in the reinforcing system.

5.2.3 *Material sampling and testing*—Typical materials testing to be performed may include:

a) Chloride-ion content testing to determine the depth and intensity of chloride penetration, and to estimate the chloride content of the original concrete. Chloride contamination can promote corrosion of the steel portions of the PT system that are in direct contact with concrete as well as conventional reinforcement and hardware whose corrosion may damage the concrete protection of the post-tensioning.

b) Evaluation of depth of carbonation into the concrete. This reaction decreases the pH of the concrete, thus making corrosion more likely in the presence of moisture and oxygen.

c) Obtaining measurements of concrete cover over tendons and reinforcing steel and, if appropriate, correlating these measurements with the chloride contamination profiles or depth of carbonation.

d) Copper-copper sulfate half-cell testing to determine whether corrosion activity is presently occurring in the mild steel-reinforcing system. Although these tests may not provide meaningful information on tendons due to lack of electrical continuity, they might indicate the likelihood of corrosion activity on anchorage hardware and reinforcement where electrical continuity of steel and contact with concrete are present.

e) After locating tendons, extracting concrete cores for compressive strength testing and for petrographic examination to include air-void system analysis.

5.3—Condition surveys of tendons

It is possible to obtain some information about the likely condition of the PT system by means of a visual inspection. A visual inspection can help to identify specific external evidence of possible internal distress. The visual inspection can also help identify if there is a pattern to the distress, and may point to the possible extent of a problem.

5.3.1 *Cracking*—Cracking of concrete in post-tensioned structures can be the result of problems with the PT system, but all cracks should be evaluated for their impact on future durability. Following are some examples of cracking patterns that may provide information about the PT system:

a) Cracks that are vertical in beams or perpendicular to the direction of span in slabs may indicate a loss of post-tensioning force. A structural analysis to consider possible losses of post-tensioning precompression from built-in restraint can provide an indication of whether such cracking could be anticipated. If a large discrepancy is found between the calculated service load stresses and the actual observed extent of cracking, then a loss of post-tensioning may be indicated.

b) Longitudinal cracks in beams, either on the sides or across the bottom in line with the tendons, may be an indication of water infiltration into the tendons. This water can then fill the sheathing and, in unheated structures in colder climates, freeze during winter months. The expansion due to formation of ice can cause splitting of the concrete cover over the tendons.

c) An unintended reversal in the curvature of the tendons at any point in the beam can cause splitting of the beam due to local internal tension in the concrete around the reversal, produced when the tendons are tensioned. Although these cracks can form without affecting the condition of the PT tendons, they can be a warning of a potential spall.

d) Cracks in the area of beam-column connections may be due to structural action, a deficiency of reinforcement intended to control bursting stresses in the tendon anchorage zone, restrained shrinkage, or a combination of these effects. Water entering these cracks may have easy access to the ends of the tendons and cause further deterioration (Nawy 1989).

5.3.2 *Staining*—Stains on the surface of the concrete can also provide external evidence of possible internal distress due to corrosion of the PT system. PT coating stains on the soffits of slabs, especially at low points of tendon profiles, can be an indication of unrepaired damage to the tendon sheathing, as well as shallow concrete cover over the tendons. Such PT coating staining may be accompanied by water stains or evidence of leaching, indicating water infiltration into the slab tendons. Rust staining may indicate a compromise in the corrosion protection of the post-tensioning and would warrant further testing and exploratory concrete removal, as discussed in 5.5.

5.3.3 *End anchorages*—Visual inspection of exposed end anchorage grout pockets should be performed, especially where exposure to moisture is evident and the findings correlated with any signs of distress such as those described previously. Evidence of shrinkage, cracking, debonding, freezing-and-thawing damage, or PT coating or rust staining coming from the grout pocket may indicate a potential breach in the corrosion protection of the anchorage and PT tendon.

5.3.4 *Strand failures*—The most obvious external evidence of corrosion damage is the presence of a section of strand protruding from the structure (Fig. 4.5.4a and 4.5.4b). An exposed loop of strand results when the strand breaks and the elastic energy is released suddenly. The strands typically will erupt from the slab at high points or low points in the tendon profile where concrete cover may be shallow and there is some curvature in the tendon profile; occasionally only a single wire will burst through the surface of the concrete. Loops formed by this phenomenon can be anywhere from 1 to 24 in. (25 to 600 mm) high. Rather than bursting from the structure at some point midway between anchorages, the tendons may also protrude out of the structure at the location of exterior anchorages a distance of several inches (millimeters) or several feet (meters).

Strand breakage can occur without visible disturbance to the concrete, so the absence of strand loops or projections is not necessarily an indication of the absence of broken tendons. Most post-tensioned structures use higher-strength concrete, with higher cracking strength, and may incorporate—perhaps unintentionally—a significant degree of restraint or redundancy. For example, below-grade construction, two-way slab construction, or both, make it possible to have a significant percentage of the tendons broken in a beam or in an area of slab without obvious distress. Strand breakage results in the release of a significant amount of energy, which is often loud enough to be heard by building occupants.

The location at which the tendon or wire has erupted out of the structure is usually some distance away from the location of actual failure. Exploratory concrete removals, combined with the removal of a portion of the broken tendon, are required to identify the location, nature, and the possible causes of the failure.

A visual review of the structure can determine the distribution of such eruptions in the structure, correlating them to detailing problems and inadequate corrosion protection that, in turn, help determine possible appropriate repair and protection measures. It might be useful to determine the distribution of eruptions over time, in addition to identifying any apparent trends in the frequency of tendon failures over time. Unfortunately, delaying repairs to monitor tendon failures might allow acceleration in the rate of deterioration. Usually, it is better to begin repairs as soon as a reasonable condition study has been completed rather than to wait for months or years to gather more information. In a few cases, the rate of corrosion activity has been observed to increase over time, indicating that the rate of tendon failures could increase with time as well.

5.4—Nondestructive testing

Currently there are no dependable nondestructive testing methods for evaluating the existing condition of tendons inside a structure. The main shortcomings of present test methods include an inability to provide information about severity or location of pitting, gross section loss, location of preexisting cracks or breaks, or breaks and section loss of individual wires in a strand. Several common procedures are reviewed in the following with comments about their limitations for application to unbonded tendons.

a) Reinforcing steel locators can locate coated and sheathed tendons, as well as bonded reinforcement. They are useful in estimating the thickness of concrete cover over anchorages and along strands, and the thickness of grout protection of strand stubs at stressing pockets. Readings for anchorages and strand stubs using these devices, as well as along strands, may require calibration by comparing against actual physical measurements of concrete cover thickness.

b) Half-cell potential (ASTM C876) can determine the presence of corrosion activity in bonded reinforcing bars and in anchors and strands that are in contact with concrete. Half-cell testing does not reliably locate corrosion along the length of a tendon because the PT coating and sheathing will electrically insulate the strand. This method is redundant where visible evidence of corrosion is present.

c) Ultrasonic pulse velocity (ASTM C597) can determine certain properties of concrete, but has not proven to be adaptable in determining the integrity of tendons.

d) Ground-penetrating radar (GPR) (ASTM D4748) can locate embedded reinforcement and tendons in concrete, including scanning large areas to locate reinforcement and tendons. The images produced by radar, however, generally do not have sufficient resolution to obtain information about pitting, breaks, or loss of cross section of individual wires. In addition, GPR generally will not be suitable to locate individual wire breaks, but will allow for identification of completely failed tendons provided a distinct physical gap exists at the break location. Also, other reinforcement can shield the tendon being checked.

e) Impact-echo testing can be used to supplement a delamination survey to more accurately determine the delaminated areas. It can also determine crack depths, voids, and honeycombing. Impact-echo provides no information regarding the strand condition.

f) Acoustic emission testing (not to be confused with continuous acoustic monitoring discussed in 6.6) is a technique for locating corrosion based on detection of high-frequency acoustic wavers emitted at sites of active corrosion. This technique has been used successfully on steel structures, such as storage tanks, and more recently has been applied to post-tensioned structures. For detection of corrosion damage in post-tensioning, acoustic emission remains a developing technique that, to date, has not be positively correlated with specific corrosion damage on a tendon. With acoustic emission testing, it is not possible to determine the nature of the corrosion damage (surface attack versus pitting), the location of the damage (strand versus anchorage or location on the strand), or whether wires or the entire strand are broken.

g) Vacuum testing of grouted stressing pockets can assess the quality of the grout and its bond to the sides of the blockout (Schupack 1991a). Such testing can help quantify the effectiveness of the grout plug in preventing water from infiltrating the PT tendon.

h) X-ray techniques are often used to locate embedded reinforcement in concrete, and are generally applied to limited areas due to the expense and environmental issues associated with the gamma rays emitted during testing. X-ray cannot provide information about pitting or loss of cross section of individual wires, but may be able to locate strand breaks, provided a distinct physical gap exists at the break location. However, it is usually impractical to use this method along the entire length of a tendon to find strand breaks due to the complexity and costs associated with the testing. Note that existing mild reinforcement may shield PT tendons and obscure the imaging process.

i) The gas purge technique removes moisture from unbonded PT tendons by pressurizing and passing a dry, noncorrosive gas, such as air or nitrogen, through the interior of the sheathing. This technique can evaluate how much moisture is present in a tendon.

5.5—Exploratory openings

5.5.1 *Concrete removal*—Usually it is necessary to chip into the concrete to inspect the strand, PT coating, sheathing, and anchorages, and to assess the extent of corrosion activity of the bonded reinforcement. This procedure is slow and is limited to parts of the structure that are accessible. For example, anchorages for beam tendons are frequently embedded in columns and are not readily exposed. Occupied structures present special problems because concrete removal is noisy, dusty, and dirty. The number of inspection opportunities may be small compared with the total number and linear footage of tendons in a structure, so the information obtained may not be statistically meaningful. Exploratory openings can be used to calibrate or confirm results from nondestructive tests. Repair of concrete removals should restore PT coating, sheathing, and concrete protection (ACI 546R).

5.5.2 *Exposing tendons*—Exposing parts of tendons is most convenient where concrete cover is shallowest, such as at low points or high points in the tendon profile (Fig 5.5.2). In most structures, it is preferable to expose the strand at low points in the profile for the following reasons:

a) Removal of the upper surface often is more disruptive to the use of the building.

b) The upper surface is often more exposed to sources of water or other aggressive agents; achieving high quality and timely repairs to reestablish protection may be difficult at the time the exploratory work is being performed, or it may be desirable to leave the test areas open for some period for repeated viewing.

c) If water is present in the tendons, it tends to collect at low points in the profile and is more likely to be detected there, especially for loosely sheathed strand.

Care should be taken in all cases not to damage the tendons because they are usually fully stressed. Consider, too, that an unblemished, properly coated strand uncovered in a convenient area can be seriously corroded at another location along the tendon. Once broken, the tendon has lost

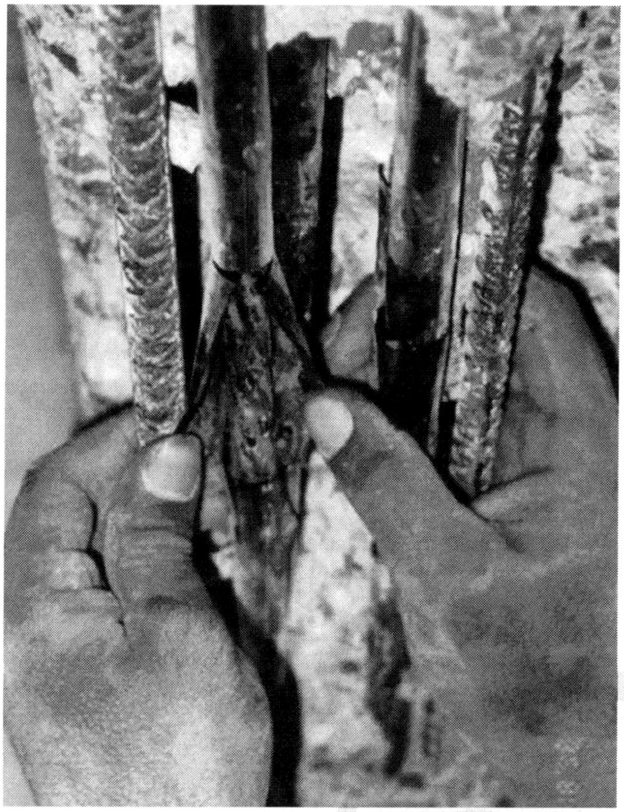

Fig. 5.5.2—Typical tendon inspection procedure.

its structural value for its entire length between anchorages, so it is important to take as many observations as possible and to locate them in the areas of suspected corrosion. Engineering judgment should be applied to estimate the existing condition and future performance of the structure.

As previously discussed (5.3.2), PT coating stains or water stains on the soffit of the slabs and beams can help identify locations where exploratory removals may be desirable. Locations that appear to be free from distress should also be examined to provide a more representative picture of the general condition of the tendons.

Exposing selected tendons for a length of at least 12 in. (300 mm), for as much of their circumference as practical, can provide the following information:

a) The type of sheathing: paper-wrapped, push-through, heat-sealed, or extruded; its tightness to the strand; and its thickness

b) Unrepaired damage to the sheathing

c) Water within the sheathing

d) The condition of the PT coating

e) Corrosion on the strand

f) Verification of strand size

g) Loose strand or wire indicating prestress loss

Exposed strand or wire is commonly checked for the presence of post-tensioning force by means of pry bars or screwdrivers. If the strand can be displaced a significant amount with a pry bar, it is likely there is little or no tension remaining in the strand. This testing is subjective, because the length of strand exposed affects the amount of displacement that can be obtained. The screwdriver penetration test can also be used to identify loose or broken wires. The inspector attempts to drive a standard (flat blade) screwdriver between the wires of the strand at an inspection port. Penetration of the screwdriver indicates little or no force in the wire(s). If sufficient length of the strand is exposed, the test can be performed between each of the six outer wires, thus identifying individual wire breaks. The test is also subjective and will not necessarily detect all wire breaks, especially if they are at some distance from the inspection location. Intermediate anchorages and bonding of the tendon by build-up of corrosion by-products may maintain sufficient force in the tendon to mask strand or wire breaks.

Exploratory concrete removals undertaken in the area of intermediate anchorages are useful, particularly where leakage has occurred through construction joints. The stressing side of the joint can be exposed without compromising the integrity of the tendon; the concrete on the bearing side may be under compression and should not be disturbed.

Exploratory concrete removals should not begin without a detailed plan of action. A specification should be prepared that outlines safety precautions and acceptable procedures for demolition, encapsulation repairs, and concrete patching. Individuals performing the chipping must be educated about tendons so they are knowledgeable about what to look for and understand the consequences of breaking a strand. They should use slow, careful means of concrete removal, and should use tools that are appropriate to the task (ACI 546R). As previously discussed, broken strands can erupt vertically through the top or bottom surfaces of the concrete, or horizontally through the anchors.

5.6—Strand removal

Strands can be removed for inspection, but it is unusual to cut a stressed tendon for this purpose. It is more common to extract part of a strand that is known to be broken, usually one that has erupted from the concrete. It is usually possible to remove the strand between the eruption and the break, but considerable effort is sometimes required. Laying the strand piece on the deck after removal is a simple way to find the location of the break. Once removed, the strand can be examined to determine the condition of the PT coating, the frequency and severity of corrosion, and the type of failure of the wires.

Signs of corrosion are sometimes subtle and look more like stains than rust, so the strand should be examined closely by someone experienced in this type of evaluation. The failed end of a strand can be submitted to a metallurgical testing laboratory for an analysis of the broken wires. Microscopic examination of the external and internal wires of the strand can be performed to look for cracking, pitting, and corrosion.

Chemical residue present on the wires at the failure location can sometimes be identified and related to the failure. Testing of physical and mechanical properties of the strand can be performed on strand taken some distance from the break to determine tensile strength, elongation, and bend testing.

5.7—Other testing and investigative procedures

Lift-off testing, which is pulling the strand loose from an anchor to measure its actual force, can be used to check tension in the strands, but is difficult to perform with single-strand tendons in a completed structure. Because the strand has been cut close to the wedges, special tools are required to grip the strand stub tightly enough to pull the strand and loosen the wedges. Corrosion increases the locking force of the wedges and weakens the stub, making the process difficult to execute. Welding an extension onto the strand is not feasible because of the likelihood of permanently damaging the anchorage and the strand. When barrel anchorages have been used, it may be possible to clamp onto the barrel and perform a lift-off of the tendon in this manner. Care should be taken, however, because barrels are generally heat-treated, making them hard, difficult to grip, and nonweldable.

Lift-off tests will not provide detailed information about the amount or distribution of corrosion on the strand, nor will they give an indication of the loss of ultimate strength. They will, however, indicate the strand's ability to sustain the force held by the jack after the wedges came loose.

An alternative to lift-off testing is to check for tension in the strand using equipment that analyzes the vibration characteristics of the strand under tension. This equipment is adapted from that used to check tendons in cable-stayed bridges and in similar applications. A portion of the strand is exposed, typically approximately 2 ft (600 mm), and the ends of the exposed length are chocked to provide node points. An accelerometer is attached to the exposed strand, and the strand excited by means of tapping. The vibration is transmitted from the accelerometer to computing equipment to determine the tension in the strand as a function of the frequency of vibration. As with lift-off testing, checking tension with an accelerometer does not provide detailed information about the amount or distribution of corrosion on the strand.

Another testing technique involves measuring the force necessary to pull a tendon at a predetermined transverse distance from its existing profile. The force required to introduce the translation can be correlated to the tensile force in the tendon. This technique requires exposing a portion of the strand, typically approximately 2 ft (600 mm), at an intermediate location along the tendon, and use of a hydraulic jack and testing frame. As with lift-off testing, checking tension in this manner does not provide detailed information about the amount or distribution of corrosion on the strand.

Load testing of slabs and beams in accordance with Chapter 20 of ACI 318-11 provides no detailed information about the condition of the individual tendons. A significant number of tendons could have failed without being detected by a load test. The period of time subsequent to the testing for which the test results are valid is subject to a great deal of uncertainty, and ACI 318 provides no guidance for estimating the remaining service life.

Load testing is expensive and disruptive. A load test can confirm the tested part of the structure has adequate strength at the time of the test, but this method has limitations when applied to structures with suspected or known corrosion damage to unbonded tendons.

CHAPTER 6—REPAIR METHODS

6.1—General

Before undertaking a repair effort of any kind, it is essential to determine the causes of corrosion damage and to identify possible preventative measures that may be taken. Incorporating measures that address causes of deterioration is essential to a repair program that is effective for extending the useful life of the structure. Whether the need for tendon repair is critical should be assessed by analyzing the structure to determine if sufficient reserve strength exists to adequately support the design loads after a few tendons are lost. In some cases, it may not be necessary to replace every broken or damaged tendon, or it could be reasonable to use smaller-diameter strand to replace the strands that were removed. The smaller-diameter tendon may be of a higher strength if required to maintain adequate precompression or ultimate strength. Where the type and condition of the strand and sheathing permits, it is possible to replace the old strand with one of the same diameter. If analysis indicates that a significant reduction in calculated nominal strength has occurred, then evacuating or shoring the affected portions of the structure may be required in advance of repairs.

When tendon corrosion has progressed to the point where the strength of any part of a structure has been reduced to an unacceptable level, repairs should be considered to correct the deficiency. The type of repair will depend on the extent, or assumed extent, of the corrosion damage and the resulting loss of strength. If the damage is known to be minor and its location easily identified, the remedy may consist of simply repairing and protecting the existing tendons. It is likely that a variety of conditions will be encountered, some that might require the removal and replacement of entire tendons. Engineering assessment of the proposed repair methods and sequence is needed to determine how the repair process may further affect the integrity of the structure.

When the actual or assumed extent of damage makes localized repairs impractical, the structure may need to be strengthened with an externally applied structural system consisting of PT tendons, structural steel sections, carbon fiber reinforcement, or concrete members. External PT systems have been designed to completely replace the original tendons in some structures. These various approaches are discussed in the following sections. When external strengthening is not feasible, partial or complete demolition and replacement may be necessary.

6.2—Tendons with observable corrosion

Visible evidence of tendon corrosion, such as what might be seen on exposed tendon end anchorages where the stressing pocket has not been grouted, should be repaired as quickly as possible. The exposed part of the anchorage should be cleaned by abrasive blasting or equivalent, coated with a rust-preventive paint or other corrosion protection, and the stressing pocket properly filled with a cementitious grout in accordance with Post-Tensioning Institute (1985).

Tendon corrosion may also occur at locations where it is not directly exposed, such as leaking cracks, and intermediate anchorages and end anchorages where leakage has occurred through the adjacent construction joint. If the strand has not experienced significant corrosion-related section loss, the strand and anchorage can be addressed in a manner similar to that described previously. This requires exposing the strand and anchorage on the stressing side of the joint, cleaning the exposed anchorage and strand by abrasive blasting or the equivalent, and coating the exposed anchor with a rust-preventive paint or other corrosion protection. The exposed portion of the strand should be protected by applying new PT coating and restoring the sheathing. Sheathing repair provisions are found in PTI M10.3.

In a structure where tendon corrosion has been diagnosed, appropriate means of slowing the rate of corrosion in the existing tendons should be applied. Eliminating water intrusion is of primary importance, so concrete repairs should be made and cracks should be routed and sealed. If extensive cracking is present or if there is widespread deficiency of protective concrete cover throughout all or a portion of the structure, consideration should be given to the application of a waterproofing membrane, possibly incorporating a wearing surface as appropriate (ACI 224.1R).

Tendons that have moisture in their sheathing but have not yet experienced significant corrosion-related cross-sectional area loss can be dried by using the gas purge technique discussed previously in the nondestructive testing section of this report. In this case, the technique is continued until the dry, noncorrosive gas (air or nitrogen) pressurizing and passing through the interior of the sheathing has removed as much moisture as desired from the unbonded PT tendon. The time required to dry a tendon will depend on the length and amount of moisture in the sheathing. Following the initial drying process, the system can be maintained and monitored to keep a tendon dry.

Alternatively, new PT coating can be injected into the tendon sheathing to protect the tendon.

The injection of PT coating into an existing sheathing is used occasionally for tendons subjected to drying, particularly where moisture could reenter the tendon sheathing like in an exterior structure such as a parking garage, and is intended to restore the tendon environment to a like-new condition. The effectiveness of PT coating injection is limited by the volume of the interstitial space inside the sheathing, and requires multiple ports along the length of the tendon to allow air to escape while the PT coating is being injected. Additionally, the injection process requires the use of a PT coating material that is less viscous than typical to facilitate injection.

6.3—Spliced repair of strand

When a strand has been inadvertently cut or damaged, or when corrosion damage is known or believed to be localized, repairs are often made by replacement of only part of the strand. When the damage occurs away from the tendon ends, the existing anchors can be reused, in which case the existing wedges may not require replacement. After detensioning any remaining stressed wires, the damaged section of strand is cut away and a new piece of strand is spliced onto the existing tendon. This technique requires that the sheathing be restored at all inspection and repair locations to provide encapsulation of the strand. Sheathing repair provisions are found in PTI M10.3.

Fig. 6.3—Repair procedure using splice couplers.

At the splice locations, couplers are used to connect the new strand section to the remaining original strand. The simplest couplers are short tubes with two sets of wedges (Fig. 6.3), whereas others allow the strand to be stressed at the splice location. The original anchorages do not need to be exposed when an in-line stressing coupler is used and the stressing process is a limited form of proof test. Some designers specify that for this type of repair the strand must be stressed, as usual, to 80 percent of ultimate before lock-off. Others acknowledge that long-term losses have already occurred and allow a lower initial tension to be applied. The repaired tendon stressed in this manner will have approximately the same effective stress as the original tendons.

When the damage to a tendon occurs at or adjacent to an end anchor, it is often easier to re-tension at the end anchor. In that case, new strand is used from the end anchor to a coupler installed beyond the extent of tendon damage. This type of repair often requires replacement of the anchor casting and wedges to ensure that the wedges and anchor are compatible and that the wedges will properly seat during stressing.

6.4—Strand replacement

Although replacing a strand for its full length and using the original anchors is also possible, dislodging the old wedges is sometimes difficult and the anchors can be damaged in the process. It is usually advisable to replace the anchors with new ones because this gives the opportunity to improve the system's durability. Replacement of the wedges is required to ensure that the new strand can be gripped properly. The new wedges should be compatible with the existing anchor casting. If not found to be compatible, the anchor will need to be replaced.

To replace a strand for its full length, strand extraction can be performed once the strand is free of its anchorage. Strand extraction is normally not difficult. In some cases, a jack can be used to pull the strand out—a reliable yet slow process. Usually the loose strand is pulled out by hand or with the assistance of a come-along or vehicle. If binding is a problem and jacking is required, the applied stress should not exceed $0.80f_{pu}$. If the tendon cannot be moved, then the cause of the binding must be located and removed, which usually requires additional openings in the concrete. The distance back to the bound location is calculated using the achieved elongation.

Before inserting the new strand into the old sheathing that remains in the structure, moisture inside the sheathing should be removed. Pulling rags through the sheathing until dry or passing dry air or a moisture-absorbent gas through the empty sheathing are effective techniques of drying out the existing sheathing. Once dry, a quantity of PT coating should be pumped into the sheathing ahead of the new strand. The goal of this procedure, or any alternative procedure, should be to fill all voids inside the sheathing so that water cannot again enter into the sheathing and start a new cycle of corrosion. The new strand used can be either bare or precoated and encased in a plastic sheathing from which it is pulled out to be inserted. In any tendon, only one type of PT coating should be used. In general, precoated strand, preferably one covered with an extruded sheathing, is desirable for use as replacement strand. The post-tensioned coating on a factory-coated strand is applied under pressure during fabrication so the coverage and penetration between the wires is better than can be achieved if the PT coating is applied in the field to a bare strand.

The replacement strand is often of the same size and grade as the original strand, depending on the condition of the existing sheathing. Alternatively, epoxy-coated strand meeting ASTM A822/A822M may be considered for strand replacement, but a smaller-diameter strand must be used to accommodate the thickness of the coating (30 to 40 mils [0.7

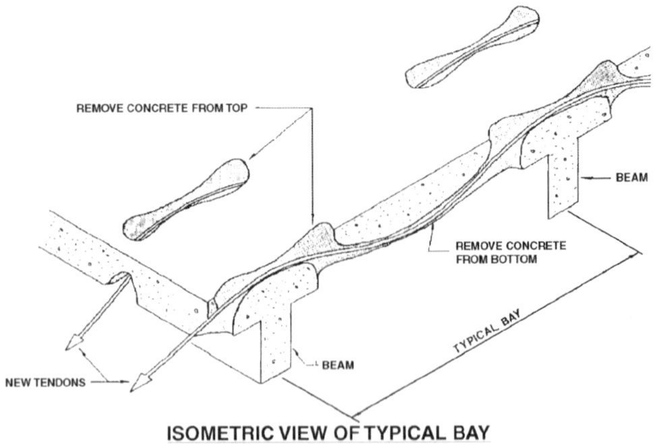

Fig. 6.5—Tendon replacement by trenching.

to 1.0 mm]). Special anchorages and wedges are required for use with epoxy-coated strand, so existing anchorages should be replaced. There have been problems in the past with slipping of strand through anchorages due to an inadequate bite of standard wedges through the epoxy coating.

New strand is either pulled into the existing sheathing, usually by first threading through a smaller-diameter strand that has been welded to the end of the new strand, or pushed through by means of a jack that has been reversed and tied off to provide the necessary reactive force. Although the strand buckles more readily at the jack as the embedment friction increases, making shorter strokes a necessity, most strands can be replaced by this means. For replacing longer lengths of new strand, intermediate openings where the strand can be gripped may be necessary.

Replacing strands is relatively expensive, intrusive, and time-consuming. However, it can be an effective remedy when architectural considerations are important because there is no change in the appearance of the structure. This work should be performed by experienced contractors who understand all aspects of proper corrosion protection for tendons.

6.5—Tendon replacement

In some structures or portions of structures where many strands are damaged in many places, it may be more cost effective to abandon some or all of the existing tendons and install new ones, rather than to replace individual strand and anchorages. This is generally practical only for slabs or wide beams because it requires that either intermittent or full-length slots be cut in the concrete. The procedure, termed "trenching," as shown in Fig. 6.5, is a patented process. Detailing at new anchorages and along the length of the new tendon should ensure that the compression, uplift, and downward forces are transferred from the new concrete, in which the new tendons are encased, to the existing concrete.

6.6—External post-tensioning

Externally applied PT systems can be effectively used to strengthen large portions of existing structures. External PT systems have found application in the retrofit of a wide

Fig. 6.6a—Example of external post-tensioning. (Courtesy of Seneca Structural Engineering, Inc.)

Fig. 6.6b—Example of external post-tensioning. (Courtesy of Wiss, Janney, Elstner Associates, Inc.)

variety of framing systems, including two-way flat plates, flat slabs, one-way slabs, beams, and girders.

Typically these systems consist of adding straight tendons along the tensile zone of existing slabs or beams, or building a tendon truss under or alongside the existing slab or beam. The lower (tension) chord of the truss is the PT tendon or tendon group. The vertical truss element is usually a structural steel shape, normally a tube with a bearing plate that bears on the soffit of the existing structure and applies the beneficial upward force. The top, or compression chord of the truss, is the existing structure itself to which the PT anchors are attached (Fig. 6.6a and 6.6b).

Advantages of externally post-tensioned retrofits include:

a) The ability to apply large upward forces to the existing structure with minimal headroom requirements

b) Little or no interference with existing utilities

c) Minimal disruption of the existing function of the building

d) Possible cost savings over methods such as tendon replacement

One possible disadvantage of externally applied post-tensioning is that it can change the architectural appearance of the structure when it is not hidden by a ceiling or other feature. Fire protection of the external PT system should be in accordance with the requirements of the governing building code. For additional information regarding fireproofing of external post-tensioning, refer to PTI DC80.3-12/ICRI 320.6. Corrosion protection may be required in accordance with the project specifications.

6.7—Continuous acoustic monitoring

Acoustic monitoring systems can provide continuous monitoring for post-tensioned structures (Nawy 1989). When a strand or wire fails, there is a sudden release of energy. In this monitoring system, sensors mounted at selected locations on the structure detect the acoustic response transmitted from the location of the break. Information collected by the sensors is recorded by terminals at the site and downloaded to a central computer where the information is analyzed to determine the probable location of the break (Fig. 6.7).

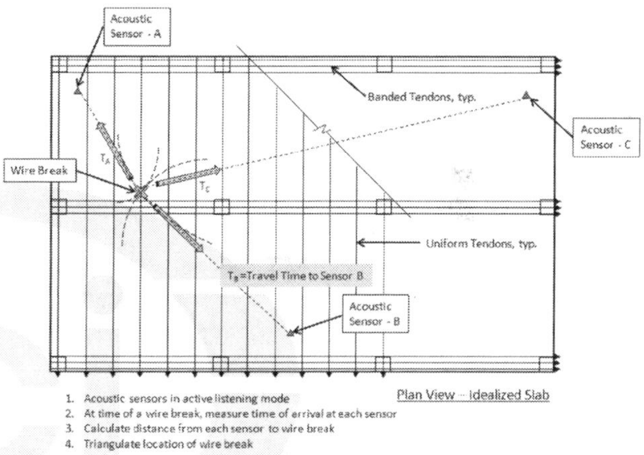

Fig. 6.7—Schematic of continuous monitoring systems.

The monitoring system does not determine the condition of the PT system at the outset of installation, but rather provides a record of the number and location of strand breaks in the structure subsequent to installation. The continuous acoustic monitoring system is intended to assist with managing the evaluation and repair of structures with known or suspected tendon corrosion problems. The goal is to minimize repair expenditures and still ensure structural integrity.

6.8—External nonprestressed reinforcement or support

Strengthening can be accomplished with nonprestressed elements, such as structural steel or reinforced concrete beams and columns. If architectural requirements permit, the most cost-effective method to strengthen an existing structure may be to add a permanent vertical support, such as a column. Most structures, however, cannot accommodate additional columns and require some form of flexural retrofit, such as beams and girders or bonded external reinforcement, that carry loads to existing columns. If nonprestressed retrofits are designed to share loads with the existing floor system (that is, the retrofit is designed only for the strength deficiency, not as a total replacement for the flexural and shear strength of the floor framing), then analysis and design should be performed based on deflection compat-

Fig. 6.8a—Example of strengthening a PT structure with mechanically anchored steel plates. (Courtesy of Wiss, Janney, Elstner Associates, Inc.)

Fig. 6.8b—Example of external strengthening of a PT structure using carbon fiber strips. (Courtesy of Wiss, Janney, Elstner Associates, Inc.)

Fig. 6.8c—Example of strengthening a post tensioned structure using reinforcing bar embedded in strips in the concrete surface. (Courtesy of Wiss, Janney, Elstner Associates, Inc.)

ibility to confirm that stresses in the existing concrete remain within allowable limits and that deflections remain within an acceptable range. Jacks placed between the existing structure and the retrofit element can be used to load the retrofit element to ensure that the retrofit partially supports existing dead loads and applied stressing-end loads, and can reduce deflections in the system. Composite action between the retrofit and existing structure can also be used to promote load sharing.

Installation of external nonprestressed reinforcement or supports can involve major disruption in use of the building and may require rerouting of existing utilities, depending on the type and details of the external reinforcing system. External reinforcement could consist of mechanically anchored structural steel shapes or plates (Fig. 6.8a), adhered carbon fiber strips on the surface of the structural element being reinforced (Fig. 6.8b), or reinforcing bar or carbon fiber strips embedded in strips cut into the concrete (Fig. 6.8c).

6.9—Total demolition and replacement

Demolition procedures for post-tensioned buildings, in their entirety, are similar to those for nonprestressed concrete buildings, except for different collapse mechanisms. Although the development of detailed demolition plans is outside the scope of this report, demolition of structures with unbonded tendons has some important considerations. For instance, the tendons cause significant catenary action to develop, which can be helpful or detrimental depending on the procedures and equipment used (Chacos 1991). Also, cutting tendons before beginning demolition is neither necessary nor advisable because the catenary action from the tendons can assist in controlling potential collapse. Furthermore, precautions may be required during concrete removal to prevent broken tendons from exiting the edges of the structure and damaging adjacent property, although some studies have shown this to be an unlikely event for single-strand systems (Aalami and Barth 1989). The amount of projection will depend on many factors, including type of sheathing, condition and quantity of PT coating, and amount of drape. The potential for tendons exiting the structure should be evaluated for each project.

CHAPTER 7—SUMMARY

During the past 50 years, unbonded tendons have become a significant form of reinforcement for concrete building structures. Tendon types have changed in response to an industry-wide recognition of the vulnerability of the earlier tendons to corrosion damage, as well as market forces that reflected the convenience and reduced labor costs associated with single-strand systems. As the causes of tendon corrosion became known, specifications were changed to implement requirements for the materials, fabrication, installation, and design details to protect the tendons from deterioration. Structures using single-strand unbonded tendons that have low-quality PT coating, insufficient quantity of PT coating, water in the sheathing, and poor corrosion protection details are more likely to suffer strand failures. Concern for the evaluation and repair of these unprotected and deteriorating structures is the primary impetus for preparing this report.

Advances in the materials, manufacturing processes, installation procedures, and specifications have significantly reduced the potential for many of the corrosion types reviewed. Proper use of fully encapsulated tendon systems meeting current specifications can substantially eliminate the potential for corrosion. PTI M10.2 requires that all tendons used in structures governed by ACI 318 be encapsulated.

The most common type of unbonded tendon to be used for buildings during the past 30 years has a single seven-wire strand with cast anchors and two-piece wedges. Techniques for repairing these tendons and the cost of performing this work are understood by many contractors and licensed design professionals. While there are many methods available to evaluate the condition of concrete and bonded reinforcement in a post-tensioned structure, there is no method available, without destructive testing, that can determine the amount that an unbonded tendon has been weakened by corrosion.

Not every building can be economically repaired by replacing all, or parts of, the tendons. Sometimes it is best to disregard the existing tendons and replace them with a new externally applied system. This method is adaptable to many circumstances and can be used for beams as well as slabs.

Total demolition is the last resort for a deteriorated structure of any type. The decision to demolish a post-tensioned structure should be made after a realistic assessment has been made of the condition of the concrete and the bonded reinforcement, and not just on an estimate of the condition of the tendons.

CHAPTER 8—REFERENCES

Committee documents are listed first by document number and year of publication followed by authored documents listed alphabetically.

American Concrete Institute

222.2R-01(10)—Corrosion of Prestressing Steels

224.1R-07—Causes, Evaluation, and Repair of Cracks in Concrete Structures

318-47—Building Code Requirements for Reinforced Concrete

318-51—Building Code Requirements for Reinforced Concrete

318-56—Building Code Requirements for Reinforced Concrete

318-63—Building Code Requirements for Reinforced Concrete

318-71—Building Code Requirements for Reinforced Concrete

318-77—Building Code Requirements for Reinforced Concrete

318-83—Building Code Requirements for Reinforced Concrete and Commentary

318-89—Building Code Requirements for Reinforced Concrete and Commentary

318-02—Building Code Requirements for Structural Concrete and Commentary

318-05—Building Code Requirements for Structural Concrete and Commentary

318-08—Building Code Requirements for Structural Concrete and Commentary

318-11—Building Code Requirements for Structural Concrete and Commentary

349-13—Code Requirements for Nuclear Safety-Related Concrete Structures and Commentary

364.1R-07—Guide for Evaluation of Concrete Structures before Rehabilitation

423.7-07—Specification for Unbonded Single-Strand Tendon Materials and Commentary

437R-03—Strength Evaluation of Existing Concrete Buildings

546R-04—Concrete Repair Guide

ASTM International

A822/A822M-04(2010)—Standard Specification for Seamless Cold-Drawn Carbon Steel Tubing for Hydraulic System Service

B117-11—Standard Practice for Operating Salt Spray (Fog) Apparatus

C597-09—Standard Test Method for Pulse Velocity Through Concrete

C876-09—Standard Test Method for Corrosion Potentials of Uncoated Reinforcing Steel in Concrete

D4748-10—Standard Test Method for Determining the Thickness of Bound Pavement Layers Using Short-Pulse Radar

Post-Tensioning Institute

DC80.3-12/ICRI 320.6-12—Guide for Evaluation and Repair of Unbonded Post-Tensioned Concrete Structures

M10.2-00—Specifications for Unbonded Single Strand Tendons

M10.3-00—Field Procedures Manual for Unbonded Single Strand Tendons

Authored documents

Aalami, B. O., and Barth, F. G., 1989, "Controlled Demolition of an Unbonded Post-Tensioned Concrete Slab," Post-Tensioning Institute, Farmington Hills, MI, 30 pp.

Bureau of Public Roads, 1954, "Criteria for Prestressed Concrete Bridges," U.S. Department of Commerce, Washington, DC, 25 pp.

Chacos, G., 1991, "Demolishing a Post-Tensioned Parking Garage," *Concrete International*, V. 13, No. 10, Oct., pp. 44-46.

Freyermuth, C. L., 1989, "Structural Integrity of Buildings Constructed with Unbonded Tendons," *Concrete International*, V. 11, No. 3, Mar., pp. 56-63.

Joint ACI-ASCE Committee 323, 1958, "Tentative Recommendations for Prestressed Concrete (JL54-30)," *ACI Journal Proceedings*, V. 54, No. 1, Jan., pp. 545-578.

Nawy, E. G., 1989, "Stresses, Strains, and Bursting Cracks in Anchorage Zones of Post-Tensioned Beams," *Cracking in Prestressed Concrete Structures*, SP-113, American Concrete Institute, Farmington Hills, MI, pp. 109-132.

Nehil, T. E., 1991, "Rehabilitating Parking Structures with Corrosion-Damaged Button-Headed Post-Tensioning Tendons," *Concrete International*, V. 13, No. 10, Oct., pp. 66-73.

Peterson, C., and Braunfield, A. H., 1957, "Our Experience with Prestressed Lift Slabs," World Conference on Prestressed Concrete, San Francisco, CA.

Post-Tensioning Institute, 1972, *Post-Tensioning Manual*, first edition, Post-Tensioning Institute, Farmington Hills, MI.

Post-Tensioning Institute, 1976, *Post-Tensioning Manual*, second edition, Post-Tensioning Institute, Farmington Hills, MI.

Post-Tensioning Institute, 1981, *Post-Tensioning Manual*, third edition, Post-Tensioning Institute, Farmington Hills, MI.

Post-Tensioning Institute, 1985, "Specification for Unbonded Single Strand Tendons," *PCI Journal*, V. 30, No. 2, Mar.-Apr., pp. 23-39.

Post-Tensioning Institute, 1993, "Specifications for Unbonded Single Strand Tendons," Post-Tensioning Institute, Farmington Hills, MI, 16 pp.

Post-Tensioning Institute, 2006, *Post-Tensioning Manual*, sixth edition, Post-Tensioning Institute, Farmington Hills, MI.

Schupack, M., 1991a, "Corrosion Protection for Unbonded Tendons," *Concrete International*, V. 13, No. 2, Feb., pp. 51-57.

Schupack, M., 1991b, "Evaluating Buildings with Unbonded Tendons," *Concrete International*, V. 13, No. 10, Oct., pp. 52-57.

Schupack, M., and Suarez, M. G., 1982, "Some Recent Corrosion Embrittlement Failures of Prestressing Systems in the United States," *PCI Journal*, V. Mar.-Apr., V. 27, No. 2, pp. 38-55.

Structural Engineers Association of Northern California (SEA/NC), 1959, "Building Code Requirements for Prestressed Concrete," *PCI Journal*, V. 4, No. 1, June, pp. 72-78.

Suarez, M. G., and Schupack, M., 1988, "Evaluation of the Performance of Single Strand Unbonded Tendons in the Collapse of the L'Ambiance Plaza Lift Slab Building," *Post-Tensioning Institute Project Report*, Apr.

Specification for Unbonded Single-Strand Tendon Materials

An ACI Standard

Reported by Joint ACI-ASCE Committee 423

Carin L. Roberts-Wollmann, Chair

Amy M. Reineke Trygestad, Secretary

Theresa M. Ahlborn
Robert W. Barnes
Florian G. Barth
Asit N. Baxi
Roger J. Becker
Kenneth B. Bondy
Charles W. Dolan
James P. Donnelly
Pierre Esselinck
Martin J. Fradua
William L. Gamble
Harry A. Gleich
Shawn P. Gross
Pawan R. Gupta
William M. Hale
H. R. Trey Hamilton III
Carol Hayek
Mohammad Iqbal
Donald P. Kline
Larry B. Krauser
Jason J. Krohn
Mark E. Moore*
Theodore L. Neff†
Sami H. Rizkalla
James Rogers
Bruce W. Russell
David H. Sanders
Thomas C. Schaeffer
Morris Schupack*
Richard W. Stone
Miroslav F. Vejvoda
Jeffrey S. Volz
H. Carl Walker
Zuming Xia
Paul Zia
Chunsheng 'Steve' Cai
Steven R. Close
Henry J. Cronin Jr.
Ward N. Marianos Jr.
Hani Melhem
Antoine E. Naaman
Thomas E. Nehil
Andrea J. Schokker

Consulting Members
Robert N. Bruce Jr.
Ned H. Burns

*Deceased
†Chair of subcommittee responsible for preparation of specification.

This material specification provides materials criteria and fabrication requirements for unbonded single-strand tendons.

Keywords: fabrication; post-tensioning; PT coating; tendon; unbonded.

CONTENTS

ACI 423.7-14 supersedes ACI 423.7-07, became effective October 22, 2014, and was adopted and published November 2014.

1—SCOPE

1.1
This specification provides materials criteria and fabrication requirements for unbonded single-strand tendons.

1.2
This specification shall not apply to post-tensioned slab-on-ground, mat, or raft foundations on expansive soils.

NOTE 1—For membrane-type structures primarily under tensile forces resulting from temperature effects and concrete shrinkage, the provisions apply where deemed appropriate by Architect/Engineer.

1.3
The text of this standard references notes and footnotes that provide explanatory material. These notes and footnotes (excluding those in tables and figures) shall not be considered as requirements of the standard.

1.4
Values stated in inch-pound units are to be regarded as standard. Values given in parentheses are mathematical conversion to SI Units that are provided for information only, and are not considered standard.

2—DEFINITIONS

2.1
The following definitions govern in this specification. See "ACI Concrete Terminology" for additional definitions. http://www.concrete.org/Tools/ConcreteTerminology.aspx

anchorage—a device used to maintain elongation in prestressing strand by transferring compression force to concrete.

Architect/Engineer—the architect, engineer, architectural firm, or engineering firm developing Contract Documents or administering the Work under Contract Documents, or both.

Contract Documents—a set of documents supplied by Owner to Contractor as the basis for construction; these documents contain contract forms, contract conditions, specifications, drawings, addenda, and contract changes.

coupler—a mechanical device that connects ends of two strands together, thereby creating a continuous tendon by transferring the post-tensioning force from one strand to the other strand.

encapsulated tendon—a tendon that is completely enclosed in a watertight covering from end to end, including anchorages, sheathing, post-tensioning coating, sleeves, and an encapsulation cap over the strand tail at each end.

encapsulation cap—plastic cap filled with post-tensioning coating that provides a watertight connection to all anchorages protecting the wedges and the tendon tail from moisture infiltration. The watertight connection shall include a mechanical interlock of a type that does not rely on friction alone to be held in place.

f_{pu}—specified tensile strength of prestressing steel, psi (MPa).

fixed anchorage—anchorage located at end of tendon where stressing of tendon is not required (also known as dead-end anchorage).

installation drawings—drawings showing information about the specifics of the post-tensioning system and tendon placement such as the number, size, length, marking, location, elongation, and tendon profiles (also referred to as shop drawings).

intermediate anchorage—anchorage located between the ends of the tendon for application of post-tensioning force.

local zone—rectangular prism (or equivalent rectangular prism for circular or oval anchorages) of concrete immediately surrounding the anchorage device and any confining reinforcement.

non-encapsulated tendon—a tendon that has bare metallic anchorages, and sheathing that is continuous between anchorages but not connected to the anchorages.

post-tensioning—method of prestressing in which prestressing steel is tensioned after concrete has hardened.

post-tensioning coating—material used to protect the prestressing steel against corrosion and reduce friction between prestressing steel and sheathing.

prestressed concrete—structural concrete in which internal stresses are introduced to reduce potential concrete tensile stresses resulting from loads.

prestressing steel—high-strength steel, most commonly a seven-wire strand, used to impart prestress forces to concrete.

sheathing—an extruded high-density polyethylene or polypropylene covering that encases prestressing steel to prevent bonding of the prestressing steel with the surrounding concrete, to provide corrosion protection, and to contain the post-tensioning coating.

sleeve—device for connecting sheathing to anchorage.

strand tail—protruding length of strand outside of the wedges that remains in place after the tendon tail has been cut off.

stressing anchorage—anchorage at one end or both ends of tendon where stressing of tendon is required (also known as live-end anchorage)

tendon—in post-tensioned applications, the tendon is a complete assembly consisting of anchorages, prestressing steel, post-tensioning coating, and sheathing.

tendon tail—protruding length of strand outside of the stressing anchorage needed temporarily for stressing of the tendons.

unbonded tendon—tendon in which prestressing steel is prevented from bonding to concrete, and is free to move relative to concrete. The longitudinal prestressing force is permanently transferred to concrete by the anchorages only.

wedges—pieces of tapered high-strength steel with serrations that bite into and grip the prestressing steel to transfer the prestressing force to the anchorage.

wedge cavity—opening in anchorage designed to accommodate the prestressing steel passing through and housing the wedges.

Work—the entire construction or separately identifiable parts thereof required to be furnished under Contract Documents.

3—REFERENCED STANDARDS

3.1—American Concrete Institute

ACI 318-11—Building Code Requirements for Structural Concrete

ACI 350-06—Code Requirements for Environmental Engineering Concrete Structures

3.2—ASTM International

ASTM A416/A416M-12—Standard Specification for Steel Strand, Uncoated Seven-Wire for Prestressed Concrete

ASTM B117-11—Standard Practice for Operating Salt Spray (Fog) Apparatus

ASTM C1077-14—Standard Practice for Laboratories Testing Concrete and Concrete Aggregates for Use in Construction and Criteria for Laboratory Evaluation

ASTM D92-12—Standard Test Method for Flash and Fire Points by Cleveland Open Cup Tester

ASTM D95-13—Standard Test Method for Water in Petroleum Products and Bituminous Materials by Distillation

ASTM D512-12—Standard Test Methods for Chloride Ion in Water

ASTM D566-02(2009)—Standard Test Method for Dropping Point of Lubricating Grease

ASTM D610-08(2012)—Standard Test Method for Evaluating Degree of Rusting on Painted Steel Surfaces

ASTM D638-10—Standard Test Method for Tensile Properties of Plastics

ASTM D792-13—Standard Test Methods for Density and Specific Gravity (Relative Density) of Plastics by Displacement

ASTM D2265-06—Standard Test Method for Dropping Point of Lubricating Grease Over Wide Temperature Range

ASTM D3867-09—Standard Test Methods for Nitrite-Nitrate in Water

ASTM D4289-13—Standard Test Method for Elastomer Compatibility of Lubricating Greases and Fluids

ASTM D4658-09—Standard Test Method for Sulfide Ion in Water

ASTM D6184-98(2005)—Standard Test Method for Oil Separation from Lubricating Grease

3.3—International Organization for Standardization

ISO/IEC Guide 65:1996—General requirements for Bodies Operating Product Certification Systems

ISO/IEC 17065:2012—Conformity Assessment – Requirements for Bodies Certifying Products, Processes and Services

3.4—Federal Test Method Standard

FTMS 791B Method 321.2

4—ORDERING INFORMATION

4.1

It shall be the responsibility of the purchaser to specify all requirements that are necessary for the manufacture and delivery of the unbonded single-strand tendon under this specification. Such requirements to be considered include, but are not limited to, the submittal items in 4.1.1 through 4.1.6. Submittal item 4.1.1.1 shall be required unless specifically excluded by the purchaser.

4.1.1 *Prestressing steel*

4.1.1.1 Certified mill test reports and typical load-elongation curves for each coil or pack of strand used in fabrication of the tendons containing the following information:

a) Heat number and identification

b) Tensile strength

c) Yield stress at 1 percent extension under load

d) Elongation at failure

e) Modulus of elasticity

f) Diameter and net area of strand

g) Type of material (normal or low-relaxation)

4.1.1.2 Relaxation test results.

4.1.1.3 Samples from each heat (or manufacturer's length, in the case of strands), properly marked, for verification of prestressing steel properties.

4.1.1.4 Strand records related to material production containing the following information:

a) Purchasing records showing the purchase of appropriate base materials used in production

b) Product traceability through production and shipping

c) Testing results for tests required under 7.1 of this specification, conformities (or nonconformities), and resultant actions

d) Calibration records for all testing and control devices

e) Statistical records of quality performance evidencing the occasion, frequency, and percentage of accepted and rejected final product; records shall include internal and external occurrences such as on-site lab results and customer responses

f) Suitability and testing of raw materials, including quality reports from wire or rod suppliers

g) Procedure for the quarantine and disposal of noncompliant product and records of same

Strand records shall be maintained for a minimum period of 5 years.

4.1.2 *Post-tensioning (PT) coating*—Test results on PT coating, tested in accordance with 7.2.1.

4.1.3 *Sheathing*—A sheathing material report containing type, thickness, and density of material; information about added ultraviolet light stabilizers (if used); and supporting test data demonstrating compliance with all requirements of 5.3.

4.1.4 *Anchorages and couplers*—Static and fatigue test reports of representative assemblies for each different assembly to be used on the project.

4.1.5 *Fabrication plant certification*—A copy of the tendon fabrication plant certification. If the certification is not from a nationally accredited organization, provide documentation demonstrating compliance with 11.1.

4.1.6 *Stressing jack calibration*—Calibration certificates for every jack and gauge set supplied to the project.

5—MATERIALS

5.1—Prestressing steel

Prestressing steel used in unbonded single-strand post-tensioning tendons shall conform to:

a) ASTM A416/A416M

b) Strand not specifically covered in ASTM A416/A416M that conforms to minimum requirements of this specification and has properties meeting requirements of ASTM A416/A416M.

5.2—Post-tensioning (PT) coating

PT coating shall meet the criteria of 7.2 and demonstrate the following performance requirements:

a) Provide lubrication between the strand and sheathing

b) Resist flow caused by gravity within anticipated temperature range of exposure

c) Provide continuous nonbrittle coating at lowest anticipated temperature of exposure

d) Be chemically stable and nonreactive with prestressing steel, reinforcing steel, sheathing material, and concrete

5.3—Sheathing

Tendon sheathing material shall be high-density polyethylene or polypropylene that meets the criteria of 8.2 and that demonstrates the following performance requirements:

a) Sufficient strength and durability to withstand damage during fabrication, handling, transport, installation, concrete placement, and stressing

b) Watertight and impermeable to water vapor over entire sheathing length

c) Chemically stable, without embrittlement or softening over the anticipated exposure temperature range and service life of the structure; free chloride ions shall not be extractable from the sheathing material

d) Nonreactive with concrete, prestressing steel, reinforcing steel, and PT coating

5.4—Anchorages and couplers

Castings shall be nonporous and free of sand, blow holes, and voids.

5.5—Connecting components

Any component used to connect the sheathing to any anchorage or coupler enclosure in encapsulated systems shall conform to the following:

a) Be watertight in conformance with 9.3

b) Meet or exceed the same requirements as the sheathing for durability during fabrication, transportation, handling, storage, and installation

c) Be translucent or have another method of verifying compliance with 5.5(a) and 6.5.2(c)

d) Be translucent or have other method of verifying overlap with sheathing

5.6—Tape

Tape used as a component for enclosing couplers shall:

a) Be self-adhesive and moisture-proof

b) Be nonreactive with sheathing, PT coating, or prestressing steel

c) Have elastic properties

d) Have a minimum width of 2 in. (50 mm)

e) Have a contrasting color to the tendon sheathing

6—MANUFACTURE

6.1—Minimum quantity of post-tensioning (PT) coating

The minimum amount of PT coating on the prestressing strand shall be at least 2.5 lb (1.14 kg) of PT coating per 100 ft (30.5 m) for 0.5 in. (12.7 mm) diameter strand, and 3.0 lb (1.36 kg) of PT coating per 100 ft (30.5 m) for 0.6 in. (15.3 mm) diameter strand. The minimum quantity of PT coating for other strand sizes can be determined by linear extrapolation.

The PT coating shall completely fill the annular space between the strand and sheathing. The PT coating shall extend over the entire tendon length.

6.2—Manufacturing processes

Sheathing shall be manufactured by an extrusion process that provides watertight encasement of the PT coating.

6.3—Sheathing coverage

Sheathing shall be continuous between anchorages and shall prevent intrusion of cement paste or loss of coating materials. For nonencapsulated applications, a maximum of 1 in. (25 mm) of unsheathed strand shall be permitted at stressing anchorages, and up to 12 in. (300 mm) shall be permitted at fixed anchorages. Coupler components shall be protected with the same PT coating used on the strand, and shall be enclosed in sleeving with adequate length to permit necessary movements during stressing.

6.4—Nonencapsulated systems

Unless otherwise specified, the use of nonencapsulated systems shall be permitted for applications not governed by ACI 318 or ACI 350.

6.5—Encapsulated systems

6.5.1 *Encapsulation*—For all tendons used in applications governed by either ACI 318 or ACI 350, protect all anchorage assemblies by encapsulation, including the anchorage, wedges, and prestressing steel, against corrosion. The encapsulation shall include a watertight connection of the sheathing to the anchorage, and a watertight closure of the wedge cavity and prestressing steel.

Encapsulation systems shall employ plastic coating to protect the metallic components from corrosion. The use of bare metallic anchorages produced from a material that is subject to corrosion shall not be permitted.

6.5.2 *Connection component*—Any component used to connect the sheathing to any anchorage or coupler enclosure shall conform to the following:

a) Have a watertight, mechanical connection to the anchorage protection or coupler enclosure and a watertight connection at the tendon sheathing

b) Have a minimum 4 in. (100 mm) overlap between the end of the extruded sheathing covering the prestressing steel and the end of the sleeve

c) Within the connecting component or enclosure, either the prestressing steel shall be covered by sheathing for its full length, or the annular space between the sleeve and the strand shall be filled with PT coating in conformance with 5.2.

7—MECHANICAL PROPERTIES

7.1—Prestressing steel

7.1.1 *Mechanical properties*—Tests shall be performed in accordance with the methods and frequency prescribed in ASTM A416/A416M.

7.1.2 *Relaxation properties*—For low-relaxation strand, finished strand shall be tested during production for relaxation at least annually, but also on any occasion of change in the type of raw material or manufacturing process. The relaxation test shall be performed according to the requirements of ASTM A416/A416M and the requirements herein.

The relaxation test shall be performed as a full 1000-hour test at initial production and every third year thereafter. Interim annual relaxation tests may be performed as 200-hour tests with results extrapolated to 1000 hours, provided that the previous full 1000-hour test exhibits satisfactory results.

7.1.3 *Premature failure*—For tests required in this specification for both normal and low-relaxation prestressing steel, failure of prestressing steel due to notch or pinching effects from wedges is not acceptable.

7.1.4 *Acceptance criteria for surface condition*—Strand used for tendon manufacture shall be dry. Surface rust, if any is present, shall be removable with a fine steel wool pad or by vigorous rubbing with a cloth. Pits on steel surface shall not exceed 0.002 in. (0.05 mm) in diameter or length.

NOTE 2—Guidance for evaluating the degree of rusting on strand is presented in Sason (1992).*

7.2—Post-tensioning (PT) coating

7.2.1 *Mechanical properties and acceptance*—The PT coating shall be a compound that complies with the tests and associated acceptance criteria specified in Table 7.2.1. PT coating shall be tested whenever any change is made to its chemical composition and at least once every 5 years.

8—DIMENSIONS, DENSITY, AND PERMISSIBLE VARIATIONS

8.1—Prestressing steel

All samples tested shall be checked to ensure that the strand meets the dimensional requirements listed in ASTM A416/A416M.

8.2—Sheathing

8.2.1 *Thickness*—Minimum thickness of sheathing shall be 0.050 in. (1.27 mm). Variations in sheathing thickness of up to 10 percent shall be acceptable, provided an average of four equidistant readings along the circumference equals or exceeds the required thickness.

8.2.1 *Density*—Minimum density of sheathing shall be 0.034 lb/in.^{3} (941 kg/m^3).

8.2.2 *Diameter*—Sheathing shall have an inside diameter at least 0.030 in. (0.76 mm) greater than the maximum diameter of the strand.

8.2.3 *Surface*—Sheathing shall provide a smooth circular outside surface and shall not visibly reveal lay of the strand.

8.3—Anchorages

8.3.1 *Bearing stresses*—Anchorage dimensions shall be such that average bearing stresses on concrete created by an anchorage shall not exceed values computed by the following equations unless testing by a certified independent laboratory indicates anchorage performance equivalent or superior to anchorages satisfying the requirements of this section.

a) At transfer load

$$f_{cp} = 0.75 f'_{ci} \sqrt{\frac{A'_b}{A_b}} \quad (8.3.1a)$$

*Sason, A. S., 1992, "Evaluation of Degree of Rusting on Prestressed Concrete Strand," *PCI Journal*, V. 37, No. 3, May-June, pp. 25-30.

Table 7.2.1—Testing of PT coating

Test no.	Test description	Test method	Acceptance criteria
1	Dropping point	ASTM D566 or D2265	Minimum 300°F (149°C)
2	Oil separation at 160°F (71°C)	FTMS 791B Method 321.2 (ASTM D6184)	0.5 percent maximum by mass
3	Water content	ASTM D95	0.1 percent maximum
4	Flash point (refers to oil component)	ASTM D92	Minimum 300°F (149°C)
5	Corrosion test (5 percent salt fog at 100°F [38°C], 0.005 in. [0.127 mm], minimum hours, Q Panel Type S)	ASTM B117	Rust Grade 7 or better after 1000 hours of exposure according to ASTM D610
6	Water-soluble ions:* a. Chlorides b. Nitrates c. Sulfides	 ASTM D512 ASTM D3867 ASTM D4658	 10 ppm maximum 10 ppm maximum 10 ppm maximum
7	Soak test (5 percent salt fog at 100°F [38°C], 0.005 in. [0.127 mm] coating, Q Panel Type S) Immerse panels 50 percent in a 5 percent salt solution and expose to salt fog)	ASTM B117 (modified)	No emulsification of the coating after 720 hours of exposure.
8	Compatibility with sheathing: a. Hardness and volume change of polymer after exposure to grease, 40 days at 150°F (66°C) b. Tensile strength change of polymer after exposure to grease, 40 days at 150°F (66°C)	 ASTM D4289 (ASTM D792 for density) ASTM D638	 Permissible change in hardness: 15 percent volume: 10 percent Permissible change in tensile strength: 30 percent

*Procedure for Test 6: The inside (bottom and sides) of a 1.06 quart (1 L) glass beaker (approximate outside diameter 4.13 in. [105 mm], height 5.71 in. [145 mm]) is thoroughly coated with 3.53 ± 0.35 oz (100 ± 10 g) of corrosion-inhibiting coating material. The coated beaker is filled with approximately 30 oz (900 cc) of distilled water and heated in an oven at a controlled temperature of 100°F ± 2°F (37.8°C ± 1.1°C) for 4 hours. The water extraction is tested by the noted test procedures for the appropriate water-soluble ions. Results are reported as ppm in the extracted water.

Notes:

Tests 1 and 2—Limiting the dropping point to 300°F (149°C) minimum is intended to ensure product stability under the elevated temperatures that are possible during tendon fabrication and installation. Together, Tests 1 and 2 minimize bleeding of the lighter components from the coating.

Test 3—Water content is limited to exclude the presence of free water in the coating material.

Test 4—This test refers to the oil component in the coating material. Too low a flash point indicates higher content of volatile derivatives, which affect the long-term stability and consistency of the coating material.

Test 5—This test provides a method to determine the effectiveness of the corrosion-inhibiting properties of the coating. The method is a standard test used for corrosion-inhibiting coatings such as paints. The acceptance criterion of Grade 7 or better (according to ASTM D610) after 1000 hours of exposure requires that only 0.3 percent of the area exposed can have indications of corrosion. The test is conducted on a 3 x 6 in. (76 x 152 mm) steel panel with a coating thickness of 0.005 in. (0.127 mm). When determining the percent of area corroded, only the area inside 1/4 in. (6 mm) from the edges of the panel is evaluated.

Test 6—Water-soluble ions known to cause corrosion are limited by this requirement.

Test 7—The soak test is designed to determine the ability of the coating to provide corrosion protection after having been exposed to standing water for a period of time. Certain coatings will absorb water to the extent that they will emulsify and thus not be an effective barrier against moisture reaching the steel. This test will guard against use of such coatings.

Test 8—Certain petroleum derivatives react with polyethylene or polypropylene, changing their physical properties to the point where they are no longer usable as sheathing materials. This test is required to preclude the use of coatings with such derivatives.

but not greater than $1.25f_{ci}'$

b) At service load

$$f_{cp} = 0.6f_c'\sqrt{\frac{A_b'}{A_b}} \qquad (8.3.1b)$$

but not greater than f_c'

where f_{cp} is the permissible concrete compressive stress; f_c' is the specified concrete compressive strength; f_{ci}' is the specified concrete compressive strength at time of initial prestress; A_b' is the maximum area of the portion of the concrete anchorage surface that is geometrically similar to and concentric with the area of the anchorage; and A_b is the net bearing area of anchorage.

The average bearing stress P/A_b in the concrete shall be computed by dividing the force P of the prestressing steel by the net bearing area A_b between concrete and bearing plate or other structural element of the anchorage that has the function of transferring force to the concrete.

8.4—Connecting components

8.4.1 Any component used to connect the sheathing to any anchorage or coupler enclosure in encapsulated systems shall have 0.050 in. (1.27 mm) minimum thickness.

8.4.2 Tape used as a component for enclosing couplers shall have a minimum width of 2 in. (50 mm).

9—ANCHORAGE ASSEMBLY TESTING

9.1—Validation

The adequacy of a tendon system shall be confirmed by static and fatigue conformance tests in accordance with the minimum requirements outlined in 9.4 and 9.5. Separate specimens shall be used for static and fatigue tests. Retesting is required whenever a component of an assembly changes.

9.2—Components

Component parts from different manufacturers shall not be used in an anchorage assembly unless the combination has been tested in accordance with 9.1.

9.3—Encapsulated systems

The encapsulation assembly shall remain watertight when tested and subjected to a uniform hydrostatic pressure for a period of 24 hours in accordance with 9.6. The following minimum uniform hydrostatic pressure shall be used in the test:

a) For applications governed by ACI 318: 1.25 psi (8.6 kPa)

b) For applications governed by ACI 350: 10 psi (68.8 kPa)

9.4—Static test

9.4.1 Static tensile tests of tendons shall be performed. The test assembly, consisting of standard production quality components and tendons, shall be at least 3.5 ft (1.1 m) long between anchorages. The test shall provide determination of the yield stress, tensile strength, and percent elongation of the complete tendon.

9.4.1.1 *Strength test criteria*—Anchorages and couplers of unbonded tendons shall develop at least 95 percent f_{pu} when tested in accordance with 9.4.

9.4.1.2 *Ductility criteria*—Total elongation under breaking load when tested in accordance with 9.4.1 shall not be less than 2 percent. Elongation shall be measured in a minimum gauge length of 3 ft (915 mm) between two points at least 3 in. (75 mm) from each anchorage. Tendon couplers shall not reduce elongation at rupture below that required for anchorages.

9.5—Fatigue test

9.5.1 Fatigue tests shall be performed on tendon specimens with representative samples from production runs and with a minimum length of 3 ft (1 m) between anchorages. In the first test, the tendon shall be subjected to 500,000 cycles between 60 and 66 percent f_{pu}. In the second test, the tendon shall be subjected to 50 cycles between 40 and 85 percent f_{pu}. One complete cycle involves change from the lower stress level to the upper stress level and back to the lower.

9.6—Hydrostatic test

9.6.1 Representative couplers and anchorages shall be tested to verify a watertight encapsulation of prestressing steel and all connections in conformance with 9.6.1.1 through 9.6.1.3. Three tests are required for each assembly with all three passing for the system to pass.

9.6.1.1 Anchorages shall remain watertight when tested and subjected to a hydrostatic pressure of no less than that specified in 9.3 for a period of 24 hours.

9.6.1.2 Representative samples from production runs, selected and assembled by the manufacturer, shall be used in testing. Stressing, intermediate, and fixed anchorage assemblies shall each be tested.

9.6.1.3 During the testing procedure, the following steps are required to detect the presence of moisture:

a) Add white pigment to the post-tensioning coating

b) Use a colored dye in the water that will contrast with the white color of the post-tensioning coating

c) After 24 hours, the encapsulation system shall be removed and the color of the coating shall be noted

No colored dye staining inside the encapsulation system anywhere on the white PT coating is permissible.

10—TESTING SERVICES

10.1—Prestressing steel

Mechanical property and dimensional tests may be performed in-house by the manufacturer. Relaxation tests shall be performed, or observed and confirmed, by an independent laboratory accredited under ASTM C1077. Any testing facility used, whether in-house or other, shall be identified completely including physical address and contact information.

10.2—Anchorage assembly

Static, fatigue, and hydrostatic tests shall be performed by an independent testing laboratory selected by the system manufacturer. The independent testing laboratory shall be accredited under ASTM C1077.

11—CERTIFICATION

11.1

Unbonded single-strand tendons shall be fabricated in a plant audited and certified by an external program accredited by a national accreditation body such as the American National Standards Institute (ANSI), International Accreditation Service (IAS), American Association for Laboratory Accreditation (A2LA), or equivalent.

Nonaccredited certification programs shall meet the essential requirements of ISO/IES Guide 65 or ISO/IEC 17065, including, at a minimum, the following:

a) Have a documented structure which safeguards impartiality including provisions to ensure the impartiality of the operations of the certification body; this structure shall enable the participation of all parties concerned in the development of policies and principles regarding the content and functioning of the certification system

b) Be free of commercial/financial pressures that might influence results and be independent from the entity they are certifying

c) Have personnel that are competent and who meet defined minimum relevant criteria as defined by the certifying agency

d) Have criteria outlined in specified standards

e) Have a procedure for making rules and procedures available to the public

12—PACKAGING AND IDENTIFICATION

12.1—Bundling and banding

Individual tendons shall be secured in bundles using a tying process that does not damage the sheathing. The tendon sheathing shall be protected from damage by banding materials using padding material between banding and tendons.

12.2—Identification of components

12.2.1 *Wedges and anchorages*—Wedges and anchorages shall be identified by individual concrete placement area, floor sequence, or both. In the event that materials intended for one concrete placement area are exchanged into another concrete placement area, the transaction shall be noted for traceability purposes.

12.2.2 *Prestressing steel*—The strand manufacturing process shall be controlled and documented in a manner providing identification and traceability with regard to coil(s) of strand, wire rod heat number, and wire coil(s) used to produce the strand.

12.2.3 *Encapsulated systems*—Identification of all component parts of the encapsulation system and assembly instructions shall be provided to the field.

13—HANDLING, STORAGE, AND SHIPPING

13.1—Handling

13.1.1 Tendons shall not be damaged during handling, loading, moving, or unloading.

13.1.2 Smooth forklift booms or padded forks shall be used to handle tendons.

13.1.3 Slings used to lift tendons shall be nonmetallic and not choked when handling tendon coils. Metal chokers, chains, or hooks shall not be used to lift tendons or tendon bundles.

13.2—Storage before shipping

13.2.1 Fabricated tendons shall be stored on a paved surface with proper drainage away from tendons or on proper dunnage.

13.2.2 Tendons that are exposed to sunlight (with possible ultraviolet ray degradation) longer than 2 weeks (or 1 month maximum when ultraviolet light stabilizers are added to the sheathing per manufacturer's recommendations) shall be protected by tenting or tarping with adequate ventilation.

13.3—Shipping

13.3.1 Nonmetallic tie-downs shall be used to secure tendon bundles to bed of the transport vehicle. Metal strapping or chains shall not be used.

13.3.2 Protection shall be provided between bed of the transport vehicle and bundles to prevent damage to the sheathing during shipping.

13.3.3 Tendons, accessories, equipment, and encapsulation materials shall be protected from exposure to moisture, deicing salts, and other corrosive elements during transportation by shipping inside of enclosed trailers, by covering with tarps, by shrink-wrapping the tendon bundles, or by other methods accepted by the Architect/Engineer. Sheathing and anchorages shall not be damaged during loading, transportation, and unloading.

13.4—Protection from moisture

13.4.1 Tendons shall be protected from exposure to moisture, deicing salts, and other corrosive elements.

14—INSTALLATION DRAWINGS

14.1 Nonprestressed reinforcement required for anchorages in the local zone shall be indicated on installation drawings.

14.2 Couplers shall be used only at locations approved by the Architect/Engineer. The location of the couplers shall be specified to maintain proper concrete cover and to allow longitudinal displacement of the couplers during stressing. Couplers shall not be used at points where tendon radius of curvature is less than 480 strand diameters.

Report on Corrosion and Repair of Grouted Multistrand and Bar Tendon Systems

Reported by Joint ACI-ASCE Committee 423

Bruce W. Russell, Chair

Carin L. Roberts-Wollmann, Secretary

Theresa M. Ahlborn	Henry J. Cronin Jr.	Mohammad Iqbal	Thomas C. Schaeffer
Robert W. Barnes	Charles W. Dolan	L. S. Paul Johal	Andrea J. Schokker
Florian G. Barth	James P. Donnelly	Donald P. Kline	Morris Schupack
Sarah L. Billington	Martin J. Fradua	Mark E. Moore	Richard W. Stone
Allan P. Bommer	William L. Gamble	Antoine E. Naaman	Amy M. Reinke Trygestad
Kenneth B. Bondy	Shawn P. Gross	Theodore L. Neff	H. Carl Walker
Ned H. Burns	Pawan R. Gupta	Sami H. Rizkalla	Paul Zia
Steven R. Close	H. R. Trey Hamilton III	David H. Sanders	

This report gives general information regarding the evaluation of corrosion damage in structures reinforced with grouted multistrand and bar tendons. A review of current practice is included along with the historical background. Specific potential problem areas for grouted tendons are discussed for each critical part of the tendon. Current methods for evaluating corrosion damage and typical repair schemes are included.

Keywords: anchorage; bonded; corrosion; duct; durability; grout; post-tensioning; prestressed; repair; strand; tendon.

CONTENTS

ACI 423.8R-10 was adopted and published October 2010.

CHAPTER 1—INTRODUCTION

1.1—General

This report is intended to provide general information on the corrosion and repair of grouted multistrand and bar tendon systems. The information is intended to represent the state of knowledge of the committee in the areas of corrosion prevention in grouted tendons, as well as evaluation and repair. It is not intended as a standard or recommended practice. A strong background in durability of grouted post-tensioning systems, including practical experience and training, is recommended before designing, constructing, evaluating, or repairing these structures.

This report focuses on the unique materials and hardware associated with grouted post-tensioning tendons. General corrosion theory for metals in structural concrete systems is not within the scope of this report. Information on corrosion theory and mechanisms can be found in ACI 222R-01 and ACI 222.2R-01.

1.2—Background

The use of grouted high-strength wire, strand, and bar post-tensioning tendons as a viable construction technique began in North America in the 1950s. Components of the standard system have changed significantly over the last half-century from shimmed button-head wire-type systems to bundles of seven-wire strand systems with wedges (Post-Tensioning Institute 2006). Post-tensioning bars may also be used in grouted systems, although they have not changed significantly since their first introduction. The latest generation of post-tensioning systems includes multiple layers of corrosion protection due to an increased awareness of the importance of a durable system.

Grout material has also undergone significant changes over the past 50 years. For the majority of this time, a standard mixture of water and portland cement was used. Sometimes chemical admixtures were included to increase pumpability or to introduce expansion in an attempt to counteract settlement and voids. As early as the 1960s (Schupack 1971), it was recognized that bleed was a problem in grouts, which could cause significant voids in the tendon. In the 1970s, an anti-bleed admixture (Schupack 1973) was developed to reduce the bleeding tendencies of grouts. This formulation was used successfully in a number of major structures, but did not see widespread use in the post-tensioning bridge and building industries. During the late 1990s and early 2000s, prepackaged grout formulations containing anti-bleed admixtures gained significant popularity in the U.S. and continue to see widespread use. A sudden change in grouting materials in the late 1990s was primarily due to the discovery of several post-tensioned bridges in Florida that had tendon corrosion (Corven Engineering, Inc. 2001). Past performance of grouted post-tensioning systems is covered in more detail in Section 4.1.

Most layers of corrosion protection for post-tensioning steel have seen significant advances in the last 5 to 10 years. The discovery of corrosion problems has also led to more standardized methods for the evaluation of corrosion damage and repair methods.

1.3—Scope

This report includes a review and summary of the following:

- Guide specifications and training as applied to grouted post-tensioning tendons;
- Potential problem areas in grouted systems;
- Durability enhancements for each component of the tendon;
- Evaluation of corrosion problems; and
- Repair methods for grouted systems.

1.4—Limitations

This report presents a summary of the typical problems that may occur in grouted post-tensioning systems and includes guidelines for evaluation and repair. The material is general in nature and may not apply to every situation. Due to the uniqueness of each post-tensioned structure, engineering judgment must be used to determine the appropriate action based on experience and the general guidance provided in this document.

This report is not intended to be included as part of a specification document for materials, evaluations, or repairs. Evaluation of grouted post-tensioned tendons is a difficult undertaking and no single method can be used to fully describe the damage or life expectancy of a tendon. Due to the many changes in materials and equipment over the last 50-plus years, the evaluator may encounter a wide variation in the types of systems encountered. Varying climates, environments, and the structure's usage and design also play a role in possible deterioration. The evaluator must be knowledgeable about the performance of similar structures under similar conditions to get an indication, if any, of internal damage of the structure.

CHAPTER 2—DEFINITIONS

ACI provides a comprehensive list of definitions through an online resource, "ACI Concrete Terminology," http://terminology.concrete.org. Definitions provided herein complement that resource.

admixture—a material other than water, aggregates, cementitious materials, and fiber reinforcement, used as an ingredient in a cementitious mixture to modify its freshly mixed, setting, or hardened properties and that is added to the batch before or during its mixing.

admixture, water-reducing—an admixture that either increases slump of freshly mixed mortar or concrete without increasing water content or maintains slump with a reduced amount of water, the effect being due to factors other than air entrainment.

backflow—a condition whereby the grout front reverses direction and flows back on top of the grout layer in the bottom of the duct.

bleed—the autogenous flow of mixing water within, or its emergence from, a newly placed cementitious mixture caused by the settlement of solid materials within the mass and filtering action of strands and wires.

cementitious materials—materials having cementing properties.

corrosion inhibitor—a chemical compound, either liquid or powder, usually intermixed in concrete and sometimes applied to concrete, and that effectively decreases corrosion of steel reinforcement.

duct—a conduit (plain or corrugated) to accommodate prestressing steel for post-tensioned concrete.

fluidity—a measure of time, expressed in seconds, necessary for a stated quantity of grout to pass through the orifice of the flow cone.

grout—a mixture of cementitious material and water, with or without aggregate, proportioned to produce a pourable consistency without segregation of the constituents; injected into the duct to fill the space around the tendon.

high-range water-reducing admixture—see **admixture, water-reducing**.

inlet—tubing or duct used for injection of grout into duct.

outlet—tubing or duct to allow the escape of air, water, grout, and bleed water from the duct.

post-tensioning—method of prestressing in which prestressing steel is tensioned after concrete has hardened.

set—the condition reached by a cement paste, mortar, or concrete when it has lost plasticity to an arbitrary degree, usually measured in terms of resistance to penetration or deformation; initial set refers to first stiffening; final set refers to attainment of significant rigidity; also, strain remaining after removal of stress.

set time—the lapsed time from the addition of mixing water to a cementitious mixture until the mixture reaches a specified degree of rigidity as measured by a specific procedure.

steel, prestressing—high-strength steel used to prestress concrete, commonly seven-wire strands, single wires, bars, rods, or groups of wires or strands.

tendon—an assembly consisting of a tensioned element (such as a wire, bar, rod, strand, or a bundle of these elements) used to impart compressive stress in concrete, along with any associated components used to enclose and anchor the tensioned element.

thixotropic—a reversible, time-dependent decrease in viscosity when a fluid is subjected to increased shear stress or shear rate.

vacuum grouting—a process by which a partial vacuum is generated within a void to create a pressure differential that allows grout to flow into and substantially fill the void through a single grout inlet.

volume change—an increase or decrease in volume due to any cause.

CHAPTER 3—REVIEW OF CURRENT PRACTICE

3.1—General

Code requirements in the area of post-tensioning durability were very minimal prior to publication of the Post-Tensioning Institute's (PTI), "Specification for Grouting of Post-Tensioned Structures," in 2001 (current edition published in 2003 [PTI Committee on Grouting Specifications 2003]). This comprehensive document focuses on grouting, including materials, construction, and testing, with emphasis on new construction.

The American Segmental Bridge Institute (ASBI) established a Grouting Certification Program in 2001. The program format, which includes both lecture and live demonstrations of grout testing and pumping, culminates in a certification exam. General detailing for durability in post-tensioning systems is covered in addition to grouting. Grouting best practices taught at this course follow the PTI Specification (PTI Committee on Grouting Specifications 2003).

Certain state departments of transportation (DOT) have developed their own detailed specifications for durability of post-tensioned bridges. The Florida Department of Transportation (FDOT), for example, maintains a comprehensive state specification (FDOT 2007).

3.2—Grouting requirements

The PTI Specification (PTI Committee on Grouting Specifications 2003) is divided into four main sections: materials, design, testing, and construction. This specification should be consulted for a full description of requirements in each area. Basic concepts are summarized in Sections 3.2.1 through 3.2.4.

3.2.1 *Materials*—Grout consists of cement and water and may also include chemical or mineral admixtures as well as fine sand. Pozzolans used in concrete may also be used in grout. Silica fume and fly ash, both Class C and Class F, are the most commonly used. Because fly ash may vary considerably from different sources, a consistent source should be used to achieve similar field properties to the ones found during mixture development in the lab. Fly ash can potentially reduce bleed, increase workability, and reduce permeability. Silica fume has been shown to improve corrosion protection in concrete by reducing the permeability of the concrete. In grouts, however, the particles may agglomerate because the coarse aggregate that aids in mixing is absent. Recent testing has shown better performance with interground silica fume cements (Brooks 2003).

Chemical admixtures used in grout have included high-range water-reducing admixtures, anti-bleed admixtures, expansive admixtures, and corrosion inhibitors. High-range water-reducing admixtures help achieve workability at low water-cementitious material ratios. For constant water content, adding only water reducers or high-range water-reducing admixtures usually increases bleed, so the admixture should not be added beyond the mixture design amount.

Anti-bleed admixtures, also called gelling agents or stabilizers, are necessary for applications requiring strong bleed resistance or water retention in grout. Silica fume and fly ash mildly improve bleed resistance, but are not adequate alone for the high water-retentive properties needed for many applications. Bleed in grouts is discussed in Section 4.3.

Expansive agents were used in grouts for many years with the premise of counteracting bleed and voids with expansion of the grout. A small amount of expansion to offset settlement may be desirable, but the use of foaming-type expansion admixtures should be avoided. These admixtures may have widely varying levels of expansion when combined with different cements. They also may cause an interconnected air void system, leading to higher permeability. The perceived benefit of counteracting bleeding does not occur because bleeding continues until the grout is set.

Corrosion inhibitors have not been shown to be effective in grouts. Accelerated testing has indicated a reduction in corrosion protection (Pacheco et al. 2007). A well-designed anti-bleed grout with low water content is preferable over the addition of corrosion inhibitors in grout.

For any grout design, care should be taken to use consistent materials between lab testing, mixture development, and field mixing. Materials should be stored per manufacturer's directions and care should be taken to make sure materials are not past their expiration date.

3.2.2 *Design*—The design section of the PTI Specification (PTI Committee on Grouting Specifications 2003) covers requirements for grouts divided by classes. Class A grouts are for nonaggressive exposure applications, Class B grouts are for aggressive exposure applications, Class C grouts are prepackaged grouts suitable for aggressive environments, and Class D grouts are for special applications. For example, a stay cable application with grouting lifts greater than 100 ft (30.5 m) would be considered a Class D grout. The PTI Specification (PTI Committee on Grouting Specifications 2003) provides performance requirements for each grout class.

3.2.3 *Testing*—Laboratory testing must be performed at a constant water content for a given grout design. Prepackaged materials must state a maximum and minimum water-to-cementitious material ratio on the bag and laboratory testing must be performed at both water contents. Property testing includes: set time, strength, permeability, volume change, fluidity, bleed resistance, corrosion resistance, and wet density. Fluidity testing gives an indication of the pumpability of the grout. A flow cone test is used to estimate fluidity. A modification is made for thixotropic grouts because these grouts require additional energy to flow. Bleed testing includes a wick-induced bleed test that indicates a grout's ability to resist the filtering effects of the strand in low vertical rise bleed situations. For more severe bleed situations, the "Test Method for Bleed Stability of Cementitious Post-Tensioning Tendon Grout (ACI 423.9M-10)" is more appropriate (Joint ACI-ASCE Committee 423 2010). This test requires pressurization of a cylinder of grout against a filter to separate the water. Various pressure levels are used to give an indication of the grout's bleed resistance at different vertical heights (columns of grout). Field testing may include fluidity, wet density, sampling for strength, and bleed resistance. Field mock-up tests may be required by the engineer.

3.2.4 *Construction*—The construction section of the PTI Specification (PTI Committee on Grouting Specifications 2003) covers material storage, preparation for grouting, the grouting operation, and post-grouting procedures. Chapter 4 reviews post-tensioning materials and construction practices related to grouting.

3.3—Tendon protection requirements

Specific tendon protection requirements for grouted post-tensioned structures are difficult to find in one comprehensive document. The FDOT specifications (FDOT 2007) contain many items in this area related to bridges, including specific types of plastic ducts. Protection levels are also covered during the Grouting Certification Program. Each post-tensioned structure may be unique, making standard durability detailing difficult. Engineering judgment is required for each given situation, with a focus on keeping aggressive agents such as moisture and chlorides away from the prestressing steel and anchorages. These protection levels may include anchor end caps, pourbacks, membranes, and a robust duct. Chapter 4 covers each of these items in more detail.

CHAPTER 4—GROUTED TENDONS

4.1—Evolution of grouted tendons

Early grouted post-tensioning tendons consisted of button-head wires with shims at the anchors. Seven-wire strand tendons anchored by toothed wedges were later introduced and are still in use today. The most significant changes in the grouting and anchor protection area have come about since the late 1990s. Durability measures such as plastic ducts, permanent plastic end caps, reinforced concrete cover over anchorages, and water-proofing membranes in the anchor area have seen increased popularity, particularly in aggressive environments.

In the late 1990s, prepackaged grouts were rarely used. A decade later, they are the most common types of grout used in bridge structures in the U.S. and are becoming more common in building structures. A large number of prepackaged products are now on the market. Approved products undergo extensive testing to meet the requirements of the PTI Specification (PTI Committee on Grouting Specifications 2003) as well as specific state DOT requirements. The majority of older structures used a plain water-cement grout, possibly with an expansive admixture, rather than a specially formulated anti-bleed grout. Although most of these structures are not showing signs of corrosion deterioration, some have grout voids that could lead to corrosion. Anti-bleed grouts have the advantage of reducing or eliminating voids caused by bleed. The past performance that instigated the widespread changes in grouting is discussed in Section 4.8.

4.2—Types of grouted tendons

Grouted tendons may include either strand, typically seven-wire, or bar. Grouted strand tendons are most commonly multi-strand tendons. Slab-type tendons, including transverse bridge tendons, may be encased in oval ducts, while other multi-strand tendons are typically found in circular ducts. Single-strand grouted tendons are also used in certain applications.

While grouted tendons are commonly referred to as bonded tendons, this is not always the case. Internal grouted tendons found within the concrete are considered bonded tendons. Tendons may also be external to the concrete (although they may be in the void of a box girder) and only bonded to the structure at discrete locations such as the anchors and deviators. These tendons may be grouted, but are considered unbonded for design purposes. External grouted tendons are most commonly used in segmental box girder construction and in repair/retrofit applications.

Grouted tendons are used in many different structural components and may be vertically or horizontally oriented. Tendons may be essentially straight or may have significant curvature with more complex geometry. The vertical rise and tendon geometry can make the grouted area a very complex space to fill without voids if the proper grout material is not used.

4.3—Potential problem areas

Grout properties such as fluidity, bleed resistance, low permeability, reasonable set time, stability, and others, are all desirable in a quality grout for post-tensioning. The property that has caused the most problems and has the biggest potential for allowing corrosion in the tendon is bleed resistance. Bleed water from the grout is trapped in the duct and tends to rise to the highest vertical point. In a draped tendon, this may be at a high-end anchor or an intermediate crest. In a vertical tendon, the bleed water may only migrate to a certain intermediate height prior to grout set and thus bleed lenses may form at several locations along the height of the tendon in addition to the top of the tendon. In a horizontal tendon, the bleed lens may be shallow and extend over a long horizontal distance. The tendency of a grout to bleed is accentuated by the filtering effect of seven-wire strand, particularly in large, multi-strand tendons. The pressure head developed over vertical distances further accentuates the bleed. Bleed should be a consideration in all grouted tendons, but the most severe bleed situations arise from large multi-strand tendons and tall vertical tendons (such as in bridge piers).

After the grout sets, the bleed lenses leave an ungrouted area full of bleed water. This water may reabsorb into the grout in some cases, or may partially remain for long periods of time in the void. The strand in the void has no protection from the high-pH grout environment and may corrode even without outside ingress of chlorides (Bricker and Schokker 2004). If chlorides or outside moisture are able to reach the duct, the voided area provides a collection point and a potential hot spot for corrosion.

To combat the bleed problem, an anti-bleed admixture, also known as a gelling agent or stabilizer, is often added to the grout formulation. Fly ash and silica fume can reduce bleed to some extent, but typically a chemical anti-bleed agent is needed for multi-strand tendons and tendons with vertical rises (including draped tendons). The amount of anti-bleed admixture required is dependent on the vertical pressure head and number of strands in the duct. Prepackaged grouts that have bleed-resistant properties typically contain some form of anti-bleed agent.

Many anti-bleed or water-retentive grouts also exhibit thixotropic properties. Thixotropic grouts will be fluid when agitated, but will appear gel-like at rest. These grouts may appear too stiff to those unaccustomed to their behavior, but a good thixotropic grout will have stable fluid behavior when mixed or agitated. Thixotropic grouts tend to flow through the duct with a steep front, which should fill all cavities as the grout front advances (Fig. 4.1(a)). This type of flow can push out air and moisture as the front passes. In contrast, water-cement grout is generally more fluid and self-leveling than thixotropic grout during injection. Consequently, water-cement grout may fill the bottom of the duct first as illustrated in Fig. 4.1(b). In some cases, backflow may result, which is when the grout front turns and flows back on top of the grout layer in the bottom of the duct. Backflow can entrap air in the duct, causing voids if it is not expelled through proper venting. Draped tendon configurations are particularly susceptible to backflow.

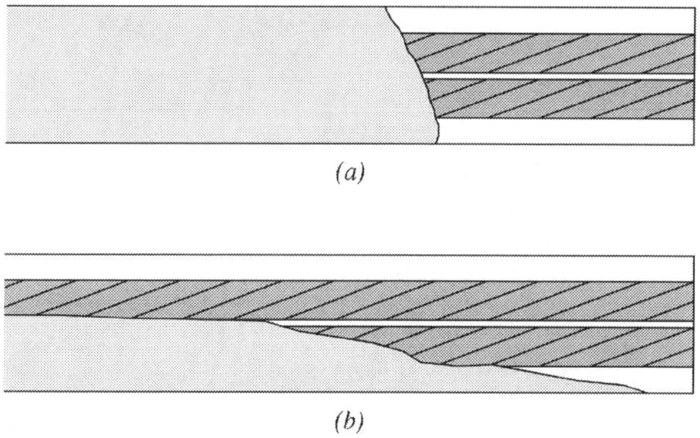

Fig. 4.1—Typical flow patterns for: (a) thixotropic grout in high viscosity state; and (b) plain water-cement grout (Schokker et al. 2002).

Voids may also result from poor construction practices and lack of quality assurance. Examples of poor construction practices that may result in voids are listed in the following:

- Lack of high point vents in ducts, or improper placement of the high point vents;
- Use of improper mixing equipment for the grout leading to poor grout quality and fluidity;
- Inaccurate measurement of water for mixing of the grout;
- Lack of shut-off valves that can hold pressure on the grout when grouting is complete; and
- Blockages within the duct or grout that leaks out of the duct.

Other potential problem areas are related primarily to providing multiple robust layers of protection between the environment and the strand. These barriers include concrete cover, membranes, anchor protection, and the duct; they are discussed in Sections 4.4 and 4.7.

4.4—Duct

Traditionally, post-tensioning ducts have been fabricated of galvanized steel. Often the seams in the metal duct are easily opened, particularly when bent. This type of duct has not proven to be an effective barrier. In addition, once the sacrificial zinc layer is gone, the underlying steel duct has no corrosion protection and may corrode rapidly. The corrosion product from the duct may also cause cracking in the surrounding concrete, providing an additional direct path for moisture and chlorides. In an aggressive environment, a ribbed plastic duct is preferred for internal tendons. The duct itself can provide an effective barrier against ingress of aggressive agents to the grout and strand. Support of the duct prior to concreting should be adequate to ensure the duct holds the proper profile and concrete cover.

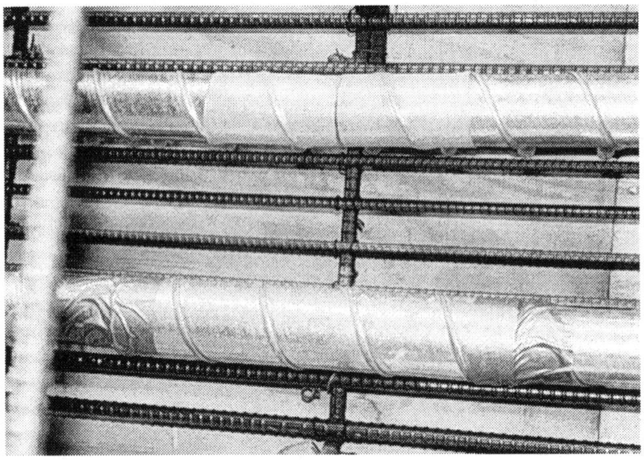

Fig. 4.2—Galvanized duct. Top: heat-shrink splice; and bottom: duct-type splice (West et al. 1999).

External ducts are typically smooth black plastic. Care must be taken that the material used in these ducts meets specific environmental stress ratings. Poor lots of plastic duct may be prone to splitting, effectively removing the duct as a protection barrier. Care must be taken with duct splicing to ensure that the splice does not become an entrance point for corrodents. A splice made up of ethylene propylene diene monomer (EPDM) marine exhaust hose with stainless steel bands used to clamp the hose to the duct has been used as an effective splice for plastic duct. Duct tape wrapping is not an adequate splice method for plastic duct systems.

Although galvanized steel duct is not recommended in aggressive environments, it is often used in nonaggressive environments. The typical splice for metallic duct is a piece of oversized duct over the splice area with each end of the splice wrapped in duct tape. Another option is a heat-shrink splice. Both splice types are shown in Fig. 4.2. Research has shown problems with the heat-shrink splices on metallic duct (Salas et al. 2004) due to trapped moisture causing corrosion of the splice.

4.4.1 *Duct testing*—Ducts should be leak tested with compressed air prior to grouting. This testing is done with oil-free air to assure the continuity and pressure tightness of the duct system. Guidance for pressure testing of ducts can be found in the FDOT Specification (FDOT 2007) and the *fib*'s *Bulletin* 7 article (*fib* 2000). Note that methods for leak testing of assembled components for system approval differ from methods used in field testing on in-place ducts. Primarily, the difference relates to the air pressure used and the acceptable values for pressure loss. If it is determined that cross-grouting is likely between two tendons and it cannot be repaired, the affected tendons should be grouted in a single operation.

4.4.2 *Flushing of ducts*—Flushing of ducts is not recommended. Ducts should be sealed from debris prior to tendon stressing and then again after stressing, prior to grouting. Water flushing is not effective in removing debris and often results in pockets of entrapped water in the strand bundle. While flushing may help increase grout flow, it is an inappropriate way to do so. The grout will often mix with the remaining water, except where strong thixotropic grouts are used. The resulting grout can have a significantly higher water-cement ratio and, thus, higher bleed than intended. Pockets of water may also be trapped in the duct.

Flushing of ducts is sometimes used as a means to remove water-soluble oils applied to the strand for temporary corrosion protection. Studies have proved this not to be an effective means of removing water-soluble oils, and this practice should be avoided (Kittleman et al. 1993). Past practice included flushing ducts with water to remove debris, removing temporary corrosion protection products, or cooling or lubricating the ducts prior to pumping.

4.5—Grout

Grout material properties and grouting practices should be in conformance with PTI specifications to ensure a successful grouting operation. The grout must have appropriate fluidity to pump and retain this fluidity in a wide range of temperatures over the time frame of pumping. Bleed resistance, which is discussed in Section 4.3, is critical in preventing voided areas. Set time, chloride penetration resistance, and strength are also important properties and are covered by the PTI Specification (PTI Committee on Grouting Specifications 2003).

Mixing procedures and equipment can have a significant effect on grout behavior. Thixotropic grouts tend to be mixed much more effectively with a high shear or colloidal mixer. A holding tank is then needed to keep the grout agitated. All grouts will have different properties when mixed with various types of mixers and different mixing times. Both over-mixing and under-mixing can be problematic. The same equipment that will be used in the field should be used for property testing and any field mockup tests. Backup equipment should be on hand in case of equipment malfunction during pumping.

Proper inlets, outlets, and vents are necessary to facilitate complete filling of the duct. Vents are placed at intermediate high points and at locations slightly downstream of the high points to ensure that air can be expelled, particularly in a backflow situation. A vent should also be included in the end cap at both the fixed and stressing ends of the tendon. Vents included at low points will aid in draining out moisture. This is particularly critical in situations where entrapped water may freeze. All inlets, outlets, and vents should have a positive shut-off valve. Past practice of bending over vent tubes and tying with tie wire is not sufficient. Vents should be shut off in successive downstream order, with the exception of the vents located slightly downstream of the intermediate high points, which should be closed prior to their associated high point vent. Grout should be allowed to flow from the vents until all visible air and water pockets are gone from the flow stream. At least 1 gal. (3.8 L) of grout should be expelled from each vent and 5 gal. (19 L) from the outlet. The vents should not be reopened prior to grout set. Vents should be cut flush with the concrete surface and sealed to make sure that they do not provide a method of ingress for chloride and moisture.

If grouting of a tendon is interrupted, the continuation of the grouting process should be done in a manner that ensures complete filling of the duct. This may require injecting grout into the last vent from which grout has flowed, tapping a new inlet adjacent to the blockage, and/or vacuum grouting.

4.6—Strand and bar

Post-tensioning tendons typically contain seven-wire strand of 0.5 or 0.6 in. (12.7 or 15.2 mm) in diameter or high-strength bar. Although strand coatings are uncommon, coating options include epoxy-coated or epoxy-coated and filled. Epoxy-coated strand has a coating of epoxy that is applied to the outside surface of the strand. Epoxy-coated and filled strand has epoxy that fills the interstices between the seven wires in addition to the coating on the outside of the strand, resulting in a coating around each of the seven wires, including the center wire. In-depth discussion about post-tensioning coatings can be found in ACI 222.2R-01.

Tendons should be sealed from the environment after insertion and prior to grouting. Occasionally, the period of time between insertion and grouting may be much longer than desired and additional protection measures are considered. Vapor phase inhibitors (VPI) can provide temporary corrosion protection for the prestressing steel. To be effective, the tendons must be completely sealed once the VPI powder is added. Another option is the use of water-soluble oils or other coatings sprayed on the strand prior to insertion of the strand in the duct. These products can provide corrosion protection, but reduce bond. The effect of bond reduction in the post-tensioning system that is permanently anchored is not likely to be critical. This bond reduction may cause the system to behave more like an unbonded system than a bonded one.

A research study on temporary corrosion protection found that several products show promising corrosion protection with minimal bond reduction (Salcedo et al. 2004). Some products, however, actually increased corrosion of the strand from biologically-induced corrosion from the components in the product or reduced bond by a large percentage. A fully sealed tendon is the first source of protection during the interim phase between stressing and grouting.

4.7—Anchorage hardware and protection

The anchorage area can be particularly vulnerable to ingress of aggressive agents. Through the 1990s, anchorage protection typically consisted of a cap, mastic, or a concrete cover over anchorage. It has been recognized that a design for durability relying on a single layer of protection cannot guarantee reliable protection of the post-tensioning anchorages. Multiple levels of anchorage protection are now recommended (PTI Committee on Grouting Specifications 2003). These include a corrosion-resistant permanent grout cap, adequate concrete cover over anchorage, and a waterproofing membrane.

Anchorages with grout ports suitable for inspection from either the top or front are recommended. The grout ports serve a dual function of grout inlet/outlet and post-grouting inspection access. The geometry of the grout ports must facilitate drilling and inspection of the grout directly behind the anchor head using a borescope.

4.8—Performance record

4.8.1 *Performance in the UK*—Problems with several bridges led to a moratorium on post-tensioned bridge construction in the UK from 1992 to 1996. Once a thorough grouting and durability document was in place (Concrete Society 1996), the ban was lifted. Because of the differences in design and construction between U.S. bridges and the bridges that failed in the UK, there was little concern in the U.S. at that time about tendon corrosion problems. This changed less than a decade later with the discovery of corrosion problems in Florida bridges (Corven Engineering, Inc. 2002).

4.8.2 *Performance in the U.S.*—While there have been incidents of corroded tendons and grout voids in some cases, there have been no failures of grouted post-tensioned structures in the U.S. due to corrosion. In 1999, tendon corrosion was discovered in the Niles Channel post-tensioned segmental bridge in Florida. This was thought to be an isolated incident until the following year when, during a routine inspection, tendon corrosion was found in the Mid-Bay Bridge in Florida. The bridge had only been open 7 years at the time. Problems included duct splitting, significant grout voids, and corrosion of the steel strand. One tendon had failed at the anchorage and had pulled out of the expansion joint diaphragm. Later that year, corrosion was discovered in the piers of the approach spans on the Sunshine Skyway (Parsons Brinckerhoff Quade and Douglas, Inc. 2001). The piers are post-tensioned segmental concrete and vertical tendon corrosion was traced back to a problem with saltwater ingress into the duct during construction. Duct splitting was also discovered.

After the discovery of tendon corrosion in the Florida bridges, other state DOTs began investigating their post-tensioned bridges in detail for signs of problems. Caltrans and many other state DOTs have reported excellent performance in their bridges (ASBI Grouting Committee 2000). Some bridges were found with grout voids, but without the severe environment of Florida, corrosion was limited. After the initial problems were discovered in Florida, the American Segmental Bridge Institute (ASBI) implemented a grout training program that is now required by many states.

CHAPTER 5—EVALUATING CORROSION DAMAGE

5.1—General

The multiple levels of protection in a grouted post-tensioning system can make inspection difficult. While duct, grout caps at anchorages, and concrete cover represent physical barriers to visual inspection of the prestressing steel, they can also hinder the performance of nondestructive evaluation (NDE) methods. Methods of inspection are outlined in Sections 5.3 and 5.4.

Visual inspection may reveal corrosion, cracking, or spalling of concrete. Substantial corrosion may be present in the duct, even without visual signs. Visual inspection should include a focus on potential problem areas, such as runoff from a roadway onto the anchorage area, or locations where voids may exist such as tendon high points.

Visual inspection of external tendons can reveal split duct or some indication of voided areas through hammer sounding. It is difficult to determine the difference between a small area where the duct has separated from the grout, versus a voided area with hammer sounding. Breaching the duct with nails to determine whether a void exists is counter-productive, as it allows oxygen into the duct and provides a path for ingress of contaminants. If the duct is opened or breached in any way, it should be inspected and then repaired immediately.

5.2—Monitoring

Electrically isolated systems have been implemented in Europe for enhanced corrosion protection and for monitoring corrosion in post-tensioned structures (Matt 2001). In this type of system, an electrical contact is made with the tendon and used to monitor changes in potential (voltage), indicating corrosion activity. An isolated system requires a nonmetallic duct as well as special anchorage details that electrically insulate the anchorage from the surrounding concrete. Corrosion measurements can be made at any time during the life of the structure and can be valuable even without continuous monitoring.

Acoustic emission has been used in some cases to detect prestressing steel breakage in post-tensioning tendons. While these systems can be effective, they entail continuous monitoring of the structure to provide beneficial information.

5.3—Nondestructive methods

Although many nondestructive evaluation (NDE) methods have been successful on reinforced concrete structures, none have proven to be comprehensive for locating voids and corrosion in post-tensioned structures. Post-tensioned structures are typically very complex with congested reinforcement and multi-strand tendons encased in concrete. Evaluation at the critical anchorage zones is further complicated by congested spiral reinforcement and often large cover concrete. NDE methods can be divided into several categories, although many permutations of these exist.

Ultrasonic methods have been successful in the medical industry as well as in the aerospace industry. A large number of methods are included under the heading of ultrasonics, including both two-dimensional and now three-dimensional imaging. All methods are based on high-frequency sound waves and their echoes. Ultrasonic techniques have been evaluated in laboratory testing for use in void detection in post-tensioning tendons (Jansohn and Scherzer 2002), including C-Scan imaging (Iyer et al. 2003). Although the method shows promise in preliminary investigations, many items need to be addressed, such as transducer type and compensation for various tendon duct cross-section curvatures, transducing medium, limitations on depth, use in multi-strand systems, signal processing, and advanced pattern recognition.

The impact-echo technique has been useful in flaw detection in concrete (Sansalone and Street 1995; Kesner and Poston 1997; Carino 2001) including promising results for detecting voids in post-tensioned structures (Jaeger et al. 1996; Abraham and Cote 2002). A major drawback of the impact-echo method for use in post-tensioned structures is that it cannot detect voids inside plastic ducts.

Radiography or x-ray has been effective at detecting voids in both concrete and grouted tendons (Shaw et al. 2000; Corven Engineering, Inc. 2001), but radiation safety concerns and expense have limited its use (Matt 2001). The equipment also needs access from both sides of the tested area.

Ground-penetrating radar is effective in locating embedded objects in concrete, including nonprestressed reinforcement, tendons, and voids. Although promising for detecting voids in plastic duct, it cannot penetrate steel duct to detect voids. Duct detection is also complicated by the presence of nonprestressed reinforcement (Corven Engineering, Inc. 2002).

Remanent magnetism has met success in detecting fractures in prestressing steel strands (Scheel and Hillemeier 1995), while the magnetic flux method has been used to detect small flaws and corrosion in post-tensioning tendons (Ghorbanpoor et al. 2000; Corven Engineering, Inc. 2002). The magnetic flux equipment must run along the duct itself, so it is limited to use in the free length of external tendons. Both methods have limitations for potential use in internal tendons due to difficulties with interference from non-prestressed reinforcement and tie wire.

Reflectometrical impulse measurement (RIMT) and time domain reflectometry (TDR) methods send an electrical signal along the tendon from the anchor to detect flaws in the tendon. The anchor must be exposed to use these methods. TDR has been used successfully for many years to detect defects in transmission lines, but only recently evaluated for post-tensioning applications (Chajes et al. 2003). Lab testing indicated promise for detecting both corrosion and voids, but testing focused on specimens with a sensor wire running alongside the steel strand. The method shows promise, but at this time requires that the sensor wire be in place at the time of construction or destructive methods be used to place the sensor and make contact with the tendon.

Electrochemical methods, such as half-cell potential monitoring and linear polarization for corrosion rate measurement, have been used successfully to help understand the corrosion state of the reinforcement within a concrete structure (West et al. 1999). Other methods common in the corrosion industry, such as electrochemical impedance spectroscopy (EIS) also have been considered for reinforced concrete. Due to the many variables affecting each of these methods, they must be used in coordination with other data such as visual inspection, chloride penetration samples, and concrete resistivity to establish the likelihood and severity of corrosion. Temperature, humidity, concrete type, bar coatings, duct type and other factors can significantly affect readings.

5.4—Exploratory methods

Because NDE methods have met only limited success in evaluation of post-tensioning tendons, exploratory methods are often necessary. The NDE methods may be useful for focusing the invasive methods in areas where corrosion is most likely. Drilling may be required to determine where voids exist in the tendon and to evaluate the extent of corrosion. A less invasive procedure involves drilling into the anchorage area through the grout port (not into the anchorage itself) and inserting a borescope to investigate the status of the tendon. Post-tensioning anchorage manufacturers have recently adapted their anchorages so that the grout port provides easier access for the borescope into the tendon. The borescope allows the user to see inside the tendon with a video camera. From this perspective, even small amounts of corrosion may appear disproportionately large, so an operator experienced with the procedure is needed. Numerous borescope evaluations have been done by the Florida DOT and full information and procedures are detailed in a report (Corven Engineering, Inc. 2001).

CHAPTER 6—REPAIR SCHEMES AND METHODS

6.1—General

Because grouted post-tensioned structures have generally performed well, there is limited experience with their repair. The most comprehensive document in this area is available from the Florida DOT (Corven Engineering, Inc. 2002). Sections 6.2 and 6.3 briefly describe several repair methods available.

6.2—Grouting of voids

Repair of grout voids is the most common type of repair encountered in grouted post-tensioned structures. Even if no corrosion is evident, eliminating voids that expose strand is desirable. The voids can either be filled with a cement-based grout or a resin type of product. Although a cement-based product can restore the alkaline environment, a resin product can also be effective if it fully coats the strand.

One method of filling the void is gravity filling with a tube and funnel. This method can effectively cover the strand bundle depending on the location of the void, but may not be able to completely fill the void due to trapped air.

Another method is vacuum grouting. Special equipment is needed for this method. A vacuum measuring technique is used to determine the volume of the void after which a vacuum is pulled on the tendon and the same volume of grout is injected into the void. This method is effective in filling the void. It does require that the void is sealed off so that the vacuum is effective.

A third method for repairing a grout void is pressure grouting. This method is effective when the void is very large and an inlet and outlet can be drilled into the duct. This method allows the use of conventional pressure grouting methods for the repair.

6.3—Tendon repair

Split or breached ducts should be repaired to restore the protective envelope for the tendon. In case of heavy corrosion resulting in strand failure, the entire tendon may need to be replaced. This is a very specialized operation that has only been completed in a few instances. The FDOT has performed tendon replacements and has detailed information on this process (Corven Engineering, Inc. 2001). Strengthening options may include external post-tensioning to counteract cracking, deflections, and to increase capacity. Detailed tendon replacement and strengthening strategies are beyond the scope of this report as they tend to be project specific. Repair design should not be undertaken without significant knowledge and experience with grouted multistrand and bar tendon systems.

CHAPTER 7—REFERENCES

7.1—Cited references

Abraham, O., and Cote, P., 2002, "Impact-Echo Thickness Frequency Profiles for Detection of Voids in Tendon Ducts," *ACI Structural Journal,* V. 99, No. 3, May-June, pp. 239-247.

ACI Committee 222, 2001, "Corrosion of Prestressing Steels (222.2R-01)," American Concrete Institute, Farmington Hills, MI, 43 pp.

ACI Committee 222, 2001, "Protection of Metals in Concrete (222R-01)," American Concrete Institute, Farmington Hills, MI, 41 pp.

American Segmental Bridge Institute (ASBI) Grouting Committee, 2000, "Interim Statement on Grouting Practices," ASBI, Dec., 6 pp.

Bricker, M. D., and Schokker, A. J., 2004, "Corrosion from Bleed Water in Grouted Post-Tensioned Tendons," PCA R&D Report 2547.

Brooks, M. A., 2003, "Evaluation of the Accelerated Corrosion Test Method for Use in Grouts for Post-Tensioning," MS thesis, Pennsylvania State University, University Park, PA, 106 pp.

Carino, N. J., 2001, "Impact-Echo Method: An Overview," *Proceedings of the Structures Congress and Exposition*, American Society of Civil Engineers, May 21-23, Washington, DC, pp. 1-18.

Chajes, M.; Hunsperger, R.; Liu, W.; Li, J.; and Kunz, E., 2003, "Void Detection in Grouted Post-Tensioned Bridges Using Time Domain Reflectometry," *Proceedings of the Transportation Research Board (TRB) Annual Meeting,* Washington, DC, http://www.ee.udel.edu/~hunsperg/corrosion/TRB2003.

Concrete Society, 1996, "Durable Bonded Post-Tensioned Concrete Bridges," *Concrete Society Technical Report* No. 47, second edition, Aug.

Corven Engineering, Inc., 2001, "Mid-Bay Bridge Post-Tensioning Evaluation," Florida Department of Transportation Report, 2622 pp.

Corven Engineering, Inc., 2002, "New Directions for Florida Post-Tensioned Bridges," Florida Department of Transportation Report, 68 pp.

Florida Department of Transportation (FDOT), 2007, "Standard Specifications for Road and Bridge Construction," Tallahassee, FL, 934 pp.

fib, 2000, "Corrugated Plastic Ducts for Internal Bonded Post-Tensioning," *fib Bulletin* No. 7, 50 pp.

Ghorbanpoor, A.; Orchelt, R.; Edwards, M.; and Abdel Salam, E., 2000, "Magnetic-Based NDE of Prestressed and Post-Tensioned Concrete Members—The MFL System," *Publication* No. FHWA-RD-00-026, Federal Highway Administration, Research, Development, and Technology, U.S. Department of Transportation, McLean, VA, May, 4 pp.

Iyer, S.; Schokker, A. J.; and Sinha, S., 2003, "Ultrasonic C-Scan Imaging—Preliminary Evaluation For Detecting Corrosion and Voids in Post-Tensioning Tendons," *Proceedings of the Transportation Research Board Annual Meeting*, Washington, DC, 24 pp.

Jaeger, B. J.; Sansalone, M. J.; and Poston, R. W., 1996, "Detecting Voids in Grouted Tendon Ducts of Post-Tensioned Concrete Structures Using the Impact-Echo Method," *ACI Structural Journal*, V. 93, No. 4, July-Aug., pp. 462-473.

Jansohn, R., and Scherzer, J., 2002, "Improper Filled Ducts Detected by Ultrasound Reflection," *Proceedings of the 8th European Conference on Nondestructive Testing*, Barcelona, Spain, http://www.ndt.net/article/ecndt02/470/470.htm.

Joint ACI-ASCE Committee 423, 2010, "Test Method for Bleed Stability of Cementitious Post-Tensioning Tendon Grout (ACI 423.9M-10)," American Concrete Institute, Farmington Hills, MI, 3 pp.

Kesner, K., and Poston, R. W., 1997, "Condition Assessment of Concrete Structures Using Impact-Echo," *Concrete Repair Bulletin*, Jan.-Feb., pp. 6-11.

Kittleman, W. M.; Davis, R. T.; Hamilton, H. R.; Frank, K. H.; and Breen, J. E., 1993, "Evaluation of Agents for Lubrication and Temporary Corrosion Protection of Post-Tension Tendons," *Research Report* No. 1264-1, Center for Transportation Research, Bureau of Engineering Research, University of Texas at Austin, Austin, TX, Aug., 124 pp.

Matt, P., 2001, "Non-Destructive Evaluation and Monitoring of Post-Tensioning Tendons," *Durability of Post-Tensioning Tendons, fib Bulletin* No. 15, pp. 103-108.

Pacheco, A. P.; Schokker, A. J.; and Hamilton, H. R.; 2007, "Revisions to Accelerated Corrosion Test Method for Post-Tensioning Grout," *ACI Materials Journal*, V. 104, No. 2, Mar.-Apr., pp. 123-128.

Parsons Brinckerhoff Quade and Douglas, Inc., 2001, "Sunshine Skyway Bridge Post-Tensioned Tendons Investigation," Florida Department of Transportation, Oct.

Post-Tensioning Institute, 2006, *Post-Tensioning Manual*, sixth edition, PTI, Farmington Hills, MI, 354 pp.

PTI Committee on Grouting Specifications, 2003, "Specification for Grouting of Post-Tensioned Structures," second edition, Post-Tensioning Institute, Farmington Hills, MI, 60 pp.

Salas, R. M.; Schokker, A. J.; West, J. S.; Breen, J. E.; and Kreger, M. E., 2004, "Durability Design of Post-Tensioned Bridge Substructure Elements: Summary," *Research Report* 1405-S, University of Texas Center for Transportation Research, Sept., 4 pp.

Salcedo, E.; Schokker, A. J.; Kreger, M. E.; and Breen, J. E., 2004, "Bond and Corrosion Evaluation of Emulsifiable Oils for Temporary Corrosion Protection in Post-Tensioned Tendons," *PTI Journal*, V. 2, No. 1, Jan., pp. 30-38.

Sansalone, M., and Street, W., 1995, "Use of the Impact-Echo Method and Field Instrument for Non-Destructive Testing of Concrete Structures," *Proceeding of the International Symposium on Non-Destructive Testing in Civil Engineering*, Berlin, V. 1, pp. 494-502.

Schokker, A. J.; Breen, J. E.; and Kreger, M. E., 2002, "Simulated Field Testing of High Performance Grouts for Post-Tensioning," *Journal of Bridge Engineering*, ASCE, V. 7, No. 2, Mar.-Apr., pp. 127-133.

Scheel, H., and Hillemeier, B., 1995, "The Capacity of the Remanent Magnetism Methods to Detect Fractures of Steel in Tendons Embedded in Prestressed Concrete," *Proceedings of the International Symposium on Non-Destructive Testing in Civil Engineering*, Berlin, V. 1, pp. 211-218.

Schupack, M., 1971, "Grouting Tests on Large Post-Tensioning Tendons for Secondary Nuclear Containment Structures," *PCI Journal*, Mar.-Apr., pp. 85-97.

Schupack, M., 1973, "Tendon Grouting Means," U.S. Patent 3,762,937, Oct. 2.

Shaw, P.; Pedersen, T. K.; and Serluppens, R., 2000, "High Energy Radiography Combined with the Agfa Strukturix DPS-Imaging System and Comparison with other NDE-Method for Inspection of Thick Reinforced Concrete Structures," *NDT*, .net-V. 5, No. 8, Aug. http://www.ndt.net/article/v05n08/shaw/shaw.htm.

West, J. S.; Schokker, A. J.; Larosche, C. L.; Breen, J. E.; and Kreger, M. E., 1999, "Long Term Post-Tensioned Beam and Column Exposure Test Specimens: Experimental Program," *Research Report* 1405-3, University of Texas Center for Transportation Research, Austin, TX, Oct., 282 pp.

Guide to Estimating Prestress Losses

Reported by Joint ACI-ASCE Committee 423

Carin L. Roberts-Wollmann*, Chair

Amy M. Reineke Trygestad, Secretary

Theresa M. Ahlborn*
Robert W. Barnes†
Florian G. Barth
Asit N. Baxi†
Roger J. Becker
Kenneth B. Bondy†
Charles W. Dolan†
James P. Donnelly
Pierre Esselinck†

Martin J. Fradua
William L. Gamble
Harry A. Gleich†
Shawn P. Gross
Pawan R. Gupta
William M. Hale
H. R. Trey Hamilton III†
Carol Hayek
Mohammad Iqbal

Donald P. Kline
Larry B. Krauser
Jason J. Krohn†
Mark E. Moore‡
Theodore L. Neff
Sami H. Rizkalla
James Rogers
Bruce W. Russell
David H. Sanders

Thomas C. Schaeffer
Morris Schupack‡
Richard W. Stone
Miroslav F. Vejvoda†
Jeffery S. Volz
H. Carl Walker
Zuming Xia†
Paul Zia*

*Co-chairs of subcommittee responsible for preparation of report.
†Subcommittee members.
‡Deceased

Consulting Members

Robert N. Bruce Jr.
Ned H. Burns
Chunsheng "Steve" Cai

Steven R. Close
Henry J. Cronin Jr.
Ward N. Marianos Jr.

Hani Melhem
Antoine E. Naaman
Thomas E. Nehil

Andrea J. Schokker

This guide is intended for estimation of prestress losses in concrete structures. Methods presented include lump sum, simplified approaches addressing individual source of loss, and additional estimation methods. They address losses in pretensioned and post-tensioned members, including bonded, unbonded, and external tendons. Note that these estimation methods have not been evaluated for relative merits. A discussion of the variability of prestress losses caused by the variability in concrete properties is also presented. Several example problems are included.

Keywords: creep; friction; post-tensioning; prestress loss; prestressed concrete; relaxation; shrinkage.

CONTENTS

ACI 423.10R-16 was adopted and published August 2016.

CHAPTER 1—INTRODUCTION

1.1—Introduction

Estimating prestress loss at any given time during the life of a prestressed concrete member is a complex issue. In pretensioned and post-tensioned members, applying prestressing force causes shortening of the concrete member that, in turn, causes a loss of tendon stress. Over time, concrete creep, concrete shrinkage, and steel relaxation result in additional reductions of tendon stress. In post-tensioned members, losses occur during the stressing operation due to friction between the tendon and sheathing or duct, which is caused by the intended and unintended tendon curvature. There are also losses due to seating of the wedges or nuts as the jacking force is transferred into the anchorage device. These and other sources of prestress loss are examined by the licensed design professional to get an estimate of the total prestress loss and resulting effective prestressing force.

Losses have inherent variability due to variations of material properties and environmental and curing conditions. Some losses may affect others. Time-dependent concrete properties are particularly difficult to estimate accurately, so losses due to creep and shrinkage are expected to be variable. Friction between the tendon and sheathing or duct, movement of wedges within the anchorage device, and modulus of elasticity of concrete are also variables. The variability within each component and the interdependence among the components make it understandable that studies comparing measured prestress losses to predictions have shown that accurate and consistent calculation of prestress loss is difficult to achieve.

The best effort to calculate prestress loss is only an estimate and, therefore, the licensed design professional should consider the consequences of actual losses being higher or lower than the estimated value. Estimation of prestress loss is an important factor for evaluating the serviceability of all types of prestressed members and the calculation of flexural strength of members with unbonded tendons. The estimation of prestress loss, however, is not a significant factor in determination of flexural strength of bonded prestressed members. When computing the shear strength of prestressed members with little or no transverse reinforcement, a conservative estimate of the effective prestressing force is warranted.

1.2—Scope

ACI 318-11 requires that the design of prestressed concrete members allow for prestress loss; however, the required level of detail for calculating losses is unspecified. The friction loss provisions for post-tensioned construction that first appeared in ACI 318-63 were removed from ACI 318-11. Although ACI 318-11 Commentary indicates that the lump sum method is obsolete, the licensed design professional's requirement to choose a method to compute losses remains. This guide is intended to aid the designer in this choice by providing an overview of the various methods available.

Many participants in the design and construction process need information on prestress losses. The licensed design professional, precasters, and post-tensioners all need an understanding of, and method to estimate, aspects of losses. To which entity is responsible for calculation of each type of loss has to be clearly defined in the contract documents.

Total losses, Δf_{pT}, are losses due to friction and seating Δf_{pFS}, elastic shortening Δf_{pES}, creep of concrete Δf_{pCR}, shrinkage of concrete Δf_{pSH}, and relaxation of tendons Δf_{pRE}. This can be expressed as Eq. (1.2)

$$\Delta f_{pT} = \Delta f_{pFS} + \Delta f_{pES} + \Delta f_{pCR} + \Delta f_{pSH} + \Delta f_{pRE} \qquad (1.2)$$

This guide presents background information and methods to calculate each type of loss.

Following the introduction and a list of notation and definitions, Chapter 3 includes a historical account of the lump sum method, currently recommended values for preliminary design, and a summary of losses that have been measured in field and laboratory studies.

Chapter 4 discusses the different types of initial losses and addresses the differences between pretensioned and post-tensioned members.

Chapter 5 presents a simplified approach to estimate long-term losses due to creep, shrinkage, and relaxation for pretensioned and post-tensioned concrete members.

Detailed approaches to estimate long-term losses are presented in Chapter 6, which also addresses changes in prestressing force caused by differential shrinkage and hydration of the concrete deck in composite members. The approaches can be used for pretensioned or post-tensioned members.

Chapter 7 discusses the variability of prestress loss calculations caused by concrete material properties, including compressive strength at transfer, modulus of elasticity, and creep and shrinkage.

Chapter 8 presents example problems and compares solutions from different methods.

1.3—Historical development

The concept of prestressing concrete dates back to the late 1800s (Naaman 2012). The performance of early prestressed concrete structures was adversely affected by time-dependent strains in the concrete—for example, creep and shrinkage, which were nearly as large as the initial steel strain due to prestressing. Before 1940, the initial steel strain induced by prestressing was limited by the low yield strength of steel. French engineer Eugene Freyssinet recognized the significance of prestress losses and the need for steels with high yield strength for prestressed applications. By 1945, higher strength steel became available, making it possible to produce the initial prestressing strain large enough so that the time-dependent strains developed in the concrete would not overcome the initial prestressing strain. As a result, the remaining prestressing force in the steel would be sufficiently large to be effective.

Prestress losses were first addressed by ACI 318 in 1963. Although the provisions catalogued the different causes of prestress loss, they only provided specific instruction on determining friction losses. These code provisions were based on an earlier committee publication that provided similar, slightly more detailed guidance on prestress loss (ACI-ASCE Committee 323 1958).

In the 1970s, the Precast/Prestressed Concrete Institute (PCI Committee on Prestress Losses 1975) and Zia et al. (1979) provided more detailed methods to estimate prestress losses. Since the 1970s, others have developed methods to estimate prestress losses (Tadros et al. 2003; Seguirant and Anderson 1985; Youakim et al. 2007; Garber et al. 2013). Gilbert and Ranzi (2011) and Branson (1977) provide general approaches to the calculation of a variety of time-dependent effects in concrete structures, including prestress losses. Computer programs have been developed to perform the tedious calculations required for stepwise analyses of prestress loss. However, due to the inherent uncertainties associated with material properties, construction practices, and in-service conditions, even the most refined calculations result in prestress loss predictions that differ from measured values.

1.3.1 *Currently available guidance on estimating prestress losses*—For pretensioned building products, the *PCI Design Handbook* (PCI 2010) presents a method to estimate prestress losses based on the method developed by Zia et al. (1979). This method is widely used for building structures and is referenced in the R18.6.1 commentary of ACI 318-11, and presented in this guide in Chapter 5. For bridge beams, the "AASHTO LRFD Bridge Design Specification" (AASHTO 2012) presents two methods. One is an approximate method and the other a refined method based on several parameters to estimate prestress losses. The refined method could be applied to building products as well. These methods are presented in Chapters 5 and 6.

1.4—Guide organization and use

This guide presents a variety of approaches for estimating prestress losses in pretensioned and post-tensioned members. This section identifies relevant sections of interest in the guide, depending on member type (pretensioned or post-tensioned) and level of effort (lump sum, simplified, or detailed). The lump sum method is only recommended for preliminary designs. The simplified method is appropriate for most typical designs. Detailed methods are most often used for more complex structures, which may have staged construction and prestressing operations.

1.4.1 *Pretensioned members*—Losses for pretensioned members are classified as initial or long-term. One group of initial loss occurs during stressing and before transfer of prestress due to friction, seating losses, and temperature effects. It is the precaster's responsibility to understand the magnitude of these losses and account for them to provide the specified strand stress before transfer. Information on these types of losses is found in:

(a) Anchorage seating—4.2.1
(b) Form and abutment deformations—4.2.2
(c) Thermal effects—4.2.4
(d) Steel relaxation—4.2.5

Another initial loss is elastic shortening of the member that occurs at the time of transfer. As the prestress force is transferred to concrete, the member shortens. The steel and concrete are fully bonded, so the steel shortens with the concrete. This shortening causes a loss in stress in the prestressing steel, known as the elastic shortening loss, which should be accounted for by the designer. Long-term losses occur due to concrete creep and shrinkage and prestressing steel relaxation. Other changes of tendon force can occur due to temperature effects and external loads placed on the member at the time of casting or in service.

1.4.1.1 *Pretensioned members/lump sum method*—The lump sum method presented in Chapter 3 is often used for preliminary design. The values presented in sources referenced in Chapter 3 typically include all losses, both initial and long-term.

1.4.1.2 *Pretensioned members/simplified method*—The simplified method is a commonly employed approach to estimate prestress losses in typical pretensioned members. The designer needs to calculate four components of loss and add them together for the total prestress loss. Components and applicable sections are:

(a) Elastic shortening—4.3.2
(b) Creep—5.2
(c) Shrinkage—5.3
(d) Relaxation—5.4

1.4.1.3 *Pretensioned members/detailed method*—This guide provides information on more detailed methods of prestress loss estimation. Two alternate methods for a more detailed calculation of elastic shortening losses in pretensioned members are:

(a) Transformed section method—4.3.1
(b) Iterative gross section method with iteration—4.3.3

More detailed approaches to calculate long-term losses are presented in Chapter 6. These methods are used with a variety of creep and shrinkage models, as opposed to the simplified method, which uses a single model. Detailed methods also allow the designer to consider the influence of a cast-in-place composite deck if needed, whereas the simplified method only accounts for the weight of the deck, but not other factors such as differential shrinkage and internal stress redistributions between the beam and the deck, if acting compositely. The detailed methods are:

(a) AASHTO LRFD refined method (AASHTO 2012)—6.3.2
(b) General age-adjusted effective modulus method (Menn 1990)—6.3.3
(c) Incremental time-step method (Nilson 1987)—6.4

Chapter 6 (6.6) also provides information on the approximation of changes due to thermal effects of deck casting.

1.4.2 *Post-tensioned members*—Several approaches can be used to approximate prestress losses in post-tensioned members. Initial losses encompass all prestress loss during the stressing operation, including friction due to wobble and curvature, seating losses, and elastic shortening losses. The estimation of long-term losses for bonded post-tensioned members is essentially the same as for pretensioned members. Calculation of long-term losses in unbonded post-tensioned members is different, because losses are related to the overall change in tendon length, rather than the change in strain at a specific section.

1.4.2.1 *Post-tensioned members/lump sum method*—The lump sum method, presented in Chapter 3, is typically used only for preliminary designs. Before adopting a value for use in preliminary design, the licensed design professional should determine if the presented value includes friction and seating losses.

1.4.2.2 *Post-tensioned members/simplified method*—The simplified method can be used to estimate prestress losses in typical post-tensioned members. The designer needs to calculate five components of loss and add them together for the total prestress loss. Components and applicable sections are:

(a) Friction and seating loss—4.4
(b) Elastic shortening loss—4.5
(c) Creep loss—5.2.1 (bonded)
(d) Creep loss—5.2.2 (unbonded)
(e) Shrinkage loss—5.3
(f) Relaxation loss—5.4

Note that elastic shortening losses only occur in post-tensioned members with multiple tendons when the tendons are stressed sequentially. Tendons stressed first will incur losses as the concrete shortens due to the stressing of subsequent tendons.

1.4.2.3 *Post-tensioned members/detailed methods*—As with pretensioned members, long-term prestress loss in post-tensioned members are estimated using more detailed methods presented in Chapter 6, with a detailed description in 1.4.1.3. Initial losses are calculated per 4.3.3 and 4.4.

CHAPTER 2—NOTATION AND DEFINITIONS

2.1—Notation

a = constant
ab = eccentricity from the centroid of the beam (gross section) to the centroid of the deck (also centroid of the deck reinforcing steel), in. (mm)
A_c = area of concrete, in.^{2} (mm^2)
A_{comp} = transformed area of the composite section, in.^{2} (mm^2)
A_d = area of composite concrete deck, in.^{2} (mm^2)
A_g = area of gross concrete section at the cross section considered, in.^{2} (mm^2)
A_{ps} = area of prestressing steel, in.^{2} (mm^2)
A_{sd} = area of deck steel, in.^{2} (mm^2)
A_{tr} = transformed cross-sectional area, in.^{2} (mm^2)
$b(y)$ = width of cross-section at depth y relative to centroid of section, in. (mm)
C = factor in calculation of prestress loss due to relaxation according to the *PCI Design Handbook* (PCI 2010) method
C_c = creep coefficient
d = friction loss over length L, psi (MPa)
e = base of Naperian logarithms
e_b = basic elongation, in. (mm)
e_c = eccentricity of centroid of tendons with respect to the centroid of the gross cross section at the center of the beam, in. (mm)
e_d = distance between centroid of deck and centroid of composite beam, in. (mm)
e_e = eccentricity of centroid of tendons with respect to the centroid of the gross cross section at the ends of the beam, in. (mm)
e_p = eccentricity of centroid of tendons with respect to the centroid of the gross concrete at the cross section considered, in. (mm)
e_{pc} = eccentricity of centroid of tendons with respect to the centroid of the composite cross section, in. (mm)

e_{tr} = eccentricity of centroid of tendons with respect to the centroid of the f_{bpt} transformed concrete at the cross section considered, in. (mm)
$E(t)$ = modulus of elasticity at any time t, psi (MPa)
E_c = modulus of elasticity of concrete, psi (MPa)
E_c' = effective modulus of elasticity of concrete, psi (MPa)
E_c'' = age-adjusted effective modulus of elasticity of concrete, psi (MPa)
E_{cd} = modulus of elasticity of the composite deck, psi (MPa)
E_{ci} = modulus of elasticity of concrete at time of application of prestress, psi (MPa)
$E_c(t_i)$ = modulus of elasticity of concrete at time t_i, psi (MPa)
E_p = modulus of elasticity of the prestressing steel, psi (MPa)
f_{anchor} = strand stress at the anchorage device after seating, psi (MPa
f_c = concrete compressive stress, psi (MPa)
f_c' = specified compressive strength of concrete, psi (MPa)
f_{cd} = concrete stress at center of gravity of prestressing force due to all superimposed permanent loads that are applied to the member after it has been prestressed, psi (MPa)
f_{ci}' = specified compressive strength of concrete at transfer of prestress, psi (MPa)
f_{ci} = concrete compressive stress immediately after transfer at fiber under investigation, psi (MPa)
f_{cir} = net compressive concrete stress at center of gravity of prestressing force immediately after the prestress has been applied to the concrete, psi (MPa)
f_{cpa} = average compressive concrete stress at the center of gravity of the tendons immediately after the prestress has been applied to the concrete, psi (MPa)
f_{cps} = concrete stress at center of gravity of prestressing force due to all prestress and applied loads, psi (MPa)
f_{dead} = tendon stress at nonstressing end, psi (MPa)
f_{jack} = jacking stress, psi (MPa)
f_L = stress in prestressing steel at a distance L from jacking end, psi (MPa)
$f_{L/2}$ = stress in prestressing steel at a distance $L/2$ from the jacking end, psi (MPa)
f_{max} = maximum stress in the prestressing steel along the tendon length, psi (MPa)
f_{pbt} = stress in prestressing steel immediately before transfer, psi (MPa)
f_{pi} = prestressing steel stress immediately following transfer, psi (MPa)
f_{po} = prestressing steel stress after jacking and seating, psi (MPa)
$f_{ps}(t)$ = stress in prestressing steel at time t, psi (MPa)
f_{pt} = stress in prestressing steel immediately after transfer, psi (MPa)
f_{pu} = specified tensile strength of prestressing steel, psi (MPa)
f_{py} = specified yield strength of prestressing steel, psi (MPa)
f_x = stress in prestressing steel at a distance x from the jacking end, psi (MPa)
I_c = moment of inertia of the composite cross section, in.^{4} (mm^4)
I_d = moment of inertia of the deck, in.^{4} (mm^4)
I_g = moment of inertia of gross concrete section about centroidal axis, neglecting reinforcement, in.^{4} (mm^4)
I_{tr} = moment of inertia of transformed concrete section about centroidal axis, including reinforcement, in.^{4} (mm^4)
J = factor in calculation of prestress loss due to relaxation according to the *PCI Design Handbook* (PCI 2010) method
k = wobble friction coefficient per unit length of tendon, per ft (per m)
k_f = factor for the effect of concrete strength
k_{hc} = humidity factor for creep
k_s = factor for the effect of volume-to-surface ratio
k_{td} = time development factor
K_{cir} = modification factor in *PCI Design Handbook* (PCI 2010) method in calculation of concrete stress due to prestressing force immediately after the prestress has been applied to the concrete
K_{cr} = coefficient in the *PCI Design Handbook* (PCI 2010) method to account for loss due to creep
K_{df} = transformed section coefficient
K_{es} = factor in calculation of elastic shortening losses in Zia et al. (1979) method
K_{id} = section modification factor from AASHTO (2012) prestress loss method
K_{re} = factor in calculation of prestress loss due to relaxation in *PCI Design Handbook* (PCI 2010) method
K_{sh} = factor in calculation of prestress losses due to shrinkage in *PCI Design Handbook* (PCI 2010) method
L = strand length from anchorage to anchorage, ft or in. (m or mm)
L_{beam} = length of beam, ft or in. (m or mm)
L_{free} = length of strand outside of beam, ft or in. (m or mm)
M = bending moment experienced by cross section immediately after transfer (usually due to self-weight), in.-lb (N-mm)
M_b^o = initial creep-producing moment in the girder, in-lb (N-mm)
M_{deck} = moment in beam due to the weight of the deck, in.-lb (N-mm)
M_g = bending moment due to dead weight of prestressed member and any other permanent loads in place at the time of prestressing, in.-lb (N-mm)
M_{sd} = moment due to all superimposed permanent loads applied after prestressing, in.-lb (N-mm)
n_p = modular ratio; modulus of prestressing steel divided by modulus of concrete
N = number of sequentially stressed tendons
N_b^o = initial creep-producing force in the girder, lb (N)
P = applied tension force, lb (N)
P_{avg} = average force in the tendon, lb (N)
P_i = initial prestress force after anchorage seating loss, lb (N)

P_j = maximum prestress force during jacking operation, lb (N)
P_o = prestress force before release, lb (N)
P_{shd} = force to fully restrain the shrinkage of the composite deck, lb (N)
RH = average ambient relative humidity in percent
s = slope of stress in prestress versus distance line, psi/ft (MPa/m)
S_b = section modulus with respect to the bottom fiber of the beam, in.^{3} (mm^3)
t = time under consideration from time of release, days
t_c = age of concrete, days
t_d = time since end of cure to time of deck placement, days
t_f = final time under consideration from time of release, days
t_i = 1-day steam cured
t_o = time of initial loading from time of release, days
T_0 = prestressing force at stressing end, lb (N)
T_x = prestressing force at point x, lb (N)
$T(y)$ = temperature of cross section at distance y from centroid, °F (°C)
V/S = ratio of volume to surface area of concrete element, in.^{3}/in.^{2} (mm^3/mm^2)
w_c = unit weight of normalweight concrete or equilibrium density of lightweight concrete, lb/ft^3 (kg/m^3)
w_t = weight of topping, lb/ft^2 (N/m^2)
x = length of tendon from stressing end to point x, ft or in. (m or mm)
x_s = length influenced by anchor set, ft or in. (m or mm)
y = distance from centroid of cross section to location under consideration, in. (mm)
y_{bott} = the distance from the centroid of the composite section to the bottom of the section, in. (mm)
y_{tr} = distance from centroid of transformed section to concrete fiber under investigation, in. (mm)
α = total angular change from jacking end to point x, radians
α_c = coefficient of thermal expansion of concrete, /°F (/°C)
α_{ps} = coefficient of thermal expansion of prestress, /°F (/°C)
β = constant such that $\beta = (28 - a)/28$
γ_h = correction factor for ambient relative humidity
γ_{st} = correction factor for specified concrete compressive strength at transfer
Δ = tendon elongation, in. (mm)
Δf_{cdf} = change in concrete stress at the center of gravity of the prestressing force due to the differential shrinkage force, psi (MPa)
Δf_{cgp} = change in concrete stress at the center of gravity of the prestressing force due to application of superimposed load, psi (MPa)
Δf_{pA} = change in stress due to anchor set, psi (MPa)
Δf_{pCD} = prestress loss due to creep of girder concrete between time of deck placement and final time, psi (MPa)
Δf_{pCR} = prestress loss due to creep, psi (MPa)
Δf_{pCR1} = prestress loss due to creep of girder concrete between transfer and deck placement, psi (MPa); (AASHTO 2012)
Δf_{pEG} = increase in prestress (elastic gain) due to addition of superimposed permanent loads, psi (MPa)
Δf_{pES} = prestress loss due to elastic shortening, psi (MPa)
Δf_{pFS} = prestress loss due to friction and seating, psi (MPa)
Δf_{pLT} = long-term prestress loss, psi (MPa)
Δf_{pR1} = prestress loss due to relaxation of prestressing strands between time of transfer and deck placement, psi (MPa)
Δf_{pR2} = prestress loss due to relaxation of prestressing strands in composite section between time of deck placement and final time, psi (MPa)
Δf_{pRE} = prestress loss due to relaxation, psi (MPa)
Δf_{pSD} = prestress loss due to shrinkage of girder concrete between time of deck placement and final time, psi (MPa)
Δf_{pSH} = prestress loss due to shrinkage, psi (MPa)
Δf_{pSR} = prestress loss due to shrinkage of girder concrete between transfer and deck placement, psi (MPa)
Δf_{pSS} = prestress gain due to shrinkage of deck in composite section, psi (MPa)
Δf_{pT} = total prestress loss, psi (MPa)
ΔL_{jack} = elongation within the jack, in. (mm)
ΔM_b = change in moment in the beam, in.-lb (N-mm)
ΔM_d = change in moment in the deck, in.-lb (N-mm)
ΔN_b = change in force in the beam, lb (N)
ΔN_d = change in force in the deck, lb (N)
ΔN_{ps} = change in prestress force, lb (N)
ΔN_{relax} = change in prestress force due to relaxation (no associated strain), lb (N)
ΔN_{sd} = change in the force in the deck reinforcement, lb (N)
ΔP = change in force, lb (N)
ΔP_c = change in force in concrete, lb (N)
ΔP_{ps} = change in force in prestressing steel, lb (N)
Δs = anchor set, in. (mm)
ΔT_1 = temperature rise, °F (°C)
ΔT_2 = temperature change, °F (°C)
$\Delta\varepsilon_b$ = change in strain in the beam at the gross section centroid
$\Delta\varepsilon_c$ = change in strain in concrete at the center of gravity of the prestressing force
$\Delta\varepsilon_{cr}$ = change in strain in concrete due to creep
$\Delta\varepsilon_d$ = change in strain at the centroid of the deck
$\Delta\varepsilon_p$ = change in strain in prestressing steel
$\Delta\varepsilon_{pA}$ = change in strain in prestressed reinforcement due to anchor set
$\Delta\varepsilon_{free}$ = change in strain in free length of prestressing steel
$\Delta\varepsilon_{shb}$ = change in strain in the beam due to shrinkage
$\Delta\varepsilon_{shb}(t_d)$ = shrinkage strain in the beam concrete at the time the deck is placed
$\Delta\varepsilon_{shb}(t_f)$ = shrinkage strain in the beam concrete at the final time considered
$\Delta\varepsilon_{sd}$ = change in strain in the deck steel
$\Delta\kappa$ = change in curvature, /in. (/mm)
$\Delta\sigma(t_i)$ = change in stress at time t_i, psi (MPa)

$\varepsilon(t)$ = concrete strain at time t
$\varepsilon(t_o)$ = initial concrete strain
$\varepsilon_{shb}(t)$ = shrinkage strain in the prestressed beam at time t following the end of cure
$\varepsilon_{shb}(t_d)$= shrinkage strain of the girder at the time the deck is placed
$\varepsilon_{shb}(t_f)$= total shrinkage strain of the beam at the time of analysis
$\varepsilon_{shd}(t)$ = shrinkage strain in the composite deck at time t following the end of cure
$\varepsilon_{shd}(t_f)$= shrinkage of the deck concrete at final time under consideration
μ = curvature friction coefficient
$\sigma(t_o)$ = stress at time t_o, psi (MPa)
$\phi(t,t_i)$ = creep coefficient at time t for loads applied at time t_i
$\phi(t,t_o)$ = creep coefficient at time t for loads applied at time t_o
$\phi(t_f,t_d)$= creep coefficient at time t_f for loads applied at time t_d
$\phi(t_f,t_o)$= creep coefficient at time t_f for loads applied at time t_o
$\phi_d(t_f,t_d)$= creep coefficient of the deck concrete at time t_f for loads applied at time t_d
χ = aging coefficient

2.2—Definitions

ACI provides a comprehensive list of definitions through an online resource, "ACI Concrete Terminology," https://www.concrete.org/store/productdetail.aspx?ItemID=CT16. Definitions provided herein complement this source.

anchorage—in post-tensioning, a device used to anchor the tendon to the concrete member; in pretensioning, a device used to maintain the elongation of a tendon during the time interval between stressing and release.

modulus of elasticity—the ratio of normal stress to corresponding strain for tensile or compressive stress below the proportional limit of the material.

post-tensioning—method of prestressing reinforced concrete in which tendons are tensioned after concrete has attained a specified minimum strength or specified minimum age.

CHAPTER 3—LUMP-SUM METHOD

3.1—Scope

In some cases, use of a simple lump-sum value for prestress loss may be adequate instead of detailed calculations. This chapter summarizes lump-sum values from past codes and literature, and discusses their use in design. In addition, data from the measurement of prestress losses are presented and summarized.

3.2—Historical code requirements

The Federal Highway Administration (FHWA), known as the Bureau of Public Roads (1954) at the time, published design criteria for prestressed concrete bridges in which losses due to creep, relaxation, shrinkage, and elastic deformation were specified as

$$\text{pretensioned concrete} = 6000 + 16f_{cps} + 0.04f_{pi} \text{ (psi)}$$
$$\text{pretensioned concrete} = 41.4 + 16f_{cps} + 0.04f_{pi} \text{ (MPa)} \quad (3.2a)$$

$$\text{post-tensioned concrete} = 3000 + 11f_{cps} + 0.04f_{pi} \text{ (psi)}$$
$$\text{post-tensioned concrete} = 20.7 + 11f_{cps} + 0.04f_{pi} \text{ (MPa)} \quad (3.2b)$$

where f_{cps} is the concrete stress at the center of gravity of the prestressing force due to all prestress and applied loads, psi (MPa); and f_{pi} is prestressing steel stress immediately following transfer, psi (MPa).

In formulating the criteria, it was assumed that the efficiency of the anchorage was 100 percent with no seating loss and applicable to normalweight concrete only. The amount of losses was to be determined by tests for lightweight concrete. In Eq. (3.2a) and (3.2b), the first term represents loss due to shrinkage, the second represents combined losses due to elastic shortening and creep, and the third represents loss due to steel relaxation. For post-tensioned concrete, elastic shortening was ignored. In the early 1950s, only lower strengths of concrete and steel were available for prestressed concrete applications. For typical designs, the value of f_{cps} was approximately 1300 psi (8.96 MPa) and the value of f_{pi} was a maximum of 200,000 psi (1380 MPa). Using these typical values for f_{cps} and f_{pi}, the criteria stated (Eq. (3.2a) and Eq. (3.2b)) for losses equates to 34,800 psi (240 MPa) for pretensioned concrete and 25,300 psi (174 MPa) for post-tensioned concrete.

Requirements for evaluating prestress loss first appeared in ACI 318-63, in new Chapter 29 on prestressed concrete. ACI 318-63 commentary recommended the following lump-sum losses, excluding friction and seating losses, based on the ACI-ASCE Committee 323 (1958) report: pretensioning = 35,000 psi (241 MPa); and post-tensioning = 25,000 psi (172 MPa).

In published discussions, Abeles (1958) argued that these recommended lump-sum values should be eliminated or increased to allow designers to assess the losses for specific conditions. If the lump-sum values were to remain, he suggested the following ranges of values: pretensioning = 30,000 to 40,000 psi (207 to 276 MPa); and post-tensioning = 20,000 to 30,000 psi (138 to 207 MPa).

Similarly, the Precast/Prestressed Concrete Institute (PCI) published building code requirements for prestressed concrete (Structural Engineers Association of Northern California 1959) specifying the following lump-sum values for estimating prestress losses, excluding friction prestressing steel loss: pretensioning = 25,000 to 35,000 psi (172 to 241 MPa); and post-tensioning = 15,000 to 25,000 psi (103 to 172 MPa).

In ACI 318-83, specific lump sum recommendations in the commentary were replaced by the following statement: "Lump-sum values of prestress losses for both pretensioned and post-tensioned members that were indicated before the 1983 Commentary are considered obsolete." The commentary also provides several references that can be used to compute prestress loss (ACI-ASCE Committee 323 1958; ACI 435R; PCI Committee on Prestress Losses 1975; Zia et al. 1979). Although ACI 318-11 does not explicitly prohibit use of lump-sum losses, the commentary is clear in warning that previously cited lump-sum losses are considered obsolete.

3.3—Industry practice

The Post-Tensioning Institute's manual (PTI 2006) provides a discussion of long-term losses and a detailed discussion of friction losses, which are addressed elsewhere in this guide, including 4.4.2. While PTI refers to Zia et al. (1979), it is noted that precise determination of losses in post-tensioned members is not critical. The discussion is closed with the statement, "A 100% variance in the estimate for total long-term losses generally results in less than a 10% difference in stress in the prestressing steel at nominal strength." Kelley (2000) indicates that for approximating prestress losses, tendons are typically assumed to be stressed to $0.8f_{pu}$; for ASTM A416/A416M Grade 270 (1860 MPa), this equates to a tendon stress of 216,000 psi (1490 MPa). In addition, it is typically assumed that the average tendon stress along a tendon after seating is $0.7f_{pu}$. Based on an assumed final effective prestress of 175,000 psi (1210 MPa), the total prestress loss is 41,000 psi (282 MPa), or 19 percent of the initial jacking stress. This results in a force of 26,800 lb (119,000 N), often rounded to 27,000 lb (120,100 N), for a single 0.5 in. (12.7 mm) diameter strand.

Before low-relaxation strand gained popularity in the 1970s, use of stress-relieved strand resulted in higher losses due to relaxation of the prestressing strand. At that time, a total loss of 52,000 to 57,000 psi (359 to 393 MPa) was typically assumed, resulting in a final effective prestress force of approximately 24,000 to 25,000 lb (106,800 to 111,200 N) per strand.

The *PCI Design Handbook* (PCI 2010) indicates that losses have no effect on the ultimate strength of a flexural component unless the tendon is unbonded or has an effective prestress of less than $0.5f_{pu}$. In addition, the handbook indicates that overestimated or underestimated losses may impact service limit states such as camber, deflection, or cracking. PCI (2010) also provides an estimate of total losses for typical concrete components:

(a) Normalweight concrete: 25,000 to 50,000 psi (172 to 345 MPa) (12 percent to 25 percent)

(b) Sand lightweight concrete: 30,000 to 55,000 psi (207 to 379 MPa) (15 percent to 27 percent)

These values are presented as total losses for typical components. Losses in terms of the percentage of an assumed initial jacking force of $0.75f_{pu}$ are also given. For refined estimation of losses, PCI (2010) refers to Zia et al. (1979).

3.4—Measured losses

Measurement of prestressing losses is challenging because it requires that concrete strain be determined from the time when the prestress is transferred until the end of the measurement period. This typically excludes use of bonded electrical resistance foil gauges because of drift due to temperature change and lack of durability when exposed to concrete and the surrounding environment. Consequently, losses are typically measured using vibrating wire strain gauges embedded in concrete or mechanical gauges, which measure relative movement of gauge points adhered to the concrete surface. These movements are used to calculate average concrete strain between the points and to track change in concrete strain over time. If perfect bond is assumed between prestressing steel and concrete, then change in concrete strain is equal to change in steel strain.

If steel strain has been tracked from the initial value, which is typically the initial stressing, then the measured strain is used to compute the change in steel stress due to change in steel strain. The conversion from strain to stress requires that the elastic modulus of the prestressing steel (E_p) be known. While determining E_p for single wires and bar is relatively straightforward, the nature of the seven-wire strand makes accurate and precise measurement of E_p more difficult. Additionally, strain measurement methods cannot account for time-dependent losses due to relaxation. For low-relaxation prestressing strand, losses due to relaxation are usually minor relative to other time-dependent losses. For other materials more susceptible to relaxation, a force-measuring device is required, which is not typically a practical alternative. The measured change in steel stress is equal to the prestress loss from the time when the initial measurement was taken to when the measurements were terminated. The ultimate value of long-term losses is practically unattainable due to the time over which measurements are taken. Fortunately, most of the long-term losses occur in the first 6 months after prestress transfer. Consequently, relatively short-term measurements of losses, which range from 6 months to 2 years, can provide useful information regarding ultimate long-term losses.

Measurement techniques described previously require that instrumentation be installed during fabrication; readings are then taken before prestress transfer and regularly during the monitoring period. For in-service girders not previously instrumented, measurement of crack opening during loading may be used to estimate the effective prestress (Pessiki et al. 1996). While not as reliable or accurate as vibrating wire or mechanical gauges, it does provide a method for estimating the effective prestress in noninstrumented, in-service girders.

Table 3.4 presents measured losses reported in the literature. Nearly all data are from measurements on precast, pretensioned, I-shaped bridge girders. Spans range from less than 10 ft (3 m) to more than 270 ft (82.3 m). In addition, concrete compressive strengths vary from 4000 psi (27.6 MPa) to more than 15,000 psi (103 MPa) with an average of approximately 9000 psi (62.0 MPa). While little data are available on prestressed concrete elements used specifically in building construction, all or part of the bridge data presented herein is useful as a benchmark against which, for future evaluation, either bridge or building components can be compared.

Measured total losses were gathered from each reference and include instantaneous and long-term losses for the associated measurement period indicated in the table. No adjustments for final total losses have been made. Losses are reported in terms of stress in the prestressing steel and as a percentage of an assumed initial prestress force equal to $0.75f_{pu}$. Overall average prestress losses were approximately 40,700 psi (281 MPa), or 21 percent, with a coefficient of variation (COV) of 40 percent that indicates a wide data scatter. While the data are quite scattered, average total loss

Table 3.4—Measured losses from literature

Reference	Total loss, ksi	Percent loss	f_c', psi	Age at f_c' test	Span, ft	Age at loss test	Girder type*	Instrumentation†	Prestressing steel type‡	Comments§
Greuel et al. (2000)	35.4	17.5	11,800	56 days	115.5	6 to 9 months	ODOT B42-48	c	270LL	
Pessiki et al. (1996)		17.2	8760	28 years	89	28 years	PennDOT 24 x 60 I	c	Unknown	
		18.1	8180	28 years	89	28 years	PennDOT 24 x 60 I	c	Unknown	
Mossiossian and Gamble (1972)	32.5	16.0	Data unavailable							
	35.1	17.3								
Kebraei et al. (1997)	17.9	8.8								
	36.8	18.2								
Shenoy and Frantz (1991)	30.5	15.1	7100	27 years	54	27 years	36W x 27D box	c	250SR	
Stanton et al. (1999)	34	16.8	Data unavailable							
	65.6	32.4								
	55.1	27.2								
	69.3	34.2								
	36.1	17.8								
Seguirant and Anderson (1985)	41.7	20.6								
	35	17.3								
	35.7	17.6								
Gross and Burns (2000) *Louetta Bridge*	43.1	21.3	13,630	56 days	132.9	761 days	TxDOT U54	v	270LL	2
	37.9	18.7	14,320	56 days	118.2	748 days	TxDOT U54	v	270LL	2
	40.3	19.9	13,290	56 days	119.7	1262 days	TxDOT U54	v	270LL	2
	33.8	16.7	13,410	56 days	132.3	1221 days	TxDOT U54	v	270LL	2
Gross and Burns (2000) *San Angelo Eastbound*	50.6	25.0	13,700	56 days	127.8	422 days	AASHTO IV	v	270LL	2
	57.2	28.3	13,700	56 days	127.8	422 days	AASHTO IV	v	270LL	2
	51.5	25.4	14,240	56 days	152.2	404 days	AASHTO IV	v	270LL	2
	52.0	25.7	14,830	56 days	152.2	746 days	AASHTO IV	v	270LL	2
	57.4	28.4	13,750	28 days	145.2	316 days	AASHTO IV	v	270LL	2
	58.2	28.7	14,490	56 days	145.2	309 days	AASHTO IV	v	270LL	2
	55.6	27.5	14,550	56 days	144.5	305 days	AASHTO IV	v	270LL	2
Gross and Burns (2000) *San Angelo Westbound*	34.7	17.1	10,130	28 days	127.8	771 days	AASHTO IV	v	270LL	
	34.4	17.0	10,130	28 days	127.8	771 days	AASHTO IV	v	270LL	
	32.7	16.1	10,130	28 days	127.8	771 days	AASHTO IV	v	270LL	
	30.6	15.1	10,260	28 days	127.8	766 days	AASHTO IV	v	270LL	
Al-Omaishi et al. (2009)	32	15.8	9025	28 days	127	468 days	NU2000	v	270LL	
	35.7	17.6	9025	28 days	127	469 days	NU2000	v	270LL	
	43.5	21.5	10,050	28 days	110	490 days	NE1400BT	v	270LL	
	42.3	20.9	10,050	28 days	110	491 days	NE1400BT	v	270LL	
	25.4	12.5	10,670	28 days	129.2	398 days	U54B	v	270LL	
	42.1	20.8	10,280	28 days	159	385 days	W83G	v	270LL	
	40	19.8	10,280	28 days	159.8	385 days	W83G	v	270LL	
Dwairi et al. (2010)	26	12.8	10,740	100 days	57.4	<1 year	AASHTO III	v	270LL	2,12
	38	18.8	9750	100 days	91.9	<1 year	AASHTO IV	v	270LL	2,12

Reference	Total loss, ksi	Percent loss	f_c', psi	Age at f_c' test	Span, ft	Age at loss test	Girder type*	Instrumentation†	Prestressing steel type‡	Comments§
Czaderski and Motavalli (2006)		19	8100	38 years	56	38 years	450W I	d	unknown	1,13
		39	8100	38 years	56	38 years	450W I	d	unknown	1,13
		20	8100	38 years	56	38 years	450W I	d	unknown	1,13
Yang and Myers (2006)	41.9	21	10,160	56 days	55	2 years	MoDOT II	v	270LL	2
Roller et al. (1993, 1995)	23	11.4	9660	660 days	69	660 days	AASHTO BT-54	1	270LL	10
Kahn and Lopez (2005)	37.2	18.4	10,229	103 days	40	100 days	AASHTO II	v	270LL	9,14
	29.6	14.6	10,868	144 days	40	100 days	AASHTO II	v	270LL	9,14
Labia et al. (1997)	62	30.6	7864	20 years	70	20 years	1213W x 838D box	c	270LL	15
	62	30.6	8265	20 years	70	20 years	1213W x 838D box	c	270LL	15
Stallings et al. (2003)	44.8	22.1	9950	56 days	112	10 months	AASHTO BT-54	v	270LL	2
Ahlborn et al. (2000)	55.1	27.2	12,600	480 days	132.8	840 days	Mn/DOT 45M	v	270LL	
	53.3	26.3	11,100	480 days	132.8	860 days	Mn/DOT 45M	v	270LL	
Gamble (1970)	31	20.0	7730	667 days	71	1212 days	48D I-beam	d	250SR	
Houdeshell et al. (1972)	33	19.6	6920	420 days	70	782 days	48D I-beam	d	250SR	
	31	18.5	6920	420 days	67.3	783 days	48D I-beam	d	250SR	
Anderson et al. (1972)	57.3	28.3	6405	28 days	9	500 days	6D I-beam	d	16	17
	66.2	32.7	6278	28 days	9	490 days	6D I-beam	d	16	17
	72.6	35.8	6052	28 days	9	480 days	6D I-beam	d	16	17
	76.2	37.6	6170	28 days	9	420 days	6D I-beam	d	16	17
	68.2	33.7	6250	28 days	9	410 days	6D I-beam	d	16	17
	76.7	37.9	6303	28 days	9	400 days	6D I-beam	d	16	17
Gamble (1979)	28.1	17.8	7030	367 days	46	400 days	42D I-beam	d	250SR	
	28.1	17.8	7030	367 days	46	400 days	42D I-beam	d	250SR	
Hamilton and O'Neill (2009)	20.2	10.0	8400	<1 years	15	<1 years	24W x 20D I.T.	c	270LL	
	45.8	22.6	8400	<1 years	16	<1 years	24W x 20D I.T.	c	270LL	1
	17.8	9	8000	4 months	18	4 months	14W x 20D I-beam	v	270LL	
	20.9	10	8000	5 months	18	5 months	14W x 20D I-beam	v	270LL	
	22.7	11	8000	6 months	18	6 months	14W x 20D I-beam	v	270LL	
	26.1	13	8000	3.5 months	18	3.5 months	14W x 20D I-beam	v	270LL	
	24.5	12	8000	3.5 months	18	3.5 months	14W x 20D I-beam	v	270LL	
	34.2	17	8000	3.5 months	18	3.5 months	14W x 20D I-beam	v	270LL	
Barr et al. (2000)	57.6	28	10,000	56 days	133	3 years	W74MG	v	270LL	2,18
	49.7	25	10,000	56 days	133	3 years	W74MG	v	270LL	2,18
	60.6	30	10,000	56 days	133	3 years	W74MG	v	270LL	2,18
	31.9	16	10,000	56 days	76.4	3 years	W74MG	v	270LL	2,18
	32.6	16	10,000	56 days	76.4	3 years	W74MG	v	270LL	2,18

Reference	Total loss, ksi	Percent loss	f_c', psi	Age at f_c' test	Span, ft	Age at loss test	Girder type*	Instrumentation†	Prestressing steel type‡	Comments§
Osborn et al. (2012)	45.3	27.1	7100	42 years	22.5	42 years	AASHTO II	c	250SR	
	55.3	33.0	7100	42 years	22.5	42 years	AASHTO II	c	250SR	
	31.6	18.8	7100	42 years	22.5	42 years	AASHTO II	c	250SR	
	34.0	20.3	7100	42 years	22.5	42 years	AASHTO II	c	250SR	
	40.1	24.0	7100	42 years	22.5	42 years	AASHTO II	c	250SR	
Attanayake and Aktan (2010)	40.6	23.2	8300	48 years	48.5	48 years	915W x 533D box	c	250SR	19
Azizinamini et al. (1996)	36.2	20.7	4330	25 years	55	25 years	Nebraska III	c	250SR	
Boone (2008)	29.8	14.7	13,785	56 days	132	10 months	AASHTO VI	v	270LL	8
	16.1	8.0	13,785	56 days	82	7 months	AASHTO VI	v	270LL	8
Brewe and Myers (2010)	66.5	32.8	8210	243 days	15	196 days	12D T-beam	d	270LL	6
	70.7	34.9	8210	243 days	15	197 days	12D T-beam	d	270LL	6
	64.5	31.9	8210	243 days	15	198 days	12D T-beam	d	270LL	6
	62.9	31.1	8210	243 days	15	199 days	12D T-beam	d	270LL	6
	67.4	33.3	8210	243 days	15	200 days	12D T-beam	d	270LL	6
	57.7	28.5	8210	243 days	15	201 days	12D T-beam	d	270LL	6
Canfield (2005)	37	18.3	15,287	160 days	89	161 days	AASHTO IV	v	270LL	2
	40.3	19.9	15,145	184 days	89	182 days	AASHTO BT-56	v	270LL	2
Furr et al. (1968)	26.3	14	6600	140 days	50	140 days	TxDOT B	d	250SR	7
	25.3	13.5	7291	140 days	50	140 days	TxDOT B	d	250SR	
	30.9	16.5	6360	140 days	50	140 days	TxDOT B	d	250SR	7
	30.9	16.5	7080	140 days	50	140 days	TxDOT B	d	250SR	
	29.9	15.9	5160	28 days	40	660 days	TxDOT B	d	250SR	7
	28.0	14.9	5160	28 days	40	660 days	TxDOT B	d	250SR	7
	40.9	21.8	5160	28 days	56	660 days	TxDOT B	d	250SR	7
	39.1	20.8	5160	28 days	56	660 days	TxDOT B	d	250SR	7
	27.3	14.6	3990	7 days	56	660 days	TxDOT B	d	250SR	
Hale and Russell (2006)	53.4	26.4	15,610	360 days	24	360 days	24D I-beam	d	270LL	
	74	36.5	11,460	360 days	24	360 days	24D I-beam	d	270LL	
	58.4	28.8	11,460	360 days	24	360 days	24D I-beam	d	270LL	
	51.8	25.6	15,610	360 days	24	360 days	24D I-beam	d	270LL	
Halsey and Miller (1996)	29.3	19.5	11,790	25 years	27	25 years	12D inverted-T	c	Unknown	
	35	23.3	11,790	25 years	27	25 years	12D inverted-T	c	Unknown	
Idriss and Solano (2002)	26.5	13.1	10,151	56 days	97	1 years	BT 1600	f	270LL	2,19
Lewis and Karbhari (2006)	8.6	4.2	6050	28 days	264.5	363 days	multiple cell box	v	270LL	1
	9	4.4	5600	28 days	271.7	303 days	multiple cell box	v	270LL	1
Kukay (2008)	48.5	27.7	9300	40 years	34.5	40 years	AASHTO II	c	250SR	
	49.1	28.0	9300	40 years	34.5	40 years	AASHTO II	c	250SR	
	42.1	24.1	9300	40 years	34.5	40 years	AASHTO II	c	250SR	
	54.9	31.4	9300	40 years	34.5	40 years	AASHTO II	c	250SR	
	56.9	32.5	9300	40 years	34.5	40 years	AASHTO II	c	250SR	
	45.6	26.0	9300	40 years	34.5	40 years	AASHTO II	c	250SR	
	30.0	17.2	9300	40 years	34.5	40 years	AASHTO II	c	250SR	
	38.9	22.2	9300	40 years	34.5	40 years	AASHTO II	c	250SR	

Reference	Total loss, ksi	Percent loss	f_c', psi	Age at f_c' test	Span, ft	Age at loss test	Girder type*	Instrumentation†	Prestressing steel type‡	Comments§
Roller et al. (2011)	28.3	14.0	11,670	90 days	131.2	650 days	BT-78	v	270LL	19
Ruiz et al. (2008)	28.6	14.1	12,520	28 days	18	265 days	6.5W x 12D	v	270LL	11
	22.8	11.3	10,700	28 days	18	258 days	6.5W x 12D	v	270LL	11
	23.8	11.8	13,100	28 days	18	258 days	6.5W x 12D	v	270LL	11
	28.2	13.9	10,340	28 days	18	270 days	6.5W x 12D	v	270LL	5
	28.2	13.9	12,890	28 days	18	255 days	6.5W x 12D	v	270LL	5
	25.1	12.4	11,320	28 days	18	290 days	6.5W x 12D	v	270LL	4
	21.3	10.5	11,420	28 days	18	286 days	6.5W x 12D	v	270LL	4
	24.1	11.9	11,740	28 days	18	286 days	6.5W x 12D	v	270LL	4
	24.6	12.1	11,000	28 days	18	274 days	6.5W x 12D	v	270LL	4
	23.5	11.6	12,030	28 days	18	274 days	6.5W x 12D	v	270LL	4
Saiidi et al. (1998)	33	16.3	5430	28 days	150	300 days	multiple cell box	e	270LL	
Saiidi et al. (1996)	36	17.8	6100	28 days	155	30 months	multiple cell box	e	270LL	
Tabatabai and Dickson (1993)	29	17.5	10,800	34 years	42.3	34 years	AASHTO II	c	240W	1
Wang et al. (2011)	76.7	37.9	6930	28 days	9.8	260 days	200W x 300D	e	270LL	
	70.7	34.9	6930	28 days	9.8	260 days	200W x 300D	e	270LL	
	77.9	38.5	6930	28 days	9.8	260 days	200W x 300D	e	270LL	
	72	35.6	6930	28 days	9.8	260 days	200W x 300D	e	270LL	
Average	41.2	21%	9460							
COV	39%	38%								
Maximum	77.9	39%	15,610							
Minimum	8.6	4%	3990							

*Notes on girder specifications: 45D indicates 45 in. deep; 533W indicates 533 mm wide.

†Notes on instrumentation: c = crack reopening; v = vibrating wire gauge; e = electrical resistance gauge; f = fiber optic gauge; d = demountable mechanical gauge (DEMEC); and l = Carleson strain gauge.

‡Notes on prestressing steel type: 270LL = ASTM A416/A416M Gr270 low-relaxation seven-wire strand; 250SR = ASTM A416/A416M Gr250 stress-relieved seven-wire strand; 240W = Grade 240 prestressing wire.

§Comments: 1. Post-tensioned member; 2. High-performance concrete; 3. Lightweight concrete; 4. Self-consolidating concrete (SCC) with Type I cement; 5. Self-consolidating concrete with Type III cement; 6. High-strength SSC; 7. Lightweight concrete; 8. Self-consolidating high-performance concrete; 9. High-performance lightweight concrete; 10. Steam cured; 11. High-strength concrete; 12. Possible gauge malfunction; 13. Direct measurement of strain in tendons; 14. Creep (CR) and shrinkage (SH), predicted at 40 years; 15. Losses averaged for both girders, possible excessive SH due to temperature; 16. 0.25 in. (6.4 mm) diameter three-strand AMGAL galvanized oceanographic rope (f_s = 277,000 psi [1910 MPa] at failure); 17. Relaxation not included, high shrinkage due to high surface area to cross section area ratio; 18. Specified concrete strength; 19. Losses averaged for four beams.

is within the range of estimated lump-sum losses of 25,000 to 50,000 psi (172 to 345 MPa) given in PCI (2010).

Wide scatter of these measurements is partially due to variation in measurement techniques and the length of time over which measurements were taken. It is also an indication that variations in material properties, design characteristics, and exposure conditions ultimately lead to prestress losses that are naturally variable and almost impossible to predict with either great precision or accuracy.

CHAPTER 4—INITIAL LOSSES

4.1—Scope

Chapter 4 addresses initial losses, sometimes called short-term, instantaneous, or immediate losses, for pretensioned and post-tensioned members. These losses include friction and seating losses during tensioning operations, and elastic shortening losses that occur when the prestressing force is transferred to concrete. Instantaneous changes in prestress force caused by applied loads are discussed later in this chapter.

Friction and seating losses are discussed for pretensioned and post-tensioned members. Basic friction concepts due to curvature and wobble are covered. The friction and seating losses that occur during the stressing operations in a precast plant are the responsibility of the precaster. Precasters understand the sources and magnitudes of the losses and compensate for them during initial stressing, as described in 4.2. The designer does not typically need to consider these losses for pretensioned members. For post-tensioned members, friction and seating loss calculations are required. The differences between unbonded, bonded, single-strand, and multi-strand tendons are shown in 4.4, along with recommended friction and wobble coefficients specific to the type of duct used. Seating losses for wedge anchorages and bar nuts are also discussed in 4.4.

Also addressed are changes in prestress force that result from elastic deformations of a prestressed concrete member. The discussion is limited to ranges of member behavior in which the concrete and prestressed and nonprestressed reinforcement exhibit linearly elastic material behavior. For the purposes of estimating initial prestress losses, this is assumed when the member is designed to satisfy the ACI 318 prestressing steel stress limits and concrete stress limits at prestress transfer.

Sections 4.2 and 4.3 discuss losses before and at transfer in pretensioned members, while 4.4 and 4.5 present these losses for post-tensioned members. Section 4.6 discusses subsequent changes to the effective prestress due to externally applied loads.

4.2—Pretensioning losses before transfer

In pretensioned construction, the prestressing takes place in a precasting facility and the producer is usually responsible for providing the correct prestress force immediately before transfer. The licensed design professional has only limited information regarding fabrication procedures.

During stressing operations and concrete curing, strand stress varies. Sources of stress variation are discussed herein, along with strand elongation, which is typically measured to ensure that the strands have the expected level of stress before prestress transfer.

4.2.1 *Anchor set*—Anchor set is the movement of strands within the anchors as the strands are stressed and anchored. Wedges in the anchorage typically grip the strand and move forward slightly as the jaws seat in the tapered wedge cavity.

Stressing-end anchor set occurs during single-strand tensioning as the final jacking force is transferred from the stressing device to the anchorage. A measurable decrease in strand elongation occurs as the wedges move into the tapered wedge cavity, and the strand stress decreases accordingly. The change in prestress due to anchor set Δf_{pA} is computed

$$\Delta f_{pA} = E_p \Delta\varepsilon_{pA} = E_p \frac{\Delta s}{L} \tag{4.2.1}$$

where E_p is modulus of elasticity of prestressed steel, psi (MPa); $\Delta\varepsilon_{pA}$ is change in strain in prestressed reinforcement due to anchor set, in./in. (mm/mm); Δs is anchor set, in. (mm); and L is strand length anchorage to anchorage, ft or in. (m or mm).

The decrease in elongation associated with this anchor set typically ranges from 1/8 to 3/8 in. (3 to 10 mm). For a specific anchor set value, the magnitude of the prestress loss due to anchor set decreases with increasing length between anchorages. In some pretensioning operations, the prestressing beds are long enough, which is commonly 300 to 500 ft (90 to 150 m), that the anchorage seating losses are relatively small. Seating losses, however, will be larger in shorter prestressing beds. Additional forces are applied during stressing to compensate for the anticipated seating loss effect.

In the case of multiple-strand tensioning operations, the stressing-end wedges are seated in the anchorage as the force is applied. Therefore, no loss of prestress force occurs relative to the measured value. Likewise, seating loss for fixed-end anchorages and splicing anchorages also occurs during tensioning, and there is no loss in prestress force relative to the measured value.

Fig. 4.2.2.1a—Self-stressing form for a double tee.

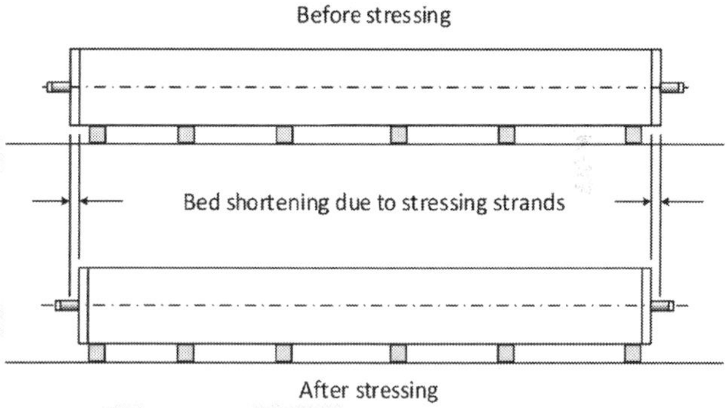

Fig. 4.2.2.1b—Self-stressing form shortening (adapted from the Precast/Prestressed Concrete Institute (PCI 2013).

4.2.2 *Form or abutment deformations*—Two strand-restraint systems—self-stressing forms and fixed-abutment beds—are considered separately in this section.

4.2.2.1 *Self-stressing forms*—Self-stressing steel forms have strand anchorages bearing against plates that transfer the jacking force directly to the forms. Therefore, form shortening occurs during the tensioning operation. Figure 4.2.2.1a shows a self-stressing form.

The form shortens incrementally as each strand is stressed. This shortening, which is illustrated in Fig. 4.2.2.1b, might not be uniform for each strand due to flexure of the form, stressing sequence, and friction between the form and its supports.

If strands are stressed sequentially, the stress in each strand will decrease when subsequent strands are stressed. The amount of form shortening that takes place should be measured periodically for different numbers of strands in common configurations. The initial tensioning force may be increased to compensate for the anticipated form-shortening effect on the average strand.

Fig. 4.2.2.2a—Fixed abutments.

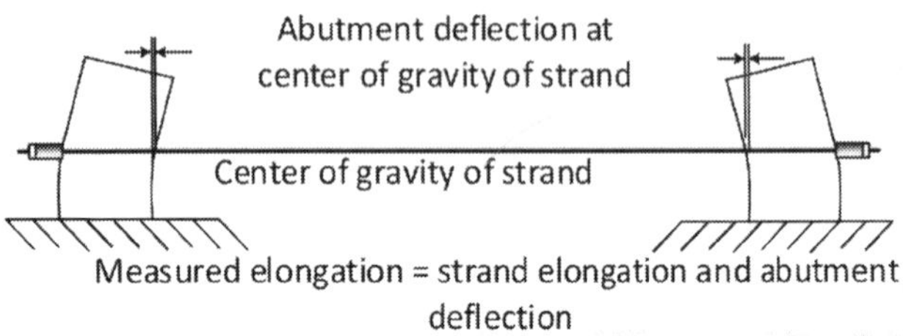

Fig. 4.2.2.2b—Fixed-abutment deformations.

4.2.2.2 *Fixed-abutment beds*—Abutment movement or movement of anchorages occurs during tensioning of fixed-abutment beds. Figure 4.2.2.2a shows two fixed abutments.

Even large, well-designed abutments will rotate and deflect slightly when loaded, as illustrated in Fig. 4.2.2.2b. Although the movement is small, it should be monitored and evaluated for strand patterns commonly used. The influence of this movement decreases with increasing bed length. As with self-stressing forms, if strands are tensioned sequentially, the tensioning force may need to be increased to compensate for the effects of abutment movement on previously stressed strands.

4.2.3 *Elongation calculation and correction*—A prestressing strand's length increases in proportion to the applied tension. This increase in length is called elongation. Measured strand elongations are typically compared to expected values to verify the prestress force before transfer. Elongation is commonly discussed using three terms: basic, gross, and net.

The basic elongation e_b is the length the strand is expected to elongate based on the following factors only:

a) Applied tension force P, lb (N)

b) Strand length from anchorage to anchorage, L, in. (mm)

c) Strand cross-sectional area A_{ps}, in.^{2} (mm^2)

d) Strand modulus of elasticity, E_p, psi (MPa)

ACI 318-11 and AASHTO (2012) design specification stress limits ensure that the prestressing reinforcement remains in the elastic range during tensioning. Therefore, Eq. (4.2.3) for calculating the basic elongation of straight strands is

$$e_b = \frac{PL}{A_{ps}E_p} \qquad (4.2.3)$$

A more complex formulation may be warranted for harped (draped) strands.

In a pretensioning operation, an initial force removes slack from the strands during installation, and the elongation measurements follow after the initial force is applied. The applied tension force for elongation computation, therefore, is determined by subtracting the initial force from the final force measured during tensioning. Strand length is the length between the stressing plates on which the strand chucks bear at each end of the casting bed. Strand cross-sectional area and modulus of elasticity are taken from the strand manufacturer's mill report for each pack of strand.

Strand area and modulus of elasticity may vary for different heat numbers, which identify specific production runs as well as manufacturers. Average values of area and modulus of elasticity for different packs of strand may be used for elongation calculations, provided they vary less than 2.5 percent for all strands in use. Otherwise, the elongation is recomputed using the actual strand properties.

Elongation is typically measured as the displacement of a mark on the strand relative to a reference point on the stressing-end prestressing chuck or anchor. The elongation measured before stressing-end anchorage seating is known as gross elongation. Net elongation is the elongation measured after anchorage seating. Gross and net elongation are used to check strand tension before and after seating loss, respectively.

Because elongation is measured at the stressing-end anchorage, measured elongations may include operational components that do not reflect actual strand elongation, such as strand seating displacements at fixed-end or splicing anchorages, or deformation of the abutments or self-stressing forms during stressing. The characteristics of each pretensioning setup are determined and appropriate compensations made when calculating the expected elongation values.

4.2.4 *Thermal effects*

4.2.4.1 *Stress changes from thermal effects*—When strands are anchored between fixed abutments, changes in strand temperature after tensioning change the strand force. Strands tensioned on a cold morning will lose force when warmer concrete raises the strand temperature because the length between anchorages is not affected by the concrete temperature. The opposite effect could occur for strands tensioned on a hot afternoon. The strand stress can decrease approximately 1000 psi (6.89 MPa) for a differential temperature increase of 5°F (2.8°C). Thermal stress changes will not occur with self-stressing forms because the form will expand and contract with temperature changes similar to the strand.

Several researchers have noted that the heat of hydration during curing of a pretensioned beam may affect the force in the prestressing steel (Gross 1998; Barr et al. 2000).

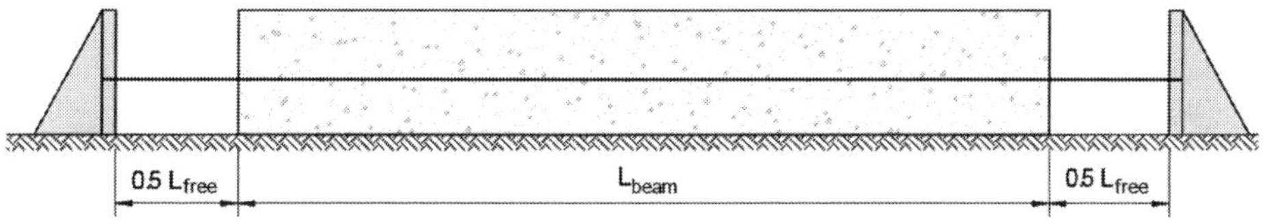

Fig. 4.2.4.2—Beam in casting bed.

Because the strands are anchored in place at the abutments, they have a constant length from the time they are anchored until they are released. If the strands, under a constant strain, are heated by the hydrating concrete, they will experience a loss in stress. The loss is difficult to quantify because it depends on when the concrete and steel begin to bond. After the concrete has reached its peak temperature, the concrete and steel begin to cool. However, because the strand contraction is restrained by the abutments, increased tensile stresses will develop in the strand, and tension may develop in the concrete beam. The gain in tension in the strand outside the beam upon cooling is not the same as the loss of tension during heating, because the steel is bonded to the concrete. However, all changes in stresses that occur due to temperature changes after the concrete and steel are fully bonded are recovered after the strands are cut. After transfer of the prestress, temperature changes will have a small influence on the prestress force because the concrete and steel will expand and contract as a unit. There is a very small difference in the coefficient of thermal expansion between the concrete and steel. The example problem given in 8.4 indicates that these thermal effects are not a significant source of loss.

4.2.4.2 *Example of beam with concentric tendon*—Consider the simple case of a concentric tendon in a pretensioned beam as shown in Fig. 4.2.4.2. For this discussion, assume that the beam is steam cured overnight, with the steam shut off at some point to allow the beam to cool before the strands are cut.

Assuming the abutments are completely fixed, if the temperature rises ΔT_1 along the entire tendon length while there is no bond between the strand and the concrete, the tension in the tendon will relax by $\Delta T_1 \alpha_{ps} E_p$, where α_{ps} is the coefficient of thermal expansion of the prestressing steel and E_p is the modulus of elasticity.

To simplify the analysis, assume that when the concrete reaches its peak temperature, which is typically maintained for several hours, perfect bond develops between the concrete and steel, and the concrete develops strength and stiffness, with a modulus of elasticity of E_c. For this discussion, another simplifying assumption is that the strand segments outside the beam and inside the beam are the same temperature.

After the steam is shut off, the tarp is removed, or forms are stripped, the beam and strand are exposed to ambient temperature, and the steel and concrete will cool before prestress transfer by some amount, ΔT_2. As the concrete and steel cool and try to shorten, the contraction is restrained, so the strand stress becomes more tensile. There will be a change in force, ΔP, which is the same in the free length of strand as it is in the beam. Within the beam, however, a portion of this force is carried by the concrete, ΔP_c, and the remainder is carried by the prestressing steel, ΔP_{ps}. Strain compatibility should be maintained between the concrete and prestressing steel within the beam, so the following Eq. (4.2.4.2a) and (4.2.4.2b) are written.

$$\Delta\varepsilon_p = \Delta\varepsilon_c \quad (4.2.4.2a)$$

$$\Delta T_2 \alpha_{ps} - \frac{\Delta P_{ps}}{E_p A_{ps}} = \Delta T_2 \alpha_c - \frac{\Delta P_c}{E_c A_c} \quad (4.2.4.2b)$$

Equation (4.2.4.2b) is written assuming that the force that develops is tensile. The other relationship that can be written is the total length change of the tendon is zero

$$\Delta\varepsilon_p L_{beam} + \Delta\varepsilon_{free} L_{free} = 0 \quad (4.2.4.2c)$$

where $\Delta\varepsilon_{free}$ is the strain in the free tendon length, and L_{free} is the total free tendon length. The strain in the free length is calculated as Eq. (4.2.4.2d)

$$\Delta\varepsilon_{free} = \Delta T_2 \alpha_{ps} - \frac{\Delta P}{E_p A_{ps}} \quad (4.2.4.2d)$$

When performing example calculations, assuming that the temperature rise and temperature fall are the same, the strand outside the beam will experience a net increase in tension and the concrete will have a net tensile stress as the composite beam resists the restraint forces. Upon release, the restraint forces are removed, and the prestress force at transfer is lower than the original jacking force due to the loss caused by increased temperatures between tensioning and bond on the fixed length bed. Precasters have reported the effects of the tensile stress in the concrete prior to release as causing occasional vertical cracking in prestressed girders before release of the strands (Gross 1998). Typically, these cracks close at prestress transfer.

Based on the prior analysis, concrete has a tensile stress before transfer, and the strand outside the beam has a higher tensile force than the original jacking force.

With these relationships, a rough idea of strand and concrete stresses are established. Many factors, however, complicate the quantification of these stresses. They include the variable nature of early-age concrete modulus of elasticity and the coefficient of thermal expansion of the young concrete. While the cooling occurs over a period of several hours, the modulus and coefficient of thermal expansion are changing and there will be restraint of thermal movements by the formwork; both are difficult to quantify. Finally, prior discussion used a concentric tendon. The tendon's eccentricity will also affect result.

In an analysis, Gross (1998) used a similar method to determine effects of temperature during casting. However, rather than trying to quantify the effect of the developing stiffness of the concrete, a factor was calibrated that accounted for the concrete's contribution to the stiffness of the embedded strand.

4.2.5 *Steel relaxation*—Relaxation is a time-dependent property of the prestressing steel, which results in a loss of stress under a constant strain. Relaxation begins immediately after stressing, so it should be considered in the calculation of losses prior to transfer. The following equation, Eq. (4.2.5), is widely accepted for calculating relaxation losses for low-relaxation strands

$$f_{ps}(t) = f_{po}\left[1 - \frac{\log t}{45}\left(\frac{f_{po}}{f_{py}} - 0.55\right)\right] \quad (4.2.5)$$

where t is duration of load, in hours; f_{po} is initial strand stress after jacking and seating, psi (MPa); and f_{py} is yield stress of strand, psi (MPa).

If concrete is cast soon after stressing, and the force is released within 24 hours, the relaxation loss will be relatively small (~1 percent) and can be compensated with an initial overstress. Producers might account for this, but designers would generally not consider this component of the relaxation loss.

4.3—Elastic shortening losses in pretensioned members

Any length change of prestressed reinforcement results in a corresponding change in the effective prestress force. During prestress transfer in a pretensioned member, the prestressing strands shorten, as illustrated in Fig. 4.3, as the surrounding concrete is compressed until equilibrium is achieved. The strain decrease in the strand is accompanied by a corresponding decrease in the strand tension. The resulting change in effective prestress is known as the elastic shortening loss.

Reasonably accurate estimation of elastic shortening prestress loss is important for two reasons:

1. Stresses in the concrete and prestressed reinforcement immediately after transfer should satisfy code-prescribed limits. These limits effectively constrain the member size, prestress amount, and concrete strength at transfer.

2. The instantaneous and time-dependent creep responses of prestressed concrete members are strongly dependent on the initial elastic shortening and state of stress at transfer. Thus, an error in the estimation of the elastic shortening loss can result in magnified errors in predicted camber and long-term prestress losses.

Members with larger concrete precompression stresses, such as bridge girders, usually experience larger elastic shortening losses than more lightly prestressed members. Prestress losses attributable to elastic shortening typically range from 4 to 10 percent of the prestress force before

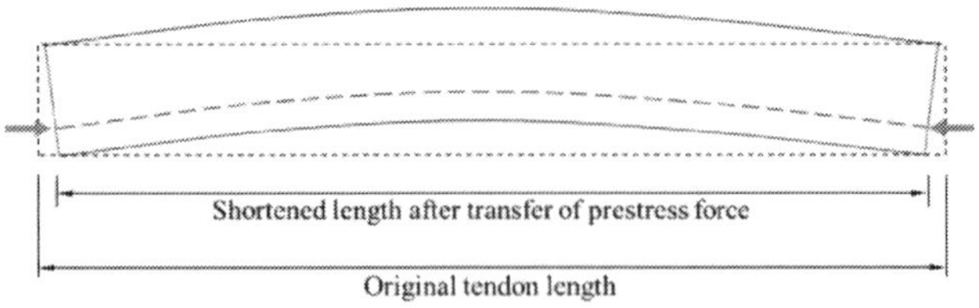

Fig. 4.3—Elastic shortening of prestressed reinforcement during prestress transfer.

transfer in fully pretensioned members—for example, ACI 318 Class U or Class T flexural members. Higher losses are possible in members with lower concrete stiffness, such as those constructed with lightweight concrete. Average elastic shortening losses in post-tensioned members are usually smaller and depend heavily on the sequence of stressing operations.

4.3.1 *Transformed-section approach*—Computation of stresses in concrete and prestressing steel immediately after transfer is straightforward using the transformed-section approach. Assuming that: a) there is a linearly elastic material behavior; b) there is a perfect bond between concrete and reinforcement; and c) the plane sections remain plane, the following relationships (Eq. (4.3.1a) and (4.3.1b)) result from satisfying axial force and moment equilibrium on a cross section immediately after transfer.

$$f_{ci} = -\frac{f_{pbt}A_{ps}}{A_{tr}} - \frac{f_{pbt}A_{ps}e_{tr}y_{tr}}{I_{tr}} + \frac{My_{tr}}{I_{tr}} \quad (4.3.1a)$$

and

$$f_{pt} = f_{pbt} - n_p\left(\frac{A_{ps}}{A_{tr}} + \frac{e_{tr}^2 A_{ps}}{I_{tr}}\right)f_{pbt} + n_p\frac{Me_{tr}}{I_{tr}} \quad (4.3.1b)$$

where f_{ci} is concrete compressive stress immediately after transfer at fiber under investigation, psi (MPa); f_{pbt} is stress in prestressing steel immediately before transfer, psi (MPa); f_{pt} is stress in prestressing steel immediately after transfer, psi (MPa); A_{ps} is area of prestressed reinforcement, in.^{2} (mm^2); A_{tr} is transformed area of cross section (transforming reinforcement area to concrete area of equivalent stiffness), in.^{2} (mm^2) (4.3.2); I_{tr} is second moment of transformed area of cross section (moment of inertia), in.^{4} (mm^4); E_{tr} is eccentricity of prestress force with respect to centroid of transformed area, in. (mm); Y_{tr} is distance from centroid of transformed section to concrete fiber under investigation, in. (mm); $n_p = E_p/E_{ci}$ = modular ratio of prestressing reinforcement with respect to concrete at transfer; E_p is modulus of elasticity of prestressed reinforcement, psi (MPa); E_{ci} is modulus of elasticity of concrete at time of prestress transfer, psi (MPa); and M is bending moment experienced by cross section immediately after transfer (usually due to self-weight), in.-lb (N-mm).

For continuous members, the total prestress moment (primary plus secondary) replaces $f_{pbt}A_{ps}e_{tr}$ in the second term of Eq. (4.3.1a).

Equation (4.3.1a) computes the concrete stress at the level of the prestressing steel immediately after transfer, f_{cir}, by setting y_{tr} equal to e_{tr}. Equation (4.3.1b) can then be simplified to yield Eq. (4.3.1c)

$$f_{pt} = f_{pbt} + n_p f_{cir} \tag{4.3.1c}$$

where f_{cir} is concrete stress at the level of the prestressing steel immediately after transfer, psi (MPa).

Although these equations use the prestress level immediately before transfer, the stresses that result from Eq. (4.3.1a) through (4.3.1c) represent stresses immediately after transfer. Thus, when using the transformed-section approach, the post-transfer concrete stresses are computed directly from the pretransfer prestress force. Accordingly, the prestress loss attributable to elastic shortening, Δf_{pES}, is simply the difference between f_{pt} and f_{pbt}, as shown in Eq. (4.3.1d) and (4.3.1e).

$$\Delta f_{pES} = -n_p\left(\frac{A_{ps}}{A_{tr}} + \frac{e_{tr}^2 A_{ps}}{I_{tr}}\right) f_{pbt} + n_p \frac{Me_{tr}}{I_{tr}} \tag{4.3.1d}$$

or

$$\Delta f_{pES} = n_p f_{cir} \tag{4.3.1e}$$

The first two terms in Eq. (4.3.1d) represent the effect of strand shortening as the concrete is compressed; the third term represents the opposing effect of the self-weight moment on the cross section.

4.3.2 *Gross-section approximation*—If computation of transformed-section properties is considered overly burdensome, the elastic shortening loss can be estimated with reasonable accuracy using gross-section properties. This is specifically true for lightly prestressed members. For linearly elastic material behavior and perfect bond between reinforcement and concrete, the change in prestress during transfer is computed by multiplying concrete stress at the level of the prestressing steel after transfer (f_{cir}) by the modular ratio (n_p), as indicated in Eq. (4.3.1e).

The concrete stress f_{cir} is computed using gross-section properties, but prior knowledge of the effective prestress force after transfer is required. The *PCI Design Handbook* (PCI 2010) simplifies the method using an approximate version of this technique. A single iteration of the calculation in Eq. (4.3.1e) is performed after an initial assumption of a 10 percent prestress loss due to elastic shortening. This assumption is embodied in the handbook's value of $K_{cir} = 0.9$ for pretensioned members. Equations (4.3.2a) and (4.3.2b) are used to calculate elastic shortening loss

$$\Delta f_{pES} = \frac{E_p}{E_{ci}} f_{cir} \tag{4.3.2a}$$

$$f_{cir} - K_{cir}\left(\frac{P_i}{A_g} + \frac{P_i e_p^2}{I_g}\right) - \frac{M_g e_p}{I_g} \tag{4.3.2b}$$

where P_i is initial prestress force (after anchorage seating loss), lb (N); e_p is eccentricity of prestressed reinforcement with respect to the gross section, in. (mm); A_g is area of the gross cross section, in.^{2} (mm^2); I_g is second moment of area of gross cross section (moment of inertia), in.^{4} (mm^4); and M_g is bending moment due to dead weight of prestressed member and any other permanent loads in place at the time of stressing, in.-lb (N-mm).

4.3.3 *Iterative gross-section approach*—A more exact value of the elastic shortening loss is obtained using gross-section properties when an iterative solution technique is employed. A trial value of elastic shortening loss is assumed and used to compute f_{cir}. After Eq. (4.3.1e) is used to update elastic shortening loss, the designer can evaluate the accuracy of the initial estimate and reiterate if desired. Two or three iterations are usually adequate to get a reasonable level of accuracy. Alternatively, a closed-form solution for directly computing the elastic shortening loss with gross-section properties is employed to condense repeated iterations to a single step.

$$\Delta f_{pES} = \frac{e_p M_g A_g - A_{ps} f_{pbt}\left(I_g + e_p{}^2 A_g\right)}{A_{ps}\left(I_g + e_p{}^2 A_g\right) + \dfrac{A_g I_g}{n_p}} \tag{4.3.3}$$

Equation (4.3.3) is functionally equivalent to the expression recommended in the AASHTO LRFD (2011) commentary.

Analysis with gross-section properties results in an elastic shortening approximation that is close to the transformed-section value if: a) the closed-form solution of Eq. (4.3.3) is employed; or b) an iterative approach is pursued until convergence is achieved. If the iterative approach is used with net section properties, the solution will nearly match the transformed section properties solution. However, accuracy of the single-iteration approximation of the *PCI Design Handbook* (PCI 2010) depends on how closely the assumed elastic shortening loss of 10 percent matches the final calculated value.

The example pretensioned double-tee beam in 8.1 provides a basis for comparison between the methods. Transformed-section analysis (8.1.2) results in a computed elastic shortening loss of 15,700 psi (108 MPa), which is approximately 7.8 percent of the pretransfer stress. Application of Eq. (4.3.3) using gross-section properties yields a computed loss of 15,550 psi (108 MPa) (7.7 percent). The single-iteration approximation from the *PCI Design Handbook* (PCI 2010) simplified method using $K_{cir} = 0.9$ is 14,900 psi (103 MPa) (7.4 percent). For this example, the relative error among these three computational techniques is less than 0.5 percent of the initial prestress level, which is small compared to the uncertainty associated with the actual material properties. In general, the small relative error between transformed-section (and iterative with net section) and gross-section analyses will increase as the ratio of bonded reinforcement to member cross-sectional area increases.

4.3.4 *Variation of elastic shortening loss along the member length*—The bending moment due to self-weight varies along the member length. Therefore, the magnitude of the elastic shortening loss is not uniform along the pretensioned member. Variation in the eccentricity or quantity of the bonded prestressed reinforcement is often employed to control concrete stresses after transfer. ACI 318 requires that member sections satisfy the prescribed stress limits. It is usually adequate to compute elastic shortening losses at cross sections that are likely to be critical for concrete stress checks immediately after transfer or under service loads. If little variation in the effective prestress force is expected along the member, it is common practice to use the elastic shortening loss computed for the midspan cross section. In members with single-point strand draping and a parabolic moment diagram from gravity loads, the cross section located at 40 percent of the simple span is often selected. End regions are often critical for concrete stress checks immediately after transfer.

4.3.5 *Sensitivity to concrete stiffness*—Inspection of Eq. (4.3.1d) indicates that the accuracy of computed elastic shortening loss is dependent on accurate estimation of the concrete modulus of elasticity at prestress transfer (E_{ci}). The elastic shortening loss depends directly on concrete material stiffness, a property that is also one of the most uncertain of the parameters involved in computing this loss. As discussed in 7.4, actual measured E_{ci} values can vary greatly from values predicted in accordance with simple, code-prescribed equations; therefore, codes often allow the use of mixture-specific stiffness values if they are known. Knowledge of the stiffness characteristics of commonly used aggregates and concrete mixtures in a region are used to improve the accuracy of the E_{ci} value assumed for design. Even when the relationship between concrete compressive strength and E_{ci} is well established, variability is often experienced because measured concrete compressive strengths usually exceed specified values to reduce the time that the prestressed element remains in the casting bed.

4.4—Post-tensioning losses during tensioning and transfer

4.4.1 *Initial losses and tendon types*—The initial losses in post-tensioning (PT) tendons are friction, anchor set, and elastic shortening. Although there are many types of PT tendons, only the most commonly used ones are discussed herein. The stressing sequence affects the initial losses of all tendons.

Unbonded single-strand tendons are coated with PT coating and sheathed with extruded high-density polyethylene (HDPE) sheathing. The PT coating lubricates the strand that can move freely within the sheathing, which reduces friction. These tendons are stressed with a single-strand jack and the elongations measured after anchor set and after the stressing jack has been removed from the tendon.

Multistrand tendons typically consist of multiple bare strands in a metal or plastic duct. The strands are stressed together with a center-hole multistrand jack; however, the elongations are measured before anchor set. Multistrand tendons are normally grouted with cementitious grout after stressing and elongation verification, and are often called bonded tendons.

Bar tendons typically consist of a single bar in a metal or plastic duct. The bar is stressed with a bar jack using a coupler and a pulling bar. The elongation is typically measured after anchor set, or the jacking force is verified with a lift-off test. Bar tendons are grouted after stressing.

4.4.2 *Friction losses*—During stressing, the prestressing force along a tendon decreases from its maximum value at the stressing end to a minimum at the fixed end. This decrease is due to friction losses between the prestressing strand and its sheathing or duct due to the intended tendon curvature and unintended tendon curvature, which is also called wobble. Once the jacking force reaches its prescribed value, the tendon is released and anchored at the stressing end. Locking the tendon is achieved in most cases by the engagement of serrated wedges, which are activated as the strand retracts and draws the wedges into a conical wedge cavity in the anchor or a wedge plate. Therefore, during stressing, there are two components of initial stress loss: one from friction and the other from anchor set. An example of a final stress diagram of a tendon with one stressing end immediately after anchor set is shown in Fig. 4.4.2a. The slope of the curve, $-s$, between f_{jack} and f_L, represents the average friction intensity. The triangular area between f_{jack}, f_{anchor}, and f_{max} divided by E_p represents the loss of elongation due to the anchor set. The curve between f_{anchor}, f_{max}, and f_L represents the strand stress along the length after anchor set. The area under this curve divided by E_p is the net tendon elongation after anchor set.

The difference in strand stress between the stressing end and the fixed end increases with increasing tendon length, greater friction coefficient, and larger angular change in the tendon profile.

In typical building construction with unbonded single-strand tendons, the slope of the stress curve will be relatively flat as the friction is small. For normal-length tendons, the average tendon stress is assumed to act at all critical sections along the length. For long tendons, over 120 ft (37 m) in length when stressed from one end only, or over 240 ft (73 m) in length when stressed from both sides, additional tendons may be required to provide the total prestressing force required at a given location in the member. For short tendons less than 40 ft (12 m) in length, the effect of the anchor set can be significant and should be considered as discussed in 4.4.3.

The stress distribution for a tendon stressed from both ends is illustrated in Fig. 4.4.2b. It is not necessary to stress the tendon at both ends simultaneously. The stressing at the second end typically takes place after stressing of the first end is completed.

As shown in Fig. 4.4.2b, when the resulting stress distribution is symmetrical when the tendon is stressed from both ends, the average friction is also symmetrical. Figure 4.4.2b also illustrates that the elongation before anchor set obtained from the first-end stressing, proportional to the

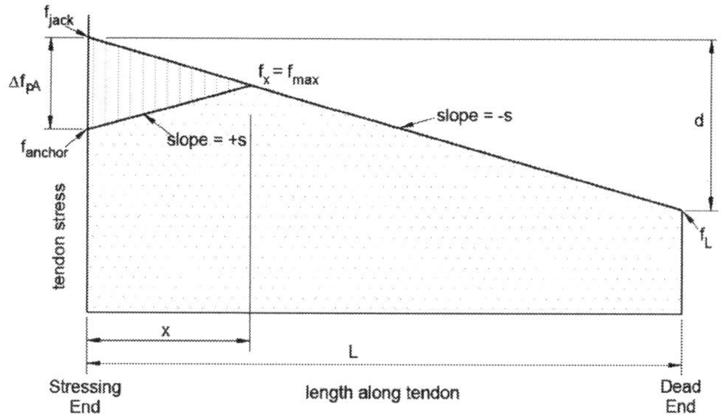

Fig. 4.4.2a—Stress distribution in tendons with one stressing end.

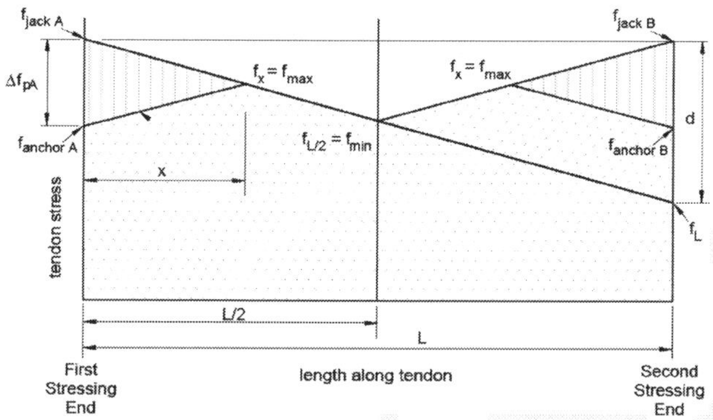

Fig. 4.4.2b—Stress distribution in tendons with two stressing ends.

area under the stress curve, f_{jackA} and f_L, will be much larger than the remaining elongation obtained from the second-end stressing, proportional to the area between $f_{L/2}$, f_{jackB}, and f_L.

Bonded multistrand tendons are usually used in bridges with deeper members, resulting in a greater angular deviation. They are also used in buildings, in transfer girders, in beams, in slabs, and also in other structures such as water tanks. As the friction increases over the supports where the angular deviation is larger, each tendon is analyzed separately so that the actual force in each span and at each critical section is evaluated. Figure 4.4.2c shows the stress diagram with variable friction intensity. In bonded tendons, no force equalization occurs, as the tendons are typically grouted shortly after stressing and an average force for the tendons is typically not used.

Friction losses are subdivided into intended tendon curvature, represented by a curvature friction coefficient μ, and the unintended tendon curvature, represented by a wobble friction coefficient k. The stiffness of the duct contributes to a lower wobble friction, and a more flexible duct could contribute to increased wobble friction. The typically stiffer metal ducts have a larger maximum support spacing requirement than the more flexible plastic ducts. The effect of friction on the force along the tendon is calculated according to the formula

$$T_x = T_o e^{-(\mu\alpha + kx)} \tag{4.4.2}$$

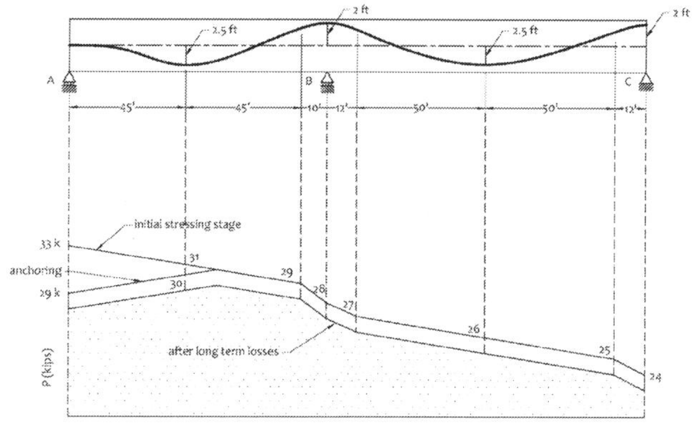

Fig. 4.4.2c—Stress distribution in tendons reflecting variable friction intensity. (Note: 1 ft = 0.305 m; 1 kip = 4.45 kN.)

where T_x is prestressing force at point x, lb (N); T_o is prestressing force at stressing end, lb (N); e is base of Naperian logarithms; μ is curvature friction coefficient (Table 4.4.2); α is total angular change in radians from jacking end to point x, rad; k is wobble friction coefficient per unit length of tendon (Table 4.4.2); and x is length of tendon from stressing end to point x, ft (m).

When different friction intensities exist along the tendon, tendon lengths with constant friction between two points are considered separately. Examples include high point to inflection point and inflection point to low point. Computer software is available to generate prestressing forces at all locations along the tendon path. The effect of the anchor set is discussed in 4.4.3.

Friction coefficients are specific to tendon type and configuration. For unbonded single-strand tendons, the type of PT coating and characteristics of the extruded plastic sheathing determine the friction values. As the PT coating and extruded plastic sheathing materials improve in quality and consistency, predictability of the friction values increases in accuracy. For bonded multistrand tendons, the type of duct is of great importance. For example, plastic ducts typically have lower friction values than corrugated metal ducts. Recommended friction coefficients are shown in Table 4.4.2 from the Post-Tensioning Institute's guide specifications (PTI M50.3-12 Table 5.1).

Bar tendons are typically straight and relatively short with no friction along the tendon length as long as the bar is not touching the duct.

4.4.3 *Anchor set*—A typical value for the anchor set is 1/4 in. (6 mm) when stressing equipment with a power seating device is used. When single-strand tendons are stressed with jacks without a power seating device, anchor setup to 3/8 in. (19 mm) is common

The effect of anchor set on tendon stresses is calculated in accordance with the formulas that follow. The anchor set is computed as the integration of the stress from the jacking point to the anchor set influence distance divided by the modulus of elasticity.

$$\Delta s = \frac{\int(\text{final stress} - \text{initial stress})dx}{E_p} \tag{4.4.3a}$$

Table 4.4.2—Recommended friction coefficients

Type of prestressing steel	Corrugated metal duct		Corrugated plastic duct		Smooth steel pipe		Smooth plastic pipe		No duct plastic sheathing	
	μ	k, ft^{-1} (m^{-1})	μ	k, ft^{-1} (m^{-1})	μ	k, ft^{-1} (m^{-1})	μ	k, ft^{-1} (m^{-1})	μ	k, ft^{-1} (m^{-1})
Strand	0.15 to 0.25	0.00005 to 0.0003 (0.0002 to 0.0010)	0.10 to 0.14	0.00005to 0.0003 (0.0002 to 0.0010)	0.25 to 0.30	0	0.10 to 0.14	0		
Strand in precast elements and constant curvature tendons	0.15 to 0.25	0.00005 to 0.0003 (0.0002 to 0.0010)	0.10 to 0.14	0.00005to 0.0003 (0.0002 to 0.0010)						
External tendons, bare dry strand					0.25 to 0.30	0	0.12 to 0.15	0		
Lubricated strand	0.12 to 0.18	0.00005 to 0.0003 (0.0002 to 0.0010)			0.20 to 0.25	0				
Strand coated and extruded*	0.01 to 0.05	0.00005 to 0.0003 (0.0002 to 0.0010)	0.01 to 0.05	0.00005 to 0.0003 (0.0002 to 0.0010)	0.01 to 0.05	0	0.01 to 0.05	0	0.01 to 0.07	0.00005 to 0.0003 (0.0002 to 0.0010)
Bars, deformed, smooth, and round	0.30	0 to 0.0002 (0 to 0.0007)	0.30	0 to 0.0002 (0 to 0.0007)						

*Post-tensioned coating in accordance with the performance specification (PTI M10.3-00).

$$\Delta s = \frac{\Delta f_{pA} x_s}{2E_p} \quad (4.4.3b)$$

The slope of the stress line represents the friction intensity or the friction over the length. The slope of the stress line after anchor set is equal and opposite to the slope of the jacking stress line. Equation (4.4.3c) only applies when the friction is linear between the two ends (or points under consideration).

$$s = \frac{d}{L} = \frac{(f_{jack} - f_L)}{L} \quad (4.4.3c)$$

$$x_s = \frac{\Delta f_{pA}}{2s} = \frac{\Delta f_{pA} L}{2d} \quad (4.4.3d)$$

$$\Delta f_{pA} = \frac{2 \cdot d \cdot x_s}{L} \quad (4.4.3e)$$

Substituting Eq. (4.4.3e) into Eq. (4.4.3b), and solving for x_s gives

$$x_s = \sqrt{\frac{E_p \Delta s L}{d}} \quad (4.4.3f)$$

If $x > L$, then the loss transfers all the way to the fixed end, as shown in Fig. 4.4.3.

Or, Eq. (4.4.3e) can be substituted into Eq. (4.4.3b) to arrive at Eq. (4.4.3g) for Δf_{pA}

$$\Delta f_{pA} = 2\sqrt{\frac{E_p d \Delta s}{L}} \quad (4.4.3g)$$

where Δf_{pA} is change in stress due to anchor set at the stressing anchor, psi (MPa); d is friction loss in length L, psi (MPa); x_s is length influenced by anchor set, in. (mm); L is strand length from anchorage to anchorage, in. (mm); Δs is anchor set, in. (mm); E_p is modulus of elasticity of prestressed reinforcement, psi (MPa); and s is the slope of the stress-versus-distance line, psi/ft (MPa/m).

When measuring anchor set, the tendon elongation within the jack should be considered.

In shorter tendons, especially when the friction is relatively low, the anchor set can affect the tendon force all the way to the fixed end. Figure 4.4.3 illustrates the stress distribution in such short tendons. The anchor set reduces the strand stress at the fixed end from f_{L1} to f_{L2}. The area under the stress curve in bold (f_{anchor}, f_{L2}) represents the elongation after anchor set. The trapezoidal area between f_{jack}, f_{anchor}, f_{L2}, and f_{L1} represents the loss in elongation due to the anchor set.

The anchor set in bar tendons is typically small and often not measured.

4.4.4 *Elongation*

4.4.4.1 *Elongation calculations*—To calculate the expected elongation, the area under the stress-versus-length plot (Fig. 4.4.2a, 4.4.2b, or 4.4.3) is divided by E_p. Elongations are computed for unbonded, bonded, and bar tendons with the same basic formula

$$\Delta = \frac{P_{avg} L}{A_{ps} E_p} = \frac{f_{p\_avg} L}{E_p} \quad (4.4.4.1)$$

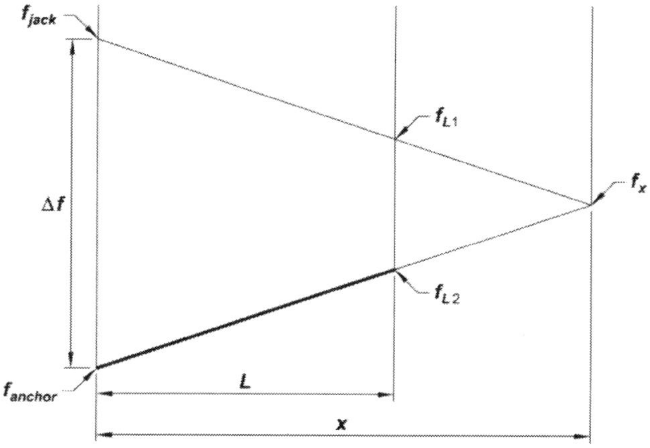

Fig. 4.4.3—Stress distribution in short tendons.

where Δ is elongation, in. (mm); P_{avg} is average force in the tendon, lb (N); f_{p_avg} is the average stress in the tendon, psi (MPa); L is strand length from anchorage to anchorage, in. (mm); A_{ps} is area of prestressing steel, in.^{2} (mm^2); and E_p is modulus of elasticity of prestressed reinforcement, psi (MPa).

The average prestressing force depends on friction losses caused by tendon curvature, intended as well as wobble, and the anchor set.

Values for strand area and modulus of elasticity are verified based on mill certificates of the steel used for the particular tendon if elongation variations are greater than permissible.

4.4.4.2 *Elongation measurements*—The procedures for tendon stressing and field measurements of elongations are in industry standard documents (PTI M10.3-00; PTI C30.4-07).

4.4.4.3 *Items considered for elongations*—In cases of elongations outside of the specified tolerances, several factors should be verified or considered:

(a) Problems, inconsistencies, or both, in tendon marking and measurement of the elongation

(b) Excessive friction due to greater angular deviation, larger friction, or both, than assumed; or excessive wobble caused by placement variations

(c) Excessive anchor set caused by cement paste in the wedge cavity or improperly functioning stressing equipment or power seating device

(d) Variations in steel area, modulus of elasticity, or both, that can be verified on mill certificates for strand supplied to the job

(e) Error in calculations; shorter tendons are affected more than longer ones by inaccuracy in measurements and in a variation in anchor set as these values represent a larger percentage of the overall elongation. The anchor set in short tendons affects the tendon stress to the fixed end

4.4.4.4 *Elongations*—The jacking force is the primary concern when stressing tendons. Calibrated stressing jacks and gauges should be used while the elongation measurements serve as a verification of the jacking force. Because of the many factors affecting elongations, a reasonable tolerance between the jacking force and the measured elongation should be used.

Different codes require specific tolerances for the correlation between the expected and measured elongations.

Table 4.4.4.4—Tolerances for the correlation between the expected and measured elongations in selected codes

Code	Tolerance in elongation Expected versus measured, percent
ACI 318-11	± 7 percent
AASHTO (1989)	± 5 percent on individual tendon based on friction coefficients confirmed with liftoff tests, and material properties of the actual materials used
AASHTO (2011)	± 5 percent for tendon length over 50 ft (15 m) ± 7 percent for tendon length 50 ft (15 m) or less
Eurocode EN 13670:2009	± 5 percent for the total force at a section ± 15 percent for an individual tendon
CEB-FIP model code (Comité Euro-International du Béton 2010)	Tendon L > 50 ft (15 m): ±10 percent for a particular tendon ±5 percent for all tendons in a section Tendon L ≤ 50 ft (15 m): ±15 percent for a particular tendon ±7 percent for all tendons in a section

Table 4.4.4.4 summarizes such tolerances. These tolerance approaches are used as guidance in resolving elongation readings outside the specified tolerances.

As indicated in Table 4.4.4.4, there are different tolerance percentages in codes for:

(a) Long and short tendons

(b) Particular tendons and all tendons in a section or a member

(c) Under elongation and over elongation

4.5—Elastic shortening loss in post-tensioned members

In post-tensioned members, a stressed tendon is elongated at the same time the surrounding concrete is being compressed. Thus, this tendon experiences no elastic shortening during its own stressing operation. Previously anchored tendons, however, shorten with the surrounding concrete during the subsequent stressing of other tendons and, therefore, experience prestress loss due to elastic shortening. The elastic shortening loss in a post-tensioned member depends on the sequence of stressing operations and is usually taken as an average of the losses experienced by the individual tendons.

At the time of stressing, all post-tensioned tendons are unbonded. Because an unbonded tendon can move within its sheathing or duct, it does not undergo the same stress-induced strain changes as the concrete surrounding it for most flexural members. For this reason, the average compressive stress, f_{cpa}, in the concrete is suggested for use in evaluating prestress losses due to elastic shortening. This procedure relates the elastic shortening prestress loss for unbonded tendons to the average strain along the length of the member rather than the strain at the point of maximum moment.

In Zia et al. (1979), elastic shortening losses are calculated as

$$\Delta f_{pES} = K_{es} E_p f_{cpa} / E_{ci} \quad (4.5.1)$$

where f_{cpa} is average compressive concrete stress at the center of gravity of the tendons immediately after the prestress has been applied to the concrete (considerations for computing this average stress are discussed later in this section), psi (MPa). K_{es} = 0.5 for post-tensioned components when tendons are tensioned in sequential order to the same tension and 0 if all tendons are tensioned simultaneously; with other PT procedures, the value for K_{es} may vary from 0 to 0.5. E_p is modulus of elasticity of prestressed reinforcement (approximately 28,500,000 psi [196,500 MPa]).

Computation of elastic shortening losses in post-tensioned members can differ relative to similar computation for pretensioned members. As discussed previously, the elastic shortening loss is estimated as the average loss experienced by all tendons due to all stressing operations. This is accomplished by using a K_{es} value of 0.5 as an averaging factor for post-tensioned members when tendons are tensioned in sequential order to the same tension (Zia et al. 1979). Similarly, the AASHTO (2011) specification recommends an averaging factor of $(N - 1)/2N$, where N is the number of sequentially stressed tendons. In cases where large tendon spacing results in minimal elastic interaction between adjacent tendons, such as some slab systems, the elastic shortening losses in post-tensioned tendons may be further reduced. AASHTO (2011) recommends a further 75 percent reduction of the computed elastic shortening loss in this case; the Zia et al. (1979) simplified method leaves the selection of a reduced K_{es} value somewhere between 0 and 0.5, which is based on the discretion of the practitioner.

When averaging the elastic shortening losses among sequentially stressed tendons, the influence of self-weight bending moment should be considered. For the case of a post-tensioned beam, the self-weight moments are likely to be induced simultaneously with the first tendon stressing operation that lifts the beam from its supports; therefore, they are already compensated for when stressing of this tendon is completed. In this case, the elastic shortening losses that occur due to subsequent stressing operations will be due to prestress forces only.

A second major difference related to computation of elastic shortening effects in post-tensioned members is that the tendons are unbonded at the time of prestress transfer. For this reason, net-section properties are used in the computation of elastic shortening loss and the accompanying concrete stresses after transfer. Net-section properties are calculated by deducting the cross-sectional area of the PT ducts. If a large amount of bonded reinforcement—prestressed or nonprestressed—is included, transformed-section analysis using net-section concrete properties can improve accuracy of elastic computations. This involves transforming the area of bonded reinforcement into an equivalently stiff area of concrete, and deducting the area of open ducts.

Also, because the tendons are unbonded at prestress transfer, shortening of the tendon is not directly related to the concrete strain at the cross section of interest. The elastic shortening of a tendon is a direct function of the relative movement of its two anchors. Thus, for a tendon with concentric anchors, the elastic shortening deformation can be computed directly from the average prestress on the cross section—as is done in the post-tensioned slab example in 8.2. Note that for draped or nonconcentric straight tendons, the average concrete stress along the full length of the tendon at its center of gravity has to be calculated to determine the elastic shortening loss.

Finally, as discussed in 4.4, friction losses result in a prestress force that varies along the length of a post-tensioned member. Therefore, the average prestress force over the length of the member after friction losses can be used when estimating elastic shortening losses in these members.

4.6—Elastic gain under superimposed loads

When superimposed dead loads are added at a later age, the prestress reinforcement can elongate and cause an increase in effective prestress force. This increase is effectively a partial recovery of prestress loss that is sometimes called elastic gain. Superimposed dead loads that may result in elastic gain include cast-in-place toppings on flooring members and decks and traffic barriers on bridge girders. For pretensioned members and post-tensioned members with bonded tendons, the elastic gain is calculated by multiplying the resulting change in concrete stress at the center of gravity of the prestress force by the modular ratio

$$\Delta f_{pEG} = n_p \Delta f_{cgp} \tag{4.6}$$

where Δf_{pEG} is increase in prestress (elastic gain) due to addition of superimposed dead load, psi (MPa); $n_p = E_p/E_c$ = modular ratio of prestressing reinforcement with respect to concrete at age when superimposed load is applied; and Δf_{cgp} is change in concrete stress at center of gravity of the prestressing force due to application of superimposed load, psi (MPa).

Little to no increase in effective prestress force is expected to accompany the addition of superimposed dead loads to PT members with unbonded tendons because the tendon length will change only slightly between anchorages.

CHAPTER 5—LONG-TERM LOSSES: SIMPLIFIED METHOD

5.1—Scope

Long-term prestress losses are caused by the time-dependent properties of concrete and steel, namely, concrete creep and shrinkage and steel relaxation. Determination of long-term prestress loss can involve complicated and laborious procedures because the rate of loss due to one factor, such as relaxation of tendons, is continually altered by changes in stress due to other factors, such as shrinkage and creep of concrete. Rate of creep is, in turn, altered by the change in tendon stress. Many of these factors are further dependent on uncertain material properties, loading time, concrete curing method, environmental conditions, and construction details.

The equations presented within this chapter are intended to reasonably estimate prestress loss from the various time-dependent sources. They are applicable for prestressed members of normal designs with: a) an extreme fiber compressive stress in the precompressed tensile zone under the full dead load condition ranging from 350 to 1750 psi (2.41 and 12.1 MPa); b) minimum concrete compressive strength f_c' of 4000 psi (27.6 MPa); and c) a unit weight of concrete at least 115 lb/ft^3 (1842 kg/m^3) (Zia et al. 1979). For unusual design conditions, consider a more detailed procedure.

Chapter 5 is based on a task group report sponsored by ACI-ASCE Committee 423, Prestressed Concrete (Zia et al. 1979). Simple equations are given to enable the designer to estimate the various sources of prestress loss individually as opposed to using a lump-sum value. These equations are intended for practical design applications and provide fairly realistic values for normal design conditions. Refer to Chapter 6 for unusual design situations and special structures where more detailed analyses may be warranted.

Total losses Δf_{pT} are losses due to friction and seating (Δf_{pFS}), elastic shortening (Δf_{pES}), creep of concrete (Δf_{pCR}), shrinkage of concrete (Δf_{pSH}), and relaxation of tendons (Δf_{pRE}). This can be expressed as Eq. (5.1)

$$\Delta f_{pT} = \Delta f_{pFS} + \Delta f_{pES} + \Delta f_{pCR} + \Delta f_{pSH} + \Delta f_{pRE} \quad (5.1)$$

Friction and seating, as well as elastic shortening, are presented in Chapter 4. Chapter 5 deals only with the long-term losses: Δf_{pCR}, Δf_{pSH}, and Δf_{pRE}.

5.2—Creep of concrete (Δf_{pCR})

Creep of concrete complicates stress-loss calculations. The rate of loss due to creep changes when the concrete stress level changes. Because the stress level is changing constantly throughout the life of the structure, the rate of loss due to creep is also constantly changing.

Part of the initial compressive strain induced in the concrete immediately after transfer is reduced by the tensile strain resulting from the superimposed permanent dead load. Prestress loss due to concrete creep can be assumed to be proportional to the net permanent compressive strain in the concrete at the level of the reinforcement.

For prestressed members made of sand lightweight concrete, there is a significantly larger amount of loss due to elastic shortening of concrete because of its lower modulus of elasticity. This results in an overall reduction in loss due to creep. This effect is accounted for by a 20 percent reduction of the creep coefficient. For members made of all lightweight concrete, special consideration should be given to the properties of the particular lightweight aggregate used.

5.2.1 *Bonded tendons*—Losses due to creep of concrete are calculated as (Zia et al. 1979)

$$\Delta f_{pCR} = K_{cr}(E_p/E_c)(f_{cir} - f_{cds}) \quad (5.2.1a)$$

where K_{cr} is 2.0 for pretensioned normalweight components, 1.6 for pretensioned sand-lightweight components, 1.6 for post-tensioned normalweight components, 1.28 for post-tensioned sand-lightweight components; f_{cir} is net compressive concrete stress at center of gravity of prestressing force immediately after the prestress has been applied to the concrete, psi (MPa) (Eq. 4.3.2b); and f_{cds} is concrete stress at center of gravity of prestressing force due to all superimposed, permanent dead loads that are applied to the member after it has been prestressed, psi (MPa)

$$f_{cds} = \frac{M_{sd} - e_p}{I_g} \quad (5.2.1b)$$

E_c is modulus of elasticity of concrete at 28 days, psi (MPa); M_{sd} is moment due to all superimposed, permanent dead, and sustained loads applied after prestressing, lb-in. (N-mm); and e_p is eccentricity of the center of gravity of the tendons with respect to the centroid of the concrete at the cross section considered, in. (mm).

5.2.2 *Unbonded tendons*—Because an unbonded tendon can move within its sheathing, it does not undergo the same stress-induced strain changes as the concrete surrounding it. For this reason, the average compressive concrete stress at the center of gravity of the tendon, f_{cpa}, is suggested for use in evaluating prestress losses due to creep of concrete. This procedure relates the prestress loss for unbonded tendons due to creep to the average member strain at the center of gravity of the tendon, rather than the strain at the center of gravity of the tendon at the section of maximum moment. The somewhat higher residual tensile stress in an unbonded tendon logically results in somewhat higher loss due to steel relaxation.

$$\Delta f_{pCR} = K_{cr}(E_p/E_c)f_{cpa} \quad (5.2.2)$$

where f_{cpa} is average compressive concrete stress along the member length at the center of gravity of the tendons immediately after the prestress has been applied to the concrete, psi (MPa).

5.3—Concrete shrinkage (Δf_{pSH})

Loss of stress in the tendon due to concrete shrinkage surrounding it is proportional to that part of the shrinkage that takes place after the transfer of prestress force to the concrete.

Shrinkage strain developed in a concrete member is influenced, among other factors, by its volume/surface ratio and the ambient relative humidity. Prestress loss due to shrinkage is the product of the shrinkage that occurs after transfer and the modulus of elasticity of prestressing steel. The factor K_{sh} accounts for the reduction in shrinkage due to delaying the application of the prestressing force.

Note that for some lightweight concrete, the coefficient for the estimation of shrinkage strain may be greater than the value 8.2×10^{-6} used in Eq. (5.3).

Losses due to shrinkage of concrete are calculated:

$$\Delta f_{pSH} = 8.2 \times 10^{-6} K_{sh} E_p (1 = 0.06V/S)(100 - RH) \text{ (for } V/S \text{ in in.)}$$
$$\Delta f_{pSH} = 8.2 \times 10^{-6} K_{sh} E_p (1 = 0.0024V/S)(100 - RH) \text{ (for } V/S \text{ in mm)} \quad (5.3)$$

Table 5.3—Values of K_{sh} for post-tensioned members (PCI 2010)

Time after end of moist curing to application of prestress, days	1	3	5	7	10	20	30	60
K_{sh}	0.92	0.85	0.80	0.77	0.73	0.64	0.58	0.45

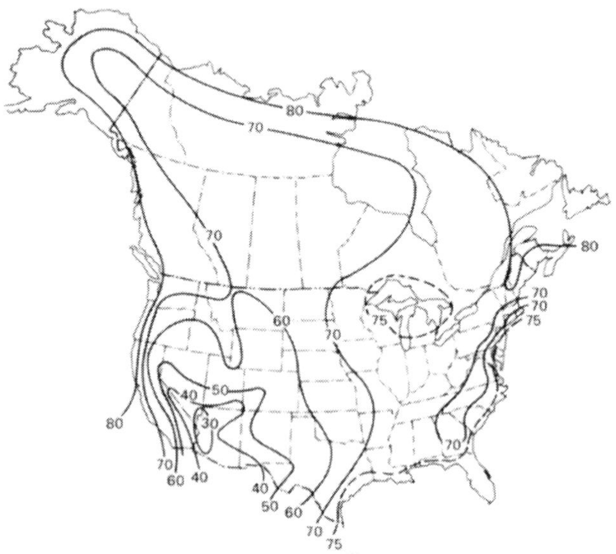

Fig. 5.3—Annual average ambient relative humidity, percent (PCI 2010).

where K_{sh} = 1.0 for pretensioned components—refer to Table 5.3 for post-tensioned components; V/S is volume-to-surface ratio, in. (mm); and RH is average ambient relative humidity, percent (Fig. 5.3).

Except for the K_{sh} term, there is no distinction between pretensioned and post-tensioned members or between those with bonded and unbonded tendons for shrinkage loss calculations.

5.4—Relaxation of tendons (Δf_{pRE})

Relaxation of tendons complicates stress-loss calculations. The rate of loss due to relaxation changes as the steel stress level changes. Because the stress level is changing constantly throughout the life of the structure, particularly in pretensioned members, the rate of loss due to relaxation is also constantly changing.

Relaxation of a prestressing tendon depends on the stress level in the tendon. Basic relaxation values K_{re} for prestressing steel are shown in Table 5.4. Because of other prestress losses, there is a continual reduction of the long-term tendon stress, thus causing a reduction in the rate of relaxation over time. The reduction in tendon stress due to elastic shortening of concrete occurs instantaneously, while the reduction due to creep and shrinkage takes place over a prolonged period of time. The factor J is specified to approximate these effects.

Losses due to relaxation of tendons are calculated in Eq. (5.4)

$$\Delta f_{pRE} = [K_{re} - J(\Delta f_{pSH} + \Delta f_{pCR} + \Delta f_{pES})]C \qquad (5.4)$$

Table 5.4—Values of K_{re} and J (PCI 2010)

Tendon type	K_{re}, psi (MPa)	J
Grade 270, stress-relieved strand or wire	20,000 (138)	0.150
Grade 250, stress-relieved strand or wire	18,500 (128)	0.140
Grade 240 or 235, stress-relieved wire	17,600 (121)	0.130
Grade 270, low-relaxation strand	5000 (34)	0.040
Grade 250, low-relaxation wire	4630 (32)	0.037
Grade 240 or 235, low-relaxation wire	4400 (30)	0.035
Grade 145 or 160, stress-relieved bar	6000 (41)	0.050

where values for K_{re} and J are taken from Table 5.4.

The following sections provide equations for the calculation of the coefficient C.

5.4.1 *Stress-relieved strand*—Historically, stress-relieved strands were widely employed in the United States and elsewhere into the mid-1980s. Since then, most of the prestressing strands manufactured and used throughout the world qualify as low-relaxation strands. Stress-relieved strand, also known as normal-relaxation strand, is not currently manufactured in the United States or Canada. Because some historical or older structures may contain stress-relieved strands and it is possible that stress-relieved strand be supplied by manufacturers outside of North America, related equations are presented. If normal-relaxation strand is ordered, manufacturers typically supply low-relaxation strand, which meets all of the requirements of normal-relaxation strand.

In the unlikely case that stress-relieved strands are specified by the licensed design professional, the relaxation properties of the strand supplied will, most likely, more closely conform to those of low-relaxation strand. The installation of strand that behaves similarly to low-relaxation strand, when normal-relaxation strand properties were used for design, results in the oversupply of prestressing force. This could cause increased creep and excessive camber.

Therefore, unless the strand is specified and supplied as low-relaxation, the relaxation properties are measured independently. The equations for stress-relieved strand that are published in this guide and elsewhere (PCI 2010) are not recommended for current applications.

For stress-relieved strand, the coefficient C in Eq. (5.4) is computed as follows.

$$\text{If } 0.75 \geq f_{pi}/f_{pu} \geq 0.70,\ C = 1 + 9(f_{pi}/f_{pu} - 0.7) \qquad (5.4.1a)$$

$$\text{If } 0.70 > f_{pi}/f_{pu} \geq 0.51,\ C = [(f_{pi}/f_{pu})/0.19][(f_{pi}/f_{pu})/0.85 - 0.55] \qquad (5.4.1b)$$

$$\text{If } f_{pi}/f_{pu} < 0.51,\ C = (f_{pi}/f_{pu})/3.83 \qquad (5.4.1c)$$

5.4.2 *Low-relaxation strand*—The coefficient C in Eq. (5.4) for low-relaxation strand is computed as follows.

$$\text{If } f_{pi}/f_{pu} \geq 0.54,\ C = [(f_{pi}/f_{pu})/0.21][(f_{pi}/f_{pu})/0.9 - 0.55] \qquad (5.4.2a)$$

$$\text{If } f_{pi}/f_{pu} \leq 0.54,\ C = (f_{pi}/f_{pu})/4.25 \qquad (5.4.2b)$$

where $f_{pi} = P_i/A_{ps}$, psi (MPa); and f_{pu} is ultimate strength of prestressing steel, psi (MPa).

5.5—AASHTO LRFD approximate estimate of time-dependent losses

A simple equation to estimate the time-dependent prestress loss due to creep of concrete, shrinkage of concrete, and relaxation of steel in typical bridge beams is presented in the "AASHTO LRFD Bridge Design Specifications" (AASHTO 2012). The following criteria should be met to use the equation:

(a) Members are made of normalweight concrete

(b) The concrete is either steam- or moist-cured

(c) Prestressing is provided by strands or bars with either stress-relieved or low-relaxation properties

(d) Average exposure conditions and temperatures characterize the site

Section C5.9.5.3 commentary of AASHTO (2012) notes that the equation was derived for sections with composite decks and bonded tendons. AASHTO does not define average conditions. In the derivation (Tadros et al. 2003), it was assumed that at the bottom fibers of a composite bridge beam, the individual stresses due to effective prestress, girder weight, deck weight, superimposed dead and live load sum to zero at time infinity. The concrete stress due to external loads at the center of gravity of the strands was assumed to be approximately equally caused by girder weight, deck weight, and live load. Further assumptions stated by Tadros et al. (2003) include:

(a) Cross section is subjected to positive bending moment (composite deck in compression)

(b) Volume/surface ratio of precast section is between 3 and 4 in. (76.2 and 101.6 mm)

(c) No nonprestressed reinforcement is present in the cross section

(d) Prestress is transferred at 1 day of concrete age under accelerated plant curing conditions

(e) Deck weight is applied to unshored precast section at least 28 days after prestress transfer

With these assumptions, the detailed method (6.3.2) is simplified to

$$\Delta f_{pLT} = 10.0\frac{f_{pi}A_{ps}}{A_g}\gamma_h\gamma_{st} + 12{,}000\gamma_h\gamma_{st} + \Delta f_{pRE}$$

(for stresses in psi) (5.5a)

$$\Delta f_{pLT} = 10.0\frac{f_{pi}A_{ps}}{A_g}\gamma_h\gamma_{st} + 83\gamma_h\gamma_{st} + \Delta f_{pRE}$$

(for stresses in MPa)

where

$$\gamma_h = 1.7 - 0.01RH \quad (5.5b)$$

$$\gamma_{st} = \frac{5000}{1000 + f'_{ci}} \text{ (for } f'_{ci} \text{ in psi)}$$

$$\gamma_{st} = \frac{34.5}{6.9 + f'_{ci}} \text{ (for } f'_{ci} \text{ in MPa)} \quad (5.5c)$$

where f_{pi} is prestressing steel stress immediately following transfer, psi (MPa); A_{ps} is area of prestressing steel, in.^{2} (mm^2); A_g is gross cross-sectional area of beam, in.^{2} (mm^2); Δf_{pRE} is estimate of relaxation loss = 2400 psi (16.5 MPa) for low-relaxation strand and 10,000 psi (68.95 MPa) for stress-relieved strand; γ_h is correction factor for ambient relative humidity; γ_{st} is correction factor for specified concrete compressive strength at transfer; RH is average annual ambient relative humidity, percent; and f_{ci}' is specified concrete compressive strength at transfer of prestress, psi (MPa).

The simplified equation was found to be a conservative estimate of prestress losses (Tadros et al. 2003) compared to full-scale test results and the results of the refined method. The simplified method, however, should not be used with noncomposite members, members with uncommon shapes or atypical levels of prestress, or unusual construction staging.

CHAPTER 6—LONG-TERM LOSSES: DETAILED METHODS

6.1—Scope

There are a variety of methods to estimate prestress losses when more detailed calculations are desired. For designs using assumed material properties, it is not always necessary to use detailed prestress loss calculations. Detailed methods can be categorized as time-step methods or age-adjusted effective modulus methods. It is also possible to account for additional effects on long-term tendon stresses due to differential shrinkage of concretes in composite systems, and thermal effects when casting composite decks or toppings.

Detailed methods are normally used for unusual design cases, such as very long span members or segmental bridges, to obtain a more accurate estimate of long-term losses using measured material properties. Detailed methods may also be a preferred option when unusual construction sequencing is used, such as spliced girders, which are initially pretensioned and then assembled and post-tensioned at the erection site. They can also be used to estimate prestress loss at any time during the life of the structure, as opposed to the simplified method, which can only provide an estimated loss at the end of service life.

6.2—Creep and shrinkage models

The ability of a detailed method to accurately estimate prestress loss is dependent on accurate models of creep and shrinkage of which several are widely used. ACI 209.2R presents four models for prediction of creep and shrinkage: 1) ACI 209R; 2) Bažant and Baweja (1995); 3) CEB-FIP model code (Comité Euro-International du Béton 1999); and 4) Gardner and Lockman (2001). Other models, such as Sakata (1993), Wendner et al. (2013), and AASHTO (2012), provide alternative calculation methods. There is no consensus on which model is the best predictor of time-dependent behavior and all have significant variability. ACI 209.2R provides comparisons of the predictions of the four presented models with a database of measurements, but does not endorse any one model.

The detailed models for estimation of prestress loss require the calculation of shrinkage strains and creep coefficients for the prestressed member concrete and, if the member has a composite deck or topping, the deck concrete. This guide does not review the models, but assumes the licensed design professional selects an appropriate model for the application.

6.3—Age-adjusted effective modulus approaches

6.3.1 *Age-adjusted effective modulus of elasticity*—One approach to modeling the effects of time-dependent strains caused by creep is to adjust the concrete modulus of elasticity to account for both the elastic and time-dependent creep strains. Figure 6.3.1 illustrates this concept. The initial strain $\varepsilon(t_0)$ and the initial stress $\sigma(t_0)$ can be related by the initial elastic modulus E_c. When stress is sustained over time, additional strain will develop as a function of the creep coefficient $\phi(t,t_0)\varepsilon(t_0)$. Creep is a function of the time under consideration (t) and the time when the load was first applied (t_0). Thus, the final strain can be expressed as a function of the creep coefficient $\phi \cdot (t,t_0)$ (Ghali et al. 2012)

$$\varepsilon(t) = \varepsilon(t_0) + \phi(t,t_0)\varepsilon(t_0) = \varepsilon(t_0)(1+\phi(t,t_0))$$
$$= \frac{\sigma(t_0)}{E_c}(1+\phi(t,t_0)) = \frac{\sigma(t_0)}{\dfrac{E_c}{(1+\phi(t,t_0))}} \quad (6.3.1a)$$

Another way to express this is to define the effective modulus of elasticity E_c'

$$E_c' = \frac{E_c}{1+\phi(t,t_0)} \quad (6.3.1b)$$

This modulus is appropriate for calculating strains due to loads that are instantaneously applied, and then left in place for some time period.

Some stresses, such as those from prestress loss or the self-equilibrating stresses from differential shrinkage, develop slowly over time. One approach to this problem is to divide the stress into several increments and sum the elastic and creep strains for each increment of stress

$$\varepsilon(t) = \sum_{i=1}^{n} \frac{\Delta\sigma(t_i)}{E_c(t_i)}\left[1+\phi(t,t_i)\right] \quad (6.3.1c)$$

where the modulus of elasticity and creep coefficient are appropriate for the concrete age at the time each new increment of stress is applied. This type of summation can be tedious and often impossible if the stress history is not well known. This summation can be approximated using an aging coefficient χ originally developed by Trost (1967). In this case, the total change in stress is applied in one step and the creep coefficient is modified to reflect that, as the stress slowly develops, the concrete is maturing and later increments of stress have less associated creep than earlier increments of stress. In this way, the total strain is expressed

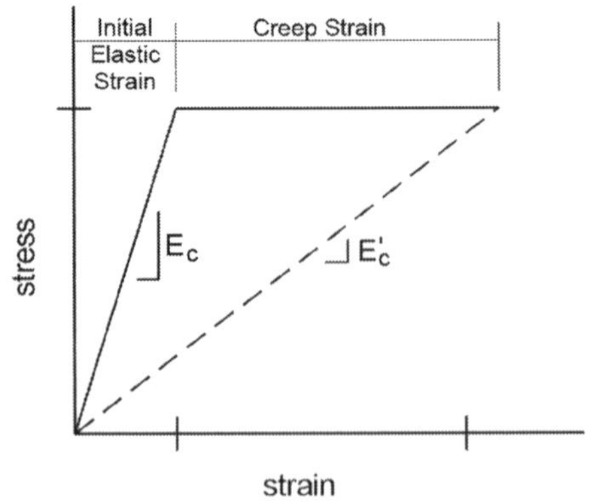

Fig 6.3.1—Graphical definition of effective modulus.

$$\varepsilon(t) = \left(\sum_{i=1}^{n} \Delta\sigma(t_i)\right)\frac{1+\chi\phi(t,t_0)}{E_c} = \frac{\sum_{1}^{n}\Delta\sigma(t_i)}{\dfrac{E_c}{1+\chi\phi(t,t_0)}} \quad (6.3.1d)$$

where the modulus of elasticity and the creep coefficient are appropriate for the concrete age at the time of the initial increment of stress. The aging coefficient is typically taken as a number between 0.7 and 0.9. ACI 209R provides guidance in determining the appropriate aging coefficient as a function of the age when the stress begins to develop and the time over which the stress develops.

The age-adjusted effective modulus of elasticity, E_c'', is defined (Ghali et al. 2012)

$$E_c'' = \frac{E_c}{1+\chi\phi(t,t_0)} \quad (6.3.1e)$$

This modulus is appropriate for calculating strains due to stresses that develop slowly over time. Two methods for estimating prestress loss outlined in this section use the age-adjusted effective modulus of elasticity in their formulations.

6.3.2 *AASHTO LRFD refined method*—The refined method for estimating prestress losses in the "AASHTO LRFD Bridge Design Specifications" (AASHTO 2012) was developed as part of National Cooperative Highway Research Program (NCHRP) Project 18-07 (Tadros et al. 2003). The method splits the estimation of long-term prestress losses into two time periods: 1) from the time of initial transfer of prestress to the time the composite deck is placed (subscript *id*); and 2) from deck placement to end of service (subscript *df*). For systems without a cast-in-place deck or topping, only one step, from initial to final (subscript *if*), is required. Although the method was developed for bridge girders, it is a general age-adjusted effective modulus method and can be applied to any pretensioned member.

The equation for time-dependent prestress loss is

$$\Delta f_{pLT} = (\Delta f_{pSR} + \Delta f_{pCR1} + \Delta f_{pR1})_{id} + (\Delta f_{pSD} + \Delta f_{pCD} + \Delta f_{pR2} + \Delta f_{pSS})_{df} \quad (6.3.2a)$$

where Δf_{pLT} is long-term prestress loss, psi (MPa); Δf_{pSR} is prestress loss due to shrinkage of beam concrete between transfer and deck placement, psi (MPa); Δf_{pCR1} is prestress loss due to creep of beam concrete between transfer and deck placement, psi (MPa); Δf_{pR1} is prestress loss due to relaxation of prestressing strands between time of transfer and deck placement, psi (MPa); Δf_{pSD} is prestress loss due to shrinkage of beam concrete between time of deck placement and final time, psi (MPa); Δf_{pCD} is prestress loss due to creep of beam concrete between time of deck placement and final time, psi (MPa); Δf_{pR2} is prestress loss due to relaxation of prestressing strands in composite section between time of deck placement and final time, psi (MPa); and Δf_{pSS} is prestress gain due to shrinkage of deck in composite section, taken as a negative value in Eq. (6.3.2a), psi (MPa).

In Section 5.9.5.4 of AASHTO (2012), an equation is presented for each component of loss. NCHRP Report 496 (Tadros et al. 2003) presents the derivation of the equations. Each equation begins with the same assumption that there is perfect bond between the concrete and the steel; therefore, over time, the concrete at the strand level undergoes the same change in strain as the strand

$$\Delta\varepsilon_p = \Delta\varepsilon_c \qquad (6.3.2b)$$

where $\Delta\varepsilon_p$ is change in strain in the prestressing steel; and $\Delta\varepsilon_c$ is change in strain in the concrete at the level of the prestressing steel.

The change in the strand strain is simply the change in prestress force (ΔP_{ps}) divided by the strand area and modulus of elasticity

$$\Delta\varepsilon_p = \frac{\Delta P_{ps}}{A_{ps}E_p} \qquad (6.3.2c)$$

The change in the concrete strain is the time-dependent change, such as creep strain or shrinkage strain, minus the change in strain associated with prestress loss, ΔP_{ps}. Prestress loss results in elastic and creep strain. Because prestress loss develops slowly over time, the creep strain is the elastic strain multiplied by the creep coefficient times the aging coefficient.

For example, for the loss due to shrinkage the total change in strain at the strand level is calculated (Tadros et al. 2003)

$$\Delta\varepsilon_c = \varepsilon_{shb}(t_d) - \left(\frac{\Delta P_{ps}}{A_g} + \frac{\Delta P_{ps}e_p^{\,2}}{I_g}\right)\frac{(1+\chi\phi(t_f,t_0))}{E_{ci}} \qquad (6.3.2d)$$

where $\varepsilon_{shb}(t_d)$ is shrinkage strain of the girder at the time the deck is placed; t_d is time since end of cure to time of deck placement, days; t_f is final time under consideration, days; and t_0 is time of initial prestress transfer, days.

Setting the change in concrete strain equal to the change in steel strain results in the equation

$$\frac{\Delta P_{ps}}{A_{ps}E_p} = \varepsilon_{shb}(t_d) - \left(\frac{\Delta P_{ps}}{A_g} + \frac{\Delta P_{ps}e_p^{\,2}}{I_g}\right)\frac{(1+\chi\phi(t_f,t_0))}{E_{ci}} \qquad (6.3.2e)$$

To determine the prestress loss, all terms with ΔP_{ps} are collected to the left side of the equation and the change in force is divided by the strand area to arrive at a change in stress. The aging coefficient is set to 0.7 and the final form of the equation is

$$\Delta f_{pSR} = \frac{\varepsilon_{shb}(t_d)E_p}{1+\dfrac{E_p}{E_{ci}}\dfrac{A_{ps}}{A_g}\left(1+\dfrac{A_g e_p^{\,2}}{I_g}\right)\left(1+0.7\phi(t_f,t_0)\right)} \qquad (6.3.2f)$$

For simplification, a new term called a section modification factor is defined

$$K_{id} = \frac{1}{1+\dfrac{E_p}{E_{ci}}\dfrac{A_{ps}}{A_g}\left(1+\dfrac{A_g e_p^{\,2}}{I_g}\right)\left(1+0.7\phi(t_f,t_0)\right)} \qquad (6.3.2g)$$

The equation for change in prestress due to shrinkage that occurs from end of cure to time of deck placement can then be written as

$$\Delta f_{pSR} = \varepsilon_{shb}(t_d)E_pK_{id} \qquad (6.3.2h)$$

This approach is used for each type of prestress loss over each time interval.

The refined method includes the effect of deck shrinkage on changes in strand stress. When the deck concrete shrinks, after composite behavior has been established, the top of the girder restrains the shrinkage. In this way, the deck exerts a compressive force on the top of the girder, and the girder exerts an equal tensile force on the deck. This results in the girder deflecting downward and tension developing in the bottom of the girder. This also results in an increase in strand tension at the bottom of the beam. This is illustrated in Fig. 6.3.2a and 6.3.2b.

In the AASHTO (2012) approach, changes in prestress from deck and girder shrinkage are evaluated separately. For determining the increase in prestress due to deck shrinkage, the force exerted by the deck on the composite girder is determined

$$P_{shd} = \varepsilon_{shd}(t_f)A_dE_{cd} \qquad (6.3.2i)$$

where P_{shd} is force to fully restrain deck shrinkage, lb (N); $\varepsilon_{shd}(t_f)$ is shrinkage of the deck concrete at final time under consideration; A_d is area of composite concrete deck, in.^{2} (mm^2); and E_{cd} is modulus of elasticity of the composite deck, psi (MPa).

This force is assumed to be applied at the centroid of the deck on the composite girder. Knowing the force and location of application, the stress at the center of gravity of the prestress force can be calculated. Because this is a slowly developing force, the modulus of elasticity is divided by (1 + $\chi\phi_d(t_f,t_d)$) to reflect the relaxation of the force due to creep. The stress at the center of gravity of the prestress force is calculated

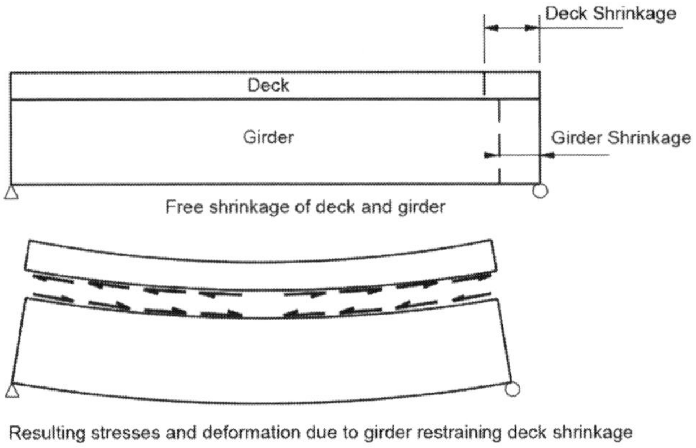

Fig 6.3.2a—Effect of differential shrinkage on girder deflection.

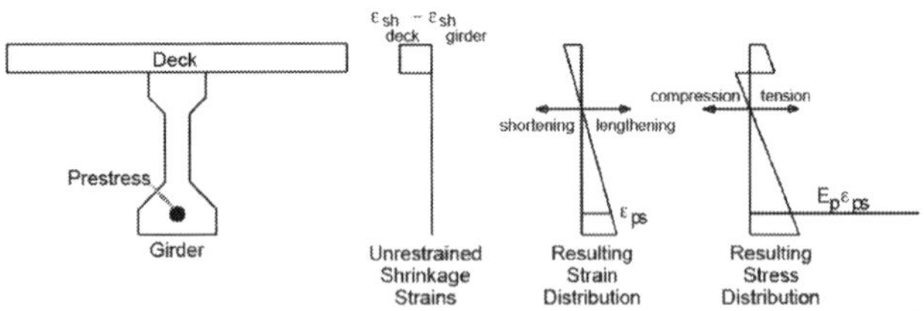

Fig. 6.3.2b—Self-equilibrating stresses due to differential shrinkage.

$$\Delta f_{cdf} = \frac{\varepsilon_{shd}(t_f) A_d E_{cd}}{\left(1+\chi\phi_d(t_f,t_d)\right)}\left(\frac{1}{A_{comp}} - \frac{e_{pc}e_d}{I_c}\right) \quad (6.3.2j)$$

where Δf_{cdf} is concrete stress at the center of gravity of the prestress force due to the differential shrinkage force, psi (MPa); A_{comp} is transformed area of the composite section, in.^{2} (mm^2); I_c is transformed moment of inertia of the composite section, in.^{4} (mm^4); e_{pc} is eccentricity of the prestress relative to the composite section centroid, in. (mm); e_d is eccentricity of the centroid of the deck relative to the composite section centroid, in. (mm); and $\phi_d(t_f,t_d)$ is creep coefficient of the deck concrete at final time due to loads placed immediately following the end of moist cure.

Then, using the same approach as with other losses, the change in steel strain and the change in concrete strain at the level of the prestress are equated, and the gain in prestress is determined. The full equation is

$$\Delta f_{pSS} = \frac{\frac{E_p}{E_c}\Delta f_{cdf}\left(1+0.7\phi(t_f,t_d)\right)}{1+\frac{E_p}{E_c}\frac{A_{ps}}{A_{comp}}\left(1+\frac{A_c e_c^2}{I_c}\right)\left(1+0.7\phi(t_f,t_0)\right)} \quad (6.3.2k)$$

where $\phi(t_f,t_d)$ is creep coefficient of the beam concrete at the final time under consideration, for loads applied at the time of deck placement.

If a designer chooses to calculate this increase in prestress force, the increase in tension at the bottom of the girder should also be considered when checking allowable stresses. Equation (6.3.2j) can be altered to calculate the concrete tension at the bottom of the beam. The term e_{pc} is replaced by y_{bott}, the distance from the centroid of the composite section to the bottom of the section, to arrive at the tensile stress at the bottom of the beam due to deck shrinkage.

Typically, the tensile stresses at the bottom of the beam due to differential shrinkage are ignored. The stresses tend to be relatively small depending on several parameters, such as girder age at the time the deck is placed, relative cross-sectional areas of the deck and girder, shrinkage properties of the deck, and amount of deck reinforcing steel. Therefore, ignoring these stresses does not typically lead to any obvious distress in the beam.

However, if a designer chooses to take advantage of the increase in prestress force due to differential shrinkage, the detrimental effects, which may be greater than the advantages, should also be considered.

6.3.3 *A general age-adjusted effective modulus approach*—A second refined method, which can be used to evaluate prestress loss, is similar to the AASHTO (2012) refined method. Both methods use an age-adjusted effective modulus approach to account for the effect of the slowly changing force in the prestress. In the AASHTO approach, the effects of creep, shrinkage, and relaxation are investigated independently. In this approach (Menn 1990), all time-dependent sources of loss are considered in a single formulation. Figure 6.3.3a shows the basic model to predict prestress losses before deck placement. There are six unknowns and six equations that can be written to solve for the unknowns. The equations are shown.

Equations of Internal Equilibrium:

$$\Delta N_b + \Delta N_{pc} = 0 \quad (6.3.3a)$$

$$\Delta M_b + \Delta N_{pc} \cdot e_p = 0 \quad (6.3.3b)$$

Constitutive Equations:

$$\Delta\varepsilon_p = \frac{(\Delta N_{ps} - \Delta N_{relax})}{A_{ps}E_p} \quad (6.3.3c)$$

$$\Delta\varepsilon_b = \frac{N_b^o\phi(t,t_o)}{A_g E_c} + \frac{\Delta Nb}{A_g E_c}\left(1+\chi\phi(t,t_0)+\varepsilon_{shb}(t)\right) \quad (6.3.3d)$$

Compatibility Equations:

$$\Delta\kappa = \frac{M_b^o\phi(t,t_0)}{I_g E_c} + \frac{\Delta M_b}{I_g E_c}\left(1+\chi\phi(t,t_0)\right) \quad (6.3.3e)$$

$$\Delta\varepsilon_b = \Delta\varepsilon_p - \Delta\kappa \cdot e_p \quad (6.3.3f)$$

where ΔN_b is change in force in the beam, lb (N); ΔN_{ps} is change in prestress force, lb (N); ΔM_b is change in moment in the beam, in.-lb (N-mm); $\Delta\varepsilon_p$ is change in prestress strain; $\Delta\varepsilon_b$ is change in strain in the beam at the gross section centroid; $\Delta\kappa$ is change in curvature, rad./in. (rad./mm);

M_b^o is initial creep-producing moment in the girder, in.-lb (N-mm); N_b^o is initial creep-producing force in the girder, lb (N); and ΔN_{relax} is change in prestress force due to relaxation (no associated strain), lb (N).

All unknowns are assumed to be positive as the equations are written. A positive notation means tension or lengthening, and a negative notation means compression or shortening. A positive curvature or moment causes compression at the top of the beam. When all unknowns are determined, a positive value will reflect the correct assumption. The cross-sectional properties of the beam are based on the net section; however, the gross properties are an acceptable approximation. The creep coefficient and shrinkage strain can be calculated with any acceptable model and reflect the time interval from prestress release to deck placement.

For a member with a composite deck, the same approach is used for the time interval from deck placement to final time. Figure 6.3.3b presents the unknowns for this interval. There are now 11 unknowns, so 11 equations are written.

Equations of Internal Equilibrium:

$$\Delta N_b + \Delta N_d + \Delta N_{ps} + \Delta N_{sd} = 0 \quad (6.3.3g)$$

$$\Delta M_d + \Delta M_b - \Delta N_d \cdot a_b + \Delta N_{ps} \cdot e_p - \Delta N_{sd} \cdot a_b = 0 \quad (6.3.3h)$$

Constitutive Equations:

$$\Delta\varepsilon_d = \frac{\Delta N_d}{E_d A_d}\left(1+\chi\phi_d(t_f,t_d)\right)+\varepsilon_{shd}(t_f) \quad (6.3.3i)$$

$$\Delta\varepsilon_b = \frac{N_b^o}{E_c A_g}(\phi(t_f,t_o)-\phi(t_d,t_o)) + \frac{\Delta N_b}{E_c A_g}(1+\chi\phi(t_f,t_d))+(\varepsilon_{shb}(t_f)-\varepsilon_{shb}(t_f)) \quad (6.3.3j)$$

$$\Delta\varepsilon_p = \frac{(\Delta N_{ps}-\Delta N_{relax})}{A_{ps}E_p} \quad (6.3.3k)$$

$$\Delta\varepsilon_{sd} = \frac{(\Delta N_{sd})}{A_{sd}E_s} \quad (6.3.3l)$$

$$\Delta\kappa = \frac{M_b^o}{E_c I_g}(\phi(t_f,t_o)-\phi(t_d,t_o)) + \frac{M_{deck}}{E_c I_g}(\phi(t_f,t_d)) + \frac{\Delta M_b}{E_c I_g}(1+\chi\phi(t_f,t_d)) \quad (6.3.3m)$$

$$\Delta\kappa = \frac{\Delta M_d}{E_d I_d}\left(1+\chi\phi_d\left(t_f,t_d\right)\right) \quad (6.3.3n)$$

Compatibility Equations:

$$\Delta\varepsilon_b = \Delta\varepsilon_p - \Delta\kappa \cdot e_p \quad (6.3.3o)$$

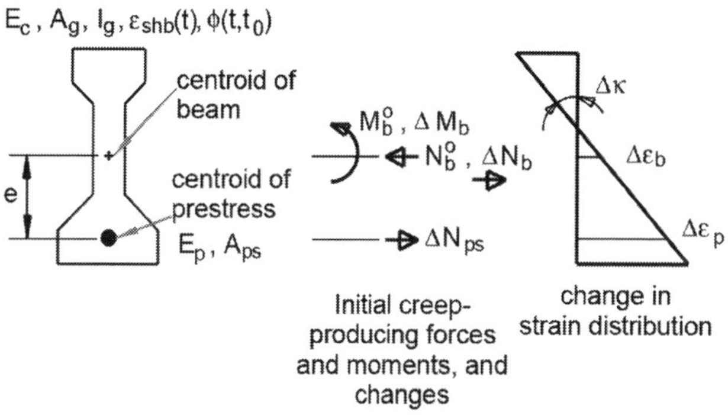

Fig. 6.3.3a—Model for estimating prestress loss.

$$\Delta\varepsilon_b = \Delta\varepsilon_d - \Delta\kappa \cdot a_b \quad (6.3.3p)$$

$$\Delta\varepsilon_d = \Delta\varepsilon_{sd} \quad (6.3.3q)$$

where e_p is eccentricity from the centroid of the beam (gross section) to the centroid of the prestress, in. (mm); a_b is eccentricity from the centroid of the beam (gross section) to the centroid of the deck (also the centroid of the deck reinforcing steel), in. (mm); ΔN_{sd} is change in the force in the deck reinforcement, lb (N); ΔN_d is change in the force in the deck concrete, lb (N); ΔM_d is change in moment in the deck concrete, in.-lb (N-mm); E_d is modulus of elasticity of the deck concrete, psi (MPa); A_d is area of the composite concrete deck, in.^{2} (mm^2); $\phi_d(t_f,t_d)$ is creep coefficient of the deck concrete at final time due to loads at time of deck placement; χ is aging coefficient; $\varepsilon_{shd}(t_f)$ is shrinkage strain in the deck at final time; $\phi(t_f,t_d)$ is creep coefficient of the beam at final time for loads placed at time of deck casting; $\phi(t_f,t_0)$ is creep coefficient of the beam at final time for loads placed at time of prestress transfer; $\phi(t_d,t_0)$ is creep coefficient in the beam at the time the deck was placed for loads placed at time of prestress transfer; N_b^o is initial creep-producing force in the girder, lb (N); M_b^o is moment in the beam just before deck placement, in.-lb (N-mm); M_{deck} is moment in the beam due to deck weight, in.-lb (N-mm); $\varepsilon_{shb}(t_f)$ is total shrinkage strain of the beam at the time of analysis; $\varepsilon_{shb}(t_d)$ is shrinkage strain of the girder at the time the deck is placed; $\Delta\varepsilon_{sd}$ is change in strain in the deck steel; I_d is moment of inertia of the deck, in.^{4} (mm^4); E_s is modulus of elasticity of the deck steel, psi (MPa); and A_{sd} is area of deck steel, in.^{2} (mm^2).

When these equations are solved simultaneously, the time-dependent prestress loss considering the effect of deck reinforcement and deck shrinkage can be determined. Note that the given system of equations is based on the centroid of the deck reinforcement at the centroid of the deck. If this is not the case, the equations can be easily altered to reflect the actual location of the reinforcement. Instantaneous changes, such as elastic shortening losses and increases in strain at the time of deck placement, are calculated using the transformed cross-sectional properties of the girder. This general method, which is adaptable to all types of composite construction, can evaluate the prestress losses in precast, post-tensioned deck panels after they have been made composite with either a steel or concrete beam. Other reinforcement layers, such

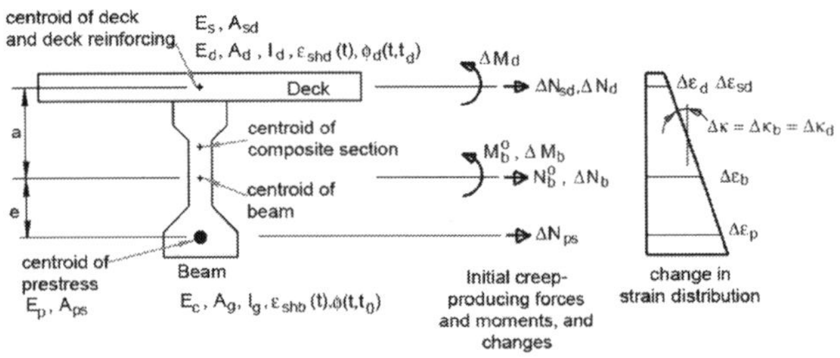

Fig. 6.3.3b—Unknowns from deck placement to final time.

as girder compression reinforcement, can be added. For any layer added, a new constitutive and compatibility equation is written and the equations of internal equilibrium altered.

6.4—Incremental time-step method

Incremental time-step approaches to prestress loss calculations are based on superposition of elastic and creep strains resulting from increments of stress. Figure 6.4 presents the concept graphically. A varying load history is represented as a series of load increments that are left in place over time. Each load increment has associated elastic and creep strains. At any point in time, the total strain can be calculated as the summation of the elastic and creep strains for all loads plus shrinkage strains.

Guidance for time-step analyses for estimating prestress losses is found in textbooks and other sources (Nilson 1987; Naaman 2012; PCI Committee on Prestress Losses 1975; Gilbert and Ranzi 2011). The life of the structure is a summation of time increments that includes important loading stages expected to occur over the structure's life. Typically, early-age increments are relatively short and later increments are longer because larger changes occur earlier in the life of the structure. For each time increment, the following five-step calculations assume all tendons are grouped together at their centroid. The subscript for the beginning of the time step is "n-1" and the end of the time step is "n."

Step 1: *Calculate the increment of creep and shrinkage for the time step*—Use the selected model to determine the creep coefficient and the shrinkage strain at the beginning and end of the time step: $\phi(t_{n-1},t_0)$, $\phi(t_n,t_0)$, $\varepsilon_{shb}(t_{n-1})$, $\varepsilon_{shb}(t_n)$, then calculate the increment of creep and shrinkage for the time step as the differences between the starting and ending values.

Step 2: *Calculate the initial elastic strains on the cross section*—Based on the axial forces and moments on the cross section at the beginning of the time step, the cross-sectional properties, and the modulus of elasticity, calculate the load-induced strain distribution through the depth of the section. The concrete strain at the center of gravity of the prestressing force is of primary importance; however, the method can also be used to determine changes in curvature and deflection.

Step 3: *Calculate time-dependent strain at the center of gravity of the prestressing force*—Multiply the elastic strain at the center of gravity of the prestressing force by the increment of the creep coefficient $\phi(t_n,t_0) - \phi(t_{n-1},t_0)$ to determine the increment of creep strain, $\Delta\varepsilon_{cr}$. Add to that the increment of shrinkage strain, $\Delta\varepsilon_{shb} = \varepsilon_{shb}(t_n) - \varepsilon_{shb}(t_{n-1})$. Multiply the sum of the time-dependent strains by the modulus of elasticity of the prestressing steel and add to this the increment of relaxation loss for the time step. This is the gross increment of prestress loss for the time step.

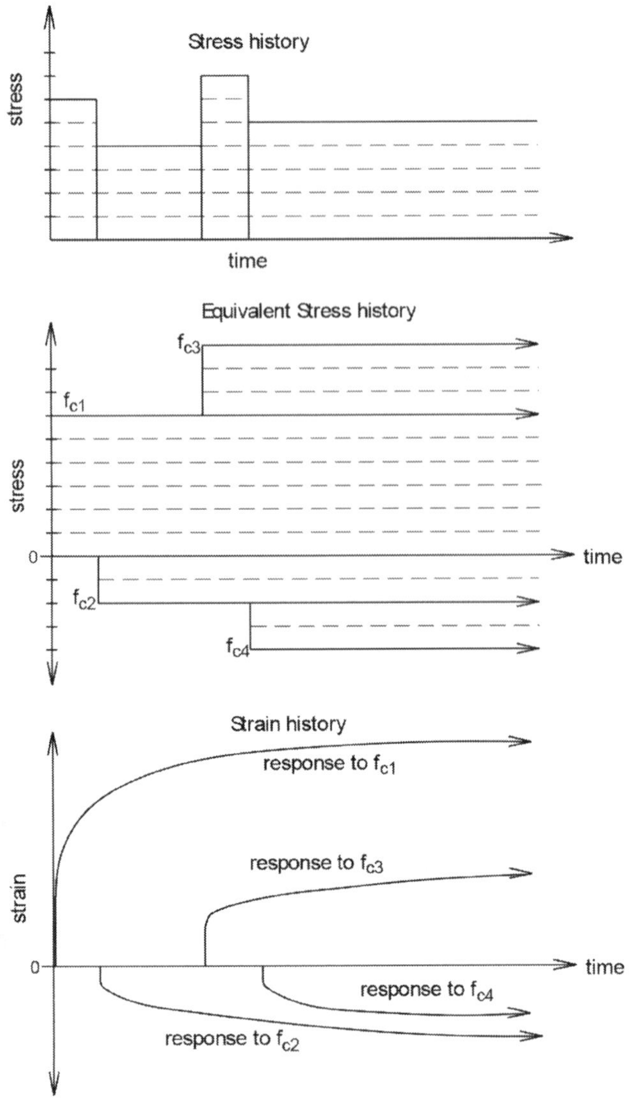

Fig. 6.4—Superposition of stresses and strains.

Step 4: *Calculate elastic rebound due to prestress loss*—Calculate the change in stress at the center of gravity of the prestressing force due to the prestress loss, and divide by the concrete modulus of elasticity to determine the associated change in strain. Multiply this strain by the steel modulus to determine the change in prestress. This will be a tensile stress, which is also called prestress recovery or elastic rebound. Subtract this stress from the loss computed in Step 3 to arrive at the net increment of prestress loss for the time step.

Step 5: *Determine creep-producing forces for the start of the next time step*—Adjust the prestress force and moment to reflect the prestress loss over the previous time step. This new force and moment are used to calculate the elastic strains at the beginning of the next time step.

6.4.1 *Variations in approach*—A variety of approaches to the time-step method have been proposed (Waldron 2004;

ACI 435R; Moustafa 1986; Gutierrez et al. 1996; Tadros et al. 1975). The simplest, especially for beams that are not made composite, is to use only one creep model throughout the calculations. This means the ultimate creep coefficient is based on the assumption that all loads are applied at one point in time. The initial elastic strains are based on the modulus of elasticity at the time the first loads are applied.

Another approach is to treat each increment of prestress loss as a newly applied load. The creep coefficient for this load is calculated based on the maturity of the concrete at the end of the time step. Then, the elastic strains at the beginning of the next time step are the sum of the initial elastic strains based on the modulus of elasticity at release and the strains associated with the prestress loss based on the concrete modulus of elasticity at the start of the time step. This approach is theoretically more correct, but also more time-consuming.

6.5—Computer programs

Several commercially available software programs include calculation of prestress losses or general time-dependent effects in concrete. The designer should determine which method of loss calculation and creep and shrinkage functions are being used in the program to know the strengths and weaknesses of the method for the individual design application.

6.6—Effects of deck temperature during casting of composite deck or topping

When fresh concrete is placed and begins to cure, the concrete heats up due to the hydration reaction between the water and cement. When deck concrete is placed on top of a beam, this temperature increase causes the top flange of the beam to expand relative to the bottom flange. The thermal gradient induces self-equilibrating stresses through the depth of the beam. Hypothetically, if the concrete did not harden once it achieved its peak temperature, and the beam was allowed to return to its original length, the stresses would be eliminated. However, as the deck concrete hardens, it can begin to take on load that locks in stresses produced by the deck heating.

The effect of deck curing on prestress losses in composite beams can be examined qualitatively; however, there are unknown or uncertain values. They include early-age concrete properties such as modulus and coefficient of thermal expansion, thermal gradient in the top of the beam, and development of composite action through bonding of the deck concrete to the top of the beam. As with the evaluation of thermal effects during beam casting, a quantitative evaluation is difficult.

The first assumption to make is with the thermal gradient. As reported by Roberts-Wollmann et al. (1995), based on studies of thermal gradients in match cast segments of segmental bridges, the temperature gradient dissipates exponentially over the first 12 in. (305 mm) of the heated element. A linear approximation is made to simulate this heat dissipation for ease of calculation. Depending on casting conditions, a reasonable assumption is that the concrete temperature will rise to approximately 110 to 120°F (43 to 49°C), and the beam will be at approximately ambient temperature.

The self-equilibrating stresses from this gradient are calculated using the common approach of calculating moment and axial forces required to fully restrain thermal movements.

$$P = \int E_c \alpha_c b(y) T(y) dy \tag{6.6a}$$

$$M = \int E_c \alpha_c b(y) T(y) y dy \tag{6.6b}$$

where α_c is coefficient of thermal expansion of the concrete; y is distance from the centroid of the cross section, in. (mm); $T(y)$ is temperature at depth y, °F (°C); and $b(y)$ is width of cross section at depth y, in. (mm).

The stresses at any location y through the depth of the beam are calculated as

$$\sigma(y) = T(y)\alpha_c E_c - \frac{P}{A} - \frac{My}{I} \tag{6.6c}$$

To calculate the effect of temperature gradient on the beam, the deck concrete modulus of elasticity is calculated over a 2-day period. Measurements have shown a 2-day period is the time required for the concrete to reach ambient temperature, according to an unpublished thesis by Marston (2010). For a simple analysis, it is assumed that when the deck concrete reaches its maximum temperature, the deck has begun to harden and the modulus of elasticity increases from that point until it reaches its 28-day modulus. If measured values of the modulus are available, a curve can be fit to represent the modulus at any time after maximum temperature. An equation that fits the data (ACI 209R) is

$$E(t_c) = \frac{E_{c28}}{(a + \beta t)} \tag{6.6d}$$

where: $E(t)$ is modulus of elasticity at any time t, psi (MPa); a is constant; and t_c is age of concrete, in days; and β is constant such that

$$\beta = \frac{(28 - a)}{28} \tag{6.6e}$$

If data are not available, ACI 209R provides typical values for the constants so a curve can be approximated.

Once the equation for the modulus is determined, the modulus is calculated at several intervals over the 2-day cooling period every 6 hours for the first and second day. This determines the curvature of the girder, the stress induced in the girder, and the camber of the girder over this time period.

After the modulus of elasticity has been determined for the appropriate times, the temperature gradient is determined for the same times. A constant temperature for the deck and linear gradient through the first 12 in. (305 mm) of the beam is a simple and reasonable assumption.

After the deck concrete has reached its maximum temperature, the deck begins to harden and the temperature cools. A composite section is created and the reduction in temperature causes the girder camber to reduce. The deck begins resisting load because it began to harden at the maximum girder camber, inhibiting the girder from returning to its original camber immediately after the deck was placed. This causes permanent stresses to be locked in the deck and girder. As the deck gets stiffer over time, more load is resisted by the deck, which further inhibits camber reduction.

As the deck cools, the stress calculation becomes dependent on the change in temperature from the maximum. At this point, the composite section properties of the girder with the deck should be calculated using the modulus of elasticity of each at the time of analysis, resulting in section properties that vary at each time interval. Because the deck is composite, it is included in the moment and axial force equations.

By calculating the moment and axial force on the girder during deck cooling, the locked-in stresses at the bottom of the composite section are calculated. This analysis predicts the tensile stress induced at the bottom of the girder, which reduces the cracking moment when the girder is loaded. The change in the prestress force can also be determined.

The analysis shows that there is additional tension at the bottom of the beam and a slight increase in prestress tension. This effect is accounted for by AASHTO (2012) and ACI 318-11 by using a somewhat smaller cracking stress $6\sqrt{f_c'}$ (for f_c' in psi) ($0.5\sqrt{f_c'}$ [for f_c' in MPa]) for the V_{ci} term in shear calculations for prestressed beams as opposed to $7.5\sqrt{f_c'}$ (for f_c' in psi) ($0.63\sqrt{f_c'}$ [for f_c' in MPa]), which is used to calculate the cracking moment for non-prestressed beams.

The effect of deck curing temperature on camber changes and prestress changes can be investigated using the methods as outlined in this section; however, the effect is small and the effort unnecessary for typical prestressed beams.

CHAPTER 7—VARIABILITY OF LOSS CALCULATIONS

7.1—Objective

Chapter 7 explores the range of calculated prestress losses based on the material properties and the modulus of elasticity values in ACI 318. The *PCI Design Handbook* (PCI 2010) is used as a baseline for calculated prestress losses and assessing the difference between calculated and measured losses. Chapter 7 is intended to be a general study on variation in losses, which enables the designer to appreciate the range of variation associated with the loss calculations.

7.2—Scope

This chapter addresses three variations of prestress losses:

(1) The variation of the basic material properties, specifically modulus of elasticity and creep, from theoretical calculations is addressed

(2) Parametric studies of selected losses on a double-tee beam and a hollow core slab are completed to evaluate the possible range of variation from standard calculations

(3) A case study of shrinkage variations is presented. These studies provide an indication of the range of measured prestress losses as they may differ from calculated losses. These variations are common to all prestressed members. Variation from anchor seating and friction are more associated with post-tensioned members and relaxation, while common to all members, has a smaller contribution to total losses.

7.3—Contributions to prestress loss

Prestress losses are inherently variable and are derived from numerous factors as discussed in 7.2. Figure 7.3a illustrates this variability and is based on the assessment of losses in a pretensioned double-tee beam, a post-tensioned slab, and a post-tensioned beam presented in Chapter 8. In particular, note the variation among the various effects. For example, shrinkage varies from 15 percent of the losses in the double-tee beam to 53 percent of the losses in the post-tensioned beam, although the difference in the total magnitude of the shrinkage losses between the members is only 800 psi (5.50 MPa). Thus, the magnitude of the shrinkage losses is only slightly greater in the post-tensioned beam, while the percentage of total loss due to shrinkage in the post-tensioned beams is much greater. The total losses of post-tensioned slab and beam are much less than that of the pretensioned double-tee beam, primarily due to elastic shortening and creep losses. The loss due to elastic shortening is the largest of all losses in the pretensioned beam, but is practically negligible for both the post-tensioned slab and beam. The stress at the centroid of the prestressing force—f_{cir} for the bonded examples and f_{cpa} for the unbonded slab example—is much larger in the double-tee (1250 psi [8.62 MPa]) than in the post-tensioned slab (160 psi [1.10 MPa]) or beam (300 psi [2.07 MPa]). The higher concrete compressive stress leads to much higher losses due to creep.

Based on these examples, it can be inferred that variability in shrinkage properties will have a larger influence as a percentage on the post-tensioned members than the pretensioned beam. However, variability in the modulus of elasticity and creep properties will have a larger influence on a pretensioned beam as compared to a post-tensioned member.

Studies have been conducted on the variation of losses and probabilistic studies on the variation are available (Steinberg 1995; Gilbertson and Ahlborn 2004). Of particular interest is the Gilbertson and Ahlborn study, as it compares alternative computation approaches and the potential variation in results. Figure 7.3b shows the variation in prestress loss for an I-girder. The figure illustrates two important issues: 1) The results of calculation of losses vary with the method selected; and 2) Even within one methodology, there is a variation of results. The details of each method are discussed in the Gilbertson and Ahlborn paper.

Gilbertson and Ahlborn (2004) examine the variation in all sources of prestress loss. In this chapter, two sources of prestress loss are considered: elastic shortening and creep. Shrinkage variations are examined later in a short case study. Losses due to anchor seating are omitted because they are often compensated by overstressing in pretensioning opera-

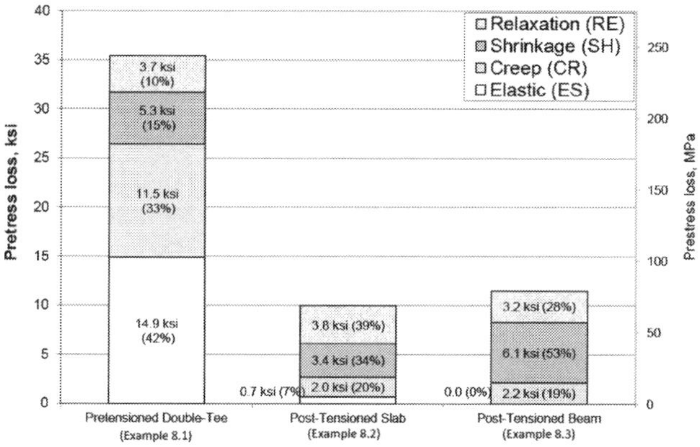

Fig. 7.3a—Example of the contribution to losses for different concrete elements.

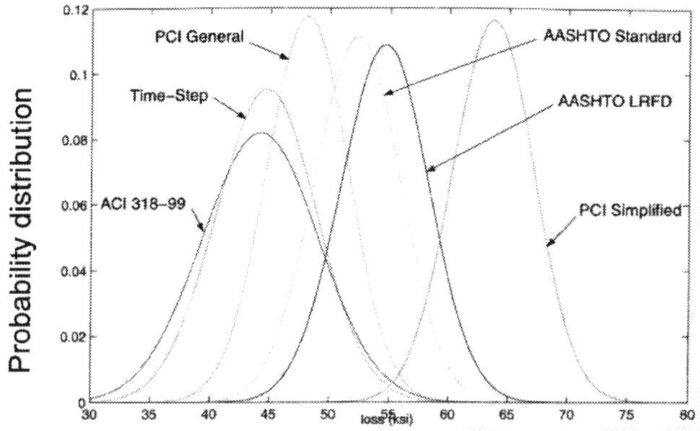

Fig. 7.3b—Variation in prestress losses for an I-girder (Gilbertson and Ahlborn 2004). (Note: 1 ksi = 6.89 MPa.)

tions, although they can directly affect some post-tensioning (PT) applications. Friction and wobble losses vary with the duct type and duct placement in PT applications, a variation given in Table 4.4.2 of suggested values. ACI 318 requires reconciliation if the difference between the predicted and measured elongation exceeds 7 percent.

7.4—Modulus of elasticity

Modulus of elasticity directly impacts the initial elastic shortening of a member and indirectly affects the creep contribution to loss. The following discussion explores variation in modulus in detail and the corresponding impact of variation on the loss calculations.

7.4.1 *Factors affecting modulus of elasticity*—Concrete modulus of elasticity, E_c, is one of the predominant factors affecting prestress loss calculation. Modulus of elasticity varies with concrete constituent materials and varies with time. Accordingly, significant variations in estimated losses are rooted in the variation of modulus of elasticity. The accurate estimate of prestress losses is dependent on an accurate valuation of the modulus of elasticity. Factors that influence the modulus of elasticity are the type and volume fraction of coarse and fine aggregate, moisture state of the specimen, loading conditions, and stiffness and porosity of the cement paste. Knowledge of these factors has long been in the literature (Hirsch 1962). Because all the factors are usually related to concrete compressive strength and unit weight, empirical equations to calculate E_c have been developed in terms of f_c' and unit weight. Different modulus of elasticity equations are considered because the value of E_c varies with the equation used.

7.4.2 *Equations for modulus of elasticity*—A multitude of equations for concrete modulus of elasticity are available in the literature. Comparisons of six different equations of E_c are examined with respect to compressive strength (Mehta and Monteiro 1993). The values for E_c based on concrete strength are summarized in Tables 7.4.2a and 7.4.2b, and Fig. 7.4.2a.

Each equation gives different modulus of elasticity values for a given compressive strength. Selecting the ACI 318-08 Eq. (1) for modulus places the user in the midrange of these predictive equations.

Each empirical equation was developed by correlating a large database of measured concrete strength with modulus test data. To apply any of the equations, measured concrete strength at a given age should be used for f_c'. However, for either design or investigation, the licensed design professional generally has no information on in-place concrete strength. Thus, common practice uses the specified concrete compressive strength of concrete. This practice can lead to considerable discrepancy in the corresponding value for the modulus of elasticity E_c, because the concrete actually produced in the field or in a prestressing plant will typically exceed the specified strength by a substantial margin to meet the acceptance criteria for a given project. Thus, while the specified properties may be well established, properties at the time of load application vary.

Storm (2011) studied the properties of plant-produced concrete. They conducted tests on multiple sets of specimens for each of 382 prestressed bridge girders of different types produced by nine prestressing plants in six mid-Atlantic and southeastern states. They found that, at prestress transfer, the measured concrete compressive strength was approximately 25 percent higher than the specified transfer strength (Fig. 7.4.2b). Similarly, the measured concrete compressive strength was approximately 45 percent higher than the specified strength at 28 days. Accordingly, Storm (2011) recommends that the specified concrete strength f_{ci}' should be multiplied by 1.25 to compute the modulus of elasticity at prestress transfer, and f_c' be multiplied by 1.45 to compute the modulus of elasticity at 28 days.

7.4.3 *Aggregate influence*—The modulus of elasticity equations do not account for factors other than unit weight and compressive strength. Other factors that influence the value of modulus of elasticity include proportion of coarse aggregate in the concrete and aggregate properties. Mehta and Monteiro (1993) as well as Mindess et al. (2003) show that the modulus of elasticity value is affected by the coarse aggregate content and type. The modulus of elasticity equations are compared using quartzite aggregate concrete as the baseline. For other types of aggregate, the modulus of elasticity is obtained by multiplying E_c with a correction factor as shown in Table 7.4.3.

Table 7.4.2a—Equations for modulus of elasticity*

Source	Equation		Equation no.
	E_c and f_c' in psi and w_c in lb/ft³	E_c and f_c' in MPa and w_c in kg/m³	
ACI 318-08	$E_c = 33w^{1.5}\sqrt{f_c'}$	$E_c = 0.043w^{1.5}\sqrt{f_c'}$	(1)
Empirical Eq. (1) (Mehta and Monteiro 1993)	$E_c = 6{,}000{,}000/(1 + (2000/f_c'))$	$E_c = 41{,}370/(1 + (13.8/f_c'))$	(2)
Empirical Eq. (2) (Mehta and Monteiro 1993)	$E_c = 1{,}800{,}000 + 460f_c'$	$E_c = 12{,}400 + 460f_c'$	(3)
ACI 363R-92	$E_c = 40{,}000\sqrt{f_c'} + 10^6(w_c/145)^{1.5}$	$E_c = 3320\sqrt{f_c'} + 6895(w_c/2330)^{1.5}$	(4)
Ahmad and Shah (1982)	$E_c = w^{2.5}(f_c')^{0.325}$	$E_c = 3.37 \times 10^{-5}w^{2.5}(f_c')^{0.325}$	(5)
CEB-FIP (2010)	$E_c = 3{,}120{,}000(f_c'/1450)^{(1/3)}$	$E_c = 21{,}500(f_c'/10)^{(1/3)}$	(6)

*Note: f_c' is the compressive strength and w the unit weight of concrete.

Table 7.4.2b—Comparison of different modulus of elasticity equations*

f_c'	Eq. (1) ACI 318-08	Eq. (2) Empirical (Mehta and Monteiro 1993)	Eq. (3) Empirical (Mehta and Monteiro 1993)	Eq. (4) ACI 363R-92	Eq. (5) Ahmad and Shah (1982)	Eq. (6) CEB-FIP (2010)
3000	3,122,000	3,600,000	3,180,000	3,357,000	3,718,000	4,000,000
4000	3,605,000	4,000,000	3,640,000	3,714,000	4,082,000	4,300,000
5000	4,031,000	4,286,000	4,100,000	4,028,000	4,389,000	4,600,000
6000	4,415,000	4,500,000	4,560,000	4,312,000	4,657,000	4,900,000
7000	4,769,000	4,667,000	5,020,000	4,573,000	4,896,000	5,200,000
8000	5,098,000	4,800,000	5,480,000	4,817,000	5,114,000	5,500,000
9000	5,407,000	4,909,000	5,940,000	5,045,000	5,313,000	5,800,000
10,000	5,700,000	5,000,000	6,400,000	5,261,000	5,498,000	6,100,000

*Note: All units are psi, concrete unit weight is 145 lb/ft³, 1000 psi = 6.896 MPa.

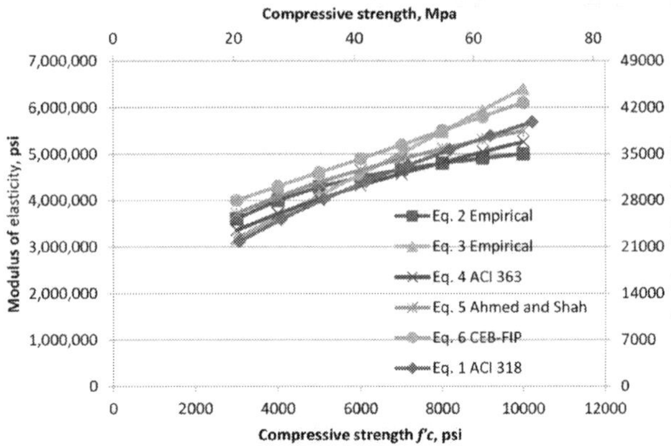

Fig. 7.4.2a—Comparison of different modulus of elasticity equations.

In Fig. 7.4.3, elastic moduli are plotted versus f_c' after taking aggregate type factor into account by applying the correction factor to the E_c as computed by the ACI 318 equation. Note that the value of E_c is highest for the concrete with basalt as a coarse aggregate, which is followed by quartz and limestone, and at the least with sandstone.

7.4.4 *Volume of coarse aggregate*—Modulus of elasticity values are also affected by the volume fraction of the coarse aggregate. Figure 7.4.4 presents the variation of modulus of elasticity with respect to variation in the volume content of various aggregates. Examining the aggregates typically used in concrete, the modulus tends to increase with increasing

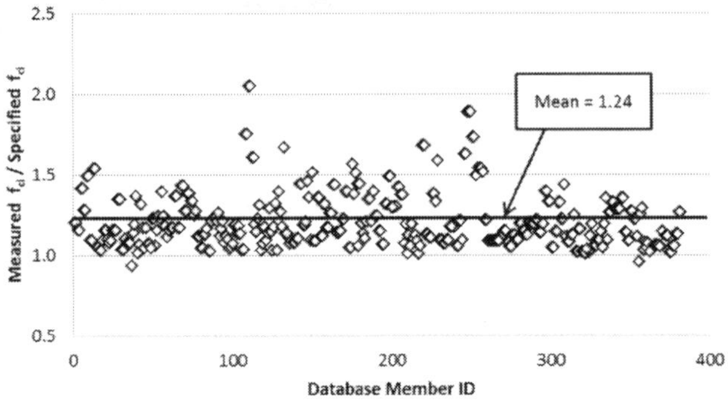

Fig. 7.4.2b—Comparison of measured concrete strength to specified strength f_{ci}' at prestress transfer (Storm 2011).

Table 7.4.3—Correction factor for different aggregate types (adapted from Myers and Carrasquillo (1999))

Aggregate type	Correction factor
Basalt, dense limestone	1.2
Quartzite	1.0
Limestone	0.9
Sandstone	0.7

coarse aggregate volume. This effect is not reflected in any of the modulus equations.

7.4.5 *Weight of aggregate*—Unit weight effects generally follow the strength corrections for aggregate types and

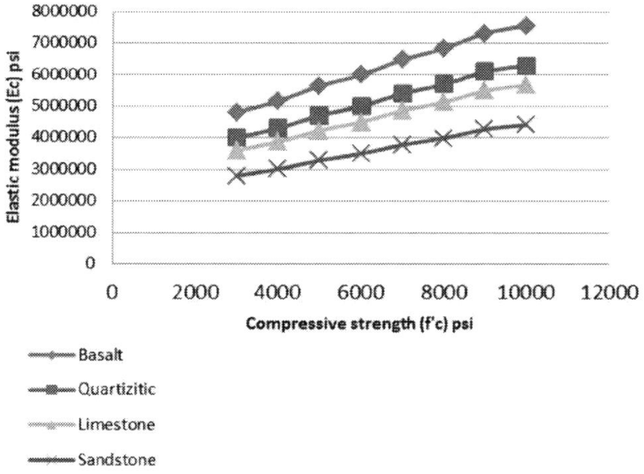

Fig. 7.4.3—Comparison of modulus of elasticity for different aggregate types (adapted from Myers and Carrasquillo (1999)). (Note: 1000 psi = 6.896 MPa.)

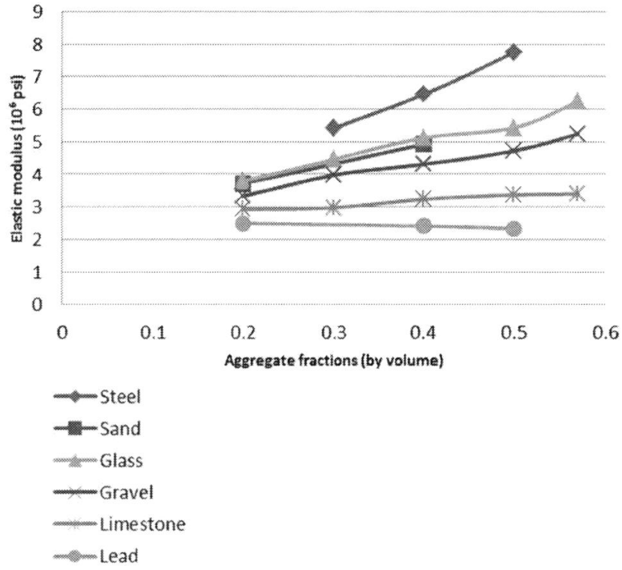

Fig. 7.4.4—Comparison of modulus of elasticity for different aggregate volume (adapted from Mindess et al. (2003)). (Note: 1000 psi = 6.896 MPa.)

therefore are not cumulative effects. The term of $33w^{1.5}$ used in Eq. (1) (Table 7.4.2a) provides the correction for unit weight.

7.4.6 *Conclusions for modulus of elasticity*—The following conclusions are based on selecting the equation in ACI 318-11 as the "correct" baseline. Empirical Eq. (1) (Table 7.4.2a) overestimates the modulus of elasticity value for f_c' range up to 6000 psi (41.4 MPa), while Eq. (2) (Table 7.4.2b) overestimates the value for f_c' above 6000 psi (41.4 MPa). Ahmad and Shah (1982) and the CEB-FIP (2010) equations in this guide give higher E_c values for the entire range of f_c'.

A correction factor to account for aggregate type in the modulus of elasticity equation is close to the difference due to the weight of coarse aggregate; therefore, mixture weight plays a prominent role in the determination of E_c. However, just accounting for the aggregate weight instead of the volume fraction or stiffness of coarse aggregate can overestimate the modulus calculation. From the discussion in 7.3 and 7.4, the licensed design professional can reasonably expect to see variations of up to 30 percent between calculated and measured modulus.

7.5—Creep

7.5.1 *Factors affecting creep*—Concrete creep is a time-dependent phenomenon of continued deformation under sustained load. It depends on several factors, including the volume of cement paste, water-cementitious material ratio (*w/cm*), type and volume of aggregates, concrete age at the time of loading, prestress level, age at which the concrete is prestressed, and member geometry. A full discussion of creep is found in ACI 209.1R and ACI 209.2R. Among these factors, creep is most affected by paste volume. Many of the interactions are similar to those affecting the modulus of elasticity (Glucklich and Ishai 1962). For example, higher aggregate volume, harder aggregates, and higher strength tend to reduce creep. Rather than reiterate all of the interactions within the concrete mixture that affect creep, licensed design professionals typically reduce a complex problem to a single creep coefficient.

7.5.2 *Creep coefficient*—For the purpose of computing losses, the effect of creep is often computed using a creep coefficient C_c, which is the ratio of the long-term creep deformation to the initial elastic deformation. Thus, creep is directly tied to the modulus of elasticity. Typical creep coefficients range between 1 and 3.5 with a value of 2.35 recommended for many normalweight concrete mixtures.

7.6—Variational analysis

To appreciate how losses can vary from calculated values, a series of loss calculations are made varying only the modulus of elasticity and the creep coefficient. The loss calculations follow the recommendations given earlier in this report. Two prestressed members are examined: a 32 in. (813 mm) deep, 10 ft (3.05 m) wide double-tee beam with a 2 in. (51 mm) topping and an 8 in. (203 mm) deep hollow core slab. Section properties are taken from the *PCI Design Handbook* (PCI 2010). Calculations consider that the number of strands varies with span length.

The study first examines a 60 ft (18.3 m) long double-tee beam with the only variable being the modulus of elasticity. Figure 7.6a shows the range of losses possible by using alternate equations for modulus. The horizontal axis is the modulus of elasticity rather than the concrete strength, and the equations used to calculate the modulus are indicated.

The assessment of modulus of elasticity indicated that a variation of up to 30 percent was possible using existing data. That range of variation translates to a ±10 percent variation in losses compared to the losses calculated using Eq. (1) from ACI 318-08.

The effect of creep on the loss calculation is more pronounced. Figures 7.6b and 7.6c indicate the loss calculations for variation of creep coefficient for both double-tee

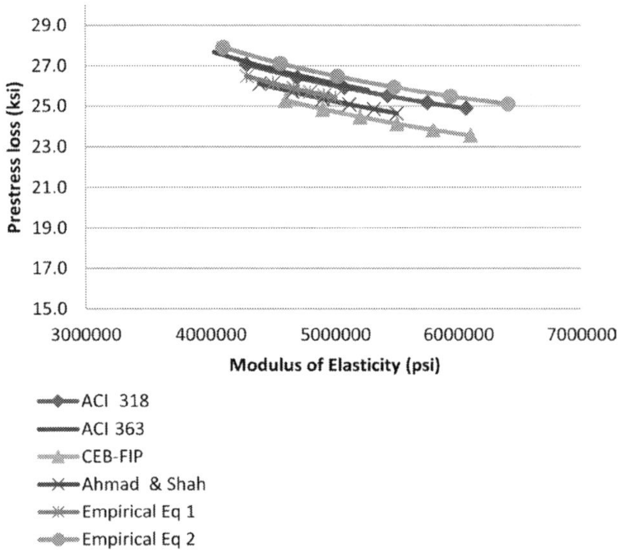

Fig. 7.6a—Prestress losses for double tee with variation in modulus of elasticity. (Note: 1000 psi = 6.896 MPa.)

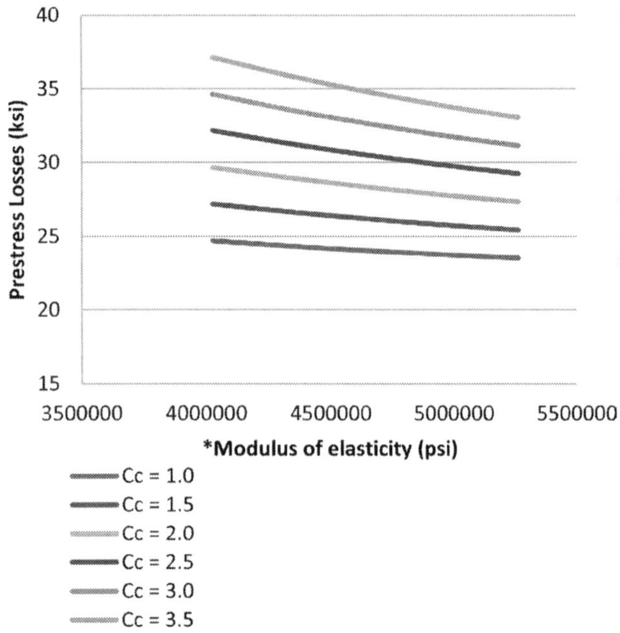

Fig. 7.6b—Double-tee beam losses for various creep coefficients. (Note: 1000 psi = 6.896 MPa.)

beam and hollow core members with f_c' varying from 5000 to 10,000 psi (34.5 to 69.0 MPa). Using a creep coefficient of 2.0 as a baseline and an increment of ±0.5 for the creep coefficient, the variation in losses for double-tee beams range from ±23 percent with the low modulus to ±17 percent for the high modulus. The loss variations for hollow core slabs are similar (Fig. 7.6c).

While the examples are based on Eq. (1) (ACI 318-08) from Table 7.4.2a, it is instructive to consider the effect of a change in the source data for the calculations. If the CEB-FIP (2010) modulus equation (Eq. (6) in Table 7.4.2a) is used for the modulus of elasticity and a normal creep coefficient is used to estimate the creep strains, then the initial

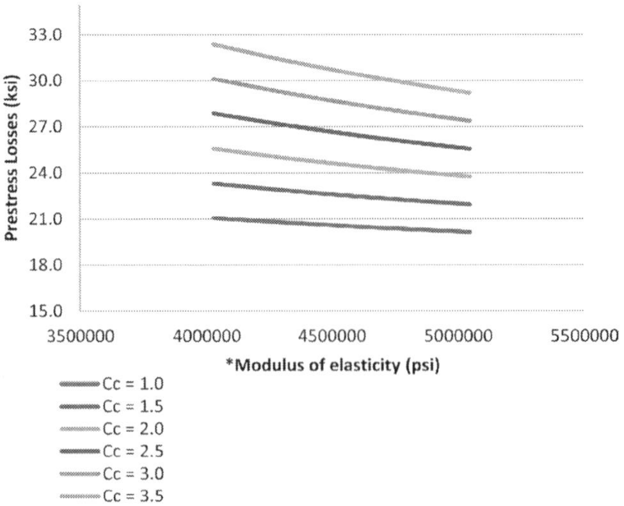

Fig. 7.6c—Hollow core slab losses for various creep coefficients. (Note: 1000 psi = 6.896 MPa.)

modulus of elasticity would have been overestimated when compared to a prediction using the ACI equation. The initial elastic shortening would then be lower. When multiplied by the creep coefficient, the result would predict a lower total deformation and correspondingly lower losses.

7.7—Shrinkage case study

The variational analysis examines only modulus of elasticity and creep. Figure 7.3a indicated that shrinkage has a large impact on losses. Shrinkage variation is not as easy to quantify as a variation in creep coefficient, so a case study is used to illustrate variations due to shrinkage. Lark et al. (2004) observed the short- and long-term behavior of two bridge girders on the Cogan and Grangetown Viaducts in the United Kingdom. Both were cast during summer, but Grangetown Viaduct was cast when the humidity level was 23 percent higher than when Cogan was cast. The difference in the strain behavior of these two bridges is shown in Fig. 7.7.

Figure 7.7 indicates substantial differences between predicted and measured strains in the structure, even when sophisticated modeling tools are used. A nearly 3-to-1 difference in strains is observed between the Cogan and Grangetown Viaducts at an age of 80 days, as noted in the figure.

Bridge girders have a smaller surface-to-volume ratio than flat slabs. Thus, it is reasonable to expect that variation of shrinkage strains in flat slab construction will be at least as large as beams, as is reflected in Fig. 7.3a.

7.8—Self-consolidating concrete

Self-consolidating concrete (SCC) is coming into wider use and its properties, as they affect losses, are changing with newer mixture designs. Because most SCC mixtures sacrifice coarse aggregate content in preference for fine materials, including both fine aggregate and cementitious materials, the modulus of elasticity of SCC is expected to be less than that of normal concrete of similar strength. Accordingly, the licensed design professional may expect that SCC will be affected by creep strains to a much larger extent than

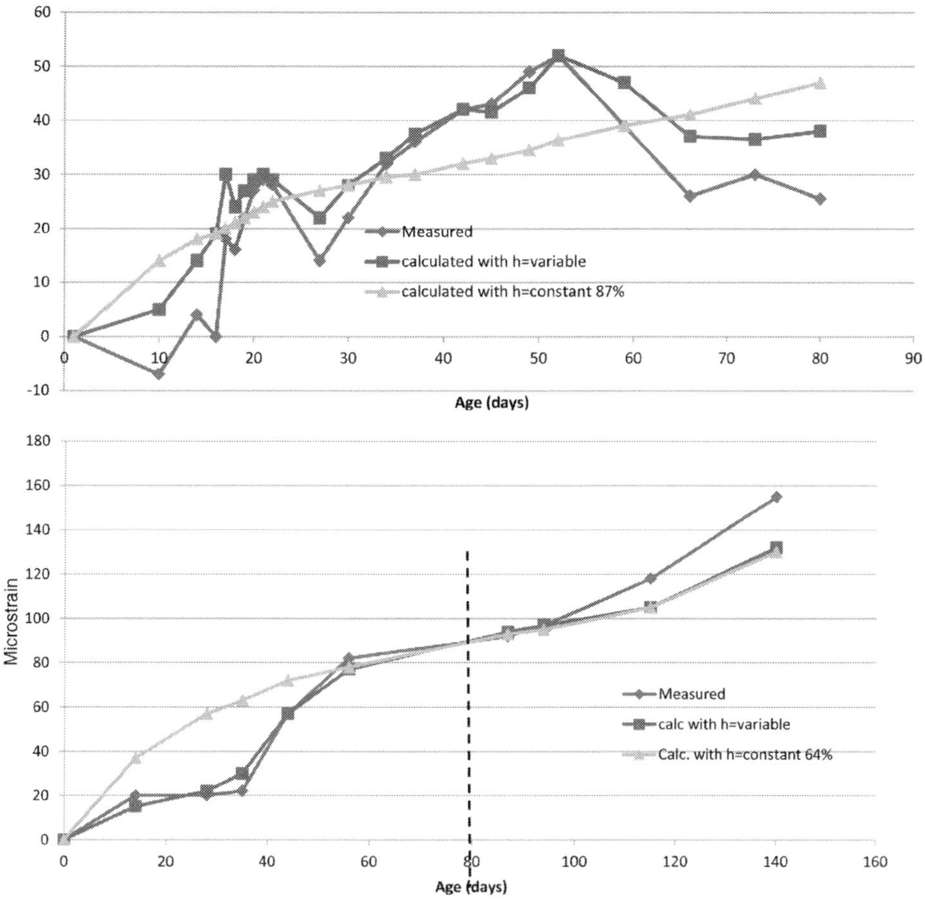

Cogan Viaduct

Fig. 7.7—Comparison of theoretical and measured strains in the Grangetown and Cogan Viaducts (adapted from Lark et al. (2004)).

normal concrete. A full discussion of the effects of SCC, their components, and the impact on variation of losses is beyond the scope of this section. An awareness of the impact of SCC on modulus, creep, and shrinkage is important to understand loss calculations. Additional information on losses in SCC is found in Levy et al. (2010), Schindler et al. (2007), Long and Khayat (2011), Gross et al. (2007), and Kavanaugh et al. (2008).

7.9—Conclusions

This chapter examines the range of a small number of factors affecting loss computations. The computations suggest that variations in losses between a baseline calculation and performance do exist. Figure 7.9, from Gilbertson and Ahlborn (2004), shows variations also exist for different calculation methods for a box girder. Variation in losses resulting from variations in modulus of elasticity and creep alone can approach ±45 percent if each factor is considered separately and then combined. From Fig. 7.3b and 7.9, the variation in losses from the mean ranges from 15 to 30 percent for the PCI general and simplified approaches discussed earlier. Further examples of loss calculations are shown in Chapter 8.

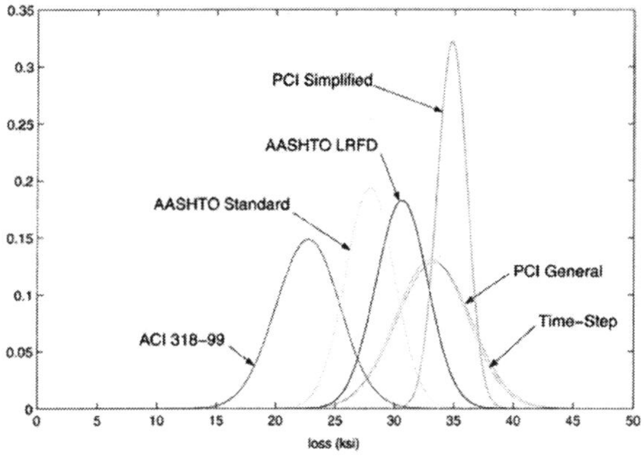

Fig. 7.9—Estimated losses for box girder from all sources (Gilbertson and Ahlborn 2004). (Note: 1 ksi = 6.895 MPa.)

Note that the maximum variation of all factors occurring simultaneously is unlikely. While variations for each factor might seem large, the variation in total losses ranges from 8 to 15 percent of the typical calculated losses. The variation between theoretical and measured loss decreases with improved knowledge of the concrete properties.

CHAPTER 8—EXAMPLES

This chapter presents several examples of prestress loss calculations. The first example is a pretensioned double-tee beam, and several methods are used to calculate losses. This example illustrates the differences in the various methods in terms of result and the computational effort required to perform the calculations. Two post-tensioned examples are then presented, an unbonded slab and a beam with bonded tendons. Only the simplified approach is used for these examples. Finally, an example is presented illustrating the effects of temperature changes prior to prestress transfer in a pretensioned beam.

8.1—Pretensioned double-tee beam

Example 8.1 presents several approaches to the calculation of prestress losses in a typical double-tee beam. The elastic shortening losses are calculated with four methods and the long-term losses are calculated with four methods. At the end of the section, results are compared.

8.1.1 *Givens and problem statement*—The beam in this problem is a double-tee beam with the designation 10LDT 32 + 2; dimensions as shown in Fig. 8.1.1.

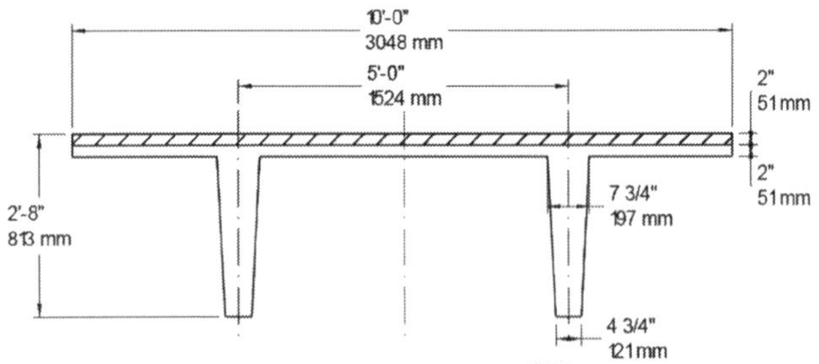

Fig. 8.1.1—Double-tee beam.

Span ℓ = 70 ft (21.3 m)
No superimposed dead load except topping
RH = 75 percent

Section properties (untopped):
A_g = 615 in.^{2} (3.97 × 10^5 mm^2)
I_g = 59,720 in.^{4} (2.49 × 10^{10} mm^4)
y_{bott} = 21.98 in. (558.3 mm)
S_b = 2717 in.^{3} (4.45 × 10^7 mm^3)
V/S = 1.69 in. (42.9 mm)

Precast concrete—sand-lightweight with unit weight:
w_c = 115 lb/ft^3 (1842 kg/m^3)

Self-weight of the double-tee beam:
w_{self} = 491 lb/ft (7170 N/m)

Weight of the normalweight concrete topping placed after erection (assume age of double-tee is 90 days):
w_t = 250 lb/ft (3650 N/m)

Properties of the sand-lightweight concrete:
f_c' = 5000 psi (34.5 MPa)
E_c = 2,900,000 psi (20.0 × 10^3 MPa)
f_{ci}' = 3500 psi (24.1 MPa)
E_{ci} = 2,400,000 psi (16.5 × 10^3 MPa)

Properties of the prestressing steel:
(12) 1/2-in. diameter, 270,000 psi, low-relaxation strands (12.7 mm dia., 1860 MPa)
A_{ps} = 12(0.153) = 1.836 in.^{2} (12 (98.7) = 1184 mm^2)
E_p = 28,500,000 psi (197 × 10^3 MPa)
Jacking stress = $0.75f_{pu}$

Depressed at midspan with the following eccentricities:
At the ends of the beam:
$e_e = 12.81$ in. (325.4 mm)
At midspan, single drape point:
$e_c = 18.73$ in. (475.7 mm)

Problem statement: Determine total prestress loss.

8.1.2 *Elastic shortening losses*—This section presents four approaches to calculate elastic shortening.

8.1.2.1 *Gross section approximation*—For depressed strands with a single drape point, the critical section is approximately at 0.4ℓ. The moments, eccentricity, and prestress force at this location are first calculated.

M at $0.4\ell = wx(\ell - x)/2 = w(0.4\ell)(\ell - 0.4\ell)/2 = 0.120w\ell^2$

The moment due to the self-weight of the beam at the 0.4ℓ point is

$M_g = 0.12(491)(70)^2 = 289{,}000$ ft-lb $= 3{,}468{,}000$ in-lb (391.8×10^6 N-mm)

The eccentricity e at $0.4\ell = 12.81 + 0.8(18.73 - 12.81) = 17.55$ in. (445.8 mm)
Assume compensation for anchorage seating loss during prestressing, so the force in the strand immediately before release is

$P_j = 0.75A_{ps}f_{pu} = 0.75(1.836)(270{,}000) = 371{,}800$ lb (1,654,000 N)

Determine the concrete stress at the tendon level, f_{cir}, making the assumption that the force after elastic shortening is 90 percent of the jacking force, as suggested in Zia et al. (1979).

$$f_{cir} = K_{cir}\left(\frac{P_j}{A_g} + \frac{P_j e_p^{\,2}}{I_g}\right) - \frac{M_g e_p}{I_g}$$

$$f_{cir} = 0.9\left(\frac{371{,}800 \text{ lb}}{615 \text{ in.}^2} + \frac{371{,}800 \text{ lb} \times (17.55 \text{ in.})^2}{59{,}720 \text{ in.}^4}\right) - \frac{3{,}468{,}000 \text{ in.-lb} \times 17.55 \text{ in.}}{59{,}720 \text{ in.}}$$

$$f_{cir} = 1251 \text{ psi } (8.625 \text{ MPa})$$

$$\Delta f_{pES} = f_{cir}\frac{E_p}{E_c'} = 1251 \text{ ksi}\frac{28{,}500{,}000 \text{ psi}}{2{,}400{,}000 \text{ psi}} = \underline{\underline{14{,}900 \text{ psi}}} \text{ (103 MPa)}$$

8.1.2.2 *Gross section with iteration*—Begin with the elastic shortening loss calculated in (8.1.2.1), and determine what fraction of the original force the loss represents; calculate a new value of K_{cir}

$$K_{cir} = \frac{\text{stress after transfer}}{\text{jacking stress}} = \frac{202{,}500 \text{ psi} - 14{,}900 \text{ psi}}{202{,}500 \text{ psi}} = 0.926$$

Use the new value of K_{cir} to calculate a new value of f_{cir}

$$f_{cir} = 0.926\left(\frac{371{,}800 \text{ lb}}{615 \text{ in.}^2} + \frac{371{,}800 \text{ lb} \times (17.55 \text{ in.})^2}{59{,}720 \text{ in.}^4}\right) - \frac{3{,}468{,}000 \text{ in.-lb} \times 17.55 \text{ in.}}{59{,}720 \text{ in.}}$$

$$f_{cir} = 1316 \text{ psi } (9.07 \text{ MPa})$$

Use the new f_{cir} to calculate a new elastic shortening loss:

Check if this is consistent with $K_{cir} = 0.926$

$$K_{cir} = \frac{\text{stress after transfer}}{\text{jacking stress}} = \frac{202{,}500 \text{ psi} - 15{,}540 \text{ psi}}{202{,}500 \text{ psi}} = 0.923$$

The new K_{cir} is within 1 percent of the starting value, so the iteration can be concluded, or an additional iteration can be performed with the new K_{cir}

$$f_{cir} = 0.923\left(\frac{371{,}800 \text{ lb}}{615 \text{ in.}^2} + \frac{371{,}800 \text{ lb} \times (17.55 \text{ in.})^2}{59{,}720 \text{ in.}^4}\right) - \frac{3{,}468{,}000 \text{ in.-lb} \times 17.55 \text{ in.}}{59{,}720 \text{ in.}}$$

$$f_{cir} = 1309 \text{ psi (9.03 MPa)}$$

$$\Delta f_{pES} = f_{cir}\frac{E_p}{E_c'} = 1309 \text{ psi}\frac{28{,}500{,}000 \text{ psi}}{2{,}400{,}000 \text{ psi}} = \underline{\underline{15{,}540 \text{ psi}}} \text{ (107.1 MPa)}$$

$$K_{cir} = \frac{\text{stress after transfer}}{\text{jacking stress}} = \frac{202{,}500 \text{ psi} - 15{,}540 \text{ psi}}{202{,}500 \text{ psi}} = 0.923$$

The new K_{cir} matches the starting value, so the iteration is complete.

8.1.2.3 *Closed form of iterative method*—Elastic shortening losses are calculated using Eq. (4.3.3), which is

$$\Delta f_{pES} = \frac{e_p M_g A_g - A_{ps} f_{pbt}\left(I_g + e_p^2 A_g\right)}{A_{ps}\left(I_g + e_p^2 A_g\right) + \dfrac{A_g I_g}{n_p}}$$

$$\Delta f_{pES} = \frac{17.55\text{in.}(3{,}468{,}000 \text{ in.-lb})(615 \text{ in.}^2) - 1.836 \text{ in.}^2(202{,}500 \text{ psi})(59{,}720 \text{ in.}^4 + (17.55 \text{ in.})^2(615 \text{ in.}^2))}{1.836 \text{ in.}^2(59{,}720 \text{ in.}^4 + (17.55 \text{ in.})^2(615 \text{ in.}^2)) + \dfrac{615 \text{ in.}^2(59{,}720 \text{ in.}^4)}{\dfrac{28{,}500{,}000 \text{ psi}}{2{,}400{,}000 \text{ psi}}}}$$

$$\Delta f_{pES} = \underline{\underline{-15{,}550 \text{ psi}}} \text{ (107.2 MPa)}$$

Note that the result is a negative number indicating a prestress loss.

8.1.2.4 *Transformed section method*—First the transformed area and moment of inertia are calculated:

$$A_t = A_g + A_{ps}(n_p - 1)$$

$$n_p = \frac{E_p}{E_{ci}'} = \frac{28{,}500{,}000 \text{ psi}}{2{,}400{,}000 \text{ psi}} = 11.88$$

$$A_t = 615 \text{ in.}^2 + 1.836 \text{ in.}^2(11.88 - 1) = 635.0 \text{ in.}^2 \text{ } (4.10 \times 10^5 \text{ mm}^2)$$

$$cg_t = \frac{A_g y_g + A_{ps}(n_p - 1)y_{ps}}{A_t}$$

$$cg_t = \frac{615 \text{ in.}^2(21.98 \text{ in.}) + (1.836 \text{ in.}^2)(11.88 - 1)(4.43 \text{ in.})}{635.0 \text{ in.}^2} = 21.43\text{in. (544.3 mm)}$$

$$I_t = I_g + A_g(y_g - cg_t)^2 + A_{ps}(n_p - 1)(y_{ps} - cg_t)^2$$

$$I_t = 59{,}720 \text{ in.}^4 + 615 \text{ in.}^2(21.98 \text{ in.} - 21.43 \text{ in.})^2 + (1.836 \text{ in.}^2)(11.88 - 1)(21.43 \text{ in.} - 4.43 \text{ in.})^2$$

$$I_t = 65{,}676 \text{ in.}^4 \text{ } (2.73 \times 10^{10} \text{ mm}^4)$$

Calculate the concrete stress at the strand level using the full jacking force, self-weight moment, and transformed section properties

$$f_{cir} = \frac{371{,}800 \text{ lb}}{635.0 \text{ in.}^2} + \frac{371{,}800 \text{ lb}(21.43 \text{ in.} - 4.43 \text{ in.})^2}{65{,}676 \text{ in.}^4} - \frac{3{,}468{,}000 \text{ in.-lb } (21.43 \text{ in.} - 4.43 \text{ in.})}{65{,}676 \text{ in.}^4}$$

$$f_{cir} = 1324 \text{ psi (9.30 MPa)}$$

$$\Delta f_{pES} = 1324\text{ psi}(11.88) = \underline{\underline{15{,}730\text{ psi}}}\ (108.5\text{ MPa})$$

Note that, using the iterative method in 8.1.2.2, with the net section properties of the beam, will give the same elastic shortening result as the transformed section method.

8.1.2.5 *Summary of elastic shortening calculations for double-tee beam*—Table 8.1.2.5 presents the results of the elastic shortening calculations using the four methods. Note that all losses are considered positive.

Table 8.1.2.5—Comparison of elastic shortening calculations

Method	Section	Elastic shortening loss	
Gross section approximation	8.1.2.1	14,900 psi	103.0 MPa
Gross section approximation with iteration	8.1.2.2	15,500 psi	107.1 MPa
Closed form solution to iteration	8.1.2.3	15,550 psi	107.2 MPa
Transformed section method	8.1.2.4	15,700 psi	108.5 MPa

The gross section approximation method predicts less loss than the other three methods, which predict similar losses. This is due to the approximation of K_{cir} as 0.9, which is less than the actual value. Depending on the problem, this approximation may be larger or smaller than the more-refined methods. This is a source of error in the approximate method; however, the effort in the calculation is less than the other three. A designer can weigh the implications of the loss of accuracy compared to the lower level of effort required. Also recall, the accuracy of all of the methods is dependent on the modulus of elasticity, which is difficult to predict accurately.

8.1.3 *Long-term losses*—In this section, four approaches to the calculation of long-term losses are presented:
(a) Simplified method (Chapter 5)
(b) AASHTO (2012) refined method (6.3.2)
(c) General age-adjusted effective modulus method (6.3.3)
(d) Incremental time-step method (6.4)
The AASHTO (2012) simplified method (5.5) is not applicable because the topping is not composite.

8.1.3.1 *Simplified method*—The starting point for the simplified method is the simplest approach to the calculation of elastic shortening loss, the gross section approximation. So the values of f_{cir} and Δf_{pES} in the calculations are 1251 psi (8.63 MPa) and 14,900 psi (103.0 MPa), respectively.

First, the moment and stress at the tendon level due to the topping are calculated:

$$M_{sd} = 0.12 \times 250\text{ lb/ft}(70\text{ ft})^2 = 147{,}000\text{ ft-lb} = 1{,}764{,}000\text{ in.-lb}\ (199 \times 10^6\text{ N-mm})$$

$$f_{cds} = \frac{M_{sd}e_p}{I_g} = \frac{1{,}764{,}000\text{ in.-lb} \times 17.55\text{ in.}}{59{,}720\text{ in.}^4} = 518\text{ psi}\ (3.57\text{ MPa})$$

Calculate the prestress loss due to creep:

$$\Delta f_{pCR} = K_{cr}(f_{cir} - f_{cds})\frac{E_p}{E_c} = 1.6(1251\text{ psi} - 518\text{ psi})\frac{28{,}500{,}000\text{ psi}}{2{,}900{,}000\text{ psi}} = \underline{\underline{11{,}500\text{ psi}}}\ (79.3\text{ MPa})$$

Calculate the loss in prestress due to shrinkage:

$$\Delta f_{pSH} = 8.2\times10^{-6}K_{sh}E_p\left(1-0.06\frac{V}{S}\right)(100-RH)$$

$$\Delta f_{pSH} = 8.2\times10^{-6}\times1.0\times28{,}500{,}000\text{ psi}(1-0.06\times1.69)(100-75) = \underline{\underline{5300\text{ psi}}}\ (36.5\text{ MPa})$$

Calculate the prestress loss due to relaxation:

$$\Delta f_{pRE} = [K_{re} - J(\Delta f_{p,SH} + \Delta f_{p,CR} + \Delta f_{p,ES})]C$$

From Table 5.4:

K_{re} = 5000 psi (34.5 MPa)

J = 0.04

Using Eq. (5.4.2a) with:

$f_{pi}/f_{pu} = 0.75$

$C = 1.0$

Δf_{pRE} = [5000 psi – 0.04(5300 psi + 11,500 psi + 14,900 psi)] 1.0 = 3700 psi (25.5 MPa)

Sum the three losses to arrive at the long-term loss:

$\Delta f_{pLT} = \Delta f_{pCR} + \Delta f_{pSH} = \Delta f_{pRE}$ = 11,500 psi + 5300 psi + 3700 psi = 20,500 psi (141.0 MPa)

The total change in prestress from just before release to end of service, not including elastic gains from topping is:

$\Delta f_p = \Delta f_{pES} + \Delta f_{pLT}$ = 14,900 psi + 20,500 psi = 35,400 psi (244.0 MPa)

The effective prestress at end of service is:

$f_{pe} = f_{pJACK} - \Delta f_p$ = 202,500 psi – 35,400 psi = 167,100 psi (1152 MPa)

To calculate the stresses in the beam at the end of service, this tendon stress is used along with gross section properties. It is common practice to not include the increase in tendon stress associated with applied loads such as the topping or live loads in the calculation of f_{pe}.

8.1.3.2 *AASHTO LRFD (2012) refined method*—The AASHTO LRFD refined method uses the elastic shortening losses and associated concrete compressive stresses as calculated with the transformed section method, so the values of f_{cir}, and $\Delta f_{p,ES}$ in the calculations are 1324 psi (9.13 MPa) and 15,700 psi (108.2 MPa), respectively. AASHTO's notation for f_{cir} is f_{cgp}.

Assume topping is noncomposite.
Calculate creep coefficients (using the AASHTO LRFD (2012) Model)
$\phi(t,t_i) = 1.9 k_s k_{hc} k_f k_{td} t_i^{-0.118}$

For the creep due to initially applied loads considered at the time of topping placement:
t_i = 1-day steam cured
t = 90 days
$k_s = 1.45 - 0.13(V/S) \geq 1.0$
$k_s = 1.45 - 0.13(1.69) = 1.23$
$k_{hc} = 1.56 - 0.008RH$
$k_{hc} = 1.56 - 0.008(75) = 0.96$

$$k_f = \frac{5}{1+f'_{ci}} = \frac{5}{1+3.5} = 1.11$$

At 90 days:

$$k_{td} = \frac{t}{61-4f'_{ci}+t} = \frac{90}{61+4(3.5)+90} = 0.657$$

$\phi(90,1) = (1.9)(1.23)(0.96)(1.11)(0.657)(1) = 1.636$

For the creep due to initially applied loads considered at the end of service:
All factors remain the same except the k_{td} term is 1.0
$\phi(\infty,1) = (1.9)(1.23)(0.96)(1.11)(1.0)(1) = 2.49$

For the creep due to the topping, applied at 90 days, considered at the end of service:
There is no change in the k_s and k_{hc} terms, and k_{td} term is 1.0

$$k_f = \frac{5}{1+f'_{ci}} = \frac{5}{1+5} = 0.833$$

$$\phi(\infty,90) = (1.9)(1.23)(0.96)(0.833)(1.0)(90^{-0.118}) = 1.10$$

Calculate shrinkage strains:

$$\varepsilon_{sh} = k_s k_{hs} k_f k_{td}(0.48 \times 10^{-3})$$

Shrinkage at 90 days:
The k_s, k_f, and k_{td} terms are the same as the first creep calculation
$k_{hs} = 2.00 - 0.014RH = 2.00 - 0.014(75) = 0.95$
$\varepsilon_{sh}(90) = (1.23)(0.95)(1.11)(0.657)0.48 \times 10^{-3} = 0.409 \times 10^{-3}$

Shrinkage at end of service:
All terms are the same as 90-day calculation, except $k_{td} = 1.0$
$\varepsilon_{sh}(\infty) = (1.23)(0.95)(1.11)(1.0)0.48 \times 10^{-3} = 0.623 \times 10^{-3}$

The AASHTO method separates the losses into two time periods: release to deck casting (*id*), and deck casting to end of service (*df*). The total time-dependent losses can be split into the following components:

$$\Delta f_{pLT} = (\Delta f_{pSR} + \Delta f_{pCR1} + \Delta f_{pR1})_{id} + (\Delta f_{pSD} + \Delta f_{pCD} + \Delta f_{pR2} - \Delta f_{pSS})_{df}$$

The following sections illustrate the calculation of each component.

Losses due to shrinkage from release to deck placement, Δf_{pSR}
First, the section modification term, K_{id}, is calculated:

$$K_{id} = \frac{1}{1+\dfrac{E_p}{E_{ci}}\dfrac{A_{ps}}{A_g}\left(1+\dfrac{A_g e_{pg}^2}{I_g}\right)\left(1+0.7\phi_b(t_f,t_i)\right)}$$

$$K_{id} = \frac{1}{1+\dfrac{28{,}500{,}000}{2{,}400{,}000}\dfrac{1.836}{615}\left(1+\dfrac{615(17.55)^2}{59{,}720}\right)(1+0.7(2.49))} = 0.72$$

The AASHTO equation for losses due to shrinkage from release to placement is:

$\Delta f_{pSR} = \varepsilon_{sh}(90)E_p K_{id} = 0.409 \times 10^{-3}(28{,}500{,}000 \text{ psi})(0.72) =$ 8400 psi (57.9 MPa)

Losses due to creep from release to deck placement, Δf_{pCR1}

$$\Delta f_{pCR1} = \frac{E_p}{E_{ci}} f_{cgp}\phi(t_d,t_i)K_{id} = \frac{28{,}500{,}000}{2{,}400{,}000}(1324)(1.636)(0.72) = 18{,}500 \text{ psi } (127.6 \text{ MPa})$$

Losses due to relaxation from release to deck placement, Δf_{pR1}
Calculate strand stress after elastic shortening losses:

$f_{pt} = f_{pJACK} - \Delta f_{pES} = 202{,}500 \text{ psi} - 15{,}700 \text{ psi} = 186{,}000 \text{ psi } (1288 \text{ MPa})$

$K_L = 30$ for low-relaxation strands

$$\Delta f_{pR1} = \frac{f_{pt}}{K_L}\left(\frac{f_{pt}}{f_{py}} - 0.55\right) = \frac{186{,}800}{30}\left(\frac{186{,}800}{243{,}000} - 0.55\right) = 1400 \text{ psi } (9.65 \text{ MPa})$$

Losses due to shrinkage from deck placement to end of service, Δf_{pSD}

Because the topping is noncomposite, the section modification term is the same for both time intervals:

$$K_{id} = K_{df} = 0.72$$

Calculate the shrinkage from time of deck placement to end of service:

$$\varepsilon_{bdf} = \varepsilon_{sh}(\infty) - \varepsilon_{sh}(90) = (0.623 - 0.409) \times 10^{-3} = 0.214 \times 10^{-3}$$

$$\Delta f_{pSD} = \varepsilon_{bdf} E_p K_{df} = (0.214 \times 10^{-3})(28{,}500{,}000)0.72 = \underline{4400\text{ psi}}\ (30.3\text{ MPa})$$

Losses due to creep from deck placement to end of service, Δf_{pCD}

There are two causes of creep during this time interval. First, there will be continued creep due to the initially applied loads: prestress and self-weight. In most cases, this creep will be compressive creep at the bottom of the beam, which will continue to reduce the strand stress. Second, there will be creep from the addition of the permanent load of the topping. This creep will be tensile, increasing the strand tension. The AASHTO (2012) method also includes applying the prestress loss in the first interval as a tensile force, and accounting for the tensile creep associated with the prestress loss. It can be argued that this is unnecessary, because using the age-adjusted modulus has already accounted for it. The method now combines a stepwise analysis with an age-adjusted approach, which is double-counting the creep associated with prestress loss. This example, however, presents the calculations as dictated by the AASHTO method.

Calculate the long-term loss from release to deck placement:

$$\Delta f_{pid} = \Delta f_{pSR} + \Delta f_{pCR1} + \Delta f_{pR1} = 8400\text{ psi} + 18{,}500\text{ psi} + 1400\text{ psi} = 28{,}300\text{ psi}\ (195.1\text{ MPa})$$

Calculate the moment at 0.4ℓ due to the topping (8.1.1)

$$M_{SDL} = 147{,}000\text{ ft-lb} = 1{,}764{,}000\text{ in-lb}\ (199{,}300\text{ N-m})$$

Apply the topping dead load moment and the prestress loss to the section and calculate the concrete stress at the strand level. Use the transformed section properties for the dead load, and the gross section properties for the prestress loss.

$$\Delta f_{cd} = \frac{(-28{,}300\text{ psi})(1.836\text{ in.}^2)}{615\text{ in.}^2} + \frac{(-28{,}300\text{ psi})(1.836\text{ in.}^2)(17.55\text{ in.})^2}{59{,}720\text{ in.}^4}$$
$$-\frac{1{,}764{,}000\text{ in.-lb}(17.02\text{ in.})}{65{,}423\text{ in.}^4} = -811\text{ psi}\ (-5.59\text{ MPa})$$

This is a tensile stress. Now calculate total change in prestress due to creep from deck placement to final time:

$$\Delta f_{pCD} = \frac{E_p}{E_{ci}} f_{cgp}\left[\phi_b(t_f,t_i) - \phi_b(t_d,t_i)\right]K_{df} + \frac{E_p}{E_c}\Delta f_{cd}\phi(t_f,t_d)K_{df}$$

$$\Delta f_{pCD} = \frac{28{,}500{,}000\text{ psi}}{2{,}400{,}000\text{ psi}}(1324\text{ psi})(2.49 - 1.636)0.72 + \frac{28{,}500{,}000\text{ psi}}{2{,}900{,}000\text{ psi}}(-811\text{ psi})(1.1)0.72$$

$$\Delta f_{pCD} = \underline{\underline{3400\text{ psi}}}\ (23.4\text{ MPa})$$

Loss due to relaxation from deck placement to end of service, Δf_{R2}

AASHTO allows the simplifying assumption that the relaxation loss from initial to deck placement is equal to the loss from deck placement to final.

$$\Delta f_{R2} = \Delta f_{R1} = \underline{1400\text{ psi}}\ (9.65\text{ MPa})$$

Gain in prestress force due to differential shrinkage of topping relative to girder

Because the topping is considered to be noncomposite, this calculation is not required. There is some interaction between the topping and the girder, which results in downward displacement and tension in the concrete at the strand level. Common practice is to ignore this interaction, because there is no reinforcement to develop composite action between the topping and the beam. If the cohesion between the topping and the beam is broken, any stresses that may have built up due to differential shrinkage will be relieved. Also, if the topping is thin, the differential shrinkage stresses will be small.

Elastic gain in prestress force due to topping

To determine the effective prestress at end of service, all elastic gains are calculated. The concrete stress at the strand level due to topping is calculated:

$$\Delta f_{cSDL} = -\frac{1{,}764{,}000 \text{ in.-lb } (17.02 \text{ in.})}{65{,}423 \text{ in.}^4} = -459 \text{ psi } (-3.16 \text{ MPa})$$

Calculate the change in prestress corresponding to the prior change in concrete stress:

$$\Delta f_{pSDL} = \frac{E_p}{E_c}\Delta f_{cSDL} = \frac{28{,}500{,}000 \text{ psi}}{2{,}900{,}000 \text{ psi}}(-459 \text{ psi}) = \underline{\underline{-4500 \text{ psi}}} \ (-31.0 \text{ MPa})$$

Summary

The total long-term change in prestress force is calculated as:

$$\Delta f_{pLT} = (\Delta f_{pSR} + \Delta f_{pCR1} + \Delta f_{pR1})_{id} + (\Delta f_{pSD} + \Delta f_{pCD} + \Delta f_{pR2} + \Delta f_{pSS})_{df}$$
$$\Delta f_{pLT} = (8400 + 18{,}500 + 1400) + (4400 + 3400 + 1400 + 0) = \underline{37{,}500 \text{ psi}} \ (258 \text{ MPa})$$

The total change in prestress from just before release to end of service, including elastic losses and gains is:

$$\Delta f_p = \Delta f_{pES} + \Delta f_{pLT} + \Delta f_{pSDL} = 15{,}700 \text{ psi} + 37{,}500 \text{ psi} - 4500 \text{ psi} = 48{,}700 \text{ psi } (336 \text{ MPa})$$

The effective prestress at end of service is:

$$f_{pe} = f_{pJACK} - \Delta f_p = 202{,}500 \text{ psi} - 48{,}700 \text{ psi} = 153{,}800 \text{ psi } (1060 \text{ MPa})$$

8.1.4 *Double-tee beam with general age-adjusted effective modulus method*

Using creep and shrinkage values from the previous example, six equations are written to solve for six unknowns. One of the unknowns is the change in the force in the prestressing strands.

Equations of internal equilibrium

$$\Delta N_b + \Delta N_{ps} = 0$$
$$\Delta M_b + \Delta N_{ps} \cdot e_p = 0$$

Using gross cross-sectional properties, e_p = 17.55 in. (445.8 mm), so:

$$\Delta M_b + 17.55\Delta N_{ps} = 0$$

Constitutive equations

$$\Delta\varepsilon_p = \frac{\Delta N_{ps} - \Delta N_{relax}}{A_{ps}E_p}$$

The total relaxation is assumed to be 2800 psi (19.3 MPa) for 1.836 in.^{2} (1184 mm^2) of prestress, –5100 lb (–22,700 N), so the equation becomes:

$$\Delta\varepsilon_p = \frac{(\Delta N_{ps} - (-5100 \text{ lb}))}{(1.836 \text{ in.}^2)(28{,}500{,}000 \text{ psi})} = \frac{\Delta N_{ps}}{52{,}326{,}000} + 97.5\times10^{-6}$$

$$\Delta\varepsilon_b = \frac{N_b^o\phi(t,t_0)}{A_gE_c} + \frac{\Delta N_b}{A_gE_c}(1+\chi\phi(t,t_0)) + \varepsilon_{shb}(t)$$

The initial force in the beam is the jacking force minus the elastic shortening loss, the cross-sectional properties are the gross properties (as a close approximation of net), and the shrinkage term is the total shrinkage from release to end of service:

$$\Delta\varepsilon_b = \frac{-(202{,}500 - 15{,}700 \text{ psi})(1.836 \text{ in.}^2)2.49}{(615 \text{ in.}^2)(2{,}400{,}000 \text{ psi})} + \frac{\Delta N_b}{(615 \text{ in.}^2)(2{,}400{,}000 \text{ psi})}(1 + 0.7(2.49)) + (-623 \times 10^{-6})$$

$$\Delta\kappa = \frac{M_b^o \phi(\infty, t_0)}{I_g E_{ci}} + \frac{M^{SDL}\phi(\infty, t_d)}{I_t E_c} + \frac{\Delta M_b}{I_g E_{ci}}\left(1 + \chi\phi(\infty, t_0)\right)$$

Note that three creep-producing moments are considered: 1) the initially applied prestress and self-weight moment; 2) the moment from superimposed dead load; and 3) the changing moment due to creep.

M_b^o = (–(202,500 psi – 15,200 psi)(1.836 in.^{2})(17.55 in.)) + 3,436,000 = –2,599,000 in.-lb (294,000 N-m)

$$\Delta\kappa = \frac{(-2{,}599{,}000 \text{ in,-lb})2.49}{(59{,}720 \text{ in.}^4)(2{,}400{,}000 \text{ psi})} + \frac{1{,}764{,}000 \text{ in.-lb}(1.1)}{(65{,}423 \text{ in.}^4)(2{,}900{,}000 \text{ psi})} + \frac{\Delta M_b}{(59{,}720 \text{ in.}^4)(2{,}400{,}000 \text{ psi})}(1 + 0.7(2.49))$$

Equation of strain compatibility

$\Delta\varepsilon_b = \Delta\varepsilon_p - \Delta\kappa \cdot e_p$

$\Delta\varepsilon_b = \Delta\varepsilon_p - \Delta\kappa \cdot 17.55$ in.

These equations are put into a matrix form to solve for the six unknowns. The resulting values are:

$\Delta\varepsilon_b = -1069$ με

$\Delta\kappa = -11$ με/in. (0.43 με/mm)

$\Delta\varepsilon_p = -1263$ με

ΔN_b = 71,200 lb (317,000 N)

ΔM_b = 1,250,000 in.-lb (141,000 N-m)

ΔN_{ps} = –71,200 lb (–317,000 N)

The total change in prestress is:

Δf_{psLT} = 71,200lb/1.836 in.^{2} = 38,800 psi (268 MPa)

This does not, however, include the elastic change in stress due to the superimposed dead load, which was previously calculated and is 4500 psi (31.0 MPa) tension. So the total change in stress from release is:

Δf_p = 15,700 psi + 38,800 psi – 4500 psi = 50,000 psi (345 MPa)

The effective prestress at end of service is:

$f_{pe} = f_{pJACK} - \Delta f_p$ = 202,500 psi – 50,000 psi = 152,500 psi (1051 MPa)

8.1.5 *Double-tee beam with stepwise method*—Time steps were selected, with smaller initial intervals and longer intervals later in the life of the beam. The days chosen were 10, 30, 90, 100, 5000, and 20,000 days. The AASHTO creep and shrinkage model, which was used for the previous examples, was also used to compute creep coefficients and shrinkage strains at each time step. Creep coefficients were determined for the initially applied loads, and the overlay load applied at 90 days. The creep coefficients and shrinkage strains are presented in Table 8.1.5a.

Table 8.1.5a—Creep coefficients and shrinkage strains for double-tee beam

Days since casting	Total shrinkage strain, microstrain	Creep coefficient for loads applied at Day 1	Creep coefficient for loads applied at Day 90
10	109	0.437	NA
30	243	0.970	NA
90	409	1.636	NA
100	424	1.694	0.216
5000	617	2.467	1.091
20,000	622	2.490	1.098

The first step is to establish the initial conditions. Based on previous calculations using the iterative method with gross cross-sectional properties for elastic shortening losses, the initial prestress stress, after elastic shortening losses, is 186,900 psi (1289 MPa), and the initial compression at the 0.4ℓ location at the prestress level is 1309 psi (9.03 MPa) compression.

The following steps are made for each time step, based on the initial conditions:

Calculate creep strain for time step

$$\varepsilon_{cr} = \frac{f_{cgp}}{E_{ci}}\phi = \frac{1309 \text{ psi}}{2{,}400{,}000 \text{ psi}}0.437 = 238\ \mu\varepsilon$$

Calculate reduction of strand tension due to creep and shrinkage

$$\Delta f_{pCR+SH} = E_{ps}(\varepsilon_{cr} + \varepsilon_{sh}) = 28{,}500{,}000 \text{ psi}(0.000238 + 0.000109) = 9900 \text{ psi (68.3 MPa)}$$

Calculate relaxation loss for the time step

$$\Delta f_{RE} = f_{pi}\frac{\log(t\cdot 24)}{45}\left(\frac{f_{pi}}{f_{py}} - 0.55\right) = 186{,}900 \text{ psi}\left(\frac{\log(10\cdot 24)}{45}\right)\left(\frac{186{,}900 \text{ psi}}{243{,}000 \text{ psi}} - 0.55\right) = 2200 \text{psi (15.2 MPa)}$$

Calculate gross loss for this step

$$\Delta f_{psgross1} = 9900 \text{ psi} + 2200 \text{ psi} = 12{,}100 \text{ psi (83.4 MPa)}$$

Determine change in concrete stress at centroid of strand due to prestress loss

$$\Delta f_{closs} = \frac{12{,}100 \text{ psi} \times 1.836 \text{ in.}^2}{615 \text{ in.}^2} + \frac{12{,}100 \text{ psi}\times 1.836 \text{ in.}^2 (17.55 \text{ in.})^2}{59{,}720 \text{ in.}^4} = 151 \text{ psi (1.04 MPa) (tension)}$$

Calculate the associated increase in tendon tension

$$\Delta f_{p\,rebound} = 0.151 \text{ ksi}\cdot\frac{28{,}500{,}000 \text{ psi}}{2{,}400{,}000 \text{ psi}} = 1800 \text{ psi (12.4 MPa)}$$

Calculate net loss for this step

$$\Delta f_{ps1} = 12{,}000 \text{ psi} - 1800 \text{ psi} = 10{,}300 \text{ psi (71.0 MPa)}$$

This loss is subtracted from the initial prestress to arrive at the prestress at the end of the time step: 186,900 psi – 10,300 psi = 176,600 psi (1218 MPa). With this steel stress, the associated concrete stress is determined. These are the starting values for the next time step. The calculations are repeated using the increment of creep and shrinkage for each time step. After the topping is placed, the creep associated with the new load is also included in the loss calculations. The results of the calculations are presented in Table 8.1.5b.

Table 8.1.5b—Time-step prestress losses

	Day 10	Day 30	Day 90	Day 100	Day 5000	Day 20,000
Start of time step f_{ps}, psi (MPa)*	186,900 (1289)	176,600 (1218)	166,700 (1149)	155,400 (1071)	155,300 (1071)	146,300 (1009)
Start of time step f_{cgp}, psi (MPa)*	–1309 (–9.03)	–1181 (–8.14)	–1057 (–7.29)	–916 (–6.32)	–915 (–6.31)	–803 (–5.54)
Creep strain, με	–238	–262	–293	12	–157	–7
Shrinkage strain, με	–109	–133	–167	–15	–193	–4
Loss due to creep (CR) + shrinkage (SH), psi (MPa)	–9900 (–68.3)	–11,300 (–77.9)	–13,100 (–90.3)	–100 (–0.69)	–10,000 (–68.9)	–300 (–2.07)
Relaxation loss, psi (MPa)	–2200 (–15.2)	–300 (–2.07)	-200 (–1.38)	–0 (–0.0)	–500 (–3.45)	–100 (–0.69)
Gross loss for time step, psi (MPa)	–12,100 (–83.4)	–11,600 (–80.0)	–13,300 (–91.7)	–100 (–0.69)	–10,500 (–72.4)	–400 (–2.76)
Elastic rebound, psi (MPa)	1800 (12.4)	1700 (11.7)	2000 (13.8)	0 (0)	1600 (11.0)	100 (0.7)
Net loss for time step, psi (MPa)	–10,300 (–71.0)	–9900 (–68.3)	–11,300 (–77.9)	–100 (–0.7)	–8900 (–61.4)	–300 (–2.1)
End of time step f_{ps}, psi (MPa)	176,600 (1218)	166,700 (1149)	155,400 (1071)	155,300 (1071)	146,300 (1009)	146,000 (1007)
End of time step f_{cgp}, psi (MPa)	–1181 (–8.14)	–1057 (–7.29)	–916 (–6.32)	–915 (–6.31)	–803 (–5.54)	–799 (–5.51)
Concrete stress at level of strand due to topping,psi (MPa)			460 (3.17)			

*Does not include stress changes due to topping.

The total time-dependent prestress loss is 40,900 psi (282. MPa), with elastic shortening losses and gain due to topping, the effective prestress is:

f_{pe} = 202,500 – 15,600 – 40,900 + 4500 = 150,900 psi (1040 MPa)

Table 8.1.5c presents a summary of the results of the various methods to estimate prestress losses, and compares to the simplified approach presented in 8.1.1.

Table 8.1.5c—Summary of results of loss calculations for example

Method	Elastic shortening, psi (MPa)	Time-dependent, psi (MPa)	Gain from topping, psi (MPa)	Total change, psi (MPa)	Effective prestress, psi (MPa)
Simplified method	–14,900 (–102.7)	–20,500 (–141.3)	—	–35,400 (–244.1)	167,100 (1152)
AASHTO LRFD (2012) method	–15,700 (–108.2)	–37,500 (–258.0)	4500 (31.0)	–48,700 (–335.8)	153,800 (1060)
Age-adjusted effective modulus	–15,700 (–108.2)	–38,800 (–267.5)	4500 (31.0)	–50,000 (–355.7)	152,500 (1051)
Stepwise	–15,600 (–107.6)	–40,900 (–282.0)	4500 (31.0)	–51,600 (–355.8)	150,900 (1040)

Although the range of difference is on the order of 10 percent, the simplified method predicts considerably smaller loss than the other three methods. This is primarily because the creep loss calculation uses the 28-day modulus rather than the modulus at release, and the elastic shortening loss is smaller because no iteration was performed.

8.2—Post-tensioned slab with unbonded tendons

This example presents calculation of prestress losses in a typical one-way post-tensioned flat slab with unbonded monostrand tendons. The initial losses are calculated based on the methods in Chapter 4 and long-term losses are calculated using the simplified method described in Chapter 5.

8.2.1 *Givens and problem statement*—This problem considers a six-span, one-way slab spanning 18 ft (5.49 m) between 14 in. (356 mm) x 34 in. (864 mm) cast-in-place concrete beams, as shown in Fig. 8.2.1

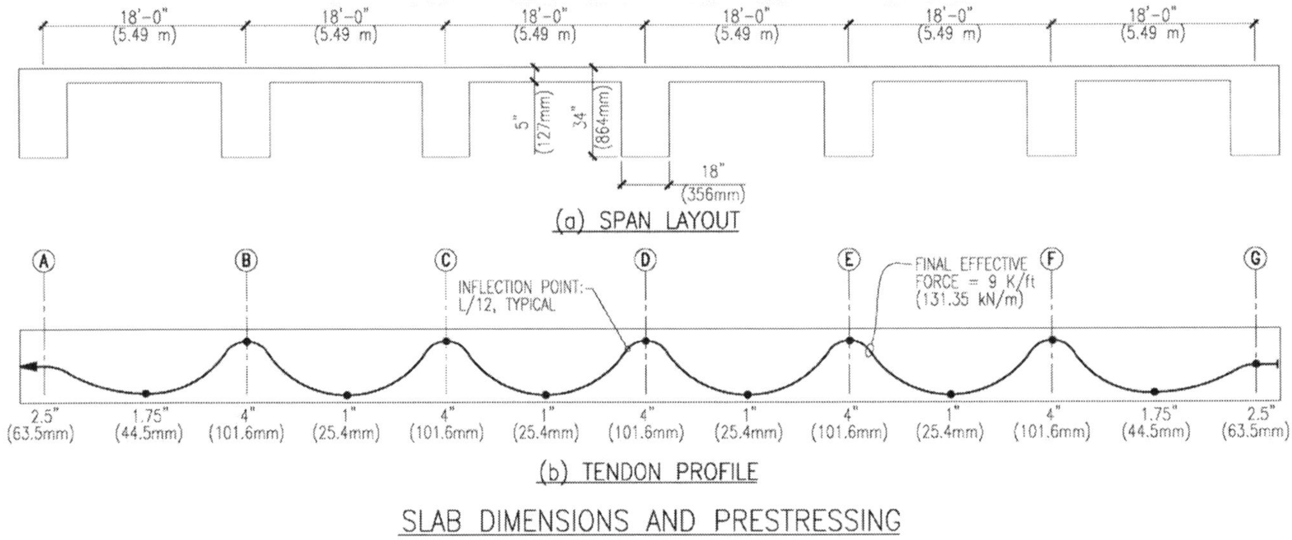

Fig. 8.2.1—Slab dimensions and prestressing.

Slab thickness = 5 in. (127 mm)
RH = 80 percent

Concrete properties: Normalweight
f_{ci}' = 3000 psi (20.68 MPa) (at transfer)
f_c' = 4000 psi (27.58 MPa)
w_c = 150 lb/ft^3 (2403 kg/m^3)
E_{ci} = 3,122,000 psi (21,530 MPa)
E_c = 3,605,000 psi (24,860 MPa)
Concrete age at stressing = 3 days

Strand properties:
Low-relaxation, unbonded system
Strand diameter = 1/2 in. (12.7 mm)
Required uniform tendon final effective force in slab = 9000 lb/ft (131,300 N/m) or one tendon spaced at 3 ft on center (0.91 m) assuming a final effective force of 27,000 lb (120,100 N) per tendon.
Strand area = 0.153 in.^{2} (98.7 mm^2)
E_p = 28,500,000 psi (196,500 MPa)
f_{pu} = 270,000 psi (1862 MPa)
Jacking stress = 0.8 × 270,000 psi (1862 MPa) = 216,000 psi (1489 MPa)
Angular friction coefficient = 0.07 (rad^{-1})
Wobble friction coefficient = 0.001 (ft^{-1}) (0.00328 m^{-1})
Anchor set = 0.25 in (6.4 mm)
Reverse parabola at L/12
Assume stressing from one end (left end)

Problem statement: Determine initial friction, anchor set and elastic shortening losses and long-term creep, shrinkage, and relaxation losses.

8.2.2 *Friction losses and elongation*—Friction and anchor set calculations are calculated using the approach described in Section 4.4 where the, "entire tendon is idealized as several individual parabolic segments." The eccentricity at each parabolic vertex is calculated with the following expression (Eq. (8.2.2a)) (variables used shown in Fig. 8.2.2a)

$$e_i = \frac{e_{i-1}(x_{i+1}-x_i)+e_{i+1}(x_i-x_{i-1})}{(x_{i+1}-x_{i-1})} \tag{8.2.2a}$$

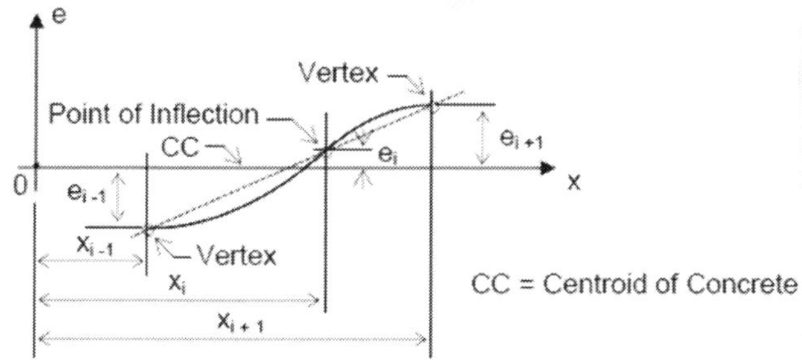

Fig. 8.2.2a—Location of inflection points.

The angular deviation is then computed for each segment (Eq. (8.2.2b)) and (Fig. 8.2.2b)

$$\theta_j = \tan^{-1}\left(2\frac{e_j-e_v}{x_v-x_j}\right) \tag{8.2.2b}$$

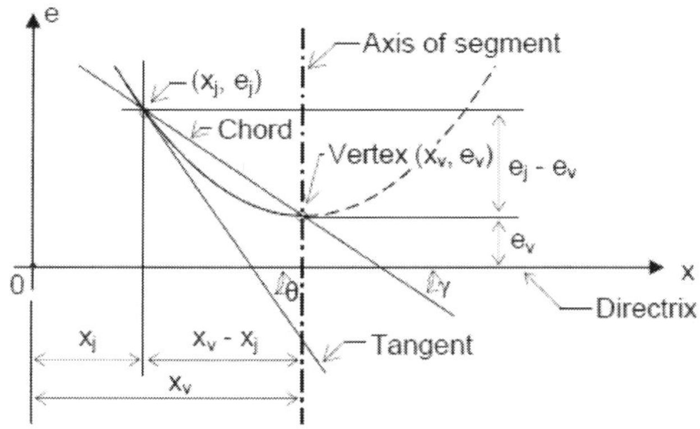

Fig. 8.2.2b—Tangent at a point on a parabolic segment.

The exponential in Eq. (8.2.2c) is computed. Multiplied by the jacking stress, it gives the tendon stress at the end of each segment before anchor set:

$$\sigma_{beforeset}(x) = \sigma_{jacking} e^{-(\mu\theta+kx)} \tag{8.2.2c}$$

Elongation of each segment before anchor set is computed with the average stress along the segment

$$elongation_i = 12 \times \frac{1}{2}\frac{\sigma_i + \sigma_{i-1}}{E_{ps}} \times (x_i - x_{i-1}) + elongation_{i-1} \tag{8.2.2d}$$

Table 8.2.2 provides the friction loss parameters calculated at various locations along the tendon length. The variables shown in the table are as defined in this section.

Table 8.2.2— Friction and anchor set losses, and elongations

	Vertex i	e_i, in. eccentricity	$\lvert e_i - e_v \rvert$, in.	x_i, ft distance from stressing end	$\lvert x_i - x_v \rvert$, in.	θ_i, rad segment angular deviation	α_i, rad angular deviation	$\exp(-k \times l_i - \alpha_i \times \mu)$ stress in percent of stressing end	Stress before set, ksi	Stress after set, ksi	Elongation, in.
Stressing end	1	2.50	0.00	0.0	0.0	0.000	0.000	1.000	216.0	188.1	0.00
	2	2.38	0.13	1.5	1.5	0.014	0.014	0.998	215.5	188.6	0.14
Midspan 1	3	1.75	0.63	9.0	7.5	0.014	0.028	0.989	213.6	190.4	0.81
	4	3.63	1.88	16.5	7.5	0.042	0.069	0.979	211.4	192.6	1.48
Support B	5	4.00	0.38	18.0	1.5	0.042	0.111	0.975	210.5	193.6	1.62
	6	3.50	0.50	19.5	1.5	0.055	0.167	0.969	209.4	194.7	1.75
Midspan 2	7	1.00	2.50	27.0	7.5	0.055	0.222	0.958	207.0	197.1	2.41
	8	3.50	2.50	34.5	7.5	0.055	0.278	0.947	204.7	199.4	3.06
Support C	9	4.00	0.50	36.0	1.5	0.055	0.333	0.942	203.6	200.5	3.19
	10	3.50	0.50	37.5	1.5	0.055	0.389	0.937	202.5	201.6	3.32
Midspan 3	11	1.00	2.50	45.0	7.5	0.055	0.444	0.927	200.2	200.2	3.95
	12	3.50	2.50	52.5	7.5	0.055	0.500	0.916	197.9	197.9	4.58
Support D	13	4.00	0.50	54.0	1.5	0.055	0.555	0.911	196.8	196.8	4.70
	14	3.50	0.50	55.5	1.5	0.055	0.611	0.906	195.8	195.8	4.83
Midspan 4	15	1.00	2.50	63.0	7.5	0.055	0.666	0.896	193.6	193.6	5.44
	16	3.50	2.50	70.5	7.5	0.055	0.722	0.886	191.4	191.4	6.05
Support E	17	4.00	0.50	72.0	1.5	0.055	0.777	0.881	190.4	190.4	6.17
	18	3.50	0.50	73.5	1.5	0.055	0.833	0.877	189.3	189.3	6.29
Midspan 5	19	1.00	2.50	81.0	7.5	0.055	0.888	0.867	187.2	187.2	6.89
	20	3.50	2.50	88.5	7.5	0.055	0.944	0.857	185.1	185.1	7.47
Support F	21	4.00	0.50	90.0	1.5	0.055	0.999	0.852	184.1	184.1	7.59
	22	3.63	0.38	91.5	1.5	0.042	1.041	0.848	183.3	183.3	7.71
Midspan 6	23	1.75	1.88	99.0	7.5	0.042	1.082	0.840	181.4	181.4	8.28
	24	2.38	0.63	106.5	7.5	0.014	1.096	0.833	179.8	179.8	8.85
Fixed end	25	2.50	0.13	108.0	1.5	0.014	1.110	0.831	179.4	179.4	8.97

	Vertex i	e_i, cm eccentricity	$\lvert e_i - e_v \rvert$, cm	x_i, m distance from stressing end	$\lvert x_i - x_v \rvert$, m	θ_i, rad segment angular deviation	α_i, rad angular deviation	$\exp(-k \times l_i - \alpha_i \times \mu)$ stress in percent of stressing end	Stress before set, MPa	Stress after set, MPa	Elongation, cm
Stressing end	1	6.4	0.0	0.00	0.00	0.000	0.000	1.000	1489.3	1296.8	0.0
	2	6.0	0.3	0.46	0.46	0.014	0.014	0.998	1485.6	1300.5	0.3
Midspan 1	3	4.4	1.6	2.75	2.29	0.014	0.028	0.989	1473.1	1313.0	2.1
	4	9.2	4.8	5.03	2.29	0.042	0.069	0.979	1457.8	1328.3	3.8
Support B	5	10.2	1.0	5.49	0.46	0.042	0.111	0.975	1451.4	1334.7	4.1
	6	8.9	1.3	5.95	0.46	0.055	0.167	0.969	1443.6	1342.5	4.4
Midspan 2	7	2.5	6.4	8.24	2.29	0.055	0.222	0.958	1427.3	1358.8	6.1
	8	8.9	6.4	10.52	2.29	0.055	0.278	0.947	1411.1	1375.0	7.8
Support C	9	10.2	1.3	10.98	0.46	0.055	0.333	0.942	1403.6	1382.6	8.1
	10	8.9	1.3	11.44	0.46	0.055	0.389	0.937	1396.0	1390.1	8.4

Midspan 3	11	2.5	6.4	13.73	2.29	0.055	0.444	0.927	1380.2	1380.2	10.0
	12	8.9	6.4	16.01	2.29	0.055	0.500	0.916	1364.6	1364.6	11.6
Support D	13	10.2	1.3	16.47	0.46	0.055	0.555	0.911	1357.3	1357.3	11.9
	14	8.9	1.3	16.93	0.46	0.055	0.611	0.906	1350.0	1350.0	12.3
Midspan 4	15	2.5	6.4	19.22	2.29	0.055	0.666	0.896	1334.7	1334.7	13.8
	16	8.9	6.4	21.50	2.29	0.055	0.722	0.886	1319.6	1319.6	15.4
Support E	17	10.2	1.3	21.96	0.46	0.055	0.777	0.881	1312.5	1312.5	15.7
	18	8.9	1.3	22.42	0.46	0.055	0.833	0.877	1305.4	1305.4	16.0
Midspan 5	19	2.5	6.4	24.71	2.29	0.055	0.888	0.867	1290.7	1290.7	17.5
	20	8.9	6.4	26.99	2.29	0.055	0.944	0.857	1276.1	1276.1	19.0
Support F	21	10.2	1.3	27.45	0.46	0.055	0.999	0.852	1269.2	1269.2	19.3
	22	9.2	1.0	27.91	0.46	0.042	1.041	0.848	1263.6	1263.6	19.6
Midspan 6	23	4.4	4.8	30.20	2.29	0.042	1.082	0.840	1250.5	1250.5	21.0
	24	6.0	1.6	32.48	2.29	0.014	1.096	0.833	1240.0	1240.0	22.5
Fixed end	25	6.4	0.3	32.94	0.46	0.014	1.110	0.831	1236.9	1236.9	22.8

8.2.3 *Anchor set losses*—The force lost in the tendon due to anchor set is represented in Fig. 8.2.3 by the area between the stress curves before and after anchor set. Knowing that the stress slopes are opposite, the anchor set length and the maximum stress f_{max} are determined numerically. First, the segment with the maximum stress is identified, then the position is determined inside that segment. In this example, the anchor set length is 38.4 ft (11.7 m) and the maximum stress f_{max} in the tendon is 202,200 psi (1394 MPa).

For tendons with many curvature reversals and low friction, such as slabs, a hand calculation is cumbersome. The simplified method shown in 4.4.2 can be used. For this example, the simplified method will give an anchor set length of 41.5 ft (12.7 m) and a maximum stress of 201,900 psi (1392 MPa). The anchor set is longer because the assumption of linear friction neglects that most friction occurs on supports.

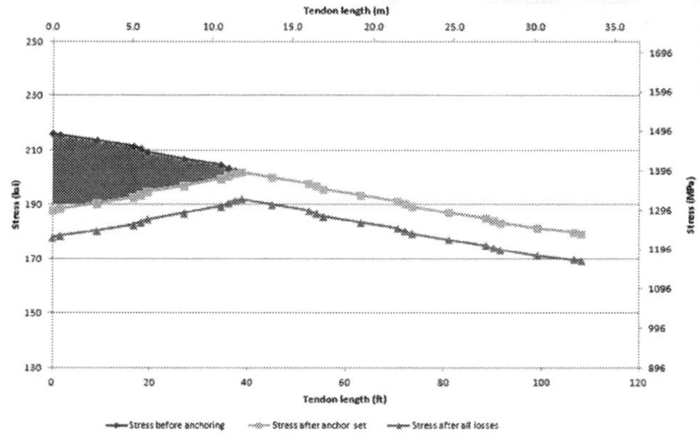

Fig. 8.2.3—Tendon stress diagram along tendon length.

8.2.4 *Elastic shortening*—The elastic shortening loss is calculated using the simplified method described in Section 4.5.

$\Delta f_{pES} = K_{es}E_p f_{cpa}/E_{ci}$

Required final effective force = 9000 lb/ft (131,300 N/m) or a tendon spacing of one tendon per 3 ft (0.91 m) on center.

The average initial stress after friction and seating loss in this example = 191,600 psi (1321 MPa). Based on a tendon spacing of one tendon per 3 ft (0.91 m), the area of prestressing steel = 0.153/3 = 0.051 in.^{2} (32.9 mm^2) per foot width of the slab.
K_{es} = 0.5 for post-tensioned members
f_{cpa} = (191,600 × 0.153/3)/(12 in. × 5 in.) = 163 psi (1.12 MPa)
Δf_{pES} = (0.5)(28,500,000 psi)(163 psi)/(3,122,000 psi) = 744 psi (5.13 MPa)

8.2.5 *Creep*—The long-term losses such as creep, shrinkage, and relaxation are calculated using the simplified method described in Chapter 5.
$\Delta f_{pCR} = K_{cr}(E_p/E_c)f_{cpa}$
K_{cr} = 1.6 (post-tensioned normalweight)

$\Delta f_{pCR} = (1.6)(28{,}500{,}000/3{,}605{,}000)(163 \text{ psi}) = 2062 \text{ psi } (14.22 \text{ MPa})$

8.2.6 *Shrinkage*

$\Delta f_{pSH} = (8.2 \times 10^{-6})K_{sh}E_p(1 - 0.06V/S)(100 - RH)$

$K_{sh} = 0.85$

$V/S = 2.5$ in. (63.5 mm) $= (8.2 \times 10^{-6})(0.85)(28{,}500{,}000)[1 - 0.06(2.5)](100 - 80) = 3377$ psi (23.28 MPa)

8.2.7 *Relaxation*

$\Delta f_{pRE} = [K_{re} - J(\Delta f_{pSH} + \Delta f_{pCR}R + \Delta f_{pES})]C$

From Table 5.4:

$K_{re} = 5000$ psi (34.47 MPa)

$J = 0.04$

Average initial tendon stress $f_{pi} = 191{,}600$ psi (1321 MPa)

$f_{pi}/f_{pu} = 191{,}600 \text{ psi}/270{,}000 \text{ psi} = 0.71$

$C = [0.71/0.21][0.71/0.9 - 0.55] = 0.807$

$\Delta f_{pRE} = [5000 - 0.04(3377 + 2062 + 744)](0.807) = 3835$ psi (26.44 MPa)

8.2.8 *Total long-term losses*—Total losses are:

$\Delta f_{pT} = \Delta f_{pES} + \Delta f_{pCR} + \Delta f_{pSH} + \Delta f_{pRE} = 744 + 2062 + 3377 + 3835 = 10{,}000$ psi (69.0 MPa)

Final effective tendon stresses after all losses are calculated by subtracting the total long-term losses from the tendon stress after anchoring (Fig. 8.2.3). The final effective force in the tendon per foot is (191,600 – 10,000)(0.153/3) = 9260 lb/ft (135.1 kN/m) compared to required 9000 lb/ft (131.4 kN/m).

8.3—Post-tensioned beam with bonded tendons

Example 8.3 presents calculation of prestress losses in a two-span post-tensioned beam with one 12-strand bonded tendon with 1/2 in. (12.7 mm) strands in corrugated metal ducts. The initial losses are calculated based on the methods in Chapter 4 and long-term losses are calculated using the simplified method described in Chapter 5.

8.3.1 *Givens and problem statement*—This problem considers a continuous two-span post-tensioned beam with 60 and 30 ft spans (18.3 m and 9.1 m) (Fig. 8.3.1) and a tributary bay width of 18 ft (5.49 m). The beam size is 18 x 34 in. (457 x 864 mm), assuming an effective flange width of 98 in. (2489 mm) and the slab is a cast-in-place 5 in. (127 mm) thick post-tensioned slab.

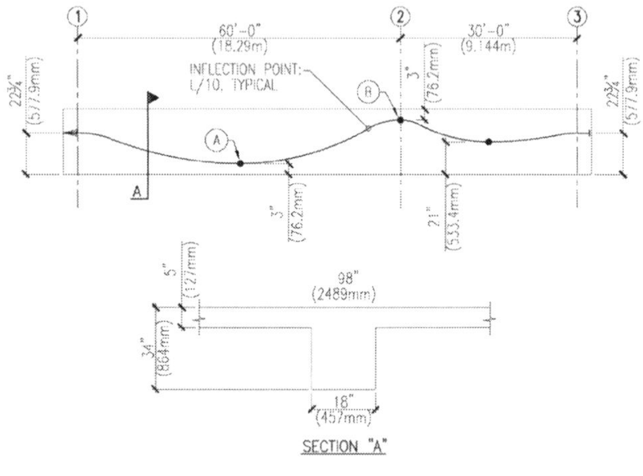

Fig. 8.3.1—Beam dimensions and prestressing.

Concrete properties: Normalweight

$f_{ci}' = 3000$ psi (20.68 MPa) (at transfer)

$f_c' = 5000$ psi (34.47 MPa)

$w = 150$ lb/ft^3 (2403 kg/m^3)

$E_{ci} = 3{,}122{,}000$ psi (21,530 MPa)

$E_c = 4{,}030{,}000$ psi (27,790 MPa)

Loads:
SW = (18 ft × 5/12 + 18/12 × (34 – 5)/12) (150 lb/ft^3) = 1670 lb/ft (24.37 kN/m)
SDL = 10 lb/ft^2 × 18 ft = 180 lb/ft (2.63 kN/m) (assume 10 lb/ft^2 [0.048 kN/m^2] superimposed sustained load after beam is placed in service)
RH = 60 percent

Section properties:
Effective flange width = 40 in. + 18 in. + 40 in. = 98 in. (2.49 m)
T-beam cross section = 5 × 98 + 18 × (34 – 5) = 1012 in.^{2} (0.653 m^2)
Moment of inertia of T-beam = 110,648 in.^{4} (0.0461 m^4)
Distance of neutral axis from top = 11.27 in. (286.3 mm)
Distance of neutral axis from bottom = 22.73 in. (577.3 mm)

Tendon properties:
Low-relaxation, 12-strand bonded system
Strand diameter = 1/2 in. (12.7 mm)
Strand area = 0.153 in.^{2} (98.7 mm^2)

Tendon area:
A_{ps} = 12 × 0.153 in.^{2} = 1.836 in.^{2} (1184 mm^2)
E_p = 28,500,000 psi (196.5 GPa)
f_{pu} = 270,000 psi (1862 MPa)
Jacking stress = 0.8 × 270,000 psi = 216,000 psi (1489 MPa)
Angular friction coefficient μ = 0.25 (rad^{-1}) (Table 4.4.2)
Wobble friction coefficient κ = 0.0002 ft^{-1} (0.000656 m^{-1})
Anchor set = 0.375 in. (9.53 mm) (assumes jack without power seating)
Reverse parabola at L/10
Stressing from one end (left end)

Slab is also prestressed in the temperature tendon direction (parallel to the beams); however, its contribution is neglected in this analysis.

Problem statement: Determine friction, anchor set, elastic, creep, shrinkage and relaxation losses at tendon location A and B as shown in Fig. 8.3.1.

8.3.2 *Friction losses and elongation*—Similar to Section 8.2.2, the tendon eccentricity and the angular deviation at each parabolic vertex is determined to compute $e^{-(\mu\theta+kx)}$. Multiplying by the jacking stress gives the tendon stress at the end of each segment before anchor set. Elongations before anchor set are computed for each segment taking the average stress along the segment.

Table 8.3.2—Friction and anchor set losses, and elongations

	Vertex i	e_i, in. eccentricity	$\|e_i - e_v\|$, in.	x_i, ft distance from stressing end	$\|x_i - x_v\|$, in.	θi, rad segment angular deviation	α_i, rad angular deviation	$\exp(-k \times l_i - \alpha_i \times \mu)$ stress in percent of stressing end	Stress before set, ksi	Stress after set, ksi	Elongation, in.
Stressing end	1	22.75	0.00	0.0	0.0	0.000	0.0000	1.000	216.0	175.4	0.00
	2	18.80	3.95	6.0	6.0	0.109	0.1093	0.972	209.9	181.5	0.54
Midspan 1 / A	3	3.00	15.80	30.0	24.0	0.109	0.2186	0.941	203.3	188.1	2.63
	4	25.40	22.40	54.0	24.0	0.154	0.3729	0.901	194.7	194.7	4.64
Support 2 / B	5	31.00	5.60	60.0	6.0	0.154	0.5272	0.866	187.1	187.1	5.12
	6	29.00	2.00	63.0	3.0	0.111	0.6379	0.842	181.9	181.9	5.35
Midspan 2	7	21.00	8.00	75.0	12.0	0.111	0.7485	0.817	176.5	176.5	6.26
	8	22.40	1.40	87.0	12.0	0.019	0.7680	0.811	175.2	175.2	7.15
Support 3	9	22.75	0.35	90.0	3.0	0.019	0.7874	0.807	174.2	174.2	7.37

	Vertex i	e_i, cm eccentricity	$\|e_i - e_v\|$, cm	x_i, m distance from stressing end	$\|x_i - x_v\|$, m	θi, rad segment angular deviation	α_i, rad angular deviation	$\exp(-k \times l_i - \alpha_i \times \mu)$ stress in percent of stressing end	Stress before set, MPa	Stress after set, MPa	Elongation, cm
Stressing end	1	57.8	0.0	0.00	0.00	0.000	0.0000	1.000	1489.3	1209.4	0.00
	2	47.8	10.0	1.83	1.83	0.109	0.1093	0.972	1447.4	1251.3	1.37
Midspan 1 / A	3	7.6	40.1	9.15	7.32	0.109	0.2186	0.941	1401.7	1297.1	6.67
	4	64.5	56.9	16.47	7.32	0.154	0.3729	0.901	1342.2	1342.2	11.78
Support 2 / B	5	78.7	14.2	18.30	1.83	0.154	0.5272	0.866	1289.8	1289.8	13.00
	6	73.7	5.1	19.22	0.92	0.111	0.6379	0.842	1253.9	1253.9	13.59
Midspan 2	7	53.3	20.3	22.88	3.66	0.111	0.7485	0.817	1216.8	1216.8	15.89
	8	56.9	3.6	26.54	3.66	0.019	0.7680	0.811	1208.0	1208.0	18.15
Support 3	9	57.8	0.9	27.45	0.92	0.019	0.7874	0.807	1201.4	1201.4	18.71

Figure 8.3.2 shows that a linear friction approximation is not considered for this tendon; rather, the specified tendon curvature is used for more accuracy.

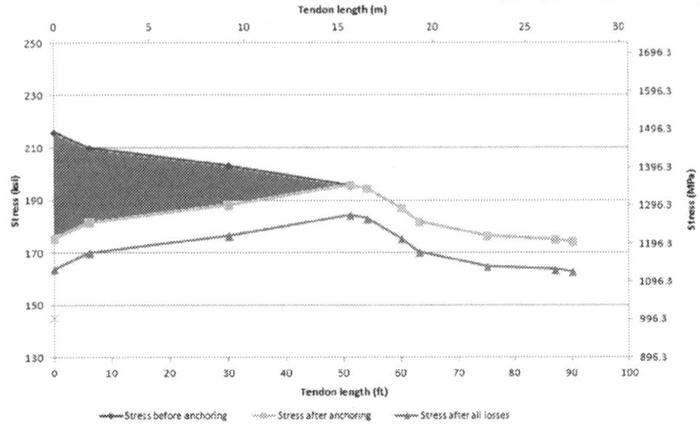

Fig. 8.3.2—Tendon tension diagram.

8.3.3 *Anchor set losses*—Anchor set losses are computed similarly to those in 8.2.3.

For this example,

Anchor set length = 51.1 ft (15.6 m)

Maximum stress = 195,700 ksi (1349 MPa)

Average initial stress = 184,500 ksi (1272 MPa)

8.3.4 *Computing concrete stresses*—A structural analysis is performed to obtain moments presented in Table 8.3.4 assuming uncracked sections.

Table 8.3.4—Moments in beam

	Point A Midspan 1	Point B Support 2
Post-tensioning (PT) moment kip-ft (kN-m)	–454.36 (–616.03)	471.5 (639.27)
SW moment kip-ft (kN-m)	469.87 (637.06)	–563.25 (–763.67)
SDL moment kip-ft (kN-m)	50.65 (68.67)	–60.71 (–82.31)

To compute creep and elastic shortening losses, f_{cpi}, f_g and f_{cds} stresses are needed at both points: f_{cpi} is concrete stress at center of gravity of tendon due to P_i; f_g isconcrete stress at center of gravity of tendon due to member self-weight; f_{cds} is concrete stress at center of gravity of tendon due to all superimposed permanent dead loads; and f_{cir} is net concrete stress at center of gravity of tendons at transfer.

As a simplification, f_{cir} is calculated with the average tendon stress after transfer. It could also be calculated with the stress determined for each location (given in Table 8.3.2)

At Point A:

$$f_{cpi} = \frac{184.500 \text{ psi} \times 1.836 \text{ in.}^2}{1012 \text{ in.}^2} + \frac{-454,360 \times 12 \times (-22.73 \text{ in.} + 3 \text{ in.})}{110,648 \text{ in.}^4} = 1307 \text{ psi}(C) \text{ (9.01 MPa)}$$

(Note: compression stress is positive and tension stress is negative)

$$f_g = \frac{469,870 \times 12 \times (-19.73 \text{ in.})}{110,648 \text{ in.}^4} = -1005 \text{ psi}(T) \text{ (6.93 MPa)}$$

$$f_{cds} = \frac{50,650 \times 12 \times (-19.73 \text{ in.})}{110,648 \text{ in.}^4} = -108 \text{ psi}(T) \text{ (0.745 MPa)}$$

$$f_{cir} = K_{cir} \times f_{cpi} - f_g = 1.0 \times 1307 \text{ psi} - 1005 \text{ psi} = 302 \text{ psi}(C) \text{ (2.08 MPa)}$$

At Point B:

$$f_{cpi} = \frac{184,500 \text{ psi} \times 1.836 \text{ in.}^2}{1012 \text{ in.}^2} + \frac{471,500 \times 12 \times (-11.27 \text{ in.} - 3 \text{ in.})}{110,648 \text{ in.}^4} = 758 \text{psi}(C) \text{ (5.23 MPa)}$$

$$f_g = \frac{563,250 \times 12 \times (-8.27 \text{ in.})}{110,648 \text{ in.}^4} = -505 \text{ psi}(T) \text{ (3.48 MPa)}$$

$$f_{cds} = \frac{-60,710 \times 12 \times (8.27 \text{ in.})}{110,648 \text{ in.}^4} = -54 \text{ psi}(T) \text{ (0.372 MPa)}$$

$$f_{cir} = K_{cir} \times f_{cpi} - f_g = 1.0 \times 758 \text{ psi} - 505 \text{ psi} = 253 \text{ psi}(C) \text{ (1.74 MPa)}$$

8.3.5 *Elastic shortening*—The elastic shortening loss is calculated using the simplified method described in Section 4.5.

$\Delta f_{pES} = K_{es}E_{ps}f_{cir}/E_{ci} - K_{es} = 0$ because all strands are tensioned simultaneously

At Point A:

Δf_{pES} = (0)(28,500,000 psi)(302 psi)/(3,122,000 psi) = 0 psi (0 MPa)

At Point B:

Δf_{pES} = (0)(28,500,000 psi)(253 psi)/(3,122,000 psi) = 0 psi (0 MPa)

8.3.6 *Creep*—The long-term losses such as creep, shrinkage and relaxation are calculated using the simplified method described in Chapter 5. For bonded tendons, creep shortening losses are computed at each point.

$\Delta f_{pCR} = K_{cr}(E_{ps}/E_c)(f_{cir} - f_{cds})$

K_{cr} = 1.6 (post-tensioned normalweight)

At Point A:

Δf_{pCR} = (1.6)(28,500,000/4,030,000)(302 psi – 108 psi) = 2195 psi (15.13 MPa)
(caution is warranted with signs; f_{cds} is subtracted if in tension)

At Point B:

Δf_{pCR} = (1.6)(28,500,000/4,030,000)(253 psi – 54 psi) = 2252 psi (15.53 MPa)

8.3.7 *Shrinkage*

$\Delta f_{pSH} = (8.2 \times 10^{-6})K_{sh}E_{ps}(1 - 0.06V/S)(100 - RH)$

K_{sh} = 0.85

V/S = 1012 in.^{2}/(2 × 98 + 2 × 29) = 3.984 in. (101.2 mm)

At both points:

Δf_{pSH} = (8.2 × 10^{-6})(0.85)(28,500,000)[1 – 0.06(3.984)](100 – 60) = 6046 psi (41.69 MPa)

8.3.8 *Relaxation*

$\Delta f_{pRE} = [K_{re} - J(\Delta f_{pSH} + \Delta f_{pCR}R + \Delta f_{pES})]C$

Relaxation losses depend on creep, elastic shortening and shrinkage losses, and are calculated at both points.

From Table 5.4:
K_{re} = 5000 psi (34.47 MPa)
J = 0.04
Average initial tendon stress f_{pi} = 184,500 psi (1272 MPa)
f_{pi}/f_{pu} = 184,500 ksi/270,000 psi = 0.683
C = [0.683/0.21][0.683/0.9 – 0.55] = 0.681

At Point A:

Δf_{pRE} = [5000 – 0.04(6046 + 2195 + 0)](0.681) = 3181 psi (21.93 MPa)

At Point B:

Δf_{pRE} = [5000 – 0.04(6046 + 2252 + 0)](0.681) = 3179 psi (21.92 MPa)

Difference is negligible in this case; relaxation losses can be computed with an average value of creep shortening losses.

8.3.9 *Total long-term losses*

At Point A:

$\Delta f_{pT} = \Delta f_{pES} + \Delta f_{pCR} + \Delta f_{pSH} + \Delta f_{pRE}$ = 0 + 2195 + 6046 + 3181 = 11,422 psi (78.8 MPa)

At Point B:

$\Delta f_{pT} = \Delta f_{pES} + \Delta f_{pCR} + \Delta f_{pSH} + \Delta f_{pRE}$ = 0 + 2252 + 6046 + 3179 = 11,477 psi (79.1 MPa)

Because long-term losses at each point of interest are close, it is sufficient to subtract the mean long-term losses (11,450 psi [78.9 MPa]) from the tendon stresses after anchor set to compute tendon stresses after all losses along the tendon (Fig. 8.3.2). The final effective force in the beam assuming the mean long-term loss of 11,450 psi (78.9 MPa) = (184,500 – 11,450) × 1.836 in.^{2} = 317,700 lb (1413.2 kN).

8.4—Example with heat of hydration during casting

The effect on prestress loss of the heat of hydration in pretensioned concrete members is presented with this example. The method presented in 4.2.4 is used in the example.

8.4.1 *Description of lab beam and instrumentation*—Small test beams were cast in the laboratory to investigate transfer and development length of prestressing strands in lightweight concrete (Cross 2012). Instrumentation included load cells on each of the three strands outside of the beam, which were monitored from the stressing operations until just before cutting the strands. In addition, a vibrating wire gauge was placed at the centroid of the strand pattern at midspan each of the test beams. Each vibrating wire gauge contains a thermistor, so temperatures in the beam were recorded from just before casting the concrete until just before testing the beams several months later. Figure 8.4.1a presents the temperature history and Fig. 8.4.1b presents the average stress calculated from the loads recorded by the load cells on the three strands from before casting to just before transfer. These beams were not steam cured, so they were left in the stressing beds for 7 days before release strength was achieved.

Figure 8.4.1c shows the cross section of the beam. There were two 24 ft (7.32 m) long beams cast in a 60 ft (18.3 m) long bed. The following calculations determine the changes in stresses in the strand and in the concrete during the heating and cooling cycles.

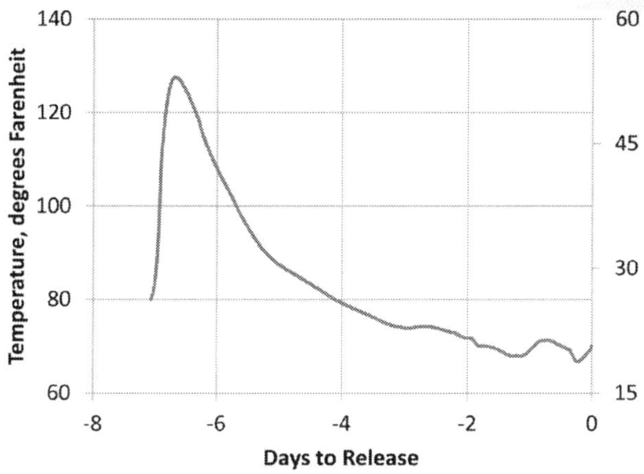

Fig. 8.4.1a—Temperature history for beam.

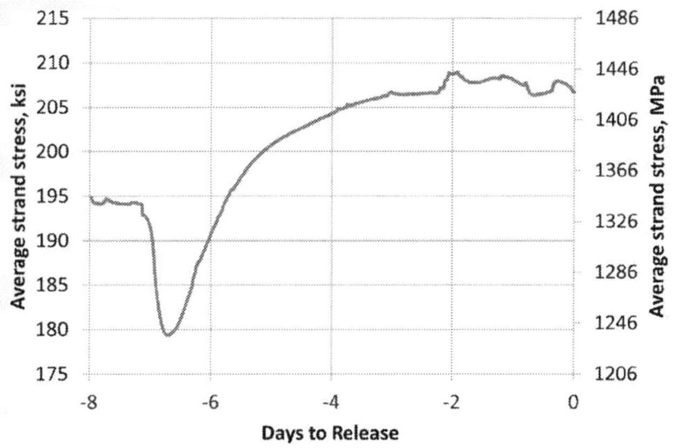

Fig. 8.4.1b—Average stress in strands from stressing to just before transfer based on load cell.

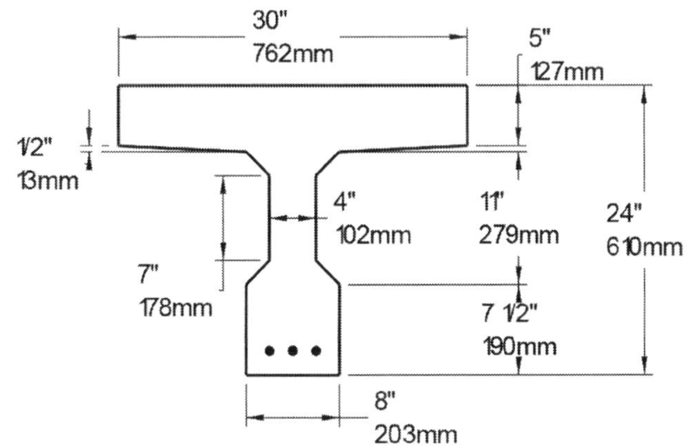

Fig. 8.4.1c—Cross section of beam in lab study.

8.4.2—*Calculations of stress changes during curing*

Given information for this problem is:

Temperature rise = 48°F (26.7°C) (Fig. 8.4.1a)

Temperature fall = 58°F (32.2°C) (Fig. 8.4.1a)

Free strand length = 12 ft = 144 in. (3658 mm)
Strand length in two beams = 48 ft = 576 in. (14,630 mm)
Coefficient of thermal expansion of steel = 6 με/°F (10.8 με/°C)
Area of strand = 3 × 0.217 in. ^{2} = 0.651 in.^{2} (420 mm^2)
Modulus of strand = 28,500,000 psi (196,500 MPa)
Coefficient of thermal expansion of concrete = 5 με/°F (9.0 με/°C)
Net area of concrete = 272 in.^{2} (175,483 mm^2)
Net moment of inertia of concrete = 15,194 in.^{4} (6.324 × 10^9 mm^4)
Modulus of concrete = 4,800,000 psi (33,095 MPa)

During the temperature rise, it is assumed that the concrete has no stiffness and there is no bond between the concrete and the steel. It is assumed that the temperature of the strands is the same as that of the concrete along the full length. Therefore, the change in the stress in the prestress is assumed to be

$$\Delta f_{temp\_rise} = \Delta T\alpha_s E_p = 48 \times 6 \times 10^{-6}(28{,}500{,}000 \text{ psi}) = 8200 \text{ psi (56.5 MPa}$$

As is seen in Fig. 8.4.1b, this is considerably less than the 15,000 psi (103 MPa) loss recorded by the load cells. This can partly be explained by higher relaxation losses at elevated temperatures, which is discussed in Rostásy et al. (1991).

Next, the calculation of stress changes due to cooling is performed.

First three changes in force are defined: ΔP_{free} is change in the force in the free strand length, lb (N); ΔP_c is change in force in the concrete cross section, lb (N); and ΔP_{ps} is change in force in the strand within the concrete beam, lb (N).

The sum of the beam forces are equal to the change in force in the free strand length:

$$\Delta P_{ps} + \Delta P_c + \Delta P_{free}$$

The total change in length of the free strand and the beams are equal to zero. The temperature was assumed to be constant along the strand to simplify the analysis. The individual changes are calculated as:

$$\Delta_{free} = -\Delta T\alpha_s L_{free} + \frac{\Delta P_{free} L_{free}}{A_{ps} E_p}$$

$$\Delta_c = -\Delta T\alpha_c L_{beam} + \frac{\Delta P_c L_{beam}}{A_c E_c} + \frac{\Delta P_c e_p^{\,2} L_{beam}}{I_c E_c}$$

$$\Delta_{ps} = -\Delta T\alpha_s L_{beam} + \frac{\Delta P_{ps} L_{beam}}{A_{ps} E_p}$$

Two compatibility equations are written:

$$\Delta_{ps} = \Delta_c$$

$$\Delta_{ps} + \Delta_{free} = 0$$

We now have a system of six equations and six unknowns. For this particular problem, the three equations for length change are:

$$\Delta_{free} = -58 \times 6 \times 10^{-6} \times 144 + \frac{\Delta P_{free} 144}{0.651 \times 28{,}500} = -0.050 + \Delta P_{free} \times 0.0078$$

$$\Delta_c = -58 \times 5 \times 10^{-6} \times 576 + \frac{\Delta P_c 576}{272 \times 4800} + \frac{\Delta P_c 13.9^2 \cdot 576}{15{,}194 \times 4800} = -0.167 + \Delta P_c \times 0.00197$$

$$\Delta_{ps} = -58 \times 6 \times 10^{-6} \times 576 + \frac{\Delta P_{ps} 576}{0.651 \times 28{,}500} = -0.200 + \Delta P_{ps} \times 0.031$$

Solving the equations simultaneously results in the following:

ΔP_{free} = 22,800 lb (101,400 N) (tension)

ΔP_c = 20,400 lb (90,700 N) (tension)

$\Delta P_{ps} = 2400$ lb (10,700 N) (tension)

$\Delta_{free} = 0.127$ in. (3.23 mm) (lengthening)

$\Delta_c = \Delta_{ps} = -0.127$ in. (–3.23 mm) (shortening)

The stress increase in the free strand length would be 35,000 psi (241.3 MPa). According to Fig. 8.4.1b, the stress increase due to cooling was approximately 28,000 psi (193.0 MPa), so the method overestimates the change. The stress increase of the strand within the beam is calculated to be 3600 psi (24.8 MPa), compared to a measured change of approximately 3000 psi (20.7 MPa) (Fig. 8.4.3). Before release, the concrete has a calculated stress of 336 psi (2.3 MPa) in tension at the level of the center of gravity of the strand.

Stress in concrete prior to transfer

$$\text{stress in concrete prior to transfer} = \frac{20.4\text{ kip}}{272\text{ in.}^2} + \frac{20.4\text{ kip}(13.9\text{ in.})^2}{15{,}194\text{ in.}^4} = 336\text{ psi (2.32 MPa)}$$

Using the original jacking force and the transformed cross-sectional properties, the compressive stress in the concrete at the level of the strand is:

$$f_c = \frac{-189\text{ ksi} \times 0.651\text{ in.}^2}{276\text{ in.}^2} + \frac{-189\text{ ksi} \times 0.651\text{ in.}^2 \times (13.7\text{ in.})^2}{15{,}934\text{ in.}^4}$$
$$+\frac{246\text{ in.-kip} \times 13.7\text{ in}}{15{,}934\text{ in.}^2} = -1685\text{ psi (11.6 MPa)}$$

The compressive stress in the concrete at transfer can be calculated using the stress that at the time the concrete initially bonded to the concrete. All changes occurring thereafter are recoverable.

stress in strand bonding = 189 ksi – 8.2 ksi = 180.8 ksi

$$f_c = -\frac{180.8\text{ ksi} \times 0.651\text{ in.}^2}{276\text{ in.}^2} - \frac{180.8\text{ ksi} \times 0.651\text{ in.}^2 \times (13.7\text{ in.})^2}{15{,}934\text{ in.}^4} + \frac{246\text{ in.-kip} \times 13.7\text{in.}}{15{,}934\text{ in.}^4}$$
$$f_c = -1600\text{ psi (11.03 MPa)}$$

The apparent compressive stress in the concrete at release would be 336 psi + 1600 psi =1936 psi (2.32 MPa + 11.03 MPa = 13.35 MPa) due to the removal of the restraint forces at transfer. Whereas, the actual stress in the concrete of 1600 psi represents a 5 percent reduction due to the loss of prestress between initial tensioning and bond. As mentioned previously, some simplifying assumptions are made in this analysis; however, it provides a general idea of the stress changes in the steel and concrete due to heating and cooling of the beam.

8.4.3 *Comparisons of model and measurements*—The graphs (Fig 8.4.1b and 8.4.3) confirm the behavior predicted by the model. As the concrete heats up, strand stress decreases as shown in Fig. 8.4.1b. The decrease, measured by load cells, is larger than would be expected. As calculated, the expected change in stress is 8200 psi (56.54 MPa), but the change measured with the load cells is approximately 15,000 psi (103.4 MPa)—almost twice the expected change. Part of the loss could be relaxation; tests have shown that relaxation can be significantly increased by elevated temperature (Rostásy et al. 1991).

As the concrete cools, the strand stress as measured by both the load cells and the internal gauges increases. Figure 8.4.3 shows the strand stress as measured by the vibrating wire gauge embedded in the concrete. As the model predicts, the increase in strand stress external to the beam is greater than that inside the beam. Assuming a concrete modulus of elasticity of 4,800,000 psi (33 GPa), the model predicted an increase of 35,000 psi (241.3 MPa) in the strand external to the concrete, while the measured increase was approximately 28,000 psi (193.0 MPa). The model predicted an increase in strand stress within the concrete to be approximately 3600 psi (24.8 MPa), which is similar to the measurement with the vibrating wire gauges (VWGs).

8.4.4 *Conclusions from temperature example*—In conclusion, the effect of temperature changes during casting can be examined qualitatively. A quantitative evaluation is difficult due to the uncertainty in determining the early-age concrete properties, the influence of temperature on relaxation, and the effects of formwork restraint. Thermal effects are considered to be partly responsible for measurements of elastic shortening, which are higher than theoretical. Because the net influence on the beam is small, for most applications it is acceptable to ignore the thermal effects during casting.

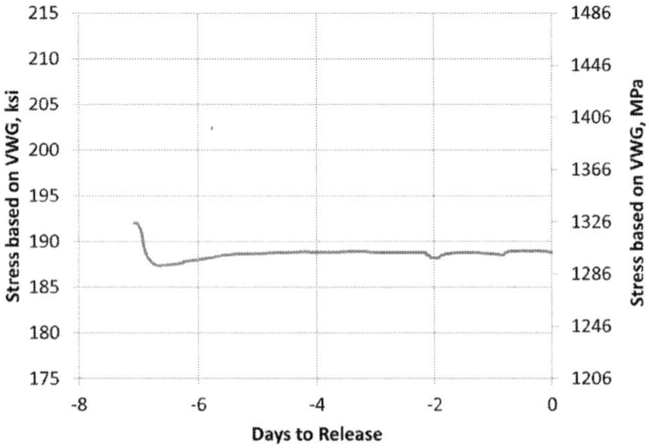

Fig. 8.4.3—Strand stress based on vibrating wire gauges (VWGs).

CHAPTER 9—REFERENCES

ACI committee documents and documents published by other organizations are listed first by document number, full title, and year of publication followed by authored documents listed alphabetically.

American Concrete Institute

ACI 209R-92(08)—Prediction of Creep, Shrinkage, and Temperature Effects in Concrete Structures

ACI 209.1R-05—Report on Factors Affecting Shrinkage and Creep of Hardened Concrete

ACI 209.2R-08—Guide for Modeling and Calculating Shrinkage and Creep in Hardened Concrete

ACI 318-63—Building Code Requirements for Reinforced Concrete

ACI 318-83—Building Code Requirements for Reinforced Concrete

ACI 318-99—Building Code Requirements for Structural Concrete and Commentary

ACI 318-08—Building Code Requirements for Structural Concrete and Commentary

ACI 318-11—Building Code Requirements for Structural Concrete and Commentary

ACI 363R-92(97)—Report on High Strength Concrete

ACI 435R-95(00)—Control of Deflection in Concrete Structures (Appendix B added 2003)

ASTM International

ASTM A416/A416M-12—Standard Specification for Steel Strand, Uncoated Seven-Wire for Prestressed Concrete

(CEN) The European Committee for Standardization

EN 13670-09—Execution of Concrete Structures

Post-Tensioning Institute

PTI C30.4-07—Training and Certification of Field Personnel for Bonded Post-Tensioning

PTI M10.3-00—Field Procedures Manual for Unbonded Single Strand Tendons

PTI M50.3-12—Guide Specification for Grouted Post-Tensioning

Authored documents

AASHTO, 1989, "Guide Specifications for Design and Construction of Segmental Concrete Bridges," first edition, American Association of State Highway and Transportation Officials, Washington, DC.

AASHTO, 2011, "LRFD Bridge Construction Specifications," third edition, Interim Revisions, American Association of State Highway and Transportation Officials, Washington, DC.

AASHTO, 2012, "AASHTO LRFD Bridge Design Specifications," American Association of State Highway and Transportation Officials, Washington, DC.

Abeles, P. W., 1958, "Losses of Prestressing Force," *Concrete and Constructional Engineering*, V. 53, Part 1, No. 9, Aug., pp. 331-340.

ACI-ASCE Committee 323, 1958, "Tentative Recommendations for Prestressed Concrete," *ACI Journal Proceedings*, V. 55, No. 7, Jan., pp. 545-578.

Ahlborn, T. M.; French, C. E.; and Shield, C. K., 2000, "High-Performance Concrete Prestressed Bridge Girders: Long-term and Flexural Behavior," *Report* 2000-32, Minnesota Department of Transportation, St. Paul, MN, Nov., 390 pp.

Ahmad, S. H., and Shah, S. P., 1982, "Complete Triaxial Stress-Strain Curves for Concrete," *Journal of the Structural Division*, V. 108, No. 4, Apr., pp. 728-742.

Al-Omaishi, N.; Tadros, M. K.; and Seguirant, S. J., 2009, "Estimating Prestress Loss in Pretensioned, High-Strength Concrete Members," *PCI Journal*, V. 54, No. 4, pp. 132-159. doi: 10.15554/pcij.09012009.132.159

Anderson, T. C.; Houdeshell, D. M.; and Gamble, W. L., 1972, "Construction and Long-term Behavior of 1/8th Scale Prestressed Concrete Bridge Components," *Structural Research Series* No. 384, Civil Engineering Studies, University of Illinois, Champaign, IL, Oct., 115 pp.

Attanayake, U., and Aktan, H., 2010, "Capacity Evaluation of a Severely Distressed and Deteriorated 50-Year Old Box-Beam with Limited Data," *Journal of Performance of Constructed Facilities*, V. 25, No. 4, Aug., pp. 229-308.

Azizinamini, A.; Keeler, B. J.; Rohde, J.; and Mehrabi, A. B., 1996, "Application of a New Nondestructive Evaluation Technique to a 25-Year-Old Prestressed Concrete Girder," *PCI Journal*, V. 41, No. 3, May-June, pp. 82-95. doi: 10.15554/pcij.05011996.82.95

Barr, P.; Eberhard, M.; Stanton, J.; Khaleghi, B.; and Hsieh, J. C., 2000, "High Performance Concrete in Washington State SR18/SR516 Overcrossing: Final Report on Girder Monitoring," Washington State Transportation Center, Dec., 153 pp.

Bažant, Z. P., and Baweja, S., 1995, "Creep and Shrinkage Prediction Model for Analysis and Design of Concrete Structures – Model B3," *RILEM Journal of Materials and Structures*, V. 28, No. 180, pp. 357-365, 415-430, 488-495.

Boone, S. D., 2008, "A Collection of New Studies Using Existing and Proposed Techniques and Instrumentation for Nondestructive and Analysis of Concrete Materials and Structures," *All Graduate Theses and Dissertations*, Paper 125, Utah State University, Logan, UT, May, 225 pp.

Branson, D. E., 1977, *Deformation of Concrete Structures*, McGraw-Hill, New York, 540 pp.

Brewe, J., and Myers, J. J., 2010, "High-Strength Self-Consolidating Concrete Girders Subjected to Elevated Compressive Fiber Stresses, Part 1: Prestress Loss and Camber Behavior," *PCI Journal*, V. 55, No. 4, pp. 59-77. doi: 10.15554/pcij.09012010.59.77

Bureau of Public Roads, 1954, "Criteria for Prestressed Concrete Bridges," U.S. Department of Commerce, Washington, DC, 25 pp.

Canfield, S. R., 2005, "Full Scale Testing of Prestressed, High Performance Concrete, Composite Bridge Girders," master's thesis, Georgia Institute of Technology, Atlanta, GA, Aug., 332 pp.

Comité Euro-International du Béton (CEB), 1999, "Structural Concrete – Textbook on Behaviour, Design and Performance, Updated Knowledge of the CEB/FIP Model Code 1990," *fib Bulletin 2*, V. 2, pp. 37-52.

Comité Euro-International du Béton (CEB), 2010, "CEB-FIP Model Code for Concrete Structures," Federation International du Beton, Lausanne, Switzerland.

Cross, B. T., 2012, "Structural Performance of High Strength Lightweight Concrete Pretensioned Bridge Girders," doctoral dissertation, Virginia Polytechnic Institute and State University, Feb., 413 pp.

Czaderski, C., and Motavalli, M., 2006, "Determining the Remaining Tendon Force of a Large-Scale, 38-Year-Old Prestressed Concrete Bridge Girder," *PCI Journal*, V. 51, No. 4, July-Aug., pp. 56-68. doi: 10.15554/pcij.07012006.56.68

Dwairi, H. M.; Wagner, M. C.; Kowalsky, M. J.; and Zia, P., 2010, "Behavior of Instrumented Prestressed High Performance Concrete Bridge Girders," *Construction & Building Materials*, V. 24, No. 11, Nov., pp. 2294-2311. doi: 10.1016/j.conbuildmat.2010.04.026

Furr, H. L.; Sinno, R.; and Ingram, L. L., 1968, "Prestress Loss and Creep Camber in a Highway Bridge with Reinforced Concrete Slab on Pretensioned Prestressed Concrete Beams," *Research Report* 69-3, Texas Transportation Institute, Texas A & M University, Oct., 152 pp.

Gamble, W. L., 1970, "Field Investigation of a Continuous Composite Prestressed I-Beam Highway Bridge Located in Jefferson County, Illinois," *Structural Research Series* No. 360, Civil Engineering Studies, University of Illinois, Champaign, IL, 136 pp.

Gamble, W. L., 1979, "Long-term Behavior of a Prestressed I-Girder Highway Bridge in Champaign County, Illinois," *Structural Research Series* No. 470, Civil Engineering Studies, University of Illinois, Champaign, IL, 120 pp.

Garber, D.; Gallardo, J.; Deschenes, D.; Dunkman, D.; and Bayrak, O., 2013, "Effect of New Prestress Loss Estimates on Pretensioned Concrete Bridge Girder Design," *Technical Report No. FHWA/TX-12/0-6374-2*, June, 280 pp.

Gardner, N. J., and Lockman, M. J., 2001, "Design Provisions for Drying Shrinkage and Creep of Normal Strength Concrete," *ACI Materials Journal*, V. 98, No. 2, Mar.-Apr., pp. 159-167.

Ghali, A.; Favre, R.; and Elbadry, M., 2012, *Concrete Structures: Stresses and Deformations: Analysis and Design for Serviceability*, fourth edition, Spon Press, New York, 646 pp.

Gilbert, R. I., and Ranzi, G., 2011, *Time-Dependent Behaviour of Concrete Structures*, Spon Press, Taylor and Francis Group, London and New York, 428 pp.

Gilbertson, C. G., and Ahlborn, T. M. T., 2004, "A Probabilistic Comparison of Prestress Loss Methods in Prestressed Concrete Beams," *PCI Journal*, V. 49, No. 5, Sept.-Oct., pp. 52-69. doi: 10.15554/pcij.09012004.52.69

Glucklich, J., and Ishai, O., 1962, "Creep Mechanism in Cement Mortar," *ACI Journal Proceedings*, V. 59, No. 7, July, pp. 923-948.

Greuel, A.; Rogers, B. T.; Miller, R. A.; Shahrooz, B. M.; and Baseheart, T. M., 2000, "Evaluation of a High Performance Concrete Box Girder Bridge," *PCI Journal*, V. 45, No. 6, Nov.-Dec., pp. 60-71. doi: 10.15554/pcij.11012000.60.71

Gross, S. P., 1998, "Field Performance of Prestressed High Performance Concrete Highway Bridges in Texas," PhD dissertation, the University of Texas at Austin, Austin, TX.

Gross, S. P., and Burns, N. H., 2000, "Field Performance of Prestressed High Performance Highway Bridges in Texas," *Research Report* 9-580/589-2, Center for Transportation Research, Bureau of Engineering Research, University of Texas at Austin, TX, V. 1, Feb., 668 pp.

Gross, S. P.; Yost, J. R.; and Gaynor, E., 2007, "Experimental Study of Prestress Loss and Camber in High-Strength SCC Beams," *Self-Consolidating Concrete for Precast Prestressed Applications*, SP-247, A. K. Schindler, D. Trejo, and R. W. Barnes, eds., American Concrete Institute, Farmington Hills, MI, pp. 77-92.

Gutierrez, S. E.; Cudmani, R. O.; and Danesi, R. F., 1996, "Time-Dependent Analysis of Reinforced and Prestressed Concrete Members," *ACI Structural Journal*, V. 93, No. 4, July-Aug., pp. 420-427.

Hale, W. M., and Russell, B. W., 2006, "Effect of Allowable Compressive Stress at Release on Prestress Loss and the Performance of Precast, Prestressed Concrete Bridge Girders," *PCI Journal*, V. 51, No. 2, Mar.-Apr., pp. 14-25. doi: 10.15554/pcij.03012006.14.25

Halsey, J. T., and Miller, R., 1996, "Destructive Testing of Two Forty-Year-Old Prestressed Concrete Bridge Beams," *PCI Journal*, V. 41, No. 5, Sept.-Oct., pp. 84-93. doi: 10.15554/pcij.09011996.84.93

Hamilton, H. R., and O'Neill, C., 2009, "Determination of Service Stresses in Pretensioned Beams," *Project BD 545-78*, Florida Department of Transportation, Department of Civil and Coastal Engineering, University of Florida, Research Management Center, Gainesville, FL, Dec., 102 pp.

Hirsch, T. J., 1962, "Modulus of Elasticity of Concrete Affected by Elastic Moduli of Cement Paste Matrix and Aggregate," *ACI Journal Proceedings*, V. 59, No. 3, Mar., pp. 427-450.

Houdeshell, D. M.; Anderson, T. C.; and Gamble, W. L., 1972, "Field Investigation of a Prestressed Concrete Highway Bridge Located in Douglas County, Illinois," *Structural Research Series* No. 375, Civil Engineering Studies, University of Illinois, Champaign, IL, Feb., 150 pp.

Idriss, R. L., and Solano, A., 2002, "Effects of Steam Curing Temperature on Early Prestress Losses in High-Performance Concrete Beams," *Transportation Research Record: Journal of the Transportation Research Board*, V. 1813, Paper No. 02-3087, pp. 218-228.

Kahn, L. F., and Lopez, M., 2005, "Prestress Losses in High Performance Lightweight Concrete Pretensioned Bridge Girders," *PCI Journal*, V. 50, No. 5, Sept-Oct, pp. 84-94. doi: 10.15554/pcij.09012005.84.94

Kavanaugh, B. P.; Schindler, A. K.; and Barnes, R. W., 2008, "Creep Behavior of Self-Consolidating Concrete (SCC)," *Proceedings of The Third North American Conference on the Design and Use of Self-Consolidating Concrete, Challenges and Barriers to Application*, Chicago, IL, V. 1, Nov. 10-12.

Kebraei, M.; Luedke, J.; and Azizinamini, A. A., 1997, "High Performance Concrete in 120th and Giles Bridge,

Sarpy County, Nebraska," Department of Roads, University of Nebraska, Lincoln, NE, V. 24, No. 6, Nov./Dec., 300 pp.

Kelley, G. S., 2000, "Prestress Losses in Post-Tensioned Structures," *PTI Technical Note*, No. 10, Sept., 6 pp.

Kukay, B. M., 2008, "Bridge Instrumentation and the Development of Non-Destructive and Destructive Techniques Used to Estimate Residual Tendon Stress in Prestressed Girders," *All Graduate Theses and Dissertations,* Paper 128, Utah State University, Logan, UT, 231 pp.

Labia, Y.; Saiidi, M. S.; and Douglas, B., 1997, "Full-Scale Testing and Analysis of 20-Year-Old Pretensioned Concrete Box Girders," *ACI Structural Journal*, V. 94, No. 5, Sept.-Oct., pp. 471-482.

Lark, R. J.; Howells, R. W.; and Barr, B. I. G., 2004, "Behaviour of Post-Tensioned Concrete Box Girders," *Proceedings of the Institution of Civil Engineers: Bridge Engineering*, V. 157, No. 2, June, pp. 71-81.

Levy, K. R.; Barnes, R. W.; and Schindler, A. K., 2010, "Time-Dependent Deformations of Pretensioned, Self-Consolidating Concrete," Proceedings of the Third International *fib* Congress, Washington DC, Mar., 14 pp.

Lewis, M., and Karbhari, V., 2006, "Experimental Verification of the Influence of Time-Dependent Material Properties on Long-Term Bridge Characteristics," *Report* SSRP 06/20, Structural Systems Research Project, University of California, San Diego, CA, Aug., 498 pp.

Long, W.-J., and Khayat, K. H., 2011, "Creep of Prestressed Self-Consolidating Concrete," *ACI Materials Journal*, V. 108, No. 5, Sept-Oct., pp. 476-484.

Marston, J., 2010, "Camber Change and Prestress Loss in Lightweight Prestressed Girders," Unpublished Masters Project and Report, Virginia Tech, Blacksburg, VA.

Mehta, P. K., and Monteiro, P. J. M., 1993, *Concrete Structure, Properties and Materials,* second edition, Prentice Hall, Englewood Cliffs, NJ.

Menn, C., 1990, *Prestressed Concrete Bridges*, Birkhauser Verlag, Basel, Boston.

Mindess, S.; Young, J. F.; and Darwin, D., 2003, *Concrete*, second edition, Prentice Hall, New York.

Mossiossian, V., and Gamble, W. L., 1972, "Time-Dependent Behavior of Noncomposite and Composite Prestressed Concrete under Field and Laboratory Conditions," *Structural Research Series* No. 385, Civil Engineering Studies, University of Illinois, Champaign, IL, May, 517 pp.

Moustafa, S. E., 1986, "Nonlinear Analysis of Reinforced and Prestressed Concrete Members," *PCI Journal*, V. 31, No. 5, Sept.-Oct., pp. 126-147. doi: 10.15554/pcij.09011986.126.147

Myers, J. J., and Carrasquillo, R. L., 1999, "Mixture Proportioning for High-Strength High-Performance Concrete Bridge Beams," *High-Performance Concrete: Research to Practice*, SP-189, American Concrete Institute, Farmington Hills, MI, Jan., pp. 37-56.

Naaman, A. E., 2012, *Prestressed Concrete Analysis and Design,* third edition, Techno Press 3000, Ann Arbor, MI, 1176 pp.

Nilson, A. H., 1987, *Design of Prestressed Concrete*, second edition, John Wiley and Sons, New York.

Osborn, G. P.; Barr, P. J.; Petty, D. A.; Halling, M. W.; and Brackus, T. R., 2012, "Residual Prestress Forces and Shear Capacity of Salvaged Prestress Concrete Bridge Girders," *Journal of Bridge Engineering*, V. 17, No. 2, Dec., pp. 302-309. doi: 10.1061/(ASCE)BE.1943-5592.0000212

PCI, 2010, *PCI Design Handbook*, seventh edition, Precast/Prestressed Concrete Institute, Chicago, IL.

PCI, 2013, *Quality Control Technician/Inspector Level I & II Training Manual,* fourth edition, TM 101-13 Precast/Prestressed Concrete Institute, Chicago, IL.

PCI Committee on Prestress Losses, 1975, "Recommendations for Estimating Prestress Losses," *PCI Journal*, V. 20, No. 4, July-Aug., pp. 44-75.

Pessiki, S. M.; Kaczinski, M.; and Wescott, H. H., 1996, "Evaluation of Effective Prestress Forces in 28-Year-Old Prestressed Concrete Bridge Beams," *PCI Journal*, V. 41, No. 6, Nov.-Dec., pp. 78-89.

PTI, 2006, *Post-Tensioning Manual*, sixth edition, Post-Tensioning Institute, Farmington Hills, MI, Skokie, IL, 354 pp.

Roberts-Wollmann, C. L.; Breen, J. E.; and Kreger, M. E., 1995, "Temperature Induced Deformations in Match Cast Segments," *PCI Journal*, V. 40, No. 4, July-Aug., pp. 62-71. doi: 10.15554/pcij.07011995.62.71

Roller, J. J.; Martin, B. T.; Russell, H. G.; and Bruce Jr., R. N., 1993, "Performance of Prestressed High Strength Concrete Bridge Girders," *PCI Journal*, V. 38, No. 3, May-June, pp. 34-45. doi: 10.15554/pcij.05011993.34.45

Roller, J. J.; Russell, H. G.; Bruce, R. N.; and Alaywan, W. R., 2011, "Evaluation of Prestress Losses in High-Strength Concrete Bulb-Tee Girders for the Rigolets Pass Bridge," *PCI Journal*, V. 56, No. 1, Winter, pp. 110-134. doi: 10.15554/pcij.01012011.110.134

Roller, J. J.; Russell, H. G.; Bruce, R. N.; and Martin, B. T., 1995, "Long-Term Performance of Prestressed, Pretensioned High Strength Concrete Bridge Girders," *PCI Journal*, V. 40, No. 6, Nov.-Dec., pp. 48-59. doi: 10.15554/pcij.11011995.48.59

Rostásy, F. S.; Thienel, K. C.; and Schütt, K., 1991, "On Prediction of Relaxation of Colddrawn Prestressing Wire under Constant and Variable Elevated-Temperature," *Nuclear Engineering and Design*, V. 130, No. 2, Sept., pp. 221-227. doi: 10.1016/0029-5493(91)90130-A

Ruiz, E. D.; Floyd, R. W.; Stanton, B. W.; Do, N. H.; and Hale, W. M., 2008, "Prestress Losses in Prestressed Bridge Girders Cast with Self-Consolidating Concrete," *Research Report* MBTC 2071, Mack-Blackwell Transportation Center, University of Arkansas, Fayetteville, AR, July, 84 pp.

Saiidi, M.; Hutchens, E.; and Gardella, D., 1998, "Bridge Prestress Losses in Dry Climate," *Journal of Bridge Engineering*, V. 3, No. 3, Aug., pp. 111-116. doi: 10.1061/(ASCE)1084-0702(1998)3:3(111)

Saiidi, M. S.; Shields, J.; O'Connor, D.; and Hutchens, E., 1996, "Variation of Prestress Force in a Prestressed Concrete Bridge during the First 30 Months," *PCI Journal*, V. 41, No. 5, Sept.-Oct., pp. 66-72. doi: 10.15554/pcij.09011996.66.72

Sakata, K., 1993, "Prediction of Concrete Creep and Shrinkage, Creep and Shrinkage of Concrete," Proceedings of the Fifth International RILEM Symposium.

Schindler, A.; Barnes, R.; Roberts, J.; and Rodriguez, S., 2007, "Properties of Self-Consolidating Concrete for Prestressed Members," *ACI Materials Journal*, V. 104, No. 1, Jan.-Feb., pp. 53-61.

Seguirant, S. J., and Anderson, R. G., 1985, "Prestress Losses (Phase I)," *Technical Bulletin 84-B2*, Concrete Technology Associates, Tacoma, WA, Apr.

Shenoy, C. V., and Frantz, G. C., 1991, "Structural Test of 27-Year-Old Prestressed Concrete Bridge Beams," *PCI Journal*, V. 36, No. 6, Sept.-Oct., pp. 80-90. doi: 10.15554/pcij.09011991.80.90

Stallings, J. M.; Barnes, R. W.; and Eskildsen, S., 2003, "Camber and Prestress Losses in Alabama HPC Bridge Girders," *PCI Journal*, V. 48, No. 5, Sept.-Oct., pp. 90-104. doi: 10.15554/pcij.09012003.90.104

Stanton, J. F.; Eberhard, M. O.; and Barr, P., 1999, "Behavior of High-Strength, High-Performance Concrete Bridge Girders," *High-Performance Concrete: Research to Practice*, SP-189, American Concrete Institute, Farmington Hills, MI, pp. 71-92.

Steinberg, E. P., 1995, "Probabilistic Assessment of Prestress Loss in Pretensioned Prestressed Concrete," *PCI Journal*, V. 40, No. 6, Nov.-Dec., pp. 76-85. doi: 10.15554/pcij.11011995.76.85

Storm, T., 2011, "Predicting Prestress Losses, Camber, and Deflection in Prestressed Concrete," MS thesis, North Carolina State University, Raleigh, NC.

Structural Engineers Association of Northern California, 1959, "Building Code Requirements for Prestressed Concrete," *PCI Journal*, V. 4, No. 1, June, pp. 72-78.

Tabatabai, H., and Dickson, T., 1993, "Structural Evaluation of a 34-Year-Old Precast Post-Tensioned Concrete Girder," *PCI Journal*, V. 38, No. 5, Sept.-Oct, pp. 50-63. doi: 10.15554/pcij.09011993.50.63

Tadros, M. K.; Al-Omaishi, N.; Seguirant, S. J.; and Gallt, J. G., 2003, "Prestress Losses in Pretensioned High-Strength Concrete Bridge Girders," *NCHRP Report 496*, Transportation Research Board, Washington, DC.

Tadros, M. K.; Ghali, A.; and Dilger, W. H., 1975, "Time-Dependent Prestress Loss and Deflection in Prestressed Concrete Members," *PCI Journal*, V. 20, No. 3, pp. 86-98. doi: 10.15554/pcij.05011975.86.98

Trost, H., 1967, "Implications of the Superposition Principle in Creep and Relaxation Problems for Concrete and Prestressed Concrete," *Beton und Stahlbetonbau (West Berlin)*, V. 62, pp. 230-238, 261-269.

Wang, W.; Dai, J.; Li, G.; and Huang, C., 2011, "Long-Term Behavior of Prestressed Old-New Concrete Composite Beams," *Journal of Bridge Engineering*, V. 16, No. 2, Mar.-Apr., pp. 275-285. doi: 10.1061/(ASCE)BE.1943-5592.0000152

Waldron, C. J., 2004, "Investigation of Long-Term Prestress Losses in Pretensioned High Performance Concrete Girders," PhD dissertation, Virginia Polytechnic Institute and State University, Blacksburg, VA, Nov., 207 pp.

Wendner, R.; Hubler, M. H.; and Bažant, Z. P., 2013, "The B4 Model for Multi-decade Creep and Shrinkage Prediction," sponsored by CONCREEP-9@ MIT, *Proceedings of the International Conference on Creep, Shrinkage and Durability Mechanics of Concrete and Concrete Structures*, Massachusetts Institute of Technology, Cambridge, MA, Sept.

Yang, Y. M., and Myers, J. J., 2006, "Prestress Loss Measurements in Missouri's First Fully Instrumented High-Performance Concrete Bridge," *Transportation Research Record*, Transportation Research Board, Washington, DC, V. 1928, pp. 118-125.

Youakim, S. A.; Ghali, A.; Hida, S. E.; and Karbhari, V. M., 2007, "Prediction of Long-Term Prestress Losses," *PCI Journal*, V. 52, No. 2, Mar-Apr., pp. 116-130. doi: 10.15554/pcij.03012007.116.130

Zia, P.; Preston, H. K.; Scott, N. L.; and Workman, E. B., 1979, "Estimating Prestress Losses," *Concrete International*, V. 1, No. 6, June, pp. 32-38.

Report on Deflection of Nonprestressed Concrete Structures

Reported by ACI Committee 435

Eric S. Musselman, Chair

Dylan Freytag, Secretary

Peter H. Bischoff
Allan P. Bommer
Flora A. Calabrese
Eamonn F. Connolly
Norbert J. Delatte
Mamdouh M. El-Badry
Amin Ghali
Mayrai Gindy
Shawn P. Gross
Young Hak Lee
Adam S. Lubell
Faris A. Malhas
Michael C. Mota
Hani H. Nassif
Edward G. Nawy
Debrethann R. Orsak
Maria A. Polak
Mahmoud M. Reda Taha
Andrew Scanlon
Richard H. Scott

Consulting Members

Alex Aswad
Finley A. Charney
Satyendra Ghosh
Peter Lenkei
Bernard L. Meyers
Vilas S. Mujumdar
Himat T. Solanki
Susanto Teng

This report presents a consolidated treatment of initial and time-dependent deflection of nonprestressed reinforced concrete members such as simple and continuous beams and one-way and two-way slab systems. It presents the current state of practice of deflection prediction as well as analytical methods for computer use in deflection estimation. Topics include material properties, deflection of reinforced concrete one-way flexural members, deflection of two-way slab systems, and reducing deflection of concrete members.

Keywords: camber; cracking; creep; curvature; deflection; modulus of rupture; moments of inertia; serviceability; shrinkage; time-dependent deflection.

CONTENTS

ACI 435R-20 supersedes ACI 435R-95(03) and became effective November 2020.

CHAPTER 1—INTRODUCTION AND SCOPE

1.1—Introduction

Design for serviceability is central to the work of structural engineers and code-writing bodies. It is also essential to users of the designed structures. Increased use of high-strength concrete and higher-strength reinforcing bars, coupled with more detailed computer-aided designs, has resulted in lighter and more material-efficient and, thus, more flexible structural members and systems. This in turn has necessitated better prediction and control of short-term and long-term behavior of concrete structures at service loads.

This report presents a consolidated treatment of initial and time-dependent deflection of nonprestressed reinforced concrete members such as simple and continuous beams and one- and two-way slab systems. It presents current engineering practice in design for control of deformation and deflection of concrete members and includes methods presented in ACI 318 plus selected other approaches suitable for computer-based use in deflection computation. Design examples are given at the end of one- and two-way framing chapters showing how to evaluate deflection and, thus, control it through adequate design for serviceability. The content of the report as well as the step-by-step examples are intended to familiarize practitioners with the current methods for estimating deflections in buildings as well as analytical methods suitable for computer-based application. The examples apply ACI 318 requirements and a recommended alternative approach with a lower cracking moment (to account for shrinkage restraint). Methods for predicting initial and time-dependent deflections of prestressed concrete are not addressed in this document, although prestressing can be an effective tool for controlling both short- and long-term deflections.

1.2—Scope

The principal causes of deflections taken into account in this report are those due to elastic deformation, flexural cracking, creep, shrinkage, and temperature effects. This document is composed of two introductory chapters and four main chapters that provide information on calculating and controlling deflections of members constructed using reinforced concrete. The organization of the report is:

a) **Chapter 1—Introduction and Scope** provides background information on the document.

b) **Chapter 2—Notation and Definitions** provides a listing of the notation used throughout the document.

c) **Chapter 3—Material Properties** discusses material properties that affect deflections.

d) **Chapter 4—Deflection of Reinforced Concrete One-Way Flexural Members** discusses behavior of uncracked and cracked members, and time-dependent effects. It also includes the relevant code procedures and expressions for deflection computation in reinforced concrete beams. Numerical examples are included to illustrate the standard calculation methods for simply supported and continuous concrete beams.

e) **Chapter 5—Deflection of Two-Way Slab Systems** covers the deflection behavior of reinforced two-way-action slabs and plates. This chapter gives an overview of classical and other methods of deflection estimation, such as the crossing beam analogy and the finite element method for immediate deflection computation. It also discusses approaches for determining the minimum thickness requirements for two-way slabs and plates and gives a detailed computational example for evaluating the long-term deflection of a two-way reinforced concrete slab. The chapter emphasizes the uncertainties inherent in estimating deflections of two-way slab systems.

f) **Chapter 6—Reducing Deflection of Concrete Members** gives practical and remedial guidelines for improving and controlling the deflection of reinforced concrete members, hence enhancing their overall long-term serviceability.

It should be emphasized that the magnitude of actual deflection in concrete structural members, particularly in buildings, which are the emphasis and the intent of this report, can only be estimated with limited accuracy. This is because of the large variability in the properties of the constituent materials of these members, the quality control exercised in their construction, and the construction methods used. Therefore, for practical considerations, the computed deflection values in the illustrative examples at the end of each chapter should be interpreted with this in mind.

In summary, this single document gives design engineers the key tools for estimating, and thereby controlling through design, the expected deflection in nonprestressed reinforced concrete building structures. The material presented and the design examples will help to enhance serviceability when used judiciously by the engineer. Designers, constructors, and codifying bodies can draw on the material presented in this document to achieve serviceable deflection of constructed facilities.

CHAPTER 2—NOTATION AND DEFINITIONS

2.1—Notation

A_s = area of nonprestressed tension steel, in.^{2} (mm^2)
A_s' = area of nonprestressed steel in compression zone, in.^{2} (mm^2)
b = width of the section, in. (mm)
b_w = web width, in. (mm)
C_t = creep coefficient of concrete at time t, days
C_u = ultimate creep coefficient of concrete
c = depth of centroidal axis, in. (mm)
d = distance from the extreme compression fiber to centroid of tension reinforcement, in. (mm)
d' = distance from the extreme compression fiber to centroid of compression reinforcement, in. (mm)
d_b = bar diameter, in. (mm)
E = modulus of elasticity, psi (MPa)
E_c = modulus of elasticity of concrete, psi (MPa)
f_c = stress in concrete, psi (MPa)
f_c' = specified compressive strength of concrete, psi (MPa)
f_{cr} = stress to cause cracking in concrete, psi (MPa)
f_r = modulus of rupture of concrete, psi (MPa)
f_{res} = stress from restraint to shrinkage, psi (MPa)
f_s = stress in nonprestressed steel, psi (MPa)
f_y = specified yield strength of nonprestressed reinforcing steel, psi (MPa)
h = overall thickness of a member, in. (mm)
h_f = flange thickness, in. (mm)
I = moment of inertia, in.^{4} (mm^4)
I_{cr} = moment of inertia of the cracked section transformed to concrete, in.^{4} (mm^4)
I_e = effective moment of inertia for computation of deflection, in.^{4} (mm^4)
I_g = moment of inertia for gross concrete section about centroidal axis, neglecting reinforcement, in.^{4} (mm^4)
k = depth of compression zone divided by d
ℓ = span length, in. (mm)
ℓ_n = distance from the inside of the support to the inside of support, clear span, in. (mm)
M = bending moment, lb-in. (N-mm)
M_1 = moment at End 1 of a continuous member, lb-in. (N-mm)
M_2 = moment at End 2 of a continuous member, lb-in. (N-mm)
M_a = maximum service load moment (unfactored) at stage deflection is computed, lb-in. (N-mm)
M_{cr} = cracking moment, lb-in. (N-mm)
M_D = moment due to dead load, lb-in. (N-mm)
M_L = moment due to live load, lb-in. (N-mm)
M_m = midspan moment, lb-in. (N-mm)
N = axial member load, lb (N)
n = modular ratio E_s/E_c
T = temperature, °F (°C)
t = time, s
w = uniformly distributed load (load per unit length), lb/in. (N/mm)
w_c = unit weight of normalweight concrete or equilibrium density of lightweight concrete, lb/ft^3 (kg/m^3)
y_t = distance from centroidal axis of gross section, neglecting reinforcement, to extreme fiber in tension, in. (mm)
Δ = elastic deflection of a beam or slab, in. (mm)
Δ_{cr} = additional deflection due to creep, in. (mm)
Δ_{inc} = incremental deflection that occurs after attachment on nonstructural elements (includes long-term deflection Δ_{LT} from sustained loads and immediate deflection from the remaining part of live load that is not sustained, in. (mm)
Δ_L = initial (immediate) deflection due to live load, in. (mm)
Δ_{LT} = deflection from long-term effects, in. (mm)
Δ_{sh} = additional deflection due to shrinkage, in. (mm)
Δ_{sus} = initial (immediate) deflection due to sustained load, in. (mm)
ε_{cf} = strain due to stress in the concrete
ε_o = free strain, such as unrestrained shrinkage
ε_{sf} = strain due to stress in nonprestressed steel
ε_{sh} = shrinkage strain of concrete
$(\varepsilon_{sh})_t$ = shrinkage strain of concrete at time t, days
$(\varepsilon_{sh})_u$ = ultimate shrinkage strain of concrete
ε_t = total strain
ζ = distribution coefficient
κ = cross section curvature, in.$^{-1}$ (mm^{-1})
κ_{sh} = shrinkage curvature, in.$^{-1}$ (mm^{-1})
λ_c = creep multiplier for long-term deflection
λ_{sh} = shrinkage warping multiplier for long-term deflection
λ_t = total multiplier for long-term deflection
λ_Δ = time-dependent multiplier for long-term deflection
ν = Poisson's ratio
ρ = nonprestressed tension reinforcement ratio (A_s/bd)
ρ' = reinforcement ratio for nonprestressed compression steel (A_s'/bd)
ξ = time-dependent multiplier for deflection

2.2—Definitions

Please refer to the latest version of ACI Concrete Terminology for a comprehensive list of definitions.

CHAPTER 3—MATERIAL PROPERTIES

3.1—Objective

Deflections in reinforced concrete structures are affected significantly by numerous material properties, including concrete and reinforcement moduli of elasticity, concrete modulus of rupture, creep, and shrinkage. The purposes of this chapter are to: 1) briefly address how each of these material properties affects deflection; and 2) provide brief guidance on the most common expressions recommended by various ACI committees for estimation of these parameters during the design process. This chapter is not intended to provide a comprehensive review of all the material property expressions that have been suggested by researchers, committees, and design codes.

3.2—Material properties affecting deflection

The primary material properties that affect the deflection of reinforced concrete structures are identified in the following.

3.2.1 *Concrete modulus of elasticity*—The elastic modulus of concrete (E_c) has a significant impact on deflections in reinforced concrete structures, as it is a direct measure of material stiffness. Based on simple mechanics principles developed for elastic theory, deflection is inversely related to the elastic modulus of the concrete. However, the concrete elastic modulus also affects the sectional moment of inertia at the cracked section (I_{cr}) and, thus, the relationship between concrete elastic modulus and deflection is far more complex.

In indeterminate structures, the member forces and moments themselves (and thus resulting deflections) will be affected by the modulus of elasticity if it varies across the different components of the structure.

3.2.2 *Concrete modulus of rupture*—The concrete modulus of rupture (f_r) affects deflection primarily in that it establishes the stress threshold at which a member cracks. Upon reaching the modulus of rupture, the member (or a portion of the member) transitions from an uncracked to a cracked state and there is a significant reduction in stiffness accompanied by an increase in deflection. While most reinforced concrete members are designed to be cracked at service loads, the extent of cracking along the length of the member will have an impact on the member deflection.

For indeterminate structures, the extent of cracking has a significant effect on member stiffness, which in turn affects internal member forces and moments used in evaluating deflections.

3.2.3 *Time-dependent concrete properties*—Creep and shrinkage of concrete have a pronounced effect on the deflection behavior of reinforced concrete members. Creep of concrete will cause additional deformation over time, resulting in increased deflection of flexural members under sustained load. In many cases, the additional deflection will exceed the instantaneous deflection. The consideration of creep is not as simple as using a multiplier based solely on the creep coefficient of the concrete material because creep also causes a redistribution of internal stresses over time. This redistribution of internal stresses can be considered to cause a variation in effective sectional properties (at any point in the member) with time because the neutral axis and moment of inertia will change as the compressive strain in the concrete increases with age due to creep.

Similarly, shrinkage of concrete will directly lead to time-dependent deflections whenever warping is present due to differential shortening within the depth of the member. Differential shortening may occur due to an unsymmetric arrangement of internal reinforcement or external environmental effects. Restraint to shrinkage, either due to external restraint or internal restraint of embedded reinforcement, can also have a significant effect on deflection by inducing tensile stresses in the concrete that effectively reduce the external load required to cause cracking and thereby increase deflection. The most common approach for calculating deflection is to use a lower cracking moment computed with a reduced modulus of rupture to account for the tensile stresses that develop from restraint to shrinkage. For indeterminate structures, the member forces and moments themselves (and thus resulting deflections) may be significantly affected by creep and shrinkage. Creep effects can lead to a redistribution of forces among the members in a structure. Shrinkage restraint can also develop among the members in a structure, leading to cracking and a subsequent loss in stiffness.

3.2.4 *Reinforcement modulus of elasticity*—The effect of reinforcement modulus of elasticity on deflections can be significant but is limited primarily to the effect on sectional properties and the restraint of concrete shrinkage. The use of a lower modulus reinforcement will lead to smaller sectional moments of inertia for a constant reinforcement ratio, particularly for the cracked section, which in turn will lead to larger deflections. On the other hand, a lower modulus reinforcement will offer less internal restraint to concrete shrinkage and thus will not have as significant an effect in reducing the effective cracking moment.

3.3—Concrete material properties

Expressions recommended by ACI committees for estimation of concrete material properties during the design process are presented in this section. In general, these properties vary widely among different concretes or even among different batches of concrete produced using a single set of mixture proportions. Therefore, these expressions should only be expected to provide rough estimates for these concrete material properties, with a relatively low degree of accuracy. When a higher level of accuracy in deflection calculations is desired, concrete material properties should be measured using appropriate ASTM standards on test specimens produced using field trial batches.

3.3.1 *Concrete modulus of elasticity*—The modulus of elasticity is strongly influenced by the constitutive materials and proportions used and is especially affected by the properties and quantity of the coarse aggregate. An increase in the modulus of elasticity is commonly associated with an increase in compressive strength because the slope of the ascending branch of the stress-strain diagram becomes steeper for higher-strength concretes, but at a lower rate than the compressive strength. The value of the secant modulus of elasticity at $0.45f_c'$ for normal-strength concretes at 28 days is usually approximately 4×10^6 psi (28,000 MPa), whereas for higher-strength concretes with compressive strengths in excess of 8000 psi (56 MPa), values as high as 8×10^6 psi (56,000 MPa) have been reported (Myers and Yang 2004).

ACI 318 recommends Eq. (3.3.1) for computing the modulus of elasticity of concretes with densities in the range of 90 to 150 lb/ft^3 (1445 to 2325 kg/m^3) based on the secant modulus at the $0.45f_c'$ intercept.

$$\begin{aligned} E_c &= 33w_c^{1.5}\sqrt{f_c'} \quad \text{(psi)} \\ E_c &= 0.043w_c^{1.5}\sqrt{f_c'} \quad \text{(MPa)} \end{aligned} \qquad (3.3.1)$$

Table 3.3.3—Creep and shrinkage ratios from age 60 days to the indicated concrete age (Branson 1977)

Creep, shrinkage ratios	Concrete age					
	2 months	3 months	6 months	1 year	2 years	>5 years
C_t/C_u	0.48	0.56	0.68	0.77	0.84	1.00
$(\varepsilon_{sh})_t/(\varepsilon_{sh})_u$—MC	0.46	0.60	0.77	0.88	0.94	1.00
$(\varepsilon_{sh})_t/(\varepsilon_{sh})_u$—SC	0.36	0.49	0.69	0.82	0.91	1.00

Notes: MC is moist-cured; and SC is steam-cured.

Although Eq. (3.3.1) is considered to provide reasonable predictions for concretes in the strength range up to 6000 psi (42 MPa), significant variation is still expected between predicted and measured values. ACI 318 commentary suggests that differences of up to 20 percent between predicted and measured values is to be expected.

3.3.2 *Concrete modulus of rupture*—ACI 318 recommends Eq. (3.3.2a) for computing the modulus of rupture of concrete with different densities

$$f_r = 7.5\lambda\sqrt{f_c'} \quad \text{(psi)}$$
$$f_r = 0.62\lambda\sqrt{f_c'} \quad \text{(MPa)} \qquad (3.3.2a)$$

where

λ = 1.0 for normal-density concrete (145 to 150 lb/ft^3 [2325 to 2400 kg/m^3])
= 0.85 for normalweight fine aggregate and lightweight coarse aggregate concrete
= 0.75 for all-lightweight aggregate concrete

Alternatively, ACI Committee 435 (1978) recommended using Eq. (3.3.2b) for computing the modulus of rupture of concrete with densities w_c in the range of 90 to 145 lb/ft^3 (1445 to 2325 kg/m^3). This equation provides a more continuous consideration of concrete density than Eq. (3.3.2a) and yields slightly higher values of f_r.

$$f_r = 0.65\sqrt{w_c f_c'} \quad \text{(psi)}$$
$$f_r = 0.013\sqrt{w_c f_c'} \quad \text{(MPa)} \qquad (3.3.2b)$$

A large degree of scatter can be expected in modulus of rupture test results, so only limited improvement in the accuracy should be expected when using the aforementioned equations in design calculations. Furthermore, these expressions are based on small prism specimens and do not reflect the presence of tensile stresses due to shrinkage restraint caused by embedded reinforcing bars and supporting elements.

3.3.3 *Time-dependent concrete properties*—ACI 318 does not specify explicit values for the concrete ultimate creep coefficient C_u and ultimate shrinkage strain $(\varepsilon_{sh})_u$. However, they can be evaluated from several equations available in the literature (ACI Committee 209 1982; Bažant and Panula 1980; Branson 1977). ACI Committee 435 (1978) suggested that the average values for C_u and $(\varepsilon_{sh})_u$ can be estimated as 1.60 and 400 × 10^{-6}, respectively. These values correspond to the following conditions:

a) 70 percent average relative humidity
b) Age of loading: 20 days for both moist and steam-cured concrete
c) Minimum thickness of component: 6 in. (150 mm)

Table 3.3.3 includes creep and shrinkage ratios at different times after loading, where the ratio indicates the percentage of ultimate creep or shrinkage that can be expected to have occurred at a certain age. Most of the time-dependent deformation can be expected to occur within the first year.

ACI Committee 209 (1971, 1982) and ACI 209R recommend a time-dependent model for creep and shrinkage under standard conditions as developed by Branson (1971, 1977). The term "standard conditions" is defined for a number of variables related to material properties, the ambient temperature, humidity, and size of members. The standard conditions for both creep and shrinkage are:

a) Age of concrete at load application = 3 days (steam), 7 days (moist)
b) Ambient relative humidity = 40 percent
c) Minimum member thickness = 6 in. (150 mm)
d) Concrete slump = 3 in. (75 mm)
e) Fine aggregate content = 50 percent
f) Air content = 6 percent

The coefficient for creep at time t (days) after load application is given by Eq. (3.3.3a)

$$C_t = \left(\frac{t^{0.6}}{10 + t^{0.6}}\right)C_u \qquad (3.3.3a)$$

where $C_u = 2.35K_h^cK_d^cK_s^cK_f^cK_{ac}^cK_{to}^c = 2.35$ for standard conditions.

Each K coefficient is a correction factor for conditions other than standard as follows:

K_h^c = relative humidity factor
K_d^c = minimum member thickness factor
K_s^c = concrete consistency factor
K_f^c = fine aggregate content factor
K_{ac}^c = air content factor
K_{to}^c = age of concrete at load application factor

Graphic representations and general equations for the modification factors (K-values) for nonstandard conditions are given in Fig. 3.3.3a (Meyers and Thomas 1983).

For concrete moist-cured for 7 days, the free shrinkage strain can be estimated using Eq. (3.3.3b) at any time t, where t is the time in days from the end of initial curing

$$(\varepsilon_{sh})_t = \left(\frac{t}{35 + t}\right)(\varepsilon_{sh})_u \qquad (3.3.3b)$$

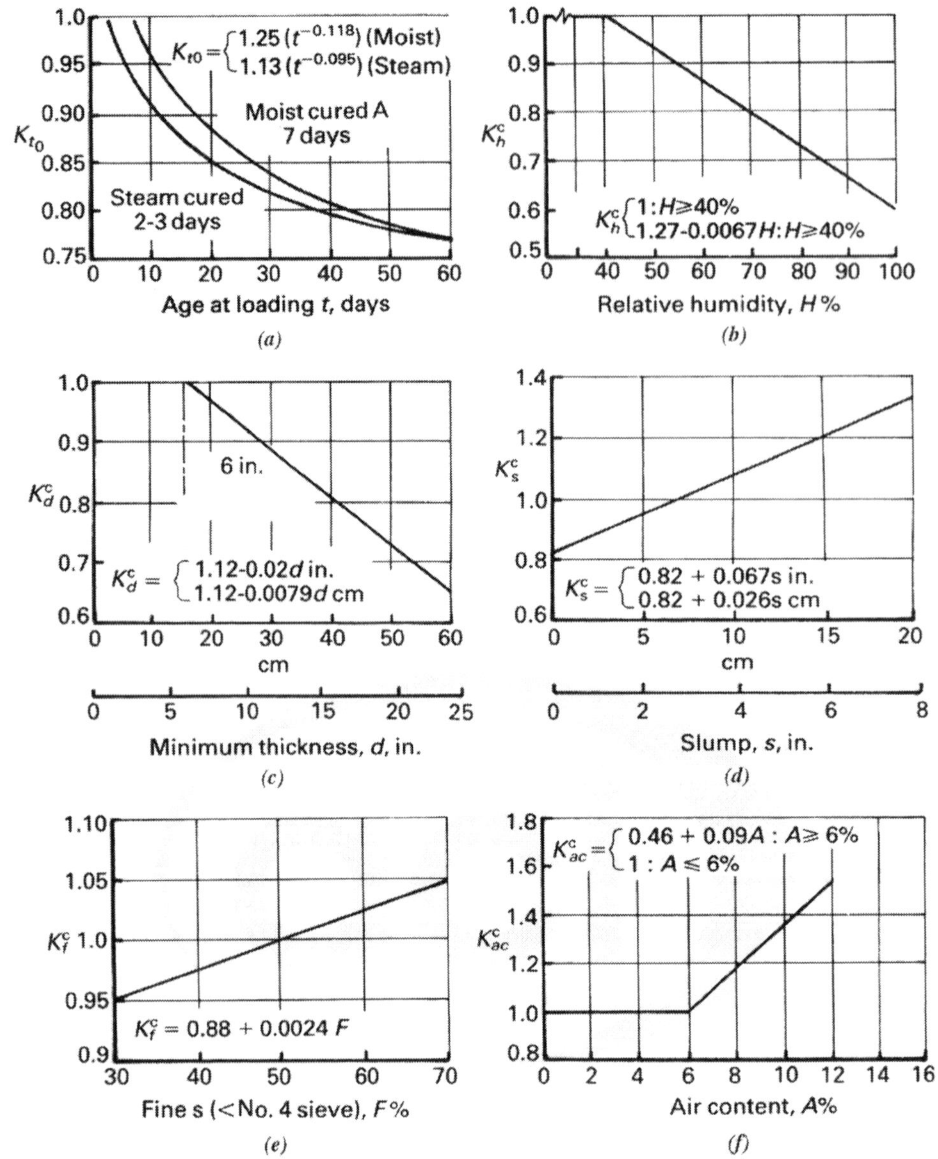

Fig. 3.3.3a—Creep correction factors for nonstandard conditions, ACI 209R method (Meyers and Thomas 1983).

and for concrete steam-cured for 1 to 3 days, the shrinkage strain can be estimated using Eq. (3.3.3c) at any time t, where t is the time in days from the end of steam curing

$$(\varepsilon_{sh})_t = \left(\frac{t}{55+t}\right)(\varepsilon_{sh})_u \qquad (3.3.3c)$$

where $(\varepsilon_{sh})_u = 780 \times 10^{-6}\, K_h^s K_d^s K_f^s K_B^s K_{ac}^s = 780 \times 10^{-6}$ for standard conditions.

Each K coefficient is a correction factor for other than standard conditions. All coefficients are the same as defined for creep except K_B^s, which is a coefficient for cement content. Graphic representation and general equations for the modification factors for nonstandard conditions are given in Fig. 3.3.3b (Meyers and Thomas 1983). The aforementioned procedure, using standard and correction equations and extensive experimental comparisons, is detailed in Branson (1977). The presence of chemical admixtures, specifically water reducers, has been shown to decrease creep and shrinkage (Brooks and Al-Qarra 1999), a factor that was not included in the development of the ACI Committee 209 (1971, 1982) and ACI 209R prediction models. ACI 209.2R outlines other creep and shrinkage functions in addition to the ACI 209R function based on Branson's work.

One approach that is employed to include the influence of creep in deflection calculations is to use an age-adjusted modulus of elasticity of the concrete, $\bar{E}_c(t,t_0)$, which can be calculated using Eq. (3.3.3d).

$$\bar{E}_c(t,t_0) = \frac{E_c(t_0)}{1+\chi C_t(t,t_0)} \qquad (3.3.3d)$$

where $E_c(t_0)$ is the modulus of elasticity of concrete at time t_0. $C_t(t,t_0)$ is the creep coefficient, which is equal to the creep

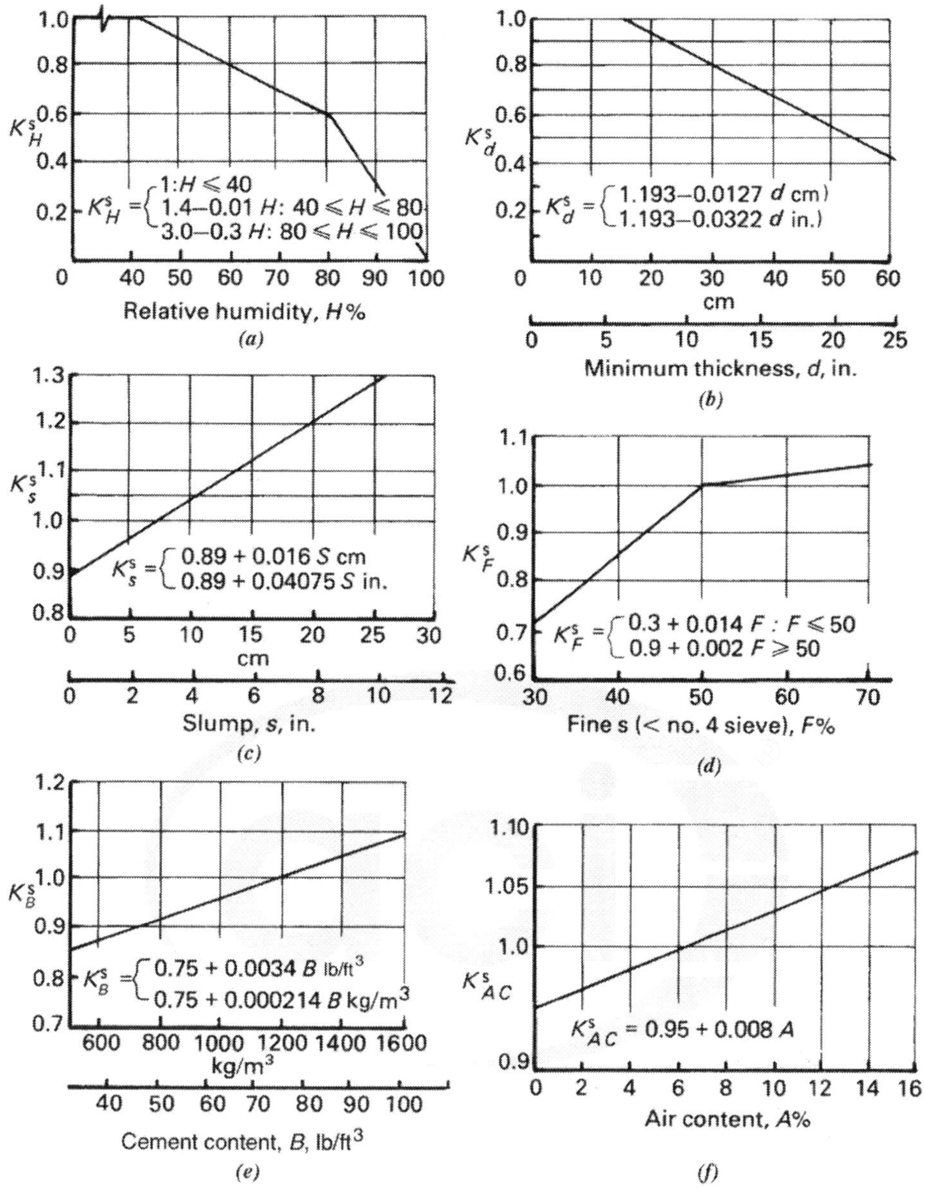

Fig. 3.3.3b—Shrinkage correction factors for nonstandard conditions, ACI 209R method (Meyers and Thomas 1983).

strain in the period t_0 to t divided by the instantaneous strain. The aging coefficient χ was introduced by Trost (1967) and Bažant and Kim (1979) and is mainly a function of t_0 and t (Ghali et al. 2011). A value of 0.8 can be used in most cases because χ varies between 0.7 and 0.9. Tabulated values of χ are given in ACI 209R.

3.3.4 *Special concretes*—In this section, brief guidance is provided on the material properties for special concretes.

3.3.4.1 *Lightweight concrete*—Lightweight concrete is structural concrete made with lightweight aggregates with bulk density less than 70 lb/ft³ (1120 kg/m³) (ACI 213R). Extensive information on the material properties of lightweight concrete is provided in ACI 213R.

The modulus of elasticity of lightweight concrete is lower than for normal-density concrete because the moduli of lightweight aggregates are significantly lower than the moduli of sand, stone, and gravel used in normal-density concretes (ACI 213R). The modulus of elasticity for lightweight concrete may be estimated using Eq. (3.3.1), which directly considers the weight difference via the parameter w_c. It is important to note however, that the simplified form of Eq. (3.3.1) often used in practice: $E_c = 57{,}000\sqrt{f_c'}$ (psi) ($4700\sqrt{f_c'}$ [MPa]) is only valid for normal-density concrete and should never be used for lightweight concrete.

The modulus of rupture is also proportionally lower for lightweight concrete as compared to normal-density concrete of the same compressive strength. Equations (3.3.2a) or (3.3.2b) may be used to estimate the modulus of elasticity of lightweight concrete. In the case of Eq. (3.3.2a), the equation should be used as presented when the tensile splitting strength f_{ct} is not specified. Otherwise, the equation should be modified by substituting $f_{ct}/6.7$ for $\sqrt{f_c'}$ (psi) ($1.8f_{ct}$ for $\sqrt{f_c'}$ [MPa]) but the value of $f_{ct}/6.7$ (psi) ($1.8f_{ct}$ [MPa]) should not exceed $\sqrt{f_c'}$. Work by Nassar (2002) indicates that Eq. (3.3.2a) may be unconservative for lightweight concretes with higher compressive strengths.

ACI 213R reports that while there is a large variation in measured values of drying shrinkage for lightweight concrete, it appears that low-strength lightweight concrete generally has greater drying shrinkage than comparable normal-density concretes. At higher strengths, some lightweight concrete has been found to exhibit lower shrinkage. Creep coefficients reported in the literature are generally higher for lightweight concrete as compared to comparable normal-density concrete.

3.3.4.2 *High-strength concrete*—High-strength concrete is defined by ACI 363R as concrete with specified compressive strengths exceeding 8000 psi (56 MPa). Extensive information on the material properties of high-strength concrete is provided in ACI 363R.

The modulus of elasticity is generally higher for high-strength concrete as compared to normal-strength concrete. Several equations have been proposed by various researchers for the modulus of elasticity of high-strength concrete, but a study by ACI 363R found that no single empirical expression estimated the modulus of elasticity to a high degree of accuracy for all concretes considered in the study. However, Eq. (3.3.4.2a) was found by ACI 363R to be a relatively reliable lower-bound expression for normal-density high-strength concrete based on most of the test data considered

$$E_c = 40{,}000\sqrt{f_c'} + 1{,}000{,}000 \quad \text{(psi)}$$
$$E_c = 3320\sqrt{f_c'} + 6900 \quad \text{(MPa)} \qquad (3.3.4.2a)$$

The values reported by various investigators (ACI 363R) for the modulus of rupture of both low-density and normal-density high-strength concretes range between $7.5\sqrt{f_c'}$ and $12\sqrt{f_c'}$ (psi) ($0.62\sqrt{f_c'}$ and $1.0\sqrt{f_c'}$ [MPa]). Therefore, Eq. (3.3.2a) can underestimate the modulus of rupture of normal-density high-strength concrete, and ACI 363R recommends the use of Eq. (3.3.4.2b) for high-strength concrete instead

$$f_r = 11.7\sqrt{f_c'} \quad \text{(psi)}$$
$$f_r = 0.97\sqrt{f_c'} \quad \text{(MPa)} \qquad (3.3.4.2b)$$

For high-strength concretes with low water-cementitious materials ratio (*w/cm*) and high binder content, the mechanisms of shrinkage may be more complex than for normal-strength concretes, with chemical and autogenous shrinkage supplementing the effects of drying shrinkage. There are considerably less data available on these mechanisms of shrinkage due to the difficulty in making such measurements. According to ACI 363R, there is no clear trend with respect to drying shrinkage of high-strength concrete, except that it is generally similar to the shrinkage of normal-strength concretes. Some researchers have reported a relatively high initial rate of shrinkage for high-strength concrete (Swamy and Anand 1973), but ACI 363R reports that there is general agreement that the drying rates in high-strength concrete will be slower when only drying shrinkage is considered. Nagataki and Yonekura (1978) reported that the shrinkage of high-strength concrete containing high-range water reducers was less than for lower-strength concrete.

On the other hand, there is general agreement that the creep coefficients for high-strength concretes are lower than for normal-strength concrete (ACI 363R). The creep coefficient (ratio of creep strain to initial elastic strain under sustained axial compression) for high-strength concrete may be as low as one-half that generally associated with low-strength concrete (Ngab et al. 1981; Nilson 1985). However, high-strength concretes are often exposed to a higher level of sustained stress such that the total creep strain may be similar to that for normal strength concrete.

3.3.4.3 *Self-consolidating concrete*—Self-consolidating concrete is concrete proportioned such that it may be consolidated without mechanical vibration. Many self-consolidating concrete mixtures use admixtures and, in some cases, a reduced quantity of coarse aggregate to maximize stability and minimize segregation and bleeding, which can lead to differences in hardened material properties relative to conventionally-placed concrete. However, self-consolidating mixtures can also be engineered to have properties similar to, or better than, those of conventionally placed concrete (ACI 237R). Information on the material properties of self-consolidating concrete is provided in ACI 237R.

In some studies, self-consolidating concrete mixtures with high sand-to-total aggregate ratios (*s/a*) have been observed to have a modulus of elasticity of approximately 10 to 15 percent lower than that of conventional concrete of similar compressive strength, whereas, in other studies, there has been little effect observed on the modulus of elasticity (ACI 237R). ACI 237R recommends the use of Eq. (3.3.1) for the modulus of elasticity of self-consolidating concrete in applications where the modulus is not critical, and recommends testing to determine the modulus of elasticity when greater accuracy is required.

ACI 237R does not recommend any unique considerations for the estimation of the modulus of rupture for self-consolidating concrete. As such, Eq. (3.3.2a) or (3.3.2b) may be used to estimate the modulus of rupture. A study by Sonebi and Bartos (2001) found that the modulus of rupture of self-consolidating concrete may actually be higher than that of conventional concrete with similar proportions.

ACI 237R notes that several studies have reported the drying shrinkage of self-consolidating concrete to be similar to or lower than that of conventional concrete with similar compressive strength. When mixture proportions are similar to those for conventionally placed concrete, self-consolidating concrete has been found to have similar creep characteristics to those for conventional concrete. When a greater paste volume is used for stability considerations, a higher creep in comparison to conventionally placed concrete may occur (ACI 237R). ACI 237R recommends testing to determine the creep coefficient when creep is an important parameter in design.

3.3.4.4 *Fiber-reinforced concrete*—Fiber-reinforced concrete consists primarily of hydraulic cements, aggregates, and discrete reinforcing fibers. Common materials used as fiber reinforcement include steel, glass, and synthetic materials, including polypropylene and polyethylene. ACI 544.1R provides detailed information on the characteristics and behaviors of the common types of fiber reinforcement, and their effect on concrete properties.

The primary contribution of fibers to concrete is to improve the pre- and post-cracking behavior of the concrete in tensile regions. Some fibers have the ability to increase the tensile capacity of the concrete, effectively increasing the modulus of rupture. No general equation has been proposed to quantify this increase as the effect is highly dependent on the fiber type, shape, and application. In addition to affecting the precracking behavior, fibers also allow for stress transfer across a crack. This typically results in a decrease in crack width, and a decrease in crack spacing. This behavior serves to decrease the immediate deflections of the structural member by reducing the rise of the neutral axis upon cracking, thereby reducing the change in the post-cracking moment of inertia relative to the gross moment of inertia. As with the modulus of rupture, no general equation has been proposed to quantify this response due to the variation in performance of various fiber types and geometries.

Fibers can also affect the creep properties of concrete. Compression creep is not significantly affected; however, long-term flexural creep is affected by the presence of fibers (ACI 544.5R). The long-term deflection response is dependent on the characteristics of the fibers used in the concrete. Longer fibers with higher moduli of elasticity tend to improve long-term deflection behavior. Specifically, fiber-reinforced concrete beams with steel fibers have shown less long-term deflections than reinforced concrete beams without fibers (Tan et al. 1994). However, fibers with lower moduli of elasticity tend to exhibit more creep behavior over time (Bernard 2004).

The shrinkage response of concrete is generally not affected by the presence of fibers. However, the potential increase in the apparent modulus of rupture and the ability of the fibers to distribute cracking may decrease the effect of shrinkage on the deflection behavior of the concrete.

3.4—Reinforcement material properties

Deflections are not as sensitive to variation in reinforcement material properties in comparison to concrete properties, so the use of specified or manufacturer-supplied values is usually adequate and additional testing on field specimens is not required. The property of interest is the reinforcement modulus of elasticity. Typical values are provided in the following.

3.4.1 *Steel reinforcement*—ACI 318 specifies using the value $E_s = 29 \times 10^6$ psi (200,000 MPa) for the modulus of elasticity of nonprestressed reinforcing steel.

3.4.2 *Fiber-reinforced polymer (FRP) reinforcement*—The modulus of elasticity of FRP reinforcement (E_f) varies widely, depending on the type and volume fraction of the fiber used in the composite. In general, elastic modulus values provided by the reinforcement manufacturer should be used when available. Ranges of elastic moduli for different commercially-available FRPs are provided by ACI 440.1R:

a) Glass-fiber-reinforced polymer (GFRP): $E_f = 5.1$ to 7.4×10^6 psi (35,000 to 51,000 MPa)
b) Carbon-fiber-reinforced polymer (CFRP): $E_f = 15.9$ to 84.0×10^6 psi (110,000 to 580,000 MPa)
c) Aramid-fiber-reinforced polymer (AFRP): $E_f = 6.0$ to 18.2×10^6 psi (41,000 to 125,000 MPa)

CHAPTER 4—DEFLECTION OF REINFORCED CONCRETE ONE-WAY FLEXURAL MEMBERS

4.1—General

4.1.1 *Introduction*—In many structures, deflection rather than strength is the controlling factor. For these cases, deflection computations will determine the proportions and configuration of the structural system members. Member stiffness is a function of short- and long-term mechanical properties of the concrete and the extent of flexural cracking. Hence, expressions defining the modulus of rupture, modulus of elasticity, creep, shrinkage, and temperature effects are important parameters in predicting the deflection of reinforced concrete members (refer to Chapter 3 for a detailed explanation of these properties). The stiffness and amount of reinforcement also control the member stiffness after cracking.

The designer may choose to calculate deflections and check that these computed values are less than specified permissible limits. Alternatively, design codes such as ACI 318 give specified maximum span/depth ratios, with limits of applicability, for which serviceability is assumed to be satisfied and deflections do not need to be explicitly calculated.

4.1.2 *Objectives*—This chapter covers the initial (immediate) and time-dependent deflections at service load levels under static conditions for one-way nonprestressed flexural concrete members. Details are also provided on deemed-to-comply limiting span-thickness ratios for which deflections do not need to be calculated. This chapter is intended to give the designer enough basic background to design concrete members that perform adequately under service loads, taking into account cracking and both short- and long-term effects.

Information is presented on the effective moment of inertia method prescribed in ACI 318 to evaluate deflection. ACI 318 permits use of an expression for the effective moment of inertia (I_e) to compute deflection, unless a more comprehensive analysis is carried out to determine the member stiffness. The existing approach for I_e used by ACI 318-14 and earlier editions can underestimate deflection in certain conditions (Bischoff and Scanlon 2007); hence, an alternative expression for the effective moment of inertia that works well over a wide range of conditions is proposed. Other modifications and more refined methods that require detailed calculations for estimating deflection are also discussed. This includes the effects of preloading from construction loads and reduced flexural stiffness that can result when tensile stresses develop in the concrete from restraint to shrinkage.

Table 4.2.1—Recommended tension reinforcement ratios for nonprestressed one-way members so that deflection will normally be within acceptable limits (ACI Committee 435 1978)

Members	Cross section	Normalweight concrete	Lightweight concrete
Not supporting or not attached to nonstructural elements likely to be damaged by large deflections	Rectangular, T, or box	$\rho \leq 35$ percent ρ_b $\rho_w \leq 40$ percent ρ_b	$\rho \leq 30$ percent ρ_b $\rho_w \leq 35$ percent ρ_b
Supporting or attached to nonstructural elements likely to be damaged by large deflections	Rectangular, T, or box	$\rho \leq 25$ percent ρ_b $\rho_w \leq 30$ percent ρ_b	$\rho \leq 20$ percent ρ_b $\rho_w \leq 25$ percent ρ_b

Notes: For continuous members, the positive region steel ratios only may be used. ρ_b refers to the balanced steel ratio based on ultimate strength.

Table 4.2.2a—Minimum thickness of nonprestressed beams and one-way slabs unless deflections are computed (from ACI 318)

		Minimum thickness, *h*		
	Simply supported	One end continuous	Both ends continuous	Cantilever
Member	Members not supporting or attached to partitions or other construction likely to be damaged by large deflections			
Solid one-way slabs	$\ell/20$	$\ell/24$	$\ell/28$	$\ell/10$
Beams or ribbed one-way slabs	$\ell/16$	$\ell/18.5$	$\ell/21$	$\ell/8$

Notes: Values given should be used directly for members with normalweight concrete and Grade 60 (Grade 420) reinforcement. For other conditions, the values shall be modified as follows:

a) For f_y other than 60,000 psi (420 MPa), the values shall be multiplied by $0.4 + f_y/100{,}000$ $(0.4 + f_y/700)$.

b) For lightweight concrete having equilibrium density w_c in the range 90 to 115 lb/ft³ (1440 to 1840 kg/m³), the values should be multiplied by $1.65 - 0.005w_c$ $(1.65 - 0.0003w_c)$ but not less than 1.09.

Table 4.2.2b—Minimum thickness of beams and one-way slabs used in roof and floor construction (ACI Committee 435 1978)

	Members not supporting or not attached to nonstructural elements likely to be damaged by large deflections				Members supporting or attached to nonstructural elements likely to be damaged by large deflection			
Member	Simply supported	One end continuous	Both ends continuous	Cantilever	Simply supported	One end continuous	Both ends continuous	Cantilever
Roof slab	$\ell/22$	$\ell/28$	$\ell/35$	$\ell/9$	$\ell/14$	$\ell/18$	$\ell/22$	$\ell/5.5$
Floor slab and roof beam or ribbed roof slab	$\ell/18$	$\ell/23$	$\ell/28$	$\ell/7$	$\ell/12$	$\ell/15$	$\ell/19$	$\ell/5$
Floor beam or ribbed floor slab	$\ell/14$	$\ell/18$	$\ell/21$	$\ell/5.5$	$\ell/10$	$\ell/13$	$\ell/16$	$\ell/4$

4.2—Control of deflection

Beams and one-way slabs are rarely built as isolated members but are part of a monolithic integrated system. Excessive deflection of a floor slab may cause dislocations in the partitions it supports or difficulty in leveling furniture or fixtures. Excessive deflection of a beam can damage a partition below, and excessive deflection of a spandrel beam or perimeter slab above a window opening could crack the glass panels. In the case of roofs or open floors, such as top floors of parking garages, ponding of water can result.

During design, deflection is controlled indirectly through reinforcement ratio limitations or minimum thickness requirements, or is controlled directly through calculating a deflection and limiting its value to a maximum based on functional requirements usually expressed as a fraction of the span length.

4.2.1 *Tension steel reinforcement ratio limitations*—One method to minimize deflection of a concrete member in flexure is by using a relatively small reinforcement ratio. When smaller reinforcement ratios are used, larger (and stiffer) sections are required to meet strength requirements. Limiting values of the reinforcement ratio ρ, ranging from $0.25\rho_b$ to $0.40\rho_b$ were recommended by ACI Committee 435 (1978), as shown in Table 4.2.1. Other methods for reducing deflection are presented in Chapter 6.

4.2.2 *Minimum thickness limitations*—Deflections of beams and one-way slabs of normal proportions and supporting typical loads in buildings are normally satisfactory when the minimum thickness provisions in Table 4.2.2a are met or exceeded. This table (ACI 318) applies only to members that are not supporting or not attached to partitions or other construction likely to be damaged by excessive deflections. Values in Table 4.2.2a were modified by ACI Committee 435 (1978) and are expanded in Table 4.2.2b to include members that are supporting or attached to nonstructural elements likely to be damaged by excessive deflections. The thickness limits in Tables 4.2.2a and 4.2.2b may be ignored when computed deflections are shown to be satisfactory. The values in Tables 4.2.2a and 4.2.2b should be used with care, as they may not be appropriate for certain situations such as sensitive window systems, long spans, and high live loads.

Due to limitations on the code provisions for minimum thickness, several authors have proposed alternative minimum thickness approaches based on computer simulations or engineering beam equations for deflection calculation (Grossman 1981, 1987; Rangan 1982; Gardner 2001; Scanlon and Lee 2006). Grossman (1981, 1987) carried out

Table 4.2.2c—Maximum permissible computed deflections (ACI 318)

Member	Condition	Deflection to be considered	Deflection limitation
Flat roofs	Not supporting or attached to nonstructural elements likely to be damaged by large deflections	Immediate deflection due to maximum of roof live load, snow and rain	$\ell/180^*$
Floors		Immediate deflection due to live load	$\ell/360$
Roof or floors	Supporting or attached to nonstructural elements	That part of the total deflection occurring after attachment of nonstructural elements, which is the sum of the time-dependent deflection due to all sustained loads and the immediate deflection due to any additional live load†	$\ell/480$‡
			$\ell/240$§

*Limit not intended to safeguard against ponding. Ponding should be checked by calculations of deflection, including added deflections due to ponded water, and considering long-term effects of all sustained loads, camber, construction tolerances, and reliability of provisions for drainage.

†Time-dependent deflection should be determined in accordance with 24.2.4 of ACI 318-14 but should be permitted to be reduced by amount of deflection calculated to occur before attachment of nonstructural elements. This amount should be calculated on basis of accepted engineering data relating to time-deflection characteristics of members similar to those being considered.

‡Limit should be permitted to be exceeded if measures are taken to prevent damage to supported or attached elements.

§Limit should not exceed tolerance provided for nonstructural elements.

many computer studies to develop a simplified expression for the minimum thickness needed to satisfy serviceability requirements (Eq. (5.3e) in Chapter 5). Work by Bischoff and Scanlon (2009) indicates that heavily loaded slabs (with a service load greater than approximately 200 lb/ft^2 [9.6 kPa]) designed using the ACI 318 minimum thickness requirements in Table 4.2.2a are not likely to satisfy the corresponding ACI 318 deflection requirements outlined in Table 4.2.2c for that part of the total (incremental) deflection occurring after attachment of the nonstructural elements; while beams with the minimum thickness requirements from Table 4.2.2a are also not likely to satisfy the deflection requirements in Table 4.2.2c. Subsequent work by Bischoff and Veysey (2011) provides recommended values of minimum member thickness needed to satisfy the $\ell/240$ and $\ell/480$ deflection limits for one-way slabs and beams. As the ACI 318 minimum thickness requirements are not always conservative, it is recommended that deflections be computed and compared with the appropriate deflection limits.

Other codes and standards such as BS EN 1992-1-1 give limiting span/effective depth ratios for flexural members and provide limits on computed values of deflection. Gardner (2011) provides an extensive comparison of requirements from numerous code or standard authorities.

4.2.3 *Computed deflection limitations*—The permissible computed deflections specified in ACI 318 for one-way systems are given in Table 4.2.2c, where the span-deflection ratios provide for a simple set of permissible deflections. Where excessive deflection may cause damage to the nonstructural or other structural elements, only that part of the deflection occurring after construction of the affected elements (such as partitions) needs to be considered. The most stringent deflection limit of $\ell/480$ in Table 4.2.2c is an example of such a case. Where excessive deflection may result in a functional problem, such as visual sagging or ponding of water, the total deflection should be considered. Project specific requirements, such as for sensitive equipment or curtain wall systems, may require more stringent limits than those listed in Table 4.2.2c.

Deflection limits can vary from code to code and represent the consensus of experience for typical structures. BS EN 1992-1-1 limits the total deflection to $\ell/250$, which should ensure that the part of deflection occurring after construction of finishes and partitions will be limited to $\ell/500$ or 0.78 in. (20 mm), whichever is less, for spans up to 32.8 ft (10 m). Provision is made in the code for longer spans and for the amount of tension and compression reinforcement. Serviceability requirements of AS 3600 limit the calculated deflection to a value appropriate to the structure and its intended use, which cannot exceed $\ell/250$ for the total deflection and $\ell/500$ for the incremental deflection that occurs after the addition or attachment of partitions when provision is made to minimize the effect of movement; otherwise the limit is $\ell/1000$. BS EN 1992-1-1 limits deflections under quasi-permanent (sustained) loads to $\ell/250$ to avoid impairment of the appearance or general utility of a structure. A degree of precamber is permitted. For deflections after construction, a limit of $\ell/500$ is deemed appropriate for quasi-permanent loads and this should avoid damage occurring to adjacent parts of the structure.

4.3—Short-term deflection calculation

The short-term or immediate deflection Δ_i from flexural curvatures can be computed using a generalized elastic equation for a prismatic member under uniformly distributed load w.

$$\Delta_i = C\frac{5}{384}\frac{w\ell^4}{E_cI_e} = K\frac{5}{48}\frac{M\ell^2}{E_cI_e} \tag{4.3}$$

where M is the moment at the critical section due to service loads at the stage deflection is computed. The critical section is usually taken at midspan for simple and continuous members and at the support face for a cantilever. K and C are restraint factors that depend on the support fixity and loading conditions. For continuous members with uniform distributed loading, $K = (1.2M_m - 0.2M_o)/M_m$, where M_m is the moment at midspan and M_o is the total static moment $w\ell^2/8$. Similarly, $C = 1.2M_o/M_m - 0.2$. K equals 2.4 and C equals 9.6 for cantilevers. Restraint factor values are given in Table 4.3 for a flexural member subjected to a uniformly distributed load w. Effects of cracking and reinforcement on member stiffness are taken into consideration with an effective moment of inertia (I_e). The modulus of elasticity

Table 4.3—Restraint factors *K* and *C* for a uniformly distributed load

Member type (support conditions)	*K*	*C*
Cantilever (fixed end)*	2.4	9.6
Simple span	1.0	1.0
One end continuous with the discontinuous end unrestrained ($K = 1.2 - 0.2M_o/M_m$)	0.925 for $M_m = w\ell^2/11$	0.673
One end continuous with the discontinuous end integral with the end support ($K = 1.2 - 0.2M_o/M_m$)	0.85 for $M_m = w\ell^2/14$	0.486
Fixed-hinged	0.80	
Both ends continuous ($K = 1.2 - 0.2M_o/M_m$)	0.80 for $M_m = w\ell^2/16$	0.40
Fixed-fixed	0.60	0.20

*Additional deflection from rotation at support of cantilever should be included.

(E_c) can be obtained from Eq. (3.3.1) for normal-strength concrete or Eq. (3.3.4.2a) for high-strength concrete.

The beam deflection Eq. (4.3) can also be written in terms of the midspan and member end moments (or midspan and member end curvatures), which is a convenient form for calculating deflection of two way-slabs. For example, $\Delta_i = (10M_m - M_1 - M_2) \times \ell^2/(96E_cI_e)$, where M_1 and M_2 are the two end moments. Textbooks on reinforced concrete (Nawy 2009; Wight and MacGregor 2009) also provides information on the different methods used to calculate deflection.

The approach described previously using Eq. (4.3) to compute deflection is most suitable for hand calculations. Use of Eq. (4.3) implies that deflection of continuous members depends on the member stiffness at the critical section, with no consideration given to the stiffness at the end supports unless a weighted average is taken of stiffness at the midspan and end supports. Other approaches involving more detailed calculations that may be amenable for implementation into computer software include integration of section curvature M/E_cI_e (4.3.3) to account for changes in stiffness along the member span.

4.3.1 *Uncracked members: gross moment of inertia* I_g—When the maximum flexural moment at service load in a beam or a slab causes a tensile stress less than the modulus of rupture (f_r), no flexural tension cracks develop at the tension side of the concrete element if the member is not restrained or the shrinkage and temperature tensile stresses are negligible. In such a case, the effective moment of inertia of the uncracked transformed section (I_t) is applicable for deflection computations. For design purposes, however, the gross moment of inertia I_g, neglecting the reinforcement contribution, can be used with negligible loss of accuracy. The combination of service loads with shrinkage and temperature effects due to end restraint or restraint from the internal reinforcement may cause cracking if the tensile stress in the concrete exceeds the modulus of rupture. In such cases, 4.3.2 applies.

4.3.2 *Cracked members: effective moment of inertia* I_e—Tension cracks occur when the imposed loads cause bending moments in excess of the cracking moment, thus resulting in tensile stresses in the concrete that are higher than its modulus of rupture. The cracking moment M_{cr} may be computed as follows

$$M_{cr} = \frac{f_r I_g}{y_t} \tag{4.3.2}$$

where y_t is the distance from the centroidal axis to the tension face of the member, and f_r is the modulus of rupture of the concrete, as expressed by Eq. (3.3.2a) or (3.3.2b). Use of a lower cracking moment to account for shrinkage induced tension is discussed in 4.3.2.3 and in 5.2.3 for two-way slabs.

Cracks develop at several sections along the member length. While the cracked moment of inertia I_{cr} applies to the cracked sections, the gross moment of inertia I_g applies to the uncracked concrete between these sections. Several methods have been developed to estimate the variations in stiffness caused by cracking along the span. These methods either modify the flexural rigidity of the member by using an average or effective moment of inertia (Yu and Winter 1960; Branson 1963), make adjustments to the curvature along the span and at critical sections (Beeby 1968), compute an average curvature or the ratio M/EI at each location along the member span (*fib* MC2010), or use an incremental evaluation of section-curvature (Ghali 1989; Ghali et al. 2011). Deflection can be conservatively estimated by using the effective stiffness or curvature at the critical section where the member stiffness is lowest.

Studies by Branson (1977, 1985) have shown that the initial deflections Δ_i occurring in a beam or a slab after the maximum moment M_a has exceeded the cracking moment M_{cr} can be evaluated using an effective moment of inertia I_e that models a gradual transition from the gross (uncracked) moment of inertia I_g to the cracked transformed moment of inertia I_{cr}.

4.3.2.1 *Branson's* I_e—With the exception of slender walls (as defined in 11.8 of ACI 318-14), ACI 318 provides an expression for the effective moment of inertia I_e originally proposed by Branson (1963). This approach was selected as being sufficiently accurate to control deflections in reinforced concrete unless stiffness values are obtained by a more comprehensive analysis. Branson's approach is known to work well for beams that typically have a reinforcement ratio greater than 1 percent, but significantly overestimates I_e for lightly reinforced slabs where the in-service moment is often only just greater than the cracking moment (Bischoff and Scanlon 2007; Gilbert 2007).

Branson's equation for the effective moment of inertia I_e for short-term deflections is as follows

$$I_e = \left(\frac{M_{cr}}{M_a}\right)^3 I_g + \left[1 - \left(\frac{M_{cr}}{M_a}\right)^3\right] I_{cr} \leq I_g \tag{4.3.2.1a}$$

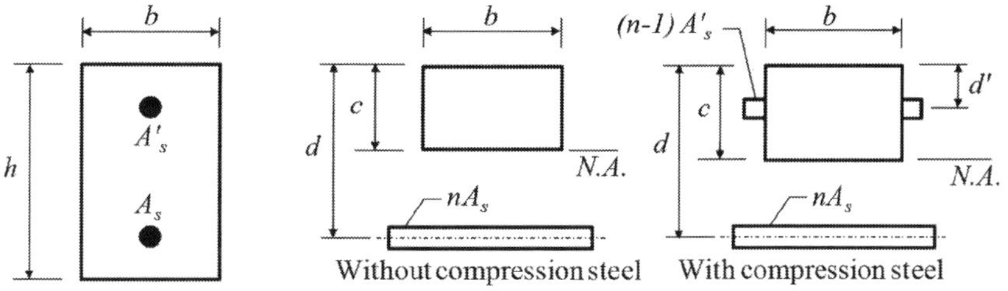

$$I_g = bh^3/12;\quad c = kd;\quad \rho = A_s/bd;\quad \rho' = A_s'/bd$$

Without compression steel:

$$k = \sqrt{(n\rho)^2 + 2n\rho} - n\rho$$

$$I_{cr} = bc^3/3 + nA_s(d-c)^2$$

With compression steel:

$$k = \sqrt{[n\rho + (n-1)\rho']^2 + 2[n\rho + (n-1)\rho'(d'/d)]} - [n\rho + (n-1)\rho']$$

$$I_{cr} = bc^3/3 + nA_s(d-c)^2 + (n-1)A_s'(c-d')^2$$

(a) Rectangular Sections

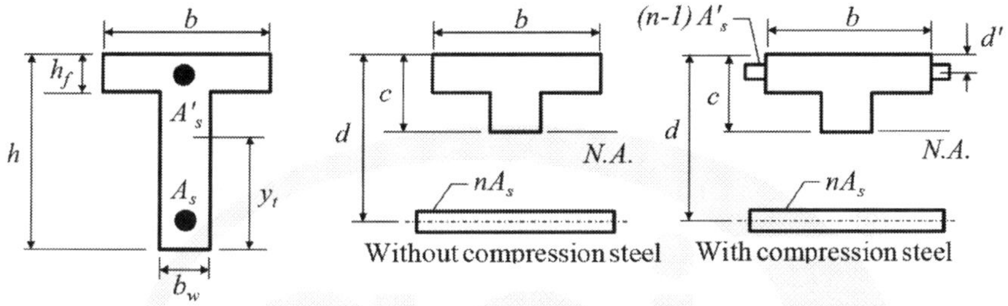

$$y_t = h - \frac{1}{2}\frac{\zeta h_f^2 + b_w h^2}{\zeta h_f + b_w h};\quad \zeta = b_f/b_w - 1;\quad \rho_w = A_s/b_w d;\quad \rho_w' = A_s'/b_w d \text{ and } c = kd$$

$$I_g = (b_f - b_w)h_f^3/12 + b_w h^3/12 + (b_f - b_w)h_f(h - h_f/2 - y_t)^2 + b_w h(y_t - h/2)^2$$

Without compression steel:

$$k = \sqrt{[n\rho_w + \zeta(h_f/d)]^2 + 2[n\rho_w + 0.5\zeta(h_f/d)^2]} - [n\rho_w + \zeta(h_f/d)]$$

$$I_{cr} = (b_f - b_w)h_f^3/12 + b_w c^3/3 + (b_f - b_w)h_f(c - h_f/2)^2 + nA_s(d-c)^2$$

With compression steel:

$$k = \sqrt{[n\rho_w + (n-1)\rho_w' + \zeta(h_f/d)]^2 + 2[n\rho_w + (n-1)\rho_w'(d'/d) + 0.5\zeta(h_f/d)^2]} - [n\rho_w + (n -$$

$$I_{cr} = (b_f - b_w)h_f^3/12 + b_w c^3/3 + (b_f - b_w)h_f(c - h_f/2)^2 + nA_s(d-c)^2 + (n-1)A_s'(c-d')^2$$

(b) Flanged Sections

Fig. 4.3.2.1—Moments of inertia of uncracked and cracked transformed sections.

or in a slightly rearranged form as

$$I_e = I_{cr} + \left(\frac{M_{cr}}{M_a}\right)^3 (I_g - I_{cr}) \le I_g \qquad (4.3.2.1b)$$

where M_{cr} is the cracking moment; M_a is the maximum (unfactored) service load moment at the stage for which deflections are being considered; I_g is the gross moment of inertia of section; and I_{cr} is the moment of inertia of cracked transformed section.

The value I_e provides a transition between the upper and the lower bounds of I_g and I_{cr}, respectively, as a function of the level of cracking, expressed as M_{cr}/M_a. The use of I_e as a weighted average of I_g and I_{cr} is assumed to give deflection values close to those obtained using a bilinear load-deflection response after cracking. Details for computing the cracked moment of inertia I_{cr} are provided in Fig. 4.3.2.1.

Deflections have traditionally been computed for each load level using either form of Eq. (4.3.2.1a) or (4.3.2.1b), such as dead load and dead load plus live load. Thus, the incremental deflection such as that due to live load alone is computed as the difference between these values at the two load levels. Another school of thought, however, assumes the structural member has been preloaded under the full (dead plus live) service load such that subsequent calculations of both the dead and live load deflection are computed using the same value of I_e corresponding to full dead plus live load (Grossman 1981). The latter approach is generally assumed to be more realistic for slab construction and provides a conservative estimate of long-term deflection. The former approach, on the other hand, provides a conservative estimate of immediate deflection from live load, although this requirement typically does not control design.

The value of I_e may be determined using M_a at the support for cantilevers, and at the midspan for simple spans. Equation (4.3.2.1a) shows that I_e is an interpolation between the well-defined limits of I_g and I_{cr}. This equation was recommended by ACI Committee 435 (1966) and has been used in ACI 318 since 1971 and the AASHTO Bridge Design Specifications since 1973. Detailed numerical examples using this method for simple and continuous beams and unshored and shored composite beams are available in Branson (1977). Textbooks on reinforced concrete such as those by Wang and Salmon (1992) and Nawy (2009) also provide an extensive treatment of the subject.

Heavily reinforced members will have an I_e approximately equal to I_{cr}, which may in some cases (flanged members) be larger than I_g of the concrete section alone. For most practical cases, the calculated I_e will be less than I_g. In both cases, I_e should be taken as the lesser of the calculated value or I_g when designing for deflection control, unless a justification can be made for rigorous transformed section computations.

Equation (4.3.2.1a) or (4.3.2.1b) is intended to represent an averaged value of I_e that accounts for the variation in stiffness along the member. Replacing the cubic power in Eq. (4.3.2.1a) with a fourth power is meant to give the effective moment of inertia at a section and is useful when integrating curvatures at multiple sections along the span length.

$$I_e = \left(\frac{M_{cr}}{M_a}\right)^4 I_g + \left[1 - \left(\frac{M_{cr}}{M_a}\right)^4\right] I_{cr} \le I_g \quad (4.3.2.1c)$$

Branson's equation (Eq. (4.3.2.1a) or (4.3.2.1b)) is an empirically derived formula that was calibrated to work well for typical steel-reinforced concrete beams having reinforcing ratios greater than approximately 1 percent (Bischoff 2005). However, this equation underestimates deflection for most members reinforced with FRP bars (ACI 440.1R), for certain types of slender tilt-up wall panels (Bischoff and Scanlon 2007), and for steel-reinforced concrete beams and slabs with reinforcing ratios below 1 percent (Bischoff 2005; Gilbert 2006). ACI 440.1R-06 recommends applying a correction factor to Eq. (4.3.2.1a) or (4.3.2.1b) for computing deflection of FRP-reinforced concrete members, while the problem with lightly reinforced slabs has not yet been addressed by ACI 318. However, ACI 318 recognizes the problem with Branson's equation for tilt-up wall panels, and now requires use of a bilinear equation taken from the *Uniform Building Code* (ICBO 1997).

The accuracy of Branson's equation for I_e is dependent on the ratio of I_g to I_{cr} (in addition to the level of loading with respect to the cracking moment), and only works well for flexural members when $I_g/I_{cr} < 3$ (corresponding to steel reinforced concrete members with a reinforcing ratio greater than approximately 1 percent). Member stiffness using Branson's approach is overestimated when $I_g/I_{cr} > 3$. The ratio I_g/I_{cr} for lightly reinforced concrete slabs (0.2 percent $< \rho <$ 0.6 percent) can be as high as 12, varies between 7 to 25 for FRP-reinforced concrete beams and slabs, and ranges from 15 to 25 for walls with a central layer of reinforcement (Bischoff and Scanlon 2007). Hence, Branson's equation (Eq. (4.3.2.1a) or (4.3.2.1b)) can significantly underestimate deflection in these three cases.

Both the member (cubic power) and section (fourth power) based expressions for I_e were calibrated from tests (Branson 1963) of uniformly loaded beams where the value of I_e was within 12 percent of I_{cr} for nearly 85 percent of the specimens assessed. The assumed level of service load ranged from two to four times the cracking moment. There was practically no difference (3 to 6 percent at most) between the member I_e and section based I_e for nearly all the 56 specimens considered. Hence, it is difficult to draw any definite conclusions about the applicability of using either equation for other types of members and loading conditions.

4.3.2.2 *Bischoff's* I_e—Bischoff (2005, 2007) applied basic concepts of tension stiffening to flexural members to develop a rational expression for I_e given by

$$I_e = \frac{I_{cr}}{1 - \beta\left(\frac{M_{cr}}{M_a}\right)\left(1 - \frac{I_{cr}}{I_g}\right)} = \frac{I_{cr}}{1 - \left(\frac{M_{cr}}{M_a}\right)^2\left(1 - \frac{I_{cr}}{I_g}\right)} \le I_g \quad (4.3.2.2a)$$

with the tension stiffening factor $\beta = M_{cr}/M_a$ that varies between 1 (at cracking) and 0 when $M_a >> M_{cr}$. Setting β equal to 1 (instead of M_{cr}/M_a) gives a bilinear response with constant tension stiffening that represents an upper bound on member stiffness as shown in Fig. 4.3.2.2a, whereas β set equal to 0 (instead of M_{cr}/M_a) provides a lower bound on member stiffness with no tension stiffening to give the I_{cr} response. Bischoff's expression for I_e gives a reasonable estimate of deflection provided the correct cracking moment is used (Scanlon and Bischoff 2008).

Equation (4.3.2.2a) can be rearranged into Eq. (4.3.2.2b) that incorporates similar ratios to Eq. (4.3.2.1a). However, comparing the arrangement of terms in the two equations highlights the fundamental difference between the two approaches. Equation (4.3.2.2b) uses a weighted average of flexibilities whereas Eq. (4.3.2.1a) uses a weighted average of stiffnesses. Because deflection is proportional to flexibility, Bischoff's approach (Eq. (4.3.2.2a) or (4.3.2.2b)) is applicable over a wider range of situations than Branson's approach (Eq. (4.3.2.1a)). The other primary difference is that the cubic power in Branson's equation (Eq. (4.3.2.1a)) has been replaced by a squared power in Bischoff's equation (Eq. (4.3.2.2a and b)) (refer to Bischoff [2005] for motivations for this).

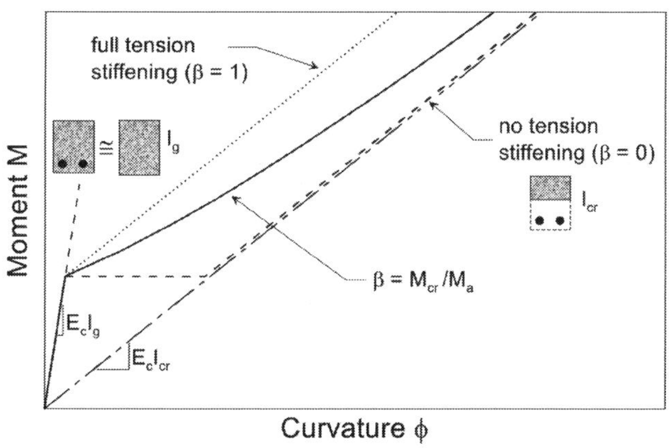

Fig. 4.3.2.2a—Moment-curvature response for different tension stiffening values with Eq. (4.3.2.2a).

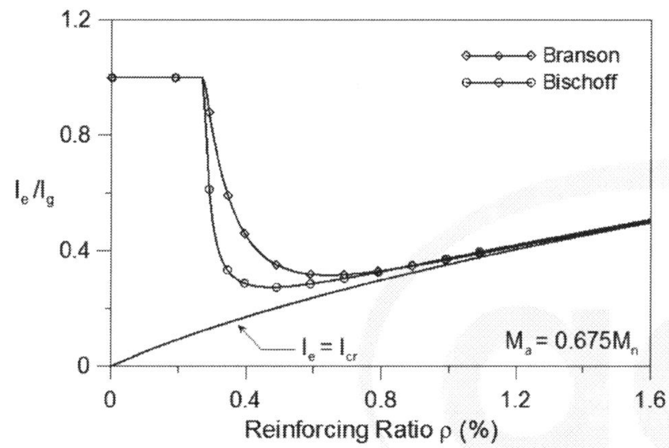

Fig. 4.3.2.2b—Comparison of I_e *using Branson and Bischoff's approach with* M_{cr}.

$$\frac{1}{I_e} = \left(\frac{M_{cr}}{M_a}\right)^2 \frac{1}{I_g} + \left[1 - \left(\frac{M_{cr}}{M_a}\right)^2\right] \frac{1}{I_{cr}} \geq \frac{1}{I_g} \qquad (4.3.2.2b)$$

Bischoff's I_e (defined by Eq. (4.3.2.2a) or (4.3.2.2b)) is simple and works equally well for concrete beams and slabs reinforced with either steel or FRP bars over a wide range of reinforcing ratios (Bischoff and Scanlon 2007). Bischoff's equation (Eq. (4.3.2.2a)) and Branson's equation (Eq. (4.3.2.1a)) give similar computed values of deflection for beams reinforced with steel at reinforcing ratios greater than 1 percent (where tension stiffening is not significant) as shown in Fig. 4.3.2.2b. Differences only become significant for flexural members with an I_g/I_{cr} greater than 3 (typically for flexural members with $\rho < 1$ percent for steel-reinforced concrete). For the reasons given previously, Bischoff's equation (Eq. (4.3.2.2a)) is recommended for computing deflection.

4.3.2.3 *Shrinkage restraint and preloading*—Restraint to shrinkage from either the internal reinforcement or adjoining members and supports induces tensile stresses in the concrete that decrease the cracking moment and reduce the flexural stiffness of the member. Decreasing the cracking moment has a significant effect on computed values of deflection for lightly reinforced members such as slabs. Deflection of moderate to more heavily reinforced members is not significantly affected at full-service load as the effective moment of inertia I_e is closely approximated by I_{cr} at the full service level of load for these members (where $M_a/M_{cr} > 3$).

CSA A23.3 uses a reduced cracking moment for computing deflection (based on the work of Scanlon and Murray [1982]) whereas AS 3600 reduces the cracking moment with a computed value of the shrinkage restraint stress (Gilbert 1999). BS EN 1992-1-1 provisions for computing deflection can also account for a reduced cracking moment. The use of a lower cracking moment also accounts for cases where the service load moment is only slightly less than the unrestrained cracking moment as factors such as shrinkage and temperature can still cause the section to crack over time.

To account for this reduction in cracking load, Gilbert (1999) recommends reducing the modulus of rupture f_r by an amount equal to the restraint stress $f_{res} = 2.5\rho E_s \varepsilon_{sh}/(1 + 50\rho)$, where ε_{sh} is the shrinkage strain in the concrete. AS 3600 uses a modified version of Gilbert's (1999) expression that includes the effect of compression steel. For design, Scanlon and Bischoff (2008) recommend using a reduced cracking moment equal to two-thirds the value of M_{cr} (computed using Eq. (4.3.2) with Eq. (3.3.2a)) when computing deflections for steel-reinforced concrete using Eq. (4.3.2.2a) for I_e. ACI 318 already uses $(2/3)M_{cr}$ for computing out-of-plane deflection of slender walls. For FRP-reinforced concrete, the cracking moment should only be reduced by 20 percent to reflect a lower degree of internal restraint when using the more flexible FRP reinforcement (Bischoff and Gross 2011b). Further reductions in the cracking moment may be required to account for axial tension induced by restraint to shrinkage from the supports and adjacent parts of the structure. For shrinkage stresses approaching f_r, or other causes of large net axial forces, it may be more appropriate and numerically stable to modify M_a instead of M_{cr}.

Preloading from construction loads prior to installation of the nonstructural elements can cause additional cracking that reduces the member stiffness and increases deflection arising from the sustained portion of loading. Scanlon and Bischoff (2008) recommend computing the immediate deflection under sustained loads with an effective moment of inertia corresponding to the full (dead plus live) service load moment to account for the reduced stiffness. This simplifies the calculation procedure considerably as only one value of I_e is now considered in deflection calculations.

4.3.2.4 *Variation in stiffness along member length*—Equation (4.3.2.2a) or (4.3.2.2b) effectively provides a section-based expression for the effective moment of inertia. For members with uniform cross section and typical reinforcement patterns, using either of these equations based on the service load moment at the critical section (where the moment is greatest) will provide a conservative estimate of deflection as the member stiffness is lowest at this location.

If warranted, a more accurate calculation of deflection can be carried out by integrating curvatures along the length of the member based on the calculated I_e value at each section (Bischoff and Gross 2011a). Using this approach gives a modified expression for Eq. (4.3.2.2a) that includes an additional factor γ to account for the variation in stiffness along the length of member. ACI 440.1R has adopted this approach for design of FRP-reinforced concrete.

$$I_e = \frac{I_{cr}}{1-\gamma\left(\frac{M_{cr}}{M_a}\right)^2\left(1-\frac{I_{cr}}{I_g}\right)} \leq I_g \qquad (4.3.2.4)$$

The integration factor $\gamma \geq 1$ accounts for the length of the uncracked regions of the member in addition to the change in stiffness along the cracked regions. The value of γ depends on the type of loading, boundary conditions, and reinforcement pattern. Earlier work by Al-Zaid et al. (1991) has also recognized that the I_e value can be affected by the type of loading (concentrated or distributed) on the member. Expressions for the factor γ in Table 4.3.2.4, provided for different types of beams and loading cases, can be used in place of numerical integration of curvature. Closed-form solutions for deflection of continuous members with moments at the end supports are not readily available and would typically require numerical integration unless simplifying assumptions are made.

The increase in member stiffness with Eq. (4.3.2.4) depends on the type of load, boundary conditions, reinforcing pattern, reinforcing material, reinforcement ratio, and level of service load relative to the cracking moment. Setting γ = 1 results in Eq. (4.3.2.2a), which provides a reasonably conservative estimate of deflection in many cases, particularly for steel reinforced concrete members subjected to a uniformly distributed load (Bischoff and Gross 2011a).

4.3.2.5 *Continuous members*—For continuous beams and one-way slabs, ACI 318 stipulates that I_e may be taken as the average of values obtained for the critical positive and negative moment sections. For prismatic members, I_e may be taken as the value obtained at midspan for continuous spans. The use of midspan section properties for continuous prismatic members is considered satisfactory in approximate calculations primarily because the midspan rigidity, including the effect of cracking, has a dominant effect on deflections (ACI Committee 435 1978).

ACI 318 suggests a simple averaging of I_e values for the positive and negative moment section if the designer chooses to average the effective moment of inertia I_e.

$$I_e = 0.50I_{e(m)} + 0.25(I_{e(1)} + I_{e(2)}) \qquad (4.3.2.5a)$$

where the subscripts m and 1 and 2 refer to midspan and the two beam ends, respectively.

Improved results for continuous prismatic members can, however, be obtained using a weighted average, as presented in the following equations.

For beams continuous on both ends

$$I_e = 0.70I_{e(m)} + 0.15(I_{e(1)} + I_{e(2)}) \qquad (4.3.2.5b)$$

For beams continuous on one end only

$$I_e = 0.85I_{e(m)} + 0.15I_{e(1)} \qquad (4.3.2.5c)$$

Work by Bischoff (2007), however, indicates that deflection of continuous members can be underestimated when using a weighted average of I_e, for cases where the member stiffness is greater at the end supports than at midspan. Once again, deflection would be reflected more correctly by taking a weighted average of the inverse stiffness (flexibility) values ($1/E_cI_e$). Reasonable values of deflection are obtained by using the midspan value of the effective moment of inertia provided that: 1) the end support moments do not exceed twice the midspan value; and 2) the value of I_e at the end supports is between one-half and three times the value at midspan.

When I_e is calculated as indicated in the previous discussion, the deflection can be obtained using a generalized form of the elastic equation such as Eq. (4.3), or any other approach used to compute elastic deflection such as the moment-area method or conjugate beam method. Numerical integration procedures can also be used. The moment-curvature (rotation) response of the member is taken into consideration with each approach.

4.3.3 *Section-based moment-curvature approach*—The calculation methods for member deflection described in 4.3.1 and 4.3.2 rely on the use of an average member stiffness between the supports. For uncracked members (4.3.1), the gross moment of inertia, I_g, is used, while an average effective moment of inertia, I_e, can be used for cracked members (4.3.2). In the latter case, the parameter I_e can account for variations in the amount of cracking along the member length as well as the tension stiffening of concrete between the cracks by using integration factors or other simplifications to represent the variation in bending moment. In general, the mathematical complexity in these techniques is intentionally low, using linear-elastic material properties to facilitate hand calculations or simple computer-based techniques, such as spreadsheets. However, with the availability of personal computers in engineering design offices, more accurate analytical procedures can be routinely considered for the calculation of deflections under complex loading scenarios, time dependent effects (4.4.3.2), or to directly consider the nonlinear material behavior.

A numerical section-based approach can directly consider the compressive stress-strain response characteristics of different concrete types within a cross section as well as the calculation of deflections before and after yielding of the reinforcement at different locations along the member. Sectional-based analytical methods can also be used to estimate the influence of nonprismatic geometry or variable reinforcement detailing on the overall deflection of members.

Numerical modeling is performed by first determining the nonlinear moment-curvature response of representative cross sections along the member span that satisfy compatibility of axial strains and equilibrium of section forces. The member deflection is then computed by numerically integrating the curvatures (Park and Paulay 1975; Collins and Mitchell 1997; Ghali et al. 2011). Such calculations can be performed

Table 4.3.2.4—Integration factor to account for changes in stiffness along the member (Bischoff and Gross 2011a)

Beam and loading type	Moment and elastic deflection equations	$I_e = I_{cr}/[1 - \gamma\eta(M_{cr}/M_a)^2]$ with $\eta = 1 - I_{cr}/I_g$
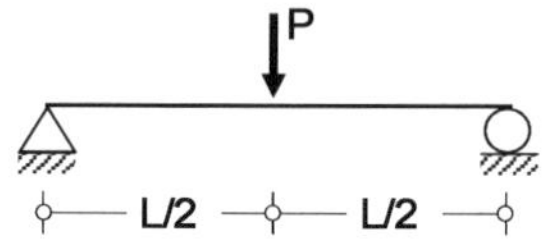	$M_a = PL/4$ $\Delta = PL^3/48E_cI_e$	$\gamma = 3 - 2(M_{cr}/M_a)$
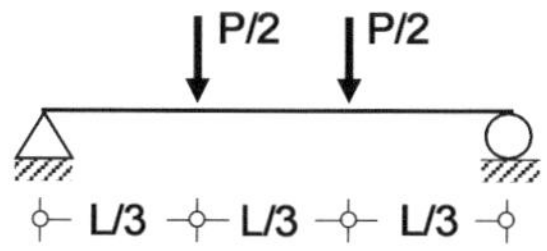	$M_a = PL/6$ $\Delta = 23PL^3/1296E_cI_e$	$\gamma = 1.7 - 0.7(M_{cr}/M_a)$
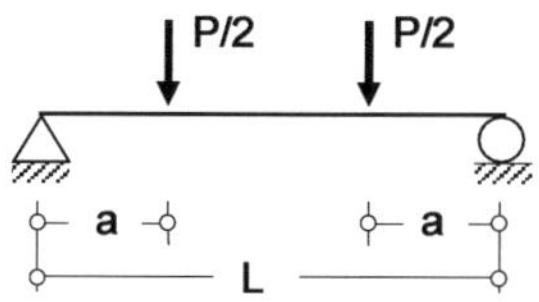	$M_a = Pa/2$ $\Delta = \alpha' PL^3/48E_cI_e$ $\alpha' = 3(a/L) - 4(a/L)^3$	$\gamma = \dfrac{3(a/L) - 4\zeta(a/L)^3}{3(a/L) - 4(a/L)^3}$ $\zeta = 4(M_{cr}/M_a) - 3$
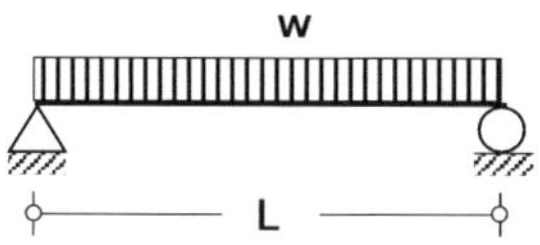	$M_a = wL^2/8$ $\Delta = 5wL^4/384E_cI_e$	$\gamma = \dfrac{1.6\zeta^3 - 0.6\zeta^4}{(M_{cr}/M_a)^2} + 2.4\ln(2-\zeta) = 1.72 - 0.72(M_{cr}/M_a)$ $\zeta = 1 - \sqrt{1 - M_{cr}/M_a}$
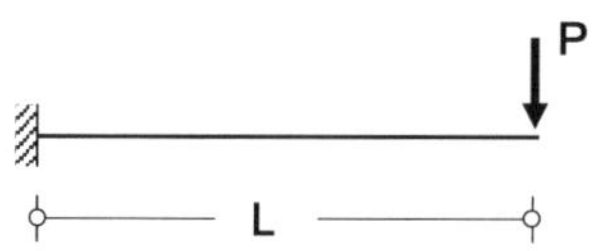	$M_a = PL$ $\Delta = PL^3/3E_cI_e$	$\gamma = 3 - 2(M_{cr}/M_a)$
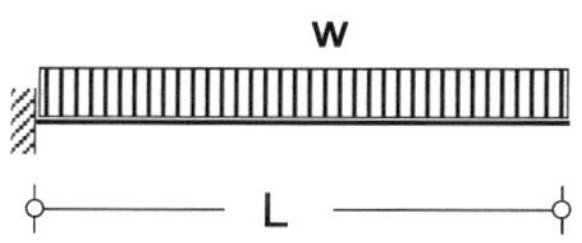	$M_a = wL^2/2$ $\Delta = wL^4/8E_cI_e$	$\gamma = 1 - 2\ln(M_{cr}/M_a)$

using a piece-wise numerical integration approach described herein, or a finite-element-based approach.

One simple method to obtain the variation in curvature along the member axis is to solve the relationship for curvature $\kappa = M_a/E_cI_e$ at selected cross sections, where M_a is the moment at the cross section and I_e is the effective moment of inertia from Eq. (4.3.2.1c) or (4.3.2.2a), depending on use of Branson's or Bischoff's estimate of I_e, respectively.

Alternatively, direct consideration of the nonlinear mechanical properties of the materials in combination with the cross section configuration can be made. For a given loading condition at each selected cross section, the axial strains are established using the well-known hypothesis that plane sections before bending remain plane after bending. Thus, a linear variation of strains occurs over the height of the cross section, with subscripts including t to denote total (Fig. 4.3.3a). For short-term loading with no time-dependent effects, the strain due to stress in the concrete, denoted as ε_{cf}, is determined at each elevation in the cross section. Local bending of the reinforcing bars is ignored and perfect bond is assumed between the concrete and the bars. Thus, the total strain in each reinforcing bar is identical to the total axial strain, ε_{tf}. For a reinforced member without prestressing, the strain due to stress in the reinforcement, ε_{sf}, is identical to ε_{tf}. With the strain distribution defined, the stress-strain response of the concrete and reinforcement is used to establish the corresponding stresses f_c and f_s. To approximate tension stiffening of the concrete, an average tensile stress to average tensile strain relationship (Vecchio and Collins 1986) can be used to estimate the tensile stress after cracking. Finally, for this overall stress condition over the cross section, the equilibrium conditions by Eq. (4.3.3a) and (4.3.3b) are checked.

$$N = \int_A f_c \cdot dA + \int_{A_s} f_s \cdot dA_s \tag{4.3.3a}$$

$$M = -\int_A f_c \cdot y \cdot dA - \int_{A_s} f_s \cdot y \cdot dA_s \tag{4.3.3b}$$

where N and M are the internal axial force and internal bending moment on the cross section due to external loading, and y is the vertical distance measured up from the centroidal axis. The corresponding curvature κ is the slope of the total strain diagram.

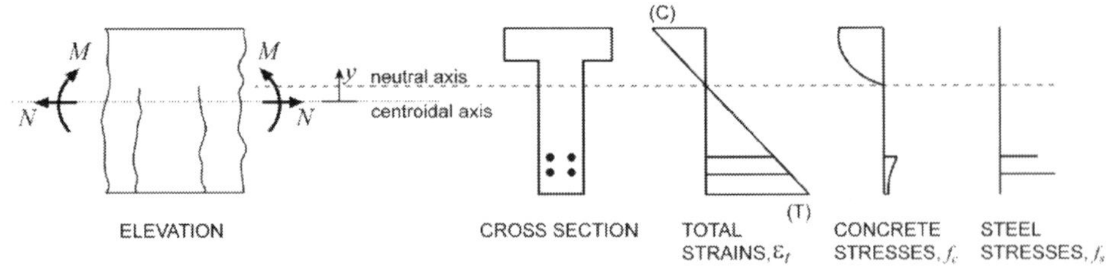

Fig. 4.3.3a—Sectional response.

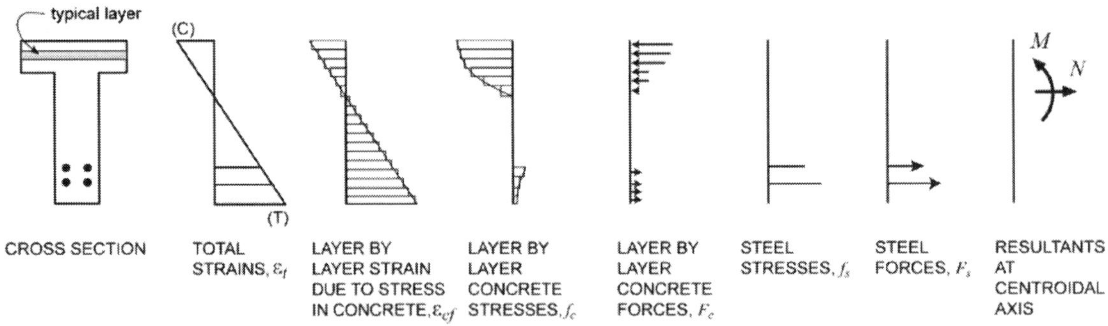

Fig. 4.3.3b—Layer-by-layer of sectional forces.

To simplify the calculations in Eq. (4.3.3a) and (4.3.3b), the calculations can be performed numerically using a layer-by-layer approach. As shown in Fig. 4.3.3b, the cross section is discretized into n layers of either constant or unequal height. From the linear distribution of total strains, the representative strain due to stress and the stress in each concrete layer can be computed based on the average strain in the layer or through further refinement based on the shape of concrete stress-strain response. Corresponding layer-by-layer forces in the concrete are then solved by considering the concrete stresses and the cross section geometry of the layer. The reinforcement strains due to stress and forces are computed from the overall strain distribution using a similar procedure, noting that the concrete and reinforcement strains are identical at the reinforcement level due to the perfect bond assumption. Typical stress-strain relationships for concrete and reinforcement are given in Wight and MacGregor (2009). In place of a nonlinear stress-strain model for concrete in tension, Collins and Mitchell (1997) suggest that tension stiffening can be approximated as a uniform tensile stress of $0.5f_{cr}$ acting over the part of the concrete within $7.5d_b$ of each reinforcing bar. Setting the concrete stress to zero for concrete in tension surrounding the reinforcement gives the section response at a crack but is not useful for computing deflection as this does not take tension stiffening into account. Collins and Mitchell (1997) also note that the layer-by-layer approach for concrete in compression can be modeled using an equivalent stress-block approach if the concrete stress-strain response has a parabolic shape.

The moment-curvature relationships for different cross sections along the member can be used to establish the approximate curvature distribution (Fig. 4.3.3c). The corresponding slopes and deflections can be obtained based on numerical integration using a number of different approaches including virtual work, conjugate beam methods, or the use of moment-area theorems.

Appendix B of ACI 435R-95(03) provides a detailed approach for computing deflection of reinforced concrete members, with or without prestressing, having a variable cross -section. Section analysis is carried out to determine curvature of an uncracked and cracked section (ignoring concrete in tension). Curvature of the uncracked and cracked section is then interpolated using *fib* MC2010 to obtain an interpolated value of average curvature at each cross -section that takes tension stiffening of the concrete into account. The average curvature obtained with this approach at sections along the member span is used to obtain deflection.

4.4—Long-term deflection calculation

4.4.1 *Traditional ACI method*—Unless values are obtained by a more comprehensive analysis, the additional long-term deflection due to the combined effects of creep and shrinkage can be calculated for nonprestressed one-way flexural members by applying a multiplier λ_Δ to the computed short-term (immediate) deflections caused by the sustained load. Hence, the long-term deflection from the sustained loads is obtained by

$$\Delta_{LT} = \lambda_\Delta \Delta_{sus} \quad (4.4.1a)$$

where Δ_{sus} is the immediate deflection due to sustained loads (deflection from dead load Δ_D plus any deflection from the sustained portion of the live load $\Delta_{L,sus}$), and λ_Δ is the time-dependent multiplier for a defined duration of time t.

$$\lambda_\Delta = \frac{\xi}{1+50\rho'} \quad (4.4.1b)$$

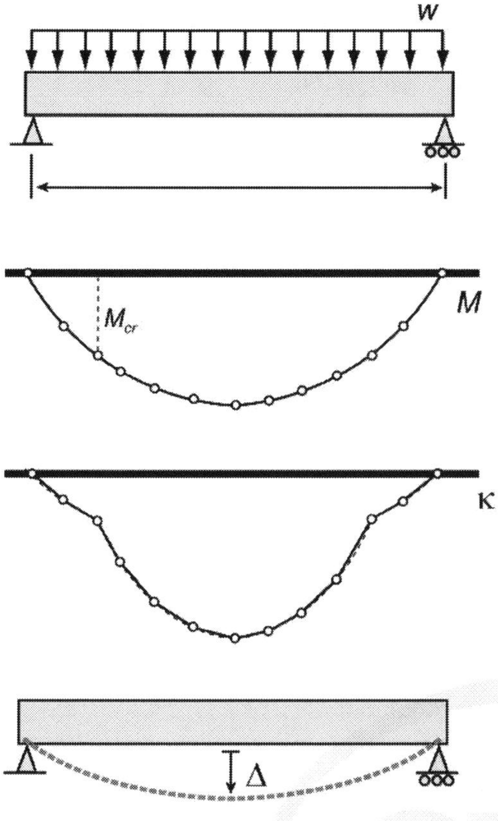

Fig. 4.3.3c—Typical curvature distribution.

where ρ′ is the reinforcement ratio for nonprestressed compression steel reinforcement; and ξ is the time-dependent factor from Fig. 4.4.1 (ACI 318).

It is permitted to assume that ξ is equal to:

5 years or more: 2.0

1 year: 1.4

6 months: 1.2

3 months: 1.0

Caution may need to be exercised with this method, particularly for lightly reinforced slabs, as the ratio of long-term deflection to short-term sustained load deflection can vary significantly from the ACI 318 λ_Δ values (Walkup et al. 2019).

The incremental deflection Δ_{inc} that occurs after attachment of the nonstructural elements includes the long-term deflection Δ_{LT} from the sustained loads plus the immediate deflection from the remaining part of live load ($\Delta_{L,add} = \Delta_L - \Delta_{L,sus}$) that is not sustained. The portion of long-term deflection occurring before attachment of the nonstructural elements may be deducted.

$$\Delta_{inc} = \lambda_\Delta \Delta_{sus} + \Delta_{L,add} \quad (4.4.1c)$$

Note that ACI 440.1R recommends using 60 percent of the time-dependent factor to give $\lambda_\Delta = 0.6\xi$ for FRP reinforcement (compression reinforcement is not considered for FRP-reinforced concrete members).

Research has shown that high-strength concrete members exhibit less sustained-load deflections than low-strength concrete members (Luebkeman et al. 1985; Nilson 1985).

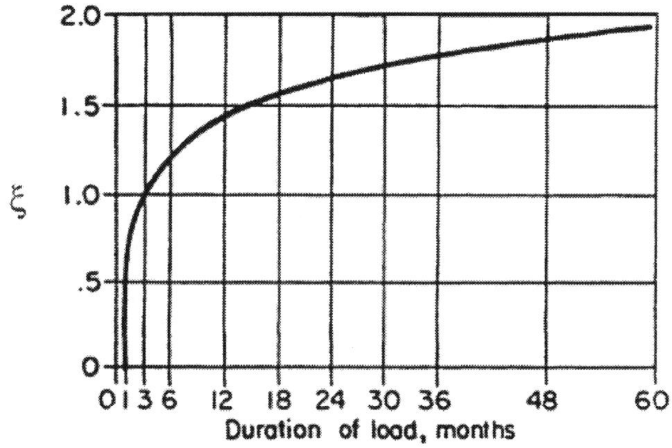

Fig. 4.4.1—ACI 318 multipliers for long-term deflections.

This behavior is mainly due to lower creep strain characteristics. However, the influence of compression steel reinforcement is less pronounced in high-strength concrete members. This is because the substantial force transfer from the compression concrete to compression reinforcement is greatly reduced for high-strength concrete members, for which creep is lower than normal-strength concrete. Nilson (1985) suggested that two modifying factors should be introduced into Eq. (4.4.1b). The first is a material modifier μ_m with values equal to or less than 1.0, applied to ξ to account for the lower creep coefficient. The second is a section modifier μ_s also having values equal to or less than 1.0, to be applied to ρ′ to account for the decreasing importance of compression steel in high-strength concrete members. Nilson (1985) introduced a single modifier μ to account for both effects simultaneously, leading to the following simplified equation

$$\lambda_\Delta = \frac{\mu\xi}{1+50\mu\rho'} \quad (4.4.1d)$$

where $0.7 \leq \mu = 1.3 - 0.00005f'_c \leq 1.0$ with f_c' in psi ($0.7 \leq \mu = 1.3 - 0.0073f_c' \leq 1.0$ with f_c' in MPa).

This equation results in μ = 1.0 for concrete strengths less than 6000 psi (42 MPa) and provides a reasonable fit of experimental data for higher concrete strengths. More data is needed, however, particularly for strengths between 9000 and 12,000 psi (62 and 83 MPa) and beyond before a definitive statement can be made. Gardner and Zhang (1996) have also proposed alternative long-term deflection multipliers based on the relative load level, tensile steel ratio, and compression steel ratio.

4.4.2 *ACI Committee 435 modified method (Branson 1963, 1977)*—For computing creep and shrinkage deflections separately, Branson's (1963, 1977) Eq. (4.4.2a) and (4.4.2b) are recommended by ACI Committee 435 (1966, 1978).

$$\Delta_{cr} = \lambda_t \Delta_{sus} \quad (4.4.2a)$$

where $\lambda_t = 0.85C_t/(1 + 50\rho')$.

$$\Delta_{sh} = k_{sh}\kappa_{sh}\ell^2 = k_{sh}\left(A_{sh}\frac{(\varepsilon_{sh})_t}{h}\right)\ell^2 \quad (4.4.2b)$$

C_t and $(\varepsilon_{sh})_t$ may be determined from Eq. (3.3.3a) through (3.3.3c) and Table 3.3.3.

$$A_{sh} = 0.7(\rho-\rho')\left(\frac{\rho-\rho'}{\rho}\right)^{1/2} \quad \text{for } \rho-\rho' \leq 3.0 \text{ percent}$$
$$= 0.7\rho^{1/3} \text{ for } \rho' = 0$$
$$= 1.0 \text{ for } \rho-\rho' > 3.0 \text{ percent}$$

The reinforcement ratios ρ and ρ' are computed at the support section for cantilevers and at the midspan sections for simple and continuous spans.

The shrinkage deflection constant k_{sh} is as follows:

Cantilevers = 0.50
Simple beams = 0.13
Spans with one end continuous (multi spans) = 0.09
Spans with one end continuous (two spans) = 0.08
Spans with both ends continuous = 0.07

Separate computations of creep and shrinkage are preferable when part of the live load is considered as a sustained load.

4.4.3 *Other methods*—Other methods for time-dependent deflection calculation in reinforced concrete beams and one-way slabs are available in the literature. They include several methods listed in ACI Committee 435 (1966), Appendix B of ACI 435R-95(03) described by Ghali et al. (2011), *fib* MC2010, and BS EN 1992-1-1. Other methods described in 5.2.4 are also suitable for calculation of time-dependent deflection. This section highlights BS EN 1992-1-1 (Eurocode 2), and briefly describes the section curvature approach, referring the reader to the literature for details.

4.4.3.1 *BS EN 1992-1-1 (Eurocode 2) method*—BS EN 1992-1-1 uses the concept of the partially cracked section. According to BS EN 1992-1-1, members that are expected to crack, but may not be fully cracked, will behave in a manner intermediate between the uncracked and fully cracked conditions. For members subjected mainly to flexure, an adequate prediction of behavior is given by Eq. (4.4.3.1a).

$$\alpha = \zeta\alpha_{II} + (1-\zeta)\alpha_I \quad (4.4.3.1a)$$

where α is the deformation parameter considered. This parameter may be, for example, a strain, a curvature, or a rotation but, as a simplification, α may also be taken as a deflection; α_I and α_{II} are the values of the parameter calculated for the uncracked and fully cracked conditions, respectively; and ζ is a distribution coefficient (allowing for tensioning stiffening at a section) given by Eq. (4.4.3.1b)

$$\zeta = 1 - \beta(\sigma_{sr}/\sigma_s)^2 \quad (4.4.3.1b)$$

($\zeta = 0$ for uncracked sections)
where β is a coefficient taking account of the influence of the duration of the loading or of repeated loading on the average strain (β = 1.0 for a single short-term loading and β = 0.5 for sustained loads or many cycles of repeated loading); σ_s is the stress in the tension reinforcement calculated on the basis of a cracked section; and σ_{sr} is the stress in the tension reinforcement calculated on the basis of a cracked section under the loading conditions causing first cracking.

Note that σ_{sr}/σ_s may be replaced by M_{cr}/M for flexure or N_{cr}/N for pure tension, where M_{cr} is the cracking moment and N_{cr} is the cracking force. Research (Scott and Beeby 2005) recommends the modification of using β = 0.5 when calculating short-term deflections because loss of tension stiffening occurs much more rapidly than was previously thought to be the case.

Deformations due to loading may be assessed using the tensile strength and the effective modulus of elasticity of the concrete. In general, it is recommended that using the mean value of the axial tensile strength of the concrete is likely to give the best estimate of the behavior.

The effects of creep may be included by using an effective modulus of elasticity for the concrete given by Eq. (4.4.3.1c)

$$E_{c,eff} = E_{cm}/(1 + C_u) \quad (4.4.3.1c)$$

where C_u is the final value of the creep coefficient. This is dependent on the age of the concrete at the time of loading and the relative humidity of the exposure conditions.

Shrinkage curvatures are assessed using Eq. (4.4.3.1d)

$$\kappa_{sh} = \varepsilon_{cs}n_e(S/I) \quad (4.4.3.1d)$$

where κ_{sh} is the curvature due to shrinkage; ε_{cs} is the free shrinkage of concrete; S is the first moment of area of the reinforcement about the centroid of the section (section modulus); I is the second moment of area of the section (moment of inertia); and n_e is the effective modular ratio equal to $E_s/E_{c,eff}$ (which allows for the effects of creep). S and I should be calculated for the uncracked and cracked conditions and the final curvature assessed using Eq. (4.4.3.1a).

BS EN 1992-1-1 observes that the most rigorous method for assessing deflections is to compute curvatures at frequent sections along the member and then calculate the deflection by numerical integration. However, it will frequently be acceptable to compute the deflection twice, assuming the whole member to be uncracked and then fully cracked, and then interpolate using Eq. (4.4.3.1a).

4.4.3.2 *Time-dependent change in strain and curvature at a section*—As described in 4.3.3, the section-based moment-curvature method is a general approach to determine the member deflection by considering the calculated response at representative cross sections. The cross section response directly considers the stress-strain characteristics of the materials and can account for differences between the total strain ε_t and the strain due to stress of the concrete and reinforcement, ε_{cf} and ε_{sf}, respectively. Therefore, this approach lends itself to consider the time-dependent influences on deflection including shrinkage, creep, and modeling the deflection of members at different construction stages.

The time-dependent action of creep in concrete results in gradual changes of deflection under sustained loading. Equation (3.3.3d) presents a relationship to determine the age-adjusted modulus of concrete in compression by means of a creep coefficient. The resulting modulus can be used to determine an effective stress-strain response of the concrete in compression for direct use within the section moment-curvature calculation methods from Section 4.3. Note that while there may be small sustained tensile loads in the concrete before cracking, or between cracks after cracking, any tension-induced creep will have a negligible influence on the bending response and can typically be ignored. The deflection at any given period should account for the sequence of loading and the amount of creep that results over each time interval of a given loading scenario.

Other time-dependent effects of concrete can result in deflections but do not rely on sustained loading. As one example, uniform shrinkage over the cross section of an unrestrained (statically determinate) member will result in a shortening of the member without external forces. However, even though this is externally free shrinkage, only the concrete shrinks while the reinforcement does not. To maintain strain compatibility between the materials, internal stresses will result, and these should be accounted for in the force equilibrium at the cross section. This can result in changes to the cracking moment M_{cr} and can also induce deflections that will depend in part on the reinforcement configuration. One approach to calculate these effects is to separate the total strain at each point in the cross section (ε_t) into a strain due to stress ε_{cf} and the other free strains ε_o. This is illustrated conceptually in Fig. 4.4.3.2 where the total strains are used to determine the member curvature κ. The part of Fig. 4.4.3.2 representing the strains due to stress are used within the force equilibrium Eq. (4.3.3a) and (4.3.3b).

4.4.4 *Finite element method*—Many finite element models have been developed to account for time-dependent deflections of reinforced concrete members. The sophistication and complexity of these methods varies widely. Some approaches are similar to section curvature methods (Hirsch 2009) whereas other approaches attempt to model the full nonlinear time-dependent complexity of reinforced concrete as a heterogeneous three-dimensional continuum. The more complex approaches are only justifiable when a high degree of precision is required for special structures and only when substantially accurate creep and shrinkage data are available. In special cases, such information on material properties is warranted and may be obtained experimentally from tests of actual materials to be used. Even with tests of the actual materials to be used, assumptions regarding tension stiffening and other behavior still need to be made to get reasonable predictions of deflection.

4.5—Temperature-induced deflections

Variations in ambient temperature can significantly affect deformations of reinforced concrete structures. Deflections occur in flexural members when a temperature gradient occurs between its opposite faces. It has become standard practice to evaluate thermal stresses and displacements in tall building structures. Movements of bridge superstructures and precast concrete members are also computed for the purpose of design of support bearings and expansion

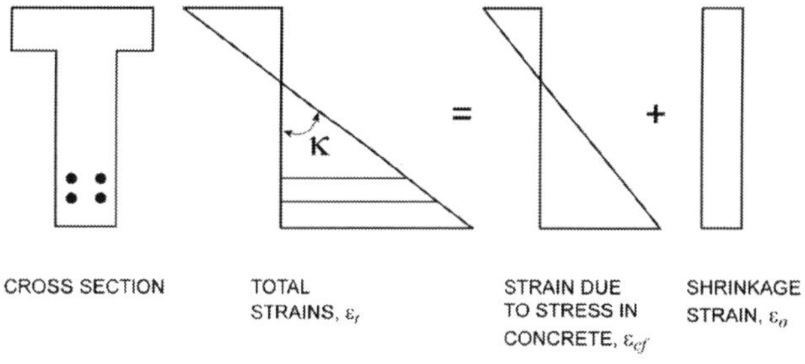

Fig. 4.4.3.2—Consideration of shrinkage strains.

joints. Before performing an analysis for temperature effects, it is necessary to select design temperature gradients. Martin (1971) summarizes design temperatures that are provided in various national and foreign codes.

ACI 435.7R outlined procedures for estimating changes in stiffness and temperature-induced deflections for reinforced concrete members. The following expressions are taken from that report.

4.5.1 *Temperature gradient on statically determinate (unrestrained) member*—With temperature distribution $T(y)$ on the cross section, thermal strain at a distance y from the bottom of the section can be expressed by

$$\varepsilon_{th}(y) = \alpha T(y) \tag{4.5.1a}$$

where α is the coefficient of thermal expansion.

To restrain the movement due to temperature $T(y)$, a stress is applied in the opposite direction to $\varepsilon_{th}(y)$

$$f(y) = E_c \alpha T(y) \tag{4.5.1b}$$

The net restraining axial force and moment are obtained by integrating over the cross-section depth

$$P = \int_A f dA = \int_0^h \left[\alpha E_c T(y) b(y)\right] dy \tag{4.5.1c}$$

$$M = \int_A f(y - y_t) dA = \int_0^h \left[\alpha E_c T(y) b(y)(y - y_t)\right] dy \tag{4.5.1d}$$

where y_t represents the distance to the centroidal axis measured from the bottom of the section. To obtain the total strains on the unrestrained cross section, P and M are applied in the opposite direction to the restraining force and moment. Assuming plane sections remain plane, axial strain ε_a and curvature κ are given by

$$\varepsilon_a = \frac{P}{AE_c} = \frac{\alpha}{A}\int_0^h \left[T(y) b(y)\right] dy \tag{4.5.1e}$$

$$\kappa = \frac{M}{E_c I} = \frac{\alpha}{I}\int_0^h \left[T(y) b(y)(y - y_t)\right] dy \tag{4.5.1f}$$

The net stress distribution on the cross section is given by

$$f_n(y) = \frac{P}{A} \pm \frac{M(y-y_t)}{I} - E_c \alpha T(y) \quad (4.5.1g)$$

and assumes the member remains uncracked.

For a linear temperature gradient varying from 0 to ΔT, the curvature is given by

$$\kappa = \frac{\alpha \Delta T}{h} \quad (4.5.1h)$$

In the case of a uniform vertical temperature gradient constant along the length of a member, deflections for simply supported (Δ_{ss}) and cantilever (Δ_{cant}) beams are calculated as

$$\Delta_{ss} = \frac{\kappa \ell^2}{8} = \frac{\alpha \Delta T}{h} \frac{\ell^2}{8} \quad (4.5.1i)$$

$$\Delta_{cant} = \frac{\kappa \ell^2}{2} = \frac{\alpha \Delta T}{h} \frac{\ell^2}{2} \quad (4.5.1j)$$

The deflection-to-span ratio is given by

$$\frac{\Delta}{\ell} = \frac{\alpha \Delta T}{h} \frac{\ell}{k} \quad (4.5.1k)$$

where $k = 8$ for simply supported beams and 2 for cantilever beams.

4.5.2 *Effect of restraint on thermal movement*—If a member is restrained from deforming under the action of temperature changes, internal stresses are developed. Cracking that occurs when tensile stresses exceed the concrete tensile strength reduces the flexural stiffness of the member and results in increased deflections under subsequent loading. Consequently, significant temperature effects should be taken into account in determining member stiffness for deflection calculation. The calculation of the effective moment of inertia should be based on maximum moment conditions.

In cases where stresses are developed in the member due to restraint of axial deformations, the induced stress due to axial restraint should be included in the calculation of the cracking moment in a manner analogous to that for including the prestressing force in prestressed concrete beams.

4.5.3 *Section moment-curvature method for temperature-induced deflections*—As noted in 4.4.3.2, the moment-curvature method can be used to predict the deflection of members that experience deflection due to internal restraint. Thus, it is a suitable method to consider deflections caused by temperature-induced deflections, including thermal gradients.

Figure 4.5.3 illustrates the case of applying a linear thermal gradient over the height of a member, where the temperature at the top surface is increased while the bottom surface is kept constant. The member is also subjected to applied external loads, resulting in bending moment M at the cross section considered. A distribution of total strains ε_t is given in Fig. 4.5.3. The free thermal strains and the strain due to stress are also illustrated in Fig. 4.5.3. Note that the internal

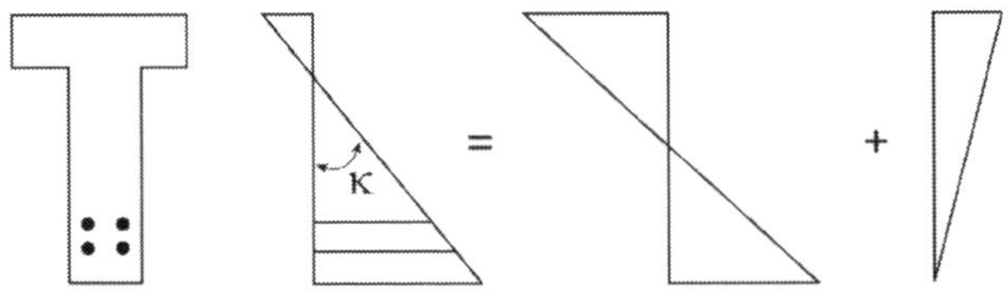

Fig. 4.5.3—Consideration of thermal gradients.

stresses in the concrete and the reinforcing steel are solved based on the ε_{cf} values. To ensure force equilibrium between the external loading causing M and the internal forces using Eq. (4.3.3a) and (4.3.3b), the total strain diagram should be solved iteratively by varying the curvature κ and the strain at one elevation, normally the extreme fiber strain on the flexural compression side. Refer to Collins and Mitchell (1997) for techniques appropriate for spreadsheet implementation and Ghali et al. (2011) for techniques suitable for finite element software.

CHAPTER 5—DEFLECTION OF A TWO-WAY SLAB SYSTEM

5.1—Introduction

Two-way slab systems are widely used for roof and floor construction in reinforced concrete buildings. Because slabs typically have large span-depth ratios, the resulting flexibility means that deflection control is usually a major factor in design. Efforts to minimize slab thickness to reduce construction costs can lead to problems associated with deflection of slab systems.

ACI 318 specifies minimum thickness requirements for control of two-way slab deflections. If the slab thickness equals or exceeds the specified minimum thickness, deflections need not be computed and serviceability in terms of deflection control is deemed to be satisfied.

Experience has shown that, in most cases, the use of ACI minimum thickness equations produces slabs that behave in a satisfactory manner under service loads. ACI 318 does permit a smaller-than-minimum thickness to be used if deflections are calculated and shown to be within limits specified by the standard. However, the calculation of slab deflections requires careful consideration of many factors if realistic estimates of deflections are to be obtained. ACI 318 provides limited guidance on deflection calculation. This chapter provides a detailed discussion of the many factors that need to be considered.

This chapter reviews calculation of two-way slab deflections. Methods of calculating slab deflections are presented. Effects of two-way action, cracking, creep, and shrinkage are considered. The minimum thickness equations in ACI 318 are examined and proposals for alternative methods of specifying minimum thickness are reviewed.

5.2—Deflection calculation methods for two-way slab systems

5.2.1 *Immediate deflection of uncracked slabs*—In this section, calculation procedures are given for immediate deflections based on three approaches, namely classical solutions, simplified crossing beam analogies, and finite element analysis.

5.2.1.1 *Classical solutions*—Immediate deflection of uncracked two-way slab systems loaded uniformly can be determined using plate bending theory for elastic thin plates. Load-deflection response is governed by the plate equation (Timoshenko and Woinowsky-Krieger 1959)

$$\frac{\partial^4\Delta}{\partial x^4}+2\frac{\partial^4\Delta}{\partial x^2\partial y^4}+\frac{\partial^4\Delta}{\partial y^4}=\frac{w}{d} \qquad (5.2.1.1a)$$

where x, y are the orthogonal coordinate axes of the middle surface; Δ is the deflection of the plate; w is the transverse load per unit area; D is the flexural rigidity per unit width, $Eh^3/12(1-\nu^2)$; E is the modulus of elasticity; h is the plate thickness; and ν is Poisson's ratio.

The relationship between moments and curvatures is given by

$$\begin{Bmatrix} M_x \\ M_y \\ M_{xy} \end{Bmatrix}=\frac{Eh^3}{12(1-\nu^2)}\begin{bmatrix} 1 & \nu & 0 \\ \nu & 1 & 0 \\ 0 & 0 & 2(1+\nu) \end{bmatrix}\begin{Bmatrix} \kappa_x \\ \kappa_y \\ \kappa_{xy} \end{Bmatrix} \qquad (5.2.1.1b)$$

For rectangular plates with uniformly distributed loads, the solution of Eq. (5.2.1.1b) leads to an expression for maximum deflection in the form

$$\Delta=\frac{\alpha w\ell^4}{D}=\frac{\alpha w\ell^4}{Eh^3/12(1-\nu^2)} \qquad (5.2.1.1c)$$

where ℓ is the longer span length; w is the uniform transverse load; and α is a coefficient depending on the boundary conditions and aspect ratio.

It is noted from Eq. (5.2.1.1c) that the influence of Poisson's ratio ν on deflections is quite small. Typical values of ν for concrete fall in the range between 0.15 and 0.25. The term $(1-\nu^2)$, in the flexural rigidity D, falls in the range 0.94 to 0.98. The error involved in neglecting Poisson's ratio is, therefore, approximately 2 to 6 percent.

Solutions of the plate equation for various geometries and support conditions have been given by Timoshenko and Woinowsky-Krieger (1959) and by Jensen (1938). Because closed-form solutions of the plate equation are available for only a limited number of cases, alternative solution procedures are required for most practical situations.

5.2.1.2 *Crossing beam methods*—Several approaches have been developed in which the two-way slab system is considered an orthogonal one-way system, thus allowing deflection calculations by beam analogy. Some of the earlier approaches were summarized in ACI 435.6R.

Rangan (1976) and Scanlon and Murray (1982) described calculation procedures in which the column and middle strips are treated as continuous beams; the middle strip is considered to be supported at its ends by column strips that run perpendicular to the middle strip. Results obtained by the crossing beam approach have been found to compare favorably with results obtained by finite element analyses (Scanlon and Murray 1982; Chang and Hwang 1996).

For two-way systems supported on a rectangular layout of columns, the ACI 318 method of design for strength involves dividing the slab into column strips and middle strips in each of the two orthogonal directions. The total static moment for each span $M_o = w_u\ell_2\ell_n^2/8$, is divided between positive and negative moment regions and then between column and middle strips, using either the direct design method or the equivalent frame method, where w_u is the intensity of load per unit area, ℓ_n is the effective span, and ℓ_2 is the dimension in the perpendicular direction. The distribution of moments approximates the elastic distribution for the given loading and therefore can be used to obtain an estimate of immediate deflections. However, the applied moments for strength design use factored loads whereas the moments for calculation of immediate deflections use service loads that are unfactored. The bending moments required to calculate deflections or curvatures of column and middle strips may be taken as the bending moments determined for factored loads according to ACI 318 multiplied by the ratio of service load to factored load.

Figure 5.2.1.2 shows a rectangular panel in a column-supported two-way slab system. The dotted areas represent a set of crossing beams from which column strip deflection Δ_c and middle strip deflection Δ_m can be obtained. Each beam can be treated as a strip of unit width for which end moments, midspan moment, and flexural rigidity properties can be obtained. Note that, by definition, end moments are those at the faces of supports, such as column or column capital faces, and that the beam span is the clear span between the faces of such supports.

Once the end moments and midspan moment have been obtained for a column or middle strip, the deflection for the strip can be calculated using the elastic beam deflection equation

$$\Delta=\frac{5}{48}\frac{\ell_n^2}{EI}\left[M_m+0.1(M_1+M_2)\right] \qquad (5.2.1.2a)$$

where ℓ_n is the clear span; M_1 and M_2 are the end moments per unit width; and M_m is the midspan moment per unit width. Positive M_m, M_1, or M_2 produce tension at bottom fiber.

Using this procedure, the deflection of each column strip (Δ_c) and of each middle strip (Δ_m) can be calculated. The midpanel deflection Δ_{mp} is obtained by adding the column and middle strip deflections.

$$\Delta_{mp} = \Delta_c + \Delta_{mp} \qquad (5.2.1.2b)$$

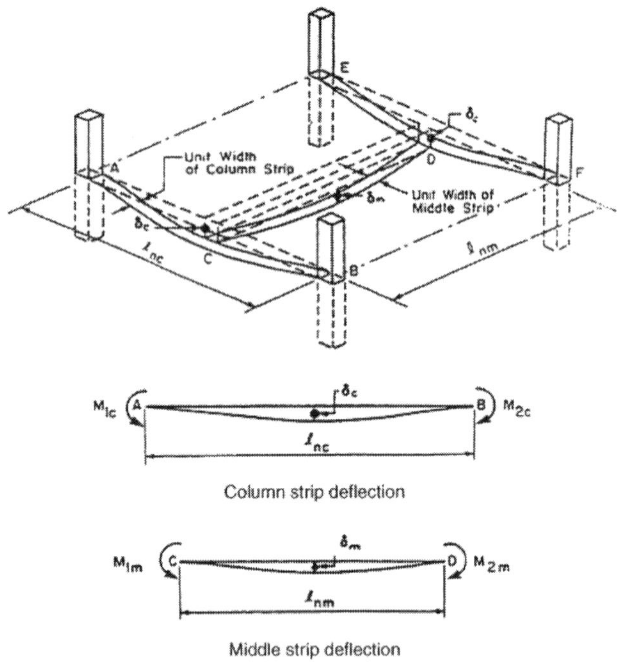

Fig. 5.2.1.2—Crossing beam approach.

For cantilever slabs, the deflection equation for cantilevers should be used and the rotation at the support should be included.

An earlier version of the equivalent frame method for calculating deflections proposed by Vanderbilt et al. (1965) considers the midpanel deflection as the sum of a column strip deflection, cantilever deflection of a portion of the middle strip extending from the column strip, and the midpanel deflection of a simply supported rectangular plate. The procedure developed by Nilson and Walters (1975), based on the equivalent frame method, is similar to the method outlined previously, except that a reference deflection is calculated for the total panel width. Deflections for column and middle strips are then obtained from this reference deflection using lateral distribution factors based on relative *M/EI* values. A numerical example (Nawy 2009) calculating the expected deflection limits using this procedure is given in Appendix B. The resulting values are applicable instead of the minimum thickness values from ACI 318-14 Table 8.3.1.1.

Ghali (1989) calculates the deflection at midspan of a column or middle strip from values of curvature calculated on the basis of compatibility and equilibrium at the midspan and supports using the relationship

$$\Delta = \frac{\ell^2}{96}(\kappa_L + 10\kappa_c + \kappa_R) \qquad (5.2.1.2c)$$

where κ_L, κ_c, and κ_R are the curvatures calculated from analysis of sections at the left end, center, and right end of column and middle strips, respectively, and ℓ is the distance between the two ends. This relationship is based on the assumption that variation of curvature over the length ℓ is parabolic. The effects of cracking, creep, and shrinkage are accounted for in the analysis for κ at each section. In the absence of prestressing, simplification can be made by use of multipliers and graphs (Ghali 1989; Ghali et al. 2011) that also account for the cracking, creep, and shrinkage effects.

5.2.1.3 *Finite element method*—The finite element method can be used to analyze plates with irregular support and loading conditions. Effects of beams and columns can be included, and general-purpose computer programs are available for elastic analysis of plate systems. Computer programs that account for cracking, creep, and shrinkage have been under development since the 1970s (Scanlon and Murray 1974; Gilbert and Warner 1978; ASCE Task Committee 1982; Hirsch et al. 2012) and commercial software is now available.

Although the method is becoming increasingly popular in engineering practice, some skill is required in selecting appropriate finite element and material models, developing an appropriate mesh, preparing computer input data, and interpreting the results.

5.2.2 *Effect of cracking*—The procedures outlined previously are applicable to linear elastic isotropic plate systems and should be modified for concrete slabs to include the effect of cracking on flexural stiffness. An early application of classical anisotropic theory to analysis of two-way reinforced concrete slabs is given in the text by Timoshenko and Woinowsky-Krieger (1959). More recently, procedures have been proposed for including cracking in finite element analysis and in the crossing beam analogies for two-way slabs.

The effective moment of inertia I_e concept, developed originally by Branson (1963) for beams, can be applied directly to the column and middle strips in the crossing beam analogies described in 5.2.1.2 for elastic uncracked plates. In Eq. (5.2.1.2a), the cross section stiffness *EI* becomes E_cI_e using the usual averaging procedures given in ACI 318 for I_e, calculated at both positive and negative moment locations. Kripanarayanan and Branson (1976) presented an extension of the Nilson and Walters (1975) equivalent frame procedure to include cracking using the I_e procedure. An alternative expression for effective moment of inertia proposed by Bischoff (2005, 2007) is discussed in 4.3.2.2.

Finite element models for cracking in two-way slabs based on layered plate modeling and concrete stress-strain relationships accounting for tension stiffening have been available since the early 1970s (Scanlon and Murray 1974; Gilbert and Warner 1978). More recently, general purpose computer programs accounting for tension stiffening have become commercially available. A simple generalization of the effective moment of inertia concept to two-way systems was suggested by Scanlon and Murray (1982) and implemented in a modified version of a linear elastic plate bending finite element by Graham and Scanlon (1986). An extension of this approach has recently been presented by Vollum (2011).

5.2.3 *Restraint cracking*—In two-way reinforced concrete slabs built monolithically with supporting column and wall elements, in-plane shortening due to shrinkage, and thermal effects is restrained. The restraint is provided by a combination of factors, including embedded reinforcement, attachment to structural supports, and lower shrinkage rates of previously placed adjacent panels when slab panels are placed at different times. Nonlinear distribution of free shrinkage strains across the cross section may also be a contributing factor.

Service load moments in two-way slabs are often of the same order of magnitude as the code-specified cracking moment M_{cr}. Deflection calculations made using the code-specified modulus of rupture will often result in an uncracked section being used when cracking may actually be present due to a combination of flexural stress and restraint stress.

ACI 318 specifies the modulus of rupture for deflection calculations as $7.5\sqrt{f_c'}$ (psi) ($0.62\sqrt{f_c'}$ [MPa]). Laboratory test data, summarized in ACI 209R, indicate values ranging from 6 to $12\sqrt{f_c'}$ (psi) (0.5 to $1.0\sqrt{f_c'}$ [MPa]).

For slab sections with low reinforcement ratios, approaching minimum reinforcement, the difference between cracked and uncracked flexural stiffness is significant. It is important, therefore, to account for effects of any restraint cracking that may be present. Unfortunately, the extent of restraint cracking is difficult to predict. To account for restraint cracking in two-way slabs, Rangan (1976) suggested that column strip deflections be based on the moment of inertia of a fully cracked section I_{cr} and that middle strip deflections be based on $(I_g + I_{cr})/2$. Good agreement was reported between calculated and field measured deflections.

A more general approach was proposed by Scanlon and Murray (1982). They suggested that the effect of restraint cracking be included by introducing a restraint stress f_{res} that effectively reduces the modulus of rupture for calculating M_{cr}—that is

$$M_{cr} = \frac{f_e I_g}{y_t} \qquad (5.2.3)$$

where $f_e = f_r - f_{res}$.

A value of $4\sqrt{f_c'}$ (psi) ($0.33\sqrt{f_c'}$ [MPa]), or approximately half of the value specified in ACI 318, was proposed for the modulus of rupture used to compute the reduced cracking moment. This approach was investigated by Tam and Scanlon (1986) and has produced good correlation between calculated deflection and reported mean field-measured deflections (Jokinen and Scanlon 1985; Graham and Scanlon 1985).

Ghali (1989) has also used the idea of a reduced modulus of rupture and demonstrated the calculation of restraint stress due to reinforcement in the presence of uniform shrinkage.

An additional consideration is that the moments used in design for strength are based on some redistribution of moments. The distribution of design moments does not reflect the high peaks of moment adjacent to columns that occur in uncracked slabs. Deflection calculations based on moment distributions used for design therefore tend to underpredict the effects of cracking on flexural stiffness.

For slab systems in which significant restraint to in-plane deformations may be present, Scanlon and Murray (1982) recommended that a reduced effective modulus of $4\sqrt{f_c'}$ (psi) ($0.33\sqrt{f_c'}$ [MPa]) be used to compute the effective moment of inertia I_e based on Branson's equation. Scanlon and Bischoff (2008) recommended a value of $5\sqrt{f_c'}$ (psi) ($0.41\sqrt{f_c'}$ [MPa]) (equivalent to $2/3M_{cr}$) along with Bischoff's equation for the effective moment of inertia provided in Eq. (4.3.2.2a).

5.2.4 *Long-term deflections*—Long-term deflections can be estimated by applying a long-term multiplier to the calculated immediate deflection. Values for the long-term multiplier are specified in design codes such as ACI 318, where a value of $2/(1 + 50\rho')$ is applied to the immediate deflection caused by the sustained load considered. As discussed in 5.2.4.2, numerous authors have suggested that the ACI 318 long-term multiplier is too low for application to two-way slab systems, being based on poor correlation between reported calculated long-term deflections and field-measured deflections. It may be that much of the discrepancy between calculated and measured deflections is due to the effect of restraint cracking described previously. Additionally, shrinkage warping is more significant in shallow slab systems than in deeper beam sections. Shrinkage warping occurs where the centroids of the steel reinforcement and the uncracked portion of the concrete do not coincide, and the shrinkage of concrete, combined with the dimensional stability of the reinforcement, warps the member causing deflections to occur.

Two approaches are presented next for estimating long-term deflections—namely, by detailed computations and by the ACI multiplier methods.

5.2.4.1 *Detailed calculations*—Effects of creep deflection and shrinkage warping may be considered separately using procedures outlined in ACI 209R, based on the work of Branson et al. (1970) and Branson and Christiason (1971).

Deflection due to creep is obtained from

$$\Delta_{cr} = k_r C_t \Delta_I \qquad (5.2.4.1a)$$

where C_t is the time-dependent creep coefficient representing creep strain at any time t, in days, after load application; k_r is the factor to account for compression reinforcement and neutral axis shift; and Δ_i is the immediate deflection due to dead load plus sustained live load, including effects of cracking.

Equation (3.3.3a) gives the ACI 209R equation for C_t. ACI 209R provides typical values of factors applying to moist-cured concrete loaded at 7 days or later (refer to Chapter 3 for details). For slabs loaded before 7 days, these values may be used for first approximations.

In a two-way slab, shrinkage occurs in all directions. The shrinkage deflection should therefore be calculated for orthogonal column and middle strips, and the results added to give the total midpanel shrinkage deflection. Although there may be a contribution to shrinkage warping from nonuniform shrinkage strains through the slab cross section,

Table 5.2.4.2—Multipliers recommended by different authors

Source	Modulus of rupture, psi	Immediate	Creep λ_c	Shrinkage λ_{sh}	Total λ_t
Sbarounis (1984)	$7.5\sqrt{f_c'}$	1.0	2.8	1.2	5.0
Branson (1977)	$7.5\sqrt{f_c'}$	1.0	2.0	1.0	4.0
Graham and Scanlon (1986)	$7.5\sqrt{f_c'}$	1.0	2.0	2.0	5.0
	$4\sqrt{f_c'}$	1.0	1.5	1.0	3.5
Hossain et al. (2011)	$7.5\sqrt{f_c'}$	1.0	3.0		4.0
ACI 318	$7.5\sqrt{f_c'}$	1.0	2.0		3.0

Note: $7.5\sqrt{f_c'}$ (psi) = $0.62\sqrt{f_c'}$ (MPa) and $4\sqrt{f_c'}$ (psi) = $0.33\sqrt{f_c'}$ (MPa).

there are insufficient experimental data available to make specific recommendations for deflection calculations.

Shrinkage warping deflection for a beam is given by

$$\Delta_{sh} = k_{sh}\kappa_{sh}\ell^2 \quad (5.2.4.1b)$$

where k_{sh} is the coefficient depending on end conditions:
= 11/128 (one end continuous)
= 1/16 (both ends continuous)
= 1/8 (simple beam)
= 1/2 (cantilever)
κ_{sh} is the shrinkage curvature
= $0.7\varepsilon_{sh}(t)\rho^{1/3}/h$, singly reinforced member
= $0.7\varepsilon_{sh}(t)(\rho - \rho')^{1/3}((\rho - \rho')/\rho)^{1/2}/h$, doubly reinforced member

Shrinkage warping deflections can also be determined using the equivalent tension force method outlined in ACI 209R.

The total deflection at any time is obtained by adding immediate deflection due to sustained load, creep deflection due to sustained load, shrinkage warping deflection, and deflection due to the part of the live load that is transient.

Sophisticated finite element models have been developed (ASCE Task Committee 1982) to account for time-dependent deformations of two-way slabs caused by creep and shrinkage. These models are generally used for research purposes and are considered too complex for normal design applications, particularly when the high variability of creep and shrinkage properties is considered.

5.2.4.2 *ACI multiplier*—While deflection calculations can be made for long-term effects of creep and shrinkage, as outlined previously, the use of a multiplier applied to the immediate deflection provides a simple calculation procedure that is adequate for most purposes. This approach is used in ACI 318, in which a sustained-load multiplier of up to 2 is applied to the immediate deflection of a member with no compression reinforcement. Several authors have recommended increasing this factor for two-way slabs, as indicated in Table 5.2.4.2. The variation among the proposals indicates that there is considerable uncertainty in predicting long-time deflections. The total long-time multiplier is expressed as

$$\lambda_t = 1 + \lambda_c + \lambda_{sh} \quad (5.2.4.2a)$$

where λ_c is the multiplier for creep and λ_{sh} is the multiplier for shrinkage warping. Expressing shrinkage warping as a multiplier is not the most accurate method for predicting the deflections associated with shrinkage warping as they are not directly proportional to the immediate deflection. However, this approach has been adopted by numerous researchers and codes (including ACI 318) due to its simplicity and convenience.

As a first approximation, the additional deflection at intermediate time intervals due to sustained loads can be calculated using the values for ξ (Table 24.2.4.1.3 of ACI 318-14) multiplied by the factor $(\lambda_t - 1)/2$.

It is recommended that, in cases where restraint stresses are expected to have an insignificant effect, the multiplier for sustained-load deflection be increased from 2 to 4, as recommended by Sbarounis (1984) and Graham and Scanlon (1986). In this case, the ACI 318 value for modulus of rupture would be used. In cases where restraint stresses are likely to have a significant effect on cracking, for example, large slab areas and stiff lateral restraint elements such as structure walls and columns, it is recommended that a reduced modulus of rupture given by $f_r = 4\sqrt{f_c'}$ (psi) ($0.33\sqrt{f_c'}$ [MPa]) be used along with a long-term sustained-load multiplier of 2.5.

Values recommended in ACI 209R for ultimate creep and shrinkage coefficients are $C_u = 2.35$ and $(\varepsilon_{sh})_u = 780 \times 10^{-6}$, respectively, at standard conditions as discussed in Chapter 3. Sbarounis (1984) has suggested that, at standard conditions, the long-term multipliers be modified if the concrete properties are known, and better estimates of ultimate creep $\bar{C}_u$ and shrinkage $\overline{(\varepsilon_{sh})}_u$ are available. Thus,

$$\lambda_t = (1+\lambda_c)\frac{\overline{C_u}}{C_u} + \lambda_{sh}\frac{\overline{(\varepsilon_{sh})_u}}{(\varepsilon_{sh})_u} \quad (5.2.4.2b)$$

5.3—Minimum thickness requirements

Because of the complexities involved in calculating two-way slab deflections, engineers have preferred to control deflections by giving minimum slab thickness as a function of span length. Equations such as those in 8.3 of ACI 318-14, as shown in Table 5.3, are based on experience gained over many years. The ACI 318 equations express minimum thickness in terms of clear span between columns, steel yield strength, and flexural stiffness of edge beams. The minimum thickness values are modified for the effects

Table 5.3—Minimum thickness of nonprestressed two-way slabs without interior beams, in. (ACI 318-14 Table 8.3.1.1)*

	Without drop panels†			With drop panels†		
	Exterior panels		Interior panels	Exterior panels		Interior panels
f_y‡	Without edge beams	With edge beams§		Without edge beams	With edge beams§	
40,000 psi (280 MPa)	$\ell_n/33$	$\ell_n/36$	$\ell_n/36$	$\ell_n/36$	$\ell_n/40$	$\ell_n/40$
60,000 psi (420 MPa)	$\ell_n/30$	$\ell_n/33$	$\ell_n/33$	$\ell_n/33$	$\ell_n/36$	$\ell_n/36$
75,000 psi (520 MPa)	$\ell_n/28$	$\ell_n/31$	$\ell_n/31$	$\ell_n/31$	$\ell_n/34$	$\ell_n/34$

*For two-way construction, ℓ_n is the length of clear span in the long direction, measured face-to-face of supports (in.).

†Drop panels as given in ACI 318-14 Section 8.2.4.

‡For f_y between the values given in the table, minimum thickness shall be determined by linear interpolation.

§Slabs with beams between columns along exterior edges. Exterior panels shall be considered to be without edge beams if α_f is less than 0.8. The value of α_f for the edge beams shall be calculated in accordance with ACI 318-14 Section 8.10.2.7.

Note: 1 in. = 25.4 mm.

of drop panels and discontinuous edges. ACI 318 permits the use of thinner slabs if deflections are computed and found to satisfy the specified maximum permissible values.

An extensive evaluation of the current ACI minimum thickness equations was reported by Thompson and Scanlon (1988). The study was based on finite element analyses of more than 300 slabs covering a range of thickness values, aspect ratios, edge beam dimensions, construction loads, and other parameters. The main conclusions of the study were as follows:

a) Calculated deflections for slabs designed according to the minimum thickness requirements of ACI 318 were within the permissible limits, when the calculations were based on the ACI 318-specified value of $7.5\sqrt{f_c'}$ (psi) ($0.62\sqrt{f_c'}$ [MPa]) for modulus of rupture, and the ACI 209R equations for creep and shrinkage deflection. Construction loads due to shoring and reshoring were also considered.

b) When the calculations were based on a reduced modulus of rupture to account for restraint cracking, the ACI 318 limit of $\ell/480$ on incremental deflection was exceeded for slab panels with aspect ratios less than 1.5. An increase of 10 percent over the current minimum thickness value for square panels was suggested to obtain calculated deflections within the allowable limits. The suggested increase in minimum thickness decreases linearly to zero for a panel with an aspect ratio equal to 1.5.

The results of this study suggest that the ACI minimum thickness equations will provide satisfactory serviceability in most cases, confirming the satisfactory performance of slabs designed and built according to the requirements in ACI 318 prior to the 1989 edition. When more stringent than normal deflection limits are required, or when nonstandard loads are applied, a more detailed analysis should be used. Other means to increase the slab stiffness, such as the addition of beams, can also be considered. Attempts have been made to develop criteria for span-depth ratios or minimum thickness of slabs that explicitly include the effects of such parameters as live-to-dead load ratio, permissible deflection-to-span ratio, effect of cracking, sustained load level, and time between construction and installation of nonstructural elements. Two such approaches are described in the following paragraphs.

Gilbert (1985) extended to two-way slabs an expression developed by Rangan (1982) for maximum allowable span-depth ratio for beams. Rangan's equation involves rearranging the basic equation for beam deflection calculations

$$\Delta = \Delta_{L,add} + \Delta_{sus} \tag{5.3a}$$

where $\Delta_{L,add}$ is the deflection due to the variable part of live load $(kw_{L,add}\ell^4)/E_cI_e$; Δ_{sus} is the total deflection due to sustained load (w_s) including sustained part of live load (= $\lambda_\Delta(kw_s\ell^4/E_cI_e)$); k is a constant based on support and load conditions; and λ_Δ is the long-term multiplier.

Replacing I_e by αbd^3, where the term α gives an approximation for I_e/bd^3 as a function of the reinforcement ratio ρ, Eq. (5.3a) can be rewritten as

$$\frac{\Delta}{\ell} = \frac{k(w_{L,add} + \lambda_\Delta w_s)\ell^3}{E_c\alpha bd^3} \tag{5.3b}$$

If Δ/ℓ is given as the maximum permissible deflection-to-span ratio, the corresponding maximum span-to-effective depth ratio can be obtained from

$$\frac{\ell}{d} < k_1\frac{\Delta}{\ell}\left(\frac{\alpha bE_c}{w_{L,add} + \lambda_\Delta w_s}\right)^{1/3} \tag{5.3c}$$

where k_1 is a combination of factors to account for support conditions and effect of beam flanges.

Gilbert extended Eq. (5.3c) by adding a slab system factor k_2 to account for two-way action—that is

$$\frac{\ell}{d} < k_1k_2\frac{\Delta}{\ell}\left(\frac{\alpha bE_c}{w_{L,add} + \lambda_\Delta w_s}\right)^{1/3} \tag{5.3d}$$

The factor k_2 was developed for a variety of conditions from parameter studies using a sophisticated finite element model. Equation (5.3d) involves an iterative procedure because the reinforcement ratio required to determine α and the dead load is initially unknown.

A somewhat simpler expression for beams was developed by Grossman (1981, 1987); it was based on numerous computer-generated beam deflection calculations. Grossman's minimum thickness equation is given by

$$h_{min} > \frac{\ell}{c}\frac{\ell}{\Delta}C_d \geq \frac{\ell}{24} \quad (5.3e)$$

Correction factors are given for variations in support conditions, d/h, f_y, and concrete density. The term c was developed from the computer analyses and depends on the load levels and construction methods used. For heavily-loaded members, a limiting value of $c = 4320$ was proposed by Grossman (1981) for heavily loaded members. Smaller thicknesses can be obtained if the required reinforcement ratio for less heavily loaded members is known and is used to obtain a larger revised value of c from Grossman's data.

The term C_d, given by

$$C_d = \frac{\lambda' D + L}{D + L} \quad (5.3f)$$

accounts for both the live-load-to-dead-load ratio, L/D, and the net long-term multiplier λ' for deflections occurring after installation of partitions in buildings.

Although developed for beams only, the equation could be extended to two-way systems using a slab system factor similar to that given by Gilbert (1985).

Scanlon and Lee (2006) extended Rangan's (1982) original proposal, incorporating an approximation for effective moment of inertia, to produce a unified minimum thickness equation that can be applied to beams, one-way slabs, edge-supported slabs, flat plates, and flat slabs. The general equation is given by

$$\frac{\ell_n}{h} = \beta_{ES}\left[\left(\frac{\Delta_{inc}}{\ell}\right)_{allow}\frac{2400k_{DP}E_c(b/12)}{Kk_{AR}k_{SS}(\lambda_\Delta w_s + w_{L,add})}\right]^{\frac{1}{3}} \quad \text{(in.-lb)}$$

$$\frac{\ell_n}{h} = \beta_{ES}\left[\left(\frac{\Delta_{inc}}{\ell}\right)_{allow}\frac{16{,}640k_{DP}E_c b}{Kk_{AR}k_{SS}(\lambda_\Delta w_s + w_{L,add})}\right]^{\frac{1}{3}} \quad \text{(SI)} \quad (5.3g)$$

where w_s is sustained load (psf, Pa [slabs]; plf, N/m [beams])
$w_{L,add}$ is additional live load (psf, Pa [slabs]; plf, N/m [beams])
$\beta_{ES} = 1$ except β_{ES} = (long span)/(short span) ≤ 2.0 for edge-supported slabs
K is deflection coefficient depending on support conditions (= 5 for simply supported, = 1.4 for both ends continuous, = 2 for one end continuous, and = 48 for fixed end cantilver)
$k_{DP} = 1$ except $k_{DP} = 1.35$ for slab with drop panels
$k_{ss} = 1$ except $k_{ss} = 1.35$ for column supported two-way slab systems
$k_{AR} = 1$ except $k_{AR} = 0.2 + 0.4\beta$ for edge-supported slabs
b = 12 in. (1000 mm) for one-way and two-way slabs
= beam width (= web width, b_w for T-beams), in. (mm)
$(\Delta_{inc})_{allow}$ is required incremental deflection limit

This equation can be specialized for flat plates as

$$\frac{\ell_n}{h} = \left[\left(\frac{\Delta_{inc}}{\ell}\right)_{allow}\frac{1200E_c}{\lambda_\Delta w_s + w_{L,add}}\right]^{\frac{1}{3}} \quad \text{(in.-lb)}$$

$$\frac{\ell_n}{h} = \left[\left(\frac{\Delta_{inc}}{\ell}\right)_{allow}\frac{8300 \times 10^3 E_c}{\lambda_\Delta w_s + w_{L,add}}\right]^{\frac{1}{3}} \quad \text{(SI)} \quad (5.3h)$$

An example taken from Scanlon and Lee (2006) is presented in Section B.3 to illustrate the application of the proposed equation.

Gardner (2011) compared span-depth approaches from several codes and standards as well as those proposed by various authors. Lee and Scanlon (2007) presented a framework based on utility theory to compute minimum thickness considering deflection variability and uncertainty in establishing deflection limits.

5.4—Loads for deflection calculation

ACI 318 stipulates that calculated deflections must not exceed certain permissible values, expressed as fractions of span length. Components of deflections to be considered are immediate live load deflection and incremental deflection, including that due to live load, after installation of nonstructural elements. The live load component of deflection is normally considered as that due to total load minus that due to dead load. Under monotonic loading, two effective moment of inertia values should be used to calculate the deflections at the two different load levels, as shown in Fig. 5.4a.

For multistory slab construction, the loads imposed on the slab during construction may be the largest use of the slab's capacity throughout the life of the structure (Grundy and Kabaila 1963). For these slabs, the extent of cracking is determined by the construction loads resulting from shoring and reshoring procedures. Under these conditions, all values of immediate deflection should be calculated using the effective moment of inertia corresponding to the construction load level, as illustrated in Fig. 5.4b. Because the construction procedure is usually unknown during design, the full specified dead load plus live load is recommended as a first approximation to the maximum load during construction. This calculation procedure usually results in a smaller live load deflection and larger dead load deflection, with correspondingly larger sustained load deflection.

A typical load-time history is shown in Fig. 5.4c for a slab in a multistory structure. During construction, where multiple levels of shoring equipment is used, the load on the slab increases as new slabs are placed above. When construction above is no longer supported by the slab under consideration, the load decreases to a value corresponding to the slab self-weight plus an allowance for superimposed dead load and sustained portion of live load (load level at t_1, in Fig. 5.4c).

A simple procedure to determine slab loads during construction was proposed by Grundy and Kabaila (1963). More refined analysis procedures reported subsequently (for

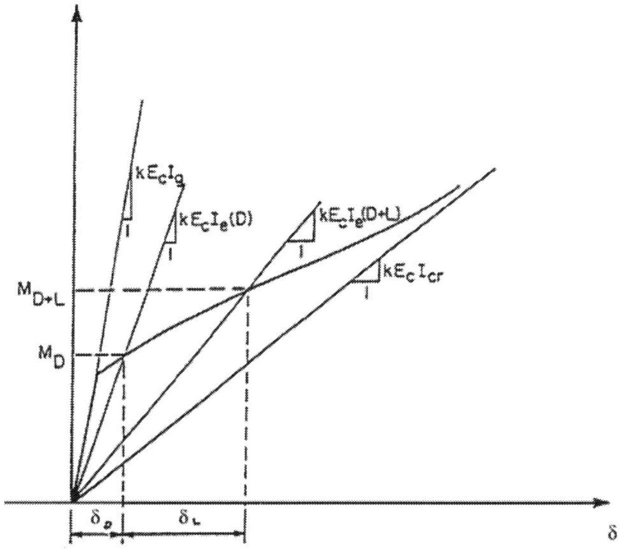

Fig. 5.4a—Definition of δ_D and δ_L for monotonic moment-deflection curve.

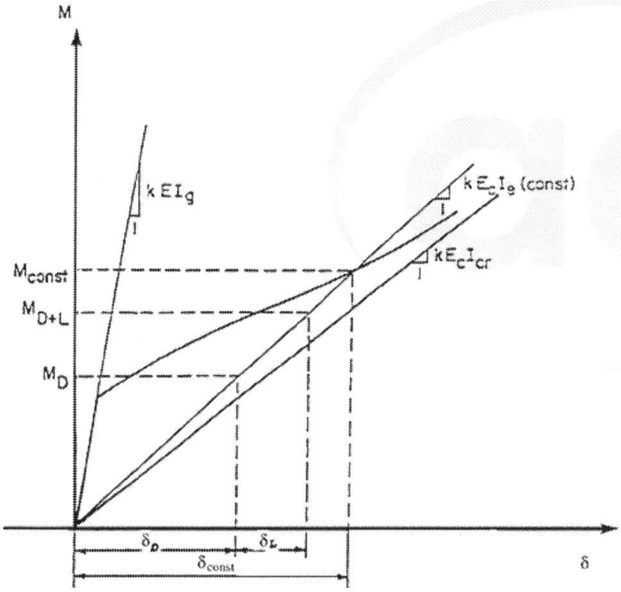

Fig. 5.4b—Definition of δ_D, δ_L, and δ_{const} when construction loads exceed specified dead plus live load.

example, Liu et al. [1985] and Aguinaga-Zapata and Bažant [1986]) give results that are similar to the original Grundy and Kabaila procedure. The maximum load during construction, including loads due to shoring and reshoring plus an allowance for construction live load, can be estimated using the following relationship

$$w = k_1k_2Rw_{slab} + w_{CL}/N \quad (5.4a)$$

where k_1 is the allowance for error in theoretical load ratio R; k_2 is the allowance for weight of formwork; R is the applied load/slab dead load ratio and the load ratio calculated by the Grundy and Kabaila procedure; w_{slab} is the slab dead load; w_{CL} is the construction live load; and N is the number of shored and reshored levels.

Gardner (1985) recommends $k_1 = k_2 = 1.1$. The construction live load may be taken as 50 lb/ft^2 (2.4 kPa) as recommended by ACI 347R. The factor k_1 accounts for errors in computing R due to variations in stiffnesses between the stories in the supporting system. The factor R has been shown to vary from 1.2 to 2.2, depending primarily on the number of stories of shores and reshores in the system. If the shoring system to be used is unknown, a value of R = 1.5 can be used in the calculation as recommended by ACI 347.2R. Instead of multiplying by a factor k_2 for formwork weight, the formwork weight can be additive to the equation. A value of 10 lb/ft^2 (0.48 kPa) is considered a reasonable allowance for most formwork systems, but this may need to be confirmed against the actual weight of formwork if deflection calculations need to be more precise. At time t_2 in Fig. 5.4c, a slight increase in the sustained load occurs as nonstructural elements are installed. The variable portion of live load may be considered as applied intermittently thereafter. One application of live load is shown at time t_3.

An analysis procedure based on this type of loading history and ACI 209R creep and shrinkage functions was developed by Graham and Scanlon (1986), using the principle of superposition. Effects of partial creep recovery were considered. Analyses were also made for the simplified load-time history shown in Fig. 5.4d with the corresponding displacement-time history. Long-term sustained load deflections were obtained using multipliers calibrated with the more complex history of Fig. 5.4c. Resulting multipliers are included in Table 5.2.4.2.

Based on the procedures suggested by Sbarounis (1984) and Graham and Scanlon (1986), the following approach

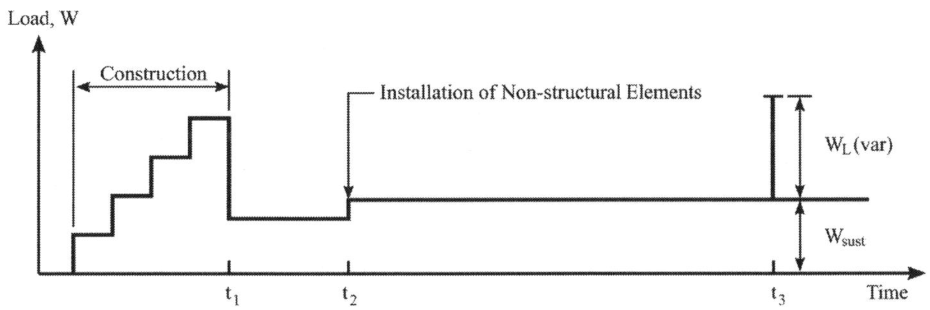

Fig. 5.4c—Schematic load-time history.

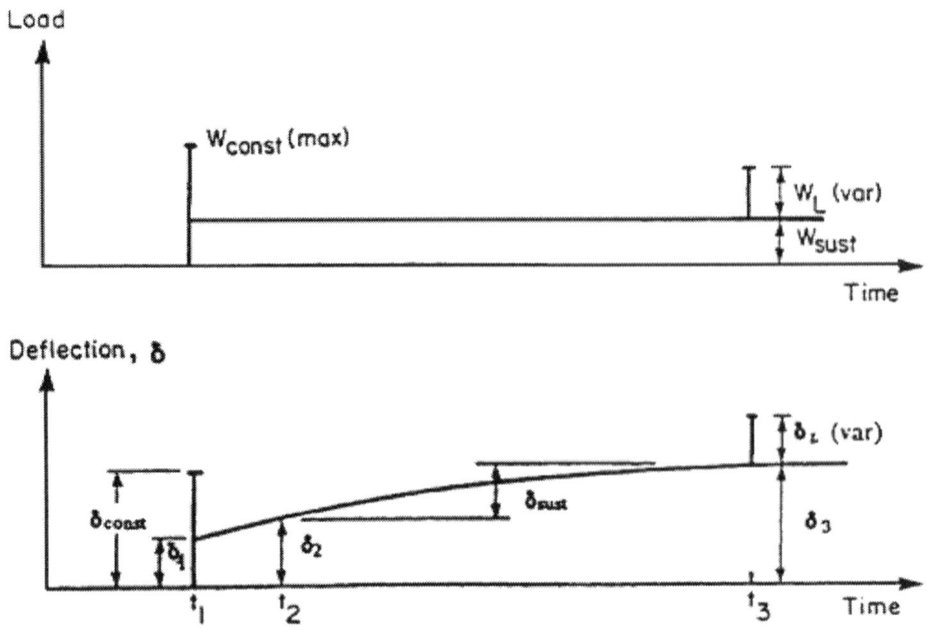

Fig. 5.4d—Simplified load-time history and corresponding deflection-time history.

based on the simplified load-time history can be used to estimate long-term deflections in multistory slab systems.

a) Estimate the maximum construction load expected, $w_{const}(max)$, based on usual procedures for multistory construction.

b) Calculate the corresponding immediate construction load deflection Δ_{const}.

c) Calculate the live load deflection by scaling the construction load deflection.

$$\Delta_L = \frac{w_L}{w_{const}(max)} \cdot \Delta_{const} \cdot \frac{E_c(Const)}{E_c(L)} \tag{5.4b}$$

where $E_c(Const)$ and $E_c(L)$ are modulus of elasticity values at application of construction load and live load, respectively, which can be calculated using the equations in 3.3.1.

d) Scale the construction load deflection to the sustained load level. Sustained load includes dead load plus any portion of the live load assumed to be sustained.

$$\Delta_{sus} = \frac{w_{sus}}{w_{const}(max)} \Delta_{const} \frac{E_c(Const)}{E_c(Sust)} \tag{5.4c}$$

where $E_c(Sust)$ is the modulus of elasticity at the time sustained load is applied (for example, at end of construction period).

e) Calculate sustained load deflection at time of installation of nonstructural elements.

$$\Delta_1 = \lambda_1 \Delta_{sus} \tag{5.4d}$$

where λ_1 is multiplier corresponding to time interval t_1 to t_2. The time function given in Eq. (3.3.3a) can be used to determine λ—that is,

$$\lambda_1 = \frac{(t_2 - t_1)^{0.60}}{10 + (t_2 - t_1)^{0.60}} \lambda_t \tag{5.4e}$$

f) Calculate ultimate sustained load deflection.

$$\Delta_2 = \lambda_t \Delta_{sus} \tag{5.4f}$$

where λ_t is long-term multiplier (Eq. 4.4.1b).

g) Calculate the deflection due to the variable portion of live load, that is, that portion of live load not assumed as sustained.

$$\Delta_{L,add} = K\Delta_L \text{ (from Step 3)} \tag{5.4g}$$

where

$$K = \frac{\text{variable live load}}{\text{total live load}} = \frac{w_{L,add}}{w_L}$$

h) Calculate increment in deflection after installation of nonstructural elements.

$$\Delta_{inc} = \Delta_2 + \Delta_{L,add} - \Delta_1 \tag{5.4h}$$

i) Compare calculated deflections with appropriate permissible values.

Scanlon and Suprenant (2011) presented a simplified method for estimating time-dependent deflections under a prescribed construction loading time history. The method is based on a generalization of creep superposition applied to time-dependent deflections calculated using the crossing beam method. Hirsch et al. (2012) showed that this approach produces results that compare favorably with available experimental data and results obtained by more complex finite element analysis.

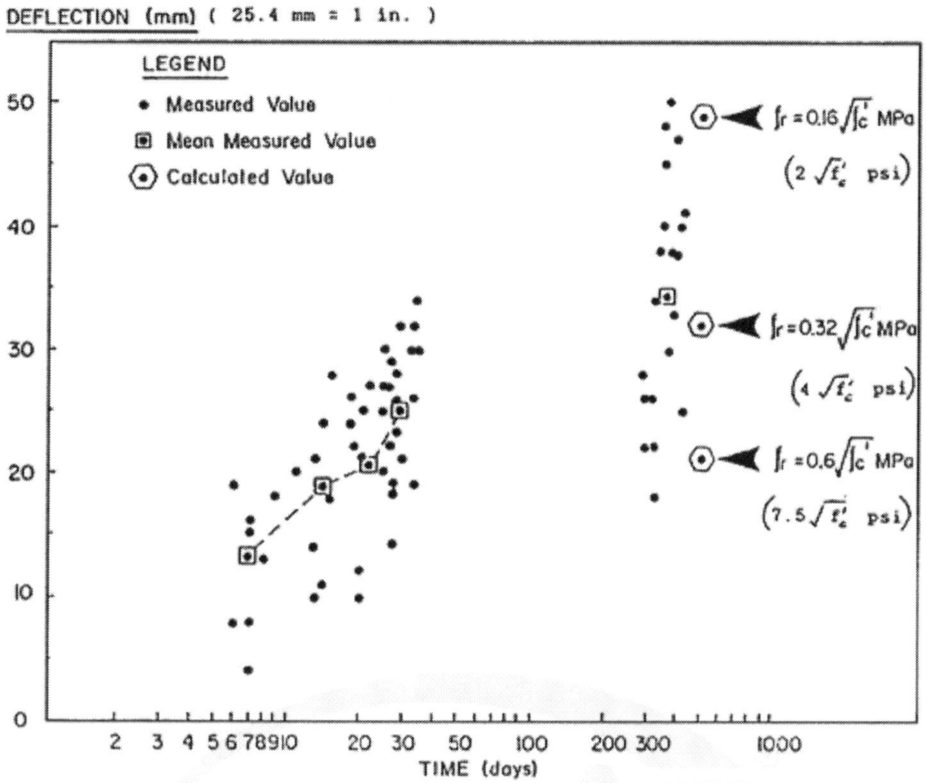

Fig. 5.5a—Field-measured deflections for 40 nominally identical slab panels in 28-story building (Jokinen and Scanlon 1985).

5.5—Variability of deflections

ACI 435.4R reported that the variability of actual deflections under nominally identical conditions is often large. For simply supported beams under laboratory conditions, it was reported that, using Branson's I-effective procedure, there is approximately a 90 percent probability that actual deflections of a particular beam will range between 80 and 130 percent of the calculated value. The variability of deflections in the field can be even greater.

Based on Monte Carlo simulation, Ramsay et al. (1979) indicated that the coefficient of variation for immediate deflection of beams ranged from 25 to 50 percent. The major source of variation was found to be flexural stiffness and tensile strength of concrete, particularly when the service load moment is close to the calculated cracking moment. Other sources of variability include variations of slab thickness and effective depth of reinforcing steel, and creep and shrinkage properties. Choi et al. (2004) extended the work of Ramsay et al. (1979) to include variability of time-dependent concrete properties and demonstrated that time-dependent deflections have higher variability than immediate deflections, as expected.

Jokinen and Scanlon (1985) presented results of an analysis of field-measured two-way slab deflections for a 28-story office tower in Edmonton, Alberta, Canada. Figure 5.5a shows a plot of deflection versus time measurements for 40 nominally identical slab panels. The high variability is evident, both during the construction period (first 35 days) and at approximately 1 year thereafter.

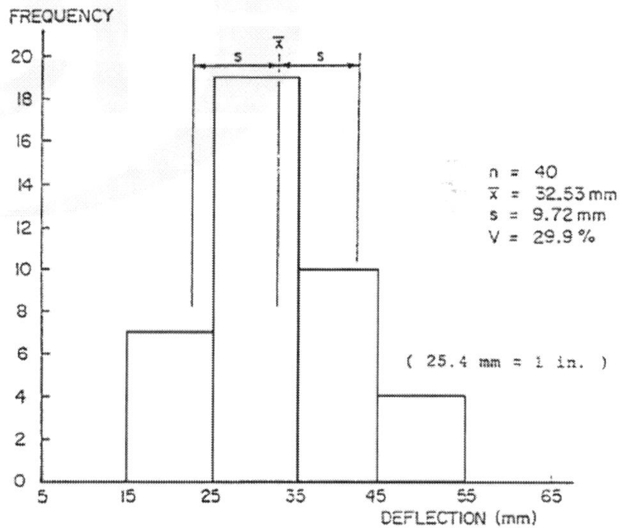

Fig. 5.5b—Histogram of 1-year deflection (Jokinen and Scanlon 1985).

A histogram of 1-year deflections, shown in Fig. 5.5b, indicates a coefficient of variation of 29.9 percent for these slabs and a range of deflections from approximately the mean minus 50 percent to the mean plus 70 percent. Calculated deflections at 1 year based on three assumed values of modulus of rupture, and long-term multipliers proposed by Jokinen and Scanlon (1985), are shown in Fig. 5.5b. These results indicate that the best estimate of the mean deflection was obtained using a reduced modulus of rupture of $4\sqrt{f_c'}$

(psi) (0.33 $\sqrt{f_c'}$ [MPa]). The calculated deflection based on the ACI 318 specified value, 7.5 $\sqrt{f_c'}$ (psi) (0.62 $\sqrt{f_c'}$ [MPa]), was found to lie at the low end of the range of measured deflections. The calculations included effects of construction loads.

Sbarounis (1984) reported on deflection measurements taken after 1 year on 175 bays of a multistory building in Chicago, IL. Measured deflections had a mean value of 1.35 in. (34.3 mm) and a coefficient of variation of 21.2 percent. The range in measured deflections was 0.53 to 2.16 in. (13.5 to 54.9 mm), that is, from the mean minus 60 percent to the mean plus 60 percent. Calculated values were close to the mean deflection.

Numerous case studies of large deflections reported in the literature have been summarized in ACI 435.8R. These case studies, including examples from Australia, Scotland, Sweden, and the United States, highlight the large number of factors that can cause variability in in-place slab deflections.

5.6—Allowable deflections

ACI 318 specifies limits on maximum permissible computed deflections for live load deflection, and incremental deflection occurring after installation of nonstructural elements. No limit is specified for total deflection. However, on the assumption that nonstructural elements will be installed shortly after construction of the slab, the specified limit on incremental deflection indirectly controls the total deflection.

The specified deflection limits apply to calculated deflections. By calculating deflections based on unfactored design loads and expected material properties, the calculated deflection should be interpreted as an estimate of the mean deflection. Recognizing the variability of actual deflections as measured in the field, some variation from the calculated deflection is to be expected. If the calculated deflection is close to the allowable deflection, there is a high probability that the actual deflection will exceed the allowable limit on computed deflection.

Because the deflection limits are based on experience and past practice resulting in generally satisfactory behavior, it should be assumed that the variability of actual deflections is accounted for indirectly in the specified limits. The specified limits in ACI 318 are based on previous experience. The primary concern of ACI 318 is public safety. The serviceability provisions are of a general nature intended to provide adequate serviceability for most design situations. Individual cases may require more stringent requirements than the limited treatment given in ACI 318. Guidance on appropriate deflection limits for a range of applications was given in ACI 435.3R.

CHAPTER 6—REDUCING DEFLECTION OF CONCRETE MEMBERS

6.1—Introduction

Building structures designed by limit states approach may have adequate strength but unsatisfactory serviceability response. Namely, they may exhibit excessive deflection. Thus, the size of flexural members is, in many cases, determined by deflection considerations rather than by strength. This chapter proposes design procedures for reducing the expected deflection that enable design engineers to proportion building structures to meet both strength and serviceability requirements. The result could be more economical structures compared to those designed with unnecessarily conservative deflection-control approaches. The discussion assumes that a competent design is prepared in accordance with ACI 318 and construction follows good practices.

To properly evaluate options for reducing deflection, a design engineer should know the level of stress in the member under consideration, particularly whether the member is uncracked, partially cracked, or fully cracked. Heavily reinforced members tend to be fully cracked because of the heavy loads they are subjected to. In this chapter, only two limiting conditions are considered: uncracked members and fully cracked members. If the maximum applied moment in the positive region is more than twice the cracking moment, the member may be considered as fully cracked ($I_e = I_{cr}$) with little loss of accuracy. Frequently, a member is only partially cracked (I_e significantly greater than I_{cr}) and the statements about both limiting conditions are not strictly applicable. In these situations, engineering judgement and appropriate calculations should be made to assess the actual serviceability conditions of the member. Chapters 4 and 5 of this report discuss methods for computing deflections of partially cracked members.

In addition to the stress conditions, there may be physical or nonstructural constraints on the use of some options, such as limits on increasing concrete dimensions. All options should be evaluated in terms of cost because some may increase the cost, and some may have offsetting considerations that reduce the cost, while still others may have little effect on cost. For each option presented, there is a discussion on the effect of implementation on deflection, the approximate range of potential reduction of deflection, and appropriate situations in which the option should be considered. The options are arranged in three groups: design techniques, construction techniques, and materials selection.

6.2—Design techniques

6.2.1 *Increasing section depth*—Increasing the depth may not be possible after schematic design of the building has been established because such dimensional changes may affect the architectural and mechanical work. There are many instances, however, where beam depth can be increased. For both uncracked and cracked sections, the increase in section stiffness from an increase in depth is more than proportional.

The moment of inertia of an uncracked rectangular section is proportional to the section height cubed, and is given by Eq. (6.2.1a). The moment of inertia of a cracked rectangular section with no axial loads is proportional to the depth squared, and is given by Eq. (6.2.1b).

$$\text{uncracked } I_g = bh^3/12 \qquad (6.2.1a)$$

$$\text{cracked (nonprestressed) } I_{cr} = nA_s(1-k)jd^2 \qquad (6.2.1b)$$

For preliminary design purposes, the cracked moment of inertia can be assumed to be proportional to the area of the

Table 6.2.1.1—Effect of increasing *d* and A_s on cracked moment of inertia

	A. Initial	B. Increase *d*	C. Increase *d* and decrease A_s
f_c'	4000 psi (28 MPa)	4000 psi (28 MPa)	4000 psi (28 MPa)
f_y	60,000 psi (410 MPa)	60,000 psi (410 MPa)	60,000 psi (410 MPa)
n	10	10	10
b	10 in. (250 mm)	10 in. (250 mm)	10 in. (250 mm)
h	18 in. (460 mm)	20 in. (510 mm)	20 in. (510 mm)
d	15.5 in. (390 mm)	17.5 in. (440 mm)	17.5 in. (440 mm)
A_s	1.0 in.^{2} (650 mm^2)	1.0 in.^{2} (650 mm^2)	0.88 in.^{2} (570 mm^2)
M_u	790 kip-in. (89 kN-m)	900 kip-in. (100 kN-m)	790 kip-in. (89 kN-m)
I_{cr}	1610 in.^{4} (6.7 × 10^8 mm^4)	2100 in.^{4} (8.7 × 10^8 mm^4)	1890 in.^{4} (7.9 × 10^8 mm^4)

reinforcement times the depth of the reinforcement squared (A_sd^2). For the cross sections in the example in 6.2.1.1, using this assumption leads to errors of less than 5 percent.

If, by increasing the depth, the concrete tensile stress in a member is reduced sufficiently so that it changes from a cracked or partially cracked member to an uncracked member, the stiffness could increase dramatically. The uncracked stiffness can be as much as three times the cracked stiffness (Grossman 1981). Caution should be exercised when assuming a member is uncracked, as shrinkage and thermal stresses may lessen the load that is required to crack the member.

6.2.1.1 *Example: Effect of depth on cracked moment of inertia*—This example shows the effect of increasing the depth of a member on the cracked moment of inertia of a cross section. Cross Section A is the initial cross section before changes. In Cross Section B, the depth is increased, but all other parameters are kept constant (and moment capacity is increased); in Cross Section C, the depth is increased the same as Cross Section B, but the area of reinforcement is decreased to give the same moment capacity as Cross Section A (Table 6.2.1.1).

6.2.2 *Increasing section width*—This option is not applicable to slabs or other members with physical constraints on their width. Where beams cannot be made deeper because of floor-to-floor height limitations, but can be made wider, the increase in stiffness is proportional to the increase in width if the member is uncracked. If the member is cracked and remains cracked after increasing the width, the increase in stiffness is very small. However, if a cracked member becomes uncracked because the width is increased, its stiffness increases appreciably, possibly by as much as a factor of three (Grossman 1981).

6.2.3 *Addition of compression reinforcement*—Compression reinforcement has some effect on immediate deflection. It can influence I_{cr}, and as a result I_e, and the corresponding initial deflection, although the influence may be small. It can, however, significantly reduce the incremental deflection that occurs due to creep and shrinkage.

The increases in deflection that occur over time have two components: creep deflection and shrinkage warping. Compression reinforcement reduces deflection because concrete creep tends to transfer the compression force from the concrete to the compression reinforcement, which does not itself creep. The closer the reinforcement is to the compression face of the member relative to the neutral axis depth, the more effective reinforcement is in reducing creep deflection. Thus, compression reinforcement is more effective in deeper than in shallower beams or slabs if the concrete cover to the compression face of the member is of constant value. For some very shallow members, due to the requirements of minimum bar cover, compression reinforcement could be at or near the neutral axis and be almost totally ineffective in reducing long-term creep deflection.

Shrinkage warping occurs where the centroids of the steel reinforcement and the uncracked portion of the concrete do not coincide, and the shrinkage of concrete, combined with the dimensional stability of the reinforcement, warps the member in a fashion similar to a piece of bimetal subject to temperature variations. Compression reinforcement reduces shrinkage warping because it brings the centroid of the tension and compression reinforcement closer to the centroid of the uncracked concrete. While compression reinforcement reduces shrinkage and warping of all flexural members, it is especially effective for T-beams where the neutral axis is close to the compression face and far from the tension reinforcement. If the T-beam has a thin slab subject to higher-than-normal shrinkage because of its high surface-to-volume ratio, then compression reinforcement will be more effective than for a rectangular beam. This will be true for ribbed slabs or joist systems as well.

6.2.4 *Addition of tension reinforcement*—For uncracked members, addition of tension reinforcement has hardly any effect on deflection. For fully cracked members, the addition of tension reinforcement reduces both immediate and long-term deflection almost in proportion to the increase in the steel reinforcement area. This can be seen from the cracked moment of inertia I_{cr}, defined in Eq. (6.2.1b). For all practical purposes, $I_{cr} = 0.9A_snd^2$ because the variation in the term $(1 - k)j$ is usually small. For example, if the total deflection of a cracked member is 1.5 in. (38 mm), as in the previous example, increasing the tension reinforcement by 50 percent will reduce the deflection to approximately 1.1 in. (28 mm). However, the increased reinforcement area should still be less than the maximum permitted by code to ensure ductile behavior. This option is most effective for lightly reinforced solid and ribbed slabs. The option of adding more tension reinforcement is not available or is limited for heavily reinforced beams unless compression reinforcement is also added to balance the increase in tension bar area and ensure code ductility requirements are satisfied.

6.2.5 *Revision of structure geometry*—Common solutions to reduce deflections include increasing the number of columns to reduce the length of the spans, adding cross members to create two-way systems, and increasing the size of columns to provide more end restraint to flexural members.

6.2.6 *Revision of deflection limit criteria*—If the predicted deflection of a member is excessive, the deflection limits may be reexamined to determine if they are unnecessarily restrictive. If experience or analysis indicates that those limits (refer to Chapter 4) can be relaxed, then other action might not be required. Many building codes do not set absolute limits on deflection. An engineer may be able to determine that the building occupancy, or construction conditions, such as a sloping roof, do not require the normal deflection limits.

6.3—Construction techniques

6.3.1 *Concrete curing to allow gain in strength*—Deflection response is affected more by concrete strength at the time of loading, than by final concrete strength. If the construction schedule makes early loading of the concrete likely or desirable, then measures to ensure high strength at first loading or construction loading can be effective. Consider for example a concrete member with design compressive strength f_c' of 4000 psi (28 MPa), that would be uncracked as designed. If loaded when the concrete's actual strength is 2500 psi (17.2 MPa), it could be excessively cracked due to a lower modulus of rupture at the time of loading. Even though its final load-carrying capacity may be deemed satisfactory, the cracked member may still deflect several times more than a similar uncracked one. Furthermore, the modulus of elasticity of a 4000 psi (28 MPa) concrete is higher than that of 2500 psi (17.2 MPa) concrete (refer to 6.4 for the effects of material selection on these parameters).

6.3.2 *Concrete curing to reduce shrinkage and creep*—Immediate deflection will not be greatly affected by concrete curing, but additional time-dependent deflection will be reduced. Assuming the time-dependent component of deflection is evenly divided between shrinkage and creep, if shrinkage is reduced 20 percent by adequate curing, the additional time dependent deflection due to shrinkage and creep will be reduced by 10 percent. The effect will be most pronounced on members subject to high shrinkage such as those with a high surface/volume ratio (smaller members), those with thin flanges, and structures in arid atmospheres or members that are restrained. The effect of adequate curing on creep is similar to its effect on shrinkage.

Curing may also reduce shrinkage warping due to nonuniform shrinkage through the depth of the member. If shrinkage at the top of a slab is higher than at the bottom due to premature drying, the slab will deflect downward.

6.3.3 *Control of shoring and reshoring procedures*—The shoring loads may be imposed on the floor slabs before the concrete has reached its design strength (refer to 6.3.1). Construction loads should never be allowed to exceed design loads, with applicable reductions included, without proper prior engineering evaluation. Also, the total construction load should not be permitted to exceed the maximum load-carrying capacities of the early age concrete slabs. ACI 347.2R provides information on construction load analysis and shoring/reshoring design.

Construction of formwork and shoring should ensure that a sag or negative camber is not built into the slab. Experience indicates that frequently the apparent deflection varies widely between slabs of identical design and construction. Some reasons for this may be that such slabs were not all built level or at the specified grade or the method and timing of form stripping was not uniformly applied. Also, construction loads may not have been applied uniformly.

6.3.4 *Delay of the first loading*—This allows the concrete to gain more strength before loading or helps to reach its design strength. Both the modulus of elasticity E_c and the modulus of rupture f_r will be increased. An increase in E_c increases the flexural stiffness. An increase in the modulus of rupture value f_r reduces the amount of cracking or even allows the member to remain uncracked with an increase in flexural stiffness EI as noted in the following section.

6.3.5 *Delay in installation of deflection-sensitive elements or equipment*—A delay in equipment installation will have no effect on immediate or total deflection (except as previously noted in 6.3.1), but incremental deflection occurring after installation of deflection-sensitive components will be reduced. The deflection occurring before a deflection-sensitive component is installed will have no effect on the component. For example, if the additional long-term deflection is 1 in. (25 mm), and installation of partitions is delayed for 3 months, the incremental deflection will be approximately 0.50 in. (13 mm), or approximately one-half as much as the total deflection.

6.3.6 *Location of deflection-sensitive equipment to avoid deflection problems*—Equipment such as printing presses, scientific equipment, and the like should remain level and should be located at midspan where the change in slope is very small with the increase in deflection. On the other hand, because the amplitude of vibration is highest at midspan, vibration-sensitive equipment may be best located near the supports.

6.3.7 *Provision of architectural details to accommodate expected deflection*—Partitions that abut columns, as an example, may show the effect of deflection by separating horizontally from the column near the top even though the partition is not cracked or otherwise damaged. Architectural details should accommodate such movements. Likewise, windows, walls, partitions, and other nonstructural elements supported by or located under deflecting concrete members should be provided with slip joints to accommodate the expected deflections or differential deflections between concrete members above and below the nonstructural elements.

6.3.8 *Building camber into floor slabs*—Built-in camber has no effect on the computed deflection of a slab.

Cambering, however, is effective for the installation of partitions and equipment, if the objective is to have a level floor slab after deflection takes place. For best results, deflection should be carefully calculated using the appropriate modulus of concrete E_c value and the correct moment of inertia I. Overestimating the deflection value can lead the designer to specify unreasonable overcambering. Hence, the pattern and value of cambering at several locations should be specified and the results monitored during construction. Procedures can be revised as necessary for slabs that are to be constructed at a later date.

6.3.9 *Ensuring that top bars are not displaced downward*—Downward displacement of top bars always reduces strength. The effect on deflection in uncracked members is minimal, but its effect on cracked members, namely, those that are heavily loaded, is in proportion to the square of the ratio of change in effective depths for cantilevers but much less for continuous spans. This reduced effect in continuous members is because the flexural stiffness and resulting deflection of the member is determined primarily by member stiffness at the midspan section. Thus, the deflection of cantilevers is particularly sensitive to misplacement of the top reinforcing bars. Deflection could increase in continuous members if the reduction in strength at negative moment regions results in redistribution of moments.

6.4—Materials selection

6.4.1 *Selection of materials for mixture design that reduce shrinkage and creep or increase the moduli of elasticity and rupture*—Materials having an effect on these properties include aggregates, cement, silica fume, and admixtures. Lower water-cement ratio (w/c), a lower slump, and changes in other materials proportions can reduce shrinkage and creep or increase the moduli of elasticity or rupture. Shrinkage-reducing admixtures may also be used.

6.4.2 *Use of concretes with a higher modulus of elasticity*—Using ACI 318 procedures, the stiffness of an uncracked member increases in proportion to the elastic modulus, which varies in proportion to the square root of the cylinder strength (ACI 318-14 Section 19.2.2). The stiffness of a cracked section is affected little by a change in the modulus of elasticity.

Table 6.5—Deflection reducing options

Option	Effect on section stiffness: Uncracked	Effect on section stiffness: Cracked
Design techniques		
1. Deeper members	$(h^*/h)^3$ for rectangular beams; less for T-beams	$(d^*/d)^2$ or (d^*/d) If change to uncracked section, up to 300 percent
2. Wider members	(b^*/b)	Small unless changed to uncracked section
3. Add A_s'	Up to 50 percent for Δ_{LT}. No effect for Δ_i.	Up to 50 percent for Δ_{LT}. No effect for Δ_i.
4. Add A_s	No effect	A_s^*/A_s
5. Add prestress	Reduces dead load deflection to nearly zero	Reduces dead load deflection to nearly zero and member to uncracked
6. Structural geometry	Large effect	Large effect
7. Revise criteria	See text	See text
Construction techniques		
8. Cure: f'_c	Same as higher E_c and f_r	Same as high E_c and f_r and could change to uncracked section.
9. Cure ε_{sh} and ε_{cr}	For long-term deflection $(\varepsilon_{sh}*/\varepsilon_{sh})$ and $(\varepsilon_{cr}*/\varepsilon_{cr})$	For long-term deflection $(\varepsilon_{sh}*/\varepsilon_{sh})$ and $(\varepsilon_{cr}*/\varepsilon_{cr})$.
10. Shoring	Large effect, see text	Large effect, see text
11. Delay first loading	Same as higher E_c and f_r	Same as high E_c and f_r and could change to uncracked section.
12. Delay installation	Up to 50+ percent depending on time delay	Up to 50+ percent depending on time delay
13. Locate equipment	See text	See text
14. Architectural details	See text	See text
15. Camber	See text	See text
16. Top bars	No effect	Up to $(d^*/d)^2$ for cantilevers
Materials		
17. Materials	See 6.4	See 6.4
18. Mixture design	See 6.3.2 and 6.4.3	See 6.3.2 and 6.4.3
19. Higher E_c	(E_c^*/E_c) or $(f_c'^*/f_c')^{0.5}$	Significant
20. Higher f_r	None	Significant
21. Use fiber reinforcement	See 6.3.2 and 6.4.4	See 6.3.2 and 6.4.4

*Parameters that have been changed to reduce deflection.

Because the stiffness of a flexural member is *EI*, both the modulus of elasticity and the moment of inertia affect stiffness significantly. Concrete with a higher modulus of elasticity will reduce deflections. The actual modulus of elasticity may vary significantly from the ACI 318 code prediction. If historic mixture data is not available, modulus of elasticity tests can be performed on trial batches.

6.4.3 *Use of concretes with a higher modulus of rupture*—Concrete with a higher modulus of rupture does not necessarily increase the stiffness of uncracked members and highly cracked members. However, stiffness of partially cracked members increases because of the reduction of the degree of cracking. The increase in stiffness (decrease in deflection) depends on steel reinforcement percentage, the increase in modulus of rupture, and the magnitude of applied moment.

6.4.4 *Addition of short discrete fibers to concrete mixture*—Adding short discrete fibers to the concrete mixture (fiber-reinforced concrete) has been reported to reduce shrinkage and increase the cracking strength, both of which might reduce deflection (Alsayed 1993). Changes in the material behavior depend on the type and amount of fibers used.

6.5—Summary

Table 6.5 summarizes some of the preventive measures needed to reduce or control deflection. This table can serve as a general guide to the design engineer but is not all inclusive, and engineering judgement should be exercised in the choice of the most effective parameters that control deflection behavior.

CHAPTER 7—REFERENCES

Committee documents are listed first by document number and year of publication followed by authored documents listed alphabetically.

American Concrete Institute

ACI 209R-92(08)—Prediction of Creep, Shrinkage, and Temperature Effects in Concrete Structures

ACI 209.2R-08—Guide for Modeling and Calculating Shrinkage and Creep in Hardened Concrete

ACI 213R-14—Guide for Structural Lightweight-Aggregate Concrete

ACI 237R-07(19)—Self-Consolidating Concrete

ACI 318-14—Building Code Requirements for Structural Concrete and Commentary

ACI 347R-14—Guide to Formwork for Concrete

ACI 347.2R-17—Guide for Shoring/Reshoring of Concrete Multistory Buildings

ACI 363R-10—Report on High-Strength Concrete

ACI 435R-95(03)—Control of Deflection in Concrete Structures

ACI 435.3R-68(89)—Allowable Deflections (no longer in print)

ACI 435.4R-72(89)—Variability of Deflections of Simply Supported Reinforced Concrete Beams (no longer in print)

ACI 435.6R-74(89)—Deflection of Two-way Reinforced Concrete Floor Systems: State-of-the-Art Report (no longer in print)

ACI 435.7R-85(97)—State-of-the-Art Report on Temperature-Induced Deflections of Reinforced Concrete Members (no longer in print)

ACI 435.8R-85(97)—Observed Deflection of Reinforced Concrete Slab Systems, and Causes of Large Deflections

ACI 440.1R-15—Guide for the Design and Construction of Concrete Reinforced with FRP Bars

ACI 544.1R-96(09)—Report on Fiber Reinforced Concrete

ACI 544.5R-10—Report on the Physical Properties and Durability of Fiber Reinforced Concrete

British Standards Institution

BS EN 1992-1-1:2004+A1:2014—Eurocode 2: Design of Concrete Structures. General Rules and Rules for Buildings

Canadian Standards Association

CSA A23.3:2014—Design of Concrete Structures

International Conference of Building Officials (merged into International Code Council)

ICBO 1997—Uniform Building Code

International Federation for Structural Concrete (fib)

fib MC2010—Model Code for Concrete Structures 2010

Standards Australia

AS 3600-2009—Design and Construction of Concrete Building Structures

Authored documents

ACI Committee 209, 1971, "Effects of Creep, Shrinkage, and Temperature Effects in Concrete Structures," *Designing for Effects of Creep Shrinkage Temperature in Concrete Structures*, SP-27, American Concrete Institute, Farmington Hills, MI, pp. 51-93.

ACI Committee 209, 1982, "Prediction of Creep, Shrinkage, and Temperature Effects in Concrete Structures," *Designing for Creep & Shrinkage in Concrete Structures: A Tribute to Adrian Pauw*, SP-76, B. B. Goyal, ed., American Concrete Institute, Farmington Hills, MI, pp. 193-300.

ACI Committee 435, 1966, "Deflections of Reinforced Concrete Flexural Members," *ACI Journal Proceedings*, V. 63, No. 6, June, pp. 637-674.

ACI Committee 435, 1978, "Proposed Revisions by Committee 435 to ACI Building Code and Commentary Provisions on Deflections," *ACI Journal Proceedings*, V. 75, No. 6, June, pp. 229-238.

Aguinaga-Zapata, M., and Bažant, Z. P., 1986, "Creep Deflections in Slab Buildings and Forces in Shores During Construction," *ACI Journal Proceedings*, V. 83, No. 5, Sept.-Oct., pp. 719-726.

Al-Zaid, R.; Al-Shaikh, A. H.; and Abu-Hussein, M., 1991, "Effect of Loading Type on the Effective Moment of Inertia of Reinforced Concrete Beams," *ACI Structural Journal*, V. 88, No. 2, Mar.-Apr., pp. 184-190.

Alsayed, S. H., 1993, "Flexural Deflection of Reinforced Fibrous Concrete Beams," *ACI Structural Journal*, V. 96, No. 1, Jan.-Feb., pp. 72-76.

ASCE Task Committee, 1982, *Finite Element Analysis of Reinforced Concrete*, ASCE, New York, 553 pp.

Bažant, Z. P., and Kim, S., 1979, "Approximate Relaxation Function for Concrete," *Journal of the Structural Division*, V. 105, No. 12, Dec., pp. 2695-2705.

Bažant, Z. P., and Panula, L., 1980, "Creep and Shrinkage Characterization for Analyzing Prestressed Concrete Structures," *PCI Journal*, V. 25, No. 3, May-June, pp. 86-122. doi: 10.15554/pcij.05011980.86.122

Beeby, A. W., 1968, "Short Term Deformations of Reinforced Concrete Members," *Cement and Concrete Association Technical Report* TRA 408, London, UK, Mar.

Bernard, E. S., 2004, "Creep of Cracked Fibre Reinforced Shotcrete Panels," *Shotcrete: More Engineering Developments*, Taylor & Francis Group, London, UK, pp. 47-57.

Bischoff, P. H., 2005, "Reevaluation of Deflection Prediction for Concrete Beams Reinforced with Steel and Fiber Reinforced Polymer Bars," *Journal of Structural Engineering*, V. 131, No. 5, pp. 752-767. doi: 10.1061/(ASCE)0733-9445(2005)131:5(752)

Bischoff, P. H., 2007, "Rational Model for Calculating Deflection of Reinforced Concrete Beams and Slabs," *Canadian Journal of Civil Engineering*, V. 34, No. 8, pp. 992-1002. doi: 10.1139/l07-020

Bischoff, P. H., and Gross, S. P., 2011a, "Equivalent Moment of Inertia Based on Integration of Curvature," *Journal of Composites for Construction*, V. 15, No. 3, pp. 263-273. doi: 10.1061/(ASCE)CC.1943-5614.0000164

Bischoff, P. H., and Gross, S. P., 2011b, "Design Approach for Calculating Deflection of FRP Reinforced Concrete," *Journal of Composites for Construction*, V. 15, No. 4, pp. 490-499. doi: 10.1061/(ASCE)CC.1943-5614.0000195

Bischoff, P. H., and Scanlon, A., 2007, "Effective Moment of Inertia for Calculating Deflections of Concrete Members Containing Steel Reinforcement and FRP Reinforcement," *ACI Structural Journal*, V. 104, No. 1, Jan.-Feb., pp. 68-75.

Bischoff, P. H., and Scanlon, A., 2009, "Span-Depth Ratios for One-Way Members Based on ACI 318 Deflection Limits," *ACI Structural Journal*, V. 106, No. 5, Sept.-Oct., pp. 617-626.

Bischoff, P. H., and Veysey, S., 2011, "Minimum Thickness Requirements Needed to Satisfy Deflection Limits for One-Way Reinforced Concrete Construction," CSCE 2011 General Conference, Ottawa, ON, Canada.

Branson, D. E., 1963, "Instantaneous and Time-Dependent Deflections of Simple and Continuous Reinforced Concrete Beams," *HPR Publication* No. 7, Part 1, Alabama Highway Department, U.S. Bureau of Public Roads, Aug., pp. 1-78.

Branson, D. E., 1971, "Compression Steel Effect on Long-Time Deflections," *ACI Journal Proceedings*, V. 68, No. 8, Aug., pp. 555-559.

Branson, D. E., 1977, *Deformation of Concrete Structures*, McGraw-Hill Book Co., Advanced Book Program, New York, 546 pp.

Branson, D. E., 1985, *Handbook of Concrete Engineering*, second edition, M. Fintel, ed., Van Nortrand Reinhold Co., New York, pp. 53-78.

Branson, D. E., and Christiason, M. L., 1971, "Time-Dependent Concrete Properties Related to Design-Strength and Elastic Properties, Creep and Shrinkage," *Symposium on Creep, Shrinkage, and Temperature Effects*, SP-27, American Concrete Institute, Farmington Hills, MI, pp. 257-277.

Branson, D. E.; Meyers, B. L.; and Kripanarayanan, K. M., 1970, "Time-Dependent Deformation of Noncomposite and Composite Prestressed Concrete Structures," *Highway Research Record*, No. 324, pp. 15-43.

Brooks, J. J., and Al-Qarra, H., 1999, "Assessment of Creep and Shrinkage of Concrete for the Flintshire Bridge," *Structural Engineering*, V. 77, No. 5, Mar., pp. 21-26.

Chang, K.-Y., and Hwang, S.-J., 1996, "Practical Estimation of Two-way Slab Deflections," *Journal of Structural Engineering*, V. 122, No. 2, pp. 150-159. doi: 10.1061/(ASCE)0733-9445(1996)122:2(150)

Choi, B.-S.; Scanlon, A.; and Johnson, P. A., 2004, "Monte Carlo Simulation of Immediate and Time-dependent Deflections of Reinforced Concrete Beams and Slabs," *ACI Structural Journal*, V. 101, No. 5, Sept.-Oct., pp. 633-641.

Collins, M. P., and Mitchell, D., 1997, *Prestressed Concrete Structures*, Prentice-Hall Inc., Englewood Cliffs, NJ, 766 pp.

Gardner, N. J., 1985, "Shoring, Reshoring, and Safety," *Concrete International*, V. 7, No. 4, Apr., pp. 28-34.

Gardner, N. J., 2001, "Span/Thickness Limitations for Deflection Control," *Code Provisions for Deflection Control in Concrete Structures*, SP-203, E. G. Nawy and A. Scanlon, eds., American Concrete Institute, Farmington Hills, MI, pp. 95-114.

Gardner, N. J., 2011, "Span/Thickness Limits for Deflection Control," *ACI Structural Journal*, V. 108, No. 4, July-Aug., pp. 453-460.

Gardner, N. J., and Zhang, J., 1996, "Controlling Deflection Serviceability by Span/Depth Limits and Long-Term Deflection Multipliers for Reinforced Concrete Beams," *Recent Developments in Deflection Evaluation of Concrete*, SP-161, American Concrete Institute, Farmington Hills, MI, pp. 165-196.

Ghali, A., 1989, "Prediction of Deflections of Two-Way Floor Systems," *ACI Structural Journal*, V. 86, No. 5, Sept.-Oct., pp. 551-562.

Ghali, A.; Favre, R.; and Elbadry, M., 2011, *Concrete Structures: Stresses and Deformations: Analysis and Design for Serviceability*, fourth edition, CRC Press, Boca Raton, FL, 646 pp.

Gilbert, R. I., 1985, "Deflection Control of Slabs Using Allowable Span to Depth Ratios," *ACI Journal Proceedings*, V. 82, No. 1, Jan.-Feb., pp. 67-77.

Gilbert, R. I., 1999, "Deflection Calculation for Reinforced Concrete Structures—Why We Sometimes Get It Wrong," *ACI Structural Journal*, V. 96, No. 6, Nov.-Dec., pp. 1027-1032.

Gilbert, R. I., 2006, "Discussion of "Reevaluation of Deflection Prediction for Concrete Beams Reinforced with Steel and Fiber Reinforced Polymer Bars," by Peter H. Bischoff," *Journal of Structural Engineering*, V. 132, No. 8, pp. 1328-1330. doi: 10.1061/(ASCE)0733-9445(2006)132:8(1328)

Gilbert, R. I., 2007, "Tension Stiffening of Lightly Reinforced Concrete Slabs," *Journal of Structural Engineering*, V. 133, No. 6, pp. 899-903. doi: 10.1061/(ASCE)0733-9445(2007)133:6(899)

Gilbert, R. I., and Warner, R. F., 1978, "Tension Stiffening in Reinforced Concrete Slabs," *Journal of the Structural Division*, V. 104, No. 12, pp. 1885-1900.

Graham, C. J., and Scanlon, A., 1985, "Deflection of Concrete Slabs under Construction Loading," *Deflections of Concrete Structures*, SP-86, G. Sabnis, ed., American Concrete Institute, Farmington Hills, MI, pp. 215-230.

Graham, C. J., and Scanlon, A., 1986, "Long-Time Multipliers for Estimating Two-Way Slab Deflections," *ACI Journal Proceedings*, V. 83, No. 5, Sept.-Oct., pp. 899-908.

Grossman, J. S., 1981, "Simplified Computations for Effective Moment of Inertia (I_e) and Minimum Thickness to Avoid Deflection Computations," *ACI Journal Proceedings*, V. 78, No. 6, Nov.-Dec., pp. 423-439, and author's closure, V. 79, No. 5, Sept.-Oct. 1982, pp. 414-419.

Grossman, J. S., 1987, *Building Structural Design Handbook*, R. N. White and C. G. Salmon, eds., John Wiley & Sons, New York, pp. 699-786.

Grundy, P., and Kabaila, A., 1963, "Construction Loads on Slabs with Shored Formwork in Multi-Story Buildings," *ACI Journal Proceedings*, V. 60, No. 12, Dec., pp. 1729-1738.

Hirsch, J., 2009, "Accurate Long-Term Deflection Prediction in Flat Slabs Using Linear Elastic Global Analysis," *24th Biennial Conference of the Concrete Institute of Australia*, Sydney, Australia, 2009, 11 pp.

Hirsch, J.; Calabrese, F.; Connolly, E.; and Bommer, A., 2012, "Practical Deflection Prediction of Concrete Slabs," SP-284, American Concrete Institute, Farmington Hills, MI. (CD-ROM)

Hossain, T. R.; Vollum, R.; and Ahmed, S. U., 2011, "Deflection Estimation of Reinforced Concrete Flat Plates Using ACI Method," *ACI Structural Journal*, V. 108, No. 4, July-Aug., pp. 405-413.

Jensen, V. P., 1938, "Solutions for Certain Rectangular Slabs Continuous Over Flexible Supports," *Bulletin* No. 303, University of Illinois Engineering Experiment Station, Urbana, IL.

Jokinen, E. P., and Scanlon, A., 1985, "Field-Measured Two-Way Slab Deflections," *Proceedings*, Annual Conference, CSCE, SK, Canada, May.

Kripanarayanan, K. M., and Branson, D. E., 1976, "Short-Time Deflections of Flat Plates, Flat Slabs, and Two-Way Slabs," *ACI Journal Proceedings*, V. 73, No. 12, Dec., pp. 686-690.

Lee, Y. H., and Scanlon, A., 2007, "Deflection Control of Concrete Structures Based on Utility Theory," *ACI Structural Journal*, V. 104, No. 1, Jan.-Feb., pp. 60-67.

Liu, X.; Chen, W. F.; and Bowman, M. D., 1985, "Construction Load Analysis for Concrete Structures," *Journal of Structural Engineering*, V. 111, No. 5, pp. 1019-1036. doi: 10.1061/(ASCE)0733-9445(1985)111:5(1019)

Luebkeman, C. H.; Nilson, A. H.; and Slate, F. O., 1985, "Sustained Load Deflection of High Strength Concrete Beams," *Research Report* No. 85-2, Department of Structural Engineering, Cornell University, Ithaca, NY, Feb.

Martin, I., 1971, "Effects of Environmental Conditions on Thermal Variations and Shrinkage of Concrete Structures in the United States," *Designing for Effects of Creep Shrinkage Temperature in Concrete Structures*, SP-27, American Concrete Institute, Farmington Hills, MI, pp. 279-300.

Meyers, B. L., and Thomas, E. W., 1983, *Handbook of Structural Concrete*, McGraw-Hill, New York, pp. 11.1-11.33.

Myers, J. J., and Yang, Y., 2004, "Field and Laboratory Performance of Prestressed High Performance Concrete Girders for Missouri Bridge Structures," *Report* No. 04-49, Center for Infrastructure Engineering Studies, Missouri University of Science and Technology, Rolla, MO, June, pp. 96-98, 102.

Nagataki, S., and Yonekura, A., 1978, "Studies of the Volume Changes of High-Strength Concretes with Superplasticizer," *Journal*, Japan Prestressed Concrete Engineering Association, V. 20, pp. 26-33.

Nassar, A. J., 2002, "Investigation of Transfer Length, Development Length, Flexural Strength and Prestress Loss Trend in Fully Bonded High-Strength Lightweight Prestressed Girders," Virginia Polytechnic Institute and State University, May 15, 136 pp.

Nawy, E. G., 2009, *Reinforced Concrete: A Fundamental Approach*, sixth edition, Pearson Education, NJ, 915 pp.

Ngab, A. S.; Nilson, A. H.; and Slate, F. O., 1981, "Shrinkage and Creep of High-Strength Concrete," *ACI Journal Proceedings*, V. 78, No. 4, July-Aug., pp. 255-261.

Nilson, A. H., 1985, "Design Implications of Current Research on High-Strength Concrete," *High-Strength Concrete*, SP-87, H. G. Russell, ed., American Concrete Institute, Farmington Hills, MI, pp. 85-117.

Nilson, A. H., and Walters Jr., D. B., 1975, "Deflection of Two-Way Floor Systems by the Equivalent Frame Method," *ACI Journal Proceedings*, V. 72, No. 5, May, pp. 210-218.

Park, R., and Paulay, T., 1975, *Reinforced Concrete Structures*, John Wiley & Sons, Inc., New York, 769 pp.

Ramsay, R. J.; Mina, S. A.; and Macgregor, J. G., 1979, "Monte Carlo Study of Short Time Deflections of Reinforced Concrete Beams," *ACI Journal Proceedings*, V. 76, No. 8, Aug., pp. 897-918.

Rangan, B. V., 1976, "Prediction of Long-Term Deflections of Flat Plates and Slabs," *ACI Journal Proceedings*, V. 73, No. 4, Apr., pp. 223-226.

Rangan, B. V., 1982, "Control of Beam Deflections by Allowable Span-Depth Ratios," *ACI Journal Proceedings*, V. 79, No. 5, Sept.-Oct., pp. 372-377.

Sbarounis, J. A., 1984, "Multi-Story Flat Plate Buildings: Construction Loads and Immediate Deflections," *Concrete International*, V. 6, No. 2, Feb., pp. 70-77.

Scanlon, A., and Bischoff, P. H., 2008, "Shrinkage Restraint and Loading History Effects on Deflection of Flexural Members," *ACI Structural Journal*, V. 105, No. 4, July-Aug., pp. 498-506.

Scanlon, A., and Lee, Y. H., 2006, "Unified Span-to-Depth Ratio Equation for Nonprestressed Concrete Beams and

Slabs," *ACI Structural Journal*, V. 103, No. 1, Jan.-Feb., pp. 142-148.

Scanlon, A., and Murray, D. W., 1974, "Time Dependent Reinforced Concrete Slab Deflections," *Journal of the Structural Division*, V. 100, pp. 1911-1924.

Scanlon, A., and Murray, D. W., 1982, "Practical Calculation of Two-way Slab Deflections," *Concrete International*, V. 4, No. 11, Nov., pp. 43-50.

Scanlon, A., and Suprenant, B., 2011, "Spreadsheet for Estimating Two-way Slab Deflections Including Construction Load Effects," *Concrete International*, V. 33, No. 7, July, pp. 29-34.

Scott, R. H., and Beeby, A. W., 2005, "Long Term Tension Stiffening Effects in Concrete," *ACI Structural Journal*, V. 102, No. 1, Jan.-Feb., pp. 31-39.

Sonebi, M., and Bartos, P. J. M., 2001, "Performance of Reinforced Columns Cast with Self Compacting Concrete," *Fifth CANMET/ACI Conference on Recent Advances in Concrete Technology-Proceeding*, SP-200, V. M. Malhotra, ed., American Concrete Institute, Farmington Hills, MI, pp. 415-432.

Swamy, R. N., and Anand, K. L., 1973, "Shrinkage and Creep of High Strength Concrete," *Civil Engineering and Public Works Review* (London), V. 68, No. 807, Oct., pp. 859-865, 867-868.

Tam, K. S. S., and Scanlon, A., 1986, "Deflection of Two-Way Slabs Subjected to Restrained Volume Change and Transverse Loads," *ACI Journal Proceedings*, V. 83, No. 5, Sept.-Oct., pp. 737-744.

Tan, K. H.; Paramasivam, P.; and Tan, K. C., 1994, "Instantaneous and Long-Term Deflections of Steel Fiber Reinforced Concrete Beams," *ACI Structural Journal*, V. 91, No. 4, July-Aug., pp. 364-393.

Thompson, D. P., and Scanlon, A., 1988, "Minimum Thickness Requirements for Control of Two-Way Slab Deflections," *ACI Structural Journal*, V. 85, No. 1, Jan.-Feb., pp. 12-22.

Timoshenko, S., and Woinowsky-Krieger, S., 1959, *Theory of Plates and Shells*, McGraw-Hill Book Co., New York, NY.

Trost, H., 1967, "Auswirkungen des Superpositionsprinzips auf Kriech- und Relaxationsprobleme bei Beton und Spann beton," *Beton- und Stahlbetonbau*, V. 61, pp. 230-238.

Vanderbilt, M. D.; Sozen, M. A.; and Siess, C. P., 1965, "Deflections of Multiple-Panel Reinforced Concrete Floor Slabs," *ASCE Proceedings*, V. 91, No. ST4, Aug., pp. 77-101.

Vecchio, F. J., and Collins, M. P., 1986, "The Modified Compression-Field Theory for Reinforced Concrete Elements Subjected to Shear," *ACI Journal Proceedings*, V. 83, 2, Mar.-Apr., pp. 219-231.

Vollum, R. L., 2011, "Influence of Construction Loading on Long-Term Slab Deflection," *Andy Scanlon Symposium on Serviceability and Safety of Concrete Structures*, SP-284, American Concrete Institute, Farmington Hills, MI, pp. 1-22.

Walkup, S.; Musselman, E.; and Gross, S., 2019, "Effect of Service Load Levels on Long-Term Deflection Multiplier," *ACI Structural Journal*, V. 116, No. 2, Mar.-Apr., pp. 89-100.

Wang, C. K., and Salmon, C. G., 1992, *Reinforced Concrete Design*, fifth edition, Harper Collins, 1030 pp.

Wight, J. K., and MacGregor, J. G., 2009, *Reinforced Concrete Mechanics and Design*, fifth edition, Pearson Prentice Hall, Upper Saddle River, NJ, 1112 pp.

Yu, W. W., and Winter, G., 1960, "Instantaneous and Long-Term Deflections of Reinforced Concrete Beams Under Working Loads," *ACI Journal Proceedings*, V. 57, No. 1, Jan., pp. 29-50.

APPENDIX A—DEFLECTION DESIGN EXAMPLES

Numerical examples are presented to demonstrate some of the calculation procedures outlined in Chapter 4. Examples are presented for a simply supported slab and for a four-span continuous beam, as well as an example for temperature-induced deflections.

Example A.1—Deflection of a simply supported slab

This example compares short- and long-term deflection computed using both Branson's (1963) and Bischoff's (2005, 2007) expression for I_e. Results are presented for a simply supported one-way slab with an 18 ft (5.5 m) span. The slab has a thickness of 10 in. (254 mm) that does not quite satisfy the ACI 318 minimum thickness requirements in Table 4.2.2a ($\ell/20$ for a simply supported slab).

Loading consists of a dead load w_D = 125 lb/ft^2 (6 kPa) from the member self-weight and live load w_L = 100 lb/ft^2 (4.8 kPa). Part of the live load $w_{L,sus}$ = 20 lb/ft^2 (1 kPa) is sustained. The concrete has a specified compressive strength of 4000 psi (27.6 MPa) and is reinforced with Grade 60 (Grade 420) steel bars. Strength requirements are satisfied using No. 4 (12.7 mm diameter) bars spaced at 7 in. (178 mm) on center to give a reinforcing ratio $\rho = 0.32$ percent.

Maximum deflection is computed using the traditional ACI 318 approach (4.4.1) and assumes the slab has been preloaded up to the full dead plus live load during the construction period. Restraint to shrinkage is also considered by using a reduced cracking moment equal to two-thirds M_{cr} (4.3.2.3).

Solution:

Material properties and bending moment values:

$$E_c = 57{,}000\sqrt{f_c'} = 57{,}000\sqrt{4000} = 3.6\times10^6 \text{ psi (24,900 MPa)}$$

$E_s = 29 \times 10^6$ psi (200,000 MPa)

modular ratio $n = E_s/E_c = (29 \times 10^6)/(3.6 \times 10^6) = 8.04$

modulus of rupture $f_r = 7.5\sqrt{f_c'} = 7.5\sqrt{4000} = 474$ psi (3.3 MPa)

Positive moment at midspan = $w\ell^2/8$ for a 1 ft (305 mm) wide strip of slab:

$M_D = 125 \times 18^2/8 = 5062.5$ ft-lb = 60,800 in.-lb (6.87 kN-m)
$M_L = 100 \times 18^2/8 = 4050$ ft-lb = 48,600 in.-lb (5.49 kN-m)
$M_D + M_L = 9112.5$ ft-lb = 109,400 in.-lb (12.37 kN-m)

Gross section properties:
$y_t = h/2 = 10/2 = 5$ in. (127 mm)
$I_g = bh^3/12 = 12 \times 10^3/12 = 1000$ in.^{4} (416.2×10^6 mm^4)
$M_{cr} = f_r I_g/y_t = 474 \times 1000/5 = 94{,}800$ in.-lb (10.7 kN-m)

Cracked section properties:
$A_s = (12/7) \times 0.20 = 0.343$ in.^{2}/ft (726 mm^2/m) width of slab
(effective number of No. 4 [12.7 mm] bars in a 1 ft [305 mm] wide strip of slab = 12/7=1.71)
$d = h - \text{cover} - d_b/2 = 10 - 0.75 - 0.5/2 = 9$ in. (229 mm)
$\rho = A_s/bd = 0.343/(12 \times 9) = 0.317$ percent
$n\rho = 8.04 \times 0.317/100 = 0.0255$

$$c = \left(\sqrt{(n\rho)^2 + 2n\rho} - n\rho\right)d = \left(\sqrt{(0.0255)^2 + 2\times 0.0255} - 0.0255\right)9 = 1.82 \text{ in. (46 mm)}$$

$$I_{cr} = \frac{bc^3}{3} + nA_s(d-c)^2 = 12\times\frac{1.82^3}{3} + 8.04\times 0.343(9-1.82)^2 = 166 \text{ in.}^4\ (69.1 \times 10^6 \text{ mm}^4)$$

$I_g/I_{cr} = 6.0$

Effective moment of inertia I_e for full dead plus live load (using M_{cr}):
$M_a = M_D + M_L = 109{,}400$ in.-lb (12.37 kN-m)
$M_{cr}/M_a = 94{,}800/109{,}400 = 0.867$
$(M_{cr}/M_a)^2 = (0.867)^2 = 0.752$
$(M_{cr}/M_a)^3 = (0.867)^3 = 0.652$
$1 - I_{cr}/I_g = 1 - 166/1000 = 0.834$
$I_{e,Branson} = (M_{cr}/M_a)^3 I_g + [1 - (M_{cr}/M_a)^3]\, I_{cr} = 0.652 \times 1000 + (1 - 0.652)166 = 710$ in.^{4} (296×10^6 mm^4)
$I_{e,Bischoff} = I_{cr}/[1 - (M_{cr}/M_a)^2(1 - I_{cr}/I_g)] = 166/(1 - 0.752 \times 0.834) = 445$ in.^{4} (185×10^6 mm^4)

The value for I_e is approximately 40 percent less when computed using Bischoff's expression.

Effective moment of inertia I_e for full dead plus live load (using $0.67M_{cr}$ to account for shrinkage restraint):
$M_a = M_D + M_L = 109{,}400$ in.-lb (12.37 kN-m)
$M_{cr}/M_a = 0.67 \times 0.867 = 0.581$
$(M_{cr}/M_a)^2 = (0.581)^2 = 0.338$
$(M_{cr}/M_a)^3 = (0.581)^3 = 0.196$
$I_{e,Branson} = (M_{cr}/M_a)^3 I_g + [1 - (M_{cr}/M_a)^3] I_{cr} = 0.196 \times 1000 + (1 - 0.196)166 = 329$ in.^{4} (137×10^6 mm^4)
$I_{e,Bischoff} = I_{cr}/[1 - (M_{cr}/M_a)^2(1 - I_{cr}/I_g)] = 166/(1 - 0.338 \times 0.834) = 231$ in.^{4} (96.2×10^6 mm^4)

I_e decreases by approximately one-half when using the reduced cracking moment. This is expected because tension stiffening has a significant influence on member stiffness whenever $M_a \leq 2M_{cr}$.

Calculation of deflection using M_{cr}:

Immediate deflection due to live load (using M_{cr}):

Using Branson's expression for I_e gives

$$\Delta_L = \frac{5M_L\ell^2}{48E_cI_e} = \frac{5\times 48{,}600\times(18\times 12)^2}{48\times 3.6\times 10^6\times 710} = 0.092 \text{ in. (2.34 mm)}$$

$$\Delta_L = 0.092 \text{ in. (2.34 mm)} \leq \frac{\ell}{360} = \frac{18\times 12}{360} = 0.6 \text{ in. (15.2 mm) for floors} \therefore\ \textbf{OK}$$

Using Bischoff's expression for I_e gives

$$\Delta_L = \frac{5M_L\ell^2}{48E_cI_e} = \frac{5\times 48{,}600\times(18\times 12)^2}{48\times 3.6\times 10^6\times 445} = 0.147 \text{ in. } (3.73 \text{ mm})$$

$$\Delta_L = 0.147 \text{ in.}(3.73 \text{ mm}) \le \frac{\ell}{360} = \frac{18\times 12}{360} = 0.6 \text{ in. } (15.2 \text{ mm}) \text{ for floors } \therefore \textbf{ OK}$$

Immediate (short-term) deflection under sustained loads (using M_{cr}):

$M_{sus} = M_D + 0.2M_L = 60{,}800 + 0.2 \times 48{,}600 = 70{,}520$ in.-lb (7.97 kN-m)

Using Branson's expression for I_e gives

$$\Delta_{sus} = \frac{5M_{sus}\ell^2}{48E_cI_e} = \frac{5\times 70{,}520\times(18\times 12)^2}{48\times 3.6\times 10^6\times 710} = 0.134 \text{ in. } (3.40 \text{ mm})$$

Using Bischoff's expression for I_e gives

$$\Delta_{sus} = \frac{5M_{sus}\ell^2}{48E_cI_e} = \frac{5\times 70{,}520\times(18\times 12)^2}{48\times 3.6\times 10^6\times 445} = 0.213 \text{ in. } (5.41 \text{ mm})$$

Long-term and incremental deflection (using $\lambda_\Delta = 2$ and M_{cr}):

Using Branson's expression for I_e gives

$\Delta_{inc} = \lambda_\Delta\Delta_{sus} + \Delta_{L,add} = 2.0 \times 0.134 + 0.8 \times 0.092 = 0.341$ in. (8.66 mm)

$$\Delta_{inc} = 0.341 \text{ in.}(8.66 \text{ mm}) \le \frac{\ell}{240} = \frac{18\times 12}{240} = 0.9 \text{ in. } (22.9 \text{ mm}) \therefore \textbf{ OK}$$

Using Bischoff's expression for I_e gives

$\Delta_{inc} = \lambda_\Delta\Delta_{sus} + \Delta_{L,add} = 2.0 \times 0.213 + 0.8 \times 0.147 = 0.543$ in. (13.8 mm)

$$\Delta_{inc} = 0.543 \text{ in.}(13.8 \text{ mm}) \le \frac{\ell}{240} = \frac{18\times 12}{240} = 0.9 \text{ in. } (22.9 \text{ mm}) \therefore \textbf{ OK}$$

Deflection calculated using Bischoff's expression for I_e gives approximately 60 percent more deflection than the value computed with Branson's approach. The immediate (ℓ/360) and long-term (ℓ /240) deflection requirements are satisfied with both approaches for this example when the full cracking moment M_{cr} is used in deflection calculations.

Calculation of deflection using $0.67M_{cr}$:

Immediate deflection due to live load (using $0.67M_{cr}$):

Using Branson's expression for I_e gives

$$\Delta_L = \frac{5M_L\ell^2}{48E_cI_e} = \frac{5\times 48{,}600\times(18\times 12)^2}{48\times 3.6\times 10^6\times 329} = 0.199 \text{ in. } (5.05 \text{ mm})$$

$$\Delta_L = 0.199 \text{ in.}(5.05 \text{ mm}) \le \frac{\ell}{360} = \frac{18\times 12}{360} = 0.6 \text{ in. } (15.2 \text{ mm}) \text{ for floors } \therefore \textbf{ OK}$$

Using Bischoff's expression for I_e gives

$$\Delta_L = \frac{5M_L\ell^2}{48E_cI_e} = \frac{5\times 48{,}600\times(18\times 12)^2}{48\times 3.6\times 10^6\times 231} = 0.283 \text{ in. } (7.19 \text{ mm})$$

$$\Delta_L = 0.283 \text{ in.}(7.19 \text{ mm}) \le \frac{\ell}{360} = \frac{18\times 12}{360} = 0.6 \text{ in. } (15.2 \text{ mm}) \text{ for floors } \therefore \textbf{ OK}$$

Short-term deflection under sustained loads (using $0.67M_{cr}$):

Using Branson's expression for I_e gives

$$\Delta_{sus} = \frac{5M_{sus}\ell^2}{48E_cI_e} = \frac{5 \times 70,520 \times (18 \times 12)^2}{48 \times 3.6 \times 10^6 \times 329} = 0.289 \text{ in. (7.35 mm)}$$

Using Bischoff's expression for I_e gives

$$\Delta_{sus} = \frac{5M_{sus}\ell^2}{48E_cI_e} = \frac{5 \times 70,520 \times (18 \times 12)^2}{48 \times 3.6 \times 10^6 \times 231} = 0.412 \text{ in. (10.47 mm)}$$

Long-term and incremental deflection (using $\lambda_\Delta = 2$ and $0.67M_{cr}$):

Using Branson's expression for I_e gives

$$\Delta_{inc} = \lambda_\Delta\Delta_{sus} + \Delta_{L,add} = 2.0 \times 0.289 + 0.8 \times 0.199 = 0.738 \text{ in. (18.75 mm)}$$

$$\Delta_{inc} = 0.738 \text{ in.(18.75 mm)} \leq \frac{\ell}{240} = \frac{18 \times 12}{240} = 0.9 \text{ in. (22.9 mm)} \therefore \textbf{OK}$$

Using Bischoff's expression for I_e gives

$$\Delta_{inc} = \lambda_\Delta\Delta_{sus} + \Delta_{L,add} = 2.0 \times 0.412 + 0.8 \times 0.283 = 1.05 \text{ in. (26.7 mm)}$$

$$\Delta_{inc} = 1.05 \text{ in.(26.7 mm)} \leq \frac{\ell}{240} = \frac{18 \times 12}{240} = 0.9 \text{ in. (22.9 mm)} \therefore \textbf{N.G.}$$

Using a reduced cracking moment equal to $0.67M_{cr}$ doubles the computed value of deflection for this example. Deflection calculated using Bischoff's expression for I_e gives approximately 40 percent more deflection than the value computed with Branson's approach. Long-term deflection requirements for this slab are no longer satisfied using Bischoff's expression for I_e, and a slightly thicker slab of 10.5 in. (267 mm) thickness is required.

Example A.2—Age-adjusted deflection of simply supported slab

Deflection calculations for the slab in Example A.1 are repeated to estimate the long-term age-adjusted values due to creep of the concrete. For this example, assume that the slab remains fully shored (that is, completed unloaded) until an age of 28 days after casting, at which time the loading becomes the member self-weight and the full service live load of $w_L = 100$ lb/ft^2 (4.8 kPa), which is sustained. The concrete has achieved a compressive strength of 4000 psi (27.6 MPa) at 28 days and, for simplicity in this example, further maturity of the concrete is assumed to have ceased at this same age. The total long-term deflection at an age 5 years later is to be determined using the age-adjusted modulus method described in 3.3.3. Use the approximation for I_e using Bischoff's expression (Eq. (4.3.2.2a)) and a cracking moment based on $(2/3)M_{cr}$ to account for restrained shrinkage.

Solution:

Similar to Example A.1, the relevant loads and section properties can be determined as follows:

Short-term material properties:

$E_c = 57{,}000\sqrt{f_c'} = 57{,}000\sqrt{4000} = 3.6 \times 10^6$ psi (24,900 MPa)

$E_s = 29 \times 10^6$ psi (200,000 MPa)

The total positive moment at midspan = $w\ell^2/8$ for a 1 ft (300 mm) wide strip of slab

$M_D + M_L = 9112.5$ ft-lb = 109,400 in.-lb (12.37 kN-m)

Creep coefficient and age-adjusted modulus

From Eq. (3.3.3a), the creep coefficient under standard conditions described in 3.3 at $t = 1825$ (5 years) can be estimated as

$$C_t = \left(\frac{t^{0.6}}{10 + t^{0.6}}\right) \times 2.35 \times 1.0 = 2.116$$

Using a typical aging coefficient value of $\chi = 0.8$, the age adjusted modulus from Eq. (3.3.3d) becomes

$$\bar{E}_c(1825, 28) = \frac{E_c(28)}{1 + \chi C_t(1825, 28)} = \frac{3.6 \times 10^6}{1 + 0.8 \times 2.116} = 1.34 \times 10^6 \text{psi (9220 MPa)}$$

The modular ratio n and I_{cr} associated with the aged adjusted modulus are

$$n = E_s/\bar{E}_c = (29 \times 10^6)/(1.34 \times 10^6) = 21.7$$

$$n\rho = 21.7 \times 0.317/100 = 0.069$$

$$c = \left(\sqrt{(n\rho)^2 + 2n\rho} - n\rho\right)d = \left(\sqrt{(0.069)^2 + 2 \times 0.069} - 0.069\right)9 = 2.78 \text{ in. (70.6 mm)}$$

$I_{cr} = bc^3/3 + nA_s(d - c)^2 = 12 \times 2.78^3/3 + 21.7 \times 0.343(9 - 2.78)^2 = 373$ in.^{4} (155.3×10^6 mm^4)

$I_g/I_{cr} = 2.68$

The age-adjusted effective moment of inertia I_e for full dead plus live load using Bischoff's expression and using $0.67M_{cr}$ to account for shrinkage restraint

$$I_{e,Bischoff} = \frac{I_{cr}}{\left[1 - \left(\frac{M_{cr}}{M_a}\right)^2\left(1 - \frac{I_{cr}}{I_g}\right)\right]} = \frac{373}{1 - 0.338 \times 0.627} = 473 \text{ in.}^4 \ (196.9 \times 10^6 \text{ mm}^4)$$

The corresponding total deflection at midspan becomes

$$\Delta_{D+L} = \frac{5}{48} \cdot \frac{M_{D+L} \cdot \ell^2}{(\bar{E}_c I_e)} = \frac{5}{48} \cdot \frac{109{,}400 \cdot (18 \times 12)^2}{(1.34 \times 10^6 \times 473)} = 0.84 \text{in. (21.3 mm)}$$

Thus, the age adjusted deflection is approximately 1.3 times larger than the short-term (instantaneous) deflection value of 0.64 in. (16.2 mm) from Example A.1.

Example A.3—Short- and long-term deflection of a four-span continuous beam

A reinforced concrete beam supporting a 4 in. (100 mm) slab is continuous over four equal spans $\ell = 36$ ft (11 m) as shown in Fig. A.3a (Nawy 2009). It is subjected to a uniformly distributed load $w_D = 700$ lb/ft (10.22 kN/m), including its self-weight and a service load $w_L = 1200$ lb/ft (17.52 kN/m). The beam has the dimensions $b = 14$ in. (356 mm), $d = 18.25$ in. (464 mm) at midspan, and a total thickness $h = 21.0$ in. (533 mm). The end span is reinforced with four No. 9 (28.6 mm diameter) bars at midspan at the bottom fibers and six No. 9 (28.6 mm diameter) bars at the top fibers of the support section (at the first interior support). Use the traditional ACI 318 method for long term effect discussed in 4.4.1.

Calculate the maximum deflection of the continuous beam using the ACI 318 method.

Given:
$f_c' = 4000$ psi (27.6 MPa), normalweight concrete
$f_y = 60{,}000$ psi (414 MPa)

50 percent of the live load is sustained for 36 months on the structure.

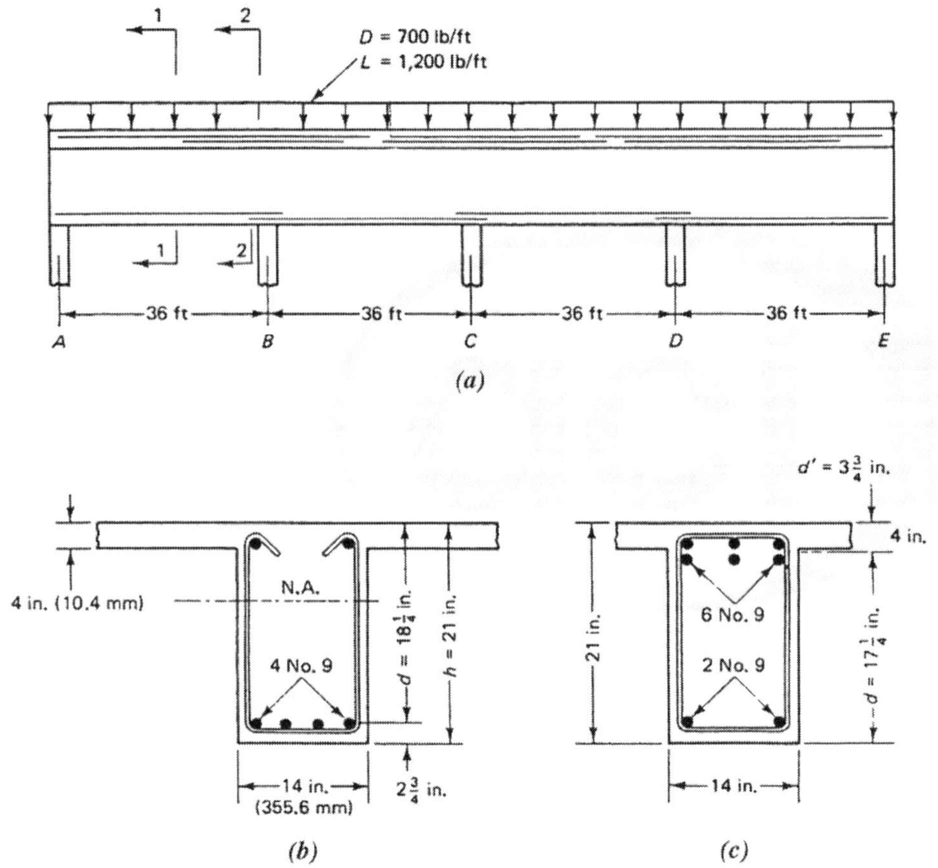

*Fig. A.3a—Details of continuous beam (*Nawy 2009; *image courtesy of Pearson/Prentiss Hall). (Note: 1 in. = 25.4 mm; 1 lb/ft = 14.6 N/m.)*

Solution—ACI method:

Note: All calculations are rounded to three significant figures.

Material properties and bending moment values

$E_c = 57{,}000\sqrt{f_c'} = 57{,}000\sqrt{4000} = 3.6 \times 10^6$ psi (24,900 MPa)
$E_s = 29 \times 10^6$ psi (200,000 MPa)
modular ratio $n = E_s/E_c = (29 \times 10^6)/(3.6 \times 10^6) = 8.1$
modulus of rupture $f_r = 7.5\sqrt{f_c'} = 7.5\sqrt{4000} = 474$ psi (3.3 MPa)
For the end span, the positive moment $= 0.0772w\ell^2$

$+M_D = 0.0772 \times 700(36.0)^2 \times 12 = 840{,}000$ in.-lb (94.9 kN-m)

$+M_L = 0.0772 \times 1200(36.0)^2 \times 12 = 1{,}440{,}000$ in.-lb (162.7 kN-m)
$+(M_D + M_L) = 0.0772 \times 1900(36.0)^2 \times 12 = 2{,}280{,}000$ in.-lb (257.7 kN-m)
For the negative moment at the first interior support = $0.1071w\ell^2$
$-M_D = 0.1071 \times 700(36.0)^2 \times 12 = 1{,}170{,}000$ in.-lb (132.2 kN-m)
$-M_L = 0.1071 \times 1200(36.0)^2 \times 12 = 2{,}000{,}000$ in.-lb (226.1 kN-m)
$-(M_D + M_L) = 0.1071 \times 1900(36.0)^2 \times 12 = 3{,}170{,}000$ in.-lb (358.3 kN-m)

Effective moment of inertia I_e

Figure A.3b shows the theoretical midspan and support cross sections to be used for calculating the gross moment of inertia I_g.

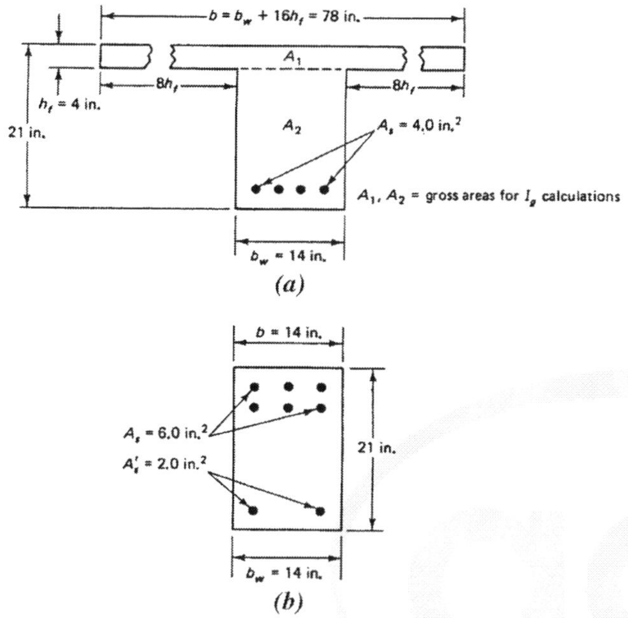

Fig. A.3b—Gross moment of interia I_g cross sections. (Note: 1 in. = 25.4 mm; 1 in.² = 645 mm².)

1. Midspan section:

width of T-beam flange = $b_w + 16h_f = 14.0 + 16 \times 4.0 = 78$ in. (1981 mm)

Uncracked section properties

Depth from compression flange to the elastic centroid is

$$\bar{y} = (A_1y_1 + A_2y_2)/(A_1 + A_2)\frac{78(4\times 2)+14\times(21-4)\times 12.5}{78\times 4+14\times 17} = 6.54 \text{ in. (166.1 mm)}$$

$y_t = h - \bar{y} = 21.0 - 6.54 = 14.5$ in. (368.3 mm)
$I_g = [78(4)^3]/12 + 78 \times 4(6.54 - 4/2)^2 + [14(21 - 4)^3]/12] + 14(21 - 4)(14.5 - [21 - 4]/2)^2 = 21{,}000$ in.⁴ (8.741×10^9 mm⁴)
$M_{cr} = f_rI_g/y_t = (474 \times 21{,}000)/14.5 = 686{,}000$ in.-lb (77.5 kN-m)

Cracked section properties

Depth of centroidal axis (A_s = four No. 9 [28.6 mm] bars = 4.0 in.² [2580 mm²]):

The location c of the centroidal axis for the cracked section is obtained by taking the moment of area of the transformed section about the top (compression) face. Assuming c lies within the flange thickness,
$b_fc^2/2 + nA_sc - nA_sd = 0$
$78c^2/2 + 8.1 \times 4 \times c - 8.1 \times 4 \times 18.25 = 0$

This gives $c = 3.5$ in. $\leq h_f = 4$ in. (89 mm $\leq$ 102 mm)

Hence, the centroidal axis is inside the flange and the flange section is analyzed as a cracked rectangular section.

$I_{cr} = [78(3.5)^3/3] + 8.1 \times 4(18.25 - 3.5)^2 = 8160$ in.⁴ (3.40×10^9 mm⁴)

M_{cr}/M_a ratio for dead load, dead plus sustained live load, and dead plus live load:
$M_{cr}/M_D = 686{,}000/840{,}000 = 0.82$
$M_{cr}/M_{D+0.5L} = 686{,}000/(840{,}000 + 0.5 \times 1{,}440{,}000) = 0.44$
$M_{cr}/M_{D+L} = 686{,}000/2{,}280{,}000 = 0.30$

Effective moment of inertia for midspan section:
$I_e = (M_{cr}/M_a)^3 I_g + [1 - (M_{cr}/M_a)^3] I_{cr} \leq I_g$
$I_{e,D} = 0.55 \times 21{,}000 + 0.45 \times 8160 = 15{,}200$ in.^{4} (6.33×10^9 mm^4)
$I_{e,D+0.5L} = 0.085 \times 21{,}000 + 0.915 \times 8160 = 9250$ in.^{4} (3.85×10^9 mm^4)
$I_{e,D+L} = 0.027 \times 21{,}000 + 0.973 \times 8160 = 8500$ in.^{4} (3.54×10^9 mm^4)

2. Support section:

Uncracked section properties

$I_g = b_w h^3/12 = 14(21)^3/12 = 10{,}800$ in.^{4} (4.50×10^9 mm^4)
$y_t = 21/2 = 10.5$ in. (266.7 mm)

$$M_{cr} = f_r I_g / y_t = \frac{474 \times 10.80}{10.5} = 488{,}000 \text{ in.-lb (55.2 kN-m)}$$

Cracked section properties

Depth of centroidal axis:
A_s = six No. 9 (28.6 mm) bars = 6.0 in.^{2} (3870 mm^2)
A_s' = two No. 9 (28.6 mm) bars = 2.0 in.^{2} (1290 mm^2)
$d = 21.0 - 3.75 = 17.25$ in. (438 mm)

A similar calculation for the centroidal axis depth c gives a value $c = 7.58$ in. (192.5 mm)

$I_{cr} = bc^3/3 + nA_s(d - c)^2 + (n - 1)A_s'(c - d')^2 = 6900$ in.^{4} (2.87×10^9 mm^4)

M_{cr}/M_a ratio for dead load, dead plus sustained live load, and dead plus live load
$M_{cr}/M_D = 488{,}000/1{,}170{,}000 = 0.42$
$M_{cr}/M_{D+0.5L} = 488{,}000/(1{,}170{,}000 + 0.5 \times 2{,}000{,}000) = 0.225$
$M_{cr}/M_{D+L} = 488{,}000 / 3{,}170{,}000 = 0.154$

Effective moment of inertia for support section:
$I_{e,D} = 0.07 \times 10{,}800 + 0.93 \times 6900 = 7170$ in.^{4} (2.98×10^9 mm^4)
$I_{e,D+0.5L} = 0.01 \times 10{,}800 + 0.99 \times 6900 = 6940$ in.^{4} (2.89×10^9 mm^4)
$I_{e,D+L} = 0.003 \times 10{,}800 + 0.997 \times 6900 = 6910$ in.^{4} (2.88×10^9 mm^4)

3. Average effective I_e for continuous span:

average: $I_e = 0.85 I_m + 0.15 I_{ec}$
$I_{e,D} = 0.85 \times 15{,}200 + 0.15 \times 7170 = 14{,}000$ in.^{4} (5.83×10^9 mm^4)
$I_{e,D+0.5L} = 0.85 \times 9250 + 0.15 \times 6940 = 8900$ in.^{4} (3.70×10^9 mm^4)
$I_{e,D+L} = 0.85 \times 8500 + 0.15 \times 6910 = 8260$ in.^{4} (3.44×10^9 mm^4)

4. Deflection calculations

Short-term deflection (maximum deflection of end span):

$\Delta = 0.0065 w \ell^4 / EI$

For all practical purposes, assume the span length ℓ equals the clear span ℓ_n

$$\Delta = \frac{0.065 w (36 \times 12)^4}{3.6 \times 10^6 \times I_e} \times \frac{1}{12} = 5.240 \frac{w}{I_e} \text{in.} \left(3.822 \times 10^9 \frac{w}{I_e} \text{ mm}\right)$$

Initial dead-load deflection:

$\Delta_D = 5.240(700)/14{,}000 = 0.26$ in. (6.6 mm)

Initial live-load deflection ($\Delta_L = \Delta_{L+D} - \Delta_D$):

$$\Delta_L = \frac{5.240(1900)}{8260} - \frac{5.240(700)}{14,000} = 1.21 - 0.26 = 0.95 \text{ in. (24.1 mm)}$$

Initial 50 percent sustained live-load deflection ($\Delta_{L,S} = \Delta_{0.5L+D} - \Delta_D$):

$$\Delta_{L,S} = \frac{5.240(1300)}{8900} - \frac{5.240(700)}{14,000} = 0.77 - 0.26 = 0.51 \text{ in. (12.9 mm)}$$

Long-term deflection:
$\rho' = A_s'/bd = 0$ (at midspan in this case)
multiplier $\lambda_\Delta = \xi/(1 + 50\rho')$

From Fig. 4.4.1,
$\xi = 1.75$ for 36-month sustained load
$\xi = 2.0$ for 5-year loading

Therefore, $\lambda_\infty = 2.0$ and $\lambda_t = 1.75$

The incremental deflection ($\Delta_{inc} = \lambda_\infty\Delta_D + \lambda_t\Delta_{LS} + \Delta_L$) is

$\Delta_{inc} = 2.0 \times 0.26 + 1.75 \times 0.51 + 0.95 = 1.41 + 0.95 = 2.36$ in. (59.9 mm)

Deflection requirements (Table 4.2.2c)
$\ell/180 = (36 \times 12)/180 = 2.4$ in. (61.0 mm) $> \Delta_L = 0.95$ in. (24.1 mm) **(OK)**
$\ell/360 = 1.2$ in. (30.5 mm) $> \Delta_L = 0.95$ in. (24.1 mm) **(OK)**
$\ell/480 = 0.9$ in. (22.9 mm) $< \Delta_{LT} = 2.36$ in. (59.9 mm) **(N.G.)**
$\ell/240 = 1.8$ in. (45.7 mm) $< \Delta_{LT} = 2.36$ in. (59.9 mm) **(N.G.)**

Hence, the continuous beam is limited to floors or roofs not supporting or attached to nonstructural elements such as partitions. For the alternative, compute the effective moment of inertia I_e as described in 4.3.2.2:

$$I_e = \frac{I_{cr}}{1-(M_{cr}/M_a)^2(1-I_{cr}/I_g)} \le I_g$$

Midspan section:
$1 - I_{cr}/I_g = 1 - 8160/21,000 = 0.61$
$I_{e,D} = 8160/(1 - 0.67 \times 0.61) = 13,800$ in.^{4} (5.74×10^9 mm^4) (9 percent smaller)
$I_{e,D+0.5L} = 8160/(1 - 0.19 \times 0.61) = 9230$ in.^{4} (3.84×10^9 mm^4) (within 1 percent)
$I_{e,D+L} = 8160/(1 - 0.09 \times 0.61) = 8630$ in.^{4} (3.59×10^9 mm^4) (1.5 percent larger)

Support section:
$1 - I_{cr}/I_g = 1 - 6900/10,800 = 0.36$
$I_{e,D} = 6900/(1 - 0.18 \times 0.36) = 7380$ in.^{4} (3.07×10^9 mm^4) (3 percent larger)
$I_{e,D+0.5L} = 6900/(1 - 0.05 \times 0.36) = 7020$ in.^{4} (2.92×10^9 mm^4) (1 percent difference)
$I_{e,D+L} = 6900/(1 - 0.02 \times 0.36) = 6950$ in.^{4} (2.89×10^9 mm^4) (0.5 percent difference)

Average effective I_e for continuous span ($I_e = 0.85I_m + 0.15I_{ec}$):
$I_{e,D} = (0.85 \times 13,800) + (0.15 \times 7380) = 12,800$ in.^{4} (5.33×10^9 mm^4)
$I_{e,D+0.5L} = (0.85 \times 9230) + (0.15 \times 7020) = 8900$ in.^{4} (3.70×10^9 mm^4)
$I_{e,D+L} = (0.85 \times 8630) + (0.15 \times 6950) = 8380$ in.^{4} (3.49×10^9 mm^4)

Midspan deflection:
$\Delta_D = 5.240(700)/12,800 = 0.29$ in. (7.4 mm)
$\Delta_L = 5.240(1900)/8380 - 5.240(700)/12,800 = 1.19 - 0.29 = 0.90$ in. (22.9 mm)
$\Delta_{L,S} = 5.240(1300)/8900 - 5.240(700)/12,800 = 0.77 - 0.29 = 0.48$ in. (12.2 mm)
$\Delta_{inc} = 2.0 \times 0.29 + 1.75 \times 0.48 + 0.90 = 1.42 + 0.90 = 2.32$ in. (58.9 mm)

There is very little difference between the two approaches for I_e used to compute deflection of the beam end span. Simply using the cracked moment of inertia at midspan (I_{cr} = 8160 in.^{4} [3.40 × 10^9 mm^4]) for this example gives Δ_{inc} = 2.35 in. (59.7 mm), which is also not very different from the computed value using I_e.

Example A.4—Temperature-induced deflections

These design examples illustrate the calculation procedures for temperature-induced deflections.

Example (a): Simply supported vertical wall panel with linear temperature gradient through the wall thickness

ΔT = 40°F (22.2°C)
α = 5.5 × 10^{-6} in./in./°F (9.9 × 10^{-6} mm/mm/°C)
h = 4 in. (101 mm)

a) Single-story span: ℓ = 12 ft (3.66 m)
Δ = (5.5 × 10^{-6} × 40 × 144^2)/(4 × 8) = 0.14 in. (3.6 mm)

b) Two-story span: ℓ = 24 ft (7.32 m)
Δ = (5.5 × 10^{-6} × 40 × 288^2)/(4 × 8) = 0.57 in. (14.5 mm)

Example (b): Simply supported T-section with linear temperature gradient over depth

ΔT = 40°F (22.2°C)
α = 5.5 × 10^{-6} in./in./°F (9.9 × 10^{-6} mm/mm/°C)
h = 36 in. (914 mm)
ℓ = 60 ft (18.3 m) simply supported
Δ = (5.5 × 10^{-6} × 40 × 720^2)/(36 × 8) = 0.40 in. (10 mm)

Example (c): Simply supported T-section with constant temperature increase over flange depth, web temperature remains constant

I = 69,319 in.^{4} (2.88 × 10^{10} mm^4)
y_t = 26.86 in. (682 mm)
ΔT = 40°F (22.2°C)
α = 5.5 × 10^{-6} in./in./°F (9.9 × 10^{-6} mm/mm/°C)
h = 36 in. (914 mm)
h_f = 3 in. (76 mm)
b = 96 in. (2438 mm)
ℓ = 60 ft (18.3 m)

$$\kappa = (\alpha/I)\int_0^{36} T(y)b(y)(y - y_t)dy = (\alpha/I)\int_{33}^{36}(40\times 96)(y - 26.86)dy$$

= (88,013 × 5.5 × 10^{-6})/69,319 = 6.98 × 10^{-6} 1/in. (2.75 × 10^{-7} 1/mm)
Δ = ($\kappa\ell^2$)/8 = (6.98 × 10^{-6} × 720^2)/8 = 0.45 in. (11.4 mm)

APPENDIX B—TWO-WAY SLAB DEFLECTION EXAMPLES

Example B.1—Deflection design example for long-term deflection of a two-way slab

The following example (Nawy 2009) illustrates the application of the equivalent frame approach developed by Nilson and Walters (1975), along with the modulus of rupture and long-term multiplier given in ACI 318.

A 7 in. (178 mm) slab of a five panel by five panel floor system spanning 25 ft (7.62 m) in the E-W direction and 20 ft (6.10 m) in the N-S direction is shown in Fig. B.1. The panel is monolithically supported by beams 15 x 27 in. (381 x 686 mm) in the E-W direction and 15 x 24 in. (381 x 610 mm) in the N-S direction. The floor is subjected to a time-dependent deflection due to an equivalent uniform working (service) load intensity w = 450 lb/ft^2 (21.5 kPa).

Material properties of the floor are:
f_c' = 4000 psi (27.6 MPa)
f_y = 60,000 psi (414 MPa)
E_c = 3.6 × 10^6 psi (24,900 MPa)

Assume:

1. Net moment from adjacent spans

E-W: Support 1: 20 × 10^3 (27.1 × 10^3); Support 2: 5 × 10^3 ft-lb (6.8 × 10^3 N-m)

N-S: Support 1: 40 × 10^3 (54.2 × 10^3); Support 4: 20 × 10^3 ft-lb (27.1 × 10^3 N-m)

2. Equivalent column stiffness, $K_{ec} = 400E_c$ lb-in. per radian ($6555E_c$ N-m/rad) in both directions.

3. Cracked moment of inertia:

E-W: I_{cr} = 45,500 in.^{4} (18.94 × 10^9 mm^4)

N-S: I_{cr} = 32,500 in.^{4} (13.53 × 10^9 mm^4)

Find the maximum central deflection of the panel due to the long-term loading and determine if its magnitude is acceptable if the floor supports sensitive equipment, which can be damaged by large deflections.

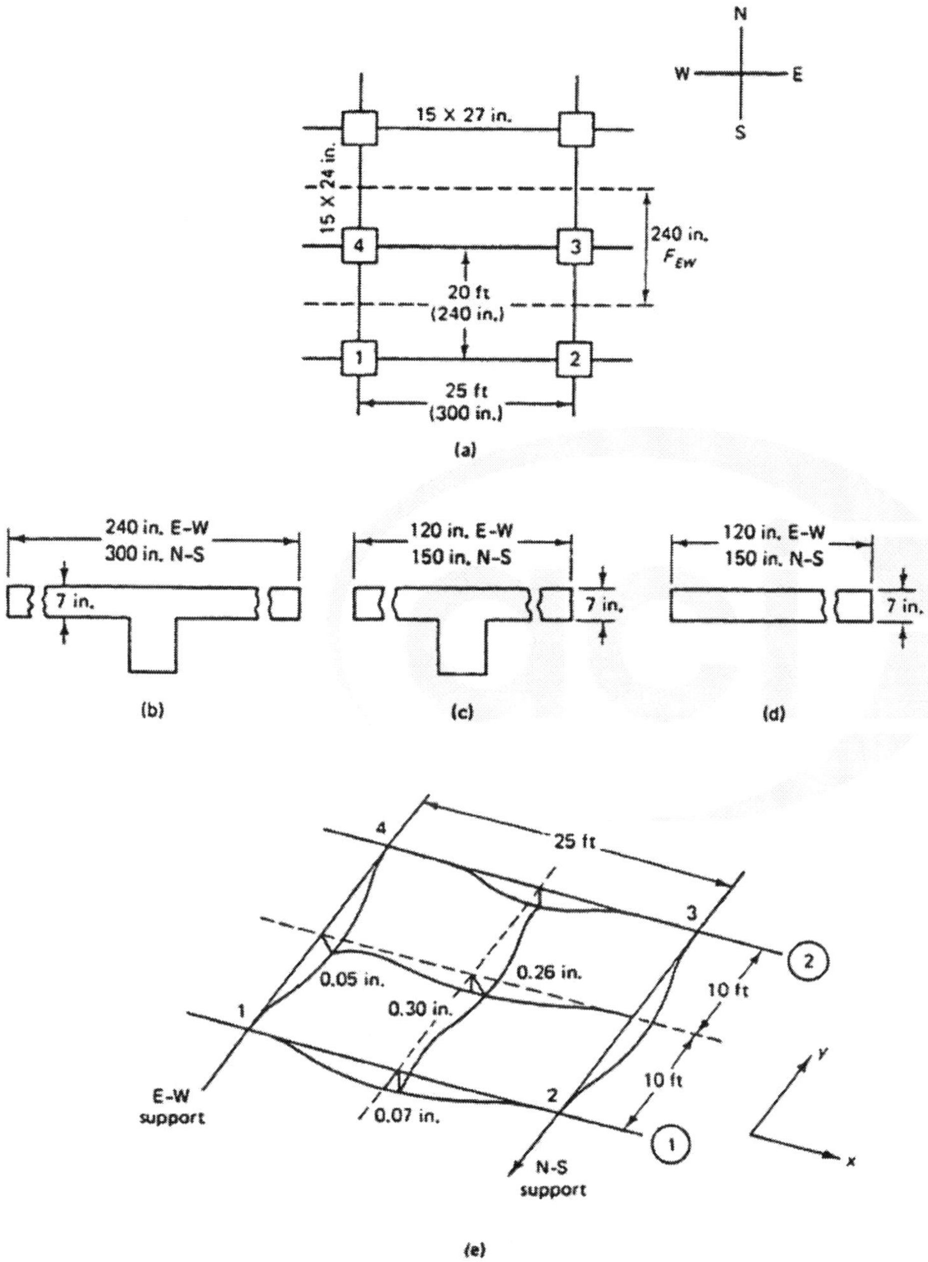

*Fig. B.1—Long-term deflection of two-way multi-panel slab on beams, equivalent frame calculation method (*Nawy 2009*; image courtesy of Prentice Hall). (Note: 1 in. = 25.4 mm; 1 ft = 0.305 m.)*

Solution:

Note: All calculations are rounded to three significant figures.

Calculate the gross moments of inertia (in.^{4} [mm^4]) of the sections in Fig. B.1, namely, the total equivalent frame I_{cs} in Part (b), the column strip beam I_c in Part (c), and the middle strip slab I_s in Part (d). These variables are:

	I_{cs}, in.[4]	I_c, in.[4]	I_s, in.[4]	I_{cr}, in.[4]
E-W	63,600	53,700	3430	45,500
N-S	47,000	40,000	4290	32,500

Note: 1 in.^{4} = 416,230 mm^4.

Next, calculate factors $\alpha_1\ell_2/\ell_1$ and $\alpha_2\ell_1/\ell_2$. In both cases, they are greater than 1.0. Hence, the factored moments coefficients (percent) obtained from the tables in 8.10 of ACI 318-14.

	Column strip (+ and –)	Middle strip (+ and –)
E-W	81.0	19.0
N-S	67.5	32.5

E-W direction deflections (span = 25 ft [7.62 m])

Long-term w = 450 lb/ft^2 (21.55 kPa)

$$\Delta'_{25} = \frac{450 \times 20(25)^4 \times 12^3}{384 \times 3.6 \times 10^6 \times 63{,}600} = 0.069 \text{ in. (1.8 mm)}$$

$$\Delta_c = 0.069 \times 0.81 \frac{63{,}600}{53{,}700} = 0.066 \text{ in. (1.7 mm)}$$

$$\Delta_s = 0.069 \times 0.19 \times \frac{63{,}600}{3430} = 0.243 \text{ in. (6.2 mm)}$$

Rotation at End 1 is

$$\theta_1 = \frac{M_1}{K_{ec}} = \frac{20 \times 10^3 \times 12}{400 \times 3.6 \times 10^6} = 1.67 \times 10^{-4} \text{rad}$$

and the rotation at End 2 is

$$\theta_2 = \frac{M_2}{K_{ec}} = \frac{5 \times 10^3 \times 12}{400 \times 3.6 \times 10^6} = 0.42 \times 10^{-4} \text{rad}$$

where θ is the rotation at one end if the other end is fixed.

Δ'' is the deflection adjustment due to rotation at supports 1 and 2 = $\theta\ell/8$
Δ'' = [(1.67 + 0.42) × 10^{-4} × 300]/8 = 0.008 in. (0.2 mm)

Therefore,
net Δ_{cx} = 0.066 + 0.008 = 0.074, say, 0.07 in. (1.8 mm)
net Δ_{sx} = 0.243 + 0.008 = 0.251, say, 0.25 in. (6.4 mm)

N-S direction deflections (span = 20 ft [6.10 m])

$$\Delta'_{20} = \frac{450 \times 25(20)^4 \times 12^3}{384 \times 3.6 \times 10^6 \times 47{,}000} = 0.048 \text{ in. (1.2 mm)}$$

$$\Delta_c = 0.048 \times 0.675 \times \frac{47{,}000}{40{,}000} = 0.038 \text{ in. (0.96 mm)}$$

$$\Delta_s = 0.048 \times 0.325 \times \frac{47{,}000}{4290} = 0.171 \text{ in. (4.3 mm)}$$

$$\text{rotation } \theta_1 = \frac{M_1}{K_{ec}} = \frac{40 \times 10^3 \times 12}{400 \times 3.6 \times 10^6} = 3.3 \times 10^{-4} \text{rad}$$

$$\text{rotation } \theta_4 = \frac{M_4}{K_{ec}} = \frac{20 \times 10^3 \times 12}{400 \times 3.6 \times 10^6} = 1.67 \times 10^{-4} \text{rad}$$

$$\Delta'' = \frac{\theta\ell_2}{8} = \frac{(3.3+1.67)10^{-4}\times 240}{8} = 0.015 \text{ in. } (0.38 \text{ mm})$$

Therefore,

net $\Delta_{cy} = 0.038 + 0.015 = 0.053$ say 0.05 in. (1.3 mm)
net $\Delta_{sy} = 0.171 + 0.015 = 0.186$ say 0.19 in. (4.8 mm)
total central deflection $\Delta = \Delta_{sx} + \Delta_{cy} + \Delta_{cx}$
$\Delta_{E\text{-}W} = \Delta_{sx} + \Delta_{cy} = 0.25 + 0.05 = 0.30$ in. (7.6 mm)
$\Delta_{N\text{-}S} = \Delta_{sy} + \Delta_{cx} = 0.19 + 0.07 = 0.26$ in. (6.6 mm)

Hence, the average deflection at the center of the interior panel $1/2(\Delta_{E\text{-}W} + \Delta_{N\text{-}S}) = 0.28$ in. (7.1 mm).

Adjustment for cracked section:

Use Branson's effective moment of inertia equation,

$$I_e = \left(\frac{M_{cr}}{M_a}\right)^3 I_g + \left[1-\left(\frac{M_{cr}}{M_a}\right)^3\right] I_{cr}$$

Calculation of ratio M_{cr}/M_a

$M_{cr} = f_r I_g / y_t$

where f_r is the modulus of rupture of concrete; y_t is the distance of center of gravity of section from outer tension fibers; E-W (240 in. [6096 mm] flange width): $y_t = 21.5$ in. (546 mm); N-S (300 in. [7620 mm] flange width): $y_t = 19.2$ in. (488 mm).

$f_r = 7.5\sqrt{f_c'} = 7.5\sqrt{4000} = 474$ psi (3.3 MPa)

Hence,

$$M_{cr}(\text{E-W}) = \frac{474\times 63{,}600}{21.5}\times\frac{1}{12} = 1.17\times 10^5 \text{ ft-lb } (159 \text{ kN-m})$$

$$M_{cr}(\text{N-S}) = \frac{474\times 47{,}000}{19.2}\times\frac{1}{12} = 0.97\times 10^5 \text{ ft-lb } (132 \text{ kN-m})$$

interior panel $M_a = w_s\ell^2/16 = [20 \times 450(25)^2]/16$ for E-W
$= 3.52 \times 10^5$ ft-lb (477 kN-m)
$= [25 \times 450(20)^2]/16$ for N-S
$= 2.81 \times 10^5$ ft-lb (381 kN-m)

Note that the moment factor 1/16 is used to be on the safe side, although the actual moment coefficients for two-way action would have been smaller.

E-W effective moment of inertia I_e
$M_{cr}/M_a = (1.17 \times 10^5)/(3.52 \times 10^5) = 0.332$
$(M_{cr}/M_a)^3 = 0.037$
$I_e = 0.037 \times 63{,}600 + (1 - 0.037)45{,}500 = 46{,}200$ in.^{4} (19.2×10^9 mm^4)

N-S effective moment of inertia I_e
$M_{cr}/M_a = (0.97 \times 10^5)/(2.81 \times 10^5) = 0.345$
$(M_{cr}/M_a)^3 = 0.041$
$I_e = 0.041 \times 47{,}000 + (1 - 0.041)32{,}500 = 33{,}100$ in.^{4} (13.8×10^9 mm^4)

Adjusted central deflection for cracked section effect $= 1.40 \times 0.28 = 0.39$ in. (10 mm)
$\ell/\Delta = [25 \times 12]/0.39 = 769 > 480$ allowed in Table 4.2.2c.

Hence, the long-term central deflection is acceptable.

Example B.2—Deflection calculation for a flat plate using the crossing beam method

An edge panel of 6 in. (150 mm) flat plate with multiple panels in each direction is shown in Fig. B.2.

The plate is supported on 16 × 16 in. (406 × 406 mm) columns. The slab is designed for an unfactored live load of 60 lb/ft² (2.87 kPa) in addition to its self-weight of 75 lb/ft² (3.59 kPa). Assume that the slab is subjected to significant in-plane restraint. Check the live load deflection and incremental deflection at midpanel if nonstructural components are installed 1 month after removal of shoring.

Material properties are:

f_c' = 3000 psi (20.7 MPa)
f_y = 60,000 psi (414 MPa)
$E_c = 3.12 \times 10^6$ psi (21,500 MPa)

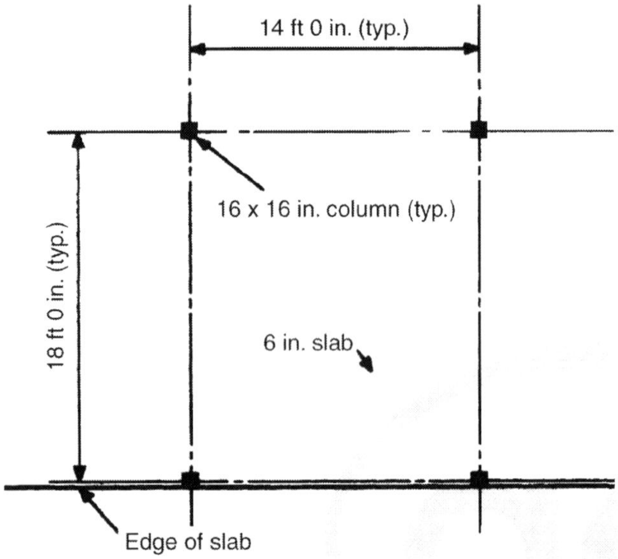

Fig. B.2—Plan of flat plate edge panel, beam crossing calculation method. (Note: 1 in. = 25.4 mm; 1 ft = 0.305 m.)

Using Eq. (5.2.1.2a), deflection of column and middle strips can be obtained from

$$\delta = \frac{5}{48}\frac{\ell_n^2}{E_c I_e}[M_m + 0.1(M_1 + M_2)]$$

in which moments and I_e are computed for a strip of unit width.

The midpanel deflection is computed as the sum of the column strip deflection in the N-S direction and the middle strip deflection in the E-W direction. Moments at unfactored load level due to dead plus live load are given in Table B.2.

Table B.2—Calcualtion of I_e for column and middle strips

	N-S column strip			E-W middle strip		
	Ext. Neg. (M_1)	Pos. (M_m)	Int. Neg. (M_2)	Neg. (M_1)	Pos. (M_m)	Neg. (M_2)
Moments (M_a), ft·kip/ft	2.45	2.93	4.94	0.72 (<M_{cr})	0.63 (<M_{cr})	0.72 (<M_{cr})
$(M_{cr}/M_a)^3$	0.154	0.09	0.02	—	—	—
$1 - (M_{cr}/M_a)^3$	0.846	0.91	0.98	—	—	—
I_{cr}, in.⁴	26.8	34.3	46.7	—	—	—
I_e, in.⁴	56.0	50.6	50.1	216 (= I_g)	216 (= I_g)	216 (= I_g)
I_e, average		52.2			216	

Note: 1 in.⁴ = 416,230 mm⁴; 1 ft·kip/ft = 4.448 kN-m/m.

Cracking moment (M_{cr})

$f_r = 4\sqrt{f_c'}$ (significant restraint) = 219 psi (1.51 MPa)
$M_{cr} = f_r I_g / y_t$
$I_g = 1/12(12)(6)^3 = 216$ in.⁴ (89.9×10^6 mm⁴)
$y_t = 6/2 = 3$ in. (76 mm)

M_{cr} = [(219)(216)/3](1/12,000) = 1.314 ft·kip/ft (5.844 kN-m/m)

Effective moment of inertia I_e

$$I_e = \left(\frac{M_{cr}}{M_a}\right) I_g + \left[1 - \left(\frac{M_{cr}}{M_a}\right)^3\right] I_{cr}$$

(Refer to Table B.2 for tabulated values)

Average I_e for column strip = 52.2 in.^{4} (21.7×10^6 mm^4)

Average I_e for middle strip = 216 in.^{4} (89.9×10^6 mm^4)

Column strip deflection

$$\Delta_c = \frac{(5)}{(48)}\frac{(16.7 \times 12)^2}{(3{,}120{,}000)(52.2)} \times [2.93 + 0.1(-2.45 - 4.94)](12{,}000) = 0.67 \text{ in. } (17.0 \text{ mm})$$

Middle strip deflection

$$\Delta_m = \frac{(5)}{(48)}\frac{(12.7 \times 12)^2}{(3{,}120{,}000)(216)} \times [0.63 + 0.1(-0.72 - 0.72)](12{,}000) = 0.02 \text{ in. } (0.51 \text{ mm})$$

Mid-panel deflection

$\Delta = \Delta_c + \Delta_m$ = 0.69 in. (17.5 mm)

Live load deflection

$$\Delta_L = \left(\frac{w_L}{w_D + w_L}\right) 0.69 = \frac{(60)}{(135)}(0.69) = 0.31 \text{ in. } (7.9 \text{ mm})$$

Span length on diagonal

$$= \sqrt{16.7^2 + 12.7^2} = 20.94 \text{ ft} = 251 \text{ in. } (6380 \text{ mm})$$

Permissible live load deflection
= ℓ/360 = 251/360
= 0.70 in. (17.8 mm)
> 0.31 in. (7.9 mm) … **OK** for short-term deflection

Incremental deflection
Use long-term multiplier = 2.5 applied to sustained load deflection.
Assume sustained load = 75 + 20 = 95 lb/ft^2 (4.55 kPa)
Additional long-term deflection = (2.5)(0.49) = 1.23 in. (31.2 mm)
Long-term deflection at 1 month = (0.5/2.0)(1.23) = 0.31 in. (7.9 mm)
Incremental deflection = 1.23 – 0.31 = 0.92 in. (23.4 mm)
Additional live load deflection = (40/135)(0.69) = 0.20 in. (5.1 mm)
Total = 1.12 in. (28.4 mm)
Permissible deflection = ℓ/480 = 251/480 = 0.53 in. (13.5 mm) < 1.12 in. (28.4 mm) ... **N.G.** for long-term deflection.

Hence, camber the slab or revise the design if nonstructural components are supported.

Example B.3—Minimum thickness calculation

The following example (Scanlon and Lee 2006) presents an alternative to the ACI 318 provisions for minimum thickness of a two-way system.

An interior panel of a floor system is shown in Fig. B.3. Equation (5.3g) is applied to six options, as summarized in Table B.3. The compressive strength of the concrete is 4000 psi (27.6 MPa) and the modulus of elasticity is 3.605×10^6 psi (24,900 MPa).

Table B.3—Factors and member dimensions: design examples

Case	Equation factors				Slab thickness, h_s, in.	Beam	
	β_{ES}	k_{DP}	k_{AR}	k_{SS}		b_w, in.	h_b, in.
1. One-way slab, short direction, narrow beam	1.0	1.0	1.0	1.0	7	12	24
2. One-way slab, short direction, wide beam	1.0	1.0	1.0	1.0	6	36	18
3. One-way slab, long direction, narrow beam	1.0	1.0	1.0	1.0	9	12	22
4. Flat plate	1.0	1.0	1.0	1.35	10	—	—
5. Flat slab	1.0	1.35	1.0	1.35	9	—	—
6. Two-way slab on beams	1.27	1.0	0.708	1.0	6	16	22

Note: 1 in = 25.4 mm.

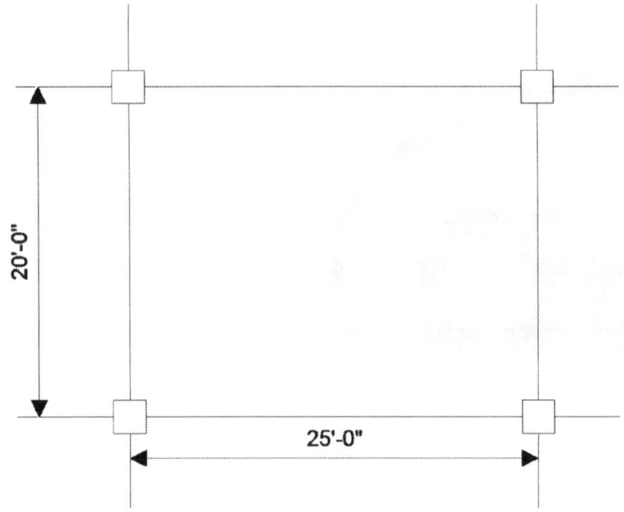

Fig. B.3—Interior panel of a two-way system. (Note: 1 ft = 0.305 m.)

Case 1: One-way slab, short direction, narrow beam (b_w = 12 in. [305 mm])

Estimate slab thickness: $h = \ell/28 = 20 \times 12/28 = 8.5$ in. (216 mm)
Slab self-weight = $(8.5/12) \times 150 = 106$ psf (5075 Pa)
Clear span $\ell_n = 20 - 1 = 19$ ft (5.79 m)
Sustained load $W_s = 106 + 15 + 30 = 151$ psf (7230 Pa)
Additional live load, $w_{L(add)} = 40$ psf (1915 Pa)

Slab:
Factors: $\beta_{ES} = 1.0$, $k_{DP} = 1.0$, $k_{AR} = 1.0$, $k_{SS} = 1.0$, $K = 1.4$

$$\frac{\ell_n}{h} = 1 \cdot \left[\frac{1}{480} \frac{2400 \times 3605000\left(\dfrac{12}{12}\right)}{(1.4)(2 \times 151 + 40)} \right]^{\frac{1}{3}} = 33.5$$

$h = 19 \times 12/33.5 = 6.8$ in. → Use $h = 7$ in. (178 mm)

(Note: a second iteration with the new depth gives $h = 6.51$ in. [165 mm])
(ACI 318 gives $h = \ell/28 = 8.5$ in. [216 mm] (for slab not supporting partitions)

Beam:
Clear span = 25 – 1 = 24 ft (7.32 m)
Tributary width = 20 ft (6.10 m)
Slab self-weight = (7/12) × 150 = 87.5 psf (4190 Pa)
Estimate weight of beam web = 400 plf (5840 N/m)
Sustained load = (87.5 + 15 + 30) × 20 + 400 = 3050 plf (44,500 N/m)
Additional live load = 40 × 20 = 800 plf (11,700 N/m)

$$\frac{\ell_n}{h} = 1 \cdot \left[\frac{1}{480} \frac{2400 \times 3605000\left(\frac{12}{12}\right)}{(1.4)(2 \times 3050 + 800)} \right]^{\frac{1}{3}} = 12.3$$

$h = 24 \times 12/12.3 = 23.4$ in. (594 mm) → Use $h = 24$ in. (610 mm)

(ACI 318 gives $h = \ell/21 = 11$ in. [279 mm] (for member not supporting partitions)

Cases 2 through 5

The calculation procedure is essentially the same for each case. A first estimate of slab thickness can be taken as $\ell/28$ in all cases. At most one iteration will be required for convergence.

Case 6: Two-way slab on beams (b_w = 16 in. [406 mm])
Clear span (long) = 25 – 1.33 = 23.67 ft (7.21 m)
Clear span (short) = 20 – 1.33 = 18.67 ft (5.69 m)
$\beta_{ES} = 23.67/18.67 = 1.27$
$k_{AR} = 0.2 + 0.4 \times 1.27 = 0.708$
$k_{DP} = 1.0$, $k_{SS} = 1.0$, $K = 1.4$

Slab:
Estimate slab depth = 24 × 12/30 = 9.6 in. (244 mm)
Slab self-weight = (9.6/12) × 150 = 120 psf (5750 Pa)
Sustained load w_s = 120 + 15 + 30 = 165 psf (7900 Pa)
Additional live load, $W_{L(add)}$ = 40 psf (1915 Pa)

$$\frac{\ell_n}{h} = 1.27 \cdot \left[\frac{1}{480} \frac{2400 \times 3605000\left(\frac{12}{12}\right)}{(0.708)(1.4)(2 \times 165 + 40)} \right]^{\frac{1}{3}} = 46.50$$

$h = (23.67 \times 12)/46.50 = 6.1$ in. (155 mm)

Try $h = 6$ in. (152 mm). Second iteration gives $h = 5.7$ in. (145 mm) → Use $h = 6$ in. (152 mm)

Beam (long span):
Clear span = 25 – 1 = 24 ft (7.32 m)
Tributary width = 20 ft (6.10 m)
Estimate beam web weight = 400 plf (5840 N/m)
Sustained load, W_S = (75 + 15 + 30) × 20 + 400 = 2800 plf (40900 N/m)
Additional live load, $W_{L(add)}$ = 40 × 20 = 800 plf (11700 N/m)

$$\frac{\ell_n}{h} = 1 \cdot \left[\frac{1}{480} \frac{2400 \times 3,605,000\left(\frac{16}{12}\right)}{(1.4)(2 \times 2800 + 800)} \right]^{\frac{1}{3}} = 13.9$$

$h = 24 \times 12/13.9 = 20.7$ in. (526 mm) → Use $h = 22$ in. (559 mm)

Beam (short span):
Clear span = 20 – 1 = 19 ft (5.79 m)
Tributary width = 25 ft (7.62 m)

$$\frac{\ell_n}{h} = 1 \cdot \left[\frac{1}{480} \frac{2400 \times 3{,}605{,}000(12/12)}{(1.4)(2 \times 3400 + 1000)} \right]^{\frac{1}{3}} = 11.8$$

$h = 19 \times 12/11.8 = 19.3$ in. (490 mm) → Use $h = 22$ in. (559 mm) (to match long span)

Observed Deflections of Reinforced Concrete Slab Systems, and Causes of Large Deflections

Reported by ACI Committee 435

D. R. Buettner*
Chairman

S. K. Ghosh*
Chairman, Subcommittee on Field Measurements

D. E. Branson
R. G. Drysdale*
A. Farah
A. B. Gogate
J. Grossman
C-T. T. Hsu
S. V. Kulkarni
M. S. Mirza*
E. G. Nawy*
M. V. Pregnoff
G. M. Sabnis*
C. G. Salmon
A. Scanlon*
J. M. Spang
M. K. Tadros
A. F. Shaikh
S. Zundelevich

*Members of the Subcommittee on Field Measurements which prepared this report.

This report is in two distinct parts.

Part I is a summary of published studies on slab deflections (3 from Australia, 1 from Scotland, 1 from Sweden, 2 from U. S.). The summary focuses on construction practices and materials quality. Comparison of deflections calculated by various methods with actual long-term deflections is made in some cases.

Part II summarizes several construction problems and material deficiencies which can contribute to large long-term deflections. Focusing on large construction loads, the authors show that construction loads may be considerably higher than design loads and that high construction loads cause high initial deflections because concrete has a lower modulus of elasticity when loaded at an early age. Furthermore, concrete creeps more when it is loaded at an early age, thereby causing additional high long-term deflections, even when construction loads are sustained only for a few days.

The authors then suggest a method of form removal and reshoring that has proved successful in the New York City area in preventing large slab deflections. Essentially, no more than an 8-foot slab span is left unsupported until a slab is mature.

A reader interested only in the Committee's findings and recommendations may proceed straight to Part II of the report.

Keywords: concrete construction; concrete slabs; creep properties; **deflection;** flat concrete plates; **form removal; loads (forces);** modulus of elasticity; reinforced concrete; **shoring;** shrinkage; two-way slabs.

This document is available as part of *Deflections of Concrete Structures*, SP-86. An abstract of SP-86 was published in the ACI JOURNAL, *Proceedings* V. 82, No. 4, July–August 1985.

Strength Evaluation of Existing Concrete Buildings

Reported by ACI Committee 437

Paul H. Ziehl*, Chair

Aaron K. Larosche*, Secretary

Nicholas J. Carino*
Paolo Casadei
John A. Frauenhoffer*
Nestore Galati*
Zareh B. Gregorian
Pawan R. Gupta
Frederick D. Heidbrink*
Jason S. Johnson*
Ashok M. Kakade
Danielle D. Kleinhans
Andrew T. Krauklis
Carl J. Larosche*
Daniel J. McCarthy*
Antonio Nanni*
Thomas E. Nehil
K. Nam Shiu
J. Gustavo Tumialan*
Jeffrey S. West

Consulting Members

Marco Arduini
Grant T. Halvorsen
Habib M. Zeim Alabideen

*Members of the committee who prepared this report.

The strength of existing concrete buildings and structures can be evaluated analytically and supplemented where necessary with load testing. The recommendations in this report indicate when such an evaluation may be needed, establish criteria for selecting the evaluation method, and indicate the data and background information necessary for an evaluation. Methods of determining material properties used in the analytical and load test investigations are described in detail. Analytical investigations should follow the principles of strength design. Working stress analysis can supplement the analytical investigations by relating the actual state of stress in structural components to the observed conditions.

Keywords: cracking; deflection; deformation; deterioration; gravity load; load; load test; reinforced concrete; strength; strength evaluation; test.

CONTENTS

ACI 437R-19 supersedes 437R-03 and was adopted and published May 2019.

CHAPTER 1—INTRODUCTION AND SCOPE

1.1—Introduction

This report defines the process of structural evaluation to determine the structural adequacy of existing concrete structures as defined by ACI 562. The procedures can be applied generally to new concrete structures, provided that appropriate evaluation criteria are agreed upon before the start of the investigation. This report covers structural concrete, including conventionally reinforced cast-in-place concrete, precast-prestressed concrete, precast post-tensioned concrete, and post-tensioned cast-in-place concrete.

1.2—Scope

1.2.1 *Background and limitations*—Engineering judgment based on rational, scientific principles is critical in the strength evaluation of concrete structures. Such judgment of a qualified licensed design professional is needed for proper application of relevant code provisions to the case being studied. An assessment of structural safety can be achieved with the information and data from field investigations, scientific computations based on sound principles, as well as subjective engineering judgment from the licensed design professional. This is particularly true for structures deteriorated from prolonged exposure to the environment or damaged in an extreme event, such as a fire, earthquake, or explosion.

Similarly, there are no generally recognized criteria for evaluating serviceability of an existing concrete structure. Such evaluation requires engineering judgment based on scientific principles, and close consultation with the owner regarding the intended use of the structure and expected level of performance.

ACI 562 discusses potential conclusions resulting from a strength evaluation; however, the determination of one or more of the following conclusions regarding the integrity of a concrete structure is possible from a strength evaluation:

a) The structure or structural element has an adequate margin of safety according to the provisions of the applicable building code.

b) The strength determined by evaluation is less than that required for factored loads but greater than required for service loads (load factors equal to or greater than 1.0 for all load cases). In this case, the structure or structural element is not adequate. In some cases, restricted use of the structure that limits the applied loads in recognition of the computed strength may be permitted.

c) The design strength of the structure is less than required for service loads under the applicable building code. In such cases, the owner should be notified and consideration given to the installation of shoring, severe restriction of use, or evacuation of the structure until remedial work can be done.

1.2.2 *Applications*—The procedures recommended in this report apply to strength evaluation of existing concrete buildings or other structures, including the following circumstances:

a) Structures that show damage from excess or improper loading, explosions, vibrations, fire, or other causes

b) Structures where there is evidence of deterioration or structural weakness, such as excessive cracking or spalling of the concrete, reinforcing bar corrosion, excessive member deflection or rotation, or other signs of distress

c) Structures that are suspected of not satisfying building code requirements in terms of design, materials, or construction

d) Structures where there is doubt as to the structural adequacy and the original design criteria are not known

e) Structures undergoing expansion or a change in use or occupancy and where the new design criteria exceed the original design criteria

f) Structures that require performance testing following remediation (repair or strengthening)

g) Structures that require testing by order of the building official

1.2.3 *Exceptions*—This report does not address the following conditions:

a) Performance testing of structures with unusual design concepts

b) Product development testing where load tests are carried out for quality control or approval of mass-produced elements

c) Evaluation of soil conditions

d) Load assessment for strength evaluation of environmental engineering concrete structures (refer to ACI 350 for additional information)

e) Liquefied gas containment structures (refer to ACI 376 for additional information)

1.2.4 *Categories of structural evaluation*—There are numerous different characteristics or levels of performance of an existing concrete structure that can be evaluated. These include:

a) Stability of the entire structure

b) Stability of individual components of the structure

c) Strength and safety of individual structural elements

d) Stiffness of the entire structure

e) Stiffness of individual structural elements

f) Susceptibility of individual structural elements to excessive long-term deformation

g) Dynamic response of individual structural elements

h) Fire resistance of the structure

i) Serviceability of the structure

j) Durability of the structure

This report deals with the evaluation of an existing concrete structure for stability, strength, and safety. Although not intended to be an in-depth review of durability, this report addresses durability-related aspects and notes significant features that could compromise structural performance, either at the time of the investigation or later.

1.2.5 *Procedure for a strength evaluation*—Most strength evaluations have many basic steps in common. Each evaluation, however, should address the unique characteristics of the structure in question and the specific concerns that have arisen regarding its structural integrity. Generally, the evaluation will consist of:

a) Defining the existing condition of the structure, including:

i. Reviewing available information

ii. Conducting a condition survey

iii. Determining the cause and rate of progression of existing distress

iv. Performing preliminary structural analysis
v. Determining the degree of repair to precede the evaluation
b) Selecting the structural elements that require detailed evaluation
c) Assessing past, present, and future loading conditions to which the structure has and will be exposed under anticipated use
d) Conducting the evaluation
e) Evaluating the results
f) Preparing a comprehensive report including description of procedure and findings of the previous steps

CHAPTER 2—DEFINITIONS

Please refer to the latest version of ACI Concrete Terminology for a comprehensive list of definitions.

CHAPTER 3—PRELIMINARY INVESTIGATION

This chapter describes the initial work that should be performed during a strength evaluation of an existing concrete structure. The objective of the preliminary investigation is to establish the existing condition of the structure to obtain a reliable assessment of the available structural capacity. This requires estimating the condition and strength of the concrete and the condition, location, strength, and area of reinforcement. Sources of information that should be reviewed before carrying out the condition survey are discussed. Available techniques for conducting a condition survey are described. Refer to ACI 562 for code requirements for the preliminary assessment.

3.1—Review of existing information

To learn as much as possible about the structure, the licensed design professional should research the history of the structure related to the design, construction, and service record. A thorough knowledge of the original design criteria minimizes the number of assumptions necessary to perform an analytical evaluation. The following list of possible information sources is intended as a guide. Not all of them need to be considered in a strength evaluation. The licensed design professional should exercise judgment in determining which sources need to be consulted for the specific strength evaluation being conducted.

3.1.1 *The original design*—Many sources of information are helpful in defining the parameters used in the original design, such as:
a) Architectural, structural, mechanical, electrical, and plumbing contract drawings and specifications
b) Structural design calculations
c) Change orders to the original contract drawings and specifications
d) Project communication records such as faxes, transcripts of telephone conversations, e-mails, and memoranda between the engineer of record and other consultants for the project
e) Records of the local building department
f) Geotechnical investigation reports including anticipated structure settlements
g) The structural design standards referenced by the local code at the time of design

3.1.2 *Construction materials*—Project documents should be checked to understand the types of materials that were specified and used for the structure, including:
a) Reports on the proportions and properties of the concrete mixtures, including information on the types of admixtures used and whether they contained more than negligible amounts of chlorides
b) Reinforcing steel mill test reports
c) Material shop drawings, including placing drawings prepared by suppliers that were used to place their products, bars, welded wire fabric, and prestressing steel; formwork drawings; and mechanical, electrical, and plumbing equipment drawings
d) Thickness and properties of any stay-in-place formwork, whether composite or noncomposite by design; such materials could include steel sheet metal and clay tile

3.1.3 *Construction records*—Documentation dating from original construction may be available such as:
a) Correspondence records of the design team, owner, general contractor, specialty subcontractors, and material suppliers and fabricators
b) Field inspection reports
c) Contractor and subcontractor daily records
d) Job progress photographs, films, and videos
e) Concrete cylinder compressive strength test reports
f) Field slump and air-content test reports
g) Delivery tickets from concrete trucks
h) As-built drawings
i) Survey notes and records
j) Reports filed by local building inspectors
k) Drawings and specifications kept in the trailers or offices of the contractor and the subcontractors during the construction period
l) Records of accounting departments that may indicate materials used in construction

3.1.4 *Design and construction personnel*—Another source of information concerning the design and construction of the structure under investigation is the individuals involved in those processes. Interviews often yield relevant information for a strength evaluation. This information can reveal problems, changes, or anomalies that occurred during design and construction.

3.1.5 *Service history of the structure*—This includes documents that define the history of the structure such as:
a) Records of current and former owners/occupants, their legal representatives, and their insurers
b) Maintenance records
c) Documents and records concerning previous repair and remodeling, including summaries of condition assessments and reports associated with the changes made
d) Records maintained by owners of adjacent structures
e) Weather records
f) Logs of seismic activity and activity or records of other extreme weather events, such as hurricanes (where applicable)
g) Photographs of the structure, including aerial photographs

3.2—Condition survey of structure

Areas of deterioration and distress in structural elements should be identified, inspected, and recorded as to type, location, and degree of severity. Procedures for performing condition surveys are described in this section. The reader should also refer to ACI 201.1R and ACI 364.1R. Engineering judgment should be exercised in performing a condition survey. All the steps outlined in the following may not be required in a particular strength evaluation. The engineer performing the evaluation should decide what information will be needed to determine the existing condition of structural elements of the particular structure being evaluated.

3.2.1 *Recognition of abnormalities*—A broad knowledge of the fundamental characteristics of structural concrete and the types of distress and defects that can be observed in a concrete structure is essential for a successful strength evaluation. Additional information on the causes and evaluation of concrete structural distress may be found in ACI 201.1R, ACI 207.3R, ACI 222R, ACI 222.2R, ACI 222.3R, ACI 224R, ACI 224.1R, ACI 224.4R, ACI 228.2R, ACI 309.2R, ACI 349.3R, ACI 362.2R, ACI 364.1R, ACI 423.4R, and ACI 423.8R, as well as documents of other organizations such as the International Concrete Repair Institute (ICRI).

3.2.2 *Visual examination*—Visual distress, deterioration, and damage existing in the structure should be located by means of a thorough visual inspection of the critical and representative structural components. Liberal use of photographs, notes, and sketches to document this examination is recommended. Abnormalities should be recorded as to type, magnitude, location, and severity.

If the engineer conducting the visual examination finds defects that render a portion or all of the structure unsafe, the condition should be reported immediately to the owner or building official. Appropriate temporary measures should be undertaken immediately to secure the structure before it is placed back into service and the survey continued.

To employ the analytical method of strength evaluation discussed in Chapter 6, it is necessary to obtain accurate information on the member properties, dimensions, and positioning of the structural components in the structure. If this information is incomplete or questionable, the missing information should be determined through a field survey. Verification of geometry and member dimensions by field measurement should be made for all critical members.

3.2.3 *In-place tests for estimating concrete compressive strength*—Numerous standard test methods are available for estimating the in-place concrete compressive strength or for determining relative concrete strengths within the structure. Traditionally, these have been called nondestructive tests to contrast them with drilling and testing core samples. A more descriptive term for these tests is in-place tests. Additional information on these methods can be found in ACI 228.1R, Malhotra (1976), Malhotra and Carino (2004), and Bungey et al. (2006).

The common feature of in-place tests is that they do not directly measure compressive strength of concrete. Rather, they measure some other property that has been found to have an empirical correlation with compressive strength. These methods are used to estimate compressive strength or to compare relative compressive strength at different locations in the structure.

If in-place tests are to be used for estimating in-place compressive strength, a strength relationship that correlates compressive strength and the test measurement should be developed by testing cores that have been drilled from areas adjacent to the in-place test locations. An attempt should be made to obtain paired data (core strength and in-place test results) from different parts of the structure to obtain the full range of in-place compressive strength. Regression analysis of the correlation data can be used to develop a prediction equation along with the confidence limits for the estimated strength. For a given test method, the strength relationship is influenced to different degrees by the specific constituents of the concrete. For accurate estimates of concrete strength, general correlation curves supplied with test equipment or developed from concrete other than that in the structure being evaluated should not be used unless they have been verified by comparison of estimated strengths with measured core strengths. Therefore, in-place testing can reduce the number of cores taken but cannot eliminate the need for drilling cores from the structure.

If in-place tests are to be used only to compare relative concrete strength in different parts of the structure, it is not necessary to develop the strength relationships. If the user is not aware of the factors that can influence the in-place test results, it is possible to draw erroneous conclusions concerning the relative in-place strength.

Sections 3.2.3.1 through 3.2.3.5 summarize a number of currently available in-place tests that have been adopted as ASTM test methods and highlight some factors that have a significant influence on test results. ACI 228.1R has detailed information on developing strength relationships and on the statistical methods that should be used to interpret the results.

3.2.3.1 *Rebound number*—Procedures for conducting this test are given in ASTM C805/C805M. The test instrument consists of a metal housing, a spring-loaded mass (the hammer), and a steel rod (the plunger). To perform a test, the plunger is placed perpendicular to the concrete surface and the instrument housing is pushed toward the concrete. This action causes the extension of a spring connected to the hammer. When the instrument is pushed to its limit, a catch is released and the hammer is propelled toward the concrete where it impacts a shoulder on the plunger. The hammer rebounds, and the rebound distance is measured on a scale numbered from 10 to 100. The rebound distance is recorded as the rebound number indicated on the scale, which represents the percentage of the original stretched length of the spring. A new instrument has been developed that measures a rebound index as the ratio of the speed of the hammer at rebound to the speed at impact. The rebound index determined in this manner is not affected by the orientation of the instrument during testing. For the same concrete, the two types of instruments do not result in the same value of the rebound index.

The rebound distance or speed depends on how much of the initial hammer energy is absorbed by the interaction of the plunger with the concrete. The greater the absorbed energy, the lower the rebound number will be. A simple, direct relationship between rebound number and compressive strength does not exist. It has been shown empirically, however, that for a given concrete mixture, there is good correlation between the gain in compressive strength and the increase in the rebound number.

The concrete in the immediate vicinity of the plunger has the greatest effect on a measured rebound number. For example, a test performed directly above a hard particle of coarse aggregate will result in a higher rebound number than a test over mortar. To account for the variations in local conditions, ASTM C805/C805M requires averaging 10 rebound readings for a test result. Procedures for discarding abnormally high or low individual rebound values are also given.

The rebound number reflects the properties of the concrete near the surface and may not be representative of the rebound value of the interior concrete. A surface layer of carbonated or deteriorated concrete results in a rebound number that does not represent interior concrete properties. Carbonation densifies the surface and will result in high rebound values. Heavily textured, soft, or surfaces with loose mortar require surface grinding before testing. Rebound number increases as the moisture content of concrete decreases, and test results on a dry surface will not be representative of interior concrete that is moist. For instruments based on measuring rebound distance, the direction of the instrument (sideward, upward, downward) affects the rebound distance, so this should be considered if comparing readings or using correlation relationships. Manufacturers provide correction factors to account for varying hammer positions, but it is good practice to verify their accuracy if possible.

The rebound number is a simple and economical method for quickly obtaining information about the near-surface concrete properties of a structural member. Factors identified in ASTM C805/C805M and ACI 228.1R should be considered when evaluating rebound number results. Unless a project-specific correlation is developed using cores, this method is recommended only for assessing uniformity and locating regions with abnormal rebound values.

3.2.3.2 *Probe penetration*—The procedures for this test method are given in ASTM C803/C803M. The device is known commercially as the Windsor probe. The test involves the use of a special powder-actuated gun to drive a hardened steel rod (probe) into the surface of a concrete member. The penetration of the probe into the concrete is taken as an indicator of concrete strength.

The probe penetration test is similar to the rebound number test, except that the probe impacts the concrete with a much higher energy level. A theoretical analysis of this test is complex. Qualitatively, it involves the initial kinetic energy of the probe and absorption of that kinetic energy by friction and failure of the concrete. As the probe penetrates the concrete, crushing of mortar and aggregate occurs along the penetration path and extensive fracturing occurs within a conical region around the probe. Hence, the strength properties of aggregates and mortar influence penetration depth. This contrasts with the behavior of ordinary strength concrete in a compression test, in which aggregate strength plays a secondary role compared with mortar strength. Thus, an important characteristic of the probe penetration test is that the type of coarse aggregate strongly affects the relationship between compressive strength and probe penetration.

Because the probe penetrates concrete, test results are not highly sensitive to local surface conditions such as texture and moisture content. The exposed lengths of the probes are measured, and a test result is the average of three probes located within 7 in. (180 mm) of each other. The probe penetration system has provisions to use a lower power level or a probe with larger tip diameter for testing relatively weak (less than 3000 psi [20 MPa]) or low-density (lightweight) concrete. Relationships between probe penetration and compressive strength are only valid for that specific power level and probe type.

In a manner similar to the rebound number test, this method is useful for comparing relative compressive strength at different locations in a structure. Strengths of cores taken from the structure and the statistical procedures detailed in ACI 228.1R are required to develop the correlation to permit compressive strength estimation on the basis of probe penetration results.

3.2.3.3 *Pulse velocity*—The procedures for this method are given in ASTM C597. The test equipment includes a transmitter, receiver, and electronic instrumentation. The test consists of measuring the time required for a pulse of ultrasonic stress-wave energy to travel through a concrete member. The ultrasonic energy is introduced into the concrete by the transmitting transducer, which is coupled to the surface with viscous acoustic couplant, such as petroleum jelly, water-soluble jelly, vacuum grease, or automotive grease. The pulse travels through the member and is detected by the receiving transducer, which is coupled to the opposite surface. The pulse transit time is measured by and displayed on associated instrumentation. The distance between the transducers is divided by the transit time to obtain the pulse velocity through the concrete under test. Most instruments allow the user to input the measured path length, and the pulse velocity is displayed along with the transit time.

The pulse velocity is proportional to the square root of the elastic modulus and inversely proportional to the square root of concrete density. The elastic modulus of concrete varies approximately in proportion to the square root of compressive strength. Hence, as concrete matures, large changes in compressive strength are accompanied by only minor changes in pulse velocity (ACI 228.1R). In addition, other factors affect pulse velocity, and these factors can easily overshadow changes due to strength. One such factor is moisture content. An increase in moisture content increases the pulse velocity, and this could be incorrectly interpreted as an increase in compressive strength. The presence of reinforcing steel aligned with the pulse travel path can also significantly increase pulse velocity. The user needs to be aware of the factors that can affect the measured pulse velocity and needs to ensure proper coupling to the concrete to obtain accurate values of the pulse velocity.

Under laboratory conditions, excellent correlations have been reported between pulse velocity and compressive strength development for a given concrete. These findings, however, should not be interpreted to mean that highly reliable estimates of in-place strength can be made routinely. Reliable strength estimates are possible only if correlation relationships include those characteristics of the in-place concrete that have a bearing on pulse velocity. Of paramount importance is the type and content of aggregate in the concrete, which have a strong effect on pulse velocity but generally have little effect on compressive strength for typical normal-density concrete. It is for this reason that the pulse velocity method is not generally recommended for estimating in-place strength. It is suitable for locating regions in a structure where the concrete is of a different quality or where there may be internal defects, such as cracking and honeycombing. It is not possible, however, to determine the nature of the defect based solely on the measured pulse velocity (3.2.5.2).

3.2.3.4 *Pullout test*—The pullout test consists of measuring the force required to pull an embedded metal insert out of a concrete member (refer to ACI 228.1R for illustration of this method). The force is applied by a jack that bears against the concrete surface through a reaction ring concentric with the insert. As the insert is extracted, a conical fragment of the concrete is also removed. The test produces a well-defined failure in the concrete and measures a static strength property. There is, however, no consensus on which strength property is measured and so a strength relationship should be developed between compressive strength and pullout strength (Stone and Carino 1983; Carino 2004a). The relationship is valid only for the particular test configuration and concrete materials used in the correlation testing. Compared with other in-place tests, strength relationships for the pullout test are affected little by details of the concrete proportions. The strength relationship, however, depends on aggregate density (lightweight or normal density) and maximum aggregate size if greater than 1-1/2 in. (40 mm).

ASTM C900 describes two procedures for performing pullout tests. In one procedure, the inserts are cast into the concrete during construction and the pullout strength is used to assess early-age in-place strength for timing of construction operations. The second procedure deals with post-installed inserts that can be used in existing construction. A commercial system is available for performing post-installed pullout tests (Petersen 1997), and the use of the system is described in ACI 228.1R.

Other types of pullout-type test configurations are available for existing construction (Mailhot et al. 1979; Chabowski and Bryden-Smith 1979; Domone and Castro 1987). These typically involve drilling a hole and inserting an anchorage device that will engage in the concrete and cause fracture in the concrete when the device is extracted. These methods, however, do not have the same failure mechanism as in the standard pullout test, and they have not been standardized by ASTM.

3.2.3.5 *Pull-off test for assessing in-place tensile strength of concrete and bond of overlay materials*—The procedures for the pull-off test are given in ASTM C1583/C1583M, and additional guidance on the use and interpretation of tests results may be found in ICRI 210.3. The test can be used to determine the near-surface tensile strength of concrete, the bond strength of a repair or overlay material to a concrete substrate, or the tensile strength of a repair or overlay material. This is a stand-alone test method and does not require the use of a preestablished strength relationship. The test is performed by drilling a shallow core perpendicular to the surface, leaving the intact core attached to the concrete substrate (Bungey and Madandoust 1992). After bonding a steel disk to the top surface of the core, a tensile load is applied to the disk until failure occurs. Alternatively, the disk can be bonded to the surface first and then the partial-depth core can be drilled. The tensile failure will occur in the weakest of four planes: in the concrete substrate; the interface between concrete and overlay; in the overlay material (if present); or at the interface of the steel disk and test surface. In preparing for the test, it is important that the disk axis be aligned with the direction of the applied tensile load and that the loading system does not introduce a bending moment during loading.

3.2.4 *In-place tests for locating reinforcing steel*—The size, number, and location of steel reinforcing bars need to be established to make an accurate assessment of structural capacity. In addition, embedded reinforcement needs to be located before drilling cores. A variety of electromagnetic devices, known as cover meters, are used for these purposes.

These devices have inherent limitations and it may be necessary to resort to radiographic methods for a reliable assessment of the reinforcement layout. Ground-penetrating radar (3.2.5.6) is also capable of locating embedded metallic objects, such as reinforcement. The following sections summarize these tools. Additional information can be found in ACI 228.2R, Malhotra and Carino (2004), and Bungey et al. (2006).

3.2.4.1 *Electromagnetic devices*—There are two general types of electromagnetic devices for locating reinforcement in concrete. One type is based on the principle of magnetic reluctance, which refers to the resistance in creating magnetic flux in a material and is analogous to electrical resistance in an electric circuit. These devices incorporate a U-shaped search head (yoke) that includes two electrical coils wound around an iron core. One coil supplies a low-frequency alternating current that results in an alternating magnetic field and an alternating magnetic flux flowing through the bar between the ends of the yoke. The other coil senses the magnitude of the flux. If a steel bar is located within the path of the flux, the reluctance decreases and the magnetic flux is increased. The sensing coil monitors the increase in flux. Thus, as the yoke is scanned over the surface of a concrete member, a maximum signal is noted on the meter display when the yoke lies directly over a steel bar. ACI 228.2R provides additional discussion for these types of meters. With proper correlation, these meters can estimate the depth of a bar if its size is known or estimate the bar size if the depth of cover is known. Bar size can also be estimated by a dual measurement technique (Tam et al. 1977). Dixon

(1987) and Snell et al. (1988) report additional details on using cover meters. Magnetic reluctance meters are affected by the presence of iron-bearing aggregates or the presence of strong magnetic fields from nearby electrical equipment.

The other type of cover meter is based on the principle of eddy currents. This type of cover meter employs a probe that includes a coil excited by a high-frequency electrical current. The alternating current sets up an alternating magnetic field. If this magnetic field encounters a metallic object, circulating currents are created in the surface of the metal. These are known as eddy currents. The alternating eddy currents, in turn, give rise to an alternating magnetic field that opposes the field created by the probe. As a result, the current through the coil decreases. By monitoring the current through the coil, the presence of a metal object can be detected. These devices are similar to a recreational metal detector. More advanced eddy-current instruments are based on the pulse-induction technique. In this case, a voltage pulse is applied to the coil. The decaying magnetic field, which is created when the voltage pulse is turned off, induces decaying eddy currents in the surface of the bar. The decaying eddy currents, in turn, generate a decaying magnetic field that induces a current in the coil. The amplitude of the measured induced current depends primarily on the depth of the bar. Smaller probes are available to discriminate individual closely-spaced bars, and larger probes are used to increase the maximum cover that can be measured, which is typically approximately 4 in. (100 mm). Bar sizes can be estimated, generally to within ±1 bar size, using special diameter probes or by making dual measurements over a bar with the sensor oriented alternately in two orthogonal directions. In the latter case, the ratio of the signal amplitudes from the two measurements can be related to the bar size.

An important distinction between these two types of meters is that reluctance meters detect only ferromagnetic objects, whereas eddy-current meters detect any type of electrically conductive metal. Cover meters are limited to detecting reinforcement located within approximately 4 in. (100 mm) of the exposed concrete surface. They are usually not effective in heavily reinforced sections, particularly sections with two or more adjacent or nearly adjacent layers of reinforcement. The ability to detect individual closely spaced bars depends on the design of the probe. Probes that can detect individual closely spaced bars, however, have limited depth of penetration. It is advisable to create a specimen composed of a bar embedded in a nonmagnetic and nonconductive material, such as dry sand, to verify that the device is operating correctly.

The accuracy of cover meters depends on the meter design, bar spacing, and thickness of concrete cover. The ratio of cover to bar spacing is an important parameter in terms of the measurement accuracy, and the manufacturer's instructions should be followed. It may be necessary to make a mockup of the member being tested to understand the limitations of the device, especially if more than one layer of reinforcement is present. Such mockups can be made by supporting bars in a plywood box or embedding bars in sand.

Results from cover meter surveys should be verified by drilling or chipping a selected area or areas as deemed necessary to confirm the measured concrete cover and bar size or develop improved project-specific correlations (3.2.4.4).

3.2.4.2 *Radiography*—By using penetrating radiation, such as X-rays or gamma rays, radiography can determine the position and configuration of embedded reinforcing steel, post-tensioning strands, and electrical wires (ACI 228.2R). As the radiation passes through the member, its intensity is reduced according to the thickness, density, and absorption characteristics of the material. The quantity of radiation passing through the member is recorded on film similar to that used in medical applications, or radiation amplitude can be recorded using special image plates that produce a two-dimensional digital array of the radiation amplitude (Mariscotti et al. 2009). The length of exposure is determined by the sensitivity of the recording media, strength of radiation, distance from the source to detector, and thickness of concrete. Reinforcing bars absorb more energy than the surrounding concrete and show up as light areas on the exposed film. Cracks and voids, on the other hand, absorb less radiation and show up as dark zones on the film. Crack planes parallel to the radiation direction are detected more readily than crack planes perpendicular to the radiation direction. With digitally recorded data, signal-processing tools can be used to extract quantitative information such as loss in cross-sectional area due to reinforcement corrosion (Mariscotti et al. 2009). If multiple images are obtained with the source at different locations, three-dimensional (3-D) images of the internal structure can be reconstructed (Mariscotti et al. 2009).

Due to the size and large electrical power requirements of X-ray units to penetrate concrete, the use of X-ray units in the field is limited. Therefore, radiography of concrete is generally performed using the man-made isotopes, such as Iridium 192 or Cobalt 60. Gamma rays result from the radioactive decay of unstable isotopes. As a result, a gamma ray source cannot be turned off, and extensive shielding is needed to contain the radiation when not in use for inspection. The shielding requirements make gamma ray sources heavy and bulky, especially if high penetrating ability is required.

The penetrating ability of gamma rays depends on the type and activity (age) of the isotope source. Iridium 192 is practical up to 8 in. (200 mm) and can be used on concrete up to 12 in. (300 mm) thick, if time and safety permit. Cobalt 60 is practical up to approximately 20 in. (0.5 m) thickness. Additional penetration depth up to approximately 24 in. (0.6 m) can be obtained by the use of intensifying screens next to the film. For thicker structural elements, such as beams and columns, a hole may be drilled and the source placed inside the member. The thickness that can be penetrated is a function of the time available to conduct the test. The area to be radiographed needs access from both sides.

Radiographic inspection poses health hazards and needs to be performed only by licensed and trained personnel. One drawback to radiography is that it can interrupt tenant or construction activities should the exposure area need to be evacuated during testing. Because of the high cost and safety concerns, radiography is usually the method of last resort if other methods are not successful.

As with other methods, results from radiographic tests should be verified by drilling or chipping selected areas as deemed necessary to confirm locations of reinforcing steel.

3.2.4.3 *Ground-penetrating radar*—Pulsed radar systems (3.2.5.6) can be used to locate embedded reinforcement. This method offers advantages over magnetic methods as a result of its greater penetration, which depends on the antenna frequency and the concrete moisture content. Access to one side of a member is all that is generally needed to perform an investigation. It is difficult to estimate bar size unless advanced processing of data is carried out (Zhan and Xie 2009). For a given antenna, the ability to discern individual bars in the received signal depends on the depth of the bars and their spacing. As cover increases, the bars have to be further apart to be discerned (Bungey et al. 1994). Interpretation of the results of a radar survey requires an experienced operator and should always be correlated to actual field measurements made by selected drilling or chipping.

3.2.4.4 *Removal of concrete cover*—This method removes the concrete cover to locate and determine the depth and size of embedded reinforcing steel, either by chipping or power drilling. This method is used primarily for verification of test results or for developing correlations with the results of the nondestructive methods outlined previously. Removal of concrete cover is the only reliable technique available to determine the condition of embedded reinforcing steel in deteriorated structures.

3.2.5 *Nondestructive tests for identifying internal abnormalities*—A strength evaluation may also involve determining if internal abnormalities exist that may reduce structural capacity, such as internal voids, cracks, or regions of inferior concrete quality. Compared with methods of strength estimation, some techniques for locating internal defects require complex instrumentation and specialized expertise to perform the tests and interpret the results. Refer to ACI 228.2R, Malhotra and Carino (2004), and Bungey et al. (2006) for additional information.

3.2.5.1 *Sounding*—Hollow areas or planes of delamination below the concrete surface can be detected by striking the surface with a hammer or a steel bar. A hollow or drum-like sound is heard when the surface over a hollow region, a delamination, or a region with shallow depth is struck. This compares with a higher-frequency ringing sound when the surface of undamaged and relatively thick concrete is struck. For slabs, such areas can be detected by dragging short lengths of steel chain over the concrete surface, unless the slab has a smooth, hard finish, in which case inadequate vibration is set up by the chain segments. Sounding is a simple and effective method for locating regions with subsurface fracture planes, but the sensitivity and reliability of the method decreases as the depth of the defect increases. For overhead applications, there are commercially available devices that use rotating sprockets on the end of a pole as a sounding method to detect delaminations. Procedures for using sounding in pavements and slabs can be found in ASTM D4580/D4580M.

3.2.5.2 *Pulse velocity*—The principle of pulse velocity is described in 3.2.3.3. Pulse travel time between the transmitting and receiving transducers is affected by the concrete properties along the travel path and the actual travel path distance. If there is a region of low-quality concrete between the transducers, the travel time increases and a lower velocity value is computed. If there is a relatively small void between the transducers, the pulse diffracts around the void as it travels through the concrete. This increases the actual path length and a lower pulse velocity is computed. While the pulse velocity method can be used to locate abnormal regions, it cannot identify the depth or the nature of the abnormality. Cores are often taken to determine the nature of the indicated abnormality.

3.2.5.3 *Impact-echo*—In the impact-echo method, a short-duration mechanical impact is applied to the concrete surface (Sansalone and Carino 1986). The impact generates stress waves that propagate away from the point of impact. The stress wave that propagates into the concrete is reflected if it encounters an interface between the concrete and a material with different acoustic properties. If the interface is between concrete and air, almost complete reflection occurs. The reflected stress wave travels back to the surface, where it is again reflected into the concrete and the cycle repeats. A receiving transducer located near the impact point monitors the surface movement resulting from the periodic arrival of the reflected stress wave. The transducer signal is recorded as a function of time from which the depth of the reflecting interface can be determined. If there is no defect, the thickness of the member can be determined, provided the thickness is small compared with the other dimensions. In general, the thickness has to be less than 20 percent of the smallest lateral dimension for the response to be dominated by reflections from the back wall. If this condition is not satisfied, reflections from the side boundaries will interfere with reflections from the back wall and interpretation of results becomes complicated.

Because the stress wave undergoes multiple reflections between the test surface and the internal reflecting interface, the recorded waveform is periodic. If the waveform is transformed into the frequency domain, the periodic nature of the waveform appears as a dominant peak in the amplitude spectrum (Carino et al. 1986). The frequency of that peak can be related to the depth of the reflecting interface by a simple relationship (Sansalone and Streett 1997).

Impact-echo can be used to measure the thickness of plate-like elements if there is access to only one face (ASTM C1383). A plate-like element is one in which the smallest lateral dimension is at least five times the thickness to be measured. Two procedures are required to measure the thickness. The first is to determine the stress-wave speed in the concrete, and the second is to measure the thickness frequency of the plate and calculate thickness. The wave speed can be established by a surface measurement technique using two transducers (Sansalone and Streett 1997), or it can be determined by measuring the thickness frequency, drilling a hole at the test point, measuring the thickness at the hole, and calculating the wave speed from the measured frequency and thickness.

The impact-echo method can be used to detect internal abnormalities and defects, such as delaminations, regions of honeycombing, voids in grouted tendon ducts, subgrade voids, and the quality of interfaces in bonded overlays (Sansalone and Carino 1988, 1989; Jaeger et al. 1996; Wouters et al. 1999; Lin and Sansalone 1996). The test provides information on the conditions in the region directly below the receiving transducer and impact point. Thus, an impact-echo survey typically comprises many tests on a predefined grid. Care is required to establish the optimal spacing between test points (Kesner et al. 1999). The degree of success in a specific application depends on factors such as the shape of the member, the nature of the defect, and the experience of the operator. It is important that the operator understands how to select the impact duration and how to recognize invalid waveforms that result from improper seating of the transducer or improper impact (Sansalone and Streett 1997). Standardized test methods (ASTM) have not been developed for internal defect detection using the impact-echo method.

Devices are available, or are being developed, that will speed up the data acquisition process so that many points can be tested in a short period of time. Some devices incorporate a rolling transducer and an electro-mechanical impactor whereas others use a support frame with an electric motor to move a transducer-impactor assembly along a straight line and make measurements at selected points along the line. Software uses the closely-spaced, impact-echo data to reconstruct images of reflecting interfaces.

3.2.5.4 *Impulse-response*—The impulse-response method is similar to the impact-echo method, except that a longer-duration impact is used, and the time history of the impact force is measured. The method measures the vibrational response of the portion of the structure surrounding the impact point (Davis et al. 1997). Measured response and the force history are used to calculate the impulse response spectrum of the structure (Carino 2004b). Depending on the quantity (displacement, velocity, or acceleration) measured by the transducer, the response spectrum has different meanings. Typically, the velocity of the surface is measured and the response spectrum represents the mobility (velocity/force) of the structure, which is affected by the geometry of the structure, the support conditions, and defects that affect the dynamic stiffness of the structure. An experienced user can extract several measures of the vibrational response that can be used to compare conditions at different test points (Davis and Dunn 1974; ACI 228.2R; Davis and Hertlein 1995).

ASTM C1740 provides a standard practice for impulse response testing. Parameters extracted from the impulse response test are plotted as contour plots from which it is possible to identify testing locations that have a different response from the rest of the structure. Those anomalous points can be subjected to more detailed investigation and the rest of the structure with similar and acceptable response can be assumed to be sound. Coring or other forms of invasive probing should be performed at the good and flawed locations to confirm the interpretations.

An impulse-response test result is affected by a larger volume of the structure surrounding the test point than the impact-echo method, but the test result cannot define the exact location or depth of a hidden defect. As a result, impulse-response testing is often used as a rapid screening method in conjunction with impact-echo or ultrasonic-echo testing for detailed investigation at identified anomalous locations.

3.2.5.5 *Ultrasonic-echo*—The ultrasonic-echo method is a time-domain, reflection-based, stress-wave method for locating reflecting interfaces within a concrete member. It is based on the pitch-catch principle in which one transducer sends out a stress-wave pulse and another receives the reflected pulse, with both transducers being located on the same surface (ACI 228.2R). The travel time from the transmitter to the receiver is measured and, based on the wave speed and distance between the transducers, the depth of the reflecting interface can be calculated as explained in ACI 228.2R.

In one commercial device that has been developed, a computer-controlled antenna composed of a 4 x 12 array of transducers permits measurements of travel times between different pairs of transducers to be made in a few seconds. The transducers are point shear-wave transducers that do not require a viscous coupling fluid. The antenna array measures 3 x 13 in. (75 x 330 mm) and, in effect, it looks into the concrete beneath the antenna. The system can measure the concrete shear-wave speed or the user can provide an approximate value. An error in the wave speed value will only affect the indicated depth of the reflecting interfaces.

The computer in the antenna controls the operation of the 12 rows of transducers so that each row functions sequentially as a transmitter while the other rows act as receivers. The multiple travel time measurements from the various transducer pairs are used as input to a signal-processing technique called synthetic aperture focusing. The end result of the signal processing is an averaged two-dimensional (2-D) image of the reflecting interfaces in the volume of concrete below the antenna.

To test a concrete member, a grid is marked on the test surface and the grid spacing is entered into the instrument. At each test location, a 2-D cross-sectional image of the reflecting interface is computed. After the entire grid has been tested, the 2-D images are transferred to a computer along with the grid geometry. The computer uses 3-D visualization software to stitch together the 2-D images and create a 3-D representation of the reflecting interfaces in the concrete below the test grid. The process is analogous to the method used in medical imaging to reconstruct 3-D images of internal organs from 2-D images at different cross sections (tomograms).

The ultrasonic-echo instrument is able to detect the same types of defects as the impact-echo method, but it can do so more rapidly. It has proven useful for locating voids in grouted tendon ducts of post-tensioned members. Compared with impact-echo, the antenna of the ultrasonic-echo system does not have to be directly over the tendon duct to obtain a clear signal. The center frequency of the transmitted pulse in the commercial system can be changed from 25 to 85 kHz, which provides control over resolution and penetration depth.

3.2.5.6 *Ground-penetrating radar*—This method is similar in principle to the other echo techniques except that electromagnetic energy is introduced into the material rather than mechanical energy. An antenna placed on the concrete surface (surface coupled antenna) sends out an extremely short-duration radio frequency pulse. The receiver in the antenna registers a strong signal due to the direct coupling of the transmitter and receiver through the near-surface concrete. The pulse travels into the concrete, and if the member contains boundaries between materials with different electrical properties, some of the energy in the pulse is reflected back to the antenna. The electrical property of interest is called the dielectric constant, which is related to the ability of a material to store electrical charge and it affects the speed of the pulse in the material (ACI 228.2R). Knowing the speed of the electromagnetic pulse in the concrete, the depth of the reflecting interface can be determined from the arrival time of the reflected pulse (ACI 228.2R). A digital recording system generates a profile view of the reflecting interfaces within the member as the antenna is moved over the surface. Changes in the reflection patterns indicate buried items, voids, and thickness of the member. Interpretation of the recorded profiles is the most difficult aspect of using commercially available radar systems. This method has been used successfully to locate embedded items, such as reinforcing steel and ducts, to locate regions of deterioration and voids or honeycombing, and to measure member thickness if access is limited to one side. The penetrating ability of the pulse depends on the electrical conductivity of the material and the frequency of the electromagnetic radiation. As electrical conductivity increases, pulse penetration decreases. In testing concrete, higher moisture content increases conductivity and reduces pulse penetration.

There are two ASTM standards on the use of ground-penetrating radar (GPR), both of which have been developed for highway applications. ASTM D4748 measures the thickness of bound pavement layers, and ASTM D6087 identifies the presence of delaminations in asphalt-covered bridge decks. With proper adaptation, these standards can be applicable to condition assessment in building structures. The Federal Communications Commission (FCC 2002) published rules that regulate the purchase and use of GPR equipment. Purchasers of GPR equipment are required to register their devices with the FCC and indicate where the devices will be used.

Ground-penetrating radar is especially helpful in locating reinforcing bars so that cores can, perhaps, be obtained free of reinforcement. As described in 3.2.4.3, GPR is an option for locating reinforcing bars. This application can be used to determine the reinforcement layout so that cores can be obtained free of reinforcement. In the simplest form of this application, GPR scans are performed in the region of interest and the detected reinforcement locations are marked on the surface of the concrete. Where more complete information is required, some GPR equipment can be used to create a 3-D model of the reinforcing steel in a section of the structure. This is accomplished by placing a plastic sheet with a permanently marked grid on the concrete surface. The antenna, with a distance-measuring transducer, is scanned along the gridlines in perpendicular directions. The acquired reflection data are manipulated by software to create a 3-D model of the portion of the structure below the test grid. The 3-D model can be manipulated to show the locations of the centerlines of the reinforcing bars. The sizes of the bars shown on the reconstructed images are, however, not to be interpreted as actual bar sizes. The image provides only the centerlines of the bars. Sophisticated signal processing is needed to estimate bar sizes from GPR data (Zhan and Xie 2009). Some GPR systems include an additional sensor that measures the magnetic field associated with alternating current in a metallic conductor. Such systems are useful to avoid cutting into live electrical conductors in the process of sampling concrete.

3.2.5.7 *Infrared thermography*—A surface having a temperature above absolute zero emits electromagnetic energy. At room temperature, the wavelength of this radiation is in the infrared region of the electromagnetic spectrum. The rate of energy emission from the surface depends on its temperature, so by using infrared detectors it is possible to notice differences in surface temperature. If a concrete member contains an internal defect, such as a large crack or void, and there is heat flow through the member, the presence of the defect can influence the temperature of the surface above the defect. A picture of the surface temperature can be created by using an infrared camera, thereby enabling the location of hot or cold spots on the surface. The locations of these hot and cold spots serve as indications of the locations of internal near-surface defects in the concrete. The technique has been used successfully to locate regions of delamination in concrete pavements and bridge decks (ASTM D4788).

Heat flow through the member must be present to use infrared thermography. This can be achieved by the natural heating from sunlight or by applying a heat source to one side of the member. In addition, the member surface should be of one material and have a uniform value of a property known as emissivity, which is a measure of the efficiency of energy radiation by the surface. Changes in emissivity cause changes in the rate of energy radiation, which can be incorrectly interpreted as changes in surface temperature. The presence of foreign material on the surface, such as paint or grease, will affect the results of infrared thermography by changing the emissivity and, therefore, the apparent temperature of the surface. It is often useful to take a photographic or video record of the areas of the concrete surface being imaged by infrared photography. By comparing the two images, surface defects with different emissivity can be eliminated from consideration as internal defects in the concrete.

3.2.5.8 *Radiography*—As discussed in 3.2.4.2, radiography can be used to determine the position and location of embedded reinforcing steel. Radiography can also be used to determine the internal condition of a structural member. As described previously, reinforcing bars absorb more energy than the surrounding concrete and show up as light areas on exposed film or digital images. Cracks and voids, on the other hand, absorb less radiation and show up as dark areas in the images. Crack planes parallel to the radiation direction are detected more readily than cracks perpendicular to the radiation direction.

CHAPTER 4—METHODS FOR MATERIAL EVALUATION

This chapter describes procedures to assess the quality and mechanical properties of the concrete and reinforcing steel in a structure. These procedures are often used to corroborate the results of in-place or nondestructive methods mentioned in Chapter 3. Sampling techniques, petrographic and chemical analyses of concrete, and test methods are discussed. When material properties are not available or not found on drawings or specifications, values specified in ACI 562 may be considered. If material properties obtained from project documents are used, ACI 562 notes that additional testing may be required to confirm the information if degradation has occurred.

4.1—Concrete

The compressive strength of concrete is the most significant concrete property with regards to the strength evaluation of concrete structures. In-place concrete strength is a function of several factors, including the concrete mixture proportions, curing conditions, degree of consolidation, location within the structure, and deterioration over time. The following sections describe the physical sampling and direct testing of concrete to assess concrete strength. The condition of the concrete and extent of distress is indirectly assessed by strength testing because deterioration results in a strength reduction. An evaluation of the condition of the concrete and causes of deterioration may be obtained directly from petrographic and chemical analyses of the concrete.

4.1.1 *Guidelines on sampling concrete*—The licensed design professional is responsible for determining the locations and the number of samples to be taken for establishing the in-place characteristics of the concrete (ACI 562). It is essential that the concrete samples be obtained, handled, identified (labeled), and stored properly to prevent damage or contamination. Sampling techniques are discussed in this section.

Guidance on developing an appropriate sampling program is provided by ASTM C823/C823M. Samples are usually taken to obtain statistical information about the properties of concrete in the entire structure, for correlation with in-place tests covered in Chapter 3, or to characterize some unusual or extreme conditions in specific portions of the structure (Bartlett and MacGregor 1996, 1997). For statistical information, sample locations should be randomly distributed throughout the structure. The number and size of samples depends on the necessary laboratory tests and the degree of confidence desired in the average values obtained from the tests.

The type of sampling plan that is required on a particular project depends on whether the concrete is believed to be uniform or if there are likely to be two or more regions that are different in composition, condition, or quality. In general, a preliminary investigation should be performed and other sources of information should be considered before a detailed sampling plan is prepared. Where a property is believed to be uniform, sampling locations should be distributed randomly throughout the area of interest and all data treated as one group. Otherwise, the study area should be subdivided into regions believed to be relatively uniform, with each region sampled and analyzed separately.

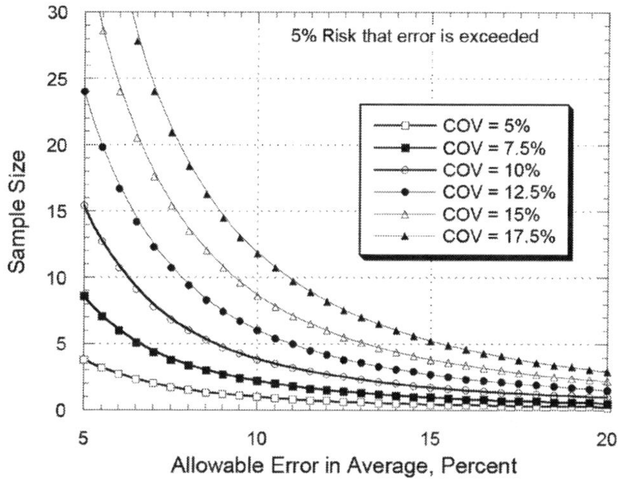

Fig. 4.1.1—Sample size based on ASTM E122; risk = 5 percent.

There are different approaches for choosing sample locations. Williams et al. (2006a,b) discuss the use of random, stratified, and adaptive sampling methods for planning nondestructive testing investigations of large concrete structures to determine the proportion of the structure that is flawed. Adaptive sampling is a method in which the next sampling location is based on the results from previous locations. From limited case studies, Williams et al. (2006a,b) determined that random sampling produced the narrowest confidence intervals, but noted that access and cost considerations may not always allow random sampling.

For tests intended to measure the average value of a concrete property, such as strength, elastic modulus, or air content, the number of samples generally depends on:

a) The maximum allowable difference (or error) between the sample average and the true average
b) The variability of the test results
c) The acceptable risk that the maximum allowable difference is exceeded

These factors can be taken into account by using ASTM E122 to estimate the sample size, as shown in Fig. 4.1.1. The vertical axis gives the number of samples needed as a function of:

a) The maximum allowable difference (as a percentage of the true average) between the sample average and true average
b) The coefficient of variation of the test results

In Fig. 4.1.1, the risk that the maximum allowable error will be exceeded is 5 percent, but other levels can be used. Because the variability of test results is usually not known in advance, an initial estimate can be made and adjusted as test results become available.

Economy should also be considered in the selection of sample sizes. In general, uncertainty in an average value is related to the inverse of the square root of the number of results used to compute that average. For large sample sizes, an increase in the sample size will result in only a small decrease in the risk that the acceptable error is exceeded. The cost of additional sampling and testing would not be justified in these situations.

Concrete is neither isotropic nor homogenous, and so its properties will vary depending on the direction that samples are taken and the position within a member. Close attention should be given to vertical concrete members, such as columns, walls, and deep beams, because concrete properties will vary with elevation due to differences in placing and compaction procedures, segregation, and bleeding. Typically, the strength of concrete at the bottom of a placement will be greater than at the top of the placement (Bartlett and MacGregor 1999).

4.1.1.1 *Core sampling*—The procedures for removing concrete samples by core drilling are described in ASTM C42/C42M, and ACI 214.4R provides additional guidance on factors to consider in obtaining cores. The following guidelines are of particular importance in core sampling:

a) Cores should be taken using water-cooled, diamond-studded core bits. Drills should be in good operating condition and supported rigidly so that the cut surfaces of the cores will be as straight as possible.

b) The number, size, and location of core samples should be selected to permit all necessary laboratory tests. If possible, use separate cores for different tests so that there will be no influence from prior tests.

c) Cores to be tested for a strength property should have a minimum diameter of at least twice the maximum nominal size of the coarse aggregate, or 3.70 in. (94 mm), whichever is greater. The use of small-diameter cores results in lower and more erratic strengths (Bungey 1979; Bartlett and MacGregor 1994a).

d) If possible, cores to be tested for a strength property should have a length of twice their diameter.

e) Embedded reinforcing steel should be avoided in a core to be tested for compressive strength. If cores cannot be obtained without embedded steel, ASTM C42/C42M permits testing cores with pieces of reinforcing steel, but engineering judgment may be needed to interpret the test results (Gaynor 1965). The criteria that will be used to evaluate measured strengths of cores with embedded reinforcement should be agreed to by the interested parties before beginning the coring program.

f) Avoid cutting electrical conduits or prestressing steel. Use cover meters (3.2.4.1) or ground-penetrating radar (3.2.4.3) to locate embedded metal items prior to drilling.

g) Select core locations to have the least effect on member strength.

h) If possible, core drilling should completely penetrate the concrete section to avoid having to break off the core to facilitate removal. If through-drilling is not feasible, the core should be drilled approximately 2 in. (50 mm) longer than required allowing for possible damage at the base of the core.

i) If cores are taken to determine strength, the number of cores should be based on the expected uniformity of the concrete and the desired confidence level in the average strength, as discussed in 4.1.1. The strength value should be taken as the average of the cores. A single core should not be used to evaluate or diagnose a particular problem.

4.1.1.2 *Random sampling of broken concrete*—Sampling of broken concrete generally should not be used if strength of concrete is in question. Broken concrete samples, however, can be used in some situations for petrographic and chemical analyses in the evaluation of deteriorated concrete members.

4.1.2 *Petrographic and chemical analyses*—Petrographic and chemical analyses of concrete are important tools for the strength evaluation of existing structures, providing valuable information related to the concrete composition, present condition, and potential for future deterioration. The concrete characteristics and properties determined by these analyses can provide insight into the nature and forms of the distress.

4.1.2.1 *Petrography*—The techniques used for a petrographic examination of concrete or concrete aggregates are based on those developed in petrology and geology to classify rocks and minerals. The examination is generally performed in a laboratory using cores removed from the structure. The cores are cut into sections and polished before microscopic examination. Petrography may also involve analytical techniques, such as scanning electron microscopy (SEM), X-ray diffraction (XRD), infrared spectroscopy, and differential thermal analysis. ASTM C1723 provides guidance on the use of energy-dispersive X-ray spectroscopy (EDX) to obtain information on the elemental composition of the specimens placed in the SEM.

A petrographic analysis is normally performed to determine the composition of concrete, assess the adequacy of the mixture proportions, and determine the cause(s) of deterioration. A petrographic analysis performed in accordance with ASTM C856 can provide some of the following information about the concrete:

a) Density of the cement paste and color of the cement
b) Type of cement used
c) Proportion of unhydrated cement
d) Presence of pozzolans or slag cement
e) Volumetric proportions of aggregates, cement paste, and air voids
f) Homogeneity of the concrete
g) Presence and type of fibers (fiber-reinforced concrete)
h) Presence of foreign materials, including debris or organic materials
i) Aggregate shape, size distribution, and composition
j) Nature of interface between aggregates and cement paste
k) Extent to which aggregate particles are coated and the nature of the coating substance
l) Potential for deleterious reactions between the aggregate and cement alkalis
m) Presence of unsound aggregates (fractured or porous)

n) Air content and various dimensional characteristics of the air-void system, including entrained and entrapped air
o) Characteristics and distribution of voids
p) Occurrence of settlement and bleeding in fresh concrete
q) Degree of consolidation
r) Presence of surface treatments

Petrography can also provide information on the following items to aid in the determination of causes of concrete deterioration:
a) Occurrence and distribution of fractures
b) Presence of contaminating substances
c) Surface-finish-related problems
d) Curing-related problems
e) Presence of deterioration caused by exposure to freezing and thawing
f) Presence of reaction products in cracks or around aggregates, indicating deleterious alkali-aggregate reactions
g) Presence of ettringite within cement paste (other than in pore system or voids) and in cracks indicating sulfate attack
h) Presence of corrosion products
i) Presence of deterioration due to abrasion or fire exposure
j) Weathering patterns from surface-to-bottom

The standard procedures for the petrographic examination of samples of hardened concrete are addressed by ASTM C856. Procedures for a microscopic assessment of the concrete air-void system, including the air content of hardened concrete and of the specific surface, void frequency, spacing factor, and paste-air ratio of the air-void system, are provided in ASTM C457/C457M. ASTM C295/C295M contains procedures specific to petrographic analysis of aggregates. Powers (2002), Mailvaganam (1992), and Erlin (1994) provide additional information on petrographic examination of hardened concrete. Mielenz (1994) describes petrographic examination of concrete aggregates in detail.

Concrete samples for petrographic analysis should be collected as described in 4.1.1 and following ASTM C823/C823M. If possible, a qualified petrographer who is familiar with problems commonly encountered with concrete should be consulted before the removal of samples from an existing structure. If the petrographic analysis is being used to assess observed concrete distress or deterioration in a structure, samples for analysis should be collected from locations in the structure exhibiting distress, rather than in a random manner as used in a general assessment (4.1.1).

The petrographer should be provided with information regarding the preconstruction, construction, and post-construction history and performance of the structure. Specific items of interest include:
a) Original concrete mixture proportions, including information on chemical admixtures and supplementary cementitious materials
b) Concrete surface treatments or coatings
c) Curing conditions
d) Placement conditions, including concrete temperature, air temperature, ambient humidity, and wind conditions
e) Placement and finishing techniques
f) Location and orientation of core or sample in structure

Fig. 4.1.2.2—Depth of carbonation as indicated by color change using a pH indicator solution.

g) Exposure conditions during service
h) Description of distressed or deteriorated locations in structure, including photographs

4.1.2.2 *Chemical tests*—Chemical testing of concrete samples can provide information on the presence or absence of various compounds and on forms of deterioration. In addition, chemical tests can be used to gauge the severity of various forms of deterioration and, in some cases, to predict the potential for future deterioration if exposure conditions remain unchanged. Examples of chemical testing for concrete include determination of cement content, chemical composition of cementitious materials, presence of chemical admixtures, content of soluble salts, detection of alkali-silica reactions, depth of carbonation, and chloride content. One of the more common uses of chemical testing is to measure the depth of carbonation and chloride concentration to assess the risk of reinforcement corrosion (corrosion mechanisms and factors for corrosion are discussed in detail in ACI 222R, ACI 222.2R, and ACI 222.3R).

Carbonation contributes to the risk of reinforcing steel corrosion by disrupting the passivity of the steel. More specifically, concrete carbonation exists if the pH of the concrete is reduced to approximately 9 or less (ACI 222R). Chemical testing to determine the depth of carbonation can be accomplished by splitting a core lengthwise and applying a solution of phenolphthalein indicator to the freshly fractured core surface. The indicator changes from colorless to a magenta color above a pH of 9. Thus, the depth of carbonation can be measured by determining the depth of material not undergoing a color change to magenta upon application of phenolphthalein indicator. Alternatively, a pH indicator solution can be applied to the freshly fractured surface. Such a solution displays different colors depending on the pH value. For example, Fig. 4.1.2.2 shows the carbonation front on a concrete core as evidenced by the color variation after applying a pH indicator solution. The abrupt color change from yellow (light shade) to purple (dark shade) indicates the depth of carbonation. Any steel within this depth could be vulnerable to carbonation-induced corrosion.

The presence of chloride ions in the concrete at the level of the reinforcement is the most common cause of reinforcement corrosion. Chlorides can be present in the concrete from the mixture constituents or due to external sources, including exposure to a marine environment or chloride-based deicing chemicals. If the chloride concentration reaches a threshold level at the reinforcement surface, corrosion of the reinforcement may begin in the presence of adequate oxygen and moisture. Thus, testing to determine chloride ion concentration is used to determine whether chloride levels are above the corrosion threshold and to predict the time to corrosion initiation (information on service-life prediction is provided in ACI 365.1R). A full assessment of corrosion risk will include the development of a chloride concentration profile of the concrete by collecting and testing samples at multiple depths from near the surface of the concrete at or below the level of reinforcement. Chemical analysis for chloride concentration is performed on powdered samples of concrete. Samples may be collected using a rotary impact drill or using cores. In the first method, concrete powder from the drilling operation is carefully collected at several depths. If cores are used, the core is cut into 0.5 in. (13 mm) thick slices at the depths of interest, and the concrete is crushed to powder for analysis. Guidance on both collection techniques is provided in ASTM C1152/C1152M, ASTM C1218/C1218M, and AASHTO T 260. Alternatively, powder samples at carefully controlled depth increments can be obtained by a device known as a profile grinder. At each depth increment, a portion of the surface with a diameter of 2.9 in. (73 mm) is removed.

Depending on the evaluation objective(s) and criteria, the samples are tested for water-soluble or acid-soluble chloride concentration (ACI 222R provides detailed information on water- and acid-soluble chlorides). Sample preparation for water-soluble and acid-soluble chloride levels is addressed in ASTM C1218/C1218M and ASTM C1152/C1152M, respectively. The chloride concentration is determined by potentiometric titration of the prepared sample with silver nitrate, as described in ASTM C114. Commercial kits for rapid (acid-soluble) chloride concentration testing using a calibrated chloride ion probe are also available. AASHTO T 260 addresses this field method for determining acid-soluble or total chloride content. ACI 222R provides more information on chloride thresholds for corrosion and chloride testing. Also, testing for the presence of corrosion inhibitors can be important in assessing the likely effect of chloride contamination on the anticipated performance of the structure.

4.1.3 *Testing concrete for compressive strength*—Direct measurement of the concrete compressive strength in an existing structure can only be achieved through removal and testing of cores. In-place or nondestructive test methods can be used to estimate compressive strength if used in conjunction with core testing.

4.1.3.1 *Testing cores*—Compressive strength of concrete cores taken from an existing structure and conditioned in accordance with ASTM C42/C42M should be determined in accordance with ASTM C39/C39M. Key points in this procedure are:

a) The mass of each core tested for compressive strength should be measured in accordance with ASTM C42/C42M. The mass and measured core dimensions are used to calculate the approximate density of the core, which may be of value in the event of unexpected compressive strength results.

b) For core length-diameter ratios less than 1.75, the appropriate strength correction factors given in ASTM C42/C42M should be applied to obtain the core strength for a length-diameter ratio of 2. These correction factors are approximate and engineering judgment should be exercised in evaluating the corrected core strength (Bartlett and MacGregor 1994b).

c) Unless specified otherwise, cores should be moisture-conditioned in accordance with the default procedure given in ASTM C42/C42M. This procedure involves keeping the cores in watertight container(s) for a specified time after they were last wetted. The time of last wetting is either when the core was drilled or at the completion of the last end treatment that involved wetting the core. The intent of the conditioning procedure is to preserve the in-place moisture content and reduce moisture gradients. Excessive moisture gradients in the cores will reduce the measured compressive strength (Bartlett and MacGregor 1994c). Additional discussion on the importance of moisture conditioning is provided by Neville (2001). For core testing related to the evaluation of concrete due to low strength test results for standard-cured cylinders during construction, ACI 318 requires that the cores be tested between 48 hours and 7 days after they were obtained, unless otherwise approved by the licensed design professional. For core testing related to the strength evaluation of an existing concrete structure, careful consideration should be given to whether procedures for moisture conditioning of cores should differ from the default procedure specified in ASTM C42/C42M.

d) Care should be exercised in end preparation of cores before testing for compressive strength. If capping compound is used, its thickness is limited by ASTM C617/C617M. This is especially critical for high-strength concrete. If cores are tested using unbonded caps in accordance with ASTM C1231/C1231M, the inner diameter of the retaining rings should be between 102 and 107 percent of the core diameter. There are limitations on the flatness of the core ends to use one of these capping procedures; therefore, saw cutting of one or both ends may be required. If ends of cores are ground, verification of flatness is required to ensure that the requirements of ASTM C39/C39M are met.

e) Core compressive strengths may be expected to be lower for cores removed from the upper portions of slabs, beams, footings, walls, and columns than from lower portions of such members (Bartlett and MacGregor 1999).

f) The interpretation of core strengths is not a simple matter. Involved parties should agree on the evaluation criteria before sampling begins (Neville 2001).

4.1.3.2 *Equivalent specified strength*—Depending on age, temperature history, moisture condition during curing, and strength level, compressive strength values obtained from core tests can either be lower or higher than those obtained from tests of standard-cured cylinders molded from samples of concrete taken during construction. It has been reported that for mature concrete, the core strength varies from 100 percent of the of the standard-cylinder strength for 3000 psi (20 MPa) concrete to 70 percent for 9000 psi (60 MPa) concrete (Mindess and Young 1981). It has been practice to divide measured core strengths by 0.85 and consider the resulting value as the equivalent cylinder strength (Hanson 2007). The basis for this practice is the work of Bloem (1965, 1968) in which it was found that, for well-cured concrete slabs, the strengths of cores drilled from the slabs were approximately 85 percent of the strength of companion standard-cured cylinders. There are many factors that affect the relationship between core strengths and the strengths of companion standard-cured cylinders; therefore, a single ratio is not applicable to all situations. In addition, for a strength evaluation, the objective is not to determine equivalent cylinder strength, but to estimate the equivalent in-place specified strength based on the strength variation measured by cores. Thus, the inappropriate practice of dividing core strengths by 0.85 is discouraged because rational procedures exist for making more reliable estimates of the equivalent in-place strength by considering the major factors that affect measured core strengths (Bartlett and MacGregor 1995).

In evaluating structural capacity, the equivalent specified strength for the concrete in the structure should be established on the basis of core strengths. The equivalent in-place specified strength represents the in-place compressive strength that is expected to be exceeded by approximately 90 percent of the concrete in the structure and is used in calculating nominal member capacities. The commentary to Chapter 27 of ACI 318-14 refers to the methods in ACI 214.4R as an acceptable procedure for determining the equivalent specified strength of the in-place concrete on the basis of cores. Section 6.4.3 of ACI 562-16, on the other hand, provides a specific statistical relationship to calculate the equivalent specified strength from the measured core strengths, which have been modified to account for core diameter and moisture condition. These procedures for estimating the equivalent in-place specified strength are based largely on work by Bartlett and MacGregor (1995).

The value of the calculated equivalent specified strength depends on the number of cores tested and the variability of the core strengths. Figure 4.1.3.2 shows the value of the equivalent specified strength as a fraction of the average in-place strength for different numbers of cores and coefficients of variation of the core test results based on Chapter 6 of ACI 562-16. The figure shows that the equivalent specified strength is a smaller fraction of the average strength with increasing variability of test results and with lower number of tests.

4.1.3.3 *In-place tests*—Currently, there are no in-place tests that provide direct measurements of compressive strength of concrete in an existing structure. In-place tests, however, can be used to estimate in-place compressive strength, provided correlations that are applicable to the concrete in the structure have been established. In-place tests are commonly used in conjunction with tests of drilled cores to reduce the amount of coring required to estimate compressive strengths throughout a large structure. These indirect tests

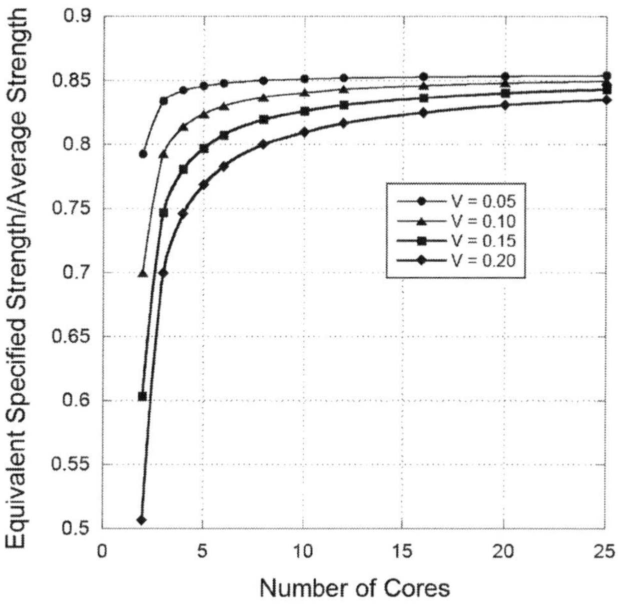

Fig. 4.1.3.2—Ratio of equivalent specified strength to average in-place compressive strength as a function of the number of cores and coefficient of variation of core strengths based on ACI 562.

do not have the same degree of uncertainty, and appropriate statistical methods are necessary to establish valid estimates of compressive strength based on indirect test results. Refer to ACI 228.1R and 3.2.3 for further information.

4.2—Reinforcing steel

4.2.1 *Determination of yield strength*—The yield strength of the reinforcing steel can be established by one of three methods as discussed in Chapter 6 of ACI 562-16. Information from project documents, including drawings; specifications; or previous testing, including mill test reports, may be used to determine yield strength. If material properties from original construction reports are used, the commentary of ACI 562 suggests that additional testing may be required to confirm these properties if degradation has occurred. If mill test reports furnished by the manufacturer of the reinforcing steel are used, the design professional and the building official should be in agreement. Yield strengths from mill test reports tend to be greater than those obtained from tests of field samples. If existing documentation of yield strength is not available or desirable, yield strength may be determined either by the historical values presented in Chapter 6 of ACI 562 or by sampling and destructive testing of specimens taken from the structure. Guidelines for sampling are provided in 4.2.2. ACI 562 allows the design professional to use a yield strength value of 27,000 psi (186 MPa) in place of testing if historic information is not available.

The Concrete Reinforcing Steel Institute (2014) provides information on reinforcing systems in older structures, including reinforcing bar specifications, yield strengths, sizes, and allowable stresses. Note that the hardness reported in historic standards may not necessarily correlate to current metallurgical hardness measures.

4.2.2 *Sampling techniques*—If the yield strength of embedded reinforcing steel is determined by testing, the recommendations listed in the following should be followed:

a) An equivalent yield strength should be calculated using Eq. (6.4.8) presented in ACI 562.

b) A minimum of three test samples should be used to compute an equivalent yield strength.

c) Tension test specimens should be the full section of the bar (ASTM A370-18 Annex 9). Requirements for specimen length, preparation, testing, and determination of the yield strength are provided by ASTM A370.

d) If bar samples meeting the length requirements of ASTM A370-18 Annex 9, cannot be obtained, samples may be prepared (machined) according to the general requirements of ASTM A370 for testing and determination of mechanical properties.

e) Samples should be removed at locations of minimum stress in the reinforcement.

f) To avoid excessive reduction in member strength, no two samples should be removed from the same cross section (location) of a structural member.

g) Locations of samples in continuous concrete construction should be separated by at least the development length of the reinforcement to avoid excessive weakening of the member.

h) For single structural elements having a span of less than 25 ft (7.5 m) or a loaded area of less than 625 ft^2 (60 m^2), at least one sample should be taken from the main longitudinal reinforcement (not stirrups or ties).

i) For longer spans or larger loaded areas, more samples should be taken from locations well distributed through the portion being investigated.

j) Sample locations should be chosen to determine whether the same strength of steel was used throughout the structure.

Sampling of prestressed reinforcement, whether from bonded or unbonded systems, is a complex undertaking and beyond the scope of this report. Discussion of extraction of unbonded single-strand tendons for testing can be found in ACI 423.4R.

4.2.3 *Additional considerations*—The strength evaluation of concrete structures can require consideration of several reinforcement-related factors in addition to the yield strength, such as development length, anchorage, and reduction in cross section or bond due to corrosion.

Reinforcing bars manufactured before 1947 are sometimes smooth or have deformation patterns not meeting modern requirements. As a result, the bond and development of these bars can be significantly different from those of modern reinforcement (Concrete Reinforcing Steel Institute 2014). Similarly, changes to details and assumptions for standard hooks can affect the development of hooked bars in older structures. For structures with reinforcing bars manufactured before 1947, the Concrete Reinforcing Steel Institute (2014) conservatively recommends assuming that the required development length is twice that based on current code provisions. Concrete deterioration, especially spalling of concrete due to corrosion or other damage mechanism, will adversely affect the development of reinforcing steel.

Corrosion of reinforcement can lead to reduction in member capacity and ductility as a result of reinforcement section loss or disruption of bond. Because reinforcement corrosion normally results in disruption and cracking of the concrete surrounding the bar, bond to the concrete will be negatively affected in addition to loss of cross-sectional area. As a result, where bond is important, the reduction in structural capacity can be higher than that based solely on the reduction of the cross-sectional area of the bar. A conservative approach should be used in assessing the residual capacity of damaged or corroded reinforcement. Special consideration should be given to situations where corrosion of prestressing steel is suspected (ACI 222.2R). Tests for determining corrosion activity include measuring half-cell potential (ASTM C876) and linear polarization resistance. Refer to ACI 222R and ACI 228.2R for additional information on these types of tests.

4.3—Fiber-reinforced polymer reinforcement systems

Fiber-reinforced polymer (FRP) systems used for strengthening of existing concrete structures come in a variety of forms, including wet layup systems; pultruded (shop-fabricated) systems, commonly referred to as laminates; and near-surface-mounted (NSM) FRP systems. ACI 440.2R provides information on different commercially available FRP systems for strengthening of concrete structures. Wet layup FRP systems consist of fiber sheets or fabrics (typically made of carbon and glass) impregnated with a saturating resin on site and bonded to the concrete surfaces. The fiber sheets can be bound together in multiple plies. Pultruded FRP systems are typically thin composite shapes manufactured off site and are also bonded to the concrete surfaces with resins. Near-surface-mounted FRP systems consist of circular or rectangular bars or plates installed and bonded into grooves cut in the concrete surface.

Material properties of installed FRP used to strengthen concrete elements can be obtained from project documents such as drawings, specifications, or results of material testing performed as quality control during construction, which can include FRP properties from tests of field sample panels or witness panels. It is not practical to extract samples to determine mechanical properties such as tensile strength and modulus of elasticity because FRP systems are adhered to the concrete substrate with resins. Therefore, samples cannot be easily retrieved without damaging the FRP itself. Small core samples, typically 0.5 in. (13 mm) in diameter, can be taken to measure the thickness of cured laminate thickness or number of plies. The licensed design professional should specify the sampling frequency. ICRI 330.2 provides guidance on sampling frequency.

The performance and effectiveness of FRP over time can be evaluated as one or more of the following procedures:

a) Visual survey of the FRP system for signs of debonding of FRP laminates or near-surface-mounted FRP bars, peeling or blistering of the FRP laminates, changes in color, and cracking of the resin used to bond the FRP to concrete

b) Visual survey of the strengthened concrete element for signs of distress such as cracking or excessive deflections, and underlying damage such as corrosion of the internal steel reinforcement

c) Sounding (hammer tap) of laminates for hollow-sounding areas to identify areas of debonding (delaminations) from the concrete surfaces. The licensed design professional should evaluate the effect of delaminations on the structural integrity and durability of the FRP system. The size, locations, and quantity of delaminations should be considered in the evaluation. ACI 440.2R provides guidance for acceptance guidelines. In general:

i. Small delaminations, less than 2 in.^{2} (1290 mm^2), are acceptable if the delaminated area is less than 5 percent of the total laminate area and there are no more than 10 such delaminations per 10 ft^2 (0.93 m^2).

ii. Large delaminations, greater than 25 in.^{2} (16,130 mm^2), should be repaired by removing delaminated areas and installing and overlapping a laminate patch of equivalent strength.

iii. Delaminations less than 25 in.^{2} (16,130 mm^2) may be repaired by epoxy injection or laminate replacement, depending on the size and number of delaminations and their locations.

d) Pull-off tension tests evaluate the bond strength of FRP laminates to the concrete substrate. ASTM D7522/D7522M describes test procedures for evaluating the pull-off strength of FRP laminates bonded to concrete surfaces. ICRI 210.3 provides additional guidance regarding pull-off tests. Testing should not be performed at critical areas (for example, area of maximum flexural demand for FRP flexural strengthening) and splice areas should be avoided. Typical sample areas are at the ends of the FRP strengthening. The licensed design professional should specify the testing frequency (ICRI 330.2). Tension adhesion strengths should exceed 200 psi (1380 kPa) and should fail in the concrete substrate. Lower strengths or failure between the FRP laminate and the concrete or between laminate plies should be evaluated by the licensed design professional for acceptance. Cored samples from bond testing can also be used to measure the laminate thickness or number of plies.

e) Ultrasonic or thermographic tests for indications of progressive delamination.

f) For near-surface-mounted strengthening, cores can be extracted to measure the bar size and visually verify the consolidation of the adhesive around the FRP bar. As in the case of laminates, the sample location should not be at critical areas.

g) Load testing to evaluate the overall performance of the element strengthened with FRP. ACI 437.2 provides requirements for load testing of concrete elements.

CHAPTER 5—ASSESSMENT OF LOADING CONDITIONS AND SELECTION OF EVALUATION METHOD

5.1—Assessment of loading and environmental conditions

A fundamental aspect of any strength evaluation is the assessment of the loads and environmental conditions, past, present, and future. These should be accurately defined so that the results of the strength evaluation process will be realistic.

5.1.1 *Dead loads*—Dead loads consist of the self-weight of the structure and any superimposed dead loads.

5.1.1.1 *Self-weight of structure*—The self-weight of the structure can be estimated using field-measured dimensions of the structure and material densities as presented in ASCE 7. Dimensions obtained solely from design drawings should be used with caution because significant differences can exist between dimensions shown on design drawings and actual, as-built dimensions. Similarly, differences can exist between material densities obtained from ASCE 7 and actual in-place densities due to variations in moisture content, material constituents, and other reasons. If differences in densities are suspected, field samples should be analyzed.

5.1.1.2 *Superimposed dead loads*—Superimposed dead loads include the weight of all materials incorporated into the structure, exclusive of the self-weight of the structure. Examples include the weight of architectural floor and ceiling finishes; nonstructural topping slabs or overlays; partitions; mechanical systems; fixed service equipment such as cranes and exterior cladding; and landscaping such as fixed planters, soils, and plantings. The magnitude of superimposed dead loads can be estimated by performing a field survey of the structure for such items and using appropriate values for loads as presented in ASCE 7 or other reference sources. Consideration should be given to superimposed dead loads that may not be present at the time of the evaluation but may be applied over the life of the structure.

5.1.2 *Live loads*—The magnitude, location, and orientation of live loads on a structural component depend on the intended use of the structure. Past, present, and future usage conditions should be established accurately so that appropriate assumptions can be made for the selection of live loads. For evaluation, refer to Chapters 1, 4, and 5, and Appendix A of ACI 562-16 regarding which building code edition should be used in establishing the live loads. In cases where the structure's occupancy does not align with the prescribed live load categories, the live load can be established if a rational assessment of past, present, or future usage is prudent and reasonable.

If the serviceability of a structure is to be evaluated in addition to strength, the live loads that will be present during normal occupancy of the structure should be estimated. Estimates of live loads can be obtained from field surveys and measurements of loads in other structures with similar occupancies. In many instances, the day-to-day live loads are much lower than the design live loads prescribed in the local building code. Data from surveys of live loads in buildings are presented in the commentary to ASCE 7. Data from surveys of live loads in parking structures are presented in Wen and Yeo (2001).

5.1.3 *Wind loads*—ASCE 7 provides guidance to determine wind loads. Site-specific historical wind-speed information can be obtained from the National Climatic Data Center of the National Oceanic and Atmospheric Administration (NOAA).

5.1.4 *Rain loads*—When evaluating roofs, loads that result from ponding or pooling of rainwater due to the nature of the roof profile, deflections of framing members, or improper roof drainage should be considered.

5.1.5 *Snow and ice loads*—Consider the possibility of partial snow loading, unbalanced roof snow loads, drifting snow loads, and sliding snow loads as defined in ASCE 7. When estimating ground snow loads, consider local and regional geographical locations. In the absence of specific requirements in the local building code, refer to ASCE 7 and information available from NOAA. Special attention is required if the structure is located near a large body of water that could be the source of moisture leading to ultra-heavy snow storms.

5.1.6 *Seismic loads*—Seismic loading conditions are presented in local building codes. In addition, detailed seismic load information is presented in ASCE 7 and ASCE 41. If the ability of the structure to resist seismic loads is of concern, the evaluation of the structure should also follow criteria contained in appropriate Building Seismic Safety Council and Federal Emergency Management Agency documents.

5.1.7 *Thermal effects*—Where restraint exists, expansion and contraction of a concrete structure due to daily and seasonal variations in ambient temperature can cause significant forces in the structural elements. The engineer should consult local weather records or NOAA to determine the range of temperatures that the structure has experienced. Approximate data regarding seasonal temperature variations are available in the *PCI Design Handbook* (Prestressed/Precast Concrete Institute 2004).

If there is a sudden change in ambient temperature, large concrete sections will respond more slowly than smaller sections. Therefore, effects of rate of heat transfer in individual concrete elements can also be important. It may also be appropriate to consider the effect of absorption of radiant heat due to the reflective properties of concrete coatings exposed to direct sunlight.

Variations in the temperature within a building can influence the magnitude of thermal effect forces. Consider conditions such as areas of the building where heating or cooling is turned off at night, inadequately or overly insulated areas, and existence of cold rooms.

5.1.8 *Creep and shrinkage*—The effects of long-term creep and shrinkage are important considerations for concrete elements (ACI 209R). Cracks or other distress can be caused by restrained shrinkage (ACI 224R). In a concrete structure, internal stresses result from restrained shrinkage and long-term creep of concrete elements. These stresses, when combined with other stresses produced by applied loading, prestressing forces, or restraint of deformations caused by prestressing forces, can be significant. An example of this effect is a reinforced concrete column under sustained loading where stresses in the embedded reinforcing steel can increase over time due to creep of the concrete. Another example occurs in unrestrained prestressed structures, such as pretensioned beams, where creep and shrinkage will reduce the tensile force in the prestressing steel over time. In restrained prestressed structures, such as post-tensioned floor systems, restrained shrinkage may result in significant tension stresses that counteract the initial concrete precompression due to post-tensioning. The complex mechanisms associated with creep and shrinkage often makes quantifying their effects with precision difficult. Guidance for estimating the effects of creep and shrinkage can be found in ACI 209R, ACI 224R, and the *PCI Design Handbook* (Prestressed/Precast Concrete Institute 2004).

5.1.9 *Soil and hydrostatic pressure*—Significant loads can be imposed on a structure from soil and hydrostatic pressure. Soil densities and the lateral soil pressure vary significantly. It is often prudent to sample and establish actual soil densities and properties such as the internal angle of friction. Variations in water table and moisture content can result in large variations in the lateral pressure. Overall stability should be checked in structures that are built on a slope, due to unbalanced soil pressure. Hydrostatic uplift forces can occur due to elevated groundwater conditions, defects or failures in pressurized water piping, and at high flood elevations. Consider possible loads or damage caused by frost heaving of soil, soil shrinkage or swelling, differential soil settlement, and improper drainage. The loads imposed on the structure due to these conditions can be determined through collaboration with a geotechnical engineer.

5.1.10 *Fire*—If the structure being evaluated has been exposed to fire, consider the effects of localized damage caused by the heat of the fire or by the firefighting efforts. Volume changes of concrete elements during a fire can cause significant damage. Restrained thermal expansion may lead to high internal stresses, which can result in concrete section loss by spalling and loss of bond. Additionally, concrete material properties, including compressive strength and elastic modulus, may deteriorate and can be permanently altered as a result of fire exposure. Potential damage to reinforcing steel or prestressing tendons should also be considered in the evaluation process. Additional information on damage due to fire is found in ACI 216.1. Petrographic analysis and in-place tests can be used to assess the extent of fire damage.

5.1.11 *Loading combinations*—For purposes of analytical strength evaluation, load combinations should conform to the provisions of ACI 562. If serviceability is to be evaluated, load factors equal to 1.0 may be appropriate. Multiple load combinations may be necessary to fully assess the performance of the structure.

5.2—Selecting the proper method of evaluation

The evaluation method selected depends on factors such as the structural framing system, information known about its existing condition, and logistical and economic considerations. The typical choices are:

a) Evaluation solely by analysis
b) Evaluation by analysis and in-place load testing

5.2.1 *Evaluation solely by analysis*—Evaluation solely by analysis is recommended if:

a) Sufficient information is available, or obtainable by field investigation, about the physical characteristics, material properties, and anticipated loadings and structural behavior
b) There is no evidence of latent defects

c) The number of unknowns is small

Analytical evaluation is appropriate if all the following conditions are satisfied:

a) There exists an accepted methodology for analyzing the type of structural system under consideration

b) Characteristics of the structural elements can be determined and modeled within acceptable limits of error

c) The distress is limited in magnitude or nature, so that the uncertainties introduced into the analysis do not render the application of the theory excessively difficult

d) Nonlinear behavior in materials and systems, if present under the loading conditions imposed, is adequately modeled. Examples of nonlinear behavior include concrete cracking, bond slip, and reinforcement yielding. Impact or blast loads can also induce nonlinear behavior.

5.2.2 *Evaluation by analysis and in-place load testing*—Considerable experience has been assembled and reported on the subject of in-place load tests of existing structures. Refer to ACI 437.2, Tumialan et al. (2012), Ziehl et al. (2008), Galati et al. (2008), Nehil et al. (2007), Anderson and Popovic (1988), Barboni et al. (1997), Bares and FitzSimons (1975), Bungey (1989), Concrete Innovation Appraisal Service (2000), Elstner et al. (1987), FitzSimons and Longinow (1975), Fling et al. (1989), Guedelhoefer and Janney (1980), Hall and Tsai (1989), Ivanyi (1976), Kaminetzky (1991), Mettemeyer et al. (1999), Nanni and Gold (1998a,b), Nanni and Mettemeyer (2001), Nanni et al. (1998), Popovic et al. (1991), and Raths and Guedelhoefer (1980). Two load testing methods, cyclic and monotonic, are described in ACI 437.1R and ACI 437.2.

Evaluation by analysis and in-place load testing is recommended in the following cases:

a) The complexity of the design concept and lack of experience with the types of structural elements present make evaluation solely by analytical methods impractical or uncertain.

b) The loading and material characteristics of the structural element(s) cannot be readily determined.

c) The existing distress introduces significant uncertainties into the parameters necessary to perform an analytical evaluation.

d) The degree of suspected deficiencies in design, material, or construction cannot be readily determined.

e) If there is doubt concerning adequacy of structural elements for new loading that exceeds the allowable stresses calculated using the original design.

CHAPTER 6—EVALUATION AND INTERPRETATION OF RESULTS

This chapter provides guidelines for performing and interpreting results of the evaluation. The evaluation should be designed with sufficient breadth and scope to allow meaningful conclusions to be developed regarding the suitability of the structure for its intended use. The evaluation may be performed solely by analytical methods or by a combination of analytical and in-place load testing methods.

Regardless of the method of evaluation, it is essential that the evaluation include all suspected defects detected in the preliminary investigation. More than one portion of the structure may need to be evaluated if multiple defects are suspected or if large areas of a structure are being evaluated. The following items should be considered for determining the extent of the evaluation:

a) Variations in the condition of the structure and material properties

b) Variation in type of structural framing systems

c) Differences in loading intensity required by intended use

d) Presence of other conditions that can affect load-carrying capacity, such as large floor openings or atypical bay sizes

Economic, schedule, and logistical considerations limit the number of specific members or the portions of the structure that can be evaluated in detail. Therefore, it is important to identify the specific critical members or portions of the structure in assessing the overall structural performance before undertaking the evaluation. Further information related to structural assessment, structural analysis, structural serviceability, and strength evaluation by load testing can be found in ACI 562-16 Sections 6.1 through 6.8.

6.1—Analytical evaluation

The information gathered from the preliminary investigation and material evaluations should be used in the analysis to determine the safe load-carrying capacity of the structure or portion of the structure being evaluated.

6.1.1 *Forms of analysis*—In the evaluation of concrete structures by analytical methods, analysis has two different meanings. One deals with finding the values of forces and moments that exist in the structure. The second uses the characteristics of the structure or member to predict how it will respond to the existing load effects.

A structure should be analyzed to determine the bending moments, torsional moments, shear forces, and axial forces at the critical sections. Many engineers will limit this part of the analysis to linear-elastic response even though this is generally not realistic for reinforced concrete, particularly for loads in excess of service level.

The alternative, nonlinear analysis may require special capabilities not found in most engineering offices. An analysis done by elastic methods, however, often provides a reasonable estimate for the values of important load effects.

For nonlinear analysis, an assumption is made about the behavior of structures. For an evaluation of structural performance at service loads, it may be reasonable to assume that concrete and reinforcing steel behave in a linearly-elastic manner. It is necessary, however, to account for the low tensile strength of concrete, and cracked section properties are often used. Understanding the working stress properties of a structure can be valuable for assessing conditions between incremental stages of loading. A working stress analysis can be beneficial when relating observed conditions (such as cracking, deflection, or camber) to the actual state of stress in the structural components.

If structural safety is the principal concern, the strength of the member or structure needs to be established. The principles of strength design, as applied in ACI 318, provide a basis for establishing a nominal strength for structural members (ACI 562).

6.1.2 *Levels of analysis*

6.1.2.1 *Rigorous analysis*—Analysis based on experimentally verified theories of structural mechanics is useful under the following conditions:

a) Loading conditions for the structure are known with a high degree of certainty after examining existing data.

b) Detailed structural engineering drawings and material specifications are available, and are believed to be reliable or have been confirmed or supplemented with data obtained by the condition survey, for example:

i. Dimensions of the structure and its members can be determined by field measurements and are used to establish dead loads.

ii. The location, size, and depth of concrete cover of embedded reinforcing steel can be determined by field investigation.

iii. Material characteristics essential to the analysis can be determined, or estimated reasonably, by the use of invasive or nondestructive tests.

iv. Estimates of the strength of the foundations can be obtained by conducting appropriate geotechnical explorations and soil tests.

c) Sufficient data can be collected to make an adequate assessment of the existing physical condition of the structure, including estimation of the effects of distress, deterioration, and damage.

6.1.2.2 *Finite-element analysis*—Linear finite element analysis and nonlinear finite element analysis provide a solution for cases where conventional methods of analysis are not sufficient. The latter method can be used to evaluate the effects of nonlinear material properties on structural response to levels of loading that produce inelastic behavior, such as concrete cracking, bond slip, and yielding of reinforcement. Nonlinear finite element analysis should be performed by experienced structural engineers competent in verifying and documenting the results.

6.1.2.3 *Approximate analysis*—Use of approximate methods of analysis requires considerable experience with the type of structural system under evaluation and its behavior. Most importantly, approximate methods require the exercise of sound engineering judgment. Two basic guidelines should be followed:

1. All assumptions necessary for performing the structural analyses should be clearly documented. If inelastic models embedded in finite element analysis software are used, the models should be well understood, properly verified, and clearly documented. Care should be taken to describe those assumptions made in the strength evaluation by accounting for existing distress, deterioration, or damage.

2. All assumptions necessary to conduct the theoretical structural analysis should provide a conservative lower-bound value for the safe load-carrying capacity of the structure.

6.1.3 *General considerations*—The assumed behavior of the structure and the results of the theoretical analyses should be consistent with the observed behavior of the structure. The analysis should model characteristics of the structure such as:

a) The effects of nonprismatic members on the relative stiffness of components in the structure

b) Torsional characteristics of structural members

c) Two-way load response in slab systems

d) Column support and structural fixities in terms of moment-rotation characteristics

e) Column base characteristics as influenced by soil conditions

6.1.4 *Acceptance criteria*—The structure or structural component being evaluated is deemed to have sufficient strength if the analytical evaluation demonstrates that the predicted design capacity of the elements satisfies the requirements and the intent of ACI 562.

Uncertainty about the structure is reduced where field work has established the material strengths of steel and concrete; the size, location, and configuration of reinforcement; and structural dimensions. This supporting work can serve as justification for using a different strength-reduction factor for evaluation as opposed to design (ACI 562-16 Chapter 5), if allowed by the local building code. Experience and engineering judgment are important in this case.

If the analytical evaluation indicates that the structure does not satisfy the intent of ACI 562, the building official may approve a lower load rating for the structure based on the results of such evaluation if allowed by the local building code.

6.1.5 *Findings of analytical evaluation*—An analytical strength evaluation has three possible findings:

1. Analyses show that the structure or structural element has an adequate margin of safety according to the provisions of the applicable building code. In this case, the design strength (nominal strength multiplied by strength-reduction factor) exceeds that required for factored loads.

2. Analyses show that the design strength is less than that required for factored loads but greater than required for service loads (load factors equal to or greater than 1.0 for all load cases). In this case, the structure or structural element is not adequate. In some cases, restricted use of the structure that limits the applied loads in recognition of the computed strength may be permitted.

3. Analyses show that the design strength of the structure is less than required for service loads under the applicable building code. In such cases, the owner should be notified and consideration given to the installation of shoring, severe restriction of use, or evacuation of the structure until remedial work can be done.

6.2—Supplementing the analytical evaluation with load tests

6.2.1 *Conditions for use*—In-place load testing is recommended if the following conditions are met:

a) The test results will permit rational interpretation of the structural strength of the element to be tested.

b) The influence of adjacent structural members, components, or entire structures can be accounted for during the load test and when evaluating the results of the tests. This influence includes full accounting of alternative load paths that are available in the structure.

c) The structure can be monitored adequately and safely by appropriate instrumentation to provide the necessary data to make an evaluation of the structural strength and, where appropriate, serviceability.
d) All participants in the test and all passersby are safe during setup and performance of the test.

An analysis should always be done before conducting a load test. This analysis can employ approximate methods. The analysis should be performed to allow for a reasonable prediction of the performance of the structure during the load test. Calculated deflections of concrete structural elements can, in many cases, be inaccurate. Care and engineering judgment are required when comparing calculated deflections with those measured during a load test. Reports are available to assist the engineer in calculating deflections of reinforced concrete structures (ACI 435R; ACI 435.8R).

ACI 423.4R describes the limitations of full-scale load testing when evaluating structures with unbonded post-tensioned tendons damaged by corrosion and highlights the need for caution related to load testing for structural systems with unbonded post-tensioned tendons. Further information on the corrosion protection of bonded and unbonded prestressing materials and prestressing system components is addressed in Chapter 8 of ACI 562-16.

6.2.2 *Identifying the form of test to be conducted*—Evaluation of structural adequacy may be aided by one or both of the following forms of load testing:
a) Static tests
b) Dynamic tests, using special test procedures developed specifically for the characteristics of the structure to be tested; dynamic test procedures are beyond the scope of this report.

6.2.3 *General requirements*—The following general requirements are applicable when conducting a load test:
a) A qualified licensed design professional, acceptable to the building official, should design and directly supervise the tests.
b) The structure should be loaded to adequately test the suspected source of weakness.
c) On environmentally exposed structures, load tests should be conducted at a time when the effects of temperature variations, wind, and sunlight on the structure and the monitoring devices are minimized; for example, early morning, late evening, or at night.
d) Load tests on exposed concrete structures should preferably be conducted at temperatures above 32°F (0°C).
e) On environmentally exposed structures, the environmental conditions, especially the ambient temperatures and wind, should be recorded at frequent intervals during the load test.

6.2.4 *Test loads*—The following guidelines may be useful for selecting the type of test load or loading device in conducting a load test of a concrete structure:
a) If the test load is applied by using separate elements, such as iron bars, bricks, sandbags, or concrete blocks, the elements should be arranged throughout the duration of the test to prevent arching action. The largest base dimension of the separate elements or stacks of elements should be less than one-sixth of the span of the structural element being tested. These elements or stacks should be separated by a clear lateral distance of at least 4 in. (100 mm).
b) Separate pieces should be of uniform shape, and the weight of each piece should not differ by more than 5 percent from the average weight. The average weight should be determined by weighing at least 20 pieces taken at random.
c) If nonuniform loading elements are used, each element should be measured to determine surface contact area, weighed, and marked appropriately.
d) The load devices should be readily removable.
e) Materials that readily absorb moisture should not be used as loading elements.
f) Test load devices applied to sloping surfaces should be securely anchored to prevent shifting. Load components in all directions should be accounted for to prevent movements.
g) Water, loose sand, or other similar materials should be contained within small compartments to prevent ponding effects or shifting during significant deformation of the structure that may occur during the test.
h) If using hydraulic or pneumatic load-application systems, adequate supports should be provided to transfer the reactions, except where these reactions are part of the loading scheme. These loading devices should continue to function in a uniform fashion, even under significant deformation of the structure.

The total accumulated test load should be within 5 percent of the intended value. Arrangement of the test load should consider the following:
a) Care should be taken in the loading scheme not to unintentionally damage any other element of the structure that is not part of the test.
b) The test load should be arranged as close as possible to the load arrangement that the structure is intended to support.
c) If the test load cannot be arranged as described previously, it should be arranged to produce load effects similar to those that would be produced by the design load.
d) If uniform loads are approximated with concentrated loads, stress concentrations at the points of load application should not be significant.
e) The test load should be designed to produce the maximum load effect in the area being tested. This includes use of checkerboard or similar pattern loads, if required by the applicable building code.

6.2.5 *Instrumentation*—The following guidelines are applicable to instrumentation systems for monitoring a load test:
a) Instrumentation should monitor deflections, lateral deformations, support rotations, and support settlement or shifting during application of the test load.
b) Measurement devices should be mounted to determine relative changes in the shape of the structure or structural element during the test.
c) During the load test, instrumentation should be protected from environmental influences such as direct sunlight, significant temperature variations, and wind.
d) Before the start of the load test, instrumentation should be installed to determine the effects of thermal changes on the deformations of the structure and on the instruments. If

necessary, compensation factors should be developed for application to the data obtained from the load test.
e) Strain measurements, if applicable, should be made at critical locations.
f) Deflection and strain measurement devices should be duplicated in critical areas.
g) Resolution of deflection measurement devices should not exceed 1/100 of the expected deflection (ACI 437.2).
h) Deflection of structural members can be measured with electronic or mechanical devices or with conventional surveying equipment. As an example, deflections can be measured using linear variable differential transformers (LVDTs).
i) Displacement transducers and resistance strain gauges can allow rapid electronic collection of data from numerous points when connected to a data acquisition system. Their installation, however, can be time-consuming, particularly if instrumentation has to be protected from the weather.
j) Inclinometers can be used to measure the rotation or slope of a test member. Because values of slope can be correlated to deflections, these instruments can be a good resource if displacement transducers are difficult to mount. Inclinometers can be mounted on a variety of vertical and horizontal surfaces.
k) Mechanical devices, such as dial gauges, are typically sturdy and simple to operate, but collection of data can be slow and often requires that someone enter the structure during performance of the test, which can be dangerous. These devices are valuable for measuring small deflections in stiff structures.
l) Large deflections can be measured easily by suspending graduated scales from critical points and reading them with a surveyor's level from a remote location.
m) Deflection measurement devices should be placed at the point(s) of maximum expected deflection. Devices should also be placed at the supports to detect column shortening, if deemed appropriate by the engineer.
n) Crack width can be measured by using graduated magnifying glasses or crack comparators. Their use during a load test is often restricted for safety reasons. If they are used, marks should be placed at each point on the cracks where readings are to be taken so that subsequent readings are taken at the same positions.
o) Crack movement (opening or closing) can be measured with dial gauges and displacement transducers. Crack movement can also be measured accurately by using gauge points and an extensometer.
p) Crack extension and the formation of new cracking can be tracked with acoustic emission sensors in combination with appropriate data acquisition systems and software (El Batanouny et al. 2014).
q) In deteriorated structures, cracks are often present either on the top or bottom surfaces of slabs and beams or the sides of columns. Some of these cracks may have meaning with respect to the structural behavior, while others may simply be the result of deterioration. For example, cracking caused by corrosion of embedded reinforcement may not directly relate to movement of structural elements during a load test. Engineering judgment should be exercised when monitoring and measuring crack movement, particularly if the structure contains numerous existing cracks or exhibits deterioration.
r) Thermometers or thermocouples should be used to measure the ambient temperature during a load test. Temperature readings should be taken in all areas of a structure that are affected by the load test. For structural slabs, thermometers should be placed above and below the slab surface. Records of variations of sunlight should be maintained for roof slabs and other areas of the structure that are exposed to direct sunlight during performance of a load test.
s) In a variety of shapes, sizes, and capacities, load cells are used to measure the load applied by hydraulic or pneumatic jacks. Pressure transducers can also be used to measure fluid pressures in the hydraulic system, which can be calibrated to a specific level of load.
t) Data acquisition systems can be used for simultaneously collecting readings from several devices as the load is being applied. Such devices include pressure transducers, load cells, LVDTs, inclinometers, extensometers, acoustic emission sensors, and strain gauges. This allows for real-time monitoring of the measured structural response. If acoustic emission sensors are used, high-speed data acquisition equipment is needed.

6.2.6 *Shoring*—Shoring should be provided before a load test, whether the entire structure or only a portion is involved, to support the structure in case of failure during the test. The shoring should be designed to carry the existing dead load and all additional superimposed test loads on the portion of the structure for which collapse is possible. The effects of impact loading on the shoring, which is likely if a structure or member fails suddenly during the test, should be considered in the selection of shoring elements. Refer to the Commentary of ACI 562-16 Chapter 9 for guidance on shoring design. This may be accomplished by designing the shoring to support at least twice the total test load plus the existing dead load. Some engineers design the shoring to support at least twice the total test load plus the existing dead load. This approach should be evaluated for appropriateness on a case-by-case basis. Additional guidance on how to address shoring loads is available in ACI 562-16 Chapter 5.

Shoring more than one level to prevent progressive collapse in the event of failure should be considered. For example, if all floors below the test floor cannot support the weight of the test element, the loads it supports, and the imposed test loads, then the shoring should be extended to the foundation level. For horizontal members, shoring should clear the underside of the structure by not more than the maximum expected deflection plus an allowance not to exceed 2 in. (50 mm). Similar arrangements should be made for other types of members. Shoring should not influence or interfere with the free movements of the structure under the test load and should be designed and constructed to protect all people working on, below, or beside the structure to be tested in case of excessive deformation or collapse.

6.2.7 *Static load tests of flexural members*—Static load testing of flexural members is addressed in ACI 437.2. Two loading protocols—monotonic and cyclic—are described.

The monotonic protocol involves the application of load for a period of 24 hours followed by a recovery period under no applied load. The cyclic protocol involves loading and unloading in incremental steps and generally requires less time to perform than the monotonic protocol. The cyclic protocol also provides the opportunity to better understand end fixity and load transfer characteristics of the tested component by comparing actual with calculated responses as the load test progresses. In addition to the information provided in ACI 437.2, the following guidelines are presented:

a) Shoring and instrumentation should be installed before any test load is applied. A series of base readings should be taken immediately before the application of the test load to serve as a datum for making measurements on the various elements of the structure during the load test.

b) No portion of the load that represents live loads should be applied before the deflections due to the dead load and superimposed dead load have effectively reached constant values.

c) After total dead load deflections have stabilized, existing cracks and other defects should be observed, marked, and recorded.

d) The licensed design professional should closely inspect the structure following application of each load increment for the formation or worsening of cracking and distress, as well as for the presence of excessive deformations or rotations. The licensed design professional should analyze the significance of any distress and determine whether it is safe to continue with the test.

e) If the cyclic protocol is followed, load-deflection curves should be developed for all critical points of deflection measurements during the load test. Electronic data-gathering and plotting equipment is available to automatically plot such curves. These curves should be closely monitored during the load test. They are a valuable tool in determining the load-deflection response of the structure and for determining if the structure is behaving elastically as the test load is increased. If the monotonic protocol is followed, deflection measurements should be recorded at each load increment, at the beginning and end of the 24-hour sustained load period, and at least 24 hours after the removal of the load. If the structure has met the acceptance criteria at the conclusion of the loading cycle, it is not necessary to wait the full 24-hour period to make the final set of response measurements.

6.2.8 *Static load tests of elements in shear*—Load testing in direct shear to proof test the shear capacity of structural elements (such as beams, two-way flat plates and slabs, and corbels) is not recommended because only limited deflections are expected before failure, providing limited warning. A great deal of reliance is placed on the judgment of a licensed design professional conducting a load test for shear capacity. If possible, analysis of shear-flexure interaction, such as at a beam-column joint, is recommended. Each test is unique in terms of the characteristics of the structural elements being evaluated. Therefore, specific guidelines for conducting such tests cannot be listed as for load tests of flexural members. The following guidelines are presented for consideration by a licensed design professional who determines that a load test for evaluation of shear capacity should be conducted:

a) The structure should be thoroughly examined before and during the test. It is important to establish the concrete strength, aggregate type, and the shear reinforcement details as constructed. These parameters significantly impact the shear capacity of a structural element.

b) The load test should be preceded by a structural analysis to predict the performance of the structure.

c) Shoring of the structure is imperative. Provide shoring similar to that discussed for testing flexural members. Shoring should be designed for impact loading in the event of a sudden failure during the load test.

d) Instrumentation of the structure should concentrate on shear crack-width monitoring and shear crack extension in addition to deflections. Electronic instrumentation should be used to monitor crack widths to avoid the need for workers to read mechanical instruments during the load test.

e) The critical components of the structure should be monitored continuously during the test.

f) If load testing is planned for two-way slab systems, attention should be paid to the effect that the transfer of unbalanced moments has on punching shear at columns supporting unequal spans or unequally loaded spans, particularly for structures designed before modern code requirements.

g) If testing an element that is likely to fail in shear, but has significant flexural contribution (for example, the shear-flexure interaction at a beam-column joint), the loading to produce the expected failure should be calculated. Shear strength is typically, although not always, assessed at a distance d away from a support. Load placed less than a distance d away from a support may provide confinement and increase the shear capacity.

h) If minimum requirements are not specified by the code, acceptance criteria for the load test should be developed based on the judgment of a qualified licensed design professional with concurrence of the building official. Such acceptance criteria may be based on crack formation and movements at and along existing crack planes.

6.2.9 *Interpretation of load test results*—Engineering judgment should be exercised in developing an appropriate interpretation of the results of a load test conducted on a concrete structure or elements within the structure. Sometimes, a concrete structure is believed to be deficient but passes a load test. This behavior can be the result of one or more of the following reasons:

a) The concrete structure has been designed conservatively. There are numerous reasons for a high degree of conservatism in reinforced concrete construction. These include the use of supplemental reinforcing steel placed arbitrarily in the structure to minimize cracking, using bar layouts with larger areas than required by calculation, use of conservative design theories, overestimation of dead loads, and inaccurate modeling of boundary and support conditions.

b) Actual concrete compressive strengths may exceed the specified design strengths.

c) The structural analyses do not accurately model the load-sharing characteristics of the structure.
d) Membrane forces may play a significant role in increasing the load capacity of reinforced and prestressed concrete slabs (Vecchio and Collins 1990).

CHAPTER 7—REFERENCES

Committee documents are listed first by document number and year of publication followed by authored documents listed alphabetically.

American Association of State Highway and Transportation Officials

AASHTO T 260-11—Sampling and Testing for Total Chloride Ion Content in Concrete and Concrete Raw Materials

American Concrete Institute

ACI 201.1R-08—Guide for Conducting a Visual Inspection of Concrete in Service

ACI 207.3R-18—Report on Practices for Evaluation of Concrete in Existing Massive Structures for Service Conditions

ACI 209R-92(08)—Prediction of Creep, Shrinkage, and Temperature Effects in Concrete Structures

ACI 214.4R-10(16)—Guide for Obtaining Cores and Interpreting Compressive Strength Results

ACI 216.1-14—Code Requirements for Determining Fire Resistance of Concrete and Masonry Construction Assemblies

ACI 222R-01(10)—Protection of Metals in Concrete against Corrosion

ACI 222.2R-14—Report on Corrosion of Prestressing Steels

ACI 222.3R-11—Guide to Design and Construction Practices to Mitigate Corrosion of Reinforcement in Concrete Structures

ACI 224R-01(08)—Control of Cracking in Concrete Structures

ACI 224.1R-07—Causes, Evaluation, and Repair of Cracks in Concrete Structures

ACI 224.4R-13—Guide to Design Detailing to Mitigate Cracking

ACI 228.1R-19—Report on Methods for Estimating In-Place Concrete Strength

ACI 228.2R-13—Report on Nondestructive Test Methods for Evaluation of Concrete in Structures

ACI 309.2R-15—Guide to Identification and Control of Visible Effects of Consolidation on Formed Concrete Surfaces

ACI 318-14—Building Code Requirements for Structural Concrete

ACI 349.3R-18—Report on Evaluation and Repair of Existing Nuclear Safety-Related Concrete Structures

ACI 350-06—Code Requirements for Environmental Engineering Concrete Structures

ACI 362.2R-00(13)—Guide for Structural Maintenance of Parking Structures

ACI 364.1R-19—Guide for Evaluation of Concrete Structures Before Rehabilitation

ACI 365.1R-17—Report on Service-Life Prediction

ACI 376-11—Code Requirements for Design and Construction of Concrete Structures for the Containment of Refrigerated Liquified Gases and Commentary

ACI 423.4R-14—Report on Corrosion and Repair of Unbonded Single-Strand Tendons

ACI 423.8R-10—Report on Corrosion and Repair of Grouted Multistrand and Bar Tendon Systems

ACI 435R-95(00)—Control of Deflections in Concrete Structures

ACI 435.8R-85(97)—Observed Deflections of Reinforced Concrete Slab Systems, and Causes of Large Deflections

ACI 437.1R-07—Load Tests of Concrete Structures: Methods, Magnitude, Protocols, and Acceptance Criteria

ACI 437.2-13—Code Requirements for Load Testing of Existing Concrete Structures and Commentary

ACI 440.2R-17—Guide for the Design and Construction of Externally Bonded FRP Systems for Strengthening Concrete Structures

ACI 562-16—Code Requirements for Assessment, Repair, and Rehabilitation of Existing Concrete Structures and Commentary

American Society of Civil Engineers

ASCE 7-16—Minimum Design Loads for Buildings and Other Structures

ASCE 41-17—Seismic Evaluation and Retrofit of Existing Buildings

ASTM International

ASTM A370-18—Standard Test Methods and Definitions for Mechanical Testing of Steel Products

ASTM C39/C39M-18—Standard Test Method for Compressive Strength of Cylindrical Concrete Specimens

ASTM C42/C42M-18—Standard Test Method for Obtaining and Testing Drilled Cores and Sawed Beams of Concrete

ASTM C114-18—Standard Test Methods for Chemical Analysis of Hydraulic Cement

ASTM C295/C295M-18—Standard Guide for Petrographic Examination of Aggregates for Concrete

ASTM C457/C457M-16—Standard Test Method for Microscopical Determination of Parameters of the Air-Void System in Hardened Concrete

ASTM C597-16—Standard Test Method for Pulse Velocity through Concrete

ASTM C617/C617M-15—Standard Practice for Capping Cylindrical Concrete Specimens

ASTM C803/C803M-18—Standard Test Method for Penetration Resistance of Hardened Concrete

ASTM C805/C805M-18—Standard Test Method for Rebound Number of Hardened Concrete

ASTM C823/C823M-12(2017)—Standard Practice for Examination and Sampling of Hardened Concrete in Constructions

ASTM C856-18—Standard Practice for Petrographic Examination of Hardened Concrete

ASTM C876-15—Standard Test Method for Half-Cell Potentials of Uncoated Reinforcing Steel

ASTM C900-15—Standard Test Method for Pullout Strength of Hardened Concrete

ASTM C1152/C1152M-04(2012)—Standard Test Method for Acid-Soluble Chloride in Mortar and Concrete

ASTM C1218/C1218M-17—Standard Test Method for Water-Soluble Chloride in Mortar and Concrete

ASTM C1231/C1231M-15—Standard Practice for Use of Unbonded Caps in Determination of Compressive Strength of Hardened Concrete Cylinders

ASTM C1383-15—Standard Test Method for Measuring P-Wave Speed and the Thickness of Concrete Plates Using the Impact-Echo Method

ASTM C1583/C1583M-13—Standard Test Method for Tensile Strength of Concrete Surfaces and the Bond Strength or Tensile Strength of Concrete Repair and Overlay Materials by Direct Tension (Pull-off Method)

ASTM C1723-16—Standard Guide for Examination of Hardened Concrete Using Scanning Electron Microscopy

ASTM C1740-16—Standard Practice for Evaluating the Condition of Concrete Plates Using the Impulse-Response Method

ASTM D4580/D4580M-12(2018)—Standard Practice for Measuring Delaminations in Concrete Bridge Decks by Sounding

ASTM D4748-10(2015)—Standard Test Method for Determining the Thickness of Bound Pavement Layers Using Short-Pulse Radar

ASTM D4788-03(2013)—Standard Test Method for Detecting Delaminations in Bridge Decks Using Infrared Thermography

ASTM D6087-08(2015)—Standard Test Method for Evaluating Asphalt-Covered Concrete Bridge Decks Using Ground Penetrating Radar

ASTM D7522/D7522M-15—Standard Test Method of Pull-Off Strength for FRP Laminate Systems Bonded to Concrete Substrate

ASTM E122-17—Standard Practice for Choice of Sample Size to Estimate a Measure of Quality for a Lot or Process

International Concrete Repair Institute (ICRI)

ICRI 210.3-13—Guide for Using In-Situ Tensile Pull-off Tests to Evaluate Bond of Concrete Surface Materials

ICRI 330.2-16—Guide Specification for Externally Bonded FRP Fabric Systems for Strengthening Concrete Structures

Authored documents

Anderson, N., and Popovic, P., 1988, "Load Testing of Long Span Prestressed Single Tee Beams," *PCI Journal*, V. 33, No. 2, Mar.-Apr., pp. 42-51. doi: 10.15554/pcij.03011988.42.51

Barboni, M.; Benedetti, A.; and Nanni, A., 1997, "Carbon FRP Strengthening of Doubly Curved Precast PC Shell," *Journal of Composites for Construction*, V. 1, No. 4, Nov., pp. 168-174. doi: 10.1061/(ASCE)1090-0268(1997)1:4(168)

Bares, R., and FitzSimons, N., 1975, "Load Tests of Building Structures," *Journal of the Structural Division*, V. 101, May, pp. 1111-1123.

Bartlett, F. M., and MacGregor, J. G., 1994a, "Effect of Core Diameter on Concrete Core Strengths," *ACI Materials Journal*, V. 91, No. 5, Sept.-Oct., pp. 460-470. doi: 10.14359/4160

Bartlett, F. M., and MacGregor, J. G., 1994b, "Effect of Core Length-to-Diameter Ratio on Concrete Core Strengths," *ACI Materials Journal*, V. 91, No. 4, July-Aug., pp. 339-348. doi: 10.14359/4160

Bartlett, F. M., and MacGregor, J. G., 1994c, "Effect of Moisture Condition on Concrete Core Strengths," *ACI Materials Journal*, V. 91, No. 3, May-June, pp. 227-236. doi: 10.14359/4328

Bartlett, F. M., and MacGregor, J. G., 1995, "Equivalent Specified Concrete Strength from Core Test Data," *Concrete International*, V. 17, No. 5, May, pp. 52-58.

Bartlett, F. M., and MacGregor, J. G., 1996, "Statistical Analysis of the Compressive Strength of Concrete in Structures," *ACI Materials Journal*, V. 93, No. 2, Mar.-Apr., pp. 158-168. doi: 10.14359/1353

Bartlett, F. M., and MacGregor, J. G., 1997, "Precision of In-Place Concrete Strengths Predicted Using Core Strength Correction Factors Obtained by Weighted Regression Analysis," *Structural Safety*, V. 19, No. 4, pp. 397-410. doi: 10.1016/S0167-4730(97)00020-9

Bartlett, F. M., and MacGregor, J. G., 1999, "Variation of In-Place Concrete Strength in Structures," *ACI Materials Journal*, V. 96, No. 2, Mar.-Apr., pp. 261-269. doi: 10.14359/454

Bloem, D. L., 1965, "Concrete Strength Measurement—Cores vs. Cylinders," *Proceedings of ASTM*, V. 65, pp. 668-696.

Bloem, D. L., 1968, "Concrete Strength in Structures," *ACI Journal Proceedings*, V. 65, No. 3, Mar., pp. 176-187.

Bungey, J. H., 1979, "Determining Concrete Strength by Using Small-Diameter Cores," *Magazine of Concrete Research*, V. 31, No. 107, June, pp. 91-98. doi: 10.1680/macr.1979.31.107.91

Bungey, J. H., 1989, *The Testing of Concrete in Structures*, second edition, Chapman and Hall, New York, 228 pp.

Bungey, J. H., and Madandoust, R., 1992, "Factors Influencing Pull-off Tests on Concrete," *Magazine of Concrete Research*, V. 44, No. 158, Mar., pp. 21-30. doi: 10.1680/macr.1992.44.158.21

Bungey, J. H.; Millard, S. G.; and Grantham, M., 2006, *Testing of Concrete in Structures*, fourth edition, Taylor & Francis, 339 pp.

Bungey, J. H.; Millard, S. G.; and Shaw, M. R., 1994, "The Influence of Reinforcing Steel on Radar Surveys of Concrete Structures," *Construction & Building Materials*, V. 8, No. 2, pp. 119-126. doi: 10.1016/S0950-0618(09)90021-6

Carino, N. J., 2004a, "Pullout Test," *Handbook on Nondestructive Testing of Concrete*, second edition, V. M. Malhotra, and N. J. Carino, eds., CRC Press Inc., Boca Raton, FL.

Carino, N. J., 2004b, "Stress Wave Propagation Methods," *Handbook on Nondestructive Testing of Concrete*, V. M. Malhotra and N. J. Carino, eds., CRC Press, Boca Raton, FL.

Carino, N. J.; Sansalone, M.; and Hsu, N. N., 1986, "Flaw Detection in Concrete by Frequency Spectrum Analysis of Impact-Echo Waveforms," *International Advances in Nondestructive Testing*, W. J. McGonnagle, ed., Gordon & Breach Science Publishers, New York, pp. 117-146.

Chabowski, A. J., and Bryden-Smith, D., 1979, "A Simple Pullout Test to Assess the In-Situ Strength of Concrete," *Concrete International*, V. 1, No. 12, Dec., pp. 35-40.

Concrete Innovation Appraisal Service, 2000, "Guidelines for the Rapid Load Testing of Concrete Structural Members," *CIAS Report* 00-1, Farmington Hills, MI, 97 pp.

Concrete Reinforcing Steel Institute, 2014, *Vintage Steel Reinforcement in Concrete Structures*, Schaumburg, IL, 352 pp.

Davis, A., and Dunn, C., 1974, "From Theory to Field Experience with the Nondestructive Vibration Testing of Piles," *Proceedings – Institution of Civil Engineers*, V. 57, No. 4, Dec., pp. 571-593. doi: 10.1680/iicep.1974.3895

Davis, A. G.; Evans, J. G.; and Hertlein, B. H., 1997, "Nondestructive Evaluation of Concrete Radioactive Waste Tanks," *Journal of Performance of Constructed Facilities*, V. 11, No. 4, pp. 161-167. doi: 10.1061/(ASCE)0887-3828(1997)11:4(161)

Davis, A. G., and Hertlein, B. H., 1995, "Nondestructive Testing of Concrete Chimneys and Other Structures," *Proceedings of Conference on Nondestructive Evaluation of Aging Structures and Dams*, SPIE, V. 2456, pp. 129-136.

Dixon, D., 1987, "Analysis of Cover Meter Data," *Computer Use for Statistical Analysis of Concrete Test Data*, SP-101, P. Balaguru and V. Ramakrishnan, eds., American Concrete Institute, Farmington Hills, MI, pp. 47-64.

Domone, P. L., and Castro, P. F., 1987, "An Expanding Sleeve Test for In-Situ Concrete and Mortar Strength Evaluation," *Proceedings*, Structural Faults and Repairs 87, Engineering Technics Press, Edinburgh, UK.

El Batanouny, M. K.; Ziehl, P. H.; Larosche, A.; Mangual, J.; Matta, F.; and Nanni, A., 2014, "Acoustic Emission Monitoring for Assessment of Prestressed Concrete Beams," *Construction & Building Materials*, V. 58, pp. 46-53. doi: 10.1016/j.conbuildmat.2014.01.100

Elstner, R. C.; Gustafson, D. P.; Hanson, J. M.; and Rice, P. F., 1987, "Application of ACI 318 Load Test Requirements," *CRSI Professional Members' Bulletin* No. 16, Concrete Reinforcing Steel Institute, Schaumburg, IL, 11 pp.

Erlin, B., 1994, "Petrographic Examination," *Significance of Tests and Properties of Concrete and Concrete-Making Materials, Part III, Tests and Properties of Concrete Aggregates,* ASTM STP 169B, P. Klieger and J. F. Lamond, eds., ASTM International, West Conshohocken, PA, pp. 210-218.

Federal Communications Commission (FCC), 2002, "DA 02-1658 Revision of Part 15 of the Commission's Rules Regarding Ultra-Wideband Transmission Systems," Citation – 17 FCC Rcd 13522 (19), July.

FitzSimons, N., and Longinow, A., 1975, "Guidance for Load Tests of Buildings," *Journal of the Structural Division*, V. 101, July, pp. 1367-1380.

Fling, R. S.; McCrate, T. E.; and Doncaster, C. W., 1989, "Load Test Compared to Earlier Structure Failure," *Concrete International*, V. 18, No. 11, Nov., pp. 22-27.

Galati, N.; Nanni, A.; Tumialan, J. G.; and Ziehl, P. H., 2008, "In Situ Evaluation of Two Concrete Slab Systems I: Load Determination and Loading Procedure," *Journal of Performance of Constructed Facilities*, V. 22, No. 4, Aug., pp. 207-216. doi: 10.1061/(ASCE)0887-3828(2008)22:4(207)

Gaynor, R. D., 1965, "Effect of Horizontal Reinforcing Steel on the Strength of Molded Cylinders," *ACI Journal Proceedings*, V. 62, No. 7, July, pp. 837-840. doi: 10.14359/51685622

Guedelhoefer, O. C., and Janney, J. R., 1980, "Evaluation of Performance by Full Scale Testing," *Full-Scale Load Testing of Structures*, ASTM STP-702, W. R. Schriever, ed., ASTM International, West Conshohocken, PA, pp. 5-24.

Hall, W. B., and Tsai, M., 1989, "Load Testing, Structural Reliability and Test Evaluation," *Structural Safety*, V. 6, No. 2-4, pp. 285-302. doi: 10.1016/0167-4730(89)90028-3

Hanson, J. M., 2007, "Survey of Practice to Determine Strength of In Situ Concrete from Core Tests," *Journal of Performance of Constructed Facilities*, V. 21, No. 1, Feb., pp. 22-25. doi: 10.1061/(ASCE)0887-3828(2007)21:1(22)

Ivanyi, M., 1976, "Load Tests of Building Structures," *Journal of the Structural Division*, V. 102, June, pp. 1260-1261.

Jaeger, B. J.; Sansalone, M. J.; and Poston, R. W., 1996, "Detecting Voids in Grouted Tendon Ducts of Post-Tensioned Concrete Structures Using the Impact-Echo Method," *ACI Structural Journal*, V. 93, No. 4, July-Aug., pp. 462-472. doi: 10.14359/9705

Kaminetzky, D., 1991, *Design and Construction Failures Lessons from Forensic Investigations*, McGraw-Hill, 600 pp.

Kesner, K.; Poston, R. W.; Salmassian, K.; and Fulton, G. R., 1999, "Repair of Marina del Ray Seawall," *Concrete International*, V. 21, No. 12, Dec., pp. 43-50.

Lin, J. M., and Sansalone, M. J., 1996, "Impact-Echo Studies of Interfacial Bond Quality in Concrete: Part I—Effects of Unbonded Fraction of Area," *ACI Materials Journal*, V. 93, No. 3, May-June, pp. 223-232.

Mailhot, G.; Bisaillon, A.; Carette, G. G.; and Malhotra, V. M., 1979, "In-Place Concrete Strength: New Pullout Methods," *ACI Journal Proceedings*, V. 76, No. 12, Dec., pp. 1267-1282.

Mailvaganam, N. P., 1992, *Repair and Protection of Concrete Structures*, CRC Press, Boca Raton, FL, 473 pp.

Malhotra, V. M., 1976, "Testing Hardened Concrete: Non-Destructive Methods," *ACI Monograph* No. 9, American Concrete Institute, Farmington Hills, MI, 204 pp.

Malhotra, V. M., and Carino, N. J., eds., 2004, *Handbook on Nondestructive Testing of Concrete*, second edition, CRC Press Inc., Boca Raton, FL.

Mariscotti, M. A. J.; Jalinoos, F.; Frigerio, T.; Ruffolo, M.; and Thieberger, P., 2009, "Gamma-Ray Imaging for Void and Corrosion Assessment," *Concrete International*, V. 31, No. 11, Nov., pp. 48-53.

Mettemeyer, M.; Serra, P.; Wuerthele, M.; Schuster, G.; and Nanni, A., 1999, "Shear Load Testing of CFRP Strengthened Double-Tee Beams in Precast Parking Garage," *Fiber Reinforced Polymer Reinforcement for Reinforced Concrete Structures*, Fourth International Symposium, SP-188, C. W. Dolan, S. H. Rizkalla, and A. Nanni, eds., American Concrete Institute, Farmington Hills, MI, pp. 1063-1072.

Mielenz, R. C., 1994, "Petrographic Evaluation of Concrete Aggregates," *Significance of Tests and Properties of Concrete and Concrete-Making Materials, Part IV, Tests and Properties of Concrete Aggregates*, ASTM STP 169B, P. Klieger and J. F. Lamond, eds., ASTM International, West Conshohocken, PA, pp. 341-364.

Mindess, S., and Young, F., 1981, *Concrete*, Prentice Hall, Inc., 671 pp.

Nanni, A., and Gold, W. J., 1998a, "Evaluating CFRP Strengthening Systems In-Situ," *Concrete Repair Bulletin*, V. 11, No. 1, Jan.-Feb., pp. 12-14.

Nanni, A., and Gold, W. J., 1998b, "Strength Assessment of External FRP Reinforcement," *Concrete International*, V. 20, No. 6, June, pp. 39-42.

Nanni, A.; Gold, W. J.; Thomas, J.; and Vossoughi, H., 1998, "FRP Strengthening and On-Site Evaluation of a PC Slab," *Proceedings of the Second International Conference on Composites in Infrastructure (ICCI-98)*, V. 1, H. Saadatmanesh and M. R. Ehsani, eds., Tucson, AZ, pp. 202-212.

Nanni, A., and Mettemeyer, M., 2001, "Diagnostic Load Testing of a Two-Way Post-Tensioned Concrete Slab," *Practice Periodical on Structural Design and Construction*, V. 6, No. 2, May, pp. 73-82. doi: 10.1061/(ASCE)1084-0680(2001)6:2(73)

Nehil, T. E.; Nanni, A.; and Masetti, F., 2007, "Strength Evaluation by Load Testing," *Concrete International*, V. 29, No. 3, Mar., pp. 53-58.

Neville, A., 2001, "Core Tests: Easy to Perform, Not Easy to Interpret," *Concrete International*, V. 23, No. 11, Nov., pp. 59-68.

Petersen, C. G., 1997, "LOK-Test and CAPO-Test Pullout Testing, Twenty Years' Experience," *Proceedings of Conference on Non-Destructive Testing in Civil Engineering*, J. H. Bungey, ed., British Institute of Non-Destructive Testing, pp. 77-96.

Popovic, P.; Stork, P.; and Arnold, R., 1991, "Use of Hydraulics for Load Testing of Prestressed Concrete Inverted Tee Girder," *PCI Journal*, V. 36, No. 6, Sept.-Oct., pp. 72-78. doi: 10.15554/pcij.09011991.72.78

Powers, L. J., 2002, "Petrography as a Concrete Repair Tool," *Concrete Repair Bulletin*, V. 15, No. 1, pp. 22-25.

Prestressed/Precast Concrete Institute, 2004, *PCI Design Handbook,* sixth edition, Chicago, IL, 728 pp.

Raths, C. H., and Guedelhoefer, O. C., 1980, "Correlation of Load Testing with Design," *Full-Scale Load Testing of Structures,* ASTM STP 702, W. R. Schriever, ed., ASTM International, West Conshohocken, PA, pp. 91-113.

Sansalone, M., and Carino, N. J., 1986, "Impact Echo, A Method for Flaw Detection in Concrete Using Transient Stress Waves," *Report* No. NBSIR 86-3452, National Bureau of Standards, NTIS, Springfield, VA, 234 pp.

Sansalone, M., and Carino, N. J., 1988, "Impact-Echo Method: Detecting Honeycombing, the Depth of Surface-Opening Cracks, and Ungrouted Ducts," *Concrete International*, V. 10, No. 4, Apr., pp. 38-46.

Sansalone, M., and Carino, N. J., 1989, "Detecting Delaminations in Concrete Slabs with and without Overlays Using the Impact Echo Method," *ACI Materials Journal*, V. 86, No. 2, Mar.-Apr., pp. 175-184. doi: 10.14359/2350

Sansalone, M., and Streett, W. B., 1997, *Impact-Echo: Nondestructive Testing of Concrete and Masonry*, Bullbrier Press, Jersey Shore, PA.

Snell, L. M.; Wallace, N.; and Rutledge, R. B., 1988, "Locating Reinforcement in Concrete," *Concrete International*, V. 10, No. 4, Apr., pp. 19-23.

Stone, W. C., and Carino, N. J., 1983, "Deformation and Failure in Large-Scale Pullout Tests," *ACI Journal Proceedings*, V. 80, No. 6, Nov.-Dec., pp. 501-513. doi: 10.14359/10871

Tam, C. T.; Lai, L. H.; and Lam, P. W., 1977, "Orthogonal Detection Technique for Determination of Size and Cover of Embedded Reinforcement," *Journal of Institution of Engineers (Malaysia)*, V. 22, June, pp. 6-16.

Tumialan, J. G.; Galati, N.; and Nanni, A., 2012, *Inspection, Testing, and Monitoring of Buildings and Bridges*, The National Council of Structural Engineers Associations (NCSEA), Chicago, IL.

Vecchio, F. J., and Collins, M. P., 1990, "Investigating the Collapse of a Warehouse," *Concrete International*, V. 12, No. 3, Mar., pp. 72-78.

Wen, Y. K., and Yeo, G. L., 2001, "Design Live Loads for Passenger Cars Parking Garages," *Journal of Structural Engineering*, V. 127, No. 3, pp. 280-289. doi: 10.1061/(ASCE)0733-9445(2001)127:3(280)

Williams, T. J.; Nozick, L. K.; Sansalone, M. J.; and Poston, R. W., 2006a, "Sampling Techniques for Evaluating Large Concrete Structures: Part I," *ACI Structural Journal*, V. 103, No. 3, May-June, pp. 399-408.

Williams, T. J.; Nozick, L. K.; Sansalone, M. J.; Poston, R. W.; and Kesner, K., 2006b, "Sampling Techniques for Evaluating Large Concrete Structures: Part II," *ACI Structural Journal*, V. 103, No. 4, July-Aug., pp. 505-512.

Wouters, J. P.; Peterson, J. E.; Kesner, K.; and Poston, R. W., 1999, "Impact-Echo Locates Subgrade Voids," *Concrete International*, V. 21, No. 5, May, pp. 23-27.

Zhan, R., and Xie, H., 2009, "GPR Measurement of the Diameter of Steel Bars in Concrete Specimens Based on the Stationary Wavelet Transform," *Insight—Non-Destructive Testing and Condition Monitoring*, V. 51, No. 3, pp. 151-155.

Ziehl, P. H.; Galati, N.; Nanni, A.; and Tumialan, J. G., 2008, "In-Situ Evaluation of Two Concrete Slab Systems II: Evaluation Criteria and Outcomes," *Journal of Performance of Constructed Facilities*, V. 22, No. 4, pp. 217-227. doi: 10.1061/(ASCE)0887-3828(2008)22:4(217)

Load Tests of Concrete Structures: Methods, Magnitude, Protocols, and Acceptance Criteria

Reported by ACI Committee 437

Antonio Nanni*
Chair

Jeffrey S. West
Secretary

Tarek Alkhrdaji	Ashok M. Kakade	Javeed Munshi	Thomas Rewerts*
Joseph A. Amon*	Dov Kaminetzky	Thomas E. Nehil†	K. Nam Shiu
Nicholas J. Carino	Andrew T. Krauklis	Renato Parretti	Avanti C. Shroff
Paolo Casadei	Chuck J. Larosche	Brian J. Pashina	Jay Thomas
Ufuk Dilek	Michael W. Lee	Stephen Pessiki	Jeffrey A. Travis
John Frauenhoffer*	Daniel J. McCarthy*	Predrag L. Popovic	Fernando V. Ulloa
Zareh B. Gregorian	Patrick R. McCormick	Guillermo Ramirez*	Paul H. Ziehl*
Pawan R. Gupta	Matthew A. Mettemeyer*		

*Member of subcommittee that prepared this report.
†Chair of subcommittee that prepared this report.

This report provides the recommendations of Committee 437 regarding selection of test load magnitudes, protocol, and acceptance criteria to be used when performing load testing as a means of evaluating safety and serviceability of concrete structural members and systems. The history of load factors and acceptance criteria as found in the ACI 318 building code is provided along with a review of other load test practice. Recommended revisions to load factors to be used at this time, additions to load testing protocol, and revisions to acceptance criteria used to evaluate the findings of load testing are provided.

Keywords: acceptance criteria; cyclic load test; deflection; deterioration; load test factors; load test protocol; monotonic load test; reinforced concrete; strength evaluation.

CONTENTS

ACI 437.1R-07 was adopted and published March 2007.

CHAPTER 1—INTRODUCTION

1.1—Background

Significant revisions were made in Chapter 9 of ACI 318-02 to the load factors to be used for determining required strength. The load factor for dead load was reduced from 1.4 to 1.2, and the load factor for live load was reduced from 1.7 to 1.6; other changes were also made as given in equations for required strength in Chapter 9. The strength-reduction factors (ϕ-factors) were also modified. The ϕ-factor for shear and torsion was reduced from 0.85 to 0.75, while the ϕ-factor for compression-controlled members was reduced from 0.70 to 0.65 unless spiral reinforcement is provided. The ϕ-factor for tension-controlled members (most flexural members) was not reduced, and remains 0.9.

The load factors and load combinations of ACI 318-05 match those of ASCE 7-02 (American Society of Civil Engineers 2002). The changes were made to unify the load factors used to design concrete structures with those generally used to design structures constructed of other materials, such as structural steel. The changes also facilitated the design of concrete structures that included members of materials other than concrete.

Chapter 20 (Strength Evaluation of Existing Structures) of 318-02 and 318-05 was not changed from the previous code with regard to load test procedures. Section 20.3.2 (Load Intensity) of ACI 318-02 was not changed from the 1999 edition; that is, the total test load (including dead load already in place) was still defined to be not less than $0.85(1.4D + 1.7L)$, with live load permitted to be reduced in accordance with the applicable building code.

The reduction in load factors used for computing required strength without a corresponding reduction in the test load intensity resulted in two effects. First, the test load was no longer a fixed percentage of the required strength. Second, the test load was now in the range of 93 to 98% of the required strength for tension-controlled sections rather than 85% of the required strength as was the case in ACI 318-71 through 318-99.

ACI Committee 318 requested that Committee 437 review and report on the load intensity requirements of Chapter 20. In the process, Committee 437 has undertaken a thorough review of the historical background of load testing and developed not only recommendations for revisions to the test load magnitude (TLM), but also to the protocol for load testing and the acceptance criteria used to evaluate the results.

1.2—Introduction

The provisions of Chapter 20 of ACI 318 have remained essentially unchanged for an unprecedented period of time since the publication of ACI 318-71, when the code was changed from working stress design to ultimate strength design. Before the 1971 code, the test load requirements or acceptance criteria were revised with almost every new edition of the code dating back to 1920. Chapter 3 and Appendix B of this report provide a detailed review of the history of the load test requirements and acceptance criteria in ACI 318. They also provide a discussion of other international standards and of significant research and reporting of other organizations on the subject of load testing.

The changes made in the load factors and load combinations of ACI 318-05 require a re-examination of the load test requirements of Chapter 20 of ACI 318-05. This report presents the recommendations of Committee 437 for revisions to the requirements of Chapter 20. Three key areas are addressed: load factors to be used in defining the TLM; the load test protocol; and acceptance criteria.

As will be discussed further in Chapter 4, the purposes of the recommended revisions to the TLM definition are twofold. The first purpose is to define a test load that will demonstrate a consistent safe margin of capacity over code-required service live load levels. Secondly, the definition of the test load primarily in terms of service live load rather than required (ultimate) strength is meant to emphasize the fact that load testing is (typically) a proof loading. In the experience of the committee members, most structures being load tested pass with small deflections. Load testing does not typically provide an indication of the ultimate strength of the structure, and that indication usually is not the goal of load testing.

Since 1920, the acceptance criteria used with load testing have incorporated a limit on measured maximum deflections after a 24-hour holding period of the total test load. The current criteria have not changed since ACI 318-63. Currently, the deflection limit is described by the formula $\Delta_{max} \leq l_t^2/20{,}000h$. The theoretical basis for this formula had its origins in the first decades of the 20th century. The committee has researched the origins of the formula and re-evaluated its appropriateness. The committee recommends adopting other more meaningful deflection acceptance criteria.

Chapters 5 and 6 of the report discuss selection of a load test protocol and recommended changes to the acceptance criteria used in strength evaluation and load testing. Committee 437 in its report 437R-03, "Strength Evaluation of Existing Concrete Buildings," has discussed a cyclic load test method that offers advantages in terms of reliability and understanding of structural response to load when compared with the conventional static load test. Chapter 6 presents recommended acceptance criteria for both the 24-hour static test and for the cyclic test. Acceptance criteria for serviceability are also given.

1.3—Limitations

Procedures and recommendations provided in this report are intended for structures and buildings using concretes of normal strengths. The methods are not intended for bridges, structures with unusual design concepts, or other special structures. The methods are not intended to be used for product development testing where load testing is used for quality control or approval of mass-produced members. Testing for resistance to wind and seismic loads is not discussed. AASHTO provisions for load testing of bridge structures are outside the scope of this report. Load testing to determine ultimate strength is also outside the scope of this report.

CHAPTER 2—NOTATION AND TERMINOLOGY

2.1—Notation

The notations reported in this section refer to the symbols used in the numbered chapters.

h = overall thickness of member, in. (mm)
l_t = span of member under load test; units depend on structural member considered (ACI 318)
s = average spacing between cracks, in. (mm)
D = total dead load: $D_w + D_s$; units depend on structural member considered
D_s = superimposed dead load; units depend on structural member considered
D_w = dead load due to self-weight; units depend on structural member considered
F = loads due to weight and pressure of fluids with well-defined densities and controllable maximum heights; units depend on structural member considered
I_{DL} = deviation from linearity index, dimensionless
I_P = permanency index, dimensionless
I_R = repeatability index, dimensionless
L = live loads produced by use and occupancy of the building not including construction, environmental loads, and superimposed dead loads; units depend on structural member considered
L_r = roof live loads produced during maintenance by workers, equipment, and materials or during life of structure by moveable objects such as planters and people; units depend on structural member considered
P = applied load during load test (Fig. 6.1 and 6.2)
P_i = load of point i in load-deflection envelope for computation of I_{DL} acceptance criterion (Fig. 6.2)
P_{min} = minimum load to be maintained during load test (typically 10% of total test load)
P_{ref} = reference load for computation of I_{DL} acceptance criterion (Fig. 6.2)
R = rain load, or related internal moments and forces; units depend on structural member considered
S = snow load; units depend on structural member considered
TL = test load per ACI 318 before 1971; units depend on structural member considered
TL_{05} = TL_{99} = test load per ACI 318-71 through ACI 318-05 = $0.85(1.4D + 1.7L) = 1.19D + 1.44L$; units depend on structural member considered
TLM = test load magnitude (including dead load already in place); units depend on structural member considered
U = required strength to resist factored loads
U_{99} = required strength per ACI 318-99 = $1.4D + 1.7L$
U_{05} = required strength per ACI 318-05 = $1.2D + 1.6L$
α_i = slope of secant line of point i in load-deflection envelope, degrees
α_{ref} = slope of reference secant line in load-deflection envelope, degrees
$\Delta\varepsilon_s$ = strain difference in longitudinal reinforcement
Δ_i = deflection of point i in load-deflection envelope for computation of I_{DL} acceptance criterion (Fig. 6.2)
Δ_{max} = measured maximum deflection, in. (mm)
Δ_{ref} = reference deflection for computation of I_{DL} acceptance criterion (Fig. 6.2)
$\Delta_{r\,max}$ = measured residual maximum deflection, in. (mm)
Δ^A_{max} = maximum deflection in Cycle A under maximum test load, in. (mm)
Δ^A_r = residual deflection after Cycle A under minimum test load, in. (mm)
Δ^B_{max} = maximum deflection in Cycle B under maximum load, in. (mm)
Δ^B_r = residual deflection after Cycle B under minimum test load, in. (mm)
ϕ = strength-reduction factor as per ACI 318

2.2—Terminology

The following definitions are important to the understanding of this report.

acceptance criteria—a set of explicit and quantitative rules to determine whether or not a structure (or a portion of it) passes a load test.

dead load (*D*), total—in this report, a distinction is made between dead load due to self-weight and superimposed dead loads. Total dead load D will include both dead load due to self-weight and superimposed dead loads; that is, $D = D_{w} + D_{s}$. This definition creates a distinction not used in ACI 318 or the International Building Code (IBC).

dead load (D_w), self-weight—dead load due to self-weight D_w is to include the weight of the concrete structural system only.

dead load (D_s), superimposed—this report uses superimposed dead load to designate all other weight of materials of construction incorporated into a building other than self-weight of the concrete structural system. Such loads include, but are not limited to, partitions, floor finishes, nonstructural topping slabs and overlays, roofing materials, ceiling finishes, cladding, stairways, fixed service equipment, and landscaping, including fixed planters, soils, and plantings.

failure—when referred to the performance of a structure (or a portion of it) under load test, it indicates that one or more acceptance criteria are not met.

proof load and proof load ratio—proof load is used to describe a load applied to a structure with intent to prove a safe margin of satisfactory performance beyond code-required service live and dead loads. For this reason, proof load is defined in terms of service loads and not in terms of required or ultimate strength. A proof load is generally not intended to provide an indication of the ultimate strength of the structure. Arithmetically, the proof load ratio is defined as the TLM minus the total dead load divided by the service live load; that is, proof load ratio = $(TLM - D)/L$.

strip or patch test load—a test load distributed over a limited portion of the tributary area of the structure or member to be tested and typically applied by means of hydraulic jacks.

test load magnitude (TLM)—TLM is defined as all existing dead load due to self-weight and existing superimposed dead load plus additional test loads used to simulate effects of factored service live loads and factored superimposed dead loads. The factors to be applied to live loads and superimposed dead loads to establish the TLM are provided in Chapter 4. The factor for superimposed dead loads is to be applied to both existing superimposed dead loads and those not already in place.

CHAPTER 3—HISTORY OF LOAD TEST, LOAD FACTORS, AND ACCEPTANCE CRITERIA

3.1—Scope of historical review

An extensive review of the existing literature has been done to develop a history of load testing of reinforced concrete structures. The results of this work are reported in detail in Appendix B. The focus of this literature search has been in the following areas that are under consideration for revision in ACI 318:

- The purpose or goal of load testing, and the types of load tests that should be used;
- Development of appropriate superimposed loads to be used in a load test; and
- Establishment of appropriate acceptance criteria for structural response to those test loads.

Appendix B begins with a history of the development of load testing within the United States and development of ACI building code requirements for load testing. This section of the appendix is followed by a section presenting general discussion of work done by various organizations in the United States and around the world in the area of load testing of concrete structures. The purpose of Appendix B is to provide a historical perspective of changes to ACI 318 recommended by Committee 437. It serves to show the origins of the present state of practice and why changes are considered appropriate. It provides a discussion of research on and practices for load testing outside the United States.

3.2—Summary and conclusions

The key points drawn from the literature survey and derived conclusions are provided herein.

3.2.1 *Purpose of load testing*

1. Load testing originated in the late 1800s as proof (or acceptance) testing to show that a structure could resist specified service loads with a reasonable margin of safety against failure. It was generally not employed to determine the ultimate strength of a concrete member;

2. Provisions for load testing in ACI 318 and prevailing industry interpretations of those provisions have, over time, blurred to imply that the purpose of load testing is: 1) to ensure that the structure being tested meets the requirements of ACI 318; and 2) to assess the ultimate strength of that concrete structure; and

3. Consideration of historical information and data suggests that the purpose of load testing should be divided into three distinct categories:

   a. Proof testing to show that a structure can safely resist intended design loads with an adequate factor of safety against failure;
   b. Proof testing to show that a structure can resist the working design loads in a serviceable fashion where deflections and cracking are within limits considered acceptable by ACI 318; and
   c. Testing to failure to show the ultimate capacity of a structural member either in the field or as a model in a laboratory setting.

3.2.2 *Test load magnitude*

1. The test load magnitude used in U.S. load testing practice generally originated as two times the live load. This criterion has been found in the oldest references reviewed, including those dating into the late 1890s. The exact origin of this test load has not been found. It is believed to be a rule of thumb that was adopted in that era;

2. This test load was used for structures designed using allowable stress design techniques that are generally no longer used in the United States;

3. The criterion for using a superimposed test load of two times the live load was abandoned by ACI in 1963, although it continued to persist in various local and state building codes well beyond that time;

4. Load test practice in ACI did not change to any appreciable degree when ultimate strength design was introduced to the ACI 318 code in 1963 and 1971. Ultimate strength design

methods generally resulted in a lower factor of safety against failure than allowable stress design methods, and the resulting designs were often more flexible than those of the earlier methods. The TLM was scaled back approximately 10%; however, the deflection criteria remained unchanged;

5. Over time, the TLM has been modified in ACI 318 from a high of $TL = 1.5D + 2.0L$ to the current low of $TL = 0.85(1.4D + 1.7L)$, which equates to $TL = 1.19D + 1.44L$. As shown in Table B.4, no agreement exists regarding load factors for defining the test load magnitude in similar documents throughout the world. Ideally, a minimum factor of safety should be explicitly agreed upon in terms of TLM;

6. It is suggested that a load level consisting of the service load equal to $1.0D + 1.0L$ should be included in the load test procedure to provide for assessment of the serviceability of the structure. Deflections and crack widths should be compared with maximum allowable, code-defined, or desirable values; and

7. More specific criteria should be developed to define what constitutes visible evidence of failure.

3.2.3 *Protocol for application of the load test*

1. Modern practice for load testing seems to be turning in the direction of applying the test load in increments that include multiple cycles of incremental loading and unloading until the full desired test load is attained. This appears to have benefits relative to ensuring that the structure is adequately and properly responding to the desired test load in terms of deflection and deflection recovery;

2. Load test practice should include application of one or more preliminary load tests at values well below the full desired superimposed test load to assess the conditions of end restraint and fixity acting in the portion of the structure being tested and to identify the degree of load sharing that is occurring from the member being loaded to the surrounding monolithic or structurally attached members; and

3. Duration of the application of the full desired test load has historically been set at 24 hours. Because a sufficient correlation of shorter-term tests with 24-hour tests has not been found, the 24-hour holding period at full TLM should be retained in the code to take creep of concrete into consideration (even if to a limited extent) and to allow the structure to properly respond and adjust to the maximum test load.

3.2.4 *Acceptance criteria for load testing*

3.2.4.1 *Use of maximum deflection*

1. The current acceptance criterion for maximum allowable deflection (that is, $\Delta_{max} = l_t^2/20{,}000h$) in a load test was developed for simple span members and does not adequately reflect any variations in end fixity of structural members from that condition. Further, that equation was developed during the era of allowable stress design methods. The equation is based on concepts of uncracked sections and maximum allowable stress in concrete. The allowable stress and elastic modulus built into the equation were derived for lower-strength concrete than is often employed in design today. The equation does not take into account the actual strength and stiffness of the concrete in the member being tested;

2. No correlation exists between structural response to a test load of $TL = 0.85(1.4D + 1.7L)$ and the deflection criteria that are currently being used in ACI load test practice;

3. The maximum deflection of a structure following application of a test load should be compared, where possible, against calculated values using the best available calculation methods that are based on thorough and comprehensive field investigation of the physical and mechanical properties of the concrete in the area of the structure under investigation; and

4. It is the current provision of IBC 2003 to limit deflections during load tests to values established as simple percentages of the span (for example, $l_t/360$) relating to serviceability criteria.

3.2.4.2 *Use of deflection recovery*

1. With the single exception of work done and reported in Israel in 1950 (Arnan et al. 1950), historical load test practice suggests that deflection recovery can be properly used as an acceptance criterion for load testing of concrete structures. The concerns expressed in the 1950 Israeli report regarding deflection recovery can be addressed through implementation of a load test practice that includes preliminary load testing or application of the test load in several cycles of loading and unloading of the structure in increasing increments until the full test load is in place;

2. Historical practice suggests that the deflection recovery after 24 hours in a static load test, without incremental loading and unloading of the structure as suggested previously, should be at least 75%. The Israeli research and more current work with cyclic load testing suggest that the deflection recovery requirement should be significantly higher, on the order of 90%, when using the cyclic load test method or when retesting a structure using the static load test method; and

3. Alternative methods of analyzing deflection recovery data to establish new criteria for acceptance have been introduced recently to accompany the cyclic load test method. If cyclic load testing is to be incorporated into ACI 318, then the appropriate accompanying deflection recovery acceptance criteria need to be defined.

CHAPTER 4—LOAD FACTORS

4.1—Introduction

A revised definition of TLM should be developed to address the change of load factors and load combinations used in ACI 318-05 for defining required strength compared with load factors used in ACI 318-71 through 318-99. The new definition should address concerns regarding whether structures designed by earlier codes should have different TLMs than structures designed in accordance with ACI 318-05. The new definition should also address whether the load test will be performed on all suspect portions of a structure or only on selected limited areas.

This chapter presents recommendations for revisions to the definition of test load magnitude (TLM). The TLM is intended for proof testing; that is, load testing to show that a structure can safely support code-required service loads. Load testing to determine ultimate strength is outside the scope of this report.

Table 4.1—Design strength and test load comparison: full load test*

Type of facility	D_w, lb/ft²† (kN/m²) (1)	D_s, lb/ft² (kN/m²) (2)	L, lb/ft² (kN/m²) (3)	$\frac{L}{D}$ (4)	U_{99}, lb/ft² (kN/m²) (5)	U_{05}, lb/ft² (kN/m²) (6)	$\frac{U_{05}}{U_{99}}$ (7)	TL_{05}, lb/ft² (kN/m²) (8)	$\frac{TL_{05}}{U_{05}}$ (9)	$\frac{TL_{05}}{D+L}$ (10)	$\frac{TL_{05}-D}{L}$ (11)	TLM, lb/ft² (kN/m²) (12)	$\frac{TLM}{TL_{05}}$ (13)	$\frac{TLM}{U_{05}}$ (14)	$\frac{TLM}{U_{99}}$ (15)	$\frac{TLM-D}{L}$ (16)
Parking slab, unreduced live load	65 (3.11)	—	50 (2.39)	0.77	176 (8.43)	158 (7.57)	0.90	150 (7.18)	0.95	1.30	1.69	135 (6.46)	0.90	0.85	0.77	1.40
Parking beam, reduced live load	100 (4.79)	—	30 (1.44)	0.30	191 (9.15)	168 (8.04)	0.88	162 (7.76)	0.97	1.25	2.08	142 (6.80)	0.87	0.85	0.74	1.40
Office slab, unreduced live load	65 (3.11)	20 (0.96)	50 (2.39)	0.59	204 (9.77)	182 (8.71)	0.89	173 (8.28)	0.95	1.28	1.77	157 (7.52)	0.91	0.86	0.77	1.44
Storage, light	110 (5.27)	—	125 (5.99)	1.14	367 (17.57)	332 (15.90)	0.91	312 (14.94)	0.94	1.33	1.61	285 (13.65)	0.91	0.86	0.78	1.40
Storage, light with heavier structure	150 (7.18)	—	125 (5.99)	0.83	423 (20.25)	380 (18.19)	0.90	359 (17.19)	0.95	1.31	1.67	325 (15.56)	0.90	0.86	0.77	1.40
Storage, heavy	150 (7.18)	—	250 (11.97)	1.67	635 (30.40)	580 (27.77)	0.91	540 (25.86)	0.93	1.35	1.56	500 (23.94)	0.93	0.86	0.79	1.40
Manufacturing, very heavy	175 (8.38)	—	400 (19.15)	2.29	925 (44.29)	850 (40.70)	0.92	786 (37.63)	0.93	1.37	1.53	735 (35.19)	0.93	0.86	0.79	1.40
Landscaped pedestrian plaza	200 (9.58)	300‡ (14.36)	100 (4.79)	0.20	870 (41.66)	760 (36.39)	0.87	740 (35.43)	0.97	1.23	2.40	670 (32.08)	0.91	0.88	0.77	1.70
Plaza, truck dock	200 (9.58)	—	250 (11.97)	1.25	705 (33.76)	640 (30.64)	0.91	599 (28.68)	0.94	1.33	1.60	550 (26.33)	0.92	0.86	0.78	1.40
Average	—	—	—	—	—	—	0.90	—	0.95	1.31	—	—	0.91	0.86	0.77	1.44

*TLM definition for testing all suspect portions of structure.

†1 lb/ft² = 47.88 N/m².

‡Landscaped pedestrian plaza value of 300 lb/ft² (14.36 kN/m²) is not defined by ASCE-7, but is selected herein for illustrative purposes to represent 2.5 ft (0.76 m) of uniformly distributed saturated soil weighing 120 lb/ft³ (1922 kg/m³) such as might be encountered in a large fixed planter containing trees.

Definitions:

D_w = dead load to self-weight; D_s = superimposed dead load; $D = D_w + D_s$ = total dead load; and L = live load.

U_{99} = required strength per 318-99 = $1.4D + 1.7L$.

U_{05} = required strength per 318-05 = $1.2D + 1.6L$.

$TL_{05} = TL_{99}$ = test load per 318-71 through 318-05 = $0.85(1.4D + 1.7L) = 1.19D + 1.44L$.

$TL_{99}/U_{99} = 0.85$ for any value of D and L.

TLM = proposed test load magnitude = $1.0D_w + 1.1D_s + 1.4L$ (simplified by assuming F, L_r, S, and R equal to 0).

4.2—Load factors for various components of service load

4.2.1 *Reasons for change*—The required strength U (and design strength) of tension-controlled members of structures designed in accordance with ACI 318-02 and 318-05 has been reduced compared with the required strength per previous editions of ACI 318. As a result, the test load as defined in Chapter 20 of ACI 318-02 and 318-05 is not a fixed percentage of the required strength.

Table 4.1 provides a comparison of required strengths as defined in ACI 318-99 and 318-05 for a variety of structures. The table assumes that the members being considered (slabs and beams) are not over-reinforced and therefore qualify as tension-controlled members, which is usually the case in most concrete structures. Representative values for dead and live loads as shown in Columns 1, 2, and 3 are taken from typical buildings. Column 4 shows that the live load to total dead load ratio varies from 0.20 to 2.29. Columns 5 and 6 show the total factored demands (or minimum required strengths) according to ACI 318-99 and 318-05, while Column 7 shows their ratios. Column 8 shows the test load computed according to ACI 318-05. Note that while the ratio of test load to required strength in ACI 318-99 was 0.85, the ratio of test load (TL_{05}) to required strength (U_{05}) defined by ACI 318-05 varies from 0.93 to 0.97 for the selected examples as shown in Column 9.

In Table 4.1, Columns 9 and 10 provide a comparison of the test loads as defined in ACI 318-05 with required strength and total service loads. Note that the ratio of test load to total service loads varies from 1.23 to 1.37 for the examples provided, which is a reasonably close range. The table also provides in Column 11 a comparison of the test load minus the total dead load divided by the live load (the proof load ratio). Note that this ratio varies from 1.53 to 2.40, which is a considerably wider spread.

A consequence of defining the test load as a constant percentage of the required design strength is that the relationship between the proof load applied to the structure and the service live load is not apparent and is not a reasonably constant ratio. The variation in this ratio is among the reasons the TLM should be redefined, the goal being more consistent proof testing of structures.

It is recommended that the TLM be redefined in terms of proof loading rather than as a percentage of required strength. As discussed in Chapter 3, proof loading has historically been the purpose of load testing. The proof load ratio readily reveals the factor of safety of test load over service loads, and therefore adds clarity to the intent of load testing.

As noted in Chapter 3, ACI 318 has wavered on whether some additional percentage of the design dead load should be included in the test load. Defining the test load as a combination of factored design dead and live loads is not unique to ACI. Introducing a factor other than 1.0 for dead loads in defining the TLM makes the relationship between the TLM and the service live loads variable (that is, a function of the relative magnitude of the dead loads and live loads). As shown in Table 4.1, when the ratio of live load to dead plus superimposed dead loads is small (Column 4), the test load as defined in ACI 318-05 approaches the required strength (Column 9). This relationship tends to penalize structures that are heavy compared with the live loads they support even though calculation of a substantially accurate dead load is achievable on existing structures. This aspect of the current test load definition is another reason modifications to the definition of the TLM are recommended.

4.2.2 *Recommended changes to test load magnitude*—As defined in Section 2.2, a proof load is a load applied to a structure to prove a safe margin of satisfactory performance beyond code-required service live and dead loads. It is proposed that the proof load be defined in terms of those parts of the total load a structure will likely be subjected to that are variable. Therefore, when defining proof load, unlike when defining required strength, there is a need to separate the components of dead load that do not vary from those that do. For this reason, dead load is separated into two categories: dead load due to self-weight (D_w) and dead load due to weight of construction and other building materials (D_s). This latter category is defined as superimposed dead loads and, as noted in Section 1.3, includes weights of finishes, cladding, partitions, and fixed landscaping elements.

Dead load due to self-weight should be based on the as-constructed dimensions of those portions of the structure to be tested or dimensions of the structural members that are considered to be representative of the as-built structure, if different. Because this is a known and existing load, there is no need to apply a factor greater than unity to this self-weight when defining the test load as a proof load.

Superimposed dead loads may be defined by the local building code or may be defined in the design documents for the structure. Because these loads represent a variable that may change over time depending on the owner's use of the facility and construction and maintenance means and methods, a factor greater than 1.0 is suggested for superimposed dead loads. The actual factor used will depend on the degree of variability anticipated by the engineer defining the load test or by the building official. A load factor of 1.1 is recommended for superimposed dead loads except as discussed herein.

For partial load testing (when only portions of the suspect areas of a structure are to be tested), a higher test load is recommended to improve the level of confidence that significant flaws or weaknesses in the design, construction, or current condition of the structure are made evident by the load test. This recommendation reinstitutes the format of ACI 437R-67, in which two different test load definitions were provided. The exception in these current recommendations is when the members to be tested are determinate (for example, cantilevers or simple span members) and the possibility exists of producing an inelastic response in the members if the test load approaches the design strength too closely. While the new strength-reduction factors of ACI 318-05 provide for a higher nominal strength with respect to design or required strength than did the factors of ACI 318-99, the new factors are still based not only on desired reliability, but also on probable inaccuracies in design or construction; for an existing structure, these latter concerns mean that it is not possible to know how great the buffer between design strength and nominal strength is. Therefore, for determinate members, the lower TLM is recommended.

Where the suspected shortcoming or weakness among structural members is highly variable throughout the structure (for example, corrosion and debonding of embedded reinforcing steel), it is critical that the engineer select areas for testing that represent conditions believed to be severe with respect to the safety and performance of the structure. It is important to note that it is not only the severity of damage to the structural member, but rather the combination of severity with the location of minimum strength reserve that is of most interest. The percentage increase in TLM recommended as follows for partial tests will not significantly improve probability that the tested structure can safely support code loads if the tested areas are not well chosen.

It is recommended that the load intensity as provided in Section 20.3.2 of ACI 318-05 be defined as follows. The equations are proposed to be consistent with the load combinations of Chapter 9.

Load intensity—When all suspect portions of a structure are to be load tested or when the members to be tested are determinate and the suspect flaw or weakness is controlled by flexural tension, the test load magnitude, ***TLM***, (including dead load already in place) shall not be less than

$$TLM = 1.2(D_w + D_s) \tag{20-1}$$

or

$$TLM = 1.0D_w + 1.1D_s + 1.4L + 0.4(L_r \text{ or } S \text{ or } R) \tag{20-2}$$

or

$$TLM = 1.0D_w + 1.1D_s + 1.4(L_r \text{ or } S \text{ or } R) + 0.9L \tag{20-3}$$

where

D_s = superimposed dead load;
D_w = dead load due to self-weight;
L = live loads, or related internal moments and forces;
L_r = roof live load, or related internal moments and forces;
R = rain load, or related internal moments and forces; and
S = snow load, or related internal moments and forces.

Table 4.2—Design strength and test load comparison: partial load test*

Type of facility	TLM, lb/ft^2 (kN/m^2) (12)	$\frac{TLM}{TL_{05}}$ (13)	$\frac{TLM}{U_{05}}$ (14)	$\frac{TLM}{U_{99}}$ (15)	$\frac{TLM-D}{L}$ (16)
Parking slab, unreduced live load	145 (6.94)	0.97	0.92	0.82	1.60
Parking beam, reduced live load	148 (7.09)	0.91	0.88	0.77	1.60
Office slab, unreduced live load	167 (7.99)	0.96	0.92	0.82	1.64
Storage, light	310 (14.84)	1.00	0.93	0.85	1.60
Storage, light with heavier structure	350 (16.76)	0.97	0.92	0.83	1.60
Storage, heavy	550 (26.33)	1.02	0.95	0.87	1.60
Manufacturing, very heavy	815 (39.02)	1.04	0.96	0.88	1.60
Landscape pedestrian plaza	690 (33.04)	0.93	0.91	0.79	1.90
Plaza, truck dock	600 (28.73)	1.00	0.94	0.85	1.60
Average	—	0.98	0.92	0.83	1.64

*TLM definition for testing only part of suspect portions of structure.

Definitions:
TLM = proposed test load magnitude = $1.0D_w + 1.1D_s + 1.6L$ (simplified by assuming F, L_r, S, and R equal to 0).

When only part of suspect portions of a structure is to be load tested and members to be tested are indeterminate, the ***TLM*** (including dead load already in place) shall not be less than

$$TLM = 1.3(D_w + D_s) \quad (20\text{-}4)$$

or

$$TLM = 1.0D_w + 1.1D_s + 1.6L + 0.5(L_r \text{ or } S \text{ or } R) \quad (20\text{-}5)$$

or

$$TLM = 1.0D_w + 1.1D_s + 1.6(L_r \text{ or } S \text{ or } R) + 1.0L \quad (20\text{-}6)$$

D_s = superimposed dead load;
D_w = dead load due to self-weight;
L = live loads, or related internal moments and forces;
L_r = roof live load, or related internal moments and forces;
R = rain load, or related internal moments and forces; and
S = snow load, or related internal moments and forces.

In Eq. (20-2), the coefficient of the live load shall be permitted to be reduced in accordance with the requirements of the applicable Model Code or General Building Code. If impact factors have been used for the live load in design of the structure, then the same impact factor should be included in the above equations.

The total dead load shall include all superimposed dead loads, $\boldsymbol{D_s}$, considered in design or considered by the engineer or building official to be relevant to the proposed load test. Where superimposed dead loads represent a significant portion of the total service loads, are not already in place on the structure, and/or may not be of controllable intensity, a factor greater than 1.1 shall be considered for the superimposed dead load in the above equations for calculating the test load magnitude.

The commentary to this section in the building code could provide further explanatory discussion on this paragraph; for example, the possible variability of soil loading intensity and construction equipment loads on a landscaped structure. For this example, if soil loads are not already in place on the structure to be tested, then it will likely be appropriate to increase the test load magnitude by using a factor such as 1.4 or 1.6 to account for the variability of the loads the structure will be subjected to during installation of the soils and other landscaping features.

Commentary language should be provided in the building code to caution users when testing structures designed according to Chapter 9 of ACI 318-02 or 318-05 that, for some structures, the test load may induce bilinear elastic (cracked) or inelastic behavior. Discussion is provided in Chapter 5 regarding linearity of response as part of acceptance criteria recommended for adoption in ACI 318.

When testing members not meeting the minimum shear reinforcement requirements of ACI 318-05, Section 11.5.6.1 but meeting strength requirements on the basis of Section 11.5.6.2, an assessment of the test load at which significant cracking or damage in the web-shear region will occur is recommended. Significant cracking that does not close after removal of the test load may result if nonprestressed reinforcement yields during the load test or if the web shear region has no nonprestressed reinforcement. An appropriate adjustment of the proof load may be required to prevent permanent damage (that is, permanent open cracking) to such members. Equations (20-1) through (20-3) are recommended for determining TLM for such cases.

Tables 4.1 and 4.2, Column 12, provide the value of the proposed TLM for the example structures selected for full and partial load tests, respectively. Comparisons of the TLM with the total test load and required strength defined by ACI 318-05 are given in Columns 13 and 14, respectively. As shown in Table 4.1, the proposed TLM definition for full load tests has the effect of reducing the test load by approximately 10% compared with the test load of ACI 318-05 (Column 8), and so also reduces the TLM relative to required strength. In fact, the TLM is typically about 86% of the required strength per ACI 318-05 (Column 14) and about 77% of required strength per ACI 318-99 (Column 15). No examples have been provided of structures supporting fluid loads; however, the 1.2 factor recommended is 86% of the load factor for fluid loads F provided in Chapter 9 of ACI 318-05 for defining required strength U, and thus would produce a TLM versus required strength ratio consistent with the ratio for structures with live loads L, L_r, R, and S.

The proposed TLM definition for partial load tests where only parts of the suspect areas are to be tested results in a test load close in magnitude to the test load of ACI 318-05, varying from 91 to 104% of the current test load for the example structures as shown in Column 13 of Table 4.2.

Proposing a ratio of the TLM to the required strength of approximately 85% for full load testing is, of course, not accidental. The ratio of test load to required strength was explicitly set at 85% in 1971. Calculations made by members of Committee 437 also indicate that the ratio of the TLM to ultimate strength appears generally to have been on the order of 80 to 85% in previous allowable stress design versions of the code. That is to say, one can design a slab or beam using the allowable stress design methods and typical materials strengths of the 1940s and 1950s, and then calculate the resulting nominal strength using current principles. If one then calculates the TLM defined in earlier editions of ACI 318 (for example, ACI 318-51 and 318-56) and compares that with the nominal strength of the designs that resulted from those code provisions, it turns out that the ratio is often approximately 80 to 85%. Thus, having an upper limit to the TLM of about 85% of required strength has considerable sustained history in ACI. This limit is furthermore considered prudent to avoid possibly causing excessive inelastic deformations in a structure as a result of load testing.

A concern, but unavoidable consequence, of maintaining the ratio of TLM to required strength at 85% is that with the reduced load factors of ACI 318-05, the proven factor of safety resulting from load testing would now be lower than at any time in the history of ACI. The proof load ratio that resulted from the TLM defined in ACI 318-71 through 318-05 has typically been on the order of 1.7 (Column 11). The proof load ratio resulting from the new TLM would typically be 1.4 when all suspect portions of a structure are to be tested, or 1.6 when only part of the suspect portions are to be tested. With respect to international standards, however, this remains about average. In addition, as a practical matter, because most load tests involve testing only part of the suspect portions of a structure, the proposed Eq. (20-4) through (20-6) will generally control and provide a TLM that is roughly 90 to 95% of the required strength and, for most of the examples presented, is close to the TLM of ACI 318-05.

The recommended new TLM provides a rational balance between providing an adequate factor of safety, but not causing damage to the structure in the process. Refer also to Section 4.3 of this report regarding modifications to load factors.

4.2.3 *Applicability of TLM to structures designed per earlier codes*—The new TLM should be considered applicable for existing structures regardless of the code under which they were designed. The nominal strength of tension-controlled members designed in accordance with the provisions of ACI 318-71 through 318-99 was approximately 10% greater than those designed per 318-05, but generally at least 10% less than members designed according to the allowable stress method of earlier codes. Members designed according to the earlier allowable stress methods would have been subjected to higher TLMs using the test loads of ACI 318-51 and 318-56. As discussed previously, the ratio of these TLMs to the members' nominal strength would have been on the order of 80 to 85%. Therefore, applying test loads defined by 318-71 through 318-05 to structures designed according to earlier codes tests them to a lower percentage of their nominal strength. This method has become accepted practice.

Model building codes such as IBC provide that the strength of structures designed per earlier codes is to be calculated according to the current code. Committee 437, in its reports ACI 437R-67 through 437R-03, has stated that strength evaluation of existing structures by analytical means is to be based on principles of strength design as applied in ACI 318 (using current principles).

Similarly, the proposed modified definition of the TLM should be considered appropriate for strength evaluation of structures designed per earlier editions of ACI 318. If the proof load recommended herein provides an acceptable margin of safety over maximum anticipated service loads for a structure designed in accordance with 318-05, then the same factor of safety should be considered adequate for structures designed in accordance with earlier codes. The proposed TLM will be less than the test loads defined in earlier editions of ACI 318. Therefore, no inherent danger exists of overloading such structures when using the proposed TLM.

4.3—Load factors for extreme ratios of live load to total dead load

Service conditions where the ratios of live load to total dead load are considered outside the normal range are defined as follows

$$\begin{cases} \dfrac{L}{D_w + D_s} < 0.50, \text{ where 0.50 is lower limit of normal range} \\ \dfrac{L}{D_w + D_s} > 2.0, \text{ where 2.0 is upper limit of normal range} \end{cases} \quad (4\text{-}1)$$

For structures where $L/(D_w + D_s) < 0.50$, the load factors applied to the dead load due to self-weight and superimposed dead load in the recommended new TLM definition achieve two ends. First, they remove the potential penalty against structures with large self-weight compared with the live loads they carry by eliminating the extra dead load component of the test load. They also reduce the TLM as a percentage of the required strength per ACI 318-05 compared with the test load defined in ACI 318-05 versus required strength. As can be seen in Table 4.1, Column 14, the ratio of the proposed new TLM to required strength remains nearly constant, regardless of the L/D, whereas Column 9 shows the penalty assigned to structures with low L/D by the current test load definition. For partial load testing, the ratio is not as constant, and Column 14 of Table 4.2 shows that structures with higher L/D ratios also have larger TLMs relative to their required strength, but the TLMs are not significantly different than the current test load.

It is recommended that the load factor for the live load component of the service loads for such structures with L/D less than 0.50 be the same as for structures falling in the normal range of L/D. The minimum TLM given by Eq. (20-1) and (20-4), however, provides an additional lower bound to the test load that will apply in those cases where the live-dead load ratio is very small (L/D less than 0.15), where the factored live load does not provide a sufficiently large proof load with respect to the self-weight and superimposed dead loads.

For structures with large live loads compared with the structure's self-weight and weight of other superimposed dead loads, that is, $L/(D_w + D_s) > 2.0$, the committee sees conflicting concerns. As noted in Chapter 3, the RILEM document TBS-2 recommends increasing the test load if the live load exceeds twice the dead load, although that document does not provide further explanation of why an increased factor of safety is considered appropriate nor what the magnitude of that increased factor of safety should be. On the other hand, this approach could result in situations where otherwise adequate structures are loaded into the inelastic range during the load test, inducing permanent deformations. This could occur, for example, when testing a structure prestressed for a lower, more typical service load condition but reinforced with bonded reinforcement to provide adequate ultimate strength for full code-required live load.

If the engineer and building official are of the opinion that the service live loads for a structure to be evaluated by load testing are known, controllable, and free from dynamic magnification effects, it is recommended that the load factor to be used on the live load portion of the service loads be reduced to 1.2 and 1.3, respectively, for full and partial load tests when $L/(D_w + D_s) > 2.0$.

The following text is proposed for inclusion in the commentary for R20.3.2 of ACI 318:

> For structures where the ratio of live load to total dead load (L/D) is larger than 2.0, the multiplier of the live load, L, can be reduced from 1.4 to 1.2 in Eq. (20-2), and from 1.6 to 1.3 in Eq. (20-5) when the engineer determines that the magnitude of the live load is known and controllable and free from dynamic magnification effects.

CHAPTER 5—LOAD TEST PROTOCOL

5.1—Introduction

To apply test loads to a structure or portion of a structure in a systematic fashion for purposes of evaluating safety and serviceability, a number of items should be considered. They include, but are not limited to: test load configuration, the means by which the test load is applied, the procedure for application of the test load, and the duration of application of the test load. These items are discussed in this chapter. In addition, two common test methods are defined and discussed in general terms.

5.2—Test load configuration

According to Chapter 20 of ACI 318-05, the test load must be arranged to maximize the deflection and stresses in the critical regions of the structural members under investigation. There are no other requirements for the configuration of the test load. Several possible options could be used to satisfy the Chapter 20 requirements. The test load could be applied so as to replicate the uniformly distributed load used for design, or the test load could be applied with a series of concentrated loads to simulate the effects of a uniformly distributed load.

5.2.1 *Uniformly distributed load pattern*—Perhaps the most obvious way to determine if a structure is capable of carrying the loads for which it is designed is to apply those loads in the same load pattern that is assumed in the design. To simulate a uniformly distributed load condition, test loads are commonly applied by means of dead weights, which is discussed in another section of this chapter. When test loads are applied in a uniform pattern over the full structure or over a large enough area to fully load the critical member being investigated as well as surrounding structural members that could contribute to supporting the load, then concerns such as load sharing and end fixity need not be as thoroughly investigated as when a small number of concentrated loads are applied.

5.2.2 *Patch or strip equivalent loads*—Chapter 20 of ACI 318-05 does not indicate the specific load distribution to be used; therefore, it is acceptable to apply equivalent concentrated (or patch) loads by means of hydraulic jacks or other methods. When using point loads applied by hydraulic jacks, it is difficult to determine the equivalent forces that will produce the same effects, including bending moments and shear forces, as the uniformly distributed load used in design. When planning a load test to determine the magnitude of the concentrated equivalent loads, the engineer may model the structural behavior of the members through the following methods:

- Numerical approaches (for example, finite element method) (Vatovec et al. 2002; Galati et al. 2004). Appropriate modeling is only possible given knowledge of material properties, internal reinforcement location, and overall geometry;
- Simplified models that analyze a portion of statically indeterminate structures. In this instance, it is necessary to have knowledge of the degree of fixity at the supports and the load sharing offered by adjacent members;
- Trial tests. For those situations where no information is available on the construction, and budget constraints disallow invasive and nondestructive testing before conducting a load test, a load-unload cycle could be used for calibration of actual member fixities and load transfer characteristics. Current practice in Europe (Lombardo and Mirabella 2004) shows that an equivalent force to substitute for uniformly distributed loads may be calibrated based on the knowledge of the deflection response of the member(s) and the surrounding structure. To this end, Appendix A presents a brief explanation of the methodologies to be used to establish service load and TLM in the case of a strip test load and patch test load(s).

5.3—Load application method

5.3.1 *Dead weights*—To simulate a uniformly distributed load condition, loads are commonly applied by means of dead weight such as masonry block, sand bags, and water, either ponded or in barrels. Test loads can typically be applied with rather unsophisticated technology, and do not require specialized equipment. Such procedures, however, lead to laborious and time-consuming activities for site preparation, affecting the overall cost of the load test. In addition, when test loads are applied by means of dead

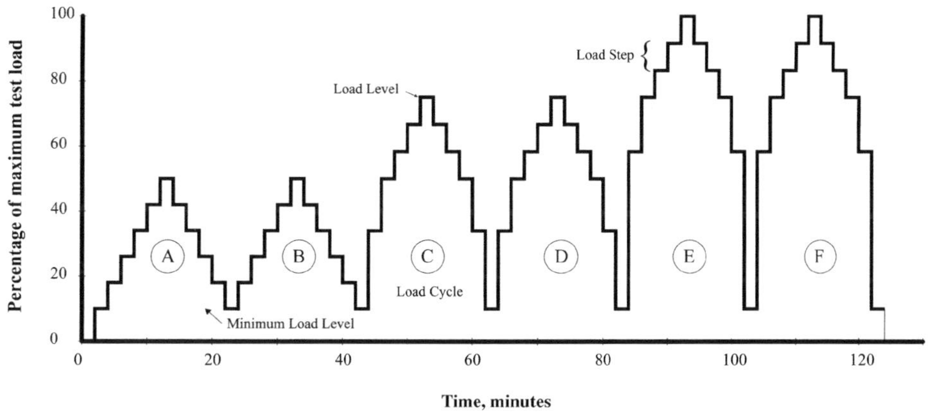

Fig. 5.1—Load tests and cycles for a cyclic load test.

weights, there is generally no feasible way to rapidly remove the load. In case of failure, adequately designed shoring becomes a critical safety measure.

5.3.2 *Hydraulic jacks*—The application of test loads using hydraulic jacks, rather than uniformly distributed dead loads, allows for faster and more controlled application of test loads. When a structure that is loaded by displacement-controlled hydraulic jacks experiences a softening postpeak behavior, the applied load decreases in a stable manner because the displacement rate remains constant. An added benefit of applying test loads with hydraulic jacks is that the test load can be removed almost instantaneously in case of impending failure. The use of hydraulics in the proper configuration may also create less of a disturbance to the occupants and finishes of the area being tested, thus resulting in a reduction of inconvenience to the users. While loading by means of hydraulic jacks may provide benefits during a load test, there is a need to create a reaction system for the hydraulic jacks that requires design and could be expensive and time consuming to implement. There are several ways to provide reactions to the hydraulic jacks that depend on the characteristics of the member to be tested and the overall site conditions. Several methods are defined in ACI 437R.

5.4—Loading procedures

Two procedures are currently in use for the application of test loads to buildings. The first has been used for many years, and involves applying loads in a monotonic fashion. The other, more recent, procedure applies test loads in a series of zero to maximum load cycles that increase incrementally (Fig. 5.1).

5.4.1 *Monotonic loading*—In current practice, monotonic loading is the standard loading procedure because of practical considerations and cost of placing and removing test loads that are commonly in the form of sand bags, water barrels, and other similar materials. Typically, loads are applied in not less than four approximately equal increments up to a predetermined maximum test load level. Data readings are usually taken at each loading stage. The time it takes to get to the maximum load depends on the test load configuration and the load application method as previously discussed. Monotonic loading is almost always used when the loads are being applied with dead weights because of the time it takes to apply and remove the loads. Monotonic loading can also be used when applying test loads with hydraulic jacks.

5.4.2 *Cyclic loading*—In the cyclic loading procedure, the loads are applied in loading-unloading cycles of increasing magnitude using hydraulic jacks that are controlled by hand or electric pumps. Using a sequence of loading and unloading cycles up to the predetermined maximum load level provides the opportunity to work the structure and assess potential changes in response to repeated loading and to increasing load levels. The load sequence is intended to identify, in an explicit manner, any undesirable response. In recent work (Mettemeyer 1999; Casadei et al. 2005), the response has been characterized by monitoring parameters such as: linearity of structural deflection response, repeatability of load-deflection response, and permanency of deflections. Because the structure is initially loaded and unloaded at low levels, the engineer has the ability to better understand end fixity and load transfer characteristics of the tested member by comparing actual deflection responses with calculated deflection responses. For statically indeterminate structures in particular, this ability allows checking the accuracy of the assumptions made regarding fixity and load sharing used to plan the load test. The advantages of cyclic loading are not yet fully understood because the data base and experience obtained using this procedure are limited, so additional validation is desirable.

5.5—Loading duration

Once the maximum test load has been reached, it is held in place for a given amount of time. Depending on the test method that is used, this may be a short duration (approximately 2 minutes) or up to as long as 24 hours.

5.5.1 *Twenty-four hours at maximum load*—For more than 80 years, the maximum test load has been held for at least 24 hours according to ACI 318 requirements. The strength of concrete under sustained load is known to be lower than the strength under short-term load. The strength under sustained load is closely related to the stress at which cracks develop in the concrete paste. These are unstable cracks that can grow under a sustained stress. Thus, the 24-hour sustained load duration is used to verify that the concrete is not stressed too close to its ultimate strength. In addition, successfully holding a test load for 24 hours has a very positive effect on the level of comfort in those who will use and occupy the structure after the load test is completed. It is generally understood, however, that this relatively brief load duration cannot demonstrate most time-dependent effects.

5.5.2 *Stability at maximum load*—Another approach has recently been introduced that significantly decreases the amount of time the maximum test load is sustained on a tested structure. The reasons for the shorter duration of sustained load are simple—economic implications and minimizing disruption for the building occupants—but the justification for not holding the test load for an extended amount of time is complex. The idea is that by studying other behavioral characteristics of the tested member (that is, deviation from linearity, repeatability, and permanency), one can determine if the tested structure is approaching its ultimate strength without maintaining the test load for a sustained duration. The drawback of the relatively shorter duration of loading is that it does not create the same level of comfort as holding the load for 24 hours in those who will use the structure after the load test is completed. The level of experience with using a shorter duration cyclic test is limited, and additional data are needed to solidify the evaluation criteria.

5.6—Load testing procedure

A variety of combinations of the aforementioned procedures have been used over the last 100 years in international load testing practice. Two load test procedures are described in the following sections. The first is the 24-hour monotonic uniform load test that has been used for many years and is prescribed by ACI 318. The second is the relatively new cyclic load test as discussed in Appendix A of ACI 437R.

5.6.1 *Twenty-four-hour monotonic uniform load test*—Once a structure has been selected to undergo a load test, a preliminary evaluation is conducted. The evaluation is meant to determine, if possible, material and section properties, loading history, and levels of deterioration of the structure. Because the test load is applied in a uniformly distributed manner similar to the design load pattern, certain characteristics of the structure may or may not be investigated. When several adjacent spans or bays are simultaneously loaded, characteristics, such as load sharing and fixity of supports, need not be fully understood before the load test begins because the structure will behave just as it would under design loading, and its ability to hold the design load will be determined directly by the load test. Preliminary calculations are typically done to determine some anticipated results; however, without fully understanding the structure's behavior, these calculations are used only as a rough guide as to how the structure will perform under the test loads and to locate instrumentation to determine maximum responses during the test. Once the structure is adequately instrumented at the locations where the maximum response is expected, initial values of each instrument are recorded not more than 1 hour before application of the first load increment. After the test is started, the uniformly distributed load is applied in not less than four approximately equal increments. If the measurements are not recorded continuously, a set of response readings are taken at each of the four load increments until the total test load has been reached and again after the test load has been applied on the structure for at least 24 hours. Once the last readings under sustained load have been taken, the test load is removed, and a set of final readings is taken 24 hours after the test load is removed. The measured deflections and deflection recovery are compared with code-specified acceptance criteria (Table B.1 and Section 6.1). In case the structure does not meet the acceptance criteria, Chapter 20 of ACI 318-05 allows the test to be repeated 72 hours after the removal of the first test load.

This test method takes advantage of one very important factor in load testing—consideration of how load is distributed in the structure. Because the load is applied in the same pattern as designed, factors such as load sharing and end fixity are inherently considered during the load test and thus do not require a full understanding of their contributions to the overall strength of the structure. By demonstrating that the structure can sustain the applied design load for a 24-hour period without deflection or permanent deformation exceeding the preset limits, the results of the load test are relatively straightforward. This method, however, does have some drawbacks. The application of a uniformly distributed load can be time consuming and laborious. The overall duration of the test is at least 3 days (half a day to set up, 24 hours at maximum load, 24 hours unloaded, and half a day to disassemble), assuming that retesting is not necessary. This amount of time with a continuous presence on a job site is costly to an owner as well as disruptive to the tenants. Testing large areas of a structure or performing multiple tests within a structure may be too time consuming and expensive to provide a thorough evaluation of the overall performance of the entire structure under design loads.

5.6.2 *Cyclic load test*—Appendix A of the ACI 437R-03 reports the protocol for conducting a cyclic load test.

Following the preliminary investigation, the initial steps for planning a cyclic load test include structural analysis and load intensity definitions, which require considerable engineering effort as compared with the 24-hour monotonic uniform load test described previously. The predetermined test load is applied to discrete areas on the tested member that have been selected to maximize specific responses that are being investigated in the member. To determine the required magnitude, quantity, and location of applied concentrated loads, one must have a thorough understanding of the structure's behavioral characteristics, including the effects of load sharing and end fixity. These normally cannot be accurately determined with simple hand calculations. Relatively complex models may be required to fully understand the structural responses to the applied test loads.

The procedure of a cyclic load test consists of the application of concentrated loads in a quasi-static manner (that is, sufficiently slow to avoid strain rate effect) to the structural member in at least six loading/unloading cycles. Even though the number of cycles and the number of steps within each cycle (five loading plus five unloading) should be considered as minimum requirements, in most cases they provide for an adequate assessment of structural performance. For this minimum test protocol, the total load test duration should be approximately 2 hours, with each loading/unloading cycle lasting approximately 20 minutes. With reference to Fig. 5.1, the protocol description is given as follows:

- **Benchmark**—The initial reading of the instrumentation should be taken no more than 30 minutes before beginning the load test and any load being applied.
- **Cycle A**—The first load cycle consists of five load steps, each increased by no more than 10% of the total test load expected in the cyclic load test. The load is increased in steps, typically until the service level of the member is reached, but no more than 50% of the total test load. The maximum load level for each cycle should be maintained until the structural response parameters have stabilized.* During each unloading phase (using similar steps as the loading phase), a minimum load P_{min} of at least 10% of the total test load should be maintained to keep the test devices engaged. Response measurements are taken during both the loading and the unloading phases. The duration of a complete loading/unloading cycle is set to a minimum of 20 minutes, which implies that each loading/unloading step including the sustained phase is 2 minutes long;
- **Cycle B**—A repeat of Cycle A that provides a check of the repeatability of the structural response parameters obtained in the first cycle. Monitoring the repeatability of load-deflection response is of relevance at any load level, including the relatively lower load Cycles A and C. For example, this allows the engineer to determine if a change in stiffness (that greatly affects linearity) is the result of cracking within the elastic range of the member;
- **Cycles C and D**—Load Cycles C and D are identical in load magnitude and achieve a maximum load level that is typically halfway between the maximum load level achieved in Cycle A and B and 100% of the total test load. The loading procedure is similar to that of Cycle A and B. For Cycle C and D, it is suggested that the load of the first of five steps be at the load level of the third step of Cycle A, and the load of the second step be at the level of maximum load attained in Cycle A. The remaining three steps should be of equal magnitude to attain the maximum load level for Cycles C and D;
- **Cycles E and F**—The fifth and sixth load cycles, E and F, respectively, should be identical in load magnitude, and they should reach the total test load. For Cycles E and F, it is suggested that the load of the first of five steps be at the load level of the third step of Cycle C, and the load of the second step be at the level of maximum load attained in Cycle C. The remaining three steps should be of equal magnitude to attain the maximum load level for Cycle E and F; and
- **Final step**—At the conclusion of Cycle F, the test load should be decreased to zero. A final reading should be taken no sooner than 2 minutes after the total test load, not including the equipment used to apply the load, has been removed.

*For each load cycle, maximum load level needs to remain approximately constant for at least 2 minutes. During this time interval, the measurands, such as deflection or strain, have to remain stable before proceeding with unloading. Stability is defined herein as a change in the measurable not exceeding 5% of the initial value over a period of 2 minutes.

The main differences between the two protocols is that, for the latter, the loads are applied in loading-unloading cycles of increasing magnitude using hydraulic jacks, and the maximum test load is maintained for a shorter duration of time. Using a sequence of loading and unloading cycles up to the predetermined maximum load level allows the engineer a real-time assessment of member performance. The load sequence is intended to identify, in an explicit manner, any undesirable response. The response can be characterized by monitoring parameters such as linearity of structural deflection response, repeatability of load-deflection response, and permanency of deflections (Chapter 6). An additional advantage is that the duration of the maximum applied load in the cyclic load test may be considerably reduced from that of the 24-hour monotonic uniform load test described previously, which has economic implications and minimizes disruption for the building occupants. The main drawbacks with the cyclic load-testing method are the amount of engineering that is required to properly determine the appropriate test loads and the relatively small amount of supporting data used to determine evaluation criteria.

CHAPTER 6—ACCEPTANCE CRITERIA

6.1—Criteria for 24-hour monotonic load test

Section 20.5 of ACI 318-05 defines acceptance criteria for interpreting the results of the 24-hour monotonic load test.

The evaluation of the member/structure is based on two different sets of acceptance criteria to certify whether or not the load test is passed: a set of visual parameters (such as no spalling or crushing of compressed concrete is evident), and the measured maximum deflections (must satisfy one of the following two equations)

$$\Delta_{max} \leq \frac{l_t^2}{20{,}000h} \quad (6\text{-}1)$$

$$\Delta_{r\ max} \leq \frac{\Delta_{max}}{4} \quad (6\text{-}2)$$

Defining an acceptable deflection criterion by the formula given in Eq. (6-1) makes it difficult to establish a relationship with typical deflection limits such as $l_t/240$, $l_t/360$, and so on. Also, the theoretical basis for Eq. (6-1), as discussed in Chapter 3, is unrelated to modern material strengths, deflection limits, degree of fixity that may be present in the structural member being tested, and current reinforced concrete construction practice. Most members/structures pass the acceptance criteria of the current monotonic load test, showing very small deflections.

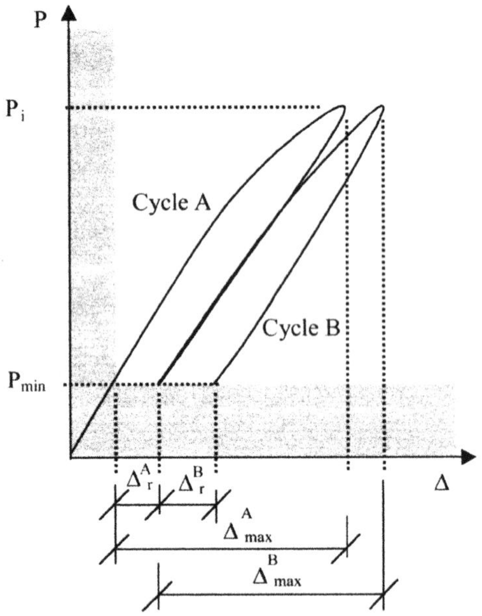

Fig. 6.1—Example of load-versus-deflection curve for two cycles at same load level.

Chapter 20 of ACI 318-05 requires that response measurements are to be made after each load increment is applied as well as after the total load has been on the structure for at least 24 hours. No commentary, however, is offered regarding the purpose of the intermediate deflection readings. These measurements clearly provide an opportunity to verify the linear response of the structure and to discontinue the test if a pronounced change in linearity is noted, as evidenced by a large increase in deflection observed after a loading increment. The concept of "deviation from linearity," discussed in more detail in the following section, could be applied to the intermediate readings of the 24-hour monotonic load test and provide an explicit guideline for interpretation of deflection readings taken during the sequence of load application steps.

Chapter 20 of ACI 318-05 does not define acceptance criteria for establishing satisfactory behavior at service load level. Even though it is recognized that calculations regarding deflection and crack width may not be sufficiently developed or accurate to justify using them as mandatory accept/reject criteria at this load level, the engineer should include the assessment under service load as an integral part of the structural performance evaluation process.

In summary, new deflection acceptance criteria must be developed. These deflection acceptance criteria should generally be based on the following principles of engineering mechanics under the assumption that accurate deflection readings are attained:

- Maximum deflection under full test load compared with calculated theoretical maximum deflection at that load level;
- Recovery of deflection upon full removal of load; and
- Linearity of deflection response during loading and unloading.

6.2—Criteria for cyclic load test

Appendix A of ACI 437R-03 describes the cyclic load test method. This alternative load test method appears to offer some advantages in terms of reliability and understanding of structural response to load. Three distinct measures of performance are proposed for the cyclic load test method (CLT method): repeatability, permanency (that is related to deflection recovery), and deviation from linearity. The acceptance criteria are based on limited testing as described in Chapter 3 of this report. The three criteria may be related to any response (for example, deflection, rotation, and strain); however, deflection appears to be the most convenient (CIAS 2000). As such, performance measures and acceptance criteria are described in this section in terms of deflection.

- Repeatability is a measure of the similarity of behavior of the member/structure during two twin load cycles (Fig. 6.1) at the same load level, and is calculated according to the following equation

$$I_R = \text{repeatability index} = \frac{\Delta^B_{max} - \Delta^B_r}{\Delta^A_{max} - \Delta^A_r} \times 100\% \qquad (6\text{-}3)$$

  Repeatability as defined herein is not an indicator of the quality of the testing technique, but rather an indicator of structural performance related to recoverable (elastic) deflection and load-deflection response in general. Experience (Mettemeyer 1999) has shown that a repeatability index I_R in the range of 95 to 105% is a satisfactory. For values of I_R inside this range, the member/structure can be considered to pass the load test;
- Permanency is the relative value of the residual deflection compared with the corresponding maximum deflection during the second of two twin load cycles at the same load level. It should be less than 10% (Mettemeyer 1999) for the member/structure to be considered passing the load test. The permanency index I_P is computed using the following equation (Fig. 6.1, Cycle B)

$$I_P = \text{permanency index} = \frac{\Delta^B_r}{\Delta^B_{max}} \times 100\% \qquad (6\text{-}4)$$

  If the level of permanency is higher than the aforementioned 10%, it may be an indication that load application has damaged the member/structure and that nonlinear effects are taking place; and
- Deviation from linearity represents the measure of the nonlinear behavior of a member/structure being tested at any time after a given threshold that typically corresponds to its service load level. To define deviation from linearity, linearity is defined first as the ratio of the slopes of two secant lines intersecting the load-deflection envelope (Fig. 6.2). Figure 6.2 shows the schematic load-deflection curves obtained by a total of six loading cycles (A through F), which consisted of three pairs of twin cycles with each pair at the same load level. The load-deflection envelope is the curve

constructed by connecting the points corresponding to only those loads that are greater than or equal to any previously applied load. As expressed by Eq. (6-5), the linearity of any point i on the load-deflection envelope is the percent ratio of the slope of the secant line* to point i, expressed by $\tan(\alpha_i)$, to the slope of the reference secant line,† expressed by $\tan(\alpha_{ref})$

$$\text{Linearity}_i = \frac{\tan(\alpha_i)}{\tan(\alpha_{ref})} \times 100\% \quad (6\text{-}5)$$

The deviation from linearity of any point i on the load-deflection envelope is the complement of the linearity of that point, as given in the following equation

$$I_{DL} = \text{deviation from Linearity}_i \text{ index} = 100\% - \text{Linearity}_i \quad (6\text{-}6)$$

Once the level of load corresponding to the reference point has been achieved, deviation from linearity should be monitored continuously until the conclusion of the cyclic load test. Experience (Mettemeyer 1999) has shown that I_{DL} values less than 25% indicate that the structure has passed the load test.

If a member/structure is initially uncracked and becomes cracked during the load test, the change in flexural stiffness as a result of a drastic change in moment of inertia at the crack location(s) can produce a very high deviation from linearity that is not necessarily related to degradation in strength (Masetti 2005). For such a member/structure, repeatability and permanency may be better indicators of damage occurrence, or I_{DL} should be only computed for the member/structure under cracked conditions.

While additional research and field testing of structures are required to verify the overall suitability of the CLT method, adoption of these measures of performance and the recommended threshold levels appear justifiable.

6.2.1 *Determination of member/structure capacity (load rating)*—The cyclic load test could also be used to determine the capacity of a given member/structure based on the three-index acceptance criteria if the load test is not terminated when the TLM level is reached (Casadei 2004). In fact, as real-time measurements and assessment are possible, the engineer can apply a number of twin load cycles at increasing load levels until one of the three acceptance criteria fails (that is, attainment of the critical load). Given the critical load and after subtracting the factored dead load, the engineer can establish the safe live load level. The validity of this load rating protocol rests on the reliability of the acceptance criteria and their threshold values to correctly predict the necessary strength reserve in the structure.

*Secant is the line that connects the origin to the point of interest on the load-deflection envelope.

†The reference point usually coincides with the peak load of the first cycle.

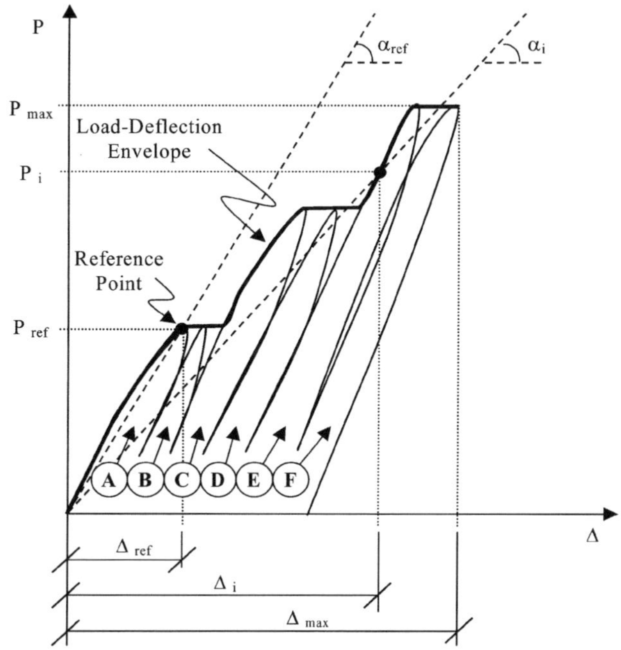

Fig. 6.2—Schematic load-versus-deflection curve for six cycles.

6.3—Considerations of performance assessment at service load level

Irrespective of the loading procedure (that is, monotonic or cyclic load) and type of load (that is, uniformly distributed load over the entire tributary area, strip load, or patch load(s)), measurements of flexural deflection and crack spacing and width under the test load equivalent to the service condition (that is, $1.0D_w + 1.0D_s + 1.0L$) should be recorded and checked against limit values established by the engineer.

When applicable, if the measured deflection or crack width exceed their respective limits set by the engineer, careful consideration should be given to continuing the load test to higher load levels. It is recognized that the variable nature of cracking and the challenges in accurately measuring and predicting crack width make the corresponding limits difficult to implement. The intent of the provision, however, is to caution the engineer that the occurrence or growth of excessive cracks under "immediate" service loads may be a signal of structural deficiencies. Influence of crack width is of particular significance for some members/structures, such as those exposed to aggressive environments. If crack widths for watertight structures or those exposed to aggressive environments exceed the preset limits, the structure need not be considered to have failed the load test with respect to safety. Provided that the structure meets the requirements for performance under full TLM, it may still be considered satisfactory if additional protective measures can be taken to prevent or retard future deterioration.

Guidance for establishing possible limit values for deflection and crack width at service load are as follows:

- Maximum measured deflection should not exceed the permissible values given in Table 9.5(b) of ACI 318-05 Chapter 9 for the various types of members. This criterion is only applicable if the load distribution pattern reflects the one used for design, which is typically not the case for test loads of the strip or patch type. Furthermore, the first two values in Table 9.5(b) are intended for immediate live load deflections, while the third and fourth deflection limits are for the additional deflection occurring after attachment of nonstructural members due to long-term deflection caused by all sustained loads plus any immediate live load deflection. This makes these limits difficult to apply in the setting of a load test where only the immediate deflection due to applied loads can be measured. Long-term deflection due to sustained loads can be calculated and then added to the load test deflection results for live loading to arrive at a value that can be compared with the latter two limits of Table 9.5(b); and

- The maximum width of new flexural cracks formed during the course of the load test or the change in width of existing flexural cracks should not exceed a limiting width determined by the engineer, owner, or building official before the load test. Consideration should be given to the intended use and exposure conditions for the structure or member. Limiting crack widths should be selected based on the following:

  1. Suggested tolerable crack widths as reported by ACI Committee 224 (ACI 224R); and

  2. The value of the analytical width computed as the product s times $\Delta\varepsilon_s$, where s is the average spacing between cracks, and $\Delta\varepsilon_s$ represents the difference in strain in the longitudinal steel reinforcement when the cross section of interest is considered cracked and uncracked, respectively, and subject to an applied moment at that location resulting from the service load.

6.4—Recommendations for acceptance criteria at test load magnitude level

Adoption of the acceptance criteria for both monotonic and cyclic load tests is recommended as described in the following sections. In contrast to service condition, acceptance criteria at the TLM level are mandatory pass-fail requirements and are established based on the load procedure adopted (that is, monotonic or cyclic load).

6.4.1 *Twenty-four-hour monotonic load test procedure*—The acceptance criteria listed as follows need to be checked:

1. While increasing the load from service to TLM and while holding the maximum load constant for 24 hours, the structure should show no signs of impending failure, such as concrete crushing in the compressive zone or concrete cracking exceeding a preset limit. This criterion is of a qualitative nature;

2. The maximum absolute deflection recorded at the 24th hour of sustained TLM should be less than the member deflection computed analytically in accordance with Sections 9.5.2.2 through 9.5.2.5 of ACI 318-05. This criterion requires that the engineer carefully considers the load distribution pattern during computations. It is recognized that a load test is typically undertaken when insufficient information is available to perform a strictly analytical evaluation. The objective of this provision is to make sure that the engineer has made a prediction, given the available information and that such prediction be used to interpret the experimental results. There should be an upper limit to the measured absolute deflection that, if exceeded, rules out the option of using deflection recovery as an acceptance criterion as well as retesting. Such a limit is suggested to be equal to $l_t/180$; and

3. The residual deflection of the member should be less than 25% of the corresponding absolute maximum deflection immediately upon unloading or 24 hours afterward, respectively.

   a. If the member/structure is sufficiently stiff, deflection recovery is not relevant. In fact, it may even be unfeasible to compute the deflection recovery due to limitations in the precision/accuracy of the deflection measurement equipment. No check on deflection recovery is required if the absolute deflection is lower than 0.05 in. (1.3 mm) or the deflection as a percentage of span length is less than $l_t/2000$; and

   b. If the member/structure fails the deflection recovery criterion on the first test, retesting should be permitted, with the stipulation that the engineer establishes that deflection does not represent a serviceability problem. An upper limit on residual deflection after the retest equal to 10% of the maximum deflection recorded during the retest is recommended.

If any one of the three aforementioned criteria listed is not met, the member/structure should be considered having failed the load test. No retesting is permitted except for the stipulation in Item 3.

6.4.2 *Cyclic load test procedure*—The following acceptance criteria need to be checked:

1. While increasing the load from service to TLM and any time during the load test, the structure should show no signs of impending failure, such as concrete crushing in the compressive zone or concrete cracking exceeding a preset limit. This criterion is of a qualitative nature;

2. The maximum deflection recorded at the second load cycle that reaches TLM should be less than the member deflection computed analytically in accordance with Sections 9.5.2.2 through 9.5.2.5 of ACI 318-05. This criterion requires that the engineer carefully considers the load distribution pattern during computations. It is recognized that a load test is typically undertaken when insufficient information is available to perform a strictly analytical evaluation. The objective of this provision is to make sure that the engineer has made a prediction given the available information and that such a prediction be used to interpret the experimental results;

3. The repeatability index I_R, a measure of the similarity of behavior of the member/structure during two equal-level load cycles, should never be outside the range of 95 to 105%;

4. The deviation from linearity index I_{DLi}, a measure of the nonlinear behavior of the member/structure being tested, should be monitored continuously during the cyclic

load test until its conclusion, and never exceed the threshold value of 25%. Special consideration should be given to a structure/member that cracks during the load test if cracking is not considered detrimental to the serviceability of the structure; and

5. The permanency index I_P, the relative value of the residual deflection compared with the corresponding maximum deflection during the second of two equal-level twin load cycles, should never exceed the threshold value of 10%.

If any one of the five aforementioned criteria listed is not met, the member/structure should be considered having failed the load test intended to reach the selected TLM. No retesting is permitted.

6.5—Strength reserve beyond load test acceptance criteria

Irrespective of the test method, it is important to understand the strength reserve that likely remains in the member/structure after it passes the load test.

Results from load tests conducted on five different structures using either the 24-hour monotonic load test or the cyclic load test followed by loading to failure were used to establish threshold values for repeatability, permanency, and deviation from linearity (Mettemeyer 1999). These threshold values were set at limits that would ensure some reserve capacity in the member once one of the threshold values was surpassed. Because the structures were loaded monotonically up to failure, after the 24-hour or cyclic load tests were concluded, the only criterion that could be calculated up to failure was deviation from linearity. A threshold value of 25% for deviation from linearity was set because it allowed for at least a 40% strength reserve in the members before collapse. The threshold values for repeatability and permanency were selected based on the extreme values experienced during the 24-hour and cyclic load tests conducted on the five members.

Additional work (Casadei et al. 2005) on nearly identical reinforced concrete one-way slabs that were loaded to ultimate failure allowed for the determination of the strength reserve before collapse after the slabs had failed either the 24-hour monotonic or the cyclic load test. The criterion that became critical during the load tests was deviation from linearity. In this project, the margin of safety (that is, strength reserve) with respect to ultimate failure was found to be approximately 20% of the maximum load applied during the test for all slabs that also collapsed with the same failure mode. Obviously, a large database including different construction systems and structural configurations would be necessary to arrive at more definite conclusions.

CHAPTER 7—SUMMARY

The TLM and acceptance criteria as currently defined in Chapter 20 of ACI 318-05 should be revised.

The purpose of revising the TLM is twofold. The first purpose is to define a test load that demonstrates an acceptable, safe margin of capacity over design service dead and live load levels and to be as consistent as possible, regardless of the self-weight of the structure or the code used for the original design. The proposed TLM puts more emphasis on the variable portions of the service loads (the live loads and superimposed dead loads) and in so doing provides a more consistent proof load than does the ACI 318-05. Second, the recommended equations given in Chapter 4 of this report to define the test load magnitude are parallel with the equations for required strength given in Chapter 9 of ACI 318-05 and so provide a consistent format and logic within the code.

While the 24-hour monotonic load test has been part of the ACI code since the early part of the last century, it is recommended that the cyclic load test method described in Chapter 5 be considered for use in the code to supplement the current test method. The cyclic method provides a technique to more thoroughly evaluate structural response than does the monotonic load test method.

The current maximum deflection and deflection recovery criteria need to be revised because the theoretical bases for the criteria are considered inapplicable to most structural systems and modern materials, and are unrelated to design criteria. A clearer rationale and explanation of deflection criteria have been provided in Chapter 6 of this report. For performance assessment at service load levels, the proposed deflection limits for evaluating test results are related to the deflection limits of Chapter 9 of ACI 318-05, and for the 24-hour monotonic load test protocol, the predicted deflection used to establish an acceptable upper bound is to be calculated using the deflection prediction equations of Chapter 9. This consistency within the code would dispel some of the mystery associated with the current deflection limit for load testing. An additional set of acceptance criteria has been proposed when cyclic load testing is used. Additional testing and verification of the appropriate values for acceptance criteria for cyclic load testing are needed to make the test interpretation more meaningful.

As discussed in Chapter 4, this report recommends that the TLM be redefined in terms of service loads rather than required strength; however, there is the acknowledgement that the intent of the new definition is to limit the test load to approximately 85 to 90% of the required strength as defined in ACI 318-05 when testing all suspect areas of a structure, or 90 to 95% of the required strength when testing only a portion of the suspect areas. The load test provisions in Chapter 20 of ACI 318-05 should be reviewed any time there is a change in the definition of required strength, strength-reduction factors (ϕ-factors), or both.

This report uses as a reference the provisions on load testing outlined in ACI 318-05, and will have to be modified if future editions of the building code change such provisions. From a legal standpoint, ACI 318 sets the binding requirements. The recommendations provided in this report have the purpose of integrating and enriching the understanding and practice of load testing and its acceptance criteria, but do not replace ACI 318 provisions.

CHAPTER 8—REFERENCES

8.1—Referenced standards and reports

The documents of the various standards-producing organizations referred to in this document are listed with their serial designations. Because some of these documents are revised frequently, the user of this report should check for the most recent version.

American Concrete Institute (ACI)
224R Control of Cracking in Concrete Structures
318 Building Code Requirements for Structural Concrete and Commentary
437R Strength Evaluation of Existing Concrete Buildings

ASTM International
E 196 Standard Practice for Gravity Load Testing of Floors and Low Slope Roofs
F 914 Standard Test Method for Acoustic Emission for Insulated and Non-Insulated Aerial Personnel Devices Without Supplemental Load Handling Attachments

International Code Council
International Building Code (IBC)

8.2—Cited references

ACI Committee 318, 1947, "Building Code Requirements for Reinforced Concrete (ACI 318-47)," American Concrete Institute, Farmington Hills, Mich., 64 pp.

ACI Committee 318, 1951, "Building Code Requirements for Reinforced Concrete (ACI 318-51)," ACI JOURNAL, *Proceedings* V. 47, No. 8, Apr., pp. 589-652.

ACI Committee 318, 1956, "Building Code Requirements for Reinforced Concrete (ACI 318-56)," ACI JOURNAL, *Proceedings* V. 52, No. 9, May, pp. 913-986.

ACI Committee 318, 1963, "Building Code Requirements for Reinforced Concrete (ACI 318-63)," American Concrete Institute, Farmington Hills, Mich., 144 pp.

ACI Committee 318, 1971, "Building Code Requirements for Reinforced Concrete (ACI 318-71)," American Concrete Institute, Farmington Hills, Mich., 78 pp.

ACI Committee 318, 1999, "Building Code Requirements for Structural Concrete (ACI 318-99) and Commentary (318R-99)," American Concrete Institute, Farmington Hills, Mich., 391 pp.

ACI Committee 318, 2002, "Building Code Requirements for Structural Concrete (ACI 318-02) and Commentary (318R-02)," American Concrete Institute, Farmington Hills, Mich., 443 pp.

ACI Committee 318, 2005, "Building Code Requirements for Structural Concrete (ACI 318-05) and Commentary (318R-05)," American Concrete Institute, Farmington Hills, Mich., 430 pp.

ACI Committee 437, 1967, "Strength Evaluation of Existing Concrete Buildings (ACI 437R-67)," American Concrete Institute, Farmington Hills, Mich., 6 pp.

ACI Committee 437, 1982, "Strength Evaluation of Existing Concrete Buildings (ACI 437R-67) (Revised 1982)," American Concrete Institute, Farmington Hills, Mich., 7 pp.

ACI Committee 437, 2003, "Strength Evaluation of Existing Concrete Buildings (ACI 437R-03)," American Concrete Institute, Farmington Hills, Mich., 28 pp.

ACI Committee E-1, 1928, "Joint Code Building Regulations for Reinforced Concrete," Report on Reinforced Concrete Building Design and Specifications Amended and Adopted as a Tentative Standard at the Twenty-Fourth Annual Convention of the American Concrete Institute, Feb. 28.

American Concrete Institute, 1920, "Standard Specification No. 23—Standard Building Regulations for the Use of Reinforced Concrete," American Concrete Institute, Farmington Hills, Mich.

American Concrete Institute, 1936, "Building Code Regulations for Reinforced Concrete," ACI 501-36T, American Concrete Institute, Farmington Hills, Mich.

American Society of Civil Engineers (ASCE), 2002, "Minimum Design Loads for Buildings and Other Structures," ASCE, Reston, Va. (CD-ROM)

American Society of Mechanical Engineers (ASME), 2005, "Reinforced Thermoset Plastic Corrosion Resistant Equipment," ASME, New York, 340 pp.

American Society of Mechanical Engineers (ASME), 2004, "BPVC Section X—Fiber-Reinforced Plastic Pressure Vessels," ASME, New York.

Arnan, M. A.; Reiner, M.; and Teinowitz, M., 1950, "Research on Loading Tests of Reinforced Concrete Structures," *Report*, Standards Institution of Israel, Jerusalem, 52 pp.

Bares, R., and FitzSimons, N., 1975, "Load Tests of Building Structures," *Journal of the Structural Division*, ASCE, May, pp. 1111-1123.

Birkmire, W. H., 1894, *Skeleton Construction in Buildings*, John Wiley & Sons, New York, 80 pp.

BRE Information Paper 2/95, 1995, "Guidance for Engineers Conducting Static Load Tests on Building Structures," Building Research Establishment, England, 4 pp.

Canadian Standards Association, 1994, "Design of Concrete Structures, Chapter 20—Strength Evaluation Procedures," Standard A23.3.

Casadei, P., 2004, "Assessment and Improvement of Capacity of Concrete Members: A Case for In-Situ Load Testing and Composite Materials" PhD dissertation, Department of Architecture and Civil Engineering, University of Missouri-Rolla, Rolla, Mo.

Casadei, P.; Parretti, R.; Heinze, T.; and Nanni, A., 2005, "In-Situ Load Testing of Parking Garage RC Slabs: Comparison Between Cyclic and 24 Hrs Load Testing," *Practice Periodical on Structural Design and Construction*, ASCE, V. 10, No. 1, Feb., pp. 40-48.

Chicago Building Ordinance, 1910.

Committee on Reinforced Concrete and Building Laws, 1916, "Proposed Revised Standards Building Regulations for the Use of Reinforced Concrete," *Proceedings of the Twelfth Annual Convention of the American Concrete Institute*, p. 172.

Committee on Reinforced Concrete and Buildings Laws, 1917, "Proposed Standard Building Regulation for the Use of Reinforced Concrete," *Proceedings of the Thirteenth Annual Convention of the American Concrete Institute*, p. 410.

Concrete Innovation Appraisal Service (CIAS), 2000, "Guidelines for the Rapid Load Testing of Concrete Structural Members," *CIAS Report 00-1*, American Concrete Institute, Farmington Hills, Mich., 97 pp.

Condron, T. L., 1917, "Principles of Design and Results of Tests on Girderless Floor Construction of Reinforced Concrete," *Proceedings of the Ninth Annual Convention of the National Association of Cement Users*, pp. 116-126.

Czechoslovak State Standard CSN 73 2030, 1977, "Loading Tests of Building Structures, Common Regulations," Publishers for the Office for Standardization and Measurement, Prague, Czechoslovakia, p. 38.

FitzSimons, N., and Longinow, A., 1975, "Guidance for Load Tests of Buildings," *Journal of the Structural Division*, ASCE, pp. 1367-1380.

Galati, N.; Casadei, P.; Lopez, A.; and Nanni, A., 2004, "Load Test Evaluation of Augspurger Ramp Parking Garage, Buffalo, NY," *Report* 04-50, University of Missouri-Rolla, Rolla, Mo.

Genel, M., 1955a, "Ripartizione Laterale dei Carichi in Seguito alla Monoliticita' del Cemento Armato," *Il Cemento*, V. 52, June, pp. 6-15.

Genel, M., 1955b, "Ripartizione Laterale dei Carichi in Seguito alla Monoliticita' del Cemento Armato (continuazione)," *Il Cemento*, V. 52, July, pp. 6-13.

Hennebique, F., 1909, *Ferro-Concrete Theory and Practice, A Handbook for Engineers and Architects*, L. G. Mouchel & Partners, Ltd., London, 359 pp.

Institution of Structural Engineers, 1964, "Report of a Committee on the Testing of Structures," London, 24 pp.

International Code Council, 2003, *International Building Code*, International Code Council, Inc., 358 pp.

Japanese Society for Nondestructive Inspection (JSNDI), 1991, "Methods for Absolute Calibration of Acoustic Emission Transducers by Reciprocity Technique," NDIS 2109.

Japanese Society for Nondestructive Inspection (JSNDI), 1997, "Evaluation of Performance Characteristics of Acoustic Emission Testing Equipment," NDIS 2106.

Japanese Society for Nondestructive Inspection (JSNDI), 1997, "Evaluation Method for the Deterioration of Acoustic Emission Sensor Sensitivity," NDIS 2110.

Japanese Society for Nondestructive Inspection (JSNDI), 1997, "Recommended Practice for the Continuous Acoustic Emission Monitoring of Pressure Vessels," NDIS 2419.

Japanese Society for Nondestructive Inspection (JSNDI), 2000, "Recommended Practice for In-Situ Monitoring of Concrete Structures by Acoustic Emission," NDIS 2421, 6 pp.

Joint Committee on Concrete and Reinforced Concrete, 1913, "Second Report of Joint Committee on Concrete and Reinforced Concrete," *1913 Proceedings of the American Society of Civil Engineers*, 45 pp.

Kramer, E. W., and Raafat, A. A., 1961, "The Ward House: a Pioneer Structure of Reinforced Concrete," *Journal of the Society of Architectural Historians*, V. 20, No. 1, Mar., pp. 34-37.

Lombardo, S., and Mirabella, G., 2004, "Il Collaudo Tecnico Amministrativo dei Lavori Pubblici" Dario Flaccovio Editore s.r.l. (in Italian)

Masetti, F., 2005, "Structural Implications of Field Load Testing Using Patch-Loads," MS dissertation, Department of Architecture and Civil Engineering, University of Missouri-Rolla, Rolla, Mo.

Masetti, F.; Galati, N.; Nehil, T.; and Nanni, A., 2006, "In-Situ Load Test: a Case Study," *Paper* 16-9, fib Second Congress, June 4-8, Naples, Italy, 11 pp. (CD-ROM)

Mettemeyer, M., 1999, "In Situ Rapid Load Testing of Concrete Structures," Master's thesis, Department of Civil Engineering, University of Missouri-Rolla, Rolla, Mo.

National Association of Cement Users (NACU), 1908, "Report of the Committee on Laws and Ordinances," National Association of Cement Users, pp. 233-239.

National Association of Cement Users (NACU), 1910, "Standard Building Regulations for the Use of Reinforced Concrete," *NACU Standard* No. 4, National Association of Cement Users, pp. 349-361.

Nehil, T.; Masetti, F.; and Nanni, A., 2006, "Test Load Magnitude and Acceptance Criteria For Strength Evaluation by Means of Load Testing: Current Recommendations of American Concrete Institute Committee 437—Strength Evaluation," *Paper* 16-24, fib Second Congress, June 4-8, Naples, Italy, 9 pp. (CD-ROM)

Perrot, E. G., 1911, "Analysis of Results of Load Test on Panels of Reinforced Concrete Buildings," *Proceedings of the Seventh Annual Convention of the National Association of Cement Users*, p. 216.

RILEM, 1980, "General Recommendation for Statical Load Test of Load-Bearing Concrete Structure In Situ," TBS-2.

Slater, W. S., 1912, "The Testing of Reinforced Concrete Buildings Under Load," *Proceedings of the Eighth Annual Convention of the National Association of Cement Users*, p. 165.

Turner, C. A. P., 1912, *Examples of the Mushroom System of Reinforced Concrete Construction*, 68 pp.

Urquhart, L. C., and O'Rourke, C. E., 1926, *Design of Concrete Structures*, McGraw-Hill, New York, 482 pp.

Vatovec, M.; Kelley, P.; Alkhrdaji, T.; and Nanni, A., 2002, "Evaluation and Carbon Fiber Reinforced Polymer Strengthening of an Existing Garage: Case Study," *Journal of Composites for Construction*, V. 6, No. 3, pp. 184-193.

Wright, F. L., 1906, "Specifications for the Construction of Unity Temple," 38 pp.

APPENDIX A—DETERMINATION OF EQUIVALENT PATCH LOAD

A.1—Notation

The selection of notations reported in this section only refers to the symbols used in this appendix.

a = dimension of patch load in longitudinal direction, in. (mm)

b = dimension of patch load in transverse direction, in. (mm)

c = comprehensive coefficient for determination of equivalent test load

c_1 = coefficient for determination of equivalent test load, longitudinal direction

c_2 = coefficient for determination of equivalent test load, transverse direction

E = concrete modulus of elasticity, psi (N/mm^2)

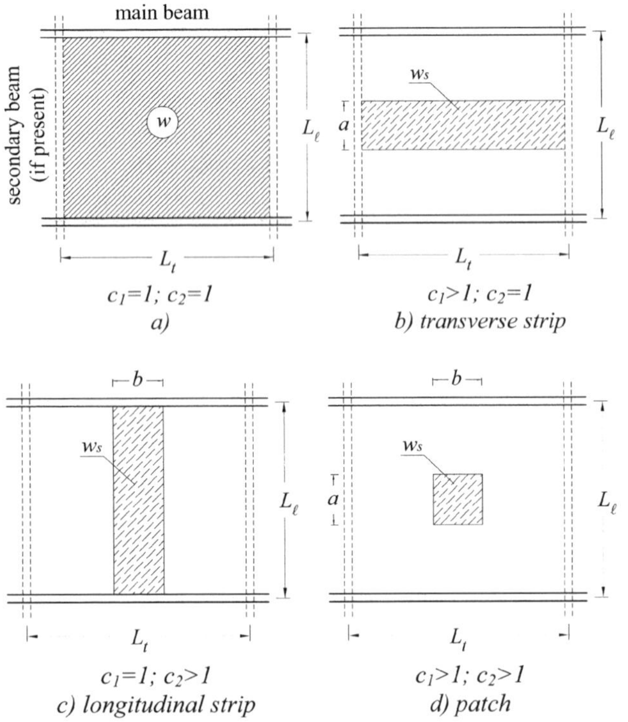

Fig. A.1—Loading configuration.

- I_s = moment of inertia of slab, in.^{4} (mm^4)
- k_{s1} = rotational stiffness at left span-end location, kip-ft (kN-m)
- k_{s2} = rotational stiffness at right span-end location, kip-ft (kN-m)
- L_ℓ = slab span in longitudinal direction, ft (m)
- L_t = slab span in transverse direction, ft (m)
- M_{int1} = moment at center-span for System 1 in Fig. A.3, kip-ft (kN-m)
- M_{int2} = moment at center-span for System 2 in Fig. A.3, kip-ft (kN-m)
- M_{L1} = moment at left span-end for System 1 in Fig. A.3, kip-ft (kN-m)
- M_{L2} = moment at left span-end for System 2 in Fig. A.3, kip-ft (kN-m)
- M_{R1} = moment at right span-end for System 1 in Fig. A.3, kip-ft (kN-m)
- M_{R2} = moment at right span-end for System 2 in Fig. A.3, kip-ft (kN-m)
- P = force corresponding to $w \times L_\ell \times L_t$, kip (kN)
- P_s = force corresponding to $w_s \times a \times b$, kip (kN)
- t = slab thickness, in. (mm)
- $v(x)$ = analytically computed deflected shape, in. (mm)
- w = magnitude of total uniformly distributed test load (TLM), lb/ft^2 (kN/m^2)
- w_s = magnitude of equivalent patch test load, lb/ft^2 (kN/m^2)
- w_{scal} = magnitude of patch test load used to calibrate coefficient c, lb/ft^2 (kN/m^2)
- x = coordinate along the longitudinal axis, in. (mm)

A.2—Introduction

A 24-hour load test or a cyclic load test conducted with hydraulic jacks that apply concentrated or patch loads (w_s) as an alternative to the distributed load ($TLM = w$) offers significant advantages. A disadvantage is the computational complexity associated with the determination of the patch load(s) that generates the internal forces (that is, shear or bending moment) at a critical location identical to that for the distributed load.

This appendix intends to provide a concise explanation of the analytical steps necessary for the determination of w_s given the value of w. To accomplish this objective, only the concepts are presented, leaving the details to available literature (Masetti 2005; Masetti et al. 2006; Nehil et al. 2006) and using as an example the case of a one-way slab (for which the positive moment at midspan is the force at the location of interest). Other cases can be approached using the methodology shown.

A.3—One-way slab system

To load-test a one-way reinforced concrete slab for positive moment at midspan, a uniformly distributed test load, $TLM = w$, should be considered as shown in Fig. A.1(a). The equivalent load test can be performed by applying the load on a restricted area of the slab of interest. For example, the applied load may consist of transverse or longitudinal load strips (Fig. A.1(b) and (c) or a patch load as shown in Fig. A.1(d)). The equivalent test load w_s, irrespective of the pattern, has to be selected to cause, at a given location, the same internal force caused by w. The relationship between the two load values is described by the following equation

$$w_s = c_1 \times c_2 \times w = c \times w \qquad \text{(A-1)}$$

where c, c_1, and c_2 are coefficients such that: $c = c_1 \times c_2$ is greater than 1.0.

The coefficient c_1 depends on both the degree of fixity of the slab restraints at the main beam locations and the loading strip length a. Its value approaches 1.0 when the strip length a approaches the slab length L_ℓ.

The coefficient c_2 is a function of the transverse stiffness of the slab; it reflects the fact that the portions of the slab to which the load is not applied participate in the load sharing. The coefficient c_2 increases with the decreasing of the loading strip width b and its value approaches 1.0 when the strip width b approaches the slab width L_t.

The determination of the coefficient $c = c_1 \times c_2$ is not trivial, and several approaches can be adopted. In the following sections, preliminary calculations and an experimental method for its refinement are described for a load test to be conducted with a patch load, as shown in Fig. A.1(d).

A.4—Procedure and preliminary calculations

The engineer in charge of the load test must estimate w_s in advance, after the pattern of patch load application and the magnitude w have been established. It is important to recognize that any structural analysis must treat fixity and stiffness as preliminary assumptions that could be refined based on actual behavior once the structure is loaded and its

Table A.1—Numerical example: geometry and loads (Fig. A.1)

Geometry			Loads				
Slab	L_ℓ, ft (m)	16 (4.88)	TLM	w, lb/ft² (kN/m²)	100 (4.79)	$P = w \times L_\ell \times L_t$, kip (kN)	28.8 (128.11)
	L_t, ft (m)	18 (5.49)	Preliminary	c_1	6.25		
	t, in. (mm)	7 (177)		c_2	9.28		
Beams, in. (mm)		18 x 24 (457 x 610)		w_s, lb/ft² (kN/m²)	5777 (276.60)	$P_s = w_s \times a \times b$, kip (kN)	8.67 (38.57)
Columns, in. (mm)		18 x 18 (457 x 457)		w_{scal}, lb/ft² (kN/m²)	575 (27.53)		
Patch load	a, in. (mm)	12 (305)	After calibration cycle	c_1	5.95		
	b, in. (mm)	18 (457)		c_2	7.83		
				w_s, lb/ft² (kN/m²)	4662 (223.22)	$P_s = w_s \times a \times b$, kip (kN)	7.01 (31.18)

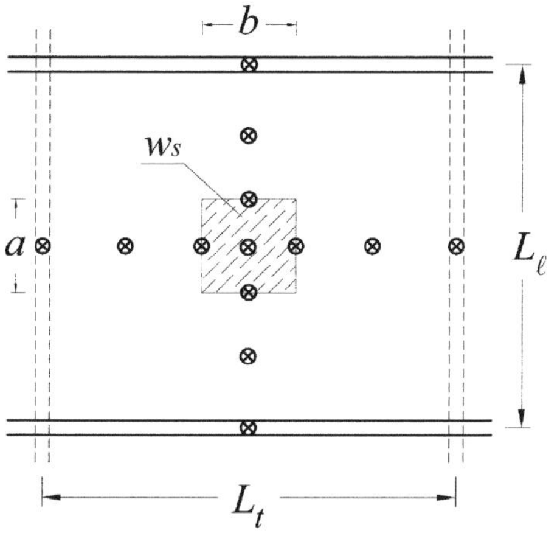

Fig. A.2—LVDT locations.

response is measured. With these refinements, the induced internal forces can be determined with a much higher degree of accuracy.

The preliminary analysis, given w and the patch-load configuration, should consist of three main steps:

1. Estimate the stiffness of every structural member in the system (that is, columns and beams);

2. Perform a calculation of the critical internal force (that is, positive moment) at the selected location due to w; and

3. Calculate w_s using the target force and the degree of fixity obtained from Steps 1 and 2.

The strength of the system subjected to w_s should be checked to ensure safety. For example, if the test is meant to produce a critical flexural response, the shear capacity of the structure is to be checked to prevent shear failure. In addition, if members within the structure are used to supply the reaction to w_s, the capacity of those members should be checked as well.

For the purpose of an example, a one-way reinforced concrete slab system with characteristics given in Table A.1 is given (Fig. A.1(d)). First, a preliminary analysis is performed to estimate w_s, and then a real load cycle at a percentage of the estimated w_s value allows for calibration. In the preliminary phase, the span-end fixities are estimated following the Commentary R13.7.4 of ACI 318-05, and the coefficient c_1 is calculated using traditional structural analysis methods imposing the patch load w_s to produce the same maximum positive moment at center-span caused by the uniformly distributed load w. The coefficient c_2 is calculated by means of the evaluation of the slab width that effectively participates in sharing the load along the transverse direction (Genel 1955a,b). As shown in Table A.1, both c_1 and c_2 assume values greater than one in the preliminary phase because the load is applied by strips of width a and b smaller than L_ℓ and L_t. After the value w_s is estimated, a pretest cycle can be performed to refine the calculation of c_1 and c_2. For the pretest, a load magnitude $w_{scal} = 10\%\ w_s$ is deemed reasonable because the structure is linear elastic in this range. Following the procedure described in Section A.5, the values of c_1 and c_2 are recomputed and the final value of w_s is obtained, as shown in Table A.1.

A.5—Calculations after calibration cycle

The objective of this section is to show the procedure to compute the values of c_1 and c_2 once a calibration cycle at a magnitude w_{scal} has been performed. The load w_{scal} should be selected taking into consideration the following aspects:

- It cannot exceed the linear elastic threshold of the structure; and
- It has to be large enough to cause a deflected shape to be read with adequate accuracy by the sensors used in the test setup.

The coefficients c_1 and c_2 are computed separately, making reference to the sketches of Fig. A.1(b) and (c). In case of a patch load, the separation is only applicable if the principle of superposition is valid (elastic and linear behavior). This separation is only for the purpose of the presentation. In reality, because the structure is subject to a patch load instead of a strip load, c_1 and c_2 are interrelated. In the given example, seven sensors (linear variable differential transducers [LVDTs]) were used along both the longitudinal and transverse directions (Fig. A.2 shows their locations), for a total of 13 devices.

A.5.1 *Determination of* c_1—Referring to Fig. A.3, the load w_s producing the same maximum positive moment in System 2 as produced by w in System 1 will be determined. The relationship between w and w_s is expressed by

$$w_s = c_1 w \tag{A-2}$$

taken from Eq. (A-1), when $c_2 = 1$.

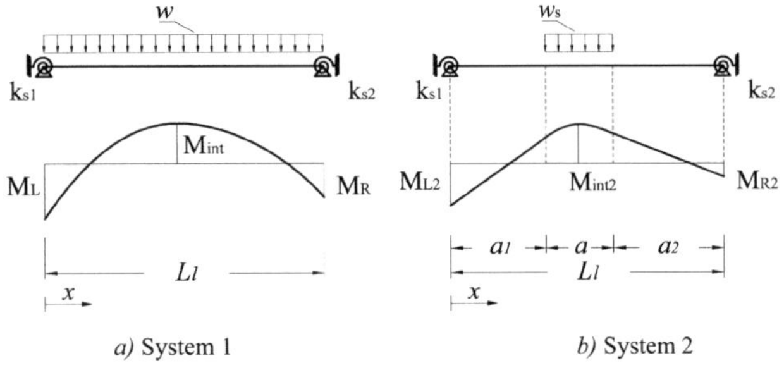

Fig. A.3—Structural models representing real slab.

The procedure presented herein allows solving Eq. (A-2) using the experimentally determined value of c_1 obtained after the calibration cycle.

Three simplifications are necessary to develop a manageable model. First, in Fig. A.3, the supports are assumed to be points, while in reality they have a finite width equal to that of the main beams; second, k_{s1} and k_{s2} are the springs representing the rotational stiffness of the slab connection to the main beams; and third, the structural system is assumed to be linear. M_{int1} and M_{int2} are the maximum positive moments, while M_{L1}, M_{R1}, M_{L2}, and M_{R2} are the moments acting at the supports in Systems 1 and 2 of Fig. A.3. For the structures presented in Fig. A.3, considering linear elastic behavior, the following equations can be obtained for System 1

$$\begin{cases} M_{L1} = M_{L1}(w, L_t, E, I_s, k_{s1}, k_{s2}) \\ M_{R1} = M_{R1}(w, L_t, E, I_s, k_{s1}, k_{s2}) \\ M_{int1} = M_{int1}(w, L_t, E, I_s, k_{s1}, k_{s2}) \end{cases} \quad \text{(A-3)}$$

and for System 2

$$\begin{cases} M_{L2} = M_{L2}(w_s, a_1, a_2, a, E, I_s, k_{s1}, k_{s2}) \\ M_{R2} = M_{R2}(w_s, a_1, a_2, a, E, I_s, k_{s1}, k_{s2}) \\ M_{int2} = M_{int2}(w_s, a_1, a_2, a, E, I_s, k_{s1}, k_{s2}) \end{cases} \quad \text{(A-4)}$$

To determine w_s as a function of w, it is necessary to impose that $M_{int1} = M_{int2}$, and solve for w_s. The values of k_{s1} and k_{s2} are unknown and need to be determined experimentally by means of the preliminary test. The degree of fixity at the slab ends can be calculated by means of the procedure outlined below consisting of eight sequential steps:

Step 1	A load w_{scal} is applied using the load pattern of System 2, and deflections are recorded at a number of locations given in Step 4.
Step 2	The moment diagram $M(x)$ has the shape shown in Fig. A.3(b); its values are unknown, but its equation has the following form: $$M(x) = \begin{cases} Ax + B & \text{if } 0 \le x < a_1 \\ Cx^2 + Dx + E & \text{if } a_1 \le x \le a_1 + a \\ Fx + G & \text{if } a_1 + a < x \le L_t \end{cases} \quad \text{(A-5)}$$ where A, B, C, D, E, F, and G are constants that depend on a_1, a_2, and a; w_{scal}; and k_{s1} and k_{s2}.
Step 3	Considering the Bernoulli's equation of the elastic line (neglecting deformations due to shear) $$v''(x) = -\frac{M(x)}{EI_s} \quad \text{(A-6)}$$ and integrating twice with respect to x (assuming EI_s constant with respect to x), the deflected shape assumes the form $$v(x) = \begin{cases} C_1x^3 + C_2x^2 + C_3x + C_4 & \text{if } 0 \le x < a_1 \\ C_5x^4 + C_6x^3 + C_7x^2 + C_8x + C_9 & \text{if } a_1 \le x \le a_1 + a \\ C_{10}x^3 + C_{11}x^2 + C_{12}x + C_{13} & \text{if } a_1 + a < x \le L_t \end{cases} \quad \text{(A-7)}$$ The 13 constants C_1, C_2, ..., C_{13}, related to k_{s1} and k_{s2}, can be determined by means of six compatibility relationships (displacement and rotation at $x = 0$, $x = a_1$, $x = a_1 + a$, and $x = a_1 + a + a_2$) and by measurement of at least seven displacements.
Step 4	The absolute displacements d_1, d_2, ..., d_7 have to be measured at the positions shown in Fig. A.4. Because zero displacement is assumed at the supports, the measures d_1, d_2, ..., d_7 should be transformed into displacements relative to the slab movement, namely, δ_1, δ_2, ..., δ_7 (where $\delta_1 = \delta_7 = 0$). The displacements δ_i can be determined as $$\delta_i = d_i - \frac{d_7 - d_1}{a_1 + a_2 + a}x_i - d_i \text{ with } i = 1,2, ..., 7$$ 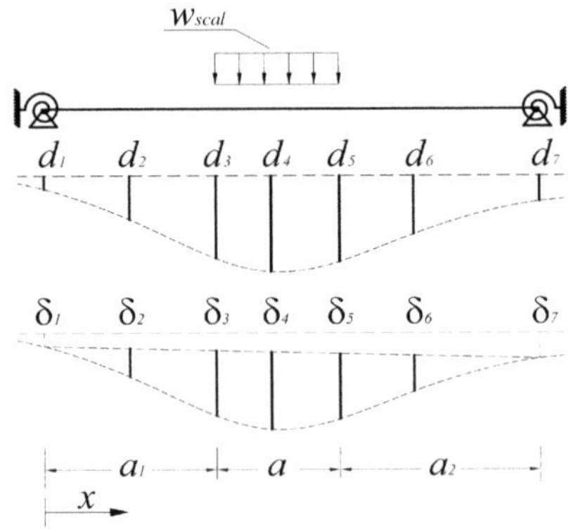 *Fig. A.4—Derivation of relative displacements δ.*
Step 5	Knowing C_1, C_2, ..., C_{13}, the form of the measured shape can be approximated by Eq. (A-7).
Step 6	Using C_1, C_2, ..., C_{13}, taking the second derivative of Eq. (A-7), plugging it into Eq. (A-6), the constants A, B, C, D, E, F, and G in Eq. (A-5) are found.
Step 7	Using A, B, C, D, E, F, and G, the approximate shape of the moment diagram due to w_{scal} is found. Therefore, M_{int2}, M_{L2}, M_{R2}, and EI_s can be derived, where EI_s represents the slab constant flexural stiffness.
Step 8	Using M_{int2}, M_{L2}, M_{R2}, and EI_s, the linear system described in Eq. (A-4), when $w_s = w_{scal}$, can be solved for k_{s1} and k_{s2}.

The suggested method can be applied in the case of a very small strip width a (that is, concentrated load), that theoretically could become a line load.

A.5.2 *Determination of* c_2—The coefficient c_2 takes into account the loading limited to a width b rather than the total slab width L_t. It can be determined by applying Betti's theorem that states, "in a system, applying two sets of forces P_i and Q_j that cause two sets of displacements δp and δq respectively, the work of the forces P_i on the displacements δq_i at the locations i is equal to the work of the forces Q_j on the displacements δp_j at the locations j." The application of this theorem allows one to determine the relationship between the calculated maximum deflection in System 1 (that is, design configuration: Fig. A.1(a)) and the measured maximum deflection in System 2 (that is, test configuration: Fig. A.1(c)), as shown in Fig. A.5. In addition, b_1, ..., b_6 represent distances between the sensors in the transverse direction; f_{I0} represents the displacement at the center in System 1; and $f_{II0}, \ldots, f_{II6}$ represent the displacements at the sensor locations in System 2.

The value of c_2 can be obtained as

$$c_2 = \frac{b_1 \times (f_{II1} + f_{II2}) + b_2(f_{II2} + f_{II3}) + b_3(f_{II3} + f_{II0}) + \varsigma}{2 \cdot f_{I0} \cdot b} \qquad \text{(A-8)}$$

where $\varsigma = b_4(f_{II0} + f_{II4}) + b_5(f_{II4} + f_{II5}) + b_6(f_{II5} + f_{II6})$.

Thus, c_2 is defined only by in-place measured displacements under a load cycle at a load level w_{scal}.

A.6—Conclusions

The following conclusions can be derived:

- Given w, it is possible to compute w_s based on a, b, and the location of the force of interest; w_s is related to w by c_1 and c_2;
- c_1 and c_2 can be estimated first by classical analysis;
- c_1 and c_2 can be calibrated during a preliminary load test at a magnitude w_{scal} under which the slab behavior is linear elastic; and
- For the one-way slab used as an example, the difference between preliminary and calibrated load values to attain the same maximum positive moment at center span was approximately 20%, indicating that calibration is an important step.

APPENDIX B—HISTORY OF LOAD TEST, LOAD FACTORS, AND ACCEPTANCE CRITERIA

B.1—Notation

The selection of notations reported in this section only refers to the symbols used in this appendix. Because this Appendix reports direct quotes from cited references, some of the symbols reported herein may conflict with more commonly accepted symbols reported elsewhere in the document and in this appendix itself.

c = distance of external fiber of section from neutral axis, in. (mm)
f_c = allowable compressive stress of concrete, psi (N/mm^2)
f'_c = specified compressive strength of concrete, psi (N/mm^2)
h = overall height of member, in. (mm)
k = coefficient used to determined value of maximum deflection
l_t = span of member under load test; units (in. or ft) depend on structural member considered (ACI 318)
t = thickness of slab, in. (mm)
w = unit weight of concrete, lb/ft^3 (kg/m^3)
n = coefficient reflecting maximum moment in structural member, dimensionless
m = coefficient reflecting maximum deflection in structural member, dimensionless

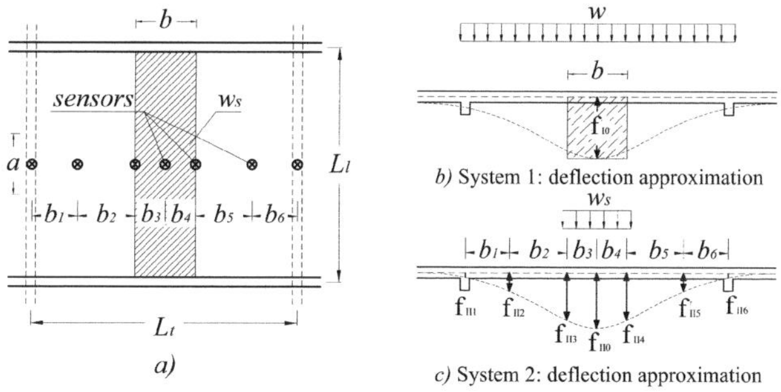

Fig. A.5—Application of Betti's theorem (approximate solution).

D = total dead load; units (lb or kips) depend on structural member considered
D = maximum deflection in the structure; units (in. or ft) depend on structural member considered
E = modulus of elasticity of concrete, psi (N/mm^2)
I = moment of inertia of the section, in.^{4} (mm^4)
L = live loads produced by use and occupancy of the building not including construction, environmental loads, and superimposed dead loads; units (lb or kips) depend on structural member considered
L = span of structural member; units (in. or ft) depend on structural member considered
M_f = factored load generating bending moment according to the Canadian Standard Association
N = number of total events
P_f = factored load generating normal force according to the Canadian Standard Association
TL = test load per ACI 318 before 1971; units (in. or ft) depend on structural member considered
TL_{05} = test load per ACI 318-71 through 318-05; units (lb or kip) depend on structural member considered
V_f = factored load generating shear force according to the Canadian Standard Association
W = applied load; units (lb or kip) depend on structural member considered
Δ = maximum deflection in structure; units (in. or ft) depend on structural member considered
Δ_{max} = measured maximum deflection, in. (mm)

B.2—Historical load test practice in the United States and according to ACI

B.2.1 *Early history in the United States from 1890 to 1920*—The practice of load testing concrete structures in the U.S. began late in the 1890s as a method of proof testing newly constructed concrete systems and structures. Development of reinforced concrete structures in this country, as well as abroad, was fostered by the development of numerous proprietary reinforcement systems. Examples include the Hennebique system developed in France in the 1870s and patented in 1892 by Francois Hennebique (Hennebique 1909); the Ransome system patented by Ernest L. Ransome in 1884 in the United States; the Kahn system that was developed by Julius Kahn and used extensively by his brother, Albert Kahn, between 1900 and 1920; and the Turner system of reinforced concrete construction, developed in the early 1900s by C. A. P. Turner. Proof load testing of those systems and many others is widely reported in the literature between 1890 and 1920.

The Ward House, built in Port Chester, N.Y., from 1873 to 1876 and constructed by William E. Ward, is generally recognized as the first reinforced concrete structure in the United States. Ward employed load testing during the construction of this residence as a means of proving the viability of this new and unique method of construction, as reported by Kramer and Raafat (1961):

> ...Ward undertook numerous field tests of his new system from which acceptable results for deflection and strength were obtained. Before constructing any of the floors, he subjected a sample beam to a weight far beyond its normal load carrying capacity. After the parlor floor had been laid for 1 year, he piled a weight of 26 tons between the two central beams, leaving it there over the winter; at the end of the testing period, the amount of deflection was only one hundredth of an inch. Ward was an eminently practical man who believed in putting his theories to the most rigid tests. He also carried out experiments on flat slabs of concrete supported on all four sides.

Another early example of proof load testing of a new concrete structure was reported by Birkmire in 1894:

> A section of a flat floor in the California Academy of Science, 15 x 22 ft, was tested in 1890 with a uniform load of 415 lb per square foot, and the load left on for 1 month. The deflection at the center of the 22 ft space was only 1/8 in.

An early example of the incorporation of load testing into the design of reinforced concrete buildings is Frank Lloyd Wright's 1906 design of the Unity Temple in Oak Park, Ill. (Wright 1906) Wright prepared design documents that contained the following performance specification for the structural design of the concrete structure of this remarkable historic building:

> Throughout floors shall be constructed to carry safely a uniformly distributed superimposed live load of 100 pounds per square foot with a maximum deflection of 1/800 of the span.

This was immediately followed with a provision requiring a full-scale load test to ensure that all work met this performance specification.

> Floor shall be tested in approved manner, at expense of this contractor at any point after cement has set 15 days. They shall be subjected to twice the loading specified for live load within the limits of the deflection specified and after removal of loading shall resume position previous to test. Any work not passing test shall be replaced and brought to test requirements.

This early specification for proof load testing of a newly constructed concrete structure contained several of the basic elements of load testing that are found in our modern building codes:

- Specification of a minimum age of the structure before testing; in this case, 15 days;
- Specification of a minimum test load; in this case, two times the live load;
- Specification of a maximum acceptable deflection; in this case, the span length divided by 800 (l_t/800); and
- Specification of a required deflection recovery; in this case, apparently 100% as the maximum deflection during the load test was to be recovered upon removal of the superimposed test load.

The earliest building code reference for load testing of reinforced concrete structures that this committee has found is the 1903 New York City Building Regulations (Urquhart and O'Rouke 1926) that contained the following:

> The Contractor must be prepared to make load tests on any portion of a concrete-steel-construction, within a reasonable time after erection, as often as may be required by the Superintendent of Buildings. The tests must show that the construction will sustain a load of three times that for which it is designed without any sign of failure.

Another early building code reference to load testing of concrete structures can be found in the Chicago Building Ordinance adopted in December of 1910 that states the following:

> Tests shall be made by the owner upon the demand of the Commissioner of Buildings on all forms of construction involving spans of over 8 ft. Such tests shall be made to the approval of the Commissioner of Buildings and must show that the construction will sustain a load equal to twice the sum of the live and dead loads, for which it was designed, without any indication of failure. Each test load shall remain in place at least 24 hours.
>
> Each test load shall cover two or more panels and shall remain in place at least 24 hours. The deflection under the full test load at the expiration of 24 hours shall not exceed 1/800th of the span. These tests shall be considered as tests of workmanship only.

The first code requirement for load testing of concrete structures in this country by a national organization is that contained in the National Association of Cement Users' (NACU) 1908 document entitled "Report of the Committee on Laws and Ordinances":

> The contractor must be prepared to make load tests on any portion of a reinforced concrete constructed building within a reasonable time after erection as often as may be required by the commissioner of buildings. The tests must show that the construction will sustain a load with a factor of safety for floors and structural members as required by Section 126 of this code.

The NACU was the forerunner to the American Concrete Institute. The NACU's 1910 "Standard Building Regulations for the Use of Reinforced Concrete" did not include any guidance on load testing.

In 1912, W. S. Slater published what may have been the first state-of-the-art report on load testing of reinforced concrete buildings in the United States. That document contains the following statements regarding the intent of load tests:

> Load tests have been required by city building departments and as a condition of acceptance of reinforced concrete buildings and have been used by construction companies and engineers to demonstrate the adequacy of various designs. Such load tests are never continued to destruction, the applied load being generally twice the design live load, and emphasis is placed upon measurement of deflections and recovery. No measurement of stresses is made in such tests and under these conditions the safe load cannot be fixed upon as a definite ratio of the ultimate load.

This document clearly shows the importance of the deflection response in evaluating a load test.

At about the same time, Emile G. Perrot summarized the prevailing mood regarding the intent of load tests among contemporaries (Perrot 1911).

> These load tests are made, not with a view of obtaining scientific data, but more particularly of satisfying both the architects and the owners that the work of the contractor had been properly performed, and that the building would sustain the loads for which it was designed.

Perrot also provided insight in that same document relating to the possible origin of the use of a test load of 2.0 times the superimposed live load, as follows:

> The practice now is to require a floor to be tested to double the live load without sign of fracture and that after the load is removed the floor must recover its normal position. It is the writer's belief that many specifications require a too rigid test by imposing the requirement of loading the floor to three or more times the live load. A little consideration will show the fallacy of this requirement, because it is not desirable to test an actual floor of a building so as to stress the reinforcement to a point equal to or greater than its elastic limit, as this would permanently weaken the section of the floor so test. Take for example a floor designed for a live load of 200 lb per sq ft; assume the test load to be three times the live load; also assume the dead weight of the construction to be 75 per sq ft. Then as usually computed with a factor of safety of 4, the breaking load of the floor would be 1100 lb per sq ft. If from this is deducted the dead weight of 75 lb per sq ft, the load to break the floor is 1025 lb per sq ft. If a test load of three times 200 lb, or 600 lb per sq ft is applied, there is likelihood of the reinforcement being stretched beyond its yield point, or elastic limit, because the average elastic limit of medium steel is one-half its ultimate strength. Hence a test load that exceeds more than one-half of 1025 lb, or about 500 lb, should not be applied. This, it will be noticed, is about 2-1/2 times the live load. The requirement of the Bureau of Building Inspection of Philadelphia for a test load is two times the live load.

T. L. Condron presented an important paper in 1917 at the 9th Annual Conference of the NACU entitled, "Principles of Design and Results of Test on Girderless Floor Construction of Reinforced Concrete" (Condron 1917). The following discussion relevant to the issues of appropriate test loads and the corresponding maximum allowable deflection in a load test is contained in the written paper published in the proceedings of that conference:

> A test load equal to twice the live load (where the live load is greater than the dead load) seems to me to be the maximum test load that should be called for. With regard to the proper amount of deflection, which should be considered satisfactory, under test load, this can only be arrived at by careful study of the many tests that have been made on various types of construction. The permissible deflection, under test load, should be less than the deflection that produces visible cracks in the finished concrete surface of the floor or ceiling. Reinforced concrete structures should not be subjected to loadings that will produce visible cracks, and certainly structures should be so designed that working loads will not produce cracks.
>
> ....a limit of 1/800th of the diagonal span for a single panel test of double the live load would be entirely reasonable, but is apparently too severe a limitation where the test load is made to cover two panels and is equal to twice the live, plus the dead load. For such a test, the permissible deflection should be at least 1/600th of the diagonal span.

The "Second Report of the Joint Committee on Concrete and Reinforced Concrete" was published in the 1913 proceedings of the American Society of Civil Engineers, and contained the following guidance on load testing:

> Load tests on portions of the finished structure shall be made where there is reasonable suspicion that the work has not been properly performed, or that, through influences of some kind, the strength has been impaired. Loading shall be carried to such a point that one and three-quarters times the calculated working stresses in critical parts are reached, and such loads shall cause no injurious permanent deformations. Load tests shall not be made until after 60 days of hardening.

The load testing requirements in this document were aimed at defective or questionable new structures, rather than proof testing, which proved to be somewhat of an anomaly at that time.

The 1916 proceedings of the 12th Annual Conference of ACI contained the following guidelines for proof load testing (Committee on Reinforced Concrete and Buildings Laws 1916):

> The Superintendent of Buildings may require a load test on a floor within reasonable time of the erection. The test shall be made under the supervision of the Superintendent of Buildings and shall show that the construction will sustain safely an applied load of twice the total live load, but in no case less than one and one-half times the total live and dead load.

Table B.1—Summary of ACI code requirements for load testing practice

Year	Minimum age	Total test load		Duration of load	Maximum deflection	Deflection recovery	Notes
		D	*L*				
1916	—	1.00	2.00	—	—	—	—
1917	60 days	1.00	2.00	—	—	80% at 3 days	—
1920	60 days	1.50	1.50	—	$l/800$	80% at 7 days	—
1924	30 days	1.50	1.50	24 hours	None	75% at 24 hours	—
1928	60 days	1.50	1.50	24 hours	None	75% at 24 hours	—
1936	None	1.50	1.50	24 hours	$l_t^2/12{,}000h$	75% at 24 hours	—
1941	None	1.50	1.50	24 hours	$l_t^2/12{,}000h$	75% at 24 hours	—
1947	None	1.50	2.00	24 hours	$l_t^2/12{,}000h$	60% at 24 hours	Structure fails if deflection exceeds three times $l_t^2/12{,}000h$
1951	None	1.50	2.00	24 hours	$l_t^2/12{,}000h$	60% at 24 hours	Structure fails if deflection exceeds three times $l_t^2/12{,}000h$
1956	56 days	1.00	2.00	24 hours	$l_t^2/12{,}000h$	75% at 24 hours	Multiple limits on maximum acceptable deflections added
1963	56 days	1.30	1.70	24 hours	$l_t^2/20{,}000h$	75% at 24 hours	—
1971 through 2005	56 days	1.19	1.45	24 hours	$l_t^2/20{,}000h$	75% at 24 hours	—

Provisions for load testing progressed a little further in the 1917 proceedings of the 13th Annual Conference of ACI (Committee on Reinforced Concrete and Buildings Laws 1917). This report included the first appearance of deflection recovery as an acceptance criterion in an ACI document and of a requirement for the minimum age of a structure at time of test:

> The Building Department may require the Owner to make load tests on portions of the finished structure where there is a reasonable suspicion that the work has not been properly performed, or that, through influences of the same kind, the strength has been impaired, or where there is any doubt as to the sufficiency of the design. The test shall show that, with a load of twice the design live load, the permanent deflection several days after load is removed to be not more than 20% of the total deflection under the test load. Load tests shall not be made before the concrete has been in place 60 days.

B.2.2 *1920 ACI regulations for reinforced concrete*—ACI issued "Standard Specifications No. 23—Standard Building Regulations for the Use of Reinforced Concrete" in 1920 (American Concrete Institute 1920). That document contained the following basic provisions for proof load testing of structures:

- Establishment of a specific applied superimposed test load of two times the live load; that is, $TL = 1.0D + 2.0L$;
- Establishment of a maximum acceptable total deflection for flexural members of $l_t/800$;
- Use of deflection recovery as an acceptance criterion; that is, recovery to be equal to or more than 80% of maximum deflection at 7 days after load removal;
- Specification of 56 days as the minimum age of structure before testing would be allowed;
- Allowance for retesting a structure that failed a load test; and
- Provision for reducing the allowable live load when a structure did not pass a load test.

This set of criteria formed the foundation for the load testing provisions contained in all future ACI codes through the ACI 318-05. Various changes were made to the ACI 1920 criteria with nearly each subsequent issuance of ACI regulations or requirements, as summarized in Table B.1 and as discussed as follows.

B.2.3 *1928 ACI tentative regulations*—In 1928, ACI issued "Tentative Building Regulations for Reinforced Concrete" (ACI Committee E-1 1928). The superimposed test load was modified to include 1.5 times the live load plus one half of the dead load added to the self-weight; that is, $TL = 1.5D + 1.5L$. No maximum deflection criterion was prescribed. The structure was considered to have failed the test if, within 24 hours after the removal of the load, the slabs or beams did not show a recovery of at least 75% of the maximum deflection recorded after the 24-hour holding period.

B.2.4 *1936 and 1941 ACI building regulations*—In 1936, ACI issued the "Building Regulations for Reinforced Concrete (ACI 501-36T)" ("T" indicates this was a tentative standard) (American Concrete Institute 1936). The TLM was not changed; however, an important change was introduced. The maximum acceptable deflection Δ was defined as the following

$$\Delta = 0.001L^2/12t \quad \text{(B-1)}$$

where L = the span (l_t in this report and ACI 318-05), and t = total depth of the slab or beam (h in the notation of this report and ACI 318-05), expressed in the same units as span and deflection.

This general form of the deflection acceptance criterion has been in the ACI Building Code ever since. Because the origin of this important and lasting criterion in ACI 318 has become lost to most current practitioners, it is discussed in some detail herein, and the derivation of Eq. (B-1) is shown.

In 1909, C.A.P. Turner (Turner 1912) discussed the fundamental principle of engineering mechanics behind this equation, indicating that according to elastic theory, the deflection due to the test load would vary roughly as the square of the span divided by the depth for a fixed maximum stress, assuming a fixed ratio of the thickness of a slab to its span. Following is his discussion on this topic; but first, the symbols are defined:

c = distance of extreme section fiber from neutral axis;
f_c = allowable compressive stress;
h = height of section;
E = modulus of elasticity of concrete;
I = moment of inertia of section;
L = span of structural member;
W = applied load; and
Δ = maximum deflection in structure.

From the theory of elasticity for flexural members

$$W = \frac{n \cdot f_c \cdot I}{L \cdot c}$$

where $c = h/2$ for an uncracked section, and n = 4, 8, 8, or 12 for the four different cases, namely: simple and restrained beams loaded with W concentrated at center and W uniformly distributed.

The requirement of stiffness (that is, a given maximum deflection) limits the load by a different formula

$$W = m\frac{E \cdot I \cdot \Delta}{L^3}$$

where m = 48, 76.8 (that is, 384/5), 192, or 384 for the same four cases.

By equating these values of W the relation between Δ and f_c is obtained

$$\Delta = \frac{n \cdot L^2 \cdot f_c}{m \cdot c \cdot E}$$

This shows that the maximum deflection for the same unit stress varies as L^2/c for beams of the same material, while coefficients n and m result in additional variation. Of course, such variations make it impossible to limit the permissible deflection to a fixed percent of the span.

ACI 501-36T (American Concrete Institute 1936) included the following regarding the allowable compressive stress and the modulus of elasticity of concrete:

Allowable compressive stress f_c	$= 0.4f_c'$ (Section 305)
Modulus of elasticity E	$= 1000f_c'$ (Section 601)

Thus, the ratio of f_c/E = 1/2500, and for the case where $c = h/2$ (for an uncracked section), the equation for maximum Δ becomes

$$\Delta = \frac{n \cdot l_t^2 \cdot f_c}{m \cdot c \cdot E} = \frac{n \cdot l_t^2}{m \cdot 1250 \cdot h} \qquad \text{(B-2)}$$

Table B.2—Maximum deflection

Type of beam	Type of load distribution	Δ_{max}
Simple span beam	Uniform load	$l_t^2/12{,}000h$
	Point load at midspan	$l_t^2/15{,}000h$
Fixed end beam	Uniform load	$l_t^2/40{,}000h$
	Point load at midspan	$l_t^2/30{,}000h$

Then, for concrete with f_c' of 2000 psi (13.8 MPa)—a typical strength in the early 1900s—the $f_c = 0.4f_c'$ = 800 psi (5.52 MPa), and the $E = 1000f_c'$ = 2,000,000 psi (13.8 GPa). The maximum deflections as a function of beam type and load distribution are listed in Table B.2.

These values of Δ_{max} for each condition of end restraint are constant with variations in concrete strength because the ratio f_c/E is constant, at least as defined in ACI 501-36T.

This derivation of the equation for maximum deflection of a uniformly loaded simply supported beam, $\Delta_{max} = l_t^2/12{,}000h$, is also confirmed in the 1950 publication in Israel, "Research on Loading Tests of Reinforced Concrete Structures" (Arnan et al. 1950).

It is evident that this original equation for maximum allowable deflection is based on simple span conditions, uncracked concrete sections, and concrete strengths and elastic moduli significantly below those used today.

B.2.5 *ACI 318-47 and 318-51*—In ACI 318-47, the following changes were made to the guidelines for load testing:

- The superimposed test load was increased to half the dead load plus twice the live load added to the self-weight; that is, $TL = 1.5D + 2.0L$;
- The maximum allowable deflection was maintained at $l_t^2/12{,}000h$ despite the increase in test load;
- The deflection recovery requirement was reduced from 75% of maximum deflection to 60%; and
- The provision was added that the structure fails the load test and no retesting is allowed if maximum deflection is greater than 3 times $l_t^2/12{,}000h$.

No changes were made to these load test provisions in ACI 318-51.

B.2.6 *ACI 318-56*—Substantial changes were made to load testing criteria in ACI 318-56. The additional dead load requirement was dropped, and the test load was returned to two times the live load only. The maximum deflection criterion was significantly expanded, as shown in the direct quotes from ACI 318-56, Section 203 (where symbols were defined as t = height of the section, D = maximum deflection in the structural member, and L = span of the structural member):

> (a) If the structure shows evident failure, the changes or modifications needed to make the structure adequate for the rated capacity shall be made; or a lower rating may be established;
>
> (b) Floor and roof construction shall be considered to conform to the load test requirements if there is no evidence of failure and the maximum deflection does not exceed
>
> $D = L^2/12{,}000t$.....................(1)

Table B.3—*k* coefficient

		Values of coefficient *k* in deflection equation for various values of concrete compressive strength f_c', psi (MPa)					
Type of beam	Type of load	2000 (13.8)	2500 (17.2)	3000 (20.7)	3750 (25.9)	4000 (27.6)	5000 (34.5)
Simple span	Uniform load	13,800	12,300	11,200	10,000	9700	8700
	Point load at midspan	17,200	15,400	14,000	12,500	12,200	10,900
Fixed ends	Uniform load	45,800	40,800	37,400	33,500	32,400	29,000
	Point load at midspan	34,400	30,700	28,000	25,100	24,300	21,700

In which all terms are in the same units. Constructions with greater deflections shall meet the requirements of subsections (c), (d), and (e);

(c) The maximum deflection of a floor or roof construction shall not exceed the limit in Table 203(c) considered by the Building Official to be appropriate for the construction;

Table 203(c)—Maximum allowable deflection

Construction	Deflection
1. Cantilevered beams and slabs	$L^2/1800t$
2. Simple beams and slabs	$L^2/1800t$
3. Beams continuous at one support and slabs continuous at one support for the direction of the principal movement	$L^2/9000t$
4. Flat slabs (L = the longer span)	$L^2/10{,}000t$
5. Beams and slabs continuous at the supports for the direction of the principal reinforcement	$L^2/10{,}000t$

(d) The maximum deflection shall not exceed $L/180$ for a floor construction intended to support or be attached to partitions or other construction likely to be damaged by large deflections of the floor; and

Deflection recovery and provisions for retesting were included as follows:

(e) Within 24 hours after the removal of the test load the recovery of deflection caused by the application of the test load shall be at least 75% of the maximum deflection if this exceeds $L^2/12{,}000t$. However, constructions failing to show 75% recovery of the deflection may be retested. The second test loading shall not be made until at least 72 hours after the removal of the test load for the first test. The maximum deflection in the retest shall conform to the requirements of Sections 203(c) and (d) and the recovery of deflection shall be at least 75%.

B.2.7 *ACI 318-63*—In ACI 318-63, ultimate strength design was introduced as an alternate to working stress design. The test load in a load test was redefined as superimposed 30% of the dead load plus 1.7 times the live load added to the self-weight; that is, $TL = 1.3D + 1.7L$. The extent to which this test load would vary from the requirement of ACI 318-56 depended on the relative magnitudes of dead and live loads and could be either higher or lower than the ACI 318-56 test load. The 24-hour holding period for the test load was reaffirmed. The acceptance criteria, however, were made more restrictive. The maximum allowable deflection at the end of the 24-hour holding period was reduced significantly to $\Delta_{max} = l_t^2/20{,}000h$. If that deflection limit was to be exceeded, then recovery of deflection within 24 hours after removal of the test load was to be at least 75% of the maximum deflection to pass the test. The maximum allowable deflections provided in Table 203(c) of ACI 318-56 were dropped from the 1963 code.

The rationale behind the change of the maximum allowable deflection from $l_t^2/12{,}000h$ to $l_t^2/20{,}000h$ is unknown to Committee 437. In an attempt to understand why this change was made, one should note that the values for allowable compressive stress in the extreme fiber of a flexural member in bending f_c and the relationship for the modulus of elasticity E were changed in ACI 318-63 as follows:

Allowable compressive stress $f_c = 0.45f_c'$ (Section 1002)
Modulus of elasticity $E = w^{1.5}33\sqrt{f_c'}$ (Section 1102)

where w = unit weight of concrete.

Table B.3 provides a summary of the values in the coefficient k used for computing the maximum deflection according to the equation $\Delta_{max} = l_t^2/kh$, where $k = mE/2nf_c'$ for various values of f_c' and various conditions of loading and end fixity, based on these 318-63 parameters and using Eq. (B-2) developed in Section B.2.4 of this report.

Table B.3 shows that the k values and, therefore, the corresponding beam deflections that result from variations in the end fixity and load type, vary by more than 300%. This is a clear illustration of the inadequacies of using a single value such as 12,000 or 20,000 for computing the maximum acceptable deflection during a load test for all conditions of end restraint and different concrete strengths.

B.2.8 *ACI 437-67*—ACI Committee 437 published the first version of "Strength Evaluation of Existing Concrete Buildings" in 1967 (ACI Committee 437 1967). That document provided extensive guidance for load testing of existing concrete buildings. The following specific criteria were included:

- Test load:
  - Where the strength of a whole structure is under investigation, test load $TL = 1.25D + 1.50L$, or $TL = 1.50D$, whichever is greater; and
  - Where the strength of only a portion of a structure is under investigation, test load $TL = 1.30D + 1.70L$, or $TL = 1.60D$, whichever is greater.
- Duration of test load = 24 hours;
- Maximum allowable deflection = $l_t^2/20{,}000h$; and
- Deflection recovery = 75% at 24 hours after load removal.

When Committee 437 revised its report in 1982 (ACI Committee 437 1982), the test load was no longer separately defined for tests on portions of a structure versus tests on a whole structure. Instead, the single definition for the TLM as given in ACI 318-71 and later editions was recommended.

Of particular interest, it was indicated in this report that if the maximum deflection of a reinforced concrete flexural member is smaller than $l_t^2/20{,}000h$, elastic behavior may be assumed, and the recovery of deflection requirement stated above may be waived. This is the first and only reference to a correlation between the equation for maximum allowable deflection under test load and the assumption of linear elastic (uncracked) behavior. No technical basis is given for stating that there is a correlation between this maximum deflection and the assumption of elastic or inelastic behavior of a reinforced concrete structure or structural component.

Finally, the following provision is included in the ACI 437R-67 that dealt with serviceability (where L = live load):

> If serviceability is also a criterion in the evaluation of the structure, the deflection at the superimposed load level of 1.0L, in addition to the simulated dead load, for any part of the structure should not exceed that stipulated by the authority, and the significance of any cracks should be duly considered.

This is believed by Committee 437 to be the first reference in any historical ACI document to consider serviceability in a load test of a concrete structure. This provision was maintained in subsequent editions of ACI 437R.

B.2.9 *ACI 318-71*—In ACI 318-71, the test load was again redefined, this time as equivalent to $0.85(1.4D + 1.7L)$; that is, $TL_{05} = 1.19D + 1.45L$. The acceptance criteria for a load test remained essentially unchanged from ACI 318-63, despite this reduction in test load intensity. The Commentary to ACI 318-71 acknowledged that the new test load represented a reduction of approximately 8 to 15% from the previous code, depending on the ratio of live load to dead load. The commentary (ACI 318R-71) noted the following:

> The new procedure has the advantage, however, that the test load is a constant percentage of the theoretical design strength. This reduction in testing load avoids possible problems in testing of prestressed members where load values stated in ACI 318-63 might induce inelastic behavior even in a member, which proves to have adequate strength capacity.

The maximum deflection criterion, $\Delta = l_t^2/20{,}000h$, was not modified despite the reduction in TLM.

No changes have been made to the provisions in Chapter 20 of ACI 318 in any subsequent edition of that document since 1971 in the areas of magnitude of test load or the maximum deflection and deflection recovery acceptance criteria. Table B.1 provides a summary of ACI code requirements as they relate to load testing practice.

B.2.10 *Cyclic load testing of concrete structures*—Mettemeyer (1999) provided a summary of the cyclic load test (CLT) method, (Section 6.2). It included a description of the general concepts, objectives, planning, evaluation of the structure, selection of members to be evaluated, methods of load application, TLM, prediction of structural responses, equipment that may be used, analysis during testing, interpretation of results, and descriptions of commercial applications.

This document (Mettemeyer 1999) also presented the use of the CLT procedure that was carried out in five case studies. All five of the case studies involved beam specimens that were strengthened with externally applied carbon fiber-reinforced polymer (CFRP), and each specimen was loaded so that it would fail in shear. The five case studies are summarized as follows:

Case study No. 1: Prestressed double T-beam—This study was conducted in a controlled laboratory environment. The dapped ends of the double T-beam were strengthened with CFRP. This dapped end was the area under investigation; therefore, the double T-beam was loaded in shear in such a way that the reactions were greater than 85% of the factored design loads. The CLT performance measures (explained in detail in Section 5.2) of repeatability, permanency, and maximum deviation from linearity were 98, 4 (maximum), and 12%, respectively. These values were indicative of acceptable behavior as discussed in Section 6.2. For the 24-hour load test method, when the performance measures of permanency and maximum deviation from linearity were applied, the values of 3 and 22% were obtained. Repeatability could not be calculated with the 24-hour load test procedure because this performance measure requires repeated cycling. Loading to failure after completion of the tests indicated that the maximum test load was approximately 50% of the ultimate capacity of the specimen.

The following studies (Cases 2 to 5) were conducted in the field. In the entire study, 20 reinforced concrete ceiling joists were tested using the CLT technique, and were also taken to failure to determine their true capacities. Sixteen of the members were strengthened with CFRP, with the remaining four serving as control specimens. The net deflection achieved during the 24-hour load test was essentially the same as that achieved during the CLT for all cases.

Case study No. 2: Short span ceiling joist (shear capacity, strengthened with CFRP)—This ceiling joist was strengthened with CFRP and was tested in shear-to-load levels that were calculated to exceed 85% of the factored design loads. The CLT performance measures of repeatability, permanency, and maximum deviation from linearity were 104, 5 (maximum), and 21%, respectively. Loading to failure after completion of the tests indicated that the maximum test load was approximately 45% of the ultimate capacity of the specimen. From the load versus deflection behavior, it was determined that the level of load to achieve a 25% deviation from linearity (threshold value explained in Section 6.2) was approximately 52% of ultimate capacity.

Case study No. 3: Short-span ceiling joist—This specimen was similar to that for case study No. 2 with the exception that two plies of CFRP reinforcement were used as opposed to only one ply for the joist in case study No. 2. The CLT performance measures of repeatability, permanency, and maximum deviation from linearity were 102, 2 (maximum), and 21%, respectively. Loading to failure after completion of the tests indicated that the maximum test load was again approximately 45% of the ultimate capacity of the specimen. From the load-versus-deflection behavior, it was determined that the level of load to achieve a 25% deviation from linearity was again approximately 52% of ultimate capacity.

Case study No. 4: Long-span ceiling joist—This specimen was strengthened with one ply of CFRP without an end anchor. The CLT performance measures of repeatability, permanency, and maximum deviation from linearity were 100, 3 (maximum), and 20%, respectively. Loading to failure after completion of the tests indicated that the maximum test load was approximately 17% of the ultimate capacity of the specimen. From the load-versus-deflection behavior, it was determined that the level of load to achieve a 25% deviation from linearity was approximately 27% of ultimate capacity.

Case study No. 5: Long-span ceiling joist (shear capacity)—This specimen was similar to the joist in case study No. 4 with the exception that an end anchor was used with the CFRP strengthening. The CLT performance measures of repeatability, permanency, and maximum deviation from linearity were 103, 5 (maximum), and 24%, respectively. Some change in deflection was noticed during the 24-hour load test, and this was attributed to temperature. Loading to failure after completion of the tests indicated that the maximum test load was approximately 35% of the ultimate capacity of the specimen. From the load-versus-deflection behavior, it was determined that the level of load to achieve a 25% deviation from linearity was approximately 41% of ultimate capacity.

B.2.11 *CIAS Report 2000 and ACI 437R-03*—In 2000, the Concrete Innovations Appraisal Service (CIAS), a subsidiary of the Concrete Research and Education Foundation, issued its appraisal report, "Guidelines for Rapid Load Testing of Concrete Structural Members," (CIAS 2000). The report discussed the CLT method as an alternative method for evaluating structures by load testing. The report was reviewed by a panel consisting of ACI members, some of whom were members of ACI Committee 437. The panel's appraisal of the information submitted stated the following:

> The panel's opinion is that the proposed Rapid Load Test protocol has great potential value to the construction industry. The method has the potential for making load testing of new structures, deteriorated structures, and repaired structures more practical and more meaningful than the present 24-hour static load test presented in the American Concrete Institute "Building Code Requirements for Structural Concrete (ACI 318-99)," Chapter 20. The controlled cyclic loading and continuous monitoring and evaluation of measured responses are seen by the panel as having real value... the panel feels that the method's potential advantages make it worthy of further development and submission to ACI Committee 437—Strength Evaluation of Existing Structures, and ACI Committee 318—Building Code. The Rapid Load Test method could supplement or form the basis for a revision to the current Chapter 20 strength evaluation provisions.

The information contained in the CIAS report was subsequently reviewed and discussed in Committee 437. The cyclic load test method was reported in Appendix A of ACI 437R-03, which provided details on the procedure and a suggested protocol.

B.3—Other historical load test practices

The following is a discussion of load test practices in various parts of the world dating back to 1903. This discussion is not presented as an all-encompassing summary; it merely represents information the Committee has been able to uncover to date. The major components of this discussion are summarized in Table B.4.

B.3.1 *Proceedings of 5th Annual Convention, National Association of Cement Users (NACU 1908)*—This document includes one of the earliest summaries of existing building regulations from around the world relating to load testing of concrete structures. Pertinent portions of a tabulated summary of these regulations are quoted in the following sections.

B.3.1.1 *Swiss Society of Engineers and Architects: 1903 recommendations*—

- "Test load to be at least 50% greater than working loads allowed in calculations."
- "Test loads not to be put on until 45 days have been allowed for setting."
- "If possible deflections of different stages of loading to be noted."

B.3.1.2 *Prussian government regulations: 1904*—

- "If loading tests are necessary, they are to be carried out under instruction of representative of building authority."
- "When a strip is cut from a floor or decking is tested, the load shall be uniformly distributed and shall not exceed the weight of the strip and twice the working load. If a strip is tested in-situ the above loading shall be increased by one-half."

B.3.1.3 *French government rules: 1907*—

- "Conditions of test and time that shall elapse before structures are brought into use must be inserted in contract, and also, the maximum deflection as far as practicable."
- "The time to elapse before use of structures must be 90 days for structures of primary importance, 45 days for ordinary constructions, and 30 days for floors."
- "Measurements to be taken during test, which are likely to be of scientific interest to engineers."
- "Test loads on floors shall be the dead and superimposed loads acting over the whole area of the floor, or at least upon a complete panel."
- "The loads to be left on for at least 24 hours, and deflection to cease after 15 hours."

B.3.1.4 *British Reinforced Concrete Committee: 1907 recommendations*—

- "Loading tests not to be made till 2 months after completion."
- "Test load not to exceed 1-1/2 times superimposed loading."
- "Consideration to be given to adjoining parts of a structure in case of partial loading."
- "No test load to be applied, which would cause metal to be stressed more than 2/3 of its elastic limit."

B.3.1.4 *Austrian Ministry of Interior Rules: 1908*—

- "Breaking tests of whole or part to be made on request.

Table B.4—Sampling of load test requirements other than those from ACI

Year and document	Minimum age	Total test load D	Total test load L	Duration of load	Maximum deflection	Deflection recovery	Notes
1903: New York City Building Regulations	—	1.00	3.00	—	—	—	—
1903: Swiss Society of Engineers and Architects (NACU 1908)	45 days	1.50	1.50	—	—	—	—
1904: Prussian Government Regulations (NACU 1908)	—	1.00	2.00	—	—	—	Test load for an isolated member
		1.00	3.00				Test load for a nonisolated member
1907: French Government Rules (NACU 1908)	90 days	—	—	24 hours	No deflections after 15 hours	—	Structures of primary importance
	65 days						Ordinary construction
	30 days						Floors
1907: Great Britain (NACU 1908)	2 months	1.00	1.50	—	—	—	No test load to be applied that would cause metal to be stressed more than 2/3 of its elastic limit
1908: Austrian Ministry of the Interior Rules (NACU 1908)	6 weeks	1.00	1.50	—	—	No permanent deflections	Acceptance criteria: no cracks or permanent deflections
1908: Borough of Manhattan, N.Y.	—	1.00	3.00	24 hours	—	—	Acceptance criteria: no sign of failure
1908: Borough of Brooklyn, N.Y.	—	1.00	2.00	—	$L/700$	—	—
1908: Buffalo, N.Y.	—	1.00	3.00	—	—	—	Acceptance criteria: no sign of failure
1908: Chicago, Denver, San Francisco	—	1.00	2.00	—	$L/700$	—	—
1908: Toledo, Ohio	—	1.00	3.00	—	—	—	Acceptance criteria: no sign of failure
1908:Baltimore, Md.	—	1.00	2.00	—	—	—	Acceptance criteria: no sign of failure
1910: Chicago Building Ordinance	—	1.00	2.00	24 hours	—	—	—
1926: Russia	—	—	—	—	$L/750$	67%	No further information available on test procedures
1934: Building Research Board, Code of Practice for RC	56 days	1.00	1.50	24 hours	—	75%	Repeat load test if deflection recovery is not met
1957: CP 114, The Structural Use of RC in Buildings, British Standard Code of Practice	56 days	1.00	1.25	24 hours	Not undue	75%	—
1959: CP 115, The Structural Use of PC in Buildings, British Standard Code of Practice	56 days	1.00	1.25	24 hours	Not undue	85%	For prestressed concrete
1963: Australian AS CA-2	56 days	1.00	2.00	24 hours	L^2/cd	75%	Coefficient c has values from 1800 to 10,000, depending on type of construction
1964: European Concrete Committee	(a)	1.00	1.00 to 1.10	(b)	(b)	(b)	(a) Concrete to have reached strength specified by engineer (b) To be decided by engineer before test
1964: Institution of Structural Engineers	—	1.10	1.25	8 hours	$L/360^*$ $L/250^\dagger$	75%	Crack width acceptance criteria also: *for live load only †for dead plus live
1975: RILEM TBS-2, General Recommendation for Loading Tests	56 days	1.20	1.40	16 hours	—	75%* 87.5%† 80%‡	*For new structures †For structures already used and exposed to load ‡Precast concrete structures
1977: Czechoslovakia State Standard—CSN 73 2030	3 months	1.00	1.05 to 1.25	24 hours	Code values × 1.1 to 1.2	75%	Crack width acceptance criteria also
1994: CSA A23.3 (CSA 1994)	28 days	1.125	1.35	—	—	60%	For whole structures under investigation
		1.25	1.50				For portions of structures under investigation
1995: Building Research Establishment (BRE 1995)	—	1.25	1.25	24 hours	Code values	90%	—

No test before expiration of six weeks after completion of ramming."

- "Loading to be such that effect is same as dead load plus 1-1/2 specified superimposed load. No cracks or permanent deflections."
- "For breaking tests load to be gradually increased."
- "Breaking load not to be less than 3-1/2 times the total dead and superimposed loads, less the weight of the member."

This 1910 NACU document also contained a tabularized summary of requirements for load testing in the time frame of 1908 in the United States. Pertinent excerpts from that summary appear in Table B.4.

B.3.2 *Research on loading tests of reinforced concrete structures: report of work carried out at standards Institution of Israel (Arnan et al.) in 1950*—This report outlines the findings of interesting research performed at this organization before 1950 in the area of load testing. The introduction of this report includes the following noteworthy text:

> The loading test is generally regarded as an indication of the safety of the structure. Whereas on one hand it has been known for years that this test is entirely unreliable, on the other hand, the scientific foundation of this test has been questioned in recent years. The board for Scientific and Industrial Research had therefore authorized a research project, the object of which was to clarify the physical foundations of the loading tests.

The document also contains the following statements:

> The significance of the load test has often been questioned. At the laboratories of the Institution, about 25 such tests have been performed since 1945, and it is notable that, although in all cases the strength of the structure was under suspicion; in no case failure to comply with the load test was found. This peculiarity of the load test to favour the builder is well known to contractors.

> The history of the load test and the theory underlying it are difficult to trace…all specifications in the various countries where a load test is prescribed, from the USSR to the USA, are extremely similar. It is difficult to avoid the conclusion that somebody, upon the beginning of the process of reinforced concrete construction one or two generations ago, introduced the loading test as a rule of thumb, and that its provisions were then copied by others for lack of something better.

This study included load tests of slabs in nine existing buildings and constructing 18 new concrete slab specimens and testing them in the laboratory. The laboratory tests included specimens intentionally constructed with defective concrete. The test methods included a method of alternate loading, in which the load was applied and immediately removed three times in succession. This was done up to the total load required by either British or American practice.

The following findings of this research were outlined in the report.

> 3.6 Discussion of Results
> (i) First Criterion
> … in the laboratory load test the first criterion [that is, the use of the maximum deflection acceptance criterion ($l_t^2/12{,}000h$)] proved to be entirely safe, while, in the load tests on existing buildings, … it proved to be entirely misleading. It was, however, consistently misleading, being always on the "unsafe" side. In other words, the actual deflections were always smaller than those calculated in accordance with formula 3.4.1 [i.e. $l_t^2/12{,}000h$]. We need not look far for the reason. The laboratory floor slabs were all freely supported, whereas the slabs in the buildings are all fixed. Formula 3.4.1 [$L^2/20{,}000h$] was established at a time when fixing reinforcing concrete slabs was hardly known. Formula 3.4.1 therefore provides an adequate criterion for freely supported slabs, but where the slabs are fixed, actual deflections are much smaller. This points to the necessity and possibility for establishing a formula analogous to 3.4.1 one which would take into account modern methods of reinforced concrete construction…
>
> (ii) Second Criterion
> …the application of the second criterion in the building tests [i.e. use of deflection recovery criterion in the tests of slabs in existing buildings] gave entirely erratic results, sometimes good and sometimes bad.

In the laboratory testing, only one slab showed on first loading a recovery of more than 75%. This slab was deliberately constructed with low-strength concrete. All other slabs failed this criterion on first loading. All slabs, whether up to or below strength requirements, showed adequate recoveries on second loading, while on third loading, the recovery was practically complete.

The report goes on to state that it can therefore be said that the second criterion (that is, the deflection recovery requirement) does not provide a means of judging the quality of the concrete.

> Generally speaking, even good concrete will fail on first loading, while even bad concrete will pass on second and further loading.

This report contains an interesting commentary on the use of deflection recovery in load testing practice. The conclusions section includes the following statements:

> Even the worst concrete attains practically 100% recovery on repeated loading... also, the data from our testing shows that the permanent deformation due to creep and non-recovered elastic after-effect is negligible. Therefore the criterion of 'recovery' has no meaning.

Regarding use of maximum deflection as an acceptance criterion, the following commentary is included in the conclusions of the report:

> We have found that the criterion of 'maximum deflection,' if calculated in accordance with a formula applicable in the case of freely supported beams, is suitable, and that the quality of concrete can thus be judged in accordance with the magnitude of *E*. If this is so, however, we must know the correct instantaneous elastic deflection, which is most pronounced when the maximum percentage of recovery is attained. In accordance with the results of our research, we are, therefore, of the opinion that it might be possible to find a correct procedure for the loading test by performing 'alternate' loadings up to the maximum load required by the test, measuring the last recovered deflection and comparing this with a calculated deflection taking into account the actual proportion of steel and concrete and the conditions of fixing. It will then be necessary to specify a maximum deflection, and the criterion would be whether this maximum deflection is exceeded or not. Further research is required in connection with this procedure in order to specify the maximum deflection for different cases of design, and especially taking into account the use of high grade steel.

The information contained in this report, although over 50 years old, is still pertinent to nearly all of the lingering worldwide concerns regarding load testing practice.

B.3.3 *Report of Committee on Testing of Structures (Institution of Structural Engineers 1964)*—This report identified two types of load tests. The first was an acceptance test to check the behavior of a structure or part of a structure under a load equal to or greater than the known working load, so as to assess its adequacy for service; the second was a test to destruction to determine ultimate strength.

Regarding test load for an acceptance test, this report indicates that the estimated dead load should be arbitrarily increased by an amount that should not exceed 10%. It is further indicated that the imposed load (that is, live load) should also be arbitrarily increased by not more than 25%. This would equate to a total test load $TL = 1.10D + 1.25L$.

The report contained the following statements relative to these guidelines:

> The above recommended increases of loadings are not intended to ensure some particular load factor against failure nor to test the structure to a specified proportional increase in stress. The acceptance test is intended merely to demonstrate that the behaviour of the structure at working loads, or at slightly higher loads that might for some reason be applied during the life of the structure, is likely to be satisfactory.

The following additional guidelines are contained in this document relative to load tests of reinforced concrete structures:

1. Duration of test loading is to be 8 hours.

2. Guidance for maximum allowable deflection during test:
   a. Proof of serviceability is the object of an acceptance test, and the only requirement in regard to any deflection or deformation is that it shall not exceed the appropriate permissible amount that is either specified in the design standard or established by the engineer.
   b. In buildings where finishes are to be applied after the deformations due to dead load are substantially complete, deformations due only to imposed load need be considered when establishing permissible limits from the viewpoint of possible damage to finishes; ($L/360$)
   c. In buildings it is also necessary to limit the total deflection (due to both dead and imposed loads) from the viewpoint of appearance; ($L/250$).
   d. In structures where damage to finishes has not to be considered, and where appearance is not as critical as in buildings, deflections due both to dead and to imposed loads may be greater than for buildings, and any limit should be a matter for the engineer's decision.

3. Recovery, measured 8 hours after load removal, equal to or greater than 75%.

4. Provision for retesting; after removal of the test imposed load, the recovery of deformation should be at least 75%. If this requirement is not met, but the recovery is nevertheless not less than 50%, the structure should be regarded as satisfactory if, on re-test, the recovery is at least 75%.

This document also introduces allowable crack width acceptance criteria:

> The maximum widths of cracks at the working load should not exceed 0.3 mm (0.012 in) for internal construction and normal conditions of exposure; 0.2 mm (0.008 in) for external constructions and normal conditions of exposure; or 0.1 mm (0.004 in) for aggressive conditions of exposure, whether for internal or external construction. Lower limits may be desirable for prestressed concrete structures.

B.3.4 *ASTM Committee E6, Performance of Buildings: 1965 to 1995*—A special task group of ASTM Committee E6 was formed in 1965 and worked for 8 years to address the issues associated with providing guidance for load testing of existing structures. The first results of that task force were outlined in a paper in 1975 (FitzSimons and Longinow 1975). While a good reference on the general aspects of load testing of all types of structures, little specific guidance was provided.

ASTM subsequently published ASTM E 196, "Standard Practice for Gravity Load Testing of Floors and Flat Roofs (ASTM E 196-95)," but that document also provided little in the way of specific guidance for load testing of existing structures.

B.3.5 *Czechoslovak State Standard CSN 73 2030 (1977)*—This document was produced as a result of a comprehensive project directed by Richard Bares, with the Institute of Theoretical and Applied Mechanics, at the Czechoslovak Academy of Science in Prague, Czechoslovakia. Bares contacted experts in many different countries, and he is considered to have produced one of the most comprehensive documents on load testing of structures at that time. This document was originally issued in 1969, but an English translation from 1977 was made available to the committee. Bares and FitzSimons (1975) provided a summary of the document.

Relevant highlights of this document include:

1. The purpose of a load test is to assess the actual behavior of a structure or member through determination of its load-bearing capacity or usability in terms of magnitude of deflection and cracking under applied loads;

2. Types of load tests:
   a. Proof tests to demonstrate the ability of a member or structure to satisfy the given purpose in accordance with design requirements, the suitability of a new construction method, or new construction materials used;
   b. Control tests to demonstrate the ability of a member or structure to satisfy the given purpose in accordance with design requirements that already have been approved; and
   c. All other tests not intended exclusively for the assessment of a single member or structure;

3. Each load test can be executed as follows:
   a. To failure of the structure or a portion of it to determine its ultimate load-bearing capacity; or
   b. By test loads specified to prove the usability of the structure or a portion of it with reference to its: (a) load-bearing ability; (b) rigidity (deflection); or (c) cracks (deformation);

4. Age of structure—the test is carried out only after the structure has attained the required properties, particularly the full strength of the materials used, or after the termination of significant creep or settlement of the structure, or both. In the case of concrete structures, it is recommended that the tests begin after 3 months;

5. Magnitude of superimposed test load—Specified as the quantity $Z = 0.5L(1 + n)$, where n is a factor varying for different structural materials from 1.1 to 1.4. For concrete the superimposed test load thus varies from $1.05L$ to $1.2L$. In testing for serviceability, the value of the test load is specified to be $1.0L$.

6. Duration of loading: 24 hours for normal weight concrete structures;

7. Acceptance criteria:
   a. The load under which the structure has failed is the load under which the structure has lost its ability for further use due to one of the following causes:
      i. Complete failure of the structure or its part or section or the rupture of reinforcement;
      ii. Loss of stability of the structure or its part or member;
      iii. Local failure that continues to grow without any increase in load;
      iv. The deformation increments under the same load measured three times in succession at identical intervals do not decrease;
      v. The deformation increment due to the last loading state equals the sum of the deformations due to the first five equally high loading stages or exceeds it;
      vi. The deflection equals or exceeds 1/50th of the span length;
      vii. In deformed concrete structures, the width of cracks equals or exceeds 1.5 mm (0.060 in.) provided these cracks are at least 200 mm (8 in.) long;
      viii. The failure of concrete structures by slanting cracks in the proximity of supports or point loads; or
      ix. Loss of bond between reinforcement and concrete.
   b. The tested building structure is considered usable with reference to its load-bearing ability if it has fulfilled simultaneously the following conditions:
      i. The magnitude of permanent deformations does not exceed 25% for reinforced concrete structures;
      ii. The state of failure due to the design load is stabilized, while the width of cracks in concrete structures does not exceed 0.3 mm (0.012 in.), if they are protected against weather, and 0.2 mm (0.008 in.) if they are exposed to weather.
   c. The tested building structure is considered usable with reference to its rigidity (deformation), if it satisfies simultaneously the following conditions:
      i. The measured elastic deformations under test load must not exceed the k multiple of the theoretically determined value (varies between 1.1 and 1.2 for reinforced concrete);
      ii. Total deflections or other total deformations under standard live load must not exceed the limit deflections or deformations given in the respective standards for the design and erection of building structures and reduced according to the magnitude and period of application of the load;
      iii. Total deflections or other total deformations under the test load must not exceed the limit deflections or deformations more than k times (varies between 1.1 and 1.2 for reinforced concrete);
      iv. A test structure of reinforced concrete is considered usable with reference to the origin and development of cracks, if it satisfies simultaneously the following conditions:
         1. The crack width under standard live load must not exceed the values stipulated by the standards for the design of structures;
         2. The distance between cracks under standard live load must not exceed the values stipulated by the standards for the design of structures;
         3. The cracks do not appear under loads less than 0.9 of the theoretically determined load for the original of the first crack according to the theory of elasticity; and
         4. After the removal of the load, the cracks close to a width less than 1/3 of the prescribed value.

B.3.6 *General Recommendation for Statical Loading Test of Load-Bearing Concrete Structures In Situ (RILEM TBS-2) (RILEM 1980)*—The following are pertinent sections of this document:

1. Loading tests are investigation processes to be carried out on buildings, load-bearing structures or parts with the aim of obtaining empirical experimental data concerning their load-bearing capacity or suitability for the purpose intended.

2. The referenced and maximum value of the test load is to be determined according to the purpose of the test. Such purposes might be:
   a. To check serviceability by safety test;
   b. To define load-bearing capacity reserve (for example, for structures of unknown load-bearing capacity);
   c. To check load-bearing capacity.

3. When testing serviceability, the reference value of test load is to be based on:
   a. The useful and expected loads specified in standards;
   b. The useful and expected loads as indicated by the designer;
   c. The values calculated from the load-bearing capacity limit, deducting dead load;
   d. The load which gives rise to the deformation limits permitted for the structure or the crack widths, reduced by the dead load.

4. It is suggested to apply a 1.4 load increasing factor for variable useful loads, while for permanent useful loads a 1.2 factor is most advisable.

5. The value of the test load should be increased if: the use of the building requires an unusually high safety factor; a decrease in load-bearing capacity with time is anticipated due to such factors as corrosion or deterioration of material properties; the effects of shrinkage, creep and relaxation are important and should be considered; the structure will be exposed to extreme environmental factors such as large

temperature variations; the effects of dynamic loads are important; the service live load exceeds twice the dead load. No quantitative guidance is provided in the document, however, on what additional increase is appropriate.

6. For tests to define the load-bearing capacity, the test load is to be determined by continuous processing and evaluating of the loading test results during the test. In this case, the maximum test load will be calculated from the load values that determine the permitted ratio between permanent and total deformation, stabilization of deformations as well as the deformation limits of the structure and the permitted crack width.

7. If the aim of the load test is to establish the load-bearing capacity (ultimate strength) of the structure, test loads will be determined from the ultimate strength calculated from the nominal yield strength of the steel in the structure as well as the nominal concrete strength, taking into consideration strength variability.

8. Minimum age of structure at time of test = 56 days.

9. Duration of maximum loading to be 16 hours.

10. If the aim of the load test is to define the actual load-bearing capacity limit of this structure, it is suggested that the structure be subjected to a number of load and unloading cycles, with ever increasing load magnitude, with half-hour increments for loading and unloading, and half-hour waiting periods between loading periods.

11. Failure criteria:
   a. The structure or any part of it has collapsed, fallen into two or even more parts;
   b. The structure or any part of it has lost its stability;
   c. The local damage of the structure increases, spreads without essential increase of loading;
   d. The deformation of the structure shows no decrease at unchanged load, measured three times, after equal intervals;
   e. At the last loading stage the extent of deformation reached or exceeded the full deformation that occurred during previous stages when the same loads were applied;
   f. Deflection equals or exceeds $L/50$;
   g. On the structure exposed to bending load, the crack width, measured at 200 mm (7.87 in.) distances, amounts to a total of 1.5 mm (0.059 in.);
   h. The extent of the diagonal cracks near the supports of the concrete structure reaches or exceeds the value in item 'g' above;
   i. The structure is separated from the stiffening.

12. Acceptance criteria—In the absence of more rigorous provisions, the structure tested is found suitable for service, if the following conditions are fulfilled:
   a. None of the failure criteria exist;
   b. The residual deflections and deformations do not exceed the following percentage values of total deflection.
      i. For new structures, at least 56 days old, when loaded for the first time:
         1. For prestressed concrete structures 20%.
         2. For reinforced concrete structures 25%.
      ii. For structures already used, previously exposed to full loading, half of the above values are used.
      iii. Structures from 28 to 35 days old may have values higher by 1.25:
         1. The remaining deflections reach 1.5 of the former values as a maximum and the deflections that remain after the unloading following the second load-bearing test are not greater than half the percentages given for the first loading test and in the case of a possible third test 1/3 of the deflections remaining after the second test;
         2. The measured deflections and deformations do not exceed 1.2 of the calculated values, provided that the true value of the elastic modulus of the structure was used for the calculating of deformations;
         3. The curves showing deformation due to loading are with good approximation linear or discontinuously linear, with definite break points and the deformations became stabilized during the loading test.
   c. The maximum crack width is within the limits stipulated for the materials and purpose of the structure;
   d. The maximum deflections and other deformations remain within the limits stipulated for the structure according to its intended use.

B.3.7 *CSA Standard A23.3 (Canadian Standards Association 1994)*—Section 20.3 of this document addresses "General Requirements for Load Tests." It contains the following relevant items:

1. Age of structure should be 28 days or more at time of testing.

2. Superimposed dead loads: A load to simulate the effect of the portion of the dead loads not already present shall be applied 24 hours before the application of the test load and shall remain in place until all testing has been completed.

3. Test load:
   a. When an entire structural system in doubt is load tested or an entire questionable portion of a system is load tested, the test load shall be 90% of the factored loads, M_f, V_f, and P_f. When only a portion of a structural system in doubt is tested and the results of the tests are taken as representative of the structural adequacy of other untested portions of the system, the test load shall be equal to the factored loads M_f, V_f, and P_f.
   b. The superimposed test load shall be applied in not less than four approximately equal increments without shock to the structure and in a manner to avoid arching of the load materials.
   c. The test load shall be left in place for 24 hours.

4. Acceptance criteria:
   a. If the portion of the structure tested fails or shows visible indications of impending failure, it shall be considered to have failed the test.
   b. Deflection recovery: For load tests of flexural systems or members for moment resistance, the required deflection recovery values are specified as follows:
      i. Nonprestressed members:

1. First test 60%
2. Retest 75%

ii. Prestressed members: 80%

This document has considerable similarity to ACI 318-05 requirements for load testing. Examination of the CSA test loads and the CSA factored loads reveals that the ACI and CSA factored loads are not substantially different. The CSA test loads are as follows:

$$TL = 0.9(1.25D + 1.5L) = 1.125D + 1.35L$$

when the entire structural system or entire portion of a system is tested.

$$TL = 1.25D + 1.5L$$

when only a portion of a structural system is tested and is intended to be representative of the untested portion of the structure.

In addition, the CSA provisions do not include a maximum acceptable deflection for a load test. Only deflection recovery criteria are included.

B.3.8 *Guidance for Engineers Conducting Static Load Tests on Building Structures (BRE 1995)*— This document contains the following guidance for testing concrete structures for serviceability.

1. The purpose of this type of testing is to establish whether the structure is likely to perform satisfactorily in service. Both long-term deformations under permanent dead load and short-term deformations due to imposed loads need to be within acceptable limits, and the structure must be able to carry its full service loading safely.

2. Where the test load is not specified in the relevant code of practice or is not applicable to the particular circumstances of the structure being considered, the test load may be chosen as the maximum the structure should sustain without suffering permanent deformation or damage, or $TL = 1.25D + 1.25L$, whichever is less.

3. The maximum applied load should be left on the structure until it has in effect come to rest. A period of 24 hours is likely to be sufficient for most structures.

4. Acceptance criteria:
   a. The maximum deflection recorded does not exceed that given in the relevant code of practice or that specified for the structure.
   b. For concrete structures that have been bedded-in by reapplying the full test load several times, the recovery 24 hours after removal of the load would be expected to exceed 90%.
   c. Existing cracking and deformation do not extend significantly during the test.
   d. The structure shows no other signs of damage or distress as a result of the test cycle.

Regarding the interesting aforementioned concept of "bedding-in," the following information is presented in this report:

1. Depending on the magnitude of the full test load, load history, type of construction, and structural material, bedding-in loads may be desirable. The object of applying bedding-in loads is to settle the structure on its supports and release any frictional restraints incorporated during construction;

2. Bedding-in loads should be applied and removed in at least five increments, with deformations being monitored. In general, the magnitude of bedding-in load should not exceed the intended future service loading. The structure can be considered to be satisfactorily bedded-in when it has recovered to its original position (+/– 10%) after a loading cycle;

3. For concrete components to be taken beyond their service loading, the full test load is itself likely to produce slight degradation of the component. In these circumstances, it may be necessary to reapply the full test load several times until a repeatable response is obtained between successive loadings.

This information echoes findings of the Israeli research report and is of significance in regard to proposals to adopt or allow cyclic load testing of concrete structures under ACI 318.

B.3.9 *2003 International Building Code (International Code Council 2003)*—The 2003 International Building Code contains guidance on conducting load tests on existing building structures. The following is a summary of the major provisions of this key building code:

1. Whenever there is a reasonable doubt as to the stability or load-bearing capacity of a completed building, structure or portion thereof for the expected loads, an engineering assessment shall be required. The engineering assessment shall involve either a structural analysis or an in-situ load test, or both.

2. The IBC refers to material standards for provisions in conducting load tests. This includes ACI 318-05.

3. For structures not covered by ACI 318-05 or any other material standard listed in the IBC, the following minimum test criteria are outlined:
   a. The test load shall be equal to two times the unfactored design loads;
   b. Under the design load, the deflection shall not exceed the limitations specified in Section 1604.3. Those deflection limits, for a superimposed live load only, vary from $L/180$ to $L/360$, and are based on serviceability limit states only;
   c. Within 24 hours after removal of the test load, the structure shall have recovered not less than 75% of the maximum deflection; and
   d. During and immediately after the test, the structure shall not show evidence of failure.

B.3.10 *Italian codes*—According to the Italian building code, the design and construction of a new building must be verified by an independent professional engineer who has been licensed for at least 10 years. In most cases, as part of the "threshold inspection," the engineer requires that a load test be conducted before the public use of the structure. The same stipulation applies to existing buildings when there is a need to assess their structural performance with respect to building code changes or changes in use.

In contrast to the U.S. construction practice where floors are generally made of cast-in-place reinforced concrete slabs with likely uniform and known properties in both directions, it is common practice in Italy to have floors made of cast-in-place or precast reinforced concrete joists spaced by voided clay tiles, then topped with a thin overlay of concrete reinforced with a steel mesh to redistribute the load. It is clear that in such a structural system it is rather difficult to determine how loads distribute because of uncertainties in the boundary conditions as well as in the transverse stiffness. For these reasons, through the years researchers and practitioners have developed methods to determine how to compute equivalent test patch loads to simulate uniformly distributed loads.

Current practice in Italy (Lombardo and Mirabella 2004) shows that an equivalent force to substitute for the uniformly distributed load may be calibrated based on the knowledge of the deflection response of the member(s) and the surrounding structure. The most common method to determine an equivalent patch load is to determine two coefficients k_1 and k_2 that take into account the transverse and longitudinal redistribution of the load, respectively. Before conducting a load test, such coefficients can be computed experimentally and then used to determine the appropriate value of the concentrated test load to simulate the uniformly distributed loads used for design. The method is based on the deflection response in the two perpendicular directions of the flooring system to a small concentrated load, usually lower than the service loads (pilot test load). Based on the longitudinal deflection response, it is possible to calibrate the coefficient that accounts for the degree of fixity at the slab boundaries. Such a coefficient is usually 1.0 when the fixity of the restraints at the supporting beam locations is that of a perfect clamp. The transverse deflection response accounts for the participation of neighboring joists.

B.3.11 *Recommended Practice for In-Situ Monitoring of Concrete Structures by Acoustic Emission (NDIS 2421) (Japanese Society for Nondestructive Inspection [JSNDI] 2000)*—This document is thought to be unique in that it is the first standardized document that makes use of the acoustic emission (AE) technique for the inspection and evaluation of reinforced concrete structures. The document incorporates four related codes: description of functions and performance on AE devices (NDIS 2106 [JSNDI 1997]); calibration of AE sensors by the reciprocal method (NDIS 2109 [JSNDI 1991]); evaluation method for the deterioration of AE sensor sensitivity (NDIS 2110 [JSNDI 1997]); and recommended practice for the continuous AE monitoring of pressure vessels (NDIS 2419 [JSNDI 1997]). The recommended practice came about in Japan because of the large number of bridges that are reaching their intended service lives coupled with the need for evaluation of structures after extreme events, such as earthquakes. The practice notes aging, fatigue, heavy traffic loads, chemical reactions, and disasters as events or environments that lead to a need for evaluation before repair and rehabilitation. Before this document, the only standardized application of the AE technique to in-service structures had been to pressure vessels.

The stated purpose of this document (NDIS 2421) is to standardize existing techniques to estimate the degree of damage through in-place monitoring. The document addresses both long-term monitoring and short-term monitoring through load testing. The recommended practice consists of 11 chapters. The chapters entitled "Monitoring System" and "Tests and Evaluation" are described as follows:

Monitoring system—This chapter addresses amplification, parameters to be measured, duration of measurement and analysis, and the type of analysis to be used including trend analysis, distribution analysis, correlation analysis, and location analysis. The signal-to-noise ratio is noted as important to the analysis, and acceptable levels are established. The treatment of noise, selection of an appropriate threshold level, postanalysis of the data, dimensions of the sensor array to be used, and the frequency range of the sensors to be used are described. A frequency range of 20 to 100 kHz is recommended to limit attenuation.

Test and evaluation—This chapter addresses the differentiation between AE signals because of service level loadings and those that are representative of damage and not observed in service conditions. It further discusses monitoring that is performed continuously or routinely, and sometimes temporarily after disasters such as earthquakes.

The deterioration process of the structure is estimated through the following AE parameters:

1. Sudden increase of AE activity normally detected by counts, hits, and events;
2. Variation of such AE parameters as RMS, energy, and amplitude distribution;
3. Clustering and concentration of AE locations; and
4. AE activity under cyclic loading.

The document notes that through sudden increases in AE activity, the deterioration process, and often impending failure, can be estimated. One example of deterioration and its relation to AE activity through the freezing-and-thawing process is given.

In regard to loading of structures, the rate process theory is described. The probability function of AE occurrence from a stress level is formulated as a hyperbolic function. A relationship between the number of total AE events (N) and the stress level is given. The change of amplitude distribution is also noted to be useful. By applying AE location procedures, moment tensor analysis is used to define tensile or shear cracks and to determine crack orientation. In direct relation to controlled load testing for the evaluation of reinforced concrete structures, the Kaiser effect (a lack of or significantly reduced acoustic emission before the previously applied maximum load) is described in detail. The relationship between crack opening and the presence of the Kaiser effect in reinforced concrete beams has been reported previously. Further documentation of this effect has been reported under truck loading of harbor structures. In relation to the Kaiser effect, two parameters are proposed:

1. *Ratio of load at the onset of AE activity to previous load*:

load ratio = load at onset of AE activity under the repeated loading/previous load

2. *Ratio of cumulative AE activity under unloading to that of previous maximum loading cycle*:

calm ratio = the number of cumulative AE activity during unloading/total AE activity at the previous maximum loading cycle

Based on these parameters, a criterion to evaluate damage is plotted schematically as calm ratio versus load ratio; and the damage is divided into heavy, intermediate, and minor damage. This approach has been applied to laboratory specimens with crack mouth opening displacement gages for correlation with the AE activity.

In regard to load testing and evaluation with acoustic emission in the United States, there are currently codes related to: 1) tanks and pressure vessels—ASME RTP-1 [ASME 2004] and ASME Section X [ASME 2004]); and, 2) aerial devices (that is, manlifts) (ASTM F 914). In the field of civil structures, AE-based techniques are currently gaining favor, but are not widely used in practice. Until some standard guidelines are developed in the United States, it will difficult for AE to become accepted.

ACI 437.2-13

Code Requirements for Load Testing of Existing Concrete Structures (ACI 437.2-13) and Commentary

Reported by ACI Committee 437

Carl J. Larosche*, Chair

J. Gustavo Tumialan*, Secretary

Joseph A. Amon
Nicholas J. Carino
Paolo Casadei
John A. Frauenhoffer
Nestore Galati*
Zareh B. Gregorian
Pawan R. Gupta
Frederick D. Heidbrink
Ashok M. Kakade

Danielle D. Kleinhans
Andrew T. Krauklis
Daniel J. McCarthy
Javeed Munshi
Antonio Nanni
Thomas E. Nehil*
Renato Parretti
K. Nam Shiu
Jeffrey S. West*

Paul H. Ziehl*

Consulting Members
Marco Arduini
Habib M. Zein Alabideen

*Member of the subcommittee that prepared this report

This code provides requirements for test load magnitudes, test protocols, and acceptance criteria for conducting a load test as a means of evaluating the safety and serviceability of concrete structural members and systems for existing buildings as provided for by ACI 562-13. A load test may be conducted as part of a structural evaluation to determine whether an existing building requires repair and rehabilitation, or to verify the adequacy of repair and rehabilitation measures applied to an existing building, or both. This code contains provisions for both a cyclic load test and a monotonic load test procedure.

Keywords: acceptance criteria; cyclic loading; load test; monotonic loading; test load magnitude; test protocol.

CONTENTS

ACI 437.2-13 replaces ACI 437-12, was adopted October 23, 2013, and published January 2014.

CODE

CHAPTER 1—GENERAL

1.1—General

1.1.1 The scope, purpose, applicability, limitations, interpretation principles, and units of measure are defined in this chapter.

1.1.2 This code supplements and is part of the "Code Requirements for Evaluation, Repair, and Rehabilitation of Existing Concrete Buildings (ACI 562-13) and Commentary" through reference.

1.2—Scope

1.2.1 The requirements of this code shall govern for the evaluation of safety and serviceability of members in existing concrete structures by load testing.

1.2.2 The requirements of this code shall apply to reinforced concrete with prestressed reinforcement, nonprestressed reinforcement, or both.

1.2.3 Procedures and requirements provided in this code are not applicable to existing structures having concretes with compressive strengths above 8000 psi unless permitted by the licensed design professional.

1.2.4 For structures being erected under the requirements of a legally adopted general building code, of which ACI 318-11 forms a part, this standard shall not replace Chapter 20 of ACI 318-11.

1.3—Purpose

1.3.1 The purpose of this code is to establish the minimum requirements for the test load magnitudes, load test procedures, and acceptance criteria applied to existing concrete structures as part of an evaluation of safety and serviceability to determine whether an existing structure requires repair and rehabilitation, to verify the adequacy of repair and rehabilitation measures applied to an existing structure, or to provide for public health and safety through structural safety and serviceability.

1.3.2 Load tests shall be conducted according to load test procedures as described in Chapter 5.

COMMENTARY

R1—GENERAL

R1.1—General

R1.1.2 Through its reference by ACI 562-13, this code is to be used only when ACI 562-13 governs. In those cases in which the general building code governs, then the applicable provisions of Chapter 20 of ACI 318-11 dealing with load testing are to be used.

R1.2—Scope

R1.2.1 The determination of situations where a load test is required is outside the scope of this code. Furthermore, this code does not address procedures for analytical strength evaluations, condition evaluation of existing structures, or assessment of structural deterioration and its consequences. The licensed design professional should advise the owner and parties participating in the load testing of a structure of the potential for damage or even failure of the portions of the structure to be load tested in accordance with the procedures of this code. Refer to 5.1 for provisions relating to safety and shoring.

R1.2.4 Experience is lacking in the application of the procedures and requirements in this code to structures having high-strength concrete (concrete compressive strengths over 8000 psi). Structures with high-strength concrete may exhibit a more linear response as the structure approaches its load-carrying limit and may experience a more brittle failure in comparison to structures with strengths less than 8000 psi.

CODE

1.3.3 All references in this code to the licensed design professional shall be understood to mean persons who are licensed to practice structural design in the jurisdiction where this code is being used. The licensed design professional for a project is responsible for and in charge of the evaluation of safety and serviceability.

1.4—Limitations

1.4.1 This code does not apply to foundations, slabs-on-ground, structures whose responses are dominated by in-plane actions, or structures having unusual design concepts as determined by the licensed design professional.

1.4.2 This code shall not be used for product development testing where load testing is used for quality control or approval of mass-produced members.

1.4.3 Testing for resistance to load effects of fatigue, wind, and earthquakes is outside the scope of this code.

1.4.4 Load testing to determine design strength or load-carrying limit at failure is outside the scope of this code.

1.5—Interpretation

1.5.1 The official version of the code is the English language version published by the American Concrete Institute using inch-pound units.

1.5.2 In case of conflict between the official version of the code and other versions of the code, the official version governs.

1.6—Language and units of measurement

Values in this code are stated in inch-pound units. A companion code in SI units is also available.

COMMENTARY

R1.4—Limitations

R1.4.1 The procedures and requirements in this code have been developed and applied to structural members subjected to gravity loads and where the structural response is dominated by flexure, such as typical elements of the gravity load-resisting system in a building. This code is not applicable to structures where the response involves significant in-plane or membrane action, where soil-structure interaction is present, or in situations where the structural behavior is not typical of building structural elements. Structures where the response may be inappropriate for use with the requirements of this code include shells, arches, bins and silos, blast-resistant structures, and chimneys.

R1.4.4 The procedures and requirements of this code are not intended to provide any indication of the design strength or load-carrying limit of a member or structure at failure. The term "design strength" refers to the strength of a member or cross section as used in the context of the basic requirement for strength design, expressed as "design strength ≥ required strength" (based on factored load combinations of the general existing building code). The design strength is calculated in accordance with the provisions and assumptions of the strength design method of the general existing building code and multiplied by the appropriate strength reduction factor ϕ.

CODE

COMMENTARY

CHAPTER 2—NOTATION AND DEFINITIONS

R2—NOTATION AND DEFINITIONS

2.1—Notation

D = total dead load: $D_w + D_s$, or related internal moments or forces

D_s = superimposed dead load, or related internal moments or forces

D_w = load due to self-weight of the concrete structural system

F = load due to weight and pressure of fluids with well-defined densities and controllable maximum heights, or related internal moments or forces

I_{DL} = deviation from linearity index

I_{Pi} = permanency index for the i-th cycle at specified load level of a cyclic load test

$I_{P(i+1)}$ = permanency index for the $(i + 1)^{th}$ cycle at specified load level of a cyclic load test

I_{pr} = permanency ratio index

L = live load due to use and occupancy of the building not including environmental load and superimposed dead load, or related internal moments or forces

L_r = roof live load produced during maintenance by workers, equipment, and materials or during life of structure by moveable objects such as planters and people, or related internal moments or forces

l_t = span of member under load test and taken as the smaller of: (a) distance between centers of supports; and (b) clear distance between supports plus thickness h of member; for a cantilever, it shall be taken as twice the distance from face of support to cantilever end, in.

P_i = applied load corresponding to point i in load-deflection envelope in a cyclic load test, used for calculation of α_i in I_{DL} acceptance criterion, lb

P_{max} = applied load corresponding to 100 percent of test load magnitude, lb

P_{min} = minimum load to be maintained during a cyclic load test, lb

P_{ref} = reference load corresponding to the peak load of the first cycle (reference point) in a cyclic load test, used for computation of α_{ref} in I_{DL} acceptance criterion, lb

R = rain load, or related internal moments and forces

S = snow load, or related internal moments or forces

α_i = slope of secant line of point i in load-deflection envelope, where secant line passes through origin and point of interest, i, in a cyclic load test, degrees

α_{ref} = slope of reference secant line in load-deflection envelope, where reference secant line passes through origin and peak load of the first cycle (reference point) in a cyclic load test, degrees

Δ_i = measured deflection corresponding to point i in load-deflection envelope in a cyclic load test, used for computation of α_i in I_{DL} acceptance criterion, in.

CODE

Δ_l = maximum measured deflection (cyclic or monotonic load test), in.

Δ_{l2} = maximum deflection measured during a retest relative to the position of the structure at the beginning of the retest, in.

Δ_r = residual deflection measured after the 24-hour recovery period following complete removal of the load at the completion of the load test, in.

Δ_{ref} = measured reference deflection corresponding to the peak load of the first cycle (reference point) in a cyclic load test, used for computation of α_{ref} in I_{DL} acceptance criterion, in.

Δ_{rrt} = permanent or residual deflection measured after the 24-hour recovery period following complete removal of the load at the completion of the repeat load test; residual deflection during the repeat test is relative to the position of the structure at the beginning of the repeat test, in.

Δ^1_{max} = measured maximum deflection at the end of the i-th cycle at a given load level of a cyclic load test, in.

Δ^1_r = measured residual deflection after unloading to the minimum load, P_{min}, at the completion of the i-th cycle at a given load level of a cyclic load test, in.

Δ^2_{max} = measured maximum deflection at the end of the $(i + 1)^{th}$ cycle at a given load level of a cyclic load test, in.

Δ^2_r = measured residual deflection after unloading to the minimum load, P_{min}, at the completion of the $(i + 1)^{th}$ cycle at a given load level of a cyclic load test, in.

2.2—Definitions

acceptance criteria—set of explicit and quantitative rules to establish whether or not a structure (or a portion of it) passes a load test.

applied test load (ATL)—load applied to the structure, in addition to dead load already in place, to produce test load magnitude.

cyclic load test—application of test load magnitude using a defined set of increasing, repeated load cycles and then maintaining the test load magnitude for at least 24 hours.

failure—when referred to the performance of a structure (or a portion of it) under load test, it indicates that one or more acceptance criteria are not met.

licensed design professional—engineer or architect who is licensed to practice structural design as defined by the statutory requirements of the professional licensing laws of a state or jurisdiction; or the architect or engineer, licensed as described, who is responsible for the structural design of a particular project.

load-deflection envelope—curve enclosing the points on a plot of applied load versus measured deflection from a cyclic load test. Only those loads that are greater than or equal to any previously applied load cycle are included.

load level—maximum load attained during a loading cycle in a cyclic load test or an intermediate load before the full test load magnitude is applied in a monotonic load test.

COMMENTARY

R2.2—Definitions

CODE

load step—load increment applied during a loading cycle to reach the load level for that cycle or the load decrement during the unloading portion of the cycle.

monotonic load test—application of test load magnitude in at least four approximately equal steps and then maintaining the test load magnitude for at least 24 hours.

reference point—point on a plot of applied cyclic load versus measured deflection corresponding to the peak load of the first load cycle.

reference secant line—line passing through the origin of a plot of applied cyclic load versus measured deflection and the reference point.

self-weight dead load, D_w—load due to self-weight of the concrete structural system only.

strip or patch test load—test load distributed over a limited portion of the tributary area of the structure or member to be tested and typically applied by means of hydraulic jacks.

superimposed dead load, D_s—load due to weight of construction materials other than self-weight of the structural system that are already incorporated into the building or will be incorporated in the future.

test load magnitude (TLM)—total load to be used in conjunction with acceptance criteria in this code to determine whether or not a structure, or a portion of it, passes a load test.

total dead load, D—total load due to self-weight, including superimposed loads.

COMMENTARY

superimposed dead load, D_s—this includes, but is not limited to, walls, roofs, ceilings, stairways, built-in partitions, finishes, cladding, and other similarly incorporated architectural and structural items; the weight of fixed service equipment such as cranes, plumbing stacks and risers, electrical, heating, ventilating and air-conditioning systems, and fire sprinkler systems.

test load magnitude (TLM)—this includes factored dead, live, snow, and rain loads, and is established using the load factors and combinations defined in Chapter 4 of this code. It includes dead load that is already in place in the structure, including self-weight and super-imposed dead load, plus additional applied test loads to reach the desired TLM.

total dead load, D—a distinction is made between dead load due to self-weight and superimposed dead loads in this code. The total dead load, D, includes both dead load due to self-weight and superimposed dead loads; that is, $D = D_w + D_s$.

CODE

COMMENTARY

CHAPTER 3—REFERENCED STANDARDS

American Concrete Institute

ACI 318-11—Building Code Requirements for Structural Concrete and Commentary

ACI 562-13—Code Requirements for Evaluation, Repair, and Rehabilitation of Existing Concrete Buildings and Commentary

American Society of Civil Engineers (ASCE)

ASCE/SEI 7-10—Minimum Design Loads for Buildings and Other Structures

CODE

CHAPTER 4—TEST LOAD

4.1—Service loads

4.1.1 Service loads shall be determined in accordance with the governing general building code requirements. The dead load portion of the service loads shall be separated into components D_w and D_s.

4.1.2 Unless documentation or tests are available to define the density of concrete used in the structure, the density of normalweight reinforced concrete shall be taken as 150 lb/ft^3 for calculation of D_w.

4.1.3 Fluid loads F, due to weight and pressure of fluids with well-defined densities and controllable maximum heights, are permitted to be included in the superimposed dead load, D_s, for defining the test load magnitude in 4.2. If fluid loads F are due to weight and pressure of fluids that do not have well-defined densities or controllable maximum heights, then F shall be included with live load L for defining the test load magnitude in 4.2.

4.2—Test load magnitude

4.2.1 Test load magnitude (TLM) shall be as defined in 4.2.2 and 4.2.3 except as modified by 4.2.4 and as permitted or required by the local building authority. For load testing and evaluation, live loads are permitted to be reduced in accordance with the general building code. If the general building code requires application of impact factors to the live loads, the same impact factors shall be included in the live load used in 4.2.2 through 4.2.4.

4.2.2 When load testing will evaluate only part of the portions of a structure that are suspected of containing deficiencies or that have been repaired or rehabilitated and whose adequacy is to be verified, and members to be tested are statically indeterminate, then Eq. (4.2.2a) through (4.2.2c) shall be used to determine the TLM.

$$\text{TLM} = 1.3(D_W + D_S) \tag{4.2.2a}$$

$$\text{TLM} = 1.0D_W + 1.1D_S + 1.6L + 0.5(L_r \text{ or } S \text{ or } R) \tag{4.2.2b}$$

$$\text{TLM} = 1.0D_W + 1.1D_S + 1.6(L_r \text{ or } S \text{ or } R) + 1.0L \tag{4.2.2c}$$

4.2.3 When all suspect portions of a structure are to be load tested, or when the elements to be tested are determinate and the suspect flaw or weakness is controlled by flexural tension, the TLM (including dead load already in place) shall not be less than the largest value from Eq. (4.2.3a) through (4.2.3c). If the deficiency in statically determinate members is not tension-controlled, then Eq. (4.2.2a) through (4.2.2c) shall be used to determine the TLM.

$$\text{TLM} = 1.2(D_W + D_S) \tag{4.2.3a}$$

COMMENTARY

R4—TEST LOAD

R4.2—Test load magnitude

The required TLM follows the recommendations of ACI 437.1R. The value of the TLM is intended to be consistent with the factored load combinations of ASCE/SEI 7. The TLM is appropriate for evaluating concrete structures designed in accordance with the current or previous editions of ACI 318.

R4.2.3 Smaller load factors are permitted for determinate tension-controlled sections to reduce the potential for inelastic deformations as a result of the load test and because of the extra ductility and associated warning expected to develop before failure or collapse. Tension-controlled sections are also usually less sensitive to variations in concrete strength.

When testing structures designed using the load factors and ϕ-factors in the 2002 and later editions of ACI 318, wherein load factors and ϕ-factors are changed from earlier editions, the test load may induce bilinear elastic (cracked)

CODE

$$\text{TLM} = 1.0D_W + 1.1D_S + 1.4L + 0.4(L_r \text{ or } S \text{ or } R) \quad (4.2.3b)$$

$$\text{TLM} = 1.0D_W + 1.1D_S + 1.4(L_r \text{ or } S \text{ or } R) + 0.9L \quad (4.2.3c)$$

4.2.4 If serviceability is a criterion in the evaluation of the structure, a test load level equivalent to $1.0D + 1.0L + 1.0(L_r$ or S or $R)$ shall be included in the loading cycles so that the behavior of the structure at service load level can be evaluated.

4.2.5 It shall be permitted to modify the load factors used in Eq. (4.2.2a) through (4.2.2c) and (4.2.3a) through (4.2.3c) in accordance with 4.2.5.1.

4.2.5.1 If the ratio of service live loads to service dead loads exceeds 2.0, that is, if $(L + (L_r$ or S or $R))/(D_W + D_S) > 2.0$ and if the suspect deficiency is tension-controlled, it shall be permitted to reduce the load factor applied to the live load L in Eq. (4.2.2b) to 1.4 and in Eq. (4.2.3b) to 1.3, and to reduce the load factor applied to roof live loads, snow loads, or rain loads (L_r or S or R) in Eq. (4.2.2c) to 1.4 and in Eq. (4.2.3c) to 1.3.

COMMENTARY

or inelastic behavior for some structures. Discussion is provided in Chapter 6 regarding linearity of response as part of acceptance criteria.

When testing members not meeting the minimum shear reinforcement requirements on the basis of 11.4.6.1 of ACI 318-11, but meeting strength requirements on the basis of 11.4.6.2 through 11.4.6.4, an assessment of the test load at which cracking in the web-shear region will occur is recommended. Section R11.3.3 of ACI 318-11 indicates that Eq. (11-12) can be used to determine the shear force that will cause web-shear cracking in prestressed concrete members. Cracking that does not close after removal of the test load may occur where nonprestressed reinforcement yields during the load test. Precast/prestressed members with thin webs, such as double tees, can be a concern in this regard because they may be prone to developing permanent damage from full test loads.

R4.2.5.1 For buildings with large live loads compared with the structure's self-weight and weight of superimposed dead loads, there may be a concern that an otherwise adequate structure could be loaded into the inelastic range during the load test, inducing permanent deformations. This could occur, for example, when testing a structure prestressed for a lower, more typical, service load condition but reinforced with bonded reinforcement to provide adequate design strength for full factored code-required live load. Section 4.2.5.1 provides some flexibility for moderating the TLM in such cases. This provision in effect permits reducing the TLM to approximately 85 percent of demand as defined in Eq. (9-2) of ACI 318-11 for the referenced equations of 4.2.2, and reducing the TLM to approximately 80 percent of demand for the referenced equations of 4.2.3. These reductions are permitted only for those cases where the limiting strength being tested is tension-controlled, implying a ductile failure mode with large accompanying deflections.

CODE

CHAPTER 5—LOADING PROTOCOL

5.1—Safety

5.1.1 Load tests shall be conducted in a manner to provide safety to people and the structure during the test. Shoring shall be designed by a licensed design professional and installed prior to the start of the load test.

5.1.2 Safety measures shall be implemented in such a way as not to interfere with load test procedures or affect results.

5.2—Load arrangement

5.2.1 The areas loaded shall be selected to maximize the response characteristics due to applied test load (ATL) specified in Chapter 4. More than one test load arrangement shall be used if a single arrangement will not simultaneously result in maximum values of the effects necessary to demonstrate the adequacy of the structure.

5.2.2 Either uniform or concentrated loads are permitted. In the case of uniform loading, arching of the applied load shall be avoided and concentrated load effects investigated. If concentrated loads are used to simulate distributed loading, they shall be of such magnitude and so located as to produce equivalent maximum internal moments and forces at critical cross sections as would be produced by uniform loads.

5.2.3 The load shall be applied at locations so that its effects on the structural elements for which safety or serviceability is in doubt are maximized and the probability of unloaded members sharing the applied load is minimized. If it is shown by analysis that adjoining unloaded elements will help resist some of the load, the contribution of these members shall be taken into account.

COMMENTARY

R5—LOADING PROTOCOL

R5.1—Safety

R5.1.1 Shoring should be provided before a load test, whether the entire structure or only a portion is involved, to support the structure in case of failure during the test. Shoring should be designed to support dead loads, the ATL, and all additional superimposed loads on the portion of the building for which collapse is possible. The effects of impact loading on the shoring, which is likely if a structure or member fails during the test, should be considered in the selection of shoring elements. This may be accomplished by designing the shoring to support at least twice the total test load plus the existing dead load.

For multistory structures, shoring more than one level to prevent progressive collapse in the event of failure should be considered. For example, if all floors below the test floor cannot support the weight of the test element, the loads it supports, and the imposed test loads, then the shoring should be extended to suitable bearing surface.

R5.1.2 For horizontal members, shoring should clear the underside of the structure by not more than the maximum expected deflection plus an allowance not to exceed 2 in. Similar arrangements should be made for other types of members. Shoring should not influence or interfere with the free movement of the structure under the test load.

R5.2—Load arrangement

R5.2.2 Arching can be caused by the stacking of masses and refers to the tendency for the load to be transmitted nonuniformly to the flexural element being tested. For example, if a slab is loaded by a uniform arrangement of bricks with the bricks contacting each other, arching can result in reduction of the load on the slab near the midspan of the slab.

The internal moment and force diagrams due to concentrated loads cannot exactly reproduce the ones due to uniform loads.

R5.2.3 The load arrangement should cause the maximum demands at the critical sections. In cases where it is shown by analysis that adjoining unloaded elements will help support some of the load, the load should be placed with consideration of the expected load sharing.

CODE

5.3—Monotonic loading protocol

5.3.1 Applied test load (ATL) shall be applied in a predetermined pattern in at least four approximately equal increments. If serviceability is a criterion in the evaluation of the structure, a load level equivalent to $1.0D + 1.0L + 1.0(L_r$ or S or $R)$ shall be included so that the behavior at service load level can be evaluated. The tolerance for the applied sustained load shall be ±5 percent of the full ATL.

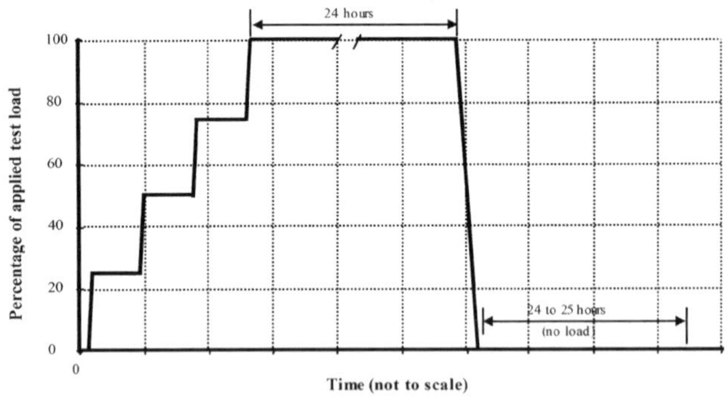

Fig. 5.3.1—Loading protocol for monotonic load test procedure.

5.3.2 Load level and load steps are shown in Fig. 5.3.1. After applying each load level, deflection measurements shall be made at equal time intervals until the deflections stabilize. Deflections shall be considered stabilized when the difference between successive deflection readings taken no less than 2 minutes apart does not exceed 10 percent of the initial deflection recorded for the current load step. Each load step shall be held for at least 2 minutes. The full ATL shall be maintained for at least 24 hours.

5.3.3 A set of response measurements shall be made at the beginning and the end of the 24-hour sustained load period.

5.3.4 After response measurements are made at the end of the 24-hour sustained load period, the test load shall be removed as quickly as practicable.

5.3.5 A final set of response measurements shall be made at least 24 hours after removal of the load. If the structure has met the acceptance criteria at the conclusion of the loading cycle, it is not necessary to wait the full 24-hour period to make the final set of response measurements.

5.4—Cyclic loading protocol

5.4.1 Applied test load (ATL) shall be applied in repeated loading and unloading cycles of increasing magnitude. At least six repeated loading cycles shall be used with the following magnitudes:

a) Cycles A and B—Service load level if serviceability is a criterion, otherwise 50 percent of ATL

b) Cycles C and D—Halfway between Cycle A and 100 percent of ATL

c) Cycles E and F—100 percent of ATL

COMMENTARY

R5.3—Monotonic loading protocol

R5.3.4 Intermediate response measurements should be made at selected intervals to determine whether significant creep is occurring during the hold period.

R5.4—Cyclic loading protocol

In the cyclic loading procedure, the loads are applied in loading-unloading cycles of increasing magnitude using hydraulic jacks. Using a sequence of loading and unloading cycles up to the predetermined maximum load level provides the opportunity to evaluate the structure at different load levels. The load sequence is intended to identify, in an explicit manner, any undesirable response. Because the structure is initially loaded and unloaded at low levels, the licensed design professional has the ability to better under-

CODE

The tolerance for the applied load shall be ±5 percent for each load cycle.

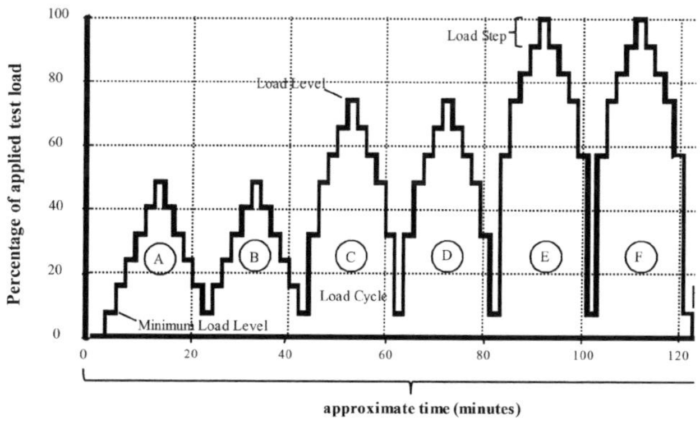

Fig. 5.4.1—Loading protocol for cyclic load test procedure.

5.4.2 Load levels and load steps for each loading cycle are shown in Fig. 5.4.1 and are described in 5.4.4 through 5.4.6. After applying each load step, deflection measurements shall be made at equal time intervals until the deflections stabilize. Deflections shall be considered stabilized when the difference between successive deflection readings taken no less than 2 minutes apart does not exceed 10 percent of the initial deflection recorded for the current load step. Each load step shall be held for at least 2 minutes.

5.4.3 At the end of each unloading phase, which shall use similar steps as the loading phase, a load P_{min} of at least 10 percent of ATL shall be maintained to keep the test devices engaged. P_{min} shall not exceed 15 percent of ATL.

5.4.4 *Load Cycles A and B*—For Load Cycle A, the load steps shall be initiated at P_{min}. The first load cycle consists of five approximately equal load steps beyond P_{min}. Load Cycle B is a repeat of Load Cycle A.

5.4.5 *Load Cycles C and D*—For Cycles C and D, the load at the first of five steps shall be at the same level as the third step of Cycle A, and the load at the second step shall be at the level of maximum load attained in Cycle A. The remaining three steps are of approximately equal magnitude.

5.4.6 *Load Cycles E and F*—For Cycles E and F, the load at the first of five steps shall be at the same level as the third step of Cycle C, and the load at the second step shall be at the level of maximum load attained in Cycle C. The remaining three steps shall be of approximately equal magnitude.

5.5—Load application

5.5.1 Load shall be applied in accordance with either 5.3 or 5.4, as determined by the licensed design professional.

COMMENTARY

stand end fixity and load transfer characteristics of the tested member by comparing actual deflection responses with calculated deflection responses. For statically indeterminate structures in particular, this ability allows checking the accuracy of the assumptions on fixity and load sharing made in the structural analysis.

R5.5—Load application

R5.5.1 The cyclic loading protocol allows performing a real-time assessment of member characteristics such as linearity and permanency of deformations at different load levels. Hydraulic loading devices are generally required

CODE

5.5.2 Load shall be applied without impact and without causing vibration of the structure.

5.5.3 The loading differential for the twin cycles in the cyclic load test shall not exceed a maximum range of 10 percent.

5.6—Response measurements

5.6.1 Response measurements shall be taken beginning 10 minutes before the application of the first load level and continued throughout the loading and unloading cycles. Deflections shall be measured. The resolution of deflection measurement devices shall not exceed 1/100 of the expected deflection.

5.6.2 Measurements shall be made at locations where maximum response is expected as established by the licensed design professional.

5.7—Visual inspection

5.7.1 The structure shall be inspected following application of each load level for the formation, widening, or lengthening of cracks and other forms of distress; as well as for the presence of excessive deformations or rotations.

5.7.2 The licensed design professional shall assess the significance of any distress and determine whether it is safe to continue with the test.

COMMENTARY

to complete the cyclic loading protocol within a reasonable period of time. In some instances, the use of hydraulic loading devices may not be practicable. Examples of such instances may include thin shell structures and structures having inadequate existing members to be used as reaction members.

R5.6—Response measurements

R5.6.1 In addition to deflection, response measurements may include rotation, strain, and crack widths. If electronic data acquisition is used, the measurements can generally be taken at a rate of once each second. The data acquisition rate can be reduced to once each minute during the 24-hour sustained load period. In some cases, mechanical instrumentation could be used, such as analog dial gauges. In such cases, the measurements should be made after each load step.

CODE

CHAPTER 6—ACCEPTANCE CRITERIA

6.1—Distress caused by load test

6.1.1 The portions or sections of the structure tested using the cyclic loading protocol or monotonic loading protocol shall show no evidence of failure.

6.1.2 Deflections due to the test loads that exceed precalculated deflections shall be evaluated by the licensed design professional. Based on this evaluation, a decision on whether to continue the load test shall be made by the licensed design professional.

6.1.3 Cracking shall be evaluated by the licensed design professional who shall decide on whether to continue the load test.

6.1.4 If at any time during the load test a member develops cracks indicating imminent shear failure, it shall be considered as having failed the load test. Retesting of the failed member is not permitted.

6.1.5 In regions of anchorage and lap splices, the appearance of a series of short cracks inclined to the axis of the reinforcement or cracks parallel to the axis reinforcement shall be evaluated by the licensed design professional who shall decide on whether to continue the load test.

COMMENTARY

R6—ACCEPTANCE CRITERIA

R6.1—Distress caused by load test

R6.1.1 A general acceptance criterion to assess the behavior of a structure under test load is that no evidence of failure is observed. Evidence of failure includes cracking, crushing in compression zones, and deflections that are incompatible with the serviceability or safety requirements of the structure. No simple rules have been developed for application to all types of structures and conditions.

R6.1.2 Load tests are typically undertaken when it is not possible to evaluate by analytical means the performance of a structure with reasonable certainty due to the degree of deterioration or uncertainties in analysis. Thereby, calculations to estimate deflections under the test loads may not be sufficiently accurate to be used as mandatory acceptance criteria. The licensed design professional should, however, make calculations to estimate deflections due to the test loads, considering as-built information for the structure. If the measured deflections exceed the calculated deflections, careful consideration should be given before continuing the load test to higher load levels. The calculated deflections need to be used to assess the performance during the load test and to interpret the load test results. Recommendations for the calculation of deflections are available in ACI 435R.

R6.1.3 Even though it is recognized that calculations regarding crack widths may not be sufficiently accurate to justify using them as mandatory acceptance criteria, the licensed design professional is to include their assessment under service load and other load levels as an integral part of the structural performance evaluation process.

The purpose of this recommendation is to caution the licensed design professional conducting the load test that the occurrence or excessive growth of cracks under service loads may be an indication of structural deficiencies.

R6.1.4 Shear forces are transmitted across a shear crack plane and are resisted by a mechanism combining the effects of the transverse reinforcement crossing the crack, if present; aggregate interlock; and dowel action of the reinforcement crossing the crack. As crack lengths increase to approach a horizontal projected length equal to the depth of the member and concurrently widen to the extent that aggregate interlock cannot occur, and as transverse stirrups, if present, begin to yield or display loss of anchorage, the member is assumed to be approaching imminent shear failure.

R6.1.5 Cracking along the axis of the reinforcement, at lap splices, and in anchorage zones may be related to high stresses associated with the transfer of forces between the reinforcement and the concrete. These cracks may be indicators of pending brittle failure of the element. The licensed design professional is required to evaluate the consequences of the observed cracking.

CODE

6.2—Performance assessment at service load level

6.2.1 If serviceability is a criterion in the evaluation of the structure, deflections, crack spacing, and crack widths under a test load equivalent to the service condition shall be recorded and checked against limit values that are established by the licensed design professional before conducting the load test. If the limit values are exceeded at the service load level, the licensed design professional shall determine whether to terminate the load test.

COMMENTARY

R6.2—Performance assessment at service load level

R6.2.1 Irrespective of the loading procedure (that is, monotonic or cyclic load) and type of load (that is, uniformly distributed load over the entire tributary area, or concentrated loads), if the measured deflection or crack width exceed limits set before the load test by the licensed design professional, careful consideration should be given as to whether to continue the load test to higher load levels. It is recognized that the variable nature of cracking and the challenges in accurately measuring and predicting crack width make the corresponding limits difficult to implement. The intent of the provision is to caution the licensed design professional that the occurrence or growth of excessive cracks under service loads may be an indication of structural deficiencies. Influence of crack width is of particular significance for some members or structures, such as those exposed to aggressive environments.

Guidance for establishing possible limit values for deflection and crack width at service load is as follows:

1. Maximum measured deflection should not exceed the permissible values given in Table 9.5(b) of ACI 318-11 for the various types of members. This recommendation is only applicable if the load distribution pattern reflects the one used for design, which is typically not the case for test loads of the strip or patch type. Furthermore, the first two values in Table 9.5(b) are intended for immediate live load deflections, while the third and fourth deflection limits are for the additional deflection occurring after attachment of nonstructural members due to long-term deflection caused by all sustained loads plus any immediate live load deflection. This makes these limits difficult to apply in the setting of a load test where only the immediate deflection due to applied loads can be measured. Additional long-term deflection due to sustained loads can be calculated and then added to the load test deflection results for live loading to arrive at a value that can be compared with the latter two limits of Table 9.5(b).
2. The maximum width of new flexural cracks formed during the course of the load test or the change in width of existing flexural cracks should not exceed a limiting width determined by the licensed design professional before the load test. Consideration should be given to the intended use and exposure conditions for the structure or member. Limiting crack widths should be selected based on the following:
   a) Suggested tolerable crack widths as reported in ACI 224R
   b) The value of the analytical width calculated as the product s times $\Delta\varepsilon_s$ (where s is the average spacing between cracks and $\Delta\varepsilon_s$ represents the difference in strain in the longitudinal steel reinforcement when the cross section of interest is considered cracked and uncracked and subject to an applied moment at that location resulting from the service load)

CODE

6.3—Acceptance criteria for monotonic loading protocol

6.3.1 *Deflection limits*

6.3.1.1 The member or structure shall be considered to have passed the load test if the residual deflection, Δ_r, satisfies Eq. (6.3.1) and the maximum deflection, Δ_l, measured during the test satisfies Eq. (6.3.2).

$$\Delta_r \leq \frac{\Delta_l}{4} \quad (6.3.1)$$

$$\Delta_l \leq \frac{l_t}{180} \quad (6.3.2)$$

6.3.1.2 If the maximum deflection measured during the test, Δ_l, is less than 0.05 in. or the deflection as a fraction of the span length, l_t, is less than $l_t/2000$, the residual deflection requirements given by Eq. (6.3.1) shall be permitted to be waived.

6.3.2 *Retesting*

6.3.2.1 If the measured residual deflection Δ_r does not satisfy Eq. (6.3.1), it shall be permitted to repeat the load test if the maximum deflection Δ_l satisfies Eq. (6.3.2).

6.3.2.2 The repeat test shall be conducted not earlier than 72 hours after removal of the test load for the first test.

6.3.2.3 The portion of the structure tested in the repeat test shall be considered acceptable if the residual deflection Δ_r, measured at least 24 hours after removal of the load, satisfies Eq. (6.3.3), where deflections are measured relative to the position of the structure at the start of the repeat test.

$$\Delta_{rrt} \leq \frac{\Delta_{l2}}{10} \quad (6.3.3)$$

where Δ_{l2} is the maximum deflection measured during the second test relative to the position of the structure at the beginning of the second test.

6.4—Acceptance criteria for cyclic loading protocol

6.4.1 *Deviation from linearity index*—The deviation from linearity index, I_{DL}, calculated in accordance with Eq. (6.4.1), shall be monitored during all the cycles of the cyclic load test.

$$I_{DL} = 1 - \frac{\tan(\alpha_i)}{\tan(\alpha_{ref})} \quad (6.4.1)$$

where $\tan(\alpha_i)$ is the secant stiffness of any point i on the increasing loading portion of the load-deflection envelope, and $\tan(\alpha_{ref})$ is the slope of the reference secant line for the load-deflection envelope, as shown in Fig. 6.4.1.

6.4.1.1 The deviation from linearity index shall be considered acceptable if it does not exceed 0.25 for any point on the loading portion of the load-deflection envelope.

COMMENTARY

R6.3—Acceptance criteria for monotonic loading protocol

R6.3.1 The residual deflection for the monotonic load test limit follows past practice. If the structure shows no evidence of failure, recovery of deflection after removal of the test load is used to determine whether the performance of the structure is satisfactory. As a departure from past practice, Eq. (6.3.2) sets an upper limit to the absolute measured deflection that, if exceeded, rules out retesting or the option of using deflection recovery as an acceptance criterion (ACI 437.1R). If Eq. (6.3.2) is satisfied and the member or structure fails the residual deflection criterion on the first test, retesting is permitted. All deflections measured during retesting are relative to the position of the structure at the start of the repeat test.

If the member or structure is sufficiently stiff, deflection recovery is not relevant. In fact, it may even be unfeasible to measure the deflection recovery due to limitations in the resolution or accuracy of the deflection measurement equipment. To avoid penalizing a satisfactory structure in such a case, recovery measurements are waived and no check on deflection recovery is required if the absolute deflection is lower than 0.05 in. or the deflection as a fraction of span length is less than $l_t/2000$ (ACI 437.1R).

R6.4—Acceptance criteria for cyclic loading protocol

R6.4.1 The deviation from linearity index should be checked at the critical locations of the member being load tested. The deviation from linearity index represents the measure of the nonlinear behavior of a member or structure being tested. As a member becomes increasingly more damaged, its behavior may become more nonlinear, and its deviation from linearity increases. To define deviation from linearity, linearity is defined first as the ratio of the slopes of two secant lines intersecting the load-deflection envelope (Fig. 6.4.1). The load-deflection envelope is the curve constructed by connecting the points of the load-deflection data in the first of the two cycles of each load cycle.

If a member or structure is initially uncracked and becomes cracked during the load test, the change in flexural stiffness as a result of a drastic change in moment of inertia at the crack location(s) can produce a high deviation

CODE

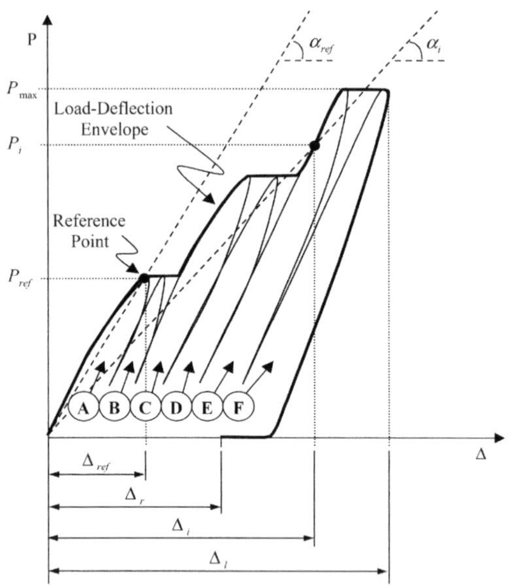

Fig. 6.4.1—Schematic load-versus-deflection curve for cyclic load test.

6.4.2 *Permanency ratio*—The permanency ratio I_{pr}, calculated in accordance with Eq. (6.4.2a), shall be monitored during all the cycles of the cyclic load test.

$$I_{pr} = \frac{I_{p(i+1)}}{I_{pi}} \tag{6.4.2a}$$

where I_{pi} and $I_{p(i+1)}$ are the permanency indexes calculated for the i-th and $(i + 1)^{th}$ load cycles, respectively, at the same load level using Eq. (6.4.2b) and (6.4.2c) and deflections as defined in Fig. 6.4.2.

$$I_{pi} = \frac{\Delta_r^i}{\Delta_{max}^i} \tag{6.4.2b}$$

$$I_{p(i+1)} = \frac{\Delta_r^{(i+1)}}{\Delta_{max}^{(i+1)}} \tag{6.4.2c}$$

6.4.2.1 The permanency ratio shall be considered acceptable if it does not exceed 0.50 for all pairs of load cycles.

6.4.3 *Residual deflection*—The residual deflection Δ_r, measured at least 24 hours after removal of the load at the completion of the load test, shall be considered acceptable if it satisfies Eq. (6.3.1), where Δ_l is the maximum deflection measured during the test as shown in Fig. 6.4.1.

COMMENTARY

from linearity that is not necessarily related to degradation in strength (Nehil et al. 2007).

If the member/structure remains uncracked during the load test, the deviation from linearity should be computed under uncracked conditions.

If the transition between uncracked and cracked condition is experienced at a load cycle other than Cycle A, the load test may be restarted and Cycles A to E performed.

R6.4.2 Permanency ratio is evaluated considering the maximum deflection and the residual deflection for two cycles at each load level. If the permanency ratio exceeds the specified limit, it may be an indication that load application has damaged the member or structure and that inelastic behavior is taking place. High permanency ratios at low load levels (generally below service loads) might not be indicative of inelastic behavior when the measuring instrument resolutions are of the same order of magnitude as the actual deflections and deflection recovery. The permanency ratio might not be of significance if the absolute deflection measured for the twin cycles is lower than 0.05 in. or, as a fraction of span length, is less than l_t/2000.

R6.4.3 If the structure shows no evidence of failure, recovery of deflection after removal of the test load is used to determine whether the performance of the structure is satisfactory. The residual deflection Δ_r is the difference between the initial and final (after the 24-hour recovery period following complete load removal) deflections for the load test. The residual deflection limit for concrete structures with nonprestressed reinforcement follows past practice. Bares and FitzSimons (1975) suggest that 25 percent residual deflection is acceptable for reinforced concrete

CODE

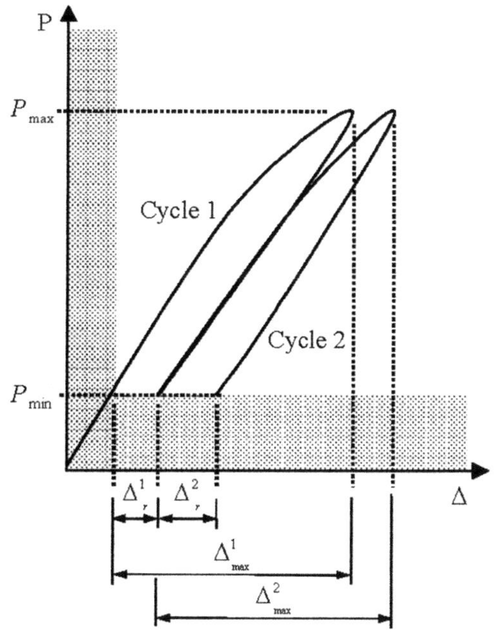

Fig. 6.4.2—Schematic of load-versus-deflection curve for two cycles at same load level.

6.4.4 *Retesting*

6.4.4.1 If any of the acceptance criteria in 6.4.1 to 6.4.3 is not satisfied, the member or structure shall be considered to have failed the load test except as provided in 6.4.4.2.

6.4.4.2 It shall be permitted to retest the member or structure if the maximum deflection, Δ_l, measured during the test satisfies Eq. (6.3.2).

6.4.4.3 The repeat test shall be conducted not earlier than 24 hours after complete removal of the test load for the first test.

6.4.4.4 The portion of the structure that is retested shall be considered acceptable if the residual deflection Δ_{rrt}, measured at least 24 hours after removal of the load, satisfies Eq. (6.3.3), where Δ_{l2} is the maximum deflection measured during the second test relative to the position of the structure at the beginning of the second test.

6.5—Provision for lower load rating

6.5.1 If the structure under investigation does not satisfy conditions of 6.1 and 6.2 and the criteria of 6.3 or 6.4, the structure shall be permitted for use at a lower load rating based on the results of the load test or structural analysis, if approved by the building official.

COMMENTARY

structures and 20 percent is acceptable for prestressed concrete structures.

R6.5—Provision for lower load rating

R6.5.1 Except for load-tested members that have failed or show extensive distress due to the load test, the building official may permit the use of a structure or member at a lower load rating that is judged to be safe and appropriate on the basis of the test results. When the lower load rating is achieved through structural analysis, the structural model used to analyze the structure should represent the actual behavior and, therefore, it should be calibrated to replicate the load test.

CODE

COMMENTARY

6.5.2 If the cyclic loading protocol is used, a number of twin load cycles at increasing load levels shall be applied until either the deviation from linearity index or permanency ratio fails. The safe live load shall be established by subtracting the factored dead load from the critical load at which one of the acceptance criteria was not met, and dividing that difference by the live load factor. Load factors used for the determination of the safe live load shall be determined in accordance with the general building code. After unloading, the residual deflection shall satisfy Eq. (6.3.3).

6.5.3 If the structure or member has undergone significant damage, as determined by the licensed design professional, the distressed structure or members shall not be put into service, even at a lower load rating.

CODE

COMMENTARY

R7—COMMENTARY REFERENCES

American Concrete Institute

ACI 224R-01—Control of Cracking in Concrete Structures (Reapproved 2008)

ACI 318-11—Building Code Requirements for Structural Concrete and Commentary

ACI 435R-95—Control of Deflection in Concrete Structures (Reapproved 2000)

ACI 437.1R-07—Load Tests of Concrete Structures: Methods, Magnitude, Protocols, and Acceptance Criteria

ACI 562-13—Code Requirements for Evaluation, Repair, and Rehabilitation of Existing Concrete Buildings and Commentary

American Society of Civil Engineers (ASCE)

ASCE/SEI 7-10—Minimum Design Loads for Buildings and Other Structures

Cited references

Bares, R., and FitzSimons, N., 1975, "Load Tests of Building Structures," *Journal of the Structural Division*, V. 101, No. 5, May, pp. 1111-1123.

Nehil, T.; Nanni, A.; and Masetti, F., 2007, "Strength Evaluation by Load Testing," *Concrete International*, V. 29, No. 3, Mar., pp. 53-58.

Types of Mechanical Splices for Reinforcing Bars

Reported by ACI Committee 439

Conrad Paulson
Chair

Mark D. Marvin
Secretary

Joseph A. Bohinsky
Jean-Jacques Braun
James L. Caldwell
Domingo J. Carreira
Louis J. Colarusso
Christian L. Dahl
David H. DeValve
Alvin C. Ericson
Salem S. Faza

Gustav G. Erlemann*
Anthony L. Felder
Paul S. Fredrickson
William C. Gallenz
Steven E. Holdsworth
Allen J. Hulshizer
James M. LaFave
Harry B. Lancelot

Murray A. Lount*
Kenneth A. Luttrell
Leroy A. Lutz
Adolfo B. Matamoros
Peter Meza
Theodore A. Mize
Richard A. Ramsey
Roy H. Reiterman

Robert C. Richardson
Mario E. Rodriguez
Philip E. Ross
Clifford A. Sabo
Robert G. Smith
Cloyd E. (Joe) Warnes
Ken Williamson
William H. Zehrt

*Deceased.

Reinforcing bar splices are key components in reinforced concrete construction. Because mechanical splices do not rely on the surrounding concrete to transfer the force between bars, splitting failures are not a concern, and the strength of the splice is not sensitive to the compressive strength of the concrete or the amount of cover. Splices in reinforcing steel are grouped in three categories: lapped bars, mechanical, or welded; the focus of this report is mechanical splices. Strength capacities established in ACI 318 are higher for mechanical splices than for lap splices. Several types of mechanical splices have been qualified to sustain inelastic reinforcing bar strain excursions representative of those that might be brought about by overload due to seismic activity, wind, or blast.

This report provides engineers and contractors with updated information about bar-to-bar mechanical splices and the types of proprietary mechanical splices currently available. There is no attempt to state conditions of acceptance or to endorse or rate one particular mechanical splice over another.

The information on mechanical splices and their installation was provided by mechanical splice manufacturers. Although it was compiled by the committee, none of the information was directly verified by the committee. An attempt was made to include generic descriptions of all types of mechanical splices generally available and sold in the North American market at the beginning of 2004; however, some mechanical splices new in the market may not be included due to lack of knowledge of their existence or because no information was submitted for inclusion at the time this report was written.

Reasons for using mechanical splices, as well as various engineering considerations that should be made when specifying mechanical splices, are discussed. Mechanical splices are described in terms of seismic type, configuration, installation procedure, clearance requirements, and other characteristics. Illustrations of the various mechanical splices are included.

Keywords: dowel; mechanical connection; mechanical splice; reinforced concrete; sleeve; splice.

CONTENTS

ACI 439.3R-07 supersedes ACI 439.3R-91 (Reapproved 1999) was adopted and published March 2007.

CHAPTER 1—GENERAL

1.1—Introduction

In reinforced concrete design, the engineer is faced with sizing individual members and establishing reinforcing steel types, placement requirements, and types of splices that will be used throughout the structure. Design drawings or specifications should clearly show or describe all splice locations and the required performance.

The importance and necessity of clearly prescribing splice requirements is evident in ACI 318. Chapter 1 describes specific items to be included on the design drawings, details, and specifications. These items include showing the location and length of lap splices, reinforcement anchorage lengths, and the type and location of mechanical and welded splices. Chapter 12 also addresses this subject, and states: "Splices of reinforcement shall be made only as required or permitted on design drawings, or in specifications, or as authorized by the engineer."

In beams, columns, walls and slabs, mechanical, welded and lap splices are usually permitted. The merits of each type of splice should always be considered. The selection of the preferred mechanical or welded splice typically must address project- and location-specific problems in addition to cost, installation ease, and other factors.

1.1.1 *Report background and scope*—This report is a revision of ACI 439.3R-91, and provides updated information about types of proprietary bar-to-bar mechanical splices generally available and sold in the North American market at the beginning of 2004. Those mechanical splice systems that have become obsolete or inactive have been removed from this report. Additionally, references to accessories and devices that connect reinforcing bars to structural steel or to headed end anchorage devices are not included because they are outside the scope of this report.

Usage and design requirements of mechanical splices, as well as size capabilities and features of selected mechanical splices, are described in Chapters 1, 2, and 3 of this report.

1.1.2 *Reinforcing bar designations and grades*—In accordance with the current practice of ASTM International, reinforcing bar sizes shown in this report are stated as number designations with inch-pound sizes shown first and SI sizes shown in parentheses. Likewise, grades of reinforcing bars, unless otherwise described, are referred to in terms of minimum yield strength; Grade 40 (280) has a specified yield strength of 40,000 psi (280 MPa), Grade 60 (420) has a specified yield strength of 60,000 psi (420 MPa), and Grade 75 (520) has a specified yield strength of 75,000 psi (520 MPa).

1.1.3 *Type 1 and Type 2 mechanical splices*—There are two primary classifications of mechanical splices, whose use is dictated by design loads in the constructed structure. Type 1 splices are used in elements where there is little concern for inelastic deformations and elevated tensile stresses from seismic events. Type 2 splices have proven, through accepted industry testing, the ability to develop the specified tensile strength of the spliced bars for resistance to elevated tensile stresses.

In accordance with ACI 318, Chapter 21, Type 2 mechanical splices are required to develop the specified tensile strength of the bars being spliced. Type 2 mechanical splices used with ASTM A 706/A 706M are required to develop 80,000 psi (550 MPa) minimum, and also not less than 1.25 times the actual yield strength of the bar. Type 2 mechanical splices used with ASTM A 615/A 615M Grade 60 (420) are required to develop 90,000 psi (620 MPa) minimum, and Type 2 mechanical splices used with ASTM A 615/A 615M Grade 40 (280) are required to develop 60,000 psi (420 MPa) minimum.

The requirements for Type 2 mechanical splices are intended to avoid a splice failure in locations where the reinforcement is subjected to stress levels that are expected to approach the tensile strength of the reinforcement. Lap splices are not recommended in locations where inelastic yielding could occur because such splices are not reliable under conditions of cyclic loading into the inelastic range. Under such circumstances, the engineer is encouraged to specify Type 2 mechanical splices.

In accordance with ACI 318, Type 1 mechanical splices should develop in tension or compression, as required, a minimum of 125% of the specified yield strength of the bar. Type 1 mechanical splices are not required to satisfy the more stringent requirements for Type 2 splices. As a result, Type 1 mechanical splices may not be capable of resisting the stress and strain demands that can take place in sections where yielding of the reinforcement is likely to occur during a seismic event.

For any mechanical splicing system under consideration for use, the reader should obtain supporting test information directly from the manufacturer pertaining to the particular mechanical splice device.

1.1.4 *Categorization of mechanical splices*—Four basic categories of mechanical splices are considered in this report. They are:

1. Compression-only mechanical splices, also known as end-bearing mechanical splices;
2. Tension-compression mechanical splices that can resist both tensile and compressive forces;
3. Dowel bar mechanical splices; and
4. Mechanical lap splices.

1.1.5 *Definitions*—In this report, specific terms are defined as follows.

bar-end check—checking the ends of reinforcing bars to determine whether they fit the devices intended for splicing the bars.

coupler—threaded device for joining reinforcing bars for the purpose of providing transfer of either axial compression, axial tension, or both, from one bar to the other.

position coupler—threaded device for joining reinforcing bars where bars are not rotated. A component is rotated to engage coupler threads or reinforcing bar threads.

coupling sleeve—nonthreaded device fitting over the ends of two reinforcing bars for the purpose of providing transfer of axial compression, axial tension, or both, from one bar to the other.

end-bearing sleeve—device fitting over the abutting ends of two reinforcing bars for the purpose of ensuring transfer of only axial compression from one bar to the other.

mechanical lap splice—complete assembly of a coupling sleeve device and any additional components configured to accomplish positive splicing of overlapping reinforcing bars.

mechanical splice—complete assembly of a coupler, a coupling sleeve, or an end-bearing sleeve, including any additional intervening material or other components required to accomplish the splicing of reinforcing bars. (Note: "mechanical splice" was formerly called "mechanical connection" in ACI 318-95 and earlier editions.)

1.1.6 *Welding*—With the exception of friction welding, which is a solid-state joint process that is used in some mechanical splice systems, it is beyond the scope of this report to provide comprehensive information about welded splices or other currently available special proprietary splicing systems. Engineers are referred to the most current AWS code for welding reinforcing steel (ANSI/AWS D1.4) and the ASTM specifications for reinforcing bars, such as ASTM A 615/A 615M, A 706/A 706M, and A 996/A 996M. Information about welding of reinforcing bars is also presented in the CRSI *Manual of Standard Practice*.

1.2—Usage

There are numerous situations that dictate the use of mechanical splices or that make their incorporation more practical and cost effective than lap splices. Some of the most common conditions are:

1. **Where size No. 14 and 18 (No. 43 and 57) bars are used.** This occurs most often in columns, raft mat foundations, and other heavily reinforced concrete structures. ACI 318 does not permit size No. 14 and 18 (No. 43 and 57) bars to be lap spliced. Lap splices are only allowed by ACI 318 in the case of size No. 11 (No. 36) and smaller bars;

2. **Where spacing of the reinforcing bars is insufficient to permit lap splicing the bars.** This generally occurs in situations requiring large amounts of reinforcement and the use of larger bars, as in heavily loaded columns;

3. **When requirements in current codes and specifications for tension lap splices result in long lap splice lengths.** This is the case especially for bar sizes such as size No. 9, 10, and 11 (No. 29, 32, and 36) bars in Grade 60 (420) steel or in epoxy-coated reinforcing bars;

4. **In tension tie members.** Tension lap splices of reinforcing bars in tension tie members are not permitted by the code;

5. **When the location of construction joints and provisions for future construction dictate the use of mechanical splices to provide tensile continuity.** Mechanical splices are often preferable to having long bar lengths projecting from existing concrete construction. A minimum of a 12 in. (305 mm) bar extension provides sufficient length for application of most mechanical splices without damaging the existing concrete during installation. If mechanical splices must be staggered, the projecting bar extension should be greater to allow for the stagger pattern;

6. **To satisfy a steel-concrete ratio of 8% maximum for reinforced concrete columns.** This limit is considered in Chapter 10 of ACI 318 to be a practical maximum for reinforcement in terms of economy and requirements for placing. By using mechanical splices instead of lap splices, the amount of longitudinal steel may be significantly reduced in the splice region;

7. **When existing reinforced concrete structures are renovated, modified or repaired and new reinforcing bars must be spliced to those already in concrete.** Mechanical splicing requires less removal of concrete compared to lap splicing; and

8. **When members may be subjected to inelastic deformations due to earthquake.** Strength capacities established in ACI 318 are higher for mechanical splices than for lap splices. Lap splices may not be reliable under conditions of cyclic loading into the inelastic range. Type 2 mechanical splices are qualified by tests to sustain inelastic reinforcing bar strain excursions.

1.3—General considerations

All mechanical splices have particular physical features, in the splice components or in the required installation equipment or procedure, that can influence design and construction. Selection of splice type, components and installation requirements, installed cost, and available options should be considered in the design process. The following subsections identify considerations for splice selection.

1.3.1 *Spacing and cover requirements*—Minimum clear distances between adjacent reinforcing bars are specified in codes and design specifications. For example, Chapter 7 of ACI 318 sets an absolute minimum clear distance between parallel bars in a layer of not less than the nominal diameter of the bar or 1 in. (25 mm), except that for columns, the minimum clear distance is not less than 1-1/2 times the nominal bar diameter or 1-1/2 in. (38 mm). Chapter 7 states that the clear distance limits also apply to adjacent bars in lapped splices, but does not address clearance limits for mechanical splices.

Clearance limits for mechanical splices may be a factor in the selection and positioning of the appropriate mechanical splice. The outside diameter of the mechanical splice should be known. Up-to-date dimensional data should be obtained from the splice manufacturer. By knowing the diameter of the mechanical splice, the engineer/specifier can decide whether the mechanical splices need to be staggered on the basis of the clearance required. For constructibility reasons, mechanical splices are usually best located in the same plane or elevation. There is little justification for staggering mechanical splices solely to prevent undesirable failure modes that are associated with lap splices because, unlike lap splices, the ability of mechanical splices to carry and transfer load from bar-to-bar is not affected by concrete cover or the compressive strength of concrete. Nevertheless, it has been the practice of some designers in the past to stagger mechanical splices as if they were lap splices. Pending any future code revisions, however, the minimum stagger length should be checked and specified by the engineer only when required to be consistent with an applicable code section, such as the provisions in Chapter 12 of ACI 318 related to splices that do not meet Type 1 or 2 requirements.

The size and operation of field equipment, if any, required for the installation of mechanical splices can sometimes dictate a minimum spacing or a stagger pattern. Some information to evaluate these requirements is available in Chapter 3 of this report.

Some proprietary mechanical splices currently available have outside diameters that are substantially larger than that of the reinforcing bars. Special consideration should be given to the minimum concrete cover of the stirrups, ties, or spirals at these splice locations. In many cases, the stirrup or tie patterns adjacent to the mechanical splices, or the location of the splice itself, can be adjusted to avoid a reduced concrete cover. Codes do not mandate the placing of stirrups and ties over mechanical splices. Where designs involving close spacing of stirrups and ties necessitates their placement over couplers or coupling sleeves, however, it may be necessary to require greater cover over the splices and longitudinal reinforcement so that stirrups or ties encompassing the mechanical splices have adequate cover. Ideally, the dimensions of hoops or stirrups at the splice location should be adjusted to account for the larger diameter of the splice, but that is not generally the case in practice. All structural components of mechanical splices should meet the code-prescribed minimum concrete cover dimension. In the case of dowel bar mechanical splices with flanges, it should be noted that flanges are nonstructural in nature. Therefore, the designer need only consider cover over flanges with respect to durability.

1.3.2 *Matching of end alignments, end preparation of deformed bars, use of plain round bars, special bar deformations, and equipment*—The engineer/specifier should be aware of any special end preparation of bars required for a particular mechanical splice. For example, for end-bearing splices, Chapter 12 of ACI 318 requires that the ends of the bars be cut to within 1-1/2 degrees of square with respect to the longitudinal axis.

By definition, couplers are threaded, and some require matching threads on the bar ends. One mechanical splice requires reinforcing bars with special thread-like deformations. In most cases, for field assembly purposes, the mechanical splice should be concentrically aligned with the longitudinal bar axis. This is especially important for couplers that require reinforcing bar rotation, so that the bar is not likely to experience large oscillations when being rotated during field assembly. In the structure, linear alignment from bar to bar can generally be preserved across the splice by using bars with straight ends and securing the continuation bar in place at the time of installation. Splice tolerances for angular misalignment and misinsertion should be obtained from the splice manufacturer. In situations where the axes of the reinforcing bars to be spliced are not concentrically aligned, such as with curved bars, mispositioned dowel bars, and precast-to-precast connections, coupling sleeve systems tolerant to misalignment should be selected.

The reinforcing steel fabricator should be made aware of any special end preparation or threading requirement for the reinforcing bars by design drawings and specifications. Special requirements may entail the use of end-threading machines or special tools and equipment at the construction site. End threading is typically completed in the fabrication shop, although automated threading machines for field use are now becoming available.

Most reinforcing bars are deformed bars with lugs or protrusions that limit longitudinal movement or slip of the bar relative to the surrounding concrete; however, ACI allows limited use of plain reinforcing bars, and ASTM standard specifications exist which define material requirements for such, in sizes up to and including 2 in. (50.8 mm). Several types of mechanical splices rely on mechanical interlock with reinforcing bar deformations and are therefore not suitable for use with plain round bars. The mechanical splice manufacturer can advise accordingly.

The contractor and fabricator should coordinate to ensure that proper tooling and equipment exists at the site for installing selected mechanical splices on the specified reinforcing steel within placing tolerance. Related to this, a bar-end check is sometimes advisable before proceeding with certain mechanical splices. The mechanical splice manufacturer can advise accordingly.

1.3.3 *Coated reinforcing bars*—Reinforcing bars can be epoxy coated to ASTM A 775/A 775M or A 934/A 934M, or zinc coated (galvanized) to ASTM A 767/A 767M as a means of protection from corrosion. Mechanical splices of coated and uncoated reinforcing bars are made in similar ways. To properly install some types of mechanical splices on coated bars, the bar ends should be shielded at the time of coating, or the coating has to be completely removed over the length of the splice device and a short distance, approximately 2 in. (50 mm), beyond the ends of the splice device.

After installation of mechanical splices on coated reinforcing bars, the splice components and any damaged coating on the bars adjacent to the mechanical splices should be touched up with the appropriate compatible patching material. Project specifications should define the requirements for touch-up of the mechanical splices and repair of damaged coatings (refer to Chapter 5 of ACI 301). Flame-cutting of coated bars in any location should be avoided, when possible, because of damage to the coating. Where mechanical splices are used, flame-damaged epoxy coating may be more difficult to remove.

1.3.4 *Stainless steel reinforcing bars*—Stainless steel bars for concrete reinforcement, such as ASTM A 955/A 955M, can be mechanically spliced by stainless steel couplers that have chemical composition and corrosion resistance similar to that of the bars. Information on availability should be obtained from the supplier or splice manufacturer. If stainless steel bars are mechanically spliced to dissimilar steels such as carbon steel reinforcing bars, the potential for galvanic corrosion due to dissimilar metals should first be assessed.

1.3.5 *Field erection*—In many applications, mechanical splices may be staggered for clearance, access, and code requirements. If staggered mechanical splices are used in columns, for example, free-standing erection and assembly of the reinforcement may be required rather than preassembled cages, thus necessitating use of external bracing or formwork to hold bars while completing splices.

There is a considerable difference in the time and equipment required to install different mechanical splices. Therefore, the field erection procedure and schedule should be coordinated with the selection and installation procedure of the mechanical splices. If special equipment is required, particular information regarding its size, weight, operation, and availability should be obtained from the supplier or splice manufacturer.

For some projects, the engineer may find it appropriate to specify the types of mechanical splices to be used. Also, the method of construction may determine the types of mechanical splices that can be most readily used. Any unique requirements of a selected mechanical splice should be considered by all parties before beginning construction. Study of the subsequent descriptions will assist in determining these requirements. The supplier or manufacturer should be able to provide information related to unique requirements, if required.

1.3.6 *Inspection and quality assurance*—For all of the splices described in Sections 3.3, 3.4, and 3.5, some type of installation inspection or quality assurance is needed to ensure satisfactory performance. Quality system program statements, certificates of compliance, or certified material test reports should be requested from mechanical splice suppliers. Mechanical splice manufacturers can provide installation acceptance criteria for use by engineers and field inspectors. Depending on jurisdiction, special inspection may be required in accordance with a building code or design specification.

Destructive tensile testing of sister splices (also known as companion splices), or a proportion of production pieces, can be specified by the engineer to verify that a particular splice will not prematurely slip or otherwise fail, particularly in critical locations where Type 2 splices are used. The frequency and manner of tensile testing may form part of the project specifications or may be prescribed in building codes. When mechanical splices are used in safety-related structures, such as nuclear containments, it may be mandatory to perform destructive tensile tests on as many as 1-in-50 splices. In some jurisdictions, two or three test specimens may be prepared and tested at the start of the splicing phase. Batch testing (or ongoing testing) is performed by some agencies on 1-in-500 splices or in accordance with instructions issued by the project engineer. ACI 318 makes no recommendations on the testing frequency of installed splices on a project.

CHAPTER 2—DESIGN REQUIREMENTS FOR MECHANICAL SPLICES

2.1—Codes and specifications

Design requirements of the applicable code (for example, ACI 318, 349, 350 or 359) or specification (for example, AASHTO *Standard Specifications for Highway Bridges* or AASHTO *LRFD Bridge Design Specifications*) for reinforcing bar mechanical splices are not reproduced or discussed in detail in this report. Codes generally specify a minimum splice strength. For example, ACI 318-05, Chapter 12, requires that a (Type 1) full mechanical splice develop in tension or compression, as required, at least 125% of the specified yield strength f_y of the bar, so some yielding will occur in the reinforcing bar adjacent to the mechanical splice before failure in the splice itself. ACI 318-05, Chapter 21, requires that Type 2 mechanical splices develop the specified tensile strength of the spliced bar. For further background information and some of the considerations made in developing the ACI 318 mechanical splice provisions, the reader is encouraged to review the commentary provided in Chapters 12 and 21 of ACI 318-05.

ASTM A 1034/A 1034M describes the methods for testing mechanical splices, such as monotonic tension tests, monotonic compression tests, and cyclic load tests, but does not quantify the parameters for testing nor acceptance criteria.

Information regarding the claimed qualifications of mechanical splices is limited in this edition of the report to the Type 1 or Type 2 criteria of ACI 318, as summarized in Table 2.1. Supporting test results for a particular proprietary mechanical splice should be obtained directly from the splice manufacturer.

Table 2.1 includes a complete listing of tension-compression mechanical splices and shows the performance claimed by the manufacturers as Types 1 and 2 when used in conjunction with Grade 60 (420) deformed reinforcing bars. It should be noted that some manufacturers have more than one version of a specific mechanical splice. For example, there might be a version that is specifically Type 1, and another version that is capable of Type 2 performance. Other splice manufacturers may offer a single version that is rated as both Type 1 and Type 2. Type 2 splices inherently satisfy Type 1 splice performance requirements because the required tensile strength of a Type 2 splice is greater than that of a Type 1 splice.

For repair, renovation, and structural modification, the strength of the mechanical splice is generally dictated by the condition and attributes of the existing reinforcement. For example, if ASTM A 615/A 615M Grade 60 (420) bars are mechanically spliced to 80-year-old structural grade bars that have a characteristic yield strength of 33,000 psi (230 MPa) with an unknown specified tensile strength, it may be unrealistic to specify a particular mechanical splice without first testing the mechanical splice with the old bar. Only certain types of mechanical splices are suitable for these kinds of applications. It is not feasible, for example, to use threaded couplers because of the impracticality of forming

Table 2.1—Tension-compression mechanical splices: availability, applications, and suitability of Types 1 and 2 splices for Grade 60 (420) uncoated deformed reinforcing bars

	Bar designation no.		Applications ■ = primary; □ = secondary A = cast-in-place B = precast to precast C = new construction D = in-place renovation/repair				Suitability ■ = primary; □ = secondary E = shop or preinstallation F = field installation G = for deformed bars H = plain round bars			
Report section and mechanical splice generic name	Type 1 splice	Type 2 splice	A	B	C	D	E	F	G	H
3.3.1 Cold-swaged steel coupling sleeve	3 to 18 (10 to 57)		■	—	■	■	□	■	■	—
3.3.2 Cold-swaged coupler with taper-threaded ends	3 to 18 (10 to 57)		■	—	■	□	■	□	■	—
3.3.3 Coupler for thread-like deformed reinforcing bars	6 to 18 (19 to 57)		■	—	■	—	□	■	■*	—
3.3.4 Extruded steel coupler with parallel threaded ends	4 to 18 (13 to 57)		■	—	■	—	■	—	■	—
3.3.5 Friction-welded bar coupler with parallel threads	6 to 18 (19 to 57)		■	—	■	—	■	—	■	□
3.3.6 Friction-welded coupler with taper-threaded ends	5 to 18 (16 to 57)		■	—	■	—	■	—	■	□
3.3.7 Grout-filled coupling sleeve	3 to 18 (10 to 57)		□	■	■	□	■	□	■	—
3.3.8 Grout-filled coupling sleeve with parallel thread	4 to 18 (13 to 57)		□	■	■	□	■	□	■	—
3.3.9 Grout-filled coupling sleeve with taper thread	6 to 18 (19 to 57)		□	■	■	□	■	□	■	—
3.3.10 Shear screw and rail coupling sleeve	4 to 18 (13 to 57)	4 to 14 (13 to 43)	■	—	■	■	—	■	■	□
3.3.11 Shear screw and wedge coupling sleeve	4 to 18 (13 to 57)	3 to 14 (10 to 43)	■	—	■	■	—	■	■	□
3.3.12 Steel-filled coupling sleeve	4 to 18 (13 to 57)		■	—	■	■	—	■	■	—
3.3.13 Taper-threaded steel coupler	3 to 18 (10 to 57)		■	—	■	—	■	—	■	□
3.3.14 Threaded coupler with standard national coarse threads	4 to 11 (13 to 36)	4 to 8 (13 to 25)	■	—	■	—	■	—	■	□
3.3.15 Threaded coupler with upsized bar threads, cold-forged	4 to 18 (13 to 57)		■	—	■	—	■	—	■	□
3.3.16 Threaded coupler with upsized bar threads, hot-forged	4 to 11 (13 to 36)		■	—	■	—	■	—	■	□
3.3.17 Upset bar and coupling sleeve with straight threads	5 to 14 (16 to 43)		■	—	■	■	■	■	■	□
3.4.1 Cold-swaged steel flanged coupler with taper-threaded ends	3 to 18 (10 to 57)		■	—	■	—	■	—	■	—
3.4.2 Coupler with standard national coarse threads and flange	4 to 11 (13 to 36)	4 to 8 (13 to 25)	■	—	■	—	■	—	■	□
3.4.3 Coupler with taper threads and mounting plate	3 to 18 (10 to 57)		■	—	■	□	■	□	■	□
3.4.4 Integrally forged coupler with flange	4 to 11 (13 to 36)		■	—	■	—	■	—	■	□

*Coupler must be used with compatible thread-like deformed reinforcing bars.

Notes: **IMPORTANT**—Contact the manufacturer for capabilities in splicing ASTM A 615/A 615M Grade 75 (520) bar. Each splice type classification is based solely on a manufacturer's claimed performance with uncoated ASTM A 615/ASTM A 615M Grade 60 (420) or ASTM A 706/A 706M bar. Splice type classifications may be different for coated bars and other grades or specifications. **Neither ACI, nor the committee, tests or approves any manufacturer's product or process, and consequently do not warrant the splice type classifications.** Splice performance, availability, suitability, limitations, and exclusions should be verified directly with the mechanical splice manufacturer. All information is subject to change.

the required threads on the exposed ends of the reinforcing bars that extend from the concrete. For this reason, Table 2.1 includes some information pertaining to applications and suitability.

2.2—Design considerations

Due to the minimum mechanical splice strength requirement, it is generally assumed in design that the presence of a mechanical splice of two reinforcing bars does not result in a reduction of the anticipated structural strength, the stiffness, or the ductility of the reinforcing steel. Design codes cover strength requirements for welded and mechanical splices, but generally do not specify how to avoid potential weaknesses that may be directly attributed to the specific details or materials of a mechanical splice.

- In a flexural member, the mechanical splice should not result in a low effective longitudinal stiffness of the reinforcement that violates the strain conditions assumed in the member design; and

- In sections where inelastic strains are anticipated, such as in potential plastic hinge sections, the assembly of the mechanical splice and the reinforcing bars should possess adequate ductility so that premature failure of the spliced reinforcement is avoided. A Type 2 splice should be used in accordance with the provisions of ACI 318-05, Chapter 21. The use of welded splices, Type 1 mechanical splices, and lap splices in these sections is not permitted by Chapter 21 of ACI 318-05.

2.3—Drawings and specifications

Design drawings and specifications should clearly define splice type(s), locations, and force requirements (for example, lapped versus mechanical, tension splice versus tension-compression splice), as well as typical installation requirements. Design drawings and specifications should define typical splices, approved manufacturers (as applicable), and available options, as well as further information on materials, installation, and post-installation testing to demonstrate proper installation (as required). Placement of the reinforcing steel should consider constructibility and, where possible, the engineer should seek input from the contractor regarding the use, installation, and location of splices.

CHAPTER 3—MECHANICAL SPLICES AND INSTALLATION

3.1—General

A variety of proprietary mechanical splices are currently available. In this chapter, the relevant features, characteristics, and installation procedure of various available mechanical splices are generally described. No attempt has been made to include dimensional data because information varies from each manufacturer and may change.

Descriptions of the mechanical splices are presented in alphabetical order by generic name. The committee does not endorse or rate one particular mechanical splice over another. An attempt was made to include all mechanical splice types available and sold in the North American market at the beginning of 2004. The descriptions of the mechanical splices are based on information furnished by the manufacturers. When no information was submitted by a manufacturer, no generic description of their product(s) was included in this report.

3.2—Compression-only mechanical splices

When considering the use of a compression-only mechanical splice, the engineer, as a safeguard, should confirm that the splice will not be subjected to stress reversal. Temporary stress reversals might be brought about by a dynamic condition such as seismic activity, wind loading, or impact. When loading is less certain, tension-compression splices rather than compression-only splices are recommended.

In compression-only end-bearing mechanical splices of the type described in Chapter 12 of ACI 318-05, compressive stress is transferred by vertical concentric bearing from one bar to the other bar. Except for a steel-filled coupling sleeve, ACI 318 requires that the ends of the bars be sawcut, or cut by some other means, within 1-1/2 degrees of square to the bar longitudinal axis. ACI 318 also requires that an end-bearing splice device be capable of holding the bars in concentric bearing contact.

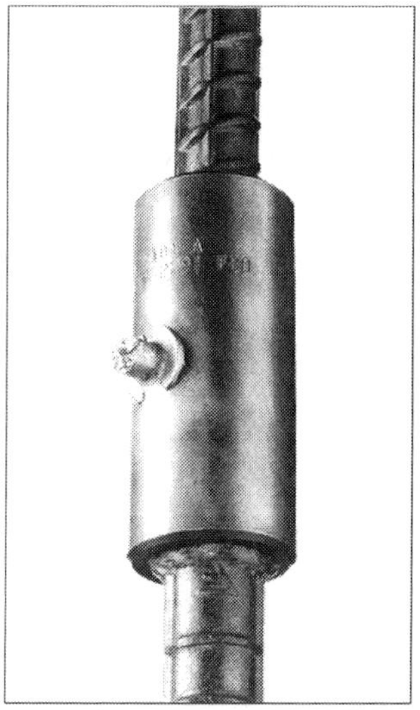

Fig. 3.1—Steel-filled coupling sleeve.

Two commercially available compression-only mechanical splice devices that remain on the market are described in terms of configuration, bar sizes that can be spliced, capability of splicing bars of different sizes, and installation procedure.

3.2.1 *Steel-filled coupling sleeve*—A steel-filled coupling sleeve, where the connection is made by introducing molten steel filler material within the interior of a steel tube or sleeve, is available for compression-only applications. By means of an exothermic process, molten filler solidifies around the deformations on the bar and internal grooves of the sleeve, creating a mechanical interlock between the bar, sleeve, and filler material (Fig. 3.1). The configuration of this coupling sleeve and its installation procedure are the same as the description of the steel-filled coupling sleeve that is given under tension-compression mechanical splices (Section 3.3.12), and will not be repeated herein.

Steel-filled coupling sleeves are available to splice bar sizes No. 11, 14, and 18 (No. 36, 43, and 57). Transition mechanical splices for different bar sizes can be made with special inserts, varying up to two sizes, such as No. 11 to 18 (No. 36 to 57). Bar ends do not require special end preparation; however, a bar-end check is recommended to prevent fitting problems in the field between the inside of the sleeve and the deformations of the larger bar. The ends can be shear cut or flame-cut because the molten steel material fills the space between the ends of the bars to enhance bearing. When used with epoxy- or zinc-coated bars, the bar ends should be shielded at the time of coating application, or the coating should be removed.

Although intended for compression-only applications, this coupling sleeve, by its nature, is capable of developing some tensile strength, but it is not intended to resist significant tensile loading. Bars can be connected in any orientation. A clear spacing of 1-1/2 nominal bar diameters is required to provide clearance for the connecting equipment. The difference

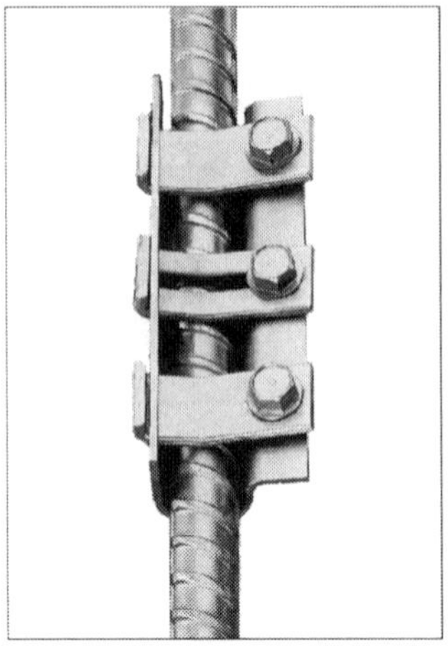

Fig. 3.2—Strap-type steel coupling sleeve.

between this steel-filled coupling sleeve for compression-only applications and the tension-compression coupling sleeve is in the length of the sleeve. For example, a 3 in. (76 mm) long coupling sleeve is used for connecting size No. 18 (No. 57) bars in compression only, whereas a 9 in. (229 mm) long coupling sleeve is used for the tension-compression mechanical splice.

3.2.2 *Strap-type steel coupling sleeve*—Strap-type steel coupling sleeves are the end-bearing sleeve type described in Chapter 12 of ACI 318-05, designed for compression-only applications. These types of sleeves are approximately a half-cylindrical shell, with a bent flange at one side and slots along the other side. L-shaped straps clip through the slots of the coupling sleeve and are bolted to the flange (Fig. 3.2).

Installation of the strap-type steel coupling sleeve consists of joining two sawcut sections of reinforcing steel at the midpoint of the sleeve. The straps are then tightened using a standard hex-shaped socket wrench. Strap-type steel coupling sleeves are available in North America to splice bar sizes No. 7 through 18 (No. 22 through 57). When used with epoxy- or zinc-coated (galvanized) bars, there is no need to shield the bar ends at the time of coating application or to remove the coating from the bar ends. Special adapter wedges are available for insertion into the coupling sleeve to allow for mechanically splicing bars of different sizes.

3.3—Tension-compression mechanical splices

Seventeen types of commercially available tension-compression mechanical splices are described in this section. Information made available to the committee is reported in terms of configuration, capability of connecting bars of different sizes, requirements for preparation of bar ends, requirements for equipment, tools, materials for installation, and installation procedure, as relevant.

All mechanical splice manufacturers claim Type 1 tension-compression capability with Grade 40 (280) and Grade 60 (420) bars in accordance with Chapter 12 of ACI 318-05. Table 2.1 summarizes the claimed conformance of the listed mechanical splices to the Type 1 or 2 criteria of ACI 318 with Grade 60 (420) bars. Certain mechanical splice manufacturers also claim to meet Type 1 tension-compression capability with Grade 75 (520) bars.

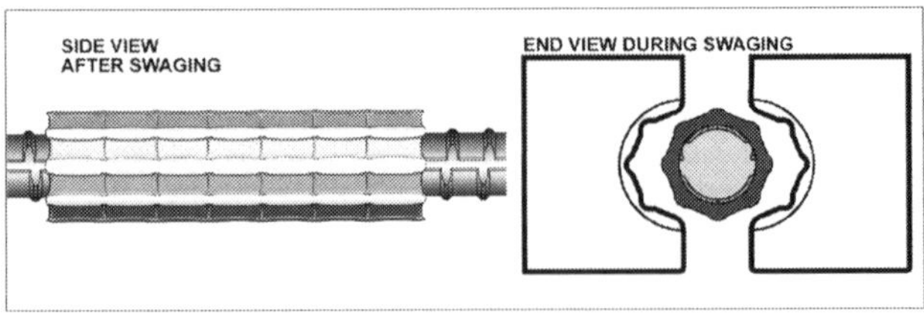

Fig. 3.3—Cold-swaged steel coupling sleeve.

3.3.1 *Cold-swaged steel coupling sleeve*—A cold-swaged steel coupling sleeve consists of a seamless steel sleeve that slips over the ends of two reinforcing bars and is deformed onto the reinforcing bar profile to produce mechanical interlock (Fig. 3.3).

Bar sizes No. 3 through 18 (No. 10 through 57) can be spliced by this method, including bars of different sizes. Coupling sleeves are available for use with reinforcing bar Grades 40 (280), 60 (420), and 75 (520). When used with epoxy-coated bars that comply with ASTM A 775/A 775M, it is not necessary to shield the bar ends at the time of coating application or to remove the coating from the bar ends. Hot-dipped galvanized coupling sleeves can be ordered for galvanized bars that comply with ASTM A 767/A 767M. The cold-swaged splicing method is suitable for new construction, field repair, and the splicing of older types of reinforcing bars, provided that such bars have deformations for mechanical interlock.

The first reinforcing bar is marked and inserted halfway into the sleeve. A hydraulic press fitted with a removable two-piece die set is used for field installation. The die set deforms the first half of the coupling sleeve in a radial direction onto the reinforcing bar in a series of overlapping pressings. The second bar is inserted to abut against the first, and the remaining unswaged portion of the sleeve is swaged onto the second bar. Field-type presses, including dies, weigh between 20 and 230 lb (9 and 105 kg). They can be supported for use in any orientation.

Larger bench-type presses with adjustable stops and insertion gauges are available for field or shop use. These machines, which weigh approximately 1600 lb (726 kg) or more, can be used to swage a coupling sleeve onto the end of a reinforcing bar before placing it onto the other bar. Adapter kits allow the aforementioned field presses to be used in the same way.

No special bar end preparation is required, so ends can be sheared, sawed, or flame-cut; however, a bar-end check is recommended. Bars can be connected from any orientation because special positioning of the press around the bar is not required. For very closely spaced bars, access of the equipment to install the mechanical splices should be checked with the manufacturer.

3.3.2 *Cold-swaged coupler with taper-threaded ends*—This type of coupler consists of two pieces, both of which are manufactured from seamless steel tubing or carbon steel bar. The internally threaded coupler has a precut taper thread

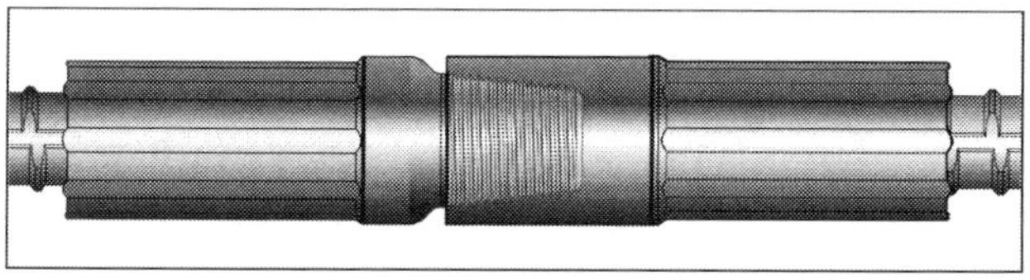

Fig. 3.4—Cold-swaged steel coupler with taper-threaded ends (partial section).

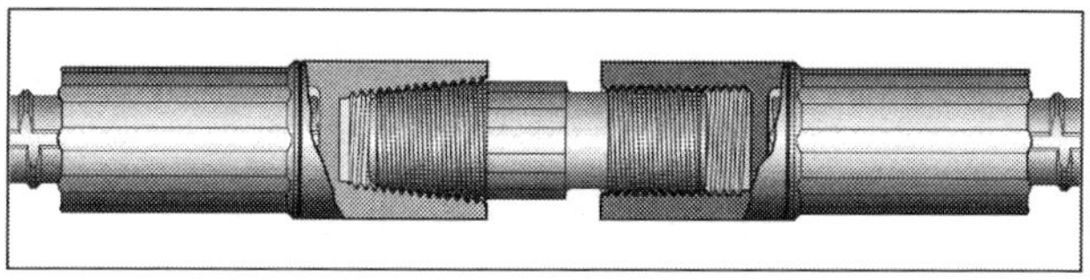

Fig. 3.5—Cold-swaged position coupler with threaded stud (partial section).

inside one end; a preformed externally threaded coupler has taper thread to match (Fig. 3.4).

The bar ends should be straight and can be sheared, sawed, or flame-cut; however, a bar-end check is recommended. Reinforcing bars are correctly positioned inside the couplers by means of internal stops. The couplers are deformed onto the bar profile to produce mechanical interlock by a swaging process using a bench press or a field-type press, as described in Section 3.3.1. The swaging can take place on the job site or at a fabricator's plant. A single pressing per coupler half is needed to install most bar sizes when using a bench press.

The tapered thread on the coupler helps to align the threads before rotating one of the two bars. The manufacturer requires that the threads be fully assembled. No special equipment or torque wrench is needed for this operation. A pipe wrench may be needed to overcome bar weight. Because threads are formed on the couplers and not on the reinforcing bar, the cross-sectional area of the bar is not reduced.

A variation of this is the position coupler, which consists of two internally threaded ends and an interconnecting externally threaded steel stud. This three-piece variation is generally used where neither reinforcing bar can be rotated to engage the threads (for example, where each bar is bent). The stud has a right-hand taper thread on one end and a right-hand parallel thread on the other. Rotating the stud connects the two internally threaded couplers together (Fig. 3.5). No torque wrench is required for this operation.

Bar Grades 40 (280), 60 (420), and 75 (520) in sizes No. 3 through 18 (No. 10 through 57), including transition sizes (Fig. 3.6), can be spliced by either the two-piece or three-piece type couplers. When used with epoxy-coated bars that comply to ASTM A 775/A 775M, it is not necessary to shield the bar ends at the time of coating application or to remove the coating from the bar ends. Hot-dipped galvanized splice components can be ordered for mechanically splicing galvanized bars that comply with ASTM A 767/A 767M.

3.3.3 *Coupler for thread-like deformed reinforcing bars*—A reinforcing bar conforming to ASTM A 615/A 615M, except for markings, with rolled-on deformations having a profile similar to the shape of a stub-acme type of thread is used. The reinforcing bars can be mechanically spliced using matching threaded couplers and nuts available from the reinforcing bar manufacturer. This mechanical splice is available for all grades up to Grade 75 (520) in bar sizes No. 6 through 18 (No. 19 through 57) and a special bar size No. 20 (No. 63), Grade 80 (550).

The bars are mechanically spliced together using one of two methods:

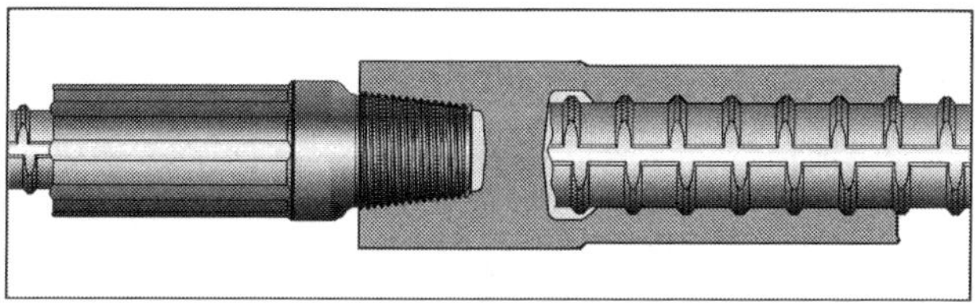

Fig. 3.6—Transition cold-swaged coupler with taper-threaded ends (section).

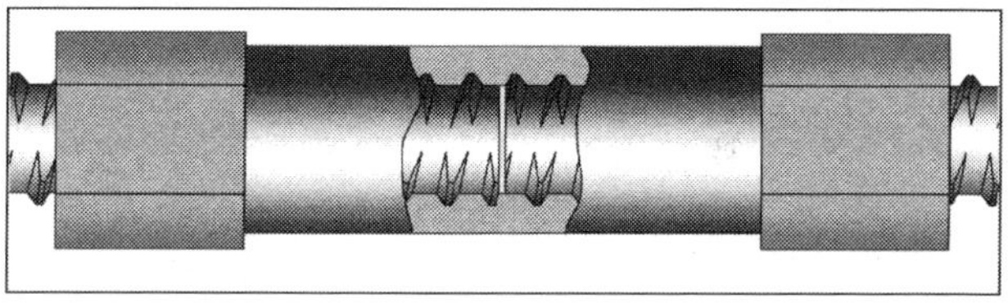

Fig. 3.7—Coupler for thread-like deformed reinforcing bars (partial section).

1. In installations where one of the two bars can be rotated, a coupler is engaged on the ends of the two opposing bars, and the two bars are tightened against each other. For mechanical splices working in compression, the bar ends should be perpendicular to within 1-1/2 degrees and may be either sawcut or abrasion wheel-cut;

2. In installations where neither of the bars to be connected can be rotated, a coupler is engaged on the end of the two opposing bars and a jam nut is tightened against each end of the coupler, as shown in Fig. 3.7. In this case, the ends will not have to be cut to the 1-1/2 degree perpendicularity tolerance.

In all mechanical splices described in Section 3.3.3, the assembly components are torqued together. Torque values range from 150 to 750 ft·lb (203 to 1016 N·m) for the No. 6 through 11 (No. 19 through 36) sizes, and are 1500 and 3000 ft·lb (2032 and 4065 N·m) for the No. 14 and 18 (No. 43 and 57) sizes, respectively. Hydraulic torque wrenches are available to achieve the larger torque values. Required torque values may be less, depending on specific project requirements.

No special cleanliness requirements for the bar ends apply. Different size bars can be spliced. Bars should have a clear spacing of 1-1/2 bar diameters to provide room for tightening. As required, the manufacturer can supply compatible bars, threaded couplers, and nuts that have been epoxy coated or galvanized. It is not necessary to shield the bar ends at the time of coating application or to remove the coating from the bar ends.

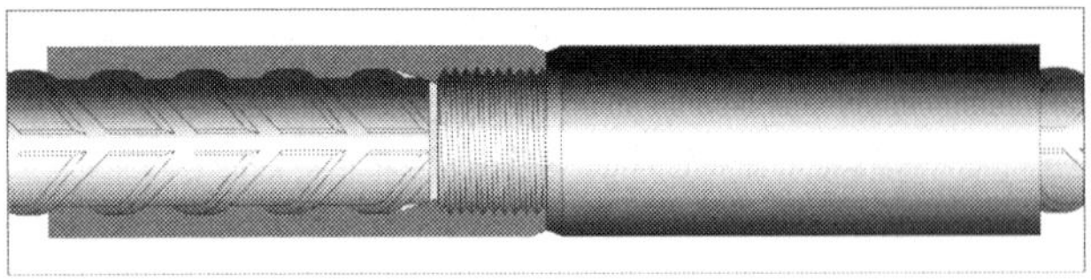

Fig. 3.8—Extruded steel coupler with parallel threaded ends (partial section).

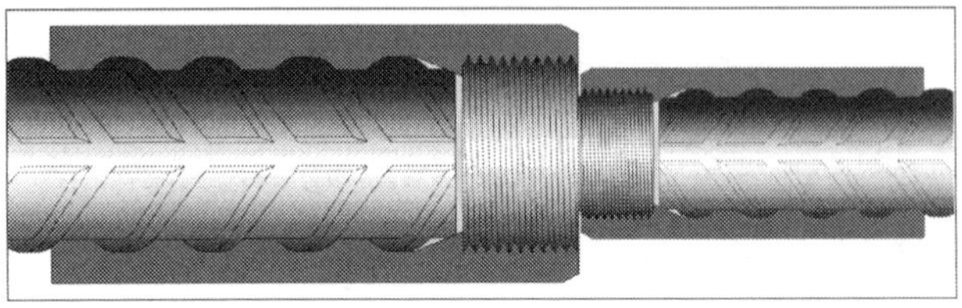

Fig. 3.9—Extruded steel transition coupler with parallel threaded ends (section).

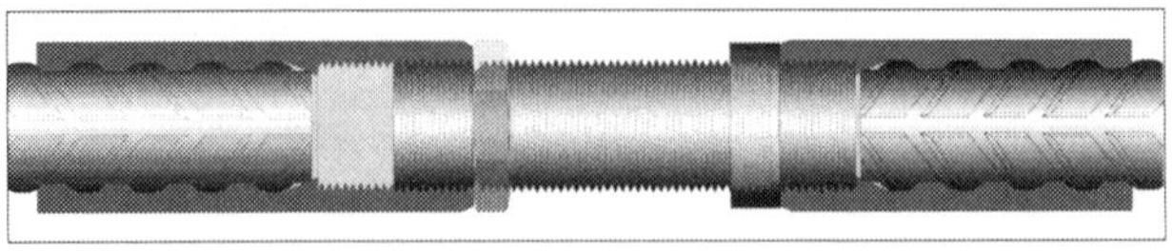

Fig. 3.10—Extruded steel positional coupler with parallel threaded ends (section).

3.3.4 *Extruded steel coupler with parallel threaded ends*—This type of splice consists of two pieces: an internally threaded sleeve on the first-stage bar, and an externally threaded sleeve on the second-stage bar. Both sleeves have a hollow portion on one end that receives the reinforcing bar, and a thread on the other end that is used to achieve the connection. Reinforcing bars with any deformation pattern can be spliced (Fig. 3.8).

Before installation, the bar ends should be individually sheared or sawcut square; a bar-end check is recommended. After the bar is inserted inside the coupler, the coupler is extruded onto the reinforcing bar using an automatic hydraulic press or a field-type press to produce mechanical interlock with the bar profile. The extrusion process can take place at a fabricator plant or on the job site. During the extrusion process, each connection is proof-tested by the application of a tensile force, and values are automatically recorded. Final assembly is done manually on site by rotating either bar. Chamfers on the coupler components help to align the threads. No torque wrench is required for this operation. Because threads are formed on the coupler components and not on the reinforcing bar, the cross-sectional area of the bar is not reduced.

Connection of reinforcing bars of different sizes is made by use of standard internally threaded couplers extruded on both bars; a transitional steel stud with matching external threads is used to complete the connection. Using standard internally threaded couplers on each side makes it unnecessary to plan for which bar size will be used for a transition. Studs can be produced with external threads for any combination of bar sizes No. 4 through 18 (No. 13 through 57) (Fig. 3.9).

When neither reinforcing bar can be rotated, a position splice is made by use of one standard internally threaded coupler and one longer internally threaded coupler in conjunction with an external thread positional steel stud that is pre-engaged into the longer internally threaded coupler. During the final stage of assembly at the job site, the position stud is unscrewed from the longer internally threaded coupler component into the standard internally threaded coupler. A locknut is provided if the second bar should be held in a specific position. Connection of reinforcing bars of different sizes can also be made by this method. A further variation uses a longer position stud to allow for a certain gap between the two bars to be spliced. Marks on the stud are provided for the installer to verify that thread engagement is sufficient on both ends of the stud (Fig. 3.10).

3.3.5 *Friction-welded bar coupler with parallel threads*—This type of splice uses two parallel threaded steel bars and an internally threaded coupler. The threaded steel bars have a stress area equal to or greater than the nominal area of the reinforcing bars being spliced, and are shop-attached to the reinforcing bar ends by an inertia friction-welding process at the factory of the mechanical splice manufacturer. In this process, a flywheel containing the threaded bar element is accelerated to a predetermined rotational speed. The threaded bar is brought into contact with a stationary restrained reinforcing bar while the drive motor is simultaneously disengaged. The components are forced together under pressure, and the energy that is stored in the flywheel is dissipated through friction in the faying surfaces as they rub together and heat is generated to create a weld interface. As the flywheel stops, an axial compressive force is maintained until the materials coalesce. Characteristically, some weld flash is produced as material is displaced from the faying surfaces. Because it is a solid-state process, friction welds are generally free from porosity and similar defects. When used with ASTM A 706/A 706M or A 615/ A 615M reinforcing bars, the strength of friction welds are comparable to that of the materials being welded.

The internally threaded coupler is fully screwed onto one of the threaded splice bars. In the field, the second threaded splice bar is butted to the end of the first. The coupler is unwound from the first threaded splice bar onto the second threaded splice bar. Some adjustment may be necessary at this stage to align the threads. An optional installation tool available from the mechanical splice manufacturer has been designed to rotate the internally threaded coupler. To complete the connection, one of the bars is rotated and snugly tightened. No torque wrench is required for this operation. End-to-end splicing of bar sizes No. 6 through 18 (No. 19 through 57), including transitions, can be accomplished by this method (Fig 3.11). Epoxy-coated bars can be spliced by shielding the bar ends at the time of epoxy application or removing the epoxy coating from the bar.

In a variation of this, where neither bar can be rotated, the system can be used as a position coupler. A locknut and the internally threaded coupler are fully screwed onto one of the threaded splice bars. A second locknut is prethreaded onto the second bar. The coupler is unscrewed from the first

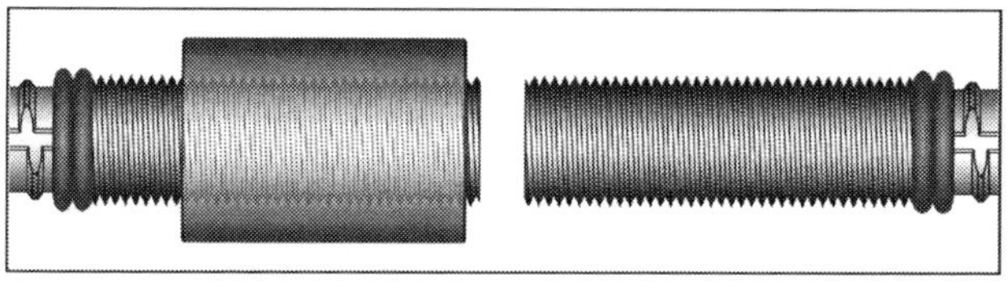

Fig. 3.11—Friction-welded bar coupler with parallel threaded ends.

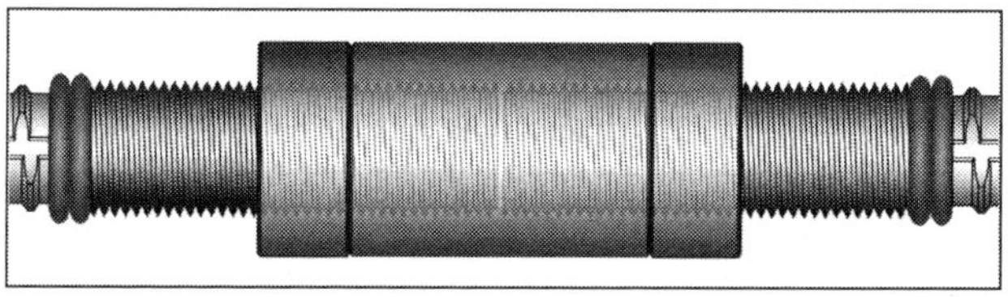

Fig. 3.12—Friction-welded bar position coupler with parallel threaded ends.

threaded splice bar onto the second threaded splice bar and centralized. The locknuts are then tightened against the coupler to a snug, tight condition. No torque wrench is required for this operation (Fig 3.12).

3.3.6 *Friction-welded coupler with taper-threaded ends*—This type of splice consists of matching external and internal taper-threaded ends and coupling sleeves designed to transfer both tensile and compressive stresses. The components are attached to reinforcing bar sizes No. 5 through 18 (No. 16 through 57) using a shop friction-welded fabrication process. Transitions to different reinforcing bar sizes are also possible. The coupler components have stress areas equal to or greater than the nominal area of the reinforcing bars being spliced. The strength of friction welds with ASTM A 706/A 706M bars has been evaluated by static and cyclic tests. Splice installation is done by positioning and turning the reinforcing bars (Fig. 3.13). The use of a heat-shrink sleeve is recommended by the splice manufacturer to properly seal the splice when the bars are epoxy coated.

A variation of this system consists of an interconnecting internally threaded sleeve that unwinds from a parallel-threaded end onto a taper-threaded end. This type of product is used where neither reinforcing bar can be rotated; for example, where each bar is bent (Fig. 3.14).

3.3.7 *Grout-filled coupling sleeve*—The grout-filled coupling sleeve is a frusto-conical* ductile iron casting with deformations similar to reinforcing bar patterns on the inner wall. A nonshrink, high-strength, proprietary grout fills the sleeve around the bars. The grout is either pumped or poured into the chambers with a low-pressure hand-grout pump. The bars do not require any special end preparation (Fig. 3.15).

This coupling sleeve may be epoxy coated to splice epoxy-coated bars, or hot-dipped galvanized for zinc-coated bars. The coating on the bars does not have to be removed for the bars to be connected.

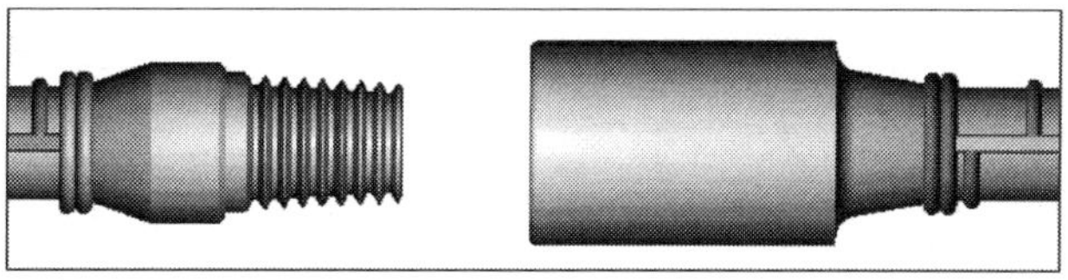

Fig. 3.13—Friction-welded coupler with taper-threaded ends.

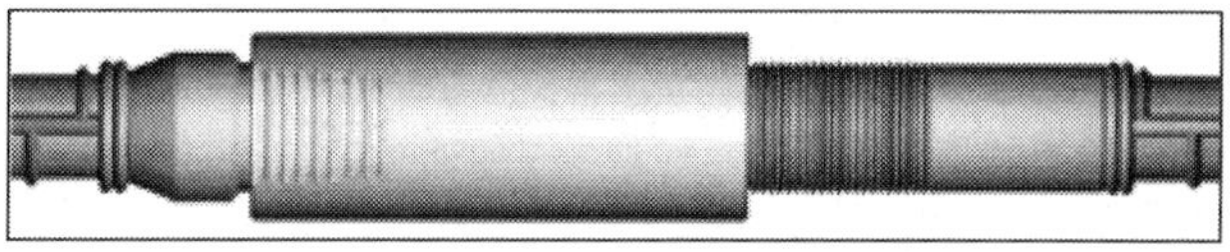

Fig. 3.14—Friction-welded bar ends with interconnecting threaded sleeve.

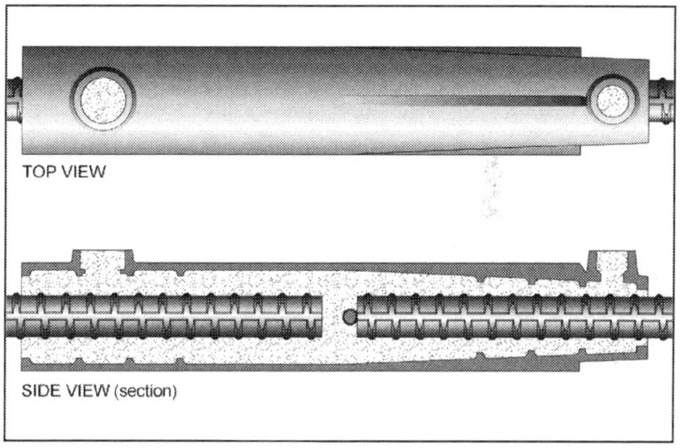

Fig. 3.15—Grout-filled coupling sleeve.

Installed splices are designed with sufficient development length of bars inside the sleeve to meet the strength requirements of ACI 318 when used with either ASTM A 706/A 706M or A 615/A 615M Grade 40 (280) and 60 (420) reinforcing bars. Grade 75 (520) bars may also be spliced. This type of mechanical splice can connect bars of varying sizes.

The grout-filled coupling sleeve is particularly suited to precast concrete construction. The sleeve is cast-in at the end of a precast member with the reinforcing bar inserted in the narrow end to the bar stop. The wide end opening is held against the forms during casting. Later in the field, it will be ready to receive a bar from the foundation or another precast member.

The coupling sleeve can be used above the construction joint, which is referred to as post-grout orientation because it is grouted after the precast members are erected and braced. It can also be used below the construction joint, which is referred to as pre-grout, because the sleeve must be filled with grout before the next precast member is erected. The former is more flexible for all weather conditions. The latter results in a connection that is more architecturally attractive because there are no visible grout ports.

The grouted coupling sleeve can also be used horizontally or at any angle in a closure pour by sliding the sleeve all the

*Frusto-conical—a geometric net shape, being that of a cone intersected by two parallel planes, usually perpendicular to the longitudinal axis of the cone.

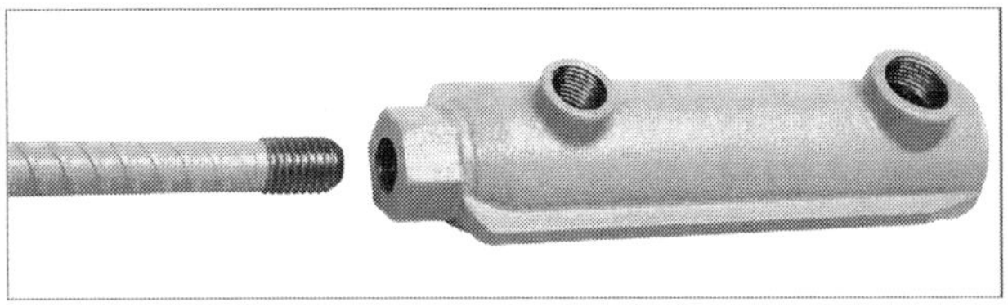

Fig. 3.16—Grout-filled coupling sleeve with parallel thread at upset bar ends.

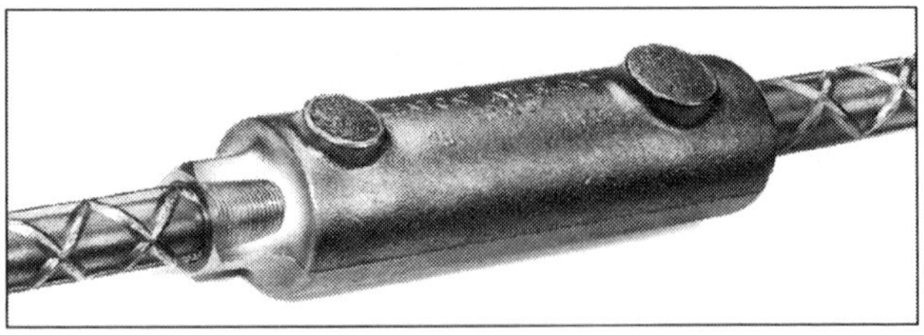

Fig. 3.17—Grout-filled coupling sleeve with taper thread.

way over onto one bar and then sliding it back over the other bar to be connected such that the ends meet in the center of the sleeve with a 1 to 2 in. (25 to 50 mm) gap, depending on bar size. The ends are sealed, and the sleeves filled through the grout ports with the hand-grout pump.

Special precautions should be taken to ensure that movement of the bars within the sleeves does not take place after grouting until the grout achieves a minimum compressive strength of 3000 to 5000 psi (21 to 34 MPa) within the initial 24-hour period. Braces, shores, and cables are typical supports used to stabilize the precast members and sleeves.

Precast members with vertical reinforcing bars may be assembled without any closure pours or formwork. This method of detailing is more fully discussed in ACI 550.1R.

3.3.8 *Grout-filled coupling sleeve with parallel thread at upset bar ends*—This mechanical splice system consists of a grout sleeve and a proprietary, metallic, nonshrink cementitious grout. The system is for use with size No. 4 through 18 (No. 13 through 57) uncoated Grade 60 (420) reinforcing bars conforming to ASTM A 615/A 615M, and is used to splice equal-size bars and different-sized bars end-to-end. Typical applications include precast and cast-in-place construction (Fig. 3.16).

The grout sleeves are produced as castings from ductile iron complying with ASTM A 536, with yield and ultimate tensile strengths of 55,000 and 80,000 psi (379 and 552 MPa), respectively. The grout sleeves are straight cylinders with one end threaded to receive a parallel threaded reinforcing bar. The bars are required to be processed in a shop having suitable automatic equipment to upset the bar end and form threads on the upset end. Because threads are formed on the upset end, the nominal specified diameter of the reinforcing bar is not compromised by the threading operations. Threaded reinforcing bar sizes No. 4 through 6 (No. 13 through 19) may have a machined integral shoulder or lock nut to facilitate torquing. Threaded reinforcing bar sizes No. 7 through 18 (No. 22 through 57) may have an optional locking nut instead of the integral shoulder. The grout sleeve is screwed onto the threaded bar using a wrench.

The opposite end of the grout sleeve has a cavity to receive a nonthreaded bar. The entrance to the cavity includes internal threads that are pitched approximately 0.286 in. (7.3 mm) on center for optional use with proprietary lifting bolts. The wide opening of the cavity allows for transitioning between bars of different sizes. The sides of the sleeve have inlet and outlet ports to receive grout. The grout is pumped into the inlet tube located at the sleeve bottom until it is visible through the outlet at the top, all entrapped air is removed, and all the spaces within the sleeve are full of grout. Adjoining bars are not permitted to be misaligned by more than 7 degrees. All spliced joints are required to be braced and supported to prevent movement until the grout gains sufficient strength; the braces are left in place for at least 24 hours or until the grout has attained a minimum compressive strength of 3000 psi (20.7 MPa). The required 28-day compressive strength of the grout is 8500 psi (59 MPa). The compressive strength of field-mixed grout can be determined by performing compressive tests on 2 in. (50 mm) cubes of the material cured in accordance with ASTM C 109/C 109M.

3.3.9 *Grout-filled coupling sleeve with taper thread*—Specifically designed for precast construction, the taper-threaded grout-filled coupling sleeve is designed to maintain structural integrity between load-bearing precast members such as columns, beams, and shearwalls.

This type of mechanical reinforcing steel splice is designed to join size No. 6 through 18 (No. 19 through 57) reinforcing bars conforming to ASTM A 615/A 615M or A 706/A 706M. Epoxy-coated bars can be spliced without the need for shielding at the time of coating application or removing the epoxy coating from the bar ends. The sleeve uses a tapered thread in conjunction with a proprietary grout. The sleeve is fastened to the first bar by means of a taper-threaded end. The assembly (bar and sleeve) is positioned within the formwork at the precaster's plant before placement of the concrete. The splice is completed at the job site, where the exposed dowel of the adjoining structural member is inserted within the sleeve and filled with the proprietary grout. Once hardened, the grout interlocks between the deformations of the reinforcing bar and internal grooves of the coupler (Fig. 3.17).

This type of system eliminates the use of anchor bolts or embedded weld plates in precast construction, and is not dependent on the condition of the surrounding concrete.

The coupling sleeves are produced as castings from ductile iron complying with ASTM A 536. Each sleeve is identified with the reinforcing bar size, part number, and manufacturer's name. Premixed and prepackaged grout is purchased from the splice manufacturer in 50 lb bags. Bags of grout are imprinted with the manufacturer's name and address, product designation, and water mixing instructions.

3.3.10 *Shear screw and rail coupling sleeve*—This type of device uses shear-head-type screws threaded into the side of seamless tubing and two serrated rails that are the same length as the coupling sleeve. The serrated rails cradle the reinforcing bar, and are embedded into the bar as the screws are

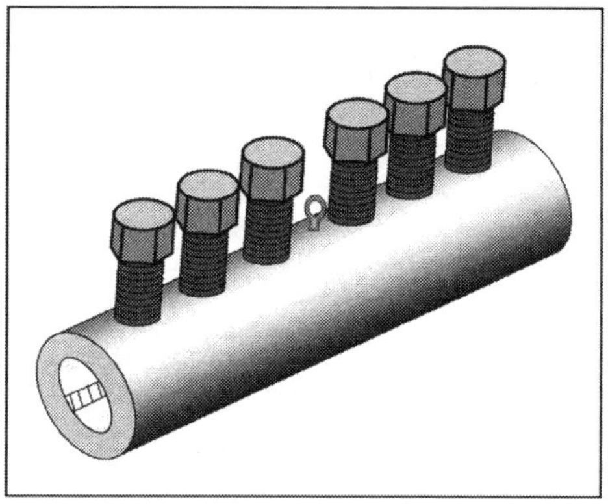

Fig. 3.18—Shear screw and rail coupling sleeve.

tightened. The heads of the screws are designed to shear off at a prescribed torque to accomplish proper installation (Fig. 3.18).

The reinforcing bar is inserted halfway into the coupler, and the coupler is held in place while the corresponding screws are tightened finger-tight. The other bar end is inserted into the coupler until the two bar ends meet and the corresponding screws are tightened finger-tight. The screws are then tightened using a wrench in an alternating pattern until the heads shear off. The system may be used with plain or deformed bars in sizes No. 4 through 18 (No. 13 through 57). Epoxy-coated bars can be spliced without the need for shielding at the time of coating application or removing the epoxy coating from the bar ends.

3.3.11 *Shear screw and wedge coupling sleeve*—This mechanical splice consists of a smooth, seamless shaped steel sleeve with two converging sides. A series of cone-pointed hex-head screws are arranged along the sleeve length (Fig. 3.19).

Reinforcing bars are inserted from each end to a center stop. No special bar-end preparation is required, so ends can be sheared, sawed, or flame-cut. As the screws are tightened, they indent into the surface of the bars, which in turn wedge into the converging sides of the sleeve. Screws can be tightened using suitable impact wrenches or hand-held ratchet wrenches. The heads of the screws shear off at a prescribed tightening torque.

When making splices between fixed points, a sleeve without a center stop can be slipped entirely onto one bar and subsequently repositioned over the two bar ends being spliced. Bar sizes No. 3 through 14 (No. 10 through 43) can be spliced as can bars of different sizes. Epoxy-coated bars that comply with ASTM A 775/A 775M can be spliced with epoxy-coated coupling sleeves without the need for shielding of the bar ends at the time of coating application or removing the epoxy coating from the bar ends. Galvanized bars can be spliced with galvanized coupling sleeves. This mechanical splice is suitable for new construction, field repair applications, and the splicing of older types of reinforcing bars.

3.3.12 *Steel-filled coupling sleeve*—The steel-filled coupling sleeve is a mechanical means of butt-splicing deformed reinforcing bars that works by introducing molten-steel filler material within the interior of a steel tube or sleeve. By means of an exothermic process, molten filler solidifies around the deformations on the bar and internal

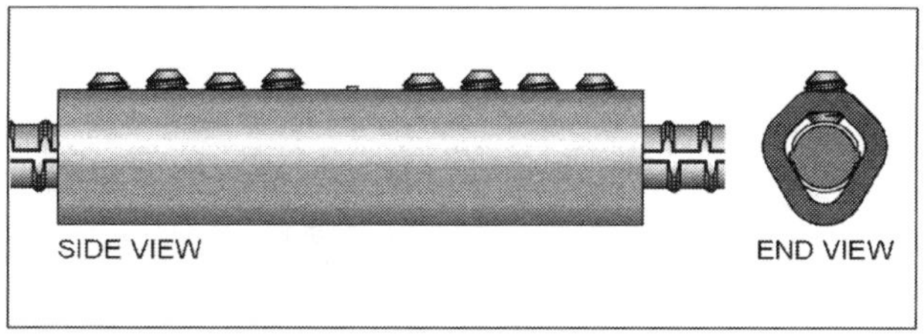

Fig. 3.19—Shear screw and wedge coupling sleeve.

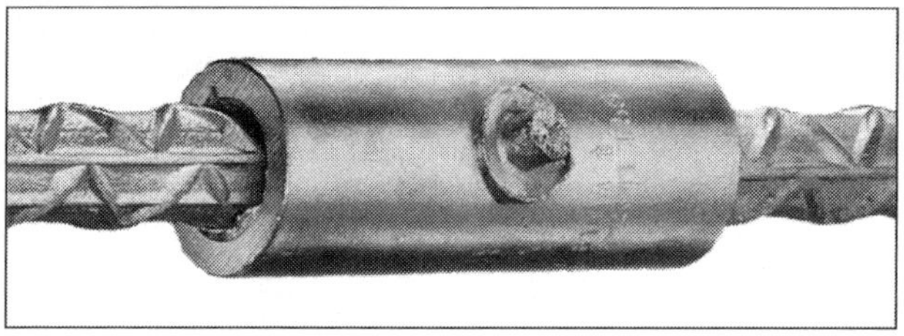

Fig. 3.20—Steel-filled coupling sleeve.

grooves of the sleeve, creating a mechanical interlock between the bar, sleeve, and filler material (Fig. 3.20).

No special bar end preparation is required, so ends can be sheared, sawed, or flame-cut. The steel-filled coupling sleeve is capable of splicing misaligned or circumferentially positioned bars due to the oversized inside diameter of the sleeve.

Field installation procedures consist of sliding the steel sleeve over one bar, inserting the second bar, and then centering the sleeve over the bar ends. Expendable graphite components are temporarily attached to the side of the sleeve. The premixed filler material bag is emptied within the graphite crucible. An exothermic reaction produces a molten metal that flows into the sleeve, filling the void inside the sleeve and surrounding the deformations of the reinforcing bars. Inspection of the completed splice consists of viewing both ends of the sleeve to ensure that filler material is visible.

Available in bar sizes No. 4 through 18 (No. 13 through 57), the system is designed to be used on all grades of ASTM A 615/A 615M reinforcing steel [Grades 40 (280), 60 (420), and 75 (520)], in addition to ASTM A 706/A 706M. To install this type of mechanical splice on epoxy- or zinc-coated bars, the bar ends should be shielded at the time of coating application, or the coating has to be completely removed. The steel-filled coupling sleeve is suited to new construction, field repair and renovation applications, and the splicing of older types of reinforcing bars, provided that such bars have deformations for mechanical interlock.

3.3.13 *Taper-threaded steel coupler*—Taper-threaded steel couplers have internal tapered threads. The use of tapered threads helps to reduce field damage or cross-threading before full thread engagement is achieved. The taper also helps to align and self-center the reinforcing steel. This system requires approximately four revolutions to assemble. Due to the proprietary nature of the taper-threaded steel coupler systems, the tapered threads and products are not interchangeable between manufacturers.

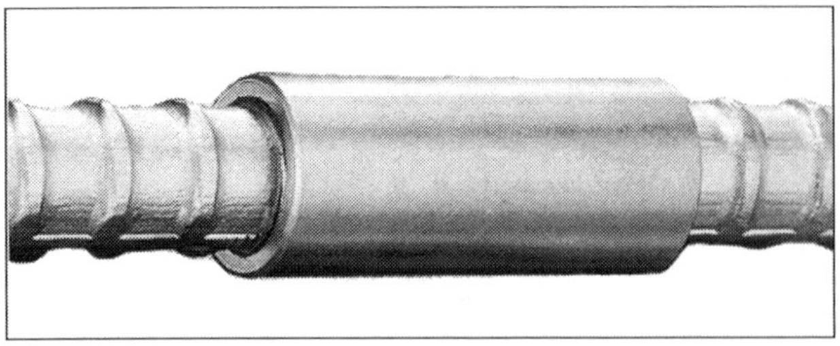

Fig. 3.21—Taper-threaded coupler assembled.

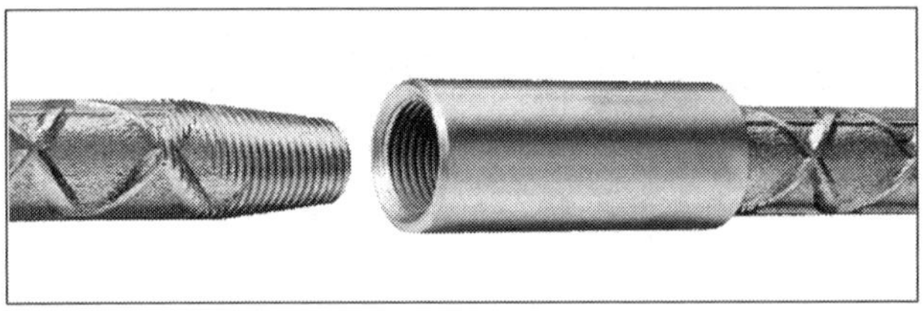

Fig. 3.22—Taper-threaded bar and coupler.

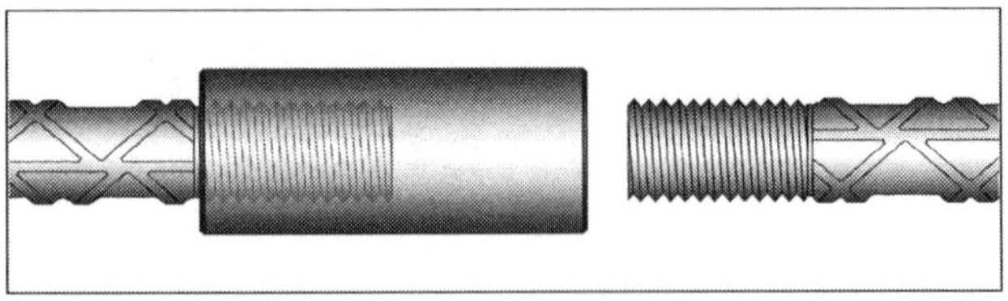

Fig. 3.23—Threaded coupler with standard national coarse threads.

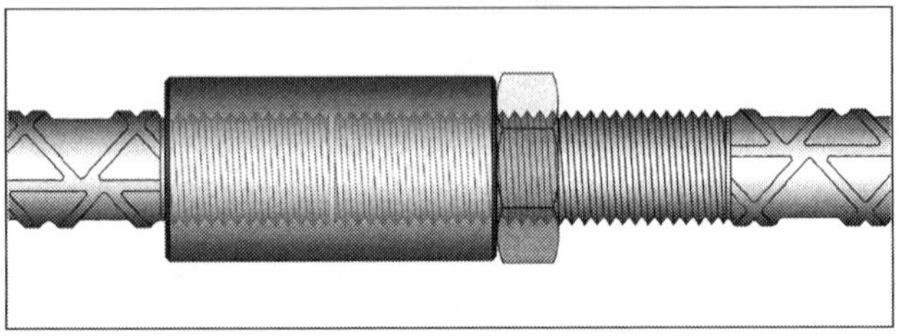

Fig. 3.24—Position coupler with standard national coarse threads.

Machining of the reinforcing bar ends is performed by reinforcing steel fabrication shops on automatic equipment supplied by the splice manufacturer. Before shipment to the job site, the threaded ends of the bars are protected by installing plastic thread protectors.

Field assembly of standard couplers, which are used where at least one bar is free to rotate (Fig. 3.21 and 3.22), consists of removing the thread protector from the end of the bar. The taper-threaded steel coupler is then fastened to the threaded end of the first bar until it is hand tight (in some cases, the coupler is installed on the bar at the fabrication shop). The second bar is inserted into the coupler and is rotated until hand tight. A pipe wrench is positioned on the second bar and rotated to complete the connection. Transition-style couplers, for splicing bars of different sizes, are assembled in the same manner as standard couplers.

Position couplers are used to join curved or bent bars that must be held in a predetermined position during the assembly process. The couplers can also be used for straight bars where neither bar is free to rotate. Position coupler installation is different from standard and transition couplers.

All styles of taper-threaded steel couplers are designed to join size No. 3 through 18 (No. 10 through 57) reinforcing bars conforming to ASTM A 615/A 615M or A 706/A 706M. Epoxy-coated couplers with taper threads are available for splicing epoxy-coated bars without the need for shielding of the bar ends at the time of coating application or removing the epoxy coating from the bar ends. Hot-dipped galvanized couplers can be produced for mechanically splicing galvanized bars. Stainless steel threaded couplers can be manufactured for splicing stainless steel reinforcing bars.

3.3.14 *Threaded coupler with standard national coarse threads*—These types of mechanical splices, which are normally used with bar sizes No. 4 through 11 (No. 13 through 36) ASTM A 615/A 615M Grade 60 (420), use couplers with internally machined threads and matching exterior threads cut or rolled directly onto the ends of the reinforcing bars. There is no requirement for preforging or for a bar end upsizing operation (Fig. 3.23).

Threading a bar end with a standard national coarse series of thread will reduce the nominal cross-sectional area of the bar by approximately 15 to 25% depending on the thread size, but the effect of the area reduction on splice strength may be offset by specifying a higher-grade material, by imposing strength controls over the mechanical properties of the specified grade, or by increasing the specified bar diameter.

Threading of the bar ends with standard national coarse threads can be performed by coupler suppliers or by reinforcing steel fabricators with suitable equipment. Precautions should be taken to protect the threaded ends of the bars during shipping and handling. A common installation practice is to connect the coupler to at least one bar end before reinforcing bar placement. The coupler may be manufactured with a thread stop at its center or the thread length on the reinforcing bars should be controlled to ensure proper thread engagement. The outside diameter of the coupler for Grade 60 (420) bars generally varies from 7/8 in. (22 mm) for the size No. 4 (No. 13) bar, to 2 in. (51 mm) for the size No. 11 (No. 36) bar. The use of a pipe wrench may be necessary to abut the bar end against the thread stop or otherwise secure the splice assembly. No torque wrench is required for this operation. A position splice is made by machining a longer thread on one of the bars to be connected. This method is used when neither reinforcing bar can be rotated. Assembly is completed in the field by unwinding a preassembled coupler from the long thread onto the adjoining threaded bar. A locknut may be used if one of the bars needs to be orientated in a certain direction (Fig. 3.24).

Epoxy-coated couplers with standard national coarse threads are available for splicing epoxy-coated bars that comply with ASTM A 775/A 775M, without the need for shielding at the time of coating application or removing the epoxy coating from the bar ends. Hot-dipped galvanized couplers can be produced for mechanically splicing galvanized

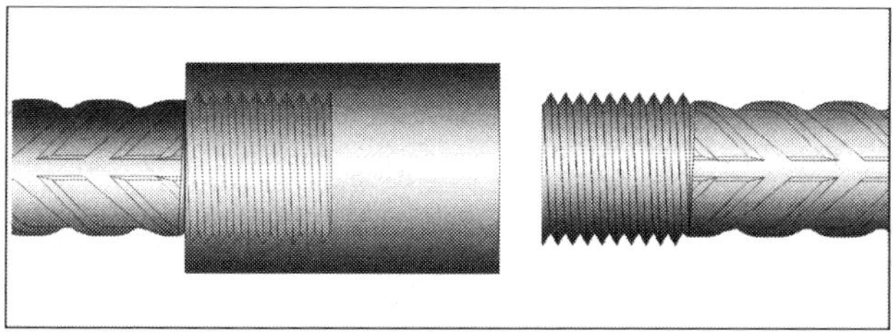

Fig. 3.25—Threaded coupler with upsized bar threads, cold-forged.

bars that comply with ASTM A 767/A 767M. Stainless steel threaded couplers can be manufactured for splicing stainless steel reinforcing bars.

3.3.15 *Threaded coupler with upsized bar threads, cold-forged*—This type of splice, for bar sizes No. 4 through 18 (No. 13 through 57), consists of two reinforcing bars with upsized parallel threaded ends that are connected by a coupler with internally machined threads (Fig. 3.25).

Each reinforcing bar with a sawcut end is inserted into a hydraulically actuated bench press that enlarges the bar end by a cold-forging operation. Thereafter, threads are formed on the upsized section of bar by use of a thread-cutting machine or a thread-rolling machine. By virtue of the upsizing method, the nominal cross-sectional area of the bar in the thread region is not reduced.

Bar end preparation can take place at a fabricator's plant or on the job site. A chamfer on the coupler helps locate the threaded ends for proper bar alignment. Assembly is completed by rotating either bar. No torque wrench is required for this operation. Connection of reinforcing bars of different sizes is made by use of a transition coupler with threads that match the corresponding threads on the reinforcing bars. Using standard threads on each bar makes it unnecessary to plan for which bar size will be used for a transition (Fig. 3.26).

When neither reinforcing bar can be rotated, a position coupler is used by machining a longer thread on one of the bars to be connected. The threads extend from the upsized section of bar into some of the reinforcing bar deformations. Assembly is completed by unwinding a preassembled coupler from the long thread onto the adjoining threaded bar. An optional locknut is provided if one of the bars must be held in a specific orientation (Fig. 3.27).

3.3.16 *Threaded coupler with upsized bar threads, hot-forged*—In this system, the mechanical splice is manufactured directly from two reinforcing bars. Each reinforcing bar is altered by a hot-forging operation that produces an integral internal or external threaded portion suitable for connecting the two bars. One reinforcing bar is hot-forged to provide a cavity that is subsequently machined by the manufacturer with internal threads. A second reinforcing bar is hot-forged to provide an upsized section of bar that is then machined by the manufacturer with external threads (Fig. 3.28).

The system uses standard thread sizes larger than the nominal diameter of the reinforcing bar. The thread size is such that the system is capable of achieving the minimum tensile strength of Grade 60 (420) reinforcing bars. For size No. 4 through 7 (No. 13 through 22) reinforcing bars, the threads are standard

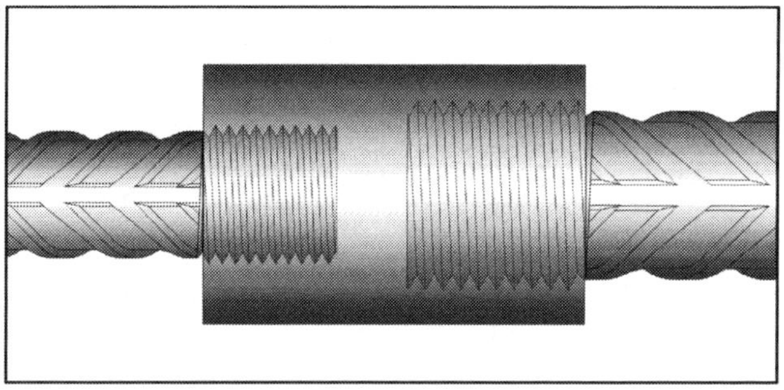

Fig. 3.26—Transition coupler with upsized bar threads, cold-forged.

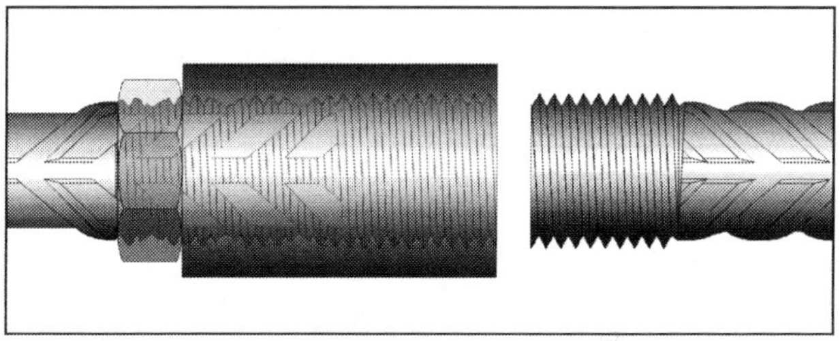

Fig. 3.27—Position coupler with upsized bar threads, cold-forged.

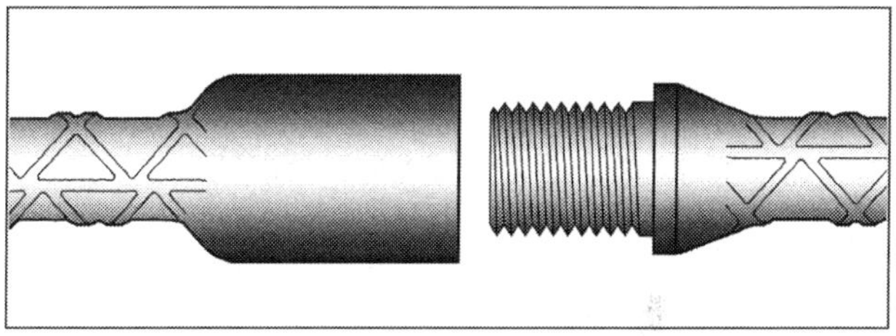

Fig. 3.28—Threaded coupler with upsized threads, hot-forged.

national coarse series, 1/8 in. (3 mm) larger than the nominal bar diameter. For size No. 8 through 11 (No. 25 through 36) bars, the threads have a pitch of eight threads per inch with thread diameters of 1-1/8, 1-1/4, 1-7/16, and 1-9/16 in., respectively.

The manufacturer can supply this type of splicing system epoxy-coated, if required. In the event a transition mechanical splice is required, the socket can be formed on a larger bar with appropriately matching internal threads to accommodate the external threads of the smaller bar end. No torque wrench is required during assembly.

A similar system to the above, for reinforcing bar sizes No. 7 through 14 (No. 22 through 43), comprises two reinforcing bars with hot-forged upsized sections of bar that are machined by the manufacturer with external threads and connected by a coupler with internally machined threads. The threads have a pitch of eight threads per inch with thread diameters as detailed previously for bar sizes No. 7 through 11 (No. 22 through 36), and with a thread diameter of 1-7/8 in. in the case of size No. 14 (No. 43).

3.3.17 *Upset bar and coupling sleeve with straight threads*—This mechanical splice consists of threaded

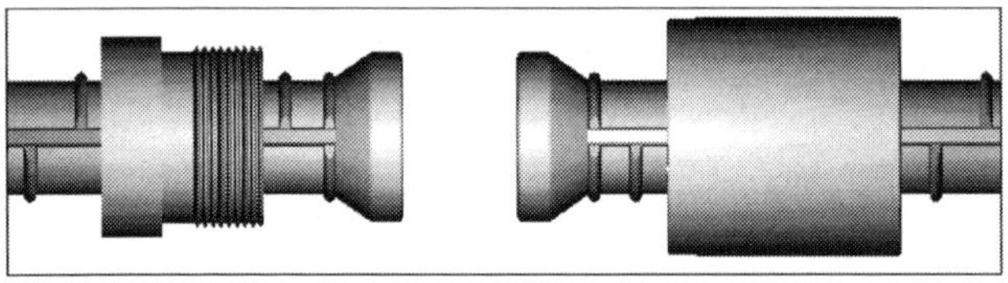

Fig. 3.29—Upset bar and coupling sleeve with straight threads.

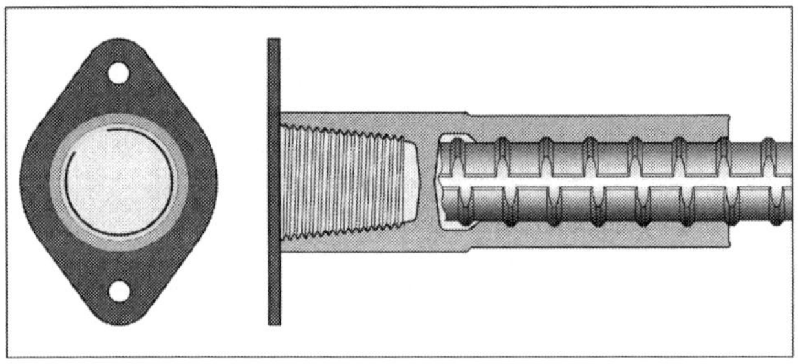

Fig. 3.30—Cold-swaged flanged coupler with taper-threaded ends.

components and two reinforcing bars with ends that have been upset by a portable hydraulic machine either on or off site. It is available zinc-plated from bar sizes No. 5 through 14 (No. 16 through 43); it is designed to be used in both retrofit and new construction. Transitions to different reinforcing bar sizes are also possible. There is no need to turn bars, which makes it suitable for bent and curved (hoop) bars. The installation procedure consists of placing threaded components over the reinforcement and heating the reinforcing bar ends to a temperature of 1650 °F (900 °C) to upset heads on reinforcing bars by hot-forging. Next, the installer hand-tightens the two threaded-component parts and locks the assembly with a torque wrench (Fig. 3.29).

For the splicing of epoxy-coated bars, the splice manufacturer recommends using a heat-shrink sleeve to seal the splice. It is necessary to either remove the coating on bar ends or to protect the operator from possible toxic fumes resulting from heating epoxy.

3.4—Dowel bar mechanical splices

Dowel bars are used at construction joints to transfer tension loads, compression loads, or both, across the joints between reinforcing bars on both sides of the joints.

Several manufacturers make dowel bar mechanical splices. All include flanged couplers for new construction that are designed to achieve the same tension-compression capability as the tension-compression splices described in the previous section.

Flanged couplers have holes that permit nailing of the flange to the sides or ends of the formwork. In each case, the coupler has internal threads that permit the attachment of a dowel splice bar or long length of reinforcing bar. These systems have the advantage of permitting continuity of reinforcement across construction joints without formwork penetration. Because flanged couplers are typically fully embedded in concrete, their use eliminates projecting bars, which can interfere with construction and can be a cause of accidents. Industry terms for these devices include dowel bar replacement and dowel bar substitute.

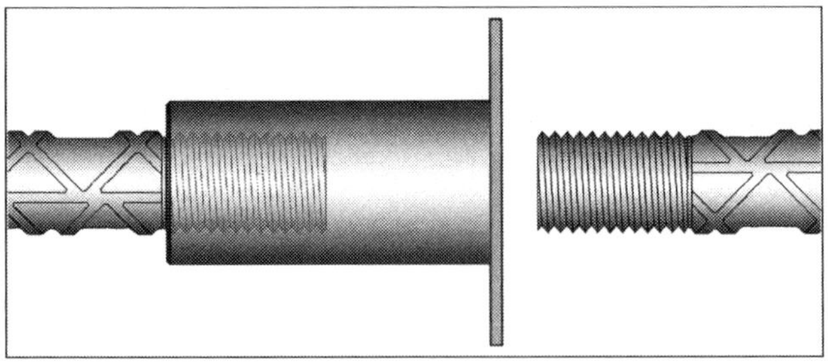

Fig. 3.31—Coupler with standard national coarse threads and flange.

Four different dowel-bar mechanical splice systems are described in this section. Although each is specifically designed for splicing reinforcing bars to dowel splice bars, nothing precludes their use as ordinary tension-compression mechanical splices subject to the general considerations in Section 1.3 and the design requirements in Chapter 2 of this report. Table 2.1 includes a listing of the four dowel bar mechanical splices and shows the performance claimed by the manufacturers when used in conjunction with Grade 60 (420) deformed reinforcing bars.

3.4.1 *Cold-swaged steel flanged coupler with taper-threaded ends*—The cold-swaged coupler (Fig. 3.30) is basically the same as the coupler with internal threaded ends described in Section 3.3.2. These couplers have flanges with holes for nailing to wooden forms, as do the other dowel bar splices. Steel studs with nuts can be used to attach these couplers to a steel bulkhead.

Epoxy-coated reinforcing bars that comply with ASTM A 775/A 775M can be spliced without the need for shielding at the time of coating application or removing the epoxy coating from the bar ends. Hot-dipped galvanized splice components can be provided for mechanically splicing galvanized bars that comply with ASTM A 767/A 767M.

3.4.2 *Coupler with standard national coarse threads and flange*—In this system, a flanged threaded coupler is used. This flanged coupler is a variation of the coupler described in Section 3.3.14. The coupler is fabricated from material conforming to ASTM A 108 and has an optional internal positive stop at midlength. The stop ensures that the reinforcing bar threaded from each end is installed to the proper depth. Alternatively, the thread lengths on the reinforcing bars are controlled to ensure proper thread engagement. A square flange plate is attached to one end of the coupler (Fig. 3.31).

Grade 60 (420) reinforcing bars conforming to ASTM A 615/A 615M are threaded with a standard national coarse thread of the same or similar size as the nominal bar diameter. Because threading of standard reinforcing bars is employed, the threaded bars may be fabricated locally or may be purchased from the coupler manufacturer. A common practice is to attach the coupler to one of the reinforcing bar ends before shipment.

Straight or bent reinforcing bars are cast in concrete with the flanged coupler attached to the form face. After the form-

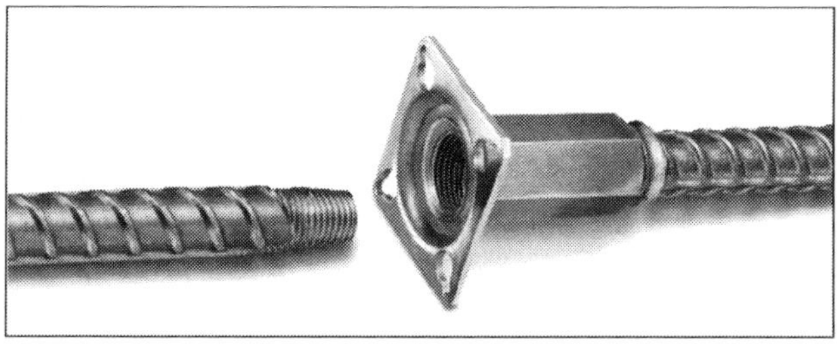

Fig. 3.32—Coupler with taper threads and mounting plate.

work is removed, straight or bent splice bars are installed into the coupler to extend the reinforcing bar. These couplers are available for size No. 4 through 11 (No. 13 through 36) bars.

Because the cross section at the threads is less than that of the reinforcing bar, the strength of the system is limited by the cross section at the threads or by the coupler itself. The effect of the area reduction on splice strength may be offset by specifying a higher-grade material, by imposing strength controls over the mechanical properties of the specified grade, or by increasing the specified bar diameter.

Epoxy-coated couplers with standard national coarse threads are available for splicing epoxy-coated bars that comply with ASTM A 775/A 775M, without the need for shielding at the time of coating application or removing the epoxy coating from the bar ends. Hot-dipped galvanized couplers can be produced for mechanically splicing galvanized bars that comply with ASTM A 767/A 767M. Stainless steel threaded couplers can be manufactured for splicing stainless steel reinforcing bars.

3.4.3 *Coupler with taper threads and mounting plate*—Taper-threaded dowel bar substitutes are a two-piece assembly designed to splice size No. 4 through 11 (No. 13 through 36) reinforcing bars where concrete formwork prevents the use of one continuous section of bar. Assemblies consist of a reinforcing bar factory-attached to a coupler, and a separate reinforcing bar with external taper threads that subsequently fastens into the coupler. Assemblies can be produced by either friction welding in bar sizes No. 4 through 7 (No. 13 through 22) (Fig. 3.32) or by taper threading both ends of a flanged-style coupler in bar sizes No. 4 through 11 (No. 13 through 36). Assemblies can be produced from bars conforming to ASTM A 615/A 615M or ASTM A 706/A 706M. Completed assemblies are shipped to the job site with factory-installed thread protectors on the coupler and externally threaded bar. Mounting plates are also factory-installed to the coupler. The mounting plates have keyed holes to assist in positioning the assemblies against the formwork.

Coated assemblies are available to meet either ASTM A 775/A 775M for epoxy or ASTM A 767/A 767M for zinc coating (galvanized). Taper-threaded stainless steel couplers can be manufactured for splicing stainless steel reinforcing bars.

3.4.4 *Integrally forged coupler with flange*—In this system, the mechanical splice is manufactured directly from two reinforcing bars. This is an extension of the integrally hot-forged threaded coupler system with upsized bar threads

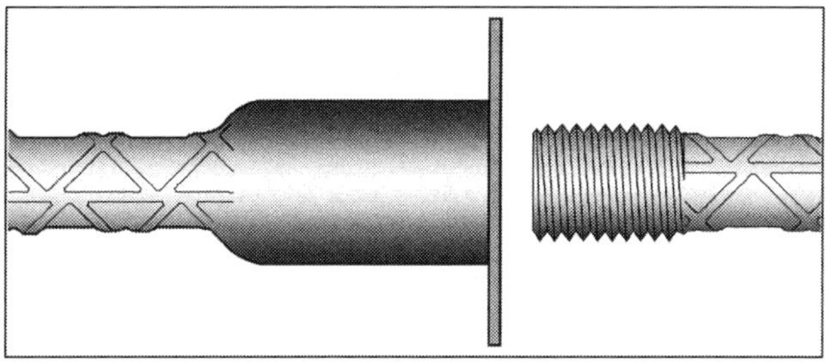

Fig. 3.33—Integrally forged coupler with flange.

described in Section 3.3.16. Each reinforcing bar is altered by a hot-forging operation that produces an internally or externally threaded portion suitable for connecting the two bars. One reinforcing bar is forged and machined to provide one end with an integrally flanged coupler with internal threads. Another reinforcing bar has a hot-forged upset end with external threads (Fig. 3.33).

The system uses standard thread sizes larger than the nominal diameter of the reinforcing bar so that splices are capable of achieving the minimum specified tensile strength of Grade 60 (420) reinforcing bars. For size No. 4 through 7 (No. 13 through 22) reinforcing bars, the threads are standard national coarse series, 1/8 in. (3 mm) larger than the nominal bar diameter. For size No. 8 through 11 (No. 25 through 36) bars, the threads have a pitch of eight threads per inch with thread diameters of 1-1/8, 1-1/4, 1-7/16, and 1-9/16 in., respectively. The internally threaded flanged coupler is hot-forged on a reinforcing bar of the same size as the hot-forged upset end with matching external threads. The internally threaded and externally threaded pieces can be manufactured with straight bars or with 90-degree hooked bars having the desired dimensions. A bar with a coupler at each end can transfer continuity of reinforcement across construction joints at two opposite faces of a member. In the event a transition mechanical splice is required, the internally threaded piece can be formed on a larger bar with appropriately matching threads to accommodate the smaller, externally threaded step-down bar. The manufacturer can supply this type of splicing system epoxy-coated, if required.

3.5—Mechanical lap splices

Two mechanical lap splices with a claimed Type 1 strength capability, as defined by ACI 318-05, Chapter 12, are described in this section. These types of mechanical splices are normally used in situations where the reinforcement is required only for tension, such as column ties, stirrup ties, hoops, spirals, flexural reinforcement in bridge decks and slabs-on-grade, or shrinkage-temperature reinforcement; however, they are also capable of transferring compressive loads.

Compared with lap splices, which rely on the surrounding concrete to transfer force between bars, mechanical lap splices are much shorter because coupling sleeves and components are used to positively connect reinforcing bars in an overlapping configuration. This makes them suited for repair and renovation applications because significantly less concrete must be removed to splice bars in concrete.

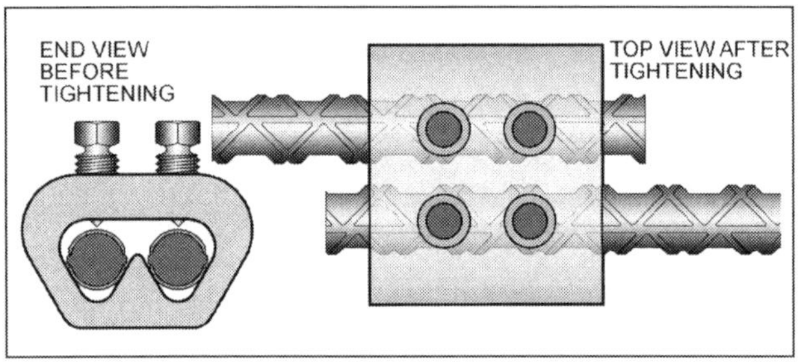

Fig. 3.34—Shear screw and double-wedge coupling sleeve.

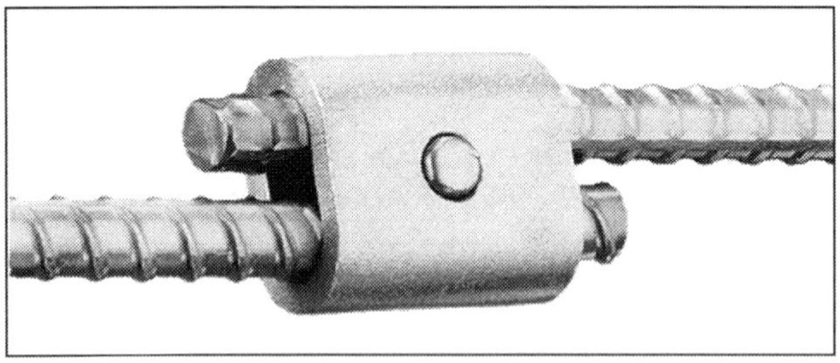

Fig. 3.35—Steel coupling sleeve with wedge.

3.5.1 *Shear screw and double-wedge coupling sleeve*—This mechanical lap splice consists of a ductile iron sleeve with two converging wedges. Two series of cone-pointed screws are arranged along the sleeve length opposite of the wedges (Fig. 3.34).

Each reinforcing bar extends out of the sleeve by approximately one bar diameter. No special bar end preparation is required. As the screws are tightened, they indent into the surface of the bars, which in turn wedge into the converging sides of the sleeve. Screws can be tightened using suitable impact wrenches or hand-held ratchet wrenches. The heads of the screws shear off at a prescribed tightening torque. The system is capable of meeting the requirements of ACI 318-05, Chapter 12, in both tension and compression.

Bar sizes No. 3 through 7 (No. 10 through 22) plus certain bars of different sizes—either uncoated or epoxy-coated—can be spliced using this coupling sleeve. There is no need to shield the bar ends at the time of coating application or remove the coating on the bars ends when using this type of splice device with ASTM A 775/A 775M reinforcing bars. Hot-dipped galvanized sleeves can be produced for mechanical lap splicing of galvanized bars that comply with ASTM A 767/A 767M. The device is suited to new construction, field repair, and renovation applications.

3.5.2 *Steel coupling sleeve with wedge*—This mechanical lap splice consists of a steel sleeve with a hydraulically driven wedge, which is designed for reinforcing bars in sizes No. 4 through 6 (No. 13 through 19). The sleeve is oval-shaped in cross section, permitting the overlapping of two reinforcing bars of the same diameter (Fig. 3.35). After the coupling sleeve is correctly positioned, a hardened wedge-shaped pin is driven with a portable hand-held hydraulic tool purchased from the manufacturer. The wedge is installed between the reinforcing bars and extends through the hole on the opposite side of the coupler. The tapered wedge pin forces the deformation of the bars into the interior side walls of the coupler. The system is capable of meeting the requirements of ACI 318-05, Chapter 12, in both tension and compression. There is no need to shield epoxy-coated bar ends at the time of coating application or remove the coating when using this type of splice device with ASTM A 775/A 775M reinforcing bars. Hot-dipped galvanized coupling sleeves can be produced for mechanical lap splicing of galvanized bars.

A 10,000 psi (69 MPa) automatic unloading pump can be used to drive the hydraulic hand-tool, which weighs less than 30 lb (13.6 kg).

The steel coupling sleeve with wedge is suited to new construction, field repair, and renovation applications.

CHAPTER 4—SUMMARY

Updated information about mechanical splices has been presented together with information on different types of mechanical splice devices generally available and sold in the North American market at the beginning of 2004. The structural engineer should also consider these points:

- It may be necessary to determine whether particular mechanical splices are acceptable to local code or building officials. The mechanical splice manufacturer should normally be the principal party to obtain the necessary recognition or approval of the splicing products it manufactures;
- The engineer should approve all splices on the project that are appropriate for certain conditions of inelastic behavior or repeated reverse loading, which should be considered in seismic design;
- Where no special requirements or job conditions exist that either favor or preclude the use of a particular mechanical splice, as in certain Type 1 splice applications, project specifications should be left open to ensure competition. Otherwise, the engineer may suggest a splice based on design and offer options (as possible) where experience dictates such;
- Availability of required materials and equipment and service capabilities of the suppliers should be reviewed before specifying any particular mechanical splice;
- Installation instructions and performance data should be secured directly from the manufacturers of the mechanical splices. Manufacturer's requirements should be referenced in design drawings and specifications to provide greater assurance that the devices will be installed properly;
- Inspection or quality assurance is needed to ensure satisfactory performance. Splice manufacturers can provide installation acceptance criteria for field inspection; and
- Testing can be performed to verify that a particular splice type will meet specified performance requirements, particularly in critical locations where Type 2 splices are used.

CHAPTER 5—REFERENCES

5.1—Referenced standards and reports

The documents of the various standards-producing organizations referred to in this report are listed below with their serial designations and the latest editions at the time this report was prepared. Because these documents are revised frequently, the reader is advised to contact the proper sponsoring group if it is desired to refer to the latest version.

American Association of State Highway & Transportation Officials (AASHTO)

Standard Specifications for Highway Bridges

LRFD Bridge Design Specifications

American Concrete Institute (ACI)

301	Specifications for Structural Concrete
318	Building Code Requirements for Structural Concrete
349	Code Requirements for Nuclear Safety Related Concrete Structures
350	Code Requirements for Environmental Engineering Concrete Structures
359	Code for Concrete Containments (published as Section III, Division 2, of the ASME Boiler and Pressure Vessel Code)
550.1R	Emulating Cast-in-Place Detailing in Precast Concrete Structures

ASTM International

A 108	Specification for Steel Bar, Carbon and Alloy, Cold-Finished
A 536	Specification for Ductile Iron Castings
A 615/A 615M	Specification for Deformed and Plain Carbon-Steel Bars for Concrete Reinforcement
A 706/A 706M	Specification for Low-Alloy Steel Deformed and Plain Bars for Concrete Reinforcement
A 767/A 767M	Specification for Zinc-Coated (Galvanized) Steel Bars for Concrete Reinforcement
A 775/A 775M	Specification for Epoxy-Coated Steel Reinforcing Bars
A 934/A 934M	Specification for Epoxy-Coated Prefabricated Steel Reinforcing Bars
A 955/A 955M	Specification for Deformed and Plain Stainless-Steel Bars for Concrete Reinforcement
A 996/A 996M	Specification for Rail-Steel and Axle-Steel Deformed Bars for Concrete Reinforcement
A 1034/A 1034M	Standard Test Methods for Testing Mechanical Splices for Steel Reinforcing Bars
C 109/C 109M	Test Method for Compressive Strength of Hydraulic Cement Mortars (Using 2-in. or [50-mm] Cube Specimens)

American Welding Society (AWS)

ANSI/AWS D1.4	Structural Welding Code—Reinforcing Steel

These publications may be obtained from the following organizations:

American Association of State Highway & Transportation Officials
444 North Capitol Street NW
Suite 249
Washington, DC 20001
www.aashto.org

American Concrete Institute
38800 Country Club Drive
Farmington Hills, MI 48331
www.concrete.org

American Society of Civil Engineers
Three Park Avenue
New York, NY 10016-5990
www.asme.org

ASTM International
100 Barr Harbor Drive
West Conshohocken, PA 19428
www.astm.org

American Welding Society
550 NW LeJeune Road
Miami, FL 33126
www.aws.org

5.2—Cited references

ACI Committee 318, 2005, "Building Code Requirements for Structural Concrete (ACI 318-05) and Commentary (318R-05)," American Concrete Institute, Farmington Hills, Mich., 430 pp.

CRSI, *Manual of Standard Practice (MSP-2-01),* Concrete Reinforcing Steel Institute, Schaumburg, Ill.

ACKNOWLEDGMENTS

The following manufacturers are to be thanked for permitting the committee to exhibit a picture or graphic of their product in this report.

Barsplice Products, Inc., Dayton, Ohio: www.barsplice.com

Dayton/Richmond Concrete Accessories, Miamisburg, Ohio: www.daytonrichmond.com

Dextra Manufacturing Co., Bangkok, Thailand: www.dextragroup.com

Dywidag Systems International, U.S.A. Inc., Bolingbrook, Ill.: www.dywidag-systems.com

Erico, Inc., Solon, Ohio: www.erico.com

Headed Reinforcement Corp., Fountain Valley, Calif.: www.hrc-usa.com

Splice Sleeve North America, Inc., Irvine, Calif.: www.splicesleeve.com

NOTICES AND DISCLAIMERS

While generic descriptions of known types of mechanical splices are presented, it is beyond the scope of this report to provide a comprehensive directory of mechanical splice manufacturers or a directory of those reinforcing steel fabricators and suppliers who buy and sell mechanical splice devices. References in the Acknowledgments Section to trade associations and organizations that provide manufacturer names are provided for reader convenience and information only. Those who wish to contribute information, pictures, or graphics for consideration in any future edition of this report are encouraged to do so by active and voluntary participation in committee meetings, by correspondence with the committee Chair, or both. Neither ACI, nor the committee itself, endorses, tests, approves, or warrants any manufacturer, product, or process described in this report.

Report on Steel Reinforcement—Material Properties and U.S. Availability

Reported by ACI Committee 439

Mark. D. Marvin
Chair

William C. Gallenz
Secretary

Joseph A. Bohinsky	Paul S. Fredrickson	Peter Meza	Mario A. Rodriguez
Domingo J. Carreira*	Todd R. Hawkinson	Theodore A. Mize*	Philip E. Ross
Augusta Carroll	Steven H. Holdsworth	Conrad Paulson	Clifford A. Sabo
Louis J. Colarusso	Allen J. Hulshizer	Ryan W. Pelter	Robert G. Smith
David H. DeValve	Harry B. Lancelot, III	Richard A. Ramsey	Dyke W. Starnes
Gustav G. Erlemann†	Kenneth A. Luttrell	Roy H. Reiterman	Kenneth W. Williamson
Salem S. Faza	LeRoy A. Lutz*	Robert C. Richardson	William H. Zehrt, Jr.
Anthony L. Felder*	A. Murray Lount†		

*Subcommittee members who produced this report.
†Deceased.

The material properties of the various types of steel reinforcement produced for use in the United States are described. Types of steel reinforcement defined in this report include deformed reinforcing bars, plain and deformed wire, welded wire reinforcement, bar mats, and prestressing reinforcement. The requirements, restrictions, and testing information of the pertinent ASTM specifications are reviewed. The availability of the various types and sizes of reinforcement in the U.S. is discussed.

Keywords: bend tests; bending (reinforcing steels); deformed reinforcement; deformed reinforcing bars; ductility; material properties; prestressing steels; reinforced concrete; reinforcing steels; specifications; spiral reinforcement; tensile strength; welded reinforcement grids; welded wire fabric; welded wire reinforcement; yield strength.

CONTENTS

ACI 439.4R-09 supersedes ACI 439.4R-89 and was adopted and published October 2009.

CHAPTER 1—INTRODUCTION AND SCOPE

1.1—Introduction

The material properties of the various types of steel reinforcement produced for use in the U.S. are described. Types of steel reinforcement defined in this report include deformed reinforcing bars, plain and deformed wire, welded wire reinforcement, bar mats, and prestressing reinforcement. The requirements, restrictions, and testing information of pertinent ASTM specifications are reviewed. The availability of the various types and sizes of reinforcement in the U.S. is discussed.

1.2—Scope

Requirements for reinforcement are stated in Section 3.5.1 of ACI 318-08 as: "Reinforcement shall be deformed reinforcement, except that plain reinforcement shall be permitted for spirals or prestressing steel..." Deformed reinforcement is defined in Section 2.2 of ACI 318-08 as: "Deformed reinforcing bars, bar mats, deformed wire, welded wire reinforcement conforming to Section 3.5.3." This report excludes discussion of structural steel, steel pipe, steel tubing, steel fibers, expanded metal, and fiber-reinforced polymers (FRP) and other nonmetallic reinforcement.

This report describes these types of steel reinforcement in terms of availability and material properties, expressed in inch-pound units and metric units. The special requirements of ACI 349 and 359 are also discussed. American Association of State Highway and Transportation Officials (AASHTO) requirements (2002, 2004) are not discussed in this report because they generally follow ACI 318.

CHAPTER 2—NOTATION AND DEFINITIONS

2.1—Notation

There is no notation used in this document.

2.2—Definitions

compacted strand—prestressing strand that is drawn through a circular die deforming the wires to produce a strand with a smaller circular shape.

indented strand—prestressing strand in which the outer wires have small indentations to permit more rapid transfer of strand force to the concrete.

See ACI CT, 'Concrete Terminology,' for definitions at the ACI Web site, http://www.concrete.org/Technical/CCT/FlashHelp/ACI_Terminology.htm.

CHAPTER 3—REINFORCING BARS

3.1—Introduction

Reinforcing bars are the most popular type of nonprestressed reinforcing steel used in concrete construction. Approximately 10 million tons of reinforcing bars are used in the U.S. every year. Nearly all reinforced concrete construction, including precast and prestressed structures, requires some reinforcing bars.

Most of the reinforcing bars used in construction are deformed bars; that is, the bars have deformations to enhance the bond of the steel to the surrounding concrete. Limits on the deformation dimensions, such as the minimum deformation height and maximum deformation spacing, are prescribed in ASTM specifications. Plain reinforcing bars, which do not have deformations, are only permitted by ACI 318 for spiral reinforcement as used in reinforced concrete columns.

3.2—Material properties

The mechanical property requirements, chemical composition restrictions, and other requirements of the ASTM specifications for reinforcing bars are summarized in Tables 3.1 through 3.3. Reinforcing bars are primarily a hot-rolled product. The bars are manufactured by a steel mill using electric-arc furnaces to melt scrap steel to produce billets. These billets are then heated and passed through a series of rolls to make reinforcing bars. The deformations are formed on the bars during the final pass through the rolls.

Most of the properties of reinforcing bars of interest to the Architect/Engineer are defined by ASTM specifications. The significance of certain properties as defined by ASTM, however, is not readily apparent, and some special properties related to structural design are not defined by ASTM. Therefore, a brief review of the significance of various mechanical properties on structural design is appropriate. Plots of idealized stress-strain curves for ASTM A615 and ASTM A706 reinforcing bars are shown for two strain ranges in Fig. 3.1.

Of prime importance in structural design is the yield strength of nonprestressed reinforcement. ACI 318 requires determination of yield strengths over 60 ksi (420 MPa) at a stress that corresponds to a strain of 0.35% and limits yield

Table 3.1—Summary of minimum ASTM strength requirements for reinforcing steel

Material	ASTM specification	Grade designation	Minimum yield strength, psi (MPa)	Minimum tensile strength, psi (MPa)
Reinforcing bars	A615/A615M	Grade 40 (280) Grade 60 (420) Grade 75 (520)	40,000 (280) 60,000 (420) 75,000 (520)	60,000 (420) 90,000 (620) 100,000 (690)
	A706/A706M	Grade 60 (420)	60,000 (420) [78,000 (540) maximum]	80,000 (550)*
Stainless steel reinforcing bars	A955/A955M	Grade 40 (300) Grade 60 (420) Grade 75 (520)	40,000 (300) 60,000 (420) 75,000 (520)	70,000 (500) 90,000 (620) 100,000 (690)
Low-carbon chromium steel reinforcing bars	A1035/A1035M	Grade 100 (690) Grade 120 (830)	100,000 (690) 120,000 (830)	150,000 (1030) 150,000 (1030)
Deformed bar mats	A184/A184M	Same as reinforcing bars		
Zinc-coated (galvanized) bars	A767/A767M	Same as reinforcing bars		
Epoxy-coated bars	A775/A775M A934/A934M	Same as reinforcing bars		
Wire—plain	A82/A82M		70,000 (485)	80,000 (550)
Wire—deformed	A496/A496M		75,000 (515)	85,000 (585)
Welded wire Plain (W1.2 (MW8) and larger)	A185/A185M		65,000 (450)	75,000 (515)
(smaller than W1.2 (MW8))			56,000 (385)	70,000 (485)
Deformed	A497/A497M		70,000 (485)	80,000 (550)
Epoxy-coated wire/welded wire	A884/A884M	Same as wire or welded wire		
Stainless steel wire/welded wire	A1022/A1022M	Same as wire or welded wire		
Seven-wire strand for prestressing	A416/A416M	Grade 250 (stress-relieved)	212,500 (1465)	250,000 (1725)
		Grade 250 (low-relaxation)	225,000 (1555)	250,000 (1725)
		Grade 270 (stress-relieved)	229,500 (1580)	270,000 (1860)
		Grade 270 (low-relaxation)	243,000 (1675)	270,000 (1860)
Wire for prestressing	A421/A421M	Stress-relieved	199,750 to 212,500 (1377 to 1465)†	235,000 to 250,000 (1620 to 1725)†
		Low-relaxation	211,500 to 225,000 (1460 to 1550)†	235,000 to 250,000 (1620 to 1725)†
Bars for prestressing	A722/A722M	Type I (plain)	127,500 (880)	150,000 (1035)
		Type II (deformed)	120,000 (830)	150,000 (1035)
Compacted strand for prestressing‡	A779/A779M	Grade 245 (stress-relieved)	213,200 (1480)	245,000 (1700)
		Grade 245 (low-relaxation)	220,500 (1530)	245,000 (1700)
		Grade 260 (stress-relieved)	226,200 (1565)	260,000 (1800)
		Grade 260 (low-relaxation)	234,000 (1620)	260,000 (1800)
		Grade 270 (stress-relieved)	234,900 (1620)	270,000 (1860)
		Grade 270 (low-relaxation)	243,000 (1675)	270,000 (1860)
Epoxy-coated seven-wire strand for prestressing‡	A882/A882M	Same as seven-wire strand		
Seven-wire indented strand for prestressing‡	A886/A886M	Grade 250I (1725I) (stress-relieved)	212,500 (1465)	250,000 (1725)
		Grade 250I (1725I) (low-relaxation)	225,000 (1550)	250,000 (1725)
		Grade 270I (1860I) (stress-relieved)	229,500 (1580)	270,000 (1860)
		Grade 270I (1860I) (low-relaxation)	243,000 (1675)	270,000 (1860)
Two- and three-wire strand for prestressing	A910/A910M	Grade 250 (stress-relieved)	212,500 (1465)	250,000 (1725)
		Grade 250 (low-relaxation)	225,000 (1550)	250,000 (1725)
		Grade 270 (stress-relieved)	229,500 (1580)	270,000 (1860)
		Grade 270 (low-relaxation)	243,000 (1675)	270,000 (1860)

*But not less than 1.25 times the actual yield strength.
†Minimum strength depends on wire sizes and type.
‡Not listed in ACI 318.

Table 3.2—ASTM mechanical property requirements for reinforcing bars

Type of steel and ASTM designation	Grade designation	Minimum yield strength, psi (MPa)*	Minimum tensile strength, psi (MPa)	Range of bar size designation†	Minimum percentage of elongation in 8 in. (203.2 mm)	Pin diameter for bend test‡ (d_b = nominal diameter of bar)
Carbon, A615/A615M	40 (280)	40,000 (280)	60,000 (420)	3 (10)	11	3-1/2d_b
				4 (13)	12	
				5 (16)		
				6 (19)		5d_b
	60 (420)	60,000 (420)	90,000 (620)	3 (10)	9	3-1/2d_b
				4 (13)		
				5 (16)		
				6 (19)		5d_b
				7 (22)	8	
				8 (25)		
				9 (29)	7	7d_b
				10 (32)		
				11 (36)		
				14 (43)		(90 degrees) 9d_b
				18 (57)		
	75 (520)	75,000 (520)	100,000 (690)	6 (19)	7	5d_b
				7 (22)		
				8 (25)		
				9 (29)	6	7d_b
				10 (32)		
				11 (36)		
				14 (43)		(90 degrees) 9d_b
				18 (57)		
Low-alloy, A706/A706M	60 (420)	60,000 minimum (420) 78,000 maximum (540)	80,000§ (550)	3 (10)	14	3d_b
				4 (13)		
				5 (16)		
				6 (19)		4d_b
				7 (22)	12	
				8 (25)		
				9 (29)		6d_b
				10 (32)		
				11 (36)		
				14 (43)	10	8d_b
				18 (57)		
Stainless, A955/A955M	40 (300)	40,000 (300)	70,000 (500)	3 (10)	20	3-1/2d_b
				4 (13)		
				5 (16)		
				6 (19)		5d_b
	60 (420)	60,000 (420)	90,000 (620)	3 (10)	20	3-1/2d_b
				4 (13)		
				5 (16)		
				6 (19)		5d_b
				7 (22)		
				8 (25)		
				9 (29)		7d_b
				10 (32)		
				11 (36)		
				14 (43)		(90 degrees) 9d_b
				18 (57)		
	75 (520)	75,000 (520)	100,000 (690)	6 (19)	20	5d_b
				7 (22)		
				8 (25)		
				9 (29)		7d_b
				10 (32)		
				11 (36)		
				14 (43)		(90 degrees) 9d_b
				18 (57)		

Table 3.2—ASTM mechanical property requirements for reinforcing bars (cont.)

Low-carbon chromium, A1035/A1035M	100 (690)	100,000 (690) 80,000 (550) at 0.35% strain	150,000 (1030)	3 (10)	7	3-1/2d_b
				4 (13)		
				5 (16)		
				6 (19)		5d_b
				7 (22)		
				8 (25)		
				9 (29)		7d_b
				10 (32)		
				11 (36)		
				14 (43)	6	(90 degrees) 9d_b
				18 (57)		
	120 (830)	120,000 (830) 90,000 (620) at 0.35% strain	150,000 (1030)	3 (10)	7	3-1/2d_b
				4 (13)		
				5 (16)		
				6 (19)		5d_b
				7 (22)		
				8 (25)		
				9 (29)		7d_b
				10 (32)		
				11 (36)		
				14 (43)	6	(90 degrees) 9d_b
				18 (57)		

*Yield point or yield strength; refer to ASTM specifications.
†Bar numbers are based on the number of eighths of an inch included in the nominal diameter of the bars (bar numbers approximate the number of millimeters of the nominal diameter of the bar).
‡Test bends over 180 degrees unless noted otherwise.
§Tensile strength should not be less than 1.25 times the actual yield strength.

strength used in design to a maximum of 80 ksi (550 MPa). An exception to this maximum is made for transverse spirals that can be used to a maximum of 100 ksi (690 MPa). The purpose of these requirements is to provide for reliable application of the strength design approach. ASTM specifications require that the yield strength be determined at a prescribed strain when the steel does not have a well-defined yield point. Variability in mechanical properties was investigated by Mirza and MacGregor (1979). For information on bend tests, see Kudder and Gustafson (1983).

3.2.1 *Ductility*—Tensile-test ductility requirements (after-fracture measurement of the elongation of a tensile specimen) as specified by ASTM resulted in the production of reinforcing bars that have generally experienced very few fractures during fabrication or in service in reinforced concrete structures, even in the case of structures subjected to seismic, blast, or wind forces. It is important in any inelastic analysis to realize that useful ductility is limited to the strain corresponding to the maximum stress on the stress-strain curve, and this may be less than half the ultimate ductility. In addition, the ductility may be reduced if the reinforcement buckles under load reversals.

Ductility of reinforcing bars is not an important parameter in reinforced concrete members subjected primarily to compression, but ductility is important in slabs and beams or in columns with significant bending or axial tension. An example would be a structure subjected to seismic, blast, or wind forces where the percentage of reinforcing steel is low and there is a possibility of steel rupture before concrete crushing (McDermott 1998). Rupture of longitudinal steel has generally been precluded in flexural members by using steels with ductility as defined by ASTM and by imposing lower limits on steel percentage, such as given in Section 10.5

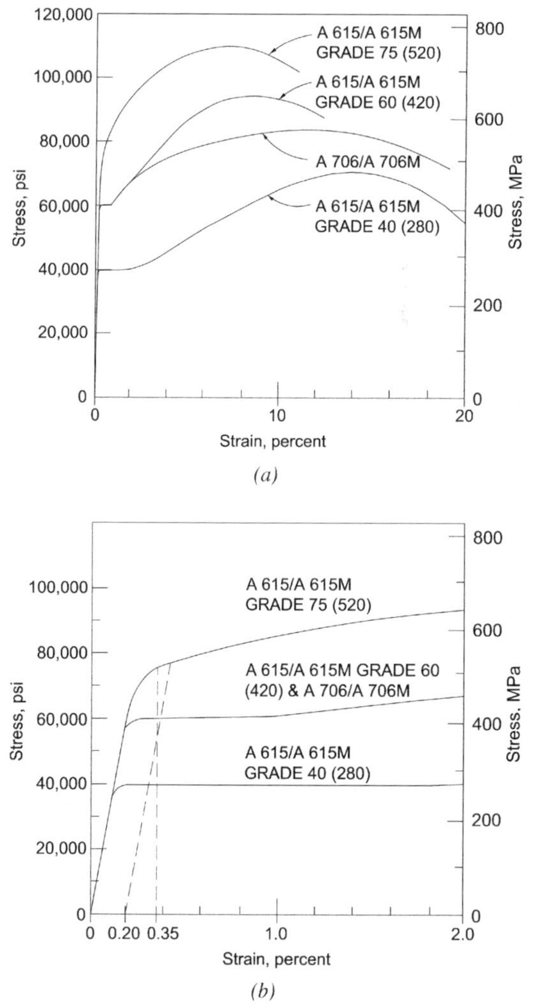

Fig. 3.1—Idealized stress-strain curves for reinforcing bars.

Table 3.3—Reinforcing bars: ASTM chemical composition restrictions and other requirements

ASTM specification	Chemical composition restrictions	Other requirements
A615/A615M	0.06% maximum phosphorus (heat) 0.075% maximum phosphorus (finished bar)	Tensile retests permitted on two random specimens when: a) Yield strength of original specimen is within 1000 psi (7 MPa) of specification minimum; b) Tensile strength of original specimen is within 2000 psi (14 MPa) of specification minimum; or c) Elongation of original specimen is within two percentage units of specification minimum. Bend retests permitted on two random specimens. Bar mark for type of steel: S
A706/A706M	Chemical composition limited to: (see below)	Same retest provisions as described previously for A615/A615M. The manufacturer should report the chemical composition and carbon equivalent for each heat of steel to the purchaser or his representative. Bar mark for type of steel: W ("S" and "W" for bars meeting both A615/A615M and A706/A706M)
A955/A955M	Chemical composition, % (see below)	Same retest provisions as described previously for A615/A615M. In addition, report: 1) Heat treat condition; 2) Magnetic permeability (if appropriate); 3) Hardness (if appropriate); and 4) Corrosion test (if appropriate).
A1035/A1035M	Chemical composition limited to: (see below)	Same retest provisions as described previously for A615/A615M Bar mark for type of steel: CS

A706/A706M — Chemical composition limited to:

Element	Maximum% (heat)	Maximum % (finished bar)
Carbon	0.30	0.33
Manganese	1.50	1.56
Phosphorus	0.035	0.043
Sulfur	0.045	0.053
Silicon	0.50	0.55

Carbon equivalent (C.E.) limited to 0.55% calculated by

$$C.E. = \%C + \frac{\%Mn}{6} + \frac{\%Cu}{40} + \frac{\%Ni}{20} + \frac{\%Cr}{10} - \frac{\%Mo}{50} - \frac{\%V}{10}$$

A955/A955M — Chemical composition, %

UNS designation	S24000	S24100	S30400	S31603	S31653	S31803
Type	XM-29	XM-28	304	316L	316LN	....
Carbon	0.08	0.15	0.08	0.03	0.08	0.03
Manganese	11.50-14.50	11.00-14.00	2.00	2.00	2.00	2.00
Phosphorus	0.060	0.060	0.045	0.045	0.045	0.045
Sulfur	0.030	0.030	0.030	0.030	0.030	0.020
Silicon	1.0	1.0	1.0	1.0	1.0	1.0
Chromium	17.00-19.00	16.50-19.00	18.00-20.00	16.00-18.00	16.00-18.00	21.00-23.00
Nickel	2.25-3.75	0.50-2.50	8.00-10.50	10.00-14.00	10.00-14.00	4.50-6.50
Molybdenum	....	....	....	2.00-3.00	2.00-3.00	2.50-3.50
Nitrogen	0.20-0.40	0.20-0.45	0.10	0.10	0.10-0.16	0.08-0.20

A1035/A1035M — Chemical composition limited to:

Element	Maximum %
Carbon	0.15
Chromium	8.0 to 10.9
Manganese	1.5
Nitrogen	0.05
Phosphorus	0.035
Sulfur	0.045
Silicon	0.50

of ACI 318-08 (McDermott 1974). Ductility is important in members subjected to membrane tension, as in some structural components, such as shearwalls or diaphragms that may be subjected to seismic, blast, or wind forces.

3.2.2 *Properties not defined by ASTM*—ASTM specifications for reinforcing bars do not include restrictions regarding strength under fatigue or impact properties when loading introduces high strain rates. Reinforcing bars are not subject to fatigue or impact failures in most structures. Refer to Helgason and Hansen (1974) for information on fatigue strength of reinforcing bars. No practical methods for testing impact properties of reinforcing bars have been devised; Charpy tests (ASTM D5942, withdrawn 1998) on machined specimens do not reflect impact properties of a deformed bar. High strain rates result in higher yield strengths and, to a lesser extent, higher tensile strengths. Ductility is generally not sensitive to the strain rate (Cowell 1965; Siess 1962; Crum 1959; McDermott 1974; Malvar 1998; Malvar and Crawford 1998).

3.3—Availability

Reinforcing bars that conform to ASTM A615/A615M for carbon steel (previously called billet-steel) are the most commonly produced and specified type. The vast majority of construction uses carbon-steel bars. Carbon-steel bars are readily available throughout the U.S. Most of the producers roll reinforcing bars in 60 ft (18 m) stock lengths. Obtaining lengths longer than 60 ft (18 m) normally requires making special arrangements with the supplier having that capability. Alternatively, longer lengths can be obtained by using mechanical splices that develop the full strength of the reinforcing bars.

Reinforcing bars are typically furnished as shop-fabricated to a construction project. Shop fabrication means the bars have been cut to lengths, bent, bundled, and tagged for identification. Companies that furnish shop-fabricated reinforcing bars are commonly called fabricators. Reinforcing bars are fabricated from information developed by the fabricator from the project drawings. Fabricators obtain stock-length and coiled reinforcing bars directly from steel mills.

In 1997, steel mills that produced reinforcing bars began to phase in the production of bars conforming to soft metric sizes. Within a few years, the shift to exclusive production of soft metric reinforcing bars was essentially achieved. Virtually all reinforcing bars currently produced in the U.S. are soft metric. Soft metric conversion enables the industry to furnish the same reinforcing bars to inch-pound and metric construction projects, eliminating the need for the steel mills and fabricators to maintain a dual inventory. The sizes of soft metric reinforcing bars are physically the same as the corresponding sizes of inch-pound bars. Soft metric bar sizes, which are designated No. 10, No. 13, No. 16, and so on, correspond to inch-pound bar sizes No. 3, No. 4, No. 5, and so on. The metric bar designations are simply a relabeling of the inch-pound bar designations. Table 3.2 shows the one-to-one correspondence of the soft metric bar designation numbers to the inch-pound bar designation numbers.

Producers may require orders in large quantities—usually heat lots—to roll lengths over 60 ft (18 m). Stock material in 20, 30, and 40 ft (6, 9, and 12 m) lengths is normally available, usually in the smaller bar sizes (No. 3 through No. 6 [No. 10 through No. 19]). Stock material in Grade 60 (420) for No. 3 and No. 4 (No. 10 and No. 13) bars is available in 1500 and 3000 lb (700 and 1400 kg) coils (approximate weight or mass). Coil stock is commonly used by fabricators with automatic bending equipment for the fabrication of ties and stirrups. Coils of reinforcing bars in sizes No. 3 through No. 5 (No. 10 through No. 16) are also used by fabricators that specialize in fabricating spirals.

In general, ASTM A615/A615M Grade 40 (280) (minimum specified yield strength = 40,000 psi [280 MPa]) in bar sizes No. 3 through No. 6 (No. 10 through No. 19) and Grade 60 (420) (minimum specified yield strength = 60,000 psi [420 MPa]) in bar sizes No. 3 through No. 11 (No. 10 through No. 36) are readily available in lengths up to 60 ft (18 m) in all parts of the country. Grade 60 (420) in bar sizes No. 14 and No. 18 (No. 43 and No. 57) is generally available, but these bar sizes are not usually kept in a fabricator's inventory.

ASTM A615/A615M also includes provisions for Grade 75 (520) in bar sizes No. 6 through No. 11 (No. 19 through No. 36), No. 14 (No. 43), and No. 18 (No. 57). The minimum yield strength-strain requirement in A615/A615M is compatible with ACI 318; that is, the Grade 75 (520) minimum yield strength of 75,000 psi (520 MPa) corresponds to a strain of 0.35%. Typically, Grade 75 (520) reinforcing bars are used in columns made with high-strength concrete.

ASTM A706/A706M for low-alloy steel reinforcing bars was issued in 1974 to meet specific needs for controlled tensile properties for seismic-resistant design and construction, and controlled chemical composition for weldability (Gustafson and Felder 1991). During the 20 years following issuance of ASTM A706/A706M, the demand, and consequently, the availability, of A706/A706M bars was not clearly defined. A trend appears to have started toward increased and exclusive use of low-alloy steel reinforcing bars. In 1995, for example, the California Department of Transportation (Caltrans) revised its Standard Specifications (Caltrans 1995) to require low-alloy steel reinforcing bars to conform to ASTM A706/A706M for virtually all reinforced concrete structures. The only exceptions to the revised requirements issued by Caltrans that allow the use of carbon-steel bars that conform to ASTM A615/A615M are slope and channel paving, certain types of concrete barriers, and temporary railing. Several other public agencies within the State of California have followed the lead of Caltrans by requiring low-alloy steel bars for construction projects. In 1996, the Nevada Department of Transportation implemented a policy that requires A706/A706M bars in the construction of essentially all reinforced concrete structures. The Illinois Department of Transportation followed suit in 2005 by permitting only A706/A706M bars for all reinforced concrete structures. Usage of low-alloy reinforcing bars has also increased in other sectors, such as other government agencies and privately-financed building construction.

Rail-steel and axle-steel reinforcing bars are covered by ASTM A996/A996M. Because rail-steel and axle-steel bars are only available in a few areas of the country, these types of bars are not discussed in detail herein. Requirements for use of rail-steel or axle-steel bars may be found in ASTM A996/A996M and in Section 3.5.3 of ACI 318-08.

Reinforcing bars with special chemical or physical properties may be produced for particular applications, such as nuclear power plants and anchor rods for transmission towers. Reinforcing bars with special material properties require large-quantity orders.

3.4—Welding

Provisions for welding are specifically excluded from ASTM A615/A615M. Low-alloy steel reinforcing bars conforming to ASTM A706/A706M are intended for welding; hence, the specification contains restrictions on chemical composition and carbon equivalent.

If it is necessary to make welded splices of ASTM A615/A615M reinforcing bars, it is important that the provisions of AWS D1.4/AWS D1.4M be followed to obtain a ductile connection. AWS D1.4/AWS D1.4M requires the calculation of the carbon equivalent, which in turn requires that the chemical composition of the bars to be welded be known either from information provided by the producer or by testing a sample. Certified Mill Test Reports should be requested in the project specifications when the construction project contains welded reinforcing bars. ASTM A615/

A615M reinforcing bars with higher carbon equivalent percentages require a prescribed amount of preheat. Traceability of reinforcing bars to their Certified Mill Test Reports after fabrication is an important quality-control issue when welding ASTM A615/A615M bars. Welding of reinforcing bars conforming to ASTM A706/A706M should also be performed in accordance with AWS D1.4/AWS D1.4M. ASTM A706/A706M limits the carbon equivalent to 0.55%, and AWS D1.4/AWS D1.4M requires little or no preheat for bars at or below this limit.

Connection of crossing bars by small arc welds, known as tack welds, is not recommended. These welds can cause stress concentrations that may lead to reduced strength and ductility. If tack welding is authorized by the Architect/Engineer, the welds should be made in conformance with all the requirements of AWS D1.4/AWS D1.4M (refer to ACI 318-08, Section 7.5.4.).

3.5—Material testing

Acceptance tests for compliance with ASTM specifications require that the yield strength, tensile strength, and percentage of elongation as well as the nominal weight (mass), bend test, deformation requirements, and phosphorous content be determined for a representative specimen of each size from each heat of reinforcing bar rolled. ASTM A370 prescribes these test methods and procedures.

For reinforced concrete building projects designed in accordance with ACI 318, the Architect/Engineer should specify the reinforcing bars in accordance with ASTM A615/A615M, A706/A706M, or A996/A996M. The special provisions for seismic design in Chapter 21 of ACI 318-08 require reinforcing bars to comply with ASTM A706/A706M. Chapter 21 of 318-08, however, permits the use of Grades 40 and 60 (300 and 420) ASTM A615/A615M bars with the following limitations:

1. The actual yield strength cannot exceed the specified yield strength by more than 18,000 psi (120 MPa); and
2. The actual tensile strength cannot be less than 1.25 times the actual yield strength.

Table 3.4 shows the standard fabricated bar bends and end hooks used in construction. These standard bends and end hooks have been established by ACI 315 in conjunction with the Concrete Reinforcing Steel Institute, based on provisions in ACI 318. For some bar sizes, the diameters of the bends and hooks are larger than the minimum values specified by ACI 318. The dimensions of the bar bends and hooks are the finished bend diameters, which are measured on the inside of the bend. Pin sizes used by fabricators are based on Section 7.2 of ACI 318-08. The slight straightening of the bent bar, known as springback, results in a larger finished bend diameter.

Note in Table 3.4 that the pin diameters for the ASTM bend tests are slightly smaller (typically one-half or one bar diameter smaller) than the finished bend diameters used in construction. The basis for the smaller pin diameters is to have a specimen tested to more severe requirements than those used in construction.

Additional testing criteria and restrictions on reinforcing bars have evolved to meet the special design and quality assurance/quality control requirements for safety-related portions of nuclear power plants. ACI 359 permits only the use of reinforcing bars meeting ASTM A615/A615M or A706/A706M. In addition, ACI 359 requires more frequent tensile testing than the ASTM specifications, that is, one tension test at a maximum frequency of 50 tons (45 t) per heat per bar size versus ASTM's one test per heat per bar size. The basis for this exception is that ACI 359 considers the ASTM test frequency inadequate for its special design requirements because commercial heat sizes may be as much as 250 tons (230 t) or more.

Another difference in ACI 359 concerns retest criteria for tension tests. ASTM A615/A615M and A706/A706M permit retests on two random specimens if the results of an original tension specimen are within specified limits (Table 3.3). ACI 359 requires retests on two random specimens, but no limits on the original test specimen results are imposed.

ACI 349 is similar to ACI 359 in that it permits only ASTM A615/A 615M or A706/A706M bars. ACI 349 also requires one tension test for each 50 tons (45 t) of each bar size produced from each heat of steel.

3.6—Spirals and mats

3.6.1 *Spirals*—Spiral reinforcement for columns is fabricated from reinforcing bars conforming to ASTM A615/A615M or A706/A706M. Spirals may also be made from plain wire that complies with ASTM A82/A82M or deformed steel wire that complies with ASTM A496/A496M.

The fabrication of spirals (with the accompanying spacing devices, if required) is generally done by reinforcing bar fabricators. Not all fabricators of reinforcing bars produce spirals; consequently, lead times may be extended in some locations during times of peak demand.

3.6.2 *Bar mats*—This type of reinforcement uses deformed reinforcing bars that conform to ASTM A615/A615M or A706/A706M. Bar mats are defined in ASTM A184/A184M as two layers of bars that are assembled at right angles to each other by welding. ASTM A184/A184M outlines the requirements for welding, the testing of the welds, tolerances, marking, and inspection.

The use of bar mats is advantageous when there is significant repetition of identical mats; however, due to low usage, bar mats may not be readily available in some geographic areas. Many reinforcing bar fabricating shops do not normally produce bar mats.

3.7—Corrosion protection products

ACI 318 permits the use of epoxy-coated and zinc-coated (galvanized) reinforcing bars as corrosion-protection systems in reinforced concrete structures. The bars to be coated should conform to Section 3.5.3.1 of ACI 318-08. ASTM A955/A955M covers stainless steel reinforcing bars.

Epoxy-coated reinforcing bars should conform to ASTM A775/A775M or A934/A934M. Two industry practices exist: one is to coat straight bars and then fabricate the coated bars (ASTM A775/A775M); the other is to prefabricate the bars and then coat them (ASTM A934/A934M). The

Table 3.4—ACI 318 standard minimum bends and ASTM bend requirements

ASTM specification	Bar designation number	Grades designations	Bend pin diameter (d_b = nominal bar diameter) ASTM*	ACI Standard†	ACI Stirrup/tie‡
A615/A615M	3, 4, 5 (10, 13, 16)	40, 60 (280, 420)	3-1/2d_b	6d_b	4d_b
	6 (19)	40, 60, 75 (280, 420, 520)	5d_b	6d_b	6d_b
	7, 8 (22, 25)	60, 75 (420, 520)	5d_b	6d_b	6d_b
	9, 10, 11 (29, 32, 36)	60, 75 (420, 520)	7d_b	8d_b	N/A
	14, 18 (43, 57)	60, 75 (420, 520)	9d_b (90 degrees)	10d_b	N/A
A706/A706M	3, 4, 5 (10, 13, 16)	60 (420)	3d_b	6d_b	4d_b
	6, 7, 8 (19, 22, 25)	60 (420)	4d_b	6d_b	6d_b
	9, 10, 11 (29, 32, 36)	60 (420)	6d_b	8d_b	N/A
	14, 18 (43, 57)	60 (420)	8d_b	10d_b	N/A
A955/A955M	3, 4, 5 (10, 13, 16)	40, 60 (300, 420)	3-1/2d_b	6d_b	4d_b
	6 (19)	40, 60, 75 (300, 420, 520)	5d_b	6d_b	6d_b
	7, 8 (22, 25)	60, 75 (420, 520)	5d_b	6d_b	6d_b
	9, 10, 11 (29, 32, 36)	60, 75 (420, 520)	7d_b	8d_b	N/A
	14, 18 (43, 57)	60, 75 (420, 520)	9d_b (90 degrees)	10d_b	N/A
A1035/A1035M	3, 4, 5 (10, 13, 16)	100, 120 (690, 830)	3-1/2d_b	6d_b	6d_b
	6, 7, 8 (19, 22, 25)	100, 120 (690, 830)	5d_b	6d_b	6d_b
	9, 10, 11 (29, 32, 36)	100, 120 (690, 830)	7d_b	8d_b	N/A
	14, 18 (43, 57)	100, 120 (690, 830)	9d_b (90 degrees)	10d_b	N/A

*ASTM bend tests 180 degrees unless otherwise specified.
†ACI standard bends 90 and 180 degrees.
‡ACI stirrup/tie bends 90 and 135 degrees for bars No. 3 to No. 8 (No. 10 to No. 25) only.

Concrete Reinforcing Steel Institute has a voluntary certification program for epoxy coating applicator plants.

Zinc-coated reinforcing bars should conform to ASTM A767/A767M. The Architect/Engineer should be aware of optional provisions in ASTM A767/A767M and specify these requirements when appropriate. For instance, bars are usually galvanized after fabrication. In this case, the specification establishes minimum finished bend diameters for bars that are fabricated before galvanizing, except that the bars may be bent tighter if they are stress-relieved by a prescribed heat treatment process. Hence, the Architect/Engineer should specify which bars require special finished bend diameters; these are usually the smaller bar sizes for stirrups and ties. ASTM A767/A767M includes two classes with different weights of zinc coating. Class I (3.5 oz/ft^2 [1070 g/m^2]) is normally specified for general construction. ACI 349 does not permit the use of zinc-coated reinforcing bars.

No uncoated reinforcing steel, or any other embedded metal dissimilar to zinc, should be permitted in the same structural concrete member with or in close proximity to galvanized reinforcing bars, except as part of a cathodic protection system. Galvanized bars should not be electrically coupled to uncoated bars.

Standards ASTM A767/A767M, A775/A775M, and A934/A934M provide nonmandatory information regarding field handling of coated bars. The specifier should prescribe requirements in the project specification for: handling and placing; job-site storage; tie wire; bar supports; touching up the ends of bars cut at the job site; prohibiting flame-cutting of epoxy-coated bars; limits on the amount of repairable damaged coating; the repair of all damaged coating; and internal vibrators for consolidating concrete. These field requirements are provided in ACI 301.

Epoxy-coated bars are generally available throughout the U.S. Zinc-coated (galvanized) reinforcing bars are available in limited areas. Because of inspection and the coating application process, delivery times might be extended compared with uncoated reinforcing bars.

ASTM A955/A955M includes three minimum yield strengths for stainless steel reinforcing bars: Grades 40, 60, and 75 (Grades 300, 420, and 520). The bar sizes are the same as those in A615/A615M. Typically, stainless steel bars lack a well-defined yield point. As a result, ASTM A955/A955M requires that the yield strength be determined as the stress corresponding to a prescribed strain of 0.35%. Interest is developing, mainly among several State Departments of Transportation, in using stainless steel bars in corrosive environments. Although the initial cost of stainless steel bars is relatively high, proponents contend the use of such bars becomes attractive when a long service life and life-cycle costs are factored into the economics of a project. Stainless steel bars have also been used in reinforced concrete structures where nonmagnetic materials were required.

CHAPTER 4—PLAIN AND DEFORMED WIRE

4.1—Introduction

Low-carbon cold-worked wire is used extensively in construction. Wire for concrete reinforcement may be plain, as defined by ASTM A82/A82M, or deformed, as defined by ASTM A496/A496M. The ASTM mechanical requirements are given in Table 4.1. Plain or deformed wire is used as concrete reinforcement in building and bridge structures. This reinforcement is also used in sewer pipes, precast components, and slabs-on-ground.

Plain steel wire used as concrete reinforcement is permitted by ACI 318 for use in caisson, column, or pier spirals and in welded wire reinforcement (WWR) where the welded intersections provide the bond strength. Deformed steel wire is permitted in ACI 318 for use as general concrete reinforcement. The deformations along the surface of the wire provide the bond strength. For deformed wire used in WWR, the deformations and the welded intersections combine to provide the bond strength.

4.2—Material properties

Low-carbon wire is cold-worked by either drawing or rolling the wire from hot-rolled rods. Thus, wire is different from reinforcing bars, which are a hot-rolled product. The rods have a low carbon content—under 0.30% and are usually in the range of 0.08 to 0.16% carbon.

Low-carbon hot-rolled rod is drawn through dies to produce plain wire of the desired size, or rolled through deforming rolls to produce deformed wire of the desired size. The drawing and rolling processes increase the yield and tensile strengths. Most low-carbon plain wire used in concrete reinforcement is drawn through one or two dies, and most low-carbon deformed wire is rolled through at least two sets of rolls. Because the dies and rolls wear during the manufacturing process, dimensional tolerances within the ASTM standards provide for an allowable size range above and below the exact wire size. Plain wire size is determined by measuring wire diameter, and deformed wire size is determined by measuring wire weight per unit length.

Because the wire has a low carbon content, it generally has good welding characteristics and excellent bendability properties. There are no specific requirements for welding wire to wire in the field. The strength of cold-worked wire may be reduced when field-welded.

ASTM A82/A82M and A496/A496M require that yield strength for plain or deformed wire be determined at a strain of 0.50%. In accordance with Sections 3.5.3 and 3.5.4 of ACI 318-08, however, the yield strength should be determined at a stress corresponding to a strain of 0.35% for all material with a yield strength greater than 60,000 psi (420 MPa) up to 80,000 psi (550 MPa). Plots of idealized stress-strain curves for various yield strength wires are shown in Fig. 4.1. A plot of the A615/A615M reinforcing bar is included as a reference.

Research at the University of Nebraska (Amorn et al. 2007) has shown that deformed wire, in the form of WWR with no welds in the high stress region, has similar fatigue resistance to that of reinforcing bars. There is presently no known data available on impact resistance of low-carbon steel wire.

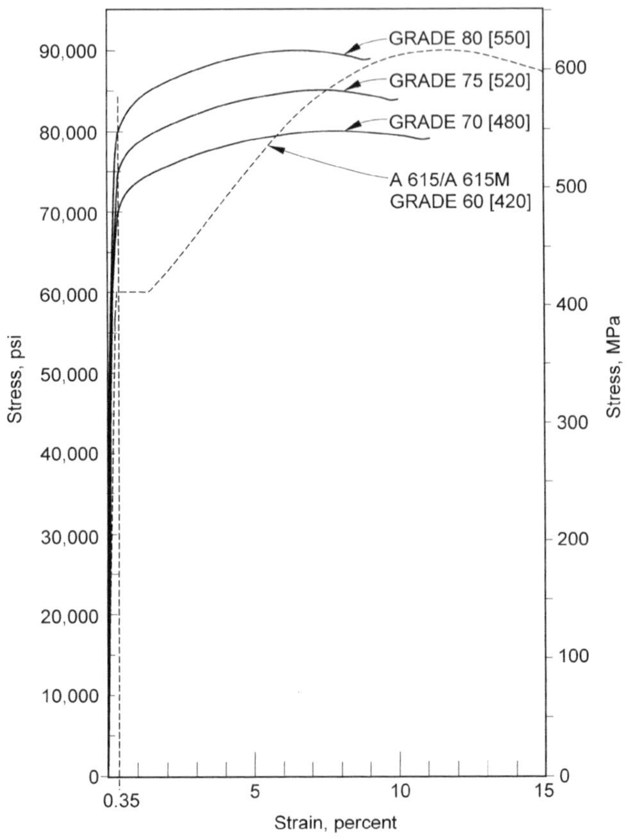

Fig. 4.1—Idealized stress-strain curves for cold-rolled ASTM A82/A82M and ASTM A496/A496M wire.

Table 4.1—ASTM mechanical property requirements for plain and deformed wire

Type of wire	Minimum yield strength, psi (MPa)	Minimum tensile strength, psi (MPa)	Pin diameter for bend tests*
Plain (ASTM A82/A82M)	70,000 (485)	80,000 (550)	W7 (MW45) and smaller......1d_b Larger than W7 (MW45)......2d_b
Deformed (ASTM A496/A496M)	75,000 (515)	85,000 (585)	D6 (MD39) and smaller......2d_b Larger than D6 (MD39)......4d_b
Plain stainless steel (ASTM 1022/A1022M)	70,000 (485)	80,000 (550)	Same as for ASTM A82/A82M
Deformed stainless steel (ASTM 1022/A1022M)	75,000 (515)	85,000 (585)	Same as for ASTM A496/A496M

*d_b = nominal diameter of specimen. Bend tests through 180 degrees for plain wire (ASTM A82/A82M) and 90 degrees for deformed wire (A496/A496M).

4.3—Wire size designation

The designation of wire sizes involves a letter-number combination. ASTM uses "W" or "MW" to designate plain wire, and "D" or "MD" for deformed wire. The number following the W or D gives the cross-sectional area of the wire in hundredths of a square inch, and MW or MD gives the cross-sectional area in square millimeters. For instance, a W5.0 (MW32) wire is a plain wire with a cross-sectional area of 0.05 in.^{2} (32 mm^2). D31.0 (MD200) indicates a deformed wire with a cross-sectional area of 0.31 in.^{2} (200 mm^2).

4.4—Availability

Plain wire (ASTM A82/A82M) used for WWR is generally available in wire sizes from W0.5 (MW3) (0.080 in. [2.03 mm] diameter) through W45 (MW290) (0.757 in. [19.22 mm] diameter). Plain wire is used in the manufacture of welded wire for concrete reinforcement and is generally limited to a minimum size of W1.4 (MW9) (0.135 in. [3.4 mm] diameter). Wire sizes smaller than W1.4 (MW9) are typically used in masonry reinforcement, and are normally galvanized.

Spirals fabricated from wire are subject to the same availability and lead time requirements as spirals using ASTM A615/A615M or A706/A706M hot-rolled plain bars.

Deformed wire (ASTM A496/A496M), most commonly used in the manufacture of deformed WWR, is normally available in wire sizes from D4 through D20 (MD26 through MD129) (larger wire sizes from some producers are available up to D45 [MD290]). Although some deformed welded wire is available in smaller sizes, Section 3.5.3.5 of ACI 318-08 limits the minimum deformed wire size to D4 (MD26).

4.5—Corrosion protection products

ASTM A884/A884M covers epoxy-coated wire and WWR. Epoxy-coated wire is used as a corrosion protection system in reinforced concrete structures or components exposed to chlorides. Availability varies depending on geographic location; local manufacturers should be contacted for additional information.

Cold-worked wire can be galvanized to conform to ASTM A641/A641M. ASTM A123/A123M covers hot-dip galvanizing of wire as well as other reinforcement. Local manufacturers should be contacted for additional information.

ASTM A933/A933M covers vinyl-coated wire and WWR. Local manufacturers should be contacted for additional information.

ASTM A1022/A1022M covers stainless steel wire and welded wire for concrete reinforcement. Stainless steel wire and WWR are used in applications requiring resistance to corrosion, controlled magnetic permeability, or both. Local manufacturers should be contacted for additional information.

CHAPTER 5—WELDED WIRE REINFORCEMENT

5.1—Introduction

Welded wire reinforcement, the material and its manufacturing process, are covered by ASTM A185/A185M for WWR made from plain wire, and ASTM A497/A497M for WWR made from deformed wire or a combination of deformed and plain wire. Welded wire reinforcement was formerly identified as welded wire fabric. Applications include reinforcement for footings, foundations, slabs-on-ground, slabs on metal deck, two-way structural slabs, and walls. Welded wire reinforcement is manufactured in flat sheets and rolls. Welded wire reinforcement sheets can be bent into various shapes such as stirrup baskets for beams, and closed shear and confinement cages for beams and columns. Welded wire reinforcement rolls made of small-diameter wire, up to W2.9 (MW19), are used for slabs-on-ground in some geographic regions.

Welded wire reinforcement is used extensively in the precast concrete industry. Rolls are used for reinforced concrete pipe, and sheets are used for box culverts, wall panels, and many other precast components.

Plain WWR uses the welded intersections to provide bond strength and anchorage in the concrete. ACI 318 limits wire spacing in plain WWR to a maximum of 16 in. (400 mm) except for shear reinforcement where the maximum spacing is 24 in. (610 mm). Deformed WWR uses the combination of deformations along the surface of the wire and the welded intersections to provide bond strength and anchorage in the concrete.

Table 5.1—ASTM mechanical property requirements for steel wires in welded wire reinforcement

Wire size	Minimum yield strength,* psi (MPa)	Minimum tensile strength, psi (MPa)	Minimum weld shear strength,† lb (N)
Welded plain wire (ASTM A185/A185M)			
W1.4 (MW9) to W45 (MW290)	65,000 (450)	75,000 (515)	$35{,}000A_w$ ($241A_w$)
Welded deformed wire (ASTM A497/A497M)			
D4 (MD26) to D45 (MD290)	70,000 (485)	80,000 (550)	$35{,}000A_w$ ($241A_w$)
Welded plain and deformed stainless steel wire (ASTM A1022/A1022M)			
W1.2 (MW8) to W45 (MW290)	65,000 (450)	75,000 (515)	$35{,}000A_w$ ($241A_w$)
D1 (MD4.5) to D45 (MD290)	70,000 (485)	80,000 (550)	$35{,}000\ A_w$ ($241A_w$)

*The minimum yield strength is determined at a strain of 0.005.

†A_w = nominal area of larger wire in in.^{2} (mm^2). The ratio of the area of the smaller wire to the area of the larger wire must be at least 0.40.

5.2—Material properties

ASTM A185/A185M and A497/A497M define: the minimum weld shear strength; tension, bend, and weld shear test methods and apparatus; and permissible variations, packaging, and identification requirements. ASTM-specified minimum mechanical property requirements for steel wires in WWR are summarized in Table 5.1.

ACI 318 requires a yield strength of 60,000 psi (420 MPa) for design unless the WWR is specified and furnished with a minimum yield strength measured at 0.35% strain. In this case, the tested minimum yield strength can be used as noted in the following section.

Recent research at the University of Nebraska (Amorn et al. 2007) showed that deformed wire, in the form of WWR with no welds in the high stress region, has similar fatigue resistance to that of reinforcing bars. The fatigue resistance is slightly reduced when welds are present in the high-stress

Table 5.2—Representative styles of metric welded wire reinforcement with equivalent in.-lb units

SI styles* (MW = plain wire)†	Area, mm²/m	Mass, kg/m²	Inch-pound styles* (W = plain wire)†	Area, in.²/ft	Weight, lb/CSF‡
102 x 102 – MW9 x MW9	88.9	1.51	4 x 4 – W1.4 x W1.4	0.042	31
102 x 102 – MW13 x MW13	127.0	2.15	4 x 4 – W2.0 x W2.0	0.060	44
102 x 102 – MW19 x MW19	184.2	3.03	4 x 4 – W2.9 x W2.9	0.087	62
102 x 102 – MW26 x MW26	254.0	4.30	4 x 4 – W4.0 x W4.0	0.120	88
152 x 152 – MW9 x MW9	59.3	1.03	6 x 6 – W1.4 x W1.4	0.028	21
152 x 152 – MW13 x MW13	84.7	1.46	6 x 6 – W2.0 x W2.0	0.040	30
152 x 152 – MW19 x MW19	122.8	2.05	6 x 6 – W2.9 x W2.9	0.058	42
152 x 152 – MW26 x MW26	169.4	2.83	6 x 6 – W4.0 x W4.0	0.080	58
305 x 305 – MW54 x MW54	175.7	3.08	12 x 12 – W8.3 x W8.3	0.083	63
305 x 305 – MW61 x MW61	199.0	3.47	12 x 12 – W9.4 x W9.4	0.094	71
305 x 305 – MW97 x MW97	317.5	5.52	12 x 12 – W15 x W15	0.150	113
305 x 305 – MW110 x MW110	362.0	6.25	12 x 12 – W17.1 x W17.1	0.171	128

*Welded wire reinforcement styles listed in this table are representative examples of potential styles or spacings. There are an unlimited number of styles that design engineers can use. This listing does not restrict other wire sizes and styles from being specified. For other available styles or wire sizes, consult WRI publications or discuss with WWR manufacturers.
†Wires may also be deformed. Prefix D (MD) should be used, except where only W (MW) is required by building codes (usually less than a W4 [MW26]). Wire sizes can be specified in 0.001 in.² increments (1 mm² increments for metric wire).
‡CSF = 100 ft².

region. There are presently no known data available on impact resistance of WWR.

5.3—Higher-strength welded wire reinforcement

The yield strengths in ASTM A82/A82M for plain wire and in A496/A496M for deformed wire are based on a strain of 0.5%. ACI 318 permits the use of design yield strengths up to 80,000 psi (550 MPa) for flexural, shear, and confinement reinforcement if the yield strength is measured at a strain of 0.35% (ACI Committee 439 1999). Section 11.4.2 of ACI 318-08 permits the design yield strength of up to 80,000 psi (550 MPa) for deformed WWR used for shear reinforcement.

The manufacturer should certify that the wire used in the WWR meets the more restrictive strain requirements of ACI 318. The Architect/Engineer uses this provision to specify higher yield strengths in designs. Materials with yield strengths higher than the minimum (65,000 psi [450 MPa] for ASTM A185/A185M and 70,000 [485 MPa] for ASTM A497/A497M) up to 80,000 psi (550 MPa) are available. Certified Mill Test Reports are available from manufacturers for all materials.

5.4—Welded wire reinforcement styles

Representative styles of WWR are given in Table 5.2. An illustrative example of a style is as follows:

6 x 12 – W16 x W8 (in-lb.)

152 x 305 – MW104 x MW52 (SI)

This denotes a WWR style made of plain wire in which:
Spacing of longitudinal wires = 6 in. (152 mm);
Spacing of transverse wires = 12 in. (305 mm);
Size of longitudinal wires =
W16 (MW104) (area = 0.16 in.^{2} [104 mm^2]); and
Size of transverse wires =
W8 (MW52) (area = 0.08 in.^{2} [52 mm^2]).

A welded deformed wire style would be noted in the same manner by substituting a D(MD) prefix to the wire sizes for the W(MW) prefix.

5.5—Availability

As noted in previous sections on plain and deformed wire, welded wire for concrete reinforcement is available with minimum wire sizes of W1.4 (MW9) for plain WWR and D4 (MD26) for deformed WWR. The majority of manufacturers produce WWR that uses wire sizes up to W20 (MW129) or D20 (MD129), with some capable of producing wire sizes up to W45 (MW290) or D45 (MD290). Most manufacturers, by special order, are capable of providing styles with variable wire spacings in both directions.

5.6—Minimum quantity requirements

Economy can be achieved by specifying standard styles. Welded wire manufacturers usually stock the typical standard styles that most users are aware of, as well as heavier styles that the industry has listed in publications, such as the Wire Reinforcement Institute's WWR-500, and in technical fact sheets on current product information.

5.6.1 *Stock styles*—Certain styles of plain WWR are carried in stock either at the producing mills or warehouses. While practice varies somewhat with different manufacturers and localities, the representative styles listed in Table 5.2 are usually available.

Some styles of plain WWR may be obtained in roll form. Typical roll widths and lengths of plain WWR are:

- U.S. (except West Coast): 60 in. x 150 ft (1524 mm x 45.7 m); and
- West Coast of U.S.: 84 in. x 150 ft or 200 ft (2134 mm x 45.7 m or 61.0 m).

It is recommended that rolls be straightened and cut to size before placement.

5.6.2 *Nonstock styles*—It is often desirable to order WWR sheets or rolls specifically produced to meet the reinforcement requirements and dimensions for individual projects.

The minimum quantity requirements for specific nonstock styles are governed by the manufacturing process requirements, which vary with different manufacturers. The following examples are merely a guide to illustrate the general requirements.

1. Longitudinal spacing, wire size, and sheet or roll width changes usually require quantities of 10 to 20 tons (9 to 18 t) per style;
2. Transverse wire spacing and size, side and end overhangs, and length changes usually require quantities of 2 to 5 tons (2 to 5 t) per style; and
3. The average weight for the total quantity in an order for any style should be approximately 20 tons (18 t). There are exceptions for less than truckload quantities, and these are handled on individual project requests with the manufacturer's acceptance. Before selecting nonstock styles, the Architect/Engineer should check with local manufacturers for both maximum and minimum quantities available.

For nonstock WWR styles, the following guidelines lead to the greatest production economies:

1. The most important factor affecting economy is keeping the longitudinal wire size and spacing the same for a single production run. It is possible, however, to vary wire sizes and spacings to obtain required areas of steel. Some manufacturers can produce variable-stepped spacing of longitudinal wires;
2. The second most important factor is that the transverse wire size and spacing are the same for a single production run. Spacing of transverse wires is typically 3, 4, 6, 8, 12, 16, and 18 in. (75, 100, 150, 200, 300, 400, and 460 mm). For applications other than structural, particularly for slab-on-ground installations, wire spacing can be as large as 18 in. (460 mm) or greater, depending on the manufacturer; and
3. As specified in the dimensions section of the corresponding ASTM specification for welded wire, a maximum allowable size differential of wires being welded together is maintained to ensure adequate weld shear strength. For both plain and deformed wire reinforcement, ASTM requires the smaller wire to have an area of 40% or more of the larger wire and with a minimum weld shear strength in pounds force of 35,000 times the area of the larger wire in square inches (in Newtons of 241 times the area of the larger wire in square millimeters).

Although it is possible to manufacture rolls using wire in the W12 (MW77) or D12 (MD77) and larger, some manufacturers recommend that sheets be specified essentially for all construction project applications as well as precast/prestressed structural members. Welded wire reinforcement for pipe, representing a major part of the welded wire industry, however, is manufactured in rolls using wire sizes of W12 (MW77) and larger.

Welded wire reinforcement may also be used as stirrup and tie reinforcement in beams and columns (Furlong et al. 1991; Guimaraes et al. 1992). Most manufacturers have hydraulic benders available to form sheets of WWR into many stirrup and tie configurations. Refer to Griezic et al. (1994) for tests of deformed welded wire stirrups.

5.7—Corrosion protection products

ASTM A884/A884M covers epoxy-coated wire and WWR. Epoxy-coated welded wire is used as a corrosion protection system in reinforced concrete structures or components exposed to chlorides. Availability varies depending on geographic location; local manufacturers should be contacted for additional information.

Cold-worked wire can be galvanized to conform to ASTM A641/A641M for use in producing WWR. Alternatively, WWR can be galvanized after fabrication. ASTM A123/A123M covers the hot-dip galvanizing of WWR as well as other reinforcement. A new ASTM standard specifically covering galvanized WWR is in process.

ASTM A933/A933M covers vinyl-coated wire and WWR. Local manufacturers should be contacted for additional information. ASTM A1022/A1022M covers stainless steel wire and welded wire for concrete reinforcement. Stainless steel wire and WWR is used in applications requiring resistance to corrosion, controlled magnetic permeability, or both. Local manufacturers should be contacted for additional information.

CHAPTER 6—PRESTRESSING REINFORCEMENT

6.1—Introduction

The previous sections of this report addressed passive reinforcement. Prestressing steel is an active reinforcement. The term "active" describes a prestressing system that applies a sustained force to a structural member, regardless of the other applied loads on that member. Because the magnitude of force that the prestressing system delivers to a structural member is of prime importance, the mechanism of transferring that force from the prestressing steel to the structural member is a major consideration. The transfer mechanism selected for any prestressing technique has a direct influence on material property requirements.

In pretensioned prestressed concrete, the force is first applied to the prestressing steel and then maintained while concrete is placed around the prestressing steel and cured. After the concrete has reached specified strength, the prestressing force is transferred to the concrete by bond to the prestressing steel. Prestressing reinforcement for prestressed concrete consists primarily of strands.

In post-tensioning systems, after the concrete has reached a prescribed compressive strength, the force is applied from the prestressing steel to anchors and bearing plates, and then to the concrete member being prestressed. Prestressing steel for post-tensioned members includes a wide spectrum of materials. Wires, bars, or strands are used interchangeably in post-tensioned members, depending upon the economics of the prestressing steel, the anchor system, the availability of the post-tensioning systems, and the considerations for the members being post-tensioned. In recent years, wire tendons have rarely been used, especially after the cessation of the post-tensioning used in nuclear power plant construction. Wire tendons are available if detensioning for inspection or retensioning of unbonded tendons to compensate for losses in the post-tensioning force is required, or if no slip anchorages are needed. At present, their most common use is for tendon replacement in nuclear containments.

Table 6.1—ASTM mechanical property requirements for prestressing wire and strand

ASTM A421/A421M wire

	Minimum yield strength, psi (MPa)				Minimum tensile strength, psi (MPa)		Minimum elongation, percent in a 10 in. (250 mm) gauge length
	Stress-relieved		Low-relaxation				
Size, in. (mm)	Type BA	Type WA	Type BA	Type WA	Type BA	Type WA	
0.192 (4.88)	*	212,500 (1465)	*	225,000 (1550)	*	250,000 (1725)	4.0
0.196 (4.98)	204,000 (1407)	212,500 (1465)	216,000 (1490)	225,000 (1550)	240,000 (1655)	250,000 (1725)	4.0
0.250 (6.35)	204,000 (1407)	204,000 (1407)	216,000 (1490)	216,000 (1490)	240,000 (1655)	240,000 (1655)	4.0
0.276 (7.01)	199,750 (1377)	199,750 (1377)	211,500 (1460)	211,500 (1460)	235,000 (1620)	235,000 (1620)	4.0

ASTM A416/A416M strand

		Minimum yield strength, psi (MPa)		Minimum tensile strength, psi (MPa)	Minimum elongation, percent in a 24 in. (600 mm) gauge length
Size, in. (mm)	Grade	Stress-relieved	Low-relaxation		
All sizes†	270 (1860)	229,500 (1580)	243,000 (1675)	270,000 (1860)	3.5
All sizes†	250 (1725)	212,500 (1465)	225,000 (1550)	250,000 (1725)	3.5

ASTM A779/A779M compacted strand‡

		Minimum yield strength, psi (MPa)		Minimum tensile strength, psi (MPa)	Minimum elongation, percent in a 24 in. (600 mm) gauge length
Size, in. (mm)	Grade	Stress-relieved	Low-relaxation		
0.5 (12.7)	270 (1860)	234,900 (1620)	243,000 (1675)	270,000 (1860)	3.5
0.6 (15.2)	260 (1800)	226,200 (1565)	234,000 (1620)	260,000 (1800)	3.5
0.7 (18.0)	245 (1700)	213,200 (1480)	220,500 (1530)	245,000 (1700)	3.5

ASTM A886/A886M indented strand‡

		Minimum yield strength, psi (MPa)		Minimum tensile strength, psi (MPa)	Minimum elongation, percent in a 24 in. (600 mm) gauge length
Size, in. (mm)	Grade	Stress-relieved	Low-relaxation		
All sizes†	270I (1860I)	229,500 (1580)	243,000 (1675)	270,000 (1860)	3.5
All sizes†	250I (1725I)	212,500 (1465)	225,000 (1550)	250,000 (1725)	3.5

ASTM A910/A910M two- and three-wire strand‡

		Minimum yield strength, psi (MPa)		Minimum tensile strength, psi (MPa)	Minimum elongation, percent in a 24 in. (600 mm) gauge length
Size, in. (mm)	Grade	Stress-relieved	Low-relaxation		
All sizes†	270 (1860)	229,500 (1580)	243,000 (1675)	270,000 (1860)	3.5
All sizes†	250 (1725)	212,500 (1465)	225,000 (1550)	250,000 (1725)	3.5

*Not commonly furnished in Type BA wire.
†Refer to Table 6.3 for sizes.
‡Not listed in ACI 318.

A wide variety of mechanisms is used to transfer the prestress force from the prestressing steel in post-tensioned construction. Most commonly used in the U.S. are wedge grips for strands, threaded connectors for bars, or upset ends (button heads) on the prestressing wires. The performance of the anchorage and bearing plate system and its effect on the prestressing steel may affect the performance of the post-tensioning system to a greater degree than variables in the prestressing steel. The prestressing steel may be bonded to the surrounding concrete after the prestressing force is transferred by injecting grout into the duct around the tendon, or it may be left unbonded and protected with a corrosion inhibitor. When the tendons are external to the member, though grouted, they are considered unbonded. The prestressing steel also imparts force to the member at points of curvature of the tendon, such as at tendon deflectors in pretensioned girders (harped or draped strands) or along lengths of curved tendons in post-tensioned members.

6.2—Material properties

Reinforcement for prestressed concrete may consist of strands as defined by ASTM A416/A416M, bars as defined by ASTM A722/A722M, or high-strength wires as defined by ASTM A421/A421M. Compacted strands (ASTM A779/A779M), indented strands (ASTM A886/A886M), and two- and three-wire strands (ASTM A 910/A 910M) are not listed in ACI 318. The lack of listing in ACI 318 does not, however, preclude their use (refer to Sections 3.5.6.2 and 1.4 in ACI 318-08). Mechanical properties for various types of prestressing steel are summarized in Tables 6.1 and 6.2. Idealized stress-strain curves for several prestressing steels are shown in Fig. 6.1.

Table 6.2—Summary of ASTM A722/A722M mechanical properties and other requirements for uncoated high-strength steel bar for prestressing concrete

Mechanical properties*	Type I (plain bars)	Type II (deformed bars)
Minimum tensile strength	150,000 psi (1035 MPa)	150,000 psi (1035 MPa)
Minimum yield strength	127,500 psi (880 MPa)	120,000 psi (830 MPa)
Strain at yield strength (extension under load method)	0.7%	0.7%
Minimum percentage of elongation: Gauge length of 20 bar diameters, or	4%	4%
Gauge length of 10 bar diameters	7%	7%
Nominal diameter ranges	3/4 to 1-3/8 in. (19 to 35 mm)	5/8 in. to 2-1/2 in. (15 to 65 mm)
Chemical requirements	Maximum % (heat): 0.040 phosphorus, 0.050 sulfur Maximum % (finished bar): 0.048 phosphorus, 0.058 sulfur	
Supplementary requirements (apply only when specified by purchaser)		
Bend test, Type I and II (135 degrees)	Nominal diameter, in. (mm)	Bend pin diameter (d_b = nominal bar diameter)
	5/8, 3/4, 1 (15, 20, 26)	$6d_b$
	1-1/4, 1-3/8 (32, 36)	$8d_b$
	1-3/4, 2-1/2 (46, 65)	$10d_b$
Reduction of area	20% minimum	—

*Bars to be cold-stressed to not less than 80% minimum tensile strength and then stress-relieved to produce prescribed mechanical properties.

Seven-wire strand for prestressing applications consists of one center wire and six helically formed outer wires with a uniform pitch of not less than 12 nor more than 16 times the nominal strand diameter. The most widely used type of strand, which is regarded as the standard type, is low-relaxation strand. Because the use of stress-relieved (normal relaxation) strand under ASTM A416/A416M is largely unavailable, this type of strand is not discussed in this report. Low-relaxation strand specified under ASTM A416/A416M is produced in two grades, Grade 250 (1725) and Grade 270 (1860), with minimum tensile strengths of 250,000 and 270,000 psi (1725 and 1860 MPa), respectively, based on nominal area of the strand.

Compacted seven-wire strand (compacted by drawing through a die and subsequently stress-relieved) produced under ASTM A779/A779M has a different strength level specified for each of three different sizes. ASTM A421/A421M, A779/A779M, and A886/A886M also provide (under Supplement I of each specification) for low-relaxation strand and relaxation testing.

For low-relaxation strand or wire, relaxation losses after 1000 hours should not be more than 2.5% when initially loaded to 70% of specified minimum tensile strength, nor more than 3.5% when loaded to 80% of specified minimum tensile strength at 68°F (20°C), as specified in ASTM A416/A416M and A910/A910M and in each supplement of ASTM A421/A421M, A779/A779M, and A886/A886M. For a given loading over a given time period, the relaxation loss for low-relaxation strand is about one-fourth that of a stress-relieved strand.

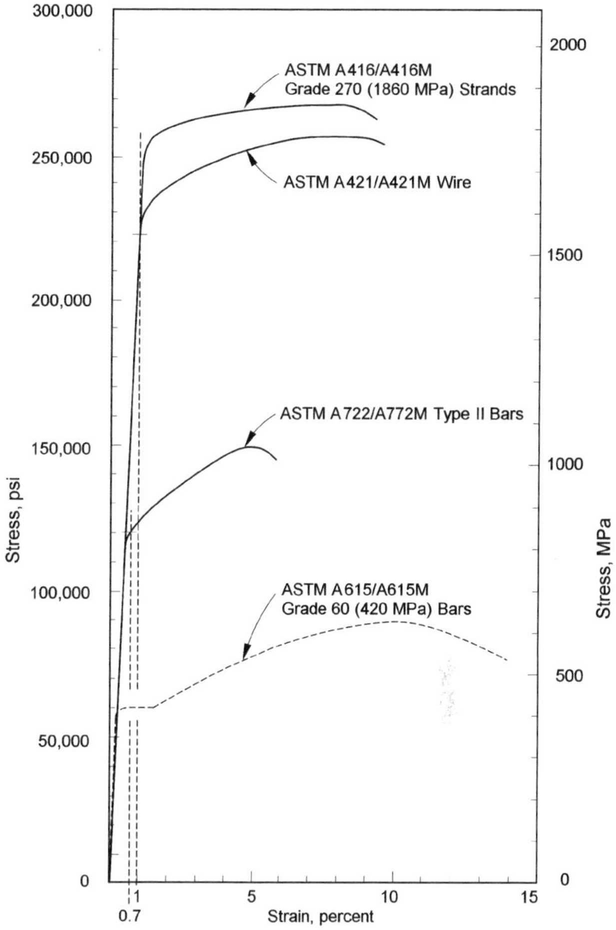

Fig. 6.1—Idealized stress-strain curves of prestressing steel.

As required in ASTM A416/A416M and A910/A910M, and in each supplement of A421/A421M, A779/A779M, and A886/A886M, the minimum yield strength for low-relaxation strand and wire measured at 1% extension under load should not be less than 90% of specified minimum tensile strength.

Material such as high-strength alloy steel bar is used as a prestressing reinforcement. ASTM A722/A722M specifies that the mechanical properties of such bar (Table 6.2) be obtained from tests on full-size bar test specimens. Supplementary requirements for ASTM A722/A722M provide for bend tests, minimum reduction-of-area, and report of chemical composition.

Wire produced under ASTM A421/A421M varies in minimum tensile strength depending on wire size. It may be used in applications that require cold-end deformations for anchorage, Type BA (button anchorage), and, for applications that require wedge anchorage of the ends, Type WA (wedge anchorage) with no cold-end deformation. ASTM A421/A421M also provides (under Supplement I) for low-relaxation wire and relaxation testing for the product.

ACI 318 references ASTM A416/A416M, A421/A421M, and A722/A722M with no additional material requirements. ACI 318, however, does require that post-tensioning anchorages and couplers for bonded and unbonded prestressing

Table 6.3—Available sizes of prestressing strand

ASTM A416/A416M strand and ASTM A886/A886M indented strand			
Nominal diameter, in. (mm)	ASTM A416/ A416M strand designation	Nominal area, in.^{2} (mm^2)	
		Grade 250 (1725)	Grade 270 (1860)
1/4 (6.35)	6	0.036 (23.2)	Not available
5/16 (7.94)	8	0.058 (37.4)	0.061 (39.4)*
3/8 (9.53)	9	0.080 (51.6)	0.085 (54.8)
7/16 (11.11)	11	0.108 (69.7)	0.115 (74.2)
1/2 (12.70)	13	0.144 (92.9)	0.153 (98.7)
6/10 (15.24)	15	0.216 (139.4)	0.217 (140.0)

ASTM A910/A910M two- and three-wire strand	
Nominal diameter, in. (mm)	Strand designation
0.228 (5.8)	2 x 0.114 (2 x 2.90)
0.189 (4.8)	3 x 0.089 (3 x 2.25)
0.205 (5.2)	3 x 0.095 (3 x 2.40)
0.244 (6.2)	3 x 0.114 (3 x 2.90)
0.256 (6.5)	3 x 0.118 (3 x 3.00)
0.295 (7.5)	3 x 0.138 (3 x 3.50)
0.340 (8.6)	3 x 0.158 (3 x 4.00)

*For ASTM A886/886M; not available for ASTM A416/A416M.

tendons develop at least 95% of the tensile strength of the tendons, when tested in an unbonded condition, without exceeding anticipated set. For bonded tendons, anchorages and couplers should be located so that 100% of the specified tensile strength of the tendons could be developed at critical sections after the tendons are bonded in the member.

ACI 423.3R has additional recommendations for unbonded tendons. Test assemblies for these additional recommended tests should consist of standard production quality components, and the tendons should be at least 10 ft (3.05 m) long. The assembly for the static test should be tested in a manner to allow accurate determination of the yield strength, tensile strength, and percent elongation of the complete tendon. Assemblies for the dynamic tests should withstand, without failure, 500,000 cycles from 60 to 66% of minimum specified tensile strength and 50 cycles from 40 to 80% of minimum specified tensile strength. The specimen used for the second fatigue test need not be the same as that used for the first fatigue test.

Some applications may require more severe dynamic testing. These dynamic/fatigue test requirements and other detailed requirements for post-tensioning systems are presented in the Post-Tensioning Institute's *Post-Tensioning Manual* and ACI 423.7.

Requirements for prestressing application on safety-related portions of nuclear power plants are covered by ACI 349 and 359. ACI 349 essentially parallels the requirements in ACI 318. ACI 359, however, identifies performance test and production test requirements. Performance tests require tendon assemblies to be statically tested with their yield strength, tensile strength, and elongation reported. The tensile strength of the assembly should be at least the minimum specified tensile strength of the prestressing steel and the elongation at least 2% over a 100 in. (2540 mm) gage length.

ACI 349 and 359 require one high-cycle dynamic test performed without failure, as also recommended in ACI 423.3R. The stress relaxation properties of the prestressing steel should also be determined by test in accordance with Supplement I of ASTM A421/A421M, A779/A779M, or A886/A886M. ACI 359 requires a minimum of three 1000-hour relaxation tests. Production testing should be performed in accordance with the applicable ASTM specification.

6.3—Availability

Low-relaxation prestressing strands (ASTM A416/A416M), high-strength steel bars (ASTM A722/A722M), and prestressing wires (ASTM A421/A421M) are generally available as part of prestressing systems that include complete tendons with their anchorage devices, ducts, and jacking equipment.

Low-relaxation prestressing strand (ASTM A416/A416M) is generally available throughout the U.S. in the sizes noted in Table 6.3. Type II uncoated high-strength steel bars for prestressing concrete (ASTM A722/A722M, deformed bar) are available throughout the U.S. in the sizes noted in Table 6.2.

Stress-relieved strands (ASTM A416/A416M), Type I high-strength steel plain bars (ASTM A722/A722M), and prestressing wires (ASTM A421/A421M) may be purchased by special order. Stress-relieved strands (ASTM A416/A416M) and Type I high-strength steel plain bars (ASTM A722/A722M), however, are generally not available in the U.S. The post-tensioning systems in the market no longer use them because of their much higher relaxation losses compared to the low-relaxation strands.

6.4—Corrosion protection products

Besides the ASTM standard prestressing reinforcement discussed previously, a variety of strands for corrosion protection is also available. These special applications are epoxy-coated seven-wire strand, galvanized strand, and stainless steel strand.

Epoxy-coated strand has been used in prestressing applications where corrosion control is of paramount importance, such as piles, bridge decks and girders, stay cables, parking structures, replacement tendons in parking structures, ground anchors (tiebacks), fender piles, and floating docks in both pretensioned and post-tensioned systems. ASTM A882/A882M prescribes requirements for filled epoxy-coated seven-wire strand. The strand is coated with epoxy using the electrostatic fusion-bonded process similar to that used for coating nonprestressed reinforcing bars, wire, and WWR. In filled strands, which are the standard type, the interstices (that is, the space in between the wires) are filled with epoxy to minimize migration of corrosive media. The use of filled epoxy-coated strand prevents water from penetrating into the interior of the strand through the interstices between the wires, thereby preventing internal corrosion. Grit may be applied to the surface of the epoxy to improve bond strength.

The strand coating is sufficiently flexible to allow stretching to high levels of strain without the formation of holidays (pinholes not discernible to the unaided eye). Unlike galvanized strand, mechanical properties such as strength and ductility are not affected by the coating process. Relaxation for unfilled epoxy-coated strand is 4%, 6.5% for filled epoxy-coated strand, and only 2.5% for uncoated strand. Tests conducted in 1985 at Fritz Engineering Laboratory of Lehigh University and by Wollmann et al. (1996) concluded that fatigue resistance is improved by reducing the fretting (damage from rubbing) between individual strands and between the strands and metal ducts using epoxy-coated strands. Filled epoxy-coated strand is also available for use in prestressed and post-tensioned concrete.

Galvanized strand has been used in prestressed circular tanks and as barrier cables for parking garages. It should be noted that due to the thermal effects during hot-dip galvanizing and the reduced steel area for the same overall strand diameter, galvanized strands have lower strengths and higher ductility than uncoated prestressing strands of the same nominal diameter.

Stainless steel strands have been used in special applications that require, in addition to moderate corrosion resistance, a low magnetic permeability, such as prestressed concrete used in piles and piers for demagnetizing stations. There is also some interest in barrier cable applications, but the high cost of stainless steel can be a factor. Although there is no current ASTM specification for stainless steel strand for prestressed concrete, ASTM A368 for stainless steel strand is intended for use as guy wires, overhead ground wires, and similar purposes.

6.5—Other prestressing reinforcement

Strands up to 300 ksi (2070 MPa) tensile strength with 270 ksi (1860 MPa) yield strength are now produced, although availability is limited. All other properties conform to ASTM A416/A416M. Due to the 11% increase in tensile strength of this steel over steel with 270 ksi (1860 MPa) tensile strength, an economic advantage results from the use of this steel. A noteworthy application is the use of 300 ksi (2070 MPa) epoxy-coated strands to replace corroded tendons in parking structures; for example, 15/32 in. (11.9 mm), 300 ksi (2070 MPa) epoxy-coated strand could replace 1/2 in. (12.70 mm), 270 ksi (1860 MPa) uncoated strand.

Many other sizes of prestressing strand not noted in Table 6.3 are made by manufacturers as special orders. For these and the other types discussed in this section, it is recommended that the suppliers be contacted directly.

CHAPTER 7—DEVELOPMENTS IN REINFORCING STEEL

There have been several developments regarding reinforcing steel. ACI adopted ACI 408.3R, which permits the use of reinforcing bars with increased height and closer spacing of the ribs or transverse deformations than used on currently produced reinforcing bars. An ASTM specification for low-carbon chromium steel reinforcing bar has also been introduced.

7.1—High relative rib area reinforcing bars

Based on bond research and specimen tests conducted primarily at the University of Kansas, ACI 408.3R prescribes tensile development and splice length requirements when and if commercial production of high relative rib area reinforcing bars is initiated. These reinforcing bars have a ratio of rib bearing area to bar surface area that is 50 to 100% more than typical reinforcing bars currently in production. This high relative rib area provides for a reduction in tension development and lap splice lengths when the bars are confined by transverse reinforcement.

7.2—Low-carbon, chromium steel reinforcing bars

In 2004, ASTM issued A1035/A1035M for low-carbon, chromium steel reinforcing bars with a single minimum yield strength of 100,000 psi (690 MPa), designated Grade 100 [690]. In 2007, bars with a minimum yield strength of 120,000 psi (830 MPa), designated Grade 120 [830], were added to this specification. Bar sizes and deformation requirements are the same as in the other ASTM reinforcing bar specifications. Idealized stress-strain curves for these two grades of A1035/A1035M steel are shown in Fig. 7.1. These bars lack a well-defined yield point. Designers should note that ASTM A1035/A1035M requires the yield strengths of 100,000 and 120,000 psi, designated Grade 100 [690] and Grade 120 [830] respectively, be determined using the 0.2% offset method. ACI 318 had required the yield strength be determined at 0.35% strain for yield strengths over 60,000 psi (420 MPa). ASTM A1035/A1035M requires a minimum yield strength of 80,000 psi (550 MPa) for Grade 100 [690] bars, and 90,000 psi (620 MPa) for Grade 120 [830] bars when yield strength is

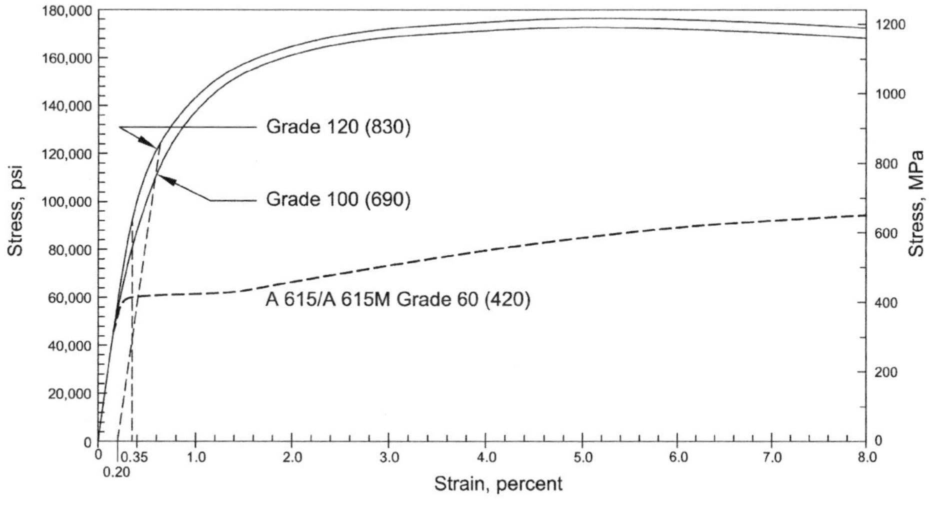

Fig. 7.1—Idealized stress-strain curves for ASTM A1035/A1035M reinforcing bars.

determined at 0.35% strain. Design should address proper development and lap splice lengths for this higher yield strength bar as well as the potential for increased cracking and deflections due use of higher bar stresses. A Summary Report by Seliem et al. (2009) addresses the bond behavior of this steel.

CHAPTER 8—REFERENCES

8.1—Referenced standards and reports

The various documents used in the development of this report follow. Because some of these documents are revised frequently, the reader is advised to contact the sponsoring organization to obtain the latest revision.

American Concrete Institute

301	Specifications for Structural Concrete
315	Details and Detailing of Concrete Reinforcement
318	Building Code Requirements for Structural Concrete
349	Code Requirements for Nuclear Safety-Related Concrete Structures
359	Code for Concrete Reactor Vessels and Containments, Section III, Division 2 of the ASME Boiler and Pressure Vessel Code
408.3R	Guide for Splice and Development Length of High Relative Rib Area Reinforcing Bars in Tension
423.3R	Recommendations for Concrete Members Prestressed with Unbonded Tendons
423.7	Specification for Unbonded Single-Strand Tendon Materials and Commentary

ASTM International

A82/A82M	Standard Specification for Steel Wire, Plain, for Concrete Reinforcement
A123/A123M	Standard Specification for Zinc (Hot-Dip Galvanized) Coatings on Iron and Steel Products
A184/A184M	Standard Specification for Welded Deformed Steel Bar Mats for Concrete Reinforcement
A185/A185M	Standard Specification for Steel Welded Wire Reinforcement, Plain, for Concrete
A368	Standard Specification for Stainless Steel Wire Strand
A370	Standard Test Methods and Definitions for Mechanical Testing of Steel Products
A416/A416M	Standard Specification for Steel Strand, Uncoated Seven-Wire for Prestressed Concrete
A421/A421M	Standard Specification for Uncoated Stress-Relieved Steel Wire for Prestressed Concrete
A496/A496M	Standard Specification for Steel Wire, Deformed, for Concrete Reinforcement
A497/A497M	Standard Specification for Steel Welded Wire Reinforcement, Deformed, for Concrete
A615/A615M	Standard Specification for Deformed and Plain Carbon-Steel Bars for Concrete Reinforcement
A641/A641M	Standard Specification for Zinc-Coated (Galvanized) Carbon Steel Wire
A706/A706M	Standard Specification for Low-Alloy Steel Deformed and Plain Bars for Concrete Reinforcement
A722/A722M	Standard Specification for Uncoated High-Strength Steel Bar for Prestressing Concrete
A767/A767M	Standard Specification for Zinc-Coated (Galvanized) Steel Bars for Concrete Reinforcement
A775/A775M	Standard Specification for Epoxy-Coated Steel Reinforcing Bars
A779/A779M	Standard Specification for Steel Strand, Seven-Wire, Uncoated, Compacted, Stress-Relieved for Prestressed Concrete
A882/A882M	Standard Specification for Filled Epoxy-Coated Seven-Wire Prestressing Steel Strand
A884/A884M	Standard Specification for Epoxy-Coated Steel Wire and Welded Wire Fabric for Reinforcement
A886/A886M	Standard Specification for Steel Strand, Indented, Seven-Wire Stress-Relieved for Prestressed Concrete
A910/A910M	Standard Specification for Uncoated, Weldless, 2- and 3-Wire Steel Strand for Prestressed Concrete
A933/A933M	Standard Specification for Vinyl (PVC) Coated Steel Wire and Welded Wire Fabric for Reinforcement
A934/A934M	Standard Specification for Epoxy-Coated Prefabricated Steel Reinforcing Bars
A955/A955M	Standard Specification for Deformed and Plain Stainless-Steel Bars for Concrete Reinforcement
A996/A996M	Standard Specification for Rail-Steel and Axle-Steel Deformed Bars for Concrete Reinforcement
A1022/A1022M	Standard Specification for Deformed and Plain Stainless Steel Wire and Welded Wire for Concrete Reinforcement
A1035/A1035M	Standard Specification for Deformed and Plain, Low-carbon, Chromium, Steel Bars for Concrete Reinforcement
D5942	Withdrawn Standard: ASTM D5942-96 Test Method for Determination of Charpy Impact Strength (withdrawn 1998)

American Welding Society

AWS D1.4/ AWS D1.4M	Structural Welding Code—Reinforcing Steel

Concrete Reinforcing Steel Institute

Manual of Standard Practice

Post-Tensioning Institute

Post-Tensioning Manual

Wire Reinforcement Institute, Inc.

WWR-500	Manual of Standard Practice—Structural Welded Wire Reinforcement

The previously mentioned publications may be obtained from the following organizations:

American Concrete Institute
P.O. Box 9094
Farmington Hills, MI 48333-9094
www.concrete.org

ASTM International
100 Barr Harbor Dr.
West Conshohocken, PA 19428-2959
www.astm.org

American Welding Society
550 NW Lejune Rd.
Miami, FL 33135
www.aws.org

Concrete Reinforcing Steel Institute
933 North Plum Grove Rd.
Schaumburg, IL 60173-4758
www.crsi.org

Post-Tensioning Institute
38800 Country Club Dr.
Farmington Hills, MI 48331
www.post-tensioning.org

Wire Reinforcement Institute, Inc.
942 Main St., Suite 300
Hartford, CT 06103
www.wirereinforcementinstitute.org

8.2—Cited references

AASHTO, 2002, "LRFD Bridge Design Specifications," second edition, American Association of State Highway and Transportation Officials, Washington, DC.

AASHTO, 2004, "Standard Specifications for Highway Bridges," 22nd edition, American Association of State Highway and Transportation Officials, Washington, DC.

ACI Committee 439, 1999, *Development of Seismic Steel Reinforcement Products and Systems*, SP-184, American Concrete Institute, Farmington Hills, MI, 97 pp.

Amorn, W.; Bowers, J.; Girgas, A.; and Tadros, M. K., 2007, "Fatigue of Deformed Welded-Wire Reinforcement," *PCI Journal*, V. 52, No. 1, Jan.-Feb. 2007, pp. 106-120.

Caltrans, 1995, *Standard Specifications*, State of California, Department of Transportation, Sacramento, CA, July.

Cowell, W. L., 1965, "Dynamic Tests of Concrete Reinforcing Steels," *Technical Report* No. R394, U.S. Naval Civil Engineering Laboratory, Port Hueneme, 34 pp.

Crum, R. G., 1959, "Tensile Impact Tests for Concrete Reinforcing Steels," ACI JOURNAL, *Proceedings* V. 56, No. 1, July, pp. 59-61.

Furlong, R.; Fenves, G.; and Kasl, E., 1991, "Welded Structural Wire Reinforcement for Columns," *ACI Structural Journal*, V. 88, No. 5, Sept.-Oct., pp. 585-591.

Griezic, A.; Cook, W.; and Mitchell, D., 1994, "Tests to Determine Performance of Deformed Welded Wire Stirrups," *ACI Structural Journal*, V. 91, No. 2, Mar.-Apr., pp. 211-220.

Guimaraes, G. N.; Kreger, M. E.; and Jirsa, J. O., 1992, "Evaluation of Joint-Shear Provisions for Interior Beam-Column-Slab Connections Using High-Strength Materials," *ACI Structural Journal*, V. 89, No. 1, Jan.-Feb., pp. 89-98.

Gustafson, D. P., and Felder, A. L., 1991, "Questions and Answers on ASTM A 706 Reinforcing Bars," *Concrete International*, V. 13, No. 7, July, pp. 54-57.

Helgason, T., and Hanson, J. M., 1974, "Investigation of Design Factors Affecting Fatigue Strength of Reinforcing Bars—Statistical Analysis," *Abels Symposium on Fatigue of Concrete*, SP-41, American Concrete Institute, Farmington Hills, MI, pp. 107-137.

Kudder, R. J., and Gustafson, D. P., 1983, "Bend Tests of Grade 60 Reinforcing Bars," ACI JOURNAL, *Proceedings* V. 80, No. 3, May, pp. 202-209.

Malvar, L. J., 1998, "Review of Static and Dynamic Properties of Steel Reinforcing Bars," *ACI Materials Journal*, V. 95, No. 5, Sept.-Oct., pp. 609-616.

Malvar, L. J., and Crawford, J. E., 1998, "Dynamic Increase Factors for Steel Reinforcing Bars," Twenty-Eighth DOD Explosive Safety Seminar, Orlando, FL, Aug.

McDermott, J. F., 1974, "Reinforcing Steel," *Reinforced Concrete Engineering*, John Wiley & Sons, New York, pp. 42-64.

McDermott, J. F., 1998, "Interrelationships between Reinforcing Bar Physical Properties and Seismic Demands," *ACI Structural Journal*, V. 95, No. 2, Mar.-Apr., pp. 175-182.

Mirza, S. A., and MacGregor, G., 1979, "Variability of Mechanical Properties of Reinforcing Bars," *Journal of the Structural Division*, ASCE, V. 105, No. ST5, May, pp. 921-937.

Seliem, H. M.; Hosny, A.; Rizkalla, S.; Zia, P. [North Carolina State University]; Briggs, M.; Miller, S.; Darwin, D.; Browning, J. [The University of Kansas]; Glass, G. M.; Hoyt, K.; Donnelly, K.; and Jirsa, J. O. [The University of Texas at Austin], 2009, "Bond Behavior of ASTM A1035 Reinforcing Bars," *ACI Structural Journal*, V. 106, No. 4, July-Aug., pp. 530-539.

Siess, C. P., 1962, "Behavior of High Strength Deformed Reinforcing Bars Under Rapid Loading," Department of Civil Engineering, University of Illinois, Urbana, IL.

U. S. Army Corps of Engineers, 1957, "Design of Structures to Resist the Effects of Atomic Weapons—Strength of Materials and Structural Elements," *Manual* No. EM 1110-345-414, 64 pp.

Wiss, Janney, Elstner Associates, Inc., 1970, "Final Report on Bar Tests," Committee of Concrete Reinforcing Bar Producers and American Iron and Steel Institute, Northbrook, IL, Apr.

Wollmann, G. P.; Yates, D. L.; Breen, J. E; and Kreger, M. E., 1996, "Fretting Fatigue in Post-Tensioned Concrete Beams," *ACI Structural Journal*, V. 93, No. 2, Mar.-Apr., pp. 172-179.

ACI 439.5R-18

Guide for the Specification, Manufacture, and Construction Use of Welded Wire Reinforcement

Reported by ACI Committee 439

Todd R. Hawkinson, Chair

David H. DeValve, Secretary

Paul B. Aubee
Brian L. Barrows
Richard H. Birley
George T. Biro
Joseph A. Bohinsky
Domingo J. Carreira
Louis J. Colarusso
Salem S. Faza
Anthony L. Felder
Augusta Gaertner
William C. Gallenz
Steven E. Holdsworth
Allen J. Hulshizer
Richard Huza
Josh Ison
Jason Koehler
Harry B. Lancelot III
Kenneth A. Luttrell
LeRoy A. Lutz
Carl D. Maki
Mark D. Marvin
David R. Maul
David B. McDonald
Theodore A. Mize
Conrad Paulson
Ryan W. Pelter
Richard A. Ramsey
Harold E. Reed
Roy H. Reiterman
Mario E. Rodriguez
Clifford A. Sabo
William H. Zehrt Jr.
Phil J. Zivich

Consulting Members

Jose Bagg
Jean-Jacques Braun
Luis E. Garcia
Kent A. Harries
Douglas D. Lee

Welded wire reinforcement (WWR) is prefabricated reinforcement consisting of high-strength cold-worked steel wires that are resistance-welded together in square or rectangular grids by continuous automatic welders. This report provides WWR product information, material specifications, and design/detailing recommendations, with an overview of manufacturing, shipping, and construction use in various applications of concrete construction.

Keywords: cast-in-place; cold-working; constructability; deformed reinforcement; mesh; precast; post-tensioned; prestressed; reinforcement; welded wire.

CONTENTS

ACI 439.5R-18 was adopted and published August 2018.

CHAPTER 1—INTRODUCTION AND SCOPE

1.1—Introduction

Welded wire reinforcement (WWR) is a prefabricated reinforcement consisting of a series of cold-worked steel wires that are resistance-welded together in square or rectangular grids by continuous automatic welders, and is one of the recognized and established forms of concrete reinforcement. United States patents covering the production of wires welded in this method were first issued in 1901 (Richardson 2000).

Plain or deformed wires, or a combination of both, may be used in WWR. Plain WWR bonds to concrete by mechanical anchorage at each wire intersection, whereas deformed WWR uses surface deformations plus welded intersections for bond and anchorage. The small-diameter, closely-spaced wires provide uniformly distributed reinforcement and effective crack control in footings, slabs, walls, paving, and roofs, in addition to providing shear resistance and confinement in columns, beams, and girders. WWR is prohibited in special seismic systems where the weld is required to resist stresses in response to confinement, lateral support of longitudinal bars, shear, or other actions in accordance with ACI 318. A wide range of wire sizes and spacings makes it possible to furnish more precise requirements for steel area. The placement of WWR can result in cost savings over nonprefabricated reinforcement. Material savings are obtainable by specifying WWR with high yield strengths as recognized by ACI 318 and ASTM A884/A884M, A1060/A1060M, and A1064/A1064M.

Extensive use was made of WWR in the Empire State Building and World Trade Center Towers in New York, and the Marina City Towers and Standard Oil Building in Chicago. The interstate highway system was built using an extensive amount of WWR in the 1950s and 1960s, and WWR continues to be used today in many transportation structural applications.

1.2—Scope

WWR is prefabricated reinforcement consisting of high-strength cold-worked steel wires that are resistance-welded together in square or rectangular grids by continuous automatic welders. This guide provides WWR product information, material specifications, and design/detailing recommendations, with an overview of manufacturing, shipping, and construction use in various applications of concrete construction.

1.3—W and D designations

Individual wire (plain and deformed) size designations are based on the cross-sectional area of a given wire. Gauge numbers were used exclusively for many years but were eliminated in the 1970s to reduce the confusion caused from misunderstanding the specified gauge size. The prefixes W and D are used in combination with a number. The letter W designates a plain wire, and the letter D denotes a deformed wire. The number following the letter gives the cross-sectional area in hundredths of a square inch (square millimeter). When describing metric wire, the prefix M is added; MW describes metric plain wire and MD metric deformed wire. The wire spacing in metric WWR is given in millimeters (mm) and the cross-sectional area of the wire is in square millimeters (mm^2).

For instance, wire designation W4 (MW26) would indicate a plain wire with a cross-sectional area of 0.04 $in.^2$ (26 mm^2) and a D10 (MD65) wire would indicate a deformed wire with a cross-sectional area of 0.10 $in.^2$ (65 mm^2). The size of wires in WWR is designated in the same manner. This system provides many advantages. Because the design professional knows the cross-sectional area of a wire and the spacing, the total cross-sectional area per unit width can easily be determined. For instance, a D6 (MD39) wire on 4 in. (102 mm) centers would provide three wires per 1 ft (305 mm) with a total cross-sectional area of 0.18 $in.^2$/ft (381 mm^2/m).

Many common wire sizes were developed to reduce peak-season lead times and aid in overall plant efficiency. The most readily available wire sizes along with the nominal diameter, area, and weight per unit length are shown in Table 1.3a (plain wire) and Table 1.3b (deformed wire). Plain wire is smooth cold-worked wire from a hot-rolled steel rod into the size or sizes needed for processing into welded sheets. Deformed wire can be indented or raised ribbed cold-worked wire from a hot-rolled rod into the size or sizes needed for processing into welded sheets. Areas of wire should be checked with the most efficient and readily available material from manufacturers. Other wire sizes are available and many manufacturers are able to produce them in 0.0015 $in.^2$ (1 mm^2) increments.

1.4—Product description

Spacing and size of wire in WWR are identified by style. The following subsections explain the orientation and definition of each part in the makeup of a WWR sheet. When the WWR sheet is designated with uniform or fixed spacing for quotation or order entry, it should be described as follows,

In in-lb units:

6 x 12 – D20.0 x D15.0 (Grade 80) – 78 in. (+12, +6) x 20 ft (24, 12) – 263.53 lb/sheet – 150 sheets needed

In SI units:

Table 1.3a—Plain wire

Size number*	Nominal diameter, in. (mm)	Nominal area,† in.^{2} (mm^2)	Nominal weight (mass), lb/ft (kg/m)
W 0.5 (MW3)	0.080 (2.03)	0.005 (3.23)	0.017 (0.025)
W 1.2 (MW8)	0.124 (3.14)	0.012 (7.74)	0.041 (0.061)
W 1.4 (MW9)	0.134 (3.39)	0.014 (9.03)	0.048 (0.071)
W 2 (MW13)	0.160 (4.05)	0.020 (12.9)	0.068 (0.101)
W 2.5 (MW16)	0.178 (4.53)	0.025 (16.1)	0.085 (0.126)
W 2.9 (MW19)	0.192 (4.88)	0.029 (18.7)	0.099 (0.147)
W 3.5 (MW23)	0.211 (5.36)	0.035 (22.6)	0.119 (0.177)
W 4 (MW26)	0.226 (5.73)	0.040 (25.8)	0.136 (0.202)
W 4.5 (MW29)	0.239 (6.08)	0.045 (29.0)	0.153 (0.228)
W 5 (MW32)	0.252 (6.41)	0.050 (32.3)	0.170 (0.253)
W 5.5 (MW36)	0.265 (6.72)	0.055 (35.5)	0.187 (0.278)
W 6 (MW39)	0.276 (7.02)	0.060 (38.7)	0.204 (0.304)
W 8 (MW52)	0.319 (8.11)	0.080 (51.6)	0.272 (0.405)
W 10 (MW65)	0.357 (9.06)	0.100 (64.5)	0.340 (0.506)
W 11 (MW71)	0.374 (9.50)	0.110 (71.0)	0.374 (0.557)
W 12 (MW77)	0.391 (9.93)	0.120 (77.4)	0.408 (0.607)
W 14 (MW90)	0.422 (10.7)	0.140 (90.3)	0.476 (0.708)
W 16 (MD103)	0.451 (11.5)	0.160 (103)	0.544 (0.810)
W 18 (MW116)	0.479 (12.2)	0.180 (116)	0.612 (0.911)
W 20 (MW129)	0.505 (12.8)	0.200 (129)	0.680 (1.01)
W 22 (MW142)	0.529 (13.4)	0.220 (142)	0.748 (1.11)
W 24 (MW155)	0.533 (14.0)	0.240 (155)	0.816 (1.21)
W 26 (MW168)	0.575 (14.6)	0.260 (168)	0.884 (1.32)
W 28 (MW181)	0.597 (15.2)	0.280 (181)	0.952 (1.42)
W 30 (MW194)	0.618 (15.7)	0.300 (194)	1.02 (1.52)
W 31 (MW200)	0.628 (16.0)	0.310 (200)	1.05 (1.57)
W 45 (MW290)	0.757 (19.2)	0.450 (290)	1.53 (2.28)

*These sizes represent the most readily available sizes in the WWR industry. Other wire sizes are available and manufacturers can produce them in 0.0015 in.^{2} (1 mm^2) increments. For sizes other than those shown, the wire size is the number of one-hundredths of a square inch (mm^2) in the nominal area of the wire cross section prefixed by the letter W (MW).

†The cross-sectional area is based on the weight of the wire. The area in square inches may be calculated by dividing the unit weight per linear inch in pounds by 0.2833, or by dividing the unit weight per linear foot of specimen in pounds by 3.4. (The area in square millimeters may be calculated by dividing the unit weight per lineal meter by 0.00784.)

152 x 305 – MD129 x MD97 (550 MPa) – 1981 mm (+305, +152) x 6.10 m (610, 305) – (119.53 kg/sheet) – 150 sheets needed

where

6 (152) = spacing of longitudinal wires, in. (mm)

12 (305) = spacing of transverse wires, in. (mm)

D20.0 (MD129) = type (deformed) and size of longitudinal wires

D15.0 (MD97) = type (deformed) and size of transverse wires

Grade 80 (550 MPa) = yield strength of specified wire sizes—80 ksi (550 MPa)

Sheet width (plus overhangs) 78 in. (+12, +6) x sheet length (includes overhangs) 20 ft (24,12); 1981 mm (+305, +152) x 6.10 m (610, 305) (refer to 1.4.1 through 1.4.3)

Total sheet weight (mass) = 263.53 lb (119.53 kg) (refer to 1.4.4).

Quantity ordered (usually in truckloads or portions thereof): 150 sheets needed (refer to 1.4.5).

This particular sheet is illustrated in Fig. 1.4. If a specified yield strength is not given in a purchase order, the manufacturer should follow the requirements of the applicable section of ASTM A1022/A1022M and A1064/A1064M.

For sheets with variable wire spacing or variable length, the sheet should be described as follows.

In in.-lb units:

V x V – D20.0 x W15.0 (Grade 80)

78 in. (+12, +6) x Var. Length (24, 12)

LS = 12 in. OH, 1 at 8 in., 7 at 10 in., 6 in. OH

TS = 24 in. OH, 2 at 6 in., 15 at 12 in., 2 at 6 in., 12 in. OH

Var. Length = variable length-alternate every other wire 20 ft- 15 ft

Sheet weight = 190.17 lb

215 sheets needed

In SI units:

V x V – MD129 x MW97 (550 MPa)

1981 mm (+305, +152) x Var. Length (610, 305)

LS = 305 mm OH, 1 at 203 mm, 7 at 254mm, 152 mm OH

TS = 610 mm OH, 2 at 152 mm, 15 at 305 mm, 2 at 152mm, 305 mm OH

Table 1.3b—Deformed wire

	Nominal dimensions				Deformation requirements*
Wire size†	Unit weight, lb/ft (kg/m)	Nominal diameter,‡ in. (mm)	Cross-sectional area,§ in.^{2} (mm^2)	Perimeter, in. (mm)	Minimum average height of deformations, in. (mm)
D1 (MD6)	0.034 (0.051)	0.113 (2.87)	0.010 (6.45)	0.354 (9.00)	0.0045 (0.114)
D2 (MD13)	0.068 (0.101)	0.160 (4.05)	0.020 (12.9)	0.501 (12.7)	0.0063 (0.160)
D3 (MD19)	0.102 (0.152)	0.195 (4.96)	0.030 (19.4)	0.614 (15.6)	0.0078 (0.198)
D4 (MD26)	0.136 (0.202)	0.226 (5.73)	0.040 (25.8)	0.709 (18.0)	0.0101 (0.257)
D5 (MD32)	0.170 (0.253)	0.252 (6.41)	0.050 (32.3)	0.793 (20.1)	0.0113 (0.287)
D6 (MD39)	0.204 (0.304)	0.276 (7.02)	0.060 (38.7)	0.868 (22.1)	0.0124 (0.315)
D7 (MD45)	0.238 (0.354)	0.299 (7.58)	0.070 (45.2)	0.938 (23.8)	0.0134 (0.304)
D8 (MD52)	0.272 (0.405)	0.319 (8.11)	0.080 (51.6)	1.00 (25.5)	0.0143 (0.363)
D9 (MD58)	0.306 (0.455)	0.339 (8.60)	0.090 (58.1)	1.06 (27.0)	0.0152 (0.386)
D10 (MD65)	0.340 (0.506)	0.357 (9.06)	0.100 (64.5)	1.12 (28.5)	0.0160 (0.406)
D11 (MD71)	0.374 (0.557)	0.374 (9.51)	0.110 (71.0)	1.18 (29.9)	0.0187 (0.475)
D12 (MD77)	0.408 (0.607)	0.391 (9.93)	0.120 (77.4)	1.23 (31.2)	0.0195 (0.495)
D13 (MD84)	0.442 (0.658)	0.407 (10.3)	0.130 (83.9)	1.28 (32.5)	0.0203 (0.516)
D14 (MD90)	0.476 (0.708)	0.422 (10.7)	0.140 (90.3)	1.33 (33.7)	0.0211 (0.536)
D15 (MD97)	0.510 (0.759)	0.437 (11.1)	0.150 (96.8)	1.37 (34.9)	0.0218 (0.554)
D16 (MD103)	0.544 (0.810)	0.451 (11.5)	0.160 (103)	1.42 (36.0)	0.0225(0.572)
D17 (MD110)	0.578 (0.860)	0.465 (11.8)	0.170 (110)	1.46 (37.1)	0.0232 (0.589)
D18 (MD116)	0.612 (0.911)	0.479 (12.2)	0.180 (116)	1.50 (38.2)	0.0239 (0.607)
D19 (MD122)	0.646 (0.961)	0.492 (12.5)	0.190 (122)	1.55 (39.2)	0.0245 (0.622)
D20 (MD129)	0.680 (1.01)	0.505 (12.8)	0.200 (129)	1.59 (40.3)	0.0252 (0.604)
D21 (MD135)	0.714 (1.06)	0.517 (13.1)	0.210 (135)	1.62 (41.3)	0.0259 (0.658)
D22 (MD141)	0.748 (1.11)	0.529 (13.4)	0.220 (141)	1.66 (42.2)	0.0265 (0.673)
D23 (MD148)	0.782 (1.16)	0.541 (13.7)	0.230 (148)	1.70 (43.2)	0.0271 (0.688)
D24 (MD154)	0.816 (1.21)	0.553 (14.0)	0.240 (154)	1.74 (44.1)	0.0277 (0.704)
D25 (MD161)	0.850 (1.26)	0.564 (14.3)	0.250 (161)	1.77 (45.0)	0.0282 (0.716)
D26 (MD167)	0.884 (1.32)	0.575 (14.6)	0.260 (167)	1.81 (45.9)	0.0288 (0.732)
D27 (MD174)	0.918 (1.37)	0.586 (14.9)	0.270 (174)	1.84 (46.8)	0.0293 (0.744)
D28 (MD180)	0.952 (1.42)	0.597 (15.2)	0.280 (180)	1.88 (47.6)	0.0299 (0.759)
D29 (MD187)	0.986 (1.47)	0.608 (15.4)	0.290 (187)	1.91 (48.5)	0.0304 (0.772)
D30 (MD193)	1.02 (1.52)	0.618 (15.7)	0.300 (193)	1.94 (49.3)	0.0309 (0.785)
D31 (MD200)	1.05 (1.57)	0.628 (16.0)	0.310 (200)	1.97 (50.1)	0.0314 (0.798)
D45 (MD290)	1.53 (2.28)	0.757 (19.2)	0.450 (290)	2.38 (60.4)	0.0379 (0.961)

*The maximum deformation spacing is 0.285 in. (7.24 mm) and the minimum deformation spacing is 0.182 in. (4.62 mm) for all wire sizes.
†These sizes represent the most readily available sizes in the WWR industry. Other wire sizes are available and manufacturers can produce them in 0.0015 in.^{2} (1 mm^2) increments. For sizes other than those shown, the wire size is the number of one-hundredths of a square inch (mm^2) in the nominal area of the deformed wire cross section prefixed by the letter D (MD).
‡The nominal diameter of a deformed wire is equivalent to the diameter of a plain wire having the same weight per unit length as the deformed wire.
§The cross-sectional area is based on the weight of the wire. The area in square inches may be calculated by dividing the unit weight per linear inch in pounds by 0.2833, or by dividing the unit weight per linear foot of specimen in pounds by 3.4. (The area in square millimeters may be calculated by dividing the unit weight per lineal meter by 0.00784.)

Var. Length = variable length-alternate every other wire 6.10 m, 4.57 m
Sheet weight = 86.26 kg/sheet
215 sheets needed
where
V = variable spacing of longitudinal wires, in. (mm), and variable spacing of transverse wires, in. (mm)
D20 (MD129) = type (deformed) and size of longitudinal wires
W15 (MW97) = type (plain) and size of transverse wires
Grade 80 (550 MPa) = yield strength of specified wire sizes, 80 ksi (550 MPa)
Sheet width (plus overhangs)78 in. (+12, +6) x Variable Length (includes overhangs) (24, 12); [1981 mm (+305, +152) x Variable Length (610, 305)] (includes overhangs) (refer to 1.4.1 through 1.4.3).
OH = overhang
LS = description of variable longitudinal spacing, in. (mm)
TS = description of variable transverse spacing, in. (mm).
Var. Length = description of variable length longitudinal wires (refer to 1.4.2)
Total sheet weight = 190.17 lb (86.26 kg) (refer to 1.4.4)
Quantity ordered (usually in truckloads or parts thereof) 215 sheets needed (refer to 1.4.5).

Plain wire can often be combined in a sheet style with deformed wire, especially where main reinforcement may be required in only one direction. Some machines can also produce sheets with variable line wire, crosswire sizes, or both. Check with local manufacturer for capabilities.

1.4.1 *Width*—Width is the center-to-center distance between outside longitudinal wires. This dimension does not include overhangs, as described in 1.4.3. Overall width is the width plus the side overhangs, or the tip-to-tip dimension of the transverse wires. The width should be specified in inches (millimeters) as the first number on the second line of the sheet style designation. Figure 1.4 demonstrates graphically the definition of width and overall width. Maximum sheet

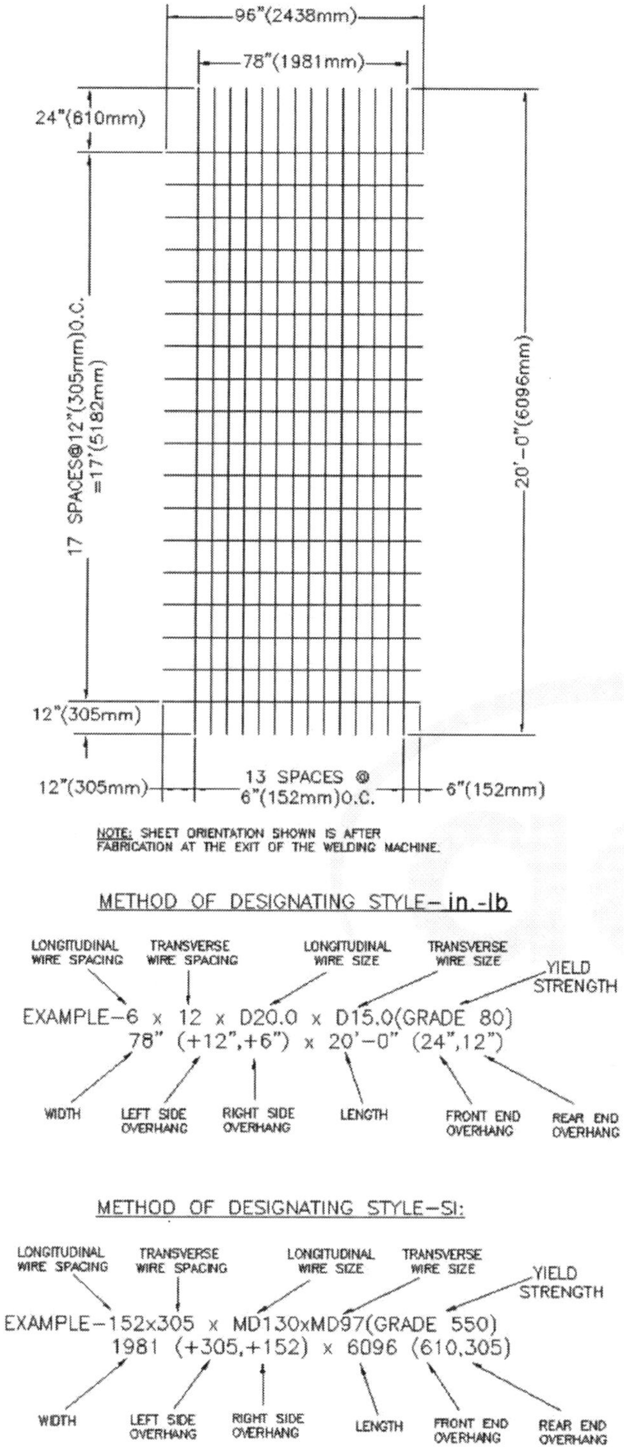

Fig. 1.4—Nomenclature.

widths are determined by machine, job site, or state shipping constraints.

1.4.2 *Length*—Length is the tip-to-tip dimension of the longitudinal wires. The length dimension is usually inclusive of the end overhangs in the sheet style designation. The length should always be specified as the fourth entry on the second line of the sheet style designation. Figure 1.4 demonstrates graphically the definition of length.

1.4.3 *Overhang*—There are two types of overhangs that define a sheet style. The side overhang is the extension of transverse wires beyond the centerline of outside longitudinal wires. The side overhangs are added to the defined sheet width to equal the overall sheet width. As described in 1.4.1, the defined sheet width is the center-to-center distance between the outermost longitudinal wires. Side overhangs should be specified and defined in inches (millimeters) as the second and third entry on the second line of the sheet style designation. The second entry on the second line refers to the left side overhang as the sheet is oriented entering the welding machine. The third entry on the second line refers to the right side overhang as the sheet is oriented entering the welding machine.

The end overhang is the extension of longitudinal wires beyond the centerline of outside transverse wires. The end overhang should be included in the overall length, which is the fourth entry on the second line of the sheet style designation. The length can also be shown as the center-to center of transverse wires with plus dimensions inside the parentheses to indicate the overall length. The end overhangs are shown in inches (millimeters) as the fifth and sixth entries on the second line of the sheet style designation. The fifth entry refers to the front of the sheet as it is entering the machine for welding. The sixth entry refers to the rear of the sheet as it is entering the machine for welding. Figure 1.4 demonstrates graphically the definition of side and end overhang.

1.4.4 *Weight*—Weight (mass) is calculated by computing the weights of longitudinal and transverse wires separately and adding the two results for the total. The weight should always be entered after the width and length listing on line three of the sheet style designation. If variable spacing or length is used, the weight should be listed on the next line after the spacing or length reference. The calculation that was used to determine weight need not be shown on the sheet style designation. The sheet weight is calculated as follows:

calculated actual weight = longitudinal weight + transverse weight

longitudinal weight = $Wt_l \times N_l \times L$ (round to two decimal places)

transverse weight = $Wt_t \times N_t \times OW$ (round to two decimal places)

1.4.5 *Sheet quantity*—Sheet quantity should always be listed on the line after the weight listing in the sheet style designation.

CHAPTER 2—NOTATION AND DEFINITIONS

2.1—Notation

L = length of sheet, ft (m)
N_l = number of longitudinal wires
N_t = number of transverse wires
OW = overall width, ft (m)
Wt_l = unit weight of one longitudinal wire, lb/ft (kg/m)

Wt_t = unit weight of one transverse wire, lb/ft (kg/m)

2.2—Definitions

ACI provides a comprehensive list of definitions through an online resource, "ACI Concrete Terminology." Definitions provided herein complement that resource.

style—spacing and size of wires arranged approximately at right angles and electrical resistance welded together to form welded-wire reinforcement (WWR).

CHAPTER 3—CODES AND STANDARDS

Three organizations govern the design, manufacture, and testing of concrete reinforcement and are discussed in the following.

3.1—AASHTO

American Association of State Highway and Transportation Officials (AASHTO) is the main governing body for transportation-related design and construction. Public road projects such as bridge decks, bridge beams, roadway or airport paving, or state-funded box culvert work usually fall under the appropriate AASHTO specification. Many AASHTO material standards reference or mirror ASTM specifications for manufacture and testing of wire and WWR products. The AASHTO standards governing wire and WWR are as follows:

a) AASHTO M 32M/32
b) AASHTO M 55M/55
c) AASHTO M 221M/221
d) AASHTO M 225M/225

3.2—American Concrete Institute

ACI 318 is the main governing code for structural concrete design in the United States. Check with building officials for locations outside the United States.

3.3—ASTM International

ASTM International provides procedures for all testing of WWR. The following specifications govern the testing of WWR as well as requirements for the application and testing of corrosion-inhibiting materials. The specifications governing WWR are as follows:

a) ASTM A884/A884M
b) ASTM A1022/A1022M
c) ASTM A1060/A1060M
d) ASTM A1064/A1064M

CHAPTER 4—MANUFACTURING PROCESS

4.1—Sustainability

WWR is a sustainable product with a high percentage of steel that is composed of recycled material that is 100 percent recyclable. Steel mills melt reclaimed steel into billets (square sections of varying lengths) with certain specified compositions for use in making different steel products. Continuous cast or reheat mills roll steel billets into various diameter rods. A welded wire manufacturer cold-works the rods into the wire size and strength required to produce WWR for use in reinforced concrete.

4.2—Cold-working

The basic material in the manufacture of wire is a hot-rolled steel rod. It is approximately circular in cross section and is produced and shipped in coils. If a rod is subjected to a tensile test, the stress-strain curve will be similar to the line for hot-rolled rod, as shown in Fig. 4.2a. The stress-strain curve has a distinct transition in behavior at which the strain increases without a corresponding linear increase in stress, and there is a considerable extension between that point and rupture. The relationships between stress and strain primarily depends on the chemical composition of the steel rods. Because the rod is hot-rolled, these properties are independent of the size of the rod.

To transform the rod into wire, it is passed through a series of reducing rollers or dies. In addition to decreasing the diameter and increasing the length, this operation has some very important effects on the mechanical properties, as is shown by the stress-strain curve for Grade 70 (480 MPa) wire in Fig. 4.2a. Here, there is no definite yield point in the conventional sense. Instead, once the steel has reached its proportional limit, the plot is a continuous curve to ultimate strength or rupture. The ultimate tensile strength is increased above that of the rod, but the strain at rupture is smaller. These changes are characteristic of any steel on which cold-working is done. The amount of change depends on the amount of cold-working. There is no heat in the process except that caused by the friction of the rod against the rollers or dies. This change is usually measured by the reduction in area. Large reductions in diameter result in higher tensile strength and smaller ultimate strain than in the hot-rolled rod used to make it. This is shown by comparing the stress-strain curve for the hot-rolled rod to Grade 75 (520 MPa) wire, also in Fig. 4.2a, which is for a wire rolled from the same rod but with a smaller finished diameter. Grade 80 (550 MPa) wire has a still larger amount of cold-working, and so the effect is even greater than the Grade 70 (480) wire. A cold-rolling line is shown in Fig. 4.2b.

In summary, the mechanical properties of a wire are functions of two things: the composition of the rod and the amount of cold-working done. Wires are made in many sizes, but rods are available in a limited number of diameters. Therefore, several different wire sizes are manufactured from one rod size, so each wire size has a different amount of cold-working.

Although there is no true yield plateau for cold-worked wire, the yield strength concept is so important in structural analysis and design that a nominal value is used. This is generally stated as the stress at some specified strain. According to ASTM A1022/A1022M and A1064/A1064M governing the manufacture of reinforcing wire, the yield strength is defined as the stress measured at a strain of 0.005 in./in. (mm/mm), or by the 0.2 percent offset method for wire tested to minimum yield strength requirements.

ACI 318 states that if a yield strength greater than or equal to 60,000 psi (420 MPa) is used for design of concrete struc-

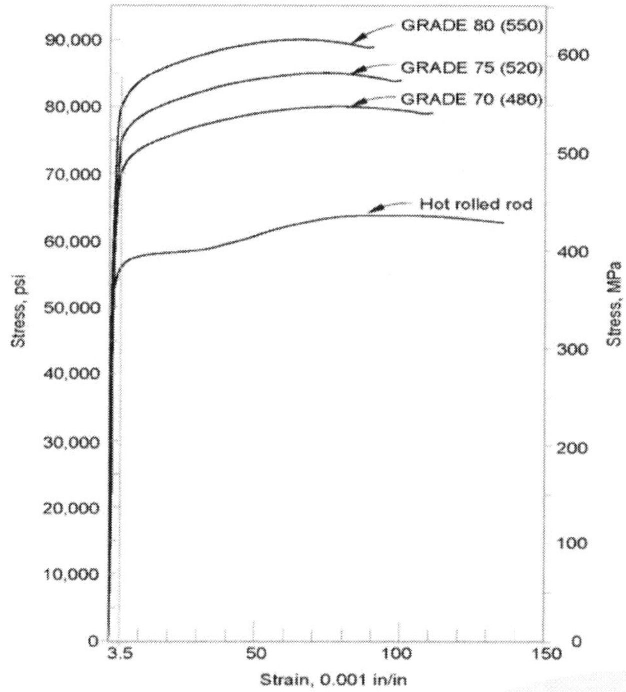

Fig. 4.2a—Idealized stress-strain curves for ASTM A1064/A1064M wire. (Note: in./in. = mm/mm.)

Fig. 4.2b—Cold-rolling line.

tures, the measured yield strength is to be determined at a strain of 0.0035 in./in. (mm/mm) or by the 0.2 percent offset method. It is vital that, when specifying WWR for reinforced concrete structures, a yield strength be specified to ensure the design requirements are met.

4.3—Straightening and cutting

The straightening and cutting operation is the second step in the manufacturing process of engineered WWR. This is where the rolled wire is fed from spools into a straightening machine and then cut to length per the manufacturing order

Fig. 4.3—Straightening and cutting lines.

(Fig. 4.3). This step is introduced into the process to enable both batch processing and flatness of the finished sheets.

4.4—Welding

Welding is the third step in the manufacturing process. High-strength cold-rolled wires are welded together into square or rectangular grids. This process uses a resistance weld where the longitudinal wire is fed through the welding machine (Fig. 4.4a) from a field of wire spools or a feed table of straightened and cut wires (Fig. 4.4b), and the cross-wire is dropped onto the longitudinal wire by an overhead hopper or fed from a spool (Fig. 4.4c). The introduction of adequate pressure and electric current creates the heat necessary for the fusing of wires for welded intersections (Fig. 4.4d). Welders fed by wire spools may produce rolls or sheets of WWR, whereas straightened- and cut-wire-fed welders produce only WWR sheets (4.4e). This completes the process of manufacturing WWR. This material is banded and prepared for delivery, or can be sent to post-weld fabrication (4.5).

4.5—Fabrication

4.5.1 *Bending*—Bending is a cold-work, post-weld process. The cold-work done to wire in the bending process is limited by minimum bend diameters prescribed in various industry standards and documents. The WWR can be bent at the manufacturing facility, project site, or precast yard. Field bending of reinforcement partially embedded in concrete is not permitted, except as shown in the construction documents or permitted by the licensed design professional.

Fig. 4.4—*Resistance welding machines and resistance welding heads.*

This process uses a bending apparatus that places the wire under anvils for resistance. The moveable arm is then lifted through a hydraulically controlled foot or hand pedal to gain the appropriate bend angle and dimension (Fig. 4.5.1). Virtually any angle or bend scenario can be performed for projects, ranging from column cages to simple 90-degree bends for footing dowels.

4.5.2 *Rolling*—Rolling is another post-weld process. The finished welded sheet is passed through a conventional three-roll roller to form a curved shape for products such as tunnel linings or circular tanks. The roller can be set to accommodate virtually any wire size combination as well as any desired radius (Fig. 4.5.2).

4.5.3 *Cutting and shearing*—Cutting and shearing of sheets are other post-weld processes. These processes can be either manual or by automatic hydraulic guillotine method. Hand-cutting is typically used to cut out various types of openings such as doors and windows, whereas automatic cutting is used to cut various size sheets from a single production run (Fig. 4.5.3).

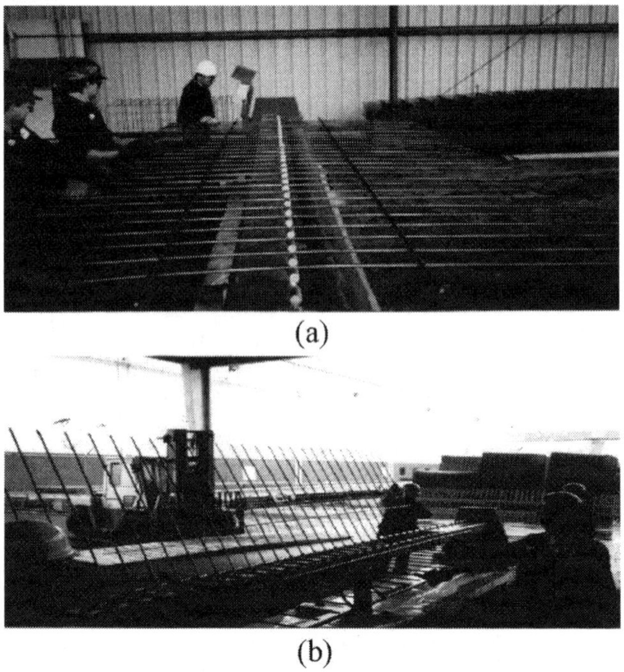

Fig. 4.5.1—(a) Bending machine; and (b) first bend.

4.6—Shipping and handling

Prior to shipment, several quality assurance tests required by the applicable ASTM standard are conducted in accordance with that standard. When required by the purchaser, material certification is provided with the shipment in accordance with the applicable ASTM standard.

The final component of the manufacturing process is shipping. This step becomes important when various sizes and scenarios require multiple shipments and sheet style mark numbers.

4.6.1 *Packaging and tagging*—Bundling instructions should be in accordance with the customer's job-site requirements and manufacturing needs. This step should be preplanned and adhered to by the shipping department to assure proper project continuity.

Fig. 4.5.2—Mercer Street Tunnel rolled WWR sheets.

Fig. 4.5.3—Shearing machine.

4.6.2 *Bundling instructions and parameters*—The following information outlines the basic bundling requirements of a typical manufacturer. WWR sheets are packaged in bundles that are secured with steel banding. The banding is generally not load-rated and should not be used for lifting the bundles. Sheets are commonly delivered to distribution centers; fabricator shops, to incorporate the welded sheets along with bundled bars or wires into a load; or directly to the construction site or precaster, on flat-bed open trailers.

The purchaser may request that the loads be covered to protect the steel from road salts during the winter months; however, this requirement should be conveyed to the WWR manufacturer at the time of project bidding.

Wooden blocking is usually placed between stacked bundles, or between stacked multiple bundles. The manufacturer will seek to fill the loads to maximum legal height or weight limits. When the height of the stacked bundles is the load-limiting parameter and packaging instructions are not specified by the purchaser, the WWR manufacturer can optimize the sheet quantities on a load by creating bundles using a method known as nested sheets. If the sheet size, weight, and wire projections are favorable, alternate sheets are flipped over as they exit the welding area of the production machine. This process results in a shorter stack of sheets than if the sheets are all stacked or all flipped. Refer to Fig. 4.6.2 for an illustration of nesting. Most welding machines will produce WWR with the crosswire always on the upper surface or always on the bottom surface.

CHAPTER 5—CONSTRUCTION APPLICATIONS

5.1—Cast-in-place reinforced concrete

Cast-in-place reinforced concrete is defined as concrete that is deposited and allowed to harden in the place where it is required to be in the completed structure. WWR has been used successfully in many applications pertaining to cast-in-place reinforced concrete.

5.1.1 *Footings*—Footings are typically designed as isolated or continuous strips. Isolated footings offer a greater potential for use of WWR because they are generally square or rectangular, and usually require simple reinforcement patterns in a single or double layer (Fig. 5.1.1).

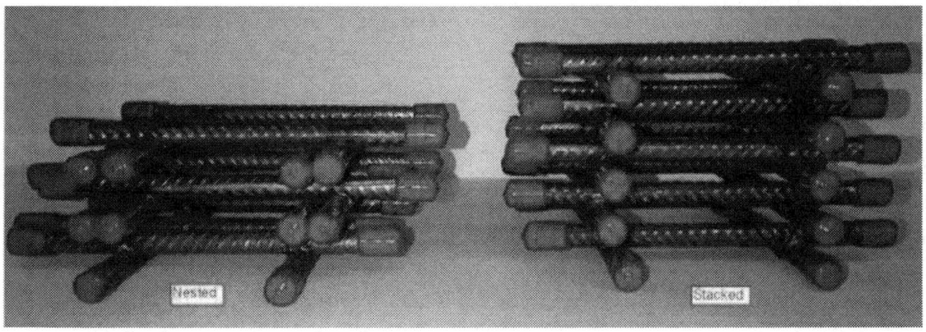

Fig. 4.6.2—Bundles showing nested and stacked methods.

Fig. 5.1.1—WWR in an isolated footing.

5.1.2 *Columns and beams*—Column ties and beam stirrups can be designed using WWR in a variety of ways. Depending on code provisions and product application, the column ties can be designed and bent as a single closed cage unit or a multi-piece unit with overlapping ties, spliced according to ACI 318. It may also be possible to incorporate the main reinforcement as part of the cage formed by the WWR (Fig. 5.1.2). WWR can include the reinforcement required for the longitudinal, confinement, transverse, and shear reinforcement, or all these.

5.1.3 *Slabs*—WWR is more often used for slabs that are cast directly on ground. Different WWR patterns are used to reinforce suspended slabs due to different design parameters and slab types. The steel reinforcement for suspended slabs is needed to carry tensile stresses caused by bending from applied loads, whereas slabs-on-ground are principally reinforced to control cracking caused by shrinkage, temperature changes, and subgrade support from nonuniform loading. With the versatility of new and existing machinery, more efficient configurations offer alternatives that at one time were unachievable. For example, by extending overhangs to account for lap splices as governed by ACI 318, WWR sheets can provide lap conditions similar to reinforcing bars. This lap is often called a finger lap splice or finger splice in the wire industry (Fig. 5.1.3). Refer to 6.5 on lap splicing and anchorage. Further, this machinery allows the licensed design professional to offer various sizes of mats within one project depending on the joint spacing rather than relying on the contractor to waste valuable time cutting sheets to fit an area. The contractor may have to cut sheets to suit isolated or small areas or elevation steps from one of the mats provided for the project.

5.1.4 *Walls*—Wall design can be classified in four primary types: nonbearing walls, bearing walls, retaining walls, and shear walls. It is difficult to divide the functions of these walls precisely, but, in general, a bearing wall is designed primarily to support vertical loads, a retaining wall to carry horizontal forces perpendicular to the face of the wall, and a shear wall to resist horizontal forces parallel to the wall. Many walls, however, serve more than one of these functions.

Steel reinforcement is used in concrete walls for two reasons: to resist tensile, shear, and compressive forces resulting from horizontal and vertical loading on the wall, and to minimize and control cracking caused by shrinkage, temperature changes, and differential settlement.

Fig. 5.1.2—Rolled WWR column retrofit.

Fig. 5.1.3—WWR sheets for airport runway.

Multi-story buildings with exterior and interior walls of the same height offer an opportunity for the use of WWR. Normally, the design documents will specify the type of reinforcement required to carry the loading. The designer should know the type of wall being constructed and the construction methods that will be used to assure an adequate WWR design that will work in tandem with the construction parameters.

5.2—Precast concrete

WWR is a widely used form of reinforcement in many types of precast concrete. Precast concrete is formed and cured at a location other than where it is finally used. With the exception of a few small items that may be cast at the job site, precast concrete is usually made at a location remote from the point of use, generally in an established precast plant.

5.2.1 *Building, housing, landscape, and specialty products*—These products are used to construct commercial and residential buildings and improve the appearance or usefulness of the physical site around a building or residential site. Systems such as precast steps, walls, and slabs are examples of some products that fall under this category. Each one of these products typically uses WWR in a prefabricated and customized manner to provide for the exact dimensions and layout of the home or office being built.

Architectural panels can be produced with special finishes ranging from exposed aggregate to colored or patterned designs cast into the product during fabrication. WWR is widely used in this product, as it produces good crack control on surfaces that require it. Further, it provides ease of installation in a repetitive product.

WWR provides for good constructability or fabrication of retaining wall systems and modular paving products. Because most of these designs do not require heavy reinforcement, the multitude of wire sizes available makes it easier for the designer and precaster to provide the areas of steel required, thus providing additional savings.

5.2.2 *Sanitary and stormwater products*—These products are typically used to convey or direct stormwater or sanitary water from the initial source to a final destination. Such products include, but are not limited to, manholes, catch basins, curb inlets, precast pipe, and pipe chambers. A primary reinforcement for the aforementioned products is WWR. The use of WWR, usually in commonly stocked rolls, provides an efficient layout for reinforcing these products.

5.2.3 *Water and wastewater products*—These products are used to hold or contain water, oil, or other liquids for the purpose of further processing into noncontaminating liquids and soil products. The typical products that fall into this category are septic tanks, wet wells, and distribution boxes. Many federal, state, and local agencies already reference the use of WWR for these products. The wire size is typically small and is bent and formed into the required shape for the product being cast.

5.2.4 *Transportation products*—These products are used in the construction, safety, and site protection of road, airport, and transportation systems. Many of these products exclusively use WWR as their primary method of reinforcement.

Traffic and median barriers use many different styles and shapes of WWR, depending on the barrier size and shape. When the barrier requires a sheet to be bent or shaped, it is usually performed at the manufacturing plant and shipped to the precast plant as a prebent product. Sound barrier wall panels have been used for many years to reduce the traffic noise from an adjacent highway into a nearby residential neighborhood. Because these panels are typically rectangular and repetitive in quantity, WWR is typically used as reinforcement. It is shipped to the precast plant in sheets that are custom designed and fabricated to the dimensions required by the various heights and widths of the precast sound wall panel.

Another increasingly popular use for this type of reinforcement is in the precast tunnel segment. This product is precast into forms with a predetermined radius. Because of this radius, it is necessary to provide a sheet of reinforcement that is prerolled and will fit into the form with no apparent adjustments (4.5.2). These segments use many different types of reinforcement configurations.

One of the more popular uses of WWR is for the box culvert precast member. The typical WWR provides for sufficient crack control and flexural resistance after cracking in both the circumferential and longitudinal areas of the box culvert. Also, the WWR reinforcement or styles are typically custom designed to match the span and rise of the box culvert being cast.

5.2.5 *Communication, utility, and industrial products*—This category of precast items includes precast products that are incorporated into electrical, gas, telephone, and other utility segments of construction. These structures are very similar to the distribution boxes discussed in 5.3.2. These products include junction boxes, also referred to as distribution boxes, and can typically be specified for use with WWR.

5.3—Prestressed and post-tensioned concrete

5.3.1 *Commercial products*—WWR is typically used in prestressed and post-tensioned products as shear reinforcement, such as the dead load of a bridge deck bearing on the prestressed girders and vehicular loads on bridge decks where the prestressed girders provided dead, live, and vehicular load support. Prestressing strands provide the compression force needed under service load conditions to offset the tension forces in the concrete from applied dead and live loads, and provide the tensile strength required under factored load conditions or nominal strength. The additional reinforcement provided by the WWR may be considered confinement reinforcement that resists splitting stresses at the ends of beams during the application of the prestressing force.

A post-tensioned structural member is typically reinforced with prestressing strands and nonprestressed steel reinforcement. The WWR will aid in resisting stresses that occur during transportation and erection as well as help redistribute induced load stresses on the designed member through the welded wire intersections. This type of product is used in the construction of buildings such as mixed use/retail, office buildings, and schools.

5.3.2 *Industrial products*—These products span the manufacturing and food-processing plants and the large warehouse and distribution centers. For many years, the typical wall construction for these buildings was double-tee. This type of construction uses a member that looks very much like its namesake. It is composed of a single monolithic top slab cast with two vertical legs or stems. Two different types of WWR are used extensively in these products: one for the stems of the product and one for the top slab or flange. Two additional products that typically use WWR are prestressed: precast columns and beams. These columns and beams typically use prestressing strand running longitudinally through the member with reinforcement stirrups to confine the main longitudinal steel. Often, WWR is provided as a prebent unit for the confinement stirrups around the prestressing strand.

5.3.3 *Transportation products*—These products are primarily used for the construction of bridges and marine projects as well as sound wall panels. One of the most popular methods for bridge building is the use of prestressed girder members. These range from the bulb-tees to the AASHTO I-beams as well as the trapezoidal members developed by many state departments of transportation, and the box beam widely used in many areas of the country.

In the bulb-tee and I-beam types of girders, WWR is used in the top and bottom flange for shear reinforcement and confinement reinforcement to resist bursting stresses that may occur after the application of the prestressing force on the beam. WWR is also used as shear reinforcement in the web of the beam and is provided in flat or bent sheets to accommodate the height and length of the bulb-tees and I-beams being cast.

5.3.4 *Housing products*—An increasingly popular option for constructing housing projects is through the use of prestressed modular units. These include, but are not limited to, dormitories, hotels, healthcare facilities, single- and multi-family houses, and retirement facilities.

The aforementioned units are cast at the precasting/prestressing plant as a modular unit typically containing the floor, walls, and roof or ceiling. The WWR is provided in multiple styles for the floor, walls, and roof. The window openings are typically cut out at the plant to ensure proper location of the opening.

5.3.5 *Specialized products*—These products have a wide range of possibilities for the use of WWR and the versatility that it allows. They typically range from correctional facilities, to stadiums and parking structures, to potentially any integrated precast/prestressed structure. These products use some of the aforementioned prestressed members such as the column and beams and the double-tee system. They may also employ the modular type construction described in 5.3.4.

One use under the category of specialized product would be for stadium risers. These members have the typical look of a stair unit and are precast and prestressed members with nonprestressed reinforcement to resist loads and control shrinkage and temperature cracking. WWR is commonly used in this product and is either prebent at the manufacturer's location or shipped in flat sheets and bent by the prestressed concrete producer. In either event, the bent sheet is placed in the form after the initial tensioning of the strand and then concrete is cast into the form. The need for quality control inspection is kept to a minimum, as the sheets are prebent to industry standards.

5.4—Tilt-up

Tilt-up construction with only a few open areas provides an excellent example of long-run wall panels where WWR can be used (Fig. 5.4). Dead, live, wind, or even seismic loads do not always govern the design of tilt-up panels. Lifting the panels may govern the design of the panel and its reinforcement, depending on panel dimensions. Furthermore, panel deflections during lifting may actually govern the amount of reinforcement required. In any case, it is the designer's responsibility to understand the design and construction parameters of tilt-up panels to assure an appropriate design with WWR.

Fig. 5.4—WWR sheets for tilt-up wall panels.

CHAPTER 6—CONSTRUCTIBILITY

6.1—Field welding

Machine and resistance welding, as used in the manufacture of WWR, is covered by ASTM A1064/A1064M. The provisions for field welding do not apply to the resistance welds in WWR. Special procedures are followed in plant production of WWR, and it is difficult to perform the same welding techniques in the field. There are portable hand-held electric resistance welders, but they are not used on job sites; they are only used in WWR plants where proper electrical power is provided and the correct safety procedures are followed.

Engineers have approved field welding of WWR, but should not allow welds in the regions of high stress; for example, in the middle of spans or close to columns. The engineer of record should specify requirements or performance criteria for field welding if it is desired over other methods of splicing, such as lapping sheets of WWR together. ANSI/AWS D1.4/D1.4M should be used as a reference for specifying field welding.

6.2—Field bending

Bending WWR in the field is common. Benders can be placed on the truck with bundles of WWR sheets and trucked to the job site. Many general concrete contractors use this method to obtain the correct bend measurements to fit the project. Fabricators have benders in their yard or plant. Sometimes, sheets are partially bent at the plant site, then final bending is completed on the site to suit exact dimensions. Partially bent sheets can be nested to provide space and maximum weight for shipment. Bending machines can be made to suit any length or width of WWR sheet, usually 15, 20, or 30 ft (4.5, 6, or 9 m) lengths or longer. Various size mandrels are used to provide the correct bend in accordance with building codes and ASTM standards (WWR 600).

6.3—Cutting

Sheets of WWR can be cut to suit project layout requirements. It is more desirable and economical to provide larger quantities of the same-size sheets than smaller quantities of various-size sheets. If sheets need to be cut, they can be cut with overhangs that are 1/2, 1, 2 in. (12.5, 25, 51 mm), or up to one-half the wire spacing. The latter is most common with WWR manufacturers. Manufacturers can provide flush cuts on the transverse or longitudinal length of the sheet. End overhangs of sheets are usually one-half the wire spacing, unless finger splices are required. Refer to 6.5 for more details on laps and splicing.

Field cutting is simple with the use of bolt cutters. Sheets can be supplied to the site without openings. Sheets can be cut on site to obtain the exact desired location for openings.

6.4—Lap splicing and anchorage

Lap splicing welded wire sheets can be accomplished in multiple ways. Plain and deformed WWR are spliced differently, and lap splice methods and calculations are provided in ACI 318.

Plain WWR is spliced by overlapping welded intersections as illustrated in Fig. 6.4a. Two welded intersections are overlapped to accomplish the splice, and the lap length is defined by the total distance between the two outermost welded intersections.

Deformed WWR can be spliced by overlapping welded intersections similarly to plain WWR, as in Fig. 6.4b, or can be spliced using the deformed wire splice criteria as illustrated in Fig. 6.4c. Because deformed wire can be developed without welded intersections, the deformed WWR can be spliced by relying only on the deformed wire length extending from the sides or ends of the sheets. This method keeps the lap splice in a single plane and does not require the overlap of welded intersections. When the extended overhangs are used in both directions, single-plane lap spices can be achieved in both directions and concrete clear cover is consistent across all lap splice locations. The extended overhangs of the deformed wires in Fig. 1.4 are intended to illustrate how this might be done in a sheet.

Refer to WWR 500 for building code requirements.

6.5—Placing tolerances

The dimensional tolerances of WWR are vital when placing sheets within screed rails to maintain required concrete cover around the forms. Tolerances for WWR are the same for other types of reinforcing and are covered in ACI 318 and ACI 117. A key to controlling cracking and to keep crack widths to a minimum to achieve aggregate interlock is to adequately support sheets of WWR. The sheets should be placed on sturdy supports to maintain uniform cover in accordance with building codes (WWR 600; WRI TF 702-08).

6.6—Detailing

Many WWR manufacturers have the capability of computer-aided detailing of WWR sheets into projects they are awarded. Many offer this service when WWR is specified and purchased by reinforcing steel fabricators (WWR 600).

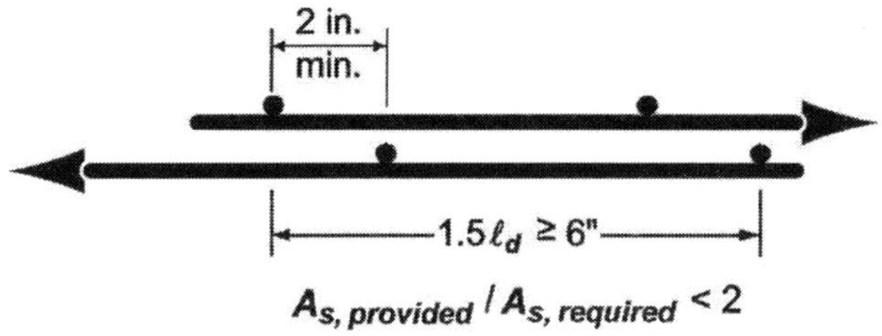

Fig. 6.4a—Lap splices of plain WWR.

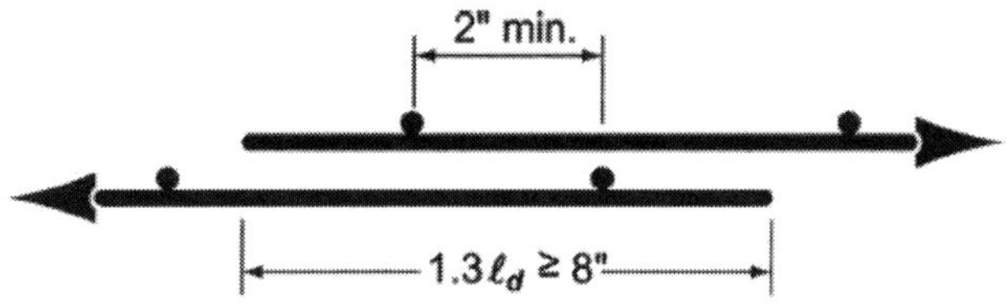

Fig. 6.4b—Lap splices of deformed WWR.

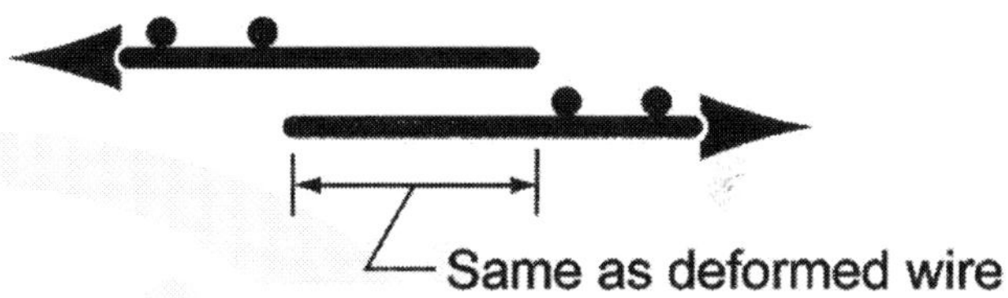

Fig. 6.4c—Lap splices of deformed WWR.

CHAPTER 7—REFERENCES

Committee documents are listed first by document number and year of publication followed by authored documents listed alphabetically.

American Association of State Highway and Transportation Officials (AASHTO)

AASHTO M 32M/32-09—Standard Specification for Steel Wire, Plain, for Concrete Reinforcement

AASHTO M 55M/55-09—Standard Method of Test for Steel Welded Wire Reinforcement, Plain, for Concrete

AASHTO M 221M/221-09(2013)—Standard Specification for Steel Welded Wire Reinforcement, Deformed, for Concrete

AASHTO M 225M/225-09(2013)—Standard Specification for Steel Wire, Deformed, for Concrete Reinforcement

American Concrete Institute (ACI)

ACI 318-14—Building Code Requirements for Structural Concrete and Commentary.

ACI 117-10(15)—Specification for Tolerances for Concrete Construction and Materials and Commentary

American National Standards Institute/American Welding Society (ANSI/AWS)

ANSI/AWS D1.4/D1.4M:2011—Structural Welding Code-Reinforced Steel

ASTM International

ASTM A884/A884M-14—Standard Specification for Epoxy-Coated Steel Wire and Welded Wire Reinforcement

ASTM A1022/A1022M-16—Standard Specification for Deformed Stainless Steel Wire and Welded Wire for Concrete Reinforcement

ASTM A1060/A1060M-16—Standard Specification for Zinc-Coated (Galvanized) Steel Welded Wire Reinforcement, Plain and Deformed, for Concrete

ASTM A1064/A1064M-17—Standard Specification for Carbon-Steel Wire and Welded Wire Reinforcement, Plain and Deformed, for Concrete

Wire Reinforcement Institute (WRI)

WRI TF 702R-08—Tech Facts: Supports Are Needed for Long-Term Performance of Welded Wire Reinforcement in Slabs-On-Grade.

WWR 500-16—Manual of Standard Practice: Structural Welded Wire Reinforcement

WWR 600-06—Structural Detailing Manual

Authored documents

Richardson, R. C., 2000, "Men of Steel—The Story of Welded Wire Fabric and the Wire Reinforcement Institute—50 Years Old and Growing," Wire Reinforcement Institute.

Guide for the Use of ASTM A1035/A1035M Type CS Grade 100 (690) Steel Bars for Structural Concrete

Reported by ACI Committee 439

Todd R. Hawkinson, Chair

David H. DeValve, Secretary

Paul B. Aubee
Brian L. Barrows
Richard H. Birley
George T. Biro
Domingo J. Carreira
Timothy L. Cartwright
Louis J. Colarusso
Salem S. Faza
Anthony L. Felder
Augusta Gaertner
William C. Gallenz
Allen J. Hulshizer
Richard Huza
Josh Ison
Jason Koehler
Harry B. Lancelot III
Kenneth A. Luttrell
LeRoy A. Lutz
Carl D. Maki
Mark D. Marvin
Theodore A. Mize
Conrad Paulson
Ryan W. Pelter
Shannon Pole
Richard A. Ramsey
Harold E. Reed
Mario E. Rodriguez
Clifford A. Sabo
William H. Zehrt Jr.
Phil J. Zivich

Consulting Members

Jose Bagg
Jean-Jacques Braun
Luis E. Garcia
Kent A. Harries
Douglas D. Lee

This guide provides recommendations on design provisions for the use of ASTM A1035/ASTM A1035M Type CS Grade 100 (690) deformed steel bars for reinforced concrete members. The recommendations address only those requirements of ACI 318-14 that limit efficient use of such steel bars. Other code requirements are not affected. Any other ACI 318 versions will be explicitly specified. Although there are limiting ACI 318 requirements, ACI 318-14 Section 1.10 would allow the use of high-strength reinforcement. "Sponsors...shall have the right to present the data on which their design is based to the building official or to a board of examiners appointed by the building official."

The International Building Code (IBC 2012) would allow the same under Section 104.11, "Alternative materials, design and methods of construction and equipment". To approve an alternative material under this section, a building department would typically require an ICC Evaluation Service (ICC-ES) Evaluation Report, which would be based on an ICC-ES Acceptance Criteria (AC) document. An AC document (ICC-ES AC429) and an Evaluation Report (ICC-ES ESR-2107) exist, permitting the use of ASTM A1035/A1035M Grade 100 reinforcement.

This guide includes a discussion of the material characteristics of Grade 100 (690) ASTM A1035/A1035M (CS) deformed steel bars and recommends design criteria for beams, columns, slab, systems, walls, and footings for Seismic Design Category (SDC) A, B, or C, and for structural components not designated as part of the seismic-force-resisting system for SDC D, E, or F.

A structure assigned to SDC A, B, or C is required to be designed for all applicable gravity and environmental loads. In the case of SDC A structures, seismic forces are notional structural integrity forces. This guide addresses all design required for SDC A, B, and C structures.

Because the modulus of elasticity for ASTM A1035/A1035M (CS) is similar to that of carbon steel (ASTM A615/A615M) using higher specified minimum yield strength f_y may result in higher steel stress at service load condition and potentially cause wider cracks and larger deflections, which may be objectionable if aesthetics and water-tightness are critical design requirements. Higher deflection can also contribute to serviceability issues. Also, with higher f_y, the required development length will be longer.

Keywords: bar; design; guide; high-strength steel; structural.

ACI 439.6R-19 supersedes ACI ITG-6R-10 and was adopted and published February 2019.

CONTENTS

CHAPTER 1—INTRODUCTION

1.1—Objective

This guide is based on ACI ITG-6R-10, "Design Guide for the Use of ASTM A1035/A1035M Grade 100 (690) Steel Bars for Structural Concrete," reported by ACI Innovation Task Group 6. The ACI ITG-6R guide provides design provisions for the use of ASTM A1035/A1035M Type CS Grade 100 (690) deformed steel bars in reinforced structural members. This guide, which is based on ACI ITG-6R-10, is a stand-alone document, references and addresses only those requirements in ACI 318-14 that limit the use of such steel bars, and should not affect the application of other code requirements. Any other ACI 318 version will be explicitly specified. This guide includes a discussion of the material characteristics of Grade 100 (690) ASTM A1035/A1035M Type CS (Chromium content 8.0 to 10.9 percent) deformed steel bars, and the design provisions are based on the specific material properties and stress-strain behavior of these bars.

Although there are limiting ACI 318 requirements, ACI 318-14 Section 1.10 would allow the use of high-strength reinforcement. "Sponsors...shall have the right to present the data on which their design is based to the building official or to a board of examiners appointed by the building official."

The IBC International Building Code (IBC 2012) would allow the same under Section 104.11, "Alternative materials, design and methods of construction and equipment". To approve an alternative material under this section, a building department would typically require an ICC Evaluation Service (ICC-ES) Evaluation Report that would be based on an ICC-ES Acceptance Criteria (AC) document. An AC document (ICC-ES AC429) and an Evaluation Report (ICC-ES ESR-2107) exist, permitting the use of ASTM A1035/A1035M Grade 100 reinforcement.

Since publication of ACI ITG-6R, additional Grade 100 (690) ASTM A1035/A1035M Types CM (Chromium content 4.0 percent to 7.9 percent) and CL (chromium content 2.0 percent to 3.9) were introduced to the market. The mechanical properties and stress-strain characteristics of these steels may affect the design criteria herein and limit the applicability of the guide. Additionally, ASTM A615/A615M Grade 100 (690) was introduced to the market. These bars will be addressed in future editions when research on concrete members reinforced with these bar types become available and the impact of the mechanical properties have been documented. Grade 120 (830) ASTM A1035/A1035M Type CS, CM and CL are available, but are not addressed under this guide.

1.2—Scope

This guide presents the material characteristics of ASTM A1035/A1035M (CS) steel bars and recommends design criteria for beams, columns, slab systems, walls, and footings for Seismic Design Category (SDC) A, B, or C. A structure assigned to SDC A, B or C is required to be designed for all applicable gravity and environmental loads. In the case of SDC A structures, seismic forces are notional structural integrity forces. This guide addresses all design required for SDC A, B, and C structures. Due to lack of adequate data, the application of this guide for SDC D, E, or F is limited to slab systems, foundations, and structural components not designated as part of the seismic-force-resisting system, but explicitly checked for the induced effects of the design displacements. The only exception is the use of transverse reinforcement for concrete confinement with a specified minimum yield strength f_y up to 100,000 psi (690 MPa) in special moment frames, special structural walls, and all components of special structural walls including coupling beams and wall piers as permitted by ACI 318. Refer to 10.1 of this guide for more information on seismic design considerations. Shells, folded plate members, and prestressed concrete are beyond the scope of this guide. However, ASTM A1035/A1035M Type CS can be used as reinforcement in prestressed concrete but not as the prestressing strands. Design examples are included to illustrate design procedures and proper application of the design criteria. Modifications to these design criteria may be justified where the design adequacy within the scope of this guide is demonstrated by successful use, analysis, or test.

1.3—Historical perspective and background

For several decades prior to ACI 318-71, the design of structural concrete was restricted to using specified minimum yield strength f_y of 60,000 psi (420 MPa) or less for reinforcing bars. Section A603(e) of ACI 318-56 specified that "Stress in tensile and compressive reinforcement at ultimate load shall not be assumed greater than the yield point or 60,000 psi, whichever is smaller."

Section 1505 of ACI 318-63, specified two requirements:

"(a) When reinforcement is used that has a yield strength, f_y, in excess of 60,000 psi (420 MPa), the yield strength to be used in design shall be reduced to $0.85f_y$ or 60,000 psi (420 MPa), whichever is greater, unless it is shown by tension tests that at a proof stress equal to the specified minimum yield strength, f_y, the strain does not exceed 0.003;

(b) Designs shall not be based on a yield strength, f_y, in excess of 75,000 psi (520 MPa). Design of tension reinforcement shall not be based on a yield strength, f_y, in excess of 60,000 psi (420 MPa) unless tests are made in compliance with Section 1508(b)."

The Commentary on Section 1505 of ACI 318-63 states that:

> This section provides limitations on the use of high strength steels to assure safety and satisfactory performance. High strength steels frequently have a strain at yield strength or yield point in excess of the 0.003 assumed for the concrete at ultimate. The requirements of Section 1505(a) are to adjust to this condition.
>
> The maximum stress in tension of 60,000 psi (420 MPa) without test is to control cracking. The absolute maximum is specified as 75,000 psi (520 MPa) to agree with present ASTM specifications and as a safeguard until there is adequate experience with the high stresses.

Then, the Commentary on Section 1508 of ACI 318-63 states that:

> When the design yield point of tension reinforcement exceeds 60,000 psi (420 MPa), detailing for crack control becomes even more important. Entirely acceptable structures have been built, particularly in Sweden, with a design yield strength approaching 100,000 psi (690 MPa) but more design criteria for crack control and considerable American practical experience with 60,000 psi (420 MPa) yield strength tension reinforcement are needed before higher yield strengths are approved for general use. The Code, therefore limits *tension* reinforcement to 60,000 psi (420 MPa) yield strength, unless special full-scale tests are made. It was thought that 75,000 psi (520 MPa) yield strength tension reinforcement should be permitted where full-scale testing is economically feasible, such as in precast members. The crack width criteria are not too difficult to meet by proper attention to reinforcing details.

When the use of Grade 40 (280) reinforcing bars in the 1930s and 1940s was replaced by the use of Grade 60 (420) bars in the 1950s and 1960s, there were concerns about fatigue resistance of the higher-strength steel bars. Similar concerns were expressed about the use of ASTM A1035/A1035M (CS) steel bars when they were introduced. El-Hacha and Rizkalla (2002) and DeJong et al. (2006) conducted studies on the fatigue behavior of ASTM A1035/A1035M (CS) steel bars. Their results indicated that a fatigue life of 1×10^6 cycles was observed at a stress range of approximately 44,000 psi (310 MPa) for ASTM A1035/A1035M (CS) steel bars. The ASTM A1035/A1035M (CS) steel bars showed comparable fatigue resistance to Grade 60 (420) reinforcing bars even though their stress at service would be higher than that of Grade 60 (420) bars.

1.4—Reinforcing steel grades availability

The most widely used deformed reinforcing bars conform to ASTM A615/A615M, which include Grade 40 (280), Grade 60 (420), Grade 75 (520), and Grade 80 (550). The Grade 60 (420) reinforcement exhibits minimum yield strength of 60,000 psi (420 MPa) with a distinct yield plateau. ACI 318 permits use of reinforcing bars with a specified minimum yield strength f_y exceeding 60,000 psi (420 MPa), but f_y is limited to the lesser of 80,000 psi (550 MPa) or the stress corresponding to a strain of 0.0035, except as follows. ACI 318 limits the specified minimum yield strength for deformed bars used as shear reinforcement to 60,000 psi (420 MPa). For deformed bars used as confinement reinforcement (ties or spirals) in compression members, ACI 318 permits the use of specified minimum yield strength of up to 100,000 psi (690 MPa).

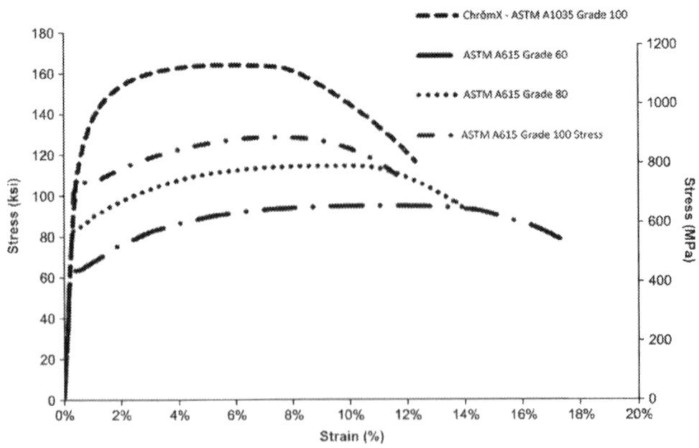

Fig. 1.5—Comparison of typical stress-strain curves for ASTM A615/A615M and ASTM A1035/A1035M (CS) reinforcing bars.

1.5—Introduction of ASTM A1035/A1035M Type CS Grade 100

The introduction of higher-strength steel reinforcing bars with a specified minimum yield strength, $f_y = 100$, the designers might be able to reduce the total cross-sectional area of required reinforcement. The reduced area of reinforcement could result in fewer bars and reduce reinforcement congestion often encountered in mat foundations, shear walls, beam-column joints, and many precast concrete elements. The reduction in reinforcement congestion facilitates concrete placement and consolidation.

The ASTM A1035/A1035M (CS) bar Grade 100 (690) (Fig. 1.5) exhibits a linear stress-strain relationship up to a proportional limit ranging from 60,000 to 80,000 psi (420 to 550 MPa), without a well-defined yield plateau. Refer to Appendix C of this guide for a discussion on how the lack of a well-defined yield plateau affects the flexural behavior of beams. Actual yield strength, determined by the 0.2 percent offset method, typically exceeds 115,000 psi (790 MPa) for Grade 100 (690) bars. The tensile strength typically exceeds 155,000 psi (1070 MPa) for Grade 100 (690) bars. The corresponding strain at the peak of the stress-strain curve ranges from 0.04 to 0.06. Refer to Chapter 3 of this guide for more details on the material characteristics of ASTM A1035/A1035M (CS) reinforcing bars.

CHAPTER 2—NOTATION AND DEFINITIONS

2.1—Notation

A_s = area of nonprestressed longitudinal tension reinforcement, in.^{2} (mm^2)

A_s' = area of nonprestressed longitudinal compression reinforcement, in.^{2} (mm^2)

A_{tr} = total cross-sectional area of all transverse reinforcement that is within the spacing s and crosses the potential plane of splitting through the reinforcement being developed or lap spliced, in.^{2} (mm^2); refer to Eq. (10.2b)

a = depth of equivalent rectangular stress block as defined in ACI 318, in. (mm)

b = width of compression face of member, in. (mm)
b_w = web width, in. (mm)
C = force in the compression zone of a beam, lbf (N)
c = distance from extreme compression fiber to neutral axis, in. (mm)
c_b = $c_{min} + 0.5d_b$, in. (mm); refer to Eq. (10.2a)
c_{bb} = clear cover of reinforcement being developed or lap spliced, measured to tension face of member, in. (mm)
c_c = clear cover to tension steel, in. (mm)
c_{max} = maximum value of c_s or c_{bb}, in. (mm)
c_{min} = minimum value of c_s or c_{bb}, in. (mm)
c_s = minimum value of c_{si} + 0.25 in. (6.35 mm) or c_{so}, in. (mm); c_{si} may be used instead of c_{si} + 0.25 in. (6.35 mm)
c_{si} = one-half of average clear spacing between bars or lap splices in a single layer, in. (mm)
c_{so} = clear cover of reinforcement being developed or lap spliced, measured to side face of member, in. (mm)
d = distance from extreme compression fiber to centroid of longitudinal tension reinforcement, in. (mm)
d' = distance from extreme compression fiber to centroid of longitudinal compression reinforcement, in. (mm)
d_b = bar diameter, in. (mm)
d_t = distance from extreme compression fiber to centroid of extreme layer of longitudinal tension reinforcement, in. (mm)
E_c = modulus of elasticity of concrete, psi (MPa)
E_s = modulus of elasticity of reinforcement, psi (MPa)
f_c' = specified compressive strength of concrete, psi (MPa)
f_s = calculated tensile stress in reinforcement, psi (MPa)
f_u = specified ultimate strength of reinforcement, psi (MPa)
f_y = specified minimum yield strength of reinforcement, psi (MPa)
f_{yt} = specified minimum yield strength of transverse reinforcement, psi (MPa)
h = overall height or thickness of member, in. (mm)
h_f = flange thickness of T-beam, in. (mm)
I_{cr} = moment of inertia of cracked section transformed to concrete, in.^{4} (mm^4)
I_e = effective moment of inertia for computation of deflection, in.^{4} (mm^4)
I_g = moment of inertia of gross concrete section about centroidal axis, neglecting reinforcement, in.^{4} (mm^4)
K_{tr} = transverse reinforcement index; refer to Eq. (10.2b)
ℓ_d = development length (also splice length), in. (mm)
M = applied moment at critical section, in.-lbf (N·mm)
M_a = maximum moment in member due to service loads at stage deflection is computed, in.-lbf (N·mm)
M_{cr} = cracking moment, in.-lbf (N·mm)
M_n = nominal flexural strength at section, in.-lbf (N·mm)
n = number of bars being developed or lap spliced along plane of splitting; refer to Eq. (10.2b)
P_n = nominal axial strength of cross section, lbf (N)
R_r = relative rib area of the reinforcement = ratio of projected rib area normal to bar axis to the product of the nominal bar diameter and the center-to-center rib spacing, may be taken conservatively as 0.07 for design
s = spacing of transverse (shear) reinforcement, in. (mm)
T = force in tension reinforcement of a beam, lbf (N)
T_s = additional bond strength provided by the transverse steel, lbf (N)
t_d = bar diameter factor = $0.78d_b + 0.22$ in. (mm); refer to Eq. (10.2d)
t_r = term representing the effect of relative rib area on T_s = $9.6R_r + 0.28 \leq 1.72$; refer to Eq. (10.2c)
V = applied shear at critical section, lbf (N)
V_c = nominal shear strength provided by concrete, lbf (N)
w_u = factored load per unit length of beam or one-way slab, lb/ft^3 (kg/m^3)
α = reinforcement location factor
β = ratio of the distance from the neutral axis to the extreme tension face to the distance from the neutral axis to the center of the tension reinforcement factor relating depth of equivalent rectangular compressive stress block to neutral axis depth, as defined in ACI 318
β_c = coating factor; refer to Eq. (10.2a)
Δ = deflection, in. (mm)
ε_s = strain in reinforcement
ε_t = net tensile strain in extreme layer of longitudinal tension steel at nominal strength, excluding strains due to creep, shrinkage, and temperature
ϕ = strength reduction factor
λ = lightweight aggregate concrete factor
ρ_b = bias factor = ratio of the average actual value to the specified minimum value of a property being analyzed
ρ = reinforcement ratio = A_s/bd
ρ_b = reinforcement ratio producing balanced strain conditions as defined in ACI 318
ρ_s = volumetric spiral reinforcement ratio
ψ = curvature
ω = factor reflecting benefit of large cover/spacing perpendicular to controlling cover/spacing = $0.1(c_{max}/c_{min}) + 0.9 \leq 1.25$; refer to Eq. (10.2a)

2.2—Definitions

ACI provides a comprehensive list of definitions through an online resource, ACI Concrete Terminology. Definitions provided herein complement that resource.

0.2% offset method—method for determining a yield strength value for a material that does not exhibit a distinct yield plateau. The yield strength is the stress on the engineering stress-strain curve at its intersection with a line having a slope equal to the initial modulus of elasticity and offset from the linear elastic portion of the engineering stress-strain curve by a strain of 0.2%.

bias factor—ratio of the average actual value to the specified minimum value of a property being considered in risk analysis.

coefficient of variation (COV)—the standard deviation divided by the mean value of a variable.

cumulative distribution function (CDF)—a function or graph describing the probability distribution of a real-valued random variable taking on a value less than or equal to a particular value.

Seismic Design Category (SDC)—classification assigned to a structure based on its occupancy category, and the severity of the design earthquake ground motion. The category assignment can range from A to F.

CHAPTER 3—MATERIAL PROPERTIES

3.1—Introduction

This guide addresses high-strength deformed reinforcing bars of Type CS as defined by ASTM A1035/A1035M, "Standard Specification for Deformed and Plain, Low-Carbon, Chromium, Steel Bars for Concrete Reinforcement." As with other specifications for steel reinforcing bars, this standard includes requirements for nominal weights and dimensions, tensile properties, chemical composition, and deformations. This chapter reviews properties of primary interest to the structural designer who may specify the use of ASTM A1035/A1035M bars.

Since the publication of ACI ITG-6R, additional Grade 100 (690) reinforcing steels were introduced to the market, including ASTM A1035/A1035M Types CM and CL and ASTM A615/A615M Grade 100 (690). The mechanical properties and stress-strain characteristics of these steels may impact the applicability of this design guide. These bars will be addressed in future editions when research on concrete members reinforced with these bar types become available and the impact of the mechanical properties have been documented.

3.2—Weights, dimensions, and deformations

ASTM A1035/A1035M (CS) specifies nominal weights (mass), areas, and dimensions. Deformation requirements are the same as those specified by ASTM A615/A615M for carbon-steel bars and ASTM A706/A706M for low-alloy steel bars. Consequently, this section does not discuss bar deformations and related aspects.

3.3—Specified tensile properties

Tensile properties are paramount for structural design. Tables 3.3a and 3.3b contain a summary of the specified tensile properties for ASTM A1035/A1035M (CS) and other reinforcement (ACI 439.4R). Table 3.3a shows that the tensile strength properties for ASTM A1035/A1035M (CS) bars are significantly greater than those for ASTM A615/A615M Grade 60 (420), Grade 80 (550), and Grade 100 (690) bars. The requirements for elongation in 8 in. (200 mm) across the fracture, as shown in Table 3.3b, are comparable to those for ASTM A615/A615M Grade 80 (550) and Grade 100 (690) reinforcement, and in some cases lower than those for ASTM A615/A615M Grade 60 (420) reinforcement. As described in 1.3 of this guide, ASTM A1035/A1035M (CS) reinforcing steel does not exhibit a distinct yield plateau (Fig. 1.5). Consequently, ASTM A1035/A1035M (CS) specifies minimum yield strength according to the 0.2 percent offset method.

Table 3.3a—Specified tensile and yield strength

Bar type	Tensile strength, minimum, psi (MPa)	Yield strength*	
		Minimum, psi (MPa)	Maximum, psi (MPa)
ASTM A615/A615M Grade 60 (420)	90,000 (620)	60,000 (420)	—
ASTM A615/A615M Grade 80 (550)	105,000 (725)	80,000 (550)	—
ASTM A615/A615M Grade 100 (690)	115,000 (790)†	100,000 (690)	—
ASTM A706/A706M Grade 60 (520)	80,000‡ (550)‡	60,000 (420)	78,000 (540)
ASTM A706/A706M Grade 80 (520)	100,000‡ (690)‡	80,000 (550)	98,000 (675)
ASTM A1035/A1035M (CS, CM, and CL) Grade 100 (690)	150,000 (1030)	100,000 (690)	—

*Observed yield point for ASTM A615/A615M and ASTM A706/A706M bars, and yield strength according to 0.2 percent offset method for ASTM A1035/A1035M bars, which is applicable to ASTM A615/A615M and ASTM A706/A706M bars only when steel bar tested does not exhibit a well-defined yield point.

†Grade 100 (690) reinforcing bars have a ratio of specified tensile strength to specified yield strength of 1.15. Designers should be aware that there will, therefore, be a lower margin of safety and reduced warning of failure following yielding when Grade 100 (690) bars are used in structural members.

‡Tensile strength for ASTM A706/A706M bars should also be not less than 1.25 times measured yield strength.

Table 3.3b—Specified elongation in 8 in. (200 mm) across fracture

Bar type	Bar size no.			
	3, 4, 5, 6 (10, 13, 16, 19)	7, 8 (22, 25)	9, 10, 11 (29, 32, 36)	14, 18 (43, 57)
	Elongation in 8 in. (200 mm) across fracture, minimum, percent			
ASTM A615/A615M Grade 60 (420)	9	8	7	7
ASTM A615/A615M Grade 80 (550), 100 (690)	7	7	6	6
ASTM A706/A706M Grade 60 (420)	14	12	12	10
ASTM A706/A706M Grade 80 (550)	12	12	12	10
ASTM A1035/A1035M (CS, CM, and CL) Grade 100 (690)	7	7	7	6

3.4—Measured tensile properties

Understanding the differences in tensile behavior of higher strength steels is essential for the safe and serviceable design

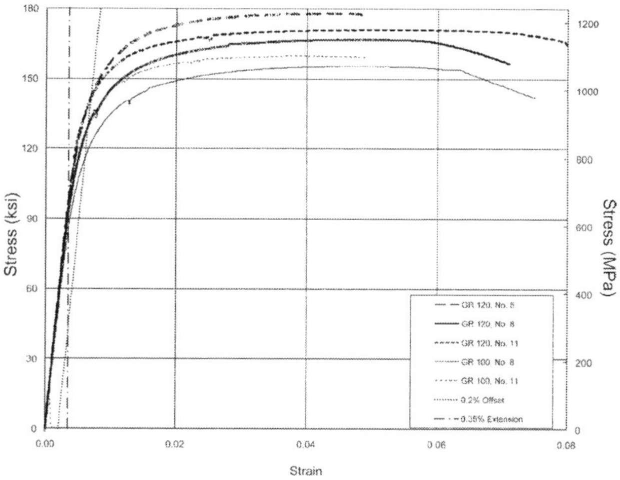

Fig. 3.4a—Actual stress-strain curves for ASTM A1035/A1035M Type CS reinforcing bars of different grades and sizes (WJE 2008).

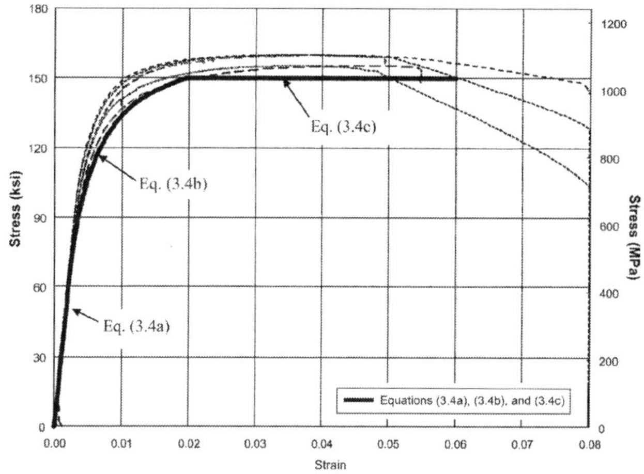

Fig. 3.4b—Equations (3.4a), (3.4b), and (3.4c) compared with actual stress-strain curves from samples of No. 8 (25) and No. 11 (36) bars of ASTM A1035/A1035M Grade 100 (690) Type CS.

of concrete members. Tensile properties are necessary in the probabilistic studies that establish the strength reduction factors used in reinforced concrete design.

Figure 1.5 shows comparisons of stress-strain curves recorded for samples of ASTM A1035/A1035M (CS) Grade 100 (690) bars to similar curves for samples of ASTM A615/A615M bars in Grades 60 (420) and 80 (550). ASTM A1035/A1035M (CS) bars have a greater tensile strength and lack a well-defined yield point and yield plateau. ASTM A1035/A1035M (CS) bars reach a proportional limit at a stress from 60,000 to 80,000 psi (420 to 550 MPa), which is similar to the yield stress of ASTM A615/A615M Grade 60 (420) and ASTM A706/A706M Grade 60 (420) bars (WJE 2008) and the strain at the peak tensile stress in the bar ranges from 0.04 to 0.06. By comparison, strains at peak tensile stress for ASTM A615/A615M Grade 60 (420) bars range from 0.07 to 0.10, and those of ASTM A706/A706M Grade 60 (420) bars range from 0.10 to 0.14.

For ASTM A1035/A1035M (CS) bars, the elongation in 8 in. (200 mm) across the fracture ranges from 0.08 to 0.13, whereas the elongation in 8 in. (200 mm) across the fracture for ASTM A615/A615M Grade 60 (420) and ASTM A706/A706M Grade 60 (420) bars range from 0.09 to 0.12 and 0.14 to 0.20, respectively. The modulus of elasticity value of 29,000,000 psi (200,000 MPa) as defined in ACI 318 is applicable to define the elastic modulus for ASTM A1035/A1035M (CS) Grade 100 (690) bars.

Figure 3.4a shows actual stress-strain curves recorded for samples of ASTM A1035/A1035M reinforcing bars (WJE 2008). Yield strength of ASTM A1035/A1035M (CS) bars, determined by the 0.2 percent offset method, exceeds 115,000 psi (790 MPa). The tensile strength for ASTM A1035/A1035M (CS) Grade 100 (690) bar exceeds 155,000 psi (1070 MPa).

An approximate lower bound for the stress-strain curves of Grade 100 (690) bars can be represented by the following three equations, as shown in Fig. 3.4b.

Equations (3.4a) through (3.4c) represent a lower bound to the stress-strain behavior of ASTM A1035/A1035M (CS) Grade 100 (690) bar. The equations are based on an assumed proportional limit of 70,000 psi (480 MPa) and an assumed tensile strength of 150,000 psi (1030 MPa), which is reached at the strain of 0.02, the upper limit for Eq. (3.4b).

$$f_s = 29{,}000\varepsilon_s \text{ (ksi) for } \varepsilon_s \leq 0.0024 \text{ (in.-lb)}$$

$$f_s = 200{,}000\varepsilon_s \text{ (MPa) for } \varepsilon_s \leq 0.0024 \text{ (SI)} \quad (3.4a)$$

$$f_s = 170 - \frac{0.4317}{\varepsilon_s + 0.0019} \text{ (ksi) for } 0.0024 < \varepsilon_s \leq 0.02 \text{ (in.-lb)}$$

$$f_s = 1170 - \frac{2.9670}{\varepsilon_s + 0.0019} \text{ (MPa) for } 0.0024 < \varepsilon_s \leq 0.02 \text{ (SI)}$$

(3.4b)

$$f_s = 150 \text{ (ksi) for } 0.02 < \varepsilon_s \leq 0.06 \text{ (in.-lb)}$$

$$f_s = 1040 \text{ (MPa) for } 0.02 < \varepsilon_s \leq 0.06 \text{ (SI)} \quad (3.4c)$$

Accordingly, an analysis was performed on the results of 137 mill tests on ASTM A1035/A1035M (CS) Grade 100 (690) bars produced after 2004 that were provided by a manufacturer. Based on the analysis of the mill test data, it was shown that Eq. (3.4b) and (3.4c) would provide a lower tolerance limit on actual stress corresponding to a strain of 0.0035, actual yield strength, and actual tensile strength for ASTM A1035/A1035M (CS) Grade 100 (690) bars such that at least 95 percent of the data are greater than the corresponding values calculated by these equations with a confidence level of 90 percent.

Cumulative distribution functions (CDFs) for the yield strength by the 0.2 percent offset method and the tensile strength were developed from the ASTM A1035/A1035M

Table 3.4—Statistical analysis of mill test data for ASTM A1035/A1035M (CS) Grade 100 (690) steel reinforcing bars*

		0.2% offset yield strength			Tensile strength	
Bar designation	No. of samples	Mean, ksi (MPa)	Bias factor λ_b	COV	Mean, ksi (MPa)	COV
No. 3 (10)	10	139.8 (964)		0.044	174.3 (1202)	0.046
No. 4 (13)	20	129.9 (895)		0.026	162.1 (1118)	0.027
No. 5 (16)	28	130.9 (903)		0.036	164.7 (1135)	0.035
No. 6 (19)	16	129.9 (895)		0.067	164.1 (1132)	0.045
No. 7 (22)	15	124.2 (857)		0.061	162.5 (1120)	0.028
No. 8 (25)	9	128.4 (886)		0.032	161.6 (1114)	0.019
No. 9 (29)	8	127.1 (877)		0.029	161.7 (1115)	0.039
No. 10 (32)	3	133.7 (922)		0.050	169.5 (1169)	0.051
No. 11 (36)	28	132.8 (916)		0.040	168.5 (1162)	0.031
All sizes	137	130.6 (901)		0.051	165.2 (1139)	0.040
Lower tail of data for all sizes	—	—	1.159	0.043	—	—
Lower tail of data for ASTM A615/A615M Grade 60 (420)*	—	—	1.145	0.055	—	—

*Statistical determination used for establishing strength reduction factors ϕ as reported by Nowak and Szerszen (2003).

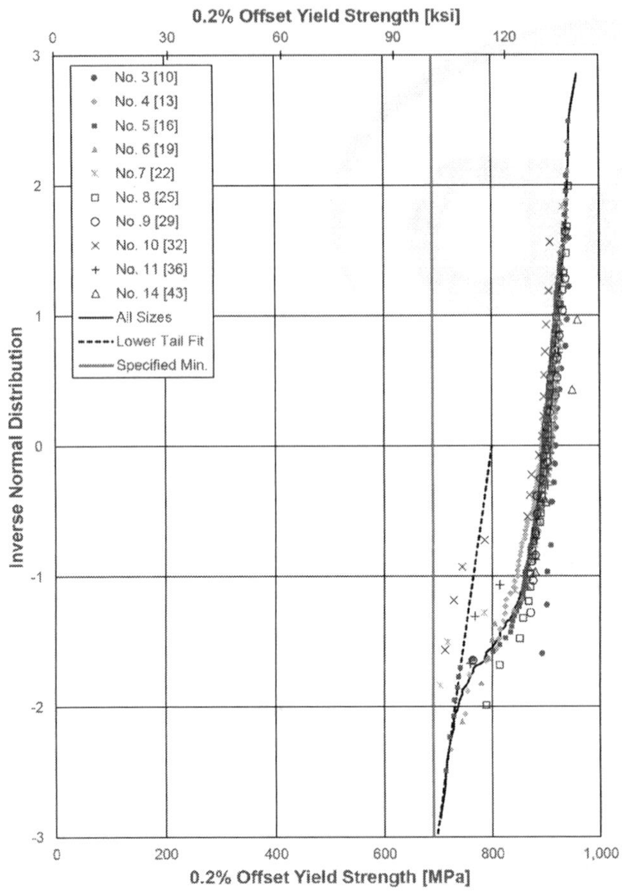

Fig. 3.4c—Cumulative distribution function for yield strength of ASTM A1035/A1035M (CS) Grade 100 (690) bars by 0.2 percent offset method. (Note: 1 MPa = 145 psi.)

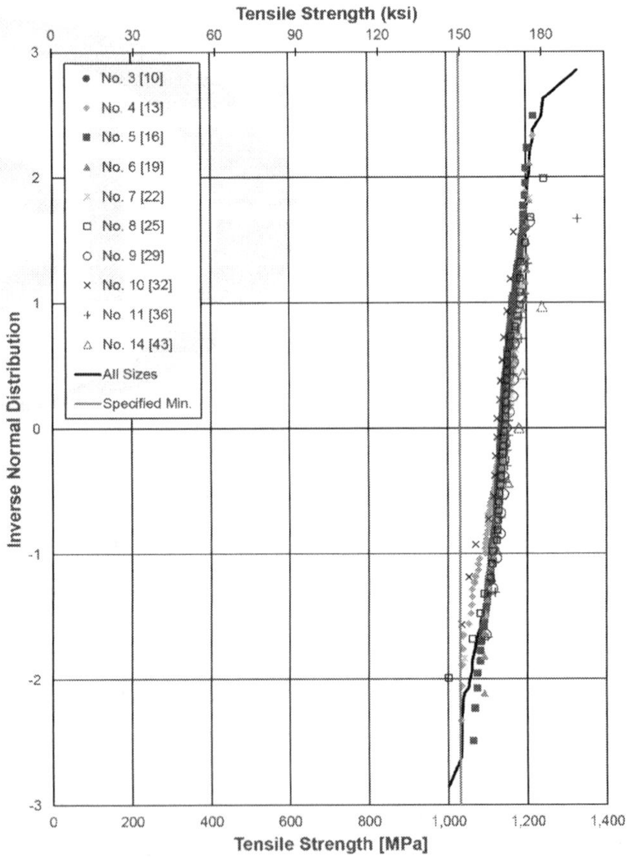

Fig. 3.4d—Cumulative distribution function for tensile strength of ASTM A1035/A1035M (CS) Grade 100 (690) bars. (Note: 1 MPa = 145 psi.)

(CS) Grade 100 (690) bar mill test data. The results are summarized in Table 3.4, Fig. 3.4c, and Fig. 3.4d. The summarized statistical data include the bias factor λ_b and the coefficient of variation (COV).

Nowak and Szerszen (2003) reported the comparable statistical analysis for yield strength of ASTM A615/A615M Grade 60 (420) bars as used for code-calibration reliability analysis. They recommended using a bias factor λ_b of 1.145

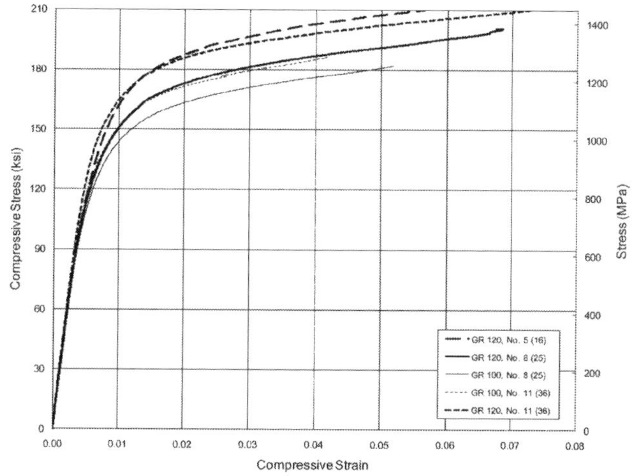

Fig. 3.5—Actual stress-strain curves in compression for ASTM A1035/A1035M (CS) reinforcing bars (WJE 2008).

and a COV of 0.05, based on the lower tail of distribution of the data used for their analysis, in the development of strength reduction factors for ACI 318. The summarized statistical data in Table 3.4 include the bias factor λ_b of 1.159 and COV of 0.043 for ASTM A1035/A1035M (CS) Grade 100 (690) reinforcement, based on a curve fit to the lower tail of the pooled data. The lower tail fit curve, shown in Fig. 3.4c, was determined in the same manner as used by Nowak and Szerszen (2003) for ASTM. The values of the bias factors for ASTM A1035/A1035M (CS) Grade 100 (690) bar is larger than 1.145 established for ASTM A615/A615M Grade 60 (420) bars, whereas the value for COV is less than the recommended value of 0.05 for ASTM A615/A615M Grade 60 (420) bars. Use of the same strength reduction factors with ASTM A1035/A1035M (CS) Grade 100 (690) bars as used for ASTM A615/A615M Grade 60 (420) bars is therefore recommended.

3.5—Actual compressive properties

WJE (2008) studied the stress-strain behavior in compression of representative samples of ASTM A1035/A1035M (CS) bars in size No. 5 (16), 8 (25), and 11 (36) in both Grades 100 (690) and 120 (830). The typical length of a compression test specimen was twice its nominal diameter. Figure 3.5 shows recorded compressive stress-strain curves. The proportional limit ranges from 60,000 to 80,000 psi (420 to 550 MPa) for both grades. Yield strength in compression by the 0.2 percent offset method is approximately 130,000 psi (900 MPa) for Grade 100 (690) bars and ranges from 135,000 to 140,000 psi (930 to 970 MPa) for Grade 120 (830) bars. The initial slopes of the stress-strain curves in compression are consistent with 29,000,000 psi (200,000 MPa) defined by ACI 318 for the modulus of elasticity.

3.6—Chemical composition

Table 3.6 shows the comparison of the specified chemical composition for ASTM A1035/A1035M (CS) low-carbon, chromium-steel bars with that of ASTM A615/A615M carbon-steel bars and ASTM A706/A706M low-alloy steel bars. The specified chemical content of ASTM A1035/A1035M (CS) bars differs from the other two primarily in that there is a required amount of chromium (8.0 to 10.9 percent). In addition, carbon content is limited to 0.15 percent and a maximum limit is specified for nitrogen (0.05 percent). Other than the chromium requirement, the chemical composition specified by ASTM A1035/A1035M (CS) is similar to ASTM A706/A706M. ASTM A1035/A1035M (CS) bars are not readily weldable because of metallurgical reasons, as discussed in 10.4 of this guide.

Table 3.6—Specified chemical composition by heat analysis

	Bar type		
	ASTM A1035/ A1035M (CS)	ASTM A615/ A615M	ASTM A706/ A706M
Element	Maximum content, percent		
Carbon	0.15	†	0.30
Chromium	8.0 to 10.9*	—	—‡
Manganese	1.50	†	1.50‡
Nitrogen	0.05	—	—
Phosphorus	0.035	0.06	0.035
Sulfur	0.045	†	0.045
Silicon	0.50	—	0.50

*Chromium content should also be a minimum of 8.0 percent and maximum of 10.9 percent.

†Content should be reported but no limit is established.

‡Regulated by maximum carbon equivalent of 0.55 percent according to the following formula: CE = %C + %Mn/6 + %Cu/40 + %Ni/20 + %Cr/10 – %Mo/50 – %V/10.

CHAPTER 4—BEAMS

4.1—Introduction

Although the previous chapters address Grade 100 (690) and Grade 120 (830) of ASTM A1035/A1035M (CS) bars, the procedures in 4.2 through 4.11 of this guide apply only to the use of ASTM A1035/A1035M (CS) Grade 100 (690) bars. Research data and experience are insufficient to allow proper evaluations of the procedures for Grade 120 (830) bars.

With the simplified strength design procedures in 4.2 and 4.3, flexural strength design using ASTM A1035/A1035M (CS) Grade 100 (690) bars is similar to that using ASTM A615/A615M Grade 60 (420) bars, whether by manual calculations or by using design software. Only certain inputs and limitations are modified.

4.2—Flexural strength

As discussed in 3.4 of this guide, Eq. (3.4a) through (3.4c) represent a lower bound to the stress-strain behavior of ASTM A1035/A1035M (CS) Grade 100 (690) bar.

For a given reinforced concrete section, the designer may perform a nonlinear flexural analysis by using the stress-strain relationship defined by Eq. (3.4a) and (3.4b) and satisfying the requirements of force equilibrium and strain compatibility. However, the analysis procedure requires an iterative process of trial and error. If such a nonlinear analysis is used for flexural design, the tensile stress in the reinforcement under service load should be determined to evaluate satisfaction of serviceability criteria such as crack control or control of member deflections. Furthermore, evaluation procedures for other strength criteria at ultimate limit state, such as shear strength, should consider nonlinear behavior of the reinforcement. Refer to Appendix B of this guide for detailed discussions of a nonlinear flexural analysis approach and for design examples comparing the nonlinear flexural analysis approach with elastic-plastic approach.

A simplified design method using ASTM A1035/A1035M (CS) bars based on the study by Mast et al. (2008) uses an idealized elastic-plastic stress-strain curve, similar to that used for ASTM A615/A615M Grades 60 (420) and 75 (520) bars (Fig. 4.2). Limitations are set on the range of applicability for this method to protect against harmful effects resulting from the actual steel stress being higher than assumed, and effects from larger steel strains at service load. Refer to 4.3 of this guide for a discussion of these limitations.

The idealized elastic-plastic stress-strain curve shown in Fig. 4.2 has an elastic portion with a modulus E_s of 29,000,000 psi (200,000 MPa) (the same as for ASTM A615/A615M bar), and a perfectly plastic behavior after reaching f_y equal to 100,000 psi (690 MPa). These parameters allow computation of flexural strength with f_y set at 100,000 psi (690 MPa) for a beam that is tension-controlled. Appendix A illustrates a series of design examples using the idealized elastic-plastic stress-strain curve of ASTM A1035/A1035M (CS) Grade 100 (690) reinforcement.

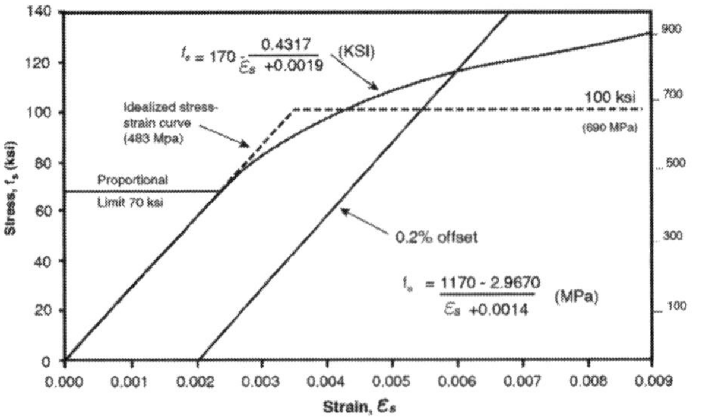

Fig. 4.2—Approximated nonlinear stress-strain relationship of ASTM A1035/A1035M (CS) Grade 100 (690) steel and idealized bilinear elastic-plastic stress-strain relationship for simplified design. (Note: 1 ksi = 6.9 MPa.)

4.3—Tension- and compression-controlled limits

4.3.1 *Historical*—It has long been accepted that reinforced concrete flexural members should behave elastically at service load but have the capability to deform inelastically before reaching maximum capacity. This is accomplished by limiting the reinforcement ratio so that the reinforcement yields before the concrete crushes. Beginning with ACI 318-63, flexural member reinforcement ratio was limited to 75 percent of the balanced reinforcement ratio ρ_b. By 1995, this criterion had been used for over 30 years, and flexural member behavior was judged to be satisfactory.

4.3.2 *Tension- and compression-controlled strain limits*—Beginning with ACI 318-95, the tension-controlled criterion based on tensile strain in the reinforcement was selected to provide behavior similar to that experienced under the $0.75\rho_b$ criterion. Nonlinear analyses of the behavior between service load and nominal strength were made for the old maximum reinforcement ratio and the new tension-controlled strain criteria. The ratio of deformation at nominal strength to that at service load for steel strain ε_s, curvature ψ, and deflection Δ were calculated and compared. Setting the tension-controlled strain limit at 0.005 for Grade 60 (420) steel provided more deformation than the $\rho = 0.75\rho_b$ limit. For Grade 60 (420) steel, the compression-controlled limit was set at 0.002, which is the strain (rounded) at the balanced condition (Mast 1992).

4.3.3 *Strain limits for sections with ASTM A1035/A1035M (CS) bar*—The definition of tension-controlled sections in ACI 318 is based on ASTM A615/A615M Grades 60 (420) and 75 (520) bars. These were the reinforcing bars available when ACI 318 strain limits were originally developed. For ASTM A1035/A1035M (CS) bar, the process to determine ACI 318 strain limits was repeated using equations similar to Eq. (3.4a) and (3.4b) of this guide and the idealized elastic-plastic stress-strain relationship for ASTM A1035/A1035M (CS) bar. It was found that a tension-controlled strain limit of 0.0066 (corresponding to c/d = 5/16) produced behavior similar to members designed by ACI 318 with Grades 60 (420) and 75 (520) bars and a strain limit of 0.005 (Mast et al. 2008). The strain limit for compression-controlled sections is discussed in the next section.

4.3.4 *Simplified design strain limits*—The tension- and compression-controlled strain limits need to be modified for simplified design using the idealized elastic-plastic stress-strain relationship described in 4.2. The tension-controlled strain limit should be adjusted to 0.009 (c/d = 0.25) to compensate for the fact that the actual steel stress at nominal strength is higher than the assumed stress of 100,000 psi (690 MPa). This adjustment ensures ductility and deformability are comparable to designs based on ACI 318 using ASTM A615/A615M Grade 60 (420) bars. Table 4.3.4 and Fig. 4.3.4 show that the simplified method using ε_t = 0.009 and f_y = 100,000 psi (690 MPa) results in a value of nominal flexural strength 18 percent less than the same section designed using ε_t = 0.0066 and the corresponding f_s = 119,000 psi (820 MPa) according to Eq. (3.4b). Using the simplified method avoids having to perform an analysis using the nonlinear stress-strain relationship of the steel.

ACI 318 defines the compression-controlled strain limit as the net tensile strain at balanced strain conditions. For f_y of 100,000 psi (690 MPa) and E_s = 29,000,000 psi (200,000 MPa), this is a strain of 0.00345. It is simpler and conservative to round the compression-controlled strain limit for the A1035/A1035M (CS) reinforcement to 0.004.

Using the limits defined previously for the simplified method, the tension-controlled limit occurs at a c/d = 0.25 (instead of 0.375 for Grade 60 [420] steel), and the compression-controlled limit is at a c/d = 3/7, or 0.43 (instead of 0.600 for Grade 60 [420] steel).

Table 4.3.4—Comparison of design methods using ASTM A1035/A1035M (CS) Grade 100 (690) steel

	Using Eq. (3.4a) and (3.4b)	Simplified method
Tension-controlled strain limit	0.0066	0.009
Steel tensile stress f_s, ksi (MPa)	119 (820)	100 (690)
Neutral axis depth c, in. (mm)	$0.3125d$	$0.25d$
Stress block depth $a = \beta_1 c$, in. (mm)	$0.3125\beta_1 d$	$0.25\beta_1 d$
Compression force C, kip (N)	$0.85f_c'ab$	$0.85f_c'ab$
Steel area $A_s = C/f_s$, in.^{2} (mm^2)	$0.85(f_c'/f_s)(0.3125\beta_1\delta)b$	$0.85(f_c'/f_s)(0.25\beta_1\delta)b$
Tension-controlled reinforcement ratio $\rho_t = A_s/bd$	$0.002232f_c'\beta_1$ $(0.0003239f_c'\beta_1)$	$0.002125f_c'\beta_1$ $(0.0003079f_c'\beta_1)$
$T = C = A_s f_s = \rho_t bdf_s$, kip (N)	$0.2656f_c'\beta_1 bd$	$0.2125f_c'\beta_1 bd$
Lever arm $= d - a/2$, in. (mm)	$d(1 - 0.156\beta_1)$	$d(1 - 0.125\beta_1)$
M_n for $f_c' = 5$ ksi (34.5 MPa); $\beta_1 = 0.8$, in.-kip (mm-N)	$0.186f_c'bd^2$	$0.153f_c'bd^2$

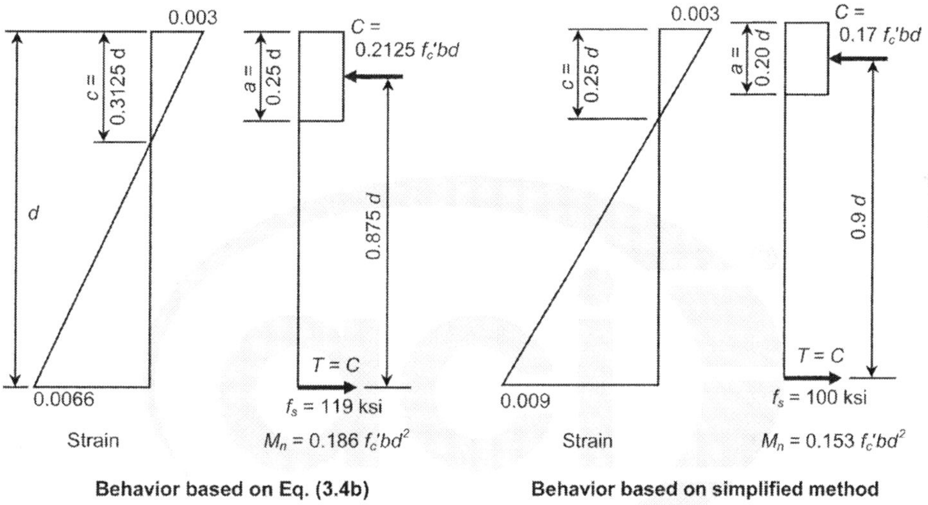

Fig. 4.3.4—Tension-controlled strain limits with $f_c' = 5$ *ksi and* $\beta_1 = 0.8$. *(Note: 1 ksi = 6.9 MPa.)*

4.4—Strength reduction factor ϕ

With the aforementioned modified tension- and compression-controlled limits for use with the simplified method, it is necessary to revise the equation for the transition in ϕ between these strain limits. The revised equation is given as follows

$$0.65 \leq (\phi = 0.45 + 50\varepsilon_t) \leq 0.9 \quad (4.4)$$

Note that for $\varepsilon_t \geq 0.009$ (tension-controlled section), $\phi = 0.9$, and for $\varepsilon_t \leq 0.004$ (compression-controlled section), $\phi = 0.65$. Figure 4.4 illustrates this relationship.

If the designer uses nonlinear analysis for flexural design, Eq. (4.4) should be modified for the value of ϕ in the transition between the tension- and compression-controlled strain limits. The modified equation for ϕ is given in Appendix B.

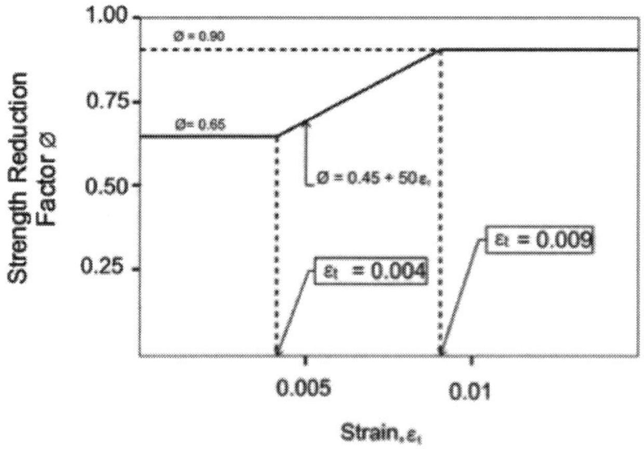

Fig. 4.4—Relationship between strength reduction factor ϕ *and strain limits for use only with simplified design based on idealized elastic-plastic stress-strain relationship for A1035/A1035M (CS) Grade 100 (690) steel reinforcement.*

4.5—Stress in steel due to flexure

For an efficient use of flexural reinforcement, it is advisable to proportion all flexural members as tension-controlled members. Figure 4.5 shows steel strain and stress diagrams for a member at the tension-controlled limit. The steel stress diagram shows both the idealized elastic-plastic and nonlinear stress distributions. The stress diagrams below the neutral axis are the same as in Fig. 4.2. There are two regions in the steel stress diagram where the idealized elastic-plastic design stress may exceed the nonlinear design stress given

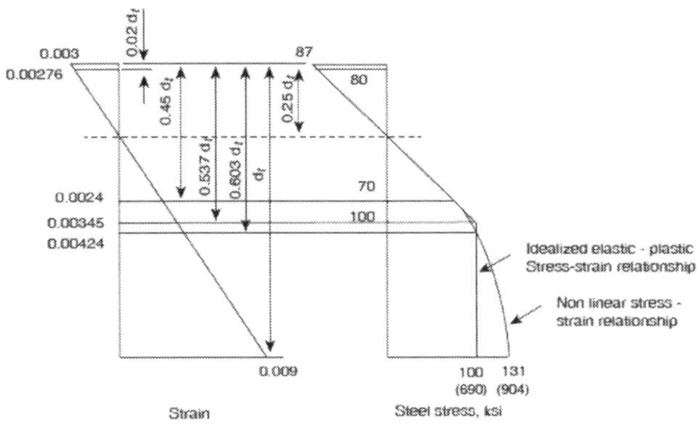

Fig. 4.5—Strain and steel stress at tension-controlled limits. (Note: 1 ksi = 6.9 MPa.)

by Eq. (3.4a) and (3.4b) of this guide. On the tension side, this occurs between $0.45d_t$ and $0.603d_t$ below the extreme compression fiber. Ordinarily, beams do not have primary tension steel at this location, but if there is, underestimating steel stress near the tension face when using the idealized elastic-plastic stress-strain model offsets strength overestimation between $0.45d_t$ and $0.603d_t$. On the compression side, in relation to a maximum usable compression steel stress of 80,000 psi (550 MPa), the steel stress is overestimated for steel with strains from 0.00276 to 0.003. For a section at the tension-controlled limit, this occurs only for steel located within $0.02d_t$ measured from the compression face. The influence of this overestimation is typically insignificant.

4.6—Compression stress limit

ACI 318 limits the maximum usable compression strain in concrete to 0.003. For this reason, ACI 318 limits the compression steel stress to 80,000 psi (550 MPa), and this limit should also apply for ASTM A1035/A1035M (CS) bars in compression.

Although most design software accepts only a single value of f_y for tension and compression, using f_y = 100,000 psi (690 MPa) as input for design software should not be a problem. For a maximum concrete compressive strain of 0.003, the steel stress in compression does not exceed 80,000 psi (550 MPa) unless the depth from the extreme compression fiber to the compression steel, d', is less than $0.08c$, where c is the depth of the compression zone. This is unlikely to happen and can be easily checked.

4.7—Moment redistribution

The moment redistribution technique in ACI 318 allows moments in continuous members to be increased or decreased by prescribed amounts relative to the magnitudes determined from the elastic theory, where sufficient rotational capacity is provided at plastic hinge locations. Corresponding adjustments are made to the moment demand envelopes in the span. As of the writing of this guide, no data are available to judge if moment redistribution is applicable to members with ASTM A1035/A1035M (CS) bars. In particular, the simplified strength design procedures in 4.2 and 4.3 of this guide may not provide adequate estimates of the actual loading conditions at which flexural hinges form. Furthermore, the inelastic rotation capacity of high-moment regions reinforced with ASTM A1035/A1035M (CS) bars has not been evaluated. Therefore, do not use moment redistribution for members containing ASTM A1035/A1035M (CS) bars until further data are available.

When determining the design forces in members with ASTM A1035/A1035M (CS) bars or other reinforcement constructed integrally with other members or elements that contain ASTM A1035/A1035M (CS) longitudinal bars, the actual forces transferred from the member containing ASTM A1035/A1035M (CS) bars may be difficult to determine at the strength limit state because ASTM A1035/A1035M (CS) bars lack a well-defined yield plateau. This situation could occur with beam-column joints of frame structures, the interface between slabs and spandrel beams, two-way slab construction, or continuous members where ASTM A1035/A1035M (CS) bars and other reinforcement types are used in different regions.

4.8—Deflection

For one-way members, ACI 318 provides two methods for controlling deflections at the service load level:

(1) Compute the expected deflections and compare against appropriate limits

(2) Implicitly control deflection through minimum thicknesses

When deflections are not computed, ACI 318 provides the minimum height or thickness h of one-way concrete members to implicitly control deflections. The minimum thickness values apply only to members not supporting or attached to construction likely to be damaged by deflections, and are given as functions of the span length, member type, and support condition. The table is based on normalweight concrete and ASTM A615/A615M Grade 60 (420) bar. For f_y other than 60,000 psi (420 MPa), footnote (b) of the table permits modification of the tabulated values by the multiplier $(0.4 + f_y/100{,}000)$. This table was first presented in ACI 318-71 where accompanying commentary stated:

> The modification for yield strength in Footnote (b) is based on judgment, experience, and studies of the results of tests and of unpublished analyses. The simple expression given is approximate but should yield conservative results for the types of members considered in the table, for typical reinforcement ratios, and for values of f_y between 40 and 80 ksi.

The range of f_y between 40,000 and 80,000 psi (280 and 550 MPa) corresponded to the range typically available in 1971.

When using ASTM A1035/A1035M (CS) longitudinal bars in designs based on the method described in 4.2 and 4.3 of this guide, the strain in the reinforcement at the service condition is higher than that in comparable members designed with Grade 60 (420) reinforcement. Under service load condition, the steel stress is usually taken as 0.67 of f_y. Using the multiplier $(0.4 + f_y/100{,}000)$ as a member thickness modification factor results in similar maximum member curvature for members with different reinforcement strengths. The guide by Mast (2006) shows that for f_y = 100,000 psi (690 MPa), the deflection at service load is 1.4 times that for members with f_y = 60,000 psi (420 MPa). This confirms the applicability of the multiplier of (0.4 + f_y/100,000) = 1.4 on the minimum height or thickness h in ACI 318 minimum thickness recommendations of Chapter 8 when using ASTM A1035/A1035M (CS) bars with f_y = 100,000 psi (690 MPa). Desalegne and Lubell (2012) recommended that direct deflection calculations should be used for the design of concrete slabs reinforced with ASTM A1035/A1035M (CS), which is discussed in Chapter 6.

For direct deflection calculation, the ACI 318 method uses an effective moment of inertia proposed by Branson (1977) to account for variable cracks at different sections along the member length. Bischoff (2005), however, showed that using Branson's equation for I_e underestimates deflections, especially when the reinforcement ratio is less than 1 percent. Bischoff (2005) proposed an alternate formulation of I_e

$$I_e = \frac{I_{cr}}{1-\left(1-\frac{I_{cr}}{I_g}\right)\left(\frac{M_{cr}}{M_a}\right)^2} \leq I_g \tag{4.8}$$

Calculated deflections using Eq. (4.8) correlate well with deflections measured in laboratory tests for members with a range of reinforcement ratios (Bischoff and Scanlon 2007). Because slabs and beams using ASTM A1035/A1035M (CS) longitudinal bars may result in lightly reinforced members, Eq. (4.8) may be used in place of ACI 318-14 Eq. (24.2.3.5a). This method is suitable when the steel stress at the load condition checked for deflection is below the proportional limit.

When ASTM A1035/A1035M (CS) steel is used as longitudinal reinforcement, the designer should check that the member deflection limits are not exceeded for lightly reinforced members.

4.9—Crack control

4.9.1 *Historical*—In the early 1960s, the developments of ultimate strength design and ASTM A615/A615M Grade 60 (420) reinforcement caused concern about crack widths. These two changes in design practice resulted in steel stress in flexural members at the service load of approximately 36,000 psi (250 MPa), almost twice the 20,000 psi (140 MPa) allowed previously for ASTM A615/A615M Grade 40 (280) steel under working stress design. Numerous research papers led to the crack control requirements of Chapter 10 in ACI 318-71. These requirements remained mostly unchanged until 1999.

Researchers agree that strain in the tension steel is one of the primary variables for controlling crack width. Because the elastic relationship between strain and stress for steel is a constant for strains smaller than the proportional limit, most crack width formulas use f_s instead of ε_s. In addition, the interaction of bond to the reinforcement and tensile stress in the concrete surrounding the reinforcement also affect crack spacing and crack width. Numerous equations have been proposed for predicting crack spacings and widths including the Gergely-Lutz equation (Gergely and Lutz 1968), which was adopted by ACI 318-71.

4.9.2 *Crack widths*—There are three reasons for limiting crack width: 1) appearance; 2) control of reinforcement corrosion; and 3) water-tightness. Aesthetic considerations by owners and designers often demand tight control of crack width. For liquid-retaining structures and building envelopes, water-tightness is essential. Control of reinforcement corrosion is paramount to achieve durability.

4.9.3 *ACI 318-99 revisions*—While the z-factor determined from Eq. (10-2) of ACI 318-71 provided adequate reinforcement distribution for crack control of normal-size beams and one-way slabs with standard concrete clear covers not exceeding 1.5 in. (38 mm), the z-factor method based on the Gergely-Lutz (Gergely and Lutz 1968) equation produced unrealistic bar spacings when larger concrete covers were used. This led designers to either ignore the provisions when larger covers were used or simply use the code-minimum cover or a standard value—typically 2.5 in. (63 mm)—rather than the actual cover in the calculations. Additionally, some building inspectors continued the old practice of comparing measured crack widths in the field against the limiting crack widths according to the z-factor as reasons for rejecting members or projects, even though research showed that there was little if any correlation between corrosion and crack width for the range of crack widths being considered. Further, the limiting crack widths implied by the z-factor—0.016 in. (0.41 mm) for interior exposure and 0.013 in. (0.33 mm) for exterior exposure (refer to Commentary on Section 10.6.4 of ACI 318-71)—represent a difference of 0.003 in. (0.08 mm), which implies an unrealistic accuracy. To find a simpler, more direct way of defining bar spacing to control cracking than Eq. (10-2), ACI Committee 318 examined the work of Kaar and Mattock (1963), Gergely and Lutz (1968), Beeby (1979), and Frosch (1999). A simple bilinear relationship between concrete clear cover and bar spacing for a maximum crack width of 0.016 in. (0.41 mm) was adopted as Eq. (10-5) in ACI 318-99. Figure 4.9.3 illustrates this bilinear relationship and three other nonlinear relationships. In Fig. 4.9.3, f_s is the bar stress at service load, assumed as 60 percent of specified minimum yield strength, and is the ratio of the distance from neutral axis to the extreme tension face to the distance from neutral axis to the center of the tension reinforcement. The bilinear relationship (ACI 318-99 Eq. (10-5)) is independent of β and bar size, whereas the other nonlinear relationships depend on these variables. A modified version of Eq. (10-5) of ACI 318-99—valid for service load steel stresses ranging from 24,000 to 48,000 psi (170 to 330 MPa)—became Eq. (10-4) of ACI 318-05.

4.9.4 *Applicability with ASTM A1035/A1035M (CS) bars*—Figure 4.9.4 is similar to the bilinear equation shown in Fig. 4.9.3, but with f_s taken as 67,000 psi (460 MPa), which is two-thirds of the specified minimum yield strength f_y. ACI

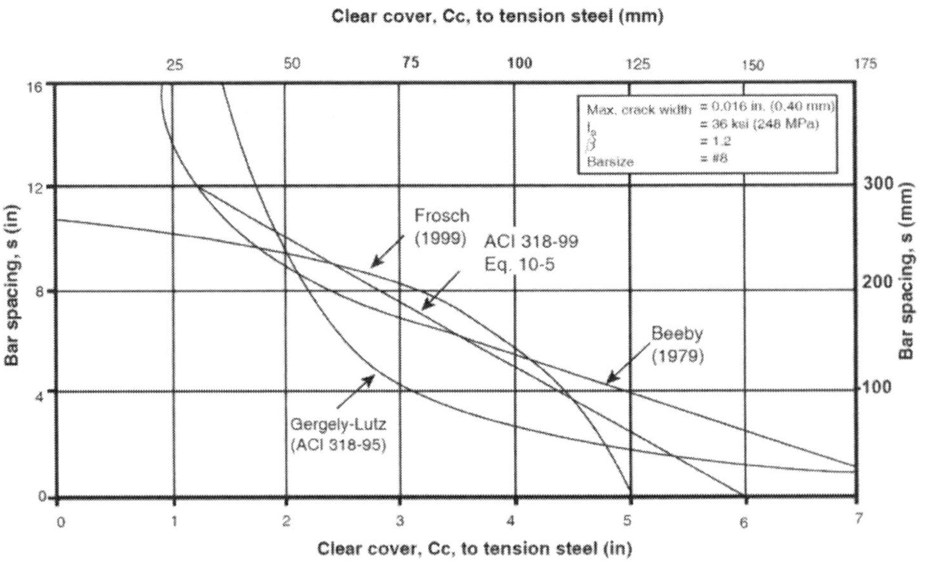

Fig. 4.9.3—Bar spacing versus clear cover for crack control. (Note: 1 in. = 25 mm; 1 ksi = 6.9 MPa.)

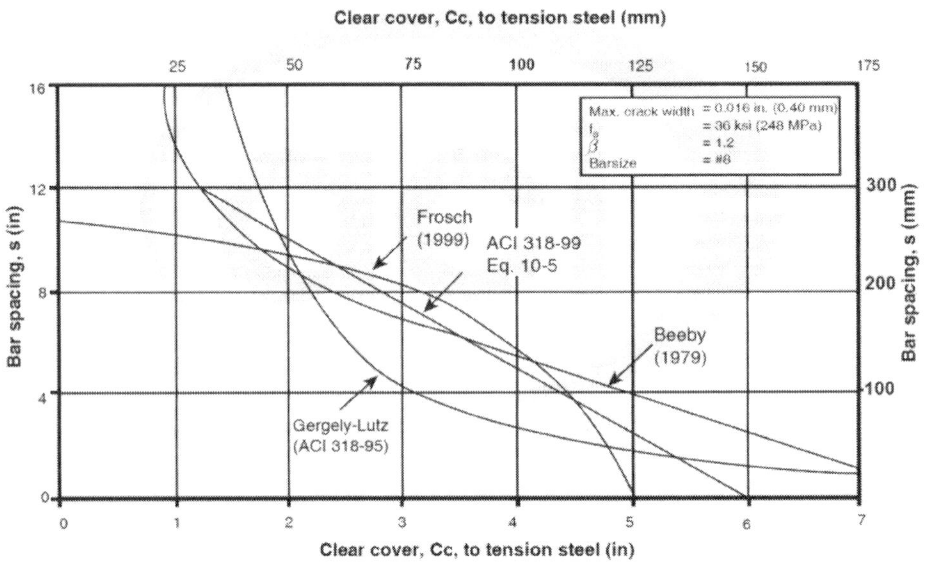

Fig 4.9.4—Bar spacing versus clear cover for crack control (f_s = 67 ksi). (Note: 1 in. = 25 mm; 1 ksi = 6.9 MPa.)

318 gives more conservative results compared with those shown in Fig. 4.9.3, but the results are restrictive in design choices. For example, the bar spacing s cannot exceed 6.5 in. (165 mm) at 1 in. (25 mm) cover c_c, and smaller bar spacing is required at greater cover to meet the 0.016 in. (0.41 mm) crack width criterion. Any combination of c_c and s above the bilinear curve leads to larger crack widths.

When crack control is critical, use of ACI 318 is appropriate. For members with minimal cover, such as slabs and walls, this is feasible when designing for f_y = 100,000 psi (690 MPa) and f_s = 67,000 psi (460 MPa). To provide crack control at a reasonable bar spacing for members with increased cover, it will be necessary to limit the steel stress at service load to less than 67,000 psi (460 MPa).

4.10—Minimum reinforcement

A minimum area of flexural reinforcement in accordance to ACI 318, As_{min}, shall be provided at every section where tension reinforcement is required by analysis with f_y = 100,000 psi (690 MPa). When the minimum reinforcing requirements for temperature and shrinkage of ACI 318 are applied, the minimum reinforcing ratio of 0.0014 will control.

4.11—Strength design for shear

Design for shear and torsion in reinforced concrete beams using ASTM A1035/A1035M (CS) bars as reinforcement should follow ACI 318, with a few exceptions noted as follows.

The shear failure of reinforced concrete beams without shear reinforcement is brittle and the shear stress at failure is influenced by several design parameters, including member depth, longitudinal reinforcement ratio, and concrete strength (Rajagopalan and Ferguson 1968; Collins and Kuchma 1999; Reineck et al. 2003; Lubell et al. 2004). Unsafe predictions of shear capacity can result from the use of ACI 318 for lightly reinforced beams that are generally with ρ less than 1 percent. Laboratory tests (Hassan et al. 2008; Desalegne and Lubell 2010) demonstrate that these trends also occur for members containing ASTM A1035/A1035M (CS) bars as longitudinal reinforcement; therefore, all lightly reinforced beams should contain minimum shear reinforcement.

For members designed according to the simplified method for flexural design in 4.2 of this guide, the shear capacity can be determined from ACI 318. Such members should include shear reinforcement that satisfies the spacing and quantity requirements of ACI 318. ACI 318 limits the design yield strength f_{yt} for shear reinforcement to 60,000 psi (420 MPa), but the value can be increased to 80,000 psi (550 MPa) for welded deformed wire reinforcement. Similarly, ACI 318 limits the design yield strength f_{yt} for torsional reinforcement to 60,000 psi (420 MPa). The limitation of 60,000 psi (420 MPa) has been specified since ACI 318-71, in part to provide a control on diagonal crack width at service load levels.

ACI 318-95 raised the f_{yt} limit to 80,000 psi (550 MPa) for welded deformed wire reinforcement. Full-scale beam tests (Griezic et al. 1994) indicated that inclined shear crack widths at service load levels were less for beams with welded deformed wire reinforcement cages designed with a yield strength of 75,000 psi (520 MPa) than for beams reinforced with deformed Grade 60 (420) stirrups. For lack of comparative data, the limitation of 60,000 psi (420 MPa) remains for torsional reinforcement even when using welded deformed wire reinforcement.

Tests of full-scale beams by Munikrishna (2008) and Sumpter et al. (2009) compared the behavior of beams reinforced with ASTM A1035/A1035M (CS) Grade 100 (690) bars as stirrups with 135-degree hooks and designed with f_{yt} of 80,000 psi (550 MPa) with the behavior of similar beams reinforced with deformed Grade 60 (420) stirrups. The results indicated that the behavior of the two groups of beams was similar at failure. At service load levels, inclined shear crack widths were larger, as expected, for beams reinforced with ASTM A1035/A1035M (CS) Grade 100 (690) stirrups designed with a yield strength of 80,000 psi (550 MPa). In all cases, the crack widths were less than the commonly accepted limit of 0.016 in. (0.41 mm).

A design yield strength f_{yt} of 80,000 psi (550 MPa) for ASTM A1035/A1035M (CS) Grade 100 (690) stirrups as shear reinforcement is appropriate if appearance and serviceability due to shear cracking is not a critical design consideration. Otherwise, f_{yt} should be limited to 60,000 psi (420 MPa). For lack of research data, f_{yt} should also be limited to 60,000 psi (420 MPa) for shear reinforcement designed for torsion.

CHAPTER 5—COLUMNS

5.1—Introduction

Column design using ASTM A1035/A1035M (CS) bars should follow ACI 318, with modifications made to the specified minimum yield strength, f_y, for the longitudinal reinforcement and f_{yt} for the lateral reinforcement.

5.2—Specified minimum yield strength for longitudinal reinforcement

For members under axial load, the maximum compressive stress of the longitudinal reinforcement should be limited to 80,000 psi (550 MPa) when computing the design axial strength ϕP_n using ACI 318.

The steel strain ε_s corresponding to this stress is approximately 0.0028, nearly equal to the maximum usable strain (0.003) for concrete in compression assumed by ACI 318.

For members subjected to axial load combined with moment, f_y should be 80,000 psi (550 MPa) for the longitudinal reinforcement in compression for the reason given previously, and 100,000 psi (690 MPa) for the longitudinal reinforcement in tension as in the case of flexural design discussed in Chapter 4 of this guide.

5.3—Specified minimum yield strength for transverse reinforcement

Transverse reinforcement in a column serves three functions:

(1) Lateral support to the longitudinal reinforcement

(2) Confinement to the core concrete when the load supported by the column approaches its axial strength

(3) Shear reinforcement when the column is subjected to shear

The transverse reinforcement is most commonly rectangular or circular ties and spirals.

ACI 318 provides minimum size and maximum spacing of ties in compression members, irrespective of the bar strength. These requirements apply for ASTM A1035/A1035M (CS) bars.

When a column is required to resist shear, provide ties for the required shear strength as discussed in 4.11 of this guide.

For columns reinforced with ASTM A1035/A1035M (CS) spirals, determine the volumetric spiral reinforcement ratio ρ_s of ACI 318 with the specified minimum yield strength f_{yt} as 100,000 psi (690 MPa). Spirals should be continuous and lap splices should not be used.

5.4—Slenderness effect

When slenderness effect is considered, design the column by using the moment magnification procedure of ACI 318. Use the magnified moment to design the column according to the procedures outlined in Chapter 4 of this guide.

CHAPTER 6—SLAB SYSTEMS

6.1—One-way slabs

Slab depth is typically determined by deflection control considerations including ACI 318 depth limits table for one-way slabs modified for f_y = 100,000 psi (690 MPa), or by the depth required to provide adequate shear strength without shear reinforcement. Design for flexure and deflection in one-way slabs, with or without joists, is based on the same principles as for beams. Thus, the recommendations in Chapter 4 of this guide apply. To control flexural cracking, the distribution of flexural reinforcement should conform to ACI 318, as discussed in 4.9 of this guide. In addition, shrinkage and temperature reinforcement should be provided to meet the requirements of ACI 318.

An analytical study of one-way slab deflection at service load level (Tang and Lubell 2008) using Eq. (4.8) for the effective moment of inertia, I_e, indicated that minimum slab thickness for deflection control was less sensitive to reinforcement strength than suggested by ACI 318 and more heavily influenced by applied loading, span length, and the assumptions for cracking moment. Desalegne and Lubell (2012) showed similar trends for members with A1035/A1035M (CS) Grade 100 reinforcement.

6.2—Shear design of one-way slabs

Shear design of one-way slabs is the same as shear design of beams and should follow the provisions of ACI 318. Shear reinforcement, however, is rarely used in one-way slabs. As indicated in 4.11 of this guide, research has demonstrated that for members without shear reinforcement, the shear stress at failure decreases as the member depth increases and as the reinforcement ratio decreases.

Bentz et al. (2006) and Bentz and Collins (2006) developed a simplified shear capacity model that accounts for the influence of member depth, concrete strength, aggregate size, crack width, and longitudinal reinforcement strains on shear strength. Hoult et al. (2008) presented an enhanced version of the model to better predict shear strengths for members with larger reinforcement strains.

For one-way slabs containing ASTM A1035/A1035M (CS) longitudinal bars designed in accordance with Chapter 4 of this guide, a simplification to the Hoult et al. (2008) model for shear strength was developed by Desalegne and Lubell (2010). This simplification assumed a longitudinal reinforcement strain of 0.0042, which corresponds to a stress of 100,000 psi (690 MPa) according to Eq. (3.4b) of this guide. For lightly reinforced members (generally with ρ less than 1 percent) without significant axial load and not containing shear reinforcement, the shear strength may be taken as

$$V_c = \frac{73}{(39+2.1d)}\sqrt{f_c'}b_w d \quad \text{(in.-lb)}$$

$$V_c = \frac{154}{(1000+2.1d)}\sqrt{f_c'}b_w d \quad \text{(SI)} \qquad (6.2)$$

Because Eq. (6.2) is based on the simplified flexural model from Mast et al. (2008) (4.2), it is not appropriate for members designed according to Appendix B of this guide. Desalegne and Lubell (2010) provide a generalized shear capacity model that can also be applied to members designed for flexure using Appendix B of this guide.

6.3—Two-way slabs

ACI 318 governs design of two-way slab systems. These provisions should be applicable when using ASTM A1035/A1035M (CS) bars with specified minimum yield strength of 100,000 psi (690 MPa) for tension and 80,000 psi (550 MPa) for compression. ACI 318 permits design of two-way slabs by the Direct Design Method (Section 8.10 of ACI 318-14) or the Equivalent Frame Method (Section 8.11 of ACI 318-14). Both methods are suitable for analysis of two-way slabs containing ASTM A1035/A1035M (CS) steel. As noted in 4.7 of this guide, however, the use of moment redistribution should not be used for members containing ASTM A1035/A1035M (CS) bars until further test data are available. Therefore, allowing up to 10 percent moment redistribution in accordance to ACI 318-14 in slabs with Grade 60 (420), steel should not be followed.

ACI 318 provides guidance on slab reinforcement details and illustrates the minimum extensions of reinforcement that are required in two-way slabs without beams. These provisions should also apply to slabs with ASTM A1035/A1035M (CS) steel if the designer directly checks for the impact on flexural capacity due to the longer bar development lengths compared with steel with f_y = 80,000 psi (550 MPa).

Only limited test data are available that examine the punching shear strength of two-way slab systems containing ASTM A1035/A1035M (CS) steel as flexural reinforcement without any shear reinforcement (Seliem et al. 2008). The punching shear strength of two specimens containing ASTM A1035/A1035M (CS) steel were similar to that of a third specimen containing Grade 60 (420) steel, and the capacities were in agreement with the existing ACI 318 punching shear analytical model. One-way shear should also be checked in accordance with 6.2 of this guide. Because of a lack of adequate research data, the specified minimum yield strength f_{yt} for shear reinforcement should be limited to 60,000 psi (420 MPa), where shear reinforcement is required to provide sufficient shear strength for a two-way slab system. Yang et al. (2010) reported that direct replacement of conventional steel bars with high-strength steel bars, having the same area, resulted in a 27 percent increase of the punching shear strength. This increase of punching shear resistance is because the higher-strength bars did not yield prior to punching failure.

When f_y = 100,000 psi (690 MPa) is used in the design of two-way slab systems, it is recommended that the minimum slab thickness without interior beams be determined according to ACI 318. When using ACI 318 depth limits table, direct extrapolation from f_y = 75,000 psi (520 MPa) to f_y = 100,000 psi (690 MPa) is recommended. When using ASTM A1035/A1035M (CS) bars as longitudinal reinforcement and members are lightly reinforced (generally with ρ less than 1 percent), direct calculations of deflection is recommended.

Due to the higher reinforcement stresses of ASTM A1035/A1035M (CS) steel at the service load level, it is essential to check for adequate control of cracking. Additional guidance on crack control in two-way structures is provided by ACI 224R.

CHAPTER 7—WALLS

7.1—Introduction

Design of walls using ASTM A1035/A1035M (CS) Grade 100 (690) bars should follow ACI 318, except that f_y for flexural reinforcement and for shear reinforcement should be limited as indicated in 7.2 through 7.4 of this guide.

7.2—Vertical reinforcement

Limit specified minimum yield strength f_y to 80,000 psi (550 MPa) for vertical reinforcement in compression, and to 100,000 psi (690 MPa) for vertical reinforcement in tension due to out-of-plane flexure or in-plane overturning forces.

7.3—Horizontal reinforcement

Limit the specified minimum yield strength f_y to 80,000 psi (550 MPa) for horizontal reinforcement in compression, and to 100,000 psi (690 MPa) for horizontal reinforcement in tension due to flexure.

7.4—Shear reinforcement

Limit the specified minimum yield strength f_y to 60,000 psi (420 MPa) where the vertical or horizontal reinforcement is required to resist shear. If appearance and serviceability due to shear cracking are not critical design considerations, f_y may be taken as 80,000 psi (550 MPa). Larger diagonal crack widths under service load are expected if f_y is taken as 80,000 psi (550 MPa). Refer to the ACI 318 Commentary on crack width and shear reinforcement.

7.5—Minimum reinforcement

Minimum reinforcement for the control of shrinkage cracking should meet the requirements of ACI 318.

CHAPTER 8—FOOTINGS AND PILE CAPS

8.1—Design

The design of footings and pile caps is governed by the provisions of ACI 318. The depth of footings is generally determined by its required shear strength.

For flexural design of shallow footings and pile caps, follow the design principles from Chapter 4 of this guide. For reinforcement in tension due to flexural forces, f_y may be taken up to 100,000 psi (690 MPa). For reinforcement in compression due to flexural forces, f_y should not be taken greater than 80,000 psi (550 MPa). For shear design of shallow footings, follow the principles in 4.11 and 6.2 of this guide.

For deep footings or regions of foundation structures meeting the criteria of ACI 318, the strut-and-tie modeling procedure in ACI 318 can be applied. Hassan et al. (2008) and Garay-Moran and Lubell (2008, 2016) reported tests of deep beams with ASTM A1035/A1035M (CS) bars as the primary tension ties, but without web reinforcement. In all cases, the failure mode was brittle and the ACI 318 provisions over-predicted the strength for beams with higher concrete strengths and larger shear span-depth ratios. Therefore, all members with ASTM A1035/A1035M (CS) ties should contain minimum web reinforcement in accordance with ACI 318 until additional test data become available.

Garay-Moran and Lubell (2008, 2016) reported tests of deep beams containing ASTM A1035/A1035M (CS) ties and Grade 60 (420) web reinforcement. Ductile failure modes were achieved, and ACI 318 strut-and-tie modeling provisions predicted strengths safely when the maximum tension tie stress was limited to 100,000 psi (690 MPa). No test data are available for the case of deep members with ASTM A1035/A1035M (CS) bars as the distributed web reinforcement. The strut-and-tie models should not be used in cases where the elastic flow of forces is not followed and also where considerable deformation capacity redistribution is needed based on the model geometry.

CHAPTER 9—MAT FOUNDATIONS

9.1—Design

ACI 318 governs the design of mat foundations. For design with ASTM A1035/A1035M (CS) bars, limit steel stress to 80,000 psi (550 MPa) for reinforcement in compression and 100,000 psi (690 MPa) for reinforcement in tension due to flexure.

Provide minimum reinforcement in each principal direction, with maximum spacing of 18 in. (460 mm), in accordance with ACI 318. Distribute reinforcement near the top or bottom of the section or allocate between the two section faces for specific conditions, such that the total area of continuous reinforcing steel satisfies ACI 318.

For shear design of mat foundations, follow the principles in 4.11, 6.2, and 6.3 of this guide.

CHAPTER 10—OTHER DESIGN CONSIDERATIONS

10.1—Seismic design limitations

ACI 318 provides design requirements for earthquake resistance but does not apply to structures assigned to Seismic Design Category (SDC) A. Structures assigned to SDC B through F should satisfy the applicable provisions of ACI 318. The recommendations presented in previous sections are applicable to structures assigned to SDC A, B, and C. For structures assigned to SDC D, E, or F, the application of this guide is limited to slab systems, foundations, and structural components not designated as part of the seismic-force-resisting system, but explicitly checked for the induced effects of the design displacements. The exception to this is the use of transverse reinforcement for concrete confinement with f_{yt} up to 100,000 psi (690 MPa) for special moment frames and special structural walls as permitted by ACI 318.

Numerous studies (Muguruma and Watanabe 1990; Muguruma et al. 1991; Sugano et al. 1990; Budek et al. 2002; Stephan et al. 2003) conducted on the use of high-strength reinforcement (f_{yt} up to and greater than 120,000 psi

[827 MPa]) show that there is no detriment to using high-strength transverse reinforcement as column confinement. The test program by Lepage et al. (2008) has demonstrated successful use of high-strength steel (HSS) in beams and columns subjected to reverse cyclic loadings (Rautenberg et al. 2010). With the exception of using f_{yt} up to 100,000 psi (690 MPa) for confinement, the following two recommendations are appropriate until comprehensive and systematic tests performed using ASTM A1035/A1035M (CS) bars justify its use in structural members of buildings assigned to SDC D and higher:

1. ASTM A1035/A1035M (CS) bars should not be used as longitudinal reinforcement in a structural member that is part of the seismic-force-resisting system of a building assigned to SDC D, E. or F. The use of ASTM A1035/A1035M (CS) bars as transverse reinforcement is permitted, provided f_{yt} is limited to 60,000 psi (420 MPa) for computing shear strength.
2. If Grade 60 (420) bar is used for column longitudinal reinforcement and ASTM A1035/A1035M (CS) bar is used for beam longitudinal reinforcement in an intermediate moment frame, it should be ensured that the beam-column joints exhibit strong column-weak beam behavior because the actual flexural strengths of beams using ASTM A1035/A1035M (CS) reinforcement exceed the nominal flexural strengths computed by the simplified method in Chapter 4 of this guide. In addition, the required shear strength of the beam should be determined based on the nominal moment M_n obtained from nonlinear analysis, as given in Appendix B of this guide.

10.2—Development and lap splice length

Design for tension development and splices of ASTM A1035/A1035M (CS) reinforcement should follow ACI 318. Some of the provisions require additional testing and evaluation for validation, as noted in the following.

There have been several investigations on bond and development length of ASTM A1035/A1035M bars of different sizes and with different concrete strengths. Earlier studies include 130 bond tests of beam-end specimens (Ahlborn and DenHartigh 2002), 15 pullout tests (El-Hacha and Rizkalla 2002), and six beam tests with unconfined spliced bars (Peterfreund 2003). Although these studies used pre-standardization high-strength bars, the bar deformations have not changed and the research results are still relevant. The results of these investigations indicated that the development length computed by Eq. (12-1) of ACI 318-99 and ACI 318-02 was sufficient for pre-standardized ASTM A1035/A1035M (CS) bars to develop the 0.2 percent offset yield strength of 120,000 psi (830 MPa).

However, a coordinated research program of bond tests of 69 large-scale specimens using ASTM A1035/A1035M bars (Seliem et al. 2009) indicated that ACI 318 should be used to predict only the strength of confined splices, whereas a similar equation proposed by ACI 408R provides a better strength estimate for both unconfined and confined splices using a strength reduction factor ϕ of 0.82.

Based on the aforementioned studies of ACI 318, it is found applicable for calculating the development and splice lengths of ASTM A1035/A1035M (CS) bars to develop the specified minimum yield strength of 100,000 psi (690 MPa) if the bars are confined. ACI 318 should not be applied to unconfined bars, as discussed in the following.

Alternatively, for both confined and unconfined spliced bars, the equation in ACI 408R with a revised strength reduction factor ϕ of 0.80, instead of 0.82, is recommended. The equations in ACI 408R for development length in the customary in.-lbf units are shown as Eq. (10.2a) through (10.2e).

$$\ell_d = \frac{\left(\dfrac{f_y}{f_c'^{\,1/4}} - \phi 2400\omega\right)\alpha\beta_c\lambda}{\phi 76.3\left(\dfrac{c_b\omega + K_{tr}}{d_b}\right)} d_b \quad \text{(in.-lb)}$$

$$\ell_d = \frac{\left(\dfrac{f_y}{f_c'^{\,1/4}} - \phi 57.4\omega\right)\alpha\beta_c\lambda}{\phi 1.83\left(\dfrac{c_b\omega + K_{tr}}{d_b}\right)} d_b \quad \text{(SI)} \tag{10.2a}$$

where $\alpha = 1.3$ for top cast bars, $\beta_c = 1.0$ for uncoated bars, $\lambda = 1.0$ for normalweight concrete, and

$$K_{tr} = (0.52t_r t_d A_{tr}/sn)\sqrt{f_c'} \quad \text{(in.-lb)}$$

$$K_{tr} = (6.26t_r t_d A_{tr}/sn)\sqrt{f_c'} \quad \text{(SI)} \tag{10.2b}$$

$$t_r = 9.6R_r + 0.28 \leq 1.72 \tag{10.2c}$$

$$t_d = 0.78d_b + 0.22 \quad \text{(in.-lb)}$$

$$t_d = 0.03d_b + 0.22 \quad \text{(SI)} \tag{10.2d}$$

$$\left(\frac{c_b\omega + K_{tr}}{d_b}\right) \leq 4.0 \tag{10.2e}$$

Although no test data are available, design for development and splices of ASTM A1035/A1035M (CS) Grade 100 (690) bars in compression may follow ACI 318, but the specified minimum yield strength f_y in compression is limited to 80,000 psi (550 MPa).

Tests by Harries et al. (2010) have shown that ASTM A1035/A1035M (CS) bars with 90- and 180-degree standard hooks can develop tensile bar stresses of at least 140,000 psi (965 MPa) and, in many cases, the strength of the bar. Instead of hooked bars, headed bars may be considered as an alternative. Section 10.3 of this guide discusses development of headed ASTM A1035/A1035M (CS) bars.

10.3—Mechanically spliced bars and headed bars

According to 10.2 of this guide, designs with f_y of 100,000 psi (690 MPa) for ASTM A1035/A1035M (CS) bars require relatively long development or splice lengths. Because these lengths may be uneconomical or impractical, the designer may consider using mechanical splices (mechanical connections) and mechanically headed bars.

ACI 318 provides minimum tensile strength requirements for mechanical splices and specifies these requirements as $1.25f_y$ for the Type 1 splice and $1.0f_u$ for the Type 2 splice. Accordingly, the requirements for Type 1 mechanical splices are intended to avoid a splice failure when the reinforcement is subject to expected stress levels in yielding regions. The additional requirement for a Type 2 mechanical splice is intended to result in a mechanical splice capable of sustaining inelastic strains through multiple cycles.

These requirements are functions of the specified strengths of the reinforcing bars being spliced (not actual bar strengths) and do not include any ductility requirements. These functions of the specified strengths were historically developed based on reinforcement with stress-strain behavior that is sharply yielding with a definite yield plateau. For gradually yielding reinforcement, such as ASTM A1035/A1035M (CS) reinforcement, strain-based requirements are more appropriate. Equivalent strain criteria for the typical Type 1 and Type 2 mechanical splices (ACI 318) can be developed for ASTM A1035/A1035M (CS) reinforcement by examination of the actual stress-strain behavior for sharply-yielding reinforcement, and establishing a strain that corresponds to the specified minimum strengths for the mechanical splice. On this basis, the Type 1 mechanical splice minimum strength corresponds to a strain in the spliced bars that is beyond the yield plateau and into the onset of strain hardening; the onset of strain hardening occurs at strains in the range of 1 to 2 percent for ASTM A615/A615M Grade 60 and ASTM A706/A706M Grade 60 reinforcement. The Type 2 mechanical splice minimum strength requirement corresponds to a strain in the spliced bars that is even larger and beyond the onset of strain hardening and into the strain hardening range, corresponding to strains in the range of 2.5 to 4 percent for ASTM A615/A615M Grade 60 and ASTM A706/A706M Grade 60 reinforcement. Consequently, to achieve stress-strain performance similar to those of the strength-based requirements for mechanical splices specified in ACI 318, new designated Type A and B mechanical splices are recommended by this guide with monotonic strain limits based on strain of the bars.

Based on the historical statistical analysis of ASTM A1035/A1035M (CS) bars' test data, it is recommended that bars be mechanically spliced to develop 2 percent static tensile strain for Type A connections and 3 percent static tensile strain for Type B connections. Testing of the mechanical splices for strain capacity should be performed under static tensile test loading, as any multiple inelastic cyclic excursions beyond the actual yield point of the bar (measured through 0.2% offset), such as defined in ICC-ES AC133, ICC-ES AC429, ASTM A1034/A1034M, or ISO 15835-2, will strain harden the reinforcing bar and reduce the strain capacity of the reinforcing steel. Inelastic cyclic testing is therefore not recommended for use in evaluation of strain capacity of mechanically spliced A1035/A1035 M (CS) Grade 100 (690) bars. Only test samples cut from original straight lengths of reinforcing steel, not from coil or post-fabricated steel, should be used for tensile strain evaluation.

ASTM A1035/A1035M (CS) bar can be developed in tension by using a mechanically attached head, as is the case with other ASTM types and grades of deformed reinforcing bars. Provisions for computing development length of headed deformed reinforcing bars were introduced in ACI 318, but they are currently limited to reinforcing bars with f_y not exceeding 60,000 psi (420 MPa) and, thus, precludes the use of these provisions with ASTM A1035/A1035M (CS) bars and other reinforcing bars with f_y in excess of 60,000 psi (420 MPa), such as ASTM A615/A615M Grade 75 (520) and Grade 80 (550) bars. However, headed ASTM A1035/A1035M (CS) bars may be used in accordance with ACI 318 "...provided that test results showing the adequacy of such attachment or device are approved by the building official."

ACI 439.3R summarizes available mechanical splices. Some manufacturers use certain methods to produce mechanical splices or headed bars that excessively heat the reinforcing bar, causing an unfavorable alteration to the microstructure of ASTM A1035/A1035M (CS) steel. The applicability of a mechanical splice or headed bar should be determined by consulting the manufacturer of the mechanical splices for the applicability with ASTM A1035/A1035M (CS) bar.

Mechanical splices and headed bars should be placed in accordance with the design drawings or as approved by the licensed design professional.

10.4—Bending and welding of bars

Experiments and field applications have shown that the ASTM A1035/A1035M (CS) reinforcing bars can be bent into standard hooks by following the requirements of ACI 318. Note, however, that because of the strain-hardening characteristics of the material, it is usually necessary to overbend the bar to achieve the desired bent angle. As it is with other reinforcing bars straightening of a bent bar embedded in concrete should be avoided, as it may cause the bar to break.

Although the chemistry (other than the chromium content) of ASTM A1035/A1035M (CS) steel is similar to that of ASTM A706/A706M steel, ASTM A1035/A1035M bars should not be welded. The heat of welding alters the microstructure of the steel. Welding of ASTM A1035 bars should be approached with caution because no specific provisions have been included in the ASTM A1035 to enhance weldability of the bars.

10.5—Use of ASTM A1035/A1035M (CS) bars with ASTM A615/A615M bars

ASTM A1035/A1035M (CS) bars may be used in direct contact with ASTM A615/A615M bars in a concrete structure because the difference in corrosion potential between these dissimilar metals is approximately 100 mV, which is well within the acceptable value of 250 mV (Hartt 2009).

Table 11—Specified minimum yield strengths for design of members using ASTM A1035/A1035M (CS) reinforcement

	Longitudinal reinforcement		Transverse reinforcement		
Type of member	Tension, psi (MPa)	Compression, psi (MPa)	Shear, psi (MPa)	Torsion, psi (MPa)	Confinement, psi (MPa)
Beams and one-way slabs	100,000 (690)	80,000 (550)	80,000 (550)	60,000 (420)	N/A
Columns	100,000 (690)	80,000 (550)	80,000 (550)	60,000 (420)	100,000 (690)
Tension ties	80,000 (550)	N/A	N/A	N/A	N/A
Compression struts	N/A	80,000 (550)	N/A	N/A	N/A
Two-way slabs	100,000 (690)	80,000 (550)	60,000 (420)	60,000 (420)	N/A
Walls	100,000 (690)	80,000 (550)	80,000 (550)	N/A	100,000 (690)
Footings and pile caps	100,000 (690)	80,000 (550)	80,000 (550)	60,000 (420)	N/A
Mat foundations	100,000 (690)	80,000 (550)	80,000 (550)	N/A	N/A

CHAPTER 11—SUMMARY

Presented in this document are guidelines for using ASTM A1035/A1035M (CS) Grade 100 (690) deformed reinforcing bars in reinforced concrete members, including beams, columns, slab systems, walls, footings, and mat foundations, for Seismic Design Category (SDC) A, B, or C. A structure assigned to SDC A, B, or C is required to be designed for all applicable gravity and environmental loads. In the case of SDC A structures, seismic forces are notional structural integrity forces. This document addresses all design required for SDC A, B, and C structures. Application of these guidelines for SDC D, E, or F is limited to slab systems, foundations, and structural components that are not designed as part of seismic-force-resisting system. Both advantages and disadvantages of using ASTM A1035/A1035M (CS) deformed bars are discussed in Chapter 1 of this guide. Design examples included in Appendixes A and B of this guide illustrate design procedures and proper applications of the recommended design criteria.

Tests indicated that ASTM A1035/A1035M (CS) Grade 100 (690) reinforcing bars exhibit a linear stress-strain relationship up to a proportional limit ranging from 60,000 to 80,000 psi (420 to 550 MPa) with a modulus of elasticity of 29,000,000 psi (200,000 MPa), but without a well-defined yield point. The yield strength determined by the 0.2 percent offset method exceeds 115,000 psi (790 MPa). The tensile strength exceeds 155,000 psi (1070 MPa) with corresponding strain ranging from 0.04 to 0.06. The elongation in 8 in. (200 mm) ranges from 0.08 to 0.13. An approximate lower-bound representation of the stress-strain curve of the Grade 100 (690) bar is given in Eq. (3.4a) through (3.4c) of this guide.

A simplified design method for flexural tension is developed based on an idealized "elastic-perfectly plastic" stress-strain curve with a yield plateau at f_y = 100 ksi (690 MPa). Because the yield plateau neglects the effect of strain hardening of the steel, the tension-controlled and compression-controlled strain limits and the strength reduction factor ϕ are adjusted to ensure satisfactory member behavior. The adjusted tension and compression strain limits are 0.009 and 0.004, respectively, and the adjusted strength reduction factor ϕ is given by Eq. (4.4) of this guide. For flexural and direct compression, f_y is taken as 80,000 psi (550 MPa) so that the compatible strain in concrete under compression will not exceed the maximum value of 0.003. Due to a lack of research data and experience, moment redistribution for flexural design using ASTM A1035/A1035M (CS) reinforcement is not recommended.

Flexural design using ASTM A1035/A1035M (CS) Grade 100 (690) bars with f_y = 100,000 psi (690 MPa) would result in higher steel stress at service load conditions, so the designer should exercise caution in applying the ACI 318 provisions on crack and deflection controls, especially for shallow members with reinforcement ratio less than 1 percent and where diagonal shear cracking is expected at the service load level.

For shear design of beams, ASTM A1035/A1035M (CS) Grade 100 (690) bars can be used as shear reinforcement with f_{yt} = 80,000 psi (550 MPa), but at service load levels, slightly increased shear crack widths should be expected. In addition, the concrete contribution to shear strength, V_c, of lightly reinforced deep beams with longitudinal reinforcement ratio less than 1 percent may be less than predicted by ACI 318. It is advisable that such beams should contain minimum shear reinforcement as determined by ACI 318.

Development and splice length for ASTM A1035/A1035M (CS) reinforcement in tension may be determined by ACI 318, provided the splice is confined. Better predictions for both unconfined and confined splices, however, can be obtained by using Eq. (10.2a) through (10.2e) of this guide, which are recommended in ACI 408R. Splice lengths shall comply with ACI 318 if ACI 408R-modified equations provide shorter lengths.

Design of slab systems using ASTM A1035/A1035M (CS) Grade 100 (690) steel bars should follow ACI 318 provisions for shear in one-way slabs and for two-way slabs. Using f_y = 100,000 psi (690 MPa) for tension reinforcement in one-way slabs, the section may become lightly reinforced. If ρ is less than 1 percent, it is recommended that the reduced concrete contribution to shear strength, V_c, as given in Eq. (6.2) of this guide, be considered. For two-way slabs, the slab thickness should be determined according to

ACI 318 based on actual deflection calculations. When the member is lightly reinforced (generally with ρ less than 1 percent), it is recommended that direct deflection calculations be used (Yang et al. 2010). If shear reinforcement is required to provide sufficient shear strength for a two-way slab system, the specified minimum yield strength, f_{yt}, for shear reinforcement should be limited to 60,000 psi (420 MPa) because adequate research data are not available to justify a higher value.

Designs for columns, walls, footings and pile caps, and mat foundations using ASTM A1035/A1035M (CS) reinforcement are discussed in Chapters 5, 7, 8, and 9, respectively, of this guide. There are no exceptions from the requirements of ACI 318 for these structural members other than the adjustments of specified minimum yield strength f_y for tension and compression reinforcements, and f_{yt} for shear reinforcement. Some other design considerations, including seismic design limitations, spliced and headed bars, bending of bars, and use of ASTM A1035/A1035M (CS) bars together with ASTM A615/A615M bars, are discussed in Chapter 10 of this guide.

The recommended specified minimum yield strengths for design of various structural members using ASTM A1035/A1035M (CS) reinforcement are summarized for convenient reference in Table 11.

CHAPTER 12—REFERENCES

ACI committee documents and documents published by other organizations are listed first by document number, full title, and year of publication followed by authored documents listed alphabetically.

American Concrete Institute (ACI)

ACI 224R-01—Control of Cracking in Concrete Structures

ACI 318-56—Building Code Requirements for Reinforced Concrete

ACI 318-63—Building Code Requirements for Reinforced Concrete

ACI 318-71—Building Code Requirements for Reinforced Concrete

ACI 318-95—Building Code Requirements for Structural Concrete and Commentary

ACI 318-99—Building Code Requirements for Structural Concrete and Commentary

ACI 318-02—Building Code Requirements for Structural Concrete and Commentary

ACI 318-05—Building Code Requirements for Structural Concrete and Commentary

ACI 318-08—Building Code Requirements for Structural Concrete and Commentary

ACI 318-11—Building Code Requirements for Structural Concrete and Commentary

ACI 318-14—Building Code Requirements for Structural Concrete and Commentary

ACI 408R-03—Bond and Development of Straight Reinforcing Bars in Tension

ACI 439.3R-07—Types of Mechanical Splices for Reinforcing Bars

ACI 439.4R-09(17)—Report on Steel Reinforcement—Material Properties and U.S. Availability

ACI ITG-6R-10—Design Guide for the Use of ASTM A1035/A1035M Grade 100 (690) Steel Bars for Structural Concrete

ASTM International

A615/A615M-18—Standard Specification for Deformed and Plain Carbon-Steel Bars for Concrete Reinforcement

A706/A706M-16—Standard Specification for Deformed and Plain Low-Alloy Steel Bars for Concrete Reinforcement

A1034/A1034M-10a(2015)—Standard Test Methods for Testing Mechanical Splices for Steel Reinforcing Bars

A1035/A1035M-16b—Standard Specification for Deformed and Plain, Low-Carbon, Chromium, Steel Bars for Concrete Reinforcement

ICC Evaluation Service, Inc.

ICC-ES AC133-1212-R1—Proposed Revisions to the Acceptance Criteria for Mechanical Connector Systems for Steel Reinforcing Bars

ICC-ES AC429-0612-R1—New Acceptance Criteria for High-strength Steel Reinforcing Bars International Organization for Standardization

ICC-ES ESR-2107-2018—Evaluation Report for ChrōmX®9100, 4100 and 2100, Grade 100 Steel Reinforcing BIBars

International Code Council

IBC 2012—International Building Code

International Organization for Standardization (ISO)

ISO 15835-2:2009—Steels for the Reinforcement of Concrete – Reinforcement Couplers for Mechanical Splices of Bars – Part 2: Test methods

Authored documents

Ahlborn, T. M., and DenHartigh, T. C., 2002, "A Comparative Bond Study of MMFX Reinforcing Steel in Concrete," *Final Report* No. CSD 2002–03, Department of Civil and Environmental Engineering, Center for Structural Durability, Michigan Tech Transportation Institute, Houghton, MI, July, 25 pp.

Ansley, M., 2002, "Investigation into the Structural Performance of MMFX Reinforcing," Preliminary Draft Report, Structures Research Center, Florida Department of Transportation, Tallahassee, FL, June, 22 pp.

Beeby, A. W., 1979, "The Prediction of Crack Widths in Hardened Concrete," *The Structural Engineer*, V. 57A, No. 1, Jan., pp. 9-17.

Bentz, E. C., and Collins, M. P., 2006, "Development of the 2004 Canadian Standards Association (CSA) A23.3 Shear Provisions for Reinforced Concrete," *Canadian Journal of Civil Engineering*, V. 33, No. 5, pp. 521-534. doi: 10.1139/l06-005

Bentz, E. C.; Vecchio, F. J.; and Collins, M. P., 2006, "Simplified Modified Compression Field Theory for Calculating Shear Strength of Reinforced Concrete Elements," *ACI Structural Journal*, V. 103, No. 4, July-Aug., pp. 614-624.

Bischoff, P. H., 2005, "Reevaluation of Deflection Prediction for Concrete Beams Reinforced with Steel and Fiber Reinforced Polymer Bars," *Journal of Structural Engineering*, V. 131, No. 5, pp. 752-767. doi: 10.1061/(ASCE)0733-9445(2005)131:5(752) (ASCE)

Bischoff, P. H., and Scanlon, A., 2007, "Effective Moment of Inertia for Calculating Deflections of Concrete Members Containing Steel Reinforcement and Fiber-Reinforced Polymer Reinforcement," *ACI Structural Journal*, V. 104, No. 1, Jan.-Feb., pp. 68-75.

Branson, D. E., 1977, *Deformation of Concrete Structures*, McGraw-Hill, New York, 577 pp.

Budek, A.; Priestley, M.; and Lee, C., 2002, "Seismic Design of Columns with High-Strength Wire and Strand as Spiral Reinforcement," *ACI Structural Journal*, V. 99, No. 5, Sept.-Oct., pp. 660-670.

Collins, M. P., and Kuchma, D., 1999, "How Safe Are our Large, Lightly Reinforced Concrete Beams, Slabs, and Footings?" *ACI Structural Journal*, V. 96, No. 4, July-Aug., pp. 482-490.

DeJong, S. J.; MacDougall, C.; and Heffernan, P. J., 2006, "Fatigue Behaviour of MMFX Corrosion-Resistant Reinforcing Steel," *Advanced Materials in Bridges*, Proceedings of the 7th International Conference on Short and Medium Span Bridges, Montreal, QC, Canada, 11 pp.

Desalegne, A. S., and Lubell, A. S., 2010, "Shear Behavior of Concrete Slabs Longitudinally Reinforced with High-Performance Steel," *ACI Structural Journal*, V. 107, No. 2, Mar.-Apr., pp. 228-236.

Desalegne, A. S., and Lubell, A. S., 2012, "Deflection Control of Concrete Slabs Longitudinally Reinforced with ASTM A1035/A1035M-07 Steel," *ACI Structural Journal*, V. 109, No. 6, Nov.-Dec., pp. 867-877.

Desalegne, A. S., and Lubell, A. S., 2015, "Shear in Concrete Beams Reinforced with High Performance Steel," *ACI Structural Journal*, V. 112, No. 6, Nov-Dec., pp. 783-792. doi: 10.14359/51687798

El-Hacha, R., and Rizkalla, S. H., 2002, "Fundamental Material Properties of MMFX Steel Rebars," *Report* No. 02-04, Constructed Facilities Laboratory, Department of Civil, Construction, and Environmental Engineering, North Carolina State University, Raleigh, NC, July, 59 pp.

Frosch, R. J., 1999, "Another Look at Cracking and Crack Control in Reinforced Concrete," *ACI Structural Journal*, V. 96, No. 3, May-June, pp. 437-442.

Garay-Moran, J.-D., and Lubell, A. S., 2008, "Behavior of Concrete Deep Beams with High Strength Reinforcement," *Proceedings of ASCE Structures Conference: Crossing Borders*, Structural Engineering Institute (SEI)/ASCE Structures Congress, Vancouver, BC, Canada, 10 pp.

Garay-Moran, J.-D., and Lubell, A. S., 2016, "Behavior of Deep Beams Containing High-Strength Longitudinal Reinforcement," *ACI Structural Journal*, V. 113, No. 1, Jan.-Feb., pp. 17-28. doi: 10.14359/51687910

Gergely, P., and Lutz, L. A., 1968, "Maximum Crack Width in Reinforced Concrete Flexural Members," *Causes, Mechanism, and Control of Cracking in Concrete*, SP-20, American Concrete Institute, Farmington Hills, MI, pp. 87-117.

Griezic, A.; Cook, W. D.; and Mitchell, D., 1994, "Tests to Determine Performance of Deformed Welded-Wire Fabric Stirrups," *ACI Structural Journal*, V. 91, No. 2, Mar.-Apr., pp. 211-220.

Harries, K. A.; Shahrooz, B.; Saltani, A.; Reis, J. M.; Wells, E. L.; Miller, R.; and Russell, H. G., 2010, "Bond and Anchorage of High Strength Reinforcing Steel," Transportation Research Board Annual Meeting 2010, Washington, DC, 18 pp.

Hartt, W., 2009, "Risk of Macro-Cell Corrosion Associated with Black Bar – MMFX 2 (ASTM A1035) Combinations in Concrete," Research Report Submitted to MMFX Technologies Corporation, Irvine, CA, 10 pp.

Hassan, T. K.; Seliem, H. M.; Dwairi, H.; Rizkalla, S. H.; and Zia, P., 2008, "Shear Behavior of Large Concrete Beams Reinforced with High-Strength Steel," *ACI Structural Journal*, V. 105, No. 2, Mar.-Apr., pp. 173-179.

Hoult, N.; Sherwood, E. G.; Bentz, E.; and Collins, M. P., 2008, "Does the Use of FRP Reinforcement Change the One-Way Shear Behavior of Reinforced Concrete Slabs?" *Journal of Composites for Construction*, V. 12, No. 2, Apr., pp. 125-133. doi: 10.1061/(ASCE)1090-0268(2008)12:2(125)

Kaar, P. H., and Mattock, A. H., 1963, "High Strength Bars as Concrete Reinforcement: Part 4—Control of Cracking," *Research and Development Bulletin*, V. 5, No. 1, Jan., pp. 15-38.

Lepage, A.; Tavallali, H.; Pujol, S.; and Rautenberg, J., 2008, "Towards Earthquake-Resistant Concrete Structures with Ultra High-Strength Steel Reinforcement," *Proceedings of the 14th World Conference on Earthquake Engineering*, Beijing, China, Oct. 12-17, 8 pp.

Lubell, A.; Sherwood, T.; Bentz, E.; and Collins, M. P., 2004, "Safe Shear Design of Large, Wide Beams," *Concrete International*, V. 26, No. 1, Jan., pp. 66-78.

Malhas, F. A., 2002, "Preliminary Experimental Investigation of the Flexural Behavior of Reinforced Concrete Beams Using MMFX Steel," *Final Report*, Contract No. 904311021, Department of Civil Engineering, University of North Florida, Jacksonville, FL, July, 55 pp.

Mast, R. F., 1992, "Unified Design Provisions for Reinforced and Prestressed Concrete Flexural and Compression Members," *ACI Structural Journal*, V. 89, No. 2, Mar.-Apr., pp. 185-199.

Mast, R. F., 2006, "Memorandum: Behavior of Flexural Members Reinforced with MMFX Steel," Jan., 15 pp.

Mast, R. F.; Dawood, M.; Rizkalla, S. H.; and Zia, P., 2008, "Flexural Strength Design of Concrete Beams Reinforced with High-Strength Bars," *ACI Structural Journal*, V. 105, No. 5, Sept.-Oct., pp. 570-577.

Muguruma, H.; Nishiyama, M.; Watanabe, F.; and Tanaka, H., 1991, "Ductile Behavior of High-Strength Concrete Columns Confined by High-Strength Transverse Reinforcement," *Evaluation and Rehabilitation of Concrete Structures and Innovations in Design*, SP-128, American Concrete Institute, Farmington Hills, MI, pp. 877-891.

Muguruma, H., and Watanabe, F., 1990, "Ductility Improvement of High-Strength Concrete Columns with Lateral Confinement," *Proceedings of the Second International Symposium on High-Strength Concrete*, SP-121, American Concrete Institute, Farmington Hills, MI, pp. 47-60.

Munikrishna, A., 2008, "Shear Behavior of Concrete Beams Reinforced with High Performance Steel Shear Reinforcement," MSCE thesis, Department of Civil, Construction, and Environmental Engineering, North Carolina State University, Raleigh, NC, Dec., 152 pp.

Nowak, A. S., and Szerszen, M. M., 2003, "Calibration of Design Code for Buildings (ACI 318): Part 1—Statistical Models for Resistance," *ACI Structural Journal*, V. 100, No. 3, May-June, pp. 377-382.

Peterfreund, S. W., 2003, "Development Length of Micro-Composite (MMFX) Steel Reinforcing Bars Used in Bridge Deck Applications," Senior Honors Report, Department of Civil and Environmental Engineering, University of Massachusetts, Amherst, MA, June, 59 pp.

Portland Cement Association (PCA), 2005, *PCA Notes on ACI 318-05 Building Code Requirements for Structural Concrete,* M. E. Kamara, and B. G. Rabbat, eds., Portland Cement Association, Skokie, IL, 1008 pp.

Rajagopalan, K. S., and Ferguson, P. M., 1968, "Exploratory Shear Tests Emphasizing Percentage of Lightweight Steel," *ACI Journal Proceedings*, V. 65, No. 8, Aug., pp. 634-638 and author's closure, V. 66, No. 2, pp. 153-154.

Rautenberg, J.; Pujol, S.; Tavallali, H.; and Lepage, A., 2010, "Cyclic Response of Concrete Columns Reinforced with High Strength Steel," *Proceedings of the Ninth US National and Tenth Canadian Conference on Earthquake Engineering,* Toronto, ON, Canada, July 25-29, 9 pp.

Reineck, K.-H.; Kuchma, D. A.; Kim, K. S.; and Marx, S., 2003, "Shear Database for Reinforced Concrete Members without Shear Reinforcement," *ACI Structural Journal*, V. 100, No. 2, Mar.-Apr., pp. 240-249.

Seliem, H. M.; Hosny, A.; Rizkalla, S.; Zia, P.; Briggs, M.; Miller, S.; Darwin, D.; Browning, J.; Glass, G. M.; Hoyt, K.; Donnelly, K.; and Jirsa, J. O., 2009, "Bond Characteristics of ASTM A1035 Steel Reinforcing Bars," *ACI Structural Journal*, V. 106, No. 4, July-Aug., pp. 530-539.

Seliem, H. M.; Lucier, G.; Rizkalla, S. H.; and Zia, P., 2008, "Behavior of Concrete Bridge Decks Reinforced with High-Performance Steel," *ACI Structural Journal*, V. 105, No. 1, Jan.-Feb., pp. 78-86.

Stephan, B.; Restrepo, J.; and Seible, F., 2003, "Seismic Behavior of Bridge Columns Built Incorporating MMFX Steel," *Report* No. SSRP-2003/09, Department of Structural Engineering, University of California, San Diego, La Jolla, CA, Oct., 37 pp.

Sugano, S.; Nagashima, T.; Kimura, H.; Tamura, A.; and Ichikawa, A., 1990, "Experimental Studies on Seismic Behavior of Reinforced Concrete Members of High Strength Concrete," *Proceedings of the Second International Symposium on High-Strength Concrete,* SP-121, American Concrete Institute, Farmington Hills, MI, pp. 61-87.

Sumpter, M.; Rizkalla, S. H.; and Zia, P., 2009, "Behavior of High Performance Steel as Shear Reinforcement for Concrete Beams," *ACI Structural Journal*, V. 106, No. 2, Mar.-Apr., pp. 171-177.

Tang, J., and Lubell, A. S., 2008, "Influence of Longitudinal Reinforcement Strength on One-Way Slab Deflection," *Canadian Journal of Civil Engineering*, V. 35, No. 10, Oct., pp. 1076-1087. doi: 10.1139/L08-050

WJE, 2008, "Mechanical Properties of ASTM A1035/A1035M High Strength Steel Bar Reinforcement," *Final Report*, WJE No. 2008.9901.0, Wiss, Janney, Elstner Associates, Inc., Chicago, IL, Dec. 19, 47 pp.

Yang, J.; Yoon, Y.; Cook, W.; and Mitchell, D., 2010, "2010, "Punching Shear Behavior of Two-Way Slabs Reinforced with High-Strength Steel," *ACI Structural Journal*, V. 107, No. 4, July-Aug., pp. 468-475.

Yotakhong, P., 2003, "Flexural Performance of MMFX Reinforcing Bars in Concrete Structures," MSCE thesis, Department of Civil, Construction, and Environmental Engineering, North Carolina State University, Raleigh, NC, Dec., 145 pp.

APPENDIX A—DESIGN EXAMPLES

A.1—Introduction

The following design examples illustrate the various design procedures recommended in this guide. The problems selected for these examples are from a Portland Cement Association (PCA) publication, *PCA Notes on ACI 318-05 Building Code Requirements for Structural Concrete* (PCA 2005). The term "CS" used in the examples refers to ASTM A1035/A1035M (CS) Grade 100 steel reinforcement.

Examples are numbered as the chapter number of this guide. Examples are grouped to illustrate the various aspects of the recommended design procedures, not in ascending numerical order. Examples appear in customary in.-lb units except Examples 6.1(SI) and 4.6(SI), which are the same as Example 6.1 and Example 4.6, but in SI units for purposes of illustration.

Where appropriate, commentaries are added to indicate the differences in design when using ASTM A1035/A1035M (CS) bars as opposed to the ASTM A615/A615M Grade 40 and Grade 60 steel bars.

A.2—Design examples

Example 4.1—Moment strength in beams using equivalent rectangular stress distribution

Similar to Example 6.1 in the *PCA Notes*, this example represents direct bar-for-bar replacement of Grade 60 reinforcing bars by CS bars.

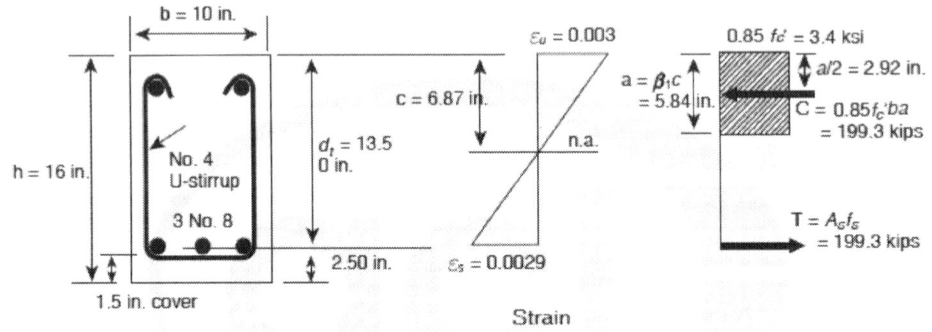

(a) For the beam section shown, calculate moment strength based on static equilibrium using the ACI equivalent rectangular stress distribution. Assume $f_c' = 4000$ psi and $f_y = 100{,}000$ psi. For simplicity, neglect hanger bars in the compression zone.

(b) Calculate the moment strength for the same concrete section reinforced with two No. 8 high-strength bars ($f_y = 100{,}000$ psi) instead of three No. 8 Grade 60 bars. For simplicity, neglect hanger bars in the compression zone.

This example represents replacement of the longitudinal tension force, A_sf_y, provided by the conventional reinforcement, which results in the use of fewer high-strength reinforcing bars.

Calculations and discussion

(a) Section reinforced with three No. 8, Grade 100 tension bars

1. Define rectangular concrete stress distribution.

$d = d_t = 16 - 2.5 = 13.50$ in.

$A_s = 3 \times 0.79 = 2.37$ in.^{2}

Assuming $\varepsilon_s < \varepsilon_y$ ($\varepsilon_y = 100/29{,}000 = 0.00345$)

$T = A_sE_s\varepsilon_s = 2.37 \times 29{,}000 \times \varepsilon_s = 68{,}730\varepsilon_s$

$C = 0.85f_c'ba = 0.85 \times 4 \times 10 \times a = 34a$

$$\varepsilon_s = \left(\frac{d_t - c}{c}\right)0.003 = \left(\frac{d_t - a/\beta_1}{a/\beta_1}\right)0.003$$

Satisfying equilibrium $T = C$

$$68,730\left(\frac{13.50 - a/0.85}{a/0.85}\right)0.003 = 34a$$

$$40a^2 + 242.58a - 2783.57 = 0$$

$$a = 5.84 \text{ in.}$$

2. *Determine net tensile strain ε_s and ϕ.*
$c = a/\beta_1 = 5.84/0.85 = 6.87$ in.

	c, in.	ϕ, curvature
Grade 60, PCA	4.92	0.00061
Grade 100, CS	6.87	0.00044

$$\varepsilon_s = \left(\frac{d_t - c}{c}\right)0.003 = \left(\frac{13.50 - 6.87}{6.87}\right)0.003 = 0.0029 < 0.004$$

(Refer to Section 4.4 of this guide.)

Therefore, the section is compression-controlled. This also confirms that $\varepsilon_s < \varepsilon_y = 0.00345$.

$\phi = 0.65$

(Refer to Section 4.4 of this guide.)

3. *Determine nominal moment strength M_n and design moment strength ϕM_n.*
$M_n = A_s E_s \varepsilon_s (d_t - (a/2)) = 68{,}730 \times 0.0029 \times (13.50 - 2.92) = 2108.8$ in.-kip = 175.7 ft-kip

$\phi M_n = 0.65(175.7) = 114.2$ ft-kip

	M_n, ft-kip	ϕM_n, ft-kip
PCA, Grade 60	135.2	121.9
CS, Grade 100	175.7	114.2

4. *Minimum reinforcement.*

$$A_{s,min} = \frac{3\sqrt{f_c'}}{f_y} b_w d \geq \frac{200 b_w d}{f_y}$$

Because $f_c' = 4000$ psi, $3\sqrt{4000} = 189.7 < 200$. Therefore, $200b_w d/f_y$ governs

$$\frac{200\,b_w d}{f_y} = \frac{200 \times 10 \times 13.50}{100{,}000} = 0.27 \text{ in.}^2$$

A_s (provided) = 2.37 in.^{2} > $A_{s,min}$ = 0.27 in.^{2} **OK**

The nominal moment strength of the beam reinforced with CS is 30 percent higher than the nominal moment strength of 135.2 ft-kip for the same beam reinforced with Grade 60 steel (*PCA Notes*, Example 6.1). The change of failure mode from tension-controlled to compression-controlled, however, reduces the ϕ value, resulting in the beam design moment with CS bars 6 percent lower than the beam reinforced with Grade 60 steel bars.

(b) Section reinforced with two No. 8, Grade 100 tension bars.

1. *Define rectangular concrete stress distribution.*
$d = d_t = 16 - 2.5 = 13.50$ in.

$A_s = 2 \times 0.79 = 1.58$ in.^{2}

Assuming $\varepsilon_s > \varepsilon_y$

$T = A_s f_y = 1.58 \times 100 = 158$ kip

$$a = \frac{A_s f_y}{0.85 f'_b b} = \frac{158}{0.85 \times 4 \times 10} = 4.65 \text{ in.}$$

2. *Determine net tensile strain* ε_s *and* ϕ.
$c = a/\beta_1 = 4.65/0.85 = 5.47$ in.

	c, in.	ϕ, Curvature
PCA, 3 No. 8 Grade 60	4.92	0.00061
CS, 3 No. 8 Grade 100	5.47	0.00055

$$\varepsilon_s = \left(\frac{d_t - c}{c}\right)0.003 = \left(\frac{13.50 - 5.47}{5.47}\right)0.003 = 0.0044 > 0.004$$

Therefore, the section is in the transition zone. This also confirms that $\varepsilon_s > \varepsilon_y = 0.00345$

$\phi = 0.45 + 50\varepsilon_s = 0.45 + 50 \times 0.0044 = 0.67$

(Refer to Section 4.4 of this guide.)

3. *Determine nominal moment strength* M_n *and design moment strength* ϕM_n.

$$M_n = A_s f_y\left(d_t - \frac{a}{2}\right) = 158 \times (13.50 - 2.33) = 1764.9 \text{ in.-kip} = 147.1 \text{ ft-kip}$$

$\phi M_n = 0.67(147.1) = 98.5$ ft-kip

	M_n, ft-kip	ϕM_n, ft-kip
PCA, 3 No. 8 Grade 60	135.2	121.9
CS, 2 No.8 Grade 100	147.1	98.5

4. *Minimum reinforcement.*

A_s (provided) = 1.58 in.^{2} > $A_{s,min}$ = 0.27 in.^{2} **OK**

In this case, even though there is a reduction of one-third of the longitudinal steel area, the failure mode is in the transition region between compression control and tension control. The corresponding reduction of the ϕ value causes the beam design moment with two No. 8 CS bars to be 19 percent lower than for the beam reinforced with three No. 8 Grade 60 steel bars.

Example 4.2—Design of beam with compression reinforcement

Similar to Example 6.2 in *PCA Notes*, this example represents replacement of Grade 60 reinforcing bars by high-strength bars. This example demonstrates the benefit of using high-strength concrete with high-strength bars.

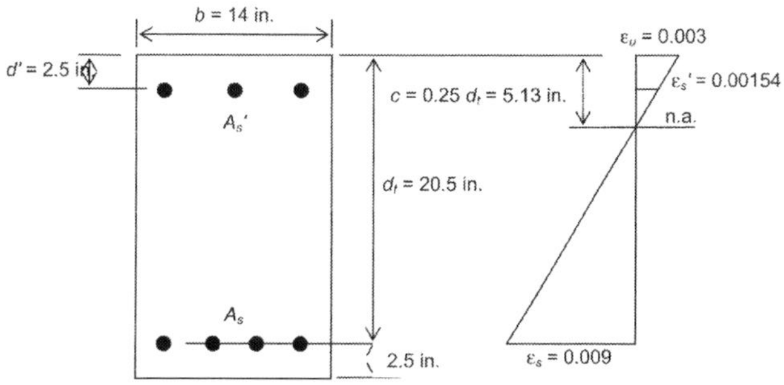

(a) A beam cross section is limited to the maximum size shown. Determine the required area of reinforcement for a factored moment M_u = 516 ft-kip; f_c' = 4000 psi; f_y = 100,000 psi (tension); and f_y = 80,000 psi (compression).

(b) Determine the required area of reinforcement for the same section using f_c' = 8000 psi.

(c) Determine the required area of reinforcement for the same section using f_c' = 8000 psi and f_y = 60,000 psi for comparison.

Calculations and discussion

(a) Grade 100 steel and f_c' = 4000 psi.

1. Determine the required nominal strength.

Design the section to be tension controlled, $\varepsilon_s = 0.009$ and $\phi = 0.9$.

(Refer to Section 4.4 of this guide.)

$M_n = M_u/\phi = 516/0.9 = 573$ ft-kip

2. Determine maximum moment without compression reinforcement.

$c = 0.25d = 0.25 \times 20.5 = 5.13$ in.

f_c' = 4000 psi	c, in.	ϕ, curvature
Grade 60, PCA	7.69	0.00040
Grade 100, CS	5.13	0.00058

$a = \beta_1 c = 0.85 \times 5.13 = 4.36$ in.

$C = T = 0.85f_c'ab = 0.85 \times 4 \times 4.36 \times 14 = 208$ kip

$$M_{nt} = T\left(d_t - \frac{a}{2}\right) = 208\left(20.5 - \frac{4.36}{2}\right) = 3811 \text{ in.-kip} = 318 \text{ ft-kip}$$

3. Required area of tension steel to develop M_{nt}.

$A_{s,nt} = 208/100 = 2.08$ in.^{2}

4. Additional moment (573 – 318 = 255 ft-kip) should be developed in T-C *couple between tension steel and compression steel.*

$$\text{Additional tension steel required: } \Delta A_s = \frac{255(12)}{(20.5-2.5)(100)} = 1.70 \text{ in.}^2$$

Total tension steel required: $A_s = 2.08 + 1.70 = 3.78$ in.^{2}

The strain in the compression steel is 0.00154, as shown in the aforementioned strain diagram. The corresponding steel stress is 44.6 ksi.

Compression steel required: $A_s' = \dfrac{255(12)}{(20.5-2.5)(44.6)} = 3.81 \text{ in.}^2$

PCA Notes solution: using Grade 60 steel, $A_s = 6.59$ in.^{2} and $A_s' = 1.43$ in.^{2}

Total area of reinforcement required for this design, including the tension steel of 3.78 in.^{2} and the compression steel of 3.81 in.^{2}, is 7.59 in.^{2} The original design in the *PCA Notes*, which uses Grade 60 steel, required a tension steel of 6.59 in.^{2} and a compression steel of 1.43 in.^{2} for a total 8.02 in.^{2} of reinforcing steel. The use of CS results in a reinforcing steel reduction of 5 percent. The total reduction is small because the compression steel increases to offset the smaller compressive force provided by the normal-strength concrete. High-strength steel (HSS) for flexural capacity is most advantageous when used with high-strength concrete, as illustrated in the following section.

(b) Grade 100 steel and $f_c' = 8000$ psi

1. Determine the required nominal strength.

$M_n = M_u/\phi = 516/0.9 = 573$ ft-kip

2. Determine maximum moment without compression reinforcement.

$c = 0.25d = 0.25 \times 20.5 = 5.13$ in.

$f_c' = 8000$ psi	c, in.	φ, curvature
Grade 60, PCA	5.98	0.00050
Grade 100, CS	5.13	0.00058

$a = \beta_1 c = 0.65 \times 5.13 = 3.33$ in.

$C = T = 0.85 f_c' ab = 0.85 \times 8 \times 3.33 \times 14 = 317$ kip

$M_{nt} = 5971$ in.-kip $= 498$ ft-kip

3. Required area of tension steel to develop M_{nt}.

$A_{s,nt} = 317/100 = 3.17$ in.^{2}

4. Additional moment (573 – 498 = 75 ft-kip) should be developed in T-C *couple between tension steel and compression steel.*

Additional tension steel required: $\Delta A_s = \dfrac{75(12)}{(20.5-2.5)(100)} = 0.05 \text{ in.}^2$

Total tension steel required: $A_s = 3.17 + 0.50 = 3.67$ in.^{2}

Compression steel required: $A_s' = \dfrac{75(12)}{(20.5-2.5)(44.6)} = 1.12 \text{ in.}^2$

By using Grade 100 steel with 8000 psi concrete, both the tension and the compression steels are reduced.

(c) $f_c' = 8000$ psi and $f_y = 60{,}000$ psi

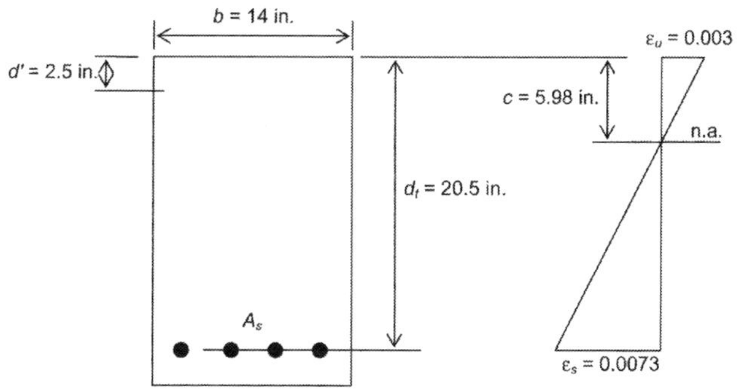

1. Determine the required nominal strength.

$M_n = M_u/\phi = 516/0.9 = 573$ ft-kip $= 6876$ in.-kip

2. Define rectangular concrete stress distribution.

$C = T = 0.85f_c'ab$

$$M_n = T\left(d_t - \frac{a}{2}\right) = 0.85 f_c' ab\left(d_t - \frac{a}{2}\right)$$

$6876 = 0.85 \times 8 \times a \times 14 \times (20.5 - a/2)$

$47.6\, a^2 - 1951.6a + 6876 = 0$

$a = 3.89$ in.

3. Determine required area of tension steel.

$c = a/\beta_1 = 3.89 / 0.65 = 5.98$ in.

$$\varepsilon_s = \left(\frac{d_t - c}{c}\right)(0.003) = \left(\frac{20.5 - 5.98}{5.98}\right)0.003 = 0.0073 > 0.00207 = e_y$$

$A_s = T/f_y = 0.85f_c'ab/f_y = 0.85 \times 8 \times 3.89 \times 14/60 = 6.17$ in.^{2} No compression steel is required.

PCA Notes solution: using Grade 60 steel with 4000 psi concrete, $A_s = 6.59$ in.^{2} and $A_s' = 1.43$ in.^{2}

Using CS for flexural capacity is most beneficial when used with high-strength concrete, as illustrated in this example. Using CS requires a total area of reinforcement, including tension and compression steel, of 4.79 in.^{2} Grade 60 reinforcement requires a total of 6.17 in.^{2} of steel. The design using CS results in a 25 percent reduction of the required flexural reinforcement.

The design using Grade 60 steel does not require compression steel but using CS does. The tension-controlled strain limit for CS is 0.009 instead of 0.005 for Grade 60 steel. This high strain level results in a smaller compression zone depth c. Consequently, the high-strength compression reinforcement should provide additional compression force if $c > d'$.

	A_s, in.^{2}	A_s', in.^{2}	Total A_s, in.^{2}
Grade 60, PCA, f_c' = 4000 psi	6.59	1.43	8.02
Grade 100, CS, f_c' = 4000 psi	3.78	3.81	7.59
Grade 60, PCA, f_c' = 8000 psi	6.17	Not needed	6.17
Grade 100, CS, f_c' = 8000 psi	3.67	1.12	4.79

Example 5.1—Axial load-moment strength, P_n and M_n, for given strain conditions

Similar to Example 6.4 in *PCA Notes*, this example represents direct bar-for-bar replacement of Grade 60 reinforcing bars by high-strength bars.

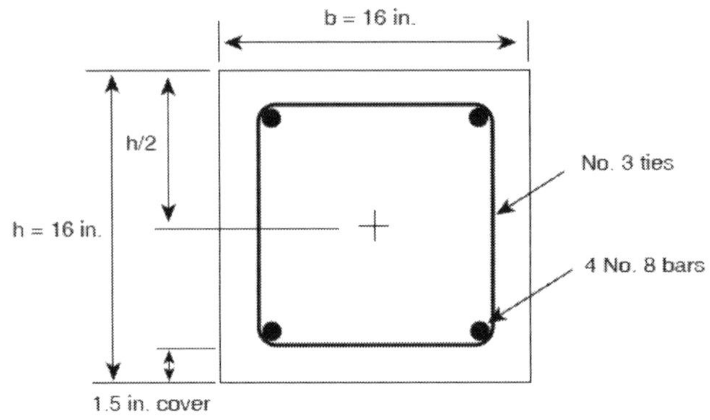

For the column section shown, calculate the load-moment strength, P_n and M_n, for four strain conditions.

(1) Bar stress near tension face of member equal to zero, $f_s = 0$

(2) Bar stress near tension face of member equal to $0.5f_y$ ($f_s = 0.5f_y$)

(3) At limit for compression-controlled section ($\varepsilon_t = 0.004$)

(4) At limit for tension-controlled section ($\varepsilon_t = 0.009$)

Use $f_c' = 4000$ psi, $f_y = 100{,}000$ psi in tension and $f_y = 80{,}000$ psi in compression.

Calculations and discussion

(a) Axial load-moment strength, P_n and M_n, for Strain Condition 1: $\varepsilon_s = 0$.

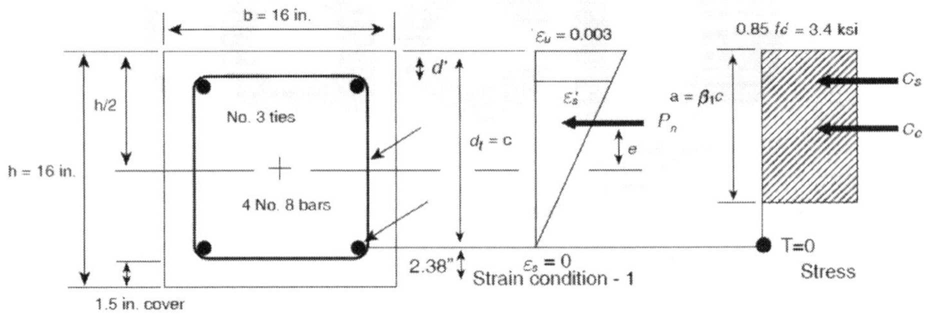

1. Define stress distribution and determine force values.

$d' = \text{Cover} + \text{No. 3 tie diameter} + d_b/2 = 1.5 + 0.375 + 0.5 = 2.38$ in.

$d_t = 16 - 2.38 = 13.62$ in.

Because $\varepsilon_s = 0$, $c = d_t = 13.62$ in.

$a = \beta_1 c = 0.85 \times 13.62 = 11.58$ in., where $\beta_1 = 0.85$ for $f_c' = 4000$ psi.

$C_c = 0.85 f_c'ba = 0.85 \times 4 \times 16 \times 11.58 = 630$ kip

Compression: $\varepsilon_y' = f_y'/E_s = 80/29{,}000 = 0.00276$

Tension: $\varepsilon_y = f_y/E_s = 100/29{,}000 = 0.00345$

From strain compatibility

$$\varepsilon_s' = \varepsilon_u\left(\frac{c-d'}{c}\right) = 0.003\left(\frac{13.62-2.38}{13.62}\right) = 0.00248 < \varepsilon_y' = 0.00276$$

Compression steel has not yielded

$C_s = A_s'E_s\varepsilon_s' = 1.58 \times 29{,}000 \times 0.00248 = 113.6$ kips

2. Determine P_n *and* M_n *from static equilibrium.*

$P_n = C_c + C_s = 630 + 113.6 = 743.6$ kip

$M_n = P_n e = C_c(ha/2 - a/2) + C_s(ha/2 - d') = 630(8 - 5.79) + 113.6(8 - 2.38) = 2030.7$ in.-kip $= 169.2$ ft-kip

$e = M_n/P_n = 2030.7/743.6 = 2.73$ in.

$\phi = 0.65$

(Refer to Section 4.4 of this guide.)

Therefore, for strain condition $\varepsilon_s = 0$; the design axial load strength, $\phi P_n = 0.65(743.6) = 483.3$ kip; and the design moment strength, $\phi M_n = 0.65(169.2) = 110.0$ ft-kip.

$\varepsilon_s = 0$	ϕP_n, kip	ϕM_n, ft-kip
Grade 60, PCA	471.1	104.3
Grade 100, CS	471.1	110.0

(b) Axial load-moment strength, P_n and M_n, for Strain Condition 2: $\varepsilon_s = 0.5\varepsilon_y$.

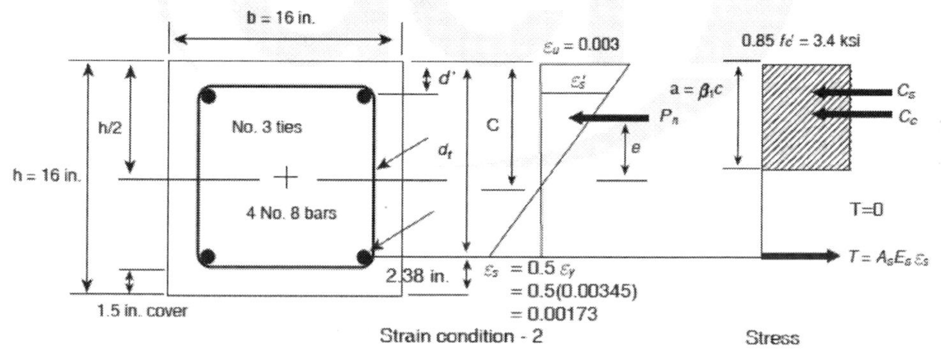

1. Define stress distribution and determine force values.

$d' = 2.38$ in. $d_t = 13.62$ in.

From strain compatibility

$$\left(\frac{c}{0.003}\right) = \left(\frac{d^t}{0.5\varepsilon_y} + 0.003\right)$$

$$c = \left(\frac{0.003 d_t}{0.5\varepsilon_Y + 0.003}\right) = \frac{0.003 \times 13.62}{0.00173 + 0.003} = 8.64 \text{ in.}$$

Strain in compression reinforcement

$$\varepsilon_s' = \varepsilon_u\left(\frac{c-d'}{c}\right) = 0.003\left(\frac{8.64-2.38}{8.64}\right) = 0.00218 < \varepsilon_y' = 0.00276$$

Compression steel has not yielded

$a = \beta_1 c = 0.85 \times 8.64 = 7.34$ in.

$C_c = 0.85 f_c' ba = 0.85 \times 4 \times 16 \times 7.34 = 399.3$ kip

$C_s = A_s' E_s \varepsilon_s' = 1.58 \times 29{,}000 \times 0.00218 = 99.9$ kip

$T = A_s E_s \varepsilon_s = 1.58 \times 29{,}000 \times 0.00173 = 79.3$ kip

2. *Determine* P_n *and* M_n *from static equilibrium.*

$P_n = C_c + C_s - T = 399.3 + 99.9 - 79.3 = 419.9$ kip

$M_n = P_n e = C_c(h/2 - a/2) + C_s(h/2 - d') + T(d_t - h/2)$
$= 399.3(8 - 3.67) + 99.9(8 - 2.38) + 79.3(13.62 - 8) = 2736.1$ in.-kip $= 228$ ft-kip

$e = M_n/P_n = 2736.1/419.9 = 6.52$ in.

$\phi = 0.65$

(Refer to Section 4.4 of this guide.)

Therefore, for strain condition $\varepsilon_s = 0.5\varepsilon_y = 0.00173$; design axial load strength, $\phi P_n = 0.65(419.9) = 272.9$ kip; and design moment strength $\phi M_n = 0.65\ (228) = 148.2$ ft-kip.

$\varepsilon_s = 0.5\varepsilon_y$	ϕP_n, kip	ϕM_n, ft-kip
Grade 60, PCA, $\varepsilon_s = 0.00104$	335.3	136.9
Grade 100, CS, $\varepsilon_s = 0.00173$	272.9	148.2

(c) Axial load-moment strength, P_n and M_n, for Strain Condition 3: $\varepsilon_s = 0.004$.

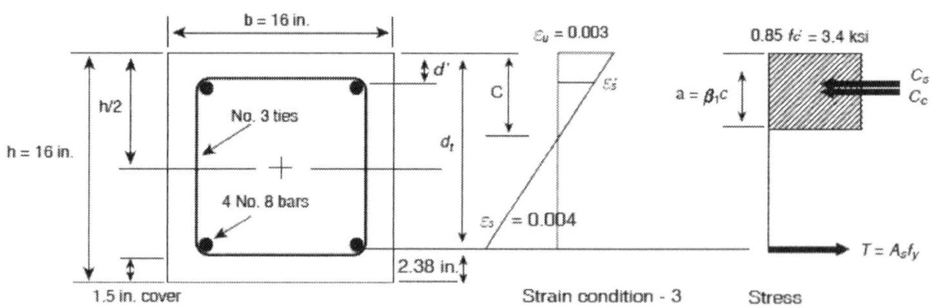

1. *Define stress distribution and determine force values.*

$d' = 2.38$ in.

$d_t = 13.62$ in.

From strain compatibility

$$\left(\frac{c}{0.003}\right) = \left(\frac{d_t}{0.004+0.003}\right)$$

$$c = \left(\frac{0.003d_t}{0.004+0.003}\right) = \frac{0.003\times 13.62}{0.004+0.003} = 5.84 \text{ in.}$$

Strain in compression reinforcement

$$\varepsilon_s' = \varepsilon_u\left(\frac{c-d'}{c}\right) = 0.003\left(\frac{5.84-2.38}{5.84}\right) = 0.00178 < \varepsilon_y' = 0.00276$$

Compression steel has not yielded

$a = \beta_1 c = 0.85 \times 5.84 = 4.96$ in.

$C_c = 0.85f_c'ba = 0.85 \times 4 \times 16 \times 4.96 = 269.8$ kip

$C_s = A_s'E_s'\varepsilon_s' = 1.58 \times 0.00178 \times 29{,}000 = 81.6$ kips

$T = A_sf_y = 1.58 \times 100 = 158.0$ kip

2. Determine P_n *and* M_n *from static equilibrium.*

$P_n = C_c + C_s - T = 269.8 + 81.6 - 158.0 = 193.4$ kip

$M_n = P_ne = C_c(h/2 - a/2) + C_s(h/2 - d') + T(d_t - h/2)$
$= 269.8(8 - 2.48) + 81.6(8 - 2.38) + 158.0(13.62 - 8) = 2835.8$ in.-kip $= 236.3$ ft-kip

$e = M_n/P_n = 2835.8/193.4 = 14.66$ in.

$\phi = 0.65$

(Refer to Section 4.4 of this guide.)

Therefore, for strain condition $\varepsilon_s = 0.004$; design axial load strength $\phi P_n = 0.65(193.4) = 125.7$ kip; and design moment strength, $\phi M_n = 0.65(236.3) = 153.6$ ft-kip.

ε_s	ϕP_n, kip	ϕM_n, ft-kip
Grade 60, PCA, $\varepsilon_s = 0.00207$	242.3	150.1
Grade 100, CS, $\varepsilon_s = 0.004$	125.7	153.6

(d) Axial load-moment strength, P_n and M_n, for Strain Condition 4: $\varepsilon_s = 0.009$.

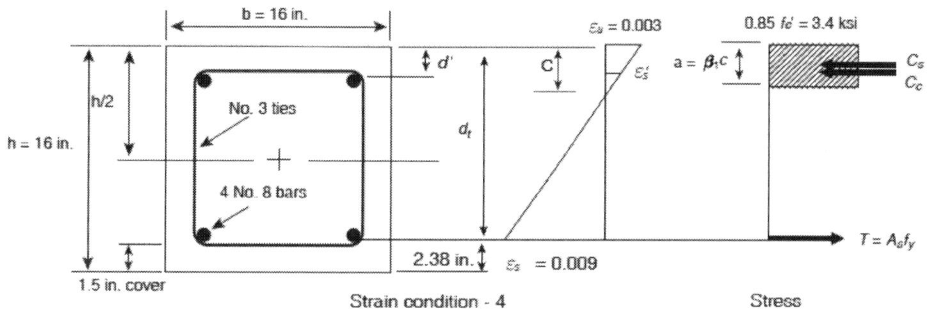

1. Define stress distribution and determine force values.

$d' = 2.38$ in. $d_t = 13.62$ in.

From strain compatibility

$$\left(\frac{c}{0.003}\right)=\left(\frac{d_t}{0.009+0.003}\right)$$

$$c=\left(\frac{0.003d'}{0.009+0.003}\right)=\frac{0.003\times 13.62}{0.009+0.003}= 3.41\text{ in.}$$

Strain in compression reinforcement

$$\varepsilon_s{}'=\varepsilon_u\left(\frac{c-d'}{c}\right)=0.003\left(\frac{3.41-2.38}{3.41}\right)=0.00091<\varepsilon_y{}'=0.00276$$

Compression steel has not yielded

$a = \beta_1 c = 0.85 \times 3.41 = 2.90$ in.

$C_c = 0.85f_c'ba = 0.85 \times 4 \times 16 \times 2.90 = 157.8$ kip

$C_s = A_s'E_s\varepsilon_s' = 1.58 \times 29{,}000 \times 0.00091 = 41.7$ kip

$T = A_s f_y = 1.58 \times 100 = 158.0$ kips

2. *Determine* P_n *and* M_n *from static equilibrium.*

$P_n = C_c + C_s - T = 157.8 + 41.7 - 158.0 = 41.5$ kip

$M_n = P_n e = C_c(h/2 - a/2) + C_s(h/2 - d') + T(d_t - h/2)$
$= 157.8(8 - 1.45) + 41.7(8 - 2.38) + 158.0(13.62 - 8) = 2155.9$ in.-kips $= 179.7$ ft-kip

$e = M_n/P_n = 2155.9/41.5 = 51.9$ in.

$\phi = 0.9$

(Refer to Section 4.4 of this guide.)

Therefore, for strain condition $\varepsilon_s = 0.009$; the design axial load strength, $\phi P_n = 0.9\ (41.5) = 37.4$ kips; and the design moment strength, $\phi M_n = 0.9\ (179.7) = 161.7$ ft-kips.

ε_s	ϕP_n, kip	ϕM_n, ft-kip
Grade 60, PCA, $\varepsilon_s = 0.005$	193.4	174.2
Grade 100, CS, $\varepsilon_s = 0.009$	179.7	161.7

The following figure shows a complete interaction diagram for this column. Grade 100, CS versus PCA Figure 6-25 for Grade 60:

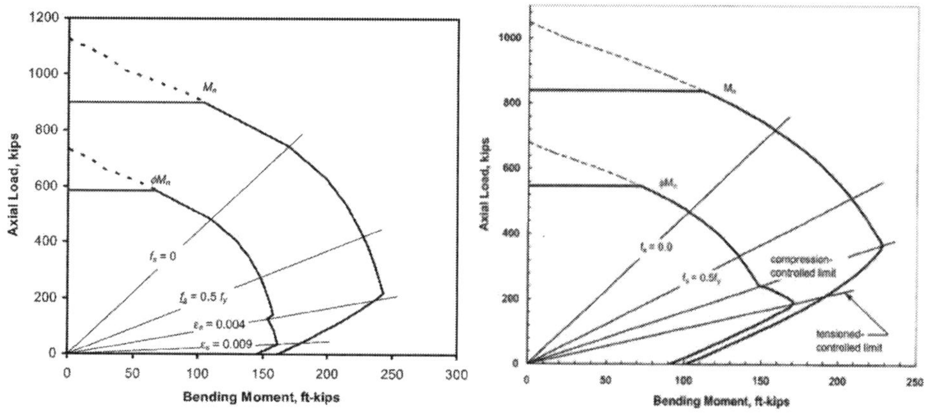

Example 6.4 in the *PCA Notes* shows the axial load-bending moment interaction diagram for the same section reinforced with Grade 60 steel. Comparing the behavior of the two columns indicates that, for a given axial load level, using CS reinforcement provides higher nominal bending moment strength than using Grade 60 steel. This difference is greater for tension-controlled sections. For the case of no axial load, the nominal bending moment strength of the column reinforced with CS is 60 percent greater than the column reinforced with Grade 60 steel. For larger axial load levels, the difference between the bending-moment strengths is less. For compression-controlled sections, the capacity of the concrete controls the capacity of the section. The type of reinforcement does not significantly affect the strength of the member.

A similar trend is observed for the axial compression design strength of the columns. For axial load levels between 150 and 250 kip, however, the design bending moment strength of the member reinforced with Grade 60 steel is higher than the column reinforced with CS. These loading conditions represent the transition region from tension-controlled failure to compression-controlled failure. The difference of the design bending moment strength is due to the lower strength reduction factor ϕ used for the high-strength column in this region.

Example 4.3—Design of rectangular beam with tension reinforcement only

Similar to Example 7.1 in the *PCA Notes*, this example highlights differences between beams designed with CS and Grade 60 reinforcement. Select a rectangular beam size and required reinforcement A_s to carry service load moments.

M_D = 56 ft-kip and M_L = 35 ft-kip. Select reinforcement to control flexural cracking.

Use f_c' = 4000 psi and f_y = 100,000 psi.

Calculations and discussion

(a) To illustrate a complete design procedure for rectangular sections with tension reinforcement only, compute a minimum beam depth using the maximum reinforcement permitted for tension-controlled flexural members. The design procedure follows the method in the *PCA Notes* and the recommendations in this guide.

1. Determine maximum tension-controlled reinforcement ratio for material strengths.

$$\rho_t = 0.002125\beta_1 f_c' = 0.002125 \times 0.85 \times 4 = 0.00723$$

(Refer to Table 4.3.4 of this guide.)

2. Compute bd² *required.*

Required moment strength

$$M_u = (1.2 \times 56) + (1.6 \times 35) = 123.2 \text{ ft-kip}$$

$$R_n = \frac{M_n}{bd^2} = \rho f_y\left(1 - \frac{0.5\rho f_y}{0.85 f_c'}\right) = 0.00723 \times 100{,}000 \times \left(1 - \frac{0.5 \times 0.00723}{0.85}\frac{100}{4}\right) = 646.1 \text{ psi}$$

$$bd^2 \text{ (required)} = \frac{M_u}{\phi R_n} = \frac{123.2 \times 12 \times 1000}{0.9 \times 646.1} = 2542.4 \text{ in.}^3$$

3. Size members so that bd² *provided* ≥ bd² *required.*

Set b = 10 in. (column width)

$$d = \sqrt{\frac{2542.4}{10}} = 15.9 \text{ in.}$$

Minimum beam depth ≈ 15.9 + 2.5 = 18.4 in.

For moment strength, a 10 x 19 in. beam size is adequate. (*PCA Notes*: a 10 x 16 in. beam size is adequate.) However, deflection is an essential consideration in designing beams by the Strength Design Method. Section 4.8 of this guide discusses deflection control, but it is not part of this example.

4. Using the 19 in. beam depth, compute a revised value of ρ.

$$d = 19 - 2.5 = 16.5 \text{ in.}$$

$$R_n = \frac{M_u}{\phi bd^2} = \frac{123.2 \times 12 \times 1000}{0.9 \times 10 \times 16.5^2} = 603.4 \text{ psi}$$

$$\rho = \frac{0.85 f_c'}{f_y}\left(1 - \sqrt{1 - \frac{2R_n}{0.85 f_c'}}\right) = \frac{0.85 \times 4}{100}\left(1 - \sqrt{\frac{2 \times 603.4}{0.85 \times 4000}}\right) = 0.00669$$

5. Compute A_s *required.*

$$A_s = (\text{revised } \rho)(bd \text{ provided}) = 0.00669 \times 10 \times 16.5 = 1.10 \text{ in.}^2$$

(b) Review the correctness of the computations by considering statics.

$T = A_s f_y = 1.10 \times 100 = 110$ kip

$$a = \frac{A_s f_y}{0.85 f_c' b} = \frac{110}{0.85 \times 4 \times 10} = 3.24 \text{ in.}$$

$c = a/\beta_1 = 3.24/0.85 = 3.81$ in.

$$\varepsilon_s = \left(\frac{d-c}{c}\right)0.003 = \left(\frac{16.5-3.81}{3.81}\right)0.003 = 0.01 > 0.009 \quad \textbf{OK}$$

Design moment strength

$$\phi M_n = \phi\left[A_s f_y\left(d - \frac{a}{2}\right)\right] = 0.9\left[110\left(16.5 - \frac{3.24}{2}\right)\right]$$

$$= 1473.1 \text{ in.-kip} = 122.8 \text{ ft-kip} \approx \text{required } M_u = 123.2 \text{ ft-kip} \quad \textbf{OK}$$

(c) Select reinforcement to satisfy distribution of flexural reinforcement requirements of ACI 318.

A_s required = 1.10 in.^{2} (*PCA Notes*: A_s required = 2.40 in.^{2})

For illustrative purposes, select two No. 6 and one No. 4 bars (A_s = 1.08). For practical design and detailing, use one bar size for total A_s.

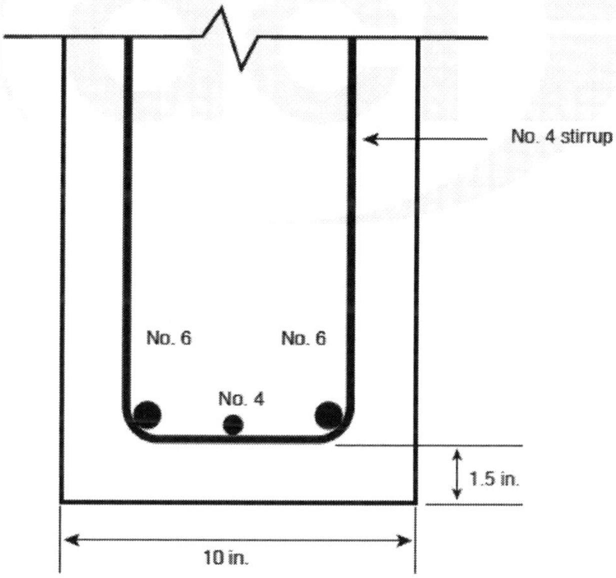

$c_c = 1.5 + 0.5 = 2.0$ in.

Maximum spacing allowed by ACI 318

$$s = 15\left(\frac{40{,}000}{f_s}\right) - 2.5c_c \le 12\left(\frac{40{,}000}{f_s}\right)$$

Use $f_s = 0.67 f_y = 67$ ksi

$$s = \frac{600}{67} - 2.5 \times 2.0 = 8.96 - 5 = 3.96 \text{ in. (governs)}$$

$$s = 12\left(\frac{40}{f_s}\right) = 12\left(\frac{40}{67}\right) = 7.16 \text{ in.}$$

$$\text{spacing provided} = \frac{1}{2}(10 - 2[1.5 + 0.5 + 3/8]) = 2.63 \text{ in.} < 3.96 \text{ in.} \quad \textbf{OK}$$

	Beam size, in.	**A_s, in.^{2}**
Grade 60, PCA	10 x 16	2.40
Grade 100, CS	10 x 19	1.10

The total area of reinforcement required for this design is 1.10 in.^{2} The original design in the *PCA Notes*, which uses Grade 60 steel, requires 2.40 in.^{2} of steel. The use of CS results in a 55 percent reduction of the required flexural reinforcement. The required beam depth, however, is 19 in.—a 20 percent increase compared with the original design. The increased beam depth provides sufficient depth for the compression zone to develop an internal moment within the beam.

Example 6.1—Design of one-way solid slab

Similar to Example 7.2 in the *PCA Notes*, this example highlights differences between slabs designed with CS and Grade 60 reinforcement. Determine required thickness and reinforcement for a one-way slab continuous over more than two equal spans. Clear span $\ell_n = 18$ ft. Use $f_c' = 4000$ psi and $f_y = 100{,}000$ psi. Service loads: $w_d = 75$ psf (assume 6 in. slab), $w_\ell = 50$ lb/ft^2.

Calculations and discussion

1. Compute required moment strengths using approximate moment analysis permitted by ACI 318 base design on end span.

Factored load $w_u = (1.2 \times 75) + (1.6 \times 50) = 170$ lb/ft^2

Positive moment at discontinuous end integral with support

$+M_u = w_u \ell_n^2/14 = 0.170 \times 18^2/14 = 3.93$ ft-kip/ft

Negative moment at exterior face of first interior support

$-M_u = w_u \ell_n^2/10 = 0.170 \times 18^2/10 = 5.51$ ft-kip/ft

2. Determine required slab thickness.

For deflection control, choose a reinforcement ratio ρ equal to $0.5\rho_t$, or half of the maximum permitted for tension-controlled sections.

$\rho_t = 0.002125\beta_1 f_c' = 0.002125 \times 0.85 \times 4 = 0.00723$

(Refer to Table 4.3.4 of this guide.)

Set $\rho = 0.5(0.00723) = 0.00362$

Design procedure follows the method outlined previously.

$$R_n = \frac{M_n}{bd^2} = \rho f_y\left(1 - \frac{0.5\rho}{0.85}\frac{f_y}{f_c'}\right) = 0.00362 \times 100{,}000 \times \left(1 - \frac{0.5 \times 0.00362}{0.85}\frac{100}{4}\right) = 342.7 \text{ psi}$$

$$\text{required } d = \sqrt{\frac{M_u}{\phi R_n b}} = \sqrt{\frac{5.51 \times 12 \times 1000}{0.9 \times 342.7 \times 12}} = 4.23 \text{ in.}$$

Assuming No. 3 bars, required $h = 4.23 + 0.375/2 + 0.75$ in. (cover) = 5.17 in.

These calculations indicate a slab thickness of 5.5 in. is adequate. Table 9.5(a) in ACI 318-11, however, indicates a minimum thickness of $(\ell/24) \times (0.4 + f_y/100{,}000) \geq 12.6$ in., unless deflections are computed. Also, note that Table 9.5(a) is applicable only to "members in one-way construction not supporting or attached to partitions or other construction likely to be damaged by large deflections." Otherwise, deflections should be computed.

For purposes of illustration, the required reinforcement is computed using $h = 5.5$ in., $d = 4.56$ in., and deflection is not considered. (*PCA Notes*: use $h = 4.5$ in. and $d = 3.59$ in.)

3. Compute required negative moment reinforcement.

$$\rho = \frac{0.85 f_c'}{f_y}\left(1 - \sqrt{1 - \frac{2R_n}{0.85 f_c'}}\right) = \frac{0.85 \times 4}{100}\left(1 - \sqrt{\frac{2 \times 294.4}{0.85 \times 4000}}\right) = 0.00308$$

Alternatively, by approximate proportion

revised ρ = (revised R_n/original R_n)(original ρ) = 294.4/342.7 × 0.00362 = 0.00311

$-A_s$ (required) = ρbd = 0.00308 × 12 × 4.56 = 0.17 in.^{2}/ft

Use No. 3 at 8 in. (A_s = 0.165 in.^{2}/ft ~ 0.17 in.^{2}/ft)

(*PCA Notes*: use No. 5 at 10 in. [A_s = 0.37 in.^{2}/ft].)

Check that the beam is tension controlled: $\rho = 0.00308 < \rho_t = 0.00723$ **OK**

4. Compute required positive moment reinforcement.

$$R_n = \frac{M_u}{\phi bd^2} = \frac{3.93 \times 12 \times 1000}{0.9 \times 12 \times 4.56^2} = 210.0 \text{ psi}$$

revised ρ = (revised R_n/original R_n)(original ρ) = 210.0/342.7 × 0.00362 = 0.00222

$+A_s$ (required) = ρbd = 0.00222 × 12 × 4.56 = 0.12 in.^{2}/ft

Use No. 3 at 12 in. (A_s = 0.11 in.^{2}/ft ~ 0.12 in.^{2}/ft)

Maximum spacing allowed by ACI 318

$$s = 15\left(\frac{40{,}000}{f_s}\right) - 2.5c_c \le 12\left(\frac{40{,}000}{f_s}\right)$$

Use $f_s = 0.67f_y$ = 67 ksi

$$s = \frac{600}{f_s} - 2.5c_c = \frac{600}{67} - 2.5 \times 0.75 = 7.08 \text{ in. (governs)}$$

$$s = 12\left(\frac{40}{f_s}\right) = 12\left(\frac{40}{67}\right) = 7.16 \text{ in.}$$

Use No. 3 at 7 in. (A_s = 0.189 in.^{2}/ft) to satisfy spacing requirements (ρ = 0.00345). (*PCA Notes*: use No. 4 at 9 in. [A_s = 0.27 in.^{2}/ft.])

Check that the beam is tension controlled: $\rho = 0.00345 < \rho_t$ **OK**

	Total area of reinforcement	Slab thickness, in.
Grade 60, PCA	0.64 in.^{2}	4.5
Grade 100, CS	0.38 in.^{2}	5.5

The total area of reinforcement provided for this design, including positive and negative moment region reinforcement, is 0.38 in.^{2}/ft. The original design in *PCA Notes*, which uses Grade 60 steel, requires a total of 0.64 in.^{2}/ft of steel. The slab thickness in this design is 5.5 in. as compared with 4.5 in. in the original design. Both designs fail to meet the minimum slab thickness for deflection control.

For a complete design, check serviceability requirements such as cracking and deflection, especially for lightly reinforced flexural members with CS reinforcement. Verify adequate shear capacity of the member.

Example 6.1 (SI)—Design of one-way solid slab (SI units)

This example is identical to Example 6.1, but presented in SI units for purposes of illustration. Determine required thickness and reinforcement for a one-way slab continuous over more than two equal spans. Clear span ℓ_n = 5.49 m. Use f_c' = 27.6 MPa and f_y = 690 MPa. Service loads: w_d = 3.59 kN/m^2 (assume 150 mm slab), w_ℓ = 2.39 kN/m^2.

Calculations and discussion

1. Compute required moment strengths using approximate moment analysis permitted by ACI 318M base design on end span.

Factored load $w_u = (1.2 \times 3.59) + (1.6 \times 2.39) = 8.13$ kN/m²

Positive moment at discontinuous end integral with support

$+M_u = w_u\ell_n^2/14 = 8.13 \times (5.49)^2/14 = 17.5$ kN-m/m

Negative moment at exterior face of first interior support

$-M_u = w_u\ell_n^2/10 = 8.13 \times (5.49)^2/10 = 24.5$ kN-m/m

2. Determine required slab thickness.

For deflection control, choose a reinforcement ratio ρ equal to approximately $0.5\rho_t$, or half of the maximum permitted for tension-controlled sections.

$\rho_t = 0.0003079\beta_1 f_c' = 0.0003079 \times 0.85 \times 27.6 = 0.00722$

(Refer to Table 4.3.4 of this guide.)

Set $\rho = 0.5(0.00722) = 0.00361$

Design procedure follows the method outlined previously.

$$R_n = \frac{M_n}{bd^2} = \rho f_y\left(1 - \frac{0.5\rho f_y}{0.85 f_c'}\right) = 0.00361 \times 690 \times \left(1 - \frac{0.5 \times 0.00361}{0.85}\frac{690}{27.6}\right) = 2.36 \text{ MPa}$$

Assuming 10M bars, required $h = 107 + 11.3/2 + 19$ mm (cover) $= 132$ mm

These calculations indicate a slab thickness of 140 mm is adequate. Table 9.5(a) of ACI 318-11, however, indicates a minimum thickness of $(\ell/24) \times (0.4 + f_y/690) \geq 320$ mm, unless deflections are computed. Also, note that Table 9.5(a) is applicable only to "members in one-way construction not supporting or attached to partitions or other construction likely to be damaged by large deflections." Otherwise, compute deflections.

For purposes of illustration, the required reinforcement will be computed using $h = 140$ mm, $d = 116$ mm, and deflection is not considered. (The design in the *PCA Notes* uses $h = 114$ mm and $d = 91$ mm.)

3. Compute required negative moment reinforcement.

$$R_n = \frac{M_u}{\phi bd^2} = \sqrt{\frac{24.5 \times 1000 \times 1000}{0.9 \times 1000 \times (116)^2}} = 2.02 \text{ MPa}$$

$$\rho = \frac{0.85 f_c'}{f_y}\left(1 - \sqrt{1 - \frac{2R_n}{0.85 f_c'}}\right) = \frac{0.85 \times 27.6}{690}\left(1 - \sqrt{1 - \frac{2 \times 2.02}{0.85 \times 27.6}}\right) = 0.00307$$

Alternatively, by approximate proportion

revised ρ = (revised R_n/original R_n)(original ρ) = (2.02/2.36) × 0.00361 = 0.00309

$-A_s$ (required) $= \rho bd = 0.00307 \times 1000 \times 116 = 356$ mm²/m

Use 10M bars at 250 mm ($A_s = 400$ mm²/m > 356 mm²/m).

Check that the beam is tension controlled: $\rho = 0.00345 < \rho_t = 0.00722$ **OK**

4. Compute required positive moment reinforcement.

$$R_n = \frac{M_u}{\phi b d^2} = \frac{17.5 \times 1000 \times 1000}{0.9 \times 1000 \times (116)^2} = 1.44 \text{ MPa}$$

revised ρ = (revised R_n/original R_n)(original ρ) = (1.44/2.36) × 0.00361 = 0.0022

$+A_s$ (required) = ρbd = 0.0022 × 1000 × 116 = 255 mm²/m

Use 10M bars at 350 mm (A_s = 286 mm²/m > 255 mm²/m)

Maximum spacing allowed by ACI 318M

$$s = 380\left(\frac{280}{f_s}\right) = 2.5c_c \leq 300\left(\frac{280}{f_s}\right)$$

Use $f_s = (2/3)f_y = 460$ MPa

$$s = 380\left(\frac{280}{f_s}\right) = 2.5c_c = \frac{106,400}{460} - 2.5 \times 19 = 184 \text{ mm}$$

$$s = 300\left(\frac{280}{f_s}\right) = 300\left(\frac{280}{460}\right) = 183 \text{ mm (governs)}$$

Use 10M bars at 180 mm (A_s = 556 mm²/m) to satisfy spacing requirements (ρ = 0.00479).

Check that the beam is tension controlled: $\rho = 0.00479 < \rho_t = 0.00722$ **OK**

	Total area of reinforcement	Slab thickness, mm
Grade 60 (420), PCA	1312 mm²/m	114
Grade 100 (690), CS	956 mm²/m	140

The total area of reinforcement provided for this design, including positive and negative moment region reinforcement, is 956 mm²/m. The original design in *PCA Notes*, which uses Grade 420 steel, requires a total of 1312 mm²/m of steel. The slab thickness in this design is 140 mm as compared with 114 mm in the original design. Both designs fail to meet the minimum slab thickness for deflection control.

For a complete design, check serviceability requirements such as cracking and deflection, especially for lightly reinforced flexural members with CS reinforcement. Verify adequate shear capacity of the member.

Example 6.1A—Design of one-way solid slab (alternate solution)

This example presents an alternative to Example 6.1 of this guide by considering the deflection criterion neglected in Example 6.1. To meet the minimum thickness requirement of Table 9.5(a) in ACI 318-11, the slab should be at least 12.6 in. thickness (refer to Example 6.1). A slab of 13 in. thick is selected.

Determine required thickness and reinforcement for a one-way slab continuous over more than two equal spans. Clear span ℓ_n = 18 ft. Use f_c' = 4000 psi and f_y = 100,000 psi. Service loads: w_d = 162.5 lb/ft² (use 13 in. slab), w_l = 50 lb/ft².

Calculations and discussion

1. Compute required moment strengths using approximate moment analysis permitted by ACI 318 base design on end span.

Factored load w_u = (1.2 × 162.5) + (1.6 × 50) = 275 lb/ft²

Positive moment at discontinuous end integral with support

$+M_u = w_u\ell_n^2/14 = 0.275 \times 18^2/14 = 6.36$ ft-kips/ft

Negative moment at exterior face of first interior support

$-M_u = w_u\ell_n^2/10 = 0.275 \times 18^2/10 = 8.91$ ft-kips/ft

2. Compute required negative moment reinforcement.

For 13 in. thick slab, use $d = 11$ in.

$$R_n = \frac{M_u}{\phi bd^2} = \sqrt{\frac{8.91\times12\times1000}{0.9\times12\times(11)^2}} = 81.8 \text{ psi}$$

$$\rho = \frac{0.85f_c'}{f_y}\left(1-\sqrt{1-\frac{2R_n}{0.85f_c'}}\right) = \frac{0.85\times4}{100}\left(1-\sqrt{1-\frac{2\times81.8}{0.85\times4000}}\right) = 0.00083 << \rho_t = 0.00723 \quad \textbf{OK}$$

So the slab is tension-controlled.

(Refer to Table 4.3.4 of this guide.)

$-A_s$ (required) $= \rho bd = 0.00083 \times 12 \times 11 = 0.11$ in.^{2}/ft

Check the minimum reinforcement requirement by ACI 318

$$-A_{s,min} = \frac{3\sqrt{f_c'}}{f_y}bd = \frac{3\sqrt{4000}}{100,000}(12)(11) = 0.25 \text{ in.}^2/\text{ft}$$

$$\text{or } -A_{s,min} = \frac{200bd}{f_y} = \frac{200(12)(11)}{100,000} = 0.264 \text{ in.}^2/\text{ft (governs)}$$

Use No. 3 at 5 in. ($A_s = 0.264$ in.^{2}/ft)

3. Compute required positive moment reinforcement.

By inspection, the positive moment reinforcement should be No. 3 at 5 in.

4. Check maximum reinforcement spacing allowed by ACI 318.

$$s = 15\left(\frac{40,000}{f_s}\right) = 2.5c_c = 12\left(\frac{40,000}{f_s}\right)$$

$c_c = 13 - 11 - 0.375/2 = 1.81$ in.

Use $f_s = 0.67f_y = 67$ ksi

$$s = \frac{600}{f_s} = 2.5c_c = \frac{600}{67} - 2.5\times1.81 = 4.43 \text{ in. (governs)}$$

$$s = 12\left(\frac{40}{f_s}\right) = 12\left(\frac{40}{67}\right) = 7.16 \text{ in.}$$

Use No. 3 at 4.25 in. for positive and negative reinforcements to satisfy the spacing requirements.

5. Check shear strength at the face of the first interior support.

$V_u = 1.15w_u\ell_n/2 = 1.15 \times 0.275 \times 18/2 = 2.85$ kip/ft wide of slab

$$V_u = 2\sqrt{f_c'}bd = 2\sqrt{4000}(12)(11)/1000 = 16.7 \text{ kip/ft wide of slab} > V_u/\phi$$

If Eq. (6.2) of this guide is used

$$V_u = \frac{73}{38+2.1d}\sqrt{f_c'}bd = \frac{73}{38+2.1(11)}\sqrt{4000}(12)(11)/1000 = 9.98 \text{ kip/ft wide of slab} >> V_u/\phi$$

No shear reinforcement required.

6. Calculate deflections.

Slab thickness has been selected to meet the minimum requirement in Table 9.5(a) of ACI 318-11, but the slab deflections are calculated for purposes of illustration. The deflection of the end span is considered.

Compute modulus of rupture, modulus of elasticity, and modular ratio.

$$f_r = 7.5\sqrt{f_c'} = 7.5\sqrt{4000} = 474 \text{ psi}$$

$$E_c = w_c^{1.5}33\sqrt{f_c'} = 150^{1.5}33\sqrt{4000} = 3.83 \times 10^6 \text{ psi}$$

$N = E_s/E_c = 29 \times 10^6/3.83 \times 10^6 = 7.6$

Compute moment of inertia of gross section and cracking moment.

$I_g = bh^3/12 = 12 \times (13)^3/12 = 2197$ in.^{4}

$M_{cr} = f_rI_g/(h/2) = 474 \times 2197/6.5 = 160{,}212$ in.-lb $= 13.35$ ft-kip

Compute dead load and live load moments.

Negative moments at interior face of first support

$-M_D = 162.5(18)^2/10 = 5265$ ft-lb $= 5.265$ ft-kip

$-M_L = 50(18)^2/10 = 1620$ ft-lb $= 1.62$ ft-kip

Positive moments at discontinuous end integral with support

$M_D = 162.5(18)^2/14 = 3761$ ft-lb $= 3.761$ ft-kip

$M_L = 50(18)^2/14 = 1157$ ft-lb $= 1.157$ ft-kip

Because $M_D + M_L$ in both cases is much less than M_{cr}, use

I_e (equivalent moment of inertia) $= I_g = 2197$ in.^{4}

Based on the commentary of ACI 318, compute the midspan deflection of the end span as

$\Delta = K(5/48)M_a\ell_n^2/E_cI_e$

where $K = 1.2 - 0.2(M_o/M_a)$; $M_o = w\ell^2/8$ (w = load on the span, ℓ = span length); and M_a = positive moment in the span. Therefore,

Under dead load

$M_o = (162.5)(18)^2/8 = 6581$ ft-lb, $M_a = 3761$ ft-lb

$K = 1.2 - 0.2(6581/3761) = 0.85$

Δ_{Di} = immediate deflection due to dead load

$= 0.85(5/48)(3761)(18)^2(1728)/(3.83 \times 10^6)(2197) = 0.022$ in.

Under live load

Δ_{Li} = immediate deflection due to live load by direct proportion

$= 0.022(1157)/(3761) = 0.007$ in. $<< \ell_n/360 = 0.6$ in.

So the immediate deflection under dead and live loads $= 0.022 + 0.007 = 0.029$ in. $<< \ell_n / 360$.

Consider the effect of long-term deflection based on ACI 318.

$$\lambda_\Delta = \frac{\xi}{1+50\rho'}$$

$\xi = 2.0$ for 5 years or more, $\rho' = 0.264/12 \times 11 = 0.002$

$\lambda_\Delta = 2/(1 + 50 \times 0.002) = 1.82$

Long-term deflection of the slab under dead load $= 0.022 \times 1.82 = 0.04$ in.

Long-term deflection plus live load deflection $= 0.04 + 0.007 = 0.047$ in. $<< \ell_n/480 = 0.45$ in.

This example demonstrates that the ACI 318 minimum depth table results were conservative. Based on the deflection control criteria of ACI 318, the slab thickness in this example requires little reinforcement to meet the minimum requirement. Further, the slab remains uncracked under the service load ($D + L$), and the immediate and long-term deflections of the slab are well within the permissible limits. The nominal shear strength of the slab exceeds the factored shear load, so no shear reinforcement is required.

Example 4.4—Design of rectangular beam with compression reinforcement

Similar to Example 7.3 in *PCA Notes*, this example illustrates the benefits of using CS reinforcing bars for design of heavily-reinforced flexural members. A beam cross section is limited to the size shown. Determine the required area of reinforcement for service load moments.

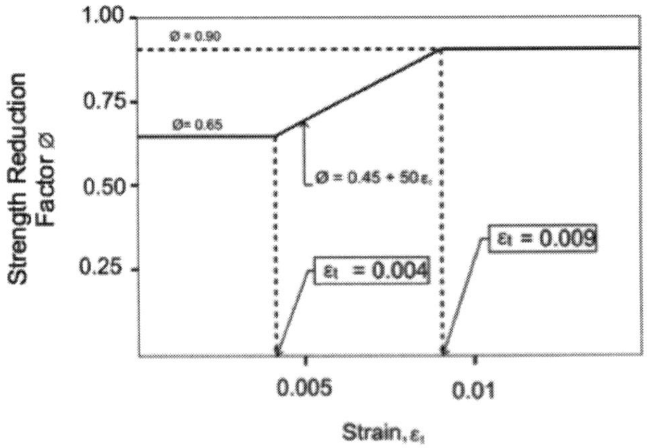

M_D = 430 ft-kip and M_L = 175 ft-kip
Check crack control requirements of Section 10.6.4 of ACI 318-08.
Use f_c' = 4000 psi, f_y = 100,000 psi (tension), and f_y' = 80,000 psi (compression)

Calculations and discussion

(a) Determine required reinforcement.

1. Determine if compression reinforcement is needed.

$$M_u = 1.2M_D + 1.6M_L = (1.2 \times 430) + (1.6 \times 175) = 796 \text{ ft-kip}$$

$$M_n = M_u/\phi = 796/0.9 = 884 \text{ ft-kip}$$

$$R_n = \frac{M_u}{bd^2} = \sqrt{\frac{884 \times 12 \times 1000}{12 \times 30^2}} = 982 \text{ psi}$$

Determine the maximum value of R_n without compression reinforcement at the tension-controlled strain limit.

$$\rho_t = 0.2125\beta_1 \frac{f_c'}{f_y} = 0.2125(0.85)\frac{4}{100} = 0.00723$$

(Refer to Table 4.3.4 of this guide.)

$$\int \omega_t = \rho_t \frac{f_y}{f_c'} = 0.00723\frac{100}{4} = 0.1808$$

$$\frac{M_{nt}}{f_c'bd^2} = 0.1614 \text{ from Table 7-1 of PCA Notes}$$

$$R_{nt} = \frac{M_{nt}}{bd^2} = 646 \text{ psi} < 982 \text{ psi}$$

Therefore, compression reinforcement is required.

2. Find the nominal strength moment resisted by the concrete section, without compression reinforcement.

$M_{nt} = R_{nt}bd^2 = 646 \times 12 \times 30^2/(12 \times 1000) = 581$ ft-kip resisted by the concrete

Required moment strength resisted by the compression reinforcement

$M_n' = 884 - 581 = 303$ ft-kip

3. *Determine the compression steel stress* f_s'.

Check yielding of the compression steel. Because the section was designed at the tension-controlled strain limit $\varepsilon_t = 0.009$, $c = 0.25d$

$c = 0.25d = 7.5$ in.

$$\varepsilon_s' = \left(\frac{c-d'}{c}\right)(0.003) = \left(\frac{7.5-2.5}{7.5}\right)(0.003) = 0.002 < \varepsilon_y' = \frac{f_y'}{E_s} = \frac{80}{29,000} = 0.00276$$

Compression reinforcement is elastic at the nominal strength.

$f_s' = E_s\varepsilon_s' = 29,000 \times 0.002 = 58$ ksi

4. *Determine the total required reinforcement.*

$$A_s' = \frac{M_n'}{f_s'(d-d')} = \frac{303 \times 12}{58(30-2.5)} = 2.28 \text{ in.}^2$$

$A_s = f_s'/f_yA_s' + \rho_t bd = 58/100 \times 2.28 + 0.00723 \times 12 \times 30 = 3.93$ in.^{2}

	A_s, in.^{2}	A_s', in.^{2}
Grade 60, PCA	7.29	0.79
Grade 100, CS	3.93	2.28

5. *Check moment capacity.*

When the compression reinforcement is elastic.

$$a = \frac{A_sf_y - A_s'f_s'}{0.85f_c'b} = \frac{3.93 \times 100 - 2.28 \times 58}{0.85 \times 4 \times 12} = 6.36 \text{ in.}$$

$\phi M_n = \phi(\rho bdf_y[d - a/2] + A_s'f_s[d - d'])$
$= 0.9[0.00723 \times 12 \times 30 \times 100 \times (30 - 3.18) + 2.30 \times 58 \times (30 - 2.5)]$
$= 9584$ in.-kip $= 799$ ft-kip $> M_u = 796$ ft-kip **OK**

(b) Select reinforcement to satisfy ACI 318 for control of flexural cracking.

Select three No. 8 bars ($A_s' = 2.37$ in.$^2 > 2.28$ in.^{2}) for compression reinforcement.

Select four No. 9 bars ($A_s = 4.00$ in.$^2 > 3.93$ in.^{2}) for tension reinforcement.

	Tension reinforcement	Compression reinforcement
Grade 60, PCA	Two No. 6 (0.88 in.^{2})	Six No. 10 (7.62 in.^{2})
Grade 100, CS	Three No. 8 (2.37 in.^{2})	Four No. 9 (4.0 in.^{2})

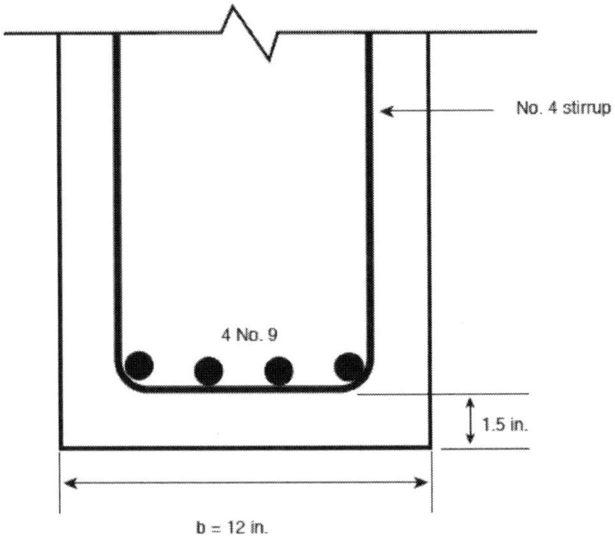

Maximum spacing allowed by ACI 318

$$s = \frac{600}{f_s} - 2.5c_c \leq 12\left(\frac{40}{f_s}\right)$$

$c_c = 1.5 + 0.5 = 2.0$ in.

Use $f_s = 0.67f_y = 67$ ksi

$$s = \frac{600}{f_s} - 2.5c_c = \frac{600}{67} - 2.5 \times 2.0 = 3.96 \text{ in. (governs)}$$

$$s = 12\left(\frac{40}{f_s}\right) = 12\left(\frac{40}{67}\right) = 7.16 \text{ in.}$$

Spacing provided = (1/3)(12 – 2[1.5 + 0.5 + 9/16]) = 2.29 in. < 3.96 in. **OK**

(c) Stirrups or ties are required throughout the distance where compression reinforcement is required for strength.

maximum spacing = 16 × longitudinal bar diameter = 16 × 1.125 = 18 in.
= 48 × tie bar diameter = 48 × 0.5 = 24 in. = least dimension of member = 12 in. (governs)

Use s_{max} = 12 in. for No. 4 stirrups

The total area of reinforcement used for this design, including tension and compression reinforcement is 6.37 in.^{2} The original design in the *PCA Notes*, which uses Grade 60 steel, uses a total of 8.5 in.^{2} of reinforcing steel. The use of CS results in a 25 percent reduction of the required flexural reinforcement. The selected high-strength tension reinforcement can be placed as one layer at the beam bottom compared with two layers of conventional steel required by the original design. The design using CS requires less labor to place the reinforcement and tie the reinforcing cages.

Example 4.5—Design of flanged section with tension reinforcement only

Similar to Example 7.4 in *PCA Notes*, this example illustrates the design procedure for flanged beams reinforced with CS. Select reinforcement for the flanged section shown to carry service dead and live load moments of

$M_D = 72$ ft-kip and $M_L = 88$ ft-kip

Use $f_c' = 4000$ psi and $f_y = 100,000$ psi

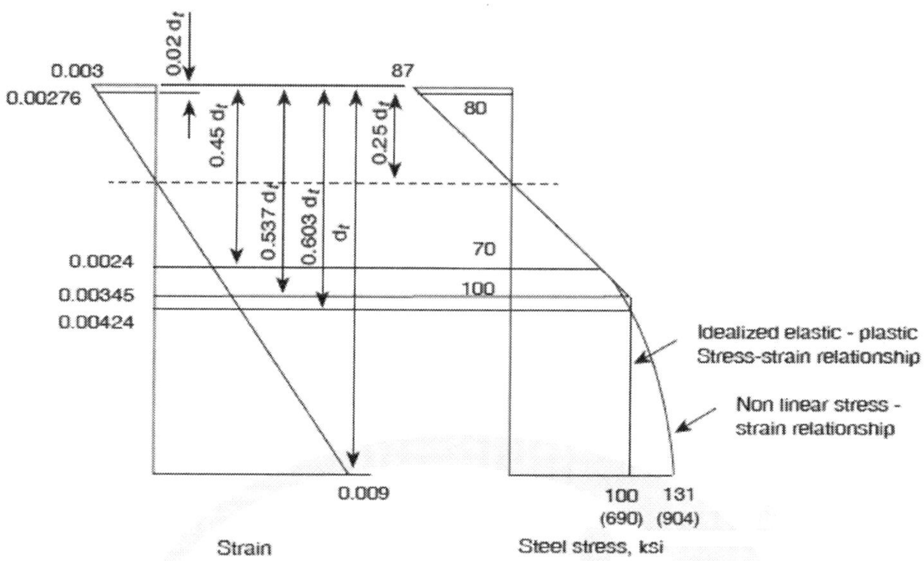

Calculations and discussion

1. Determine the required flexural strength.

$M_u = 1.2M_D + 1.6M_L = (1.2 \times 72) + (1.6 \times 88) = 227$ ft-kip

2. Determine depth of equivalent stress block, a, *for a rectangular section. Assume* $\phi = 0.9$.

$M_u = \phi C(d - a/2) = \phi 0.85 f_c' ab(d - a/2)$

$227 \times 12 = 0.9 \times 0.85 \times 4 \times a \times 30 \times (19 - a/2)$

$45.9a^2 - 1744.2a + 2724 = 0$

$a = 1.63$ in. < 2.5

with $a < h_f$, determine A_s for a rectangular section (refer to Example 4.6 for the case when $a > h_f$)

Check ϕ

$c = a/\beta_1 = 1.63/0.85 = 1.92$ in.

$$\varepsilon_s = \left(\frac{d-c}{c}\right)(0.003) = \left(\frac{19-1.92}{1.92}\right)(0.003) = 0.0267 > 0.009$$

Section is tension-controlled, and $\phi = 0.9$

3. Compute A_s *required.*

$$A_s = \frac{0.85 f_c' ab}{f_y} = \frac{0.85 \times 4 \times 1.63 \times 30}{100} = 1.66 \text{ in.}^2$$

Use three No. 7 bars (A_s = 1.8 in.^{2} > 1.66 in.^{2})

	Tension reinforcement	A_s, in.^{2}
Grade 60, PCA	Three No. 9	3.0 in.^{2}
Grade 100, CS	Three No. 7	1.8 in.^{2}

4. Check minimum required reinforcement.

For f_c' = 4000 psi, the governing requirement is

$$A_{s,min} = \frac{200}{f_y} b_w d = \frac{200}{100{,}000} \times 10 \times 19 = 0.38 \text{ in.}^2 < A_s \text{ provided} = 1.8 \text{ in.}^2 \quad \textbf{OK}$$

5. Check distribution of reinforcement.

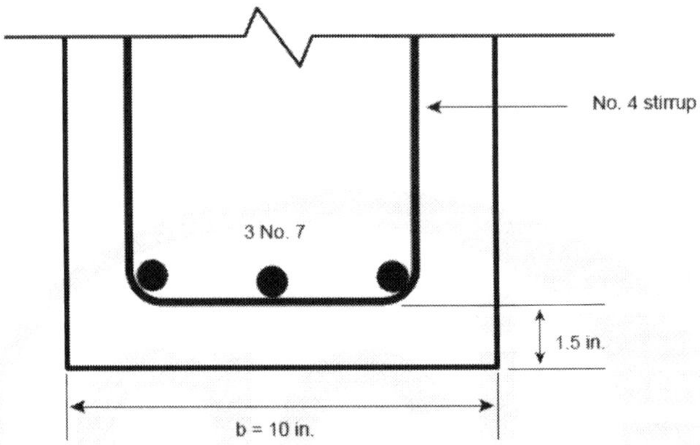

Maximum spacing allowed by ACI 318

$$s = \frac{600}{f_s} - 2.5c_c \leq 12\left(\frac{40}{f_s}\right)$$

c_c = 1.5 + 0.5 = 2.0 in.

Use $f_s = 0.67 f_y$ = 67 ksi

$$s = \frac{60}{f_s} - 2.5c_c = \frac{600}{67} - 2.5 \times 2.0 = 3.96 \text{ in. (governs)}$$

$$s = 12\left(\frac{40}{f_s}\right) = 12\left(\frac{40}{67}\right) = 7.16 \text{ in.}$$

Spacing provided = (1/2)(10 – 2([1.5 + 0.5 + 7/16]) = 2.56 in. < 3.96 in. **OK**

The total area of flexural reinforcement provided by this design is 1.8 in.^{2} The original design in *PCA Notes*, which uses Grade 60 steel, uses 3.0 in.^{2} of reinforcing steel. This represents a 40 percent reduction of the required flexural reinforcement. Flanged sections provide a wide compression block compared with rectangular sections. The wide flange provides adequate compression force to balance the tension force provided by the CS without the need for additional compression reinforcement.

Example 4.6—Design of flanged section with tension reinforcement only

Similar to Example 7.5 in *PCA Notes*, this example illustrates the design procedure for flanged beams reinforced with CS with a neutral axis that lies in the beam web. Select reinforcement for the flanged section shown to carry a factored moment of $M_u = 400$ ft-kip.

Use $f_c' = 4000$ psi and $f_y = 100{,}000$ psi.

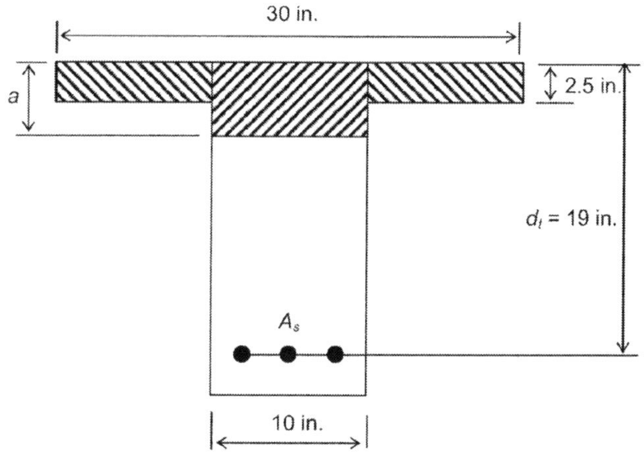

Calculations and discussion

(a) Determine required reinforcement.

1. Determine the depth of equivalent stress block, a, *as for a rectangular section.*

Assume tension-controlled section, $\phi = 0.9$

$M_u = \phi 0.85 f_c' a b_w (d - a/2)$

$400 \times 12 = 0.9 \times 0.85 \times 4 \times a \times 30 \times (19 - a/2)$

$45.9a^2 - 1744.2a + 4800 = 0$

$a = 2.98$ in. > 2.5 in.

Because the value of a as a rectangular section exceeds the flange thickness, the equivalent stress block extends into the web, and the design should be based on T-section behavior. (Refer to Example 4.5 when a is less than the flange depth.)

2. Compute required reinforcement A_{sf} *and nominal moment strength* M_{nf} *corresponding to the overhanging beam flange in compression.*

Compressive strength of flange

$C_f = 0.85 f_c'(b - b_w)h_f = 0.85 \times 4 \times (30 - 10) \times 2.5 = 170$ kip

Required A_{sf} to equilibrate C_f

$A_{sf} = C_f/f_y = 170/100 = 1.70$ in.^{2}

Nominal moment strength of flange

$$M_{nf} = A_{sf} f_y \left(d - \frac{h_f}{2} \right) = 1.70 \times 100 \times (19 - 1.25) = 3018 \text{ in.-kip} = 252 \text{ ft-kip}$$

3. Required nominal moment strength carried by beam web.

$M_{nw} = M_u/\phi - M_{nf} = 400/0.9 - 252 = 193$ ft-kip

4. *Compute reinforcement* A_{sw} *required to develop moment strength to be carried by the web.*

$M_{nw} = 0.85 f_c' a_w b_w (d - a_w/2)$

$193 \times 12 = 0.85 \times 4 \times a_w \times 10 \times (19 - a_w/2)$

$17a_w^2 - 646a_w + 2316 = 0$

$a_w = 4.01$ in. $> h_f = 2.5$ in.

5. *Check to see if section is tension-controlled.*

$c_w = a_w/\beta_1 = 4.01/0.85 = 4.72$ in.

$$\varepsilon_s = \left(\frac{d - c_w}{c_w}\right)(0.003) = \left(\frac{19 - 4.72}{4.72}\right)(0.003) = 0.0091 > 0.009$$

Therefore, the section is tension-controlled and $\phi = 0.9$

(Refer to Section 4.4 of this guide.)

$$A_{sw} = \frac{0.85 f_c' a_w b_w}{f_y} = \frac{0.85 \times 4 \times 4.01 \times 10}{100} = 1.36 \text{ in.}^2$$

6. *Total reinforcement required to carry factored moment* $M_u = 400$ *ft-kip.*

$A_s = A_{sf} + A_{sw} = 1.70 + 1.36 = 3.06$ in.^{2}

	A_{sf}	A_{sw}
Grade 60, PCA (5.10 in.^{2})	2.83 in.^{2}	2.27 in.^{2}
Grade 100, CS (3.06 in.^{2})	1.70 in.^{2}	1.36 in.^{2}

7. *Check moment capacity.*

$$\phi M_n = \phi\left(A_{sf} f_y \left[d - \frac{h_f}{2}\right] + A_{sw} f_y \left[d - \frac{a_w}{2}\right]\right) = 0.9\left(1.70 \times 100\left[19 - \frac{2.5}{2}\right] + 1.36 \times 100\left[19 - \frac{4.01}{2}\right]\right)$$

$= 4796$ in.-kip $= 400$ ft-kip **OK**

(b) Select reinforcement to satisfy crack control criteria.

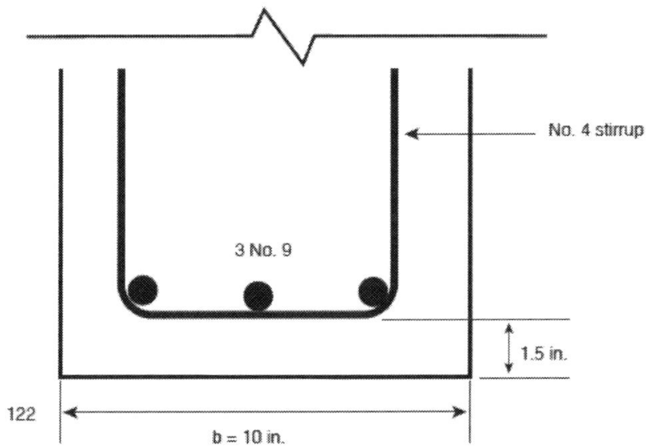

Use three No. 9 bars (A_s = 3.0 in.^{2}, 2 percent less than required, assumed sufficient)

Maximum spacing allowed by ACI 318

$c_c = 1.5 + 0.5 = 2.0$ in.

Use $f_s = 0.67, f_y = 67$ ksi

$$s = \frac{600}{f_s} - 2.5c_c = \frac{600}{67} - 2.5 \times 2.0 = 3.96 \text{ in. (governs)}$$

$$s = 12\left(\frac{40}{f_s}\right) = 12\left(\frac{40}{67}\right) = 7.16 \text{ in.}$$

Spacing provided = (1/2)(10 – 2[1.5 + 0.5 + 9/16]) = 2.44 in. < 3.96 in. **OK**

The total area of flexural reinforcement provided in this design is 3.0 in.^{2} The original design in *PCA Notes*, which uses Grade 60 steel, uses 5.0 in.^{2} of reinforcing steel. This represents a 40 percent reduction of the required flexural reinforcement. For members exhibiting T-beam behavior at nominal strength, reduce the required flexural reinforcement by using CS with a suitable design approach.

Example 4.6 (SI)—Design of flanged section with tension reinforcement only (SI units)

This example is identical to Example 4.6 but presented in SI units. Select reinforcement for the flanged section shown, to carry a factored moment of M_u = 542.4 kN-m. Use f_c' = 27.6 MPa, and f_y = 690 MPa.

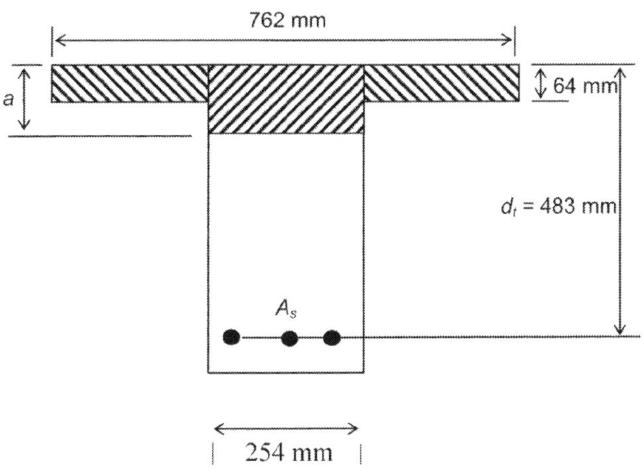

Calculations and discussion

(a) Determine required reinforcement.

1. Determine the depth of equivalent stress block, a, *as for a rectangular section.*

Assume tension-controlled section, $\phi = 0.9$

$M_u = \phi 0.85 f_c' a b_w (d - a/2)$

$542.4 \times 1000 = 0.9 \times 0.85 \times 27.6 \times a \times 0.762 \times (483 - a/2)$

$8.04a^2 - 7770a + 542{,}400 = 0$

$a = 75.8 \text{ mm} > 64 \text{ mm}$

Because the value of a as a rectangular section exceeds the flange thickness, the equivalent stress block extends into the web, and the design should be based on T-section behavior. (Refer to Example 4.5 when a is less than the flange depth.)

2. Compute required reinforcement A_{sf} *and nominal moment strength* M_{nf} *corresponding to the overhanging beam flange in compression.*

Compressive strength of flange

$C_f = 0.85 f_c'(b - b_w)h_f = 0.85 \times 27.6 \times (762 - 254) \times 64 = 762{,}730 \text{ N}$

Required A_{sf} to equilibrate C_f

$A_{sf} = C_f/f_y = 762{,}730/690 = 1105 \text{ mm}^2$

Nominal moment strength of flange

$$M_{nf} = A_{sf} f_y \left(d_t - \frac{h_f}{2} \right) = 1105 \times 690 \times (483 - 32) = 343{,}865{,}000 \text{ N-mm} = 343.9 \text{ kN-m}$$

3. Required nominal moment strength carried by beam web.

$M_{nw} = M_u/\phi - M_{nf} = 542.4/0.9 - 343.9 = 258.8 \text{ kN-m}$

4. Compute reinforcement A_{sw} *required to develop moment strength to be carried by the web.*

$M_{nw} = 0.85 f_c' a_w b_w (d - a_w/2)$

$258.8 \times 1000 = 0.85 \times 27.6 \times a_w \times 0.254 \times (483 - a_w/2)$

$2.98a_w^2 - 966a_w + 86{,}846 = 0$

$a_w = 100 \text{ mm} > h_f = 64 \text{ mm}$

5. Check to see if section is tension-controlled.

$c_w = a_w/\beta_1 = 100/0.85 = 118 \text{ mm}$

$$\varepsilon_s = \left(\frac{d - c_w}{c_w} \right)(0.003) = \left(\frac{483 - 118}{118} \right)(0.003) = 0.0093 > 0.009$$

Therefore, the section is tension-controlled and $\phi = 0.9$

(Refer to Section 4.4 of this guide.)

$$A_{sw} = \frac{0.85 f_c' a_w b_w}{f_y} = \frac{0.85 \times 27.6 \times 100 \times 254}{690} = 864 \text{ mm}^2$$

6. *Total reinforcement required to carry factored moment* $M_u = 542.4$ *kN-m.*

$A_s = A_{sf} + A_{sw} = 1105 + 864 = 1969$ mm²

7. *Check moment capacity.*

$$\phi M_n = \phi\left(A_{sf} f_y \left[d - \frac{h_f}{2}\right] + A_{sw} f_y \left[d - \frac{a_w}{2}\right]\right) = 0.9\left(1105 \times 690\left[483 - \frac{64}{2}\right] + 864 \times 690\left[483 - \frac{100}{2}\right]\right)$$

$= 541{,}802{,}000$ N·mm $= 542$ kN·m **OK**

(b) Select reinforcement to satisfy crack control criteria.

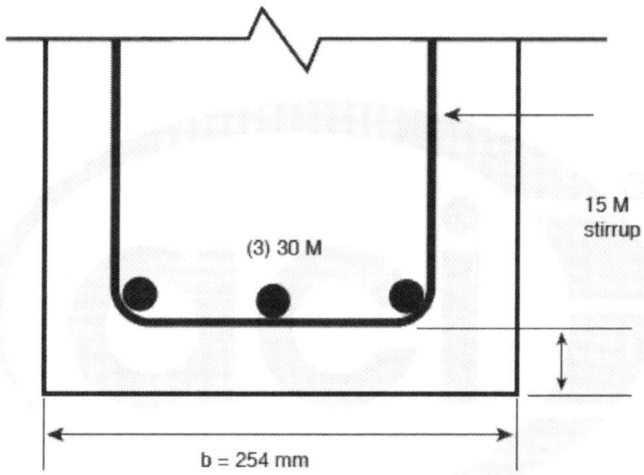

Use three 30M bars ($A_s = 2100$ mm² > 1969 mm² required).

Maximum spacing allowed by ACI 318M

$$s = 380\left(\frac{280}{f_s}\right) - 2.5c_c \leq 300\left(\frac{280}{f_s}\right)$$

$c_c = 38 + 16 = 54$ mm

Use $f_s = 0.67 f_y = 462$ MPa

$$s = \frac{106{,}400}{f_s} - 2.5c_c = \frac{106{,}400}{462} - 2.5 \times 54 = 95 \text{ mm (governs)}$$

$$s = 300\left(\frac{280}{f_s}\right) = 300\left(\frac{280}{462}\right) = 182 \text{ mm}$$

Spacing provided = (1/2)(254 – 2[38 + 16 + 29.9/2]) = 58 mm < 95 mm **OK**

The total area of flexural reinforcement provided in this design is 1969 mm^2. The original design in the *PCA Notes*, which uses Grade 420 steel, uses 3226 mm^2 of reinforcing steel. This represents a 39 percent reduction of the required flexural reinforcement. For members exhibiting T-beam behavior at nominal strength, reduce the required flexural reinforcement by using CS with a suitable design approach.

Example 6.2—Design of one-way joist

Similar to Example 7.6 in *PCA Notes*, this example illustrates the design procedure for continuous one-way joist systems reinforced with CS. Determine the required depth and reinforcement for the one-way joist system shown in the following. The joists are 6 in. wide and are spaced 36 in. on center. The slab is 3.5 in. thick. Use f_c' = 4000 psi and f_y = 100,000 psi.

Service DL = 130 lb/ft^2 (assumed total for joists and beams plus superimposed dead loads)
Service LL = 60 lb/ft^2
Width of spandrel beam = 20 in., width of interior beams = 36 in.
Columns: interior = 18 x 18 in., exterior = 16 x 16 in.
Story height (typ.) = 13 ft

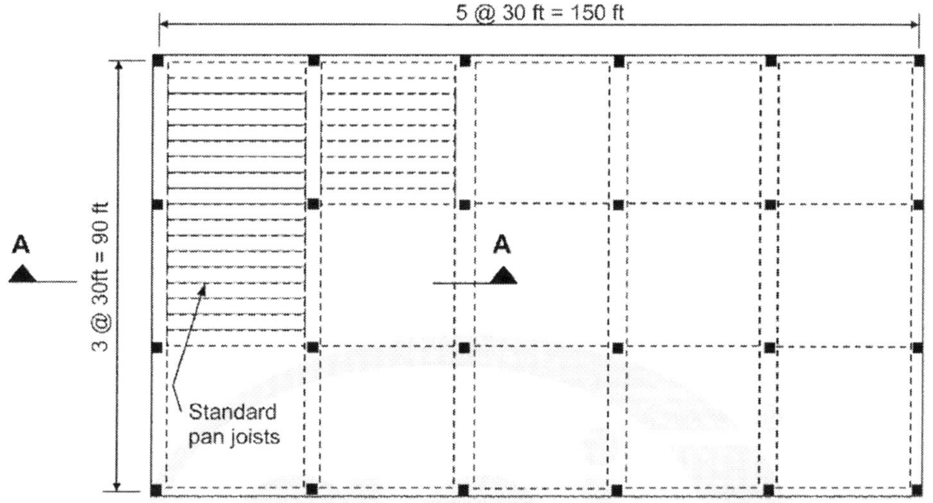

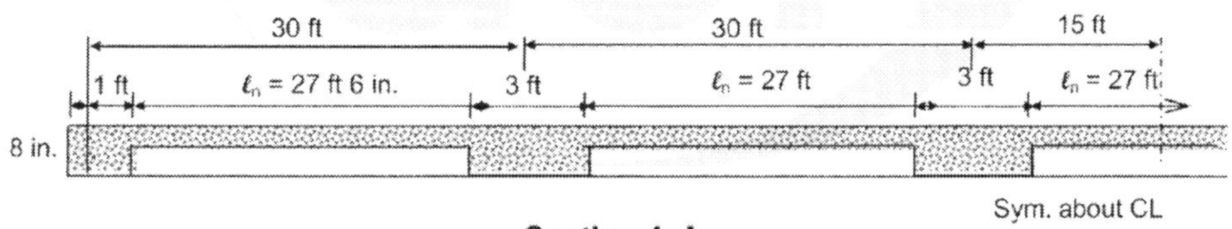

Section A-A

Calculations and discussion

(a) Compute the factored moments at the faces of the supports and determine the depth of the joists.

w_u = [(1.2 × 0.13) + (1.6 × 0.06)] × 3 = 0.756 ft-kip

Using the approximate coefficients, the following table summarizes the factored moments along the span.

Location	M_u, ft-kip
	End span
Exterior negative	$w_u\ell_n^2/24 = 0.756 \times 27.5^2/24 = 23.8$
Positive	$w_u\ell_n^2/14 = 0.756 \times 27.5^2/14 = 40.8$
Interior negative	$w_u\ell_n^2/10 = 0.756 \times 27.25^2/10 = 56.1$
	Interior span
Positive	$w_u\ell_n^2/16 = 0.756 \times 27^2/16 = 34.4$
Negative	$w_u\ell_n^2/11 = 0.756 \times 27^2/11 = 50.1$

For reasonable deflection control, choose a reinforcement ratio ρ equal to approximately one-half ρ_t.

$\rho_t = 0.002125\beta_1 f_c' = 0.002125 \times 0.85 \times 4 = 0.00723$

(Refer to Table 4.3.4 of this guide.)

Set $\rho = 0.5 \times 0.00723 = 0.00362$

Determine the required depth of the joist based on $M_u = 56.1$ ft-kip

$$\omega = \frac{\rho f_y}{f_c'} = \frac{0.00362 \times 100}{4} = 0.0905$$

From Table 7-1 in the *PCA Notes*, $M_u/\phi f_c' bd^2 = 0.0855$

$$d = \sqrt{\frac{M_u}{\phi f_c' b_w (0.0855)}} = \sqrt{\frac{56.1 \times 12}{0.9 \times 4 \times 6 \times 0.0855}} = 19.1 \text{ in.}$$

Allowing 1.25 in. for concrete cover and half bar diameter, then $h \approx 19.1 + 1.25 = 20.4$ in.

To satisfy the requirements for joist construction in ACI 318, $h_{max} = 3.5 \times b_w = 3.5 \times 6 = 21$ in.

These calculations indicate a 21 in. joist depth is adequate. ACI 318, however, indicates a minimum thickness of $(\ell/18.5) \times (0.4 + f_y/100{,}000) = 27$ in., unless deflections are computed. This is applicable only to "members in one-way construction not supporting or attached to partitions or other construction likely to be damaged by large deflections." Otherwise, deflections must be computed.

For purposes of illustration, compute the required reinforcement for a 21 in. deep joist without considering deflection further. Assume $d = 21$ in. – 1.25 in. = 19.75 in.

(b) Compute required reinforcement.

1. End span, exterior negative.

$$\frac{M_u}{\phi f_c' bd^2} = \frac{23.8 \times 12}{0.9 \times 4 \times 6 \times 19.75^2} = 0.0339$$

From Table 7-1 in *PCA Notes*, $\omega \approx 0.0346$

$$A_s = \frac{\omega bd f_c'}{f_y} = \frac{0.0346 \times 6 \times 19.75 \times 4}{100} = 0.16 \text{ in.}^2$$

For $f_c' = 4000$ psi, use

$$A_{s,min} = \frac{200 b_w d}{f_y} = \frac{200 \times 6 \times 19.75}{100{,}000} = 0.24 \text{ in.}^2 > 0.16 \text{ in.}^2$$

Use $A_s = 0.24$ in.^{2}

Distribute bars uniformly in top of slab

$A_s = 0.24/3 = 0.08$ in.^{2}/ft

Maximum spacing allowed by ACI 318

$$s = \frac{600}{f_s} - 2.5c_c \leq 12\left(\frac{40}{f_s}\right)$$

Assuming No. 3 bars

$c_c = h - d - 3/16 = 21 - 19.75 - 3/16 = 1.1$ in.

Use $f_s = 0.67f_y = 67$ ksi

$$s = \frac{600}{f_s} - 2.5c_c = \frac{600}{67} - 2.5 \times 1.1 = 6.21 \text{ in. (governs)}$$

$$s = 12\left(\frac{40}{f_s}\right) = 12\left(\frac{40}{67}\right) = 7.16 \text{ in.}$$

Use No. 3 at 6 in. ($A_s = 0.22$ in.^{2}/ft). $A_s = 6 \times 0.11$ in.^{2} = 0.66 in.^{2} in 36 in. for each joist.

(*PCA Notes*: use No. 3 at 10 in.^{2})

Check if the joist is tension-controlled

$$\rho = \frac{A_s}{b_w d} = \frac{0.66}{6 \times 19.75} = 0.0056 < \rho_t = 0.00723 \quad \textbf{OK}$$

2. *End span, positive.*

$$\frac{M_u}{\phi f_c' bd^2} = \frac{40.8 \times 12}{0.9 \times 4 \times 36 \times 19.75^2} = 0.0097$$

From Table 7-1 in *PCA Notes*, $\omega \approx 0.01$

$$A_s = \frac{\omega bdf_c'}{f_y} = \frac{0.01 \times 36 \times 19.75 \times 4}{100} = 0.28 \text{ in.}^2$$

Check rectangular section behavior

$$a = \frac{A_s f_y}{0.85 f_c' b} = \frac{0.28 \times 100}{0.85 \times 4 \times 36} = 0.22 < 3.5 \text{ in.} \quad \textbf{OK}$$

Use one No. 5 bar ($A_s = 0.31$ in.^{2}) (*PCA Notes*: use two No. 5 bars)

3. *End span, interior negative.*

$$\frac{M_u}{f_c' bd^2} = \frac{56.1 \times 12}{0.9 \times 4 \times 6 \times 19.75^2} = 0.0799$$

From Table 7-1 in *PCA Notes*, $\omega \approx 0.084$

$$A_s = \frac{\omega bd\, f_c'}{f_y} = \frac{0.084 \times 6 \times 19.75 \times 4}{100} = 0.40 \text{ in.}^2$$

Distribute bars uniformly in top of slab

$A_s = 0.40/3 = 0.13$ in.^{2}/ft

Use No. 3 at 6 in. for crack control considerations as previously described.

(*PCA Notes*: use No. 5 at 12 in.^{2})

4. Obtain reinforcement for the other sections in similar fashion. The following table summarizes the results. Note that at all sections, the requirements in ACI 318 for crack control are satisfied.

Location	M_u, ft-kip	A_s, in.^{2} CS 100 ksi	A_s, in.^{2} PCA 60 ksi	Details CS 100 ksi	Details PCA 60 ksi
End span exterior negative	23.8	0.24	0.37	No.3 @ 6 in.	No.3 @ 10 in.
End span positive	40.8	0.28	0.53	1 No.5	2 No.5
End span interior negative	56.1	0.40	0.73	No.3 @ 6 in.	No.5 @ 12 in.
Interior span positive	34.4	0.23	0.42	2 No.3	2 No.5
Interior span negative	50.1	0.35	0.65	No.3 @ 6 in.	No.5 @ 6 in.

5. The slab reinforcement normal to the joists is located at mid-depth of the slab to resist positive and negative moments.

Use $M_u = \dfrac{w_u \ell^2}{12} = \dfrac{0.185 \times 2.5^2}{12} = 0.096$ ft-kip , where $w_u = 1.2(44 + 30) + 1.6(60) = 185$ lb/ft^2 = 0.185 kip/ft^2.

$$\frac{M_u}{\phi f'_c bd^2} = \frac{0.096 \times 12}{0.9 \times 4 \times 12 \times 1.75^2} = 0.0087$$

From Table 7-1 in *PCA Notes*, $\omega \approx 0.0087$.

$$A_s = \frac{\omega bd f'_c}{f_y} = \frac{0.087 \times 12 \times 1.75 \times 4}{100} = 0.01 \text{ in.}^2$$

ACI 318 governs minimum slab reinforcement.

$$A_{s,min} = 0.0018 \times 12 \times 3.5 \times \frac{60{,}000}{100{,}000} = 0.05 \text{ in.}^2/\text{ft}$$

$s_{max} = 5h = 5 \times 3.5 = 17.5$ in. but not more than 18 in.

Use No. 3 at 16 in. (A_s = 0.08 in.^{2}/ft)

(*PCA Notes*: Use No. 3 at 16 in.)

(c) Shear at supports should be checked, but it is beyond the scope of this example.

Based on flexural strength requirements, using CS results in a reduction of the required reinforcement between 30 to 65 percent within the joist system as compared with using Grade 60 steel (PCA 2005). One notable exception is at the exterior end of the end span. This corresponds to the location of lowest negative moment. Due to the low magnitude of the applied moment at this location, cracking requirements for the design govern the required steel. Due to the higher level of stress in the reinforcement at the service load level, the maximum allowable spacing for high-strength reinforcement is lower than the spacing used for conventional steel. Consequently, at this location, the amount of CS provided is 67 percent greater than that of the design using Grade 60 steel. The required joist depth for the design using CS is 10 percent greater than using Grade 60 steel. This is due to the higher tension-controlled strain limit of the high-strength reinforcement. The deflection of the joist should be calculated to determine if it is within acceptable limits.

Example 6.3—Flexural design of support beam for one-way joist

Similar to Example 7.7 in *PCA Notes*, this example illustrates the design procedure for continuous beams reinforced with CS. Determine the required depth and reinforcement for the support beams along the interior column line in Example 6.2. The width of the beams is 36 in. Use $f_c' = 4000$ psi and $f_y = 100{,}000$ psi.

Service DL = 130 lb/ft² (assumed total for joists and beams plus superimposed dead loads)
Service LL = 60 lb/ft²
Width of spandrel beam = 20 in.
Width of interior beams = 36 in.
Columns: interior = 18 x 18 in.
exterior = 16 x 16 in.
Story height (typ.) = 13 ft

Calculations and discussion

(a) Compute the factored moments at the faces of the supports and determine the beam depth.

$w_u = [(1.2 \times 0.13) + (1.6 \times 0.06)] \times 30 = 7.56$ kip/ft

Using the approximate coefficients, the following table summarizes the factored moments along the span.

Location	M_u, ft-kip
	End span
Exterior negative	$w_u\ell_n^2/16 = 7.56 \times 28.58^2/16 = 385.9$
Positive	$w_u\ell_n^2/14 = 7.56 \times 28.58^2/14 = 441.1$
Interior negative	$w_u\ell_n^2/10 = 7.56 \times 28.54^2/10 = 615.8$
	Interior span
Positive	$w_u\ell_n^2/16 = 7.56 \times 28.5^2/16 = 383.8$

For overall economy, choose a beam depth equal to the joist depth used in Example 6.2.

Check the 21-in. depth for $M_u = 615.8$ ft-kip.

$\omega_t = 0.2125\beta_1 = 0.2125 \times 0.85 = 0.1806$

$\phi R_{nt} = \omega_t f_c'(1 - 0.59\omega_t) = 0.1806 \times 4 \times (1 - 0.59 \times 0.1806) \times 1000 = 645$ psi

$M_{ut} = \phi R_{nt} bd^2 = 645 \times 36 \times 18.5^2/(12 \times 1000) = 662$ ft-kip $> M_{u,max} = 615.8$ ft-kip **OK**

Section will be tension-controlled without compression reinforcement.

These calculations indicate a beam depth of 21 in. is adequate. ACI 318, however, indicates a minimum thickness of $\ell/18.5 \times (0.4 + f_y/100{,}000) = 27$ in., unless deflections are computed. This is applicable only to "members in one-way construction not supporting or attached to partitions or other construction likely to be damaged by large deflections." Otherwise, deflections must be computed.

For purposes of illustration, compute the required reinforcement for a 21 in. deep beam without considering deflection further.

(b) Compute required reinforcement.

1. End span, exterior negative.

$$\frac{M_u}{\phi f_c' bd^2} = \frac{385.9 \times 12}{0.9 \times 4 \times 36 \times 18.5^2} = 0.1044$$

From Table 7-1 in *PCA Notes*, $\omega \approx 0.112$.

$$A_s = \frac{\omega bd\, f_c'}{f_y} = \frac{0.112 \times 36 \times 18.5 \times 4}{100} = 2.98 \text{ in.}^2$$

Check for minimum required reinforcement according to ACI 318.

For $f_c' = 4000$ psi, use

$$A_{s,min} = \frac{200 b_w d}{f_y} = \frac{200 \times 36 \times 18.5}{100,000} = 1.33 \text{ in.}^2 < A_s \quad \textbf{OK}$$

Use 10 No. 5 bars ($A_s = 3.10$ in.^{2}).

(*PCA Notes*: use seven No. 8 bars [$A_s = 5.53$ in.^{2}].)

Check distribution of flexural reinforcement requirements of ACI 318

Maximum spacing allowed by ACI 318.

$$s = \frac{600}{f_s} - 2.5c_c \le 12\left(\frac{40}{f_s}\right)$$

$c_c = 1.5 + 0.5 = 2.0$ in.

Use $f_s = 0.67 f_y = 67$ ksi

$$s = \frac{600}{f_s} - 2.5c_c = \frac{600}{67} - 2.5 \times 2.0 = 3.96 \text{ in. (governs)}$$

$$s = 12\left(\frac{40}{f_s}\right) = 12\left(\frac{40}{67}\right) = 7.16 \text{ in.}$$

$$\text{spacing provided} = \frac{1}{9}\left(36 - 2\left[1.5 + 0.5 + \frac{5}{16}\right]\right) = 3.49 \text{ in.} < 3.96 \text{ in.} \quad \textbf{OK}$$

2. End span, positive.

$$\frac{M_u}{\phi f_c' bd^2} = \frac{441.1 \times 12}{0.9 \times 4 \times 36 \times 18.5^2} = 0.1193$$

From Table 7-1 in *PCA Notes*, $\omega \approx 0.129$.

$$A_s = \frac{\omega bd\, f_c'}{f_y} = \frac{0.129 \times 36 \times 18.5 \times 4}{100} = 3.44 \text{ in.}^2$$

Use 11 No. 5 bars ($A_s = 3.41$ in.^{2}) (one percent less than required, assumed sufficient).

(*PCA Notes*: use 11 No. 7 bars [$A_s = 6.60$ in.^{2}].)

Note that this reinforcement satisfies the cracking requirements in ACI 318 and fits within the beam width. It also can be used conservatively at the midspan section of the interior span.

3. End span, interior negative.

$$\frac{M_u}{\phi f_c' bd^2} = \frac{615.8 \times 12}{0.9 \times 4 \times 36 \times 18.5^2} = 0.1666$$

From Table 7-1 in *PCA Notes*, $\omega \approx 0.187$.

$$A_s = \frac{\omega bd\, f_c'}{f_y} = \frac{0.187 \times 36 \times 18.5 \times 4}{100} = 4.98 \text{ in.}^2$$

Use nine No. 7 bars (A_s = 5.40 in.^{2}). (*PCA Notes*: use 10 No. 9 bars [A_s = 10.0 in.^{2}].)

$$\text{spacing provided } = \frac{1}{8}\left(36 - 2\left[1.5 + 0.5 + \frac{7}{2 \times 8}\right]\right) = 3.89 \text{ in.} < 3.96 \text{ in.} \quad \textbf{OK}$$

	PCA Grade 60 A_s, in.^{2}	**PCA Grade 60 details**	**CS Grade 100 A_s, in.^{2}**	**CS Grade 100 details**
End span, exterior negative	3.10	10 No.5	5.53	7 No.8
End span, positive	3.41	11 No.5	6.60	11 No.7
End span, interior negative	5.40	9 No.7	10.00	10 No.9

High-strength steel (HSS) reinforcement was 3.10 in.^{2} to 5.40 in.^{2} at locations throughout the beam. The original design in *PCA Notes* provided 5.53 in.^{2} to 10.0 in.^{2} of Grade 60 reinforcement. This is a reduction of 45 to 50 percent at the corresponding locations in the beam. The beam depth was 10 percent greater using CS. Calculate beam deflection to determine if it is within acceptable limits.

Example 10.1—Development of bars in tension

Similar to Example 4.3 in *PCA Notes*, this example illustrates the calculation of the development for CS reinforcing bars using ACI 408R equations.

Calculate required development length for the inner two No. 8 bars in the beam shown in the following. Make the two No. 8 outer bars are continuous along the full length of beam. Use $f_c' = 4000$ psi (normalweight concrete) and $f_y = 100{,}000$ psi, and uncoated bars. Stirrups provide the minimum code requirements for beam shear reinforcement.

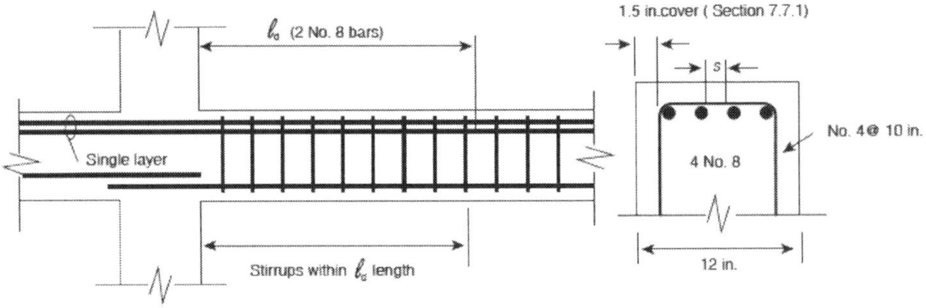

Calculations and discussion

Determine the development length of No. 8 bar with $f_y = 100{,}000$ psi.

Nominal bar diameter of No. 8 bar = 1.0 in.

$$\ell_d = \frac{\left(\dfrac{f_y}{(f_c')^{1/4}} - \phi 2400\omega\right)\alpha\beta_c\lambda}{\phi 76.3\left(\dfrac{c\omega + K_{tr}}{d_b}\right)} d_b$$

(Refer to Section 10.2 of this guide.)

$\alpha = 1.3$ for top cast bars; $\beta_c = 1.0$ for uncoated bars.

$\lambda = 1.0$ for normalweight concrete.

Clear spacing = [12 – 2(cover) – two (No. 4 stirrups) – four No. 8 bars)]/three spaces
= [12 – 2(1.5) – 2(0.5) – 4(1.0)]/3 = 1.33 in.

c_{si} = 1/2 clear spacing = 0.67 in.

$c_{so} = c_b$ = cover + No. 4 stirrup = 1.5 + 0.5 = 2.0 in.

$c_s = \min(c_{so}, c_{si} + 0.25) = \min(2.0, 0.92) = 0.92$ in.

$c_{min} = \min(c_s, c_b) = \min(0.92, 2.0) = 0.92$ in.

$c_{max} = \max(c_s, c_b) = \max(0.92, 2.0) = 2.0$ in.

$c = c_{min} + 0.5d_b = 0.92 + 0.5(1.0) = 1.42$ in.

$$\omega = 0.1\left(\frac{c_{max}}{c_{min}}\right) + 0.9 \le 1.25 = 0.1\left(\frac{2.0}{0.92}\right) + 0.9 = 1.12 < 1.25$$

$$K_{tr} = \frac{0.52 t_r t_d A_{tr}}{sn}\sqrt{f_c'}$$

$t_r = 9.6R_r + 0.28 \leq 1.72 = 9.6(0.07) + 0.28 = 0.95 \leq 1.72$

$t_d = 0.78d_b + 0.22 = 0.78(1.0) + 0.22 = 1.0$ in.

$s = 10$ in. spacing of stirrups

$n =$ two bars being developed

A_{tr} (two No. 4) = 2(0.2) = 0.4 in.^{2}

$$K_{tr} = \frac{(0.52)(0.94)(1.0)(0.4)}{(10)(2)}\sqrt{4000} = 0.62 \text{ in.}$$

$$\frac{c\omega + K_{tr}}{d_b} = \frac{(1.42)(1.12) + 0.62}{1.0} = 2.21 < 4$$

$$\ell_d = \frac{\left(\frac{100{,}000}{(4000)^{1/4}} - (0.8)(2400)(1.12)\right)(1.3)(1.0)(1.0)}{(0.8)(76.3)(2.21)}(1.0) = 100.5 \text{ in.}$$

The development length ℓ_d for two conventional No. 8 Grade 60 bars (*PCA Notes*, Example 4.3) was calculated using two different methods allowed in ACI 318 as 61.7 in. and 47.0 in. The calculated development length for the high-strength reinforcing bars represents an increase of 63 and 114 percent. This indicates that with a yield strength of 100,000 psi, CS bars require increased development length.

Example 10.2—Development of flexural reinforcement

Similar to Example 4.4 in *PCA Notes*, this example illustrates the calculation of the required bar lengths for continuous beams reinforced with CS. This example also highlights design differences between continuous beams reinforced with CS as compared with Grade 60 reinforcement.

Determine the lengths of the top and bottom bars for the exterior span of the continuous beam shown in the following. The concrete is normal weight and the bars are Grade 100. The total uniformly distributed factored gravity load on the beam is w_u = 6.0 kip/ft (including weight of the beam). Use f_c' = 4000 psi and f_y = 100,000 psi; b = 16 in.; and h = 22 in. Concrete cover = 1.5 in.

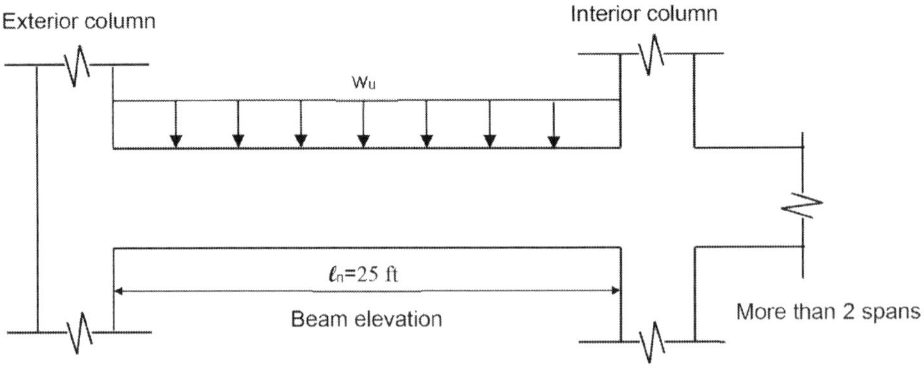

Calculations and discussion

(a) Preliminary design for moment and shear reinforcement

1. Use approximate analysis for moment and shear.

Location	Factored moments and shears
Interior face of exterior support	$-M_u = w_u\ell_n^2/16 = 6(25^2)/16 = -234.4$ ft-kip
End span positive	$+M_u = w_u\ell_n^2/14 = 6(25^2)/14 = 267.9$ ft-kip
Exterior face of first interior support	$-M_u = w_u\ell_n^2/10 = 6(25^2)/10 = -375.0$ ft-kip
Exterior face of first interior support	$V_u = 1.15w_u\ell_n/2 = 1.15(6)(25)/2 = 86.3$ kip

2. Determine required flexural reinforcement using procedures of Chapter 4 of this guide. With 1.5 in. cover, No. 4 bar stirrups, and No. 6 or 7 flexural bars, d ≈ *19.6 in.*

M_u	A_s required	Bars	A_s provided
–234.4 ft-kip	1.74 in.^{2}	Four No. 6	1.76 in.^{2}
267.9 ft-kip	2.02 in.^{2}	Two No. 6 and Two No. 7	2.08 in.^{2}
–375 ft-kip	4.64 in.^{2}	Four No. 10	5.08 in.^{2}

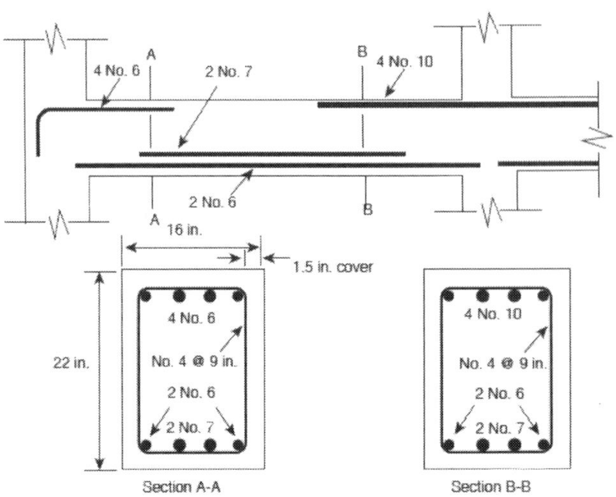

Using the reinforcement outlined in the previous table, the beam exhibits a compression-controlled failure at the exterior face of the first interior support and therefore $\phi = 0.65$ is used. Due to the higher tension-controlled strain limit for the high-strength reinforcing bars, the section cannot be designed as tension-controlled without including additional compression reinforcement. The bars on the bottom of the beam do not have adequate length to develop the full yield strength in compression at the exterior face of the first interior support. Make the bars continuous across the support to provide the necessary anchorage. This reduces the required top reinforcement at this section by approximately 18 percent ($A_s = 3.35$ in.^{2}). Check the bar spacing of the proposed reinforcing to satisfy crack control limits. For illustration purposes, continue with the reinforcement as outlined.

3. Determine required shear reinforcement using V_c *per ACI 318.*

V_u at d, distance from face of support

$$V_u = 86.3 - 6(19.6/12) = 76.5 \text{ kip}$$

$$\phi V_c = \phi(2\sqrt{f'_c}b_w d) = 0.75(2)(\sqrt{4000})(16)(19.6)/1000 = 29.8 \text{ kip}$$

Use No. 4 stirrups at 9 in. spacing $< s_{max} = d/2 = 9.8$ in.

$$\phi V_s = \frac{\phi A_v f_y d}{s} = \frac{0.75(0.4)(80)(19.6)}{9} = 52.3 \text{ kip}$$

$$\phi V_n = \phi V_c + \phi V_s = 29.8 + 52.3 = 82.1 \text{ kip} > 76.6 \text{ kip} \quad \textbf{OK}$$

Distance from support where stirrups not required

$$V_u < 1/2\phi V_c = 29.8 / 2 = 14.9 \text{ kip}$$

$$V_u = 86.3 - 6x = 14.9 \text{ kip}$$

$$x = 11.9 \text{ ft} \approx 1/2 \text{ span}$$

Use No. 4 U stirrups at 9 in. (entire span).

(*PCA Notes*: use No. 4 U stirrups at 7 in. [entire span].)

(b) Bar lengths for bottom reinforcement

1. Required number of bars to be extended into supports.

Extend 1/4 of (+A_s) at least 6 in. into the supports. With a longitudinal bar required at each corner of the stirrups of ACI 318, at least two bars should be extended full length. Extend the two No. 6 bars full span length (plus 6 in. into the supports) and cut off the two No. 7 bars within the span.

2. Determine cut-off locations for the two No. 7 bars and check other development requirements.

Shear and moment diagrams for loading condition causing maximum factored positive moment are shown in the following figure.

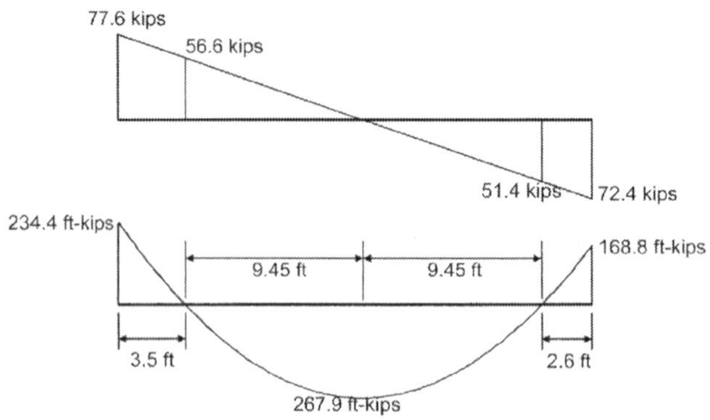

The positive moment portion of the M_u diagram is shown in the following at a larger scale, including the design moment strengths ϕM_n for the total A_s (two No. 6 and two No. 7) and for two No. 6 bars separately. For two No. 6 and two No. 7, ϕM_n = 276.0 ft-kip. For two No. 6, ϕM_n = 124 ft-kip.

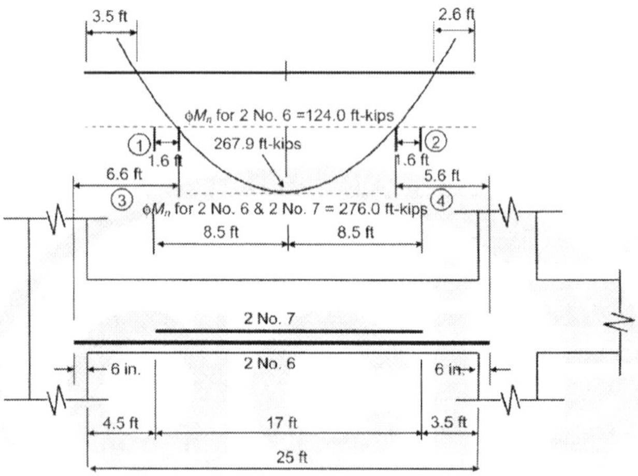

The two No. 6 bars extend full span length plus 6 in. into the supports. Cut the two No. 7 bars at 4.5 ft and 3.5 ft from the exterior and interior supports. Determine these cutoff locations as follows.

Dimensions (1) and (2) should be the larger of d or $12d_b$

$d = 19.6 \text{ in.}^2 = 1.6$ ft (governs)

$12d_b = 12(0.88) = 10.5$ in.

Within the development length ℓ_d, only two No. 6 bars are developed (two No. 7 bars are already developed in the length of 8.5 ft).

Development for No. 6 corner bars (refer to *PCA Notes* Example 4.3 for a detailed calculation of development length for CS reinforcing bars) is as follows.

$\ell_d = 31$ in. = 2.6 ft

Dimension (3): 6.6 ft > 2.6 ft **OK**

Dimension (4): 5.6 ft > 2.6 ft **OK**

Check required development length ℓ_d for two No. 7 bars. Two No. 6 bars are already developed in length 2.6 ft from bar end.

$\ell_d = 44$ in. = 3.7 ft < 8.5 ft **OK**

For No. 6 bars, check development requirements at point of inflection (PI):

$$\ell_d \le \frac{M_n}{V_u} + \ell_a$$

For two No. 6 bars, M_n = 124.0/0.9 = 137.8 ft-kip

At the left PI, V_u = 77.6 – 6(3.5) = 56.6 kip

ℓ_a = larger of $12d_b$ = 9 in. or d = 19.6 in. (governs)

$$\ell_d \le \frac{137.8 \times 12}{56.6} + 19.6 = 48.8 \text{ in.}$$

For No. 6 bars, ℓ_d = 31 in. < 48.8 in. **OK**

At the right PI, V_u = 51.4 kip; by inspection, the development requirements for the No. 6 bars are satisfactory.

With both tentative cutoff points located in a zone of flexural tension, one of the three conditions of ACI 318 should be satisfied. The applicability of this provision to high-strength reinforcing bars has not been demonstrated experimentally. The code provisions are checked herein for illustrative purposes.

At left cutoff point (4.5 ft from support).

V_u = 77.6 – (6)(4.5) = 50.6 kip

ϕV_n = 82.1 kip (No. 4 U-stirrups at 9 in.)

2/3(82.1) = 54.7 kip > 50.6 kip **OK**

For illustrative purposes, determine if the bar cutoff point of Section 12.10.5.3 of ACI 318-11 is also satisfied:

M_u = 54.1 ft-kip at 4.5 ft from support

A_s = 0.38 in.^{2}

For two No. 6 bars, A_s provided = 0.88 in.^{2}

0.88 in.^{2} > 2(0.38) = 0.76 in.^{2} **OK**

3/4(82.1) = 61.6 kip > 50.6 kip **OK**

Therefore, ACI 318 is also satisfied at cutoff location.

At right cutoff point (3.5 ft from support)

V_u = 72.4 – (3.5)(6) = 51.4 kip

2/3(ϕV_n) = 54.7 kip > 51.4 kip **OK**

Summary: The tentative cutoff locations for the bottom reinforcement meet all proposed development requirements. Place the two No. 7 bars by 17 ft unsymmetrically within the span. Ensure proper placement of the No. 7 bars by specifying an 18 ft length for symmetrical bar placement within the span, 3.5 ft from each support. The recommended bar arrangement is shown at the end of this example.

(c) Bar lengths for top reinforcement

Shear and moment diagrams for loading condition causing maximum factored negative moments are shown in the following.

The negative moment portions of the M_u diagram are shown in the following at a larger scale, including the design moment strengths ϕM_n for the total negative A_s at each support (four No. 6 at exterior support and four No. 10 at interior support). For four No. 6, $\phi M_n = 237.4$ ft-kip ($\phi = 0.9$). For four No. 10, $\phi M_n = 376.7$ ft-kip ($\phi = 0.65$).

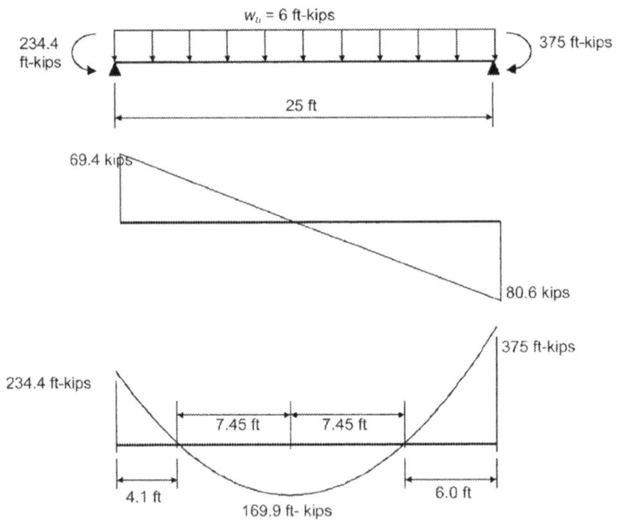

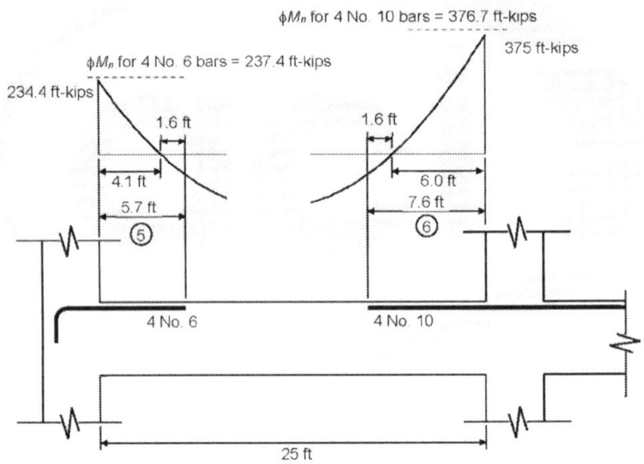

(d) Development requirements for four No. 6 bars at exterior support

1. Required number of bars to be extended.

Extend 1/3 of ($-A_s$) provided at supports beyond the inflection point a distance equal to the greater of d, $12d_b$, or $\ell_n/16$.

$d = 19.6 = 1.6$ ft (governs)

$12d_b = 12(0.75) = 9.0$ in.

$\ell_n/16 = 25 \times 12/16 = 18.8$ in.

Because the inflection point is 4.1 ft from the support, the total length of the No. 6 bars will be short even with the required 1.6 ft extension beyond the inflection point. Check the required development length ℓ_d for a cutoff location at 5.75 ft (5 ft 9 in.) from the support face.

Dimension (5) in the aforementioned figure should be at least equal to ℓ_d.

For No. 6 bars, $\ell_d = 31$ in.

With four No. 6 bars developed at same location (support face)

Including top bar effect, $\ell_d = 1.3(31) = 40.3$ in.

For No. 6 top bars, $\ell_d = 40.3$ in. = 3.35 ft < 5.75 ft **OK**

2. Anchorage into exterior column.

The No. 6 bars can be anchored into the column with a standard end hook following the provisions of ACI 318. Recent tests have shown that ASTM A1035/A1035M bars with standard hooks can develop tensile bar stresses of at least 140,000 psi (965 MPa) and, in many cases, the strength of the bar. (Refer to Section 10.2 of this guide.)

$$\ell_{dh} = \left(0.02\psi_e \frac{f_y}{\lambda}\sqrt{f'_c}\right)d_b$$

$\psi_e = 1.0$ for uncoated bars

$\lambda = 1.0$ for normalweight concrete

$$\ell_{dh} = (0.02)(1.0)\left(\frac{100,000}{1.0}\right)\sqrt{4000}(0.75) = 23.7 \text{ in.}$$

The required ℓ_{dh} for the hook is reduced if excess reinforcement is considered. Because A_s (provided) ≈ A_s (required), however, this effect is not considered in this example.

Overall depth of column required is 23.7 + 2 = 25.7 in.

(e) Development requirements for four No. 10 bars at interior column.
1. Required extension for one-third of (–A_s)

$d = 19.6$ in. = 1.6 ft (governs)

$12d_b = 12(1.27) = 15.2$ in.

$\ell_n/16 = 18.8$ in.

For No. 10 bars, $\ell_d = 129.6$ in. = 10.8 ft

Dimension (6) in the aforementioned figure = 7.6 ft < ℓ_d = 10.8 ft **NG**

Extend the No. 10 bars to provide the required ℓ_d beyond the face of the column.

(f) Summary: selected bar lengths for the top and bottom reinforcement is shown in the following.

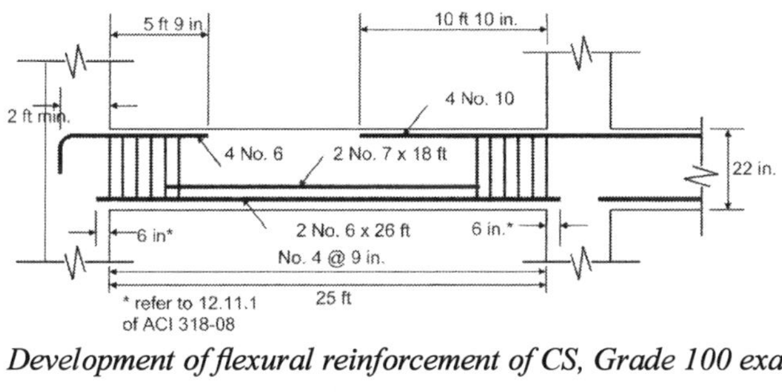

Development of flexural reinforcement of CS, Grade 100 example

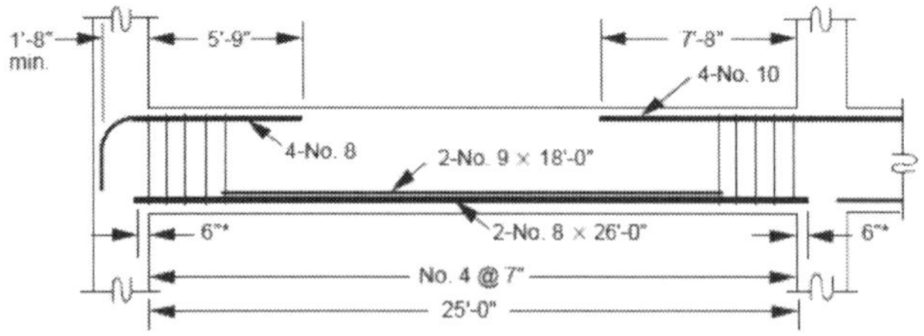

Development of flexural reinforcement of PCA Grade 60 example

(g) Supplementary requirements

If the beam was part of a primary lateral load-resisting system, the two No. 6 bottom bars extending into the supports would have to be anchored to develop the bar yield strength at the face of supports. At the exterior column, anchorage should be provided by using end hooks or headed bars.

At the interior column, the two No. 6 bars could be extended a distance ℓ_d beyond the support face into the adjacent span, mechanically coupled or lap spliced with extended bars from the adjacent span. A Class A lap splice or a mechanical splice may be considered to satisfy the intent of ACI 318.

The design using CS requires an equivalent number of smaller-diameter bars as compared with a similar design using conventional steel reinforcement (PCA 2005). The development length required for the smaller-diameter high-strength reinforcing bars is similar to that required for the larger-diameter conventional steel bars.

In this example, using high-strength reinforcing bars results in an overall savings of the required amount of flexural steel of approximately 20 percent as compared with the design using Grade 60 steel bars. Wider spacing of the CS stirrups provides a 20 percent reduction of the shear reinforcement.

Example 4.7—Deflections of simple-span rectangular beam

(Similar to Example 10.1 in *PCA Notes*) This example illustrates the calculation of short- and long-term deflections for a rectangular beam reinforced with CS bars. The section was designed to have similar flexural strength and the same failure mode as the beam in the original example, which uses Grade 40 reinforcing bars. The current design represents a 60 percent reduction of the tension steel provided and a 45 percent reduction of the compression steel as compared with the original design using Grade 40 reinforcement.

Required: Analysis of short-term deflections, and long-term deflections at 3 months and 5 years (ultimate value). (Refer to Section 4.8 of this guide.)

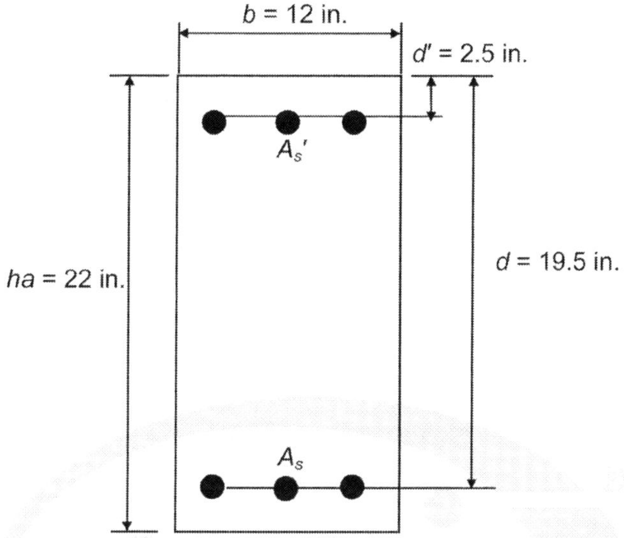

f_c' = 3000 psi (normalweight concrete)
f_y = 100,000 psi
f_y' = 80,000 psi
A_s = two No. 5 and one No. 3 = 0.73 in.^{2}
E_s = 29,000,000 psi
$\rho = A_s/bd = 0.0031$
A_s' = one No. 3 = 0.33 in.^{2}
(A_s' not required for strength)
$\rho' = A_s'/bd = 0.0014$
Superimposed dead load = 120 lb/ft (not including beam weight)
Live load = 300 lb/ft (50 percent sustained)
Span = 25 ft

Calculations and discussion

(a) Minimum beam thickness for members not supporting or attached to partitions or other construction likely to be damaged by large deflections

According to Table 9.5(a) of ACI 318-11, footnotes for CS

h_{min} = span/16 multiplied by 1.4 factor for 100,000 psi steel

$$= \frac{25 \times 12}{16} \times 1.4 = 26 \text{ in.} > 22 \text{ in.}$$

Therefore, calculate the beam deflection to confirm that the beam depth is satisfactory.

(b) Moments

$w_d = 0.12 + (12)(22)(0.150)/144 = 0.395$ kip/ft

$$M_d = \frac{w_d \ell^2}{8} = \frac{(0.395)(25)^2}{8} = 30 \text{ ft-kip}$$

$$M_\ell = \frac{w_\ell \ell^2}{8} = \frac{(0.300)(25)^2}{8} = 23.4 \text{ ft-kip}$$

$M_{d+\ell} = 54.3$ ft-kip

$M_{sus} = M_d + 0.5M_l = 30.9 + (0.5)(23.4) = 42.6$ ft-kip

(c) Modulus of rupture, modulus of elasticity, modular ratio

$$f_r = 7.5\sqrt{f_c'} = 7.5\sqrt{3000} = 411 \text{ psi}$$

$$E_c = \omega_c^{1.5} 33\sqrt{f_c'} = (150)^{1.5} 33\sqrt{3000} = 3.32 \times 10^6 \text{ psi}$$

$$n_s = \frac{E_s}{E_c} = \frac{29 \times 10^6}{3.32 \times 10^6} = 8.7$$

(d) Gross and cracked section moments of inertia, using Table 10-2 from *PCA Notes*

$$I_g = \frac{bh^3}{12} = \frac{(12)(22)^3}{12} = 10{,}650 \text{ in.}^4$$

$$B = \frac{b}{nA_s} = \frac{12}{(8.7)(0.73)} = 1.89 \text{ in.}$$

$$r = \frac{(n-1)A_s'}{(nA_s)} = \frac{(7.7)(0.33)}{(8.7)(0.73)} = 0.400$$

$$kd = \left[\sqrt{2dB(1 + rd'/d) + (1+r)^2} - (1+r)\right] / B$$
$$= \left[\sqrt{2(19.5)(1.89)(1 + (0.400)(2.5)/(19.5)) + (1+0.400)^2} - (1+0.400)\right] / 1.89 = 3.98$$

$$I_{cr} = \frac{bk^3d^3}{3} + nA_s(d-kd)^2 + (n-1)A_s'(kd-d')^2$$
$$= \frac{(12)(3.98)^3}{3} + (8.7)(0.73)(19.5-3.98)^2 + (7.2)(0.33)(3.98-2.5)^2$$
$$= 1788 \text{ in.}^4$$

$I_g/I_{cr} = 5.96$

(e) Effective moment of inertia, I_{eff}, per ACI 318, and as per Eq. (4.8) of this guide.

$$M_{cr} = \frac{f_r I_g}{y_t} = \frac{(411)(10{,}650)}{(11)} / 12{,}000 = 33.2 \text{ ft-kip}$$

1. Under dead load only.

$$\frac{M_{cr}}{M_d} = \frac{33.2}{30.9} > 1$$

Hence, $(I_e)_d = I_g = 10{,}650 \text{ in.}^4$

2. Under sustained load.

$$\left(\frac{M_{cr}}{M_{sus}}\right)^3 = \left(\frac{33.2}{42.6}\right)^3 = 0.473$$

$$\begin{aligned}(I_e)_{sus} &= (M_{cr}/M_a)^3 I_g + \left(1 - [M_{cr}/M_a]^3\right) I_{cr} \le I_g \\ &= (0.473)(10{,}650) + (1 - 0.473)(1788) \\ &= 5980 \text{ in.}^4\end{aligned}$$

$(I_e)_{sus} = 3612 \text{ in.}^4$ (Eq. (4.8))

3. Under dead + live load.

$$\left(\frac{M_{cr}}{M_{d+\ell}}\right)^3 = \left(\frac{33.2}{54.3}\right)^3 = 0.229$$

$$\begin{aligned}(I_e)_{d+\ell} &= (0.229)(10{,}650) + (1 - 0.229)(1788) \\ &= 3817 \text{ in.}^4\end{aligned}$$

$(I_e)_{d+\ell} = 2595 \text{ in.}^4$ (Eq. (4.8))

(f) Initial or short-time deflections, using Eq. (3) of *PCA Notes*

$$(\Delta_i)_d = \frac{K(5/48)M_d\ell^2}{E_c(I_e)_d} = \frac{(1)(5/48)(30.9)(25)^2(12)^3}{(3320)(10{,}650)} = 0.098 \text{ in.}$$

$K = 1$ for simple spans (refer to Table 8-3 of *PCA Notes*)

$$(\Delta_i)_{sus} = \frac{K(5/48)M_{sus}\ell^2}{E_c(I_e)_d} = \frac{(1)(5/48)(42.6)(25)^2(12)^3}{(3320)(5980)} = 0.241 \text{ in.} \quad \text{(0.495 in. per this guide)}$$

$$(\Delta_i)_{d+l} = \frac{K(5/48)M_{d+l}\ell^2}{E_c(I_e)_d} = \frac{(1)(5/48)(54.3)(25)^2(12)^3}{(3320)(3817)} = 0.482 \text{ in.} \quad \text{(0.689 in. per this guide)}$$

$(\Delta_i)_\ell = (\Delta_i)_{d+\ell} - (\Delta_i)_d = 0.482 - 0.098 = 0.384$ in. (0.591 in. per this guide)

Allowable deflections (Table 9.5(b) of ACI 318-08).

Flat roofs not supporting and not attached to nonstructural elements likely to be damaged by large deflections

$(\Delta_i)_\ell \le$ span/180 = 300/180 = 1.67 in. > 0.384 in. **OK** (> 0.591 in. **OK** per this guide)

Floors not supporting and not attached to nonstructural elements likely to be damaged by large deflections

$(\Delta_i)_\ell \le$ span/360 = 300/360 = 0.83 in. > 0.384 in. **OK** (> 0.591 in. **OK** per this guide)

(g) Additional long-term deflections at 3 months and 5 years (ultimate value).

Combined creep and shrinkage deflections of ACI 318 and Eq. (4) of *PCA Notes*: $\Delta_{cp} + \Delta_{sh} = \lambda(\Delta_i)_{sus}$

Duration	ξ	$\lambda = \dfrac{\xi}{1+50\rho'}$	$(\Delta_i)_{sus}$, in.	$(\Delta_i)_l$, in.	$\Delta_{cp} + \Delta_{sh} = \lambda(\Delta_i)_{sus}$, in.	$\Delta_{cp} + \Delta_{sh} + (\Delta_i)_l$, in.
5 years	2.0	1.87	0.241	0.384	0.451	0.84
3 months	1.0	0.93	0.241	0.384	0.224	0.61

Separate creep and shrinkage deflections, using Eq. (5) and (6) of *PCA Notes*

For $\rho = 0.0031$; $\rho' = 0.0014$

For $\rho = 100$, $\rho = 0.31$ and $\rho' = 100$, $\rho' = 0.14$, read $A_{sh} = 0.287$ (Fig. 10.3 of *PCA Notes*) and $K_{sh} = 0.125$ for simple spans (Table 10-5 of *PCA Notes*).

Duration	C_t	$\lambda_{cp} = \dfrac{0.85C_t}{1+50\rho'}$	$\Delta_{cp} = \lambda_{cp}(\Delta_i)_{sus}$, in.	ε_{sh}, in./in.	$\phi_{sh} = \dfrac{A_{sh}\mu_{sh}}{h}$, 1/in.	$\Delta_{sh} = K_{sh}\phi_{sh}l^2$, in.	$\Delta_{cp} + \Delta_{sh} + (\Delta_i)_l$, in.
5 years	1.6 (ultimate)	1.27	0.31	400×10^{-6}	$\dfrac{0.287 \times 400 \times 10^{-6}}{22} = 5.22 \times 10^{-6}$	$\dfrac{1}{8} \times 5.22 \times 10^{-6} \times (25 \times 12)^2 = 0.059$	0.31 + 0.059 + 0.384 = 0.753
3 months	$0.56 \times 1.6 = 0.9$	0.71	0.17	$0.6 \times 400 \times 10^{-6} = 240 \times 10^{-6}$	3.13×10^{-6}	= 0.035	0.17 + 0.035 + 0.384 = 0.59

Allowable deflection of ACI 318.

Roof or floor construction supporting or attached to nonstructural elements are likely to be damaged by large deflections (very stringent limitation).

$\Delta_{cp} + \Delta_{sh} + (\Delta_i)_l \leq$ Span/480 = 300/480 = 0.63 in. **NG** by both methods

Roof or floor construction supporting or attached to nonstructural elements not likely to be damaged by large deflections

$\Delta_{cp} + \Delta_{sh} + (\Delta_i)_l \leq$ Span/240 = 300/240 = 1.25 in. **OK** by both methods and this guide.

For beams reinforced with high-strength reinforcing steel, the calculated short-term and long-term deflections are 25 to 40 percent higher than the corresponding deflections of an equivalent beam reinforced with conventional steel reinforcement. This is due to the reduced area of CS required to achieve equal moment capacity of the section. The reduction of the required steel area results in a corresponding reduction of the section's effective moment of inertia, which increases the calculated beam deflections. Deflection limitations, rather than strength requirements, may govern the design of concrete beams reinforced with CS bars. Deflection calculations are based on ACI 318 using an effective moment of inertia and a long-term multiplier and are used for illustrative purposes. The applicability of these provisions to high-strength reinforcing bars has not been demonstrated experimentally.

Example 4.8—Design for shear: members subject to shear and flexure only

Similar to Example 12.1 in *PCA Notes*, this example illustrates the calculation of the required high-strength shear reinforcement to resist a given level of applied load. This example considers the allowable increase of the shear reinforcement yield strength.

Determine required size and spacing for vertical U-stirrups for a 30 ft span, simply-supported beam. (Refer to Section 4.11 of this guide.)

$b_w = 13$ in.
$d = 20$ in.
$f_c' = 3000$ psi
$f_{yt} = 80{,}000$ psi
$w_u = 4.5$ kip/ft

Calculations and discussion

Assume the live load is present on the full span so that the design shear at the centerline of the span is zero. (Obtain a design shear greater than zero at midspan by using partial live loading of the span.) Using design procedure for shear reinforcement outlined in this part.

1. Determine factored shear forces.

At support: $V_u = 4.5(15) = 67.5$ kip

At distance d from support: $V_u = 67.5 - 4.5(20/12) = 60$ kip

2. Determine shear strength provided by concrete.

$$\phi V_c = \phi 2\sqrt{f_c'}b_w d$$

$$\phi = 0.75$$

$$\phi V_c = (0.75)2\sqrt{3000}(13)(20) = 21.4 \text{ kip}$$

$V_u = 60$ kip $> \phi V_c = 21.4$ kip

Therefore, shear reinforcement is required.

3. Compute $V_u - \phi V_c$ *at critical section.*

$$V_u - \phi V_c = 60 - 21.4 = 38.6 \text{ kip} < \phi 8\sqrt{f_c'}b_w d = 85.4 \text{ kip} \quad \textbf{OK}$$

4. Determine distance x_c *from support beyond which minimum shear reinforcement is required* ($V_u = \phi V_c$).

$$x_c = \frac{V_u \text{ at support} - \phi V_c}{w_u} = \frac{67.5 - 21.4}{4.5} = 10.2 \text{ ft}$$

Determine distance x_m from support beyond which concrete can carry total shear force ($V_u = \phi V_c/2$).

$$x_m = \frac{V_u \text{ at support} - (\phi V_c/2)}{w_u} = \frac{67.5 - (21.4/2)}{4.5} = 12.6 \text{ ft}$$

5. Determine required spacing of vertical U-stirrups.

At critical section, $V_u = 60$ kip $> \phi V_c = 21.4$ kip

$$s \text{ (required)} = \frac{\phi A_v f_{yt} d}{V_u - \phi V_c}$$

Assuming No. 4 U-stirrups (A_v = 0.40 in.^{2})

$$s \text{ (required)} = \frac{0.75(0.4)(80)(20)}{38.6} = 12.4 \text{ in.}$$

Check maximum permissible spacing of stirrups.

$$s \text{ (max)} \leq d/2 = 20/2 = 10 \text{ in. (governs)}$$

$$\leq 24 \text{ in. because } V_u - \phi V_c = 38.6 < \phi 4\sqrt{f_c'} b_w d = 42.7 \text{ kip}$$

Maximum stirrup spacing based on minimum shear reinforcement.

$$s \text{ (max)} \leq \frac{A_v f_{yt}}{0.75\sqrt{f_c'} b_w} = \frac{0.4(80,000)}{0.75\sqrt{3000}(13)} = 60 \text{ in.}$$

$$\leq \frac{A_v f_{yt}}{50 b_w} = \frac{0.4(80,000)}{50(13)} = 49$$

Because s (required) = 12 in. > s (max) = 10 in., provide stirrups at 10 in. spacing.

Stirrup spacing using No. 4 U-stirrups

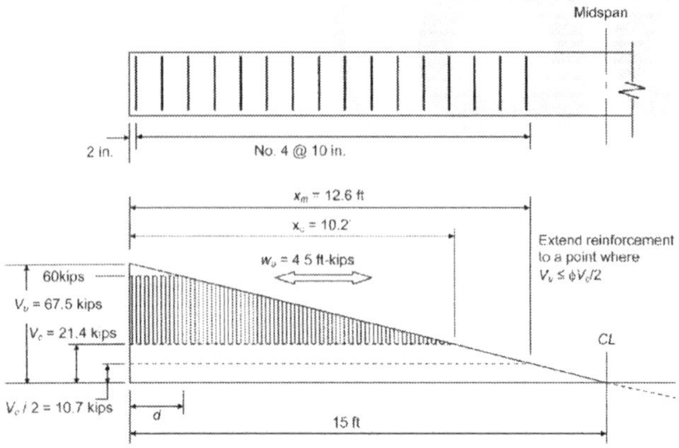

The design with CS resulted in a 25 percent reduction of the shear reinforcement as compared with the design with Grade 40 steel (*PCA Notes*). Due to the increased yield strength of the CS shear reinforcement, minimum spacing or area requirements, rather than strength requirements, may govern the spacing of the shear reinforcement. The design with high-strength shear reinforcement results in constant spacing along the region requiring shear reinforcement. This minimizes construction errors, which can occur from different spacing requirements along the beam length.

Example 7.1—Shear design of wall

Similar to Example 21.4 in *PCA Notes*, this example illustrates the design of a reinforced concrete wall using CS reinforcing bars to carry lateral loads.

Determine the shear and flexural reinforcement for the wall shown.

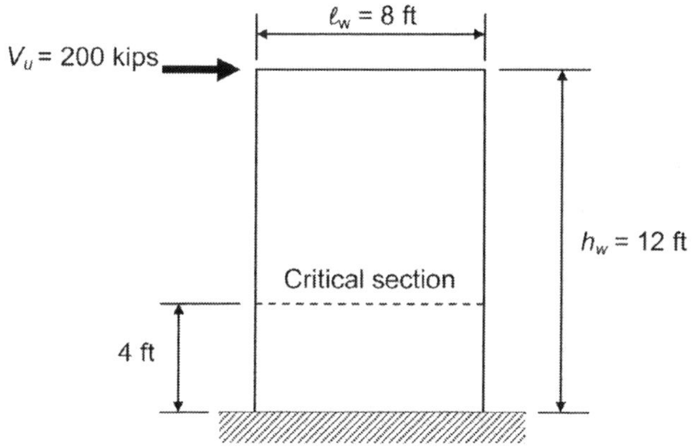

h (wall thickness) = 8 in.
$f_c' = 3000$ psi
$f_y = 100{,}000$ psi
$f_{yt} = 80{,}000$ psi
$f_y' = 80{,}000$ psi

Calculations and discussion

1. Check maximum shear strength permitted.

$$\phi V_n = \phi 10\sqrt{f_c'}hd$$

where $d = 0.8\ell_w = 0.8 \times 8 \times 12 = 76.8$ in.

$$\phi V_n = 0.75 \times 10\sqrt{3000} \times 8 \times 76.8/1000 = 252.4 \text{ kip} > 200 \text{ kip} \quad \textbf{OK}$$

2. Calculate shear strength provided by concrete V_c.

Critical section for shear

$\ell_w/2 = 8/2 = 4$ ft (governs)

or

$h_w/2 = 12/2 = 6$ ft

$$V_c = 3.3\sqrt{f_c'}hd + \frac{N_u d}{4\ell_w}$$
$$= 3.3\sqrt{3000}(8)(76.8)/1000 + 0 = 111 \text{ kips}$$

or

$$V_c = \left[0.6\sqrt{f'_c} + \frac{\ell_w\left(1.25\sqrt{f'_c} + \dfrac{0.2N_u}{\ell_w h}\right)}{\dfrac{M_u}{V_u} - \dfrac{\ell_w}{2}}\right] hd$$

$$= \left[0.6\sqrt{3000} + \frac{96(1.25\sqrt{3000}+0)}{96-48}\right]\left(\frac{8\times 76.8}{1000}\right) = 104 \text{ kips (governs)}$$

where $M_u = (12 - 4)V_u = 8V_u$ ft-kip $= 96V_u$ in.-kip

3. Determine required horizontal shear reinforcement.

$V_u = 200$ kip $> \phi V_c/2 = 0.75(104)/2 = 39.0$ kip

Provide shear reinforcement in accordance with ACI 318.

$$\begin{aligned} V_u &\le \phi V_n \\ &\le \phi(V_c + V_s) \\ &\le \phi V_c + \phi A_v f_y d/s_2 \end{aligned}$$

$$\frac{A_v}{s_2} = \frac{(V_u - \phi V_c)}{\phi f_y d} = \frac{[200-(0.75\times 104)]}{0.75\times 80\times 76.8} = 0.0265$$

For two No. 3: $s_2 = \dfrac{2\times 0.11}{0.0265} = 8.3$ in.

For two No. 4: $s_2 = \dfrac{2\times 0.20}{0.0265} = 15.1$ in.

For two No. 5: $s_2 = \dfrac{2\times 0.31}{0.0265} = 23.4$ in.

Use two No. 4 at 15 in.

$$\rho_t = \frac{A_v}{A_g} = \frac{2\times 0.2}{8\times 15} = 0.0033 > 0.0025$$

OK

$$\text{maximum spacing} = \begin{cases} \ell_w/5 = 96/5 = 19.2 \text{ in.} \\ 3h = 3\times 8 = 24 \text{ in.} \\ 18 \text{ in. (governs)} \end{cases}$$

Use two No. 4 at 15 in.

4. Determine vertical shear reinforcement.

$$\rho_\ell = 0.0025 + 0.5\left(2.5 - \frac{h_w}{\ell_w}\right)(\rho_t - 0.0025) \ge 0.0025$$

$$= 0.0025 + 0.5\left(2.5 - \frac{12}{8}\right)(0.0033 - 0.0025) \ge 0.0025$$

$$= 0.0029$$

$$\text{maximum spacing} = \begin{cases} \ell_w/3 = 96/3 = 32 \text{ in.} \\ 3h = 3 \times 8 = 24 \text{ in.} \\ 18 \text{ in. (governs)} \end{cases}$$

Use two No. 4 at 15 in. ($\rho_t = 0.0033$).

5. *Design for flexure.*

$M_u = V_u h_w = 200 \times 12 = 2400$ ft-kip

Assume tension-controlled section ($\phi = 0.9$)

with $d = 0.8\ell_w = 0.8 \times 96 = 76.8$ in.

(Note: An exact value of d is determined by a strain compatibility analysis in the following.)

$$R_n = \frac{M_u}{\phi b d^2} = \frac{2400 \times 12{,}000}{0.9 \times 8 \times 76.8^2} = 678 \text{ psi}$$

$$\rho = \frac{0.85 f'_c}{f_y}\left(1 - \sqrt{1 - \frac{2R_n}{0.85 f'_c}}\right)$$
$$= \frac{0.85 \times 3}{100}\left(1 - \sqrt{1 - \frac{2 \times 678}{0.85 \times 3000}}\right) = 0.00805$$

$A_s = \rho b d = 0.00805 \times 8 \times 76.8 = 4.95$ in.^{2}

Use nine No. 6 ($A_s = 3.96$ in.^{2}) at each end of wall, which provides less area of steel than that determined based on $d = 0.8\ell_w$.

Check moment strength of the wall with nine No. 6 bars using a strain compatibility analysis (refer to the following figure for reinforcement layout) based on the idealized elastic-plastic stress-strain curve of steel of the simplified method discussed in Chapter 4 of this guide.

From strain compatibility analysis (including No. 4 vertical bars)

$c = 20$ in.

$d \approx 90$ in. (to centroid of No. 6 bars)

$\varepsilon_t = 0.011$ (at extreme tension steel)

$M_n = 3215$ ft-kip

$\phi M_n = 0.9 \times 3215 = 2894$ ft-kip > 2400 ft-kip **OK**

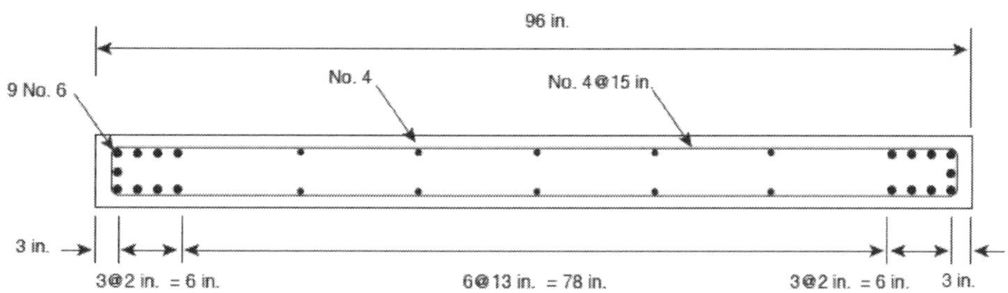

Use nine No. 6 bars each side ($A_s = 3.96$ in.^{2})

Including the main flexural reinforcement and the vertical shear reinforcement, this design provides 9.92 in.^{2} of vertical CS reinforcement. The design using conventional steel reinforcement requires 16.22 in.^{2} of vertical reinforcement. The use of CS reinforcement results in a 40 percent reduction of the amount of vertical steel. The use of CS results in a 33 percent reduction of the required horizontal shear reinforcement.

No research data are available to validate the use of the equations for V_c in Step 2 of the solution when using CS reinforcement.

Example 8.1—Design for flexural reinforcement of footing

Similar to Example 22.3 in *PCA Notes*, this example illustrates the design of the main flexural reinforcement of a typical footing using CS bars. Determine the required reinforcement for the footing shown per linear ft.

$f_c' = 3000$ psi
$f_y = 100{,}000$ psi
$P_u = 860$ kip
$q_s = 5.10$ kip/ft^2
Footing width $b = 13$ ft, column dimensions = 12 x 30 in.

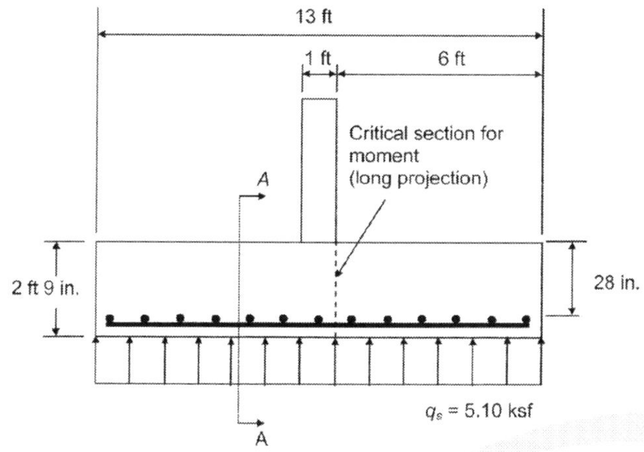

Calculations and discussion

1. Critical section for moment is at column face.

$M_u = 5.10 \times 13 \times 6^2/2 = 1193$ ft-kip

2. Compute required A_s *assuming tension-controlled section* ($\phi = 0.9$).

$$\text{required } R_n = \frac{M_u}{\phi b d^2} = \frac{1193 \times 12 \times 1000}{0.9 \times 156 \times 28^2} = 130 \text{ psi}$$

$$\rho = \frac{0.85 f_c'}{f_y}\left(1 - \sqrt{1 - \frac{2R_n}{0.85 f_c'}}\right) = \frac{0.85 \times 3}{100}\left(1 - \sqrt{1 - \frac{2 \times 130}{0.85 \times 3000}}\right) = 0.0013$$

Based on gross area, $\rho = (d/h) \times 0.0013 = (28/33) \times 0.0013 = 0.0011$

Check minimum A_s required for footings of uniform thickness; for Grade 100 reinforcement

$$\rho_{min} = \frac{0.0018 \times 60{,}000}{f_y} = \frac{0.0018 \times 60{,}000}{100{,}000} = 0.0011 < 0.0014 \text{ (required)}$$

$\rho_{min} = 0.0014$ (controls)

Therefore, required $A_s = \rho_{min} bh = 0.0014 \times 156 \times 33 = 7.21$ in.^{2}

Use 12 No. 7 bars ($A_s = 7.20$ in.^{2}) each way.

A lesser amount of reinforcement is required in the perpendicular direction due to lesser M_u, but for ease of placement, use the same uniformly distributed reinforcement each way. For perpendicular direction, $d_t = 27.1$ in.

3. Check net tensile strain (ε_t).

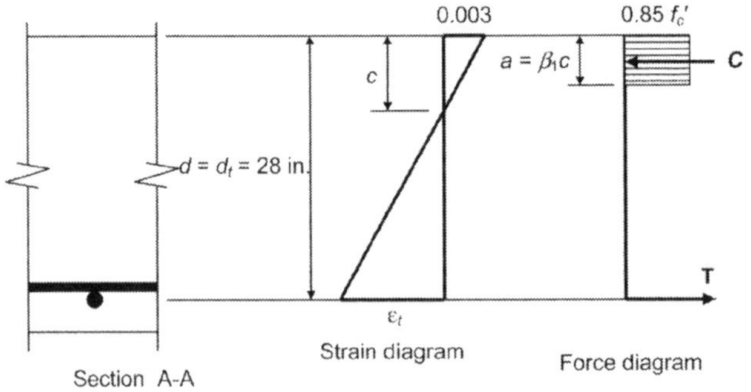

$$a = \frac{A_s f_y}{0.85 f_c' b} = \frac{7.20 \times 100}{0.85 \times 3 \times 156} = 1.81 \text{ in.}$$

$$c = \frac{a}{\beta_1} = \frac{1.81}{0.85} = 2.13 \text{ in.}$$

$$\varepsilon_t = \left(\frac{0.003}{c}\right) d_t - 0.003 = \frac{0.003}{2.13} \times 28 - 0.003 = 0.036 > 0.009$$

Therefore, section is tension-controlled and initial assumption is valid. **OK**

Use 12 No. 7 bars each way.

4. *Check development of reinforcement.*

Critical section for development is the same as that for moment (at column face)

$$\ell_d = \frac{\left(\frac{f_y}{(f_c')^{1/4}} - \phi 2400\omega\right)\alpha\beta_c\lambda}{\phi 76.3\left(\frac{c\omega + K_{tr}}{d_b}\right)} d_b$$

$\alpha = 1.0$ for bottom cast bars

$\beta_c = 1.0$ for uncoated bars

$\lambda = 1.0$ for normalweight concrete

Clear spacing = [156 – 2(cover) – 12 (No. 7 bars)]/11 spaces

= [156 – 2(3.0) – 12(7/8)]/11 = 12.7 in.

c_{si} = 1/2 clear spacing = 6.35 in.

$c_{so} = c_b$ = cover = 3.0 in.

$c_s = \min(c_{so}, c_{si} + 0.25) = \min(3.0, 6.6) = 3.0$ in.

$c_{min} = \min(c_s, c_b) = \min(3.0, 3.0) = 3.0$ in.

$c_{max} = \max(c_s, c_b) = \max(3.0, 3.0) = 3.0$ in.

$c = c_{min} + 0.5d_b = 3.0 + 0.5(7/8) = 3.44$ in.

$$\omega = 0.1\left(\frac{c_{max}}{c_{min}}\right) + 0.9 \leq 1.25$$

$$= 0.1\left(\frac{3.0}{3.0}\right) + 0.9 = 1.0 < 1.25$$

$K_{tr} = 0$ (no transverse reinforcement)

$$\frac{c\omega + K_{tr}}{d_b} = \frac{(3.44)(1.0) + 0}{7/8} = 3.93 < 4$$

$$\ell_d = \frac{\left(\frac{100{,}000}{3000^{1/4}} - 0.82(2400)(1.0)\right)(1.0)(1.0)(1.0)}{(0.82)(76.3)(3.93)}(7/8) = 41.1 \text{ in.} > 12 \text{ in.}$$

OK

Because $\ell_d = 41.4$ in. is less than the available embedment length in the short direction

$$\left(\frac{156}{2} - \frac{30}{2} - 3 = 60 \text{ in.}\right)$$

the No. 7 bars can be fully developed.

Use 12 No. 7 each way.

The current design using CS reinforcing bars represents a 30 percent reduction of the total area of flexural reinforcing steel provided in the footing compared with using Grade 60 steel (PCA 2005). Minimum reinforcement requirements for shrinkage and temperature govern the use of high-strength bars. Designers need to check for shear and punching shear. Strength requirements govern design using conventional steel.

APPENDIX B—FLEXURAL ANALYSIS USING NONLINEAR STRESS-STRAIN CURVE OF ASTM A1035/A1035M (CS) GRADE 100 (690) REINFORCEMENT

B.1—Introduction

The flexural strength of members reinforced with ASTM A1035/A1035M (CS) Grade 100 steel may be determined by using the nonlinear stress-strain curve of the reinforcement described in Chapter 3 of this guide.

A general solution is adopted for calculating the flexural strength of beam sections by a method of successive approximations for the neutral axis depth. This solution may be implemented by using an electronic spreadsheet, "439.6R-18 Flexural Analysis Spreadsheet.xlsx," which can be accessed online through the ACI Bookstore. Figure B.1 shows a flowchart for the solution by successive approximation for the general case of a doubly-reinforced beam. All variables with a prime correspond to the top reinforcement.

When using the nonlinear stress-strain curve of reinforcement in flexural design, the tensile stress in the reinforcement under service load may exceed the proportional limit. The impact of this higher tensile stress on member deflection and crack control should be considered.

Furthermore, when the longitudinal reinforcement has been designed according to the method in this appendix, the possible nonlinear response and higher strains of the longitudinal reinforcement may have significant impact on shear capacity and development length. Due to a lack of experimental data, it is recommended that members designed by this method should contain at least the minimum shear reinforcement specified by ACI 318. Desalegne and Lubell (2010, 2015) present shear design models for members with and without shear reinforcement where the longitudinal reinforcement is designed according to Appendix B. To fully develop the higher tensile stress f_s in the longitudinal reinforcement, the shear reinforcement may also be used as confinement reinforcement when determining the required development length of the longitudinal bars.

B.2—Design assumptions

The strength design of members for flexure and axial loads is based on ACI 318. However, when using ASTM A1035/A1035M (CS) Grade 100 steel bars, use Eq. (3.4a) to (3.4c) of this guide instead of ACI 318. In addition, the stress in compression reinforcement should not be taken greater than 80 ksi.

The design moment strength determined by using the nonlinear stress-strain curve of ASTM A1035/A1035M (CS) Grade 100 steel uses a strength reduction factor ϕ from 0.65 to 0.90, similar to the range given in ACI 318 and in Section 4.3 of this guide. ACI 318 defines the compression-controlled strain limit using the balanced strain conditions, which refers to the strain in the reinforcement at first yield. Similarly, consider the 0.2 percent offset yield strength of 100 ksi as the strain in the reinforcement at first yield. This strain occurs at 0.00424 according to Eq. (3.4b) and defines the compression-controlled strain limit. The tension-controlled strain limit is 0.0066 as discussed in Section 4.3 of this guide.

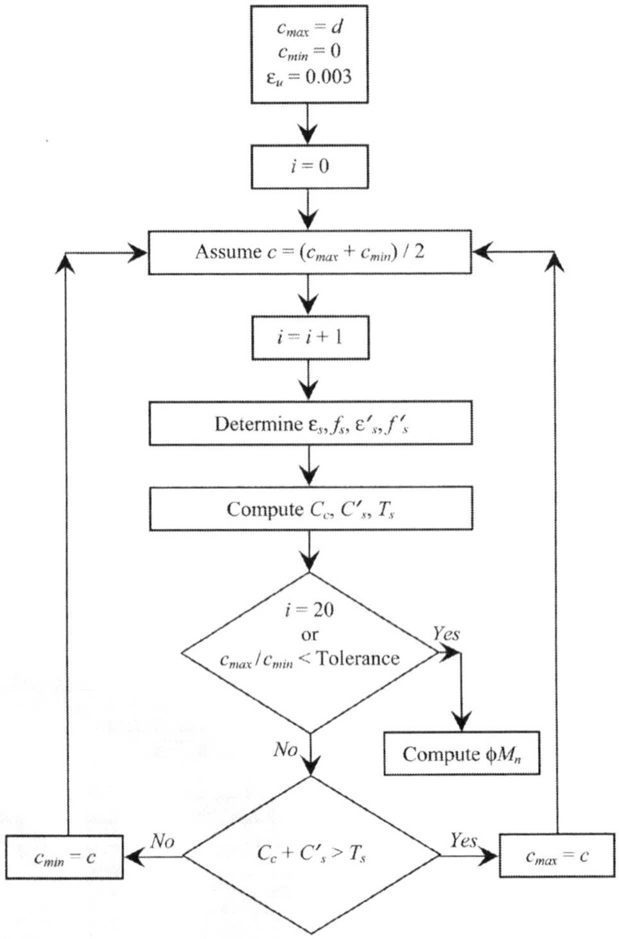

Fig. B.1—Flowchart for successive approximation solution for neutral axis, c.

In this appendix, a simple and conservative expression of ϕ is adopted after assigning $\phi = 0.65$ corresponding to a tensile strain of 0.0042 (f_s =100 ksi) and assigning $\phi = 0.90$ corresponding to a tensile strain of 0.0067 (f_s =120 ksi).

$$0.65 \leq (\phi = 0.23 + 100\varepsilon_t) \leq 0.9 \quad \text{(B.2)}$$

where the value of ε_t is positive for tension.

B.3—Spreadsheet implementation

An electronic spreadsheet is programmed to compute the flexural strength of a doubly-reinforced T-beam section as shown in Fig. B.3, where a singly-reinforced rectangular section corresponds to setting h_f and A_s' to zero. The spreadsheet, "439.6R-18 Flexural Analysis Spreadsheet.xlsx," can be accessed online through the ACI Bookstore. The spreadsheet is organized by examples with flexural strength calculations as shown in Figs. B.4a to B.4j, where all cell values are easily reproduced using hand calculations following the equations given in the following.

The spreadsheet contains two blocks of cells. Use the top block, Cells A1 to J19, to highlight the main input variables. These variables, except for f_c', correspond to the general T-beam cross section of Fig. B.3. Use the bottom block,

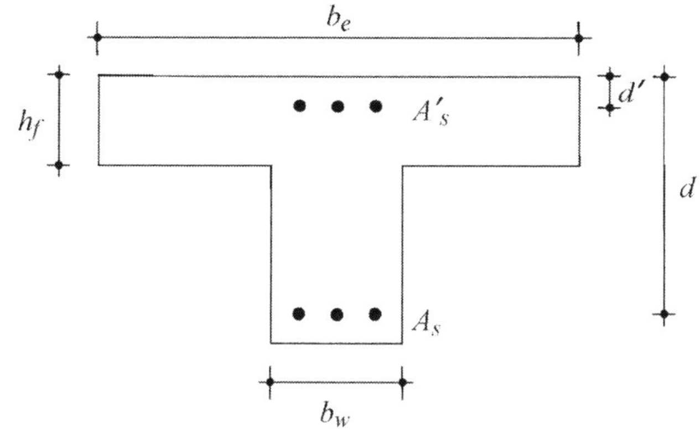

Fig. B.3—Doubly-reinforced T-beam section.

Cells A20 to J50, to perform the calculations that support the value of ϕM_n in Cell B47.

Trial values of c in Cells B25 to B44 correspond to the midpoint between c_{max} and c_{min} of Cells I25 and J25, to I44 and J44. The number of iterations is fixed to 20 even though solutions of reasonable accuracy are obtained with less iterations. After each iteration, update the values of c_{max} and c_{min} depending on the value for the summation of internal forces associated with the assumed c, as recorded by Cells G25 to G44 using Eq. (B.3a)

$$\sum \text{forces} = C_c + C_s' + T_s \tag{B.3a}$$

where

$$C_c = 0.85 f_c' \times b_e \times \beta_1 \times c \text{ (for } \beta_1 \times c \leq h_f) \tag{B.3b(a)}$$

$$C_c = 0.85 f_c'[(b_e - b_w)h_f + b_w \times \beta_1 \times c] \text{ (for } \beta_1 \times c > h_f) \tag{B.3b(b)}$$

$$C_s' = A_s' \times f_s' \tag{B.3c}$$

$$T_s = A_s \times f_s \tag{B.3d}$$

The variables are illustrated in Fig. B.3 and explained in Fig. B.4a to B.4j. Apply Eq. (3.4a), (3.4b) and (3.4c), as stated in Section B.2, to obtain the stress in the top reinforcement, f_s', and the stress in the bottom reinforcement, f_s. The strains ε_s' and ε_s to determine the stresses f_s' and f_s are based on Eq. (B.3e) and (B.3f)

$$\varepsilon_s' = \frac{\varepsilon_u}{c}(c - d') \tag{B.3e}$$

$$\varepsilon_s = \frac{\varepsilon_u}{c}(c - d) \tag{B.3f}$$

where ε_u is the usable compressive strain of concrete, defined as 0.003 in accordance with ACI 318. Compressive strains are taken as positive, therefore in Eq. (B.3a) compression forces are taken as positive. If the summation in Eq. (B.3a) results positive, reduce the value of c (and its value in the current iteration defines the new high bound c_{max}) in Cells I25 to I44. If the summation in Eq. (B.3a) results negative, increase the value of c (and its value in the current iteration defines the new low bound c_{min}) in Cells J25 to J44.

Use Eq. (B.2) and (B.3g(a)) to obtain the design flexural strength ϕM_n computed in Cell B47.

For $\beta_1 \times c \leq h_f$

$$M_n = 0.85 f_c' b_e \beta_1 c\left(d - \frac{\beta_1 c}{2}\right) + A_s' f_s'(d - d') \tag{B.3g(a)}$$

For $\beta_1 \times c > h_f$

$$M_n = 0.85 f_c'(b_e - b_w)h_f\left(d - \frac{h_f}{2}\right) + 0.85 f_c' b_w \beta_1 c\left(d - \frac{\beta_1 c}{2}\right) + A_s' f_s'(d - d') \tag{B.3g(b)}$$

B.4—Design examples

The spreadsheet described in Section B.3 is applied to several of the design examples in Appendix A. The spreadsheet, "439.6R-18 Flexural Analysis Spreadsheet.xlsx," can be accessed online through the ACI Bookstore. Figures B.4a to B.4j includes solutions to examples involving flexural strength calculations. Brief comments are made after each example. The results are summarized in Table B.4. It can be seen that by using the nonlinear analysis presented in this appendix, a saving of over 20 percent of the amount of reinforcement is realized in each case.

Example 4.1a

	A	B	C	D	E	F	G	H	I	J
1										
2		Analysis of doubly-reinforced beam sections:								
3										
4										
5	*units*	s		(*s* for standard or *m* for metric)						
6										
7	b_e =	0.0	in	width of effective compression flange (use 0 in rectangular sections)						
8	b_w =	10.0	in	web width						
9	h_f =	0.0	in	flange thickness (use 0 in rectangular sections)						
10	d =	13.50	in	effective depth, tension reinforcement						
11	d' =	0.00	in	effective depth, compression reinforcement						
12										
13	f'_c =	4.0	ksi	concrete strength						
14	$f_{y,A1035}$ =	100	ksi	yield strength of reinforcement (0.2% offset)						
15	E_s =	29000	ksi	steel modulus						
16										
17	A'_s =	0.00	in^2	area of compression reinforcement						
18	A_s =	2.37	in^2	area of tension reinforcement						
19										
20										
21	β_1 =	0.85			ACI 318-08 Section 10.2.7.3					
22	ε_u =	0.003			ACI 318-08 Section 10.2.3					
23										
24	*Iteration*	c	ε_s	f_s	ε'_s	f'_s	Σ *forces*		c_{max}	c_{min}
25	1	6.75	-0.0030	-82.2	0.0030	80.0	0.2		13.50	0.00
44	20	6.75	-0.0030	-82.3	0.0030	80.0	0.0		6.75	6.75
45										
46	ϕ =	0.65								
47	ϕM_n =	112.3	k-ft							
48										
49										
50										

Fig. B.4a—Example 4.1a.

The compression-controlled section has the tension reinforcement at a strain of approximately 0.0030 for an actual steel stress lower than the stress based on a linear stress-strain relationship, as assumed in Appendix A. The computed ϕM_n following Appendix B is 112.3 ft-kip compared with 114.2 ft-kip following Appendix A, a difference of less than 2 percent.

Example 4.1b

	A	B	C	D	E	F	G	H	I	J
1										
2		Analysis of doubly-reinforced beam sections:								
3										
4										
5	*units*	s		(*s* for standard or *m* for metric)						
6										
7	b_e =	0.0	in	width of effective compression flange (use 0 in rectangular sections)						
8	b_w =	10.0	in	web width						
9	h_f =	0.0	in	flange thickness (use 0 in rectangular sections)						
10	d =	13.50	in	effective depth, tension reinforcement						
11	d' =	0.00	in	effective depth, compression reinforcement						
12										
13	f'_c =	4.0	ksi	concrete strength						
14	$f_{y,A1035}$ =	100	ksi	yield strength of reinforcement (0.2% offset)						
15	E_s =	29000	ksi	steel modulus						
16										
17	A'_s =	0.00	in^2	area of compression reinforcement						
18	A_s =	1.58	in^2	area of tension reinforcement						
19										
20										
21	β_1 =	0.85			ACI 318-08 Section 10.2.7.3					
22	ε_u =	0.003			ACI 318-08 Section 10.2.3					
23										
24	*Iteration*	c	ε_s	f_s	ε'_s	f'_s	Σ *forces*		c_{max}	c_{min}
25	1	6.75	-0.0030	-82.2	0.0030	80.0	65.1		13.50	0.00
44	20	5.52	-0.0043	-101.0	0.0030	80.0	0.0		5.52	5.52
45										
46	ϕ =	0.66								
47	ϕM_n =	98.4	k-ft							
48										
49										
50										

Fig. B.4b—Example 4.1b.

A steel tensile stress of 101.0 ksi at a strain of 0.00433 balances the concrete compression forces. In this case, Appendix B gives a flexural strength of ϕM_n = 98.4 ft-kip, nearly identical to that obtained in Appendix A, where ϕM_n = 98.5 ft-kip. The solution in Appendix B uses a slightly lower ϕ factor.

Example 4.2a

	A	B	C	D	E	F	G	H	I	J
1										
2		Analysis of doubly-reinforced beam sections:								
3										
4										
5	*units*	s		(*s* for standard or *m* for metric)						
6										
7	b_e =	0.0	in	width of effective compression flange (use 0 in rectangular sections)						
8	b_w =	14.0	in	web width						
9	h_f =	0.0	in	flange thickness (use 0 in rectangular sections)						
10	d =	20.50	in	effective depth, tension reinforcement						
11	d' =	2.50	in	effective depth, compression reinforcement						
12										
13	f'_c =	4.0	ksi	concrete strength						
14	$f_{y,A1035}$ =	100	ksi	yield strength of reinforcement (0.2% offset)						
15	E_s =	29000	ksi	steel modulus						
16										
17	A'_s =	2.44	in^2	area of compression reinforcement						
18	A_s =	3.21	in^2	area of tension reinforcement						
19										
20										
21	β_1 =	0.85			ACI 318-08 Section 10.2.7.3					
22	ε_u =	0.003			ACI 318-08 Section 10.2.3					
23										
24	*Iteration*	c	ε_s	f_s	ε'_s	f'_s	Σ *forces*		c_{max}	c_{min}
25	1	10.25	-0.0030	-82.2	0.0023	65.8	311.4		20.50	0.00
44	20	6.34	-0.0067	-120.0	0.0018	52.7	0.0		6.34	6.34
45										
46	ϕ =	0.90								
47	ϕM_n =	516.0 k-ft								
48										
49										
50										

Fig. B.4c—Example 4.2a.

The doubly-reinforced rectangular section is designed so that the tension reinforcement is at a strain of 0.0067 with a corresponding steel stress of 120 ksi instead of the 100 ksi assumed in Appendix A. The solution in Appendix B uses compression reinforcement at a strain of 0.00182 for a stress of 52.7 ksi, whereas Appendix A uses a strain of 0.00154 for a stress of 44.6 ksi. The total area of steel, $A_s' + A_s$, of 5.65 in.^{2} obtained in Appendix B represents a 26 percent reduction compared with the 7.59 in.^{2} obtained in Appendix A.

Example 4.2b

	A	B	C	D	E	F	G	H	I	J
1										
2		Analysis of doubly-reinforced beam sections:								
3										
4										
5	*units*	s		(s for standard or m for metric)						
6										
7	b_e =	0.0	in	width of effective compression flange (use 0 in rectangular sections)						
8	b_w =	14.0	in	web width						
9	h_f =	0.0	in	flange thickness (use 0 in rectangular sections)						
10	d =	20.50	in	effective depth, tension reinforcement						
11	d' =	2.50	in	effective depth, compression reinforcement						
12										
13	f'_c =	8.0	ksi	concrete strength						
14	$f_{y,A1035}$ =	100	ksi	yield strength of reinforcement (0.2% offset)						
15	E_s =	29000	ksi	steel modulus						
16										
17	A'_s =	0.00	in^2	area of compression reinforcement						
18	A_s =	3.01	in^2	area of tension reinforcement						
19										
20										
21	β_1 =	0.65			ACI 318-08 Section 10.2.7.3					
22	ε_u =	0.003			ACI 318-08 Section 10.2.3					
23										
24	*Iteration*	c	ε_s	f_s	ε'_s	f'_s	Σ *forces*		c_{max}	c_{min}
25	1	10.25	-0.0030	-82.2	0.0023	65.8	386.7		20.50	0.00
44	20	5.99	-0.0073	-123.1	0.0017	50.7	0.0		5.99	5.99
45										
46	ϕ =	0.90								
47	ϕM_n =	515.6 k-ft								
48										
49										
50										

Fig. B.4d—Example 4.2b.

This example uses 8 ksi concrete on the same cross section of Example 4.2a. Appendix A requires compression reinforcement for a total amount of steel, $A_s' + A_s$, of 4.79 in^2. The solution in Appendix B, where A_s = 3.01 in.^{2} without need of compression reinforcement, offers a steel reduction of 37 percent.

Example 4.3

	A	B	C	D	E	F	G	H	I	J
1										
2		Analysis of doubly-reinforced beam sections:								
3										
4										
5	*units*	*s*		(*s* for standard or *m* for metric)						
6										
7	b_e =	0.0	in	width of effective compression flange (use 0 in rectangular sections)						
8	b_w =	10.0	in	web width						
9	h_f =	0.0	in	flange thickness (use 0 in rectangular sections)						
10	d =	16.50	in	effective depth, tension reinforcement						
11	d' =	0.00	in	effective depth, compression reinforcement						
12										
13	f'_c =	4.0	ksi	concrete strength						
14	$f_{y,A1035}$ =	100	ksi	yield strength of reinforcement (0.2% offset)						
15	E_s =	29000	ksi	steel modulus						
16										
17	A'_s =	0.00	in^2	area of compression reinforcement						
18	A_s =	0.82	in^2	area of tension reinforcement						
19										
20										
21	β_1 =	0.85			ACI 318-08 Section 10.2.7.3					
22	ε_u =	0.003			ACI 318-08 Section 10.2.3					
23										
24	*Iteration*	c	ε_s	f_s	ε'_s	f'_s	Σ *forces*		c_{max}	c_{min}
25	1	8.25	-0.0030	-82.2	0.0030	80.0	171.0		16.50	0.00
44	20	3.80	-0.0100	-133.9	0.0030	80.0	0.0		3.80	3.80
45										
46	ϕ =	0.90								
47	ϕM_n =	122.6 k-ft								
48										
49										
50										

Fig. B.4e—Example 4.3.

For a target flexural strength of ϕM_n = 123 ft-kip, Appendix A requires A_s = 1.10 in.^{2} Using the same beam dimensions, Appendix B requires A_s = 0.82 in.^{2}, a 25 percent reduction.

Example 4.4

	A	B	C	D	E	F	G	H	I	J
1										
2		Analysis of doubly-reinforced beam sections:								
3										
4										
5	*units*	*s*		(*s* for standard or *m* for metric)						
6										
7	b_e =	0.0	in	width of effective compression flange (use 0 in rectangular sections)						
8	b_w =	12.0	in	web width						
9	h_f =	0.0	in	flange thickness (use 0 in rectangular sections)						
10	d =	30.00	in	effective depth, tension reinforcement						
11	d' =	2.50	in	effective depth, compression reinforcement						
12										
13	f'_c =	4.0	ksi	concrete strength						
14	$f_{y,A1035}$ =	100	ksi	yield strength of reinforcement (0.2% offset)						
15	E_s =	29000	ksi	steel modulus						
16										
17	A'_s =	1.28	in^2	area of compression reinforcement						
18	A_s =	3.36	in^2	area of tension reinforcement						
19										
20										
21	β_1 =	0.85			ACI 318-08 Section 10.2.7.3					
22	ε_u =	0.003			ACI 318-08 Section 10.2.3					
23										
24	*Iteration*	c	ε_s	f_s	ε'_s	f'_s	Σ *forces*		c_{max}	c_{min}
25	1	15.00	-0.0030	-82.2	0.0025	72.3	336.4		30.00	0.00
44	20	9.28	-0.0067	-120.0	0.0022	63.6	0.0		9.28	9.28
45										
46	ϕ =	0.90								
47	ϕM_n =	796.6 k-ft								
48										
49										
50										

Fig. B.4f—Example 4.4.

For the doubly-reinforced rectangular section designed for ϕM_n = 796 ft-kip while attaining a tension-controlled strain limit, Appendix A requires a total steel area, $A_s' + A_s$, of 6.21 in.^{2} with a tensile steel strain of 0.009 and a compression steel strain of 0.002. The solution in Appendix B requires a total steel area, $A_s' + A_s$, of 4.64 in.^{2} with a tensile steel strain of 0.0067 and a compression steel strain of 0.00219. This represents a 25 percent reduction of steel with respect to Appendix A.

Example 4.5

	A	B	C	D	E	F	G	H	I	J
1										
2		Analysis of doubly-reinforced beam sections:								
3										
4										
5	*units*	s		(*s* for standard or *m* for metric)						
6										
7	b_e =	30.0	in	width of effective compression flange (use 0 in rectangular sections)						
8	b_w =	10.0	in	web width						
9	h_f =	2.5	in	flange thickness (use 0 in rectangular sections)						
10	d =	19.00	in	effective depth, tension reinforcement						
11	d' =	0.00	in	effective depth, compression reinforcement						
12										
13	f'_c =	4.0	ksi	concrete strength						
14	$f_{y,A1035}$ =	100	ksi	yield strength of reinforcement (0.2% offset)						
15	E_s =	29000	ksi	steel modulus						
16										
17	A'_s =	0.00	in^2	area of compression reinforcement						
18	A_s =	1.11	in^2	area of tension reinforcement						
19										
20										
21	β_1 =	0.85			ACI 318-08 Section 10.2.7.3					
22	ε_u =	0.003			ACI 318-08 Section 10.2.3					
23										
24	*Iteration*	c	ε_s	f_s	ε'_s	f'_s	Σ *forces*		c_{max}	c_{min}
25	1	9.50	-0.0030	-82.2	0.0030	80.0	353.3		19.00	0.00
44	20	1.92	-0.0267	-150.0	0.0030	80.0	0.0		1.92	1.92
45										
46	ϕ =	0.90								
47	ϕM_n =	227.1	k-ft							
48										
49										
50										

Fig. B.4g—Example 4.5.

A singly-reinforced T-beam section is designed for $\phi M_n = 227$ ft-kip. The beam requires 1.66 in.^{2} according to Appendix A. The steel area of 1.11 in.^{2} obtained in Appendix B, with a tensile steel strain of 0.0267 and a steel stress of 150 ksi, represents a 33 percent reduction in steel area. The strain exceeds the 0.015 threshold indicated in Section 4.2 of this design guide. Cracking and deflections may be critical and should be investigated.

Example 4.6

	A	B	C	D	E	F	G	H	I	J
1										
2		Analysis of doubly-reinforced beam sections:								
3										
4										
5	*units*	*s*		(*s* for standard or *m* for metric)						
6										
7	b_e =	30.0	in	width of effective compression flange (use 0 in rectangular sections)						
8	b_w =	10.0	in	web width						
9	h_f =	2.5	in	flange thickness (use 0 in rectangular sections)						
10	d =	19.00	in	effective depth, tension reinforcement						
11	d' =	0.00	in	effective depth, compression reinforcement						
12										
13	f'_c =	4.0	ksi	concrete strength						
14	$f_{y,A1035}$ =	100	ksi	yield strength of reinforcement (0.2% offset)						
15	E_s =	29000	ksi	steel modulus						
16										
17	A'_s =	0.00	in^2	area of compression reinforcement						
18	A_s =	2.34	in^2	area of tension reinforcement						
19										
20										
21	β_1 =	0.85			ACI 318-08 Section 10.2.7.3					
22	ε_u =	0.003			ACI 318-08 Section 10.2.3					
23										
24	*Iteration*	c	ε_s	f_s	ε'_s	f'_s	Σ *forces*		c_{max}	c_{min}
25	1	9.50	-0.0030	-82.2	0.0030	80.0	252.1		19.00	0.00
44	20	4.71	-0.0091	-130.9	0.0030	80.0	0.0		4.71	4.71
45										
46	ϕ =	0.90								
47	ϕM_n =	400.0 k-ft								
48										
49										
50										

Fig. B.4h—Example 4.6.

The steel area required for ϕM_n = 400 ft-kip is 2.34 in.^{2} according to Appendix B, whereas Appendix A requires 3.06 in.^{2}, a reduction of 24 percent.

Example 4.6(SI)

	A	B	C	D	E	F	G	H	I	J
1										
2		Analysis of doubly-reinforced beam sections:								
3										
4										
5	*units*	*m*		(*s* for standard or *m* for metric)						
6										
7	b_e =	762.0	mm	width of effective compression flange (use 0 in rectangular sections)						
8	b_w =	254.0	mm	web width						
9	h_f =	63.5	mm	flange thickness (use 0 in rectangular sections)						
10	d =	482.6	mm	effective depth, tension reinforcement						
11	d' =	0.0	mm	effective depth, compression reinforcement						
12										
13	f'_c =	27.6	MPa	concrete strength						
14	$f_{y,A1035}$ =	690	MPa	yield strength of reinforcement (0.2% offset)						
15	E_s =	200000	MPa	steel modulus						
16										
17	A'_s =	0.00	mm^2	area of compression reinforcement						
18	A_s =	1510	mm^2	area of tension reinforcement						
19										
20										
21	β_1 =	0.85			ACI 318-08 Section 10.2.7.3					
22	ε_u =	0.003			ACI 318-08 Section 10.2.3					
23										
24	*Iteration*	c	ε_s	f_s	ε'_s	f'_s	Σ *forces*		c_{max}	c_{min}
25	1	241.30	-0.0030	-567.2	0.0030	551.7	1121.7		482.60	0.00
44	20	119.75	-0.0091	-902.6	0.0030	551.7	0.0		119.75	119.75
45										
46	ϕ =	0.90								
47	ϕM_n =	542.5 kN-m								
48										
49										
50										

Fig. B.4i—Example 4.6(SI).

The steel area required for ϕM_n = 542 kN-m is 1510 mm^2 according to Appendix B, whereas Appendix A requires 1970 mm^2, a reduction of 24 percent.

Example 6.3

	A	B	C	D	E	F	G	H	I	J
1										
2		Analysis of doubly-reinforced beam sections:								
3										
4										
5	*units*	*s*		(*s* for standard or *m* for metric)						
6										
7	b_e =	0.0	in	width of effective compression flange (use 0 in rectangular sections)						
8	b_w =	36.0	in	web width						
9	h_f =	0.0	in	flange thickness (use 0 in rectangular sections)						
10	d =	18.50	in	effective depth, tension reinforcement						
11	d' =	0.00	in	effective depth, compression reinforcement						
12										
13	f'_c =	4.0	ksi	concrete strength						
14	$f_{y,A1035}$ =	100	ksi	yield strength of reinforcement (0.2% offset)						
15	E_s =	29000	ksi	steel modulus						
16										
17	A'_s =	0.00	in^2	area of compression reinforcement						
18	A_s =	3.87	in^2	area of tension reinforcement						
19										
20										
21	β_1 =	0.85		ACI 318-08 Section 10.2.7.3						
22	ε_u =	0.003		ACI 318-08 Section 10.2.3						
23										
24	*Iteration*	c	ε_s	f_s	ε'_s	f'_s	Σ *forces*		c_{max}	c_{min}
25	1	9.25	-0.0030	-82.2	0.0030	80.0	644.1		18.50	0.00
44	20	4.80	-0.0086	-128.9	0.0030	80.0	0.0		4.80	4.80
45										
46	ϕ =	0.90								
47	ϕM_n =	616.1 k-ft								
48										
49										
50										

Fig. B.4j—Example 6.3.

The negative moment on the one-way joist needs a flexural strength of ϕM_n = 616 ft-kip. Appendix A gives a solution with A_s = 4.98 in.^{2}, whereas Appendix B requires A_s = 3.87 in.^{2} for 22 percent reduction in steel area.

Table B.4—Summary of required flexural reinforcement in selected examples

Example	Appendix A, in.^{2}	Appendix B, in.^{2}	Ratio B/A
4.2a	7.59	5.65	0.74
4.2b	4.79	3.01	0.63
4.3	1.10	0.82	0.75
4.4	6.21	4.64	0.75
4.5	1.66	1.11	0.67
4.6	3.06	2.34	0.76
6.3	4.98	3.87	0.78

APPENDIX C—FLEXURAL BEHAVIOR OF BEAMS REINFORCED WITH ASTM A1035/A1035M BARS

The stress-strain curve of ASTM A1035/A1035M (CS) Grade 100 (690) bar differs from that of ASTM A615/A615M Grade 60 (420) bar in that the former is without a well-defined yield plateau. The Grade 100 (690) bar has a specified minimum yield strength of 100,000 psi (690 MPa) as determined by the 0.2% offset method, whereas the Grade 60 (420) bar has a specified minimum yield strength of 60,000 psi (410 MPa), usually determined by observation of a distinct yield point or knee in the stress-strain curve. These differences affect the flexural behavior of beams significantly.

Figure C.1 shows the load-deflection curves obtained by Yotakhong (2003) from flexural tests of three 12 x 18 in. (300 x 450 mm) beams over a simple span of 15 ft (4.6 m) with two-point loading producing a constant moment region at midspan. The number of bars and the grade of steel of the tension reinforcement are shown in the figure. Each beam contained two No. 4 Grade 60 (420) bars as compression reinforcement. The beams also contained No. 4 Grade 60 (420) stirrups at 8 in. (200 mm) on center within the shear spans and at 9 in. (225 mm) on center in the constant moment region.

Before the initial flexural cracking occurred at an applied load of approximately 8 kip (36 kN), the behavior of all three beams was virtually identical. After the initial cracking, the load-deflection response of all beams was nearly linear. Beams A and C showed similar stiffness after cracking because both contained three No. 6 bottom bars. Beam B, however, deflected more than Beams A and C at a similar load because it contained only two No. 6 bottom bars and, therefore, the stiffness was less, the neutral axis depth was smaller, and the bar stress was 50 percent higher.

At the load level of 33 kip (147 kN), the reinforcement in Beam A reached its yield strength and the beam deflection continued to increase with only a slight increase in loading. Failure of Beam A occurred when the concrete reached the maximum measured strain of 0.004 in the extreme compression fiber. At this same load level of 33 kip (147 kN), however, Beam B had deflected more than Beam A because Beam B contained only two No. 6 bottom bars and the bar was stressed to 96,000 psi (662 MPa) compared with 64,000 psi (441 MPa) for Beam A. Being reinforced with Grade 100 (690) bars that lack a well-defined yield plateau, both Beams B and C continued to carry more load with increasing deflections and decreasing stiffness (as the reinforcement stress progressed into the inelastic range) until failure occurred when the concrete reached the measured maximum strain of 0.003 in the extreme compression fiber. All three beams failed in flexure with crushing of the concrete near midspan.

Yotakhong (2003) compared the test results with the theoretical predictions based on the actual stress-strain relationships of the reinforcements. The comparisons are close, as shown in Table C.1.

It is worth noting that both Beams B and C showed considerable deflection before failure, even though the deflection at failure was less than that of Beam A. Furthermore, both beams carried proportionally more load than Beam A, which was reinforced with Grade 60 (420) bars.

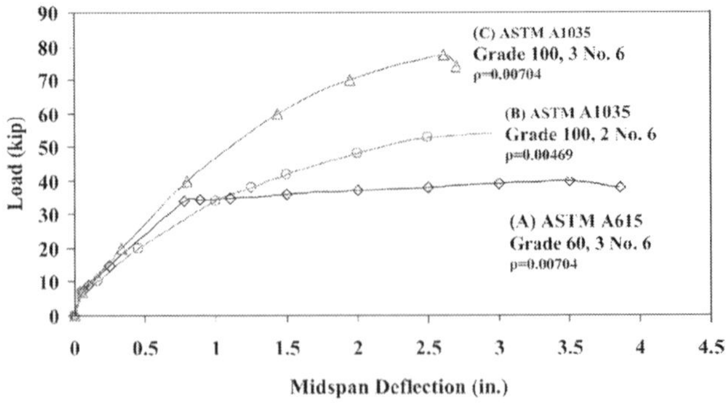

Fig. C.1—Load-deflection curves of beams reinforced with ASTM A1035/A1035M (CS) and ASTM A615/A615M reinforcing bars (Yotakhong 2003). (Note: 1 kip = 4.45 kN; 1 in. = 25.4 mm.).

The flexural behavior of the beams described previously also indicates the difference in the internal moment-resisting mechanism between Beam A reinforced with Grade 60 (420) bars and Beams B and C reinforced with Grade 100 (690) bars. Initially, under increasing load, the internal forces T (tension in the reinforcement) and C (compression in the concrete) of all three beams gradually increase and the lever arm between T and C also slowly increases as the neutral axis moves slowly toward the extreme compression fiber, gradually reducing the area of the compression zone. At the higher load near failure, the internal forces T and C of Beam A must remain constant because T has reached its yield value, so any increase in moment is accompanied with an increase in the lever arm between T and C, causing the neutral axis to move quickly closer to the extreme compression fiber, reducing the area of compression block, and producing larger curvature and deflection. On the other hand, for Beams B and C under increasing load toward failure, the internal forces T and C continue to increase (because the reinforcement lacks a yield plateau) without the need of a pronounced increase in the lever arm between T and C, thus producing smaller curvatures and deflections for a given applied load.

Figure C.2 shows the relationship for a singly reinforced rectangular beam between the nominal flexural strength M_n and reinforcement ratio ρ for three different reinforcement yield strengths f_y: 60,000 psi (420 MPa) for Grade 60 (420) bar, 80,000 psi (550 MPa) for the upper limit permitted by ACI 318, and 100,000 psi (690 MPa) for the simplified design method (Mast et al. 2008) recommended in this guide. Also shown in the figure is the theoretical prediction obtained by using a nonlinear stress-strain relationship for ASTM A1035/A1035M (CS) Grade 100 (690) reinforcement closely similar to Eq. (3.4a), (3.4b), and (3.4c) recommended in this guide.

It is important to note that each curve consists of two segments connected at the point corresponding to the balanced reinforcement ratio ρ_b. Tension-controlled failure of a beam occurs when its reinforcement ratio is less than ρ_b, whereas compression-controlled failure of a beam occurs when its reinforcement ratio is greater than ρ_b. As the reinforcement yield strength f_y increases, the balanced reinforcement ratio is greatly reduced. For beams with tension-

Table C.1—Comparisons between measured and predicted ultimate load and the corresponding deflection of beams tested by Yotakhong (2003)

Beam ID	Ultimate load			Deflection at ultimate load		
	Measured, kip (kN) (a)	Calculated, kip (kN) (b)	(a)/(b)	Measured, in. (mm) (c)	Calculated, in. (mm) (d)	(c)/(d)
Beam A	40.7 (181)	41 (182)	0.99	3.87 (98)	3.1 (79)	1.25
Beam B	54.7 (243)	58 (258)	0.94	3.1 (79)	3.1 (79)	1.0
Beam C	77.9 (347)	80 (356)	0.97	2.7 (69)	2.5 (64)	1.08

Table C.2—Details of flexural tests by Malhas (2002) and Ansley (2002)

Beam ID	Bottom bars	Top bars	Stirrups	f_c', psi (MPa)	Measured M_n, ft-kip (m-kN) (a)	Calculated M_n, ft-kip (m-kN) (b)	(a)/(b)
Malhas (2002) tests							
A1	Two No. 8	Two No. 4	No. 4 at 6 in.	6200 (43)	227 (309)	194 (264)	1.17
A2	Four No. 4	Two No. 4	No. 4 at 6 in.	6200 (43)	145 (197)	102 (139)	1.42
A3	Two No. 4	Two No. 4	No. 4 at 6 in.	6200 (43)	73 (99)	54 (73)	1.35
Ansley (2002) test							
1	Two No. 6	None	None	6683 (46)	202.5 (275)	112.6 (153)	1.80

controlled failure, the nominal flexural strength is governed by the reinforcement yield strength; therefore, as the yield strength increases, the flexural strength increases substantially. On the other hand, for beams with compression-controlled failure, the nominal flexural strength is governed by the concrete strength; therefore, the three curves merge into a single curve for large values of ρ.

Also shown in Fig. C.2 are flexural test results obtained by Malhas (2002) of the University of North Florida and by Ansley (2002) of the Florida Department of Transportation, both using 12 x 18 in. (300 x 450 mm) beams reinforced with ASTM A1035/A1035M (CS) bars. The details of these test beams are shown in Table C.2. The calculated nominal flexural strength M_n given in the table is based on the simplified method recommended in this guide.

It can be seen from Table C.2 that the test results exceeded the predictions using the simplified method by a substantial margin. Furthermore, Fig. C.2 shows that the test results compared closely with the theoretical predictions from nonlinear analysis. These comparisons suggest that it is justifiable to use the ϕ factor specified by ACI 318 for the simplified design method. Additionally, because the failure is dominated by yielding of tension reinforcement, reliability of the calculated flexural strength is directly related to reliability of the steel properties of ASTM A1035/A1035M (CS) bars, which are comparable to reliability of the steel properties of ASTM A615/A615M bars.

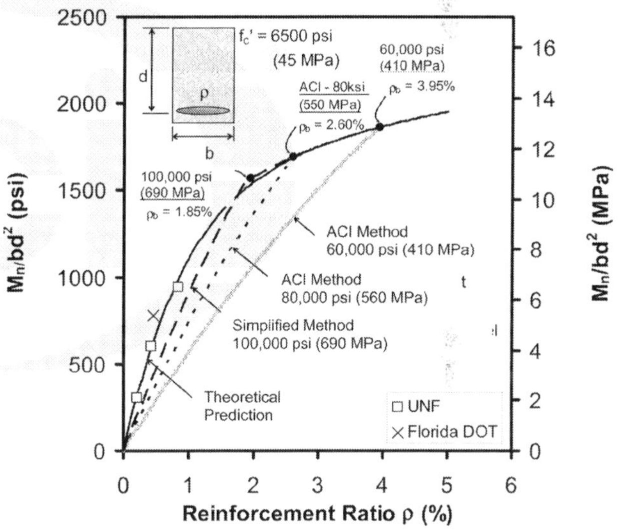

Fig. C.2—Comparisons of nominal flexural strength versus reinforcement ratio for different reinforcement yield strength (Mast et al. 2008).

Designers using ASTM A1035/A1035M (CS) should check for cracking and deflection of members being designed.

Report on Fiber-Reinforced Polymer (FRP) Reinforcement for Concrete Structures

Reported by ACI Committee 440

John P. Busel Chair	Carol K. Shield Secretary

Tarek Alkhrdaji	Russell Gentry	James G. Korff	Hayder A. Rasheed
Charles E. Bakis	Janos Gergely	Michael W. Lee	Sami H. Rizkalla
P. N. Balaguru	William J. Gold	Ibrahim M. Mahfouz	Morris Schupack
Lawrence C. Bank	Nabil F. Grace	Orange S. Marshall	Rajan Sen
Abdeldjelil Belarbi	Mark F. Green*	Amir Mirmiran	Khaled A. Soudki
Brahim Benmokrane	Zareh B. Gregorian	Ayman S. Mosallam	Samuel A. Steere III
Gregg J. Blaszak	Doug D. Gremel	John J. Myers	Gamil S. Tadros
Timothy E. Bradberry	H. R. Trey Hamilton III	Antonio Nanni	Jay Thomas
Gordon L. Brown, Jr.	Isaam E. Harik	Kenneth Neale	Houssam A. Toutanji*
Vicki L. Brown	Kent A. Harries	John P. Newhook	J. Gustavo Tumialan
Raafat El-Hacha	Mark P. Henderson	Max L. Porter	Milan Vatovec
Garth J. Fallis	Bohdan N. Horeczko	Mark A. Postma	Stephanie L. Walkup
Amir Z. Fam	Vistasp M. Karbhari	Andrea Prota	David White
Edward R. Fyfe			

*Subcommittee co-chairs responsible for preparing the report.

Several voting members of the committee contributed chapters or made other substantial contributions to the report. In addition, the committee would like to acknowledge the contribution of associate members T. Ivan Campbell and Luke A. Bisby.

Applications of fiber-reinforced polymer (FRP) composites as reinforcement for concrete structures have been growing rapidly in recent years. ACI Committee 440 has published design guidelines for internal FRP reinforcement, externally bonded FRP reinforcement for strengthening, prestressed FRP reinforcement, and test methods for FRP products. Although these guidelines exist, new products and applications continue to be developed. Thus, this report summarizes the current state of knowledge on these materials and their application to concrete and masonry structures. The purpose of this report is to act as an introduction to FRP materials in areas where ACI guides exist, and to provide information on the properties and behavior of concrete structures containing FRP in areas where guides are not currently available. If an ACI guide is available, the guide document supersedes information in this report, and the guide should always be followed for design and application purposes. ACI Committee 440 is also in the process of developing new guides and thus the current availability of guides should be checked by the reader. In addition to the material properties of the constituent materials (that is, resins and fibers) and products, current knowledge of FRP applications, such as internal reinforcement including prestressing, external strengthening of concrete and masonry structures, and structural systems, is discussed in detail. The document also addresses durability issues and the effects of extreme events, such as fire and blast. A summary of some examples of field applications is presented.

Keywords: aramid fibers; blast; bridges; buildings; carbon fibers; composite materials; corrosion; design; dowels; ductility; durability; external reinforcement; fatigue; fiber-reinforced polymer (FRP); fibers; fire; glass fiber; masonry; mechanical properties; polymer resin; prestressed concrete; seismic; stay-in-place forms; structural systems; test methods.

CONTENTS

ACI 440R-07 supersedes ACI 440R-96 and was adopted and published September 2007.

CHAPTER 1—INTRODUCTION AND SCOPE

1.1—Introduction

The purpose of this report is to present the current state of knowledge with regard to applications of fiber-reinforced polymer (FRP) materials in concrete. This report summarizes the fundamental behavior, the most current research, design codes, and practical applications of concrete and masonry structures containing FRP. This document is intended to complement other reports (for example, standards and design guidelines) produced by ACI Committee 440, either by summarizing the research that supports those documents or by providing information on future developments of those documents. If an ACI guide is available, the guide document supersedes information in this report, and the guide should always be followed for design and application purposes. ACI Committee 440 is also in the process of developing new guides; thus, the current availability of guides should be checked by the reader.

FRP materials are composite materials that typically consist of strong fibers embedded in a resin matrix. The fibers provide strength and stiffness to the composite and generally carry most of the applied loads. The matrix acts to bond and protect the fibers and to provide for transfer of stress from fiber to fiber through shear stresses. The most common fibers are glass, carbon, and aramid. Matrixes are typically epoxies, polyesters, vinylesters, or phenolics.

1.2—Historical perspective of FRP composites

While the concept of composites has been in existence for several millennia (for example, bricks made from mud and straw), the incorporation of FRP composite technology into the industrial world is less than a century old. The age of plastics emerged just after 1900, with chemists and industrialists taking bold steps to have plastics (vinyl, polystyrene, and Plexiglas) mimic and outdo natural materials. Spurred on by the needs of electronics, defense, and eventually space technologies, researchers created materials with properties that seemed to defy known principles, such as bullet-stopping Kevlar. The first known FRP product was a boat hull manufactured in the mid-1930s as part of a manufacturing experiment using a fiberglass fabric and polyester resin laid in a foam mold (ACMA MDA 2006). From this modest beginning, FRP composite applications have revolutionized entire industries, including aerospace, marine, electrical, corrosion resistance, and transportation.

FRP composite materials date back to the early 1940s in the defense industry, particularly for use in aerospace and naval applications. The U.S. Air Force and Navy capitalized on FRP composites' high strength-weight ratio and inherent resistance to the corrosive effects of weather, salt air, and the sea. Soon the benefits of FRP composites, especially its corrosion resistance capabilities, were communicated to the public sector. Fiberglass pipe, for instance, was first introduced in 1948 (ACMA MDA 2006) for what has become one of its widest use areas within the corrosion market, the oil industry. FRP composites proved to be a worthy alternative to other traditional materials even in the high-pressure, large-diameter situations of chemical processing. Besides superior corrosion resistance, FRP pipe offered both durability and strength, thus eliminating the need for interior linings, exterior coatings, and cathodic protection. Since the early 1950s, FRP composites have been used extensively for equipment in the chemical processing, pulp and paper, power, waste treatment, metal refining, and other manufacturing industries (ACMA MDA 2006). Myriads of products and FRP installations help build a baseline of proven performance in the field.

The decades after the 1940s brought new, and often revolutionary, applications for FRP composites (ACMA MDA 2006). The same technology that produced the reinforced plastic hoops required for the Manhattan nuclear project in World War II spawned the development of high-performance composite materials for solid rocket motor cases and tanks in the 1960s and 1970s. In fact, fiberglass wall tanks were used on the Skylab orbiting laboratory to provide oxygen for the astronauts. In 1953, the first Chevrolet Corvette with fiberglass body panels rolled off the assembly line (ACMA MDA 2006). Now, high-performance race cars are the proving ground for technology transfer to passenger vehicles. In the 1960s, the British and U.S. Navies were simultaneously developing FRP-based minesweeper ships because FRP composites are not only superior to other materials in harsh marine environments, they are also nonmagnetic. It was also noticed at that time that one of the features of FRP is the ability of the materials to reduce the radar signature of the structure, such as a ship or an aircraft. High-performance composite materials have been demonstrated in advanced technology aircraft such as the F-117 Stealth Fighter and B-2 Bomber. Currently, FRP composites are being used for space applications and are involved in several NASA test initiatives (ACMA MDA 2006).

While the majority of the historical and durability data of FRP composite installations comes from the aerospace, marine, and corrosion-resistance industries (ACMA MDA 2006), FRP composites have been used as a construction material for several decades. FRP composite products were first demonstrated to reinforce concrete structures in the mid-1950s (ACMA MDA 2006). In the 1980s, a resurgence in interest arose when new developments were launched to apply FRP reinforcing bars in concrete that required special performance requirements such as nonmagnetic properties or in areas that were subjected to severe chemical attack.

Composites have evolved since the 1950s, starting with temporary structures and continuing with restoration of historic buildings and structural applications. Typical products developed were domes, shrouds, translucent sheet panels, and exterior building panels. A major development of FRP for civil engineering has been the application of externally bonded FRP for rehabilitation and strengthening of concrete structures.

During the late 1970s and early 1980s, many applications of composite reinforcing products were demonstrated in Europe and Asia. In 1986, the world's first highway bridge using composite reinforcing tendons was built in Germany. The first all-composite bridge deck was demonstrated in China. The first all-composite pedestrian bridge was installed in 1992 in Aberfeldy, Scotland. In the U.S., the first FRP-reinforced concrete bridge deck was built in 1996 at McKinleyville, West Virginia, followed by the first all-composite vehicular bridge deck (The No-Name Creek Bridge in 1996) in Russell, Kansas. Numerous composite pedestrian bridges have been installed in U.S. state and national parks in remote locations not accessible by heavy construction equipment, or for spanning over roadways and railways (ACMA MDA 2006).

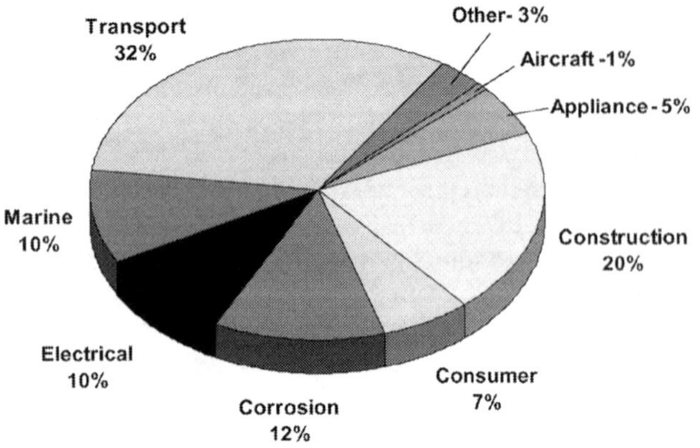

Fig. 1.1—U.S. composites shipments 2004 market share by end-use application (ACMA 2005). Total estimated volume: 1.8 billion kg (4.0 billion lb) (ACMA 2005 and PPG).

Composite fabricators and suppliers are actively developing products for the civil infrastructure, which is considered to be the largest potential market for FRP composites (ACMA MDA 2006). Concrete repair and reinforcement, bridge deck repair and new installation, composite-hybrid technology (the marriage of composites with concrete, wood, and steel), marine piling, and pier upgrade programs are just some of the areas that are currently being explored. This document describes all aspects of applications of FRP composites for concrete and masonry structures including internal reinforcement, strengthening, prestressing, and integrated stay-in-place forms.

1.2.1 *Industry statistics*—The composites industry associations and producers have traditionally tracked FRP market growth in several primary markets: aircraft/aerospace, appliance/business equipment, construction, consumer products, corrosion-resistant equipment, electrical, marine, transportation, and other applications. The American Composites Manufacturers Association (ACMA 2005) and PPG Industries reported that estimates of composites shipments in 2004 reached 1.8 billion kilograms (4.0 billion pounds). Figure 1.1 shows the distribution of FRP composites materials shipped in 2004. According to the *Composites News International* (ACMA 2005), the estimated size of the composites industry in North America is approximately $9 billion. The composites industry has shown considerable growth over the past 10 years, and is projected to increase as FRP composites are accepted in new markets.

1.2.2 *Product and benefits for construction applications*—FRP composites provide many solutions to the needs of the owner and civil engineer. FRP products for civil infrastructure/construction applications are more resistant to corrosion than reinforcing steel, thus the service life of the structure may be increased. FRP products have high strength-to-weight ratios and strength properties greater than those of steel. In repair and rehabilitation, the light weight and ease of application of the materials can result in saving in labor costs. The main drawback of the materials is their relatively high material cost.

Currently, many FRP products are available to build or repair civil engineering structures. These products have been extensively demonstrated and used around the world. Examples of FRP composite products include:

- FRP composite systems for repair, strengthening, and seismic retrofit for beams, columns, slabs, and walls;
- FRP reinforcing bars, grids, and tendons for concrete reinforcement;
- Bridge deck panels and pedestrian bridge systems;
- New structural shapes;
- Piling products and systems for marine waterfront structures;
- FRP dowel bars for durable long-term service in concrete highway pavements; and
- FRP tie connectors and FRP grid shear connectors for concrete sandwich wall construction.

Several examples of such applications are described in Chapter 13 of this report.

CHAPTER 2—NOTATION AND DEFINITIONS

2.1—Notation

A_f = area of FRP reinforcement, mm^2 (in.^{2})
a = depth of equivalent rectangular stress block; length of shear span, mm (in.)
b = width of carbon FRP plate, mm (in.)
c = neutral axis depth, mm (in.)
D = diameter, mm (in.)
D_i = inner diameter of tube, mm (in.)
D_o = outer diameter of tube, mm (in.)
d = distance from extreme compression fiber to centroid of tensile reinforcement, mm (in.)
d_b = bar diameter, mm (in.)
E_c = modulus of concrete core, MPa (psi)
E_{FRP} = modulus of elasticity of FRP reinforcement, MPa (psi)
E_f = equivalent orthotropic elastic modulus of tube in hoop direction, MPa (psi)
E_s = elastic modulus of steel reinforcement, MPa (psi)
f'_c = specified concrete compressive strength, MPa (psi)
f'_{cc} = compressive strength of confined concrete, MPa (psi)
f_{fu} = ultimate strength of FRP material, MPa (psi)
f_{pi} = initial prestressing stress, MPa (psi)
h = depth of reinforced concrete beam, mm (in.)
I_{cr} = cracked moment of inertia of the section of a reinforced concrete beam, mm^4 (in.^{4})
I_e = effective moment of inertia, mm^4 (in.^{4})
I_{ey} = overall effective beam moment of inertia at yielding, mm^4 (in.^{4})
L = length of CFFT, mm (in.)
l_d = development length, mm (in.)
l_t = transfer length for prestressing tendons, mm (in.)
M_u = factored moment, kNm (lb-in.)
N = axial load, kN (lb)
R = radius of FRP tube, mm (in.)
T_g = glass transition temperature, °C (°F)
t = thickness of FRP plate, mm (in.)
V_c = nominal shear strength provided by concrete, kN (lb)
V_{exp} = experimentally obtained shear force, kN (lb)
Δ = deflection of beam subjected to load, mm (in.)
Δf_p = stress range in fatigue loading, MPa (psi)

ε_{cc} = axial strain to calculate confinement pressure in CFFT
ε_{FRP} = strain level in FRP reinforcement
$\varepsilon_{FRP\text{-}service}$ = strain level in FRP reinforcement in service
ε_{fe} = effective strain in FRP laminate
ε_{fu} = design rupture strain of FRP
ε_s = strain of steel reinforcement
ϕ = strength-reduction factor
ϕ_u = curvature of beam at failure, 1/mm (1/in.)
ϕ_y = curvature of beam at yielding of internal steel, 1/mm (1/in.)
θ = rotation of a section of a beam, radian or degree; or direction of fibers
κ_m = coefficient to calculate effective strain in FRP laminate
μ = ductility index
ρ = reinforcement ratio
ρ_{FRP} = FRP longitudinal reinforcement ratio
ρ_{fc} = fiber composite reinforcement ratio
σ_R = confinement pressure in circular CFFT, MPa (psi)
σ_x = hoop tensile strength, MPa (psi)
σ_y = axial compressive strength, MPa (psi)
τ_m = bond strength of FRP bar, MPa (psi)
ν_c = Poisson's ratio of concrete
ν_f = longitudinal Poisson's ratio of tube

2.2—Definitions

AFRP—aramid fiber-reinforced polymer.

aging—the process of exposing materials to an environment for an interval of time.

aramid fiber—highly oriented organic fiber derived from polyamide incorporating aromatic ring structure.

balanced FRP reinforcement ratio—the reinforcement ratio in a flexural member that causes the ultimate strain of FRP bars and the ultimate compressive strain of concrete (assumed to be 0.003) to be simultaneously attained.

bar—a composite material formed into a long, slender structural shape suitable for the internal reinforcement of concrete and consisting of primarily longitudinal unidirectional fibers bound and shaped by a rigid polymer resin material. The bar may have a cross section of variable shape (commonly circular or rectangular), and may have a deformed or roughened surface to enhance bonding with concrete.

binder—chemical treatment applied to the random arrangement of glass fibers to give integrity to mats. Specific binders are used to promote chemical compatibility with the various laminating resins used.

BMC—bulk molding compound.

bond-critical applications—applications of FRP systems for strengthening structural members that rely on bond to the concrete substrate; flexural and shear strengthening of beams and slabs are examples of bond-critical applications.

braided string or rope—string or rope made by braiding continuous fibers or strands.

braiding—a process whereby two or more systems of yarns are wound together in the bias direction to form an integrated structure. Braided material differs from woven and knitted fabrics in the method of yarn introduction into the fabric and the manner by which the yarns are interlaced.

b-stage—intermediate stage in the polymerization reaction of thermosets, following which material will soften with heat and is plastic and fusible. The resin of an uncured prepreg or premix is usually in b-stage.

carbon fiber—fiber produced by pyrolysis of organic precursor fibers. Used interchangeably with graphite. Types of carbon fibers include mesophase pitch carbon and pan carbon (polyacrylonitrile).

catalyst—organic peroxide used to activate the polymerization.

CET—coefficient of thermal expansion; change in linear dimension per unit length due to change in temperature.

CFFT—concrete-filled FRP tube.

CFRP—carbon fiber-reinforced polymer (includes graphite fiber-reinforced polymer).

composite—a combination of one or more materials differing in form or composition on a macroscale. Note: The constituents retain their identities; that is, they do not dissolve or merge completely into one another, although they act together. Normally, the components can be physically identified and exhibit an interface between one another.

concrete substrate—the original concrete or any cementitious repair material used to repair or replace the original concrete; the substrate can consist entirely of original concrete, entirely of repair materials, or of a combination of original concrete and repair materials; the substrate includes the surface to which the FRP system is adhered.

contact-critical applications—applications of FRP systems that rely on continuous intimate contact between the concrete substrate and the FRP system. In general, contact-critical applications consist of FRP systems that completely wrap around the perimeter of the section. For most contact-critical applications, the FRP system is bonded to the concrete to facilitate installation, but does not rely on that bond to perform as intended. Confinement of columns for seismic retrofit is an example of a contact-critical application.

continuous fiber reinforcement—any construction of resin-bound continuous fibers used to reinforce a concrete matrix. The construction may be in the shape of continuous fiber bars, tendons, or other shapes.

continuous filament—fiber that is made by spinning or drawing into one long continuous entity.

continuous filament tow—parallel filaments coated with sizing, drawn together into single or multiple strands, and wound into a cylindrical package.

continuous filament yarn—yarn that is formed by twisting two or more continuous filaments into a single continuous strand.

coupling agent—part of a surface treatment or finish that is designed to provide a bonding link between the fiber surface and the laminating resin.

cps—centipoises, unit of viscosity. The standard unit is poise. For relatively low viscosities, the units are often referred to as centipoises (cps) or 0.01 poise. Water is the standard at 1 cps.

crimp—waviness of a fiber, a measure of the difference between the length of the unstraightened and straightened fibers.

cross-link—a chemical bond between polymer molecules. Note: An increased number of cross-links per polymer molecule increases strength and modulus at the expense of ductility.

cure of FRP systems—the process of causing the irreversible change in the properties of a thermosetting resin by chemical reaction. Cure is typically accomplished by addition of curing (cross-linking) agents or initiators, with or without heat and pressure. Full cure is the point at which a resin reaches the specified properties. Undercure is a condition where specified properties have not been reached.

curing agent—a catalytic or reactive agent that, when added to a resin, causes polymerization. Also called hardener or initiator.

deformability factor—the ratio of energy absorption (area under the moment-curvature curve) at ultimate strength of the section to the energy absorption at service level.

denier—measure of fiber diameter, taken as the weight in grams of 9000 m of the fiber.

durability—ability to resist cracking, oxidation, chemical degradation, delamination, wear, fatigue, and/or the effects of foreign object damage for a specified period of time, under the appropriate load conditions, and under specified environmental conditions.

E-glass—a family of glass with a calcium alumina borosilicate composition and a maximum alkali content of 2.0%.

epoxy resin—resin formed by the chemical reaction of epoxide groups with amines, alcohols, phenols, and others.

extrusion—process by which a molten resin is forced through a die of a desired shape.

fabric—arrangement of fibers held together in two dimensions. A fabric may be woven, nonwoven, or stitched.

fabric, nonwoven—material formed from fibers or yarns without interlacing. This can be stitched, knit, or bonded.

fabric, woven—material constructed of interlaced yarns, fibers, or filaments.

fiber—general term for a filamentary material. Any material whose length is at least 100 times its diameter, typically 0.10 to 0.13 mm.

fiber content—the amount of fiber present in a composite. Note: This usually is expressed as a percentage volume fraction or weight fraction of the composite.

fiberglass—a composite material consisting of glass fibers in resin.

fiber-reinforced polymer (FRP)—composite material consisting of continuous fibers impregnated with a fiber-binding polymer then molded and hardened in the intended shape.

fiber volume fraction—the ratio of the volume of fibers to the volume of the composite.

fiber weight fraction—the ratio of the weight of fibers to the weight of the composite.

filament—smallest unit of a fibrous material. A fiber made by spinning or drawing into one long continuous entity.

filament winding—process for forming FRP parts by winding continuous rovings onto a rotating mandrel. The rovings may be dry or saturated with resin.

filler—a relatively inert substance added to a resin to alter its properties or to lower cost or density. Sometimes the term is used specifically to mean particulate additives.

fire retardant—chemicals that are used to reduce the tendency of a resin to burn; these can be added to the resin or coated on the surface of the FRP.

FRC—fiber-reinforced composite.

GFRP—glass fiber-reinforced polymer.

glass fiber—fiber drawn from an inorganic product of fusion that has cooled without crystallizing. Types of glass fiber include alkali-resistant (AR-glass); general purpose (E-glass); and high-strength (S-glass).

glass transition temperature—the midpoint of the temperature range over which an amphorous material (such as glass or a high polymer) changes from (or to) brittle, vitreous state to (or from) a rubbery state.

graphite fiber—fiber containing more than 99% crystalline carbon made from a precursor by oxidation.

grating—a planar FRP form. Gratings may be manufactured using molding methods or by mechanically assembling pultruded FRP elements (bars, I-shapes, or rods) together in two orthogonal directions to produce sheets.

grid—a planar FRP form in which continuous fibers are aligned in two orthogonal directions and combined with resin to produce a open mesh-like structure. Grids may be made using continuous manufacturing methods and supplied on rolls, or sheets made by molding methods and supplied in sheets.

GRP—glass-reinforced plastic.

hand lay-up—an open-mold manufacturing process in which resin is applied by hand, brush, or roller on to dry fiber reinforcements and exposed to the atmosphere for cure. This process can be done in a mold or performed on an object.

hardener—substance used to cure epoxy resins.

HFRP—hybrid FRP.

hybrid—a combination of two or more different fibers, such as carbon and glass or carbon and aramid, into a structure.

impregnation—saturation of voids and interstices of a reinforcement with a resin.

initiator—a substance, usually peroxide, that speeds up the curing of a resin.

interface—the boundary or surface between two different, physically distinguishable media. On fibers, the contact area between fibers and coating/sizing.

interlaminar shear—shearing force tending to produce a relative displacement between two laminae in a laminate along the plane of their interface.

isophthalic polyester resin—product of isophthalic acid, glycol, and maleic anhydride (Ashland Specialty Chemical 2006).

laminate—two or more layers of fibers, bound together in a resin matrix.

lay-up—the process of placing the FRP reinforcing material in position for molding.

LEED—Leadership in Energy and Environment Design—a certification program for sustainable building.

mat—a fibrous material for reinforced polymer, consisting of randomly oriented chopped filaments, short fibers (with or without a carrier fabric), or long random filaments loosely held together with a binder.

matrix—in the case of FRPs, the materials that serve to bind the fibers together, transfer load to the fibers, and protect them against environmental attack and damage due to handling.

MF-FRP—mechanically fastened FRP.

monomer—an organic molecule of relatively low molecular weight that creates a solid polymer by reacting with itself or other compounds of low molecular weight or both.

multifilament—yarn consisting of many continuous filaments.

NDI—nondestructive inspection.

NEFMAC—new fiber composite material for advanced concrete.

NSM—near-surface-mounted.

nylon—polyamide polymer that is thermoplastic in nature.

PAN—polyacrylonitrile, a precursor fiber used to make carbon fiber.

PAN carbon fiber—carbon fiber made from polyacrylonitrile (PAN) fiber.

phenolic resin—thermoset resin produced by condensation of aromatic alcohol.

pitch—a black residue from the distillation of petroleum.

pitch carbon fiber—carbon fiber made from petroleum pitch.

plastisol—a plastisol is a liquid dispersion of polyvinyl chloride resin in a plasticizer along with materials such as stabilizers, colorants, fillers, and other additives.

ply—a single layer of fabric or mat; multiple plies, when molded together, make up the laminate.

PMC—polymer matrix composite.

polyester—one of a large group of synthetic resins, mainly produced by reaction of dibasic acids with dihydroxy alcohols; commonly prepared for application by mixing with a vinyl-group monomer and free-radical catalysts at ambient temperatures and used as binders for resin mortars and concretes, fiber laminates (mainly glass), adhesives, and the like. Commonly referred to as unsaturated polyester.

polyester resin—resin produced by the polycondensation of dihydroxy derivatives and dibasic organic acids or anhydrides yielding resins that can be compounded with styrol monomers to give highly cross-linked thermoset resins.

polymer—a high-molecular-weight organic compound, natural or synthetic, containing repeating units.

polymerization—the chemical reaction in which two or more molecules of the same substance combine to form a compound containing the same elements and in the same proportions but of higher molecular weight.

polyurethane—reaction product of an isocyanate with any of a wide variety of other compounds containing an active hydrogen group; used to formulate tough, abrasion-resistant coatings and matrixes.

postcuring, FRP—additional elevated-temperature curing that increases the level of polymer cross-linking; final properties of the laminate or polymer are enhanced.

pot life—time interval after preparation during which a liquid or plastic mixture is to be used.

precusor—the rayon, PAN, or pitch fibers from which carbon fibers are derived.

prepreg—semi-hardened fiber-matrix construction made by soaking strands or roving with resin or resin precursors.

pultrusion—process by which a molten or curable resin and continuous fibers are pulled through a die of a desired structural shape of constant cross section, usually to form a rod or tendon.

PVC—polyvinyl chloride.

reinforcement—material, ranging from short fibers through complex textile forms, that is combined with a resin to provide it with enhanced mechanical properties.

resin—polymeric material that is rigid or semi-rigid at room temperature, usually with a melting-point or glass transition temperature above room temperature.

resin content—the amount of resin in a laminate, expressed as either a percentage of total mass or total volume.

resistance factor—factor applied to a specified material property or to the resistance of a member for the limit state under considerations, which takes into account the variability of dimensions, material properties, workmanship, type of failure, and uncertainty in the prediction of resistance.

roving—a number of yarns, strands, tows, or ends of fibers collected into a parallel bundle with little or no twist.

RTM—resin transfer molding.

SCRIMP—Seemans composite reinforcement infusion molding process—a vacuum process to combine resin and reinforcement in an open mold.

SFRP—steel FRP.

shape—construction made of continuous fibers in a shape other than used to reinforce concrete monoaxially, or in the specific shape of a grid or mesh. Generally not a bar, tendon, grid, or mesh, although may be used generically to include one or more of these.

sheet, FRP—a dry, flexible ply used in wet lay-up FRP systems. Unidirectional FRP sheets consist of continuous fibers aligned in one direction and held together in-plane to create a ply of finite width and length. Fabrics are also referred to as sheets.

shelf life—the length of time packaged materials can be stored under specified conditions and remain usable.

SIP—Structurally integrated stay-in-place. Used to describe SIP form systems described in Chapter 9.

sizing—surface treatment or coating applied to filaments to improve the filament-to-resin bond and to impart processing and durability attributes.

SMC—sheet molding compound.

spray-up—method of contact molding wherein resin and chopped strands of continuous filament roving are deposited on the mold directly from a chopper gun.

spun yarn—yarn made by entangling crimped staple.

staple—short fibers of uniform length usually made by cutting continuous filaments. Staple may be crimped or uncrimped.

strand—bundle of filaments bonded with sizing.

synthetic fiber, types—polyacrylonitrile (PAN, acrylic); polyamide: nylon (aliphatic) and aramid (aromatic); polyvinyl alcohol; polyvinyl chloride (PVC); polyethylene (PE) (olefin).

textile—fabric, usually woven.

thermoplastic—polymer that is not cross-linked. Thermoplastic polymer generally can be remelted and recycled.

thermoset—resin that is formed by cross-linking polymer chains. A thermoset cannot be melted and recycled.

tow—bundle of fibers, usually a large number of spun yarns.

twisted string or rope—string or rope made by twisting continuous fibers or strands.

uncrimped—fibers with no crimp.

unsaturated polyester—product of a condensation reaction between dysfunctional acids and alcohols, one of which, generally the acid, contributes olefinic unsaturation.

URM—unreinforced masonry.

UV—ultraviolet.

vinylester resin—resin characterized by reactive unsaturation located primarily in terminal positions that can be compounded with styrol monomers to give highly cross-linked thermoset copolymers.

VARTM—vacuum resin transfer molding—a vacuum process to combine resin and reinforcement in an open mold.

volume fraction—the proportion from 0.0 to 1.0 of a component within the composite, measured on a volume basis, such as fiber-volume fraction.

weaving—a multidirectional arrangement of fibers. For example, polar weaves have reinforcement yarns in the circumferential, radial, and axial (longitudinal) directions; orthogonal weaves have reinforcement yarns arranged in the orthogonal (Cartesian) geometry, with all yarns intersecting at 90 degrees.

wet lay-up—a method of making a laminate product by applying the resin system as a liquid when the fabric or mat is put in place.

wet-out—the process of coating or impregnating roving, yarn, or fabric in which all voids between the strands and filaments are filled with resin; it is also the condition at which this state is achieved.

yarn—group of fibers held together to form a string or rope.

CHAPTER 3—CODES AND STANDARDS

This chapter provides an overview of available design documents addressing applications of FRP composite materials either as internal reinforcement or as an external repair material for concrete structures. For design guidance, the referenced design guidelines or codes should be consulted and take precedence over information in this report. With the exception of materials testing standards, the Canadian code documents (CSA S6 and S806), and the Egyptian code (Egyptian Ministry of Housing, Utilities, and Urban Development 2005), the available documents are intended to provide guidance for the use of FRP materials and do not carry the weight of design standards. In some cases, the issuing agency has chosen to term the documents Bulletins (fib 2001), Interim Guidance (IStructE 1999) or, in the case of ACI, Emerging Technology Documents (ACI 440.2R and 440.4R). Nonetheless, all are consensus documents and represent the current state of the practice. ACI removed the Emerging Technology designation from ACI 440.1R.

3.1—Materials

Most major standards-writing organizations have a number of available materials testing standards appropriate for determining a variety of FRP material properties and characteristics. Materials test standards, however, are always generic and are intended to set a standard of performance based on laboratory tests; therefore, they are not necessarily well suited to determine the in-place properties of interest to the concrete designer.

ACI Committee 440 has initiated the development of test methods specifically intended for FRP products and materials used with reinforced concrete. These ACI-developed standards are aimed at generating design values useful to the concrete practitioner. Once accepted by ACI, these test methods are submitted to ASTM Committee D30.05, "Composite Materials—Structural Test Methods" to be developed as formal ASTM standards. The first collection of ACI test methods is compiled in ACI 440.3R. ASTM D30.05 is beginning the process of revising and adopting these initial standards.

Most international standards organizations issue standards for determining basic physical properties of FRP materials. Many of these have been adapted by various industry and research interests for products with applications for concrete structures. Chapter 5 of this report discusses FRP materials testing more completely.

3.2—Internal FRP reinforcement

Fundamental design methodologies for FRP-reinforced concrete are similar to those of conventional steel-reinforced concrete. Cross-sectional equilibrium, strain compatibility, and constitutive material behavior form the basis of all code approaches to designing reinforced concrete, regardless of the reinforcing material. The nonductile and anisotropic natures of FRP reinforcing products, however, need to be addressed in design guidelines. In flexural design, for instance, the ultimate limit state may be defined either by FRP rupture or concrete crushing, provided that the strength and serviceability criteria are met. Because of the lack of ductility of such failure modes, a higher reserve of strength is required. Thus, strength-reduction factors or material-resistance factors are generally lower for FRP-reinforced concrete members than for steel-reinforced concrete members. In all cases, FRP design guidelines and codes are consistent with the applicable reinforced concrete design codes. No attempt is made to adjust load factors; thus, only strength-reduction factors or material-resistance factors are adjusted to reflect the use of FRP. Chapters 6 and 7 of this document discuss, in more detail, internally FRP-reinforced and FRP-prestressed concrete members, respectively.

ACI 440.1R uses a strength design approach for FRP-reinforced concrete members that is consistent with ACI

318. In this approach, a member is designed based on its required strength and checked for fatigue and creep rupture endurance, and serviceability criteria. In many cases, these latter criteria may control the design of members, particularly where lower stiffness aramid or glass reinforcement is used. ACI 440.1R, consistent with ACI 318, provides strength-reduction factors related to the ductility of the failure mode considered. Guidance, in the form of additional reduction factors, is also provided for all durability-related criteria with the exception of fire resistance. ACI 440.4R provides similar design guidance for prestressing structures with FRP tendons.

The CSA S806 represents the first stand-alone formalized design code (rather than guide) addressing FRP-reinforced concrete. Consistent with CSA A23.3, CSA S806 prescribes partial material-resistance factors rather than strength-reduction factors. Additionally, CSA S806 does not consider FRP rupture to be a valid failure, permitting only concrete crushing. The CSA S806 does not address durability directly, but rather references CSA S478 instead. CSA S806 requires the designer to consider fire resistance, but offers no provisions or guidance in this regard. CSA S806 also provides provisions for prestressing concrete members with aramid or carbon bars.

The CSA S6 also includes formalized provisions for the application of FRP reinforcement or prestressing. Material-resistance factors and stress limits associated with serviceability, pretensioning, and post-tensioning are provided. Durability is addressed through material-resistance factors for FRP type and application, and the effects of creep and fatigue are assumed to be included in the factors and limits provided.

The ISIS *Design Manual 3—Reinforcing Concrete Structures with Fibre Reinforced Polymers* (Rizkalla and Mufti 2001) takes an approach very similar to ACI 440.1R in permitting both FRP rupture and concrete crushing failures. *Design Manual 3* provides material-resistance factors, though it does not differentiate between expected failure modes. *Design Manual 3* discusses durability, but does not provide additional factors to account for environmental exposure. Creep and fatigue are addressed in the same manner as in CSA S6. Finally, *Design Manual 3* provides listings of product-specific material properties.

In Europe, the Federation Internationale de Beton (fib) Task Group 9.3, FRP Reinforcement for Concrete Structures, is preparing a technical bulletin addressing the use of internal FRP reinforcement. As of early 2007, this document is not yet available, but several European countries have developed their own guidelines for such applications.

In Great Britain, the *Interim Guidance on the Design of Reinforced Concrete Structures using Fibre Composite Reinforcement* (IStructE 1999) takes the form of suggested changes to the British design codes BS8110, "Structural Use of Concrete, Parts 1 and 2" (British Standards Institution [BSI] 1997) and BS5400, "Part 4 Code of Practice for the Design of Concrete Bridges" (BSI 1990). The approach taken by the IStructE document is consistent with that taken by CSA S6.

The Japanese "Recommendations for the Design and Construction of Concrete Structures using Continuous Fiber Reinforced Materials" (JSCE 1997) is intended to supplement the "JSCE Standard Specification for Concrete Structures" (JSCE 1986) with respect to the use of FRP reinforcement for structures other than buildings. The JSCE recommendations (1997) are based on limit states design with verification of the ultimate, serviceability, and fatigue limits states. Although limits associated with durability phenomena are prescribed, these are typically given in reference to the results of a standard material property test for the FRP bar or strand considered. "Design Guidelines of FRP Reinforced Concrete Building Structures" (Sonobe et al. 1997), provides similar guidelines for building structures. A third guide, "Design Methods for Prestressed FRP-Reinforced Concrete Building Structures," is reported (Bakis et al. 2002) although it does not appear to be available in English.

The Egyptian FRP code (Egyptian Ministry of Housing, Utilities, and Urban Development 2005) represents a stand-alone formalized design code addressing FRP-reinforced concrete that includes provisions for the design and construction of concrete reinforced with FRP bars. The Egyptian document uses the limit states design approach for FRP-reinforced concrete members that is consistent with the formalized Egyptian "Code for the Design and Construction of Concrete Structures" (2001). In this approach, a member is designed based on its required strength and checked for fatigue and creep rupture endurance and serviceability criteria. The Egyptian FRP code prescribes a material-resistance factors approach related to the ductility of the failure mode considered in addition to all durability-related criteria with the exception of fire resistance.

In North America, two approaches for evaluating the resistance of members are used. In the ACI approach, the strength-reduction factor depends on the action being resisted (for example, the reduction factor is different for flexure, shear, and axial loads). The strength-reduction factor (less than 1.0) is applied to the total nominal resistance, and depends on the mode of failure. On the other hand, Canadian codes are based on a unified limit state design philosophy in which material-resistance factors, always less than 1.0, are used to reflect uncertainties in determining the nominal resistance structural components. In CSA S6, the material-resistance factor for each FRP type changes with application and was arrived at by considering both material variability and environmental factors for those specific applications.

3.3—External FRP reinforcement

Recommended design methodologies of externally bonded FRP strengthening schemes are universally consistent with concrete design methodologies. Currently, the three formalized documents addressing applications of externally bonded FRP are CSA S806, CSA S6, and the Egyptian FRP code (Egyptian Ministry of Housing, Utilities, and Urban Development 2005). All other available documents are guides. The approaches taken, in each case, are consistent with local concrete codes and related guidance for internal FRP reinforcement (Section 3.2). Similar to internal FRP reinforcement, failure modes may be associated with concrete crushing or FRP rupture or loss of bond. Additionally, the initial strain conditions of the existing internal steel reinforcement should also be considered. Load-reduction factors and material-resistance factors are consistent with those recommended for internal FRP reinforcement. Chapter 8 of this document discusses externally bonded FRP reinforcement used for structural repair or retrofit in more detail.

Beyond simply providing the required capacity, three significant design issues, associated with applications of external FRP for repair and strengthening, are:

1. Quality of bond (or anchorage, or both) between FRP and concrete substrate;

2. The durability and long-term performance of FRP materials; and

3. Strengthening limits.

The consensus of available design guides is that each of these issues requires special attention and additional research. Additionally, most guides for design of externally bonded composites address only reinforced concrete rectangular sections. ACI Committee 440 is currently preparing documents that address concrete members having pre- or post-tensioned steel reinforcement, irregular geometry, compression FRP reinforcement, near-surface-mounted (NSM) FRP applications, or upgrade of masonry structures.

ACI 440.2R provides strength-reduction factors based on ductility of the expected failure mode (as measured by reinforcing steel strain) consistent with ACI 318. ACI 440.2R specifically addresses environmental exposure with environmental reduction factors associated with various typical exposures. Fatigue and creep effects are addressed through prescribed FRP stress limits. Cover delamination and debonding of the FRP are mitigated through strain limits applied to the FRP. These limits are believed to reduce the shear transfer demand of the cover concrete and adhesive layers sufficiently to prevent bond failure. Finally, *de facto* strengthening limits are determined by stipulating that the ultimate capacity of the structure without the FRP reinforcement should exceed a minimum level that corresponds to the expected loads under such situations. Additional guidance is provided for fire event loading requirements. These strengthening limits are based on the premise that the FRP retrofit is rendered entirely ineffective under particular conditions, such as fire or vandalism, and the member (without FRP) should be able to carry service loads without collapse. A revision of ACI 440.2R to include NSM reinforcement and to update load factors to those adopted in the ACI 318-05 (ACI Committee 318 2005) is underway. Similarly, a separate document addressing the repair of masonry using externally bonded FRP reinforcement is in preparation.

According to ACI 440.2R, the shear strength can be improved by wrapping the FRP system around three sides of the member (U-wrap), bonding to the two sides of the member, or completely wrapping the member. The FRP system can be installed continuously along the span length of a member or placed as discrete strips. The nominal shear capacity of an FRP-strengthened concrete member can be determined by adding contributions from the reinforcing steel (stirrups, ties, or spirals) and concrete and the contribution of the FRP reinforcement based on a maximum FRP strain at ultimate that is no greater than 0.004. FRP contribution for shear is further reduced using a reduction factor that is less than 1.0.

For compression members, ACI 440.2R allows FRP systems to be used to increase the axial compression strength of a concrete member by providing confinement with lateral (hoop) fibers. The hoop fibers provide confining effects similar to conventional spiral or tie reinforcing steel. ACI 440.2R uses a simplified approach to determine the peak confined concrete strength as a function of the FRP confining pressure. The effectiveness of FRP confinement is determined as a function of the geometry of the section of the confined member. An efficiency factor is introduced to address the shape effects. For circular sections, the efficiency factor can be taken equal to 1.0.

ACI 440.2R gives some guidelines for detailing FRP sheets or laminates to avoid bond-related failures. The details depend on the geometry of the structure, the soundness and quality of substrate, and levels of load to be sustained by the FRP sheets or laminates.

For fatigue, ACI 440.2R suggests fatigue stress limits on FRP laminates to avoid fatigue failures of the FRP. The stress limits are lowest for glass FRP (GFRP), and highest for carbon FRP (CFRP). Existing recommendations for stress range limits for reinforced concrete beams (ACI 215R) may be applied to the internal reinforcing steel, but these typically limit the service stresses in the steel and concrete to levels that may be exceeded in cases where the structure is strengthened. Indeed, exceeding these stress limits may be the reason for requiring strengthening. Finally, it should also be taken into account that by the time strengthening takes place, the structure may have been in service for a significant number of years, exhausting some of its fatigue life.

To calculate the design strength of a strengthened member, the ACI approach uses strength-reduction factors applied to the nominal resistance, while other codes use material-resistance factors applied to the strength properties of the materials themselves (including FRPs). In addition to an overall strength-reduction factor, ACI 440.2R applies an additional partial resistance factor to the portion of the strength contributed by the FRP reinforcement. To account for durability, ACI 440.2R also recommends environmental reduction factors to reduce the strength of the FRP that can be used in design. These factors depend on the type of FRP and the exposure conditions for the element to be strengthened.

The International Code Council (ICC) Evaluation Service, Inc. has acceptance criteria AC-125 and AC-178 (ICC 2003a,b). Although these criteria are not codes or standards, they are documents intended to evaluate FRP materials as alternatives to code-approved materials. They are developed in an open forum at public meetings presided by building officials.

The Canadian CSA S806 represents the first stand-alone formalized design code (rather than guide) addressing externally bonded FRP reinforcement for concrete. For strengthening applications, CSA S806 considers all possible failure modes rather than just concrete crushing (as it does for internal reinforcement). Durability is addressed through reference to CSA S478. Although a single limiting FRP tensile strain (0.007) is provided, delamination and debonding are addressed by directing the designer to "currently available information appropriate to the combination of sheets and adhesive." Strengthening limits are based on the requirement that the structure should support its nominal service load in the event that the FRP becomes ineffectual. CSA S806 addresses the repair of masonry structures.

ISIS *Design Manual 4—Strengthening Reinforced Concrete Structures with Externally-Bonded Fibre Reinforced Polymers* (Neale 2001) provides considerable guidance and a number of design examples for the use of externally bonded FRP. This document, however, is written as a state-of-the-art report, referencing the recommendations of others rather than making its own clear design recommendations. Generally, ISIS *Design Manual 4* references the recommendations of ACI 440.2R (which was in draft form when ISIS *Design Manual 4* was released).

The European Fédération Internationale de Béton (fib) *Bulletin* 14 (fib 2001), produced by fib Task Group 9.3, also represents a combination guide and state-of-the-art report. Significantly, fib *Bulletin* 14 recognizes the difference in expected performance, not only between FRP material types, but between preformed and wet layed-up FRP systems in the form of different material safety factors. Delamination and debonding are extensively addressed in fib *Bulletin* 14 using a simplified bilinear bond model and by also addressing the effects of the loss of composite action between the FRP and concrete substrate. Durability effects are discussed, but no clear design guidance is provided. FRP-strengthened members are expected to withstand accidental loads in the case where the FRP no longer functions. In this case, reference is made to "Eurocode 1" (CEN 1994) for the loads to be considered.

In Great Britain, the Concrete Society published *Technical Report* 55 (Concrete Society 2000, 2004). This document is similar to ISIS *Design Manual 4* (Neale 2001) and fib *Bulletin* 14 (2001) in its approach and scope. Importantly, *Technical Report* 55 has been followed by *Technical Report* 57 (Concrete Society 2003). This report addresses more practical construction issues associated with externally bonded FRP materials.

The Japanese "Recommendations for the Upgrading of Concrete Structures with Use of Continuous Fiber Sheets" (JSCE 2001) adopts a performance-based approach to the design of externally bonded FRP materials. Flexure and shear capacity, column deformation capacity (jacketing applications), flexural crack width, and protection of the concrete substrate from chloride ion penetration are all considered explicitly in the JSCE recommendations. A variety of guidelines produced by various Japanese authorities are also available primarily for column and/or pier retrofit using FRP jacketing techniques (Bakis et al. 2002).

The Egyptian FRP code (Egyptian Ministry of Housing, Utilities, and Urban Development 2005) also addresses externally bonded FRP reinforcement for concrete. The approach is very similar to that of ACI 440.2R.

The Italian "Guide for the Design and Construction of Externally Bonded FRP Systems for Strengthening Existing Structures" (CNR 2004) includes the application of externally bonded FRP materials to reinforced or prestressed concrete structures as well as to masonry structures. Delamination failure mechanisms are explained, and a fracture mechanics approach is adopted to predict such failure modes; based on that approach, an ultimate design strength for laminate/sheet is provided. This document describes a detailed design example of strengthening a two-story building with CFRPs.

CHAPTER 4—COMPOSITE MATERIALS AND PROCESSES

4.1—Introduction

FRP composites are defined as a polymer matrix, either thermoset or thermoplastic, that is reinforced with a fiber or other reinforcing material with a sufficient aspect ratio (length-to-thickness) to provide a discernible reinforcing function in one or more directions. FRP composites are different from traditional construction materials such as steel or aluminum. FRP composites are anisotropic (properties vary with the direction), whereas steel or aluminum are isotropic (uniform properties in all directions, independent of applied load). Therefore, FRP composite properties are directional, and typically the most favorable mechanical properties are in the direction of the fiber placement.

Many terms have been used to define FRP composites. Modifiers have been used to identify a specific fiber such as glass FRP (GFRP), carbon FRP (CFRP), aramid FRP (AFRP), steel FRP (SFRP), and hybrid FRP (HFRP) for composites containing different types of fibers. In addition, other acronyms were developed over the years; their use depended on geographical location or market use. For example, fiber-reinforced composites (FRCs), glass-reinforced plastics (GRPs), and polymer matrix composites (PMCs) are found in many references.

Although these composites are defined as a polymer matrix that is reinforced with fibers, this definition should be further refined when describing composites in structural applications. In the case of structural applications such as FRP composite reinforced concrete, at least one of the constituent materials should be a continuous reinforcement phase supported by a stabilizing matrix material. For the special class of matrix materials discussed herein (that is, thermosetting polymers), the continuous fibers will usually be stiffer and stronger than the matrix.

Composite materials, in the sense that they will be discussed in this chapter, will be at the macrostructural level. This chapter will address the gross structural forms and constituents of composites, including the matrix resins and reinforcing fibers. This chapter will also briefly address additives and fillers, as well as process considerations.

The performance of any composite depends on the materials of which the composite is made, the arrangement of the primary load-bearing portion of the composite (reinforcing fibers), and the interaction between the materials (fibers and matrix).

Each of the constituent materials or ingredients plays an important role in the processing and final performance of the end product. The resin or polymer holds the fibers in place and influences the physical properties of the end product. The reinforcement provides the mechanical strength. The fillers and additives are used as process or performance aids to impart special properties to the end product.

The mechanical properties and composition of FRP composites can be engineered for their intended use. The type and quantity of materials selected, in addition to the manufacturing process to fabricate the product, will affect the mechanical properties and performance. Important considerations for the design of composite products include:

- Type and percentage of fiber or fiber volume;
- Orientation of fiber (0, 90, ±45 degrees, or a combination of these);
- Type of resin;
- Cost of product;
- Volume of production (to help determine the best manufacturing method); and
- Service conditions.

4.2—Polymer matrix: resins

Resins are divided into two major groups: thermoset and thermoplastic. Thermoplastic resins become soft when heated, and may be shaped or molded while in a heated semifluid state. They become rigid when cooled. Thermoset resins, on the other hand, are usually liquids or low-melting-point solids in their initial form. When used to produce finished goods, these thermosetting resins are cured with a catalyst, heat, or a combination of the two. Once cured, solid thermoset resins cannot be converted back to their original liquid form. Unlike thermoplastic resins, cured thermosets will not melt and flow but will soften when heated (and lose hardness). Once formed, they cannot be reshaped. Heat distortion temperature and the glass transition temperature T_g are used to measure the softening of a cured resin. Test methods for both heat distortion temperature and T_g measure the approximate temperature where the cured resin will soften significantly to yield (bend or sag) under load.

The most common thermosetting resins used in the composites industry are unsaturated polyesters, epoxies, vinylesters, and phenolics.

4.2.1 *Polyester*—Unsaturated thermoset polyester resins are the workhorses of the composites industry, and they represent approximately 75% of the total resins used (ACMA MDA 2006). To avoid any confusion in terms, readers should be aware that there is an unrelated family of thermoplastic polyesters that are best known for their use as fibers for textiles and clothing. Polyesters are produced by the condensation polymerization of dicarboxylic acids and difunctional alcohols (glycols). In addition, unsaturated polyesters contain an unsaturated material, such as maleic anhydride or fumaric acid, as part of the dicarboxylic acid component. The finished polymer is dissolved in a reactive monomer such as styrene to give a low-viscosity liquid. When this resin is cured, the monomer reacts with the unsaturated sites on the polymer, converting it to a solid thermoset structure.

Polyesters are versatile because of their capacity to be modified or tailored during the building of the polymer chains. They have been found to have almost unlimited usefulness in all segments of the composites industry. The principal advantage of these resins is a balance of properties (including mechanical, chemical, and electrical), dimensional stability, cost, and ease of handling or processing.

Unsaturated polyesters are divided into classes depending upon the structure of their basic building blocks. Some common examples would be orthophthalic (ortho), isophthalic (iso), dicyclopentadiene, and bisphenol fumarate resins. In addition, polyester resins are classified according to end-use application as either general purpose or specialty polyesters.

The general-purpose polyesters define products that are relatively low in cost, offer good mechanical and electrical performance, and have a well-defined set of processing/fabricating characteristics. Almost all ortho and dicyclopentadiene resins and some iso resins fall into the general-purpose category.

Because polyesters can be chemically tailored to meet the requirements of a wide range of applications, a number of specialty polyesters that address specific performance, such as flexibility, electrical insulation, corrosion resistance, heat resistance, fire retardancy, and optical translucence, are available.

4.2.2 *Epoxy*—Epoxy resins have a well-established record in a wide range of composite parts, structures, and concrete repair. The structure of the resin can be engineered to yield a number of different products with varying levels of performance. A major benefit of epoxy resin over unsaturated polyester resins is its lower shrinkage. Epoxy resin can also be formulated with different materials or blended with other epoxy resins to achieve specific performance features. Cure rates can be controlled to match process requirements through the proper selection of hardeners and/or catalyst systems. Generally, epoxies are cured by addition of an anhydride or an amine hardener as a two-part system. Different hardeners, or different quantities of hardener, produce different cure profiles and give different properties to the finished composite.

Epoxies are primarily used for fabricating high-performance composites with superior mechanical properties, resistance to corrosive liquids and environments, superior electrical properties, good performance at elevated temperatures, and excellent adhesion to a substrate. Epoxy resins do not, however, have particularly good ultraviolet (UV) resistance. Because the viscosity of epoxy is much higher than that of most polyester resins, they require a post-cure (elevated heat) to obtain ultimate mechanical properties. Epoxies have little or no odor as compared with polyesters.

Epoxy resins are used with a number of fibrous reinforcing materials, including glass, carbon, and aramid. Epoxies are compatible with most composite manufacturing processes, particularly vacuum-bag molding, autoclave molding, pressure-bag molding, compression molding, filament winding, and hand lay-up. Currently, epoxy is the predominant resin used in external repair of concrete using FRP sheet and fabric products.

4.2.3 *Vinylester*—Vinylesters were developed to combine the advantages of epoxy resins with those of unsaturated polyester resins. These resins are produced by reacting epoxy resin with acrylic or methacrylic acid. This provides an unsaturated site, much like that produced in polyester resins when maleic anhydride is used. The resulting material is dissolved in styrene to yield a liquid that is similar to polyester resin. Vinylesters are also cured with the conventional organic peroxides used with polyester resins. Vinylesters offer mechanical toughness and excellent corrosion resistance. These enhanced properties are obtained without complex processing, handling, or special shop fabricating practices that are typical with epoxy resins.

4.2.4 *Phenolic*—Phenolics are a class of resins commonly based on phenol (carbolic acid) and formaldehyde, and cure through a condensation reaction producing water that should be removed during processing. Phenolic composites have many desirable performance qualities, including high temperature resistance, creep resistance, excellent thermal insulation and sound-damping properties, corrosion resistance, and excellent fire/smoke/smoke toxicity properties. Phenolics are applied as adhesives or matrix binders in engineered woods (plywood), brake linings, clutch plates, and circuit boards. Phenolic composites are not currently widely available for the construction industry.

4.2.5 *Structural engineered polyvinyl chloride (PVC) plastisol*—Structural engineered PVC plastisol resins are, by definition, thermoplastic resins with many performance characteristics of thermosets but with many more processing-friendly attributes than traditional thermoset resins. This new resin is a single-component material with no volatiles. This resin has the potential of impacting FRP composites products; however, to date, no products have been made or tested. It features a viscosity of approximately 1.5 to 3 Pa·S (1500 to 3000 cps) with a shelf life in excess of 6 months with essentially no settling of solids. To fully fuse (or cure) this material, it is subjected to temperatures between 177 and 199 °C (350 and 390 °F) for between 10 seconds and several minutes depending on the thickness of the part and the speed of the operation. Wetout of traditional glass and carbon fibers is effective, with chemical bonding occurring between the silanes present in the sizings and the fully fused PVC plastisol.

From a processing standpoint, PVC plastisol resins are efficient in handling liquid resins and minimizing waste associated with startup and shutdown. Structural engineered PVC plastisol resins are formulated with no styrene, solvent, or volatiles. Additive packages offer additional benefits of excellent UV resistance and fire retardancy. Additionally, there is virtually no microvoiding in the finished part, which ensures much better electrical properties and significantly lower water absorption compared with traditional thermosets. The resin exhibits a T_g of approximately 63 °C (145 °F).

4.3—Reinforcing fibers

The primary function of fibers or reinforcements is to carry load along the length of the composite to provide strength and stiffness in one direction. Reinforcements can be oriented to provide tailored properties in various directions. Reinforcements can be both natural and synthetic. Most commercial reinforcements, however, are synthetic. The principal types of fibers in commercial use for civil engineering applications are glass, carbon (or graphite), and aramid.

Of these, glass fiber has by far the largest volume of reinforcement measured either in quantity consumed or in product sales. Other composite reinforcing materials include ultra-high molecular weight polyethylene, polypropylene, polyester, and nylon.

The most common form of FRPs used in structural applications is called a laminate. Laminates are made by stacking a number of thin layers (lamina) of fibers and matrixes and consolidating them into the desired thickness. Fiber orientation in each layer, as well as the stacking sequence of the various layers, can be controlled to generate a range of physical and mechanical properties.

A unidirectional or one-dimensional fiber arrangement is transversely isotropic. This fiber orientation results in a maximum strength and modulus in the direction of the fiber axis. A planar arrangement of fibers is two-dimensional and has different strengths at all angles of fiber orientation (orthotropic). A three-dimensional array has substantially reduced strengths over the one-dimensional arrangement. Mechanical properties in any one direction are proportional to the amount of fiber by volume oriented in that direction.

Most reinforcement for either thermosetting or thermoplastic resins receives some form of surface treatment, either during fiber manufacture or as a subsequent treatment. Other materials applied to fibers as they are produced include resinous binders to hold fibers together in bundles and lubricants to protect fibers from degradation caused by process abrasion.

4.3.1 *Glass fibers*—Glass has been the predominant fiber for many civil engineering applications because of an economical balance of cost and specific strength properties. Glass fibers are commercially available in E-glass formulation (for electrical grade), the most widely used general-purpose form of composite reinforcement, and other formulations for high strength (S-2® glass), improved acid resistance (ECR glass), and alkali resistance (AR glass).

Based on an alumina-lime-borosilicate composition, E-glass fibers are considered the predominant reinforcement for polymer matrix composites because of their high electrical insulating properties, low susceptibility to moisture, and high mechanical properties.

Glass fibers used for reinforcing composites generally range in diameter from 9 to 23 microns. Fibers are drawn at high speeds through small holes in electrically heated bushings. These bushings form the individual filaments. The filaments are gathered into groups or bundles called strands or tows. The filaments are water and air cooled and then coated with a proprietary chemical binder or sizing to protect the filaments and enhance the composite laminate properties. The sizing also determines the processing characteristics of the glass fiber and the conditions at the fiber-matrix interface in the composite.

Glass is generally a good impact-resistant fiber, but is denser than carbon or aramid. Composites made from this material exhibit very good electrical and thermal insulation properties. Glass fibers are also transparent to radio frequency radiation and are used in radar antenna applications.

4.3.2 *Carbon fibers*—Carbon fiber is made from polyacrylonitrile (PAN), pitch, or rayon fiber precursors. The properties of carbon fiber are controlled by molecular structure and degree of freedom from defects. The formation of carbon fibers requires processing temperatures above 1000 °C (1830 °F).

There are two basic types of carbon fiber: high modulus and high strength. The difference in properties between them is a result of the differences in fiber microstructure, derived from the precursor type and the processing temperature (1000 to 3000 °C [1830 to 5430 °F]).

PAN-based carbon fiber is the predominant form used in the civil engineering venue because of its very high strength (up to about 5000 MPa [720 ksi] in unidirectional form) and relatively high modulus (up to about 140 GPa [20 msi] in unicomposite form or 420 GPa [60 msi] in fiber form). Pitch-based carbon fibers have extremely high modulus values (up to 970 GPa [140 msi]) with lower strengths (up to about 690 MPa [100 ksi] in unicomposite form). They are used primarily in space and satellite applications. Carbon fiber is about five to 10 times more expensive than glass fiber; however, it has about twice the usable strength and four times the modulus of glass.

Carbon fibers are supplied in a number of different forms, from continuous filament tows to chopped fibers and mats. The highest strength and modulus are obtained from unidirectional continuous reinforcement. Carbon fiber composites have a lower strain capacity than do glass or aramid composites. Carbon fiber is highly resistant to alkali or acid attack. Nevertheless, carbon fibers can cause galvanic corrosion when in contact with metals. A barrier material, such as glass and resin, is used to prevent this occurrence. Also, carbon fibers can conduct electric current, and are thus vulnerable to strikes by lightening.

Rayon and isotropic pitch precursors are used to produce low-modulus carbon fibers (50 GPa [7000 ksi]). Both PAN and liquid crystalline pitch precursors are made into higher modulus carbon fibers by carbonizing above 800 °C (1400 °F). Fiber modulus increases with heat treatment from 1000 to 3000 °C (1830 to 5430 °F). The results vary with the precursor selected. Fiber strength appears to maximize at a lower temperature of 1500 °C (2730 °F) for PAN and some pitch precursor fibers, but increases for most mesophase (anisotropic) pitch precursor fibers.

The axial-preferred orientation of graphene layers in carbon fibers determines the modulus of the fiber. Both axial and radial textures and flaws affect the fiber strength. Orientation of graphene layers at the fiber surface affects wetting and strength of the interfacial bond to the matrix.

Carbon fibers are not easily wet by resins, particularly the higher-modulus fibers. Surface treatments that increase the number of active chemical groups (and sometimes roughen the fiber surface) have been developed for some resin matrix materials. Carbon fibers are frequently shipped with an epoxy size treatment applied to prevent fiber abrasion, improve handling, and provide an epoxy resin matrix compatible interface. Fiber and matrix interfacial bond strength approaches the strength of the resin matrix for lower-modulus carbon fibers. Higher-modulus PAN-based fibers show substantially lower interfacial bond strengths. Failure in high-modulus fiber occurs in its surface layer in much the same way as with aramids.

Carbon fibers are available as tows or bundles of parallel fibers. The range of individual filaments in the tow is normally from 1000 to 200,000 fibers. These tows can be woven or knitted into fabrics. Carbon fiber is also available as a prepreg where the fibers are preimpregnated with resin, as well as in the form of unidirectional tow sheets.

4.3.3 *Aramid fibers*—Aramid fiber is an aromatic polyamide organic fiber for composite reinforcement. Aramid fibers offer good mechanical properties at a low density with the added advantage of toughness or impact resistance. They are characterized as having reasonably high tensile strength, a medium modulus, and a very low density as compared with glass and carbon. The tensile strength of aramid fibers is higher than that of glass fibers, and the modulus is approximately 50% higher than that of glass. These fibers increase the impact resistance of composites and provide products with higher tensile strengths. Aramid fibers are insulators of both electricity and heat. They are resistant to organic solvents, fuels, and lubricants. Aramid composites have poor compressive strength. Dry aramid fibers are tough, have been used as cables or ropes, and are frequently used in ballistic applications.

4.3.4 *Steel fibers*—A new type of FRP composite strengthening system has emerged that uses high-strength steel fibers and is commonly known as SFRP. The high-strength steel fibers demonstrate a linear elastic stress-strain relationship that is similar to carbon and glass fibers. The steel fibers (also referred to as wires) have a tensile strength in the range of 2400 to 3100 MPa (350 to 450 ksi) and an elastic modulus of 200 MPa (30,000 ksi). The high-strength steel wires are twisted together to form steel cords, each containing five to 13 wires. Typical wire diameters range from 0.5 to 1.3 mm (0.02 to 0.05 in.). Because of the unwinding effects of the cords at higher tensile loads, ultimate tensile strains in the range of 0.02 to 0.05 can be achieved. The twisted steel wires have a deformed surface that allows for adequate bond to the surrounding matrix. SFRP is currently being applied for strengthening of concrete structures in a similar manner to other externally bonded FRP materials (Casadei et al. 2005a,b; Prota et al. 2006a,b).

Typical properties of glass, carbon, and aramid fibers are reported in Table 4.1.

4.4—Types of reinforcement

Regardless of the material, reinforcements are available in forms to serve a wide range of processes and end-product requirements. Fibers supplied as reinforcement include (untwisted) roving; (twisted) yarns; milled fiber; chopped strands; and continuous, chopped, or thermoformable mats. Reinforcement materials can be designed with unique fiber architectures and be preformed or shaped depending on the product requirements and manufacturing process.

4.4.1 *Multi-end and single-end rovings*—Multi-end rovings consist of many individual strands or bundles of filaments, which are then chopped and randomly deposited into the resin matrix. They can also be woven into fabric reinforcement. Processes such as sheet molding compound (SMC) and spray-up use multi-end rovings. Multi-end rovings can also be used in some filament winding and

Table 4.1—Typical material properties of fibers

Fiber	Typical diameter, microns	Density, g/cm³ (lb/in.³)	Tensile modulus, GPa (10⁶ psi)	Tensile strength, MPa (ksi)	Strain to failure, %	Coefficient of thermal expansion, 10^{-6}/°C	Poisson's ratio
Commercial composite reinforcing fibers (constructed from Mallick [1988])							
Glass							
E-glass	10	2.54 (0.092)	72.4 (10.5)	3450 (500.0)	4.8	5.0	0.2
S-glass	10	2.49 (0.090)	86.9 (12.6)	4300 (625.0)	5.0	2.9	0.22
Carbon PAN-carbon							
T-300	7	1.76 (0.064)	231 (34)	3650 (530)	1.4	–0.1 to –0.5 (longitudinal), 7 to 12 (radial)	–0.2
AS	7	1.77 (0.064)	220 (32)	3100 (450)	1.2	–0.5 to –1.2 (longitudinal), 7 to 12 (radial)	—
t-40	6	1.81 (0.065)	276 (40)	5650 (820)	2.0	—	—
HSB	7	1.85 (0.067)	345 (50)	2340 (340)	0.58	—	—
Fortafil 3™	7	1.80 (0.065)	227 (33)	3800 (550)	1.7	–0.1	—
Fortafil 5™	7	1.80 (0.065)	345 (50)	2760 (400)	0.8	—	—
Toray M40J	—	1.77 (0.064)	377 (55)	4410 (640)	1.2	—	—
Zoltek (2006)	7	1.81 (0.065)	242 (35)	3800 (550)	—	—	—
Pitch-carbon							
P-555	10	2.00 (0.072)	380 (55)	1900 (275)	0.5	–0.9 (longitudinal)	—
P-100	10	2.16 (0.078)	758 (110)	2410 (350)	0.32	–1.6 (longitudinal)	—
Aramid							
Kevlar™ 49	12	1.45 (0.052)	131 (19)	3620 (525)	2.8	–2.0 (longitudinal), +59 (radial)	0.35
Twaron™ 1055*	12	1.45 (0.052)	127 (18)	3600 (533)	2.5	–2.0 (longitudinal), +59 (radial)	0.35

*Mechanical properties: single filament at 22 °C (72 °F) per ASTM D 2101.

pultrusion applications. The single-end roving consists of many individual filaments wound into a single strand. This product is generally used in processes that use a unidirectional reinforcement such as filament winding or pultrusion. These manufacturing processes are described in Section 4.8.

4.4.2 *Mats*—These are two-dimensional random arrays of chopped strands. The fiber strands are deposited onto a continuous conveyor and then passed through a region where thermosetting resin is dusted on them. This resin is heat set and holds the mat together. The binder resin dissolves in the polyester or vinylester matrix, thereby allowing the mat to conform to the shape of the mold.

4.4.3 *Woven, stitched, and braided fabrics*—Many types of fabrics can be used to reinforce resins in a composite. Multidirectional reinforcements are produced by weaving, knitting, stitching, or braiding continuous fibers into a fabric from twisted and plied yarns. Fabrics refer to all flat-sheet, rolled goods. Fabrics can be manufactured with almost any reinforcing fiber. The most common fabrics are constructed with glass, carbon, or aramid fibers. Fabrics are available in several weave constructions and thicknesses from 0.025 to 10.2 mm (0.0010 to 0.40 in.). They offer oriented strengths and high reinforcement loadings often found in high-performance applications.

Fabrics are typically supplied on rolls of 23 to 274 m (25 to 300 yd) in length and 25 to 3050 mm (1 to 120 in.) in width. The fabric should be inherently stable enough to be handled, cut, and transported to the mold, but pliable enough to conform to the mold shape and contours. Properly designed, the fabric will allow for quick wet-out and wet-through of the resin, and will stay in place once the resin is applied. Fabrics, like rovings and chopped strands, come with specific sizings or binder systems that promote adhesion to the resin system.

Woven fabrics are fabricated on looms in a variety of weights, weaves, and widths. In a plain weave, each fill yarn or roving is alternately crossed over and under each fiber in the other direction, allowing the fabric to conform to curved surfaces. Woven fabrics are manufactured where half of the fiber strands are laid at right angles to the other half (0 to 90 degrees). Woven fabrics are commonly used in boat manufacturing.

Stitched fabrics, also known as nonwoven, noncrimped, stitched, or knitted fabrics, have optimized strength properties because of the fiber architecture. Stitched fabrics are produced by assembling successive layers of aligned fibers. Typically, the available fiber orientations include the 0-degree direction (warp), 90-degree direction (weft or fill), and ±45-degree direction (bias). The assembly of each layer is then sewn together. These fabrics have been traditionally used in boat hulls. Other applications include light poles, wind turbine blades, trucks, busses, and underground tanks. These fabrics are currently used in bridge decks and column repair systems. Multiple orientations provide a quasi-isotropic reinforcement.

Braided fabrics are engineered with a system of two or more yarns intertwined in such a way that all of the yarns are interlocked for optimum load distribution. Biaxial braids provide reinforcement in the bias direction only with fiber angles ranging from ±15 to ±95 degrees. Triaxial braids provide reinforcement in the bias direction with fiber angles ranging from ±10 to ±80 degrees and axial (0-degree) direction.

4.4.4 *Unidirectional*—Unidirectional reinforcements include tapes, tows, tow sheets, rovings (which are collections of fibers or strands), grids, and pultruded bars. Fibers in this form are all aligned parallel. The aligned fibers are held together with a very light crossweave or tacked in place with a thermoplastic cross yarn or other light carrier. Composites using unidirectional tapes or sheets have high strength in the direction of the fiber. Unidirectional sheets are thin, and multiple layers are required for most structural applications.

Unidirectional sheets of carbon, glass, and aramid fibers are used extensively for wrapping of concrete beams and columns, while unidirectional reinforcing bars or prestressing tendons are applied for reinforcing or prestressing concrete structures, respectively.

4.5—Additives and fillers

4.5.1 *Additives and modifiers*—A wide variety of additives are used in composites to modify material properties and tailor the performance of the laminate. Although these materials are generally used in relatively low quantity by weight compared with resins, reinforcements, and fillers, they perform critical functions. These include affecting shrinkage, fire resistance, emissions, viscosity, mold release, and coloration.

4.5.2 *Fillers*—Fillers reduce the cost of composites; they also frequently impart performance improvements that might not otherwise be achieved by the reinforcement and resin ingredients alone. Fillers can improve mechanical properties, including fire and smoke performance, by reducing organic content in composite laminates. Also, filled resins shrink less than unfilled resins, thereby improving the dimensional control of molded parts. Important properties, including water resistance, weathering, surface smoothness, stiffness, dimensional stability, and temperature resistance, can all be improved through the proper use of fillers.

Commonly used fillers include calcium carbonate, clay, alumina trihydrate, and calcium sulfate. Others include mica, feldspar, wollastonite, silica, glass microspheres, and flake glass.

When used in composite laminates, inorganic fillers can account for 40 to 65% of the composite by weight. In comparison to resins and reinforcements, fillers are the least expensive of the major ingredients.

4.6—Core materials for sandwich structures

Bonded sandwich structures have been a basic component of the composites industry for over 45 years. The concept of using relatively thin, strong face sheets bonded to thicker, lightweight core materials has allowed the industry to build strong, stiff, light, and highly durable structures that otherwise would not be practical. This technology has been demonstrated in boats, trucks, building panels, and, more recently, in bridge decks.

Face sheets can be of almost any material. In the composites industry, the most common face sheet laminates are glass and carbon. The common core materials are foam, syntactic foam, honeycomb, and balsa wood (Fig. 4.1). Some core materials can be shaped, such as a waffle pattern or corrugation, to achieve the desired mechanical properties.

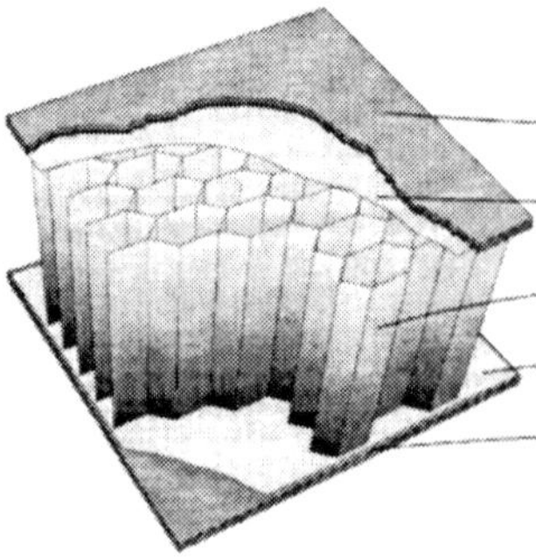

Fig. 4.1—Honeycomb sandwich construction (Hohe et al. 2001).

4.7—Adhesives

Adhesives are used to attach composites to themselves as well as to other surfaces such as concrete. Adhesive bonding is the method of choice for bonding thermoset composites. Several considerations are involved in applying adhesives effectively. The joint or interface connection should be engineered to select the proper adhesive and application method to ensure bond strength. Careful surface preparation and cure are critical to bond performance.

The most common adhesives are acrylics, epoxies, and urethanes. Epoxies provide a high-strength bond with high temperature resistance, whereas acrylics provide moderate temperature resistance with good strength and rapid curing. For applications where toughness is needed, urethane may be selected.

4.8—FRP manufacturing processes

Unique to the composites industry is the ability to create a product from many different manufacturing processes. There are a wide variety of processes available to the composites manufacturer to produce cost-efficient products. Each of the fabrication processes has characteristics that define the type of products to be produced.

4.8.1 *Pultrusion*—Pultrusion is a continuous molding process that combines fiber reinforcements and thermosetting resin. The pultrusion process is used in the fabrication of composite parts that have a constant cross-sectional profile. Typical examples include various rods and bar sections, bridge beams, and decks. Most pultruded laminates are formed using rovings aligned along the major axis of the part. Various continuous strand mats, fabrics (braided, woven, and knitted), and texturized or bulked rovings are used to obtain strength in the cross axis or transverse direction.

The process is normally continuous and highly automated (Fig. 4.2). Reinforcement materials, such as rovings, mats, or fabrics, are positioned in a specific location using preforming shapers or guides to form the profile. The reinforcements are drawn through a resin bath or wet-out where the material is thoroughly coated or impregnated with a liquid thermosetting resin. The resin-saturated reinforcements enter a heated metal pultrusion die. The dimensions and shape of the die will define the finished part being fabricated. Inside the metal die, heat is transferred under precise temperature control to the reinforcements and liquid resin. The heat activates the

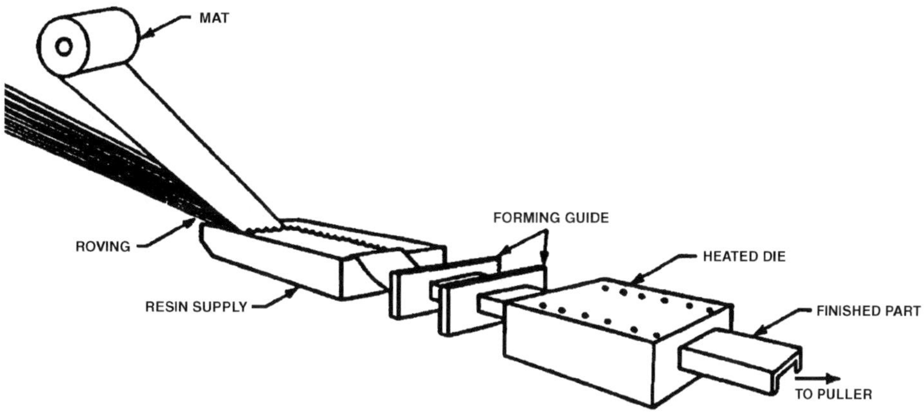

Fig. 4.2—Pultrusion process (Creative Pultrusions, Inc. 1994).

curing or polymerization of the thermoset resin, changing it from a liquid to a solid. The solid laminate emerges from the pultrusion die to the exact shape of the die cavity. The laminate solidifies when cooled, and it is continuously pulled through the pultrusion machine and cut to the desired length. The process is driven by a system of caterpillar or tandem pullers located between the die exit and the cutoff mechanism.

The process provides flexibility in the design of pultruded profiles. Currently, profiles up to 1830 mm (72 in.) wide and 530 mm (21 in.) high are possible. Pultrusion can manufacture both simple and complex profiles, eliminating the need for extensive post-production assembly of components. Because the process is continuous, length variations are limited to shipping capabilities. This process allows for optimized fiber architectures with uniform color, eliminating the need for many painting requirements.

4.8.2 *Filament winding*—This process takes continuous fibers in the form of parallel strands (rovings), impregnates them with matrix resin, and winds them on a rotating cylinder. The resin-impregnated rovings traverse back and forth along the length of the cylinder. The moving relationship between the rotating surface and the roving/matrix is usually controlled by a computer. A controlled thickness, wind angle, and fiber volume fraction laminate is thereby created. The material is cured on the cylinder and then removed (Fig. 4.3).

Pipes, torsion tubes, rocket cases, pressure bottles, storage tanks, and airplane fuselages are made by this process. For the construction industry, pipes are after-filled with concrete for piles and offer structural systems as described in Chapter 9. Filament winding is also used for concrete column repair (Chapter 8). There can be additional add-on fiber/matrix placement systems to add chopped short-length fibers or particulate materials to increase thickness at low cost. Polyester, vinylester, and epoxy matrix materials are used.

4.8.3 *Compression molding*—Compression molding is the most common method of molding thermosetting materials such as sheet molding compound (SMC) and bulk molding compound (BMC). This molding technique involves compressing materials containing a temperature-activated catalyst in a heated matched metal die using a vertical press.

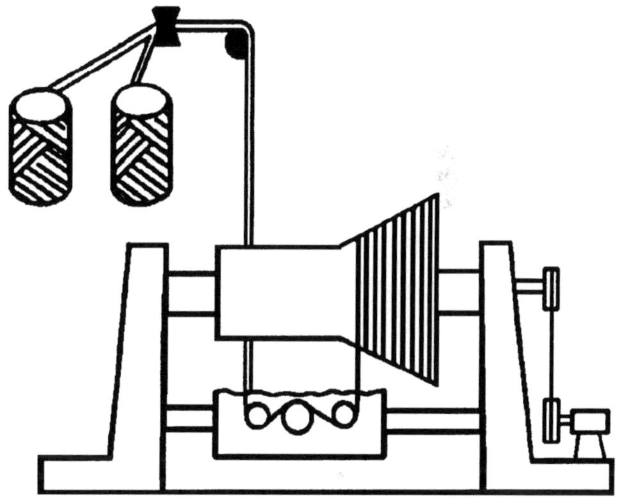

Fig. 4.3—Filament winding process (Mettes 1969).

The molding process begins with the delivery of a high-viscosity uncured composite material to the mold. Mold temperatures typically are in the range of 177 to 204 °C (350 to 400 °F). As the mold closes, composite viscosity is reduced under heat and pressure at approximately 7 MPa (1000 psi). The resin and reinforcements flow to fill the mold cavity.

While the mold remains closed, the thermoset material undergoes a chemical change or cure that permanently hardens it into the shape of the mold cavity. Mold closure times vary from 30 seconds up to several minutes depending on the design and material formulation.

When the mold opens, parts are ready for finishing operations such as deflashing, painting, bonding, and installation of inserts for fasteners. By varying the formulation of the thermoset material and the reinforcements, parts can be molded to meet applications ranging from automotive Class A exterior body panels to structural members such as automobile bumper beams.

4.8.4 *Resin transfer molding*—In resin transfer molding (RTM) (Fig. 4.4), a closed mold process, reinforcement material is placed between two matching mold surfaces that are then closed and clamped. A low-viscosity thermoset

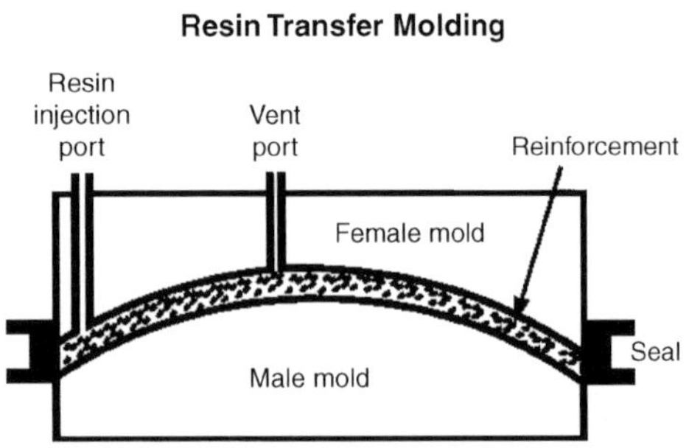

Fig. 4.4—Resin transfer molding.

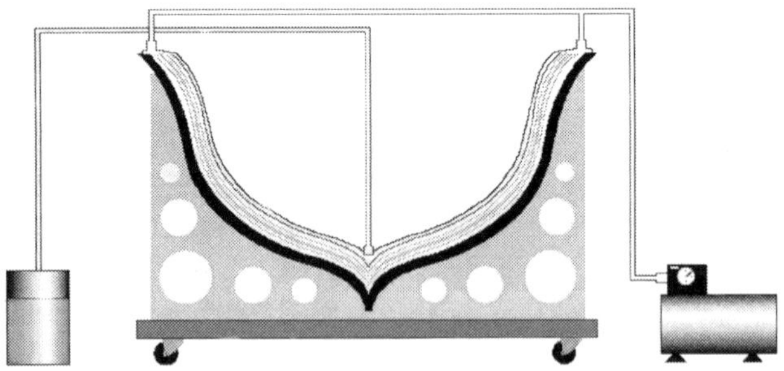

Fig. 4.5—Diagram of VARTM process.

resin is injected under moderate pressure 350 to 690 kPa (50 to 100 psi) into the mold cavity through a port or series of ports within the mold. The resin is injected to fill all voids within the mold and to penetrate all surfaces of the reinforcing materials. The reinforcements may include a variety of fiber types in various forms, such as continuous fibers or mats. The part is typically cured with heat.

RTM is compatible with a variety of resin systems including polyester, vinylester, epoxy, phenolic, modified acrylic, and hybrid resins such as polyester and urethane.

4.8.5 *Vacuum-assisted resin transfer molding*—Vacuum-assisted resin transfer molding (VARTM) parts are made by placing dry fiber reinforcing fabrics into a mold, applying a vacuum bag to the open surface, and pulling a vacuum, while at the same time infusing a resin to saturate the fibers until the part is fully cured (Fig. 4.5).

This process has been used in many new and large applications ranging from turbine blades to bridge decks. Unique to this process is the manufacturing method that allows the efficient processing of VARTM to produce large structural shapes that are generally void-free. This process has been used to make both thin and very thick laminates. In addition, complex shapes with unique fiber architectures allow the fabrication of large parts that have a high structural performance.

A ribbed distribution layer is placed under the vacuum bag, and the resin flow is initially along the surface of the part before being pulled by the vacuum through the dry fabrics.

4.8.6 *Hand lay-up*—Hand lay-up is the oldest and simplest method for producing FRP laminates. Capital investment for hand lay-up processes is relatively low. The most expensive piece of equipment is typically a spray gun for resin and gel coat application. Some fabricators pour or brush the resin into the molds so that a spray gun is not required for this step. There is virtually no limit to the size of the part that can be made. The molds can be made of wood, sheet metal, plaster, or FRP composites.

In the hand lay-up (or wet lay-up) process, highly soluble resin is sprayed, poured, or brushed into a mold. The reinforcement is then wet-out with resin. The reinforcement is placed in the mold. Depending upon the thickness or density of the reinforcement, it may receive additional resin to improve wet-out and allow better drapeability into the mold surface. The reinforcement is then rolled, brushed, or applied using a squeegee to remove entrapped air and to compact it against the mold surface. Curing may be at room temperature or elevated temperatures using roll-in ovens or lamps.

Hand lay-up is also widely applied in the field for strengthening structures with FRP sheets.

4.8.7 *Centrifugal casting*—Centrifugal casting is suited to the production of hollow parts such as pipes. Fiber reinforcement in the form of fabric is placed inside a mandrel, and the resin is sprayed onto it from inside the mandrel rotating at high speed. The centrifugal force pushes the resin outwards, impregnating the reinforcement while minimizing the void ratio.

CHAPTER 5—PROPERTIES, TEST METHODS, AND NONDESTRUCTIVE EVALUATION

5.1—Introduction

This chapter lists some of the reported test methods that have been used to evaluate various properties of FRP materials used as internal or external reinforcement for reinforced and prestressed concrete. Also included are discussions of typical FRP products and properties used in applications involving concrete, as well as a brief description of nondestructive evaluation techniques for FRP systems. ACI 440.3R should be consulted for information on generally accepted test methods for various properties of FRP reinforcement or strengthening materials. A number of relevant ASTM standards also exist for the determination of physical properties of FRP materials such as specific gravity, coefficient of thermal expansion, and fiber volume fraction. These are not discussed in detail in this document.

5.2—Typical properties of currently available products

Advances in FRP technology, increased use of FRP systems in the construction industry, and steadily decreasing costs of FRP materials have led to the development of a wide range of available FRP products for a variety of specific applications. It is beyond the scope of this document to list all currently available FRP products and systems.

The properties of currently available FRP systems vary significantly depending on their specific formulation, constituents, and manufacturing method. The reader is reminded that the material properties of FRPs are highly directionally dependent, and the properties typically quoted are in the direction of the principal reinforcement. Off-axis properties are typically significantly different than those in the fiber direction. In the fiber direction, the products typically display linear elastic behavior to failure (Fig. 5.1), but hybrid combinations, such as carbon and glass or carbon and aramid,

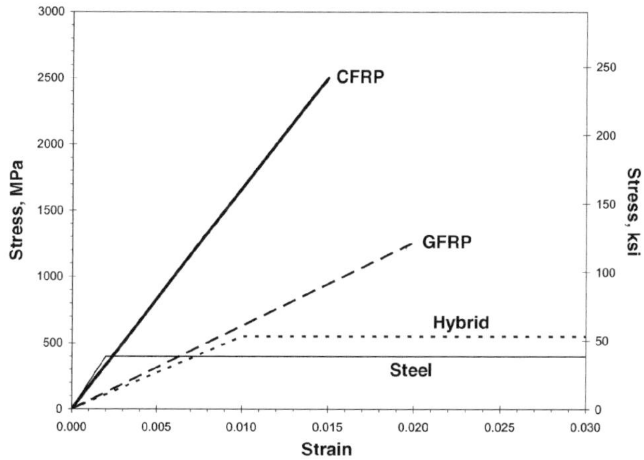

Fig. 5.1—Typical stress-strain curves for FRP products. Based on data from Teng et al. (2002), Tanužs et al. (1996), and Apinis et al. (1998).

display monotonic stress-strain behavior (Fig. 5.1) similar to that of steel (Tamužs et. al. 1996; Apinis et. al. 1998).

Tables 5.1 and 5.2 present overviews of the properties of some typical FRP products for internal reinforcement and external strengthening, respectively, of reinforced concrete structures. Included in the tables is information on FRP bars, rods, sheets, plates, and tapes. The tabulated information is that reported by the manufacturer. In Table 5.2, the material properties are reported using one of two methods specified by ACI 440.3R. Method 1 is based on composite area and uses the measured width and thickness of the specimen to determine the cross-sectional area of the specimen for calculation of the tensile strength and modulus of elasticity. Method 2 is based on equivalent fiber area and uses the equivalent thickness of a fiber layer without resin and the measured width for calculation of the tensile strength and modulus of elasticity. Specific information on the potential applications of each of these systems can be obtained by contacting the FRP manufacturers (also included in the tables). These tables should not be considered as endorsements for a particular FRP product.

TechFab LLC (2006) produces a very light carbon/epoxy grid with strengths ranging from 14 kN/m (1000 lb/ft) to over 220 kN/m (15,000 lb/ft) with grid openings from approximately 12 mm (0.5 in.) square to over 50 mm (2 in.) square. Because of the flexible nature of the textile process used to manufacture grids, grid opening sizes and strengths can be customized to suit the many various concrete product uses.

Table 5.1—Selected properties of typical currently available FRP reinforcing products as reported by the manufacturer

Reinforcement type	Designation	Diameter, mm (in.)	Area, mm^2 ($in.^2$)	Tensile strength, MPa (ksi)	Elastic modulus, GPa (ksi)
Deformed steel	No. 3	9.5 (0.375)	71 (0.11)	420* (60)	208 (30,000)
Pultrall Inc. (2005)					
V-ROD CFRP rod	3/8	9.5 (0.375)	71 (0.11)	1596 (230)	120 (17,400)
V-ROD GFRP rod	3/8	9.5 (0.375)	71 (0.11)	852 (123)	43 (6700)
Autocon Composites Inc. (2006)					
NEFMAC GFRP rod	G10	N/A	79 (0.12)	600 (87)	30 (4400)
NEFMAC CFRP rod	C16	N/A	100 (0.16)	1200 (174)	100 (14,500)
Hughes Brothers (2005)					
Aslan 100 GFRP rod	No. 3	9.5 (0.375)	84 (0.13)	760 (110)	41 (5900)
Aslan 200 CFRP rod	No. 3	9.0 (0.360)	65 (0.10)	2068 (300)	124 (18,000)
Mitsubishi (2005)					
LEADLINE™ CFRP rod	Round	12 (0.470)	113 (0.18)	2255 (327)	147 (21,300)
Sika Corp. (2007)					
CarboDur CFRP rod	3/8	9.5 (0.375)	71 (0.11)	2800 (406)	155 (22,500)

*Specified yield strength.

5.3—Test methods for mechanical properties

The basic mechanical properties of FRP materials can be determined by applying tensile, compressive, and shear loadings. Flexural testing is also used in some cases. Static material properties are usually of primary interest.

5.3.1 *Tension test methods*—Axial tension testing of high-strength unidirectional composites is often a challenge because load should be transmitted from the testing apparatus to the specimen via shear, and the shear strength of a unidirectional composite is typically much lower than its axial tensile strength. Further, shear gripping will load the external fibers more than the internal ones causing shear lag and progressive fiber failure. To avoid these problems, end tabs are required when testing flat laminates, as described in detail in ACI 440.3R. Special anchors are required for testing of FRP rods and bars by inserting their ends into steel cylinders that are subsequently filled with either a polymer resin or a cement-based grout (also described in ACI 440.4R). For grid-type FRP reinforcements, linear test specimens can be prepared by cutting away extraneous material in such a way as not to affect the performance of the part to be tested. ASTM D 618 (Porter and Barnes 1991; ACI 440.3R) recommends leaving a minimum 2 mm (0.08 in.) projection of crossbars.

5.3.2 *Compression test methods*—It has been said that there is no true axial compression test for composites. The mode of failure is buckling, ranging from buckling of the entire specimen cross section or local microbuckling of individual fibers. Thus, the greater resistance to buckling the test fixture provides, the higher the compressive strength values obtained. For flat laminate FRP composites, many axial compression test methods in current use are some variation of the Celanese compression test as in ASTM D 3410. This test uses a thin, straight-sided specimen that looks very much like an axial tension specimen except that the distance between tabs is much smaller. For FRP rods, tests can be carried out according to ASTM D 695. Common problems associated with compression testing of FRPs are discussed in detail by Harper et al. (1995).

5.3.3 *Shear test methods*—Several ASTM standards are available for determining the shear properties of composite

Table 5.2—Selected properties of typical currently available FRP strengthening systems as reported by the manufacturer

FRP system	Fiber type	Weight, g/m^2 (lb/ft^2)	Design thickness, mm (in.)	Tensile strength, MPa (ksi)	Tensile elastic modulus, GPa (ksi)	ACI 440.3R test reporting method
Fyfe Co. LLC (2005)						
Tyfo SEH51 sheet	Glass	915 (0.19)	1.3 (0.052)	575 (83.4)	26.1 (3785)	Method 1
Tyfo SCH41 sheet	Carbon	644 (0.14)	1.0 (0.040)	985 (143)	95.8 (13,900)	Method 1
Sika Corp. (2007)						
SikaWrap Hex 100G sheet	Glass	913 (0.19)	1.0 (0.040)	531 (77)	23.6 (3430)	Method 1
SikaWrap Hex 103C sheet	Carbon	618 (0.13)	1.0 (0.040)	717 (104)	65.1 (9450)	Method 1
CarboDur S plate	Carbon	1800 (0.37)	1.2 to 1.4 (0.048 to 0.055)	2800 (406)	165 (23,900)	Method 1
CarboDur M plate	Carbon	1900 (0.39)	1.2 (0.048)	2400 (348)	210 (30,500)	Method 1
CarboDur H plate	Carbon	1900 (0.39)	1.2 (0.048)	1300 (189)	300 (43,500)	Method 1
BASF (2006)						
MBrace EG 900 sheet	Glass	900 (0.19)	0.37 (0.015)	1517 (220)	72.4 (10,500)	Method 2
MBrace AK 60 sheet	Aramid	600 (0.12)	0.28 (0.011)	2000 (290)	120 (17,400)	Method 2
MBrace CF 130	Carbon	300 (0.062)	0.17 (0.007)	3800 (550)	227 (33,000)	Method 2
MBrace CF 160	Carbon	600 (0.124)	0.33 (0.013)	3800 (550)	227 (33,000)	Method 2
S&P 100/1.4	Carbon	—	1.4 (0.055)	2700 (390)	159 (23,000)	Method 1
Hughes Brothers (2005)						
Aslan 400 plate	Carbon	—	1.4 (0.055)	2400 (350)	131 (19,000)	Method 1
Aslan 500 tape	Carbon	—	2.0 (0.079)	2068 (300)	124 (18,000)	Method 1
Aslan 500 tape	Carbon	—	4.5 (0.177)	1965 (285)	124 (18,000)	Method 1

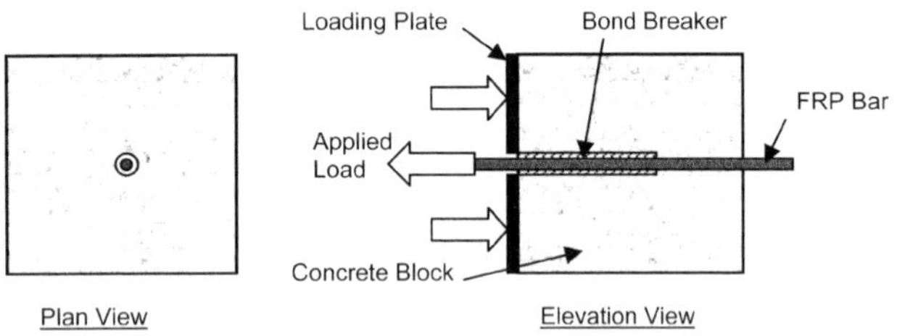

Fig. 5.2—Pullout test specimen.

materials. For flat FRP laminates, direct in-plane shear test methods include the V-notched beam test (ASTM D 5379/D 5379M), the two-rail and three-rail shear tests (ASTM D 4255/D 4255M), and the 45-degree laminate tensile shear test (ASTM D 3518/D 3518M). For FRP rods, shear test devices have been constructed so that a rod-shaped test specimen is sheared on two planes as in ACI 440.3R.

5.3.4 *Flexural test methods*—Flexural tests are relatively easy to perform; thus, they are fairly popular. The difficulty is that flexural testing does not directly provide information on basic material properties. The stress state on the loaded side of the specimen is compression, on the other side is tension, and at the neutral axis is pure shear. Usually, the shear component is minimized by making the test specimen long relative to its thickness (a ratio of approximately 32:1 is commonly used) (Adams 1992). Although flexural tests do not directly provide basic design data, their use is often justified on the basis that the material is subjected to flexural loading in service.

For testing FRP rods in flexure, ASTM D 4476 describes a standard test method for flexural properties of fiber-reinforced pultruded rods. This method has been used for FRP reinforcing bars with some modifications, as in Wang et al. (2002). FRP laminates can also be tested in flexure in three- or four-point bending as described in Adams (1992).

5.3.5 *Bond test methods for internal FRP reinforcement*—Bond characteristics influence the mechanism of load transfer between FRP reinforcement and concrete, and therefore control the concrete crack spacing, crack width, required concrete cover to the reinforcement, and the reinforcement development length. The behavior of reinforced concrete structures thus depends on the integrity of the bond. Many test methods have been developed and carried out to determine the bond strength of FRP bars and rods in concrete.

5.3.5.1 *Pullout tests*—The concentric pullout test is a popular test method adopted by researchers for comparative bond assessment for FRP reinforcement. In this method, an FRP bar specimen is embedded in a concrete block, with the embedded length of the FRP rod typically being five times the rod diameter, and the bar is pulled in tension as shown in Fig. 5.2. The average bond stress and bond stress versus slip at the loaded end can thus be obtained. The pullout test is used for comparative tests, but is not considered a valid test to determine the development that is needed for the design embedment length for flexural beam design. Additional information and specimen fabrication details are given in ACI 440.3R.

The ring test (Tepfers and Olsson 1992) is carried out as a pullout test, where the specimen is a concrete cylinder with the studied reinforcing bar embedded along the center axis of the cylinder. This test allows for determination of the splitting tendency of the bar in all stages of loading up to bond failure by determination of the angle between bond forces and the bar axis. The bond length corresponds to the height of the cylinder, which is chosen to be three diameters of the bar, so the bond stress along the bond length becomes practically evenly distributed. The cylinder is cast in a thin steel tube (the ring), which becomes a part of the test specimen.

5.3.5.2 *Flexural bond tests*—Pullout tests, while extremely useful in many circumstances, are not representative of stress field conditions in an FRP reinforced concrete beam. Flexural beam tests, such as the beam-end test, solve some of the stress field discrepancies that are present in pullout tests, and thus offer the advantages of representing the beam stress field more closely.

The beam-end test method is described in detail for steel reinforcing bars in ASTM A 944, and is specified as a test for the relative bond strength of reinforcing bars in concrete. In this method, a bar is embedded in a concrete block, and a tensile load is introduced into the bar using a special test setup as shown schematically in Fig. 5.3. This test setup has the advantages of simulating the stress field condition in a reinforced concrete member in flexure.

Various additional flexural test methods exist and have been used to study the bond behavior of FRP bars embedded in concrete. Several of these are shown schematically in Fig. 5.4, including (a) the simple beam test (Larralde et al. 1994); (b) the notched beam test (Daniali 1992); (c) the hinged beam test (Magnusson 1997); (d) the spliced beam test (Tighiouart et al. 1999); (e) the cantilever beam test (Fish 1992); and (f) the trussed beam test. If these tests are done upside down, then the effects of self-weight are practically eliminated.

5.3.5.3 *Direct axial tension test*—Two types of direct axial tension tests have been used with a concrete specimen having a continuous bar embedded in its center (Hasau et al. 1990). In the first type, load is directly applied to the bar protruding from the concrete ends. In the second type, load is applied to the concrete. A concrete specimen is cast with an FRP bar embedded at its center and a notch at midlength to allow for the formation of the first crack at a preset location (Fig. 5.5). Bond strength is studied indirectly through the observation of crack spacing and width in the concrete block.

5.3.6 *Bond test methods for externally bonded FRP reinforcement*—Flexural and shear strengthening techniques of bonding FRP plates or sheets to reinforced concrete elements is now a method of choice for structural rehabilitation. Failure of these bonded systems before achieving their FRP capacity, however, can result from a number of mechanisms collectively referred to as delamination or debonding. Teng et al. (2002) provide a thorough review of bond-related failure mechanisms and the present state of the art for modeling these. The behavior of the FRP-adhesive-concrete interface region, which should include the region of cover concrete adjacent to the interface, is collectively interpreted as bond behavior. Transfer of stresses through the cover concrete, for instance, can lead to failures through the cover concrete that are nonetheless affected by the properties of the bonding adhesive. Complicating interface region behavior is the mixed mode behavior of the debonding process and the heterogeneous material properties in this region.

Bond behavior is a critical parameter in strengthening a concrete structure with externally bonded FRP, and numerous test methods have been presented. Test methods

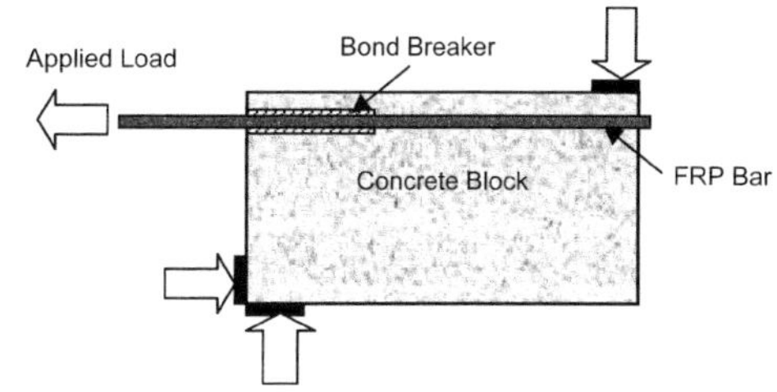

Fig. 5.3—Beam-end test method.

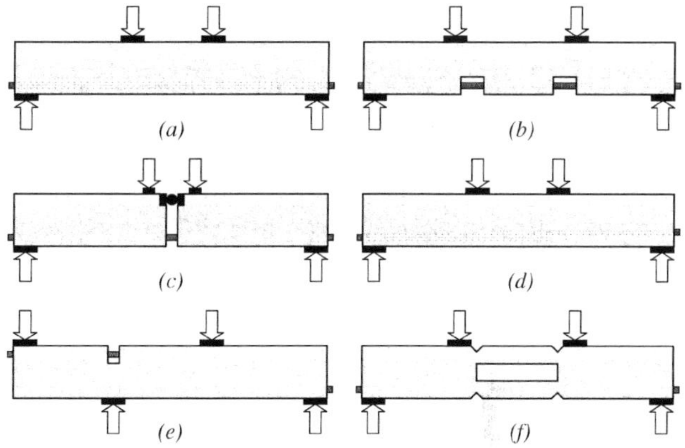

Fig. 5.4—Additional flexural bond test method configurations: (a) simple beam test; (b) notched beam test; (c) hinged beam test; (d) spliced beam test; (e) cantilever beam test; and (f) trussed beam test.

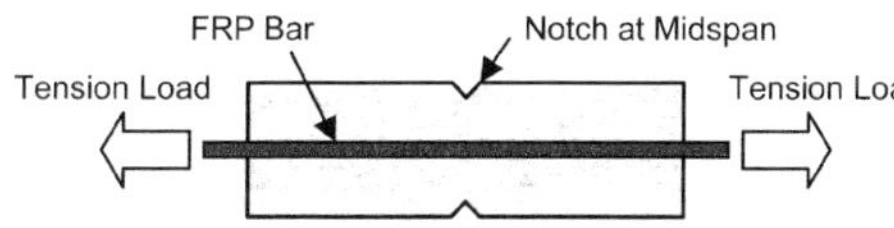

Fig. 5.5—Direct axial tension test.

can be broadly classified into three types: shear bond, tension, or mixed mode. Recent research has focused largely on mixed mode bond test methods in an effort to accurately characterize the FRP-concrete bond, which in practice experiences a combination of shear and normal stresses. A summary of bond test methods for externally-bonded FRP systems is provided by Ueda and Dai (2005).

5.3.6.1 *Shear bond type tests*—Because externally bonded FRP materials are most often used in configurations where stress transfer across the interface is dominated by shear (Mode II) stresses, various shear bond test methods are available in the literature. The various methods can be used to study the local interfacial shear bond behaviors by studying the bond strain distribution (Ueda and Dai 2005). Schematics showing the configurations of various types of shear bond tests are given in Fig. 5.6, including: (a) the single-lap shear bond test (Bizindavyi and Neale 1999; Chajes et al. 1996); (b) the push-apart; (c) the pull-apart double-lap shear bond test (Horiguchi and Saeki 1997; Brosens and Gemert 1997); (d) the bending-type shear bond test (Horiguchi and Saeki 1997); and (e) the inserted-type shear bond test (Fukuzawa et al. 1997).

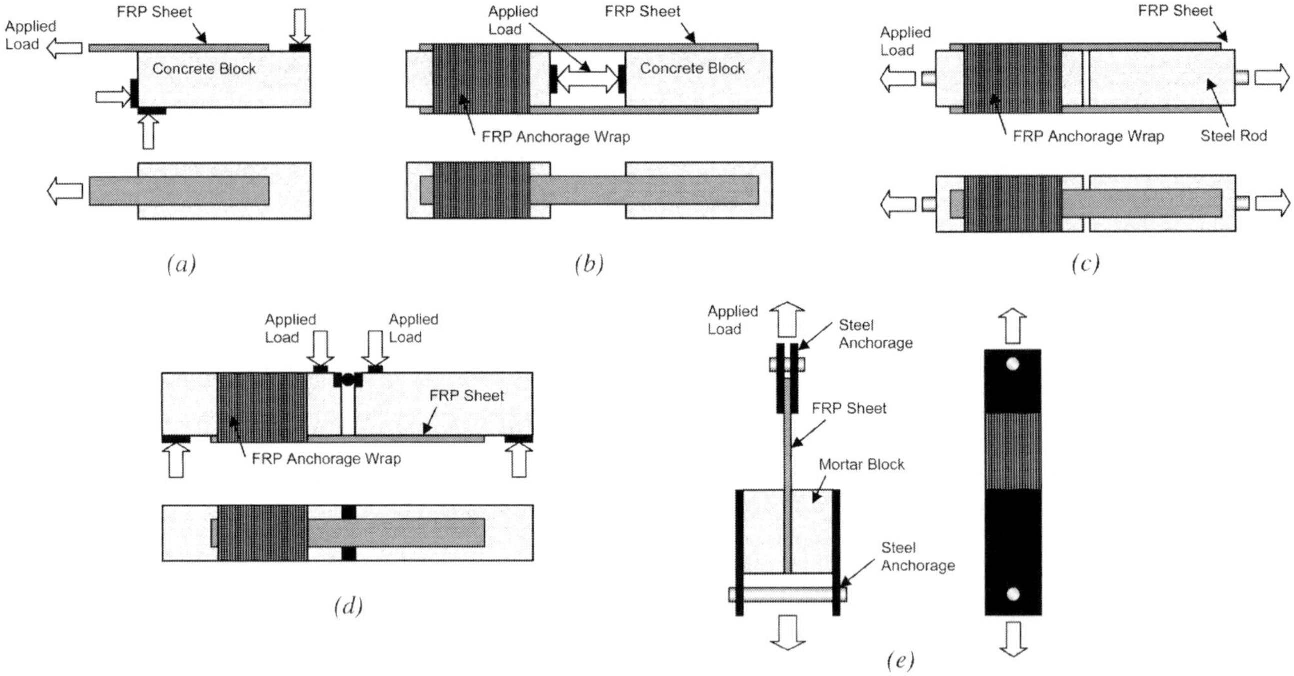

Fig. 5.6—Various shear-bond test method configurations: (a) single-lap shear bond test; (b) push-apart; (c) pull-apart double-lap shear bond test; (d) bending-type shear bond test; and (e) inserted-type shear bond test.

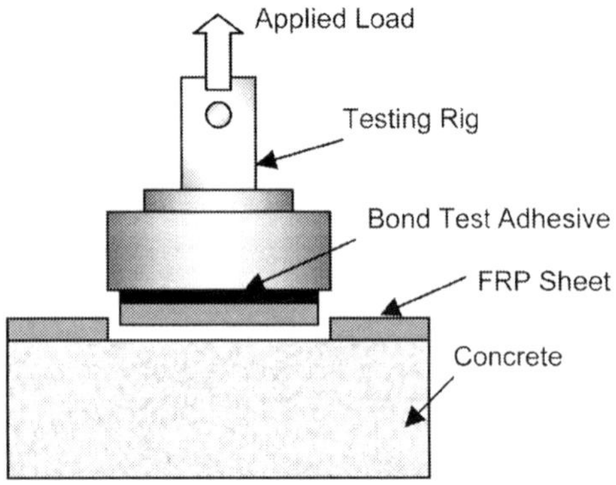

Fig. 5.7—Direct tension pulloff test method (Xu et al. 2002).

5.3.6.2 *Tension-type bond tests*—At least three test methods are available in the literature to evaluate the strength of the FRP-concrete bond subject to pure normal (Mode I) stresses. Figure 5.7 shows a schematic of the direct tension pulloff test method (Horiguchi and Saeki 1997), which is also presented in detail in ACI 440.3R. In this method, a special testing rig is used to pull a circular disc of bonded FRP away from the surface of the concrete in direct tension. The circular area is typically created using a core drill, and an adhesion disc is used, with an appropriate adhesive, to apply an increasing tensile load to the disc until bond failure occurs. For well-bonded FRP materials, the failure surface is in (and thus limited by) the substrate concrete. An alternate shear test arrangement uses a similar test setup, but twists the adhesion disc off of the substrate; this torsional pulloff test is not included in ACI 440.3R. Other tension-type bond test methods not discussed in detail herein, or in ACI 440.3R, include the three-point bending pulloff test (Xu et al. 2002) and the wedge splitting test (Dai et al. 2003).

5.3.6.3 *Mixed-mode bond tests*—Mixed mode bond test methods are more representative of the interface bond behavior for concrete structures retrofitted with FRP sheets (Ueda and Dai 2005). Mixed-mode testing methods include the variable angle peel test (Karbhari and Engineer 1996), the beam-type and slab-type dowel tests (Ueda and Dai 2005), the single contoured cantilever beam test (Boyajian et al. 2002), the double cantilever beam test (Kim et al. 2002; Wan et al. 2002), and the modified double cantilever beam test (Wan et al. 2004). These bond test methods are considerably more complex and difficult to perform than those discussed previously, and they are typically used to study the fracture behavior of FRP-concrete interfaces. The reader is encouraged to consult the listed references for more information on mixed-mode bond testing.

Test methods are also available to study tension stiffening and secondary crack formation in applications of externally bonded FRP materials (Nanni et al. 1997).

5.3.6.4 *Bond overlap tests*—Test methods are available to evaluate the strength of FRP to FRP bond overlaps or splices (ACI 440.3R), and can be conducted in single-lap or double-lap configurations.

5.3.7 *Concrete surface preparation for bond testing*—The effects of surface preparation of the concrete on the bond between concrete and FRP are very important, and have been studied by several authors. Chajes et al. (1996) found that mechanical abrasion increased the interfacial bond strength. Shen et al. (2002) also studied the effect of surface roughness on the bond performance between FRP laminates and concrete. Surface preparation was by rotary water jetting to produce different roughness grades, which were classified using a laser profilometer. Specimens were subjected to flexural bond testing and direct tension and torsional pulloff tests. They concluded that moderate surface roughness was necessary for adequate bond properties to be developed and that different CFRP systems tested had different optimal surface roughnesses.

5.4—Durability testing methods

The reduction of strength and the subsequent failure of materials subjected to cyclic or sustained loading combined with environmental effects have emerged as some of the most fundamental problems of engineering materials. Because composite materials are regarded as having good long-term performance, they are used in applications where the degradation of strength and life expectancy by fatigue, creep, and environmental effects would otherwise be a concern. Thus, testing methods to evaluate the susceptibility of FRP materials to these conditions is required.

5.4.1 *Tensile fatigue testing methods*—Fatigue tests on FRPs can be carried out using several specimen configurations, including direct tests on rods using various kinds of gripping mechanisms (ACI 440.3R), and tests on concrete beam specimens reinforced with FRP reinforcements. In the case of performing direct tests on rods, problems are often encountered in gripping the rod, because some grips perform poorly when used for cyclic tests, and failure may occur in the anchoring zone. When performing fatigue tests on concrete beam specimens, the rod is continuously in contact with concrete, which may affect the fatigue behavior of the FRP rod through fretting or friction heating (Adimi et al. 2000).

5.4.1.1 *Concrete prism tensile fatigue test method*—Because of the relatively low modulus of elasticity of FRP reinforcement combined with its high strength, it is very difficult to develop high levels of stress in FRP rods used as tensile reinforcement in concrete beams. A reinforced concrete prism specimen for tensile fatigue testing has been developed (Adimi et al. 2000). Figure 5.8 shows the details of the specimen, where top and bottom blocks of concrete serve as anchors for the FRP rod while the middle block provides the concrete environment for the FRP rod at the part where the failure is expected. The middle block is separated from the end blocks by a bond breaker during casting. This specimen is loaded in tension by collars at the shoulders of the end blocks, and can be tested under any stress amplitude and with the maximum stress almost equal to the tensile strength of the FRP rod. To avoid stress concentrations near the rod and splitting of concrete end blocks during the test, a bond breaker is included, as shown, of approximately five bar diameters, and the end blocks are precompressed near the constriction with a steel compression ring (not shown in the figure).

5.4.1.2 *Direct fatigue tests*—Fatigue tests can also be conducted directly on FRP specimens. These specimens can be rods, flat laminates, or sections of FRP grids. A suggested test method for tensile fatigue of FRP bars is discussed in detail in ACI 440.3R.

5.4.2 *Creep test methods*—FRP materials may fail by creep at strengths below the maximum static strength when subjected to a significant sustained stress for long periods. This creep-rupture strength varies according to the type of FRP, and so the creep-rupture strength should be evaluated when determining the allowable level of sustained tension in FRP used as reinforcement. A suggested test method for creep testing of FRP bars is presented in ACI 440.3R. Traditional methods of performing creep tests, including direct gravity loading, compressed spring loading, or gravity loading using a lever arm (so called creep frame) to permit greater loads, have all been adapted for testing FRP materials.

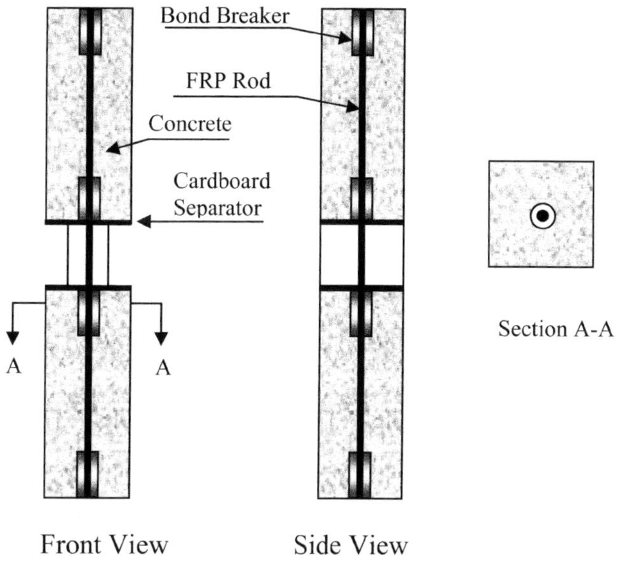

Fig. 5.8—Schematic view of test specimen (adapted from Adimi et al. 2000).

5.5—Nondestructive inspection techniques for FRP materials

Nondestructive inspection is needed to assure the integrity of FRP materials and FRP-strengthened structures over their service life (U.S. Department of Defense 2002). Structures made of traditional materials are all susceptible to damage and environmental degradation in service. As a result, they are periodically inspected. Although decades of experience have shown composites to be reliable structural materials, they are also susceptible to damage and possible environmental degradation. Consequently, structures employing composite reinforcement should also be subjected to periodic nondestructive inspection in service. Test methods should be designed to evaluate a material system economically and quickly. This section discusses quality assurance and nondestructive testing of composites.

5.5.1 *Nondestructive inspection*—Nondestructive inspection (NDI) tests are used to find flaws and damage such as resin starvation, resin richness, wrinkles, ply bridging, discoloration due to overheating or lightning strike, impact damage by any cause, foreign matter, blisters, and disbanding. Some of the discrepancies are readily discernible by a visual inspection (Crane 2000; Pethrick 2000; U.S. Department of Defense 2002).

Composite NDI techniques commonly used in production of composites include visual, ultrasonic, thermography, laser shearography, tap, and x-ray (Crane 2000; U.S. Department of Defense 2002). Advantages of thermography and shearography are that they are large area, noncontact methods that create permanent records. This is useful for monitoring possible damage growth over the life of a structure.

Despite the recognized importance of NDI to ensure reliability of composite structures, there has not been a great deal of research on NDI methods for FRP-strengthened concrete structures (Newman and Zweben 2002a). This is a significant issue considering the long service life of civil engineering structures.

5.5.1.1 *Visual inspection*—NDI by visual means is by far the oldest and most economical NDI method (U.S. Department of Defense 2002). Consequently, visual inspection is performed routinely as a means of quality control and damage assessment.

Once detected, the affected area becomes a candidate for closer inspection. Flashlights, magnifying glasses, mirrors, and borescopes are employed as aids in the visual inspection of composites. Visual inspection cannot find internal flaws in the composite, such as delaminations, disbands, or matrix crazing. In addition, impact damage is frequently not visible to the naked eye, and tight surface cracks and edge delaminations may not be detected visually. Therefore, visual inspection techniques need to be supplemented by other methods of nondestructive testing.

5.5.1.2 *Tap testing*—Sometimes referred to as audio, sonic, or coin tap, this technique makes use of frequencies in the audible range (10 to 20 Hz) to detect damage (U.S. Department of Defense 2002). Although it can be an accurate method in the hands of skilled and experienced personnel, tap testing is highly subjective, time consuming, and does not create a permanent test record. Despite these limitations, tap testing is a common technique used for the detection of delamination or disbonding. The method is performed by tapping the inspection area with a solid round disk or lightweight hammer and listening to the audible response of the structure. A clear, sharp, ringing sound is indicative of a well-bonded solid structure, while a dull sound indicates a damaged area.

5.5.1.3 *Ultrasonics*—Ultrasonic inspection has proven to be a useful tool for the detection of internal delaminations, voids, or inconsistencies in composite components not otherwise discernible using visual or tap testing (Tittmann and Crane 2000; U.S. Department of Defense 2002).

A high-frequency sound wave is introduced into the structure, and may be directed to travel normally to its surface, along the surface, or at some predefined angle to the surface. Different directions are often used because flaws may not be noticeable from only one direction.

When an ultrasonic wave strikes an interrupting object, part of the wave or energy is transmitted through the object, and the rest of the energy is reflected back to the surface. The disrupted or diminished sonic energy is then picked up by a receiving transducer and converted into a display on an oscilloscope or a chart recorder. The display allows the operator to comparatively evaluate the discrepant indications against those areas known to be good.

5.5.1.4 *Radiography*—Radiography, or x-ray as it is often referred to, allows a view into the interior of the part (Crane 2000; U.S. Department of Defense 2002) by passing x-rays through the part while recording the absorption of the rays onto an x-ray sensitive film. Because the method records changes in total density through its thickness, it is not a preferred method for detecting defects, such as delaminations, that are in a plane normal to the ray direction.

Although radiography may be useful in a factory environment, it is expensive and not practical for field application. Furthermore, most composites are nearly transparent to x-rays, so low-energy rays should be used. Opaque penetrant can be used to enhance the visibility of surface breaking defects, although this is generally not available for in-service inspections.

Operators should be protected by sufficient lead shields because the possibility of exposure exists either from the x-ray tube or from scattered radiation. Maintaining a distance from the x-ray source is essential.

5.5.1.5 *Shearography*—Shearography is an optical NDI technique that detects defects by measuring variations in reflected light (speckle pattern) from the surface of the object when stressed (Krishnaswamy 2000; U.S. Department of Defense 2002; Newman and Zweben 2002a,b). Using a laser light source, an original image of the illuminated surface is recorded via a video image. The part is subsequently stressed by heating, mechanical loading, changes in pressure, or acoustic vibrations during which a second video image is made. Changes in the surface contour caused by debonding or delaminations become visible on the video display. Shearography is being used in production environments for rapid inspection of bonded composite structures.

5.5.1.6 *Thermography*—Thermal inspection includes all methods in which heat-sensing devices are used to measure temperature variations (Thomas 2000; U.S. Department of Defense 2002; Newman and Zweben 2002a,b). The basic principle of thermal inspection consists of measuring or mapping surface temperatures when heat flows from, to, or through a test object.

All thermographic techniques rely on differentials in thermal conductivity between normal, defect-free areas and those having a defect. Normally, a heat source is used to elevate the temperature of the specimen being examined while the surface heating effects are observed. Because defect-free areas conduct heat more efficiently than areas with defects, the amount of heat that is absorbed or reflected indicates the quality of the bond. The type of defects that affect the thermal properties include debonding, cracks, impact damage, panel thinning, and water ingress. Thermal methods are most effective for thin laminates or for defects near the surface.

CHAPTER 6—PERFORMANCE OF CONCRETE MEMBERS WITH INTERNAL FRP REINFORCEMENT

This chapter summarizes research findings regarding the performance of FRP as the main structural reinforcement for non-prestressed concrete members. FRP internal reinforcement is widely used in commercial applications. Much of the development of design guidelines for internal FRP reinforcement dates back to the pioneering work of Nanni (1993). Additional work by others facilitated and contributed to the design guidelines in their current format (Jaeger et al. 1997; GangaRao and Vijay 1997; Theriault and Benmokrane 1998). The widespread implementation of FRP as reinforcement for concrete structural members requires: 1) a comprehensive understanding of how these two materials behave together as a structural system; and 2) analytical techniques that reliably predict the composite behavior. In this regard, three important physical characteristics of FRP materials should be considered: 1) high tensile strength; 2) low modulus of elasticity; and 3) linear-elastic brittle behavior to failure. Substitution of FRP for steel on an equal area basis typically results in significantly higher deflections with wider crack widths and greater flexural strength. As a consequence, serviceability criteria are often governing parameters in design considerations.

6.1—Strength

Flexural failure of concrete members reinforced with currently available FRP materials is governed by either concrete crushing or FRP tensile rupture. Such failure modes are brittle and differ from the behavior of concrete beams under-reinforced with steel. In addition, the shear capacity of concrete beams is reduced as a result of increased crack width and reduced size of compressive stress blocks.

6.1.1 *Flexural strength*—Nawy and Neuwerth (1971) monotonically tested 20 simply supported rectangular beams reinforced with GFRP and steel reinforcing bars. The beams failed by compression of the concrete; hence, they did not develop the full capacity of the FRP bars. The authors suggested that because the modulus of FRP bars was only slightly higher than that of concrete, limited tensile stresses could be transmitted from the concrete to the FRP reinforcement. In a second study, Nawy and Neuwerth (1977) found that the behavior of the beams with respect to cracking, ultimate load, and deflection could be predicted with the same degree of accuracy as for steel reinforced concrete beams. Satoh et al. (1991) reached similar conclusions.

Faza and GangaRao (1992) investigated the flexural performance of simply supported rectangular concrete beams reinforced with GFRP reinforcing bars. They concluded that, to take advantage of the high FRP reinforcing bar ultimate strength, use of high-strength concrete instead of normal-strength concrete (69 versus 27.6 MPa [10 versus 4 ksi]) is essential. The ultimate moment capacity of high-strength concrete beams increased by 90% when an equal area of FRP reinforcing bars of ultimate tensile strength of 896 MPa (130 ksi) was used instead of mild steel reinforcing bars. Sand-coated FRP reinforcing bars, in addition to high-strength concrete, were found to increase the cracking moment of the beams and to reduce the crack widths in addition to eliminating the sudden propagation of cracks toward the compression zone.

Bank and Xi (1992) tested nine simply supported slabs having shear-span-to-effective-depth ratios of approximately three and reinforced with a variety of molded and pultruded GFRP gratings. Following initial cracking, flexural cracks developed in the constant moment region at regular spacing of approximately 76 mm (3 in.). With increasing load, diagonal tension shear cracking developed in the shear span. Flexural compression failure occurred in three of the first six slabs, and the remaining slabs failed in shear. The slabs that failed in compression had the lowest concrete strength. In several of the shear failures, the concrete below the reinforcement in the shear span was completely separated from the slab. In all slabs, the experimental shear force V_{exp} was significantly larger than V_c as predicted by ACI 318.

Nanni et al. (1992) tested five concrete beams reinforced with hybrid reinforcing bars, steel deformed bars, and FRP reinforcing bars. Load-deflection behavior for the different reinforcing bar types was characterized as follows: 1) for steel reinforcing bars, a typical three-stage behavior of an under-reinforced concrete beam consisting of uncracked-section, cracked-section linear elastic to yield, and post-yield of reinforcement; 2) for FRP reinforcing bars, a two-stage behavior consisting of uncracked section behavior followed by cracked-section linear-elastic behavior to failure; and 3) for hybrid reinforcing bars, a three-stage behavior typical of under-reinforced steel beam characterized by an uncracked section and linear-elastic response followed by steel core yielding before ultimate failure. Test results showed that, relative to ultimate flexural capacity, coating the FRP and hybrid reinforcing bars with sand increased flexural capacity by approximately 25%. Smaller crack widths and higher postcrack flexural rigidity were also reported for the sand-coated reinforcing bars as compared with the corresponding uncoated reinforcing bars.

Matthys and Taerwe (2000) carried out four-point bending tests on eight reinforced concrete slabs. The materials used to reinforce the slabs were steel, carbon, and hybrid carbon-glass NEFMAC grids and CFRP bars. The authors found that the design of concrete structures with FRP reinforcement was mainly governed by serviceability criteria. In general, crack control was less restrictive than deflection control. A deformability index was defined as the ratio of the deflection at ultimate load to deflection at service load. For the FRP-reinforced slabs of this study, the deformability index was between 10 and 110% higher than the deformability index obtained for steel-reinforced slabs. The margin between the ultimate load and the service load of the FRP-reinforced slabs was on the order of 2.5 to 4.5 times higher than for steel-reinforced structures.

Nanni and Faza (2002) outlined design and construction using FRP bars with an objective to provide a meaningful overview of the design protocol for FRP-reinforced concrete members. Thiagarajan (2003) presented an analytical and experimental study on the flexural behavior of concrete beams reinforced with sand-coated CFRP rods and found that the analytical capacities, predicted using equilibrium equations; strain compatibility condition; and finding the concrete compressive forces by integrating Hognestad's equations, were within 8% of the experimental capacities.

6.1.2 *Beam shear*—Larralde (1992) tested a series of eight FRP-grating/concrete composite slabs in which the shear span-depth ratio and concrete deck thickness was varied in an attempt to force different types of failures. Test results showed that, for samples with a shear span-depth ratio of 7.7 or greater, failure occurred by crushing of the concrete. For these samples, the calculated flexural capacity was very close to the test results. For shear span-depth ratios of 5 or less, failure occurred as a result of diagonal tension cracking.

Shehata (1999) investigated the effect of bend radius, the crack angle, the stirrup anchorage, the stirrup spacing, and the material type of flexural reinforcement in tests on 52 panel specimens. The following limitations were proposed for detailing FRP stirrups to achieve a capacity of at least 50% of the guaranteed strength parallel to the fibers: 1) the bend radius should not be less than 50 mm (2 in.) or four times the effective bar diameter, whichever is greater; and 2) the tail length should not be less than 70 mm (2.75 in.), or six times the effective bar diameter, whichever is greater. Beams reinforced with CFRP strands for flexure showed less concrete contribution to shear strength than did beams reinforced with steel strands. This observation was attributed to the wide cracks, the small depth of the compression zone, and poor dowel action associated with the use of FRP as longitudinal reinforcement. Beams with GFRP stirrups showed better performance than those with CFRP stirrups.

Razaqpur and Mostofinejad (1999) tested four continuous beams that were reinforced with FRP grids as shear reinforcement. The beams did not collapse, and retained nearly 80% of their strength when they reached their load capacity in the negative moment regions. CFRP stirrups experienced strains of 4000 $\mu\varepsilon$ without rupture, which was well above the limit of 2000 $\mu\varepsilon$ set by the CSA S6. The authors observed the same tendency in the behavior of beams reinforced with CFRP stirrups and beams reinforced with steel stirrups. The use of CFRP for longitudinal reinforcement reduced the shear strength of the beams due to a lower contribution of aggregate interlock to shear resistance. The bars in dowel action carried low shear force due to the relatively poor strength and stiffness of FRP bars in the transverse direction. After failure, the concrete was removed from some of the beams to inspect the CFRP grid stirrups. Researchers noticed that, even in the most distressed parts of the beam, no CFRP stirrup had ruptured.

Deitz et al. (1999) tested 12 concrete deck panels reinforced with three reinforcing schemes. One scheme used GFRP reinforcement at the top and bottom of the deck. Another scheme used all epoxy-coated steel reinforcement, and the third scheme used GFRP bars at the top and steel at the bottom. Two modes of failure were observed: the specimens reinforced with steel failed in flexure, while the specimens with the other two reinforcement schemes failed in diagonal tension. The slabs with GFRP failed in diagonal tension because the lower modulus of the GFRP reinforcement compared with steel allowed wider diagonal cracks leading to shear failure before flexural failure. The diagonal tension modes provided adequate warning of failure because large crack widths and displacements developed.

In 2006, ACI 440.1R adopted the beam shear model of Tureyen and Frosch (2003) in which the concrete shear capacity V_c of flexural members using FRP bars as main reinforcement is based on the depth of the elastic cracked section neutral axis. The formulation is independent of reinforcing bar material properties but accounts for the axial stiffness of the FRP reinforcement indirectly through the elastic neutral axis depth. This method has shown good agreement with test data from a variety of bar stiffnesses and concrete strengths (Tureyen and Frosch 2003). Additional testing by El-Sayed et al. (2005, 2006) found that the method proposed by Tureyen and Frosch (2003) gave reasonably conservative results.

6.1.3 *Punching shear*—El-Ghandour et al. (1999) carried out tests on flat slabs reinforced with FRP bars. The authors found problems of bond slip and crack localization over the main flexural bars. As a result, the slabs failed at loads lower than their expected flexural and punching shear capacities. According to the authors' observations, the ACI 318-95 (ACI Committee 318 1995) equation for punching shear capacity ignored the influence of tension flexural reinforcement when calculating the concrete shear resistance, which was heavily dependent on the concrete strength. The authors estimated that the ACI equation for punching shear was not conservative, and proposed an alternative equation.

Matthys and Taerwe (2000) conducted tests on 17 FRP circular reinforced concrete simply supported slabs reinforced with FRP grids. The authors found that most slabs had a punching shear failure mode. The slabs with FRP reinforcement with a similar strength as the reference steel reinforced slabs required a lower load to produce cracking and had lower postcracking stiffness than the reference slabs. Evaluation of their test results showed that the punching shear design provisions in the European, Japanese, and ACI 318 codes need to be modified to accurately predict the punching capacity of slabs with FRP reinforcement. The computed punching failure load was lower than the measured failure load for slabs reinforced with FRP bars with a low modulus. The authors proposed to modify the value in the punching shear capacity code equations that account for the reinforcement ratio effect with the factor E_{frp}/E_s.

Ospina et al. (2003) tested four interior column two-way slab specimens subjected to uniform gravity loading, two of which were reinforced with GFRP bars and one with a GFRP grid. Results show that the axial stiffness of the FRP reinforcement, as well as the concrete strength f_c', significantly affect the concentric punching shear response of FRP-reinforced slabs. Punching shear failure in slabs reinforced with FRP bars was sudden and brittle; however, two-way slabs reinforced with FRP grids exhibited a much more ductile failure (Bank and Xi 1992; Jacobson et al. 2006; Ospina et al. 2003). El-Gamal et al. (2005) tested six full-scale bridge deck slabs reinforced with GFRP, CFRP, or steel bars. They found little to no correlation between reinforcement ratio and punching shear capacity for the GFRP and CFRP reinforced slabs.

Ospina (2005) performed a statistical evaluation of available punching shear test results from slabs with either FRP reinforcement or steel and concluded that the one-way shear design model of Tureyen and Frosch (2003) can be modified to predict the punching capacity of two-way slabs with FRP reinforcement. This modified equation forms the basis of the punching shear equation adopted by ACI 440.1R.

6.2—Serviceability

Serviceability of FRP-reinforced flexural members is described in terms of deflection and crack width limitations.

6.2.1 *Deflection considerations*—Nawy and Neuwerth (1971) determined that deflection of FRP-beams at ultimate load was approximately three times greater than that of the corresponding steel-reinforced beams. Larralde et al. (1988) found that theoretical deflection predictions as used for steel-reinforced concrete beams underestimated test results for loads above 50% of ultimate; deflection values were fairly well predicted at load levels up to approximately 30% of ultimate. The study suggested a procedure in which values of curvature calculated at different sections of the beam should be used to obtain a better estimate of deflection values.

Faza and GangaRao (1992) found predicted deflections of FRP-reinforced beams to be underestimated using the effective moment of inertia I_e as prescribed by ACI 318-89 (ACI Committee 318 1989). The authors introduced a new method of calculating the effective moment of inertia of concrete beams reinforced with FRP reinforcement. The new expression was based on the assumption that the concrete section between the point loads was fully cracked, while the end sections were assumed to be partially cracked.

Benmokrane et al. (1994) compared the flexural behavior of concrete beams reinforced with GFRP bars to conventionally reinforced ones. At 25% of M_u, the crack pattern and spacing in FRP-reinforced beams were similar to those in conventional steel-reinforced beams. At service (50% of M_u) and ultimate (90% of M_u) loads, there were more and wider cracks in the FRP-reinforced beams than in the steel-reinforced beams. At service and ultimate loads, FRP-reinforced beams experienced maximum deflections three times higher than steel-reinforced beams. Thus, predicted deflections using the Branson expression for effective moment of inertia I_e, as prescribed in ACI 318-89 (ACI Committee 318 1989), resulted in underestimated deflections. This was attributed to the width, depth, and spacing of the cracks.

A study by Engel et al. (1999) proposed modified ACI 318 equations to compute the deflection of concrete beams reinforced with FRP grids. The researchers tested beams with different grid configurations. Grids with fibers that crossed the grid intersections and grids with fibers turned at a right angle to the grid intersections were tested. For all grid joint designs, the modified ACI code design predictions were in good agreement with the measured deflection response for the midpoint deflection of FRP grid-reinforced concrete beams. The stiffness of the grids can be calculated from a grid section in a stand-alone tension test.

Joh et al. (1999) tested 17 beams reinforced with nine types of FRP bars. The beams were reinforced with GFRP, AFRP, CFRP, and steel. The average ratio of long-term deflection, measured at 9.5 months, to short-term deflection was 0.34 for beams with GFRP bars, 0.65 for beams with AFRP bars, and 0.56 for beams with CFRP bars. The ratio for the steel-reinforced control beam was 0.60. Although the AFRP and CFRP reinforced beams had ratios similar to the control beam, the GFRP-reinforced beams' ratio averaged only 56% of the control beam ratio. This finding is in agreement with Brown (1997), who performed a parametric study of GFRP-reinforced rectangular beams and concluded that the ratio of long-term deflection to short-term deflection was, on average, 55% of the same ratio for steel-reinforced beams.

Toutanji and Saafi (1999) tested concrete beams reinforced with GFRP bars. The authors found that the deflections predicted using the ACI 318 equation were lower than the measured deflections.

A method for determining deflections of FRP-reinforced concrete beams was proposed by Razaqpur et al. (2000), assuming that the moment-curvature relation of FRP sections was linear in the precracked and postcracked stages, and that the tension-stiffening effect could be ignored. The predicted deflections agreed within 5%, with deflections obtained in a number of flexure tests on reinforced concrete beams and slabs measured at service load (50 to 60% of ultimate). The method presented did not account for shear deformations.

Rasheed et al. (2004) developed closed-form expressions for deflection calculation by exactly integrating the bilinear moment-curvature response along the beam span. The results were in excellent agreement with experiments. A modified version of ACI 318 was proposed.

Kassem et al. (2002) tested 12 full-scale concrete beams reinforced with carbon FRP bars. The test parameters included the type of the CFRP bars (mechanical properties and surface finishing) and the reinforcement ratio. They concluded that the deflection predictions based on beams reinforced with GFRP bars may underestimate the experimental deflection values for CFRP-reinforced beams, especially at load levels less than twice the cracking moment.

Because of the variable stiffness, brittle-elastic nature, and particular bond features of FRP reinforcement, deflections of FRP-reinforced concrete members are more sensitive to the variables affecting deflection than steel-reinforced members of identical size and reinforcement layout (Ospina et al. 2001). Deflections in members with FRP reinforcement tend to be greater in magnitude because of the lower stiffness of most commercially available FRP reinforcements. An indirect deflection-control procedure in which deflections in beams and one-way slabs with FRP reinforcement are controlled by the selection of a minimum member thickness for preliminary design (Ospina and Gross 2006) is presented in ACI 440.1R.

6.2.2 *Crack width and patterns*—As with steel reinforcement, the maximum crack width at the tension face of a reinforced concrete member is heavily influenced by the strain level in the reinforcement, the concrete cover, the bar spacing, and the mechanical properties of the reinforcement (Frosch 1999). ACI 440.1R includes a variation of Frosch's (1999) crack control equation found in ACI 318, with a recommendation to limit the predicted crack widths to 0.5 and 0.7 mm (0.020 to 0.028 in.) for exterior and interior exposure conditions, respectively. These values, adopted from CSA S806, are larger than the crack width limits traditionally accepted in steel-reinforced concrete design due to the superior corrosion resistance of FRP.

Nawy and Neuwerth (1971) found that beams reinforced with steel had fewer cracks than corresponding FRP-reinforced beams. The large number of well-distributed cracks in the FRP-reinforced beams indicated that good mechanical bond was developed between the FRP bar and surrounding concrete.

Faza and GangaRao (1992) determined that concrete beams reinforced with spirally deformed FRP reinforcing bars using normal-strength concrete (27.6 MPa [4000 psi]) exhibited crack formation that was sudden and propagated toward the compression zone soon after the concrete stress reached its tensile strength. Crack spacing was very close to the stirrup spacing, and cracks formed at or near the stirrups. This sudden propagation of cracks and wider crack widths decreased when higher-strength concrete 51.7 to 69 MPa (7.5 to 10 ksi) and sand-coated FRP reinforcing bars were employed. Based on the assumption that maximum crack width can be approximated by an average strain in FRP reinforcing bar multiplied by expected crack spacing, an expression for maximum crack spacing governed by the following parameters was developed: 1) bond strength of the FRP reinforcing bar; 2) splitting tensile strength of the concrete; 3) area of the concrete cross section in tension; 4) number of reinforcing bars in tension; 5) size of reinforcing bar; and 6) working stress of the FRP reinforcing bar.

In a study by Toutanji and Saafi (2000), the authors suggest that crack width depends primarily on average crack spacing and on average reinforcement stress, which are related to the mechanical properties of the concrete, the reinforcement, and the bond between them. They proposed an equation for crack width prediction.

6.3—Bond and development of reinforcement

The evaluation of bond characteristics of FRP reinforcement is of prime importance in the design of FRP-reinforced concrete members. Bond characteristics are influenced by factors such as: 1) size and type of reinforcement (wires or strands); 2) surface conditions (smooth, deformed, and sand-coated); 3) Poisson's ratio; 4) concrete strength; 5) concrete confinement (helix or stirrups); 6) type of loading (static, cyclic, and impact); 7) time-dependent effects; 8) amount of concrete cover; and 9) type and volume of fiber and matrix.

A study on bond of GFRP reinforcing bars was conducted by Tao (1994) on 102 straight and 90-degree hook specimens. New limits for allowable slip were suggested as 0.064 mm (0.0025 in.) at the free end, or 0.38 mm (0.015 in.) at the loaded end. According to this study, the basic development length of straight and hooked GFRP reinforcing bars could be computed knowing the bar size, ultimate strength of the reinforcement, and concrete strength.

Shield et al. (1999) tested 72 inverted half-beam specimens. The tests did not use stirrups in the beams to simulate the lack of confinement present in bars of bridge decks. GFRP rods of 19.1 mm (0.75 in.) diameter and No. 5 bars were tested, and the development length was determined to be between $42d_b$ and $63d_b$. Nonuniformly spaced cracks perpendicular to the bar developed.

Katz et al. (1999) studied the bond behavior of five reinforcing bar types at room and high temperatures. One bar had lugs on the surface to provide the desired bond to concrete. Three other bar types had an extra layer of resin on the surface and a coat of sand. Also, a steel bar was included in the study for comparison purposes. The bars were tested at temperatures of 20, 130, and at 250 °C (68, 266, and 482 °F). Two main bond failure mechanisms were observed. At room temperature, portions of the concrete surface in contact with the FRP bars pulled out in some specimens. At high temperatures, however, all the slip took place at the surface of the FRP bar because the shear properties of the resin on the surface of the FRP bar deteriorated. At 250 °C (482 °F), all FRP bars lost 80% or more of their room temperature bond strength, while the steel bars lost only 40% of their room temperature bond strength. The bond strength of all FRP bars decreased with increases in temperature, and it appeared to level off at a temperature of approximately 200 °C (392 °F). The authors agreed with a model proposed by Greszczuk (1969) that assumed the controlling parameters for bond strength were the shear modulus and the thickness of the outer layer of resin on the FRP reinforcing bar. The bond stiffness (initial slope of the bond-slip curve at the loaded end) decreased with increasing temperature. FRP bars with helical fiber wrapping and sand coating tended to exhibit better bond behavior than was seen in bars that relied on lugs of polymer for bond strength. The FRP bars with lugs of resin showed abrupt drops in the bond-slip curves after the peak load was reached.

Several weaknesses in the use of pullout tests (cylinders or pullout specimens) have been identified in discussions in ACI Committees 408 and 440 because such tests do not sufficiently account for all types of mechanical behavior, such as the flexural curvature and the combinations of shear and flexure. Many attempts have been made to find a better standard test method.

Malvar (1995) conducted tension and pullout bond tests on four types of reinforcing bars and studied the effects of confining pressure on bond. All bars had a diameter of 19 mm (3/4 in.) and were composed of E-glass fibers with a fiber volume fraction of 45%, embedded in a vinylester or polyester resin.

Five different confining pressure levels were studied: 3.45, 10.3, 17.2, 24.1, and 31 MPa (500, 1500, 2500, 3500, and 4500 psi). The measured adhesion (bond stress at zero average slip) was between 0.69 and 2.1 MPa (100 and 300 psi). Beyond this adhesion, the slope of the bond-slip curve appeared to increase with higher confinement. For an identical amount of confinement, the bond strength for a steel bar was, on average, 1.2 to 1.5 times higher than that for an FRP bar. Large variations in indentation depths of the bars resulted in a large variation of bond strength. Bond strength can be increased threefold by increasing confining pressure by a factor of 7.

Cosenza et al. (1997) reviewed the current knowledge of bond performance of FRP bars embedded in concrete. They found that the longitudinal shear modulus of the FRP bar was an important parameter. Coating the surface of bars with sand enhanced the bond strength and bond stiffness, but led to more brittle bond behavior. Bars with a surface coated with sand had excellent bond strength ($\tau_m > 10$ MPa [1.45 ksi]), even higher than steel bars. The bond strength of smooth FRP bars was poor ($\tau_m < 2.37$ MPa [0.34 ksi]). The bond strength of spirally glued bars had values slightly higher than those of smooth bars (τ_m = 4.50 MPa [0.65 ksi]).

Wambeke and Shield (2004) consolidated a database of 269 beam bond tests. The database was limited to beam-end tests, notch-beam tests, and splice tests. Two equations for development length were developed; one based on a splitting mode of failure, and the other on a pullout mode of failure. These equations form the basis for the design recommendations in ACI 440.1R.

Research on spliced FRP bars conducted by Aly (2005), Aly and Benmokrane (2005), and Aly et al. (2005, 2006) found that the development length of FRP bars was dependent on the square root of the longitudinal modulus of the bars.

6.4—Fatigue performance

Benmokrane et al. (1999) tested concrete slabs 3200 mm (126 in.) long, 1000 mm (39.3 in.) wide, and 260 mm (10.2 in.) thick, reinforced with CFRP gratings. The beams were subjected to 4 million cycles of loading, with load amplitude of 10 to 100 kN (2.25 to 22.5 kips). The first 2 million cycles were applied at a frequency of 2 Hz, and the remaining cycles at 3 Hz. After 4 million cycles, the maximum deflection was 12 mm (0.47 in.) (span/250). The same slab had a deflection of 8 mm (0.34 in.) under a 100 kN (22.5 kips) load applied statically before the fatigue test.

Rahman et al. (2000) conducted a 4 million cycle fatigue test on a full-scale model of a bridge deck slab. The slab was a composite structure supported by steel I-beams. The deck was reinforced using CFRP grating size C10 bars with 46 mm^2 (0.071 in.^{2}) cross-sectional area and 100 mm (4 in.) grid spacing. The FRP reinforcement ratio was 0.3%. The researchers found that the deflection after 4 million cycles increased by 26% when compared with the deflection under static 100 kN (22.5 kip) loading. A deflection increase of 32% was observed for the 125 kN (28.1 kip) loading. Crack widths increased 42% after 4 million cycles under the 100 kN (22.5 kip) load. Stresses in the reinforcement were also increased by the fatigue loading by approximately 10%. The failure modes of the slab were punching shear in five loading cases and flexure in one loading case. The minimum capacity of the slab was over five times the design service load.

Kumar and GangaRao (1998) found that the rate of fatigue degradation in bridge decks reinforced with GFRP bars was similar to that of bridge decks reinforced with steel for a range of stringer stiffnesses and amount of composite action.

El-Ragaby et al. (2006) tested four full-size concrete deck slabs reinforced with GFRP bars to failure under a concentric fatigue loading. They concluded that the GFRP reinforced concrete bridge decks had better fatigue performance and longer fatigue life than equivalent steel reinforced decks.

6.5—Members reinforced with FRP grating systems

Bank et al. (1991) tested seven full-size concrete bridge deck slabs, six of which were reinforced with pultruded GFRP gratings, and one with steel reinforcing bars. The slabs were designed for a live load moment designated by AASHTO (1989) using a nominal HS-25 loading with a live load impact factor of 30%. Failure was the result of concrete crushing followed immediately by propagation of a flexural-shear crack in a diagonal path towards the outer support. This crack was intercepted by the top surface of the FRP grating and redirected horizontally along the top surface of the grating to the free end. No failure of the FRP grating was observed. The steel-reinforced slab failed by yielding of the reinforcing bars and subsequent crushing of the concrete. The service load midspan deflections for all FRP-reinforced slabs were close to the allowable limit of 4.9 mm (0.192 in.). Deflection was found to stabilize after a limited number of cycles. All slabs failed at loads in excess of three times the service load.

Bank and Xi (1992) tested the performance of four full-scale two-span continuous concrete slabs, doubly reinforced (top and bottom) with 51 mm (2 in.) deep, commercially produced, pultruded FRP gratings having longitudinal bar spacings of 76 and 51 mm (3 and 2 in.) on center, respectively. Slabs were loaded as follows: first, under a monotonically increasing load to 115 kN (26 kips), then subjected to 10 loading/unloading cycles of 0 to 115 kN (0 to 26 kips), and finally loaded monotonically to failure. Flexural cracking developed early in both the positive and negative moment regions and were in line with the transverse bar locations. All slabs experienced shear failure in the short shear span between the middle support and the load point. The ratios of failure to service load for FRP-reinforced slabs were 4.26, 3.89, 4.17, and 4.16. For the steel-reinforced slab, this ratio was 3.34. No evidence of shear cracking was observed before failure. The local radius of curvature in the positive moment region generally satisfied AASHTO serviceability specifications. In the negative moment region, however, this criterion was violated. Service load deflections were well below the *L*/500 limit, where *L* is the length of the beam.

6.6—Members reinforced with FRP grids

Since 2004, carbon fiber/epoxy resin grids have been used to reinforce a wide variety of precast concrete structures including insulated wall panels, architectural cladding panels, double tees, and residential members. In addition to being used a secondary reinforcement for crack control, the grids have also been used as primary reinforcement.

Most grids being used as internal reinforcement for concrete parts are constructed using continuous carbon fibers impregnated with thermoset epoxy resins. Glass and basalt fibers may also be used. Grids can be produced by a number of manufacturing processes. Grids can be woven and then impregnated with a thermoset or thermoplastic resin.

Grids further enhance the potential for more effective reinforcement of concrete structures using composite materials because they can be used near either surface of the concrete structure (but embedded within the concrete) to control cracking, reduce the thickness of the concrete structure, and to provide tension reinforcement. Grids may be used as the only reinforcement material or in combination with traditional reinforcement such as reinforcing bars and welded wire fabric. Because of differences in the physical and mechanical behavior of FRP materials versus steel, combined with the fact that grids offer the unique opportunity to position a noncorrosive structural reinforcement very close to the surface, unique guidance on the engineering and construction of concrete structures reinforced with FRP grids is needed and is being developed as an activity of ACI Committee 440.

6.7—Pavement applications

FRP dowel bars have been extensively investigated by Davis and Porter (1999) and others. These dowel bars have been used primarily in highway concrete pavement construction. The results of this research have shown that GFRP dowels are able to carry the shear transfer across the pavement joint. For example, 44 mm (1.75 in.) diameter GRFP dowels spaced at 200 and 300 mm (8 and 12 in.) were shown to give favorable results compared with the typical 38 mm (1.5 in.) diameter steel dowel for up to 10 million cycles of repeated loading simulating 40 kN (9 kip) wheel loads on each side of the joint for full-scale pavement slabs.

Choi and Chen (2003, 2005) studied the concrete stresses induced in a GFRP-reinforced concrete slab due to concrete shrinkage and temperature variations. The results showed that the stress level in concrete was reduced with GFRP reinforcing bars because of the low Young's modulus of GFRP. The stress distribution and crack width in the GFRP-reinforced continuously reinforced concrete pavement section were predicted under various design considerations, such as the coefficient of thermal expansion (CTE) of concrete, the friction from the subbase of the pavement, and the bond-slip between concrete and reinforcement. The stress levels in the GFRP reinforcement, the crack widths, and the crack spacing of the proposed pavements were shown to be within the allowable AASHTO pavement design requirements. Other work on FRP-reinforced concrete pavements has been conducted by Walton and Bradberry (2005).

CHAPTER 7—PRESTRESSED CONCRETE MEMBERS

Much of the available research has concentrated on the short-term static behavior of bonded FRP prestressed beams. Both pretensioned and post-tensioned systems have been considered. ACI 440.4R provides details of prestressing systems for FRP and guidelines for the design of FRP prestressed concrete structures.

7.1—FRP tendons

This section examines commercially available tendons and anchorage systems. For more detailed information, the reader is referred to ACI 440.4R. Properties and characteristics of selected available tendon/anchorage systems are summarized in Table 7.1 and are based on the manufacturer's published data. The trade names of the products are used for clarity. The properties of the tendons may change as the manufacturers modify their products. More details about the manufacturers are provided in ACI 440.4R. Variations in the properties of the tendons (for example, modulus and ultimate stress) are generally due to differences in the fiber volume fraction or to the orientation of the fibers in the tendon. For example, Arapree tendons have a volume fraction of 0.45, whereas Parafil rope has no polymer at all. Thus, the strength and modulus values for Parafil are higher. Also, the CFCC tendons are twisted seven-wire strands; thus, not all the fibers are aligned along the longitudinal axis. Therefore, the modulus and strength properties are lower than the Leadline tendons that have all the fibers oriented longitudinally.

7.2—Anchorages

Conventional anchorage systems for steel prestressing reinforcement have hardened teeth that bite into the reinforcement to enhance transfer of stress. With FRP reinforcement, these teeth cause premature failure of the reinforcement at the grip location. Different anchorage systems have been used with FRP tendons, such as clamp, plug and cone, resin sleeve, resin potted, metal overlay, and split wedge anchorages, as shown in Fig. 7.1. These anchorages and their gripping mechanisms are briefly described herein and in more detail in ACI 440.4R.

7.2.1 *Clamp anchorage*—A clamp anchorage (Fig. 7.1(a)) consists of two or four grooved steel plates sandwiching the FRP tendon and held together by pretensioned bolts and springs. The force is transferred by a shear-friction mechanism (Malvar and Bish 1995).

7.2.2 *Plug and cone anchorage*—The plug and cone (or barrel and spike) anchorage (Fig. 7.1(b)) is made of a socket housing and a conical spike (Burgoyne 1992). Such a system is suited to anchor Parafil ropes where the fibers are not encased in resin media but are held only by an outer protection sheath.

7.2.3 *Resin sleeve anchorage*—In this anchorage system, the FRP tendon is embedded in resin filling a metallic housing such as a steel or copper tube (Fig. 7.1(c)). The resin used ranges from nonshrink cement with or without sand to epoxy-based material. The load transfer mechanism depends completely on the bonding and interlocking between the anchorage components. This type of anchorage has been used for both round and flat tendons.

7.2.4 *Resin potted anchorage*—The potted anchorage (Fig. 7.1(d)) varies depending on the internal configuration of the socket: straight, linearly tapered, or parabolically tapered. The load transfer mechanism from the tendon to the sleeve is by interfacial shear stress that is a function of bonding and normal stress produced by the variation of the resin profile (Kim and Meier 1991). The practical drawbacks include precutting the tendons to length and the curing time for the resin or grout.

7.2.5 *Metal overlaying*—The die-cast wedge system (Fig. 7.1(e)) for the carbon fiber composite cable (CFCC) requires that the tendon length be predefined so that metal can be cast onto the tendon during fabrication with the result that adjustment on site is not possible. The tendon and the metal material are compressed by a typical wedge anchorage similar to those used on steel tendons. The load transfer of this anchorage is achieved by shear friction, which is a function of compressive stress and friction at the contact surfaces.

7.2.6 *Split wedge anchorage*—Split wedge anchorages (Fig. 7.1(f)) are generally preferred because of their compactness, ease of assembly, reusability, and reliability. The wedge system is widely used in anchoring steel tendons. The anchorages are modified for use with FRP tendons by

Table 7.1—Manufacturer's material properties (Nanni et al. 1996)

Tendon	Fiber type	Type	Fiber volume ratio	Nominal diameter, mm (in.)	Cross section, mm^2 (in.^{2})	Young's modulus, GPa (ksi)	Ultimate stress, MPa (ksi)	Ultimate strain, %	Density, g/cm^3 (lb/ft^3)	Poisson's ratio	Anchorage type	Reference
Arapree	Aramid	—	0.45	7.5 (0.30)	44 (0.071)	60 (8700)	1400 (200)	2.4	1.25 (78)	0.38	Wedge	Gerritse and Werner 1988; Sireg 2007
Parafil	Aramid	Type G	1.00	13.5 (0.53)	87 (0.135)	120 (17,400)	1200 to 1900 (170 to 270)	1.5	1.44 (90)	N/A	Plug and cone	Burgoyne 1988; Linear Composites 2007
Leadline	Carbon	Indented	0.65	7.9 (0.31)	46 (0.071)	150 (21,800)	2250 (325)	1.3	1.53 (95)	0.27	Wedge	Mitsubishi 2005
CFCC	Carbon	Twisted seven wire	0.65	12.5 (0.49)	76 (0.118)	137 (19,900)	1900 (270)	1.6	1.50 (94)	N/A	Resin-filled die-cast/ wedge	Santoh 1993

increasing their length and placing a soft metal sleeve around the tendon to prevent notching. The mechanism of gripping relies on friction and clamping force between the wedges, barrel, and tendon.

A metallic anchorage was developed (Sayed-Ahmed and Shrive 1998; Campbell et al. 2000) for 8 mm (0.31 in.) CFRP. This anchorage has also been studied experimentally and analytically by Al-Mayah et al. (2001). The researchers also developed a new anchor system that allowed the maximum load capacity of the CFRP rod to be achieved and required no presetting of the wedges (Al-Mayah 2004; Al-Mayah et al. 2003). Nonmetallic versions of this anchorage have also been developed consisting of either ultra-high-performance concrete wrapped with CFRP sheets (Reda Taha and Shrive 2003a,b) or carbon fiber-reinforced reactive powder concrete (Shaheen 2004).

7.3—Flexural behavior

The load-deflection response of FRP prestressed beams is generally bilinear due to the linear elastic nature of the FRP tendons. The beams exhibit a linear load-deflection response up to cracking and with a reduced stiffness after cracking up to failure (Abdelrahman et al. 1995; Abdelrahman and Rizkalla 1997; Arockiasamy et al. 1995).

Two modes of flexural failure have been reported for FRP prestressed beams: 1) rupture of prestressing reinforcement; and 2) crushing of concrete in compression. These failure modes are dependent on whether or not the cross section has sufficient rotation capacity to develop the ultimate tensile stress of FRP reinforcement. When the concrete section has enough rotation capacity (enough concrete area in compression), failure by rupture of the tendon occurs. Rupture of the tendon is normally accompanied by horizontal cracks at the level of the prestressing reinforcement, causing spalling of the concrete cover. According to a number of researchers (Abdelrahman et al. 1995; Naaman and Jeong 1995; Currier 1995), rupture of FRP prestressed reinforcement is accompanied by a release of elastic strain energy that is partly absorbed by the concrete, thus causing extensive cracking and spalling. When failure of a prestressed beam is caused by crushing of concrete (when the cross section has insufficient rotation capacity), cracking is less extensive compared with those beams that fail by rupture of tendons. Some researchers have reported that failure of FRP prestressed beams by crushing of concrete results in increased ductility. This behavior is attributed to the progressive failure of concrete in compression (Grace and Abdel-Sayed 1998a).

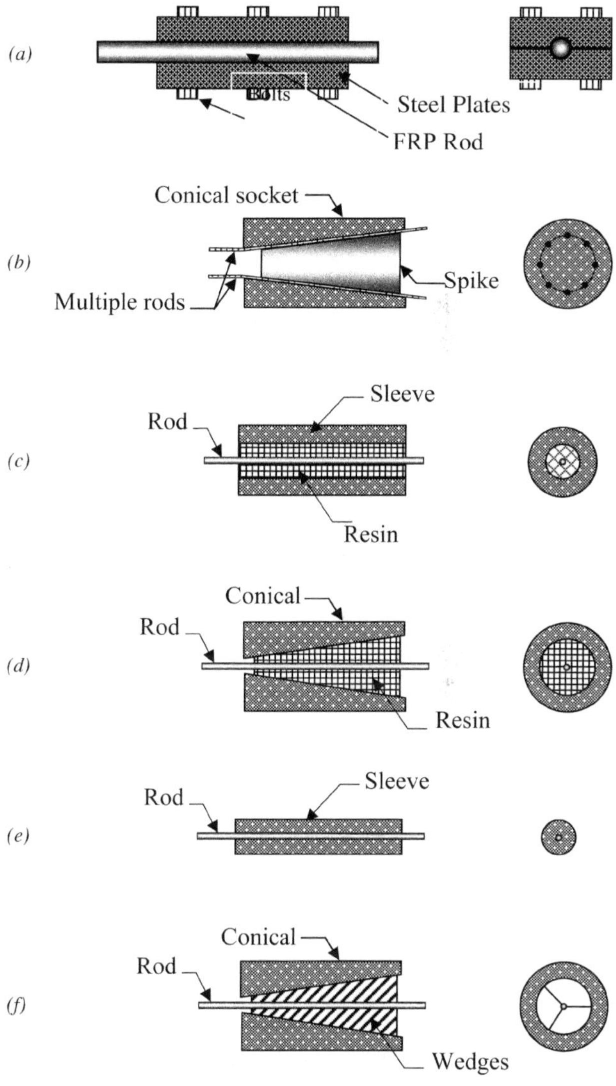

Fig. 7.1—(a) Clamp; (b) plug and cone; (c) resin sleeve; (d) resin potted; (e) metal overlay; and (f) split wedge anchorage systems.

Comparison of flexural behavior of similar concrete beams prestressed with FRP and steel reinforcement shows that the beams possess very similar precracking stiffnesses to each other, but the postcracking stiffnesses are markedly different. Abdelrahman and Rizkalla (1997) found that beams pretensioned with FRP reinforcement had a much lower postcracking stiffness resulting in larger postcracking deflections. The ultimate deflection in the steel prestressed beam was, however, much greater due to yielding of steel reinforcement. The ultimate load for the CFRP beams, however, was higher compared with steel prestressed beams.

For conventional steel prestressing, bonded tendons are often preferred due to concerns over corrosion of unbonded tendons. Because FRP tendons are resistant to corrosion, bonding may not be required. This would save on the extra expenditure required for grouting. Unbonded tendons also facilitate replacement of prestressing tendons in case of strength inadequacy or increase in the design loads. Unbonded FRP prestressed members are also more deformable than similar bonded prestressed members. Problems that arise with unbonded FRP tendons in beams include anchorage of FRP reinforcement and the fatigue strength of these beams, especially at the tendon anchorage junction.

Post-tensioned beams with unbonded FRP tendons exhibit less cracking under short-term loading conditions compared with beams with bonded prestressed reinforcement (Maissen and De Smet 1995), but the cracking is normally concentrated in a few wide cracks. The load-deflection response of unbonded post-tensioned beams is bilinear, but compared with pretensioned and bonded post-tensioned beams, the postcracking stiffness is much less.

Research on crack widths of flexural members prestressed with bonded FRP was performed by Dolan et al. (2000). Both monotonic and cyclic loadings were studied. Crack widths were larger than for comparable steel prestressed beams because of the lower modulus of the FRP. Fatigue loading increased the crack widths slightly.

Bryan and Green (1996) studied the low temperature behavior of beams prestressed with bonded CFRP. No significant differences in behavior were observed for short-term exposure to low temperature.

Braimah et al. (1998) and Marshe and Green (1999) investigated the feasibility of transversely prestressing concrete bridge deck slabs with bonded CFRP tendons. They found that this approach was effective in improving the serviceability and strength of the deck slabs. CFRP prestressed slabs performed as well, if not better than, comparable steel prestressed deck slabs in terms of serviceability and strength requirements.

Parallel-lay aramid ropes have been developed for use as unbonded prestressing cables at the University of Cambridge (Burgoyne 1992, 1993, 2001). Parafil is not strictly an FRP because no resin (polymer) binds the fibers together. The ropes have an external polyethylene sheath that serves to retain the shape of the rope, and also to shield the aramid fibers from UV light, to which they are susceptible. The intended uses for the ropes are as external prestressing tendons, or as unbonded tendons inside ducts in the section. This arrangement avoids any problems associated with the fibers coming into contact with the alkaline concrete. Stresses are limited by stress-rupture considerations, typically approximately 1000 MPa (145 ksi) at initial stressing. Several prototype structures have been built with these tendons in Japan (as external strengthening tendons), in England (for prestressed brickwork), in Canada (prestressed timber), and in Scotland (as stay cables to the all-composite bridge at Aberfeldy).

7.4—Fatigue behavior

Fatigue resistance of beams is typically investigated by calculating the stress range Δf_p produced in the prestressing tendon during cyclic loading, and comparing this stress range with that obtained from the S-N curves for the prestressing tendon.

If the level of precompression in a prestressed concrete member is sufficient to ensure an uncracked section throughout the service life of the member, the fatigue characteristics of the prestressing reinforcement and anchorages are not likely to be critical design factors. In cracked prestressed concrete members, however, localized stress increases in prestressing reinforcement across cracks in pretensioned beams, fretting fatigue of prestressing reinforcement in post-tensioned beams, and the fatigue of the tendon-anchorage assembly become very significant. ACI Committee 215 and the Post-Tensioning Institute have recommended that tendon-anchorage assemblies consisting of prestressing steel and anchorage systems be able to withstand, without failure, 500,000 cycles of stress varying from 60 to 66% of the specified ultimate strength of the assembly (ACI 215R; Post-Tensioning Institute 1990).

Iwamoto et al. (1993) conducted tests on concrete beams reinforced with two types of bonded prestressing tendons: AFRP or steel wire. Three levels of initial tension of the tendon were selected at 40, 60, and 70% of the tensile strength of the FRP bar. The failure modes of the beams with aramid tendons included tendon rupture, development of bond cracks, and shear compression failure. The fatigue strength at 2 million cycles was not less than 65% of the static ultimate strength of the beams. The bond between aramid tendons and concrete deteriorated more than that of the specimens containing prestressing wire because of rubbing between aramid fibers and concrete. Therefore, the fatigue strength of beams with aramid fiber tendons cannot be predicted by fatigue tests on isolated aramid fiber bars. The rigidity of the beam did not decrease much under cyclic loading.

Abdelrahman et al. (1995) tested four prestressed beams using two bonded CFRP reinforcement types (CFCC and Leadline™). The fatigue loading was performed with a maximum load equal to the cracking load and the minimum equal to 70% of the cracking load. The authors concluded that the beams survived 2 million cycles of fatigue loading without measurable effects on the beam stiffness.

Dolan et al. (2000) reported that cracked CFRP pretensioned beams showed no fatigue failure after 3 million flexural cycles with nominal tensile stresses of $0.5\sqrt{f_c'}$ MPa ($6\sqrt{f_c'}$ ksi) at the extreme fiber of the beams. Gradual softening of the girders was observed, but the beams indicated no loss of strength due to the fatigue loading.

Braimah (2000) completed tests on unbonded CFRP prestressed concrete beams. Only one CFRP tendon survived the full 2 million cycles. Most of the tendons ruptured at the anchorage/tendon interface when the fatigue stress range in the tendon was greater than 100 MPa (15 ksi). Braimah concluded that totally unbonded tendons may not be appropriate for fatigue loading due to stress concentrations at the tendon/anchorage junction.

7.5—Time-dependent behavior

Material research has shown that FRP reinforcement has different relaxation properties than steel reinforcement (Erki and Rizkalla 1993) leading to the need to evaluate the long-term behavior of FRP prestressed members. Long-term prestress losses are computed based on creep and shrinkage of the concrete and relaxation of the tendons. Losses for FRP tendons due to creep and shrinkage are typically less than the corresponding losses for steel tendons due to the lower modulus of elasticity of the FRP. Relaxation losses are more problematic because there is little data that describes relaxation loss profiles. Relaxation characteristics vary with the fiber type, and are generally less than 12% over the life of the structure (JSCE 1993).

Sen et al. (1999) examined AFRP pretensioned beams under simulated tidal and temperature changes and found that the combined effects of temperature change and tidal simulation were detrimental to the performance of AFRP pretensioned beams. The loss of moment capacity of the beams increased with length of exposure, varying between 43 and 55% within the first 21 months.

Early research results from long-term tests showed that FRP prestressed beams exhibited similar behavior to those prestressed with steel. The long-term deformations depended on the fiber type, and were often higher than those of steel prestressed beams. Currier (1995) found that long-term deformations of FRP prestressed beams could be predicted by using conventional methods developed for steel prestressed beams, albeit with minor modifications.

Matthys and Taerwe (1998) tested three series of pretensioned concrete slabs using steel wires or AFRP tendons as prestressing reinforcement. They found that the long-term deformations of the AFRP pretensioned slabs were higher than those of the steel pretensioned slabs due to the lower elastic modulus and higher relaxation characteristics of the AFRP tendons.

Braimah et al. (1999) and Braimah (2000) tested three pretensioned CFRP beams and one steel prestressed beam under sustained loading for up to 2 years. The long-term behavior of the CFRP beams was very similar to that of the steel prestressed beam. A model for predicting the long-term deflections of the beams was also developed and was shown to predict the experimental measurements with reasonable accuracy.

7.6—Ductility and deformability

Ductility describes the ability of a structural member to sustain large inelastic deformations in comparison with elastic ones before collapse without significant loss in resistance. FRPs have no ductility as is commonly associated with yielding of steel because they exhibit linear elastic behavior until brittle failure. The lower strain at failure is of significant concern to design engineers. A portion of the strain is used for prestressing and flexural strain, leaving very little strain reserve for safety against brittle failure.

The most common ductility ratios or indexes are given in terms of curvature, rotation, or deflection as follows

$$\mu_\phi = \frac{\phi_u}{\phi_y},\ \mu_\theta = \frac{\theta_u}{\theta_y},\ \text{or}\ \mu_\Delta = \frac{\Delta_u}{\Delta_y} \tag{7-1}$$

where μ = ductility index; ϕ = curvature, θ = rotation, and Δ = deflection. The subscript y denotes yielding of the reinforcement, while u denotes ultimate limit state. Because FRP reinforcements do not yield, the conventional definition of ductility is not applicable. A number of alternative definitions have been suggested in the literature.

Abdelrahman et al. (1995) proposed deformability ratio criteria that relied on the equivalent deformations of the uncracked section and deformations at ultimate. Curvature and displacement-based deformation indexes were defined as the ratio of the curvature of deflection of the beam at failure to the equivalent elastic curvature or displacement of the uncracked beam at a load equal to the ultimate load.

Naaman and Jeong (1995) proposed an energy-based definition of the ductility index for perfectly elasto-plastic behavior. According to the proposed definition, the ductility ratio was expressed as the ratio of the total energy to the elastic energy at the failure state of a beam.

The energy-based definition of the ductility index of concrete structures appears to be the most elegant, and a number of researchers including Grace and Abdel-Sayed (1998a,b) have tried to improve flexural ductility by using different techniques to achieve an increase in the inelastic component of the total energy in FRP prestressed or reinforced structures. Most of these efforts have, however, achieved increased deflections at ultimate without a significant increase in the inelastic energy.

According to Lees and Burgoyne (1999), the high rotation capacity in FRP prestressed beams is essential to warn of incipient failure. These large rotations are most often achieved in unbonded or partially bonded beams where the beam rotations are concentrated in a few cracks. If the tendons are partially bonded (for example, where the bond is chosen to be of a known low shear strength or the tendon is bonded intermittently along the length), then a system can be designed where a controlled amount of debonding takes place at the crack locations. In this manner, both a high ultimate moment capacity and a large rotation capacity concentrated at the few cracks can be achieved.

Another approach is to use a ratio of the curvatures under ultimate and service loads (Dolan and Burke 1996).

Deformability indexes based on this rationale were calculated for harped prestressed beams from published data (Table 7.2). The service loads were considered as the load to produce a tensile stress in the concrete of $0.25\sqrt{f_c'}$ MPa ($3\sqrt{f_c'}$ in psi). The index is effectively a function of the ratio of ultimate strain to the prestressing strain, with slight modification due to the differences in the neutral axis of

Table 7.2—Indexes and ratios to evaluate deformability

Reference	Condition	Deformability index	Ultimate/ initial prestress	Maximum deflection/ span	*c*/*d* ratio
Abdelrahman (1997)	Under-reinforced	3.6	3.6	1/32	0.115
Abdelrahman (1997)	Over-reinforced	4.4	3.6	1/32	0.345
Currier (1995)	Over-reinforced	2.4	2.0	1/32	0.307

elastic and inelastic behavior. Abdelrahman's indexes are higher because his prestress level was lower than the other tests.

The data indicate that the most efficient method to obtain high deformability is to reduce the initial prestress. Reducing the initial prestress strain provides more tendon strain reserve, greater curvature or deflection capacity, and a higher index.

One other method for improving the deformability, ductility, or both of beams prestressed with FRP is combining FRP prestressed reinforcement with non-prestressed stainless steel reinforcement (Tung and Campbell 2002).

7.7—Transfer and development length

The transfer and development length of an FRP tendon is a function of the perimeter configuration area and surface condition of the FRP, the stress in the FRP, and the method used to transfer the FRP force to the concrete. The mechanism of bond differs between FRP and steel strands due to the large variation of FRP bars in terms of shapes, surface treatments, and elastic moduli.

7.7.1 *Transfer length*—The transfer length in pretensioned concrete is the length required to transfer the full prestressing force to the concrete. In general, most FRP tendons exhibit transfer lengths shorter than that of steel.

Nanni et al. (1992) examined the transfer length of braided epoxy-impregnated AFRPs and found that friction was the predominant bonding mechanism in aramid fibers and that these fibers showed little tendon slippage compared with steel. Taerwe and Pallemans (1995) also studied AFRP tendons and suggested a transfer length of 16 times the nominal diameter of the tendons. Ehsani et al. (1997b) found the transfer length to be between 33 and 50 bar diameters, depending on the type of tendon.

Mahmoud and Rizkalla (1996) studied the bond characteristics of CFRP tendons in 24 pretensioned concrete beams and found that the measured transfer length varied from 450 to 650 mm (17.7 to 25.6 in., or 56 to 81 bar diameters) for Leadline™ tendons, and from 300 to 425 mm (11.8 to 16.7 in.) for CFCC strands. After further study and comparison with an analytical model, they recommended a formula for transfer length dependent on the tendon prestress, diameter and type, and the concrete strength (Mahmoud et al. 1997).

Ehsani et al. (1997b) also conducted tests on Leadline™ (8 mm [0.31 in.]) and CFCC (8.3 mm [0.33 in.]) carbon tendons and found the transfer length to be 54 bar diameters for Leadline™ and 50 bar diameters for CFCC.

Soudki et. al. (1997) reported the transfer length for Leadline™ CFRP tendons as 80 times the bar diameter, and found

Table 7.3—Typical transfer length and development length based on tests of various FRP tendons

Material	Type	Diameter, mm (in.)	f_{pi}/f_{fu}	l_t/d_b	l_d/d_b
Aramid	Arapree	10.0 (0.39)	0.5 to 0.7	16 to 50	100
Carbon	Leadline	7.9 (0.31)	0.5 to 0.7	50 to 80	175
	CFCC	8.3 (0.33)	0.5 to 0.7	50	N/A
Steel	Seven-wire	12.7 (0.50)	0.75	50 to 60	106

that existing models for steel may give unconservative transfer lengths for the CFRP tendon.

7.7.2 *Flexural bond length*—Nanni et al. (1992) examined the development and flexural bond lengths in 25 AFRP pretensioned concrete beams. The development length was determined to range from 80 to 120 times the nominal diameter of the tendon.

Ehsani et al. (1997b) examined the transfer and flexural bond lengths of CFRP and AFRP prestressing tendons and compared the results with those of steel strands. Two types of CFRP strands were tested: Leadline™ and CFCC. Three AFRP types were tested and they concluded that the ACI development length requirements for steel strands were conservative for AFRP tendons, but were not adequate for Leadline™ CFRP tendons.

7.7.3 *Summary*—The bond and development for prestressing FRP tendons made of carbon, aramid, or glass fibers are intended to provide bond integrity for the strength of the member. Typical values for transfer length and development lengths of various FRP tendons are given in Table 7.3.

7.8—Shear behavior

One of the earliest applications of FRP as shear reinforcement was demonstrated in the one-third scale model beams for the AASHTO girders of the Taylor Bridge in Canada, which were tested by Fam et al. (1997). In this study, five I-girders, each 9.3 m (30.5 ft) long, were exclusively reinforced for shear and prestressing using CFRP, including both CFCC and Leadline™ reinforcement.

To simulate the composite action of the bridge deck, the CFRP stirrups were projected from the girder into the slab, which was cast after fabrication of the girders. Various web reinforcement ratios were used for each type of CFRP reinforcement. This was controlled by the stirrup size and number of legs. The web reinforcement ratio was shown to affect the induced stress level in stirrups and the diagonal crack width; however, the effect was not linearly proportional to the web reinforcement ratio. Also, no slip was measured between the deck slab and girder, which suggested that the stirrups had adequate dowel strength. Because these beams were controlled by flexural capacity, the web reinforcement ratio had virtually no effect on the overall performance.

The study demonstrated the feasibility of using CFRP harped prestressed reinforcement and shear reinforcement in bridge girders, and led to successful detailing of stirrups at the junction between the web and bottom flange to avoid premature shear failure. The study also resulted in a successful design and construction of the Taylor Bridge in Canada, which was the first to use FRP shear reinforcement of this nature.

A comprehensive study was later performed on shear strength of beams reinforced by GFRP and CFRP stirrups (Shehata 1999) and showed that the effective capacity of FRP stirrups might be as low as 50% of the guaranteed strength as described in detail in Section 6.1.2.

7.9—External tendons

External unbonded tendons can be an effective method to exploit the properties of FRP. The unbonded FRP tendon is strained much less than a bonded FRP tendon, and its brittle behavior may be less critical than for a bonded FRP tendon. Nevertheless, the overall behavior is dependent on the performance of the anchorage, and the tendons are fully exposed to potential damage. Furthermore, the contribution of the unbonded tendon to the flexural strength of the member is significantly less than that of a bonded tendon.

CSA S6 permits FRP tendons as the main components of an external post-tensioned strengthening system. The stress limits for external tendons are slightly lower than those for bonded tendons, but GFRP tendons are allowed in nonalkaline grout.

In external prestressing, special attention should be paid to the anchorage that grips the tendon and the interaction of the tendon with the deviation saddles. A slight reduction in the cable strength of CFCC is observed when the angle of curvature at the saddles is 10 degrees or less (Santoh 1993). To reduce the friction between the tendons and the saddle, a cushion material or Teflon sheaths may be used at the saddle point.

In a comparative study between the behavior of concrete members prestressed with internal unbonded and external prestressed FRP tendons, Burgoyne (1992) tested two beams prestressed with parallel-lay aramid (Parafil) ropes. The beam with the internal unbonded tendon did not collapse completely, but the externally prestressed beam failed suddenly and completely, with a total loss of prestressing. Based on this test, Burgoyne proposed that if the tendons were external to the concrete, they should pass through loose rings so that, in the event of failure, the tendons would be forced to deflect with the beam.

Other studies on beams with external tendons were conducted by Mutsuyoshi and Machida (1993), Tsuchida et al. (1993), Miyamoto et al. (1994a,b), El-Hacha (1997), and Jerrett et al. (1996). Saeki et al. (1993, 1995) found that beams prestressed with external tendons could withstand 2 million cycles of fatigue loading.

Horiguchi et al. (1995) studied the effects of temperature on the fatigue properties of externally prestressed concrete beams. They found that the ultimate flexural strength of the beams improved, while their fracture toughness decreased at low temperature. In the fatigue tests, only tension failure of steel reinforcement was observed, but the fatigue strength of external prestressing cables decreased at low temperature.

Grace and Abdel-Sayed (1996a,b) suggested a new system consisting of precast prestressed concrete double-tee girders covered with a deck slab. The double-tee girders were reinforced with GFRP bars and prestressed internally and externally with harped CFRP tendons. The test results indicated that the system was a promising approach to produce a ductile, crack-free and long-lasting bridge system. Also, the tested double-tee bridge had confirmed its ability to sustain a repeated load of 60% of its ultimate load-carrying capacity for 7 million cycles without any significant changes in its dynamic and static characteristics.

El-Hacha et al. (1996) conducted an analytical study on externally prestressed beams and compared the results with available experimental results (that is, strength and stress in prestress at ultimate). They concluded that the equation proposed by Harajli and Kanj (1991) provided the closest results to the experimental values on the conservative side, while the equation proposed by Naaman and Alkhairy (1991) overestimated the strength of beams prestressed with FRP.

Miyamoto et al. (1998) carried out an analytical and experimental study. They applied external prestressing to strengthen the Misaka Bridge in Japan to evaluate the performance of the bridge such as load-carrying capacity and deflection control. The bridge was 35.3 m (116 ft) in length, 6.0 m (19.7 ft) in width, and composed of three simply supported composite steel girders. The external tendons were placed in a queen post arrangement through a rigid concrete diaphragm. Static load tests and moving load tests were carried out before and after completion of the strengthening for comparison purposes. They compared the calculated results of finite element analysis with the measured values, and good agreement was obtained. Their static load test results showed that the strengthening reduced deformation to approximately 50% of the prestrengthening level. The moving vehicle load test results showed that the dynamic deformation was not significant.

7.10—Prestressed poles

Road environments are known to be very aggressive, causing corrosion at the fixture of plain steel, steel-reinforced, or steel-prestressed concrete lighting posts. When steel tendons are used, in many applications, a significant concrete cover (40 to 50 mm [1.5 to 2.0 in.]) is required to protect the prestressing steel from aggressive internal or external environments, or both. In contrast, when durable CFRP tendons are used, only a relatively small concrete cover (15 to 20 mm [0.6 to 0.8 in.]) is required. Therefore, CFRP-prestressed poles can be significantly lighter than equivalent steel-prestressed structures. Recent work has considered the design and installation of 8 m (26 ft) nominal height CFRP prestressed concrete lighting columns fulfilling the requirements of the relevant European Standards (Terrasi and Lees 2003).

CHAPTER 8—REPAIR, STRENGTHENING, AND RETROFITTING

Applications of FRP materials for retrofitting and strengthening existing concrete structures have been rapidly growing in North America (Labossière et al. 1997; Hazen et al. 1998; Grace and Abdel-Sayed 2003), Europe (Meier et al. 1992; Steiner 1996; Nanni 1997; Matthys et al. 2004; Blasi et al. 2004; Rostasy et al. 2004), and Japan (Ichimasu et al. 1993; Katsumata et al. 2001). This strengthening technique provides an efficient, noncorroding alternative to externally bonded steel plates. The purpose of retrofitting by FRP composite strengthening systems is to strengthen or improve the flexural capacity, shear capacity, axial capacity, and ductility, or any combination of them. The high durability of FRP is valuable in environments that may cause steel corrosion. FRP is gaining popularity in view of its many advantages such as low unit weight, ease of handling and application, and low installation and maintenance costs. Some of the disadvantages of FRP include their linear stress-strain behavior to failure, low strain at failure compared with steel, negligible dowel action at flexural and shear cracks, and relatively high material cost. In certain situations, however, FRP provides the most economical solution to strengthening problems because labor costs can be dramatically reduced (Meier and Erki 1997). ACI Committee 440 has developed guidelines for external strengthening applications such as ACI 440.2R. This chapter is introduction to the subject of repair, and the guide should always be followed for design and application purposes.

FRP can be used to strengthen concrete beams in flexure and shear; slabs in flexure; and columns in compression, shear, and flexure. FRP sheets, plates, or NSM reinforcement can be attached to the tensile face of beams or slabs for flexural strengthening. For shear strengthening of beams, FRP sheets can be bonded to the sides of the beams. For columns, FRP sheets or precured shells can be placed on the exterior of columns to increase confinement, ductility, and strength.

The process for the field installation of externally bonded FRP reinforcement consists of the following basic steps: concrete surface preparation (that is, cleaning, sealing cracks, rustproofing of existing steel reinforcement, and smoothing), application of a primer coat, application of a resin undercoat, adhesion of the sheets, curing, and application of the finish coat.

8.1—Flexural strengthening with non-prestressed FRP

The initial development of the FRP-strengthening technique with prefabricated laminates took place in Switzerland (Meier and Kaiser 1991) and Germany. The initial work on strengthened full-scale reinforced concrete beams (Meier and Kaiser 1991) validated the strain compatibility method in the analysis of cross sections and suggested that inclined cracking (shear cracks) may lead to premature failure by peeling-off of the strengthening laminate at the crack. An analytical model for the composite plate anchoring, which agreed with the test results, was developed (Meier and Kaiser 1991). A comprehensive analytical and experimental study of the short-term flexural behavior of strengthened FRP reinforced concrete beams was carried out by Triantafillou and Plevris (1990 and 1992). They concluded that the flexural behavior of reinforced concrete beams strengthened with FRP laminates can be adequately described by the classical theory of plane-sectional analysis to predict the moment-curvature of the load-deflection response at a specific section when premature peeling or debonding failure of the FRP is avoided. Other researchers have subsequently validated these results (Assih et al. 1997).

8.1.1 *Static behavior in flexure*—Meier and Kaiser (1991) conducted tests on 26 flexural reinforced concrete beams

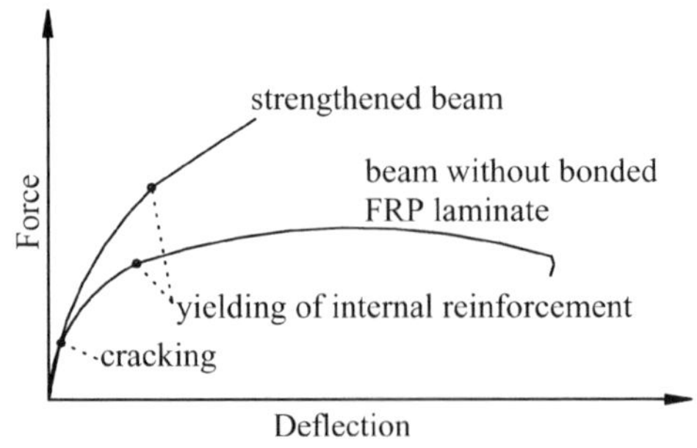

Fig. 8.1—Typical load-deflection curve of strengthened and unstrengthened concrete beams (adapted from Meier and Kaiser [1991]).

with a 2.4 m (8 ft) span strengthened with CFRP laminates. A typical load-deflection curve is shown in Fig. 8.1, which shows a doubling of strength for the strengthened beam, but with reduced deflections, and hence ductility, at failure. The laminate also caused a more distributed cracking pattern in the strengthened beam with reduced crack widths. Other researchers have subsequently found similar results (Hefferman and Erki 1996; Swamy et al. 1996a,b; Beber et al. 1999; Naaman 1999; Kachlakev and Barnes 1999; Jonaitis et al. 1999).

Several failure modes controlling the ultimate strength in concrete beams strengthened with FRP laminates have been observed and reported in the literature by many researchers (Meier and Kaiser 1991; Triantafillou and Plevris 1992; Triantafillou 1998a). These studies show that the main failure modes are:

- Crushing of the concrete in compression before yielding of the reinforcing steel;
- Yielding of the steel in tension followed by rupture of the FRP laminate;
- Yielding of the steel in tension followed by concrete crushing;
- Shear/tension delamination of the concrete cover (cover delamination); and
- Debonding of the FRP from the concrete substrate (FRP debonding).

An et al. (1991) performed a parametric study of reinforced concrete beams strengthened with FRP plates. They found that for beams with high internal reinforcement ratios, a stiffer plate in combination with higher-strength concrete is more effective than a plate with a lower stiffness in combination with lower-strength concrete. For beams with the same internal reinforcement ratios and concrete compressive strength, they found that, as the plate strength and stiffness increased, the ultimate moment capacity increased until the failure mode changed to crushing of the concrete in compression. Cha et al. (1999) found similar results.

Triantafillou and Plevris (1992) developed charts to relate the failure mode of strengthened beams to the level of external FRP and internal steel reinforcement (Fig. 8.2). For a given amount of internal steel reinforcement, the failure mode changed from steel yielding to concrete crushing as the amount of FRP was increased. FRP fracture only occurred for beams with low amounts of internal steel reinforcement.

They validated these charts with tests on small-scale beams (1.2 m [4 ft] span) and found that peeling failures placed an upper limit on the amount of FRP that could be effective for strengthening.

Triantafillou and Plevris (1995) developed a reliability-based design procedure for strengthening with CFRP sheets or plates. Two strength-reduction factors were derived to achieve a reliability index of approximately 3 over a broad spectrum of design conditions. From the analysis, they proposed a general strength-reduction factor, $\phi = 0.85$, and a partial reduction factor, $\phi_{fc} = 0.95$, for the fiber composite strength.

White et al. (1998) investigated the effects of loading rates on the behavior of 3.0 m (9.8 ft) span reinforced concrete beams strengthened with CFRP laminates. Their test results showed that service and ultimate flexural capacity increased as the rate of loading increased. Cracking and failure modes were not affected by the rate of loading.

Maalej and Bonacci (1998) tested two 3.7 m (12.1 ft) span reinforced concrete beams with external CFRP sheets to study the effect of strengthening under a loaded condition. The beam strengthened under load was approximately 6% weaker than a similar beam strengthened without an applied load. The difference occurred because the force in the sheet was lower in the beam strengthened under load. This result demonstrated the importance of considering the initial strain condition in the beam when strengthening occurs and was confirmed by the research of Tan and Mathivoli (1999). On the other hand, David et al. (1999) studied the influence of precracking and found that it did not affect the structural capacity but did reduce the stiffness of the strengthened member.

Aboutaha (1999) experimentally investigated the application of CFRP sheets for strengthening a damaged AASHTO Type II prestressed concrete girder (9.4 m [31 ft] span). To simulate the damage, the undamaged girder was loaded to its ultimate flexural capacity, resulting in severe flexural cracks under the loading points and bond deterioration between the steel strands and concrete. The response of the girder in its damaged condition showed a serious deterioration of stiffness, approximately 50% lower than that of the original girder. The stiffness of the repaired girder was approximately 25% lower than that of the original girder. The ultimate strength of the repaired girder was much higher than that of the original girder.

Rasheed and Pervaiz (2003) developed a bond-slip analytical solution for externally strengthened beams with FRP plates using beam theory with a deformable adhesive layer. The solution provided closed-form deflection expressions for beams with long and short FRP plates subjected to three-point bending, four-point bending, or uniform load. The results showed that the FRP tension force cannot be fully developed when the adhesive shear modulus is below 65 MPa/mm (239 ksi/in.) of the adhesive layer thickness regardless of the length of the plate.

Charkas et al. (2003) developed an analytical solution for the deflection calculation of simply supported reinforced concrete beams strengthened with FRP at any load stage. The solution assumes a trilinear moment-curvature response. It incorporates some tension-stiffening effects and assumes the section to be fully cracked only upon or near steel yielding, depending on the concrete nonlinearity in compression. Comparisons with experiments indicate the effectiveness of the procedure for properly anchored plates. An extensive parametric study was conducted to reveal a single linear relationship between the cracked moment of inertia of the section I_{cr} and the overall effective beam moment of inertia at yielding I_{ey} for a wide range of geometric and material parameters.

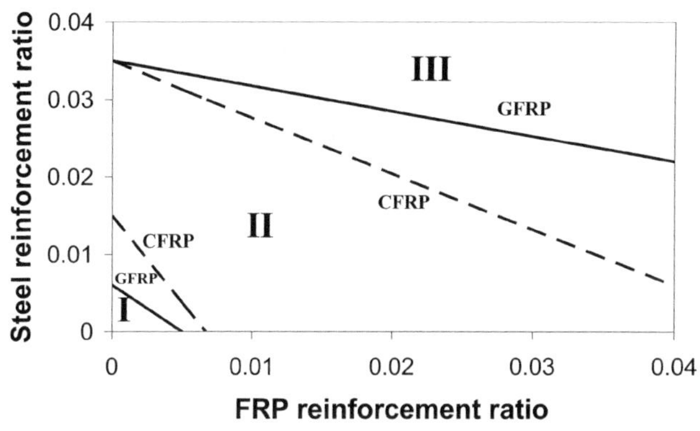

Fig. 8.2—Failure diagram for different composite sheets: CFRP and GFRP. (I = steel yield-FRP fracture [below lower lines]; II = steel yield-concrete crushing [between bounding lines]; and III = compression failure [above upper lines].) (Adapted from Triantafillou and Plevris [1992]).

Other work has been conducted by Kotynia (1999, 2005a) and Kaminska and Kotynia (2000).

8.1.2 *Debonding failures*—Most methods of externally bonding FRP materials to a concrete substrate have exhibited some degree of failure due to debonding. Flexure and shear retrofit measures are typically bond critical, and a number of debonding modes have been observed. Debonding has been characterized in a wide variety of ways in attempts to formulate rational approaches to debonding mitigation. Oehlers (2005) provides a summary of the five generic debonding modes observed in bond critical FRP applications:

8.1.2.1 Axial intermediate crack debonding describes the debonding observed in direct pull tests where the FRP is in tension and there is no curvature present in the concrete substrate. Axial intermediate crack debonding is a fundamental bond property occurring only in laboratory tests.

8.1.2.2 Plate end debonding results from large normal stresses that occur at the curtailment of an FRP plate or sheet. Plate end debonding typically propagates through the cover concrete, often at the level of the internal reinforcement. Plate end debonding may be effectively mitigated through the use of transverse clamping reinforcement (U-stirrups of FRP) (Garden and Hollaway 1998; Garden et al. 1997; Spadea et al. 2001), mechanical anchorage (Shahrooz et al. 2002), or by ensuring that the FRP curtailment extends into the uncracked region of a beam (Roberts 1989). Additionally, using a more flexible adhesive or tapering the FRP plate (width, thickness, or both) can reduce the stress concentration at the end of the plate (Roberts 1989).

8.1.2.3 Intermediate crack-induced debonding initiates at flexure or flexure-shear cracks, or both, located in the shear span of a member, and propagates in the direction of decreasing moment. Such cracks result from the small degree of prying action resulting from the relative distortion present at the toe of a crack in the shear span of a member. Marginal improvement in intermediate crack-induced debonding behavior has been observed when continuous transverse clamping reinforcement is present (Kotynia 2005b). Intermediate crack-induced debonding has been observed to initiate at multiple cracks over the same shear span (Harries et al. 2006).

8.1.2.4 Critical diagonal crack-induced debonding is similar to intermediate crack-induced debonding but is associated with a single critical shear crack in the member. Mitigation of critical diagonal crack-induced debonding is affected by controlling the formation of the shear crack itself.

8.1.2.5 Interfacial shear-induced debonding is associated with the elementary interface shear between stress concentrations (for example, between cracks in a constant moment region). Such debonding is rarely observed in conventional bonded applications but may control the behavior of FRP bonded to prestressed members where cracking is suppressed or in cases where large FRP thicknesses are used to control deflections at service load (and thus, largely uncracked) conditions (Oehlers 2005).

8.1.3 *Plate and sheet anchors*—Examination of the various premature failure modes in concrete leads to the conclusion that premature failure usually initiates at the end of the plate or sheet. This is due to the geometric singularity created by termination of the plate, and it implies that proper detailing of the plate end and additional plate anchorage techniques should be used.

Several approaches have been investigated to avoid peeling failures. Some of these include mechanical anchorages at the ends of the sheet, wrapped sheets around the web of the beam over the longitudinal FRP sheet, or changes in the geometry of the sheets in the anchorage zones as suggested by Karam (1992).

Swamy et al. (1987) showed that premature debonding of steel plates can effectively be avoided by ensuring that the width-thickness ratio of the plate is not less than 50. Swamy and Mukopadhyaya (1995) have shown that this recommendation holds true for FRP plates when glass, glass-carbon, and aramid fibers are used. In the case of CFRP, the plates are generally so thin that the criterion is automatically satisfied. Another technique can be used if multiple layers are being prestressed. The layers can then be terminated (and the prestress transferred) at different locations along the beam. This technique is effective at reducing the magnitude of shear and normal stress concentrations that occur at the ends of the plate (Wight 1998).

By using bonded angle plates or transverse FRP wraps, longitudinal FRP strips can be effectively anchored to the tensile face of the beam. The effectiveness of this technique has been commented on by Arduini et al. (1995) who found that a great deal of ductility could be observed until rupture of the plate or shear failure of the wrap or angle plates.

Deblois et al. (1992) compared bonded unidirectional and bidirectional GFRP sheets with bolted unidirectional GFRP sheets using 1.0 and 4.1 m (3.3 and 13.5 ft) specimens. They found that bolted sheets or bidirectional sheets were more effective than unbolted unidirectional sheets.

Sharif et al. (1994) studied three different anchorage schemes for GFRP plates using small-scale beams: 1) bolting the plates to the tension face; 2) bolting combined with FRP plates bonded to the sides of the beams; and 3) a special one-piece I-jacket plate that was glued to the bottom along the whole span and to the sides of the beam in the shear span. The I-jacket was the most effective anchorage scheme because it prevented all types of peeling failures. Bolts were not very effective because the beams failed by shear peeling.

Lamothe et al. (1998), Spadea et al. (1998), Grace et al. (1999), and Van Gemert (1999) also found that wrapping the sides of the beams with vertical FRP sheets provided effective anchorage for the flexural sheets. Quantrill et al. (1996) found that GFRP angles attached to the sides of the beams were effective anchors. On the other hand, Naaman (1999) tested 3.0 m (9.8 ft) span T-beams and did not find any improvement in the strength of the beams when U-shaped anchors were used.

Garden et al. (1997) tested concrete beams strengthened with CFRP plates to study the effects of three parameters: plate aspect ratio (b/t plate width divided by plate thickness) at constant cross-sectional area, shear span-depth ratio of beams (a/h), and the form of plate end anchorage. The plates were anchored by extending them under the supports or attaching them to GFRP angles bonded to the sides of the beams. The ultimate capacities of the strengthened beams decreased with reducing plate width-thickness ratios. Failure was always accompanied by concrete cover separation from internal reinforcement. Increasing the shear span-depth ratios resulted in improved ultimate capacities. Anchoring the sheets increased the strength of the beams.

Teng et al. (1999) conducted an experimental study into strengthening deficient cantilever concrete slabs by bonding GFRP strips on the top surface. Different anchorage systems were used, and the most effective method was to anchor the GFRP strips into the walls through horizontal slots and into the slab with fiber anchors. This method allowed the full strength of the strips to be developed, and the strength was almost four times that of the unstrengthened beam.

8.1.4 *Flexural strengthening using inorganic matrix*—An inorganic matrix is a low-viscosity resin prepared by blending an aluminosilicate powder with a water-based activator. It is suitable for penetrating carbon or glass fiber sheets and fabrics (Kurtz and Balaguru 2001; Foden et al. 1996; Toutanji et al. 2001). One disadvantage of organic matrix composites is their lack of fire resistance (Chapter 12). Inorganic matrixes have a number of advantages such as high resistance to fire and high temperature, resistance to UV radiation, and easy handling because the inorganic resin is water-based and emits no odor and toxins during construction or curing. Temperature exposure tests show that, by using an inorganic matrix, only 37% of the composite's initial flexural strength is lost after 1 hour of exposure at 800 °C (1472 °F) (Foden et al. 1996).

Kurtz and Balaguru (2001) compared the performance of an inorganic matrix with that of an organic one when used to externally strengthen reinforced concrete beams with carbon fiber sheets. Strength, stiffness, ductility, failure pattern, and cracking of beams strengthened with the two systems (organic and inorganic) were compared. The results showed that the inorganic matrix was as effective in increasing the strength and stiffness of the reinforced concrete beams as the organic matrix, with a minor reduction in ductility. The failure mechanism changed from sheet delamination for the organic system to sheet rupture for the inorganic system.

Toutanji et al. (2001) confirmed these results with further tests and studied the fatigue behavior (Toutanji et al. 2006). Large-scale reinforced concrete beams were strengthened with three layers of carbon fiber sheets and tested under fatigue. The relationships between fatigue strength, crack width, and the number of cycles were studied and analyzed. Results showed that both load capacity and the number of fatigue cycles of reinforced concrete beams were significantly increased with carbon fiber sheets.

8.1.5 *Slabs*—Arockiasamy et al. (1995) conducted a test on the flexural capacity of a structurally damaged solid slab bridge model strengthened with bonded thin CFRP laminates. The unstrengthened slab was tested to failure and then retrofitted with CFRP. The retrofitted slab showed a 90% recovery in the ultimate moment, but with larger deflections.

Malvar et al. (1996) tested six 1/5-scale under-reinforced concrete slabs. Both one-way and two-way strengthening schemes were evaluated. Their test results showed that slabs with one-way CFRP sheets exhibited an increase in the ultimate load of up to 32%, and a decrease in the ultimate deflection of up to 42%. On the other hand, the slabs with two-way CFRP sheets exhibited an increase in ultimate load of 47% and a decrease in ultimate deflection of 27% when compared with the control specimen. Half-scale slabs strengthened with three longitudinal and one transverse CFRP layer exhibited a 20% increase in the ultimate load. The typical failure mode for all the specimens was punching shear.

Ospina et al. (2001) examined the punching shear behavior of damaged two-way slabs with FRP sheets. They tested two full-scale 4.2 m (13.8 ft) square slabs under concentric uniform gravity loading. The slabs were strengthened with 250 mm (10 in.) wide CFRP strips bonded to the top surface in a cruciform pattern. Results showed that the bonded FRP strips tended to stiffen the slab, but no punching shear strength increase was obtained compared with the results of a control virgin slab specimen.

8.2—Flexural strengthening with prestressed FRP

Another method for FRP flexural strengthening is achieved using prestressed FRP laminates. This specialized application combines the benefits of passive bonded FRP laminate systems with the advantages associated with external prestressing. By applying a prestress to the laminate, the material may be used more efficiently because a greater portion of its tensile capacity is engaged. Prestressed FRP

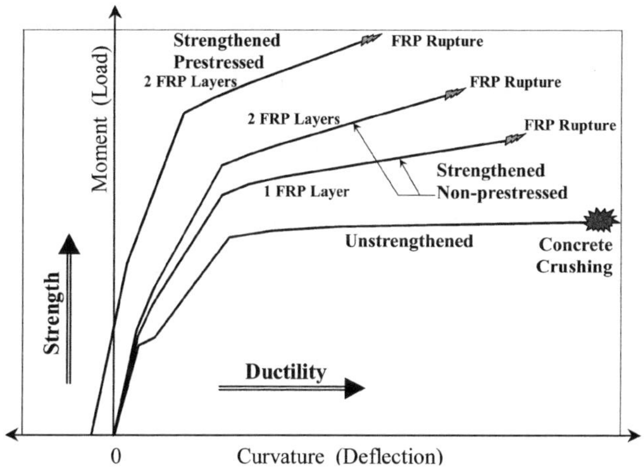

Fig. 8.3—Comparison of prestressed and non-prestressed laminates (El-Hacha 2000).

laminates, when applied to the tension face of a concrete beam, can improve the serviceability of a beam and provide control for cracks by closing existing cracks completely and delaying the formation of new ones. The benefit of prestressed FRP laminates in delaying the cracking and closing of existing cracks may help to restore durability of the structure because the cracks provide additional access for moisture into the concrete and the adhesive layer.

Anchoring the ends of prestressed FRP laminates may further increase the ultimate strength by preventing premature failure modes (Meier 1995; Wight et al. 2001). Prestressed CFRP laminates are likely to show fewer long-term losses than external unbonded tendons (Hollaway and Leeming 1999). Prestress levels of at least 25% of the laminate strength may be necessary to achieve a significant improvement in terms of structural stiffness and load-carrying capacity of the concrete (Garden et al. 1998). Meier et al. (1992) suggested that a prestress level as high as 50% of the plate strength might be necessary to increase ultimate strength by delaying the peeling failure of the FRP plate.

Highly tensioned prestressed laminates will tend to have high shear stresses where the laminates terminate, and this may cause premature shear peeling failure (Triantafillou and Deskovic 1991). To prevent this failure, the shear strength of the concrete may be increased by the application of transverse reinforcement or anchors (Wight et al. 2001; Lees et al. 1999).

Typical load-deflection curves for a reinforced concrete beam, and similar concrete beams strengthened with non-prestressed and prestressed FRP laminates, are shown in Fig. 8.3 (El-Hacha 2000). As shown in the figure, the prestressed laminates are able to increase both the strength and stiffness of the beam. Several systems have been developed to induce a prestress in the laminate for flexural strengthening, and these generally fall into three categories: cambered beam systems, systems that tension the FRP laminate against an independent external reaction frame, and systems that tension the FRP laminate against the strengthened beam itself. The literature concerning this technology is very limited when compared with that for conventional nonprestressed applications. El-Hacha et al. (2001) recently summarized much of the pertinent literature.

A new concept for prestressing the CFRP laminates was outlined recently by Stoecklin and Meier (2003). Once the FRP is prestressed, the central portion is cured and bonded to the beam using heating elements. The prestress level is then reduced progressively as the curing moves towards the ends of the laminates. Using this method, the prestress profile can be designed for the specific beam with the maximum prestress in the middle and the lower prestress towards the ends of the laminate. The shear stresses in the concrete cover at the end of the laminates are also reduced, and anchors are not required to transfer the prestress at release of jacking.

The fatigue of beams strengthened by prestressed FRP laminates has been investigated by Wight and Erki (2001) and Ford (2004).

8.3—Shear strengthening

Javor et al. (1999) conducted tests on 1.6 m (5.2 ft) span reinforced concrete beams that were deficient in shear. The beams were strengthened with CFRP plates bonded on the sides of the beams. Two arrangements were considered: vertical plates and plates at a 45-degree angle. Both arrangements increased the shear capacity of the beams by approximately 40%.

Naaman (1999) considered shear strengthening with either plates or sheets bonded vertically to the sides of rectangular beams (1.3 m [4.3 ft] span) and T-beams (1.9 m [6.2 ft] span). Both the plates and the sheets were found to increase the shear strength by at least 30%. The sheets were found to be more effective than the plates because they could be anchored around the bottom of the beams.

Variations in the number of layers, as well as the fiber orientation and layout, were studied (Lamothe et al. 1998; Naaman 1999; Triantafillou 1998a; Javor et al. 1999; Kachlakev and Barnes 1999; Hutchinson and Rizkalla 1999; Fanning and Kelly 1999; Khalifa et al. 2000; Triantafillou and Antonopoulos 2000; Deniaud and Cheng 2000; Lombard et al. 2000). It was observed that the shear strength increased with increasing number of laminate layers. The maximum shear strength was obtained for beams that were wrapped over their full depth by sheets having fibers perpendicular to the beam's longitudinal axis.

8.4—Axial strengthening of columns

Wrapping concrete columns with FRP provides confinement to the specimen. Confinement is highly dependent upon the cross-sectional shape and aspect ratio of the column (Prota et al. 2006b; Mukherjee et al. 2004), and is most effective for circular sections (Demers and Neale 1994). Transverse strain in the column section caused by longitudinal cracking and section dilation results in hoop strains in the FRP that produce confining pressure. This increase in confining pressure can cause an increase in axial strength and ductility. Concrete confined with fiber composite materials behaves differently than concrete confined with steel. Harmon et al. (1995) found that FRP-wrapped composite concrete failed at higher axial and radial strains than steel-wrapped concrete. Another difference in behavior between steel- and fiber-wrapped specimens is the ductility of the failure. Steel, in general, results in a ductile failure with adequate warning, whereas fibers, such as carbon, have violent explosive failures with little to no warning.

Demers and Neale (1994) studied small-scale (circular and square) unreinforced specimens, wrapped with carbon or glass fiber unidirectional sheets. They concluded that FRPs produced an increase in strength of up to 70% over unwrapped specimens and improved ductility. Demers and Neale (1994) also examined a few confinement models and concluded that they tend to overestimate the specimen capacity.

Howie and Karbhari (1995) and Karbhari and Howie (1997) investigated the effects of various composite wrap architectures on the confining effects and overall performance enhancement of concrete cylinders. They altered the wrap architecture by changing the orientation of the entire composite layer. The wraps varied in configuration from along the longitudinal axis, 45 degrees off the axis, and in the hoop direction. They found that the failure strength increased with the percentage of composite reinforcement in the hoop direction.

Toutanji (1999) investigated the performance of concrete columns confined with FRP composite sheets. Concrete columns were wrapped with three different types of FRP composites. Axial load and axial lateral strains were obtained to evaluate stress-strain behavior, ultimate strength, stiffness, and ductility. The results showed that both the strength and ductility of tested specimens were significantly enhanced over unwrapped specimens. In addition, an analytical model was developed to predict the entire stress-strain relationship of the wrapped specimens. Results from a series of experimental tests on concrete confined with FRP sheets compared favorably with the results obtained by the proposed model.

Confinement effectiveness of FRP jackets in concrete columns was studied by a number of researchers (Mirmiran and Shahawy 1996; Saafi et al. 1999). The improvement in mechanical properties depends on several parameters, including concrete strength, types of fibers and resin, fiber volume fraction and fiber orientation in the jacket, jacket thickness, shape of cross section, length-diameter ratio of the column, and the interface bond between the core and the jacket.

Spoelstra and Monti (1999) developed a uniaxial model for concrete confined with FRP. The model explicitly accounted for the continuous interaction with the FRP wrap due to the lateral strain of concrete through an incremental-iterative approach. The relation between the axial and lateral strains was implicitly derived through equilibrium between the dilating confined concrete and the wrap. The model was compared with a set of experimental tests, and showed very good agreement in both the axial stress-strain and the stress-lateral strain response.

Seible et al. (1999a) conducted a large-scale test on one as-built and four composite-wrapped rectangular flexural bridge spandrel columns to assess the effectiveness of different retrofit schemes using FRP composite jackets. The tests showed that FRP composite jacketing systems clearly can be installed without affecting the overall geometry or appearance of the structure. The authors emphasized the importance of designing retrofit strategies to control the mode of failure. Retrofitting one weakness without considering other potential modes of failure could lead to ineffective and poor designs (Seible et al. 1999a).

Monti et al. (2001) proposed a design equation to determine the optimal thickness of FRP jackets and to enhance the ductility of existing reinforced concrete bridge piers with circular cross sections. The design procedure stems from the definition of an upgrading index given as the ratio of the target to the available ductility at the pier base section attained through FRP jacketing; it results in a simple expression in terms of easily computable quantities, such as the ultimate strain and the peak strength of concrete before and after upgrading. The index, despite its simplicity, yields excellent predictions of the ductility increase obtained through FRP wrapping, and is used to develop a design equation. The equation allows the design of the optimal thickness of FRP jackets in terms of the desired upgrading index, mechanical characteristics of the selected composite material, and quantities defining the initial state of the pier section.

The experimental and analytical behavior of axially loaded large-scale columns confined with FRP wrapping reinforcement was investigated by Matthys et al. (2005, 2006). Different wrapping orientations and different types of FRP reinforcement, consisting of CFRP sheets, GFRP fabrics, and hybrid FRP (HFRP) fabric were used to confine the concrete columns. The effective circumferential FRP failure strain and the effect of an increasing confining action on the stress-strain behavior of reinforced concrete columns confined with FRP were examined. Different models for the prediction of the stress-strain behavior of FRP-confined concrete were reviewed. The results showed that few of the available models that were derived based on small cylinders could accurately predict the ultimate strength and the stress-strain response of large-scale columns.

8.5—Seismic strengthening and retrofitting

The various applications of external FRP composite retrofit in buildings as well as bridges have shown that, in most cases, the implementation of composite materials provides adequate seismic detailing to the structure, and improves ductility and seismic performance. This provides an economic alternative to rebuilding. Seismic rehabilitation is not covered in 440.2R, but is planned to be for future revisions.

The flexural behavior of earthquake-damaged reinforced concrete columns repaired with FRP wraps has been studied by Saadatmanesh et al. (1997a). The columns were repaired with prefabricated FRP wraps and retested under simulated earthquake loading. Seismic performance of repaired columns in terms of their hysteretic response was evaluated and compared with those of the original and unretrofitted columns. The results indicate that the proposed repair technique was highly effective. Both flexural strength and displacement ductility of the repaired columns were higher than those of the original columns.

The behavior of rectangular bridge columns with substandard design details for seismic forces was investigated by Saadatmanesh et al. (1997b). High-strength FRP straps were wrapped around the column in the potential plastic hinge region to increase confinement and to improve the behavior under seismic forces. Results indicated that significant improvement in ductility and energy-absorption capacity could be achieved due to this retrofitting technique.

8.5.1 *Seismic retrofit design*—In the aftermath of the 1995 Hyogo-Ken Nanbu earthquake in which many failures of bridges and piers occurred, numerous studies were conducted on ways to retrofit existing reinforced concrete columns and piers. Mutsuyoshi et al. (1999) found that continuous fiber sheets offered a feasible means of retrofitting. Consequently, several design guidelines on the use of FRP sheets for retrofitting highway, railway, and subway structures have been proposed in recent years. The JSCE Concrete Committee on FRP Sheets has been commissioned to establish a new design method for seismic retrofitting of bridge columns and piers. It seeks to unify all the existing guidelines on a performance-based design concept. The new design method for shear strengthening and ductility enhancement of retrofitted reinforced concrete structures using FRP sheets is described in the study.

Seible et al. (1997a,b) and Seible (2001) described jacket design criteria for various seismic column failure modes and provided detailed examples of their application to retrofit columns with circular and rectangular geometry, different reinforcement ratios, and detailing. The carbon jacket designs were validated through large-scale bridge column model tests and were found to be just as effective as steel shell jacketing in providing desired inelastic design deformation capacity levels.

Sheikh (2001) tested columns retrofitted with CFRP and GFRP composites to improve the seismic resistance of concrete columns. The results showed that retrofitting with FRP for both circular and square columns could improve their brittle behavior, thus significantly improving their seismic resistance.

Pantelides and Gergely (2002) presented analysis and design procedures for the CFRP composite seismic retrofit of a reinforced concrete three-column bridge bent. The seismic retrofit was successful, and the bridge bent retrofitted with CFRP composites reached a displacement ductility level in excess of the target ductility and double the hysteretic energy dissipation of the as-built bent. Descriptions of the CFRP composite layout and validation of the design assumptions from the experimental results were presented. Recommendations for improvement of the original CFRP composite seismic retrofit design were offered based on the lessons learned from the in-place tests.

Saatcioglu and Grira (2001) carried out an experimental investigation to verify the use of FRP grids as transverse reinforcement in concrete structures, placing emphasis on concrete confinement and seismic performance. Large-scale column specimens were tested under simulated seismic loading. Test parameters included grid spacing and pattern, the volumetric ratio of grid reinforcement, and the level of

Fig. 8.4—Attaching FRP strips with power-actuated fasteners (Bank et al. 2003).

axial load. Results showed improved deformability of columns when confined by properly designed grids.

Sause et al. (2004) proposed a design procedure for seismic retrofit of columns using FRP jackets that was based on curvature capacity and concrete strain demand. The design recommendations were verified with experimental data of four full-scale square columns tested under reverse cyclic loading. One column was tested without retrofitting, while the remaining three were retrofitted with various plies of carbon jackets. The authors concluded that a simple fiber model approach, accounting for regions of confined and unconfined concrete, appropriately estimated a conservative extreme concrete fiber compression strain required to achieve a desired deformation capacity. Further, adequate structural response of building columns may be achieved using lower levels of confinement than those previously presented for highway bridge pier columns with larger cross sections.

8.6—Mechanically fastened fiber-reinforced polymer (MF-FRP) laminates

A method for strengthening reinforced concrete members using pultruded FRP laminates and mechanical fasteners has been developed (Lamanna et al. 2001, 2004; Bank et al. 2003; Rizzo et al. 2004). The strengthening is obtained by attaching pultruded FRP laminates with high bearing and longitudinal strengths to the concrete using several, closely spaced steel power-actuated fastening pins or steel expansion anchors (Fig. 8.4). The MF-FRP method is rapid and uses conventional, typically available hand-tools and lightweight materials. In addition, unlike the bonded method, the MF-FRP method requires minimal surface preparation and permits immediate use of the strengthened structure. Previous research on this method on a variety of beam sizes has shown promising results in terms of installation efficiency, the level of strengthening achieved, and the prevention of laminate delamination before concrete crushing.

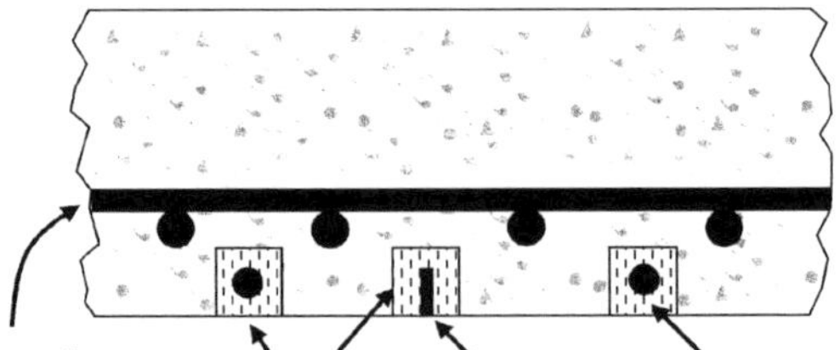

Fig. 8.5—Cross section of a slab strengthened with NSM FRP reinforcement.

8.7—Strengthening using near-surface-mounted FRP reinforcement

Near-surface-mounted (NSM) FRP bars and laminates are being used as an alternative technology to externally bonded FRP laminates. Figure 8.5 shows a cross section of a slab strengthened with NSM. The NSM reinforcement technique involves cutting a groove in the surface of the member, roughening and cleaning the groove, filling the groove halfway with a structural adhesive, installing the FRP reinforcing bar or laminate, filling the groove completely with structural adhesive, and leveling the surface. The NSM FRP technique may be more suitable for cases in which the concrete surface is very rough, weak, or requires significant surface preparation. NSM systems can also be used for shear and flexural strengthening of unreinforced masonry walls (Tumialan et al. 2003a,b,c).

8.7.1 *Flexural strengthening*—Alkhrdaji (2001) strengthened a reinforced concrete bridge deck span with NSM CFRP bars that was tested in place to failure. Compared with an unstrengthened span, they achieved approximately a 30% increase in flexural capacity. Failure was due to the rupture of CFRP bars.

Nordin et al. (2001) conducted a pilot study on concrete beams strengthened in flexure with prestressed NSM CFRP strips. Test results showed a substantial increase in cracking and failure loads for the strengthened specimens. Prestressing the strips did not influence the mode of failure. Compared with unstrengthened specimens, the prestressed beams had considerably smaller deflections at failure.

De Lorenzis et al. (2000) presented the results of tests on large T-beams (with steel reinforcement ratio equal to approximately 25% of the balanced amount) and rectangular beams (with two steel reinforcement ratios, equal to approximately 35 and 59% of the balanced amount). Debonding of the FRP bars was the dominant failure mode for the T-beams and the rectangular beams with the lower reinforcement ratio. The rectangular beams with the higher reinforcement ratio failed by concrete crushing. The authors proposed a system for post-tensioning of the NSM system in cases where access to the ends of the beam was not feasible.

Hassan and Rizkalla (2002) investigated the feasibility of using different techniques for strengthening prestressed concrete bridge decks in flexure. Large-scale models of a prestressed concrete bridge deck were strengthened with

FRP systems and tested to failure. Test results showed that the efficiency of NSM CFRP strips was three times greater than that of the externally bonded strips.

Rosenboom et al. (2004) strengthened 12 prestressed concrete girders with various CFRP systems and tested them under static and fatigue loading. The NSM CFRP bars and strips strengthened girders and achieved a 20% increase in ultimate flexural capacity compared with the control girder when monotonically loaded to failure. The NSM-strengthened girders also performed well under fatigue loading conditions, surviving over 2 million cycles of increased service loading with little degradation and reduced crack widths.

8.7.2 *Shear strengthening*—De Lorenzis and Nanni (2001a,b) investigated the structural performance of simply supported reinforced concrete beams strengthened in shear with NSM CFRP rods. Test results showed NSM FRP rods were effective at enhancing the shear capacity of reinforced concrete beams. The beams strengthened in shear showed an increase in capacity from 28 to 41% over the control specimen.

Nanni et al. (2004) applied NSM CFRP shear strengthening as well as externally bonded precured CFRP laminate flexural strengthening to a prestressed concrete girder taken off an overloaded bridge. The girder was tested in four-point bending, and the use of the rectangular NSM CFRP bars increased the shear capacity of the prestressed girder by at least 53%.

8.7.3 *Development length and bond*—Hassan and Rizkalla (2003, 2004) studied the bond characteristics of NSM CFRP bars and strips. The influence of the groove dimensions, groove spacing, and the limited adhesive cover was investigated. They concluded that the tensile stresses at the concrete-adhesive interface as well as at the FRP-adhesive interface were highly dependent on the groove dimensions and controlled the mode of failure of NSM FRP bars and strips. They recommended widening of the groove to minimize the induced tensile stresses at the concrete-epoxy interface and increase the debonding loads of NSM bars. A mathematical model to find the development length of NSM FRP bars was presented along with design charts correlated to experimental results.

De Lorenzis and Nanni (2002) and De Lorenzis et al. (2004) investigated the bond characteristics between NSM FRP rods and concrete. The influence of several parameters, such as bond length, diameter and surface configurations of the rods, surface condition of the groove, groove size, and type of groove-filling material, was examined. These tests showed that bars with some surface deformation should be used to avoid interfacial failure between the bar and the epoxy, and that the degree of groove roughness as normally obtained by saw-cutting was able to prevent pure interfacial failure at the epoxy-to-concrete interface. Debonding was then controlled by failure within the epoxy or the concrete, or a combination of the two. The effects of bar surface deformation and a groove size-to-bar diameter ratio on failure mode and bond strength were analyzed. A groove size-to-bar diameter ratio equal to 2.0 was recommended. A less satisfactory performance of cement paste as a groove-filling material was also reported. De Lorenzis (2004) proposed equations to describe the local bond-slip curves of different NSM systems and analytical curves giving the bond failure load as a function of the bond length, outlining a design methodology for anchorage. A simplified analytical model of splitting bond failure and a three-dimensional finite element model of the NSM joint were also reported.

Sena Cruz and Barros (2004) analyzed the influence of bond length and concrete strength on NSM CFRP reinforcement in reinforced concrete by conducting pullout tests. A local bond stress-slip relationship was calibrated with the experimental results. They concluded that the concrete strength had marginal influence on the pullout behavior and that various parameters defining the local bond stress-slip relationship were dependent on the slip at peak bond stress.

Other work has been conducted by Alkhrdaji et al. (1999), De Lorenzis et al. (2000), Arduini et al. (2001), Carolin et al. (2001), Täljsten and Carolin (2001), El-Hacha and Rizkalla (2004), and Kotynia (2005b).

8.8—Design procedures

Interest in developing flexural design procedures for FRP-strengthened sections was evident in the mid-1990s. Picard et al. (1995) developed analysis equations to compute the ultimate flexural capacity of FRP-strengthened beam sections using ACI or CSA-type formulations. Chaallal et al. (1998) and Saadatmanesh and Malek (1998) independently presented comprehensive design approaches for reinforced concrete flexural members strengthened with externally bonded FRP. Iterations were required to determine the FRP area corresponding to a certain strengthening level. Rasheed and Pervaiz (2003) formulated closed form design equations allowing for a direct procedure without iterations.

Efforts to develop shear strengthening design models appeared in early 1990s. Triantafillou (1998a) reviewed the earlier models, concluding that they yield contradictory results in most cases. He extended the design model of the Eurocode 2 to account for the contribution of FRP shear reinforcement. Khalifa et al. (1998) modified Triantafillou's equation to describe combined shear failure and FRP rupture. Triantafillou and Antonopoulos (2000) further refined these models by considering three limit states of effective FRP strains.

CHAPTER 9—STRUCTURALLY INTEGRATED STAY-IN-PLACE FRP FORMS

9.1—Introduction

The construction industry is expressing great demand for innovative and durable structural members such as bridge decks and piers, piling, utility poles, and highway overhead sign structures. FRPs have been widely used for structural retrofit in the form of bars and tendons for concrete structures as replacements for steel reinforcement. FRP for structurally integrated stay-in-place (SIP) formwork maximizes the advantages of both FRP and concrete, while simplifying the construction procedure and reducing construction time. In the last several years, a number of such systems have been developed. One of the first published studies on SIP forms proposed a rectangular FRP box section with one open side to be filled with concrete for beam applications as shown in Fig. 9.1(a) (Fardis and Khalili 1981). Deskovic et al. (1995)

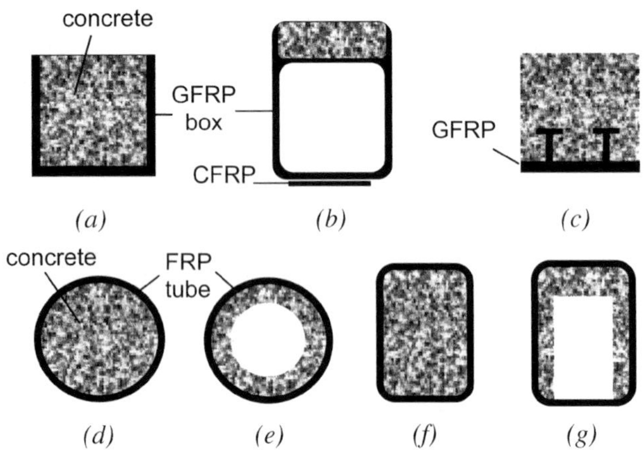

Fig. 9.1—Examples of different configurations of SIP FRP form systems for structural concrete members (Fam 2000).

proposed a novel beam system that consisted of a GFRP box beam with a CFRP laminate bonded to its tension side to take the tension forces and a concrete layer cast on its compression side as shown in Fig. 9.1(b). Hall and Mottram (1998) proposed a novel slab system that consisted of a concrete slab cast onto a pultruded FRP panel with a flat continuous base and two T-up stands as shear studs. The FRP panels serve as permanent forms and reinforcement for the slab as shown in Fig. 9.1(c). Concrete-filled FRP tubes (CFFT) are another promising system, shown in Fig. 9.1(d), that could be used as piles in corrosive marine environments and as bridge girders, piers, and columns (Mirmiran and Shahawy 1996; Fam and Rizkalla 2001a; Seible 1996). The tubes could also be rectangular, as shown in Fig. 9.1(f) and (g) for beam applications (Fam et al. 2003a; Mirmiran et al. 1999). Furthermore, in flexural applications, providing a central hole inside the tube as shown in Fig. 9.1(e) and (g) could substantially reduce the self-weight of the member (Fam and Rizkalla 2002).

9.2—Advantages and limitations of system

The most common characteristics of the (SIP) FRP form systems are as follows:

1. The FRP shape acts as a permanent form for the concrete; hence, it will save the cost of formwork involved in conventional cast-in-place or precast industries;

2. Main reinforcement for concrete is provided externally by the FRP shape, even though additional reinforcement of other materials, such as steel or FRP or both, may be provided internally. In CFFT, the tube provides both shear and flexural reinforcement using the multidirectional fiber orientation. This saves the time and cost of assembling longitudinal bars together with stirrups in conventional construction. The closed composite tubes provide passive confinement to the concrete, which significantly improves the strength and ductility;

3. Depending on the nature of loading, the capacity and performance of the system may depend on the composite action between the concrete and FRP shape. For example, under axial loads, it is desired to unbond the FRP tube from the concrete core, while under bending, it is essential to provide full composite action;

4. The SIP FRP form system lends itself to optimization based on material properties of each component. A hybrid system provides the designers with several flexible parameters that can be controlled to achieve optimum design in individual applications including type of fibers, orientation of the fibers, number of layers in the composite shell, and the composite shell wall thickness; and

5. The confined concrete, in the case of CFFT, is protected from intrusion of moisture with corrosive agents that could otherwise deteriorate the concrete core.

It is also important to note the limitations of the system. Fire is a concern in some structural applications, and measures should be taken to provide fire protection. Chapter 12 discusses this issue in more detail. Vandalism is also a concern; thus, measures should be taken either to limit access to the FRP composite or to provide a paint or texture similar to concrete. Finally, when the FRP form is in direct contact with water, it should be protected against moisture intrusion through selection of an appropriate resin. Such durability issues are discussed in Chapter 11.

9.3—Structural composition of FRP forms

The FRP component of the SIP system accommodates the concrete, acts as reinforcement in one or more directions, and, in some cases, provides confinement for the concrete at the same time. The FRP form is typically a thin shell (laminate) that could be a closed form, such as circular and rectangular tubes, or open forms. The FRP laminate typically includes a number of layers, each consisting of unidirectional fibers or woven fabric in a polymeric matrix. Glass fibers are the most economical and commonly used fibers in such applications. The direction of the fibers in each layer θ with respect to a global longitudinal axis (usually defined along the member length) could vary to provide the required strength and stiffness in more than one direction. The in-plane laminate behavior can be treated as an orthotropic membrane, mainly subjected to in-plane normal and shear forces. The configuration of the laminate, indicating its layer ply composition and sequence of various plies, is referred to as stacking sequence. A laminate is considered symmetric when, for each layer on one side of the middle surface, there is a corresponding layer at an equal distance from that reference plane on the other side with identical thickness, orientation, and properties. A laminate is considered balanced when it consists of pairs of layers of identical thickness and elastic properties, but with similar $+\theta$ and $-\theta$ orientations. Given the structure of the laminate in terms of number of layers, fiber orientation in each layer, thickness of individual layers, fiber and resin types, fiber volume fraction, and mechanical properties of a single unidirectional layer, classical lamination theory is typically used to analyze FRP laminates (Daniel and Ishai 1994). The theory provides equivalent mechanical properties of the laminate in the major directions and allows the prediction of strength under different loading conditions. Mechanical properties of a single layer can be obtained experimentally using coupon tests following ASTM specifications (ASTM 1990).

9.4—Fabrication processes of FRP structural forms

The most commonly used fabrication processes of FRP forms are filament winding, pultrusion, centrifugal casting, hand lay-up, and vacuum infusion processing (resin infusion). Filament winding and centrifugal casting are generally used to produce hollow closed FRP forms, such as tubes. Details on these processes are provided in Section 4.8 of this report.

9.5—Concrete component

Concrete may be plain, reinforced, or prestressed. The internal reinforcement could be steel or FRP and either material could be prestressed. Hybrid combinations are also possible. While any conventional concrete mixture may be used, it is possible to use a shrinkage-compensating admixture to improve bond with FRP in confined sections. On the other hand, recent studies have found shrinkage of concrete cores in CFFT to be quite negligible because the core is essentially sealed from the drying environment (Naguib and Mirmiran 2001). Additional measures, such as shear connectors or sand-coating the FRP surface, may be taken to improve bond between FRP and concrete.

9.6—Construction considerations

9.6.1 *Concreting*—FRP tubes may be filled with concrete in a vertical, inclined, or horizontal position. Care should be taken to support the weight of fresh concrete until it hardens to avoid sagging. Concrete may be pumped or poured in place. Some fabricators prefer to pressure-pump the concrete fill uphill against gravity as a means of eliminating air voids. Concrete may be consolidated using electric vibrators that may be hand-held or mounted on the FRP form. Plugs or caps may be necessary to contain the wet concrete at the ends. Figure 9.2 shows a CFFT cast in an inclined position (Fam et al. 2003c).

9.6.2 *Bond between concrete and FRP*—Composite action and load transfer between the FRP composite and concrete depend on their interface bond. Push-off tests have been used to obtain the average interfacial bond strength in concrete-filled FRP tubes (Helmi et al. 2005). The following methods have been proposed to achieve the required bond strength:

1. Mechanical interlock can be provided by forming circumferential ridges or a grid of pockets on the inside of the FRP form. Alternatively, extensions of the FRP shell into the concrete could be provided. Mechanical interlock can also be achieved by coating the FRP surface with adhesive, followed by applying a thin layer of coarse silica sand;
2. Chemical bond between concrete and FRP may be achieved using special concrete adhesives. The drawbacks of this method include the cost and difficulty of application; and
3. Shrinkage-reducing admixtures can also be used in concrete mixtures to keep the concrete engaged with the FRP form in tubular sections.

9.6.3 *Protective coating of FRP*—A durable finish may be required for FRP forms exposed to UV rays or to maintain color and gloss (Lancaster Composite, Inc. 2006). The success and longevity of the coating depend on the coating material itself and on the quality of the surface preparation. Gentle sanding or blasting with sand or metal may be sufficient to roughen the surface and promote mechanical lock with the coating material. Care should be taken, however, not to penetrate the resin-rich surface and expose the fibers. The service environment often determines the most appropriate coating material for use. Sufficient curing time should be allowed before shipping and handling. Special care should be taken not to scratch the finished surface during handling and transportation.

Fig. 9.2—Filling FRP tubes with concrete at an inclined position (Fam et al. 2003c).

9.7—Behavior of axial members

The most common forms of axial members are circular and rectangular CFFT (Fig. 9.1(d), (e), and (f)). The FRP tube should be able to withstand the lateral pressure of fresh concrete. The tube should also have adequate stiffness and strength in the circumferential direction to provide confinement for the concrete core under axial loads. The following sections are focused primarily on axial members with circular FRP tubes.

9.7.1 *Background*—Attempts to confine concrete with nonmetallic shells started in the late 1970s, using PVC pipes (Kurt 1978) with limited success. Fardis and Khalili (1981) used GFRP jackets to confine concrete cylinders. They pointed out that fibers in the hoop direction confine the concrete by controlling its lateral expansion and microcracking, whereas fibers in the axial direction resist tension caused by bending and improve buckling resistance. They also established that failure of the system was governed by failure of the GFRP shell. Mirmiran and Shahawy (1996) reported the results of CFFTs of standard cylinder size with three different shell thicknesses controlled by the number of GFRP layers. The shells were made from filament-wound E-glass/polyester, with all the fibers oriented at winding angles of ±85 degrees. In loading the specimens axially, care

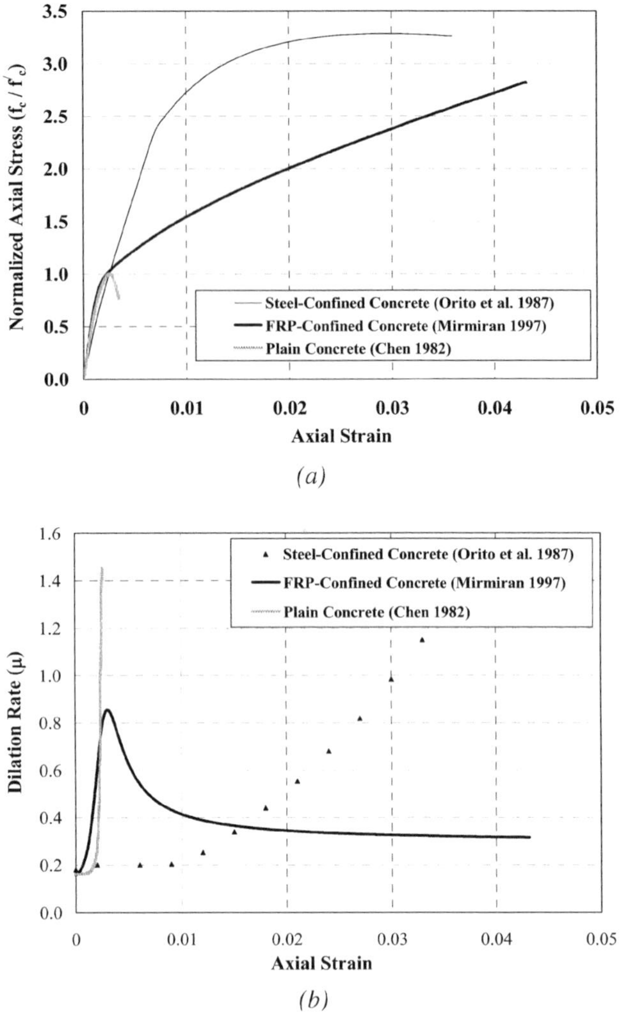

Fig. 9.3—FRP-confined concrete versus steel-confined concrete: (a) stress-strain curves; and (b) dilation behavior (Samaan et al. 1998).

was taken to only load the concrete and to not load the tube axially. Significant improvement in strength and ductility, compared with unconfined concrete, was observed. The stress-strain curve was almost bilinear, with a curved transition near the point of unconfined concrete strength, as shown in Fig. 9.3(a). Fam and Rizkalla (2001a) tested concrete-filled filament-wound GFRP tubes with 2/3 of the fibers oriented in the hoop direction and 1/3 in the axial direction. Axial load was applied to both the tube and concrete. Specimens with and without central holes of different sizes and a tube-in-tube system with concrete filling in between were tested. The study showed that the larger the size of the inner hole, the lower the confinement effect. Also, loading the tube axially reduced the confinement effect. Kanathrana and Lu (1998) and Fam and Rizkalla (2001a) showed that FRP pultruded tubes, with fibers in the axial direction, would split before adequate confinement occurred. This effect was also recognized by Davol (1998), who tested concrete-filled filament-wound CFRP tubes with 80% of the fibers oriented at ±10 degrees and 20% oriented mainly in the hoop direction.

9.7.2 *FRP confinement versus steel confinement*—The stress-strain behavior of concrete-filled GFRP tubes is compared with that of concrete-filled steel tubes in Fig. 9.3(a) (Samaan et al. 1998). The steel jacket confines concrete at low load levels if not loaded axially, due to its high Young's modulus. FRP-confined concrete is insensitive to small lateral expansion due to its lower Young's modulus, particularly for GFRP. As the unconfined strength is approached, high lateral expansion occurs due to major microcracking, which activates the FRP jacket. Passive confining pressure, continuously increasing due to the linear characteristics of the FRP, is induced. Once the jacket reaches its hoop strength, it ruptures, and the concrete fails. On the other hand, steel jackets induce constant confining pressure once the steel yields. Samaan et al. (1998) attributed the difference in behavior to the distinctly different dilation behavior of concrete as shown in Fig. 9.3(b). The dilation response of FRP-confined concrete is initially similar to unconfined concrete. It reaches a peak value, after which it decreases and finally stabilizes. Steel jackets effectively confine the concrete and control dilation before yielding. After yielding of the confining steel, the lateral confining pressure from steel confinement becomes essentially constant, thus allowing for increasing lateral dilation of confined concrete with increasing axial strains.

9.7.3 *Critical factors affecting confinement*—The following equation summarizes the critical parameters affecting the confinement pressure σ_R in circular CFFT, which directly contributes to the enhanced axial strength of the concrete (Fam and Rizkalla 2001b)

$$\sigma_R = \frac{\nu_c}{\frac{R}{E_f t} + \frac{1 - \nu_c}{E_c}} \varepsilon_{cc} \qquad (9\text{-}1)$$

At any axial strain level ε_{cc}, σ_R is increased as the stiffness of the tube in the hoop direction ($E_f t/R$) is increased, where E_f is the equivalent orthotropic elastic modulus of the tube in the hoop direction, t is the thickness of the tube, and R is the radius. The confinement pressure σ_R also increases as the dilation of concrete increases as reflected by its Poisson's ratio ν_c. E_c is the modulus of the concrete core.

Equation (9-1) represents an FRP tube fully utilized in the circumferential direction (debonded from concrete and not loaded axially), as shown in Fig. 9.4(a). Figure 9.5 shows the effect of the stiffness of the tube ($E_f t/R$) on the confinement effectiveness (f'_{cc}/f'_c), where f'_{cc} and f'_c are the strengths of the confined and unconfined concrete, respectively (Fam and Rizkalla 2003).

9.7.4 *Effect of loading tube axially*—Axial loading of the tubes is often unavoidable due to friction, adhesion, and surface irregularities, which provide some axial load transfer to the shell, as shown in Fig. 9.4(b). In this case, the confinement pressure σ_R can be given by an expression similar to Eq. (9-1), except that the numerator is ($\nu_c - \nu_f$) instead of ν_c, where ν_f is the longitudinal Poisson's ratio of the tube defined as the expansion in the hoop direction due to loading in the axial direction. This indicates that, as the tube is loaded axially, it expands radially and tends to separate from the concrete, which reduces the confinement effect (Davol et al. 2001; Fam and Rizkalla 2001b). Therefore, the FRP laminate should be designed to provide a small longitudinal Poisson's ratio ν_f. Unlike steel, FRP laminates can be engineered to control such variables. Loading the FRP tube also results in development of a biaxial state of stress as shown in Fig. 9.4(b), which reduces the strength of the FRP shell in the hoop direction. In this case, the biaxial failure envelope of the laminate needs to be established and the combined state of stress (axial compression and hoop tension) compared with the failure envelope to detect failure, as shown in Fig. 9.4(c) (Fam and Rizkalla 2001b).

9.7.5 *Effect of central holes*—Providing a central hole inside the concrete core as shown in Fig. 9.1(e) reduces the self-weight of the member, but it also reduces the confinement effect by reducing the confining pressure as shown in Fig. 9.5, which shows the effect of inner-outer diameter ratio (D_i/D_o) on the confinement ratio (Fam and Rizkalla 2003). It was also shown that providing an additional FRP tube inside the hole would improve the confinement efficiency of these members.

9.7.6 *Slenderness effect*—Mirmiran et al. (1998) reported a 20% reduction in axial strength of CFFTs with length-diameter ratios (L/D) of 5:1 as compared with specimens with (L/D) of 2:1. More recent studies (Mirmiran et al. 2001; Yuan and Mirmiran 2001) concluded that for CFFT columns not braced against sidesway with slenderness ratio (as defined in ACI 318) of 11 or less, the effect of slenderness may be neglected. For CFFT columns with slenderness ratio higher than 11, however, the maginified sway moment should be taken into account. It should be noted that this limit is lower than that specified in ACI 318 for conventional reinforced concrete columns, which is typically 22. To calculate the maginified sway moment, the study further recommended that the stiffness reduction factor in the moment magnification method of ACI 318 for CFFTs be reduced according to the equivalent orthotropic axial modulus of the FRP tube.

9.7.7 *Effect of geometry of cross section*—Square CFFTs with a corner radius equal to 4.2% of the width of the columns were tested by Mirmiran et al. (1998). Filament-wound GFRP tubes with different wall thicknesses were used. The ultimate axial strength was lower than the peak strength and stabilized at 70% of the unconfined peak strength, irrespective of the tube thickness. The specimens failed due to stress concentration at the corners (Mirmiran 2003). Picher et al. (1996) tested specimens with corner radius varying from 3.3 to 50% of the width of the specimen. Although the specimens were wrapped with CFRP sheets instead of CFRP tube, the effect of corner radius was clearly shown as in Fig. 9.6. The figure shows that increasing the radius causes the behavior of the square columns to become gradually similar to that of circular specimens. This concept was implemented in the proposal of confined-rectangular sections that feature a corner radius of 1/3 of the section width (Burgueño et al. 1996). The behavior of concrete-filled confined-rectangular CFRP tubes was investigated by Davol (1998).

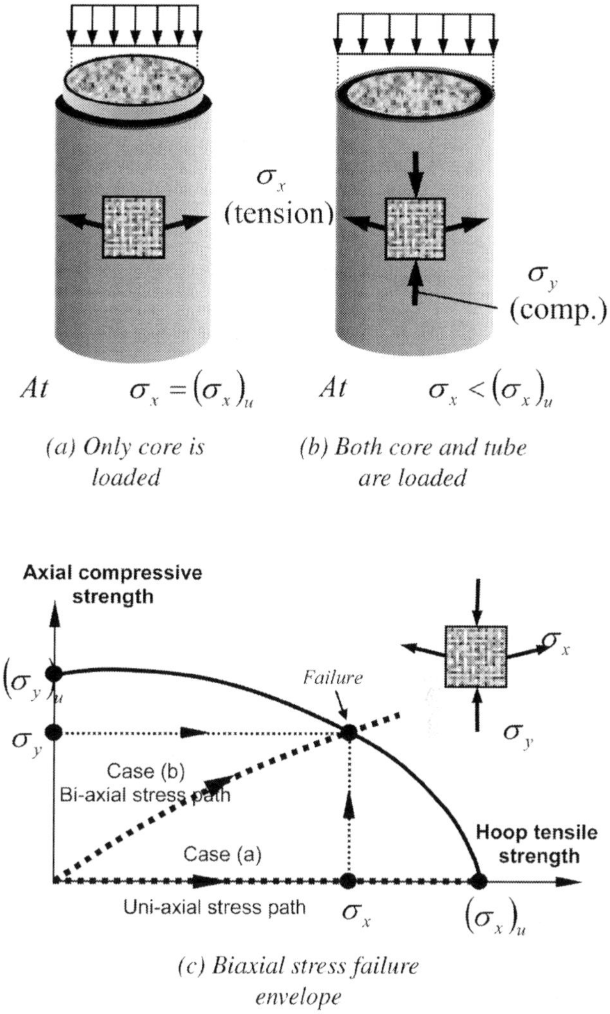

Fig. 9.4—Effect of loading FRP tube axially (Fam 2000).

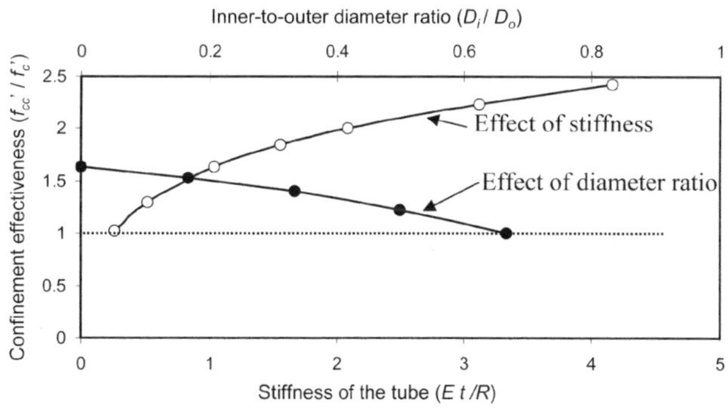

Fig. 9.5—Effect of stiffness of tube and inner hole size on confinement (Fam and Rizkalla 2003).

9.7.8 *Effect of sustained loading*—Creep of concrete in CFFT is much lower than that estimated by ACI 209R models due to sealing of concrete by the tube, lateral confinement, and stress redistribution that takes place between concrete and the FRP tube in the axial direction (Naguib and

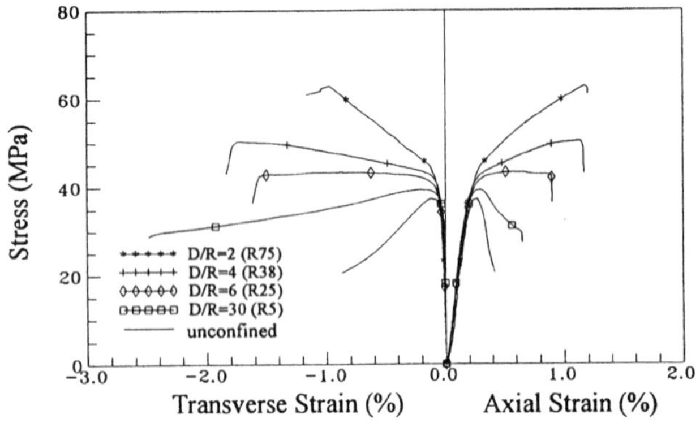

Fig. 9.6—Effect of corner radius on confinement of rectangular members (Picher et al. 1996).

Mirmiran 2001). As the stiffness of the tube increases relative to that of concrete core, larger stress redistribution occurs over time, yielding a lower creep coefficient for the concrete core.

9.7.9 *Bond effects*—Bond between the FRP jacket and the concrete core is more important for flexural members than it is for axial members, and can be achieved by using adhesive or mechanical shear connectors. Fam and Rizkalla (2001b) have shown that, for circular CFFTs under axial compression, the confinement effect is higher if the tube is not loaded axially, which can be achieved by debonding and loading the core only. Mirmiran et al. (1998) tested square CFFTs, with shear connector ribs in the axial and transverse directions, under axial compression. Although the specimens did not achieve the same effectiveness of circular tubes, it was reported that the ribs helped improve the load-carrying capacity by distributing the confining pressure more effectively around the circumference of the tube and minimizing the stress concentrations at the corners. Longitudinal ribs increased the buckling and compressive resistance. The horizontal ribs helped to maintain the cross-sectional shape of the skin and aided the shell in resisting hoop stresses.

9.7.10 *Confinement models*—Different analytical models have been developed to predict the stress-strain response of confined concrete using FRP round shells. The models recognize that confining pressure is variable and is a function of the interaction between the dilating concrete core and the FRP tube with linear characteristics (that is, passive confinement), which is different from steel-confined concrete, where the confining pressure is stable once the steel yields. Some of the available models can predict the entire stress-strain relationship for confined concrete, whereas others may only predict the peak strength (Harries et al. 1998). Some models are based on dilation relationship for confined concrete (Davol 1998; Samaan et al. 1998), modified Ahmad and Shah's model (Saafi et al. 1999; Toutanji 1999), or modified Mander's model (Mander et al. 1988) using different relationships between lateral and axial strains (Mirmiran and Shahawy 1995; Spoelstra and Monti 1999; Burgueño 1999; Fam and Rizkalla 2001b), the energy-balanced technique (Parent and Labossière 1997), or octahedral stress theory for triaxial stress state (Becque 2000). Some of the models account for the biaxial state of stress in the FRP shell under axial loading as well as the effect of a central hole in the concrete core (Fam and Rizkalla 2001b; Davol et al. 2001). A recent study (Naguib and Mirmiran 2001) showed the significant effect of strain rate on the level of confinement, which could potentially affect the comparison of various confinement tests. Several researchers (Picher et al. 1996; Mirmiran et al. 2000a) used plasticity models, such as Drucker-Prager, in finite element analysis to predict the behavior of FRP-confined concrete. A comparative study and survey of some of the confinement models that were published before June 2001 can be found in De Lorenzis and Tepfers (2003).

9.8—Behavior of flexural and axial/flexural members

Hybrid structural members, such as those shown in Fig. 9.1, have great potential for flexural and combined loading applications. Although the confinement of concrete is significantly reduced in bending, even in CFFTs, the advantages of using FRP as permanent formwork and ease of fabrication still make this system attractive. The role of concrete is to resist the compressive stresses, while that of the FRP component is to resist the tensile stresses. In CFFTs, however, concrete has another role to support the FRP shell in the compression side to avoid local buckling failure. Davol (1998), Burgueño (1999), and Fam and Rizkalla (2002) have reported increases in flexural strength of CFFTs of over two times that of the hollow tube and have also shown that, the thicker (or stiffer) the tube, the lower the contribution of concrete to strength.

9.8.1 *Background of closed form systems*—Seible et al. (1995) introduced a CFFT using carbon shells for bridge piers and girders. Two different connection designs were introduced for bridge piers in seismic zones: a ductile design concept with steel bars connecting the CFFT to the footing, and a strength design concept with the CFFT embedded inside the footing. A carbon/epoxy CFFT system was also developed and proposed as the main element for entire bridges such as a dual tied arch, a space truss with post-tensioned members, and cable-stayed bridges (Seible et al. 1998). Conventional reinforced concrete and FRP decks were also proposed to be connected to the CFFT girders using steel or FRP dowels (Burgueño 1999). A special mandrel was used to produce helical ribs on the inside of the shell to provide mechanical bond to the concrete (Burgueño 1999). Studies by Seible et al. (1997a), Burgueño et al. (1998), and Davol et al. (2001) have shown that the design of carbon CFFT structures is stiffness driven. Application of carbon CFFT to bridge systems for the California DOT was investigated through a building-block research program, including full-scale testing of CFFT components (columns and girders), connections, and beam-slab assemblies (Karbhari et al. 1998; Davol 1998; Burgueño 1999; Wernli 1999; Zhao 1999). Tests on 40% scale test units were used to develop a design philosophy for carbon CFFT bridge columns under seismic loads. Flexural behavior was investigated on full-scale carbon CFFT girders for beam-slab bridges through four-point bending tests (Davol 1998; Burgueño 1999). The application of carbon CFFT to beam-slab bridges was validated through the full-scale testing of single and multiple beam/slab assemblies with conventional reinforced concrete decks and modular E-glass/vinylester FRP decks as shown in Fig. 9.7 (Burgueño 1999; Zhao 1999). These later experimental studies addressed not only short-term behavior, but also long-term damage tolerance through simulated cyclic and fatigue loading.

Fam (2000) conducted a large experimental program to study the flexural behavior of hollow and concrete-filled tubes ranging from 90 to 940 mm (3.5 to 37 in.) in diameter with different laminate structure, central holes, and with different diameter-to-wall thickness ratios to model fender piles. The spans of the beams ranged from 1.07 to 10.4 m (3.5 to 34 ft) as shown in Fig. 9.8. Fam et al. (2003b) also studied the behavior of CFFTs under combined bending and axial loads by applying eccentric axial loads. The full interaction curves were developed including tension, compression, and balanced failure regions. Behavior of rectangular CFFTs was studied by Davol (1998), Mirmiran et al. (1999), and by Fam et al. (2003a) using configurations similar to those in Fig. 9.1(f) and (g), respectively.

9.8.2 *Background of open form systems*—Fardis and Khalili (1981) proposed casting concrete into FRP boxes as shown in Fig. 9.1(a). They pointed out that FRP resisted the tensile forces, provided partial confinement of concrete, and carried part of the shear force. They also showed that adhesion between the concrete and FRP was not necessary provided that the FRP box was closed at the two ends. The strength of all the beams tested was greater than that of typical conventional reinforced concrete beams of the same dimensions. Triantafillou and Meier (1992) and Deskovic et al. (1995) presented a hybrid box section, shown in Fig. 9.1(b), consisting of a GFRP box section to resist shear, with an upper layer of concrete in the compression side and a thin layer of CFRP in the tension side. They observed gradual failure, starting by fracture of CFRP before the GFRP, accompanied by the sudden drop of load and reduced stiffness. This provided adequate warning before complete failure (pseudoductility). Seible et al. (1998) proposed a similar hybrid system of prefabricated hybrid carbon-E-glass/vinylester box sections compositely connected to a fiber-reinforced concrete deck. Hall and Mottram (1998) presented a hybrid section combining GFRP off-the-shelf pultruded sections, commercially available as floor panels, with concrete as shown in Fig. 9.1(c). The GFRP system had two T-up stands and a continuous base. Beam tests concluded that adhesive bond was essential and that failure was often by concrete in shear. The beams were over-reinforced in flexure.

Dieter et al. (2002) reported results of laboratory tests conducted on a concrete bridge deck constructed using FRP stay-in-place forms. The pultruded FRP form spanned between, but was not continuous over the girders. Each 457 mm (18 in.) wide FRP form was stiffened by two 76 mm (3 in.) square, hollow corrugations. To ensure composite action through horizontal shear transfer, coarse aggregates were bonded to the FRP surface using epoxy. Additional FRP grid and reinforcing bar reinforcement was placed in the top of the slab and in the longitudinal directions. The deck failed by

Fig. 9.7—Testing of CFFT with concrete and GFRP deck for bridge applications (Burgueño 1999).

Fig. 9.8—Beam test of large CFFT for marine pile applications (Fam 2000).

punching shear. The tested system was used to construct a 66.5 x 13.7 m (218 x 45 ft) bridge deck on prestressed concrete girders on State Highway 151 over State Route 26 in Waupun, WI, in 2003. The bridge is currently in service, carrying an average daily traffic of 18,000 vehicles. The construction and cost analysis of the bridge were described in Berg et al. (2004).

Ulloa et al. (2004) reported the use of a U-shape GFRP beam section in composite action with a concrete deck for the San Patricio County bridge in Texas. The connection consisted of horizontal steel bars passing through holes in the GFRP webs and embedded in the concrete deck. The research demonstrated that composite action from a reinforced concrete deck could overcome deflection limitations inherent in FRP beams.

9.8.3 *Effect of reinforcement ratio and laminate structure in CFFT flexural members*—Fam and Rizkalla (2002) showed that flexural strength and stiffness of CFFTs were

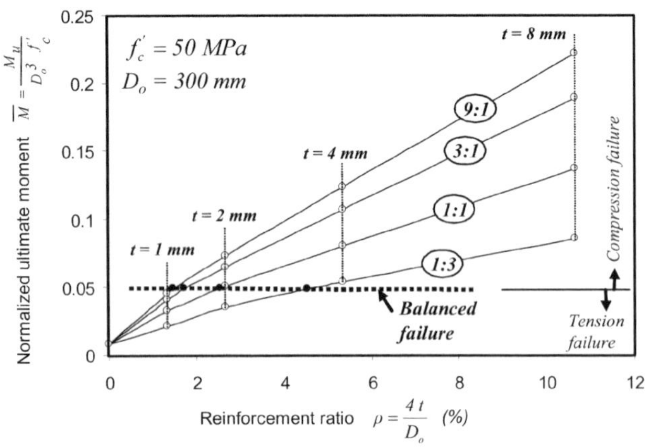

Fig. 9.9—Variation of flexural strength with reinforcement ratio for different laminate structures (Fam and Rizkalla 2002).

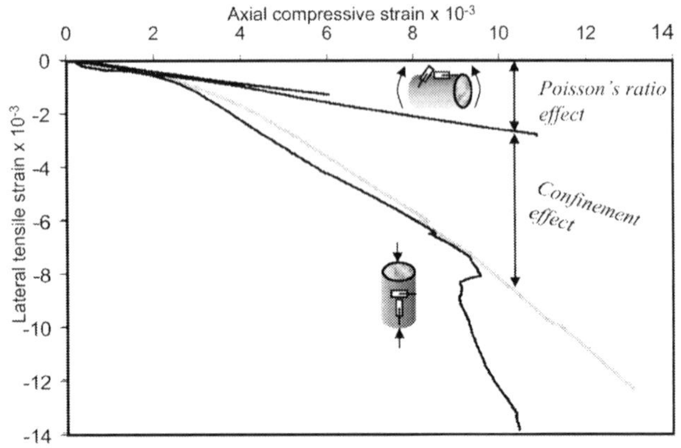

Fig. 9.10—Comparison between axial-lateral strain behavior in beams and columns (Fam and Rizkalla 2002).

governed by both the reinforcement ratio ρ and the laminate structure of the tube, where $\rho = 4t/D$, D is the diameter, and t is the thickness. Figure 9.9 shows the variation of the flexural strength with ρ for different laminate structures of the FRP tube, which had fibers oriented in the axial and hoop directions with various proportions designated as 1:3, 1:1, 3:1, and 9:1. A 1:3 laminate indicates that only 25% of the fibers are oriented in the axial direction. For a given laminate structure, increasing the wall thickness could change the failure mode from tension to compression. Similarly, for a given ρ, changing the laminate structure by increasing the stiffness in the axial direction could change the failure mode from tension to compression. Figure 9.9 also shows that the balanced ρ is reduced as the tube becomes stiffer (or thicker) in the axial direction.

9.8.4 *Confinement effect in CFFTs in bending*—When CFFT round members are subjected to bending, experimental studies (Burgueño 1999; Davol et al. 2001; Fam and Rizkalla 2002) have shown that the effect of confinement of concrete is insignificant. Figure 9.10 shows the axial strain versus the lateral strain behavior of the FRP tube in the compression zone of a beam, tested by Fam and Rizkalla (2002), versus that of a column of the same type. The figure shows that the behavior is bilinear for columns, with significant increase in

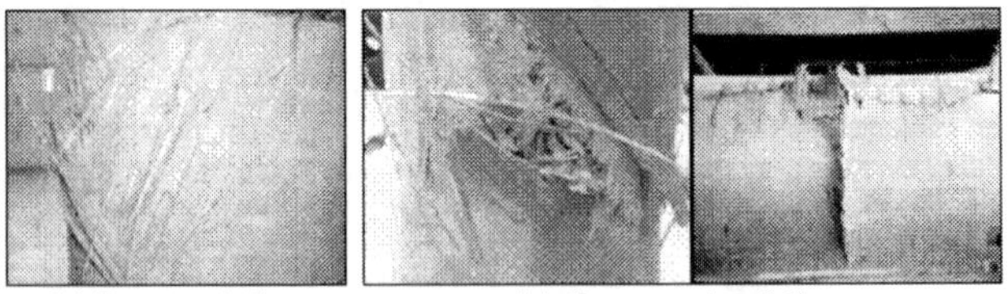

(a) Tension failure *(b) Compression failure*

Fig. 9.11—Failure modes of CFFT: (a) tension failure; and (b) compression failure (Fam et al. 2003b).

lateral strains due to confinement. For beams, however, the behavior is linear, with a slope proportional to the longitudinal Poisson's ratio of the tube, which indicates lack of confinement. This is attributed to the strain gradient, where most of the cross section of the beam is in tension.

9.8.5 *CFFTs subjected to combined bending and axial loads*—Round CFFTs have been tested under constant axial loads and increasing bending (Seible et al. 1995; Mirmiran et al. 2000b) and under increasing eccentric axial loads (Fam et al. 2003b). Compression and tension failures were achieved as shown in Fig. 9.11. Fam et al. (2002) studied the effect of both the wall thickness and laminate structure (different proportions of fibres in the axial and hoop directions) on the interaction curves. The study showed that, for thin tubes, increasing the ratio of fibers in the hoop direction would increase the axial strength and reduce the flexural strength as evident from the curves in Fig. 9.12(a), which intersect at the optimal points for laminate design for each eccentricity. For thick tubes, increasing the amount of fibers in the axial direction increased both the axial and flexural strength, as shown in Fig. 9.12(b). Additionally, for thick tubes, the entire interaction curve could be governed by compression failure.

9.8.6 *Splices and joints in CFFTs*—Because of the limited lengths of CFFTs, splices could be needed. Parvathaneni et al. (1996) produced a 13.7 m (45 ft) long CFFT pile using three 4.57 m (15 ft) long units, spliced using short steel tubing 0.6 m (2 ft) long, which matched with the inside of the GFRP tubes. Ductile joints have been proposed between CFFT bridge columns and footings using short steel dowels (Seible et al. 1998). Pseudoductile plastic hinges have also been proposed for girders (Wernli and Seible 1998; Wernli 1999) using CFRP dowels that provide ductility through gradual slip between the concrete and the bars. The achieved ductility, however, is only in one direction, and the deformation and damage cannot be reversed. The load-slip characteristics for CFRP dowels with varying anchorage details were determined through numerous pullout tests. The behavior of the connection concepts was validated through full-scale flexural testing of longitudinally spliced girders (Wernli 1999). Seible et al. (1998) have also introduced connections between CFFT beams and deck slabs using steel dowels. The connections were studied through pushout tests and full-scale testing of CFFT beam/slab assemblies, which led to the development of design and analysis recommendations (Zhao 1999). Steel dowels for CFFT columns in seismic zones were introduced by Seible et al. (1995) and Burgueño (1999).

9.8.7 *Prestressed members*—Parvathaneni et al. (1996) proposed using filament-wound CFFTs prestressed in the

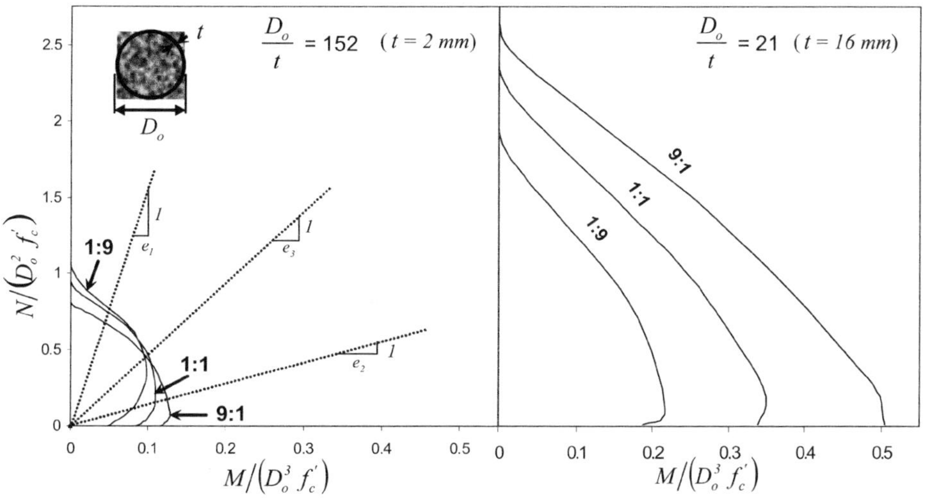

Fig. 9.12—Effect of thickness and laminate structure of tubes on interaction curves of CFFT (Fam et al. 2003b).

axial direction to produce alternative mooring piles. It was decided to take advantage of the high confined strength by prestressing the concrete to a high compressive stress, which was described as super-prestressing. Three 35 mm (1.4 in.) diameter steel Dywidag bars were pretensioned inside the tubes; 35 MPa (5 ksi) concrete was cast and cured; and finally, the bars were destressed, producing a 31 MPa (4.5 ksi) compressive stress in concrete. The conventional method of driving the pile was used. The maximum recorded dynamic strain in concrete was 1360 microstrains in compression, and no tension stresses were induced.

9.8.8 *Hysteretic behavior of CFFTs*—Seible et al. (1995) have tested carbon shell CFFTs with and without mild steel reinforcement anchorage bars as the CFFT-to-footing connection to study the response of CFFT bridge columns under simulated seismic loads. Shao (2003) has modeled and tested CFFTs under low-cycle fatigue, including the effect of loading and unloading on FRP-confined concrete and the seismic behavior of CFFTs with and without internal reinforcement, as shown in Fig. 9.13. Fan et al. (2000) have reported a ductility factor of 10 for CFFTs with internal mild steel reinforcement.

9.8.9 *Sustained loading*—Recent experimental and analytical investigations by Naguib and Mirmiran (2001) have shown that creep effects reduce the flexural stiffness of CFFT. Ultimate strength, however, is not significantly altered. A slow rate of loading and short-term creep at 70% of static capacity may cause premature rupture of the tube. Fiber analysis of CFFT beam-columns by discretizing the section into filled and hollow FRP tubes can adequately simulate the flexural creep behavior. Isochronous sustained stress-creep strain curves may be used as a constitutive nonlinear relationship for creep analysis in flexure. Creep deflection of beam-columns is much less than that of beams, mainly because axial compressive loads tend to retard cracking of concrete and tensile creep of the FRP tube. The axial stiffness ratio of the FRP tube with respect to the concrete core has a pronounced effect on creep deflection of CFFT beam-columns. As the stiffness ratio increases, creep deflections decrease. There exists, however, a threshold beyond which stiffer tubes do not provide additional benefit.

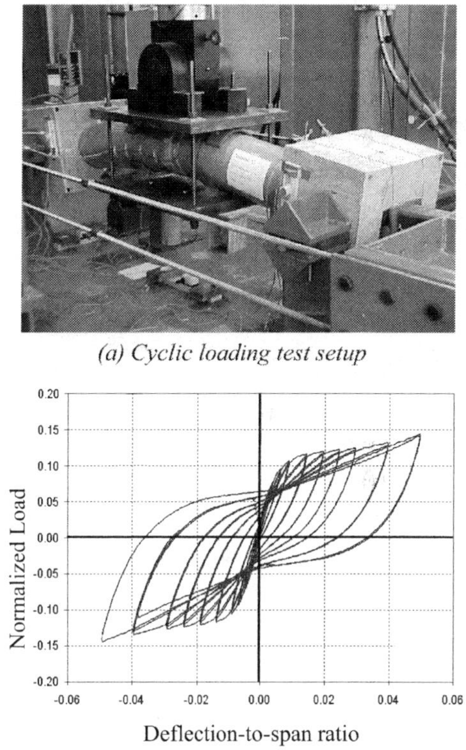

(a) Cyclic loading test setup

(b) Cyclic loading test results

Fig. 9.13—Cyclic loading test of CFFT (Shao 2003).

CHAPTER 10—MASONRY APPLICATIONS

10.1—Introduction

"Building Code Requirements for Masonry Structures" (ACI 530/ASCE 5/TMS 402) covers the design and construction of masonry structures. The code provides requirements for design and construction of new structures. Repair, retrofitting, and rehabilitation of masonry structures are not included in the document.

FRP composites using various polymer and cementitious matrixes, and fiber reinforcement of treated and untreated glass, carbon, and aramid fibers have all been applied for strengthening of masonry. This chapter summarizes work that has been focused on FRP composite systems for strengthening of masonry structures.

Potential advantages of retrofitting masonry using FRP composites include low installation costs, flexibility of use, and minimum changes in the member size after repair. Disturbance to occupants and loss of usable space are also minimized. From a structural point of view, the dynamic properties of the existing structure remain unchanged because there is little addition of weight. If stiffness change is required, it may be engineered on a case-by-case basis by properly designing the composite retrofit.

Even though most of the research on FRP composites and field applications has focused on strengthening reinforced concrete members, available literature shows high potential for reinforcing and strengthening masonry. A research project between the U.S. Army Corps of Engineers and the Market Development Alliance (MDA) of the FRP Composites Industry (Marshall et al. 1999) tested over 100 clay and concrete masonry walls under in-plane loading, and produced a wealth of data on the increased strength and ductility of these walls. The efforts ended with a seismic simulation test in which four FRP composite systems were used to seismically retrofit a half-scale two-story brick building with dimensions of 3.66 x 3.66 x 3.66 m (12 x 12 x 12 ft). The main conclusions were:

- FRP composites can be applied to increase the strength of masonry walls in shear;
- FRP composites enable greater wall drift before failure occurs;
- For shear, glass fiber is preferred over carbon fiber because of the lower stiffness of glass; and
- The failure mode of masonry wall sections can be changed by the application of FRP. By the proper placement and selection of FRP composites on an unreinforced or a lightly reinforced masonry wall, failure modes such as x-cracking can be prevented while transferring the failure to a more ductile mode such as bed joint sliding or rocking before toe crushing.

The U.S. Army Corps of Engineers has published guidelines on the specification and construction of masonry repaired with FRP composites (UFGS 2004a,b). Furthermore, ACI Committee 440 and the Existing Masonry Committee of The Masonry Society (TMS) have established a joint task group for the development of design provisions.

10.2—FRP strengthening techniques

FRP composite products, in the form of externally bonded laminates and grids, and NSM bars are the typical approaches used to strengthen masonry structures. FRP composites have been primarily investigated for enhancing the structural capacity of masonry walls and columns (Masia and Shrive 2003).

10.3—FRP repair and strengthening of masonry

10.3.1 *Flexural strengthening*—Many research projects have been conducted to study FRP systems for flexural strengthening of masonry walls. Ehsani and Saadatmanesh (1996) investigated the flexural behavior of small-scale unreinforced masonry (URM) walls strengthened with GFRP sheets and found that the flexural capacity was increased up to 24 times compared with the unreinforced control specimen. According to the test results, the effect of the mortar strength appeared to be negligible, and both specimens failed by crushing of the masonry.

Velazquez-Dimas et al. (2000) reported test results of half-scale URM walls tested under out-of-plane cyclic loading. Two of the walls were strengthened on both faces with GFRP strips. Substantial increases in strength and deformation capability were achieved. The retrofitted walls resisted pressures up to 24 times the weight of the wall and deflected as much as 5% of the wall height. To avoid very stiff behavior and improve the hysteretic response, the authors recommended limiting the reinforcement ratio to two times the balanced condition. The balanced condition is defined as the point at which failure of the masonry in compression and rupture of the composite in tension occur at the same time. Although the brittle URM walls were retrofitted with a linear elastic material, the combination resulted in a system capable of dissipating some energy representing system nonlinearity.

Hamilton and Dolan (2001) investigated the flexural behavior of small-scale URM walls strengthened with different composite materials. Strengthening with high-strength composite materials such as CFRP and AFRP (with vertical fiber orientation) led to modes of failure such as delamination and shear in the masonry. To use the composite material more efficiently, two alternatives were recommended: first, to increase the spacing of the material until rupture of the laminate governed failure, and second, to use less expensive materials such as GFRP. These more efficient alternatives resulted in four failure modes: debonding, laminate rupture, masonry shear, and face shell pullout. They reported that debonding from the masonry substrate caused the failure of most of the test specimens.

The successful use of NSM bars for improving the flexural capacity of reinforced concrete members (De Lorenzis et al. 2000) led to extending this technique to URM walls. As an example, masonry panels of concrete blocks were tested by Tumialan et al. (2002). One specimen was strengthened with one No. 3 GFRP bar (9.5 mm [0.375 in.] nominal diameter), the second with two No. 3 GFRP bars, and the third was strengthened with an externally bonded GFRP laminate (width = 76 mm [3 in.]). For comparison purposes, specimens one and three had an equivalent axial stiffness $E_{FRP} \times A_f$ (modulus of elasticity × area) to each other. The wall strengthened with one GFRP bar failed due to debonding of the paste from the masonry. Initial flexural cracks formed at the mortar bed joints perpendicular to the reinforcement, and caused secondary cracks at the epoxy paste-masonry interface resulting in debonding and subsequent wall failure. The wall strengthened with two bars failed due to masonry shear, while the specimen with the GFRP laminate failed due to debonding. This experimental program was used as a validation for the strengthening of two URM concrete walls at an educational facility in Kansas City, Missouri, where the walls exhibited cracking in the bed joints at the midheight region.

By using epoxy strengthened with short fibers, Bajpai and Duthinh (2003) were able to prevent debonding of NSM GFRP bars and consistently rupture the bars in flexural tests of masonry walls. This method resulted in higher wall strength and a more brittle behavior.

The capacity of flexural walls strengthened with FRP laminates is a function of the axial load level (Triantafillou 1998b). Moreover, FRP composites are highly effective in the case of walls that can be treated as simply supported (that is, walls exhibiting a large slenderness ratio). For a wall with a low slenderness ratio built between rigid supports, FRP is less effective because arching action of the wall dominates over the effect of the FRP because crushing of the masonry units at the boundary regions controls ultimate behavior (Tumialan et al. 2002; Galati 2002).

In summary, available literature indicates that URM walls strengthened with FRP exhibit the following modes of failure: 1) debonding of the FRP laminate from the masonry substrate; 2) flexural failure (that is, rupture of the FRP laminate in tension or crushing of the masonry in compression); or 3) shear failure in the masonry. Of these three modes of failure, the literature has shown that the controlling mode is mostly debonding of the FRP laminate. Thus, the quantity of FRP reinforcement should be balanced against the masonry shear strength; if a large amount of FRP reinforcement is provided, a brittle masonry shear failure may result. Proper masonry design philosophy dictates that brittle shear failure should be avoided by ensuring masonry flexural capacity is exceeded by its shear capacity.

Debonding is directly related to substrate surface characteristics such as roughness, soundness, and porosity. For instance, Roko et al. (1999) observed that absorption of epoxy is limited in extruded brick units as compared with that in molded bricks, leading to a reduction of the bond strength at the FRP laminate-masonry interface.

Tumialan et al. (2002) suggested that, rather than attempting to predict bond failure, the strain in the FRP laminates could be limited. The effectiveness of the FRP reinforcement depends on the bond of the FRP laminate to the masonry substrate. Because the flexural capacity is dependent on the strain developed in the laminate, effective strain in the laminate ε_{fe} can be expressed as the product $\kappa_m \varepsilon_{fu}$, where κ_m is the bond-dependent coefficient, and ε_{fu} is the design rupture strain of FRP. Tumialan et al. (2002) concluded that for nonputtied surfaces, κ_m can be assumed to be 0.45, and for puttied surfaces, κ_m can be 0.65.

Luciano et al. (2001) investigated the possibility of reinforcing masonry arches using FRP composite materials and found that the FRP laminates greatly enhanced the capacity of masonry arches.

To enhance the out-of-plane seismic resistance of the facades of historic masonry buildings, unobtrusive FRP rehabilitation techniques that incorporate intermittently bonded NSM carbon fiber rope and unbonded and intermittently bonded NSM carbon fiber composite cables (CFCC) were developed at McMaster University (Korany 2004). Ten full-size clay brick wall panels were retrofitted and tested under both monotonic loading and quasi-static cyclic loading using an airbag. Korany and Drysdale (2004) reported significant increases in ultimate capacities, energy absorption, and deformability compared with the behavior of the unreinforced walls.

10.3.2 *Shear strengthening*—Schwegler (1995) investigated strengthening methods for masonry shearwalls with FRP laminates. CFRP laminates were bonded diagonally to the masonry walls and mechanically anchored to the adjoining reinforced concrete slabs. The test results showed that the strengthened walls exhibited 50 and 300% increases in ultimate capacity and displacement, respectively, as compared with unstrengthened walls.

Cracked URM concrete block walls were repaired by Gergely and Young (2001) using CFRP laminates attached to both sides of the specimens and subjected to cyclic out-of-plane loads and in-plane loads. The symmetric laminates significantly increased the flexural and shear capacity of damaged walls. The specimens failed as a result of severe shear damage in the concrete masonry blocks.

Concrete masonry walls strengthened with FRP laminates in the horizontal direction only and tested with in-plane loading along the wall diagonal were observed to fail due to sliding shear along an unstrengthened joint (Tumialan et al. 2001; Morbin 2001). This mode of failure, which is undesirable if there are adjacent columns such as in the case of infill walls, may be controlled by placing FRP bars in the vertical direction to act as dowels.

As in the case of URM walls strengthened for flexure with FRP laminates, the type of masonry has been observed to be one of the factors influencing the in-plane wall behavior. Thus, in the case of clay brick masonry walls strengthened with laminates for shear, FRP strengthening has been observed to be more efficient than in the case of concrete masonry (Grando et al. 2003). This can be attributed to characteristics of the parent material such as height of masonry courses (that is, smaller in the case of brick masonry) and better mortar-masonry unit bond characteristics. Grando et al. (2003) also reported that the in-plane capacity of clay masonry walls strengthened on one and two faces doubled when the amount of FRP reinforcement was doubled.

Valluzzi et al. (2002) reported experimental results on small-scale clay brick masonry specimens using variables such as the type of FRP laminates (CFRP and GFRP) and strengthening configurations (single-side versus double-side strengthening, and square grid and diagonal). Double-sided strengthened specimens were more effective than single-sided specimens. In general, the diagonal strengthening configuration was observed to be more effective than the grid configuration; also, GFRP laminates were more effective than CFRP at increasing shear capacity.

Bastidas et al. (2002) investigated the strengthening with GFRP laminates of small-scale nonstructural masonry walls built with clay tiles. In addition, a full-scale wall was tested to validate the technology. The strengthening configurations included vertical and horizontal laminate strips, combinations of both, and diagonal laminates (cross-pattern). The test results showed the efficiency of the GFRP reinforcement for increasing the shear strength as well as the ductility of the system. The cross-pattern layout on both sides of the wall proved to be the most effective configuration. A significant global reduction of damage levels was observed for the strengthened masonry wall when compared with results reported by the same authors on similar URM walls. Also, global stability and overall seismic behavior were greatly improved with the GFRP reinforcement for in-plane loading.

Strengthening by FRP structural repointing can also remarkably increase the shear capacity of URM walls. Repointing is a technique used in masonry to repair and replace the mortar in the joint. With FRP repointing, the mortar is cleaned out, epoxy is placed into the groove, and an FRP bar is placed into the same groove into the epoxy. This was evident from the results of tests conducted on concrete masonry walls loaded along the diagonal (Tumialan et al. 2001; Morbin 2001). The maximum increase in shear capacity was 80% for walls strengthened with GFRP bars placed at every bed joint. Walls with reinforcement staggered on both wall faces exhibited the largest pseudoductility.

FRP structural repointing was also used to improve the in-plane structural performance of masonry infill walls (Tumialan et al. 2003a,b,c). Full-scale specimens were subjected to in-plane cyclic load. The specimens were surrounded by reinforced concrete frame and a stand-alone reinforced concrete support. The results indicated that FRP-strengthened specimens could reach lateral drifts of 0.7% without losing lateral carrying capacity, and, that for this drift level, the degradation of lateral stiffness in the strengthened walls did not implicate degradation of lateral carrying capacity.

With the objective to find alternative embedding materials to epoxy-based paste, Turco et al. (2003) investigated the in-plane behavior of concrete masonry walls strengthened with GFRP bars embedded in two different materials: epoxy-based paste and latex-modified cementitious paste. The in-plane test results showed that the performance of walls with both materials yielded similar results. The use of less expensive pastes, such as the latex-modified ones, makes the FRP structural repointing technique more appealing because the structural performance is not reduced and the appearance of the filled joints is similar to conventional mortar joints.

The effectiveness of increasing the shear strength of brick masonry by epoxy-bonding FRP overlays to the exterior surfaces was evaluated by Ehsani et al. (1997a). The variables in the test included the strength of the composite fabric, fiber orientation, and anchorage length. Specimens were tested under static loading. The results showed that both the strength and ductility of tested specimens were significantly enhanced. The orientation of the angle of fibers with respect to the plane of loading had a major effect on the stiffness of the retrofitted system, but did not affect the ultimate strength significantly.

The experimental results of three half-scale unreinforced brick walls retrofitted with vertical composite strips were presented by Ehsani et al. (1999). The specimens were subjected to cyclic out-of-plane loading. Five reinforcement ratios involving two different glass fabric composite densities were investigated. The mode of failure was controlled by tensile failure when wider and lighter composite fabrics were used and by delamination when stronger ones were used. The strengthened specimens were able to support a lateral load up to 32 times the weight of the wall. A deflection as much as 2% of the wall height was measured.

Avorio and Borri (2001) studied the problem of seismic strengthening of monumental arches and vaults. The interest in this technique came about from the examination of the types of collapse involving arches and vaults and from problems shown by structures strengthened with traditional methods. In formulating the criteria, attention was paid to the behavior of the vaults according to their constructional type and the type of texture and pattern.

Moon et al. (2003) and Moon (2004) tested, under lateral loads, a full-size two-story URM brick building that was strengthened using several FRP techniques. On one three-wythe wall, GFRP was epoxy-bonded vertically on the inside face, while NSM glass rods were epoxy-bonded into horizontal bed joints on the exterior face. This two-way retrofit increased the lateral strength, caused cracks to be well distributed, and produced a ductile-type failure mode with broad hysteresis loops and considerable energy dissipation. The four FRP systems used in this project included: precured structural grids embedded in trowel-applied epoxy adhesive (Fig. 10.1); wet lay-up unidirectional glass fabrics with an epoxy matrix (Fig. 10.2); epoxy adhesive-applied NSM GFRP rods (Fig. 10.3); and glass grids in cementitious trowel-applied matrix (Fig. 10.4). Application of GFRP systems on the other multiwythe walls worked well in in-plane shear retrofit because header bricks every sixth course generally maintained continuity between wythes.

In-place tests were performed by Corradi et al. (2002) on FRP retrofitted masonry walls damaged by recent earthquakes. Both CFRP and GFRP unidirectional laminates were used to retrofit the masonry panels, followed by in-plane tests. The tests confirmed that the shear capacity of the masonry panels was significantly increased by the FRP materials.

10.3.3 *Settlement repair*—Hartley et al. (1996) tested two full-sized 200 mm (8 in.) concrete block walls, 2.4 m (8 ft) high and 6.0 m (20 ft) long to investigate the feasibility of using unidirectional CFRP sheets to repair settlement damage. In the study, settlement loads were first applied to induce characteristic step cracking. CFRP was then applied to one surface, and the wall retested. Strength gains of over 50% were recorded. The results suggested that CFRP was suitable for rehabilitating concrete block walls damaged by foundation settlement.

10.4—Design and application considerations

10.4.1 *FRP system selection requirements*—Several suitable FRP systems are currently available to repair or retrofit masonry structures. To select the proper FRP system for a particular project, several factors have to be considered by the design engineer and building owner. Some of these factors are:

- Types of masonry construction: non-load bearing, load bearing, or retaining walls, and parent material (that is, concrete or clay masonry unit);

Fig. 10.1—Precured structural grids embedded in trowel-applied epoxy adhesive (Moon 2004).

Fig. 10.3—Epoxy adhesive-applied NSM GFRP rods (Moon 2004).

Fig. 10.2—Wet lay-up unidirectional glass fabrics with epoxy matrix (Moon 2004).

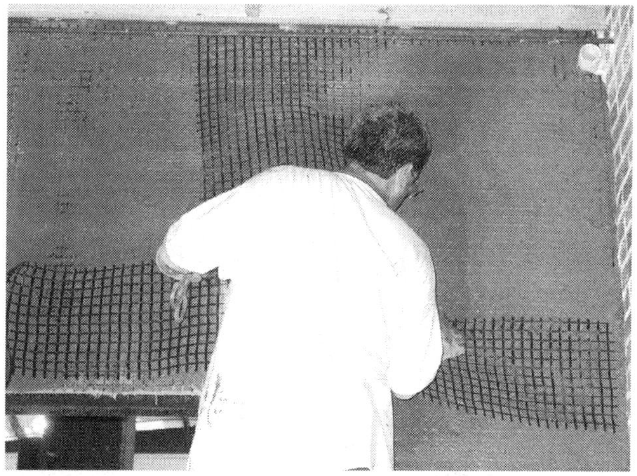

Fig. 10.4—Glass grid in cementitious trowel-applied matrix (Moon 2004).

- Overall building condition (damage level, presence of cracks, surface coatings, accessibility), with particular emphasis on the condition of the substrate. The condition and strength of the masonry substrate is an important parameter for bond-critical applications, including flexure or shear strengthening. The existing masonry substrate should possess the necessary strength to develop the design stresses of the FRP system through bond;
- Repair or retrofit impact on building operation;
- Building occupancy and use;
- Architectural considerations; and
- Building code and fire code requirements. Coatings can be used to limit smoke and flame spread.

Many of these factors have been addressed in several publications. Saadatmanesh (1994) provided an overview of the FRP applications for existing structures, including seismic retrofit of URM buildings. Christensen et al. (1996) studied the architectural implications of reinforcing existing masonry walls with FRP composite materials, the problems associated with the various substrates, and building code issues related to smoke and fire hazards. In the study, they also evaluated FRP-compatible architectural finishes.

10.4.2 *Detailing requirements*—In addition to strength and stiffness requirements, FRP retrofit or strengthening of masonry walls should also address application specific requirements. Among these are FRP composite reinforcement strength development, anchoring systems, and connections for the composite systems and between structural elements (Hamilton and Dolan 2001).

Proper FRP reinforcement detailing at wall boundaries is necessary to ensure expected performance and avoid premature failure due to debonding. For externally bonded laminates, this may be attained with anchorage systems that include steel angles, steel bolts, and FRP bars. Different systems offer their own advantages and disadvantages. Steel angles are easy to install, but aesthetically problematic. Angles in direct contact with the masonry surface may locally fracture the wall due to displacement and rotation restraint. Steel bolts have shown high effectiveness, but require a demanding installation effort (Schwegler 1995).

A technique similar to the one used for anchoring laminates in reinforced concrete joists strengthened in shear (Annaiah 2000) can be used for anchoring laminates in masonry applications. The installation technique consists of grooving a slot in the upper and lower boundary members. The ply is wrapped around an FRP bar that is placed and bonded in the slot with a suitable epoxy-based paste (Carney and Myers 2003a,b). For NSM or structural repointing construction, bars can be easily anchored into adjacent concrete members by drilling and embedding their extended terminations (Tumialan et al. 2003a,b,c).

NSM FRP bars can also be used to improve the anchorage of masonry walls to boundary reinforced concrete beams or foundations (Tumialan et al. 2002; Carney and Myers 2003a,b). Multiwythe steel reinforced masonry parapets built using clay units with standard dimensions were tested in-place. Several steel bars were missing or irregularly placed in the original construction. Before installing the GFRP bars, the holes in the reinforced concrete beam and slots in the masonry were filled with an epoxy-based paste. The design rationale was to provide enough flexural reinforcement to force the occurrence of shear failure. The masonry walls were loaded in-plane as cantilever walls. The control wall lost carrying capacity due to the crack growth caused by rocking. In the retrofitted wall, the opening of the horizontal crack in the strengthened side was controlled. Because of the eccentricity of the GFRP anchors, the wall tilted, preventing the development of the full flexural capacity. The improvement in capacity with respect to the unanchored parapet, however, was over 100%.

10.4.3 *Surface preparation*—Once the adequate FRP system has been selected and designed for the repair or retrofit project, the masonry surface to which the FRP system will be applied should be properly prepared. Surface preparation is necessary to adequately transfer the forces between masonry elements and surface-bonded FRP composite overlays. This preparation consists of complete removal of all mortar droppings, dust, dirt, oil, existing paint or coatings, and efflorescence from the masonry surface. Smooth-faced epoxy-coated or glazed units should first be roughened by grinding or sandblasting.

For unspoiled new clay or concrete masonry surfaces, wire brushing proved to be adequate to remove any loose particles or dust. The surface preparation of older clay or concrete masonry structural members, however, may require more intrusive techniques such as water blasting, grinding, or wire brushing with power tools. Concrete masonry units may be lightly sandblasted (Hamoush et al. 2001), but this method should be used with caution for clay units.

10.4.4 *Installation of FRP system*—For the installation of an FRP system, the recommendations of the system manufacturer should be precisely followed, with deviations requiring approval of the manufacturer and the design engineer. Typical masonry details such as weep holes, concrete masonry control joints, and clay masonry expansion joints should be maintained in their original condition. For example, no resin or FRP laminate should cover weep holes, and no FRP material should bridge masonry movement joints.

CHAPTER 11—DURABILITY OF FRP USED IN CONCRETE

A significant design issue for FRP composites is the consideration of overall durability of these materials, especially as related to their capacity for sustained performance under harsh and changing environmental conditions under load.

Although FRP composites have been successfully used in the automotive, marine, industrial, and aerospace sectors, critical differences exist in loading, environmental exposure, and the types of materials and processes used in these applications as compared with those likely to be used in civil infrastructure applications. Anecdotal evidence provides substantial reason to believe that, if appropriately designed and fabricated, these systems can provide longer lifetimes and lower maintenance than equivalent structures fabricated from conventional materials. Experimental data on durability, however, is sparse, not well documented, and not readily available. Additionally, some evidence has been found of rapid degradation of specific types of FRP composites exposed to certain environmental conditions found in civil engineering.

11.1—Definition of durability

In the context of this report, the durability of FRP or structural members using FRP is their ability to resist cracking, oxidation, chemical degradation, delamination, wear, fatigue, the effects of foreign object damage, or a combination of these for a specified period of time, under the appropriate load conditions, and under specified environmental conditions.

11.2—Durability of FRP composites

11.2.1 *Materials*—Without the protection of the appropriate resin, E-glass fibers are the most susceptible to degradation due to moisture and alkalinity. Similarly, aramid fibers are resistant to abrasion and impact, but show a propensity to creep, absorb moisture, and degrade under ultraviolet exposure. Carbon fibers are relatively inert to the environment. In composite design, however, the individual fibers are encapsulated in a suitable resin system to form the composite. Thus, the fibers are protected from the environment by the resin. The durability of the resin system is dependent on several factors, including the resin components and proportions as well as curing time and conditions. The composites industry has many resin systems that are designed for specific end-use applications and should be chosen based on the mechanical, physical, chemical, electrical, or other considerations in the operating environment. Properties of materials typically used in FRP composites used with concrete can be found Chapters 4 and 5 of this document or in 440.1R and 440.2R.

Both the fiber system and the resin system should be chosen based on requirements of structural performance, constructibility, and durability. Different fibers and resin systems provide differing degrees of resistance to environmental conditions such as moisture, alkaline solution, UV radiation, or extreme temperatures. Thus, the constituents need to be chosen based on both performance and durability requirements. The processing method and quality assurance and quality control used during processing are important predicators of quality and durability of the fabricated composite component. A summary of the significant factors affecting durability was presented by Porter at a SAMPE Symposium (Porter 1999).

11.2.2 *Overview of ASCE/CERF document*—The ASCE/CERF (Karbhari et al. 2003) document "Durability Gap Analysis for Fiber Reinforced Polymer Composites in Civil Infrastructure" provides the results of a gap analysis identifying critical areas in which data are needed. In this document, the use of FRP composites in the form of reinforcing bars, external reinforcement of concrete structures, seismic retrofit of concrete and masonry structures, replacement and new bridge deck systems, and wall panels and profiles were identified to be of the maximum interest. Therefore, the evaluation of durability issues with these applications was assigned the highest priority in the report. The following durability issues were addressed: moisture or aqueous solutions, alkaline environment, thermal effects, creep and relaxation, fatigue, and UV radiation. An analysis of the existing data was performed to rank the importance and availability of data in each of the topic areas.

11.2.3 *Environments*—The intent of this section is to outline the environments that FRP composites used with concrete and masonry are likely to encounter. Each section gives a brief summary of the effect that each of these environments has on the constituents usually found in FRP composites used with concrete.

11.2.3.1 *Moisture (water and salt solution)*—All resins absorb moisture, with the percentage of moisture sorption depending on the resin structure, degree of cure, and temperature. The two primary effects of moisture uptake are plasticization and a reduction in glass transition temperature. In general, moisture effects over the short term cause more pronounced degradation in strength as opposed to stiffness of the composite. Salt solutions can cause blistering due to osmotic effects. In some cases, moisture has been observed to wick along the fiber-matrix interface, resulting in a loss of structural integrity. Moisture can also affect the fracture toughness of FRP composites, with reported results being somewhat contradictory (Karbhari et al. 2003).

In the case of glass fibers, degradation is initiated by moisture-extracting ions from the fiber. Fibers need protection of the resin to avoid such deterioration. Aramid fibers absorb moisture, causing loss of transverse and compressive strength (Karbhari et al. 2003).

11.2.3.2 *Chemical solutions*—In most cases, the effect of chemical solutions is on the resin system, with the absorption following a diffusion-based process similar to that of water (Karbhari et al. 2003). The presence of specific salts or other chemicals in the solution can accelerate deterioration in the presence of inappropriate resin systems. A large number of specialty resin systems are available that are resistant to varying levels of chemical attack and exposure.

11.2.3.3 *Alkaline environment*—Alkaline solutions, such as the pore water of concrete, have a high pH and a high concentration of alkali ions (Neville 1995). Carbon fibers are resistant to alkaline solutions. Resin damage via alkali attack is generally more severe than that due to moisture. E-glass fiber systems should be properly designed and fabricated with the appropriate resin system to protect the reinforcement from alkali attack. Alkali-resistant glass fibers are available and can decrease the rate of deterioration substantially. A significant amount of extremely high pH testing has been conducted at Iowa State University, resulting in several investigations, such as by Mehus (1995). These tests show that, under a loading resulting in a stress of 40% of ultimate, some FRP reinforcing bars can fail while submerged in a solution with a pH of 12.8. Tests from the same source also show that the reinforcing bar that had been submerged in the high pH solution lost as much as 60% of its tensile strength. Later tests from the same location indicated that when improved resins were used, the results improved significantly; thus, durable resins need to be used for these environments (Boris and Porter 1999).

11.2.3.4 *Extreme temperature and thermal cycling*—The primary effects of temperature are on the viscoelastic response of the resin, and hence, of the composite. As temperature increases, the modulus of the resin will decrease. If the temperature exceeds the glass transition temperature T_g, FRP composite performance will decrease substantially. Thermal cycling below T_g generally does not cause deleterious effects, although extended thermal cycling of brittle resin systems can result in microcrack formation (Karbhari et al. 2003).

The coefficients of thermal expansion (CTEs) for FRP composites are generally quite different from those of steel and concrete. For the case of glass FRPs, the CTE is generally higher than that of steel and concrete. For carbon and aramid FRPs, the CTE is generally lower than that of steel and concrete in the direction of fibers (Hollaway and Leeming 1999). The CTE will vary considerably with fiber and resin type as well as fiber orientation and constituent volume fractions. The difference in CTE should be considered when composites are used in direct combination with steel and concrete systems.

11.2.3.5 *Low temperature and freezing and thawing*—In general, low temperature and freezing-and-thawing exposures do not affect fibers, although they can affect the resin and the fiber-resin interface. Polymeric resin systems are known to embrittle, resulting in increased strength and stiffness under sub-zero (but noncryogenic) conditions (Chawla 1998). Freezing-and-thawing effects can be more severe due to moisture-initiated effects, causing microcrack growth and coalescence because of cycling. The presence of road salts in wet conditions with subsequent freezing and thawing can cause microcrack formation and gradual degradation due to crystal formation and increased salt concentration.

11.2.3.6 *Creep and relaxation*—Polymer resins generally exhibit creep and relaxation behavior. The addition of fibers increases the creep resistance of the resins. Consequently, creep and relaxation behavior are more pronounced when load is applied transverse to fibers or when the composite has a low fiber volume fraction (Karbhari et al. 2003). Typically, thermosetting resins (unsaturated polyesters, vinylesters, epoxies, and phenolics) are more resistant to creep than are thermoplastics (polypropylene, nylons, and polycarbonates).

Carbon fibers are the least susceptible to creep rupture; aramid fibers are moderately susceptible; and glass fibers are most susceptible to creep rupture (Hollaway and Leeming 1999). Extrapolations of short-term creep data to longer service lifetimes in room-temperature air suggest rupture strengths of 29 to 55%, 47 to 66%, and 79 to 93% of the initial strength for essentially unidirectional GFRP, AFRP, and CFRP materials, respectively (Yamaguchi et al. 1997; Ando et al. 1997; Seki et al. 1997; Greenwood 2002).

E-glass fibers are also susceptible to environmental stress corrosion cracking. This is a delayed brittle fracture effect that is caused by synergism between stress and the environment (Jones 1999).

Stress limits for FRP composites under sustained load to avoid premature failure due to stress rupture can be found in ACI 440.1R and 440.2R.

11.2.3.7 *Fatigue*—The fatigue performance of FRP composite materials depends on the matrix composition and, to some extent, on the type of fiber (Curtis 1989). The individual fibers within unidirectional composites have few defects, and are consequently resistant to crack initiation. Additionally, any crack that does form travels through the matrix and is not transmitted through adjacent fibers. These toughness and crack-arresting properties contribute to the good fatigue performance of FRP materials.

11.2.3.8 *UV radiation*—Polymeric materials undergo degradation when exposed to UV radiation between 290 and 400 nm due to dissociation of chemical bonds (Karbhari et al. 2003). The subsequent reaction with oxygen can lead to oxidation, chain scission, or cross-linking. In general, effects are rarely severe in terms of mechanical performance, although some resins can show significant embrittlement and surface erosion. The most deleterious effect of UV exposure is generally not the UV-related damage, which is surface limited, but the potential for increased penetration of moisture and other agents via the damaged region. In some cases, degradation at the surface has been found to affect mechanical properties disproportionately because flaws can serve as stress concentrations (Chawla 1998). FRP composites can be protected from UV-related degradation with appropriate additives in the resin, appropriate coatings, or both.

11.3—Internal reinforcement

11.3.1 *Introduction*—This section covers the degradation process and mechanisms affecting hygro-thermo-mechanical properties of FRP reinforcing rods under exposure to alkaline environments, alternate wet and dry cycles (in corrosive and noncorrosive mediums), freezing-and-thawing conditions, temperature and humidity variations, and loads (creep and fatigue).

11.3.2 *Moisture*—Sen et al. (1998) conducted a 45-month study on the long-term performance of AFRP and CFRP pretensioned elements used to reinforce piles driven in tidal waters, based on destructive tests. Results indicated that bond degradation adversely affected the ultimate capacity of AFRP-reinforced piles, but the CFRP-reinforced piles were largely unaffected.

11.3.3 *Alkaline environment*—The reaction of FRP composites to alkaline conditions in concrete is a major design consideration. The internal concrete environment initially has high alkalinity, with the pH between 12 and 13. This alkaline environment can have an effect on fibers, such as glass and aramid, as discussed in the previous section.

Although an appropriate resin matrix (vinylester, epoxy) provides a high level of protection to fibers from this degradation, migration of high pH solutions and alkali salts through the resin (at voids, cracks, and interface between the fiber and matrix) to the fiber surface is possible (ACI 440.1R). In addition, the application of special surface coatings or fillers, selection of suitable chemistry, and improvement in manufacturing processes can all improve the durability of FRP composite reinforcement in an alkaline environment. For instance, Shah et al. (2002) observed a 50% reduction in the room-temperature diffusion rate of water into vinylester resin filled with montmorillonite nano-clay, although the equilibrium moisture content increased compared with the same resin without filler.

Aqueous solutions with high pH are known to degrade the tensile strength and stiffness of GFRP bars (Porter and Barnes 1998; Rostasy 1997; Sen et al. 1998; Takewaka and Khin 1996; Sheard et al. 1997; GangaRao and Vijay 1997). On the other hand, Devalapura et al. (1998) concluded that GFRP reinforcement exposed to both alkaline and acidic environments retained significant load-bearing capacities for extended life cycles under conditions harsher than expected in field service. Al-Dulaijan et al. (1996) detected a considerable reduction in bond strength of bars immersed in a high-pH solution for 28 days. This reduction appeared to be a result of degradation of the resin.

Arockiasamy and Sandepudi (1994) concluded from experimental studies that the Young's modulus of CFRP composite reinforcement in a combined seawater and alkaline solution with sustained tension was reduced by approximately 12% over exposure periods from 3 to 9 months. The same exposure, however, did not affect the ultimate strength.

In 2004, ISIS Canada approved a project to study the performance of GFRP reinforcement that has been used in many demonstration concrete structures across Canada (Mufti et al. 2005). The objective of the study was to provide the engineering community with the results of the performance of GFRP reinforcing bars that have been exposed to a concrete environment in built structures and to calibrate the Canadian Highway Bridge Design Code (CHBDC) performance factors on the GFRP reinforcement. Core specimens of GFRP reinforcement were collected from five field demonstration projects across Canada. Analytical methods, such as scanning electron microscopy and energy-dispersive x-ray, optical microscopy, differential scanning calorimetry, and infrared spectroscopy, were used to determine the degradation of GFRP in concrete structures.

Based on the results of the aforementioned analyses described, Mufti et al. (2005) found no visible degradation of the GFRP reinforcement (rods and grids) in the concrete environment in real engineering structures exposed to natural environmental conditions for 5 to 8 years. The results from scanning electron microscopy and x-ray analyses suggest no degradation of the GFRP reinforcement materials in the demonstration concrete structures. The x-ray analyses indicate no alkali ingress in the GFRP reinforcement from the concrete pore solution. The conclusion of the project was that the GFRP reinforcement is durable and highly compatible with the concrete material. Also, the team concluded that the CHBDC is conservative; therefore, the material performance factors of the GFRP bars should be increased. This change has been incorporated in the new addendum for the CHBDC.

11.3.4 *Low temperature and freezing and thawing*—At low temperatures, complex residual stresses arise within FRP composites as a result of matrix stiffening and mismatch of CTEs of matrix and resin as well as FRP and concrete (Chawla 1998). Residual stresses can cause microcracks in the matrix and fiber-to-matrix interface, which can grow under low-temperature thermal cycling and coalesce to form transverse matrix cracks and debonding between the fibers and the matrix.

In general, the reported literature shows that unidirectional tensile strength decreases when exposed to temperatures between –10 to –40 °C (14 to –40 °F), whereas the off-axis and transverse strengths may increase due to matrix hardening. Increasing freezing-and-thawing cycles have been shown to accentuate residual stresses, resulting in increased severity and density of cracks. An apparent increase in matrix brittleness and decrease in tensile strength has also been reported (Lord and Dutta 1988).

Mashima and Iwamoto (1993) determined that bond strength of GFRP and CFRP rods was not influenced by freezing and thawing, but AFRP (both braided and coiled) rods showed a gradual reduction of bond strength up to about 20% with continued freezing and thawing.

11.3.5 *Temperature*—Researchers reported that extremely elevated temperatures (above the glass transition temperature) have a detrimental effect on bond, probably because of lower shear stiffness in the FRP. The GFRP achieved the highest residual bond strength, while the AFRP achieved the lowest, but slip increased dramatically with increases in temperature in all the materials (Nanni et al. 1995; Katz et al. 1999).

11.3.6 *Creep and relaxation*—Odagiri et al. (1997) investigated relaxation characteristics of 6 and 7.4 mm (0.24 and 0.29 in.) diameter AFRP tendons with anchorages. Overall, the relaxation rates for AFRP rods were found to be approximately 11% at 1000 hours, and 15% at 17,700 hours (2 years).

Creep rupture is another important concern when FRP reinforcing bars are subjected to long-term loading. Creep-rupture is the tensile rupture of a material subjected to sustained high stress levels over a period of time. The creep-rupture behavior of 3 mm (0.12 in.) diameter FRP composite circular rods of glass, aramid, and carbon fiber was evaluated by Dolan et al. (1997) by nonaccelerated techniques. Aramid specimens were more susceptible to stress concentrations at anchor points. Substantial decay was found in the long-term resistance of the glass tendons, especially when in direct contact with cementitious material.

Nikurunziza et al. (2002) conducted stress rupture tests on GFRP bars. After 60 days of exposure, the loss of tensile strength was 4% for a water exposure and 11% for alkaline exposure at temperatures of 65 to 75 °C (149 to 167 °F).

Almusallam et al. (2002) conducted stress rupture tests on GFRP bars embedded in concrete beams and exposed to water. The strength losses were less than 5% for unstressed specimens, but as high as 30% for stressed specimens.

Den Uijl (1991) predicted the long-term performance of aramid bars in an alkaline environment using temperature as the varying parameter. The tests indicated that the time to failure under constant load at 60 °C (140 °F) was between 10 and 15 times shorter than when tested at 20 °C (68 °F). Ando et al. (1997) found similar results.

Sheard et al. (1997) studied durability of FRP-reinforced concrete in aggressive alkali and wet and dry environments at different temperatures and stress levels. Based on the test results, the authors suggested a 100-year life threshold stress limit of about 25% for GFRP, 50% for AFRP, and 75% for CFRP.

Creep experiments performed by Apinis et. al. (2000) showed that carbon fibers do not creep at strain levels as high as 0.69% for 17,700 hours. Aramid fibers, starting at a strain level of 1.38%, creep by 69% after 16,800 hours. Glass fibers creep by 5% from a starting strain level of 0.78% after 16,600 hours. Experiments and analysis performed by Tamužs et. al. (2001) concluded that hybrid composites, CFRP+AFRP and CFRP+GFRP, have a considerably higher creep resistance compared with pure aramid and glass composites. If brittle components such as carbon fibers are used in the hybrid composite, however, a certain risk of overloading the component exists due to the large creep of aramid fibers. After the breakage of carbon fibers, the creep will accelerate and may lead to the failure of the whole composite if the load level for the remaining fibers is high.

11.3.7 *Fatigue*—As described in Section 11.2.3.7, FRP composites generally have very good resistance to fatigue. In the case of internal reinforcement, fatigue loading is more likely to affect the concrete and the bond between the concrete and the FRP reinforcement than the FRP itself. It should, however, be noted that cracking in concrete could cause abrasion related damage to the FRP surface, which could result in higher environmental exposure to moisture and solutions.

11.3.8 *UV exposure*—In the case of internal FRP reinforcement, there is no direct exposure of the FRP to UV and hence this is not of concern except through any deterioration that UV may cause to concrete at the systems level.

11.4—External reinforcement

This section is devoted to the performance of external FRP reinforcement for concrete that has been adhesively bonded to the surface of the concrete. Included in this classification are both confinement systems such as column wrapping and flexural and shear reinforcement. External reinforcement is likely to see a more varied environmental exposure than that of internal reinforcement including moisture cycling, chemical solutions, and UV radiation.

In adhesively bonded FRP systems, one face of the material is adhered to the concrete and one is exposed to the environment. Consequently, the exposure conditions (related to moisture) for the FRP composite are affected simultaneously by the local environment and the underlying concrete. Highly porous concrete will readily transmit any available moisture to the bond line. Accumulation of salts at the bond line may cause large pore pressures to develop, thus reducing the bond capacity.

Bonded FRP systems that are not anchored rely entirely on the bond between the FRP composite and the underlying concrete and masonry. The nature of the bond between the substrate and FRP composite is tied to the FRP strength and durability as well as that of the concrete.

Failure of the external systems is generally governed by the tensile strength of the concrete. Also, cracking in the beam (due to flexure, shear, or other causes) will have a profound effect on the failure load of beams plated with FRP (Teng et al. 2002). The consequence of this key structural aspect of external FRP reinforcement is that the durability of the system may have little to do with the durability of the components of the FRP reinforcement, and have much to do with the durability of the underlying substrate.

11.4.1 *Moisture (water and salt solution)*—Prolonged exposure of FRP composites to moisture can cause problems. Moisture can gain access to the resin and fibers from the exposed face. In addition, cracks and porosity in the underlying concrete or masonry can allow moisture transport by diffusion and capillary action. Thus, both faces of externally applied composites may be exposed to moisture and salt solution.

Karbhari et al. (1997) studied the effect of water, seawater, and freezing and thawing on bonded FRP systems using a peel test. They found that the nature of the bond changed and reduced in strength when exposed to water. Wan et al. (2003) and Wan (2002) found similar results even though the FRP was not directly exposed to moisture. In the same study, the presence of moisture on the FRP substrate during FRP application was also investigated. Once again, bond behavior deteriorated with increased moisture present during application.

Chajes et al. (1995) found similar results for 48 small-scale beams strengthened with glass, aramid, or carbon fiber composites. The beams were subjected to wetting and drying or freezing and thawing, and then tested in flexure. The results indicated that the beams strengthened with carbon fibers maintained their strength better than glass or aramid fiber composites. They also found that the wetting and drying caused slightly greater degradation than that of freezing and thawing. They also found that the exposure changed the failure mode to debonding, indicating a loss in adhesion due to exposure.

Moisture damage can occur either in the adhesive at the bond line or just below the bond line in the concrete and masonry substrate (cracking). Beaudoin et al. (1998) tested beams that had been strengthened with CFRP laminates and then subjected to moisture cycling. Although there was strength reduction noted in one series, the loss of capacity was attributed to the adhesive and not to failure of the substrate.

Murthy et al. (2002) conducted environmental tests combined with sustained loading and found significant decreases in the bond capacity.

In summary, it appears that long-term exposure of externally bonded FRP composites to moisture can cause deterioration of the adhesion between the composite and concrete or masonry substrate. Furthermore, the bond may initially appear adequate, and yet deteriorate with time and exposure to moisture, freezing and thawing, or both.

11.4.2 *Alkaline environment*—Concrete has a pore water pH level as high as 13.5. Research has shown that alkaline solutions and ions can cause severe degradation to bare glass fibers and some polymer systems (Karbhari et al. 2003). Resins are known to be more or less susceptible to alkali attacks because alkaline solutions can accelerate the degradation of bond and of resins themselves, especially if they are not fully cured. An important role of the resin, in addition to transferring load between the fibers, is to protect the fibers against alkali and other harmful agents.

In addition, a number of FRP systems may be in contact with soil, which could also have a high alkaline content. Thus, the determination of durability of FRP composite systems in contact with alkalis is essential. Although a number of these effects could be in the form of solutions and could conceivably be considered under the first topic, this condition is considered separately due to its relationship to concrete.

Alkali environments can also have adverse effects on the fibers, with glass fibers being the most susceptible to damage. Aramid fibers are generally considered more resistant than glass to alkali attacks. Carbon fibers are noted to have high resistance to alkali attack to the point of being almost unaffected (Meier 1995).

11.4.3 *Extreme temperature and thermal cycling*—The effects of extreme temperature and thermal cycling on FRP materials in general is discussed in Section 11.2.3.4. An additional concern for externally bonded FRP is the effect on the bond of the FRP to the substrate. Not only may the adhesive layer be affected by temperature, but a mismatch of CTE between the FRP, adhesive, and substrate may result in unexpected thermally induced stresses. In general, stresses due to differential thermal behavior are not likely to be significant for GFRP, with a CTE relatively close to that of concrete and steel. CFRP and AFRP, on the other hand, need more consideration because their CTE values differ significantly from those of concrete and steel.

11.4.4 *Freezing and thawing*—The effects of freezing and thawing on FRP materials in general is discussed in Section 11.2.3.5. Again, the bond to the substrate may also be affected by freezing and thawing. Experience has shown that voids—regions between the concrete and FRP not completely filled with adhesive and or resin—exist in most external FRP applications. The expansive nature of most resin systems tends to minimize these voids, but poor surface preparation, latent material, or moisture presence during FRP application, among other factors, can create significant voids along the FRP-substrate interface. If liquid water can get to these voids and freeze, a lensing effect is likely. The interface region will deteriorate, and the voids will grow with continued freezing-and-thawing cycles. Good surface preparation techniques and quality-control procedures should minimize this concern. Some specific research on freezing-and-thawing effects is summarized in this section.

Karbhari and Eckel (1993) tested FRP-wrapped cylinders at a low temperature (–18°C [0 °F]) and found increased brittleness of FRP fibers at a low temperature. Soudki and Green (1996) found good performance of CFRP-strengthened columns when subjected to up to 200 cycles of freezing and thawing consisting of 16 hours of freezing (–18 °C [0 °F]) followed by 8 hours of thawing in a water bath (+15 °C [59 °F]). Toutanji and Balaguru (1998) exposed FRP-wrapped concrete cylinders to 300 freezing-and-thawing cycles. They found some deterioration due to freezing and thawing with CFRP performing better than GFRP.

Chajes et al. (1995) tested FRP-strengthened concrete beams exposed to 100 freezing-and-thawing cycles. They found strength losses of 21% for CFRP and 27% for GFRP. In similar tests, Bisby and Green (2002) and Green et al. (2003) found little strength loss and no noticeable bond deterioration with FRP-strengthened beams exposed to up to 200 cycles of freezing and thawing. Lopez et al. (1999) tested small-scale beams strengthened with two types of CFRP and subsequently exposed to up to 300 freezing-and-thawing cycles. Their results showed, for one type of FRP, that precracking of the beams before strengthening could result in significant deterioration (up to 40% loss in strength) due to freezing and thawing. This result indicates that the combination of load and cold region exposure may be more severe than cold region exposure alone.

Arockiasamy and Thayer (1998) and Lopez et al. (2003) considered fatigue loading of CFRP-strengthened beams at low temperature and found no observable difference in behavior caused by the exposure to low temperature.

11.4.5 *Creep*—Most externally bonded FRP systems are intended to strengthen an existing structure. Consequently, the level of sustained load carried by the FRP will likely be minimal. As long as the sustained load is kept below the limits prescribed in ACI 440.2R, the remaining capacity of the FRP system will still be available to resist live load. If, however, a change of use is proposed for a structure previously strengthened with FRP, then creep rupture limits should be re-examined as a potential limiting factor.

Plevris and Triantafillou (1994) analytically and experimentally studied the time-dependent behavior of concrete beams strengthened with FRP plates. Three 1.5 m (4.9 ft) span beams were tested under sustained load. The analytical model was based on the age-adjusted effective-modulus method for concrete. Their model predicted the test results of deflection to within 15%. From parametric studies, they found that CFRP and GFRP laminates bonded to concrete beams improved the long-term behavior of the strengthened beams by controlling creep strain. AFRP was not as effective because of significant creeping of the AFRP itself.

11.4.6 *Fatigue*—Relatively little investigation of the fatigue behavior of plain or reinforced concrete beams with bonded FRP composites has been conducted. Meier et al. (1993), Heffernan (1997), Barnes and Mays (1999), Shahawy and Beitelman (1999), Papakonstantinou et al. (2001), Masoud et al. (2001), Breña et al. (2002), Aidoo et al. (2004), and Quattlebaum et al. (2005) all report fatigue testing of concrete beams with bonded FRP. In all cases, the eventual failure was fatigue fracture of the longitudinal steel reinforcement, which was similar to that of the unstrengthened companion specimens. Increases in fatigue life observed in these tests were all attributed to the applied FRP reducing the stress range in the existing steel reinforcement.

In some studies, some extent of debonding of the FRP material from the concrete substrate was observed. Once debonding occurred, stress carried by the FRP is redistributed back to the internal reinforcing steel. Thus, improvement in fatigue performance was only attained as long as the FRP was adequately bonded to the concrete (Aidoo et al. 2004).

Meier (1995) performed fatigue tests on a 6.0 m (19.7 ft) T-beam strengthened with a CFRP laminate. The fatigue loading caused stresses in the FRP to vary from 102 to 210 MPa (14.8 to 30.5 ksi). The laminate survived 14 million cycles and failed in shear peeling after failure of the internal reinforcing steel.

Muszynski and Sierakowski (1996) carried out fatigue tests on reinforced concrete beams with external CFRP reinforcement. Two million cycles of load were applied, and fatigue damage was measured using two nondestructive methods. The endurance limit of beams reinforced with CFRP was greater than 250% of their static flexural strength, which showed the excellent fatigue properties of CFRP strengthening.

Toutanji et al. (2006) conducted fatigue tests on reinforced concrete beams externally strengthened with CFRP reinforcement. The beams, with a span of approximately 2 m (6.5 ft), were subjected to 2 million cycles. Results show that, due to FRP strengthening, the fatigue strength increased by 55%.

Larson et al. (2005) tested five CFRP-strengthened pretensioned prestressed concrete T-girders to examine the effect of increased live loads on fatigue. CFRP strengthening was designed to allow for stress ranges of 124 and 248 MPa (18 and 36 ksi) in the steel strands under the upgraded live load levels. They found that the girders successfully sustained 1 million fatigue cycles with no loss of ultimate capacity, and 3 million cycles with 19% loss of ultimate capacity.

11.4.7 *UV exposure*—The effects of ultraviolet exposure on externally bonded FRP materials in general is discussed in Section 11.2.3.8. Generally, when UV exposure is a concern, an appropriate admixture in the resin or barrier coating may be specified.

11.4.8 *Condition of existing structural members*—In some cases, FRP reinforcement is applied externally to replace strength that has been lost due to deterioration of the existing structure. Common elements include corrosion-damaged piers, piles, and beams. Corrosion damage may be caused by chlorides available from either a marine environment or deicing salts. ACI 440.2R clearly addresses "corrosion-related damage" and that the "cause of corrosion should be addressed" before application of the FRP system. This statement implies that the chloride-contaminated concrete should be replaced with sound concrete or other repair material before application of an FRP system. ACI 440.2R further indicates that FRP systems should not be applied over concrete that is suspected of containing corroded reinforcement.

For bonded FRP composites to remain effective over the remaining service life of the structure, the underlying concrete substrate should remain sound and capable of transferring load to the externally applied composite. This is particularly important in the case of bond-critical applications (flexural and shear strengthening).

Placement of any externally bonded FRP system will greatly affect the movement of moisture both into and out of the underlying concrete. For interior applications, this can serve to enhance the overall durability of the structure by preventing the ingress of moisture and other deleterious chemicals. If, however, the structure is exposed to a humid or wet environment, then the FRP can actually trap unwanted moisture inside the structure and cause damage to the underlying concrete (fib 2001). Before the placement of the FRP repair system, the existing moisture flow characteristics of the structure should be evaluated and then maintained. For example, external shear reinforcement can be applied in thin strips rather than full coverage to allow moisture to move freely across the concrete surface.

Another potentially damaging effect that an FRP system can have on a structure relates to corrosion and galvanic action. Aramid and glass fibers are considered strong insulators, and do not allow for movement of electrical current, but carbon is electrically conductive and can potentially facilitate the corrosion process. As a result, carbon fiber-based FRP systems should never be placed in direct contact with reinforcing steel (fib 2001).

Notwithstanding these comments, recent research has suggested that, under the right conditions, FRP wraps can slow, or even stop, steel reinforcement corrosion in chloride-contaminated concrete columns (Debaiky 2002; Debaiky et al. 2002; Pantazopoulou et al. 2001; Teng et al. 2003; Sen 2003; Chauvin et al. 2000). Pantazopoulou et al. (2001) subjected small concrete columns containing steel reinforcement to accelerated corrosion conditions and then repaired them with GFRP wraps. The repaired specimens were then subjected to continued accelerated corrosion conditions, and there appeared to be some reduction in the corrosion rates with the application of the wraps.

Debaiky et al. (2002) conducted similar tests on small concrete cylinders; corrosion data were also taken, including half-cell potential and linear polarization. They found that under strict control and monitoring, CFRP wraps reduced the corrosion activity in the reinforcing steel. This protection did not vary with the number of layers, so it was concluded that the resin provided the protection (not the fibers). They also found that partial wrapping led to macrocell corrosion in the steel reinforcement at the extremities of the wrap. These results were also confirmed with tests on larger-scale columns (Debaiky 2002). The researchers recommended more research into the effect of partial coverage on full-scale columns. Teng et al. (2003) found similar results.

For FRP-wrapped beams undergoing active corrosion, improved structural performance may be exhibited by a combination of the following two mechanisms: 1) confinement of the concrete section, thereby lessening corrosion cracking and bond splitting cracks; and 2) increased structural resistance.

Sherwood and Soudki (1999) studied the effects of reinforcing bar corrosion on the behavior of CFRP-strengthened reinforced concrete beams. During corrosion, test results showed that the CFRP laminates adequately confined corrosion cracking up to only 6% corrosion. Beyond this level, extensive cracking and CFRP laminate delamination were observed. To provide better corrosion confinement, it was suggested to extend the transverse laminate up to the full height of the beam instead of half the beam height. Load test results indicated that the strengthened corroded specimen retained 92% of the control beam strength, while the unstrengthened corroded beam retained only 79% of the strength.

The bond enhancement of FRP repaired and concrete members with corroded reinforcing bars were investigated by Soudki and Sherwood (2003). Following corrosion, the specimens were tested by bar pullout to determine the bond strength versus slipping of the reinforcing bars. The bond strength of the reinforcing bars in the concrete prisms increased at low levels of corrosion, but decreased as the degree of corrosion increased. The failure mode was typically bond splitting in unwrapped specimens. Strengthened specimens exhibited increased bond strength; they also exhibited failure by bar pullout due to the confining effects of the CFRP strengthening.

Masoud et al. (2001) investigated the structural behavior of corroded beams strengthened by CFRP sheets. The results showed that the use of CFRP sheets for strengthening reinforced concrete beams experiencing steel reinforcement corrosion was an efficient technique that could maintain the structural integrity and enhance the structural behavior of such beams. The fatigue life of the CFRP strengthened-corroded specimens was increased within a range of 2.5 to 6.0 times that of a similar unstrengthened-corroded specimen.

Although under strictly controlled conditions, FRP wraps and jackets have been demonstrated to mitigate corrosion damage; the opposite has also been observed to be true (Pantazopoulou et al. 2001). In particular, some field studies have shown accelerated corrosion rates under a wrap or jacket. Until further evidence becomes available that bonded FRP composites will effectively slow corrosion rates of existing steel reinforcement, it is advisable to repair existing corrosion damage and mitigate the underlying cause of the corrosion. An alternative would be to monitor the continuing corrosion of the wrapped column.

11.5—Structurally integrated stay-in-place (SIP) forms

Stay-in-place (SIP) FRP forms have shown promise in a variety of applications, such as piles (Mirmiran 1995; Saafi et al. 1999), girders (Zhao 1999), and bridge decks (Thompson and Iyer 2001). This type of construction has its specific durability issues, primarily because FRP is exposed and acts as the first line of defense against environmental factors. Moisture absorption, UV degradation, and freezing and thawing are the most critical issues with these systems. The durability issues are thus similar to those for externally bonded FRP.

Durability studies on SIP systems, however, are rare and limited. Lopez-Anido et al. (1999) studied the fatigue of SIP deck systems on steel girders at high temperatures, and found no significant damage after 1 million fatigue cycles at high temperature.

Fam et al. (2002) studied the effect of moisture absorption on the behavior of concrete-filled GFRP tubes of two different laminate structures submerged in fresh water for a period of up to 1 year. The ends of specimens were not sealed; therefore, moisture was allowed to penetrate the specimens from both ends. The study showed strength degradations on the order of 13 to 20% for the two different laminate structures within 1 year. The rate of degradation was reduced over time, and strength recovery was observed for one of the specimens after 1 year. The authors noted that the gain in strength of submerged concrete might have offset the loss of strength in the FRP tube. Additional studies, however, are needed to verify these findings.

CHAPTER 12—FIRE AND BLAST EFFECTS

12.1—Introduction

Recent research into applications of FRP materials for reinforcement and strengthening of reinforced concrete infrastructure has highlighted the need for information in two specific and specialized research areas: fire and blast. This chapter discusses fire and blast effects on FRP systems for concrete separately in two sections.

12.2—Fire

To date, FRP reinforcement and strengthening systems have been used predominantly in structures either where fire resistance considerations are typically not critical factors in design or where the existing structure is capable without FRP on supporting design loads during a fire event. In cases where strict fire code regulations apply, FRP systems are not being used to their full potential, largely because of their unknown behavior in the event of fire. This gap in knowledge is a primary factor preventing the widespread application of FRP materials in buildings, and it has thus been recognized as a critical research need (Sorathia et al. 2001; Harries et al. 2003; Karbhari et al. 2003). This chapter presents a brief overview of the state of the art with respect to the high temperature and fire performance of FRP reinforcement strengthening and systems for concrete structures. More complete discussions are available elsewhere (Bisby et al. 2005a).

12.2.1 *Fire safety in structures*—Fire safety engineering is concerned with limiting the likelihood of loss of life, injury, and property damage in an unwanted fire (ASCE 1992; Buchanan 2001). Most current codes focus their attention on life safety objectives. Prevention of property loss is left largely to building owners. In general, life safety is protected by preventing fires, alerting people to the presence of a fire, and providing means of escape for an adequate period of time. Design for fire safety should also ensure that building occupants are not overcome by flames, heat, or smoke, that the fire is contained to as small an area as possible, and that the structure does not collapse.

Structural fire protection is concerned with providing adequate structural fire endurance (preventing collapse) and ensuring that fire barriers remain intact (compartmentalizing the fire). These goals are typically accomplished using proven structural materials and systems, such as reinforced concrete, masonry, or structural steel, whose ability to withstand fires has been demonstrated through full-scale standard fire tests, such as ASTM E 119 guidelines. Before new construction materials and systems, such as FRP, can be used in buildings, their ability to meet fire resistance criteria set out in building codes should be demonstrated.

12.2.2 *FRPs and fire*—While the structural performance of FRP reinforced or strengthened concrete members during fire has only recently begun to receive significant research attention, a relatively large amount of research has been performed to characterize the overall behavior of FRP materials when subjected to fire (Sorathia et al. 1992; Sorathia 2004; Davies et al. 2004). Because fire safety concerns not only structural performance, but also issues of fuel load, flame spread, smoke generation, and smoke toxicity, many factors have been studied with a view to FRP applications in various specific structural and nonstructural applications (mostly in marine, aerospace, and offshore conditions). Nevertheless, because a wide variety of different FRP products and formulations are currently available, properties can vary widely, and generalizations regarding their fire performance are difficult to make.

When thick FRPs are exposed to fire, pyrolysis of the polymer matrix near the exposed surface will occur, leading to the formation of a protective (insulating) char. The thickness of the char layer increases with increasing fire exposure and forms a thermal barrier that insulates the interior of the FRP component from the fire. The char eventually degrades after prolonged exposure to high temperatures (for example, above approximately 500 °C [932 °F] for a typical polyester resin), and beyond this point, only the fibers, which are much less sensitive to elevated temperatures, remain (Davies et al. 2004). Thus, compared with unreinforced polymers, thick FRPs have advantages with regard to their behavior in fire in that the noncombustible fibers displace polymer resin, making less fuel available for the fire. Also, when the outermost layers of a composite lose their resin because of combustion, the remaining fibers act as an insulating layer for the underlying material, thus providing thermal protection to the FRP (Sorathia et al. 1992).

The aforementioned advantages are not particularly beneficial when FRP materials are used as reinforcement or strengthening for concrete or masonry because charring is not effective with thin laminates or internal reinforcement. In internal FRP reinforcing applications, where the FRPs are protected against combustion and insulated by the concrete cover, loss of bond at temperatures exceeding the glass transition temperature of the polymer matrix is likely to be the critical factor during fire. Concerns associated with flame spread and smoke generation, while important, are obviously less critical for internal reinforcement because the concrete protects the bars. In typical strengthening applications involving externally bonded FRP materials, the FRPs are generally too thin for a protective char to form, and thus the bond between the substrate and the FRP would likely be lost very early in a fire, well before any protective char would have

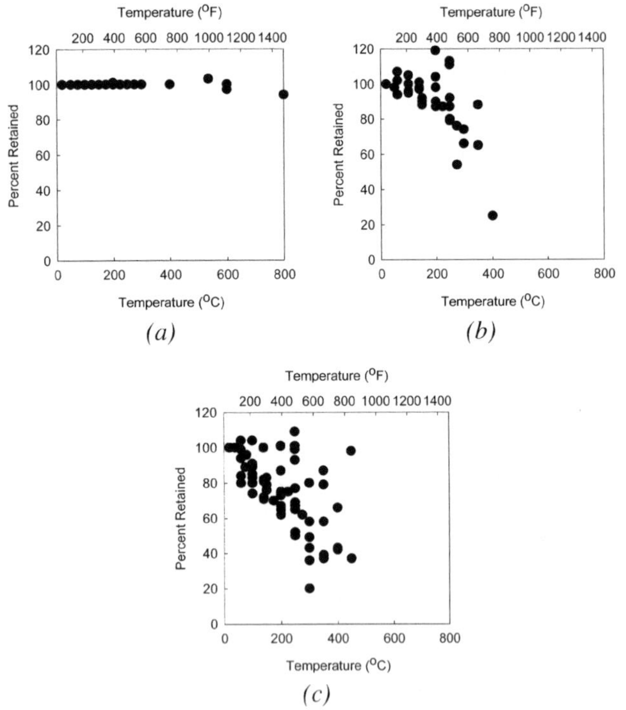

Fig. 12.1—Variation in tensile strength for: (a) carbon fibers; (b) unidirectional carbon/epoxy FRP; and (c) variation in tensile elastic modulus for unidirectional carbon/epoxy FRP. All plots show data assembled by Bisby (2003) and Williams (2004) from a variety of sources and on a variety of different fiber-matrix combinations, some of which may have incorporated elevated temperature cure matrix materials.

time to form. Flame spread, smoke generation, and smoke toxicity are clearly serious concerns that should be addressed when unprotected externally bonded FRP wraps are used in buildings.

The various high temperature properties are clearly important for the fire performance of FRP-reinforced or strengthened concrete and masonry members. The following sections present overviews of the strength, stiffness, bond, and combustion properties of FRP during fire.

12.2.2.1 *Strength and stiffness at elevated temperature*—Virtually all structural materials display reduced strength and stiffness properties at elevated temperature, and FRPs are no exception. Reductions in strength and stiffness at elevated temperature are important considerations when FRP materials are used to reinforce or strengthen structures. Relatively little information is available in the literature on the high temperature performance of FRP materials used in concrete or masonry reinforcing or strengthening applications. Data have been presented by several authors studying the variation in both strength and stiffness of various specific FRP formulations (used in applications ranging from aerospace to infrastructure).

The available test data have been assembled by Blontrock et al. (1999), Bisby (2003), and Williams (2004), and used to develop generalized semi-empirical models to describe strength and stiffness degradation with temperature for various types of FRP materials used in concrete reinforcing applications. As an example of the data available in the literature, Fig. 12.1 shows the assembled data for the variation in strength and stiffness of carbon/epoxy FRPs with temperature. While the carbon fibers themselves are relatively insensitive to the effects of elevated temperatures (Fig. 12.1(a)), the strength and stiffness of the composites are significantly degraded at temperatures of more than about 200 °C (392 °F) (Fig. 12.1(b) and (c)). The aforementioned data is provided for the purposes of illustration only, and tests should be conducted on any specific FRP material being contemplated for use, because FRP formulations vary widely. Plots similar to Fig. 12.1 for glass/epoxy and aramid/epoxy FRPs have been presented previously by Bisby et al. (2005a).

12.2.2.2 *Bond properties at elevated temperature*—The bond between concrete or masonry and FRPs is a critical factor in the performance of most FRP-reinforcing or strengthening systems for concrete and masonry. The bond, which relies heavily on the mechanical (shear) properties of the polymer matrix or adhesive, can be expected to be severely reduced at temperatures exceeding the glass transition temperature T_g of the matrix or the adhesive. Indeed, for the case of internal FRP reinforcement, testing by Katz et al. (1999) has shown that bond strength reductions in the order of 80 to 90% can be expected at temperatures between 100 and 200 °C (212 and 392 °F) (that is, in the range of T_g). Very little information is available on the post-heating residual strength of the bond between FRP bars and concrete, and further research is needed in this area.

Essentially no information is currently available on the specific behavior of the bond between unprotected externally bonded FRP materials and concrete or masonry at high temperature. The bond will likely be lost very quickly under fire exposure, because externally bonded systems are typically very thin (<2 mm [0.08 in.]), and the T_g of the resin will thus be exceeded almost instantaneously during a standard fire exposure. For insulated FRP systems, however, it is not clear exactly how long the bond between externally bonded FRPs and substrate can be maintained during fire. This is due in part to the type of concrete or masonry used, the effectiveness and thickness of the insulation, and various other factors. In addition, post-fire residual behavior of these systems is unknown. Further research is needed.

12.2.2.3 *Flame spread, smoke generation, and toxicity*—As stated previously, various important concerns, namely fuel load, flame spread, smoke generation, and smoke toxicity, are associated with combustion of organic polymer matrixes for FRPs. The likelihood of polymer ignition and combustion is dependent on the specific polymer formulation, and can vary widely from polymer to polymer (Nelson 1995). Sorathia et al. (1992) studied the smoke generation and toxicity characteristics of a variety of FRP formulations and demonstrated that thermoset resins commonly used in structural FRPs generate unacceptable quantities of smoke and have relatively poor flame spread characteristics. Furthermore, burning these materials generates varying quantities of potentially harmful gases such as carbon monoxide, hydrogen fluoride, hydrogen chloride, hydrogen sulfide, and hydrogen cyanide. While resin additives exist that can enhance the ability of polymer materials to resist ignition and combustion, these tend to unacceptably diminish mechanical properties and, in some cases, to increase the production of smoke upon eventual combustion (Nelson 1995).

When FRPs are contemplated for interior building applications, some means of preventing ignition and minimizing their involvement in fire is required. The ignition and flame spread characteristics of conventional FRP systems can be significantly improved by applying barrier treatments or coatings. Ceramic fabrics, ceramic or latex coatings, and intumescent coatings have been used to improve the ignition temperatures and flame spread characteristics of FRP systems (Sorathia et al. 1992; Apicella and Imbrogno 1999). These thin coatings, however, cannot be expected to maintain the structural effectiveness of FRP materials for significant durations of fire exposure. Thicker spray-applied and board insulation systems have also been used to protect fire-exposed FRPs, with varying degrees of success (Deuring 1994; Blontrock et al. 2000, 2001; Bisby et al. 2004a; Sorathia 2004; Williams 2004).

12.2.3 *Fire tests on FRP-reinforced or strengthened concrete*—The behavior in fire of conventional construction materials such as reinforced concrete, steel, timber, and masonry has been researched extensively, and is now reasonably well documented and understood. Data are available from numerous full-scale fire tests on conventional building systems and structural members, and, in many cases, complex numerical fire simulation models are also available to conduct parametric studies and consider specific fire scenarios on a case-by-case basis. This is not true for structural members that incorporate FRP materials, because relatively few fire tests have been reported in the literature (Deuring 1994; Blontrock et al. 2000, 2001; Bisby et al. 2004a; Williams 2004).

12.2.3.1 *FRP-reinforced concrete or FRP prestressed structural concrete members*—Several test programs have been reported in the literature that present data from fire tests on concrete members internally reinforced with FRP materials. The available data, however, are from tests using specific FRP-reinforcing systems, and the conclusions drawn are not necessarily applicable to other FRP materials or systems. Early work was performed in Japan to demonstrate the adequacy of specific FRP-reinforced curtain wall elements (Fujisaki et al. 1993; Nakagawa et al. 1993) and FRP-reinforced or prestressed beams and slabs (Okamoto et al. 1993; Sakashita 1997; NEFCOM Corp. 1998). More recently, fire tests and numerical studies on FRP-reinforced beams and slabs have been performed in North America (Kodur and Baingo 1998; Saafi 2002; Kodur and Bisby 2005) and the United Kingdom (Abassi and Hogg 2004). Essentially, it has been shown that FRP-reinforced concrete structures behave similarly to steel-reinforced concrete structures in fire, but that larger concrete covers are required to maintain FRP temperatures below acceptable limits. FRP bars appear to be more sensitive than steel bars to elevated temperature.

12.2.3.2 *FRP-strengthened structural concrete members*—Very few tests on concrete members reinforced with externally bonded FRP materials are available in the literature. Initial research in this area was performed in Europe by Deuring (1994), who demonstrated the extreme susceptibility of externally bonded FRP flexural strengthening systems to fire. Deuring also demonstrated the need for thermal insulation of the externally bonded FRPs to maintain their effectiveness during fire. Fire tests followed on FRP-strengthened concrete beams (Blontrock et al. 2000) and slabs (Blontrock et al. 2001) using a variety of supplemental fire insulation schemes. These tests confirmed the need for thermal insulation of the externally bonded FRP, but also demonstrated that the bond between the FRP and the concrete was lost in less than 1 hour even for well-insulated systems.

More recently, an extensive experimental and numerical research program aimed at investigating the fire performance of externally bonded FRP strengthening systems for concrete has been conducted through a joint research study between ISIS Canada, the National Research Council of Canada (NRC), Queen's University, and industry partners. To date, intermediate and full-scale fire tests have been performed on FRP-strengthened concrete columns (Bisby et al. 2004a), slabs (Williams et al. 2006), and beam-slab assemblies (Williams 2004). In addition, numerical models have been developed to simulate the fire behavior of these types of members (Bisby et al. 2005b), and these models have been used to conduct parametric studies (Bisby et al. 2004b). This research program has demonstrated that while FRP materials are sensitive to elevated temperatures, appropriately designed and insulated FRP-strengthened reinforced concrete beams, slabs, and columns are capable of achieving fire ratings of greater than 4 hours according to ASTM E 119. The specimens in these tests were designed in accordance with 440.2R. External coatings of gypsum-based or cementitious fire protection were applied (Kodur et al. 2006).

Preliminary studies have also been presented on the residual performance of FRP-confined concrete cylinders after exposure to elevated temperatures by Saafi and Romine (2002) and Cleary et al. (2003).

12.2.4 *Current treatment in codes and guidelines*—Thanks to an enormous research effort over the past 15 years, design codes and guidelines now exist for the application of FRP materials as internal (Rizkalla and Mufti 2001; CSA S6 and S806; ACI 440.3R and 440.4R) and externally bonded (Neale 2001; ACI 440.2R; CSA S806; ICC Evaluation Service 2003a,b; Concrete Society 2004) reinforcement for concrete structures. All of these codes and guidelines comment on the susceptibility of FRP materials to elevated temperatures. Because information on the fire performance of FRP-reinforced or strengthened concrete remains scarce, few of the existing design documents provide specific guidance or procedures for safe fire design of FRP systems. Instead, existing design documents tend to limit the use of FRP systems for reinforcing and strengthening concrete members.

With respect to fire-safe design of internal FRP reinforcement for concrete structures, CSA S806 provides guidance through a series of illustrative design charts (based on the work of Kodur and Baingo (1998)) that can be used to estimate the required concrete cover to achieve a specified fire resistance for FRP reinforcement. The difficulty in using these charts in

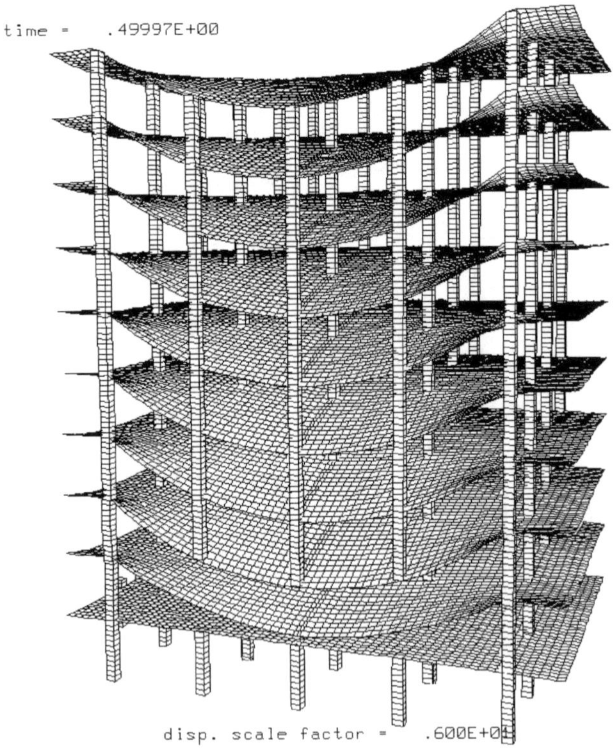

Fig. 12.2—Progressive collapse from loss of first-floor columns (Crawford et al. 1997).

practice is that they require that the critical temperature for the FRP reinforcement be known. The critical temperature for conventional steel reinforcement is 593 °C (1100 °F) in North America, which is the temperature at which it loses about 50% of its room temperature yield strength (ASCE 1992). Critical temperatures for most currently available FRP bars and tendons are not known, and additional research is required in this area. It is worth noting that the critical temperatures for FRP reinforcements will likely be governed not by tensile strength, but by bond strength, because bond strength is severely reduced at only mildly elevated temperatures (as discussed in Section 12.2.2.2).

With respect to fire-safe design of externally bonded FRP-strengthening systems for concrete, a common approach taken in existing codes and guidelines is to require that FRP materials be completely ignored during fire. This requirement is based on the assumption that unprotected FRPs will be rendered completely ineffective within minutes of fire exposure. Strengthening limits are also often imposed (ACI 440.2R).

A similar rationale is used by The Concrete Society's *Technical Report* TR55 (Concrete Society 2004). Because fire is an accidental load, *Technical Report 55* states that the safety factors used in design can be reduced and, in some cases, the FRP system can fail completely without risking failure of the structure.

The effect of the FRP composite system on fire-resistance construction is evaluated according to the appropriate section of the governing building code (ICC Evaluation Service, Inc. 2003b).

12.3—Blast effects

Structural components, such as columns, beams, and slabs, are vulnerable to the large dynamic loads from blasts. Such loads can lead to the shear failure of main load-bearing columns and result in the collapse of whole buildings, as in the case of the Oklahoma City Alfred P. Murrah building, resulting in a large number of casualties. In this building, approximately 87% of the people in the collapsed portion of the building died (153 out of 175); whereas only 5% of the people in the uncollapsed portion died (10 out of 186) (ASCE 1996). Thus, had the structure resisted the blast loads, the casualties could have been reduced significantly. Even if the columns survive, such loads can also lead to the disintegration of walls, creating secondary fragments that could cause fatalities in adjacent rooms. FRP composite materials can be used to reinforce concrete and masonry structures to prevent such disasters.

12.3.1 *Blast strengthening of reinforced concrete columns*—Numerical models can predict the shear failure of main load bearing columns, and the resulting progressive building collapse (Fig. 12.2) (Crawford et al. 1995, 1996, 1997). The models also predicted that composite wraps, or steel jackets, could increase the shear capacity of the column, thus preventing column failure and building collapse. This concept had also been reported by the Federal Emergency Management Agency (ASCE 1996).

Once the potential to resist blast loads using FRP composites had been demonstrated, it was necessary to develop both a simple design procedure and a test series to demonstrate the expected capabilities. Such a program was undertaken by the Defense Threat Reduction Agency, which sponsored the development of a simplified single-degree-of-freedom model to assess the survivability of the columns, and sponsored a series of full-scale tests to demonstrate the concepts (Morrill et al. 1999, 2004). A retrofit design procedure was developed using steel jackets or FRP wraps that were able to prevent column damage and building collapse, for circular, square, and rectangular columns. This simple column blast analysis and retrofit design procedure was based on first-principle calculations, and has been verified and validated against many full-scale blast tests (Crawford et al. 2003; Morrill et al. 2004). This code has been validated for reinforced concrete columns against several full-scale blast tests, using full-scale buildings (Fig. 12.3), or full-scale components in a testing frame that provides the proper end restraints (including the vertical load from the building) (Fig. 12.4).

Figure 12.3(a) and 12.4(a) show that blast loads will often cause unretrofitted columns to fail in diagonal shear, whereas Fig. 12.3(b), 12.3(c), and 12.4(b) show that composite wraps or steel jackets can provide sufficient diagonal shear enhancement to make the column survive. After blast testing, if the column had been properly retrofitted, its remaining axial capacity typically significantly exceeded its required axial resistance. Retrofitted and unretrofitted column deflection predictions using the column blast analysis and retrofit design software compared well with test results.

In addition, quasi-static laboratory tests with loading conditions simulating blast loads were conducted to measure

(a) Unretrofitted column (b) Steel jacketed column (c) Wrapped column

Fig. 12.3—Full-scale blast testing of reinforced concrete column in building (Crawford 2002b).

the flexural resistance of the columns (Fig. 12.5(a)). The measured flexural response is shown in Fig. 12.5(b) (Morrill et al. 2001).

While retrofitting columns with steel jackets and composite wraps is reminiscent of seismic retrofits, there are many basic differences. Besides the highly transient dynamics included into the single-degree-of-freedom design procedure, the deformed shape of first-floor building columns under blast loads usually includes two inflection points versus a single one for seismic loading (Morrill et al. 2001). This is an important difference because the deformed shape with two inflection points allows for the development of a compression membrane, which can increase the lateral resistance of the column tenfold. In turn, this requires increased shear capacity, often to be provided by composite wraps.

These building column retrofit concepts are currently being included into Department of Defense Unified Facilities Criteria to harden structures and to prevent progressive collapse (Malvar 2005). This technology has also been applied to bridge columns (Lan et al. 2003).

12.3.2 *Blast strengthening of walls*

12.3.2.1 *Reinforced concrete walls*—Research conducted by Muszynski (1998) investigated the performance of unreinforced and CFRP- or AFRP-reinforced concrete walls subjected to blast loadings at two different distances from the blast source. The reinforced walls were damaged due to either delamination of the composite material, or composite failure in tension at wall midheight. The CFRP- or AFRP-strengthened walls had less residual displacement than the bare control walls. In a follow-up test series, Muszynski et al. (2003a,b) used an autoclave-cured, three-ply, carbon fiber epoxy laminate and a knitted biaxial E-glass fabric, and later a Kevlar/glass FRP to reinforce concrete walls. The externally reinforced walls suffered high displacements, but did not fail.

Knox et al. (2000) sprayed the back of concrete walls with various unreinforced polymers, particularly polyurethane and polyurea. These polymers showed exceptional ductility, and were able to prevent deadly wall fragmentation.

12.3.2.2 *Unreinforced masonry walls*—Muszynski (1998) used CFRP and Muszynski et al. (2003b) used CFRP

(a) Unretrofitted column (b) Composite wrapped column

Fig. 12.4—Full-scale blast testing of reinforced concrete column in loading frame (Morrill et al. 2001).

(a) Quasi-static test setup

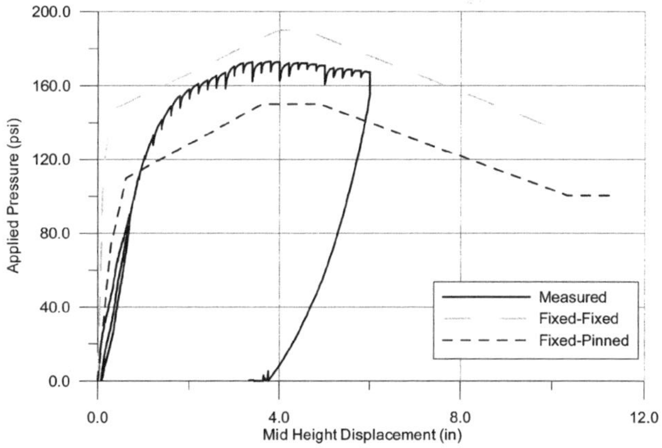

(b) Measured-versus-predicted resistance

Fig. 12.5—Quasi-static laboratory tests under simulated blast loads (Morrill et al. 2001).

and Kevlar/glass FRP to strengthen masonry walls. The reinforcements reduced the residual displacements (compared with the displacements of the unstrengthened walls). The Kevlar/glass FRP seemed to provide additional ductility and further reduced the displacements.

Hammons et al. (2002) expanded the application of polyurea to concrete masonry walls and found it very successful in preventing deadly wall fragmentation. These polymers can be further strengthened by including continuous fibers (Crawford 2002a,b). Tests were recently conducted to further enhance wall response by including aramid fiber meshes within the unreinforced polymer (Morrill et al. 2005).

Baylot et al. (2002) conducted experiments on concrete masonry unit walls using quarter-scale models of typical 200 mm (8 in.) wide concrete masonry units. Three types of retrofitting schemes were evaluated for ungrouted and partially grouted walls. The first retrofit method consisted of 1 mm (0.04 in.) thick GFRP fabric bonded to the back face of the wall. The second retrofit technique consisted of a two-part sprayed polyurea product applied to the back face of the wall. The third retrofit method used galvanized steel sheet metal attached to the back of the wall. All the walls failed during the blast testing, but the GFRP retrofitted walls were successful in preventing loose fragments from penetrating within the structure, and were, therefore, considered successful retrofits.

Myers et al. (2003) tested eight masonry walls to investigate the feasibility of using FRP laminates and rods to improve the blast resistance of masonry walls. Two series of walls with different slenderness ratios and different strengthening schemes were tested at varying blast charge weights and standoff distances. Results indicated that shear capacity controlled the blast behavior of the strengthened walls, emphasizing that any retrofit scheme should not only examine the flexural strength of the system, but also the shear strength. The use of composites in the form of bonded laminates also demonstrated that the failure mode of the system was different from the control walls; in the case of bonded laminates on the interior face of the wall, very limited scatter of debris was noted—an important safety factor for building occupants.

In a follow-up study by Carney and Myers (2003b), the continuity between infill walls strengthened with FRP and concrete boundary elements was studied. Twelve walls were tested in the lab using a pressure bag with various strengthening schemes and connection detailing. Two validation blast tests in the field were performed to verify the observed static test results. Two connection details were examined, including an embedded NSM bar and an embedded sheet with an NSM bar in addition to a control without continuity. Increases in strength and strain energy on the order of two to three times were observed for the FRP-strengthened walls with the boundary connection compared with the control walls. The study again highlighted the control of debris scatter using FRP retrofits.

Crawford et al. (2002) and Crawford and Morrill (2003) used numerical models to study various retrofit techniques for stud and masonry walls.

12.3.2.3 *Wall design*—As for the case of columns, a simple software program was developed to assess the capacity of existing walls and determine the effectiveness of the retrofit (Morrill et al. 2004). Compression membrane action was also shown to provide walls with a significant lateral capacity that was often overlooked.

CHAPTER 13—FIELD APPLICATIONS

Remarkable progress has been made in field applications of FRP for internal reinforcement, prestressing and post-tensioning tendons, external reinforcement, and piles. Many reinforced concrete structures have been built and strengthened with FRP reinforcement in North America, Europe, Japan, and several other countries. The following projects have been selected to demonstrate the wide range of applications of FRP in new construction and in retrofitting and rehabilitation of existing structures.

13.1—FRP as internal reinforcement

13.1.1 *FRP for concrete bridge decks*—The Morristown Bridge (Benmokrane et al. 2006) is a single span, slab-on-girder bridge that spans over Ryder Brook in Vermont. The bridge has a span of 43.0 m (141 ft). The 230 mm (9 in.) thick concrete slab is supported by five girders spaced at 2.4 m (7.8 ft) on center, with a 915 mm (3 ft) cantilever on each side. This concrete bridge deck slab is totally reinforced with GFRP bars. The top and the bottom mesh consist of No. 6 GFRP bars that are spaced at 100 and 150 mm (4 and 6 in.) in the transverse and the longitudinal directions, respectively. The bridge was opened to traffic in 2002.

Other bridges have been constructed in Quebec (Benmokrane et al. 2004; El-Salakawy et al. 2005), and Texas (Bradberry and Wallace 2003) (Fig. 13.1).

13.1.2 *Hall's Harbour Wharf, (2000) (Nova Scotia, Canada) GFRP bar reinforcement*—Hall's Harbour (originally constructed in 1904) is a small fishing harbor in rural Nova Scotia. The Fundy shore, where the wharf is situated, is known for its large tidal range that averages about 10 m (33 ft). The tidal surge, when whipped up by high winds, can have devastating effects on the wharf. A large section of the wharf collapsed after a severe winter storm in 1998. The destroyed wharf needed to be replaced with a new and durable structure (Fig. 13.2).

The rehabilitation involved the construction of a pile-supported concrete deck structure. The concrete deck panels were precast archpanels (Newhook et al. 2000) that used a top layer of GFRP reinforcement to resist the effects of uplift caused by wave action on the soffit of the panels. The panels were supported on deep reinforced concrete beams spaced at 4.0 m (13.1 ft) centers and supported by steel piles. The beams were reinforced with a hybrid scheme of an internal layer of minimal steel longitudinal bars with an outer layer of longitudinal and transverse GFRP reinforcement. Many sensors, consisting of conventional electric resistance strain gauges and fiber optic sensors, were embedded within the critical parts of the wharf to monitor its performance (Newhook et al. 2000).

13.1.3 *FRP barrier walls*—Several slab-on-girder bridges have been constructed in Canada employing steel-free bridge deck technology that takes advantage of the internal arching action that develops in a well-confined deck slab. In these cases, FRP bars (glass and carbon) are used as tensile reinforcement in select locations, including barrier walls and cantilever sections. Four such projects and their respective usage of FRP are: Salmon River Bridge (1996) (Nova Scotia,

Fig. 13.1—GFRP deck slab reinforcement during concrete placing of Sierrita de la Cruz Creek Bridge (Bradberry and Wallace 2003).

Fig. 13.2—Wharf at Hall's Harbour.

Canada), GFRP grid in concrete curbs and parapet walls; Crowchild Trail Bridge (1997) (Alberta, Canada), GFRP bars in cantilever slabs and as longitudinal continuity reinforcement over piers, and GFRP grid in the concrete barrier walls (Tadros et al. 1998); Chatham Bridge (1996) (Ontario, Canada), CFRP grid in cantilever section of slabs and GFRP grid reinforcement in barrier walls; and Waterloo Creek Bridge (1996) (British Columbia, Canada), GFRP reinforcement in barrier walls and as continuity reinforcement with external abutments. A view of the cantilever reinforcement terminating over the interior girders for the Chatham Bridge is shown in Fig. 13.3.

13.1.4 *Specialty applications*—Conventionally reinforced underground hydroelectric chambers are prone to deterioration due to several factors, including attack from the deicing salts and the chemicals present in the ground water. Many chambers have to be replaced annually. Concrete chambers reinforced with GFRP reinforcement have been constructed and installed in the field as part of a joint study undertaken by

Fig. 13.3—Cantilever reinforcement for Chatham Bridge (photo courtesy of B. Bakht).

Fig. 13.4—GFRP reinforcement cage (ISIS Canada 2006).

Universite de Sherbrooke and Hydro Quebec (ISIS Canada, 2006; Fig. 13.4). The chambers have been instrumented and monitored for their performance.

FRP reinforcement is often chosen for its magnetic transparency. One example is the expansion of the magnetic resonance imaging center in Halifax, Nova Scotia, Canada. One of the challenging aspects of the performance criteria was that only a very minimal amount of steel was permitted within a predetermined distance from the magnetic resonance imaging unit. Because of the close proximity of the magnetic resonance imaging unit to its cast-in-place concrete foundation, GFRP reinforcing bars were used as the primary reinforcement for the foundation. The FRP reinforcement (Fig. 13.5) allowed the foundation to meet the structural requirements, as well as the magnetic transparency needed to meet performance criteria.

Toll highways use loop detectors to detect the characteristics of the vehicles passing through a toll plaza. These detectors are embedded within the continuously reinforced concrete pavement of the ramps and the main lanes. Metallic reinforcement can introduce noise to the signals from the loop detectors and lead to unreliable information about the passing vehicles. For loop detectors to function properly, the concrete in their immediate vicinity should be free of any metallic reinforcement up to a depth of 180 mm (7 in.). GFRP reinforcement was

Fig. 13.5—Reinforcement for magnetic resonance imaging base.

(a) Parking garage

(b) Grids in double tee flange

Fig. 13.6—Precast double tees containing CFRP grids, Milwaukee, WI (Carboncast 2006).

applied in a concrete pavement for Loop 1-SH 45 Williams County toll plaza in Texas (Bradberry and Wallace 2003). The design of the deck slab in the in the area housings the toll plaza was amended to a 254 mm (10 in.) thick slab with GFRP bars as top reinforcement and conventional steel bars as bottom reinforcement.

13.1.5 *Applications of CFRP grids in precast concrete structures*—Carbon fiber grids (Fig. 13.6) are routinely used to reinforce precast concrete members (Carboncast 2006). Their strength and resistance to corrosion make them ideal

(a) New Jersey Aquarium

(b) Budweiser Cold Storage

Fig. 13.7—Carbon fiber grids for shear connectors in sandwich walls (Carboncast 2006).

for reducing the thickness and, hence, the weight of precast concrete structures. They have been used in a variety of precast concrete products including double tees, architectural cladding panels, and insulated concrete sandwich panels. To date, over 5 million square feet of carbon grid reinforced concrete products have been erected or are being produced (TechFab LLC 2006).

The new parking garage for General Electric Healthcare in Wisconsin was constructed of precast concrete double tees. The flanges of the double tees for the roof were reinforced with a single layer of carbon fiber grid. The fire rating required for this project was satisfied by full-scale assembly testing.

Carbon fiber grids have also been used as shear connectors in a variety of insulated sandwich wall panel projects (Carboncast 2006). The carbon fiber grids resist horizontal shear between the concrete wythes of a sandwich panel, making the panel behave as a fully composite member. In addition, the relatively low thermal conductivity of the carbon fiber grids help to increase the thermal and energy efficiency of the panels. The New Jersey Aquarium and a Budweiser cold storage distribution facility are shown in Fig. 13.7.

13.1.6 *Applications of CFRP grids in concrete repair*—Carbon fiber grids are also applied for repair and strengthening of concrete structures. The restoration of the Naumburg Bandshell in Central Park (Fig. 13.8) used carbon grids in rebuilding the roof shell. The first phase of restoration involved repairing the leaking concrete roof. After removing the old roof, a new waterproofing membrane was installed, ribs constructed, and a thin concrete shell formed with three layers of carbon grid as reinforcement for the thin lifts of concrete placed by hand troweling. The carbon grid easily conformed to the hemispherical shape of the roof.

(a) Overview of bandshell

(b) Carbon grid reinforcement

Fig. 13.8—Restoration of the Naumburg Bandshell in Central Park, New York (Carboncast 2006).

Fig. 13.9—Beddington Trail Bridge (ISIS Canada 2006).

Fig. 13.10—Taylor Bridge, Manitoba, Canada (ISIS Canada 2006).

13.2—Prestressing applications

13.2.1 *Internal pretensioned reinforcement*—Two bridges were constructed in Canada in 1993 and 1998 with this technology. The Beddington Trail Bridge is a two-span, continuous skew bridge (Fig. 13.9). Each span consists of 13 bulb-T precast, prestressed concrete girders. Carbon fiber cables were used for four girders, while two other girders were prestressed using two rod tendons. Prestressing of CFRP was adapted to the practice by coupling the carbon fiber composite cables and Leadline™ rods to conventional steel strands. Couplers helped to minimize the length of CFRP tendons, and were staggered to allow use of same spacing for the conventional steel-reinforcing tendons. The girders using CFRP prestressing cables and rods have been instrumented with Bragg grating fiber optic sensors and are being monitored by ISIS Canada (2006).

The Taylor Bridge (Fig. 13.10) is situated in Winnipeg, Manitoba, Canada. This is a multispan bridge with a total length of 165 m (542 ft). Four, from a total of 40 prestressed precast concrete girders, were prestressed with CFRP reinforcement. FRP stirrups were also provided for shear resistance. This field application was one of the first in which bridge girders were reinforced totally with FRP (ISIS Canada 2006).

13.2.2 *Post-tensioning applications*—A continuous concrete box girder bridge over the Verdasio River in Switzerland (Keller 2003) had two spans of 31.4 and 37.6 m (103 and 123 ft), with a width of 6.0 m (19.7 ft). Both the webs of the box girder were fully prestressed. The concrete in the web had accumulated excessive amounts of chlorides due to the effects of deicing salts. Some of the prestressing cables were corroded, and there was up to 100% loss of cross section of non-prestressed reinforcement. The external prestressing system consisted of 19 pultruded carbon wires (5 mm [0.20 in.] diameter), with an outer diameter of 32 mm (1.3 in.). The cables were stressed to 65% of their nominal ultimate capacity to restore the initial structural resistance.

The BASF bridge in Ludwigshafen, Germany is a two-lane prestressed concrete bridge with a total length of 85 m (279 ft). Four CFRP cables were used in conjunction with 16 conventional steel cables as internal unbonded post-tensioned cables (Zoch et al. 1991). Each cable consisted of 19 CFRP strands of 12.5 mm (0.50 in.) diameter, with a prestressing force of 70 kN (15.7 kips) applied on each strand. All cables were guided through tubes embedded in the concrete. Steel tendon tubes were filled with concrete after tensioning, while the CFRP tendon tubes were not filled to allow inspection, exchange of cables, and data collection.

The Bridge Street Bridge in Southfield, Michigan (Fig. 13.11), completed in 2001, is a three-span prestressed concrete bridge, 63 m (69 yd) long and 17 m (18.6 yd) wide. Both bonded and unbonded CFRP tendons were used in the longitudinal as well as transverse directions (Grace and Abdel-Sayed 2003).

13.2.3 *CFRP tendons for lighting poles*—Prestressed concrete lighting poles were used for a parking garage in Switzerland (Terrasi and Lees 2003). The overall height of

(a) Bridge

(b) External Tendons

Fig. 13.11—Bridge Street Bridge (Grace et al. 2004).

Fig. 13.12—CFRP strip application to girder (Labossière et al. 2000).

the poles was 9.2 m (30.2 ft), from which 1.2 m (4 ft) was buried in the ground, leaving a clear height of 8.0 m (26.2 ft). The poles had an outer diameter of 120 mm (4.7 in.) at the tip and 220 mm (8.7 in.) at the base. These poles were made from centrifugally cast, high-strength concrete. Each pole contained six CFRP tendons that were prestressed to 60% of their ultimate tendon stress. The concrete had a minimum 28-day cube strength of over 90 MPa (13 ksi), a minimum tensile strength of 5 MPa (725 psi), and a Young's modulus of 38,550 MPa (5590 ksi). The poles were tested successfully under torsional and bending loads.

13.3—External reinforcement

13.3.1 *Beam and girder repair*—Sainte-Emile-de-l'Energie Bridge is a single-span bridge originally built in 1951. The four-girder bridge has a span of 21.3 m (70 ft). The vehicle weights have increased significantly since the bridge was constructed. Analysis showed that to sustain

Fig. 13.13—Maryland Street Bridge CFRP wrap on external girder (ISIS Canada 2006).

the increased loads, the flexural and shear strength of the bridge had to be increased by 35 and 20%, respectively. Carbon fiber reinforcing strips were used for increasing the flexural strength (Labossière et al. 2000) (Fig. 13.12). Glass-fiber reinforcement strips were used for shear strength enhancement. The performance of the strengthening scheme was verified by load testing.

The City of Winnipeg implemented an application of CFRP sheets as a first step in upgrading the shear capacity of the Maryland Street Bridge in Winnipeg (Fig. 13.13). The twin five-span continuous precast, prestressed concrete structures were designed and constructed in 1969. Analysis using current codes, however, indicated that the shear strength of the I-shaped girders was not sufficient to withstand the increased truck loads. An experimental study was conducted to examine the use of CFRP sheets on this particular girder shape. Four girders have since been strengthened using the sheets that were placed vertically with a horizontal layer placed across the top and bottom of the web for anchorage (Hutchinson et al. 2003).

The Farm-to-Market 1927 Bridge in Texas was originally constructed in 1964. This simply supported bridge has two 13.7 m (45 ft) end spans and two 18.3 m (60 ft) interior spans, with four prestressed concrete girders spaced 2.4 m (8 ft) center in each span. In January 2002, an overheight vehicle struck the inner span of the bridge, severely damaging an external girder between two diaphragms. The bottom flange and the web of the girder were fractured into several pieces. The damage, however, was mainly limited to concrete. The prestressing strands were found to have retained most of their preimpact tension. The concrete deck, the top flange, and the rest of the girder were in good shape. The restoration work consisted of repairing the damaged concrete by conventional methods and then wrapping with CFRP composite wrap (Bradberry and Wallace 2003). The CFRP wrap was provided to enhance the integrity of the

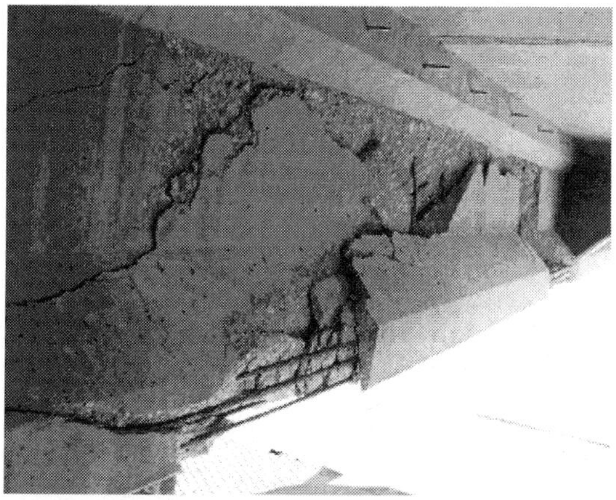

Fig. 13.14—Damaged and rehabilitated girder, Farm-to-Market Bridge, TX (Bradberry and Wallace 2003).

Fig. 13.15—CFRP application to girders, Sue Creek Bridge, TX (Bradberry and Wallace 2003).

Fig. 13.16—Portage Creek Bridge (ISIS Canada 2006).

repair work. The concrete was repaired using rapid-set, non-shrink grout and concrete, and epoxy injection (Fig. 13.14). The composite wrap was unidirectional carbon-fiber fabric and a compatible epoxy adhesive.

The Sue Creek Bridge was originally constructed before 1964. This simply supported bridge has two 9.1 m (30 ft) spans. The planned widening of the bridge necessitated the strengthening of the bridge. The two spans of the bridge were strengthened with two different types of CFRP strengthening systems (Bradberry and Wallace 2003). The first span was strengthened with longitudinally oriented CFRP fabric as the primary strengthening reinforcement and transverse CFRP fabric straps as the secondary reinforcement to control debonding of the longitudinal reinforcement. The second span was strengthened with longitudinally oriented CFRP pultruded laminates as the primary strengthening reinforcement and transverse CFRP fabric straps as the secondary reinforcement to control the debonding of longitudinal reinforcement (Fig. 13.15).

Nine two-story single-bay frames of a five-story beach resort in Egypt were found to be structurally deficient. Each bay spanned 12 m (39.4 ft) and had a 1.2 m (4 ft) deep beam. The frames lost between 20 and 30% of their flexural reinforcement and between 10 and 15% of their shear reinforcement because of corrosion of the reinforcement. The preparatory work involved removal of the rust and the repair of the concrete substrate. The flexural strengthening was applied as four 100 mm (4 in.) wide CFRP laminates bonded to the bottom fiber of the girder, and two 100 mm (4 in.) wide laminates attached to the sides of the beam (Mahfouz and Rizk 2003). U-shaped CFRP sheets, 100 mm (4 in.) wide, were used to enhance the shear capacity of the beam as well as to prevent the delamination of CFRP laminates.

13.3.2 *Column wrapping*—The Portage Creek Bridge (Fig. 13.16) in Victoria, British Columbia, Canada, was built before current seismic design codes and construction practices and would not resist potential earthquake forces as required by the current standards. Analysis showed that two short columns could fail catastrophically in shear during a large earthquake. The seismic upgrade included a retrofit scheme using FRP wraps to strengthen the short columns and a system for structural health monitoring (ISIS Canada 2006).

The structural system of the Faisal Bank's newly acquired office building in downtown Cairo, Egypt was analyzed to ascertain its adequacy for its intended use by the bank. The load-carrying capacity of 121 of its columns had to be increased by 25%. The columns, circular in cross section, with diameters varying from 300 to 750 mm (12 to 30 in.), were all 3 m (10 ft) high. Because of the premium on the office space in Cairo, steel or concrete jacketing was not an option. In addition, larger columns in the basement would have made parking of cars impossible. From the earthquake point of view, the steel or concrete jacketing would also have resulted in increase in the stiffness of the columns. The time

Fig. 13.17—Ring beam during rehabilitation (Demers et al. 2003).

(a) Koerschtal Bridge, Germany *(b) Sung San Bridge, Seoul, Korea*

(c) Main Street Bridge, MB, Canada

Fig. 13.18—Site applications with prestressed CFRP plates or sheets: (a) the Koerschtal Bridge, Germany (photo courtesy of Sika Services AG, Basler et al. 2004); (b) The Sung San Bridge, Seoul, Korea (photo courtesy of Sika Services AG, Basler et al. 2004); and (c) the Main Street Bridge, Manitoba, Canada (Kim et al. 2006).

required to complete the upgrade was also an important consideration. Wrapping of the columns was done with one or more layers of CFRP with the orientation of fibers being circumferential, thus avoiding any increase in the flexural stiffness (Mahfouz and Rizk 2003). The columns were strengthened either by continuous or discontinuous CFRP wrapping, depending on the strength requirements.

13.3.3 *Nuclear reactor containment structure*—Severe deterioration of the concrete in a containment structure occurred due to harsh environmental conditions. The deteriorated concrete was repaired, and the ring beam of the containment structure was strengthened (Fig. 13.17). The repair scheme involved the removal and replacement of deteriorated concrete, and the application of over 700 m^2 (7500 ft^2) of GFRP fabric (Demers et al. 2003).

13.3.4 *Prestressed FRP plates or sheets*—The application of prestressed FRP sheets or plates is mentioned in Section 8.2, including its structural benefits and techniques for prestressing. Basler et al. (2004) reported three reinforced concrete bridges strengthened with prestressed CFRP plates, including particular attention on various anchor systems. The first North American site application using prestressed CFRP sheets was conducted in the Main Street Bridge Overpass No. 4, Winnipeg, Manitoba, Canada (Kim et al. 2006). The bridge, 56 m long and 26 m wide (183 x 85 ft), was frequently stuck with heavy trucks, resulting in rupture of internal prestressing strands. The insufficient flexural capacity was fully recovered after strengthening. Serviceability was also improved to meet code requirements. Figure 13.18 shows photos of some of these applications.

13.4—Masonry applications

The cathedral church of Arequipa, an example of one of the most important neoclassical monuments of Peru, was first built in 1629. Its two 28 m (92 ft) high slender towers were damaged as a result of an earthquake in 2001. The upper part of one of the towers collapsed, and the second tower was severely damaged. The rehabilitation scheme involved CFRP strips to strengthen the towers (Tumialan et al. 2003d). The reinforcing consisted of 250 mm (10 in.) wide CFRP strips, spaced at 550 mm (22 in.) centers, around the circumference of the member. A concrete ring beam was provided in the second segment where the vertical CFRP laminate strips from the first segment were anchored. The second segment was strengthened by providing circumferential CFRP laminates. Because of the porosity of the masonry surface, the high-viscosity saturant was directly applied to the masonry surface instead of the primer conventionally used on concrete members. A masonry wall was later built over the CFRP laminates to retain the original appearance of the towers.

Masonry domes at El-Aini, Egypt have diameters of 12 m (39 ft) and rest over cylindrical brick walls (Mahfouz and Rizk 2003). The domes in the cracked condition behaved as a set of isolated arches connected at the apex and supported by the cylindrical walls. First, the cylindrical walls were strengthened providing 200 mm (8 in.) wide circumferential CFRP strips spaced at 500 mm (20 in.). Subsequently, the domes were also strengthened by providing CFRP circumferential wraps. The surface of the wraps was coated with brick powder to blend with the existing brickwork.

13.5—Stay-in-place FRP forms

Stay-in-place FRP forms that are structurally integrated in the system have been used in a number of applications such as end-bearing or friction piles, fender piles, and structural supports for pier decks over the last 15 years. FRP forms provide a solution for deteriorated piers and waterfront structures subjected to the harsh marine environment. The population growth of marine borers, strict environmental laws that limit toxic treatments for wood, and prohibitions on traditional maintenance practices (lead-based primers, sandblasting, and solvent-based paints) have resulted in higher maintenance and replacement costs for traditional materials. Composite materials offer both long-term and low maintenance solutions. In this section, three available systems and

Fig. 13.19—New Castle Pier.

Fig. 13.20—Hardcore monopile.

Fig. 13.21—Bridge over Nottoway River (Pando et al. 2003).

Fig. 13.22—Naval pier at Ingleside (Fam et al. 2003d).

their case studies or field applications are summarized. In addition, a novel bridge girder application using stay-in-place forms is also described.

13.5.1 *Marine pile systems and applications*—New Castle Pier in Delaware was reconstructed using 300 and 450 mm (12 and 18 in.) diameter hollow FRP composite tubes, driven over the remains of the original wood piling, in 2002. The hollow tubes were then filled with concrete and used to anchor the new pressure-treated timber deck as shown in Fig. 13.19.

Another field application was a monopile of composite tube as a single 24.4 m (80 ft) long, 1.5 m (5 ft) diameter pile, weighing only 6350 kg (13,990 lb). The pile was installed for the Delaware River and Bay Authority at the Lewes, Delaware, ferry terminal as shown in Fig. 13.20. The pile driving was performed using a vibratory hammer, and was faster than a typical steel monopile because the composite pile was fabricated in one continuous length, eliminating the need for field welding or splices. The pile was driven hollow and was not filled with concrete. The monopile was tested under a force of 1070 kN (240 kips), with a deflection of 790 mm (31 in.).

In a recent TEA-21 project, the Federal Highway Administration and Virginia DOT used CFFT piles in one bent of Route 40 Bridge over Nottoway River, as shown in Fig. 13.21 (Pando et al. 2003). The tubes were 14 m (46 ft) long, with a 0.6 m (2 ft.) diameter and a wall thickness of 6.4 mm (0.25 in.). The piles had a design capacity of 690 kN (155 kips).

At the U.S. Naval Facility pier in Ingleside, TX, CFFT composite piles with 330 to 355 mm (13 to 14 in.) diameters were used as vertical and batter piles for fender applications as shown in Fig. 13.22 (Fam et al. 2003d). The piles were designed with a factor of safety of 10 at a design capacity of 178 kN.

As part of a pilot study for the Florida DOT, empty or concrete-filled FRP tubes, with or without field splices, were installed in March 2000 as shown in Fig. 13.23 (Mirmiran and Shahawy 2003). Field tests showed composite piles to be a feasible alternative for bridge substructure. No damage was observed at the top of the filled tube, nor in the tube-concrete interface, under more severe driving conditions than those allowed by current specifications for concrete piles (Mirmiran et al. 2002).

13.5.2 *Bridge girders*—The research and development efforts on concrete-filled CFRP tubes at the University of California, San Diego, have been implemented in a bridge demonstration project by the California DOT (Caltrans). One of the two parallel structures for the Kings Stormwater Channel Bridge (Fig. 13.24) of Route 86 in Riverside County was designed with the modular concrete-filled carbon shell system (CSS) construction concept (Karbhari et al. 2000). The bridge consists of a two-span, 20 m (66 ft) long and 13 m (43 ft) wide continuous system with a beam and slab superstructure and a multicolumn intermediate pier. The superstructure is composed of six carbon epoxy girders with 10 mm (0.40 in.) wall thickness and 0.34 m (13.4 in.)

Fig. 13.23—Pile being driven as load-bearing pile in Florida (Mirmiran and Shahawy 2003).

Fig. 13.24—Kings Stormwater Channel Bridge, San Diego, CA (Karbhari et al. 2000).

inside diameter filled on site with lightweight concrete. The girders are connected compositely with cellular E-glass vinylester FRP deck panels (Seible et al. 1999b). Continuing the carbon shell into the reinforced concrete elements and providing a conventional steel reinforcement anchorage allows connection of the carbon CFFT girders to the abutment end diaphragms and the center cap-beam. The pier columns are conventionally reinforced concrete pile extensions. After proof testing, the bridge was placed in service in July 2000. Long-term health monitoring of the Kings Stormwater Channel Bridge is being conducted by using permanent field monitoring instrumentation and a remote communication and data collection system (Zhao et al. 2001).

CHAPTER 14—RESEARCH NEEDS

14.1—Introduction

The research community has made great progress identifying and quantifying the characteristics of FRP composite materials for infrastructure applications. A growing number of demonstration projects and commercial applications contribute to the knowledge base in this area. The use of FRP materials in infrastructure is still in its infancy, however, and a number of issues remain to be adequately addressed. In some cases, the lack of understanding or design guidance represents a significant limitation to the broader implementation of FRP materials in infrastructure applications. This chapter identifies research areas and issues that industry, practitioners, and academia have identified as requiring further attention to improve the understanding of the behavior of FRP materials and their applications to concrete structures.

The following discussion of research needs should not be construed as representing an impediment to the broader adoption of FRP materials for concrete applications. Rather, this discussion of research needs acknowledges that, as with any emerging technology, limitations and gaps in knowledge exist.

As the various ACI Committee 440 subcommittees develop their respective reports and guidelines, five general needs are continually repeated. These critical needs are identified as:

- Data on the durability of FRP materials and FRP-reinforced concrete systems;
- Design and construction guidelines and specifications;
- Standardized material test methods;
- New materials and systems; and
- Future research directions.

These five themes will be repeated in more specific terms in the following sections.

14.2—Key research needs

In October 2004, the National Science Foundation sponsored a workshop (Porter and Harries 2005), and the primary objective was to identify critical research needs affecting the implementation of FRP composites in construction applications and to develop a consensus on the priority of these needs. This workshop was developed out of a 2002 survey of research needs (Harries et al. 2003). This section is based on the consensus of the workshop.

To address the immediate needs of the use of FRP in concrete construction, the following research topics are highly recommended for investigation. These topics address needs based on widely accepted applications of FRP in civil infrastructure.

14.2.1 *Durability and performance-related topics*

14.2.1.1 *Identification of appropriate environments for durability testing*—Considerable disagreement exists among researchers and practitioners as to exactly what environmental parameters need to be considered when using FRP materials.

Additionally, the intended use, regional climates, and maintenance practices will significantly impact which parameters affect a particular application. Such research studies should include participation from all primary climatic regions.

14.2.1.2 *Durability studies of externally bonded FRP repair or retrofit measures*—Time-dependent effects and factors affecting the durability of concrete structures with FRP (including fatigue) need to be identified. Clearly, the durability of the adhesive bond or interface between the FRP and the substrate is of primary concern.

14.2.1.3 *Durability studies of internal FRP reinforcement*—The same time-dependent effects identified for externally bonded FRP apply equally to internal FRP reinforcement. The critical issue for internal reinforcement is the behavior of FRP in-place; thus, studies should account for the concrete environment in which the FRP is embedded, the expected behavior (such as cracking) of that environment (which may differ from steel-reinforced members), and the environmental factors of importance (which also differ somewhat from those of importance for steel-reinforced members).

14.2.1.4 *Service life prediction of structures using FRP*—Models need to be developed to extrapolate short-term test results to long-term service life predictions. This topic also deals primarily with parameters related to durability, and should involve models of degradation processes. Fatigue life of bonded FRP has been shown to be of particular concern, and predictive models of this behavior are required.

14.2.1.5 *Fire resistance and protection of FRP*—The behavior of FRP materials, whether embedded in concrete or externally applied, subject to fire loading is largely unknown. Modeling techniques should be developed and verified for predicting the performance of structures in fire.

14.2.1.6 *Seismic and blast resistance of FRP systems*—FRP is very often used for structural retrofit, including efforts to mitigate the effects of earthquake or blast loads. Beyond pseudostatic testing, which does not capture strain rate effects, little is known of the performance of these retrofit systems under such extreme loads. Methods of assessing the appropriateness of existing and innovative FRP systems for mitigating the effects of extreme loading need to be developed.

14.2.2 *Development of standardized test methods*—A consensus on accelerated environmental conditioning techniques and subsequent durability test methods is required. Methods are required for both external FRP applications and internal FRP reinforcement applications.

14.2.3 *Design and construction guidelines and specifications*—Reduction and synthesis of data and development of analysis techniques and design guidelines, which are verified by experimental data, should form a major portion of future research programs. The inherent differences between the behavior of FRP materials and steel could have major implications on design requirements. Therefore, extensive evaluation of existing design philosophies and their merits with regard to application to FRP-reinforced concrete structures is also necessary. These philosophies and axioms can then be modified to become suitable for application to FRP-reinforced and prestressed concrete structures. Furthermore, with the continued acceptance of limit states design, reliability and probabilistic studies should be conducted to develop appropriate load and resistance factors for FRP materials and systems for incorporation into codes and design specifications. Construction guidelines and specifications are also needed for implementation in the field.

14.2.4 *New materials and systems*

14.2.4.1 *Innovative and hybrid materials*—FRP materials are often not mechanically or hygrothermally suited to applications in concrete infrastructure. Research aimed at developing new and hybrid FRP materials with properties better matched to concrete is necessary. Such systems may be as simple as combined CFRP, GFRP, and AFRP products, or as innovative as polymer-free chemically prestressed systems.

14.2.4.2 *Innovative reinforcing schemes*—Taking proper advantage of the FRP materials should involve more than simply replacing steel with FRP, and should move toward the development of innovative reinforcing schemes. Such advancement should make both FRP reinforcement and concrete construction more cost effective. One role that concrete plays in reinforced concrete systems is to protect the reinforcing system. If FRP systems can be made more robust and durable, this role for concrete becomes obsolete, and should result in savings.

14.2.4.3 *Self-sensing FRP structural health-monitoring systems*—FRP materials are unique in terms of their properties and their fabrication, which lends itself well to the development of integrated sensor systems. Such systems facilitate improved structural health monitoring and may be developed to allow the structure to interact with its occupants.

14.2.5 *Future research directions*—FRP materials represent a new paradigm in structural materials, and thus, innovation in this direction is required. Research aimed at developing innovative and appropriate applications of FRP should be encouraged.

14.2.5.1 *Leveraging properties of FRP in infrastructure*—Many properties of FRP materials are not currently exploited in the construction industry. For example, glass transmits light and carbon conducts electricity. Additionally, FRP materials may be tailored to retain stiffness or exhibit particular desired stiffness properties throughout their loading history. The inherent properties of FRP materials, as described, could be leveraged in smart structures that would be able to interact with their occupants, or offer stabilizing or self-centering behavior, respectively.

14.2.5.2 *Innovative material properties*—The tailorable nature of FRP materials makes them well suited to developing a wide spectrum of unique material properties. As an example, chemical or shape memory prestressing systems may represent an appropriate use of FRP materials. The exceptional fatigue performance of CFRP materials may make them well suited for active structural systems.

14.2.5.3 *FRP in sustainable construction*—FRPs are believed to represent a very sustainable construction tool. FRP materials have a relatively large impact-to-cost ratio (that is, few resources are engaged to result in a significant structural benefit). Furthermore, FRP materials and systems may offer otherwise obsolete structures a prolonged useful life. Integrating FRP materials into Leadership in Energy and

Environmental Design (LEED) certification is a recommended beginning in this regard.

14.2.5.4 *Research partnerships*—FRP materials and systems are not uniform, and are typically proprietary. For seminal research to be conducted, a close collaboration between academe and industry is essential. The corollary to this, of course, is the need to make trade secrets available. For these competing reasons, innovative industry-academe-government partnerships should be developed to facilitate seminal research. Such partnerships are even more critical considering the large number of disparate FRP systems being developed. Nonetheless, research involving fundamental interaction between industry and academe should take the highest priority, and is likely to yield the most fruitful results.

As has been made clear, long-term performance and durability of FRP materials and systems are considered to be of critical importance. Because of the great differences in environmental exposure of civil infrastructure in North America, a need exists for a coordinated effort involving research institutions located in different climatic regions to develop standardized environmental exposure test methods and protocols.

Educational partnerships are equally important. Few civil engineering departments teach courses addressing FRP materials. Broader impact criteria for assessing research proposals should help to address this need by favoring activities that will also incorporate FRP materials into the university classroom.

Finally, although FRP is becoming accepted in infrastructure engineering, it is still not well understood by the lay engineer. For example, the 2004 roof collapse at Charles de Gaulle airport in Paris, France was first blamed on an FRP detail that was later cleared of having any contribution to the collapse. For this reason, site visits with the intent of forensic investigation of structural failures of FRP systems should be supported. The findings of such visits could help to address concerns associated with FRP and to significantly augment the existing knowledge base of its use.

14.3—Conclusions

As various organizations and institutions, including ACI Committee 440 Subcommittees, develop their respective reports and guidelines, five general needs are continually repeated. These critical needs are identified as:

- Data on the durability of FRP materials and FRP-reinforced concrete systems;
- Design and construction guidelines and specifications;
- Standardized material test methods;
- New materials and systems; and
- Future research directions.

The presentation of extensive research needs herein is not an indication of an immature technology, but rather a technology that is rapidly maturing. FRPs for applications in concrete are limited only by the innovation of the practitioners and researchers involved in its development. This discussion presupposes that FRP materials offer a number of advantages to the concrete construction community. A key responsibility, however, is to apply this new technology in appropriate ways; this requires a more thorough case-specific understanding of FRP materials and their interaction with concrete materials and structures than is currently available. Thus, this chapter is intended as a guide to researchers and funding agencies alike to help them to prioritize their activities.

CHAPTER 15—REFERENCES

15.1—Referenced standards and reports

The standards and reports listed below were the latest editions at the time this document was prepared. Because these documents are revised frequently, the reader is advised to contact the proper sponsoring group if it is desired to refer to the latest version.

American Concrete Institute

116R Cement and Concrete Terminology

209R Prediction of Creep, Shrinkage, and Temperature Effects in Concrete Structures

215R Considerations for Design of Concrete Structures Subjected to Fatigue Loading

318 Building Code Requirements for Structural Concrete

440.1R Guide for the Design and Construction of Structural Concrete Reinforced with FRP Bars

440.2R Guide for the Design and Construction of Externally Bonded FRP Systems for Strengthening Concrete Structures

440.3R Guide Test Methods for Fiber-Reinforced Polymers (FRPs) for Reinforcing or Strengthening Concrete Structures

440.4R Prestressing Concrete Structures with FRP Tendons

530 (ASCE 5/TMS 402) Building Code Requirements for Masonry Structures

ASTM International

A 944 Standard Test Method for Comparing Bond Strength of Steel Reinforcing Bars to Concrete Using Beam-End Specimens

D 618 Standard Practice for Conditioning Plastics for Testing

D 695 Standard Test Method for Compressive Properties of Rigid Plastics

D 2101 Test Method for Tensile Properties of Single Man-Made Textile Fibers Taken from Yarns and Tows (Withdrawn 1995)

D 3410/ D 3410 Standard Test Method for Compressive Properties of Polymer Matrix Composite Materials with Unsupported Gage Section by Shear Loading

D 3518/ D 3518M Standard Test Method for In-Plane Shear Response of Polymer Matrix Composite Materials by Tensile Test of a ±45° Laminate

D 4255/ D 4225M Standard Test Method for In-Plane Shear Properties of Polymer Matrix Composite Materials by the Rail Shear Method

D 4476 Standard Test Method for Flexural Properties of Fiber Reinforced Pultruded Plastic Rods

D 5379/ D 5379M Standard Test Method for Shear Properties of Composite Materials by the V-Notched Beam Method

E 119 Standard Test Methods of Fire Test of Building Construction and Materials

Canadian Standards Association

A23.3	Design of Concrete Structures
S6	Canadian Highway Bridge Design Code
S478	Guideline of Durability in Buildings
S806	Design and Construction of Building Components with Reinforced Polymers

These publications may be obtained from these organizations:

American Concrete Institute
P.O. Box 9094
Farmington Hills, MI 48333
www.concrete.org

ASTM International
100 Barr Harbor Drive, PO Box C700,
West Conshohocken, PA, 19428-2959
www.astm.org

Canadian Standards Association
178 Rexdale Blvd.
Toronto, ON
M9W 1R3 Canada
www.csa.ca

15.2—Cited references

AASHTO, 1989, "Standard Specifications for Highway Bridges," American Association of State Highway and Transportation Officials, Washington D.C.

Abbasi, A., and Hogg, P. J., 2004, "Fire Testing of Concrete Beams with Fibre Reinforced Plastic Rebar," *Proceedings of the Advanced Polymer Composites for Structural Applications in Construction,* Woodhead Publishing Ltd., Cambridge, UK, pp. 445-456.

Abdelrahman, A. A., and Rizkalla, S. H., 1997, "Serviceability of Concrete Beams Prestressed by Carbon Fiber-Reinforced Plastic Bars," *ACI Structural Journal*, V. 94, No. 4, July-Aug., pp. 447-457.

Abdelrahman, A. A.; Tadros, G.; and Rizkalla, S. H., 1995, "Test Model for the First Canadian Smart Highway Bridge," *ACI Structural Journal*, V. 92, No. 4, July-Aug., pp. 451-458.

Aboutaha, R. S., 1999, "Structural Rehab of Prestressed Concrete Bridge Girders Using CFRP Composites," *Proceeding of Structural Faults and Repair 99, The 8th International Conference on Extending the Life of Bridges, Civil and Building Structures*, London, 9 pp.

ACI Committee 318, 1989, "Building Code Requirements for Structural Concrete (ACI 318-89) and Commentary (318R-89)," American Concrete Institute, Farmington Hills, MI, 353 pp.

ACI Committee 318, 1995, "Building Code Requirements for Structural Concrete (ACI 318-95) and Commentary (318R-95)," American Concrete Institute, Farmington Hills, MI, 369 pp.

ACI Committee 318, 2005, "Building Code Requirements for Structural Concrete (ACI 318-05) and Commentary (318R-05)," American Concrete Institute, Farmington Hills, MI, 430 pp.

ACMA, 2005, Composites Industry Statistics, Arlington, VA, 35 pp., http://www.acmanet.org/professionals/2005-Composites-Industry-Statistics.pdf (accessed July 2006).

ACMA, 2006, http://www.mdacomposites.org (accessed 10 July 2006).

Adams, D. F., 1992, "Test Methods for Composite Materials: Seminar Notes," *Technical Report,* University of Wyoming, Laramie, WY, 340 pp.

Adimi, M. R.; Rahman, A. H.; and Benmokrane, B., 2000, "New Method for Testing Fibre Reinforced Polymer Rods Under Fatigue," *Journal Composites for Construction*, V. 4, No. 4, Nov., pp. 206-213.

Aidoo, J.; Harries, K. A.; and Petrou, M. F., 2004, "Fatigue Behavior of CFRP-Strengthened Reinforced Concrete Bridge Girders," *Journal of Composites for Construction*, ASCE, V. 8, No. 6, Nov.-Dec., pp. 501-509.

Al-Dulaijan, S.; Nanni, A.; Al-Zaharani, M.; Bakis, C.; and Boothby, T., 1996, "Bond Evaluation of Environmentally Conditioned GFRP/Concrete Systems," *Proceedings, 2nd Conference on Advanced Composite Materials in Bridge Structures*, M. El-Badry, ed., Canadian Society for Civil Engineering, Montréal, QC, Canada, pp. 845-852.

Alkhrdaji, T., 2001, "Destructive Testing of a Highway Bridge Strengthened with FRP Systems," PhD thesis, University of Missouri-Rolla, Rolla, MO, 235 pp.

Alkhrdaji, T.; Nanni, A.; Chen, G.; and Barker, M., 1999, "Upgrading the Transportation Infrastructure: Solid RC Decks Strengthened with FRP," *Concrete International*, V. 21, No. 10, Oct., pp. 37-41.

Al-Mayah, A., 2004, "Interfacial Behaviour of CFRP-Metal Couples for Wedge Anchor Systems," PhD thesis, University of Waterloo, Waterloo, ON, Canada, 199 pp.

Al-Mayah, A.; Soudki, K. A.; and Plumtree, A., 2001, "Experimental and Analytical Investigation of a Stainless Steel Anchorage for CFRP Pretensioning Tendons," *PCI Journal*, V. 46, No. 2, pp. 88-100.

Al-Mayah, A.; Soudki, K.; and Plumtree, A., 2003, *Tension Anchorage*, Patent No. PCT/CA03/01469.

Almusallam, T. H.; Al-Salloum, Y. A.; Alsayed, S. H.; and Albozaimy, A. M., 2002, "Tensile Strength of GFRP Bars in Concrete Beams under Sustained Loads at Different Environments," *Proceedings of the Second International Conference on Durability of FRP Composites for Construction (CDCC2002)*, B. Benmokrane and E. El-Salakawy, eds., Montreal, QC, Canada, pp. 523-534.

Aly, R. S., 2005, "Experimental and Analytical Studies on Bond Behaviour of Tensile Lap Spliced FRP Reinforcement Bars in Concrete," PhD thesis, Department of Civil Engineering, University of Sherbrooke, Sherbrooke, QC, Canada, 189 pp.

Aly, R. S., and Benmokrane, B., 2005, "Flexural Behaviour of Lab Splicing CFRP Bars in Concrete: Bundled Bars," *Proceedings of the Third International Conference on Composites in Constructions*, V. 2, P. Hamelin, D. Biguaud, E. Ferrier, and E. Jacquelin, eds., Lyon, France, July 11-13, pp. 807-814.

Aly, R. S.; Benmokrane, B.; and Ebead, U., 2005, "Bond Splitting Strength of Lab-Spliced GFRP Bars in Concrete," *Proceedings*, 33rd Annual General Conference of the

Canadian Society for Civil Engineering, Toronto, ON, Canada, pp. 1-10. (CD-ROM)

Aly, R. S.; Benmokrane, B.; and Ebead, U., 2006, "Tensile Lab Splicing Bars of Bundled CFRP Reinforcing Bars in Concrete," *Journal of Composites for Construction*, ASCE, V. 10, No. 4, July-Aug., pp. 287-294.

An, W.; Saadatmanesh H.; and Ehsani M., 1991, "RC Beams Strengthened with GFRP Plates: Part II: Analysis and Parametric Study," *Journal of Structural Engineering*, ASCE, V. 117, No. 11, Nov., pp. 3433-3455.

Ando, N.; Matsukawa, H.; Hattori, A.; and Mashima, A., 1997, "Experimental Studies on the Long-Term Tensile Properties of FRP Tendons," *Proceeding of the Third International Symposium on Non-Metallic (FRP) Reinforcement for Concrete Structures (FRPRCS-3)*, V. 2, Japan Concrete Institute, Sapporo, Japan, pp. 203-210.

Annaiah, H. R., 2000, "Shear Performance of Reinforced Concrete Beams Strengthened in situ with Composites," Master's thesis, Department of Civil Engineering, University of Missouri-Rolla, Rolla, MO, 136 pp.

Apicella, F., and Imbrogno, M., 1999, "Fire Performance of CFRP-Composites Used for Repairing and Strengthening Concrete. Materials and Construction: Exploring the Connection," *Proceedings of the Fifth ASCE Materials Engineering Congress*, American Society of Civil Engineers, New York, pp. 260-266.

Apinis, R.; Modniks, J.; Tamužs, V.; and Tepfers, R., 1998, "Ductility of Hybrid Fiber Composite Reinforcement FRP for Concrete," *8th European Conference on Composite Materials (ECCM8)*, V. 2, June 3-5, Naples, pp. 89-96.

Apinis, R.; Modniks, J.; Tamužs, V.; and Tepfers, R., 2000, "Creep Tests using Hybrid Composite Rods for Reinforcement in Concrete," *3rd International Conference on Advanced Composite Materials in Bridges and Structures*, Aug. 15-18, Ottawa, Canada, pp. 21-28.

Arduini, M.; Di Tommaso, A.; and Manfroni, O., 1995, "Fracture Mechanisms of Concrete Beams Bonded with Composite Plates," *Proceedings of the Non-Metallic (FRP) Reinforcement for Concrete Structures*, L. Taerwe, ed., RILEM, pp. 483-491.

Arduini, M.; Gottardo, R.; and De Riva, F., 2001, "FRP Rods for Flexural Reinforcement of Existing Beams: Experimental Research and Applications," *Proceedings of the International Conference on FRP Composites in Civil Engineering*, V. II, CICE, Hong Kong, China, pp. 1051-1058.

Arockiasamy, M., and Sandepudi, K., 1994, "Active Deformation Control of Bridge Prestressed with Aramid Fiber Reinforced Plastic (AFRP) Cables," *WPI* No. 0510614, Contract No. C-4131, Florida Department of Transportation, 277 pp.

Arockiasamy, M.; Sowrirajan, R.; Shahawy, M.; and Beitlman, T. E., 1995, "Repair of Damaged Pretensioned Solid Slab Using CFRP Laminates," *Proceedings of the Second International RILEM Symposium Non-Metallic (FRP) Reinforcement for Concrete Structures (FRPRCS-2)*, L. Taerwe, ed., Ghent, Belgium, pp. 492-500.

Arockiasamy, M., and Thayer, C. C., 1998, "Damage Process of CFRP Composite-Concrete Interface Under Fatigue Loading at Low Temperatures," *Proceedings of the Eighth International Offshore and Polar Engineering Conference*, V. 4, Montreal, QC, Canada, pp. 211-218.

ASCE, 1992, "Structural Fire Protection," *American Society of Civil Engineers Manuals and Reports on Engineering Practice* No. 78, ASCE, New York, p. 141.

ASCE, 1996, "The Oklahoma City Bombing: Improving Building Performance through Multi-Hazard Mitigation," FEMA 277, Federal Emergency Management Agency, American Society of Civil Engineers, New York, 185 pp.

Ashland Specialty Chemical, 2006, www.ashchem.com/ascc/compoly/Isophthalic.asp (accessed June 2006)

Assih, T.; Li, Y.; and Delmas, Y., 1997, "Repairing of Reinforced Concrete Beams by Bonded Carbon Fiber Composite Sheets," *Proceedings of the U.S.-Europe Workshop on Recent Advances in Bridge Engineering: Advanced Rehabilitation, Durable Materials, Non-destructive Evaluation, and Management*, U. Meier and R. Betti, eds., Dubendorf and Zurich, pp. 119-126.

ASTM International, 1990, *Standards and Literature References for Composite Materials*, 2nd Edition, West Conshohocken, PA, 553 pp.

Autocon Composites, 2006, http://www.autoconcomposites.com/ (accessed Dec. 2006).

Avorio, A., and Borri, A., 2001, "Strengthening Monumental Arches and Caults with FRP: Laboratory Tests and First Applications," *Proceedings of the International Conference FRP Composites in Civil Engineering*, V. 2, Elsevier Science, New York, pp. 985-992.

Bajpai, K., and Duthinh, D., 2003, "Flexural Strengthening of Masonry Walls with External Composite Bars," *The Masonry Society Journal*, Aug., pp. 9-20.

Bakis, C. E.; Bank, L. C.; Brown, V. L.; Cosenza, E.; Davalos, J. F.; Lesko, J. J.; Machida, A.; Rizkalla, S. H.; and Triantafillou, T. C., 2002, "Fiber-Reinforced Polymer Composites for Construction—State-of-the-Art Review," *Journal of Composites for Construction*, ASCE, V. 6, No. 2, pp. 73-86.

Bank, L. C.; Oliva, M. G; Arora, D.; and Borowicz, D.T., 2003, "Rapid Strengthening of Reinforced Concrete Bridges," *Report* WHRP 03-06, Wisconsin Highway Research Program, 167 pp.

Bank, L. C., and Xi, Z., 1992, "Performance of Doubly-Reinforced Pultruded Grating/Concrete Slabs," *Advanced Composite Materials in Bridges and Structures*, Proceedings of the First International Conference, Sherbrooke, QC, Canada, pp. 351-360.

Bank, L. C.; Xi, Z.; and Mosallam, A., 1991, "Experimental Study of FRP Grating Reinforced Concrete Slabs," *Advanced Composite Materials in Civil Engineering Structures, Proceedings of the Specialty Conference*, Las Vegas, NV, Feb., pp. 111-122.

Barnes, A. R., and Mays, C. G., 1999, "Fatigue Performance of Concrete Beams Strengthened with CFRP Plates," *Journal of Composites for Construction*, ASCE, V. 3, No. 2, pp. 63-72.

BASF Building Systems, 2006, http://www.chemrex.com/ (accessed Sept. 2006)

Basler, M.; Clenin, R.; and Desroches, M., 2004, "Bridge Strengthening with Prestressed CFRP Plate Systems," *4th International Conference on Advanced Composite Materials and Structures (ACMBS-IV)*, Calgary, AB, Canada, 8 pp. (CD-ROM)

Bastidas, E. D.; Molina, L. F.; Yamin, L.; Sarria, A.; Ortega, J. P.; and Tumialan, J. G., 2002, "Non-Structural Masonry Walls Strengthened with GFRP Laminates," *Ninth North American Masonry Conference,* June 1-4, Clemson, SC, 11 pp. (CD-ROM)

Baylot, J. T.; Dennis, S. T.; and Woodson, S. C., 2002, "Response of 1/4 Scale Concrete Masonry Unit (CMU) Walls to Blast," *Journal of Engineering Mechanics,* V. 128, No. 2, Feb., pp. 134-142.

Beaudoin, Y.; Labossière, P.; and Neale, K. W., 1998, "Wet-Dry Action on the Bond Between Composite Materials and Reinforced Concrete Beams," *Proceedings of CDCC 1998*, B. Benmokrane and H. Rahman, eds., Sherbrooke, QC, Canada, pp. 537-546.

Beber, A. J.; Filloh, A. C.; and Campagnolo, J. L., 1999, "Flexural Strengthening of R/C Beams with CFRP Sheets," *Structural Faults and Repair 99, The Eighth International Conference on Extending the Life of Bridges*, Civil and Building Structures, London, 9 pp.

Becque, J., 2000, "Analytical Modeling of Concrete Columns Confined by FRP," Master's thesis, The University of Manitoba, Winnipeg, MB, Canada, 222 pp.

Benmokrane, B.; El-Salakawy, E. F.; Desgagné, G.; and Lackey, T., 2004, "Building a New Generation of Concrete Bridge Decks Using FRP Bars," *Concrete International*, V. 26, No. 8, Aug., pp. 84-90.

Benmokrane, B.; El-Salakawy, E. F.; El-Ragaby, A.; and Lackey, T., 2006, "Designing and Testing of Concrete Bridge Decks Reinforced with Glass FRP Bars," *Journal of Bridge Engineering*, ASCE, V. 11, No. 2, pp. 217-229.

Benmokrane, B.; Masmoudi, R.; Chekired, M.; Rahman, H.; Debbache, Z.; and Tadros, G., 1999, "Design, Construction, and Monitoring of Fiber Reinforced Polymer Reinforced Concrete Bridge Deck," *Fourth International Symposium on Fiber Reinforced Polymer Reinforcement for Concrete Structures*, SP-188, C. W. Dolan, S. H. Rizkalla, and A. Nanni, eds., American Concrete Institute, Farmington Hills, MI, pp. 87-101.

Benmokrane, B.; Tighiouart, B.; and Chaallal, O., 1994, "Investigation on Bond Performance of FRP Rebars," *Summary during Technical Session on Bond of FRP Reinforcing Bars and Tendons*, ACI Spring Convention, San Francisco, CA, Mar., pp. 1-4.

Berg, A. C.; Bank, L. C.; Oliva, M. G.; and Russell, J. S., 2004, "Construction of a FRP Reinforced Bridge Deck on U.S. Highway 151 in Wisconsin," *Proceedings of the Eighty Third Annual Transportation Research Board Meeting*, Washington, DC, Jan. 11-14, 11 pp. (CD-ROM)

Bisby, L. A., 2003, "Fire Behaviour of FRP Reinforced or Confined Concrete," PhD thesis, Department of Civil Engineering, Queen's University, Kingston, ON, Canada, 371 pp.

Bisby, L. A., and Green, M. F., 2002, "Resistance to Freezing and Thawing of Fiber-Reinforced Polymer-Concrete Bond," *ACI Structural Journal*, V. 99, No. 2, Mar.-Apr., pp. 215-223.

Bisby, L. A.; Green, M. F.; and Kodur, V. K. R., 2004a, "Performance in Fire of FRP-Confined Reinforced Concrete Columns," *Proceeding of ACMBS-IV*, July 20-23, Calgary, AB, Canada, 8 pp. (CD-ROM)

Bisby, L. A.; Green, M. F.; and Kodur, V. K. R., 2004b, "Numerical Parametric Studies on the Fire Endurance of Fibre-Reinforced-Polymer-Confined Concrete Columns," *Canadian Journal of Civil Engineering*, V. 31, pp. 1090-1100.

Bisby, L. A.; Green, M. F.; and Kodur, V. K. R., 2005a, "Response to Fire of Concrete Structures that Incorporate FRP," *Progress in Structural Engineering and Materials*, V. 7, No. 3, pp. 136-149.

Bisby, L. A.; Green, M. F.; and Kodur, V. K. R., 2005b, "Modeling the Behavior of Fiber Reinforced Polymer-Confined Concrete Columns Exposed to Fire," *Journal of Composites for Construction*, V. 9, No. 1, pp. 15-24.

Bizindavyi, L., and Neale, K. W., 1999, "Transfer Lengths and Bond Strength for Composites Bonded to Concrete," *Journal Composites for Construction*, Nov., pp. 153-160.

Blasi, C.; Coisson, E.; and Ferretti, D., 2004, "The Use of Carbon Fibers in Restoring the Stables of the Medici Villa of Poggio a Caiano-Florence," *Proceedings of First International Conference on Innovative Materials and Technologies for Construction and Restoration*, Lecca, Italy, V. 2, pp. 643-652.

Blontrock, H.; Taerwe, L.; and Matthys, S., 1999, "Properties of Fibre Reinforced Plastics at Elevated Temperatures with Regard to Fire Resistance of Reinforced Concrete Members," *Fourth International Symposium on Fiber Reinforced Polymer Reinforcement for Concrete Structures*, SP-188, C. W. Dolan, S. H. Rizkalla, and A. Nanni, eds., American Concrete Institute, Farmington Hills, MI, pp. 43-54.

Blontrock, H.; Taerwe, L.; and Vandevelde, P., 2000, "Fire Tests on Concrete Beams Strengthened with Fibre Composite Laminates," *Third PhD Symposium,* Vienna, Austria, 10 pp.

Blontrock, H.; Taerwe, L.; and Vandevelde, P., 2001, "Fire Testing of Concrete Slabs Strengthened with Fibre Composite Laminates," *Proceedings of the Fiftieth Annual Symposium on Fibre-Reinforced-Plastic Reinforcement for Concrete Structures*, Thomas Telford, London, pp. 547-556.

Boris, T, and Porter, M. L., 1999, "Advancements in GFRP Materials Provide Improved Durability for Reinforced Concrete," *Proceedings 1999 International Composites Expo*, Cincinnati, OH, Published by the Composite Institute, Harrison, NY, May, pp. 10A1-6.

Boyajian, D. M.; Davalos, J. F.; Ray, I.; and Qiao, P., 2002, "Evaluation of Interface Fracture of Concrete Externally Reinforced with FRP," *Proceedings of the Second International Conference on Durability of Fibre Reinforced Polymers for Construction,* Montreal, QC, Canada, pp. 309-320.

Bradberry, T. E., and Wallace, S., 2003, "FRP Reinforced Concrete in Texas Transportation—Past, Present, Future," *Field Applications of FRP Reinforcement: Case Studies*, SP-215, S. Rizkalla and A. Nanni, eds., American Concrete Institute, Farmington Hills, MI, pp. 3-20.

Braimah, A., 2000, "Long-Term and Fatigue Behaviour of Carbon Fibre Reinforced Polymer Prestressed Concrete Beams," PhD thesis, Queen's University, Kingston, ON, Canada, 288 pp.

Braimah, A.; Green, M. F.; and Soudki, K. A., 1998, "Polypropylene FRC Bridge Deck Slabs Transversely Prestressed with CFRP Tendons," *Journal of Composites for Construction*, ASCE, V. 2, No. 4, Nov., pp. 149-157.

Braimah, A.; Green, M. F.; Soudki, K. A.; and Clapp, F. D., 1999, "Long-Term Behavior of Concrete Beams Prestressed with Carbon Fiber Tendons," *Proceedings of the Fourth International Conference on FRP in Reinforced Concrete Structures*, Baltimore, MD, pp. 547-557.

Breña, S. F.; Wood, S. L.; and Kreger M. L., 2002, "Fatigue Tests of Reinforced Concrete Beams Strengthened using Carbon Fiber Reinforced Polymer Composites," *Proceedings of the Second International Conference on Durability of Fibre Reinforced Polymer (FRP) Composites for Construction*, Sherbrooke, QC, Canada, pp. 575-586.

British Standards Institution, 1990, "Part 4 Code of Practice for the Design of Concrete Bridges," BS5400, London, 66 pp.

British Standards Institution, 1997, "Structural Use of Concrete," BS8110, London, 172 pp.

Brosens, K., and Gemert, D. V., 1997, "Anchoring Stress between Concrete and Carbon Fibre Reinforced Laminates," *Proceedings of the Third International Conference on Non-Metallic FRP Reinforcement for Concrete Structures*, Japan, V. 1, pp. 271-278.

Brown, V. L., 1997, "Sustained Load Deflections in GFRP-Reinforced Concrete Beams," *Proceedings of the Third International Symposium on Non-Metallic (FRP) Reinforcement for Concrete Structures,* V. 2, Japan Concrete Institute, Sapporo, Japan, pp. 495-502.

Bryan, P. E., and Green, M. F., 1996, "Low Temperature Behaviour of CFRP Prestressed Concrete Beams," *Canadian Journal of Civil Engineering*, V. 23, No. 2, Apr., pp. 464-470.

Buchanan, A. H., 2001, *Structural Design for Fire Safety*, John Wiley and Sons, Ltd., Chichester, UK, 444 pp.

Burgoyne, C. J., 1988, "Engineering Applications of Parafil Rope," *Symposium on Engineering Applications of Parafil Rope*, Imperial College, London, pp. 39-47.

Burgoyne, C. J., 1992, "Tests on Beams Prestressed with Polyaramid Tendons," *Proceedings of the First International Conference of Advanced Composite Materials in Bridges and Structures*, K. W. Neale, and P. Labossière, eds., CSCE Publication, Sherbrooke, QC, Canada, pp. 231-239.

Burgoyne, C. J., 1993, "Should FRP Be Bonded to Concrete?" *Fiber-Reinforced-Plastic Reinforcement for Concrete Structures*, SP-138, A. Nanni and C. W. Dolan, eds., American Concrete Institute, Farmington Hills, MI, pp. 367-380.

Burgoyne, C. J., 2001, "Rational Use of Advanced Composites in Concrete," *Proceedings of the Institute of Civil Engineers, Structures and Buildings*, V. 146, Aug., pp. 253-262.

Burgueño, R., 1999, "System Characterization and Design of Modular Fiber Reinforced Polymer (FRP) Short- and Medium-Span Bridges," PhD thesis, University of California, San Diego, La Jolla, CA, 587 pp.

Burgueño, R.; Davol, A.; and Seible F., 1998, "The Carbon Shell System for Modular Bridge Components," *Proceeding of the First International Conference on Composites in Infrastructure ICCI'96*, H. Saadatmanesh and M.R. Ehsani, eds., Tucson, AZ, Jan., pp. 341-354.

Burgueño, R.; Davol, A.; Zhao, L.; Seible, F.; and Karbhari, V. K., 1996, "Flexural Behavior of a Hybrid FRP/Concrete Beam/Slab Bridge Component," *ACI Structural Journal*, V. 101, No. 2, Mar.-Apr., pp. 228-236.

Campbell, T. I.; Shrive, N. G.; Soudki, K. A.; Al-Mayah, A.; Keatley, J. P.; and Reda, M. M., 2000, "Design and Evaluation of a Wedge-Type Anchor for Fiber Reinforced Polymer Tendons," *Canadian Journal of Civil Engineering*, V. 27, pp. 985-992.

Carboncast, 2006, http://altusprecast.blade6.donet.com/uploads/1130941009/pdf/carboncast_brochure.pdf (accessed July 2006)

Carney, P., and Myers, J. J., 2003a, "Static and Blast Resistance of FRP Strengthened Connections for Unreinforced Masonry Walls," *Report* No. 03-46, Center for Infrastructure Engineering Studies, July, 129 pp.

Carney, P., and Myers, J. J., 2003b, "Shear and Flexural Strengthening of Masonry Infill Walls with FRP for Extreme Out-of-Plane Loading," *Proceedings for the Architectural Engineering Institute 2003 Annual Meeting,* Austin, TX, Sept., 7 pp.

Carolin A.; Hordin, H.; and Täljsten, B., 2001, "Concrete Beams Strengthened with Near Surface Mounted Reinforcement of CFRP," *Proceedings of the International Conference on FRP Composites in Civil Engineering*, V. II, CICE, Hong Kong, China, pp. 1059-1066.

Casadei, P.; Nanni, A.; and Alkhrdaji, T., 2005a, "Steel-Reinforced Polymer: an Innovative and Promising Material for Strengthening the Infrastructures," *Concrete Engineering International*, V. 9, No. 1, Spring, pp. 54-56.

Casadei, P.; Nanni, A.; Alkhrdaji, T.; and Thomas, J., 2005b, "Performance of Double-T Prestressed Concrete Beams Strengthened with Steel Reinforced Polymer," *Journal of Advances in Structural Engineering*, pp. 427-442.

CEN, 1994, "Eurocode 1: Basis for Design and Actions on Structures—Part 1: Basis of Design," EMV 1991-1, Comité Européen de Normalisation, Brussels, 129 pp.

Cha, J. Y.; Balaguru, P.; and Chung, L., 1999, "Experimental and Analytical Investigation of Partially Prestressed Concrete Beams Strengthened with Carbon Reinforcement," *Proceedings of the Fourth International Symposium on Fiber Reinforced Polymer Reinforcement for Reinforced Concrete Structures (FRPRCS-4)*, Baltimore, MD, pp. 625-633.

Chaallal, O.; Nollet, M. J.; and Perraton, D., 1998, "Strengthening of Reinforced Concrete Beams with Externally Bonded Fiber-Reinforced-Plastic-Plates: Design Guidelines for Shear and Flexure," *Canadian Journal of Civil Engineering*, V. 25, pp. 692-704.

Chajes, M. J.; Finch, W. W.; Januszka, T. F.; and Thomson, T. A., 1996, "Bond and Force Transfer of Composite Material Plates Bonded to Concrete," *ACI Structural Journal,* V. 93, No. 2, Mar.-Apr., pp. 208-217.

Chajes, M. J.; Thomson, T. A.; and Farshman, C. A., 1995, "Durability of Concrete Beams Externally Reinforced with Composite Fabrics," *Construction and Building Materials*, V. 9, No. 3, pp. 141-148.

Charkas, H.; Rasheed, H.; and Mehlem, H., 2003, "Rigorous Procedure for Calculating Deflections of Fiber Reinforced Polymers Strengthened Reinforced Concrete Beams," *ACI Structural Journal*, V. 100, No. 4, July-Aug., pp. 529-539.

Chauvin, M.; Shield C.; French, C.; and Smyrl, W., 2000, "Evaluation of Electrochemical Chloride Extraction (ECE) and Fibre-Reinforced Polymer (FRP) Wrap Technology," *Final Report* 2000-24, Minnesota Department of Transportation, June, 250 pp.

Chawla, K. K., 1998, *Composite Materials: Science and Engineering*, Springer, NY, 483 pp.

Choi, J. H., and Chen, H. L., 2002, "Design Considerations of GFRP-Reinforced CRCP," *Field Applications of FRP Reinforcement: Case Studies*, SP-215, S. Rizkalla and A. Nanni, eds., American Concrete Institute, Farmington Hills, MI, pp. 139-158.

Choi, J. H., and Chen, H. L., 2005, "Design of Continuously Reinforced Concrete Pavements Using Glass Fiber Reinforced Polymer Rebars," *Publication* No. FHWA-HRT-05-081, Federal Highway Administration, U.S. Department of Transportation, McLean, VA, 79 pp.

Christensen, J. B.; Gilstrap, J.; and Dolan, C. W., 1996, "Composite Materials Reinforcement of Existing Masonry Walls," *Journal of Architectural Engineering*, ASCE, V. 2, No. 2, pp. 63-70.

Cleary, D. B.; Cassino, C. D.; and Tortorice, R., 2003, "Effect of Elevated Temperature on a Fiber Composite Used to Strengthen Concrete Columns," *Journal of Reinforced Plastics and Composites*, V. 22, No. 10, pp. 881-895.

CNR (National Research Council), 2004, "Guide for the Design and Construction of Externally Bonded FRP Systems for Strengthening Existing Structures," Advisory Committee on Technical Recommendations for Construction of National Research Council, Rome, Italy, 154 pp.

Concrete Society, 2000, "Design Guidance on Strengthening Concrete Structures Using Fibre Composite Materials," *Technical Report 55* (TR55), The Concrete Society, London, UK, 70 pp.

Concrete Society, 2003, "Strengthening Concrete Structures with Fibre Composite Materials: Acceptance, Inspection and Monitoring," *Technical Report 57* (TR57), The Concrete Society, London, UK, 49 pp.

Concrete Society, 2004, "Design Guidance for Strengthening Concrete Structures Using Fibre Composite Materials," *Technical Report* 55 (TR55), The Concrete Society, 128 pp.

Corradi, M.; Borri, A.; and Vignoli, A., 2002, "Strengthening Techniques Tested on Masonry Structures Struck by the Umbria-Marche Earthquake of 1997-1998," *Construction and Building Materials Journal*, V. 16, pp. 229-239.

Cosenza, E.; Manfredi, G.; and Realfonzo, R., 1997, "Behavior and Modeling of Bond of FRP Rebars to Concrete," *Journal of Composites for Construction*, ASCE, May, pp. 40-51.

Crane, R. L., 2000, "Nondestructive Inspection of Composites," *Comprehensive Composite Materials, Volume 5: Test Methods, Nondestructive Evaluation, and Smart Materials*, L. Carlsson, R. L. Crane, K. Uchino, A. Kelly, and C. Zweben, eds., Pergamon Press, Elsevier Science Ltd., Oxford, pp. 227-233.

Crawford, J. E., and Morrill, K. B., 2003, "Retrofit Techniques Using Polymers and Fiber Reinforced Polymers for Preventing Injurious Wall Debris," *Proceedings of the 6th International Symposium on Fibre-Reinforced Polymer (FRP) Reinforcement for Concrete Structures (FRPRCS-6)*, Singapore, pp. 1199-1208.

Crawford, J. E., 2002a, "Retrofit Methods to Mitigate Progressive Collapse," *Proceedings of the Workshop on Prevention of Progressive Collapse*, National Institute of Building Sciences, Rosemont, IL, 55 pp.

Crawford, J. E., 2002b, "Modeling Blast-Resistant Protection Systems Composed of Polymers and Fabric," *TMS 2000 Fall Meeting*, The Minerals, Metals and Materials Society, Columbus, Ohio, 16 pp.

Crawford, J. E.; Malvar, L. J.; Dunn, B. W.; and Gee, D. J., 1996, "Retrofit of Reinforced Concrete Columns Using Composite Wraps to Resist Blast Effects," *27th DOD Explosive Safety Seminar*, Las Vegas, NV, 17 pp. (CD-ROM)

Crawford, J. E.; Malvar, L. J.; Ferritto, J. M.; Morrill, K. B.; Dunn, B. W.; and Bong, P., 2003, "Description of Design Software for Retrofitting Reinforced Concrete Columns to Improve their Resistance to Blast," *Report* TR-01-16.2, Karagozian and Case Structural Engineers, Burbank, CA, 380 pp.

Crawford, J. E.; Malvar, L. J.; Wesevich, J. W.; Valancius, J.; and Reynolds, A. D., 1997, "Retrofit of Reinforced Concrete Structures to Resist Blast Effects," *ACI Structural Journal*, V. 94, No. 4, July-Aug., pp. 371-377.

Crawford, J. E.; Wesevich, J.; Valancius, J.; and Reynolds, A., 1995, "Evaluation of Jacketed Columns as a Means to Improve the Resistance of Conventional Structures to Blast Effects," *Proceedings of the 66th Shock and Vibration Symposium*, Biloxi, MS, 9 pp. (CD-ROM)

Crawford, J. E.; Xin, X.; and Bogosian, D. D., 2002, "Retrofit Techniques for Conventional Walls to Prevent Their Causing Injurious Debris Due to a Terrorist Bombing," *Proceedings of the 17th International Symposium on Military Aspects of Blast and Shock*, Las Vegas, NV, 21 pp. (CD-ROM)

Creative Pultrusions Inc., 1994, *Design Guide*, Alum Bank, PA, pp. 1-3.

Currier, J., 1995, "Deformation of Prestressed Concrete Beams with FRP Tendons," Master's thesis, Department of Civil and Architectural Engineering, University of Wyoming, Laramie, WY, 113 pp.

Curtis, P. T., 1989, "The Fatigue Behavior of Fibrous Composite Materials," *Journal of Strain Analysis*, London, V. 24, No. 4, pp. 235-244.

Dai, J. G.; Sato, Y.; Ueda, T.; and Muttaqin, H., 2003, "Mode I Fracture Behaviours of FRP-Concrete Interfaces," *Proceedings of the Japan Concrete Institute*, V. 25, pp. 1577-1582.

Daniali, S., 1992, "Development Length for Fiber-Reinforced Plastic Bars," *Proceedings of Advanced Composite*

Materials in Bridges and Structures, K. W. Neale and P. Labossière, eds., Sherbrooke, QC, Canada, pp. 179-188.

Daniel, I., and Ishai, O., 1994, *Engineering Mechanics of Composite Materials*, Oxford University Press, 395 pp.

David, E.; Djelal, C.; Ragneau, E.; and Buyle-Bodin, F., 1999, "Use of FRP to Strengthen and Repair RC Beams: Experimental Study and Numerical Simulations," *Structural Faults and Repair 99*, The 8th International Conference on Extending the Life of Bridges, Civil and Building Structures, London, July 13-15, 10 pp.

Davies, J. M.; Wang, Y. C.; and Wong, P. M. H., 2004, "Polymer Composites in Fire," *Proceedings of the Advanced Polymer Composites for Structural Applications in Construction*, Woodhead Publishing Ltd., Cambridge, UK, pp. 3-17.

Davis, D., and Porter, M. L., 1999, "Glass Fiber Reinforced Polymer Dowel Bars for Transverse Pavement Joints," *Fiber Reinforced Polymer Reinforcement for Reinforced Concrete Structures*, SP-188, C. W. Dolan, S. H. Rizkalla, and A. Nanni, eds., American Concrete Institute, Farmington Hills, MI, pp. 297-304.

Davol, A., 1998, "Structural Characterization of Concrete Filled Fiber Reinforced Shells," PhD thesis, University of California, San Diego, La Jolla, CA, 240 pp.

Davol, A.; Burgueño, R.; and Seible, F., 2001, "Flexural Behavior of Circular Concrete Filled FRP Shell," *Journal of Structural Engineering,* V. 127, No. 7, pp. 810-817.

Debaiky, A. S., 2002, "Rehabilitation of Corrosion-Damaged Reinforced Concrete Columns Using Carbon Fibre Reinforced Polymer (CFRP) Wraps," PhD thesis, Queen's University, Kingston, ON, Canada, 306 pp.

Debaiky, A. S.; Green, M. F.; and Hope, B. B., 2002, "Carbon Fiber-Reinforced Polymer Wraps for Corrosion Control and Rehabilitation of Reinforced Concrete Columns," *ACI Materials Journal*, V. 99, No. 2, Mar.-Apr., pp. 129-137.

Deblois, A. P.; Picard, A.; and Beaulieu, D., 1992, "Reinforcement de Poutres en Beton Armé a l'aide de Materiaux Composites: Etudes Theorique et Experimentale," *First International Conference, Advanced Composite Materials in Bridges and Structures*, K. W. Neale and P. Labossière, eds., Sherbrooke, QC, Canada, pp. 265-275.

Deitz, D.; Harik, I.; and Gesund, H., 1999, "One-Way Slabs Reinforced with Glass Fiber Reinforced Polymer Reinforcing Bars," *Fiber Reinforced Polymer Reinforcement for Reinforced Concrete Structures*, SP-188, C. W. Dolan, S. H. Rizkalla, and A. Nanni, eds., American Concrete Institute, Farmington Hills, MI, pp. 279-286.

De Lorenzis, L., 2004, "Anchorage Length of Near-Surface Mounted FRP Bars for Concrete Strengthening—Analytical Modeling," *ACI Structural Journal*, V. 101, No. 3, May-June, pp. 375-386.

De Lorenzis, L.; Lundgren, K.; and Rizzo, A., 2004, "Anchorage Length of Near-Surface Mounted Fiber Reinforced Polymer Bars for Concrete Strengthening—Experimental Investigation and Numerical Modeling," *ACI Structural Journal*, V. 101, No. 2, Mar.-Apr., pp. 269-278.

De Lorenzis, L., and Nanni, A., 2001a, "Shear Strengthening of Reinforced Concrete Beams with Near-Surface-Mounted Fiber Reinforced Polymer Rods," *ACI Structural Journal*, V. 98, No. 1, Jan.-Feb., pp. 60-68.

De Lorenzis, L., and Nanni, A., 2001b, "Characterization of NSM FRP Rods as Near-Surface Mounted Reinforcement," *Journal of Composites for Construction*, V. 5, No. 2, May, pp. 114-121.

De Lorenzis, L., and Nanni, A., 2002, "Bond between Near-Surface Mounted Fiber-Reinforced Polymer Rods and Concrete in Structural Strengthening," *ACI Structural Journal*, V. 99 No. 2, Mar.-Apr., pp. 123-132.

De Lorenzis, L.; Nanni A.; and La Tegola, A., 2000, "Flexural and Shear Strengthening of Reinforced Concrete Structures with Near Surface Mounted FRP Rods," *Proceedings of Third International Conference on Advanced Composite Materials in Bridges and Structures*, Ottawa, Canada, pp. 521-528.

De Lorenzis, L., and Tepfers, R., 2003, "Comparative Study of Models on Confinement of Concrete Cylinders with Fiber-Reinforced Polymer Composites," *Journal of Composites for Construction*, ASCE, V. 7, No. 3, Aug., pp. 219-237.

Demers, M., and Neale, K. W., 1994, "Strengthening of Concrete Columns with Unidirectional Composite Sheets," *Fourth International Conference on Short and Medium Span Bridges*, Halifax, NS, Canada, pp. 895-905.

Demers, M.; Popovic, A.; Neale, K. W.; Rizkalla, S.; and Tadros, G., 2003, "FRP Retrofit of the Ring-Beam of a Nuclear Reactor Containment Structure," *Field Applications of FRP Reinforcement: Case Studies*, SP-215, S. Rizkalla and A. Nanni, eds., American Concrete Institute, Farmington Hills, MI, pp. 303-316.

Den Uijl, J. A., 1991, *Mechanical Properties of Arapree: Part 4, Creep and Stress Rupture*, TU Delft, Report 25-87-31 for HBG N.V. and AKZO, Stevin Laboratory, TU Delft, The Netherlands, 143 pp.

Deniaud, C., and Cheng, J. J. R., 2000, "Shear Rehabilitation of G-Girder Bridges in Alberta Using Fiber Reinforced Polymer Sheets," *Canadian Journal of Civil Engineering*, V. 27, No. 5, pp. 960-971.

Deskovic, N.; Triantafillou, T. C.; and Meier, U., 1995, "Innovative Design of FRP Combined with Concrete: Short-Term Behavior," *Journal of Structural Engineering*, V. 121, No. 7, July, pp. 1069-1078.

Deuring, M., 1994, "Brandversuche an nachtraglich verstarkten tragern aus beton," *Research Report EMPA* No. 148'795, Swiss Federal Laboratories for Materials Testing and Research, Dubendorf, 97 pp.

Devalapura, R. K.; Greenwood, M. E.; Gauchel, J. V.; and Humphrey, T. J., 1998, "Evaluation of GFRP Performance Using Accelerated Test Methods," *Durability of Fiber Reinforced Polymer Composites for Construction*, B. Benmokrane and H. Rahman, eds., pp. 107-116.

Dieter, D. A.; Dietsche, J. S.; Bank, L.; Oliva, M. G.; and Russell, J. S., 2002, "Concrete Bridge Decks Constructed with Fiber-Reinforced Polymer Stay-in-Place Forms and Grid Reinforcement," *Journal of the Transportation Research Board*, No. 1814, pp. 219-226.

Dolan, C. W., and Burke, C. R., 1996, "Flexural Strength and Design of FRP Prestressed Beams," *Advanced Composite Materials in Bridges and Structures—2nd International Symposium*, Canadian Society of Civil Engineers, Montreal, QC, Canada, pp. 383-390.

Dolan, C. W.; Hamilton, H. R.; Bakis, C. E.; and Nanni, A., 2000, "Design Recommendations for Concrete Structures Prestressed with FRP Tendons," *Department of Civil and Architectural Engineering Report* DTFH61-96-C-00019, V. 1, University of Wyoming, Laramie, WY, 133 pp.

Dolan, C.; Leu, B. L.; and Hundley, A., 1997, "Creep-Rupture of Fiber Reinforced Plastic (FRP) Reinforcement for Concrete Structures: Properties and Applications," *Developments in Civil Engineering*, V. 42, pp. 129-163.

Egyptian Ministry of Housing, Utilities, and Urban Development, 2001, "Egyptian Code for Design and Construction of Concrete Structures," Giza, Egypt, 217 pp.

Egyptian Ministry of Housing, Utilities, and Urban Development, 2005, "Egyptian Code for the Use of Fiber Reinforced Polymers (FRP) in the Construction Fields," Egyptian Housing and Building National Research Center, Giza, Egypt.

Ehsani, M. R., and Saadatmanesh, H., 1996, "Seismic Retrofit of URM Walls with Fiber Composites," *The Masonry Society Journal*, V. 14, No. 2, pp. 63-72.

Ehsani, M. R.; Saadatmanesh, H.; and Al-Saidy, A., 1997a, "Shear Behavior of URM Retrofitted with FRP Overlays," *Journal of Composites for Construction*, ASCE, V. 1, No. 1, pp. 17-25.

Ehsani, M. R.; Saadatmanesh, H.; and Nelson, C. T., 1997b, "Transfer and Flexural Bond Performance of Aramid and Carbon FRP Tendons," *PCI Journal*, V. 42, No. 1, Jan.-Feb., pp. 76-86.

Ehsani, M. R.; Saadatmanesh, H.; and Velasquez-Dimas, J. I., 1999, "Behavior of Retrofitted URM Walls under Simulated Earthquake Loading," *Journal of Composites for Construction*, ASCE, V. 3, No. 3, pp. 134-142.

El-Gamal, S.; El-Salakawy, E.; and Benmokrane, B., 2005a, "Behavior of Concrete Bridge Deck Slabs Reinforced with FRP Bars under Concentrated Loads," *ACI Structural Journal*, V. 102, No. 5, Sept.-Oct., pp. 727-735.

El-Ghandour, A.; Pilakoutas, K.; and Waldron, P., 1999, "New Approach of Punching Shear Capacity Prediction of Fiber Reinforced Polymer Reinforced Concrete Flat Slabs," *Fiber Reinforced Polymer Reinforcement for Reinforced Concrete Structures*, SP-188, C. W. Dolan, S. H. Rizkalla, and A. Nanni, eds., American Concrete Institute, Farmington Hills, MI, pp. 135-144.

El-Hacha, R., 1997, "Strengthening of Concrete Members with Advanced Composite Materials," Master's thesis, Concordia University, Department of Civil Engineering, Montréal, QC, Canada, 324 pp.

El-Hacha, R., 2000, "Prestressed CFRP Sheets for Strengthening Concrete Beams at Room and Low Temperatures," PhD thesis, Queen's University, Department of Civil Engineering, Kingston, ON, Canada, 525 pp.

El-Hacha, R.; El-Badry, M.; and Abdalla, H., 1996, "Behaviour of Concrete Beams Externally Prestressed with Fiber Reinforced Plastic Cables," *The XXIVth IAHS World Housing Congress— How to House a Nation: the Challenge for the XXIst Century*, Ankara, Turkey, May 27-31, pp. 814-825.

El-Hacha, R., and Rizkalla, S. H., 2004, "Near-Surface-Mounted FRP Reinforcements for Flexural Strengthening of Concrete Structures," *ACI Structural Journal*, V. 101, No. 5, Sept.-Oct., pp.717-726.

El-Hacha, R.; Wight, R. G.; and Green, M. F., 2001, "Prestressed Fiber-Reinforced Polymer Laminates for Strengthening Structures," *Progress in Structural Engineering and Materials,* V. 3, pp. 111-121.

El-Ragaby, A.; El-Salakawy, E. F.; and Benmokrane, B., 2006, "Experimental Investigation on the Fatigue Behaviour of GFRP Reinforced Concrete Bridge Deck Slabs," *Proceedings, 1st International Structural Specialty Conference*, ST-126: 1-10, Calgary, AB, Canada. (CD-ROM)

El-Salakawy, E. F.; Benmokrane, B.; El-Ragaby, A.; and Nadeau, D., 2005, "Field Investigation on the First Bridge Deck Slab Reinforced with Glass FRP Bars Constructed in Canada," *Journal of Composites for Construction*, ASCE, V. 9, No. 6, Nov.-Dec., pp. 470-479.

El-Sayed, A. K.; El-Salakawy, E. F.; and Benmokrane, B., 2005, "Shear Strength of One-way Concrete Slabs Reinforced with FRP Composite Bars," *Journal of Composites for Construction*, ASCE, V. 9, No. 2, pp. 147-157.

El-Sayed, A. K.; El-Salakawy, E. F.; and Benmokrane, B., 2006, "Shear Strength of FRP-Reinforced Concrete Beams without Transverse Reinforcement," *ACI Structural Journal*, V. 103, No. 2, Mar.-Apr., pp. 235-243.

Engel, R.; Croyle, M.; Bakis, C.; and Nanni, A., 1999, "Deflection of Reinforced Concrete Beams Reinforced by Fiber Reinforced Polymer Grids with Various Joint Designs," *Fiber Reinforced Polymer Reinforcement for Reinforced Concrete Structures*, SP-188, C. W. Dolan, S. H. Rizkalla, and A. Nanni, eds., American Concrete Institute, Farmington Hills, MI, pp. 75-85.

Erki, M. A., and Rizkalla S. H., 1993, "FRP Reinforcement for Concrete Structures," *Concrete International*, V. 15, No. 6, June, pp. 48-53.

Fam, A. Z., 2000, "Concrete-Filled Fiber Reinforced Polymer Tubes For Axial and Flexural Structural Members," PhD thesis, The University of Manitoba, Winnipeg, MB, Canada, 261 pp.

Fam, A. Z., and Rizkalla, S. H., 2001a, "Behavior of Axially Loaded Concrete-Filled Circular Fiber Reinforced Polymer Tubes," *ACI Structural Journal*, V. 98, No. 3, May-June, pp. 280-289.

Fam, A. Z., and Rizkalla, S. H., 2001b, "Confinement Model for Axially Loaded Concrete Confined by FRP Tubes," *ACI Structural Journal*, V. 98, No. 4, July-Aug., pp. 451-461.

Fam, A. Z., and Rizkalla, S. H., 2002, "Flexural Behavior of Concrete-Filled Fiber-Reinforced Polymer Circular Tubes," *Journal of Composites for Construction*, ASCE, V. 6, No. 2, pp. 123-132.

Fam, A. Z., and Rizkalla, S., 2003, "Large Scale Testing and Analysis of Hybrid Concrete/Composite Tubes for Circular Beam-Column Applications," *Construction and Building Materials*, V. 17, No. 6-7, pp. 507-516.

Fam, A. Z.; Rizkalla, S. H.; and Tadros, G., 1997, "Behavior of CFRP for Prestressing and Shear Reinforcements of Concrete Highway Bridges," *ACI Structural Journal*, V. 94, No. 1, Jan.-Feb., pp. 77-86.

Fam, A. Z.; Schnerch, D.; and Rizkalla, S. H., 2002, "Moisture Effect on Durability of Concrete-Filled GFRP Tubular Piles," *NCSU-CFL Report* No. 02-04, Constructed Facilities Laboratory, North Carolina State University, Raleigh, NC, 15 pp.

Fam, A. Z.; Schnerch, D.; and Rizkalla, S. H., 2003a, "Innovative FRP/Concrete Hybrid Rectangular Beams," *6th International Symposium on FRP Reinforcement for Concrete Structures (FRPRCS-6)*, Singapore, July 8-10, pp. 685-694.

Fam, A. Z.; Flisak, B.; and Rizkalla, S., 2003b, "Experimental and Analytical Investigations of Beam-Column Behavior of Concrete-Filled FRP Tubes," *ACI Structural Journal*, V. 100, No. 4, July-Aug., pp. 499-509.

Fam, A. Z.; Pando, M.; Filtz, G.; and Rizkalla, S., 2003c, "Precast Composite Piles for the Route 40 Bridge in Virginia," *PCI Journal*, V. 48, No. 3, May-June, pp. 32-45.

Fam, A.; Greene, R.; and Rizkalla, S., 2003d, "Field Applications of Concrete-Filled FRP Tubes for Marine Piles," *Field Application of FRP Reinforcement: Case Studies*, SP-215, S. H. Rizkalla and A. Nanni, eds., American Concrete Institute, Farmington Hills, MI, pp. 161-180.

Fan, L.; Zhou, W.; and Xue, Y., 2000, "Preliminary Study of Seismic Performance of FRP Tube Confined Column Short Column," *Proceedings of the 1st Chinese National Conference of Application Technology of FFRP Materials in Civil Engineering*, National Diagnosis and Rehabilitation of Industrial Building Research Center, Beijing, China, pp. 1113-1177.

Fanning, P., and Kelly, O., 1999, "Shear Strengthening of Reinforced Concrete Beams: An Experimental Study Using CFRP Plates," *Proceedings of the 8th International Conference on Structural Faults and Repair 99,* London, July 13-15, 7 pp.

Fardis, M. N., and Khalili, H., 1981, "Concrete Encased in Fiberglass-Reinforced Plastic," ACI JOURNAL, *Proceedings* V. 78, No. 6, Nov.-Dec., pp. 440-446.

Faza, S. S., and GangaRao, H. V. S., 1992, "Bending and Bond Behavior of Concrete Beams Reinforced with Fiber Reinforced Plastic Rebars," *WVODOH-RP-83 Phase I Report*, West Virginia University, Morgantown, WV, pp. 128-173.

fib, 2001, "Externally Bonded FRP Reinforcement for RC Structures," *Bulletin* 14, Fédération Internationale de Béton, 130 pp.

Fish, K., 1992, "Development Length of Fiber-Composite Concrete Reinforcement," Master's thesis, Department of Civil and Construction Engineering, Iowa State University, Ames, IA, 129 pp.

Foden, A.; Lyon, R.; Balaguru, P.; and Davidovitz, J., 1996, "High Temperature Inorganic Resin for Use in Fiber Reinforced Composites," *Proceedings: 1st International Conference on Composite in Infrastructure, ICCI'96*, University of Arizona, Tucson, AZ, pp. 166-177.

Ford, J., 2004, "Effects of Freeze/Thaw Cycling on the Fatigue Behaviour of Prestressed Concrete Beams Strengthened with Prestressed CFRP Sheets," Master's thesis, Queen's University, Kingston, ON, Canada, 162 pp.

Frosch, R. J., 1999, "Another Look at Cracking and Crack Control in Reinforced Concrete," *ACI Structural Journal*, V. 96, No. 3, May-June, pp. 437-442.

Fujisaki, T.; Nakatsuji, T.; and Sugita, M., 1993, "Research and Development of Grid Shaped FRP Reinforcement," *Fiber-Reinforced Plastic Reinforcement for Concrete Structures*, SP-138, A. Nanni and C. W. Dolan, eds., American Concrete Institute, Farmington Hills, MI, pp. 177-192.

Fukuzawa, K.; Numao, T.; Wu, Z.; Yoshizawa, H.; and Mitsui, M., 1997, "Critical Strain Energy Release Rate of Interface Debonding Between Carbon Fibre Sheet and Mortar," *Proceedings of the Third International Conference on Non-Metallic FRP Reinforcement for Concrete Structures*, Japan, V. 1, pp. 295-302.

Fyfe Co. LLC, 2005, http://www.fyfeco.com (accessed May 2005)

Galati, N., 2002, "Arching Effect on Masonry Walls Strengthened with FRP Materials," Master's thesis, Department of Engineering Mechanics, University of Missouri-Rolla, Rolla, MO, 124 pp.

GangaRao, H., and Vijay, P. V., 1997, "Aging of Structural Composites under Varying Environmental Conditions," *Non-Metallic (FRP) Reinforcement for Concrete Structure: Proceeding of the Third International Symposium*, V. 2, Japan Concrete Institute, Sapporo, Japan, pp. 91-98.

Garden, H. N., and Hollaway, L. C., 1998, "An Experimental Study of the Failure Modes of Reinforced Concrete Beams Strengthened with Prestressed Carbon Composite Plates," *Composites Part B*, V. 29, No. 4, pp. 411-424.

Garden, H. N.; Hollaway, L. C.; and Thorne, A. M., 1997, "A Preliminary Evaluation of Carbon Fiber Reinforced Polymer Plates for Strengthening Reinforced Concrete Members," *Structures and Buildings*, May, pp. 127-142.

Garden, H. N.; Hollaway, L. C.; and Thorne, A. M., 1998, "The Strengthening and Deformation Behavior of Reinforced Concrete Beams Upgraded Using Prestressed Composite Plates," *Materials and Structures*, V. 31, No. 208, May, pp. 247-258.

Gergely, J., and Young, D. T., 2001, "Masonry Wall Retrofitted with CFRP Materials," *Proceedings of the International Conference Composites in Construction*, Portugal, Oct. 10-12, pp. 565-569.

Gerritse, A., and Werner, J., 1988, *ARAPREE: The Prestressing Element Composed of Resin Bonded Twaron Fibers*, Holland Beton Group, The Netherlands, June, 8 pp.

Grace, N. F., and Abdel-Sayed, G., 1996a, "Behavior of CFRP/GFRP Bridge System," *Second International Conference, Advanced Composite Materials in Bridges and Structures, ACMBS II*, M. M. El-Badry, ed., CSCE Publication, Montreal, QC, Canada, Aug. 11-14, pp. 289-296.

Grace, N. F., and Abdel-Sayed, G., 1996b, "Double Tee and CFRP/GFRP Bridge System," *Concrete International*, V. 18, No. 2, Feb., pp. 39-44.

Grace, N. F., and Abdel-Sayed, G., 1998a, "Ductility of Prestressed Bridges Using CFRP Strands," *Concrete International*, V. 20, No. 6, June, pp. 25-30.

Grace, N. F., and Abdel-Sayed, G., 1998b, "Behavior of Externally Draped CFRP Tendons in Prestressed Concrete Bridges," *PCI Journal*, V. 43, No. 5, Sept.-Oct., pp. 88-101.

Grace, N. F., and Abdel-Sayed, G., 2003, "Construction and Evaluation of Full-Scale CFRP Prestressed Concrete DT-Girder," *Proceedings of Fiber-Reinforced Polymer Reinforcement for Concrete Structures*, Singapore, July 8-10, pp. 1281-1290.

Grace, N. F.; Abdel-Sayed, G.; Soliman, A. K.; and Saleh, K. R., 1999, "Strengthening Reinforced Concrete Beams Using Fiber Reinforced Polymer (FRP) Laminates," *ACI Structural Journal*, V. 96, No. 5, Sept.-Oct., pp. 865-874.

Grace, N. F.; Roller, J. J.; Navarre, F. C.; Nacey, R. B.; and Bonus, W., 2004, "Load Testing a CFRP-Reinforced Bridge," *Concrete International*, V. 26, No. 7, July, pp. 51-57.

Grando, S.; Valluzzi, M. R.; Tumialan, J. G.; and Nanni, A., 2003, "Shear Strengthening of URM Clay Walls with FRP Systems," *Proceedings of the International Symposium FRPRCS-6*, Singapore, July, pp. 1229-1238.

Green, M. F.; Dent, A.; and Bisby, L. A., 2003, "Effect of Freeze-Thaw Cycling on the Behaviour of Reinforced Concrete Beams Strengthened in Flexure with FRP Sheets," *Canadian Journal of Civil Engineering,* V. 30, No. 6, pp. 1081-1088.

Greenwood, M., 2002, "Creep-Rupture Testing to Predict Long-Term Performance," *Proceedings of the 2nd International Conference on Durability of Fibre Reinforced Polymer (FRP) Composites for Construction*, University of Sherbrooke, Sherbrooke, QC, Canada, 212 pp.

Greszczuk, L., 1969, "Theoretical Studies on the Mechanics of the Fiber-Matrix Interface in Composites, Interfaces in Composites," ASTM STP 452, ASTM International, West Conshohocken, PA, 42-58 pp.

Hall, J. E., and Mottram, J. T., 1998, "Combined FRP Reinforcement and Permanent Formwork for Concrete Members," *Journal of Composites for Construction*, ASCE, V. 2, No. 2, May, pp. 78-86.

Hamilton III, H. R., and Dolan, C. W., 2001, "Flexural Capacity of Glass FRP Strengthened Concrete Masonry Walls," *Journal of Composites for Construction*, ASCE, V. 5, No. 3, pp. 170-178.

Hammons, M. I.; Knox, K. J.; and Porter, J. R., 2002, "Polymer Materials for Blast Retrofit of Masonry Construction," *Proceedings, 30th Department of Defense Explosives Safety Seminar*, Atlanta, GA, 11 pp. (CD-ROM)

Hamoush, S. A.; McGinley, M. W.; Mlakar, P.; Scott, D.; and Murray, K., 2001, "Out-of-Plane Strengthening of Masonry Walls with Reinforced Composites," *Journal of Composites for Construction*, ASCE, V. 5, No. 3, pp. 139-145.

Harajli, M. H., and Kanj, M. Y., 1991, "Ultimate Flexural Strength of Concrete Members Prestressed with Unbonded Tendons," *ACI Structural Journal*, V. 91, No. 6, Nov.-Dec., pp. 663-673.

Harmon, T. K.; Slattery, K.; and Ramakrishnan, S., 1995, "The Effect of Confinement Stiffness on Confined Concrete," *Non-Metallic (FRP) Reinforcement for Concrete Structures*, L. Taerwe, ed., St. Louis, MO, pp. 584-592.

Harper, J. F.; Miller, N. A.; and Yap, S. C., 1995, "Problems Associated with the Compression Testing of Fibre Reinforced Plastic Composites," *Polymer Testing*, V. 12, pp. 15-29.

Harries, K. A.; Kestner, J.; Pessiki, S.; Sause, R.; and Ricles, J., 1998, "Axial Behavior of Reinforced Concrete Columns Retrofit with FRPC Jackets," *Second International Conference on Composites in Infrastructure*, Tuscon, AZ, pp. 411-425.

Harries, K. A.; Porter, M.; and Busel, J., 2003, "FRP Materials and Concrete—Research Needs, *Concrete International*, V. 25, No. 10, Oct., pp. 69-74.

Harries, K. A.; Reeve, B.; and Zorn, A., 2006, "Effect of Adhesive-line Properties on the Monotonic and Fatigue Behaviour of Externally Bonded CFRP Systems," *Proceedings of 11th International Conference on Structural Faults and Repair*, Edinburgh, June, 10 pp. (CD-ROM)

Hartley, A.; Mullins, G.; and Sen, R., 1996, "Repair of Concrete Masonry Block Walls Using Carbon Fiber," *Advanced Composite Materials in Bridges and Structures*, M. El-Badry, ed., Canadian Society of Civil Engineers, Montreal, QC, Canada, pp. 795-802.

Hasau, K.; Okamoto, T.; and Tanigai, M., 1990, "Study on Braided Aramid Fiber Rods," *Architectural Institute of Japan Convention,* V. C, pp. 991-992. (in Japanese)

Hassan, T., and Rizkalla, S., 2002, "Flexural Strengthening of Prestressed Bridge Slabs with FRP Systems," *PCI Journal*, V. 47, No. 1, pp. 76-93.

Hassan, T., and Rizkalla, S., 2003, "Investigation of Bond in Concrete Structures Strengthened with Near Surface Mounted CFRP Strips," *Journal of Composites for Construction*, ASCE, V. 7, No. 3, pp. 248-257.

Hassan, T., and Rizkalla, S., 2004, "Bond Mechanism of NSM FRP Bars for Flexural Strengthening of Concrete Structures," *ACI Structural Journal*, V. 101, No. 6, Nov.-Dec., pp. 830-839.

Hazen, J. R.; Bassett, S.; McDoneel, V. P.; Dawson, D.; Morals, L.; Reque, K.; and Comar, G., 1998, *Composites for Infrastructures: A Guide for Engineers*, Ray Publishing, Wheat Ridge, CO, pp. 64-65.

Heffernan, P. J., 1997, "Fatigue Behavior of Reinforced Concrete Beams Strengthened with CFRP Laminates," PhD thesis, Department of Civil Engineering, Royal Military College of Canada, Kingston, ON, Canada, 157 pp.

Hefferman, P. J., and Erki, M. A., 1996, "Equivalent Capacity and Efficiency of Reinforced Concrete Beams Strengthened with Carbon Fiber Reinforced Plastic Sheets," *Canadian Journal of Civil Engineering*, V. 23, No. 1, Feb., pp. 21-29.

Helmi, K.; Fam, A.; and Mufti, A., 2005, "Field Installation, Splicing, and Flexural Testing of Hybrid FRP/Concrete Piles," *7th International Symposium on Fiber-Reinforced Polymer (FRP) Reinforcement for Concrete Structures*, SP-230, C. K. Shield, J. P. Busel, S. L. Walkup, and D. D. Gremel, eds., American Concrete Institute, Farmington Hills, MI, pp. 1103-1120.

Hohe, J.; Becker, W.; and Goswami, S., 2001, "Singular Stress Fields in Cellular Cores for Structural Sandwich Panels," *Composite Structures*, V. 53, pp. 19-91.

Hollaway, L. C., and Leeming, M. B., 1999, *Strengthening of Reinforced Concrete Structures Using Externally-Bonded FRP Composites in Structural and Civil Engineering*, CRC Press, England, 352 pp.

Horiguchi, T., and Saeki, N., 1997, "Effect of Test Methods and Quality of Concrete on Bond Strength of CFRP Sheets," *Proceedings of the Third International Conference on Non-Metallic FRP Reinforcement for Concrete Structures*, Japan, V. 1, pp. 265-270.

Horiguchi, T.; Saeki, N.; and Hata, S., 1995, "Fatigue of Reinforced Beams Externally Prestressed with Aramid Fiber Cables," *Proceedings of the Second International RILEM Symposium (FRPRCS-2), RILEM Proceedings 29 Non-Metallic (FRP) Reinforcement for Concrete Structures*, L. Taerwe, ed., Ghent, Belgium, pp. 601-607.

Howie, I., and Karbhari, V. M., 1995, "Effect of Tow Sheet Cinoisute Wrap Architecture on Strengthening of Concrete Due to Confinement: I—Experimental Studies," *Journal of Reinforced Plastics and Composites*, V. 14, Sept., pp. 1008-1029.

Hughes Brothers, 2005, http://www.hughesbros.com (accessed May 2005)

Hutchinson, R. L., and Rizkalla, S. H., 1999, "Shear Strengthening of AASHTO Bridge Girders Using Carbon Fiber Reinforced Polymer Sheets," *Fiber Reinforced Polymer Reinforcement for Reinforced Concrete Structures*, SP-188, C. W. Dolan, S. H. Rizkalla, and A. Nanni, eds., American Concrete Institute, Farmington Hills, MI, pp. 945-958.

Hutchinson, R.; Tadros, G.; Kroman, J.; and Rizkalla, S., 2003, "Use of Externally Bonded FRP Systems for Rehabilitation of Bridges in Western Canada," *Field Applications of FRP Reinforcement: Case Studies*, SP-215, S. Rizkalla and A. Nanni, eds., American Concrete Institute, Farmington Hills, MI, pp. 239-248.

ICC Evaluation Service, Inc., 2003a, "Interim Criteria for Inspection and Verification of Concrete and Reinforced and Unreinforced Masonry Strengthening Using Fiber-Reinforced Polymer (FRP) Composite Systems," AC-178, 7 pp.

ICC Evaluation Service, Inc., 2003b, "Interim Criteria for Concrete and Reinforced and Unreinforced Masonry Strengthening Using Fiber-Reinforced Polymer (FRP) Composite Systems," AC-125, 10 pp.

Ichimasu, H.; Maruyama, M.; Watanabe, H.; and Hirose, T., 1993, "RC Slabs Strengthened by Bonded Carbon FRP Plates Part 2—Application," *International Symposium on Fiber-Reinforced-Polymer Reinforcement for Concrete Structures*, SP-138, A. Nanni and C. W. Dolan, eds., American Concrete Institute, Farmington Hills, MI, pp. 957-970.

ISIS Canada, 2006, http://www.isiscanada.com (accessed Sept. 2006)

IStructE, 1999, *Interim Guidance on the Design of Reinforced Concrete Structures using Fibre Composite Reinforcement*, Institution of Structural Engineers, London, 116 pp.

Iwamoto, K.; Uchita, Y.; Takagi, N.; and Kojima, T., 1993, "Flexural Fatigue Behavior of Prestressed Concrete Beams Using Aramid-Fiber Tendons," *Fiber-Reinforced-Plastic Reinforcement for Concrete Structures*, SP-138, A. Nanni and C. W. Dolan, eds., American Concrete Institute, Farmington Hills, MI, pp. 509-524.

Jacobson, D. A.; Bank, L. C.; Oliva, M. G.; and Russell, J. S., 2006, "Punching Shear Capacity of Double Layer FRP Grid Reinforced Slabs," *7th International Symposium on Fiber-Reinforced (FRP) Polymer Reinforcement for Concrete Structures*, SP-230, C. K. Shield, J. P. Busel, S. L. Walkup, and D. D. Gremel, eds., American Concrete Institute, Farmington Hills, MI, pp. 857-876.

Jaeger, L. G.; Mufti, A. A.; and Tadros, G., 1997, "The Concept of the Overall Performance Factor in Rectangular-Section Reinforced Concrete Beams," *Proceedings of the Third International Symposium on Non-Metallic (FRP) Reinforcement for Concrete Structures (FRPRCS-3)*, V. 2, Japan Concrete Institute, Sapporo, Japan, pp. 551-558.

Javor, T.; Prigans, S.; and Alarashi, A., 1999, "Repair of Deficient Shear Strength Beams by CFRP," *Proceedings of the International Conference on Extending Performance of Concrete Structures,* R. K. Dhir and P. A. J. Tittle, eds., Dundee, Scotland, pp. 131-136.

Jerrett, C. V.; Ahmad, S. H.; and Scotti, G., 1996, "Behavior of Prestressed Concrete Beams Strengthened by External FRP Post-Tensioned Tendons," *Second International Conference, Advanced Composite Materials in Bridges and Structures, ACMBS II*, M. M. El-Badry, ed., CSCE Publication, Montreal, QC, Canada, Aug. 11-14, pp. 305-312.

Joh, O.; Wang, Z.; and Goto, Y., 1999, "Long-Term Deflection of Fiber Reinforced Polymer Concrete Beams," *Fiber Reinforced Polymer Reinforcement for Reinforced Concrete Structures*, SP-188, C. W. Dolan, S. H. Rizkalla, and A. Nanni, eds., American Concrete Institute, Farmington Hills, MI, pp. 577-590.

Jonaitis, B.; Papinigis, V.; and Kamaitis, Z., 1999, "Behavior of Reinforced Concrete Flexural Members Strengthened by Bonded Steel Plates and Glass Fiber Plates," *Proceedings of the International Conference on Extending Performance of Concrete Structures*, R. K. Dhir and P. A. J. Tittle, eds., Sept. 7, Dundee, Scotland, pp. 33-38.

Jones, F. R., 1999, "Durability of Reinforced Plastics in Liquid Environments," *Reinforced Plastics Durability*, Woodhead Publishing, pp. 70-110.

JSCE, 1986, "Standard Specification for Concrete Structures," Japanese Society of Civil Engineers, Tokyo, 274 pp. (English translation 1996)

JSCE, 1993, "State-of-the Art Report on Continuous Fiber Reinforcing Materials," *Concrete Engineering Series* No. 3, A. Machida, ed., Research Committee on Continuous Fiber Reinforcing Material, Japanese Society of Civil Engineers, Tokyo, Oct., 164 pp.

JSCE, 1997, "Recommendations for the Design and Construction of Concrete Structures using Continuous Fiber Reinforced Materials," *Concrete Engineering Series* No. 23, Japanese Society of Civil Engineers, Tokyo, 325 pp. (available in English)

JSCE, 2001, "Recommendations for the Upgrading of Concrete Structures with use of Continuous Fiber Sheets," *Concrete Engineering Series* No. 41, Japanese Society of Civil Engineers, Tokyo, 250 pp. (available in English)

Kachlakev, D. I., and Barnes, W. A., 1999, "Flexural and Shear Performance of Concrete Beams Strengthened with Fiber Reinforced Polymer Laminates," *Fiber Reinforced Polymer Reinforcement for Reinforced Concrete Structures*, SP-188, C. W. Dolan, S. H., Rizkalla, and A. Nanni, eds., American Concrete Institute, Farmington Hills, MI, pp. 959-972.

Kaminska, M. E., and Kotynia, R., 2000, "Experimental Research on RC Beams Strengthened with CFRP Strips," *Report* No. 9, Department of Concrete Structures, Technical University of Lodz, Poland, 55 pp.

Kanatharana, J., and Lu, L. W., 1998, "Strength and Ductility of Concrete Columns Reinforced by FRP Tubes," *Proceeding of the First International Conference on Composites in Infrastructure ICCI'96*, H. Saadatmanesh and M. R. Ehsani, eds., Tucson, AZ, Jan., pp. 370-384.

Karam, G. N., 1992, "Optimal Design for Prestressing with FRP Sheets in Structural Members," *Proceedings of the First International Conference on Advanced Composite Materials in Bridges and Structures (ACMBS I)*, K. W. Neale and P. Labossière, eds., QC, Canada, pp. 277-285.

Karbhari, V. M.; Chin, J. W.; Hunston, D.; Benmokrane, B.; Juska, T.; Morgan, R.; Lesko, J. J.; Sorathia, U.; and Reynaud, D., 2003, "Durability Gap Analysis for Fiber-Reinforced Composites in Civil Infrastructure," *Journal of Composites for Construction*, V. 7, No. 3, Aug., pp. 238-247.

Karbhari, V. M., and Eckel II, D. A., 1993, "Effect of a Cold-Regions-Type Climate on the Strengthening Efficiency of Composite Wraps for Columns," *Technical Report*, University of Delaware Center for Composite Materials, Newark, DE, 21 pp.

Karbhari, V. M., and Engineer, M., 1996, "Investigation of Bond Between Concrete and Composites: Use of a Peeling Test," *Journal of Reinforced Plastics and Composites*, V. 15, pp. 208-227.

Karbhari, V. M.; Engineer, M.; and Eckel II, D. A., 1997, "On the Durability of Composite Rehabilitation Schemes for Concrete: Use of a Peel Test," *Journal of Materials Science*, V. 32, pp. 147-156.

Karbhari, V. M., and Howie, I., 1997, "Effect of Composite Wrap Architecture on Strengthening of Concrete Due to Confinement: II—Strain and Damage Effects," *Journal of Reinforced Plastics and Composites*, V. 16, No. 11, pp. 1039-1063.

Karbhari, V. M.; Seible, F.; Burgueño, R.; Davol, A.; Wernli, M.; and Zhao, L., 2000, "Structural Characterization of Fiber-Reinforced Composite Short- and Medium-Span Bridge Systems," *Applied Composite Materials*, V. 7, Issue 2/3, May, pp. 151-182.

Karbhari, V. M.; Seible, F.; Burgueño, R.; Davol, A.; and Zhao, L., 1998, "Structural Characterization of Fiber-Reinforced Composite Short-and Medium-Span Bridge System," *Proceeding of ECCM-8, European Conference on Composite Materials*, V. 2, I. Crivelli Visconti, ed., pp. 35-42.

Kassem, C.; El-Salakawy, E. F.; and Benmokrane, B., 2002, "Deflection Behaviour of Concrete Beams Reinforced with Carbon FRP Composite Bars," *Deflection Control for the Future*, SP-210, N. J. Gardner, ed., American Concrete Institute, Farmington Hills, MI, pp. 173-190.

Katsumata, H.; Kimura, K.; and Murahashi, H., 2001, "Experience of FRP Strengthening for Japanese Historical Structures," *Proceedings International Conference FRP Composites in Civil Engineering*, V. 2, Elsevier Science, New York, pp. 1001-1008.

Katz, A.; Berman, N.; and Bank, L. C., 1999, "Effect of High Temperature on the Bond Strength of FRP Rebars," *Journal of Composites for Construction*, ASCE, V. 3, No. 2, pp. 73-81.

Keller, T., 2003, "Strengthening of Concrete Bridges with Carbon Cables and Strips," *Proceedings FRPRCS-6*, pp. 1331-1340.

Khalifa, A.; De Lorenzis, L.; and Nanni, A., 2000, "FRP Composites for Shear Strengthening of RC Beams," *Proceedings of the 3rd International Conference on Advanced Composite Materials in Bridges and Structures, (ACMBS III)*, J. Humar and A. Razaqpur, eds., Ottawa, ON, Canada, Aug. 15-18, pp. 137-144.

Khalifa, A.; Gold, W. J.; Nanni, A.; and Aziz, A., 1998, "Contribution of Externally Bonded FRP to Shear Capacity of RC Flexural Members," *Journal of Composites for Construction*, V. 2, No. 4, pp. 195-202.

Kim, J. K.; Gao, B.; and Leung, C. K., 2002, "Effect of Toughened Resin on FRP Bond and Durability of RC Beams Strengthened with FRP," *Proceedings of the Second International Conference on Durability of Fibre Reinforced Polymers for Construction,* Montreal, QC, Canada, pp. 441-452.

Kim, P., and Meier, U., 1991, "CFRP Cables for Large Structures," *Proceedings of the Speciality Conference Advanced Composites Materials in Civil Engineering Structures*, L. Srinivasa and R. Sen, eds., Las Vegas, NV., American Society of Civil Engineers, New York, pp. 233-244.

Kim, Y. J.; Green, M. F.; Fallis, G. J.; Wight, G. R.; and Eden, R., 2006, "Damaged Bridge Girder Strengthening," *Concrete International*, V. 28, No. 11, Nov., pp. 61-66.

Knox, K. J.; Hammonds, M. I.; Porter, J. R.; and Lewis, T. T., 2000, "Polymer Materials For Structural Retrofit," *Proceedings, 29th Department of Defense Explosives Safety Seminar*, New Orleans, LA, 17 pp. (CD-ROM)

Kodur, V. K. R., and Baingo, D., 1998, "Fire Resistance of FRP Reinforced Concrete Slabs," *IRC Internal Report* No. 758, National Research Council of Canada, Ottawa, ON, Canada, 39 pp.

Kodur, V. K. R., and Bisby, L. A., 2005, "Evaluation of Fire Endurance of Concrete Slabs Reinforced with FRP Bars," *Journal of Structural Engineering*, ASCE, V. 131, No. 1, pp. 34-43.

Kodur, V. K. R.; Bisby, L. A.; and Green, M. F., 2006, "FRP Retrofitted Concrete Under Fire Conditions," *Concrete International,* V. 28, No. 12, Dec., pp. 37-44.

Korany, Y., 2004, "Rehabilitation of Masonry Walls Using Unobtrusive FRP Techniques for Enhanced Out-of-Plane Seismic Resistance," PhD thesis, McMaster University, Hamilton, ON, Canada, 276 pp.

Korany, Y., and Drysdale, R., 2004, "Enhancing Flexural Resistance of Historic Masonry Walls Using Carbon Fibre Ropes," *The Masonry Society Journal*, V. 22, No. 1, pp. 27-38.

Kotynia, R., 1999, "Ductility and Load Capacity of Reinforced Concrete Members Strengthened with CFRP

Strips," PhD thesis, Department of Civil Engineering, Architecture and Environmental Engineering, University of Lodz, Poland, 215 pp.

Kotynia R., 2005a, "Strain Efficiency of Near-Surface Mounted CFRP-Strengthened Reinforced Concrete Beams," *Proceedings of the Third International Conference Composites in Construction*, Lyon, France, pp. 35-42.

Kotynia, R., 2005b, "Debonding Failures of RC Beams Strengthened with Externally Bonded Strips," *Proceedings of the International Symposium on Bond Behaviour of FRP in Structures*, Hong Kong, pp. 247-252.

Krishnaswamy, S., 2000, "Optical Methods of Inspecting Composites (Holography and Shearography)," *Comprehensive Composite Materials, Test Methods, V. 5: Nondestructive Evaluation, and Smart Materials*, L. Carlsson, R. L. Crane, and K. Uchino, eds., Pergamon Press, Elsevier Science Ltd., Oxford, pp. 447-468.

Kumar, S., and GangaRao, H., 1998, "Fatigue Response of Concrete Decks Reinforced with FRP Rebars," *Journal of Structural Engineering*, V. 124, No. 1, pp. 11-16.

Kurt, C. E., 1978, "Concrete Filled Structural Plastic Columns," *Journal of the Structural Division*, ASCE, V. 104, No. ST1, Jan., pp. 55-63.

Kurtz, S., and Balaguru, P., 2001, "Comparison of Inorganic and Organic Matrices for Strengthening of RC Beams with Carbon Sheets," *Journal of Structural Engineering*, ASCE, V. 127, No. 1, pp. 35-42.

Labossière, P.; Neale, K. W.; and Martel, S., 1997, "Strengthening with Composite Materials: Practical Applications in Quebec," *Proceedings of the U.S.-Europe Workshop on Recent Advances in Bridge Engineering: Advanced Rehabilitation, Durable Materials, Nondestructive Evaluation, and Management*, V. 11, U. Meier and R. Betti, eds., Dubendorf and Zurich, July 14-15, pp. 89-96.

Labossière, P.; Neale, K. W.; Rochette, P.; Demers, M.; Lamothe, P.; Lapierre, P.; and Desgangne, G., 2000, "Fiber Reinforced Polymer Strengthening of Sainte-Emilie-de-l'Energie Bridge: Design, Instrumentation, and Field Testing," *Canadian Journal of Civil Engineering*, V. 27, No. 5, Oct., pp. 916-927.

Lamanna, A. J.; Bank, L. C.; and Scott, D. W., 2001, "Flexural Strengthening of RC Beams using Fasteners and FRP Strips," *ACI Structural Journal*, V. 98, No. 3, May-June, pp. 368-376.

Lamanna, A. J.; Bank, L. C.; and Scott, D. W., 2004, "Flexural Strengthening of RC Beams by Mechanically Attaching FRP Strips," *Journal of Composites for Construction*, ASCE, V. 8, No. 3, pp. 203-210.

Lamothe, P.; Labossière, P.; and Neale, K. W., 1998, "Post-Strengthening of Reinforced Concrete T-Beams with Composite Materials," *Proceedings of the 1998 CSCE Annual Conference*, V. 3b, Halifax, NS, Canada, June 10-13, pp. 623-631.

Lan, S.; Crawford, J. E.; and Xin, X., 2003, "Assessment of Blast Resistance of Large Bridge Columns and the Effectiveness of Retrofitting Them with CFRP," *Proceedings, 3rd Asia-Pacific Specialty Conference on Fiber Reinforced Materials Incorporating FRC, FRP, and Others*, Changsha, China, 8 pp.

Lancaster Composite, Inc., 2006, "Fiberglass/Concrete Composite Marine Piling," *CP40 Pile Design Guide*, 32 pp. (CD-ROM)

Larralde, J., 1992, "Feasibility of FRP Molded Grating-Concrete Composite for One-Way Slab System," *Materials Performance and Prevention of Deficiencies and Failures, Proceedings of the Materials Engineering Congress*, Atlanta, GA, Aug., pp. 645-654.

Larralde, J.; Renbaum, L.; and Morsi, A., 1988, "Fiber-Glass Reinforced Plastic Rebars in Lieu of Steel Rebars," *Proceedings of the Structures Congress, TRB Annual Meeting, Task Force A2C51—Structural Applications of Fiber Reinforced Plastics*, Aug. pp. 261-269.

Larralde, J.; Silva-Rodrigues, R.; Burdette, J.; and Harris, B., 1994, "Bond Tests of Fiberglass-Reinforced Plastic Bars in Concrete," *Journal of Testing and Evaluation*, V. 22, No. 4, July, pp. 351-359.

Larson, K. H.; Peterman, R. J.; and Rasheed, H. A., 2005, "Strength-Fatigue Behavior of FRP Strengthened Prestressed Concrete T Beams," *Journal of Composites for Construction*, ASCE, V. 9, No. 4, July-Aug., pp. 313-326.

Lees, J. M., and Burgoyne, C. J., 1999, "Experimental Study of the Influence of Bond on the Flexural Behaviour of Concrete Beams Pretensioned with AFRPs," *ACI Structural Journal*, V. 96, No. 3, May-June, pp. 377-385.

Lees, J. M.; Winistoerfer, A. U.; and Meier, U., 1999, "Nonlaminated Pin-Loaded Carbon Fiber Reinforced Plastic Straps For Shear Enhancement of Concrete," *Fiber Reinforced Polymer Reinforcement for Reinforced Concrete Structures*, SP-188, C. W. Dolan, S. H. Rizkalla, and A. Nanni, eds., American Concrete Institute, Farmington Hills, MI, pp. 985-993.

Linear Composites, 2007, http://www.linearcomposites.co.uk/prod_parafil.htm (accessed Feb 2007)

Lombard, J. C.; Lau, D. T.; Humar, J. L.; Cheung, M. S.; and Foo, S., 2000, "Seismic Repair and Strengthening of Reinforced Concrete Shear Walls for Flexure and Shear Using Carbon Fiber Sheets," *Proceedings of the 3rd International Conference on Advanced Composite Materials in Bridges and Structures, (ACMBS III)*, J. Humar and A. G. Razaqpur, eds., Ottawa, ON, Canada, Aug. 15-18, pp. 645-652.

Lopez, M. M.; Naaman, A. E.; Pinkerton, L.; and Till, R. D., 2003, "Behavior of RC Beams Strengthened with FRP Laminates and Tested Under Cyclic Loading at Low Temperatures," *International Journal of Materials and Product Technology*, V. 19, No. 1-2, pp. 108-117.

Lopez, M. M.; Naaman, A. E.; and Till, R., 1999, "Bending Behavior of RC Beams Strengthened with CFRP Laminates and Subjected to Freeze-Thaw Cycles," *Fiber Reinforced Polymer Reinforcement for Reinforced Concrete Structures*, SP-188, C. W. Dolan, S. H. Rizkalla, and A. Nanni, eds., American Concrete Institute, Farmington Hills, MI, pp. 559-576.

Lopez-Anido, R.; Dutta, P.; Bouzon, J.; Morton, S.; Shahrooz, B.; and Harik, I., 1999, "Fatigue Evaluation of FRP-Concrete Bridge Deck on Steel Girders at High Temperature," *Proceedings of the 44th International SAMPE Symposium.*

Lord, H. W., and Dutta, P. K., 1988, "On the Design of Polymeric Composite Structures for Cold Regions Applications," *Journal of Reinforced Plastics and Composites*, V. 7, pp. 435-450.

Luciano, R.; Marfia, S.; and Sacco, E., 2001, "Reinforcement of Masonry Arches by FRP Materials: Experimental Tests and Numerical Investigations," *Proceedings of the International Conference Composites in Construction*, Portugal, Oct. 10-12. (CD-ROM)

Maalej, M., and Bonacci, J. F., 1998, "Repair and Strengthening of Beams Using FRP Reinforcement," *Proceedings of the 1998 CSCE Annual Conference*, V. 3b, Halifax, NS, Canada, June 10-13, pp. 643-652.

Magnusson, J., 1997, "Bond and Anchorage of Deformed Bars in High-Strength Concrete," Licentiate thesis, Chalmers University of Technology, Division of Concrete Structures, 234 pp.

Mahfouz, I., and Rizk, T., 2003, "Field Application of FRP Strengthening in Egypt," *Field Applications of FRP Reinforcement: Case Studies*, SP-215, S. Rizkalla and A. Nanni, eds., American Concrete Institute, Farmington Hills, MI, pp. 279-290.

Mahmoud, Z. I., and Rizkalla, S. H., 1996, "Bond of CFRP Prestressing Reinforcement," *Advanced Composite Materials in Bridges and Structures (ACMBS-II)*, Montreal, QC, Canada, Aug., pp. 877-884.

Mahmoud, Z. I.; Rizkalla, S. H.; and Zaghloul, E., 1997, "Transfer and Development Length of CFRP Reinforcement," *Proceedings of the 1997 CSCE Annual Conference*, Sherbrooke, QC, Canada, May, pp. 101-110.

Maissen, A., and De Smet, C. A. M., 1995, "Comparison of Concrete Beams Prestressed with Carbon Fibre Reinforced Plastic and Steel Strands," *Non-Metallic (FRP) Reinforcement for Concrete Structures, Proceedings of the Second International RILEM Symposium (FRPRCS-2)*, Ghent, Belgium, Aug. 23-25, pp. 379-386.

Mallick, P. K., 1988, *Fiber-Reinforced Composites—Materials, Manufacturing, and Design*, Marcel Dekker, Inc., New York, pp. 18-19.

Malvar, L. J., 1995, "Tensile and Bond Properties of GFRP Reinforcing Bars," *ACI Materials Journal*, V. 92, No. 3, May-June, pp. 276-285.

Malvar, L. J., 2005, "Specific Local Resistance Concepts," *Concrete International*, V. 27, No. 12, Dec., pp. 23-27.

Malvar, L. J., and Bish, J., 1995, "Grip Effects in Tensile Testing of FRP Bars," *Proceedings of the Second International RILEM Symposium (FRPRCS-2), Non-Metallic (FRP) Reinforcement for Concrete Structures,* Ghent, Belgium, Aug., pp. 108-115.

Malvar, L. J.; Warren, G. E.; and Inaba, C., 1996, "Large Scale Tests on NAVY Reinforced Concrete Pier Decks Strengthened with CFRP Sheets," *Second International Conference on Advanced Composite Materials in Bridges and Structures (ACMBS II)*, M. M. El-Badry, ed., Montreal, QC, Canada, Aug. 11-14, pp. 497-504.

Mander, J. B.; Priestley, M. J. N.; and Park, R., 1988, "Theoretical Stress-Strain Model for Confined Concrete," *Journal of Structural Engineering*, V. 114 No. 8, Aug., pp. 1805-1826.

Marshall, O. S.; Sweeney, S. C.; and Trovillion, J. C., 1999, "Seismic Rehabilitation of Unreinforced Masonry Walls," *Fourth International Symposium on Fiber Reinforced Polymer Reinforcement for Reinforced Concrete Structures*, SP-138, A. Nanni and C. W. Dolan, eds., American Concrete Institute, Farmington Hills, MI, pp. 287-295.

Marshe, S., and Green, M. F., 1999, "Punching Behaviour of Composite Bridge Decks Transversely Prestressed with Carbon Fibre Reinforced Plastic Tendons," *Canadian Journal of Civil Engineering,* V. 26, No. 5, pp. 618-630.

Mashima, M., and Iwamoto, K., 1993, "Bond Characteristics of FRP Rod and Concrete After Freeze-Thaw Deterioration," *International Symposium on Fiber Reinforced Plastic Reinforcement for Concrete Structures*, SP-138, A. Nanni and C. Dolan, eds., American Concrete Institute, Farmington Hills, MI, pp. 51-69.

Masia, M. J., and Shrive, N. G., 2003, "Carbon Fibre Reinforced Polymer Wrapping for the Rehabilitation of Masonry Columns," *Canadian Journal of Civil Engineering*, V. 30, No. 4, pp. 734-744.

Masoud, S.; Soudki, K.; and Topper, T., 2001, "CFRP-Strengthened and Corroded RC Beams under Monotonic and Fatigue Loads," *Journal of Composites for Construction*, ASCE, V. 5, No. 4, pp. 228-236.

Matthys, S., and Taerwe, L., 1998, "Long-Term Behaviour of Concrete Slabs Pre-Tensioned with AFRP or Prestressing Steel," *Durability of Fibre Reinforced Polymer (FRP) Composites for Construction,* B. Benmokrane and H. Rahman, eds., pp. 95-106.

Matthys, S., and Taerwe, L., 2000, "Concrete Slabs Reinforced with FRP Grids: One-Way Bending," *Journal of Composites for Construction*, V. 4, Aug., pp. 145-153.

Matthys, S.; Taerwe, L.; and Janssens, J., 2004, "Repair and Strengthening of Reptile Building of the Antwerp Aoo," *Proceedings of 1st International Conference on Innovative Materials and Technologies for Construction and Restoration*, Lecca, Italy, V. 2, June 6-9, pp. 653-663.

Matthys, S.; Toutanji, H.; and Taerwe, L., 2005, "Axial Load Behavior of Large-Scale Columns Confined with FRP Composites," *ACI Structural Journal*, V. 102, No. 2, Mar.-Apr., pp. 258-267.

Matthys, S.; Toutanji, H.; and Taerwe, L., 2006, "Stress-Strain Behavior of Large-Scale Circular Columns Confined with RP Composites," *Journal of Structural Engineering*, ASCE, V. 132, No. 1, Jan., pp. 123-133.

Mehus, J., 1995, "Long Term Durability of Fiber Composite Reinforcement for Concrete," MS thesis, Iowa State University, Ames, IA, 1995, 203 pp.

Meier, U., 1995, "Strengthening of Structures Using Carbon Fiber/Epoxy Composites," *Construction and Building Materials*, V. 9, No. 6, pp. 341-351.

Meier, U.; Deuring, M.; Meier, H.; and Schwegler, G., 1992, "Strengthening of Structures with CFRP Laminates: Research and Applications in Switzerland," *Proceedings of the First International Conference on Advanced Composite Materials in Bridges and Structures*, K. W. Neale and P. Labossière, eds., CSCE Publication, Sherbrooke, QC, Canada, Oct., pp. 243-251.

Meier, U.; Deuring, M.; Meier, H.; and Schwegler, G., 1993, "Strengthening of Structures with Advanced Composites," *Alternate Materials for the Reinforcement and Prestressing of Concrete*, J. L. Clarke, ed., Blackie Academic and Professional, Glasgow, pp. 153-171.

Meier, U., and Erki, M. A., 1997, "Advantages of Composite Materials in the Post-Strengthening Technique for Developing Countries," *Proceedings of the 6th International Colloquium on Concrete in Developing Countries,* Lahore, Pakistan, Jan. 4-7, pp. 1-11.

Meier, U., and Kaiser, H., 1991, "Strengthening of Structures with CFRP Laminates," *Advanced Composite Materials in Civil Engineering Structures, Proceedings of the Speciality Conference Advanced Composites Materials in Civil Engineering Structures*, S. L. Iyer, and R. Sen, eds., Las Vegas, NV, American Society of Civil Engineers, New York, pp. 224-232.

Mettes, D. G., 1969, *Handbook of Fiberglass and Advanced Plastics Composites*, G. Lubin, ed., Van Nostrand Reinhold Co., New York, 912 pp.

Mirmiran, A., 1995, "Concrete Composite Construction for Durability and Strength," *Symposium on Extending Lifespan of Structures*, International Association for Bridge and Structural Engineering (IABSE), San Francisco, CA, pp. 1155-1160.

Mirmiran, A., 2003, "Stay-In-Place FRP Form for Concrete Columns," *Advances in Structural Engineering—An International Journal*, Essex, UK, V. 6, No. 3, pp. 231-241.

Mirmiran, A., and Shahawy, M., 1995, "A Novel FRP-Concrete Composite Construction for the Infrastructure," *Proceedings of the ASCE Structures Congress XIII*, Boston, MA, pp. 1663-1666.

Mirmiran, A., and Shahawy, M., 1996, "A New Concrete-Filled Hollow FRP Composite Column," *Composites Part B: Engineering*, V. 27B, No. 3-4, June, pp. 263-268.

Mirmiran, A., and Shahawy, M., 2003, "Composite Pile: A Successful Drive," *Concrete International*, V. 25, No. 3, Mar., pp. 89-94.

Mirmiran, A.; Shahawy, M.; and Beitleman, T., 2001, "Slenderness Limit for Hybrid FRP-Concrete Columns," *Journal of Composites for Construction*, ASCE, V. 5, No. 1, pp. 26-34.

Mirmiran, A.; Shahawy, M.; El Khoury, C.; and Naguib, W., 2000b, "Large Beam-Column Tests on Concrete-Filled Composite Tubes," *ACI Structural Journal*, V. 97, No. 2, Mar.-Apr., pp. 268-276.

Mirmiran, A.; Shahawy, M.; and Samaan, M., 1999, "Strength and Ductility of Hybrid FRP-Concrete Beam-Columns," *Journal of Structural Engineering*, ASCE, V. 125, No. 10, pp. 1085-1093.

Mirmiran, A.; Shahawy, M.; Samaan, M.; El Echary, H.; Mastrapa, J. C.; and Pico, O., 1998, "Effect of Column Parameters on FRP-Confined Concrete," *Journal of Composites for Construction*, ASCE, V. 2, No. 4, Nov., pp. 175-185.

Mirmiran, A.; Shao, Y.; and Shahawy, M., 2002, "Analysis and Field Tests on the Performance of Composite Tubes under Pile Driving Impact," *Composite Structures*, Elsevier, V. 55, No. 2, pp. 127-135.

Mirmiran, A.; Zagers, K.; and Wenqing, Y., 2000a, "Nonlinear Finite Element Modeling of Concrete Confined by Fiber Composites," *Finite Elements in Analysis and Design*, Elsevier, V. 35, No. 1, pp. 79-96.

Mitsubishi, 2005, www.mitsubishichemical.com (accessed May 2005)

Miyamoto, A.; Inoshita, Y.; Takuya, M.; and Katsunori, Y., 1994a, "Mechanical Behavior and Design Concept of Prestressed Composite Girder with FRP Tendons," *Fourth International Conference on Short and Medium Span Bridges*, A. Mufti, B. Bakht, and L. G. Jaeger, eds., CSCE Publication, Halifax, NS, Canada, Aug. 8-11, pp. 931-946.

Miyamoto, A.; Maki, A.; Yagishita, F.; and Kobukata, Y., 1994b, "Application of Prestressing Technique for Bridge Strengthening," *Fourth International Conference on Short and Medium Span Bridges*, A. Mufti, B. Bakht, and L. G. Jaeger, eds., CSCE Publication, Halifax, NS, Canada, Aug. 8-11, pp. 1231-1245.

Miyamoto, A.; Tamada, H.; Kato, T.; and Kugo, M., 1998, "Performance Evaluation of Existing Composite Girder Bridge Strengthened with External Prestressing Technique," *The 5th International Conference on Short and Medium Span Bridges (SMSB-V)*, Calgary, AB, Canada, July 13-16, 12 pp.

Monti, G.; Nistico, N.; and Santini, S., 2001, "Design of FRP Jackets for Upgrade of Circular Bridge Piers," *Journal of Composites for Construction*, ASCE, V. 5, No. 2, pp. 94-101.

Moon, F. L., 2004, "Seismic Strengthening of Low-Rise Unreinforced Masonry Structures with Flexible Diaphragms," PhD thesis, Georgia Institute of Technology, Atlanta, GA, 649 pp.

Moon, F. L.; Yi, T.; Leon, R. T.; and Kahn, L. F., 2003, "Large-Scale Tests of an Unreinforced Masonry Low-Rise Building," *9th North American Masonry Conference*, Clemson, SC, June 1-4, 12 pp.

Morbin, A., 2001, *Strengthening of Masonry Elements with FRP Composites*, Tesi di Laurea, Dipartimento di Ingegneria Edile, Università di Padova, Italy, 190 pp.

Morrill, K. B.; Malvar, L. J.; and Crawford, J. E., 1999, "Retrofit Design Procedure For Existing RC Buildings to Increase their Resistance to Terrorist Bombs," *9th International Symposium on Interaction of the Effects of Munitions with Structures (IEMS)*, Berlin, Germany, pp. 587-594.

Morrill, K. B.; Malvar, L. J.; Crawford, J. E.; and Ferritto, J. M., 2004, "Blast Resistant Design and Retrofit of Reinforced Concrete Columns and Walls," *ASCE Structures Congress*, Nashville, TN, pp. 1-8.

Morrill, K. B.; Malvar, L. J.; Crawford, J. E.; Hegemier, G.; and Seible, F., 2001, "Full-Scale Testing of Reinforced Concrete Column Retrofits to Resist Blast Loads," *10th International Symposium on Interaction of the Effects of Munitions with Structures*, San Diego, CA, 8 pp. (CD-ROM)

Morrill, K. B.; Nicolaisen, K.; and Hutchinson, T., 2005, "Laboratory Testing of Kevlar® Reinforced Polyurea©," *Report* IR-04-52.1, Karagozian and Case, Burbank, CA, 53 pp.

Mufti, A. A.; Onofrei, M.; Benmokrane, B.; Banthia, N.; Boulfiza., M.; Newhook, J.; Baidar, B.; Tadros, G.; and Brett, P., 2005, "Durability of GFRP Reinforced Concrete in Field Structures," *Proceedings of the Third International Conference on Composites in Constructions*, V. 2, P. Hamelin, D. Biguaud, E. Ferrier, and E. Jacquelin, eds., Lyon, France, pp. 889-895.

Mukherjee, A.; Boothby, T. E.; Bakis, C. E.; Joshi, M. V.; and Maitra, S. R., 2004, "Mechanical Behavior of Fiber-Reinforced Polymer-Wrapped Concrete Columns-Complicating Effects," *Journal of Composites for Construction*, ASCE, V. 8, No. 2, pp. 97-103.

Murthy, S. S.; Myers, J. J.; and Micelli, F., 2002, "Environmental Effects of Concrete-FRP Bond under Various Degrees of Sustained Loading," *Proceedings of CDCC 2002*, B. Benmokrane and E. El-Salakawy, eds., Montreal, QC, Canada, pp. 333-346.

Muszynski, L. C., 1998, "Explosive Field Tests to Evaluate Composite Reinforcement of Concrete and Masonry Walls," *Fiber Composites in Infrastructure: Proceedings of the Second International Conference on Composites in Infrastructure*, Tucson, AZ, pp. 276-284.

Muszynski, L. C., and Purcell, M. R., 2003a, "Composite Reinforcement to Strengthen Existing Concrete Structures against Air Blast," *Journal of Composites for Construction*, ASCE, V. 7, No. 2, pp. 93-97.

Muszynski, L. C., and Purcell, M. R., 2003b, "Use of Composite Reinforcement to Strengthen Concrete and Air-Entrained Concrete Masonry Walls against Air Blast," *Journal of Composites for Construction*, ASCE, V. 7, No. 2, pp. 98-108.

Muszynski, L. C., and Sierakowski, R. L., 1996, "Fatigue Strength of Externally Reinforced Concrete Beams," *Materials for the New Millennium: Proceedings of the Fourth Materials Engineering Conference*, Washington D.C., Nov. 10-14, pp. 648-656.

Mutsuyoshi, H.; Ishibashi, T.; Okano, M.; and Katsuki, F., 1999, "New Design Method for Seismic Retrofit of Bridge Columns with Continuous Fiber Sheet Performance-Based Design," *Fiber Reinforced Polymer Reinforcement for Reinforced Concrete Structures*, SP-188, C. W. Dolan, S. H. Rizkalla, and A. Nanni, eds., American Concrete Institute, Farmington Hills, MI, pp. 229-241.

Mutsuyoshi, H., and Machida, A., 1993, "Behaviour of Prestressed Concrete Beams using FRP as External Cable," *International Symposium on Fiber-Reinforced-Polymer Reinforcement for Concrete Structures*, SP-138, A. Nanni and C. W. Dolan, eds., American Concrete Institute, Farmington Hills, MI, pp. 401-418.

Myers, J. J.; Belarbi, A.; and El-Domiaty, K. A., 2003, "Blast Resistance of Unreinforced Masonry Walls Retrofitted with Fiber Reinforced Polymers," *Proceedings for the North American Masonry Society 2003 Annual Meeting*, Clemson, SC, 12 pp. (CD-ROM)

Naaman, A. E., 1999, "Repair and Strengthening of Reinforced Concrete Beams Using CFRP Laminates," *Research Project* No. RC-1372, Testing and Research Section, Construction and Technology Division, Department of Civil and Environmental Engineering, University of Michigan, Ann Arbor, MI, 23 pp.

Naaman, A. E., and Alkhairy, F. M., 1991, "Stress at Ultimate in Unbonded Post-Tensioning Tendons: Part 1—Evaluation of the State of the Art," *ACI Structural Journal*, V. 88, No. 5, Sept.-Oct., pp. 641-651.

Naaman, A. E., and Jeong, S. M., 1995, "Structural Ductility of Concrete Beams Prestressed with FRP," *Proceedings of the Second International RILEM Symposium (FRPRCS-2), RILEM Proceedings 29 Non-Metallic (FRP) Reinforcement for Concrete Structures*, L. Taerwe, ed., Ghent, Belgium, pp. 379-386.

Naguib, W., and Mirmiran, A., 2001, "Time-Dependent Behavior of FRP-Confined Concrete Columns: I. Experiments and II. Modeling," *ACI Structural Journal*, V. 99, No. 2, Mar.-Apr., pp. 142-148.

Nakagawa, H.; Kobayashi, M.; Suenaga, T.; Ouchi, T.; Watanabe, S.; and Satoyama, K., 1993, "Application of Three-Dimensional Fabric Reinforced Concrete to Building Panels," *Fiber-Reinforced Plastic Reinforcement for Concrete Structures*, SP-138, A. Nanni and C. W. Dolan, eds., American Concrete Institute, Farmington Hills, MI, pp. 211-232.

Nanni, A., 1993, "Flexural Behavior and Design of Reinforced Concrete Using FRP Rods," *Journal of Structural Engineering*, ASCE, V. 119, No. 11, pp. 3344-3359.

Nanni, A., 1997, "CFRP Strengthening," *Concrete International*, V. 19, No. 6, June, pp. 19-23.

Nanni, A.; Bakis, C. E.; and Boothby, T. E., 1995, "Test Methods for FRP-Concrete Systems Subjected to Mechanical Loads: State of the Art Review," *Journal of Reinforced Plastics and Composites*, V. 14, pp. 524-558.

Nanni, A.; Bakis, C. E.; and Boothby, T. E., 1997, "Externally Bonded FRP Composites for Repair of RC Structures," *Proceedings of the Third International Conference on Non-Metallic FRP Reinforcement for Concrete Structures*, Japan, V. 2, pp. 187-194.

Nanni, A.; Bakis, C. E.; O'Neil, E.; and Dixon, T. O., 1996, "Performance of FRP Tendon-Anchor Systems for Prestressed Concrete Structures," *PCI Journal*, V. 41, No. 1, Jan.-Feb., pp. 34-44.

Nanni, A.; Di Ludovico, M.; and Parretti, R., 2004, "Shear Strengthening of a PC Bridge Girder with NSM CFRP Rectangular Bars," *Advances in Structural Engineering*, V. 7, No. 4, Aug., pp. 97-109.

Nanni, A., and Faza, S., 2002, "Designing and Constructing with FRP Bars: An Emerging Technology," *Concrete International*, V. 24, No. 11, Nov., pp. 53-58.

Nanni, A.; Utsunomiya, T.; Yonekura, H.; and Tanigaki, M., 1992, "Transmission of Prestressing Force to Concrete by Bonded Fiber Reinforced Plastic Tendons," *ACI Structural Journal*, V. 89, No. 3, May-June, pp. 335-344.

Nawy, E. G., and Neuwerth, G. E., 1971, "Behavior of Fiber Glass Reinforced Concrete Beams," *Journal of the Structural Division*, ASCE, V. 97, Sept., pp. 2203-2215.

Nawy, E. G., and Neuwerth G. E., 1977, "Fiberglass Reinforced Concrete Slabs and Beams," *Journal of Structural Division*, ASCE, V. 103, No. 2, pp. 421-440.

Neale, K., 2001, *Design Manual 4—Strengthening Reinforced Concrete Structures with Externally-Bonded Fibre Reinforced Polymers*, ISIS Canada, Winnipeg, MB, Canada, 209 pp.

NEFCOM Corp., 1998, "Technical Leaflet 3: Fire Resistance of Concrete Labs Reinforced by NEFMAC," Tokyo, 11 pp.

Nelson G. L., 1995, "Fire and Polymers: An Overview," *Proceedings, Fire and Polymers II. ACS Symposium Series 599*, Washington, DC, American Chemical Society, pp. 1-26.

Neville, A. M., 1995, *Properties of Concrete*, Longman Group Ltd., Essex, England, 844 pp.

Newhook, J. P.; Bakht, B.; Tadros, G.; and Mufti, A. A., 2000, "Design and Construction of a Concrete Marine Structure using Innovative Technologies," *Proceedings: The Third International Conference on Advanced Composite Materials in Bridges and Structures*, Ottawa, ON, Canada, pp. 777-784.

Newman, J., and Zweben, C., 2002a, "Nondestructive Inspection of FRP-Reinforced Concrete," *Proceedings, CDCC 2002, Second International Conference on Durability of Fiber Reinforced Polymer (FRP) Composites for Construction*, Montreal, QC, Canada, 12 pp. (CD-ROM)

Newman, J., and Zweben, C., 2002b, "Nondestructive Inspection of Composite-Strengthened Concrete Structures," *Proceedings, SAMPE 2002*, Long Beach, CA, May 12-16, pp. 691-702.

Nikurunziza, G.; Masmoudi, R.; and Benmokrane, B., 2002, "Effect of Sustained Tensile Stress and Temperature on Residual Strength of GFRP Bars," *Proceedings of the Second International Conference on Durability of FRP Composites for Construction (CDCC2002)*, Montreal, QC, Canada, B. Benmokrane and E. El-Salakawy, eds., pp. 347-358.

Nordin, H.; Taljsten, B.; and Carolin, A., 2001, "Concrete Beams Strengthened with Prestressed Near Surface Mounted Reinforcement," *Proceedings of the International Conference on FRP Composites in Civil Engineering*, Hong Kong, China, pp. 1067-1075.

Odagiri, T.; Matsumoto, K.; and Nakai, H., 1997, "Fatigue and Relaxation Characteristics of Continuous Aramid Fiber Reinforced Plastic Rods," *Third International Symposium on Non-Metallic (FRP) Reinforcement for Concretes Structures (FRPRCS-3)*, V. 2, Japan Concrete Institute, Sapporo, Japan, pp. 227-234.

Oehlers, D., 2005, "Generic Debonding Mechanisms in FRP Plated Beams and Slabs," *Proceedings of Bond Behaviour of FRP in Structures (BBFS 2005)*, Hong Kong, Dec. 7-9, pp. 35-44.

Okamoto, T.; Matsubara, S.; Tanigaki, M.; and Hasuo, K., 1993, "Practical Application and Performance of PPC Beams Reinforced with Braided FRP Bars," *Fiber-Reinforced Plastic Reinforcement for Concrete Structures*, SP-138, A. Nanni and C. W. Dolan, eds., American Concrete Institute, Farmington Hills, MI, pp. 875-894.

Ospina, C. E., 2005, "Alternative Model for Concentric Punching Capacity Evaluation of Reinforced Concrete Two-Way Slabs, *Concrete International*, V. 27, No. 9, Sept., pp. 53-57.

Ospina, C. E.; Alexander, S.; and Cheng, J. J., 2001, "Behaviour of Concrete Slabs with Fibre-Reinforced Polymer Reinforcement," *Structural Engineering Report* No. 242, Department of Civil and Environmental Engineering, University of Alberta, AB, Canada, 355 pp.

Ospina, C. E.; Alexander, S. D. B.; and Cheng, J. J. R., 2003, "Punching of Two-Way Concrete Slabs with Fiber-Reinforced Polymer Reinforcing Bars or Grids," *ACI Structural Journal*, V. 100, No. 5, Sept.-Oct., pp. 589-598.

Ospina, C. E., and Gross, S. P., 2006, "Rationale for the ACI 440.1R-06 Indirect Deflection Control Design Provisions," *7th International Symposium on Fiber-Reinforced Polymer (FRP) Reinforcement for Concrete Structures*, SP-230, C. K. Shield, J. P. Busel, S. L. Walkup, and D. D. Gremel, eds., American Concrete Institute, Farmington Hills, MI, pp. 651-670.

Pando, M.; Fam, A.; Lesko, J.; and Filz, G., 2003, "New Bridge Piers Using Load Bearing Concrete-Filled GFRP Tubes," *Field Application of FRP Reinforcement: Case Studies*, SP-215, S. H. Rizkalla and A. Nanni, eds., American Concrete Institute, Farmington Hills, MI, pp. 181-200.

Pantazopoulou, S. J.; Bonacci, J. F.; Sheikh, S.; Thomas, M. D. A.; and Hearn, N., 2001, "Repair of Corrosion-Damaged Columns with FRP Wraps," *Journal of Composites for Construction*, ASCE, V. 5, No. 1, pp. 3-11.

Pantelides, C. P., and Gergely, J., 2002, "Carbon-Fiber-Reinforced Polymer Seismic Retrofit of RC Bridge Bent: Design and In Situ Validation," *Journal of Composites for Construction*, ASCE, V. 6, No. 1, pp. 52-60.

Papakonstantinou, C. G.; Petrou, M. F.; and Harries, K. A., 2001, "Fatigue of Reinforced Concrete Beams Strengthened with GFRP Sheets," *Journal of Composites in Construction*, ASCE, V. 5, No. 4, pp. 246-253.

Parent, S., and Labossière, P., 1997, "Finite Element Analysis of Reinforced Concrete Columns Confined with Composite Materials," *Paper*, University of Sherbrooke, Sherbrooke, QC, Canada, pp. 400-411.

Parvathaneni, H. K.; Iyer, S., L.; and Greenwood, M., 1996, "Design and Construction of Test Mooring Pile Using Superprestressing," *Proceedings of Advanced Composite Materials in Bridges and Structures*, Montreal, QC, Canada, pp. 313-324.

Pethrick, R. A., 2000, "Bond Inspection in Composite Structures," *Comprehensive Composite Materials, V. 5: Test Methods, Nondestructive Evaluation, and Smart Materials*, L. Carlsson, R. L., Crane, and K. Uchino, eds., Pergamon Press, Elsevier Science Ltd., Oxford, pp. 359-392.

Picard, A.; Massicotte, B.; and Boucher, E., 1995, "Strengthening of Reinforced Concrete Beams with Composite Materials: Theoretical Study," *Composite Structures*, V. 33, pp. 63-75.

Picher, F.; Rochette, P.; and Labossière, P., 1996, "Confinement of Concrete Cylinders with CFRP," *Proceedings of the First International Conference on Composites in Infrastructure (ICCI'96)*, Tucson, AZ, Jan., pp. 829-841.

Plevris, N., and Triantafillou, T. C., 1994, "Time-Dependent Behavior of RC Members Strengthened with FRP

Laminates," *Journal of Structural Engineering*, ASCE, V. 120, No. 3, Mar., pp. 1016-1042.

Porter, M. L., 1999, "Durability Issues for FRP Reinforcement to Concrete Structures," *44th International SAMPE Symposium and Exhibition*, V. 44, Book 2, Society for the Advancement of Material in Process Engineering, Corina, CA, pp. 2246-2254.

Porter, M. L., and Barnes, B. A., 1991, "Tensile Testing of Glass Fiber Composite Rods," *Proceedings of the Conference of Advanced Composites in Civil Engineering Structures*, Las Vegas, NV, ASCE, New York, Feb., pp. 123-131.

Porter, M. L., and Barnes, B. A., 1998, "Accelerated Aging Degradation of Glass Fiber Composites," *2nd International Conference on Composites in Infrastructure*, V. 2, H. Saadatmanesh and M. R. Eshanni, eds., University of Arizona, Tucson, AZ, pp. 446-459.

Porter, M. L., and Harries, K. A., 2005, "Workshop on Research in FRP Composites in Concrete Construction," *Final Report*, submitted to the National Science Foundation, 35 pp.

Post-Tensioning Institute (PTI), 1990, *Post-Tensioning Manual*, 5th Edition, Post-Tensioning Institute, Phoenix, AZ, 406 pp.

Prota, A.; Tan, K. Y.; Nanni, A.; Pecce, M.; and Manfredi, G., 2006a, "Performance of Shallow Reinforced Concrete Beams with Externally Bonded Steel-Reinforced Polymers," *ACI Structural Journal*, V. 103, No. 2, Mar.-Apr., pp. 163-170.

Prota, A.; Manfredi, G.; and Cosenza, E., 2006b, "Ultimate Behavior of Axially Loaded RC Wall-Like Columns Confined with GFRP," *Composites Part B: Engineering*, V. 37, pp. 670-678.

Pultrall Inc., 2005, www.pultrall.com (accessed May 2005)

Quantrill, R. J.; Hollaway, L. C.; and Thorne, A. M., 1996, "Experimental and Analytical Investigation of FRP Strengthened Beam Response: Part I," *Magazine of Concrete Research*, V. 48, No. 177, Dec., pp. 331-342.

Quattlebaum, J.; Harries, K. A.; and Petrou, M. F., 2005, "Comparison of Three CFRP Flexural Retrofit Systems under Monotonic and Fatigue Loads," *Journal of Bridge Engineering*, ASCE, V. 10, No. 6, pp. 731-740.

Rahman, A. H.; Kingsley, C. Y.; and Kobayashi, K., 2000, "Service and Ultimate Load Behavior of Bridge Deck Reinforced with Carbon FRP Grid," *Journal of Composites for Construction*, pp. 16-23.

Rasheed, H. A.; Nayal, R.; and Melhem, H. G., 2004, "Response Prediction of Concrete Beams Reinforced with FRP Bars," *Composite Structures*, V. 65, pp. 193-204.

Rasheed, H. A., and Pervaiz S., 2003, "Closed Form Design Equations for Flexural Strengthening of RC Beams," *Composites Part B: Engineering*, V. 34, No. 6, Sept., pp. 539-550.

Razaqpur, A. G., and Mostofinejad, D., 1999, "Experimental Study of Shear Behavior of Continuous Beams Reinforced with Carbon Fiber Reinforced Polymer," *Fiber Reinforced Polymer Reinforcement for Reinforced Concrete Structures*, SP-188, C. W. Dolan, S. H. Rizkalla, and A. Nanni, eds., American Concrete Institute, Farmington Hills, MI, pp. 169-178.

Razaqpur, A. G.; Svecova, D.; and Cheung, M. S., 2000, "Rational Method for Calculating Deflection of Fiber-Reinforced Polymer Reinforced Beams," *ACI Structural Journal*, V. 97, No. 1, Jan.-Feb., pp. 175-184.

Reda Taha, M. M., and Shrive, N. C., 2003a, "New Concrete Anchors for Carbon Fiber-Reinforced Post-Tensioned Tendons—Part 1: State-of-the-Art Review/ Design," *ACI Structural Journal*, V. 100, No. 1, Jan.-Feb., pp. 86-95.

Reda Taha, M. M., and Shrive, N. C., 2003b, "New Concrete Anchors for Carbon Reinforced-Reinforced Post-Tensioned Tendons—Part 2: Development/Experimental Investigation," *ACI Structural Journal*, V. 100, No. 1, Jan.-Feb., pp. 96-104.

Rizkalla, S., and Mufti, A., 2001, *Design Manual 3—Reinforcing Concrete Structures with Fibre Reinforced Polymers*, ISIS Canada, Winnipeg, MB, Canada, 158 pp.

Rizzo, A.; Galati, N.; and Nanni, A., 2004, "Design and In-Situ Load Testing of Bridge No. 1330005 Route 3560—Phelps County, MO," *CIES Report* 04-055-a, University of Missouri-Rolla, Rolla, MO, 116 pp.

Roberts, T. M., 1989, "Approximate Analysis of Shear and Normal Stress Concentrations in the Adhesive Layer of Plated RC Beams," *The Structural Engineer*, V. 67, No. 12, pp. 229-233.

Roko, K.; Boothby, T. E.; and Bakis, C. E., 1999, "Failure Modes of Sheet Bonded Fiber Reinforced Polymer Applied to Brick Masonry," *Fourth International Symposium on Fiber Reinforced Polymer (FRP) for Reinforced Concrete Structures*, Baltimore, MD, Nov., pp. 305-311.

Rosenboom, O.; Hassan, T.; Mirmiran, A.; and Rizkalla, S., 2004, "Static and Fatigue Performance of 40-Year-Old Prestressed Concrete Girders Strengthened with Various CFRP Systems," *International Conference Future Vision and Challenges for Urban Development*, Cairo, Egypt, Dec. 20-22.

Rostasy, F. S., 1997, "Durability of FRP in Aggressive Environments," *Non-Metallic (FRP) Reinforcement for Concrete Structures, FRPRCS-3*, V. 2, pp. 113-128.

Rostasy, F. S.; Neubauer, U.; and Nothnagel, R., 2004, "Strengthening of Historic Cast Iron Girders with Bonded CFRP Plates," *Proceedings of the FRP Composites in Civil Engineering—CICE 2004*, Adelaide, Australia, July 8-10, pp. 725-731.

Saadatmanesh, H., 1994, "Fiber Composites for New and Existing Structures," *ACI Structural Journal*, V. 91, No. 3, May-June, pp. 346-354.

Saadatmanesh, H.; Ehsani, M. R.; and Limin, J., 1997a, "Repair of Earthquake-Damaged RC Columns with FRP Wraps," *ACI Structural Journal*, V. 94, No. 2, Mar.-Apr., pp. 206-215.

Saadatmanesh, H.; Ehsani, M. R.; and Jin, L., 1997b, "Seismic Retrofitting of Rectangular Bridge Columns with Composite Straps," *Earthquake Spectra*, V. 13, No. 2, pp. 281-304.

Saadatmanesh, H., and Malek, A. M., 1998, "Design Guidelines for Flexural Strengthening of RC Beams with

FRP Plates," *Journal of Composites for Construction*, V. 2, No. 4, pp. 158-164.

Saafi, M., and Romine, P., 2002, "Effect of Fire on Concrete Cylinders Confined with GFRP," *Proceedings, 2nd International Conference on Durability of Fibre Reinforced Polymer (FRP) Composites for Construction*, Advantage Inc., Sherbrooke, QC, Canada, pp. 512-521.

Saafi, M., 2002, "Effect of Fire on FRP Reinforced Concrete Members," *Composite Structures*, V. 58, pp. 11-20.

Saafi, M.; Toutanji, H.; and Li, Z., 1999, "Behavior of Concrete Columns Confined with Fiber-Reinforced Polymer Tubes," *ACI Materials Journal*, V. 96, No. 4, July-Aug., pp. 500-509.

Saatcioglu, M., and Grira, M., 2001, "Seismic Performance and Design of Concrete Columns Confined with CFRP Grids," *Proceedings, International Conference on FRP Composites in Civil Engineering*, V. 2, Elsevier Science, New York, pp. 1227-1234.

Saeki, N.; Horiguchi, T.; and Hata, S., 1995, "Plastic Performance of Reinforced Beams by Using External Prestressing of Aramid Fiber Cables," *Proceedings of the Second International RILEM Symposium (FRPRCS-2), RILEM Proceedings 29 Non-Metallic (FRP) Reinforcement for Concrete Structures*, Ghent, Belgium, L. Taerwe, ed., Aug., pp. 608-615.

Saeki, N.; Horiguchi, T.; Inomata, M.; Hata, S.; and Ikeda, T., 1993, "Strengthening of Damaged Concrete Beams by External Prestressing of Aramid Fiber Cable," *International Symposium on Fiber-Reinforced-Polymer Reinforcement for Concrete Structures*, SP-138, A. Nanni and C. W. Dolan, eds., American Concrete Institute, Farmington Hills, MI, pp. 913-932.

Sakashita, M., 1997, "Deflection of Continuous Fiber Reinforced Concrete Beams Subjected to Loaded Heating," *Proceedings, Non-Metallic (FRP) Reinforcement for Concrete Structures*, V. 2, Japan Concrete Institute, pp. 51-58.

Samaan, M.; Miamian, A.; and Seahawk, M., 1998, "Model of Concrete Confined by Fiber Composites." *Journal of Structural Engineering*, ASCE, V. 124, No. 9, pp. 1025-1031.

Santoh, N., 1993, "CFCC (Carbon Fiber Composite Cable)," *Fiber-Reinforced-Plastic (FRP) Reinforcement for Concrete Structures: Properties and Applications*, V. 42, A. Nanni, ed., Developments in Civil Engineering, Elsevier Science Publishers, pp. 223-248.

Satoh, K.; Kodama, K.; and Ohki, H., 1991, "A Study on the Bending Behavior of Repaired Reinforced Concrete Beams Using Fiber Reinforced Plastic (FRP) and Polymer Mortar," *Evaluation and Rehabilitation of Concrete Structures and Innovations in Design*, SP-128, V. M. Malthotra, ed., American Concrete Institute, Farmington Hills, MI, pp. 1017-1031.

Sause, R.; Harries, K. A.; Walkup, S. L.; Pessiki, S. P.; and Ricles, J. M., 2004, "Flexural Behavior of Concrete Columns Retrofitted with Carbon Fiber-Reinforced Polymer Jackets," *ACI Structural Journal*, V. 101, No. 5, Sept.-Oct., pp. 708-716.

Sayed-Ahmed, E. Y., and Shrive, N. G., 1998, "A New Steel Anchorage System for Post-Tensioning Applications using Carbon Fibre Reinforced Plastic Tendon," *Canadian Journal of Civil Engineering*, V. 25, No. 1, Feb., pp. 113-127.

Schwegler, G., 1995, "Masonry Construction Strengthened with Fiber Composites in Seismically Endangered Zones," *Proceedings of the Tenth European Conference on Earthquake Engineering*, Rotterdam, Netherlands, pp. 2299-2303.

Seible, F., 1996, "Advanced Composites Materials for Bridges in the 21st Century," *Proceedings of Advanced Composite Materials in Bridges and Structures*, Montreal, pp. 17-30.

Seible, F., 2001, "Seismic Retrofit of RC Columns with Continuous Carbon Fiber Jackets," *Proceedings International Conference FRP Composites in Civil Engineering*, V. 1, Elsevier Science, New York, pp. 817-824.

Seible, F.; Burgueño, R.; Abdallah, M. G.; and Nuismer, R., 1995, "Advanced Composite Carbon Shell System for Bridge Columns Under Seismic Loads," *Proceedings from the National Seismic Conference on Bridges and Highways*, San Diego, CA, 16 pp. (CD-ROM)

Seible, F.; Innamorato, D.; Baumgartner, J.; Karbhari, V.; and Sheng, L. H., 1999a, "Seismic Retrofit of Flexural Bridge Spandrel Columns Using Fiber Reinforced Polymer Composite Jackets," *Fiber Reinforced Polymer Reinforcement for Reinforced Concrete Structures*, SP-188, C. W. Dolan, S. H. Rizkalla, and A. Nanni, eds., American Concrete Institute, Farmington Hills, MI, pp. 919-931.

Seible, F.; Karbhari, V. M.; and Burgueño, R., 1999b, "Kings Stormwater Channel and I-5/Gilman Bridges," *Structural Engineering International: Journal of the International Association for Bridge and Structural Engineering*, V. 9, No. 4, Nov., pp. 250-253.

Seible, F.; Karbhari, V.; Burgueño, R.; and Seaberg, E., 1998, "Modular Advanced Composite Bridge System for Short and Medium Span Bridges," *Proceedings of Developments in Short and Medium Span Bridge Engineering*, Calgary, AB, Canada, pp. 431-441.

Seible, F.; Karbhari, V.; Burgueño, R.; Davol, A.; and Wernli M., 1997a, "Design Issues with Carbon Fiber Composite Reinforced Concrete Structures," *Proceedings of the Third International Symposium, Non-Metallic (FRP) Reinforcement for Concrete Structures*, Japan, V. 2, Oct. pp. 89-112.

Seible, F.; Priestley, M. J. N.; Hegemier, G. A.; and Innamorato, D., 1997b, "Seismic Retrofit of RC Columns with Continuous Carbon Fiber Jackets," *Journal of Composites for Construction*, ASCE, V. 1, No. 2, pp. 52-62.

Seki, H.; Sekijima, K.; and Konno, T., 1997, "Test Method on Creep of Continuous Fiber Reinforcing Materials," *Proceedings of the Third International Symposium on Non-Metallic (FRP) Reinforcement for Concrete Structures (FRPRCS-3)*, Sapporo, Japan, V. 2, pp. 195-202.

Sen, R., 2003, "Advances in the Application of FRP for Repairing Corrosion Damage," *Progress in Structural Engineering Materials*, V. 5, pp. 99-113.

Sen, R.; Shahawy, M.; Rosas, J.; and Sukumar, S., 1999, "Durability of Aramid Fiber Reinforced Plastic Pretensioned Elements Under Tidal/Thermal Cycles," *ACI Structural Journal*, V. 96, No. 1, Jan.-Feb., pp. 95-104.

Sen, R.; Shahawy, M.; Sukumar, S.; and Rosas, J., 1998, "Durability of Carbon Pretensioned Elements in a Marine Environment," *ACI Structural Journal*, V. 95, No. 6, Nov.-Dec., pp. 716-724.

Sena Cruz, J. M., and Barros, J. A. O., 2004, "Bond Between Near-Surface Mounted Carbon-Fiber-Reinforced Polymer Laminate Strips and Concrete," *Journal of Composites for Construction*, Nov.-Dec., pp. 519-527.

Shah, A. P.; Gupta, R. K.; GangaRao, H. V. S.; and Powell, C. E., 2002, "Moisture Diffusion Through Vinyl-ester Nanocomposites Made with Montmorillonite Clay," *Polymer Engineering & Science*, V. 42, No. 9, pp. 1852-1863.

Shahawy, M., and Beitelman, E. T., 1999, "Static and Fatigue Performance of RC Beams Strengthened with CFRP Laminates," *Journal of Structural Engineering*, ASCE, V. 125, No. 6, pp. 613-621.

Shaheen, E., 2004, "Carbon Fiber Reinforced Reactive Powder Concrete Anchorage System," PhD thesis, University of Calgary, Calgary, AB, Canada, 292 pp.

Shahrooz, B. M.; Boy, S.; and Baseheart, T. M., 2002, "Flexural Testing of Four 76-Year-Old T-Beams with Various Fiber-Reinforced Polymer Systems: Testing and Analysis," *ACI Structural Journal*, V. 99, No. 5, Sept.-Oct., pp. 681-691.

Shao, Y., 2003, "Behavior of FRP-Concrete Beam-Columns under Cyclic Loading," PhD thesis, North Carolina State University, Raleigh, NC, 237 pp.

Sharif, A.; Al-Sulaimani, G. J.; Basunbul, I. A.; Baluch, M. H.; and Ghaleb, B. N., 1994, "Strengthening of Initially Loaded Reinforced Concrete Beams Using FRP Plates," *ACI Structural Journal*, V. 91, No. 2, Mar.-Apr., pp. 160-168.

Sheard, P. A.; Clarke, J. L.; Dill, M. J. U.; Hammersley, G. P.; and Richardson, D. M., 1997, "Eurocrete—Taking Account of Durability for Design of FRP Reinforced Concrete Structures," *Proceedings of the Third International Symposium on Non-Metallic (FRP) Reinforcement for Concrete Structures*, V. 2, Japan Concrete Institute, Oct., pp. 75-82.

Shehata, E., 1999, "Fibre-Reinforced Polymer (FRP) for Shear Reinforcement in Concrete Structures," PhD thesis, University of Manitoba, Winnipeg, MB, Canada, 412 pp.

Sheikh, S. A., 2001, "Use of FRP Composites for Seismic Upgrade of Concrete Columns," *Proceedings, International Conference on FRP Composites in Civil Engineering*, V. 1, Elsevier Science, New York, pp. 901-908.

Shen, X.; Myers, J.; Maerz, N.; and Galicki, G., 2002, "Effect of Surface Roughness on the Bond Performance Between FRP Laminates and Concrete," *Proceedings of the Second International Conference on Durability of Fibre Reinforced Polymers for Construction*, Montreal, QC, Canada, pp. 607-616.

Sherwood, E. G., and Soudki, K. A., 1999, "Confinement of Corrosion Cracking in Reinforced Concrete Beams with CFRP Laminates," *Fiber Reinforced Polymer Reinforcement for Reinforced Concrete Structures*, SP-188, C. W. Dolan, S. H. Rizkalla, and A. Nanni, eds., American Concrete Institute, Farmington Hills, MI, pp. 591-603.

Shield, C.; French, C.; and Hanus, J., 1999, "Bond of Glass Fiber Reinforced Plastic Reinforcing Bar for Consideration in Bridge Decks," *Fiber Reinforced Polymer Reinforcement for Reinforced Concrete Structures*, SP-188, C. W. Dolan, S. H. Rizkalla, and A. Nanni, eds., American Concrete Institute, Farmington Hills, MI, pp. 393-406.

Sika Corp., 2007, http://www.sika.com (accessed May 2007)

Sireg, 2007, http://www.sireg.net (accessed Feb. 2007)

Sonobe, Y.; Fukuyama, H.; Okamoto, T.; Kani, N.; Kimura, K.; Kobatashi, K.; Masuda, Y.; Matsuzaki, Y.; Mochizuki, S.; Nagasaka, T.; Shimizu, A.; Tanano, H.; Tanigaki, M.; and Teshigawara, M., 1997, "Design Guidelines of FRP Reinforced Concrete Building Structures," *Journal of Composites for Construction*, ASCE, V. 1, No. 3, pp. 90-115.

Sorathia, U., 2004, "Chapter 9: Materials in Military Applications," *Handbook of Building Materials for Fire Protection*, McGraw-Hill, 76 pp.

Sorathia, U.; Dapp, T.; and Beck, C., 1992, "Fire Performance of Composites," *Materials Engineering*, V. 109, No. 9, pp. 10-12.

Sorathia, U.; Ohlemiller, T.; Lyon, R.; Riffle, J.; and Schultz, N., 2001, "Chapter 9: Effects of Fire," *Gap Analysis for Durability of Fibre Reinforced Polymer Composites in Civil Infrastructure*, American Society of Civil Engineers, New York, pp. 100-121.

Soudki, K. A., and Green, M. F., 1996, "Performance of CFRP Retrofitted Concrete Columns at Low Temperatures," *2nd International Conference on Advanced Composite Materials in Bridges and Structures*, Montreal, QC, Canada, Aug. 11-14, pp. 427-434.

Soudki, K. A.; Green, M. F.; and Clapp, F., 1997, "Transfer Length of Carbon Fiber Rods in Precast Pretensioned Concrete Beams," *PCI Journal*, V. 42, No. 4, Sept.-Oct., pp. 78-87.

Soudki, K. A., and Sherwood, E. G., 2003, "Bond Behaviour of Corroded Steel Reinforcement in Concrete Wrapped with CFRP Sheets," *Journal of Materials in Civil Engineering*, ASCE, V. 15, No. 4, Aug., pp. 358-370.

Spadea, G.; Bencardino, F.; and Swamy, R. N., 1998, "Structural Behavior of Composite RC Beams with Externally Bonded CFRP," *Journal of Composites for Construction*, ASCE, V. 2, No. 3, pp. 132-137.

Spadea, G.; Swamy, R. N.; and Bencardino, F., 2001, "Strength and Ductility of RC Beams Repaired with Bonded CFRP Laminates," *Journal of Bridge Engineering*, ASCE, V. 6, No. 5, pp. 349-355.

Spoelstra, M., and Monti, G., 1999, "FRP-Confined Concrete Model," *Journal of Composites for Construction*, ASCE, V. 3, No. 3, Aug., pp. 143-150.

Steiner, W., 1996, "Strengthening of Structures with CFRP Strips," *Second International Conference on Advanced Composite Materials in Bridges and Structures, ACMBS II*, M. M. El-Badry, ed., Montréal, QC, Canada, Aug. 11-14, pp. 407-417.

Stoecklin, I., and Meier, U., 2003, "Strengthening of Concrete Structures with Prestressed and Gradually-Anchored C-FRP Strips," *Proceedings of the Sixth International Conference on FRP Reinforcement for Concrete Structures*, World Scientific, Singapore, V. 2, pp. 1321-1330.

Swamy, R. N., and Mukhopadhyaya, P., 1995, "Role and Effectiveness of Non-Metallic Pates in Strengthening and Upgrading of Structures," *Non-Metallic (FRP) Reinforcement for Concrete Structures*, L. Taerwe, ed., RILEM, pp. 473-482.

Swamy, R. N.; Jones, R.; and Bloxham, J. W., 1987, "Structural Behavior of Reinforced Concrete Beams Strengthened by Epoxy-Bonded Steel Plates," *The Structural Engineer (London)*, V. 65A, No. 2, Feb., pp. 59-68.

Swamy, R. N.; Jones, R.; and Charif, A., 1996a, "Contribution of Externally Bonded Steel Plate Reinforcement to the Shear Resistance of Reinforced Concrete Beams," *Repair and Strengthening of Concrete Members with Adhesive Bonded Plates*, SP-165, R. N. Swamy and R. Gaul, eds., American Concrete Institute, Farmington Hills, MI, pp. 1-24.

Swamy, R. N.; Lynsdale, C. J.; and Mukhopadhaya, P., 1996b, "Effective Strengthening with Ductility: Use of Externally Bonded Plates of Non-Metallic Composite Materials," *Second International Conference on Advanced Composite Materials in Bridges and Structures, ACMBS II,* M. M. El-Badry, ed., Montréal, QC, Canada, Aug. 11-14, pp. 481-488.

Tadros, G.; Trompech, E.; and Mufti, A. A., 1998, "University Drive/Crowchild Trail Bridge Superstructure Replacement," *Proceedings, Second International Conference on Composites in Infrastructure*, Tucson, AZ, V. 1, pp. 693-704.

Taerwe, L., and Pallemans, I., 1995, "Force Transfer of AFRP Bars in Concrete Prisms," *Proceedings of the 2nd International RILEM Symposium*, Ghent, Belgium, pp. 154-163.

Takewaka, K., and Khin, M., 1996, "Deterioration of Stress-Rupture of FRP Rods in Alkaline Solution Simulating as Concrete Environment," *Advanced Composite Materials in Bridges and Structures*, M. M. El-Badry, ed., Canadian Society for Civil Engineering, Montreal, QC, Canada, pp. 649-664.

Täljsten, B., and Carolin, A., 2001, "Concrete Beams Strengthened with Near Surface Mounted CFRP Laminates," *Proceedings of the 5th International Conference on Fibre-Reinforced Plastics for Reinforced Concrete Structures (FRPRCS-5)*, V. 1, Cambridge, UK, pp. 107-116.

Tamužs, V.; Apinis, R.; Vilks, U.; Modniks, J.; Tepfers, R., 1996, "Ductility of Hybrid Fiber Composite Reinforcement FRP for Concrete," *First International Conference on Composites in Infrastructure, ICCI '96,* Tucson, AZ, Jan. 15-17, pp. 109-122.

Tamužs, V.; Maksimovs, R.; and Modniks, J., 2001, "Long-Term Creep of Hybrid FRP Bars," *Fifth International Conference on Non-Metallic Reinforcement for Concrete Structures—FRPRCS-5*, T. Telford, London, July 16-18, pp. 527-534.

Tan, K. H., and Mathivoli, M., 1999, "Behavior of Preloaded Reinforced Concrete Beams Strengthened with Carbon Fiber Sheets," *Proceedings of the Fourth International Symposium on Fiber Reinforced Polymer Reinforcement for Reinforced Concrete Structures (FRPRCS-4)*, Baltimore, MD, pp. 159-170.

Tao, S., 1994, "Bond of Glass-Fiber-Reinforced-Plastic Reinforcing Bars to Concrete," PhD thesis, Department of Civil Engineering, University of Arizona, Tucson, AZ, 197 pp.

TechFab LLC, 2006, *TechFab Press Releases* (www.techfabllc.com/News.htm), TechFab LLC, Anderson, SC (accessed 10 July 2006)

Teng, J. G.; Chen, J. F.; Smith, S. T.; and Lam, L., 2002, *FRP-Strengthened RC Structures*, John Wiley and Sons, 245 pp.

Teng, J. G.; Lam, L.; Chan, W.; and Wang, J., 1999, "Strengthening of Cantilever Concrete Slabs Using GFRP Strips," *Proceedings of the International Conference on Extending Performance of Concrete Structures*, R. K. Dhir and P. A. J. Tittle, eds., Dundee, Scotland, Sept. 7, pp. 189-200.

Teng, M. H.; Sotelino, E. D.; and Chen, W. F., 2003, "Performance Evaluation of Reinforced Concrete Bridge Columns Wrapped with Fiber Reinforced Polymers," *Journal of Composites for Construction*, V. 7, No. 2, pp. 83-92.

Tepfers, R., and Olsson, P. Å., 1992, "Ring Test for Evaluation of Bond Properties of Reinforcing Bars," *International Conference: Bond in Concrete—From Research to Practice*, Riga, Latvia, Oct., pp. 1-99.

Terrasi, G. P., and Lees, J. M., 2003, "CFRP Prestressed Concrete Lighting Columns," *Field Applications of FRP Reinforcement: Case Studies*, SP-215, S. H. Rizkalla and A. Nanni, eds., American Concrete Institute, Farmington Hills, MI, pp. 55-74.

Theriault, M., and Benmokrane, B., 1998, "Effects of FRP Reinforcement Ratio and Concrete Strength on Flexural Behavior of Concrete Beams," *Journal of Composites in Construction*, ASCE, V. 2, No. 1, pp. 7-16.

Thiagarajan G., 2003, "Experimental and Analytical Behavior of Carbon Fiber-Based Rods as Flexural Reinforcement," *Journal of Composites for Construction*, ASCE, V. 7, No. 1, pp. 64-72.

Thomas, R. L., 2000, "Thermal Methods Used in Composite Inspection," *Comprehensive Composite Materials, V. 5: Test Methods, Nondestructive Evaluation, and Smart Materials*, L. Carlsson, R. L. Crane, K. Uchino, A. Kelly, and C. Zweben, eds., Pergamon Press, Elsevier Science Ltd., Oxford, pp. 427-446.

Thompson, R., and Iyer, S. L., 2001, "FRP Deck System Used for the Columbus Bridge in Ohio," *Proceedings, 33rd International SAMPE Technical Conference*, Seattle, WA, 8 pp.

Tighiouart, B.; Benmokrane, B.; and Mukhopadhyaya, P., 1999, "Bond Strength of Glass FRP Rebar Splices in Beams Under Static Loading," *Construction and Building Materials*, V. 13, No. 7, pp. 383-392.

Tittmann, B. R., and Crane, R. L., 2000, "Ultrasonic Inspection of Composites," *Comprehensive Composite Materials, V. 5: Test Methods, Nondestructive Evaluation, and Smart Materials*, L. Carlsson, R. L. Crane, K. Uchino, A. Kelly and C. Zweben, eds., Pergamon Press, Elsevier Science Ltd., Oxford, pp. 259-320.

Toutanji, H., and Balaguru, P., 1998, "Durability Characteristics of Concrete Columns Wrapped with FRP Tow Sheets," *Journal of Materials in Civil Engineering*, ASCE, V. 10, No. 1, pp. 52-57.

Toutanji, H., and Saafi, M., 1999, "Deflection and Crack Width Predictions of Concrete Beams Reinforced with Fiber Reinforced Polymer Bars," *Fiber Reinforced Polymer Reinforcement for Reinforced Concrete Structures*, SP-188,

C. W. Dolan, S. H. Rizkalla, and A. Nanni, eds., American Concrete Institute, Farmington Hills, MI, pp. 1023-1034.

Toutanji, H., 1999, "Stress-Strain Characteristics of Concrete Columns Externally Confined with Advanced Fiber Composite Sheets," *ACI Materials Journal*, V. 96, No. 3, May-June, pp. 397-404.

Toutanji, H.; Deng, Y.; Zhang, Y.; and Balaguru, P., 2001, "The Use of Inorganic Matrix for Strengthening of RC Beams with Carbon Sheets," *Proceedings: CICE 2001—International Conference on FRP Composites in Civil Engineering*, Hong Kong, pp. 1117-1126.

Toutanji H., and Saafi M., 2000, "Flexural Behavior of Concrete Beams Reinforced with Glass Fiber Reinforced Polymer," *ACI Structural Journal*, V. 97, No. 5, Sept.-Oct., pp. 712-719.

Toutanji, H.; Zhao, L.; Deng, Y.; Zhang, Y.; and Balaguru, P., 2006, "Cyclic Behavior of RC Beams Strengthened with Carbon-Fiber Sheets Bonded with an Inorganic Matrix," *Journal of Materials in Civil Engineering*, ASCE, V. 18, No. 1, Jan.-Feb., pp. 28-35.

Triantafillou, T. C., 1998a, "Shear Strengthening of Reinforced Concrete Beams Using Epoxy-Bonded FRP Composites," *ACI Structural Journal*, V. 95, No. 2, Mar.-Apr., pp. 107-115.

Triantafillou, T. C., 1998b, "Strengthening of Masonry Structures Using Epoxy-Bonded FRP Laminates," *Journal of Composites for Construction*, ASCE, V. 2, No. 2, pp. 96-104.

Triantafillou, T. C., and Antonopoulos, C. P., 2000, "Design of Concrete Flexural Members Strengthened in Shear with FRP," *Journal of Composites for Construction*, ASCE, V. 4, No. 4, pp. 198-205.

Triantafillou, T. C., and Deskovic, N., 1991, "Innovative Prestressing with FRP Sheets: Mechanics of Short-Term Behavior," *Journal of Engineering Mechanics*, ASCE, V. 117, No. 7, pp. 1652-1672.

Triantafillou, T. C., and Meier, U., 1992, "Innovative Design of FRP Combined with Concrete." *Proceedings: Advanced Composite Materials in Bridges and Structures*, CSCE, K.W. Neale and P. Labossière, eds., pp. 491-500.

Triantafillou, T. C., and Plevris, N., 1990, "Flexural Behavior of Concrete Structures Strengthened with Epoxy-Bonded Fiber Reinforced Plastics," *International Seminar on Structural Repairs/Strengthening by the Plate Bonding Technique*, University of Sheffield, England, Sept., pp. 152-161.

Triantafillou, T. C., and Plevris, N., 1992, "Strengthening of RC Beams with Epoxy-Bonded Fiber-Composite Materials," *Materials and Structures Journal*, V. 25, May, pp. 201-211.

Triantafillou, T. C., and Plevris, N., 1995, "Reliability Analysis of Reinforced Concrete Beams Strengthened with CFRP Laminates," *Proceedings of the Second International RILEM Symposium (FRPRCS-2), RILEM Proceedings 29 Non-Metallic (FRP) Reinforcement for Concrete Structures*, L. Taerwe, ed., Ghent, Belgium, pp. 576-583.

Tsuchida, K.; Mutsuyoshi, H.; Taniguchi, H.; and Kita, T., 1993, "PC Beams Prestressed with Aramid Rope as External Cables," *Transactions of the Japan Concrete Institute*, V. 15, pp. 339-410.

Tumialan, J. G.; Galati, N.; Namboorimadathil, S. M.; and Nanni, A., 2002, "Strengthening of Masonry with FRP Bars," *Third International Conference on Composites in Infrastructure (ICCI 2002)*, San Francisco, CA, June 10-12, 11 pp. (CD-ROM)

Tumialan, J. G.; Galati, N.; and Nanni, A., 2003a, "Field Assessment of URM Walls Strengthened with FRP Laminates," *Journal of Structural Engineering*, ASCE, V. 129, No. 8, Sept., pp. 1047-1056.

Tumialan, J. G.; Galati, N.; and Nanni, A., 2003b, "Fiber-Reinforced Polymer Strengthening of Unreinforced Masonry Walls Subject to Out-of-Plane Loads," *ACI Structural Journal*, V. 100, No. 3, May-June, pp. 321-329.

Tumialan, J. G.; Morbin, A.; Nanni, A.; and Modena, C., 2001, "Shear Strengthening of Masonry Walls with FRP Composites," *COMPOSITES 2001 Convention and Trade Show, Composites Fabricators Association*, Tampa, FL, Oct. 3-6, 6 pp. (CD-ROM)

Tumialan, J. G.; San Bartolome, A.; and Nanni, A., 2003c, "Strengthening of URM Infill Walls by FRP Structural Repointing," *9th North American Masonry Conference*, Clemson, SC, June 1-4, 12 pp. (CD-ROM)

Tumialan, J. G.; Torealva, D.; Pasquel, E.; and Nanni, A., 2003d, "Reinforcing of Cathedral Tower with CFRP Laminates," *Field Applications of FRP Reinforcement: Case Studies*, SP-215, S. Rizkalla and A. Nanni, eds., American Concrete Institute, Farmington Hills, MI, pp. 401-412.

Tung, D. P., and Campbell, T. I., 2002, "Ductility of Pretensioned Concrete Beams with Hybrid FRP/Stainless Steel Reinforcements," *4th Structural Specialty Conference*, CSCE, Montreal, QC, Canada, 10 pp. (CD-ROM)

Turco, V.; Galati, N.; Tumialan, J. G.; and Nanni, A., 2003, "Flexural Strengthening of URM Walls with FRP Systems," *Proceedings of the International Symposium FRPRCS-6*, Singapore, V. 2, July, pp. 1219-1228.

Tureyen, A. K., and Frosch, R. J., 2003, "Concrete Shear Strength: Another Perspective," *ACI Structural Journal*, V. 100, No. 5, Sept.-Oct., pp. 609-615.

Ueda, T., and Dai, J., 2005, "Interface Bond between FRP Sheets and Concrete Substrates: Properties, Numerical Modelling and Roles in Member Behaviour," *Progress in Structural Engineering Materials*, V. 7, pp. 27-43.

UFGS, 2004a, "Rehabilitation of Reinforced and Unreinforced Masonry Walls Using Surface Applied FRP Composites," *Department of Defense Unified Facility Guide Specifications* (UFGS)-UFGS 04920, NIBS, Washington, DC, June, 29 pp.

UFGS, 2004b, "Rehabilitation of Reinforced and Unreinforced Masonry Walls Using FRP Composite Structural Repointing," *Department of Defense Unified Facility Guide Specifications* (UFGS)-UFGS 04910, NIBS, Washington, DC, June, 24 pp.

Ulloa, F.; Ziehl, P.; Medlock, R.; and Fowler, T., 2004, "Hybrid Bridges in Texas," *Concrete International*, V. 26, No. 5, May, pp. 38-43.

U.S. Department of Defense, 2002, *Composite Materials Handbook*, MIL-HDBK-17-1F, June, 586 pp.

Valluzzi, M. R.; Tinazzi, D.; and Modena, C., 2002, "Shear Behavior of Masonry Panels Strengthened by FRP Laminates," *Construction and Building Materials Journal*, Elsevier, V. 16, No. 7, Oct., pp. 409-416.

Van Gemert, D., 1999, "Extending Structural Performance of Concrete Constructions," *Proceedings Extending of the International Conference on Performance of Concrete Structures*, R. K. Dhir and P. A. J. Tittle, eds., Dundee, Scotland, Sept. 7, pp. 1-16.

Velazquez-Dimas, J.; Ehsani, M.; and Saadatmanesh, H., 2000, "Out-of-Plane Behavior of Brick Masonry Walls Strengthened with Fiber Composites," *ACI Structural Journal*, V. 97, No. 3, May-June, pp. 377-387.

Walton, S., and Bradberry, T., 2005, "Feasibility of a Concrete Pavement Continuously Reinforced by Glass Fiber Reinforced Polymer Bars," *Proceedings of 3rd International Conference on Construction Materials: Performance, Innovations and Structural Implications (ConMat05)*, Vancouver, BC, Canada, 10 pp. (CD-ROM)

Wambeke, B., and Shield, C., 2004, "Development Length of Glass Fiber Reinforcing Polymer Bars in Concrete," *ACI Structural Journal*, V. 103, No. 1, Jan.-Feb., pp. 11-17.

Wan, B., 2002, "Study of the Bond Between FRP Composites and Concrete," PhD thesis, Department of Civil and Environmental Engineering, University of South Carolina, Columbia, SC, 205 pp.

Wan, B.; Harries, K. A.; Petrou, M. F.; Sutton, M. A.; and Li, N., 2003, "Experimental Investigation of Bond between FRP and Concrete," *Proceedings of 2003 SEM Annual Conference*, Charlotte, NC, 6 pp. (CD-ROM)

Wan, B.; Petrou, M. F.; Harries, K. A.; Sutton, M. A.; and Yang, B., 2002, "Experimental Investigation of Bond Between FRP and Concrete," *Proceedings of Third International Conference on Composites in Infrastructure*, San Francisco, CA, June, 10 pp. (CD-ROM)

Wan, B.; Sutton, M. A.; Petrou, M. F.; Harries, K. A.; and Li, N., 2004, "Investigation of Bond between Fiber Reinforced Polymer and Concrete Undergoing Global Mixed-Mode I/II Loading," *Journal of Engineering Mechanics*, V. 130, No. 12, pp. 1467-1475.

Wang, P.; Masmoudi, R.; and Benmokrane, B., 2002, "Durability of GFRP Bars: Assessment and Improvement," *Proceedings of the Second International Conference on Durability of Fibre Reinforced Polymers for Construction*, Montreal, QC, Canada, pp. 153-163.

Wernli, M., 1999, "The Anchorage of Tension Members Made of Advanced Composite Materials," PhD thesis, University of California, San Diego, La Jolla, CA, 268 pp.

Wernli, M., and Seible, F., 1998, "Ductile Connections of Modular Bridge Girders Made of Concrete Filled FRP-Tubes," *Proceeding of the First International Conference on Composites in Infrastructure ICCI'96*, H. Saadatmanesh and M. R. Ehsani, eds., Tucson, AZ., Jan., pp. 356-367.

White, T.; Soudki, K. S.; and Erki, M. A., 1998, "Loading Rate Effects on Reinforced Concrete Beams Strengthened with Carbon-Fibre Reinforced Polymer Laminates," *Proceedings of the 1998 Annual Conference of the Canadian Society for Civil Engineering*, V. IIIb, Halifax, NS, Canada, June, pp. 633-642.

Wight, R. G., 1998, "Strengthening Concrete Beams with Prestressed Fiber Reinforced Polymer Sheets," PhD thesis, Queen's University, Kingston, ON, Canada, 189 pp.

Wight, R. G. and Erki, M. A., 2001, "Prestressed CFRP Sheets for Strengthening Concrete Slabs in Fatigue," *1st International Conference on Composite in Civil Engineering (CICE' 01)*, Hong Kong, China, pp. 1093-1110.

Wight, R. G.; Green M. F.; and Erki M. A., 2001, "Strengthening Concrete Beams with Prestressed FRP Sheets," *Journal of Composites for Construction*, ASCE, V. 5, No. 4, pp. 214-220.

Williams, B. K., 2004, "Fire Performance of FRP-Strengthened Reinforced Concrete Flexural Members," PhD thesis, Department of Civil Engineering, Queen's University, Kingston, ON, Canada, 389 pp.

Williams, B. K.; Bisby, L. A.; Kodur, V. K. R.; Green, M. F.; and Chowdhury, E., 2006, "Fire Insulation Schemes for FRP-Strengthened Concrete Slabs," *Composites: Part A*, V. 37, pp. 1151-1160.

Xu, Y. W.; Qiao, P. Z.; and Davalos, J. F., 2002, "Exploratory Evaluation of Mode-I Fracture Toughness of Concrete-Composite Bonded Interfaces," *Proceedings of the 15th ASCE Engineering Mechanics Conference*, New York, pp. 1-7.

Yamaguchi, T.; Kato, Y.; Nishimura, T.; and Uomoto, T., 1997, "Creep Rupture of FRP Rods made of Aramid, Carbon, and Glass Fibers," *Non-Metallic (FRP) Reinforcement for Concrete Structure, Proceeding of the Third International Symposium*, V. 2, Japan Concrete Institute, Oct., Sapporo, Japan, pp. 179-186.

Yuan, W., and Mirmiran, A., 2001, "Buckling Analysis of Concrete-Filled FRP Tubes," *International Journal of Structural Stability and Dynamics,* World Scientific, V. 1, No. 3, pp. 367-383.

Zhao, L., 1999, "Prototyping and Characterization of Connection of Pultruded-Core Composite Bridge Decks to a Carbon Shell Girder System," PhD thesis, University of California, San Diego, La Jolla, CA, 280 pp.

Zhao, L.; Sikorsky, C. S.; Seible, F.; and Karbhari, V. M., 2001, "Field Monitoring of the Kings Stormwater Channel Advanced Composite Bridge," *6th Caltrans Seismic Research Workshop, California Department of Transportation*, Sacramento, CA, No. 7-102, June, 8 pp.

Zoch, P.; Kimura, H.; Iwasaki, T.; and Heym, M., 1991, "Carbon Fiber Composite Cables—A New Class of Prestressing Members," *Transportation Research Board*, 19 pp.

Zoltek, 2006, http://www.zoltek.com/panex_products (accessed 14 July 2006)

Guide for the Design and Construction of Structural Concrete Reinforced with Fiber-Reinforced Polymer (FRP) Bars

Reported by ACI Committee 440

Carol K. Shield, Chair	Doug. D. Gremel	Rajan Sen	Russell Gentry
William J. Gold, Secretary	Shawn P. Gross*	Rudolf Seracino	Arie Gerritse
Tarek Alkhrdaji	H. R. Trey Hamilton III	Pedro F. Silva	Srinivasa L. Iyer
Charles E. Bakis	Issam E. Harik	Khaled A. Soudki	Koichi Kishitani
Lawrence C. Bank	Kent A. Harries	Samuel A. Steere III	Howard S. Kliger
Abdeldjelil Belarbi	Mark P. Henderson	Jay Thomas	Kyuichi Maruyama
Brahim Benmokrane	Bohdan N. Horeczko	Houssam A. Toutanji	Antoine E. Naaman
Luke A. Bisby	Michael W. Lee	J. Gustavo Tumialan	Hajime Okamura
Gregg J. Blaszak	Maria Lopez de Murphy	Milan Vatovec	Mark A. Postma
Hakim Bouadi	Ibrahim M. Mahfouz	David White	Ferdinand S. Rostasy
Timothy E. Bradberry	Amir Mirmiran	Sarah E. Witt	Surendra P. Shah
Gordon L. Brown Jr.	John J. Myers		Mohsen Shahawy
Vicki L. Brown	Antonio Nanni	*Consulting Members*	Yasuhisa Sonobe
John P. Busel	Ayman M. Okeil	P. N. Balaguru	Minoru Sugita
Raafat El-Hacha	Carlos E. Ospina	Craig A. Ballinger	Luc R. Taerwe
Garth J. Fallis	Renato Paretti	Harald G. F. Budelmann	Ralejs Tepfers
Amir Z. Fam	Max L. Porter	C. J. Burgoyne	Taketo Uomoto
Nabil F. Grace	Andrea Prota	Elliot P. Douglas	Paul Zia
Mark F. Green	Hayder A. Rasheed	Rami M. Elhassan	
Zareh B. Gregorian	Sami H. Rizkalla	David M. Gale	*Subcommittee Chair.

Fiber-reinforced polymer (FRP) materials have emerged as an alternative for producing reinforcing bars for concrete structures. Fiber-reinforced polymer reinforcing bars offer advantages over steel reinforcement because they are noncorrosive. Some FRP bars are nonconductive as well. Due to other differences in the physical and mechanical behavior of FRP materials versus steel, unique guidance on the engineering and construction of concrete structures reinforced with FRP bars is necessary. Other countries and regions, such as Japan, Canada, and Europe have established design and construction guidelines specifically for the use of FRP bars as concrete reinforcement. This guide offers general information on the history and use of FRP reinforcement, a description of the unique material properties of FRP, and guidelines for the design and construction of structural concrete members reinforced with FRP bars. This guide is based on the knowledge gained from worldwide experimental research, analytical work, and field applications of FRP reinforcement.

Keywords: anchorage (structural); aramid fiber; carbon fiber; crack control; concrete construction; concrete slabs; cover; creep rupture; deflections; design examples; durability; fiber-reinforced polymer; flexural strength; glass fiber; moments; reinforced concrete; reinforcement; serviceability; shear strength; spans; strength analysis; stresses; structural concrete; structural design.

CONTENTS

ACI 440.1R-15 supersedes ACI 440.1R-06 and was adopted and published March 2015.

CHAPTER 1—INTRODUCTION AND SCOPE

1.1—Introduction

Conventional concrete structures are reinforced with nonprestressed and prestressed steel. The steel is initially protected against corrosion by the alkalinity of the concrete, usually resulting in durable and serviceable construction. For many structures subjected to aggressive environments, such as marine structures, bridges, and parking garages exposed to deicing salts, combinations of moisture, temperature, and chlorides reduce the alkalinity of the concrete and result in the corrosion of reinforcing steel. The corrosion process ultimately causes concrete deterioration and loss of serviceability.

Composite materials made of fibers embedded in a polymeric resin, also known as fiber-reinforced polymer (FRP), are an alternative to steel reinforcement for concrete structures. Fiber-reinforced polymer reinforcing materials are made of continuous aramid fiber (AFRP), carbon fiber (CFRP), or glass fiber (GFRP) embedded in a resin matrix. Examples of FRP reinforcing bars are shown in Fig. 1.1. Because FRP materials are nonmagnetic and noncorrosive, the problems of electromagnetic interference and steel corrosion can be avoided with FRP reinforcement. Additionally, FRP materials exhibit several properties, such as high tensile strength, that make them suitable for use as structural reinforcement (ACI 440R; Benmokrane and Rahman 1998; Burgoyne 2001; Cosenza et al. 2001; Dolan et al. 1999; El-Badry 1996; Figueiras et al. 2001; Humar and Razaqpur 2000; Iyer and Sen 1991; Japan Society of Civil Engineers (JSCE) 1992, 1997a; Nanni 1993a; Nanni and Dolan 1993; Neale and Labossiere 1992; Saadatmanesh and Ehsani 1998; Taerwe 1995; Teng 2001; White 1992).

The mechanical behavior of FRP reinforcement differs from the behavior of conventional steel reinforcement. Accordingly, a change in the traditional design philosophy of concrete structures is needed for FRP reinforcement. Fiber-reinforced polymer materials are anisotropic and are characterized by high tensile strength only in the direction of the reinforcing fibers. This anisotropic behavior affects the shear strength and dowel action of FRP bars as well as the bond performance. Furthermore, FRP materials do not yield; rather, they are elastic until failure. Design procedures should account for a lack of ductility in structural concrete members reinforced with FRP bars.

This guide was first developed in 2001 as a guide for the design and construction of structural concrete with FRP bars. Other countries and regions, such as Japan (Japan Society of Civil Engineers 1997b), Canada (CAN/CSA-S6-06, CAN/CSA-S806-12), and Europe (*fib* 2007, 2010) have also established similar design-related documents. There is adequate analytical and experimental information on FRP-reinforced concrete, and significant field experience implementing this knowledge. Successful applications worldwide using FRP composite reinforcing bars during the past few decades have demonstrated that it can be used successfully and practically. Research and field implementation is ongoing and design recommendations continue to evolve. When using this technology, exercise judgment as to the appropriate application of FRP reinforcement and be aware of its limitations as discussed in this guide.

Note: Any reference to ACI 318 in this document without a year designation refers to ACI 318-11. All exceptions will have a specific year designation.

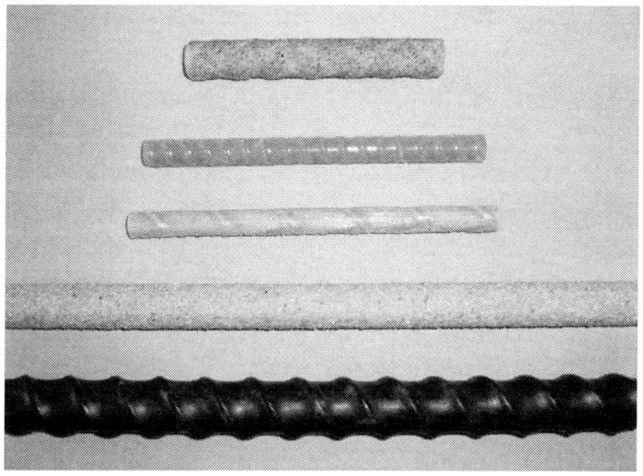

Fig. 1.1—Examples of FRP reinforcing bars.

1.2—Scope

This guide provides recommendations for the design and construction of FRP-reinforced concrete structures for nonprestressed FRP reinforcement; concrete structures prestressed with FRP tendons are covered in ACI 440.4R. The basis for this guide is knowledge gained from worldwide experimental research, analytical research work, and field applications of FRP reinforcement.

Design recommendations are based on the current knowledge and are intended to supplement existing codes and guidelines for conventionally reinforced concrete structures and to provide licensed design professionals and building officials with assistance in the design and construction of structural concrete reinforced with FRP bars.

ACI 440.3R provides a comprehensive list of test methods and material specifications to support design and construction guidelines. ACI 440.5 provides specification details for construction with FRP reinforcing bars. Material specifications for FRP bars are found in ACI 440.6.

The use of FRP reinforcement in combination with steel reinforcement for structural concrete is not addressed in this guide.

CHAPTER 2—NOTATION AND DEFINITIONS

2.1—Notation

a = depth of equivalent rectangular stress block, in. (mm)

A_f = area of fiber-reinforced polymer (FRP) reinforcement, in.^{2} (mm^2)

$A_{f,bar}$ = area of one FRP bar, in.^{2} (mm^2)

$A_{f,min}$ = minimum area of FRP reinforcement needed to prevent failure of flexural members upon cracking, in.^{2} (mm^2)

$A_{f,sh}$ = area of shrinkage and temperature FRP reinforcement per linear ft (m), in.^{2} (mm^2)

A_{fv} = amount of FRP shear reinforcement within spacing s, in.^{2} (mm^2)

$A_{fv,min}$ = minimum amount of FRP shear reinforcement within spacing s, in.^{2} (mm^2)

A_s = area of tension steel reinforcement, in.^{2} (mm^2)

b = width of rectangular cross section, in. (mm)
b_o = perimeter of critical section for slabs and footings, in. (mm)
b_w = width of the web, in. (mm)
c = distance from extreme compression fiber to the neutral axis, in. (mm)
c_b = distance from extreme compression fiber to neutral axis at balanced strain condition, in. (mm)
c_c = clear cover, in. (mm)
C = spacing or cover dimension, in. (mm)
C_E = environmental reduction factor for various fiber type and exposure conditions
d = distance from extreme compression fiber to centroid of tension reinforcement, in. (mm)
d_b = diameter of reinforcing bar, in. (mm)
d_c = thickness of concrete cover measured from extreme tension fiber to center of bar or wire location closest thereto, in. (mm)
$d_{c,side}$ = thickness of concrete cover measured from side face of member to center of longitudinal bar or wire location closest thereto, in. (mm)
E_c = modulus of elasticity of concrete, psi (MPa)
E_f = design or guaranteed modulus of elasticity of FRP defined as mean modulus of sample of test specimens ($E_f = E_{f,ave}$), psi (MPa)
$E_{f,ave}$ = average modulus of elasticity of FRP, psi (MPa)
E_s = modulus of elasticity of steel, psi (MPa)
f_c' = specified compressive strength of concrete, psi (MPa)
$\sqrt{f_c'}$ = square root of specified compressive strength of concrete, psi (MPa)
f_f = stress in FRP reinforcement in tension, psi (MPa)
f_{fb} = strength of bent portion of FRP bar, psi (MPa)
f_{fe} = bar stress that can be developed for embedment length ℓ_e, psi (MPa)
f_{fr} = required bar stress, psi (MPa)
f_{fs} = stress level induced in FRP at service loads, psi (MPa)
$f_{fs,sus}$ = stress level induced in FRP by sustained service loads, psi (MPa)
f_{fu} = design tensile strength of FRP, defined as the guaranteed tensile strength multiplied by the environmental reduction factor ($f_{fu} = C_E f_{fu}^*$), psi (MPa)
f_{fu}^* = guaranteed tensile strength of FRP bar, defined as mean tensile strength of sample of test specimens minus three times standard deviation ($f_{fu}^* = f_{fu,ave} - 3\sigma$), psi (MPa)
f_{fv} = tensile strength of FRP for shear design, taken as smallest of design tensile strength f_{fu}, strength of bent portion of FRP stirrups f_{fb}, or stress corresponding to $0.004E_f$, psi (MPa)
f_s = service stress in steel reinforcement, psi (MPa)
$f_{s,allow}$ = allowable service stress in steel reinforcement, psi (MPa)
$f_{u,ave}$ = mean tensile strength of sample of test specimens, psi (MPa)
f_y = specified yield stress of nonprestressed steel reinforcement, psi (MPa)
h = overall height of flexural member, in. (mm)
I = moment of inertia, in.^{4} (mm^4)
I_{cr} = moment of inertia of transformed cracked section, in.^{4} (mm^4)
I_e = effective moment of inertia, in.^{4} (mm^4)
I_{e+} = effective moment of inertia at location of maximum positive moment, in.^{4} (mm^4)
I_{e1-} = effective moment of inertia at location of maximum negative moment at near end of span, in.^{4} (mm^4)
I_{e2-} = effective moment of inertia at location of maximum negative moment at far end of span, in.^{4} (mm^4)
I_g = gross moment of inertia, in.^{4} (mm^4)
k = ratio of depth of neutral axis to reinforcement depth
k_b = bond-dependent coefficient
K_1 = parameter accounting for boundary conditions
K_4 = coefficient used in computing development length of bent bar
ℓ = span length of member, ft (m)
ℓ_a = additional embedment length at support or at point of inflection, in. (mm)
ℓ_{bhf} = basic development length of FRP standard hook in tension, in. (mm)
ℓ_d = development length, in. (mm)
ℓ_e = embedded length of reinforcing bar, in. (mm)
ℓ_{thf} = length of tail beyond hook in FRP bar, in. (mm)
L = distance between joints in slab-on-grade, ft (m)
M_a = maximum service load moment in member, lb-in. (N-mm)
M_{cr} = cracking moment, lb-in. (N-mm)
M_n = nominal moment capacity, lb-in. (N-mm)
$M_{s,sus}$ = moment due to sustained service loads, lb-in. (N-mm)
M_{serv} = service level moment
M_u = factored moment at section, lb-in. (N-mm)
n_f = ratio of modulus of elasticity of FRP bars to modulus of elasticity of concrete
r_b = internal radius of bend in FRP reinforcement, in. (mm)
s = stirrup spacing or pitch of continuous spirals, in. (mm)
s_{max} = maximum permissible center-to-center bar spacing for flexural crack control, in. (mm)
T_g = glass transition temperature, °F (°C)
u = average bond stress acting on the surface of FRP bar, psi (MPa)
V_c = nominal shear strength provided by concrete, lb (N)
V_f = shear resistance provided by FRP stirrups, lb (N)
V_n = nominal shear strength at section, lb (N)
V_s = shear resistance provided by steel stirrups, lb (N)
V_u = factored shear force at section, lb (N)
w = maximum allowable crack width, in. (mm)
w_{slab} = dead weight of slab, lb/ft^2 (N/m^2)

y_t = distance from centroidal axis of gross section, neglecting reinforcement, to tension face, in. (mm)
α = top bar modification factor
α_L = longitudinal coefficient of thermal expansion, 1/°F (1/°C)
α_T = transverse coefficient of thermal expansion, 1/°F (1/°C)
α_1 = ratio of average stress of equivalent rectangular stress block to f_c'
β = ratio of distance from neutral axis to extreme tension fiber to distance from neutral axis to center of tensile reinforcement
β_1 = factor taken as 0.85 for concrete strength f_c' up to and including 4000 psi (28 MPa). For strength above 4000 psi (28 MPa), this factor is reduced continuously at a rate of 0.05 per each 1000 psi (7 MPa) of strength in excess of 4000 psi (28 MPa), but is not taken less than 0.65
$\Delta_{(cp+sh)}$ = additional deflection due to creep and shrinkage under sustained loads, in. (mm)
$(\Delta_i)_{sus}$ = immediate deflection due to sustained loads, in. (mm)
$(\Delta/\ell)_{max}$ = limiting deflection-span ratio
ε_c = strain in concrete
ε_{cu} = ultimate strain in concrete
ε_f = strain in FRP reinforcement
$\varepsilon_{f,ave}$ = mean tensile strain at rupture of sample of test specimens
ε_{fs} = strain level induced in FRP at service loads
ε_{fu} = design rupture strain of FRP reinforcement, defined as the guaranteed tensile rupture strain multiplied by the environmental reduction factor ($\varepsilon_{fu} = C_E\varepsilon_{fu}$*)
ε_{fu}* = guaranteed rupture strain of FRP reinforcement defined as the mean tensile strain at failure of sample of test specimens minus three times standard deviation ($\varepsilon_{fu}* = \varepsilon_{f,ave} - 3\sigma$), in./in. (mm/mm)
ϕ = strength reduction factor
γ = parameter to account for the variation in stiffness along the length of the member
η = ratio of distance from extreme compression fiber to centroid of tension reinforcement (d) to overall height of flexural member (h)
λ = modification factor reflecting the reduced mechanical properties of lightweight concrete
λ_Δ = multiplier for additional long-term deflection
μ = coefficient of subgrade friction for calculation of shrinkage and temperature reinforcement
θ = angle of inclination of stirrups or spirals
ρ' = ratio of steel compression reinforcement; $\rho' = A_s'/bd$
ρ_b = steel reinforcement ratio producing balanced strain conditions
ρ_f = fiber-reinforced polymer reinforcement ratio
ρ_f' = ratio of FRP compression reinforcement
$\rho_{f,ts}$ = reinforcement ratio for temperature and shrinkage FRP reinforcement
ρ_{fb} = fiber-reinforced polymer reinforcement ratio producing balanced strain conditions
ρ_{fv} = ratio of FRP shear reinforcement
ρ_{min} = minimum reinforcement ratio for steel
σ = standard deviation
ξ = time-dependent factor for sustained load

2.2—Definitions

ACI provides a comprehensive list of definitions through an online resource, "ACI Concrete Terminology," http://www.concrete.org/tools/concreteterminology.aspx. Definitions provided herein compliment that source.

alkalinity—the condition of having or containing hydroxyl (OH^-) ions; containing alkaline substances.

aramid fiber—highly oriented organic fiber derived from polyamide incorporating into an aromatic ring structure.

balanced fiber-reinforced polymer reinforcement ratio—an amount and distribution of reinforcement in a flexural member such that in strength design, the tensile fiber-reinforced polymer (FRP) reinforcement reaches its ultimate design strain simultaneously with the concrete in compression reaching its assumed ultimate strain of 0.003.

braiding—a process whereby two or more systems of yarns are intertwined in the bias direction to form an integrated structure; braided material differs from woven and knitted fabrics in the method of yarn introduction into the fabric and the manner by which the yarns are interlaced.

carbon fiber—fiber produced by heating organic precursor materials containing a substantial amount of carbon, such as rayon, polyacrylonitrile (PAN), or pitch in an inert environment.

cross-link—a chemical bond between polymer molecules; increased number of cross-links per polymer molecule increases strength and modulus at the expense of ductility.

deformability factor—ratio of energy absorption (area under the moment-curvature curve) at ultimate strength of the section to the energy absorption at service level.

degradation—deleterious change in the chemical structure, physical properties, or appearance of a FRP composite.

E-glass—a family of glass with a calcium alumina borosilicate composition and a maximum alkali content of 2.0 percent.

endurance limit—the number of cycles of deformation or load that causes a material, test specimen, or structural member to fail.

fiber-reinforced polymer (FRP) bar—composite material formed into a long, slender structural shape suitable for the internal reinforcement of concrete and consisting of primarily longitudinal unidirectional fibers bound and shaped by a rigid polymer resin material.

fiber volume fraction—the ratio of the volume of fibers to the volume of the composite.

fiber weight fraction—the ratio of the weight of fibers to the weight of the composite.

glass fiber—fiber drawn from an inorganic product of fusion that has cooled without crystallizing.

grid—a two-dimensional (planar) or three-dimensional (spatial) rigid array of interconnected FRP bars that form a contiguous lattice that can be used to reinforce concrete.

precursor—for carbon or graphite fiber, the rayon, PAN, or pitch fibers from which carbon and graphite fibers are derived.

pultrusion—continuous process for manufacturing composites that have a uniform cross-sectional shape; process consists of pulling a fiber-reinforcing material through a resin impregnation bath then through a shaping die where the resin is subsequently cured.

vinyl esters—class of thermosetting resins containing ester of acrylic, methacrylic acids, or both, many of which have been made from epoxy resin.

CHAPTER 3—BACKGROUND

3.1—Historical development

The development of fiber-reinforced polymer (FRP) reinforcement can be traced to the expanded use of composites after World War II in the 1940s. The aerospace industry had long recognized the advantages of the high strength and light weight of composite materials and, during the Cold War, advancements in the aerospace and defense industry caused an increase in their use. Furthermore, the rapidly expanding economy of the United States demanded inexpensive materials to meet consumer demands. Pultrusion offered a fast and economical method of forming constant profile parts, and pultruded composites were being used to make golf clubs and fishing poles. It was not until the 1960s, however, that these materials were seriously considered for use as reinforcement in concrete.

Expansion of the national highway systems in the 1950s increased the need to provide year-round maintenance. It became common to apply deicing salts on highway bridges and, as a result, reinforcing steel in these structures and those subject to marine salt experienced extensive corrosion, becoming a major concern and leading to high maintenance cost. Various solutions were investigated, including galvanized coatings, electro-static-spray fusion-bonded (powder resin) coatings, polymer-impregnated concrete, epoxy coatings, alloyed steel bars, and glass FRP reinforcing bars (ACI 440R). Of these options, epoxy-coated steel reinforcement appeared to be the best solution and was therefore implemented in aggressive corrosion environments. Fiber-reinforced polymer reinforcing bar was not considered a viable solution and not commercially available until the late 1970s.

Initially, GFRP bars were considered a viable alternative to steel as reinforcement for polymer concrete because their use eliminated the need to address the incompatibility of thermal expansion characteristics between polymer concrete and steel. The 1980s market demanded nonmetallic reinforcement for specific advanced technology. The largest demand for electrically nonconductive reinforcement was in facilities for MRI medical equipment. FRP reinforcement became the standard in this type of construction. Other uses developed as the advantages of FRP reinforcement became better known and desired, specifically in seawall construction, substation reactor bases, airport runways, and electronics laboratories (Brown and Bartholomew 1996).

Concern for the deterioration of bridges due to chloride-ion-induced corrosion dates back to the 1970s, and its effects on aging bridges in the United States has become apparent (Boyle and Karbhari 1994). Additionally, detection of corrosion in the commonly used epoxy-coated reinforcing bars increased interest in alternative methods of avoiding corrosion. Once again, FRP reinforcement began to be considered as a general solution to address problems of corrosion in bridge decks and other structures (Benmokrane et al. 1996).

3.2—History of use

Up to the mid-1990s, the Japanese had the most FRP reinforcement applications, with more than 100 demonstration or commercial projects. Fiber-reinforced polymer design provisions were included in the design and construction recommendations of the Japanese Society of Civil Engineering (1997b). In the 2000s, China became the largest user of composite reinforcement for new construction in applications that span from bridge decks to underground works (Ye et al. 2003).

The use of FRP reinforcement in Europe began in Germany with the construction of a prestressed FRP highway bridge in 1986 (Meier 1992). Since the construction of this bridge, programs have been implemented to increase the research and use of FRP reinforcement in Europe (Taerwe 1997).

Canadian civil engineers have developed provisions for FRP reinforcement in the Canadian Highway Bridge Design Code (CAN/CSA-S6-06) and have constructed a number of FRP-reinforced concrete structures. The Headingley Bridge in Manitoba included both CFRP and GFRP reinforcement (Rizkalla 1997). The Floodway Bridge over the Red River in Winnipeg, MB, Canada, was completed in 2006. The bridge comprises 16 spans, each approximately 50 x 143 ft (15.3 x 43.5 m). All concrete elements above the girders are reinforced with GFRP bars. The project consumed 310,000 lb (140,000 kg) of GFRP reinforcing bar, making it the largest nonmetallic-reinforced concrete bridge in the world. Moreover, several bridges have been built in Quebec using GFRP bars in the decks, such as the Wotton Bridge in Wotton, the Magog Bridge on Highway 55 North, the Cookshire-Eaton Bridge on Route 108, and the Val-Alain Bridge on Highway 20 East (El-Salakawy and Benmokrane 2003; El-Salakawy et al. 2003, 2005; Benmokrane et al. 2004, 2007). Some of these bridges have been in service for more than 10 years without any signs of deterioration of the GFRP reinforcement (Mufti et al. 2007, 2011). Consequently, there is a remarkable increase in the use of GFRP bars in Canada where more than 200 bridge structures have been successfully constructed. Straight and bent FRP bars were used for the deck slab, for the concrete barriers and girders of the bridges, or both (Drouin et al. 2011). In addition, GFRP bars have been used in Canada in other concrete structures such as parking garages (Benmokrane et al. 2012), highway concrete pavement (Benmokrane et al. 2007), water tanks (Benmokrane and Mohamed 2014), and incinerators (Beaulieu-Michaud et al. 2013). In the United States, typical uses

Fig. 3.2a—FRP-reinforced deck constructed in Lima, OH (Pierce Street Bridge), in 1999.

Fig. 3.2b—GFRP bars used in the redecking of Dayton, OH, Salem Avenue Bridge in 1999.

Fig. 3.2c—Transverse view of GFRP bars in Sierrita de la Cruz Creek Bridge deck near Amarillo, TX, in 2000.

of FRP reinforcement have been previously reported (ACI 440R). Figures 3.2a, 3.2b, and 3.2c show applications in bridge deck construction. The use of GFRP bars in MRI hospital room additions is becoming commonplace. Other applications, such as waterfront construction, top mat reinforcing for bridge decks, various precast applications, and ornamental and architectural concrete, are also becoming more frequent. Some of the largest projects include the Gonda Building at the Mayo Clinic in Rochester, MN; the

Fig. 3.2d—Emma Park Bridge deck panel with GFRP reinforcing bars, top and bottom mat.

Fig. 3.2e—GFRP bars used in rail plinths.

National Institute of Health in Bethesda, MD, for MRI applications; the bridge on RM 1061 at Sierrita de la Cruz Creek in Potter County, TX, and the bridge at 53rd Avenue in Bettendorf, IA, for deck reinforcement applications (Nanni 2001). Glass FRP bars are making great strides to support accelerated bridge construction with application in precast concrete deck panels. One example is the Emma Park Bridge in Utah constructed in 2009. Glass FRP bars were used in the top and bottom mat, as shown in Fig. 3.2d. In 2011 in Miami, FL, 2.4 miles (3.9 km) of GFRP bar were used in the rail plinths (Fig. 3.2e) for AirportLink, which connects the existing Earlington Heights Station to the new Miami Intermodal Center (MIC). Glass FRP bars were selected, as they provided electrical insolation in the rail bed. Designed by the Florida Department of Transportation, the MIC is a major transportation hub that serves as a central transfer point to different modes of transportation, including Metrorail,

Fig. 3.2f—Construction of seawall with GFRP bars.

Fig. 3.2g—I-635 Bridge over State Ave. (Kansas City, KS).

Metrobus, Tri-Rail, Amtrak, intercity buses, tour buses, taxi cabs, rental cars, and the Automated People Mover (APM) connection to the Miami International Airport (Gremel 2012).

Seawalls constructed with GFRP reinforcing bar are being used in aggressive salt environments to provide long-term service. An example is the seawall that protects the Honoapiilani Highway, Maui, HI. First constructed in 2001 with steel reinforcing bar, the seawall required replacement in 2012, as shown in Fig. 3.2f.

Glass FRP reinforcing bar is also found on larger and high-volume traffic bridges. In 2013, the I-635 bridge deck over State Avenue in Kansas City, KS was replaced with cast-in-place GFRP reinforcing bar. Figure 3.2g shows GFRP reinforcing bar in the top and bottom mat of the deck panels used in the 32 x 232 ft (9.8 x 70.7 m) bridge. Construction bids for this project were the same for installed cost of epoxy-coated as well as GFRP reinforcement.

Tunnel construction, where GFRP reinforcement is used in the portion of the concrete wall to be excavated by the tunnel-boring machine, has become common in many major metropolitan areas of the world.

3.3—Material characteristics

The material characteristics of fiber-reinforced polymer (FRP) reinforcement should be considered when determining whether FRP reinforcement is suitable or necessary in a particular structure. The material characteristics are described in detail in Chapter 5. Table 3.3 provides some guidance on appropriate conditions for the use of FRP bars.

Table 3.3—Considerations for the use of fiber-reinforced polymer (FRP) reinforcing bars

Why should FRP bars be considered?
a) Impervious to chloride ion and chemical attack
b) Tensile strength is greater than steel
c) Light weight – one-fourth to one-fifth the weight of steel reinforcing bar
d) Transparent to magnetic fields, radio frequencies (glass FRP only)
e) Thermally and electrically nonconductive (glass FRP only)
f) Less concrete cover is possible
g) Admixtures to reduce corrosion are not needed
h) High fatigue endurance
i) Easily "consumed" by excavation equipment when used in temporary structures
j) In corrosive environments, service life much greater than that of steel
k) Better field handling damage tolerance than epoxy coated steel and no touch-ups required
Differences between FRP and steel
a) FRP is linear elastic to failure whereas steel yields
b) FRP is anisotropic whereas steel is isotropic
c) Due to lower modulus of FRP bars, design for serviceability often controls
d) FRP bars have a lower creep-rupture threshold than steel
e) Coefficient of thermal expansion different in longitudinal and radial directions
f) Endurance time in fire and elevated temperature applications less than that of steel
g) Should degradation of FRP bars occur, the degradation mechanism is benign to the surrounding concrete unlike steel that expands and causes failure of the member
Where should FRP bars considered?
a) Any concrete member susceptible to corrosion by chloride ions or chemicals
b) Any concrete member requiring nonferrous reinforcement due to electromagnetic considerations
c) As a cost-effective alternative to epoxy-coated and galvanized
d) As a first cost savings over stainless steel bar
e) Where machinery will "consume" the reinforced member, mainly in tunneling and mining
f) Applications requiring thermal nonconductivity
g) In mass concrete exposed to marine chlorides near chloride exposure in hybrid applications with steel

The corrosion resistance of FRP reinforcement is a significant benefit for structures in highly corrosive environments such as seawalls and other marine structures, bridge decks and superstructures exposed to deicing salts, and pavements treated with deicing salts. In structures supporting magnetic resonance imaging (MRI) units or other equipment sensitive to electromagnetic fields, the nonmagnetic properties of FRP reinforcement are of principal importance. Fiber-reinforced polymer reinforcement has a high tensile strength, significant elongation, and exhibits linear stress-strain behavior to failure. The use of FRP reinforcement should be limited to structures that will significantly benefit from other properties, such as the noncorrosive or nonconductive behavior of its materials. Due to lack of experience in its use, FRP reinforcement is not recommended for moment frames or zones where moment redistribution is required.

Research on circular and rectangular columns reinforced with glass FRP (GFRP) or carbon FRP (CFRP) bars and stirrups (De Luca et al. 2010; Tobbi et al. 2012; Pantelides et al. 2013; Afifi et al. 2013) indicates that FRP circular hoops and spirals were found to be efficient in confining concrete and that the GFRP and CFRP bars can develop up to 0.4 percent and 0.7 percent compressive strain, confirming that the FRP bars were effective in resisting compression until after crushing of the concrete (Afifi et al. 2013). However, the average load carried by the longitudinal FRP bars in the GFRP- and CFRP-reinforced columns ranged between only 5 percent and 12 percent of the maximum load. Due to the combined effect, the relatively lower elastic modulus of FRP compared with steel and the lower elastic modulus of FRP bars in compression as compared to tension, the maximum contribution of compression FRP reinforcement calculated at ultimate concrete strain (typically at ε_{cu} = 0.003) is expected to be small. Therefore, the contribution of FRP reinforcement should be neglected when used as reinforcement in columns, in compression members, or as compression reinforcement in flexural members. It is acceptable for FRP tension reinforcement to experience compression due to moment reversals or changes in load pattern.

There are advantages and disadvantages regarding sustainability when comparing FRP reinforcement to steel reinforcement. Because of the high temperatures required for production, steel reinforcement has a higher carbon footprint than FRP reinforcement (Pearson et al. 2011). Recycling FRP reinforcement, however, is not as easy or commonplace as recycling steel reinforcement. Because FRP does not corrode, life cycle costs associated with FRP reinforced structures are likely to be lower than those for steel-reinforced structures where steel corrosion is a concern (Pearson et al. 2011). Katz (2004) compared the environmental impact of FRP-reinforced pavements to steel-reinforced pavements and determined that the FRP reinforcement had a significantly smaller environmental impact than steel-reinforced pavement over the pavement lifetime.

CHAPTER 4—MATERIAL CHARACTERISTICS

This chapter presents physical and mechanical properties of fiber-reinforced polymer (FRP) reinforcing bars to provide a fundamental understanding of the behavior of these bars and the properties that affect their use in concrete structures. Furthermore, the effects of factors, such as loading history and duration, temperature, and moisture, on the properties of FRP bars are discussed.

Fiber-reinforced polymer bars are anisotropic in nature and can be manufactured using a variety of techniques such as pultrusion, braiding, and weaving (Bank 1993; Bakis 1993). Factors such as fiber volume, fiber type, resin type, fiber orientation, dimensional effects, and quality control during manufacturing all play a major role in defining the characteristics of an FRP bar. The material characteristics described in this chapter may not apply to all commercially available products. Therefore, the manufacturer's material data should be consulted for specific product properties. In addition, several consensus-based test methods for FRP reinforcement for use in structural concrete are presented (ACI 440.3R; ASTM D7205; ASTM D7337; ASTM D7617; ASTM D7705; Japan Society of Civil Engineers 1997b).

Table 4.1.1—Typical densities of reinforcing bars, lb/ft³ (g/cm³)

Steel	GFRP	CFRP	AFRP
493.00 (7.90)	77.8 to 131.00 (1.25 to 2.10)	93.3 to 100.00 (1.50 to 1.60)	77.80 to 88.10 (1.25 to 1.40)

Table 4.1.2—Typical coefficients of thermal expansion for reinforcing bars*

	CTE, × 10⁻⁶/°F (× 10⁻⁶/°C)				
Direction	Steel	GFRP	CFRP	AFRP	Concrete†
Longitudinal, α_L	6.5 (11.7)	3.3 to 5.6 (6.0 to 10.0)	–4.0 to 0.0 (–9.0 to 0.0)	–3.3 to –1.1 (–6 to –2)	4 to 6 (7.2 to 10.8)
Transverse, α_T	6.5 (11.7)	11.7 to 12.8 (21.0 to 23.0)	41 to 58 (74.0 to 104.0)	33.3 to 44.4 (60.0 to 80.0)	4 to 6 (7.2 to 10.8)

*Typical values for fiber volume fraction ranging from 0.5 to 0.7.

†Mindess et al. (2003).

4.1—Physical properties

4.1.1 *Density*—Fiber-reinforced polymer bars have a density ranging from 77.8 to 131.3 lb/ft³ (1.25 to 2.1 g/cm³), one-sixth to one-fourth that of steel (Table 4.1.1). Reduced weight lowers transportation costs and eases handling of the bars on the project site.

4.1.2 *Coefficients of thermal expansion*—The coefficients of thermal expansion of FRP bars vary in the longitudinal and transverse directions, depending on the types of fiber, resin, and volume fraction of fiber. The longitudinal coefficient of thermal expansion is dominated by the properties of the fibers whereas the transverse coefficient is dominated by the resin (Bank 1993). Table 4.1.2 lists the longitudinal and transverse coefficients of thermal expansion for typical FRP and steel bars. Note that a negative coefficient of thermal expansion indicates that the material contracts with increased temperature and expands with decreased temperature. Fiber-reinforced polymers have a thermal expansion in the transverse direction much higher than in the longitudinal direction and also higher than the thermal expansion of hardened concrete (Masmoudi et al. 2005). The difference between the transverse coefficient of thermal expansion of FRP bars and concrete may cause splitting cracks within the concrete under temperature increase and, ultimately, failure of the concrete cover if the confining action of concrete is insufficient (Gentry and Husain 1999; Bellakehal et al. 2013; Zaidi et al. 2013). Experimental results show that the transverse coefficient of thermal expansion of glass FRP (GFRP) bars is 12×10^{-6}/°F (22×10^{-6}/°C) on average, and the ratio between the transverse and longitudinal coefficients of thermal expansion of these FRP bars is equal to 4. A ratio of concrete cover thickness to bar diameter, c/d_b, greater than 1.6 is sufficient to avoid cracking of concrete under high temperature up to 175°F (80°C) (Masmoudi et al. 2005; Zaidi and Masmoudi 2008; Zaidi et al. 2013). Experimental results, however, showed no significant reduction in bond strength for temperatures up to 140°F (60 °C) (Masmoudi et al. 2011). For reference, concrete has a coefficient of thermal expansion that varies from 4×10^{-6} to 6×10^{-6}/°F (7.2×10^{-6}

Table 4.2.1—Typical tensile properties of reinforcing bars*

	Steel	GFRP	CFRP	AFRP
Nominal yield stress, ksi (MPa)	40 to 75 (276 to 517)	NA	NA	NA
Tensile strength, ksi (MPa)	70 to 100 (483 to 1600)	70 to 230 (483 to 690)	87 to 535 (600 to 3690)	250 to 368 (1720 to 2540)
Elastic modulus, × 10^3 ksi (GPa)	29.0 (200.0)	5.1 to 7.4 (35.0 to 51.0)	15.9 to 84.0 (120.0 to 580.0)	6.0 to 18.2 (41.0 to 125.0)
Yield strain, percent	0.14 to 0.25	NA	NA	NA
Rupture strain, percent	6.0 to 12.0	1.2 to 3.1	0.5 to 1.7	1.9 to 4.4

*Typical values for fiber volume fraction ranging from 0.5 to 0.7.

to 10.8 × 10^{-6}/°C) and is usually assumed to be isotropic (Mindess et al. 2003).

4.2—Mechanical properties and behavior

4.2.1 *Tensile behavior*—When loaded in tension, FRP bars exhibit no plastic behavior (yielding) before rupture. The tensile behavior of FRP bars consisting of one type of fiber material is characterized by a linearly elastic stress-strain relationship until failure. The tensile properties of some commonly used FRP bars are summarized in Table 4.2.1.

The tensile strength and stiffness of an FRP bar are dependent on several factors. Because the resin has a much lower strength than the fibers, the ratio of the volume of fiber to the overall volume of the FRP (fiber-volume fraction) significantly affects the tensile properties of an FRP bar. Strength and stiffness variations will occur in bars with various fiber-volume fractions, even in bars with the same diameter, appearance, and constituents. The rate of curing, the manufacturing process, and the manufacturing quality control also affect the mechanical characteristics of the bar (Wu 1990).

Unlike steel, the unit tensile strength of an FRP bar can vary with diameter. For example, GFRP bars from three different manufacturers show tensile strength reductions of up to 40 percent as the diameter increases proportionally from 0.375 to 0.875 in. (9.5 to 22.2 mm) (Faza and GangaRao 1993b). However, similar cross section changes do not appear to affect the strength of twisted carbon FRP (CFRP) strands (Santoh 1993). The sensitivity of aramid FRP (AFRP) bars to cross section size has been shown to vary from one commercial product to another. For example, in braided AFRP bars, there is a less than 2 percent strength reduction as bars increase in diameter from 0.28 to 0.58 in. (7.3 to 14.7 mm) (Tamura 1993). The strength reduction in a unidirectionally pultruded AFRP bar with added aramid fiber surface wraps is approximately 7 percent for diameters increasing from 0.12 to 0.32 in. (3 to 8 mm) (Noritake et al. 1993). The FRP bar manufacturer should be contacted for particular strength values of differently sized FRP bars.

Determining FRP bar strength by testing is challenging because stress concentrations in and around anchorage points on the test specimen can lead to premature failure. An adequate testing grip should allow failure to occur in the middle of the test specimen. Test methods for determining the tensile strength and stiffness of FRP bars are available in ASTM D7205.

The tensile properties of a particular FRP bar should be obtained from the bar manufacturer. Usually, a normal (Gaussian) distribution is assumed to represent the strength of a population of bar specimens (Kocaoz et al. 2005). Manufacturers should report a guaranteed tensile strength f_{fu}^* ($f_{fu}^* = f_{u,ave} - 3\sigma$), and similarly report a guaranteed rupture strain, ε_{fu}^*($\varepsilon_{fu}^* = \varepsilon_{u,ave} - 3\sigma$), and a specified tensile modulus E_f ($E_f = E_{f,ave}$). These guaranteed values of strength and strain provide a 99.87 percent probability that the indicated values are exceeded by similar FRP bars, provided that at least 25 specimens are tested (Dally and Riley 1991; Mutsuyoshi et al. 1990). If fewer specimens are tested or a different distribution is used, texts and manuals on statistical analysis should be consulted to determine the confidence level of the distribution parameters. In any case, the manufacturer should provide a description of the method used to obtain the reported tensile properties.

An FRP bar cannot be bent once it has been manufactured (an exception to this would be an FRP bar with a thermoplastic resin that could be reshaped with the addition of heat and pressure). Fiber-reinforced polymer bars, however, can be fabricated with bends. In FRP bars produced with bends, a strength reduction of 40 to 50 percent, compared with the tensile strength of a straight bar, can occur in the bend portion due to fiber bending and stress concentrations (Nanni et al. 1998b).

4.2.2 *Compressive behavior*—While design of FRP bars to resist compressive stresses is not recommended, the following section is presented to fully describe the behavior of FRP bars. In the technical literature, there is no direct measurement of time-dependent creep of FRP bar coupons subject to compression. Experimental evidence obtained from FRP-pultruded shapes indicates that the ratio of creep strain to initial strain is low for stress level up to 26 percent of the ultimate compressive strength (25 ksi [170 MPa] in this research) (McClure and Mohammadi 1995). Accordingly, creep should not cause geometrical integrity problems at the stress levels typical of FRP bars in compression.

Tests on FRP bars with a length-diameter ratio from 1:1 to 2:1 have shown that the compressive strength is lower than the tensile strength (Wu 1990). The mode of failure for FRP bars subjected to longitudinal compression can include transverse tensile failure, fiber microbuckling, or shear failure. The mode of failure depends on the type of fiber, the fiber-volume fraction, and the type of resin. Compressive strengths of 55, 78, and 20 percent of the tensile strength have been reported for GFRP, CFRP, and AFRP, respectively (Mallick 1988; Wu 1990). In general, compressive strengths are higher for bars with higher tensile strengths, except in the case of AFRP, where the fibers exhibit nonlinear behavior in compression at a relatively low level of stress.

The compressive modulus of elasticity of FRP reinforcing bars appears to be smaller than its tensile modulus of elasticity. Test reports on samples containing 55 to 60 percent volume fraction of continuous E-glass fibers in a matrix of vinyl ester or isophthalic polyester resin indicate a compressive modulus of elasticity of 5000 to 7000 ksi (35 to 48 GPa) (Wu 1990). According to reports, the compressive modulus of elasticity is approximately 80 percent for GFRP, 85 percent for CFRP, and 100 percent for AFRP of the tensile modulus of elasticity for the same product (Mallick 1988; Ehsani 1993). The most widely accepted explanation for the slightly lower values of modulus of elasticity in compression is the premature failure in the test resulting from end brooming and internal fiber microbuckling under compressive loading.

Standard test methods are not yet established to characterize the compressive behavior of FRP bars. If the compressive properties of a particular FRP bar are needed, they should be obtained from the bar manufacturer. The manufacturer should provide a description of the test method used to obtain the reported compression properties.

4.2.3 *Shear behavior*—Most FRP bar composites are relatively weak in interlaminar shear where layers of unreinforced resin lie between layers of fibers. Because there is usually no reinforcement across layers, the interlaminar shear strength is governed by the relatively weak polymer matrix. Orientation of the fibers in an off-axis direction across the layers of fiber will increase the shear resistance, depending on the degree of offset. For FRP bars, this can be accomplished by braiding or winding fibers transverse to the main fibers. Off-axis fibers can also be placed in the pultrusion process by introducing a continuous strand mat in the roving/mat creel. If the shear properties of a particular FRP bar are needed, these should be obtained from the bar manufacturer. The manufacturer should provide a description of the test method used to obtain the reported shear values.

4.2.4 *Bond behavior*—Bond performance of an FRP bar is dependent on the design, manufacturing process, mechanical properties of the bar itself, and the environmental conditions (Al-Dulaijan et al. 1996; Nanni et al. 1997; Bakis et al. 1998; Bank et al. 1998; Freimanis et al. 1998). When anchoring a steel reinforcing bar in concrete, the bond force can be transferred by adhesion resistance of the interface, also known as chemical bond; frictional resistance of the interface against slip; and mechanical interlock due to irregularity of the interface.

In FRP bars, it is postulated that bond force is transferred through the resin to the reinforcement fibers, and a bond-shear failure in the resin is also possible. When a bonded deformed bar is subjected to increasing tension, the adhesion between the bar and the surrounding concrete breaks down, and deformations on the surface of the bar cause inclined contact forces between the bar and the surrounding concrete. The stress at the surface of the bar resulting from the force component in the direction of the bar can be considered the bond stress between the bar and the concrete.

The bond properties of FRP bars have been extensively investigated by numerous researchers through different types of tests, such as pullout tests, splice tests, and cantilever beams, to determine an empirical equation for embedment length (Faza and GangaRao 1990; Ehsani et al. 1996a,b; Benmokrane 1997; Shield et al. 1999; Mosley 2002; Wambeke and Shield 2006; Tighiouart et al. 1999).

4.3—Time-dependent behavior

4.3.1 *Creep rupture*—Fiber-reinforced polymer reinforcing bars subjected to a constant tension over time can suddenly fail after a time period called the endurance time. This phenomenon is known as creep rupture or static fatigue. Creep rupture is not an issue with steel bars in reinforced concrete except in extremely high temperatures, such as those encountered in a fire. As the ratio of the sustained tensile stress to the short-term strength of the FRP bar increases, endurance time decreases. The creep rupture endurance time can also irreversibly decrease under sufficiently adverse environmental conditions such as high temperature, ultraviolet radiation exposure, high alkalinity, wet and dry cycles, or freezing-and-thawing cycles. Literature on the effects of such environments exists, although the extraction of generalized design criteria is hindered by a lack of standard creep test methods and reporting as well as the diversity of constituents and processes used to make proprietary FRP products. In addition, little data are currently available for endurance times beyond 100 hours. These factors have resulted in design criteria judged to be conservative until more research has been done on this subject. Several representative examples of endurance times for bar and bar-like materials follow. No creep strain data are available in these cases.

In general, carbon fibers are the least susceptible to creep rupture, whereas aramid fibers are moderately susceptible, and glass fibers are the most susceptible. A comprehensive series of creep rupture tests was conducted on 0.25 in. (6 mm) diameter smooth FRP bars reinforced with glass, aramid, and carbon fibers (Yamaguchi et al. 1997). The bars were tested at different load levels at room temperature in laboratory conditions using split conical anchors. Results indicated that a linear relationship exists between creep rupture strength and the logarithm of time for times up to nearly 100 hours. The ratios of stress level at creep rupture to the initial strength of the GFRP, AFRP, and CFRP bars after 500,000 hours (more than 50 years) were linearly extrapolated to be 0.29, 0.47, and 0.93, respectively.

In another investigation, endurance times were determined for braided AFRP bars and twisted CFRP bars, both using epoxy resin as the matrix material (Ando et al. 1997). These commercial bars were tested at room temperature in laboratory conditions and were anchored with an expansive cementitious grout inside of friction-type grips. Bar diameters ranged from 0.26 to 0.6 in. (5 to 15 mm), but were not found to affect the results. The ratios of stress level at creep rupture to the initial strength after 50 years calculated using a linear relationship extrapolated from data available to 100 hours was found to be 0.79 for CFRP and 0.66 for AFRP.

An investigation of creep rupture in GFRP bars in room-temperature laboratory conditions was reported by Seki et al. (1997). The molded E-glass/vinyl ester bars had a small (0.0068 in.^{2} [4.4 mm^2]) rectangular cross section and integral GFRP tabs. The percentage of initial tensile strength retained followed a linear relationship with logarithmic time, reaching a value of 55 percent at an extrapolated 50-year endurance time.

Creep rupture data characteristics of a 0.5 in. (12.5 mm) diameter commercial CFRP twisted strand in an indoor environment is available (Tokyo Rope 2000). The rupture strength at a projected 100-year endurance time is reported to be 85 percent of the initial strength.

An investigation of creep deformation (not rupture) in one commercial AFRP and two commercial CFRP bars tested to 3000 hours has been reported (Saadatmanesh and Tannous 1999a,b). The bars were tested in laboratory air and in room-temperature solutions with a pH equal to 3 and 12. The bars had diameters between 0.313 to 0.375 in. (8 to 10 mm), and the applied stress was fixed at 40 percent of initial strength. The results indicated a slight trend toward higher creep strain in the larger-diameter bars and in the bars immersed in the acidic solution. Bars tested in air had the lowest creep strains of the three environments. Considering all environments and materials, the range of strains recorded after 3000 hours was 0.002 to 0.037 percent. Creep strains were slightly higher in the AFRP bar than in the CFRP bars.

For experimental characterization of creep rupture, refer to the test method proposed by Japan Society of Civil Engineers (1997b). Creep characteristics of FRP bars can also be determined from pullout test methods cited in ACI 440.3R. Recommendations on sustained stress limits imposed to avoid creep rupture are provided in 7.4.

4.3.2 *Fatigue*—A substantial amount of data for fatigue behavior and life prediction of stand-alone FRP materials exists (National Research Council 1991; Wicaksono and Chai 2013). During most of this time period, the focus of research investigations was on materials suitable for aerospace applications. Some general observations on the fatigue behavior of FRP materials can be made, even though the bulk of the data is obtained from FRP specimens intended for aerospace applications rather than construction. Unless stated otherwise, the cases that follow are based on flat, unidirectional coupons with approximately 60 percent fiber-volume fraction and subjected to tension-tension sinusoidal cyclic loading at a frequency low enough not to cause self-heating, ambient laboratory environments, a stress ratio (ratio of minimum applied stress to maximum applied stress) of 0.1, and a direction parallel to the principal fiber alignment.

Test conditions that raise the temperature and moisture content of FRP materials generally degrade the ambient environment fatigue behavior.

Of all types of current FRP composites for infrastructure applications, CFRP is generally thought to be the least prone to fatigue failure. On a plot of stress versus the logarithm of the number of cycles at failure (S-N curve), the average downward slope of CFRP data is usually approximately 5 to 8 percent of initial static strength per decade of logarithmic life. At 1 million cycles, the fatigue strength is generally between 50 and 70 percent of initial static strength and is relatively unaffected by realistic moisture and temperature exposures of concrete structures unless the resin or fiber/resin interface is substantially degraded by the environment. Some specific reports of data to 10 million cycles indicated a continued downward trend of 5 to 8 percent per decade in the S-N curve (Curtis 1989).

Individual glass fibers, such as E-glass and S-glass, are generally not prone to fatigue failure. Individual glass fibers, however, have demonstrated delayed rupture caused by the stress corrosion induced by the growth of surface flaws in the presence of even minute quantities of moisture in ambient laboratory environment tests (Mandell and Meier 1983). When many glass fibers are embedded into a matrix to form an FRP composite, a cyclic tensile fatigue effect of approximately 10 percent loss in the initial static capacity per decade of logarithmic lifetime has been observed (Mandell 1982). This fatigue effect is thought to be due to fiber-fiber interactions and is not dependent on the stress corrosion mechanism described for individual fibers. No clear fatigue limit can usually be defined. Environmental factors play an important role in the fatigue behavior of glass fibers due to their susceptibility to moisture, alkaline, and acidic solutions.

Aramid fibers, for which substantial durability data are available, appear to behave similarly to carbon and glass fibers in fatigue. The tension-tension fatigue behavior of an impregnated aramid fiber bar is excellent. Strength degradation per decade of logarithmic lifetime is approximately 5 to 6 percent (Roylance and Roylance 1981). While no distinct endurance limit is known for AFRP, 2 million cycle fatigue strengths of commercial AFRP bars for concrete applications have been reported in the range of 54 to 73 percent of initial bar strengths (Odagiri et al. 1997). Based on these findings, Odagiri et al. (1997) suggested that the maximum stress be set at 54 to 73 percent of the initial tensile strength. Because the slope of the applied stress versus logarithmic creep-rupture time of AFRP is similar to the slope of the stress versus logarithmic cyclic lifetime data, the individual fibers appear to fail by a strain-limited creep-rupture process. This failure condition in commercial AFRP bars was noted to be accelerated by exposure to moisture and elevated temperature (Roylance and Roylance 1981; Rostasy 1997).

Although the influence of moisture on the fatigue behavior of unidirectional FRP materials is generally thought to be detrimental if the resin or fiber/matrix interface is degraded, research findings are inconclusive because the performance depends on fiber and matrix types, preconditioning methods, solution content, and the environmental condition during fatigue (Hayes et al. 1998; Rahman et al. 1997). In addition, factors such as gripping and presence of concrete surrounding the bar during the fatigue test need to be considered.

Fatigue strength of CFRP bars encased in concrete has been observed to decrease when the environmental temperature increases from 68 to 104°F (20 to 40°C) (Adimi et al.

1998). In this same investigation, the endurance limit was found to be inversely proportional to the loading frequency. It was also found that higher cyclic loading frequencies in the 0.5 to 8 Hz range corresponded to higher bar temperatures due to sliding friction. Thus, an endurance limit at 1 Hz could be more than 10 times higher than that at 5 Hz. In the cited investigation, a stress ratio (minimum stress divided by maximum stress) of 0.1 and a maximum stress of 50 percent of initial strength resulted in runouts of greater than 400,000 cycles when the loading frequency was 0.5 Hz. These runout specimens had no loss of residual tensile strength.

With CFRP bars, the endurance limit also depends on the mean stress and the ratio of maximum-to-minimum cyclic stress. Higher mean stress or a lower stress ratio (minimum divided by maximum) will cause a reduction in the endurance limit (Rahman and Kingsley 1996; Saadatmanesh and Tannous 1999a).

Although GFRP is weaker than steel in shear, fatigue tests on specimens with unbonded GFRP dowel bars have shown fatigue behavior similar to that of steel dowel bars for cyclic transverse shear loading of up to 10 million cycles. The test results and the stiffness calculations have shown that an equivalent performance can be achieved between FRP and steel bars subjected to transverse shear by changing some of the parameters, such as diameter, spacing, or both (Porter et al. 1993; Hughes and Porter 1996).

The addition of ribs, wraps, and other types of deformations improve the bond behavior of FRP bars. Such deformations, however, have been shown to induce local stress concentrations that significantly affect the performance of a GFRP bar under fatigue loading situations (Katz 1998). Local stress concentrations degrade fatigue performance by imposing multi-axial stresses that serve to increase matrix-dominated damage mechanisms normally suppressed in fiber-dominated composite materials. Additional fiber-dominated damage mechanisms can be also activated near deformations, depending on the construction of the bar.

The effect of fatigue on the bond of deformed GFRP bars embedded in concrete has been investigated in detail using specialized bond tests (Sippel and Mayer 1996; Nanni et al. 1998a; Katz 2000). Different GFRP materials, environments, and test methods were followed in each cited case, and the results indicated that bond strength can increase, decrease, or remain the same following cyclic loading. Bond fatigue behavior has not been sufficiently investigated to date, and conservative design criteria based on specific materials and experimental conditions are recommended.

Design limitations on fatigue stress ranges for FRP bars ultimately depend on the manufacturing process of the FRP bar, environmental conditions, and the type of fatigue load being applied. Given the ongoing development in the manufacturing process of FRP bars, conservative design criteria should be used for all commercially available FRP bars. Design criteria are given in 7.4.2.

With regard to the fatigue characteristics of FRP bars, the designer is referred to the provisional standard test methods cited in ACI 440.3R. The designer should always consult with the bar manufacturer for fatigue response properties.

4.4—Effects of high temperatures and fire

The fire resistance of FRP-reinforced concrete flexural elements can, given due consideration to possible differences in their response to heating, be determined in a similar way as that of steel-reinforced concrete slabs (ACI 216.1). The effects of elevated temperature exposure on the shear and axial compressive capacity of FRP-reinforced concrete elements are not well known, and additional research is warranted in these areas.

The type of reinforcement, aggregate type, and concrete cover thickness will all influence the fire performance of FRP-reinforced members. The type of reinforcement is important because all FRP materials will experience significantly different reductions in mechanical and bond properties at elevated temperatures (Bisby et al. 2005). The aggregate type and concrete cover are important because these two parameters significantly influence heat transfer to the reinforcement and, hence, the temperatures experienced during fire exposure both in the bars along the exposed zone and in the anchorage zone (Kodur and Bisby 2005; Nigro et al. 2011b).

4.4.1 *Heat transfer*—Research has shown (Kodur and Bisby 2005; Nigro et al. 2011b) that the heat transfer behavior of fire-exposed FRP-reinforced concrete slabs and beams is similar to those reinforced with steel bars.

4.4.2 *Material degradation*—Because FRP materials are generally more sensitive than steel reinforcing bars to elevated temperatures, the reinforcement type influences the fire resistance of FRP-reinforced concrete elements; these may have lower fire resistances compared to steel-reinforced elements of equivalent ambient temperature capacity (Bisby and Kodur 2007). Because FRP reinforcement is embedded within concrete, the reinforcement does not contribute significantly to fire severity or toxicity; however, the fire resistance of FRP-reinforced concrete members could be less than that of steel-reinforced members due to reduction of FRP mechanical properties, the loss of bond from a softening of the resin from which the bars are fabricated, or both.

4.4.3 *Serviceability limit state*—The properties of the polymer at the surface of the bar are essential in maintaining bond between FRP and concrete. At a temperature close to its glass transition temperature T_g, the mechanical properties of the polymer begin to reduce and the polymer is less able to transfer stresses from the concrete to the fibers, resulting in considerable reductions in bond strength (Katz et al. 1999; Nigro et al. 2012). The value of T_g depends on the type of resin, but is typically in the range of 200 to 250°F (93 to 120°C) for the resins used in most FRP bars for concrete. The fibers, which generally for glass and carbon fibers exhibit better thermal properties than the resin with a melting point of 1600°F (880°C) for glass fibers and 2900°F (1600°C) for carbon, can continue to support considerable load in the longitudinal direction; however, the tensile strength and stiffness of the overall FRP composite are reduced due to a reduction in force transfer between fibers through bond to the resin (Chowdhury et al. 2011). Reduction of stiffness is generally expected later than reduction in strength. The onset glass transition temperature, obtained by dynamic mechanical analysis (DMA) testing, should generally not be exceeded to assure preservation of the mechanical and bond properties of the FRP reinforcement.

4.4.4 *Ultimate limit state*—Should an anchorage length of FRP remain outside the area exposed to high temperature during fire, loss of bond in the heated zone is less critical, as the fibers themselves are less sensitive to elevated temperature. Hence, well-anchored FRP bars can retain considerable strength and stiffness at temperatures well above T_g. For example, Robert and Benmokrane (2010) demonstrated that well-anchored GFRP bars with a T_g of 250°F (120°C) retained as much as 50 percent of their tensile strength at temperatures exceeding 570°F (300°C); Maluk et al. (2011) demonstrated that well-anchored CFRP bars with a T_g of 250°F (121°C) retained as much as 65 percent of their tensile strength at temperatures exceeding 625°F (330°C).

4.4.5 *Structural fire resistance design*—The applicable load and resistance factors, as well as the deflection and crack width serviceability criteria used during the design process, will also affect fire performance. Differences in the design philosophy between design guides and codes for steel-reinforced concrete (which is typically controlled by strength), when compared to those for FRP-reinforced concrete (which is often controlled by serviceability criteria), mean that when subjected to a fire, the flexural capacity of FRP-reinforced members may inherently have a much greater reserve strength. This issue is discussed in detail by Bisby and Kodur (2007).

Given these aforementioned issues, rational design of FRP-reinforced concrete members for fire safety should be performed with a clear understanding of the appropriate structural fire endurance criteria applicable to the buildings in which they are proposed. The designer should account for temperature-induced reductions in the bond between FRP bars and concrete; note that bond strength reductions are not typically considered important for fire-safety design with conventional deformed steel reinforcing bars. The general approach to fire endurance testing given by ASTM E119 is recommended. When conducting fire tests using ASTM E119 for FRP-reinforced concrete members, the critical temperature approach should not be taken for the FRP reinforcement unless rational consideration of holistic member performance in fire has been performed and has led to the definition of the critical temperature. In most cases, the concrete element should be tested in a loaded condition and the fire endurance should be determined based on strength and deflection requirements rather than on the temperature in the FRP. Relevant integrity and insulation fire resistance criteria still apply (ACI 216.1).

Several full-scale tests performed in accordance with ASTM E119, CAN/ULC S101-M89, DIN EN 1363 (Deutsches Institut für Normung e.V. 2012), and ISO 834-1 provisions have been performed on GFRP- and CFRP-reinforced beams and slabs (Nigro et al. 2011a, 2013). These tests have led to structural fire resistance ratings for FRP-reinforced concrete elements in excess of 3 hours for cases where the FRP bars were well anchored into cool regions of the concrete. For example, Nigro et al. (2013) showed that simply-supported concrete slabs reinforced with well-anchored GFRP bars (with T_g of 210°F [100°C] and a clear concrete cover of 1 in. [25.4 mm]) were able to carry 60 percent of their design ultimate strength for more than 3 hours of exposure to the ISO 834 standard fire. The temperature of the bars exceeded 930°F (500°C); however, the bars continued to carry load due to the presence of effective anchoring details.

CHAPTER 5—DURABILITY

Fiber-reinforced polymer (FRP) bars are susceptible to varying amounts of strength and stiffness changes in the presence of environments before, during, and after construction. These environments can include water, ultraviolet exposure, elevated temperature, alkaline or acidic solutions, and saline solutions. Strength and stiffness may increase, decrease, or remain the same, depending on the particular material and exposure conditions. Tensile and bond properties of FRP bars are the primary parameters of interest for reinforced concrete construction.

5.1—Accelerated durability testing

The environmental condition that has attracted the most interest by investigators concerned with FRP bars is the highly alkaline pore water found in outdoor concrete structures (Gerritse 1992; Takewaka and Khin 1996; Rostasy 1997; Yamaguchi et al. 1997). Methods for systematically accelerating the strength degradation of bare, unstressed, glass filaments in concrete using temperature have been successful (Litherland et al. 1981) and have also often been applied to glass FRP (GFRP) materials to predict long-term performance in alkaline solutions. Test methods for evaluating the durability of FRP bars in alkaline solutions are presented in ASTM D7705. There is no substantiation to date, however, that accelerated methods for bare glass (where only one chemical reaction controls degradation) apply to GFRP composites (where multiple reactions and degradation mechanisms may be activated at once or sequentially). Also, the effect of applied stress during exposure should be factored as well. Due to insufficient data on combined weathering and applied stress, the discussions of weathering, creep, and fatigue are kept separate in this guide. Therefore, while short-term experiments using aggressive environments enable quick comparisons of materials, extrapolation of the results to field conditions and expected lifetimes is not possible in the absence of real-time data (Barkatt et al. 1998; Clarke and Sheard 1998). In most cases available to date, bare bars were subjected to the aggressive environment under no load. The relationships between data on bare bars and data on bars embedded in concrete are affected by additional variables, such as the degree of protection offered to the bars by the concrete (Clarke and Sheard 1998; Scheibe and Rostasy 1998; Sen et al. 1998a,b). Test times are typically in the 10- to 30-month range.

5.2—Durability of FRP bars

Due to the large amount of literature on the subject (Benmokrane and Rahman 1998) and the limited space herein, some generalizations should be made at the expense of presenting particular quantitative results. With these cautions in mind, representative experimental results for a range of FRP bar materials and test conditions are reviewed in the balance of this section. Conservatism is advised in applying these results in design until additional long-term durability data are available.

5.2.1 *Alkaline solutions*—Aqueous solutions with high values of pH varying from 11.5 to 13.0 are known to degrade the tensile strength and stiffness of GFRP bars (Porter and Barnes 1998). Particular results, however, vary significantly according to differences in test methods that, in addition to pH, include composition of the chemical solution, temperature, and presence of load. Higher temperatures and longer exposure times exacerbate the problem. Most data have been generated using temperatures as low as slightly subfreezing and as high as a few degrees below the T_g of the resin. The degree to which the resin protects the glass fibers from the diffusion of deleterious hydroxyl (OH^-) ions figures prominently in the alkali resistance of GFRP bars (Bank and Puterman 1997; Coomarasamy and Goodman 1997; GangaRao and Vijay 1997b; Porter et al. 1997; Bakis et al. 1998; Tannous and Saadatmanesh 1999; Uomoto 2000). Most researchers believe that vinyl ester resins have superior resistance to moisture ingress compared with other commodity resins. The type of glass fiber also appears to be an important factor in the alkali resistance of GFRP bars (Devalapura et al. 1996). Research examining the durability of different glass-fiber-based products, including North American GFRP reinforcing bars, have been performed by several researchers (Robert and Benmokrane 2013; Belarbi and Wang 2012; Kamal and Boulfiza 2011; Robert et al. 2009; Chen et al. 2007). Tensile strength reductions in GFRP bars ranging from 0 to 75 percent of initial values have been reported in the cited literature. Tensile stiffness reductions in stressed and unstressed GFRP bars range between 0 and 20 percent in many cases. Tensile strength and stiffness of aramid FRP (AFRP) rods in elevated-temperature alkaline solutions, either with or without tensile stress applied, have been reported to decrease between 10 and 50 percent and 0 and 20 percent of initial values, respectively (Takewaka and Khin 1996; Rostasy 1997; Sen at al. 1998b). In the case of unstressed carbon FRP (CFRP), strength and stiffness have been reported to each decrease between 0 and 20 percent (Takewaka and Khin 1996).

5.2.2 *Ultraviolet radiation and moisture*—Exposure of FRP bars to ultraviolet rays and moisture before their placement in concrete could adversely affect their tensile strength due to degradation of the polymer constituents, including aramid fibers and all resins. Proper construction practices and resin additives can mitigate this type of weathering problem significantly. It is highly recommended that, before placement in concrete, FRP bars are protected from direct exposure to sunlight and moisture. Some results from combined ultraviolet and moisture exposure tests with and without applied stress applied to the bars have shown tensile strength reductions of 0 to 20 percent of initial values in CFRP, 0 to 30 percent in AFRP, and 0 to 40 percent in GFRP (Sasaki et al. 1997; Uomoto 2000). An extensive study of GFRP, AFRP, and CFRP bars kept outdoors in a rack by the ocean showed no significant change of tensile strength or modulus of any of the bars (Tomosawa and Nakatsuji 1996, 1997).

5.2.3 *Saline solutions*—It has been shown (Rahman et al. 1996) that the addition of various types of salts to the solution in which FRP bars are immersed does not necessarily make a significant difference in the strength and stiffness of many FRP bars when compared with those in a solution without salt. Most studies, however, do not separate the effects of water and salt added to water. One study found a 0 to 20 percent reduction of initial tensile strength in GFRP bars subjected to a saline solution at room temperature and cyclic freezing-and-thawing temperatures (Vijay and GangaRao 1999); another found a 15 percent reduction in the strength of AFRP bars in a marine environment (Sen et al. 1998b).

5.3—Durability of bond between FRP and concrete

Studies of the durability of bond between FRP and concrete have been mostly concerned with the moist, alkaline environment found in concrete. The bond of FRP reinforcement relies on the transfer of shear and transverse forces at the interface between bar and concrete, and between individual fibers within the bar. These resin-dominated mechanisms are in contrast to the fiber-dominated mechanisms that control properties such as longitudinal strength and stiffness of FRP bars. Environments that degrade the polymer resin or fiber/resin interface are therefore also likely to degrade the bond strength of an FRP bar.

Numerous bond test methods (pullout tests, tension tests, and beam-end tests) have been proposed for FRP bars, although the direct pullout test has been the most popular due to its simplicity and low cost despite its inability to represent the concrete stress state in most of the practical situations (Nanni et al. 1995; Tepfers 2002). Pullout specimens with CFRP and GFRP bars have been subjected to natural environmental exposures and have not indicated significant decreases in bond strength over periods of time between 1 and 2 years (Clarke and Sheard 1998; Sen et al. 1998a). Positive and negative trends in pullout strength with respect to shorter periods of time have been obtained with GFRP bars subjected to wet elevated-temperature environments in concrete, with or without artificially added alkalinity (Al-Dulaijan et al. 1996; Bakis et al. 1998; Bank et al. 1998; Porter and Barnes 1998). Similar observations on such pullout tests using specimens subjected to accelerated environmental exposure carry over to AFRP and CFRP bars (Conrad et al. 1998). Longitudinal cracking in the concrete cover can seriously degrade the bond capability of FRP bars, and sufficient measures should be taken to prevent such cracking in laboratory tests and field applications (Sen et al. 1998a). The ability of chemical agents to pass through the concrete to the FRP bar is another important factor thought to affect bond strength (Porter and Barnes 1998).

CHAPTER 6—GENERAL DESIGN CONSIDERATIONS

The general design recommendations for flexural concrete elements reinforced with FRP bars are presented in this chapter. Recommendations are based on principles of equilibrium and compatibility, and the constitutive laws of the materials. Also, the brittle behavior of both FRP reinforcement and concrete allows for consideration to be given to either compression-controlled or tension-controlled modes of flexural failure. Detrimental effects of high temperature and fire on FRP-reinforced structures are discussed in 4.4.

6.1—Design philosophy

Although strength and working stress design approaches were considered, the strength design approach of reinforced concrete members reinforced with FRP bars was preferred to ensure consistency with other ACI documents. Design recommendations are based on limit state design principles. In many instances, serviceability criteria or fatigue and creep rupture endurance limits could control the design of concrete members reinforced for flexure with FRP bars, especially aramid FRP (AFRP) and glass FRP (GFRP) that exhibit low stiffness.

The load factors given in ACI 318 are used to determine the required strength of a concrete member reinforced with FRP.

6.2—Design material properties

Material properties provided by the manufacturer, such as the guaranteed tensile strength, should be considered as initial properties that do not include the effects of long-term exposure to the environment. Because long-term exposure to various types of environments can reduce the tensile strength and creep rupture and fatigue endurance of FRP bars, the material properties used in design equations should be reduced based on the type and level of environmental exposure.

Equations (6.2a) and (6.2b) give the tensile properties that should be used in all design equations. The design tensile strength should be determined by

$$f_{fu} = C_E f_{fu}^* \tag{6.2a}$$

The design rupture strain should be determined as

$$\varepsilon_{fu} = C_E \varepsilon_{fu}^* \tag{6.2b}$$

The design modulus of elasticity will be the same as the value reported by the manufacturer as the mean elastic modulus (guaranteed value) of a sample of test specimens ($E_f = E_{f,ave}$).

The environmental reduction factors given in Table 6.2 are conservative estimates, depending on the durability of each fiber type, and are based on the consensus of ACI Committee 440. Temperature effects are included in the C_E values. Fiber-reinforced polymer bars, however, should not be used in environments with a service temperature higher than the T_g of the resin used for their manufacturing. It is expected that with continued research, these values will become more reflective of actual effects of environment. The methodology regarding the use of these factors, however, is not expected to change.

Table 6.2—Environmental reduction factor for various fibers and exposure conditions

Exposure condition	Fiber type	Environmental reduction factor C_E
Concrete not exposed to earth and weather	Carbon	1.0
	Glass	0.8
	Aramid	0.9
Concrete exposed to earth and weather	Carbon	0.9
	Glass	0.7
	Aramid	0.8

6.2.1 *Tensile strength of FRP bars at bends*—The design tensile strength of FRP bars at a bend can be determined as

$$f_{fb} = \left(0.05 \cdot \frac{r_b}{d_b} + 0.3\right) f_{fu} \leq f_{fu} \tag{6.2.1}$$

Equation (6.2.1) is adapted from design recommendations by the Japan Society of Civil Engineers (1997b). Limited research on FRP hooks (Ehsani et al. 1995) indicates that the tensile force developed by the bent portion of a GFRP bar is mainly influenced by the ratio of the bend radius to the bar diameter, r_b/d_b; the tail length; and, to a lesser extent, the concrete strength.

For an alternative determination of the reduction in tensile strength due to bending, manufacturers of bent bars may provide test results based on test methodologies cited in ACI 440.3R.

CHAPTER 7—FLEXURE

The design of FRP-reinforced concrete members for flexure is analogous to the design of steel-reinforced concrete members. Experimental data on concrete members reinforced with FRP bars show that flexural capacity can be calculated based on assumptions similar to those made for members reinforced with steel bars (Faza and GangaRao 1993a; Nanni 1993b; GangaRao and Vijay 1997a). The design of members reinforced with FRP bars should take into account the uniaxial stress-strain relationship of FRP materials.

7.1—General considerations

This chapter specifically references rectangular sections with a single layer of one type of tensile FRP reinforcement, as the experimental work has almost exclusively considered members with this cross-sectional shape and reinforcement layout. The concepts described herein, however, can also be applied to the analysis and design of members with different geometry and multiple types, multiple layers, or both, of FRP reinforcement. Although there is no evidence that the flexural theory, as developed herein, does not apply equally well to nonrectangular sections, the behavior of nonrectangular sections has yet to be confirmed by experimental results.

7.1.1 *Flexural design philosophy*—Steel-reinforced concrete sections are commonly designed to ensure tension-controlled behavior exhibited by yielding of steel before the crushing of concrete. The yielding of the steel provides ductility and a warning of failure of the member. The nonductile behavior of FRP reinforcement necessitates a reconsideration of this approach.

If FRP reinforcement ruptures, failure of the member is sudden and catastrophic (Nanni 1993b; Jaeger et al. 1997; GangaRao and Vijay 1997a; Theriault and Benmokrane 1998); however, there would be limited warning of impending failure in the form of extensive cracking and large deflection due to the significant elastic elongation that FRP reinforcement experiences before rupture. In any case, the member would not exhibit ductility as is commonly observed for tension-controlled concrete beams reinforced with steel reinforcing bars, in which the bars exhibit plastic deformation prior to concrete crushing.

Compression-controlled behavior is marginally more desirable for flexural members reinforced with FRP bars (Nanni 1993b). By experiencing concrete crushing prior to tensile rupture of the FRP reinforcement, a flexural member does exhibit some inelastic behavior before failure.

In conclusion, both compression- and tension-controlled sections are acceptable in the design of flexural members reinforced with FRP bars, provided that strength and serviceability criteria are satisfied. To compensate for the lack of ductility, the member should possess a higher reserve of strength. The margin of safety suggested by this guide against failure is therefore higher than that used in traditional steel-reinforced concrete design.

The use of high-strength concrete allows for better use of the high-strength properties of FRP bars and can increase the stiffness of the cracked section, but the brittleness of high-strength concrete, as compared with normal-strength concrete, can reduce the overall deformability of the flexural member (GangaRao and Vijay 1997a).

Figure 7.1.1 shows a comparison of the theoretical moment-curvature behavior of beam cross sections designed for the same strength ϕM_n following the principles of ultimate strength design described in this chapter, including the recommended strength reduction factors according to this guide for FRP and ACI 318 for steel. Three cases are presented in addition to the steel-reinforced cross section: two sections reinforced with glass FRP (GFRP) bars, and one reinforced with carbon FRP (CFRP) bars. For the tension-controlled GFRP-reinforced section, the concrete dimensions are larger than for the other beams to attain the same design capacity.

7.1.2 *Assumptions*—Computations of the strength of cross sections should be performed based on the following assumptions:

a) Strain in the concrete and the FRP reinforcement is proportional to the distance from the neutral axis (a plane section before loading remains plane after loading).

b) The maximum usable compressive strain in the concrete is assumed to be 0.003.

c) The tensile strength of concrete is ignored.

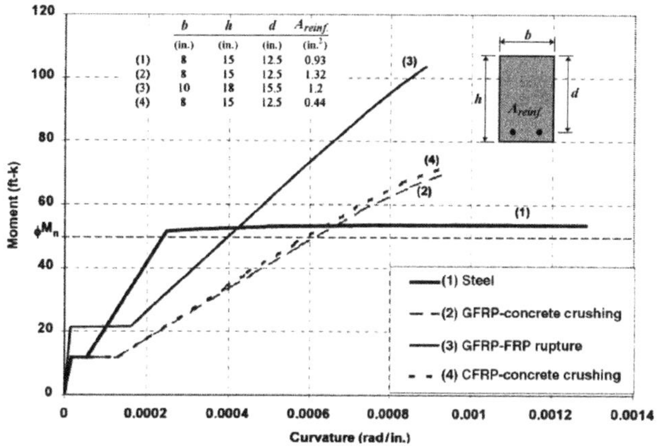

Fig. 7.1.1—Theoretical moment-curvature relationships for reinforced concrete sections using steel and FRP bars (ϕ factors of 0.9, 0.65, 0.55, and 0.65 for tension-controlled steel, compression-controlled GFRP, tension-controlled GFRP, and compression-controlled CFRP, respectively).

d) The tensile behavior of the FRP reinforcement is linearly elastic until failure.

e) A perfect bond exists between concrete and FRP reinforcement.

7.2—Flexural strength

The strength design philosophy states that the design flexural strength at a section of a member should exceed the factored moment (Eq. (7.2)). Design flexural strength refers to the nominal flexural strength of the member multiplied by a strength reduction factor (ϕ, discussed in 7.2.3). The factored moment refers to the moments calculated by the use of factored loads as prescribed in ACI 318 (for example, $1.2D + 1.6L + ...$)

$$\phi M_n \geq M_u \tag{7.2}$$

The nominal flexural strength of an FRP-reinforced concrete member can be determined based on strain compatibility, internal force equilibrium, and the controlling strength limit state (concrete crushing or FRP rupture). Figure 7.2 illustrates the stress, strain, and internal forces for the three possible cases of a rectangular section reinforced with FRP bars.

7.2.1 *Strength limit states*—The flexural capacity of an FRP-reinforced flexural member is dependent on whether it is controlled by concrete crushing or FRP rupture. The controlling limit state can be determined by comparing the FRP reinforcement ratio to the balanced reinforcement ratio, which is a ratio where concrete crushing and FRP rupture occur simultaneously. Because FRP does not yield, the balanced ratio of FRP reinforcement is computed using its design tensile strength. The FRP reinforcement ratio can be computed from Eq. (7.2.1a), and the balanced FRP reinforcement ratio can be computed from Eq. (7.2.1b)

$$\rho_f = \frac{A_f}{bd} \tag{7.2.1a}$$

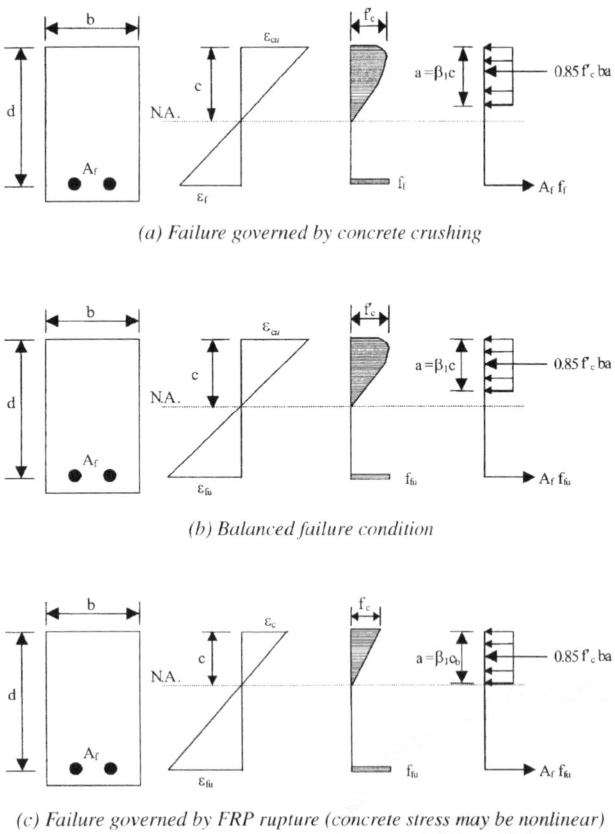

Fig. 7.2—Strain and stress distribution at ultimate conditions.

$$\rho_{fb} = 0.85\beta_1 \frac{f'_c}{f_{fu}} \frac{E_f \varepsilon_{cu}}{E_f \varepsilon_{cu} + f_{fu}} \qquad (7.2.1b)$$

If the reinforcement ratio is less than the balanced ratio ($\rho_f < \rho_{fb}$), the FRP rupture limit state controls. Otherwise, ($\rho_f > \rho_{fb}$) the concrete crushing limit state controls.

Table 7.2.1 reports some typical values for the balanced reinforcement ratio, showing that the balanced ratio for FRP reinforcement ρ_{fb} is much lower than the balanced ratio for steel reinforcement ρ_b. The balanced ratio for FRP reinforcement can be even lower than the minimum reinforcement ratio for steel ($\rho_{min} = 0.0035$ for Grade 60 steel and $f_c' = 5000$ psi [35 MPa]).

7.2.2 *Nominal flexural strength*—When $\rho_f > \rho_{fb}$, the controlling limit state is crushing of the concrete, and the stress distribution in the concrete can be approximated with the ACI rectangular stress block. Based on the equilibrium of forces and strain compatibility (shown in Fig. 7.2), the following can be derived

$$M_n = A_f f_f \left(d - \frac{a}{2}\right) \qquad (7.2.2a)$$

$$a = \frac{A_f f_f}{0.85 f_c' b} \qquad (7.2.2b)$$

$$f_f = E_f \varepsilon_{cu} \frac{\beta_1 d - a}{a} \qquad (7.2.2c)$$

Table 7.2.1—Typical values for balanced reinforcement ratio for a rectangular section with f_c' = 5000 psi (34.5 MPa)

Bar type	Yield strength f_y or tensile strength f_{fu}, ksi (MPa)	Modulus of elasticity, ksi (GPa)	ρ_b or ρ_{fb}
Steel	60 (414)	29,000 (200)	0.0335
GFRP	80 (552)	6000 (41.4)	0.0078
AFRP	170 (1172)	12,000 (82.7)	0.0035
CFRP	300 (2070)	22,000 (152)	0.0020

Substituting a from Eq. (7.2.2b) into Eq. (7.2.2c) and solving for f_f gives

$$f_f = \left(\sqrt{\frac{(E_f \varepsilon_{cu})^2}{4} + \frac{0.85\beta_1 f_c'}{\rho_f} E_f \varepsilon_{cu}} - 0.5 E_f \varepsilon_{cu}\right) \leq f_{fu} \qquad (7.2.2d)$$

The nominal flexural strength can be determined from Eq. (7.2.2a), (7.2.2b), and (7.2.2d). The FRP reinforcement is linearly elastic at the concrete crushing limit state, so the stress level in the FRP can be found from Eq. (7.2.2c) because it is less than f_{fu}.

Alternatively, the nominal flexural strength at a section can be expressed in terms of the FRP reinforcement ratio, as given in Eq. (7.2.2e) to replace Eq. (8.2.2a).

$$M_n = \rho_f f_f \left(1 - 0.59 \frac{\rho_f f_f}{f_c'}\right) b d^2 \qquad (7.2.2e)$$

When $\rho_f < \rho_{fb}$, the controlling limit state is rupture of the FRP reinforcement, and the nominal flexural strength at a section can be computed as shown in Eq. (7.2.2f)

$$M_n = A_f f_{fu} \left(d - \frac{\beta_1 c}{2}\right) \qquad (7.2.2f)$$

Although the stress in the reinforcement is known, the analysis incorporates two unknowns: the concrete compressive strain at ultimate when the FRP ruptures in tension (ε_c) and the depth to the neutral axis c. The analysis involving these unknowns becomes complex and is not easily solved by closed-form solution. The ACI equivalent rectangular stress block parameters are also not applicable because the maximum concrete strain may not be attained ($\varepsilon_c < \varepsilon_{cu}$). In this case, equivalent rectangular stress block parameters (α_1 and β_1) that approximate the centroid of the stress distribution in the concrete at the particular strain level reached would need to be used. The factor α_1 is the ratio of the average concrete stress to the concrete strength. Factor β_1 is the ratio of the depth of the equivalent rectangular stress block to the depth of the neutral axis.

For a given section, the product of $\beta_1 c$ in Eq. (7.2.2f) varies depending on material properties and FRP reinforcement ratio. For a section controlled by the limit state of FRP rupture, the maximum value for this product is equal to $\beta_1 c_b$

and is achieved when the maximum concrete strain (0.003) is attained. Therefore, a simplified and conservative lower bound calculation of the nominal flexural strength of the member can be based on Eq. (7.2.2g) and (7.2.2h) as follows

$$M_n = A_f f_{fu}\left(d - \frac{\beta_1 c_b}{2}\right) \quad (7.2.2g)$$

$$c_b = \left(\frac{\varepsilon_{cu}}{\varepsilon_{cu} + \varepsilon_{fu}}\right)d \quad (7.2.2h)$$

7.2.3 *Strength reduction factor for flexure*—Because FRP members do not exhibit ductile behavior, a conservative strength reduction factor should be adopted to provide a higher reserve of strength in the member. The Japanese recommendations for design of flexural members using FRP suggest a strength reduction factor equal to 0.77 (Japan Society of Civil Engineers 1997b). Other researchers (Benmokrane et al. 1996) suggest a value of 0.75 based on probabilistic concepts.

Based on ACI 318, the ϕ factor for design of a compression-controlled section is 0.65, with a target reliability index between 3.5 and 4.0 (Szerszen and Nowak 2003). A reliability analysis on FRP-reinforced beams in flexure using Load Combination 2 from ACI 318 for live-to-dead load ratios between 1 and 3 indicated reliability indexes between 3.5 and 4.0 when the ϕ factor was set to 0.65 for a compression-controlled section, and 0.55 for a tension-controlled section using Eq. (7.2.2g) (Shield et al. 2011). A nonlinear sectional analysis of curvatures at failure showed that the curvatures of typical FRP-reinforced beams at failure varied between $0.016/d$ and $0.018/d$ for tension-controlled failures, and between $0.011/d$ and $0.02/d$ for compression-controlled failures (Shield et al. 2011).

ACI 318 considers the section tension-controlled whenever the curvature is greater than $0.008/d$ (corresponding to a strain in the steel of 0.005). This indicates that due to the low modulus of elasticity of the reinforcement, FRP-reinforced beams will have large deflections at ultimate, and that FRP-reinforced beams with controlling limit states of FRP reinforcing bar rupture will have larger deflections at ultimate than those that are controlled by concrete crushing. Even though the curvature values of FRP-reinforced beams are larger than those of equivalent steel-reinforced beams, the committee recommends a ϕ factor of 0.55 for tension-controlled section design to maintain a minimum reliability index of 3.5.

While a concrete crushing limit state can be predicted based on calculations, the member as constructed may not fail accordingly. For example, if the concrete strength is higher than specified, the section capacity may be controlled by FRP rupture. For this reason, and to establish a transition between the two values of ϕ, a compression-controlled FRP reinforced concrete section is defined as a section in which $\rho_f \geq 1.4\rho_{fb}$, and a tension-controlled FRP reinforced concrete section is defined as one in which $\rho_f \leq \rho_{fb}$. A section with $\rho_{fb} < \rho_f < 1.4\rho_{fb}$ will theoretically be controlled by the concrete crushing limit state, but a reduced value of ϕ should be used, relative to a compression-controlled section.

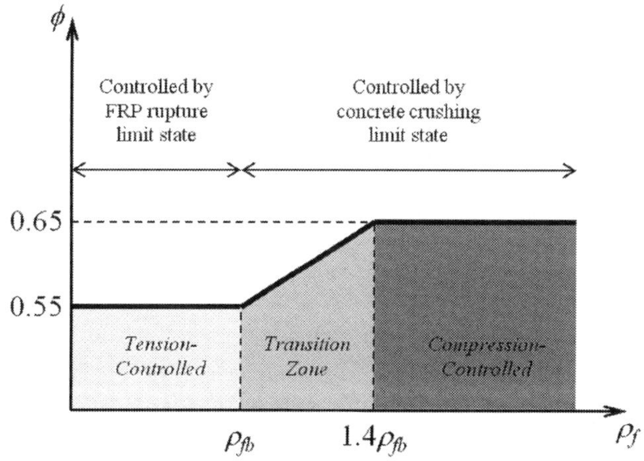

Fig. 7.2.3—Strength reduction factor as function of reinforcement ratio.

The strength reduction factor for flexure can be computed by Eq. (7.2.3). This equation is represented graphically by Fig. 7.2.3, and gives a factor of 0.65 for compression-controlled sections, 0.55 for tension-controlled sections, and provides a linear transition between the two.

$$\phi = \begin{cases} 0.55 \text{ for } \rho_f \leq \rho_{fb} \\ 0.3 + 0.25\dfrac{\rho_f}{\rho_{fb}} \text{ for } \rho_{fb} < \rho_f < 1.4\rho_{fb} \\ 0.65 \text{ for } \rho_f \geq 1.4\rho_{fb} \end{cases} \quad (7.2.3)$$

7.2.4 *Minimum FRP reinforcement*—If a section is tension controlled ($\rho_f \leq \rho_{fb}$), a minimum amount of reinforcement should be provided to prevent failure upon concrete cracking, $\phi M_n \geq M_{cr}$, where M_{cr} is the cracking moment. The provisions in ACI 318 for minimum reinforcement are based on this concept and, with modifications, are applicable to FRP-reinforced members. The modifications result from a different strength reduction factor—0.55 for tension-controlled sections instead of 0.9. The minimum reinforcement area for FRP-reinforced members is obtained by multiplying the existing ACI 318 equation for steel reinforcement by 1.64 (1.64 = 0.90/0.55). This results in Eq. (7.2.4)

$$A_{f,min} = \frac{4.9\sqrt{f_c'}}{f_{fu}} b_w d \geq \frac{330}{f_{fu}} b_w d \quad (7.2.4)$$

For SI units

$$A_{f,min} = \frac{0.41\sqrt{f_c'}}{f_{fu}} b_w d \geq \frac{2.3}{f_{fu}} b_w d$$

If a section is not tension controlled ($\rho_f > \rho_{fb}$), the minimum amount of reinforcement to prevent failure upon cracking is automatically achieved. Therefore, Eq. (7.2.4) is required as a check only if $\rho_f \leq \rho_{fb}$. The requirements of Eq. (7.2.4) need not be applied if at every section the area of tensile reinforcement provided is at least one-third greater than that required by analysis.

7.2.5 *Special considerations*

7.2.5.1 *Multiple layers of reinforcement and combinations of different FRP types*—In a tension-controlled section, all steel tension reinforcement is assumed to yield at ultimate when using the strength design method to calculate the nominal flexural strength of members with steel reinforcement arranged in multiple layers. Therefore, the tension force is assumed to act at the centroid of the reinforcement with a magnitude equal to the area of tension reinforcement times the yield strength of steel. Because FRP materials have no plastic region, the stress in each reinforcement layer will vary depending on its distance from the neutral axis. In these cases, strain in the outermost layer should be used to determine if the section is compression- or tension-controlled. The analysis of the flexural capacity should be based on a strain-compatibility approach. Similarly, if different types of FRP bars are used to reinforce the same member, the variation in the stress level in each bar type should be considered when calculating the flexural capacity.

7.2.5.2 *Moment redistribution*—Plastic hinges do not form in members reinforced with FRP bars due to their linear-elastic behavior up to rupture. Consequently, moment redistribution in continuous beams or other statically indeterminate structures should not be considered for FRP-reinforced concrete.

7.2.5.3 *Compression reinforcement*—FRP reinforcement has a significantly lower compressive strength than tensile strength, and is subject to significant variation (Kobayashi and Fujisaki 1995; Japan Society of Concrete Engineers 1997b). Therefore, the strength of any FRP bar in compression should be ignored in design calculations (Almusallam et al. 1997).

Placing FRP bars in the compression zone of flexural members, however, cannot be avoided in some cases. Examples include the supports of continuous beams or where bars secure the stirrups in place. In these cases, confinement should be considered for the FRP bars in compression regions to prevent their instability and to minimize the effect of the relatively high transverse expansion of some types of FRP bars. The transverse FRP reinforcement in the form of ties should have a spacing smaller than the least cross-sectional dimension or 16 longitudinal bar diameters or 48 tie bar diameters.

7.3—Serviceability

Fiber-reinforced polymer reinforced concrete members have a relatively small stiffness after cracking, when compared to steel-reinforced concrete members with the same reinforcement ratio. Consequently, permissible deflections under service loads can control the design. In general, designing FRP-reinforced cross sections for flexural strength may not satisfy serviceability criteria for deflection and crack control (Nanni 1993a; GangaRao and Vijay 1997a; Theriault and Benmokrane 1998; Bischoff 2005).

Serviceability can be defined as satisfactory performance under service load conditions. This, in turn, can be described in terms of two criteria:

1) Cracking—Excessive crack width is undesirable for aesthetic and other reasons (for example, to prevent water leakage) that can damage or deteriorate the structural concrete.

2) Deflection—Deflections should be within acceptable limits imposed by the use of the structure (for example, supporting attached nonstructural elements without damage).

The serviceability provisions given in ACI 318, Section 9.5 for deflection and Section 10.6 for crack control, need to be modified for FRP-reinforced members to account for the increased flexibility when using lower stiffness reinforcement. The substitution of FRP for steel on an equal area basis, for example, would typically result in larger deflections and wider crack widths (Gao et al. 1998a; Tighiouart et al. 1998).

7.3.1 *Cracking*—Fiber-reinforced polymer bars are corrosion-resistant; therefore, larger crack widths as compared to steel-reinforced concrete can be tolerated when corrosion of reinforcement is the primary reason for crack control. Other considerations with regard to acceptable crack width limits include aesthetics, creep rupture, and shear effects.

Two design methodologies exist for proportioning reinforcement to control flexural cracking: 1) a direct procedure in which crack widths are calculated; and 2) an indirect procedure in which maximum bar spacing limits are specified. Direct crack control refers to the calculation of a probable crack width and its comparison with an allowable crack width. This approach is followed by the Japan Society of Civil Engineers (1997b), which only considers aesthetics in setting a maximum allowable crack width of 0.020 in. (0.50 mm) and by CAN/CSA S6-06, which explicitly permits proportioning of the FRP reinforcement in such a way that the crack width does not exceed 0.020 in. (0.50 mm) for members subjected to aggressive environments, and 0.028 in. (0.70 mm) for other members. The ACI 318 crack control provisions do not address FRP reinforcement. For comparison purposes, the crack control provisions for steel reinforcement in ACI 318 correspond to a maximum crack width that varies approximately between 0.018 and 0.022 in. (0.46 and 0.56 mm).

Because of concerns about the adequacy of the empirically tuned model proposed by Gergely and Lutz (1968) for predicting crack widths in flexural members with large bar covers, ACI 318-99 replaced the traditional *z*-factor approach for crack control with an indirect procedure that controls flexural crack widths through a maximum reinforcing bar spacing. The dependence on exposure conditions was also eliminated. Rather than specifying a maximum allowable crack width, the objective now is to determine the maximum bar spacing necessary to achieve the required serviceability limit state of cracking based on the FRP reinforcement properties and the corresponding FRP bar stress (or strain) at service load levels. The current procedure also acknowledges the dominant effect that clear cover has on flexural cracking.

The maximum reinforcement spacing provisions of ACI 318 are derived from the crack width formulation developed by Frosch (1999), which is based on a physical model rather than being empirically derived.

To be consistent with ACI 318, flexural crack control in FRP-reinforced concrete beams and one-way slabs can be accomplished by specifying a maximum FRP bar spacing equal to

$$s_{max} = 1.15\frac{E_f w}{f_{fs} k_b} - 2.5c_c \leq 0.92\frac{E_f w}{f_{fs} k_b} \quad (7.3.1a)$$

The proposed flexural crack control procedure is said to be indirect because the maximum FRP bar spacing stipulated by Eq. (7.3.1a) would indirectly comply with a target maximum allowable crack width. The rationale behind this indirect flexural crack control approach is described by Ospina and Bakis (2007).

For the selected FRP stress level and target crack width limit, the evaluation of the maximum bar spacing per Eq. (7.3.1a) shall be based on a d_c value that complies with Eq. (7.3.1b). If a larger d_c value is required for specific durability requirements or any other reason and the maximum crack width limit cannot be relaxed, it is then necessary to reduce the stress level in the FRP reinforcement. This can be accomplished, for example, by increasing the amount of flexural reinforcement.

$$d_c \leq \frac{E_f w}{2 f_{fs} \beta k_b} \quad (7.3.1b)$$

Selection of the limiting crack width to be used in Eq. (7.3.1a) and (7.3.1b) depends on the intended use of the structure. The procedure allows for controlling different levels of flexural cracking, spanning from very narrow cracks in structures in aggressive environments or where water tightness is required, to situations where wider cracks may be acceptable due to the superior corrosion resistance of FRP reinforcement. In general, crack widths in FRP-reinforced members will be larger than those in steel-reinforced members. In situations where crack widths are limited by aesthetic reasons, limiting crack widths in the range of 0.016 to 0.028 in. (0.4 to 0.7 mm) are generally acceptable.

The FRP stress at service, f_{fs}, can be evaluated by performing a cracked-elastic section analysis. Recommendations for the bar stress at service load in ACI 318 (f_s = $0.67f_y$) are only applicable to steel-reinforced concrete when strength governs design and the steel reinforcement yields at failure. The same approach is not possible for FRP-reinforced concrete, as design is most often governed by serviceability limits related to deflection and crack width control or by fatigue and creep rupture effects.

The k_b term is a coefficient that accounts for the degree of bond between the FRP bar and the surrounding concrete. For FRP bars having bond behavior similar to uncoated steel bars, the bond coefficient k_b is assumed equal to 1.0. For FRP bars having bond behavior inferior to steel, k_b is larger than 1.0, and for FRP bars having bond behavior superior to steel, k_b is smaller than 1.0. A test method for the determination of k_b has been approved by the Canadian Standards Association in CAN/CSA S806-12. For an analysis of crack width data performed by Bakis et al. (2006) on a variety of concrete cross-sections and FRP bar manufacturers, fiber types, resin formulations, and surface treatments, average k_b values ranged from 0.60 to 1.72, with a mean of 1.10. Data for rough sand-coated FRP bar surface treatments trended toward the lower end of this range. The consensus of the committee, for the case where k_b is not known from experimental data, is that a conservative value of 1.4 should be assumed. Smooth bars and grids are specifically excluded from this recommendation. Further analysis is needed before a committee consensus can be reached on k_b for such reinforcement.

Bakis and Boothby (2004) found that crack widths in GFRP-reinforced concrete beams under sustained loads increased beyond initial values by approximately 40 percent in an indoor environment and by approximately 60 percent in an outdoor environment over a period of 3 years. Further research is needed to examine the impact of long-term effects on these flexural crack control design provisions.

7.3.2 *Deflections*—In general, ACI 318 provisions for deflection control are concerned with deflections that occur at service levels under immediate and sustained static loads, and do not apply to dynamic loads such as earthquakes, transient winds, or vibration of machinery. Two methods are given in ACI 318 for control of deflections of one-way flexural members:

1) The indirect method of mandating the minimum thickness of the member (Table 9.5(a) in ACI 318).

2) The direct method of limiting computed deflections (Table 9.5(b) in ACI 318).

Because of the variable stiffness, brittle-elastic nature, and particular bond features of FRP reinforcement, deflections of FRP-reinforced concrete members are more sensitive to the variables affecting deflection than steel-reinforced members of identical size and reinforcement layout. Deflections in members with FRP reinforcement also tend to be greater in magnitude because of the lower stiffness associated with commercially available FRP reinforcement. This guide therefore recommends the use of a direct method of deflection control, as outlined as follows in 7.3.2.2 and 7.3.2.3. Recommended minimum thicknesses for FRP-reinforced members are provided in 7.3.2.1 for convenience in establishing initial member proportions for design only. Member dimensions may need to be revised based on the limits of calculated deflections.

7.3.2.1 *Recommended minimum thicknesses for design*—Recommended minimum thicknesses for design of one-way slabs and beams are provided in Table 7.3.2.1. The table is intended only to provide guidance for initial design. Use of these recommended minimum thicknesses does not guarantee that all deflection considerations will be satisfied for a particular project.

Values in Table 7.3.2.1 are based on a generic maximum span-to-depth ratio limitation (Ospina et al. 2001) corre-

Table 7.3.2.1—Recommended minimum thickness of nonprestressed beams or one-way slabs

	Minimum thickness h			
Member	Simply supported	One end continuous	Both ends continuous	Cantilever
Solid one-way slabs	$\ell/13$	$\ell/17$	$\ell/22$	$\ell/5.5$
Beams	$\ell/10$	$\ell/12$	$\ell/16$	$\ell/4$

sponding to the limiting curvature associated with a target deflection-span ratio (Eq. (7.3.2.1)). The procedure, which is described in detail by Ospina and Gross (2005), can be applied to flexural members at service loads with any type of linear-elastic reinforcement.

$$\frac{\ell}{h} \leq \frac{48\eta}{5K_1}\left(\frac{1-k}{\varepsilon_{fs}}\right)\left(\frac{\Delta}{\ell}\right)_{max} \quad (7.3.2.1)$$

where $\eta = d/h$; k is as defined in Eq. (7.3.2.2b); and $(\Delta/\ell)_{max}$ is the limiting service load deflection-span ratio. K_1 is a parameter that accounts for boundary conditions. It may be taken as 1.0, 0.8, 0.6, and 2.4 for uniformly loaded simply supported, one end continuous, both ends continuous, and cantilevered spans, respectively. The term ε_{fs} is the strain in the FRP reinforcement under service loads, evaluated at midspan except for cantilevered spans. For cantilevers, ε_{fs} shall be evaluated at the support.

Equation (7.3.2.1) assumes no tensile contribution from concrete between cracks, also referred to as tension stiffening. To consider the effects of tension stiffening in developing Table 7.3.2.1, the values resulting from Eq. (7.3.2.1) were modified by the ratio of effective and fully cracked moments of inertia computed using Eq. (7.3.2.2c) and (7.3.2.2a), respectively. Tabulated values are based on an assumed service deflection limit of $\ell/240$ under total service load, and assumed reinforcement ratios of $2.0\rho_{fb}$ and $3.0\rho_{fb}$ for slabs and beams, respectively.

7.3.2.2 *Effective moment of inertia*—An uncracked section has a moment of inertia equal to the gross moment of inertia, I_g. Cracking occurs when the maximum service load moment M_a exceeds the cracking moment M_{cr}, and this causes a reduction in member stiffness. The moment of inertia of the cracked section, I_{cr}, for a singly-reinforced rectangular member is calculated for a cracked transformed section using an elastic analysis given by Eq. (7.3.2.2a) and (7.3.2.2b).

$$I_{cr} = \frac{bd^3}{3}k^3 + n_f A_f d^2 (1-k)^2 \quad (7.3.2.2a)$$

$$k = \sqrt{2\rho_f n_f + (\rho_f n_f)^2} - \rho_f n_f \quad (7.3.2.2b)$$

The overall flexural stiffness E_cI of a cracked member varies between E_cI_g and E_cI_{cr}, depending on the magnitude of the applied service moment and the extent of cracking along the member. Branson (1965) introduced the concept of an effective moment of inertia, I_e, to allow for a gradual transition from I_g to I_{cr}. This approach accounts for two different phenomena: the effect of concrete tension stiffening and the variation of EI along the member. Branson's (1965) equation for the effective moment of inertia, I_e, was adopted by ACI 318-71.

As demonstrated by Bischoff (2005), Branson's equation overestimates member stiffness when the I_g/I_{cr} of the member is greater than approximately 3 or 4. This corresponds to most FRP-reinforced concrete beams that typically have an I_g/I_{cr} between 5 and 25. It is for this reason that past research on deflection of FRP-reinforced concrete beams (Nawy and Neuwerth 1977; Benmokrane et al. 1996; Yost et al. 2003) has shown that Branson's equation underestimates deflection, particularly for members with a high I_g/I_{cr}.

Several authors have proposed effective moment of inertia expressions for FRP-reinforced concrete. Benmokrane et al. (1996) first incorporated an empirical correction factor in Branson's equation that was needed to reduce tension stiffening and gave reasonable estimates of computed deflection. Other modifications to Branson's equation involving various correction factors have been proposed in the literature (Gao et al. 1998a; Toutanji and Saafi 2000; Yost et al. 2003). This modified-Branson approach was adopted by past editions of ACI 440.1R. Others have proposed a variety of methods that rely on different assumptions regarding tension stiffening and variations in stiffness along the length of the member (Faza and GangaRao 1992; Razaqpur et al. 2000; Rasheed et al. 2004; Bischoff 2005). These models have displayed varying degrees of accuracy when compared to experimental databases (Mota et al. 2006; Bischoff et al. 2009).

Bischoff (2005) proposed an alternative section-based expression for the effective moment of inertia, I_e, that works equally well for both steel- and FRP-reinforced concrete members without the need for empirical correction factors. Branson's original expression represents a weighted average of the uncracked and cracked member stiffness (E_cI), whereas Bischoff's proposed approach represents a weighted average of flexibility ($1/E_cI$). The approach using a weighted average of flexibility better represents the deflection response of members with discrete cracks along their length (Bischoff and Scanlon 2007).

The section-based expression proposed by Bischoff (2005) is modified as follows to include an additional factor γ to account for the variation in stiffness along the length of the member.

$$I_e = \frac{I_{cr}}{1-\gamma\left(\frac{M_{cr}}{M_a}\right)^2\left[1-\frac{I_{cr}}{I_g}\right]} \leq I_g \text{ where } M_a \geq M_{cr} \quad (7.3.2.2c)$$

This approach provides reasonable estimates of deflection for FRP-reinforced concrete beams and one-way slabs (Bischoff et al. 2009). The factor γ is dependent on load and boundary conditions and accounts for the length of the uncracked regions of the member and for the change in stiffness in the cracked regions. In place of a more comprehensive analysis as suggested in Bischoff and Gross (2011), the factor can be taken as $\gamma = 1.72 - 0.72(M_{cr}/M_a)$, which is the result from integrating the curvature over the length of a simply-supported beam with a uniformly distributed load.

Unless stiffness values are obtained by more comprehensive analysis, immediate deflections should be computed with the effective moment of inertia given by Eq. (7.3.2.2c) using the maximum service load moment M_a in the member. Deflection of continuous members can be estimated using a weighted average of I_e computed at the critical positive and negative moment sections, as recommended by ACI 318 and ACI 435R. For spans with both ends continuous, the effective moment of inertia may be approximated as $I_e = 0.70I_{e+} + 0.15(I_{e1-} + I_{e2-})$. For spans with one end continuous, I_e may be approximated based on the location of maximum moment along the span (DeSimone 2009).

The cracking moment M_{cr} is as specified in ACI 318 and should be computed using

$$M_{cr} = \frac{7.5\lambda\sqrt{f_c'}I_g}{y_t} \quad (7.3.2.2d)$$

For SI units

$$M_{cr} = \frac{0.62\lambda\sqrt{f_c'}I_g}{y_t}$$

When $M_a \geq M_{cr}$, the effects of cracking should be considered using Eq. (7.3.2.2c). Recommended minimum thickness values in Table 7.3.2.1 assume this condition. When calculations result in $M_a < M_{cr}$ but the difference between the two values is small, inherent variability in the tensile strength of concrete and the restraint of shrinkage due to the reinforcement may still cause the section to crack. In such cases, deflections will be significantly underestimated by the use of gross section properties. The designer should exercise judgment and consider the project-specific impacts of underestimating deflections in determining whether Eq. (7.3.2.2c) should be employed with a M_{cr}/M_a less than unity to give a more conservative estimate of deflection.

7.3.2.3 *Calculation of deflection (direct method)*—When deflections are estimated by computation according to the provisions of this section, the designer should compare computed deflections to acceptable limits set as part of the design criteria for the project. In many cases, these deflection criteria are set by local building codes.

The short-term deflections (instantaneous deflection under service loads) of an FRP one-way flexural member can be calculated using the effective moment of inertia of the FRP-reinforced beam and the usual structural analysis techniques.

The magnitude of long-term deflection can be several times the short-term deflection, and both short-term and long-term deflections under service loads should be considered in the design. The long-term increase in deflection is a function of member geometry (reinforcement area and member size), load characteristics (age of concrete at the time of loading, and magnitude and duration of sustained load), and material characteristics (elastic moduli of the concrete and FRP reinforcement, creep and shrinkage of concrete, formation of new cracks, and widening of existing cracks).

Data on time-dependent deflections of FRP-reinforced members due to creep and shrinkage indicate that the time-versus-deflection curves of FRP- and steel-reinforced members have the same basic shape, indicating that the same fundamental approach for estimating the long-term deflection can be used (Brown 1997).

According to ACI 318, the long-term deflection due to creep and shrinkage $\Delta_{(cp+sh)}$ can be computed according to the following equations

$$\Delta_{(cp+sh)} = \lambda_\Delta(\Delta_i)_{sus} \quad (7.3.2.3a)$$

$$\lambda_\Delta = \frac{\xi}{1+50\rho_f'} \quad (7.3.2.3b)$$

The parameter λ_Δ in Eq. (7.3.2.3b) reduces to ξ because compression reinforcement is not considered for FRP reinforced members ($\rho_f' = 0$). Values of ξ are reported in ACI 318.

These equations can be used for FRP reinforcement with modifications to account for the differences in the axial stiffness of the reinforcement for FRP-reinforced concrete members as compared with steel-reinforced concrete members. With either FRP or steel reinforcement, concrete creep leads to an effective reduction in the flexural stiffness E_cI. For simplicity, this reduction can be considered as the superposition of two contradictory effects. The first effect is the decrease in effective elastic modulus as a direct result of the concrete creep. The second effect, which can be approximated using an elastic cross section analysis with the reduced elastic modulus for concrete, is an increase in neutral axis depth. This increased neutral axis depth leads to an effective increase in the moment of inertia of the cracked section. The increase in neutral axis depth can be shown to be proportionally more significant for FRP-reinforced members than for steel-reinforced members because of the lower axial stiffness of the reinforcement in typical FRP-reinforced concrete members. As a result, the time-dependent deflection increase for FRP-reinforced concrete can be expected to be proportionally less than for steel-reinforced concrete.

Brown (1997) observed that the time-dependent deflection of FRP-reinforced beams with no compression reinforcement over a sustained loading period of 6 months was 60 to 90 percent of the initial deflection. The measured additional time-dependent deflection was only 50 to 75 percent of the deflection suggested by Eq. (7.3.2.3a) and (7.3.2.3b). Similar results have been reported in other studies (Vijay et al. 1998; Arockiasamy et al. 1998) for both glass FRP (GFRP) and carbon FRP (CFRP).

Based on the aforementioned results, a modification factor of 0.6 is recommended to be applied to Eq. (7.3.2.3a). For typical applications, the long-term deflection of FRP-reinforced members can therefore be determined from Eq. (7.3.2.3c)

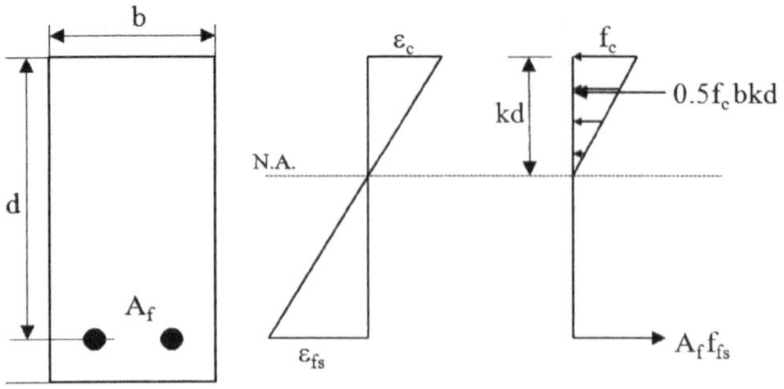

Fig. 7.4—Elastic stress and strain distribution.

$$\Delta_{(cp+sh)} = 0.6\xi(\Delta_i)_{sus} \tag{7.3.2.3c}$$

Gross et al. (2003) found that for beams that are not cracked prior to application of the sustained load, Eq. (7.3.2.3c) may significantly underestimate the time-dependent deflection multiplier. This situation may be found in members where most or all of the service load is sustained load. Gross et al. (2003) attributed this underestimation to the fact that additional flexural cracks were observed to form in the beams over time under the sustained loading. Further experimental work is necessary to validate Eq. (7.3.2.3c) for applications with high levels of sustained load.

7.4—Creep rupture and fatigue

To avoid creep rupture of the FRP reinforcement under sustained stresses or failure due to cyclic stresses and fatigue of the FRP reinforcement, the stress levels in the FRP reinforcement under these stress conditions should be limited. Because these stress levels will be within the elastic range of the member, the stresses can be computed through an elastic analysis, as depicted in Fig. 7.4.

7.4.1 *Creep rupture stress limits*—To avoid failure of an FRP-reinforced member due to creep rupture of the FRP, stress limits should be imposed on the FRP reinforcement. The stress level in the FRP reinforcement can be computed using Eq. (7.4.1), with $M_{s,sus}$ equal to the unfactored moment due to all sustained loads (dead loads and the sustained portion of the live load)

$$f_{fs,sus} = M_{s,sus}\frac{n_f d(1-k)}{I_{cr}} \tag{7.4.1}$$

The cracked moment of inertia, I_{cr}, and the ratio of the effective depth to the depth of the elastic neutral axis, k, are computed using Eq. (7.3.2.2a) and (7.3.2.2b).

Values for safe sustained stress levels are given in Table 7.4.1. These values are based on the creep rupture stress limits previously stated in 4.3.1, with an imposed safety factor of 1/0.60.

7.4.2 *Fatigue stress limits*—If the structure is subjected to fatigue regimes, the FRP stress should be limited to the values stated in Table 7.4.1. The FRP stress can be calculated using Eq. (7.4.1), by replacing $M_{s,sus}$ with the moment due to

Table 7.4.1—Creep rupture stress limits in FRP reinforcement

Fiber type	GFRP	AFRP	CFRP
Creep rupture stress limit $f_{fs,sus}$	$0.20f_{fu}$	$0.30f_{fu}$	$0.55f_{fu}$

all sustained loads plus the maximum moment induced in a fatigue loading cycle.

CHAPTER 8—SHEAR

The design of fiber-reinforced polymer (FRP)-reinforced concrete is similar to that of steel-reinforced concrete members. The different mechanical properties of FRP bars, however, affect shear strength and should be considered. This chapter addresses the shear resistance of FRP-reinforced beams and one-way slabs, the use of FRP stirrups, and the punching shear capacity of FRP-reinforced two-way slabs.

8.1—General considerations

Several issues should be considered for the shear design of FRP-reinforced members. Fiber-reinforced polymer has:

1) A relatively low modulus of elasticity;
2) A low transverse shear resistance;
3) A high tensile strength and no yield point.

Also, the tensile strength of the bent portion of an FRP bar is significantly lower than that of the straight portion.

8.1.1 *Shear design philosophy*—The design of FRP shear reinforcement is based on the strength design method. The strength reduction factor of 0.75 given by ACI 318 for reducing nominal shear capacity of steel-reinforced concrete members should also be used for FRP reinforcement. The design shear strength ϕV_n must be larger than the factored shear force V_u at the section considered. Computation of the maximum shear force V_u at beam supports can be attained following ACI 318 provisions.

8.2—Shear strength of FRP-reinforced members

According to ACI 318, the nominal shear strength of a reinforced concrete cross section, V_n, is the sum of the shear resistance provided by concrete, V_c, and the steel shear reinforcement, V_s.

Compared with a steel-reinforced section with equal areas of longitudinal reinforcement, a cross section using FRP flexural reinforcement after cracking has a smaller depth to the neutral axis because of the lower axial stiffness (product of reinforcement area and modulus of elasticity). The compression region of the cross section is reduced, and the crack widths are wider. As a result, the shear resistance provided by both aggregate interlock and compressed concrete is smaller. Research on the shear capacity of flexural members without shear reinforcement has indicated that the concrete shear strength is influenced by the stiffness of the tensile (flexural) reinforcement (Nagasaka et al. 1993; Zhao et al. 1995; Japan Society of Concrete Engineers 1997b; Sonobe et al. 1997; Michaluk et al. 1998; Tureyen and Frosch 2002, 2003).

The contribution of longitudinal FRP reinforcement in terms of dowel action has not been determined. Because of the lower strength and stiffness of FRP bars in the transverse direction, however, it is assumed that their dowel action contribution is less than that of an equivalent steel area. Further research is needed to quantify this effect.

The concrete shear capacity V_c of flexural members using FRP as main reinforcement can be evaluated according to Eq. (8.2a)

$$V_c = 5\sqrt{f_c'}b_w(kd) \quad (8.2a)$$

For SI units

$$V_c = \frac{2}{5}\sqrt{f_c'}b_w(kd)$$

The parameter k may be evaluated using Eq. (8.3.2.2b).

Equation (8.2a) accounts for the axial stiffness of the FRP reinforcement through the neutral axis depth kd, which is a function of the reinforcement ratio ρ_f and the modular ratio n_f. This equation has been shown to provide a reasonable factor of safety for FRP-reinforced specimens across the range of reinforcement ratios and concrete strengths tested to date (Tureyen and Frosch 2003). Research on lightweight concrete members reinforced with GFRP bars is limited (Pantelides et al. 2012a,b). Based on these experiments, a reduction factor $\lambda = 0.80$ in conjunction with Eq. (8.2a) has been proposed for predicting the shear capacity of sand-lightweight concrete reinforced with GFRP bars (Liu and Pantelides 2013); alternatively, the Modified Compression Field Theory could be used to predict the shear capacity (Liu and Pantelides 2012).

Equation (8.2a) may be rewritten as Eq. (8.2b). This form of the equation indicates that Eq. (8.2a) is simply the ACI 318 shear equation for steel reinforcement V_c modified by the factor ([5/2]k), which accounts for the axial stiffness of the FRP reinforcement.

$$V_c = \left(\frac{5}{2}k\right)2\sqrt{f_c'}b_w d \quad (8.2b)$$

The ACI 318 method used to calculate the shear contribution of steel stirrups is applicable when using FRP as shear reinforcement. The shear resistance provided by FRP stirrups perpendicular to the axis of the member, V_f, can be written as

$$V_f = \frac{A_{fv}f_{fv}d}{s} \quad (8.2c)$$

The stress level in the FRP shear reinforcement should be limited to control shear crack widths and maintain shear integrity of the concrete and to avoid failure at the bent portion of the FRP stirrup (Eq. (6.2.1)). Equation (8.2d) gives the stress level in the FRP shear reinforcement at ultimate for use in design

$$f_{fv} = 0.004E_f \leq f_{fb} \quad (8.2d)$$

When using shear reinforcement perpendicular to the axis of the member, the required spacing and area of shear reinforcement can be computed from Eq. (8.2e)

$$\frac{A_{fv}}{s} = \frac{(V_u - \phi V_c)}{\phi f_{fv}d} \quad (8.2e)$$

When inclined FRP stirrups are used as shear reinforcement, Eq. (8.2f) is used to calculate the contribution of the FRP stirrups

$$V_f = \frac{A_{fv}f_{fv}d}{s}(\sin\theta + \cos\theta) \quad (8.2f)$$

When continuous FRP rectangular spirals are used as shear reinforcement (in this case, s is the pitch and θ is the angle of inclination of the spiral), Eq. (8.2g) gives the contribution of the FRP spirals

$$V_f = \frac{A_{fv}f_{fv}d}{s}(\sin\theta) \quad (8.2g)$$

Shear failure modes of members with FRP as shear reinforcement can be classified into two types (Nagasaka et al. 1993): shear-tension failure mode (controlled by the rupture of FRP shear reinforcement) and shear-compression failure mode (controlled by the crushing of the concrete web). The first mode is more brittle, and the latter results in larger deflections. Experimental results have shown that the modes of failure depend on the shear reinforcement index $\rho_{fv}E_f$, where ρ_{fv} is the ratio of FRP shear reinforcement $A_{fv}/b_w s$. As the value of $\rho_{fv}E_f$ increases, the shear capacity in shear tension increases, and the mode of failure changes from shear tension to shear compression.

8.2.1 *Limits on tensile strain of shear reinforcement*—The design assumption that concrete and reinforcement capacities are added is accurate when shear cracks are adequately controlled. Therefore, the tensile strain in FRP shear reinforcement should be limited to ensure that the ACI design approach is applicable.

CAN/CSA-S6-06 limits the tensile strain in FRP shear reinforcement to 0.002 in./in (mm/mm). It is recognized that this strain value (corresponding to the yield strain of Grade 60 steel) may be very conservative. Experimental evidence indicated that higher strain values were always attained (Wang 1998; Zhao et al. 1995; Okamoto et al. 1994). The Eurocrete Project provisions limit the value of the shear strain in FRP reinforcement to 0.0025 in./in. (mm/mm) (Dowden and Dolan 1997). In no case should effective strain in FRP shear reinforcement exceed 0.004, nor should the design strength exceed the strength of the bent portion of the stirrup, f_{fb}. Ahmed et al. (2010a,b) reported that using 0.004 yields more accurate yet conservative predictions of the shear strength of concrete members reinforced with GFRP stirrups. The value of 0.004 is justified as the strain that prevents degradation of aggregate interlock and corresponding concrete shear (Priestley et al. 1996). CAN/CSA-S6S1-10 adopted the 0.004 limit in the shear design of concrete members reinforced with FRP stirrups.

8.2.2 *Minimum amount of shear reinforcement*—ACI 318 requires a minimum amount of shear reinforcement when V_u exceeds $\phi V_c/2$. This requirement is to prevent or restrain shear failure in members where the sudden formation of cracks can lead to excessive distress (Joint ACI-ASCE Committee 426 1973). To prevent brittle shear failure, adequate reserve strength should be provided to ensure a factor of safety similar to ACI 318 provisions for steel reinforcement. Equation (8.2.2) gives the recommended minimum amount of FRP shear reinforcement

$$A_{fv,min} = \frac{50 b_w s}{f_{fv}} \tag{8.2.2}$$

For SI units

$$A_{fv,min} = 0.35\frac{b_w s}{f_{fv}}$$

with b_w and s in in. (mm), and f_{fv} in psi (MPa).

The minimum amount of reinforcement given by Eq. (8.2.2) is independent of the concrete strength. If steel stirrups are used, the minimum amount of reinforcement provides a shear strength that varies from $1.50V_c$ when f_c' is 2500 psi (17 MPa) to $1.25V_c$ when f_c' is 10,000 psi (69 MPa). Equation (8.2.2), which was derived for steel-reinforced members, is more conservative when used with FRP-reinforced members. For example, when applied to a flexural member having GFRP as longitudinal reinforcement, the shear strength provided by Eq. (8.2.2) could exceed $3V_c$. The ratio of the shear strength provided by Eq. (8.2.2) to V_c will decrease as the stiffness of longitudinal reinforcement increases or as the strength of concrete increases.

8.2.3 *Shear failure due to crushing of the web*—Studies by Nagasaka et al. (1993) indicate that for FRP-reinforced sections, the transition from rupture to crushing limit states occurs at an average value of $0.3f_c'b_wd$ for V_c, but can be as low as $0.18f_c'b_wd$. When V_c is smaller than $0.18f_c'b_wd$, shear-tension can be expected, whereas when V_c exceeds $0.3f_c'b_wd$, crushing is expected. The correlation between rupture and the crushing limit states is not fully understood, and it is more conservative and recommended to use the ACI 318 limit of $8\sqrt{f_c'}\,b_wd$ rather than $0.3f_c'b_wd$. In fact, the ACI limitation is aimed at controlling excess shear crack widths and is, thus, below values corresponding to crushing of the web.

8.3—Detailing of shear stirrups

The maximum spacing of vertical steel stirrups given in ACI 318 as the smaller of $d/2$ or 24 in. (600 mm) is used for vertical FRP shear reinforcement. This limit ensures that each shear crack is intercepted by at least one stirrup.

Tests by Ehsani et al. (1995) indicated that for specimens with r_b/d_b of zero, the reinforcing bars failed in shear at very low load levels at the bends. Therefore, although manufacturing of FRP bars with sharp bends is possible, such details should be avoided. A minimum r_b/d_b ratio of 3 is recommended. In addition, FRP stirrups should be closed with 90-degree hooks.

Fig. 8.3—Required tail length for FRP stirrups.

ACI 318 provisions for bond of hooked steel bars cannot be applied directly to FRP reinforcing bars because of their different mechanical properties. The tensile force in a vertical stirrup leg is transferred to the concrete through the tail beyond the hook, as shown in Fig. 8.3. Ehsani et al. (1995) found that for a tail length ℓ_{thf} beyond $12d_b$, there is no significant slippage and no influence on the tensile strength of the stirrup leg. Therefore, a minimum tail length of $12d_b$ should be used.

8.4—Shear strength of FRP-reinforced two-way concrete slabs

Experimental evidence (Ahmad et al. 1993; Bank and Xi 1995; Banthia et al. 1995; Matthys and Taerwe 2000; El-Ghandour et al. 2003; Ospina et al. 2003) shows that the axial stiffness of the FRP reinforcement, as well as the concrete strength f_c', significantly affect the transverse shear response of interior FRP-reinforced two-way slab-column connections. Test results of isolated FRP-reinforced two-way slab specimens subjected to uniform gravity loading indicate that an increase in the top FRP mat stiffness increases punching shear capacity and decreases the ultimate slab deflection. Punching shear failure in slabs reinforced with FRP bars is sudden and brittle. Conversely, punching test results (Bank and Xi 1995; Ospina et al. 2003) show that two-way slabs reinforced with FRP grids rather than bars do not exhibit a sharp load drop at punching failure. Instead, they continue to absorb energy in a stable fashion following initial failure. Hassan et al. (2013a) reported that increasing the reinforcement ratio resulted in higher punching-shear capacity, lower reinforcement and concrete strains, and lower deflections. In addition, the use of high-strength concrete increased the punching-shear capacity, significantly reduced concrete strains, increased strains in the GFRP reinforcing bars, and reduced deflection due to the high tensile strength and modulus of elasticity.

A statistical evaluation of test results reveal that the one-way shear design model proposed by Tureyen and Frosch (2003), which accounts for reinforcement stiffness, can be modified (Ospina 2005) to account for the shear transfer in two-way concrete slabs. The modification leads to Eq. (8.4a), which can be used to calculate the concentric punching shear capacity of FRP-reinforced two-way concrete slabs that are either supported by interior columns or subjected to concentrated loads that are either square or circular in shape

$$V_c = 10\sqrt{f_c'}b_o(kd) \tag{8.4a}$$

For SI units

$$V_c = \frac{4}{5}\sqrt{f_c'}b_o(kd)$$

The parameter k may be evaluated using Eq. (7.3.2.2b).

In the evaluation of Eq. (8.4a), b_o should be computed at $d/2$ away from the column face. In addition, the shape of the critical surface should be the same as that of the column.

Equation (8.4a) can be rewritten as Eq. (8.4b). This equation is simply the basic ACI 318 concentric punching shear equation for steel-reinforced slabs V_c modified by the factor ([5/2]k) that accounts for the axial stiffness of the FRP reinforcement.

$$V_c = \left(\frac{5}{2}k\right)4\sqrt{f_c'}b_o d \tag{8.4b}$$

Equation (8.4a) provides a reasonable factor of safety for FRP-reinforced concrete two-way slabs across the range of reinforcement ratios and concrete strengths tested. Experimental results and design recommendations on shear strength of FRP-reinforced two-way slabs can be found in the literature (CAN/CSA S806-12 2012; Lee et al. 2009; Dulude et al. 2013; Hassan et al. 2013b). Further research is needed to examine the punching capacity of FRP-reinforced two-way slabs supported by edge and corner columns, as well as the effects of column rectangularity and unbalanced moment transfer on the punching capacity of FRP-reinforced concrete two-way slabs supported on interior columns.

CHAPTER 9—SHRINKAGE AND TEMPERATURE REINFORCEMENT

Shrinkage and temperature reinforcement is intended to limit crack width. The stiffness and strength of reinforcing bars control this behavior. Shrinkage cracks perpendicular to the member span are restricted by flexural reinforcement; therefore, shrinkage and temperature reinforcement are only required in the direction perpendicular to the span. ACI 318 requires a minimum steel reinforcement ratio of 0.0020 when using Grade 40 or 50 deformed steel bars, and 0.0018 when using Grade 60 deformed bars or welded reinforcement (deformed or smooth). ACI 318 also requires that the spacing of shrinkage and temperature reinforcement not exceed five times the member thickness or 18 in. (500 mm).

9.1—Minimum FRP reinforcement ratio

No experimental data are available for the minimum FRP reinforcement ratio for shrinkage and temperature. ACI 318, Section 7.12.2, states that for slabs with steel reinforcement having a yield stress exceeding 60 ksi (414 MPa) measured at a yield strain of 0.0035, the ratio of reinforcement to gross area of concrete should be at least 0.0018 × 60/f_y, where f_y is in ksi, but not less than 0.0014. The stiffness and the strength of shrinkage and temperature FRP reinforcement can be incorporated in this formula. Therefore, when deformed FRP shrinkage and temperature reinforcement is used, the amount of reinforcement should be determined by using Eq. (9.1)

$$\rho_{f,ts} = 0.0018 \times \frac{60,000}{f_{fu}}\frac{E_s}{E_f} \tag{9.1}$$

For SI units

$$\rho_{f,ts} = 0.0018 \times \frac{414}{f_{fu}}\frac{E_s}{E_f}$$

Due to limited experience, it is recommended that the ratio of temperature and shrinkage reinforcement given by Eq. (9.1) be taken not less than 0.0014, the minimum value specified by ACI 318 for steel shrinkage and temperature reinforcement. The licensed design professional may consider an upper limit for the ratio of temperature and shrinkage reinforcement equal to 0.0036, or compute the ratio based on calculated strain levels corresponding to the nominal flexural capacity rather than the strains calculated using Eq. (9.1). Spacing of shrinkage and temperature FRP reinforcement should not exceed three times the slab thickness or 12 in. (300 mm), whichever is less. The use of FRP for temperature and shrinkage reinforcement for slabs-on-ground is presented in Appendix A.

CHAPTER 10—DEVELOPMENT AND SPLICES OF REINFORCEMENT

In a reinforced concrete flexural member, the tension force carried by the reinforcement balances the compression force in the concrete. The tension force is transferred to the reinforcement through the bond between the reinforcement and the surrounding concrete. Bond stresses exist whenever the force in the tensile reinforcement changes. Bond between fiber-reinforced polymer (FRP) reinforcement and concrete is developed through a mechanism similar to that of steel reinforcement and depends on FRP type, elastic modulus, surface deformation, and the shape of the FRP bar (Al-Zahrani et al. 1996; Uppuluri et al. 1996; Gao et al. 1998b).

10.1—Development of stress in straight bar

Figure 10.1 shows the equilibrium condition of an FRP bar of length ℓ_e embedded in concrete. The force in the bar

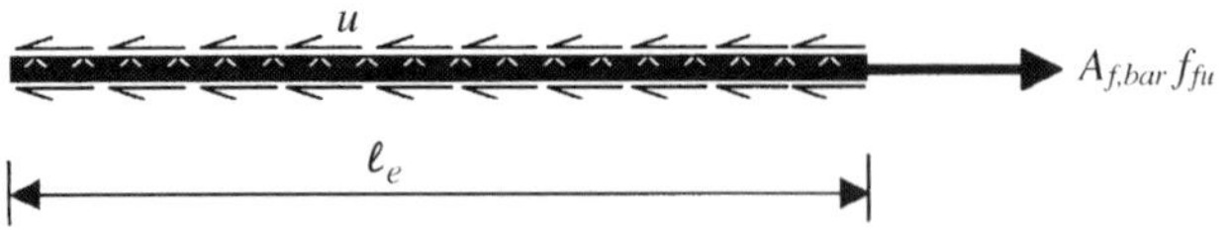

Fig. 10.1—Transfer of force through bond.

is resisted by an average bond stress u acting on the surface of the bar. Equilibrium of forces can be written as follows

$$\ell_e \pi d_b u = A_{f,bar} f_f \tag{10.1a}$$

where f_f is the stress developed in the bar at the end of the embedment length. In contrast to steel bars, the full strength of an FRP bar need not be developed, especially when flexural capacity is controlled by concrete crushing and the required stress in the bar at failure is less than its guaranteed ultimate strength. Additionally, changing the limit state from bar fracture or concrete crushing to bond failure does not significantly change the ductility associated with the failure.

The development length equation for steel reinforcing bars found in ACI 318 is based on the work done by Orangun et al. (1977). The development length equation was based on 62 unconfined splice tests and 54 confined splice tests that failed by splitting concrete. Orangun et al. (1977) developed an equation relating the average bond stress normalized by the square root of the concrete compressive strength to the normalized cover to the center of the bar C/d_b and the normalized splice length d_b/ℓ_e using linear regression. This equation served as the foundation for the development length equation for steel reinforcing bars found in ACI 318.

A similar methodology was followed by Wambeke and Shield (2006) in which a consolidated database of 269 beam bond tests was created. The database was limited to beam-end tests, notch-beam tests, and splice tests. The majority of the bars represented in the database were composed of glass FRP (GFRP). In Wambeke and Shield's database (2006), there were 82 beam tests that resulted in splitting failures based on the work of Ehsani et al. (1996a), Daniali (1992), Shield et al. (1997, 1999), and Tighiouart et al. (1999). The tests included both spiral wrap and helical lug patterned bars with and without confining reinforcement. A linear regression of the normalized average bond stress versus the normalized cover and embedment (splice) length resulted in the following relationship after rounding the coefficients

$$\frac{u}{\sqrt{f_c'}} = 4.0 + 3.0\frac{C}{d_b} + 100\frac{d_b}{\ell_e} \tag{10.1b}$$

For SI units

$$\frac{u}{0.083\sqrt{f_c'}} = 4.0 + 0.3\frac{C}{d_b} + 100\frac{d_b}{\ell_e}$$

where C is the lesser of the cover to the center of the bar (d_c or $d_{c,side}$) or one-half of the center-on-center spacing of the bars being developed. The bar surface (spiral wrap versus helical lug) did not appear to affect the results, nor surprisingly did the presence of confining reinforcement (Wambeke and Shield 2006). Darwin et al. (1996) found that confining steel used in beams that had steel reinforcing bars with a high relative rib area had more of a beneficial increase in the bond force over the same-size steel bars with moderate rib area. The counter-argument is proposed herein. The GFRP bars have a very low relative rib area and, therefore, the presence of confinement may not increase the average bond stress. Additional research into the effect of confining reinforcement on bond of GFRP bars, however, is warranted.

Equations (10.1a) and (10.1b) can be solved for the achievable bar stress given the existing embedment length and cover. A random subset of the full database developed by Wambeke and Shield (2006) was used to determine a factor of safety for use with these equations. The resulting probability of a test-predicted ratio less than 1.0 was 22 percent. This database included both splitting and pullout failures with embedment lengths of at least $19d_b$. Additionally, a limit of 3.5 was put on the C/d_b term so that the same equation could be used to predict developable bar stresses for either splitting or pullout bond failure mode. When the normalized cover was over 3.5 and the embedment length was greater than $19d_b$, the failure mode was always pullout. The resulting expression for developable bar stress is

$$f_{fe} = \frac{\sqrt{f_c'}}{\alpha}\left(13.6\frac{\ell_e}{d_b} + \frac{C}{d_b}\frac{\ell_e}{d_b} + 340\right) \le f_{fu} \tag{10.1c}$$

For SI units

$$f_{fe} = \frac{0.083\sqrt{f_c'}}{\alpha}\left(13.6\frac{\ell_e}{d_b} + \frac{C}{d_b}\frac{\ell_e}{d_b} + 340\right) \le f_{fu}$$

in which the term C/d_b should not be taken larger than 3.5, and α is a factor to account for bar location (discussed in 10.1.1). The mean of the test-predicted ratio of bar stresses using this equation was 1.14, with a coefficient of variation of 15.8 percent. The appropriateness of this equation for embedment lengths greater than $100d_b$ is questionable, as the dataset used to develop it did not include any bond failures with embedment lengths greater than $100d_b$. Embedment lengths shorter than $20d_b$ are not recommended. Additional work needs to be performed to determine the effect of the chosen factor of safety for bond on the flexural reliability of the section.

When applying Eq. (10.1c) for design purposes, it should be assumed that the maximum achievable bar stress varies linearly from 0 to the value produced by Eq. (10.1c) along the first $20d_b$ of the bar embedment. After this point, Eq. (10.1c) can be used to determine the achievable bar stress along the bar. A check should be made to determine if adequate moment capacity can be achieved at the end of the available embedment length. If not, then the embedment length should be increased, the number of bars increased so that a lower stress in each bar is required at ultimate, or the nominal moment capacity should be recalculated to include the possibility of bond failure, as described in 10.1.3. Note that increasing the number of bars may decrease the stress developed in any one bar, as the term C/d_b may decrease as the bar spacing decreases.

10.1.1 *Bar location modification factor*—The default bar location modification factor is 1.0. While placing concrete, air, water, and fine particles migrate upward through the concrete. This can cause a significant drop in bond strength under the horizontal reinforcement. The term "top reinforcement" usually refers to horizontal reinforcement with more than 12 in. (305 mm) of concrete below it at the time of embedment. The database assembled by Wambeke and Shield (2006) included 15 tests using top bars with embedment lengths greater than $16d_b$ that resulted in bond failure (Ehsani et al. 1996a; Mosley 2002). Based on these data, for bars with more than 12 in. (300 mm) of concrete cast below, α in Eq. (10.1c) should be taken as 1.5.

10.1.2 *Material modification factor*—A limited amount of bond data exists in the Wambeke and Shield (2006) database for aramid FRP (AFRP) bars (Mosley 2002). Based on these few tests, however, the development length of AFRP bars appears to be similar to that of GFRP bars. Therefore, the development length equations provided are also reasonable for AFRP bars without the addition of a material modification factor. No data exist in the database for carbon FRP (CFRP) bars; it is anticipated that the much larger stiffness of the CFRP bars will likely decrease the required development lengths and, correspondingly, its material modification factor. At this time, a material factor equal to 1.0 is recommended for CFRP bars.

10.1.3 *Nominal moment strength of bond critical sections*—Bond critical sections are defined as sections where the maximum achievable stress in the FRP bar is limited by Eq. (10.1c). For this case, the nominal moment capacity should be recalculated using a modification of the method described in 7.2. When bond limits the stress that can be developed in the bar, the two possible limits states are concrete crushing and bond failure. The capacity for a concrete crushing limit state can be calculated using Eq. (7.2.2e). This equation is applicable if the bar stress that can be developed (f_{fe} as determined from Eq. (10.1c)) is greater than or equal to the bar stress determined by Eq. (7.2.2d). When $\rho_f \leq \rho_{fb}$ or $\rho_f > \rho_{fb}$ and the bar stress required in Eq. (7.2.2d) cannot be developed, the capacity for the limit state of bond failure can be determined using Eq. (7.2.2g) with f_{fe} from Eq. (10.1c) substituted for f_{fu}, and f_{fe}/E_f substituted for ε_{fu} in Eq. (7.2.2h). A strength reduction factor of 0.55 is recommended for flexure when the limit state is bond failure.

10.2—Development length of bent bar

Limited experimental data are available on the bond behavior of hooked FRP reinforcing bars. ACI 318 provisions for development length of hooked steel bars are not applicable to FRP bars due to the differences in material characteristics.

Ehsani et al. (1996b) tested 36 specimens with hooked GFRP bars. Based on the results of the study, the expression for the development length of a 90-degree hooked bar ℓ_{bhf} was proposed as follows

$$\ell_{bhf} = K_4 \frac{d_b}{\sqrt{f_c'}} \tag{10.2a}$$

The K_4 factor for the calculation of the development length in this equation is 1820 (150 for SI units) for bars with f_{fu} less than 75,000 psi (517 MPa). This factor should be multiplied by $f_{fu}/75{,}000$ ($f_{fu}/517$) for bars having a tensile strength between 75,000 and 150,000 psi (517 and 1034 MPa).

When the side cover (normal to the plane of hook) is more than 2-1/2 in. (64 mm) and the cover extension beyond hook is not less than 2 in. (50 mm), another multiplier of 0.7 can be applied (Ehsani et al. 1996b). These modification factors are similar to those in ACI 318, Section 12.5.3, for steel hooked bars. To account for the lack of experimental data, the use of Eq. (10.2b) in calculating the development length of hooked bars is recommended

$$\ell_{bhf} = \begin{cases} 2000\dfrac{d_b}{\sqrt{f_c'}} & \text{for } f_{fu} \leq 75{,}000 \text{ psi} \\ \dfrac{f_{fu}}{37.5}\dfrac{d_b}{\sqrt{f_c'}} & \text{for } 75{,}000 < f_{fu} < 150{,}000 \text{ psi} \\ 4000\dfrac{d_b}{\sqrt{f_c'}} & \text{for } f_{fu} \geq 150{,}000 \text{ psi} \end{cases} \tag{10.2b}$$

For SI units

$$\ell_{bhf} = \begin{cases} 165\dfrac{d_b}{\sqrt{f_c'}} & \text{for } f_{fu} \leq 520 \text{ MPa} \\ \dfrac{f_{fu}}{3.1}\dfrac{d_b}{\sqrt{f_c'}} & \text{for } 520 < f_{fu} < 1040 \text{ MPa} \\ 330\dfrac{d_b}{\sqrt{f_c'}} & \text{for } f_{fu} \geq 1040 \text{ MPa} \end{cases}$$

with ℓ_{bhf} and d_b in in. (mm) and f_{fu} and f_c' in psi (MPa), in mm.

The value calculated using Eq. (10.2b) should not be less than $12d_b$ or 9 in. (230 mm). These values are based on test results reported by Ehsani et al. (1995), in which the tensile force and slippage of a hooked bar stabilized near $12d_b$. The tail length of a hooked bar, ℓ_{thf} (Fig. 8.3), should not be less than $12d_b$. Longer tail lengths were found to have an insignificant influence on the ultimate tensile force and slippage of the hook. To avoid shear failure at the bend, the radius of the bend should not be less than $3d_b$ (Ehsani et al. 1995).

10.3—Development of positive moment reinforcement

In general, the requirements of 12.10 and 12.11 of ACI 318 should be met when using FRP reinforcement with the following changes: for straight bars, the stress to be developed f_{fr} should be the minimum of f_{fu}, the stress given by Eq. (7.2.2d), and the stress given by Eq. (11.1c). The development length for straight bars is defined as the bond length required to develop f_{fr} and is given by

$$\ell_d = \frac{\alpha \dfrac{f_{fr}}{\sqrt{f'_c}} - 340}{13.6 + \dfrac{C}{d_b}} d_b \qquad (10.3a)$$

For SI units

$$\ell_d = \frac{\alpha \dfrac{f_{fr}}{0.083\sqrt{f'_c}} - 340}{13.6 + \dfrac{C}{d_b}} d_b$$

where C is the lesser of the cover to the center of the bar (d_c or $d_{c,side}$) or one-half of the center-on-center spacing of the bars being developed. Because of the reduced resistance factor compared with steel, the provision for development of positive reinforcement at points of inflection and simple supports given in ACI 318, Section 12.11.3, should be altered to

$$\ell_d \leq \frac{\phi M_n}{V_u} + \ell_a \qquad (10.3b)$$

where M_n is the nominal moment strength assuming all reinforcement at the section to be stressed to the required bar stress f_{fr}; V_u is the factored shear force at the section; and ℓ_a, at a support, is the embedment length beyond center of the support, or ℓ_a, at a point of inflection, is the larger of the effective depth of the member or $12d_b$. The value of $\phi M_n/V_u$ may be increased by 30 percent when the ends of the reinforcement are confined by a compressive reaction. This restriction on the development length need not be met if it can be shown by refined analysis that the design moment capacity is greater than the factored moment everywhere along the development length.

10.4—Tension lap splice

ACI 318, Section 12.15, distinguishes between two types of tension lap splices, depending on the fraction of the bars spliced in a given length and on the reinforcement stress in the splice. For steel reinforcement, the splice length for a Class A splice is $1.0\ell_d$, and for a Class B splice is $1.3\ell_d$. This classification for FRP applications is inappropriate, as often the full tensile strength of the bar need not be developed; hence, the assumption that all splices are Class B splices is conservative. Limited data are available for the minimum development length of FRP tension lap splices (Benmokrane 1997; Mosley 2002). Consequently, a value of $1.3\ell_d$ is recommended for all splices.

CHAPTER 11—DESIGN EXAMPLES

A series of design examples is provided to illustrate the design concepts and procedures presented earlier in this design guide. The examples have been developed to be similar to examples presented in the Portland Cement Association (PCA) publication, *Notes on ACI 318-08 Building Code Requirements for Structural Concrete* (Portland Cement Association 2008). In a few cases, the problem statements are identical. In most cases, minor modifications have been made to the original PCA (2008) *Notes* problem statements to make them more suitable for use as examples for FRP-reinforced concrete. It is anticipated that additional examples will be added, as new provisions are developed for future editions of this design guide. Examples are presented first in customary inch-pound units, and then in SI units.

Material properties are given for the concrete and reinforcement in the problem statement for each example. Material properties are generally reflective of glass FRP (GFRP) reinforcement because this is the most commonly used FRP reinforcing material for internal reinforcement today. As noted in 4.2.1, the guaranteed ultimate tensile strength (f_{fu}*) of the reinforcement typically varies with bar size for a given manufacturer, with smaller reinforcement having higher strengths. For simplicity, this variation is neglected in the examples presented herein, and the tensile strength is taken as a single value regardless of bar diameter.

Example 1—Flexural (moment) strength using equivalent rectangular concrete stress distribution (compression-controlled section)

This example is similar to Example 6.1 of PCA Notes on ACI 318-08.

For the beam section shown, calculate moment strength based on static equilibrium using the equivalent rectangular concrete stress distribution as shown in Fig. 7.2 of this guide. Assume f_c' = 4000 psi, f_{fu}* = 80,000 psi, and E_f = 6000 ksi. Assume interior exposure conditions and neglect compression reinforcement.

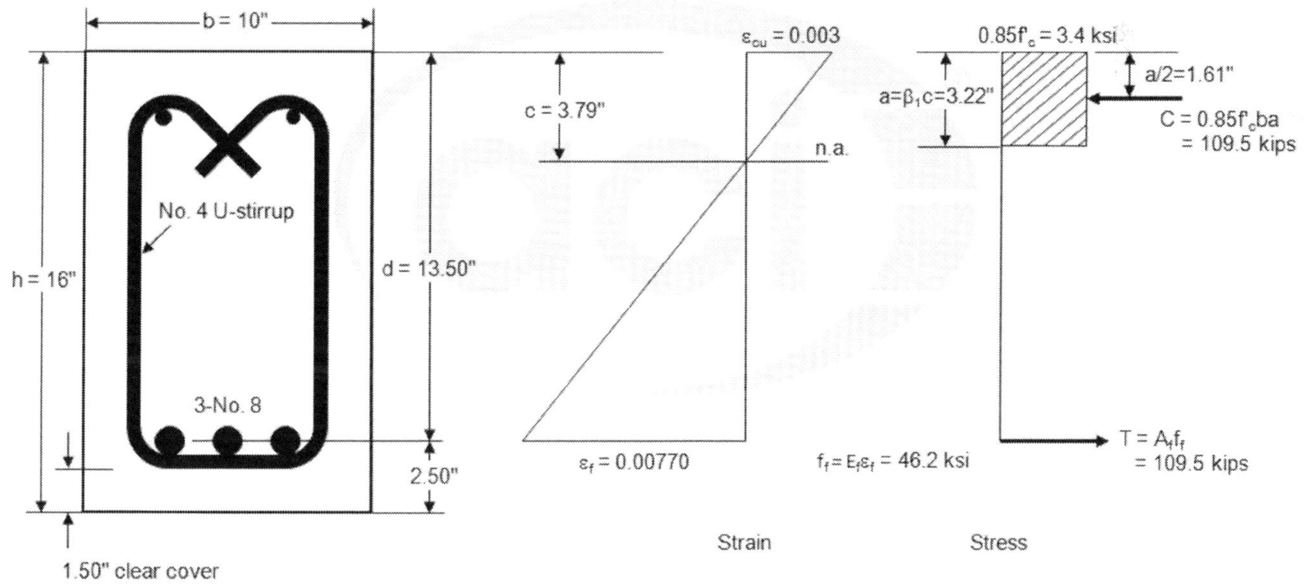

Calculations and discussion	**Reference**
No. 8 bar properties:	440.6-08, Table 7.1
d_b = 1.00 in.	
$A_{f,bar}$ = 0.79 in.^{2}	
Design material properties:	
C_E = 0.8	440.1R, Table 6.2
$f_{fu} = C_E f_{fu}* = (0.8)(80) = 64$ ksi	440.1R, Eq. (6.2a)
1. Determine the strength reduction factor.	
$d = 16 - 1.5 - 0.5 - (1.00/2) = 13.50$ in.	
$A_f = (3)(0.79\text{ in.}^2) = 2.37$ in.^{2}	
$\rho_f = \dfrac{A_f}{bd} = \dfrac{2.37}{(10)(13.50)} = 0.01756$	440.1R, Eq. (7.2.1a)

$$\rho_{fb} = 0.85\beta_1 \frac{f_c'}{f_{fu}} \frac{E_f \varepsilon_{cu}}{E_f \varepsilon_{cu} + f_{fu}}$$ 440.1R, Eq. (7.2.1b)

$E_f\varepsilon_{cu} = (6000)(0.003) = 18.00$ ksi

$$\rho_{fb} = 0.85(0.85)\frac{(4)}{(64)}\frac{(18.00)}{(18.00\ +\ 64)} = 0.00991$$

$$\frac{\rho_f}{\rho_{fb}} = \frac{0.01756}{0.00991} = 1.77$$

Because $\rho_f \geq 1.4\rho_{fb}$, the section is compression-controlled. 440.1R, Eq. (7.2.3)
$\phi = 0.65$

2. Determine stress in tensile reinforcement at ultimate conditions.

$$f_f = \sqrt{\frac{(E_f\varepsilon_{cu})^2}{4} + \frac{0.85\beta_1 f_c'}{\rho_f} E_f \varepsilon_{cu}} - 0.5E_f\varepsilon_{cu} \leq f_{fu}$$ 440.1R, Eq. (7.2.2d)

$$f_f = \sqrt{\frac{(18.00)^2}{4} + \frac{0.85(0.85)(4)}{(0.01756)}(18.00)} - 0.5(18.00) \leq 64$$

$f_f = 46.2$ ksi

3. Determine nominal flexural strength M_n and design flexural strength ϕM_n.

$$a = \frac{A_f f_f}{0.85 f_c' b} = \frac{(2.37)(46.2)}{0.85(4)(10)} = 3.22 \text{ in.}$$ 440.1R, Eq. (7.2.2b)

$$M_n = A_f f_f \left(d - \frac{a}{2}\right) = (2.37)(46.2)\left(13.50 - \frac{3.22}{2}\right) = 1302 \text{ in.-kip} = 108.5 \text{ ft-kip}$$ 440.1R, Eq. (7.2.2a)

Alternatively, compute M_n directly:

$$M_n = \rho_f f_f \left(1 - 0.59\frac{\rho_f f_f}{f_c'}\right) bd^2$$ 440.1R, Eq. (7.2.2e)

$$= (0.01756)(46.2)\left(1 - 0.59\frac{(0.01756)(46.2)}{4}\right)(10)(13.50)^2 = 1302 \text{ in.-kip} = 108.5 \text{ ft-kip}$$

$\phi M_n = (0.65)(108.5) = 70.5$ ft kip

4. Minimum reinforcement.
The minimum reinforcement provisions do not apply because the section is not tension-controlled. 440.1R, Sec. 7.2.4

Note: While the general procedure and principles used in this example are applicable for an FRP-reinforced beam of any cross-sectional shape, the specific equations used in this example are restricted to singly-reinforced rectangular cross sections (or flanged sections that exhibit rectangular section behavior) with reinforcement in a single layer.

Example 2—Flexural (moment) strength using equivalent rectangular concrete stress distribution (tension-controlled section)

This example is similar to Example 6.1 of PCA Notes on ACI 318-08.

For the beam section shown, calculate moment strength based on static equilibrium using the equivalent rectangular concrete stress distribution as shown in Fig. 7.2 of this guide. Assume $f_c' = 4000$ psi, $f_{fu}^* = 90{,}000$ psi, and $E_f = 6000$ ksi. Assume interior exposure conditions and neglect compression reinforcement.

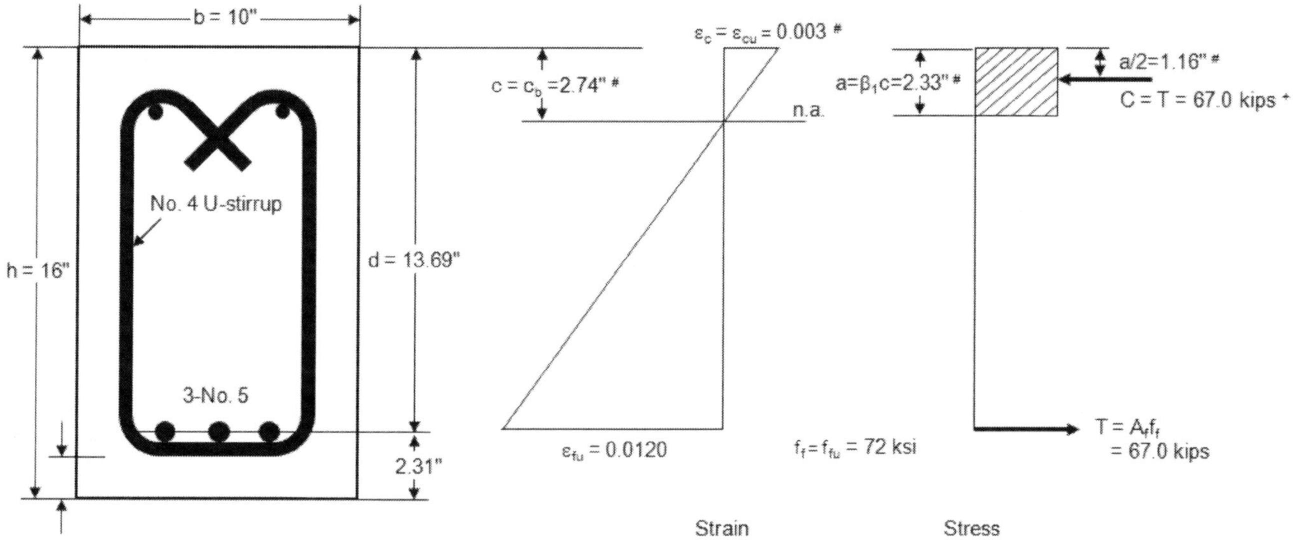

Conservatively assumed for design simplification. True values for c, a, and ε_c will be smaller, since actual NA will be higher.
+ Compression force must equal tension force. C cannot be calculated as 0.85f'_cba, since the correct value of a is not known.

Calculations and discussion	**Reference**
No. 5 bar properties: $d_b = 0.625$ in. $A_{f,bar} = 0.31$ in.^{2}	440.6-08, Table 7.1
Design material properties: $C_E = 0.8$	440.1R, Table 6.2
$f_{fu} = C_E f_{fu}^* = (0.8)(90) = 72$ ksi	440.1R, Eq. (6.2a)

1. Determine the strength reduction factor.

$d = 16 - 1.5 - 0.5 - (0.625/2) = 13.69$ in.

$A_f = (3)(0.31\text{ in.}^2) = 0.93\text{ in.}^2$

$$\rho_f = \frac{A_f}{bd} = \frac{0.93}{(10)(13.69)} = 0.00679 \qquad \text{440.1R, Eq. (7.2.1a)}$$

$$\rho_{fb} = 0.85\beta_1 \frac{f_c'}{f_{fu}} \frac{E_f \varepsilon_{cu}}{E_f \varepsilon_{cu} + f_{fu}} \qquad \text{440.1R, Eq. (7.2.1b)}$$

$E_f\varepsilon_{cu} = (6000)(0.003) = 18.00$ ksi

$$\rho_{fb} = 0.85(0.85)\frac{(4)}{(72)}\frac{(18.00)}{(18.00 + 72)} = 0.00803$$

$$\frac{\rho_f}{\rho_{fb}} = \frac{0.00679}{0.00803} = 0.85$$

Because $\rho_f \leq \rho_{fb}$, the section is tension-controlled and $\phi = 0.55$. 440.1R, Eq. (7.2.3)

2. Determine stress in tensile reinforcement at ultimate conditions.

Because section is tension-controlled, $f_f = f_{fu} = 72$ ksi

3. Determine nominal flexural strength M_n and design flexural strength ϕM_n.

$$c_b = \left(\frac{\varepsilon_{cu}}{\varepsilon_{cu} + \varepsilon_{fu}}\right)d = \left(\frac{0.003}{0.003 + \dfrac{72}{6000}}\right)(13.69) = 2.74 \text{ in.}$$ 440.1R, Eq. (7.2.2h)

$$M_n = A_f f_{fu}\left(d - \frac{\beta_1 c_b}{2}\right) = (0.93)(72)\left(13.69 - \frac{(0.85)(2.74)}{2}\right) = 839 \text{ in.-kip} = 69.9 \text{ ft-kip}$$ 440.1R, Eq. (7.2.2g)

$\phi M_n = (0.55)(69.9) = 38.4$ ft-kip

4. Minimum reinforcement

The minimum reinforcement provisions apply since the section is tension-controlled. 440.1R, Sec. 7.2.4

$$A_{f,min} = \frac{4.9\sqrt{f_c'}}{f_{fu}} b_w d \geq \frac{330}{f_{fu}} b_w d$$ 440.1R, Eq. (7.2.4)

$$A_{f,min} = \frac{4.9\sqrt{4000}}{(72)(1000)}(10)(13.69) \geq \frac{330}{(72)(1000)}(10)(13.69)$$

$$= 0.59 \geq 0.63$$

$A_{f,min} = 0.63$ in.^{2}

A_f(provided) $= 0.93$ in.$^2 > A_{f,min} = 0.63$ in.^{2} OK

Note: Whereas the general procedure and principles used in this example are applicable for an FRP-reinforced beam of any cross-sectional shape, the specific equations used in this example are restricted to singly-reinforced rectangular cross sections (or flanged sections that exhibit rectangular section behavior) with reinforcement in a single layer.

The procedure used in this example, described in 7.2.2, is approximate and conservative. A more detailed analysis of a tension-controlled cross section would require the computation of the neutral axis location based on the principles of strain compatibility and a numerical constitutive model for the concrete (for example, parabolic and Hognestad). This type of analysis can be computationally intensive and is generally not suitable for hand calculations. For problems such as the one shown in this example, differences in results (computed flexural strength) are likely to be negligible. A strain compatibility analysis is suggested for cases where there are multiple layers of reinforcement or when different FRP bar types are mixed (7.2.5.1).

Example 3—Design of a rectangular beam with tension reinforcement only

This example is similar to Example 7.1 of PCA Notes on ACI 318-08.

Design a rectangular beam of width $b = 12$ in. to have adequate flexural strength. The beam must resist service load moments $M_D = 56$ ft-kip and $M_L = 35$ ft-kip. Assume interior exposure conditions.

$f_c' = 4000$ psi
$f_{fu}{}^* = 80{,}000$ psi
$E_f = 6500$ ksi

Calculations and discussion — Reference

Design material properties:
$C_E = 0.8$ 440.1R, Table 6.2
$f_{fu} = C_E f_{fu}{}^* = (0.8)(80) = 64$ ksi 440.1R, Eq. (6.2a)

1. A common starting point for design of a reinforced concrete member of unknown dimensions is the assumption of a reinforcement ratio.
As a starting point, assume $\rho_f = 1.5\rho_{fb}$.

$$\rho_{fb} = 0.85\beta_1 \frac{f_c'}{f_{fu}} \frac{E_f \varepsilon_{cu}}{E_f \varepsilon_{cu} + f_{fu}}$$ 440.1R, Eq. (7.2.1b)

$E_f \varepsilon_{cu} = (6500)(0.003) = 19.50$ ksi

$$\rho_{fb} = 0.85(0.85)\frac{(4)}{(64)}\frac{(19.50)}{(19.50+64)} = 0.01055$$

$\rho_f = 1.5\rho_{fb} = 1.5(0.01055) = 0.01583$

Because $\rho_f \geq 1.4\rho_{fb}$, the section is compression-controlled. 440.1R, Eq. (7.2.3)
$\phi = 0.65$

2. Compute bd^2 required.

First, determine the required design moment strength:
$\phi M_{n,reqd} = M_u = 1.2M_D + 1.6M_L = 1.2(56) + 1.6(35) = 123.2$ ft-kip 318-11, Eq. (9-2)

Calculate the stress in the tensile reinforcement (f_f) at ultimate conditions for the assumed value of ρ_f.

$$f_f = \sqrt{\frac{(E_f\varepsilon_{cu})^2}{4} + \frac{0.85\beta_1 f_c'}{\rho_f}E_f\varepsilon_{cu}} - 0.5E_f\varepsilon_{cu} \leq f_{fu}$$ 440.1R, Eq. (7.2.2d)

$$f_f = \sqrt{\frac{(19.50)^2}{4} + \frac{0.85(0.85)(4)}{(0.01583)}(19.50)} - 0.5(19.50) \leq 64$$

$f_f = 50.7$ ksi

Use the moment capacity equation to determine required dimensions for the cross section.

$$M_u = \phi M_n = \phi\rho_f f_f\left(1 - 0.59\frac{\rho_f f_f}{f_c'}\right)bd^2$$ 440.1R, Eq. (7.2.2e)

$$(123.2)(12) = (0.65)(0.01583)(50.7)\left(1 - 0.59\frac{(0.01583)(50.7)}{(4)}\right)bd^2$$

$bd^2 = 3214$ in.^{3}

3. Size member so that $(bd^2)_{provided} \geq (bd^2)_{required}$.

Recall $b = 12$ in., so $d = \sqrt{\frac{3214}{12}} = 16.36$ in.

4. Now, determine the required reinforcement, select bars, and determine depth.
$A_{f,reqd} = \rho_f bd = (0.01583)(12)(16.36) = 3.11$ in.^{2}
Select four No. 8 bars ($A_f = 3.16$ in.^{2})

Note: Examining alternative designs using other bar sizes may require changing the assumed value of f_{fu}* *to a value appropriate for the selected bar size.*

No. 8 bar diameter: $d_b = 1.00$ in. 440.6-08, Table 7.1

For interior exposure, clear cover is 1.5 in. 440.5-08, Table 3.1

Assuming No. 4 stirrups:

$$h = (16.36) + (1.5) + (0.5) + \frac{(1.00)}{2} = 18.86 \text{ in.}$$

Round up to be conservative. So, select a 12 x 20 in. beam.

5. Determine capacity of cross section.

$$d = 20 - \left[(1.5) + (0.5) + \frac{(1.00)}{2}\right] = 17.50 \text{ in.}$$

$$\rho_f = \frac{A_f}{bd} = \frac{3.16}{(12)(17.50)} = 0.01505$$ 440.1R, Eq. (7.2.1a)

$$\frac{\rho_f}{\rho_{fb}} = \frac{0.01505}{0.01055} = 1.43$$

Because $\rho_f \geq 1.4\rho_{fb}$, $\phi = 0.65$. 440.1R, Eq. (7.2.3)

$$f_f = \sqrt{\frac{(E_f \varepsilon_{cu})^2}{4} + \frac{0.85\beta_1 f_c'}{\rho_f} E_f \varepsilon_{cu}} - 0.5E_f \varepsilon_{cu} \leq f_{fu}$$ 440.1R, Eq. (7.2.2d)

$$f_f = \sqrt{\frac{(19.50)^2}{4} + \frac{0.85(0.85)(4)}{(0.01505)}(19.50)} - 0.5(19.50) \leq 64$$

$f_f = 52.2$ ksi

$$M_n = \rho_f f_f \left(1 - 0.59\frac{\rho_f f_f}{f_c'}\right) bd^2$$ 440.1R, Eq. (7.2.2e)

$$= (0.01505)(52.2)\left(1 - 0.59\frac{(0.01505)(52.2)}{4}\right)(12)(17.50)^2 = 2553 \text{ in.-kip} = 212.7 \text{ ft-kip}$$

$\phi M_n = (0.65)(212.7) = 138.3$ ft-kip $> M_u = 123.2$ ft-kip OK

Note: Many designs for FRP-reinforced concrete are governed by serviceability requirements related to crack control, deflections, and creep rupture, rather than by flexural strength requirements. Calculations related to these serviceability requirements are covered in other example problems in this chapter.

Example 4—Design of one-way solid slab

This example is similar to Example 7.2 of PCA Notes on ACI 318-08.

Determine the required thickness and reinforcement for a one-way slab continuous over three or more equal spans. Center-to-center span $\ell = 19$ ft and clear span $\ell_n = 18$ ft. Assume interior exposure conditions.

$f_c' = 4000$ psi

$f_{fu}^* = 95{,}000$ psi

$E_f = 6000$ ksi

Service loads: w_D = slab self-weight (no superimposed dead load), $w_L = 50$ lb/ft^2

Calculations and discussion — Reference

Design material properties:

$C_E = 0.8$ 440.1R, Table 6.2

$f_{fu} = C_E f_{fu}^* = (0.8)(95) = 76$ ksi 440.1R, Eq. (6.2a)

1. Determine required slab thickness.
Based on minimum thickness table, consider estimated depth.
End span will control thickness:

$$h \approx \frac{\ell}{17} = \frac{(19)(12)}{17} = 13.4 \text{ in.}$$ 440.1R, Table 7.3.2.1

Table 7.3.2.1 is only intended to provide guidance for initial design, therefore, assume $h = 12$ in.

2. Compute the design moments using approximate moment analysis permitted by ACI 318-11, Sec. 8.3.3. Design will be based on the end span because it will yield the highest moments. Assume the end of the end span is integral with the support.

12 in. slab weighs $[(12)/(12)](150) = 150$ lb/ft^2

Factored load $q_u = 1.2(150) + 1.6(50) = 260$ lb/ft^2 318-11, Eq. (9-2)

Positive moment at discontinuous end integral with support:

$$+M_u = \frac{q_u \ell_n^2}{14} = \frac{(0.260)(18)^2}{14} = 6.02 \text{ ft-kip/ft}$$ 318-11, Sec. 8.3.3

Negative moment at exterior face of first interior support:

$$-M_u = \frac{q_u \ell_n^2}{10} = \frac{(0.260)(18)^2}{10} = 8.42 \text{ ft-kip/ft}$$ 318-11, Sec. 8.3.3

3. Determine required reinforcement and select bars.

Assume section is tension-controlled. For this case, $f_f = f_{fu} = 76$ ksi and $\phi = 0.55$.
For interior exposure, clear cover is 0.75 in. 440.5-08, Table 3.1

Assume No. 5 bars for flexural reinforcement.
No. 5 bar diameter: $d_b = 0.625$ in. 440.6-08, Table 7.1

$$d = 12 - \left[(0.75) + \frac{(0.625)}{2}\right] = 10.94 \text{ in.}$$

$$c_b = \left(\frac{\varepsilon_{cu}}{\varepsilon_{cu} + \varepsilon_{fu}}\right) d = \left(\frac{0.003}{0.003 + \frac{76}{6000}}\right)(10.94) = 2.09 \text{ in.}$$ 440.1R, Eq. (7.2.2h)

Use moment strength equation to solve for area of reinforcement.

Consider $-M$ because it governs.

$$M_u = \phi M_{n,reqd} = \phi A_{f,reqd} f_{fu}\left(d - \frac{\beta_1 c_b}{2}\right)$$ 440.1R, Eq. (7.2.2g)

$$A_{f,reqd} = \frac{M_u}{\phi f_{fu}\left(d - \frac{\beta_1 c_b}{2}\right)} = \frac{(8.42)(12)}{(0.55)(76)\left(10.94 - \frac{(0.85)(2.09)}{2}\right)} = 0.24 \text{ in.}^2\text{/ft}$$

Note that this requirement is well less than the minimum reinforcement, as computed in the following.

$$\rho_{f,ts} = 0.0014 \leq 0.0018 \times \frac{60{,}000}{f_{fu}} \frac{E_s}{E_f} \leq 0.0036$$ 440.1R, Eq. (9.1)

$$\rho_{f,ts} = 0.0014 \leq 0.0018 \times \frac{60{,}000}{76{,}000} \frac{29{,}000}{6000} \leq 0.0036$$

$\rho_{f,ts} = 0.0014 \leq 0.0069 \leq 0.0036$, so $\rho_{f,ts} = 0.0036$
$A_{f,min} = \rho_{f,min} bh = (0.0036)(12)(12) = 0.52$ in.^{2}/ft

Select No. 5 at 7 in. spacing ($A_f = 0.53$ in.^{2}/ft)

Verify assumption of tension-controlled behavior:

$$\rho_{fb} = 0.85\beta_1 \frac{f_c'}{f_{fu}} \frac{E_f \varepsilon_{cu}}{E_f \varepsilon_{cu} + f_{fu}}$$ 440.1R, Eq. (7.2.1b)

$E_f \varepsilon_{cu} = (6000)(0.003) = 18.00$ ksi

$$\rho_{fb} = 0.85(0.85)\frac{(4)}{(76)} \frac{(18.00)}{(18.00 + 76)} = 0.00728$$

$$\rho_f = \frac{A_f}{bd} = \frac{0.53}{(12)(10.94)} = 0.00404$$ 440.1R, Eq. (7.2.1a)

$$\frac{\rho_f}{\rho_{fb}} = \frac{0.00404}{0.00728} = 0.55 \quad \text{OK}$$

The slab may be designed to be 12 in. thick with No. 5 at 7 in. for $-M$. By observation, the same minimum reinforcement will be required for $+M$.

In addition to flexural strength, the slab should be examined for shear and the serviceability criteria of crack control, deflections, and creep rupture stress limits. Calculations related to these requirements are covered in other example problems in this chapter.

Further calculations show that this slab will be uncracked at service by a significant margin, and is even uncracked at ultimate conditions. The slab will work, but will be highly inefficient. Therefore, consider a more efficient slab design that is selected to be cracked at service.

4. Redesign the slab to be cracked at service loads.

When cracked, FRP-reinforced concrete slabs are seldom governed by flexural strength. Whereas a slab designed for flexural strength alone would have a ratio of service level moment to nominal moment strength (M_{serv}/M_n) of approximately 0.40 to 0.45, depending on the ratio of dead-to-live load, most FRP-reinforced slabs are governed by serviceability requirements and will exhibit ratios closer to 0.20 to 0.25. As a design approximation, design for a flexural strength corresponding to approximately twice the actual factored moment ($2.0M_u$).

Select a reinforcement ratio corresponding to a compression-controlled section, as this will promote the use of enough reinforcement to control cracking and reduce deflections.

Design for $2.0M_u$ and as a starting point, assume $\rho_f = 1.5\rho_{fb}$.

$\rho_f = 1.5\rho_{fb} = 1.5(0.00728) = 0.01092$

Calculate the stress in the tensile reinforcement (f_f) at ultimate conditions for the assumed value of ρ_f.

$$f_f = \sqrt{\frac{(E_f\varepsilon_{cu})^2}{4} + \frac{0.85\beta_1 f_c'}{\rho_f}E_f\varepsilon_{cu}} - 0.5E_f\varepsilon_{cu} \le f_{fu}$$ 440.1R, Eq. (7.2.2d)

$$f_f = \sqrt{\frac{(18.00)^2}{4} + \frac{0.85(0.85)(4)}{(0.01092)}(18.00)} - 0.5(18.00) \le 76$$

$f_f = 60.6$ ksi

Estimate dead load based on an assumed 8 in. slab thickness:
8 in. slab weighs $[(8)/(12)](150) = 100$ lb/ft^2
Factored load $q_u = 1.2(100) + 1.6(50) = 200$ lb/ft^2 318-11, Eq. (9-2)

Negative moment at exterior face of first interior support (governs):

$$-M_u = \frac{q_u \ell_n^2}{10} = \frac{(0.200)(18)^2}{10} = 6.48 \text{ ft-kip/ft}$$ 318-11, Sec. 8.3.3

Use the moment capacity equation to determine a depth for the slab.

$$M_u = \phi M_n = \phi\rho_f f_f\left(1 - 0.59\frac{\rho_f f_f}{f_c'}\right)bd^2$$ 440.1R, Eq. (7.2.2e)

$$2(6.48)(12) = (0.65)(0.01092)(60.6)\left(1 - 0.59\frac{(0.01092)(60.6)}{(4)}\right)(12)d^2$$

$d = 5.78$ in.

Assume No. 6 bars for flexural reinforcement.

$$h = (5.78) + (0.75) + \frac{(0.75)}{2} = 6.91 \text{ in.}$$

Round up to be conservative (7 in. thick slab [h = 7 in.]).

Correct dead load for 7 in. thickness:
7 in. slab weighs [(7)/(12)](150) = 87.5 lb/ft²
Factored load q_u = 1.2(87.5) + 1.6(50) = 185 lb/ft² — 318-11, Eq. (9-2)

The flexural capacity could be checked, but it should be satisfactory by inspection because the design was based on providing a capacity of $2.0M_u$. Thus, capacity calculations are not necessary.

Select reinforcement for the slab. Assume the same reinforcement for $-M$ and $+M$.
$A_f = \rho_f bd$ = (0.01092)(12)(5.78) = 0.76 in.²/ft — 440.1R, Eq. (7.2.1a)
Select No. 6 bars at 6 in. spacing (A_f = 0.88 in.²/ft).

5. Select temperature and shrinkage reinforcement for transverse direction.

$$\rho_{f,ts} = 0.0014 \le 0.0018 \times \frac{60{,}000}{f_{fu}}\frac{E_s}{E_f} \le 0.0036 \qquad \text{440.1R, Eq. (9.1)}$$

$$\rho_{f,ts} = 0.0014 \le 0.0018 \times \frac{60{,}000}{76{,}000}\frac{29{,}000}{6000} \le 0.0036$$

$\rho_{f,ts} = 0.0014 \le 0.0069 \le 0.0036$, so $\rho_{f,ts} = 0.0036$

$A_{f,ts} = \rho_{f,ts}bh$ = (0.0036)(12)(7.00) = 0.30 in.²/ft

Select No. 4 bars at 8 in. spacing (A_f = 0.30 in.²/ft).

Note: The slab should now be examined for shear, crack control, deflections, and creep rupture. Calculations related to these requirements are covered in other example problems in this chapter. If these design criteria are not satisfied, then the slab thickness may be increased incrementally or additional reinforcement added.

Example 5—Distribution of reinforcement for effective crack control

This example is similar to Example 9.1 of PCA Notes on ACI 318-08.

A rectangular beam is being designed for a factored moment M_u = 131 ft-kip. The beam has been sized as a 16 in. wide by 24 in. deep beam, and the required area of GFRP flexural reinforcement has been computed as $A_{f,reqd}$ = 2.27 in.² Determine an arrangement of flexural reinforcement that satisfies crack control requirements in 7.3.1. Assume bottom clear cover to the flexural reinforcement (c_c) of 2.0 in. (1.5 in. clear cover + 0.5 in. stirrup diameter). Assume side clear cover to the flexural reinforcement ($c_{c,side}$) of 2.5 in. (1.5 in. clear cover + 0.5 in. stirrup diameter + 0.5 in. to accommodate inner radius of stirrup bend). The unfactored service load moment is M_s = 90 ft-kip. Assume interior exposure conditions.

f_c' = 4000 psi
f_{fu}^* = 80,000 psi
E_f = 5800 ksi

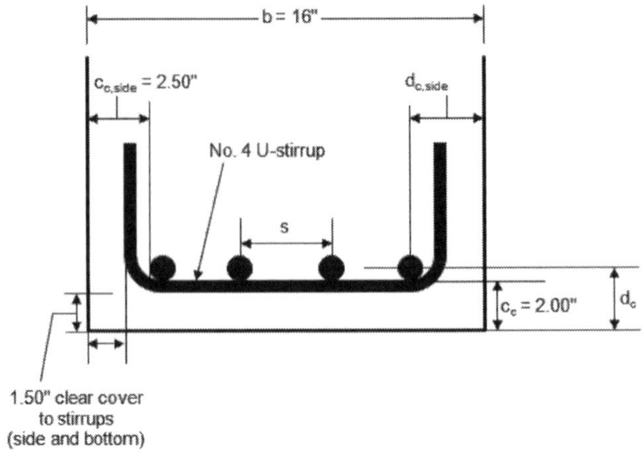

Calculations and discussion **Reference**

Design material properties:

$C_E = 0.8$ — 440.1R, Table 6.2

$f_{fu} = C_E f_{fu}^* = (0.8)(80) = 64$ ksi — 440.1R, Eq. (6.2a)

1. Select reinforcement.

Four No.7 bars ($A_f = 2.40$ in.^{2})

No. 7 bar diameter: $d_b = 0.875$ in. — 440.6-08, Table 7.1

2. Check limitation on value of d_c:

$$d_c \le \frac{E_f w}{2 f_{fs} \beta k_b}$$ 440.1R, Eq. (7.3.1b)

Assume a maximum allowable crack width w of 0.028 in. — 440.1R, Sec. 7.3.1

Take $k_b = 1.4$. — 440.1R, Sec. 7.3.1

The parameters β and f_{fs} must be determined from a linear-elastic cracked section analysis under service loads. Many of these calculations are covered in 7.3.2.2 and 7.4.1.

$$k = \sqrt{2\rho_f n_f + (\rho_f n_f)^2} - \rho_f n_f$$ 440.1R, Eq. (7.3.2b)

$$d_c = c_c + \frac{d_b}{2} = 2.00 + \frac{0.875}{2} = 2.44 \text{ in.}$$

$$d = h - d_c = 24 - 2.44 = 21.56 \text{ in.}$$

$$\rho_f = \frac{A_f}{bd} = \frac{2.40}{(16)(21.56)} = 0.00696$$ 440.1R, Eq. (7.2.1a)

$$E_c = 57{,}000\sqrt{f_c'} = 57{,}000\sqrt{4000} = 3.605 \times 10^6 \text{ psi} = 3605 \text{ ksi}$$ 318-11, Sec. 8.5.1

$$n_f = \frac{E_f}{E_c} = \frac{5800}{3605} = 1.609$$

$$\rho_f n_f = (0.00696)(1.609) = 0.01120$$

$$k = \sqrt{2\rho_f n_f + (\rho_f n_f)^2} - \rho_f n_f = \sqrt{2(0.01120) + (0.01120)^2} - 0.01120 = 0.139$$

$$I_{cr} = \frac{bd^3}{3}k^3 + n_f A_f d^2 (1-k)^2$$ 440.1R, Eq. (7.3.2.2a)

$$= \frac{(16)(21.56)^3}{3}(0.139)^3 + (1.609)(2.40)(21.56)^2(1-0.139)^2 = 1474 \text{ in.}^4$$

Service level stress in FRP reinforcement (refer to Eq. (7.4.1) for a similar calculation based on sustained service load):

$$f_{fs} = M_s \frac{n_f d(1-k)}{I_{cr}} = ((90)(12))\frac{(1.609)(21.56)(1-0.139)}{1474} = 21.9 \text{ ksi}$$ 440.1R, Eq. (7.4.1)

$$\beta = \frac{h-kd}{d-kd} = \frac{24-(0.139)(21.56)}{21.56-(0.139)(21.56)} = 1.13$$

Once parameters have been calculated, check limit on d_c

$$d_c \le \frac{E_f w}{2 f_{fs} \beta k_b}$$ 440.1R, Eq. (7.3.1b)

$$2.44 \text{ in.} \le \frac{(5800)(0.028)}{2(21.9)(1.13)(1.4)}$$

2.44 in. > 2.34 in. NG

Although the limit on d_c is not satisfied, continue to check the bar spacing for illustrative purposes in the following.

Check bar spacing against maximum spacing:

$$s_{max} = 1.15\frac{E_f w}{f_{fs} k_b} - 2.5c_c \le 0.92\frac{E_f w}{f_{fs} k_b}$$ 440.1R, Eq. (7.3.1a)

$$s_{max} = 1.15\frac{(5800)(0.028)}{(21.9)(1.4)} - 2.5(2.00) \le 0.92\frac{(5800)(0.028)}{(21.9)(1.4)}$$

$s_{max} = 1.09 \le 4.87$, so $s_{max} = 1.09$ in.

Provided center-to-center bar spacing:

$$s_{provided} = \frac{b - 2d_{c,side}}{N-1}$$

$d_{c,side}$ = side cover to center of exterior bars = 2.5 + (1/2)(0.875) = 2.94 in.
N = number of bars = 4

$$s_{provided} = \frac{16-2(2.94)}{4-1}$$

$s_{provided}$ = 3.37 in. > s_{max} = 1.09 in. NG

The selected reinforcement does not meet crack control requirements based on the calculations shown above. Both the cover (d_c) and spacing (s) limits are not satisfied.

Note that minimum clear bar spacing requirements must also be satisfied. In this case, maximum center-to-center bar spacing of 1.09 in. per Eq. (7.3.1a) is so small that minimum clear spacing (d_b, 1.0 in., or 4/3 times the maximum aggregate size) per ACI 318-11, Sec. 7.6, could not possibly be provided.

3. Choose a new reinforcement selection. Generally, if the aforementioned checks are not satisfied, then additional reinforcement will be required such that the FRP bar stress at service is reduced. Increase the amount of reinforcement and try two different possibilities:

Three No. 9 bars (A_f = 3.00 in.^{2}; d_b = 1.128 in.) 440.6-08, Table 7.1
Five No. 7 bars (A_f = 3.00 in.^{2}; d_b = 0.875 in.) 440.6-08, Table 7.1

Calculations are repeated for these two possibilities as well as the original selection checked previously. Values are presented in table format to save space and facilitate comparison.

Parameter	Unit	Four No. 7	Three No. 9	Five No.7
A_f	in.^{2}	2.40	3.00	3.00
d	in.	21.56	21.44	21.56
d_c	in.	2.44	2.56	2.44
ρ_f		0.00696	0.00875	0.00870
k		0.139	0.154	0.154
I_{cr}	in.^{4}	1474	1779	1801
f_{fs}	ksi	21.9	17.7	17.6
β		1.13	1.14	1.13
d_c (limit value)	in.	2.34	2.87	2.91
Check if $d_c \leq d_c$ (limit value)		**NG**	**OK**	**OK**
s_{max} (limit value)	in.	1.09	2.53	2.58
$d_{c,side}$	in.	2.94	3.06	2.94
$s_{provided}$	in.	3.37	4.94	2.53
Check if $s_{provided} \leq s_{max}$		**NG**	**NG**	**OK**

As can be seen in the table, the choice of three No. 9 bars provides sufficient additional reinforcement to reduce the stress in the reinforcing bars. The reinforcement, however, is spaced too far apart to meet the maximum spacing requirement, as the reinforcement is not sufficiently distributed to control crack widths.

The choice of five No. 7 bars provides the same amount of reinforcement as the choice of three No. 9 bars, but distributes the reinforcement better within the width of the cross section. The five bars would be spaced closely enough to control cracking, and far enough apart to provide minimum clear spacing required by ACI 318-11, Sec. 7.6.

Use five No. 7 bars distributed a single layer.

Example 6—Deflection of a simple-span nonprestressed rectangular beam

This example is similar to Example 10.1 of PCA Notes on ACI 318-08.

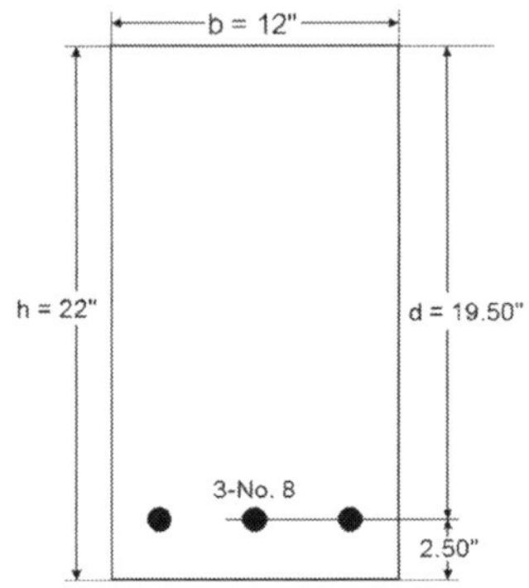

Required: analysis of deflections and comparison to ACI 318-11 allowable deflections. Consider long-term deflections at 3 months and 5 years (ultimate value).

Data:

f_c' = 3000 psi (normalweight concrete)

E_f = 6500 ksi

Superimposed dead load (not including beam weight) = 120 lb/ft

Live load = 300 lb/ft (20 percent sustained)

Span = 25 ft (simply-supported)

Calculations and discussion — **Reference**

No. 8 bar properties: — 440.6-08, Table 7.1

$d_b = 1.00$ in.
$A_{f,bar} = 0.79$ in.^{2}

Reinforcement quantities needed for calculations:
$A_f = (3)(0.79 \text{ in.}^2) = 2.37 \text{ in.}^2$

$$\rho_f = \frac{A_f}{bd} = \frac{2.37}{(12)(19.50)} = 0.01013$$ 440.1R, Eq. (7.2.1a)

1. Check recommended minimum beam thickness.
Based on minimum thickness table, consider estimated depth:

$$h \approx \frac{\ell}{10} = \frac{(25)(12)}{10} = 30 \text{ in.}$$ 440.1R, Table 7.3.2.1

Table 7.3.2.1 is intended to provide guidance for preliminary design, and does not guarantee that deflection limits will be met, so deflections should be computed.

2. Moments:

$$w_D = 0.120 + \frac{(12)(22)(0.150)}{144} = 0.395 \text{ kip/ft}$$

$$M_D = \frac{w_D \ell^2}{8} = \frac{(0.395)(25)^2}{8} = 30.86 \text{ ft-kip}$$

$$M_L = \frac{w_L \ell^2}{8} = \frac{(0.300)(25)^2}{8} = 23.44 \text{ ft-kip}$$

$M_{sus} = M_D + 0.20(M_L) = 30.86 + (0.20)(23.44) = 35.55$ ft-kip
$M_{D+L} = 30.86 + 23.44 = 54.30$ ft-kip

3. Modulus of elasticity and modular ratio:

$$E_c = 57{,}000\sqrt{f_c'} = 57{,}000\sqrt{3000} = 3.122 \times 10^6 \text{psi} = 3122 \text{ ksi}$$ 318-11, Sec. 8.5.1

$$n_f = \frac{E_f}{E_c} = \frac{6500}{3122} = 2.082$$

4. Gross and cracked section moment of inertia:

$$I_g = \frac{bh^3}{12} = \frac{(12)(22)^3}{12} = 10{,}648 \text{ in.}^4$$

$\rho_f n_f = (0.01013)(2.082) = 0.02109$

$$k = \sqrt{2\rho_f n_f + (\rho_f n_f)^2} - \rho_f n_f = \sqrt{2(0.02109) + (0.02109)^2} - 0.02109 = 0.185$$ 440.1R, Eq. (7.3.2.2b)

$$I_{cr} = \frac{bd^3}{3} k^3 + n_f A_f d^2 (1-k)^2$$ 440.1R, Eq. (7.3.2.2a)

$$= \frac{(12)(19.50)^3}{3}(0.185)^3 + (2.082)(2.37)(19.50)^2(1-0.185)^2 = 1434 \text{ in.}^4$$

$$\frac{I_g}{I_{cr}} = \frac{10{,}648}{1434} = 7.43 \qquad \frac{I_{cr}}{I_g} = \frac{1434}{10{,}648} = 0.135$$

5. Effective moments of inertia:

$$M_{cr} = \frac{7.5\lambda\sqrt{f'_c}I_g}{y_t} = \frac{7.5(1)\left(\sqrt{3000}\right)(10{,}648)}{(1/2)(22)(12{,}000)} = 33.14 \text{ ft-kip}$$ 440.1R, Eq. (7.3.2.2d)

Assuming that the member is loaded to the full dead load before any live load is applied, the effective moment of inertia may be computed on the basis of the dead load alone. (This assumption will yield a conservative estimate of live load deflection in the next step). Because $M_D < M_{cr}$, this beam is uncracked for this condition. Thus, the effective moment of inertia for dead load is I_g.

$(I_e)_D = I_g = 10{,}648$ in.^{4}

Find the effective moment of inertia due to dead plus live load.

$$\frac{M_{cr}}{M_a} = \frac{M_{cr}}{M_{D+L}} = \frac{33.14}{54.30} = 0.610$$

$\gamma = 1.72 - 0.72(M_{cr}/M_a) = 1.72 - 0.72(0.610) = 1.281$ 440.1R, Sec. 7.3.2.2

$$(I_e)_{D+L} = \frac{I_{cr}}{1-\gamma\left(\frac{M_{cr}}{M_a}\right)^2\left[1-\frac{I_{cr}}{I_g}\right]}$$ 440.1R, Eq. (7.3.2.2c)

$$= \frac{1434}{1-(1.281)(0.610)^2[1-0.135]} = 2440 \text{ in.}^4$$

6. Immediate deflections:

$K = 1$ for simple spans

$$(\Delta_i)_D = \frac{K(5)M_D\ell^2}{48E_c(I_e)_D} = \frac{(1)(5)(30.86)(25)^2(12)^3}{48(3122)(10{,}648)} = 0.104 \text{ in.}$$

$$(\Delta_i)_{D+L} = \frac{K(5)M_{D+L}\ell^2}{48E_c(I_e)_{D+L}} = \frac{(1)(5)(54.30)(25)^2(12)^3}{48(3122)(2440)} = 0.802 \text{ in.}$$

The immediate live load deflection is found by subtracting the dead load deflection from the deflection due to both dead and live load.

$(\Delta_i)_L = (\Delta_i)_{D+L} - (\Delta_i)_D = 0.802 - 0.104 = 0.698$ in.

It is conservative to find the immediate deflection due to sustained load by using the effective moment of inertia based on the full dead plus live load.

$$(\Delta_i)_{sus} = \frac{K(5)M_{sus}\ell^2}{48E_c(I_e)_{D+L}} = \frac{(1)(5)(35.55)(25)^2(12)^3}{48(3122)(2440)} = 0.525 \text{ in.}$$

$(\Delta_i)_{L,unsustained} = (\Delta_i)_{D+L} - (\Delta_i)_{sus} = 0.802 - 0.525 = 0.277$ in.

7. Incremental portion of total deflection occurring after placement of nonstructural elements (sum of the long-term deflection due to all sustained loads and the immediate deflection due to any additional live load):

a. 3 months ($\xi = 1.0$) 318-11, Sec. 9.5.2.5

$\Delta_{incr} = \Delta_{LT,sus} + \Delta_{L,unsustained} = 0.6\xi(\Delta_i)_{sus} + (\Delta_i)_{L,unsustained} = 0.6(1.0)(0.525) + 0.277 = 0.592$ in. 440.1R, Eq. (7.3.2.3c)

b. 5 years ($\xi = 2.0$) 318-11, Sec. 9.5.2.5

$\Delta_{incr} = \Delta_{LT,sus} + \Delta_{L,unsustained} = 0.6\xi(\Delta_i)_{sus} + (\Delta_i)_{L,unsustained} = 0.6(2.0)(0.525) + 0.277 = 0.907$ in. 440.1R, Eq. (7.3.2.3c)

8. Compare computed deflections to allowable deflections.

Allowable deflections (immediate live load): 318-11, Table 9.5(b)

Flat roofs not supporting and not attached to nonstructural elements likely to be damaged by large deflections:

$$\left\{(\Delta_i)_L = 0.698 \text{ in.}\right\} \leq \left\{\frac{\ell}{180} = \frac{(25)(12)}{180} = 1.67 \text{ in.}\right\} \quad \text{OK}$$

Floors not supporting and not attached to nonstructural elements likely to be damaged by large deflections:

$$\left\{(\Delta_i)_L = 0.698 \text{ in.}\right\} \leq \left\{\frac{\ell}{360} = \frac{(25)(12)}{360} = 0.83 \text{ in.}\right\} \quad \text{OK}$$

Allowable deflections (incremental after placement of nonstructural elements): 318-11, Table 9.5(b)

Roof or floor construction supporting or attached to nonstructural elements not likely to be damaged by large deflections:

$$\left\{\Delta_{incr} = 0.907 \text{ in.}\right\} \leq \left\{\frac{\ell}{240} = \frac{(25)(12)}{240} = 1.25 \text{ in.}\right\} \quad \text{OK}$$

Roof or floor construction supporting or attached to nonstructural elements likely to be damaged by large deflections (very stringent criteria):

$$\left\{\Delta_{incr} = 0.907 \text{ in.}\right\} > \left\{\frac{\ell}{480} = \frac{(25)(12)}{480} = 0.63 \text{ in.}\right\} \quad \text{NG}$$

If nonstructural elements likely to be damaged by large deflections are to be supported by this member, it will need to be redesigned to have a greater flexural stiffness. Otherwise, the member is adequate.

Example 7—Creep rupture stress check under sustained loads

This example does not have a companion in PCA Notes on ACI 318-08 *because creep rupture is an FRP phenomenon.*

The rectangular beam shown is designed to carry service load moments $M_D = 18$ ft-kip and $M_L = 18$ ft-kip. Assume 20 percent of the live load is sustained. Verify that the beam satisfies the creep rupture stress limits in 7.4.1. Assume interior exposure conditions.

$f_c' = 4000$ psi
$f_{fu}^* = 80,000$ psi
$E_f = 6000$ ksi

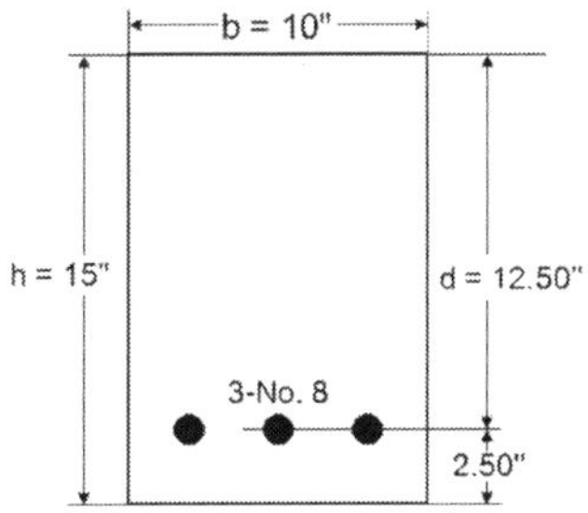

Calculations and discussion | **Reference**

No. 8 bar properties: *440.6-08, Table 7.1*
$d_b = 1.00$ in.
$A_{f,bar} = 0.79$ in.^{2}

Reinforcement quantities needed for calculations:
$A_f = (3)(0.79 \text{ in.}^2) = 2.37 \text{ in.}^2$

$$\rho_f = \frac{A_f}{bd} = \frac{2.37}{(10)(12.50)} = 0.01896$$ 440.1R, Eq. (7.2.1a)

Design material properties:
$C_E = 0.8$ 440.1R, Table 6.2
$f_{fu} = C_E f_{fu}^* = (0.8)(80) = 64$ ksi 440.1R, Eq. (6.2a)

1. Sustained load moment:
$M_{sus} = M_D + 0.20(M_L) = 18 + (0.20)(18) = 21.6$ ft-kip

2. Modulus of elasticity and modular ratio:

$$E_c = 57{,}000\sqrt{f_c'} = 57{,}000\sqrt{4000} = 3.605 \times 10^6 \text{ psi} = 3605 \text{ ksi}$$ 318-11, Sec. 8.5.1

$$n_f = \frac{E_f}{E_c} = \frac{6000}{3605} = 1.664$$

3. Cracked section moment of inertia:
$\rho_f n_f = (0.01896)(1.664) = 0.03155$

$$k = \sqrt{2\rho_f n_f + (\rho_f n_f)^2} - \rho_f n_f = \sqrt{2(0.03155) + (0.03155)^2} - 0.03155 = 0.222$$ 440.1R, Eq. (7.3.2.2b)

$$I_{cr} = \frac{bd^3}{3}k^3 + n_f A_f d^2 (1-k)^2$$ 440.1R, Eq. (7.3.2.2a)

$$= \frac{(10)(12.50)^3}{3}(0.222)^3 + (1.664)(2.37)(12.50)^2(1-0.222)^2 = 444 \text{ in.}^4$$

4. Reinforcement stress under sustained service loads:

$$f_{fs,sus} = M_{s,sus}\frac{n_f d(1-k)}{I_{cr}}$$ 440.1R, Eq. (7.4.1)

$$f_{fs,sus} = (21.6)(12)\frac{(1.664)(12.50)(1-0.222)}{444} = 9.45 \text{ ksi}$$

Allowable reinforcement stress: 440.1R, Table 7.4.1
$\{f_{fs,sus} = 9.45 \text{ ksi}\} \leq \{0.20 f_{fu} = 0.20(64) = 12.8 \text{ ksi}\}$ OK

Example 8—Design for shear (members subject to shear and flexure only)

This example is similar to Example 12.1 of PCA Notes on ACI 318-08.

Determine the required size and spacing of vertical U-stirrups for an 18 ft span, simply-supported normalweight reinforced concrete beam. Assume interior exposure conditions.

$b_w = 12$ in.
$d = 19.5$ in.
$f_c' = 4000$ psi
$f_{fu}^* = 100{,}000$ psi (stirrups)
$E_f = 6000$ ksi (stirrups)
$w_u = 4.82$ kip/ft (includes self-weight)
$\rho_f = 0.0270$ (longitudinal reinforcement)
$r_b/d_b = 4$ (assumed curvature of bent stirrup bars)

Calculations and discussion **Reference**

Design material properties:
$C_E = 0.8$ 440.1R, Table 6.2
$f_{fu} = C_E f_{fu}^* = (0.8)(100) = 80$ ksi 440.1R, Eq. (6.2a)

For the purposes of this example, the live load will be assumed to be present on the full span so that the design shear at the centerline of span is zero. (A design shear greater than zero at midspan is obtained by considering partial live loading of the span.)

1. Determine factored shear forces.

At support: $V_u = 4.82\left(\frac{18}{2}\right) = 43.38 \text{ kip}$

At distance d from support: $V_u = 43.38 - 4.82\left(\frac{19.5}{12}\right) = 35.55 \text{ kip}$

2. Determine shear strength provided by concrete.

$$\phi V_c = \phi\left(\frac{5}{2}k\right)2\sqrt{f_c'}b_w d$$ 440.1R, Eq. (8.2b)

$$E_c = 57{,}000\sqrt{f_c'} = 57{,}000\sqrt{4000} = 3.605\times10^6 \text{ psi} = 3605 \text{ ksi}$$ 318-11, Sec. 8.5.1

$$n_f = \frac{E_f}{E_c} = \frac{6000}{3605} = 1.664$$

$\rho_f n_f = (0.0270)(1.664) = 0.04493$

$$k = \sqrt{2\rho_f n_f + (\rho_f n_f)^2} - \rho_f n_f = \sqrt{2(0.04493) + (0.04493)^2} - 0.04493 = 0.258$$ 440.1R, Eq. (7.3.2.2b)

$$\phi V_c = (0.75)\left(\frac{5}{2}(0.258)\right)2\frac{\sqrt{4000}}{1000}(12)(19.5) = 14.32 \text{ kip}$$

$V_u = 35.55 \text{ kip} > \phi V_c/2 = 7.16 \text{ kip}$

Therefore, shear reinforcement is required. 440.1R, Sec. 8.2.2

3. Compute $V_u - \phi V_c$ at critical section.

$V_u - \phi V_c = 35.55 - 14.32 = 21.23 \text{ kip} < \phi 8\sqrt{f_c'}b_w d = 88.8 \text{ kip}$ OK 440.1R, Sec. 8.2.3

4. Determine distance x_c from support beyond which shear reinforcement is not required for strength ($V_u = \phi V_c$):

$$x_c = \frac{V_u\text{@support} - \phi V_c}{w_u} = \frac{43.38 - 14.32}{4.82} = 6.03 \text{ ft}$$

Determine distance x_m from support beyond which shear reinforcement is not required ($V_u = \phi V_c/2$):

$$x_m = \frac{V_u\text{@support} - (\phi V_c/2)}{w_u} = \frac{43.38 - (14.32/2)}{4.82} = 7.51 \text{ ft}$$

Therefore, only minimum shear reinforcement is required between 6.03 ft and 7.51 ft from the supports. Shear reinforcement is not required past 7.51 ft from the supports.

5. Determine design tensile stress in shear reinforcement.
Tensile strength of bent bars:

$$f_{fb} = \left(0.05\cdot\frac{r_b}{d_b} + 0.3\right)f_{fu} \le f_{fu}$$ 440.1R, Eq. (6.2.1)

$f_{fb} = (0.05 \cdot 4 + 0.3)f_{fu} \le f_{fu}$
$f_{fb} = (0.50)f_{fu} \le f_{fu}$
$f_{fb} = (0.50)(80) = 40 \text{ ksi}$

The design tensile strength is based on a strain of 0.004:
$f_{fv} = 0.004E_f \le f_{fb}$ 440.1R, Eq. (8.2d)

$f_{fv} = 0.004(6000) \leq 40$
$f_{fv} = 24 \leq 40$
$f_{fv} = 24$ ksi

6. Determine required spacing of vertical U-stirrups at the critical section.

$$\frac{A_{fv}}{s} = \frac{(V_u - \phi V_c)}{\phi f_{fv} d} = \frac{(35.55 - 14.32)}{(0.75)(24)(19.5)} = 0.0605 \text{ in.}^2/\text{in.}$$ 440.1R, Eq. (8.2e)

Assuming No. 4 U-stirrups ($A_{fv} = 0.40$ in.^{2})

$$s = \frac{0.40}{0.0605} = 6.61 \text{ in.}$$

Check maximum permissible spacing of stirrups:
$s = d/2 = 9.75$ in. ≤ 24 in. 440.1R, Sec. 8.3
because $V_u - \phi V_c = 21.23$ kip $< \phi 4\sqrt{f_c'} b_w d = 44.4$ kip OK 318-11, Sec. 11.4.5

Maximum stirrup spacing based on minimum shear reinforcement:

$$s = \frac{A_{fv,min} f_{fv}}{50 b_w} = \frac{(0.40)(24)(1000)}{50(12)} = 16 \text{ in.}$$ 440.1R, Eq. (8.2.2)

Therefore, the spacing at the critical section is restricted to the smallest of 6.61, 9.75, and 16 in. Select spacing of 6 in. at the critical section.

Determine the distance x where a transition can be made to a spacing of 9 in. (to satisfy strength and maximum spacing requirement calculated above as $d/2 = 9.75$ in.):

$$\frac{A_{fv}}{s} = \frac{(V_u - \phi V_c)}{\phi f_{fv} d}$$ 440.1R, Eq. (8.2e)

$$\frac{0.40}{9} = \frac{(V_u - \phi V_c)}{(0.75)(24)(19.5)}$$

$V_u - \phi V_c = 15.60$ kip
$V_u = 15.60 + 14.32 = 29.92$ kip

$$x = \frac{V_u@\text{support} - V_u}{w_u} = \frac{43.38 - 29.92}{4.82} = 2.79 \text{ ft}$$

Therefore, a transition may be made from 6 to 9 in. spacing at 2.79 ft from the support.

7. Select a stirrup spacing arrangement that satisfies previous calculations.

Place first stirrup at 2 in. from support. Use 6 in. spacing to 3 ft-2 in. from support. Use 9 in. spacing to 7 ft-8 in. from support.

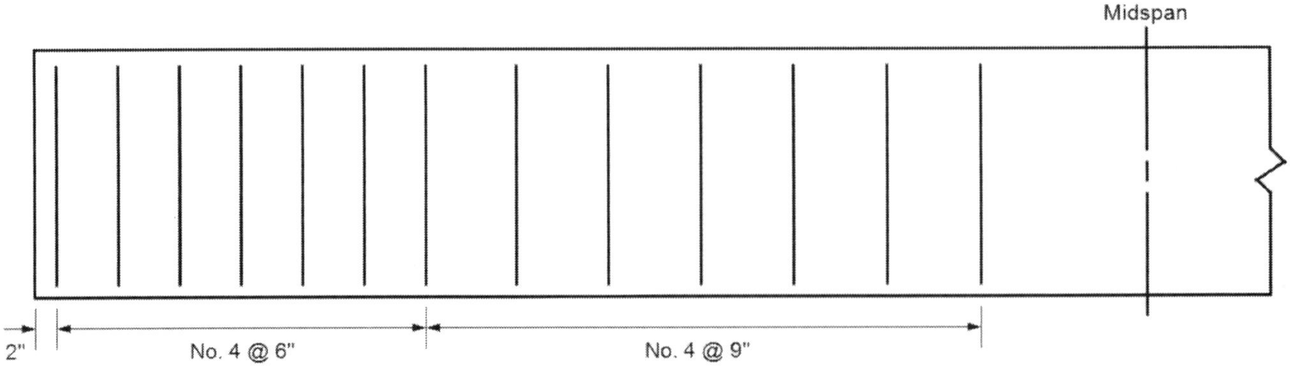

Note: Many designers would carry the reinforcement spacing of 9 in. through midspan, even though it is not required by calculations.

Example 9—Development of bars in tension (compression-controlled or transition zone section)

This example is similar to Example 4.2 of PCA Notes on ACI 318-08.

Calculate the required tension development length for the No. 8 glass FRP (GFRP) bars (alternate short bars) embedded in the normalweight concrete one-way slab shown in the following figure to develop the full moment capacity at Section A.

Assume $f_c' = 4000$ psi, $f_{fu}^* = 80{,}000$ psi, and $E_f = 6000$ ksi. Assume interior exposure conditions.

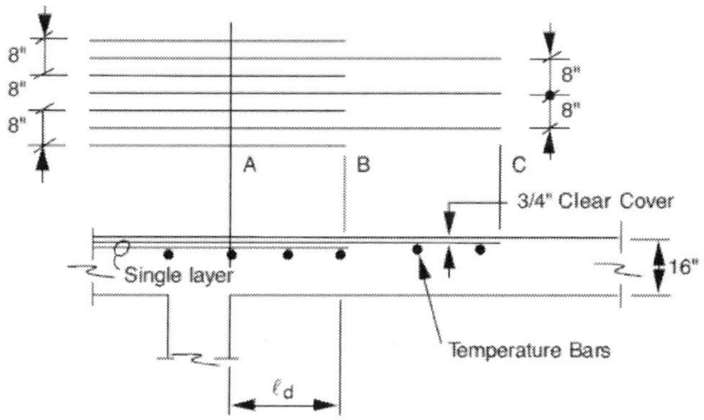

Calculations and discussion	**Reference**
Assume short bars are developed within distance AB while long bars are developed within BC.	
No. 8 bar properties:	440.6-08, Table 7.1
$d_b = 1.00$ in.	
$A_{f,bar} = 0.79$ in.^{2}	
Design material properties:	
$C_E = 0.8$	440.1R, Table 6.2
$f_{fu} = C_E f_{fu}^* = (0.8)(80) = 64$ ksi	440.1R, Eq. (6.2a)
1. Determine the type of section.	
$d = 16 - 0.75 - (1.00/2) = 14.75$ in.	
$A_f = (0.79\text{ in.}^2)/(4\text{ in. spacing}) = 0.198\text{ in.}^2/\text{in.} = 2.37\text{ in.}^2/\text{ft}$ of slab	
$\rho_f = \frac{A_f}{bd} = \frac{2.37}{(12)(14.75)} = 0.01339$	440.1R, Eq. (7.2.1a)
$\rho_{fb} = 0.85\beta_1 \frac{f_c'}{f_{fu}} \frac{E_f \varepsilon_{cu}}{E_f \varepsilon_{cu} + f_{fu}}$	440.1R, Eq. (7.2.1b)
$E_f\varepsilon_{cu} = (6000)(0.003) = 18.00$ ksi	
$\rho_{fb} = 0.85(0.85)\frac{(4)}{(64)}\frac{(18.00)}{(18.00+64)} = 0.00991$	
$\frac{\rho_f}{\rho_{fb}} = \frac{0.01339}{0.00991} = 1.35$	
Because $\rho_{fb} < \rho_f < 1.4\rho_{fb}$, the section is in the transition zone.	440.1R, Eq. (7.2.3)
2. Determine stress in tensile reinforcement at ultimate conditions.	
$f_f = \sqrt{\frac{(E_f\varepsilon_{cu})^2}{4} + \frac{0.85\beta_1 f_c'}{\rho_f} E_f\varepsilon_{cu}} - 0.5E_f\varepsilon_{cu} \le f_{fu}$	440.1R, Eq. (7.2.2d)
$f_f = \sqrt{\frac{(18.00)^2}{4} + \frac{0.85(0.85)(4)}{(0.01339)}(18.00)} - 0.5(18.00) \le 64$	
$f_f = 54.0$ ksi	

3. Determine development length.

$$\ell_d = \frac{\alpha \dfrac{f_{fr}}{\sqrt{f_c'}} - 340}{13.6 + \dfrac{C}{d_b}} d_b$$ 440.1R, Eq. (10.3a)

The bar stress that needs to be developed is the stress at the ultimate condition. There is no need to develop the full strength of the bar in this case.
$f_{fr} = f_f = 54.0$ ksi 440.1R, Sec. 10.3

Bar location modification factor should be taken as $\alpha = 1.5$ because more than 12 in. of concrete is cast below the reinforcement. 440.1R, Sec. 10.1.1

Center-to-center spacing of bars being developed = 8 in.

$$C = \min\left(d_c, \frac{\text{ctr-to-ctr spacing}}{2}\right) \leq 3.5d_b$$ 440.1R, Sec. 10.1

$$C = \min\left(0.75 + \frac{1.00}{2}, \frac{8}{2}\right) \leq 3.5(1.00) = \min(1.25, 4.00) \leq 3.50$$

$C = 1.25$ in.

$$\ell_d = \frac{\alpha \dfrac{f_{fr}}{\sqrt{f_c'}} - 340}{13.6 + \dfrac{C}{d_b}} d_b = \frac{(1.5)\dfrac{(54.0)(1000)}{\sqrt{4000}} - 340}{13.6 + \dfrac{1.25}{1.00}}(1.00) = 63.3 \text{ in.}$$

The required development length is 63.3 in.

Example 10—Development of bars in tension (tension-controlled section)

This example is similar to Example 4.3 of PCA Notes on ACI 318-08.

Calculate the length required to develop the inner two No. 8 glass FRP (GFRP) bars at the column face in the following figure. The two No. 8 outer bars are to be made continuous along full length of beam. Use $f_c' = 4000$ psi (normalweight concrete), $f_{fu}^* = 80{,}000$ psi, and $E_f = 6000$ ksi. Assume $d = 27.50$ in. Assume interior exposure conditions.

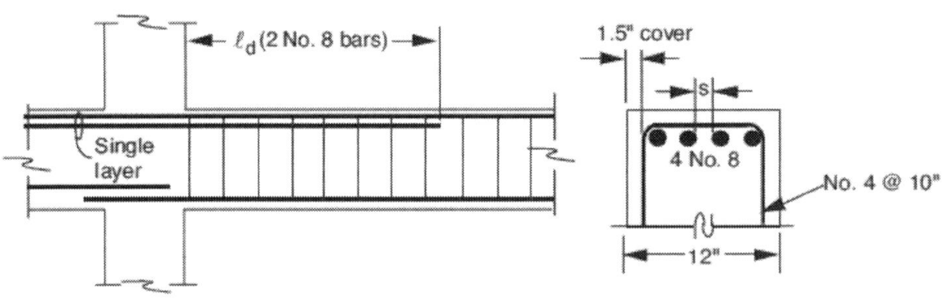

Calculations and discussion **Reference**

No. 8 bar properties: 440.6-08, Table 7.1
$d_b = 1.00$ in.
$A_{f,bar} = 0.79$ in.^{2}

Design material properties:
$C_E = 0.8$ 440.1R, Table 6.2
$f_{fu} = C_E f_{fu}^* = (0.8)(80) = 64$ ksi 440.1R, Eq. (6.2a)

1. Determine the type of section.
$A_f = (4)(0.79 \text{ in.}^2) = 3.16 \text{ in.}^2$

$$\rho_f = \frac{A_f}{bd} = \frac{3.16}{(12)(27.50)} = 0.00958$$ 440.1R, Eq. (7.2.1a)

$$\rho_{fb} = 0.85\beta_1 \frac{f_c'}{f_{fu}} \frac{E_f \varepsilon_{cu}}{E_f \varepsilon_{cu} + f_{fu}}$$ 440.1R, Eq. (7.2.1b)

$E_f\varepsilon_{cu} = (6000)(0.003) = 18.00$ ksi

$$\rho_{fb} = 0.85(0.85)\frac{(4)}{(64)}\frac{(18.00)}{(18.00+64)} = 0.00991$$

$$\frac{\rho_f}{\rho_{fb}} = \frac{0.00958}{0.00991} = 0.97$$

Because $\rho_f \le \rho_{fb}$, the section is tension-controlled. 440.1R, Eq. (7.2.3)

2. Determine stress in tensile reinforcement at ultimate conditions.
Because section is tension-controlled, $f_f = f_{fu} = 64$ ksi

3. Determine development length.

$$\ell_d = \frac{\alpha \dfrac{f_{fr}}{\sqrt{f_c'}} - 340}{13.6 + \dfrac{C}{d_b}} d_b$$ 440.1R, Eq. (10.3a)

$f_{fr} = f_f = 64.0$ ksi 440.1R, Sec. 10.3

Bar location modification factor should be taken as $\alpha = 1.5$ because more than 12 in. of concrete is cast below the reinforcement. 440.1R, Sec. 10.1.1

Clear spacing of bars being developed = [12 – 2(1.50) – 2(0.50) – 4(1.00)]/3 = 1.33 in.

$$C = \min\left(d_c, \frac{\text{ctr-to-ctr spacing}}{2}\right) \le 3.5d_b$$ 440.1R, Sec. 10.1

$$= \min\left(1.50 + 0.50 + \frac{1.00}{2}, \frac{0.50 + 1.33 + 0.50}{2}\right) \le 3.5(1.00) = \min(2.50,\ 1.17) \le 3.50$$

$C = 1.17$ in.

$$\ell_d = \frac{\alpha \dfrac{f_{fr}}{\sqrt{f_c'}} - 340}{13.6 + \dfrac{C}{d_b}} d_b = \frac{(1.5)\dfrac{(64)(1000)}{\sqrt{4000}} - 340}{13.6 + \dfrac{1.17}{1.00}}(1.00) = 79.5 \text{ in.}$$

The required development length is 79.5 in.

Example 11—Shear strength of slab at column support

This example is similar to Example 16.1 of PCA Notes on ACI 318-08.

Determine two-way action shear strength at an interior column support of a flat plate slab system for the following design conditions.

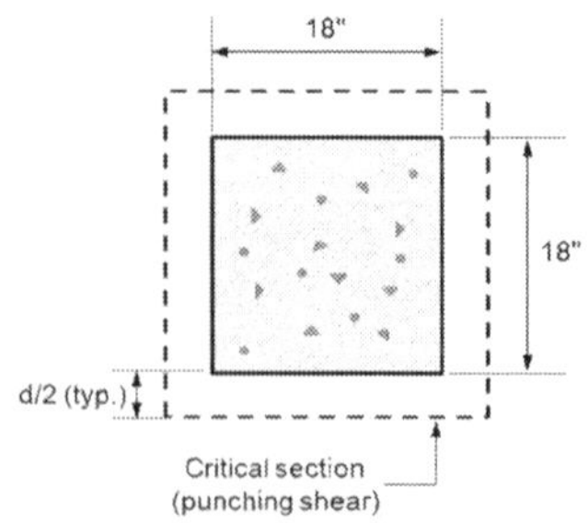

Column dimensions = 18 in. x 18 in.
Slab effective depth d = 6.5 in.
Specified concrete strength f_c' = 4000 psi (normalweight concrete)
E_f = 5800 ksi
ρ_f = 0.0120 (flexural reinforcement in column strips in both directions, top mat of reinforcement)

Calculations and discussion | Reference

1. Two-way action shear (punching shear) without shear reinforcement:

$V_u \leq \phi V_n$ — 440.1R, Sec. 8.1

$V_u \leq \phi V_c \quad (V_f = 0)$ — 440.1R, Sec. 8.2

2. Determine shear strength provided by concrete.

$$\phi V_c = \phi\left(\frac{5}{2}k\right)4\sqrt{f_c'}b_o d$$ 440.1R, Eq. (8.4a)

$$E_c = 57{,}000\sqrt{f_c'} = 57{,}000\sqrt{4000} = 3.605\times10^6 \text{ psi} = 3605 \text{ ksi}$$ 318-11, Sec. 8.5.1

$$\rho_f n_f = \rho_f \frac{E_f}{E_c} = (0.0120)\left(\frac{5800}{3605}\right) = (0.0120)(1.609) = 0.01931$$

$$k = \sqrt{2\rho_f n_f + (\rho_f n_f)^2} - \rho_f n_f = \sqrt{2(0.01931)+(0.01931)^2} - 0.01931 = 0.178$$ 440.1R, Eq. (7.3.2.2b)

b_o = 4(18 + 3.25 + 3.25) = 98 in. — 440.1R, Sec. 8.4

$$\phi V_c = (0.75)\left(\frac{5}{2}(0.178)\right)4\frac{\sqrt{4000}}{1000}(98)(6.5) = 53.83 \text{ kip}$$

Example 1M—Flexural (moment) strength using equivalent rectangular concrete stress distribution (compression-controlled section)

This example is similar to Example 6.1 of PCA Notes on ACI 318-08.

For the beam section shown, calculate moment strength based on static equilibrium using the equivalent rectangular concrete stress distribution as shown in Fig. 7.2 of this guide. Assume f_c' = 28 MPa, f_{fu}* = 550 MPa, and E_f = 41,000 MPa. Assume interior exposure conditions and neglect compression reinforcement.

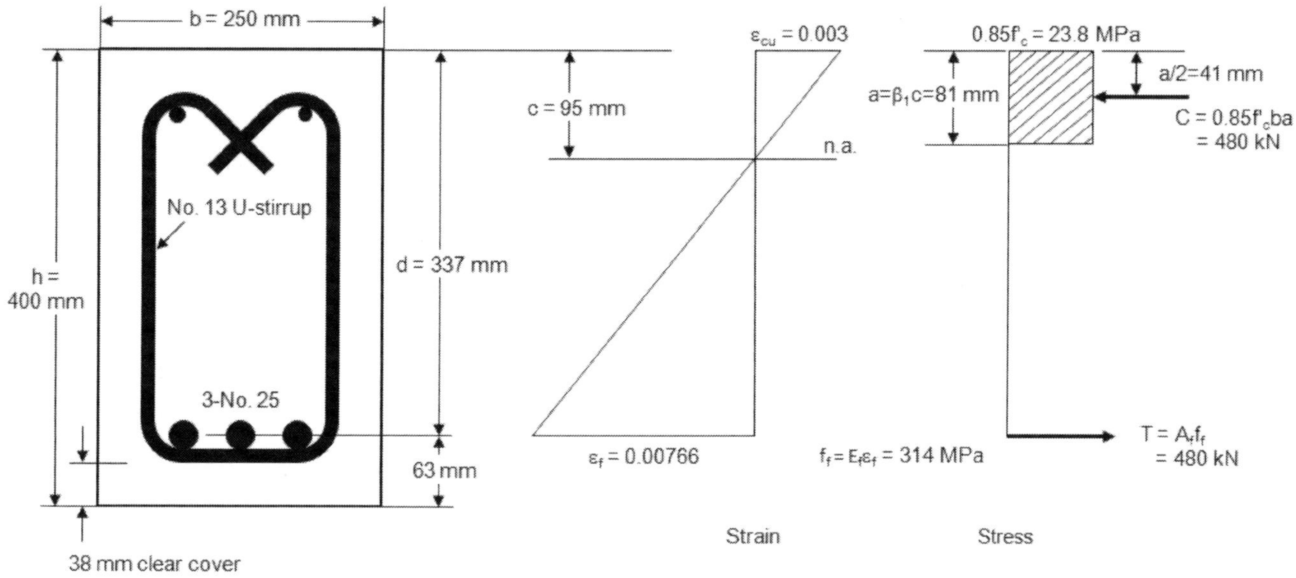

Calculations and discussion — **Reference**

No. 25 bar properties: — 440.6-08, Table 7.1

$d_b = 25.4$ mm

$A_{f,bar} = 510$ mm^2

Design material properties:

$C_E = 0.8$ — 440.1R, Table 6.2

$f_{fu} = C_E f_{fu}^* = (0.8)(550) = 440$ MPa — 440.1R, Eq. (6.2a)

1. Determine the strength reduction factor.

$d = 400 - 38 - 12.7 - (25.4/2) = 337$ mm

$A_f = (3)(510\text{ mm}^2) = 1530\text{ mm}^2$

$$\rho_f = \frac{A_f}{bd} = \frac{1530}{(250)(337)} = 0.01816$$ 440.1R, Eq. (7.2.1a)

$$\rho_{fb} = 0.85\beta_1 \frac{f_c'}{f_{fu}} \frac{E_f \varepsilon_{cu}}{E_f \varepsilon_{cu} + f_{fu}}$$ 440.1R, Eq. (7.2.1b)

$E_f\varepsilon_{cu} = (41{,}000)(0.003) = 123$ MPa

$$\rho_{fb} = 0.85(0.85)\frac{(28)}{(440)}\frac{(123)}{(123\ +\ 440)} = 0.01004$$

$$\rho_f/\rho_{fb} = \frac{0.01816}{0.01004} = 1.81$$

Because $\rho_f \geq 1.4\rho_{fb}$, the section is compression-controlled. — 440.1R, Eq. (7.2.3)

$\phi = 0.65$

2. Determine stress in tensile reinforcement at ultimate conditions.

$$f_f = \sqrt{\frac{(E_f\varepsilon_{cu})^2}{4} + \frac{0.85\beta_1 f_c'}{\rho_f}E_f\varepsilon_{cu}} - 0.5E_f\varepsilon_{cu} \ \leq f_{fu}$$ 440.1R, Eq. (7.2.2d)

$$f_f = \sqrt{\frac{(123)^2}{4} + \frac{0.85(0.85)(28)}{(0.01814)}(123)} - 0.5(123) \leq 440$$

$f_f = 314$ MPa

3. Determine nominal flexural strength M_n and design flexural strength ϕM_n.

$$a = \frac{A_f f_f}{0.85 f_c' b} = \frac{(1530)(314)}{0.85(28)(250)} = 81 \text{ mm}$$ 440.1R, Eq. (7.2.2b)

$$M_n = A_f f_f \left(d - \frac{a}{2}\right) = (1530)(314)\left(337 - \frac{81}{2}\right) = 142.4 \times 10^6 \text{ N-mm} = 142.4 \text{ kN-m}$$ 440.1R, Eq. (7.2.2a)

Alternatively, compute M_n directly:

$$M_n = \rho_f f_f \left(1 - 0.59\frac{\rho_f f_f}{f_c'}\right) bd^2$$ 440.1R, Eq. (7.2.2e)

$$= (0.01814)(314)\left(1 - 0.59\frac{(0.01814)(314)}{28}\right)(250)(337)^2 = 142.3 \times 10^6 \text{ N-mm} = 142.3 \text{ kN-m}$$

$\phi M_n = (0.65)(142.3) = 92.5$ kN-m

4. Minimum reinforcement.
The minimum reinforcement provisions do not apply because the section is not tension-controlled. 440.1R, Sec. 7.2.4

Note: While the general procedure and principles used in this example are applicable for an FRP-reinforced beam of any cross-sectional shape, the specific equations used in this example are restricted to singly-reinforced rectangular cross sections (or flanged sections that exhibit rectangular section behavior) with reinforcement in a single layer.

Example 2M—Flexural (moment) strength using equivalent rectangular concrete stress distribution (tension-controlled section)

This example is similar to Example 6.1 of PCA Notes on ACI 318-08.

For the beam section shown, calculate moment strength based on static equilibrium using the equivalent rectangular concrete stress distribution as shown in Fig. 7.2 of this guide. Assume $f_c' = 28$ MPa, $f_{fu}^* = 620$ MPa, and $E_f = 41{,}000$ MPa. Assume interior exposure conditions and neglect compression reinforcement.

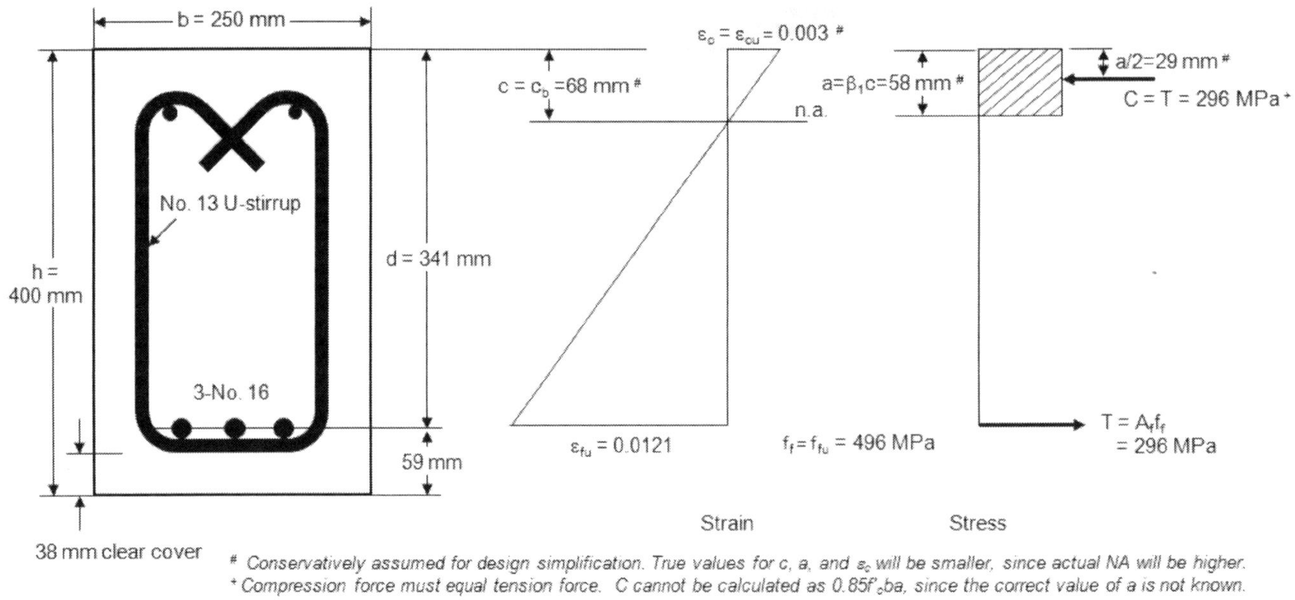

Calculations and discussion	**Reference**
No. 16 bar properties:	440.6-08, Table 7.1
$d_b = 15.9$ mm	
$A_{f,bar} = 199$ mm^2	

Design material properties:

$C_E = 0.8$ — 440.1R, Table 6.2

$f_{fu} = C_E f_{fu}^* = (0.8)(620) = 496$ MPa — 440.1R, Eq. (6.2a)

1. Determine the strength reduction factor.

$d = 400 - 38 - 12.7 - (15.9/2) = 341$ mm

$A_f = (3)(199\ \text{mm}^2) = 597\ \text{mm}^2$

$$\rho_f = \frac{A_f}{bd} = \frac{597}{(250)(341)} = 0.00700$$ 440.1R, Eq. (7.2.1a)

$$\rho_{fb} = 0.85\beta_1 \frac{f_c'}{f_{fu}} \frac{E_f \varepsilon_{cu}}{E_f \varepsilon_{cu} + f_{fu}}$$ 440.1R, Eq. (7.2.1b)

$E_f\varepsilon_{cu} = (41{,}000)(0.003) = 123$ MPa

$$\rho_{fb} = 0.85(0.85)\frac{(28)}{(496)}\frac{(123)}{(123+496)} = 0.00810$$

$$\frac{\rho_f}{\rho_{fb}} = \frac{0.00700}{0.00810} = 0.86$$

Because $\rho_f \leq \rho_{fb}$, the section is tension-controlled and $\phi = 0.55$. — 440.1R, Eq. (7.2.3)

2. Determine stress in tensile reinforcement at ultimate conditions.

Because section is tension-controlled, $f_f = f_{fu} = 496$ MPa

3. Determine nominal flexural strength M_n and design flexural strength ϕM_n.

$$c_b = \left(\frac{\varepsilon_{cu}}{\varepsilon_{cu} + \varepsilon_{fu}}\right)d = \left(\frac{0.003}{0.003 + \frac{496}{41{,}000}}\right)(341) = 68 \text{ mm}$$ 440.1R, Eq. (7.2.2h)

$$M_n = A_f f_{fu}\left(d - \frac{\beta_1 c_b}{2}\right) = (597)(496)\left(341 - \frac{(0.85)(68)}{2}\right) = 92.4 \times 10^6 \text{ N-mm} = 92.4 \text{ kN-m}$$ 440.1R, Eq. (7.2.2g)

$\phi M_n = (0.55)(92.4) = 50.8$ kN-m

4. Minimum reinforcement

The minimum reinforcement provisions apply since the section is tension-controlled. — 440.1R, Sec. 7.2.4

$$A_{f,min} = \frac{0.41\sqrt{f_c'}}{f_{fu}} b_w d \geq \frac{2.3}{f_{fu}} b_w d$$ 440.1R, Eq. (7.2.4)

$$A_{f,min} = \frac{0.41\sqrt{28}}{496}(250)(341) \geq \frac{2.3}{496}(250)(341)$$
$$= 373 \geq 395$$

$A_{f,min} = 395\ \text{mm}^2$

A_f (provided) $= 597\ \text{mm}^2 > A_{f,min} = 395\ \text{mm}^2$ OK

Note: Whereas the general procedure and principles used in this example are applicable for an FRP-reinforced beam of any cross-sectional shape, the specific equations used in this example are restricted to singly-reinforced rectangular cross sections (or flanged sections that exhibit rectangular section behavior) with reinforcement in a single layer.

The procedure used in this example, described in 7.2.2, is approximate and conservative. A more detailed analysis of a tension-controlled cross section would require the computation of the neutral axis location based on the principles of strain compatibility and a numerical constitutive model for the concrete (for example, parabolic and Hognestad). This type of analysis can be computationally intensive and is generally not suitable for hand calculations. For problems such as the one shown in this example, differences in results (computed flexural strength) are likely to be negligible. A strain compatibility analysis is suggested for cases where there are multiple layers of reinforcement or when different FRP bar types are mixed (7.2.5.1).

Example 3M—Design of a rectangular beam with tension reinforcement only

This example is similar to Example 7.1 of PCA Notes on ACI 318-08.

Design a rectangular beam of width b = 300 mm to have adequate flexural strength. The beam must resist service load moments M_D = 76 kN-m and M_L = 47 kN-m. Assume interior exposure conditions.

f_c' = 28 MPa

$f_{fu}*$ = 550 MPa

E_f = 45,000 MPa

Calculations and discussion — **Reference**

Design material properties:

$C_E = 0.8$ — 440.1R, Table 6.2

$f_{fu} = C_E f_{fu}* = (0.8)(550) = 440$ MPa — 440.1R, Eq. (6.2a)

1. A common starting point for design of a reinforced concrete member of unknown dimensions is the assumption of a reinforcement ratio.
As a starting point, assume $\rho_f = 1.5\rho_{fb}$.

$$\rho_{fb} = 0.85\beta_1 \frac{f_c'}{f_{fu}} \frac{E_f \varepsilon_{cu}}{E_f \varepsilon_{cu} + f_{fu}}$$ 440.1R, Eq. (7.2.1b)

$E_f\varepsilon_{cu} = (45,000)(0.003) = 135$ MPa

$$\rho_{fb} = 0.85(0.85)\frac{(28)}{(440)}\frac{(135)}{(135+440)} = 0.01079$$

$\rho_f = 1.5\rho_{fb} = 1.5(0.01079) = 0.01619$

Because $\rho_f \geq 1.4\rho_{fb}$, the section is compression-controlled. — 440.1R, Eq. (7.2.3)

$\phi = 0.65$

2. Compute bd^2 required.

First, determine the required design moment strength:

$\phi M_{n,reqd} = M_u = 1.2M_D + 1.6M_L = 1.2(76) + 1.6(47) = 166.4$ kN-m — 318-11, Eq. (9-2)

Calculate the stress in the tensile reinforcement (f_f) at ultimate conditions for the assumed value of ρ_f.

$$f_f = \sqrt{\frac{(E_f\varepsilon_{cu})^2}{4} + \frac{0.85\beta_1 f_c'}{\rho_f}E_f\varepsilon_{cu}} - 0.5E_f\varepsilon_{cu} \leq f_{fu}$$ 440.1R, Eq. (7.2.2d)

$$f_f = \sqrt{\frac{(135)^2}{4} + \frac{0.85(0.85)(28)}{(0.01619)}(135)} - 0.5(135) \leq 440$$

f_f = 349 MPa

Use the moment capacity equation to determine required dimensions for the cross section.

$$M_u = \phi M_n = \phi\rho_f f_f\left(1 - 0.59\frac{\rho_f f_f}{f_c'}\right)bd^2$$ 440.1R, Eq. (7.2.2e)

$$(166.4)(10^6) = (0.65)(0.01619)(349)\left(1 - 0.59\frac{(0.01619)(349)}{(28)}\right)bd^2$$

$bd^2 = 51.43 \times 10^6$ mm³

3. Size member so that $(bd^2)_{provided} \geq (bd^2)_{required}$.

Recall $b = 300$ mm, so $d = \sqrt{\dfrac{51.43\times10^6}{300}} = 414$ mm

4. Now, determine the required reinforcement, select bars, and determine depth.
$A_{f,reqd} = \rho_f bd = (0.01619)(300)(414) = 2011$ mm²
Select four No. 25 bars ($A_f = 2040$ mm²)

Note: Examining alternative designs using other bar sizes may require changing the assumed value of f_{fu}* *to a value appropriate for the selected bar size.*

No. 25 bar diameter: $d_b = 25.4$ mm — 440.6-08, Table 7.1

For interior exposure, clear cover is 38 mm — 440.5-08, Table 3.1

Assuming No. 13 stirrups:

$$h = (414) + (38) + (12.7) + \frac{(25.4)}{2} = 477 \text{ mm}$$

Round up to be conservative. So, select a 300 x 500 mm beam.

5. Determine capacity of cross section.

$$d = 500 - \left[(38) + (12.7) + \frac{(25.4)}{2}\right] = 437 \text{ mm}$$

$$\rho_f = \frac{A_f}{bd} = \frac{2040}{(300)(437)} = 0.01556$$ 440.1R, Eq. (7.2.1a)

$$\frac{\rho_f}{\rho_{fb}} = \frac{0.01556}{0.01079} = 1.44$$

Because $\rho_f \geq 1.4\rho_{fb}$, $\phi = 0.65$. — 440.1R, Eq. (7.2.3)

$$f_f = \sqrt{\frac{(E_f\varepsilon_{cu})^2}{4} + \frac{0.85\beta_1 f_c'}{\rho_f}E_f\varepsilon_{cu}} - 0.5E_f\varepsilon_{cu} \leq f_{fu}$$ 440.1R, Eq. (7.2.2d)

$$f_f = \sqrt{\frac{(135)^2}{4} + \frac{0.85(0.85)(28)}{(0.01556)}(135)} - 0.5(135) \leq 440$$

$f_f = 357$ MPa

$$M_n = \rho_f f_f\left(1 - 0.59\frac{\rho_f f_f}{f_c'}\right)bd^2$$ 440.1R, Eq. (7.2.2e)

$$= (0.01556)(357)\left(1 - 0.59\frac{(0.01556)(357)}{28}\right)(300)(437)^2 = 281.0 \times 10^6 \text{ N-mm} = 281.0 \text{ kN-m}$$

$\phi M_n = (0.65)(281.0) = 182.3$ kN-m $> M_u = 166.4$ kN-m OK

Note: Many designs for FRP-reinforced concrete are governed by serviceability requirements related to crack control, deflections, and creep rupture, rather than by flexural strength requirements. Calculations related to these serviceability requirements are covered in other example problems in this chapter.

Example 4M—Design of one-way solid slab

This example is similar to Example 7.2 of PCA Notes on ACI 318-08.

Determine the required thickness and reinforcement for a one-way slab continuous over three or more equal spans. Center-to-center span $\ell = 5.8$ m and clear span $\ell_n = 5.5$ m. Assume interior exposure conditions.

$f_c' = 28$ MPa

$f_{fu}* = 650$ MPa

$E_f = 41{,}000$ MPa

Service loads: w_D = slab self-weight (no superimposed dead load), $w_L = 24$ kN/m^2

Calculations and discussion	**Reference**
Design material properties:	
$C_E = 0.8$	440.1R, Table 6.2
$f_{fu} = C_E f_{fu}* = (0.8)(650) = 520$ MPa	440.1R, Eq. (6.2a)

1. Determine required slab thickness.
Based on minimum thickness table, consider estimated depth.
End span will control thickness:

$$h \approx \frac{\ell}{17} = \frac{(5.8)(1000)}{17} = 341 \text{ mm}$$ 440.1R, Table 7.3.2.1

Table 7.3.2.1 is only intended to provide guidance for initial design, therefore, assume $h = 300$ mm.

2. Compute the design moments using approximate moment analysis permitted by ACI 318-11, Sec. 8.3.3. Design will be based on the end span because it will yield the highest moments.

Assume the end of the end span is integral with the support.

300 mm slab weighs [(300)/(1000)](24) = 7.2 kN/m^2

Factored load $q_u = 1.2(7.2) + 1.6(2.4) = 12.5$ kN/m^2 — 318-11, Eq. (9-2)

Positive moment at discontinuous end integral with support:

$$+M_u = \frac{q_u \ell_n^2}{14} = \frac{(12.5)(5.5)^2}{14} = 27.0 \text{ kN-m/m}$$ 318-11, Sec. 8.3.3

Negative moment at exterior face of first interior support:

$$-M_u = \frac{q_u \ell_n^2}{10} = \frac{(12.5)(5.5)^2}{10} = 37.8 \text{ kN-m/m}$$ 318-11, Sec. 8.3.3

3. Determine required reinforcement and select bars.

Assume section is tension-controlled. For this case, $f_f = f_{fu} = 520$ MPa and $\phi = 0.55$.

For interior exposure, clear cover is 19 mm. — 440.5-08, Table 3.1

Assume No. 16 bars for flexural reinforcement.

No. 16 bar diameter: $d_b = 15.9$ mm 440.6-08, Table 7.1

$$d = 300 - \left[(19) + \frac{(15.9)}{2}\right] = 273 \text{ mm}$$

$$c_b = \left(\frac{\varepsilon_{cu}}{\varepsilon_{cu} + \varepsilon_{fu}}\right)d = \left(\frac{0.003}{0.003 + \dfrac{520}{41{,}000}}\right)(273) = 52 \text{ mm}$$ 440.1R, Eq. (7.2.2h)

Use moment strength equation to solve for area of reinforcement.

Consider $-M$ because it governs.

$$M_u = \phi M_{n,reqd} = \phi A_{f,reqd} f_{fu} \left(d - \frac{\beta_1 c_b}{2} \right)$$ 440.1R, Eq. (7.2.2g)

$$A_{f,reqd} = \frac{M_u}{\phi f_{fu} \left(d - \frac{\beta_1 c_b}{2} \right)} = \frac{(37.8)(10^6)}{(0.55)(520)\left(273 - \frac{(0.85)(52)}{2} \right)} = 527 \text{ mm}^2\text{/m}$$

Note that this requirement is well less than the minimum reinforcement, as computed in the following.

$$\rho_{f,ts} = 0.0014 \le 0.0018 \times \frac{414}{f_{fu}} \frac{E_s}{E_f} \le 0.0036$$ 440.1R, Eq. (9.1)

$$\rho_{f,ts} = 0.0014 \le 0.0018 \times \frac{414}{520} \frac{200{,}000}{41{,}000} \le 0.0036$$

$\rho_{f,ts} = 0.0014 \le 0.0070 \le 0.0036$, so $\rho_{f,ts} = 0.0036$
$A_{f,min} = \rho_{f,min} bh = (0.0036)(1000)(300) = 1080 \text{ mm}^2\text{/m}$

Select No. 16 at 180 mm spacing ($A_f = 1106 \text{ mm}^2\text{/m}$)

Verify assumption of tension-controlled behavior:

$$\rho_{fb} = 0.85\beta_1 \frac{f_c'}{f_{fu}} \frac{E_f \varepsilon_{cu}}{E_f \varepsilon_{cu} + f_{fu}}$$ 440.1R, Eq. (7.2.1b)

$E_f\varepsilon_{cu} = (41{,}000)(0.003) = 123$ MPa

$$\rho_{fb} = 0.85(0.85)\frac{(28)}{(520)} \frac{(123)}{(123+520)} = 0.00744$$

$$\rho_f = \frac{A_f}{bd} = \frac{1106}{(1000)(273)} = 0.00405$$ 440.1R, Eq. (7.2.1a)

$$\frac{\rho_f}{\rho_{fb}} = \frac{0.00405}{0.00744} = 0.54 \quad \text{OK}$$

The slab may be designed to be 300 mm thick with No. 16 at 180 mm for $-M$. By observation, the same minimum reinforcement will be required for $+M$.

In addition to flexural strength, the slab should be examined for shear and the serviceability criteria of crack control, deflections, and creep rupture stress limits. Calculations related to these requirements are covered in other example problems in this chapter.

Further calculations show that this slab will be uncracked at service by a significant margin, and is even uncracked at ultimate conditions. The slab will work, but will be highly inefficient. Therefore, consider a more efficient slab design that is selected to be cracked at service.

4. Redesign the slab to be cracked at service loads.

When cracked, FRP-reinforced concrete slabs are seldom governed by flexural strength. Whereas a slab designed for flexural strength alone would have a ratio of service level moment to nominal moment strength (M_{serv}/M_n) of approximately 0.40 to 0.45, depending on the ratio of dead-to-live load, most FRP-reinforced slabs are governed by serviceability requirements and will exhibit ratios closer to 0.20 to 0.25. As a design approximation, design for a flexural strength corresponding to approximately twice the actual factored moment ($2.0M_u$).

Select a reinforcement ratio corresponding to a compression-controlled section, as this will promote the use of enough reinforcement to control cracking and reduce deflections.

Design for $2.0M_u$ and as a starting point, assume $\rho_f = 1.5\rho_{fb}$.

$\rho_f = 1.5\rho_{fb} = 1.5(0.00744) = 0.01116$

Calculate the stress in the tensile reinforcement (f_f) at ultimate conditions for the assumed value of ρ_f.

$$f_f = \sqrt{\frac{(E_f\varepsilon_{cu})^2}{4} + \frac{0.85\beta_1 f_c'}{\rho_f}E_f\varepsilon_{cu}} - 0.5E_f\varepsilon_{cu} \le f_{fu}$$ 440.1R, Eq. (7.2.2d)

$$f_f = \sqrt{\frac{(123)^2}{4} + \frac{0.85(0.85)(28)}{(0.01116)}(123)} - 0.5(123) \le 520$$

$f_f = 415$ MPa

Estimate dead load based on an assumed 200 mm slab thickness:
200 mm slab weighs $[(200)/(1000)](24) = 4.8$ kN/m²
Factored load $q_u = 1.2(4.8) + 1.6(2.4) = 9.6$ kN/m² 318-11, Eq. (9-2)
Negative moment at exterior face of first interior support (governs):

$$-M_u = \frac{q_u \ell_n^2}{10} = \frac{(9.6)(5.5)^2}{10} = 29.0 \text{ kN-m/m}$$ 318-11, Sec. 8.3.3

Use the moment capacity equation to determine a depth for the slab.

$$M_u = \phi M_n = \phi\rho_f f_f\left(1 - 0.59\frac{\rho_f f_f}{f_c'}\right)bd^2$$ 440.1R, Eq. (7.2.2e)

$$2(29.0)(10^6) = (0.65)(0.01116)(415)\left(1 - 0.59\frac{(0.01116)(415)}{(28)}\right)(1000)d^2$$

$d = 146$ mm

Assume No. 19 bars for flexural reinforcement.

$$h = (146) + (19) + \frac{(19.1)}{2} = 175 \text{ mm}$$

Round up to be conservative (180 mm thick slab [$h = 180$ mm]).

Correct dead load for 180 mm thickness:
180 mm slab weighs $[(180)/(1000)](24) = 4.3$ kN/m²
Factored load $q_u = 1.2(4.3) + 1.6(2.4) = 9.0$ kN/m² 318-11, Eq. (9-2)

The flexural capacity could be checked, but it should be satisfactory by inspection because the design was based on providing a capacity of $2.0M_u$. Thus, capacity calculations are not necessary.

Select reinforcement for the slab.
Assume the same reinforcement for $-M$ and $+M$.
$A_f = \rho_f bd = (0.01116)(1000)(146) = 1629$ mm²/m 440.1R, Eq. (7.2.1a)
Select No. 19 bars at 150 mm spacing ($A_f = 1890$ mm²/m).

5. Select temperature and shrinkage reinforcement for transverse direction.

$$\rho_{f,ts} = 0.0014 \le 0.0018 \times \frac{414}{f_{fu}}\frac{E_s}{E_f} \le 0.0036$$ 440.1R, Eq. (9.1)

$$\rho_{f,ts} = 0.0014 \le 0.0018 \times \frac{414}{520}\frac{200{,}000}{41{,}000} \le 0.0036$$

$\rho_{f,ts} = 0.0014 \le 0.0070 \le 0.0036$, so $\rho_{f,ts} = 0.0036$
$A_{f,ts} = \rho_{f,ts}bh = (0.0036)(1000)(180) = 648$ mm²/m

Select No. 13 bars at 180 mm spacing (A_f = 717 mm²/m).

Note: The slab should now be examined for shear, crack control, deflections, and creep rupture. Calculations related to these requirements are covered in other example problems in this chapter. If these design criteria are not satisfied, then the slab thickness may be increased incrementally or additional reinforcement added.

Example 5M—Distribution of reinforcement for effective crack control

This example is similar to Example 9.1 of PCA Notes on ACI 318-08.

A rectangular beam is being designed for a factored moment M_u = 177 kN-m. The beam has been sized as a 400 mm wide by 600 mm deep beam, and the required area of GFRP flexural reinforcement has been computed as $A_{f,reqd}$ = 1460 mm² Determine an arrangement of flexural reinforcement that satisfies crack control requirements in 7.3.1. Assume bottom clear cover to the flexural reinforcement (c_c) of 50 mm (approximately 38 mm clear cover + 13 mm stirrup diameter). Assume side clear cover to the flexural reinforcement ($c_{c,side}$) of 65 mm (approximately 38 mm clear cover + 13 mm stirrup diameter + 13 mm to accommodate inner radius of stirrup bend). The unfactored service load moment is M_s = 122 kN-m. Assume interior exposure conditions.

f_c' = 28 MPa

f_{fu}* = 550 MPa

E_f = 40,000 MPa

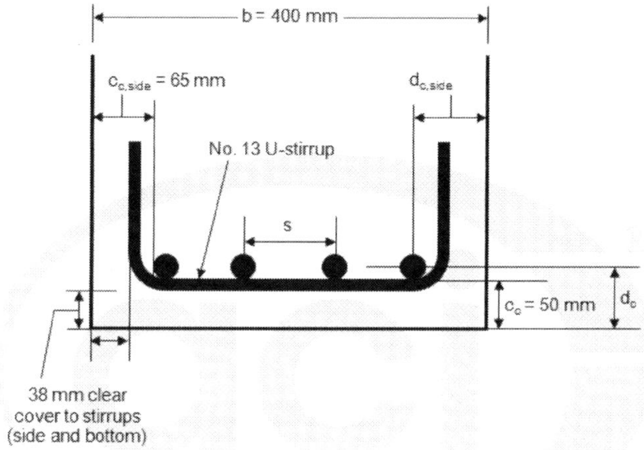

Calculations and discussion | Reference

Design material properties:

C_E = 0.8 — 440.1R, Table 6.2

$f_{fu} = C_E f_{fu}^* = (0.8)(550) = 440$ MPa — 440.1R, Eq. (6.2a)

1. Select reinforcement.

Four No. 22 bars (A_f = 1550 mm²)

No. 22 bar diameter: d_b = 22.2 mm — 440.6-08, Table 7.1

2. Check limitation on value of d_c:

$$d_c \le \frac{E_f w}{2 f_{fs} \beta k_b}$$ — 440.1R, Eq. (7.3.1b)

Assume a maximum allowable crack width w of 0.70 mm. — 440.1R, Sec. 7.3.1

Take k_b = 1.4. — 440.1R, Sec. 7.3.1

The parameters β and f_{fs} must be determined from a linear-elastic cracked section analysis under service loads. Many of these calculations are covered in 7.3.2.2 and 7.4.1.

$$k = \sqrt{2\rho_f n_f + (\rho_f n_f)^2} - \rho_f n_f$$ — 440.1R, Eq. (7.3.2b)

$$d_c = c_c + \frac{d_b}{2} = 50 + \frac{22.2}{2} = 61 \text{ mm}$$

$d = h - d_c = 600 - 61 = 539$ mm

$$\rho_f = \frac{A_f}{bd} = \frac{1550}{(400)(539)} = 0.00719$$ 440.1R, Eq. (7.2.1a)

$$E_c = 4700\sqrt{f_c'} = 4700\sqrt{28} = 24{,}900 \text{ MPa}$$ 318-11, Sec. 8.5.1

$$n_f = \frac{E_f}{E_c} = \frac{41{,}000}{24{,}900} = 1.647$$

$$\rho_f n_f = (0.00719)(1.647) = 0.01184$$

$$k = \sqrt{2\rho_f n_f + (\rho_f n_f)^2} - \rho_f n_f = \sqrt{2(0.01184) + (0.01184)^2} - 0.01184 = 0.142$$

$$I_{cr} = \frac{bd^3}{3}k^3 + n_f A_f d^2(1-k)^2$$ 440.1R, Eq. (7.3.2.2a)

$$= \frac{(400)(539)^3}{3}(0.142)^3 + (1.647)(1550)(539)^2(1-0.142)^2 = 605.7 \times 10^6 \text{ mm}^4$$

Service level stress in FRP reinforcement (refer to Eq. (7.4.1) for a similar calculation based on sustained service load)

$$f_{fs} = M_s \frac{n_f d(1-k)}{I_{cr}} = ((122)(10^6))\frac{(1.647)(539)(1-0.142)}{605.7\times 10^6} = 153 \text{ MPa}$$ 440.1R, Eq. (7.4.1)

$$\beta = \frac{h-kd}{d-kd} = \frac{600-(0.142)(539)}{539-(0.142)(539)} = 1.13$$

Once parameters have been calculated, check limit on d_c

$$d_c \le \frac{E_f w}{2 f_{fs}\beta k_b}$$ 440.1R, Eq. (7.3.1b)

$$61 \text{ mm} \le \frac{(41{,}000)(0.70)}{2(153)(1.13)(1.4)}$$

61 mm > 59 mm NG

Although the limit on d_c is not satisfied, continue to check the bar spacing for illustrative purposes in the following.

Check bar spacing against maximum spacing:

$$s_{max} = 1.15\frac{E_f w}{f_{fs}k_b} - 2.5c_c \le 0.92\frac{E_f w}{f_{fs}k_b}$$ 440.1R, Eq. (7.3.1a)

$$s_{max} = 1.15\frac{(41{,}000)(0.70)}{(153)(1.4)} - 2.5(50) \le 0.92\frac{(41{,}000)(0.70)}{(153)(1.4)}$$

$s_{max} = 29 \le 123$, so $s_{max} = 29$ mm

Provided center-to-center bar spacing:

$$s_{provided} = \frac{b - 2d_{c,side}}{N-1}$$

$d_{c,side}$ = side cover to center of exterior bars = 65 + (1/2)(22.2) = 76 mm
N = number of bars = 4

$$s_{provided} = \frac{400-2(76)}{4-1}$$

$s_{provided} = 83$ mm $> s_{max} = 29$ mm NG

The selected reinforcement does not meet crack control requirements based on the calculations shown above. Both the cover (d_c) and spacing (s) limits are not satisfied.
Note that minimum clear bar spacing requirements must also be satisfied. In this case, maximum center-to-center bar spacing of 1.09 in. per Eq. (7.3.1a) is so small that minimum clear spacing (d_b, 25 mm, or 4/3 times the maximum aggregate size) per ACI 318-11, Sec. 7.6, could not possibly be provided (318-11, Sec. 7.6).

3. Choose a new reinforcement selection. Generally, if the aforementioned checks are not satisfied, then additional reinforcement will be required such that the FRP bar stress at service is reduced. Increase the amount of reinforcement and try two different possibilities:

Three No. 29 bars (A_f = 1940 mm^2; d_b = 28.7 mm) 440.6-08, Table 7.1
Five No. 22 bars (A_f = 1940 mm^2; d_b = 22.2 mm) 440.6-08, Table 7.1

Calculations are repeated for these two possibilities as well as the original selection checked previously. Values are presented in table format to save space and facilitate comparison.

Parameter	Unit	Four No. 22	Three No. 29	Five No. 22
A_f	mm^2	1550	1940	1940
d	mm	539	536	539
d_c	mm	61	64	61
ρ_f		0.00719	0.00905	0.00900
k		0.142	0.158	0.158
I_{cr}	mm^4	605.7×10^6	731.8×10^6	740.5×10^6
f_{fs}	MPa	153	124	123
β		1.13	1.14	1.13
d_c (limit value)	mm	59	73	74
Check if $d_c \leq d_c$ (limit value)		NG	OK	OK
s_{max} (limit value)	mm	29	65	67
$d_{c,side}$	mm	76	79	76
$s_{provided}$	mm	83	121	62
Check if $s_{provided} \leq s_{max}$		NG	NG	OK

As can be seen in the table, the choice of three No. 29 bars provides sufficient additional reinforcement to reduce the stress in the reinforcing bars. The reinforcement, however, is spaced too far apart to meet the maximum spacing requirement, as the reinforcement is not sufficiently distributed to control crack widths.

The choice of five No. 22 bars provides the same amount of reinforcement as the choice of three No. 29 bars, but distributes the reinforcement better within the width of the cross section. The five bars would be spaced closely enough to control cracking, and far enough apart to provide minimum clear spacing required by ACI 318-11, Sec. 7.6.

Use five No. 22 bars distributed a single layer.

Example 6M—Deflection of a simple-span nonprestressed rectangular beam

This example is similar to Example 10.1 of PCA Notes on ACI 318-08.

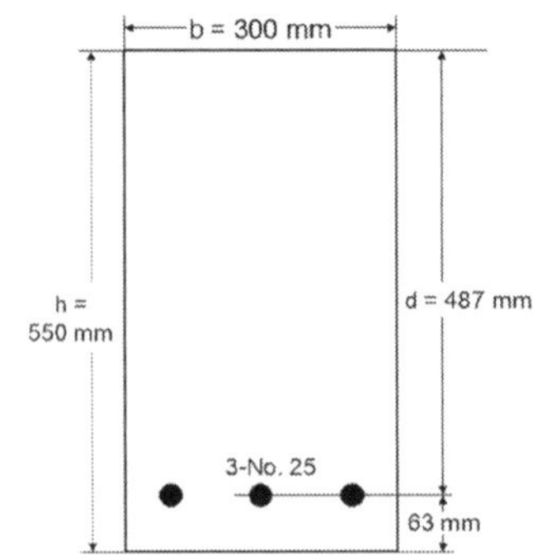

Required: analysis of deflections and comparison to ACI 318-11 allowable deflections. Consider long-term deflections at 3 months and 5 years (ultimate value).

Data:

f_c' = 21 MPa (normalweight concrete)

E_f = 45,000 MPa

Superimposed dead load (not including beam weight) = 1.8 kN/m

Live load = 4.4 kN/m (20 percent sustained)

Span = 7.5 m (simply-supported)

Calculations and discussion	**Reference**
No. 25 bar properties: d_b = 25.4 mm; $A_{f,bar}$ = 510 mm^2	440.6-08, Table 7.1

Reinforcement quantities needed for calculations:

$A_f = (3)(510\ \text{mm}^2) = 1530\ \text{mm}^2$

$$\rho_f = \frac{A_f}{bd} = \frac{1530}{(300)(487)} = 0.01047$$ 440.1R, Eq. (7.2.1a)

1. Check recommended minimum beam thickness.

Based on minimum thickness table, consider estimated depth:

$$h \approx \frac{\ell}{10} = \frac{(7.5)(1000)}{10} = 750\ \text{mm}$$ 440.1R, Table 7.3.2.1

Table 7.3.2.1 is intended to provide guidance for preliminary design, and does not guarantee that deflection limits will be met, so deflections will be computed.

2. Moments:

$$w_D = 1.8 + \frac{(300)(550)(24)}{1000} = 5.8\ \text{kN/m}$$

$$M_D = \frac{w_D \ell^2}{8} = \frac{(5.8)(7.5)^2}{8} = 40.8\ \text{kN-m}$$

$$M_L = \frac{w_L \ell^2}{8} = \frac{(4.4)(7.5)^2}{8} = 30.9\ \text{kN-m}$$

$M_{sus} = M_D + 0.20(M_L) = 40.8 + (0.20)(30.9) = 47.0$ kN-m

$M_{D+L} = 40.8 + 30.9 = 71.7$ kN-m

3. Modulus of elasticity and modular ratio:

$$E_c = 4700\sqrt{f_c'} = 4700\sqrt{21} = 21{,}500\ \text{MPa}$$ 318-11, Sec. 8.5.1

$$n_f = \frac{E_f}{E_c} = \frac{45{,}000}{21{,}500} = 2.093$$

4. Gross and cracked section moment of inertia:

$$I_g = \frac{bh^3}{12} = \frac{(300)(550)^3}{12} = 4.159\times10^9 \text{ mm}^4$$

$\rho_f n_f = (0.01047)(2.093) = 0.02191$

$$k = \sqrt{2\rho_f n_f + (\rho_f n_f)^2} - \rho_f n_f = \sqrt{2(0.02191)+(0.02191)^2} - 0.02191 = 0.189$$ 440.1R, Eq. (7.3.2.2b)

$$I_{cr} = \frac{bd^3}{3}k^3 + n_f A_f d^2(1-k)^2$$ 440.1R, Eq. (7.3.2.2a)

$$= \frac{(300)(487)^3}{3}(0.189)^3 + (2.093)(1530)(487)^2(1-0.189)^2 = 0.578\times10^9 \text{ mm}^4$$

$$\frac{I_g}{I_{cr}} = \frac{4.159\times10^9}{0.578\times10^9} = 7.20 \quad \frac{I_{cr}}{I_g} = \frac{0.578\times10^9}{4.159\times10^9} = 0.139$$

5. Effective moments of inertia:

$$M_{cr} = \frac{0.62\lambda\sqrt{f_c'}I_g}{y_t} = \frac{0.62(1)\left(\sqrt{21}\right)(4.159\times10^9)}{(1/2)(550)(10^6)} = 43.0 \text{ kN-m}$$ 440.1R, Eq. (7.3.2.2d)

Assuming that the member is loaded to the full dead load before any live load is applied, the effective moment of inertia may be computed on the basis of the dead load alone. (This assumption will yield a conservative estimate of live load deflection in the next step). Because $M_D < M_{cr}$, this beam is uncracked for this condition. Thus, the effective moment of inertia for dead load is I_g.

$(I_e)_D = I_g = 4.159\times10^9$ mm^4

Find the effective moment of inertia due to dead plus live load.

$$\frac{M_{cr}}{M_a} = \frac{M_{cr}}{M_{D+L}} = \frac{43.0}{71.7} = 0.600$$

$\gamma = 1.72 - 0.72(M_{cr}/M_a) = 1.72 - 0.72(0.600) = 1.288$ 440.1R, Sec. 7.3.2.2

$$(I_e)_{D+L} = \frac{I_{cr}}{1-\gamma\left(\frac{M_{cr}}{M_a}\right)^2\left[1-\frac{I_{cr}}{I_g}\right]}$$ 440.1R, Eq. (7.3.2.2c)

$$= \frac{0.578\times10^9}{1-(1.288)(0.600)^2[1-0.139]} = 0.962\times10^9 \text{ mm}^4$$

6. Immediate deflections:

$K = 1$ for simple spans

$$(\Delta_i)_D = \frac{K(5)M_D\ell^2}{48E_c(I_e)_D} = \frac{(1)(5)(40.8)(7.5)^2(10^{12})}{48(21{,}500)(4.159\times10^9)} = 2.7 \text{ mm}$$

$$(\Delta_i)_{D+L} = \frac{K(5)M_{D+L}\ell^2}{48E_c(I_e)_{D+L}} = \frac{(1)(5)(71.7)(7.5)^2(10^{12})}{48(21{,}500)(0.962\times10^9)} = 20.3 \text{ mm}$$

The immediate live load deflection is found by subtracting the dead load deflection from the deflection due to both dead and live load.

$(\Delta_i)_L = (\Delta_i)_{D+L} - (\Delta_i)_D = 20.3 - 2.7 = 17.6$ mm

It is conservative to find the immediate deflection due to sustained load by using the effective moment of inertia based on the full dead plus live load.

$$(\Delta_i)_{sus} = \frac{K(5)M_{sus}\ell^2}{48E_c(I_e)_{D+L}} = \frac{(1)(5)(47.0)(7.5)^2(10^{12})}{48(21{,}500)(0.962\times10^9)} = 13.3 \text{ mm}$$

$$(\Delta_i)_{L,unsustained} = (\Delta_i)_{D+L} - (\Delta_i)_{sus} = 20.3 - 13.3 = 7.0 \text{ mm}$$

7. Incremental portion of total deflection occurring after placement of nonstructural elements (sum of the long-term deflection due to all sustained loads and the immediate deflection due to any additional live load):

a. 3 months ($\xi = 1.0$) — 318-11, Sec. 9.5.2.5

$$\Delta_{incr} = \Delta_{LT,sus} + \Delta_{L,unsustained} = 0.6\xi(\Delta_i)_{sus} + (\Delta_i)_{L,unsustained} = 0.6(1.0)(13.3) + 7.0 = 15.0 \text{ mm}$$ 440.1R, Eq. (7.3.2.3c)

b. 5 years ($\xi = 2.0$) — 318-11, Sec. 9.5.2.5

$$\Delta_{incr} = \Delta_{LT,sus} + \Delta_{L,unsustained} = 0.6\xi(\Delta_i)_{sus} + (\Delta_i)_{L,unsustained} = 0.6(2.0)(13.3) + 7.0 = 23.0 \text{ mm}$$ 440.1R, Eq. (7.3.2.3c)

8. Compare computed deflections to allowable deflections.

Allowable deflections (immediate live load): — 318-11, Table 9.5(b)

Flat roofs not supporting and not attached to nonstructural elements likely to be damaged by large deflections:

$$\left\{(\Delta_i)_L = 17.6 \text{ mm}\right\} \le \left\{\frac{\ell}{180} = \frac{(7.5)(1000)}{180} = 41.7 \text{ mm}\right\} \quad \text{OK}$$

Floors not supporting and not attached to nonstructural elements likely to be damaged by large deflections:

$$\left\{(\Delta_i)_L = 17.6 \text{ mm}\right\} \le \left\{\frac{\ell}{360} = \frac{(7.5)(1000)}{360} = 20.8 \text{ mm}\right\} \quad \text{OK}$$

Allowable deflections (incremental after placement of nonstructural elements): — 318-11, Table 9.5(b)

Roof or floor construction supporting or attached to nonstructural elements not likely to be damaged by large deflections:

$$\left\{\Delta_{incr} = 23.0 \text{ mm}\right\} \le \left\{\frac{\ell}{240} = \frac{(7.5)(1000)}{240} = 31.3 \text{ mm}\right\} \quad \text{OK}$$

Roof or floor construction supporting or attached to nonstructural elements likely to be damaged by large deflections (very stringent criteria):

$$\left\{\Delta_{incr} = 23.0 \text{ mm}\right\} > \left\{\frac{\ell}{480} = \frac{(7.5)(1000)}{480} = 15.6 \text{ mm}\right\} \quad \text{NG}$$

If nonstructural elements likely to be damaged by large deflections are to be supported by this member, it will need to be redesigned to have a greater flexural stiffness. Otherwise, the member is adequate.

Example 7M—Creep rupture stress check under sustained loads

This example does not have a companion in PCA Notes on ACI 318-08 *because creep rupture is an FRP phenomenon.*

The rectangular beam shown is designed to carry service load moments M_D = 24 kN-m and M_L = 24 kN-m. Assume 20 percent of the live load is sustained. Verify that the beam satisfies the creep rupture stress limits in 7.4.1. Assume interior exposure conditions.

f_c' = 28 MPa

$f_{fu}* = 550$ MPa
$E_f = 41{,}000$ MPa

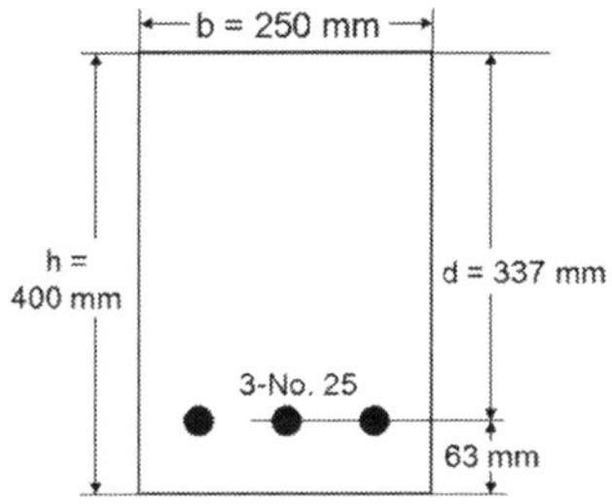

Calculations and discussion | **Reference**

No. 25 bar properties: — 440.6-08, Table 7.1
$d_b = 25.4$ mm
$A_{f,bar} = 510$ mm^2

Reinforcement quantities needed for calculations:
$A_f = (3)(510 \text{ mm}^2) = 1530 \text{ mm}^2$

$$\rho_f = \frac{A_f}{bd} = \frac{1530}{(250)(337)} = 0.01816$$ — 440.1R, Eq. (7.2.1a)

Design material properties:
$C_E = 0.8$ — 440.1R, Table 6.2
$f_{fu} = C_E f_{fu}* = (0.8)(550) = 440$ MPa — 440.1R, Eq. (6.2a)

1. Sustained load moment:
$M_{sus} = M_D + 0.20(M_L) = 24 + (0.20)(24) = 28$ kN-m

2. Modulus of elasticity and modular ratio:

$$E_c = 4700\sqrt{f_c'} = 4700\sqrt{28} = 24{,}900 \text{ MPa}$$ — 318-11, Sec. 8.5.1

$$n_f = \frac{E_f}{E_c} = \frac{41{,}000}{24{,}900} = 1.647$$

3. Cracked section moment of inertia:
$\rho_f n_f = (0.01816)(1.647) = 0.02991$

$$k = \sqrt{2\rho_f n_f + (\rho_f n_f)^2} - \rho_f n_f = \sqrt{2(0.02991) + (0.02991)^2} - 0.02991 = 0.216$$ — 440.1R, Eq. (7.3.2.2b)

$$I_{cr} = \frac{bd^3}{3}k^3 + n_f A_f d^2(1-k)^2$$ — 440.1R, Eq. (7.3.2.2a)

$$= \frac{(250)(337)^3}{3}(0.216)^3 + (1.647)(1530)(337)^2(1-0.216)^2 = 0.208 \times 10^9 \text{ mm}^4$$

4. Reinforcement stress under sustained service loads:

$$f_{fs,sus} = M_{s,sus}\frac{n_f d(1-k)}{I_{cr}}$$ — 440.1R, Eq. (7.4.1)

$$f_{fs,sus} = (28.8)(10^6)\frac{(1.647)(337)(1-0.216)}{0.208 \times 10^9} = 60 \text{ MPa}$$

Allowable reinforcement stress: — 440.1R, Table 7.4.1

$\{f_{fs,sus} = 60 \text{ MPa}\} \leq \{0.20 f_{fu} = 0.20(440) = 88 \text{ MPa}\}$ OK

Example 8M—Design for shear (members subject to shear and flexure only)

This example is similar to Example 12.1 of PCA Notes on ACI 318-08.

Determine the required size and spacing of vertical U-stirrups for an 5.5 m span, simply-supported normalweight reinforced concrete beam. Assume interior exposure conditions.

$b_w = 300$ mm
$d = 500$ mm
$f_c' = 28$ MPa
$f_{fu}^* = 700$ MPa (stirrups)
$E_f = 41{,}000$ MPa (stirrups)
$w_u = 70.8$ kN/m (includes self-weight)
$\rho_f = 0.0272$ (longitudinal reinforcement)
$r_b/d_b = 4$ (assumed curvature of bent stirrup bars)

Calculations and discussion	Reference
Design material properties:	
$C_E = 0.8$	440.1R, Table 6.2
$f_{fu} = C_E f_{fu}^* = (0.8)(700) = 560$ MPa	440.1R, Eq. (6.2a)

For the purposes of this example, the live load will be assumed to be present on the full span so that the design shear at the centerline of span is zero. (A design shear greater than zero at midspan is obtained by considering partial live loading of the span.)

1. Determine factored shear forces.

At support: $V_u = 70.8\left(\frac{5.5}{2}\right) = 194.7 \text{ kN}$

At distance d from support: $V_u = 194.7 - 70.8\left(\frac{500}{1000}\right) = 159.3 \text{ kN}$

2. Determine shear strength provided by concrete.

$$\phi V_c = \phi\left(\frac{5}{2}k\right)2\sqrt{f_c'}b_w d$$ 440.1R, Eq. (8.2b)

$$E_c = 4700\sqrt{f_c'} = 4700\sqrt{28} = 24{,}900 \text{ MPa}$$ 318-11, Sec. 8.5.1

$$n_f = \frac{E_f}{E_c} = \frac{41{,}000}{24{,}900} = 1.647$$

$\rho_f n_f = (0.0272)(1.647) = 0.04480$

$$k = \sqrt{2\rho_f n_f + (\rho_f n_f)^2} - \rho_f n_f = \sqrt{2(0.04480) + (0.04480)^2} - 0.04480 = 0.258$$ 440.1R, Eq. (7.3.2.2b)

$$\phi V_c = (0.75)\left(\frac{0.258}{5}\right)2\sqrt{28}(300)(500)\left(\frac{1}{1000}\right) = 61.4 \text{ kN}$$

$V_u = 159.3 \text{ kN} > \phi V_c/2 = 31.2 \text{ kN}$

Therefore, shear reinforcement is required. 440.1R, Sec. 8.2.2

3. Compute $V_u - \phi V_c$ at critical section.

$V_u - \phi V_c = 159.3 - 61.4 = 97.9 \text{ kN} < \phi 0.66\sqrt{f_c'}b_w d = 393 \text{ kN}$ OK 440.1R, Sec. 8.2.3

4. Determine distance x_c from support beyond which shear reinforcement is not required for strength ($V_u = \phi V_c$):

$$x_c = \frac{V_u@\text{support} - \phi V_c}{w_u} = \frac{194.7 - 61.4}{70.8} = 1.88 \text{ m}$$

Determine distance x_m from support beyond which shear reinforcement is not required ($V_u = \phi V_c/2$):

$$x_m = \frac{V_u@\text{support} - (\phi V_c/2)}{w_u} = \frac{194.7 - (61.4/2)}{70.8} = 2.32 \text{ m}$$

Therefore, only minimum shear reinforcement is required between 1.88 m and 2.32 m from the supports. Shear reinforcement is not required past 2.32 m from the supports.

5. Determine design tensile stress in shear reinforcement.
Tensile strength of bent bars:

$$f_{fb} = \left(0.05 \cdot \frac{r_b}{d_b} + 0.3\right) f_{fu} \leq f_{fu}$$ 440.1R, Eq. (6.2.1)

$f_{fb} = (0.05 \cdot 4 + 0.3) f_{fu} \leq f_{fu}$
$f_{fb} = (0.50) f_{fu} \leq f_{fu}$
$f_{fb} = (0.50)(560) = 280$ MPa

The design tensile strength is based on a strain of 0.004:
$f_{fv} = 0.004 E_f \leq f_{fb}$ 440.1R, Eq. (8.2d)
$f_{fv} = 0.004(41{,}000) \leq 280$
$f_{fv} = 164 \leq 280$
$f_{fv} = 164$ MPa

6. Determine required spacing of vertical U-stirrups at the critical section.

$$\frac{A_{fv}}{s} = \frac{(V_u - \phi V_c)}{\phi f_{fv} d} = \frac{(159.3 - 61.4)1000}{(0.75)(164)(500)} = 1.59 \text{ mm}^2/\text{mm}$$ 440.1R, Eq. (8.2e)

Assuming No. 13 U-stirrups ($A_{fv} = 258$ mm^2)

$$s = \frac{258}{1.59} = 162 \text{ mm}$$

Check maximum permissible spacing of stirrups:
$s = d/2 = 250$ mm ≤ 600 mm 440.1R, Sec. 8.3
because $V_u - \phi V_c = 97.9$ kN $< \phi 4\sqrt{f_c'} b_w d = 196$ kN OK 318-11, Sec. 11.4.5

Maximum stirrup spacing based on minimum shear reinforcement:

$$s = \frac{A_{fv,min} f_{fv}}{0.35 b_w} = \frac{(258)(164)}{0.35(300)} = 403 \text{ mm}$$ 440.1R, Eq. (8.2.2)

Therefore, the spacing at the critical section is restricted to the smallest of 162, 250, and 403 mm. Select spacing of 150 mm at the critical section.

Determine the distance x where a transition can be made to a spacing of 250 mm (to satisfy strength and maximum spacing requirement calculated above as $d/2 = 250$ mm):

$$\frac{A_{fv}}{s} = \frac{(V_u - \phi V_c)}{\phi f_{fv} d}$$ 440.1R, Eq. (8.2e)

$$\frac{258}{250} = \frac{(V_u - \phi V_c)}{(0.75)(164)(500)1000}$$

$V_u - \phi V_c = 63.5$ kN
$V_u = 63.5 + 61.4 = 124.9$ kN

$$x = \frac{V_u@\text{support} - V_u}{w_u} = \frac{194.7 - 124.9}{70.8} = 0.99 \text{ m}$$

Therefore, a transition may be made from 150 to 250 mm spacing at 0.99 m from the support.

7. Select a stirrup spacing arrangement that satisfies previous calculations.

Place first stirrup at 50 mm from support. Use 150 mm spacing to 1.10 m from support. Use 250 mm spacing to 2.3 m from support.

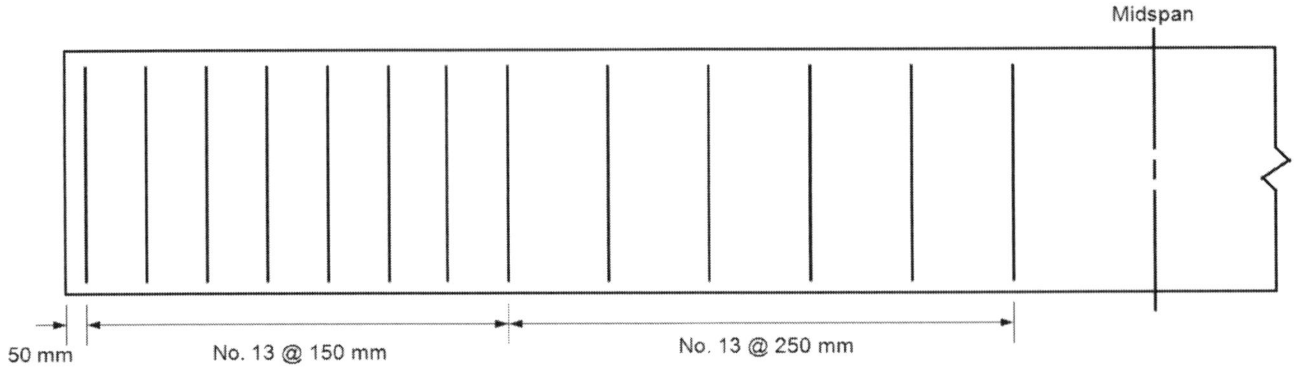

Note: Many designers would carry the reinforcement spacing of 250 mm through midspan, even though it is not required by calculations.

Example 9M—Development of bars in tension (compression-controlled or transition zone section)

This example is similar to Example 4.2 of PCA Notes on ACI 318-08.

Calculate the required tension development length for the No. 25 glass FRP (GFRP) bars (alternate short bars) embedded in the normalweight concrete one-way slab shown in the following figure to develop the full moment capacity at Section A.

Assume $f_c' = 28$ MPa, $f_{fu}^* = 550$ MPa, and $E_f = 41,000$ MPa. Assume interior exposure conditions.

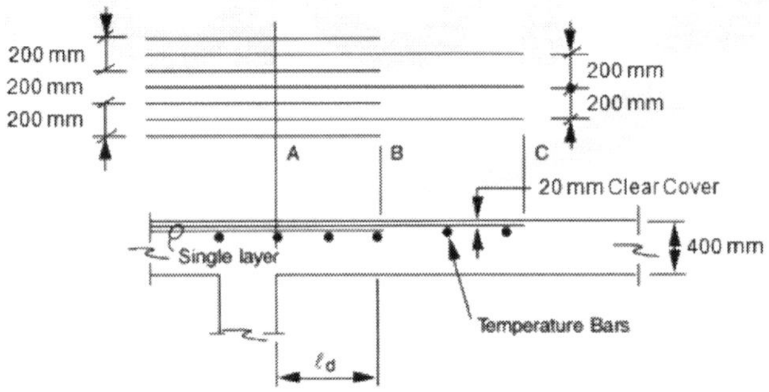

Calculations and discussion	**Reference**
Assume short bars are developed within distance AB while long bars are developed within BC.	
No. 25 bar properties:	440.6-08, Table 7.1
$d_b = 25.4$ mm	
$A_{f,bar} = 510$ mm^2	
Design material properties:	
$C_E = 0.8$	440.1R, Table 6.2
$f_{fu} = C_E f_{fu}^* = (0.8)(550) = 440$ MPa	440.1R, Eq. (6.2a)
1. Determine the type of section.	
$d = 400 - 20 - (25.4/2) = 367$ mm	
$A_f = (510\text{ mm}^2)/(100\text{ mm spacing}) = 5.10\text{ mm}^2/\text{mm} = 5100\text{ mm}^2/\text{m}$ of slab	
$\rho_f = \dfrac{A_f}{bd} = \dfrac{5100}{(1000)(367)} = 0.01390$	440.1R, Eq. (7.2.1a)

$$\rho_{fb} = 0.85\beta_1 \frac{f_c'}{f_{fu}} \frac{E_f \varepsilon_{cu}}{E_f \varepsilon_{cu} + f_{fu}}$$ 440.1R, Eq. (7.2.1b)

$E_f\varepsilon_{cu} = (41{,}000)(0.003) = 123$ MPa

$$\rho_{fb} = 0.85(0.85)\frac{(28)}{(440)}\frac{(123)}{(123+440)} = 0.01004$$

$$\frac{\rho_f}{\rho_{fb}} = \frac{0.01390}{0.01004} = 1.38$$

Because $\rho_{fb} < \rho_f < 1.4\rho_{fb}$, the section is in the transition zone. 440.1R, Eq. (7.2.3)

2. Determine stress in tensile reinforcement at ultimate conditions.

$$f_f = \sqrt{\frac{(E_f\varepsilon_{cu})^2}{4} + \frac{0.85\beta_1 f_c'}{\rho_f} E_f\varepsilon_{cu}} - 0.5E_f\varepsilon_{cu} \le f_{fu}$$ 440.1R, Eq. (7.2.2d)

$$f_f = \sqrt{\frac{(123)^2}{4} + \frac{0.85(0.85)(28)}{(0.01390)}(123)} - 0.5(123) \le 440$$

$f_f = 366$ MPa

3. Determine development length.

$$\ell_d = \frac{\alpha \dfrac{f_{fr}}{0.083\sqrt{f_c'}} - 340}{13.6 + \dfrac{C}{d_b}} d_b$$ 440.1R, Eq. (10.3a)

The bar stress that needs to be developed is the stress at the ultimate condition. There is no need to develop the full strength of the bar in this case.
$f_{fr} = f_f = 366$ MPa 440.1R, Sec. 10.3

Bar location modification factor should be taken as $\alpha = 1.5$ because more than 300 mm of concrete is cast below the reinforcement. ACI 440.1R, Sec. 10.1.1

Center-to-center spacing of bars being developed = 200 mm

$$C = \min\left(d_c, \frac{\text{ctr-to-ctr spacing}}{2}\right) \le 3.5d_b$$ 440.1R, Sec. 10.1

$$C = \min\left(20 + \frac{25.4}{2}, \frac{200}{2}\right) \le 3.5(25.4) = \min(33, 100) \le 89$$

$C = 33$ mm

$$\ell_d = \frac{\alpha \dfrac{f_{fr}}{0.083\sqrt{f_c'}} - 340}{13.6 + \dfrac{C}{d_b}} d_b = \frac{(1.5)\dfrac{(366)}{0.083\sqrt{28}} - 340}{13.6 + \dfrac{33}{25.4}}(25.4) = 1550 \text{ mm}$$

The required development length is 1550 mm.

Example 10M—Development of bars in tension (tension-controlled section)

This example is similar to Example 4.3 of PCA Notes on ACI 318-08.

Calculate the length required to develop the inner two No. 25 glass FRP (GFRP) bars at the column face in the following figure. The two No. 25 outer bars are to be made continuous along full length of beam. Use f_c' = 28 MPa (normalweight concrete), $f_{fu}{}^*$ = 550 MPa, and E_f = 41,000 MPa. Assume d = 700 mm. Assume interior exposure conditions.

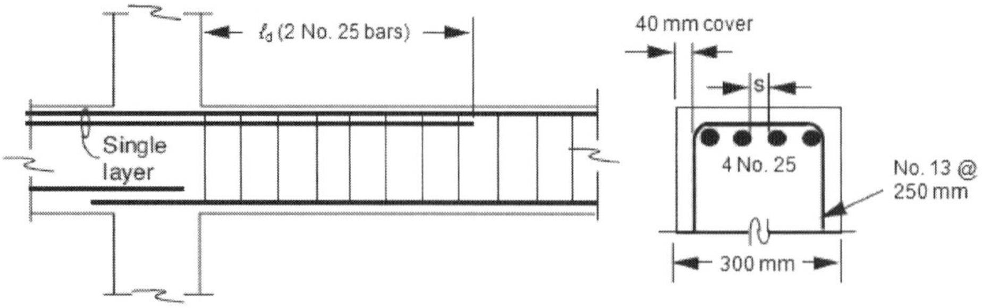

Calculations and discussion	**Reference**
No. 25 bar properties:	440.6-08, Table 7.1
d_b = 25.4 mm	
$A_{f,bar}$ = 510 mm^2	

Design material properties:

$C_E = 0.8$	440.1R, Table 6.2
$f_{fu} = C_E f_{fu}{}^* = (0.8)(550) = 440$ MPa	440.1R, Eq. (6.2a)

1. Determine the type of section.

$A_f = (4)(510\text{ mm}^2) = 2040\text{ mm}^2$

$$\rho_f = \frac{A_f}{bd} = \frac{2010}{(300)(700)} = 0.00957$$ 440.1R, Eq. (7.2.1a)

$$\rho_{fb} = 0.85\beta_1 \frac{f_c'}{f_{fu}} \frac{E_f \varepsilon_{cu}}{E_f \varepsilon_{cu} + f_{fu}}$$ 440.1R, Eq. (7.2.1b)

$E_f\varepsilon_{cu}$ = (41,000)(0.003) = 123 MPa

$$\rho_{fb} = 0.85(0.85)\frac{(28)}{(440)}\frac{(123)}{(123+440)} = 0.01004$$

$$\frac{\rho_f}{\rho_{fb}} = \frac{0.00957}{0.01004} = 0.95$$

Because $\rho_f \leq \rho_{fb}$, the section is tension-controlled. 440.1R, Eq. (7.2.3)

2. Determine stress in tensile reinforcement at ultimate conditions.

Because section is tension-controlled, $f_f = f_{fu}$ = 440 MPa

3. Determine development length.

$$\ell_d = \frac{\alpha \dfrac{f_{fr}}{0.083\sqrt{f_c'}} - 340}{13.6 + \dfrac{C}{d_b}} d_b$$ 440.1R, Eq. (10.3a)

$f_{fr} = f_f$ = 440 MPa 440.1R, Sec. 10.3

Bar location modification factor should be taken as α = 1.5 because more than 300 mm of concrete is cast below the reinforcement. ACI 440.1R, Sec. 10.1.1

Clear spacing of bars being developed = [300 – 2(40) – 2(12.7) – 4(25.4)]/3 = 31 mm

$$C = \min\left(d_c, \frac{\text{ctr-to-ctr spacing}}{2}\right) \leq 3.5d_b$$ 440.1R, Sec. 10.1

$$= \min\left(40 + 12.7 + \frac{25.4}{2}, \frac{12.7 + 31 + 12.7}{2}\right) \leq 3.5(25.4) = \min(65,28) \leq 89$$

$C = 28$ mm

$$\ell_d = \frac{\alpha\dfrac{f_{fr}}{0.083\sqrt{f_c'}} - 340}{13.6 + \dfrac{C}{d_b}} d_b = \frac{(1.5)\dfrac{(440)}{0.083\sqrt{28}} - 340}{13.6 + \dfrac{28}{25.4}}(25.4) = 2010 \text{ mm}$$

The required development length is 2010 mm.

Example 11M—Shear strength of slab at column support

This example is similar to Example 16.1 of PCA Notes on ACI 318-08.

Determine two-way action shear strength at an interior column support of a flat plate slab system for the following design conditions.

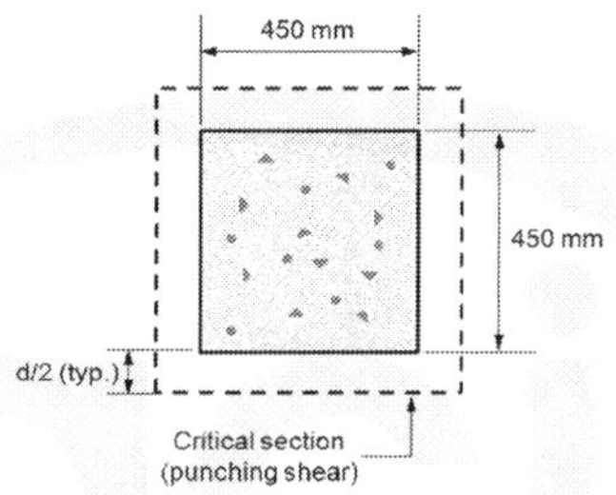

Column dimensions = 450 mm x 450 mm
Slab effective depth d = 165 mm
Specified concrete strength f_c' = 28 MPa (normalweight concrete)
E_f = 40,000 MPa
ρ_f = 0.0120 (flexural reinforcement in column strips in both directions, top mat of reinforcement)

Calculations and discussion — Reference

1. Two-way action shear (punching shear) without shear reinforcement:

$V_u \leq \phi V_n$ 440.1R, Sec. 8.1
$V_u \leq \phi V_c$ $(V_f = 0)$ 440.1R, Sec. 8.2

2. Determine shear strength provided by concrete.

$$\phi V_c = \phi\left(\frac{k}{5}\right)4\sqrt{f_c'}b_o d$$ 440.1R, Eq. (8.4b)

$$E_c = 4700\sqrt{f_c'} = 4700\sqrt{28} = 24{,}900 \text{ MPa}$$ 318-11, Sec. 8.5.1

$$\rho_f n_f = \rho_f \frac{E_f}{E_c} = (0.0120)\left(\frac{40{,}000}{24{,}900}\right) = (0.0120)(1.606) = 0.01927$$

$$k = \sqrt{2\rho_f n_f + (\rho_f n_f)^2} - \rho_f n_f = \sqrt{2(0.01927) + (0.01927)^2} - 0.01927 = 0.178$$ 440.1R, Eq. (7.3.2.2b)

$b_o = 4(450 + 82.5 + 82.5) = 2460$ mm 440.1R, Sec. 8.4

$$\phi V_c = (0.75)\left(\frac{0.178}{5}\right)4\sqrt{28}(2460)(165)\left(\frac{1}{1000}\right) = 229 \text{ kN}$$

CHAPTER 12—REFERENCES

ACI committee documents and documents published by other organizations are listed first by document number, full title, and year of publication followed by authored documents listed alphabetically.

American Concrete Institute

216.1-07/TMS-0216-07—Code Requirements for Determining Fire Resistance of Concrete and Masonry Construction Assemblies

318-71—Building Code Requirements for Reinforced Concrete and Commentary

318-99—Building Code Requirements for Structural Concrete and Commentary

318-08—Building Code Requirements for Structural Concrete and Commentary

318-11—Building Code Requirements for Structural Concrete and Commentary

360R-10—Guide to Design of Slabs-on-Ground

435R-95(00)—Control of Deflection in Concrete Structures (Appendix B added 2003)

440R-07—Report on Fiber-Reinforced Polymer (FRP) Reinforcement for Concrete Structures

440.3R-04—Guide Test Methods for Fiber-Reinforced Polymers (FRPs) for Reinforcing or Strengthening Concrete Structures

440.4R-04(11)—Prestressing Concrete Structures with FRP Tendons (Reapproved 2011)

440.5-08—Specification for Construction with Fiber-Reinforced Polymer Reinforcing Bars

440.6-08— Specification for Carbon and Glass Fiber-Reinforced Polymer Bar Materials for Concrete Reinforcement

ASTM International

ASTM D7205/D7205M-06(2011)—Standard Test Method for Tensile Properties of Fiber Reinforced Polymer Matrix Composite Bars

ASTM D7337/D733M-12—Standard Test Method for Tensile Creep Rupture of Fiber Reinforced Polymer Matrix Composite Bars

ASTM D7617/D7617M-11—Standard Test Method for Transverse Shear Strength of Fiber-reinforced Polymer Matrix Composite Bars

ASTM D7705/D7705M-12—Standard Test Method for Alkali Resistance of Fiber Reinforced Polymer (FRP) Matrix Composite Bars used in Concrete Construction

ASTM E119-12a—Standard Test Methods for Fire Tests of Building Construction and Materials.

Canadian Standards Association

CAN/CSA-S6-06—Canadian Highway Bridge Design Code

CAN/CSA S6S1-10—Canadian Highway Bridge Design Code

CAN/CSA-S806-12—Design and Construction of Building Components with Fibre-Reinforced Polymers

Canadian Standards Association/Underwriters Laboratory Canada

CAN/ULC-S101-M89—Standard Methods of Fire Endurance Tests of Building Construction and Materials

International Organization for Standardization

ISO 834-1:1999—Fire-Resistance Tests – Elements of Building Construction – Part 1: General Requirements

Authored documents

Adimi, R.; Rahman, H.; Benmokrane, B.; and Kobayashi, K., 1998, "Effect of Temperature and Loading Frequency on the Fatigue Life of a CFRP Bar in Concrete," *Proceedings of the Second International Conference on Composites in Infrastructure (ICCI-98)*, Tucson, AZ, V. 2, pp. 203-210.

Afifi, M. Z.; Mohamed, H. M.; and Benmokrane, B., 2013, "Axial Capacity of Circular Concrete Columns Reinforced with Glass-FRP Bars and Spirals," *11th International Conference on Fibre Reinforced Polymer (FRP) Reinforcement for Concrete Structures (FRPRCS-11)*, Porto, Portugal, June 26-28, 10 pp.

Ahmed, E. A.; El-Salakawy, E. F.; and Benmokrane, B., 2010a, "Shear Performance of RC Bridge Girders Reinforced with Carbon FRP Stirrups," *Journal of Bridge Engineering*, V. 15, No. 1, pp. 44-54. doi: 10.1061/(ASCE) BE.1943-5592.0000035

Ahmed, E. A.; El-Salakawy, E. F.; and Benmokrane, B., 2010b, "Performance Evaluation of GFRP Shear Reinforcement for Concrete Beams," *ACI Structural Journal*, V. 107, No. 1, Jan.-Feb., pp. 53-62.

Ahmad, S. H.; Zia, P.; Yu, T.; and Xie, Y., 1993, "Punching Shear Tests of Slabs Reinforced with 3-D Carbon Fiber Fabric," *Concrete International*, V. 16, No. 6, June, pp. 36-41.

Al-Dulaijan, S. U.; Nanni, A.; Al-Zahrani, M. M.; and Bakis, C. E., 1996, "Bond Evaluation of Environmentally Conditioned GFRP/Concrete System," *Proceedings of the Second International Conference on Advanced Composite Materials in Bridges and Structures (ACMBS-2)*, M. M. El-Badry, ed., Canadian Society for Civil Engineering, Montreal, QC, Canada, pp. 845-852.

Almusallam, T. H.; Al-Salloum, Y.; Alsayed, S.; and Amjad, M., 1997, "Behavior of Concrete Beams Doubly Reinforced by FRP Bars," *Proceedings of the Third International Symposium on Non-Metallic (FRP) Reinforcement for Concrete Structures (FRPRCS-3)*, Japan Concrete Institute, Tokyo, Japan, V. 2, pp. 471-478.

Al-Zahrani, M. M.; Nanni, A.; Al-Dulaijan, S. U.; and Bakis, C. E., 1996, "Bond of FRP to Concrete for Rods with Axisymmetric Deformations," *Proceedings of the Second International Conference on Advanced Composite Materials in Bridges and Structures (ACMBS-II)*, Montreal, QC, Canada, pp. 853-860.

Ando, N.; Matsukawa, H.; Hattori, A.; and Mashima, A., 1997, "Experimental Studies on the Long-Term Tensile Properties of FRP Tendons," *Proceedings of the Third International Symposium on Non-Metallic (FRP) Reinforcement for Concrete Structures (FRPRCS-3)*, V. 2, Japan Concrete Institute, Tokyo, Japan, pp. 203-210.

Arockiasamy, M.; Amer, A.; and Shahawy, M., 1998, "Environmental and Long-Term Studies on CFRP Cables and CFRP Reinforced Concrete Beams," *Proceedings of the First International Conference on Durability of Composites for Construction*, B. Benmokrane and H. Rahman, eds., Sherbrooke, QC, Canada, pp. 599-610.

Bakis, C. E., 1993, "FRP Composites: Materials and Manufacturing," *Fiber-Reinforced-Plastic for Concrete Structures: Properties and Applications*, A. Nanni, ed., Elsevier, Amsterdam, pp. 13-58.

Bakis, C. E., and Boothby, T. E., 2004, "Evaluation of Crack Width and Bond Strength in GFRP Reinforced Beams Subjected to Sustained Loads," *Proceedings of the 4th International Conference on Advanced Composite Materials in Bridges and Structures—ACMBS-IV*, M. El-Badry and L. Dunaszegi, eds. (CD-ROM)

Bakis, C. E.; Freimanis, A. J.; Gremel, D.; and Nanni, A., 1998, "Effect of Resin Material on Bond and Tensile Properties of Unconditioned and Conditioned FRP Reinforcement Rods," *Proceedings of the First International Conference on Durability of Composites for Construction*, B. Benmokrane and H. Rahman, eds., Sherbrooke, QC, Canada, pp. 525-535.

Bakis, C.; Ospina, C.; Bradberry, T.; Benmokrane, B.; Gross, S.; Newhook, J.; and Thiagarajan, G., 2006, "Evaluation of Crack Widths in Concrete Flexural Members Reinforced with FRP Bars," *Proceedings of the 3rd International Conference on FRP Composites in Civil Engineering*, A. Mirmiran and A. Nanni, eds., Florida International University, Miami, FL, pp. 307-310.

Bank, L. C., 1993, "Properties of FRP Reinforcement for Concrete," *Fiber-Reinforced-Plastic (FRP) Reinforcement for Concrete Structures: Properties and Applications, Developments in Civil Engineering*, V. 42, A. Nanni, ed., Elsevier, Amsterdam, pp. 59-86.

Bank, L. C., and Puterman, M., 1997, "Microscopic Study of Surface Degradation of Glass Fiber-Reinforced Polymer Rods Embedded in Concrete Castings Subjected to Environmental Conditioning," *High Temperature and Environmental Effects on Polymeric Composites*, V. 2, ASTM STP 1302, T. S. Gates and A.-H. Zureick, eds., ASTM International, West Conshohocken, PA, pp. 191-205.

Bank, L. C.; Puterman, M.; and Katz, A., 1998, "The Effect of Material Degradation on Bond Properties of FRP Reinforcing Bars in Concrete," *ACI Materials Journal*, V. 95, No. 3, May-June, pp. 232-243.

Bank, L. C., and Xi, Z., 1995, "Punching Shear Behavior of Pultruded FRP Grating Reinforced Concrete Slabs," *Proceedings of the 2nd International Symposium on Non-Metallic (FRP) Reinforcement for Concrete Structures*, Ghent, Belgium, pp. 360-367.

Banthia, N.; Al-Asaly, M.; and Ma, S., 1995, "Behavior of Concrete Slabs Reinforced with Fiber-Reinforced Plastic Grid," *Journal of Materials in Civil Engineering*, V. 7, No. 4, Nov., pp. 252-257. doi: 10.1061/(ASCE)0899-1561(1995)7:4(252)

Barkatt, A.; Bank, L. C.; Prian, L.; and Gentry, T. R.; 1998, "Accelerated Test Methods to Determine the Long-Term Behavior of Composite Highway Structures Subject to Environmental Loading," *Journal of Composites, Technology and Research*, V. 20, No. 1, Jan., pp. 38-50. doi: 10.1520/CTR10499J

Beaulieu-Michaud, M.-C. B.; Mohamed, H. M.; and Benmokrane, B., 2013, "Conception, Construction et Monitorage d'une Dalle de Grande Dimension en Béton Armé de Polymères Renforcés de Fibre (PRF)," *Proceedings of the CSCE Annual General Meeting & Conference, Montreal*, Québec, Canada, May 29-June 1, 10 pp.

Belarbi, A., and Wang, H., 2012, "Durability of FRP Bars Embedded in Fiber-Reinforced-Concrete," *Journal of Composites for Construction*, V. 16, No. 4, pp. 371-380. doi: 10.1061/(ASCE)CC.1943-5614.0000270

Bellakehal, H.; Zaidi, A.; Masmoudi, R.; and Bouhicha, M., 2013, "Combined Effect of Sustained Load and Freeze–Thaw Cycles on One-Way Concrete Slabs Reinforced with Glass Fibre-Reinforced Polymer," *Canadian Journal of Civil Engineering*, V. 40, No. 11, pp. 1060-1067. doi: 10.1139/cjce-2012-0514

Benmokrane, B., 1997, "Bond Strength of FRP Rebar Splices," *Proceedings of the Third International Symposium on Non-Metallic (FRP) Reinforcement for Concrete Structures (FRPRCS-3)*, V. 2, Japan Concrete Institute, Tokyo, Japan, pp. 405-412.

Benmokrane, B.; Ahmed, E.; Dulude, C.; and Boucher, E., 2012, "Design, Construction, and Monitoring of the First Worldwide Two Way Flat Slab Parking Garage Reinforced with GFRP Bars," *Sixth International Conference on FRP Composites in Civil Engineering*, Rome, Italy, June 13-15, 8 pp.

Benmokrane, B.; Chaallal, O.; and Masmoudi, R., 1996, "Flexural Response of Concrete Beams Reinforced with FRP Reinforcing Bars," *ACI Structural Journal*, V. 93, No. 1, Jan.-Feb., pp. 46-55.

Benmokrane, B.; El-Salakawy, E.; Desgagné, G.; and Lackey, T., 2004, "FRP Bars for Bridges," *Concrete International*, V. 26, No. 8, pp. 84-90.

Benmokrane, B.; El-Salakawy, E.; El-Gamal, S.; and Goulet, S., 2007, "Construction and Testing of an Innovative Concrete Bridge Deck Totally Reinforced with Glass FRP Bars: Val-Alain Bridge on Highway 20 East," *Journal of Bridge Engineering*, V. 12, No. 5, Sept., pp. 632-645. doi: 10.1061/(ASCE)1084-0702(2007)12:5(632)

Benmokrane, B., and Mohamed, M. H., 2014, "Extending the Service Life of Water Treatment Structures: Glass Fiber-Reinforced Polymer Bars Used in Water Treatment Plant," *Concrete International*, V. 36, No. 2, Feb., pp 40-45.

Benmokrane, B., and Rahman, H., eds., 1998, "Durability of Fiber Reinforced Polymer (FRP) Composites for Construction," *Proceedings of the First International Conference (CDCC '98)*, QC, Canada, Feb., 692 pp.

Bisby, L. A., and Kodur, V. K. R., 2007, "Evaluating the Fire Endurance of Concrete Slabs Reinforced with FRP Bars: Considerations for a Holistic Approach," *Composites. Part B, Engineering*, V. 38, No. 5-6, pp. 547-558. doi: 10.1016/j.compositesb.2006.07.013

Bisby, L. A.; Green, M. F.; and Kodur, V. K. R., 2005, "Response to Fire of Concrete Structures that Incorporate FRP," *Progress in Structural Engineering and Materials*, V. 7, No. 3, pp. 136-149. doi: 10.1002/pse.198

Bischoff, P., 2005, "Reevaluation of Deflection Prediction for Concrete Beams Reinforced with Steel and Fiber-Reinforced Polymer Bars," *Journal of Structural Engineering*, V. 131, No. 5, May, pp. 752-767. doi: 10.1061/(ASCE)0733-9445(2005)131:5(752)

Bischoff, P. H., and Gross, S. P., 2011, "Equivalent Moment of Inertia Based on Integration of Curvature," *Journal of Composites for Construction*, V. 15, No. 3, pp. 263-273. doi: 10.1061/(ASCE)CC.1943-5614.0000164

Bischoff, P. H.; Gross, S. P.; and Ospina, C. E., 2009, "The Story Behind the Proposed ACI 440 Changes for Prediction of Deflections in Reinforced Concrete," *Serviceability of Concrete Members Reinforced with Internal/External FRP Reinforcement,* ACI SP-264, T. Alkhardji, P. H. Bischoff, and C. E. Ospina, eds., American Concrete Institute, Farmington Hills, MI, pp. 53-76.

Bischoff, P. H., and Scanlon, A., 2007, "Effective Moment of Inertia for Calculating Deflections of Concrete Members Containing Steel Reinforcement and Fiber-Reinforced Polymer Reinforcement," *ACI Structural Journal*, V. 104, No. 1, pp. 68-75.

Boyle, H. C., and Karbhari, V. M., 1994, "Investigation of Bond Behavior between Glass Fiber Composite Reinforcements and Concrete," *Journal of Polymer-Plastic Technology Engineering*, V. 33, No. 6, pp. 733-753. doi: 10.1080/03602559408013105

Branson, D. E., 1965, "Instantaneous and Time-Dependent Deflections of Simple and Continuous Reinforced Concrete Beams," *HPR Report No. 7*, Alabama Highway Department, Bureau of Public Roads, Montgomery, AL, Part 1, 78 pp.

Brown, V., 1997, "Sustained Load Deflections in GFRP-Reinforced Concrete Beams," *Proceedings of the Third International Symposium on Non-Metallic (FRP) Reinforcement for Concrete Structures (FRPRCS-3)*, V. 2, Japan Concrete Institute, Tokyo, Japan, pp. 495-502.

Brown, V., and Bartholomew, C., 1996, "Long-Term Deflections of GFRP-Reinforced Concrete Beams," *Proceedings of the First International Conference on Composites in Infrastructure (ICCI-96)*, H. Saadatmanesh and M. R. Ehsani, eds., Tucson, AZ, pp. 389-400.

Burgoyne, C., ed., 2001, "Non-Metallic Reinforcement for Concrete Structures—FRPRCS-5," *Proceedings*, International Conference, Cambridge, UK.

Clarke, J., and Sheard, P., 1998, "Designing Durable FRP Reinforced Concrete Structures," *Proceedings of the First International Conference on Durability of Composites for Construction*, B. Benmokrane and H. Rahman, eds., Sherbrooke, QC, Canada, pp. 13-24.

Chen, Y.; Davalos, J. F.; Ray, I.; and Kim, H.-Y., 2007, "Accelerated Aging Tests for Evaluations of Durability Performance of FRP Reinforcing Bars for Concrete Structures," *Composite Structures*, V. 78, No. 1, Mar., pp. 101-111. doi: 10.1016/j.compstruct.2005.08.015

Chowdhury, E. U.; Eedson, R.; Bisby, L. A.; Green, M. F.; Bénichou, N.; and Kodur, V. K. R., 2011, "Mechanical Characterization of FRP Materials at High Temperature," *Fire Technology*, V. 45, No. 4, pp. 1-18.

Conrad, J. O.; Bakis, C. E.; Boothby, T. E.; and Nanni, A., 1998, "Durability of Bond of Various FRP Rods in Concrete," *Proceedings of the First International Conference on Durability of Composites for Construction*, B. Benmokrane and H. Rahman, eds., Sherbrooke, QC, Canada, pp. 299-310.

Coomarasamy, A., and Goodman, S., 1997, "Investigation of the Durability Characteristics of Fiber Reinforced Polymer (FRP) Materials in Concrete Environment," *American Society for Composites—Twelfth Technical Conference*, Dearborn, MI.

Cosenza, E.; Manfredi, G.; and Nanni, A., eds., 2001, "Composites in Construction: A Reality," *Proceedings, International Workshop*, Capri, Italy, ASCE, Reston, VA, 277 pp.

Curtis, P. T., 1989, "The Fatigue Behavior of Fibrous Composite Materials," *Journal of Strain Analysis*, V. 24, No. 4, pp. 235-244. doi: 10.1243/03093247V244235

Dally, J. W., and Riley, W. F., 1991, *Experimental Stress Analysis*, third edition, McGraw Hill, New York, 672 pp.

Daniali, S., 1992, "Development Length for Fiber-Reinforced Plastic Bars," *Advanced Composite Materials in Bridges and Structures*, K. W. Neale and P. Labossiere, eds., pp. 179-188.

Darwin, D.; Zuo, J.; Tholen, M.; and Idun, E., 1996, "Development Length Criteria for Conventional and High Relative Rib Area Reinforcing Bars," *ACI Structural Journal*, V. 93, No. 3, May-June, pp. 347-359.

De Luca, A.; Matta, F.; and Nanni, A., 2010, "Behavior of Full-Scale Glass Fiber-Reinforced Polymer Reinforced Concrete Columns under Axial Load," *ACI Structural Journal*, V. 107, No. 5, Sept-Oct., pp. 589-596.

DeSimone, M. V., 2009, "Deflections of Continuous GFRP-Reinforced Concrete Beams," MSCE thesis, Department of Civil and Environmental Engineering, Villanova University, Villanova, PA.

Deutsches Institut für Normung e.V., 2012, "German Standards for Testing of Hardened Concrete, Part 1—Impermeability to Water," DIN 1048, pp. 9-10.

Devalapura, R. K.; Greenwood, M. E.; Gauchel, J. V.; and Humphrey, T. J., 1996, *Advanced Composite Materials in Bridges and Structures*, M. M. El-Badry, ed., Canadian Society for Civil Engineering, Montreal, QC, Canada, pp. 107-116.

Dolan, C. W.; Rizkalla, S.; and Nanni, A., eds., 1999, *Fourth International Symposium on Fiber Reinforced Polymer Reinforcement for Reinforced Concrete Structures*, SP-188, C. W. Dolan, S. H. Rizkalla, and A. Nanni, eds., American Concrete Institute, Farmington Hills, MI, 1182 pp.

Dowden, D. M., and Dolan, C. W., 1997, "Comparison of Experimental Shear Data with Code Predictions for FRP Prestressed Beams," *Proceedings of the Third International Symposium on Non-Metallic (FRP) Reinforcement for Concrete Structures (FRPRCS-3)*, V. 2, Japan Concrete Institute, Tokyo, Japan, pp. 687-694.

Drouin, B.; Latour, G.; and Mohamed, H., 2011, "More than 10 Years Successful Field Applications of FRP Bars in Canada," *CDCC 2011, The Fourth International Conference on Durability & Sustainability of Fiber Reinforced Polymer*

(FRP) Composites for Construction and Rehabilitation, Quebec City, QC, Canada, July, 6 pp.

Dulude, C.; Hassan, M.; Ahmed, E. A.; and Benmokrane, B., 2013, "Punching Shear Behavior of Two-Way Flat Concrete Slabs Reinforced with GFRP Bars," *ACI Structural Journal*, V. 110, No. 5, Sept.-Oct., pp. 723-734.

Ehsani, M. R., 1993, "Glass-Fiber Reinforcing Bars," *Alternative Materials for the Reinforcement and Prestressing of Concrete*, J. L. Clarke, Blackie Academic & Professional, London, UK, pp. 35-54.

Ehsani, M. R.; Saadatmanesh, H.; and Tao, S., 1995, "Bond of Hooked Glass Fiber Reinforced Plastic (GFRP) Reinforcing Bars to Concrete," *ACI Materials Journal*, V. 92, No. 4, July-Aug., pp. 391-400.

Ehsani, M. R.; Saadatmanesh, H.; and Tao, S., 1996a, "Design Recommendation for Bond of GFRP Rebars to Concrete," *Journal of Structural Engineering*, V. 122, No. 3, pp. 247-254. doi: 10.1061/(ASCE)0733-9445(1996)122:3(247)

Ehsani, M. R.; Saadatmanesh, H.; and Tao, S., 1996b, "Bond Behavior and Design Recommendations for Fiber-Glass Reinforcing Bars," *Proceedings of the First International Conference on Composites in Infrastructure (ICCI-96)*, H. Saadatmanesh and M. R. Ehsani, eds., Tucson, AZ, pp. 466-476.

El-Badry, M., ed., 1996, "Advanced Composite Materials in Bridges and Structures (ACMBS-II)," *Proceedings of the Second International Conference*, Montreal, QC, Canada, 1027 pp.

El-Ghandour, A. W.; Pilakoutas, K.; and Waldron, P., 2003, "Punching Shear Behavior of Fiber Reinforced Polymers Reinforced Concrete Flat Slabs: Experimental Study," *Journal of Composites for Construction*, V. 7, No. 3, pp. 258-265. doi: 10.1061/(ASCE)1090-0268(2003)7:3(258)

El-Salakawy, E. F., and Benmokrane, B., 2003, "Design and Testing of a Highway Concrete Bridge Deck Reinforced with Glass and Carbon FRP Bars," *ACI Special Publication, Field Applications of FRP Reinforcement: Case Studies, SP-215-2*, Aug., pp. 37–54.

El-Salakawy, E. F.; Benmokrane, B.; and Desgagné, G., 2003, "Fibre-Reinforced Polymer Composite Bars for the Concrete Deck Slab of Wotton Bridge," *Canadian Journal of Civil Engineering*, V. 30, No. 5, pp. 861-870. doi: 10.1139/l03-055

El-Salakawy, E. F.; Benmokrane, B.; El-Ragaby, A.; and Nadeau, D., 2005, "Field Investigation on the First Bridge Deck Slab Reinforced with Glass FRP Bars Constructed in Canada," *Journal of Composites for Construction*, V. 9, No. 6, Dec., pp. 470-479. doi: 10.1061/(ASCE)1090-0268(2005)9:6(470)

Faza, S. S., and GangaRao, H. V. S., 1990, "Bending and Bond Behavior of Concrete Beams Reinforced with Plastic Rebars," *Transportation Research Record*, V. 1290, pp. 185-193.

Faza, S. S., and GangaRao, H. V. S., 1992, "Pre- and Post-Cracking Deflection Behaviour of Concrete Beams Reinforced with Fibre-Reinforced Plastic Rebars," *Advanced Composite Materials in Bridges and Structures*, K. W. Neale and P. Labossiere, eds., Canadian Society for Civil Engineering, Montreal, QC, Canada, pp. 151-160.

Faza, S. S., and GangaRao, H. V. S., 1993a, "Theoretical and Experimental Correlation of Behavior of Concrete Beams Reinforced with Fiber Reinforced Plastic Rebars," *Fiber-Reinforced-Plastic Reinforcement for Concrete Structures—International Symposium*, SP-138, A. Nanni and C. W. Dolan, eds., American Concrete Institute, Farmington Hills, MI, pp. 599-614.

Faza, S. S., and GangaRao, H. V. S., 1993b, "Glass FRP Reinforcing Bars for Concrete," *Fiber-Reinforced-Plastic (FRP) Reinforcement for Concrete Structures: Properties and Applications, Developments in Civil Engineering,* V. 42, A. Nanni, ed., Elsevier, Amsterdam, pp. 167-188.

fib, 2007, "FRP Reinforcement in RC Structures," *Bulletin No. 40*, Federal Institute of Technology, Lausanne, Switzerland.

fib, 2010, *Model Code for Concrete Structures* (MC2010), Ernst and Son, Federal Institute of Technology, Lausanne, Switzerland.

Figueiras, J.; Juvandes, L.; and Furia, R., eds., 2001, "Composites in Construction," *Proceedings of the International Conference Composites in Construction—CCC2001*, Portugal.

Freimanis, A. J.; Bakis, C. E.; Nanni, A.; and Gremel, D., 1998, "A Comparison of Pullout and Tensile Behaviors of FRP Reinforcement for Concrete," *Proceedings of the Second International Conference on Composites in Infrastructure (ICCI-98)*, V. 2, Tucson, AZ, pp. 52-65.

Frosch, R. J., 1999, "Another Look at Cracking and Crack Control in Reinforced Concrete," *ACI Structural Journal*, V. 96, No. 3, May-June, pp. 437-442.

GangaRao, H. V. S., and Vijay, P. V., 1997a, "Design of Concrete Members Reinforced with GFRP Bars," *Proceedings of the Third International Symposium on Non-Metallic (FRP) Reinforcement for Concrete Structures (FRPRCS-3)*, V. 1, Japan Concrete Institute, Tokyo, Japan, pp. 143-150.

GangaRao, H. V. S., and Vijay, P. V., 1997b, "Aging of Structural Composites under Varying Environmental Conditions," *Proceedings of the Third International Symposium on Non-Metallic (FRP) Reinforcement for Concrete Structures (FRPRCS-3)*, V. 2, Japan Concrete Institute, Tokyo, Japan, pp. 91-98.

Gao, D.; Benmokrane, B.; and Masmoudi, R., 1998a, "A Calculating Method of Flexural Properties of FRP-Reinforced Concrete Beam: Part 1: Crack Width and Deflection," *Technical Report*, Department of Civil Engineering, University of Sherbrooke, Sherbrooke, QC, Canada, 24 pp.

Gao, D.; Benmokrane, B.; and Tighiouart, B., 1998b, "Bond Properties of FRP Rebars to Concrete," *Technical Report*, Department of Civil Engineering, University of Sherbrooke, Sherbrooke, QC, Canada, 27 pp.

Gentry, T. R., and Husain, M., 1999, "Thermal Compatibility of Concrete and Composite Reinforcements," *Journal of Composites for Construction*, V. 3, No. 2, pp. 82-86. doi: 10.1061/(ASCE)1090-0268(1999)3:2(82)

Gergely, P., and Lutz, L. A., 1968, "Maximum Crack Width in Reinforced Concrete Flexural Members," *Causes,*

Mechanism, and Control of Cracking in Concrete, SP-20, American Concrete Institute, Farmington Hills, MI, pp. 87-117.

Gerritse, A., 1992, "Durability Criteria for Non-Metallic Tendons in an Alkaline Environment," *Proceedings of the First International Conference on Advance Composite Materials in Bridges and Structures (ACMBS-I)*, Canadian Society of Civil Engineers, Sherbrooke, QC, Canada, pp. 129-137.

Gremel, D., 2012, "Ensuring Electrical Isolation in Elevate Rail," *Concrete International*, V. 34, No. 2, Feb., pp. 22-25.

Gross, S.; Yost, J.; and Kevgas, G., 2003, "Time-Dependent Behavior of Normal and High Strength Concrete Beams Reinforced with GFRP Bars Under Sustained Loads," *High Performance Materials in Bridges*, ASCE, Reston, VA, pp. 451-462.

Hassan, M.; Ahmed, E. A.; and Benmokrane, B., 2013a, "Punching-Shear Strength of Normal- and High-Strength Concrete Flat Slabs Reinforced with GFRP Bars," *Journal of Composites for Construction*, V. 17, No. 6, Sept., 12 pp.

Hassan, M.; Ahmed, E. A.; and Benmokrane, B., 2013b, "Punching Shear Strength of Glass Fiber-Reinforced Polymer Reinforced Concrete Flat Slabs," *Canadian Journal of Civil Engineering*, V. 40, No. 10, pp. 951-960. doi: 10.1139/cjce-2012-0177

Hayes, M. D.; Garcia, K.; Verghese, N.; and Lesko, J., 1998, "The Effect of Moisture on the Fatigue Behavior of a Glass/Vinyl Ester Composite," *Proceedings of the Second International Conference on Composites in Infrastructure (ICCI-98)*, V. 1, Tucson, AZ, pp. 1-12.

Hughes, B. W., and Porter, M. L., 1996, "Experimental Evaluation of Non-Metallic Dowel Bars in Highway Pavements," *Proceedings of the First International Conference on Composites in Infrastructure (ICCI-96)*, H. Saadatmanesh and M. R. Ehsani, eds., Tucson, AZ, pp. 440-450.

Humar, J., and Razaqpur, A. G., eds., 2000, "Advanced Composite Materials in Bridges and Structures," *Proceedings of the third International Conference*, Ottawa, ON, Canada.

Iyer, S. L., and Sen, R., eds., 1991, "Advanced Composite Materials in Civil Engineering Structures," *Proceedings of the Specialty Conference*, ASCE, Reston, VA, 443 pp.

Jaeger, L. G.; Mufti, A.; and Tadros, G., 1997, "The Concept of the Overall Performance Factor in Rectangular-Section Reinforced Concrete Beams," *Proceedings of the Third International Symposium on Non-Metallic (FRP) Reinforcement for Concrete Structures (FRPRCS-3)*, V. 2, Japan Concrete Institute, Tokyo, Japan, pp. 551-558.

Japan Society of Civil Engineers, 1997a, *Proceedings of the Third International Symposium on Non-Metallic (FRP) Reinforcement for Concrete Structures (FRPRCS-3)*, V. 2, Japan Concrete Institute, Tokyo, Japan, pp. 511-518.

Japan Society of Civil Engineers, 1997b, "Recommendation for Design and Construction of Concrete Structures Using Continuous Fiber Reinforcing Materials," *Concrete Engineering Series* No. 23, 325 pp.

Japan Society of Civil Engineers Subcommittee on Continuous Fiber Reinforcement, 1992, *Proceedings of the Utilization of FRP-Rods for Concrete Reinforcement*, Japan Society of Civil Engineers, Tokyo, Japan, 314 pp.

Joint ACI-ASCE Committee 426, 1973, "The Shear Strength of Reinforced Concrete Members," *Journal of the Structural Division*, V. 99, No. 6, pp. 1091-1197.

Kamal, A., and Boulfiza, M., 2011, "Durability of GFRP Rebars in Simulated Concrete Solutions under Accelerated Aging Conditions," *Journal of Composites for Construction*, V. 15, No. 4, pp. 473-481. doi: 10.1061/(ASCE)CC.1943-5614.0000168

Katz, A., 1998, "Effect of Helical Wrapping on Fatigue Resistance of GFRP," *Journal of Composites for Construction*, V. 2, No. 3, pp. 121-125. doi: 10.1061/(ASCE)1090-0268(1998)2:3(121)

Katz, A., 2000, "Bond to Concrete of FRP Rebars after Cyclic Loading," *Journal of Composites for Construction*, V. 4, No. 3, pp. 137-144. doi: 10.1061/(ASCE)1090-0268(2000)4:3(137)

Katz, A., 2004, "Environmental Impact of Steel and Fiber-Reinforced Polymer Reinforced Pavements," *Journal of Composites for Construction*, V. 8, No. 6, pp. 481-488. doi: 10.1061/(ASCE)1090-0268(2004)8:6(481)

Katz, A.; Berman, N.; and Bank, L. C., 1999, "Effect of High Temperature on the Bond Strength of FRP Rebars," *Journal of Composites for Construction*, V. 3, No. 2, pp. 73-81. doi: 10.1061/(ASCE)1090-0268(1999)3:2(73)

Kobayashi, K., and Fujisaki, T., 1995, "Compressive Behavior of FRP Reinforcement in Non-Prestressed Concrete Members," *Proceedings of the Second International RILEM Symposium on Non-Metallic (FRP) Reinforcement for Concrete Structures (FRPRCS-2)*, Ghent, Belgium, pp. 267-274.

Kocaoz, S.; Samaranayake, V. A.; and Nanni, A., 2005, "Tensile Characterization of Glass FRP Bars," *Composites. Part B, Engineering*, V. 36, No. 2, Jan., pp. 127-134. doi: 10.1016/j.compositesb.2004.05.004

Kodur, V. K. R., and Bisby, L. A., 2005, "Evaluation of Fire Endurance of Concrete Slabs Reinforced with Fiber-Reinforced Polymer Bars," *Journal of Structural Engineering*, V. 131, No. 1, Jan., pp. 34-43. doi: 10.1061/(ASCE)0733-9445(2005)131:1(34)

Lee, J. H.; Yoon, Y. S.; and Mitchell, D., 2009, "Improving Punching Shear Behavior of Glass Fiber-Reinforced Polymer Reinforced Slabs," *ACI Structural Journal*, V. 106, No. 4, July-Aug., pp. 427-434.

Litherland, K. L.; Oakley, D. R.; and Proctor, B. A., 1981, "The Use of Accelerated Aging Procedures to Predict the Long Term Strength of GRC Composites," *Cement and Concrete Research*, V. 11, No. 3, pp. 455-466. doi: 10.1016/0008-8846(81)90117-4

Liu, R., and Pantelides, C. P., 2012, "Shear Capacity of Concrete Slabs Reinforced with Glass-Fiber- Reinforced Polymer Bars using the Modified Compression Field Theory," *PCI Journal*, V. 57, No. 3, pp. 83-99. doi: 10.15554/pcij.06012012.83.99

Liu, R., and Pantelides, C. P., 2013, "Shear Strength of GFRP Reinforced Precast Lightweight Concrete Panels," *Construction & Building Materials*, V. 48, No. 11, July, pp. 51-58. doi: 10.1016/j.conbuildmat.2013.06.057

Mallick, P. K., 1988, *Fiber Reinforced Composites, Materials, Manufacturing, and Design*, Marcell Dekker, Inc., New York, 469 pp.

Maluk, C.; Bisby, L. A.; Terrasi, G.; and Green, M. F., 2011, "Bond Strength of CFRP and Steel Bars in Concrete at Elevated Temperature," *Advances in Fire Design of Concrete Structures*, SP-279, American Concrete Institute, Farmington Hills, MI, pp. 1-36. doi: 10.14359/51682965

Mandell, J. F., 1982, "Fatigue Behavior of Fiber-Resin Composites," *Developments in Reinforced Plastics*, V. 2, Applied Science Publishers, London, pp. 67-107.

Mandell, J. F., and Meier, U., 1983, "Effects of Stress Ratio Frequency and Loading Time on the Tensile Fatigue of Glass-Reinforced Epoxy," *Long Term Behavior of Composites*, ASTM STP 813, ASTM International, West Conshohocken, PA, pp. 55-77.

Masmoudi, R.; Masmoudi, A.; Ouezdou, M.; and Daoud, A., 2011, "Long-Term Bond Performance of GFRP Bars in Concrete under Temperature Ranging From 20 °C to 80 °C," *Construction & Building Materials*, V. 25, No. 2, pp. 486-493. doi: 10.1016/j.conbuildmat.2009.12.040

Masmoudi, R.; Zaidi, A.; and Gérard, P., 2005, "Transverse Thermal Expansion of FRP Bars Embedded in Concrete," *Journal of Composites for Construction*, V. 9, No. 5, pp. 377-387. doi: 10.1061/(ASCE)1090-0268(2005)9:5(377)

Matthys, S., and Taerwe, L., 2000, "Concrete Slabs Reinforced with FRP Grids. II: Punching Resistance," *Journal of Composites for Construction*, V. 4, No. 3, pp. 154-161. doi: 10.1061/(ASCE)1090-0268(2000)4:3(154)

McClure, G., and Mohammadi, Y., 1995, "Compression Creep of Pultruded E-Glass-Reinforced-Plastic Angles," *Journal of Materials in Civil Engineering*, V. 7, No. 4, Nov., pp. 269-276. doi: 10.1061/(ASCE)0899-1561(1995)7:4(269)

Meier, U., 1992, "Carbon Fiber Reinforced Polymers: Modern Materials in Bridge Engineering," *Journal of the International Association for Bridge and Structural Engineering*, V. 2, No. 1, pp. 7-12. doi: 10.2749/101686692780617020

Michaluk, C. R.; Rizkalla, S.; Tadros, G.; and Benmokrane, B., 1998, "Flexural Behavior of One-Way Concrete Slabs Reinforced by Fiber Reinforced Plastic Reinforcement," *ACI Structural Journal*, V. 95, No. 3, May-June, pp. 353-364.

Mindess, S.; Young, J. F.; and Darwin, D., 2003, *Concrete*, second edition, Prentice-Hall, Upper Saddle River, NJ, 644 pp.

Mosley, C. P., 2002, "Bond Performance of Fiber Reinforced Plastic (FRP) Reinforcement in Concrete," MS thesis, Purdue University, West Lafayette, IN.

Mota, C.; Alminar, S.; and Svecova, D., 2006, "Critical Review of Deflection Formulas for FRP-RC Members," *Journal of Composites for Construction*, V. 10, No. 3, pp. 183-194. doi: 10.1061/(ASCE)1090-0268(2006)10:3(183)

Mufti, A.; Onofrei, M.; Benmokrane, B.; Banthia, N.; Boulfiza, M.; Newhook, J. P.; Bakht, B.; Tadros, G.; and Brett, P., 2007, "Field Study of Glass-Fibre-Reinforced Polymer Durability in Concrete," *Canadian Journal of Civil Engineering*, V. 34, No. 3, pp. 355-366. doi: 10.1139/l06-138

Mufti, A. A.; Newhook, J.; Benmokrane, B.; Tadros, G.; and Vogel, H. M., 2011, "*Durability of GFRP Rods in Field Demonstration Projects across Canada*," *Proceedings of the 4th International Conference on Durability and Sustainability of Fibre Reinforced Polymer (FRP) Composites for Construction and Rehabilitation, CDCC2011*, Quebec City, Canada, July 20-22, 9 pp.

Mutsuyoshi, H.; Uehara, K.; and Machida, A., 1990, "Mechanical Properties and Design Method of Concrete Beams Reinforced with Carbon Fiber Reinforced Plastics," *Transaction of the Japan Concrete Institute*, V. 12, Japan Concrete Institute, Tokyo, Japan, pp. 231-238.

Nagasaka, T.; Fukuyama, H.; and Tanigaki, M., 1993, "Shear Performance of Concrete Beams Reinforced with FRP Stirrups," *Fiber-Reinforced-Plastic Reinforcement for Concrete Structures—International Symposium*, SP-138, A. Nanni and C. W. Dolan, eds., American Concrete Institute, Farmington Hills, MI, pp. 789-811.

Nanni, A., ed., 1993a, "Fiber-Reinforced-Plastic (FRP) Reinforcement for Concrete Structures: Properties and Applications," *Developments in Civil Engineering*, V. 42, 450 pp.

Nanni, A., 1993b, "Flexural Behavior and Design of RC Members Using FRP Reinforcement," *Journal of Structural Engineering*, V. 119, No. 11, pp. 3344-3359. doi: 10.1061/(ASCE)0733-9445(1993)119:11(3344)

Nanni, A., 2001, "Relevant Field Applications of FRP Composites in Concrete Structures," *Proceedings of the International Conference Composites in Construction—CCC2001*, J. Figueiras, L. Juvandes, and R. Faria, eds., Portugal, pp. 661-670.

Nanni, A.; Bakis, C. E.; Al-Zahrani, M. M.; Al-Dulaijan, S. U.; and Boothby, T. E., 1998a, "Effect of Cyclic Loading on Bond Behavior of GFRP Rods Embedded in Concrete Beams," *Journal of Composites, Technology and Research*, V. 20, No. 1, pp. 29-37. doi: 10.1520/CTR10498J

Nanni, A.; Rizkalla, S.; Bakis, C. E.; Conrad, J. O.; and Abdelrahman, A. A., 1998b, "Characterization of GFRP Ribbed Rod Used for Reinforced Concrete Construction," *Proceedings of the International Composites Exhibition (ICE-98)*, Nashville, TN, pp. 16A/1-6.

Nanni, A.; Bakis, C. E.; and Boothby, T. E., 1995, "Test Methods for FRP-Concrete Systems Subjected to Mechanical Loads: State of the Art Review," *Journal of Reinforced Plastics and Composites*, V. 14, pp. 524-588.

Nanni, A., and Dolan, C. W., eds., 1993, *Fiber-Reinforced-Plastic Reinforcement for Concrete Structures—International Symposium*, SP-138, American Concrete Institute, Farmington Hills, MI, 977 pp.

Nanni, A.; Nenninger, J.; Ash, K.; and Liu, J., 1997, "Experimental Bond Behavior of Hybrid Rods for Concrete Reinforcement," *Structural Engineering & Mechanics*, V. 5, No. 4, pp. 339-353. doi: 10.12989/sem.1997.5.4.339

National Research Council, 1991, "Life Prediction Methodologies for Composite Materials," Committee on Life Prediction Methodologies for Composites, NMAB-460, National Materials Advisory Board, Washington, DC, 66 pp.

Nawy, E., and Neuwerth, G., 1977, "Fiberglass Reinforced Concrete Slabs and Beams," *Journal of the Structural Division*, V. 103, No. 2, Feb., pp. 421-440.

Neale, K. W., and Labossiere, P., eds., 1992, "Advanced Composite Materials in Bridges and Structures," *Proceed-*

ings of the First International Conference (ACMBS-I), Sherbrooke, QC, Canada, 705 pp.

Nigro, E.; Bilotta, A.; Cefarelli, G.; Manfredi, G.; and Cosenza, E., 2012, "Performance under Fire Situations of Concrete Members Reinforced with FRP Rods: Bond Models and Design Nomograms," *Journal for Composites in Construction*, V. 16, No. 4, pp. 395-406.

Nigro, E.; Cefarelli, G.; Bilotta, A.; Manfredi, G.; and Cosenza, E., 2011a, "Fire Resistance of Concrete Slabs Reinforced with FRP Bars. Part I: Experimental Investigations on the Mechanical Behavior," *Composites. Part B, Engineering*, V. 42, No. 6, pp. 1739-1750. doi: 10.1016/j.compositesb.2011.02.025

Nigro, E.; Cefarelli, G.; Bilotta, A.; Manfredi, G.; and Cosenza, E., 2011b, "Fire Resistance of Concrete Slabs Reinforced with FRP Bars. Part II: Experimental Results and Numerical Simulations on the Thermal Field," *Composites. Part B, Engineering*, V. 42, No. 6, pp. 1751-1763. doi: 10.1016/j.compositesb.2011.02.026

Nigro, E.; Cefarelli, G.; Bilotta, A.; Manfredi, G.; and Cosenza, E., 2013, "Adhesion at High Temperature of FRP Bars Straight or Bent at the end of Concrete Slabs," *Journal of Structural Fire Engineering*, V. 4, No. 2, June, pp. 71-86. doi: 10.1260/2040-2317.4.2.71

Noritake, K.; Kakihara, R.; Kumagai, S.; and Mizutani, J., 1993, "Technora, an Aramid FRP Rod," *Fiber-Reinforced-Plastic (FRP) Reinforcement for Concrete Structures: Properties and Applications, Developments in Civil Engineering*, V. 42, A. Nanni, ed., Elsevier, Amsterdam, pp. 267-290.

Odagiri, T.; Matsumoto, K.; and Nakai, H., 1997, "Fatigue and Relaxation Characteristics of Continuous Aramid Fiber Reinforced Plastic Rods," *Proceedings of the Third International Symposium on Non-Metallic (FRP) Reinforcement for Concrete Structures (FRPRCS-3)*, V. 2, Japan Concrete Institute, Tokyo, Japan, pp. 227-234.

Okamoto, T.; Nagasaka, T.; and Tanigaki, M., 1994, "Shear Capacity of Concrete Beams Using FRP Reinforcement," *Journal of Structural Construction Engineering*, No. 455, pp. 127-136.

Orangun, C.; Jirsa, J. O.; and Breen, J. E., 1977, "A Reevaluation of Test Data on Development Length and Splices," *ACI Journal Proceedings*, V. 74, No. 3, Mar., pp. 114-122.

Ospina, C. E., 2005, "Alternative Model for Concentric Punching Capacity Evaluation of Reinforced Concrete Two-Way Slabs," *Concrete International*, V. 27, No. 9, Sept., pp. 53-57.

Ospina, C. E.; Alexander, S.; and Cheng, J. J., 2001, "Behaviour of Concrete Slabs with Fibre-Reinforced Polymer Reinforcement," *Structural Engineering Report* No. 242, Department of Civil and Environmental Engineering, University of Alberta, AB, Canada, 355 pp.

Ospina, C. E.; Alexander, S. D. B.; and Cheng, J. J. R., 2003, "Punching of Two-way Concrete Slabs with Fiber-Reinforced Polymer Reinforcing Bars or Grids," *ACI Structural Journal*, V. 100, No. 5, Sept.-Oct., pp. 589-598.

Ospina, C. E., and Bakis, C. E., 2007, "Indirect Flexural Crack Control of Concrete Beams and One-Way Slabs Reinforced with FRP Bars," *Proceedings of the 8th International Symposium on Fiber Reinforced Polymer Reinforcement for Concrete Structures (FRPRCS-8)*, T.C. Triantafillou, ed., University of Patras, Greece. (CD-ROM).

Ospina, C. E., and Gross, S. P., 2005, "Rationale for the ACI 440.1R-06 Indirect Deflection Control Design provisions," *Proceedings of the 7th International Symposium on Fiber-Reinforced Polymer (FRP) Reinforcement for Concrete Structures (FRPRCS-7)*, SP-230, C. Shield, J. Busel, S. Walkup, and D. Gremel, eds., American Concrete Institute, Farmington Hills, MI, pp. 651-670.

Pantelides, C. P.; Besser, B. T.; and Liu, R., 2012a, "One-Way Shear Behavior of Lightweight Concrete Panels Reinforced with GFRP Bars," *Journal of Composites for Construction*, V. 16, No. 1, Feb., pp. 2-9. doi: 10.1061/(ASCE)CC.1943-5614.0000240

Pantelides, C. P.; Liu, R.; and Reaveley, L. D., 2012b, "Lightweight Concrete Precast Bridge Deck Panels Reinforced with Glass Fiber-Reinforced Polymer Bars," *ACI Structural Journal*, V. 109, No. 6, Nov.-Dec., pp. 879-888.

Pantelides, C. P.; Gibbons, M. E.; and Reaveley, L. D., 2013, "Axial Load Behavior of Concrete Columns Confined with GFRP Spirals," *Journal of Composites for Construction*, V. 17, No. 3, June, pp. 305-313. doi: 10.1061/(ASCE)CC.1943-5614.0000357

Pearson, C.; Donchev, T.; and Limbachiya, M., 2011, "An Investigation into the Sustainability of FRP Reinforcement Bars," *CDCC-11 4th International Conference on Durability and Sustainability of Fibre Reinforced Polymer (FRP) Composites for Construction and Rehabilitation*, Quebec City, Canada, July, pp. 71-80.

Porter, M. L., and Barnes, B. A., 1998, "Accelerated Aging Degradation of Glass Fiber Composites," *Second International Conference on Composites in Infrastructure*, V. II, H. Saadatmanesh and M. R. Eshani, eds., University of Arizona, Tucson, AZ, pp. 446-459.

Porter, M. L.; Hughes, B. W.; Barnes, B. A.; and Viswanath, K. P., 1993, "Non-Corrosive Tie Reinforcing and Dowel Bars for Highway Pavement Slabs," *Report* No. HR-343, Iowa Highway Research Board and Iowa Department of Transportation, Ames, IA.

Porter, M. L.; Mehus, J.; Young, K. A.; O'Neil, E. F.; and Barnes, B. A., 1997, "Aging for Fiber Reinforcement in Concrete," *Proceedings of the Third International Symposium on Non-Metallic (FRP) Reinforcement for Concrete Structures (FRPRCS-3)*, V. 2, Japan Concrete Institute, Tokyo, Japan, pp. 59-66.

Portland Cement Association, 1990, "Concrete Floors on Ground," Skokie, IL, 36 pp.

Portland Cement Association, 2008, *Notes on ACI 318-08 Building Code Requirements for Structural Concrete*, M. Kamara, L. Novak, and B. Rabbat, eds., Portland Cement Association, Skokie, IL, 1028 pp.

Priestley, M. N.; Seible, F.; and Calvi, G. M., 1996, *Seismic Design and Retrofit of Bridges*, John Wiley and Sons, New York, 704 pp.

Rahman, A. H.; Kingsley, C. Y.; and Crimi, J., 1996, "Durability of FRP Grid Reinforcement," *Advanced Composite Materials in Bridges and Structures*, M. M. El-Badry, ed.,

Canadian Society for Civil Engineering, Montreal, QC, Canada, pp. 681-690.

Rahman, A. H., and Kingsley, C. Y., 1996, "Fatigue Behavior of a Fiber-Reinforced-Plastic Grid as Reinforcement for Concrete," *Proceedings of the First International Conference on Composites in Infrastructure (ICCI-96)*, H. Saadatmanesh and M. R. Ehsani, eds., Tucson, AZ, pp. 427-439.

Rahman, H.; Adimi, R.; and Crimi, J., 1997, "Fatigue Behavior of a Carbon FRP Grid Encased in Concrete," *Proceedings of the Third International Symposium on Non-Metallic (FRP) Reinforcement for Concrete Structures (FRPRCS-3)*, V. 2, Japan Concrete Institute, Tokyo, Japan, pp. 219-226.

Rasheed, H. A.; Nayal, R.; and Melhem, H. G., 2004, "Response Prediction of Concrete Beams Reinforced with FRP Bars," *Composite Structures*, V. 65, No. 2, pp. 193-204. doi: 10.1016/j.compstruct.2003.10.016

Razaqpur, A. G.; Svecova, D.; and Cheung, M. S., 2000, "Rational Method for Calculating Deflection of Fiber-Reinforced Polymer Reinforced Beams," *ACI Structural Journal*, V. 97, No. 1, Jan.-Feb., pp. 175-185.

Rizkalla, S. H., 1997, "A New Generation of Civil Engineering Structures and Bridges," *Proceedings of the Third International Symposium on Non-Metallic (FRP) Reinforcement for Concrete Structures (FRPRCS-3)*, V. 1, Japan Concrete Institute, Tokyo, Japan, pp. 113-128.

Robert, M. and Benmokrane, B., 2010, "Behaviour of GFRP Reinforcing Bars Subjected to Extreme Temperature," *Journal of Composites for Construction*, V. 14, No. 4, pp. 353-360

Robert, M., and Benmokrane, B., 2013, "Combined Effects of Saline Solution and Moist Concrete on Long-Term Durability of GFRP Reinforcing Bars," *Journal of Construction and Building Materials*, V. 38, Jan, pp. 274-284. doi: 10.1016/j.conbuildmat.2012.08.021

Robert, M.; Cousin, P.; and Benmokrane, B., 2009, "Durability of GFRP Reinforcing Bars Embedded in Moist Concrete," *Journal of Composites for Construction*, V. 13, No. 2, pp. 66-73. doi: 10.1061/(ASCE)1090-0268(2009)13:2(66)

Rostasy, F. S., 1997, "On Durability of FRP in Aggressive Environments," *Proceedings of the Third International Symposium on Non-Metallic (FRP) Reinforcement for Concrete Structures (FRPRCS-3)*, V. 2, Japan Concrete Institute, Tokyo, Japan, pp. 107-114.

Roylance, M., and Roylance, O., 1981, "Effect of Moisture on the Fatigue Resistance of an Aramid-Epoxy Composite," *Organic Coatings and Plastics Chemistry*, American Chemical Society, Washington, DC, V. 45, pp. 784-788.

Saadatmanesh, H., and Ehsani, M. R., eds., 1998, "Fiber Composites in Infrastructure," *Proceedings of the Second International Conference on Composites in Infrastructure (ICCI-98)*, Tucson, AZ.

Saadatmanesh, H., and Tannous, F. E., 1999a, "Relaxation, Creep, and Fatigue Behavior of Carbon Fiber-Reinforced Plastic Tendons," *ACI Materials Journal*, V. 96, No. 2, Mar.-Apr., pp. 143-153.

Saadatmanesh, H., and Tannous, F. E., 1999b, "Long-Term Behavior of Aramid Fiber-Reinforced Plastic Tendons," *ACI Materials Journal*, V. 96, No. 3, May-June, pp. 297-305.

Santoh, N., 1993, "CFCC (Carbon Fiber Composite Cable)," *Fiber-Reinforced-Plastic (FRP) Reinforcement for Concrete Structures: Properties and Applications, Developments in Civil Engineering*, V. 42, A. Nanni, ed., Elsevier, Amsterdam, pp. 223-247.

Sasaki, I.; Nishizaki, I.; Sakamoto, H.; Katawaki, K.; and Kawamoto, Y., 1997, "Durability Evaluation of FRP Cables by Exposure Tests," *Proceedings of the Third International Symposium on Non-Metallic (FRP) Reinforcement for Concrete Structures (FRPRCS-3)*, V. 2, Japan Concrete Institute, Tokyo, Japan, pp. 131-137.

Scheibe, M., and Rostasy, F. S., 1998, "Stress-Rupture Behavior of AFRP Bars in Concrete and Under Natural Environment," *Second International Conference on Composites in Infrastructure*, V. II, H. Saadatmanesh and M. R. Eshani, eds., University of Arizona, Tucson, AZ, pp. 138-151.

Seki, H.; Sekijima, K.; and Konno, T., 1997, "Test Method on Creep of Continuous Fiber Reinforcing Materials," *Proceedings of the Third International Symposium on Non-Metallic (FRP) Reinforcement for Concrete Structures (FRPRCS-3)*, V. 2, Japan Concrete Institute, Tokyo, Japan, pp. 195-202.

Sen, R.; Shahawy, M.; Sukumar, S.; and Rosas, J., 1998a, "Effect of Tidal Exposure on Bond of CFRP Rods," *Second International Conference on Composites in Infrastructure*, V. II, H. Saadatmanesh and M. R. Eshani, eds., University of Arizona, Tucson, AZ, pp. 512-523.

Sen, R.; Shahawy, M.; Rosas, J.; and Sukumar, S., 1998b, "Durability of Aramid Pretensioned Elements in a Marine Environment," *ACI Structural Journal*, V. 95, No. 5, Sept.-Oct., pp. 578-587.

Shield, C.; French, C.; and Hanus, J., 1999, "Bond of GFRP Rebar for Consideration in Bridge Decks," *Fourth International Symposium on Fiber Reinforced Polymer Reinforcement for Reinforced Concrete Structures*, SP-188, C. W. Dolan, S. H. Rizkalla, and A. Nanni, eds., American Concrete Institute, Farmington Hills, MI, pp. 393-406.

Shield, C.; French, C.; and Retika, A., 1997, "Thermal and Mechanical Fatigue Effects on GFRP Rebar-Concrete Bond," *Proceedings of the Third International Symposium on Non-Metallic (FRP) Reinforcement for Concrete Structures (FRPRCS-3)*, V. 2, Japan Concrete Institute, Tokyo, Japan, pp. 381-388.

Shield, C., Galambos, T., and Gulbrandsen, P., 2011, "On the History and Reliability of the Flexural Strength of FRP Reinforced Concrete Members in ACI 440.1R," *10th International Symposium on Fiber Reinforced Polymer Reinforcement for Concrete Structures*, SP-275, R. Sen, R. Seracino, C. Shield, and W. Gold, eds., American Concrete Institute, Farmington Hills, MI, pp. 1-18.

Sippel, T. M., and Mayer, U., 1996, "Bond Behavior of FRP Strands under Short-Term, Reversed and Cyclic Loading," *Proceedings of the Second International Conference on Advanced Composite Materials in Bridges and Structures (ACMBS-2)*, M. M. El-Badry, ed., Canadian Society for Civil Engineering, Montreal, QC, Canada, pp. 837-844.

Sonobe, Y.; Fukuyama, H.; Okamoto, T.; Kani, N.; Kimura, K.; Kobayashi, K.; Masuda, Y.; Matsuzaki, Y.; Mochizuki,

S.; Nagasaka, T.; Shimizu, A.; Tanano, H.; Tanigaki, M.; and Tenshigawara, M., 1997, "Design Guidelines of FRP Reinforced Concrete Building Structures," *Journal of Composites for Construction*, V. 1, No. 3, pp. 90-115. doi: 10.1061/(ASCE)1090-0268(1997)1:3(90)

Szerszen, M., and Nowak, A., 2003, "Calibration of Design Code for Buildings (ACI 318): Part 2—Reliability Analysis and Resistance Factors," *ACI Structural Journal*, V. 100, No. 3, May-June, pp. 383-391.

Taerwe, L., ed., 1995, "Non-Metallic (FRP) Reinforcement for Concrete Structures," *Proceedings of the Second International RILEM Symposium on Non-Metallic (FRP) Reinforcement for Concrete Structures (FRPRCS-2)*, Ghent, Belgium, 714 pp.

Taerwe, L., 1997, "FRP Activities in Europe: Survey of Research and Applications," *Proceedings of the Third International Symposium on Non-Metallic (FRP) Reinforcement for Concrete Structures (FRPRCS-3)*, V. 1, Japan Concrete Institute, Tokyo, Japan, pp. 59-74.

Takewaka, K., and Khin, M., 1996, "Deterioration of Stress-Rupture of FRP Rods in Alkaline Solution Simulating as Concrete Environment," *Advanced Composite Materials in Bridges and Structures*, M. M. El-Badry, ed., Canadian Society for Civil Engineering, Montreal, QC, Canada, pp. 649-664.

Tamura, T., 1993, "FiBRA," *Fiber-Reinforced-Plastic (FRP) Reinforcement for Concrete Structures: Properties and Applications, Developments in Civil Engineering*, V. 42, A. Nanni, ed., Elsevier, Amsterdam, pp. 291-303.

Tannous, F. E., and Saadatmanesh, H., 1999, "Durability of AR-Glass Fiber Reinforced Plastic Bars," *Journal of Composites for Construction*, V. 3, No. 1, pp. 12-19. doi: 10.1061/(ASCE)1090-0268(1999)3:1(12)

Teng, J.-G., ed., 2001, "FRP Composites in Civil Engineering," *Proceedings CICE 2001*, Hong Kong, China, V. 1 and 2.

Tepfers, R., 2002, "Test System for Evaluation of Bond Properties of FRP Reinforcement in Concrete," *Proceedings of the Third International Symposium on Bond in Concrete—From Research to Standards*, Budapest, Hungary, pp. 657-666.

Theriault, M., and Benmokrane, B., 1998, "Effects of FRP Reinforcement Ratio and Concrete Strength on Flexural Behavior of Concrete Beams," *Journal of Composites for Construction*, V. 2, No. 1, pp. 7-16. doi: 10.1061/(ASCE)1090-0268(1998)2:1(7)

Tighiouart, B.; Benmokrane, B.; and Gao, D., 1998, "Investigation of Bond in Concrete Member with Fiber Reinforced Polymer (FRP) Bars," *Construction and Building Materials Journal*, Dec., 10 pp.

Tighiouart, B.; Benmokrane, B.; and Mukhopadhyaya, P., 1999, "Bond Strength of Glass FRP Rebar Splices in Beams under Static Loading," *Construction & Building Materials*, V. 13, No. 7, pp. 383-392. doi: 10.1016/S0950-0618(99)00037-9

Tobbi, H.; Farghaly, A. S.; and Benmokrane, B., 2012, "Concrete Columns Reinforced Longitudinally and Transversally with Glass Fiber-Reinforced Polymers Bars," *ACI Structural Journal*, V. 109, No. 4, July-Aug., pp. 551-558.

Tokyo Rope, 2000, "CFCC, Carbon Fiber Composite Cable," *Product Circular* No. 991-2T-SA, Tokyo Rope Manufacturing Co., Tokyo, http://www.tokyorope.co.jp/

Tomosawa, F., and Nakatsuji, T., 1996, "Evaluation of the ACM Reinforcement Durability by Exposure Test," *Advanced Composite Materials in Bridges and Structures*, M. M. El-Badry, ed., Canadian Society for Civil Engineering, Montreal, QC, Canada, pp. 699-706.

Tomosawa, F., and Nakatsuji, T., 1997, "Evaluation of the ACM Reinforcement Durability by Exposure Test," *Proceedings of the Third International Symposium on Non-Metallic (FRP) Reinforcement for Concrete Structures (FRPRCS-3)*, V. 2, Japan Concrete Institute, Tokyo, Japan, pp. 139-146.

Toutanji, H., and Saafi, M., 2000, "Flexural Behavior of Concrete Beams Reinforced with Glass Fiber-Reinforced Polymer (GFRP) Bars," *ACI Structural Journal*, V. 97, No. 5, Sept.-Oct., pp. 712-719.

Tureyen, A. K., and Frosch, R. J., 2002, "Shear Tests of FRP Reinforced Concrete Beams without Stirrups," *ACI Structural Journal*, V. 99, No. 4, July-Aug., pp. 427-434.

Tureyen, A. K., and Frosch, R. J., 2003, "Concrete Shear Strength: Another Perspective," *ACI Structural Journal*, V. 100, No. 5, Sept.-Oct., pp. 609-615.

Uomoto, T., 2000, "Durability of FRP as Reinforcement for Concrete Structures," *Proceedings of the 3rd International Conference on Advanced Composite Materials in Bridges and Structures*, ACMBS-3, J. L. Humar and A. G. Razaqpur, eds., Canadian Society for Civil Engineering, Montreal, QC, Canada, pp. 3-17.

Uppuluri, V. S.; Bakis, C. E.; Nanni, A.; and Boothby, T. E., 1996, "Analysis of the Bond Mechanism in FRP Reinforcement Rods: The Effect of Rod Design and Properties," *Proceedings of the Second International Conference on Advanced Composite Materials in Bridges and Structures (ACMBS-II)*, Montreal, QC, Canada, pp. 893-900.

Vijay, P. V., and GangaRao, H. V. S., 1999, "Accelerated and Natural Weathering of Glass Fiber Reinforced Plastic Bars," *Fourth International Symposium on Fiber Reinforced Polymer Reinforcement for Reinforced Concrete Structures*, SP-188, C. W. Dolan, S. H. Rizkalla, and A. Nanni, eds., American Concrete Institute, Farmington Hills, MI, pp. 605-614.

Vijay, P. V.; GangaRao, H. V. S.; and Kalluri, R., 1998, "Hygrothermal Response of GFRP Bars under Different Conditioning Schemes," *Proceedings of the First International Conference (CDCC 1998)*, Sherbrooke, QC, Canada, pp. 243-252.

Wambeke, B., and Shield, C., 2006, "Development Length of Glass Fiber Reinforced Polymer Bars in Concrete," *ACI Structural Journal*, V. 103, No. 1, Jan.-Feb., pp. 11-17.

Wang, J., 1998, "Determination of the Shear Resistance of Concrete Beams and Slabs Reinforced with Fibre Reinforced Plastics," MS thesis, Carleton University, Ottawa, ON, Canada.

White, T. D., ed., 1992, "Composite Materials and Structural Plastics in Civil Engineering Construction," *Proceedings of the Materials Engineering Congress*, ASCE, Reston, VA, pp. 532-718.

Wicaksono, S., and Chai, G. B., 2013, "A Review of Advances in Fatigue and Life Prediction of Fiber-Reinforced Composites," *Journal of Materials Design and Applications*, V. 227, pp. 179-195.

Wu, W. P., 1990, "Thermomechanical Properties of Fiber Reinforced Plastic (FRP) Bars," PhD dissertation, West Virginia University, Morgantown, WV, 292 pp.

Yamaguchi, T.; Kato, Y.; Nishimura, T.; and Uomoto, T., 1997, "Creep Rupture of FRP Rods Made of Aramid, Carbon and Glass Fibers," *Proceedings of the Third International Symposium on Non-Metallic (FRP) Reinforcement for Concrete Structures (FRPRCS-3)*, V. 2, Japan Concrete Institute, Tokyo, Japan, pp. 179-186.

Ye, L. P.; Feng, P.; Zhang, K.; Lin, L.; Hong, W. H.; Yue, Q. R.; Zhang, N.; and Yang, T., 2003, "FRP in Civil Engineering in China: Research and Applications," *Proceedings of the Sixth International Symposium on FRP Reinforcement for Concrete Structures (FRPRCS-6)*, K. H. Tan, ed., Singapore, 1401 pp.

Yost, J.; Gross, S.; and Dinehart, D., 2003, "Effective Moment of Inertia for GFRP Reinforced Concrete Beams," *ACI Structural Journal*, V. 100, No. 6, Nov.-Dec., pp. 732-739.

Zaidi, A., and Masmoudi, R., 2008, "Thermal Effect on Fiber Reinforced Polymer Reinforced Concrete Slabs," *Canadian Journal of Civil Engineering*, V. 35, No. 3, Mar., pp. 312-320. doi: 10.1139/L07-110

Zaidi, A.; Masmoudi, R.; and Bouhicha, M., 2013, "Numerical Analysis of Thermal Stress-Deformation in Concrete Surrounding FRP Bars in Hot Region," *Construction & Building Materials*, V. 38, Jan, pp. 204-213. doi: 10.1016/j.conbuildmat.2012.08.047

Zhao, W.; Maruyama, K.; and Suzuki, H., 1995, "Shear Behavior of Concrete Beams Reinforced by FRP Rods as Longitudinal and Shear Reinforcement," *Proceedings of the Second International RILEM Symposium on Non-Metallic (FRP) Reinforcement for Concrete Structures (FRPRCS-2)*, Ghent, Belgium, pp. 352-359.

APPENDIX A—SLABS-ON-GROUND

Two of the most common types of construction for slabs-on-ground are discussed in this appendix: plain concrete slabs and slabs reinforced with temperature and shrinkage reinforcement.

A.1—Design of plain concrete slabs

Plain concrete slabs-on-ground transmit loads to the subgrade with minimal distress and are designed to remain uncracked under service loads. To reduce shrinkage crack effects, the spacing of construction joints, contraction joints, or both, is usually limited. For details of design methods of plain concrete slabs-on-ground, refer to ACI 360R.

A.2—Design of slabs with shrinkage and temperature reinforcement

When designing a slab with shrinkage and temperature reinforcement, it should be considered a plain concrete slab without reinforcement to determine its thickness. The slab is assumed to remain uncracked when loads are placed on its surface. Shrinkage crack width and spacing are limited by a nominal amount of distributed fiber-reinforced polymer (FRP) reinforcement placed in the upper half of the slab. The primary purpose of shrinkage reinforcement is to control the width of any crack that forms between joints. Shrinkage reinforcement does not prevent cracking, nor does it significantly add to the flexural capacity of the slab. Increasing the thickness of the slab can increase the flexural capacity.

Although the slab is intended to remain uncracked under service loading, the reinforcement is used to limit crack spacing and width, permit the use of wider joint spacing, increase the ability to transfer load at joints, and provide a reserve strength after shrinkage or temperature cracking has occurred.

The subgrade drag method is frequently used to determine the amount of nonprestressed shrinkage and temperature reinforcement that is needed, but does not apply when prestressing or randomly distributed fibers are used (Portland Cement Association 1990). When using steel reinforcement, the drag equation is as follows

$$A_s = \frac{\mu L w_{slab}}{2 f_{s,allow}} \tag{A.2a}$$

where A_s is the cross-sectional area of steel per linear foot, in.^{2} (mm^2 per linear meter); $f_{s,allow}$ is the allowable stress in steel reinforcement, psi (MPa), commonly taken as two-thirds to three-fourths of f_y; μ is the coefficient of subgrade friction (1.5 is recommended for floors on ground [Portland Cement Association 1990]); L is the distance between joints, ft (m); and w is the dead weight of the slab, lb/ft^2 (N/m^2), usually assumed to be 12.5 lb/ft^2 per in. of slab thickness (24 N/m^2 per mm).

Because of the lower modulus of the FRP reinforcement, the governing equation should be based on the strain rather than the stress level when designing shrinkage and temperature FRP reinforcement. At the allowable stress, the strain in steel reinforcement is approximately 0.0012; implementing the same strain for FRP will result in a stress of $0.0012E_f$, and Eq. (A.2a) can be written as

$$A_{f,sh} = \frac{\mu L w}{2\left(0.0012 E_f\right)} \tag{A.2b}$$

Equation (A.2b) can also be used to determine joint spacing L for a set amount of reinforcement. No experimental data have been reported on FRP slab-on-ground applications; research is required to validate this approach.

Guide for the Design and Construction of Externally Bonded FRP Systems for Strengthening Concrete Structures

Reported by ACI Committee 440

Carol K. Shield, Chair

William J. Gold, Secretary

Tarek Alkhrdaji
Charles E. Bakis
Lawrence C. Bank
Abdeldjelil Belarbi
Brahim Benmokrane
Luke A. Bisby
Gregg J. Blaszak
Hakim Bouadi
Timothy E. Bradberry
Vicki L. Brown
John Busel
Raafat El-Hacha
Garth J. Fallis
Amir Z. Fam
Russell Gentry
Nabil F. Grace
Mark F. Green
Zareh B. Gregorian
Doug D. Gremel
Shawn P. Gross
H. R. Trey Hamilton III
Issam E. Harik
Kent A. Harries*
Mark P. Henderson
Ravindra Kanitkar
Yail Jimmy Kim
Michael W. Lee
Maria Lopez de Murphy
Ibrahim M. Mahfouz
Amir Mirmiran
John J. Myers
Antonio Nanni
Ayman M. Okeil
Carlos E. Ospina
Renato Parretti
Maria A. Polak
Max L. Porter
Andrea Prota
Hayder A. Rasheed
Sami H. Rizkalla
Rajan Sen
Rudolf Seracino
Venkatesh Seshappa
Pedro F. Silva
Samuel A. Steere, III
Jennifer E. Tanner
Jay Thomas
Houssam A. Toutanji
J. Gustavo Tumialan
Milan Vatovec
David White
Sarah E. Witt*

*Co-chairs of the subcommittee that prepared this document.

Consulting Members

P. N. Balaguru
Craig A. Ballinger
Harald G. F. Budelmann
C. J. Burgoyne
Rami M. Elhassan
David M. Gale
Srinivasa L. Iyer
Koichi Kishitani
Howard S. Kliger
Kyuichi Maruyama
Antoine E. Naaman
Hajime Okamura
Mark A. Postma
Ferdinand S. Rostasy
Mohsen Shahawy
Surendra P. Shah
Yasuhisa Sonobe
Minoru Sugita
Luc R. Taerwe
Ralejs Tepfers
Taketo Uomoto
Paul Zia

Fiber-reinforced polymer (FRP) systems for strengthening concrete structures are an alternative to traditional strengthening techniques such as steel plate bonding, section enlargement, and external post-tensioning. FRP strengthening systems use FRP composite materials as supplemental externally-bonded or near-surface-mounted reinforcement. FRP systems offer advantages over traditional strengthening techniques: they are lightweight, relatively easy to install, and noncorroding. Due to the characteristics of FRP materials as well as the behavior of members strengthened with FRP, specific guidance on the use of these systems is needed. This guide offers general information on the history and use of FRP strengthening systems; a description of the material properties of FRP; and recommendations on the engineering, construction, and inspection of FRP systems used to strengthen concrete structures. This guide is based on the knowledge gained from experimental research, analytical work, and field applications of FRP systems used to strengthen concrete structures.

Keywords: aramid fibers; bridges; buildings; carbon fibers; corrosion; cracking; development length; earthquake resistance; fiber-reinforced polymers; structural design.

ACI 440.2R-17 supersedes ACI 440.2R-08 and was adopted and published May 2017.

CONTENTS

CHAPTER 1—INTRODUCTION AND SCOPE

1.1—Introduction

The strengthening or retrofitting of existing concrete structures to resist higher design loads, correct strength loss due to deterioration, correct design or construction deficiencies, or increase ductility has historically been accomplished using conventional materials and construction techniques. Externally bonded steel plates, steel or concrete jackets, and external post-tensioning are some of the many traditional techniques available.

Composite materials made of fibers in a polymeric resin, also known as fiber-reinforced polymers (FRPs), have emerged as a viable option for repair and rehabilitation. For the purposes of this guide, an FRP system is defined as the fibers and resins used to create the composite laminate, all applicable resins used to bond it to the concrete substrate, and all applied coatings used to protect the constituent materials. Coatings used exclusively for aesthetic reasons are not considered part of an FRP system.

FRP materials are lightweight, noncorroding, and exhibit high tensile strength. These materials are readily available in several forms, ranging from factory-produced pultruded laminates to dry fiber sheets that can be wrapped to conform to the geometry of a structure before adding the polymer resin. The relatively thin profiles of cured FRP systems are often desirable in applications where aesthetics or access is a concern. FRP systems can also be used in areas with limited access where traditional techniques would be difficult to implement.

The basis for this document is the knowledge gained from a comprehensive review of experimental research, analytical work, and field applications of FRP strengthening systems. Areas where further research is needed are highlighted in this document and compiled in Appendix C.

1.1.1 *Use of FRP systems*—This document refers to commercially available FRP systems consisting of fibers and resins combined in a specific manner and installed by a specific method. These systems have been developed through material characterization and structural testing. Untested combinations of fibers and resins could result in an unexpected range of properties as well as potential material incompatibilities. Any FRP system considered for use should have sufficient test data to demonstrate adequate performance of the entire system in similar applications, including its method of installation. ACI 440.8 provides a specification for unidirectional carbon and glass FRP materials made using the wet layup process.

The use of FRP systems developed through material characterization and structural testing, including well-documented proprietary systems, is recommended. The use of untested combinations of fibers and resins should be avoided. A comprehensive set of test standards and guides for FRP systems has been developed by several organizations, including ASTM, ACI, ICRI, and ICC.

1.1.2 *Sustainability*—Sustainability of FRP materials may be evaluated considering environmental, economic, and social goals. These should be considered not only throughout the construction phase, but also through the service life of the structure in terms of maintenance and preservation, and for the end-of-life phase. This represents the basis for a life-cycle approach to sustainability (Menna et al. 2013). Life cycle assessment (LCA) takes into account the environmental impact of a product, starting with raw material extraction, followed by production, distribution, transportation, installation, use, and end of life. LCA for FRP composites depends on the product and market application, and results vary. FRP composite materials used to strengthen concrete elements can use both carbon fiber and glass fiber, which are derived from fossil fuels or minerals, respectively, and therefore have impacts related to raw material extraction. Although carbon and glass fibers have high embodied energies associated with production, on the order of 86,000 Btu/lb and 8600 Btu/lb (200 and 20 mJ/kg), respectively (Howarth et al. 2014), the overall weight produced and used is orders of magnitude lower than steel (having embodied energy of 5600 Btu/lb [13 mJ/kg]), concrete (430 Btu/lb [1 mJ/kg]), and reinforcing steel (3870 Btu/lb [9 mJ/kg]) (Griffin and Hsu 2010). The embodied energy and potential environmental impact of resin and adhesive systems are less studied, although the volume used is also small in comparison with conventional construction materials. In distribution and transportation, FRP composites' lower weight leads to less impact from transportation, and easier material handling allows smaller equipment during installation. For installation and use, FRP composites are characterized as having a longer service life because they are more durable and require less maintenance than conventional materials. The end-of-life options for FRP composites are more complex.

Although less than 1 percent of FRP composites are currently recycled, composites can be recycled in many ways, including mechanical grinding, incineration, and chemical separation (Howarth et al. 2014). It is difficult, however, to separate the materials, fibers, and resins without some degradation of the resulting recycled materials. The market for recycled composite materials is small, although aircraft manufacturers in particular are considering methods and programs to recycle and repurpose composite materials at the end of an aircraft's life cycle.

Apart from the FRP materials and systems, their use in the repair and retrofit of structures that may otherwise be decommissioned or demolished is inherently sustainable. In many cases, FRP composites permit extending the life or enhancing the safety or performance of existing infrastructure at a monetary and environmental cost of only a fraction of replacement. Additionally, due to the high specific strength and stiffness of FRP composites, an FRP-based repair of an existing concrete structure will often represent a less energy-intensive option than a cementitious or metallic-based repair.

Within this framework of sustainability, FRP retrofit of existing structures may lead to benefits, contributing to the longevity and safety of retrofitted structures. Thus, FRP retrofit can be regarded as a viable method for sustainable design for strengthening and rehabilitation of existing structures. The environmental advantages of FRP, as evaluated by LCA investigations, have been enumerated by Napolano et al. (2015), Moliner Santisteve et al. (2013), Zhang et al. (2012), and Das (2011).

1.2—Scope

This document provides guidance for the selection, design, and installation of FRP systems for externally strengthening concrete structures. Information on material properties, design, installation, quality control, and maintenance of FRP systems used as external reinforcement is presented. This information can be used to select an FRP system for increasing the strength, stiffness, or both, of reinforced concrete beams or the ductility of columns and other applications.

A significant body of research serves as the basis for this guide. This research, conducted since the 1980s, includes analytical studies, experimental work, and monitored field applications of FRP strengthening systems. Based on the available research, the design procedures outlined herein are considered conservative.

The durability and long-term performance of FRP materials has been the subject of much research; however, this research remains ongoing. The design guidelines in this guide account for environmental degradation and long-term durability by providing reduction factors for various environments. Long-term fatigue and creep are also addressed by stress limitations indicated in this document. These factors and limitations are considered conservative. As more research becomes available, however, these factors may be modified, and the specific environmental conditions and loading conditions to which they should apply will be better defined. Additionally, the coupling effect of environmental conditions and loading conditions requires further study. Caution is advised in applications where the FRP system is subjected simultaneously to extreme environmental and stress conditions. The factors associated with the long-term durability of the FRP system may also affect the tensile modulus of elasticity of the material used for design.

Many issues regarding bond of the FRP system to the substrate remain the focus of a great deal of research. For both flexural and shear strengthening, there are many different modes of debonding failure that can govern the strength of an FRP-strengthened member. While most of the debonding modes have been identified by researchers, more accurate methods of predicting debonding are still needed. Throughout the design procedures, significant limitations on the strain achieved in the FRP material (and thus, the stress achieved) are imposed to conservatively account for debonding failure modes. Future development of these design procedures should include more thorough methods of predicting debonding.

This document gives guidance on proper detailing and installation of FRP systems to prevent many types of debonding failure modes. Steps related to the surface preparation and proper termination of the FRP system are vital in achieving the levels of strength predicted by the procedures in this document. Research has been conducted on various methods of anchoring FRP strengthening systems, such as U-wraps, mechanical fasteners, fiber anchors, and U-anchors. Because no anchorage design guidelines are currently available, the performance of any anchorage system should be substantiated through representative physical testing that includes the specific anchorage system, installation procedure, surface preparation, and expected environmental conditions.

The design equations given in this document are the result of research primarily conducted on moderately sized and proportioned members fabricated of normalweight concrete. Caution should be given to applications involving strengthening of very large or lightweight concrete members or strengthening in disturbed regions (D-regions) of structural members such as deep beams, corbels, and dapped beam ends. When warranted, specific limitations on the size of members and the state of stress are given herein.

This guide applies only to FRP strengthening systems used as additional tensile reinforcement. These systems should not be used as compressive reinforcement. While FRP materials can support compressive stresses, there are numerous issues surrounding the use of FRP for compression. Microbuckling of fibers can occur if any resin voids are present in the laminate. Laminates themselves can buckle if not properly adhered or anchored to the substrate, and highly unreliable compressive strengths result from misaligning fibers in the field. This document does not address the construction, quality control, and maintenance issues that would be involved with the use of the material for this purpose, nor does it address the design concerns surrounding such applications.

This document does not specifically address masonry (concrete masonry units, brick, or clay tile) construction, including masonry walls. Information on the repair of unreinforced masonry using FRP can be found in ACI 440.7R.

1.2.1 *Applications and use*—FRP systems can be used to rehabilitate or restore the strength of a deteriorated structural member, retrofit or strengthen a sound structural member to resist increased loads due to changes in use of the structure, or address design or construction errors. The licensed design professional should determine if an FRP system is a suitable strengthening technique before selecting the type of FRP system.

To assess the suitability of an FRP system for a particular application, the licensed design professional should perform a condition assessment of the existing structure that includes establishing its existing load-carrying capacity, identifying deficiencies and their causes, and determining the condition of the concrete substrate. The overall evaluation should include a thorough field inspection, a review of existing design or as-built documents, and a structural analysis in accordance with ACI 364.1R. Existing construction documents for the structure should be reviewed, including the design drawings, project specifications, as-built information, field test reports, past repair documentation, and maintenance history documentation. The licensed design professional should conduct a thorough field investigation of the existing structure in accordance with ACI 437R, ACI 562, ACI 369R, and other applicable ACI documents. As a minimum, the field investigation should determine the following:

a) Existing dimensions of the structural members
b) Location, size, and cause of cracks and spalls
c) Quantity and location of existing reinforcing steel
d) Location and extent of corrosion of reinforcing steel
e) Presence of active corrosion
f) In-place compressive strength of concrete
g) Soundness of the concrete, especially the concrete cover, in all areas where the FRP system is to be bonded to the concrete

The tensile strength of the concrete on surfaces where the FRP system may be installed should be determined by conducting a pull-off adhesion test in accordance with ASTM C1583/C1583M. The in-place compressive strength of concrete should be determined using cores in accordance with ACI 562 requirements. The load-carrying capacity of the existing structure should be based on the information gathered in the field investigation, the review of design calculations and drawings, and as determined by analytical methods. Load tests or other methods can be incorporated into the overall evaluation process if deemed appropriate.

FRP systems used to increase the strength of an existing member should be designed in accordance with Chapters 9 through 15, which include a comprehensive discussion of load limitations, rational load paths, effects of temperature and environment on FRP systems, loading considerations, and effects of reinforcing steel corrosion on FRP system integrity.

1.2.1.1 *Strengthening limits*—In general, to prevent sudden failure of the member in case the FRP system is damaged, strengthening limits are imposed such that the increase in the load-carrying capacity of a member strengthened with an FRP system is limited. The philosophy is that a loss of FRP reinforcement should not cause member failure. Specific guidance, including load combinations for assessing member integrity after loss of the FRP system, is provided in Chapter 9.

1.2.1.2 *Fire and life safety*—FRP-strengthened structures should comply with applicable building and fire codes. Smoke generation and flame spread ratings in accordance with ASTM E84 should be satisfied for the installation according to applicable building codes, depending on the classification of the building. Coatings (Apicella and Imbrogno 1999) and insulation systems (Williams et al. 2006) can be used to limit smoke and flame spread.

Because of the degradation of most FRP materials at high temperature, the strength of externally bonded FRP systems is assumed to be lost completely in a fire, unless it can be demonstrated that the FRP will remain effective for the required duration of the fire. The fire resistance of FRP-strengthened concrete members may be improved through the use of certain resins, coatings, insulation systems, or other methods of fire protection (Bisby et al. 2005b). Specific guidance, including load combinations and a rational approach to calculating structural fire resistance, is given in 9.2.1.

1.2.1.3 *Maximum service temperature*—The physical and mechanical properties of the resin components of FRP systems are influenced by temperature and degrade at temperatures close to or above their glass-transition temperature T_g (Bisby et al. 2005b). The T_g for commercially available, ambient temperature-cured FRP systems typically ranges from 140 to 180°F (60 to 82°C). The T_g for a particular FRP system can be obtained from the system manufacturer or through testing by dynamic mechanical analysis (DMA) according to ASTM E1640. Reported T_g values should be accompanied by descriptions of the test configuration; sample preparation; curing conditions (time, temperature, and humidity); and size, heating rate, and frequency used. The T_g defined by this method represents the extrapolated onset temperature for the sigmoidal change in the storage modulus observed in going from a hard and brittle state to a soft and rubbery state of the material under test. This transition occurs over a temperature range of approximately 54°F (30°C) centered on the T_g. This change in state will adversely affect the mechanical and bond properties of the cured laminates. For a dry environment, it is generally recommended that the anticipated service temperature of an FRP system not exceed T_g – 27°F (T_g – 15°C) (Xian and Karbhari 2007), where T_g is taken as the lowest T_g of the components of the system comprising the load path. This recommendation is for elevated service temperatures such as those found in hot regions or certain industrial environments. In cases where the FRP will be exposed to a moist environment, the wet glass-transition temperature T_{gw} should be used (Luo and Wong 2002). Testing may be required to determine the critical service temperature for FRP in other environments. The specific case of fire is described in more detail in 9.2.1.

1.2.1.4 *Minimum concrete substrate strength*—FRP systems need to be bonded to a sound concrete substrate and should not be considered for applications on structural members containing corroded reinforcing steel or deteriorated concrete unless the substrate is repaired using the recommendations in 6.4. Concrete distress, deterioration, and corrosion of existing reinforcing steel should be evaluated and addressed before the application of the FRP system. Concrete deterioration concerns include, but are not limited to, alkali-silica reactions, delayed ettringite formation, carbonation, longitudinal cracking around corroded reinforcing steel, and laminar cracking at the location of the steel reinforcement.

The strength of the existing concrete substrate is an important parameter for bond-critical applications, including flexure or shear strengthening. The substrate should possess the necessary strength to develop the design stresses of the FRP system through bond. The substrate, including all bond surfaces between repaired areas and the original concrete, should have sufficient direct tensile and shear strength to transfer force to the FRP system. For bond-critical applications, the tensile strength should be at least 200 psi (1.4 MPa), determined by using a pull-off type adhesion test per ICRI 210.3R or ASTM C1583/C1583M. FRP systems should not be used when the concrete substrate has a compressive strength f_c' less than 2500 psi (17 MPa). Contact-critical applications, such as column wrapping for confinement that rely only on intimate contact between the FRP system and the concrete, are not governed by these minimum values. Design stresses in the FRP system are developed by deformation or dilation of the concrete section in contact-critical applications.

The application of FRP systems will not stop the ongoing corrosion of existing reinforcing steel (El-Maaddawy et al. 2006). If steel corrosion is evident or is degrading the concrete substrate, placement of FRP reinforcement is not recommended without arresting the ongoing corrosion and repairing any degradation of the substrate.

CHAPTER 2—NOTATION AND DEFINITIONS

2.1—Notation

A_c = cross-sectional area of concrete in compression member, in.^{2} (mm^2)
A_{cw} = area of concrete section of individual vertical wall, in.^{2} (mm^2)
A_e = cross-sectional area of effectively confined concrete section, in.^{2} (mm^2)
A_f = area of FRP external reinforcement, in.^{2} (mm^2)
$A_{fanchor}$ = area of transverse FRP U-wrap for anchorage of flexural FRP reinforcement, in.^{2} (mm^2)
A_{fv} = area of FRP shear reinforcement with spacing s, in.^{2} (mm^2)
A_g = gross area of concrete section, in.^{2} (mm^2)
A_p = area of prestressed reinforcement in tension zone, in.^{2} (mm^2)
A_s = area of nonprestressed steel reinforcement, in.^{2} (mm^2)
A_{sc} = area of the longitudinal reinforcement within a distance of w_f in the compression region, in.^{2} (mm^2)
A_{si} = area of i-th layer of longitudinal steel reinforcement, in.^{2} (mm^2)
A_{st} = total area of longitudinal reinforcement, in.^{2} (mm^2)
A_{sw} = area of longitudinal reinforcement in the central area of the wall, in.^{2} (mm^2)
a = depth of the equivalent concrete compression block, in. (mm)
a_b = smaller cross-sectional dimension for rectangular FRP bars, in. (mm)
b = width of compression face of member, in. (mm)
= short side dimension of compression member of prismatic cross section, in. (mm)
b_b = larger cross-sectional dimension for rectangular FRP bars, in. (mm)
b_w = web width or diameter of circular section, in. (mm)
C_E = environmental reduction factor
C_{sc} = compressive force in A_{sc}, lb (N)
c = distance from extreme compression fiber to the neutral axis, in. (mm)
c_y = distance from extreme compression fiber to the neutral axis at steel yielding, in. (mm)
D = diameter of compression member for circular cross sections or diagonal distance equal to $\sqrt{b^2 + h^2}$ for prismatic cross section (diameter of equivalent circular column), in. (mm)
d = distance from extreme compression fiber to centroid of tension reinforcement, in. (mm)
d' = distance from the extreme compression fiber to the center of A_{sc}, in. (mm)
d'' = distance from the extreme tension fiber to the center of A_{st}, in. (mm)
$d_{b\ell}$ = diameter of longitudinal steel in confined plastic hinge, in. (mm)
d_f = effective depth of FRP flexural reinforcement, in. (mm)
d_{fv} = effective depth of FRP shear reinforcement, in. (mm)
d_i = distance from centroid of i-th layer of longitudinal steel reinforcement to geometric centroid of cross section, in. (mm)
d_p = distance from extreme compression fiber to centroid of prestressed reinforcement, in. (mm)
E_2 = slope of linear portion of stress-strain model for FRP-confined concrete, psi (MPa)
E_c = modulus of elasticity of concrete, psi (MPa)
E_f = tensile modulus of elasticity of FRP, psi (MPa)
E_{ps} = modulus of elasticity of prestressing steel, psi (MPa)
E_s = modulus of elasticity of steel, psi (MPa)
e_s = eccentricity of prestressing steel with respect to centroidal axis of member at support, in. (mm)
e_m = eccentricity of prestressing steel with respect to centroidal axis of member at midspan, in. (mm)
f_c = compressive stress in concrete, psi (MPa)
f_c' = specified compressive strength of concrete, psi (MPa)
f_{cc}' = compressive strength of confined concrete, psi (MPa)
f_{co}' = compressive strength of unconfined concrete; also equal to $0.85f_c'$, psi (MPa)
$f_{c,s}$ = compressive stress in concrete at service condition, psi (MPa)
f_f = stress in FRP reinforcement, psi (MPa)

f_{fd} = design stress of externally bonded FRP reinforcement, psi (MPa)
f_{fe} = effective stress in the FRP; stress attained at section failure, psi (MPa)
$f_{f,s}$ = stress in FRP caused by a moment within elastic range of member, psi (MPa)
f_{fu} = design ultimate tensile strength of FRP, psi (MPa)
f_{fu}^* = ultimate tensile strength of the FRP material as reported by the manufacturer, psi (MPa)
f_l = maximum confining pressure due to FRP jacket, psi (MPa)
f_{ps} = stress in prestressed reinforcement at nominal strength, psi (MPa)
$f_{ps,s}$ = stress in prestressed reinforcement at service load, psi (MPa)
f_{pu} = specified tensile strength of prestressing tendons, psi (MPa)
f_s = stress in nonprestressed steel reinforcement, psi (MPa)
f_{sc} = stress in the longitudinal reinforcement corresponding to A_{sc}, psi (MPa)
f_{si} = stress in the *i*-th layer of longitudinal steel reinforcement, psi (MPa)
$f_{s,s}$ = stress in nonprestressed steel reinforcement at service loads, psi (MPa)
f_{st} = stress in the longitudinal reinforcement corresponding to A_{st}, psi (MPa)
f_{sw} = stress in the longitudinal reinforcement corresponding to A_{sw}, psi (MPa)
f_y = specified yield strength of nonprestressed steel reinforcement, psi (MPa)
g = clear gap between the FRP jacket and adjacent members, in (mm)
h = overall thickness or height of a member, in. (mm)
= long side cross-sectional dimension of rectangular compression member, in. (mm)
h_f = member flange thickness, in. (mm)
h_w = height of entire wall from base to top, or clear height of wall segment or wall pier considered, in. (mm)
I_{cr} = moment of inertia of cracked section transformed to concrete, in.^{4} (mm^4)
I_{tr} = moment of inertia of uncracked section transformed to concrete, in.^{4} (mm^4)
K = ratio of depth of neutral axis to reinforcement depth measured from extreme compression fiber
k_1 = modification factor applied to κ_v to account for concrete strength
k_2 = modification factor applied to κ_v to account for wrapping scheme
k_f = stiffness per unit width per ply of the FRP reinforcement, lb/in. (N/mm); $k_f = E_f t_f$
L_e = active bond length of FRP laminate, in. (mm)
L_p = plastic hinge length, in. (mm)
L_w = length of the shear wall, in. (mm)
ℓ_{db} = development length of near-surface-mounted FRP bar, in. (mm)
$\ell_{d,E}$ = length over which the FRP anchorage wraps are provided, in. (mm)
ℓ_{df} = development length of FRP system, in. (mm)
ℓ_o = length, measured along the member axis from the face of the joint, over which special transverse reinforcement must be provided, in. (mm)
ℓ_{prov} = length of steel lap splice, in. (mm)
M_{cr} = cracking moment, in.-lb (N-mm)
M_n = nominal flexural strength, in.-lb (N-mm)
M_{nf} = contribution of FRP reinforcement to nominal flexural strength, lb-in. (N-mm)
M_{np} = contribution of prestressing reinforcement to nominal flexural strength, lb-in. (N-mm)
M_{ns} = contribution of steel reinforcement to nominal flexural strength, lb-in. (N-mm)
M_s = service moment at section, in.-lb (N-mm)
M_{snet} = service moment at section beyond decompression, in.-lb (N-mm)
M_u = factored moment at a section, in.-lb (N-mm)
N = number of plies of FRP reinforcement
n_f = modular ratio of elasticity between FRP and concrete = E_f/E_c
n_s = modular ratio of elasticity between steel and concrete = E_s/E_c
P_e = effective force in prestressing reinforcement (after allowance for all prestress losses), lb (N)
P_n = nominal axial compressive strength of a concrete section, lb (N)
P_u = factored axial load, lb (N)
$\bar{p}_{fu}$ = mean tensile strength per unit width per ply of FRP reinforcement, lb/in. (N/mm)
p_{fu}^* = ultimate tensile strength per unit width per ply of FRP reinforcement, lb/in. (N/mm); $p_{fu}^* = f_{fu}^* t_f$
R_n = nominal strength of a member
$R_{n\phi}$ = nominal strength of a member subjected to elevated temperatures associated with a fire
R = radius of gyration of a section, in. (mm)
r_c = radius of edges of a prismatic cross section confined with FRP, in. (mm)
S_{DL} = dead load effects
S_{LL} = live load effects
s_f = center-to-center spacing of FRP strips, in. (mm)
T_f = tensile force in FRP, lb (N)
T_g = glass-transition temperature, °F (°C)
T_{gw} = wet glass-transition temperature, °F (°C)
T_{ps} = tensile force in prestressing steel, lb (N)
T_{st} = tensile force in A_{st}, lb (N)
T_{sw} = tensile force in A_{sw}, lb (N)
t_f = nominal thickness of one ply of FRP reinforcement, in. (mm)
t_w = thickness of the existing concrete shear wall, in. (mm)
V_c = nominal shear strength provided by concrete with steel flexural reinforcement, lb (N)
V_e = design shear force for load combinations including earthquake effects, lb (N)
V_f = nominal shear strength provided by FRP stirrups, lb (N)

V_n = nominal shear strength, lb (N)
V_n^* = shear strength of existing member, lb (N)
V_s = nominal shear strength provided by steel stirrups, lb (N)
w_f = width of FRP reinforcing plies, in. (mm)
y_b = distance from centroidal axis of gross section, neglecting reinforcement, to extreme bottom fiber, in./in. (mm/mm)
y_t = vertical coordinate within compression region measured from neutral axis position. It corresponds to transition strain ε_t', in. (mm)
α = angle of application of primary FRP reinforcement direction relative to longitudinal axis of member
α_1 = multiplier on f_c' to determine intensity of an equivalent rectangular stress distribution for concrete
α_L = longitudinal coefficient of thermal expansion, in./in./°F (mm/mm/°C)
α_T = transverse coefficient of thermal expansion, in./in./°F (mm/mm/°C)
β_1 = ratio of depth of equivalent rectangular stress block to depth of the neutral axis
ε_b = strain in concrete substrate developed by a given bending moment (tension is positive), in./in. (mm/mm)
ε_{bi} = strain in concrete substrate at time of FRP installation (tension is positive), in./in. (mm/mm)
ε_c = strain in concrete, in./in. (mm/mm)
ε_c' = compressive strain of unconfined concrete corresponding to f_c', in./in. (mm/mm); may be taken as 0.002
ε_{ccu} = ultimate axial compressive strain of confined concrete corresponding to $0.85f_{cc}'$ in a lightly confined member (member confined to restore its concrete design compressive strength), or ultimate axial compressive strain of confined concrete corresponding to failure in a heavily confined member
$\varepsilon_{c,s}$ = strain in concrete at service, in./in. (mm/mm)
ε_{ct} = concrete tensile strain at level of tensile force resultant in post-tensioned flexural members, in./in. (mm/mm)
ε_{cu} = ultimate axial strain of unconfined concrete corresponding to $0.85f_{co}'$ or maximum usable strain of unconfined concrete, in./in. (mm/mm), which can occur at $f_c = 0.85f_c'$ or $\varepsilon_c = 0.003$, depending on the obtained stress-strain curve
ε_f = strain in the FRP reinforcement, in./in. (mm/mm)
ε_{fd} = debonding strain of externally bonded FRP reinforcement, in./in. (mm/mm)
ε_{fe} = effective strain in FRP reinforcement attained at failure, in./in. (mm/mm)
ε_{fu} = design rupture strain of FRP reinforcement, in./in. (mm/mm)
$\overline{\varepsilon}_{fu}$ = mean rupture strain of FRP reinforcement based on a population of 20 or more tensile tests per ASTM D3039/D3039M, in./in. (mm/mm)
ε_{fu}^* = ultimate rupture strain of FRP reinforcement, in./in. (mm/mm)
ε_{pe} = effective strain in prestressing steel after losses, in./in. (mm/mm)
ε_{pi} = initial strain in prestressed steel reinforcement, in./in. (mm/mm)
ε_{pnet} = net strain in flexural prestressing steel at limit state after prestress force is discounted (excluding strains due to effective prestress force after losses), in./in. (mm/mm)
$\varepsilon_{pnet,s}$ = net strain in prestressing steel beyond decompression at service, in./in. (mm/mm)
ε_{ps} = strain in prestressed reinforcement at nominal strength, in./in. (mm/mm)
$\varepsilon_{ps,s}$ = strain in prestressing steel at service load, in./in. (mm/mm)
ε_s = strain in nonprestessed steel reinforcement, in./in. (mm/mm)
ε_{sy} = strain corresponding to yield strength of nonprestressed steel reinforcement, in./in. (mm/mm)
ε_t = net tensile strain in extreme tension steel at nominal strength, in./in. (mm/mm)
ε_t' = transition strain in stress-strain curve of FRP-confined concrete, in./in. (mm/mm)
ϕ = strength reduction factor
ϕ_D = design curvature for a confined concrete section
$\phi_{y,frp}$ = curvature of the FRP confined section at steel yielding
κ_a = efficiency factor for FRP reinforcement in determination of f_{cc}' (based on geometry of cross section)
κ_b = efficiency factor for FRP reinforcement in determination of ε_{ccu} (based on geometry of cross section)
κ_v = bond-dependent coefficient for shear
κ_ε = efficiency factor equal to 0.55 for FRP strain to account for the difference between observed rupture strain in confinement and rupture strain determined from tensile tests
θ_p = plastic hinge rotation demand
ρ_f = FRP reinforcement ratio
ρ_g = ratio of area of longitudinal steel reinforcement to cross-sectional area of a compression member (A_s/bh)
ρ_l = longitudinal reinforcement ratio
ρ_s = ratio of nonprestressed reinforcement
σ = standard deviation
τ_b = average bond strength for near-surface-mounted FRP bars, psi (MPa)
ψ_e = factor used to modify development length based on reinforcement coating
ψ_f = FRP strength reduction factor
= 0.85 for flexure (calibrated based on design material properties)
= 0.85 for shear (based on reliability analysis) for three-sided FRP U-wrap or two sided strengthening schemes
= 0.95 for shear fully wrapped sections
ψ_s = factor used to modify development length based on reinforcement size
ψ_t = factor used to modify development length based on reinforcement location

2.2—Definitions

ACI provides a comprehensive list of definitions through an online resource, "ACI Concrete Terminology," https://www.concrete.org/store/productdetail.aspx?ItemID=CT13. Definitions provided herein complement that source.

aramid fiber—fiber in which chains of aromatic polyamide molecules are oriented along the fiber axis to exploit the strength of the chemical bond.

aramid fiber-reinforced polymer—composite material comprising a polymer matrix reinforced with aramid fiber cloth, mat, or strands.

carbon fiber—fiber produced by heating organic precursor materials containing a substantial amount of carbon, such as rayon, polyacrylonitrile, or pitch in an inert environment.

carbon fiber-reinforced polymer—composite material comprising a polymer matrix reinforced with carbon fiber cloth, mat, or strands.

catalyst—substance that accelerates a chemical reaction and enables it to proceed under conditions more mild than otherwise required and that is not, itself, permanently changed by the reaction.

contact-critical application—strengthening or repair system that relies on load transfer from the substrate to the system material achieved through contact or bearing at the interface.

creep rupture—breakage of a material under sustained loading at stresses less than the tensile strength.

cross-linking—formation of covalent bonds linking one polymer molecule to another.

E-glass—family of glass fibers used in reinforced polymers with a calcium alumina borosilicate composition and a maximum alkali content of 2.0 percent.

fabric—two-dimensional network of woven, nonwoven, knitted, or stitched fibers; yarns; or tows.

fiber content—the amount of fiber present in a composite, expressed as a percentage volume fraction or mass fraction of the composite.

fiber fly—short filaments that break off dry fiber tows or yarns during handling and become airborne.

fire retardant—additive to the resin or a surface coating used to reduce the tendency of a resin to burn.

fiber volume fraction—ratio of the volume of fibers to the volume of the composite containing the fibers.

full cure—period at which components of a thermosetting resin have reacted sufficiently for the resin to produce specified properties.

glass fiber—filament drawn from an inorganic fusion typically comprising silica-based material that has cooled without crystallizing.

glass fiber-reinforced polymer—composite material comprising a polymer matrix reinforced with glass fiber cloth, mat, or strands.

glass-transition temperature—representative temperature of the temperature range over which an amorphous material (such as glass or a high polymer) changes from (or to) a brittle, vitreous state to (or from) a plastic state.

impregnate—to saturate fibers with resin or binder.

initiator—chemical used to start the curing process for unsaturated polyester and vinyl ester resins.

interlaminar shear—force tending to produce a relative displacement along the plane of the interface between two laminae.

intumescent coating—covering that swells, increasing volume and decreasing density, when exposed to fire imparting a degree of passive fire protection.

lamina—single layer of fiber reinforcement.

laminate—multiple plies or lamina molded together.

layup—process of placing reinforcing material and resin system in position for molding.

monomer—organic molecule of low molecular weight that creates a solid polymer by reacting with itself or other compounds of low molecular weight.

phenolic resin—thermosetting resin produced by the condensation reaction of an aromatic alcohol with an aldehyde (usually a phenol with formaldehyde).

pitch—viscid substance obtained as a residue of petroleum or coal tar for use as a precursor in the manufacture of some carbon fibers.

polyacrylonitrile—synthetic semi-chrystalline organic polymer-based material that is spun into a fiber form for use as a precursor in the manufacturer of some carbon fibers.

polyester—one of a large group of synthetic resins, mainly produced by reaction of dibasic acids with dihydroxy alcohols.

postcuring—application of elevated temperature to material containing thermosetting resin to increase the degree of polymer crosslinking and enhance the final material properties.

prepreg—sheet of fabric or mat preimpregnated with resin or binder that is partially cured and ready for final forming and curing.

pultrusion—continuous process for manufacturing fiber-reinforced polymer composites in which resin-impregnated fiber reinforcements (roving or mats) are pulled through a shaping and curing die to produce composites with uniform cross sections.

putty—thickened polymer-based resin used to prepare the concrete substrate.

resin content—amount of resin in a fiber-reinforced polymer composite laminate, expressed as either a percentage of total mass or total volume.

roving—parallel bundle of continuous yarns, tows, or fibers with little or no twist.

saturating resins (or saturants)—polymer-based resin used to impregnate the reinforcing fibers, fix them in place, and transfer load between fibers.

shelf life—length of time packaged materials can be stored under specified conditions and remain usable.

sizing—surface treatment applied to filaments to impart desired processing, durability, and bond attributes.

storage modulus—measure of the stored energy in a viscoelastic material undergoing cyclic deformation during dynamic mechanical analysis.

tow—untwisted bundle of continuous filaments.

vinylester resin—thermosetting reaction product of epoxy resin with a polymerizable unsaturated acid (usually methacrylic acid) that is then diluted with a reactive monomer (usually styrene).

volatile organic compound—organic compound that vaporizes under normal atmospheric conditions.

wet layup—manufacturing process where dry fabric fiber reinforcement is impregnated on site with a saturating resin matrix and then cured in place.

wet-out—process of coating or impregnating roving, yarn, or fabric to fill the voids between the strands and filaments with resin; it is also the condition at which this state is achieved.

witness panel—small mockup manufactured under conditions representative of field application, to confirm that prescribed procedures and materials will yield specified mechanical and physical properties.

yarn—twisted bundle of continuous filaments.

CHAPTER 3—BACKGROUND INFORMATION

Externally bonded fiber-reinforced polymer (FRP) systems have been used to strengthen and retrofit existing concrete structures around the world since the mid-1980s. The number of projects using FRP systems worldwide has increased dramatically, from a few in the 1980s to many thousands today. Structural elements strengthened with externally bonded FRP systems include beams, slabs, columns, walls, joints/connections, chimneys and smoke-stacks, vaults, domes, tunnels, silos, pipes, and trusses. Externally bonded FRP systems have also been used to strengthen masonry, timber, steel, and cast-iron structures. Externally bonded FRP systems were developed as alternatives to traditional external reinforcing techniques such as steel plate bonding and steel or concrete column jacketing. The initial development of externally bonded FRP systems for the retrofit of concrete structures occurred in the 1980s in Europe and Japan.

3.1—Historical development

In Europe, FRP systems were developed as alternates to steel plate bonding. Bonding steel plates to the tension zones of concrete members with adhesive resins was shown to be a viable technique for increasing flexural strength (Fleming and King 1967). This technique has been used to strengthen many bridges and buildings around the world. Because steel plates can corrode, leading to a deterioration of the bond between the steel and concrete, and because they are difficult to install, requiring the use of heavy equipment, researchers looked to FRP materials as an alternative to steel. Experimental work using FRP materials for retrofitting concrete structures was reported as early as 1978 in Germany (Wolf and Miessler 1989). Research in Switzerland led to the first applications of externally bonded FRP systems to reinforced concrete bridges for flexural strengthening (Meier 1987; Rostasy 1987).

FRP systems were first applied to reinforced concrete columns for providing additional confinement in Japan in the 1980s (Fardis and Khalili 1981; Katsumata et al. 1987). A sudden increase in the use of FRPs in Japan was observed following the 1995 Hyogoken-Nanbu earthquake (Nanni 1995).

Researchers in the United States have had a continuous interest in fiber-based reinforcement for concrete structures since the 1930s. Development and research into the use of these materials for retrofitting concrete structures, however, started in the 1980s through the initiatives of the National Science Foundation (NSF) and the Federal Highway Administration (FHWA). The research activities led to the construction of many field projects that encompassed a wide variety of environmental conditions. Previous research and field applications for FRP rehabilitation and strengthening are described in ACI 440R and conference proceedings, including those of the Fiber Reinforced Polymers for Reinforced Concrete Structures (FRPRCS), Composites in Civil Engineering (CICE), and Conference on Durability of Composites for Construction (CDCC) series.

The development of codes and standards for externally bonded FRP systems is ongoing in Europe, Japan, Canada, and the United States. The first published codes and standards appeared in Japan (Japan Society of Civil Engineers 2001) and Europe (International Federation for Structural Concrete 2001). In the United States, ACI 440.8, ICC AC125, and NCHRP Report 655 (Zureick et al. 2010) provide criteria for evaluating FRP systems.

3.2—Commercially available externally bonded FRP systems

FRP systems come in a variety of forms, including wet layup systems and precured systems. FRP system forms can be categorized based on how they are delivered to the site and installed. The FRP system and its form should be selected based on the acceptable transfer of structural loads and the ease and simplicity of application. Common FRP system forms suitable for the strengthening of structural members are listed in 3.2.1 through 3.2.4.

3.2.1 *Wet layup systems*—Wet layup FRP systems consist of dry unidirectional or multidirectional fiber sheets or fabrics impregnated with a saturating resin on site. The saturating resin, along with the compatible primer and putty, bonds the FRP sheets to the concrete surface. Wet layup systems are saturated on site and cured in place and, in this sense, are analogous to cast-in-place concrete. Three common types of wet layup systems are listed as follows:

1. Dry unidirectional fiber sheets where the fibers run predominantly in one planar direction. ACI 440.8 provides specifications for unidirectional carbon fiber-reinforced polymer (CFRP) and glass fiber-reinforced polymer (GFRP) wet layup systems.

2. Dry multidirectional fiber sheets or fabrics where the fibers are oriented in at least two planar directions.

3. Dry fiber tows that are wound or otherwise mechanically applied to the concrete surface. The dry fiber tows are impregnated with resin on site during the winding operation.

3.2.2 *Prepreg systems*—Prepreg FRP systems consist of partially cured unidirectional or multidirectional fiber sheets or fabrics that are preimpregnated with a saturating resin in the manufacturer's facility. Prepreg systems are bonded to the concrete surface with or without an additional resin application, depending on specific system requirements.

Prepreg systems are saturated off site and, like wet layup systems, cured in place. Prepreg systems usually require additional heating for curing. Prepreg system manufacturers should be consulted for storage and shelf-life recommendations and curing procedures. Three common types of prepreg FRP systems are:

1. Preimpregnated unidirectional fiber sheets where the fibers run predominantly in one planar direction
2. Preimpregnated multidirectional fiber sheets or fabrics where the fibers are oriented in at least two planar directions
3. Preimpregnated fiber tows that are wound or otherwise mechanically applied to the concrete surface

3.2.3 *Precured systems*—Precured FRP systems consist of a wide variety of composite shapes manufactured off site. Typically, an adhesive, along with the primer and putty, is used to bond the precured shapes to the concrete surface. The system manufacturer should be consulted for recommended installation procedures. Precured systems are analogous to precast concrete. Three common types of precured systems are:

1. Precured unidirectional laminate sheets typically delivered to the site in the form of large flat stock or as thin ribbon strips coiled on a roll
2. Precured multidirectional grids, typically delivered to the site coiled on a roll
3. Precured shells, typically delivered to the site in the form of shell segments cut longitudinally so they can be opened and fitted around columns or other members; multiple shell layers are bonded to the concrete and to each other to provide confinement

3.2.4 *Near-surface-mounted (NSM) systems*—Surface-embedded NSM FRP systems consist of circular or rectangular bars or plates installed and bonded into grooves made on the concrete surface. A suitable adhesive is used to bond the FRP bar into the groove, and is cured in-place. The NSM system manufacturer should be consulted for recommended adhesives. Two common FRP bar types used for NSM applications are:

1. Round bars usually manufactured using pultrusion processes, typically delivered to the site in the form of single bars or in a roll, depending on bar diameter
2. Rectangular bars and plates usually manufactured using pultrusion processes, typically delivered to the site in a roll

CHAPTER 4—CONSTITUENT MATERIALS AND PROPERTIES

The physical and mechanical properties of fiber-reinforced polymer (FRP) materials presented in this chapter explain the behavior and properties affecting their use in concrete structures. The effects of factors such as loading history and duration, temperature, and moisture on the properties of FRP are discussed.

FRP strengthening systems come in a variety of forms (wet layup, prepreg, and precured). Factors such as fiber volume, type of fiber, type of resin, fiber orientation, dimensional effects, and quality control during manufacturing all play a role in establishing the characteristics of an FRP material. The material characteristics described in this chapter are generic and do not apply to all commercially available products. Standard test methods are available to characterize certain FRP products (refer to Appendix B). ACI 440.8 provides a specification for unidirectional carbon FRP (CFRP) and glass FRP (GFRP) materials made using the wet layup process. The licensed design professional should consult with the FRP system manufacturer to obtain the relevant characteristics for a specific product and the applicability of those characteristics.

4.1—Constituent materials

The constituent materials used in commercially available FRP repair systems, including all resins, primers, putties, saturants, adhesives, and fibers, have been developed for the strengthening of structural concrete members based on materials and structural testing.

4.1.1 *Resins*—A wide range of polymeric resins, including primers, putty fillers, saturants, and adhesives, are used with FRP systems. Commonly used resin types, including epoxy, vinyl esters, and polyesters, have been formulated for use in a wide range of environmental conditions. FRP system manufacturers use resins that have:

a) Compatibility with and adhesion to the concrete substrate
b) Compatibility with and adhesion to the FRP composite system
c) Compatibility with and adhesion to the reinforcing fiber
d) Resistance to environmental effects, including, but not limited to, moisture, salt water, temperature extremes, and chemicals normally associated with exposed concrete
e) Filling ability
f) Workability
g) Pot life consistent with the application
h) Development of appropriate mechanical properties for the FRP composite

4.1.1.1 *Primer*—Primer is used to penetrate the surface of the concrete, providing an improved adhesive bond for the saturating resin or adhesive.

4.1.1.2 *Putty fillers*—Putty is used to fill small surface voids in the substrate, such as bug holes, and to provide a smooth surface to which the FRP system can bond. Filled surface voids also prevent bubbles from forming during curing of the saturating resin.

4.1.1.3 *Saturating resin*—Saturating resin is used to impregnate the reinforcing fibers, fix them in place, and provide a shear load path to effectively transfer load between fibers. The saturating resin also serves as the adhesive for wet layup systems, providing a shear load path between the previously primed concrete substrate and the FRP system.

4.1.1.4 *Adhesives*—Adhesives are used to bond precured FRP laminate and near-surface mounted (NSM) systems to the concrete substrate. The adhesive provides a shear load path between the concrete substrate and the FRP reinforcing system. Adhesives are also used to bond together multiple layers of precured FRP laminates.

4.1.2 *Fibers*—Continuous glass, aramid, and carbon fibers are common reinforcements used in FRP systems. The fibers give the FRP system its strength and stiffness. Typical ranges of the tensile properties of fibers are given in Appendix A. A more detailed description of fiber types is given in ACI 440R.

4.1.3 *Protective coatings*—The protective coating protects the bonded FRP reinforcement from potentially damaging environmental and mechanical effects. Coatings are typically applied to the exterior surface of the FRP system after some prescribed degree of adhesive or saturating resin cure. The protection systems are available in a variety of forms. These include:

a) Polymer coatings that are generally epoxy or polyurethanes.

b) Acrylic coatings that can be either straight acrylic systems or acrylic cement-based systems. The acrylic systems can also come in different textures.

c) Cementitious systems that may require roughening of the FRP surface (such as broadcasting sand into wet resin) and can be installed in the same manner as they would be installed on a concrete surface.

d) Intumescent coatings that are polymer-based coatings used to provide a degree of passive fire protection and control flame spread and smoke generation per code requirements.

There are several reasons why protection systems are used to protect FRP systems that have been installed on concrete surfaces. These include:

a) *Ultraviolet light protection*—The epoxy used as part of the FRP strengthening system will be affected over time by exposure to ultraviolet light. There are many available methods used to protect the system from ultraviolet light. These include acrylic coatings, cementitious surfacing, aliphatic polyurethane coatings, and others. Certain types of vinylester resins have higher ultraviolet light durability than epoxy resins.

b) *Fire protection*—Fire protection systems are discussed in 1.2.1.2 and 9.2.1.

c) *Vandalism*—Protective systems that are to resist vandalism should be hard and durable. There are different levels of vandalism protection, ranging from polyurethane coatings that will resist cutting and scraping to cementitious overlays that provide greater protection.

d) *Impact, abrasion, and wear*—Protection systems for impact, abrasion, and wear are similar to those used for vandalism protection; however, abrasion and wear are different than vandalism in that they result from repeated exposure rather than a one-time event, and their protection systems are usually chosen for their hardness and durability.

e) *Aesthetics*—Protective topcoats may be used to conceal the FRP system. These may be acrylic latex coatings that are gray in color to match concrete, or they may be various other colors and textures to match the existing structure.

f) *Chemical resistance*—Exposure to harsh chemicals, such as strong acids, may damage the FRP system. In such environments, coatings with better chemical resistance, such as urethanes and novolac epoxies, may be used.

g) *Submersion in potable water*—In applications where the FRP system is to be submerged in potable water, the FRP system may leach compounds into the water supply. Protective coatings that do not leach harmful chemicals into the water may be used as a barrier between the FRP system and the potable water supply.

4.2—Physical properties

4.2.1 *Density*—FRP materials have densities ranging from 75 to 130 lb/ft^3 (1.2 to 2.1 g/cm^3), which is four to six times lower than that of steel (Table 4.2.1).

Table 4.2.1—Typical densities of FRP materials, lb/ft^3 (g/cm^3)

Steel	Glass FRP (GFRP)	Carbon FRP (CFRP)	Aramid FRP (AFRP)
490 (7.9)	75 to 130 (1.2 to 2.1)	90 to 100 (1.5 to 1.6)	75 to 90 (1.2 to 1.5)

4.2.2 *Coefficient of thermal expansion*—The coefficients of thermal expansion of unidirectional FRP materials differ in the longitudinal and transverse directions, depending on the types of fiber, resin, and volume fraction of fiber. Table 4.2.2 lists the longitudinal and transverse coefficients of thermal expansion for typical unidirectional FRP materials. Note that a negative coefficient of thermal expansion indicates that the material contracts with increased temperature and expands with decreased temperature. For reference, the isotropic values of coefficient of thermal expansion for concrete and steel are also provided in Table 4.2.2. Refer to 9.3.1 for design considerations regarding thermal expansion.

4.2.3 *Effects of high temperatures*—Above the glass transition temperature T_g, the elastic modulus of a polymer is significantly reduced due to changes in its molecular structure. The value of T_g depends on the type of resin and is normally in the region of 140 to 180°F (60 to 82°C). In an FRP composite material, the fibers, which exhibit better thermal properties than the resin, can continue to support some load in the longitudinal direction until the temperature threshold of the fibers is reached. This can occur at temperatures exceeding 1800°F (1000°C) for carbon fibers, 530°F (275°C) for glass fibers, and 350°F (175°C) for aramid fibers. Due to a reduction in force transfer between fibers through bond to the resin, however, the tensile properties of the overall composite are reduced. Test results have indicated that temperatures of 480°F (250°C)—much higher than the resin T_g—will reduce the tensile strength of GFRP and CFRP materials exceeding 20 percent (Kumahara et al. 1993). Other properties affected by the shear transfer through the resin, such as bending strength, are reduced significantly at lower temperatures (Wang and Evans 1995).

For bond-critical applications of FRP systems, the properties of the polymer at the fiber-concrete interface are essential in maintaining the bond between FRP and concrete. At a temperature close to its T_g, the mechanical properties of the polymer are significantly reduced and the polymer begins to lose its ability to transfer stresses from the concrete to the fibers.

4.3—Mechanical properties

4.3.1 *Tensile behavior*—When loaded in direct tension, unidirectional fiber-reinforced polymer (FRP) materials do not exhibit any plastic behavior (yielding) before rupture.

Table 4.2.2—Typical coefficients of thermal expansion for FRP materials*

Direction	Coefficient of thermal expansion, $\times 10^{-6}$/°F ($\times 10^{-6}$/°C)				
	Concrete	Steel	GFRP	CFRP	AFRP
Longitudinal, α_L	4.0 to 6.0 (7.0 to 11.0)	6.5 (11.7)	3.3 to 5.6 (6 to 10)	–0.6 to 0 (–1 to 0)	–3.3 to –1.1 (–6 to –2)
Tranverse, α_T	4.0 to 6.0 (7.0 to 11.0)	6.5 (11.7)	10.4 to 12.6 (19 to 23)	12 to 27 (22 to 50)	33 to 44 (60 to 80)

*Typical values for fiber-volume fractions ranging from 0.5 to 0.7.

The tensile behavior of FRP materials consisting of a single type of fiber material is characterized by a linear elastic stress-strain relationship until failure, which is sudden and brittle.

The tensile strength and stiffness of an FRP material is dependent on several factors. Because the fibers in an FRP material are the main load-carrying constituents, the type of fiber, the orientation of fibers, the quantity of fibers, and method and conditions in which the composite is produced affect the tensile properties of the FRP material. Due to the primary role of the fibers and methods of application, the properties of an FRP repair system are sometimes reported based on the net-fiber area. In other instances, such as in precured laminates, the reported properties are based on the gross-laminate area.

The gross-laminate area of an FRP system is calculated using the total cross-sectional area of the cured FRP system, including all fibers and resin. The gross-laminate area is typically used for reporting precured laminate properties where the cured thickness is constant and the relative proportion of fiber and resin is controlled.

The net-fiber area of an FRP system is calculated using the known area of fiber, neglecting the total width and thickness of the cured system; thus, resin is excluded. The net-fiber area is typically used for reporting properties of wet layup systems that use manufactured fiber sheets and field-installed resins. The wet layup installation process leads to controlled fiber content and variable resin content. A method similar to net-fiber area reporting is to report the tensile force or stiffness per unit width of the FRP system as required by ASTM D7565/D7565M.

System properties reported using the gross laminate area have higher relative thickness dimensions and lower relative strength and modulus values, whereas system properties reported using the net-fiber area have lower relative thickness dimensions and higher relative strength and modulus values. Regardless of the basis for the reported values, the load-carrying strength ($f_{fu}A_f$) and axial stiffness (A_fE_f) of the composite remain constant. Properties reported based on the net-fiber area are not the properties of the bare fibers. When tested as a part of a cured composite, the measured tensile strength and ultimate rupture strain of the net-fiber are typically lower than those measured based on a dry fiber test. The properties of an FRP system should be characterized as a composite, recognizing not just the material properties of the individual fibers, but also the efficiency of the fiber-resin system, the fabric architecture, and the method used to create the composite. The mechanical properties of all FRP systems, regardless of form, should be based on the testing of laminate samples with known fiber content.

The tensile properties of some commercially available FRP strengthening systems are given in Appendix A. The tensile properties of a particular FRP system, however, should be obtained from the FRP system manufacturer or using the appropriate test method described in ASTM D3039/D3039M, D7205/D7205M, or D7565/D7565M. Manufacturers should report an ultimate tensile strength, which is defined as the mean tensile strength of a sample of test specimens minus three times the standard deviation ($f_{fu}^* = \bar{f}_{fu} - 3\sigma$) and, similarly, report an ultimate rupture strain ($\varepsilon_{fu}^* = \bar{\varepsilon}_{fu} - 3\sigma$). This approach provides a 99.87 percent probability that the actual ultimate tensile properties will exceed these statistically-based design values for a standard sample distribution (Mutsuyoshi et al. 1990). The elastic modulus should be calculated in accordance with ASTM D3039/D3039M, D7205/D7205M, or D7565/D7565M. A minimum number of 20 replicate test specimens should be used to determine the ultimate tensile properties. The manufacturer should provide a description of the method used to obtain the reported tensile properties, including the number of tests, mean values, and standard deviations.

4.3.2 *Compressive behavior*—Externally bonded FRP systems should not be used as compression reinforcement due to insufficient testing to validate its use in this type of application. The mode of failure for FRP laminates subjected to longitudinal compression can include transverse tensile failure, fiber microbuckling, or shear failure. The mode of failure depends on the type of fiber, the fiber-volume fraction, and the type of resin. In general, compressive strengths are higher for materials with higher tensile strengths, except in the case of aramid FRP (AFRP), where the fibers exhibit nonlinear behavior in compression at a relatively low level of stress (Wu 1990). The compressive modulus of elasticity is usually smaller than the tensile modulus of elasticity of FRP materials (Ehsani 1993).

4.4—Time-dependent behavior

4.4.1 *Creep rupture*—FRP materials subjected to a sustained load can suddenly fail after a time period referred to as the endurance time. This type of failure is known as creep rupture. As the ratio of the sustained tensile stress to the short-term strength of the FRP laminate increases, endurance time decreases. The endurance time also decreases under adverse environmental conditions, such as high temperature, ultraviolet-radiation exposure, high alkalinity, wet and dry cycles, or freezing-and-thawing cycles.

In general, carbon fibers are the least susceptible to creep rupture, aramid fibers are moderately susceptible, and glass fibers are most susceptible. Creep rupture tests have been conducted on 0.25 in. (6 mm) diameter FRP bars reinforced with glass, aramid, and carbon fibers. The FRP bars were tested at different load levels at room temperature. Results indicated that a linear relationship exists between creep rupture strength and the logarithm of time for all load levels. The ratios of stress to cause creep rupture after 500,000 hours (approximately 50 years) to the short-term ultimate strength of the GFRP, AFRP, and CFRP bars were extrapolated to be approximately 0.3, 0.5, and 0.9, respectively (Yamaguchi et al. 1997; Malvar 1998). Recommendations on sustained stress limits imposed to avoid creep rupture are given Chapter 9 through 15. As long as the sustained stress in the FRP is below the creep rupture stress limits, the strength of the FRP is available for nonsustained loads.

4.4.2 *Fatigue*—A substantial amount of data for fatigue behavior and life prediction of stand-alone FRP materials is available (National Research Council 1991). Most of these data were generated from materials typically used by the aerospace industry. Despite the differences in quality and consistency between aerospace and commercial-grade FRP materials, some general observations on the fatigue behavior of FRP materials can be made. Unless specifically stated otherwise, the following cases are based on a unidirectional material with approximately 60 percent fiber-volume fraction and subjected to tension-tension sinusoidal cyclic loading at:

a) A frequency low enough to not cause self heating

b) Ambient laboratory environments

c) A stress ratio (ratio of minimum applied stress to maximum applied stress) of 0.1

d) A direction parallel to the principal fiber alignment

Test conditions that raise the temperature and moisture content of FRP materials generally degrade the ambient environment fatigue behavior.

Of all types of FRP composites for infrastructure applications, CFRP is the least prone to fatigue failure. An endurance limit of 60 to 70 percent of the initial static ultimate strength of CFRP is typical. On a plot of stress versus the logarithm of the number of cycles at failure (S-N curve), the downward slope for CFRP is usually approximately 5 percent of the initial static ultimate strength per decade of logarithmic life (Curtis 1989). At 1 million cycles, the fatigue strength is generally between 60 and 70 percent of the initial static ultimate strength and is relatively unaffected by the moisture and temperature exposures of concrete structures unless the resin or fiber/resin interface is substantially degraded by the environment.

In ambient-environment laboratory tests (Mandell and Meier 1983), individual glass fibers demonstrated delayed rupture caused by stress corrosion, which had been induced by the growth of surface flaws in the presence of even minute quantities of moisture. When many glass fibers were embedded into a matrix to form an FRP composite, a cyclic tensile fatigue effect of approximately 10 percent loss in the initial static strength per decade of logarithmic lifetime was observed (Mandell 1982). This fatigue effect is thought to be due to fiber-fiber interactions and is not dependent on the stress corrosion mechanism described for individual fibers. Usually, no clear fatigue limit can be defined. Environmental factors can play an important role in the fatigue behavior of glass fibers due to their susceptibility to moisture, alkaline, or acidic solutions.

Aramid fibers, for which substantial durability data are available, appear to behave reasonably well in fatigue. Neglecting in this context the rather poor durability of all aramid fibers in compression, the tension-tension fatigue behavior of an impregnated aramid fiber strand is excellent. Strength degradation per decade of logarithmic lifetime is approximately 5 to 6 percent (Roylance and Roylance 1981). While no distinct endurance limit is known for AFRP, 2-million-cycle endurance limits of commercial AFRP tendons for concrete applications have been reported in the range of 54 to 73 percent of the ultimate tensile strength (Odagiri et al. 1997). Because the slope of the applied stress versus logarithmic endurance time of AFRP is similar to the slope of the stress versus logarithmic cyclic lifetime data, the individual fibers appear to fail by a strain-limited creep rupture process. This lifetime-limiting mechanism in commercial AFRP bars is accelerated by exposure to moisture and elevated temperature (Roylance and Roylance 1981; Rostasy 1997).

4.5—Durability

Many FRP systems exhibit reduced mechanical properties after exposure to certain environmental factors, including high temperature, humidity, and chemical exposure. The exposure environment, duration of exposure, resin type and formulation, fiber type, and resin-curing method are some of the factors that influence the extent of the reduction in mechanical properties. These factors are discussed in more detail in 9.3. The tensile properties reported by the manufacturer are based on testing conducted in a laboratory environment, and do not reflect the effects of environmental exposure. These properties should be adjusted in accordance with the recommendations in 9.4 to account for the anticipated service environment to which the FRP system may be exposed during its service life.

4.6—FRP systems qualification

FRP systems should be qualified for use on a project based on independent laboratory test data of the FRP-constituent materials and the laminates made with them, structural test data for the type of application being considered, and durability data representative of the anticipated environment. Test data provided by the FRP system manufacturer demonstrating the proposed FRP system should meet all mechanical and physical design requirements, including tensile strength, durability, resistance to creep, bond to substrate, and T_g, should be considered. ACI 440.8 provides a specification for unidirectional carbon and glass FRP materials made using the wet layup process.

FRP composite systems that have not been fully tested should not be considered for use. Mechanical properties of FRP systems should be determined from tests on laminates manufactured in a process representative of their field installation. Mechanical properties should be tested in general conformance with the procedures listed in Appendix B. Modifications of standard testing procedures may be permitted to emulate field assemblies.

The specified material-qualification programs should require sufficient laboratory testing to measure the repeatability and reliability of critical properties. Testing of multiple batches of FRP materials is recommended. Independent structural testing can be used to evaluate a system's performance for the specific application.

CHAPTER 5—SHIPPING, STORAGE, AND HANDLING

5.1—Shipping

Fiber-reinforced polymer (FRP) system constituent materials should be packaged and shipped in a manner that conforms to all applicable federal and state packaging and shipping codes and regulations. Packaging, labeling, and shipping for thermosetting resin materials are controlled by CFR 49.

5.2—Storage

5.2.1 *Storage conditions*—To preserve the properties and maintain safety in the storage of FRP system constituent materials, the materials should be stored in accordance with the manufacturer's recommendations. Certain constituent materials, such as reactive curing agents, hardeners, initiators, catalysts, and cleaning solvents, have safety-related requirements and should be stored in a manner as recommended by the manufacturer and OSHA. Catalysts and initiators (usually peroxides) should be stored separately.

5.2.2 *Shelf life*—The properties of the uncured resin components can change with time, temperature, or humidity. Such conditions can affect the reactivity of the mixed system and the uncured and cured properties. The manufacturer sets a recommended shelf life within which the properties of the resin-based materials should continue to meet or exceed stated performance criteria. Any component material that has exceeded its shelf life, has deteriorated, or has been contaminated should not be used. FRP materials deemed unusable should be disposed of in a manner specified by the manufacturer and acceptable to state and federal environmental control regulations.

5.3—Handling

5.3.1 *Safety data sheet*—Safety data sheets (SDSs) for all FRP-constituent materials and components should be obtained from the manufacturers, and should be accessible at the job site.

5.3.2 *Information sources*—Detailed information on the handling and potential hazards of FRP-constituent materials can be found in company literature and guides, OSHA guidelines, and other government informational documents.

5.3.3 *General handling hazards*—Thermosetting resins describe a generic family of products that includes unsaturated polyesters, vinyl esters, epoxy, and polyurethane resins. The materials used with them are generally described as hardeners, curing agents, peroxide initiators, isocyanates, fillers, and flexibilizers. There are precautions that should be observed when handling thermosetting resins and their component materials. Some general hazards that may be encountered when handling thermosetting resins are listed as follows:

a) Skin irritation, such as burns, rashes, and itching

b) Skin sensitization, which is an allergic reaction similar to that caused by poison ivy, building insulation, or other allergens

c) Breathing organic vapors from cleaning solvents, monomers, and dilutents

d) With a sufficient concentration in air, explosion or fire of flammable materials when exposed to heat, flames, pilot lights, sparks, static electricity, cigarettes, or other sources of ignition

e) Exothermic reactions of mixtures of materials causing fires or personal injury

f) Nuisance dust caused by grinding or handling of the cured FRP materials (manufacturer's literature should be consulted for specific hazards)

The complexity of thermosetting resins and associated materials makes it essential that labels and the SDS are read and understood by those working with these products. CFR 16 Part 1500 regulates the labeling of hazardous substances and includes thermosetting-resin materials. ANSI Z400.1/Z129.1-2010 provides further guidance regarding classification and precautions.

5.3.4 *Personnel safe handling and clothing*—Disposable suits and gloves are suitable for handling fiber and resin materials. Disposable rubber or plastic gloves are recommended and should be discarded after each use. Gloves should be resistant to resins and solvents. Safety glasses or goggles should be used when handling resin components and solvents. Respiratory protection, such as dust masks or respirators, should be used when fiber fly, dust, or organic vapors are present, or during mixing and placing of resins if required by the FRP system manufacturer.

5.3.5 *Workplace safe handling*—The workplace should be well ventilated. Surfaces should be covered as needed to protect against contamination and resin spills. Each FRP system constituent material has different handling and storage requirements to prevent damage. The material manufacturer should be consulted for guidance. Some resin systems are potentially dangerous during mixing of the components. The manufacturer's literature should be consulted for proper mixing procedures, and the SDS for specific handling hazards. Ambient cure resin formulations produce heat when curing, which in turn accelerates the reaction. Uncontrolled reactions, including fuming, fire, or violent boiling, may occur in containers holding a mixed mass of resin; therefore, containers should be monitored.

5.3.6 *Cleanup and disposal*—Cleanup can involve use of flammable solvents, and appropriate precautions should be observed. Cleanup solvents are available that do not present flammability concerns. All waste materials should be contained and disposed of as prescribed by the prevailing environmental authority.

CHAPTER 6—INSTALLATION

Procedures for installing fiber-reinforced polymer (FRP) systems have been developed by the system manufacturers and often differ between systems. In addition, installation procedures can vary within a system, depending on the type and condition of the structure. This chapter presents general guidelines for the installation of FRP systems. Contractors trained in accordance with the installation procedures developed by the system manufacturer should install FRP systems. Deviations from the procedures developed by the FRP system manufacturer should not be allowed without consulting with the manufacturer.

6.1—Contractor competency

The FRP system installation contractor should demonstrate competency for surface preparation and application of the FRP system to be installed. Contractor competency can be demonstrated by providing evidence of training and documentation of related work previously completed by the contractor or by actual surface preparation and installation of the FRP system on portions of the structure. The FRP system manufacturer or its authorized agent should train the contractor's application personnel in the installation procedures of its system and ensure they are competent to install the system.

6.2—Temperature, humidity, and moisture considerations

Temperature, relative humidity, and surface moisture at the time of installation can affect the performance of the FRP system. Conditions to be observed before and during installation include surface temperature and moisture condition of the concrete, air temperature, relative humidity, and corresponding dew point.

Primers, saturating resins, and adhesives should generally not be applied to cold or frozen surfaces. When the surface temperature of the concrete surface falls below a minimum level as specified by the FRP system manufacturer, improper saturation of the fibers and improper curing of the resin constituent materials can occur, compromising the integrity of the FRP system. An auxiliary heat source can be used to raise the ambient and surface temperature during installation and maintain proper temperatures during curing. The heat source should be clean and not contaminate the surface or the uncured FRP system.

Resins and adhesives should generally not be applied to damp or wet surfaces unless they have been formulated for such applications. FRP systems should not be applied to concrete surfaces that are subject to moisture vapor transmission. The transmission of moisture vapor from a concrete surface through the uncured resin materials typically appears as surface bubbles and can compromise the bond between the FRP system and the substrate.

6.3—Equipment

Some FRP systems have unique, often system-specific, equipment designed specifically for their application. This equipment can include resin impregnators, sprayers, lifting/positioning devices, and winding machines. All equipment should be clean and in good operating condition. The contractor should have personnel trained in the operation of all equipment. Personal protective equipment, such as gloves, masks, eye guards, and coveralls, should be chosen and worn for each employee's function. All supplies and equipment should be available in sufficient quantities to allow continuity in the installation project and quality assurance.

6.4—Substrate repair and surface preparation

The behavior of concrete members strengthened or retrofitted with FRP systems is highly dependent on a sound concrete substrate and proper preparation and profiling of the concrete surface. An improperly prepared surface can result in debonding or delamination of the FRP system before achieving the design load transfer. The general guidelines presented in this chapter should be applicable to all externally bonded FRP systems. Specific guidelines for a particular FRP system should be obtained from the FRP system manufacturer.

6.4.1 *Substrate repair*—All problems associated with the condition of the original concrete and the concrete substrate that can compromise the integrity of the FRP system should be addressed before surface preparation begins. ACI 546R and ICRI 310.2R detail methods for the repair and surface preparation of concrete. All concrete repairs should meet the requirements of the design drawings and project specifications. The FRP system manufacturer should be consulted on the compatibility of the FRP system with materials used for repairing the substrate.

6.4.1.1 *Corrosion-related deterioration*—Externally bonded FRP systems should not be applied to concrete substrates suspected of containing actively corroding reinforcing steel. The expansive forces associated with the corrosion process are difficult to determine and could compromise the structural integrity of the externally applied FRP system. The cause(s) of the corrosion should be addressed, and the corrosion-related deterioration should be repaired before the application of any externally bonded FRP system.

6.4.1.2 *Injection of cracks*—Cracks that are 0.010 in. (0.3 mm) and wider can affect the performance of the externally bonded FRP systems. Consequently, cracks wider than 0.010 in. (0.3 mm) should be pressure-injected with epoxy before FRP installation in accordance with ACI 224.1R. Smaller cracks exposed to aggressive environments may require resin injection or sealing to prevent corrosion of existing steel reinforcement. Crack-width criteria for various exposure conditions are given in ACI 224.1R.

6.4.2 *Surface preparation*—Surface preparation requirements should be based on the intended application of the FRP system. Applications can be categorized as bond-critical or contact-critical. Bond-critical applications, such as flexural or shear strengthening of beams, slabs, columns, or walls, require an adhesive bond between the FRP system and the concrete. Contact-critical applications, such as confinement of columns, only require intimate contact between the FRP system and the concrete. Contact-critical applications do not require an adhesive bond between the FRP system and the concrete substrate, although one is typically provided to facilitate installation.

6.4.2.1 *Bond-critical applications*—Surface preparation for bond-critical applications should be in accordance with recommendations of ACI 546R and ICRI 310.2R. The concrete or repaired surfaces to which the FRP system is to be applied should be freshly exposed and free of loose or unsound materials. Where fibers wrap around corners, the corners should be rounded to a minimum 0.5 in. (13 mm) radius to reduce stress concentrations in the FRP system and voids between the FRP system and the concrete. Roughened corners should be smoothed with putty. Obstructions, inside corners, concave surfaces, and embedded objects can affect the performance of the FRP system and should be addressed. Obstructions and embedded objects may need to be removed before installing the FRP system. Inside corners and concave surfaces may require special detailing to ensure that the bond of the FRP system to the substrate is maintained. Surface preparation can be accomplished using abrasive or water-blasting techniques. All laitance, dust, dirt, oil, curing compound, existing coatings, and any other matter that could interfere with the bond of the FRP system to the concrete should be removed. Bug holes and other small surface voids should be completely exposed during surface profiling. After the profiling operations are complete, the surface should be cleaned and protected before FRP installation so that no materials that can interfere with bond are redeposited on the surface.

The concrete surface should be prepared to a surface profile not less than CSP 3, as defined by ICRI 310.2R or to the tolerances recommended by the FRP system manufacturer. Localized out-of-plane variations, including form lines, should not exceed 1/32 in. (1 mm) or the tolerances recommended by the FRP system manufacturer. Localized out-of-plane variations can be removed by grinding, before abrasive or water blasting, or can be smoothed over using resin-based putty if the variations are very small. Bug holes and voids should be filled with resin-based putty.

All surfaces to receive the strengthening system should be as dry as recommended by the FRP system manufacturer. Water in the pores can inhibit resin penetration and reduce mechanical interlock. Moisture content should be evaluated in accordance with the requirements of ACI 503.4.

6.4.2.2 *Contact-critical applications*—In applications involving confinement of structural concrete members, surface preparation should promote continuous intimate contact between the concrete surface and the FRP system. Surfaces to be wrapped should, at a minimum, be flat or convex to promote proper loading of the FRP system. Large voids in the surface should be patched with a repair material compatible with the existing concrete. Materials with low compressive strength and elastic modulus, such as plaster, can reduce the effectiveness of the FRP system and should be removed.

6.4.3 *Near-surface mounted (NSM) systems*—NSM systems are typically installed in grooves cut onto the concrete surface. The existing steel reinforcement should not be damaged while cutting the groove. The soundness of the concrete surface should be checked before installing the bar. The inside faces of the groove should be cleaned to ensure adequate bond with concrete. The resulting groove should be free of laitance or other compounds that may interfere with bond. The moisture content of the parent concrete should be controlled to suit the bonding properties of the adhesive. The grooves should be completely filled with the adhesive. The adhesive should be specified by the NSM system manufacturer.

6.5—Mixing of resins

Mixing of resins should be done in accordance with the FRP system manufacturer's recommended procedure. All resin components should be at the proper temperature and mixed in the correct ratio until there is a uniform and complete mixing of components. Resin components are often contrasting colors, so full mixing is achieved when color streaks are eliminated. Resins should be mixed for the prescribed mixing time and visually inspected for uniformity of color. The material manufacturer should supply recommended batch sizes, mixture ratios, mixing methods, and mixing times.

Mixing equipment can include small electrically powered mixing blades or specialty units, or resins can be mixed by hand stirring, if needed. Resin mixing should be in quantities sufficiently small to ensure that all mixed resin can be used within the resin's pot life. Mixed resin that exceeds its pot life should not be used because the viscosity will continue to increase and will adversely affect the resin's ability to penetrate the surface or saturate the fiber sheet.

6.6—Application of FRP systems

Fumes can accompany the application of some FRP resins. FRP systems should be selected with consideration for their impact on the environment, including emission of volatile organic compounds and toxicology.

6.6.1 *Primer and putty*—Where required, primer should be applied to all areas on the concrete surface where the FRP system is to be placed. The primer should be placed uniformly on the prepared surface at the manufacturer's specified rate of coverage. The applied primer should be protected from dust, moisture, and other contaminants before applying the FRP system.

Putty should be used in an appropriate thickness and sequence with the primer as recommended by the FRP manufacturer. The system-compatible putty, which is typically a thickened resin-based paste, should be used only to fill voids and smooth surface discontinuities before the application of other materials. Rough edges or trowel lines of cured putty should be ground smooth before continuing the installation.

Before applying the saturating resin or adhesive, the primer and putty should be allowed to cure as specified by the FRP system manufacturer. If the putty and primer are fully cured, additional surface preparation may be required before the application of the saturating resin or adhesive. Surface preparation requirements should be obtained from the FRP system manufacturer.

6.6.2 *Wet layup systems*—Wet layup FRP systems are typically installed by hand using dry fiber sheets and a saturating resin, typically per the manufacturer's recommendations. The saturating resin should be applied uniformly to all prepared surfaces where the system is to be placed. The fibers can also be impregnated in a separate process using a resin-impregnating machine before placement on the concrete surface.

The reinforcing fibers should be gently pressed into the uncured saturating resin in a manner recommended by the FRP system manufacturer. Entrapped air between layers should be released or rolled out before the resin sets. Sufficient saturating resin should be applied to achieve full saturation of the fibers.

Successive layers of saturating resin and fiber materials should be placed before the complete cure of the previous layer of resin. If previous layers are cured, interlayer surface preparation, such as light sanding or solvent application as recommended by the system manufacturer, may be required.

6.6.3 *Machine-applied systems*—Machine-applied systems can use resin-preimpregnated tows or dry-fiber tows. Prepreg tows are impregnated with saturating resin off site and delivered to the jobsite as spools of prepreg tow material. Dry fibers are impregnated at the jobsite during the winding process.

Wrapping machines are primarily used for the automated wrapping of concrete columns. The tows can be wound either horizontally or at a specified angle. The wrapping machine is placed around the column and automatically wraps the tow material around the perimeter of the column while moving up and down the column.

After wrapping, prepreg systems should be cured at an elevated temperature. Usually, a heat source is placed around the column for a predetermined temperature and time schedule in accordance with the manufacturer's recommendations. Temperatures are controlled to ensure consistent quality. The resulting FRP jackets do not have any seams or welds because the tows are continuous. In all the previous application steps, the FRP system manufacturer's recommendations should be followed.

6.6.4 *Precured systems*—Precured systems include shells, strips, and open grid forms that are typically installed with an adhesive. Adhesives should be uniformly applied to the prepared surfaces where precured systems are to be placed, except in certain instances of concrete confinement where adhesion of the FRP system to the concrete substrate may not be required.

Precured laminate surfaces to be bonded should be clean and prepared in accordance with the manufacturer's recommendation. The precured sheets or curved shells should be placed on or into the wet adhesive in a manner recommended by the FRP manufacturer. Entrapped air between layers should be released or rolled out before the adhesive sets. The adhesive should be applied at a rate recommended by the FRP manufacturer.

6.6.5 *Near-surface mounted (NSM) systems*—NSM systems consist of installing rectangular or circular FRP bars in grooves cut onto the concrete surface and bonded in place using an adhesive. Grooves should be dimensioned to ensure adequate adhesive around the bars. Typical groove dimensions for NSM FRP rods and plates are found in 14.3. NSM systems can be used on the topside of structural members and for overhead applications. Adhesive type and installation method should be specified by the NSM system manufacturer.

6.6.6 *Protective coatings*—Coatings should be compatible with the FRP strengthening system and applied in accordance with the manufacturer's recommendations. Typically, the use of solvents to clean the FRP surface before installing coatings is not recommended due to the deleterious effects that solvents can have on the polymer resins. The FRP system manufacturer should approve any use of solvent wipe preparation of FRP surfaces before the application of protective coatings. The coatings should be periodically inspected and maintenance should be provided to ensure the effectiveness of the coatings.

6.7—Alignment of FRP materials

The FRP ply orientation and ply stacking sequence should be specified. Small variations in angle, as little as 5 degrees, from the intended direction of fiber alignment can cause a substantial reduction in strength and modulus. Deviations in ply orientation should only be made if approved by the licensed design professional.

Sheet and fabric materials should be handled in a manner to maintain the fiber straightness and orientation. Fabric kinks, folds, or other forms of waviness should be reported to the licensed design professional.

6.8—Multiple plies and lap splices

Multiple plies can be used, provided that all plies are fully impregnated with the resin system, the resin shear strength is sufficient to transfer the shearing load between plies, and the bond strength between the concrete and FRP system is sufficient. For long spans, multiple lengths of fiber material or precured stock can be used to continuously transfer the load by providing adequate lap splices. Lap splices should be staggered unless noted otherwise by the licensed design professional. Lap splice details, including lap length, should be based on testing and installed in accordance with the manufacturer's recommendations. Due to the characteristics of some FRP systems, multiple plies and lap splices are not always possible. Specific guidelines on lap splices are given in Chapter 14.

6.9—Curing of resins

Curing of resins is a time-temperature-dependent phenomenon. Ambient-cure resins can take several days to reach full cure. Temperature extremes or fluctuations can retard or accelerate the resin curing time. The FRP system manufacturer may offer several prequalified grades of resin to accommodate these situations.

Elevated cure systems require the resin to be heated to a specific temperature for a specified time. Various combinations of time and temperature within a defined envelope should provide full cure of the system.

All resins should be cured according to the manufacturer's recommendation. Field modification of resin chemistry should not be permitted. Cure of installed plies should be monitored before placing subsequent plies. Installation of successive layers should be halted if there is a curing anomaly.

6.10—Temporary protection

Adverse temperatures; direct contact by rain, dust, or dirt; excessive sunlight; high humidity; or vandalism can damage an FRP system during installation and cause improper cure of the resins. Temporary protection, such as tents and plastic screens, may be required during installation and until the resins have cured. If temporary shoring is required, the FRP system should be fully cured before removing the shoring and allowing the structural member to carry the design loads. In the event of suspected damage to the FRP system during installation, the licensed design professional should be notified and the FRP system manufacturer consulted.

CHAPTER 7—INSPECTION, EVALUATION, AND ACCEPTANCE

Quality assurance and quality control (QA/QC) programs and criteria are to be maintained by the fiber-reinforced polymer (FRP) system manufacturers, the installation contractors, and others associated with the project. QA is typically an owner or a licensed professional activity whereas QC is a contractor or supplier activity. The QC program should be comprehensive and cover all aspects of the strengthening project, and should be detailed in the project specifications by a licensed professional. The degree of QC and the scope of testing, inspection, and record keeping depends on the size and complexity of the project.

Quality assurance is achieved through a set of inspections and applicable tests to document the acceptability of the installation. Project specifications should include a requirement to provide a QA plan for the installation and curing of all FRP materials. The plan should include personnel safety issues, application and inspection of the FRP system, location and placement of splices, curing provisions, means to ensure dry surfaces, QA samples, cleanup, and the suggested submittals listed in 15.3.

7.1—Inspection

FRP systems and all associated work should be inspected as required by the applicable codes. In the absence of such requirements, the inspection should be conducted by or under the supervision of a licensed design professional or a qualified inspector. Inspectors should be knowledgeable of FRP systems and be trained in the installation of FRP systems. The qualified inspector should require compliance with the design drawings and project specifications. During the installation of the FRP system, daily inspection should be conducted and should include:

a) Date and time of installation

b) Ambient temperature, relative humidity, and general weather observations

c) Surface temperature of concrete

d) Surface moisture

e) Surface preparation methods and resulting profile using the ICRI surface profile chips

f) Qualitative description of surface cleanliness

g) Type of auxiliary heat source, if applicable

h) Widths of cracks not injected with epoxy

i) Fiber or precured laminate batch number(s) and approximate location in structure

j) Batch numbers; mixture ratios; mixing time; and qualitative descriptions of the appearance of all mixed resins including primers, putties, saturants, adhesives, and coatings mixed for the day

k) Observations of progress of cure of resins

l) Conformance with installation procedures

m) Pull-off test results: bond strength, failure mode, and location

n) FRP properties from tests of field sample panels or witness panels, if required

o) Location and size of any delaminations or air voids

p) General progress of work

The inspector should provide the licensed design professional or owner with the inspection records and witness panels. Records and witness panels should be retained for a minimum of 10 years or a period specified by the licensed design professional. The installation contractor should retain sample cups of mixed resin and maintain a record of the placement of each batch.

7.2—Evaluation and acceptance

FRP systems should be evaluated and accepted or rejected based on conformance or nonconformance with the design drawings and specifications. FRP system material properties, installation within specified placement tolerances, presence of delaminations, cure of resins, and adhesion to substrate should be included in the evaluation. Placement tolerances, including fiber orientation, cured thickness, ply orientation, width and spacing, corner radii, and lap splice lengths, should be evaluated.

Witness panel and pull-off tests are used to evaluate the installed FRP system. In-place load testing can also be used to confirm the installed behavior of the FRP-strengthened member (Nanni and Gold 1998).

7.2.1 *Materials*—Before starting the project, the FRP system manufacturer should submit certification of specified material properties and identification of all materials to be used. Additional material testing can be conducted if deemed necessary based on the size and complexity of the project or other factors. Evaluation of delivered FRP materials can include tests for tensile strength, T_g, gel time, pot life, and adhesive shear strength. These tests are usually performed on material samples sent to a laboratory according to the QC test plan. Tests for pot life of resins and curing hardness are usually conducted on site. Materials that do not meet the minimum requirements as specified by the licensed design professional should be rejected.

Witness panels can be used to evaluate the tensile strength and modulus, lap splice strength, hardness, and T_g of the FRP system installed and cured on site using installation procedures similar to those used to install and cure the FRP system. During installation, flat panels of predetermined dimensions and thickness can be fabricated on site according to a predetermined sampling plan. After curing on site, the panels can then be sent to a laboratory for testing. Witness panels can be retained or submitted to an approved laboratory in a timely manner for testing of strength and T_g. Strength and elastic modulus of FRP materials can be determined in accordance with the requirements of ASTM D3039/D3039M, D7205/D7205M, or D7565/D7565M. The properties to be evaluated by testing should be specified. The licensed design professional may waive or alter the frequency of testing.

Some FRP systems, including precured and machine-wound systems, do not lend themselves to the fabrication of small, flat, witness panels. For these cases, the licensed design professional can modify the requirements to include test panels or samples provided by the manufacturer. During installation, sample cups of mixed resin should be prepared according to a predetermined sampling plan and retained for testing to determine the degree of cure (7.2.4).

7.2.2 *Fiber orientation*—Fiber or precured-laminate orientation should be evaluated by visual inspection. Fiber waviness—a localized appearance of fibers that deviate from the general straight-fiber line in the form of kinks or waves—should be evaluated for wet layup systems. Fiber or precured laminate misalignment of more than 5 degrees from that specified on the design drawings (approximately 1 in./ft [80 mm/m]) should be reported to the licensed design professional for evaluation and acceptance.

7.2.3 *Delaminations*—The cured FRP system should be evaluated for delaminations or air voids between multiple plies or between the FRP system and the concrete. Inspection methods should be capable of detecting delaminations of 2 in.^{2} (1300 mm^2) or greater. Methods such as acoustic sounding (hammer sounding), ultrasonics, and thermography can be used to detect delaminations.

The effect of delaminations or other anomalies on the structural integrity and durability of the FRP system should be evaluated. Delamination size, location, and quantity relative to the overall application area should be considered in the evaluation.

General acceptance guidelines for wet layup systems are:

a) Small delaminations less than 2 in.^{2} (1300 mm^2) each are permissible as long as the delaminated area is less than 5 percent of the total laminate area and there are no more than 10 such delaminations per 10 ft^2 (1 m^2).

b) Large delaminations greater than 25 in.^{2} (16,000 mm^2) can affect the performance of the installed FRP and should be repaired by selectively cutting away the affected sheet and applying an overlapping sheet patch of equivalent plies.

c) Delaminations less than 25 in.^{2} (16,000 mm^2) may be repaired by resin injection or ply replacement, depending on the size and number of delaminations and their locations.

For precured FRP systems, each delamination should be evaluated and repaired in accordance with the licensed design professional's direction. Upon completion of the repairs, the laminate should be reinspected to verify that the repair was properly accomplished.

7.2.4 *Cure of resins*—The relative cure of FRP systems can be evaluated by laboratory testing of witness panels or resin cup samples using ASTM D3418. The relative cure of the resin can also be evaluated on the project site by physical observation of resin tackiness and hardness of work surfaces or hardness of retained resin samples. The FRP system manufacturer should be consulted to determine the specific resin-cure verification requirements. For precured systems, adhesive hardness measurements should be made in accordance with the manufacturer's recommendation.

7.2.5 *Adhesion strength*—For bond-critical applications, tension adhesion testing of cored samples should be conducted in accordance with the requirements of ASTM D7522/D7522. Such tests cannot be performed when using near-surface-mounted (NSM) systems. The sampling frequency should be specified. Tension adhesion strengths should exceed 200 psi (1.4 MPa) and should exhibit failure of the concrete substrate. Lower strengths or failure between the FRP system and the concrete or between plies should be reported to the licensed design professional for evaluation and acceptance. For NSM strengthening, sample cores may be extracted to visually verify the consolidation of the resin adhesive around the FRP bar. The location of this core should be chosen such that the continuity of the FRP reinforcement is maintained (that is, at the ends of the NSM bars).

7.2.6 *Cured thickness*—Small core samples, typically 0.5 in. (13 mm) in diameter, may be taken to visually ascertain the cured laminate thickness or number of plies. Cored samples required for adhesion testing also can be used to ascertain the laminate thickness or number of plies. The sampling frequency should be specified. Taking samples from high-stress areas or splice areas should be avoided. For aesthetic reasons, the cored hole can be filled and smoothed with a repair mortar or the FRP system putty. If required, a 4 to 8 in. (100 to 200 mm) overlapping FRP sheet patch of equivalent plies may be applied over the filled and smoothed core hole immediately after taking the core sample. The FRP sheet patch should be installed in accordance with the manufacturer's installation procedures.

CHAPTER 8—MAINTENANCE AND REPAIR

8.1—General

As with any strengthening or retrofit repair, the owner should periodically inspect and assess the performance of the fiber-reinforced polymer (FRP) system used for strengthening or retrofit repair of concrete members.

8.2—Inspection and assessment

8.2.1 *General inspection*—A visual inspection looks for changes in color, debonding, peeling, blistering, cracking, crazing, deflections, indications of reinforcing bar corrosion, and other anomalies. In addition, ultrasonic, acoustic sounding (hammer tap), or thermographic tests may indicate signs of progressive delamination.

8.2.2 *Testing*—Testing can include pull-off tension tests (7.2.5) or conventional structural loading tests (ACI 437R).

8.2.3 *Assessment*—Test data and observations are used to assess any damage and the structural integrity of the strengthening system. The assessment can include a recommendation for repairing any deficiencies and preventing recurrence of degradation.

8.3—Repair of strengthening system

The method of repair for the strengthening system depends on the causes of the damage, the type of material, the form of degradation, and the level of damage. Repairs to the fiber-reinforced polymer (FRP) system should not be undertaken without first identifying and addressing the causes of the damage.

Minor damage should be repaired, including localized FRP laminate cracking or abrasions that affect the structural integrity of the laminate. Minor damage can be repaired by bonding FRP patches over the damaged area. The FRP patches should possess the same characteristics, such as thickness or ply orientation, as the original laminate. The FRP patches should be installed in accordance with the material manufacturer's recommendation. Minor delaminations can be repaired by resin injection. Major damage, including peeling and debonding of large areas, may require removal of the affected area, reconditioning of the cover concrete, and replacement of the FRP laminate.

8.4—Repair of surface coating

In the event that the surface-protective coating should be replaced, the FRP laminate should be inspected for structural damage or deterioration. The surface coating may be replaced using a process approved by the system manufacturer.

CHAPTER 9—GENERAL DESIGN CONSIDERATIONS

General design recommendations are presented in this chapter. The recommendations presented are based on the traditional reinforced concrete design principles stated in the requirements of ACI 318 and knowledge of the specific mechanical behavior of fiber-reinforced polymer (FRP) reinforcement.

FRP strengthening systems should be designed to resist tensile forces while maintaining strain compatibility between the FRP and the concrete substrate. FRP reinforcement should not be relied on to resist compressive forces. It is acceptable, however, for FRP tension reinforcement to experience compression due to moment reversals or changes in load pattern. The compressive strength of the FRP reinforcement, however, should be neglected.

9.1—Design philosophy

These design recommendations are based on limit-states-design principles. This approach sets acceptable levels of safety for the occurrence of both serviceability limit states (excessive deflections and cracking) and ultimate limit states (failure, stress rupture, and fatigue). In assessing the nominal strength of a member, the possible failure modes and subsequent strains and stresses in each material should be assessed. For evaluating the serviceability of a member, engineering principles, such as transformed section calculations using modular ratios, can be used.

FRP strengthening systems should be designed in accordance with ACI 318 strength and serviceability requirements using the strength and load factors stated in ACI 318. Additional reduction factors applied to the contribution of the FRP reinforcement are recommended by this guide to reflect uncertainties inherent in FRP systems different from steel-reinforced and prestressed concrete. These reduction factors were determined based on statistical evaluation of variability in mechanical properties, predicted versus full-scale test results, and field applications. FRP-related reduction factors were calibrated to produce reliability indexes typically above 3.5. Reliability indexes between 3.0 and 3.5 can be encountered in cases where relatively low ratios of steel reinforcement combined with high ratios of FRP reinforcement are used. Such cases are less likely to be encountered in design because they violate the recommended strengthening limits of 9.2. Reliability indexes for FRP-strengthened members are determined based on the approach used for reinforced concrete buildings (Nowak and Szerszen 2003; Szerszen and Nowak 2003). In general, lower reliability is expected in retrofitted and repaired structures than in new structures.

9.2—Strengthening limits

Careful consideration should be given to determine reasonable strengthening limits. These limits are imposed to guard against collapse of the structure should bond or other failure of the FRP system occur due to damage, vandalism, or other causes. The unstrengthened structural member, without FRP reinforcement, should have sufficient strength to resist a certain level of load. The existing strength of the structure should be sufficient to resist a level of load as described by Eq. (9.2)

$$(\phi R_n)_{existing} \geq (1.1S_{DL} + 0.75S_{LL})_{new} \quad (9.2)$$

A dead load factor of 1.1 is used because a relatively accurate assessment of the dead loads of the structure can be determined. A live load factor of 0.75 is used to exceed the statistical mean of the yearly maximum live load factor of 0.5, as given in ASCE 7. The strengthening limit resulting from compliance with Eq. (9.2) will allow the strengthened member to maintain sufficient structural capacity until the damaged FRP is repaired.

In cases where the design live load acting on the member to be strengthened has a high likelihood of being present for a sustained period of time, a live load factor of 1.0 should be used instead of 0.75 in Eq. (9.2). Examples include library stack areas, heavy storage areas, warehouses, and other occupancies with a live load exceeding 150 lb/ft^2 (730 kg/m^2). More specific limits for structures requiring a fire resistance rating are given in 9.2.1.

9.2.1 *Structural fire resistance*—The level of strengthening that can be achieved through the use of externally bonded FRP reinforcement can be limited by the code-required fire-resistance rating of a structure. The polymer resins typically used in wet layup and prepreg FRP systems and the polymer adhesives used in precured FRP systems suffer deterioration of mechanical and bond properties at temperatures close to or exceeding the T_g of the polymer, as described in 1.2.1.3.

Although the FRP system itself is significantly affected by exposure to elevated temperature, a combination of the FRP system with an existing concrete structure may still have an adequate fire resistance. When considering the fire resistance of an FRP-strengthened concrete element, it is important to recognize that the strength of a reinforced concrete element is reduced during fire exposure due to heating of both the reinforcing steel and the concrete. Performance in fire of the existing concrete member can be enhanced by installing an insulation system, which will provide thermal protection to existing concrete and internal reinforcing steel, thus improving the overall fire rating, although the FRP system contribution may be reduced (Bisby et al. 2005a; Williams et al. 2006; Palmieri et al. 2011; Firmo et al. 2012).

By extending the methods in ACI 216.1 to FRP-strengthened reinforced concrete, limits on strengthening can be used to ensure a strengthened structure will not collapse in a fire event. A member's resistance to load effects, with reduced steel and concrete strengths and without the contribution of the FRP reinforcement, can be compared with the load demand on the member during the fire event to ensure the strengthened member can support these loads for the required fire duration (or fire rating time) without failure

$$R_{n\theta} \geq 1.0S_{DL} + 1.0S_{LL} \quad (9.2.1a)$$

Alternately, ACI 562 specifies the following

$$R_{n\theta} \geq 1.2S_{DL} + 0.5S_{LL} + 0.2S_{SL} + 1.0A_k \quad (9.2.1b)$$

where $R_{n\theta}$ is the nominal resistance of the member at an elevated temperature, and S_{DL}, S_{LL}, and S_{SL} are the specified dead, live, and snow loads, respectively, calculated for the strengthened structure. For cases where the design live load has a high likelihood of being present for a sustained period of time, a live load factor of 1.0 should be used in place of 0.5 in Eq. (9.2.1b). Due to the lack of guidance for the calculation of A_k, the load or load effect resulting from the fire event, use of Eq. (9.2.1a) is recommended.

If the FRP system is meant to allow greater load-carrying capacity, such as an increase in live load, the load effects should be computed using these greater loads. If the FRP system is meant to address a loss in strength, such as deterioration, the resistance should reflect this loss.

The nominal resistance of the member at an elevated temperature $R_{n\theta}$ may be determined using the procedure outlined in ACI 216.1 or through testing. The nominal resistance $R_{n\theta}$ should be calculated based on the reduced material properties of the existing member. The resistance should be computed for the time required by the member's fire-resistance rating—for example, a 2-hour fire rating—and should not account for the contribution of the FRP system unless the continued effectiveness of the FRP can be proven through testing. More research is needed to accurately identify temperatures at which effectiveness is lost for different types of FRP. Until better information on the properties of FRP at high temperature is available, the critical temperature can be taken as the lowest T_g of the components of the system comprising the load path.

9.2.2 *Overall structural strength*—While FRP systems are effective in strengthening members for flexure and shear and providing additional confinement, other modes of failure, such as punching shear and bearing capacity of footings, may be only marginally affected by FRP systems (Sharaf et al. 2006). All members of a structure should be capable of withstanding the anticipated increase in loads associated with the strengthened members.

Additionally, analysis should be performed on the member strengthened by the FRP system to check that, under overload conditions, the strengthened member will fail in a flexural mode rather than in a shear mode.

9.2.3 *Seismic applications*—Requirements for seismic strengthening using FRP are addressed in Chapter 13.

9.3—Selection of FRP systems

9.3.1 *Environmental considerations*—Environmental conditions uniquely affect resins and fibers of various FRP systems. The mechanical properties (for example, tensile strength, ultimate tensile strain, and elastic modulus) of some FRP systems degrade under exposure to certain environments such as alkalinity, salt water, chemicals, ultraviolet light, high temperatures, high humidity, and freezing-and-thawing cycles. The material properties used in design should account for this degradation in accordance with 9.4.

The licensed design professional should select an FRP system based on the known behavior of that system in the anticipated service conditions. Some important environmental considerations that relate to the nature of specific systems are given as follows. Specific information can be obtained from the FRP system manufacturer.

a) *Alkalinity/acidity*—The performance of an FRP system over time in an alkaline or acidic environment depends on the matrix material and the reinforcing fiber. Dry, unsaturated bare, or unprotected carbon fiber is resistant to both alkaline and acidic environments whereas bare glass fiber can degrade over time in these environments. A properly selected and applied resin matrix, however, should isolate and protect the fiber from the alkaline/acidic environment and resist deterioration. Sites with high alkalinity and high moisture or relative humidity favor the selection of carbon-fiber systems over glass-fiber systems.

b) *Thermal expansion*—FRP systems may have thermal expansion properties that are different from those of concrete. In addition, the thermal expansion properties of the fiber and polymer constituents of an FRP system can vary. Carbon fibers have a coefficient of thermal expansion near zero whereas glass fibers have a coefficient of thermal expansion similar to concrete. The polymers used in FRP strengthening systems typically have coefficients of thermal expansion roughly five times that of concrete. Calculation of thermally-induced strain differentials are complicated by variations in fiber orientation, fiber volume fraction, and thickness of adhesive layers. Experience indicates, however, that thermal expansion differences do not affect bond for small ranges of temperature change, such as ±50°F (±28°C) (Motavalli et al. 1997; Soudki and Green 1997; Green et al. 1998).

c) *Electrical conductivity*—Glass FRP (GFRP) and aramid FRP (AFRP) are effective electrical insulators, whereas carbon FRP (CFRP) is conductive. To avoid potential galvanic corrosion of steel elements, carbon-based FRP materials should not come in direct contact with steel.

9.3.2 *Loading considerations*—Loading conditions uniquely affect different fibers of FRP systems. The licensed design professional should select an FRP system based on the known behavior of that system in the anticipated service conditions. Some important loading considerations that relate to the nature of the specific systems are given in the following. Specific information should be obtained from material manufacturers.

a) *Impact tolerance*—AFRP and GFRP systems demonstrate better tolerance to impact than CFRP systems.

b) *Creep rupture and fatigue*—CFRP systems are highly resistive to creep rupture under sustained loading and fatigue failure under cyclic loading. GFRP systems are more sensitive to both loading conditions.

9.3.3 *Durability considerations*—Durability of FRP systems is the subject of considerable ongoing research (Dolan et al. 2008; Karbhari 2007). The licensed design professional should select an FRP system that has undergone durability testing consistent with the application environment. Durability testing may include hot-wet cycling, alkaline immersion, freezing-and-thawing cycling, ultraviolet exposure, dry heat, and salt water (Cromwell et al. 2011).

Any FRP system that completely encases or covers a concrete section should be investigated for the effects of a variety of environmental conditions including those of freezing and thawing, steel corrosion, alkali and silica aggregate reactions, water entrapment, vapor pressures, and moisture vapor transmission (Masoud and Soudki 2006; Soudki and Green 1997; Porter et al. 1997; Christensen et al. 1996; Toutanji 1999). Many FRP systems create a moisture-impermeable layer on the surface of the concrete. In areas where moisture vapor transmission is expected, adequate means should be provided to allow moisture to escape from the concrete structure.

9.3.4 *Protective-coating selection considerations*—A coating or insulation system can be applied to the installed FRP system to protect it from exposure to certain environmental conditions (Bisby et al. 2005a; Williams et al. 2006). The thickness and type of coating should be selected based on the requirements of the composite repair; resistance to environmental effects such as moisture, salt water, temperature extremes, fire, impact, and ultraviolet exposure; resistance to site-specific effects; and resistance to vandalism. Coatings are relied on to retard the degradation of the mechanical properties of the FRP systems. The coatings should be periodically inspected and maintained to ensure continued effectiveness.

External coatings or thickened coats of resin over fibers can protect them from damage due to impact or abrasion. In high-impact or traffic areas, additional levels of protection may be necessary. Portland cement plaster and polymer coatings are commonly used for protection where minor impact or abrasion is anticipated.

9.4—Design material properties

Unless otherwise stated, the material properties reported by manufacturers, such as the ultimate tensile strength, typically do not consider long-term exposure to environmental conditions and should be considered as initial properties. Because long-term exposure to various types of environments can reduce the tensile properties and creep-rupture and fatigue endurance of FRP laminates, the material properties used in design equations should be reduced based on the environmental exposure condition.

Equations (9.4a) through (9.4c) give the tensile properties that should be used in all design equations. The design ultimate tensile strength should be determined using the environmental reduction factor given in Table 9.4 for the appropriate fiber type and exposure condition

$$f_{fu} = C_E f_{fu}^* \quad (9.4a)$$

Similarly, the design rupture strain should also be reduced for environmental exposure conditions

$$\varepsilon_{fu} = C_E \varepsilon_{fu}^* \quad (9.4b)$$

Because FRP materials are linear elastic until failure, the design modulus of elasticity for unidirectional FRP can be determined from Hooke's law. The expression for the modulus of elasticity, given in Eq. (9.4c), recognizes that the modulus is typically unaffected by environmental conditions. The modulus given in this equation will be the same as the initial value reported by the manufacturer

$$E_f = f_{fu}/\varepsilon_{fu} \quad (9.4c)$$

The constituent materials, fibers, and resins of an FRP system affect its durability and resistance to environmental exposure. The environmental reduction factors given in Table 9.4 are conservative estimates based on the relative durability of each fiber type.

As Table 9.4 illustrates, if the FRP system is located in a relatively benign environment, such as indoors, the reduction factor is closer to unity. If the FRP system is located in an aggressive environment where prolonged exposure to high humidity, freezing-and-thawing cycles, salt water, or alkalinity is expected, a lower reduction factor should be used. The reduction factor can be modified to reflect the use of a protective coating if the coating has been shown through testing to lessen the effects of environmental exposure and the coating is maintained for the life of the FRP system.

Table 9.4—Environmental reduction factor for various FRP systems and exposure conditions

Exposure conditions	Fiber type	Environmental reduction factor C_E
Interior exposure	Carbon	0.95
	Glass	0.75
	Aramid	0.85
Exterior exposure (bridges, piers, and unenclosed parking garages)	Carbon	0.85
	Glass	0.65
	Aramid	0.75
Aggressive environment (chemical plants and wastewater treatment plants)	Carbon	0.85
	Glass	0.50
	Aramid	0.70

CHAPTER 10—FLEXURAL STRENGTHENING

Bonding fiber-reinforced polymer (FRP) reinforcement to the tension face of a concrete flexural member with fibers oriented along the length of the member will provide an increase in flexural strength. Increases in overall flexural strength from 10 to 160 percent have been documented (Meier and Kaiser 1991; Ritchie et al. 1991; Sharif et al. 1994). When taking into account the strengthening limits of 9.2 and ductility and serviceability limits, however, strength increases of up to 40 percent are more reasonable.

This chapter does not apply to FRP systems used to enhance the flexural strength of members in the expected plastic hinge regions of ductile moment frames resisting seismic loads; these are addressed in Chapter 13.

10.1—Nominal strength

The strength design approach requires that the design flexural strength of a member exceed its required factored moment, as indicated by Eq. (10.1). The design flexural strength ϕM_n refers to the nominal strength of the member multiplied by a strength reduction factor, and the factored moment M_u refers to the moment calculated from factored loads (for example, $\alpha_{DL}M_{DL} + \alpha_{LL}M_{LL} + ...$)

$$\phi M_n \geq M_u \quad (10.1)$$

This guide recommends that the factored moment M_u of a section be calculated by use of load factors as required by ACI 318. An additional strength reduction factor for FRP, ψ_f, should be applied to the flexural contribution of the FRP reinforcement alone, M_{nf}, as described in 10.2.10. The additional strength reduction factor, ψ_f, is used to improve the reliability of strength prediction and accounts for the different failure modes observed for FRP-strengthened members (delamination of FRP reinforcement).

The nominal flexural strength of FRP-strengthened concrete members with mild steel reinforcement and with bonded prestressing steel can be determined based on strain compatibility, internal force equilibrium, and the controlling mode of failure. For members with unbonded prestressed steel, strain compatibility does not apply and the stress in the unbonded tendons at failure depends on the overall deformation of the member and is assumed to be approximately the same at all sections.

10.1.1 *Failure modes*—The flexural strength of a section depends on the controlling failure mode. The following flexural failure modes should be investigated for an FRP-strengthened section (GangaRao and Vijay 1998):

a) Crushing of the concrete in compression before yielding of the reinforcing steel

b) Yielding of the steel in tension followed by rupture of the FRP laminate

c) Yielding of the steel in tension followed by concrete crushing

d) Shear/tension delamination of the concrete cover (cover delamination)

e) Debonding of the FRP from the concrete substrate (FRP debonding)

Concrete crushing is assumed to occur if the compressive strain in the concrete reaches its maximum usable strain ($\varepsilon_c = \varepsilon_{cu} = 0.003$). Rupture of the externally bonded FRP is assumed to occur if the strain in the FRP reaches its design rupture strain ($\varepsilon_f = \varepsilon_{fu}$) before the concrete reaches its maximum usable strain.

Cover delamination or FRP debonding can occur if the force in the FRP cannot be sustained by the substrate (Fig. 10.1.1a). Such behavior is generally referred to as debonding, regardless of where the failure plane propagates within the FRP-adhesive-substrate region. Guidance to avoid the cover delamination failure mode is given in Chapter 13.

Away from the section where externally bonded FRP terminates, a failure controlled by FRP debonding may govern (Fig. 10.1.1a(b)). To prevent such an intermediate crack-induced debonding failure mode, the effective strain in FRP reinforcement should be limited to the strain at which debonding may occur, ε_{fd}, as defined in Eq. (10.1.1)

$$\varepsilon_{fd} = 0.083\sqrt{\frac{f'_c}{nE_f t_f}} \leq 0.9\varepsilon_{fu} \quad \text{(in.-lb)}$$

$$\varepsilon_{fd} = 0.41\sqrt{\frac{f'_c}{nE_f t_f}} \leq 0.9\varepsilon_{fu} \quad \text{(SI)} \qquad (10.1.1)$$

Equation (10.1.1) takes a modified form of the debonding strain equation proposed by Teng et al. (2003, 2004) that was based on committee evaluation of a significant database for flexural beam tests exhibiting FRP debonding failure. The proposed equation was calibrated using average measured values of FRP strains at debonding for flexural tests experiencing intermediate crack-induced debonding to determine the best-fit coefficient of 0.083 (0.41 in SI). Reliability of the FRP contribution to flexural strength is addressed by incorporating an additional strength reduction factor for FRP, ψ_f, in addition to the strength reduction factor ϕ per ACI 318 for structural concrete. Anchorage systems such as U-wraps, mechanical fasteners, fiber anchors, and U-anchors (examples are shown schematically in Fig. 10.1.1b) have been proven successful at delaying, and sometimes preventing, debonding failure of the longitudinal FRP (Kalfat et al. 2013; Grelle and Sneed 2013). Experimental studies have shown that these anchorage systems can increase the effective strain in the flexural FRP to values up to tensile rupture (Lee et al. 2010; Orton et al. 2008).

For near-surface-mounted (NSM) FRP applications, the value of ε_{fd} may vary from $0.6\varepsilon_{fu}$ to $0.9\varepsilon_{fu}$, depending on many factors such as member dimensions, steel and FRP reinforcement ratios, and surface roughness of the FRP bar. Based on analysis of a database of existing studies (Bianco et al. 2014), the committee recommends the use of $\varepsilon_{fd} = 0.7\varepsilon_{fu}$. To achieve the debonding design strain of NSM FRP bars, ε_{fd}, the bonded length should be greater than the development length given in Chapter 13.

10.2—Reinforced concrete members

This section presents guidance on the calculation of the flexural strengthening effect of adding longitudinal FRP

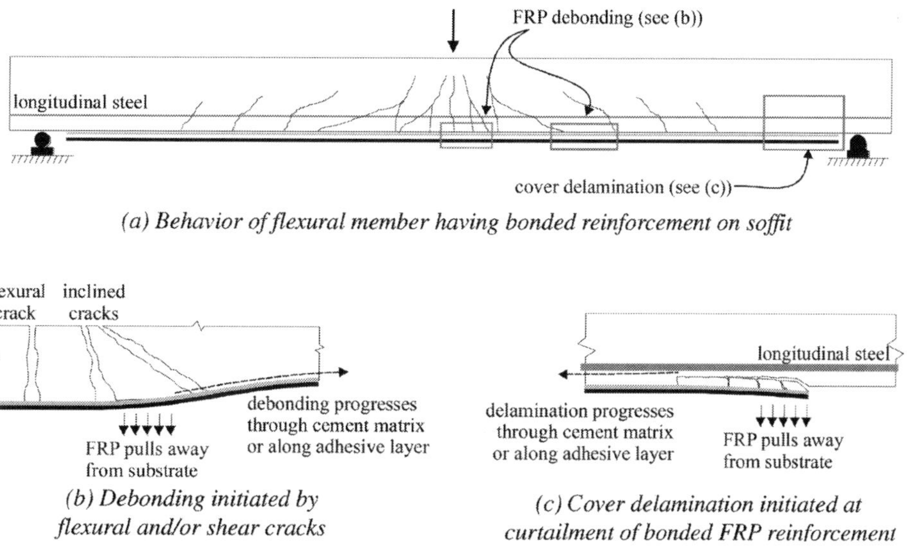

Fig. 10.1.1a—Debonding and delamination of externally bonded FRP systems.

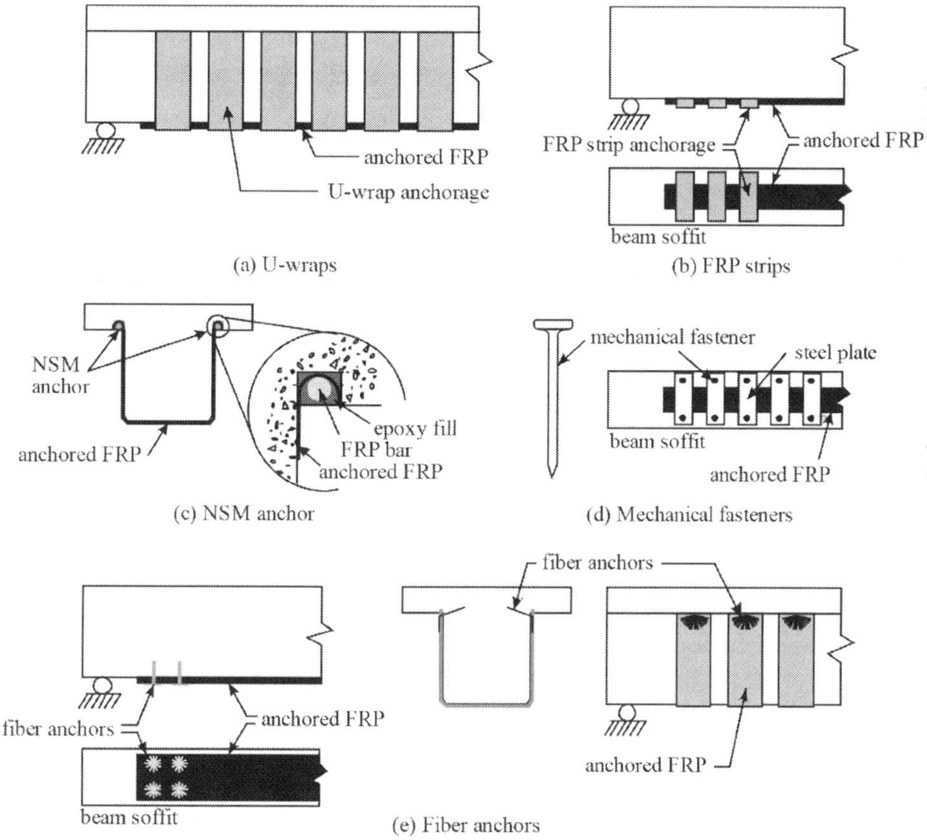

Fig. 10.1.1b—FRP anchorage systems.

reinforcement to the tension face of a reinforced concrete member. A specific illustration of the concepts in this section applied to strengthening of existing rectangular sections reinforced in the tension zone with nonprestressed steel is given. The general concepts outlined herein can, however, be extended to nonrectangular shapes (T-sections and I-sections) and to members with steel compression reinforcement.

10.2.1 *Assumptions*—The following assumptions are made in calculating the flexural resistance of a section strengthened with an externally applied FRP system:

a) Design calculations are based on the dimensions, internal reinforcing steel arrangement, and material properties of the existing member being strengthened.

b) The strains in the steel reinforcement and concrete are directly proportional to their distance from the neutral axis.

That is, a plane section before loading remains plane after loading.

c) There is no relative slip between external FRP reinforcement and the concrete.

d) The shear deformation within the adhesive layer is neglected because the adhesive layer is very thin with only slight variations in its thickness.

e) The maximum usable compressive strain in the concrete is 0.003.

f) The tensile strength of concrete is neglected.

g) The FRP reinforcement has a linear elastic stress-strain relationship to failure.

While some of these assumptions are necessary for the sake of computational ease, the assumptions do not accurately reflect the true fundamental behavior of FRP flexural reinforcement. For example, there will be shear deformation in the adhesive layer, causing relative slip between the FRP and the substrate. The inaccuracy of the assumptions will not, however, significantly affect the computed flexural strength of an FRP-strengthened member. An additional strength reduction factor (presented in 10.2.10) will conservatively compensate for any such discrepancies.

10.2.2 *Shear strength*—When FRP reinforcement is being used to increase the flexural strength of a member, the member should be capable of resisting the shear forces associated with the increased flexural strength. The potential for shear failure of the section should be considered by comparing the design shear strength of the section to the required shear strength. If additional shear strength is required, FRP laminates oriented transverse to the beam longitudinal axis can be used to resist shear forces, as described in Chapter 11.

10.2.3 *Existing substrate strain*—Unless all loads on a member, including self-weight and any prestressing forces, are removed before installation of FRP reinforcement, the substrate to which the FRP is applied will be strained. These strains should be considered initial strains and should be excluded from the strain in the FRP (Arduini and Nanni 1997; Nanni and Gold 1998). The initial strain on the bonded substrate, ε_{bi}, can be determined from an elastic analysis of the existing member, considering all loads that will be on the member during the installation of the FRP system. The elastic analysis of the existing member should be based on cracked section properties.

10.2.4 *Flexural strengthening of concave soffits*—The presence of curvature in the soffit of a concrete member may lead to the development of tensile stresses normal to the adhesive and surface to which the FRP is bonded. Such tensile stresses result when the FRP tends to straighten under load, and can promote the initiation of FRP debonding or interlaminar failures that reduce the effectiveness of the FRP flexural strengthening (Aiello et al. 2001; Eshwar et al. 2003). If the extent of the curved portion of the soffit exceeds a length of 40 in. (1.0 m) with a rise of 0.2 in. (5 mm), the surface should be made flat before strengthening. Alternately, anchorage systems such as U-wraps, mechanical fasteners, fiber anchors, or NSM anchors should be installed to mitigate delamination (Eshwar et al. 2005).

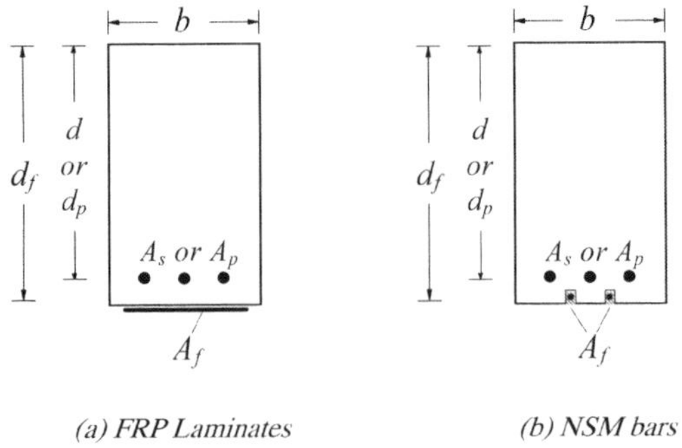

Fig. 10.2.5—Effective depth of FRP systems.

10.2.5 *Strain in FRP reinforcement*—It is important to determine the strain in the FRP reinforcement at the ultimate limit state. Because FRP materials are linear elastic until failure, the strain in the FRP will dictate the stress developed in the FRP. The maximum strain that can be achieved in the FRP reinforcement will be governed by either the strain developed in the FRP at the point at which concrete crushes, the point at which the FRP ruptures, or the point at which the FRP debonds from the substrate. The effective strain in the FRP reinforcement at the ultimate limit state can be found from Eq. (10.2.5)

$$\varepsilon_{fe} = \varepsilon_{cu}\left(\frac{d_f - c}{c}\right) - \varepsilon_{bi} \le \varepsilon_{fd} \qquad (10.2.5)$$

where ε_{bi} is the initial substrate strain as described in 10.2.3, and d_f is the effective depth of FRP reinforcement, as indicated in Fig. 10.2.5.

10.2.6 *Stress in the FRP reinforcement*—The effective stress in the FRP reinforcement is the maximum level of stress that can be developed in the FRP reinforcement before flexural failure of the section. This effective stress can be found from the strain in the FRP, assuming perfectly elastic behavior

$$f_{fe} = E_f \varepsilon_{fe} \qquad (10.2.6)$$

10.2.7 *Strength reduction factor*—The use of externally bonded FRP reinforcement for flexural strengthening will reduce the ductility of the original member. In some cases, the loss of ductility is negligible. Sections that experience a significant loss in ductility, however, should be addressed. To maintain a sufficient degree of ductility, the strain in the steel at the ultimate limit state should be checked. For reinforced concrete members with nonprestressed steel reinforcement, adequate ductility is achieved if the strain in the steel at the point of concrete crushing or failure of the FRP, including delamination or debonding, is at least 0.005, according to the definition of a tension-controlled section as given in ACI 318.

The approach taken by this guide follows the philosophy of ACI 318. A strength reduction factor given by Eq. (10.2.7) should be used, where ε_t is the net tensile strain in extreme tension steel at nominal strength, as defined in ACI 318.

$$\phi = \begin{cases} 0.90 \text{ for } \varepsilon_t \geq 0.005 \\ 0.65 + \dfrac{0.25(\varepsilon_t - \varepsilon_{sy})}{0.005 - \varepsilon_{sy}} \text{ for } \varepsilon_{sy} < \varepsilon_t < 0.005 \\ 0.65 \text{ for } \varepsilon_t \leq \varepsilon_{sy} \end{cases} \quad (10.2.7)$$

This equation sets the reduction factor at 0.90 for ductile sections and 0.65 for brittle sections where the steel does not yield, and provides a linear transition for the reduction factor between these two extremes. The use of Eq. (10.2.7) is limited to steel having a yield strength f_y less than 80 ksi (550 MPa) (ACI 318).

10.2.8 *Serviceability*—The serviceability of a member (deflections and crack widths) under service loads should satisfy applicable provisions of ACI 318. The effect of the FRP external reinforcement on the serviceability can be assessed using the transformed-section analysis.

To avoid inelastic deformations of reinforced concrete members with nonprestressed steel reinforcement strengthened with external FRP reinforcement, the existing internal steel reinforcement should be prevented from yielding under service load levels, especially for members subjected to cyclic loads (El-Tawil et al. 2001). The stress in the steel reinforcement under service load should be limited to 80 percent of the yield strength, as shown in Eq. (10.2.8a). In addition, the compressive stress in concrete under service load should be limited to 60 percent of the compressive strength, as shown in Eq. (10.2.8b)

$$f_{s,s} \leq 0.80 f_y \quad (10.2.8a)$$

$$f_{c,s} \leq 0.60 f_c' \quad (10.2.8b)$$

10.2.9 *Creep rupture and fatigue stress limits*—To avoid creep rupture of the FRP reinforcement under sustained stresses or failure due to cyclic stresses and fatigue of the FRP reinforcement, the stress in the FRP reinforcement under these stress conditions should be checked. Because this stress will be within the elastic response range of the member, the stresses can be computed by elastic analysis using cracked section properties as appropriate.

In 4.4, the creep rupture phenomenon and fatigue characteristics of FRP material were described and the resistance to its effects by various types of fibers was examined. As stated in 4.4.1, research has indicated that glass, aramid, and carbon fibers can sustain approximately 0.3, 0.5, and 0.9 times their ultimate strengths, respectively, before encountering a creep rupture problem (Yamaguchi et al. 1997; Malvar 1998). To avoid failure of an FRP-reinforced member due to creep rupture and fatigue of the FRP, stress limits for these conditions should be imposed on the FRP reinforcement. The stress in the FRP reinforcement can be computed using elastic analysis and an applied moment due to all sustained loads (dead loads and the sustained portion of the live load) plus the maximum moment induced in a fatigue loading cycle (Fig. 10.2.9). The sustained stress should be limited as expressed by Eq. (10.2.9) to maintain safety. Values for safe sustained plus cyclic stress are given in Table 10.2.9. These values are based approximately on the stress limits previously stated in 4.4.1 with an imposed safety factor of 1/0.6

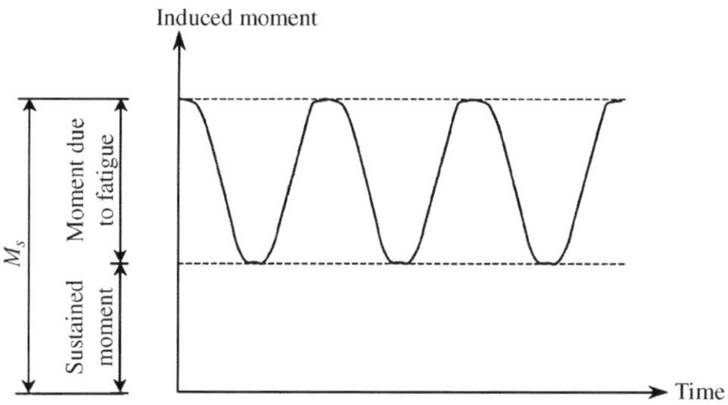

Fig. 10.2.9—Illustration of the level of applied moment to be used to check the stress limits in the FRP reinforcement.

$$f_{f,s} \leq \text{sustained plus cyclic stress limit} \quad (10.2.9)$$

Table 10.2.9—Sustained plus cyclic service load stress limits in FRP reinforcement

Stress type	Fiber type		
	GFRP	AFRP	CFRP
Sustained plus cyclic stress limit	$0.20f_{fu}$	$0.30f_{fu}$	$0.55f_{fu}$

10.2.10 *Ultimate strength of singly reinforced rectangular section*—To illustrate the concepts presented in this chapter, this section describes the application of these concepts to a nonprestressed singly-reinforced rectangular section. Figure 10.2.10 illustrates the internal strain and stress distribution for a rectangular section under flexure at the ultimate limit state.

The calculation procedure used to arrive at the ultimate strength should satisfy strain compatibility and force equilibrium, and should consider the governing mode of failure. Several calculation procedures can be derived to satisfy these conditions. The calculation procedure described herein illustrates an iterative method that involves selecting an assumed depth to the neutral axis, c, calculating the strain in each material using strain compatibility; calculating the associated stress in each material; and checking internal force equilibrium. If the internal force resultants do not equilibrate, the depth to the neutral axis should be revised and the procedure repeated.

For any assumed depth to the neutral axis, c, the strain in the FRP reinforcement can be computed from Eq. (10.2.5). This equation considers the governing mode of failure for the assumed neutral axis depth. If the left term of the inequality controls, concrete crushing controls flexural failure of the section. If the right term of the inequality controls, FRP failure (rupture or debonding) controls flexural failure of the section.

The effective stress in the FRP reinforcement can be found from the strain in the FRP, assuming perfectly elastic behavior using Eq. (10.2.6). Based on the strain in the FRP reinforcement, the strain in the nonprestressed steel rein-

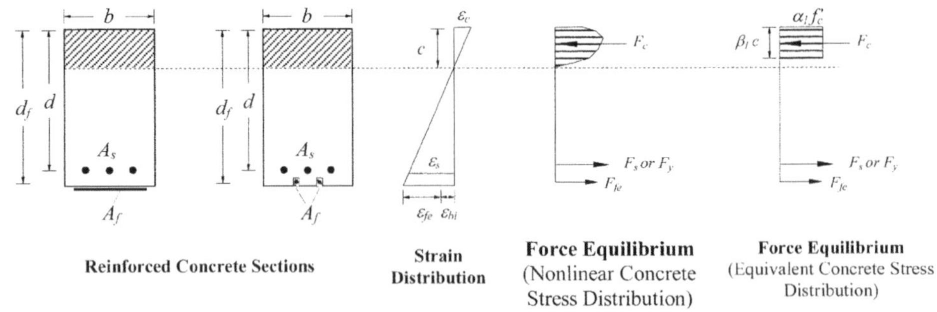

Fig. 10.2.10—Internal strain and stress distribution for a rectangular section under flexure at ultimate limit state.

forcement can be found from Eq. (10.2.10a) using strain compatibility

$$\varepsilon_s = (\varepsilon_{fe} + \varepsilon_{bi})\left(\frac{d-c}{d_f - c}\right) \qquad (10.2.10a)$$

The stress in the steel is determined from the strain in the steel using its assumed elastic-perfectly plastic stress-strain curve

$$f_s = E_s\varepsilon_s \leq f_y \qquad (10.2.10b)$$

With the stress in the FRP and steel reinforcement determined for the assumed neutral axis depth, internal force equilibrium may be checked using Eq. (10.2.10c)

$$\alpha_1 f_c'\beta_1 bc = A_s f_s + A_f f_{fe} \qquad (10.2.10c)$$

The terms α_1 and β_1 in Eq. (10.2.10c) are parameters defining a rectangular stress block in the concrete equivalent to the nonlinear distribution of stress. If concrete crushing is the controlling mode of failure (before or after steel yielding), α_1 and β_1 can be taken as the values associated with the Whitney stress block (ACI 318); that is, $\alpha_1 = 0.85$ and $\beta_1 = 0.85$ for f_c' between 2500 and 4000 psi (17 and 27 MPa), and β_1 is reduced linearly at a rate of 0.05 for each 1000 psi (7 MPa) of concrete strength exceeding 4000 psi (27 MPa). Note that β_1 shall not be taken less than 0.65. If FRP rupture, cover delamination, or FRP debonding occur, the Whitney stress block will give reasonably accurate results. A nonlinear stress distribution in the concrete or a more accurate stress block appropriate for the strain level reached in the concrete at the ultimate-limit state may be used.

The depth to the neutral axis, c, is found by simultaneously satisfying Eq. (10.2.5), (10.2.6), (10.2.10a), (10.2.10b), and (10.2.10c), thus establishing internal force equilibrium and strain compatibility. To solve for the depth of the neutral axis, c, an iterative solution procedure can be used. An initial value for c is first assumed and the strains and stresses are calculated using Eq. (10.2.5), (10.2.6), (10.2.10a), and (10.2.10b). A revised value for the depth of neutral axis, c, is then calculated from Eq. (10.2.10c). The calculated and assumed values for c are then compared. If they agree, then the proper value of c is reached. If the calculated and assumed values do not agree, another value for c is selected, and the process is repeated until convergence is attained.

The nominal flexural strength of the section with FRP external reinforcement is computed from Eq. (10.2.10d). An additional reduction factor for FRP, ψ_f, is applied to the flexural-strength contribution of the FRP reinforcement. The recommended value of ψ_f is 0.85. This reduction factor for the strength contribution of FRP reinforcement is based on the reliability analysis discussed in 9.1, which was based on the experimentally calibrated statistical properties of the flexural strength (Okeil et al. 2007)

$$M_n = A_s f_s\left(d - \frac{\beta_1 c}{2}\right) + \psi_f A_f f_{fe}\left(d_f - \frac{\beta_1 c}{2}\right) \qquad (10.2.10d)$$

10.2.10.1 *Stress in steel under service loads*—The stress in the steel reinforcement can be calculated based on a cracked-section analysis of the FRP-strengthened reinforced concrete section, as indicated by Eq. (10.2.10.1)

$$f_{s,s} = \frac{\left[M_s + \varepsilon_{bi}A_f E_f\left(d_f - \frac{kd}{3}\right)\right](d-kd)E_s}{\left[\begin{array}{l} A_s E_s\left(d - \frac{kd}{3}\right)(d-kd) \\ +A_f E_f\left(d_f - \frac{kd}{3}\right)(d_f - kd)\end{array}\right]} \qquad (10.2.10.1)$$

The distribution of strain and stress in the reinforced concrete section is shown in Fig. 10.2.10.1. Similar to conventional reinforced concrete, the depth to the neutral axis at service, kd, can be computed by taking the first moment of the areas of the transformed section. The transformed area of the FRP may be obtained by multiplying the area of FRP by the modular ratio of FRP to concrete. Although this method ignores the difference in the initial strain of the FRP, the initial strain does not greatly influence the depth to the neutral axis in the elastic response range of the member.

The stress in the steel under service loads computed from Eq. (10.2.10.1) should be compared against the limits described in 10.2.8. The value of M_s from Eq. (10.2.10.1) is equal to the moment due to all sustained loads (dead loads

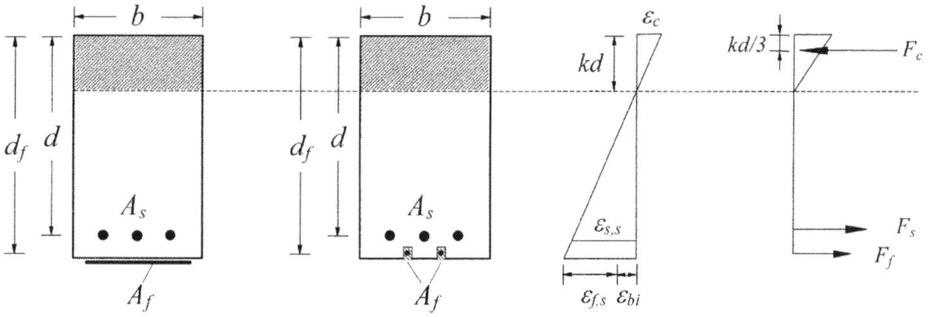

Fig. 10.2.10.1—Elastic strain and stress distribution.

and the sustained portion of the live load) plus the maximum moment induced in a fatigue loading cycle, as shown in Fig. 10.2.9.

10.2.10.2 *Stress in FRP under service loads*—The stress in the FRP reinforcement can be computed using Eq. (10.2.10.2) with $f_{s,s}$ from Eq. (10.2.10.1). Equation (10.2.10.2) gives the stress in the FRP reinforcement under an applied moment within the elastic response range of the member

$$f_{f,s} = f_{s,s}\left(\frac{E_f}{E_s}\right)\frac{d_f - kd}{d - kd} - \varepsilon_{bi}E_f \quad (10.2.10.2)$$

The stress in the FRP under service loads computed from Eq. (10.2.10.2) should be compared against the limits described in 10.2.9.

10.3—Prestressed concrete members

This section presents guidance on the effect of adding longitudinal FRP reinforcement to the tension face of a rectangular prestressed concrete member. The general concepts outlined herein can be extended to nonrectangular shapes (T-sections and I-sections) and to members with tension, compression, or both, nonprestressed steel reinforcement.

10.3.1 *Members with bonded prestressing steel*

10.3.1.1 *Assumptions*—In addition to the basic assumptions for concrete and FRP behavior for a reinforced concrete section listed in 10.2.1, the following assumptions are made in calculating the flexural resistance of a prestressed section strengthened with an externally applied FRP system:

a) Strain compatibility can be used to determine strain in the externally bonded FRP, strain in the nonprestressed steel reinforcement, and the strain or strain change in the prestressing steel.

b) Additional flexural failure mode controlled by prestressing steel rupture should be investigated.

c) For cases where the prestressing steel is draped or harped, several sections along the span of the member should be evaluated to verify strength requirements.

d) The initial strain of the concrete substrate, ε_{bi}, should be calculated and excluded from the effective strain in the FRP. The initial strain can be determined from an elastic analysis of the existing member, considering all loads that will be applied to the member at the time of FRP installation. Analysis should be based on the actual condition of the member (cracked or uncracked section) to determine the substrate initial strain.

10.3.1.2 *Strain in FRP reinforcement*—The maximum strain that can be achieved in the FRP reinforcement will be governed by strain limitations due to either concrete crushing, FRP rupture, FRP debonding, or prestressing steel rupture. The effective design strain for FRP reinforcement at the ultimate-limit state for failure controlled by concrete crushing can be calculated by the use of Eq. (10.2.5)

For failure controlled by prestressing steel rupture, Eq. (10.3.1.2a) can be used. For Grade 270 and 250 ksi (1860 and 1725 MPa) strand, the value of ε_{pu} to be used in Eq. (10.3.1.2a) is 0.035

$$\varepsilon_{fe} = (\varepsilon_{pu} - \varepsilon_{pi})\left(\frac{d_f - c}{d_p - c}\right) - \varepsilon_{bi} \le \varepsilon_{fd} \quad (10.3.1.2a)$$

in which

$$\varepsilon_{pi} = \frac{P_e}{A_p E_p} + \frac{P_e}{A_c E_c}\left(1 + \frac{e^2}{r^2}\right) \quad (10.3.1.2b)$$

10.3.1.3 *Strength reduction factor*—To maintain a sufficient degree of ductility, the strain in the prestressing steel at the nominal strength should be checked. Adequate ductility is achieved if the strain in the prestressing steel at the nominal strength is at least 0.013. Where this strain cannot be achieved, the strength reduction factor is decreased to account for a less ductile failure. The strength reduction factor for a member prestressed with standard 270 and 250 ksi (1860 and 1725 MPa) prestressing steel is given by Eq. (10.3.1.3), where ε_{ps} is the prestressing steel strain at the nominal strength

$$\phi = \begin{cases} 0.90 & \text{for } \varepsilon_{ps} \ge 0.013 \\ 0.65 + \dfrac{0.25(\varepsilon_{ps} - 0.010)}{0.013 - 0.010} & \text{for } 0.010 < \varepsilon_{ps} < 0.013 \\ 0.65 & \text{for } \varepsilon_{ps} \le 0.010 \end{cases} \quad (10.3.1.3)$$

10.3.1.4 *Serviceability*—To avoid inelastic deformations of the strengthened member, the prestressing steel should be prevented from yielding under service load levels. The stress in the steel under service load should be limited per Eq. (10.3.1.4a) and (10.3.1.4b). In addition, the compressive stress in the concrete under service load should be limited to 45 percent of the compressive strength

$$f_{ps,s} \leq 0.82 f_{py} \quad (10.3.1.4a)$$

$$f_{ps,s} \leq 0.74 f_{pu} \quad (10.3.1.4b)$$

When fatigue is a concern, the stress in the prestressing steel due to transient live loads should be limited to 18 ksi (125 MPa) when the radii of prestressing steel curvature exceeds 29 ft (9 m), or to 10 ksi (70 MPa) when the radii of prestressing-steel curvature does not exceed 12 ft (3.6 m). A linear interpolation should be used for radii between 12 and 29 ft (3.6 and 9 m) (AASHTO 2004). These limits have been verified experimentally for prestressed members with harped and straight strands strengthened with externally bonded FRP (Rosenboom and Rizkalla 2006).

10.3.1.5 *Creep rupture and fatigue stress limits*—To avoid creep rupture of the FRP reinforcement under sustained stresses or failure due to cyclic stresses and fatigue of the FRP reinforcement, the stress in the FRP reinforcement under these stress conditions should not exceed the limits provided in 10.2.9.

10.3.1.6 *Nominal strength*—The calculation procedure to compute nominal strength should satisfy strain compatibility and force equilibrium, and should consider the governing mode of failure. The calculation procedure described herein uses an iterative method similar to that discussed in 10.2.

For any assumed depth to the neutral axis, c, the effective strain and stress in the FRP reinforcement can be computed from Eq. (10.2.5) and (10.2.6), respectively. This equation considers the governing mode of failure for the assumed neutral axis depth. If the left term of the inequality in Eq. (10.2.5) controls, concrete crushing controls flexural failure of the section. If the right term of the inequality controls, FRP failure (rupture or debonding) controls flexural failure of the section.

The strain in the prestressed steel can be found from Eq. (10.3.1.6a) based on strain compatibility

$$\varepsilon_{ps} = \varepsilon_{pe} + \frac{P_e}{A_c E_c}\left(1 + \frac{e^2}{r^2}\right) + \varepsilon_{pnet} \leq 0.035 \quad (10.3.1.6a)$$

in which ε_{pe} is the effective strain in the prestressing steel after losses, and ε_{pnet} is the net tensile strain in the prestressing steel beyond decompression, at the nominal strength. The value of ε_{pnet} will depend on the mode of failure, and can be calculated using Eq. (10.3.1.6b) and 10.3.1.6c)

$$\varepsilon_{pnet} = 0.003\left(\frac{d_p - c}{c}\right) \text{ for concrete crushing failure} \quad (10.3.1.6b)$$

$$\varepsilon_{pnet} = (\varepsilon_{fe} + \varepsilon_{bi})\left(\frac{d_p - c}{d_f - c}\right) \text{ for FRP rupture or debonding failure modes} \quad (10.3.1.6c)$$

The stress in the prestressing steel is calculated using the material properties of the steel. For a typical seven-wire low-relaxation prestressing strand, the stress-strain curve may be approximated by the following equations (Prestressed/Precast Concrete Institute 2004)

For Grade 250 ksi (1725 MPa) steel

$$f_{ps} = \begin{cases} 28{,}500\varepsilon_{ps} & \text{for } \varepsilon_{ps} \leq 0.0076 \\ 250 - \dfrac{0.04}{\varepsilon_{ps} - 0.0064} & \text{for } \varepsilon_{ps} > 0.0076 \end{cases} \quad \text{(in.-lb)}$$

$$f_{ps} = \begin{cases} 196{,}500\varepsilon_{ps} & \text{for } \varepsilon_{ps} \leq 0.0076 \\ 1720 - \dfrac{0.276}{\varepsilon_{ps} - 0.0064} & \text{for } \varepsilon_{ps} > 0.0076 \end{cases} \quad \text{(SI)} \quad (10.3.1.6d)$$

For Grade 270 ksi (1860 MPa) steel

$$f_{ps} = \begin{cases} 28{,}500\varepsilon_{ps} & \text{for } \varepsilon_{ps} \leq 0.0086 \\ 270 - \dfrac{0.04}{\varepsilon_{ps} - 0.007} & \text{for } \varepsilon_{ps} > 0.0086 \end{cases} \quad \text{(in.-lb)}$$

$$f_{ps} = \begin{cases} 196{,}500\varepsilon_{ps} & \text{for } \varepsilon_{ps} \leq 0.0086 \\ 1860 - \dfrac{0.276}{\varepsilon_{ps} - 0.007} & \text{for } \varepsilon_{ps} > 0.0086 \end{cases} \quad \text{(SI)} \quad (10.3.1.6e)$$

With the strain and stress in the FRP and prestressing steel determined for the assumed neutral axis depth, internal force equilibrium may be checked using Eq. (10.3.1.6f)

$$\alpha_1 f_c' \beta_1 bc = A_p f_p + A_f f_{fe} \quad (10.3.1.6f)$$

For the concrete crushing mode of failure, the equivalent compressive stress block factor α_1 can be taken as 0.85, and β_1 can be estimated as described in 10.2.10. If FRP rupture, cover delamination, or FRP debonding failure occurs, the use of equivalent rectangular concrete stress block factors is appropriate. Methods considering a nonlinear stress distribution in the concrete can also be used.

The depth to the neutral axis, c, is found by simultaneously satisfying Eq. (10.2.5), (10.2.6), and (10.3.1.6a) to (10.3.1.6f), thus establishing internal force equilibrium and strain compatibility. To solve for the depth of the neutral axis, c, an iterative solution procedure can be used. An initial value for c is first assumed, and the strains and stresses are calculated using Eq. (10.2.5), (10.2.6), and (10.3.1.6a) to (10.3.1.6e). A revised value for the depth of neutral axis, c, is then calculated from Eq. (10.3.1.6f). The calculated and assumed values for c are then compared. If they agree, then the proper value of c is reached. If the calculated and assumed values do not agree, another value for c is selected, and the process is repeated until convergence is attained.

The nominal flexural strength of the section with FRP external reinforcement can be computed using Eq. (10.3.1.6g). The additional reduction factor $\psi_f = 0.85$ is applied to the flexural-strength contribution of the FRP reinforcement

$$M_n = A_p f_{ps}\left(d_p - \frac{\beta_1 c}{2}\right) + \psi_f A_f f_{fe}\left(d_f - \frac{\beta_1 c}{2}\right) \quad (10.3.1.6g)$$

10.3.1.7 *Stress in prestressing steel under service loads*—The stress in the prestressing steel can be calculated based on the actual condition (cracked or uncracked section) of the strengthened reinforced concrete section. The strain in prestressing steel at service, $\varepsilon_{ps,s}$, can be calculated as

$$\varepsilon_{ps,s} = \varepsilon_{pe} + \frac{P_e}{A_c E_c}\left(1 + \frac{e^2}{r^2}\right) + \varepsilon_{pnet,s} \quad (10.3.1.7a)$$

in which ε_{pe} is the effective prestressing strain, and $\varepsilon_{pnet,s}$ is the net tensile strain in the prestressing steel beyond decompression at service. The value of $\varepsilon_{pnet,s}$ depends on the effective section properties at service, and can be calculated using Eq. (10.3.1.7b) and (10.3.1.7c)

$$\varepsilon_{pnet,s} = \frac{M_s e}{E_c I_g} \text{ for uncracked section at service} \quad (10.3.1.7b)$$

$$\varepsilon_{pnet,s} = \frac{M_{snet} e}{E_c I_{cr}} \text{ for cracked section at service} \quad (10.3.1.7c)$$

where M_{snet} is the net service moment beyond decompression. The stress in the prestressing steel under service loads can then be computed from Eq. (10.3.1.6d) and (10.3.1.6e), and compared against the limits described in 10.3.1.4.

10.3.1.8 *Stress in FRP under service loads*—Equation (10.3.1.8) gives the stress in the FRP reinforcement under an applied moment within the elastic response range of the member. The calculation procedure for the initial strain ε_{bi} at the time of FRP installation will depend on the state of the concrete section at the time of FRP installation and at service condition. Prestressed sections can be uncracked at installation/uncracked at service, uncracked at installation/cracked at service, or cracked at installation/cracked at service. The initial strain on the bonded substrate, ε_{bi}, can be determined from an elastic analysis of the existing member, considering all loads that will be on the member during the installation of the FRP system. The elastic analysis of the existing member should be based on cracked or uncracked section properties, depending on existing conditions. In most cases, the initial strain before cracking is relatively small, and may conservatively be ignored

$$f_{f,s} = \left(\frac{E_f}{E_c}\right)\frac{M_s y_b}{I} - \varepsilon_{bi} E_f \quad (10.3.1.8)$$

Depending on the actual condition at service (cracked or uncracked), the moment of inertia, I, can be taken as the moment of inertia of the uncracked section transformed to concrete, I_{tr}, or the moment of inertia of the cracked section transformed to concrete, I_{cr}. The variable y_b is the distance from the centroidal axis of the gross section, neglecting reinforcement, to the extreme bottom fiber. The computed stress in the FRP under service loads should not exceed the limits provided in 10.2.9.

10.4—Moment redistribution

Moment redistribution for continuous reinforced concrete beams strengthened using externally bonded FRP can be used to decrease factored moments calculated by elastic theory at sections of maximum negative or maximum positive moment for any assumed loading arrangement by not more than $1000\varepsilon_t$ percent, to a maximum of 20 percent. Moment redistribution is only permitted when the strain in the tension steel reinforcement, ε_t, exceeds 0.0075 at the section at which moment is reduced. Moment redistribution is not permitted where approximate values of bending moments are used.

The reduced moment should be used for calculating redistributed moments at all other sections within the spans. Static equilibrium should be maintained after redistribution of moments for each loading arrangement. El-Refaie et al. (2003) demonstrated that continuous reinforced concrete beams strengthened with carbon FRP sheets can redistribute moment in the order of 6 to 31 percent. They also concluded that lower moment redistribution was achieved for beam sections retrofitted with higher amounts of carbon FRP reinforcement. Silva and Ibell (2008) demonstrated that sections that can develop a curvature ductility capacity greater than 2.0 can produce moment redistribution of at least 7.5 percent of the design moment.

CHAPTER 11—SHEAR STRENGTHENING

Fiber-reinforced polymer (FRP) systems have been shown to increase the shear strength of existing concrete beams and columns by wrapping or partially wrapping the members (Malvar et al. 1995; Chajes et al. 1995; Norris et al. 1997; Kachlakev and McCurry 2000). Orienting FRP fibers transverse to the axis of the member or perpendicular to potential shear cracks is effective in providing additional shear strength (Sato et al. 1996). An increase in shear strength may be required when flexural strengthening is implemented to ensure that flexural capacity remains critical. Flexural failures are relatively more ductile in nature compared with shear failures.

11.1—General considerations

This chapter presents guidance on the calculation of added shear strength resulting from the addition of FRP shear reinforcement to a reinforced concrete beam or column. The additional shear strength that can be provided by the FRP system is based on many factors, including geometry of the beam or column, wrapping scheme, and existing concrete strength, but should be limited in accordance with the recommendations of Chapter 9.

Shear strengthening using external FRP may be provided at locations of expected plastic hinges or stress reversal and for enhancing post-yield flexural behavior of members in moment frames resisting seismic loads, as described in Chapter 13.

11.2—Wrapping schemes

The three types of FRP wrapping schemes used to increase the shear strength of prismatic, rectangular beams, or columns are illustrated in Fig. 11.2. Completely wrapping the FRP system around the section on all four sides is the most efficient wrapping scheme and is most commonly used in column applications where access to all four sides of the column is available. In beam applications where an integral slab makes it impractical to completely wrap the member, the shear strength can be improved by wrapping the FRP system around three sides of the member (U-wrap) or bonding to two opposite sides of the member.

Although all three techniques have been shown to improve the shear strength of a rectangular member, completely wrapping the section is the most efficient, followed by the three-sided U-wrap. Bonding to two sides of a beam is the least efficient scheme.

For shear strengthening of circular members, only complete circumferential wrapping of the section in which the FRP is oriented perpendicular to the longitudinal axis of the member (that is, $\alpha = 90$ degrees) is recommended.

In all wrapping schemes, the FRP system can be installed continuously along the span of a member or placed as discrete strips. As discussed in 9.3.3, the potential effects of entrapping moisture in the substrate when using continuous reinforcement should be carefully considered. Specific means of allowing moisture vapor transmission out of the substrate should be employed where appropriate.

11.3—Nominal shear strength

The design shear strength of a concrete member strengthened with an FRP system should exceed the required shear strength (Eq. (11.3a)). The required shear strength of an FRP-strengthened concrete member should be computed with the load factors required by ACI 318. The design shear strength should be calculated by multiplying the nominal shear strength by the strength reduction factor ϕ, as specified by ACI 318

$$\phi V_n \geq V_u \quad (11.3a)$$

The nominal shear strength of an FRP-strengthened concrete member can be determined by adding the contribution of the FRP external shear reinforcement to the contributions from the reinforcing steel (stirrups, ties, or spirals) and the concrete (Eq. (11.3b)). An additional reduction factor ψ_f is applied to the contribution of the FRP system

Fig. 11.2—Typical wrapping schemes for shear strengthening using FRP laminates.

$$\phi V_n = \phi(V_c + V_s + \psi_f V_f) \quad (11.3b)$$

where V_c and V_s are the concrete and internal reinforcing steel contributions to shear capacity calculated using the provisions of ACI 318, respectively. For prestressed members, V_c is the minimum of V_{ci} and V_{cw} defined by ACI 318.

Based on a reliability analysis using data from Bousselham and Chaallal (2006), Deniaud and Cheng (2001, 2003), Funakawa et al. (1997), Matthys and Triantafillou (2001), and Pellegrino and Modena (2002), the reduction factor ψ_f of 0.85 is recommended for the three-sided FRP U-wrap or two-opposite-sides strengthening schemes. Insufficient experimental data exist to perform a reliability analysis for fully-wrapped sections; however, there should be less variability with this strengthening scheme, as it is less bond-dependent and, therefore, the reduction factor ψ_f of 0.95 is recommended. The ψ_f factor was calibrated based on design material properties. These recommendations are given in Table 11.3.

Table 11.3—Recommended additional reduction factors for FRP shear reinforcement

$\psi_f = 0.95$	Completely wrapped members
$\psi_f = 0.85$	Three-side and two-opposite-sides schemes

11.4—FRP contribution to shear strength

Figure 11.4 illustrates the dimensional variables used in shear-strengthening calculations for FRP laminates. The contribution of the FRP system to shear strength of a member is based on the fiber orientation and an assumed crack pattern (Khalifa et al. 1998). The shear strength provided by the FRP reinforcement can be determined by calculating the force resulting from the tensile stress in the FRP across the assumed crack. The shear contribution of the FRP shear reinforcement is then given by Eq. (11.4a)

$$V_f = \frac{A_{fv} f_{fe} (\sin\alpha + \cos\alpha) d_{fv}}{s_f} \quad (11.4a)$$

For rectangular sections

$$A_{fv} = 2nt_f w_f \quad (11.4b)$$

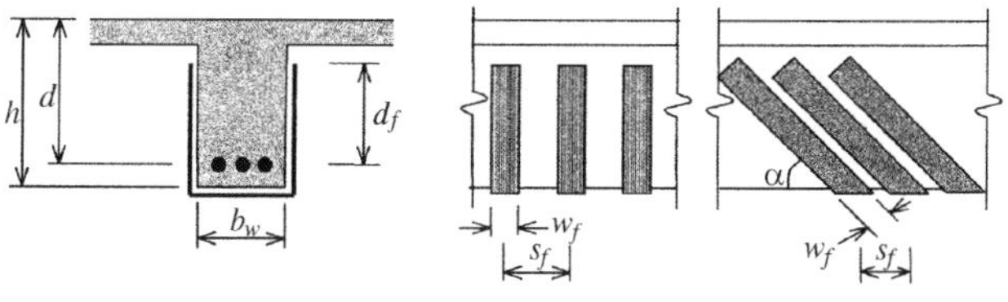

Fig. 11.4—Illustration of the dimensional variables used in shear-strengthening calculations for repair, retrofit, or strengthening using FRP laminates.

For circular sections, d_{fv} is taken as 0.8 times the diameter of the section and

$$A_{fv} = (\pi/2)nt_f w_f \quad (11.4c)$$

The tensile stress in the FRP shear reinforcement at nominal strength is directly proportional to the strain that can be developed in the FRP shear reinforcement at nominal strength

$$f_{fe} = E_f \varepsilon_{fe} \quad (11.4d)$$

11.4.1 *Effective strain in FRP laminates*—The effective strain is the maximum strain that can be achieved in the FRP system at the nominal strength and is governed by the failure mode of the FRP system and of the strengthened reinforced concrete member. The licensed design professional should consider all possible failure modes and use an effective strain representative of the critical failure mode. The following subsections provide guidance on determining this effective strain for different configurations of FRP laminates used for shear strengthening of reinforced concrete members.

11.4.1.1 *Completely wrapped members*—For reinforced concrete column and beam members completely wrapped by FRP, loss of aggregate interlock of the concrete has been observed to occur at fiber strains less than the ultimate fiber strain. To preclude this mode of failure, the maximum strain used for design should be limited to 0.4 percent for members that are completely wrapped with FRP (Eq. (11.4.1.1))

$$\varepsilon_{fe} = 0.004 \leq 0.75\varepsilon_{fu} \quad (11.4.1.1)$$

This strain limitation is based on testing (Priestley et al. 1996) and experience. Higher strains should not be used for FRP shear-strengthening applications.

11.4.1.2 *Bonded U-wraps or bonded face plies*—FRP systems that do not enclose the entire section (two- and three-sided wraps) have been observed to delaminate from the concrete before the loss of aggregate interlock of the section. For this reason, bond stresses have been analyzed to determine the efficiency of these systems and the effective strain that can be achieved (Triantafillou 1998). The effective strain is calculated using a bond-reduction coefficient κ_v applicable to shear

$$\varepsilon_{fe} = \kappa_v \varepsilon_{fu} \leq 0.004 \quad (11.4.1.2a)$$

The bond-reduction coefficient is a function of the concrete strength, the type of wrapping scheme used, and the stiffness of the laminate. The bond-reduction coefficient can be computed from Eq. (11.4.1.2b) through (11.4.1.2e) (Khalifa et al. 1998)

$$\kappa_v = \frac{k_1 k_2 L_e}{468\varepsilon_{fu}} \leq 0.75 \quad \text{(in.-lb)}$$
$$\kappa_v = \frac{k_1 k_2 L_e}{11{,}900\varepsilon_{fu}} \leq 0.75 \quad \text{(SI)} \qquad (11.4.1.2b)$$

The active bond length L_e is the length over which the majority of the bond stress is maintained. This length is given by Eq. (11.4.1.2c)

$$L_e = \frac{2500}{(nt_f E_f)^{0.58}} \quad \text{(in.-lb)}$$
$$L_e = \frac{23{,}300}{(nt_f E_f)^{0.58}} \quad \text{(SI)} \qquad (11.4.1.2c)$$

The bond-reduction coefficient also relies on two modification factors, k_1 and k_2, that account for the concrete strength and the type of wrapping scheme used, respectively. Expressions for these modification factors are given in Eq. (11.4.1.2d) and (11.4.1.2e)

$$k_1 = \left(\frac{f_c'}{4000}\right)^{2/3} \quad \text{(in.-lb)}$$
$$k_1 = \left(\frac{f_c'}{27}\right)^{2/3} \quad \text{(SI)} \qquad (11.4.1.2d)$$

$$k_2 = \begin{cases} \dfrac{d_{fv} - L_e}{d_{fv}} & \text{for U-wraps} \\ \dfrac{d_{fv} - 2L_e}{d_{fv}} & \text{for two sides bonded} \end{cases} \qquad (11.4.1.2e)$$

The methodology for determining κ_v has been validated for members in regions of high shear and low moment, such as monotonically loaded simply supported beams. Although the methodology has not been confirmed for shear strengthening in areas subjected to combined high flexural and shear stresses or in regions where the web is primarily in compression (negative moment regions), it is suggested that κ_v is sufficiently conservative for such cases. The design procedures outlined herein have been developed by a combination of analytical and empirical results (Khalifa et al. 1998).

Anchorage details have been used to develop higher strains in bonded U-wraps used in shear strengthening applications. Anchorage systems include mechanical fasteners, FRP strips, fiber anchors, and near-surface mounted (NSM) anchors; examples are shown schematically in Fig. 10.1.1b. (Khalifa et al. 1999; Kalfat et al. 2013; Grelle and Sneed 2013). Properly anchored U-wraps can be designed to fail by FRP rupture (Belarbi et al. 2011). In no case, however, should the effective strain in the anchored FRP U-wrap exceed the lesser of 0.004 or $0.75\varepsilon_{fu}$, and $\psi_f = 0.85$ remains appropriate for anchored U-wraps.

11.4.2 *Spacing*—Spaced FRP strips used for shear strengthening should be investigated to evaluate their contribution to the shear strength. Spacing should adhere to the limits prescribed by ACI 318 for internal steel shear reinforcement. The spacing of FRP strips is defined as the distance between the centerline of the strips.

11.4.3 *Reinforcement limits*—The total shear strength provided by reinforcement should be taken as the sum of the contribution of the FRP shear reinforcement and the steel shear reinforcement. The sum of the shear strengths provided by the shear reinforcement should be limited based on the criteria given for steel alone in ACI 318

$$V_s + V_f \leq 8\sqrt{f_c'}b_w d \quad \text{(in.-lb)}$$
$$V_s + V_f \leq 0.66\sqrt{f_c'}b_w d \quad \text{(SI)} \qquad (11.4.3)$$

For circular sections, $b_w d$ in Eq. (11.4.3) is taken as $0.8D^2$, where D is the member diameter.

CHAPTER 12—STRENGTHENING OF MEMBERS SUBJECTED TO AXIAL FORCE OR COMBINED AXIAL AND BENDING FORCES

Confinement of reinforced concrete columns by means of fiber-reinforced polymer (FRP) jackets can be used to enhance their strength and ductility. An increase in capacity is an immediate outcome typically expressed in terms of improved peak load resistance. Ductility enhancement, on the other hand, requires more complex calculations to determine the ability of a member to sustain rotation and drift without a substantial loss in strength. This chapter applies only to members confined with FRP systems.

12.1—Pure axial compression

FRP systems can be used to increase the axial compression strength of a concrete member by providing confinement with an FRP jacket (Nanni and Bradford 1995; Toutanji 1999). Confining a concrete member is accomplished by orienting the fibers transverse to the longitudinal axis of the member. In this orientation, the transverse or hoop fibers are similar to conventional spiral or tie reinforcing steel. Any contribution of longitudinally aligned fibers to the axial compression strength of a concrete member should be neglected.

FRP jackets provide passive confinement to the compression member, remaining unstressed until dilation and cracking of the wrapped compression member occur. For this reason, intimate contact between the FRP jacket and the concrete member is critical.

Depending on the level of confinement, the uniaxial stress-strain curve of a reinforced concrete column could be depicted by one of the curves in Fig. 12.1a, where f_c' and f_{cc}' represent the peak concrete strengths for unconfined and confined cases, respectively. These strengths are calculated as the peak load minus the contribution of the steel reinforcement, all divided by the cross-sectional area of the concrete. The ultimate strain of the unconfined member corresponding to $0.85f_c'$ (Curve (a)) is ε_{cu}. The strain ε_{ccu} corresponds to: a) $0.85f_{cc}'$ in the case of the lightly confined member (Curve (b)); and b) the failure strain in both the heavily confined-softening case (the failure stress is larger than $0.85f_{cc}'$ (Curve (c)) or in the heavily confined-hardening case (Curve (d)).

The definition of ε_{ccu} at $0.85f_{cc}'$ or less is arbitrary, although consistent with modeling of conventional concrete (Hognestad 1951) and such that the descending branch of the stress-strain curve at that level of stress ($0.85f_{cc}'$ or higher) is not as sensitive to the test procedure in terms of rate of loading and stiffness of the equipment used.

The axial compressive strength of a nonslender, normal-weight concrete member confined with an FRP jacket may be calculated using the confined concrete strength (Eq. (12.1a) and (12.1b)). The axial force acting on an FRP-strengthened concrete member should be computed using the load factors required by ACI 318, and the values of the ϕ factors as established in ACI 318 for both types of transverse reinforcing steel (spirals or ties) apply.

For nonprestressed members with existing steel spiral reinforcement

$$\phi P_n = 0.85\phi[0.85f_{cc}'(A_g - A_{st}) + f_y A_{st}] \qquad (12.1a)$$

For nonprestressed members with existing steel-tie reinforcement

$$\phi P_n = 0.8\phi[0.85f_{cc}'(A_g - A_{st}) + f_y A_{st}] \qquad (12.1b)$$

Several models that simulate the stress-strain behavior of FRP-confined compression sections are available in the literature (Teng et al. 2002; De Lorenzis and Tepfers 2003; Lam and Teng 2003a). The stress-strain model by Lam and Teng (2003a,b) for FRP-confined concrete is illustrated in Fig. 12.1b and computed using the following expressions

$$f_c = \begin{cases} E_c\varepsilon_c - \dfrac{(E_c - E_2)^2}{4f_c'}\varepsilon_c^2 & 0 \leq \varepsilon_c \leq \varepsilon_t' \\ f_c' + E_2\varepsilon_c & \varepsilon_t' \leq \varepsilon_c \leq \varepsilon_{c,max} \end{cases} \qquad (12.1c)$$

$$\varepsilon_{c,max} \leq \varepsilon_{ccu} \leq 0.01 \qquad (12.1d)$$

$$E_2 = \frac{f_{cc}' - f_c'}{\varepsilon_{ccu}} \qquad (12.1e)$$

$$\varepsilon_t' = \frac{2f_c'}{E_c - E_2} \qquad (12.1f)$$

The maximum confined concrete compressive strength, f_{cc}', and the maximum confinement pressure f_ℓ are calculated using Eq. (12.1g) and (12.1h), respectively (Lam and Teng 2003a,b) with the inclusion of an additional reduction factor $\psi_f = 0.95$

$$f'_{cc} = f'_c + \psi_f 3.3\kappa_a f_\ell \quad (12.1g)$$

$$f_\ell = \frac{2E_f n t_f \varepsilon_{fe}}{D} \quad (12.1h)$$

In Eq. (12.1g), f_c' is the unconfined cylinder compressive strength of concrete, and the efficiency factor κ_a accounts for the geometry of the section, circular and noncircular, as defined in 12.1.1 and 12.1.2. In Eq. (12.1h), the effective strain in the FRP at failure, ε_{fe}, is given by

$$\varepsilon_{fe} = \kappa_\varepsilon \varepsilon_{fu} \quad (12.1i)$$

The FRP strain efficiency factor κ_ε accounts for the premature failure of the FRP system (Pessiki et al. 2001), related primarily to stress concentration regions caused by cracking of the concrete as it dilates. Based on experimental calibration using mainly carbon FRP (CFRP)-confined concrete specimens, an average value of 0.586 was computed for κ_ε by Lam and Teng (2003a). Similarly, a database of 251 test results (Harries and Carey 2003) computed a value of $\kappa_\varepsilon = 0.58$, whereas experimental tests on medium- and large-scale columns resulted in values of $\kappa_\varepsilon = 0.57$ and 0.61, respectively (Carey and Harries 2005).

Based on tests by Lam and Teng (2003a,b), the ratio f_ℓ/f_c' should not be less than 0.08. This is the minimum level of confinement required to assure a nondescending second branch in the stress-strain performance, as shown by Curve (d) in Fig. 12.1a. This limitation was later confirmed for circular cross sections by Spoelstra and Monti (1999) using their analytical model. A strain efficiency factor κ_ε of 0.55 and a minimum confinement ratio f_ℓ/f_c' of 0.08 should be used.

The maximum compressive strain in the FRP-confined concrete, ε_{ccu}, can be found using Eq. (12.1j) (Concrete Society 2004). The maximum concrete strain, $\varepsilon_{c,max}$, used in Eq. (12.1c) should be limited to 0.01 to prevent excessive cracking and the resulting loss of concrete integrity.

$$\varepsilon_{ccu} = \varepsilon_c' \left(1.50 + 12\kappa_b \frac{f_\ell}{f_c'}\left(\frac{\varepsilon_{fe}}{\varepsilon_c'}\right)^{0.45}\right) \quad (12.1j)$$

In Eq. (12.1j), the efficiency factor κ_b accounts for the geometry of the section in the calculation of the ultimate axial strain, as defined in 12.1.1 and 12.1.2.

Strength enhancement for compression members with f_c' of 10,000 psi (70 MPa) or higher has not been experimentally verified. Enhancement of concrete having strength f_c' in excess of 10,000 psi (70 MPa) should be based on experimental testing.

12.1.1 *Circular cross sections*—FRP jackets are most effective at confining members with circular cross sections (Demers and Neale 1999; Pessiki et al. 2001; Harries and Carey 2003; Youssef 2003; Matthys et al. 2005; Rocca et al. 2006). The FRP system provides a circumferentially uniform confining pressure to the radial expansion of the compression member when the fibers are aligned transverse to the

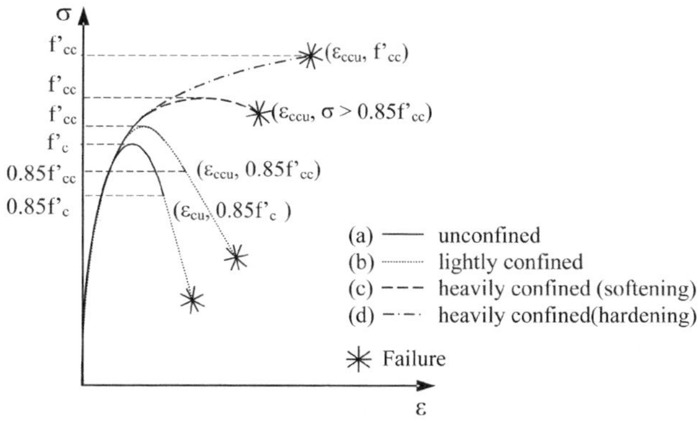

Fig. 12.1a—Schematic stress-strain behavior of unconfined and confined reinforced concrete columns (Rocca et al. 2006).

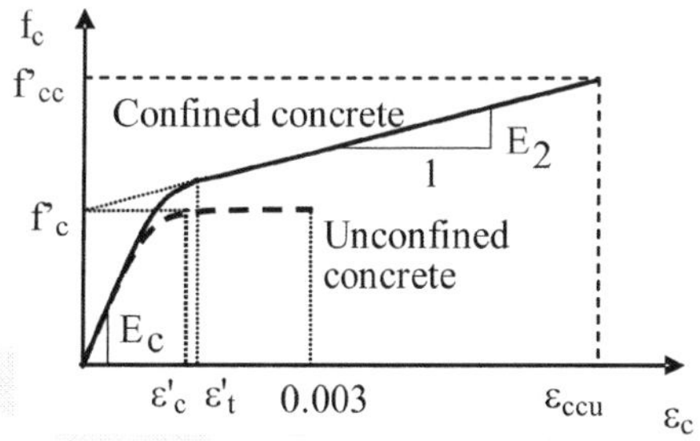

Fig. 12.1b—Stress-strain model for FRP-confined concrete (Lam and Teng 2003a).

longitudinal axis of the member. For circular cross sections, the shape factors κ_a and κ_b in Eq. (12.1g) and (12.1j), respectively, can be taken as 1.0.

12.1.2 *Noncircular cross sections*—Testing has shown that confining square and rectangular members with FRP jackets can provide marginal increases in the maximum axial compressive strength f'_{cc} of the member (Pessiki et al. 2001; Wang and Restrepo 2001; Harries and Carey 2003; Youssef 2003; Rocca et al. 2008). The provisions in this guide are not recommended for members featuring side aspect ratios h/b greater than 2.0, or face dimensions b or h exceeding 36 in. (900 mm), unless testing demonstrates their effectiveness.

For noncircular cross sections, f_ℓ in Eq. (12.1h) corresponds to the maximum confining pressure of an equivalent circular cross section with diameter D equal to the diagonal of the rectangular cross section

$$D = \sqrt{b^2 + h^2} \quad (12.1.2a)$$

The shape factors κ_a in Eq. (12.1g) and κ_b in Eq. (12.1j) depend on two parameters: the cross-sectional area of effectively confined concrete A_e, and the side-aspect ratio h/b, as shown in Eq. (12.1.2b) and (12.1.2c), respectively

$$\kappa_a = \frac{A_e}{A_c}\left(\frac{b}{h}\right)^2 \quad (12.1.2b)$$

$$\kappa_b = \frac{A_e}{A_c}\left(\frac{h}{b}\right)^{0.5} \quad (12.1.2c)$$

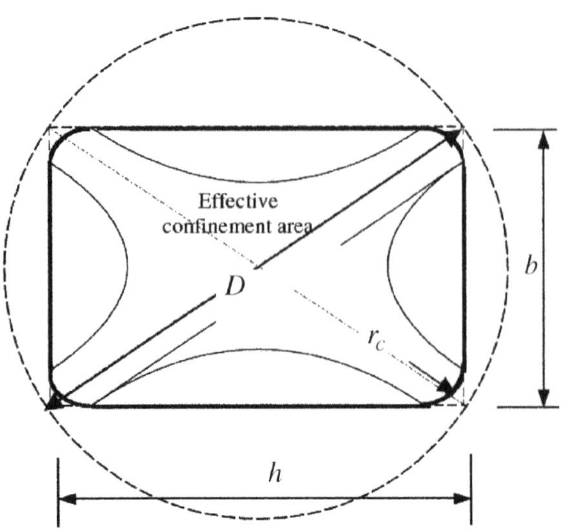

Fig. 12.1.2—Equivalent circular cross section (Lam and Teng 2003b).

The generally accepted theoretical approach for the definition of A_e consists of four parabolas within which the concrete is fully confined, outside of which negligible confinement occurs (Fig. 12.1.2). The shape of the parabolas and the resulting effective confinement area is a function of the dimensions of the column (b and h), the radius of the corners, r_c, and the longitudinal steel reinforcement ratio ρ_g, and can be expressed as

$$\frac{A_e}{A_c} = \frac{1 - \dfrac{\left[\left(\dfrac{b}{h}\right)(h-2r_c)^2 + \left(\dfrac{h}{b}\right)(b-2r_c)^2\right]}{3A_g} - \rho_g}{1-\rho_g} \qquad (12.1.2d)$$

12.1.3 *Serviceability considerations*—As loads approach factored load levels, damage to the concrete in the form of significant cracking in the radial direction might occur. The FRP jacket contains the damage and maintains the structural integrity of the column. At service load levels, however, this type of damage should be avoided. In this way, the FRP jacket will only act during overloading conditions that are temporary in nature.

To ensure that radial cracking will not occur under service loads, the transverse strain in the concrete should remain below its cracking strain at service load levels. This corresponds to limiting the compressive stress in the concrete to $0.65f_c'$. In addition, the service stress in the longitudinal steel should remain below $0.60f_y$ to avoid plastic deformation under sustained or cyclic loads. By maintaining the specified stress in the concrete at service, the stress in the FRP jacket will be relatively low. The jacket is only stressed to significant levels when the concrete is transversely strained above the cracking strain and the transverse expansion becomes large. Service load stresses in the FRP jacket should never exceed the creep rupture stress limit. In addition, axial deformations under service loads should be investigated to evaluate their effect on the performance of the structure.

12.2—Combined axial compression and bending

Wrapping with an FRP jacket can also provide strength enhancement for a member subjected to combined axial compression and flexure (Nosho 1996; Saadatmanesh et al. 1996; Chaallal and Shahawy 2000; Sheikh and Yau 2002; Iacobucci et al. 2003; Bousias et al. 2004; Elnabelsy and Saatcioglu 2004; Harajli and Rteil 2004; Sause et al. 2004; Memon and Sheikh 2005).

For predicting the effect of FRP confinement on strength enhancement, Eq. (12.1a) and (12.1b) are applicable when the eccentricity present in the member is less than or equal to $0.1h$. When the eccentricity is larger than $0.1h$, the methodology and equations presented in 12.1 can be used to determine the concrete material properties of the member cross section under compressive stress. Based on that, the axial load-moment (*P-M*) interaction diagram for the FRP-confined member can be constructed using well-established procedures (Bank 2006).

The following limitations apply for members subjected to combined axial compression and bending:

a) The effective strain in the FRP jacket should be limited to the value given in Eq. (12.2) to ensure the shear integrity of the confined concrete

$$\varepsilon_{fe} = 0.004 \leq \kappa_\varepsilon \varepsilon_{fu} \qquad (12.2)$$

b) The strength enhancement can only be considered when the applied ultimate axial force and bending moment, P_u and M_u, respectively, fall above the line connecting the origin and the balanced point in the *P-M* diagram for the unconfined member (Fig. 12.2). This limitation stems from the fact that strength enhancement is only significant for members in which compression failure is the controlling mode (Bank 2006).

P-M diagrams may be developed by satisfying strain compatibility and force equilibrium using the model for the stress-strain behavior for FRP-confined concrete presented in Eq. (12.1c) through (12.1f). For simplicity, the portion of the unconfined and confined *P-M* diagrams corresponding to compression-controlled failure can be reduced to two bilinear curves passing through three points (Fig 12.2). For values of eccentricity greater than $0.1h$ and up to the point corresponding to the balanced condition, the methodology provided in Appendix D may be used for the computation of a simplified interaction diagram. The values of the ϕ factors as established in ACI 318 for both types of transverse reinforcing steel (spirals or ties) apply.

12.3—Ductility enhancement

Increased ductility of a section results from the ability to develop greater compressive strains in the concrete before compressive failure (Seible et al. 1997). The FRP jacket can also serve to delay buckling of longitudinal steel reinforce-

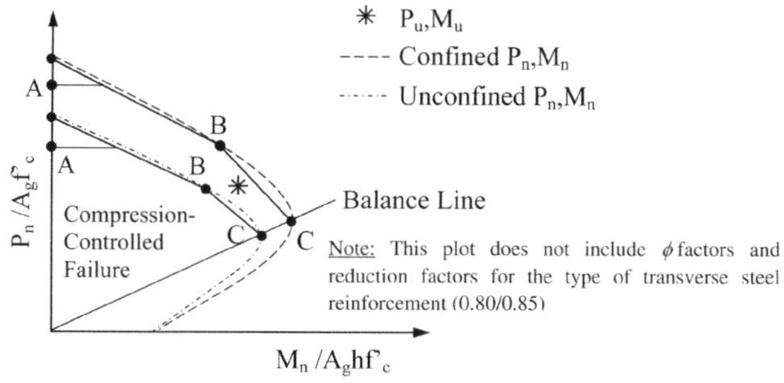

Fig. 12.2—Representative interaction diagram.

ment in compression and to clamp lap splices of longitudinal steel reinforcement.

For seismic applications, FRP jackets should be designed to provide a confining stress sufficient to develop concrete compression strains associated with the displacement demands as described in Chapter 13. Shear forces should also be evaluated in accordance with Chapter 11 to prevent brittle shear failure in accordance with ACI 318.

12.3.1 *Circular cross sections*—The maximum compressive strain for FRP-confined members with circular cross sections can be found from Eq. (12.1j) with f'_{cc} from Eq. (12.1g) and using $\kappa_b = 1.0$.

12.3.2 *Noncircular cross sections*—The maximum compressive strain for FRP-confined members with square or rectangular sections can be found from Eq. (12.1j), with f'_{cc} from Eq. (12.1g), and using κ_b as given in Eq. (12.1.2c). The confining effect of FRP jackets should be assumed to be negligible for rectangular sections with aspect ratio h/b exceeding 2.0, or face dimensions b or h exceeding 36 in. (900 mm), unless testing demonstrates their effectiveness.

12.4—Pure axial tension

FRP systems can be used to provide additional tensile strength to a concrete member. Due to the linear-elastic nature of FRP materials, the tensile contribution of the FRP system is directly related to its strain and is calculated using Hooke's Law.

The tension capacity provided by the FRP is limited by the design tensile strength of the FRP and the ability to transfer stresses into the substrate through bond (Nanni et al. 1997). The effective strain in the FRP can be determined based on the criteria given for shear strengthening in Eq. (11.4.1.1) through (11.4.1.2d). The value of k_2 in Eq. (11.4.1.2b) can be taken as 1.0. A minimum bonded length of ℓ_{df}, as calculated in 14.1.3, should be provided to develop this level of strain.

CHAPTER 13—SEISMIC STRENGTHENING

Many strengthening techniques have been developed and used for repair and rehabilitation of earthquake damaged and seismically deficient structures (Federal Emergency Management Agency 2006). Identification of an effective rehabilitation method is directly related to the outcome of a seismic evaluation of the structure and is based on consideration of many factors, including type of structure, rehabilitation objective, strengthening scheme effectiveness, constructability, and cost.

A classification of seismic rehabilitation methods for buildings in ASCE/SEI 41 and ACI 369R gives the following strategies: local modification of components, removal or lessening of existing irregularities and discontinuities, global structural stiffening, global structural strengthening, mass reduction, seismic isolation, and supplemental energy dissipation. Strengthening using FRP materials and systems allows for local modification of components and can be implemented in improving the overall seismic performance of the structure. The main advantages of FRP strengthening can be summarized as follows:

a) At the component level, FRP strengthening can be used to efficiently mitigate brittle mechanisms of failure. These may include shear failure of unconfined beam-column joints; shear failure of beams, columns, or both; and lap splice failure. FRP strengthening can also be used to increase the flexural capacity of reinforced concrete members, to resist the buckling of flexural steel bars, and to increase the inelastic rotational capacity of reinforced concrete members.

b) Implementing FRP strengthening schemes translates into an increase in the global displacement and energy dissipation capacities of the structure, thus improving the overall behavior of reinforced concrete structures subjected to seismic actions.

c) FRP shear strengthening and confinement has a small effect on the stiffness or mass of the structure. In such cases, a reevaluation of the seismic demand after strengthening is typically not required. When the structural stiffness needs to be increased, FRP strengthening of local components can be coupled with other traditional global upgrade techniques.

Many research programs have evaluated the adequacy of externally bonded FRP composites for seismic rehabilitation of concrete structures (Haroun et al. 2005; Pantelides et al. 2000; Ghobarah and Said 2002; Gergely et al. 2000; Antonopoulos and Triantafillou 2002; Hamed and Rabinovitch 2005; Pampanin et al. 2007; Di Ludovico et al. 2008a). Other research programs have confirmed the potential of FRP techniques for upgrading the seismic performance of local elements such as reinforced concrete columns (Bousias et al. 2004) and connections (Antonopoulos and Triantafillou 2002; Prota et al. 2004). Research results for FRP applied at the local element or partial structural frame level were subsequently validated on full-scale structures (Pantelides et al. 2000; 2004; Balsamo et al. 2005; Engindeniz et al. 2008a,b). In addition, several structures that include FRP-strengthened members have experienced seismic events. Failure of these members has not been reported.

This chapter presents design guidelines for the seismic strengthening of reinforced concrete elements using externally bonded FRP composites. The design guidelines described herein are intended to be used in conjunction with the fundamental concepts, analysis procedures, design philosophy, seismic rehabilitation objectives, and acceptance criteria set forth in documents such as ASCE/SEI 41 and ACI 369R. Strengthening of RC building components or structures with FRP shall follow capacity protection principles. In capacity design (Hollings 1968; Park and Paulay 1976), a desirable mechanism of inelastic response under seismic action is ensured by providing a strength hierarchy (strong column-weak beam; shear strength > flexural strength). Application of these design guidelines for the seismic rehabilitation of nonbuilding structures such as bridges, wharves, silos, and nuclear facilities warrant additional consideration.

These guidelines do not provide information required to complete a seismic evaluation of an existing structure, determine if retrofit is required, or identify the seismic deficiencies that need to be corrected to achieve the desired performance objective. These guidelines are also not meant to address post-seismic conditions or residual strength of the structure and the FRP retrofit system. After a seismic event, a structure that has been retrofitted with FRP composites could develop large displacements and excessive cracking, resulting in residual stresses or damage to the FRP system. In such cases, an investigation of the stability, ductility, and residual strength of the structure should be performed after the seismic event to assess the adequacy of the existing FRP retrofit system and to determine if additional remedial measures are needed.

13.1—Background

One of the most comprehensive documents developed to assess the need for seismic rehabilitation of reinforced concrete buildings is ASCE/SEI 41. FEMA P695 (Federal Emergency Management Agency 2009) provides further guidance in the selection of appropriate design criteria to achieve the seismic performance objectives. ACI 369R estimates the desired seismic performance of concrete components that are largely based on the format and content of ASCE/SEI 41.

FEMA (Federal Emergency Management Agency 2006) provides a complete list of references on technical design standards and analysis techniques that are available to design professionals. Other resources dealing with seismic upgrade of existing reinforced concrete structures can be obtained from Japan Building Disaster Prevention Association (2005), Eurocode 8 (2005), International Federation for Structural Concrete (2003, 2006), Italian National Research Council (2004), and Sabnis et al. (1996).

Experience gained from examining the performance of reinforced concrete structures after a seismic event indicates that many structural deficiencies result from inadequate confinement of concrete, insufficient transverse and continuity reinforcement in connections and structural members, buckling of flexural reinforcement, lap splice failures, and anchorage failures (Priestley et al. 1996; Haroun et al. 2003; Sezen et al. 2003; Pantelides et al. 1999, 2004). These deficiencies have typically led to brittle failures, soft-story failure, and large residual displacements (Moehle et al. 2002; Di Ludovico et al. 2008b; Prota et al. 2004; Pessiki et al. 1990). Experimental work has also demonstrated that externally bonded FRP systems can be effective in addressing many of the aforementioned structural deficiencies (Engindeniz et al. 2005; Pantelides et al. 2008; Silva et al. 2007).

13.2—FRP properties for seismic design

For seismic upgrades, the material environmental factors given in Table 9.4 should be used in the design of the FRP strengthening solution. The creep rupture limits in Table 10.2.9 need not be considered for seismic strengthening applications unless initial strains are imposed on the FRP as part of the retrofit scheme. Typically, when used for seismic retrofit, the FRP material will not be exposed to significant sustained service loads and creep rupture failure will not govern the design. Creep rupture limits should be considered, however, in cases where the application may impose initial or service strains that can produce sustained stresses on the FRP. Some examples include applications with expansive grouts, pretensioned FRP, or other methods that generate sustained stress in the FRP material. When this chapter is used in conjunction with ASCE/SEI 41, FRP material properties should be considered lower-bound material properties.

13.3—Confinement with FRP

Jacketing concrete structural members with FRP having the primary fibers oriented around the perimeter of the member provides confinement to plastic hinges, mitigates the splitting failure mode of poorly detailed lap splices, and prevents buckling of the main reinforcing bars.

13.3.1 *General considerations*—In seismic applications, jacketing concrete structural members with FRP is not recommended for rectangular sections with aspect ratios *h/b* greater than 1.5, or face dimensions *b* or *h* exceeding 36 in. (900 mm) (Seible et al. 1997), unless testing demonstrates the effectiveness of FRP for confinement of these members. For rectangular sections with an aspect ratio greater than 1.5, the section can be modified to be circular or oval to enhance the effectiveness of the FRP jacket (Seible et al. 1997). FRP anchors have been shown to increase the effectiveness of the FRP jacket in rectangular sections with aspect ratios greater than 1.5 (Kim et al. 2011).

13.3.2 *Plastic hinge region confinement*—FRP-jacketed reinforced concrete members achieve higher inelastic rotational capacity of the plastic hinge (Seible et al. 1997). FRP jacketing can be used to increase the concrete compressive strength when the concrete member complies with the condition in 12.3. For concrete members that do not satisfy this condition, only the ultimate concrete strains can be increased by FRP jacketing. Increase in flexural strength due to higher concrete compressive strength should be considered to verify that hinges can form prior to reaching the shear strength of members.

The design curvature ϕ_D for a confined reinforced concrete section at the plastic hinge can be calculated using Eq. (13.3.2a).

$$\phi_D = \frac{\theta_p}{L_p} + \phi_{y,frp} \le \phi_{u,frp} \tag{13.3.2a}$$

where θ_p is the plastic rotation demand, which can be determined following the analytical procedures outlined in ASCE/SEI 41. In Eq. (13.3.2a), the curvatures of the FRP-confined section at steel yielding, $\phi_{y,frp}$, and at ultimate capacity, $\phi_{u,frp}$, are determined by Eq. (13.3.2b) and (13.3.2c), and L_p is the plastic hinge length computed using Eq. (13.3.2d).

$$\phi_{y,frp} = \frac{\varepsilon_y}{d - c_{y,frp}} \quad (13.3.2b)$$

where ε_y and $c_{y,frp}$ are the steel strain and depth of the neutral axis at steel yielding, respectively, and d is the distance from the extreme compression fibers to the extreme tension steel.

$$\phi_{u,frp} = \frac{\varepsilon_{ccu}}{c_{u,frp}} \quad (13.3.2c)$$

where ε_{ccu} and $c_{u,frp}$ are the extreme compression fiber strain and depth of the neutral axis at ultimate, respectively.

For beams, the plastic hinge length L_p can be approximated as twice the beam height ($2h$). The plastic hinge length can also be determined using detailed analysis but should not be less than the beam height. In FRP-jacketed columns, the plastic hinge length L_p can be computed using Eq. (13.3.2d) (Priestley et al. 1996)

$$L_p = g + 0.0003 f_y d_{b\ell} \text{ (in.-lb)}$$
$$L_p = g + 0.044 f_y d_{b\ell} \text{ (SI)} \quad (13.3.2d)$$

where $d_{b\ell}$ and f_y are the diameter and yield stress of the flexural steel, respectively, and g is the clear gap between the FRP jacket and adjacent members, as shown in Fig. 13.3.2. The gap g should not be greater than 2 in. (50.8 mm).

In plastic hinge regions, the FRP confinement should be provided over a length not less than the larger of the plastic hinge length and ℓ_o, where ℓ_o is the length, measured along the member axis from the face of the joint, over which special transverse reinforcement must be provided as defined in Chapter 18 of ACI 318-14. It should be noted that plastic hinges may occur at locations other than the ends of the member. Complete wrapping around the perimeter of the member should be used for plastic hinge confinement. Continuous (full) coverage of the plastic hinge length with an FRP jacket is recommended. When a continuous jacket is not possible, discrete transverse FRP strips around the perimeter of the section can be used.

Once the design curvature ϕ_D has been established, the ultimate extreme compression fiber strain in the concrete at ultimate, ε_{ccu}, can be calculated using Eq. (13.3.2e)

$$\varepsilon_{ccu} = \phi_D c_u \leq 0.01 \quad (13.3.2e)$$

where c_u is the neutral axis depth at the ultimate design limit state. For members subjected to combined axial and flexural forces, ε_{ccu} should be limited to 0.01.

Once ε_{ccu} is determined, the thickness of the FRP jacket can be determined in accordance with 12.1 and 12.2. To ensure the shear integrity of the confined concrete section,

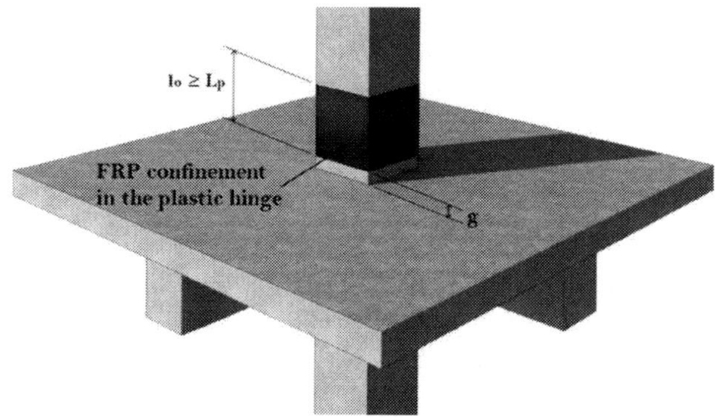

Fig. 13.3.2—Column plastic hinge confinement.

the effective design strain in the FRP jacket, ε_{fe}, should be limited to the value given by Eq. (12.2).

13.3.3 *Lap splice clamping*—The capacity of lap splices having inadequate lap length, especially those located in plastic hinge regions, can be improved by continuously confining the section over at least the length of the splice with externally bonded FRP (Seible et al. 1997; Haroun and Elsanadedy 2005). The required thickness of the FRP jacket can be calculated as follows

where D in inches and E_f in ksi,

circular sections: $nt_f = 145(D/E_f)$
rectangular sections: $nt_f = 218(D/E_f)$

where D in mm and E_f is in MPa (13.3.3a)

circular sections: $nt_f = 1000(D/E_f)$
rectangular sections: $nt_f = 1500(D/E_f)$

where n is the number of FRP plies; t_f is the thickness per ply; D is the diameter of a circular member or the greater dimension of rectangular sections (per Eq. (12.1.2a)); and E_f is the tensile modulus of the FRP jacket.

While confining the section with FRP can mitigate the splitting mode of failure, the pullout failure mode may control the capacity of the confined lap splice. Therefore, regardless of FRP retrofit, the stress in the flexural reinforcing bar, f_s, should not exceed the limit given in Eq. (13.3.3b) (Harries et al. 2006).

$$f_s \leq \frac{33\ell_{prov}\sqrt{f_c'}}{d_{b\ell}\psi_t\psi_e\psi_s} \text{ (in.-lb)}$$
$$f_s \leq \frac{2.75\ell_{prov}\sqrt{f_c'}}{d_{b\ell}\psi_t\psi_e\psi_s} \text{ (SI)} \quad (13.3.3b)$$

where ℓ_{prov} is the length of splice provided; $d_{b\ell}$ is the diameter of the flexural reinforcement; and the ψ factors are those given in Section 25.4 of ACI 318-14.

13.3.4 *Preventing buckling of flexural steel bars*—Continuous or discrete FRP strips having the primary fibers oriented around the perimeter of the member can be used to prevent buckling of the flexural steel bars (Priestley et al.

1996). For circular sections, the volumetric reinforcement ratio provided by transverse FRP, ρ_f, is

$$\rho_f = \frac{4nt_f}{D}\frac{w_f}{s_f} \qquad (13.3.4a)$$

where n is the number of FRP plies; t_f is the thickness per ply; D is the diameter of the section; w_f is the FRP strip width; and s_f is the center-to-center spacing of the FRP strips. For continuous confinement, $w_f/s_f = 1$. In rectangular sections, the volumetric reinforcement ratio provided by the FRP, ρ_f, is (Priestley et al. 1996)

$$\rho_f = 2nt_f\left(\frac{b+h}{bh}\right)\frac{w_f}{s_f} \qquad (13.3.4b)$$

where b and h are the dimensions of the rectangular section.

The amount of volumetric transverse reinforcement ratio should be at least

$$\rho_f \geq \frac{0.0052\rho_\ell D}{d_{b\ell}}\frac{f_y}{f_{fe}} \qquad (13.3.4c)$$

where ρ_ℓ is the flexural reinforcement ratio; D is the diameter of a circular section or the diagonal length of a rectangular section (Eq. (12.1.2a)); $d_{b\ell}$ and f_y are the diameter and the yield strength of the flexural reinforcement, respectively; ρ_f is the volumetric transverse reinforcement ratio computed by Eq. (13.3.4a) or (13.3.4b); and f_{fe} is the effective design stress in the FRP jacket computed by Eq. (13.3.4d)

$$f_{fe} = \varepsilon_{fe}E_f \qquad (13.3.4d)$$

where ε_{fe} is the effective design strain in the FRP jacket given by Eq. (12.2) and E_f is the tensile modulus of the FRP jacket.

When discrete FRP strips rather than a continuous jacket are used, the clear spacing between FRP strips should not exceed the limits in Eq. (13.3.4e)

$$s_f \leq \left[3-6\left(\frac{f_u}{f_y}-1\right)\right]d_{b\ell} \leq 6d_{b\ell} \qquad (13.3.4e)$$

where f_u, f_y, and $d_{b\ell}$ are the ultimate and yield strengths and the smallest diameter of the internal flexural reinforcement, respectively. The clear spacing should not exceed 6 in. (150 mm). These requirements ensure that if the cover concrete spalls in the region between strips, the FRP can provide sufficient resistance against bar buckling. This approach neglects any contribution from the existing internal transverse reinforcement because the internal ties may not coincide within the open spaces between the FRP strips, and the interaction of the internal ties and external FRP strips has not been studied.

13.4—Flexural strengthening

The flexural capacity of reinforced concrete beams and columns in expected plastic hinge regions can be enhanced using FRP only in cases where strengthening will eliminate inelastic deformations in the strengthened region and transfer inelastic deformations to other locations in the member or the structure that are able to handle the ensuing ductility demands. The required flexural strength should be calculated in accordance with the design standard being used for rehabilitation, such as ASCE/SEI 41 and ACI 369R. When this chapter is used in conjunction with ASCE/SEI 41, the strengthened reinforced concrete members with FRP should be considered force-controlled unless a deformation-controlled classification can be justified based on experimental data.

The flexural capacity of reinforced concrete beams and columns can be enhanced using the design methodology presented in Chapter 10. The flexural strength ϕM_n should satisfy the requirement of Eq. (13.4)

$$\phi M_n \geq M_u \qquad (13.4)$$

where M_u is the ultimate moment demand resulting from combined gravity and seismic demands. The flexural capacity of reinforced concrete members should be evaluated based on concrete and reinforcing steel strain limits set forth in the design standard. ASCE/SEI 41 provides a comprehensive list of concrete and reinforcing steel strain limits. In addition, the stress in the reinforcing steel should be limited to the stress that can be achieved based on the existing development lengths and lap-splice details. The strength reduction factor ϕ should be per the design standard being used for the rehabilitation. The additional strength reduction factor for FRP, ψ_f, shall be applied to the flexural contribution of the FRP reinforcement as described in 10.2.10.

13.4.2 *Development and anchorage of flexural FRP reinforcement*—This section provides conceptual methods for anchorage of flexural FRP reinforcement under seismic loads. Any anchorage method must be properly evaluated before it is selected for field implementation.

In seismic applications and within plastic hinge regions, the flexural FRP reinforcement should be confined using FRP strips that completely wrap around the perimeter of the section. Alternatively, the flexural FRP reinforcement could be confined over its entire length to provide higher resistance against debonding of the flexural FRP reinforcement. Because no anchorage design guidelines are currently available, the performance of any anchorage system should be substantiated through representative physical testing.

Such detailing provides higher resistance against debonding of the flexural FRP reinforcement. In applications involving floor systems, complete wrapping of the beam may require localized cutting of the slab to continue the FRP around the section.

Away from the plastic hinge region, transverse FRP U-wrap strips should be used to provide anchorage to the FRP flexural reinforcement. Other anchorage systems may also be used alone or in conjunction with FRP U-wrap strips.

Anchorage systems must be verified experimentally to demonstrate their effectiveness in preventing the debonding of the flexural FRP reinforcement. Several details for FRP anchorage at ends of retrofitted members are discussed in Orton et al. (2009).

The area of the transverse FRP wrap reinforcement, $A_{f,anchor}$, should be determined in accordance with Eq. (14.1.2). In addition, the length over which the FRP anchorage wraps are provided, $\ell_{d,E}$, should not be less than the value given by Eq. (13.4.2a).

$$\ell_{d,E} \geq \ell_o + \ell_{df} \quad (13.4.2a)$$

where ℓ_o is defined per Fig. 13.3.2, and ℓ_{df} is the required development length of the FRP system computed using Eq. (14.1.3).

For adequately anchored flexural FRP reinforcement, the effective design strain for FRP should be limited to

$$\varepsilon_{fd} \leq \min(0.90\varepsilon_{fu}, C_E\varepsilon_{fu}) \quad (13.4.2b)$$

Figure 13.4.2 depicts a conceptual detail for flexural strengthening of beams and columns at a joint and is intended to convey the critical elements of such a flexural strengthening. The design professional should detail the flexural FRP reinforcement to achieve continuity of the FRP across the joint. Appropriate development of the flexural FRP at both ends as well as adequate transverse reinforcement for confinement of the flexural FRP should be provided.

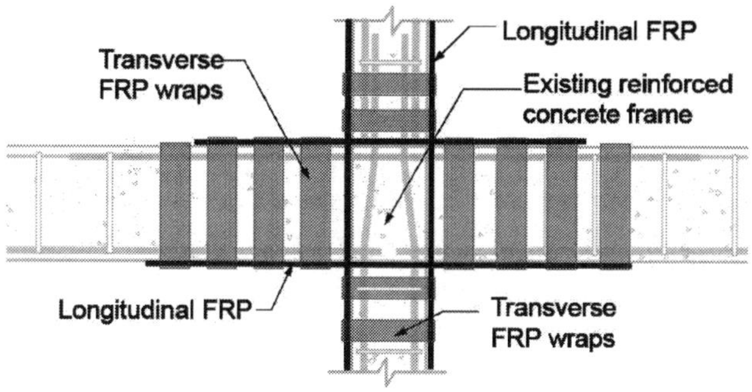

Fig. 13.4.2—Conceptual FRP strengthening detail (cross section elevation).

13.5—Shear strengthening

FRP shear strengthening can prevent brittle failures and promote the development of plastic hinges, resulting in an enhanced seismic behavior of concrete members. The design shear strength ϕV_n of a concrete member strengthened with FRP should satisfy Eq. (13.5)

$$\phi V_n \geq V_e \quad (13.5)$$

where ϕ should be per the design standard being used for the rehabilitation, and V_e is the design shear force. When this chapter is used in conjunction with ASCE/SEI 41, the shear in the strengthened member should be considered force-controlled unless a deformation-controlled classification can be justified based on experimental data.

13.5.1 *Design shear force* V_e—The design shear force should be calculated in accordance with the design standard being used for the rehabilitation, such as ASCE/SEI 41 and ACI 369R. The shear capacity should be equal to or greater than the shear corresponding to the flexural capacity of the section. For example, when the rehabilitation is based on ASCE/SEI 41, the design shear force is based on the seismic category and targeted seismic performance of the structure. When required by ASCE/SEI 41 for the determination of the design shear force, the calculation of the probable flexural strength should be based on FRP stress taken as the lesser of $1.2f_{fd}$ and f_{fu}, ϕ should be taken as unity, and the FRP strength reduction factor ψ_f should be 1.0. Other limits for FRP strain and strength specified in this document should also be considered.

13.5.2 *Nominal shear strength* V_n—The shear strength of the existing member V_n^* should be determined following the procedures described in the design standard being used for rehabilitation, such as ASCE/SEI 41 and ACI 369R. The shear strength of an FRP-strengthened concrete member is calculated using Eq. (13.5.2).

$$V_n = V_n^* + \psi_f V_f \quad (13.5.2)$$

where ψ_f is the reduction factor applied to the contribution of the FRP system in accordance with Chapter 11. The contributions of FRP to shear strength, V_f, should be determined in accordance with Chapter 11. To account for effects of stress reversal, FRP shear strengthening should be provided with complete continuity around the perimeter of the section.

13.6—Beam-column joints

Experimental tests (Bracci et al. 1992a; Prota et al. 2004; Pampanin et al. 2007) and observations of post-seismic damage (Moehle et al. 2002) in structures designed to withstand only gravity loads show that unconfined beam-column joints frequently led to brittle failures and prevented structures from achieving higher global displacements before failure. Experimental evidence (Pantelides et al. 2008; Silva et al. 2007; Pampanin et al. 2007; Bracci et al. 1992b,c) has shown that FRP systems can be effective for increasing the shear and energy dissipation capacity of unconfined joints. FRP layout and detailing will depend on the geometry of the existing joint and the number of members framing into it. FRP reinforcement in both directions is typically required at the joint to resist the cyclic loading effects of a seismic event (Engindeniz et al. 2008a). The FRP used to confine the joints should be anchored to be effective. Pantelides et al. (2008), Silva et al. (2007), and Engindeniz et al. (2008b) provide guidance on determining if FRP is a viable option for enhancing the performance of unconfined joints. Additionally, FRP reinforcement can be used to provide continuity across joints with discontinuous internal reinforcement (Orton et al. 2009).

13.7—Strengthening reinforced concrete shear walls

13.7.1 *General considerations*—This section presents design guidelines for the seismic strengthening of reinforced concrete walls. Applying horizontal FRP strips along the height of the walls can increase the shear capacity of reinforced concrete shear walls. For short or squat walls, with height-to-length ratios less than 1.5, vertical FRP strips may also be required (ACI 318). Likewise, the in-plane flexural capacity of reinforced concrete shear walls can be increased by placing vertical FRP strips at the ends or boundaries of walls (Lombard et al. 2000; Hiotakis et al. 2004).

The shear strength of walls reinforced with FRP for flexure should be evaluated and compared to the shear strength corresponding to the nominal flexural strength of the retrofitted structure to promote a flexural failure rather than a brittle shear failure. Similarly, a shear retrofit should achieve greater shear capacity than the shear corresponding to the nominal flexural capacity of the wall. When this chapter is used in conjunction with ASCE/SEI 41, the flexure and shear in the strengthened portion of the wall should be considered force-controlled action unless a deformation-controlled classification is justified based on experimental data.

13.7.2 *Flexural strengthening*—FRP reinforcement for flexural strengthening of walls may be provided on one or both sides of the wall. Figure 13.7.2 shows a wall retrofitted with FRP reinforcement placed at the extreme ends of the wall. This figure also provides a description of the main variables required for design.

13.7.2.1 *Concrete strain limits*—The concrete compressive strains ε_c should be limited by Eq. (13.7.2.1a)

$$\varepsilon_c = \varepsilon_{fc}\left(\frac{1}{L_w/c-1}\right) \leq \varepsilon_{cu} \qquad (13.7.2.1a)$$

where ε_{fd} corresponds to the strain at which debonding of the FRP may occur, per Eq. (10.1.1). In concrete shear walls, the concrete compressive strains at ultimate, ε_{cu}, should be limited to the following values (Wallace 1995)

$\varepsilon_{cu} \leq 0.010$ for confined concrete at boundaries

(13.7.2.1b)

$\varepsilon_{cu} \leq 0.003$ for unconfined concrete at boundaries

When confined boundary elements are required per Eq. (13.7.2.1b), means other than FRP may be required to meet the concrete strain requirements.

13.7.2.2 *Anchorage of flexurally strengthened walls*—Flexurally strengthened walls require anchorage to the foundations for load path continuity. Similarly, flexural FRP should be continuous through existing slabs to ensure continuity of the load path. Two conceptual methods for anchorage of a strengthened shear wall to the foundation are provided in Fig. 13.7.2.2. Any anchorage method, including the ones shown in Fig. 13.7.2.2, should be properly evaluated prior to field implementation. In shear walls, the vertical flexural FRP reinforcement does not need to be confined by transverse FRP strips or U-wraps that extend around the perimeter of the section.

13.7.3 *Shear strengthening of reinforced concrete shear walls*—Experimental investigations have demonstrated

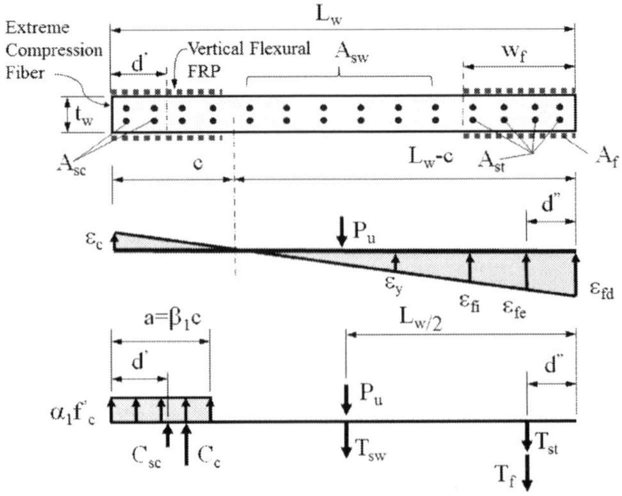

Fig. 13.7.2—FRP reinforcement for flexural strengthening.

the effectiveness of FRP for enhancing the shear performance of reinforced concrete walls subjected to seismic or cyclic loading (Haroun and Mosallam 2002; Khomwan and Foster 2005). The design shear strength ϕV_n of a reinforced concrete shear wall strengthened with FRP should satisfy Eq. (13.7.2.2a).

$$\phi V_n \geq V_u \qquad (13.7.2.2a)$$

The strength reduction factor ϕ should be per the design standard being used for the rehabilitation.

For shear walls with externally bonded FRP, the nominal shear strength V_n can be computed using Eq. (13.7.2.2b)

$$V_n = V_n^* + \psi_f V_f \qquad (13.7.2.2b)$$

where V_n^* is the nominal shear strength of the existing shear wall; ψ_f is the reduction factor applied to the contribution of the FRP in accordance with Chapter 11; and V_f is the shear strength provided by the FRP. The shear strength enhancement for a wall section of length L_w in the direction of the applied shear force, with a laminate thickness t_f on two sides or one side of the wall, can be calculated using Eq. (13.7.2.2c) (Haroun et al. 2005)

for a two-sided retrofit: $V_f = 2t_f\varepsilon_{fe}E_f d_{fv}$

(13.7.2.2c)

for a one-sided retrofit: $V_f = 0.75t_f\varepsilon_{fe}E_f d_{fv}$

where d_{fv} is the effective depth of the shear wall, as defined by Chapter 18 of ACI 318-14, but not to exceed h_w, and ε_{fe} is according to 11.4.1 of this guide. FRP should be provided on two faces of the wall if the ratio of the existing transverse steel reinforcement to gross concrete area, ρ_t, is less than 0.0015. The intent of this provision is to ensure proper shear resistance of concrete in the event of severe cracking during a seismic event. The maximum nominal shear strength of a wall segment should not exceed the value given in Eq. (13.7.2.2d)

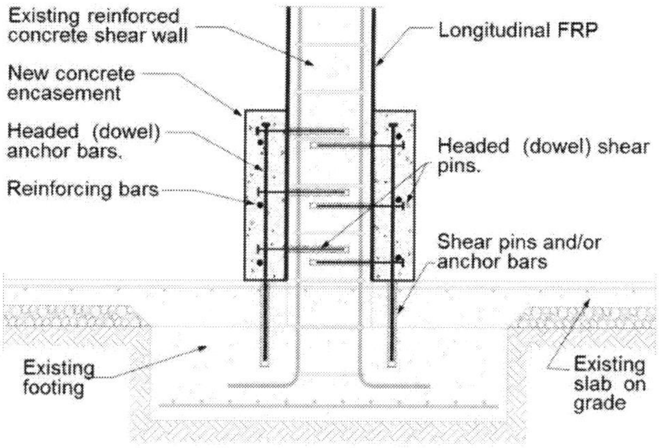

(a) Anchorage with concrete encasement

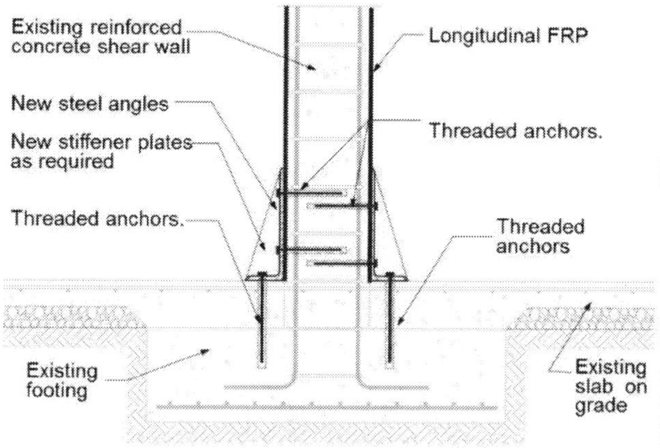

(b) Anchorage with steel angles

Fig. 13.7.2.2—Conceptual anchorage methods for strengthened shear wall.

$$V_n \leq 10\sqrt{f_c'}A_{cw} \quad (13.7.2.2d)$$

where A_{cw} is the area of the concrete section of an individual vertical wall.

13.7.3.1 *Detailing of FRP shear reinforcement*—Anchorage of shear FRP is considered good practice but it is not required to attain the shear strengths computed using the provisions of this chapter. Anchoring of the FRP shear reinforcement can be achieved by wrapping the FRP layers around the ends of the wall, by using mechanical anchorage devices such as steel anchors and steel plates (Paterson and Mitchell 2003), or by using FRP anchors (Binici and Ozcebe 2006).

The maximum clear spacing between the FRP shear strips should be limited to the minimum of one-fifth of the overall length of the wall, three times the thickness of the wall, or 18 in. (457 mm).

CHAPTER 14—FIBER-REINFORCED POLYMER REINFORCEMENT DETAILS

This chapter provides guidance for detailing externally bonded fiber-reinforced polymer (FRP) reinforcement. Detailing will typically depend on the geometry of the structure, the soundness and quality of the substrate, and the levels of load that are to be sustained by the FRP sheets or laminates. Many bond-related failures can be avoided by following these general guidelines for detailing FRP sheets or laminates:

a) Do not turn inside corners such as at the intersection of beams and joists or the underside of slabs

b) Provide a minimum 1/2 in. (13 mm) radius when the sheet is wrapped around outside corners

c) Provide adequate development length

d) Provide sufficient overlap when splicing FRP plies

14.1—Bond and delamination

The actual distribution of bond stress in an FRP laminate is complicated by cracking of the substrate concrete. The general elastic distribution of interfacial shear stress and normal stress along an FRP laminate bonded to uncracked concrete is shown in Fig. 14.1.

The weak link in the concrete/FRP interface is the concrete. The soundness and tensile strength of the concrete substrate will limit the overall effectiveness of the bonded FRP system. Design requirements to mitigate FRP debonding failure modes are discussed in 10.1.1.

14.1.1 *FRP debonding*—In reinforced concrete members having relatively long shear spans or where the end peeling (refer to 14.1.2) has been effectively mitigated, debonding may initiate at flexural cracks, flexural/shear cracks, or both, near the region of maximum moment. Under loading, these cracks open and induce high local interfacial shear stress that initiates FRP debonding that propagates across the shear span in the direction of decreasing moment. Typically, this failure does not engage the aggregate in the concrete, progressing through the thin mortar-rich layer comprising the surface of the concrete substrate. This failure mode is exacerbated in regions having a high shear-moment ratio.

Anchorage systems, such as U-wraps, mechanical fasteners, fiber anchors, and near-surface-mounted (NSM) anchors, have been proven successful at delaying, and sometimes preventing, debonding failure of the longitudinal FRP (Kalfat et al. 2013; Grelle and Sneed 2013). Numerical and experimental studies have shown that these systems can increase the effective strain in the flexural FRP to values up to tensile rupture (Lee et al. 2010; Orton et al. 2008). A few studies have proposed analytical models to predict the behavior of specific anchor systems (Kim and Smith 2010); however, no published anchorage design guidelines are currently available. Therefore, the performance of any anchorage system should be substantiated through representative physical testing.

14.1.2 *FRP end peeling*—FRP end peeling (also referred to as concrete cover delamination) can result from the normal stresses developed at the ends of externally bonded FRP reinforcement. With this type of delamination, the existing internal reinforcing steel provides a weak horizontal plane along which the concrete cover pulls away from the rest of the beam, as shown in Fig. 14.1.2a.

The tensile concrete cover splitting failure mode is controlled, in part, by stress at the termination point of

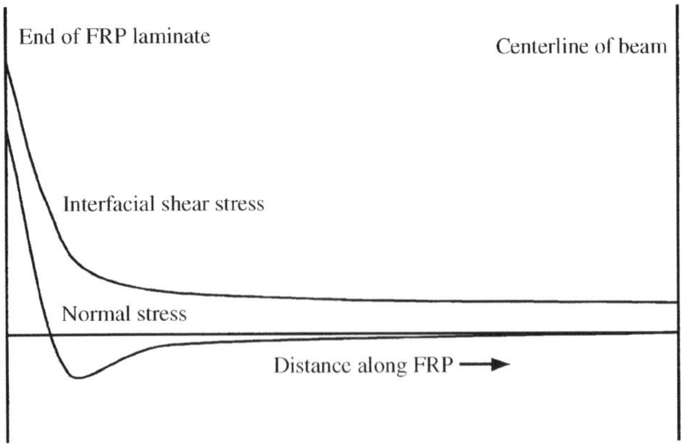

Fig. 14.1—Conceptual interfacial shear and normal stress distributions along the length of a bonded FRP laminate (Roberts and Haji-Kazemi 1989; Malek et al. 1998).

the FRP. In general, the FRP end peeling failure mode can be mitigated by using anchorage (U-wraps, mechanical fasteners, fiber anchors, or NSM anchors), by minimizing the stress at the FRP curtailment by locating the curtailment as close to the region of zero moment as possible, or by both. When the factored shear force at the termination point is greater than two-thirds of the concrete shear strength ($V_u > 0.67V_c$), the FRP laminates should be anchored with transverse reinforcement to prevent the concrete cover layer from splitting. The area of the transverse clamping FRP U-wrap reinforcement, $A_{fanchor}$, can be determined in accordance with Eq. (14.1.2) (Reed et al. 2005)

$$A_{fanchor} = \frac{(A_f f_{fe})_{longitudinal}}{(E_f \kappa_v \varepsilon_{fu})_{anchor}} \quad (14.1.2)$$

in which κ_v is calculated using Eq. (11.4.1.2b). Instead of detailed analysis, the following general guidelines for the location of cutoff points for the FRP laminate can be used to avoid end peeling failure mode:

a) For simply supported beams, a single-ply FRP laminate should be terminated at least a distance equal to ℓ_{df} past the point along the span at which the resisted moment falls below the cracking moment M_{cr}. For multiple-ply laminates, the termination points of the plies should be tapered. The outermost ply should be terminated not less than ℓ_{df} past the point along the span at which the resisted moment falls below the cracking moment. Each successive ply should be terminated not less than an additional 6 in. (150 mm) beyond the previous ply (Fig. 14.1.2b).

b) For continuous beams, a single-ply FRP laminate should be terminated at least a distance $d/2$ or 6 in. (150 mm) beyond the inflection point (point of zero moment resulting from factored loads). For multiple-ply laminates, the termination points of the plies should be tapered. The outermost ply should be terminated no less than 6 in. (150 mm) beyond the inflection point. Each successive ply should be terminated no less than an additional 6 in. (150 mm) beyond the previous ply. For example, if a three-ply laminate is required, the ply directly in contact with the concrete substrate should be terminated at least 18 in. (450 mm) past the inflection point (Fig. 14.1.2b). These guidelines apply for positive and negative moment regions.

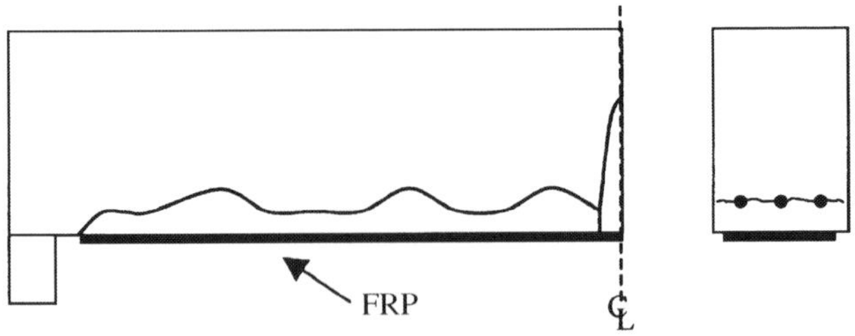

Fig. 14.1.2a—Delamination caused by tension failure of the concrete cover.

14.1.3 *Development length*—The bond capacity of FRP is developed over a critical length ℓ_{df}. To develop the effective FRP stress at a section, the available anchorage length of FRP should exceed the value given by Eq. (14.1.3) (Teng et al. 2003).

$$\ell_{df} = 0.057\sqrt{\frac{nE_f t_f}{\sqrt{f'_c}}} \quad \text{(in.-lb)}$$

$$\ell_{df} = \sqrt{\frac{nE_f t_f}{\sqrt{f'_c}}} \quad \text{(SI)} \qquad (14.1.3)$$

14.2—Detailing of laps and splices

Splices of FRP laminates should be provided only as permitted on drawings, specifications, or as authorized by the licensed design professional as recommended by the system manufacturer.

The fibers of FRP systems should be continuous and oriented in the direction of the largest tensile forces. Fiber continuity can be maintained with a lap splice. For FRP systems, a lap splice should be made by overlapping the fibers along their length. The required overlap, or lap-splice length, depends on the tensile strength and thickness of the FRP material system and on the bond strength between adjacent layers of FRP laminates. Sufficient overlap should be provided to promote the failure of the FRP laminate before debonding of the overlapped FRP laminates. The required overlap for an FRP system should be provided by the material manufacturer and substantiated through testing that is independent of the manufacturer.

Jacket-type FRP systems used for column members should provide appropriate development area at splices, joints, and termination points to ensure failure through the FRP jacket thickness rather than failure of the spliced sections.

For unidirectional FRP laminates, lap splices are required only in the direction of the fibers. Lap splices are not required in the direction transverse to the fibers. FRP laminates consisting of multiple unidirectional sheets oriented in more than one direction or multidirectional fabrics require lap splices in more than one direction to maintain the continuity of the fibers and the overall strength of the FRP laminates.

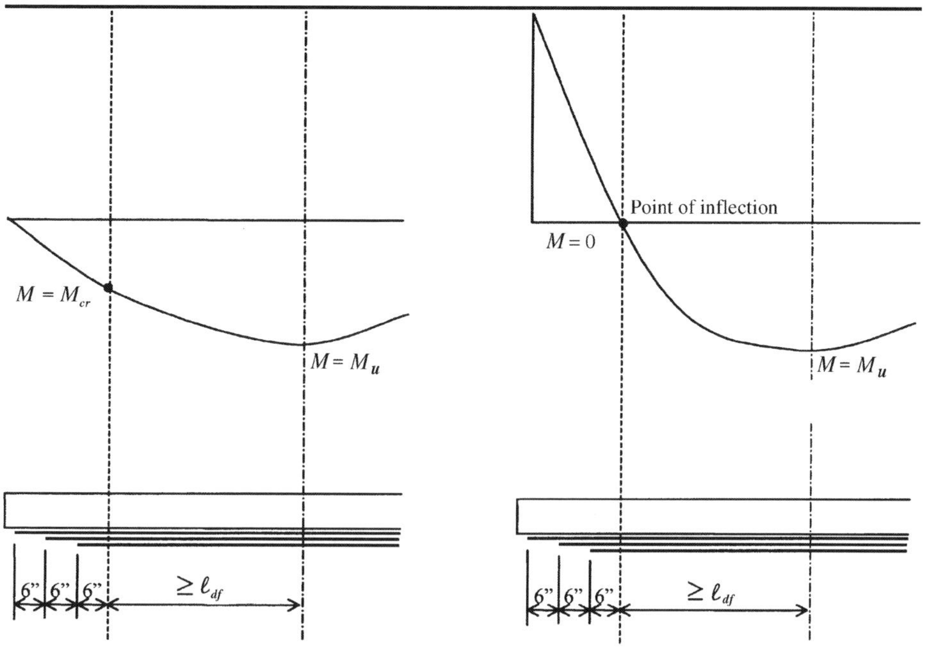

(a) Simply supported beam (b) Continuous beam

Fig. 14.1.2b—Graphical representation of the guidelines for allowable termination points of a three-ply FRP laminate.

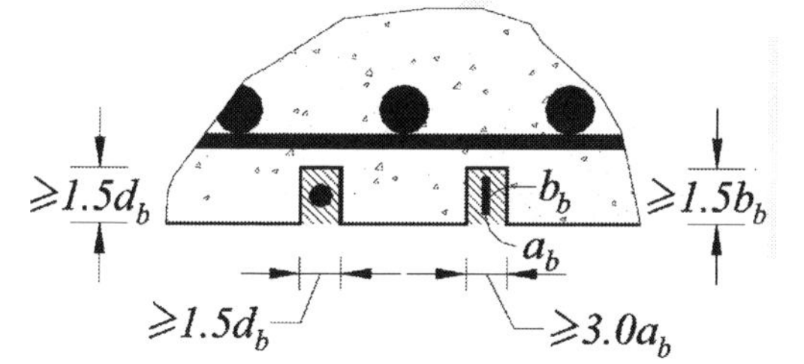

Fig. 14.3a—Minimum dimensions of grooves.

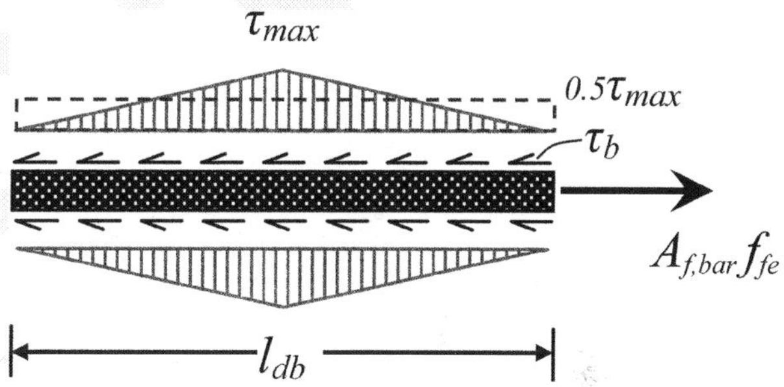

Fig. 14.3b—Transfer of force in NSM FRP bars.

14.3—Bond of near-surface-mounted systems

For near-surface-mounted (NSM) systems, the minimum dimension of the grooves should be taken at least 1.5 times the diameter of the FRP bar (De Lorenzis and Nanni 2001; Hassan and Rizkalla 2003). When a rectangular bar with a large aspect ratio is used, however, the limit may lose significance due to constructibility. In such a case, a minimum groove size of $3.0a_b$ x $1.5b_b$, as depicted in Fig. 14.3a, is suggested, where a_b is the smallest bar dimension. The minimum clear groove spacing for NSM FRP bars should be greater than twice the depth of the NSM groove to avoid overlapping of the tensile stresses around the NSM bars. Furthermore, a clear edge distance of four times the depth of the NSM groove should be provided to minimize edge effects that could accelerate debonding failure (Hassan and Rizkalla 2003).

Bond properties of NSM FRP bars depend on many factors such as cross-sectional shape and dimensions and surface properties of the FRP bar (Hassan and Rizkalla 2003; De Lorenzis et al. 2004). Figure 14.3b shows the equilibrium condition of an NSM FRP bar with an embedded length equal to its development length ℓ_{db} having a bond strength of τ_{max}. Using a triangular stress distribution, the average bond strength can be expressed as $\tau_b = 0.5\tau_{max}$. Average bond strength τ_b for NSM FRP bars in the range of 500 to 3000 psi (3.5 to 20.7 MPa) has been reported (Hassan and Rizkalla 2003; De Lorenzis et al. 2004); therefore, τ_b = 1000 psi (6.9 MPa) is recommended for calculating the bar development length. Using force equilibrium, the following equations for development length can be derived

$$\ell_{db} = \frac{d_b}{4\tau_b} f_{fd} \text{ for circular bars} \tag{14.3a}$$

$$\ell_{db} = \frac{a_b b_b}{2(a_b + b_b)(\tau_b)} f_{fd} \text{ for rectangular bars} \tag{14.3b}$$

CHAPTER 15—DRAWINGS, SPECIFICATIONS, AND SUBMITTALS

15.1—Engineering requirements

Although federal, state, and local codes for the design of externally bonded fiber-reinforced polymer (FRP) systems do not exist, other applicable code requirements may influence the selection, design, and installation of the FRP system. For example, code requirements related to fire or potable water may influence the selection of the coatings used with the FRP system. All design work should be performed under the guidance of a licensed design professional familiar with the properties and applications of FRP strengthening systems.

15.2—Drawings and specifications

The licensed design professional should document calculations summarizing the assumptions and parameters used to design the FRP strengthening system and should prepare design drawings and project specifications. The drawings and specifications should show, at a minimum, the following information specific to externally applied FRP systems:

a) FRP system to be used

b) Location of the FRP system relative to the existing structure

c) Dimensions and orientation of each ply, laminate, or near-surface-mounted (NSM) bar

d) Number of plies and bars and the sequence of installation

e) Location of splices and lap length

f) General notes listing design loads and allowable strains in the FRP laminates

g) Material properties of the FRP laminates and concrete substrate

h) Concrete surface preparation requirements, including corner preparation, groove dimensions for NSM bars, and maximum irregularity limitations

i) Installation procedures, including surface temperature and moisture limitations, and application time limits between successive plies

j) Curing procedures for FRP systems

k) Protective coatings and sealants, if required

l) Shipping, storage, handling, and shelf-life guidelines

m) Quality control and inspection procedures, including acceptance criteria

n) In-place load testing of installed FRP system, if necessary

15.3—Submittals

Specifications should require the FRP system manufacturer, installation contractor, and inspection agency (if required) to submit product information and evidence of their qualifications and experience to the licensed design professional for review.

15.3.1 *FRP system manufacturer*—Submittals required of the FRP system manufacturer should include:

a) Indication of compliance with existing specifications (such as ACI 440.8) as applicable

b) Product data sheets indicating the physical, mechanical, and chemical characteristics of the FRP system and all its constituent materials

c) Tensile properties of the FRP system, including the method of reporting properties (net fiber or gross laminate), test methods used, and the statistical basis used for determining the properties (4.3)

d) Installation instructions, maintenance instructions, and general recommendations regarding each material to be used; installation procedures should include surface preparation requirements

e) Manufacturer's Safety Data Sheets (SDSs) for all materials to be used

f) Quality control procedure for tracking FRP materials and material certifications

g) Durability test data for the FRP system in the types of environments expected

h) Structural test reports pertinent to the proposed application

i) Reference projects

15.3.2 *FRP system installation contractor*—Submittals required of the FRP system installation contractor should include:

a) Documentation from the FRP system manufacturer of having been trained to install the proposed FRP system

b) Project references, including installations similar to the proposed installation (for example, for an overhead application, the contractor should submit a list of previous installations involving the installation of the proposed FRP system in an overhead application)

c) Evidence of competency in surface preparation techniques

d) Quality control testing procedures including voids and delaminations, FRP bond to concrete, and FRP tensile properties

e) Daily log or inspection forms used by the contractor

15.3.3 *FRP system inspection agency*—If an independent inspection agency is used, submittals required of that agency should include:

a) A list of inspectors to be used on the project and their qualifications

b) Sample inspection forms

c) A list of previous projects inspected by the inspector

CHAPTER 16—DESIGN EXAMPLES

16.1—Calculation of FRP system tensile properties

This example calculation shown in Table 16.1b illustrates the derivation of material properties based on net-fiber area versus the properties based on gross-laminate area. As described in 4.3.1, both methods of determining material properties are valid. It is important, however, that any design calculations consistently use material properties based on only one of the two methods (for example, if the gross-laminate thickness is used in any calculation, the strength based on gross-laminate area should be used in the calculations as well). Reported design properties should be based on a population of 20 or more coupons tested in accordance with ASTM D3039/D3039M. Reported properties should be statistically adjusted by subtracting three standard deviations from the mean tensile stress and strain, as discussed in 4.3.1.

A test panel is fabricated from two plies of a carbon fiber/resin unidirectional fiber-reinforced polymer (FRP) system using the wet layup technique. Based on the known fiber content of this FRP system, the net-fiber area is 0.0065 in.^{2}/in. (0.165 mm^2/mm) width per ply. After the system has cured, five 2 in. (50.8 mm) wide test coupons are cut from the panel. The test coupons are tested in tension to failure in accordance with ASTM D3039/D3039M. Tabulated in Table 16.1a are the results of the tension tests.

Table 16.1a—FRP system tension test results

Coupon ID	Specimen width		Measured coupon thickness		Measured rupture load	
	in.	mm.	in.	mm.	kips	kN
T-1	2	50.8	0.055	1.40	17.8	79.2
T-2	2	50.8	0.062	1.58	16.4	72.9
T-3	2	50.8	0.069	1.75	16.7	74.3
T-4	2	50.8	0.053	1.35	16.7	74.3
T-5	2	50.8	0.061	1.55	17.4	77.4
Average	2	50.8	0.060	1.52	17.0	75.6

Table 16.1b—FRP system net fiber and gross laminate property calculations

Net-fiber area property calculations		Gross-laminate area property calculations	
Calculate A_f using the known, net-fiber area ply thickness: $A_f = nt_fw_f$	$A_f = (2)(0.0065\text{ in.}^2/\text{in.})(2\text{ in.}) = 0.026\text{ in.}^2$	Calculate A_f using the average, measured laminate thickness: $A_f = t_fw_f$	$A_f = (0.060\text{ in.})(2\text{ in.}) = 0.120\text{ in.}^2$
	$A_f = (2)(0.165\text{ mm}^2/\text{mm})(50.8\text{ mm}) = 16.8\text{ mm}^2$		$A_f = (1.52\text{ mm})(50.8\text{ mm}) = 77.4\text{ mm}^2$
Calculate the average FRP system tensile strength based on net-fiber area: $\overline{f_{fu}} = \dfrac{\text{average rupture load}}{A_f}$	$\overline{f_{fu}} = \dfrac{17\text{ kip}}{0.026\text{ in.}^2} = 650\text{ ksi}$	Calculate the average FRP system tensile strength based on gross-laminate area: $\overline{f_{fu}} = \dfrac{\text{average rupture load}}{A_f}$	$\overline{f_{fu}} = \dfrac{17\text{ kip}}{0.120\text{ in.}^2} = 140\text{ ksi}$
	$\overline{f_{fu}} = \dfrac{75.62\text{ kN}}{16.8\text{ mm}^2} = 4.5\text{ kN/mm}^2$		$\overline{f_{fu}} = \dfrac{75.62\text{ kN}}{77.4\text{ mm}^2} = 0.997\text{ kN/mm}^2$
Calculate the average FRP system tensile strength per unit width based on net-fiber area: $\overline{p_{fu}} = \dfrac{\overline{f_{fu}}A_f}{w_f}$	$\overline{p_{fu}} = \dfrac{(650\text{ ksi})(0.026\text{ in.}^2)}{2\text{ in.}} = 8.4\text{ kip/in.}$	Calculate the average FRP system tensile strength per unit width based on laminate area: $\overline{p_{fu}} = \dfrac{\overline{f_{fu}}A_f}{w_f}$	$\overline{p_{fu}} = \dfrac{(140\text{ ksi})(0.120\text{ in.}^2)}{2\text{ in.}} = 8.4\text{ kip/in.}$
	$\overline{p_{fu}} = \dfrac{(4.5\text{ kN/mm}^2)(16.8\text{ mm}^2)}{50.8\text{ mm}} = 1.49\text{ kN/mm}$		$\overline{p_{fu}} = \dfrac{(0.98\text{ kN/mm}^2)(77.4\text{ mm}^2)}{50.8\text{ mm}} = 1.49\text{ kN/mm}$

16.2—Comparison of FRP systems' tensile properties

Two FRP systems are being considered for strengthening concrete members. The mechanical properties of two FRP systems are available from respective manufacturers. System A consists of dry, carbon-fiber unidirectional sheets and is installed with an adhesive resin using the wet layup technique. System B consists of precured carbon fiber/resin laminates that are bonded to the concrete surface with an adhesive resin. Excerpts from the data sheets provided by the FRP system manufacturers are given in Table 16.2a. After reviewing the material data sheets sent by the FRP system manufacturers, the licensed design professional compares the tensile strengths of the two systems.

Because the data sheets for both systems are reporting statistically based properties, it is possible to directly compare the tensile strength and modulus of both systems, as shown in Table 16.2b.

Table 16.2a—Material properties and description of two types of FRP systems

System A (excerpts from data sheet)	System B (excerpts from data sheet)
System type: dry, unidirectional sheet Fiber type: high-strength carbon Polymer resin: epoxy System A is installed using a wet layup procedure where the dry carbon-fiber sheets are impregnated and adhered with an epoxy resin on-site.	System type: precured, unidirectional laminate Fiber type: high-strength carbon Polymer resin: epoxy System B's precured laminates are bonded to the concrete substrate using System B's epoxy paste adhesive.
Mechanical properties*†‡	Mechanical properties*†
t_f = 0.013 in. (0.33 mm)	t_f = 0.050 in. (1.27 mm)
f_{fu}^* = 550 ksi (3792 N/mm²)	f_{fu}^* = 380 ksi (2620 N/mm²)
ε_{fu}^* = 1.6%	ε_{fu}^* = 1.5%
E_f = 33,000 ksi (227,527 N/mm²)	E_f = 22,000 ksi (151,724 N/mm²)

*Reported properties are based on a population of 20 or more coupons tested in accordance with ASTM D3039/D3039M.

†Reported properties have been statistically adjusted by subtracting three standard deviations from the mean tensile stress and strain.

‡Thickness is based on the net-fiber area for one ply of the FRP system. Resin is excluded. Actual installed thickness of cured FRP is 0.04 to 0.07 in. (1.0 to 1.8 mm) per ply.

Table 16.2b—Procedure comparing two types of FRP systems

Procedure	Calculation in in.-lb units	Calculation in SI units
Step 1A—Calculate the tensile strength per unit width of System A $p_{fu}^* = f_{fu}^* t_f$	p_{fu}^* = (550 ksi)(0.013 in.) = 7.15 kip/in.	p_{fu}^* = (3.79 kN/mm²)(0.33 mm) = 1.25 kN/mm
Step 1B—Calculate the tensile strength per unit width of System B $p_{fu}^* = f_{fu}^* t_f$	p_{fu}^* = (380 ksi)(0.050 in.) = 19 kip/in.	p_{fu}^* = (2.62 kN/mm²)(1.27 mm) = 3.33 kN/mm
Step 2A—Calculate the tensile modulus per unit width of System A $k_f = E_f t_f$	k_f = (33,000 ksi)(0.013 in.) = 429 kip/in.	k_f = (227.5 kN/mm²)(0.33 mm) = 75.1 kN/mm
Step 2B—Calculate the tensile modulus per unit width of System B $k_f = E_f t_f$	k_f = (22,000 ksi)(0.050 in.) = 1100 kip/in.	k_f = (151.7 kN/mm²)(1.27 mm) = 192.7 kN/mm
Step 3—Compare the two systems Compare the tensile strengths: p_{fu}^*(System A) p_{fu}^*(System B)	$\frac{p_{fu}^*(\text{System B})}{p_{fu}^*(\text{System A})} = \frac{19\text{ kip/in.}}{7.15\text{ kip/in.}} = 2.66$ ∴ three plies of System A are required for each ply of System B for an equivalent tensile strength	$\frac{p_{fu}^*(\text{System B})}{p_{fu}^*(\text{System A})} = \frac{3.33\text{ kN/mm}}{1.25\text{ kN/mm}} = 2.66$ ∴ three plies of System A are required for each ply of System B for an equivalent tensile strength
Compare the stiffnesses: k_f(System A) k_f(System B)	$\frac{k_f(\text{System B})}{k_f(\text{System A})} = \frac{1100\text{ kip/in.}}{429\text{ kip/in.}} = 2.56$ ∴ three plies of System A are required for each ply of System B for an equivalent stiffness	$\frac{k_f(\text{System A})}{k_f(\text{System B})} = \frac{192.7\text{ kN/mm}}{75.1\text{ kN/mm}} = 2.56$ ∴ three plies of System A are required for each ply of System B for an equivalent stiffness

Because all the design procedures outlined in this document limit the strain in the FRP material, the full nominal strength of the material is not used and should not be the basis of comparison between two material systems. When considering various FRP material systems for a particular application, the FRP systems should be compared based on equivalent stiffness only. In addition, each FRP system under consideration should have the ability to develop the strain associated with the effective strain required by the application without rupturing, $\varepsilon_{fu} > \varepsilon_{fe}$.

In many instances, it may be possible to vary the width of the FRP strip as opposed to the number of plies (use larger widths for systems with lower thicknesses and vice versa). In such instances, equivalent stiffness calculations typically will not yield equivalent contributions to the strength of a member. In general, thinner (lower nt_f) and wider (higher w_f) FRP systems will provide a higher level of strength to a member due to lower bond stresses. The exact equivalency, however, can only be found by performing complete calculations (according to procedures described in Chapters 10, 11, and 12 of this guide) for each system.

16.3—Flexural strengthening of an interior reinforced concrete beam with FRP laminates

A simply supported concrete beam reinforced with three No. 9 bars (Fig. 16.3) is located in an unoccupied warehouse and is subjected to a 50 percent increase in its live-load-carrying requirements. An analysis of the existing beam indicates that the beam still has sufficient shear strength to resist the new required shear strength and meets the deflection and crack-control serviceability requirements. Its flexural strength, however, is inadequate to carry the increased live load.

Summarized in Table 16.3a are the existing and new loadings and associated midspan moments for the beam. The existing reinforced concrete beam should be strengthened with the FRP system described in Table 16.3b, specifically, two 12 in. (305 mm) wide x 23.0 ft (7 m) long plies bonded to the soffit of the beam using the wet layup technique.

By inspection, the degree of strengthening is reasonable in that it does meet the strengthening limit criteria specified in Eq. (9.2). That is, the existing moment strength without FRP, $(\phi M_n)_{w/o}$ = 266 kip-ft (361 kN-m), is greater than the unstrengthened moment limit, $(1.1M_{DL} + 0.75M_{LL})_{new}$ = 177 kip-ft (240 kN-m). The design calculations used to verify this configuration follow in Table 16.3c.

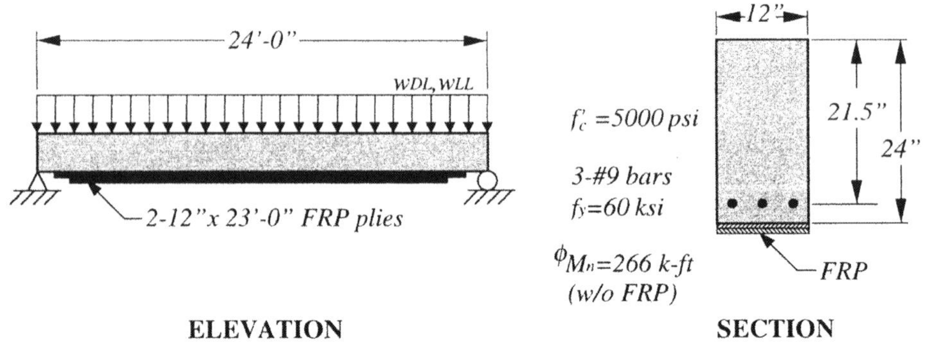

Length of the beam ℓ	24 ft	7.32 m
Width of the beam w	12 in.	305 mm
d	21.5 in.	546 mm
h	24 in.	609.6 mm
f_c'	5000 psi	34.5 N/mm^2
f_y	60 ksi	414 N/mm^2
ϕM_n without FRP	266 k-ft	361 kN-m
Bars	No. 9	ϕ = 28.6 mm

Fig. 16.3—Schematic of the idealized simply supported beam with FRP external reinforcement.

Table 16.3a—Loadings and corresponding moments

Loading/moment	Existing loads		Anticipated loads	
Dead loads w_{DL}	1.00 kip/ft	14.6 N/mm	1.00 kip/ft	14.6 N/mm
Live load w_{LL}	1.20 kip/ft	17.5 N/mm	1.80 kip/ft	26.3 N/mm
Unfactored loads ($w_{DL} + w_{LL}$)	2.20 kip/ft	32.1 N/mm	2.80 kip/ft	40.9 N/mm
Unstrengthened load limit ($1.1w_{DL} + 0.75w_{LL}$)	NA	NA	2.50 kip/ft	35.8 N/mm
Factored loads ($1.2w_{DL} + 1.6w_{LL}$)	3.12 kip/ft	45.5 N/mm	4.08 kip/ft	59.6 N/mm
Dead-load moment M_{DL}	72 kip-ft	98 kN-m	72 kip-ft	98 kN-m
Live-load moment M_{LL}	86 kip-ft	117 kN-m	130 kip-ft	176 kN-m
Service-load moment M_s	158 kip-ft	214 kN-m	202 kip-ft	274 kN-m
Unstrengthened moment limit ($1.1M_{DL} + 0.75M_{LL}$)	NA	NA	177 kip-ft	240 kN-m
Factored moment M_u	224 kip-ft	304 kN-m	294.4 kip-ft	399 kN-m

Table 16.3b—Manufacturer's reported FRP system properties

Thickness per ply t_f	0.040 in.	1.02 mm
Ultimate tensile strength f_{fu}^*	90 ksi	621 N/mm^2
Rupture strain ε_{fu}^*	0.015 in./in.	0.015 mm/mm
Modulus of elasticity of FRP laminates E_f	5360 ksi	37,000 N/mm^2

Table 16.3c—Procedure for flexural strengthening of an interior reinforced concrete beam with fiber-reinforced polymer laminates

Procedure	Calculation in in.-lb units	Calculation in SI metric units
Step 1—Calculate the FRP system design material properties		
The beam is located in an interior space and a carbon FRP (CFRP) material will be used. Therefore, per Table 9.4, an environmental reduction factor of 0.95 is suggested.		
$f_{fu} = C_E f_{fu}^*$	f_{fu} = (0.95)(90 ksi) = 85 ksi	f_{fu} = (0.95)(621 N/mm^2) = 590 N/mm^2
$\varepsilon_{fu} = C_E \varepsilon_{fu}^*$	ε_{fu} = (0.95)(0.015 in./in.) = 0.0142 in./in.	ε_{fu} = (0.95)(0.015 mm/mm) = 0.0142 mm/mm
Step 2—Preliminary calculations		
Properties of the concrete:		
β_1 from ACI 318-14, Section 22.2.2.4.3	$\beta_1 = 1.05 - 0.05(f_c'/1000) = 0.80$	$\beta_1 = 1.05 - 0.05(f_c'/1000) = 0.80$
$E_c = 57{,}000\sqrt{f_c'}$	$E_c = 57{,}000\sqrt{5000\text{ psi}} =$ 4,030,000 psi	$E_c = 4700\sqrt{34.5\text{ N/mm}^2} =$ 27,600 N/mm^2
Properties of the existing reinforcing steel:	A_s = 3(1.00 in.^{2}) = 3.00 in.^{2}	A_s = 3(645 mm^2) = 1935 mm^2
Properties of the externally bonded FRP reinforcement:		
$A_f = nt_f w_f$	A_f = (2 plies)(0.040 in./ply)(12 in.) = 0.96 in.^{2}	A_f = (2 plies)(1.02 mm/ply)(305 mm) = 619 mm^2
Step 3—Determine the existing state of strain on the soffit		
The existing state of strain is calculated assuming the beam is cracked and the only loads acting on the beam at the time of the FRP installation are dead loads. A cracked section analysis of the existing beam gives k = 0.334 and I_{cr} = 5937 in.^{4} = 2471 × 10^6 mm^4		
$\varepsilon_{bi} = \dfrac{M_{DL}(d_f - kd)}{I_{cr}E_c}$	$\varepsilon_{bi} = \dfrac{(864\text{ kip-in.})[24\text{ in.} - (0.334)(21.5\text{ in.})]}{(5937\text{ in.}^4)(4030\text{ ksi})}$ = 0.00061	$\varepsilon_{bi} = \dfrac{(97.6\text{ kN-mm})[609.6\text{ mm} - (0.334)(546.1\text{ mm})]}{(2471\times10^6\text{ mm}^4)(27.6\text{ kN/mm}^2)}$ = 0.00061
Step 4—Determine the design strain of the FRP system		
The design strain of FRP accounting for debonding failure mode ε_{fd} is calculated using Eq. (10.1.1)	$\varepsilon_{fd} = 0.083\sqrt{\dfrac{5000\text{ psi}}{2(5{,}360{,}000\text{ psi})(0.04\text{ in.})}}$ $= 0.009 \leq 0.9(0.0142) = 0.0128$	$\varepsilon_{fd} = 0.41\sqrt{\dfrac{34.5\text{ N/mm}^2}{2(37{,}000\text{ N/mm}^3)(1.02\text{ mm})}}$ $= 0.009 \leq 0.9(0.0142) = 0.0128$
Because the design strain is smaller than the rupture strain, debonding controls the design of the FRP system.		

Table 16.3c (cont.)—Procedure for flexural strengthening of an interior reinforced concrete beam with fiber-reinforced polymer laminates

Procedure	Calculation in in.-lb units	Calculation in SI metric units
Step 5—Estimate *c*, the depth to the neutral axis		
A reasonable initial estimate of c is $0.20d$. The value of the c is adjusted after checking equilibrium.		
$c = 0.20d$	$c = (0.20)(21.5\text{ in.}) = 4.30\text{ in.}$	$c = (0.20)(546.1\text{ mm}) = 109\text{ mm}$
Step 6—Determine the effective level of strain in the FRP reinforcement		
The effective strain level in the FRP may be found from Eq. (10.2.5).		
$\varepsilon_{fe} = 0.003\left(\frac{d_f - c}{c}\right) - \varepsilon_{bi} \le \varepsilon_{fd}$	$\varepsilon_{fe} = 0.003\left(\frac{24\text{ in.} - 4.3\text{ in.}}{4.3\text{ in.}}\right) - 0.00061 \le 0.009$	$\varepsilon_{fe} = 0.003\left(\frac{609.6\text{ mm} - 109.2\text{ mm}}{109.2\text{ mm}}\right) - 0.00061 \le 0.009$
Note that for the neutral axis depth selected, FRP debonding would be in the failure mode because the second expression in this equation controls. If the first expression governed, then concrete crushing would be in the failure mode. Because FRP controls the failure of the section, the concrete strain at failure ε_c may be less than 0.003 and can be calculated using similar triangles:	$\varepsilon_{fe} = 0.0131 > 0.009$ $\varepsilon_{fe} = \varepsilon_{fd} = 0.009$	$\varepsilon_{fe} = 0.0131 > 0.009$ $\varepsilon_{fe} = \varepsilon_{fd} = 0.009$
$\varepsilon_c = (\varepsilon_{fe} + \varepsilon_{bi})\left(\frac{c}{d_f - c}\right)$	$\varepsilon_c = (0.09 + 0.00061)\left(\frac{4.3\text{ in.}}{24\text{ in.} - 4.3\text{ in.}}\right) = 0.0021$	$\varepsilon_c = (0.009 + 0.00061)\left(\frac{109.2\text{ mm}}{609.6\text{ mm} - 109.2\text{ mm}}\right) = 0.0021$
Step 7—Calculate the strain in the existing reinforcing steel		
The strain in the reinforcing steel can be calculated using similar triangles according to Eq. (10.2.6).		
$\varepsilon_s = (\varepsilon_{fe} + \varepsilon_{bi})\left(\frac{c}{d_f - c}\right)$	$\varepsilon_s = (0.09 + 0.00061)\left(\frac{21.5\text{ in.} - 4.3\text{ in.}}{24\text{ in.} - 4.3\text{ in.}}\right) = 0.0084$	$\varepsilon_s = (0.09 + 0.00061)\left(\frac{546.1\text{ mm} - 109.2\text{ mm}}{609.6\text{ mm} - 109.2\text{ mm}}\right) = 0.0084$
Step 8—Calculate the stress level in the reinforcing steel and FRP		
The stresses are calculated using Eq. (10.2.10b) and Hooke's Law.		
$f_s = E_s\varepsilon_s \le f_y$	$f_s = (29{,}000\text{ ksi})(0.0084) \le 60\text{ ksi}$ $f_s = 244\text{ ksi} \le 60\text{ ksi}$ Hence, $f_s = 60\text{ ksi}$	$f_s = (200\text{ kN/mm}^2)(0.0084) \le 0.414\text{ kN/mm}^2$ $f_s = 1.68\text{ kN/mm}^2 \le 0.414\text{ kN/mm}^2$ Hence, $f_s = 0.414\text{ kN/mm}^2$
$f_{fe} = E_f\varepsilon_{fe}$	$f_{fe} = (5360\text{ ksi})(0.009) = 48.2\text{ ksi}$	$f_{fe} = (37\text{ kN/mm}^2)(0.009) = 0.33\text{ kN/mm}^2$

Table 16.3c (cont.)—Procedure for flexural strengthening of an interior reinforced concrete beam with fiber-reinforced polymer laminates

Procedure	Calculation in in.-lb units	Calculation in SI metric units
Step 9—Calculate the internal force resultants and check equilibrium Concrete stress block factors may be calculated using ACI 318. Approximate stress block factors may also be calculated based on the parabolic stress-strain relationship for concrete as follows: $\beta_1 = \frac{4\varepsilon_c' - \varepsilon_c}{6\varepsilon_c' - 2\varepsilon_c}$	$\beta_1 = \frac{4(0.0021) - 0.0021}{6(0.0021) - 2(0.0021)} = 0.749$	$\beta_1 = \frac{4(0.0021) - 0.0021}{6(0.0021) - 2(0.0021)} = 0.749$
$\alpha_1 = \frac{3\varepsilon_c'\varepsilon_c - \varepsilon_c^2}{3\beta_1\varepsilon_c'^2}$ where ε_c' is strain corresponding to f_c' calculated as	$\alpha_1 = \frac{3(0.0021)(0.0021) - (0.0021)^2}{3(0.749)(0.0021)^2} = 0.886$	$\alpha_1 = \frac{3(0.0021)(0.0021) - (0.0021)^2}{3(0.749)(0.0021)^2} = 0.886$
$\varepsilon_c' = \frac{1.7f_c'}{E_c}$ Force equilibrium is verified by checking the initial estimate of c with Eq. (10.3.1.6g).	$\varepsilon_c' = \frac{1.7(5000)}{4.03 \times 10^6} = 0.0021$	$\varepsilon_c' = \frac{1.7(34.5)}{27,600} = 0.0021$
$c = \frac{A_s f_s + A_f f_{fe}}{\alpha_1 f_c' \beta_1 b}$	$c = \frac{(3.00\text{ in.}^2)(60\text{ ksi}) + (0.96\text{ in.}^2)(48.2\text{ ksi})}{(0.886)(5\text{ ksi})(0.749)(12\text{ in.})}$ $c = 5.68\text{ in.} \neq 4.30\text{ in. n.g.}$ ∴ revise estimate of c and repeat Steps 6 through 9 until equilibrium is achieved.	$c = \frac{(1935.48\text{ mm}^2)(414\text{ N/mm}^2) + (619\text{ mm}^2)(330\text{ N/mm}^2)}{(0.886)(34.5\text{ N/mm}^2)(0.749)(304.8\text{ mm})}$ $c = 149\text{ mm} \neq 109\text{ mm n.g.}$ ∴ revise estimate of c and repeat Steps 6 through 9 until equilibrium is achieved.
Step 10—Adjust c until force equilibrium is satisfied Steps 6 through 9 were repeated several times with different values of c until equilibrium was achieved. The results of the final iteration are $c = 5.17$ in.; $\varepsilon_s = 0.0083$; $f_s = f_y = 60$ ksi; $\beta_1 = 0.786$; $\alpha_1 = 0.928$; and $f_{fd} = 48.2$ ksi	$c = \frac{(3.00\text{ in.}^2)(60\text{ ksi}) + (0.96\text{ in.}^2)(48.2\text{ ksi})}{(0.925)(5\text{ ksi})(0.786)(12\text{ in.})}$ $c = 5.17$ in. ∴ the value of c selected for the final iteration is correct.	$c = \frac{(1935.5\text{ mm}^2)(414\text{ N/mm}^2) + (619\text{ mm}^2)(330\text{ N/mm}^2)}{(0.928)(34.5\text{ N/mm}^2)(0.786)(304.8\text{ mm})}$ $c = 131$ mm ∴ the value of c selected for the final iteration is correct.
Step 11—Calculate flexural strength components The design flexural strength is calculated using Eq. (10.2.10d). An additional reduction factor, $\psi_f = 0.85$, is applied to the contribution of the FRP system. Steel contribution to bending: $M_{ns} = A_s f_s \left(d - \frac{\beta_1 c}{2}\right)$	$M_{ns} = (3.00\text{ in.}^2)(60\text{ ksi})\left(21.5\text{ in.} - \frac{0.786(5.17\text{ in.})}{2}\right)$ $M_{ns} = 3504$ kip-in. = 292 kip-ft	$M_{ns} = (1935.5\text{ mm}^2)(414\text{ N/mm}^2)\left(546.1\text{ mm} - \frac{0.786(131\text{ mm})}{2}\right)$ $M_{ns} = 3.963 \times 10^8$ N-mm = 396.3 kN-m
FRP contribution to bending: $M_{nf} = A_f f_{fe}\left(d_f - \frac{\beta_1 c}{2}\right)$	$M_{nf} = (0.96\text{ in.}^2)(48.2\text{ ksi})\left(24\text{ in.} - \frac{0.786(5.17\text{ in.})}{2}\right)$ $M_{nf} = 1020$ kip-in. = 85 kip-ft	$M_{nf} = (619\text{ mm}^2)(330\text{ N/mm}^2)\left(609.6\text{ mm} - \frac{0.786(131\text{ mm})}{2}\right)$ $M_{nf} = 1.140 \times 10^8$ N-mm = 114 kN-m

Table 16.3c (cont.)—Procedure for flexural strengthening of an interior reinforced concrete beam with fiber-reinforced polymer laminates

Procedure	Calculation in in.-lb units	Calculation in SI metric units
Step 12—Calculate design flexural strength of the section The design flexural strength is calculated using Eq. (10.1) and (10.2.10d). Because $\varepsilon_s = 0.0083 > 0.005$, a strength reduction factor of $\phi = 0.90$ is appropriate per Eq. (10.2.7). $\phi M_n = \phi[M_{ns} + \psi_f M_{nf}]$	$\phi M_n = 0.9[292 \text{ kip-ft} + 0.85(85 \text{ kip-ft})]$ $\phi M_n = 327 \text{ kip-ft} \geq M_u = 294 \text{ kip-ft}$ ∴ the strengthened section is capable of sustaining the new required moment strength.	$\phi M_n = 0.9[396.3 \text{ kN-m} + 0.85(114 \text{ kN-m})]$ $\phi M_n = 443 \text{ kN-m} \geq M_u = 399 \text{ kN-m}$ ∴ the strengthened section is capable of sustaining the new required moment strength.
Step 13—Check service stresses in the reinforcing steel and the FRP. Calculate the elastic depth to the cracked neutral axis. This can be simplified for a rectangular beam without compression reinforcement as follows: $k = \sqrt{\left(\rho_s \frac{E_s}{E_c} + \rho_f \frac{E_f}{E_c}\right)^2 + 2\left(\rho_s \frac{E_s}{E_c} + \rho_f \frac{E_f}{E_c}\left(\frac{d_f}{d}\right)\right)} - \left(\rho_s \frac{E_s}{E_c} + \rho_f \frac{E_f}{E_c}\right)$	* $k = 0.343$ $kd = (0.343)(21.5 \text{ in.}) = 7.37 \text{ in.}$ †	‡ $k = 0.343$ $kd = (0.343)(546.1 \text{ mm}) = 187 \text{ mm}$ §
Calculate the stress level in the reinforcing steel using Eq. (10.2.10.1) and verify that it is less than the recommended limit per Eq. (10.2.8a). $f_{s,s} = \frac{\left[M_s + \varepsilon_{bi} A_f E_f\left(d_f - \frac{kd}{3}\right)\right](d - kd)E_s}{A_s E_s\left(d - \frac{kd}{3}\right)(d - kd) + A_f E_f\left(d_f - \frac{kd}{3}\right)(d_f - kd)}$ $f_{s,s} \leq 0.80 f_y$	$f_{s,s} = 40.4 \text{ ksi} \leq (0.80)(60 \text{ ksi}) = 48 \text{ ksi}$ ∴ the stress level in the reinforcing steel is within the recommended limit.	$f_{s,s} = 279 \text{ N/mm}^2 \leq (0.80)(410 \text{ N/mm}^2) = 330 \text{ N/mm}^2$ ∴ the stress level in the reinforcing steel is within the recommended limit.

Table 16.3c (cont.)—Procedure for flexural strengthening of an interior reinforced concrete beam with fiber-reinforced polymer laminates

Procedure	Calculation in in.-lb units	Calculation in SI metric units
Step 14—Check creep rupture limit at service of the FRP Calculate the stress level in the FRP using Eq. (10.2.10.2) and verify that it is less than creep-rupture stress limit given in Table 10.2.9. Assume that the full service load is sustained.		
$f_{f,s} = f_{s,s}\left(\frac{E_f}{E_s}\right)\left(\frac{d_f - kd}{d - kd}\right) - \varepsilon_{bi}E_f$	$f_{f,s} = 40.4\text{ ksi}\left(\frac{5360\text{ ksi}}{29{,}000\text{ ksi}}\right)\left(\frac{24\text{ in.} - 7.37\text{ in.}}{21.5\text{ in.} - 7.37\text{ in.}}\right)$ $-(0.00061)(5360\text{ ksi})$	$f_{f,s} = 0.278\text{ kN/mm}^2\left(\frac{37\text{ kN/mm}^2}{200\text{ kN/mm}^2}\right)\left(\frac{609.6\text{ mm} - 187\text{ mm}}{546\text{ mm} - 187\text{ mm}}\right)$ $-(0.00061)(38\text{ N/mm}^2)$
For a carbon FRP system, the sustained plus cyclic stress limit is obtained from Table 10.2.9: Sustained plus cyclic stress limit = $0.55f_{fu}$	$f_{f,s} = 5.60\text{ ksi} \leq (0.55)(85\text{ ksi}) = 47\text{ ksi}$ ∴ the stress level in the FRP is within the recommended sustained plus cyclic stress limit.	$f_{f,s} = 38\text{ N/mm}^2 \leq (0.55)(590\text{ N/mm}^2) = 324\text{ N/mm}^2$ ∴ the stress level in the FRP is within the recommended sustained plus cyclic stress limit.

*
$$k = \sqrt{\left[0.0116\left(\frac{29{,}000}{4030}\right) + 0.00372\left(\frac{5360}{4030}\right)\right]^2 + 2\left[0.0116\left(\frac{29{,}000}{4030}\right) + 0.00372\left(\frac{5360}{4030}\right)\left(\frac{24\text{ in.}}{21.5\text{ in.}}\right)\right]} - \left[0.0116\left(\frac{29{,}000}{4030}\right) + 0.00372\left(\frac{5360}{4030}\right)\right]$$

†
$$f_{s,s} = \frac{\left(\left[2424\text{ kip-in.} + \left[(0.00061)(0.96\text{ in.}^3)\times(5360\text{ ksi})\left(24\text{ in.} - \frac{7.37\text{ in.}}{3}\right)\right]\right]\times\left[(21.5\text{ in.} - 7.37\text{ in.})(29{,}000\text{ ksi})\right]\right)}{\left(\left[(3.00\text{ in.}^2)(29{,}000\text{ ksi})\times\left(21.5\text{ in.} - \frac{7.37\text{ in.}}{3}\right)(21.5\text{ in.} - 7.37\text{ in.})\right] + \left[(0.96\text{ in.}^2)(5360\text{ ksi})\left(24\text{ in.} - \frac{7.37\text{ in.}}{3}\right)(24\text{ in.} - 7.37\text{ in.})\right]\right)}$$

‡
$$k = \sqrt{\left[0.0116\left(\frac{200}{27.6}\right) + 0.00372\left(\frac{37}{27.6}\right)\right]^2 + 2\left[0.0116\left(\frac{200}{27.6}\right) + 0.00372\left(\frac{37}{27.6}\right)\left(\frac{609.6\text{ mm}}{546\text{ mm}}\right)\right]} - \left[0.0116\left(\frac{200}{27.6}\right) + 0.00372\left(\frac{37}{27.6}\right)\right]$$

§
$$f_{s,s} = \frac{\left(\left[273{,}912\text{ kN-mm} + \left[(0.00061)(619\text{ mm}^2)(37\text{ kN/mm}^2)\times\left(609.6\text{ mm} - \frac{187\text{ mm}}{3}\right)\right]\right]\times\left[(546\text{ mm} - 187\text{ mm})(200\text{ kN/mm}^2)\right]\right)}{\left(\left[(1935\text{ mm}^2)(200\text{ kN/mm}^2)\times\left(546\text{ mm} - \frac{187\text{ mm}}{3}\right)(546\text{ mm} - 187\text{ mm})\right] + \left[(619\text{ mm}^2)(37\text{ kN/mm}^2)\left(607\text{ mm} - \frac{187\text{ mm}}{3}\right)(607\text{ mm} - 187\text{ mm})\right]\right)}$$

In detailing the FRP reinforcement, the FRP should be terminated a minimum of ℓ_{df}, calculated per Eq. (14.1.3), past the point on the moment diagram that represents cracking. The factored shear force at the termination should also be checked against the shear force that causes FRP end peeling, estimated as two-thirds of the concrete shear strength. If the shear force is greater than two-thirds of the concrete shear strength, the FRP strips should be extended further toward the supports. U-wraps may also be used to reinforce against cover delamination.

16.4—Flexural strengthening of an interior reinforced concrete beam with near-surface-mounted FRP bars

An existing reinforced concrete beam (Fig. 16.4) is to be strengthened using the loads given in Table 16.3a and the near-surface-mounted (NSM) FRP system described in Table 16.4a. Specifically, three No. 3 carbon FRP (CFRP) bars are to be used at a distance 23.7 in. (602.1 mm) from the extreme top fiber of the beam.

By inspection, the degree of strengthening is reasonable in that it does meet the strengthening limit criteria put forth in Eq. (10.1). That is, the existing flexural strength without FRP, $(\phi M_n)_{w/o}$ = 266 kip-ft (361 kN-m), is greater than the unstrengthened moment limit, $(1.1M_{DL} + 0.75M_{LL})_{new}$ = 177 kip-ft (240 kN-m). The design calculations used to verify this configuration follow in Table 16.4b.

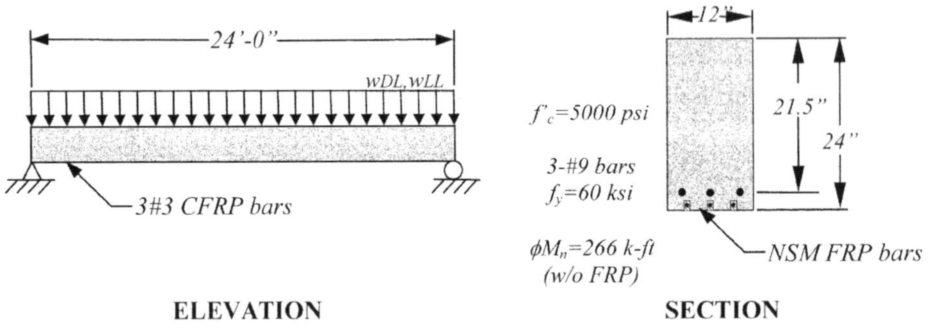

Fig. 16.4—Schematic of the idealized simply supported beam with FRP external reinforcement.

Table 16.4a—Manufacturer's reported NSM FRP system properties

Area per No. 3 bar	0.10 in.^{2}	64.5 mm^2
Ultimate tensile strength f_{fu}*	250 ksi	1725 N/mm^2
Rupture strain ε_{fu}*	0.013 in./in.	0.013 mm/mm
Modulus of elasticity of FRP laminates E_f	19,230 ksi	132,700 N/mm^2
Length of the beam ℓ	29 ft	8.84 m
Bay width l_2	30 ft	9.14 m
Width of beam w	24 in.	610 mm
d_p	22.5 in.	571 mm
h	25 in.	635 mm
Effective flange width b_f	87 in.	2210 mm
Flange thickness h_f	4 in.	102 mm
f_c'	4000 psi	27.6 N/mm^2
Strands diameter	1/2 in.	12.7 mm
f_{pe}	165 ksi	1138 N/mm^2
f_{py}	230 ksi	1586 N/mm^2
f_{pu}	270 ksi	1860 N/mm^2
E_p	28,500 ksi	1.96 × 10^5 N/mm^2
ϕM_n without FRP	336 kip-ft	455 kN-m

Table 16.4b—Procedure for flexural strengthening of an interior reinforced concrete beam with NSM FRP bars

Procedure	Calculation in in.-lb units	Calculation in SI metric units
Step 1—Calculate the FRP system design material properties The beam is located in an interior space and a CFRP material will be used. Therefore, per Table 9.4, an environmental reduction factor of 0.95 is suggested.		
$f_{fu} = C_E f_{fu}^*$	f_{fu} = (0.95)(250 ksi) = 237.5 ksi	f_{fu} = (0.95)(1725 N/mm^2) = 1639 N/mm^2
$\varepsilon_{fu} = C_E \varepsilon_{fu}^*$	ε_{fu} = (0.95)(0.013 in./in.) = 0.0123 in./in.	ε_{fu} = (0.95)(0.013 mm/mm) = 0.0123 mm/mm
Step 2—Preliminary calculations Properties of the concrete:		
β_1 from ACI 318-14, Section 22.2.2.4.3	$\beta_1 = 1.05 - 0.05\dfrac{f_c'}{1000} = 0.85$	$\beta_1 = 1.05 - 0.05\dfrac{f_c'}{6.9} = 0.85$
$E_c = 57{,}000\sqrt{f_c'}$	$E_c = 57{,}000\sqrt{5000\text{ psi}} = 4{,}030{,}00\text{ psi}$	$E_c = 4700\sqrt{34.5\text{ N/mm}^2} = 27{,}600\text{ N/mm}^2$
	A_s = 3(1.00 in.^{2}) = 3.00 in.^{2}	A_s = 3(645.2 mm^2) = 1935 mm^2
	A_f = (3 bars)(0.01 in.^{2}/bar) = 0.3 in.^{2}	A_f = (3 bars)(64.5 mm^2/bar) = 194 mm^2
Step 3—Determine the existing state of strain on the soffit The existing state of strain is calculated assuming the beam is cracked and the only loads acting on the beam at the time of the FRP installation are dead loads. A cracked section analysis of the existing beam gives k = 0.334 and I_{cr} = 5937 in.^{4} = 2471 × 10^6 mm^4		
$\varepsilon_{bi} = \dfrac{M_{DL}(d_f - kd)}{I_{cr}E_c}$	$\varepsilon_{bi} = \dfrac{(864\text{ kip-in.})[23.7\text{ in.} - (0.334)(21.5\text{ in.})]}{(5937\text{ in.}^4)(4030\text{ ksi})} = 0.00061$	$\varepsilon_{bi} = \dfrac{(97.6\text{ kN-mm})[602\text{ mm} - (0.334)(546\text{ mm})]}{(2471\times10^6\text{ mm}^4)(27.6\text{ kN/mm}^2)} = 0.00061$
Step 4—Determine the bond-dependent coefficient of the FRP system Based on the manufacturer's recommendation, the dimensionless bond-dependent coefficient for flexure κ_m is 0.7.	κ_m = 0.7	κ_m = 0.7
Step 5—Estimate *c*, the depth to the neutral axis A reasonable initial estimate of c is 0.20d. The value of the c is adjusted after checking equilibrium.		
$c = 0.20d$	c = (0.20)(21.5 in.) = 4.30 in.	c = (0.20)(546 mm) = 109 mm

Table 16.4b (cont.)—Procedure for flexural strengthening of an interior reinforced concrete beam with NSM FRP bars

Procedure	Calculation in in.-lb units	Calculation in SI metric units
Step 6—Determine the effective level of strain in the FRP reinforcement		
The effective strain level in the FRP may be found from Eq. (10.2.5).		
$\varepsilon_{fe} = \left(\frac{d_f - c}{c}\right) - \varepsilon_{bi} \leq \kappa_m \varepsilon_{fd}$	$\varepsilon_{fe} = 0.003\left(\frac{23.7\text{ in.} - 4.3\text{ in.}}{4.3\text{ in.}}\right) - 0.00061 = 0.0129$	$\varepsilon_{fe} = 0.003\left(\frac{602\text{ mm} - 109\text{ mm}}{109\text{ mm}}\right) - 0.00061 = 0.0129$
Note that for the neutral axis depth selected, FRP debonding would be the failure mode because the second expression in this equation controls. If the first expression governed, then concrete crushing would be the failure mode. Because FRP controls the failure of the section, the concrete strain at failure, ε_c, may be less than 0.003 and can be calculated using similar triangles:	$\kappa_m\varepsilon_{fd} = 0.7(0.0123) = 0.00865$ Hence, $\varepsilon_{fe} = 0.00865$ (Mode of failure is FRP debonding)	$\kappa_m\varepsilon_{fd} = 0.7(0.0123) = 0.00865$ Hence, $\varepsilon_{fe} = 0.00865$ (Mode of failure is FRP debonding)
$\varepsilon_c = (\varepsilon_{fd} + \varepsilon_{bi})\left(\frac{c}{d_f - c}\right)$	$\varepsilon_c = (0.00865 + 0.00061)\left(\frac{4.3}{23.7 - 4.3}\right) = 0.0020$	$\varepsilon_c = (0.00865 + 0.00061)\left(\frac{109}{602 - 109}\right) = 0.0020$
Step 7—Calculate the strain in the existing reinforcing steel		
The strain in the reinforcing steel can be calculated using similar triangles according to Eq. (10.2.10a).		
$\varepsilon_s = (\varepsilon_{fe} + \varepsilon_{bi})\left(\frac{d - c}{d_f - c}\right)$	$\varepsilon_s = (0.00865 + 0.00061)\left(\frac{21.5 - 4.3}{23.7 - 4.3}\right) = 0.0082$	$\varepsilon_s = (0.00865 + 0.00061)\left(\frac{546 - 109}{602 - 109}\right) = 0.0082$
Step 8—Calculate the stress level in the reinforcing steel and FRP.		
The stresses are calculated using Eq. (10.2.10b) and Hooke's Law.		
$f_s = E_s\varepsilon_s \leq f_y$	$f_s = (29{,}000\text{ ksi})(0.0082) \leq 60\text{ ksi}$ $f_s = 238\text{ ksi} > 60\text{ ksi}$ therefore, $f_s = 60\text{ ksi}$	$f_s = (200\text{ kN/mm}^2)(0.0082) \leq 0.414\text{ kN/mm}^2$ $f_s = 1.64\text{ kN/mm}^2 \leq 0.414\text{ kN/mm}^2$ therefore, $f_s = 0.414\text{ kN/mm}^2$
$f_{fe} = E_f\varepsilon_{fe}$	$f_{fe} = (19{,}230\text{ ksi})(0.00865) = 166\text{ ksi}$	$f_{fe} = (132{,}700\text{ N/mm}^2)(0.00865) = 1147\text{ N/mm}^2$

Table 16.4b (cont.)—Procedure for flexural strengthening of an interior reinforced concrete beam with NSM FRP bars

Procedure	Calculation in in.-lb units	Calculation in SI metric units
Step 9—Calculate the internal force resultants and check equilibrium Concrete stress block factors may be calculated using ACI 318. Approximate stress block factors may also be calculated based on the parabolic stress-strain relationship for concrete as follows:		
$\beta_1 = \frac{4\varepsilon_c' - \varepsilon_c}{6\varepsilon_c' - 2\varepsilon_c}$	$\beta_1 = \frac{4(0.0021) - 0.002}{6(0.0021) - 2(0.002)} = 0.743$	$\beta_1 = \frac{4(0.0021) - 0.002}{6(0.0021) - 2(0.002)} = 0.743$
$\alpha_1 = \frac{3\varepsilon_c'\varepsilon_c - \varepsilon_c^2}{3\beta_1\varepsilon_c'^2}$ where ε_c' is strain corresponding to f_c' calculated as	$\alpha_1 = \frac{3(0.0021)(0.002) - (0.002)^2}{3(0.743)(0.0021)^2} = 0.870$	$\alpha_1 = \frac{3(0.0021)(0.002) - (0.002)^2}{3(0.743)(0.0021)^2} = 0.870$
$\varepsilon_c' = \frac{1.7f_c'}{E_c}$	$\varepsilon_c' = \frac{1.7(5000)}{4030 \times 10^6} = 0.0021$	$\varepsilon_c' = \frac{1.7(34.5)}{27,606} = 0.0021$
Force equilibrium is verified by checking the initial estimate of c with Eq. (10.3.1.6g).		
$c = \frac{A_s f_s + A_f f_{fe}}{\alpha_1 f_c' \beta_1 b}$	$c = \frac{(3.00\text{ in.}^2)(60\text{ ksi}) + (0.3\text{ in.}^2)(166\text{ ksi})}{(0.87)(5\text{ ksi})(0.743)(12\text{ in.})}$ $c = 5.92$ in. $\neq$ 4.30 in. n.g. ∴ revise estimate of c and repeat Steps 6 through 9 until equilibrium is achieved.	$c = \frac{(1935\text{ mm}^2)(414\text{ N/mm}^2) + (194\text{ mm}^2)(1147\text{ N/mm}^2)}{(0.87)(34.5\text{ N/mm}^2)(0.743)(305\text{ mm})}$ $c = 150$ mm $\neq$ 109 in. n.g. ∴ revise estimate of c and repeat Steps 6 through 9 until equilibrium is achieved.
Step 10—Adjust *c* until force equilibrium is satisfied Steps 6 through 9 were repeated several times with different values of c until equilibrium was achieved. The results of the final iteration are		
$c = 5.26$ in.; $\varepsilon_s = 0.0082$; $f_s = f_y = 60$ ksi; $\varepsilon_{fe} = 0.00865$; $\varepsilon_c = 0.0027$; $\beta_1 = 0.786$; $\alpha_1 = 0.928$; and $f_{fe} = 166$ ksi	$c = \frac{(3.00\text{ in.}^2)(60\text{ ksi}) + (0.3\text{ in.}^2)(166\text{ ksi})}{(0.928)(5\text{ ksi})(0.786)(12\text{ in.})}$ $c = 5.25$ in. $\approx$ 5.26 in. ∴ the value of c selected for the final iteration is correct.	$c = \frac{(1935\text{ mm}^2)(414\text{ N/mm}^2) + (193\text{ mm}^2)(1147\text{ N/mm}^2)}{(0.928)(34.5\text{ N/mm}^2)(0.786)(305\text{ mm})}$ $c = 133$ mm $\approx$ 134 mm ∴ the value of c selected for the final iteration is correct.

Table 16.4b (cont.)—Procedure for flexural strengthening of an interior reinforced concrete beam with NSM FRP bars

Procedure	Calculation in in.-lb units	Calculation in SI metric units
Step 11—Calculate flexural strength components. The design flexural strength is calculated using Eq. (10.2.10d). An additional reduction factor, $\psi_f = 0.85$, is applied to the contribution of the FRP system. Steel contribution to bending: $M_{ns} = A_s f_s \left(d - \frac{\beta_1 c}{2}\right)$ FRP contribution to bending: $M_{nf} = A_f f_{fe} \left(d_f - \frac{\beta_1 c}{2}\right)$	$M_{ns} = (3.0\ \text{in.}^2)(60\ \text{ksi})\left(21.5\ \text{in.} - \frac{0.786(5.25\ \text{in.})}{2}\right)$ $M_{ns} = 3498$ kip-in. = 291 kip-ft $M_{nf} = (0.3\ \text{in.}^2)(166\ \text{ksi})\left(23.7\ \text{in.} - \frac{0.786(5.25\ \text{in.})}{2}\right)$ $M_{nf} = 1077$ kip-in. = 90 kip-ft	$M_{ns} = (1935\ \text{mm}^2)(414\ \text{N/mm}^2)\left(546\ \text{mm} - \frac{0.786(133\ \text{mm})}{2}\right)$ $M_{ns} = 394$ kN-m $M_{nf} = (194\ \text{mm}^2)(1147\ \text{N/mm}^2)\left(602.1\ \text{mm} - \frac{0.786(133\ \text{mm})}{2}\right)$ $M_{nf} = 122$ kN-m
Step 12—Calculate design flexural strength of the section The design flexural strength is calculated using Eq. (10-1) and (10.2.10d). Because $\varepsilon_s = 0.0082 > 0.005$, a strength reduction factor of $\phi = 0.90$ is appropriate per Eq. (10.2.7). $\phi M_n = \phi[M_{ns} + \psi_f M_{nf}]$	$\phi M_n = 0.9[291\ \text{kip-ft} + 0.85(90\ \text{kip-ft})]$ $\phi M_n = 331\ \text{kip-ft} \geq M_u = 294\ \text{kip-ft}$ ∴ the strengthened section is capable of sustaining the new required flexural strength.	$\phi M_n = 0.9[394\ \text{kN-m} + 0.85(122\ \text{kN-m})]$ $\phi M_n = 448\ \text{kN-m} \geq M_u = 398\ \text{kN-m}$ ∴ the strengthened section is capable of sustaining the new required flexural strength.
Step 13—Check service stresses in the reinforcing steel and the FRP Calculate the elastic depth to the cracked neutral axis. This can be simplified for a rectangular beam without compression reinforcement as follows: $k = \sqrt{\left(\rho_s \frac{E_s}{E_c} + \rho_f \frac{E_f}{E_c}\right)^2 + 2\left(\rho_s \frac{E_s}{E_c} + \rho_f \frac{E_f}{E_c}\left(\frac{d_f}{d}\right)\right)} - \left(\rho_s \frac{E_s}{E_c} + \rho_f \frac{E_f}{E_c}\right)$ Calculate the stress level in the reinforcing steel using Eq. (10.2.10.1) and verify that it is less than the recommended limit per Eq. (10.2.8a). $f_{s,s} = \frac{\left[M_s + \varepsilon_{bi} A_f E_f \left(d_f - \frac{kd}{3}\right)\right](d - kd)E_s}{A_s E_s \left(d - \frac{kd}{3}\right)(d - kd) + A_f E_f \left(d_f - \frac{kd}{3}\right)(d_f - kd)}$ $f_{s,s} \leq 0.80 f_y$	* $k = 0.345$ $kd = (0.345)(21.5\ \text{in.}) = 7.4$ in. †$f_{s,s} = 40.3\ \text{ksi} \leq (0.80)(60\ \text{ksi}) = 48\ \text{ksi}$ ∴ the stress level in the reinforcing steel is within the recommended limit.	‡ $k = 0.345$ $kd = (0.345)(546\ \text{mm}) = 188$ mm §$f_{s,s} = 278\ \text{N/mm}^2 \leq (0.80)(410\ \text{N/mm}^2) = 330\ \text{N/mm}^2$ ∴ the stress level in the reinforcing steel is within the recommended limit.

Table 16.4b (cont.)—Procedure for flexural strengthening of an interior reinforced concrete beam with NSM FRP bars

Procedure	Calculation in in.-lb units	Calculation in SI metric units
Step 14—Check creep rupture limit at service of the FRP Calculate the stress level in the FRP using Eq. (10.2.10.2) and verify that it is less than creep-rupture stress limit given in Table 10.2.9. Assume that the full service load is sustained.		
$f_{f,s} = f_{s,s}\left(\frac{E_f}{E_s}\right)\left(\frac{d_f - kd}{d - kd}\right) - \varepsilon_{bi}E_f$	$f_{f,s} = 40.3 \text{ ksi}\left(\frac{19{,}230 \text{ ksi}}{29{,}000 \text{ ksi}}\right)\left(\frac{23.7 \text{ in.} - 7.4 \text{ in.}}{21.5 \text{ in.} - 7.4 \text{ in.}}\right) - (0.00061)(19{,}230 \text{ ksi})$	$f_{f,s} = 0.278 \text{ kN/mm}^2\left(\frac{133 \text{ kN/mm}^2}{200 \text{ kN/mm}^2}\right)\left(\frac{602 \text{ mm} - 188 \text{ mm}}{546 \text{ mm} - 188 \text{ mm}}\right) - (0.00061)(133\text{N/mm}^2)$
For a carbon FRP system, the sustained plus cyclic stress limit is obtained from Table 10.2.9: Sustained plus cyclic stress limit = $0.55f_{fu}$	$f_{f,s} = 19 \text{ ksi} \leq (0.55)(85 \text{ ksi}) = 50 \text{ ksi}$ ∴ the stress level in the FRP is within the recommended sustained plus cyclic stress limit.	$f_{f,s} = 134 \text{ N/mm}^2 \leq (0.55)(590 \text{ N/mm}^2) = 324.5 \text{ N/mm}^2$ ∴ the stress level in the FRP is within the recommended sustained plus cyclic stress limit.

* $$k = \sqrt{\left[0.0116\left(\frac{29{,}000}{4030}\right) + 0.0012\left(\frac{19{,}230}{4030}\right)\right]^2 + 2\left[0.0116\left(\frac{29{,}000}{4030}\right) + 0.0012\left(\frac{19{,}230}{4030}\right)\left(\frac{23.7 \text{ in.}}{21.5 \text{ in.}}\right)\right]} - \left[0.0116\left(\frac{29{,}000}{4030}\right) + 0.0012\left(\frac{19{,}230}{4030}\right)\right]$$

† $$f_{s,s} = \frac{\left(\left[2424 \text{ kip-in.} + \left[(0.00061)(0.3 \text{ in.}^2) \times (19{,}230 \text{ ksi})\left(23.7 \text{ in.} - \frac{7.4 \text{ in.}}{3}\right)\right]\right] \times [(21.5 \text{ in.} - 7.4 \text{ in.})(29{,}000 \text{ ksi})]\right)}{\left(\left[(3.00 \text{ in.}^2)(29{,}000 \text{ ksi})\left(21.5 \text{ in.} - \frac{7.4 \text{ in.}}{3}\right) \times (21.5 \text{ in.} - 7.4 \text{ in.})\right] + \left[(0.3 \text{ in.}^2)(19{,}230 \text{ ksi})\left(23.7 \text{ in.} - \frac{7.4 \text{ in.}}{3}\right) \times (23.7 \text{ in.} - 7.4 \text{ in.})\right]\right)}$$

‡ $$k = \sqrt{\left[0.0116\left(\frac{200}{27.6}\right) + 0.0012\left(\frac{133}{27.6}\right)\right]^2 + 2\left[0.0116\left(\frac{200}{27.6}\right) + 0.0012\left(\frac{133}{27.6}\right)\left(\frac{602 \text{ mm}}{546 \text{ mm}}\right)\right]} - \left[0.0116\left(\frac{200}{27.6}\right) + 0.0012\left(\frac{133}{27.6}\right)\right]$$

§ $$f_{s,s} = \frac{\left(\left[273{,}912 \text{ kN-mm} + \left[(0.00061)(194 \text{ mm}^2) \times (132.7 \text{ kN/mm}^2) \times \left(602 \text{ mm} - \frac{188 \text{ mm}}{3}\right)\right]\right] \times [(546 \text{ mm} - 188 \text{ mm})(200 \text{ kN/mm}^2)]\right)}{\left(\left[(1935 \text{ mm}^2)(200 \text{ kN/mm}^2) \times \left(546 \text{ mm} - \frac{188 \text{ mm}}{3}\right)(546 \text{ mm} - 188 \text{ mm})\right] + \left[(194 \text{ mm}^2)(132.7 \text{ kN/mm}^2) \times \left(602 \text{ mm} - \frac{188 \text{ mm}}{3}\right)(602 \text{ mm} - 188 \text{ mm})\right]\right)}$$

In detailing the FRP reinforcement, FRP bars should be terminated at a distance equal to the bar development length past the point on the moment diagram that represents cracking.

16.5—Flexural strengthening of an interior prestressed concrete beam with FRP laminates

A number of continuous prestressed concrete beams with five 1/2 in. (12.7 mm) diameter bonded strands (Fig. 16.5) are located in a parking garage that is being converted to an office space. All prestressing strands are Grade 270 ksi (1860 N/mm^2) low-relaxation seven-wire strands. The beams require an increase in their live-load-carrying capacity from 50 to 75 lb/ft^2 (244 to 366 kg/m^2). The beams are also required to support an additional dead load of 10 lb/ft^2 (49 kg/m^2). Analysis indicates that each existing beam has adequate flexural capacity to carry the new loads in the negative moment region at the supports but is deficient in flexure at midspan and in shear at the supports. The beam meets the deflection and crack control serviceability requirements. The cast-in-place beams support a 4 in. (100 mm) slab. For bending at midspan, beams should be treated as T-sections. Summarized in Table 16.5a are the existing and new loads and associated midspan moments for the beam. FRP system properties are shown in Table 16.3b.

By inspection, the degree of strengthening is reasonable in that it does meet the strengthening limit criteria put forth in Eq. (10.1). That is, the existing flexural strength without FRP, $(\phi M_n)_{w/o}$ = 336 kip-ft (455 kN-m), is greater than the unstrengthened moment limit, $(1.1M_{DL} + 0.75M_{LL})_{new}$ = 273 kip-ft (370 kN-m). The design calculations used to verify this configuration follow. The beam is to be strengthened using the FRP system described in Table 16.3b. A one-ply, 24 in. (610 mm) wide strip of FRP is considered for this evaluation.

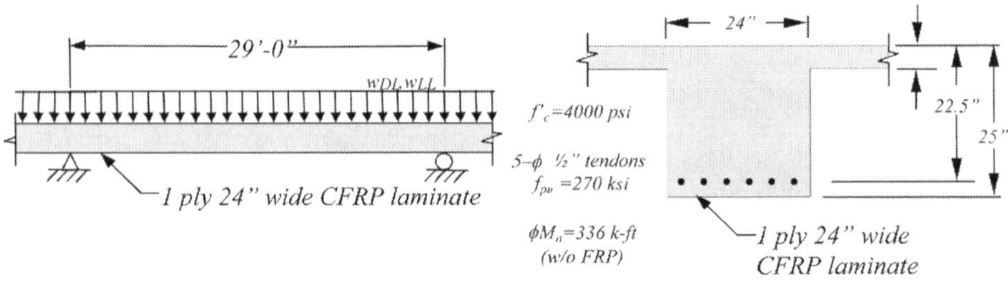

Fig. 16.5—Schematic of the idealized continuous prestressed beam with FRP external reinforcement.

Table 16.5a—Loadings and corresponding moments

Loading/moment	Existing loads		Anticipated loads	
Dead loads w_{DL}	2.77 kip/ft	40.4 N/mm	3.09 kip/ft	45.1 N/mm
Live load w_{LL}	1.60 kip/ft	23.3 N/mm	2.4 kip/ft	35 N/mm
Unfactored loads $(w_{DL} + w_{LL})$	4.37 kip/ft	63.8 N/mm	5.49 kip/ft	80.2 N/mm
Unstrengthened load limit $(1.1w_{DL} + 0.75w_{LL})$	NA	NA	5.2 kip/ft	75.9 N/mm
Factored loads $(1.2w_{DL} + 1.6w_{LL})$	5.88 kip/ft	85.9 N/mm	7.55 kip/ft	110.2 N/mm
Dead-load moment M_{DL}	147 kip-ft	199 kN-m	162 kip-ft	220.2 kN-m
Live-load moment M_{LL}	85 kip-ft	115 kN-m	126 kip-ft	171.1 kN-m
Service-load moment M_s	232 kip-ft	314 kN-m	288 kip-ft	391.3 kN-m
Unstrengthened moment limit $(1.1M_{DL} + 0.75M_{LL})_{new}$	NA	NA	273 kip-ft	371 kN-m
Factored moment M_u	312 kip-ft	423 kN-m	397 kip-ft	538 kN-m

Table 16.5b—Procedure for flexural strengthening of an interior prestressed concrete beam with FRP laminates

Procedure	Calculation in in.-lb units	Calculation in SI metric units
Step 1—Calculate the FRP-system design material properties		
The beam is located in an interior space and a CFRP material will be used. Therefore, per Table 9.4, an environmental reduction factor of 0.95 is suggested.		
$f_{fu} = C_E f_{fu}^*$	f_{fu} = (0.95)(90 ksi) = 85 ksi	f_{fu} = (0.95)(621 N/mm^2) = 590 N/mm^2
$\varepsilon_{fu} = C_E \varepsilon_{fu}^*$	ε_{fu} = (0.95)(0.015 in./in.) = 0.0142 in./in.	ε_{fu} = (0.95)(0.015 mm/mm) = 0.0142 mm/mm
Step 2—Preliminary calculations		
Properties of the concrete:		
β_1 from ACI 318-14, Section 22.2.2.4.3	$\beta_1 = 1.05 - 0.05\frac{f_c'}{1000} = 0.85$	$\beta_1 = 1.05 - 0.05\frac{f_c'}{6.9} = 0.85$
$E_c = 57{,}000\sqrt{f_c'}$	$E_c = 57{,}000\sqrt{4000 \text{ psi}} = 3{,}605{,}00 \text{ psi}$	$E_c = 4700\sqrt{27.6 \text{ N/mm}^2} = 24{,}700 \text{ N/mm}^2$
Properties of the existing prestressing steel:	A_{ps} = 5(0.153 in.^{2}) = 0.765 in.^{2}	A_{ps} = 5(99 mm^2) = 495 mm^2
Area of FRP reinforcement:		
$A_f = nt_f w_f$	A_f = (1 ply)(0.040 in./ply)(24 in.) = 0.96 in.^{2}	A_f = (1 ply)(1.0 mm/ply)(610 mm) = 610 mm^2
Cross-sectional area:		
$A_{cg} = b_f h_f + b_w(h - h_f)$	A_{cg} = (87 in.)(4 in.) + (24 in.)(25 in. – 4 in.) = 852 in.^{2}	A_{cg} = (2210 mm)(102 mm) + (610 mm)(612 mm – 102 mm) = 5.5 × 10^5 mm^2
Distance from the top fiber to the section centroid:		
$y_t = \dfrac{b_f \frac{h_f^2}{2} + b_w(h-h_f)\left(h_f + \frac{(h-h_f)}{2}\right)}{A_{cg}}$	$y_t = \dfrac{87 \text{ in.} \times \frac{4 \text{ in.}^2}{2} + 24 \text{ in.} \times 21 \times 14.5}{852} = 9.39 \text{ in.}$	$y_t = \dfrac{2210 \text{ mm} \times \frac{102 \text{ mm}^2}{2} + 610 \text{ mm} \times 533 \times 368}{5.5 \times 10^5} = 238 \text{ mm}$
Gross moment of inertia:		
$I_g = \frac{b_f h_f^3}{12} + b_f h_f\left(y_t - \frac{h_f}{2}\right)^2 + \frac{b_w(h-h_f)^3}{12} + b_w(h-h_f)\left(y_t - \frac{h-h_f}{2}\right)^2$	$I_g = \frac{87 \text{ in.} \times 4 \text{ in.}^3}{12} + 87 \text{ in.} \times 4 \text{ in.}(9.39 \text{ in.} - 2)^2 + \frac{24 \text{ in.} \times 21^3}{12} + 24 \text{ in.} \times 21(9.39 - 10.5)^2 = 38{,}610 \text{ in.}^4$	$I_g = \frac{2210 \text{ mm} \times 102 \text{ mm}^3}{12} + 2210 \text{ mm} \times 102 \text{ mm}(238 - 51)^2 + \frac{610\text{mm} \times 533^3}{12} + 610 \text{ mm} \times 533(238 - 266.7)^2 = 1.61 \times 10^{10} \text{mm}^4$
Radius of gyration:		
$r = \sqrt{\frac{I_g}{A_{cg}}}$	$r = \sqrt{\frac{38{,}610}{852}} = 6.73 \text{ in.}$	$r = \sqrt{\frac{1.61 \times 10^{10}}{5.5 \times 10^5}} = 171 \text{ mm}$
Effective prestressing strain:		
$\varepsilon_{pe} = \frac{f_{pe}}{E_p}$	$\varepsilon_{pe} = \frac{165}{28{,}500} = 0.00578$	$\varepsilon_{pe} = \frac{1138}{1.96 \times 10^5} = 0.00579$
Effective prestressing force:		
$P_e = A_{ps} f_{pe}$	P_e = 0.765 × 165 = 126.2 kip	P_e = 495 × 1138 = 563,310 N
Eccentricity of prestressing force:		
$e = d_p - y_t$	e = 22.5 – 9.39 = 13.1 in.	e = 571 – 238 = 333 mm

Table 16.5b (cont.)—Procedure for flexural strengthening of an interior prestressed concrete beam with fiber-reinforced polymer laminates

Procedure	Calculation in in.-lb units	Calculation in SI metric units
Step 3—Determine the existing state of strain on the soffit The existing state of strain is calculated assuming the beam is uncracked and the only loads acting on the beam at the time of the FRP installation are dead loads. Distance from extreme bottom fiber to the section centroid: $y_b = h - y_t$ Initial strain in the beam soffit: $\varepsilon_{bi} = \frac{-P_e}{E_c A_{cg}}\left(1+\frac{ey_b}{r^2}\right)+\frac{M_{DL}y_b}{E_c I_g}$	$y_b = 25 - 9.39 = 15.61$ in. $\varepsilon_{bi} = \frac{-126.2}{3605\times852}\left(1+\frac{13.1\times15.6}{6.73^2}\right)+\frac{147\times12\times15.6}{3605\times38,610}$ $\varepsilon_{bi} = -2.88 \times 10^{-5}$	$y_b = 635 - 238 = 397$ mm $\varepsilon_{bi} = \frac{-563,310}{24,700\times5.5\times10^5}\left(1+\frac{333\times397}{171^2}\right)+\frac{199\times10^6\times397}{24,700\times1.61\times10^{10}}$ $\varepsilon_{bi} = -2.88 \times 10^{-5}$
Step 4—Determine the design strain of the FRP system The design strain of FRP accounting for debonding failure mode ε_{fd} is calculated using Eq. (10.2.1.1) Because the design strain is smaller than the rupture strain, debonding controls the design of the FRP system.	$\varepsilon_{fd} = 0.083\sqrt{\frac{4000\text{ psi}}{1(5,360,000\text{ psi})(0.04\text{ in.})}}$ $= 0.0113 \leq 0.9(0.0142) = 0.0128$	$\varepsilon_{fd} = 0.042\sqrt{\frac{27.6\text{ N/mm}^2}{1(37,000\text{ N/mm}^2)(1.016\text{ mm})}}$ $= 0.0113 \leq 0.9(0.0142) = 0.0128$
Step 5—Estimate *c*, the depth to the neutral axis A reasonable initial estimate of c is $0.1h$. The value of the c is adjusted after checking equilibrium. $c = 0.1h$	$c = (0.1)(25\text{ in.}) = 2.50$ in.	$c = (0.1)(635\text{ mm}) = 63.5$ mm
Step 6—Determine the effective level of strain in the FRP reinforcement The effective strain level in the FRP may be found from Eq. (10.3.1.6c). $\varepsilon_{fe} = 0.003\left(\frac{d_f - c}{c}\right) - \varepsilon_{bi} \leq \varepsilon_{fd}$ Note that for the neutral axis depth selected, FRP debonding would be the failure mode because the second expression in this equation controls. If the first (limiting) expression governed, then FRP rupture would be the failure mode.	$\varepsilon_{fe} = 0.003\left(\frac{25-2.5}{2.5}\right) - 0.00003 = 0.027$ $> \varepsilon_{fd} = 0.0113$ Failure is governed by FRP debonding $\varepsilon_{fe} = \varepsilon_{fd} = 0.0113$	$\varepsilon_{fe} = 0.003\left(\frac{635-63.5}{63.5}\right) - 0.00003 = 0.027$ $> \varepsilon_{fd} = 0.0113$ Failure is governed by FRP debonding $\varepsilon_{fe} = \varepsilon_{fd} = 0.0113$

Table 16.5b (cont.)—Procedure for flexural strengthening of an interior prestressed concrete beam with fiber-reinforced polymer laminates

Procedure	Calculation in in.-lb units	Calculation in SI metric units
Step 7—Calculate the strain in the existing prestressing steel The strain in the prestressing steel can be calculated using Eq. (10.3.1.6c) and (10.3.1.6a).		
$\varepsilon_{pnet} = (\varepsilon_{fe} + \varepsilon_{bi})\left(\frac{d_p - c}{d_f - c}\right)$	$\varepsilon_{pnet} = (0.0113 + 0.00003)\left(\frac{22.5 - 2.5}{25 - 2.5}\right)$ $\varepsilon_{pnet} = 0.01$	$\varepsilon_{pnet} = (0.0113 + 0.00003)\left(\frac{571 - 63.5}{635 - 63.5}\right)$ $\varepsilon_{pnet} = 0.01$
$\varepsilon_{ps} = \varepsilon_{pe} + \frac{P_e}{A_{cg}E_c}\left(1 + \frac{e^2}{r^2}\right) + \varepsilon_{pnet} \leq 0.035$	$\varepsilon_{ps} = 0.00589 + \frac{126.2}{852 \times 3605}\left(1 + \frac{13.1^2}{6.73^2}\right) + 0.01$ $\varepsilon_{ps} = 0.016 \leq 0.035$	$\varepsilon_{ps} = 0.00589 + \frac{563,310}{5.5 \times 10^5 \times 24,700}\left(1 + \frac{333^2}{171^2}\right) + 0.01$ $\varepsilon_{ps} = 0.016 \leq 0.035$
Step 8—Calculate the stress level in the prestressing steel and FRP The stresses are calculated using Eq. (10.3.1.6e) and Hooke's Law.		
$f_{ps} = \begin{cases} 28,500\varepsilon_{ps} & \text{for } \varepsilon_{ps} \leq 0.0086 \\ 270 - \frac{0.04}{\varepsilon_{ps} - 0.007} & \text{for } \varepsilon_{ps} > 0.0086 \end{cases}$	$f_{ps} = 270 - \frac{0.04}{0.016 - 0.007} = 265.6 \text{ ksi}$	$f_{ps} = 1860 - \frac{0.276}{0.016 - 0.007} = 1831 \text{ N/mm}^2$
$f_{fe} = E_f\varepsilon_{fe}$	$f_{fe} = (5360 \text{ ksi})(0.0113) = 60.6 \text{ ksi}$	$f_{fe} = (37,000 \text{ N/mm}^2)(0.0113) = 418 \text{ N/mm}^2$
Step 9—Calculate the equivalent concrete compressive stress block parameters α_1 and β_1. The strain in concrete at failure can be calculated from strain compatibility as follows:		
$\varepsilon_c = (\varepsilon_{fe} + \varepsilon_{bi})\left(\frac{c}{d_f - c}\right)$	$\varepsilon_c = (0.0113 + 0.00003)\left(\frac{2.5}{25 - 2.5}\right) = 0.0013$	$\varepsilon_c = (0.0113 + 0.00003)\left(\frac{63.5}{635 - 63.5}\right) = 0.0013$
The strain ε_c' corresponding to f_c' is calculated as		
$\varepsilon_c' = \frac{1.7 f_c'}{E_c}$	$\varepsilon_c' = \frac{1.7(4000)}{3605 \times 10^6} = 0.0019$	$\varepsilon_c' = \frac{1.7(27.6)}{24,700} = 0.0019$
Concrete stress block factors can be estimated using ACI 318. Approximate stress block factors may be calculated from the parabolic stress-strain relationship for concrete and is expressed as follows:		
$\beta_1 = \frac{4\varepsilon_c' - \varepsilon_c}{6\varepsilon_c' - 2\varepsilon_c}$	$\beta_1 = \frac{4(0.0019) - 0.0013}{6(0.0019) - 2(0.0013)} = 0.716$	$\beta_1 = \frac{4(0.0019) - 0.0013}{6(0.0019) - 2(0.0013)} = 0.716$
$\alpha = \frac{\varepsilon_c'\varepsilon_c - \varepsilon_c}{\beta\,\varepsilon'}$	$\alpha_1 = \frac{3(0.0019)(0.0013) - (0.0013)^2}{3(0.716)(0.0019)^2} = 0.738$	$\alpha_1 = \frac{3(0.0019)(0.0013) - (0.0013)^2}{3(0.716)(0.0019)^2} = 0.738$

Table 16.5b (cont.)—Procedure for flexural strengthening of an interior prestressed concrete beam with fiber-reinforced polymer laminates

Procedure	Calculation in in.-lb units	Calculation in SI metric units
Step 10—Calculate the internal force resultants and check equilibrium Force equilibrium is verified by checking the initial estimate of c with Eq. (10.2.10c).	$c = 1.42$ in. $\neq 2.50$ in. n.g. ∴ revise estimate of c and repeat Steps 6 through 10 until equilibrium is achieved.	$c = 36$ mm $\neq 63.5$ in. n.g. ∴ revise estimate of c and repeat Steps 6 through 10 until equilibrium is achieved.
Step 11—Adjust c until force equilibrium is satisfied Steps 6 through 10 were repeated several times with different values of c until equilibrium was achieved. The results of the final iteration are $c = 1.86$ in.; $\varepsilon_{ps} = 0.016$; $f_{ps} = f_y = 265.6$ ksi; $\varepsilon_{fe} = 0.0113$; $f_{fe} = 60.6$ ksi; $\varepsilon_c = 0.00091$; $\alpha_1 = 0.577$; and $\beta_1 = 0.698$.	$c = 1.86$ in. $= 1.86$ in. ∴ the value of c selected for the final iteration is correct.	$c = 47$ mm $= 47$ mm ∴ the value of c selected for the final iteration is correct.
Step 12—Calculate flexural strength components The design flexural strength is calculated using Eq. (10.3.1.6g). An additional reduction factor, $\psi_f = 0.85$, is applied to the contribution of the FRP system. Prestressing steel contribution to bending: FRP contribution to bending:	$M_{np} = 4440$ kip-in. $= 370$ kip-ft $M_{nf} = 1417$ kip-in. $= 118$ kip-ft	$M_{np} = 501.6 \times 10^6$ N-mm $= 501.6$ kN-m $M_{nf} = 160.1 \times 10^6$ N-mm $= 160.1$ kN-m
Step 13—Calculate design flexural strength of the section The design flexural strength is calculated using Eq. (10.1) and (10.3.1.6g). Because $\varepsilon_{ps} = 0.016 > 0.015$, a strength reduction factor of $\phi = 0.90$ should be used per Eq. (10.2.7). An additional reduction factor $\psi_f = 0.85$ is used to calculate the FRP contribution to nominal capacity. $\phi M_n = \phi[M_{np} + \psi_f M_{nf}]$	$\phi M_n = 0.9[370 \text{ kip-ft} + 0.85(118 \text{ kip-ft})]$ $\phi M_n = 423 \text{ kip-ft} \geq M_u = 397 \text{ kip-ft}$ ∴ the strengthened section is capable of sustaining the new required flexural strength.	$\phi M_n = 0.9[506.1 \text{ kN-m} + 0.85(160.1 \text{ kN-m})]$ $\phi M_n = 573 \text{ kN-m} \geq M_u = 538 \text{ kN-m}$ ∴ the strengthened section is capable of sustaining the new required flexural strength.
Step 14—Check service condition of the section Calculate the cracking moment and compare the service moment:	$f_r = 7.5\sqrt{4000} = 474 \text{ psi} = 0.474 \text{ ksi}$ $M_{cr} = 3695$ kip-in. $= 308$ kip-ft $> M_s = 288$ kip-ft ∴ the strengthened section is uncracked at service.	$f_r = 0.6\sqrt{27.6} = 3.15 \text{ N/mm}^2$ $M_{cr} = 411{,}654{,}013$ N-mm $= 412$ kN-mm $> M_s = 391.3$ kN-m ∴ the strengthened section is uncracked at service.

Table 16.5b (cont.)—Procedure for flexural strengthening of an interior prestressed concrete beam with fiber-reinforced polymer laminates

Procedure	Calculation in in.-lb units	Calculation in SI metric units
Step 15—Check stress in prestressing steel at service condition		
Calculate the cracking moment and compare to service moment: $$\varepsilon_{ps,s}=\varepsilon_{pe}+\frac{P_e}{A_cE_c}\left(1+\frac{e^2}{r^2}\right)+\frac{M_se}{E_cI_g}$$	$\varepsilon_{ps,s} = 0.0063 \leq 0.0086$	$\varepsilon_{ps,s} = 0.0063 \leq 0.0086$
Calculate the steel stress using Eq. (10.3.1.6d): $$f_{ps,s}=\begin{cases}28{,}500\varepsilon_{ps,s} & \text{for } \varepsilon_{ps,s}\leq 0.0086\\ 270-\dfrac{0.04}{\varepsilon_{ps,s}-0.07} & \text{for } \varepsilon_{ps,s}\leq 0.0086\end{cases}$$	$f_{ps,s} = 28{,}500(0.0063) = 180$ ksi	$f_{ps,s} = 1.96 \times 10^5(0.0063) = 1238$ N/mm²
Check the service stress limits of Eq. (10.3.1.4a) and (10.3.1.4b):		
$f_{ps,s} \leq 0.82f_{py}$	$f_{ps,s} = 180$ ksi $< 0.82(230) = 189$ ksi OK	$f_{ps,s} = 1238$ N/mm² $< 0.82(1586) = 1300$ N/mm² OK
$f_{ps,s} \leq 0.74f_{pu}$	$f_{ps,s} = 180$ ksi $< 0.74(270) = 200$ ksi OK	$f_{ps,s} = 1238$ N/mm² $< 0.74(1860) = 1376$ N/mm² OK
Step 16—Check stress in concrete at service condition		
Calculate the cracking moment and compare to service moment: $$\varepsilon_{c,s}=\frac{-P_e}{A_cE_c}\left(1+\frac{e^2}{r^2}\right)-\frac{M_sy_t}{E_cI_g}$$	$$\varepsilon_{c,s}=\frac{-26.2}{852\times3605}\left(1+\frac{13.1^2}{7.75^2}\right)-\frac{289\times12\times9.39}{3605\times51{,}150}$$ $\varepsilon_{c,s} = 0.00016$	$$\varepsilon_{cs}=\frac{-563{,}310}{5.5\times10^5\times24{,}700}\left(1+\frac{333^2}{197^2}\right)-\frac{391.3\times10^6\times238}{24{,}700\times2.13\times10^{10}}$$ $\varepsilon_{c,s} = 0.00016$
$f_{c,s} = E_c\varepsilon_{c,s}$	$f_{c,s} = 3{,}605{,}000$ psi $(0.00016) = 577$ psi	$f_{c,s} = 24{,}700$ N/mm² $(0.00016) = 3.95$ N/mm²
$f_{c,s} \leq 0.45f_c'$	$0.45f_c' = 0.45(4000) = 1800$ psi $f_{c,s} = 577$ psi $< 0.45f_c' = 1800$ psi OK	$0.45f_c' = 0.45(27.6) = 12.42$ N/mm² $f_{c,s} = 3.95$ N/mm² $< 0.45f_c' = 12.42$ N/mm² OK
Step 17—Check service stresses in the FRP reinforcement		
The stress in the FRP at service condition can be calculated using Eq. (10.3.1.8): $$f_{f,s}=\left(\frac{E_f}{E_c}\right)\frac{M_sy_b}{I}-\varepsilon_{bi}E_f$$ Because the section is uncracked at service, the gross moment of inertia of the section must be used.	$$f_{f,s}=\left(\frac{5360\text{ ksi}}{3605\text{ ksi}}\right)\frac{289\text{ kip-ft}\times12\text{ in./ft}\times15.61\text{ in.}}{51{,}150\text{ in.}^4}$$ -0.00003×5360 ksi $f_{f,s} = 1.41$ ksi	$$f_{f,s}=\left(\frac{37{,}700\text{ N/mm}^2}{24{,}700\text{ N/mm}^2}\right)\frac{391.3\times10^6\text{ N/mm}\times397\text{ mm}}{2.13\times10^{10}\text{ mm}^4}$$ $-0.00003\times37{,}700$ N/mm² $f_{f,s} = 9.7$ N/mm²
The calculated stress in FRP should be checked against the limits in Table 10.2.9. For carbon FRP:		
$f_{f,s} \leq 0.55f_{fu}$	$0.55f_{fu} = 0.55(85) = 47$ ksi $f_{f,s} = 1.41$ ksi $< 0.55f_{fu} = 47$ ksi OK	$0.55f_{fu} = 0.55(586) = 322$ N/mm² $f_{f,s} = 9.7$ N/mm² $< 0.55f_{fu} = 322$ N/mm² OK

In detailing the FRP reinforcement, the FRP should be terminated a minimum of ℓ_{df}, calculated per Eq. (14.1.3), past the point on the moment diagram that represents cracking. The factored shear force at the termination should also be checked against the shear force that causes FRP end peeling, estimated as two-thirds of the concrete shear strength. If the shear force is greater than two-thirds of the concrete shear strength, FRP strips should be extended further toward the supports. U-wraps may also be used to reinforce against cover delamination.

16.6—Shear strengthening of an interior T-beam

A reinforced concrete T-beam (f_c' = 3000 psi [20.7 N/mm²]) located inside of an office building is subjected to an increase in its live-load-carrying requirements. An analysis of the existing beam indicates that the beam is still satisfactory for flexural strength; however, its shear strength is inadequate to carry the increased live load. Based on the analysis, the nominal shear strength provided by the concrete is V_c = 44.2 kip (196.6 kN), and the nominal shear strength provided by steel shear reinforcement is V_s = 19.6 kip (87.2 kN). Thus, the design shear strength of the existing beam is $\phi V_{n,existing}$ = 0.75(44.2 kip + 19.6 kip) = 47.85 kip (213 kN). The factored required shear strength, including the increased live load, at a distance d away from the support is V_u = 57 kip (253.5 kN). Figure 16.6a shows the shear diagram with the locations where shear strengthening is required along the length of the beam.

Supplemental FRP shear reinforcement is designed as shown in Fig. 16.6b and summarized in Table 16.6a. Each FRP strip consists of one ply (n = 1) of a flexible carbon sheet installed by wet layup. The FRP system manufacturer's reported material properties are shown in Table 16.6b.

The design calculations used to arrive at this configuration follow in Table 16.6c.

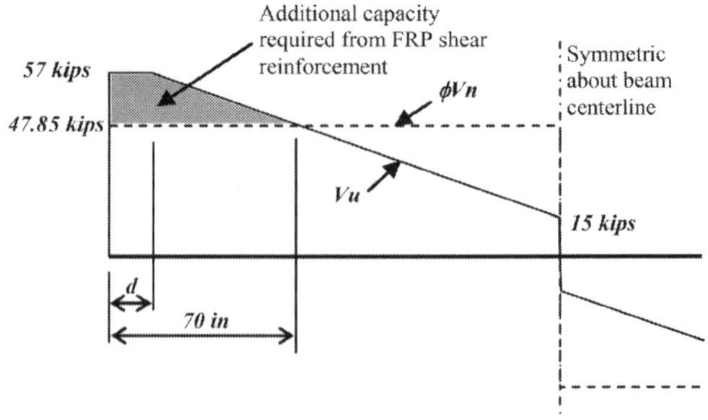

Fig. 16.6a—Shear diagram showing demand versus existing strength. The FRP reinforcement should correct the deficiency shown shaded.

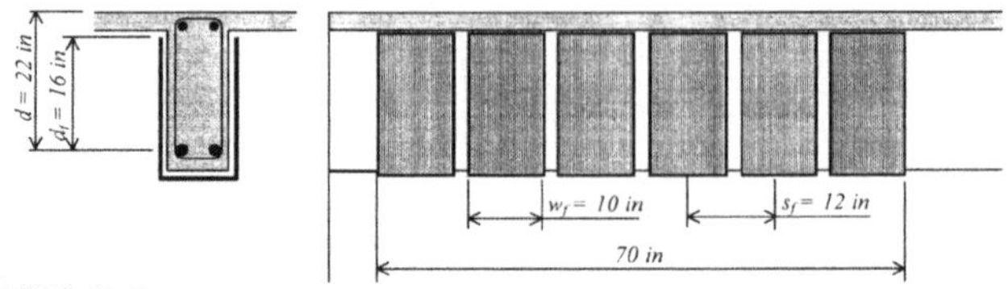

Fig. 16.6b—Configuration of the supplemental FRP shear reinforcement.

Table 16.6a—Configuration of the supplemental FRP shear reinforcement

d	22 in.	559 mm
d_{fv}	16 in.	406 mm
Width of each sheet w_f	10 in.	254 mm
Span between each sheet s_f	12 in.	305 mm
FRP strip length	70 in.	1778 mm

Table 16.6b—Manufacturer's reported FRP system properties

Thickness per ply t_f	0.0065 in.	0.165 mm
Ultimate tensile strength f_{fu}*	550,000 psi	3790 N/mm²
Rupture strain ε_{fu}*	0.017 in./in.	0.017 mm/mm
Modulus of elasticity E_f	33,000,000 psi	227,530 N/mm²

Table 16.6c—Procedure for shear strengthening of an interior T-beam

Procedure	Calculation in in.-lb units	Calculation in SI metric units
Step 1—Compute the design material properties		
The beam is located in an enclosed and conditioned space and a CFRP material will be used. Therefore, per Table 9.4, an environmental-reduction factor of 0.95 is suggested.		
$f_{fu} = C_E f_{fu}^*$	f_{fu} = (0.95)(550 ksi) = 522.5 ksi	f_{fu} = (0.95)(3.79 kN/mm^2) = 3.60 kN/mm^2
$\varepsilon_{fu} = C_E \varepsilon_{fu}^*$	ε_{fu} = (0.95)(0.017) = 0.016	ε_{fu} = (0.95)(0.017) = 0.016
Step 2—Calculate the effective strain level in the FRP shear reinforcement		
The effective strain in FRP U-wraps should be determined using the bond-reduction coefficient κ_v. This coefficient can be computed using Eq. (11.4.1.2b) through (11.4.2.1e).		
$L_e = \dfrac{2500}{(nt_f E_f)^{0.58}}$	$L_e = \dfrac{2500}{[(1)(0.0065\text{ in.})(33\times10^6\text{ psi})]^{0.58}} = 2.0$ in.	$L_e = \dfrac{416}{[(1)(0.1651\text{ mm})(227.5\times10^3\text{ kN/mm}^2)]^{0.58}} = 50.8$ mm
$k_1 = \left(\dfrac{f_c'}{4000}\right)^{2/3}$	$k_1 = \left(\dfrac{3000\text{ psi}}{4000}\right)^{2/3} = 0.825$	$k_1 = \left(\dfrac{20.7\text{ kN/mm}^2}{254}\right)^{2/3} = 0.825$
$k_2 = \left(\dfrac{d_{fv} - L_e}{d_{fv}}\right)$	$k_2 = \left(\dfrac{16\text{ in.} - 2.0\text{ in.}}{16\text{ in.}}\right) = 0.875$	$k_2 = \left(\dfrac{406\text{ mm} - 50.8\text{ mm}}{406\text{ mm}}\right) = 0.875$
$\kappa_v = \dfrac{k_1 k_2 L_e}{468\varepsilon_{fu}} \le 0.75$	$\kappa_v = \dfrac{(0.82)(0.875)(2\text{ in.})}{468(0.016)} = 0.193 \le 0.75$	$\kappa_v = \dfrac{(0.82)(0.875)(50.8\text{ mm})}{11{,}910(0.016)} = 0.193 \le 0.75$
The effective strain can then be computed using Eq. (11.4.1.2a) as follows:		
$\varepsilon_{fe} = \kappa_v \varepsilon_{fu} \le 0.004$	$\varepsilon_{fe} = 0.193(0.016) = 0.0031 \le 0.004$	$\varepsilon_{fe} = 0.193(0.016) = 0.0031 \le 0.004$
Step 3—Calculate the contribution of the FRP reinforcement to the shear strength		
The area of FRP shear reinforcement can be computed as:		
$A_{fv} = 2nt_f w_f$	A_{fv} = 2(1)(0.0065 in.)(10 in.) = 0.13 in.^{2}	A_{fv} = 2(1)(0.1651 mm)(254 mm) = 83.87 mm^2
The effective stress in the FRP can be computed from Hooke's law.		
$f_{fe} = \varepsilon_{fe} E_f$	f_{fe} = (0.0031)(33,000 ksi) = 102 ksi	f_{fe} = (0.0031)(227.6 kN/mm^2) = 0.703 kN/mm^2
The shear contribution of the FRP can be then calculated from Eq. (11.4a):		
$V_f = \dfrac{A_{fv} f_{fe} (\sin\alpha + \cos\alpha) d_{fv}}{s_f}$	$V_f = \dfrac{(0.13\text{ in.}^2)(102\text{ ksi})(1)(16\text{ in.})}{(12\text{ in.})}$ V_f = 17.7 kip	$V_f = \dfrac{(83.87\text{ mm}^2)(0.703\text{ kN/mm}^2)(1)(406\text{ mm})}{(304.8\text{ mm})}$ V_f = 78.5 kN

Table 16.6c (cont.)—Procedure for shear strengthening of an interior T-beam

Procedure	Calculation in in.-lb units	Calculation in SI metric units
Step 4—Calculate the shear strength of the section The design shear strength can be computed from Eq. (11.3b) with $\psi_f = 0.85$ for U-wraps.		
$\phi V_n = \phi(V_c + V_s + \psi_f V_f)$	$\phi V_n = 0.75[44.2 + 19.6 + (0.85)(17.7)]$ $\phi V_n = 59$ kip $> V_u = 57$ kip ∴ the strengthened section is capable of sustaining the required shear strength.	$\phi V_n = 0.75[196.6 + 87.2 + (0.85)(78.5)]$ $\phi V_n = 263$ kN $> V_u = 253.3$ kN ∴ the strengthened section is capable of sustaining the required shear strength.

16.7—Shear strengthening of an exterior column

A 24 x 24 in. (610 x 610 mm) square column requires an additional 60 kip (267 kN) of shear strength (ΔV_u = 60 kip [267 kN]). The column is located in an unenclosed parking garage and experiences a wide variation in temperature and climate. A method of strengthening the column using FRP is sought.

An E-glass-based FRP complete wrap is selected to retrofit the column. The properties of the FRP system, as reported by the manufacturer, are shown in Table 16.7. The design calculations to arrive at the number of complete wraps required follow.

Table 16.7a—Manufacturer's reported FRP system properties*

Thickness per ply t_f	0.051 in.	1.3 mm
Guaranteed ultimate tensile strength f_{fu}^*	80,000 psi	552 N/mm^2
Guaranteed rupture strain ε_{fu}^*	0.020 in./in.	0.020 mm/mm
Modulus of elasticity E_f	4,000,000 psi	27,600 N/mm^2

*The reported properties are laminate properties.

Table 16.7b—Procedure for shear strengthening of an exterior column

Procedure	Calculation in in.-lb units	Calculation in SI metric units
Step 1—Compute the design material properties The column is located in an exterior environment and a glass FRP (GFRP) material will be used. Therefore, per Table 9.4, an environmental reduction factor of 0.65 is suggested. $f_{fu} = C_E f_{fu}^*$ $\varepsilon_{fu} = C_E \varepsilon_{fu}^*$	f_{fu} = (0.65)(80 ksi) = 52 ksi ε_{fu} = (0.65)(0.020) = 0.013	f_{fu} = (0.65)(552 N/mm²) = 358.5 N/mm² ε_{fu} = (0.65)(0.020) = 0.013
Step 2—Calculate the effective strain level in the FRP shear reinforcement The effective strain in a complete FRP wrap can be determined from Eq. (11.4.1.1): $\varepsilon_{fe} = 0.004 \le 0.75\varepsilon_{fu}$	$\varepsilon_{fe} = 0.004 \le 0.75(0.013) = 0.010$ ∴ use an effective strain of ε_{fe} = 0.004.	$\varepsilon_{fe} = 0.004 \le 0.75(0.013) = 0.010$ ∴ use an effective strain of ε_{fe} = 0.004.
Step 3—Determine the area of FRP reinforcement required The required shear contribution of the FRP reinforcement can be computed based on the increase in strength needed, the strength reduction factor for shear, and a partial-reduction factor ψ_f = 0.95 for completely wrapped sections in shear. $V_{f,reqd} = \dfrac{\Delta V_u}{\phi(\psi_f)}$ The required area of FRP can be determined by reorganizing Eq. (11.4a). The required area is left in terms of the spacing. $A_{fv,reqd} = \dfrac{V_{f,reqd} s_f}{\varepsilon_{fe} E_f (\sin\alpha + \cos\alpha) d_f}$	$V_{f,reqd} = \dfrac{60\text{ kip}}{0.85(0.95)} = 74.3\text{ kip}$ $A_{fv,reqd} = \dfrac{(74.3\text{ kip})s_f}{(0.004)(4000\text{ ksi})(1)(24\text{ in.})} = 0.194 s_f$	$V_{f,reqd} = \dfrac{266.9\text{ kN}}{0.85(0.95)} = 330.5\text{ kN}$ $A_{fv,reqd} = \dfrac{(330.5\text{ kN})s_f}{(0.004)(27.6\text{ kN/mm}^2)(1)(610\text{ mm})} = 4.91 s_f$
Step 4—Determine the number of plies, and strip width and spacing. The number of plies can be determined in terms of the strip width and spacing as follows: $n = \dfrac{A_{f,reqd}}{2 t_f w_f}$	$n = \dfrac{0.194 s_f}{2(0.051\ in.)w_f} = 1.90\dfrac{s_f}{w_f}$ ∴ use two plies (n = 2) continuously along the height of the column ($s_f = w_f$).	$n = \dfrac{4.91 s_f}{2(1.3\text{ mm})w_f} = 1.90\dfrac{s_f}{w_f}$ ∴ use two plies (n = 2) continuously along the height of the column ($s_f = w_f$).

16.8—Strengthening of a noncircular concrete column for axial load increase

A 24 x 24 in. (610 x 610 mm) square column requires an additional 20 percent of axial load-carrying capacity. Concrete and steel reinforcement material properties as well as details of the cross section of the column are shown in Table 16.8a. The column is located in an interior environment, and a CFRP material will be used. A method of strengthening the column is sought.

A carbon-based FRP complete wrap is selected to retrofit the columns. The properties of the FRP system, as reported by the manufacturer, are shown in Table 16.8b. The design calculations to arrive at the number of required complete wraps follow.

Table 16.8a—Column cross section details and material properties

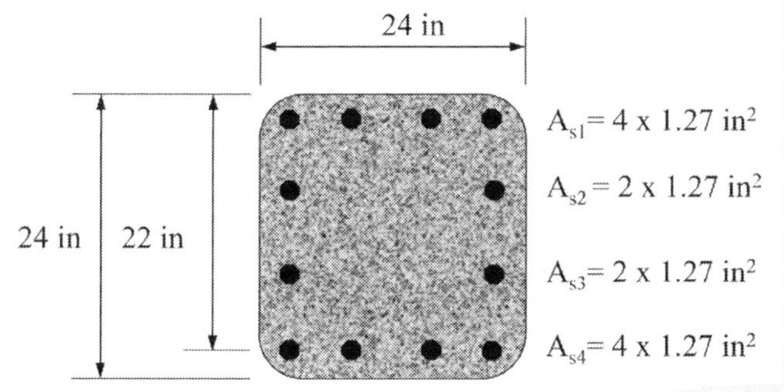

f_c'	6.5 ksi	45 MPa
f_y	60 ksi	400 MPa
r_c	1 in.	25 mm
Bars	12 No. 10	12ϕ32
A_g	576 in.^{2}	3716 cm^2
A_{st}	15.24 in.^{2}	98 cm^2
ρ_g, %	2.65	2.65
ϕP_n without FRP	2087 kip	9281 kN
$\phi P_{n(req)}$	2504 kip	11,138 kN

Note: The column features steel ties for transverse reinforcement.

Table 16.8b—Manufacturer's reported FRP system properties

Thickness per ply t_f	0.013 in.	0.33 mm
Ultimate tensile strength f_{fu}^*	550 ksi	3792 MPa
Rupture strain ε_{fu}^*	0.0167 in./in.	0.0167 mm/mm
Modulus of elasticity E_f	33,000 ksi	227,527 MPa

Table 16.8c—Procedure for strengthening of a noncircular concrete column for axial load increase

Procedure	Calculation in in.-lb units	Calculation in SI metric units
Step 1—Compute the design FRP material properties The column is located in an interior environment and a CFRP material will be used. Therefore, per Table 9.4, an environmental reduction factor of 0.95 is suggested. $f_{fu} = C_E f_{fu}^*$ $\varepsilon_{fu} = C_E \varepsilon_{fu}^*$	$f_{fu} = (0.95)(550\text{ ksi}) = 522.5\text{ ksi}$ $\varepsilon_{fu} = (0.95)(0.0167) = 0.0159\text{ in./in.}$	$f_{fu} = (0.95)(3792\text{ MPa}) = 3603\text{ MPa}$ $\varepsilon_{fu} = (0.95)(0.0167) = 0.0159\text{ mm/mm}$
Step 2—Determine the required maximum compressive strength of confined concrete f'_{cc} f'_{cc} can be obtained by reordering Eq. (12.1b): $f'_{cc} = \frac{1}{0.85(A_g - A_{st})}\left(\frac{\phi P_{n,req}}{0.80\phi} - f_y A_{st}\right)$	$f'_{cc} = \frac{1}{0.85\times(576\text{ in.}^2 - 15.24\text{ in.}^2)} \times\left(\frac{2504\text{ kip}}{0.80\times 0.65} - 60\text{ ksi}\times 15.24\text{ in.}^2\right)$ $f'_{cc} = 8.18\text{ ksi}$	$f'_{cc} = \frac{1}{0.85\times(371{,}612\text{ mm}^2 - 9832\text{ mm}^2)} \times\left(\frac{11{,}138\text{ kN}}{0.80\times 0.65} - 414\text{ MPa}\times 9832\text{ mm}^2\right)$ $f'_{cc} = 56.4\text{ MPa}$
Step 3—Determine the maximum confining pressure due to the FRP jacket f_ℓ f_ℓ can be obtained by reordering Eq. (12.1g): $f_\ell = \frac{f'_{cc} - f'_c}{3.3\kappa_a}$ where $\kappa_a = \frac{A_e}{A_c}\left(\frac{b}{h}\right)^2$ $\frac{A_e}{A_c} = \frac{1 - \frac{\left[\left(\frac{b}{h}\right)(h-2r_c)^2 + \left(\frac{h}{b}\right)(b-2r_c)^2\right]}{3A_g} - \rho_g}{1-\rho_g}$	$f_\ell = \frac{8.18\text{ ksi} - 6.5\text{ ksi}}{0.95\times 3.3\times 0.425} = 1.26\text{ ksi}$ $\kappa_a = 0.425(1)^2 = 0.425$ $\frac{A_e}{A_c} = \frac{1 - \frac{[2\times(1)(24\text{ in.} - 2\times 1\text{ in.})^2]}{3\times 576\text{ in.}^2} - 0.0265}{1 - 0.0265}$ $\frac{A_e}{A_c} = 0.425$	$f_\ell = \frac{56.4\text{ MPa} - 44.8\text{ MPa}}{0.95\times 3.3\times 0.425} = 8.7\text{ MPa}$ $\kappa_a = 0.425(1)^2 = 0.425$ $\frac{A_e}{A_c} = \frac{1 - \frac{[2\times(1)(610\text{ mm} - 2\times 25\text{ mm})^2]}{3\times 371{,}612\text{ mm}^2} - 0.0265}{1 - 0.0265}$ $\frac{A_e}{A_c} = 0.425$
Step 4—Determine the number of plies *n*. *n* can be obtained by reordering Eq. (12.1h): $n = \frac{f_\ell\sqrt{b^2+h^2}}{2E_f t_f \varepsilon_{fe}}$ $\varepsilon_{fe} = \kappa_\varepsilon \varepsilon_{fu}$ Checking the minimum confinement ratio: $\frac{f_\ell}{f'_c} \geq 0.08$	$n = \frac{1.26\text{ ksi}\sqrt{(24\text{ in.})^2 + (24\text{ in.})^2}}{2(33{,}000\text{ ksi})(0.013\text{ in.})(8.8\times 10^{-3}\text{ in./in.})}$ $n = 5.7 \approx 6\text{ plies}$ $\varepsilon_{fe} = 0.55 \times 0.0159\text{ in./in.} = 8.8\times 10^{-3}\text{ in./in.}$ $\frac{f_\ell}{f'_c} = \frac{1.26\text{ ksi}}{6.5\text{ ksi}} = 0.19 > 0.08\ \text{OK}$	$n = \frac{8.7\text{ MPa}\sqrt{(610\text{ mm})^2 + (610\text{ mm})^2}}{2(227{,}527\text{ MPa})(0.33\text{ mm})(8.8\times 10^{-3}\text{ mm/mm})}$ $n = 5.7 \approx 6\text{ plies}$ $\varepsilon_{fe} = 0.55 \times 0.0159\text{ mm/mm} = 8.8\times 10^{-3}\text{ mm/mm}$ $\frac{f_\ell}{f'_c} = \frac{8.7\text{ MPa}}{44.8\text{ MPa}} = 0.19 > 0.08\ \text{OK}$

Table 16.8c (cont.)—Procedure for strengthening of a noncircular concrete column for axial load increase

Procedure	Calculation in in.-lb units	Calculation in SI metric units
Step 5—Verify that the ultimate axial strain of the confined concrete $\varepsilon_{ccu} \leq 0.01$		
ε_{ccu} can be obtained using Eq. (12.1j):		
$\varepsilon_{ccu} = \varepsilon_c' \left(1.5 + 12\kappa_b \frac{f_\ell}{f_c'}\left(\frac{\varepsilon_{fe}}{\varepsilon_c'}\right)^{0.45}\right)$	$\varepsilon_{cc} = (0.002 \text{ in./in.}) \times \left(1.5 + 12 \times 0.425 \times \frac{1.2 \text{ ksi}}{6.5 \text{ ksi}}\left(\frac{8.8\times10^{-3} \text{ in./in.}}{0.002 \text{ in./in.}}\right)^{0.45}\right)$	$\varepsilon_{cc} = (0.002 \text{ mm/mm}) \times \left(1.5 + 12 \times 0.425 \times \frac{8.3 \text{ MPa}}{44.8 \text{ MPa}}\left(\frac{8.8\times10^{-3} \text{ mm/mm}}{0.002 \text{ mm/mm}}\right)^{0.45}\right)$
where	$\varepsilon_{cc} = 0.0067$ in./in. < 0.01 OK	$\varepsilon_{cc} = 0.0067$ mm/mm < 0.01 OK
$\kappa_b = \frac{A_e}{A_c}\left(\frac{h}{b}\right)^{0.5}$	$\kappa_b = 0.425(1)^{0.5} = 0.425$	$\kappa_b = 0.425(1)^{0.5} = 0.425$
If the case that ε_{ccu} was to be greater than 0.01, then f_{cc}' should be recalculated from the stress-strain model using Eq. (12.1c).		

16.9—Strengthening of a noncircular concrete column for increase in axial and bending forces

The column described in 16.8 is subjected to an ultimate axial compressive load P_u = 1900 kip (8451 kN) and an ultimate bending moment M_u = 380 kip-ft (515 kN-m) (e = $0.1h$). It is sought to increase load demands by 30 percent at constant eccentricity (P_u = 2470 kip, M_u = 494 kip-ft).

The calculations to determine moment-axial interaction for the FRP confined column follow in Table 16.9

Table 16.9—Procedure for strengthening of a noncircular concrete column for increase in axial and bending forces

Procedure	Calculation in in.-lb units	Calculation in SI metric units
Step 1—Determine the simplified curve for the unstrengthened column (n = 0 plies)		
Points A, B, and C can be obtained by well-known procedures, and also by using Eq. (D-1) to (D-5) considering $\psi_f = 1$, $f_{cc}' = f_c'$, $E_2 = 0$, and $\varepsilon_{ccu} = \varepsilon_{cu} = 0.003$.	$\phi P_{n(A)}$ = 2087 kip; $\phi M_{n(A)}$ = 0 kip-ft $\phi P_{n(B)}$ = 1858 kip; $\phi M_{n(B)}$ = 644 kip-ft $\phi P_{n(C)}$ = 928 kip; $\phi M_{n(C)}$ = 884 kip-ft	$\phi P_{n(A)}$ = 9283 kN; $\phi M_{n(A)}$ = 0 kN-m $\phi P_{n(B)}$ = 8265 kN; $\phi M_{n(B)}$ = 873 kN-m $\phi P_{n(C)}$ = 4128 kN; $\phi M_{n(C)}$ = 1199 kN-m

Table 16.9 (cont.)—Procedure for strengthening of a noncircular concrete column for increase in axial and bending forces

Procedure	Calculation in in.-lb units	Calculation in SI metric units
Step 2—Determine the simplified curve for a strengthened column A wrapping system composed of six plies will be the starting point to construct the bilinear Curve A-B-C and then be compared with the position of the required P_u and M_u.		
Points A, B, and C of the curve can be computed using Eq. (12-1), (D-1), and (D-2):	Point A: Nominal axial capacity:	Point A: Nominal axial capacity:
$\phi P_{n(A)} = \phi 0.8(0.85 f_{cc}'(A_g - A_{st}) + f_y A_{st})$	$\phi P_{n(A)}$ = 0.65 × 0.8(0.85 × 8.26 ksi × (576 in.^{2} – 15.24 in.^{2}) + 60 ksi × 15.24 in.^{2}) $\phi P_{n(A)}$ = 2523 kip where f_{cc}' = 6.5 ksi + 3.3(0.425)(1.26 ksi) f_{cc}' = 8.26 ksi $f_l = \dfrac{0.95 \times 2 \times 33{,}000 \text{ ksi} \times 6 \times 0.013 \text{ in.} \times \left(0.55 \times 0.0159 \dfrac{\text{in.}}{\text{in.}}\right)}{\sqrt{(24 \text{ in.})^2 + (24 \text{ in.})^2}}$ f_ℓ = 1.26 ksi	$\phi P_{n(A)}$ = 0.65 × 0.8(0.85 × 56.96 MPa × (371,612 mm^2 – 9832 mm^2) + 414 MPa + 9232 mm^2 $\phi P_{n(A)}$ = 11,223 kN where f_{cc}' = 44.8 MPa + 3.3(0.425)(8.7 MPa) f_{cc}' = 56.96 MPa $f_l = \dfrac{0.95 \times 2 \times 227{,}500 \text{ MPa} \times 6 \times 0.33 \text{ mm} \times \left(0.55 \times 0.0159 \dfrac{\text{mm}}{\text{mm}}\right)}{\sqrt{(610 \text{ mm})^2 + (610 \text{ mm})^2}}$ f_ℓ = 8.67 MPa
	Point B: Nominal axial capacity:	Point B: Nominal axial capacity:
$\phi P_{n(B,C)} = \phi(A(y_t)^3 + B(y_t)^2 + C(y_t) + D) + \sum A_{si} f_{si}]$ $\phi M_{n(B,C)} = \phi(E(y_t)^4 + F(y_t)^3 + G(y_t)^2 + H(y_t) + I) + \sum A_{si} f_{si} d_i$	$\phi P_{n(B)}$ = 0.65[–0.22 kip/in.^{3}(15.33 in.)^3 + 10.17 ksi (15.33 in.)^2 – 56 kip/in.(15.33 in.) + 3645.2 kip] + 5.08 in.^{2} (60 ksi) + 2.54 in.^{2}(60 ksi) + 2.54 in.^{2}(37.21 ksi)] $\phi P_{n(B)}$ = 2210 kip	$\phi P_{n(B)}$ = 0.65[–6.003 × 10^{-5} kN/mm^3(389 mm)^3 + 70.14 × 10^{-3} kN/mm^2(389 mm)^2 – 27.32 kN/mm(389 mm) + 16,215 kN] + 3277 mm^2(414 MPa) + 1639 mm^2(414 MPa) + 1639 mm^2(257 ksi) $\phi P_{n(B)}$ = 9892 kN
The coefficients A, B, C, D, E, F, G, H, and I of the previous expressions are given by Eq. (D-3):	where	where
$A = \dfrac{-b(E_c - E_2)^2}{12 f_c'}\left(\dfrac{\varepsilon_{ccu}}{c}\right)^2$	$A = \dfrac{-24 \text{ in.}(4595 \text{ ksi} - 190.7 \text{ ksi})^2}{12 \times 6.5 \text{ ksi}}\left(\dfrac{0.0042 \text{ in./in.}}{22 \text{ in.}}\right)^2$ $= -0.22$ kip/in.^{3}	$A = \dfrac{-610 \text{ mm}(31{,}685 \text{ MPa} - 1315 \text{ MPa})^2}{12 \times 44.8 \text{ MPa}}\left(\dfrac{0.0042 \text{ mm/mm}}{559 \text{ mm}}\right)^2$ $= -6.003 \times 10^{-5}$ kN/mm^3
$B = \dfrac{b(E_c - E_2)}{2}\left(\dfrac{\varepsilon_{ccu}}{c}\right)$ $C = b f_c'$	$B = \dfrac{24 \text{ in.}(4595 \text{ ksi} - 190.7 \text{ ksi})}{2}\left(\dfrac{0.0042 \text{ in./in.}}{22 \text{ in.}}\right)$ $= 10.17$ ksi C = –24 in. × 6.5 ksi = –156 kip/in.	$B = \dfrac{600 \text{ mm}(31{,}685 \text{ MPa} - 1315 \text{ MPa})}{2}\left(\dfrac{0.0042 \text{ mm/mm}}{559 \text{ mm}}\right)$ $= 70.14 \times 10^{-3}$ kN/mm^2 C = –610 mm × 44.84 MPa = –27.32 kN/mm
$D = b f_c' + \dfrac{bcE_2}{2}(\varepsilon_{ccu})$ $E = \dfrac{-b(E_c - E_2)^2}{16 f_c'}\left(\dfrac{\varepsilon_{ccu}}{c}\right)^2$ $F = b\left(c - \dfrac{h}{2}\right)\dfrac{(E_c - E_2)^2}{12 f_c'}\left(\dfrac{\varepsilon_{ccu}}{c}\right)^2 + \dfrac{b(E_c - E_2)}{3}\left(\dfrac{\varepsilon_{ccu}}{c}\right)$ $G = -\left(\dfrac{b}{2} f_c' + b\left(c - \dfrac{h}{2}\right)\dfrac{(E_c - E_2)}{2}\left(\dfrac{\varepsilon_{ccu}}{c}\right)\right)$ $H = b f_c'\left(c - \dfrac{h}{2}\right)$ $I = \dfrac{bc^2}{2} f_c' - bcf_c'\left(c - \dfrac{h}{2}\right) + \dfrac{bc^2 E_2}{3}(\varepsilon_{ccu}) - \dfrac{bcE_2}{2}\left(c - \dfrac{h}{2}\right)(\varepsilon_{ccu})$	$D = 24 \text{ in.} \times 22 \text{ in.} \times 6.5 \text{ ksi} + \dfrac{24 \text{ in.} \times 22 \text{ in.} \times 190.7 \text{ ksi}}{2}$ × (0.0042 in./in.) = 3645.2 kip	$D = 610 \text{ mm} \times 559 \text{ mm} \times 44.8 \text{ MPa} + \dfrac{610 \text{ mm} \times 559 \text{ mm} \times 1315 \text{ MPa}}{2}$ × (0.0042 mm/mm) = 16,215 kN

Table 16.9 (cont.)—Procedure for strengthening of a noncircular concrete column for increase in axial and bending forces

Procedure	Calculation in in.-lb units	Calculation in SI metric units
Key parameters of the stress-strain model:	For the calculation of the coefficients, it is necessary to compute key parameters from the stress-strain model:	For the calculation of the coefficients, it is necessary to compute key parameters from the stress-strain model:
$y_t = c\frac{\varepsilon_t'}{\varepsilon_{ccu}}$	$y_t = 22\text{ in.}\times\frac{0.003\text{ in./in.}}{0.0042\text{ in./in.}} = 15.33\text{ in.}$	$y_t = 559\text{ mm}\times\frac{0.003\text{ mm/mm}}{0.0042\text{ mm/mm}} = 389\text{ mm}$
$c = \begin{cases} d & \text{for Point B} \\ d\frac{\varepsilon_{ccu}}{\varepsilon_{sy}+\varepsilon_{ccu}} & \text{for Point C}\end{cases}$	$c = 22\text{ in.}$	$c = 559\text{ mm}$
$\varepsilon_t' = \frac{2f_c'}{E_c - E_2}$	$\varepsilon_t' = \frac{2\times 6.5\text{ ksi}}{4595\text{ ksi} - 190.7\text{ ksi}} = 0.003\text{ in./in.}$	$\varepsilon_t' = \frac{2\times 44.8\text{ MPa}}{31{,}685\text{ MPa} - 1315\text{ MPa}} = 0.003\text{ mm/mm}$
$E_2 = \frac{f_{cc}' - f_c'}{\varepsilon_{ccu}}$	$E_2 = \frac{7.31\text{ ksi} - 6.5\text{ ksi}}{0.0042\text{ in./in.}} = 190.7\text{ ksi}$	$E_2 = \frac{50.4\text{ MPa} - 44.8\text{ MPa}}{0.0042\text{ mm/mm}} = 1315\text{ MPa}$
$f_{cc}' = f_c' + 3.3\kappa_a f_l$	$f_{cc}' = 6.5\text{ ksi} + 3.3(0.425)(0.58\text{ ksi}) = 7.31\text{ ksi}$	$f_{cc}' = 44.8\text{ MPa} + 3.3(0.425)(3.97\text{ MPa}) = 50.4\text{ MPa}$
$\varepsilon_{ccu} = \varepsilon_c'\left(1.5 + 12\kappa_b\frac{f_l}{f_c'}\left(\frac{\varepsilon_{fe}}{\varepsilon_c'}\right)^{0.45}\right)$ $\varepsilon_{fe} = \min(0.004, \kappa_\varepsilon\varepsilon_{fu})$	$\varepsilon_{ccu} = 0.002\text{ in./in.}$ $\times\left(1.5 + 12\times 0.425\left(\frac{0.58\text{ ksi}}{6.5\text{ ksi}}\right)\left(\frac{0.004\text{ in./in.}}{0.002\text{ in./in.}}\right)^{0.45}\right)$ $\varepsilon_{ccu} = 0.0042\text{ in./in.}$	$\varepsilon_{ccu} = 0.002\text{ mm/mm}$ $\times\left(1.5 + 12\times 0.425\left(\frac{3.97\text{ MPa}}{44.8\text{ MPa}}\right)\left(\frac{0.004\text{ mm/mm}}{0.002\text{ mm/mm}}\right)^{0.45}\right)$ $\varepsilon_{ccu} = 0.0042\text{ mm/mm}$
$\kappa_a = \frac{A_e}{A_c}\left(\frac{b}{h}\right)^2$ $\kappa_b = \frac{A_e}{A_c}\left(\frac{h}{b}\right)^{0.5}$	$\kappa_a = \kappa_b = 0.425$	$\kappa_a = \kappa_b = 0.425$
$\frac{\psi_f\ E_f n t_f \varepsilon_{fe}}{\sqrt{b^2\ \ h^2}}$	$f_l = \frac{0.95\times 2\times 33{,}000\text{ ksi}\times 6\times 0.013\text{ in.}\times(0.004\text{ in./in.})}{\sqrt{(24\text{ in.})^2 + (24\text{ in.})^2}}$	$f_l = \frac{0.95\times 2\times 227{,}527\text{ MPa}\times 6\times 0.33\text{ mm}\times(0.004\text{ mm/mm})}{\sqrt{(610\text{ mm})^2 + (610\text{ mm})^2}}$
Notes: The designer should bear in mind that, for the case of pure compression, the effective strain in the FRP, ε_{fe}, is limited by $\kappa_\varepsilon\varepsilon_{fu}$ and, in the case of combined axial and bending, by $\varepsilon_{fe} = \min(0.004, \kappa_\varepsilon\varepsilon_{fu})$.	Checking the minimum confinement ratio: $f_l/f_c' = 0.58\text{ ksi}/6.5\text{ ksi} = 0.09 \geq 0.08$ OK The strains in each layer of steel are determined by similar triangles in the strain distribution. The corresponding stresses are then given by: $f_{s1} = \varepsilon_{s1}E_s = 0.0038\text{ in./in.}\times 29{,}000\text{ ksi} \rightarrow 60\text{ ksi}$ $f_{s2} = \varepsilon_{s2}E_s = 0.0026\text{ in./in.}\times 29{,}000\text{ ksi} \rightarrow 60\text{ ksi}$ $f_{s3} = \varepsilon_{s3}E_s = 0.0013\text{ in./in.}\times 29{,}000\text{ ksi} = 37.2\text{ ksi}$ $f_{s4} = \varepsilon_{s4}E_s = 0\text{ in./in.}\times 29{,}000\text{ ksi} = 0\text{ ksi}$ Nominal bending moment: $\phi M_{n(B)} = 0.65[-0.166\text{ kip/in.}^3(15.33\text{ in.})^4 + 8.99\text{ ksi}(15.33\text{ in.})^3 - 179.73\text{ kip/in.}(15.33\text{ in.})^2 + 1560\text{ kip}(15.33\text{ in.}) + 4427\text{ kip-in.}] + 5.08\text{ in.}^2(60\text{ ksi})(10\text{ in.}) + 2.54\text{ in.}^2(60\text{ ksi})(3.3\text{ in.}) - 2.54\text{ in.}^2(37.21\text{ ksi})(3.3\text{ in.})$ $\phi M_{n(B)} = 682\text{ kip-ft}$ where $E = \frac{-24\text{ in.}(4595\text{ ksi} - 190.7\text{ ksi})^2}{16\times 6.5\text{ ksi}}\left(\frac{0.0042\text{ in./in.}}{22\text{ in.}}\right)^2$ $= -0.166\text{ kip/in.}^3$	Checking the minimum confinement ratio: $f_l/f_c' = 3.97\text{ MPa}/44.8\text{ MPa} = 0.09 \geq 0.08$ OK The strains in each layer of steel are determined by similar triangles in the strain distribution. The corresponding stresses are then given by: $f_{s1} = \varepsilon_{s1}E_s = 0.0038\text{ mm/mm}\times 200{,}000\text{ MPa} \rightarrow 414\text{ MPa}$ $f_{s2} = \varepsilon_{s2}E_s = 0.0026\text{ mm/mm}\times 200{,}000\text{ MPa} \rightarrow 414\text{ MPa}$ $f_{s3} = \varepsilon_{s3}E_s = 0.0013\text{ mm/mm}\times 200{,}000\text{ MPa} = 257\text{ MPa}$ $f_{s4} = \varepsilon_{s4}E_s = 0\text{ mm/mm}\times 200{,}000\text{ MPa} = 0\text{ MPa}$ Nominal bending moment: $\phi M_{n(B)} = 0.65[-4.502\times 10^{-5}\text{ kN/mm}^3(389\text{ mm})^4 + 62.01\times 10^{-3}\text{ kN/mm}^3(389\text{ mm})^3 - 31.48\text{ kN/mm}(389\text{ mm})^2 + 6939\text{ kN}(389\text{ mm}) + 500{,}162\text{ kN-mm}] + 3277\text{ mm}^2(414\text{ MPa})(254\text{ mm}) + 1639\text{ mm}^2(414\text{ MPa})(85\text{ mm}) - 1639\text{ mm}^2(257\text{ MPa})(85\text{ mm})$ $\phi M_{n(B)} = 924\text{ kN-m}$ where $E = \frac{-610\text{ mm}(31{,}685\text{ MPa} - 1315\text{ MPa})^2}{16\times 44.8\text{ MPa}}$ $\times\left(\frac{0.0042\text{ mm/mm}}{559\text{ mm}}\right)^2 = -0.452\times 10^{-5}\text{ kN/mm}^3$

Table 16.9 (cont.)—Procedure for strengthening of a noncircular concrete column for increase in axial and bending forces

Procedure	Calculation in in.-lb units	Calculation in SI metric units
	$F = 24\text{ in.}(22\text{ in.} - 12\text{ in.}) \times \frac{(4595\text{ ksi} - 190.7\text{ ksi})^2}{12 \times 6.5\text{ ksi}} \times \left(\frac{0.0042\text{ in./in.}}{22\text{ in.}}\right)^2 + \frac{24\text{ in.}(4595\text{ ksi} - 190.7\text{ ksi})}{3} \times \left(\frac{0.0042\text{ in./in.}}{22\text{ in.}}\right) = 0.899\text{ ksi}$	$F = 610\text{ mm}(559\text{ mm} - 305\text{ mm}) \times \frac{(31{,}685\text{ MPa} - 1315\text{ MPa})^2}{12 \times 44.8\text{ MPa}} \times \left(\frac{0.0042\text{ mm/mm}}{559\text{ mm}}\right)^2 + \frac{610\text{ mm}(31{,}685\text{ MPa} - 1315\text{ MPa})}{3} \times \left(\frac{0.0042\text{ mm/mm}}{559\text{ mm}}\right) = 62.01 \times 10^{-3}\text{ kN/mm}^2$
	$G = 6.5\text{ ksi} \times 12\text{ in.} + 24\text{ in.}(22\text{ in.} - 12\text{ in.}) \times \left(\frac{4595\text{ ksi} - 190.7\text{ ksi}}{2}\right)\left(\frac{0.0042\text{ in./in.}}{22\text{ in.}}\right) = -179.73\text{ kip/in.}$	$G = 44.8\text{ MPa} \times 305\text{ mm} + 610\text{ mm}(559\text{ mm} - 305\text{ mm}) \times \left(\frac{31{,}685\text{ MPa} - 1315\text{ MPa}}{2}\right)\left(\frac{0.0042\text{ mm/mm}}{559\text{ mm}}\right) = -31.48\text{ kN/mm}$
	$H = 6.5\text{ ksi} \times 24\text{ in.}(22\text{ in.} - 12\text{ in.}) = 1560\text{ kip}$	$H = 44.8\text{ MPa} \times 610\text{ mm}(559\text{ mm} - 305\text{ mm}) = 6939\text{ kN}$
	$I = 6.5\text{ ksi} \times 24\text{ in.} \times ([22\text{ in.}]^2/2) - 6.5\text{ ksi}(22\text{ in.} - 12\text{ in.}) \times 22\text{ in.} \times 24\text{ in.} + 190.7\text{ ksi} \times 24\text{ in.} \times ([22\text{ in.}]^2/3)(0.0042\text{ in./in.}) - 190.7\text{ ksi} \times 24\text{ in.} \times (22\text{ in.}/3)(22\text{ in.} - 12\text{ in.})(0.0042\text{ in./in.}) = 4427\text{ kip-in.}$	$I = 44.8\text{ MPa} \times 610\text{ mm} \times ([559\text{ mm}]^2/2) - 44.8\text{ MPa}(559\text{ mm} - 305\text{ mm}) \times (559\text{ mm})(610\text{ mm}) + 1315\text{ MPa} \times 610\text{ mm} \times ([559\text{ mm}]^2/3)(0.0042\text{ mm/mm}) - 1315\text{ MPa} \times 610\text{ mm} \times (559\text{ mm}/3)(559\text{ mm} - 305\text{ mm})(0.0042\text{ mm/mm}) = 500{,}162\text{ kN-mm}$
	The distances from each layer of steel reinforcement to the geometric centroid of the cross section are:	The distances from each layer of steel reinforcement to the geometric centroid of the cross section are:
	$d_1 = 10\text{ in.}$ $d_2 = d_3 = 3.3\text{ in.}$	$d_1 = 254\text{ mm}$ $d_2 = d_3 = 85\text{ mm}$
	Point C: Nominal axial capacity: $\phi P_{nl} = 0.65[-0.49\text{ kip/in.}^3(10.3\text{ in.})^3 + 15.14\text{ ksi}(10.3\text{ in.})^2 - 156\text{ kip-in.}(10.3\text{ in.}) + 2448.71\text{ kip}] + 5.08\text{ in.}^2(60\text{ ksi}) + 2.54\text{ in.}^2(50.79\text{ ksi}) + 2.54\text{ in.}^2(-4.61\text{ ksi}) + 5.08\text{ in.}^2(-60\text{ ksi})$	Point C: Nominal axial capacity: $\phi P_{n(C)} = 0.65[-1.33 \times 10^{-4}\text{ kN/mm}^3(262\text{ mm})^3 + 104.41 \times 10^{-3}\text{ kN/mm}^2 \times (262\text{ mm})^2 - 27.32\text{ kN/mm}(262\text{ mm}) + 10{,}892\text{ kN}] + 3277\text{ mm}^2(414\text{ MPa}) + 1315\text{ mm}^2(350\text{ MPa}) + 1315\text{ mm}^2(-31.8\text{ MPa}) + 3277\text{ mm}^2(-414\text{ MPa})$
	$\phi P_{nl} = 1320\text{ kip}$	$\phi P_{n(C)} = 5870\text{ kN}$
	where	where
	$A = \frac{-24\text{ in.}(4595\text{ ksi} - 190.7\text{ ksi})^2}{12 \times 6.5\text{ ksi}}\left(\frac{0.0042\text{ in./in.}}{14.78\text{ in.}}\right)^2 = -0.49\text{ kip/in.}^3$	$A = \frac{-610\text{ mm}(31{,}681\text{ MPa} - 1315\text{ MPa})^2}{12 \times 44.8\text{ MPa}} \times \left(\frac{0.0042\text{ mm/mm}}{375\text{ mm}}\right)^2 = 1.33 \times 10^{-4}\text{ kN/mm}^2$
	$B = \frac{-24\text{ in.}(4595\text{ ksi} - 190.7\text{ ksi})}{2}\left(\frac{0.0042\text{ in./in.}}{14.78\text{ in.}}\right) = 15.14\text{ ksi}$	$B = \frac{-610\text{ mm}(31{,}681\text{ MPa} - 1315\text{ MPa})}{2} \times \left(\frac{0.0042\text{ mm/mm}}{375\text{ mm}}\right) = -104.41 \times 10^{-3}\text{ kN/mm}^2$
	$C = -24\text{ in.} \times 6.5\text{ ksi} = -156\text{ kip/in.}$	$C = -610\text{ mm} \times 44.8\text{ MPa} = -27.32\text{ kN/mm}$
	$D = 24\text{ in.} \times 14.78\text{ in.} \times 6.5\text{ ksi} + \frac{24\text{ in.} \times 14.78\text{ in.} \times 190.7\text{ ksi}}{2} \times (0.0042\text{ in./in.}) = 2448.71\text{ kip}$	$D = 610\text{ mm} \times 375\text{ mm} \times 44.8\text{ MPa} + \frac{610\text{ mm} \times 375\text{ mm} \times 1315\text{ MPa}}{2} \times (0.0042\text{ mm/mm}) = 10{,}892\text{ kN}$

Table 16.9 (cont.)—Procedure for strengthening of a noncircular concrete column for increase in axial and bending forces

Procedure	Calculation in in.-lb units	Calculation in SI metric units
	For the calculation of the coefficients, it is necessary to compute key parameters from the stress-strain model:	For the calculation of the coefficients, it is necessary to compute key parameters from the stress-strain model:
	$c = 22\text{ in.} \times \left(\dfrac{0.0042\text{ in./in.}}{0.0021\text{ in./in.} + 0.0042\text{ in./in.}}\right)$ $= 14.67\text{ in.}$	$c = 560\text{ mm} \times \left(\dfrac{0.0042\text{ mm/mm}}{0.0021\text{ mm/mm} + 0.0042\text{ mm/mm}}\right)$ $= 373\text{ mm}$
	$y_t = 14.64\text{ in.}\dfrac{0.003\text{ in./in.}}{0.0042\text{ in./in.}} = 10.5\text{ in.}$	$y_t = 373\text{ mm}\dfrac{0.003\text{ in./in.}}{0.0042\text{ in./in.}} = 266\text{ mm}$
	The strains in each layer of steel are determined by similar triangles in the strain distribution. The corresponding stresses are then given by: $f_{s1} = \varepsilon_{s1}E_s = 0.0037\text{ in./in.} \times 29{,}000\text{ ksi} \rightarrow 60\text{ ksi}$ $f_{s2} = \varepsilon_{s2}E_s = 0.0018\text{ in./in.} \times 29{,}000\text{ ksi} = 50.78\text{ ksi}$ $f_{s3} = \varepsilon_{s3}E_s = -1.59 \times 10^{-4}\text{ in./in.} \times 29{,}000\text{ ksi} = -4.61\text{ ksi}$ $f_{s4} = \varepsilon_{s4}E_s = -0.0021\text{ in./in.} \times 29{,}000\text{ ksi} = -60\text{ ksi}$	The strains in each layer of steel are determined by similar triangles in the strain distribution. The corresponding stresses are then given by: $f_{s1} = \varepsilon_{s1}E_s = 0.0037\text{ mm/mm} \times 200{,}000\text{ MPa} \rightarrow 414\text{ MPa}$ $f_{s2} = \varepsilon_{s2}E_s = 0.0018\text{ mm/mm} \times 200{,}000\text{ MPa} = 350\text{ MPa}$ $f_{s3} = \varepsilon_{s3}E_s = -1.59 \times 10^{-4}\text{ mm/mm} \times 200{,}000\text{ MPa} = -31.8\text{ MPa}$ $f_{s4} = \varepsilon_{s4}E_s = -0.0021\text{ mm/mm} \times 200{,}000\text{ MPa} = -414\text{ MPa}$
	Nominal bending moment: $\phi M_{n(C)} = 0.65[-0.37\text{ kip/in.}^3(10.3\text{ in.})^4 + 11.46\text{ ksi}(10.3\text{ in.})^3 - 120.08\text{ kip/in.}(10.3\text{ in.})^2 + 433.5\text{ kip}(10.3\text{ in.}) + 11{,}643\text{ kip-in.}] + 5.08\text{ in.}^2(60\text{ ksi})(10\text{ in.}) + 2.54\text{ in.}^2(50.79\text{ ksi})(3.33\text{ in.}) - 2.54\text{ in.}^2(-4.61\text{ ksi})(3.33\text{ in.}) - 5.08\text{ in.}^2(-60\text{ ksi})(10\text{ in.})$	Nominal bending moment: $\phi M_{n(C)} = 0.65[-9.98 \times 10^{-5}\text{ kN/mm}^3(262\text{ mm})^4 + 79 \times 10^{-3}\text{ kN/mm}^2(262\text{ mm})^3 - 21.03\text{ kN/mm}(262\text{ mm})^2 + 1928\text{ kN}(262\text{ mm}) + 1{,}315{,}453\text{ kN-mm}] + 3277\text{ mm}^2(414\text{ MPa})(254\text{ mm}) + 1639\text{ mm}^2(-31.8\text{ MPa})(85\text{ mm}) - 3277\text{ mm}^2(-414\text{ MPa})(254\text{ mm})$
	$\phi M_{n(C)} = 992\text{ kip-ft}$ where	$\phi M_{n(C)} = 1345\text{ kN-m}$ where
	$E = \dfrac{-24\text{ in.}(4595\text{ ksi} - 190.7\text{ ksi})^2}{16 \times 6.5\text{ ksi}}\left(\dfrac{0.0042\text{ in./in.}}{14.78\text{ in.}}\right)^2$ $= -0.37\text{ kip/in.}^3$	$E = \dfrac{-610\text{ mm}(31{,}681\text{ MPa} - 1315\text{ MPa})^2}{16 \times 44.8\text{ MPa}}\left(\dfrac{0.0042\text{ mm/mm}}{375\text{ mm}}\right)^2$ $= -9.98 \times 10^{-5}\text{ kN/mm}^3$
	$F = 24\text{ in.}(14.78\text{ in.} - 12\text{ in.}) \times \dfrac{(4595\text{ ksi} - 190.7\text{ ksi})^2}{12 \times 6.5\text{ ksi}}\left(\dfrac{0.0042\text{ in./in.}}{14.78\text{ in.}}\right)^2 + \dfrac{24\text{ in.}(4595\text{ ksi} - 190.7\text{ ksi})}{3} \times \left(\dfrac{0.0042\text{ in./in.}}{14.78\text{ in.}}\right) = 11.46\text{ ksi}$	$F = 610\text{ mm}(375\text{ mm} - 305\text{ mm}) \times \dfrac{(31{,}681\text{ MPa} - 1315\text{ MPa})^2}{12 \times 44.8\text{ MPa}}\left(\dfrac{0.0042\text{ mm/mm}}{375\text{ mm}}\right)^2 + \dfrac{610\text{ mm}(31{,}681\text{ MPa} - 1315\text{ MPa})}{3} \times \left(\dfrac{0.0042\text{ mm/mm}}{375\text{ mm}}\right) = 79 \times 10^{-3}\text{ kN/mm}^2$
	$G = -6.5\text{ ksi} \times 12\text{ in.} + 24\text{ in.}(14.78\text{ in.} - 12\text{ in.}) \times \left(\dfrac{4595\text{ ksi} - 190.7\text{ ksi}}{2}\right)\left(\dfrac{0.0042\text{ in./in.}}{14.78\text{ in.}}\right)$ $= -120.08\text{ kip/in.}$	$G = -44.8\text{ MPa} \times 305\text{ mm} + 610\text{ mm}(375\text{ mm} - 305\text{ mm}) \times \left(\dfrac{31{,}681\text{ MPa} - 1315\text{ MPa}}{2}\right)\left(\dfrac{0.0042\text{ mm/mm}}{375\text{ mm}}\right)$ $= -21.03\text{ kN/mm}$
	$H = 6.5\text{ ksi} \times 24\text{ in.}(14.78\text{ in.} - 12\text{ in.}) = 433.5\text{ kip}$	$H = 44.8\text{ MPa} \times 610\text{ mm}(375\text{ mm} - 305\text{ mm}) = 1928\text{ kN}$
	$I = 6.5\text{ ksi} \times 24\text{ in.} \times ([14.78\text{ in.}]^2/2) - 6.5\text{ ksi}(14.78\text{ in.} - 12\text{ in.})(14.78\text{ in.})(24\text{ in.}) + 190.7\text{ ksi} \times 24\text{ in.} \times ([14.78\text{ in.}]^2/3)(0.0042\text{ in./in.}) - 190.7\text{ ksi} \times 24\text{ in.} \times (14.78\text{ in.}/3)(14.78\text{ in.} - 12\text{ in.})(0.0042\text{ in./in.}) = 11{,}643\text{ kip-in.}$	$I = 44.8\text{ MPa} \times 610\text{ mm} \times ([375\text{ mm}]^2/2) - 44.8\text{ MPa}(375\text{ mm} - 305\text{ mm}) \times (375\text{ mm})(610\text{ mm}) + 1315\text{ MPa} \times 610\text{ mm} \times ([375\text{ mm}]^2/3)(0.0042\text{ mm/mm}) - 1315\text{ MPa} \times 610\text{ mm} \times (375\text{ mm}/2)(375\text{ mm} - 305\text{ mm})(0.0042\text{ mm/mm}) = 1{,}315{,}453\text{ kN-mm}$

Table 16.9 (cont.)—Procedure for strengthening of a noncircular concrete column for increase in axial and bending forces

Procedure	Calculation in in.-lb units	Calculation in SI metric units
Step 3—Comparison of simplified partial interaction diagram with required P_u and M_u	The following table summarizes the axial and bending nominal capacities (unstrengthened and strengthened) for Points A, B, and C. These points are plotted in the following figure	The following table summarizes the axial and bending nominal capacities (unstrengthened and strengthened) for Points A, B, and C. These points are plotted in the following figure.

Point	n = 0 plies (unstrengthened member)		n = 6 plies	
	ϕP_n, kip	ϕM_n, kip-ft	ϕP_n, kip	ϕM_n, kip-ft
A	2087	0	2523	0
B	1858	644	2210	682
C	928	884	1320	992

Point	n = 0 plies (unstrengthened member)		n = 6 plies	
	ϕP_n, kN	ϕM_n, kN-m	ϕP_n, kN	ϕM_n, kN-m
A	9283	0	11,223	0
B	8264	873	9829	924
C	4128	1199	5870	1345

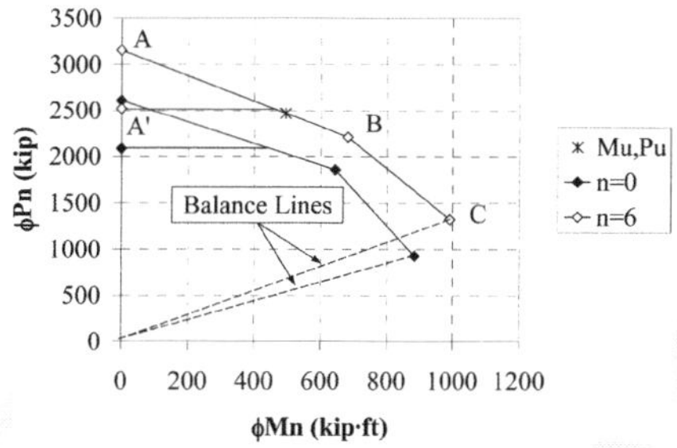

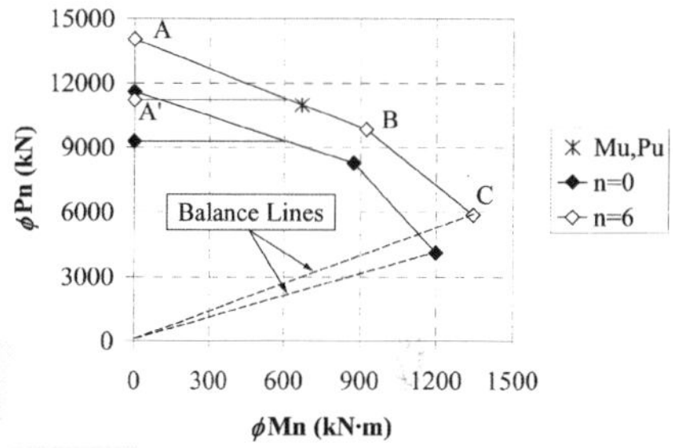

16.10—Plastic hinge confinement for seismic strengthening

This example illustrates the design of an FRP retrofit to enhance the plastic rotation capacity of a nonductile reinforced concrete column. In this example, the column cannot efficiently resist seismic loads for two reasons: (1) the tie spacing does not conform to current seismic design codes; and (2) the ties do not project into the core and under seismic loads may open once the cover concrete begins to spall. The aforementioned deficiencies indicate that the column may have inadequate shear strength, inadequate confinement of the plastic hinge region, and inadequate clamping of lap splices. In addition, the designer should ensure adequate resistance against buckling of the main longitudinal reinforcement. In this example, the deficiency under consideration is inadequate plastic rotation capacity, which can be enhanced by confinement of the plastic hinge region with FRP. A seismic analysis has already determined that the column is capable of resisting the required seismic moments. Thus, there is no need to increase the flexural capacity of the section. This example is limited in scope to the FRP design requirements, and does not cover the seismic analysis. ASCE/SEI 41 is used as the base standard for this example.

The column, which is to be part of a lateral load-resisting system, is illustrated in Fig. 16.10a. Expected material properties and other relevant information are listed in Table 16.10a. From a seismic analysis, the column should be capable of developing a plastic rotation $\theta_p = 0.025$ rad. The axial load on the column, including gravity plus seismic loads, is $P_u = 75$ kip (334 kN). This plastic rotation demand exceeds the limiting value of 0.015 stipulated in ASCE/SEI 41 numerical acceptance criteria for reinforced concrete columns that do not conform to current seismic design codes. The concrete and reinforcing steel strain limitations of ASCE/SEI 41 as listed in Table 16.10b should not be exceeded.

The column is strengthened with CFRP laminates having the composite properties listed in Table 16.10c and bonded around the column using the wet layup technique. Glass FRP (GFRP), however, can similarly be used if desired. The design process was initiated by considering a wrapping system composed of three plies. After two iterations, the final wrapping system was found to require five plies. Only the calculations used to verify the final configuration are provided. These calculations are shown in Table 16.10d.

Figure 16.10b shows the moment curvature analysis of the as-built and retrofit rectangular sections. The moment curvature analysis results show the as-built ultimate curvature capacity is significantly lower than the required curvature demand.

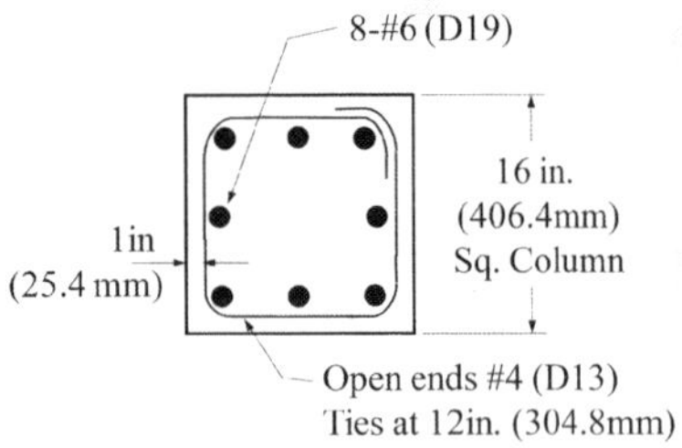

Fig. 16.10a—As-built column details.

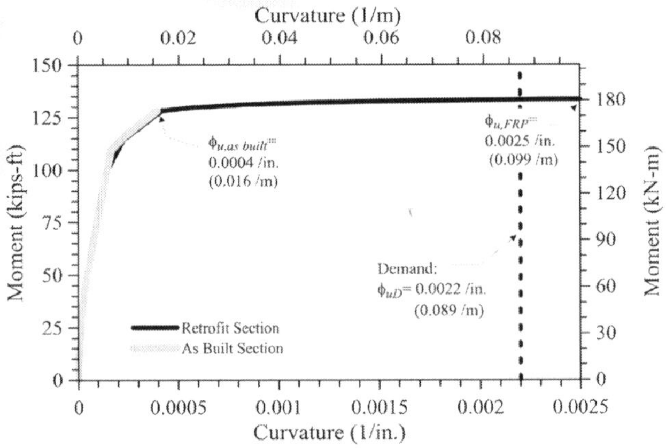

Fig. 16.10b—Moment-curvature analysis.

Table 16.10a—Column material properties

Concrete strength f_c'	4000 psi	27.6 MPa
Concrete elastic modulus $E_c = 57{,}000\sqrt{f_c'}$ (psi) $E_c = 4700\sqrt{f_c'}$ (MPa)	3605 ksi	25 GPa
Longitudinal reinforcing steel: yield strength f_y	44,000 psi	303 MPa
Modulus of elasticity of steel E_s	29,000 ksi	200 GPa
Longitudinal reinforcing steel: yield strain ε_y	0.0015	0.0015
Column height between plastic hinges L	10 ft	3.05 m
Distance to extreme tension steel d	14.625 in.	371 mm

Table 16.10b—Maximum usable strain levels (ASCE/SEI 41)

	Strain limits	
	Unconfined sections	**Confined sections per 13.3.1**
Compressive (concrete)	0.003	ε_{ccu}
Tensile (steel)	0.005	0.05
Compressive (steel)	Limited by the concrete	0.02

Table 16.10c—Manufacturer's reported composite properties

Thickness per ply t_f	0.023 in.	0.584 mm
Ultimate tensile strength f_{fu}^*	155 ksi	1072 MPa
Rupture strain ε_{fu}^*	0.015	0.015
Modulus of elasticity, E_f	9326 ksi	64.3 GPa

Table 16.10d—Procedure for designing plastic hinge confinement for seismic strengthening

Procedure	Calculation in in.-lb units	Calculation in SI units
Step 1—FRP-system design material properties		
$\varepsilon_{fu} = C_E\varepsilon_{fu}{}^*$	$\varepsilon_{fu} \leq 0.95 \times 0.015 = 0.014$	
$C_E = 0.95$		
The column is located in an interior space. Therefore, per Table 9.4, an environmental-reduction factor of 0.95 is used.	$\kappa_e = 0.55$	
$\varepsilon_{fe} = 0.004 \leq \kappa_e\varepsilon_{fu}$	$\varepsilon_{fe} = 0.004 \leq 0.55 \times 0.014 = 0.0078 \therefore \varepsilon_{fe} = 0.04$	$\therefore \varepsilon_{fe} = 0.04$
Step 2—Plastic hinge length	$L_p = 2 + 0.0003 \times 44{,}000 \times 0.75 = 12$ in.	$L_p = 50.8 + 0.044 \times 303.4 \times 19 = 305$ in.
In FRP jacketed columns, the plastic hinge length is: $L_p = g + 0.0003f_yd_{b\ell}$ (psi and in.) $L_p = g + 0.044f_yd_{b\ell}$ (MPa and mm)	Note: Use a maximum gap between the FRP and the column base of 2 in. (50.8 mm)	
Step 3—Preliminary calculations		
$b = h = 16$ in. (406 mm)		
From 12.1.2:		
$D = \sqrt{h^2 + b^2}$	$D = \sqrt{16^2 + 16^2} = 22.63$ in.	$D = \sqrt{406.4^2 + 406.4^2} = 575$ mm
	$A_e/A_c = 0.62$	$A_e/A_c = 0.62$
$\kappa_a = \frac{A_e}{A_c}\left(\frac{b}{h}\right)^2$	$\kappa_a = 0.62\left(\frac{A_e}{A_c}\right)^2 = 0.62$	$\kappa_a = 0.62$
$\kappa_b = \frac{A_e}{A_c}\sqrt{\frac{h}{b}}$	$\kappa_b = 0.62\sqrt{\frac{16}{16}} = 0.62$	$\kappa_b = 0.62$
where A_e/A_c is calculated as:	$\varepsilon_c' = 0.002$	$\varepsilon_c' = 0.002$
$\frac{A_e}{A_c} = \frac{1-\left[\frac{\left(\frac{b}{h}\right)(h-2r_c)^2+\left(\frac{h}{b}\right)(b-2r_c)^2}{3A_g}\right]-\rho_g}{1-\rho_g}$	$h/b = 1.00$ Note: For rectangular members, plastic hinge confinement by jacketing is not recommended for members featuring side aspect ratios, h/b, greater than 1.5, or face dimensions, b or h, exceeding 36 in. (900 mm), unless testing demonstrates their effectiveness (13.3.1).	
Step 4—Confining concrete model variables		
Using the trial design of five plies, compute the confined concrete model parameters listed in the following:		
$f_\ell = \frac{2E_ft_fn_f\varepsilon_{fe}}{D}$	$f_l = \frac{2 \times 9{,}326{,}000 \times 0.023 \times 5 \times 0.004}{22.63} = 380$ psi	$f_l = \frac{2 \times 64{,}300 \times 0.584 \times 5 \times 0.004}{575} = 2.61$ MPa
	$\psi_f = 095$ for fully wrapped sections	$\psi_f = 095$
$f_{cc}' = f_c' + \psi_f 3.3\kappa_a f_l$	$f_{cc}' = 4000 + 0.95 \times 3.3 \times 0.62 \times 380 = 4740$ psi	$f_{cc}' = 27.6 + 0.95 \times 3.3 \times 0.62 \times 2.61 = 32.66$ MPa
$\varepsilon_{ccu} = \varepsilon_c'\left(1.50 + 12\kappa_b\frac{f_\ell}{f_c'}\left(\frac{\varepsilon_{fe}}{\varepsilon_c'}\right)^{0.45}\right)$	$\varepsilon_{ccu} = 0.002\left(1.50 + 12 \times 0.62 \times \frac{380}{4000}\left(\frac{0.004}{0.002}\right)^{0.45}\right) = 0.0049 \leq 0.01$	$\varepsilon_{ccu} = 0.002\left(1.50 + 12 \times 0.62 \times \frac{2.61}{27.60}\left(\frac{0.004}{0.002}\right)^{0.45}\right) = 0.0049 \leq 0.01$
	ε_{ccu} is limited to 0.01 to prevent excessive cracking and the resulting loss of concrete integrity	
Note: The expressions presented previously are used in conjunction with a moment-curvature (M-ϕ) analysis program to obtain the yield and ultimate curvature. These curvatures are presented in Steps 5 and 6.		

Table 16.10d (cont.)—Procedure for designing plastic hinge confinement for seismic strengthening

Procedure	Calculation in in.-lb units	Calculation in SI units
Step 5—Output from M-ϕ analysis program obtain the neutral axis at yielding of longitudinal reinforcement, $c_{y,frp}$	Neutral axis: $c_{y,frp}$ = 5.34 in.	Neutral axis: $c_{y,frp}$ = 136 mm
	Note: Yield strain of longitudinal reinforcement: ε_y = 0.0049	
Step 6—Output from M-ϕ analysis program obtain the neutral axis at ultimate, $c_{u,frp}$	Neutral axis: $c_{u,frp}$ = 1.95 in.	Neutral axis: $c_{u,frp}$ = 50 mm
1. Concrete compression strain: $\varepsilon_{ccu} = 0.0049$	Note: The ultimate concrete compressive strain is computed per Step 4: ε_{ccu} = 0.0049	
2. Steel tension strain: $\varepsilon_s = \varepsilon_{ccu}\left(\frac{d}{c_u} - 1\right)$	$\varepsilon_s = 0.0049\left(\frac{14.625}{5.34} - 1\right) = 0.032 < 0.05$	$\varepsilon_s = 0.0049\left(\frac{371}{136} - 1\right) = 0.032 < 0.05$
Step 7—Compute the yield curvature $\phi_{y,frp} = \frac{\varepsilon_y}{d - c_{y,frp}}$	From a M-ϕ analysis the yield curvature is: $\phi_{y,frp} = \frac{0.0015}{14.625 - 5.34} = 0.000163\text{/in.}$	From a M-ϕ analysis the yield curvature is: $\phi_{y,frp} = \frac{0.0015}{371 - 136} = 0.0064\text{/m}$
Step 8—Compute the ultimate curvature capacity: $\phi_{u,frp} = \frac{\varepsilon_{ccu}}{c_{u,FRP}}$	From a M-ϕ analysis the ultimate curvature is: $\phi_{u,frp} = \frac{0.0049}{1.95} = 0.0025\text{/in.}$	From a M-ϕ analysis the ultimate curvature is: $\phi_{u,frp} = \frac{0.0049}{50} = 0.099\text{/m}$
Step 9—Compute the ultimate curvature demand $\phi_D = \frac{\theta_p}{L_p} + {}_{y,frp}$	$\phi_D = \frac{0.025}{12} + 0.000163 = 0.0022\text{/in.}$	$\phi_D = \frac{0.025}{0.305} + 0.0064 = 0.084\text{/m}$
Step 10—Verify design: $\phi_D < \phi_{u,frp}$	ϕ_D = 0.0022/in. < 0.0025/in. OK	ϕ_D = 0.084/m < 0.099/m OK
Step 11—Length of confined region, l_o Transverse reinforcement as specified per Section 18.7 of ACI 318-14 shall be provided over a length l_o	$\ell_o \geq \begin{cases} 16\text{ in.} \\ \frac{120/2}{6} = 10\text{ in.} \\ 18\text{ in.} \end{cases}$	$\ell_o \geq \begin{cases} 406.4\text{ mm} \\ \frac{3050/2}{6} = 254\text{ mm} \\ 457\text{ mm} \end{cases}$
	$l_o > L_p$ = 18 in. (457 mm)	
Design summary: Completely wrap the section with five transverse plies. Confining jacket should extend at least 18 in. (457 mm) beyond the joint interface.		

16.11—Lap-splice clamping for seismic strengthening

This example illustrates the design of an FRP retrofit to improve the seismic performance of a reinforced concrete column that is constructed with a lap splice in a region of plastic rotations. Material properties, details, and other relevant information are provided in Table 16.11a and Fig. 16.11. The column is strengthened with CFRP laminates having the composite properties listed in Table 16.11b and bonded around the column using the wet layup technique. Glass FRP, however, can similarly be used if desired.

The final wrapping system was found to require five plies. Only the calculations used to verify the final configuration are provided. These calculations are shown in Table. 16.11c.

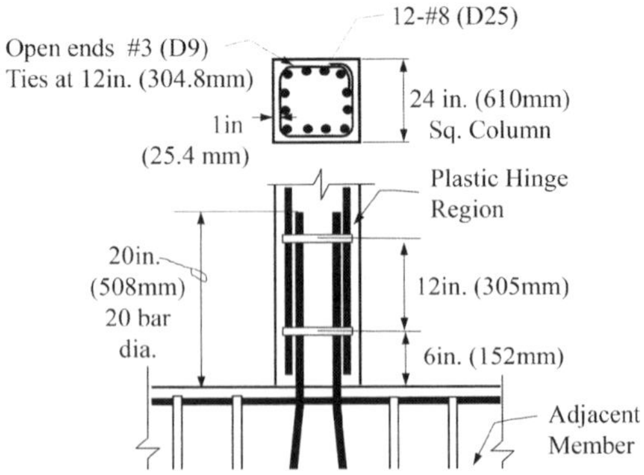

Fig. 16.11—Detail of lap splice in a plastic hinge region.

Table 16.11a—Column material properties

Concrete strength f_c'	4000 psi	27.6 MPa
Concrete elastic modulus $E_c = 57,000\sqrt{f_c'}$ (psi) $E_c = 4700\sqrt{f_c'}$ (MPa)	3605 ksi	25 GPa
Longtudinal reinforcing steel: yield strength f_y	44,000 psi	303.4 MPa
Modulus of elasticity of steel E_s	29,000 ksi	200 GPa
Longitudinal reinforcing steel: yield strain ε_y	0.0015	0.0015
Column height between plastic hinges, L	10 ft	3.05 m
Distance to extreme tension steel, d	14.625 in.	371 mm

Table 16.11b—Manufacturer's reported composite properties

Thickness per ply t_f	0.08 in.	2 mm
Ultimate tensile strength	143 ksi	986 MPa
Rupture strain	0.010	0.010
Modulus of elasticity E_f	13,900 ksi	95.8 GPa

Table 16.11c—Procedure for designing lap-splice clamping for seismic strengthening

Procedure	Calculation in in.-lb units	Calculation in SI units
Step 1—Compute the tensile stress that existing splice can develop: (ACI 318-14 Section 25.4)	$s = 12$ in. $n = 1$ Longitudinal bars in the potential plane of splitting	$s = 305$ mm $n = 1$
$f_s = \dfrac{40\ell_d\lambda\sqrt{f_c'}\left(\dfrac{c_b K_{tr}}{d_b}\right)}{3d_b\psi_t\psi_e\psi_s}$ (in. and psi)	$K_{tr} = \dfrac{40\times1\times0.11}{12\times1} = 0.37$ in.	$K_{tr} = \dfrac{40\times1\times71}{305\times1} = 9.3$ mm
$f_s = \dfrac{3.33\ell_d\lambda\sqrt{f_c'}\left(\dfrac{c_b K_{tr}}{d_b}\right)}{3d_b\psi_t\psi_e\psi_s}$ (mm and MPa)	$\dfrac{c_b + K_{tr}}{d_b} = \dfrac{(1.00+0.375+0.5)+0.37}{1.00} = 2.24 \leq 2.50$	$\dfrac{c_b + K_{tr}}{d_b} = \dfrac{(25.4+9.53+12.7)+9.3}{25.4} = 2.24 \leq 2.50$
$K_{tr} = \dfrac{40A_{tr}}{sn}$	Per Eq. (25.4.2.3a) of ACI 318-14 the value $\dfrac{c_b + K_{tr}}{d_b}$ cannot be greater than 2.5.	
$\psi_t = \psi_e = \psi_s = 1.00$ $\lambda = 1.00$ (normalweight concrete)	$f_s = \dfrac{40\times20.0\times1.0\sqrt{4000}\times2.24}{3\times1.0\times1.0\times1.0\times1.0}$ $= 37{,}779$ psi $\leq 44{,}000$ psi	$f_s = \dfrac{3.33\times508\times1.0\sqrt{27.6}\times2.24}{3\times25.4\times1.0\times1.0\times1.0}$ $= 260$ MPa ≤ 303 MPa
	Note: Computed stress f_s does not reach f_y, longitudinal bar yield strength, and as such lap splice must be clamped.	
Step 2—Stress corresponding to pullout capacity of splice	$\ell_{prov} = 20d_{b\ell} = 20 \times 1.0 = 20$ in.	$\ell_{prov} = 20d_{b\ell} = 20 \times 25.4 = 508$ mm
$f_s \leq \dfrac{33\ell_d\lambda\sqrt{f_c'}}{d_{b\ell}\psi_t\psi_e\psi_s}$ (in. and psi) $f_s \leq \dfrac{2.75\ell_d\lambda\sqrt{f_c'}}{d_{b\ell}\psi_t\psi_e\psi_s}$ (mm and MPa)	Pullout capacity of splice: $f_s = \dfrac{33\times20.0\times\sqrt{4000}}{1.0\times1.0\times1.0\times1.0} = 41{,}742$ psi	Pullout capacity of splice: $f_s = \dfrac{2.75\times508\times\sqrt{27.6}}{25.4\times1.0\times1.0\times1.0} = 289$ MPa
Step 3—Compute the required jacket thickness, t_j	$D = 24$ in. (maximum dimension in a rectangular member)	$D = 610$ mm
$nt_f = 1500\times\dfrac{D}{E_f}$ (ksi and in.) $nt_f = 1500\times\dfrac{D}{E_f}$ (MPa and mm)	$E_f = 13{,}900$ ksi $t_j = 218\times\dfrac{24}{13{,}900} = 0.38$ in.	$E_f = 95{,}800$ MPa $t_j = 1500\times\dfrac{610}{95{,}800} = 9.6$ mm
Step 4—Compute the required number of plies, n	$t_f = 0.08$ in.	$t_f = 2$ mm
$n = t_j/t_f$	$n = 0.38/0.08 =$ five plies	$n = 9.60/2 =$ five plies

Design summary:
Completely wrap the section with five transverse plies.
Confining jacket should extend at least full height of lap splice; that is, 20 in. (508 mm)
Using five plies of the FRP-specified results in a splice capacity of 41,742 psi (289 MPa), which is controlled by the maximum pull-out capacity calculated in Step 2.

16.12—Seismic shear strengthening

This example illustrates the main steps in calculating the amount of FRP required for the shear strengthening of a reinforced concrete member. The member used in this example is illustrated in Fig. 16.12. Column material properties are shown in Table 16.12a, the configuration of the FRP shear reinforcement is described in Table 16.12b, and the CFRP laminate material properties are listed in Table 16.12c. ASCE/SEI 41 is used as the standard for this example. Glass FRP (GFRP) can similarly be used if desired. Calculations for determining the FRP required as shown in Table 16.12d.

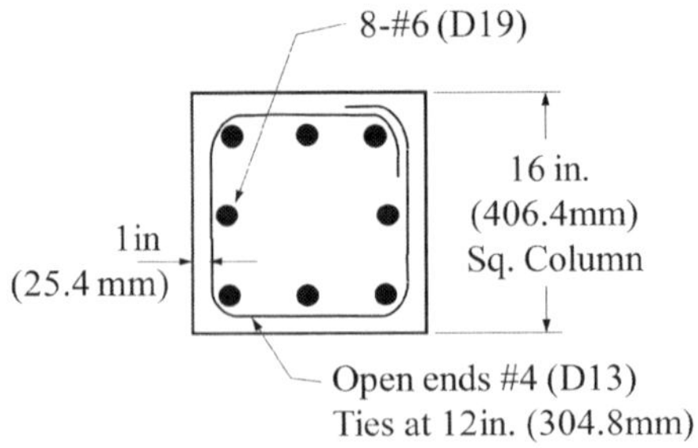

Fig. 16.12—As-built column details.

Table 16.12a—Column material properties

Concrete strength f_c'	4000 psi	27.6 MPa
Concrete elastic modulus $E_c = 57{,}000\sqrt{f_c'}$ (psi) $E_c = 4700\sqrt{f_c'}$ (MPa)	3605 ksi	25 GPa
Longtudinal reinforcing steel: yield strength f_y	44,000 psi	303.4 MPa
Modulus of elasticity of steel E_s	29,000 ksi	200 GPa
Longitudinal reinforcing steel: yield strain ε_y	0.0015	0.0015
Column height between plastic hinges, L	10 ft	3.05 m
Distance to extreme tension steel, d	14.625 in.	371 mm
Ultimate axial load P_u	75 kip	333.62 kN

Table 16.12b—Configuration of supplemental FRP shear reinforcement

Minimum section dimension	16 in.	406 mm
d_{fv} (governed by minimum section dimension)	16 in.	406 mm
Width w_f	8 in.	203 mm
Spacing s_f (full coverage)	8 in.	203 mm

Table 16.12c—Manufacturer's reported composite properties

Thickness per ply t_f	0.023 in.	0.584 mm
Ultimate tensile strength f_{fu}*	155 ksi	1072 MPa
Rupture strain ε_{fu}*	0.015	0.015
Modulus of elasticity E_f	9326 ksi	64.3 GPa

Table 16.12d—Procedure for seismic shear strengthening

Design steps	Calculation in in.-lb units	Calculation in SI units
Step 1—Compute the probable moment capacity of the member, M_{pr}	From a M-ϕ analysis, the probable moment capacity is: $M_{pr} = 322$ kip-ft	From a M-ϕ analysis, the probable moment capacity is: $M_{pr} = 437$ kN-m
Per 13.5.1; in calculating the probable moment $\phi = 1$ and $\psi_f = 1$ are used.	Notes: (1) In seismic applications, the shear design process initiates with the calculation of the probable moment capacity of the section. (2) Moment M_{pr} is computed at both top and bottom ends of the member.	
Step 2—Obtain the yield and ultimate curvature from a M-ϕ analysis	Yield curvature: $\phi_{y,frp} = 0.000163$/in. Ultimate curvature: $\phi_{u,frp} = 0.0019$/in.	Yield curvature: $\phi_{y,frp} = 0.0064$/m Ultimate curvature: $\phi_{u,frp} = 0.074$/m
Step 3—Compute the displacement ductility and reduction factor, k		
In FRP jacketed columns, the plastic hinge length is (Eq. (1)): $L_p = g + 0.0003 f_y d_b$ f_y is in ksi and d_b is in inches $L_p = g + 0.044 f_y d_b$ f_y is in MPa and d_b is in mm	Plastic hinge length: $L_p = 2 + 0.0003 \times 44{,}000 \times 0.75 = 12$ in. Note: Use a maximum gap between the FRP and the column base of 2 in.	Plastic hinge length: $L_p = 50.8 + 0.044 \times 303.4 \times 19 = 305$ mm Note: Use a maximum gap between the FRP and the column base of 50.8 mm.
In this example: $L_{eff} = \frac{L}{2} = 60$ in. (1524 mm)		
Yield deflection $\Delta_{y,frp}$: $\Delta_{y,frp} = \frac{\phi_{y,frp} L_{eff}^2}{3}$	Yield deflection: $\Delta_{y,frp} = \frac{0.000163 \times 60^2}{3} = 0.20$ in.	Yield deflection: $\Delta_y = \frac{0.0065 \times 1524^2}{3 \times 1000} = 5.1$ mm
Plastic deflection $\Delta_{p,frp}$: $\Delta_{p,frp} = (\phi_u - \phi_y) L_p \left(L_{eff} - \frac{L_p}{2}\right)$	Plastic deflection: $\Delta_{p,frp} = (0.0019 - 0.000163) \times 12 \left(60 - \frac{12}{2}\right) = 1.13$ in.	Plastic deflection: $\Delta_{p,frp} = \frac{(0.074 - 0.0064)}{1000} 305 \left(1524 - \frac{305}{2}\right) = 28.6$ mm
Displacement ductility μ_Δ: $\mu_\Delta = 1 + \frac{\Delta_p}{\Delta_y}$	Displacement ductility: $\mu_\Delta = 1 + \frac{1.13}{0.20} = 6.7$	Displacement ductility: $\mu_\Delta = 1 + \frac{28.6}{5.1} = 6.7$
Shear reduction factor per ASCE/SEI 41 $\begin{cases} \mu_\Delta \le 2; & k = 1.0 \\ 2 \le \mu_\Delta \le 6; & k = 0.7 + \frac{0.30 \times (6 - \mu_\Delta)}{4} \\ \mu_\Delta > 6; & k = 0.7 \end{cases}$	Reduction factor: $k = 0.70$	Reduction factor: $k = 0.70$
Step 4—Compute the shear design force, V_u	In this example, with the exception of the earthquake forces, there are no applied uniform loads on the member, as such: $W_u = 0.00$ From Step 1, the probable moment capacity is: $M_{pr,top} = M_{pr,bot} = 322$ kip-ft (437 kN-m)	
Per ACI 318-14 Section 18.7.6: $V_u = \frac{M_{pr,top} + M_{pr,bot}}{L} \pm \frac{w_u L}{2}$	Design shear force: $V_u = \frac{322 + 322}{120/12} \pm 0 = 64.4$ kip	Design shear force: $V_u = \frac{437 + 437}{3.05} \pm 0 = 286$ kN

Table 16.12d (cont.)—Procedure for seismic shear strengthening

Design steps	Calculation in in.-lb units	Calculation in SI units
Step 5—Calculate the concrete, V_c, and steel contribution, V_s Concrete contribution, V_c: Per ACI 318-14 Section 18.7.6.2.1, and because: $$\frac{M_{pr,r}+M_{pr,l}}{\ell_n} \geq \left(\frac{w_u \ell_n}{2}\right)\frac{1}{2} \Rightarrow$$ $V_c = 0$ Steel contribution V_s $$V_s = A_{sh} f_y \frac{d}{s}$$	Concrete contribution to shear capacity: $V_c = 0$ Steel contribution to shear capacity: $$V_s = 2\times 0.20\times 44\times\frac{14.625}{12} = 21.4 \text{ kip}$$ Combined concrete and steel contribution: $k(V_c + V_s) = 0.70(0 + 21.4) = 14.98$ kip	Concrete contribution to shear capacity: $V_c = 0$ Steel contribution to shear capacity: $$V_s = 2\times 129\times 303.4\times\frac{371.475}{305} = 95.2 \text{ kN}$$ Combined concrete and steel contribution: $k(V_c + V_s) = 0.70(0 + 95.2) = 66.6$ kN
Step 6—Calculate the required FRP force, V_f $$V_{f,R} = \frac{\left[\frac{V_u}{\phi} - k(V_c + V_s)\right]}{\psi_f}$$ Per ASCE/SEI 41, $\phi = 1.00$ For completely wrapped members, Table 11.3 recommends $\psi_f = 0.95$.	Required FRP force: $$V_{f,R} = \frac{\left[\frac{64.4}{1.00} - 14.98\right]}{0.95} = 45.26 \text{ kip}$$	Required FRP force: $$V_{f,R} = \frac{\left[\frac{286}{1.00} - 66.6\right]}{0.95} = 201.3 \text{ kN}$$
Step 7—Calculate the effective stress, f_{fe}. For fully wrapped members, the effective strain is computed using Eq. (11.4.1.1): $\varepsilon_{fe} = 0.75 C_E \varepsilon_{fu}^* \leq 0.004$ $C_E = 0.95$ The column is located in an interior space. Therefore, per Table 9.4, an environmental-reduction factor of 0.95 is used. The effective FRP stress can be computed from Hooke's law: $f_{fe} = \varepsilon_{fe} E_f$	Effective strain: $\varepsilon_{fe} = 0.75 \times 0.95 \times 0.015 = 0.11 \leq 0.004$ Effective stress: $f_{fe} = 0.004 \times 9326 = 37.3$ ksi	Effective strain: $\varepsilon_{fe} = 0.75 \times 0.95 \times 0.015 = 0.11 \leq 0.004$ Effective stress: $f_{fe} = 0.004 \times 64{,}300 = 257$ MPa
Step 8—Calculate the number of plies, n_f Area of a single ply for a fully wrapped member (Eq. (11.4a)): $A_{fv} = 2 t_f w_f$ The shear contribution of the FRP can be then calculated from Eq. (11-3). $$V_f = A_{fv} f_{fe} \frac{(\sin\alpha + \cos\alpha) d_{fv}}{s_f}$$	Area per ply: $A_{fv} = 2 \times 0.023 \times 8 = 0.37$ in.^{2} Force per ply: $$V_f = 0.37\times 37.3\frac{(1)16}{8} = 27.6 \text{ kip}$$ Number of plies: $$n_{fv} = \frac{V_{f,R}}{V_f} = \frac{45.26}{27.6} = 2 \text{ (minimum)}$$	Area per ply: $A_{fv} = 2 \times 0.584 \times 203 = 237$ mm^2 Force per ply: $$V_f = 237\times 257\frac{(1)406}{203} = 122 \text{ kN}$$ Number of plies: $$n_{fv} = \frac{V_{f,R}}{V_f} = \frac{201.3}{122} = 2 \text{ (minimum)}$$
Design summary: Completely wrap the section with two transverse plies		

16.13—Flexural and shear seismic strengthening of shear walls

This example illustrates the use of an FRP retrofit to increase the shear and flexural strength of an existing cantilevered concrete wall. The shear strengthening is achieved by installing horizontally oriented FRP on one face of the wall. The flexural strengthening is achieved by installing multiple layers of vertically oriented FRP on both faces of the wall near the wall ends. Details for anchorage of the FRP to the foundation are not addressed. Carbon FRP (CFRP) is used in this example. Glass FRP (GFRP) can similarly be used if desired.

The example incorporates the following two major phases:

1. Determine the shear and flexural capacity of the existing wall
2. Design the FRP to achieve the required strength

Details of the wall and relevant information are provided in Fig. 16.13a and Table 16.13a. The wall is assumed to be an ordinary shear wall. The wall is strengthened with FRP having the composite properties listed in Table 16.13b. A factored axial load, P_u, of 12 kip (53 kN) is assumed in addition to the lateral force. ASCE/SEI 41 is assumed to be the standard used as the basis for the rehabilitation.

This example illustrates a manual calculation approach for the design of FRP strengthening of a shear wall. A moment curvature analysis of the existing and repaired wall, shown in Fig. 16.13b, is used to assess the accuracy of the manual approach. The reasonably good agreement of the flexural results from the design example with those from a moment-curvature analysis validates the illustrative example. The design calculations are shown in Table 16.13c.

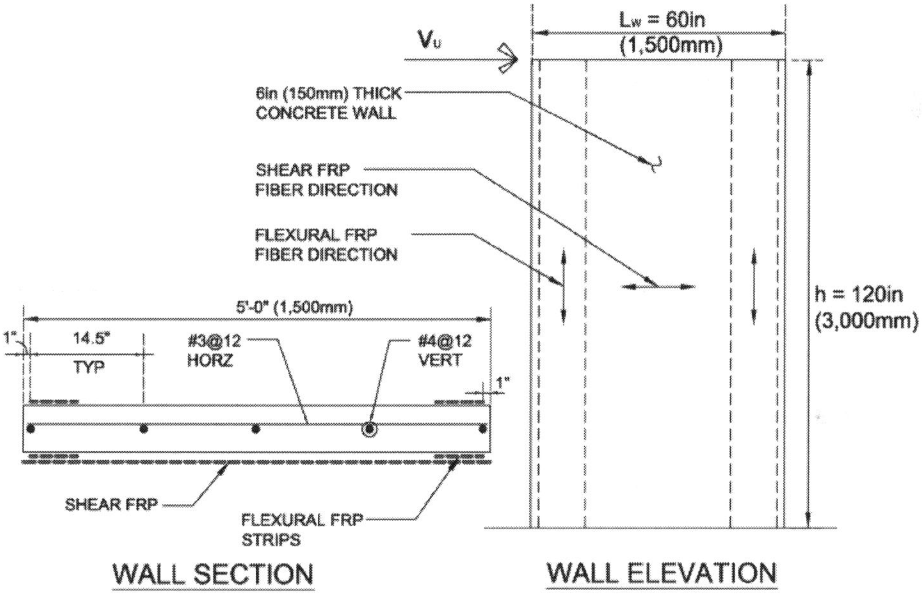

Fig. 16.13a—Concrete wall details.

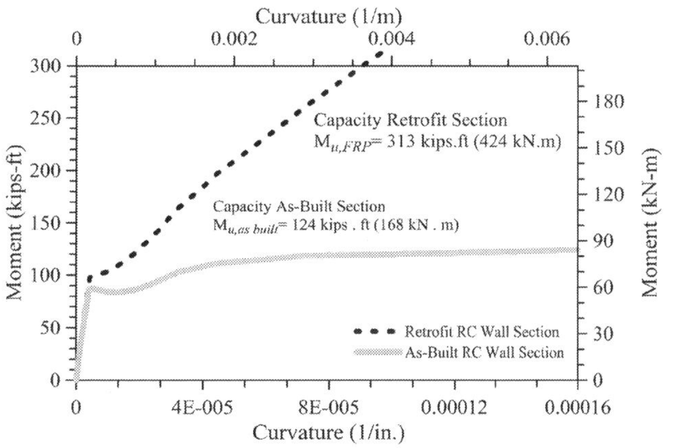

Fig. 16.13b—Moment-curvature analysis for shear wall example.

Table 16.13a—As-built shear wall properties and demands

Concrete strength	2500 psi	17.23 MPa
Longitudinal reinforcing steel yield strength	40.0 ksi	275.8 MPa
Modulus of elasticity of steel	29,000 ksi	200 GPa
Longitudinal reinforcing yield strain	0.0014 in./in.	0.0014 mm/mm
Shear wall height h	10 ft	3000 mm
Shear wall length	60.0 in.	1500 mm
Shear wall thickness	6 in.	150 mm
Existing wall reinforcement Horizontal: No. 3 at 12 in. (305 mm) on center => $\rho_t = 0.0015$ Vertical: No. 4 at 12 in. (305 mm) on center => $\rho_l = \rho_b = 0.0027$	—	—
Axial factored load P_u	12 kip	53.4 kN
Ultimate shear demand V_u	52 kip	232 kN
Ultimate moment demand at wall base M_u	260 kip-ft	348 kN-m
A_{sw} = area of wall web steel => area of four No. 4 bars (neglecting one bar within the compression zone)	0.8 in.^{2}	500 mm^2

Table 16.13b—Manufacturer's reported composite properties

Thickness per ply t_f	0.023 in.	0.575 mm
Ultimate tensile strength f_{fu}^*	140 ksi	965.3 MPa
Rupture strain ε_{fu}^*	0.012 in./in.	0.012 mm/mm
Modulus of elasticity E_f	9600 ksi	66.2 GPa

Table 16.13c—Procedure for flexural and shear seismic strengthening of shear walls

Procedure	Calculation in in.-lb units	Calculation in SI units
Step 1—Compute existing wall capacity		
Shear capacity: Shear capacity per ACI 318:		
$V_c = 2\lambda\sqrt{f'_c}t_w d_{fv}$	$V_c = \frac{2(6\times0.8\times60)\sqrt{2500}}{1000}$	$V_c = \frac{(0.167\times0.8\times1500)\sqrt{17.23}}{1000}$
$V_c = \frac{0.167 b_w d\sqrt{f'_c}}{1000}$	$V_c = 28.8$ kip	$V_c = 124.8$ kN
$V_{sw} = \frac{A_v f_v d_{fv}}{s}$	$V_{sw} = 0.11\times40{,}000\times\frac{0.8\times60}{12}$ $V_{sw} = 17.6$ kip	$V_{sw} = \frac{71\times275.8\times0.8\times\frac{1500}{305}}{1000}$ $V_{sw} = 77.0$ kN
$V_n = V_c + V_{sw}$	$V_n = 28.8$ kip + 17.6 kip $V_n = 46.4$ kip	$V_n = 124.8$ kN + 77.0 kN $V_n = 201.8$ kN $\phi V_n = 201.8$ kN $< V_u = 232$ kN
Using $\phi = 1.0$ for shear (per ASCE/SEI 41):		
Flexural capacity: A quick method to assess the flexural capacity is shown in the following. Assume that all web steel in the wall yields. This includes all the longitudinal reinforcement except for the one bar that is adjacent to the compression face, that is, four No. 4 bars can be considered to yield.	$\phi V_n = 46.4$ kip $< V_u = 52$ kip	
$a = \frac{A_{sw}f_y + P_u}{0.85 f'_c t_w}$	$a = \frac{0.8\text{ in.}^2\times40\text{ ksi}+12\text{ kip}}{0.85\times2.5\text{ ksi}\times6\text{ in.}} = 3.45$ in.	$a = \frac{\left(\frac{500\text{ mm}^2\times275.8\text{ MPa}}{1000}\right)+53.4\text{ kN}}{\left(\frac{0.85\times17.23\text{ MPa}\times150\text{ mm}}{1000}\right)} = 87.1$ mm
$M_n = (A_{sw}f_y + P_u)(d - a/2)$ where $d = L_w/2$	$d = L_w/2 = 30$ in. $M_n = 104$ kip-ft	$d = 750$ mm $M_n = 135.1$ kN-m
Using $\phi = 1.0$:	$\phi M_n = 104$ kip-ft $< M_u = 260$ kip-ft	$\phi M_n = 135.1$ kN-m $< M_u = 348$ kN-m
It can be observed that the wall does not have adequate capacity.		
Step 2—FRP design material properties	For interior exposure for carbon FRP: $C_E = 0.95$	
Use environmental reduction factors from Table 9.4.		
Per 9.4: $f_{fu} = C_E f_{fu}^*$ $\varepsilon_{fu} = C_E \varepsilon_{fu}^*$	$f_{fu} = 133$ ksi $\varepsilon_{fu} = 0.0114$ in./in.	$f_{fu} = 917$ MPa $\varepsilon_{fu} = 0.0114$ mm/mm
Step 3—Flexural strengthening		
Step 3a—Compute the debonding strain limit, ε_{fd}		
This is the limit for the effective strain in the FRP.		
Per Eq. (10.2.1.1):		
$\varepsilon_{fd} = 0.083\sqrt{\frac{f'_c}{nE_f t_f}} \le 0.9\varepsilon_{fu}$ (in.-lb) $\varepsilon_{fd} = 0.41\sqrt{\frac{f'_c}{nE_f t_f}} \le 0.9\varepsilon_{fu}$ (SI)	For one layer of FRP, $\varepsilon_{fd} = 0.0088$ For two layers of FRP, $\varepsilon_{fd} = 0.0062$ For three layers of FRP, $\varepsilon_{fd} = 0.0051$	For one layer of FRP, $\varepsilon_{fd} = 0.0088$ For two layers of FRP, $\varepsilon_{fd} = 0.0062$ For three layers of FRP, $\varepsilon_{fd} = 0.0051$

Table 16.13c (cont.)—Procedure for flexural and shear seismic strengthening of shear walls

Procedure	Calculation in in.-lb units	Calculation in SI units
Flexural design involves iteration to achieve equilibrium across the section. Usually, the first step is to assume the depth of the neutral axis c.	Use the available information to assist with an assumption for the depth of the neutral axis c.	
Assume that the flexural strengthening will require three layers of 8 in. (200 mm) wide strips on each side of the wall at each end. Assume that the effective strain is at the centroid of the FRP area. Knowing the maximum effective strain in the FRP, compute the force in the FRP	$w_f = 8$ in. $n_f = 3$ $t_f = 0.023$ in. A_f = two sides × 3(0.023)(8) $A_f = 1.104$ in.^{2}	$w_f = 200$ mm $n_f = 3$ $t_f = 0.575$ mm $A_f = 690$ mm^2
$T_f = C_E A_f \varepsilon_{fd} E_f$	$T_f = 0.85(1.104)(0.0051)(9600\text{ ksi}) = 45.94$ kip	$T_f = \frac{0.85(690)(0.0051)(66{,}200\text{ MPa})}{1000} = 198.0$ kN
Depth of corresponding compression block: $a = \frac{A_{sw}f_y + P_u + T_f}{0.85 f'_c t_w}$	$a = 7.05$ in.	$a = 177.2$ mm
Compute an estimate of the depth of the neutral axis, c:	$c = a/0.85 = 8.3$ in.	$c = a/0.85 = 208.5$ mm
Check actual strain at centroid of FRP area and corresponding force in the FRP:	$\varepsilon_{fe\,CG} = \varepsilon_{fd}\left(1 + \frac{W_f/2}{c + 1 - L_w}\right) = 0.00469$	$\varepsilon_{fe\,CG} = \varepsilon_{fd}\left(1 + \frac{W_f/2}{c + 25.4 - L_w}\right) = 0.00469$
It is observed that the force in the FRP does not agree with that based on the initial assumption. However, the above steps provide a reasonable starting point for an assumption for c.	Corresponding force in FRP, $T_f = C_E A_f \varepsilon_{feCG} E_f = 42.32$ kip Assume $c = 8.0$ in.	Corresponding force in FRP, $T_f = C_E A_f \varepsilon_{feCG} E_f = 182.1$ kN Assume $c = 200$ mm
Compute concrete strain at extreme compression surface per Eq. (13.7.2.1a): $\varepsilon_c = \varepsilon_{fd}\left(\frac{1}{(L_w/c) - 1}\right) \leq \varepsilon_{cu}$	$\varepsilon_c = 0.0008$ $\varepsilon_c < \varepsilon_{cu} = 0.003$ OK	$\varepsilon_c = 0.0008$ $\varepsilon_c < \varepsilon_{cu} = 0.003$ OK
Compute strain in the bar in the compression zone: $\varepsilon_{sc} = \left(\frac{\varepsilon_c}{c}\right)(c - d')$	$\varepsilon_{sc} = 0.007$ where $d' = 1$ in. $\varepsilon_{sc} < \varepsilon_y = 0.0014$	$\varepsilon_{sc} = 0.007$ where $d' = 25$ mm $\varepsilon_{sc} < \varepsilon_y = 0.0014$
Compute strain in the bars in the tension zone:	$\varepsilon_{st1} = \left(\frac{\varepsilon_c}{c}\right)(14.5 + 1 - 8.0)$ $\varepsilon_{st1} = 0.0075 < \varepsilon_y$	$\varepsilon_{st1} = \left(\frac{\varepsilon_c}{c}\right)(362.5 + 25 - 200)$ $\varepsilon_{st1} = 0.0075 < \varepsilon_y$
Compute strain at centroid of FRP area: Recompute total tensile force components at the above determined strain levels:	Similarly, $\varepsilon_{st2} = 0.0022 > \varepsilon_y$ $\varepsilon_{st3} = 0.0037 > \varepsilon_y$ $\varepsilon_{st4} = 0.0051 > \varepsilon_y$ $\varepsilon_{feCG} = 0.0047$	Similarly, $\varepsilon_{st2} = 0.0022 > \varepsilon_y$ $\varepsilon_{st3} = 0.0037 > \varepsilon_y$ $\varepsilon_{st4} = 0.0051 > \varepsilon_y$ $\varepsilon_{feCG} = 0.0047$
Recalculate depth of compression block and depth to neutral axis: Iterate as required to reach convergence:	$T_f = 42.34$ kip $T_{sw} = 0.2 \times 0.00075 \times 29{,}000 + 3 \times 0.2 \times 40$ $= 28.35$ kip	$T_f = 182.5$ kN $T_{sw} = 125 \times 0.00075 \times 200{,}000 + 3 \times 125 \times 275.8$ $= 122.175$ kN $= 122.2$ kN

Table 16.13c (cont.)—Procedure for flexural and shear seismic strengthening of shear walls

Procedure	Calculation in in.-lb units	Calculation in SI units
Step 3b—Calculate the strains and force components in concrete, FRP and reinforcing steel	$a = 6.48$ in. $c = 7.63$ in.	$a = 163$ mm $c = 191.8$ mm
Force in the bar in the compression zone:	Final value of the depth of the neutral axis is achieved after iteration. $c = 7.648$ in. $a = 6.5$ in.	$c = 191$ mm $a = 162.4$ mm
Strains and forces in bars in tensile zone:	$\varepsilon_c = 0.00076 < \varepsilon_{cu}$ $\varepsilon_{sc} = 0.00066 < \varepsilon_y$ $C_{sc} = 0.2 \times 0.00066 \times 29{,}000 = 3.83$ kip	$\varepsilon_c = 0.00076 < \varepsilon_{cu}$ $\varepsilon_{sc} = 0.00066 < \varepsilon_y$ $C_{sc} = 125 \times 0.00066 \times 200 = 16.5$ kN
Strain and force in FRP:	$\varepsilon_{st1} = 0.0078 < \varepsilon_y$ $\varepsilon_{st2} = 0.0022 > \varepsilon_y$ $\varepsilon_{st3} = 0.0037 > \varepsilon_y$ $\varepsilon_{st4} = 0.0051 > \varepsilon_y$	$\varepsilon_{st1} = 0.0078 < \varepsilon_y$ $\varepsilon_{st2} = 0.0022 > \varepsilon_y$ $\varepsilon_{st3} = 0.0037 > \varepsilon_y$ $\varepsilon_{st4} = 0.0051 > \varepsilon_y$
From equilibrium, compressive force in concrete:	$T_{s1} = 4.52$ kip $T_{s2} = T_{s3} = T_{s4} = 8.0$ kip $\sum T_s = 28.52$ kip	$T_{s1} = 19.5$ kN $T_{s2} = T_{s3} = T_{s4} = 34.48$ kN $\sum T_s = 122.94$ kN
Step 3c—Calculate the moment capacity of the section	$\varepsilon_{feCG} = 0.0047$ $T_f = 42.37$ kip	$\varepsilon_{feCG} = 0.0047$ $T_f = 182.5$ kN
Compute lever arm for different force components:	$C_c = (\sum T_s + T_f + P_u + C_{sc}) = 79.06$ kip Bar in compression: $d_1 = c - 1$ in. $= 7.648 - 1 = 6.65$ in. Concrete compression: $\left(c - \frac{a}{2}\right) = 7.648 - \frac{6.5}{2} = 4.4$ in. First bar in tension: (1 in. + 14.5 in – c) = 7.85 in. Second bar in tension: 22.35 in. Third bar in tension: 36.85 in. Fourth bar in tension: 51.35 in. FRP: $L_w - c - W_f/2 - 1$ in. $= 47.35$ in. P_u: $L_w/2 - c = 22.35$ in.	$C_c = (\sum T_s + T_f + P_u + C_{sc}) = 342.4$ kN Bar in compression: $d_1 = c - 25 = 166$ mm Concrete compression: $\left(c - \frac{a}{2}\right) = 109.8$ mm First bar in tension: (25 mm + 362.5 mm – c) = 196.5 mm Second bar in tension: 559 mm Third bar in tension: 921.5 mm Fourth bar in tension: 1284 mm FRP: $L_w - c - W_f/2 - 25$ mm $= 1184$ mm P_u: $L_w/2 - c = 559$ mm
Nominal moment capacity, M_n: Per Section 10.2.10, include reduction factor for FRP contribution, $\psi = 0.85$.	$M_n = 79.06 \times 4.4 + 3.83 \times 6.65 + 4.52 \times 7.85 + 8 \times 22.35 + 8 \times 36.85 + 8 \times 51.35 + 0.85 \times 42.37 \times 47.35 + 12 \times 22.35 = 272.2$ kip-ft $\phi M_n = 272.2$ kip-ft $> M_u = 260$ kip-ft $\phi M_n > M_u$ ∴ OK	$M_n = 353$ kN·m $\phi M_n = 353$ kN·m $> M_u = 348$ kN·m $\phi M_n > M_u$ ∴ OK

Table 16.13c (cont.)—Procedure for flexural and shear seismic strengthening of shear walls

Procedure	Calculation in in.-lb units	Calculation in SI units
Step 4—Shear strengthening		
Per 13.7.3, Eq. (13.7.2.2b):	From Step 1, $V_n^* = 46.4$ kip	From Step 1, $V_n^* = 1.0(201.8 + 81.5$ kN)
V_f is the shear contribution of FRP and is computed in accordance with Chapter 11.		
Step 4a—Calculate V_f		
Because the FRP is only on one side of the wall, per Eq. (13.7.2.2c):	For one layer of the FRP: $L_e = 1.99$ in. $k_1 = 0.731$ $k_2 = 0.91$ $\kappa_v = 0.2$	
Since the FRP will be face-bonded to the wall, the effective FRP strain will be: $\varepsilon_{fe} = \kappa_v \varepsilon_{fu} \leq 0.004$	$\varepsilon_{fe} = 0.2 \times 0.014 = 0.0028$ $f_{fe} = 0.0028 \times 9600 = 26.9$ ksi	$\varepsilon_{fe} = 0.2 \times 0.014 = 0.0028$ $f_{fe} = 0.0028 \times 66.2 = 0.185$ GPa
Use Eq. (11.4.1.2b7), (11.4.1.2c), (11.4.1.2d), and (11.4.2.1e) to compute κ_v.		
Corresponding tensile stress in the FRP: Per ACI 318-14, Section 11.5.4.2, d_{fv} is taken as $0.8L_w$.	It is assumed that FRP is installed over the full height of the wall and not in discrete strips. $d_{fv} = 0.8L_w = 48$ in. $V_f = 0.75 \times 0.023 \times 0.0028 \times 9600 \times 48 = 22.25$ kip	$d_{fv} = 0.8L_w = 1200$ mm $V_f = 0.75 \times 0.575 \times 0.0028 \times 66.2 \times 1200 = 95.9$ kN
	Reduction factor for FRP shear contribution, $\psi_f = 0.85$. $\psi_f V_f = 18.9$ kip	$\psi_f V_f = 81.5$ kN
Compute shear capacity of FRP strengthened wall using $\phi = 1.0$:	Shear capacity of the retrofitted wall: $\phi V_n = 1.0(46.4 + 18.9)$ kip $\phi V_n = 65.3$ kip $> V_u = 52$ kip $\phi V_n > V_u \therefore$ OK $M_n = 272.2$ kip-ft	$\phi V_n = 1.0(203.1 + 81.5)$ kN $\phi V_n = 283.3$ kN $> V_u = 232$ kN $\phi V_n > V_u \therefore$ OK $M_n = 353$ kN·m
Step 4b — Compute the shear corresponding to the nominal flexural strength. See Section 13.7.1. Note: The FRP for shear strengthening could be optimized by using horizontal strips rather than full coverage. For anchorage of shear FRP see Section 13.6.3.1.	Shear corresponding to the nominal flexural strength: $V_{M\,nom} = 272.2$ kip-ft/10 ft = 27.2 kip The shear strength of the wall is: $V_n = 65.3$ kip $V_n \geq V_{M\,nom} \therefore$ OK	$V_{M\,nom} = 353$ kN·m/3 m = 117.1 kN The shear strength of the wall is: $V_n = 284.6$ kN $V_n \geq V_{M\,nom} \therefore$ OK

CHAPTER 17—REFERENCES

Committee documents are listed first by document number and year of publication followed by authored documents listed alphabetically.

American Concrete Institute (ACI)

ACI 216.1-14—Code Requirements for Determining Fire Resistance of Concrete and Masonry Construction Assemblies

ACI 224.1R-07—Causes, Evaluation, and Repair of Cracks in Concrete Structures

ACI 318-14—Building Code Requirements for Structural Concrete and Commentary

ACI 364.1R-07—Guide for Evaluation of Concrete Structures before Rehabilitation

ACI 369R-11—Guide for Seismic Rehabilitation of Existing Concrete Frame Buildings and Commentary

ACI 437R-03—Strength Evaluation of Existing Concrete Buildings

ACI 440R-07—Report on Fiber-Reinforced Polymer (FRP) Reinforcement for Concrete Structures

ACI 440.3R-12—Guide Test Methods for Fiber-Reinforced Polymers (FRPs) for Reinforcing or Strengthening Concrete Structures

ACI 440.7R-10—Guide for the Design and Construction of Externally Bonded Fiber-Reinforced Polymer Systems for Strengthening Unreinforced Masonry Structures

ACI 440.8-13—Specification for Carbon and Glass Fiber-Reinforced Polymer (FRP) Materials Made by Wet Layup for External Strengthening of Concrete and Masonry Structures

ACI 503.4-92—Standard Specification for Repairing Concrete with Epoxy Mortars (Reapproved by 2003)

ACI 546R-14—Guide to Concrete Repair

ACI 562-13—Code Requirements for Evaluation, Repair, and Rehabilitation of Concrete Buildings and Commentary

American National Standards Institute (ANSI)

ANSI Z400.1/Z129.1-2010—Hazardous Workplace Chemicals - Hazard Evaluation and Safety Data Sheet and Precautionary Labeling Preparation

American Society of Civil Engineers (ASCE)

ASCE 7-10—Minimum Design Loads for Buildings and Other Structures

ASCE/SEI 41-13—Seismic Rehabilitation and Retrofit of Existing Buildings

ASTM International

ASTM C1583/C1583M-13—Standard Test Method for Tensile Strength of Concrete Surfaces and the Bond Strength or Tensile Strength of Concrete Repair and Overlay Materials by Direct Tension (Pull-off Method)

ASTM D648-16—Test Method for Deflection Temperature of Plastics Under Flexural Load in the Edgewise Position

ASTM D696-16—Standard Test Method for Coefficient of Linear Thermal Expansion of Plastics Between –30°C and 30°C with a Vitreous Silica Dilatometer

ASTM D790-15—Standard Test Methods for Flexural Properties of Unreinforced and Reinforced Plastics and Electrical Insulating Materials

ASTM D2240-15—Standard Test Method for Rubber Property—Durometer Hardness

ASTM D2538-02(2010)—Standard Practice for Fusion of Poly(Vinyl Chloride) (PVC) Compounds Using a Torque Rheometer

ASTM D2584-11—Standard Test Method for Ignition Loss of Cured Reinforced Resins

ASTM D2990-09—Standard Test Methods for Tensile, Compressive, and Flexural Creep and Creep-Rupture of Plastics

ASTM D3039/D3039M-14—Standard Test Method for Tensile Properties of Polymer Matrix Composite Materials

ASTM D3171-15—Standard Test Methods for Constituent Content of Composite Materials

ASTM D3418-15—Standard Test Method for Transition Temperatures and Enthalpies of Fusion and Crystallization of Polymers by Differential Scanning Calorimetry

ASTM D3479/D3479M-12—Standard Test Method for Tension-Tension Fatigue of Polymer Matrix Composite Materials

ASTM D4476/D4476M-14—Standard Test Method for Flexural Properties of Fiber Reinforced Pultruded Plastic Rods

ASTM D7205/D7205M-06(2011)—Standard Test Method for Tensile Properties of Fiber Reinforced Polymer Matrix Composite Bars

ASTM D7337/D7337M-12—Standard Test Method for Tensile Creep Rupture of Fiber Reinforced Polymer Matrix Composite Bars

ASTM D7522/D7522M-15—Standard Test Method for Pull-Off Strength for FRP Bonded to Concrete Substrate

ASTM D7565/D7565M-10(2017)—Standard Test Method for Determining Tensile Properties of Fiber Reinforced Polymer Matrix Composites Used for Strengthening of Civil Structures

ASTM D7616/D7616M-11—Standard Test Method for Determining Apparent Overlap Splice Shear Strength Properties of Wet Lay-Up Fiber-Reinforced Polymer Matrix Composites Used for Strengthening Civil Structures

ASTM D7617/D7617M-11—Standard Test Method for Transverse Shear Strength of Fiber-Reinforced Polymer Matrix Composite Bars

ASTM E84-16—Standard Test Method for Surface Burning Characteristics of Building Materials

ASTM E328-13—Standared Test Methods for Stress Relaxation Tests for Materials and Structures

ASTM E831-14—Standard Test Method for Linear Thermal Expansion of Solid Materials by Thermomechanical Analysis

ASTM E1356-08(2014)—Standard Test Method for Assignment of the Glass Transition Temperatures by Differential Scanning Calorimetry

ASTM E1640-13—Standard Test Method for Assignment of the Glass Transition Temperature by Dynamic Mechanical Analysis

ASTM E2092-13—Test Method for Distortion Temperature in Three-Point Bending by Thermomechanical Analysis

Code of Federal Regulations (CFR)

CFR 16 Part 1500-2015—Hazardous Substances and Articles; Administration and Enforcement Regulations

CFR 49-2015—Transportation

International Code Council (ICC)

ICC AC125(2013)—Acceptance Criteria for Concrete and Reinforced and Unreinforced Masonry Strengthening Using Externally Bonded Fiber-Reinforced Polymer (FRP) Composite Systems

International Concrete Repair Institute (ICRI)

ICRI 210.3R-2009—Guide for Using In-Situ Tensile Pull-Off Tests to Evaluate Bond of Concrete Surface Materials

ICRI 310.2R-2013—Selecting and Specifying Concrete Surface Preparation for Sealers, Coatings, Polymer Overlays, and Concrete Repair

Authored documents

AASHTO, 2004, "LRFD Bridge Design Specifications," third edition, American Association of State Highway and Transportation Officials Washington, DC.

Aiello, M. A.; Galati, N.; and Tegola, L. A., 2001, "Bond Analysis of Curved Structural Concrete Elements Strengthened using FRP Materials," *Fifth International Symposium on Non-Metallic (FRP) Reinforcement for Concrete Structures (FRPRCS-5)*, Cambridge-Thomas Telford, London, pp. 680-688.

Antonopoulos, C. P., and Triantafillou, T. C., 2002, "Analysis of FRP-Strengthened RC Beam-Column Joints," *Journal of Composites for Construction*, V. 6, No. 1, pp. 41-51. doi: 10.1061/(ASCE)1090-0268(2002)6:1(41)

Apicella, F., and Imbrogno, M., 1999, "Fire Performance of CFRP-Composites Used for Repairing and Strengthening Concrete," *Proceedings of the 5th ASCE Materials Engineering Congress*, Cincinnati, OH, pp. 260-266.

Arduini, M., and Nanni, A., 1997, "Behavior of Pre-Cracked RC Beams Strengthened with Carbon FRP Sheets," *Journal of Composites for Construction*, V. 1, No. 2, pp. 63-70. doi: 10.1061/(ASCE)1090-0268(1997)1:2(63)

Balsamo, A.; Manfredi, G.; Mola, E.; Negro, P.; and Prota, A., 2005, "Seismic Rehabilitation of a Full-Scale RC Structure using GFRP Laminates," *7th International Symposium on Fiber-Reinforced (FRP) Polymer Reinforcement for Concrete Structures*, SP-230, C. K. Shield, J. P. Busel, S. L. Walkup, and D. D. Gremel, eds., American Concrete Institute, Farmington Hills, MI, pp. 1325-1344.

Bank, L. C., 2006, *Composites for Construction: Structural Design with FRP Materials*, John Wiley & Sons, Hoboken, NJ, 560 pp.

Belarbi, A.; Bae, S.; Ayou, A.; Kuchma, D.; Mirmiran, A.; and Okeil, A., 2011, "Design of FRP Systems for Strengthening Concrete Girders in Shear," NCHRP Report 678, National Cooperative Highway Research Program, 130 pp.

Bianco, V.; Monti, G.; and Barros, J. A. O., 2014, "Design Formula to Evaluate the NSM FRP Strips Shear Strength Contribution to a RC Beam," *Composites. Part B, Engineering*, V. 56, pp. 960-971. doi: 10.1016/j.compositesb.2013.09.001

Binici, B., and Ozcebe, G., 2006, "Seismic Evaluation of Infilled Reinforced Concrete Frames Strengthened with FRPS," *Proceedings of the 8th U.S. National Conference on Earthquake Engineering*, Earthquake Engineering Research Center, San Francisco, CA, 10 pp.

Bisby, L. A.; Green, M. F.; and Kodur, V. K. R., 2005a, "Fire Endurance of Fiber-Reinforced Polymer-Confined Concrete Columns," *ACI Structural Journal*, V. 102, No. 6, Nov.-Dec., pp. 883-891.

Bisby, L. A.; Green, M. F.; and Kodur, V. K. R., 2005b, "Response to Fire of Concrete Structures that Incorporate FRP," *Progress in Structural Engineering and Materials*, V. 7, No. 3, pp. 136-149. doi: 10.1002/pse.198

Bousias, S.; Triantafillou, T.; Fardis, M.; Spathis, L.; and O'Regan, B., 2004, "Fiber-Reinforced Polymer Retrofitting of Rectangular Reinforced Concrete Columns with or without Corrosion," *ACI Structural Journal*, V. 101, No. 4, July-Aug., pp. 512-520.

Bousselham, A., and Chaallal, O., 2006, "Behavior of Reinforced Concrete T-Beams Strengthened in Shear with Carbon Fiber-Reinforced Polymer—An Experimental Study," *ACI Structural Journal*, V. 103, No. 3, May-June, pp. 339-347.

Bracci, J. M.; Mander, J. B.; and Reinhorn, A. M., 1992a, "Seismic Resistance of Reinforced Concrete Frame Structures Designed only for Gravity Loads: Part I - Design and Properties of a One-Third Scale Model Structure," *Technical Report* NCEER-92-0027, National Center for Earthquake Engineering Research, State University of New York, SUNY, Buffalo, New York, 186 pp.

Bracci, J. M.; Mander, J. B.; and Reinhorn, A. M., 1992b, "Seismic Resistance of Reinforced Concrete Frame Structures Designed only for Gravity Loads: Part II – Experimental Performance and Analytical Study of Retrofitted Structural Model," *Technical Report* NCEER-92-0031, National Center for Earthquake Engineering Research, State University of New York, SUNY, Buffalo, New York, 166 pp.

Bracci, J. M.; Mander, J. B.; and Reinhorn, A. M., 1992c, "Seismic Resistance of Reinforced Concrete Frame Structures Designed only for Gravity Loads: Part III - Experimental Performance and Analytical Study of Study of Structural Model," *Technical Report* NCEER-92-0029, National Center for Earthquake Engineering Research, State University of New York, SUNY, Buffalo, New York, 168 pp.

Carey, S. A., and Harries, K. A., 2005, "Axial Behavior and Modeling of Small-, Medium-, and Large-Scale Circular Sections Confined with CFRP Jackets," *ACI Structural Journal*, V. 102, No. 4, July-Aug., pp. 596-604.

Chaallal, O., and Shahawy, M., 2000, "Performance of Fiber-Reinforced Polymer-Wrapped Reinforced Concrete Column under Combined Axial-Flexural Loading," *ACI Structural Journal*, V. 97, No. 4, July-Aug., pp. 659-668.

Chajes, M.; Januska, T.; Mertz, D.; Thomson, T.; and Finch, W., 1995, "Shear Strengthening of Reinforced Concrete Beams Using Externally Applied Composite Fabrics," *ACI Structural Journal*, V. 92, No. 3, May-June, pp. 295-303.

Christensen, J. B.; Gilstrap, J. M.; and Dolan, C. W., 1996, "Composite Materials Reinforcement of Exisiting Masonry Structures," *Journal of Architectural Engineering*, V. 2, No. 2, pp. 63-70. doi: 10.1061/(ASCE)1076-0431(1996)2:2(63)

Concrete Society, 2004, "Design Guidance for Strengthening Concrete Structures Using Fibre Composite Materials," *Technical Report* No. 55 (TR55), second edition, Surrey, UK, 128 pp.

Cromwell, J. R.; Harries, K. A.; and Shahrooz, B. M., 2011, "Environmental Durability of Externally Bonded FRP Materials Intended for Repair of Concrete Structures," *Journal of Construction and Building Materials*, V. 25, No. 5, pp. 2528-2539. doi: 10.1016/j.conbuildmat.2010.11.096

Curtis, P. T., 1989, "Fatigue Behavior of Fibrous Composite Materials," *Journal of Strain Analysis*, V. 24, No. 4, pp. 235-244. doi: 10.1243/03093247V244235

Das, S., 2011, "Life Cycle Assessment of Carbon Fiber-Reinforced Polymer Composites," *The International Journal of Life Cycle Assessment*, V. 16, No. 3, pp. 268-282. doi: 10.1007/s11367-011-0264-z

De Lorenzis, L., and Nanni, A., 2001, "Characterization of FRP Rods as Near Surface Mounted Reinforcement," *Journal of Composites for Construction*, V. 5, No. 2, pp. 114-121. doi: 10.1061/(ASCE)1090-0268(2001)5:2(114)

De Lorenzis, L., and Tepfers, R., 2003, "Comparative Study of Models on Confinement of Concrete Cylinders with Fiber-Reinforced Polymer Composites," *Journal of Composites for Construction*, V. 7, No. 3, pp. 219-237. doi: 10.1061/(ASCE)1090-0268(2003)7:3(219)

De Lorenzis, L.; Lundgren, K.; and Rizzo, A., 2004, "Anchorage Length of Near-Surface-Mounted FRP Bars for Concrete Strengthening—Experimental Investigation and Numerical Modeling," *ACI Structural Journal*, V. 101, No. 2, Mar.-Apr., pp. 269-278.

Demers, M., and Neale, K., 1999, "Confinement of Reinforced Concrete Columns with Fibre Reinforced Composites Sheets—An Experimental Study," *Canadian Journal of Civil Engineering*, V. 26, No. 2, pp. 226-241. doi: 10.1139/l98-067

Deniaud, C., and Cheng, J. J. R., 2001, "Shear Behavior of Reinforced Concrete T-Beams with Externally Bonded Fiber-Reinforced Polymer Sheets," *ACI Structural Journal*, V. 98, No. 3, May-June, pp. 386-394.

Deniaud, C., and Cheng, J. J. R., 2003, "Reinforced Concrete T-Beams Strengthened in Shear with Fiber Reinforced Polymer Sheets," *Journal of Composites for Construction*, V. 7, No. 4, pp. 302-310. doi: 10.1061/(ASCE)1090-0268(2003)7:4(302)

Di Ludovico, M.; Balsamo, A.; Prota, A.; and Manfredi, G., 2008a, "Comparative Assessment of Seismic Rehabilitation Techniques on a Full Scale 3-Story RC Moment Frame Structure," *Journal of Structural Engineering and Mechanics*, V. 28, No. 6, pp. 727-747. doi: 10.12989/sem.2008.28.6.727

Di Ludovico, M.; Manfredi, G.; Mola, E.; Negro, P.; and Prota, A., 2008b, "Seismic Behavior of a Full-Scale RC Structure Retrofitted Using GFRP Laminates," *Journal of Structural Engineering*, V. 134, No. 5, pp. 810-821. doi: 10.1061/(ASCE)0733-9445(2008)134:5(810)

Dolan, C. W.; Tanner, J.; Mukai, D.; Hamilton, H. R.; and Douglas, E., 2008, "Design Guidelines for Durability of Bonded CFRP Repair/Strengthening of Concrete Beams," *NCHRP Web-Only Document 155*, http://onlinepubs.trb.org/onlinepubs/nchrp/nchrp_w155.pdf (accessed Aug. 17, 2015).

Ehsani, M. R., 1993, "Glass-Fiber Reinforcing Bars," *Alternative Materials for the Reinforcement and Prestressing of Concrete*, J. L. Clarke, Blackie Academic & Professional, London, pp. 35-54.

El-Maaddawy, T.; Chahrour, A.; and Soudki, K. A., 2006, "Effect of Fiber-Reinforced Polymer Wraps on Corrosion Activity and Concrete Cracking in Chloride-Contaminated Concrete Cylinders," *Journal of Composites for Construction*, V. 10, No. 2, pp. 139-147. doi: 10.1061/(ASCE)1090-0268(2006)10:2(139)

Elnabelsy, G., and Saatcioglu, M., 2004, "Seismic Retrofit of Circular and Square Bridge Columns with CFRP Jackets," *Advanced Composite Materials in Bridges and Structures*, Calgary, AB, Canada, 8 pp. (CD-ROM)

El-Refaie, S. A.; Ashour, A. F.; and Garrity, S. W., 2003, "Sagging and Hogging Strengthening of Continuous Reinforced Concrete Beams using CFRP sheets," *ACI Structural Journal*, V. 100, No. 4, July-Aug., pp. 446-453.

El-Tawil, S.; Ogunc, C.; Okeil, A. M.; and Shahawy, M., 2001, "Static and Fatigue Analyses of RC Beams Strengthened with CFRP Laminates," *Journal of Composites for Construction*, V. 5, No. 4, pp. 258-267. doi: 10.1061/(ASCE)1090-0268(2001)5:4(258)

Engindeniz, M.; Kahn, L. F.; and Zureick, A. H., 2005, "Repair and Strengthening of Reinforced Concrete Beam-Column Joints: State of the Art," *ACI Structural Journal*, V. 102, No. 2, Mar.-Apr., pp. 187-197.

Engindeniz, M.; Kahn, L. F.; and Zureick, A. H., 2008a, "Pre-1970 RC Corner-Beam Column-Slab Joints: Seismic Adequacy and Upgradability with CFRP Composites," *Proceedings of the Fourteenth World Conference on Earthquake Engineering*, Beijing, China, Oct.

Engindeniz, M.; Kahn, L. F.; and Zureick, A. H., 2008b, "Performance of an RC Corner-Beam Column Joint Severely Damaged under Bidirectional Loading and Rehabilitated with FRP Composites," SP-258, *Seismic Strengthening of Concrete Buildings Using FRP Composites*, Farmington Hills, MI, pp. 19-36.

Eshwar, N.; Ibell, T. J.; and Nanni, A., 2003, "CFRP Strengthening of Concrete Bridges with Curved Soffits," *International Conference Structural Faults + Repair 2003*, M. C. Forde, ed., Commonwealth Institute, London, 10 pp. (CD-ROM)

Eshwar, N.; Nanni, A.; and Ibell, T. J., 2005, "Effectiveness of CFRP Strengthening On Curved Soffit RC Beams," *Advances in Structural Engineering*, V. 8, No. 1, pp. 55-68. doi: 10.1260/1369433053749607

Eurocode 8, 2005, "Design of Structures for Earthquake Resistance, Part 3: Strengthening and Repair of Buildings," European Standard EN 1998-3, European Committee for Standardization, Brussels, Belgium, 89 pp.

Fardis, M. N., and Khalili, H., 1981, "Concrete Encased in Fiberglass Reinforced Plastic," *ACI Journal Proceedings*, V. 78, No. 6, Nov.-Dec., pp. 440-446.

Federal Emergency Management Agency, 2006, "Techniques for the Seismic Rehabilitation of Existing Buildings," FEMA-547, Building Seismic Safety Council for the Federal Emergency Management Agency, Washington, DC, 571 pp.

Federal Emergency Management Agency, 2009, "Quantification of Building Seismic Performance Factors," FEMA-P695, Applied Technology Council FEMA P695, Washington, DC, 421 pp.

Firmo, J. P.; Correia, J. R.; and França, P., 2012, "Fire Behaviour of Reinforced Concrete Beams Strengthened with CFRP Laminates: Protection Systems with Insulation of the Anchorage Zones," *Composites. Part B, Engineering*, V. 43, No. 3, pp. 1545-1556. doi: 10.1016/j.compositesb.2011.09.002

Fleming, C. J., and King, G. E. M., 1967, "The Development of Structural Adhesives for Three Original Uses in South Africa," *RILEM International Symposium, Synthetic Resins in Building Construction*, Paris, pp. 75-92.

Funakawa, I.; Shimono, K.; Watanabe, T.; Asada, S.; and Ushijima, S., 1997, "Experimental Study on Shear Strengthening with Continuous Fiber Reinforcement Sheet and Methyl Methacrylate Resion," *Third International Symposium on Non-Metallic (FRP) Reinforcement for Concrete Structures (FRPRCS-3)*, V. 1, Japan Concrete Institute, Tokyo, Japan, pp. 475-482.

GangaRao, H. V. S., and Vijay, P. V., 1998, "Bending Behavior of Concrete Beams Wrapped with Carbon Fabric," *Journal of Structural Engineering*, V. 124, No. 1, pp. 3-10. doi: 10.1061/(ASCE)0733-9445(1998)124:1(3)

Gergely, J.; Pantelides, C. P.; and Reaveley, L. D., 2000, "Shear Strengthening of RC T-Joints Using CFRP Composites," *Journal of Composites for Construction*, V. 4, No. 2, pp. 56-64. doi: 10.1061/(ASCE)1090-0268(2000)4:2(56)

Ghobarah, A., and Said, A., 2002, "Shear Strengthening of Beam-Column Joints," *The Journal of Earthquake, Wind, and Ocean Engineering*, V. 24, No. 7, pp. 881-888.

Green, M.; Bisby, L.; Beaudoin, Y.; and Labossiere, P., 1998, "Effects of Freeze-Thaw Action on the Bond of FRP Sheets to Concrete," *Proceedings of the First International Conference on Durability of Composites for Construction*, Sherbrooke, QC, Canada, pp. 179-190.

Grelle, S. V., and Sneed, L. H., 2013, "Review of Anchorage Systems for Externally-Bonded FRP Laminates," *International Journal of Concrete Structures and Materials*, V. 7, No. 1, pp. 17-33. doi: 10.1007/s40069-013-0029-0

Griffin, C. T., and Hsu, R. S., 2010, "Comparing the Embodied Energy of Structural Systems in Buildings," *ICSA 2010 – 1st International Conference on Structures & Architecture*, CRC Press, 8 pp.

Hamed, E., and Rabinovitch, O., 2005, "Dynamic Behavior of Reinforced Concrete Beams Strengthened with Composite Materials," *Journal of Composites for Construction*, V. 9, No. 5, Sept., pp. 429-440. doi: 10.1061/(ASCE)1090-0268(2005)9:5(429)

Harajli, M., and Rteil, A., 2004, "Effect of Confinement Using Fiber-Reinforced Polymer or Fiber-Reinforced Concrete on Seismic Performance of Gravity Load-Designed Columns," *ACI Structural Journal*, V. 101, No. 1, Jan.-Feb., pp. 47-56.

Haroun, M. A., and Elsanadedy, H. M., 2005, "Fiber-Reinforced Plastic Jackets for Ductility Enhancement of Reinforced Concrete Bridge Columns with Poor Lap-Splice Detailing," *Journal of Bridge Engineering*, V. 10, No. 6, pp. 749-757. doi: 10.1061/(ASCE)1084-0702(2005)10:6(749)

Haroun, M. A., and Mosallam, A., 2002, "Shear Behavior of Unreinforced Masonry Wall strengthened with FRP Composites," *Proceedings of SAMPE Conference*, Baltimore, MD, pp. 862-871.

Haroun, M. A.; Mosallam, A. S.; and Allam, K. H., 2005, "Cyclic In-Plane Shear of Concrete Masonry Walls Strengthened by FRP Laminates," SP-230, *7th International Symposium on Fiber-Reinforced Polymer FRP Reinforcement for Concrete Structures editors*: C. Shield, J. P. Busel, S. L. Walkup, and D. G. Gremel, eds., American Concrete Institute, Farmington Hills, MI, pp. 327-340.

Haroun, M. A.; Mosallam, A. S.; Feng, M. Q.; and Elsanadedy, H. M., 2003, "Experimental Investigation of Seismic Repair and Retrofit of Bridge Columns by Composite Jackets," *Journal of Reinforced Plastics and Composites*, V. 22, No. 14, Sept., pp. 1243-1268. doi: 10.1177/0731684403035573

Harries, K. A., and Carey, S. A., 2003, "Shape and 'Gap' Effects on the Behavior of Variably Confined Concrete," *Cement and Concrete Research*, V. 33, No. 6, pp. 881-890. doi: 10.1016/S0008-8846(02)01085-2

Harries, K. A.; Ricles, J. R.; Pessiki, S.; and Sause, R., 2006, "Seismic Retrofit of Lap Splices in Nonductile Square Columns Using Carbon Fiber-Reinforced Jackets," *ACI Structural Journal*, V. 103, No. 6, Nov.-Dec., pp. 874-884.

Hassan, T., and Rizkalla, S., 2003, "Investigation of Bond in Concrete Structures Strengthened with Near Surface Mounted Carbon Fiber Reinforced Polymer Strips," *Journal of Composites for Construction*, V. 7, No. 3, pp. 248-257. doi: 10.1061/(ASCE)1090-0268(2003)7:3(248)

Hiotakis, S.; Lau, D. T.; and Londono, N. L., 2004, "Research on Seismic Retrofit and Rehabilitation of Concrete Shear Walls Using FRP Materials," *NSC-NRC Taiwan-Canada Workshop on Construction Technologies National Center for Research on Earthquake Engineering*, Taipei, Taiwan, pp. 17-26.

Hognestad, E., 1951, "A Study of Combined Bending and Axial Load in Reinforced Concrete Members," *Bulletin* 399, University of Illinois Engineering Experiment Station, Urbana, IL.

Hollings, J., 1968, "Reinforced Concrete Seismic Design," *Bulletin of the New Zealand Society for Earthquake Engineering*, V. 23, pp. 217-250.

Howarth, J.; Mareddy, S. S. R.; and Mativenga, P. T., 2014, "Energy Intensity and Environmental Analysis of

Mechanical Recycling of Carbon Fibre Composite," *Journal of Cleaner Production*, V. 81, No. 15, pp. 50-64.

Iacobucci, R.; Sheikh, S.; and Bayrak, O., 2003, "Retrofit of Square Concrete Columns with Carbon Fiber-Reinforced Polymer for Seismic Resistance," *ACI Structural Journal*, V. 100, No. 6, Nov.-Dec., pp. 785-794.

International Federation for Structural Concrete, 2001, *FIB 2001: Externally Bonded FRP Reinforcement for RC Structures, fib*, Lausanne, Switzerland, 138 pp.

International Federation for Structural Concrete, 2003, "Seismic Assessment and Retrofit of Reinforced Concrete Buildings" *fib Bulletin* No. 24, *fib*, Lausanne, Switzerland, 312 pp.

International Federation for Structural Concrete, 2006, "Retrofitting of Concrete Structures by Externally Bonded FRPs with Emphasis on Seismic Applications" *fib Bulletin* No. 35, *fib*, Lausanne, Switzerland, 220 pp.

Italian National Research Council, 2004, "Guide for the Design and Construction of Externally Bonded FRP Systems for Strengthening Existing Structures: Materials, RC and PC Structures, Masonry Structures," CNR-DT 200/2004, Advisory Committee on Technical Recommendations for Construction, Rome, Italy, 154 pp.

Japan Building Disaster Prevention Association, 2005, "Recent Development of Seismic Retrofit Methods in Japan," Technical Committee on Evaluation of Building Disaster Prevention Methods, T. Kabeyasawa, ed., 94 pp.

Japan Society of Civil Engineers, 2001, "Recommendations for Upgrading of Concrete Structures with Use of Continuous Fiber Sheet," *Concrete Engineering Series 41*, JSCE.

Kachlakev, D., and McCurry, D., 2000, "Testing of Full-Size Reinforced Concrete Beams Strengthened with FRP Composites: Experimental Results and Design Methods Verification," Report No. FHWA-OR-00-19, U.S. Department of Transportation Federal Highway Administration, 109 pp.

Kalfat, R.; Al-Mahaidi, R.; and Smith, S. T., 2013, "Anchorage Devices used to Improve the Performance of Reinforced Concrete Beams Retrofitted with FRP Composites: State-of-the-Art Review," *Journal of Composites for Construction*, V. 17, No. 1, pp. 14-33. doi: 10.1061/(ASCE)CC.1943-5614.0000276

Karbhari, V., ed., 2007, *Durability of Composites for Civil Structural Applications*, Woodhead Publishing, 384 pp.

Katsumata, H.; Kobatake, Y.; and Takeda, T., 1987, "A Study on the Strengthening with Carbon Fiber for Earthquake-Resistant Capacity of Existing Concrete Columns," *Proceedings from the Workshop on Repair and Retrofit of Existing Structures, U.S.-Japan Panel on Wind and Seismic Effects*, U.S.-Japan Cooperative Program in Natural Resources, Tsukuba, Japan, pp. 1816-1823.

Khalifa, A.; Gold, W.; Nanni, A.; and Abel-Aziz, M., 1998, "Contribution of Externally Bonded FRP to the Shear Capacity of RC Flexural Members," *Journal of Composites for Construction*, V. 2, No. 4, pp. 195-203. doi: 10.1061/(ASCE)1090-0268(1998)2:4(195)

Khalifa, A.; Alkhrdaji, T.; Nanni, A.; and Lansburg, S., 1999, "Anchorage of Surface-Mounted FRP Reinforcement," *Concrete International*, V. 21, No. 10, Oct., pp. 49-54.

Khomwan, N., and Foster, S. J., 2005, "FE Modelling of FRP-Strengthened RC Shear walls Subjected to Reverse Cyclic Loading," *Proceedings of the International Symposium on Bond Behaviour of FRP in Structures 2005*, pp. 519-524.

Kim, I. S.; Jirsa, J. O.; and Bayrak, O., 2011, "Use of Carbon Fiber-Reinforced Polymer Anchors to Repair and Strengthen Lap Splices of Reinforced Concrete Columns," *ACI Structural Journal*, V. 108, No. 5, Nov.-Dec., pp. 630-640.

Kim, S. J., and Smith, S. T., 2010, "Pullout Strength Models for FRP Anchors in Uncracked Concrete," *Journal of Composites for Construction*, V. 14, No. 4, pp. 406-414. doi: 10.1061/(ASCE)CC.1943-5614.0000097

Kumahara, S.; Masuda, Y.; and Tanano, Y., 1993, "Tensile Strength of Continuous Fiber Bar Under High Temperature," *International Symposium on Fiber-Reinforced Plastic Reinforcement for Concrete Structures*, SP-138, A. Nanni and C. W. Dolan, eds., American Concrete Institute, Farmington Hills, MI, pp. 731-742.

Lam, L., and Teng, J., 2003a, "Design-Oriented Stress-Strain Model for FRP-Confined Concrete," *Construction & Building Materials*, V. 17, No. 6-7, pp. 471-489. doi: 10.1016/S0950-0618(03)00045-X

Lam, L., and Teng, J., 2003b, "Design-Oriented Stress-Strain Model for FRP-Confined Concrete in Rectangular Columns," *Journal of Reinforced Plastics and Composites*, V. 22, No. 13, pp. 1149-1186. doi: 10.1177/0731684403035429

Lee, J.-H.; Chacko, R. M.; and Lopez, M. M., 2010, "Use of Mixed Mode Fracture Interfaces for the Modeling of Large Scale FRP Strengthened Beams," *Journal of Composites for Construction*, V. 14, No. 6, pp. 845-855. doi: 10.1061/(ASCE)CC.1943-5614.0000143

Lombard, J.; Lau, D. T.; Humar, J. L.; Foo, S.; and Cheung, G. M. S., 2000, "Seismic Strengthening and Repair of Reinforced Concrete Shear Walls," *Proceedings of the 12th World Conference on Earthquake Engineering*, Auckland, New Zealand, 8 pp.

Luo, S., and Wong, C. P., 2002, "Thermo-Mechanical Properties of Epoxy Formulations with Low Glass Transition Temperatures," *Proceedings of the 8th International Symposium on Advanced Packaging Materials*, pp. 226-231.

Malek, A.; Saadatmanesh, H.; and Ehsani, M., 1998, "Prediction of Failure Load of R/C Beams Strengthened with FRP Plate Due to Stress Concentrations at the Plate End," *ACI Structural Journal*, V. 95, No. 1, Jan.-Feb., pp. 142-152.

Malvar, L., 1998, "Durability of Composites in Reinforced Concrete," *Proceedings of the First International Conference on Durability of Composites for Construction*, Sherbrooke, QC, Canada, Aug., pp. 361-372.

Malvar, L.; Warren, G.; and Inaba, C., 1995, "Rehabilitation of Navy Pier Beams with Composite Sheets," *Second FRP International Symposium on Non-Metallic (FRP) Reinforcement for Concrete Structures*, Ghent, Belgium, Aug., pp. 533-540.

Mandell, J. F., 1982, "Fatigue Behavior of Fibre-Resin Composites," *Developments in Reinforced Plastics*, V. 2, Applied Science Publishers, London, UK, pp. 67-107.

Mandell, J. F., and Meier, U., 1983, "Effects of Stress Ratio Frequency and Loading Time on the Tensile Fatigue of Glass-Reinforced Epoxy," *Long Term Behavior of Composites*, ASTM STP 813, ASTM International, West Conshohocken, PA, pp. 55-77.

Masoud, S. A., and Soudki, K. A., 2006, "Evaluation of Corrosion Activity in FRP Repaired RC Beams," *Cement and Concrete Composites*, V. 28, No. 10, pp. 969-977. doi: 10.1016/j.cemconcomp.2006.07.013

Matthys, S., and Triantafillou, T., 2001, "Shear and Torsion Strengthening with Externally Bonded FRP Reinforcement," *Proceedings of the International Workshop on: Composites in Construction: A Reality*, ASCE, Reston, VA, pp. 203-212.

Matthys, S.; Toutanji, H.; Audenaert, K.; and Taerwe, L., 2005, "Axial Load Behavior of Large-Scale Columns Confined with Fiber-Reinforced Polymer Composites," *ACI Structural Journal*, V. 102, No. 2, Mar.-Apr., pp. 258-267.

Meier, U., 1987, "Bridge Repair with High Performance Composite Materials," *Materials Technology*, V. 4, pp. 125-128. (in German)

Meier, U., and Kaiser, H., 1991, "Strengthening of Structures with CFRP Laminates," *Advanced Composite Materials in Civil Engineering Structures*, ASCE Specialty Conference, pp. 224-232.

Memon, M., and Sheikh, S., 2005, "Seismic Resistance of Square Concrete Columns Retrofitted with Glass Fiber-Reinforced Polymer," *ACI Structural Journal*, V. 102, No. 5, Sept.-Oct., pp. 774-783.

Menna, C.; Asprone, D.; Jalayer, F.; Prota, A.; and Manfredi, G., 2013, "Assessment of Ecological Sustainability of a Building Subjected to Potential Seismic Events During its Lifetime," *The International Journal of Life Cycle Assessment*, V. 18, No. 2, pp. 504-515. doi: 10.1007/s11367-012-0477-9

Moehle, J. P.; Elwood, K. J.; and Sezen, H., 2002, "Gravity Load Collapse of Building Frames during Earthquakes," SP-197, *Behavior and Design of Concrete Structures for Seismic Performance*, American Concrete Institute, Farmington Hills, MI, pp. 215-238.

Moliner Santisteve, E.; Bastida, J.; Cseh, M.; and Vidal, R., 2013, "Life Cycle Assessment of a Fibre-Reinforced Polymer Made of Glass Fibre Phenolic Resin with Brominated Flame Retardant," *1st Symposium of the Spanish LCA Network: LCA & Bioenergy*.

Motavalli, M.; Terrasi, G. P.; and Meier, U., 1997, "On the Behavior of Hybrid Aluminum/CFRP Box Beams at Low Temperatures," *Composites. Part A, Applied Science and Manufacturing*, V. 28, No. 2, pp. 121-129. doi: 10.1016/S1359-835X(96)00100-5

Mutsuyoshi, H.; Uehara, K.; and Machida, A., 1990, "Mechanical Properties and Design Method of Concrete Beams Reinforced with Carbon Fiber Reinforced Plastics," *Transaction of the Japan Concrete Institute*, V. 12, Japan Concrete Institute, Tokyo, Japan, pp. 231-238.

Nanni, A., 1995, "Concrete Repair with Externally Bonded FRP Reinforcement," *Concrete International*, V. 17, No. 6, June, pp. 22-26.

Nanni, A., and Bradford, N., 1995, "FRP Jacketed Concrete Under Uniaxial Compression," *Construction & Building Materials*, V. 9, No. 2, pp. 115-124. doi: 10.1016/0950-0618(95)00004-Y

Nanni, A., and Gold, W., 1998, "Strength Assessment of External FRP Reinforcement," *Concrete International*, V. 20, No. 6, June, pp. 39-42.

Nanni, A.; Bakis, C. E.; Boothby, T. E.; Lee, Y. J.; and Frigo, E. L., 1997, "Tensile Reinforcement by FRP Sheets Applied to RC," *9C/1-8, ICE 97 International Composites Exposition*, Nashville, TN, Jan., pp. 9C/1 to 8.

Napolano, L.; Menna, C.; Asprone, D.; Prota, A.; and Manfredi, G., 2015, "LCA-Based Study on Structural Retrofit Options for Masonry Buildings," *The International Journal of Life Cycle Assessment*, V. 20, No. 1, pp. 23-35. doi: 10.1007/s11367-014-0807-1

National Research Council, 1991, "Life Prediction Methodologies for Composite Materials," *Committee on Life Prediction Methodologies for Composites*, NMAB-460, National Materials Advisory Board, Washington, DC, 75 pp.

Norris, T.; Saadatmanesh, H.; and Ehsani, M., 1997, "Shear and Flexural Strengthening of R/C Beams with Carbon Fiber Sheets," *Journal of Structural Engineering*, V. 123, No. 7, pp. 903-911. doi: 10.1061/(ASCE)0733-9445(1997)123:7(903)

Nosho, K. J., 1996, "Retrofit of Rectangular Reinforced Concrete Columns using Carbon Fiber," MS thesis, University of Washington, Seattle, WA, 194 pp.

Nowak, A. S., and Szerszen, M. M., 2003, "Calibration of Design Code for Buildings (ACI 318): Part 1—Statistical Models for Resistance," *ACI Structural Journal*, V. 100, No. 3, May-June, pp. 377-382.

Odagiri, T.; Matsumoto, K.; and Nakai, H., 1997, "Fatigue and Relaxation Characteristics of Continuous Aramid Fiber Reinforced Plastic Rods," *Third International Symposium on Non-Metallic (FRP) Reinforcement for Concrete Structures (FRPRCS-3)*, V. 2, Japan Concrete Institute, Tokyo, Japan, pp. 227-234.

Okeil, A. M.; Bingol, Y.; and Alkhrdaji, T., 2007, "Analyzing Model Uncertainties for Concrete Beams Flexurally Strengthened with FRP Laminates," *Proceedings of the Transportation Research Board 86th Annual Meeting*, Washington, DC, 15 pp. (CD-ROM)

Orton, S. L.; Jirsa, J. O.; and Bayrak, O., 2008, "Design Considerations of Carbon Fiber Anchors," *Journal of Composites for Construction*, V. 12, No. 6, pp. 608-616. doi: 10.1061/(ASCE)1090-0268(2008)12:6(608)

Orton, S. L.; Jirsa, J. O.; and Bayrak, O., 2009, "CFRP for Continuity in Existing RC Buildings Vulnerable to Collapse," *ACI Structural Journal*, V. 106, No. 5, Sept.-Oct., pp. 608-616.

Palmieri, A.; Matthys, S.; and Taerwe, L., 2011, "Fire Testing of RC Beams Strengthened with NSM Reinforcement," *10th International Symposium on Fiber-Reinforced Polymer Reinforcement for Concrete Structures (FRPRCS-*

10), SP-275, American Concrete Institute, Farmington Hills, MI. (CD-ROM)

Pampanin, S.; Bolognini, D.; and Pavese, A., 2007, "Performance-Based Seismic Retrofit Strategy for Existing Reinforced Concrete Frame Systems Using Fiber-Reinforced Polymer Composites," *Journal of Composites for Construction*, V. 11, No. 2, pp. 211-226. doi: 10.1061/(ASCE)1090-0268(2007)11:2(211)

Pantelides, C. P.; Gergely, J.; Reaveley, L. D.; and Volnyy, V. A., 1999, "Retrofit of R/C Bridge Pier with CFRP Advanced Composites," *Journal of Structural Engineering*, V. 125, No. 10, pp. 1094-1099. doi: 10.1061/(ASCE)0733-9445(1999)125:10(1094)

Pantelides, C. P.; Clyde, C.; and Reaveley, L. D., 2000, "Rehabilitation of R/C Building Joints with FRP Composites," *12th World Conference on Earthquake Engineering*, Auckland, New Zealand. (CD-ROM)

Pantelides, C. P.; Alameddine, F.; Sardo, T.; and Imbsen, R., 2004, "Seismic Retrofit of State Street Bridge on Interstate 80," *Journal of Bridge Engineering*, V. 9, No. 4, pp. 333-342. doi: 10.1061/(ASCE)1084-0702(2004)9:4(333)

Pantelides, C. P.; Okahashi, Y.; and Reaveley, L. D., 2008, "Seismic Rehabilitation of Reinforced Concrete Frame Interior Beam-Bolumn Joints with FRP Composites," *Journal of Composites for Construction*, V. 12, No. 4, pp. 435-445. doi: 10.1061/(ASCE)1090-0268(2008)12:4(435)

Park, R., and Paulay, T., 1976, *Reinforced Concrete Structures*, Wiley, 800 pp.

Paterson, J., and Mitchell, D., 2003, "Seismic Retrofit of Shear Walls with Headed Bars and Carbon Fiber Wrap," *Journal of Structural Engineering*, V. 129, No. 5, pp. 606-614. doi: 10.1061/(ASCE)0733-9445(2003)129:5(606)

Prestressed/Precast Concrete Institute, 2004, *PCI Design Handbook Precast and Prestressed Concrete*, sixth edition, Prestressed/Precast Concrete Institute, Chicago, IL, 750 pp.

Pellegrino, C., and Modena, C., 2002, "Fiber Reinforced Polymer Shear Strengthening of Reinforced Concrete Beams with Transverse Steel Reinforcement," *Journal of Composites for Construction*, V. 6, No. 2, pp. 104-111. doi: 10.1061/(ASCE)1090-0268(2002)6:2(104)

Pessiki, S. P.; Conley, C. H.; Gergely, P.; and White, R. N., 1990, "Seismic Behavior of Lightly Reinforced Concrete Column and Beam-Column Joint Details," *NCEER Report* No. 90-0014, 184 pp.

Pessiki, S.; Harries, K. A.; Kestner, J.; Sause, R.; and Ricles, J. M., 2001, "The Axial Behavior of Concrete Confined with FRP Jackets," *Journal of Composites for Construction*, V. 5, No. 4, pp. 237-245. doi: 10.1061/(ASCE)1090-0268(2001)5:4(237)

Porter, M. L.; Mehus, J.; Young, K. A.; O'Neil, E. F.; and Barnes, B. A., 1997, "Aging for Fiber Reinforcement in Concrete," *Proceedings of the Third International Symposium on Non-Metallic (FRP) Reinforcement for Concrete Structures*, Japan Concrete Institute, Sapporo, Japan.

Priestley, M.; Seible, F.; and Calvi, G., 1996, *Seismic Design and Retrofit of Bridges*, John Wiley and Sons, New York, 704 pp.

Prota, A.; Nanni, A.; Manfredi, G.; and Cosenza, E., 2004, "Selective Upgrade of Underdesigned Reinforced Concrete Beam-Column Joints Using Carbon Fiber-Reinforced Polymers," *ACI Structural Journal*, V. 101, No. 5, Sept.-Oct., pp. 699-707.

Reed, C. E.; Peterman, R. J.; and Rasheed, H. A., 2005, "Evaluating FRP Repair Method for Cracked Prestressed Concrete Bridge Members Subjected to Repeated Loadings (Phase 1)," *KTRAN Report No. K-TRAN: KSU-01-2*, Kansas Department of Transportation, Topeka, KS, 106 pp.

Ritchie, P.; Thomas, D.; Lu, L.; and Conneley, G., 1991, "External Reinforcement of Concrete Beams Using Fiber Reinforced Plastics," *ACI Structural Journal*, V. 88, No. 4, July-Aug., pp. 490-500.

Roberts, T. M., and Haji-Kazemi, H., 1989, "Theoretical Study of the Behavior of Reinforced Concrete Beams Strengthened by Externally Bonded Steel Plates," *Proceedings of the Institute of Civil Engineers*, Part 2, V. 87, No. 9344, pp. 39-55.

Rocca, S.; Galati, N.; and Nanni, A., 2006, "Experimental Evaluation of FRP Strengthening of Large-Size Reinforced Concrete Columns," *Report* No. UTC-142, University of Missouri-Rolla, MO.

Rocca, S.; Galati, N.; and Nanni, A., 2008, "Review of Design Guidelines for FRP Confinement of Reinforced Concrete Columns of Noncircular Cross Sections," *Journal of Composites for Construction*, V. 12, No. 1, Jan.-Feb., pp. 80-92. doi: 10.1061/(ASCE)1090-0268(2008)12:1(80)

Rosenboom, O. A., and Rizkalla, S. H., 2006, "Behavior of Prestressed Concrete Strengthened with Various CFRP Systems Subjected to Fatigue Loading," *Journal of Composites for Construction*, V. 10, No. 6, Nov.-Dec., pp. 492-502. doi: 10.1061/(ASCE)1090-0268(2006)10:6(492)

Rostasy, F. S., 1987, "Bonding of Steel and GFRP Plates in the Area of Coupling Joints. Talbrucke Kattenbusch," *Research Report* No. 3126/1429, Federal Institute for Materials Testing, Braunschweig, Germany. (in German)

Rostasy, F. S., 1997, "On Durability of FRP in Aggressive Environments," *Third International Symposium on Non-Metallic (FRP) Reinforcement for Concrete Structures* (FRPRCS-3), V. 2, Japan Concrete Institute, Tokyo, Japan, pp. 107-114.

Roylance, M., and Roylance, O., 1981, "Effect of Moisture on the Fatigue Resistance of an Aramid-Epoxy Composite," *Organic Coatings and Plastics Chemistry*, V. 45, American Chemical Society, Washington, DC, pp. 784-788.

Saadatmanesh, H.; Ehsani, M. R.; and Jin, L., 1996, "Seismic Strengthening of Circular Bridge Pier Models with Fiber Composites," *ACI Structural Journal*, V. 93, No. 6, Nov.-Dec., pp. 639-647.

Sabnis, G. M.; Shroff, A. C.; and Kahn, L. F., eds., 1996, "Seismic Rehabilitation of Concrete Structures," SP-160, American Concrete Institute, Farmington Hills, MI, 318 pp.

Sato, Y.; Ueda, T.; Kakuta, Y.; and Tanaka, T., 1996, "Shear Reinforcing Effect of Carbon Fiber Sheet Attached to Side of Reinforced Concrete Beams," *Advanced Composite Materials in Bridges and Structures*, M. M. El-Badry, ed., pp. 621-627.

Sause, R.; Harries, K. A.; Walkup, S. L.; Pessiki, S.; and Ricles, J. M., 2004, "Flexural Behavior of Concrete Columns with Carbon Fiber Composite Jackets," *ACI Structural Journal*, V. 101, No. 5, Sept.-Oct., pp. 708-716.

Seible, F.; Priestley, M. J. N.; Hegemier, G. A.; and Innamorato, D., 1997, "Seismic Retrofit of RC Columns with Continuous Carbon Fiber Jackets," *Journal of Composites for Construction*, V. 1, No. 2, pp. 52-62. doi: 10.1061/(ASCE)1090-0268(1997)1:2(52)

Sezen, H.; Whittaker, A. S.; Elwood, K. J.; and Mosalam, K. M., 2003, "Performance of Reinforced Concrete Buildings during the August 17, 1999 Kocaeli, Turkey Earthquake, and Seismic Design and Construction Practice in Turkey," *Journal of Engineering Structures*, V. 25, No. 1, pp. 103-114. doi: 10.1016/S0141-0296(02)00121-9

Sharaf, M. H.; Soudki, K. A.; and Van Dusen, M., 2006, "CFRP Strengthening for Punching Shear of Interior Slab-Column Connections," *Journal of Composites for Construction*, V. 10, No. 5, pp. 410-418. doi: 10.1061/(ASCE)1090-0268(2006)10:5(410)

Sharif, A.; Al-Sulaimani, G.; Basunbul, I.; Baluch, M.; and Ghaleb, B., 1994, "Strengthening of Initially Loaded Reinforced Concrete Beams Using FRP Plates," *ACI Structural Journal*, V. 91, No. 2, Mar.-Apr., pp. 160-168.

Sheikh, S., and Yau, G., 2002, "Seismic Behavior of Concrete Columns Confined with Steel and Fiber-Reinforced Polymers," *ACI Structural Journal*, V. 99, No. 1, Jan.-Feb., pp. 72-80.

Silva, P. F., and Ibell, T. J., 2008, "Evaluation of Moment Redistribution in Continuous FRP-Strengthened Concrete Structures," *ACI Structural Journal*, V. 105, No. 6, Nov.-Dec., pp. 729-739.

Silva, P. F.; Ereckson, N. J.; and Chen, G., 2007, "Seismic Retrofit of Bridge Joints in the Central U.S. with CFRP Composites," *ACI Structural Journal*, V. 104, No. 2, Mar.-Apr., pp. 207-217.

Soudki, K. A., and Green, M. F., 1997, "Freeze-Thaw Response of CFRP Wrapped Concrete," *Concrete International*, V. 19, No. 8, Aug., pp. 64-67.

Spoelstra, M. R., and Monti, G., 1999, "FRP-Confined Concrete Model," *Journal of Composites for Construction*, V. 3, No. 3, pp. 143-150. doi: 10.1061/(ASCE)1090-0268(1999)3:3(143)

Suppliers of Advanced Composite Materials Association, 1994, *SACMA Recommended Methods (SRM) Manual*, Suppliers of Advanced Composite Materials Association, Arlington, VA.

Szerszen, M. M., and Nowak, A. S., 2003, "Calibration of Design Code for Buildings (ACI 318): Part 2—Reliability Analysis and Resistance Factors," *ACI Structural Journal*, V. 100, No. 3, May-June, pp. 383-391.

Teng, J. G.; Chen, J. F.; Smith, S. T.; and Lam, L., 2002, *FRP Strengthened RC Structures*, John Wiley & Sons, West Sussex, UK, 266 pp.

Teng, J. G.; Smith, S. T.; Yao, J.; and Chen, J. F., 2003, "Intermediate Crack Induced Debonding in RC Beams and Slabs," *Construction & Building Materials*, V. 17, No. 6-7, pp. 447-462. doi: 10.1016/S0950-0618(03)00043-6

Teng, J. G.; Lu, X. Z.; Ye, L. P.; and Jiang, J. J., 2004, "Recent Research on Intermediate Crack Induced Debonding in FRP Strengthened Beams," *Proceedings of the 4th International Conference on Advanced Composite Materials for Bridges and Structures*, Calgary, AB, Canada.

Toutanji, H., 1999, "Stress-Strain Characteristics of Concrete Columns Externally Confined with Advanced Fiber Composite Sheets," *ACI Materials Journal*, V. 96, No. 3, May-June, pp. 397-404.

Triantafillou, T. C., 1998, "Shear Strengthening of Reinforced Concrete Beams Using Epoxy-Bonded FRP Composites," *ACI Structural Journal*, V. 95, No. 2, Mar.-Apr., pp. 107-115.

Wallace, J. W., 1995, "Seismic Design of RC Structural Walls; Part I: New Code Format," *Journal of Structural Engineering*, V. 121, No. 1, pp. 75-87. doi: 10.1061/(ASCE)0733-9445(1995)121:1(75)

Wang, N., and Evans, J. T., 1995, "Collapse of Continuous Fiber Composite Beam at Elevated Temperatures," *Composites*, V. 26, No. 1, pp. 56-61. doi: 10.1016/0010-4361(94)P3630-J

Wang, Y. C., and Restrepo, J. I., 2001, "Investigation of Concentrically Loaded Reinforced Concrete Columns Confined with Glass Fiber-Reinforced Polymer Jackets," *ACI Structural Journal*, V. 98, No. 3, May-June, pp. 377-385.

Williams, B. K.; Bisby, L. A.; Kodur, V. K. R.; Green, M. F.; and Chowdhury, E., 2006, "Fire Insulation Schemes for FRP-Strengthened Concrete Slabs," *Composites. Part A, Applied Science and Manufacturing*, V. 37, No. 8, pp. 1151-1160. doi: 10.1016/j.compositesa.2005.05.028

Wolf, R., and Miessler, H. J., 1989, "HLV-Spannglieder in der Praxis," *Erfahrungen Mit Glasfaserverbundstaben Beton*, V. 2, pp. 47-51.

Wu, W., 1990, "Thermomechanical Properties of Fiber Reinforced Plastics (FRP) Bars," PhD dissertation, West Virginia University, Morgantown, WV, 292 pp.

Xian, G., and Karbhari, V. M., 2007, "Segmental Relaxation of Water-Aged Ambient Cured Epoxy," *Journal of Polymer Degradation and Stability*, V. 92, No. 9, pp. 1650-1659. doi: 10.1016/j.polymdegradstab.2007.06.015

Yamaguchi, T.; Kato, Y.; Nishimura, T.; and Uomoto, T., 1997, "Creep Rupture of FRP Rods Made of Aramid, Carbon and Glass Fibers," *Third International Symposium on Non-Metallic (FRP) Reinforcement for Concrete Structures* (FRPRCS-3), V. 2, Japan Concrete Institute, Tokyo, Japan, pp. 179-186.

Youssef, M. N., 2003, "Stress Strain Model for Concrete Confined by FRP Composites," PhD dissertation, University of California-Irvine, Irvine, CA, 310 pp.

Zhang, C.; Lin, W. X.; Abududdin, M.; and Canning, L., 2012, "Environmental Evaluation of FRP in UK Highway Bridge Deck Replacement Applications Based on a Comparative LCA Study," *Advanced Materials Research*, V. 374, pp. 43-48.

Zureick, A.-H.; Ellingwood, B. R.; Nowak, A. S.; Mertz, D. R.; and Triantafillou, T. C., 2010, "Recommended Guide Specification for the Design of Externally Bonded FRP Systems for Repair and Strengthening of Concrete Bridge Elements," *NCHRP Report 655*, Transportation Research Board, Washington, DC, 49 pp.

APPENDIX A—MATERIAL PROPERTIES OF CARBON, GLASS, AND ARAMID FIBERS

Table A.1 presents ranges of values for the tensile properties of carbon, glass, and aramid fibers. The tabulated values are based on the testing of impregnated fiber yarns or strands in accordance with suppliers of SACMA Recommended Method 16R-1994 (Suppliers of Advanced Composite Materials Association 1994). The strands or fiber yarns are impregnated with resin, cured, and then tested in tension. The tabulated properties are calculated using the area of the fibers; the resin area is ignored. Hence, the properties listed in Table A.1 are representative of unidirectional FRP systems whose properties are reported using net-fiber area (4.3.1).

Table A.2 presents ranges of tensile properties for carbon FRP (CFRP), glass FRP (GFRP), and aramid FRP (AFRP) bars with fiber volumes of approximately 50 to 70 percent. Properties are based on gross-laminate area (4.3.1).

Table A.3 presents ranges of tensile properties for CFRP, GFRP, and AFRP laminates with fiber volumes of approximately 40 to 60 percent. Properties are based on gross-laminate area (4.3.1). The properties are shown for unidirectional, bidirectional, and +45/–45-degree fabrics. Table A.3 also shows the effect of varying the fiber orientation on the 0-degree strength of the laminate.

Table A.4 gives the tensile strengths of some commercially available FRP systems. The strength of unidirectional laminates is dependent on fiber type and dry fabric weight.

These tables are not intended to provide ultimate strength values for design purposes.

Table A.1—Typical tensile properties of fibers used in FRP systems

Fiber type	Elastic modulus		Ultimate strength		Rupture strain, minimum, %
	10^3 ksi	GPa	ksi	MPa	
Carbon					
General purpose	32 to 34	220 to 240	300 to 550	2050 to 3790	1.2
High-strength	32 to 34	220 to 240	550 to 700	3790 to 4820	1.4
Ultra-high-strength	32 to 34	220 to 240	700 to 900	4820 to 6200	1.5
High-modulus	50 to 75	340 to 520	250 to 450	1720 to 3100	0.5
Ultra-high-modulus	75 to 100	520 to 690	200 to 350	1380 to 2400	0.2
Glass					
E-glass	10 to 10.5	69 to 72	270 to 390	1860 to 2680	4.5
S-glass	12.5 to 13	86 to 90	500 to 700	3440 to 4140	5.4
Aramid					
General purpose	10 to 12	69 to 83	500 to 600	3440 to 4140	2.5
High-performance	16 to 18	110 to 124	500 to 600	3440 to 4140	1.6

Table A.2—Tensile properties of FRP bars with fiber volumes of 50 to 70 percent

FRP system description	Elastic modulus, 10^3 ksi (GPa)	Ultimate tensile strength, ksi (MPa)	Rupture strain, %
High-strength carbon/epoxy	17 to 24 (115 to 165)	180 to 400 (1240 to 2760)	1.2 to 1.8
E-glass/epoxy	4 to 7 (27 to 48)	70 to 230 (480 to 1580)	1.6 to 3.0
High-performance aramid	8 to 11 (55 to 76)	130 to 280 (900 to 11,930)	2.0 to 3.0

Table A.3—Tensile properties of FRP laminates with fiber volumes of 40 to 60 percent

FRP system description (fiber orientation)	Elastic modulus		Ultimate tensile strength		Rupture strain at 0 degrees, %
	Property at 0 degrees	Property at 90 degrees	Property at 0 degrees	Property at 90 degrees	
	10^3 ksi (GPa)	10^3 ksi (GPa)	ksi (MPa)	ksi (MPa)	
High-strength carbon/epoxy, degrees					
0	15 to 21 (100 to 140)	0.3 to 1 (2 to 7)	150 to 350 (1020 to 2080)	5 to 10 (35 to 70)	1.0 to 1.5
0/90	8 to 11 (55 to 76)	8 to 11 (55 to 75)	100 to 150 (700 to 1020)	100 to 150 (700 to 1020)	1.0 to 1.5
+45/–45	2 to 4 (14 to 28)	2 to 4 (14 to 28)	25 to 40 (180 to 280)	25 to 40 (180 to 280)	1.5 to 2.5
E-glass/epoxy, degrees					
0	3 to 6 (20 to 40)	0.3 to 1 (2 to 7)	75 to 200 (520 to 1400)	5 to 10 (35 to 70)	1.5 to 3.0
0/90	2 to 5 (14 to 34)	2 to 5 (14 to 35)	75 to 150 (520 to 1020)	75 to 150 (520 to 1020)	2.0 to 3.0
+45/–45	2 to 3 (14 to 21)	2 to 3 (14 to 20)	25 to 40 (180 to 280)	25 to 40 (180 to 280)	2.5 to 3.5
High-performance aramid/epoxy, degrees					
0	7 to 10 (48 to 68)	0.3 to 1 (2 to 7)	100 to 250 (700 to 1720)	5 to 10 (35 to 70)	2.0 to 3.0
0/90	4 to 5 (28 to 34)	4 to 5 (28 to 35)	40 to 80 (280 to 550)	40 to 80 (280 to 550)	2.0 to 3.0
+45/–45	1 to 2 (7 to 14)	1 to 2 (7 to 14)	20 to 30 (140 to 210)	20 to 30 (140 to 210)	2.0 to 3.0

Notes:

FRP composite properties are based on FRP systems having an approximate fiber volume of 50 percent and a composite thickness of 0.1 in. (2.5 mm). In general, FRP bars have fiber volumes of 50 to 70 percent, precured systems have fiber volumes of 40 to 60 percent, and wet layup systems have fiber volumes of 25 to 40 percent. Because the fiber volume influences the gross-laminate properties, precured laminates usually have higher mechanical properties than laminates created using the wet layup technique.

Zero degrees represents unidirectional fiber orientation.

Zero/90 degrees (or +45/–45 degrees) represents fiber balanced in two orthogonal directions, where 0 degrees is the direction of loading, and 90 degrees is normal to the direction of loading.

Tension is applied to 0-degree direction. All FRP bar properties are in the 0-degree direction.

Table A.4—Ultimate tensile strength* of some commercially available FRP systems

FRP system description (fiber type/saturating resin/fabric type)	Fabric weight		Ultimate strength†	
	oz/yd^3	g/m^3	lb/in.	kN/mm
General purpose carbon/resin unidirectional sheet	6	200	2600	500
	12	400	3550	620
High-strength carbon/resin unidirectional sheet	7	230	1800	320
	9	300	4000	700
	18	620	5500	960
High-modulus carbon/resin unidirectional sheet	9	300	3400	600
General-purpose carbon/resin balanced sheet	9	300	1000	180
E-glass/resin unidirectional sheet	27	900	4100	720
	10	350	1300	230
E-glass/balanced fabric	9	300	680	120
Aramid/resin unidirectional sheet	12	420	4000	700
High-strength carbon/resin precured, unidirectional laminate	70‡	2380‡	19,000	3300
E-glass/vinyl ester precured, unidirectional shell	50‡	1700‡	9000	1580

*Values shown should not be used for design.

†Ultimate tensile strength per unit width of sheet or fabric.

‡Precured laminate weight.

APPENDIX B—SUMMARY OF STANDARD TEST METHODS

Table B provides a summary of test methods for the short- and long-term mechanical and durability testing of FRP rods and sheets. The recommended test methods are based on the knowledge gained from research results and literature worldwide and include those methods described in ACI 440.3R that have not yet been adopted by ASTM.

Durability-related tests use the same test methods but require application-specific preconditioning of specimens. Acceptance of the data generated by the listed test methods can be the basis for FRP material system qualification and acceptance (for example, ACI 440.8).

Table B—Test methods for FRP material systems

Property	ASTM test method(s)	ACI 440.3R test method	Summary of differences
Test methods for sheets, prepreg, and laminates			
Surface hardness	D2538 D2240 D3418	—	No ACI methods developed
Coefficient of thermal expansion	D696	—	No ACI methods developed
Glass-transition temperature	E1640	—	No ACI methods developed
Volume fraction	D3171 D2584	—	No ACI methods developed
Sheet to concrete adhesion (direct tension pulloff)	D7522/D7522M	L.1*	ACI method provides specific requirements for specimen preparation not found in the ASTM method
Tensile strength and modulus	D3039/D3039M or D7565/D7565M, as appropriate	L.2*	ACI method provides methods for calculating tensile strength and modulus on gross cross-sectional and effective fiber area basis. Section 3.3.1 is used to calculate design values.
Lap shear strength	D7616/D7616M	L.3*	ACI method provides specific requirements for specimen preparation.
Test methods for FRP bars			
Cross-sectional area	D7205/D7207M	B.1*	Two options for bar area are provided in ASTM D7205/D7205M (nominal and actual) whereas only nominal area is used in ACI 440.3R Method B.1
Longitudinal tensile strength and modulus	D7205/D7205M	B.2*	Strain limits for calculation of modulus are different in the two methods.
Shear strength	D7617/D7617M	B.4*	The ACI method focuses on dowel action of bars and does not overlap with existing ASTM methods that focus mainly on beam shearing failure modes. Bar shear strength is of specific concern for applications where FRP rods are used to cross construction joints in concrete pavements.
Durability properties	—	B.6	No existing ASTM test methods available.
Fatigue properties	D3479/D3479M	B.7	ACI methods provide specific information on anchoring bars in the test fixtures and on attaching elongation measuring devices to the bar. The ACI methods also require specific calculations that are not provided in the ASTM methods.
Creep properties	D7337/D7337M	B.8*	
Relaxation properties	D2990 E328	B.9	
Flexural tensile properties	—	B.11	No existing ASTM test methods available.
Flexural properties	D790 D4476/D4476M	—	No ACI methods developed.
Coefficient of thermal expansion	E831 D696	—	No ACI methods developed.
Glass-transition temperature	E1356 E1640 D648 E2092	—	No ACI methods developed.
Volume fraction	D3171	—	No ACI methods developed.

*Test method in ACI 440.3R is replaced by reference to appropriate ASTM method.

APPENDIX C—AREAS OF FUTURE RESEARCH

Future research is needed to provide information in areas that are still unclear or are in need of additional evidence to validate performance. The list of topics presented in this appendix provides a summary.

a) Materials

i. Methods of fireproofing FRP strengthening systems

ii. Behavior of FRP-strengthened members under elevated temperatures

iii. Behavior of FRP-strengthened members under cold temperatures

iv. Fire rating of concrete members strengthened with FRP systems

v. Effect of different coefficients of thermal expansion between FRP systems and member substrates

vi. Creep-rupture behavior and endurance times of FRP systems

vii. Strength and stiffness degradation of FRP systems in harsh environments

b) Flexure/axial force

i. Compression behavior of noncircular members wrapped with FRP systems

ii. Behavior of members strengthened with FRP systems oriented in the direction of the applied axial load

iii. Effects of high concrete strength on behavior of FRP-strengthened members

iv. Effects of lightweight concrete on behavior of FRP-strengthened members

v. Maximum crack width and deflection prediction and control of concrete reinforced with FRP systems

vi. Long-term deflection behavior of concrete flexural members strengthened with FRP systems

c) Shear

i. Effective strain of FRP systems that do not completely wrap around the section

ii. Use of FRP systems for punching shear reinforcement in two-way systems

d) Detailing

i. Anchoring of FRP systems

The design guide specifically indicates that test methods are needed to determine the following properties of FRP:

a) Bond characteristics and related bond-dependent coefficients

b) Creep rupture and endurance times

c) Fatigue characteristics

d) Coefficient of thermal expansion

e) Shear strength

f) Compressive strength

APPENDIX D—METHODOLOGY FOR COMPUTATION OF SIMPLIFIED *P-M* INTERACTION DIAGRAM FOR NONCIRCULAR COLUMNS

Axial load-moment (*P-M*) interaction diagrams may be developed by satisfying strain compatibility and force equilibrium using the model for the stress strain behavior for FRP-confined concrete presented in Eq. (12.1c) through (12.1e). For simplicity, the portion of the unconfined and confined P-M diagrams corresponding to compression-controlled failure can be reduced to two bilinear curves passing through the following points (Fig 12.2). (The following only makes reference to the confined case because the unconfined case is analogous):

a) Point A (pure compression) at a uniform axial compressive strain of confined concrete ε_{ccu}

b) Point B with a strain distribution corresponding to zero strain at the layer of longitudinal steel reinforcement nearest to the tensile face, and a compressive strain ε_{ccu} on the compression face

c) Point C with a strain distribution corresponding to balanced failure with a maximum compressive strain ε_{ccu} and a yielding tensile strain ε_{sy} at the layer of longitudinal steel reinforcement nearest to the tensile face

For confined concrete, the value of ϕP_n corresponding to Point A (ϕM_n equals zero) is given in Eq. (12.1a) and (21.1b), while the coordinates of Points B and C can be computed as:

$$\phi P_{n(B,C)} = \phi \begin{bmatrix} A(y_t)^3 + B(y_t)^2 + C(y_t) \\ +D + \sum A_{si} f_{si} \end{bmatrix} \quad \text{(D-1)}$$

$$\phi M_{n(B,C)} = \phi \begin{bmatrix} E(y_t)^4 + F(y_t)^3 + G(y_t)^2 \\ +H(y_t) + I + \sum A_{si} f_{si} d_i \end{bmatrix} \quad \text{(D-2)}$$

where

$$A = \frac{-b(E_c - E_2)^2}{12 f_c'}\left(\frac{\varepsilon_{ccu}}{c}\right)^2 \quad \text{(D-3a)}$$

$$B = \frac{b(E_c - E_2)}{2}\left(\frac{\varepsilon_{ccu}}{c}\right) \quad \text{(D-3b)}$$

$$C = -bf_c' \quad \text{(D-3c)}$$

$$D = bcf_c' + \frac{bcE_2}{2}(\varepsilon_{ccu}) \quad \text{(D-3d)}$$

$$E = \frac{-b(E_c - E_2)^2}{16 f_c'}\left(\frac{\varepsilon_{ccu}}{c}\right)^2 \quad \text{(D-3e)}$$

$$F = \begin{bmatrix} b\left(c - \dfrac{h}{2}\right)\dfrac{(E_c - E_2)^2}{12 f_c'}\left(\dfrac{\varepsilon_{ccu}}{c}\right)^2 \\ + \dfrac{b(E_c - E_2)}{3}\left(\dfrac{\varepsilon_{ccu}}{c}\right) \end{bmatrix} \quad \text{(D-3f)}$$

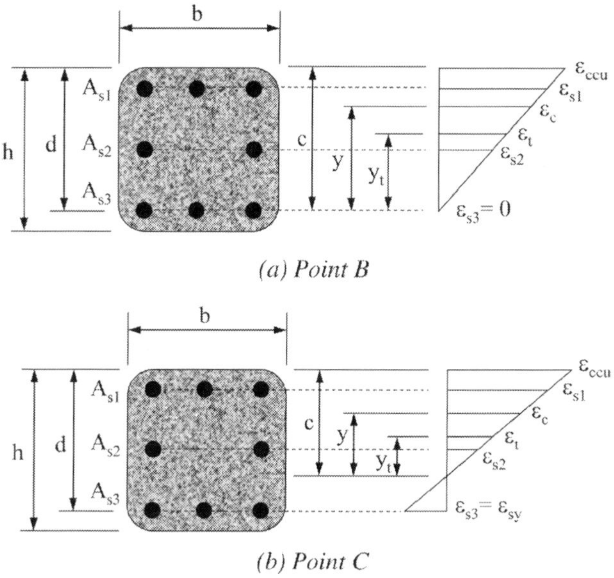

Fig. D.1—Strain distributions for Points B and C for simplified interaction diagram.

$$G = \left(\frac{b}{2} f_c' + b\left(c - \frac{h}{2}\right)\frac{(E_c - E_2)}{2}\left(\frac{\varepsilon_{ccu}}{c}\right)\right) \quad \text{(D-3g)}$$

$$H = bf_c'\left(c - \frac{h}{2}\right) \quad \text{(D-3h)}$$

$$I = \begin{bmatrix} \dfrac{bc^2}{2} f_c' - bcf_c'\left(c - \dfrac{h}{2}\right) + \dfrac{bc^2 E_2}{3}(\varepsilon_{ccu}) \\ -\dfrac{bcE_2}{2}\left(c - \dfrac{h}{2}\right)(\varepsilon_{ccu}) \end{bmatrix} \quad \text{(D-3i)}$$

In Eq. (D-3a) through (D-3i), c is the distance from the extreme compression fiber to the neutral axis (Fig D.1) and it is given by Eq. (D-4). The parameter y_t represents the vertical coordinate within the compression region measured from the neutral axis position and corresponds to the transition strain ε_t' (Eq. (D-5) [refer to Fig. D.1]).

$$c = \begin{cases} d & \text{for Point B} \\ d\dfrac{\varepsilon_{ccu}}{\varepsilon_{sy} + \varepsilon_{ccu}} & \text{for Point C} \end{cases} \quad \text{(D-4)}$$

$$y_t = c\frac{\varepsilon_t'}{\varepsilon_{ccu}} \quad \text{(D-5)}$$

where f_{si} is the stress in the i-th layer of longitudinal steel reinforcement. The values are calculated by similar triangles from the strain distribution corresponding to Points B and C. Depending on the neutral axis position c, the sign of f_{si} will be positive for compression and negative for tension. A flowchart illustrating the application of the proposed methodology is shown in Fig. D.2.

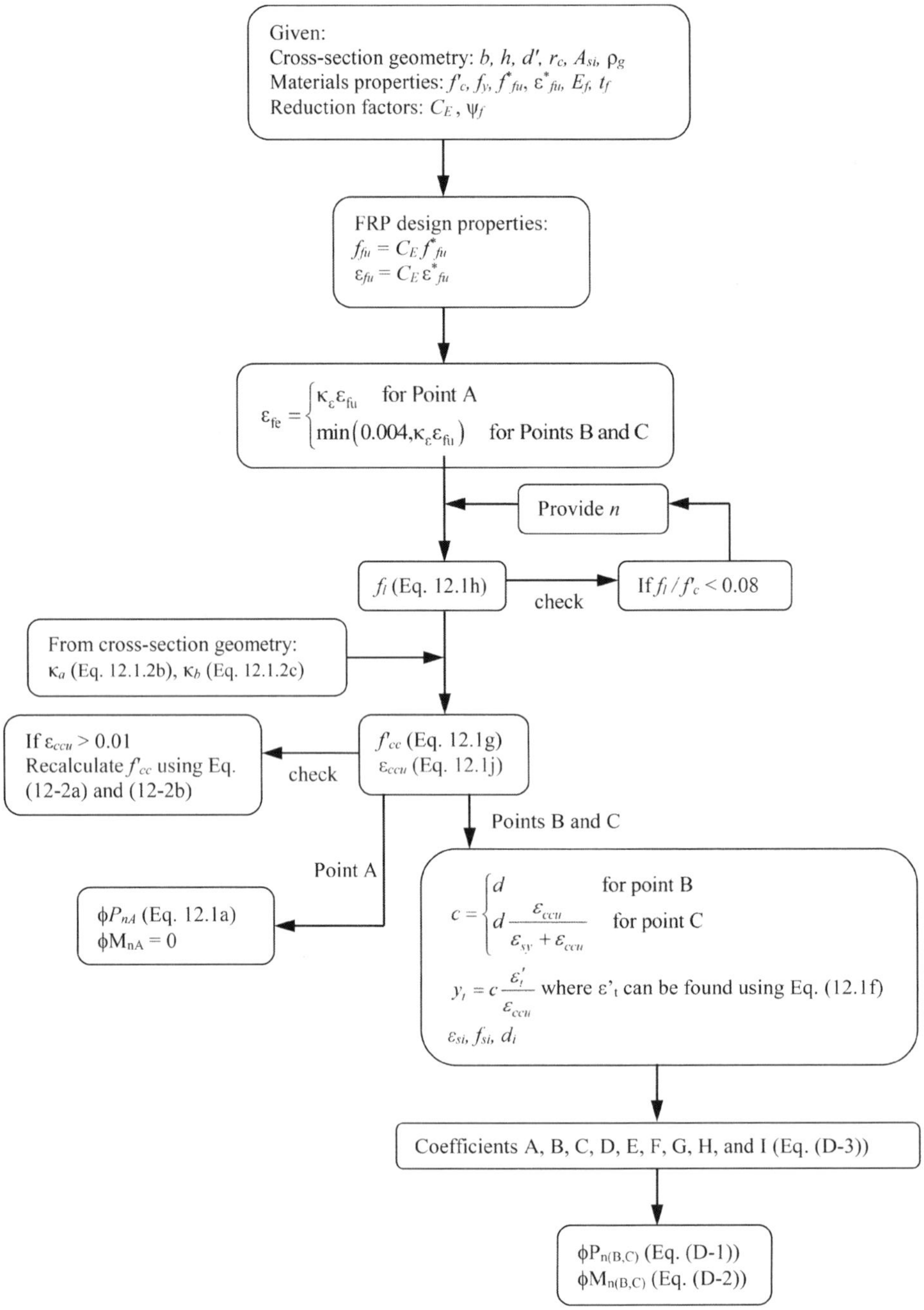

Fig. D.2—Flowchart for application of methodology.

Guide Test Methods for Fiber-Reinforced Polymer (FRP) Composites for Reinforcing or Strengthening Concrete and Masonry Structures

Reported by ACI Committee 440

Carol K. Shield, Chair

William J. Gold, Secretary

Tarek Alkhrdaji
Charles E. Bakis*
Lawrence C. Bank
Abdeldjelil Belarbi
Brahim Benmokrane
Luke A. Bisby
Gregg J. Blaszak
Hakim Bouadi
Timothy E. Bradberry
Gordon L. Brown Jr.
Vicki L. Brown
John P. Busel
Raafat El-Hacha
Garth J. Fallis
Amir Z. Fam
Nabil F. Grace
Mark F. Green
Zareh B. Gregorian
Doug D. Gremel
Shawn P. Gross
H. R. Trey Hamilton III
Issam E. Harik
Kent A. Harries
Mark P. Henderson
Bohdan N. Horeczko
Michael W. Lee
Maria Lopez de Murphy
Ibrahim M. Mahfouz
Amir Mirmiran
John J. Myers
Antonio Nanni
Ayman M. Okeil
Carlos E. Ospina
Renato Parretti
Max L. Porter
Andrea Prota
Hayder A. Rasheed
Sami H. Rizkalla
Rajan Sen
Rudolf Seracino
Pedro F. Silva
Khaled A. Soudki
Samuel A. Steere III
Jay Thomas
Houssam A. Toutanji
J. Gustavo Tumialan
Milan Vatovec
David White
Sarah E. Witt

*Chair of authoring subcommittee.

Consulting members

P. N. Balaguru
Craig A. Ballinger
Harald G. F. Budelmann
C. J. Burgoyne
Elliot P. Douglas
Rami M. Elhassan
David M. Gale
Russell Gentry
Arie Gerritse
Srinivasa L. Iyer
Koichi Kishitani
Howard S. Kliger
Kyuichi Maruyama
Antoine E. Naaman
Mark A. Postma
Hajime Okamura
Ferdinand S. Rostasy
Mohsen Shahawy
Surendra P. Shah
Yasuhisa Sonobe
Minoru Sugita
Luc R. Taerwe
Ralejs Tepfers
Taketo Uomoto
Paul Zia

ACI 440.3R-12 supersedes ACI 440.3R-04 and was adopted and published August 2012.

Due to differences in the physical and mechanical behavior of fiber-reinforced polymer (FRP) materials compared to steel, unique test methods for FRP bars and laminates are required. This guide provides model test methods for the short- and long-term mechanical, thermo-mechanical, and durability testing of FRP bars and laminates. It is anticipated that these model test methods may be considered, modified, and adopted, either in whole or in part, by a U.S. national standards-writing agency such as ASTM International or AASHTO. The publication of these test methods by ACI Committee 440 is an effort to aid in this adoption. The recommended test methods are based on the knowledge gained worldwide from review of research results and literature.

Keywords: anchorage; bond; coupler; creep; fatigue; fiber-reinforced polymer composites; modulus of elasticity; shear; splice; stirrup; strength; tendon.

CONTENTS

PART 1—GENERAL

1.1—Introduction

Conventional concrete and masonry structures are reinforced with nonprestressed steel, prestressed steel, or both. Recently, composite materials made of fibers embedded in a polymeric resin, also known as fiber-reinforced polymer (FRP) composites, have become an alternative to steel reinforcement. Because FRP materials are nonmetallic and noncorrosive, the problems of steel corrosion are avoided with FRP reinforcement. Additionally, FRP materials exhibit several properties, such as high tensile strength, that make them suitable for use as structural reinforcement. FRP materials are supplied as bars for reinforced and prestressing applications and in flat sheets or laminates for use as repair materials.

The mechanical behavior of FRP differs from the behavior of steel reinforcement. FRP materials, which are anisotropic due to the fiber orientation in the bars and laminates, are characterized by high tensile strength only in the direction of the reinforcing fibers. This anisotropic behavior affects the shear strength and dowel action of FRP bars and the bond performance of FRP bars to concrete.

FRP composites are available with a wide range of mechanical properties, including tensile strengths, bond strengths, and elastic moduli. Generally, FRP composites are not covered by national material standards. Instead, manufacturers of FRP composites provide test data and recommend design values based on these test data. Therefore, it is difficult to compare test results between product manufacturers. In addition, research has considered the durability of FRP reinforcement in environments containing moisture, high and low temperatures, and alkaline environments. Test methods that allow for the comparison of mechanical property retention in a wide range of standard environments are needed so that durable FRP-reinforced structures can be ensured.

1.2—Scope

This guide provides model test methods for determining the short- and long-term mechanical properties of FRP reinforcing bars, grids, tendons, and laminates—both prestressed and nonprestressed—for concrete and masonry structures. As noted in the individual methods, most of the methods for bars are also suitable for sections cut from grids. Where necessary, the tests consider the bars and laminates tested embedded in or bonded to concrete or masonry. For the most part, however, these tests are considered to be material tests and not component or structural tests.

These model test methods are intended for consideration, modification, and adoption, either in whole or in part, by a U.S. national standards-writing agency such as ASTM International or AASHTO. The publication of these test methods by ACI Committee 440 is an effort to aid in this adoption.

The guide contains only test methods and not material standards. The individual test methods contained in this guide do not specify minimum material properties that must be met for the materials to be considered acceptable for use. Guidance on deciding whether a material is acceptable based on test results is made in the material specifications and design provisions that complement these test methods (ACI 440.1R; ACI 440.2R; ACI 440.6).

The test methods presented in this guide are the recommendations of ACI Committee 440, and have not been adopted by ACI as standards. As such, they are, for the most part, written in nonmandatory language, using "should" and "may" rather than "shall" and "must." In keeping with the usual test method format, however, some language is imperative ("Fill a cylinder with water..." rather than "A cylinder should be filled with water..."). Although typically considered to be mandatory language, the use of imperative language in these test methods is for readability, and remains as committee recommendations only.

1.3—Existing ASTM test methods

The recommended test methods provided herein are based on the knowledge obtained worldwide from review of research results and literature. Relevant ASTM standards are referenced in the individual methods; others are listed in Table 1.3. In many cases, existing ASTM test methods are appropriate to determine material properties for FRP bars and laminates. Where such methods are completely acceptable for FRP reinforcement, no new method has been proposed. The new methods that are provided have been developed for one or more of the following reasons:

a) To provide a test method where no current method exists

b) To provide more detailed requirements that are specific to FRP reinforcing bars or laminates, such as details on how to grip the reinforcement in the test fixture

c) To adapt a test method originally developed for steel reinforcing bars to work with FRP bars

d) To provide calculated test results compatible with other ACI documents.

Table 1.3 lists specific ASTM test methods and complementary ACI 440.3R methods for various material properties. Where both ASTM and ACI 440.3R test methods exist, the differences between the methods are summarized. Hundreds of ASTM test methods are applicable to FRP composites and organic polymers. The table only describes key material properties and selected ASTM tests that can be used to determine these properties. For some properties, ASTM provides more than one test procedure. The table does not attempt to discuss the differences between various ASTM test methods.

1.4—Definitions

ACI provides a comprehensive list of definitions through an online resource, "ACI Concrete Terminology," http://terminology.concrete.org. The definitions provided herein complement that resource.

alkaline—having a pH greater than 7 (OH^- concentration greater than 1×10^{-7} M).

anchorage—device at the ends of an FRP tendon that grips the tendon, allowing a minimum of slip and transferring prestressing load from the tendon to the abutment.

anchorage reinforcement—latticed or spiral reinforcing steel or FRP bars provided as confining reinforcement for the anchorage and arranged in front of it.

anchoring section—FRP bar section embedded in the anchorage and anchorage reinforcement, including the surrounding concrete or masonry.

aramid fiber—highly oriented organic fiber derived from polyamide incorporating into an aromatic ring structure.

average load (stress)— mean value of the maximum and minimum repeated loads (stresses).

bend capacity—ultimate tensile stress carried by the FRP stirrup, provided that failure occurs in the bend.

bend radius—inside radius of the bend.

bending angle—angle formed by the straight sections of a specimen on either side of the deflector.

bending diameter ratio—ratio of the external diameter of the deflector surface in contact with the FRP bar to the diameter of the FRP bar.

bending tensile capacity—tensile capacity at failure of a specimen within the deflected section.

bonded length—length of the test bar that is in contact with concrete.

carbon fiber—fiber produced by heating organic precursor materials containing a substantial amount of carbon, such as rayon, polyacrylonitrile (PAN), or pitch, in an inert environment.

characteristic length—for bars or tendons that have a repeating surface deformation pattern, the characteristic length is the distance, in inches (mm), of this pattern; for a spiral pattern, the characteristic length is the pitch.

coefficient of thermal expansion (CTE)—measure of the relative change in linear dimension in a material based on a unit increase in temperature of that material. Note: Due to the anisotropy of FRP composites, the CTE in the longitudinal direction of the bar is likely to be different from that measured in the transverse direction.

creep rupture—material failure due to deformation (accumulated strain) caused by creep.

creep rupture load capacity—load at which failure occurs after a specified period of time from initiation of a sustained load.

cure—to irreversibly change the properties of a thermosetting resin by chemical reaction such as condensation, ring closure, or addition. Note: Cure can be accomplished by adding curing (cross-linking) agents with or without heat and pressure.

deflected section—section of an FRP bar that is bent and maintained at the required bending angle and bending diameter ratio.

deflector—device used to maintain the position, alter the bending angle, or alleviate the stress concentrations in an FRP bar; such a device may sometimes be installed in the deflected section.

equivalent bar diameter—equivalent bar diameter is determined based on the cross-sectional area of the FRP bar (ASTM D7205/D7205M).

equivalent circumference—circumference of an assumed circle with the equivalent area determined according to ASTM D7205/D7205M.

fatigue life—number of cycles of deformation or load required to bring about failure of a material, test specimen, or structural member.

fatigue strength—maximum load (stress) range at which the specimen does not fail at a prescribed number of cycles.

fiber-volume fraction—ratio of the volume of fibers to the volume of the composite.

frequency—number of loading (stressing) cycles per second.

Table 1.3—Test methods for FRP composites used in concrete and masonry structures

Property	ASTM test method(s)	ACI 440 test method	Summary of differences
Cross-sectional area	D7205/D7205M	—	ASTM D7205/D7205M replaces ACI 440.3R-04 Method B.1.
Longitudinal tensile strength and modulus	D7205/D7205M	—	ASTM D7205/D7205M replaces ACI 440.3R-04 Method B.2.
Bond properties	A944	B.3	ASTM Pullout Test C234 (ASTM C234) has been withdrawn and, as written, did not provide adequate specimen size to prevent splitting of concrete cylinder when using FRP bars. The only remaining ASTM test method for bond of steel bars to concrete is beam-end test method (A944), which has not been modified for use with FRP bars. Ongoing work by ACI Committee 440 is expected to produce beam bond test methods.
Shear strength	D7617/D7617M	—	ASTM D7617/D7617M replaces ACI 440.3R-04 Method B.4.
Bent bar capacity	—	B.5	No existing ASTM test method available.
Durability properties	—	B.6	No existing ASTM test method available.
Fatigue properties	D3479	B.7	The ACI method provides specific information on anchoring bars in the test fixtures and on attaching elongation measuring devices to the bars. The ACI method also requires specific calculations that are not provided in the ASTM method.
Creep properties	D7337/D7337M	—	ASTM D7337/D7337M replaces ACI 440.3R-04 Method B.8.
Relaxation properties	D2990 E328	B.9	The ACI method provides specific information on anchoring bars in the test fixtures and on attaching elongation measuring devices to the bars. The ACI method also requires specific calculations that are not provided in the ASTM method.
Anchorage properties	—	B.10	No existing ASTM test method available.
Tensile properties of deflected bars	—	B.11	No existing ASTM test method available.
Effect of corner radius on strength	—	B.12	No existing ASTM test method available.
Flexural properties	D790 D4476	—	No ACI method developed.
Coefficient of thermal expansion (CTE)	E831 D696	—	No ACI method developed.
Glass transition temperature	E1356 E1640 D648 E2092	—	No ACI method developed.
Volume fraction	D3171 D2584	—	No ACI method developed.
Direct tension pulloff	D7522/D7522M	—	ASTM D7522/D7522M replaces ACI 440.3R-04 Method L.1.
Tensile strength and modulus	D7565/D7565M	—	ASTM D7565/D7565M replaces ACI 440.3R-04 Method L.2.
Lap shear strength	D7616/D7616M	—	ASTM D7616/D7616M replaces ACI 440.3R-04 Method L3.
Bond strength	D4551 C882	—	No ACI method developed.

FRP bar—composite material formed into a long, slender, structural shape suitable for the internal reinforcement of concrete and consisting of primarily longitudinal unidirectional fibers bound and shaped by a rigid polymer resin material. The bar may have a cross section of variable shape, commonly circular or rectangular, and may have a deformed or roughened surface to enhance bonding with concrete.

FRP laminates—two or more layers of fiber reinforcements, such as, glass, carbon, and aramid arranged in one or more orientations—for example, 0, 90, +45, –45 degrees—and held together by a polymer matrix. Laminates come in the physical form of dry, prepreg, and precured materials.

FRP tendon—FRP element, such as a bar or strand, or a bundle of such elements primarily used in tension to impart compressive stress to concrete or masonry.

gauge length—distance between two gauge points on the test section, over which the percentage of elongation is determined.

glass fiber—fiber drawn from an inorganic fusion of silica (SiO_2) and other compounds that has cooled without crystallization.

grid—two-dimensional (planar) or three-dimensional (spatial) rigid array of interconnected FRP bars that form a contiguous lattice that can be used to reinforce concrete. Note: The lattice can be manufactured with integrally connected bars or made of mechanically connected individual bars.

impregnate—in the case of FRP composites, to saturate the fibers with resin.

load (stress) amplitude—one-half of the load (stress) range.

load (stress) range—difference between the maximum and minimum repeated loads (stress).

load (stress) ratio—minimum load (stress) divided by maximum load (stress).

maximum repeated load (stress)—maximum load (stress) during repeated loading (stressing).

minimum repeated load (stress)—minimum load (stress) during repeated loading (stressing).

number of cycles—number of times the repeated load (stress) is applied to the test specimen.

pitch—black residue from the distillation of petroleum, used as a precursor for the production of carbon fibers; the other precursor for polymer fibers is polyacrylonitrile.

polyacrylonitrile (PAN)—polymeric precursor for the production of carbon fibers; the other precursor for carbon fibers is pitch.

precured FRP—fully cured FRP that is usually made in a factory and brought to the site as a rigid solid; if used as a repair material for concrete, a precured FRP should be bonded to the surface of the concrete with an adhesive.

precursor—rayon, polyacrylonitrile, or pitch fibers from which carbon fibers are derived.

prepreg FRP—reinforcement fabrics for FRP laminates that have been preimpregnated with a resin; usually this resin is cured to an intermediate stage (B-staged) and the resulting prepreg is stored at cold temperatures; the cure restarts once the prepreg is brought to room temperature.

relaxation—reduction of stress (or load) in a material under a constant state of strain (or deformation).

relaxation rate—absolute value of the slope of the relaxation curve at a given time. In particular, the relaxation value after 1 million hours is referred to as the million-hour relaxation rate.

repeated load (stress)—load (stress) alternating cyclically between fixed maximum and minimum values.

***S-N* curve**—graphical plot of the repeated load (stress) along a vertical axis versus the number of cycles to fatigue failure on the horizontal axis.

stress—load divided by the cross-sectional area of a lot of specimens. Refer to ASTM D7205/D7205M for the calculation of cross-sectional area.

tail length—length provided beyond the bend portion.

tensile capacity—maximum tensile load carried by test specimen before failure.

tensile strength—ultimate tensile strength of FRP bars in the direction parallel to the fibers.

test section—portion of a specimen between the anchoring sections of the test specimen.

thermoset—resin that is formed by cross-linking polymer chains. Note: A thermoset cannot be melted and reshaped because the polymer chains form a three-dimensional network.

ultimate strain—change in length per unit length corresponding to the tensile capacity.

1.5—Notation

A = cross-sectional area of FRP bar (according to ASTM D7205/D7205M), mm^2
a_1 = empirical constant
b_1 = empirical constant
C_b = equivalent circumference of FRP bar, mm
D = external diameter at deflector surface position, mm
d_b = equivalent diameter of reinforcing bar, mm
E_L = axial (longitudinal) modulus of elasticity of FRP bar, MPa
F = tensile load, N
F_{u1} = tensile capacity before immersion, N
F_{u2} = tensile capacity after immersion, N
F_{ub} = ultimate load capacity according to bend test of FRP bars, N
f_u = ultimate tensile strength parallel to the fibers (according to ASTM D7205/D7205M), MPa
f_{ub} = bend capacity of the FRP stirrup, MPa
ℓ = bonded or overlap length, mm
L_c = length from the top of the embedded bar to the point of attachment of the measuring device, mm
L_t = tail length of bend bar, mm
P = tensile failure load of specimen, N
R_{et} = tensile capacity retention rate, percent
r_t = radius of bend in FRP reinforcement, mm
S = reduced tensile strength of specimen corresponding to a specific corner radius, MPa
S_c = elastic elongation, mm
T = time, h
W_0 = initial mass of the specimen before immersion, g
W_1 = mass of the specimen after immersion for a period (Time 1), g
Y_r = relaxation rate, percent
α = bending angle, degrees
χ = strength reduction factor due to bend effect
$\Delta\varepsilon$ = strain increment
ΔP = tensile load increment, N
τ = bond or shear stress, MPa

PART 2—TEST METHODS FOR FRP BARS

B.1—Test method for cross-sectional properties of FRP bars

This test method has been replaced by ASTM D7205/D7205M, Tensile Properties of Fiber Reinforced Polymer Matrix Composite Bars.

B.2—Test method for longitudinal tensile properties of FRP bars

This test method has been replaced by ASTM D7205/D7205M, Tensile Properties of Fiber Reinforced Polymer Matrix Composite Bars.

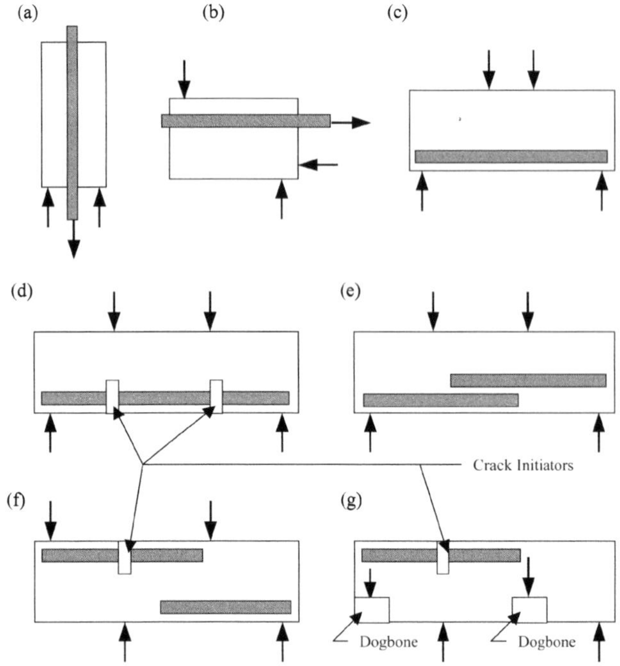

Fig. B.3.1a—Types of test methods for different bond values of FRP reinforcement in concrete: (a) pullout specimen; (b) beam-end specimen; (c) simple beam specimen; (d) hinged beam-end specimen; (e) splice specimen; (f) cantilever beam specimen (without dogbones); and (g) cantilever beam specimen (with dogbones).

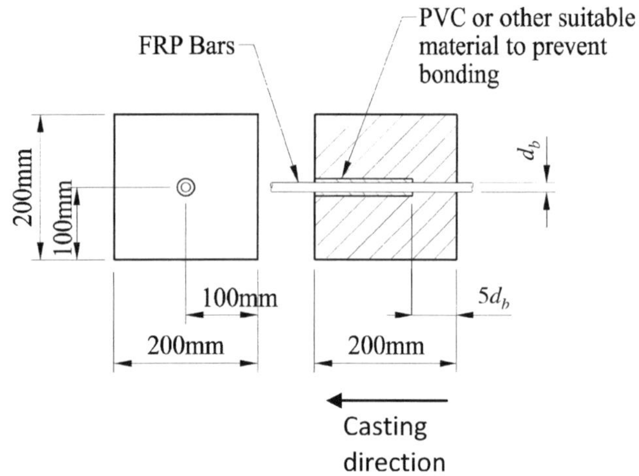

Fig. B.3.1b—Vertical bond test specimen.

B.3—Test method for bond strength of FRP bars by pullout testing

The following references are provided for historical and informational purposes only—Al-Zahrani et al. (1996); Benmokrane et al. (2002); Chaallal and Benmokrane (1993); and Freimanis et al. (1998).

B.3.1 *Scope*—This test method specifies the test requirements for determining the bond strength of FRP bars used as reinforcing bars or prestressing tendons in concrete by pullout testing. Various types of test methods are available for the determination of different bond values of FRP reinforcement in concrete structures, as shown schematically in Fig. B.3.1a. This test method describes the pullout test specimen to determine the bond strength.

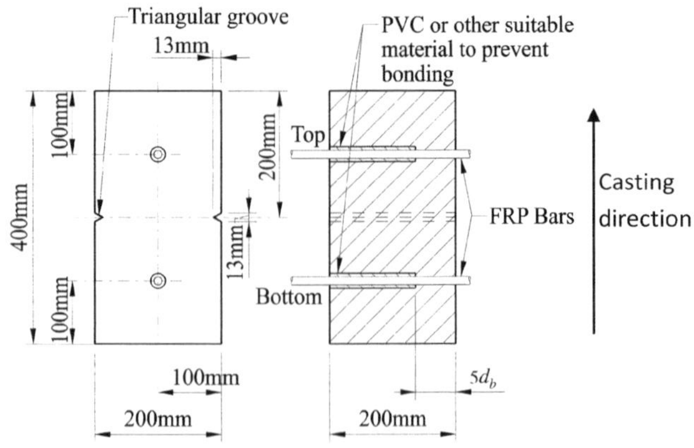

Fig. B.3.1c—Horizontal bond test.

Two methods for casting test specimens are provided. The first method aligns the bar with the concrete casting direction, similar to that of a longitudinal bar in a reinforced concrete column (Fig. B.3.1b). The second method aligns the bars transverse to the concrete casting direction, similar to that of the longitudinal bar in a reinforced concrete beam or slab (Fig. B.3.1c).

B.3.2 *Referenced documents*

ASTM A944-10—Standard Test Method for Comparing Bond Strength of Steel Reinforcing Bars to Concrete Using Beam End Specimens

ASTM C39/C39M-12—Standard Test Method for Compressive Strength of Cylindrical Concrete Specimens

ASTM C143/C143M-10—Standard Test Method for Slump of Hydraulic Cement Concrete

ASTM C192/C192M-07—Standard Practice for Making and Curing Concrete Test Specimens in the Laboratory

ASTM C293/C293M-10—Standard Test Method for Flexural Strength of Concrete (Using Simple Beam with Center Point Loading)

ASTM C511-09—Standard Specification for Mixing Rooms, Moist Cabinets, Moist Rooms, and Water Storage Tanks Used in the Testing of Hydraulic Cements and Concretes

ASTM C617/C617M-11 Standard Practice for Capping Cylindrical Concrete Specimens

ASTM D618-08—Standard Practice for Conditioning Plastics for Testing

ASTM D7205/D7205M-06(2001)—Standard Test Method for Tensile Properties of Fiber Reinforced Polymer Matrix Composite Bars

ASTM E4-10—Standard Practices for Force Verification of Testing Machines

B.3.3 *Significance and use*

B.3.3.1 This test method for measuring bond strength by pullout testing is intended for use in laboratory tests in which the principal variable is the size or type of FRP bars. The test method should not be used to establish design bond values and development lengths for FRP bars embedded in concrete.

B.3.3.2 This test method is intended to determine the bond behavior for material specifications, research and development, and quality assurance. The bond behavior will be specimen-configuration-dependent, which may affect both analysis and design. The primary test result is the bond strength of the specimen to normalweight concrete, which is an important factor to consider in the use of FRP bars as reinforcing bars or tendons.

B.3.3.3 This test method may also be used to determine the conformance of a product or a treatment to a requirement relating to its effect on the bond developed between FRP bar and concrete. The result obtained from this test method should be used only for comparative purposes to compare parameters or variables of bond strength. The method may be used to establish long-term environmental effects on bond to concrete, including environmental reduction factors for FRP bars embedded in concrete.

B.3.4 *Test equipment and requirements*

B.3.4.1 Use a testing machine with a loading capacity in excess of the tensile capacity of the test specimen and calibrated according to ASTM E4. A testing machine with either loading-rate or displacement-rate control is preferred. The load should be applied to the reinforcing bar at a rate no greater than 20 kN/minute or at the no-load speed of the testing machine head of not greater than 1.3 mm/minute, depending on the type of testing machine used and the means provided for ascertaining or controlling testing speed.

B.3.4.2 The loading plate (Fig. B.3.4.2) should be a machined steel plate at least 200 mm square and 20 mm thick, and have a hole drilled through its center of sufficient diameter to accommodate the FRP bar.

B.3.4.3 The loading end of the FRP bar should be fitted with an anchor capable of transmitting loads until the bar is pulled out of the concrete by a bond failure. The load transmission device should only transmit axial loads to the FRP bars, without applying either torsion or bending.

B.3.4.4 The displacement measuring devices fitted to both the free end and loaded end of the FRP bars should be displacement measuring devices (LVDTs) or similar apparatuses, reading accurately to 0.01 mm. Three LVDTs at 120-degree intervals at the loaded end and either one concentric gauge or two gauges at 180-degree intervals at the free end of the bar are recommended (Fig. B.3.4.4).

B.3.4.5 Two types of molds for bond test specimens are required:

1. For 200 mm concrete cubes, each containing a vertically embedded bar
2. For 200 x 200 x 400 mm prisms, each containing two horizontally embedded bars

Preferably, the molds should be made of metal no less than 6 mm thick. The molds should be watertight and constructed for easy removal without disturbing the embedded bars.

B.3.5 *Specimen preparation*

B.3.5.1 FRP bar specimens should be representative of the lot or batch being tested. Each specimen should be cut into 1200 mm long sections and assembled with an anchor

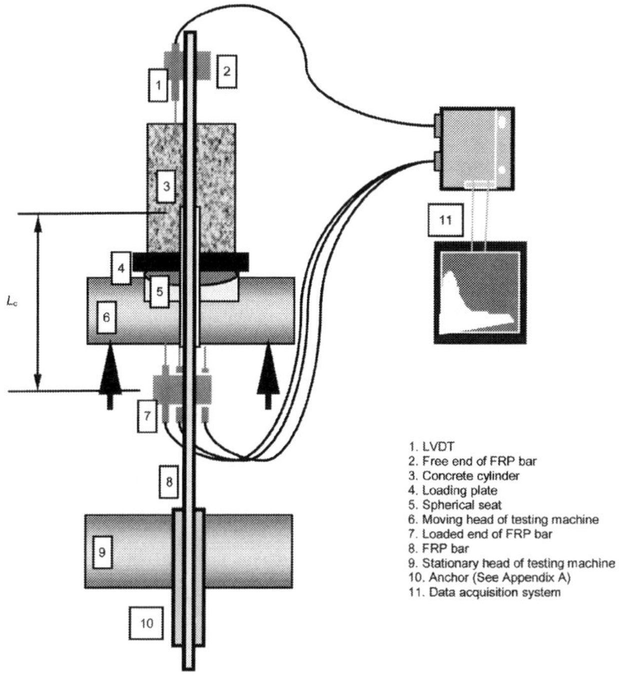

Fig. B.3.4.2—Schematic details of pullout bond test setup.

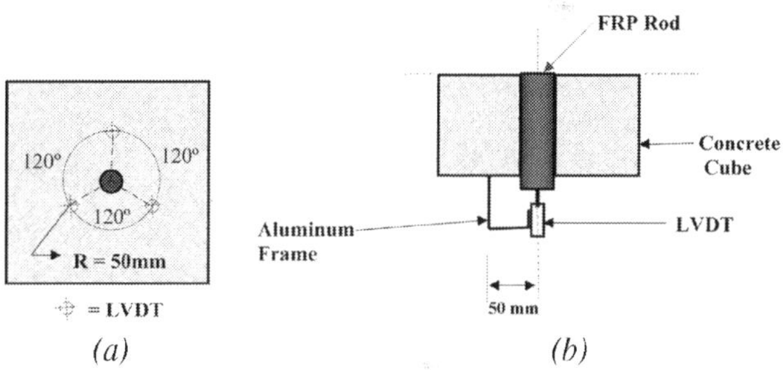

Fig. B.3.4.4—Positions of LVDTs at surface of concrete cube: (a) plan view of three LVDTs at loaded end; and (b) one LVDT at free end.

(ASTM D7205/D7205M) at one end. The test specimens should contain either one FRP bar embedded parallel to the direction of casting of the concrete (Fig. B.3.1b), or two FRP bars embedded perpendicular to the direction of casting of the concrete (Fig. B.3.1c). Five specimens of each type should constitute a set of test specimens. If a specimen has failed or slipped at the anchoring section, or split the concrete cover, an additional test should be performed on a separate specimen taken from the same lot as the failed specimen.

B.3.5.1.1 *Specimens for parallel embedded bar (Fig. B.3.1b)*—These specimens should consist of concrete cubes, 200 mm on each edge, with a single FRP bar embedded vertically along the central axis in each specimen. The bar should project upward from the top face a sufficient length to extend through the bearing blocks and the support of the testing machine, and provide an adequate length to be gripped for application of load. If splitting failure of concrete occurs consistently, a 300 mm cube is needed and new tests should be performed.

B.3.5.1.2 *Specimens for perpendicular embedded bar (*Fig. B.3.1c*)*—These specimens should consist of concrete prisms 200 x 200 x 400 mm, with the longer axis in the vertical direction. Two bars should be embedded in each specimen, perpendicular to the longer axis and parallel to and equidistant from the sides of the prism. In the vertical direction, one bar should be located with its axis 100 mm from the bottom of the prism and the other with its axis 300 mm from the bottom. The bar should project from the concrete face a sufficient length to extend through the bearing blocks and the support of the testing machine, and provide an adequate length to be gripped for application of load. A triangular groove should be formed on each of the two opposite sides of the prism parallel to the bars and at the midheight of the prism. These grooves should be at least 13 mm deep, measured perpendicular to the surface of the concrete. The grooves should facilitate breaking of the prism into two test specimens at the weakened plane before performing the bond tests.

The bonded length of the FRP bar should be five times the diameter of the FRP bar. If the bonded length, as defined previously, does not represent the bonding characteristics of the FRP bar, the bonded length may be extended as appropriate. Outside of the bonded section, the embedded bar should be sheathed with polyvinyl chloride (PVC) or other suitable material to prevent bonding. At the free end, bars should protrude from the concrete slightly so that the end of the LVDT(s) will bear on the bar (Fig. B.3.4.4).

B.3.5.2 Place the bars into the molds as follows:

a) The opening in the form through which the FRP bar is inserted should be sealed using oil, putty, or similar materials to prevent egress of water and other substances

b) The orientation of the specimen should not be changed until the form is removed

B.3.5.3 Before casting the test specimens, coat the inside surface of the molds with a thin film of mineral oil, petroleum jelly, or stearic acid paste. The following procedures are recommended for placement of concrete in the molds unless another well-established method is used.

a) For 200 x 200 x 400 mm prisms, place the concrete in four layers of approximately equal thickness and rod each layer 25 times with a 16 mm diameter tamping rod

b) For 200 mm cubes, place the concrete in four layers of approximately equal thickness and rod each layer 25 times with a 16 mm diameter tamping bar

c) After the top layer has been consolidated, strike off the surface with a trowel and protect against moisture evaporation by one of the acceptable methods described in 7.1 of ASTM C192/C192M. Care should be taken that evaporation does not take place in the area adjacent to the protruding FRP bar for vertically cast specimens

B.3.5.4 The concrete should be a standard mixture, with coarse aggregates having a maximum dimension of 20 to 25 mm. It should be batched and mixed in accordance with the applicable portions of ASTM C192/C192M. The concrete should have slump of 100 ± 20 mm in accordance with ASTM C143/C143M, and the compressive strength at 28 days should be 30 ± 3 MPa in accordance with ASTM C39/C39M. A minimum of five standard 150 x 300 mm or 100 x 200 mm control cylinders should be made for determining compressive strength from each batch of concrete.

B.3.5.5 Molds should not be removed from the specimens earlier than 20 hours after casting. Extreme care should be taken to prevent striking or otherwise disturbing the FRP bars. Immediately after removing the molds, specimens should be cured in accordance with ASTM C192/C192M until the time of testing. Specimens should be tested at an age of 28 days.

B.3.5.6 When the specimens are between 7 and 14 days old, the 200 x 200 x 400 mm prisms should be broken in half in flexure to form two 200 mm cubes. Specimens should be broken as simple beams with center-point loading in accordance with ASTM C293/C293M. The two triangular grooves in the upper and lower faces of the prisms should be located at midspan. The load should be applied to a 19 mm diameter bar laid in the upper groove until fracture occurs. Care should be taken not to strike or otherwise disturb the FRP bars during the operation.

B.3.5.7 The surface of the 200 mm cube containing the vertically embedded bar should be capped so as to use it as the bearing surface in the pullout test. The applicable portions of ASTM C617, relative to capping materials and procedures, should be followed.

B.3.6 *Conditioning*—Unless a different testing environment is specified, the pullout tests should be conducted at the standard laboratory atmosphere 23 ± 3°C and 50 ± 10 percent relative humidity.

Preconditioning of FRP bar specimens before casting in concrete, such as post-production machining, abrading, or other such processing, is permitted but should be reported.

B.3.7 *Test method*

B.3.7.1 The specimen should be mounted in the testing machine in one of the following two test setups.

B.3.7.1.1 The capped or bearing surface of the cube from which the long end of the bar projects should be in contact with the bearing block (or plaster pad) assembly. The spherically seated bearing block should rest on a support that transfers the reaction from this block to the load cell of the testing machine. The projecting FRP bar should extend through the bearing block assembly and the support, and the anchor should be gripped for tension by the jaws of the testing machine (Fig. B.5.7.1). The free end of the bar may also be potted in an anchor as described in ASTM D7205/D7205M.

B.3.7.1.2 The concrete cube should be fixed on the stationary head of the testing machine. LVDTs at the loaded end and free end are attached to measure the slip of the FRP bar, as shown in Fig. B.3.4.4. The anchor is then threaded into or gripped by the connection on the moving head of the testing machine.

B.3.7.2 Assemble the testing apparatus on the specimen. Carefully measure and record, to the nearest 0.5 mm, the distance between the top surface of the bonded length and the point of attachment of the measuring device on the FRP bar. The elongation of the FRP bar over this distance may be calculated and subtracted from the measured slip plus elongation to obtain the loaded-end slip. Moreover, free-end slip shall be measured to the nearest 0.5 mm.

B.3.7.3 Apply load to the FRP bar at a load rate no greater than 20 kN/minute, or at a testing machine head speed not greater than 1.3 mm/minute.

B.3.7.4 Read and record the applied load and the LVDT readings at a sufficient number of intervals throughout the test to provide at least 15 readings by the time a slip of 0.25 mm has occurred at the loaded end of the FRP bar. The slippage of the free end should be recorded in increments of 0.01 mm, together with the corresponding applied load.

B.3.7.5 Continue the loading and readings at appropriate intervals until rupture of the FRP bar occurs or slippage of at least 2.5 mm occurs at the loaded end of the embedded length.

B.3.7.6 For cases where a specimen is judged to have undergone a tensile failure at an anchoring section, to have slipped out of an anchoring section before the FRP bar has slipped from the concrete, or where the load is significantly reduced due to splitting or cracking of the concrete, the data should be disregarded and additional tests should be performed until the number of valid tests is not less than five.

B.3.8 *Calculations*

B.3.8.1 The average bond stress should be calculated according to Eq. (B.3.8.1) and reported with a precision to three significant digits, and the data for the pullout or bond stress versus slippage at both free-end and loaded-end displacement for each specimen should be plotted.

$$\tau = \frac{F}{C_b \ell} \qquad \text{(B.3.8.1)}$$

where C_b is π times the effective bar diameter from ASTM D7205/D7205M.

B.3.8.2 Average bond stresses causing slippage at the free end and the loaded end of 0.05 mm, 0.10 mm, and 0.25 mm, and the maximum bond stress (the bond strength) at failure, should be calculated.

B.3.8.3 At each load level, the slip at the loaded end should be calculated as the average of the readings of the LVDTs, minus the elongation S_c of the FRP bar in the length L_c between the top surface of bonded length and the point of attachment of the measuring devices on the FRP bar, the latter being calculated using Eq. (B.3.8.3)

$$S_c = \frac{FL_c}{E_L A} \qquad \text{(B.3.8.3)}$$

B.3.9 *Test report*—The test report should include the following items.

B.3.9.1 *Properties of concrete*—The following concrete properties should be included in the test report.

a) The mixture proportions of cement, fine aggregate, coarse aggregate, admixture (if used), and the *w/c*

b) Slump of freshly mixed concrete as determined in accordance with ASTM C143/C143M

c) Twenty-eight-day strength of control cylinders as determined in accordance with ASTM C39/C39M

d) Any deviation from the stipulated standards in such aspects as mixing, curing, dates of demolding, and testing of control cylinders

B.3.9.2 *Properties of FRP bar*—The following FRP bar properties should be included in the test report.

a) The trade name, shape, and date of manufacture (if available), and lot number of product tested

b) Type of fiber and fiber binding material (as reported by the manufacturer), fiber volume fraction, surface treatment, and preconditioning of FRP bar

c) Designation, equivalent diameter, and cross-sectional area, as determined according to ASTM D7205/D7205M

d) Modulus of elasticity and ultimate tensile strength, as determined in accordance with ASTM D7205/D7205M

e) A close-up photograph of the bars showing surface deformations and characteristics

B.3.9.3 *Additional items*—The following additional items should be included in the test report.

a) Numbers or identification marks on test specimens

b) Date of test, test temperature, and loading rate

c) Dimensions of test specimens and bonded length of FRP bar

d) A brief description of the gripping device

e) Average bond stress causing slippage at the free end of 0.05, 0.10, and 0.25 mm for each specimen

f) Average bond stress causing slippage at the loaded end of 0.05, 0.10, and 0.25 mm for each test specimen

g) Maximum bond stress, failure mode, and averages for each test specimen

h) Bond stress-slippage displacement (free-end and loaded-end) plots for each test specimen

B.4—Test method for transverse shear strength of FRP bars

This test method has been replaced by ASTM D7617/D7617M, Standard Test Method for Transverse Shear Strength of Fiber Reinforced Polymer Matrix Composite Bars.

B.5—Test method for strength of FRP bent bars and stirrups in bend locations

The following reference is provided for historical and informational purposes only—Morphy et al. (1997).

B.5.1 *Scope*—This test method specifies test requirements for the strength of FRP bent bars used as an anchorage for stirrups in concrete structures.

B.5.2 *Referenced documents*

ASTM International

ASTM C39/C39M-12—Standard Test Method for Compressive Strength of Cylindrical Concrete Specimens

ASTM C143/C143M-10—Standard Test Method for Slump of Hydraulic Cement Concrete

ASTM C192/C192M-07—Standard Practice for Making and Curing Concrete Test Specimens in the Laboratory

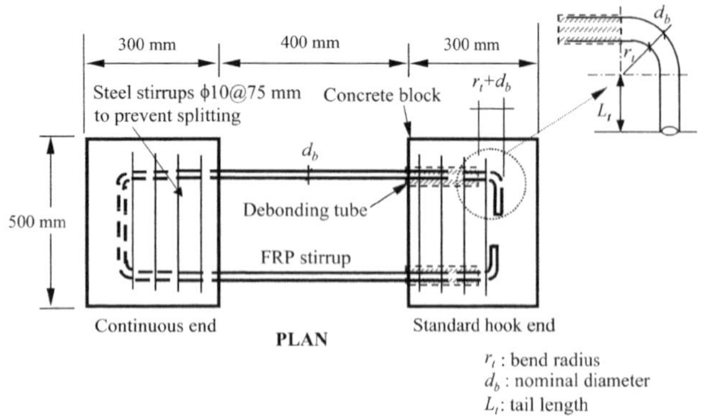

Fig. B.5.5.1—Configuration of specimen.

ASTM D7205/D7205M-06(2001)—Standard Test Method for Tensile Properties of Fiber Reinforced Polymer Matrix Composite Bars

ASTM E4-10—Standard Practices for Force Verification of Testing Machines

B.5.3 *Significance and use*

B.5.3.1 This test method is intended for use in laboratory tests to determine the strength of the bent portion provided as an anchorage in which the principal variable is the size, bend radius, or type of FRP stirrup.

B.5.3.2 Bending of FRP stirrups to develop anchorage leads to a significant reduction in the strength of the stirrups. The bend radius and tail length beyond the bend are important factors affecting the bend capacity.

B.5.3.3 This test method measures the ultimate load capacity of a single FRP stirrup subjected to tensile forces in the direction of the straight portion.

B.5.3.4 This test method is intended to determine the bend capacity and strength reduction for material specifications, research and development, quality assurance, and structural design and analysis. The behavior of bent bars and stirrups should be measured according to the method given herein, in keeping with the intended purposes.

B.5.4 *Test equipment and requirements*—The hydraulic cylinder and load cell should be calibrated according to ASTM E4, have a loading capacity in excess of the capacity of the specimen, and be capable of applying load at the required loading rate. The load cell should also be capable of giving readings accurate to within 1 percent throughout the test.

B.5.5 *Specimen preparation*

B.5.5.1 The configuration of a typical specimen is shown in Fig. B.5.5.1. The dimensions of each concrete block used to anchor the FRP stirrup may be varied according to the dimensions of the stirrup used. The free length of the stirrup between the two blocks, however, should not be less than 200 mm, with 400 mm suggested. The concrete block should be reinforced using steel stirrups, as shown in Fig. B.5.5.1, to prevent splitting of the concrete block before rupture of the FRP stirrup in the bend. The tail length L_t of the FRP stirrup tested to evaluate the bend capacity should not exceed 150 mm. The debonding tube is used to eliminate the straight-bar development of the hooked bar. The debonding tube should slip over the reinforcing bar, and the ends of the debonding tube should be filled with caulk to prevent the tubes from filling with concrete during casting.

B.5.5.2 The concrete should be a standard mixture, with coarse aggregates having a maximum dimension of 20 to 25 mm. It should be batched and mixed in accordance with the applicable portions of ASTM C192/C192M. The concrete should have slump of 100 ± 20 mm in accordance with ASTM C143/C143M, and the compressive strength at 28 days should be 30 ± 3 MPa in accordance with ASTM C39/C39M. A minimum of five standard 150 x 300 mm or 100 x 200 mm control cylinders should be made for determining compressive strength from each batch of concrete.

B.5.5.3 The number of valid test results for each test condition should not be less than five. If a specimen fails by splitting of the concrete block, the test result should be disregarded and an additional test should be performed on a separate specimen taken from the same lot as the failed specimen.

B.5.5.4 If test specimens fail due to pullout of the bent bar from the concrete, this is an indication that the bend radius and tail length are inadequate for the bar being tested. It is necessary to adjust these parameters (perhaps the size of the test blocks as well) and retest.

B.5.6 *Conditioning*—Unless a different testing environment is specified as part of the experiment, the tests should be conducted at the standard laboratory atmosphere: 23 ± 3°C and 50 ± 10 percent relative humidity.

Preconditioning of FRP bars before casting in concrete is permissible but must be reported.

B.5.7 *Test method*

B.5.7.1 The test setup, shown in Fig. B.5.7.1, consists of a hydraulic jack to apply the relative displacement between the two concrete blocks and a load cell to measure the applied load. Place the steel plates and plaster bags in front of the load cell and the hydraulic jack to distribute the applied load to the surface of the concrete. A spherical washer can also be used at the end of the ram. The block containing the test section of the bar should be placed on top of steel rollers to minimize the friction forces between the block and testing bed.

B.5.7.2 Tensile strength of straight FRP bars with the same diameter as the FRP stirrups should be evaluated according to ASTM D7205/D7205M.

B.5.7.3 The test specimens should not be subjected to any shock, vibration, or torsion during the test. Increase the force in the jack in a smooth, continuous manner until the specimen fails. Do not pause the application of load during the test. The loading rate should be selected so that the specimen fails between 1 and 10 minutes from the start of the test.

B.5.7.4 Record the failure load and failure mode for the specimen.

B.5.8 *Calculations*

B.5.8.1 The bend capacity of the FRP stirrup should only be assessed on the basis of the specimen undergoing failure in the bend. In cases where block splitting has clearly taken place, the data should be disregarded, and additional tests should be performed until the number of the test specimens failing in the bend is not less than five.

B.5.8.2 The bend capacity of the FRP stirrup should be calculated according to Eq. (B.5.8.2), and rounded to three significant digits

$$f_{ub} = \frac{F_{ub}}{2A} \quad \text{(B.5.8.2)}$$

B.5.8.3 The strength-reduction factor is calculated according to Eq. (B.5.8.3)

$$\chi = \frac{f_{ub}}{f_u} \quad \text{(B.5.8.3)}$$

B.5.9 Test *report*—The test report should include the following items.

B.5.9.1 *Properties of concrete*—The following concrete properties should be included in the test report:

a) The mixture proportions of cement, fine aggregate, coarse aggregate, admixture (if any used), and *w/c*

b) Slump of freshly mixed concrete as determined in accordance with ASTM C143/C143M

c) Twenty-eight-day strength of control cylinders as determined in accordance with ASTM C39/C39M

d) Any deviation from the stipulated standards in such aspects as mixing, curing, dates of demolding, and testing of control cylinders

B.5.9.2 *Additional items*—The following additional items should be included in the test report.

a) Trade name, shape, and date of manufacture (if available), and lot number of FRP bar tested for stirrups

b) Type of fiber and matrix used in the FRP stirrup, and fiber volume fraction

c) Process used to fabricate the stirrups, as reported by the manufacturer

d) Numbers or identification marks on test stirrups

e) Designation, diameter, and cross-sectional area

f) Dimensions of concrete block, configuration (diameter and space) of steel stirrup confinement, debonded length, bend radius, and tail length of the bent bar.

g) Preconditioning of FRP bars before casting

h) Date of test and test temperature

i) Type and capacity of load cell

j) Bend capacity and strength reduction factor for each test stirrup

k) Average bend capacity and strength reduction factor for all specimens that failed in the bend as intended

B.6—Accelerated test method for alkali resistance of FRP bars

The following references are provided for historical and informational purposes only—Benmokrane et al. (2001, 2002); Gangarao and Vijay (1997); Gentry (2001); Mehus (1995); Nkurunziza et al. (2002); and Porter et al. (1995, 1996).

B.6.1 *Scope*—This test method specifies the test requirements for evaluating alkali resistance of FRP bars used as reinforcing bars in concrete and masonry by immersion in aqueous alkaline solution. Alkali resistance is measured by subjecting the FRP bars to an alkali environment, with or

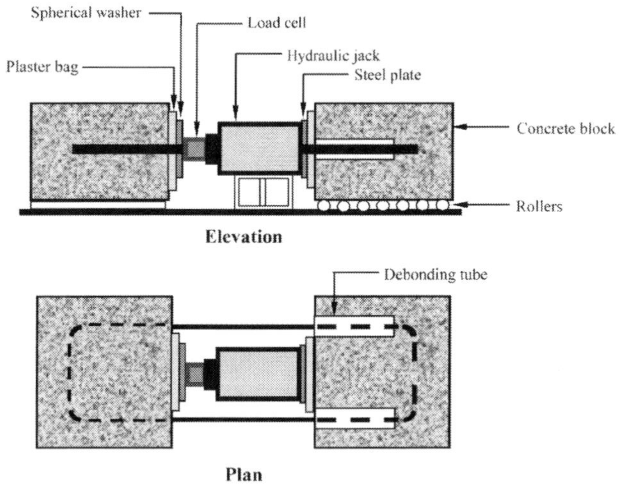

Fig. B.5.7.1—Test setup.

without stress, and then testing them in tension according to ASTM D7205/D7205M.

This test method presents three procedures conducted at a temperature of 60°C, each defining different loading conditions.

a) *Procedure A*—A system in which FRP specimens are immersed in the alkaline solution with no tensile load applied. The test control parameters are the pH value and temperature of the alkaline solution and immersion time.

b) *Procedure B*—A system in which FRP specimens are immersed in the alkaline solution under sustained tensile load. The test parameters are the sustained load level, the pH value, and temperature of the alkaline solution and immersion time.

c) *Procedure C*—A system in which FRP specimens, surrounded by moist concrete, are subjected to a sustained tensile load. The test parameters are the sustained load level, the pH value, and temperature of concrete and embedded time.

B.6.2 *Referenced documents*

ASTM International

ASTM C192/C192M-07—Standard Practice for Making and Curing Concrete Test Specimens in the Laboratory

ASTM C511-09—Standard Specification for Mixing Rooms, Moist Cabinets, Moist Rooms, and Water Storage Tanks Used in the Testing of Hydraulic Cements and Concrete

ASTM D618-08—Standard Practice for Conditioning Plastics for Testing

ASTM D5229/D5229M-92(2010)—Standard Test Method for Moisture Absorption Properties and Equilibrium Conditioning of Polymer Matrix Composite Materials

ASTM D7205/D7205M-06(2011)—Standard Test Method for Tensile Properties of Fiber Reinforced Polymer Matrix Composite Bars

ASTM E4-10—Standard Practices for Force Verification of Testing Machines

B.6.3 *Significance and use*

B.6.3.1 This test method for investigating the alkali resistance of FRP bars is intended for use in laboratory tests

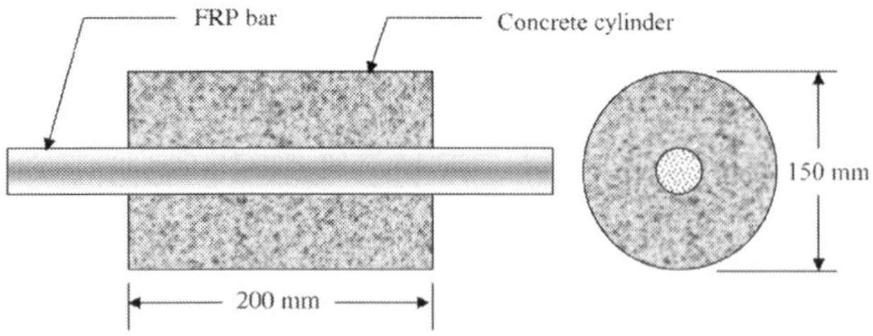

Fig. B.6.6.3—Dimensions of concrete cylinder.

in which the principal variables are the temperature and concentration of alkaline solution, the type of FRP bars, and the sustained load level.

B.6.3.2 This test method measures the mass change and tensile capacity after immersion of FRP bars in alkaline solution without stressing (Procedure A), and the tensile capacity after immersion of FRP bars in an alkaline solution and embedment of FRP bars in moist concrete under sustained load condition (Procedures B and C).

B.6.3.3 This test method is intended to determine the alkaline-resistant data for material specifications, research and development, quality assurance, and structural design and analysis. The alkaline resistance should be measured according to the method given herein. The primary test result is the mass change and tensile capacity retention of the test specimen, which are important factors to consider in the use of FRP bars.

B.6.3.4 The level of sustained loading for Procedures B and C is not specified as part of the test method. Typically, the stress in the sustained load tests should be equal to the stress caused by the dead loads, and any part of the live loads that is sustained. If service load conditions are not known, the sustained tensile stress in glass FRP bars should be set to induce a tensile strain equal to 2000 microstrain. Higher levels of sustained stress can be used as an accelerating condition. The level of sustained stress should be reported.

B.6.4 *Test equipment and requirements*—An analytical balance capable of the appropriate accuracy in accordance with Procedure A of ASTM D618 should be used. The testing machine and devices should conform to ASTM E4.

B.6.5 *Specimen preparation*

B.6.5.1 FRP bar specimens should be representative of the lot or batch being tested. In general, test specimens should not be subjected to any processing beyond manufacturing. For grid-type FRP bars, linear test specimens may be prepared by cutting away extraneous material in such a way as not to affect the performance of the tested part.

B.6.5.2 During the sampling and preparation of test specimens, all deformation, heating, outdoor exposure to ultraviolet light, and other conditions possibly causing changes to material properties of the specimen should be avoided.

B.6.5.3 The length of the specimen should be the sum of the length of the test section and the lengths of the anchoring sections. The length of the test section should not be less than 40 times the diameter of the FRP bar. For FRP bar in twisted strand form, the length of the test section should also be greater than two times the strand pitch.

B.6.5.4 The number of test specimens for pre- and post-immersion testing should not be less than five. Each specimen should be clearly labeled with identifying markings.

B.6.5.5 Coat the ends of bars and the ends and transverse elements of grids with epoxy resin to prevent the infiltration of solution via these cuts. Allow resin to cure completely before immersion.

B.6.5.6 The alkaline solution in Procedures A and B should be a composition representative of the pore water inside portland-cement concrete. The suggested composition of alkaline solution consists of 118.5 g of $Ca(OH)_2$, 0.9 g of NaOH, and 4.2 g of KOH in 1 L of deionized water. The solution should have a pH value of 12.6 to 13—a representative pH value of mature concrete pore solution. The alkaline solution should be covered before and during testing to prevent interaction with atmospheric CO_2 and to prevent evaporation.

B.6.6 *Conditioning*

B.6.6.1 Samples for Procedure A should be immersed in the alkaline solution at 60 ± 3°C for exposure times of 1, 2, 3, 4, and 6 months, unless longer exposure periods are specified. After the allotted times, the samples should be removed from the alkaline solution, thoroughly washed in deionized water, towel dried, weighed, and tested in tension to failure.

B.6.6.2 Samples for Procedure B should be installed in anchoring devices at both ends in accordance with ASTM D7205/D7205M. The test section of the specimen should be immersed in the alkaline solution inside an environmental cabinet or container holding the alkaline solutions and having a constant temperature of 60 ± 3°C. The specimen should be held in a loading fixture to subject to a constant tensile sustained load for exposure times of 1, 2, 3, 4, and 6 months, unless longer exposure periods are specified.

B.6.6.3 Samples for Procedure C should be prepared by embedding the test section of the specimens in moist concrete. Typical dimensions of the concrete cylinder are shown in Fig. B.6.6.3. The 150 mm dimension of the specimens may be increased if larger-diameter bars are used. The concrete should be a standard mixture, specified in Test Method B.3. The concrete mixture should be batched and mixed in accordance with the applicable portions of ASTM C192/C192M, and the curing procedure should conform to ASTM C511. After 28 days of curing in water, the samples should have installed anchors at both their ends in accordance with ASTM D7205/D7205M, and then positioned in the conditioning fixture for subjecting to a sustained tensile load for exposure times of 1, 2, 3, 4, and 6 months, unless longer exposure periods are specified. The concrete cylinder should be kept moist and inside an environmental cabinet having a constant temperature of 60 ± 3°C during the testing. The FRP specimens should be tested with the concrete cylinder still attached.

B.6.7 *Test method*

B.6.7.1 The pH value of the alkaline solution should be measured at the beginning of the test and after the alkali-resistance test. During immersion of the test specimens, the pH value of the alkaline solution should be monitored at least every 5 days and adjusted, if necessary, to keep the same constituents and pH value as the beginning of the test.

B.6.7.2 The external appearance of the specimen should be examined before and after the alkali resistance test for comparison of color, surface condition, and change of shape. If necessary, the specimen may be sectioned and polished, and the condition of the cross section examined under a microscope.

B.6.7.3 *Mass change testing of specimens for Procedure A*—Before immersion, the specimen should be dried until its mass is unchanged, according to Procedure D of ASTM D5229/D5229M, which is the initial mass W_0. After immersion for the prescribed period of time, the specimen should be removed from the alkaline solution, quickly washed with deionized water, dried with tissue paper, and then immediately weighed, which is the mass at Time 1, denoted as W_1. Then the specimen should have anchors installed at both ends for tensile capacity retention testing.

B.6.7.4 *Tensile capacity retention testing of specimens for Procedures A, B, and C*—Specimens should be tested in tension to failure within 24 hours after removal from the conditioning environment at the prescribed exposure length of time. The tensile test method should follow ASTM D7205/D7205M.

B.6.8 *Calculations*

B.6.8.1 The mass change of FRP bars should be calculated according to Eq. (B.6.8.1a) and (B.6.8.1b).

$$\text{mass gain (\%)} = \frac{W_1 - W_0}{W_0} \times 100 \quad \text{(B.6.8.1a)}$$

$$\text{mass loss (\%)} = \frac{W_0 - W_1}{W_0} \times 100 \quad \text{(B.6.8.1b)}$$

B.6.8.2 The material properties of FRP bars should be assessed only for those test specimens undergoing failure in the test section. In cases when tensile failure or slippage has occurred at an anchoring section, the data should be disregarded and additional tests performed from the same conditioning lot as the failed specimen. The tensile capacity retention should be calculated according to Eq. (B.6.8.2), with a precision to two significant digits.

$$R_{et} = \frac{F_{u2}}{F_{u1}} \times 100 \quad \text{(B.6.8.2)}$$

B.6.9 *Test report*—The test report should include the following items.

B.6.9.1 *Common items*—The test report should include the following common items:

a) The trade name, shape, and date of manufacture (if available), and lot number of product tested

b) Type of fiber and fiber binding material, as reported by the manufacturer, and fiber volume fraction

c) Numbers or identification marks on test specimens

d) Designation, diameter, and cross-sectional area

e) Date of start and end of immersion

B.6.9.2 *Items related to alkaline solution immersion*—The test report should include the following items related to alkaline solution immersion:

a) Composition of alkaline solution, pH, temperature, immersion period, and time

b) Sustained load level, time, and procedure of solution monitoring and adjusting

c) Record of observation of external appearance and mass change

d) Specimen mass at each time interval

e) Plots of percent mass change versus time, calculated by Eq. (B.6.8.1a) or (B.6.8.1b)

B.6.9.3 *Items related to tensile testing*—The test report should include the following items related to tensile testing:

a) Test temperature and loading

b) Tensile capacities for immersed and nonimmersed test specimens at the 1-, 2-, 3-, 4-, and 6-month intervals, with averages and standard deviations of tensile capacities and tensile strength

c) Modulus of elasticity and the average for all immersed and nonimmersed test specimens, respectively

d) Ultimate strain for all immersed and nonimmersed test specimens and average ultimate strain

e) Tensile capacity retention

f) Stress-strain curves for all immersed and nonimmersed test specimens

g) Plot of tensile capacity retention versus time of exposure

B.7—Test method for tensile fatigue of FRP bars

The following references are provided for historical and informational purposes only—Adimi et al. (1998); Odagiri et al. (1997); and Rahman and Kingsley (1996).

B.7.1 *Scope*—This test method specifies the test requirements for tensile fatigue under a constant tensile load range for FRP bars used as reinforcing bars or prestressing tendons in concrete and masonry.

The specimens should be linear or grid FRP formed from continuous fibers in such a manner as to act mechanically as a monolithic body.

Various types of fatigue testing, such as tension-tension; tension-compression; compression-compression; and various combinations of tension, compression, and shear are possible. The tension-tension test method given herein is considered to be the most basic for evaluating material characteristics. The test control parameter is the load, and the loading machine is controlled so that the specimen is subjected to repetitive constant load range. In this procedure, the test control parameter may be described using either stress or applied load as a constant amplitude fatigue variable.

B.7.2 *Referenced documents*

ASTM International

ASTM D618-08—Standard Practice for Conditioning Plastics for Testing

ASTM D3039/D3039M-08—Standard Test Method for Tensile Properties of Polymer Matrix Composite Materials

ASTM D3479/D3479M-96(2007)—Standard Test Method for Tension-Tension Fatigue of Polymer Matrix Composite Materials

ASTM D7205/D7205M-06(2011)—Standard Test Method for Tensile Properties of Fiber Reinforced Polymer Matrix Composite Bars

ASTM E4-10—Standard Practices for Force Verification of Testing Machines

ASTM E739-10—Standard Practice for Statistical Analysis of Linear or Linearized Stress Life (S-N) and Strain-Life (ε-N) Fatigue Data

B.7.3 *Significance and use*

B.7.3.1 This test method is intended for use in laboratory tests in which the principal variable is the maximum value of the repeated load.

B.7.3.2 This test method is intended to develop the tensile fatigue data for material specifications, research and development, quality assurance, and structural design and analysis. The primary test result is the fatigue life of the specimen under a specific loading and environmental condition. Guidance in statistical analysis of fatigue life data, such as determination of linear or linearized stress life (*S-N*), can be found in ASTM E739.

B.7.3.3 Fatigue properties of reinforced or prestressed concrete and masonry structures are important factors to consider in design. For FRP bars used as reinforcing bars or tendons, the fatigue behavior should be measured according to the method given herein, in keeping with the intended purposes. Factors that can affect the fatigue lifetime of an FRP bar include the maximum load, the load ratio (minimum load divided by maximum load), the wave shape (sinusoidal, triangular, square, and so on), the rate of loading (Hz), the environment (such as moisture and temperature), and the method of material conditioning and specimen preparation. These factors should be reported.

B.7.3.4 This test method determines the number of repeated loading cycles required to fail an FRP bar. Such data can be used to create *S-N* curves for a particular set of testing conditions where the principal variable is the maximum value of the repeated load. Because FRP bars often have slight variations in cross-sectional area from one axial position to another, the area of a particular lot of bars is used to compute the stress in those specimens.

B.7.3.5 This test method can be used in the study of fatigue damage in an FRP bar, such as the occurrence of microscopic cracks, fiber fractures, or delaminations. The specimen's residual strength, stiffness, or both, may change due to these damage mechanisms. The loss in strength associated with fatigue damage may be determined by discontinuing cyclic loading to obtain the quasi-static tensile strength using ASTM D7205/D7205M. The loss in stiffness may be quantified by discontinuing cyclic loading at selected cycle intervals to obtain the quasi-static axial tensile stress-strain using modulus determination described in ASTM D7205/D7205M.

B.7.4 *Test equipment and requirements*

B.7.4.1 The testing machine should generally conform to ASTM E4 and be capable of maintaining constant load amplitude, maximum and minimum repeated load, and frequency. The testing machine should be fitted with a counter capable of recording the number of cycles to failure of the specimen. The load indicator should be capable of measuring loads with an accuracy of not less than 1 percent of the load range.

B.7.4.2 Anchorages should be in accordance with ASTM D7205/D7205M. Ideally, the same type of anchorage should be used for all specimens in a given series of tests.

B.7.4.3 If strain measurements are required as the result of the fatigue tests, an extensometer capable of maintaining an accuracy of ±1 percent of the indicated value should be used. The extensometer should allow a gauge length that includes an integer number of characteristic lengths of the specimen. The integer should be at least two, but possibly greater than two to obtain a gauge length of at least 50 mm.

B.7.5 *Specimen preparation*

B.7.5.1 Specimens should be representative of the lot or batch being tested. During the sampling and preparation of test specimens, all deformation, heating, outdoor exposure to ultraviolet light, and other conditions possibly causing changes to the material properties of the specimen should be avoided.

B.7.5.2 The specimen should be prepared and handled in accordance with ASTM D7205/D7205M.

B.7.5.3 There should be a minimum of five test specimens for each load (stress) level. At least five loading levels should be used to construct a plot of stress versus load cycles to failure. If a specimen fails at or slips out of an anchoring section, an additional test should be performed on a separate specimen taken from the same lot as the failed specimen.

B.7.5.4 The total length of the specimen should be 40 bar diameters plus the total gripped length of bar. The bar diameter is found by referring to ASTM D7205/D7205M.

B.7.6 *Conditioning*

B.7.6.1 *Standard conditioning procedure*—Condition specimens in accordance with Procedure A of ASTM D618. Store and test at the standard laboratory atmosphere—23 ± 3°C and 50 ± 10 percent relative humidity—unless a different environment, such as temperature or alkaline solution, is specified as part of the experiment.

B.7.6.2 *Alternate preconditioning procedure*—If another preconditioning procedure is used, the sequence of that preconditioning during specimen preparation can be important. Report if the specimen was preconditioned before or after anchors were applied.

B.7.7 *Test method*

B.7.7.1 The mounting of test specimens should be in accordance with ASTM D7205/D7205M.

B.7.7.2 For purposes of determining an *S-N* curve, the maximum and minimum loads may be set by one of the following methods:

a) Fix the average load and vary the load amplitude

b) Fix the minimum load and vary the maximum load

c) Fix the load ratio and vary the maximum and minimum load according to this fixed ratio

The method adopted should be determined according to the purpose of the test. In any case, a minimum of five load (stress) levels should be chosen such that the range of the number of cycles to failure is between 1000 and 2×10^6. Typical *S-N* curves for FRP materials are generated using Method C and use a fixed load (stress) ratio *R* of 0.1.

B.7.7.3 The following procedure may be used where the maximum stress level for the initial test is difficult to determine:

a) Select an appropriate maximum stress in the range of 20 to 60 percent of the quasi-static tensile strength, and commence fatigue testing with this value as the repeated maximum load.

b) If the specimen does not fail after 10^4 cycles at this repeated maximum stress, add 5 percent of the quasi-static tensile strength and perform the test uninterrupted using the same specimen.

c) If failure does not occur after 10^4 cycles following Step b, a further 5 percent should be added to the repeated maximum load.

d) Repeat Step c until the specimen fails.

e) The initial maximum stress applied to an untested specimen can be set at the previously found maximum load minus 5 percent of the quasi-static tensile strength.

For tendons for prestressing applications, the stress levels of interest can be in the range of 50 to 75 percent of the quasi-static tensile strength. For bars for reinforced concrete, however, the stress levels of interest may be 15 to 30 percent of the quasi-static tensile strength.

B.7.7.4 The loading frequency should be within the range of 1 to 10 Hz, preferably 4 Hz. The temperature of the bar can increase due to internal damping. Such temperature changes can affect the fatigue performance of the bar. Thus, if temperature changes are likely to occur or if any doubts about temperature exist, temperature should be monitored with a suitable device or indicator, such as a contact thermocouple or infrared thermocouple. The temperature should be monitored for newly developed FRP materials and in any case where the loading frequency exceeds 4 Hz. If measured, temperature changes should be reported. The waveform of the load signal (sinusoidal, triangular, and square) should be held constant in any series of fatigue tests that are to be compared to each other in an *S-N* curve, unless the purpose of the tests is specifically to evaluate the effects of waveform.

B.7.7.5 Static load should be applied up to the average load, after which repeated loading should begin at the prescribed frequency. The prescribed load should be introduced as rapidly as possible without dynamic effects. The maximum and minimum repeated loads should not change for the duration of the test. Counting the number of cycles should normally commence when the load on the specimen has reached the prescribed maximum load.

B.7.7.6 Complete separation (breaking) of the specimen should be deemed to constitute failure. The number of cycles to failure should be recorded. If the specimen does not fail after 2×10^6 cycles, the test may be discontinued. A specimen that does not fail should not be reused.

B.7.7.7 Tests for each specimen should normally be conducted without interruption from the start of the test to the end of the test. When a test is interrupted, the number of cycles up to the time of interruption and the period of the interruption should be recorded.

B.7.8 *Calculations*

B.7.8.1 Five valid test specimens are required at each load level. A valid specimen is one that does not fail in the anchor, fails in the gauge length at a cycle count of between 1000 and 2×10^6 cycles, or does not fail after 2×10^6 cycles.

B.7.8.2 If an *S-N* curve is desired, the curve should be plotted in accordance with ASTM E739, with maximum repeated stress, stress range, or stress amplitude represented on a linear scale on the vertical axis, and the number of cycles to failure represented on a logarithmic scale on the horizontal axis. Where measurement points coincide, the number of coinciding points should be noted. Right-facing arrows should be added to indicate points from test results for test specimens that do not fail.

B.7.8.3 If a fatigue strength is desired, the fatigue strength at 2×10^6 cycles should be derived by interpolation on an *S-N* curve obtained by one of the three load selection methods (A, B, C). The fatigue strength should be reported with a precision of three significant digits.

B.7.9 *Test report*—The test report should include the following items:

a) The trade name, shape, and date of manufacture (if available), and lot number of product tested

b) Type of fiber and fiber binding material as reported by the manufacturer, and fiber volume content

c) Numbers or identification marks on test specimens

d) Designation, diameter, and cross-sectional area for the lot from which the specimens are taken

e) Length of specimens (total and between anchors)

f) Description of preconditioning applied to specimens, including the sequence of preconditioning during specimen preparation

g) Date of test, test temperature and humidity (from start to end of test), and type of testing machine

h) Description of anchorage

i) Description of extensometer(s) and extensometer gauge length

j) Method of selecting maximum and minimum loads for a series of tests to construct an *S-N* curve (Method A, B, or C)

k) Maximum load (stress), minimum load (stress), load (stress) range, number of cycles to failure, loading waveform, and frequency for each test specimen

l) Record of observed failure mode for each test specimen

m) *S-N* curve, if obtained in a series of tests

n) Fatigue strength at 2×10^6 cycles, if obtained from an *S-N* curve

B.8—Test method for creep rupture of FRP bars

This test method has been replaced by ASTM D7337/D7337M.

B.9—Test method for long-term relaxation of FRP bars

The following reference is provided for historical and informational purposes only—Odagiri et al. (1997).

B.9.1 *Scope*—This test method specifies the test requirements for evaluating the long-term relaxation behavior of FRP bars used as prestressing tendons in concrete and masonry under a given constant temperature and strain.

B.9.2 *Referenced documents*

ASTM International

ASTM D618-08—Standard Practice for Conditioning Plastics for Testing

ASTM D7205/D7205M-06(2011)—Standard Test Method for Tensile Properties of Fiber Reinforced Polymer Matrix Composite Bars

ASTM E4-10—Standard Practices for Force Verification of Testing Machines

B.9.3 *Significance and use*

B.9.3.1 This test method for investigating long-term relaxation of FRP bars is intended for use in laboratory tests in which the principal variables are the size or type of FRP bars, magnitude of applied stress, and duration of load application.

B.9.3.2 This test method measures the load-induced, time-dependent tensile strain at selected ages for FRP bars under an arbitrary set of controlled environmental conditions and a corresponding load ratio.

B.9.3.3 This test method is intended to determine the relaxation data for material specifications, research and development, quality assurance, and structural design and analysis. The primary test result is the relaxation rate of the specimen under a specific loading and environmental condition.

B.9.3.4 Relaxation properties of prestressed concrete and masonry structures are important to consider in design. For FRP bars used as prestressing tendons, the relaxation behavior should be measured according to the method given herein, in keeping with the intended purposes.

B.9.4 *Test equipment and requirements*

B.9.4.1 Use a testing machine with a loading capacity in excess of the relaxation load of the test specimen and calibrated according to ASTM E4. The testing machine should be capable of loading at a rate of 200 ± 50 MPa per minute and sustaining load while maintaining a constant strain. It should also control strain fluctuations to no greater than $\pm 25 \times 10^{-6}$ in the specimen throughout the test period once the strain in the specimen has been fixed. If the FRP bar slips from an anchoring section, the slippage distance should be compensated so as not to affect the test results. The accuracy of the initial load applied to the specimen should be as follows:

a) Testing machines with loading capacity equal to or less than 1 kN: ±1.0 percent of set load

b) Testing machines with loading capacity of more than 1 kN: ±2.0 percent of set load

The accuracy of readings or automatic recording of loads should be within 0.1 percent of the initial load.

B.9.4.2 The anchorage should be in accordance with ASTM D7205/D7205M.

B.9.4.3 If an extensometer or strain gauge is to be fitted to the test specimen, the extensometer or strain gauge should be in accordance with ASTM D7205/D7205M.

B.9.5 *Specimen preparation*

B.9.5.1—Specimens should be representative of the lot or batch being tested. During the sampling and preparation of test specimens, all deformation, heating, outdoor exposure to ultraviolet light, and other conditions possibly causing changes to the material properties of the specimen should be avoided.

B.9.5.2 Test specimens should be prepared and handled in accordance with ASTM D7205/D7205M.

B.9.5.3 The number of valid test results for each test condition should not be less than five. If a specimen fails at or slips entirely out of an anchoring section, the test result should be disregarded and an additional test should be performed on a separate specimen taken from the same lot as the failed specimen.

B.9.6 *Standard conditioning procedure*—Condition specimens in accordance with Procedure A of ASTM D618. Store and test at the standard laboratory atmosphere—23 ± 3°C and 50 ± 10 percent relative humidity—unless a different environment is specified as part of the experiment.

B.9.7 *Test method*

B.9.7.1 Mounting of the specimen and gauge length should be in accordance with ASTM D7205/D7205M.

B.9.7.2 If a strain gauge is to be attached to the test specimen, the specimen should be preloaded by applying a load of 10 to 40 percent of the prescribed initial load, after which the strain gauge should be attached and correctly calibrated.

B.9.7.3 The initial load should be either 70 percent of the guaranteed tensile capacity or 80 percent of the million-hour creep rupture capacity, whichever is smaller. In some cases, these conditions may result in a load that causes creep rupture but not failure due to relaxation. In such cases, it should be confirmed under actual loading conditions that the load does not result in creep rupture of the FRP specimens by increasing the initial load as necessary.

B.9.7.4 The initial load should be applied without subjecting the specimen to shock or vibration. The specified rate of loading should be 200 ± 50 MPa per minute. The strain on the specimen should be kept constant after the initial load has been applied and maintained for 120 ± 2 seconds. This time should be deemed to be the test start time.

B.9.7.5 Load reduction should generally be measured over a period of at least 1000 hours. Preferably, load reduction should be recorded automatically by a recorder attached to the testing machine. If no recorder is attached to the testing machine, stress relaxation should be measured and recorded at the following times: 1, 3, 6, 9, 15, 30, and 45 minutes; and 1, 1.5, 2, 4, 10, 24, 48, 72, 96, and 120 hours. Subsequent measurements should be taken at least once every 120 hours.

B.9.8 *Calculations*

B.9.8.1 The relaxation value should be calculated by dividing the load measured in the relaxation test by the initial load.

B.9.8.2 The relaxation curve should be plotted on a semi-logarithmic graph where the relaxation value, in percent, is represented on an arithmetic scale along the vertical axis, and test time, in hours, is represented on a logarithmic scale along the horizontal axis. An approximation line should be derived from the graph data by means of the least-squares method according to Eq. (B.9.8.2)

$$Y_r = a_1 - b_1 \log T \qquad (B.9.8.2)$$

B.9.8.3 The relaxation rate after 1 million hours should be determined from the approximation line; this value represents the million-hour relaxation rate. Where the service life of the structure in which the FRP bars are to be used is determined in advance, the relaxation rate for the number of years of service life, or service life relaxation rate, should also be determined.

B.9.9 *Test report*—The test report should include the following items:

a) The trade name, shape, and date of manufacture (if available), and lot number of product tested

b) Type of fiber and fiber binding material, as reported by the manufacturer, and fiber volume fraction

c) Numbers or identification marks on test specimens

d) Designation, diameter, and cross-sectional area

e) Date of test and test environmental conditions (humidity and temperature and their fluctuations)

f) Type and manufacturer of testing machine

g) Initial load and loading rate

h) Guaranteed tensile capacity and ratio of initial load to guaranteed tensile capacity

i) Relaxation curve for each test specimen

j) Average relaxation rates at 10, 120, and 1000 hours

k) Formula for determining the approximation line

l) Million-hour relaxation rate

m) Relaxation rate corresponding to design service life, or service life relaxation rate, where applicable

B.10—Test method for performance of anchorages of FRP bars

The following references are provided for historical and informational purposes only—Chaallal and Benmokrane (1993) and Nanni et al. (1996).

B.10.1 *Scope*—This test method specifies the test requirements for the performance of the anchorage of FRP bars used as pretensioned prestressing tendons in concrete and masonry. In this application, the anchorage is used only for a short period to transfer tension into the tendons. The long-term anchorage is via bond of the tendon to concrete.

The anchorage performance over extended periods of use, for instance, fatigue performance, is not covered in this test.

B.10.2 *Referenced documents*

ASTM International

ASTM D7205/D7205M-06(2011)—Standard Test Method for Tensile Properties of Fiber Reinforced Polymer Matrix Composite Bars

ASTM E4-10—Standard Practices for Force Verification of Testing Machines

B.10.3 *Significance and use*—This test method for investigating the performance of anchorage is intended for use in laboratory tests in which the principal variable is the size or type of FRP tendons used in pretensioned prestressing of tendons. This test method should not be used to develop tendon design values.

B.10.4 *Test equipment and requirements*—Use a testing machine or hydraulic jack and load cell with a loading capacity in excess of the tensile capacity of the test specimen and calibrated according to ASTM E4. A testing machine with either loading-rate or displacement-rate control is preferred. Tendons may be tested vertically or horizontally using a hydraulic jack.

FRP tendon should be loaded at a rate of 100 to 500 MPa/minute, and the test should finish in 1 to 10 minutes.

B.10.5 *Specimen preparation*—A specimen should be prepared by attaching the prestressing anchorage to be assessed to one end of the FRP tendon. The other end of the tendon should be anchored according to ASTM D7205/D7205M.

The length of the specimen should be the sum of the length of the tendon and the lengths of the anchoring sections at each end. The length of the tendon should not be less than 100 mm, nor should it be less than 40 times the diameter of the FRP tendon. For FRP tendons in twisted strand form, the length should also be greater than two times the strand pitch.

The number of test specimens should be no less than five.

B.10.6 *Test conditions*—Unless a different testing environment is specified as part of the experiment, the pullout anchorage tests should be conducted at the standard laboratory atmosphere of 23 ± 3°C and 50 ± 10 percent relative humidity.

Preconditioning of FRP tendons is permissible. Preconditioning may include abrasion of the surface of the tendon, environmental conditioning of the tendon, and post-curing of the tendon.

B.10.7 *Test method*—Test specimens should be mounted and supported by a tensile testing machine. The area and geometry of the surface supporting the anchorage and the manner of application of force should approximate the actual conditions within the prestressed concrete structure as closely as possible. The anchorage end of the tendon should be supported on a platen in a manner similar to that for the tensile specimen shown in Fig. A1.3 in ASTM D7205/D7205M.

Loading should be continued up to tensile failure, as determined by either failure of the FRP tendon or excessive deformation of the anchoring device.

B.10.8 *Calculations*—The tensile capacity for each specimen and average tensile capacity should be calculated. The modes of failure should also be recorded, noting any deformation or damage of the anchorage.

B.10.9 Test *report*—The test report should include the following items.

a) The trade name, shape, and date of manufacture (if available), and lot number of product tested

b) Type of testing machine

c) Type of fiber and fiber binding material, as reported by the manufacturer, and fiber volume fraction

d) Numbers or identification marks on test specimens

e) Designation, diameter, and cross-sectional area

f) Preconditioning of tendons, if any

g) Date of test, test temperature, and loading rate

h) Dimensions of test specimens and description of anchorage

i) Tensile failure capacity for each test specimen, average tensile capacity, location of failure, and failure modes

j) Records of any deformation or damage to anchorage

B.11—Test method for tensile properties of deflected FRP bars

B.11.1 *Scope*—This test method specifies the test requirements for tensile properties of deflected FRP bars used as prestressing tendons in concrete.

B.11.2 *Referenced documents*

ASTM International

ASTM D618-08—Standard Practice for Conditioning Plastics for Testing

ASTM D7205/D7205M-06(2011)—Standard Test Method for Tensile Properties of Fiber Reinforced Polymer Matrix Composite Bars

ASTM E4-10—Standard Practices for Force Verification of Testing Machines

B.11.3 *Significance and use*

B.11.3.1 This test method for obtaining the tensile properties of deflected FRP bars is intended for use in laboratory tests in which the principal variable is the size or type of FRP bars.

B.11.3.2 When FRP bars manufactured as straight bars are tensioned after being bent, the bending stress or added lateral pressure may cause strength reduction. This test method is developed for FRP bars bent as external prestressing or arranged in a curved layout as internal prestressing.

B.11.3.3 This test method is intended to determine the flexural tensile data of bent FRP bars for material specifications, research and development, quality assurance, and structural design and analysis. The primary test result is the bent tensile capacity of the specimen under a specific loading and environmental condition.

B.11.3.4 Flexural tensile properties of prestressed concrete and masonry structures are important factors to consider in design. For FRP bars used as tendons, the flexural tensile behavior should be measured according to the method given herein.

B.11.4 *Test equipment and requirements*

B.11.4.1 Use a testing machine with a loading capacity in excess of the tensile capacity of the test specimen and calibrated according to ASTM E4. A testing machine with either loading- or displacement-rate control is preferred. The test fixture should include a loading device, load indicator, anchorage holder, and deflector.

B.11.4.2 The loading device should have a loading capacity in excess of the tensile capacity of the test specimen and should be capable of applying load at the required loading rate.

B.11.4.3 The load indicator should be capable of displaying loads with an accuracy of not less than 1 percent of the failure load.

B.11.4.4 The anchorage holder should be suitable for the geometry of the test specimen and should be capable of accurately transmitting loads from the testing machine to the test specimen. The anchorage should be constructed so as to transmit only axial loads to the test specimen, without transmitting either torsion or flexural force.

B.11.4.5 The deflector should be capable of maintaining the required bending angle and bending diameter during the test until failure of the test specimen. The deflector surface in contact with the specimen should be suitable for contact with the bar's outer surface. For a round bar, a smooth rigid deflector surface may be used. For bars that are stranded or spiraled, it may be appropriate to use a deflector machined to match the surface of the bar.

B.11.5 *Specimen preparation*

B.11.5.1 Specimens should be representative of the lot or batch being tested. During the sampling and preparation of test specimens, all deformation, heating, outdoor exposure to ultraviolet light, or other conditions possibly causing changes to the material properties of the specimen should be avoided.

B.11.5.2 Test specimens should be prepared and handled in accordance with ASTM D7205/D7205M.

B.11.5.3 The length of the specimen should be the sum of the length of the test section and the lengths of the anchoring sections. The length of the test section should be such that the distance from an anchorage to the deflected section should not be less than 100 mm or less than 40 times the diameter of the FRP bar. For FRP bars in strand form, the distance from an anchorage to the deflected section should not be less than two times the strand pitch.

B.11.5.4 The number of test specimens for each test condition (combination of bending diameter and bending angle) should not be less than five. If a specimen fails at an anchoring section, or slips out of an anchoring section, an additional test should be performed on a separate specimen taken from the same lot as the failed specimen.

B.11.6 *Standard conditioning procedure*—Condition specimens in accordance with Procedure A of ASTM D618. Store and test at the standard laboratory atmosphere of 23 ± 3°C and 50 ± 10 percent relative humidity, unless a different environment (such as temperature or alkaline solution) is specified as part of the experiment.

B.11.7 *Test method*

B.11.7.1 The bending diameter and bending angle should be set appropriately for the test (Fig. B.11.7.1). This combination forms a single test condition. Each specimen should have only one deflected section.

B.11.7.2 Care should be taken when mounting the specimen in the testing machine to maintain the required bending angle and bending diameter at the deflected section during the test.

B.11.7.3 The specified rate of loading should be between 100 and 500 MPa/minute and the test should finish in 1 to 10 minutes. The deflected bar should exit the deflector on a tangent and not bend over the corner of the deflector.

B.11.7.4 Loading should be continued until failure of the test specimen. Failure load and failure location should be measured and recorded.

B.11.8 *Calculations*

B.11.8.1 The material properties of FRP bars should only be assessed on the basis of the specimen undergoing failure in the test section. In cases where tensile failure or slippage has occurred at an anchoring section, the data should be disregarded and additional tests should be performed until the number of test specimens failing in the test section is no less than five.

B.11.8.2 The maximum, minimum, average, and standard deviation of the bending tensile capacity for each set of test conditions should be calculated in accordance with ASTM D7205/D7205M.

B.11.8.3 The location and mode of failure should be observed and recorded for each test specimen.

B.11.9 *Test report*—The test report should include the following items:

a) The trade name, shape, and date of manufacture (if available), and lot number of product tested

b) Type of fiber and fiber binding material, as reported by the manufacturer, and fiber volume fraction

c) Numbers or identification marks on test specimens

d) Designation, diameter, and cross-sectional area

e) Condition of surface of FRP bar (material, thickness, configuration of any coating, and so on)

f) Date of test, test temperature, loading rate, and type and manufacturer of test machine

g) Type of anchor device

h) Bending angle, external diameter of deflector surface in contact with FRP bar, bending diameter ratio, and material and surface configuration

i) Bending tensile capacity for each test specimen

j) Location and mode of failure for each test specimen

k) Numbers of test specimens for each set of conditions, and maximum, minimum, average, and standard deviation of the bending tensile capacity

B.12—Test method for determining effect of corner radius on tensile strength of FRP bars

B.12.1 *Scope*—This method determines the effect of corner radius on the tensile properties of FRP bars used as reinforcing bars or prestressing tendons in concrete where bars are subject to tensile stresses. Tension tests are conducted using a three-component test fixture.

B.12.2 *Referenced documents*

ASTM International

ASTM D883-11—Standard Terminology Relating to Plastics

ASTM D3039/D3039M-08—Standard Test Method for Tensile Properties of Polymer Matrix Composite Materials

ASTM D7205/D7205M-06(2011)—Standard Test Method for Tensile Properties of Fiber Reinforced Polymer Matrix Composite Bars

ASTM E4-10—Standard Practices for Force Verification of Testing Machines

ASTM E6-09—Standard Terminology Relating to Methods of Mechanical Testing

B.12.3 *Significance and use*

B.12.3.1 Tension tests are conducted using a unique test fixture. The testing fixture consists of three components: upper and lower parts and interchangeable corner inserts made of any suitable metal, such as aluminum or steel (Fig. B.12.3.1a). Detailed dimensions are shown in Fig. B.12.3.1b through B.12.3.1f. The FRP specimen in the form of a bent bar is placed in the groove that runs along the sides and the top of the test fixture and corner inserts. Instrumentation is mounted depending on the variables being monitored. If modulus and strain distribution are required, strain gauges can be mounted around the corner areas. The load is applied until failure of the FRP specimen.

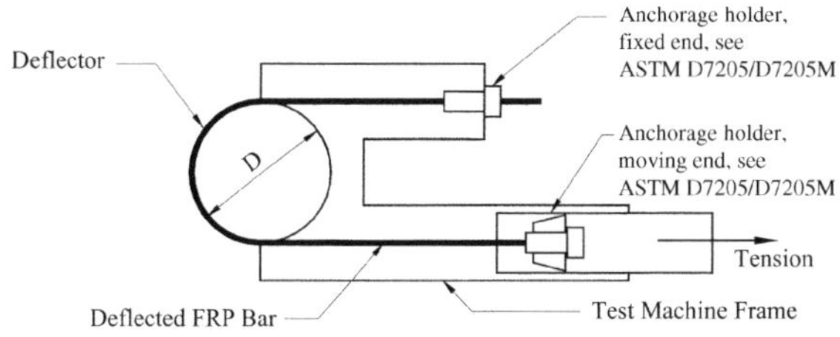

(a) Test setup, bending angle of 180°

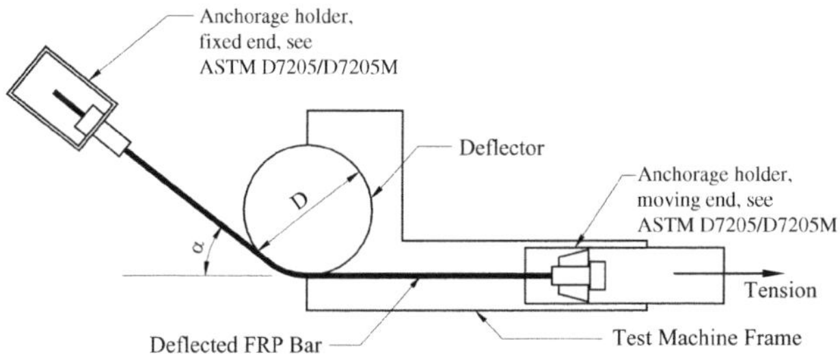

(b) Test setup, bending angle "α"

Fig. B.11.7.1—Typical bending tensile test setups.

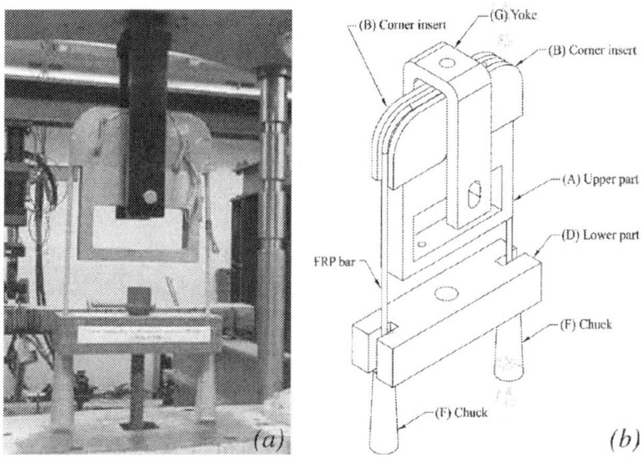

Fig. B.12.3.1a—(a) Assembled test fixture; and (b) assembled test fixture: sketch.

B.12.3.2 This tension test can provide accurate information with regard to the effect of corner radius on the tensile properties of FRP bent bars when applied under conditions with similar interaction mechanisms to those of the test method.

B.12.3.3 The method can be used for testing FRP bent bars with any fiber or resin type in rectangular or round solid shape.

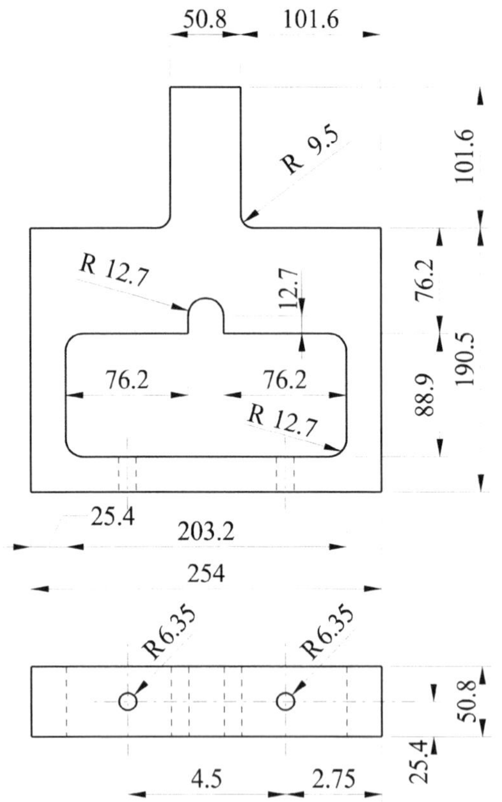

Fig. B.12.3.1b—Upper part of test fixture (in mm).

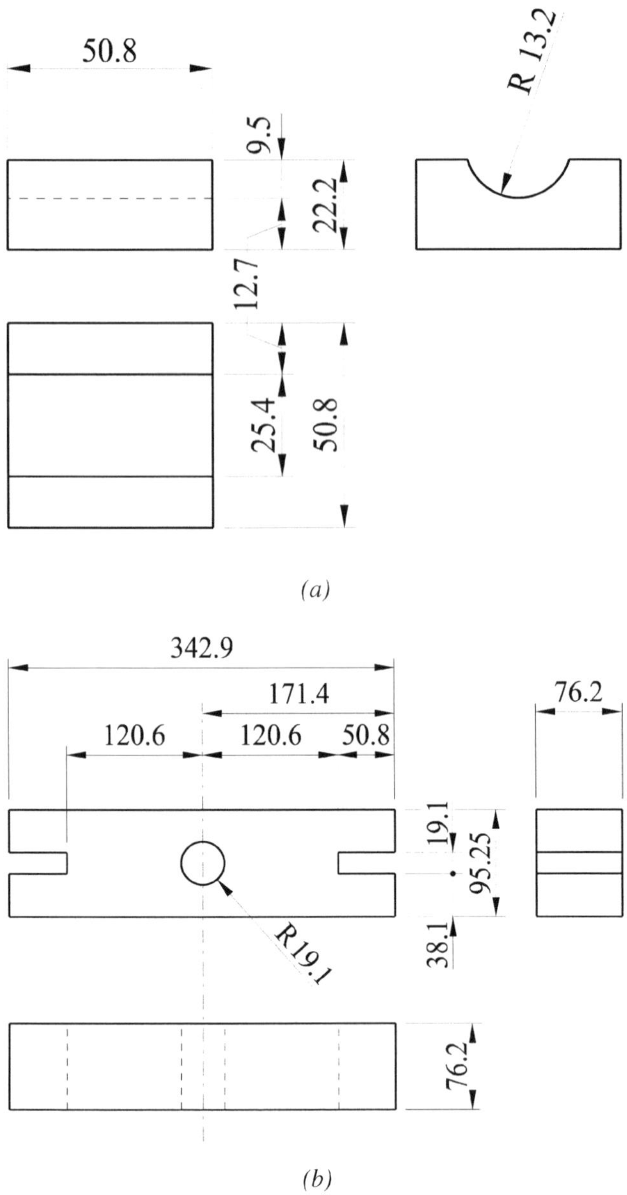

Fig. B.12.3.1c—Lower part of test fixture: (a) wedge; and (b) lower part (in mm).

B.12.3.4 The data provided by this test can be used for research and development, design, and acceptance/rejection criteria.

B.12.4 *Test equipment*

B.12.4.1 *Gauges*—Gauges should be accurate to at least 0.01 mm for measuring the width of the specimen.

B.12.4.2 *Testing machine*—The testing machine should be comprised of the following:

a) Fixed member—fixed or essentially stationary member supporting the load fixture

b) Movable member—member capable of applying a tensile load to the tensile loading fixture and transfer the load to the test fixture

c) Drive mechanism—drive for imparting to the movable member a controlled speed with respect to the stationary member

d) Load indicator—A suitable load-indicating mechanism capable of showing the total tensile load carried by the test fixture. This mechanism should indicate the load with an accuracy of 1 percent or better of the actual value. The accuracy should be verified in accordance with ASTM E4

B.12.4.3 *Strain recording*—A suitable strain-recording system is required for modulus determination and strain distribution.

B.12.4.4 *Test fixture*—The test fixture consists of the upper and lower parts around which the FRP bar is placed. The upper part is used as the testing area with interchangeable corner inserts located at its two corners. The lower part is used to anchor the specimen through anchors. Two bolts and two steel rods are used to hold the upper and lower fixture sections together during specimen installation.

B.12.4.5 *Tensile loading fixture*—The fixture is used to transfer the load from the testing machine to the test fixture (Fig. B.12.3.1e). All drilled holes in the tensile loading fixture and test fixture should be oversized to avoid any twisting and tightening when the rods and nuts are installed. All holes should be drilled 1.60 mm larger than nominal value.

B.12.5 *Specimen preparation*

B.12.5.1 *Geometry*—The FRP test specimen should conform to the dimensions of the test fixture shown in Fig. B.12.3.1b and B.12.3.1c.

B.12.5.1.1 *Anchorage*—An anchorage system at the ends of the test bar should be provided with a sufficient length

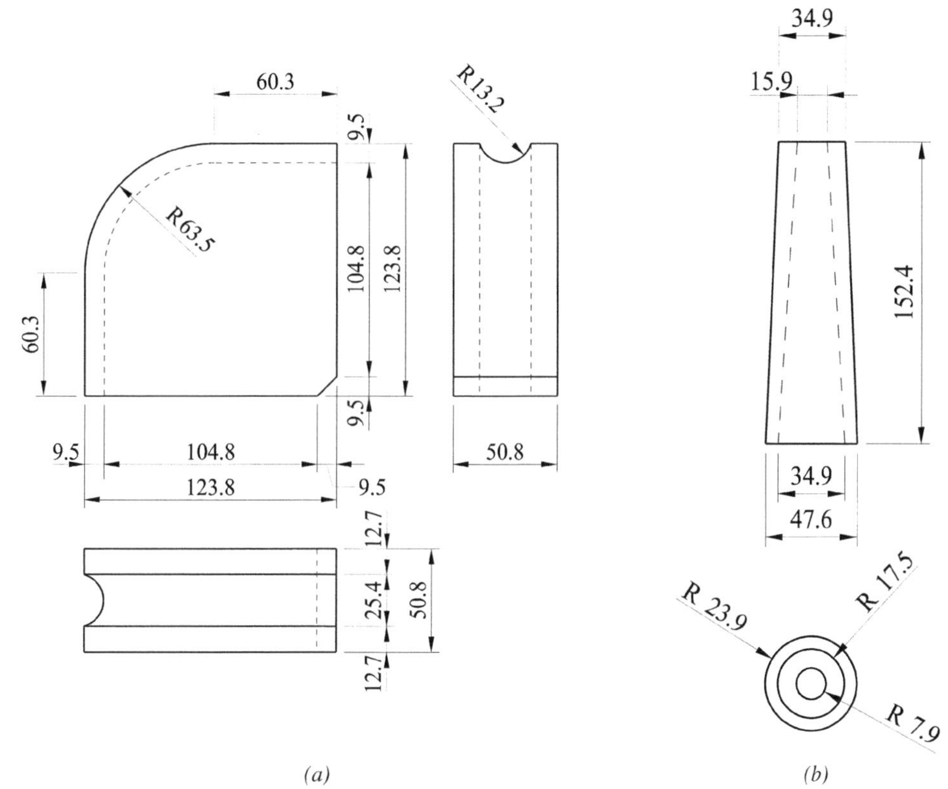

Fig. B.12.3.1d—Interchangeable corner inserts and wedge: (a) corner insert; and (b) chuck (in mm).

and size in accordance with ASTM D7205/D7205M so that failure occurs at the bent portion.

B.12.5.1.2 *Alignment*—The specimen should be centered in the grooves of the test fixture.

B.12.5.2 *Strain*—Where load-strain data are desired, the specimen may be instrumented with strain gauges. The strain gauges should be oriented parallel to the fiber direction. Strain gauges should be mounted in the straight regions of the bar in between and outside the corners. A recommended strain gauge arrangement is shown in Fig. B.12.3.1f.

B.12.5.3 *Number of specimens*—A minimum of five specimens is recommended for each material.

B.12.6 *Conditioning*—Store and test at the standard laboratory conditions of 23 ± 3°C and 50 ± 10 percent relative humidity, unless a different environment, such as temperature or alkaline solution, is specified as part of the experiment.

B.12.7 *Test method*

B.12.7.1 The tensile loading fixture is placed onto the upper cross-head of the testing machine, and the upper test fixture is connected to the tensile loading fixture using a short plain steel rod through the holes in the upper part. The lower part of the test fixture is placed onto the lower plate of the testing machine using a threaded steel bar and two nuts. The steel bar should be long enough to provide sufficient space for the anchor of the test bar when placed between the lower part and the lower cross-head.

B.12.7.2 There should be no misalignment between the test fixture and tensile loading frame. No twisting should be produced in the test fixture and specimen.

B.12.7.3 Connect data recording equipment.

B.12.7.4 Apply and release a small load on the specimen (less than 5 percent of expected failure load) to realign the test fixture. The strain gauges are zeroed in this step.

B.12.7.5 Set the recommended speed of testing. Speed of testing shall be determined by the specifications for the material being tested or by the client. When the speed of testing is not specified, however, a speed of 1.0 to 2.0 mm/minute is recommended.

B.12.7.6 Record the strain and load values continuously.

B.12.7.7 Record the maximum load from the load indicator of the testing machine.

B.12.7.8 Record the failure mode of the specimen. The failure zone is normally located around the corner.

B.12.8 *Calculations*

B.12.8.1 *Reduced tensile strength*—Calculate the reduced tensile strength using Eq. (B.12.8.1). Report the results to three significant figures

$$S = P/(2A) \qquad \text{(B.12.8.1)}$$

B.12.8.2 *Tensile modulus*—Calculate the tensile modulus of different points using Eq. (B.12.8.2). Report the results to three significant figures. The modulus should be calculated using the load and strain values corresponding to 20 and 50 percent of the reduced tensile strength *S*.

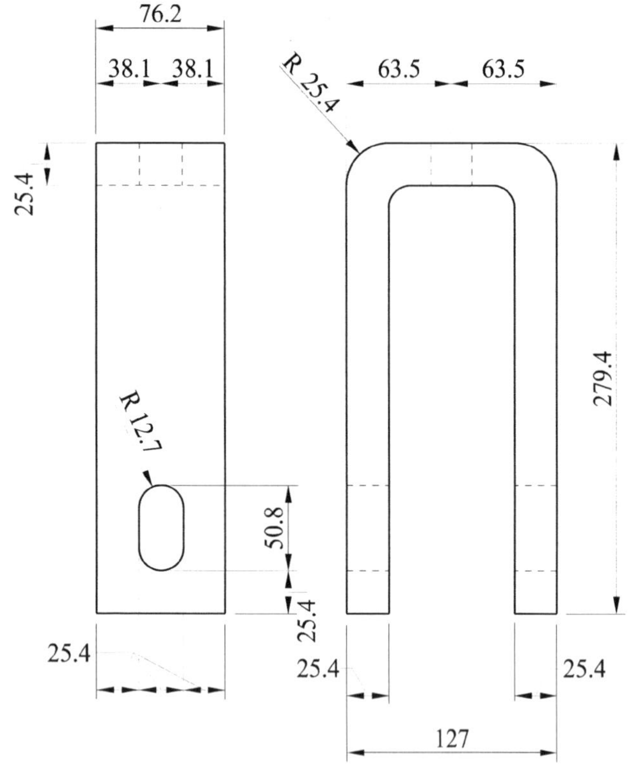

Fig. B.12.3.1e—Tensile fixture (two sets needed) (yoke) (in mm).

$$E_L = \Delta P/(2A\Delta\varepsilon) \quad \text{(B.12.8.2)}$$

B.12.9 *Test report*—The test report should include the following items:

a) Complete identification of the material tested, including source, manufacturer's code number, form, previous history, resin type, processing details, specimen quality control, description of equipment used, deviations from this standard test method, and description of the fabrication process of the specimen compared with the specified one provided by the manufacturer

b) Method of preparing test specimen and verification of quality

c) Test specimen dimensions and corner radius

d) Numbers or identification marks on test specimens

e) Speed of testing

f) Failure load

g) Individual specimen strength and average values

h) Individual specimen modulus and average values

i) Date of tests

PART 3—TEST METHODS FOR FRP LAMINATES

L.1—Test method for direct tension pulloff test

This test method has been replaced by ASTM D7522/D7522M.

L.2—Test method for tension test of flat specimen

This test method has been replaced by ASTM D7565/D7565M.

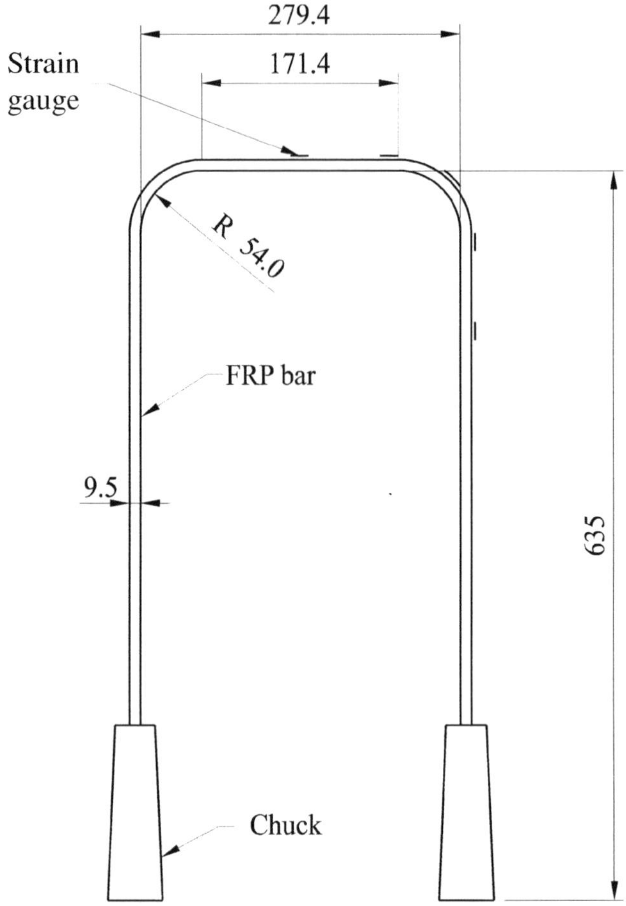

Fig. B.12.3.1f—Installation of specimen and possible strain gauge arrangement (in mm).

L.3—Test method for overlap splice tension test

This test method has been replaced by ASTM D7616/D7616M.

REFERENCES

American Concrete Institute

ACI 440.1R-06—Guide for the Design and Construction of Concrete Reinforced with FRP Bars

ACI 440.2R-08—Guide for the Design and Construction of Externally Bonded FRP Systems for Strengthening Concrete Structures

ACI 440.6-08—Specification for Carbon and Glass Fiber-Reinforced Polymer Bar Materials for Concrete Reinforcement

ASTM International

ASTM D7205/D7205M—Standard Test Method for Tensile Properties of Fiber Reinforced Polymer Matrix Composite Bars

B.3 *Test method for bond strength of FRP bars by pullout testing*

Al-Zahrani, M. M.; Nanni, A.; Al-Dulaijan, S. U.; and Bakis, C. E., 1996, "Bond of FRP to Concrete for Rods with

Axisymmetric Deformations," *Proceedings*, ACMBS-II, Montreal, QC, Canada, Aug., pp. 853-860.

Benmokrane, B.; Zhang, B.; Laoubi, K.; Tighiouart, B.; and Lord, I., 2002, "Mechanical and Bond Properties of New Generation of Carbon Fibre Reinforced Polymer Reinforcing Bars for Concrete Structures," *Canadian Journal of Civil Engineering*, V. 29, No. 2, pp. 338-343.

Chaallal, O., and Benmokrane, B., 1993, "Pullout and Bond of Glass-Fiber Rods Embedded in Concrete," *RILEM Journal of Materials and Structures*, V. 26, No. 157, pp. 167-175.

Freimanis, A. J.; Bakis, C. E.; Nanni, A.; and Gremel, D., 1998, "A Comparison of Pull-Out and Tensile Behaviors of FRP Reinforcement for Concrete," *Proceedings*, ICCI-98, Tucson, AZ, V.2, Jan. 5-7, pp. 52-65.

B.5 *Test method for strength of FRP bent bars and stirrups at bend locations*

Morphy, R.; Shehata, E.; and Rizkalla, S., 1997, "Bent Effect on Strength of CFRP Stirrups," *Proceedings of the Third International Symposium on Non-Metallic (FRP) Reinforcement for Concrete Structures*, Sapporo, Japan, Oct., pp. 19-26.

B.6 *Accelerated test method for alkali resistance of FRP bars*

Benmokrane, B.; Wang, P.; and Ton-That, T. M., 2001, "Durability of GFRP Reinforcing Bars in Alkaline Environments," *Proceedings of the CICE 2001*, International Conference on FRP Composites in Civil Engineering, Hong Kong, Dec. 12-15, pp. 1527-1534.

Benmokrane, B.; Wang, P.; Ton-That, T. M.; Rahman, H.; and Robert, J.-F., 2002, "Durability of GFRP Reinforcing Bars in Concrete Environment," *Journal of Composites for Construction*, ASCE, V. 6, No. 3, pp. 143-155.

Gangarao, H. V. S., and Vijay, P. V., 1997, "Aging of Structural Composites under Varying Environmental Conditions in Non-Metallic (FRP) Reinforcement for Concrete Structures," *Proceedings of the 3rd International Symposium*, FRPRCS-3, Sapporo, Japan, V. 2, 91 pp.

Gentry, T. R., 2001, "Life Assessment of Glass-Fiber Reinforced Composites in Portland Cement Concrete," Paper No. 077, *Proceedings*, 16th Annual Meeting of the American Society for Composites, Blacksburg, VA, Sept. 10-12.

Mehus, J., 1995, "Long Term Durability of Fiber Composite Reinforcement for Concrete," MS thesis, Iowa State University, Ames, IA, Dec.

Nkurunziza, G.; Masmoudi, R.; and Benmokrane, B., 2002, "Effect of Sustained Tensile Stress and Temperature on Residual Strength of GFRP Composite Bars," *Proceedings of the Second International Conference on Durability of Fiber Reinforced Polymer (FRP) Composites for Construction*, CDCC 2002, Montreal, QC, Canada, May 29-31, pp. 347-358.

Porter, M. L.; Mehus, J.; and Young, K. A., 1995, "Aging Degradation of Fiber Composite Reinforcements for Structural Concrete," *Final Report Submitted to Army and Navy*, Engineering Research Institute, College of Engineering, Iowa State University, Ames, IA, Sept., 203 pp.

Porter, M. L.; Mehus, J.; Young, K. A.; Barnes, B. A.; and O'Neil, E. F., 1996, "Aging Degradation of Fiber Composite Reinforcement for Structures," *Advanced Composite Materials in Bridges and Structures*, M. M. El-Badry, ed., Canadian Society for Civil Engineering, Montreal, QC, Canada, 641 pp.

B.7 *Test method for tensile fatigue of FRP bars*

Adimi, R.; Rahman, A. H.; Benmokrane, B.; and Kobayashi, K., 1998, "Effect of Temperature and Loading Frequency on the Fatigue Life of a CFRP Bar in Concrete," *Fiber Composite in Infrastructure*, Proceedings of the 2nd International Conference on Composites in Infrastructure (ICCI'98), H. Saadatmanesh and M. R. Ehsani, eds., Tucson, AZ.

Odagiri, T.; Matsumoto, K.; and Nakai, 1997, "Fatigue and Relaxation Characteristics of Continuous Fiber Reinforced Plastic Rods," *Proceedings of the Third International Symposium on Non-Metallic (FRP) Reinforcement for Concrete Structures*, FRPRCS-3, Japan Concrete Institute, Sapporo, Japan, V. 2, pp. 227-234.

Rahman, A. H., and Kingsley, C. Y., 1996, "Fatigue Behaviour of a Fiber-Reinforced-Plastic Grid as Reinforcement for Concrete," *Fiber Composite in Infrastructure*, Proceedings of the 1st International Conference on Composites in Infrastructure, H. Saadatmanesh and M. R. Ehsani, eds., Tucson, AZ.

B.9 *Test method for long-term relaxation of FRP bars*

Odagiri, T.; Matsumoto, K.; and Nakai, 1997, "Fatigue and Relaxation Characteristics of Continuous Fiber Reinforced Plastic Rods," *Proceedings of the Third International Symposium on Non-Metallic (FRP) Reinforcement for Concrete Structures*, FRPRCS-3, Japan Concrete Institute, Sapporo, V. 2, pp. 227-234.

B.10 *Test method for performance of anchorages of FRP bars*

Chaallal, O., and Benmokrane, B., 1993, "Physical and Mechanical Performance of an Innovative Glass-Fiber-Reinforced Plastic Rod for Concrete and Grouted Anchorages," *Canadian Journal of Civil Engineering*, V. 20, No. 2, pp. 254-268.

Nanni, A.; Bakis, C. E.; O'Neil, E.; and Dixon, T. O., 1996, "Performance of FRP Tendon-Anchor Systems for Prestressed Concrete Structures," *PCI Journal*, V. 41, No. 1, pp. 34-44.

ACI 440.4R-04
(Reapproved 2011)
Emerging Technology Series

Prestressing Concrete Structures with FRP Tendons

Reported by ACI Committee 440

ACI encourages the development and appropriate use of new and emerging technologies through the publication of the ***Emerging Technology Series****. This series presents information and recommendations based on available test data, technical reports, limited experience with field applications, and the opinions of committee members. The presented information and recommendations, and their basis, may be less fully developed and tested than those for more mature technologies. This report identifies areas in which information is believed to be less fully developed, and describes research needs. The professional using this document should understand the limitations of this document and exercise judgment as to the appropriate application of this emerging technology.*

Sami H. Rizkalla
Chair

John P. Busel
Secretary

Tarek Alkhrdaji	Edward R. Fyfe	Vistasp M. Karbhari	Morris Schupack
Charles E. Bakis	Ali Ganjehlou	James G. Korff	David W. Scott
P. N. Balaguru	Duane J. Gee	Michael W. Lee	Rajan Sen
Lawrence C. Bank	T. Russell Gentry	John Levar	Mohsen A. Shahawy
Abdeldjelil Belarbi	Janos Gergely	Ibrahim Mahfouz	Carol K. Shield
Brahim Benmokrane	William J. Gold	Henry N. Marsh	Khaled A. Soudki
Gregg J. Blaszak	Nabil F. Grace	Orange S. Marshall	Robert E. Steffen
Timothy E. Bradberry	Mark F. Green	Amir Mirmiran	Gamil Tadros
Gordon L. Brown	Mark Greenwood	Ayman S. Mosallam	Jay Thomas
Vicki L. Brown	Doug D. Gremel	Antonio Nanni	Houssam A. Toutanji
Thomas I. Campbell*	H. R. Hamilton	Kenneth Neale	Miroslav Vadovic
Charles W. Dolan*	Issam E. Harik	John P. Newhook	David Vanderpool
Dat Duthinh	Kent A. Harries	Max L. Porter	Milan Vatovec
Garth J. Fallis	Mark P. Henderson	Mark A. Postma	David White
Amir Fam	Bohdan Horeczko	Hayder A. Rasheed	

*Co-chairs of Subcommittee 440-I.

Note: The committee acknowledges the significant contribution of associate member Raafat El-Hacha.

Fiber-reinforced polymers (FRPs) have been proposed for use instead of steel prestressing tendons in concrete structures. The promise of FRP materials lies in their high-strength, lightweight, noncorrosive, nonconducting, and nonmagnetic properties. This document offers general information on the history and use of FRP for prestressing applications and a description of the material properties of FRP. The document focuses on the current state of design, development, and research needed to characterize and ensure the performance of FRP as prestressing reinforcement in concrete structures. The proposed guidelines are based on the knowledge gained from worldwide experimental research, analytical work, and field applications of FRPs used as prestressed reinforcement. The current development includes a basic understanding of flexure and axial prestressed members, FRP shear reinforcement, bond of FRP tendons, and unbonded or external FRP tendons for prestressing applications. The document concludes with a description of research needs.

Keywords: anchorage; bond length; crack; deflection; deformation; development length; ductility; fatigue; jacking stresses; post-tensioning; prestressed concrete; pretensioning; reinforcement ratio; shear; tendon.

CONTENTS

ACI 440.4R-04 became effective September 21, 2004.

CHAPTER 1—INTRODUCTION

Fiber-reinforced polymer (FRP) composites have been proposed for use as prestressing tendons in concrete structures. The promise of FRP materials lies in their high-strength, lightweight, noncorrosive, nonconducting, and nonmagnetic properties. In addition, FRP manufacturing, using various cross-sectional shapes and material combinations, offers unique opportunities for the development of shapes and forms that would be difficult or impossible with conventional steel materials. Lighter-weight materials and preassembly of complex shapes can boost constructibility and efficiency of construction. At present, the higher cost of FRP materials suggests that FRP use will be confined to applications where the unique characteristics of the material are most appropriate. Efficiencies in construction and reduction in fabrication costs will expand their potential market. FRP reinforcement is available in the form of bars, grids, plates, and tendons. This document examines both internal and external prestressed reinforcement in the form of tendons.

One of the principal advantages of FRP tendons for prestressing is the ability to configure the reinforcement to meet specific performance and design objectives. FRP tendons may be configured as rods, bars, and strands as shown in Fig. 1.1. The surface texture of FRP tendons may vary, resulting in bond with the surrounding concrete that varies from one tendon configuration to another. Unlike conventional steel reinforcement, there are no standardized shapes, surface configurations, fiber orientation, constituent materials, and proportions for the final products. Similarly, there is no standardization of the methods of production, such as pultrusion, braiding, filament winding, or FRP preparation for a specific application. Thus, FRP materials require considerable engineering effort to use properly. Bakis (1993) has outlined manufacturing processes.

FRP tendons are typically made from one of three basic fibers. These fibers are aramid, carbon, and glass. Aramid fibers consist of a semicrystalline polymer known as aromatic polyamide. Carbon fibers are based on the layered graphene (hexagonal) networks present in graphite, while glass generally uses either E-glass or S-glass fibers. E-glass is a low-cost calcium-aluminoborosilicate glass used where strength, low conductivity, and acid resistance are important. S-glass is a magnesium-aluminosilicate glass that has higher strength, stiffness, and ultimate strain than E-glass. S-glass costs more than E-glass, and both are susceptible to degradation in alkaline environments. Table 1.1 gives properties of typical fibers.

The selection of the fiber is primarily based on consideration of cost, strength, stiffness, and long-term stability. Within these fiber groups, different performance and material characteristics may be achieved. For example, aramids may come in low, high, and very high modulus configurations. Carbon fibers are also available with moduli ranging from below that of steel to several multiples of that of steel. Of the several fiber types, glass-based FRP reinforcement is least expensive and generally uses either E-glass or S-glass fibers.

Table 1.1—Typical fiber properties

	High-modulus carbon (Thornel 2004)	Intermediate-modulus carbon (Torayca 1996)	Standard-modulus carbon (Torayca 1996)	Aramid (Kevlar 1992)	E-glass (Hartman et al. 1994)	S-glass
Tensile strength, GPa	2.1 to 2.4	5.4 to 6.3	3.5 to 4.9	2.9 to 3.0	3.4	4.9
Tensile modulus, GPa	517 to 827	294	230	71 to 112	72.3	86.9
Ultimate strain, %	0.3	1.9 to 2.2	1.5 to 2.1	2.4 to 3.6	4.8	5.7
Mass density, g/cm^3	2.0 to 2.2	1.8	1.8	1.44	2.58	2.46
Diameter, μm	10	5	7	12	3 to 24	6 to 14
Longitudinal CTE, ppm/°C	–1.5	–0.6	–0.4	–4 to –5	5.4	1.6

Note: 1 GPa = 145 ksi; 1 g/cm^3 = 62.4 lb/ft^3; 1/°C = 1.8/°F; 1 μm = 4 × 10^{-5} in.; CTE = coefficient of thermal expansion.

The resins used for fiber impregnation are usually thermosetting and may be polyester, vinylester, epoxy, phenolic, or polyurethane. The formulation, grade, and physical-chemical characteristics of resins are practically limitless. The possible combinations of fibers, resins, additives, and fillers make generalization of the properties of FRP tendons very difficult. Additionally, FRP composites are heterogeneous and anisotropic. Final characteristics of an FRP tendon are dependent on fiber and resin properties, as well as the manufacturing process. Specific details of a particular tendon should be obtained from the manufacturer of the tendon.

1.1—Organization and limitations of document

The emphasis of this document is on flexural members in concrete buildings and bridges pretensioned with aramid or carbon FRP tendons. An FRP prestressing system consists of the tendon and the anchorage. Properties, performance, and overall behavior are dependent on the tendon/anchorage system and on the individual components. Performance of independent elements should be verified by test. Information is provided for bonded and unbonded post-tensioned applications where it is available. Only fully prestressed members are considered, with no attempt being made to address partially prestressed members. In general, to account for the brittle characteristics of FRP, unlike steel, which is ductile, the recommendations herein are conservative. Specific limitations of FRP tendons are addressed and research needs are listed in Chapter 9. The committee feels that this document is relevant to simple spans and to spans made continuous by placing steel reinforcement in the deck of a bridge structure. No recommendations are made for beams made continuous with FRP tendons or for moment-resisting frames where ductility or large deformations are required for seismic loading. The worldwide number of prestressed FRP applications is less than 100 (MDA 2004; IABSE 2003). Most are bridge structures where issues of fire were not considered critical.

1.2—Historical development and use of FRP reinforcement

The concept of using short glass fiber reinforcement in concrete was first introduced in the 1930s but was not developed into long fiber reinforcement for nearly two decades. In the 1950s and 1960s, the U.S. Army Corps of Engineers was sufficiently interested in long glass fibers for reinforcement

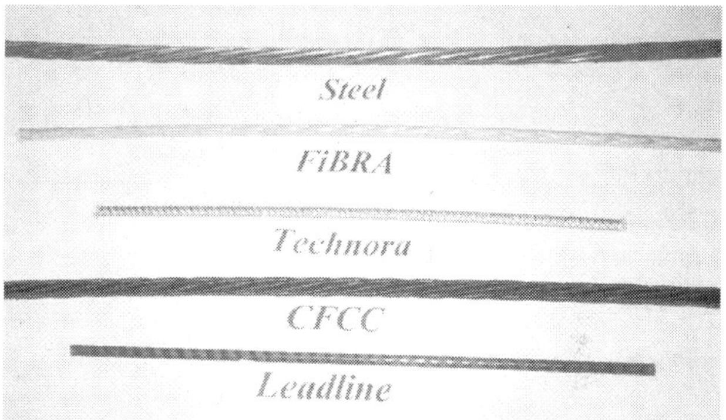

Fig. 1.1—Sample FRP reinforcement configurations.

that a series of comprehensive reports was compiled (Mather and Tye 1955; Pepper and Mather 1959; Wines et al. 1966). Although these reports were generated, research and site applications were limited. In the 1970s, corrosion-induced deterioration of concrete structures, particularly bridge decks, led to a renewed interest in design strategies that would reduce susceptibility of structures to corrosive environments.

In the 1970s, research activities started in Germany on glass FRP-based prestressing tendons. In 1978, a joint venture between German contractor Strabag-Bau and German chemical producer Bayer resulted in glass fiber-reinforced polymer (GFRP) tendons and an anchorage system for post-tensioning applications. These tendons were incorporated in several bridges in Germany and Austria. After various transition stages, however, Strabag stopped its activities in this field in the early 1990s. The National Bureau of Standards (NBS)—now renamed the National Institute of Standards and Technology (NIST)—examined nonmetallic rods for antenna guy wires. In the process, they conducted some of the first research into anchorage of composite rods that became relevant to prestressed concrete application of FRP materials (NBS 1976).

Interest in the corrosion-resistant properties of nonmetallic bars and tendons continued to grow in the 1980s. In 1983, AKZO, a chemical producer in the Netherlands, and HBG, a contractor, jointly developed aramid fiber reinforced polymer (AFRP) prestressing tendons. The Japanese have also undertaken an extensive national program to examine the use of FRP reinforcement in concrete structures. Around 1980, research and development began in Japan on production techniques for FRP reinforcement and its application to concrete structures. This research and development originally focused on the development of FRP-reinforced concrete members that used FRPs instead of steel reinforcing bars and prestressing tendons. In the United States, a new anchorage was developed for glass fiber tendons (Iyer and Kumarswamy 1988), and the prestressing use of Kevlar™ was investigated (Dolan 1989). Iyer's anchorage was supported financially by the Florida Department of Transportation (FDOT), which funded a major study to investigate the prestressing application of glass fiber tendons for bridge and marine substructures (Sen, Issa, and Mariscal 1992). This research culminated in the first conference to focus on FRP composites for civil engineering applications (Iyer and Sen 1991), and the construction of the first FRP prestressed bridge in Rapid City, South Dakota (Iyer 1993). These and similar efforts led to the development of several commercial tendon systems, many of which are discussed in the proceedings of the First International Symposium for FRP in Reinforced Concrete Structures (FRPRCS-1) (Nanni and Dolan 1993), and in a Japanese Society of Civil Engineers publication (JSCE 1996).

1.3—Design guidelines and technical committees

In 1993, the first design guidelines for FRP-reinforced and prestressed concrete buildings were established by the Japanese Society of Civil Engineers. The Japanese version of the guideline was released in 1995, while the English version (Sonobe et al. 1997) was published in 1997. The Canadian Standards Association has produced two standards—CAN/CSA S6-00 and CAN/CSA S806-02—that contain code provisions for the use of FRP prestressing tendons in bridges and buildings, respectively.

In Europe, unified design guidelines for FRP reinforcement are under development. A task group with this aim was established at the end of 1996, within the former CEB (Euro-International Concrete Committee). In December 1997, a 4-year training and mobility of researchers network project, titled "Development of Guidelines for the Design of Concrete Structures, Reinforced, Prestressed or Strengthened with Advanced Composites," started. This so-called "ConFiber-Crete Network" was comprised of 11 teams from nine different European countries.

Since the merger of CEB and FIP (*Federation Internationale de la Precontrainte*), this task group has been integrated in the new *fib* (*Federation Internationale du Béton*). Task Group 9.3 of *fib* Commission 9 is charged with developing design guidelines for concrete structures reinforced, prestressed, or strengthened with FRP, based on the design format of the CEB-FIP Model Code and Eurocode 2 (*fib* TG9.3). Recent activities in Europe have been summarized by Matthys and Taerwe (2001).

The American Society of Civil Engineers (ASCE) established a standards committee to address stand-alone FRP products. The Transportation Research Board (TRB) has formally established Committee A2C07 to examine the use of FRP in bridge structures. Other societies, including the Society for the Advancement of Material and Process Engineering (SAMPE) and the Market Development Alliance (MDA) of the FRP Composites Industry, have been active in the area of FRP for construction use.

1.4—Research efforts

Early development work on prestressing with FRP was carried out at the South Dakota School of Mines by Iyer who developed the first prestressing anchorage in the late 1980s. The first glass fiber prestressed bridge built in the United States was in Rapid City, South Dakota, in 1990. Other research was carried out at the University of South Florida, Florida Atlantic University, and the Florida Department of Transportation (Sen, Issa, and Mariscal 1992). Studies included static and fatigue testing of beams and half-scale bridges, durability studies, and full-scale testing of piles. All three materials—aramid, carbon, and glass—were evaluated.

Work relating to FRP prestressing in the United States has been documented by Dolan (1999). Much of the research on FRP reinforcement in the United States has been conducted by individual investigators at the University of Arizona (Eshani, Saadatmanesh and Nelson 1997), the University of Michigan (Naaman and Jeong 1995), the Massachusetts Institute of Technology (Triantafillou and Deskovic 1991), Pennsylvania State University (Nanni et al. 1996a,b), South Dakota School of Mines (Iyer et al. 1996), the University of California (Long Beach), West Virginia University (Vijay and GangaRao 2001), the University of Wyoming (Dolan et al. 2000), and Lawrence Technological University (Grace 2000a,b). In 1993, FHWA sponsored research into accelerated aging and standardized testing of FRP materials at Georgia Institute of Technology, Pennsylvania State University, and the Catholic University of America. In 1994, FHWA sponsored research into development of design recommendations for FRP prestressing for bridge girders that led to design specification recommendations for AASHTO (Dolan et al. 2000). Lawrence Technological University has developed a demonstration bridge using external unbonded FRP tendons (Grace 1999).

In Canada, several researchers, mostly within ISIS Canada (Canadian Network of Centers of Excellence on Intelligent Sensing for Innovative Structures), have been investigating applications of FRP prestressing tendons. Work at the University of Manitoba has emphasized carbon FRP tendons and has considered the behavior of prestressed beams under both service and ultimate conditions (Abdelrahman et al. 1995; Fam et al. 1997). Extensive studies on bond and transfer length have also been conducted. At the Royal Military College of Canada (RMC), work has been done on aramid FRP tendons (McKay and Erki 1993), and more recently on carbon FRP tendons. Researchers at Queen's University have been investigating the low temperature and long-term behavior of beams prestressed with CFRP (Bryan and Green 1996). Additionally, the possibility of using CFRP rods to prestress bridge deck slabs was investigated (Braimah et al. 1996), and work on the transfer length of beams prestressed with carbon FRP was conducted (Soudki et al. 1997). Partially prestressed, partially bonded FRP tendons have been investigated (Rizkalla et al. 2001). Noncorrosive wedge-type anchorages for FRP tendons have been developed at the University of Calgary (Sayed-Ahmed and Shrive 1998; Campbell et al. 2000; Shaheen 2004). At the University of Sherbrooke, extensive investigations are being conducted into the durability of FRP rods for reinforcing and prestressing concrete, and into the development of a bond-type anchorage (Zhang 2002). The effects of temperature on beams prestressed with FRP tendons have been the focus of research at Concordia University, and applications of unbonded FRP tendons have been considered at the University of Windsor (Salib et al. 1999). Researchers at Carleton University and the University of Waterloo are also investigating applications of FRP prestressing tendons.

The decline in the Japanese economy in the 1990s slowed the Japanese development program and has curtailed the availability of many Japanese products for evaluation and testing in North America. Developments in Japan have been addressed by Fukuyama (1999), and details of some recent projects are available from the Advanced Composites Cables (ACC) Club.

Activities in Europe relating to the use of FRP have been reported by Taerwe and Matthys (1999). Of the projects financially supported by the European community, the BRITE/EURAM project, "Fiber Composite Elements and Techniques as Non-Metallic Reinforcement for Concrete," started November 1991 and ended in 1996. The Universities of Braunschweig and Ghent, together with industrial partners from Germany and The Netherlands, investigated performance characteristics and structural aspects of FRP reinforcement for prestressed and reinforced concrete members. The EUROCRETE project, a pan-European collaborative research program with partners from the United Kingdom, the Netherlands, Switzerland, France, and Norway, began in December 1993 and ended in 1997. This project included material development, research on durability in aggressive environments, determination of structural behavior, and development of design guidance, techno-economic, and feasibility studies. The project included construction of demonstration structures.

1.5—Demonstrations and field applications

During the late 1980s and throughout the 1990s, several demonstration projects have shown the potential of FRP prestressing applications. The Marienfelde Bridge in Berlin, completed in 1988, is the first structure to be built in Germany since 1945 with external unbonded prestressing (Wolff and Miesseler 1989). It is a two-span, 27.61 m (90.5 ft) and 22.98 m (75.1 ft) continuous double-T section beam providing a pedestrian walkway and a bridle path. Seven external Polystal® tendons, each composed of 19 glass fiber bars, are located between the beam webs. Each tendon has a working load of 600 kN (135 kip), is 7.5 mm (0.3 in.) in diameter, and is covered with a 5 mm (0.2 in.) thick sheathing. Four bars out of a group of 19 in a tendon slipped out of the anchorage during prestressing. The treatment for bar slippage was not very difficult because the external prestressing system readily permitted replacement of the entire tendon, complete with the duct and anchorages, at both ends.

Fig. 1.2—Construction of the FRP prestressed girders for the Taylor Bridge in Headingly, Manitoba, Canada.

The BASF bridge in Ludwigshafen, Germany, is a prestressed roadbridge (Zoch et al. 1991). The bridge girder has width for two lanes, and the total length is 85 m (278 ft). Four CFRP tendons were used in conjunction with 16 conventional steel tendons as internal unbonded post-tensioned reinforcement. Each tendon consisted of 19 carbon fiber-reinforced polymer (CFRP) strands of 12.5 mm (0.5 in) diameter, with a prestressing force of 70 kN (16 kip) applied on each strand. The die-cast/wedge anchorage was chosen for applying prestress by post-tensioning to the bridge girder. All tendons were guided through tubes embedded in the concrete, steel tendon tubes were filled with cement grout after tensioning, while the CFRP tendon tubes were not filled to allow inspection, exchange of tendons, and data collection.

In 1993, a bridge was built in Calgary using FRP pretensioned tendons incorporating fiber optic sensors (Rizkalla and Tadros 1994). This was the first bridge of its kind in Canada and one of the first in the world. A second bridge incorporating FRP prestressing tendons was built at Headingly, Manitoba, in 1997. Figure 1.2 shows the fabrication of the prestressed girders for the Taylor Bridge in Headingly, and Fig. 1.3 shows the completed bridge.

Figures 1.4 and 1.5 show the Bridge Street Bridge in Southfield, Michigan. This bridge was completed in 2001 and is a three-span structure that used bonded and unbonded CFRP tendons in both longitudinal and transverse directions (Grace et al. 2002). The Bridge Street Bridge Deployment Project consists of two parallel, independent bridges (Structure A and Structure B) over the Rouge River in the City of Southfield, Michigan. Both structures are designed to accommodate one traffic lane and incorporate three 15-degree skewed spans. Structure A consists of new substructure and superstructure and incorporates five equally spaced conventional AASHTO-I beams in each of the three spans, with a continuous cast-in-place concrete deck slab. Structure B consists of twelve double T (DT) beams, each prestressed using pretensioned Leadline tendons and post-tensioned, in the longitudinal and transverse directions, using carbon fiber composite cable (CFCC) strands. Details of the construction of the Bridge Street Bridge can be found elsewhere (Grace et al. 2002). Bridge Structures A and B were designed

Fig. 1.3—Completed Taylor Bridge in Headingly, Manitoba, Canada.

Fig. 1.4—Erection of the Bridge Street Bridge, Southfield, Mich. (Courtesy of HRC Consulting Engineers, Inc., Mich.)

for two traffic lanes using the provisions to the extent possible of the American Association of State Highway and Transportation Officials (AASHTO) Standard Specifications for Highway Bridges (sample calculations in Grace and Singh [2003]). The superstructure dead loads for Structure B included the DT beams, composite CFRP-reinforced concrete topping, surfacing mixture, pedestrian sidewalk, barrier wall, and the bridge parapet and railing. Live load design was based on the Michigan MS-23 standard truck loading. This live load condition corresponds to 125% of the Michigan MS-18 (AASHTO HS20-44) standard truck

Fig. 1.5—Completed Bridge Street Bridge (Grace et al. 2002).

loading. The live load distribution factor used for design of the DT beams was calculated two different ways, using provisions defined in both the AASHTO Standard and AASHTO LRFD Bridge Design Specifications.

Four demonstration projects, in which FRP pretensioned concrete piles have been driven successfully, have been carried out in the United States. Further details of these projects are given in Chapter 8.

1.6—Definitions

Definitions of terms used in concrete structures are found in "Cement and Concrete Terminology (ACI 116R)." Terms and notation that have specific application to FRP reinforcement and prestressing are defined as follows.

A

AFRP—aramid fiber-reinforced polymer.

alkalinity—the condition of having or containing hydroxyl (OH^-) ions; containing alkaline substances. Note: in concrete, the alkaline environment has a pH above 12.

anchor—in prestressed concrete, to lock the stressed tendon in position so that it will retain its stressed condition.

anchorage—In post-tensioning, a device used to anchor the tendon to the concrete member; in pretensioning, a device used to maintain the elongation of a tendon during the time interval between stressing and release.

anchorage, dead-end—the anchorage at that end of a tendon that is opposite to the jacking end.

B

balanced FRP reinforcement ratio—the reinforcement ratio in a flexural member that causes the ultimate strain of FRP and the ultimate compressive strain of concrete to be simultaneously attained.

bar—a nonprestressed FRP element, made mostly of continuous fibers with nominally rectangular or circular cross section, used to reinforce a structural concrete component; the bar may have a deformed or roughened surface to enhance bonding with concrete.

bond stress—the stress between the concrete and the reinforcing element.

bond stress, average—the force in a reinforcing element divided by the product of the perimeter and the development length of the reinforcing element.

bond, transfer—in pretensioning, the bond stress resulting from the transfer of stress from the tendon to the concrete.

bonded member—a prestressed concrete member in which the tendons are bonded to the concrete either directly or through grouting.

bonded tendon—prestressing tendon that is bonded to concrete either directly or through grouting.

C

camber—a deflection that is intentionally built into a structural element or form to improve appearance or to nullify the deflection of the element under the effects of loads, shrinkage, and creep.

CFRP—carbon fiber-reinforced polymer.

coefficient of thermal expansion—change in linear dimension per unit length.

composite—a combination of one or more materials differing in form or composition on a macroscale. Note: the constituents retain their identities; that is, they do not dissolve or merge completely into one another, although they act in concert. Normally, the components can be physically identified and exhibit an interface between one another.

concrete clear cover—the least distance from the concrete surface to the nearest surface of embedded prestressing tendon or reinforcement.

creep—the inelastic strain of a material under a sustained load over a period of time.

creep-rupture—the tensile fracture of a material subjected to sustained high stress levels over a period of time.

D

deformability—ratio of deflection parameter of a member at ultimate to the same parameter at cracking.

development length—length of embedded reinforcement required to develop the nominal capacity of a member.

duct—a conduit for post-tensioning tendons.

E

effective depth of section—distance measured from the extreme compression fiber to the centroid of the tensile reinforcement.

effective prestress—stress remaining in prestressing tendons after all losses have occurred.

embedment length—length of embedded reinforcement provided beyond a critical section.

F

fiber, aramid—highly oriented organic fiber derived from polyamide incorporating an aromatic ring structure.

fiber, carbon—fiber produced by the heating and stretching for molecular alignment of organic precursor materials containing a substantial amount of carbon, such as rayon, polyacrylonitrile (PAN), or pitch, in an inert environment.

fiber content—the amount of fiber present in a composite. Note: this usually is expressed as a percentage volume fraction or weight fraction of the composite.

fiber-reinforced polymer (FRP)—composite material consisting of continuous fibers impregnated with a fiber-binding polymer, then molded and hardened in the intended shape.

fiber volume fraction—the ratio of the volume of the fibers to the total volume of the fiber-reinforced composite.

H

harped tendons—tendons that have a trajectory that is bent with respect to the center of gravity axis of the concrete member.

J

jacking force—the temporary force exerted by the device that introduces tension into prestressing tendons.

L

limit states—those conditions of a structure in which it ceases to fulfill the relevant function for which it was designed.

load factor—factor applied to a specified load that, for the limit state under consideration, takes into account the variability of the loads and load patterns, and analysis of their effects.

load, factored—product of a specified load and its load factor.

load, sustained—specified dead load plus that portion of the specified live load expected to act over a period of time sufficient to cause significant long-term deflection.

loss, elastic—in prestressed concrete, the reduction in prestressing load resulting from the elastic shortening of the member.

loss, friction—the stress loss in a prestressing tendon resulting from friction between the tendon and duct or other device during stressing.

loss, shrinkage—reduction of stress in prestressing tendon resulting from shrinkage of concrete.

loss of prestress—the reduction in the prestressing force that results from the combined effects of slip at anchorage, relaxation of stress, frictional loss, and the effects of elastic shortening, creep, and shrinkage of the concrete.

M

matrix—the material that serves to bind the fibers together, transfer loads to the fibers, and protect them against environmental attack and damage due to handling.

P

post-tensioning—a method of prestressing in which the tendons are tensioned after the concrete has hardened.

prestressed concrete—concrete in which the internal stresses have been initially introduced so that the subsequent stresses resulting from dead load and superimposed loads are counteracted to a desired degree; this may be accomplished by post-tensioning or pretensioning.

pretensioning—a method of prestressing in which the tendons are tensioned before the concrete is placed.

pultrusion—a continuous process for manufacturing composites that have a uniform cross-sectional shape; the process consists of pulling a fiber-reinforced material through a resin impregnation bath then through a shaping die where the resin is subsequently cured.

R

relaxation—the loss of tension with time in a stressed tendon maintained at constant length and temperature.

resistance factor—factor applied to a specified material property or to the resistance of a member for the limit state under consideration, which takes into account the variability of dimensions, material properties, workmanship, type of failure, and uncertainty in the prediction of resistance.

S

splitting tensile strength—tensile strength of concrete determined by a concrete cylinder splitting test.

strand—a linear component that constitutes either a part or the whole of a prestressing tendon.

T

tendon—an element such as a wire, bar, rope, a single strand, or an assembly of multiple strands used to impart prestress to a structural component. Note: Distinction is made between strands and tendons. Strands refer, for example, to pultruded FRP rods and tendons to bundled FRP rods stressed by common anchors.

thermoplastic—resin that is not cross-linked, and generally can be remelted and recycled.

thermoset—resin that is formed by cross-linking polymer chains and cannot be melted or recycled because the polymer chains form a three-dimensional network.

tow—an untwisted bundle of continuous fibers.

transfer—act of transferring force in prestressing tendons from jacks or the pretensioned anchorage to the concrete member.

transfer length—the length from the end of the member where the tendon stress is zero, to the point along the tendon where the prestress is fully effective, also called transmission length.

V

vinyl esters—a class of thermosetting resins containing esters of acrylic, methacrylic acids, or both, many of which have been made from epoxy resin.

1.7—Notation

A_p = cross-sectional area of FRP tendon, mm^2 (in.^{2})
A_{pi} = cross-sectional area of FRP tendon at level i, mm^2 (in.^{2})
$A_{p\ int}$ = area of the internal prestressed reinforcement, mm^2 (in.^{2})
$A_{p\ tot}$ = total area of internal and external prestressed reinforcement, mm^2 (in.^{2})
A_s = area of nonprestressed tensile reinforcement, mm^2 (in.^{2})
A'_s = area of nonprestressed compressive reinforcement, mm^2 (in.^{2})
A_v = amount of FRP shear reinforcement within spacing s, mm^2 (in.^{2})
$A_{v,min}$ = minimum amount of FRP shear reinforcement within spacing s, mm^2 (in.^{2})
a = depth of the equivalent compression block, mm (in.)
b = width of compression face of member, mm (in.)
b_w = width of web (width of a rectangular cross section), mm (in.)
c = depth of neutral axis, mm (in.)
c_u = depth of neutral axis at ultimate, mm (in.)
DI = deformability index
d = depth to the FRP tendon (distance from extreme compression fiber to centroid of tension reinforcement), mm (in.)
d_b = diameter of reinforcing bar, mm (in.)
d_e = effective depth, mm (in.)
d_i = depth to the FRP tendon in layer i, mm (in.)
d_p = depth from the concrete top fiber to the centroid of the prestressing tendon, mm (in.)
d_{pu} = depth of the external prestressing tendons at ultimate limit state, mm (in.)
E_c = modulus of elasticity of concrete, MPa (psi)
E_f = modulus of elasticity of the fibers, MPa (psi)
E_p = modulus of elasticity of the prestressed tendon, MPa (psi)
E_r = elastic modulus of the resin, MPa (psi)
e_{eff} = effective eccentricity, mm (in.)
f = maximum stress in tendon due to combined axial load and harping, MPa (psi)
f_c = concrete stress in the extreme compression fiber, MPa (psi)
f'_c = specified compressive strength of concrete, MPa (psi)
f'_{ci} = compressive strength of concrete at time of initial prestress, MPa (psi)
f_{fb} = stress in the bent FRP stirrup, MPa (psi)
f_{fu} = ultimate tensile strength of the FRP stirrup, MPa (psi)
f_h = stress in tendon induced by harping, MPa (psi)
f_i = FRP stress in tendon layer i, MPa (psi)
f_m = stress increase in bottom layer of FRP tendons due to flexure, MPa (psi)
f_p = tendon stress, MPa (psi)
f_{pe} = effective stress in tendon (after allowance for all prestress losses), MPa (psi)
f_{pu} = design ultimate tensile strength of prestressed FRP tendon and anchorage system, MPa (psi)
f_y = yield strength of nonprestressed tensile reinforcement, MPa (psi)
f'_y = yield strength of nonprestressed compressive reinforcement, MPa (psi)
h_f = depth of the flange in T-section, mm (in.)
I_{cr} = cracked moment of inertia, mm^4 (in.^{4})
I_e = effective moment of inertia, mm^4 (in.^{4})
I_g = gross moment of inertia, mm^4 (in.^{4})
k = ratio of neutral axis depth to FRP tendon depth
k_b = bond correction factor

k_u = ratio of neutral axis depth to FRP tendon depth for an over-reinforced section
L = effective span of beam, length of the prestressing tendon between anchorages, mm (in.)
L_d = development length, mm (in.)
L_{fb} = flexural bond development length, mm (in.)
L_t = transfer length, mm (in.)
l_{tfh} = tail length of the FRP stirrup, mm (in.)
M = unfactored live load moment, N-mm (in.-lb)
M_a = maximum moment in a member at which the deflection is being computed, N-mm (in.-lb)
M_{cr} = cracking moment, N-mm (in.-lb)
M_n = nominal moment capacity, N-mm (in.-lb)
m = number of layers of tendons
n = modular ratio = E_f/E_c
n_r = modular ratio of the resin = E_r/E_f
P_j = jacking load, N (lbf)
R = radius of curvature of the harping saddle (radius of saddle), mm (in.)
r = radius of the bend of the FRP stirrup or radius of FRP tendon, mm (in.)
REL = relaxation losses, %
R_d = depth reduction factor, %
R_f = relaxation component due to relaxation of the fibers, %
R_p = relaxation component due to the relaxation of the polymer matrix
R_s = relaxation component due to straightening of the fibers
R_t = radius of the tendon, mm (in.)
S_d = spacing of the deviators, mm
s = stirrup spacing or pitch of continuous spirals, mm (in.)
y_{cr} = distance from the extreme compression fiber to the neutral axis of the cracked section, mm (in.)
y_{eff} = distance from the extreme compression fiber to the effective neutral axis, mm (in.)
V_c = shear resistance provided by concrete, N (lbf)
V_{frp} = shear resistance provided by FRP stirrups, N (lbf)
V_n = nominal shear strength of a reinforced concrete cross section, N (lbf)
V_p = vertical component of prestressing force, N (lbf)
V_s = shear resistance provided by steel stirrups, N (lbf)
V_u = factored shear load, N (lbf)
v_f = volume of fibers in the tendon
v_r = volume of resin in the tendon
α = factor used in Eq. (3-24)
α_{fb} = factor for the flexural bond length of FRP tendon
α_t = factor for the transfer length of FRP tendon
β_1 = stress-block factor for concrete
β_b = factor for softening equivalent moment of inertia
Δf_p = change in stress in an unbonded prestressed tendon due to flexure, MPa (psi)
$\Delta\varepsilon_p$ = change in strain in prestressed tendon due to flexure
ε_{cp} = average strain in concrete at centroid of the tendon
ε_{cu} = ultimate strain in concrete in compression
ε_d = additional strain in tendon that causes the extreme precompressed fiber to reach zero strain (decompression)
ε_f = strain available in tendon for flexure after decompression
ε_p = tendon strain
ε_{pe} = effective strain in the FRP tendon after all losses
ε_{pi} = initial elastic strain in FRP tendon
ε_{pr} = loss of strain capacity due to sustained loads
ε_{ps} = strain in the FRP tendon at service
ε_{pu} = ultimate tensile strain in the prestressed FRP tendons
ϕ = strength reduction factor
ϕ_{bend} = strength reduction for bent stirrup
λ = material constant
ρ = reinforcement ratio
ρ_b = balanced reinforcement ratio
ρ_i = reinforcement ratio for reinforcement at level i
Ω = strain reduction coefficient
Ω_u = strain reduction coefficient at ultimate
ξ = ratio of effective stress to ultimate tensile strength
ψ_i = ratio of depth of tendon layer i to depth of bottom tendon layer

CHAPTER 2—FRP TENDONS AND ANCHORAGES

2.1—FRP tendon characterization

FRP tendons are produced from a wide variety of fibers, resins, shapes, and sizes. Only aramid and carbon fibers, however, are recommended in this document. Glass fibers have poor resistance to creep under sustained loads and are more susceptible to alkaline degradation than carbon and aramid fibers. An FRP tendon is identified by the type of fiber used to make the tendon. The Japanese have developed a system to define FRP tendons (Table 2.1).

The tendon and anchorage comprise a system. The tendon properties are developed for a specific anchorage. Using an alternative anchorage may not result in the same strength and performance properties of the system.

2.2—Commercial tendons

Properties and characteristics of commercially available tendons are summarized in Table 2.2 and are based on the manufacturer's published data. The trade names of the products are used for clarity and historical perspective. Not all tendons are currently available, and the properties of the tendons are subject to change. Some of these tendons are shown in Fig. 1.1.

2.3—Description of tendons

Arapree®—Arapree® was a combined development of AKZO Chemicals and Hollandsche Beton Groep in The

Netherlands. The manufacturing rights have been transferred to Sireg S.P.A., an Italian company. Arapree® consists of aramid (Twaron®) fibers embedded in epoxy resin. Two types of cross section are available in the marketplace: rectangular and circular. The former is easier to grip with a wedge-type anchorage (Gerritse and Werner 1988).

The anchorage developed for Arapree®, both flat and round rod types, consists of a tapered metal sleeve into which the tendon is either grouted (post-tensioning application) or clamped between two wedges. The wedge anchorage, designed primarily for temporary use, is comprised of a steel sleeve and two semicylindrical tapered wedges made of Polyamide PA6. The outer surface of the wedges and the inner surface of the metal socket are smooth and noncoated. The inner surface of the wedge, which holds the tendon, is similarly smooth, and gripping of the tendon relies solely on the frictional resistance provided by the plastic.

FiBRA—Mitsui Construction Company of Japan produces an FRP rod known as FiBRA. Braiding of fiber tows, followed by epoxy resin impregnation and curing, form the rod. Aramid and carbon fibers have been used, with aramid being the most common. Small variations in the manufacturing process produce two types of rod: rigid and flexible. Rigid rods are used for concrete reinforcement, whereas flexible rods, which can be coiled, are used as prestressing tendons (Tamura, Kanda, and Tsuji 1993).

FiBRA has two different types of anchorage: a resin-potted anchorage used for single-tendon anchoring and a wedge anchorage for either single- or multiple-tendon anchoring. This wedge anchorage is made of four steel wedges (held together by an O-ring) that slip inside a steel sleeve with a conical interior surface. Grit is applied to the inner surface of the wedges. The exterior surface of the wedges and the interior surface of the steel sleeve are coated with a dry lubricant to assist in seating and removal of the anchorage.

Technora®—Technora® tendon is an aramid product jointly developed by Sumitomo Construction Company and Teijin Corporation, both of Japan. Technora® is a spirally wound pultruded tendon impregnated with a vinyl ester resin (Noritake et al. 1993). Manufacturing of a spirally wound tendon begins with pultruding the impregnated straight fiber bundles through an unheated die. Identical fiber bundles are spirally wound around the tendon to produce a deformed surface. Longitudinal fiber bundles are added to the outer surface before a second spiral winding is added. The product is then cured in an oven.

Technora® tendons employ either wedge-type or potted-type anchorages. Anchorages for single rod tendons or multiple rod tendons, numbering from three to 19 rods, are

Table 2.1—Fiber identification

Fiber type	Identification
Aramid fiber	A
Carbon fiber	C
Multiple fibers	*

*In a tendon using multiple fibers, the first letter of the two fibers making up the material will be used with the dominant fiber going first (for example, CA for carbon/aramid composite with carbon having the larger volume).

Table 2.2—Properties of FRP tendons (compiled from various references)

	AFRP				CFRP	
Property	Arapree®	FiBRA	Technora®	Parafil®	Leadline™	CFCC
Fiber	Twaron	Kevlar49	Technora	Kevlar49	Carbon	Carbon
Resin	Epoxy	Epoxy	Vinyl ester	—	Epoxy	Epoxy
Fiber volume ratio	0.45	0.65	0.65	—	0.65	0.65
Density, g/cm^3	1.25	1.28	1.3	1.44	1.53	1.5
Longitudinal tensile strength, GPa	1.2 to 1.5	1.25 to 1.4	1.7 to 2.1	1.2 to 1.9	2.25 to 2.55	1.8 to 2.1
Transverse tensile strength, MPa	—	30	—	—	57	—
Longitudinal modulus, GPa	62 to 64	65 to 70	54	120 to 130	142 to 150	137
Transverse modulus, GPa	—	5.5	—	—	10.3	—
In-plane shear strength, MPa	—	4.9	—	—	71	—
In-plane shear modulus, GPa	—	2.2	—	—	7.2	—
Major Poisson's ratio	0.38	0.34 to 0.6*	0.35	—	0.27	—
Minor Poisson's ratio	—	0.02	—	—	0.02	—
Bond strength, MPa	7.7	10 to 13	10 to 16	—	4 to 20	7 to 11
Maximum longitudinal strain, %	2.4	2.0 to 3.7	3.7 to 3.8	1.5 to 1.7	1.3 to 1.5	1.57
Maximum transverse strain, %	—	—	—	—	0.6	—
Longitudinal compressive strength, MPa	—	335	—	—	1440	—
Transverse compressive strength, MPa	—	158	—	—	228	—
Longitudinal thermal expansion coefficient/°C	-2×10^{-6}	-2×10^{-6}	-3×10^{-6}	—	-0.9×10^{-6}	0.5×10^{-6}
Transverse thermal expansion coefficient/°C	50×10^{-6}	60×10^{-6}	—	—	27×10^{-6}	21×10^{-6}
Relaxation ratio at room temperature, % loss from jacking stress	11 to 14	12 at 103 h	8 to 13 at 103 h	6 to 9 at 105 h	2 to 3	0.5 to 1 at 102 h

*Pseudo Poisson's ratio due to structure of tendon.

Notes: A "—" indicates that information is not available or is not applicable. 1 MPa = 145 psi; 1 GPa = 145 ksi; 1 g/cm^3 = 62.4 lb/ft^3; and 1/°C = 1.8/°F.

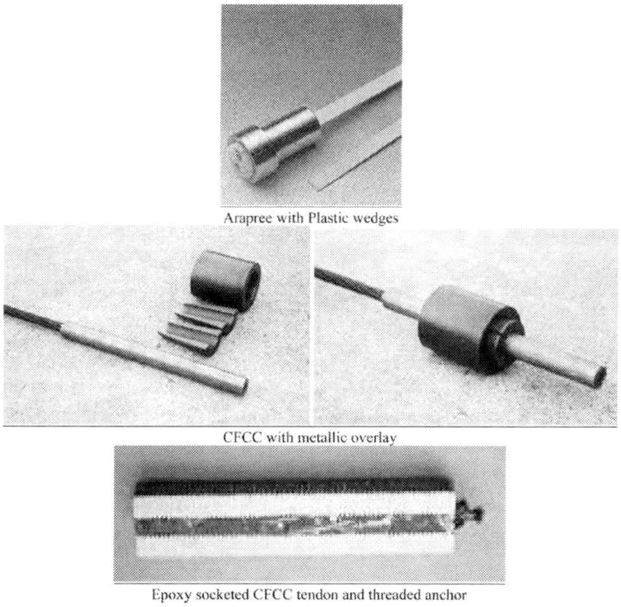

Fig. 2.1—Photos of representative anchorages.

available.A screw thread is cut into the outer surface of the housing so that the anchorage can be secured with a nut onto a bearing plate.

Parafil®—Linear Composites Limited of England is the producer of a parallel-lay tendon composed of dry fibers contained within a protective polymeric sheath (Burgoyne 1988). Parafil® was originally developed in the early 1960s to moor navigation platforms in the North Atlantic. Parafil® has several features that distinguish it from other FRP prestressing tendons: it cannot be bonded to concrete; it contains no resin; and it was not initially developed for prestressing. Nevertheless, it has been used for prestressing concrete on a number of occasions.

Parafil® tendons are anchored by means of a barrel and spike fitting, which grips the fibers between a central tapered spike and an external matching barrel. Aluminum alloy, galvanized mild steel, stainless steel, and composite materials may be used for the anchorages because this scheme takes advantage of the fibers of the tendon simply being tightly packed in the protective outer sheathing.

Leadline™— Mitsubishi Kasei Corporation of Japan has developed a CFRP tendon called Leadline™ that is pultruded and epoxy impregnated. There are several varieties of Leadline™ tendons that differ in pattern and method of fabrication of their surfaces. Smooth tendons have no surface deformations. Indented tendons have two shallow helical depressions in the surface that spiral in opposite directions. Ribbed tendons have either raised helical windings similar to the indented pattern or a circumferential winding transverse to the longitudinal axis of the tendon.

Leadline™ employs a steel wedge system to anchor the tendons with an aluminum sleeve that fits between the wedges and the tendon. The sleeve has four independent arms that extend along the length of the tendon. The wedges are placed around the sleeve so that the gap between adjacent wedges falls over the solid portion of the sleeve. The sleeve is intended to spread the stresses imposed on the tendon by the wedges. A plastic film is placed around the whole assembly to secure the multiple pieces together for insertion into the steel sleeve.

CFCC—Carbon fiber composite cable (CFCC) was developed by Tokyo Rope Mfg. Co., Ltd. and Toho Rayon Co., both of Japan. The tendon is formed by twisting a number of rods in a manner similar to a conventional stranded steel tendon (Santoh 1993). Materials used for CFCC include PAN-based carbon fiber and epoxy resin developed by Tokyo Rope. Multiple pieces of prepreg (semi-hardened tows with a resin precursor) are made into a bundle that is treated with a proprietary coating and formed into a small-diameter rod. A number of these rods are stranded and formed into a composite tendon that is heated and cured to form the finished product. The coating protects the tendon from UV radiation and mechanical damage while increasing the bond characteristics with concrete.

The manufacturer classifies CFCC anchoring methods as resin-filling and die-cast methods. The anchorages are chosen based on the intended application. The resin-filling method bonds the tendon to a steel cylinder utilizing a high-performance epoxy. These cylinders can be threaded as necessary to allow anchoring with nuts. The die-cast method attaches a die-molded alloy and steel tube to the tendon. Steel wedges then clamp the steel tube in a similar manner to steel tendon systems. Several commercial tendon/anchorage systems have been tested and evaluated by Nanni et al. (1996a,b). Some of these anchorages are shown in Fig. 2.1.

2.4—Anchorage characterization

The various types of anchorages used with FRP tendons can be classified as clamp, plug and cone, resin sleeve, resin potted, metal overlay, and split wedge anchorages, and are shown in Fig. 2.2. These anchorages are briefly described as follows.

2.4.1 *Clamp anchorage*—A clamp anchorage consists of grooved steel plates sandwiching the FRP rod and held together by bolts. The force is transferred from the tendon to the anchorage by a shear-friction mechanism and is influenced by parameters such as the roughness of the interface surfaces and the clamping force applied by the bolts. The performance of the anchorage is improved by using a sleeve material (protection media) to encase the tendon. This intermediary material with low stiffness and high ultimate elongation smooths the lateral pressure distribution on the tendon (Malvar and Bish 1995). The length of the anchorage may be varied, depending on the material chosen to ensure that the ultimate strength of the tendon is reached.

2.4.2 *Plug and cone anchorage*—The plug and cone (or barrel and spike) anchorage is made of a socket housing and a conical spike (Burgoyne 1988). Such a system is particularly well suited to anchor Parafil® tendons where the aramid fibers are not encased in resin media, but are held only by an outer protective sheath. Parafil® is unique in its ability to evenly distribute the aramid fibers around the spike, thereby achieving high anchorage efficiency. The gripping mechanism of the anchorage is similar to that of a wedge anchorage,

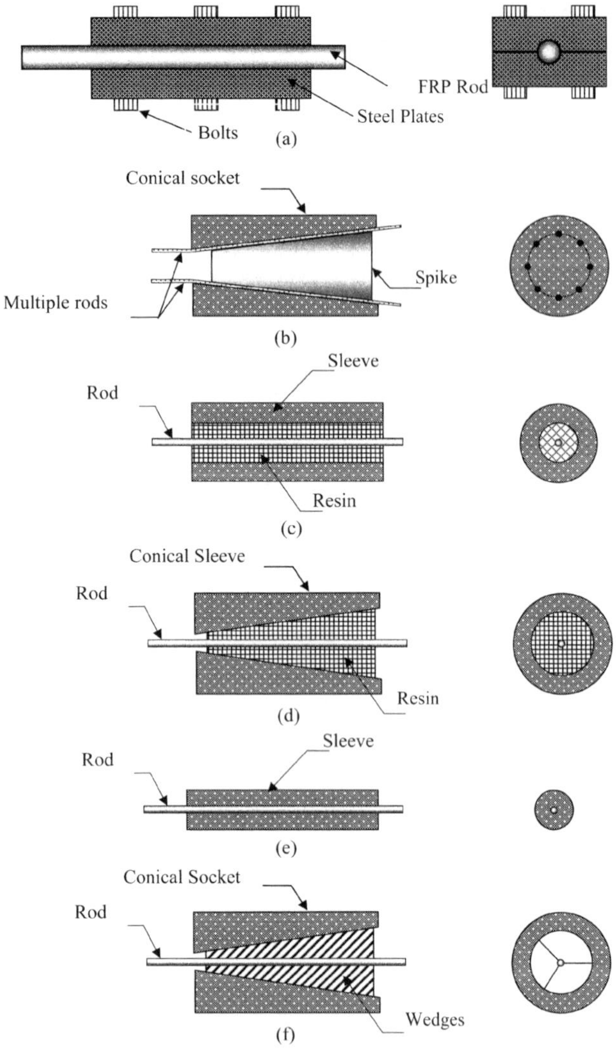

Fig. 2.2—(a) Clamp; (b) plug and cone; (c) straight sleeve; (d) contoured sleeve; (e) metal overlay; and (f) split wedge anchorages.

where the tendon is held by the compressive force applied to the fiber by inserting the spike into the barrel. This compressive stress, together with the friction between the rod material and the socket, in addition to the spike, generate a frictional stress that resists the slipping of the tendon out of the socket. The field application of this system requires the removal of the plastic sheath, combing and spreading of the individual fibers, and proper placement of the spike with a uniform distribution of fibers all around it.

2.4.3 *Straight sleeve anchorage*—In this anchorage, the FRP tendon is embedded in resin that fills a tubular metallic housing such as steel or copper. The potting material ranges from nonshrink, cement, with or without sand, to expansive cement to epoxy-based materials. In the case of nonshrink cement and polymeric potting materials, the load-transfer mechanism depends completely on bonding and interlocking between the anchorage components. The mechanism of load transfer is by bond at the interface between the rod and the filling material, and between the filling material and the metallic sleeve. To increase the bond between the anchorage components in such cases, an internally threaded sleeve is used, or a rigid filler material, such as sand, is added to the resin, or both. Filler in the resin also serves to reduce chemical shrinkage of the resin during curing. To enhance the bond between the tendon and the grout material, surface modifications such as braiding, twisting, or ribs can be applied to the tendon.

Harada and coworkers were among the earliest users of expansive cementitious materials to fill straight metallic sleeve anchorages (Harada et al. 1993). Expansive cement generates significant lateral pressure and increases the slipping resistance of the tendon. The 25 to 40 MPa (3600 to 5800 psi) internal radial pressures developed by the expansive cementitious materials are adequate for gripping FRP tendons with a wide range of surface configurations without causing undue stress concentrations at the point of egress of the tendon during quasistatic and cyclic loadings (Harada et al. 1997). Anchorages as long as 300 to 400 mm (12 to 16 in.) have been shown to be sufficient for developing the full strength of CFRP tendons without slippage at the free end of the anchorage. The effectiveness of such anchorages has also been demonstrated in elevated temperature, extended duration relaxation tests (Dye, Bakis, and Nanni 1998). At this time, the main shortcoming of this type of anchorage appears to be the 2- to 3-day curing time for expansive cementitious materials.

2.4.4 *Contoured sleeve anchorage*—The contoured sleeve anchorage has the same components as the straight sleeve anchorage. The principal difference between the two systems is the varied profile of the inner surface of the contoured sleeve, which may be linearly tapered or parabolically tapered. The load-transfer mechanism from the tendon to the sleeve is by interface shear stress, which is a function of bonding, and radial stress produced by the variation of the potting material profile. A conical profile with a constant taper angle is the most popular type of resin potted anchorage. Kim and Meier (1991) developed a variable stiffness anchorage for CFRP tendons. The work was based on concepts of a commercial anchorage developed by the Swiss company BBR. The anchorage was made of a cone filled with an epoxy matrix containing high-modulus ceramic filler. Holte, Dolan, and Schmidt (1993) described a parabolic anchorage with epoxy/sand filler.

The following parameters affect the performance of a resin potted anchorage: length of the anchorage, angle of the anchorage cone, front radius of anchorage cone, modulus of elasticity of the potted material, and length and modulus characteristic of the "soft zone" in resin filler at the front of the anchorage. Potted anchorages often fail through pullout of the tendon from the resin/grout rather than by rupture of the tendon. Practical drawbacks with this anchor include precutting the tendons to length and the curing time for the potting material.

2.4.5 *Metal overlaying*—The die-cast wedge system for CFCC requires that the tendon length be predefined so that a metal tube can be cast onto the tendon at a specific location during fabrication with the result that adjustment on site is limited. The metal overlay is added to the ends of the tendon by means of die-molding during the manufacturing process.

The die-cast molding can then be gripped at the location of the metal material using a typical wedge anchorage. The use of this system is limited because of the inflexibility in the specified length of the tendon. The load transfer in this anchorage is achieved by shear (friction) stress, which is a function of the radial compressive stress and the friction at contact surfaces.

2.4.6 *Split-wedge anchorage*—Split-wedge anchorages are generally preferred because of their compactness, ease of assembly, reusability, and reliability. This type of anchorage can be subdivided into two categories: systems with direct contact between plastic or steel wedges and the tendon, and systems using a sleeve between the wedges and the tendon. Wedge anchorages are widely used in anchoring steel tendons but should be modified for use with FRP tendons by increasing their length to reduce transverse stress on the tendon and controlling roughness in the wedge to prevent notching the tendon. The number of the wedges in the split-wedge anchorage varies from two to six wedges inserted into the barrel. The main reason for increasing the number of wedges is to provide a smoother lateral stress distribution in the radial direction of the tendon. The mechanism of gripping relies on friction and clamping force between the wedges, barrel, and tendon. Using a small taper on the wedges is of great importance to provide a smooth and uniformly distributed transverse stress.

A metallic anchorage was developed as part of the ISIS Canada program (Sayed-Ahmed and Shrive 1998; Campbell et al. 2000) for 8 mm Leadline™ CFRP tendons. The anchorage consists of three components: a stainless steel barrel with a conical socket; a four-piece stainless steel conical wedge set and a thin, soft metal sleeve that is placed between the wedges and the tendon. The distinct feature of the anchorage is that the taper angle of the wedge is 0.1 degrees greater than that of the inner surface of the barrel. The difference in angle between the barrel and the wedges helps to produce more desirable radial stress distribution on the tendon and ensures that failure of the tendon occurs outside the anchorage. An experimental and analytical investigation of this anchorage has been reported by Al-Mayah, Soudki, and Plumtree (2001). Nonmetallic versions of this anchorage, in which the elements are made either from ultra-high-performance concrete (UHPC), where the barrel is wrapped with CFRP sheet, or from carbon fiber reinforced reactive powder concrete, have been developed and tested by Reda Taha and Shrive (2003a,b) and Shaheen (2004), respectively.

2.4.7 *Failure modes of anchorages*—Various failure modes have been observed with wedge anchorages and FRP tendons, and these are summarized by separating them into two main categories: failure of the anchorage system, and rupture.

- **Failure of the anchorage system**—This kind of failure can be classified into four modes:

1. *Movement or slip of the tendon out of the anchorage caused by insufficient grip (low shear force) between the tendon and sleeve*—Grip can be increased by increasing the friction at the contact surfaces, by increasing the normal force applied, or both.

2. *Slip of the sleeve and tendon together relative to the wedges*—This indicates a high shear force between the tendon and sleeve together with a lower shear force between the sleeve and wedges. This can be overcome in the same manner as mentioned in 1.

3. *Slip of the wedges relative to barrel*—This rarely happens, mainly because of the design and geometrical configuration of the wedges and the barrel. It is often accompanied by crushing of the tendon.

4. *Rupture of the rod inside the anchorage*—High stress concentrations can be generated in the tendon inside the anchorage, causing damage of the fibers. An anchorage design that results in low stress concentration and uniform load distribution in the anchor overcomes this problem.

- **Failure of the tendon outside the anchorage**—If the tendon does not break in or within three diameters of the anchorage, then the anchorage is not contributing to the failure of the tendon and is considered a satisfactory anchorage design.

CHAPTER 3—FLEXURAL DESIGN

3.1—General considerations

Numerous flexural tests on FRP prestressed members are reported in the literature. Comprehensive testing began in the mid 1980s in Japan on beams with carbon and aramid tendons, and in Europe on beams with aramid and glass tendons (Gerritse and Werner 1991). The first tests in the United States were conducted on beams with aramid tendons (Dolan 1990). In the intervening years, several attempts have been made to combine these data into design or performance recommendations (FIP 1992; JSCE 1992). The lack of uniform testing protocols and consistent reporting procedures made completion of these criteria very difficult.

A conventional prestressed concrete beam with steel tendons will deform elastically until cracking, and then the rate of member deflection will progressively increase as the tendons yield until failure occurs by concrete crushing or tendon rupture. On the other hand, an FRP prestressed beam will deform elastically until cracking, then continue to deform in an approximate linear manner under increasing load until the tendon ruptures or the ultimate concrete compression strain is exceeded. These two behaviors are compared in Fig 3.1. The lower modulus of elasticity of FRP (Table 2.2) is reflected in the lower post-cracking behavior.

Flexural design limit states include service level stresses and flexural strength requirements. Service level stresses may be computed using techniques similar to those for conventional prestressed concrete. The loss of prestress in FRP tendons is caused by the following factors:

- Anchorage seating at transfer of prestress;
- Creep of the concrete;
- Shrinkage of the concrete;
- Elastic shortening of the concrete; and
- Relaxation of the tendons.

Anchorage seating loss is a function of the tendon system. Losses due to creep, shrinkage, and elastic shortening of concrete may be calculated using standard methods for concrete sections prestressed with steel tendons (PCI 1975,

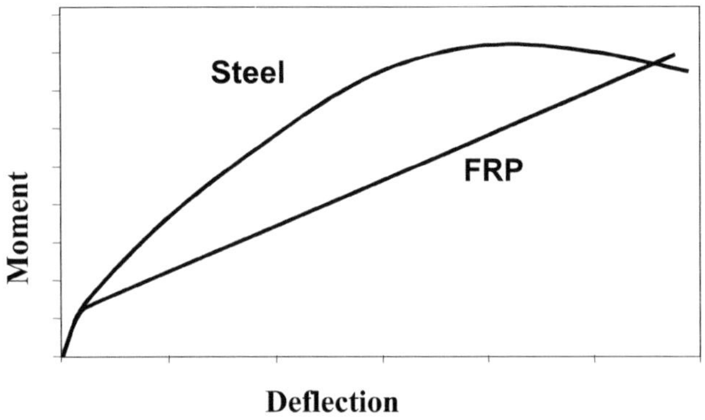

Fig. 3.1—Schematic representation of moment-deflection responses of prestressed concrete elements.

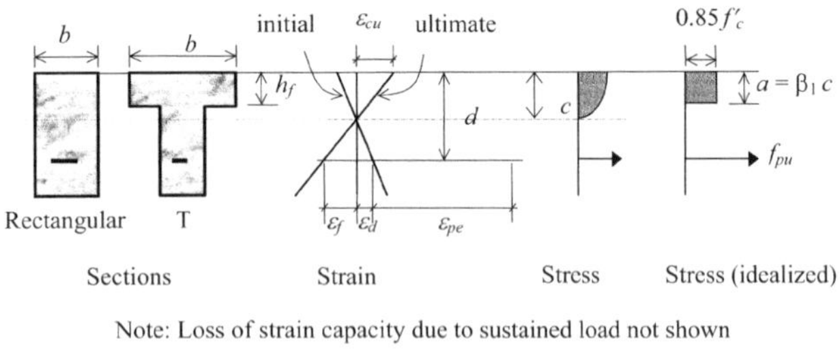

Fig. 3.2—Balanced ratio: stress and strain conditions.

2000); however, the modulus of elasticity of the FRP tendon should be used in place of the modulus of elasticity of steel. Losses for FRP tendons due to these three sources are typically less than the corresponding losses for steel tendons due to the lower modulus of elasticity of FRP tendons. Relaxation losses are more problematic and are less well understood, as there is little experimental data available that describes relaxation loss profiles for FRP tendons. Relaxation characteristics vary with the fiber type (Table 2.2) and are projected to be less than 12% of the initial level of prestress over the life of the structure (ACI 440R; JSCE 1993). More details are given in Section 3.10.

3.2—Strength design methodology

The approach to strength design of FRP prestressed beams is based on the concept of a balanced ratio, which is defined as the reinforcement ratio that simultaneously results in rupture of the tendons and crushing of the concrete. Concrete failure is taken as an extreme compression strain of ε_{cu} = 0.003. This strain limit assumes no allowance is made for confinement of concrete due to the presence of closed stirrups. A rectangular stress block is used to model the concrete behavior. Tendon failure is defined as occurring when the strain in the tendon reaches the ultimate tensile strain capacity ε_{pu}.

3.3—Balanced ratio

Figure 3.2 defines the cross section and the strain and stress conditions in a bonded section for the condition where the tendon ruptures simultaneously with compression failure of the concrete. The derivation is for a rectangular section or a T-section with a single layer of prestressed reinforcement where the compression block is within the depth of the flange, that is, $a < h_f$.

If the total strain capacity of the tendon is denoted by ε_{pu}, then the amount of strain available for flexure, ε_f, is the total strain capacity less the strain used for prestressing, ε_{pe}, the strain used to decompress the concrete, ε_d, and any loss of strain capacity due to sustained loads, ε_{pr}. This relationship is given in Eq. (3-1).

$$\varepsilon_f = \varepsilon_{pu} - \varepsilon_{pe} - \varepsilon_d - \varepsilon_{pr} \quad (3\text{-}1)$$

The strain compatibility from Fig. 3.2 allows determination of the c/d ratio in terms of the available strains. Therefore, by using similar triangles

$$\frac{c}{d} = \frac{\varepsilon_{cu}}{\varepsilon_{cu} + \varepsilon_{pu} - \varepsilon_{pe} - \varepsilon_d - \varepsilon_{pr}} \quad (3\text{-}2)$$

Equilibrium on the cross section equates the tensile force in the tendon to the compressive force on the concrete. Hence,

$$0.85 f_c' \beta_1 c b = \rho b d f_{pu} \quad (3\text{-}3)$$

where β_1 = 0.85 for concrete strengths up to 27.5 MPa (4000 psi), after which it is reduced at a rate of 0.05 for each 6.9 MPa (1000 psi) of strength in excess of 27.5 MPa (4000 psi) to a minimum value of 0.65.

Solving Eq. (3-3) for the balanced reinforcement ratio, $\rho = \rho_b$, where $\rho = A_p/bd$ is the prestress reinforcement ratio, gives

$$\rho_b = 0.85\beta_1 \frac{f_c'}{f_{pu}} \frac{c}{d} \quad (3\text{-}4)$$

Substituting the expression for c/d from Eq. (3-2) into Eq. (3-4) gives the balanced ratio in terms of material properties.

$$\rho_b = 0.85\beta_1 \frac{f_c'}{f_{pu}} \frac{\varepsilon_{cu}}{\varepsilon_{cu} + \varepsilon_{pu} - \varepsilon_{pe} - \varepsilon_d - \varepsilon_{pr}} \quad (3\text{-}5)$$

An examination of the strain components in Eq. (3-5) leads to a simplification. First, the strain loss due to sustained loads is nearly zero if the strain due to sustained load is less than 50% of the ultimate tensile strain and is recovered at the nominal strength condition. Second, the decompression strain, ε_d, is typically an order of magnitude less than the flexural strain. Setting these two strain values to zero gives the following simplified definition for ρ_b.

$$\rho_b = 0.85\beta_1 \frac{f_c'}{f_{pu}} \frac{\varepsilon_{cu}}{\varepsilon_{cu} + \varepsilon_{pu} - \varepsilon_{pe}} \quad (3\text{-}6)$$

The strain ($\varepsilon_{pu} - \varepsilon_{pe}$) represents the strain available for flexure in the tendon. FRP tendons have ultimate strains ranging from 1.3 to 3.8% (Table 2.2). Assuming that 40 to 50% of the strain capacity remains after prestressing, there is 0.006 to 0.019 strain capacity available in the FRP tendon to allow the flexural member to deform and crack before failure. This reserve provides substantial deflection, thereby giving visual warning of overload conditions, and concrete cracking occurs at 0.002 net tensile strain.

The effective strain ε_{pe} is known because the designer selects it, and the manufacturer provides the ultimate stress and strain capacity for the tendon. The designer does need to know the basis for the selection of these ultimate values because some manufacturers provide the mean ultimate stress and strain while others, typically Japanese manufacturers, provide a strength value equal to the mean less three standard deviations. Eventually, the industry will establish a standard for reporting that is consistent with the specified capacity reduction factors for FRP tendons. ACI 440.1R refers to the guaranteed tensile strength as mean strength less three standard deviations.

3.4—Flexural design and capacity prediction

3.4.1 *Bonded construction*—The flexural behavior of a beam may be described according to whether the critical section is a compression-controlled or a tension-controlled section. A compression-controlled section condition occurs when the reinforcement ratio ρ is greater than ρ_b and the concrete crushes with no failure of the tendons. When ρ is less than ρ_b, a tension-controlled section condition occurs as a result of rupture of the tendons before crushing of the concrete.

3.4.1.1 *Tension-controlled section: $\rho \le \rho_b$*—For reinforcement ratios less than ρ_b, the beam strength is governed by the tensile strength of the tendon, and the section is described as tension-controlled. In this case, the concrete strain will not reach 0.003 at failure of the beam, and, strictly speaking, the use of the rectangular stress block assumption would not be valid. For sections with $0.5\rho_b < \rho < \rho_b$, however, the concrete stress distribution will be substantially nonlinear at failure and thus an equivalent rectangular stress distribution can be assumed. Further, studies of the nominal capacity of lightly reinforced beams ($\rho < 0.5\rho_b$), where the stress in the concrete approximates a linear stress distribution, indicate use of a rectangular stress block assumption produces less than 3% error compared to an elastic analysis of a cracked section (Dolan and Burke 1996). Therefore, nominal moment capacity of a tension-controlled section with a single layer of reinforcement is developed using strength design based on a rectangular stress block.

Summing moments about the compression centroid (Fig. 3.2) defines the nominal moment capacity as

$$M_n = \rho bdf_{pu}\left(d - \frac{a}{2}\right) \tag{3-7}$$

where a is computed from equilibrium of forces on the cross section, giving

$$a = \frac{\rho d f_{pu}}{0.85 f_c'} \tag{3-8}$$

Combining the value for a with the nominal moment equation provides a combined form for prediction of nominal moment capacity

$$M_n = \rho bd^2 f_{pu}\left(1 - \frac{\rho f_{pu}}{1.7 f_c'}\right) \tag{3-9}$$

The nominal capacity of a section in which the prestressing tendons are not in a single layer is discussed in Section 3.4.2.

3.4.1.2 *Compression-controlled section: $\rho \ge \rho_b$*—In a beam with $\rho \ge \rho_b$, the concrete will fail in compression before failure of the tendon. The stress and strain compatibility is the same as in Fig. 3.2; however, the value of the tendon strain is not known. This condition is analyzed by locating a neutral axis, assuming a linear elastic tendon and a rectangular stress block for the concrete. This is done by defining the strain in the tendon, equilibrating the horizontal forces on the section, solving for the neutral axis location, and finally summing moments about the tendon location.

With this amount of reinforcement in the beam, the concrete will have developed a highly nonlinear stress-strain relation so that the use of an equivalent rectangular stress block is appropriate. The depth of the neutral axis, c, can be found by considering axial force equilibrium of the cross section

$$\rho bdf_p = 0.85 f_c' b\beta_1 c \tag{3-10}$$

Because FRP tendons are linear elastic to failure, the stress in the tendon f_p is a function only of the modulus of elasticity and the strain in the tendon

$$f_p = \varepsilon_p E_p \tag{3-11}$$

The flexural strain in the tendon can be determined from the strain diagram at failure using similar triangles

$$\varepsilon_f = \varepsilon_{cu}\frac{d-c}{c} \tag{3-12}$$

The total strain in the tendon can be expressed as the sum of the flexural strain and the effective prestressing strain ε_{pe}

$$\varepsilon_p = \varepsilon_{pe} + \varepsilon_{cu}\frac{d-c}{c} \tag{3-13}$$

Substitution of Eq. (3-13) and (3-11) into Eq. (3-10) and defining $k_u = c/d$ results in

$$\rho\left(\varepsilon_{pe} + \varepsilon_{cu}\frac{1-k_u}{k_u}\right)E_p = 0.85 f_c' \beta_1 k_u \tag{3-14}$$

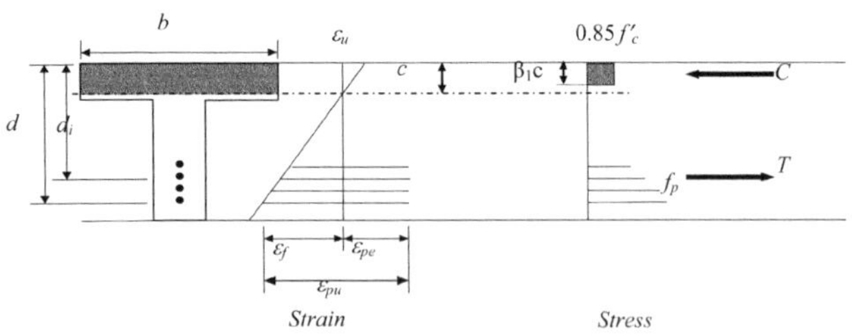

Note: Loss of strain capacity due to sustained load not shown

Fig. 3.3—Section with vertically distributed tendons.

Defining a material constant λ such that

$$\lambda = \frac{E_p \varepsilon_{cu}}{0.85 f_c' \beta_1} \tag{3-15}$$

and substituting Eq. (3-15) into Eq. (3-14) allows the resulting quadratic equation to be solved for k_u, giving

$$k_u = \sqrt{\rho\lambda + \left(\frac{\rho\lambda}{2}\left(1 - \frac{\varepsilon_{pe}}{\varepsilon_{cu}}\right)\right)^2} - \frac{\rho\lambda}{2}\left(1 - \frac{\varepsilon_{pe}}{\varepsilon_{cu}}\right) \tag{3-16}$$

The nominal moment capacity can be determined by summing moments about the tendon centroid, giving

$$M_n = 0.85 f_c' b \beta_1 k_u d^2 \left(1 - \frac{\beta_1 k_u}{2}\right) \tag{3-17}$$

3.4.2 *Development of flexural capacity for vertically distributed tendons in an under-reinforced section*—The above derivations assume that the prestressing tendons are all located at the same depth from the compression face. A method for determining the strength of vertically aligned tendons, based on the stress and the strain distribution, shown in Fig. 3.3, is presented in this section (Dolan and Swanson 2002).

The analysis is developed for a T-section with the assumption that the neutral axis is in the flange. Strain due to decompression of the concrete is small and has been neglected. Strains due to elastic and nonelastic shortening of the member are neglected as these strains are regained at the maximum curvature.

All tendons are assumed to be stressed to the same level f_{pe} and thus the stress increase at failure in the bottom tendon may be defined as $f_m = f_{pu} - f_{pe}$. Failure of the bottom tendon by rupture will lead to failure of the other tendons. The stress in each tendon can be determined from its strain, which can be expressed as a ratio of the strain at any level to the strain in the bottom tendon.

$$f_i = f_{pe} + f_m \frac{d_i - c/d}{d - c/d} = f_{pe} + f_m \left(\frac{\frac{d_i}{d} - \frac{c}{d}}{1 - c/d}\right) \tag{3-18}$$

where d_i is the depth of each individual tendon, and d is the depth of the bottom tendon.

Defining the initial prestress ratio $\xi = (f_{pe}/f_{pu})$ and $c = kd$, and assuming $\rho_i = (A_{pi}/bd)$ is the reinforcement ratio at level i, it can be shown that

$$k = \frac{\sqrt{\left(n\sum_{i=1}^{m}\rho_i\right)^2 + 2(1-\xi)n\sum_{i=1}^{m}\rho_i\left(\xi + \frac{d_i}{d}(1-\xi)\right)} - n\sum_{i=1}^{m}\rho_i}{1-\xi} \tag{3-19}$$

where m is the number of layers of tendons.

Defining the depth ratio of the tendons as $\psi_i = (d_i/d)$ and assuming uniformly prestressed tendons, the moment capacity is

$$M_n = bd^2 \sum_{i=1}^{m} \rho_i f_i \left(\psi_i - \frac{\beta_1 c}{2d}\right) \tag{3-20}$$

If the tendons through the depth of the member are prestressed to different levels, the nominal moment capacity is given as

$$M_n = bd^2 \sum_{i=1}^{m} \left\{\rho_i \left[f_{pei} + f_m \left(\frac{\psi_i - c/d}{1 - c/d}\right)\left(\psi_i - \frac{\beta_1 c}{2d}\right)\right]\right\} \tag{3-21}$$

where f_{pei} is the effective prestress at the level i.

Most beams have a relatively small number of layers of tendons. Therefore, Eq. (3-20) and (3-21) are readily solved using a spreadsheet.

3.4.3 *Design implications of vertically distributed tendons*—Equations (3-20) and (3-21) provide a means to optimize the design of FRP prestressed beams. The prestressing ratio ξ may be varied at each depth so that the final stress distribution in the tendons is uniform with depth at maximum curvature. The reinforcement ratio may also be varied with depth to account for beams with mixed strand patterns such as harped strands and a large number of straight strands in the bottom flange.

3.4.4 *Unbonded construction*—For an unbonded prestressed member where strain compatibility between the concrete and the tendons cannot be assumed, the stress in the tendons at ultimate for an under-reinforced beam may be determined using the approach suggested by Naaman et al. (2002). This approach uses a strain-reduction coefficient Ω_u to relate the strain in an unbonded tendon to that in an equivalent bonded tendon. Knowing the strain in the tendons, the moment capacity may be computed as above for a bonded section. A description of this approach is given in Section 7.1. Bakis et al. (2001) applied this approach in beams post-tensioned with unbonded CFRP tendons, and obtained acceptable correlation between predicted and measured load-deflection relationships.

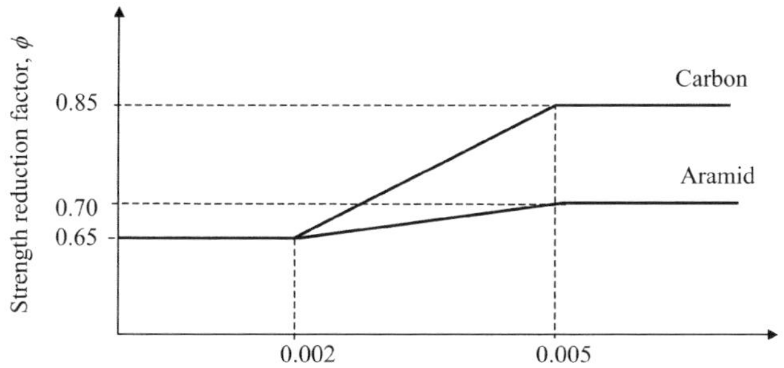

Fig. 3.4—Variation in strength reduction factor with net tensile strain.

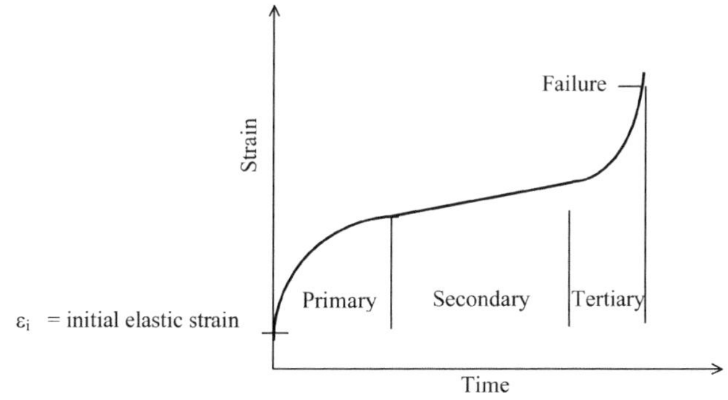

Fig. 3.5—The three stages of creep deformation.

3.5—Strength reduction factors for flexure

The experimental strength from flexural tests on members prestressed with FRP tendons reported in the literature was compared to the strength predicted by the previous equations by Dolan and Burke (1996). The analysis of strength-reduction factors examined tendon type and failure mode. The strongest correlation was to tendon type. Capacity reduction factors for aramid and carbon tendons are presented in Table 3.1. ACI 440.1R recommends different capacity reduction factors for nonprestressed reinforcement. In that case, however, the reinforcement is predominately glass fibers. Thus, there is a consistent philosophy that performance is tied to reinforcement materials.

ACI 318-02 presents a unified approach to variation in strength reduction factors based on net tensile strain in the reinforcement. In keeping with this philosophy, a transition zone between tension-controlled and compression-controlled sections is presented. Tension-controlled sections have a net tensile strain, strain in the reinforcement farthest from the compression face, greater than or equal to 0.005, while compression controlled-sections have a net tensile strain less than or equal to 0.002. The transition in strength reduction factor is indicated in Fig. 3.4.

3.6—Flexural service stresses

Concrete service load stresses are the same as specified by the AASHTO Standard Specification for Highway Bridges and are given in Table 3.2. These values are the same as ACI 318-02, except that the limitations on the tensile stresses in the concrete are more restrictive. Construction to these limits meets the requirements for a Class U member in ACI 318-02.

Table 3.1—Recommended strength reduction factors (Dolan and Burke 1996)

Tendon type	Strength reduction factor φ	Condition
Aramid	0.70	Tension-controlled behavior
Carbon	0.85	
Aramid or carbon	0.65	Compression-controlled behavior

Table 3.2—Allowable concrete stresses

Allowable stress at transfer of prestress (before losses)	Units, MPa (psi)
(a) Extreme fiber stress in compression	$0.6f'_{ci}$ $(0.6f'_{ci})$
(b) Extreme fiber stress in tension except for (c)	$0.25\sqrt{f'_{ci}}$ $(3\sqrt{f'_{ci}})$
(c) Extreme fiber stress in tendon at ends	$0.5\sqrt{f'_{ci}}$ $(6\sqrt{f'_{ci}})$
Allowable stresses under service loads (after losses)	Units, MPa (psi)
(a) Extreme fiber stress in compression due to prestress plus sustained loads	$0.45f'_c$ $(0.45f'_c)$
(b) Extreme fiber stress in compression due to prestress plus total loads	$0.6f'_c$ $(0.6f'_c)$
(c) Extreme fiber stress in precompressed tensile zone	$0.5\sqrt{f'_c}$ $(6\sqrt{f'_c})$

Table 3.3—Allowable tendon stresses at jacking

Allowable jack stresses	
Carbon	$0.65f_{pu}$
Aramid	$0.50f_{pu}$
Allowable stress immediately following transfer	
Carbon	$0.60f_{pu}$
Aramid	$0.40f_{pu}$

3.7—Jacking stresses

Steel tendons are typically stressed to 85% of their yield stress or approximately 0.005 strain. Allowable stresses in FRP tendons are typically limited to 40 to 65% of their ultimate strength due to stress-rupture limitations (Dolan et al 2000). This lower range of allowable stress actually corresponds to strains between 0.008 and 0.012, or 1.5 to 2.5 times the typical prestressing strain used in steel tendons. The recommended maximum jacking stresses for FRP tendons are given in Table 3.3. These stress limitations, which are based on creep-rupture characteristics of the tendons, should not exceed 65% of the maximum stress that can be developed when tested using the anchorage specified by the manufacturer.

3.8—Creep rupture of FRP tendons

Creep is the inelastic strain of a material under a sustained load over a period of time. Creep rupture is the tensile fracture of a material subjected to sustained high stress levels over a period of time. There are three stages of creep: primary, secondary, and tertiary, as shown in Fig. 3.5. The primary stage, characterized by a continually decreasing strain rate, confined to a short time (relative to creep rupture time) immediately following application of load, is common for FRP composites. The second stage is defined as the time period in which the strain rate is constant under constant stress. In this period, some of the weaker fibers may fail, but the friction or resin adhesion between the fibers transfers the load to adjacent fibers. If the stress level is low enough, fiber

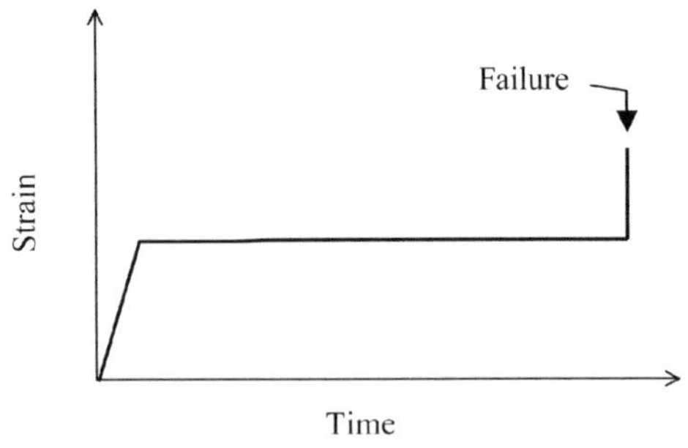

Fig. 3.6—Carbon creep-rupture curve.

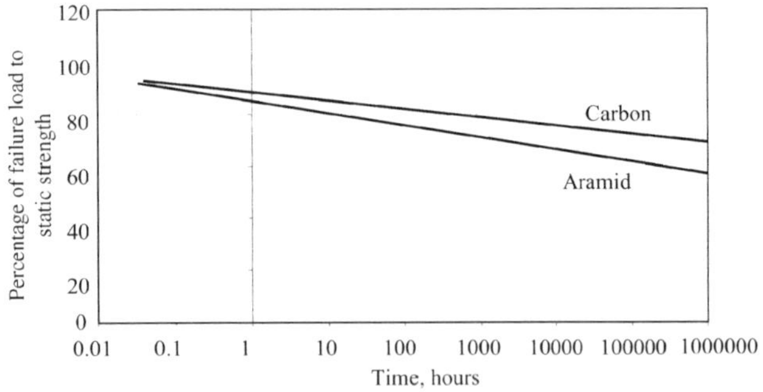

Fig. 3.7—Comparison of creep-rupture curves for aramid and carbon FRP rods under environmental exposure.

damage is confined to the secondary creep level, and the tendon has an unlimited service life. The tertiary stage, characterized by an increasing strain rate, represents rapid, progressive failure of the fibers until the tendon ruptures.

The creep strain behavior of AFRP tendons typically follows the three stages of creep: primary, secondary, and tertiary (Fig. 3.5). The strain rate of CFRP tendons during the secondary stage drops to zero. These tendons exhibit near-instantaneous failure compared to the secondary stage when the initial stress level exceeds the creep-rupture threshold (Fig 3.6).

Designers need to know the limiting percentage of ultimate load that may be applied to a tendon so that the tertiary stage is never reached. If the creep in the secondary stage is small, an equilibrium condition will be achieved and the material becomes stable.

Figure 3.7 shows the extrapolation of creep-rupture data obtained by Dolan et al. (2000) for aramid and carbon tendons encased in concrete surrounded by saltwater and subjected to a constant load. This figure shows the time to failure of the FRP tendon as a function of applied load in the tendon.

The creep-rupture curves provide allowable prestress levels for AFRP and CFRP rods for any given service life. The predicted stress level for CFRP is 70% of the tendon's short-term ultimate strength for 100 years of service. The upper limit of service load at a 100-year life for AFRP is predicted to be 55% of the tendon's short-term ultimate strength in a concrete environment.

The curves in Fig. 3.7 were extrapolated from experimental data obtained from tests conducted for over 12,000 h that used applied loads between 50 to 80% of the ultimate load. At conclusion of the tests, all unbroken tendons were tested to failure. The residual strength of the CFRP was 90% of its original static capacity, while that of the AFRP was 80% of its original capacity.

3.9—Correction of stress for harped tendons

FRP tendons behave linearly elastically to failure. Consequently, draping or harping of tendons results in a loss of tendon strength in the vicinity of the harping or draping points due to increased strain induced by curvature of the tendons. When the FRP tendon is deflected, the jacking stresses should be reduced to account for localized stress increases. Dolan et al. (2000) proposed that the stress increase due to harping, in solid and stranded tendons, can be defined by the following equation

$$f_h = \frac{E_f R_t}{R} \tag{3-22}$$

where E_f is the modulus of elasticity of the fiber; R_t is the radius of the tendon; and R is the radius of curvature of the harping saddle.

The combined stress in a tendon of cross-sectional area, A_p, at a harping saddle, due to the jacking load P_j, is given by Eq. (3-23) and should be less than the allowable stress values

$$f = \frac{P_j}{A_p} + \frac{E_f R_t}{R} \tag{3-23}$$

3.10—Relaxation and friction losses

Relaxation losses (*REL*) in FRP tendons results from three sources: relaxation of polymer R_p; straightening of fibers R_s; and relaxation of fibers R_f. The total relaxation loss, $REL = R_p + R_s + R_f$, expressed as a percentage of the stress at transfer, can be estimated by assessing these three effects separately:

Relaxation of polymer R_p—When the tendon is initially stressed, a portion of the load is carried in the resin matrix. Over time, the matrix relaxes and loses its contribution to the load-carrying capacity. This initial matrix relaxation occurs within the first 24 to 96 h and may be accelerated by heat curing of prestressed concrete beams (Dolan et al. 2000). This relaxation is affected by two characteristics of the tendon: the modular ratio of the resin to the fiber, n_r, and the volume fraction of fibers in the tendon, v_f. The modular ratio n_r is defined as the ratio of the elastic modulus of the resin, E_r, to the modulus of the fiber, E_f. The relaxation loss is the product of the volume fraction of resin, $v_r = 1 - v_f$, and the modular ratio of the resin n_r, giving

$$R_p = n_r \times v_r$$

For resins used in pultrusion operations, the modular ratio is approximately 1.5% for carbon and 3% for aramid. The resin volume fraction is typically 35 to 40% of the tendon cross section. Therefore, the total relaxation in the first phase is in the range 0.6 to 1.2% of the transfer stress. This relaxation loss can be compensated by overstressing, as long as the total stress limits given in Table 3.3 are not exceeded. Overstressing and allowing the relaxation to reduce the stress to the values given in Table 3.3 is not recommended because the loss does not occur in the fibers and the fibers would be permanently overstrained.

Straightening of fibers R_s—The fibers in a pultruded section are nearly, but not completely, parallel. Therefore, stressed fibers flow through the matrix and straighten, and this straightening appears as a relaxation loss. Straightening of the fibers is a function of the quality control of the pultrusion process. A 1 to 2% relaxation is adequate to predict this phase of the loss calculation (Dolan et al. 2000).

Relaxation of fibers R_f—Fiber relaxation is dependent on the fiber type. Carbon fibers are reported to have no relaxation; therefore, R_f for carbon may be assumed to be zero. Aramid materials creep when loaded, and this creep behavior is reflected in its relaxation behavior. The long-term relaxation for Kevlar™ fibers has been reported by DuPont to be 1 to 3% per decade on a log scale. By assuming that the relaxation count begins after the first 24 h, the total relaxation for aramids can be assumed to be 6 to 18% in a 100-year design life (Dolan 1989).

Zoch et al. (1991) performed relaxation tests on CFCC and steel strands under equivalent conditions. Strands of both materials with 12.5 mm diameter were prestressed to 80% of their proof load, and relative decrease in load over time was measured. The results showed that over a period of 100 h, the stress was reduced by 2% for CFCC, whereas steel relaxed approximately 8% from the original load. This indicates a superior relaxation resistance of CFCC to permanent stress. Santoh (1993) conducted relaxation tests on CFCC and steel strands at room temperature. They found that at 50% of the ultimate stress, the relaxation value after 100 h for CFCC was 0.48% and was 1.02% for steel; at 65% of the ultimate stress, the relaxation of CFCC was 0.81%, while that of steel was 2.28%. At 80% of the ultimate stress, it was 0.96% for CFCC and 7.35% for steel. They concluded from their tests that the relaxation of CFCC is 50% or less than that of steel.

In assessing friction loss, relevant curvature friction and wobble coefficients should be used. Such data are sparse. Burke, Nelson, and Dolan (2000) found that, for a CFRP tendon in a PVC duct, the curvature friction coefficient could range from 0.25 for stick-slip (dynamic) behavior to 0.6 for no-stick-slip (static) behavior. Because the wobble coefficient relates primarily to the type of duct, values specified for steel prestressing systems with plastic ducts may be applied to FRP tendons.

Traditional losses associated with prestressing tendons, such as friction, initial elastic shortening, and concrete creep and shrinkage, can be computed in the same manner as for prestressed steel tendons. Relevant friction and wobble coefficients should be obtained from the manufacturer of the prestressing system used.

3.11—Overall design approach

The overall design approach for FRP prestressing tendons is to use the equations for flexural behavior to establish the tendon size to meet the strength requirements of the section. A prestress level of 40 to 50% of the tendon strength is selected for the initial prestress force, and service level stresses are checked. If the section is sufficient, the flexural design is completed. If it is not, the number or size of the tendons is increased or the section size adjusted to meet service requirements, and then the strength capacity is rechecked. Nonprestressed FRP reinforcement may be used to increase the strength of a section.

3.12—Ductility or deformability

FRP tendons and concrete are both brittle materials; therefore, classical ductility, which requires plastic deformation, is difficult to obtain. Possible ways to obtain quasiductile behavior, even with brittle materials, are by confining the concrete in compression, partial prestressing, or permitting some bond slip. If ductility is defined as the energy absorbed, then the small area under a FRP prestressed load-deflection curve suggests that energy absorption ductility is limited (Naaman and Jeong 1995). Grace and Abdel-Sayed (1998a), however, have demonstrated that the ductility of a CFRP prestressed concrete beam can be enhanced by combining bonded internal and unbonded externally draped tendons. Conversely, deformability may be defined as the ratio of deflection at ultimate to deflection at cracking. Using the latter definition, FRP prestressed members can have considerable deformability.

Deformability is a key issue in determining the safety of FRP prestressed structures. Because FRP tendons do not exhibit ductility under the traditional definition, care should be taken to ensure that sufficient warning is exhibited before failure. Due to the lack of ductility, the concept of deformability as an index to measure performance provides a method of ensuring that this warning exists. Several approaches have been taken to quantify this concept into a deformability index (Mufti, Newhook, and Tadros 1996). This is usually accomplished through a ratio of deflections or curvatures under ultimate loads to those same quantities under service loads. The deformability index is intended to be a design check and not a long and complicated design procedure.

The use of an ultimate deflection-to-service-deflection ratio gives an indication of the warning to failure, but there are two main difficulties with this approach. The first issue is that deflections for point loads or uniform loads on a simple span beam are simple to calculate, but deflections for continuous structures with various loads require rigorous analysis or the use of a computer analysis program. The second issue is the difficulty in determining the deflection under ultimate loads. Traditional approaches to calculating deflections for concrete members consider the softening effect due to cracking under service loads. Near ultimate loads, this softening occurs at a faster rate and is harder to quantify.

An alternative approach is to use a ratio of the curvatures at ultimate and service loads. This is more easily accomplished by using quantities already calculated during the design process. The formula for this approach to the deformability index, DI, is shown in Eq. (3-24) (Dolan and Burke 1996).

$$\mathrm{DI} = \frac{(1-k)\ \varepsilon_{pu}}{\left(1 - \frac{\alpha}{d\beta_1}\right)\varepsilon_{ps}} \tag{3-24}$$

where $\alpha = \rho d f_{pu}/0.85 f_c'$.

If service loads are considered as the load to produce a tensile stress in the concrete of $0.25\sqrt{f_c'}$ MPa ($3\sqrt{f_c'}$ psi), the decompression strains and the strain to produce $0.25\sqrt{f_c'}$ MPa ($3\sqrt{f_c'}$ psi) are very small compared to the prestress strain. The index is effectively a function of the ratio of ultimate strain to the prestressing strain with a slight modification due to the differences in the neutral axis of elastic and inelastic behavior.

It follows from this definition that the most efficient method to obtain high deformability is to reduce the initial FRP tendon strain, which provides more tendon strain reserve, greater curvature or deflection capacity, and a higher index. A ratio of ultimate to initial prestress strain is easier to compute than the deformability index and provides similar results. Limiting the initial jacking stress to the values given in Table 3.3 results in acceptable deformability.

3.13—Minimum reinforcement

The amount of prestressed and nonprestressed reinforcement should be adequate to develop a factored flexural resistance ϕM_n greater than the smaller of 1.5 times the cracking strength or, for a tension-controlled section, 1.5 times the moment due to factored loads (CAN/CSA S806).

CHAPTER 4—SERVICABILITY

4.1—General

The allowable stresses specified in Table 3.2 ensure that the tensile strength of the concrete will not be exceeded and members will meet the ACI 318-02 requirements for Class U. It is still necessary, however, to ensure that the deflection, both short- and long-term, is acceptable for end use of the member.

4.2—Deflection

4.2.1 *Short-term deflection*—Deflections for FRP prestressed beams can be divided into two categories: short-term and long-term deflections. Before cracking, the gross moment of inertia can be used to calculate the deflections from traditional mechanics of materials. If deflection calculation following crack formation is required, methods that take into account the softening effect that cracking has on concrete members need to be used. This can be accomplished using a modified effective moment of inertia, I_e, proposed by ACI 440.1R and given by Eq. (4-1)

$$I_e = \left(\frac{M_{cr}}{M_a}\right)^3 \beta_d I_g + \left(1 - \left(\frac{M_{cr}}{M_a}\right)^3\right) I_{cr} \le I_g \tag{4-1}$$

in which β_d is a factor to soften the effective moment of inertia and is

$$\beta_d = 0.5\left[\frac{E_p}{E_s} + 1\right]$$

where E_p is the modulus of elasticity of FRP tendon; E_s is the modulus of elasticity of steel; I_g is the gross moment of inertia; M_{cr} is the cracking moment; M_a is the maximum moment in a member at which the deflection is being computed; and I_{cr} is the cracked moment of inertia, which, for a rectangular section or a flanged section with $kd < h_f$, can be calculated from Eq. (4-2).

$$I_{cr} = \frac{b(kd)^3}{3} + nA_p(d - kd)^2 \tag{4-2}$$

In Eq. (4-2), the parameter k may be found from Eq. (3-19). If the reinforcement is in a single layer $i = 1$ and $d_1 = d$. This approach is valid for loads between cracking of the concrete and up to around 50% of the ultimate load.

Subsequent to cracking, deflection due to specified live load should be calculated as the difference between the deflection due to the total service load and the deflection due to dead load. This bilinear approach is the same as recommended by PCI (2000). The deflection is not only due to the change of the effective moment of inertia I_e, but also due to the change of the eccentricity of the prestressing force after cracking. Abdelrahman and Rizkalla (1998) have suggested modifications to account for the change in effective eccentricity by finding the effective neutral axis location

$$y_{eff} = \left(\frac{M_{cr}}{M_a}\right)^2 y_g + \left(1 - \left(\frac{M_{cr}}{M_a}\right)^2\right) y_{cr} \tag{4-3}$$

where y_{eff} is the distance from the extreme compression fiber to the effective neutral axis; y_g is the distance from the extreme compression fiber to the centroid of the gross section; and y_{cr} is the distance from the extreme compression fiber to the neutral axis of the cracked section (that is, kd).

4.2.2 *Long-term deflection*—For long-term deflections, camber and deflection are separated into individual components, adjusted by a multiplier, and then superimposed to obtain final deflections in a manner similar to that for conventional steel prestressed concrete members (PCI 2000). The multipliers for FRP prestressed concrete for predicting long-term deflections, as suggested by Currier (1995), are shown in Table 4.1. These values are based on one series of experimental data consisting of three beams with each tendon type. The change in concrete strain over a period of approximately 1 year has been documented by Grace (2000a) for DT girders prestressed with CFRP tendons. Research on the long-term loss of prestress and resultant time-dependant camber/deflection effects is needed to define the time-dependant deflection relationship of concrete members prestressed by FRP tendons. Alternatively, the multiplier method of Kelly, Bradberry, and Breen (1987) may be used to calculate camber and deflection, if the time-dependant loss of prestress for a chosen FRP tendon is known, or assumed, along with the time-dependant concrete properties.

Table 4.1—Suggested multipliers for FRP tendons

		Without composite topping	
		Carbon	Aramid
At erection	Deflection due to self-weight	1.85	1.85
	Camber due to prestress	1.80	2.00
Final	Deflection due to self-weight	2.70	2.70
	Camber due to prestress	1.00	1.00
	Deflection due to applied loads	4.10	4.00

The multipliers in Table 4.1 are for members without composite toppings. The *PCI Design Handbook* (PCI 2000) indicates that multipliers relating to final deflection for topped members are smaller than for members without composite topping. Consequently, the use of values in Table 4.1 should be conservative for topped members.

4.3—Crack width and spacing

Dolan et al. (2000) studied crack widths in flexural members prestressed with CFRP. Both monotonic and cyclic loadings were studied. A linear growth in crack width with increasing load was observed after the initial precompression was overcome. The cracks were evenly spaced and occurred at the location of the stirrups, suggesting that major debonding of the carbon tendons did not occur during static or fatigue loading. Initial crack width was correlated with the Gergely-Lutz crack width equation when corrected by scaling the stress term by the ratio of the modulus of elasticity of steel to that for the FRP tendons. Crack lengths and widths increased during flexural cycling. Final crack widths were approximately three times those in a comparable steel prestressed concrete beam.

4.4—Fatigue

It is unlikely that fatigue will be a problem in uncracked members as the stress range in the tendons under repeated loading will be small (Grace 2000b; and Grace, Enomoto, and Yagi 2002). Further, Dolan et al. (2000) have shown that cracked CFRP prestressed concrete beams showed no fatigue failure after 3,000,000 flexural cycles with nominal tensile stresses of 0.5 $\sqrt{f_c'}$ MPa (6 $\sqrt{f_c'}$ psi) at the extreme fiber of the beams. Cracking was observed after the first 100,000 load cycles. Gradual softening was observed, but the beams indicated no loss of strength due to the fatigue loading.

A forensic examination of the beams after static loading to failure indicated that the tendons failed in tension, as evidenced by the "straw broom" failure shown in Fig. 4.1. Because all the carbon tendons failed in tension at close to the predicted strength, it can be assumed that the full tensile strength of the tendon was developed (the flexural cycling had no effect on the tensile strength of the tendon). An investigation of the saddle for the harping point indicated some evidence of sliding contact between the tendons and the saddle during cycling, as shown in Fig. 4.2; however, no significant fraying of the tendons was visible. There was evidence of contact near the edge of the saddle. It is recommended that a small radius be turned on the edges of any production saddle to minimize possible edge or bending effects.

Fig. 4.1—Carbon tendon after "straw broom" failure.

Fig. 4.2—Friction evidence on saddle.

CHAPTER 5—SHEAR

The use of FRP as shear reinforcement in beams prestressed with FRP tendons has not been thoroughly investigated (Fam, Rizkalla, and Tadros 1997; Shehata 1999). From the assessment of theoretical and experimental nominal shear capacities, a rational approach that conservatively modifies the ACI 318-02 expressions to account for the special characteristics of FRP stirrups is suggested.

5.1—General considerations in design of FRP stirrups

Because of differences between FRP and steel, several issues need to be addressed when using FRP as shear reinforcement, namely:

- FRP may have a relatively low modulus of elasticity;
- FRP has high tensile strength and no yield plateau;
- Tensile strength of the bent portion of an FRP bar is significantly lower than the straight portion;
- FRP has lower dowel resistance and tensile strength in any direction other than that of the fibers; and
- The bond characteristics of FRP stirrups may vary

significantly from steel reinforcement.

5.2—Shear strength with FRP stirrups

According to ACI 318-02, the nominal shear strength of any concrete cross section, V_n, is considered as the sum of the shear resistance provided by concrete, V_c, the shear resistance provided by stirrups, V_s, and the shear resistance provided by the vertical component of prestressing force, V_p.

Members prestressed with FRP tendons behave similarly to members prestressed with steel tendons. The nominal shear resistance V_n is given by

$$V_n = V_c + V_{frp} + V_p \quad (5\text{-}1)$$

The following formulation for V_c is suggested

$$V_c = 0.17\sqrt{f_c'}\,b_w d \text{ N} \quad (2.0\sqrt{f_c'}\,b_w d \text{ lb}) \quad (5\text{-}2)$$

The ACI code provides several alternative formulations for the concrete contribution to shear resistance. A search of the literature on the shear strength of FRP prestressed beams found information on only a small number of samples in which there was a significant variation in specimen size and tendon type (Dowden and Dolan 1997). This paucity of test data, together with the fact that wider shear cracks, which will probably occur in a beam prestressed with FRP rather than steel tendons, does not warrant expansion of V_c beyond the minimum value specified in Eq. (5-2) at this time.

A similar approach to account for the contribution of FRP stirrups has been taken. The shear resistance provided by FRP stirrups, V_{frp}, when the stirrups are vertical can be written as

$$V_{frp} = \frac{f_{fb} A_v d}{s} \quad (5\text{-}3)$$

and

$$f_{fb} = \phi_{bend} f_{fu} \quad (5\text{-}4)$$

where the value of ϕ_{bend} is as follows

$$\phi_{bend} = \left(0.11 + 0.05\frac{r}{d_b}\right) \text{ and } 0.25 \le \phi_{bend} \le 1.0 \quad (5\text{-}5)$$

where r is the radius of the bend.

The maximum stress in the FRP reinforcement is limited to the smaller of 0.002 times the modulus of elasticity of the stirrup, or the strength of the bent portion of the stirrups $\phi_{bend} f_{fu}$.

The form of Eq. (5-3) is the same as that given in ACI 318-02, except f_y is replaced by a stress in the FRP rod based on its r/d_b, as shown in Eq. (5-4). The parameter ϕ_{bend} (Eq. (5-5)) is modified from the original JSCE (1997) formulation so that the minimum reduction factor is limited to 0.25. This was based on the push-apart tests performed on the aramid/nylon and carbon/nylon stirrups, which resulted in a ϕ_{bend} of 0.25 (Currier 1995). Also, in the equation for the design stress of the FRP stirrups provided in JSCE (1997), the factor 0.3 has been reduced to 0.11. Moreover, the modified expression for ϕ_{bend} is more conservative than the form suggested in JSCE (1997); consequently, Eq. (5-5) is suitable for design applications until more refinement of the r/d_b response is available.

5.3—Spacing limits for shear reinforcement

ACI 318-02, Section 11.5.4, guards against excessive crack width by limiting the maximum spacing of shear reinforcement for prestressed members to $0.75h$ or 600 mm (24 in). The maximum stirrup spacing is reduced to half of this when the maximum shear V_{frp} exceeds $0.33\sqrt{f_c'}b_w d$ N $(4.0\sqrt{f_c'}b_w d$ lb). The Canadian Highway Bridge Design Code limits the tensile strain in FRP shear reinforcement to 0.002. These two conditions ensure that cracks are relatively small and uniformly distributed in the member. For FRP stirrups, the maximum spacing should be limited to $d/2$ or 600 mm (24 in.) as suggested by ACI 318-02 for nonprestressed members.

5.4—Minimum amount of shear reinforcement

ACI 318-02 requires a minimum amount of shear reinforcement when V_u, the factored shear force at a section, exceeds $\phi V_c/2$. This requirement is to prevent shear failure in members where the sudden formation of cracks can lead to excessive distress (Joint ACI-ASCE Committee 426 1973). Equation (5-6) gives the recommended minimum amount of shear reinforcement for FRP members.

$$A_{v,min} = \frac{1}{16}\sqrt{f_c'}\,\frac{b_w s}{\phi_{bend} f_{fu}} \text{ (mm}^2) \quad (5\text{-}6)$$

$$\text{or } A_{v,min} = 0.75\sqrt{f_c'}\,\frac{b_w s}{\phi_{bend} f_{fu}} \text{ (in.}^2)$$

The expression is a modification to ACI 318-02, Section 11.5.5.3. It uses the effective strength of the stirrup at the bends instead of the yield strength of reinforcing steel to conservatively estimate the minimum amount of stirrups. The value of $\phi_{bend} f_{fu}$ should not exceed 0.002 times the modulus of elasticity of the stirrup.

5.5—Detailing of shear stirrups

Limited experimental data are available on bond behavior of the bent portion of FRP stirrups. ACI 318-02 provisions for bond of hooked steel bars cannot be applied directly to FRP reinforcing bars because of their different mechanical properties. Research on FRP hooks (Ehsani, Saadatmanesh, and Tao 1995) indicates that the tensile strength of the bent portion of a GFRP bar is mainly influenced by the ratio of the bend radius to the bar diameter, r/d_b, the tail length, and to a lesser extent, the concrete strength. Tests by Ehsani, Saadatmanesh, and Tao (1995) indicated that for specimens with r/d_b of zero, the reinforcing bars failed in shear at very low load

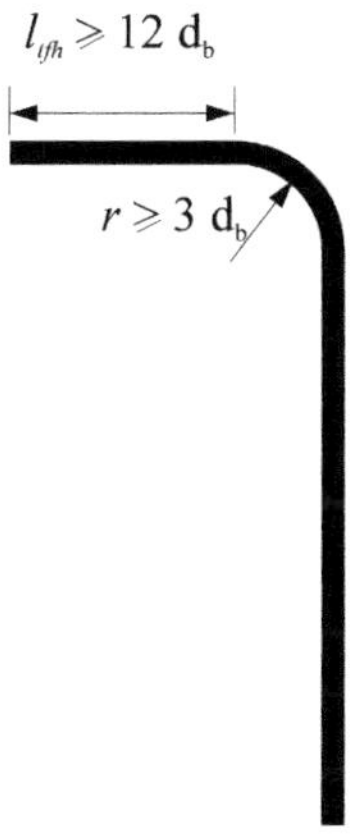

Fig. 5.1—Minimum radius and tail length of a stirrup bend.

levels at the bends. Therefore, although manufacturing of FRP bars with sharp bends is possible, such details should be avoided due to the low shear strength of FRP bars. As indicated by Ehsani, Saadatmanesh, and Tao (1995), a minimum r/d_b of 3 is recommended. In addition, FRP stirrups should be closed stirrups with 90-degree hooks. The tensile force in a vertical stirrup leg is transferred to the concrete through the tail beyond the hook as shown in Fig. 5.1. Ehsani, Saadatmanesh, and Tao (1995) found that for a tail length l_{tfh} beyond $12d_b$, there is no significant slippage and no influence on the tensile strength of the stirrup leg. Therefore, it is recommended that a minimum tail length of $12d_b$ be used.

CHAPTER 6—BOND AND DEVELOPMENT

6.1—Introduction

The transfer and development length of an FRP tendon is a function of the configuration of the perimeter and the surface condition of the FRP, the stress in the FRP, the method used to transfer the FRP force to the concrete, and the strength and cover of the concrete. The mechanism of bond differs between FRP and steel strands due to differences in shapes, surface treatments, and elastic moduli. FRP may be produced using unique manufacturing processes, which result in different characteristic surface textures. In general, a prestressing rod with a rough, irregular surface will require a shorter transfer length than a smooth one. The transfer length will be greater if the release of tension is sudden rather than gradual. A higher initial prestress in the tendon will require a larger transfer length. In general, the bond of FRP tendons could be influenced by (ACI 440R):

- Tensile strength (600 to 3000 MPa [87,000 to 435,000 psi]);
- Modulus of elasticity (40 to 170 GPa [5.8×10^6 to 25×10^6 psi]);
- Hoyer effect;
- Cross-sectional shape;
- Surface preparation (braided, deformed, smooth);
- Type and volume of fiber and matrix;
- The method of force transfer; and
- Concrete strength and cover.

Both transfer length and development length vary considerably among the different fibers and tendon configurations. This variation is discussed in the following sections.

6.2—Transfer length

The transfer length in pretensioned concrete is the minimum length required to transfer the full prestressing force to the concrete. This transfer occurs gradually, rising from zero at the location where bond is initiated and increasing gradually within the transfer length until it reaches a constant value at the effective prestress level.

6.2.1 *Aramid FRP*—Nanni et al. (1992) examined the transfer length of FiBRA tendons comprising braided epoxy-impregnated AFRP and having diameters ranging from 8 to 16 mm (5/16 to 5/8 in.). Their experimental transfer lengths ranged from 300 to 400 mm (12 to 16 in.) for low levels (25% of ultimate) of prestress and from 250 to 550 mm (10 to 22 in.) for high levels (50% of ultimate) of prestress. The transfer length values varied with the nominal diameter of the different tendons and ranged from 30 to 40 strand diameters. The authors concluded that friction was the predominant bonding mechanism in aramid fibers and that these fibers showed little tendon slippage compared to steel.

Taerwe and Pallemans (1995) carried out a study on Arapree® aramid fiber rods having diameters of 7.5 and 5.3 mm (0.3 and 0.2 in.) and different surface finishes. They suggested a transfer length of 16 times the nominal diameter for Arapree® rods.

Ehsani, Saadatmanesh, and Nelson (1997) conducted tests on Arapree® (10 mm [0.39 in.]), FiBRA (10.4 mm [0.41 in.]), and Technora (7.4 mm [0.29 in.]) tendons.The transfer length was found to be 33 bar diameters for FiBRA, 43 bar diameters for Technora®, and 50 bar diameters for Arapree.® For Arapree® tendons, the transfer length was affected significantly by the level of prestress.

Lu et al. (2000) conducted transfer length tests on Technora (7.9 mm [0.31 in.]) aramid/vinylester tendons. The transfer lengths varied between 45 to 47 diameters during tendon release, immediately after full tendon release, and 90 days following full release. In consideration of other types of FRP tendons tested by Lu et al. (2000), a minimum transfer length of 50 times the strand diameter was recommended.

6.2.2 *Carbon FRP*—Domenico (1995) investigated the transfer length and bond characteristics of CFCC strands. Variables in the study were the depth of the beams, diameter of the CFCC tendons, concrete covers and strength, and prestressing level. The measured transfer length was found to be proportional to the diameter of the CFCC strand and the prestress level, and varied from 140 to 400 mm (5.5 to 15.7 in.).

Mahmoud and Rizkalla (1996), and Mahmoud, Rizkalla, and Zaghloul (1997) studied the bond characteristics of Leadline™ and CFCC strands. The authors reported that the measured transfer length varied from 450 to 650 mm (17.7 to 25.6 in.) or 56 to 81 bar diameters for Leadline™ rods, and from 300 to 425 mm (11.8 to 16.7 in.) for CFCC strands. They recommended Eq. (6-1) for transfer length of carbon FRP.

$$L_t = \frac{f_{pe} d_b}{\alpha_t f_c'^{\,0.67}} \qquad (6\text{-}1)$$

Table 6.1—Typical transfer and development lengths for FRP tendons

Material	Type	Diameter, mm (in.)	Young's modulus, MPa (ksi)	Tensile strength, MPa (ksi)	f_{pe}/f_{pu}	L_t/d_b	L_d/d_b
Aramid	Arapree®	9.9 (0.39)	127,600 (18,500)	2450 (355)	0.5 to 0.7	16 to 50	100
	FiBRA	10.4 (0.41)	48,270 (7000)	1430 (208)	0.4 to 0.6	20 to 50	90
	Technora®	7.4 (0.291)	68,600 (9950)	1720 (250)	0.6	50	140
Carbon	Leadline™	7.9 (0.312)	149,600 (21,700)	1980 (287)	0.5 to 0.7	50 to 80	175
	CFCC	8.3 (0.327)	137,200 (19,900)	2220 (322)	0.5 to 0.7	50	N/A

where α_t is 1.9 for N-mm units (10.0 for inch-pound units) for Leadline™, and 4.8 for N-mm units (25.3 for inch-pound units) for CFCC. Grace (2000a) found that α_t is 1.95 for N-mm units (10.2 for inch-pound units) for Leadline™ and 2.12 for N-mm units (11.2 for inch-pound) for CFCC, and suggested further study on the transfer length of CFCC strands.

Ehsani, Saadatmanesh, and Nelson (1997) conducted tests on Leadline™ and CFCC carbon tendons and found the transfer length to be 54 times the diameter for Leadline™ and 50 times the diameter for CFCC.

Soudki, Green, and Clapp (1997) reported on the transfer length for Leadline™ CFRP rods in concrete T-beams. They found that the measured transfer length was approximately 80 times the rod diameter, and that existing models for steel may give unconservative transfer lengths for the CFRP rod.

Lu et al. (2000) investigated single Leadline™ and generic carbon/epoxy tendons with 7.9 mm (0.31 in.) nominal diameter. The generic tendons had a shallow waffle-type of surface deformation applied during the pultrusion manufacturing process. Transfer lengths in the beams investigated were between 52 and 53 diameters for both tendons. A conservative value of transfer length equal to 50 times diameter was recommended for the tested CFRP tendons.

6.3—Flexural bond length

6.3.1 *Aramid FRP*—Nanni and Tanigaki (1992) examined the development and flexural bond lengths of AFRP tendons. The unfactored development length of AFRP tendons in bonded pretensioned prestressed concrete members was determined as: 120, 100, and 80 times the nominal diameter for type K64, K128, and K256 tendons, respectively.

Lu et al. (2000) found that the development length of single 7.9 mm (0.31 in.) diameter Technora® tendons could not be directly measured due to the concrete crushing or tendon slip failures in the beams investigated. By using an extrapolation procedure on the linear plots of tendon force at failure versus embedment length, however, the development length that would be required to fail the tendons in tension was determined to be 141 diameters. ACI 318-02 equations were determined to be overconservative in comparison with Lu et al.'s measurements. A recommended relationship for predicting the development length of single AFRP tendons was reported.

6.3.2 *Carbon FRP*—Ehsani, Saadatmanesh, and Nelson (1997) examined the transfer and flexural bond lengths of carbon and aramid FRP prestressing tendons. Two types of CFRP strands, Leadline™ and CFCC, and three AFRP types, Technora®, FiBRA, and Arapree® were tested. It was concluded that the ACI 318 development length requirements for steel strands are conservative for AFRP tendons but are not adequate for Leadline™ tendons.

Lu et al. (2000) found that the development length of single Leadline™ carbon/epoxy tendons and generic carbon/epoxy tendons (both 7.9 mm [0.31 in.] diameter) with a shallow waffle pattern could not be determined due to compressive failure in the concrete beams tested. Using an extrapolation procedure, maximum development lengths for tensile failure of the Leadline™ and generic tendons were determined to be 138 and 146 diameters, respectively. Lu et al. (2000) recommended Eq. (6-2), which predicted the development length of these CFRP tendons with good accuracy.

$$L_d = \frac{1}{3}f_{pe}d_b + \frac{3}{4}(f_{pu} - f_{pe})d_b \quad (6\text{-}2)$$

Mahmoud and Rizkalla (1996) and Mahmoud, Rizkalla, and Zaghloul (1997) recommended the following equation for flexural bond of carbon FRP

$$L_{fb} = \frac{(f_{pu} - f_{pe})d_b}{\alpha_{fb}f_c'^{\,0.67}} \quad (6\text{-}3)$$

where α_{fb} is 1.0 for N-mm units (5.3 for inch-pound units) for Leadline™ and 2.8 for N-mm units (14.8 for inch-pound units) for CFCC.

6.4—Design considerations

Bond and development for prestressing FRP tendons made of carbon or aramid fibers are intended to provide bond integrity for the strength of the member. Typical values for transfer and development lengths of various FRP tendons are given in Table 6.1.

Many of the bond tests were conducted with direct pullout tests (Nanni et al. 1992; Nanni and Tanigaki 1992). These tests indicated that failure of the tendon surface was the principal factor limiting bond capacity. Research at the University of Minnesota (Shield, French, and Hanus 1999) recorded cracking and splitting of the concrete. The body of experimental data, especially that from pullout tests, does not provide sufficient information to define tendon spacing limitations or cover restrictions for FRP tendons. The designer should select spacing based on the specific tendon contemplated.

CHAPTER 7—UNBONDED AND EXTERNAL TENDON SYSTEMS

7.1—Unbonded prestressed members

To predict the strength of a beam post-tensioned with unbonded FRP tendons, it is necessary to determine the stress in the prestressing tendons at failure of the beam using the following relation

$$f_p = f_{pe} + \Delta f_p \tag{7-1}$$

where f_{pe} is the effective prestress in the tendon when the beam carries only the dead load after the prestress losses have occurred, and Δf_p is the stress increase above f_{pe} due to any additional applied load. The inclusion of dead load and losses is important because, when computing prestress losses needed to determine the final value of the prestressing force, the dead load moment is considered active in estimating creep losses. There is no compatibility between the strain in the unbonded prestressing FRP tendon and the concrete at every cross section in a member. The increment of stress in the tendon due to applied loading should be evaluated by considering deformation of the whole member, rather than at a section, as for a bonded member.

Many attempts have been made by investigators to estimate the stress in a tendon at different loading conditions, especially at the ultimate nominal flexural resistance for concrete members prestressed with unbonded steel tendons. The principal difference between unbonded steel and FRP tendons is the modulus of elasticity. Thus, models for unbonded steel tendons serve as a basis for developing guidelines for unbonded FRP tendons. A method based on determining the strain in the critical section (maximum moment) was proposed by Naaman (1987) for steel tendons and later extended to FRP tendons (Naaman et al. 2002). The method assumes compatibility of strains, as if the tendons were bonded, and applies a strain reduction factor Ω to account for the fact that the tendons were unbonded. The increment in strain at the critical section is as shown in Eq. (7-2).

$$\Delta\varepsilon_{average} = (\Delta\varepsilon_p)_{unbonded} = \Omega(\Delta\varepsilon_p)_{bonded} \tag{7-2}$$

The strain $(\Delta\varepsilon_p)_{bonded}$ can be computed using strain compatibility and, assuming linear elastic behavior of the tendon, the change in stress Δf_p in the unbonded tendon is given by

$$\Delta f_p = \Omega_u E_p \varepsilon_{cu}\left(\frac{d_p}{c_u} - 1\right) \tag{7-3}$$

where ε_{cu} is the strain in the extreme compression fiber at ultimate, and c_u is the depth of the neutral axis at ultimate.

Equation (7-1) can then be written as

$$f_p = f_{pe} + \Omega_u E_p \varepsilon_{cu}\left(\frac{d_p}{c_u} - 1\right) \tag{7-4}$$

Combining Eq. (7-4) with the equilibrium equation for a flanged section results in

$$0.85f_c' b_w \beta_1 c_u + 0.85f_c'(b - b_w)h_f = A_p f_p + A_s f_y - A_s' f_y' \tag{7-5}$$

where A_s' = area of nonprestressed compressive reinforcement; A_s = area of nonprestressed tensile reinforcement; f_y = yield strength of nonprestressed tensile reinforcement; f_y' = yield strength of nonprestressed compressive reinforcement; b = width of compression flange; b_w = width of web; and h_f = thickness of compression flange, gives a quadratic equation in c_u having the following root

$$c_u = \frac{-B + \sqrt{(B^2 - 4AC)}}{2A} \tag{7-6}$$

where

$A = 0.85f_c' b_w \beta_1$;

$B = A_p(E_p\varepsilon_{cu}\Omega_u - f_{pe}) + A_s'f_y' - A_s f_y + 0.85f_c'(b - b_w)h_f$; and

$C = -A_p E_p \varepsilon_{cu} \Omega_u d_p$

The strain reduction coefficient at ultimate, Ω_u, depends on a number of variables, such as loading configuration and extent of the cracks in a beam and varies theoretically between 0 and 1.0, with a value of 1.0 corresponding to a bonded tendon. The following values for Ω_u at ultimate were suggested by Alkhairi (1991) based on a regression analysis on test data from 143 beams prestressed with steel tendons

$$\Omega_u = \frac{2.6}{(L/d_p)} \text{ (for one-point loading)} \tag{7-7a}$$

$$\Omega_u = \frac{5.4}{(L/d_p)} \text{ (for two-point or uniform loading)} \tag{7-7b}$$

For design purposes, Alkhairi (1991) recalibrated the coefficient Ω_u in such a way that most of the predicted values are smaller than the experimental results and made the following recommendations

$$\Omega_u = \frac{1.5}{(L/d_p)} \text{ (for one-point loading)} \tag{7-7c}$$

$$\Omega_u = \frac{3.0}{(L/d_p)} \text{ (for two-point or uniform loading)} \tag{7-7d}$$

Recent research at Queen's University (Gangkatharan 2003) has indicated that Eq. (7-7c) and (7-7d) are applicable to beams containing CFRP tendons having an unbonded length greater than 15 times the depth of the beam.

7.2—External prestressing

7.2.1 *General*—Externally prestressed systems consist of unbonded tendons located outside the beam cross-sectional area and post-tensioned against anchorages and deviators attached to or built into the beam. The tendons may be

Fig. 7.1—Saddles used on the Bridge Street Bridge (Grace et al. 2002).

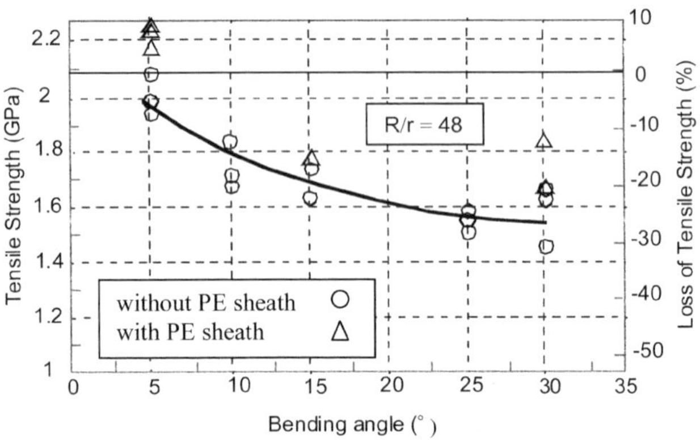

Fig. 7.2—Relationship between the tensile strength with the angle of curvature and with the radius of curvature (after Santoh [1993]).

external to the cross section or may be in the void of a box section. External prestressing has the advantage of reducing the dead load of a structure due to the reduction in web thickness as a result of eliminating internal ducts, easing inspection of tendons and replacement of old or damaged tendons, and improving shear resistance due to the elimination of shear resistance deficiencies that are present at internal duct locations. Some of the practical difficulties of this system include providing anchorage for the post-tensioning tendons and maintaining the lateral stability of the girders during post-tensioning. In addition, the tendons are exposed and therefore are susceptible to damage during construction and throughout their service life. Suitable protection should be provided for the external tendons during the life of the structure.

Straight and harped tendon profiles can be used. For harped tendons, deviation saddles are used to hold the tendons in place and transfer components of the prestressing force from the tendons into the concrete girder.

7.2.2 *Internal versus external unbonded tendons*—One of the main differences between external and internal unbonded tendons is the variation of eccentricity in the case of external tendons during deformation of the beam under load. This variation is caused by the external tendon remaining rectilinear between the deviators and the end anchorages, while the beam deformation is curvilinear. Therefore, the eccentricity of an external prestressing tendon changes as the loading progressively increases to ultimate. Thus, it is necessary to investigate the deformation behavior of externally prestressed members throughout all stages of loading.

7.2.3 *Design considerations*—The effect of tendon bending at the deviator points on the ultimate strength of the cables should be taken into consideration when using FRP as externally prestressed draped tendons. Special saddles should be designed to avoid inducing stress concentrations in the tendon by providing a system with limited curvature, friction, and sufficient area to limit the lateral stresses on the tendon. Figure 7.1 shows the saddle used on the Bridge Street bridge (Grace et al. 2002). Figure 7.2, adopted from the Tokyo Rope catalog (Santoh 1993), depicts the relationship between the tensile strength of the CFCC and the angle of curvature through which the tendon is bent. A reduction in the tendon strength is observed as the angle of curvature increases and is similar to the discussion in Section 3.9. This reduction in strength of the CFCC can be mitigated by flaring the strands at the harp points, and by reducing the friction between the CFCC tendons and the saddle by providing a cushioning material, such as low-density polyethylene (PE), on the saddle. This is particularly important when the angle of bend at the saddle is larger than 5 degrees. The effects of drape angle on the strength of Leadline™ and CFCC have been discussed by Grace and Abdel-Sayed (1998b).

External tendons are designed to ensure longitudinal prestressing of a beam and generally represent only a portion of the total flexural reinforcement. The remaining reinforcement may consist of internal prestressed reinforcement, nonprestressed reinforcement, or a combination of both, depending on the structural system and the type of construction. A minimum amount of bonded reinforcement is also necessary to control the distribution of cracks and to limit the crack widths.

7.2.4 *Stress at ultimate in external unbonded prestressed tendons*—Aravinthan and Mutsuyoshi (1997) investigated the stress in external unbonded prestressed steel tendons and noted that the variations in the eccentricity could have a significant influence on the ultimate strength of externally prestressed beams. As a result, they introduced the concept of depth reduction factor R_d to estimate the effective depth d_e of an external tendon at ultimate as follows

$$d_e = R_d d_p \qquad (7\text{-}8)$$

where

$$R_d = 1.14 - 0.005\left(\frac{L}{d_p}\right) - 0.19\left(\frac{S_d}{L}\right) \leq 1.0 \qquad (7\text{-}9a)$$

for one-point loading; and

$$R_d = 1.25 - 0.010\left(\frac{L}{d_p}\right) - 0.38\left(\frac{S_d}{L}\right) \leq 1.0 \qquad (7\text{-}9b)$$

for third-point loading;

and S_d is the spacing of the deviators, and L is the span of the beam.

Based on a parametric evaluation, Aravinthan and Mutsuyoshi (1997) proposed alternative equations for the strain reduction coefficient Ω_u. These equations, which they claim can be used to predict the behavior at ultimate of beams with external prestressing or a combination of internal and external prestressing, are as follows

$$\Omega_u = \frac{0.21}{(L/d_p)} + 0.04\left(\frac{A_{p\,int}}{A_{p\,tot}}\right) + 0.04 \qquad (7\text{-}10a)$$

for one-point loading; and

$$\Omega_u = \frac{2.31}{(L/d_p)} + 0.21\left(\frac{A_{p\,int}}{A_{p\,tot}}\right) + 0.06 \qquad (7\text{-}10b)$$

for third-point loading

where $A_{p\,int}$ is the area of the internal prestressed reinforcement, and $A_{p\,tot}$ is the total area of internal and external prestressed reinforcement.

The values of Ω_u from Eq. (7-10) can be incorporated in Eq. (7-4) with d_p replaced by d_e to determine the stress at ultimate.

The significance of change in the eccentricity of the external tendons during loading in the calculation of the stress at ultimate has been corroborated by Rao and Mathew (1996) and Tan and Ng (1997). The applicability of above equations to FRP prestressed tendons needs to be established.

CHAPTER 8—PILE DRIVING AND IN-PLACE FLEXURE

8.1—General

Piles are commonly installed by a succession of blows that are applied using a drop, diesel, steam, or compressed-air-powered hammer. Four types of damage may occur during driving of prestressed concrete piles:

- Concrete spalling at the pile head due to high compressive stresses;
- Concrete spalling at the pile tip due to hard driving resistance;
- Transverse cracking due to tensile stress reflected from the pile tip or pile head; and
- Spiral or transverse cracking due to combined torsion and tensile stress.

Though identified for steel prestressed piles, this type of damage may be anticipated in FRP prestressed piles as well. Although FRP prestressed piles may provide the same effective prestress, verification of their performance is required to allay concerns relating to the use of any new material. Specifically,

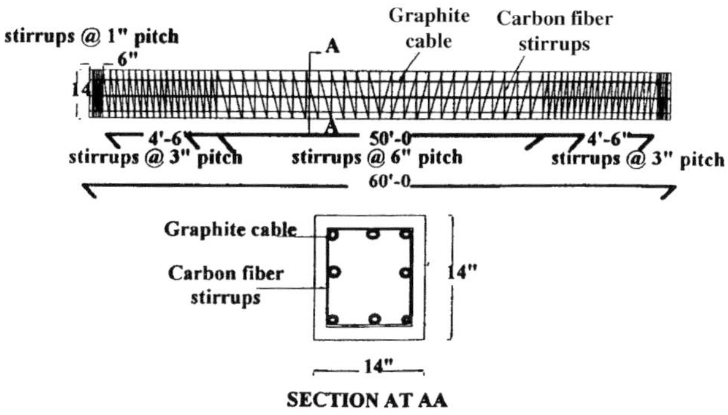

Fig. 8.1—Pile details (Iyer et al. 1996).

the effectiveness of confinement provided by FRP ties needs to be proven. Additionally, it is necessary to demonstrate that there is no debonding of the FRP strands under impact loading and consequent loss of effective prestress.

8.2—Demonstration studies

Four studies have been conducted in the United States in which full-size FRP prestressed concrete piles were driven and monitored to verify their performance. In the earliest study by Sen, Issa, and Mariscal (1992) and Sen et al. (1993), S-2 glass strands were used for prestressing. In subsequent studies, CFRP material was tested by Iyer (1995), Iyer et al. (1996), Arockiasamy and Amer (1998), and Schiebel and Nanni (2000).

8.2.1 *Study by Iyer (1995) and Iyer et al. (1996)*—Twelve CFRP prestressed piles were used to support two 6.1 m (20 ft) bays of a composite pier at the Naval Facilities Engineering Service Center, Port Hueneme, Calif. The piles were 18.3 m (60 ft) long, 356 mm (14 in.) square, and prestressed by eight 12.5 mm (1/2 in.) CFRP strands with 9.5 mm (3/8 in.) CFRP spirals. The eight strands were stressed to provide an effective prestress of 4.83 MPa (700 psi). Figure 8.1 shows the layout and cross section of the piles.

A 17.8 kN (4000 lb) hammer was used to drive the piles. The subsurface below the mudline consisted of well-graded moderately dense silty sands with intermittent silty clay and clayey silt lenses. Figure 8.2 shows the installation of the CFRP piles.

The maximum measured compressive stresses ranged from 20.7 to 23.4 MPa (3 to 3.4 ksi) and were below the recommended limit of 30.3 MPa (4.4 ksi). The pile cushion employed a fresh stack of plywood with a predriving thickness of 125 mm (5 in.). A visible surface crack was noticed after 263 blows at a location 7.62 m (25 ft) from the pile top. The installation was temporarily stopped and continued after an additional 125 mm (5 in.) plywood cushion was placed on the pile top and the hammer reset to a lower level. Maximum computed pile tensile stress reached 6.41 MPa (0.93 ksi) before the adjustment. The pile was successfully installed with no propagation of the surface crack while tensile stresses remained below 5.9 MPa (0.85 ksi).

Although the piles were 18.3 m (60 ft) long, the required bearing capacity was reached after 13.7 m (45 ft). As a result, all piles were cut to 13.7 m (45 ft) and then

Fig. 8.2—View of installation of CFRP test pile (Iyer 1995).

Fig. 8.3—Construction on CFRP prestressed piles supporting pier.

driven on site. This meant that the spiral spacing was increased to 150 mm (6 in.) from the end one spacing of 25 mm (1 in.) (Fig. 8.1). There were no cracks in the shortened installed piles. A view of the construction and installation of the CFRP piles is shown in Fig. 8.3.

8.2.2 *Study by Arockiasamy and Amer (1998)*—In this study, four 7.62 m (25 ft) long by 250 mm (10 in.) square CFRP prestressed piles were fabricated and tested. Two of the piles used CFRP spirals while two others used steel spirals. Details of the cross section of the pile and tie spacing are shown in Fig. 8.4. The all-CFRP piles used CFCC 1 x 7 7.5 mm (0.3 in.) diameter strands as spiral ties. The spacing of the spirals was the same as that used by Iyer et al. (1996) (Fig. 8.1). Steel spirals having a diameter of 5 mm (0.2 in.) were used in the two other specimens. The ultimate

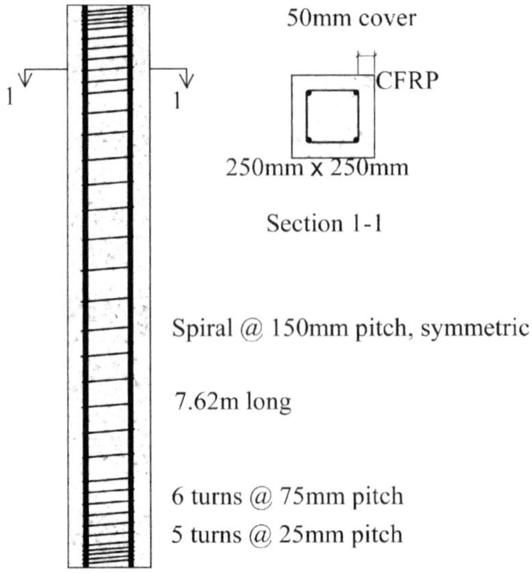

Fig. 8.4—CFRP pile details (Arockiasamy and Amer 1998).

strength of the CFRP spirals was 2117 MPa (307 ksi) versus 1034 MPa (150 ksi) for the steel spirals. Prestressing was provided by four CFCC 1 x 7 12.5 mm (0.5 in.) strands produced by Tokyo Rope. The jacking stress was designed to produce an average compressive stress of 5.5 MPa (800 psi).This effective stress was comparable to that recommended by ACI 543R. A 13.4 kN (3012 lbf) hammer was used to drive the piles.

The average specified drop height was 1.82 m (6 ft), though it averaged 2.12 m (7 ft) based on the energy transferred to the top of the piles. The stresses predicted by wave analysis varied between 0 to 7.1 MPa (0 to 1.03 ksi) tensile and 16.6 to 24.3 MPa (2.41 to 3.53 ksi) compressive strength and were in reasonable agreement with the experimental results. The tensile stresses were maximum at the initial stage of driving while the compressive stresses were higher at the end of the driving stage. All stresses were within the limits permitted by the Florida Department of Transportation (FDOT) specifications. No damage was observed in the piles.

8.2.3 *Study by Schiebel and Nanni (2000)*—Four 7.3 m (24 ft) long by 300 mm (12 in.) by 300 mm (12 in.) cross-sectional piles were fabricated. Two of these piles were CFRP prestressed and two others were prestressed using steel strands. CFRP spirals were used over a portion of the CFRP prestressed pile. Of the four piles, two—one CFRP prestressed and one steel prestressed pile—were driven. The other two were subject to flexural tests in the laboratory. CFRP spirals have a 3.62 GPa (525 ksi) guaranteed tensile strength. Two identical sections of spiral were fabricated, one for each of the two piles. Each spiral section had 21 turns (five turns at 25 mm [1 in.] pitch and 16 turns at 75 mm [3 in.] pitch for a total length of 1.325 m [53 in.] [Fig. 8.5]).The spiral sections were installed in the driven end of the piles. The rest of the pile section was confined with conventional steel spirals.

A 14.4 kN (3240 lbf) hammer was used to drive the piles (Fig. 8.6).The piles were driven into a rocky-clay fill material on the approach to a new bridge. Neither pile was driven to bedrock. Once the pile was driven, the pile head was inspected

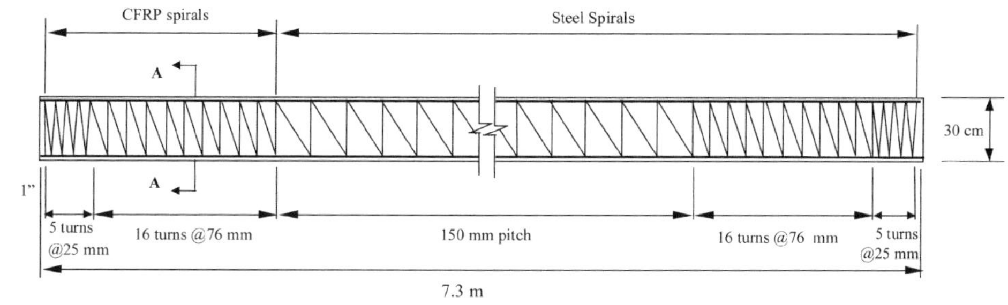

Fig. 8.5—Pile details (Schiebel and Nanni 2000).

for damage. Aside from some minor chipping at the square edges, no damage was observed. Static field-testing indicated that there was no reduction in flexural capacity.

8.3—Discussion

Piles are prestressed to withstand tensile stresses that arise when they are driven through soft soils. FRP and steel may be used to develop the same effective prestress. As FRP is stressed to a smaller fraction of its ultimate tensile strength and its interface with concrete differs from steel, transfer lengths will differ. Arockiasamy and Amer (1998) found that transfer length for the CFRP strands was 40 diameters (compared with 50 diameters for prestressed steel). For 12.5 mm (0.5 in.) strands, this amounts to 125 mm (5 in.) difference in the location where the full effective prestress is transferred to the pile. For this case, the performance of the CFRP would be marginally better.

8.4—Conclusions

The principal conclusions from the demonstration studies are:

1. The performance of FRP prestressed and steel prestressed piles during driving are similar;

2. The majority of the piles relied on friction rather than end-bearing for support;

3. FRP ties performed satisfactorily based on the absence of damage following the driving operation. The tie spacing used was identical to that in comparable steel prestressed piles;

4. The FRP prestressed piles that are properly engineered in terms of effective prestress and lateral ties provided, and employ driving parameters based on wave analysis, may be expected to perform satisfactorily; and

5. The results indicate that there are no inherent problems in driving FRP prestressed piles.

An assessment of the test reports suggest the following guidelines for FRP prestressed piles:

1. The specified strength of concrete should be at least 41.4 MPa (6000 psi) to prevent compression failure;

2. The cover for steel prestressed piles is 75 mm (3 in.). Such large covers are not needed for corrosion-resistant FRP material. Recommendations in specifications for FRP reinforcement may be used;

3. Effective prestress should be between 4.8 to 5.5 MPa (700 to 800 psi);

Fig. 8.6—View of installation of hybrid CFRP/steel (Schiebel and Nanni 2000).

4. FRP spiral ties provided should have twice the ultimate capacity as steel ties. The same spacing used in steel prestressed piles should be used for FRP prestressed piles;

5. FRP prestressed piles should be installed following the recommendations in ACI 543R;

6. Driving stresses in selected FRP test piles need to be carefully monitored using a pile dynamic analyzer (PDA); and

7. A predriving analysis is required for selection of the pile hammer, the drop height, and the thickness of the pile and helmet cushions to ensure driving stresses comply with limits set in project specifications.

CHAPTER 9—RESEARCH NEEDS

While there is a substantial volume of laboratory tests and field installations, a number of research areas remain to be investigated. The following represent the committee's major concerns.

Tendon and anchorage systems—The lack of commercial tendon and anchorage systems remains a major impediment to the industry. Without a commercial system, every project remains a research endeavor. This concern extends to the economics of FRP tendons. The current price premium makes them uneconomical for many applications.

Anchorages—Anchorages need additional development in several areas. Field and construction-friendly anchorages are needed. The fatigue effects at anchorages for unbonded systems need evaluation.

Fire protection—Even though carbon fibers may be heat resistant to 1000 °C (1800 °F), the resins are generally sensitive to heat. Fire protection of anchorages is also of concern.

Harping devices—The larger radii needed to limit stresses make commercial push-down and roller-harping devices unsuitable. New harping devices and saddles need to be developed.

Long-term bond—The FRP tendon bond depends on the resin surface. While there is some experience with epoxy-coated steel bars, the long-term durability behavior of bond in FRP prestressing tendons is not yet defined.

Galvanic action—Carbon fibers are higher on the galvanic table than steel, creating a potential corrosion problem. The resin bonding agents may preclude this effect, but research is needed to investigate this behavior and provide design guidance.

External post-tensioning for rehabilitation—FRP has potential for rehabilitation applications. The lower elastic modulus of FRP makes stressing short tendons more attractive. The lack of tendon-anchorage systems remains a key application issue.

Tendon replacement—FRP replacement tendons for corroded structures are an application area with high potential. The design of a tendon to fit in the existing duct and to function over tendon bend points in the reinforcement needs to be resolved.

Circular prestressed tanks—Design guides for prestressing circular "wire- and strand-wrapped" tanks using FRP tendons need to be developed.

Stressing procedure—Guidelines for stressing procedure/methods need to be developed.

Reliability assessment—The strength-reduction factors require recalibration to ensure conformance with the load factors in ACI 318-02.

Shear capacity—Shear design guidelines for FRP prestressed beams with FRP stirrups need further development.

Bond and development—Protocols for validation of FRP tendon transfer length and development length are needed.

CHAPTER 10—REFERENCES

10.1—Referenced standards and reports

The standards and reports listed below were the latest editions at the time this document was prepared. Because these documents are revised frequently, the reader is advised to contact the proper sponsoring group if it is desired to refer to the latest version.

American Concrete Institute

116R	Cement and Concrete Notation
440R	State-of-the-Art Report on Fiber Reinforced Plastic (FRP) Reinforcement for Concrete Structures
440.1R	Guide for the Design and Construction of Concrete Reinforced with FRP Bars
543R	Design, Manufacture, and Installation of Concrete Piles

AASHTO

AASHTO	Standard Specifications for Highway Bridges

Canadian Standards Association

CAN/CSA S6	Canadian Highway Bridge Design Code
CAN/CSA S806	Design and Construction of Building Components with Reinforced Polymers

These publications may be obtained from these organizations:

American Concrete Institute
38800 Country Club Dr.
Farmington Hills, MI 48331

AASHTO
444 North Capitol St. N.W., Suite 249
Washington, DC 20001

Canadian Standards Association
178 Rexdale Blvd.
Toronto, ON
M9W 1R3 Canada

10.2—Cited references

Abdelrahman, A. A., and Rizkalla, S. H., 1998, "Deflection Control of Concrete Beams Pretensioned by CFRP Reinforcements," *Journal of Composites for Construction*, ASCE, Feb., pp. 55-62.

Abdelrahman, A. A.; Tadros, G.; and Rizkalla, S. H., 1995, "Test Model for the First Canadian Smart Highway Bridge," *ACI Structural Journal*, V. 92, No. 4, July-Aug., pp. 451-458.

ACI Committee 318, 2002, "Building Code Requirements for Structural Concrete (ACI 318-02) and Commentary (318R-02)," American Concrete Institute, Farmington Hills, Mich., 443 pp.

Alkhairi, F. M., 1991, "On the Behavior of Concrete Beams Prestressed with Unbonded Internal and External Tendons," PhD thesis, University of Michigan, Ann Arbor, Mich., 415 pp.

Al-Mayah, A.; Soudki, K. A.; and Plumtree, A., 2001, "Experimental and Analytical Investigation of a Stainless Steel Anchorage for CFRP Pretensioning Tendons," *PCI Journal*, V. 46, No. 2, pp. 88-100.

Aravinthan, T., and Mutsuyoshi, H., 1997, "Prediction of the Ultimate Flexural Strength of Externally Prestressed PC Beams," *Transaction of the Japan Concrete Institute*, V. 19, pp. 225-230.

Arockiasamy, M., and Amer, A., 1998, "Studies on CFRP Prestressed Concrete Bridge Columns and Piles in Marine Environment," Final Report Submitted to Florida Department of Transportation, Tallahassee, Fla., July.

Bakis, C. E., 1993, "FRP Composites: Materials and Manufacturing," *Fiber-Reinforced-Plastic for Concrete Structures*, A. Nanni, ed., Elsevier, Amsterdam, Chapter 3, pp. 13-58.

Bakis, C. E.; Bhat, B. B.; Schokker, A. J.; and Boothby, T. E., 2001, "Flexure of Concrete Beams Prestressed with FRP Tendons," *Proceedings of the Fifth International Symposium on Fiber Reinforced Polymer Reinforcement for Concrete Structures (FRPRCS-5),* Cambridge, London, pp. 689-697.

Braimah, A.; Green, M. F.; and Soudki, K. A., 1996, "Eliminating Steel from Bridge Deck Slabs by Combining CFRP Tendons with Polypropylene FRC," *Proceedings of the Second International Conference on Advanced Composite Materials in Bridges and Structures (ACMBS2)*, Montreal, Quebec, Canada, Aug., pp. 735-742.

Bryan, P. E., and Green, M. F., 1996, "Low Temperature Behavior of CFRP Prestressed Concrete Beams," *Canadian Journal of Civil Engineering*, V. 23, No. 2, pp. 464-470.

Burgoyne, C. J., 1988, "Engineering Applications of Parafil Rope," *Symposium on Engineering Applications of Parafil Rope*, Imperial College, London, pp. 39-47.

Burke, B.; Nelson, J.; and Dolan, C. W., 2000, "Friction Loss of CFRP Post-Tensioned Tendons," *Proceedings of the Third International Conference on Advanced Composite Materials in Bridges and Structures (ACMBS3)*, Ottawa, Canada, Aug., pp. 249-257.

Campbell, T. I.; Shrive, N. G.; Soudki, K. A.; Al-Mayah, A.; Keatley, J. P.; and Reda, M. M., 2000, "Design and Evaluation of a Wedge-Type Anchor for Fiber Reinforced Polymer Tendons," *Canadian Journal of Civil Engineering*, V. 27, pp. 985-992.

Currier, J., 1995,"Deformation of Prestressed Concrete Beams with FRP Tendons," MS thesis, Department of Civil and Architectural Engineering, University of Wyoming, Laramie, Wy., 112 pp.

Dolan, C. W., 1989, "Prestressed Concrete Using KEVLAR Reinforced Tendons," PhD dissertation, Cornell University, Ithaca, N.Y., 204 pp.

Dolan, C. W., 1990, "Developments in Non-Metallic Prestressing Tendons," *PCI Journal*, V. 35, No. 5, Sept., p. 80.

Dolan, C. W., 1999, "FRP Prestressing in the USA," *Concrete International*, V.21, No. 10, Oct., pp. 21-24.

Dolan, C. W., and Burke, C. R., 1996, "Flexural Strength and Design of FRP Prestressed Beams," *Proceedings of the Second International Conference on Advanced Composite Materials in Bridges and Structures (ACMBS2)*, Montreal, Quebec, Canada, Aug., pp. 383-390.

Dolan, C. W.; Hamilton, H. R.; Bakis, C. E.; and Nanni, A., 2000, "Design Recommendations for Concrete Structures Prestressed with FRP Tendons, Final Report," *Department of Civil and Architectural Engineering Report DTFH61-96-C-00019*, University of Wyoming, Laramie, Wyo., May, 113 pp.

Dolan, C. W., and Swanson, D., 2002, "Development of Flexural Capacity of FRP Prestressed Beam with Vertically Distributed Tendons," *Composites Part B: Engineering*, V. 3, No. 1, Elsevier, New York, 2002, pp. 1-6.

Domenico, N., 1995, "Bond Properties of CFCC Prestressing Strands in Pretensioned Concrete Beams," MSc thesis, Department of Civil Engineering, University of Manitoba, Winnipeg, Manitoba, Canada, 160 pp.

Dowden, D. M., and Dolan, C. W., 1997, "Comparison of Experimental Shear Data for FRP Prestressed Beams with Code Provisions" *Proceedings*, FRPRCS-3, V. 2, Sapporo, Japan, Oct., 687 pp.

Dye, W. K.; Bakis, C. E.; and Nanni, A., 1998, "Accelerated Testing of Carbon FRP Tendon-Anchor Systems for Post-Stressed Concrete Applications," *Proceedings of the First International Conference on Durability of Composites for Construction*, V. 2, Sherbrooke, Quebec, Canada, pp. 463-473.

Ehsani, M. R.; Saadatmanesh, H.; and Nelson, C. T., 1997, "Transfer and Flexural Bond Performance of Aramid and Carbon FRP Tendons," *PCI Journal*, V. 42, No. 1, Jan.-Feb., pp. 76-86.

Ehsani, M. R.; Saadatmanesh, H.; and Tao, S.; 1995, "Bond of Hooked Glass Fiber Reinforced Plastic (GFRP) Reinforcing Bars to Concrete," *ACI Materials Journal*, V. 92, No. 4, July-Aug., pp. 391-400.

Fam, A. Z.; Rizkalla, S. H.; and Tadros, G., 1997, "Behavior of CFRP for Prestressing and Shear Reinforcements of Concrete Highway Bridges," *ACI Structural Journal*, V 94, No. 1, Jan.-Feb., pp. 77-86.

FIB TG9.3, http://allserv.rug.ac.be/~smatthys/fibTG9.3/

FIP, 1992, "High-Strength Fiber Composite Tensile Elements for Structural Concrete," *FIP Commission on Prestressing Materials and Systems*, Institute of Structural Engineers, London, July, 160 pp.

Fukuyama, H., 1999, "FRP Composites in Japan," *Concrete International*, V. 21, No. 10, Oct., pp. 29-32.

Gangkatharan, J., 2003, "Strength of Partially Unbonded Prestressed Concrete Beams," MSc thesis, Queen's University, Kingston, Canada, 134 pp.

Gerritse A., and Werner, J., 1988, *ARAPREE: The Prestressing Element Composed of Resin Bonded Twaron Fibers*, Holland Beton Group, The Netherlands, June, 8 pp.

Gerritse, A. and Werner, J., 1991, "ARAPREE—A Non-Metallic Tendon," *Advanced Composite Materials in Civil Engineering Structures*, ASCE, Materials Engineering Division, New York, pp. 143-154.

Grace, N. F., 1999, "Continuous CFRP Prestressed Concrete Bridges," *Concrete International,* V. 21, No. 10, Oct., pp. 42-47.

Grace, N. F., 2000a, "Transfer Length of CFRP/CFCC Strands for Double-T Girders," *PCI Journal*, V. 45, No. 5, Sept.-Oct. 2000, pp. 110-126.

Grace, N. F., 2000b, "Response of Continuous CFRP Prestressed Concrete Bridges Under Static and Repeated Loadings," *PCI Journal*, V. 45, No. 6, Nov.-Dec., pp. 84-101.

Grace, N. F., and Abdel-Sayed, G., 1998a, "Ductility of Prestressed Bridges using CFRP Strands," *Concrete International*, V. 20, No. 6, June, pp. 25-30.

Grace, N. F., and Abdel-Sayed, G., 1998b, "Behavior of Externally Draped CFRP Tendons in Prestressed Concrete Bridges," *PCI Journal*, V. 43, No. 5. Sept.-Oct., pp. 88-101.

Grace, N. F.; Enomoto, T.; and Yagi, K., 2002, "Behavior of CFCC and CFRP Leadline Prestressing Systems in Bridge Construction," *PCI Journal*, V. 47, No. 3, May-June, pp. 90-103.

Grace, N. F.; Navarre, F. C.; Nacey, R. B.; Bonus, W.; and Collavino, L., 2002, "Design-Construction of Bridge Street

Bridge—First CFRP Bridge in the United States," *PCI Journal*, V. 47, No. 5. Sept.-Oct., pp. 20-35.

Grace, N. F., and Singh, S. B., 2003, "Design Approach for Carbon Fiber-Reinforced Polymer Prestressed Concrete Bridge Beams," *ACI Structural Journal*, V. 100, No. 3, May-June, pp. 365-376.

Harada, T.; Idemitsu, T.; Watanabe, A.; Khin, M.; and Soecha, K., 1993, "New FRP Tendon Anchorage System Using Highly Expansive Material for Anchoring," *FIP Symposium 93*, Kyoto, pp. 711-718.

Harada, T.; Soeda, M.; Enomoto, T.; Tokumitsu, S.; Khin, M.; and Idemitsu, T., 1997, "Behavior of Anchorage for FRP Tendons Using Highly Expansive Material Under Cyclic Loading," *Non-Metallic Reinforcement for Concrete Structures*, Sapporo, Proceedings FRPRCS-3, V. II, pp. 719-726.

Hartman, D. R.; Greenwood, M. E.; and Miller, D. M., 1994, "High Strength Glass Fibers," *Technical Paper 1-PL-19025*, Owens Corning Fiberglas Corp., Toledo, Ohio, Mar.

Holte, L. E.; Dolan, C. W.; and Schmidt, R. J., 1993, "Epoxy Socketed Anchors for Non-Metallic Prestressed Tendons," *Fiber-Reinforced-Plastic Reinforcement of Concrete Structures*, SP-138, A. Nanni and C. W. Dolan, eds., American Concrete Institute, Farmington Hills, Mich., pp. 381-400.

IABSE, 2003, "Use of Fiber Reinforced Polymers in Bridge Construction," *Structural Engineering Documents* No. 7, IABSE-AIPC-IVBH, Zurich, Switzerland.

Iyer, S. L., 1993, "First Composite Cable Prestressed Bridge in the USA," *Proceedings of the Thirty-Eight International SAMPE Symposium*, Anaheim, Calif.

Iyer, S. L., 1995, "Demonstration of Advanced Composite Cables for use as Prestressing in Concrete Waterfront Structures," Final report submitted to U.S. Army Corps of Engineers, Construction Engineering Research Laboratory, Champaign, Ill., Nov.

Iyer, S. L., and Kumarswamy, C., 1988, "Performance Evaluation of Glass Fiber Composite Cable for Prestressing Concrete Units," *Proceedings of the Thirty Third International SAMPE Symposium*, Anaheim, Calif, Mar.

Iyer, S.; Lampo, R.; Hoy, D.; and McCarthy, N., 1996, "First Navy Pier Built in the USA Using FRP Cables for Prestressing," *Proceedings of the International Conference on FRP in Civil Engineering at IIT Madras*, Dec., pp. 490-498.

Iyer, S. L., and Sen, R., 1991, "Advanced Composite Materials in Civil Engineering Structures," *Proceedings of Specialty Conference*, ASCE, Las Vegas, Nev.

Joint ACI-ASCE Committee 426, 1973, "The Shear Strength of Reinforced Concrete Members," *Proceedings*, ASCE, V. 99, No. ST6, pp. 1091-1187.

JSCE, 1992, "Application of Continuous Fiber Reinforcing Materials to Concrete Structures," *Concrete Library International*, No. 19, Japan Society of Civil Engineers, Tokyo, p. 89.

JSCE, 1993, "State-of-the Art Report on Continuous Fiber Reinforcing Materials," *Research Committee on Continuous Fiber Reinforcing Material*, A. Machida, ed., *Concrete Engineering Series* No. 3, Japan Society of Civil Engineers, Tokyo, Oct., 164 pp.

JSCE, 1996, "Research Committee on Continuous Fiber Reinforcing Materials: Recommendation for Design and Construction of Concrete Structures Using Continuous Fiber Reinforcing Materials," *Concrete Library* No. 88, Japan Society of Civil Engineers, Tokyo, Japan, Sept.

JSCE, 1997, "Recommendation for Design and Construction of Concrete Structures Using Continuous Fiber Reinforcing Materials," *Concrete Engineering Series* No. 23, 325 pp.

Kelly, D. J.; Bradberry, T. E.; and Breen, J. E., 1987, "Time-Dependent Deflections of Pretensioned Beams," *Research Report* 381-1, Research Project 3-5-84-381, Center for Transportation Research, Bureau of Engineering Research, University of Texas at Austin, Austin, Tex.

Kevlar Technical Guide, 1992, "Du Pont Fibers," Wilmington, Del., Dec., 134 pp.

Kim, P., and Meier, U., 1991, "CFRP Cables for Large Structures," *Proceedings of the Specialty Conference on Advanced Composites Materials in Civil Engineering Structures*, ASCE, Las Vegas, Nev., pp. 233-244.

Lu, Z.; Boothby, T. E.; Bakis, C. E.; and Nanni, A., 2000, "Transfer and Development Length of FRP Prestressing Tendons," *PCI Journal*, V. 45, No. 2, pp. 84-95.

Mahmoud, Z. I., and Rizkalla, S. H., 1996, "Bond of CFRP Prestressing Reinforcement," *Proceedings of the Second International Conference on Advanced Composite Materials in Bridges and Structures (ACMBS2)*, Montreal, Quebec, Canada, Aug., pp. 877-884.

Mahmoud, Z. I.; Rizkalla, S. H.; and Zaghloul, E., 1997, "Transfer and Development Length of CFRP Reinforcement," *Proceedings of the 1997 CSCE Annual Conference*, Sherbrooke, Quebec, May, pp. 101-110.

Malvar, L. J., and Bish, J., 1995, "Grip Effects in Tensile Testing of FRP Bars," *Proceedings of the Second International RILEM Symposium (FRPRCS-2)*, Non-Metallic (FRP) Reinforcement for Concrete Structures, Ghent, Belgium, Aug., pp. 108-115.

Mather, B., and Tye, R. V., 1955, "Plastic-Glass Fiber Reinforcement for Reinforced and Prestressed Concrete: Summary of Information Available as of July 1, 1955," *Technical Memorandum* No. 6-421, Report 1, Corps of Engineers, Waterways Experiment Station, Vicksburg, Miss.

Matthys, S., and Taerwe, L., 2001, "FRP Reinforcement: Developments in Europe," *Concrete International* V. 25, No. 6, June, pp. 20-21.

McKay, K. S., and Erki, M. A., 1993, "Flexural Behavior of Concrete Beams Pretensioned with Aramid Fiber Reinforced Plastic Tendons," *Canadian Journal Civil Engineering*, V. 20. No. 4, pp. 688-695.

MDA, 2004, Market Development Alliance of the FRP Composites Industry, www.mdacomposites.org/bridge_statistics.htm.

Mitsubishi Kasei Corp., 1993, "Leadline Carbon Fiber Tendons/Bars," *Product Specification Manual*, June, 18 pp.

Mufti, A. A.; Newhook, J. P.; and Tadros, G., 1996, "Deformability Versus Ductility in Concrete Beams with FRP Reinforcement," *Proceedings of the Second International Conference on Advanced Composite Materials in*

Bridges and Structures (ACMBS2), Montreal, Quebec, Canada, Aug., pp. 189-199.

Naaman, A. E., 1987, "Partial Prestressing in the Rehabilitation of Concrete Bridges," *Proceedings of U.S.-European Workshop on Bridge Evaluation, Repair and Rehabilitation, St. Rémy-lès-Chevreux*, University of Michigan, Ann Arbor, Mich., June, pp. 392-406.

Naaman, A. E.; Burns, N.; French, C.; Gamble, W. L.; and Mattock, A. H., 2002, "Stresses in Unbonded Prestressing Tendons at Ultimate: Recommendation," *ACI Structural Journal*, V. 99, No. 4, July-Aug., pp. 520-531.

Naaman, A. E., and Jeong, S. M., 1995, "Structural Ductility of Concrete Beams Prestressed with FRP Tendons," *Non-metallic (FRP) Reinforcement for Concrete Structures*, Proceedings of the Second International RILEM Symposium (FRPRCS-2), Non-Metallic (FRP) Reinforcement for Concrete Structures, Ghent, Belgium, Aug., pp. 379-386.

Nanni, A.; Bakis, C. E.; O'Neil, E. F.; and Dixon, T. O., 1996a, "Performance of FRP Tendon-Anchor Systems for Prestressed Concrete Structures," *PCI Journal*, V. 41, pp. 34-44.

Nanni, A.; Bakis, C. E.; O'Neil, E. F.; and Dixon, T. O., 1996b, "Short-Term Sustained Loading of FRP Tendon-Anchor Systems," *Construction and Building Materials*, V. 10, pp. 255-266.

Nanni, A., and Dolan, C. W., eds., 1993, *Fiber-Reinforced-Plastic Reinforcement of Concrete Structures*, SP-138, American Concrete Institute, Farmington Hills, Mich., 977 pp.

Nanni, A., and Tanigaki, M., 1992, "Pretensioned Prestressed Concrete Members with Bonded Fiber Reinforced Plastic Tendons: Development and Flexural Bond Length (Static)," *ACI Structural Journal*, V. 89, No. 4, July-Aug., pp. 433-441.

Nanni, A.; Utsunomiya, T.; Yonekura, H.; and Tanigaki, M., 1992, "Transmission of Prestressing Force to Concrete by Bonded Fiber Reinforced Plastic Tendons," *ACI Structural Journal*, V. 89, No. 3, May-June, pp. 335-344.

NBS, 1976, "Non-Metallic Antenna Support Materials Pultruded Rods for Antenna Guys, Catenaries and Communications Structures," *Technical Report AFML-TR-76-42*, National Bureau of Standards, Washington, D.C., 126 pp.

Noritake, K.; Kakihara, R.; Kumagai, S.; and Mizutani, J., 1993, "Technora, an Aramid FRP Rod," *FRP Reinforcement for Concrete Structure: Properties and Applications*, A. Nanni, ed., Developments in Civil Engineering, V. 42. Elsevier Science Publishers B.V., Amsterdam. pp. 267-290.

PCI, 1975, "Recommendations for Estimating Prestress Losses," *PCI Journal*, V. 20, No. 4, July-Aug., pp. 43-75.

PCI, 2000, *Design Handbook*, Fifth Edition, Precast/Prestressed Concrete Institute, Chicago, Ill., 690 pp.

Pepper, L., and Mather, B., 1959, "Plastic-Glass Fiber Reinforcement for Reinforced Prestressed Concrete: Summary of Information from 1 July 1955 to 1 January 1959," *Technical Memorandum* No. 6-421 Report 2, Corps of Engineers, Waterways Experiment Station, Vicksburg, Miss.

Rao, S., and Mathew, G., 1996, "Behavior of Externally Prestressed Concrete Beams with Multiple Deviators," *ACI Structural Journal*, V. 93, No. 4, July-Aug., pp. 387-396.

Reda Taha, M. M., and Shrive, N. C., 2003a, "New Concrete Anchors for Carbon Fiber-Reinforced Post-Tensioned Tendons—Part 1: State-of-the-Art Review/Design," *ACI Structural Journal*, V. 100, No. 1, Jan.-Feb., pp. 86-95.

Reda Taha, M. M., and Shrive, N. C., 2003b, "New Concrete Anchors for Carbon Reinforced-Reinforced Post-Tensioned Tendons—Part 2: Development/Experimental Investigation," *ACI Structural Journal*, V. 100, No. 1, Jan.-Feb., pp. 96-104.

Rizkalla, N. S.; Fang, Z.; and Campbell, T. I., 2001, "Partially Bonded Partially Prestressed Pretensioned Beams with Hybrid FRP/Stainless Steel Reinforcement," *Proceedings of the Fifth International Symposium on Fiber Reinforced Polymer Reinforcement for Concrete Structures*, (FRPRCS-5), London, pp. 731-740.

Rizkalla, S. H., and Tadros, G., 1994, "A Smart Highway Bridge in Canada," *Concrete International*, V. 16, No. 6, June, pp. 42-44.

Salib, S. R.; Abdel-Sayed, G.; and Grace, N. F., 1999, "Crack Formation in Fiber Reinforced Polymers Concrete Beams," *Fiber Reinforced Polymer Reinforcement of Reinforced Concrete Structures: Selected Presentation Proceedings*, C. W. Dolan, S. H. Rizkalla, and A. Nanni, eds., American Concrete Institute, Farmington Hills, Mich., pp. 219-231.

Santoh, N., 1993, "CFCC: Carbon Fiber Composite Cable," *Fiber-Reinforced-Plastic (FRP) Reinforcement for Concrete Structures: Properties and Applications*, A. Nanni, ed., Developments in Civil Engineering, V. 42, Elsevier Science Publishers B.V., Amsterdam, pp. 223-248.

Sayed-Ahmed, E. Y., and Shrive, N. G., 1998, "A New Steel Anchorage System for Post-Tensioning Applications using Carbon Fiber Reinforced Plastic Tendons," *Canadian Journal of Civil Engineering*, V. 25, No. 1, pp. 113-127.

Schiebel, S., and Nanni, A., 2000, "Axial and Flexural Performance of Concrete Piles Prestressed with CFRP Tendons," *Proceedings of the Third International Conference on Advanced Composite Materials in Bridges and Structures (ACMBS3)*, Ottawa, Canada, Aug., pp. 471-478.

Sen, R.; Issa, M.; and Mariscal, D., 1992, "Feasibility of Fiberglass Pretensioned Piles in a Marine Environment," *Final Report No. CEM/ST/92/1* submitted to Florida Department of Transportation.

Sen, R.; Issa, M.; Wadsack, P.; and Shahawy, M., 1993, "Driving Stresses in Fiberglass Pretensioned Piles," *ACI Structural Journal*, V. 90, No. 6, Nov.-Dec., pp. 666-674.

Shaheen, E., 2004, "Carbon Fiber Reinforced Reactive Powder Concrete Anchorage System," PhD thesis, University of Calgary, Calgary, Alberta, Canada.

Shehata, E., 1999, "Fiber-Reinforced Polymer (FRP) for Shear Reinforcement in Concrete Structures," PhD thesis, Department of Civil and Environmental Engineering, University of Manitoba, Winnipeg, Canada.

Shield, C.; French, C.; and Hanus, J., 1999, "Bond of GFRP Rebar for Consideration in Bridge Decks," *Fiber Reinforced Polymer Reinforcement of Reinforced Concrete Structures*, SP-188, C. W. Dolan, S. H. Rizkalla, and A.

Nanni, eds., American Concrete Institute, Farmington Hills, Mich., pp. 393-406.

Sonobe, Y.; Fukuyama, H.; Okamoto, T.; Kani, N.; Kimura, K.; Kobayashi, K.; Masuda, Y.; Matsuzaki, Y.; Mochizuki, S.; Nagasaka, T.; Shimizu, A.; Tanano, H.; Tanigaki, M.; and Teshigawara, M., 1997, "Design Guidelines of FRP Reinforced Concrete Building Structures," *Journal of Composite for Construction*, ASCE, V. 1, No. 3, pp. 90-115.

Soudki, K. A.; Green, M. F.; and Clapp, F. D., 1997, "Transfer Length of Carbon Fiber Rods in Precast Pretensioned Concrete Beams," *PCI Journal*, V. 42, No. 5, pp. 78-87.

Taerwe, L., and Matthys, S., 1999, "FRP for Concrete Construction: Activities in Europe," *Concrete International*, V. 21, No. 10, Oct., pp. 33-36.

Taerwe, L., and Pallemans, I., 1995, "Force Transfer of AFRP Bars in Concrete Prisms," *Proceedings of the Second International RILEM Symposium (FRPRCS-2),* Non-Metallic (FRP) Reinforcement for Concrete Structures, Ghent, Belgium, Aug., 154-163.

Tamura, T.; Kanda, M.; and Tsuji, Y., 1993, "Applications of FRP Materials to Prestressed Concrete Bridges and Other Structures in Japan," *PCI Journal*, V. 38, No. 4, pp. 50-58.

Tan, K. H., and Ng, C. K., 1997, "Effects of Deviators and Tendon Configuration on Behavior of Externally Prestressed Beams," *ACI Structural Journal*, V. 94, No. 1, Jan.-Feb., pp. 13-22.

Thornel, Continuous Pitch-Based Carbon Fibers Literature, 2004, www.cytec.com, Mar.

Torayca Technical Reference Manual, 1996, Toray Industries, Inc., Carbon Fibers Dept., Tokyo, Japan, 1996.

Triantafillou, T. C., and Deskovic, N., 1991, "Innovative Prestressing with FRP Sheets: Mechanics of Short Term Behavior," *Journal of Engineering Mechanics*, V. 117, No. 7, pp. 1652-1672.

Vijay, P. V., and GangaRao, H. V. S., 2001, "Bending Behavior and Deformability of Glass Fiber-Reinforced Polymer Reinforced Concrete Members," *ACI Structural Journal*, V. 98, No. 6, Nov.-Dec., pp. 834-842.

Wines, J. C.; Dietz, R. J.; and Hawley, J. L., 1966, "Laboratory Investigation of Plastic-Glass Fiber Reinforcement for Reinforced and Prestressed Concrete," *Report 1 & 2*, U.S. Army Corps of Engineers, Waterways Experiment Station, Vicksburg, Miss.

Wolff, R., and Miesseler, H.-J., 1989, "New Materials for Prestressing and Monitoring Heavy Structures," *Concrete International*, V. 11, No. 9, Sept., pp. 86-89.

Zoch, P.; Kimura, H.; Iwasaki, T.; and Heym, M., 1991, *Carbon Fiber Composite Cables—A New Class of Prestressing Members,* Transportation Research Board, Washington, D.C., 19 pp.

Zhang, B., 2002, "Experimental and Theoretical Investigations on New Bond-Type Anchorage System for Post-Tensioning Applications with FRP Tendons," PhD thesis, University of Sherbrooke, Canada.

APPENDIX A—DESIGN EXAMPLE

Example 1—Pretensioned single T-beam

A simply supported T-beam is to be designed to carry two 15 kip (67 kN) concentrated loads spaced 6.78 ft (2.1 m) on center. A Rocky Mountain Prestress 12DT40 section, cut in half along the centerline, is to be used for the section. The beam is to be designed to carry its self-weight plus live load. Tendons are harped at midspan with an end eccentricity of zero.

Procedure	Calculation in inch-pound units	SI equivalents
Step 1—Define section properties		
A	333 in.^{2}	214,800 mm^2
I_g	51,020 in.^{4}	2.124×10^{10} mm^4
S_t	3516 in.^{3}	57.6×10^6 mm^3
S_b	2001 in.^{3}	32.8×10^6 mm^3
S_p	2378 in.^{3}	39.0×10^6 mm^3
y_t	14.51 in.	369 mm
y_b	25.49 in.	647 mm
w_g	370 lb/ft	5.40 kN/m
e (eccentricity = y_b – cover)	21.49 in.	546 mm
b_w	7.25 in.	184 mm
b	72 in.	1829 mm
h	40 in.	1016 mm
h_f	2 in.	51 mm
d	36 in.	914 mm
Step 2—Define material properties		
f_c'	6500 psi	44.8 MPa
f_{ci}'	4000 psi	27.6 MPa
E_{ci}	3.08×10^6 psi	21.2 GPa
E_c	4.60×10^6 psi	31.7 GPa
$f_{ci} = 0.6f_{ci}'$	2400 psi	16.5 MPa
$f_{ti} = 3\sqrt{f_{ci}'}$	190 psi	1.3 MPa
$f_c = 0.45f_c'$	2925 psi	20.2 MPa
$f_t = 6\sqrt{f_c'}$	484 psi	3.3 MPa
Number of tendons – $m = 6$		
f_{pu}	425,000 psi	2.93 GPa
d_p	0.375 in.	9.5 mm
A_p	0.11 in.^{2}	71.0 mm^2
E_p	18,000 ksi	124 GPa
$\varepsilon_{cu} = f_{pu}/E_p$	425/18,000 = 0.024	0.024
Stress due to harping $= R_tE_p/R$	0.12(18,000)/24 = 90 ksi	620 MPa
Initial prestress = 60% f_{pu} less 20% for harping	Percent of strength = 100(90)/425 = 21.2%	
Stress tendon to 40% of f_{pu} to allow for harping $P_i = 0.40f_{pu}mA_p$	0.4(425)6(0.11) = 112.2 kips	499 kN
Step 3—Beam span and loadings		
Length L	39.37 ft	12.0 m
Load spacing b_1	6.78 ft	2.1 m
Shear span $a_1 = 1/2(L - b_1)$	1/2(39.37 – 6.78) = 13.16 ft	4.01 m
Dead load w	370 lbf/ft	5.40 kN/m

$M_d = wL^2/8$	0.370(39.37)2/8 = 71.7 ft-kips	97.2 kN-m
Live load P	15 kips	66.7 kN
$M_1 = Pa_1$	15 × 13.16 = 197 ft-kips	267 kN-m
$M_u = 1.2M_d + 1.6M_1$ (ACI 318-02)	1.2 × 71.7 + 1.6 × 197 = 401 ft-kips	544 kN-m
$V_u = 1.2wL/2 + 1.6P$	1.2 × 0.370 × 39.37/2 + 1.6 × 15 = 32.7 kips	145 kN
Step 4—Loss Calculations		
Modular ratio $n = E_f/E_{ci}$	n = 18,000/3080 = 5.8	
Elastic shortening $\Delta f_{es} = nf_{cp} = n(P_i/A + P_ie/S_p - M_d/S_p)$	5.8(112.2/333 + 112.2 × 21.45/2387 – 71.7 × 12/2387) = 5710 psi	39.4 MPa
Creep: assume 2 times initial elastic shortening $\Delta f_{cr} = C_c\Delta f_{es}$	2.0(5710) = 11,420 psi	78.7 MPa
Shrinkage: assume 0.0006 net strain at time of testing $\Delta f_s = \varepsilon_s E_p$	(0.0006)18,000,000 = 10,800 psi	74.5 MPa
Relaxation $\Delta f_r = 0.03f_{pi}$	0.03(0.4)425,000 = 5100 psi	35.2 MPa
Total losses $\Delta f_{es} + \Delta f_c + \Delta f_s + \Delta f_r$	5710 + 11,420 + 10,800 + 5100 = 33,030 psi	228 MPa
Total losses =	19.4%	19.4%
$f_{pe} = 0.4f_{pu}$ – losses	0.4(425) – 33.0 = 137.0 ksi	944 MPa
Final prestress = mA_pf_{pe}	6(0.11)137.0 = 90.4 kips	402 kN
Step 5—Check service level stresses at midspan		
Initial stresses		
Top = $P_i/A - P_ie/S_t + M_d/S_t$	112.2/333 – 112.2 × 21.45/3516 + 71.7 × 12/3516 = –0.108 ksi	–0.74 MPa
Bottom = $P_i/A + P_ie/S_b - M_d/S_b$	112.2/333 + 112.2 × 21.45/2001 – 71.7 × 12/2001 = 1.110 ksi	7.6 MPa
Both are within stress limits.		
Final stresses		
Top = $P_e/A - P_ee/S_t + M_d/S_t + M_1/S_t$	90.4/333 – 90.4 × 21.45/3516 + 71.7 × 12/3516 + 197 × 12/3516 = 0.635 ksi tension	4.38 MPa
Bottom = $P_e/A + P_ee/S_b - M_d/S_b - M_1/S_b$	90.4/333 + 90.4 × 21.45/2001 – 71.7 × 12/2001 – 197 × 12/2001 = 0.370 ksi tension	2.55 MPa
Both are within stress limits.		
Step 6—Check strength capacity		
Effective flange width $b_{eff} = L/4$	= 39.27 × 12/4 = 119 in.	3.20 m
$= b_w + 4h_f$	use 7.25 + 4(2) = 15.25 in.	387 mm
$= b$	72 in.	1.83 m
Assume neutral axis is in the flange		
$a = f_{pu}mA_p/(0.85f_c'b_{eff})$	425(6)0.11/[(0.85)(6.5)(15.25)] = 3.33 in. > h_f, therefore need to separate forces	84.6 mm
Tendon area for flange		
$A_p' = (b_{eff} - b_w)\,0.85f_c'/f_{pu}$	(15.25 – 7.25) 0.85(6.5)/425 = 0.104 in.[2]	67.1 mm[2]
New neutral axis		
$a = (A_p - A_p')f_{pu}/0.85f_c'b_w)$	(0.66 – 0.104)425/(0.85(6.5)7.25) = 5.90 in.	150 mm
Check tension control		
$\beta_1 = 0.85 - 0.05(f_c' - 4000)/1000$	0.85 – 0.05(6500 – 4000)/1000 = 0.725	0.725
$c/d < 0.375$	= 0(590/0.725)/36 = 0.225 OK	OK
$M_n = A_p'f_{pu}(d - h_f/2) + (A_p - A_p')f_{pu}(d - a/2)$	0.104(425)(36 – 2/2) + (0.66 – 0.104)425 (36 – 5.9/2) = 9357 in.-kip = 750 ft-kip	1057 kN-m
$\phi M_n > M_u$	0.9(780) = 702 ft-kips > M_u = 481 ft-kips OK	952 kN-m > 544 kN-m
Step 7—Check shear capacity		
Shear capacity will be checked against V_c and, if insufficient, stirrups added		
$\phi V_c = \phi 2\sqrt{f_c'}\,b_wd$	(0.75)2 $\sqrt{6500}$ × (7.25 × 36)/1000 = 31.6 kips	140 kN
$\phi V_p = P_ee/L/2$	0.75(90.4)21.45/(32.79 × 12/2) = 7.4 kips	32.9 kN
$\phi V_n = \phi V_c + \phi V_p$	31.6 + 7.4 = 39.0 kips	173 kN
$\phi V_n > V_u$	39.0 kips > 32.7 kips OK	173 kN > 145 kN OK
$V_u > \phi V_c/2$, therefore, minimum shear reinforcement will be required.		
Step 8—Check live load		
Deflections against $L/480$ limit	a_1/L = 16.29/39.37 = 0.413	
$\delta = \dfrac{PL^3(a_1/L)}{24E_cI_g}(3 - 4 \times (a_1/L)^2)$	15(39.37)[3]0.413(3 – 4 × (0.413)[2])1728/(24(4.6 × 10[6]) 51,020) = 0.26 in.	6.6 mm
	δ/L = 0.26/39.37 × 12 = 1/1817 OK	1/1817 OK

Closing comment—The above beam was fabricated as designed by Rocky Mountain Prestress. It was tested to 3,000,000 load cycles producing $6\sqrt{f_c'}$ tensile stress, based on uncracked section. The beam cracked after 100,000 cycles. Following the fatigue test, the beam was loaded to failure. The experimental load capacity was 6% less than the calculated nominal capacity.

A metric version of this document (ACI 440.5M-08) is available at **www.concrete.org**

Specification for Construction with Fiber-Reinforced Polymer Reinforcing Bars

An ACI Standard

Reported by ACI Committee 440

John P. Busel
Chair

Carol K. Shield*
Secretary

Tarek Alkhrdaji	Russell Gentry	James G. Korff	Andrea Prota
Charles E. Bakis	Janos Gergely	Michael W. Lee	Hayder A. Rasheed
Lawrence C. Bank*	William J. Gold	Maria Lopez de Murphy	Sami H. Rizkalla
Abdeldjelil Belarbi	Nabil F. Grace	Ibrahim M. Mahfouz	Morris Schupack
Brahim Benmokrane	Mark F. Green	Orange S. Marshall	Rajan Sen
Luke A. Bisby	Zareh B. Gregorian	Amir Mirmiran	Khaled A. Soudki
Gregg J. Blaszak	Doug D. Gremel*	Ayman S. Mosallam	Samuel A. Steere III*
Timothy E. Bradberry*	Shawn P. Gross	John J. Myers	Gamil S. Tadros
Gordon L. Brown Jr.	H. R. (Trey) Hamilton III	Antonio Nanni†	Jay A. Thomas
Vicki L. Brown	Issam E. Harik	Kenneth Neale	Houssam A. Toutanji
Raafat El-Hacha	Kent A. Harries	John P. Newhook	J. Gustavo Tumialan
Garth J. Fallis*	Mark P. Henderson*	Ayman M. Okeil	Milan Vatovec
Amir Z. Fam	Bohdan N. Horeczko	Carlos E. Ospina*	Stephanie L. Walkup
Edward R. Fyfe	Vistasp M. Karbhari	Max L. Porter	David White

*Contributing authors.
†Task group Chair.

The committee also thanks Nick Carino, Bernard Drouin, and Renato Parretti for their contributions.

This Reference Specification covers construction using fiber-reinforced polymer reinforcing bars that the Architect/Engineer can make applicable to any construction project by citing it in the Project Specifications. The Architect/Engineer supplements the provisions of this Reference Specification as needed by designating or specifying individual project requirements.

Keywords: concrete; concrete construction; fiber-reinforced polymer reinforcement; placing; specification; tolerance.

NOTE TO SPECIFIER

This Specification is incorporated by reference in the Project Specification using the wording in P3 of the Preface and including information from the Mandatory Requirements, Optional Requirements, and Submittals Checklists following the Specifications.

PREFACE

P1. ACI Specification 440.5-08 is intended to be used by reference or incorporation in its entirety in the Project Specification. Do not copy individual Parts, Sections, Articles, or Paragraphs into the Project Specification, because taking them out of context may change their meaning.

P2. If Sections or Parts of ACI Specification 440.5-08 are copied into the Project Specification or any other document, do not refer to them as an ACI Specification, because the specification has been altered.

P3. A statement such as the following will serve to make ACI Specification 440.5-08 a part of the Project Specification: "Work on (Project Title) shall conform to all requirements of ACI 440.5-08, "Specification for Construction with Fiber-Reinforced Polymer Reinforcing Bars," published by the American Concrete Institute, Farmington Hills, Michigan, except as modified by these Contract Documents."

P4. Each technical Section of ACI Specification 440.5-08 is written in the three-part Section format of the Construction Specifications Institute, as adapted for ACI requirements. The language is imperative and terse.

ACI 440.5-08 was adopted April 30, 2008, and published June 2008.

P5. The Specification is written to the Contractor. When a provision of this Specification requires action by the Contractor, the verb “shall” is used. If the Contractor is allowed to exercise an option when limited alternatives are available, the phrasing “either...or...” is used. Statements provided in the Specification as information to the Contractor use the verbs “may” or “will.” Informational statements typically identify activities or options that “will be taken” or “may be taken” by the Owner or Architect/Engineer.

CONTENTS

PART 1—GENERAL

1.1—Scope

1.1.1 *Work specified*—This Specification covers cast-in-place structural concrete reinforced with fiber-reinforced polymer (FRP) bars except where other provisions are specified in the Contract Documents.

1.1.2 *Work not specified*—The subjects covered by ACI 301 are not in the scope of this Specification.

1.2—Definitions

acceptable or **accepted**—determined to be satisfactory by the Architect/Engineer.

acceptance—acknowledgment by Architect/Engineer that submittal or completed work is acceptable.

Architect/Engineer—the architect, engineer, architectural firm, or engineering firm issuing Contract Documents or administering the Work under Contract Documents, or both.

dielectric—a substance, such as porcelain, glass, and most plastics, that is a poor conductor of electricity but an efficient supporter of electrostatic fields.

1.3—Reference standards

1.3.1 *Reference standards*—Standards of ACI referred to in this Specification are listed with serial designation including year of adoption or revision, and are part of this Specification.

1.3.1.1 *ACI standards*

117-06	Specifications for Tolerances for Concrete Construction and Materials
301-05	Specifications for Structural Concrete
440.6-08	Specification for Carbon and Glass Fiber-Reinforced Polymer (FRP) Bar Materials for Concrete Reinforcement

1.4—Submittals

Unless otherwise required by Contract Documents, submit data and drawings specified in 1.4.1 through 1.4.3 for review and acceptance before fabrication and execution:

1.4.1 Submit the data specified in 1.4.1.1 through 1.4.1.4 unless otherwise specified:

1.4.1.1 *FRP reinforcement*—Submit FRP bar manufacturer’s certified test report.

1.4.1.2 *Placing drawings*—Submit placing drawings showing fabrication dimensions and placement locations of FRP reinforcement and reinforcement supports.

1.4.1.3 *Splices*—Submit a list of splices and request to use splices not indicated in Contract Documents.

1.4.1.4 *Supports*—Submit description of reinforcement supports and materials for fastening reinforcement not described in 2.1.3.

1.4.2 *FRP reinforcement relocation*—When alternatives are proposed, submit a request to relocate any FRP reinforcement that exceeds specified placement tolerances. Include new arrangement of reinforcement and justification for new arrangement.

1.4.3 *Inspection and quality control*—Submit quality assurance/quality control program of plant manufacturing FRP reinforcement unless otherwise specified.

1.5—Materials delivery, storage, and handling

1.5.1 Prevent bending, coating with earth, oil, or other material, or otherwise damaging the FRP reinforcement.

1.5.2 When handling FRP reinforcement, use equipment that avoids damaging or abrading the FRP bar. Lift bundles of FRP reinforcement at multiple pickup points. Do not drop or drag FRP reinforcement.

1.5.3 FRP reinforcing bars shall be stored above the surface of the ground on platforms, skids, or other supports as close as possible to the point of placement. If stored outdoors for more than 4 months, the FRP reinforcing bars shall be covered with opaque plastic or other types of cover that protect the bars from ultraviolet rays.

1.5.4 Prevent exposure of FRP reinforcing bars to temperatures above 120 °F during storage.

PART 2—PRODUCTS

2.1—Materials

2.1.1 *FRP reinforcing bars*—FRP reinforcement shall be deformed, sand coated, or both to provide bond with concrete unless otherwise stated in the Contract Documents.

FRP reinforcement shall be the types and sizes required by Contract Documents and shall conform to ACI 440.6-08.

2.1.1.1 *Permitted damage*—The maximum total unrepaired visible damage permitted on each linear foot of each FRP bar shall not exceed 2% of the surface area in that linear foot of bar. The depth of the permissible damage shall not exceed 0.04 in.

2.1.2 *Grids and gratings*—Premanufactured grids and gratings are not covered by this Specification. Mat reinforcement made of preassembled FRP bars is covered by this Specification.

2.1.3 *FRP reinforcement supports*—Use reinforcement supports made of dielectric material or, if ferrous, coated with dielectric material, including epoxy or another polymer.

2.1.4 *Precast concrete reinforcement supports*—Use concrete supports that have a surface area of not less than 4 in.^{2} and a compressive strength and durability properties equal to or greater than the specified compressive strength of the concrete being placed.

2.2—Fabrication

2.2.1 *FRP reinforcement*—No field fabrication is permitted except tying of FRP reinforcing bars, field cutting in accordance with the design or placement drawings, or both. Field cutting shall be performed as described in 3.4.

PART 3—EXECUTION

3.1—Preparation

3.1.1 When concrete is placed, FRP reinforcement shall be free of materials deleterious to mechanical bond.

3.2—Bar placement

3.2.1 *Tolerances*—Place, support, and fasten FRP reinforcement as shown on the Project Drawings. Do not exceed the placing tolerances specified in ACI 117 before concrete is placed. Placing tolerances shall not reduce cover requirements except as specified in ACI 117.

3.2.2 *FRP reinforcement relocation*—When it is necessary to move FRP reinforcement beyond the specified placing tolerances to avoid interference with other reinforcement, conduits, or embedded items, submit the resulting reinforcement arrangement for approval.

3.2.3 *Concrete cover*—Unless otherwise specified, concrete cover for FRP reinforcement shall be as indicated in Table 3.1. Table 3.1 reports specified concrete cover requirements for FRP reinforcement for all exposure conditions except those required for additional fire protection. For bundled FRP bars, specified concrete cover shall be equal to the equivalent diameter of the bundle, but need not be greater than 2 in., except the specified cover shall not be less than indicated in Table 3.1. The equivalent diameter of the bundle shall be computed based on the total area of the bundle. Tolerances on specified concrete cover shall meet the requirements of ACI 117.

3.2.4 *FRP reinforcement supports*—Unless otherwise permitted, use the reinforcement supports as given in 3.2.4.1 through 3.2.4.7.

Table 3.1—Concrete cover requirements for FRP reinforcement

	Specified cover, in.
Slabs and joists	
Top and bottom bars	
No. 10 bars and smaller	3/4
Formed concrete surfaces exposed to earth and for bottoms bearing on work mat, or slabs supporting earth cover	
No. 5 bars and smaller	1-1/2
No. 6 through 10 bars	2
Beams, formed	
Stirrups, spirals, and ties	1-1/2
Primary reinforcement	2
Exposed to earth	
Stirrups and ties	2
Primary reinforcement	2-1/2
Walls	
No. 10 bars and smaller	3/4
Formed concrete surfaces exposed to earth or in contact with ground	2
Footings and base slabs	
At formed surfaces and bottoms bearing on concrete work mat	2
At unformed surfaces and bottoms in contact with earth	3
Top of footings	Same as slabs
Over top of piles	2

3.2.4.1 Use precast concrete reinforcement supports to support reinforcement from the ground or a mud mat.

3.2.4.2 Use wire reinforcement supports that are galvanized, coated with dielectric material, or made of dielectric material (including FRP) to support FRP reinforcement.

3.2.4.3 Reinforcement and embedded steel items used with FRP reinforcement shall be zinc-coated (galvanized) or coated with nonmetallic materials, or stainless steel or stainless steel clad. In the case of carbon FRP (CFRP) bars, zinc coatings, stainless steel, or stainless steel clad shall not be used in direct contact with CFRP bars.

3.2.4.4 Use steel wires or dowels coated with dielectric material or stainless steel when precast concrete reinforcement supports with embedded steel tie wires or dowels are used with FRP reinforcement.

3.2.4.5 Reinforcement used as supports for FRP reinforcement shall be epoxy-coated or made of FRP.

3.2.4.6 In walls, use spreader bars that are epoxy-coated steel or made of FRP. Proprietary combination bar clips and spreaders used in walls with FRP reinforcement shall be made of corrosion-resistant material or coated with dielectric material.

3.2.4.7 Tie wire shall be plastic- or polymer-coated wire. For applications requiring complete electromagnetic neutrality, glass FRP (GFRP) bars shall be tied in place using nylon or plastic wire (zip) ties. Alternatively, thermoplastic injection molded clips are permitted.

3.2.5 *Mat reinforcement*—For slabs-on-ground, extend FRP mat reinforcement to within 2 in. of the concrete edge. Lap splice edges and ends of FRP mat reinforcement sheets as shown on the Project Drawings. Unless otherwise specified or permitted, do not extend FRP mat reinforcement through

Table 3.2—Minimum radius of prefabricated bends

FRP bar size	Minimum inside bend diameter
No. 3 through 8	Three bar diameters
No. 9 and 10	Four bar diameters

contraction joints. Support FRP mat reinforcement during placing of concrete to maintain specified elevation in the slab. Do not place FRP mat reinforcement on grade and subsequently raise into position in concrete.

3.2.6 *Splices*—Form lap splices as indicated on the Project Drawings. Mechanical splices for FRP reinforcement are not permitted.

3.2.7 *Field bending or straightening*—If the plans require bent FRP bars, the bends shall be incorporated during bar manufacture. Field bending or straightening of FRP bars are not permitted.

3.2.8 *Factory-formed bends*—Minimum inside bend radii shall conform to the requirements of Table 3.2. In addition, the beginning of the bend shall not be closer to the concrete surface than the minimum diameter of the bend.

3.2.9 *FRP reinforcement through expansion joint*—Do not continue FRP reinforcement or other embedded FRP items bonded to concrete through expansion joints. FRP dowels that extend through a joint or waterstop shall be unbonded or bonded to only one side of a joint or waterstop.

3.3—Repair of fiber-reinforced polymer bars

3.3.1 Damage to an FRP bar resulting in visible fibers (other than at cut ends) or a cut or defect greater than 0.04 in. deep shall be cause for rejection of a bar. Visible damage to the FRP bars exceeding 2% of surface area per foot of bar and not resulting in rejection by the Architect/Engineer shall be repaired. The remedy for a damaged FRP bar shall be to lap splice a new FRP bar adjacent to the damaged portion with the appropriate lap length on either side of the damage.

3.4—Field cutting of fiber-reinforced polymer reinforcement

3.4.1 Field cut FRP reinforcement only when specifically permitted using cutting methods specified by or acceptable to the Architect/Engineer. When FRP reinforcing bars are cut in the field, shear cutting is not permitted.

3.4.2 Cut ends shall be sealed if required by the FRP bar manufacturer or the Architect/Engineer, and all surface damage due to cutting shall be inspected by the Architect/Engineer and repaired.

3.5—Concrete placement

If FRP reinforcement is not adequately supported or tied to resist settlement, floating upward, or movement in any direction during concrete placement, concrete placement shall be halted until corrective measures are taken.

FOREWORD TO CHECKLISTS

F1. This Foreword is included for explanatory purposes only; it does not form a part of ACI Specification 440.5-08.

F2. ACI Specification 440.5-08 may be referenced by the Specifier in the Project Specification for any building project, together with supplementary requirements for the specific project. Responsibilities for project participants must be defined in the Project Specification. The ACI Specification cannot and does not address responsibilities for any project participant other than the Contractor.

F3. Checklists do not form a part of ACI Specification 440.5-08. Checklists assist the Specifier in selecting and specifying project requirements in the Project Specification.

F4. The Specifier shall make adjustments to the needs of a particular project by reviewing each of the items in the checklists and including the items the Specifier selects as mandatory requirements in the Project Specification.

F5. The Mandatory Requirements Checklist indicates work requirements regarding specific qualities, procedures, materials, and performance criteria that are not defined in ACI Specification 440.5-08.

F6. The Optional Requirements Checklist identifies Specifier choices and alternatives. The checklists identify the Sections, Parts, and Articles of the Reference Specification and the action required or available to the Specifier.

F7. The Submittal Checklist identifies Specifier choices for information or data to be provided by the Contractor before, during, or after construction.

MANDATORY REQUIREMENTS CHECKLIST

Section/Part/Article	Notes to Specifier
2.1.1 FRP reinforcing bars	State in the Contract Documents where the use of FRP reinforcing bars is required. State the types and sizes of FRP bars.
3.4 Field cutting of fiber-reinforced polymer reinforcement	Specify the permitted method of cutting FRP bars. Acceptable methods include cutting with high-speed grinder or circular blade saw. Carborundum or diamond blades are preferred.

OPTIONAL REQUIREMENTS CHECKLIST

Section/Part/Article	Notes to Specifier
1.4 Submittals	Specify which items, if any, listed in 1.4.1 through 1.4.3 are not required to be submitted.
2.1.1 FRP reinforcing bars	Specify where FRP reinforcement shall not be deformed or sand coated. Smooth FRP bars should be used as dowels where the intent is to transfer shear force while allowing bar sliding.
3.2.3 Concrete cover	Specify where concrete cover for FRP reinforcement shall be different from the values specified in Table 3.1.
3.2.4 FRP reinforcement supports	Specify permitted reinforcement supports other than those listed in 3.2.4.1 through 3.2.4.7.
3.2.5 Mat reinforcement	Specify where FRP mat reinforcement is permitted to extend through a contraction joint.

SUBMITTALS CHECKLIST

Section/Part/Article	Notes to Specifier
1.4.1.1 FRP reinforcement	FPR bar manufacturer's certified test report.
1.4.1.2 Placing drawings	Placing drawings showing fabrication dimensions and placement locations of FRP reinforcement and reinforcement supports.
1.4.1.3 Splices	A list of splices and request to use splices not indicated in Contract Documents.
1.4.1.4 Supports	Description of reinforcement supports and materials for fastening reinforcement.
1.4.2 and **3.2.2** FRP reinforcement relocation	Reinforcement arrangement when it is necessary to move FRP reinforcement beyond the specified placing tolerances to avoid interference with other reinforcement, conduits, or embedded items.
1.4.3 Inspection and quality control	Quality assurance/quality control program of plant manufacturing FRP reinforcement.

A metric version of this document (ACI 440.6M-08) is available at **www.concrete.org**

ACI 440.6-08
(Reapproved 2017)

Specification for Carbon and Glass Fiber-Reinforced Polymer Bar Materials for Concrete Reinforcement

An ACI Standard

Reported by ACI Committee 440

John P. Busel
Chair

Carol K. Shield*
Secretary

Tarek Alkhrdaji	Russell Gentry	James G. Korff	Andrea Prota
Charles E. Bakis	Janos Gergely	Michael W. Lee	Hayder A. Rasheed
Lawrence C. Bank*	William J. Gold	Maria Lopez de Murphy	Sami H. Rizkalla
Abdeldjelil Belarbi	Nabil F. Grace	Ibrahim M. Mahfouz	Morris Schupack
Brahim Benmokrane	Mark F. Green	Orange S. Marshall	Rajan Sen
Luke A. Bisby	Zareh B. Gregorian	Amir Mirmiran	Khaled A. Soudki
Gregg J. Blaszak	Doug D. Gremel*	Ayman S. Mosallam	Samuel A. Steere III*
Timothy E. Bradberry*	Shawn P. Gross	John J. Myers	Gamil S. Tadros
Gordon L. Brown Jr.	H. R. (Trey) Hamilton III	Antonio Nanni†	Jay A. Thomas
Vicki L. Brown	Issam E. Harik	Kenneth Neale	Houssam A. Toutanji
Raafat El-Hacha	Kent A. Harries	John P. Newhook	J. Gustavo Tumialan
Garth J. Fallis*	Mark P. Henderson*	Ayman M. Okeil	Milan Vatovec
Amir Z. Fam	Bohdan N. Horeczko	Carlos E. Ospina*	Stephanie L. Walkup
Edward R. Fyfe	Vistasp M. Karbhari	Max L. Porter	David White

*Contributing authors.
†Task group Chair.

The committee also thanks Nick Carino, Bernard Drouin, Jack Lesko, and Renato Parretti for their contributions.

This Material Specification covers provisions governing testing, evaluation, and acceptance of carbon and glass fiber-reinforced polymer (FRP) bars used as reinforcement for concrete.

Keywords: carbon fiber; concrete; concrete construction; FRP reinforced concrete; fiber-reinforced polymer reinforcement; glass fiber; specification.

CONTENTS

ACI 440.6-08 was adopted April 30, 2008, and published June 2008.

SECTION 1—SCOPE

1.1 This specification describes permitted constituent materials, limits on constituent volumes, and minimum performance requirements for carbon and glass fiber-reinforced polymer (FRP) bars to be used as reinforcement for nonprestressed concrete.

1.2 Only carbon and glass FRP bars are covered by this specification.

1.3 FRP bars made of more than one fiber type (hybrid FRP) are not covered by this specification.

1.4 Pultruded FRP bars with no external surface enhancement (that is, plain or smooth bars) to facilitate bond with concrete are not covered by this specification. Similarly, hollow FRP bars are not considered due to lack of documented performance as reinforcement for concrete.

1.5 Plain FRP bars used as dowels (that is, devices that transfer shear across concrete joints) where the intended function requires slip of the dowel are not covered by this specification.

1.6 This specification does not cover premanufactured grids and gratings made with FRP materials. FRP mats resulting from assembly of deformed FRP bars, however, are covered by this document.

1.7 This specification does not cover FRP bars when used for external and near-surface-mounted strengthening applications.

1.8 The text of this specification references notes and footnotes that provide explanatory material. These notes and footnotes (excluding those in tables and figures) shall not be considered as requirements of the specification.

SECTION 2—REFERENCED DOCUMENTS

2.1—ASTM standards

A615/A615M-08a	Standard Specification for Deformed and Plain Carbon-Steel Bars for Concrete Reinforcement
C904-01(2006)	Standard Terminology Relating to Chemical-Resistant Nonmetallic Materials
D570-98(2005)	Standard Test Method for Water Absorption of Plastics
D2584-02	Standard Test Method for Ignition Loss of Cured Reinforced Resins
D3171-06	Standard Test Methods for Constituent Content of Composite Materials
D4475-02(2008)	Standard Test Method for Apparent Horizontal Shear Strength of Pultruded Reinforced Plastic Rods by the Short-Beam Method
D5117-03	Standard Test Method for Dye Penetration of Solid Fiberglass Reinforced Pultruded Stock
D5229/D5229M-92(2004)	Standard Test Method for Moisture Absorption Properties and Equilibrium Conditioning of Polymer Matrix Composite Materials
D7205/D7205M-06	Standard Test Method for Tensile Properties of Fiber Reinforced Polymer Matrix Composite Bars
E1356-03	Standard Test Method for Assignment of the Glass Transition Temperatures by Differential Scanning Calorimetry
E1640-04	Standard Test Method for Assignment of the Glass Transition Temperature by Dynamic Mechanical Analysis

2.2—ACI report

The following test methods from ACI 440.3R,* "Guide Test Methods for Fiber-Reinforced Polymers (FRPs) for Reinforcing or Strengthening Concrete Structures," are referenced. Because these test methods are not written in mandatory language, purchaser and manufacturer shall agree on the protocols to be used.

- B.3 Test method for bond strength of FRP bars by pullout testing
- B.4 Test method for transverse shear strength of FRP bars
- B.5 Test method for strength of FRP bent bars and stirrups at bend locations
- B.6 Accelerated test method for alkali resistance of FRP bars
- B.12 Test method for determining the effect of corner radius on tensile strength of FRP bars

SECTION 3—TERMINOLOGY

3.1—Definitions

For definitions of terms used in this specification, refer to ASTM C904.

3.2—Definitions of terms specific to this specification

commercial-grade material—a material formulated for and used in industrial (not consumer) applications.

production lot—any lot of FRP bar produced from start to finish with the same constituent materials used in the same proportions without changing any production parameter, such as cure temperature or line speed.

property, guaranteed—a characteristic value provided by the manufacturer no greater than the mean minus three standard deviations of at least the required number of samples tested according to a specified test method. This definition is applicable to tensile strength, shear strength (perpendicular to the bar), bond strength, and strength of bent bars.

property, nominal—a value provided by the manufacturer no greater than the mean of at least the required number of samples tested according to a specified method. This definition is applicable to tensile elastic modulus, moisture absorption, and resistance to alkaline environment.

SECTION 4—CLASSIFICATION

4.1 FRP bars shall be classified according to fiber composition with the first letter of the acronym designating fiber type as follows:

- CFRP: carbon fiber-reinforced polymer bar
- GFRP: glass fiber-reinforced polymer bar

*ACI Committee 440, 2004, "Guide Test Methods for Fiber-Reinforced Polymers (FRPs) for Reinforcing or Strengthening Concrete Structures (ACI 440.3R-04)," American Concrete Institute, Farmington Hills, MI, 40 pp.

SECTION 5—ORDERING INFORMATION

5.1 The purchaser shall specify the following:

- The classification of bar (see Note 1);
- The method for creating the deformed bar surface;
- The bar size;
- The bar length; and
- For bent bars, the shape of the bend, the radius of the bend, and the length of the legs.

(Note 1—The bar classification—CFRP or GFRP—corresponds to given constituents and the minimum mechanical properties in this specification.)

SECTION 6—MATERIALS AND MANUFACTURE

6.1—Fibers

Fibers shall be in the form of unidirectional rovings (glass fibers) or unidirectional tows (carbon fibers) of given size and mass (see Note 2). Fiber sizings and coupling agents shall be compatible with the resin system used to impregnate them.

(Note 2—The fiber type and fiber amount determine the physical and mechanical properties of the FRP bar.)

6.2—Matrix resins

6.2.1 Vinylester and epoxy resin systems are permitted provided the finished product meets the physical and durability requirements of this specification. Blending of vinylester and epoxy resins is permitted.

6.2.2 The base polymer in the resin system shall not contain any polyester.

6.2.3 Styrene is permitted to be added to the polymer resin during processing. Added styrene shall be less than 10% by mass of the polymer resin. The amount of styrene, as a mass percentage of the polymer resin, added during processing shall be reported.

6.2.4 Constituent content of resin shall be determined by ASTM D3171.

6.3—Fillers and additives

6.3.1 Only commercial-grade inorganic fillers such as kaolin clay, calcium carbonate, and alumina trihydrate are permitted, and shall not exceed 20% by mass of the polymer resin constituent.

6.3.2 Only commercial-grade additives and process aids, such as release agents, low-profile shrink additives, initiators, promoters, hardeners, catalysts, pigments, fire retardants, and ultraviolet inhibitors are permitted and depend on the processing method. Shrink additives, if used, shall be less than 10% by mass of the polymer resin.

6.3.3 Only commercial-grade inorganic or organic nonwoven surfacing mats or veils are permitted.

6.4—Manufacturing process

6.4.1 The manufacturer shall produce FRP bars using variations of the pultrusion process.

6.4.2 Process or material modifications are not permitted during the production of a single lot.

6.4.3 The manufacturer shall document the process used and report the date of production and production lot size.

Table 7.1—Size designation of FRP round bars

Bar size designation	Nominal diameter, in.	Nominal area, in.^{2}
2	0.250	0.05
3	0.375	0.11
4	0.500	0.20
5	0.625	0.31
6	0.750	0.44
7	0.875	0.60
8	1.000	0.79
9	1.128	1.00
10	1.270	1.27

SECTION 7—PHYSICAL PROPERTIES

7.1—Fiber content

The fiber content shall be measured by ASTM D3171 or D2584. When ASTM D3171 is used, fiber content shall not be less than 55% by volume. When ASTM D2584 is used, fiber content shall not be less than the fraction by mass corresponding to 55% by volume. The manufacturer shall report the fiber content of the end product by volume or by mass in accordance with the method used.

7.2—Glass transition temperature

The glass transition temperature T_g of the resin shall not be less than 212 °F (see Note 3). The glass transition temperature shall be measured on a coupon cut from the as-produced bar using either the differential scanning calorimetry (DSC) method in ASTM E1356 or the dynamic mechanical analysis (DMA) method in ASTM E1640. When using the DSC method, test results for both the first scan (according to ASTM E1356) and the second scan shall be reported.

(Note 3—This temperature does not represent the maximum permitted service temperature, and is intended for purchaser's quality assurance only. ACI 440.1R* does not currently designate maximum service temperatures of FRP bars. Individual manufacturers should be consulted to determine the appropriate maximum service temperature for each product.)

7.3—Bar sizes

7.3.1 Only FRP bars of solid round or elliptical cross-section shape are allowed.

7.3.2 The size of FRP bars shall be consistent with standard sizes for steel reinforcing bars given in ASTM A615/A615M and as listed in Table 7.1.

7.3.3 The calculated diameter of an FRP bar is equivalent to that of a smooth round bar having the same area as the FRP bar measured by ASTM D7205/D7205M.

7.3.4 When the FRP bar is of elliptical shape, the minimum and maximum outside dimensions of the bar cross section shall be provided in addition to the calculated diameter. The calculated diameter of elliptical FRP bars is equivalent to that of a solid round bar having the same cross-sectional area as determined according to ASTM D7205/D7205M.

*ACI Committee 440, 2006, "Guide for the Design and Construction of Structural Concrete Reinforced with FRP Bars (ACI 440.1R-06)," American Concrete Institute, Farmington Hills, MI, 44 pp.

7.3.5 The nominal diameter of an FRP bar to be used for designation and design shall be equal to the calculated diameter. When the calculated diameter does not correspond to one of the nominal values given in Table 7.1, the next immediately smaller nominal diameter given in such table shall be used.

SECTION 8—MECHANICAL PROPERTIES

8.1—Tensile strength

The tensile strength shall be determined for two purposes: 1) overall product certification (that is, guaranteed value) from multiple production lots; and 2) manufacturer's quality control and purchaser's quality assurance for each production lot.

8.1.1 The guaranteed tensile strength for product certification shall be measured according to ASTM D7205/D7205M based on nominal dimensions and at a frequency and number of specimens, as indicated in Section 11. Minimum guaranteed tensile strength values are listed in Table 8.1. The manufacturer shall report the individual test results.

8.1.2 The tensile strength of a production lot shall be measured for the purchaser's quality assurance according to ASTM D7205/D7205M at a frequency and number of specimens as indicated in Section 11. The strength of each specimen shall be reported and be not less than the guaranteed strength reported by the manufacturer. If strength is less than the guaranteed strength, the production lot shall be rejected.

8.2—Tensile modulus of elasticity

The tensile modulus of elasticity shall be determined for the purpose of product certification (that is, nominal value). The nominal tensile modulus of elasticity of GFRP bars shall be at least 5700 ksi regardless of bar size or geometry. The tensile modulus of elasticity of CFRP bars shall be at least 18,000 ksi regardless of bar size or geometry. The tensile modulus of elasticity is derived from specimens tested in accordance with ASTM D7205/D7205M at a frequency and number of specimens as indicated in Section 11. The manufacturer shall report the individual test results.

8.3—Shear strength (perpendicular to bar)

The transverse shear strength shall be determined for the purpose of product certification (that is, guaranteed value). The guaranteed transverse shear strength of FRP bars shall be at least 18 ksi as determined by a test method submitted by the manufacturer for acceptance by the purchaser (see Note 4), at a frequency and number of specimens as indicated in Section 11. The manufacturer shall report the individual test results.

(Note 4—ACI 440.3R, Test Method B.4, may be considered by the manufacturer for this purpose.)

8.4—Ultimate tensile strain

The ultimate tensile strain shall be calculated for the purpose of product certification (that is, nominal value). The nominal ultimate tensile strain shall be calculated by dividing the guaranteed tensile strength by the nominal elastic modulus (see Note 5). The nominal ultimate tensile strain of CFRP and GFRP bars obtained by this procedure shall be at least 0.5 and 1.2%, respectively.

Table 8.1—Minimum guaranteed tensile strength for FRP bars

Bar size designation	Minimum guaranteed tensile strength	
	GFRP, ksi	CFRP, ksi
2	110	210
3	110	190
4	100	170
5	95	160
6	90	160
7	85	N/A
8	80	N/A
9	75	N/A
10	70	N/A

Note: N/A indicates that CFRP bars of these sizes are currently not available.

(Note 5—The calculation method is based on the assumption that the stress-strain behavior is linear elastic [straight line]).

8.5—Bond strength

The bond strength shall be determined for the purpose of product certification (that is, guaranteed value) for each bar size. The guaranteed bond strength of FRP bars shall be at least 1.4 ksi as determined by a test method submitted by the manufacturer for acceptance by the purchaser (see Note 6), at a frequency and number of specimens as indicated in Section 11. The manufacturer shall report the individual test results and the method used for casting the test specimens.

(Note 6—ACI 440.3R, Test Method B.3, may be considered by the manufacturer for this purpose. Such method is intended to determine the relative bond behavior for material specifications, but is not intended for design purposes. Two methods for casting the test specimens are provided in Test Method B.3.).

SECTION 9—DURABILITY PROPERTIES

9.1—Moisture absorption

Moisture absorption tests shall be determined for the purpose of product certification (that is, nominal value) in accordance with ASTM D570, Section 7.4, or D5229/D5229M, Procedure B, using a water temperature of 122 °F at a frequency and number of specimens as indicated in Section 11. The individual moisture absorption test results shall be reported, and their average shall be less than 1.0%.

9.2—Resistance to alkaline environment

Resistance to alkaline environment tests shall be determined for the purpose of product certification (that is, nominal value) in accordance with a test method submitted by the manufacturer for acceptance by the purchaser (see Note 7), at a frequency and number of specimens as indicated in Section 11. The manufacturer shall report the individual test results and the test method. Minimum strength retention values have not yet been established.

(Note 7—ACI 440.3R, Test Method B.6, may be considered by the manufacturer for this purpose.)

9.3—Longitudinal wicking

Longitudinal wicking shall be determined for the purposes of the purchaser's quality assurance. Five consecutive 1 in. long segments cut from an FRP bar shall be tested in accordance with ASTM D5117 at a frequency as indicated in Section 11. No continuous voids shall be permitted in the resin (see Note 8). A continuous void is one that appears in all five consecutive test specimens. The presence of hollow fibers is not considered a void.

(Note 8—This requirement is intended to check for continuous voids that could occur due to shrinkage of the resin during processing or as the result of poor consolidation of the fiber and resin matrix during production.)

SECTION 10—OTHER REQUIREMENTS

10.1—Bend radius

Bends shall be formed in FRP bars that are made with thermosetting resin and only while the resin is in a physical liquid state (see Note 9). The minimum inside bend radii for factory-formed FRP bar bends are specified in Table 10.1.

(Note 9—After the resin has passed the liquid state, bending or alteration of the FRP bar is not possible due to the inability of the fibers to move [or reorient] within the resin matrix. Because thermosetting polymers are highly cross linked, heating the bar on site will not be allowed as it leads to a decomposition of the resin, thus creating a loss of strength in the FRP bar.)

10.2—Strength of bends

The strength of bends shall be determined for the purpose of overall product certification (that is, guaranteed value).

10.2.1 The guaranteed strength of bends for product certification shall be measured according to a test method submitted by the manufacturer for acceptance by the purchaser (see Note 10). Testing frequency and number of specimens shall be as indicated in Section 11. The manufacturer shall report the individual test results and the test method. Minimum strength values have not yet been established.

(Note 10—ACI 440.3R, Test Method B.5 or B.12, may be considered by the manufacturer for this purpose.)

10.2.2 For a measure of the manufacturer's quality control and purchaser's quality assurance on FRP bends, one of the two test methods listed as follows shall be used at a frequency and number of specimens as indicated in Section 11 (see Note 11). The manufacturer shall report the individual test results and the test method.

(Note 11—Testing for the manufacturer's quality control does not require measurement of strength of the bends as per Section 10.2.1.)

10.2.2.1 When it is possible to extract a straight portion of the tail of the bent bar of sufficient length, such an element shall be tested according to ASTM D7205/D7205M. The strength of each specimen shall be not less than the guaranteed strength reported by the manufacturer for the bar of that diameter.

10.2.2.2 When the bend size does not allow for the tensile testing of one of its straight portions, a specimen shall be obtained from the bend and tested for interlaminar (horizontal) shear strength according to ASTM D4475 and for fiber content according to ASTM D2584 or D3171. The manufacturer shall report the individual test results and the test method. Minimum interlaminar shear strength values have not yet been established. The fiber content of each specimen shall be not less than the minimum percent by volume reported in Section 7.1.

Table 10.1—Minimum inside bend radius of bent bars

Bar size designation	Bend radius,* in.
2	3/4
3	1-1/8
4	1-1/2
5	1-7/8
6	2-1/4
7	2-5/8
8	3
9	4-1/2
10	5

*Based on three bar diameters for No. 2 through No. 8, and four bar diameters for No. 9 and No. 10.

SECTION 11—SAMPLING

11.1—Sampling frequency and number of specimens

11.1.1 For the determination of each of the mechanical and durability properties for the manufacturer's quality control tests and for the purchaser's quality assurance tests, at least five samples of sufficient length to perform the required tests shall be obtained from each production lot.

11.1.2 For the determination of each of the mechanical and durability properties for product certification (that is, guaranteed and nominal values), at least 25 samples of sufficient length to perform the required tests shall be obtained in groups of five from five different production lots. Tests for the determination of the mechanical and durability properties for product certification shall be repeated at least every 3 years.

11.1.3 For orders of bends in varying quantities that together will comprise a production lot of bends, at the discretion of the purchaser, testing of bends for the manufacturer's quality control and for the purchaser's quality assurance may be limited to a minimum of five samples of a 90-degree bend with the smallest bend radius to bar diameter ratio.

11.2—Methods of sample selection

Samples from each production lot to be used for preparing test specimens shall be selected by the manufacturer on a random basis.

SECTION 12—REJECTION

12.1 The purchaser has the option to reject material that fails to conform to the requirements of this specification. Rejection shall be reported to the manufacturer or supplier promptly and in writing.

SECTION 13—PRODUCT CERTIFICATION

13.1 When specified in the purchase order or contract, the purchaser shall be furnished with:

- Documentation showing the constituents, their quantities, and their properties as provided by the suppliers; and
- Product certification stating that samples representing each production lot have been tested and inspected as indicated in this specification and the requirements have been met.

Certifications shall bear the signature of an authorized representative of the bar manufacturer.

13.2 When specified in the purchase order or contract, a report of the results of the manufacturer's quality control tests and the purchaser's quality assurance tests shall be furnished. The test report shall include the following for a given production lot traceable to the identifying marking on the FRP straight and bent bars furnished:

- Test date and laboratory where testing was performed;
- Nominal bar diameter;
- Production lot number or identifying marking;
- Description of specimens tested, that is, specimen length, free length, and anchorage details;
- Results of individual tensile strength tests of each specimen;
- Results of dye penetration tests (wicking); and
- Description of testing apparatus, that is, load frame type and capacity, extensometer, data acquisition software, and other pertinent details.

13.3 When specified in the purchase order or contract, a report of the test results of product certification shall be furnished. The test report shall include the following for the FRP straight and bent bars furnished:

- Test date and laboratory where testing was performed;
- Resin matrix, fillers, and additives;
- Fiber manufacturer's specific product identification nomenclature used;
- Description of specimens tested, that is, specimen length, free length, and anchorage details;
- Description of the testing method when the choice of more than one test method or variation has been specified in this document;
- Geometric properties: cross-sectional area and diameter;
- Physical properties: fiber content with indication of reference by mass or volume, glass transition temperature, and bar size and dimensions of elliptical bars;
- Guaranteed mechanical properties: tensile strength, transverse shear strength (perpendicular to the bar), bond strength, and strength of bent bars;
- Nominal mechanical properties: tensile modulus of elasticity, and ultimate tensile strain (calculated);
- Nominal durability properties: moisture absorption and resistance to alkaline environment; and
- Description of testing apparatus, that is, load frame type and capacity, extensometer, data acquisition software, and other pertinent details.

SECTION 14—MARKINGS

14.1 Each bundle of bars shall be identified with a corresponding production lot number. Production lot numbers shall be printed on each bar or affixed to each bundle by means of a durable tag.

14.2 Each bundle of bars shall be labeled with the following information:

- A symbol to identify the manufacturer;
- A marking to indicate the classification of fiber (that is, G for glass and C for carbon);
- The nominal bar diameter designation (in metric);
- A marking to indicate the guaranteed tensile strength; and
- A marking to indicate the nominal modulus of elasticity.

14.3 The only marking required for each individual bar shall be the nominal bar diameter designation, and it shall be made by a durable means.

Bar markings shall take the form of permanent physical marking of the bar size in metric on the product or permanent color coding applied to the bar as shown in Table 14.1 (see Note 12).

(Note 12—Markings will be used at the construction site to verify that the specified bars are being used.)

Table 14.1—Color codes for marking bar sizes

Bar size designation	Color*
2	Black
3	Orange
4	Green
5	Red
6	Blue
7	Yellow
8	Purple
9	Gray
10	Brown

*Color coding should be at least 6 in. of the FRP at each end of the bars.

Guide for the Design and Construction of Externally Bonded Fiber-Reinforced Polymer Systems for Strengthening Unreinforced Masonry Structures

Reported by ACI Committee 440

*ACI encourages the development and appropriate use of new and emerging technologies through the publication of the **Emerging Technology Series**. This series presents information and recommendations based on available test data, technical reports, limited experience with field applications, and the opinions of committee members. The presented information and recommendations, and their basis, may be less fully developed and tested than those for more mature technologies. This report identifies areas in which information is believed to be less fully developed, and describes research needs. The professional using this document should understand the limitations of this document and exercise judgment as to the appropriate application of this emerging technology.*

John P. Busel
Chair

Carol K. Shield
Secretary

Tarek Alkhrdaji	Amir Z. Fam	Maria Lopez de Murphy	Sami H. Rizkalla
Charles E. Bakis	William J. Gold	Ibrahim M. Mahfouz	Morris Schupack
Lawrence C. Bank	Nabil F. Grace	Orange S. Marshall	Rajan Sen
Abdeldjelil Belarbi	Mark F. Green	Amir Mirmiram	Khaled A. Soudki
Brahim Benmokrane	Zareh B. Gregorian	John J. Myers	Samuel A. Steere III
Luke A. Bisby	Doug D. Gremel	Antonio Nanni	Gamil S. Tadros
Gregg J. Blaszak	Shawn P. Gross	John P. Newhook	Jay Thomas
Hakim Bouadi	H. R. Trey Hamilton III	Ayman M. Okeil	Houssam A. Toutanji
Timothy E. Bradberry	Issam E. Harik	Carlos E. Ospina	J. Gustavo Tumialan
Vicki L. Brown	Kent A. Harries	Renato Parretti	Milan Vatovec
Gordon L. Brown Jr.	Mark P. Henderson	Max L. Porter	David White
Raafat El-Hacha	Bohdan N. Horeczko	Andrea Prota*	Sarah E. Witt
Garth J. Fallis	Michael W. Lee	Hayder A. Rasheed	

*Chair of the subcommittee that prepared this report.
The committee acknowledges the significant contributions of Associate Members Nestore Galati and Rudi Seracino. The committee also thanks the members of The Masonry Society (TMS) Existing Masonry Committee for their help in reviewing this document.

Fiber-reinforced polymer (FRP) systems are an option to consider for strengthening unreinforced masonry (URM) structures. Traditional strengthening systems include external steel plates, reinforced concrete (RC) overlays, span shortening with steel subframing or bracing, and internal steel reinforcement. Relative to traditional systems, features of FRP systems include high tensile strength, light weight, ease of construction, and resistance to corrosion. This guide offers general information on FRP systems use, a description of their unique material properties, and recommendations for the design, construction, and inspection of FRP systems for strengthening URM structures. These guidelines are based on knowledge gained from a comprehensive review of experimental and analytical investigations and field applications.

Keywords: buildings; cracking; cyclic loading; detailing; earthquake resistance; fiber-reinforced polymers; fibers; flexure; masonry; shear; structural analysis;structural design; unreinforced.

ACI 440.7R-10 was adopted and published April 2010.

CONTENTS

CHAPTER 1—INTRODUCTION AND SCOPE

1.1—Introduction

Masonry is a generic term used to describe a type of construction where clay, or concrete masonry units, or natural stones are bonded together to form a load-bearing structure or a component in a structure. Non-load-bearing masonry includes partitions and veneers.

Although masonry is used in flexural applications such as retaining walls, roof and floor beams, and lintels, it is more frequently used in load-bearing walls primarily resisting compression loads. Reinforced and unreinforced masonry (URM) walls have been used in constructing structural load-bearing components. In buildings, masonry walls can serve effectively as part of the lateral load-resisting system to resist wind and earthquake loads. Infill masonry walls play a significant role in enhancing in-plane stiffness and shear resistance of both reinforced concrete (RC) and steel frames, if properly connected to the structural frame.

Unreinforced masonry structures have shown their vulnerability to major events such as earthquakes, severe wind, blast, and impact. In addition, factors such as change in occupancy, deterioration, or an increase in lateral-load demand, may create the need to perform structural retrofit. ACI 530 covers the design and construction of new masonry; the repair, retrofitting, and rehabilitation of masonry structures are not included in that code.

The repair and retrofit/rehabilitation of existing masonry structures have traditionally been accomplished using conventional materials and construction techniques. Externally bonded steel plates, RC overlays, grouted cell reinforcements, and post-tensioning are just some of the many traditional techniques available. Fiber-reinforced polymer (FRP) (ACI 440R) composites have emerged as an alternative to traditional materials for strengthening masonry structures. Fiber-reinforced polymer materials are lightweight and noncorrosive. They exhibit high tensile strength and elastic modulus (carbon FRP), and are impact resistant with electromagnetic transparency. These materials, which are available in a variety of forms, including flat sheets and plates and reinforcing bars and prestressing tendons of typically round cross section, provide the licensed design professional with flexibility in achieving desired performance. Fiber-reinforced polymer systems can be used in seismic, wind, or blast strengthening of URM structural elements.

Advantages of retrofitting masonry using FRP composites include easier handling and installation with resulting lower installation prices, and minimal changes to the structure's appearance. Disturbance to occupants and loss of usable space are also minimized. Dynamic properties of the existing structure remain unchanged because there is little weight addition or stiffness modification. Disadvantages of using FRP may include diminished performance at elevated temperatures, requirement for protective coatings, degradation of mechanical properties after long-term exposure to certain environmental conditions such as extensive moisture intrusion and frequent wet freezing-and-thawing cycles, and the relatively higher level of site supervision and inspection required.

1.2—Scope

This guideline provides information on the selection and design of FRP systems limited to externally bonded FRP laminates and near-surface-mounted FRP bars/strips for increasing the in-plane and out-of-plane strength of existing ungrouted, grouted, or partially grouted URM walls; infill walls are not included in this guide. The guide is applicable to URM structures made of clay bricks, concrete masonry units, and natural stones using conventional types of mortar.

For masonry with significant deterioration, questionable mortar bond, and cracking and/or element displacement, traditional procedures may be required as well as FRP strengthening. The evaluation for the need to apply traditional modes of strengthening is not covered in this guide.

CHAPTER 2—NOTATION AND DEFINITIONS

ACI provides a comprehensive list of notation and definitions through an online resource, "ACI Concrete Terminology," http://terminology.concrete.org. Definitions provided herein complement that resource.

2.1—Notation

A_f = cross-sectional area of FRP reinforcement, in.^{2} (mm^2)
$A_{f,bar}$ = area of one rectangular and/or circular FRP bar, in.^{2} (mm^2)
A_n = area of net mortared/grouted section, in.^{2} (mm^2)
a_b = smallest dimension of a FRP rectangular bar, in. (mm)
b_b = largest dimension of a FRP rectangular bar, in. (mm)
C_E = environmental reduction factor
c = distance from the fiber of maximum compressive strain to the neutral axis, in. (mm)
c_s = distance from the fiber of maximum compressive strain to the neutral axis at service limit state, in. (mm)
D = dead load effect
d_b = diameter of a FRP bar, in. (mm)
d_f = effective depth of FRP flexural reinforcement, in. (mm)
d_i = distance of force F_i measured from the extreme compression fiber, in. (mm)
E_f = tensile modulus of elasticity of FRP, psi (MPa)
E_s = tensile modulus of elasticity of steel, psi (MPa)
e = eccentricity of axial load, in. (mm)
F_i = force acting on the *i*-th FRP strip, lb (N)
f_a = axial compressive stress due to gravity loads, psi (MPa)
f'_{dt} = specified masonry diagonal tension strength, psi (MPa)
f_f = stress in the FRP reinforcement, psi (MPa)
f_{fe} = effective stress level in the FRP reinforcement, psi (MPa)
f_{fs} = stress in the FRP reinforcement at service, psi (MPa)
f_{fu} = design ultimate tensile strength of FRP, psi (MPa)
f^*_{fu} = ultimate tensile strength of the FRP material as reported by the manufacturer, psi (MPa)
$\overline{f_{fu}}$ = mean ultimate tensile strength of FRP reinforcement based on a population of 20 or more tensile tests per ASTM D3039/D3039M, psi (MPa)
f'_m = specified masonry compressive strength, psi (MPa)
f_{tm} = average value of the tensile strength of masonry, psi (MPa)
H = lateral earth pressure effect
h = effective height of the wall, in. (mm)
h_{eff} = height to resultant of lateral force, in. (mm)
k = coefficient accounting for the boundary conditions of the wall
L = length of the wall, in. (mm)
ℓ_d = development length, in. (mm)
M_n = nominal flexural strength, in.-lb (N-mm)
M_s = moment due to sustained loads, in.-lb (N-mm)
M_u = factored moment, in.-lb (N-mm)
n = number of plies of FRP laminates, and number of circular and rectangular bars, or both
P_n = nominal axial strength, lb (N)
P_s = axial load due to all sustained loads, lb (N)

P_u = factored axial load as a resultant of uniformly distributed load acting on the wall, lb (N)
p_{fm}, p_{fv} = force per unit width (per unit bar for NSM FRP systems), lb/in. (lb/bar) [N/mm (N/bar)]
R_n = nominal strength of a member
s_f = spacing of FRP reinforcement, in. (mm)
T_g = glass transition temperature, °F (°C)
t = nominal thickness of the wall, in. (mm)
t_b = effective depth of NSM FRP system, in. (mm)
t_f = nominal thickness of one ply of the FRP reinforcement, in. (mm)
V_{bjs} = nominal lateral strength corresponding to joint sliding, lb (N)
V_{dt} = nominal lateral strength corresponding to diagonal tension, lb (N)
V_f = FRP contribution to the nominal lateral strength for FRP-strengthened walls, lb (N)
V_n = nominal lateral strength of FRP-strengthened wall, lb (N)
V_n^{URM} = nominal lateral strength of URM, lb (N)
V_{tc} = nominal lateral strength corresponding to toe crushing, lb (N)
V_u = factored shear force, lb (N)
W = wind load effect
w_f = width of FRP laminate, in. (mm)
α = angle of inclination of diagonal strips, degrees
α_L = longitudinal coefficient of thermal expansion, in./in./°F (mm/mm/°C)
α_T = transverse coefficient of thermal expansion, in./in./°F (mm/mm/°C)
β_1 = ratio of the depth of the equivalent rectangular stress block to the depth of the neutral axis, in. (mm)
ε_{fe} = effective strain in FRP reinforcement, in./in. (mm/mm)
ε_{fu} = design rupture strain of the FRP reinforcement, in./in. (mm/mm)
ε_{fu}^* = ultimate rupture strain of the FRP reinforcement as reported by the manufacturer
$\overline{\varepsilon}_{fu}$ = mean rupture strain of FRP reinforcement based on a population of 20 or more tensile tests per D3039/D3039M, in./in. (mm/mm)
ε_{mu} = maximum usable masonry strain, in./in. (mm/mm)
ε_{bi} = strain level in the masonry substrate at the time of the FRP installation, in./in. (mm/mm)
γ = multiplier of f'_m to determine the intensity of an equivalent rectangular stress distribution for masonry
κ_m = bond-dependent coefficient for flexure
κ_v = bond-dependent coefficient for shear
ϕ = strength reduction factor
ω_f = FRP reinforcement index

2.2—Definitions

The following definitions clarify terms pertaining to the use of FRP in externally bonded systems for strengthening URM structures.

alkali-resistant (AR) glass—glass fiber formulated to resist degradation in alkaline conditions such as cementitious mixtures.

batch—quantity of material mixed at one time or in one continuous process.

binder—material forming the matrix of concretes, mortars, and sanded grouts; or chemical treatment applied to fibers to give integrity to mats, roving, and fabric.

catalyst—a substance that accelerates a chemical reaction and enables it to proceed under conditions more mild than otherwise required and that is not, itself, permanently changed by the reaction. (See also **initiator** and **hardener**.)

composite—engineering materials—for example, concrete or fiber-reinforced polymer—made from two or more constituent materials that remain distinct, but combine to form materials with properties not possessed by any of the constituent materials individually; the constituent materials are generally characterized as matrix and reinforcement or matrix and aggregate.

creep-rupture—breakage of a material under sustained loading at stresses less than the tensile strength.

cure of FRP systems—the process of causing irreversible change in the properties of a thermosetting resin by chemical reaction, typically accomplished by the addition of curing (cross-linking) agents or initiators, with or without heat and pressure.

curing agent—a catalytic or reactive agent that induces cross-linking in a thermosetting resin. (See also **hardener** and **initiator**.)

debonding—failure of cohesive or adhesive bond at the interface between a substrate and a strengthening or repair system.

degradation—a decline in the quality of the mechanical properties of a material.

delamination—a planar separation in a material that is roughly parallel to the surface of the material.

durability —the ability of a material to resist weathering action, chemical attack, abrasion, and other conditions of service for an extended period of time.

E-glass—a calcium-aluminosilicate glass that contains less than 2% alkali and may or may not contain boron. In electrical circuit board applications, the composition contains boron and less than 1% alkali. This glass is widely used as reinforcement in various resin-based components. Its high electrical resistivity makes it suitable for uses in electrical applications.

fabric—a two-dimensional network of woven, nonwoven, knitted, or stitched fibers.

fabric construction—the details of structure of fabric. Includes such information as style, width, type of knit or weave, threads per inch in warp and fill, and weight of goods.

fiber—a slender and greatly elongated solid material, generally with a length at least 100 times its diameter, that has properties making it desirable for use as reinforcement.

fiber, aramid—fiber in which chains of aromatic polyamide molecules are oriented along the fiber axis to exploit the strength of the chemical bond.

fiber, carbon—fiber produced by heating organic precursor materials containing a substantial amount of carbon, such as rayon, polyacrylonitrile (PAN), or pitch in an inert environment and at temperatures of 2700°F (1500°C) or greater.

fiber, glass—filament drawn from an inorganic fusion typically comprising silica-based material that has cooled without crystallizing. Types of glass fibers include alkali-lime (A-glass), alkali-resistant (AR-glass), general-purpose (E-glass), high-strength (S-glass), and boron-free (ECR-glass).

fiber content—the fiber present in a composite, usually expressed as a percentage volume fraction or weight fraction of the composite.

fiber fly—short filaments that break off dry fiber tows or yarns during handling and become airborne; usually classified as a nuisance dust.

fiberglass—(see **glass fiber-reinforced polymer**.)

fiber-reinforced polymer (FRP)—a general term for a composite material comprising a polymer matrix reinforced with fibers in the form of fabric, mat, strands, or any other fiber form. (See also **composite**.)

filament—see **fiber**.

filler—a finely divided, relatively inert material, such as pulverized limestone, silica, or colloidal substances, added to portland cement, paint, resin, or other materials to reduce shrinkage, improve workability, reduce price, or reduce density.

fraction, fiber volume—the ratio of the volume of fibers to the volume of the composite containing the fibers.

fraction, fiber weight—the ratio of the weight of fibers to the weight of the composite containing the fibers.

glass fiber-reinforced polymer (GFRP) —a composite material comprising a polymer matrix reinforced with glass fiber cloth, mat, or strands.

hardener—in a two-component adhesive or coating, the chemical component that causes the resin component to cure.

impregnate—to saturate fibers with resin or binder.

initiator—a chemical used to start the curing process for unsaturated polyester and vinyl ester resins. (See also **catalyst**.)

lamina—a single layer of fabric or mat reinforcing bound together in a cured resin matrix

laminate—multiple plies or lamina molded together.

layup—the process of placing reinforcing material and resin system in position for molding.

layup, wet—the process of placing the reinforcing material in the mold or its final position and applying the resin as a liquid.

length, development—the bonded length required to transfer the design strength of reinforcement at a critical section.

load, sustained—a constant load that in structures is due to dead load and long-term live load.

masonry substrate—the existing masonry, including any repair materials, to which the FRP system is applied.

matrix—the resin or binders that hold the fibers in FRP together, transfer load to the fibers, and protect them against environmental attack and damage due to handling.

monomer—an organic molecule of relatively low molecular weight that creates a solid polymer by reacting with itself or other compounds of low molecular weight.

near-surface-mounted (NSM) systems—an externally bonded FRP system in which FRP bars are bonded to grooves made in the masonry surface.

polyacrylonitrile (PAN)—a polymer fiber used to make high-modulus carbon fiber through pyrolysis.

pitch—viscid substance obtained as a residue of petroleum or coal tar and used as a precursor in the manufacture of some carbon fibers.

ply—see **lamina**.

polyester—one of a large group of synthetic resins, mainly produced by reaction of dibasic acids with dihydroxy alcohols; commonly prepared for application by mixing with a vinyl-group monomer and free-radical catalysts at ambient temperatures and used as binders for resin mortars and concretes, fiber laminates (mainly glass), adhesives, and the like. Commonly called "unsaturated polyester."

polymer—the product of polymerization; more commonly a rubber or resin consisting of large molecules formed by polymerization.

polymerization—the reaction in which two or more molecules of the same substance combine to form a compound containing the same elements and in the same proportions but of higher molecular weight.

polyurethane—reaction product of an isocyanate with any of a wide variety of other compounds containing an active hydrogen group; used to formulate tough, abrasion-resistant coatings.

pot life—time interval, after mixing of thermosetting resin and initiators, during which the mixture can be applied without degrading the final performance of the resulting polymer composite beyond specified limits.

prepreg—a sheet of fabric or mat containing resin or binder usually advanced to the B-stage and ready for final forming and cure.

resin—generally a thermosetting polymer used as the matrix and binder in FRP composites.

resin content—the resin in a fiber-reinforced polymer composite laminate, expressed as either a percentage of total mass or total volume.

resin, epoxy—a class of organic chemical bonding systems used in the preparation of special coatings or adhesives for concrete or as binders in epoxy-resin mortars, concretes, and fiber-reinforced polymer composites.

resin, thermosetting—a material that hardens by an irreversible cross-linking of monomers, typically when subjected to a chemical reaction.

roving—a parallel bundle of continuous yarns, tows, or fibers with little or no twist.

S-glass—a magnesium-aluminosilicate glass possessing tensile strength, modulus, and application temperature higher than E-glass. It is typically used as reinforcement in resin-based, high-performance and high-temperature composite applications.

sheet, FRP—a dry, flexible ply used in wet layup FRP systems. Fabrics are also called sheets.

shelf life—the length of time packaged materials can be stored under specified conditions and remain usable.

sizing—surface treatment applied to filaments to impart desired processing, durability, and bond attributes.

temperature, glass-transition—the midpoint of the temperature range over which an amorphous material (such as glass or polymer) changes from (or to) a brittle, vitreous state to (or from) a plastic state.

tow—an untwisted bundle of continuous filaments.

vinylester resin—a thermosetting reaction product of epoxy resin with a polymerizable unsaturated acid (usually methacrylic acid) that is then diluted with a reactive monomer (usually styrene).

volume fraction—see **fraction, fiber volume**.

witness panel—a small mockup manufactured under conditions representative of field application, to confirm that prescribed procedures and materials will yield specified mechanical and physical properties.

yarn—a twisted bundle of continuous filaments.

CHAPTER 3—CONSTITUENT MATERIALS AND PROPERTIES

The behavior of FRP-reinforced masonry structures depends on the physical and material properties of the existing masonry, as well as the FRP system. The physical and mechanical properties of the masonry should be investigated. The effects of factors such as loading history and duration, temperature, and moisture on the properties of FRP systems, are discussed in this chapter.

Fiber-reinforced polymer systems are available in a variety of forms such as wet layup, prepreg, and precured. Factors such as fiber volume, type of fiber, type of resin, fiber orientation, dimensional effects, and quality control during manufacturing, all play a role in establishing characteristics of the FRP system. The material characteristics described in this chapter are generic and do not apply to all commercially available products. Test methods according to ASTM standards should be used for material characterization; for test methods not included in ASTM standards, reference should be made to ACI 440.3R. Preconstruction quality assurance testing of the FRP strengthening system, hereby called FRP system or FRP reinforcement, is recommended.

3.1—Constituent materials

Constituent materials used in most commercially available FRP systems, including resins, primers, putties, saturants, adhesives, and fibers, have been developed for strengthening concrete structures. They are proven effective for strengthening masonry structures.

3.1.1 *Resins*—A wide range of polymeric resins, including primers, putty, fillers, saturants, and adhesives, are used in the manufacturing and installation of FRP systems. Common resins such as epoxies, vinyl esters, and polyesters have been formulated for use in a wide range of environmental conditions. Fiber-reinforced polymer system manufacturers use resins that have the following characteristics:

- Compatibility with and adhesion to the reinforcing fiber;
- Compatibility with and adhesion to the masonry;
- Compatibility with and adhesion to the FRP system—substrate interface;
- Resistance to environmental effects, including but not limited to moisture, salt water, temperature extremes, and chemicals normally associated with exposed masonry;
- Workability during installation;
- Resin pot life consistent with time needed for the application; and
- Development of appropriate mechanical properties for the FRP system.

3.1.1.1 *Primer*—Primer is used to penetrate the masonry surface, providing an improved adhesive bond for the saturating resin or adhesive.

3.1.1.2 *Putty fillers*—Putty is used to fill small surface voids in the substrate, such as bug holes, and to provide a smooth surface to which the FRP system can bond. Filled surface voids also prevent bubbles from forming during curing of the saturating resin.

3.1.1.3 *Saturating resin*—Saturating resin is used to impregnate the reinforcing fibers, fix them in place, and provide a shear load path to effectively transfer load between fibers. The saturating resin serves as the adhesive for wet layup systems, providing a shear load path between the previously primed substrate and the FRP system.

3.1.1.4 *Adhesives*—Adhesives are used to bond precured FRP laminate and near-surface-mounted (NSM) systems to the masonry. The adhesive provides a load path between the masonry and the FRP system. Adhesives are also used to bond together multiple layers of precured FRP laminates.

3.1.2 *Fibers*—Continuous glass, aramid, and carbon fibers are common reinforcements used in FRP systems. The fibers give the FRP system its strength and stiffness. A more detailed description of fibers is given in ACI 440R.

3.1.3 *Protective coatings*—Protective coating is used to protect the bonded FRP system from potentially damaging environmental and mechanical effects. Coatings are typically applied to the exterior surface of the FRP system after the adhesive or saturating resin has cured. Protection systems are available in a variety of forms that include:

- Polymer coatings, generally epoxies or polyurethanes;
- Acrylic coatings, either straight acrylic or acrylic cement-based; acrylic systems also come in different textures;
- Cementitious systems that may require roughening of the FRP surface by broadcasting sand into wet resin, and can be installed similarly as they would on a concrete surface; and
- Intumescent coatings that are polymer-based used, in some cases, to control flame spread and smoke generation.

Reasons for using protection systems with FRP systems installed on masonry surfaces include:

- *Ultraviolet light protection*—The resin used in the FRP system can be affected over time by exposure to ultraviolet light. To prevent degradation, there are a number of available coatings used to protect the FRP system such as acrylic coatings, cementitious surfacing, and aliphatic polyurethane coatings;
- *Vandalism*—Protective systems to resist vandalism should be hard and durable. Polyurethane coatings can offer protection against cutting and scraping. Cementitious overlays can provide more protection;

- *Impact, abrasion, and wear*—These protection systems are similar to those used for vandalism protection. Abrasion and wear, however, are different in that they result from continuous exposure rather than a one-time event and their performance is chosen for their hardness and durability;
- *Aesthetics*—Protective topcoats may be used to conceal the FRP system. These systems may be acrylic latex coatings that match the color of masonry or other colors and textures to match the existing structure;
- *Chemical resistance*—Exposure to harsh chemicals such as strong acids may damage the FRP system. In such environments, coatings with better chemical resistance such as urethanes and novolac epoxies may be used; and
- *Fire protection*—Coatings capable of reducing the spread of flames and smoke production should be employed when fire is a concern or as required by local building codes. Fiber-reinforced polymer materials are sensitive to high temperatures during a fire. When the temperature of the FRP system approaches the glass transition temperature of the resin (T_g), the strength, stiffness, and bond properties of the installed FRP system are reduced. In the case of FRP systems applied as external reinforcement to concrete or masonry members, exposure to high temperature produces a fast degradation of the bond between the FRP system and substrate. As a result, degradation of the strengthening effectiveness and debonding of FRP system may take place.

3.2—Physical properties

3.2.1 *Density*—Fiber-reinforced polymer materials have densities ranging from 75 to 130 lb/ft^3 (1.2 to 2.1 g/cm^3), which is four to six times less than that of steel, as indicated in Table 3.1 (ACI 440.2R).

3.2.2 *Coefficient of thermal expansion*—The coefficients of thermal expansion of unidirectional FRP materials differ in longitudinal and transverse directions, depending on the fiber type, resin type, and fiber content. Table 3.2 (ACI 440.2R) lists the longitudinal and transverse coefficients of thermal expansion for typical unidirectional FRP materials. Note that a negative coefficient of thermal expansion indicates the material contracts with increased temperature and expands with decreased temperature. For reference, concrete has a coefficient of thermal expansion that varies from 4×10^{-6} to 6×10^{-6}/°F (7×10^{-6} to 11×10^{-6}/°C) and is usually assumed to be isotropic (Mindess and Young 1981). Steel has an isotropic coefficient thermal expansion of 6.5×10^{-6}/°F (11.7×10^{-6}/°C).

3.2.3 *Effect of high temperature*—At temperatures near and above the T_g, the elastic modulus, creep resistance, and strength of a polymer decrease due to changes in its molecular structure. The T_g depends on a range of factors including the type, degree of cure, and moisture content of the polymer. In FRP composite materials used in masonry repair, commonly used fibers such as glass, carbon, and aramid maintain their mechanical integrity at temperatures in excess of those tolerated by polymer resins. Due to a reduction in force transfer between fibers and across adhesively bonded interfaces, however, the effectiveness of the composite strengthening system is diminished at temperatures near and above the T_g of the resin(s) used in the system.

Table 3.1—Typical densities of materials, lb/ft^3 (g/cm^3)

Steel	GFRP	CFRP	AFRP
490 (7.9)	75 to 130 (1.2 to 2.1)	90 to 100 (1.5 to 1.6)	75 to 90 (1.2 to 1.5)

Table 3.2—Typical coefficient of thermal expansion for FRP materials*

Direction	Coefficient of thermal expansion, × 10^{-6}/°F (× 10^{-6}/°C)		
	GFRP	CFRP	AFRP
Longitudinal, α_L	3.3 to 5.6 (6 to 10)	–0.6 to 0 (–1 to 0)	–3.3 to –1.1 (–6 to –2)
Transverse, α_T	10.4 to 12.6 (19 to 23)	12 to 27 (22 to 50)	33 to 44 (60 to 80)

*Typical values for fiber-volume fractions ranging from 0.5 to 0.7.

3.3—Mechanical properties

3.3.1 *Tensile behavior*—When loaded in direct tension, FRP materials do not exhibit any plastic behavior, or yielding, before rupture. The tensile behavior of FRP materials consisting of one type of fiber material is characterized by a linear elastic stress-strain relationship until failure, which is sudden and without warning.

The tensile strength and stiffness of a FRP material depends on several factors. Because the fibers in a FRP material are the main load-carrying constituent, fiber type, fiber orientation, fiber quantity, and method and conditions that the composite is manufactured affect the tensile properties of the FRP material. Due to the primary role of the fibers as tensile reinforcement and methods of application, the properties of a FRP system are sometimes reported based on the net-fiber area (Method 2 of 440.3R). In other examples, such as in pre-cured laminates, the reported properties are based on the gross-laminate area (Method 1 of 440.3R).

The gross-laminate area of a FRP system is calculated using the total cross-sectional area of the cured FRP system, including all fibers and resin. The gross-laminate area is typically used for reporting precured laminate properties where the cured thickness is constant and the relative proportion of fiber and resin is controlled.

The net-fiber area of a FRP system is calculated using the known area of fiber, neglecting the total width and thickness of the cured system; thus, resin is excluded. For example, the net-fiber area is typically used for reporting properties of wet layup FRP systems that use fiber sheets or fabrics and field-installed resins. The wet layup installation process results in controlled fiber content and a variable resin content.

Fiber-reinforced polymer system properties reported using the gross-laminate area have higher relative thickness dimensions and lower relative strength and modulus values, whereas system properties reported using the net-fiber area have lower relative thickness dimensions and higher relative strength and modulus values. Regardless of the basis for the reported values, the tensile strength ($f_{fu}A_f$) and axial stiffness (A_fE_f) remain constant. Properties reported based on the net-fiber area are not the properties of the bare fibers. The properties of a FRP system should be characterized as a composite, recognizing not only the material properties of the individual fibers, but also the efficiency of the fiber-resin system, the fabric architecture, and the method used to create the composite. The mechanical properties of all FRP systems, regardless of form, should be based on the testing of laminate samples with known fiber content.

The tensile properties of a particular FRP system, however, should be obtained from the FRP system manufacturer. Manufacturers should report an ultimate tensile strength, which is defined as the mean tensile strength of a sample of test specimens minus three times the standard deviation ($f_{fu}^* = \overline{f}_{fu} - 3\sigma$) and, similarly, report a guaranteed ultimate rupture strain ($\varepsilon_{fu}^* = \overline{\varepsilon}_{fu} - 3\sigma$). This approach provides a 99.87% probability that the ultimate tensile properties will exceed these statistically-based design values (Mutsuyoshi et al. 1990). Young's modulus should be calculated as the chord modulus between 0.003 and 0.006 strains, in accordance with ASTM D3039/D3039M. A minimum number of 20 replicate test specimens should be used to determine the mean ultimate tensile properties. The manufacturer should provide a description of the method used to obtain the reported tensile properties, including the number of tests, mean values, and standard deviations.

3.3.2 *Compressive behavior*—Externally bonded FRP systems should not be used as compression reinforcement because of insufficient test results validating its use in this type of application. While reliance on externally bonded FRP systems to resist compressive stresses is not recommended, the following section is presented to fully characterize the behavior of FRP materials.

Coupon tests on FRP laminates used for repair on concrete have shown that the compressive strength of the FRP system is lower than the tensile strength (Wu 1990). The mode of failure for FRP laminates subjected to longitudinal compression can include transverse tensile failure, fiber microbuckling, or shear failure. The mode of failure depends on the type of fiber, the fiber-volume fraction, and the type of resin. Compressive strengths of 55, 78, and 20% of the tensile strength have been reported for GFRP, carbon fiber-reinforced polymer (CFRP), and aramid fiber-reinforced polymer (AFRP), respectively (Wu 1990). In general, compressive strengths are higher for materials with higher tensile strengths, except in the case of AFRP, where the fibers exhibit nonlinear behavior in compression at a relatively low level of stress.

The compressive modulus of elasticity is usually smaller than the tensile modulus of elasticity of FRP materials. Test reports on samples containing a 55 to 60% volume fraction of continuous E-glass fibers in a matrix of vinyl ester or isophthalic polyester resin have indicated a compressive modulus of elasticity of 5000 to 7000 ksi (34,000 to 48,000 MPa) (Wu 1990). The compressive modulus of elasticity is approximately 80% for GFRP, 85% for CFRP, and 100% for AFRP of the tensile modulus of elasticity for the same product (Ehsani 1993).

3.4—Time-dependent behavior

3.4.1 *Creep rupture*—Fiber-reinforced polymer materials subjected to a constant tensile load over time can suddenly fail after a time period called the endurance time. As the ratio of the sustained tensile stress to the short-term strength of the FRP laminate increases, endurance time decreases. The creep rupture time may also decrease under adverse environmental conditions, such as high temperature, ultraviolet-radiation exposure, high alkalinity, wetting-and-drying cycles, or freezing-and-thawing cycles.

In general, carbon fibers are the least susceptible to creep rupture, aramid fibers are moderately susceptible, and glass fibers are most susceptible. Creep rupture tests have been conducted on 0.25 in. (6 mm) diameter FRP bars reinforced with glass, aramid, and carbon fibers. The FRP bars were tested at different load levels at room temperature. Results indicated that a linear relationship exists between creep-rupture strength and the logarithm of time for all load levels. The ratios of stress level at creep-rupture after 500,000 hours, which is about 50 years, to the initial ultimate strength of the GFRP, AFRP, and CFRP bars were extrapolated to be 0.3, 0.47, and 0.91, respectively (Yamaguchi et al. 1997). Similar values have been determined by Malvar (1998).

Recommendations on sustained stress limits imposed to avoid creep-rupture are given in Section 9.6 of this guide. As long as the sustained stress in the FRP is below the creep rupture stress limits, the strength of the FRP is available for nonsustained loads.

3.4.2 *Fatigue*—Fatigue of the FRP system is not an issue in URM structures that are typically strengthened with FRP systems because the systems are intended to resist loads with low cycle counts, such as earthquake, hurricane, and blast loads. Guidance on the fatigue performance for FRP composites can be found in ACI 440.2R.

3.5—Durability

Many FRP systems exhibit reduced mechanical properties after exposure to certain environmental factors, including high temperature, humidity, and chemical exposure (Karbhari 2007). The type of FRP strengthening technique may also affect the long-term durability performance of the strengthened system. The exposure environment, duration of the exposure, resin type and formulation, fiber type and volume fraction, and resin-curing method are some of the factors that influence the extent of the reduction in mechanical properties. Unless otherwise indicated, the tensile properties (ASTM D7565/D7565M) reported by the manufacturer are generally based on testing conducted in a laboratory environment and do not reflect the effects of environmental exposure.

3.6—Fiber-reinforced polymer system qualification

Fiber-reinforced polymer systems should be qualified for use on a project based on independent laboratory test data of the FRP constituent materials and the laminates made with them, structural test data for the type of application being considered, and durability data representative of the anticipated environment. Test data provided by the FRP system manufacturer demonstrating that the proposed FRP system meets all mechanical and physical design requirements including tensile strength, durability, resistance to creep, bond to substrate, and T_g, should be considered, but not used as the sole basis for qualification.

Untested FRP systems should not be considered for use. Mechanical properties of FRP systems should be determined from tests on laminates manufactured in a process representative of their field installation according to the procedures in ASTM D7565/D7565M.

The specified material-qualification programs should require laboratory testing to measure the repeatability and reliability of critical properties. Testing five batches of FRP materials is recommended. Independent structural testing can be used to evaluate a system's performance for the specific application.

CHAPTER 4—SHIPPING, STORAGE, AND HANDLING

4.1—Shipping

The user of FRP constituent materials is advised to observe federal and state packaging and shipping regulations. Packaging, labeling, and shipping for thermosetting resin materials are controlled by the Code of Federal Regulations (CFR 49). Many materials are classified as corrosive, flammable, or poisonous in Subchapter C (CFR 49) under Hazardous Materials Regulations.

4.2—Storage

4.2.1 *Storage conditions*—To preserve the properties and maintain safety in the storage of FRP system constituent materials, materials should be stored in accordance with the manufacturer's recommendations. The storage of FRP system constituent materials should consider exposure to ultraviolet light (UV), extreme temperatures, and seawater spray or other environmental conditions that can be deleterious to the FRP. Certain constituent materials, such as reactive curing agents, hardeners, initiators, catalysts, and cleaning solvents, have safety-related requirements and should be stored as recommended by the manufacturer and Occupational Safety and Health Administration (OSHA). Catalysts and initiators, which are usually peroxides, should be stored separately.

4.2.2 *Shelf life*—The properties of uncured resin components can change with time, temperature, or humidity. Such conditions can affect the reactivity of the mixed system and the uncured and cured properties. The manufacturer sets a recommended shelf life within which the properties of the resin-based materials should continue to meet or exceed stated performance criteria. Any component material that has exceeded its shelf life, deteriorated, or been contaminated should not be used. Fiber-reinforced polymer materials deemed unusable should be disposed of as specified by the manufacturer and acceptable to state and federal environmental control regulations.

4.3—Handling

4.3.1 *Material safety data sheet*—Material safety data sheets (MSDS) for all FRP constituent materials and components should be obtained from the manufacturers and accessible at the job site.

4.3.2 *Information sources*—Detailed information on the handling and potential hazards of FRP-constituent materials can be found in information sources, such as ACI and International Concrete Repair Institute (ICRI) reports, company literature and guides, OSHA guidelines, and other government documents. ACI 503R is specifically noted as a general guideline for the safe handling of epoxy and other resin adhesive compounds.

4.3.3 *General handling hazard*—Thermosetting resins describe a generic family of products that includes unsaturated polyesters, vinyl esters, epoxy, and polyurethane resins. Materials used with them are generally described as hardeners, curing agents, peroxide initiators, fillers, and flexibilizers. There are precautions to observe when handling thermosetting resins and their component materials. General hazards that may be encountered when handling thermosetting resin materials are:

- Skin irritation, such as burns, rashes, and itching;
- Skin sensitization, which is an allergic reaction similar to that caused by poison ivy, building insulation, or other allergens;
- Breathing organic vapors from cleaning solvents, monomers, and diluents;
- With a sufficient concentration in air, explosion or fire of flammable materials when exposed to heat, flames, pilot lights, sparks, static electricity, cigarettes, or other sources of ignition;
- Exothermic reactions of mixtures of materials causing fires or personal injury; and
- Nuisance dust caused by grinding or handling of the cured FRP materials (consult manufacturer's literature for specific hazards).

The complexity of thermosetting resins and associated materials makes it essential that labels and MSDS are read and understood by those working with these products. Document CFR 16, Part 1500, regulates labeling of hazardous substances and includes thermosetting-resin materials. Document ANSI Z-129.1 provides further guidance regarding classification and precautions.

4.3.4 *Personnel safe handling and clothing*—Disposable suits and gloves are suitable for handling fiber and resin materials. Disposable rubber or plastic gloves are recommended and should be discarded after each use. Gloves should be resistant to resins and solvents. Safety glasses or goggles should be used when handling resin components and solvents. Respiratory protection, such as dust masks or respirators, should be used when fiber fly, dust, or organic vapors are present, or during mixing and placing of resins if required by the FRP system manufacturer.

4.3.5 *Workplace safe handling*—The workplace should be well ventilated. Surfaces should be covered as needed to protect against contamination and resin spills. Each FRP system constituent material has different handling and storage requirements to prevent damage. Consult with the material system manufacturer for guidance. Some resin systems are potentially dangerous during the mixing of the components. Consult the system manufacturer's literature for proper mixing procedures and MSDSs for specific handling hazards. Ambient cure resin formulations produce heat when curing, which in turn accelerates the reaction. Uncontrolled reactions, including fuming, fire, or violent boiling, may occur in containers holding a mixed mass of resin; therefore, containers should be monitored.

4.3.6 *Clean-up and disposal*—Clean-up often involves the use of flammable solvents. When using flammable solvents, appropriate precautions should be observed. There are also clean-up solvents available that do not present flammability concerns. All waste materials should be contained and disposed of as prescribed by the prevailing environmental authority.

CHAPTER 5—INSTALLATION

Procedures for installing FRP systems have been developed by the system manufacturers and often differ between systems. In addition, installation procedures can vary within a system, depending on the type and condition of the structure. This chapter presents general guidelines for installation of FRP systems. Contractors trained in accordance with the installation procedures developed by the system manufacturer should install FRP systems. Deviations from the procedures developed by the FRP system manufacturer should not be allowed without getting the manufacturer's approval.

5.1—Contractor competency

The FRP system installation contractor should demonstrate competency for surface preparation and application of the FRP system to be installed. Contractor competency can be demonstrated by providing evidence of training and documentation of related work previously completed by the contractor, surface preparation and installation of the FRP system on portions or mock-ups of the structure. The FRP system manufacturer or its authorized agent should train the contractor's application personnel in the installation procedures of its system. Near-surface-mounted FRP systems require less training for surface preparation and installation than surface-mounted FRP systems.

5.2—Temperature, humidity, and moisture considerations

Temperature, relative humidity, and surface moisture at the time of installation can affect performance of the FRP system. Studies on these effects have been undertaken on concrete systems (Myers and Ekenel 2005). Conditions observed before and during installation include surface temperature of the masonry, air temperature, relative humidity, and corresponding dew point.

Primers, saturating resins, and adhesives generally should not be applied to cold or frozen surfaces. When the surface temperature of the masonry surface falls below a minimum level as specified by the FRP system or adhesive manufacturer, improper saturation of the fibers and improper curing of the resin constituent materials can occur, compromising the integrity of the FRP system. An auxiliary heat source can be used to raise the ambient and surface temperature during installation. The heat source should be clean and not contaminate the surface or the uncured FRP system.

Resins and adhesives generally should not be applied to damp or wet surfaces unless they are suitable for such applications. Fiber-reinforced polymer systems should not be applied to masonry surfaces that are subject to moisture vapor transmission. The transmission of moisture vapor from a masonry surface through the uncured resin materials typically appears as surface bubbles and can compromise the bond between the FRP system and the substrate.

5.3—Equipment

As different FRP systems may be used for strengthening URM, special equipment requirements may be specific to the selected system. In general, the advantage of FRP strengthening is associated with light weight, ease, and speed of application; therefore, special equipment requirements are limited. Equipment may include a resin impregnator, resin sprayer, and grinding and grooving tools. All equipment should be maintained, clean, and in good operating condition.

The contractor should have personnel trained in the operation of all equipment. Personal protective gear, such as gloves, masks, eye guards, and coveralls, should be worn as required by manufacturer's specifications.

Equipment and material supplies in sufficient quantities should be available to allow continuity in installation and quality control tasks. Safe and convenient access to surfaces to be strengthened to ensure proper FRP application and close inspection should be provided.

5.4—Substrate repair and surface preparation

The behavior of masonry members strengthened or retrofitted with FRP systems is highly dependent on masonry substrate and proper preparation and profiling of the masonry surface. An improperly prepared surface can result in debonding or delamination of the FRP system before achieving the design load transfer. The general guidelines presented in this chapter should be applicable to all externally bonded FRP systems. Specific guidelines for a particular FRP system should be obtained from the FRP system manufacturer.

5.4.1 *Substrate repair*—All problems associated with the condition of the original masonry and the masonry substrate that can compromise the integrity of the FRP system should be addressed before surface preparation begins. The FRP system manufacturer should be consulted to verify the compatibility of materials used for repairing the substrate with the FRP system.

5.4.2 *Surface preparation*—Surface preparation requirements depend on the FRP system used to strengthen a masonry element. Specific guidelines regarding procedures for surface preparation for each FRP system should be obtained from the system manufacturer.

5.4.2.1 *Surface preparation for externally bonded FRP laminates*—Surface preparation can involve sandblasting to roughen the surface, grinding of excess mortar in the joints, and application of epoxy primers and putty fillers. Particular care should be taken to ensure that the surface is dry and clean from dust and laitance, and to avoid unintentional damage to the substrate by using excessive force. Epoxy primers formulated for concrete substrates may not be appropriate for use on concrete masonry units (CMUs) due to their low viscosity and the porous nature of CMU. Before using epoxy primers on CMU, consult with the FRP system manufacturer regarding the specific primer that is recommended. Generally, the epoxy primers used for concrete substrates are suitable for clay brick masonry units. Masonry built with extruded brick units, having less porosity, may require the use of an epoxy primer. Putty filler applied to the entire surface has experimentally been shown to notably improve bonding between the FRP laminates and masonry (Tumialan et al. 2003a). Mortar joints should be filled with adequate putty to ensure smoothness of the surface with no unevenness between mortar joints and masonry units.

5.4.2.2 *Surface preparation for NSM FRP bars*—Near-surface-mounted systems are typically installed in grooves cut into the masonry surface. The soundness of the masonry surface should be checked before installing the bar (both circular and rectangular bars can be used). The inside faces of the groove should be cleaned to ensure adequate bond with masonry. The resulting groove should be free of laitance or other compounds that may interfere with the bond. More information on groove dimensions is reported in Section 11.2.2. The moisture content of the parent masonry should be evaluated and determine to be suitable for bonding properties of the adhesive. Care should also be taken to ensure that the grooves are completely filled with the adhesive. The adhesive should be compatible with masonry and NSM FRP bars.

5.5—Resin mixing

Mixing of resins should be done in accordance with the FRP system manufacturer's recommended procedure. All resin components should be at a proper temperature and mixed in the correct ratio until there is a uniform and complete mixing of the components. Resin components are often contrasting colors, so full mixing is achieved when color streaks are eliminated. Resins should be mixed for the prescribed mixing time and visually inspected for uniformity of color. The material manufacturer should supply recommended batch sizes, mixture ratios, mixing methods, and mixing times.

Mixing equipment can include small electrically powered mixing blades or specialty units, or resins can be mixed by hand stirring, if allowed by the manufacturer. Resin mixing should be in sufficiently small quantities to ensure that all mixed resin can be used within the resin's pot life. Mixed resin that exceeds its pot life should not be used because the viscosity will continue to increase and will adversely affect the resin's ability to penetrate the surface or saturate the fiber sheet.

5.6—Application of constituent materials

Fumes can accompany the application of some resins. Fiber-reinforced polymer system should be selected with consideration for their impact on the installation environment, including emission of volatile organic compounds and toxicology.

5.6.1 *Primer and putty*—If required, primer should be applied to all areas on the masonry surface where the FRP system is to be placed. Primer should be placed uniformly on the prepared surface at the manufacturer's specified rate of coverage. The applied primer should be protected from dust, moisture, and other contaminants before applying the FRP system.

Putty should be used in an appropriate thickness and sequence with the primer as recommended by the FRP system manufacturer. The system-compatible putty, which is typically a thickened epoxy paste, should be used only to fill voids and smooth surface discontinuities before the application of other materials. Rough edges or trowel lines of cured putty should be ground smooth before continuing the installation.

Before applying the saturating resin or adhesive, the primer and putty should be allowed to cure as per the FRP system manufacturer recommendations. If the putty and primer are fully cured, additional surface preparation may be required before application of the saturating resin or adhesive according to the procedures specified by the FRP system manufacturer.

5.6.2 *Wet layup systems*—Wet layup FRP systems are typically installed by hand using dry fiber sheets and fabrics with a saturating resin and installed as per the manufacturer's recommendations. The procedure consists of applying dry fiber sheets and fabrics and resin directly to the member being strengthened. The saturating resin is applied uniformly to all prepared surfaces where the system is placed. Alternatively, the fibers can be impregnated in a separate operation using a resin-impregnating machine before placement of the FRP system on the masonry surface.

The reinforcing fibers are gently pressed into the uncured saturating resin in a manner recommended by the FRP system manufacturer. Entrapped air between layers is released by using rib rollers and pressing the fabric against the substrate. Rib rollers are specialized hand tools that use metal or plastic rollers with combs or ribs to facilitate saturation of the fibers with resin. Sufficient saturating resin is needed to achieve full saturation of the fibers.

Successive layers of saturating resin and fiber materials are placed before the complete cure of the previous layer of resin. If previous layers are cured, interlayer surface preparation, such as light sanding or solvent application recommended by the system manufacturer, may be required.

5.6.3 *Precured systems*—Precured systems include strips, plates, bars, and open grid forms. They are typically installed with an adhesive. Adhesives are applied uniformly to the prepared surfaces where precured systems are to be placed, except in certain instances of confinement where adhesion of the FRP system to the substrate may not be required.

Precured laminate surfaces to be bonded should be clean and prepared in accordance with the system manufacturer's recommendation. The precured plates are placed on or into the wet adhesive in a manner recommended by the system manufacturer.

5.6.4 *Near-surface-mounted systems*—Near-surface-mounted systems are precured systems that consist of installing rectangular or circular FRP bars into grooves cut onto the masonry surface and bonding in place using an adhesive. Grooves should be dimensioned to ensure adequate adhesive around the bars (Chapter 11). There are many application methods and types of adhesive that have been successfully used in the field for NSM systems. Adhesive type and installation method should be selected as recommended by the system manufacturer.

5.6.5 *Protective coatings*—Coatings should be compatible with the FRP system and applied in accordance with the manufacturer's recommendations. Typically, the use of solvents to clean the FRP surface before installing coatings is not recommended due to the deleterious effects that solvents can have on the polymer resins. The FRP system manufacturer should approve any use of solvent-wipe preparation of FRP surfaces before applying the protective coatings.

Coatings should be periodically inspected and maintenance provided to ensure their effectiveness. Inspections should be performed periodically in conjunction with other regular inspections of the structure or at a frequency that is based on the exposure conditions and facility use.

5.7—Alignment of FRP materials

The FRP fiber reinforcement ply orientation and ply stacking sequence should be specified. Small variations in angle, as little as 5 degrees, from the intended direction of fiber alignment can cause a substantial reduction in strengthening performance (Yang et al. 2002). Deviations in ply orientation should only be made if approved by the licensed design professional.

Sheet and fabric materials should be handled in a manner to maintain the fiber straightness and orientation. Kinks, folds, or other forms of severe waviness in the fiber reinforcement layer should be reported to the licensed design professional; refer to Section 6.2.2 for allowable fiber misalignment.

5.8—Multiple plies and lap splices

Multiple plies can be used, provided all plies are fully impregnated with the resin system, the resin shear strength is sufficient to transfer the shearing load between plies, and the bond strength between masonry and FRP system is sufficient to transfer design forces. Lap splices should be staggered. Lap splice details, including lap length, should be based on the results of tests performed in accordance with ACI 440.3R (L.3), until ASTM tests become available. Due to the unique characteristics of some FRP systems, multiple plies and lap splices are not always possible.

5.9—Resins curing

Curing of resins is a time-temperature-dependent phenomenon. Ambient-cure resins can take several days to reach full cure. Temperature extremes or fluctuations can retard or accelerate the resin curing time. The FRP system manufacturer may offer several prequalified grades of resin to accommodate various ambient conditions.

Elevated cure systems require the resin to be heated to a specific temperature for a specified period of time. Various combinations of time and temperature within a defined envelope provide full cure of the system.

Resins should be cured according to the system manufacturer's recommendation. Field modification of resin chemistry should not be permitted without consulting the system manufacturer.

5.10—Temporary protection

Adverse temperatures, direct contact by rain, dust, dirt, excessive sunlight, high humidity, or vandalism can damage a FRP system during installation and cause improper cure of the resins. Temporary protection, such as tents and plastic screens, may be required during installation and until the resins have cured. If temporary shoring is required, the FRP system should be fully cured before removing the shoring and allowing the structural member to carry the design loads. In the event of suspected damage to the FRP system during installation, the licensed design professional should be notified and the FRP system manufacturer consulted.

CHAPTER 6—INSPECTION, EVALUATION, AND ACCEPTANCE

Quality assurance and quality control (QA/QC) programs and criteria are to be developed and maintained by the FRP system manufacturers, the installation contractors, and others associated with the project. The quality-control program should be comprehensive and cover all aspects of strengthening procedures. The degree of quality control and the scope of testing, inspection, and record-keeping depend on the size and complexity of the project.

Quality assurance is achieved through a set of inspections and applicable tests to document the acceptability of the installation. Project specifications should include a requirement to provide a quality-assurance plan for the installation and curing of all FRP materials. The plan should include:

- Personnel safety issues;
- Application and inspection of the FRP system;
- Location and placement of splices;
- Curing provisions;
- Means to ensure dry surfaces;
- Quality-assurance samples;
- Cleanup; and
- Required submittals listed in Section 12.3.

6.1—Inspection

Fiber-reinforced polymer systems and all associated work should be inspected as required by the applicable local codes. In the absence of such requirements, inspection should be conducted by or under the supervision of a licensed design professional or a qualified inspector. Inspectors should be knowledgeable of and trained in the installation of FRP systems. The qualified inspector should require compliance with design drawings and project specifications. During installation of the FRP system, the scope of the inspection should include:

- Date and time of installation;
- Ambient temperature, relative humidity, and general weather observations;
- Surface temperature of masonry;
- Surface preparation methods and resulting profile;

- Qualitative description of surface cleanliness;
- Type of auxiliary heat source, if applicable;
- Fiber or precured laminate batch number(s) and approximate location in structure;
- Batch numbers, mixture ratios, mixing times, and qualitative descriptions of the appearance of all mixed resins, including primers, putties, saturants, adhesives, and coatings mixed for the day;
- Observations of progress of cure of resins;
- Conformance with installation procedures;
- Pulloff test results according to ASTM D4541/D4541M done or supervised by a licensed design professional or owner's independent testing agency;
- Fiber-reinforced polymer properties from tests of field sample panels or witness panels, if required;
- Location and size of any delaminations or air voids; and
- General progress of work.

The inspector should provide the licensed design professional or owner with the inspection records and witness panels. The installation contractor should retain sample cups of mixed resin and maintain a record of the placement of each batch.

6.2—Evaluation and acceptance

Fiber-reinforced polymer systems should be evaluated based on conformance with the design drawings and specifications, and the manufacturer's installation recommendations. Nonconformance of the FRP system should be reported to the licensed design professional for further evaluation. The FRP system material properties, installation within specified placement tolerances, presence of delaminations, cure of resins, and adhesion to substrate should be evaluated. The evaluation should also consider fiber orientation, ply orientation, width and spacing, corner radii, and lap splice lengths of the installed FRP system.

Witness panel and pulloff tests are used to evaluate the installed FRP system. In-place load testing can also be used where applicable to confirm the installed behavior of the FRP-strengthened member (Nanni and Gold 1998).

6.2.1 *Materials*—Before starting the project, the FRP system manufacturer should submit certification of specified material properties and identification of all materials to be used. Additional material testing can be conducted if deemed necessary based on the complexity and intricacy of the project. Evaluation of delivered FRP materials can include tests for tensile strength, gel time, pot life, and adhesive strength. These tests are usually performed on material samples sent to a laboratory, according to the quality-control test plan. Tests for pot life of resins and curing hardness are usually conducted on-site. Materials that do not meet minimum requirements as specified by the licensed design professional should be rejected.

Witness panels can be used to evaluate the tensile strength and modulus, lap splice strength, and T_g of the FRP system installed and cured on-site using installation procedures similar to those used to install and cure the FRP system. During installation, flat panels of the specified dimensions and thickness can be fabricated on-site according to a predetermined sampling plan. After curing on-site, the panels can then be sent to a laboratory for testing. Witness panels can be retained or submitted to an approved laboratory for testing of tensile strength (ASTM D7205/D7205M), and T_g (ASTM D4065). Strength and elastic modulus of the FRP system is determined in accordance with the requirements of Section 3.3.1 of this guide and ACI 440.3R (L.2) or CSA-S806-02. ASTM standard tests should be used as they become available. The properties to be evaluated by testing should be specified by the licensed design professional. The licensed design professional may waive or alter the frequency of testing.

During installation, sample cups of mixed resin should be prepared according to the specified sampling plan and retained for testing to determine the level of cure (Section 6.2.4).

6.2.2 *Fiber orientations*—Fiber or precured-laminate orientation should be evaluated by inspection using a level or a straightedge. Fiber waviness—a localized appearance of fibers that deviate from the general straight-fiber line in the form of kinks or waves—should be evaluated for wet layup systems.

Fiber or precured laminate misalignment of more than 5 degrees from that specified on the design drawings (approximately 1 in./ft [80 mm/m]) should be reported to the licensed design professional for evaluation. The licensed design professional should calculate the capacity of the system considering this misalignment to determine if the design criteria are still satisfied. If the design criteria cannot be satisfied, remedial action may be warranted.

6.2.3 *Delaminations*—The cured FRP system should be evaluated for delaminations or air voids between multiple plies or between the FRP system and the masonry surface. Inspection methods should be capable of detecting delaminations of 2 in.^{2} (1300 mm^2) or greater. Methods such as acoustic sounding, for example, hammer sounding, ultrasonics, and infrared thermography can be used to detect delaminations.

The effect of delaminations or other anomalies on the structural integrity and durability of the FRP system should be evaluated by the licensed design professional. Delamination size, location, and quantity relative to the overall application area should be considered in the evaluation.

General acceptance guidelines for wet layup systems are:

- Small delaminations less than 2 in.^{2} (1300 mm^2) each are permissible as long as the delaminated area is less than 5% of the total laminate area and there are no more than 10 such delaminations per 10 ft^2 (1 m^2);
- Large delaminations, greater than 25 in.^{2} each (16,000 mm^2), compromise the structural performance of the installed FRP and should be repaired by selectively cutting away the affected sheet and applying an overlapping sheet patch of equivalent plies; and
- Delaminations less than 25 in.^{2} (16,000 mm^2) each may be repaired by resin injection or ply replacement, depending on the size and number of delaminations and their locations.

For precured FRP systems, each delamination should be evaluated and repaired in accordance with the licensed design professional's direction. Upon completion of repairs, the laminate should be re-inspected to verify the repair was properly installed.

6.2.4 *Cure of resins*—The relative cure of FRP systems can be evaluated by laboratory testing of witness panels or resin-cup samples using ASTM D3418. The relative cure of the resin can also be evaluated on the project site by physical observation of resin tackiness and hardness of work surfaces or hardness of retained resin samples. The FRP system manufacturer should be consulted to determine the specific resin-cure verification requirements. For precured systems, adhesive-hardness measurements should be made in accordance with the manufacturer's recommendation.

6.2.5 *Adhesion strength*—Tension adhesion testing of cored samples should be conducted using the method in ASTM D4541. Sampling frequency should be specified. Due to the low tensile strength of masonry units, the evaluation of tension adhesion tests should be based primarily on the failure mode. Tension adhesion tests should exhibit failure of the masonry substrate. Failure between the FRP system and the masonry substrate or interlaminar failure between layers of the FRP system should be reported to the engineer and evaluated for acceptance.

6.2.6 *Cured thickness*—Small core samples, typically 0.5 in. (13 mm) diameter, may be taken to visually ascertain the cured laminate thickness or number of plies at locations approved by the licensed design professional. Cored samples required for adhesion testing also can be used to ascertain the laminate thickness or number of plies. The sampling frequency should be specified by the licensed design professional. Taking samples from high-stress areas or splice areas should be avoided. For aesthetic reasons, the cored hole can be filled and smoothed with a repair mortar or the FRP system putty. If required by the licensed design professional, a 4 to 8 in. (100 to 200 mm) overlapping FRP sheet patch of equivalent plies may be applied over the filled and smoothed core hole immediately after taking the core sample. The FRP sheet patch should be installed in accordance with the manufacturer's installation procedures.

CHAPTER 7—MAINTENANCE AND REPAIR

7.1—General

As with any FRP system, the owner should periodically inspect and assess the performance of the FRP system used for strengthening of masonry members. Inspections should be performed periodically in conjunction with other regular inspections of the structure or at a frequency that is determined based on the exposure conditions and use of the facility. The causes of any damage or deficiencies detected during routine inspections should be identified and addressed before performing any repairs or maintenance.

7.2—Inspection and assessment

7.2.1 *General inspection*—A visual inspection should be performed to observe any changes in color, debonding, peeling, blistering, cracking, crazing, deflections, and other anomalies. In addition, ultrasonic, acoustic sounding, for example, hammer tap, or thermographic tests may reveal signs of progressive debonding and delamination. More information on inspection methods is available in ACI 440R.

7.2.2 *Assessment*—Test data and observations are used to assess any damage and the structural integrity of the strengthening system. The assessment should include repair recommendations and suggestions for reducing the incidence of future damage.

7.3—Repair of strengthening system

The repair method of the strengthening system depends on causes of the damage, the type of material, the form of degradation, and the level and extent of damage. Before repairing the FRP system, causes of the damage should be identified.

Minor damage should be repaired, including localized FRP laminate cracking or abrasions that affect the structural integrity of the laminate. Minor damage can be repaired by bonding FRP patches over the damaged area. The FRP patches should possess the same characteristics, such as thickness and ply orientation, as the original laminate. The FRP patches should be installed in accordance with the material manufacturer's recommendation. Minor delaminations can be repaired by epoxy-resin injection. Major damage, including peeling and debonding of large areas, may require removal of the affected area and replacement of the FRP laminate.

7.4—Repair of surface coating

In the event that the surface-protective coating should be replaced, the FRP laminate should be inspected for structural damage or deterioration. Consult the system manufacturer for repair of the surface coating.

CHAPTER 8—GENERAL DESIGN CONSIDERATIONS

8.1—Design philosophy

The design methodology presented in Chapters 9 and 10 is based on limit state design principles to provide acceptable safety levels.

To evaluate the nominal strength of a FRP-strengthened masonry member, all possible failure modes with associated strains and stresses in each material should be assessed.

Material properties such as compressive and tensile strength, elastic modulus in compression, and shear strength of existing masonry, should preferably be determined by in-place tests or laboratory tests of extracted samples. If these are unavailable, the material properties of existing masonry may be obtained from available guidelines, including ASCE 41-06, ACI 530, or historical records found in the literature. Dimensions of the masonry elements should be obtained from the existing drawings or field measurements.

If there are uncertainties regarding existing material strengths or substrate conditions, the engineer may wish to incorporate more conservative strength-reduction factors than those discussed in this guide.

This guide can be used to design a FRP strengthening system for existing individual masonry walls. The engineer designing the FRP strengthening system for individual wall(s) should evaluate the effect wall strengthening has on the overall structure. If appropriate, a global analysis of the overall structure should be performed (Moon et al. 2006; ASCE 41-06).

In strengthening of URM members, the relationship between each element and the entire structure is of particular relevance. Specifically, the in-plane rigidity of floor diaphragms and the effectiveness of the connections between the floor diaphragms and the walls as well as among intersecting walls themselves can be critical to the global behavior of the building. For example, post-earthquake observations of damage have shown out-of-plane wall failures due to the lack of in-plane rigidity of floor diaphragms and inadequate connectivity between walls and diaphragms. Conversely, in cases where the proper connection was ensured by anchors into the walls or by retrofit measures such as RC ring beams or steel ties at each floor level, out-of-plane failures were prevented and the seismic actions could mobilize in-plane strength of the walls (Magenes and Calvi 1997).

8.2—Strengthening limits

For concrete structural members, FRP strengthening limits are imposed to guard against failure of the strengthened element should the FRP reinforcement become ineffective due to fire, substrate degradation, vandalism, or impact damage (ACI 440.2R). These limitations ensure that the concrete member, typically a gravity-load-carrying element, can resist sustained loads typically expressed as dead loads and a portion of the live loads without the contribution of the FRP system.

In the case of masonry elements, FRP systems are usually installed to increase the strength of URM walls subjected to special loading events such as earthquake, wind, hurricane, and blast loads. These loads are rare and typically not sustained. The probability of simultaneous occurrence of damage to FRP, for example from vandalism or exposure to high temperatures and high short-term loads, like from earthquake or wind, is low; therefore, a strengthening limit for these applications is unnecessary.

There are two cases, however, in which the strengthening limits should be considered. These cases include masonry walls resisting out-of-plane loads due to earth pressure and walls that are part of the primary lateral load-carrying system resisting in-plane loads from wind. For walls resisting out-of-plane loads due to earth pressures, the strength of the existing wall should be sufficient to meet the limitation given in Eq. (8-1)

$$(\phi R_n)_{existing,\ out\text{-}of\text{-}plane} \geq (0.9D + 1.0H)_{new} \tag{8-1}$$

For cases of walls that are part of the primary lateral-load-carrying system, the in-plane strength of the wall should be sufficient to meet the limitation given in Eq.(8-2)

$$(\phi R_n)_{existing,\ in\text{-}plane} \geq (0.9D + 1.0W)_{new} \tag{8-2}$$

8.3—Design material properties

Unless otherwise stated, material properties reported by the system's manufacturer, such as tensile strength, typically do not consider long-term exposure to environmental conditions and, therefore, should be considered as initial properties. Because long-term exposure to various types of environments can reduce the tensile properties and creep-rupture and fatigue endurance of FRP laminates, the material properties used in design equations should be reduced based on the environmental exposure condition. Equations (8-3) through (8-5) give the tensile properties that are used in all design equations. The design tensile strength is determined using the environmental reduction factor given in Table 8.1 for the appropriate fiber type and exposure condition.

Table 8.1—Environmental reduction factors for various FRP systems and exposure conditions

Exposure conditions	Fiber type	Environmental reduction factor C_E
Interior exposure (for example, partitions)	Carbon	0.95
	Glass	0.75
	Aramid	0.85
Exterior exposure (including internal side of exterior walls)	Carbon	0.85
	Glass	0.65
	Aramid	0.75
Aggressive environment (basement walls)	Carbon	0.85
	Glass	0.50
	Aramid	0.70

$$f_{fu} = C_E f_{fu}^* \tag{8-3}$$

Similarly, the design rupture strain should also be reduced for environmental-exposure conditions.

$$\varepsilon_{fu} = C_E \varepsilon_{fu}^* \tag{8-4}$$

Fiber-reinforced polymer materials consisting of one type of fiber oriented predominantly in one direction are practically linearly elastic until failure. Their modulus of elasticity does not vary significantly with environmental exposure and loading history and can be computed according to Eq. (8-5).

$$E_f = \frac{f_{fu}}{\varepsilon_{fu}} \tag{8-5}$$

Constituent materials affect the durability and resistance to environmental exposure of a FRP system. The environmental-reduction factors given in Table 8.1 are estimates based on the relative durability of each fiber type. As more research information is developed and becomes available, these values may be refined. Durability test data for FRP systems with and without protective coatings may be obtained from the FRP manufacturer of the system under consideration. As Table 8.1 illustrates, if the FRP system is located in a relatively benign environment, such as indoors, the reduction factor is closer to unity. If the FRP system is located in an aggressive environment where prolonged exposure to high humidity, freezing-and-thawing cycles, salt water, or alkalinity is expected, a lower reduction factor should be used. The reduction factor can reflect the use of a protective coating if the coating has been shown through testing to lessen the effects of environmental exposure and the coating is maintained for the life of the FRP system; in such a case, the environmental reduction factor should never be larger than the values provided in Table 8.1 (ACI 440.2R) for the interior exposure conditions.

The relative durability of NSM systems versus surface-applied laminates may be better. Given the lack of data, however, this guide conservatively recommends use of the same C_E factors for both applications.

8.4—Effective strain and stress in the FRP reinforcement at the strength limit state

Debonding of the FRP system can occur if the force in the FRP system at the strength limit state cannot be sustained by the masonry substrate. For a typical FRP system that is linear elastic until failure, the level of strain in the FRP system will dictate the level of stress developed in the system. To prevent debonding, a limitation is placed on the strain level developed in the FRP laminate. The maximum strain and corresponding stress that FRP systems can attain before debonding from the masonry substrate are defined as effective strain ε_{fe} and effective stress f_{fe}.

8.4.1 *Effective strain for flexure-controlled failure modes*—The effective strain ε_{fe} and effective stress f_{fe} used for the design of flexural out-of-plane and in-plane FRP strengthening of masonry walls (Sections 9.4, 10.5, and 10.6) can be computed according to Eq. (8-6) and (8-7), respectively

$$\varepsilon_{fe} = \kappa_m \varepsilon_{fu}^* \le C_E \varepsilon_{fu}^* \quad (8\text{-}6)$$

$$f_{fe} = E_f \varepsilon_{fe} \quad (8\text{-}7)$$

where κ_m is a bond reduction coefficient calibrated using available experimental data (Albert et al. 2001; Hamilton and Dolan 2001; Tumialan 2001; Bajpai and Duthinh 2003; Tumialan et al. 2003a,b; Turco et al. 2003; Galati et al. 2006), defined as in Eq. (8-8).

$$\kappa_m = \begin{cases} 0.45 \text{ for surface-mounted FRP systems} \\ 0.35 \text{ for NSM FRP systems} \end{cases} \quad (8\text{-}8)$$

Based on current knowledge and experimental data, Eq. (8-8) is applicable only when the total force per unit width (per bar for NSM systems) that the FRP system transfers to the masonry substrate satisfies the limitation given in Eq. (8-9).

$$p_{fm} = \begin{cases} nt_f f_{fe} \le 1500 \text{ lb/in. for surface-mounted FRP systems} \\ A_{f,bar} f_{fe} \le 10{,}000 \text{ lb/bar for NSM FRP systems} \end{cases} \text{(in.-lb)}$$
$$p_{fm} = \begin{cases} nt_f f_{fe} \le 260 \text{ N/mm for surface-mounted FRP systems} \\ A_{f,bar} f_{fe} \le 44{,}500 \text{ N/bar for NSM FRP systems} \end{cases} \text{(SI)} \quad (8\text{-}9)$$

8.4.2 *Effective strain for shear-controlled failure modes*—The effective strain ε_{fe} and effective stress f_{fe} to be used for the design of shear in-plane FRP strengthening of masonry walls (Sections 10.4 and 10.6) can be computed according to Eq. (8-10) and (8-11), respectively

$$\varepsilon_{fe} = \kappa_v \varepsilon_{fu}^* \le C_E \varepsilon_{fu}^* \quad (8\text{-}10)$$

$$f_{fe} = E_f \varepsilon_{fe} \quad (8\text{-}11)$$

The bond reduction coefficient for shear-controlled failure modes κ_v depends on the FRP reinforcement index ω_f, defined in Eq. (8-12).

$$\omega_f = \frac{1}{1000} \frac{A_f E_f}{A_n \sqrt{f'_m}} \text{ for in.-lb units}$$
$$\omega_f = \frac{1}{85} \frac{A_f E_f}{A_n \sqrt{f'_m}} \text{ for SI units} \quad (8\text{-}12)$$

For shear-controlled failure modes, the bond reduction coefficient is again calibrated based on experimental data (Tinazzi and Nanni 2000; Tumialan et al. 2001b; Morbin and Nanni 2002; Valluzzi et al. 2002; Grando et al. 2003; Zhao et al. 2003; Santa Maria et al. 2004, 2006; Senescu and Mosalam 2004; Stratford et al. 2004). The coefficient for shear-controlled failure modes is equal for both FRP laminates and NSM FRP systems and is given in Eq. (8-13).

$$\kappa_v = \begin{cases} 0.40 \text{ for } \omega_f \le 0.20 \\ 0.64 - 1.2\omega_f \text{ for } 0.20 < \omega_f \le 0.45 \\ 0.10 \text{ for } \omega_f > 0.45 \end{cases} \quad (8\text{-}13)$$

Similar to flexure-controlled failure modes, Eq. (8-13) is applicable only when the force per unit width (per bar for NSM systems) that the FRP system transfers to the masonry substrate satisfies the limitation given in Eq. (8-14).

$$p_{fv} = \begin{cases} nt_f f_{fe} \le 1500 \text{ lb/in. for surface-mounted FRP systems} \\ A_{f,bar} f_{fe} \le 10{,}000 \text{ lb/bar for NSM FRP systems} \end{cases} \text{(in.-lb)}$$
$$p_{fv} = \begin{cases} nt_f f_{fe} \le 260 \text{ N/mm for surface-mounted FRP systems} \\ A_{f,bar} f_{fe} \le 44{,}500 \text{ N/bar for NSM FRP systems} \end{cases} \text{(SI)} \quad (8\text{-}14)$$

It is recognized that in Eq. (8-6) and (8-10), the κ values will always control over the C_E values. The κ values, however, have been set as a lower bound from experimental data. It is expected that further experimental data may result in higher κ values in the future. Limitation involving the C_E value is presented to establish the design philosophy that will remain consistent. Furthermore, if experimental data for a particular application is available, the designer may wish to incorporate different C_E or κ values. In this case, it is recommended to follow the limitations given by Eq. (8-6) and (8-10).

CHAPTER 9—WALL STRENGTHENING FOR OUT-OF-PLANE LOADS

9.1—Background information

A number of research projects have been conducted to study the use of FRP systems for flexural strengthening of masonry walls. Ehsani and Saadatmanesh (1996) investigated the flexural behavior of URM walls strengthened with GFRP sheets. The specimen dimensions were 8.5 in. wide, 4 in. high, and 57 in. long (216 mm wide, 100 mm high, and 1450 mm long). Specimens were subjected to four-point bending. Tension failure was observed when a low composite reinforcement ratio was used. When the number of plies was increased, the masonry failed in compression. It was observed that the flexural capacity was increased up to 24 times compared with the unreinforced control specimen. According to test results, effects of the mortar strength appeared to be negligible and both specimens with the higher composite reinforcement ratio failed by masonry crushing.

Velazquez-Dimas et al. (2000) reported test results of half-scale URM walls tested under out-of-plane cyclic loading. The test specimens had a width of 48 in. (1220 mm) and a height of 56 in. (1420 mm), with a slenderness ratio (span length divided by the specimen thickness) of 28. Two of the walls were strengthened on both faces with GFRP strips. Understanding that the balanced condition represents simultaneous failure of masonry and rupture of composite laminate, one wall had reinforcement equivalent to the balanced ratio, while the other wall had three times the balanced reinforcement. The specimen with balanced reinforcement showed extensive composite delamination at failure. The specimen with the higher reinforcement ratio failed due to high shear stresses along the lower brick course. Substantial increases in strength and deformation capability were achieved. The retrofitted walls resisted pressures up to 24 times the weight of the wall and deflected as much as 5% of the wall height. To avoid very stiff behavior and improve the hysteretic response, it was recommended to limit the reinforcement ratio to two times the balanced condition.

Hamilton and Dolan (2001) investigated the flexural behavior of URM walls strengthened with different composite materials. The walls were built with standard 8 in. (200 mm) concrete masonry units, with an overall dimension of 2 x 6 ft (0.61 x 1.83 m). The use of high-strength composite materials such as CFRP and AFRP, with vertical fiber orientation, led to modes of failure such as FRP delamination and shear in the masonry. To use the composite material more efficiently, two alternatives were recommended. The first alternative was to increase spacing of the material until observing the laminate rupture. The second alternative was to use less-expensive materials, such as GFRP. These more efficient alternatives resulted in four failure modes: debonding, laminate rupture, masonry shear, and face shell pullout. It was reported that debonding from the masonry substrate caused failure for most of the test specimens. Hamoush et al. (2001) investigated the influence of two different surface preparation methods, sand blasting and wire brush, on the behavior of 4 ft x 6 ft x 8 in. (1.2 m x 1.8 m x 0.2 m) specimens strengthened using GFRP strips placed along the horizontal and vertical directions subjected to a distributed load. They concluded that adding more than one layer of FRP reinforcement improved the structural performance of the wall. When only one layer of FRP reinforcement was used, its distance from the support had a minor influence on the wall behavior.

Successful use of NSM bars for improving the flexural capacity of RC members led to extending this technique to URM walls. When installed to masonry walls, NSM bars are typically placed in horizontal masonry bed joints. As an example, masonry panels of concrete masonry units were tested by Tumialan et al. (2002). One specimen was strengthened with one No. 10 (No. 3) GFRP bar, the second with two No. 10 (No. 3) GFRP bars, and the third was strengthened with an externally bonded GFRP laminate (width = 3 in. [7.6 mm]) to provide the same axial stiffness as the first specimen. The wall strengthened with one GFRP bar failed due to debonding of the epoxy paste from the masonry. Initial flexural cracks formed at the mortar bed joints perpendicular to the reinforcement, and originated secondary cracks at the epoxy paste-masonry interface causing debonding and subsequent wall failure. The wall strengthened with two bars failed due to masonry shear, while the specimen with the GFRP laminate failed due to debonding. As expected, Specimens 1 and 3 exhibited similar behavior in terms of load-carrying capacity and ultimate deflection. This experimental program was used as a validation for the strengthening of two URM concrete walls at an educational facility in Kansas City, MO, where the walls exhibited cracking in the bed joints at the midheight region. By using epoxy strengthened with short fibers, Bajpai and Duthinh (2003) were able to prevent debonding of NSM glass FRP bars and consistently rupture the bars in flexural tests of masonry walls. This method resulted in higher wall strength and more brittle behavior.

The load-carrying capacity of flexural walls strengthened with FRP laminates is a function of the axial load level (Triantafillou 1998). Moreover, FRP composites are highly effective in the case of walls that can be treated as simply supported—for example, walls having a large slenderness ratio. For a wall with low slenderness ratio built between rigid supports, when the out-of-plane deflection increases, the wall is restrained from free rotation at its extremities. This action induces an in-plane compressive force that, depending on the degree of support fixity, can increase the URM wall ultimate capacity by several times. This mechanism is known as arching. Due to arching, the increase of flexural capacity in walls strengthened with FRP laminates may be considerably less than expected. Laboratory and field experimentation have shown that the contribution of FRP to the flexural capacity of fixed-end walls is less than in the case of simply supported conditions, as crushing masonry units at the boundary regions controls ultimate behavior (Tumialan et al. 2003a; Galati 2002).

A wide experimental campaign has been conducted by Galati et al. (2004) on URM walls strengthened using FRP laminates and NSM bars. Twenty-five specimens had dimensions of 23.6 x 47 x 3.7 in. (600 x 1200 x 95 mm); 12 were built using concrete masonry units and 13 using clay bricks. Glass fiber-reinforced polymer and AFRP strips having widths ranging between 3 and 12 in. (75 and 300 mm) have been installed. In addition, 15 specimens were strengthened using GFRP and CFRP NSM bars; their width was 3.7 in. (95 mm) for clay specimens and ranged between 3.6 and 5.6 in.

(92 and 143 mm) for concrete specimens. All of the specimens had dimensions of 24 x 48 in. (610 x 1220 mm). Four-point bending tests have been conducted to assess the effectiveness of the strengthening method. The test results showed three types of failure modes: shear, debonding, and flexural. For large amounts of reinforcement, shear failure resulted to be the controlling mode; for other reinforcement ratios, either FRP rupture or debonding was observed. Based on the experimental data obtained in their study on FRP laminates, this guide recommends maximum usable strains equal to $0.45\varepsilon_{fu}$ for non-puttied surfaces, and $0.65\varepsilon_{fu}$ for puttied surfaces. They also proposed maximum usable strains depending on the cross section shape of NSM bars and type of adhesive. A similar investigation has been conducted by Tan and Patoary (2004) on clay brick plates simply supported on the four sides. Thirty plates with dimensions of 39 x 39 x 4 in. (1000 x 1000 x 110 mm) have been subjected to out-of-plane loads with different loading paths. The study has also analyzed the influence of surface preparation and the effectiveness of mechanical anchorages using steel bars or GFRP bolts. It was concluded that the combination of surface grinding and fiber bolt anchorage system resulted in the greatest increase of wall strength. Bidirectional GFRP fabrics provided larger strength increases than carbon or glass fiber sheets bonded with appropriate adhesive.

Parretti et al. (2004) validated a system consisting of GFRP grids embedded in polyurea resin for the strengthening of natural masonry walls. Both ultimate capacity and member ductility were improved with respect to the original, unstrengthened wall, even though the geometry and compressive strength of the masonry were such that GFRP rupture was never the obtained failure mode.

9.2—General considerations

9.2.1 *Assumptions*—The following assumptions are considered:

1. The strains in the FRP reinforcement and masonry are directly proportional to their distance from the neutral axis, that is, a plane section before loading remains plane after loading;

2. The maximum usable compressive strain in concrete masonry is 0.0025, while for clay and natural stone masonry, a value of 0.0035 can be assumed;

3. The tensile behavior of the FRP reinforcement is linear elastic up to failure;

4. The contributions of both masonry in tension and FRP reinforcement in compression are neglected;

5. There is no relative slip between external FRP reinforcement and masonry until debonding failure occurs; and

6. The wall behaves as a simply supported element or very nearly so, and the influence of wall arching mechanisms can be neglected. Arching mechanisms can potentially develop in a wall with a height-to-thickness ratio (h/t) less than 8 and built between stiff supports. In such a case, the FRP contribution should not be added to the out-of-plane capacity corresponding to arching mechanism. The influence of arching in the out-of-plane behavior decreases for walls with h/t larger than 14. This guide recommends that Tables 7-5 and 7-10 of ASCE 41-06, developed for seismic design, can be used to determine h/t for which URM walls do not need to be analyzed for out-of-plane loads and, therefore, do not require strengthening.

9.2.2 *Shear strength*—When externally bonded FRP systems are used to increase the out-of-plane flexural strength of a masonry wall, the wall should be checked to verify it has adequate shear strength (Section 9.3.2) to resist the design shear forces. If the shear strength of the existing masonry is inadequate, where shear controls the design, supplementary methods for shear strengthening could be implemented to increase the shear strength. These methods can involve grouting of hollow units near the supports, and lateral bracing.

9.3—Existing wall strength

To determine whether FRP strengthening is needed, the existing out-of-plane strength of the wall should be evaluated first. Unreinforced masonry walls should be analyzed for out-of-plane seismic forces and wind pressures, or both, and earth pressures as isolated elements spanning between floor levels and spanning horizontally, or both, between columns or pilasters (ACI 530; ASCE 41-06) and the applicable local building code, or both.

9.3.1 *Flexural and axial strength*—Assuming that the wall behaves as a simply supported element, the nominal out-of-plane flexural strength of the unreinforced and unstrengthened wall M_n can be determined as the bending moment calculated based on the masonry modulus of rupture or flexural tensile strength. The flexural tensile strength can be obtained by in-place testing following one of the test methods recommended by ASCE 41-06 (Chapter 7). In the absence of test results, the flexural tensile strength of an URM wall can be conservatively estimated using the lower-bound flexural tensile strength values shown in Table 7-1 of ASCE 41-06, which depend on the condition of the masonry. If the masonry is in good condition and built after 1960, the flexural tensile strength can be estimated using the modulus of rupture shown in Table 3.1.8.2.1 of ACI 530. The nominal axial strength P_n should be determined in accordance with the ACI 530 provisions using appropriate compressive strengths, determined by in-place testing, historical records for old masonry, or f'_m values shown in Table 7-1 of ASCE 41-06. The design strength should be obtained as the nominal strength multiplied by the strength reduction factor ϕ as determined from a reliability-based calibration study that accounts for uncertainties in materials, fabrications, construction details, and geometric properties. In the event that such a detailed reliability study is not available, it is recommended that a strength reduction factor ϕ of 0.6 as suggested by ACI 530 be used.

9.3.2 *Shear strength*—The shear strength can be obtained by in-place testing following one of the test methods recommended by ASCE 41-06 (Chapter 7). In the absence of test results, the shear strength of a URM wall can be conservatively estimated using the lower-bound shear strength values shown in Table 7-1 of ASCE 41-06, which depend on the condition of the masonry. Based on engineering judgment, the shear strength of masonry proved to be in good condition and built after 1960 can be estimated using the shear strength computed according to Section 3.2.4 of ACI 530. The design strength should be obtained as the nominal strength multiplied by the strength reduction factor ϕ, as determined from a statistical analysis of the as-built masonry materials. In the event that such a detailed materials investigation is not made, a strength reduction factor ϕ of 0.8 as suggested by ACI 530-08 should be used.

9.4—Nominal flexural strength of FRP-reinforced masonry walls subjected to out-of-plane loads

The strength design method requires that the flexural strength of the FRP-strengthened wall exceeds the factored moment, as indicated by Eq. (9-1).

$$\phi M_n \geq M_u \tag{9-1}$$

The Committee recommends the use of a strength reduction factor of ϕ = 0.6 as required by ACI 530 for URM walls subjected to flexural load, axial load, or a combination thereof. The use of a URM ϕ-factor is justified by the reduced ductility and potential for a brittle failure of FRP-strengthened walls.

Assuming that the factored axial load P_u acts at $t/2$ (t = thickness of wall), the nominal flexural strength M_n of the FRP-strengthened masonry wall can be determined from Eq. (9-2) using strain compatibility, internal force equilibrium, and the controlling mode of failure.

$$M_n = A_f f_{fe}\left(d_f - \frac{\beta_1 c}{2}\right) + P_u\left(\frac{t}{2} - \frac{\beta_1 c}{2}\right) \tag{9-2}$$

where f_{fe} represents the effective stress to be calculated according to Eq. (8-7) with the strain level ε_{fe} as defined in Eq. (9-3).

9.4.1 *Failure modes*—The following failure modes can control the out-of-plane behavior of URM walls strengthened with externally bonded FRP systems (Tumialan et al. 2003b; Galati et al. 2006):

- Crushing of the masonry in compression
- Debonding of the FRP system from the masonry substrate

Crushing of the masonry is an acceptable failure mode provided that strength and serviceability criteria are satisfied. If no mechanical anchorage is provided, FRP debonding is the most common failure mode in masonry walls strengthened with FRP systems (Tumialan et al. 2003b; Galati et al. 2005, 2006). The design protocols for flexural strengthening in this guide consider that either masonry crushing or debonding govern the flexural behavior of FRP-strengthened walls. Debonding occurs at the masonry-FRP interface and typically starts from flexural cracks at the maximum bending moment region and propagates toward the supports. As the tensile strength of masonry is lower than that of the epoxy used to bond the FRP system, the failure typically occurs in the masonry substrate. To prevent debonding, a bond-dependent coefficient is introduced to limit the strain level in the FRP reinforcement according to the provisions given in Section 8.4.1.

9.4.2 *Strain level in the FRP reinforcement*—The maximum strain level that can be achieved in the FRP reinforcement is determined by the strain developed in the FRP system at the ultimate limit state by either crushing of the masonry or FRP system debonding. The maximum strain or effective strain level in the FRP reinforcement may be calculated as

$$\varepsilon_{fe} = \varepsilon_{mu}\left(\frac{t-c}{c}\right) \leq \min(\kappa_m \varepsilon_{fu}^*,\ C_E \varepsilon_{fu}^*) \tag{9-3}$$

9.4.3 *Stress level in the FRP reinforcement*—The effective stress level in the FRP reinforcement is the maximum level of stress that can be developed in the FRP reinforcement before reaching the ultimate limit state. This effective stress can be calculated from the strain level in the FRP system, assuming perfectly elastic behavior as given in Eq. (8-7).

9.4.4 *Flexural strength computation*—Strain compatibility and equilibrium of forces are used to calculate the member flexural strength. The calculation described herein follows a simple trial-and-error method. This procedure requires the assumption of the neutral axis depth c. After assuming the neutral axis depth, the strain in the FRP can be calculated using Eq. (9-3). If the left side term in the equation controls, crushing of the masonry is the governing failure mode. If the second term controls, FRP system debonding or FRP rupture is the governing failure mode. If the equilibrium condition is not satisfied, the depth of the neutral axis should be revised and the procedure repeated. As an alternative, the flexural capacity can be computed (Section 13.1) by assuming a failure mode and calculating the corresponding neutral axis depth and area of FRP reinforcement (Bank 2006).

When FRP debonding is the governing failure mode, the effective stress in the FRP reinforcement can be computed from Eq. (8-7) assuming perfectly elastic behavior. When crushing of the masonry is the governing failure mode, the coefficients γ and β_1 defining the equivalent rectangular stress block approach, are each equal to 0.80, as reported in ACI 530. If FRP system rupture or debonding occurs before crushing of the masonry, methods considering a nonlinear stress-strain distribution can be used to estimate γ and β_1. These methods may include the use of a parabolic representation of the nonlinear compressive behavior of the masonry or other more elaborated methods such as Todeschini's method appropriately modified for the use with masonry (Galati et al. 2005). Alternative values of γ and β_1, both equal to 0.70, can provide reasonably accurate results (Tumialan et al. 2003b).

9.5—Serviceability

The serviceability of a masonry member, for example, deflections and crack widths, under service loads should satisfy applicable provisions of ACI 530. If deflection and crack width requirements are not satisfied before strengthening, supplementary strengthening or stiffening methods, or both, might need to be implemented to satisfy these

Table 9.1—Sustained plus cyclic service load stress limits in FRP reinforcement

Fiber type		
Carbon	Glass	Aramid
$0.55f_{fu}$	$0.20f_{fu}$	$0.30f_{fu}$

requirements. Fiber-reinforced polymer systems can be moderately effective in reducing deflections and crack widths; however, they are primarily effective in increasing strength.

The effect of FRP reinforcement on the in-service performance of a masonry member can be evaluated using the transformed section analysis. The strain level in the FRP reinforcement at service conditions should be compared with the limitations described in Section 9.6 for the case of sustained or cyclic load conditions.

Calculation of the stress level in the FRP reinforcement at service is required for masonry walls that resist sustained loads, such as retaining walls. The in-service stress in the FRP reinforcement can be estimated using Eq. (9-4), where M_s and P_s represent the moment and axial load due to all sustained loads, respectively. In the case of retaining walls, this load should include the entire lateral load due to earth and hydrostatic pressure and, depending on the service condition, a fraction or all lateral loads due to surcharge. The stress at service may be computed as

$$f_{fs} = \begin{cases} \dfrac{M_s - P_s\left(\frac{t}{2} - \frac{c_s}{3} \pm e\right)}{A_f\left(t - \frac{c_s}{3}\right)} & \text{for surface-mounted FRP systems} \\ \dfrac{M_s - P_s\left(\frac{t}{2} - \frac{c_s}{3} \pm e\right)}{A_f\left(t_b - \frac{c_s}{3}\right)} & \text{for NSM FRP systems} \end{cases} \quad (9\text{-}4)$$

where c_s is the neutral axis depth at service conditions.

The stress in the FRP reinforcement should not exceed the limits presented in Section 9.6.

9.6—Creep rupture stress limits

Masonry walls can be used as structural members to resist sustained lateral loads. For instance, concrete masonry walls are commonly used as retaining walls. To avoid creep-rupture of the FRP systems under sustained loads, the in-service stress calculated from Eq. (9-4) should not exceed the values shown in Table 9.1 (ACI 440.1R-06). As the stress levels will be within the elastic response range, or very nearly so, the stresses can be estimated using elastic analysis.

CHAPTER 10—WALL STRENGTHENING FOR IN-PLANE LOADS

10.1—Background information

Schwegler (1995) investigated strengthening methods for masonry shear walls with FRP laminates. The objective of this study was to determine how to increase the system ductility, generate uniform crack distribution, and increase load-carrying capacity. Carbon fiber-reinforced polymer laminates were bonded diagonally to the masonry walls and mechanically anchored to the adjoining RC slabs. The test results showed that the strengthened wall exhibited 50% to 300% increases in ultimate strength and displacement, respectively, as compared with the same wall without any strengthening system applied to the surface. By comparing walls strengthened on one versus two faces, it was observed that in the case of the one-face application, the capacity was half of the two-face strengthening, without any other major effect on the in-plane load-carrying capacity. In all the strengthened walls, fine cracks were observed perpendicular to the FRP laminates with constant spacing and small width.

Hartley et al. (1996) tested two full-sized 8 in. (203 mm) concrete block walls, 8 ft (2.44 m) high and 20 ft (6.1 m) long to investigate the feasibility of using unidirectional CFRP sheets to repair damage caused by differential foundation settlement. In the study, settlement loads were first applied to induce characteristic step cracking. Carbon fiber-reinforced polymer was then applied to one surface, and the wall retested. Strength gains of over 50% were recorded. The results suggested that CFRP was suitable for rehabilitating concrete block walls damaged by foundation settlement.

Al-Chaar and Hasan (1999) experimentally investigated the behavior of masonry bearing and shear walls subjected to seismic forces provided by an earthquake simulator. The FRP composite retrofit applied to one side of the URM walls enhanced the in-plane load capacity of these walls. Seismic rehabilitation of URM walls with FRP was investigated at the U.S. Army Corps of Engineers (Marshall et al. 1999). A large number of concrete and clay masonry walls were tested using several different FRP systems, such as laminates, grids, and NSM bars.

The U.S. Army Corps of Engineers (Marshall and Sweeney 2002) Engineer Research and Development Center, Construction Engineering Research Laboratory (ERDC-CERL), in partnership with the Market Development Alliance of the FRP Composites Industry, investigated FRP composite systems to strengthen both lightly reinforced and URM walls to better withstand damage during earthquakes. A total of 107 masonry walls, 64 clay brick walls, and 43 CMU walls, with FRP applied to one side only, were tested using a variety of different FRP configurations. The walls were tested in shear tension, in-plane shear, and in-plane rocking. Both E-glass and carbon fabrics were evaluated, as well as NSM shapes. The testing objective was to determine failure modes, mechanisms, and strengths using a variety of FRP configurations to optimize material placement. The performance of buildings in earthquakes is characterized by both the strength of the structure and its ability to survive large deformations without collapsing. In the ERDC-CERL tests, the ability to strengthen the walls using FRP was demonstrated. Placement of the FRP composites were shown to increase the strength and change the failure mode of masonry walls in shear. The configurations that performed best on clay brick walls were carbon full-coverage and both carbon and glass X-frame. The X-frame configuration consisted of an 8 in. (200 mm) wide strip of FRP placed vertically at each edge of the test specimen with two diagonal 8 in. (200 mm) strips starting from the corners to form an X across the center of the specimen. Full coverage and NSM CFRP rectangle performed best on CMU. The NSM CFRP configuration was rectangular shapes 5/8 in. (16 mm) wide by 1/8 in. (3 mm) thick with a roughened surface to enhance bond with the adhesive. The grooves were cut 4 in. (100 mm) from the outer edges and were approximately 1/4 in. (6 mm) wide and 1 in. (25 mm) deep. It was found in this investigation that because of its stiffness, carbon FRP failed in sudden bond failure.

Cracked URM concrete block walls were repaired by Gergely and Young (2001) using CFRP laminates attached to both sides of the specimens and subjected to cyclic out-of-plane loads, and in-plane loads. The symmetric laminates significantly increased the flexural and shear capacity of damaged walls. The specimens failed as a result of severe shear damage in the concrete masonry blocks. Experimental and analytical studies showed that the repaired walls behaved similarly to traditional composite sandwich panels.

Concrete masonry walls strengthened with FRP laminates in the horizontal direction only and tested with in-plane loading along the wall diagonal, have been observed to fail due to sliding shear along an unstrengthened joint (Tumialan et al. 2001a). This mode of failure is undesirable if there are adjacent columns such as in the case of infill walls, but may be controlled by placing FRP bars in the vertical direction to act as dowels.

As in the case of URM walls strengthened for flexure with FRP laminates, the masonry topology has been observed to be one of the factors influencing the in-plane wall behavior. Thus, in the case of clay brick masonry walls strengthened with laminates for shear, the FRP strengthening has been observed to be more efficient than in the case of concrete masonry (Grando et al. 2003). This can be attributed to characteristics of the parent material such as height of masonry courses—for example, smaller in the case of brick masonry—and better mortar-masonry unit bond characteristics. Grando et al. (2003) also reported that the in-plane capacity of clay masonry walls strengthened on one and two faces doubled, when doubling the FRP reinforcement.

Valluzzi et al. (2002) reported experimental results on clay brick masonry specimens having dimensions of 20 x 20 x 4.7 in. (0.51 x 0.51 x 0.12 m). The investigation considered variables such FRP laminate types (CFRP and GFRP); strengthening configurations (single-side versus double-side strengthening) and square grid and diagonal. Single-side-strengthened specimens failed due to splitting of masonry. Crack growth was larger on the unstrengthened sides due to reinforcement eccentricity. In some cases, the maximum loads were lower than those of the control panels. The double-side strengthened specimens failed due to delamination or rupture of the FRP laminate. In general, the diagonal strengthening configuration was observed to be more effective than the grid configuration; also, GFRP laminates were more effective at increasing the shear capacity.

Unreinforced masonry walls built with clay tile units are commonly used as inner walls in building envelopes. From the early 1900s to the 1960s, structural clay tiles were commonly used in infill walls throughout the U.S. This masonry typology is still widely used in other parts of the world where failure of this kind of wall has been reported to cause a large amount of loss of human life and material damage. In that context, Bastidas et al. (2002) investigated the strengthening with GFRP laminates of nonstructural masonry walls built with clay tiles. This investigation included diagonal compression tests of 26 specimens of 3.5 x 27.5 x 27.5 in. (90 x 700 x 700 mm) and 3.5 x 47 x 47 in. (90 x 1200 x 1200 mm), strengthened with different layouts of GFRP laminates. In addition, a full-scale wall was tested to validate the technology. The strengthening configurations included vertical and horizontal laminate strips, combinations of both, and diagonal laminates (cross-pattern). The test results showed that GFRP reinforcement increased the shear strength as well as the ductility of the system. The cross-pattern layout on both sides of the wall proved to be the most effective configuration. It was observed that areas of high load concentration, such as corners, should be filled with grout to avoid the fracture of the brittle clay tile units and premature failure of the wall. An infill clay tile masonry wall surrounded by an RC frame was built to validate the effectiveness of the GFRP reinforcement in larger specimens. The wall was 7.2 ft (2.2 m) high x 9.8 ft (3.0 m) long x 3.5 in. (90 mm) thick. Based on the results obtained from the testing of the small specimens, the wall was strengthened with 1.0 in. (25 mm) wide diagonal strips. The GFRP strips were applied on both sides of the wall to ensure symmetrical behavior during the in-plane cyclic loading. A significant global reduction of damage levels was observed for the strengthened masonry wall when compared with results reported by the same authors on similar URM walls. Also, global stability and overall seismic behavior were greatly improved with the GFRP reinforcement for in-plane loading.

Strengthening by FRP structural repointing, that is, placement of NSM bars in bed joints, can also remarkably increase the shear capacity of URM walls. This was evident from the results of tests conducted on concrete masonry walls loaded along the diagonal (Tumialan et al. 2001b). The maximum increase in shear capacity was 80%, registered in walls strengthened with GFRP bars placed at every bed joint. Strengthened walls showed stability after failure, that is, no loose material was observed, potentially reducing the risk of partial or total collapse. Due to reinforcement eccentricity, which caused the crack growth on the unstrengthened side to increase at a higher rate than the strengthened side, walls with FRP reinforcement on one side tilted toward the direction of the strengthened face. Walls with reinforcement staggered on both wall faces exhibited the largest displacement capacity. In-place tests were performed by Corradi et al. (2002) on FRP retrofitted masonry walls damaged by recent earthquakes. Both CFRP and GFRP unidirectional laminates were used to retrofit the masonry panels, followed by in-plane tests. The tests confirmed that the shear capacity of the masonry panels was significantly increased by the FRP materials.

Fiber-reinforced polymer structural repointing has also been used to improve the in-plane structural performance of masonry infill walls (Tumialan et al. 2003c). Full-scale specimens were subjected to in-plane cyclic load. The specimens were surrounded by an RC frame and a stand-alone RC support. The infill walls were built with concrete masonry blocks following a running bond pattern. The dimensions of the infill masonry walls were 8 ft 8 in. long and 8 ft high (2.64 x 2.44 m). The RC frames consisted of columns and a beam with cross sections of 12 x 12 in. (305 x 305 mm). The RC frame was designed to have sufficient ductility and strength; in this way, the behavior of the lateral force-resisting system would be controlled by the infill masonry wall and not by the frame. The results indicated that FRP-strengthened specimens could reach lateral drifts of 0.7% without losing lateral load-carrying capacity.

With the objective to find alternative embedding materials to the epoxy-based paste, Turco et al. (2003) investigated the in-plane behavior of concrete masonry walls strengthened with GFRP bars embedded in two different materials: epoxy-based paste and latex-modified cementitious paste. The in-plane test results showed that the performance of walls with both materials yielded similar results. The use of less expensive pastes, such as the latex-modified ones, makes the FRP structural repointing technique more appealing, as the structural performance is not reduced and the appearance of the filled joints is similar to conventional mortar joints. A study on the shear behavior of walls strengthened with FRP laminates was reported by Gu et al. (2003) confirming the effectiveness of this technique to improve both shear capacity and ductility of tested specimens.

Moon et al. (2003, 2007) tested under lateral loads a full-size two-story URM brick building that was strengthened using several FRP techniques. On one three-wythe wall, GFRP was epoxy-bonded vertically on the inside face, while NSM glass rods were epoxy-bonded into horizontal bed joints on the exterior face. This two-way retrofit increased the lateral strength, caused cracks to be well distributed, and produced a ductile-type failure mode with broad hysteresis loops and considerable energy dissipation. The four FRP systems used in this project included: precured structural thin grids embedded in trowel applied epoxy adhesive; wet layup unidirectional glass fabrics with an epoxy matrix; epoxy adhesive-applied NSM GFRP rods; and finally, glass grids in cementitious trowel-applied matrix. Application of GFRP systems on the other multi-wythe walls worked well in in-plane shear retrofit because header bricks every sixth course generally maintained continuity between wythes. Moon et al. (2006) concluded that the full three-dimensional behavior of the structural system should be considered, including flange effects, to properly design FRP retrofit systems. Each FRP system enhanced the performance of the URM building.

Stratford et al. (2004) reported the results of an experimental campaign on clay bricks and concrete masonry walls externally bonded using GFRP laminates under axial load and shear. The analysis of test results points out significant strength increases provided by the external reinforcement; the experiments have also shown that if proper anchorage is provided to FRP laminates, the wall could also attain a ductile failure mode. The use of CFRP laminates was investigated by To-Nan et al. (2006) on 5.4 x 5.2 ft (1640 x 1570 mm) walls subjected to in-plane cyclic loads; the CFRP laminates have been placed along the wall diagonals. It was pointed out that the presence of the CFRP strengthening provided a more ductile behavior of the strengthened walls compared with the URM wall and increased their energy dissipation capacity.

Marcari et al. (2007) investigated the behavior of tuff masonry panels having dimensions of 5.2 x 4.9 x 1.7 ft (1570 x 1480 x 530 mm) and subjected to shear-compression load. The main objectives were to assess the influence of different types of fibers—for example, CFRP versus GFRP—and FRP layout—for example, cross versus grid layout). The analysis of test results showed the shear strength of the wall panels was considerably improved by FRP laminates. In some cases, the original failure mode changed from shear to a shear/flexural mode; consequently, substantial gains in lateral strength were achieved up to the threshold of the value corresponding to as-built flexural capacity. In cases when such transition did not occur, strength increases proportional to the axial stiffness of the external reinforcement were measured. It was also underlined that the elastic stiffness of FRP-strengthened panels was not substantially modified by the external reinforcement and that the large increases of lateral strength were attained without significantly modifying the inelastic deformation capacity of FRP-strengthened panels. The use of GFRP laminates to improve the seismic behavior of tuff walls was also assessed by means of shake-table tests on a two story structure (Bergamo et al. 2006); shear strengthening by means of diagonal strips and flexural strengthening by means of vertical strips were designed for the tested structure. The results pointed out that the GFRP strengthening improved the seismic performance of the structure by reducing the crack opening related to rocking mechanism and allowing for a more energy dissipating behavior.

10.2—General considerations

Experimental tests have confirmed the effectiveness of FRP systems to increase the in-plane lateral strength of masonry walls (El-Gawady et al. 2005; Gergely and Young 2001; Marcari et al. 2007; Moon et al. 2007; Shrive 2006; Triantafillou 1998; Zhao et al. 2003). The design lateral strength of FRP-strengthened masonry walls should exceed the required shear strength, computed according to the load factors required by ASCE 7-05.

10.3—Existing wall strength

The behavior of URM walls under in-plane loads depends on several parameters related to geometric properties, including height, thickness, slenderness, and bond pattern; mechanical properties, for example, strength of the masonry units and mortar; and loading and support conditions. In general, the following failure modes can be recognized: joint sliding, diagonal tension, and toe crushing (Fig. 10.1).

Joint sliding and diagonal tension are shear-controlled failure modes. Joint sliding occurs on the bed joints and can develop either along a horizontal plane, generally near or at the wall base, or along a stair-stepped diagonal crack. Sliding does not reduce the axial load capacity of the wall, although the sliding distance should be controlled so that the

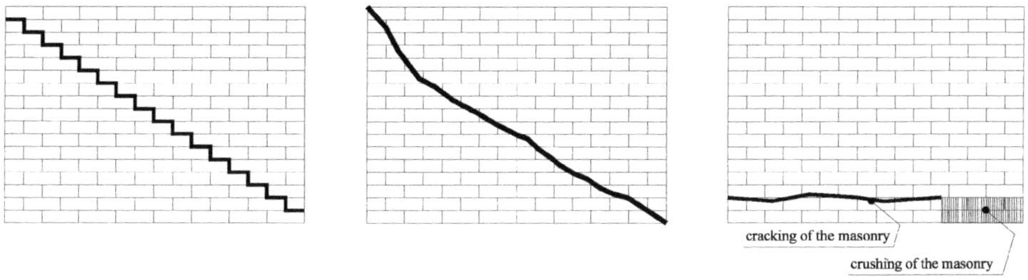

(a) Joint sliding (b) Diagonal tension (c) Toe crushing

Fig. 10.1—Failure modes.

structure does not slide off the foundation. It does, however, reduce the out-of-plane flexural load capacity of the wall. In earthquakes, sliding can be a pseudo-ductile mode, which dissipates considerable energy through friction. Bed-joint sliding is preferred to diagonal tension, as the latter is a brittle failure mode. Diagonal tension is characterized by diagonal cracks that can propagate through both mortar joints and masonry units, depending on their relative strength and on the level of axial stress.

Unlike joint sliding and diagonal tension, toe-crushing is a flexure-controlled failure mode, characterized by the formation of flexural cracks at the heel and crushing of the toe when the compression stresses reach the compressive strength of the masonry. In a steel-reinforced wall, toe crushing is often a ductile failure mode. In a URM structure, toe crushing together with a large axial load plus the increased axial force caused by steel or FRP retrofit can result in loss of vertical load capacity and wall collapse. Under such axial loads, P-Δ effects should be considered as required by ACI 530. For slender walls characterized by large material shear capacity and subjected to small compressive stresses, a rigid body rotation about the toe, that is, rocking, could occur after flexural cracking (FEMA 306). Yet, the overall response of the entire structural system should be considered (Moon et al. 2006). As the toe crushing strength generally exceeds that of the other failure modes, typically rocking does not control the design and, therefore, relevant checks will not be discussed in this guide.

The nominal lateral strength of URM walls should be computed as follows:

$$V_n^{URM} = \min(V_{bjs}, V_{dt}, V_{tc}) \tag{10-1}$$

where V_{bjs} is the nominal lateral strength corresponding to joint sliding, V_{dt} is the nominal lateral strength corresponding to diagonal tension, and V_{tc} is the nominal lateral strength corresponding to toe crushing.

To determine whether FRP strengthening is needed, the existing in-plane strength of the wall should be evaluated first. The design lateral strength should be obtained as the nominal lateral strength multiplied by the strength reduction factor ϕ, equal to 0.8 for shear-controlled failure modes (for example, joint sliding or diagonal tension) or equal to 0.6 for flexure-controlled failure mode (that is, toe crushing) in compliance with the provisions for URM walls specified by ACI 530. The shear and flexural tensile strength of masonry can be obtained by in-place testing following one of the test methods recommended by ASCE 41-06 (Chapter 7). In the absence of test results, the shear and flexural tensile strength can be conservatively estimated using the lower-bound values shown in Table 7-1 of ASCE 41-06, based on the conditions of the masonry and by consulting local building codes, or both. Alternatively, the shear and flexural tensile strength of masonry built after 1960 and proven to be in good condition may be estimated in accordance with ACI 530 provisions for URM if, in the judgment of the licensed design professional, the use of the ACI 530 provisions in place of the lower bound values from Table 7-1 of ASCE 41-06 can be justified.

10.4—Nominal shear strength of FRP-reinforced masonry walls subjected to in-plane loads

The strength design method requires that the shear strength of the FRP-strengthened wall exceeds the factored shear demand as indicated by Eq. (10-2).

$$\phi V_n \geq V_u \tag{10-2}$$

The design strength should be obtained as the nominal strength multiplied by the strength reduction factor ϕ equal to 0.8 as required by ACI 530 for URM walls subject to in-plane shear loads.

Unreinforced masonry walls requiring shear strengthening are those walls whose failure mode is due to either stepped joint sliding or diagonal tension as discussed in Section 10.3. A typical FRP strengthening scheme performed either with wet layup or NSM systems is indicated in Fig. 10.2(a) and (b), respectively. This guide discusses the most commonly used placement layouts of FRP for shear strengthening. Other layouts, including fibers placed diagonally, have been used, but they are not covered in the scope of this guide (Marcari et al. 2007).

Extending the FRP systems a distance ℓ_d (Chapter 11) beyond the wall, as shown in Fig. 10.2, or otherwise anchoring is a good detailing practice but it is not required by the shear design methodology. This is because the design method is based on the assumption that FRP debonding governs the behavior of the FRP-strengthened wall. Bonding the FRP into boundary elements such as beams or columns can help reduce the risk of wall overturning due to out-of-plane excitations. In several situations, however, extension

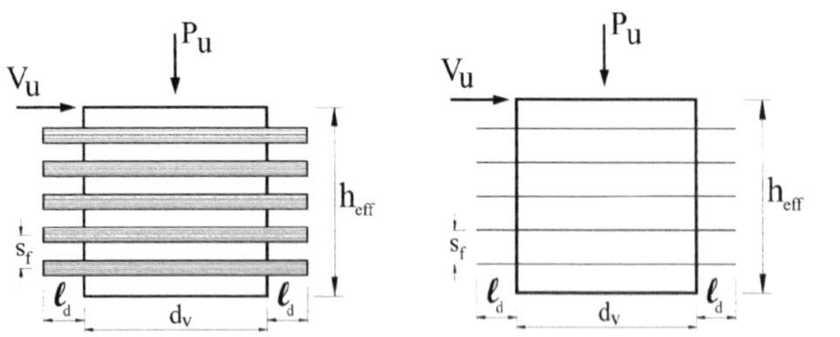

a) Horizontal strips b) Horizontal bars (NSM system)

Fig. 10.2—Fiber-reinforced polymer strengthening of shear-controlled walls.

Table 10.1—Limitations for FRP use

Masonry type	Wall construction	FRP strengthening layout
Hollow unit masonry wall	t = 8 in. (200 mm) or less Ungrouted or partially grouted walls with grouted cells spaced greater than 48 in. (1.20 m)	FRP on one face of wall is acceptable
	t = 8 in. (200 mm) or less Fully or partially grouted walls with grouted cells spaced at 48 in. (1.20 m) or less	FRP on two faces of wall is required
	t = 10 to 12 in. (250 to 300 mm) Ungrouted or partially grouted walls with grouted cells spaced greater than 60 in. (1.50 m)	FRP on two faces of wall is required
	t = 10 to 12 in. (250 to 300 mm) Fully or partially grouted walls with grouted cells spaced at 60 in. (1.50 m) or less	Use of FRP is not recommended
	t greater than 12 in. (300 mm) Ungrouted or grouted	Use of FRP is not recommended
Solid unit masonry wall	Single-wythe walls with t = 4 in. (100 mm) or less	FRP on one face of wall is acceptable
	Double-wythe walls with t = 8 in. (200 mm) or less	FRP on two faces of wall is required
	Multi-wythe walls with t > 8 in. (200 mm)	Use of FRP is not recommended

of FRP may not be practical due to existing field conditions. For example, intersecting walls, slabs, or columns with dimensions larger than the wall thickness can preclude extending the FRP beyond the wall. In such cases, other mechanical anchors should be considered.

The in-plane performance of FRP-strengthened walls is highly dependent on the type of masonry construction and the FRP strengthening layout. Experimental investigations have shown that FRP systems can significantly increase the shear capacity of URM walls when the original shear strength of the wall is not large (for example, single-wythe or ungrouted walls). Contrarily, when the shear strength of the URM wall is large (for example, multi-wythe or grouted walls), the contribution of FRP has been observed to be marginal in some situations. Test results also indicate that the FRP layout influences the wall's structural performance. For instance, in thick walls, FRP placed on the two wall sides has been shown to be more effective than FRP placed on one side. In the absence of project specific experimental evidence, this guide recommends FRP strengthening layouts based on the masonry construction as shown in Table 10.1, where the wall thicknesses are given in nominal dimensions.

The nominal shear strength of the FRP-strengthened wall can be computed by adding the FRP contribution V_f (Prota et al. 2008) to the nominal strength of the URM wall, computed according to the provisions given in Section 10.3; V_n^{URM} as

$$V_{n,s} = V_n^{URM} + V_f \tag{10-3}$$

The nominal lateral strength of the FRP-strengthened wall is the minimum of the nominal shear strength given in Eq. (10-3) and the nominal lateral strength corresponding to toe crushing of the URM wall.

The FRP contribution to the shear strength V_f can be determined as

$$V_f = \begin{cases} p_{fv} w_f \dfrac{d_v}{s_f} & \text{for surface-mounted FRP systems} \\ p_{fv} \dfrac{d_v}{s_f} & \text{for NSM FRP systems} \end{cases} \tag{10-4}$$

where p_{fv} is computed according to Eq. (8-14), w_f is the width of the FRP laminates, s_f is the center-to-center spacing between each strip, and d_v is the effective masonry depth for shear calculations given by

$$d_v = \min(H, L) \tag{10-5}$$

10.5—Nominal flexural strength of FRP-reinforced walls subjected to in-plane loads

The strength design method requires that the flexural strength of the FRP-strengthened wall exceeds the factored flexural demand, as indicated by Eq. (10-6)

$$\phi M_n \geq M_u \tag{10-6}$$

The design strength should be obtained as the nominal strength multiplied by the strength reduction factor ϕ equal to 0.6, as required by ACI 530 for URM walls under combination of axial load and bending.

Unreinforced masonry walls requiring flexural strengthening are those walls whose failure mode is due to toe crushing, as discussed in Section 10.3. A typical FRP strengthening scheme is depicted in Fig. 10.3 for a wet layup system. The effectiveness of this scheme as flexural reinforcement is highly dependent on the continuation of the load path by anchoring the wall at its base, which can be achieved either by extending or anchoring the FRP reinforcement or installing steel dowels. Suggested details for load path continuation, as well as requirements for minimum anchorage length $\ell_{d,}$ are given in Chapter 11.

Assuming that the factored axial load P_u acts at $L/2$ (L = length of the wall), the nominal moment capacity M_n of a FRP-strengthened masonry wall subjected to in-plane

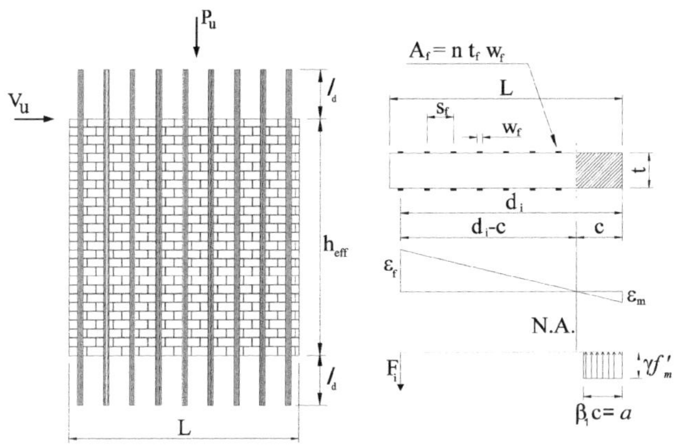

Fig. 10.3—Fiber-reinforced polymer strengthening of flexure-controlled masonry wall.

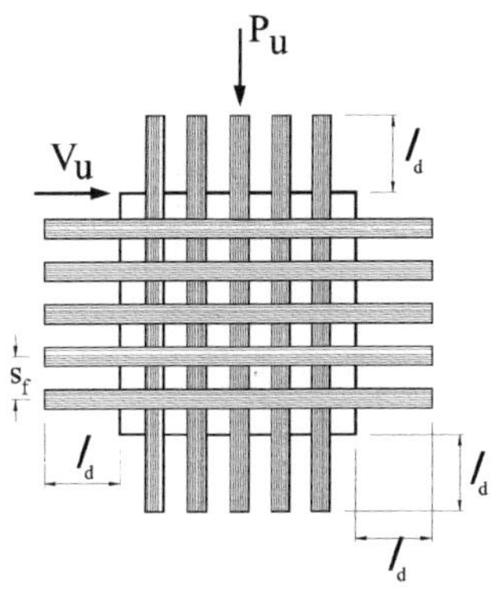

Fig. 10.4—Possible FRP scheme for both shear and flexural strengthening of URM walls.

loading can be calculated from Eq. (10-7) according to the assumptions and provisions given in Section 9.4.

$$M_n = \sum_i F_i\left(d_i - \frac{\beta_1 c}{2}\right) + P_u\left(\frac{L}{2} - \frac{\beta_1 c}{2}\right) \qquad (10\text{-}7)$$

where F_i is the force acting on the i-th FRP strip located at a distance d_i from the extreme compression fiber as indicated in Fig. 10.3. In cases where P_u does not act at $L/2$, Eq. (10-7) should account for the eccentricity of the axial load by inserting an appropriate value instead of $L/2$.

The nominal lateral strength corresponding to flexural failure of the FRP-strengthened wall can be obtained as

$$V_n = \frac{M_n}{k \cdot h_{eff}} \qquad (10\text{-}8)$$

where k is the coefficient that accounts for the boundary condition of the wall, k ($k = 0.5$ and $k = 1.0$ for a fixed-fixed and fixed-free wall, respectively), and h_{eff} is the wall height (Step 9, Section 13.2).

The nominal lateral strength of the FRP-strengthened wall is the minimum of the nominal lateral strength corresponding to flexural failure given in Eq. (10-8) and the nominal lateral strength corresponding to shear failure of the URM wall (that is, minimum between joint sliding and diagonal tension).

Similar considerations and identical procedures may be repeated in case of NSM FRP systems installed on the wall as an alternative to the wet layup system of Fig. 10.3.

10.6—Wall strengthening for shear and flexure

In some cases, the existing situation could require an increase in the lateral strength of the wall with respect to both shear and flexural failure modes. The objective can be achieved according to the principles given in Sections 10.4 and 10.5. A typical FRP scheme that could be used for this purpose is depicted in Fig. 10.4 for a wet layup system. When the URM wall is strengthened for both shear and flexure, the nominal lateral strength of the FRP-strengthened wall is the minimum of the strength corresponding to its shear (Section 10.4) and flexural (Section 10.5) failure. For the former, the nominal shear strength of the FRP-strengthened wall can be computed according to Eq. (10-3) in which the masonry contribution is that provided by ACI 530 for steel reinforced masonry.

Similar considerations may be repeated when NSM FRP systems replace the wet layup system. The installation of NSM systems in two directions, however, may not be practical from a constructibility stand point.

CHAPTER 11—DETAILING

11.1—General requirements

Chapter 11 provides guidance for detailing externally bonded and NSM FRP systems. Detailing will typically depend on the geometry of the structure, the quality of the substrate, and the levels of load that are to be sustained by the FRP system.

References relating to detailing issues presented herein are very limited and care should be taken in detailing and design of anchorages. In many cases, the following recommendations are empirical and should be implemented with caution.

11.2—Fiber-reinforced polymer debonding

11.2.1 *Externally bonded FRP systems*—For FRP systems installed according to Chapter 5, the weak link in the masonry/FRP interface is the masonry. The quality and tensile strength of the substrate will limit the overall effectiveness of the bonded FRP system.

Debonding of a properly installed FRP laminate can result from a lack of bonded area of the FRP laminate to the substrate. The masonry cannot maintain the interfacial shear and normal stresses, and the FRP laminate debonds from the substrate with a relatively thin layer of masonry attached to it.

11.2.2 *NSM FRP systems*—For NSM systems, the minimum dimension of the grooves should be at least $1.5d_b$ (d_b = FRP bar diameter) when an epoxy-based paste is used to embed the bar (De Lorenzis and Nanni 2001; Hassan and Rizkalla 2003). When a rectangular bar with a large aspect ratio is used, the limit may lose significance due to constructibility. In such a case, a minimum groove size of $3.0a_b$ x $1.5b_b$ is suggested, where a_b is the smallest bar dimension.

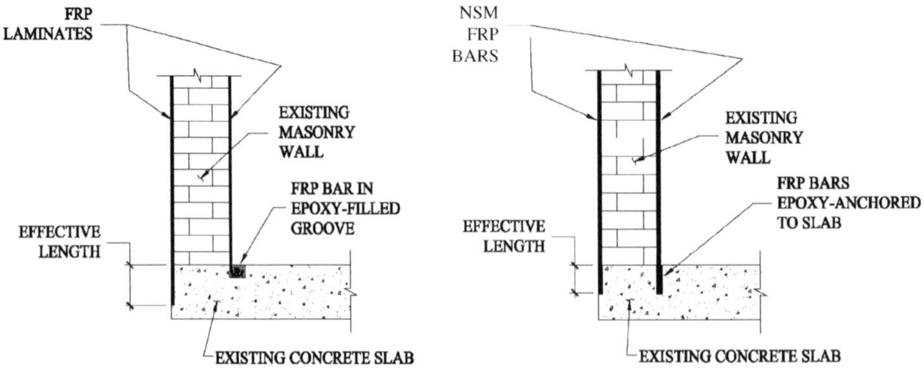

(a) Externally bonded systems *(b) NSM systems*

Fig. 11.1—Recommended anchorage details for externally bonded systems.

The minimum clear groove spacing for NSM FRP bars should be greater than twice the depth of the NSM groove to avoid overlapping of the tensile stresses around the NSM bars. Furthermore, a clear edge distance of four times the depth of the NSM groove should be provided to minimize the edge effects, which could accelerate debonding failure (Hassan and Rizkalla 2003). If final appearance of masonry requires cement-based mortar joints, the groove depth may be deeper to accommodate the cement mortar over top the resin paste adhesive. Turco et al. (2003) assessed the use of cementitious grouts to embed FRP bars. They concluded that a square groove with a least dimension of $2.5d_b$ could be adequate. Due to limited experimental data, however, the committee has no recommendation for the groove dimensions for cementitious grouts at this time.

11.2.3 *Development length*—For proper development length, ℓ_d expressed in in. (mm), of externally bonded FRP systems, the Committee recommends using Eq. (11-1) (CNR-DT 200/2004)

$$\ell_d = \begin{cases} \sqrt{\dfrac{E_f t_f}{50 f_{tm}}} & \text{in in.-lb units} \\ \sqrt{\dfrac{E_f t_f}{2 f_{tm}}} & \text{in SI units} \end{cases} \tag{11-1}$$

where t_f is the thickness the FRP system and f_{tm} is the average tensile strength of the masonry. Instead of specific data, when f_{tm} is unknown it may be assumed equal to $0.1f'_m$.

The development length of NSM FRP systems should be provided by the manufacturer and substantiated by testing independent of the manufacturer.

11.3—Spacing limits

11.3.1 *Out-of-plane flexural reinforcement*—While the horizontal clear spacing of vertical strips as high as $5.5t$ (for modern clay brick masonry walls) has been shown to be effective (Willis et al. 2009), it is recommended that the maximum clear spacing of externally bonded strips be limited to $3t$ to account for the fact that strengthening is typically performed on existing masonry. For NSM bars, the maximum center-to-center spacing should be limited to $3t$.

11.3.2 *In-plane shear reinforcement*—Based on the reinforcement spacing requirements for reinforced masonry walls subject to in-plane loading (ACI 530), the maximum clear reinforcement spacing for externally bonded strips (or the maximum center-to-center spacing for NSM bars) should be 16 in. (400 mm).

11.4—Anchorages of FRP reinforcement

Termination of the FRP reinforcement in a manner that will minimize the potential for premature debonding is recommended. Some typical termination details are shown in Fig. 10.1. Other termination details may also be possible depending on the configuration of the masonry element being strengthened. It is important to note that the details shown in Fig. 10.1 do not necessarily establish load path continuity between the masonry wall and adjoining concrete element. Additional details are discussed in Section 11.6, which may be used when continuity needs to be established.

11.4.1 *Externally bonded FRP systems*—For external faces (left face, Fig. 11.1(a)) it may be possible to extend the FRP system beyond the edge of the masonry wall by at least the effective length defined in ACI 440.2R using the appropriate concrete compressive strength. When FRP laminates are applied to interior faces (right face, Fig. 11.1(a)) it is recommended that the fabric be wrapped around a circular FRP bar embedded in the boundary element in the near-surface-mounted approach. The dimensions of the square groove should be in accordance with Section 11.2.2 considering the FRP bar diameter and thickness of the FRP sheet. The minimum recommended FRP bar diameter is 0.5 in. (12 mm). When pultruded FRP laminates are used on inside faces, it will not be possible to wrap the ends. In this case, it is recommended that a slot be cut into the boundary element and the FRP inserted as would be the case in the near-surface mounted approach (similar to right face, Fig. 11.1(b)).

Care should be taken to avoid air voids and that the epoxy resin completely fills the groove. Care should also be taken to avoid cutting or otherwise damaging embedded slab reinforcement that is critical to the performance of the overall structure.

11.4.2 *NSM FRP systems*—For exterior faces (left face, Fig. 11.1(b)) it may be possible to extend the FRP bars beyond the base of the masonry wall by at least the effective length, and groove dimensions, defined in ACI 440.2R using the appropriate concrete compressive strength. Although practically more difficult, the same is recommended for internal faces (right face, Fig. 11.1(b)). Care should be taken to avoid air voids and that the epoxy resin completely fills the groove. Care should also be taken to avoid cutting or otherwise damaging embedded slab reinforcement that is critical to performance of overall structure.

11.5—Alternate forms of anchoring

If adhesively bonded anchorage methods similar to those described in Section 10.4 are not possible, the following alternatives may be considered for externally bonded FRP systems.

11.5.1 *Embedded mechanical anchoring*—It has been shown experimentally that various forms of embedded mechanical anchors, including fiber anchors, are effective in anchoring ends of externally bonded FRP systems. Research, however, is still limited and it is recommended that the performance of any embedded mechanical anchorage system be substantiated through testing.

11.5.2 *External steel anchors*—External steel anchors similar to that illustrated in Fig. 11.2 are minimally invasive and can be used, provided that they do not interfere with external finishes. The width of the external steel anchor should at least be the same as the FRP laminates or the full width of the wall if a continuous FRP laminate is used. The external steel anchor should be bonded to the FRP using an adhesive suitable for FRP-to-steel joints, and should extend beyond the edge of the FRP by at least the anchorage length of the FRP-to-masonry interface defined by Section 12.3. Care should be taken to ensure a good bond between the steel and the FRP system and to prevent the possibility of galvanic corrosion if carbon fibers are used. For external faces (right face, Fig. 11.2) a bent steel shape (for example, plate, angle or channel) may be used and anchored to the concrete boundary with expansion or epoxy anchors. The steel member and anchors should be designed to resist the maximum axial tensile force that can be resisted by the FRP system. For internal faces (left face, Fig. 11.2) a steel angle is required. Prying of the angle due to the forces transferred by the FRP laminates should be taken into consideration when sizing the angle.

11.6—Load path continuity

The resistance of a masonry building subject to in-plane loads is defined by the inter-connectivity between the individual structural components. A structurally adequate connection between the different components creates a continuous load path for the loads. In masonry buildings, the load path is achieved by connecting the walls to the roof and floor diaphragms, and to the slabs or foundations. Thereby, load path continuity should be provided when a FRP-strengthened wall will be part of the lateral load-resisting system of a building structure. In addition, the increased loads due to the application of FRP systems may result in other failure mechanisms such as bed-joint sliding or rocking, for the case of in-plane loading, which may not be resisted by FRP systems. In such cases, internal steel bar anchors are an alternative that can be used to provide load path continuity at the wall base. Internal steel bar anchors do require disruption of the masonry wall and grouting of masonry cells, but they can be installed without impacting the exterior dimensions of the wall. These anchors should be located and spaced similarly to the FRP reinforcement and should be designed to provide adequate development length and embedment for the steel bar used.

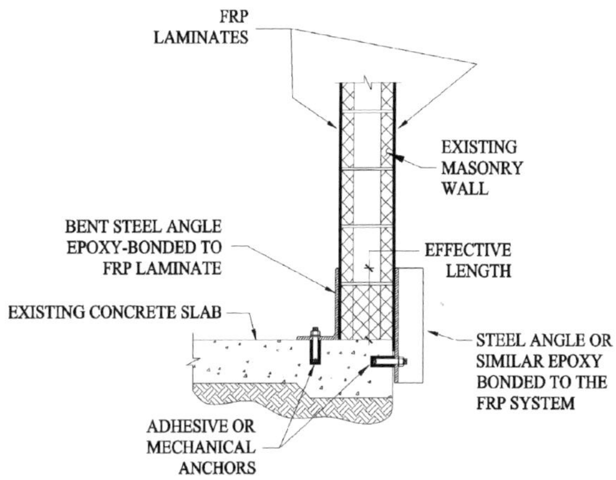

Fig. 11.2—Typical external steel anchor detail.

A variety of wall-to-floor and wall-to-roof details exist. The applicability of each of those details depends on the type of floor or roof structure and the magnitude of the loads. These details may also include the use of screw or bolt connections with angles.

Many other methods outside the scope of this guide can be used to ensure load path continuity. The use of a particular method depends on the application and it should be evaluated on a case-by-case basis.

11.7—Splices

Splices of FRP laminates should be provided only as permitted on drawings or in specifications, or as approved by the licensed design professional, following recommendations from the system manufacturer.

The fibers in FRP systems should be continuous and oriented in a direction that resists the largest tensile forces. Fiber continuity is maintained with a lap splice. For FRP systems, a lap splice should be made by overlapping the fibers along their length. The required overlap, or lap-splice length, depends on the tensile strength and thickness of the FRP system and on the bond strength between adjacent layers of FRP laminates. Sufficient overlap should be provided to promote the failure of the FRP laminate before debonding of the overlapped FRP laminates. The required overlap for a FRP system should be provided by the material manufacturer and substantiated through testing, by an independent testing agency.

For unidirectional FRP laminates, lap splices are required only in the direction of the fibers. Lap splices are not required in the direction transverse to the fibers. Fiber-reinforced polymer laminates consisting of multiple unidirectional sheets oriented in more than one direction or multidirectional fabrics require lap splices in more than one direction to maintain the continuity of the fibers and the overall strength of the FRP system.

CHAPTER 12—DRAWINGS, SPECIFICATIONS, AND SUBMITTALS

12.1—Engineering requirements

In 2009, federal, state, and local codes do not specifically address the structural design of externally bonded FRP systems. Other applicable code requirements, however, may influence the selection, design, and installation of the FRP system. For example, code requirements related to fire or potable water may influence the selection of the coatings used with the FRP system. All design work should be performed under the guidance of a licensed design professional familiar with the properties and applications of FRP systems and applicable building code requirements.

12.2—Drawings and specifications

This guide recommends the licensed design professional document calculations summarizing the assumptions and parameters used to design the FRP strengthening, prepare design drawings, and draft project specifications. The design and installation of the FRP systems requires close coordination between the licensed design professional and FRP system manufacturer. The drawings and specifications should show, at a minimum, the following information specific to externally applied FRP systems:

- Fiber-reinforced polymer system to be used;
- Location of the FRP system relative to the existing structure;
- Dimensions and orientation of each ply, laminate, or NSM bar;
- Number of plies and bars and the sequence of installation;
- Location of splices and lap length;
- General notes listing design loads and strains in the FRP systems;
- Material properties of the FRP systems and masonry substrate;
- Masonry surface preparation requirements, including groove dimensions for NSM bars, and maximum irregularity limitations;
- Installation procedures, including surface temperature and moisture limitations, and application time limits between successive installation procedures for the FRP materials such as plies for laminates;
- Curing procedures of FRP systems;
- Protective coatings and sealants if required;
- Shipping, storage, handling, and shelf-life guidelines; and
- Quality control and inspection procedures, including acceptance criteria.

12.3—Submittals

Specifications should require the FRP system manufacturer; installation contractor; inspection agency, if required; and all those involved with the project to submit product information and evidence of their qualifications and experience to the licensed design professional for review.

12.3.1 *Fiber-reinforced polymer system manufacturer*—Submittals required of the FRP system manufacturer should include:

- Product data sheets indicating the physical, mechanical, and chemical characteristics of the FRP system and all its constituent materials;
- Tensile properties of the FRP system including the method of reporting properties (net fiber or gross laminate), test methods used, and the statistical basis used for determining the properties;
- Installation instructions, maintenance instructions, and general recommendations regarding each material to be used. Installation procedures should include surface preparation requirements;
- Manufacturer's Material Safety Data Sheets (MSDS) for all materials to be used; and
- Quality-control procedure for tracking FRP materials and material certifications.

12.3.2 *Fiber-reinforced polymer system installation contractor*—Submittals required from the FRP system installation contractor include:

- Documentation from the FRP system manufacturer of having been trained to install the proposed FRP system or project references;
- Evidence of competency in surface preparation techniques; and
- Quality-control procedures including the daily log or inspection forms used by the contractor.

12.3.3 *Fiber-reinforced polymer system inspection agency*—If an independent inspection agency is used, submittals required of that agency should include:

- A list of inspectors to be used on the project and their qualifications;
- Sample inspection forms; and
- A list of previous projects inspected by the inspector or proof of training.

CHAPTER 13—DESIGN EXAMPLES

13.1—Increasing the flexural capacity of a wall subjected to out-of-plane loads

An existing warehouse has a roof built with steel open web joists and beams supported by steel interior columns and load-bearing masonry walls along the building perimeter. An elevation view of the shorter direction of the building is shown in Fig. 13.1. Determine the out-of-plane flexural strength of the wall panel shown shaded in Fig. 13.1(a) and design the required GFRP strengthening system to resist the applied loads.

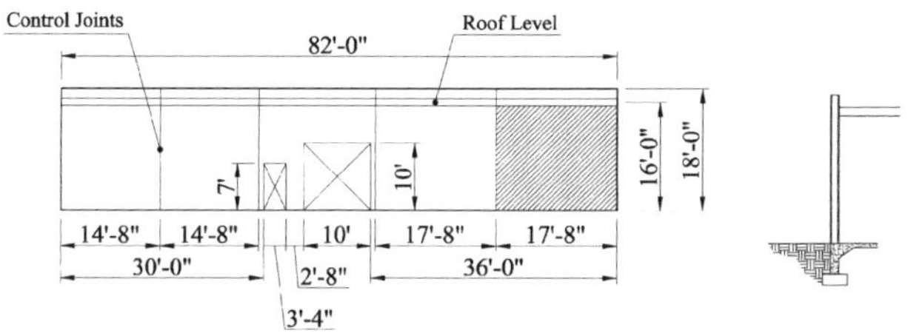

(a) Elevation view *(b) Section*

Fig. 13.1—Schematic of the masonry wall (1 in. = 25.4 mm).

The ungrouted masonry wall has a 12 in. (300 mm) nominal thickness as shown in Fig. 13.2(a); a Type N mortar was used during the construction. Factored shear and bending moment diagrams are depicted in Fig. 13.2(b). The applied factored axial load is P_u = 576 lb/ft (8406 N/m).

Table 13.1 summarizes the masonry geometrical and mechanical properties while Table 13.2 shows the FRP mechanical properties as reported by the manufacturer.

The values stated in either in-lb or SI units are to be regarded separately. The values stated in each system are not exact equivalents; therefore, each system should be used independently.

Table 13.1—Masonry geometrical and mechanical properties

Height of the wall H	18 ft	5.50 m
Wall thickness t	11.63 in.	295 mm
Section area A_n	36 in.^{2}/ft	76,000 mm^2/m
$I/(t/2)$, S	160 in.^{3}/ft	8.570 × 10^6 mm^3/m
Masonry compressive strength f'_m	1500 psi	10 MPa
Masonry compressive strain ε_{mu}	0.0025	0.0025

Table 13.2—Manufacturer's reported FRP system properties

Reinforcement type	f_{fu}^*, ksi (MPa)	E_f, ksi (MPa)	ε_{fu}^*	Thickness t_f, in. (mm)	Area $A_{f,bar}$, in.^{2} (mm^2)
GFRP laminate	220 (1517)	10,500 (72,000)	0.021	0.014 (0.356)	—
GFRP bar	120 (827)	5920 (41,000)	0.02	—	0.05 (32)

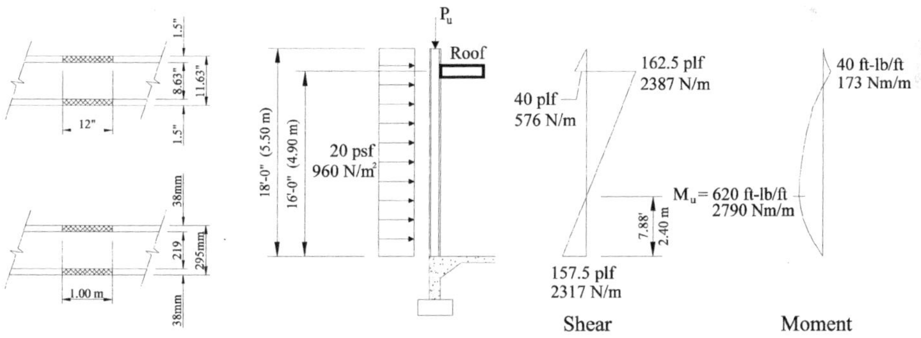

(a) Wall thickness *(b) Shear and bending moment diagrams*

Fig. 13.2—Wall thickness and external loading.

PROCEDURE	CALCULATION IN in.-lb
STEP 1: Compute the as-built tensile strength f_b.	
Determine f_r from Table 7.1 of ASCE 41 (in general, f_r can be determined according to ASCE 41 and/or the applicable local building code)	$f_r = 20$ psi
Assume $\phi = 0.6$	$\phi f_r = 0.6 \times 20 \text{ psi} = 12 \text{ psi}$
Compute $f_b = \dfrac{M_u}{S} - \dfrac{P_u}{A_n}$	$f_b = \dfrac{12 \text{ in./ft} \times 620 \text{ ft-lb/ft}}{160 \text{ in.}^3\text{/ft}} - \dfrac{576 \text{ lb/ft}}{36 \text{ in.}^2\text{/ft}} = 30.5 \text{ psi} > \phi f_r$ The masonry wall cannot sustain the applied load and needs to be strengthened. The selected strengthening system is the GFRP laminate shown in Table 13.2.
STEP 2: Compute the nominal flexural strength M_n.	
The required nominal flexural strength is calculated as follows $M_n = \dfrac{M_u}{\phi}$	$M_n = \dfrac{12 \text{ in./ft} \times 620 \text{ ft-lb/ft}}{0.6} = 12{,}400 \text{ in.-lb/ft}$
STEP 3: Compute the design mechanical properties of the FRP system.	
Assuming an environmental reduction factor C_E equal to 0.65, the FRP design tensile strength and strain can be determined from Eq. (8-3) and (8-4) $f_{fu} = C_E f_{fu}^*$ $\varepsilon_{fu} = C_E \varepsilon_{fu}^*$	$f_{fu} = 0.65 \times 220{,}000 \text{ psi} = 143{,}000 \text{ psi}$ $\varepsilon_{fu} = 0.65 \times 0.021 = 0.01365$
The bond reduction factor for flexure-controlled failure mode applicable to FRP laminates is assumed, from Eq. (8-8), equal to $\kappa_\mu = 0.45$. Therefore $f_{fe} = \kappa_m f_{fu}^* \le C_E f_{fu}^*$ $\varepsilon_{fe} = \kappa_m \varepsilon_{fu}^* \le C_E \varepsilon_{fu}^*$	$f_{fe} = 0.45 \times 220{,}000 \text{ psi} = 99{,}000 \text{ psi} < 143{,}000 \text{ psi}$ $\varepsilon_{fe} = 0.45 \times 0.021 = 0.00945 < 0.01365$
STEP 4: Determine the area of FRP reinforcement required.	
An assumption has to be made regarding the failure mode. As shown in Section 9.4.1, the failure mode can be caused by crushing of the masonry or debonding of the FRP system. Assume that the failure mode is controlled by debonding of the FRP system (such an assumption is later verified in Step 5). Assuming the stress block parameters $\gamma = \beta_1 = 0.7$ as suggested in Section 9.4.4, the two equilibrium equations can be written as follows: $\gamma \cdot f'_m \cdot \beta_1 \cdot c \cdot b - P_u = A_f \cdot f_{fe}$ $M_n = A_f \cdot f_{fe} \cdot t/2 + \gamma \cdot f'_m \cdot \beta_1 \cdot c \cdot b(t/2 - \beta_1 \cdot c/2)$ Replacing A_f obtained from the first equation $A_f = \dfrac{\gamma \cdot f'_m \cdot \beta_1 \cdot c \cdot b - P_u}{f_{fe}}$ (a) into the second equation, the following second-order equation in the unknown c (neutral axis position) can be found $\alpha \cdot c^2 + \beta \cdot c + \delta = 0$	
where $\alpha = (1/2)\gamma \cdot f'_m \cdot \beta_1^2 \cdot b$ $\beta = -\gamma \cdot f'_m + \beta_1 \cdot b \cdot t$ $\delta = M_n + (1/2)P_u \cdot t$	$\alpha = (1/2)(0.7)(1500 \text{ psi})(0.7)^2(12 \text{ in.}) = 3087 \text{ lb/in.}$ $\beta = -(0.7)(1500 \text{ psi})(0.7)(12 \text{ in.})(11.63 \text{ in.}) = -102{,}576.6 \text{ lb}$ $\delta = 12{,}400 \text{ in.-lb} + (1/2)(576 \text{ lb})(11.63 \text{ in.}) = 15{,}749.4 \text{ in.-lb}$
The neutral axis dept c can then be found by taking the only solution with physical meaning as follows $c = \dfrac{-\beta - \sqrt{\beta^2 - 4\alpha\delta}}{2\alpha}$	$c = \dfrac{102{,}576.6 \text{ lb} - \sqrt{(-102{,}576.6 \text{ lb})^2 - 4(3087 \text{ lb/in.})(15{,}749.4 \text{ in.-lb})}}{2(3087 \text{ lb/in.})} = 0.15 \text{ in.}$
The area of FRP reinforcement required can finally be computed by replacing c into Eq. (a).	$A_{f,required} = \dfrac{(0.7)(1500 \text{ psi})(0.7)(0.15 \text{ in.})(12 \text{ in.}) - (576 \text{ lb})}{99{,}000 \text{ psi}} = 0.0078 \text{ in.}^2\text{/ft}$

PROCEDURE	CALCULATION IN in.-lb
STEP 5: Failure mode.	
As we assumed that the failure mode was governed by debonding of the FRP system, we should now verify that the compressive strain in the masonry, ε_m, does not exceed, ε_{mu}, as indicated by Eq. (9-3) rewritten as follows $\varepsilon_m = \varepsilon_{fe}\left(\frac{c}{t-c}\right) \le \varepsilon_{mu}$	$\varepsilon_m = (0.00945)\left(\frac{0.15\text{ in.}}{11.63\text{ in.} - 0.15\text{ in.}}\right) = 0.00012 << \varepsilon_{mu} = 0.0025$
In the previous equation we neglected the initial strain in the masonry due to the applied loads before strengthening.	Fiber-reinforced polymer debonding is confirmed to be the governing failure mode.
STEP 6: Compute the area of FRP reinforcement to be installed.	
The width of the FRP reinforcement required can be determined as $w_{f,required} = \frac{A_{f,required}}{t_f}$ 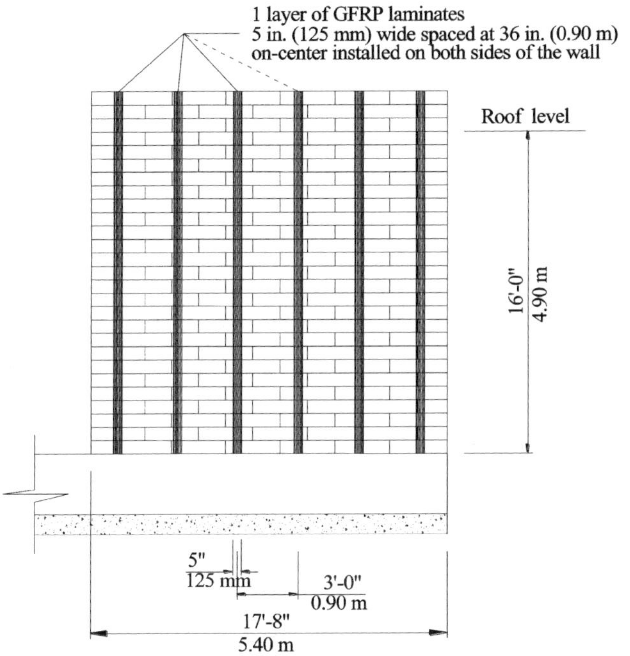 *Fig. 13.3—Fiber-reinforced polymer reinforcement layout for out-of-plane loads.*	$w_{f,required} = \frac{0.0078\text{ in.}^2/\text{ft}}{0.014\text{ in.}} = 0.56\text{ in./ft}$ Place one layer of a 5 in. wide GFRP laminate spaced at 36 in. on center (Fig. 13.3). $A_{f,installed} = \frac{12\text{ in./ft}}{36\text{ in.}}(5\text{ in.})(0.014\text{ in.}) = 0.023\text{ in.}^2/\text{ft} > A_{f,required}$
STEP 7: Maximum spacing requirement.	
For externally bonded FRP systems, the maximum spacing between FRP laminates should meet the requirement of Section 11.3.1 translated into the following equation $s_{f,max} < 3t + w_f$	$s_{f,max} < 3(11.63\text{ in.}) + 5\text{ in.} = 40\text{ in.}$ The maximum spacing requirement is satisfied.
STEP 8: Check the maximum force in the FRP system.	
According to Section 8.4.1, the maximum force per unit width in the FRP reinforcement should meet the requirement of Eq. (8-9) $p_{fm} = n \cdot t_f \cdot f_{fe} < 1500\text{ lb/in.}$	$p_{fm} = (1)(0.014\text{ in.})(99{,}000\text{ psi}) = 1386\text{ lb/in.} < 1500\text{ lb./in.}$ The maximum force per unit width requirement is satisfied.

PROCEDURE	CALCULATION IN in.-lb
STEP 9: Check the out-of-plane shear strength of the existing wall.	
The out-of-plane design shear strength ϕV_n of the existing masonry wall is calculated according to ACI 530 as $$\phi V_n = \phi \cdot \min\begin{cases} 3.8A_n\sqrt{f'_m} \\ 300A_n \\ 56A_n + 0.45P_u \end{cases}$$ larger than the factored shear force V_u equal to 162.5 lb/ft as indicated in Fig. 13.2(b).	$$\phi V_n = (0.8)\cdot\min\begin{cases} 3.8\cdot(12\text{ in.}\times 2\times 1.5\text{ in.})\sqrt{1500\text{ psi}} = 5298\text{ lb/ft} \\ 300\cdot(12\text{ in.}\times 2\times 1.5\text{ in.}) = 10{,}800\text{ lb/ft} \\ 56\cdot(12\text{ in.}\times 2\times 1.5\text{ in.}) + 0.45\cdot 576\text{ lb/ft} = 2275\text{ lb/ft} \end{cases}$$ $\phi V_n = 1820$ lb/ft $\phi V_n > V_u$ OK
STEP 10: Determine the development length to anchor the FRP system.	
The appropriate development length is calculated according to Eq. (11-1) as follows $$\ell_d = \sqrt{\frac{E_f t_f}{50 f_{tm}}}$$ On the exterior face of the wall, the FRP system is installed as indicated in Fig. 13.3. On the interior side, the FRP system is anchored as indicated in Fig. 11.2 using a steel angle epoxy bonded to the FRP laminate.	$$\ell_d = \sqrt{\frac{(10{,}500{,}000\text{ psi})(0.014\text{ in.})}{50(0.1\times 1500\text{ psi})}} = 4.42\text{ in., assume 5 in.}$$

PROCEDURE	CALCULATION IN SI UNITS
STEP 1: Compute the as-built tensile strength f_b.	
Determine f_r from Table 7.1 of ASCE 41 (in general, f_r can be determined according to ASCE 41 and/or the applicable local building code)	$f_r = 0.14$ MPa
Assume $\phi = 0.6$	$\phi f_r = 0.6 \times 0.14 \text{ MPa} = 0.084 \text{ MPa}$
Compute $f_b = \dfrac{M_u}{S} - \dfrac{P_u}{A_n}$	$f_b = \dfrac{2{,}790{,}000 \text{ N·mm/m}}{8.570 \times 10^6 \text{ mm}^3\text{/m}} - \dfrac{8406 \text{ N/m}}{76{,}000 \text{ mm}^2\text{/m}} = 0.215 \text{ MPa} > \phi f$ The masonry wall cannot sustain the applied load and needs to be strengthened. The selected strengthening system is the GFRP laminate shown in Table 13.2.
STEP 2: Compute the nominal flexural strength M_n.	
The required nominal flexural strength is calculated as follows $M_n = \dfrac{M_u}{\phi}$	$M_n = \dfrac{2790 \text{ Nm/m}}{0.60} = 4650 \text{ Nm/m}$
STEP 3: Compute the design mechanical properties of the FRP system.	
Assuming an environmental reduction factor C_E equal to 0.65, the FRP design tensile strength and strain can be determined from Eq. (8-3) and (8-4) $f_{fu} = C_E f_{fu}^*$ $\varepsilon_{fu} = C_E \varepsilon_{fu}^*$	$f_{fu} = 0.65 \times 1517 \text{ MPa} = 986 \text{ MPa}$ $\varepsilon_{fu} = 0.65 \times 0.021 = 0.01365$
The bond reduction factor for flexure-controlled failure mode applicable to FRP laminates is assumed, from Eq. (8-8), equal to $\kappa_\mu = 0.45$. Therefore $f_{fe} = \kappa_m f_{fu}^* \leq C_E f_{fu}^*$ $\varepsilon_{fe} = \kappa_m \varepsilon_{fu}^* \leq C_E \varepsilon_{fu}^*$	$f_{fe} = 0.45 \times 1517 \text{ MPa} = 683 \text{ MPa} < 986 \text{ MPa}$ $\varepsilon_{fe} = 0.45 \times 0.021 = 0.00945 < 0.001365$
STEP 4: Determine the area of FRP reinforcement required.	
An assumption has to be made regarding the failure mode. As shown in Section 9.4.1, the failure mode can be caused by crushing of the masonry or debonding of the FRP system. Assume that the failure mode is controlled by debonding of the FRP system (such an assumption is later verified in Step 5). Assuming the stress block parameters $\gamma = \beta_1 = 0.7$ as suggested in Section 9.4.4, the two equilibrium equations can be written as follows: $\gamma \cdot f'_m \cdot \beta_1 \cdot c \cdot b - P_u = A_f \cdot f_{fe}$ $M_n = A_f \cdot f_{fe} \cdot t/2 + \gamma \cdot f'_m \beta_1 \cdot c \cdot b(t/2 - \beta_1 \cdot c/2)$ Replacing A_f obtained from the first equation $A_f = \dfrac{\gamma \cdot f'_m \cdot \beta_1 \cdot c \cdot b - P_u}{f_{fe}}$ (a) into the second equation, the following second-order equation in the unknown c (neutral axis position) can be found: $\alpha \cdot c^2 + \beta \cdot c + \delta = 0$ where $\alpha = (1/2)\gamma \cdot f'_m \cdot \beta_1^2 \cdot b$ $\beta = -\gamma \cdot f'_m + \beta_1 \cdot b \cdot t$ $\delta = M_n + (1/2)P_u \cdot t$	$\alpha = (1/2)(0.7)(10 \text{ MPa})(0.7)^2(1000 \text{ mm}) = 1715 \text{ N/mm}$ $\beta = -(0.7)(10 \text{ MPa})(0.7)(1000 \text{ mm})(295 \text{ mm}) = -1{,}445{,}500 \text{ N}$ $\delta = 4{,}650{,}000 \text{ N·mm} + (1/2)(8406 \text{ N})(295 \text{ mm}) = 5{,}889{,}885 \text{ N·mm}$
The neutral axis dept c can then be found by taking the only solution with physical meaning as follows $c = \dfrac{-\beta - \sqrt{\beta^2 - 4\alpha\delta}}{2\alpha}$	$c = \dfrac{1{,}445{,}500 \text{ N} - \sqrt{(-1{,}445{,}500 \text{ N})^2 - 4(1715 \text{ N/mm})(5{,}889{,}885 \text{ N·mm})}}{2(1715 \text{ N/mm})} = 4$ mm
The area of FRP reinforcement required can finally be computed by replacing c into Eq. (a).	$A_{f,required} = \dfrac{(0.7)(10 \text{ MPa})(0.7)(4 \text{ mm})(1000 \text{ mm}) - (8406 \text{ N})}{683 \text{ MPa}} = 16 \text{ mm}^2\text{/mm}$

PROCEDURE	CALCULATION IN SI UNITS
STEP 5: Failure mode.	
As we assumed that the failure mode was governed by debonding of the FRP system, we should now verify that the compressive strain in the masonry, ε_m, does not exceed ε_{mu} as indicated by Eq. (9-3) rewritten as follows $\varepsilon_m = \varepsilon_{fe}\left(\frac{c}{t-c}\right) \le \varepsilon_{mu}$ In the previous equation we neglected the initial strain in the masonry due to the applied loads before strengthening.	$\varepsilon_m = (0.00945)\left(\frac{4\ \text{mm}}{295\ \text{mm} - 4\ \text{mm}}\right) = 0.00012 << \varepsilon_{mu} = 0.0025$ Fiber-reinforced polymer debonding is confirmed to be the governing failure mode.
STEP 6: Compute the area of FRP reinforcement to be installed.	
The width of the FRP reinforcement required can be determined as $w_{f,required} = \frac{A_{f,required}}{t_f}$ 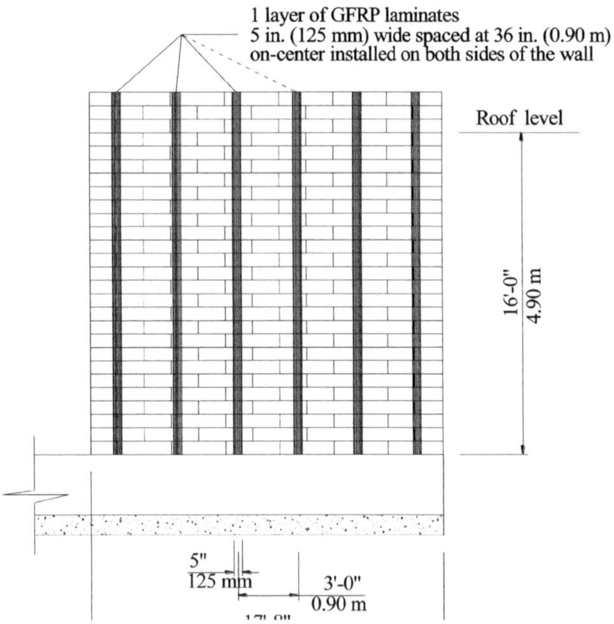 *Fig. 13.3—Fiber-reinforced polymer reinforcement layout for out-of-plane loads.*	$w_{f,required} = \frac{16\ \text{mm}^2/\text{mm}}{0.356\ \text{mm}} = 45\ \text{mm/m}$ Place one layer of a 125 mm wide GFRP laminate spaced at 900 mm on center (Fig. 13.3). $A_{f,installed} = \frac{1000\ \text{mm/m}}{900\ \text{mm}}(125\ \text{mm})(0.356\ \text{mm}) = 49\ \text{mm}^2/\text{m} > A_{f,required}$
STEP 7: Maximum spacing requirement.	
For externally bonded FRP systems, the maximum spacing between FRP laminates should meet the requirement of Section 11.3.1 translated into the following equation $s_{f,max} < 3t + w_f$	$s_{f,max} < 3(295\ \text{mm}) + 125\ \text{mm} = 1010\ \text{mm}$ The maximum spacing requirement is satisfied.
STEP 8: Check the maximum force in the FRP system.	
According to Section 8.4.1, the maximum force per unit width in the FRP reinforcement should meet the requirement of Eq. (8-9) $p_{fm} = n \cdot t_f \cdot f_{fe} < 1500$ lb/in.	$p_{fm} = (1)(0.356\ \text{mm})(683\ \text{MPa}) = 243\ \text{N/mm} < 260\ \text{N/mm}$ The maximum force per unit width requirement is satisfied.

PROCEDURE	CALCULATION IN SI UNITS
STEP 9: Check the out-of-plane shear strength of the existing wall.	
The out-of-plane design shear strength ϕV_n of the existing masonry wall is calculated according to ACI 530 as $\phi V_n = \phi \cdot \min\begin{cases} 0.3A_n\sqrt{f'_m} \\ 2A_n \\ 0.4A_n + 0.45P_u \end{cases}$ larger than the factored shear force V_u equal to 2387 N/m as indicated in Fig. 13.2(b).	$\phi V_n = (0.8) \cdot \min\begin{cases} 0.3 \cdot (76{,}000\ \text{mm}^2/\text{m})\sqrt{10\ \text{MPa}} = 72{,}100\ \text{N/m} \\ 2 \cdot (76{,}000\ \text{mm}^2/\text{m}) = 152{,}000\ \text{N/m} \\ 0.4 \cdot (76{,}000\ \text{mm}^2/\text{m}) + 0.15 \cdot 8406\ \text{N/m} = 31{,}661\ \text{N/m} \end{cases}$ $\phi V_n = 25{,}329$ N/m $\phi V_n > V_u$ OK
STEP 10: Determine the development length to anchor the FRP system.	
The appropriate development length is calculated according to Eq. (11-1) as follows $\ell_d = \sqrt{\dfrac{E_f t_f}{50 f_{tm}}}$ On the exterior face of the wall, the FRP system is installed as indicated in Fig. 13.3. On the interior side, the FRP system is anchored as indicated in Fig. 11.2 using a steel angle epoxy bonded to the FRP laminate.	$\ell_d = \sqrt{\dfrac{(41{,}000\ \text{MPa})(0.356\ \text{mm})}{2(0.1 \times 10\ \text{MPa})}} = 85$ mm, assume 120 mm

The 12 in. (300 mm) UMR wall can be strengthened to support a 620 ft-lb/ft (2790 Nm/m) moment by bonding nine 5 in. (125 mm) wide GFRP laminates at 3 ft (0.90 m) on centers (Fig. 13.3). Each strip extends 18 ft (5.50 m) to the roof level. The first of the nine strips is placed 16 in. (400 mm) from the end of the 17 ft 8 in. (5.40 m) wide wall. They should be bonded to both wall surfaces to account for load reversal under seismic loading.

13.2—Increasing the lateral capacity of a wall subjected to in-plane loads

The same warehouse analyzed in the previous example is subjected to a factored lateral in-plane force equal to $V_u = 20$ kip (90 kN) acting at the level of the roof H_{roof} = [16 in. – 0 in. (4.90 m)], as indicated in Fig. 13.1(a). Design the shear strengthening using the reinforcement type indicated as GFRP bars in Table 13.2 of Section 13.1 for the wall panel shown shaded in Fig. 13.1(a). Additional information is provided in Table 13.3. Figure 13.4 shows the strengthening systems used to enhance the wall capacity for in-plane and out-of-plane (Section 13.1) loads.

The values stated in either in.-lb or SI units are to be regarded separately. The values stated in each system are not exact equivalents; therefore, each system should be used independently of the other.

Table 13.3—Additional masonry mechanical and geometrical properties and loading information

Lower-bound masonry shear strength, v_{tl}	15 psi	0.10 MPa
Superimposed dead load, P_D	280 lb/ft	4000 N/m
Factor accounting for boundary conditions, α	1.00	1.00
Lower-bound axial compressive force due to gravity load, Q_G	400 lb/ft	5800 N/m
Spacing of every other course of masonry unit where the NSM FRP reinforcement is to be installed, s	16 in.	400 mm

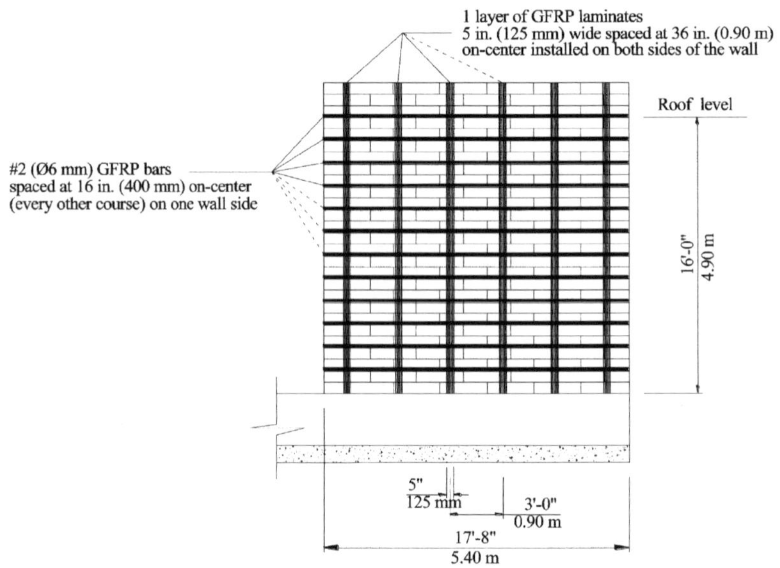

Fig. 13.4—Schematic of the strengthened wall for in-plane loads.

PROCEDURE	CALCULATION IN in.-lb
STEP 1: Compute the shear capacity due to bed-joint sliding, V_{bjs}.	
The nominal shear capacity of the unreinforced masonry wall for bed-joint sliding is found according to the provisions of ASCE 41 as follows $V_{bjs} = v_{mL}A_n$ where $v_{mL} = 0.75\dfrac{0.75v_{tL} + \dfrac{P_D}{A_n}}{1.5}$ In general, the nominal shear capacity of the URM wall for bed-joint sliding can be determined according to ASCE 41 and/or the applicable local building code.	$V_{bjs} = (9.5\text{ psi})(636\text{ in.}^2) = 6042\text{ lb}$ $v_{mL} = 0.75\dfrac{0.75(15\text{ psi}) + \dfrac{280\text{ lb/ft} \times 17.7\text{ ft}}{2(1.5\text{ in.} \times 212\text{ in.})}}{1.5} = 9.5\text{ psi}$
STEP 2: Compute the shear capacity due to diagonal tension, V_{dt}.	
The nominal shear capacity of the URM wall for diagonal tension is found according to the provisions of FEMA 356 as follows $V_{dt} = f'_{dt} \cdot A_n\left(\dfrac{L}{h_{eff}}\right)\sqrt{1 + \dfrac{f_a}{f'_{dt}}}$ where f'_{dt} may be replaced by v_{mL} calculated in the previous step; L is length of the wall assumed equal to 17 ft 8 in. (Fig. 13.1(a)); h_{eff} is the height to resultant of lateral force equal to 16 ft 0 in. (Fig. 13.1(a)); and f_a represents the axial compressive stress due to gravity load, computed as $f_a = \dfrac{P_D}{A_n}$	$V_{dt} = (9.5\text{ psi})(636\text{ in.}^2)\left(\dfrac{17.7\text{ ft}}{16\text{ ft}}\right)\sqrt{1 + \dfrac{7.8\text{ psi}}{9.5\text{ psi}}} = 9020\text{ lb}$ $f_a = \dfrac{280\text{ lb/ft} \times 17.7\text{ ft}}{636\text{ in.}^2} = 7.8\text{ psi}$
STEP 3: Compute the shear capacity due to toe crushing, V_{tc}.	
The nominal shear capacity of the URM wall for toe crushing is found according to the provisions of ASCE 41 as follows $V_{tc} = \alpha \cdot Q_G\left(\dfrac{L}{h_{eff}}\right)\left(1 - \dfrac{f_a}{0.7f'_m}\right)$ where L and h_{eff} are as defined in the previous step, and f_a is calculated as follows $f_a = \dfrac{Q_G}{A_n}$ In general, the nominal shear capacity of the URM wall for toe crushing can be determined according to ASCE 41 and/or the applicable local building code.	$V_{tc} = (1.00)(7080\text{ lb})\left(\dfrac{17.7\text{ ft}}{16\text{ ft}}\right)\left(1 - \dfrac{11.1\text{ psi}}{0.7 \times 1500\text{ psi}}\right) = 7749\text{ lb}$ $f_a = \dfrac{400\text{ lb/ft} \times 17.7\text{ ft}}{636\text{ in.}^2} = 11.1\text{ psi}$
STEP 4: Compute the as-built shear capacity, ϕV_n^{URM}.	
As indicated in Section 10.3, the nominal shear capacity or nominal lateral strength of the URM wall V_n^{URM} is obtained as the minimum of Eq. (10-1) $V_n^{URM} = \min(V_{bjs}, V_{dt}, V_{tc})$ Assuming $\phi = 0.8$, the design lateral strength can be computed as ϕV_n^{URM}.	$V_n^{URM} = \min(6042\text{ lb}, 9020\text{ lb}, 7749\text{ lb}) = 6042\text{ lb}$ $\phi V_n^{URM} = 0.8(6042\text{ lb}) = 4834\text{ lb}$ As $\phi V_n^{URM} < V_u$, the wall needs strengthening for in-plane loads
STEP 5: Compute the design mechanical properties of the NSM FRP system.	
Assuming an environmental reduction factor C_E equal to 0.65, the FRP design tensile strength and strain can be determined from Eq. (8-3) and (8-4) $f_{fu} = C_E f_{fu}^*$ $\varepsilon_{fu} = C_E \varepsilon_{fu}^*$	$f_{fu} = (0.65)(120{,}000\text{ psi}) = 78{,}000\text{ psi}$ $\varepsilon_{fu} = (0.65)(0.02) = 0.013$

PROCEDURE	CALCULATION IN in.-lb
STEP 6: Compute the effective strain and stress for shear-controlled failure mode.	
To determine the effective strain in the FRP reinforcement, compute the area of FRP reinforcement as follows	
$A_f = \frac{H_{roof}}{s}A_{f,bar}$	$A_f = \frac{12\text{ in./ft} \times 16\text{ ft}}{16\text{ in.}}0.05\text{ in.}^2 = 0.6\text{ in.}^2$
The value of ω_f is defined in Eq. (8-12)	
$\omega_f = \frac{1}{1000}\frac{A_f E_f}{A_n\sqrt{f'_m}}$	$\omega_f = \frac{1}{1000}\frac{(0.6\text{ in.}^2)(5{,}920{,}00\text{ psi})}{(636\text{ in.}^2)\sqrt{1500\text{ psi}}} = 0.144$
According to Section 8.4.2, the effective strain and stress in the FRP reinforcement can be found from Eq. (8-10) and (8-11) as a function of κ_v (Eq. (8-13))	
$\kappa_v = \begin{cases} 0.40 \text{ for } \omega_f \le 0.20 \\ 0.64 - 1.2\omega_f \text{ for } 0.20 < \omega_f \le 0.45 \\ 0.10 \text{ for } \omega_f > 0.45 \end{cases}$	$\kappa_v = 0.40$
$\varepsilon_{fe} = \kappa_v \varepsilon_{fu}^* \le C_E \varepsilon_{fu}^*$ $f_{fe} = E_f \varepsilon_{fe}$	$\varepsilon_{fe} = (0.40)(0.02) = 0.008 < 0.013$ $f_{fe} = (5{,}920{,}000\text{ psi})(0.008) = 47{,}360\text{ psi}$
STEP 7: Compute the FRP contribution to the shear strength, V_f.	
The total force per bar that the FRP system can transfer to the masonry substrate is given by Eq. (8-14) as follows: $p_{fv} = A_{f,bar}f_{fe} \le 10{,}000$ lb/bar	$p_{fv} = (0.05\text{ in.}^2)(47{,}360\text{ psi}) = 2368\text{ lb/bar} < 10{,}000\text{ lb/bar}$
The NSM FRP contribution to the shear strength can be determined from Eq. (10-4) as follows $V_f = p_{fv}\frac{d_v}{s_f}$	$V_f = (2368\text{ lb})\frac{212\text{ in.}}{16\text{ in.}} = 31{,}376\text{ lb}$
where $d_v = \min(H, L)$; and s_f represents the spacing of the NSM FRP reinforcement equal to the spacing of every other course of masonry unit, s.	$d_v = \min(18\text{ ft}, 17\text{ ft } 8\text{ in.}) = 17\text{ ft } 8\text{ in.}$
STEP 8: Compute the nominal shear strength of the FRP-strengthened wall.	
According to Eq. (10-3), the nominal shear strength of the FRP-strengthened wall is $V_{n,s} = V_n^{URM} + V_f$	$V_{n,s} = (6042\text{ lb}) + (31{,}376\text{ lb}) = 37{,}418\text{ lb}$

PROCEDURE	CALCULATION IN in.-lb
STEP 9: Compute the nominal flexural strength of the FRP-reinforced wall under in-plane loads.	
As the design lateral strength determined in the previous step is larger than the lateral strength due to toe crushing of the unreinforced wall computed in Step 3, a check needs to be performed to ensure that the FRP reinforcement designed in Section 13.1 will prevent the rupture of the wall due to toe crushing. To this extent, the FRP reinforcement designed in Section 13.1 should effectively be anchored as suggested in Section 11.4. The maximum lateral force $V_{n,f}$ that the wall can sustain before flexural failure is computed as follows $V_{n,f} = \dfrac{M_n}{k \cdot H_{roof}}$ where M_n is the nominal moment capacity computed according to Eq. (10-7) using a trial-and-error procedure that allows the determination of the neutral axis depth, $c = 12.76$ in., and gives $M_n = 3975$ in.-kip. By installing an extra layer of GFRP vertical reinforcement 5 in. wide spaced at 36 in. on-center on both sides of the wall (Fig. 13.5), the neutral axis depth results in $c = 24.06$ in. and $M_n = 7483$ in.-kip.	$V_{n,f} = \dfrac{3975 \text{ in.-kip}}{1.0 \times 192 \text{ in.}} = 20.703 \text{ kip} < V_{n,s}$ computed in Step 8 (not good) $V_{n,f} = \dfrac{7483 \text{ in.-kip}}{1.0 \times 192 \text{ in.}} = 38.974 \text{ kip} > V_{n,s}$ computed in Step 8 (OK)
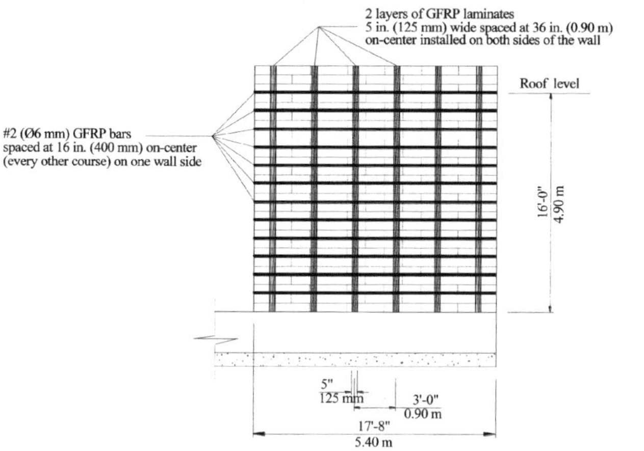 *Fig. 13.5—Schematic of the strengthened wall for in-plane loads with the additional layer of vertical reinforcement.*	
STEP 10: Compute the design lateral strength of the FRP-strengthened wall.	
The nominal lateral strength of the FRP-strengthened wall is obtained as the minimum of the nominal shear strength as determined in Step 8 and the nominal lateral strength corresponding to flexural failure of the FRP-strengthened wall as determined in Step 9 $V_n = \min(V_{n,s}, V_{n,f})$ As failure is controlled by the shear strength, the strength reduction factor is taken equal to 0.8 (Section 10.4).	$V_n = \min(37.418 \text{ kip}, 38.974 \text{ kip}) = 37.418$ kip $\phi V_n = (0.8)(37.418 \text{ kip}) = 29.934 \text{ kip} > V_u = 20$ kip

PROCEDURE	CALCULATION IN SI UNITS
STEP 1: Compute the shear capacity due to bed-joint sliding, V_{bjs}.	
The nominal shear capacity of the unreinforced masonry wall for bed-joint sliding is found according to the provisions of ASCE 41 as follows $V_{bjs} = v_{mL}A_n$ where $v_{mL} = 0.75\dfrac{0.75v_{tL} + \dfrac{P_D}{A_n}}{1.5}$ In general, the nominal shear capacity of the URM wall for bed-joint sliding can be determined according to ASCE 41 and/or the applicable local building code.	$V_{bjs} = (0.064\text{ MPa})(410{,}400\text{ mm}^2) = 26{,}266\text{ N}$ $v_{mL} = 0.75\dfrac{0.75(0.10\text{ MPa}) + \dfrac{4000\text{ N/m} \times 5.40\text{ m}}{2(38\text{ mm} \times 5400\text{ mm})}}{1.5} = 0.064\text{ MPa}$
STEP 2: Compute the shear capacity due to diagonal tension, V_{dt}.	
The nominal shear capacity of the URM wall for diagonal tension is found according to the provisions of FEMA 356 as follows $V_{dt} = f'_{dt} \cdot A_n\left(\dfrac{L}{h_{eff}}\right)\sqrt{1 + \dfrac{f_a}{f'_{dt}}}$ where f'_{dt} may be replaced by v_{mL} calculated in the previous step, L is length of the wall assumed equal to 5.4 m (Fig. 13.1(a)); h_{eff} is the height to resultant of lateral force equal to 4.88 m (Fig. 13.1(a)); and f_a represents the axial compressive stress due to gravity load, computed as $f_a = \dfrac{P_D}{A_n}$	$V_{dt} = (0.064\text{ MPa})(410{,}400\text{ mm}^2)\left(\dfrac{5.40\text{ m}}{4.90\text{ m}}\right)\sqrt{1 + \dfrac{0.053\text{ MPa}}{0.064\text{ MPa}}} = 39{,}137\text{ N}$ $f_a = \dfrac{4000\text{ N/m} \times 5.40\text{ m}}{410{,}400\text{ mm}^2} = 0.053\text{ MPa}$
STEP 3: Compute the shear capacity due to toe crushing, V_{tc}.	
The nominal shear capacity of the URM wall for toe crushing is found according to the provisions of ASCE 41 as follows $V_{tc} = \alpha \cdot Q_G\left(\dfrac{L}{h_{eff}}\right)\left(1 - \dfrac{f_a}{0.7f'_m}\right)$ where L and h_{eff} are as defined in the previous step, and f_a is calculated as follows $f_a = \dfrac{Q_G}{A_n}$ In general, the nominal shear capacity of the URM wall for toe crushing can be determined according to ASCE 41 and/or the applicable local building code.	$V_{tc} = (1.00)(5800\text{ N/m} \times 5.40\text{ m})\left(\dfrac{5.40\text{ m}}{4.90\text{ m}}\right)\left(1 - \dfrac{0.079\text{ MPa}}{0.7 \times 10\text{ MPa}}\right) = 34{,}126\text{ N}$ $f_a = \dfrac{5800\text{ N/m} \times 5.40\text{ m}}{410{,}400\text{ mm}^2} = 0.076\text{ MPa}$
STEP 4: Compute the as-built shear capacity, ϕV_n^{URM}.	
As indicated in Section 10.3, the nominal shear capacity or nominal lateral strength of the URM wall V_n^{URM} is obtained as the minimum of Eq. (10-1) $V_n^{URM} = \min(V_{bjs}, V_{dt}, V_{tc})$ Assuming $\phi = 0.8$, the design lateral strength can be computed as ϕV_n^{URM}.	$V_n^{URM} = \min(26{,}266\text{ N}, 39{,}137\text{ N}, 34{,}126\text{ N}) = 26{,}266\text{ N}$ $\phi V_n^{URM} = 0.8(26{,}266\text{ N}) = 21{,}013\text{ N}$ As $\phi V_n^{URM} < V_u$, the wall needs strengthening for in-plane loads
STEP 5: Compute the design mechanical properties of the NSM FRP system.	
Assuming an environmental reduction factor C_E equal to 0.65, the FRP design tensile strength and strain can be determined from Eq. (8-3) and (8-4) $f_{fu} = C_E f_{fu}^*$ $\varepsilon_{fu} = C_E \varepsilon_{fu}^*$	$f_{fu} = (0.65)(827\text{ MPa}) = 538\text{ MPa}$ $\varepsilon_{fu} = (0.65)(0.02) = 0.013$

PROCEDURE	CALCULATION IN SI UNITS
STEP 6: Compute the effective strain and stress for shear-controlled failure mode.	
To determine the effective strain in the FRP reinforcement, compute the area of FRP reinforcement as follows $A_f = \frac{H_{roof}}{s} A_{f,bar}$	$A_f = \frac{4900\text{ mm}}{400\text{ mm}} 32\text{ mm}^2 = 392\text{ mm}^2$
The value of ω_f is defined in Eq. (8-12) $\omega_f = \frac{1}{85} \frac{A_f E_f}{A_n \sqrt{f'_m}}$	$\omega_f = \frac{1}{85} \frac{(392\text{ mm}^2)(41{,}000\text{ MPa})}{(410{,}400\text{ mm}^2)\sqrt{10\text{ MPa}}} = 0.146$
According to Section 8.4.2, the effective strain and stress in the FRP reinforcement can be found from Eq. (8-10) and (8-11) as a function of κ_v (Eq.(8-13)) $\kappa_v = \begin{cases} 0.40 & \text{for } \omega_f \le 0.20 \\ 0.64 - 1.2\omega_f & \text{for } 0.20 < \omega_f \le 0.45 \\ 0.10 & \text{for } \omega_f > 0.45 \end{cases}$	$\kappa_v = 0.40$
$\varepsilon_{fe} = \kappa_v \varepsilon_{fu}^* \le C_E \varepsilon_{fu}^*$ $f_{fe} = E_f \varepsilon_{fe}$	$\varepsilon_{fe} = (0.40)(0.02) = 0.008 < 0.013$ $f_{fe} = (41{,}000\text{ MPa})(0.008) = 328\text{ MPa}$
STEP 7: Compute the FRP contribution to the shear strength, V_f.	
The total force per bar that the FRP system can transfer to the masonry substrate is given by Eq. (8-14) as follows: $p_{fv} = A_{f,bar} f_{fe} \le 10{,}000$ lb/bar	$p_{fv} = (32\text{ mm}^2)(328\text{ MPa}) = 10{,}496\text{ N/bar} < 44{,}500\text{ N/bar}$
The NSM FRP contribution to the shear strength can be determined from Eq. (10-4) as follows $V_f = p_{fv} \frac{d_v}{s_f}$	$V_f = (10{,}496\text{ N}) \frac{5400\text{ mm}}{400\text{ mm}} = 141{,}696\text{ N}$
where $d_v = \min(H, L)$; and s_f represents the spacing of the NSM FRP reinforcement equal to the spacing of every other course of masonry unit, s.	$d_v = \min(5.50\text{ m}, 5.40\text{ m}) = 5.40\text{ m}$
STEP 8: Compute the nominal shear strength of the FRP-strengthened wall.	
According to Eq. (10-3), the nominal shear strength of the FRP-strengthened wall is $V_{n,s} = V_n^{URM} + V_f$	$V_{n,s} = (26{,}266\text{ N}) + (141{,}696\text{ N}) = 167{,}962\text{ N}$

PROCEDURE	CALCULATION IN SI UNITS
STEP 9: Compute the nominal flexural strength of the FRP reinforced wall under in-plane loads.	
As the design lateral strength determined in the previous step is larger than the lateral strength due to toe crushing of the unreinforced wall computed in Step 3, a check needs to be performed to ensure that the FRP reinforcement designed in Section 13.1 will prevent the rupture of the wall due to toe crushing. To this extent, the FRP reinforcement designed in Section 13.1 should effectively be anchored as suggested in Section 11.4. The maximum lateral force $V_{n,f}$ that the wall can sustain before flexural failure is computed as follows $V_{n,f} = \dfrac{M_n}{k \cdot H_{roof}}$ where M_n is the nominal moment capacity computed according to Eq. (10-7) using a trial-and-error procedure that allows the determination of the neutral axis depth, c = 324 mm, and gives M_n = 449,115 N·m. By installing an extra layer of GFRP vertical reinforcement 125 mm wide spaced at 0.90 m on-center on both sides of the wall (Fig. 13.5), the neutral axis depth results in c = 611 mm and M_n = 845,465 N·m. 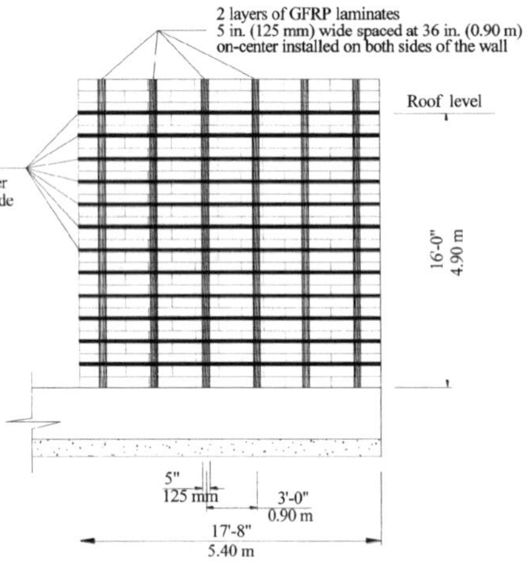 *Fig. 13.5—Schematic of the strengthened wall for in-plane loads with the additional layer of vertical reinforcement.*	$V_{n,f} = \dfrac{449{,}155\ \text{N·m}}{1.0 \times 4.90\ \text{m}}$ = 91,656 N < $V_{n,s}$ computed in Step 8 (not good) $V_{n,f} = V_{n,f} = \dfrac{845{,}465\ \text{N·m}}{1.0 \times 4.90\ \text{m}}$ = 172,544 N > $V_{n,s}$ computed in Step 8 (OK)
STEP 10: Compute the design lateral strength of the FRP-strengthened wall.	
The nominal lateral strength of the FRP-strengthened wall is obtained as the minimum of the nominal shear strength as determined in Step 8 and the nominal lateral strength corresponding to flexural failure of the FRP-strengthened wall as determined in Step 9 $V_n = \min(V_{n,s}, V_{n,f})$ As failure is controlled by the shear strength, the strength reduction factor is taken equal to 0.8 (Section 10.4).	V_n = min(167,962 N, 172,544 N) = 167,962 N ϕV_n = (0.8)(167,962 N) = 134,370 N > V_u = 90,000 N

CHAPTER 14—REFERENCES

14.1—Referenced standards and reports

The standards and reports listed below were the latest editions at the time this guide was prepared. Because these documents are revised frequently, the reader is advised to contact the proper sponsoring group if it is desired to refer to the latest version.

American Concrete Institute

440R Report on Fiber-Reinforced Polymer (FRP) Reinforcement for Concrete Structures

440.1R Guide for the Design and Construction of Structural Concrete Reinforced with FRP Bars

440.2R Guide for the Design and Construction of Externally Bonded FRP Systems for Strengthening Concrete Structures

440.3R Guide Test Methods for Fiber-Reinforced Polymers (FRPs) for Reinforcing or Strengthening Concrete Structures

503R Use of Epoxy Compounds with Concrete

530/530.1 Building Code Requirements for Masonry Structures

American Society of Civil Engineers

7-05 Minimum Design Loads for Buildings and Other Structures

41-06 Seismic Rehabilitation of Existing Buildings

ASTM International

D3039/D3039M Standard Test Method for Tensile Properties of Polymer Matrix Composite Materials

D3418 Standard Test Method for Transition Temperatures of Polymers by Differential Scanning Calorimetry

D4065 Standard Practice for Plastics: Dynamic Mechanical Properties: Determination and Report of Procedures

D4541/D4541M Standard Test Method for Pull-Off Strength of Coatings Using Portable Adhesion Testers

D7205/D7205M Standard Test Method for Tensile Properties of Fiber Reinforced Polymer Matrix Composite Bars

D7565/D7565M Standard Test Method for Determining Tensile Properties of Fiber Reinforced Polymer Matrix Composites Used for Strengthening of Civil Structures

Canadian Standards Association

S806 Design and Construction of Building Components with Fiber-Reinforced Polymers

Federal Emergency Management Agency (FEMA)

306 Evaluation of Earthquake Damaged Concrete and Masonry Wall Buildings—Basic Procedures Manual

356 Prestandard and Commentary for the Seismic Rehabilitation of Buildings

These publications may be obtained from these organizations:

American Concrete Institute
38800 Country Club Drive
Farmington Hills, MI 48331
www.concrete.org

American Society of Civil Engineers
1801 Alexander Bell Dr.
Reston, VA 20191
www.asce.org

ASTM International
100 Barr Harbor Dr.
West Conshohocken, PA 19428
www.astm.org

Canadian Standards Association
5060 Spectrum Way, Suite 100
Mississauga, ON
L4W 5N6 Canada
www.csa.ca

Federal Emergency Management Agency
FEMA Distribution Center
PO Box 430
Buckeystown, MD 21717
www.fema.gov

14.2—Cited references

Albert, L. M.; Elwi, A. E.; and Cheng, J. J., 2001, "Strengthening of Unreinforced Masonry Walls Using FRPs," *Journal of Composites for Construction*, ASCE, V. 5, No. 2, pp. 76-84.

Al-Chaar, G. K., and Hasan, H. A., 1999, "Seismic Testing and Dynamic Analysis of Masonry Bearing and Shear Walls Retrofitted with Overlay Composite," *Proceedings of the 1999 International Composites Expo*, Society of the Plastic Industry, Cincinnati, OH, May 10-12, pp. 1-10.

ANSI Z-129.1, 2006, "Hazardous Industrial Chemicals Precautionary Labeling," American National Standards Institute, Washington, DC.

Bajpai, K., and Duthinh, D., 2003, "Flexural Strengthening of Masonry Walls with External Composite Bars," *The Masonry Society Journal*, V. 21, No. 1, pp. 9-20.

Bank, L. C., 2006, *Composites for Construction: Structural Design with FRP Materials*, Wiley, New York, 236 pp.

Bastidas, E. D.; Molina, L. F.; Yamin, L.; Sarria, A.; Ortega, J. P.; and Tumialan, J. G., 2002, "Non-Structural Masonry Walls Strengthened with GFRP Laminates," *9th North American Masonry Conference*, Clemson, SC, June 1-4, 11 pp. (CD-ROM)

Bergamo, G.; Eusebio, M.; Manfredi, G.; and Prota, A., 2006, "Shake Table Test on a Tuff Masonry Building," *8th US National Conference on Earthquake Engineering*, Paper 698, Earthquake Engineering Research Institute, San Francisco, CA, Apr. 18-22, 10 pp. (CD-ROM)

CFR 16, 2008, "Part 1500: Hazardous Substances and Articles; Administration and Enforcement Regulations," Code of Federal Regulations, V. 73, No. 228, pp. 71, 545.

CFR 49, 2008, "Chapter C, Transportation," Code of Federal Regulations, 464 pp.

CNR-DT 200, 2004, "Guide for the Design and Construction of Externally Bonded FRP Systems for Strengthening Existing Structures," Italian National Research Council, Rome, Italy, 38 pp.

Corradi, M.; Borri, A.; and Vignoli, A., 2002, "Strengthening Techniques Tested on Masonry Structures Struck by the Umbria-Marche Earthquake of 1997-1998," *Construction and Building Materials*, V. 16, No. 4, pp. 229-239.

De Lorenzis, L., and Nanni, A., 2001, "Characterization of FRP Rods as Near Surface Mounted Reinforcement," *Journal of Composites for Construction*, V. 5, No. 2, pp. 114-121.

Ehsani, M. R., 1993, "Glass-Fiber Reinforcing Bars," *Alternative Materials for the Reinforcement and Prestressing of Concrete*, J. L. Clarke, Blackie Academic & Professional, London, England, pp. 35-54.

Ehsani, M. R., and Saadatmanesh, H., 1996, "Seismic Retrofitting of URM Walls with Fiber Composites," *Journal of the Masonry Society*, V. 14, No. 2, pp. 63-72.

El-Gawady, A.; Lestuzzi, P.; and Badoux, M., 2005, "In-Plane Seismic Response of URM Walls Upgraded with FRP," *Journal of Composites in Construction*, V. 9, No. 6, pp. 524-535.

Galati, N., 2002, "Arching Effect on Masonry Walls Strengthened with FRP Materials," master's thesis, University of Missouri-Rolla, Rolla, MO, 124 pp.

Galati, N.; Garbin, E.; Tumialan, J. G.; and Nanni, A., 2005, "Design Guidelines for Masonry Structures: Out of Plane Loads," *7th International Symposium on Fiber-Reinforced (FRP) Polymer Reinforcement for Concrete Structures*, SP-230, C. Shield, J. P. Busel, S. L. Walkup, and D. D. Gremel, eds., American Concrete Institute, Farmington Hills, MI, pp. 269-288.

Galati, N.; Micelli, F.; Tumialan, G. J.; La Tegola, A.; and Nanni, A., 2004, "Comparison between FRP-Strengthening Techniques on the Out-of-Plain Behaviour of URM Masonry Walls," *Proceedings of the First International Conference on Innovative Materials and Technologies for Construction and Restoration,* A. La Tegola and A. Nanni, eds., Lecce, Italy, June 6-9, 2004, V. 2, pp. 440-457.

Galati, N.; Tumialan, J. G.; and Nanni, A., 2006, "Strengthening with FRP Bars of URM Walls Subject to Out-of-Plane Loads," *Construction and Building Materials*, V. 20, No. 1-2, pp. 101-110.

Gergely, J., and Young, D. T., 2001, "Masonry Wall Retrofitted with CFRP Materials," *Proceedings of Composites in Construction*, J. Figueiras, L. Juvandes, and R. Furia, eds., Porto, Portugal, pp. 565-569.

Grando, S.; Valluzzi, M. R.; Tumialan, J. G.; and Nanni, A., 2003, "Shear Strengthening of URM Clay Walls with FRP Systems," *Proceedings of the International Symposium FRPRCS-6*, Singapore, V. 2, pp. 1229-1238.

Gu, X. L.; Ouyang, Y.; Zhang, W. P.; and Ye, F. F., 2003, "Seismic Behaviour of Masonry Structural Walls Strengthened with CFRP Plates," *Proceedings of the Sixth International Symposium on FRP Reinforcement for Concrete Structures,* K.-H. Tan, ed., Singapore, July 8-10, 2003, V. 2, pp. 1259-1268.

Hamilton III, H. R., and Dolan, C. W., 2001, "Flexural Capacity of Glass FRP Strengthened Concrete Masonry Walls," *Journal of Composites for Construction*, V. 5, No. 3, pp. 170-178.

Hamoush, S.; Mlakar, P.; McGinley, M.; and Salami, R., 2001, "Out-of-Plane Shear Strength of Masonry Walls," *Proceedings of the International Conference on FRP Composites in Civil Engineering*, J. G. Teng, ed., Hong Kong, China, V. 2, Dec. 12-15, pp. 1009-1018.

Hartley, A.; Mullins, G.; and Sen, R., 1996, "Repair of Concrete Masonry Block Walls Using Carbon Fiber," *Advanced Composite Materials in Bridges and Structures*, M. El-Badry, ed., Canadian Society of Civil Engineers, Montreal, QC, Canada, pp. 795-802.

Hassan, T., and Rizkalla, S., 2003, "Investigation of Bond in Concrete Structures Strengthened with Near Surface Mounted CFRP Strips," *Journal of Composites for Construction*, V. 7, No. 3, pp. 248-257.

Karbhari, V. M., ed., 2007, *Durability of Composites for Civil Structural Applications*, Woodhead Publishing, pp. 1-79.

Magenes, G., and Calvi, G. M., 1997, "In-Plane Seismic Response of Brick Masonry Walls," *Earthquake Engineering and Structural Dynamics*, V. 26, pp. 1091-1112.

Malvar, L., 1998, "Durability of Composites in Reinforced Concrete," *Proceedings of the First International Conference on Durability of Composites for Construction*, Sherbrooke, Canada, Aug., pp. 361-372.

Marcari, G.; Manfredi, G.; Prota, A.; and Pecce, M., 2007, "In-Plane Shear Performance of Masonry Panels Strengthened with FRP," *Composites Part B*, V. 38, No. 7-8, pp. 887-901.

Marshall, O. S., and Sweeney, S. C., 2002, "In-Plane Shear Performance of Masonry Walls Strengthened with FRP," *Proceedings of the 47th International SAMPE Symposium and Exhibition*, Long Beach, CA, May, pp. 287-295.

Marshall, O. S.; Sweeney, S. C.; and Trovillion, J. C., 1999, "Seismic Rehabilitation of Unreinforced Masonry Walls," *Fourth International Symposium on Fiber Reinforced Polymer Reinforcement for Reinforced Concrete Structures*, SP-188, American Concrete Institute, Farmington Hills, MI, pp 287-295.

Mindess, S., and Young, J., 1981, *Concrete*, Prentice-Hall, Englewood Cliffs, NJ, 671 pp.

Moon, F. L.; Yi, T.; Leon, R. T.; and Kahn, L. F., 2003, "Large-Scale Tests of an Unreinforced Masonry Low-Rise Building," *9th North American Masonry Conference*, Clemson, SC, June 1-4, 12 pp.

Moon, F. L.; Yi, T.; Leon, R. T.; and Kahn, L. F., 2006, "Recommendations for the Seismic Evaluation and Retrofit of Low-Rise URM Structures," *Journal of Structural Engineering*, V. 132, No. 5, pp. 663-672.

Moon, F. L.; Yi, T.; Leon, R. T.; and Kahn, L. F., 2007, "Testing of a Full-Scale Unreinforced Masonry Building Following Seismic Strengthening," *Journal of Structural Engineering*, V. 133, No. 9, pp. 1215-1226.

Morbin, A., and Nanni, A., 2002, "Strengthening of Masonry Elements with FRP Composites," *Report* CIES 02-23, Center for Infrastructure Engineering Studies, University of Missouri-Rolla, Rolla, MO, 264 pp.

Mutsuyoshi, H.; Uehara, K.; and Machida, A., 1990, "Mechanical Properties and Design Method of Concrete Beams Reinforced with Carbon Fiber Reinforced Plastics," *Transaction of the Japan Concrete Institute*, Japan Concrete Institute, Tokyo, Japan, V. 12, pp. 231-238.

Myers, J. J., and Ekenel, M., 2005, "Performance of FRP-Concrete Bond Subjected to Environmental Conditioning during Installation Process," *7th International Symposium on Fiber-Reinforced (FRP) Polymer Reinforcement for Concrete Structures*, SP-230, C. Shield, J. P. Busel, S. L. Walkup, and D. D. Gremel, eds., American Concrete Institute, Farmington Hills, MI, pp. 1447-1464.

Nanni, A., and Gold, W., 1998, "Strength Assessment of External FRP Reinforcement," *Concrete International*, V. 20, No. 6, June, pp. 39-42.

Parretti, R.; Nanni, A.; Prota, A.; Manfredi, G.; and Giarletta, I., 2004, "Preservation of Historical Masonry Buildings using GFRP Combined with Sprayed Polyurea," *2nd Conference on Mechanics of Masonry Structures Strengthened with FRP Materials*, Venezia, Italy, pp. 281-291.

Prota, A.; Manfredi, G.; and Nardone, F.; 2008, "Assessment of Design Formulas for In-Plane FRP Strengthening of Masonry Walls," *Journal of Composites in Construction*, V. 12, No. 6, pp. 643-649.

Santa Maria, H.; Duarte, G.; and Garib, A., 2004, "Experimental Investigation of Masonry Panels Externally Strengthened with CFRP Laminates and Fabric Subjected to In-Plane Shear Load," *13th Word Conference on Earthquake Engineering*, Vancouver, BC, Canada, 10 pp.

Santa Maria, H.; Alcaino, P.; and Luders, C., 2006, "Experimental Response of Masonry Walls Externally Reinforced with Carbon Fiber Fabrics," *Paper* No. 1402, 8th U.S. National Conference on Earthquake Engineering, San Francisco, CA, 10 pp.

Schwegler, G., 1995, "Masonry Construction Strengthened with Fiber Composites in Seismically Endangered Zones," *Proceedings of the Tenth European Conference on Earthquake Engineering*, Rotterdam, Netherlands, pp. 2299-2303.

Senescu, R., and Mosalam, K., 2004, "Retrofitting of Unreinforced Masonry Walls Using Fiber Reinforced Polymer Laminates," *Report* UCB/SEMM 2004/3, University of California, Berkeley, CA, 80 pp.

Shrive, N. G., 2006, "The Use of Fibre Reinforced Polymers to Improve Seismic Resistance of Masonry," *Construction and Building Materials*, V. 20, No. 4, pp. 269-277.

Stratford, T.; Pascale, G.; Manfroni, O.; and Bonfiglioli, B., 2004, "Shear Strengthening Masonry Panels with Sheet Glass-Fiber Reinforced Polymer," *Journal of Composites for Construction*, V. 8, No. 5, Sept.-Oct., pp. 434-443.

Tan, K. H., and Patoary, M. K. H., 2004, "Strengthening of Masonry Walls against Out-of-Plane Loads using FRP Reinforcement," *Journal of Composites for Construction*, V. 8, No. 1, Jan.-Feb., pp. 79-87.

Tinazzi, D., and Nanni, A., 2000, "Assessment of Technologies of Masonry Retrofitted with FRP," *Report* No. CIES 00-18, Center for Infrastructure Engineering Studies, University of Missouri-Rolla, Rolla, MO, 148 pp.

To-Nan, C.; Ja-Shian, C.; Yen-Chih, L.; and Yu-Chun, L., 2006, "Performance of URM Walls Strengthened with CFRP Under Horizontal Cyclic Loading," *First European Conference on Earthquake Engineering and Seismology*, Geneva, Switzerland, Sept. 3-8, 9 pp.

Triantafillou, T. C., 1998, "Strengthening of Masonry Structures Using Epoxy-Bonded FRP Laminates," *Journal of Composites for Construction*, V. 2, No. 2, pp. 96-104.

Tumialan J. G., 2001, "Strengthening of Masonry Structures with FRP Composites," doctoral dissertation, University of Missouri-Rolla, Rolla, MO, 186 pp.

Tumialan, J. G.; Morbin, A., Nanni A.; and Modena, C., 2001a, "Shear Strengthening of Masonry Walls with FRP Composites," *COMPOSITES 2001 Convention and Trade Show*, Composites Fabricators Association, Tampa, FL, Oct. 3-6, 6 pp. (CD-ROM)

Tumialan, J. G.; Huang, P. C.; Nanni, A.; and Silva, P., 2001b, "Strengthening of Masonry Walls by FRP Structural Repointing," *Non-metallic Reinforcement for Concrete Structures*, Cambridge, England, July 16-18, pp. 1033-1042.

Tumialan, J. G.; Galati, N.; Namboorimadathil, S. M.; and Nanni, A., 2002, "Strengthening of Masonry with FRP Bars," *Third International Conference on Composites in Infrastructure*, San Francisco, CA, June 10-12, 11 pp. (CD-ROM)

Tumialan, J. G.; Galati, N.; and Nanni, A., 2003a, "Field Assessment of URM Walls Strengthened with FRP Laminates," *Journal of Structural Engineering,* V. 129, No. 8. pp. 1047-1056.

Tumialan, J. G.; Galati, N.; and Nanni, A., 2003b, "Fiber-Reinforced Polymer Strengthening of Unreinforced Masonry Walls Subject to Out-of-Plane Loads," *ACI Structural Journal*, V. 100, No. 3, May-June, pp. 321-329.

Tumialan, J. G.; San Bartolome, A.; and Nanni, A., 2003c, "Strengthening of URM Infill Walls by FRP Structural Repointing," *9th North American Masonry Conference*, Clemson, SC, June 1-4, 12 pp. (CD-ROM)

Turco, V.; Galati, N.; Tumialan, J. G.; and Nanni, A., 2003, "Flexural Strengthening of URM Walls with FRP Systems," *Proceedings of the Sixth International Symposium on FRP Reinforcement for Concrete Structures (FRPRCS-6)*, Singapore, pp. 1219-1228.

Valluzzi, M. R.; Tinazzi, D.; and Modena, C., 2002, "Shear Behavior of Masonry Panels Strengthened by FRP Laminates," *Construction and Building Materials*, V. 16, No. 7, pp. 409-416.

Velazquez-Dimas, J.; Ehsani, M.; and Saadatmanesh, H., 2000, "Out-of-Plane Behavior of Brick Masonry Walls Strengthened with Fiber Composites," *ACI Structural Journal*, V. 97, No. 3, May-June, pp. 377-387.

Willis, C. R.; Yang, Q.; Seracino, R.; and Griffith, M. C., 2009, "Damaged Masonry Walls in Two-Way Bending Retrofitted with Vertical FRP Strips," *Construction and Building Materials*, V. 23, No. 4, Apr., pp. 1591-1604.

Wu, W., 1990, "Thermomechanical Properties of Fiber Reinforced Plastics (FRP) Bars," doctoral dissertation, West Virginia University, Morgantown, WV, 292 pp.

Yamaguchi, T.; Kato, Y.; Nishimura, T.; and Uomoto, T., 1997, "Creep Rupture of FRP Rods Made of Aramid, Carbon and Glass Fibers," *Third International Symposium on Non-Metallic (FRP) Reinforcement for Concrete Structures*, Japan Concrete Institute, Tokyo, Japan, V. 2, pp. 179-186.

Yang, X. B.; Nanni, A.; Haug, S., and Sun, C. L., 2002, "Strength and Modulus Degradation of Carbon Fiber-Reinforced Polymer Laminates from Fiber Misalignment," *Journal of Material in Civil Engineering*, V. 14, No. 4, pp. 320-326.

Zhao, T.; Zhang, C. J.; and Xie, J., 2003, "Experimental Study on Earthquake Strengthening of Brick Walls with Continuous Carbon Fibre Sheet," *Masonry International*, V. 16, No. 1, pp. 21-25.

Specification for Carbon and Glass Fiber-Reinforced Polymer (FRP) Materials Made by Wet Layup for External Strengthening of Concrete and Masonry Structures

An ACI Standard

Reported by ACI Committee 440

Carol K. Shield*, Chair

William J. Gold*, Secretary

Tarek Alkhrdaji
Charles E. Bakis*
Lawrence C. Bank
Abdeldjelil Belarbi
Brahim Benmokrane
Luke A. Bisby
Gregg J. Blaszak*
Hakim Bouadi
Timothy E. Bradberry
Gordon L. Brown Jr.
Vicki L. Brown
John P. Busel
Raafat El-Hacha
Garth J. Fallis
Amir Z. Fam
Nabil F. Grace
Mark F. Green
Zareh B. Gregorian
Doug D. Gremel
Shawn P. Gross
H. R. Trey Hamilton III
Issam E. Harik
Kent A. Harries
Mark P. Henderson
Bohdan N. Horeczko
Michael W. Lee
Maria E. Lopez de Murphy
Ibrahim M. Mahfouz
Amir Mirmiran
John J. Myers
Antonio Nanni*
Ayman M. Okeil
Carlos E. Ospina
Renato Parretti
Max L. Porter
Andrea Prota
Hayder A. Rasheed
Sami H. Rizkalla
Rajan Sen
Rudolf Seracino
Pedro F. Silva
Khaled A. Soudki
Samuel A. Steere III
Jay Thomas
Houssam A. Toutanji
J. Gustavo Tumialan
Milan Vatovec
David White
Sarah E. Witt†

Consulting members
P. N. Balaguru
Craig A. Ballinger
Harald G. F. Budelmann
C. J. Burgoyne
Elliot P. Douglas
Rami M. Elhassan
David M. Gale
Russell Gentry
Arie Gerritse
Srinivasa L. Iyer
Koichi Kishitani
Howard S. Kliger
Kyuichi Maruyama
Antoine E. Naaman
Hajime Okamura
Mark A. Postma
Ferdinand S. Rostasy
Surendra P. Shah
Mohsen Shahawy
Yasuhisa Sonobe
Minoru Sugita
Luc R. Taerwe
Ralejs Tepfers
Taketo Uomoto
Paul Zia

*Contributing authors.
†Task Group Chair.

This material specification covers the requirements for carbon and glass fiber-reinforced polymer (FRP) systems made by the wet layup process. These systems are externally bonded to concrete or masonry structures and intended for the structural strengthening of them.

Keywords: carbon fiber; fiber-reinforced polymer (FRP); glass fiber; masonry; wet layup.

CONTENTS

ACI 440.8 was adopted November 29, 2013 and published January 2014.

1—SCOPE

1.1 This specification describes permitted constituent materials, minimum performance requirements of those constituent materials, and minimum performance requirements for carbon fiber-reinforced polymer (CFRP) and glass fiber-reinforced polymer (GFRP) laminae made from those constituent materials using the wet layup process.

1.2 This specification only covers the fabric reinforcement and saturating resin that comprise the FRP system. Primers and putty fillers are excluded.

1.3 This specification applies to FRP laminae consisting of one ply of continuous, unidirectional fiber reinforcement and saturating resin fabricated using the wet layup process. Laminae with fibers in more than one direction that are intended to provide strength in more than one direction are excluded.

1.4 Precured FRP systems are not covered by this specification.

1.5 Hybrid FRP systems consisting of more than one type of fiber are excluded.

1.6 Only FRP laminae with epoxy resins are covered by this specification.

1.7 Values in this specification are stated in in.-lb units. A companion specification in SI units is available.

2—DEFINITIONS

2.1 The following definitions govern in this specification. For definitions not given below, refer to "ACI Concrete Terminology," www.concrete.org/Tools/ConcreteTerminology.aspx. For definitions used in but not specific to this specification, refer to ASTM C904.

fabric reinforcement—reinforcing fibers in fabric form.

fiber-reinforced polymer lamina—a single layer of composite material made by combining fabric reinforcement and saturating resin matrix.

fiber-reinforced polymer system—the product fabricated using the wet layup method with the FRP system manufacturer's fiber reinforcement and saturating resin per the FRP system manufacturer's instructions.

mean property—a property value provided by the manufacturer no greater than the mean of results obtained by a specified test method for at least the specified number of specimens. *Discussion*—This value is applicable to FRP chord tensile stiffness, resin properties, and for calculating retention values for durability.

ply—a single layer of fiber reinforcement.

roving—a parallel bundle of continuous yarns, tows, or fibers with little or no twist.

tow—an untwisted bundle of continuous filaments.

ultimate property—a property value provided by the manufacturer no greater than the mean minus three standard deviations of results obtained by a specified test method for at least the specified number of specimens. *Discussion*—This definition is applicable to maximum tensile force.

wet layup—a manufacturing process where dry fabric fiber reinforcement is impregnated on-site with a saturating resin matrix and then cured in-place.

3—REFERENCED STANDARDS

3.1 *ASTM International*

C581-03(2008)—Standard Practice for Determining Chemical Resistance of Thermosetting Resins Used in Glass-Fiber-Reinforced Structures Intended for Liquid Service

C904-01(2012)—Standard Terminology Relating to Chemical-Resistant Nonmetallic Materials

D638-10—Standard Test Method for Tensile Properties of Plastics

D790-10—Standard Test Methods for Flexural Properties of Unreinforced and Reinforced Plastics and Electrical Insulating Materials

D1141-98(2013)—Standard Practice for the Preparation of Substitute Ocean Water

D2247-11—Standard Practice for Testing Water Resistance of Coatings in 100% Relative Humidity

D3045-92(2010)—Standard Practice for Heat Aging of Plastics Without Load

D3776/D3776M-09a(2013)—Standard Test Methods for Mass Per Unit Area (Weight) of Fabric

D7565/D7565M-10—Standard Test Method for Determining Tensile Properties of Fiber Reinforced Polymer Matrix Composites Used for Strengthening of Civil Structures

E1640-13—Standard Test Method for Assignment of the Glass Transition Temperature by Dynamic Mechanical Analysis

G153-13—Standard Practice for Operating Enclosed Carbon Arc Light Apparatus for Exposure of Nonmetallic Materials.

3.2 *International Code Council—Evaluation Service (ICC-ES)*

AC125—Acceptance Criteria for Concrete and Reinforced and Unreinforced Masonry Strengthening Using Externally Bonded Fiber-Reinforced Polymer (FRP) Composite Systems

4—CLASSIFICIATION

4.1 Fiber-reinforced polymer systems shall be classified according to their fiber reinforcement composition with the first letter of the acronym designating fiber type as follows:

a) CFRP—Carbon fiber-reinforced polymer systems;

b) GFRP—Glass fiber-reinforced polymer systems

5—MATERIALS AND MANUFACTURE

5.1 *Fiber reinforcement*

5.1.1 Fiber reinforcement shall be arranged in a usable form as either continuous glass fiber rovings or carbon fiber tows of a single size and areal weight.

5.1.2 Rovings or tows shall be assembled by the fabric manufacturer into fabric reinforcement such that the rovings or tows are aligned in a single direction along the fabric length. Fiber of any material running at an angle to the rovings or tows is permitted to hold the fabric together.

5.2 *Resin*

5.2.1 Only epoxy resins shall be used for the saturating resin.

5.2.2 Resins shall not be diluted with organic solvents such as thinner.

6—PHYSICAL PROPERTIES

6.1 *Glass transition temperature*

6.1.1 Glass transition temperature of the FRP lamina shall be determined from specimens at a frequency defined in 9.1.

6.1.2 The glass transition temperature of the FRP lamina shall not be less than 140°F. The glass transition temperature shall be measured according to the dynamic mechanical analysis (DMA) method in ASTM E1640 on a coupon cut from panels produced in accordance with ASTM D7565/D7565M. Manufacturer shall report any pretreatments used.

6.2 *Areal weight*

6.2.1 Areal weight of dry fabric shall be determined from specimens at a frequency defined in 9.1.

6.2.2 The areal weight of dry fabric shall be determined by measuring the mass per unit area according to ASTM D3776/D3776M Option C and reported in oz/yd^2.

7—MECHANICAL PROPERTIES

7.1 *Mechanical properties of saturating resins*

7.1.1 Mechanical properties of the saturating resins shall be determined from specimens at a frequency defined in 9.2.

7.1.2 Saturating resin shall conform to the requirements for mechanical properties shown in Table 7.1.2.

7.2 *Mechanical properties of FRP lamina*—Mechanical properties of the FRP lamina shall be determined in accordance with ASTM D7565/D7565M on specimens taken from test panels. All properties shall be reported on a per ply basis.

7.2.1 *Ultimate tensile force per unit width*—The ultimate tensile force per unit width shall be determined at a frequency defined in 9.3. The ultimate tensile force per unit width shall be divided by the areal weight of the dry fabric as determined in 6.2 and conform to the requirements shown in Table 7.2.1.

7.2.2 *Chord tensile stiffness per unit width*—The mean chord tensile stiffness per unit width shall be determined at a frequency defined in 9.3. The chord tensile stiffness per unit width shall be divided by the areal weight of the dry fabric as determined in 6.2, and conform to the requirements shown in Table 7.2.1.

Table 7.1.2—Minimum properties for saturating resins

Property	ASTM test method	Mean value
Ultimate tensile strength	D638 Type 1	6000 psi
Tensile modulus	D638 Type 1	250,000 psi
Elongation at failure	D638 Type 1	3 percent
Ultimate flexural strength	D790	10,000 psi
Flexural modulus	D790	250,000 psi

Table 7.2.1—Tensile properties for FRP lamina

System designation	Minimum ultimate* tensile force per areal weight [kip/in./(oz/yd^2)]	Minimum mean chord tensile stiffness per areal weight [kip/in./(oz/yd^2)]
Carbon	0.170	20.0
Glass	0.0625	4.160

*Mean minus three standard deviations.

7.2.3 *Ultimate rupture strain*—The ultimate rupture strain shall be calculated as the ultimate tensile force per unit width divided by the mean chord tensile stiffness per unit width.

8—DURABILITY PROPERTIES

8.1 *FRP lamina*—The FRP lamina shall conform to the retained tensile force per unit width requirements shown in Table 8.1 for the given environmental exposure conditions. The mean tensile force per unit width shall be determined in accordance with ASTM D7565/D7565M. Mean values of test results for the unexposed and exposed specimens, tested at a frequency defined in 9.4, shall be used to calculate percent retention of tensile strength.

9—SAMPLING, FREQUENCY, AND NUMBER OF SPECIMENS

9.1 *Physical properties*

9.1.1 For determining areal weight, at least five specimens shall be tested from five fabric samples.

9.1.2 For determining glass transition temperature, at least five specimens shall be tested from five FRP single- or double-ply lamina panels made in accordance with ASTM D7565/D7565M. At least one specimen for each test shall be taken from each of the five panels.

9.1.3 Tests for physical properties shall be performed at least once every three years for each product and when the constituent materials are modified.

9.2 *Mechanical properties of saturating resins*

9.2.1 Each mechanical property shall be determined as the mean value of test results from at least five specimens.

9.2.2 Tests shall be performed at least once for each resin product and when the constituent materials are modified.

9.3 *Mechanical properties of FRP lamina*

9.3.1 For the determination of tensile stiffness and ultimate force, at least 20 specimens shall be tested from four FRP single- or double-ply lamina panels made in accordance with ASTM D7565/D7565M. At least five specimens shall be taken from each of the four panels. Tensile stiffness per unit width and tensile force per unit width shall be measured

Table 8.1—Minimum mean retained tensile force per unit width of FRP lamina after environmental conditioning*

Environmental exposure conditions	ASTM test method	Conditioning	Percent retention		
			1000 hours	2000 hours	3000 hours
Water resistance	D2247	100 percent relative humidity at 100°F ± 2°F	90	NA	85
Salt water resistance	D1141 C581	Immersion at 73°F ± 2°F	90	NA	85
Alkali resistance	C581	Immersion in $Ca(CO_3)$ at pH = 9.5 and 73°F ± 3°F	90	NA	85
Dry heat resistance	D3045	140°F ± 5°F	90	NA	85
Exterior exposure	G153	2000-hour conditioning as follows: Exposed to cycles consisting of 102 min light and 18 min light and water spray. The black-body temperature is 143°F.	NA	90	NA

*Note: Additional durability documentation, such as freezing-and-thawing resistance, is often requested by the purchaser or a governmental regulatory agency. The freezing-and-thawing test method described in ICC Evaluation Service Acceptance Criteria (AC125) may be used for this purpose.

on each specimen. Results shall be reported on a per-ply basis.

9.3.2 Ultimate tensile force per unit width shall be determined as the mean value of the test results minus three times the standard deviation.

9.3.3 Chord tensile stiffness per unit width shall be determined as the mean value of the test results.

9.3.4 Tests to determine mechanical properties shall be performed at least once every three years for each product and when the constituent materials are modified.

9.4 *Durability properties of FRP lamina*

9.4.1 For the evaluation of strength retention under prescribed exposures, specimens shall be obtained from five FRP single- or double-ply lamina panels made in accordance with ASTM D7565/D7565M. Five specimens shall be tested before exposure and five specimens shall be tested after each exposure condition. All specimens shall have the same number of plies. The mean values of the test results for the five unexposed and five exposed specimens shall be used to calculate percent retention.

9.4.2 Tests to determine the strength retention values under prescribed environments shall be performed at least once for each product and when the constituent materials are modified.

10—PRODUCT DATA SHEETS

10.1 The manufacturer's product data sheets shall include, at a minimum, the following information for the FRP single-ply lamina and the constituent materials:

a) Product name(s);
b) Physical properties of the fabric and lamina;
c) Mechanical properties of the lamina;
d) Mechanical properties of the saturating resin;
e) Any pretreatments used when determining physical or mechanical properties of the lamina;
f) Statement that lamina meets all the requirements of this specification;
g) Coverage rates of resins by weight or volume;
h) Product packaging information;
i) Resin mixing instructions;
j) Approximate pot life of saturating resin based on ambient temperature;
k) Shelf life of all constituent materials;
l) Storage requirements;
m) Safety information.

Guide to Accelerated Conditioning Protocols for Durability Assessment of Internal and External Fiber-Reinforced Polymer (FRP) Reinforcement

Reported by ACI Committee 440

Carol K. Shield, Chair

William J. Gold, Secretary

Tarek Alkhrdaji
Russell Gentry
Maria E. Lopez de Murphy
Rudolf Seracino
Charles E. Bakis
Nabil F. Grace
Ibrahim M. Mahfouz
Venkatesh Seshappa
Lawrence C. Bank
Mark F. Green
Amir Mirmiran
Pedro F. Silva
Abdeldjelil Belarbi
Zareh B. Gregorian
John J. Myers*
Samuel A. Steere III
Brahim Benmokrane
Doug D. Gremel
Antonio Nanni
Jennifer E. Tanner*

Luke A. Bisby
Shawn P. Gross
Ayman M. Okeil
Jay Thomas
Gregg J. Blaszak
H. R. Trey Hamilton, III*
Carlos E. Ospina
Houssam A. Toutanji
Hakim Bouadi
Issam E. Harik
Renato Parretti
J. Gustavo Tumialan
Timothy E. Bradberry
Kent A. Harries
Maria A. Polak
Milan Vatovec
Vicki L. Brown
Mark P. Henderson
Max L. Porter
David White

John P. Busel
Bohdan N. Horeczko†
Andrea Prota
Sarah E. Witt
Raafat El-Hacha
Ravindra Kanitkar
Hayder A. Rasheed
Garth J. Fallis
Yail Jimmy Kim
Sami H. Rizkalla
Amir Z. Fam
Michael W. Lee
Rajan Sen

Consulting Members
P. N. Balaguru
Craig A. Ballinger
Harald G. F. Budelmann
C. J. Burgoyne
Rami M. Elhassan

David M. Gale
Srinivasa L. Iyer
Koichi Kishitani
Howard S. Kliger
Kyuichi Maruyama
Antoine E. Naaman
Hajime Okamura
Mark A. Postma
Ferdinand S. Rostasy
Surendra P. Shah
Mohsen Shahawy
Yasuhisa Sonobe
Minoru Sugita
Luc R. Taerwe
Ralejs Tepfers
Taketo Uomoto
Paul Zia

*Chairs of the subcommittee who prepared this document.
†Deceased.

Fiber-reinforced polymer (FRP) composites, when designed, fabricated, and installed, provide a sustainable and durable reinforcement system for concrete. This document presents guidance for assessing the durability performance of internal and external FRP composite reinforcement using accelerated conditioning protocols (ACPs) in combination with standard test methods for mechanical properties. The objective of ACPs is to enable manufacturers to characterize the durability of their FRP composite products and encourage researchers and testing laboratories to adopt common test protocols to build a meaningful database of durability testing of FRP materials. Results of the tests conducted using the recommended ACPs are not intended to be used in the design of FRP composites as concrete reinforcement. In the future, however, when the relationship between field performance and ACPs is better understood, ACPs may be refined to allow use in quality control and design.

Keywords: accelerated conditioning; bond; durability; externally bonded; fiber-reinforced polymer composites; modulus of elasticity.

CONTENTS

ACI 440.9R-15 was adopted and published May 2015.

CHAPTER 1—INTRODUCTION AND SCOPE

1.1—Introduction

This document is a guide to the assessment of the durability performance of internal and external fiber-reinforced polymer (FRP) composite reinforcement using accelerated conditioning protocols (ACPs) in combination with standard test methods for mechanical properties. The purpose of this guide is to document ACPs so that a standardized method can be created to gather data to eventually be used as a screening or acceptance tool.

FRP composites are increasingly being used in infrastructure applications as reinforcing bars and externally bonded reinforcement for strengthening reinforced concrete elements. The use of FRP composites is predicated on performance attributes linked to their light weight, high stiffness-to-weight and strength-to-weight ratios, ease of installation in the field, potential low system cost, and potentially high overall durability.

FRP composites are used by many industries, including automotive, marine, and aerospace. They have successfully been applied in pipelines, underground storage tanks, building façades, and as architectural components. The materials, loading conditions, and environments seen in many infrastructure applications, however, are unique. Anecdotal evidence provides substantial reason to believe that, if appropriately designed and fabricated, FRP composites can provide longer service life and lower maintenance costs than steel-reinforced structures.

FRP composites have been in use as concrete reinforcement since the 1980s. Consequently, long-term performance field data are limited, making it essential that potential vulnerabilities regarding FRP durability be identified and addressed early to ensure expected long-term service. One means to identify long-term vulnerability is through the use of accelerated conditioning. Few standard protocols for conducting durability testing exist, making it difficult to draw detailed conclusions from the present database of test results generated over the past two decades. Comparing tests conducted at different laboratories is often complicated by the large number of variables among tests.

FRP composite reinforcement embedded in concrete will experience different environmental influences than those experienced by externally bonded FRP composite reinforcement. Externally bonded FRP composite reinforcement is typically exposed directly to ambient environmental conditions where embedded reinforcement is not. In many applications, the bond of externally bonded FRP composite reinforcement is critical to the short- and long-term structural performance of the system. Due to the fundamental difference in exposure conditions of internal and external FRP composite reinforcement, different ACPs and mechanical testing for internal and external FRP composite reinforcement are necessary. In either case, durability, in the context of this guide, is defined as a measure of the retention of FRP physical and mechanical properties when exposed to the ACP environments for the prescribed duration.

An overview of the evaluation process includes the following four elements:

1) Specimen fabrication and preparation—Process used to fabricate the specimen and prepare it for exposure to the ACP.

2) Accelerated conditioning protocol—Sets out the parameters for the environment and stress, including duration, to which the specimen will be exposed (Chapter 4). Additional control specimens are stored in ambient laboratory conditions.

3) Mechanical testing—Tests the accelerated conditioned (AC) and control specimens following the exposure period. Testing is completed under unexposed conditions (Chapters 5 and 6).

4) Residual mechanical property determination—The method used to evaluate the effect of ACP on mechanical properties (Chapters 5 and 6).

1.2—Scope

This document provides guidance on using ACPs and associated standard mechanical test methods to assess the durability of FRP composite reinforcement for concrete with the objective to enable manufacturers to characterize the durability of their FRP products and to encourage researchers and testing laboratories to adopt common test protocols to build a meaningful database of durability test results for FRP materials. Results of the tests conducted using the recommended protocols are not intended for use directly in the design of FRP composites. They are meant to generate a database of consistent test results that can be used in the future to refine the environmental factors recommended in ACI 440.1R, 440.2R, and 440.7R. Recommended environmental reduction factors used in the design of FRP composite reinforcement are provided in ACI 440.1R and 440.2R and are outside the scope of this guide.

The results of the conditioning and testing recommended in this guide are not intended to be useful in the prediction of service life. The use of these methods in conjunction with field evaluations of the performance of structures with FRP composite reinforcement in service conditions is encouraged. Future correlation of accelerated methods with field performance can help validate and improve confidence in their use.

Physicochemical tests that measure the changes in the physical and chemical properties of FRP materials can also be useful in assessing degradation of the material and its constituents after being subjected to the ACPs described in this guide.

CHAPTER 2—NOTATION AND DEFINITIONS

2.1—Notation

b = cross-sectional width of beam specimen

E_{L1} = average elastic modulus of unexposed control specimens, psi (MPa)

E_{L2} = average elastic modulus of accelerated conditioned specimens, psi (MPa)

f_u = ultimate tensile strength of fiber-reinforced polymer reinforcing bar, psi (MPa)

F_{p1} = average pull-off bond strength of unexposed control specimens, lbf (N)

F_{p2} = average pull-off bond strength of accelerated conditioned specimens, lbf (N)

F_{u1} = average tensile strength of unexposed control specimens, lbf (N)

F_{u2} = average tensile strength of accelerated conditioned specimens, lbf (N)

h = cross-sectional height of beam specimen

L = length of beam specimen

P_{b1} = average beam strength of unexposed control specimens, lbf (N)

P_{b2} = average beam strength of accelerated conditioned specimens, lbf (N)

R_{eb} = beam bond retention, percent

R_{em} = elastic modulus retention, percent

R_{ep} = pull-off bond retention, percent

R_{es} = ultimate strain retention, percent

R_{et} = tensile strength retention, percent

W_0 = average unexposed specimen mass, g

W_1 = average specimen mass of accelerated conditioned specimens, g

ε_{u1} = average ultimate strain of unexposed control specimens

ε_{u2} = average ultimate strain of accelerated conditioned specimens

2.2—Definitions

ACI provides a comprehensive list of definitions through an online resource, "ACI Concrete Terminology," http://concrete.org/tools/concreteterminology.aspx. Definitions provided herein complement that source.

aramid fiber—highly oriented organic fiber produced from an aromatic polyamide.

B-stage—intermediate stage in the reaction of certain thermosetting resins in which the material softens when heated and it is plastic and fusible but may not entirely dissolve or fuse; resin in an uncured prepreg is usually at this stage.

carbon fiber—fiber produced by heating organic precursor materials containing a substantial amount of carbon, such as rayon, polyacrylonitrile (PAN), or pitch, in an inert environment.

cure—inducing a reaction leading to cross-linking in a thermosetting resin using chemical initiators, catalysts, radiation, heat, or pressure.

degradation—change in the mechanical or physical properties of a material that results in a detrimental effect on performance.

durability—measure of the retention of fiber-reinforced polymer physical and mechanical properties when exposed to the accelerated conditioning protocol environment for the prescribed duration.

E-glass—family of glass fibers with a calcium alumina borosilicate composition and a maximum alkali content of 2 percent.

fiber-reinforced polymer bar—composite material formed into a long, slender, structural shape suitable for the internal reinforcement of concrete and consisting of primarily longitudinal unidirectional fibers bound and shaped by a rigid polymer resin material. The bar may have a cross section of variable shape, commonly circular or rectangular, and a deformed or roughened surface to enhance bonding with concrete.

fiber-reinforced polymer laminate—one or more layers of fiber reinforcements, such as glass, carbon, and aramid, arranged in one or more orientations—for example, 0, 90, +45, –45 degrees—and held together by a polymer matrix; FRP laminates come in the physical form of dry, prepreg, and precured materials.

glass fiber—fiber drawn from an inorganic fusion of silica (SiO_2) and other compounds that have cooled without crystallization.

impregnate—for fiber-reinforced polymer composites, to saturate the fibers with resin.

precured fiber-reinforced polymer—fully cured fiber-reinforced polymer composite material that is usually made in a factory and brought to the site as a rigid solid.

prepreg fiber-reinforced polymer—reinforcement fabrics for FRP laminates that have been preimpregnated with a resin that usually is cured to an intermediate stage (B-stage).

vinyl ester—thermosetting reaction product of epoxy resin with a polymerizable unsaturated acid that is then diluted with a reactive monomer.

wet layup—manufacturing process where dry fabric fiber reinforcement is impregnated on site with a saturating resin matrix and then cured in place.

CHAPTER 3—DURABILITY OF FIBER-REINFORCED POLYMER COMPOSITES

This chapter provides an introduction of reported factors that impact fiber-reinforced polymer (FRP) composite durability. FRP composite materials are formed through the physical combination of two or more constituent materials, of which at least one is a reinforcing material (for example, fiber) and a polymer matrix material (for example, resin). The matrix, which is typically a thermoset resin such as a vinyl ester or epoxy, binds the fibers together, allowing shear transfer between fibers; provides stability to the slender fibers; and serves as a protective barrier from environmental exposure. E-glass fibers offer high tensile strength and economical cost; however, they are susceptible to degradation due to moisture and alkalinity without the protection of an appropriate resin system. Similarly, aramid fibers are resistant to abrasion and impact, but show a propensity to creep, absorb moisture, and degrade under ultraviolet exposure. Carbon fibers are relatively inert to their environment. The arrangement of fibers in terms of their orientation typically results in FRP composites having anisotropic and heterogeneous material properties.

Durability of the resin system is dependent on several factors, including the resin components and proportions, as well as curing time and environmental exposure conditions. Resin systems available commercially are typically designed for specific end-use applications based on mechanical, physical, chemical, electrical, or other considerations associated with the intended operating environment; formulations are usually proprietary. Mechanical properties of the resins are typically published by manufacturers and are based on standard test methods. Typical properties of FRP composite materials used with concrete structures are found in ACI 440.1R and ACI 440.2R. While the important mechanical properties used for design and quality control are generally addressed within a project specification, durability properties are less so.

Currently, there is lack of consensus on standard test methods that evaluate the durability of resins and fibers. The more prominently investigated conditions affecting FRP durability include moisture (water and salt solution), other chemical solutions, alkaline environment, extreme temperature and thermal cycling, freezing and thawing, creep and relaxation, fatigue, and ultraviolet (UV) radiation. The influence of these conditions is not only dependent upon physical and environmental synergy, but also application. Internal reinforcement is likely to see a less varied environmental exposure than external reinforcement due to the protection provided by the concrete. For more-detailed discussion on FRP durability on both internal and external FRP systems, refer to ACI 440R.

CHAPTER 4—ACCELERATED CONDITIONING

Several environments presented in Chapter 3 adversely affect mechanical properties of fiber-reinforced polymer (FRP) composites over time. One of the inherent difficulties in durability testing is determining the time period required to impose changes in those properties. Ideally, to ensure that FRP composite reinforcement maintains its mechanical properties over the service life of the structure, the system should be built and observed in that environment for the specified time period—an obviously impractical approach. Consequently, as with durability testing of most materials, degradation of material properties is accelerated so that the exposure time is reasonable and practical.

4.1—Background

Bank et al. (1995) evaluated accelerated conditioning methods for determining the long-term performance of FRP composite reinforcement. They indicated that the degradation of fiber-reinforced polymer (FRP) composites could be accelerated with such environments as high or low temperature, moisture, chemicals, and gaseous mixtures, and that exposure to these environments would cause a change in mechanical or physical state. Further degradation could result by adding synergistic effects such as mechanical stress. In accelerated conditioning tests, the material is aged at a higher rate using an accelerating factor (generally temperature) according to the Arrhenius law, which assumes that the degradation process remains unchanged. The effect of temperature is more critical because the chemical degradation rate is not proportional but increases exponentially with the temperature (Robert et al. 2010).

Accelerating conditioning carries the risk that the accelerated process may also change the fundamental degradation mechanism or that the mechanism is not strictly chemical. In bonded FRP applications, the bond between FRP composite and concrete is a combination of hydrogen bonding and mechanical interlock. Consequently, the Arrhenius law is not strictly applicable to the overall degradation of the bond (Tatar et al. 2013). Because accelerated conditioning procedures are typically performed without a direct relation to the field conditions and time scale, accelerated conditions may vary from field conditions, resulting in either unconservative or overly conservative results.

Few standards are available that require evaluation of FRP composite reinforcement mechanical properties after being exposed to accelerated conditioning. AC125 specifies conditioning requirements, mechanical tests, and retention limits for resistance to moisture, saltwater, alkali, dry heat, ultraviolet (UV) light, and freezing and thawing. The specified conditioning protocols are to be applied over durations from 1000 to 10,000 hours. Many of the procedures in AC125 are similar to the accelerated conditioning protocols (ACPs) noted in this guide. One significant difference, however, is that AC125 uses an elevated temperature with water (100 ± 2°F [37.8 ± 1.1°C]) or laboratory temperature with an alkaline solution (73 ± 3°F [23.8 ± 1.7°C]). This guide suggests elevated temperature in conjunction with moisture or chemical solution to accelerate the degradation of the system.

4.2—Accelerated conditioning protocols

Although fiber rupture is a concern, the primary mechanism of degradation for fiber-reinforced polymer (FRP) reinforcing bars is the effect of alkalinity and moisture on the resin. For externally bonded FRP composites, the critical mechanism is considered to be the effect of moisture on the adhesion of FRP composite to concrete. The aim of conditioning specimens is to determine the effect of the environment on the mechanical properties. Sections 4.2.1 through 4.2.4 describe accelerated conditioning protocols (ACPs) designed to accelerate the degradation of mechanical properties. The ACP is a description of the environment and stress that the specimen should be exposed to during the conditioning period. Environment might include moisture by water or chemical solution, temperature, or a combination of them. In general, ACPs could be the environment, stress, or both, applied simultaneously. Although other exposures could be added in the future, this guide addresses environment and stress.

4.2.1 *Standard laboratory conditions*—For the purposes of comparison to accelerated conditioning (AC), standard laboratory temperature should be defined as 73.4 ± 5°F (23 ± 3°C) and standard laboratory humidity should be defined as 50 ± 10 percent relative humidity.

4.2.2 *ACP continuous immersion in water*—Continuous immersion in 122 ± 5°F (50 ± 3°C) potable water.

4.2.3 *ACP continuous exposure to humidity*—Continuous exposure to 100 percent relative humidity at 140 ± 5°F (60 ± 3°C).

4.2.4 *ACP continuous immersion in alkaline solution*—Continuous immersion in 140 ± 5°F (60 ± 3°C) alkaline solution in accordance with Section 11.3 of ASTM D7705/D7705M. Covering the alkaline solution before and during testing should prevent interaction with atmospheric CO_2 and prevent evaporation. Check the solution pH at the end of conditioning.

4.3—Mass change

The amount of moisture absorbed during exposure to the accelerated conditioning protocols (ACPs) may be important to the long-term performance of the fiber-reinforced polymer (FRP) system being tested. For many of the test procedures, mass change measurement can be performed in parallel with the ACP described in 4.2 and mechanical testing described in Chapters 5 and 6 for FRP bars and externally bonded FRP, respectively. If a measurement of the mass change is desired, measure the mass of the dry specimen according to Procedure D of ASTM D5229/D5229M (the initial mass W_0). Then, after immersion for the prescribed period of time, the specimen should be removed from the solution, quickly washed with deionized water, surface dried with tissue paper, and immediately weighed (this is the mass at Time 1, denoted as W_1).

In general, specimens are expected to absorb moisture during exposure. Consequently, a gain is expected over that of the control specimens. Mass loss would indicate a deleterious chemical reaction with the resin or fibers. Mass change should be calculated using Eq. (4.3) as follows

$$\Delta M = \frac{W_1 - W_0}{W_0} \times 100 \qquad (4.3)$$

where a positive sign indicates an increase in mass.

CHAPTER 5—FIBER-REINFORCED POLYMER REINFORCING BAR TESTS

Four test methods can be used individually or combined to determine the environmental effects on mechanical properties of fiber-reinforced polymer (FRP) reinforcing bars. Three of these methods are specified in ASTM D7705/D7705M: 1) unstressed conditioning followed by tension test; 2) conditioning with sustained tension, followed by tension test; and 3) conditioning with sustained tension inside concrete, followed by tension test. Section 5.1 presents the fourth method: the sustained bending stress test. As described in Chapter 3, durability of FRP composite reinforcement embedded in concrete can be affected by a number of environments. The primary concern, however, is the resistance of the FRP composite reinforcement to the alkaline environment caused by the surrounding concrete in combination with moisture.

5.1—Sustained bending stress test

The following procedures are intended to induce flexural stresses in fiber-reinforced polymer (FRP) reinforcing bars during conditioning. FRP bar segments are bent and secured in a curved strain jig in accordance with ASTM D543 to create a specified sustained flexural strain (Fig. 5.1). The recommended accelerated conditioning protocol (ACP) is immersion in a heated alkaline solution (Section 4.2.4) for a specified time period. The unexposed control specimens are associated with Section 4.2.1. Control parameters include solution temperature and sustained stress. Residual tensile strength, elastic modulus, and ultimate strain values are measured using ASTM D7205/D7205M and compared to those of unexposed control specimens. Final testing is conducted at room temperature rather than the elevated exposure temperature. At least five control specimens and five specimens exposed to the specified ACP should be tested.

5.1.1 *Specimen fabrication and preparation*—Specimens should be cut to a length sufficient to fit the strain jig and to be tested in accordance with ASTM D7205/D7205M.

5.1.2 *Accelerated conditioning protocol*—

(a) Accelerated conditioning—Immersion in alkaline solution as described in 4.2.4

(b) Duration—3000 hours

(c) Sustained stress—FRP specimens should be bent and secured in a curved stress jig in accordance with ASTM D543 to create a sustained flexural stress of 20 percent of the ultimate tensile strength f_u of the bar.

5.1.3 *Standard laboratory conditions*

Control specimens—Standard laboratory conditions as described in 4.2.1.

5.1.4 *Mechanical testing*—Following exposure to the specified ACP, the specimens should be carefully removed from the jig, thoroughly rinsed with water, and post-conditioned according to Procedure B of ASTM D618. Grips should be applied and accelerated conditioning specimens and control specimens should be tested in accordance with ASTM D7205/D7205M within 24 hours following completion of post-conditioning.

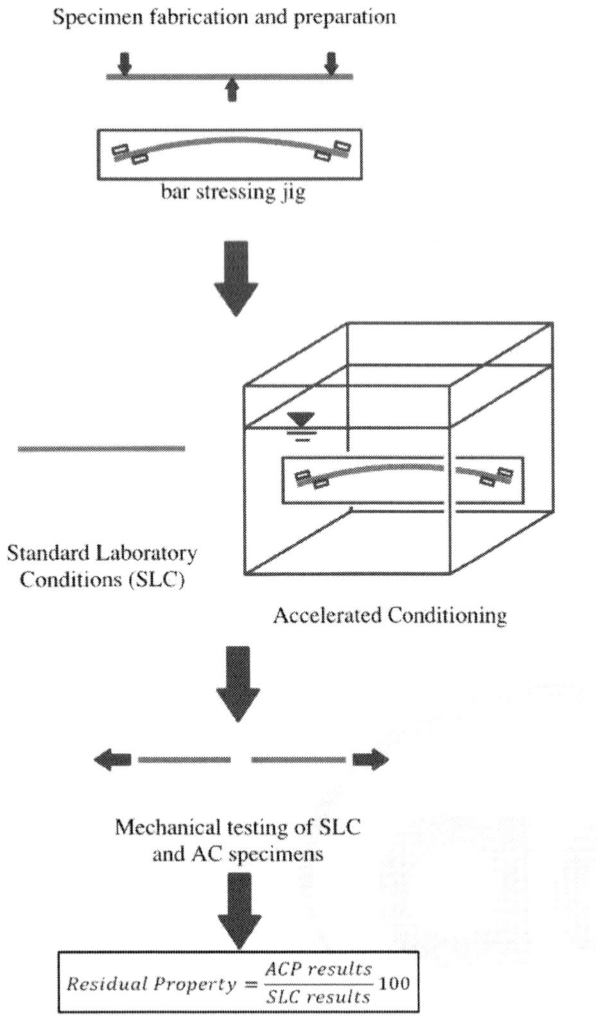

Fig. 5.1—Example procedure for combining accelerated conditioning and mechanical testing for determining residual mechanical properties for reinforcing bars.

5.1.5 *Residual mechanical properties*—The tensile strength retention should be calculated using Eq. (5.1.4a) with a precision of two significant figures.

$$R_{et} = \frac{F_{u2}}{F_{u1}} \times 100 \qquad (5.1.4a)$$

The elastic modulus retention should be calculated using Eq. (5.1.4b) with a precision of three significant figures. Elastic modulus should be calculated as directed in ASTM D7205/D7205M.

$$R_{em} = \frac{E_{L2}}{E_{L1}} \times 100 \qquad (5.1.4b)$$

The ultimate strain retention should be calculated using Eq. (5.1.4c) with a precision of three significant figures. The ultimate strain capacity of the specimen, ε_u, should be calculated as directed in ASTM D7205/D7205M.

$$R_{es} = \frac{\varepsilon_{u2}}{\varepsilon_{u1}} \times 100 \qquad (5.1.4c)$$

CHAPTER 6—EXTERNALLY BONDED FIBER-REINFORCED POLYMER TESTS

Durability of fiber-reinforced polymer (FRP) composite reinforcement externally bonded to concrete can be affected by a number of environments. This chapter presents tests that are used to determine the effect that an accelerated conditioning protocol (ACP) will have on important mechanical properties of the FRP-to-concrete bond.

The recommended ACP involves immersion in heated water. Three test procedures and the ACPs described pertain to tensile behavior of FRP laminate and bond of the FRP system to concrete for both cured-in-place and precured systems. A cured-in-place system might include fabric, saturant, primer, and putty. A precured system might include laminate, adhesive, and putty. Section 6.1 addresses specimen fabrication for the three procedures. Sections 6.2 and 6.3 address FRP externally bonded onto concrete, which is schematically outlined in Fig. 6. Section 6.4 addresses procedures for testing the FRP system alone.

6.1—Specimen fabrication and preparation

6.1.1 *Specimen fabrication and preparation for beam bond test*—Concrete described in 6.1.4 should be used to construct plain concrete beams, as shown in Fig. 6.1.1. The recommended dimensions for the beam are 4 in. (100 mm) square with a block length of 14 in. (350 mm) and a clear span of 12 in. (300 mm).

The specimen should be saw cut on a form-finished face at midspan to a depth of $h/2$. The width (kerf) of the saw cut may be cut with a composite or diamond blade. The same blade width should be used throughout a particular series of tests to allow for greater repeatability in testing. Saw cutting should be performed in a manner that minimizes spalling, microcracking, and other near-surface deterioration adjacent to the saw cut. As an alternate to saw cutting, the required weakened plane can be created by inserting a thin shim into the formwork. The notch gap width should be approximately 0.04 to 0.125 in. (1 to 3 mm). The average slot width should be reported.

Application of the externally bonded fiber-reinforced polymer (FRP) composite reinforcement should be in accordance with the manufacturer's recommendations. Surface preparation should be in accordance with 6.4.2.1 of ACI 440.2R-08. FRP composites should be cured for 7 days at standard laboratory conditions as noted in 4.2.1 before exposure to accelerated conditioning protocols (ACPs). Control specimens should be stored at standard laboratory conditions.

The FRP reinforcement should be a single layer with a minimum width of 1 in. (25 mm). Fabric reinforcement should be 0.123 lb/ft^2 (600 gsm). If a larger beam size is required to obtain an adhesive failure mode—for example, a 6 x 6 in. (150 x 150 mm)—then the length of the FRP composite reinforcement may be adjusted. In this case, a

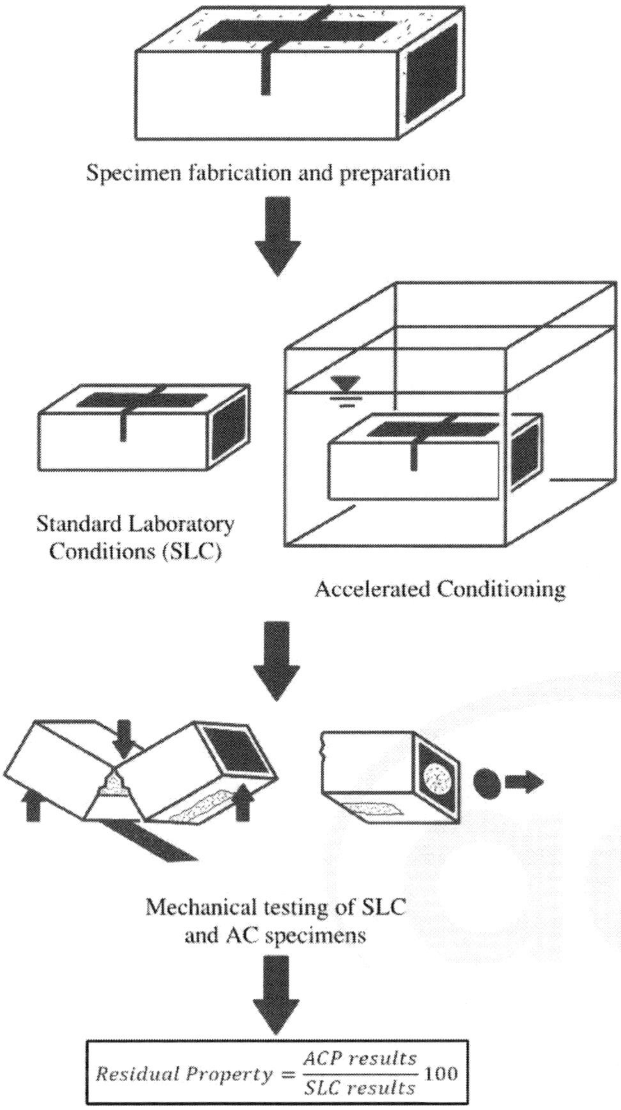

Fig. 6—Example procedure for combining accelerated conditioning and mechanical testing to determine residual mechanical properties for externally bonded reinforcement. Beam (Section 6.2) and pull-off (Section 6.3) test procedures are shown.

pilot study should be conducted without ACP to determine the FRP composite reinforcement width w and concrete strength required to give an adhesive failure mode in the control specimens. Excessive FRP composite reinforcement will generally lead to a shear failure of the concrete prism, which is an invalid failure mode for this test.

6.1.2 *Specimen fabrication and preparation for pull-off bond test*—If a direct tension pull-off test is to be conducted in conjunction with the beam test, then a test patch for this purpose should be applied to the end of the beam specimen as shown in Fig. 6.1.1. Concrete described in 6.1.4 should be used to construct control and ACP specimens.

Application of the externally bonded FRP composite should be in accordance with the manufacturer's recommendations. Surface preparation should be in accordance with 6.4.2.1 of ACI 440.2R-08. FRP composites should be cured for 7 days at standard laboratory conditions before exposure to ACP. Control specimens should be stored at standard laboratory conditions as instructed in 4.2.1. The circular saw cut and the circular loading fixture for the pull-off test should be applied after ACP exposure.

6.1.3 *Specimen fabrication and preparation for tensile stress test*—Companion tension coupons (6.4) should be fabricated at the same time as beam and pull-off bond specimens. Coupons should be fabricated to allow testing in accordance with ASTM D7565/D7565M for wet layup and ASTM D3039/D3039M for pultruded plate with the exception of conditioning according to Procedure A of ASTM D618.

6.1.4 *SLC and ACP concrete mixture*—Concrete used for accelerated conditioning tests described in 6.2 and 6.3 should be formulated as follows:

6.1.4.1 *Aggregates*—Aggregates should conform to ASTM C33/C33M, and the maximum aggregate size should be 1/2 in. (12.7 mm) for a 4 in. (100 mm) beam width.

6.1.4.2 *Cement*—Type I/II portland cement conforming to ASTM C150/C150M should be used. The concrete mixture should not include any other supplementary cementitious or pozzolonic materials, such as slag, fly ash, silica fume, and limestone powder; or chemical admixtures, such as air-entraining agents, water reducers, high-range water reducers, shrinkage-compensating admixtures, corrosion inhibitors, set retarders, or set accelerators.

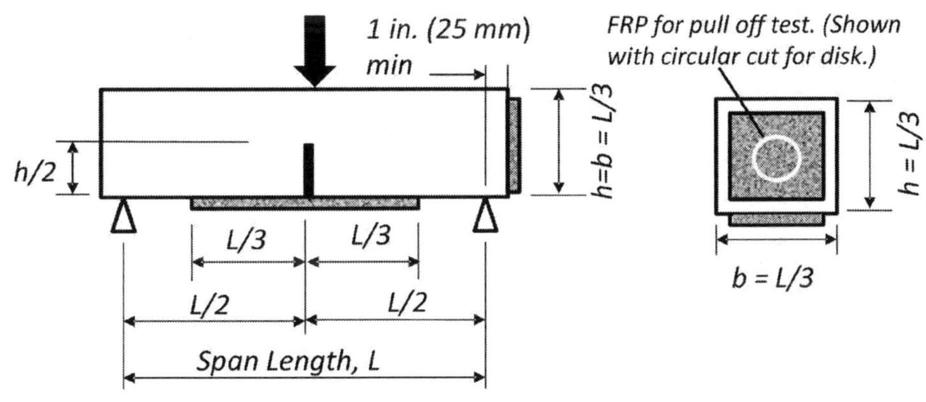

Fig. 6.1.1—Beam bond test specimen configuration.

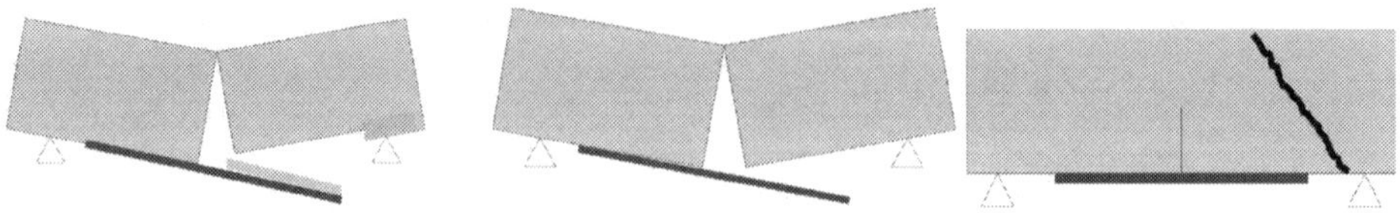

(a) (b) (c)

Fig. 6.2—FRP concrete failure modes: (a) cohesive failure mode; (b) adhesive failure mode; and (c) shear failure of concrete block.

6.1.4.3 *Curing*—Specimens should be cured for 28 days in a moist environment in accordance with ASTM C511. Cylinders should undergo the same curing regime. After curing, specimens should be dry prior to applying FRP.

6.1.4.4 *Concrete strength*—The 28-day compressive strength of the cylinders should be between 6500 and 8500 psi (46 and 60 MPa). Cylinders should be cast and tested in accordance with ASTM C31/C31M and ASTM C39/C39M.

6.2—Beam bond test

The beam bond test consists of loading small concrete beams reinforced in flexure with externally bonded fiber-reinforced polymer (FRP) reinforcement; specimens are loaded in three-point bending to failure. Residual load capacities of beams with bond failures are measured and compared to those of the unexposed control specimens. The test procedure, which was developed by Gartner et al. (2011), is similar in specimen configuration and testing procedures to those used to determine modulus of rupture of concrete (ASTM C78/C78M). For each test, at least five control specimens (standard laboratory conditions [SLCs]) and five specimens exposed to the specified accelerated conditioning protocols (ACPs) should be tested.

In bonded FRP reinforcement design, a cohesive bond failure mode is typically assumed, which is where the fracture surface passes entirely through the concrete substrate and some concrete remains adhered to the FRP (Fig. 6.2(a)). Durability of this failure mode, however, is mainly dependent on the durability of the concrete through which the fracture surface passes. The intent of the combined ACP and bond tests recommended in this section is not to test the concrete durability, but rather to test the durability of the adhesive strength of the FRP system being applied to the surface of the concrete (Fig. 6.2(b)). The failure mode of accelerated conditioning (AC) specimens should be primarily adhesive to provide any notable information on the durability of the bond of the system. If both control and AC specimens exhibit adhesive failure modes, the true degradation of the adhesive bond is determined. A mode change from cohesive in the control specimens to adhesive in the AC specimens also provides an indication of changes in the bond of the system associated to its durability and should be noted in the final data. A third failure mode, which is shear cracking of the concrete beam as indicated in Fig. 6.2(c), is considered an invalid test result. Shear failure is avoided by using concrete blocks with sufficiently high concrete compressive strength, as described in 6.1.4.

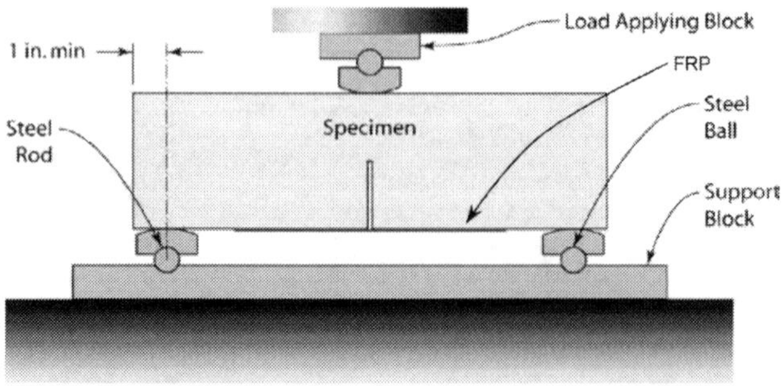

Fig. 6.2.3—Schematic view of a suitable apparatus for flexure test of concrete by three-point loading method.

6.2.1 *Accelerated conditioning protocol*—

(a) Accelerated conditioning—Immersion in water as described in 4.2.2

(b) Duration—3000 hours

(c) Sustained stress—None.

6.2.2 *Standard laboratory conditions*

Control specimens–Standard laboratory conditions as described in 4.2.1.

6.2.3 *Mechanical testing*—SLC and AC specimens should be post-conditioned for 24 hours in 100 percent relative humidity at a temperature of 73.4 ± 3.6°F (23 ± 2°C). The beam test should be conducted within 2 to 5 hours following the post conditioning. Companion direct tension pull-off tests as described in 6.3 should be conducted with the beam test. Following exposure to the specified ACP, both SLC and AC beams should be tested using ASTM C78/C78M procedures with the setup adjusted as shown in Fig. 6.2.3. Testing is conducted at room temperature rather than at the elevated exposure temperature to which some AC specimens are subjected. Concrete compressive strength, failure mode, and beam specimen ultimate load (P_b) should be reported for each test. When possible, force-displacement behavior and the concrete mixture design used should be reported. The initial fabric or laminate type, width, and thickness should be reported to ensure consistency between laboratory-prepared specimens. A final composite thickness should be reported after testing.

6.2.4 *Residual mechanical properties*—The beam bond retention should be calculated with a precision of three significant figures using Eq. (6.2.4) as follows

$$R_{eb} = \frac{P_{b2}}{P_{b1}} \times 100 \qquad (6.2.4)$$

The standard deviation, coefficient of variation, or both, should be reported with the average beam strength.

6.3—Pull-off bond test

The direct tension, pull-off test is performed in accordance with ASTM D7522/D7522M by pulling a small circular patch of externally bonded fiber-reinforced polymer (FRP) concrete composites off the concrete substrate. Residual bond strength and failure mode of the accelerated conditioning (AC) specimens are measured and compared to those of unexposed control specimens. The failure mode is also noted and may be used to characterize bond. For each test, at least five control specimens and five specimens exposed to the specified accelerated conditioned protocol (ACP) should be tested.

6.3.1 *Accelerated conditioning protocol—*

(a) Accelerated conditioning—Continuous immersion in water as described in 4.2.2

(b) Duration—3000 hours

(c) Sustained stress—None.

6.3.2 *Standard laboratory conditions—*

Control specimens—Standard laboratory conditions as described in 4.2.1.

6.3.3 *Mechanical testing*—Pull-off tests should be conducted according to ASTM D7522/D7522M. Cutting the core should occur immediately after removing the specimen from the water bath prior to post-conditioning. Specimens should be post-conditioned for 24 hours in 100 percent relative humidity at a temperature of 73.4 ± 3.6°F (23 ± 2°C). The direct tension pull-off test should be conducted between 2 and 5 hours following the post-conditioning. Testing is conducted at room temperature rather than at the elevated AC exposure temperature. Reference cylinders (ASTM C39/C39M) should be tested concurrent with control specimens to establish the concrete compressive strength. The initial fabric or laminate type along with the width and thickness should be reported to ensure consistency between laboratory-prepared specimens.

6.3.4 *Residual mechanical properties*—Pull-off bond retention should be calculated with a precision of three significant figures using Eq. (6.3.4) as follows

$$R_{ep} = \frac{F_{p2}}{F_{p1}} \times 100 \tag{6.3.4}$$

6.4—Tensile test of fiber-reinforced polymer

This test evaluates the tensile capacity of the fiber-reinforced polymer (FRP) fabric or laminate under standard laboratory conditions (SLC) and accelerated conditioning (AC). Specimen mass, residual tensile strength, elastic modulus, and ultimate strain values should be measured and compared to those of unexposed control specimens (refer to 6.1.1 and 6.1.2). The recommended accelerated conditioning protocol (ACP) parameters are water temperature and immersion time. These specimens may serve as companion specimens to either the beam and pull-off test (refer to 6.2 and 6.3).

6.4.1 *Accelerated conditioning protocol—*

Accelerated conditioning—Continuous immersion in water as described in 4.2.2

Duration—3000 hours

Sustained stress—None.

6.4.2 *Standard laboratory conditions—*

Control specimens—Standard laboratory conditions as described in 4.2.1.

6.4.3 *Mechanical testing*—Following exposure to the specified ACP, the specimens should be rinsed thoroughly with water and post-conditioned according to Procedure A of ASTM D618. Tabs should be applied and both AC and control specimens should be tested in accordance with ASTM D7565/D7565M or ASTM D3039/D3039M as appropriate. The initial fabric or laminate and final thickness of the FRP composite material should be reported to ensure consistency between laboratory-prepared specimens.

Residual mechanical properties R_{et}, R_{em}, and R_{es} should be calculated and reported as described in 5.1.4.

CHAPTER 7—FUTURE WORK AND RECOMMENDATIONS

Test methods and conditioning protocols presented in this document can be used to test fiber-reinforced polymer (FRP) composite reinforcement for durability performance when exposed to moisture and high-alkaline environments. Results of these tests can be used by manufacturers to improve product components and manufacturing processes and by researchers and testing laboratories to adopt common test protocols to build a meaningful database of results from durability testing of FRP materials. Although the use of these test protocols does not duplicate service life conditions, the data acquired will yield an improved understanding of FRP durability performance and allow accelerated conditioning protocol (ACP) test calibration between conditioned and unconditioned systems. As these tests are used more broadly and long-term field data become available, correlations between results from ACP and field observations can be studied to provide information regarding service life. As a future work, acceptable limits on the residual mechanical properties should be developed for ACP test results.

The ACP protocols do not provide recommended laminate thicknesses or bar diameters to be used for durability testing. Experience in the aerospace industry has shown that, for a given exposure environment, the moisture absorption profile for a laminate cross section is highly dependent on the laminate thickness, exposure time, and diffusion properties of the laminate (Campbell 2010). Future work should investigate the applicability of Fick's law of diffusion on the moisture absorption profile through the specimen thickness/diameter and provide recommendations regarding the specimen thickness or diameter to be used with the ACP. In addition, real-world fluctuations in temperature, humidity, and sustained stress should be considered in the expected long-term moisture profile. Effects of post-cure, including temperature and length of cure time on the composite, should be evaluated between the control and conditioned specimens.

Physicochemical tests, such as thermogravimetric analysis (with or without evolved gas analysis and Fourier transform infared spectroscopy), differential scanning calorimetry, dynamic mechanical analysis, thermomechanical analysis, X-ray diffraction, scanning electron microscopy, and transmission electron microscopy that measure the changes in the physical and chemical properties of FRP materials, are needed to fully characterize degradation of the FRP material and its constituents after subjection to ACPs described in this guide. Future research will be directed to this area to enable complete understanding of the degradation mechanisms under ACP with an emphasis toward a more scientific determination of C_E factors for related ACI 440 design guide documents (ACI 440.1R, ACI 440.2R, and ACI 440.7R).

Results of the tests conducted using the recommended ACP are not intended for use in the design of FRP composites as concrete reinforcement. In the future, however, when the relationship between field performance and accelerated conditioning is better understood and characterized, the ACPs are envisioned to be adjusted to allow for use in design. Sufficient data, however, are not yet available to calibrate ACP test results to either a particular service life or the environmental (C_E) factor used in ACI 440.1R and 440.2R.

CHAPTER 8—REFERENCES

ACI committee documents and documents published by other organizations are listed first by document number, full title, and year of publication followed by authored documents listed alphabetically.

American Concrete Institute

440R-07—Report on Fiber-Reinforced Polymer (FRP) Reinforcement for Concrete Structures

440.1R-06—Guide for the Design and Construction of Structural Concrete Reinforced with FRP Bars.

440.2R-08—Guide for the Design and Construction of Externally Bonded FRP Systems for Strengthening Concrete Structures.

440.7R-10—Guide for the Design and Construction of Externally Bonded Fiber-Reinforced Polymer Systems for Strengthening Unreinforced Masonry Structures.

ASTM International

C31/C31M-12—Standard Practice for Making and Curing Concrete Test Specimens in the Field

C33/C33M-13—Standard Specification for Concrete Aggregates

C39/C39M-14—Standard Test Method for Compressive Strength of Cylindrical Concrete Specimens

C78/C78M-10—Standard Test Method for Flexural Strength of Concrete (Using Simple Beam with Third-Point Loading)

C150/C150M-12—Standard Specification for Portland Cement

C511-13—Standard Specification for Mixing Rooms, Moist Cabinets, Moist Rooms, and Water Storage Tanks Used in the testing of Hydraulic Cements and Concrete

D543-14—Standard Practices for Evaluating the Resistance of Plastics to Chemical Reagents

D618-13—Standard Practice for Conditioning Plastics for Testing

D3039/D3039M-14—Standard Test Method for Tensile Properties of Polymer Matrix Composite Materials

D5229/D5229M-14—Standard Test Method for Moisture Absorption Properties and Equilibrium Conditioning of Polymer Matrix Composite Materials

D7205/D7205M-06(2011)—Standard Test Method for Tensile Properties of Fiber Reinforced Polymer Matrix Composite Bars

D7522/D7522M-09—Standard Test Method for Pull-Off Strength for FRP Bonded to Concrete Substrate

D7565/D7565M-10—Standard Test Method for Determining Tensile Properties of Fiber Reinforced Polymer Matrix Composites Used for Strengthening of Civil Structures

D7705/D7705M-12—Standard Test Method of Alkali Resistance of Fiber Reinforced Polymer (FRP) Matrix Composite Bars used in Concrete Construction

International Code Council

AC125-12—Concrete and Reinforced and Unreinforced Masonry Strengthening Using Fiber-Reinforced Polymer (FRP) Composite Systems

Authored documents

Bank, L. C.; Gentry, T. R.; and Barkatt, A., 1995, "Accelerated Test Methods to Determine the Long-Term Behavior of FRP Composites: Environmental Effects," *Journal of Reinforced Plastics and Composites*, V. 14, Jan, pp. 559-587.

Campbell, F. C., 2010, "Structural Composite Materials," ASM International, Society for Materials Scientists and Engineers, Materials Park, OH, 612 pp.

Gartner, A.; Douglas, E. P.; Dolan, C. W.; and Hamilton, H. R., 2011, "Small Beam Bond Test Method for CFRP Composites Applied to Concrete," *Journal of Composites for Construction*, V. 15, No. 1, Jan-Feb., pp. 52-61. doi: 10.1061/(ASCE)CC.1943-5614.0000151

Robert, M.; Wang, P.; Cousin, P.; and Benmokrane, B., 2010, "Temperature as an Accelerating Factor for Long-Term Durability Testing of FRPs: Should There Be Any Limitations?" *Journal of Composites for Construction*, V. 14, No. 4, Aug., pp. 361-367. doi: 10.1061/(ASCE) CC.1943-5614.0000102

Tatar, J.; Weston, C.; Blackburn, P.; and Hamilton, H. R., 2013, "Direct Shear Adhesive Bond Test," FRPRCS-11: 11th International Symposium on Fiber Reinforced Polymer for Reinforced Concrete Structures, 2013, 10 pp.

High-Strength Concrete Columns

Reported by Joint ACI-ASCE Committee 441

S. Ali Mirza*
Chair

Atorod Azizinamini*
Subcommittee Chair

Perry E. Adebar
Secretary

Alaa E. Elwi	Richard A. Lawrie	Patrick Paultre*	Murat Saatcioglu*
Richard W. Furlong	Douglas D. Lee	Bashkim Prishtina	Arturo E. Schultz
Roger Green	James G. MacGregor*	Franz N. Rad	Lawrence G. Selna
H. Richard Horn, Jr.	Sheng-Taur Mau	B. Vijaya Rangan	Shamim A. Sheikh
Cheng-Tzu Thomas Hsu	Robert Park	M. Ala Saadeghvaziri	

*Subcommittee members who prepared this report.

This report reviews the state of the knowledge of the behavior of high-strength concrete (HSC) columns. High-strength concrete, as used in this report, is defined as concrete with compressive strength exceeding 70 MPa (10,000 psi). The report provides highlights of research available on the performance of HSC columns under monotonically increasing concentric or eccentric compression, and with incrementally increasing lateral deformation reversals and constant axial compression.

Research results are used to discuss the effect of cover concrete and parameters related to transverse reinforcement on strength and ductility of HSC columns subjected to concentric load.

The behavior of HSC columns subjected to combined axial load and bending moment is discussed in terms of variables related to concrete and transverse reinforcement. In addition to discussion on flexural and axial capacity, this report also focuses on seismic performance of HSC columns.

Keywords: axial load; bending moment; columns; cover concrete; ductility; flexural strength; high-strength concrete; longitudinal reinforcement; seismic design; transverse reinforcement.

CONTENTS

ACI 441R-96 became effective November 25, 1996.

CHAPTER 1—INTRODUCTION

One application of high-strength concrete (HSC) has been in the columns of buildings. In 1968 the lower columns of the Lake Point Tower building in Chicago, Illinois, were constructed using 52 MPa concrete.[1] More recently, several high rise buildings[1-4] have utilized concrete with compressive strengths in excess of 100 MPa in construction of columns.

Many studies[4-9] have demonstrated the economy of using HSC in columns of high-rise buildings, as well as low to mid-rise buildings.[10] In addition to reducing column sizes and producing a more durable material, the use of HSC has been shown to be advantageous with regard to lateral stiffness and axial shortening.[11] Another advantage cited in the use of HSC columns is reduction in cost of forms. This is achieved by using HSC in the lower story columns and reducing concrete strength over the height of the building while keeping the same column size over the entire height.

The increasing use of HSC caused concern over the applicability of current building code requirements for design and detailing of HSC columns. As a result, a number of research studies have been conducted in several countries during the last few years. The purpose of this paper is to summarize major aspects of some of the reported data.

The major objectives of reported studies have been to investigate the validity of applying the current building code requirements to the case of HSC, to evaluate similarities or differences between HSC and normal-strength concrete (NSC) columns, and to identify important parameters affecting performance of HSC columns designed for seismic as well as non-seismic areas. These concerns arise from the fact that requirements for design and detailing of reinforced concrete columns in different model codes are primarily empirical and are developed based on experimental data obtained from testing column specimens having compressive strengths below 40 MPa.

The reported information can be divided into two general categories: performance of HSC columns under concentric axial load; and performance of HSC columns under combined axial load and bending moment. This report gives the highlights of the reported data in each of these categories. In this report, HSC is defined as concrete with compressive strength greater than 70 MPa.

CHAPTER 2—PERFORMANCE OF HSC COLUMNS UNDER CONCENTRIC LOADS

The majority of reported studies[12-27] in the field of HSC columns concern the behavior of columns subjected to concentric loads. Understanding the behavior of columns under concentric loads assists in quantifying the parameters affecting column performance. However, conclusions from this type of loading should not necessarily be extended to the case of combined loading, a situation most frequently encountered in columns used in buildings.

Reported data indicate that stress-strain characteristics of high-strength concrete, cover concrete, and parameters related to confining steel have the most influence on response of HSC columns subjected to concentric loads. The effect of the first parameter is discussed in Sec. 3.1. The remaining two parameters are discussed in the following sections.

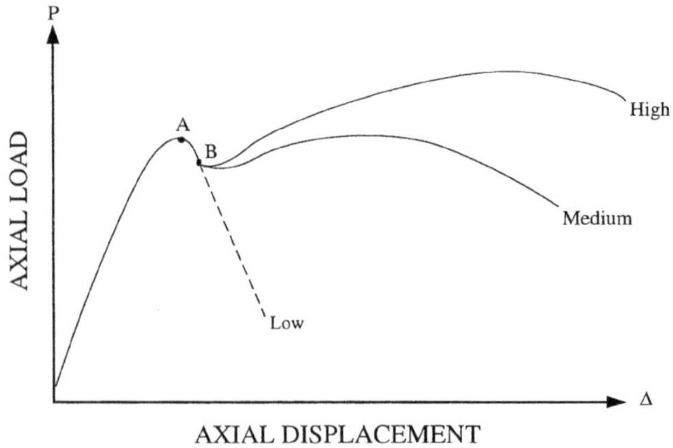

Fig. 1—Schematic behavior of HSC columns subjected to concentric axial loads, incorporating low, medium, and high amounts of transverse reinforcement.

2.1—Effect of cover concrete

Figure 1 shows a schematic load-axial deformation response under concentric loads of HSC columns with transverse reinforcement. As concrete strength increases, the ascending portion of the curve approaches a straight line. In general, spalling of the cover concrete is reported[12-27] to occur prior to achieving the axial load capacity of HSC columns, as calculated by the following equation:

$$P_o = 0.85 f_c' (A_g - A_{st}) + A_{st} f_y \quad (1)$$

where:

P_o = pure axial load capacity of columns calculated according to the nominal strength equations of ACI 318-89
f_c' = concrete compressive strength
A_g = gross cross-sectional area of column
A_{st} = area of longitudinal steel
f_y = yield strength of longitudinal steel

The 1994 edition of the Canadian Code for Design of Concrete Structures also uses this equation for computing P_o, except that the factor 0.85 is replaced by

$$\alpha_1 = (0.85 - 0.0015 f_c') \geq 0.67$$

in which f_c' is in MPa. Hence, P_o calculated by the Canadian code will be somewhat less than that calculated by ACI 318-89.

Point A in Fig. 1 indicates the loading stage at which cover concrete spalls off. The behavior of HSC columns beyond this point depends on the relative areas of the column and the core and on the amount of transverse reinforcement provided. Following spalling of the cover concrete, the load-carrying capacity of columns generally drops to point B in Fig. 1. Beyond this point, Bjerkeli et al.,[19] Cusson et al.,[25] and Nishiyama et al.[28] report that it is possible to increase the maximum axial strength of columns up to 150% of that calculated by the ACI 318-89 provisions and obtain a ductile behavior by providing sufficient transverse reinforcement. The effect of the amount of transverse reinforcement is shown schematically in Fig. 1 and will be discussed further in later sections.

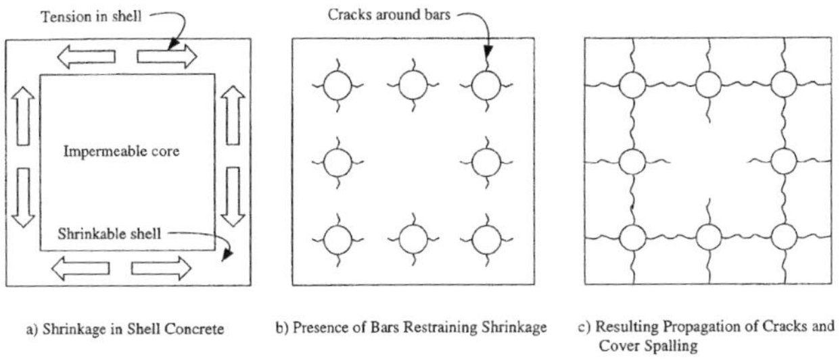

Fig. 2—Factors promoting cover spalling in high-strength concrete columns (adapted from Ref. 29).

The loss of cover concrete in HSC columns before reaching the axial capacity calculated by ACI 318-89 is contrary to the observed behavior of concrete columns made of NSC. Collins et al.[29] provide the following explanation for the factors resulting in early spalling of cover concrete in HSC columns. According to those authors, the low permeability of HSC leads to drying shrinkage strain in cover concrete, while the core remains relatively moist. As a result, tensile stresses are developed in the cover concrete as shown in Fig. 2a. Moreover, longitudinal steel, as depicted in Fig. 2b, promotes additional cracking. The combination of these two mechanisms (see Fig. 2c) then results in the formation of a cracking pattern that, according to those authors, is responsible for early loss of cover concrete, thereby preventing HSC columns from reaching their axial load capacity predicted by Eq. (1) prior to spalling of cover concrete.

Early spalling of concrete cover may also be initiated by the presence of a closely spaced reinforcement cage that separates core and cover concrete. Cusson et al.[25] attributed the spalling of the cover to planes of weakness created by the dense steel cages. They state that spalling becomes more prevalent as the concrete strength increases.

Saatcioglu and Razvi[27,30] also observed early spalling of cover concrete in their tests. Those researchers indicated that the presence of closely spaced reinforcement cage between the core and the cover concrete provided a natural plane of separation, which resulted in an instability failure of the cover concrete under high compressive stresses. The spalling in their tests occurred at a stress level below that corresponding to the crushing of plain concrete.

2.2—Effect of volumetric ratio of transverse reinforcement

In the case of NSC, an increase in the amount of transverse reinforcement has been shown to increase strength and ductility.[31] The same observation has been reported[19,25,27] for the case of HSC, though to a lesser degree. Some researchers have attributed this phenomenon to the relatively smaller increase in volume during microcracking of HSC, resulting in less lateral expansion of the core. The lower lateral expansion of core concrete delays the utilization of transverse reinforcement.

Reported data[12-27,30] indicate that in the case of HSC, little improvement in strength and ductility is obtained when the volumetric ratio of transverse reinforcement is small. For instance, Bjerkeli et al.[19] report that a volumetric ratio of

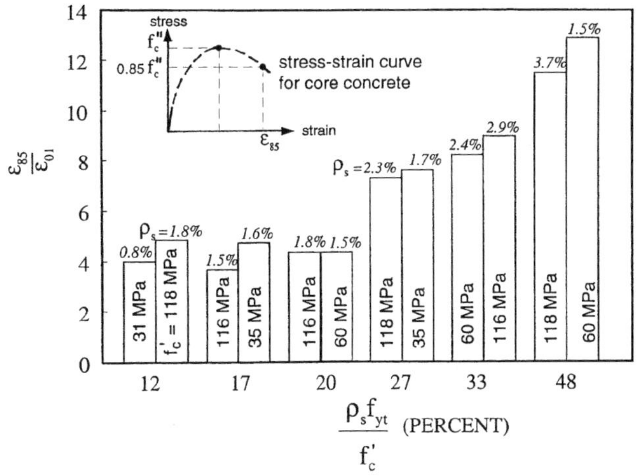

Fig. 3—Columns with different concrete strengths showing similar axial ductility ratios (f_c' = concrete compressive strength based on standard cylinder test) (adapted from Ref. 30).

1.1% was not sufficient to generate any improvement in column behavior, while the use of 3.1% resulted in columns performing in a ductile manner.

Sugano et al.,[32] Hatanaka et al.,[23] and Saatcioglu et al.[27,30] report a correlation between the non-dimensional parameter, $\rho_s f_{yt}/f_c'$, and axial ductility of HSC columns subjected to concentric loads. Figure 3 shows the relationship between this parameter and axial ductility of columns with different compressive strengths. In this figure, the axial ductility of columns is represented by the ratio $\varepsilon_{85}/\varepsilon_{01}$, where ε_{85} is the axial strain in core concrete when column load on the descending branch is reduced to 85% of the peak value and ε_{01} is the axial strain corresponding to peak stress of plain concrete. For each pair of columns compared, similar reinforcement arrangements and tie spacings were maintained. As indicated in this figure, columns of different compressive strength having the same $\rho_s f_{yt}/f_c'$ value result in almost the same axial ductility, provided that certain minimum limitations are met for the volumetric ratio and spacing of transverse reinforcement.[30]

Figure 4 shows the relationship between the parameter $\rho_s f_{yt}/f_c'$ and the ratio of experimentally obtained axial load capacity for 111 HSC columns to that predicted by Eq. 1. From this plot it could be observed that columns with a low volumetric ratio of transverse reinforcement may not achieve their strength as calculated by ACI 318-89; however, well-confined columns can result in strength in excess of that calculated by ACI 318-89. Excess strength of columns with relatively higher amounts of transverse reinforcement is generally obtained after spalling of cover concrete. This strength enhancement comes as a result of an increase in strength of the confined core concrete.

2.3—Effect of longitudinal and transverse reinforcement strength

The yield strength of the confinement steel determines the upper limit of the confining pressure. A higher confining pressure applied to the core concrete, in turn, results in higher strength and ductility. Figure 5 shows normalized

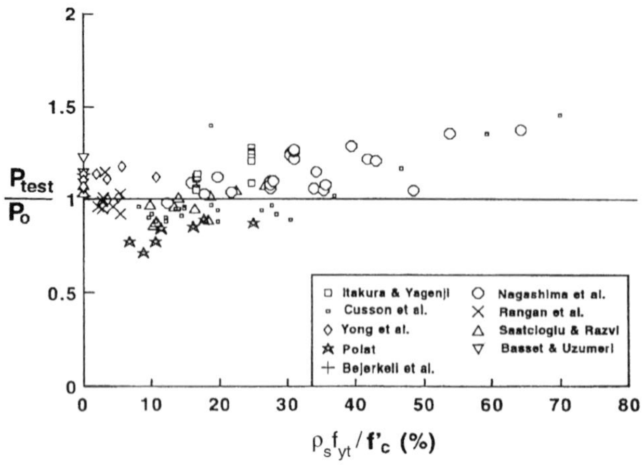

Fig. 4—Comparison of experimental and calculated concentric strengths of columns (adapted from Ref. 30).

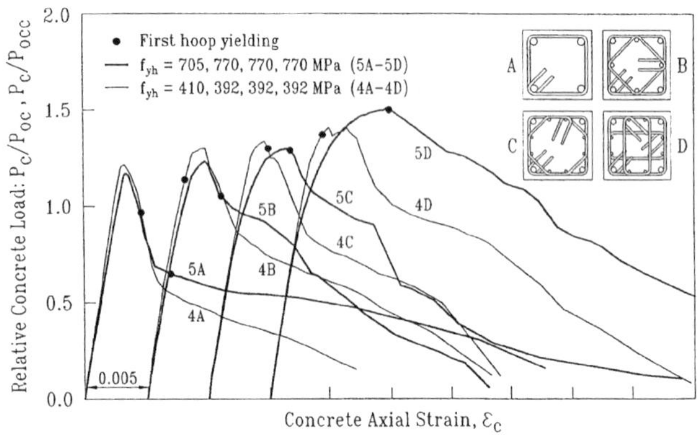

Fig. 5—Effect of transverse reinforcement yield strength (adapted from Ref. 25).

axial load-axial strain response of core concrete for four pairs of HSC columns.[25] For each pair of columns, all parameters were kept constant except the yield strength of the transverse reinforcement. The yield strength of transverse reinforcement for columns 4A, 4B, 4C, and 4D and columns 5A, 5B, 5C, and 5D was approximately 400 MPa and 700 MPa, respectively. As indicated in this figure, for well confined columns (C and D), increasing the yield strength of transverse reinforcement results in an increase in strength and ductility. However, for type A columns, where only peripheral ties are provided, the gain in strength and ductility is negligible. Reported data of HSC columns[17,25,27] indicate that when high-strength concrete is used in well-confined columns, the full yield strength of transverse reinforcement is utilized. On the other hand, in a poorly confined HSC column, tensile stresses that develop in the transverse reinforcement remain below yield strength even at the time of column failure.

2.4—Effect of longitudinal and transverse reinforcement arrangement

Well-distributed longitudinal and transverse reinforcement results in a larger effectively confined concrete area and more uniform distribution of the confining pressure,

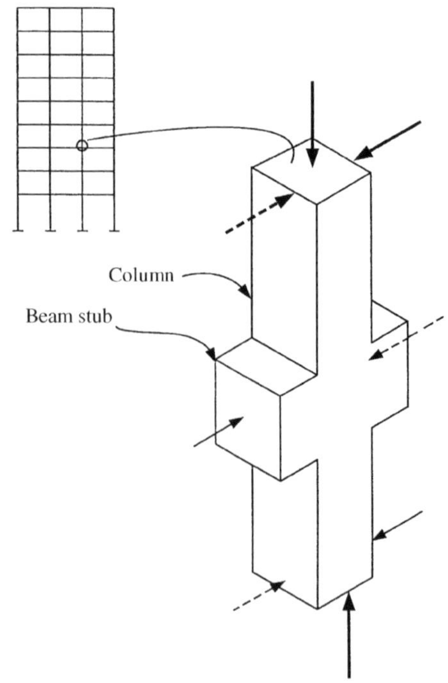

Fig. 6—Overall view of test specimens.

thereby improving the effectiveness of the confining reinforcement. In the case of NSC,[30,33] the arrangement of the transverse reinforcement and laterally supported longitudinal reinforcement has been shown to have a significant influence on strength and ductility of columns. Similar observations have been reported in the case of HSC columns.[17,27,30] Transverse reinforcement in the form of single peripheral hoops has been shown to result in very low strength and ductility of HSC columns.[17,25,27] Similar observations have also been reported for NSC columns.[34]

More detailed discussions of the behavior of HSC columns subjected to concentric axial load are presented in Refs. 25 and 30.

CHAPTER 3—PERFORMANCE OF HSC COLUMNS UNDER COMBINED AXIAL LOAD AND BENDING MOMENT

Two major questions must be addressed when designing HSC columns. First, does the rectangular stress block described in Section 10.2.7 of ACI 318-89 apply to HSC? Second, are the confinement rules given in ACI 318-89 Sections 10.9.3 and 21.4.4 adequate for HSC? In regions of high seismicity, a major concern has been the ductility of HSC columns, resulting in a reluctance to use HSC in these areas compared with regions of low seismicity. As a result, the focus of most reported investigations[32,35-43] on performance of HSC columns under combined loading has been primarily to comprehend the seismic behavior of these columns. Some of these studies also have presented data that could be used to assess the flexural capacity of HSC columns subjected to combined loading. However, available data for HSC columns subjected to combined loading are relatively limited compared with HSC columns subjected to concentric loading.

To date, most experimental research has involved testing of scaled columns. Figure 6 shows a general configuration of

Table 1—Comparison of calculated and experimental flexural strengths for specimens tested by Bing et al. (adapted from Reference 40)

Specimen number	Axial load level $P/f_c'A_g$	f_c', MPa	f_y of ties, MPa	$M_{EXP}/M_{NZS3101}$	M_{EXP}/M_{MOD}
1	0.3	98	1317	0.94	0.99
2	0.3	98	453	0.98	1.03
3	0.6	93	1317	0.87	1.08
4	0.6	93	453	0.86	1.07
5	0.6	93	1317	0.82	1.02

a typical column specimen used in most reported studies. This type of specimen represents half of the upper and lower column, together with a small portion of the floor beam. These specimens are usually subjected to constant axial load and to a repeated lateral displacement sequence similar to the one shown in Fig. 7. This type of specimen is designed so that no damage is inflicted on the beam-column joint.

3.1—Flexural strength

There is no universal agreement on the applicability of ACI 318-89 code requirements for calculating flexural strength of HSC column sections subjected to combined axial load and bending moment.

Columns are usually designed for combined axial load and bending moment using the rectangular stress block defined in ACI 318-89 Section 10.2.7. This stress block was originally derived by Mattock et al.,[44] based on tests of unreinforced concrete columns loaded with axial load and moments so as to have the neutral axis on one face of the test specimen.[45] The concrete strengths ranged up to 52.5 MPa. The stress block was defined by two parameters: the intensity of stress in the stress block, which was designated as α_1; and the ratio of the depth of the stress block to the depth of the neutral axis, which was designated as β_1. Mattock et al.[44] proposed $\alpha_1 = 0.85$ and β_1 as follows:

$$\beta_1 = 1.05 - 0.05(f_c'/6.9) \text{ but not more than } 0.85$$

for f_c' in MPa. That proposal was incorporated into Sec. 1504g of ACI 318-63.

Based on similar tests of concrete columns with concrete strengths ranging from 79 to 98 MPa, Nedderman[46] proposed a lower limit on β_1 of 0.65 for concrete strengths in excess of 55 MPa. This limit was incorporated in ACI 318-77. Similar tests were carried out by Kaar et al.[47] on concretes with compressive strength ranging from 24 to 102 MPa and by Swartz et al.[48] on concretes ranging from 58 to 77 MPa in compressive strength.

When the equation for β_1 was compared with the test data, a conservative lower bound was selected and the product $\alpha_1\beta_1$ was shown to lead to a conservative estimate for the total compression force in concrete in an eccentrically loaded column. For a rectangular stress block, the distance from the resultant compressive force in concrete to the centroid of the rectangular cross-section is $(h/2 - \beta_1 c/2)$, where h is the total depth of the cross-section. A conservative lower bound estimate of β_1 leads to an overestimation of this

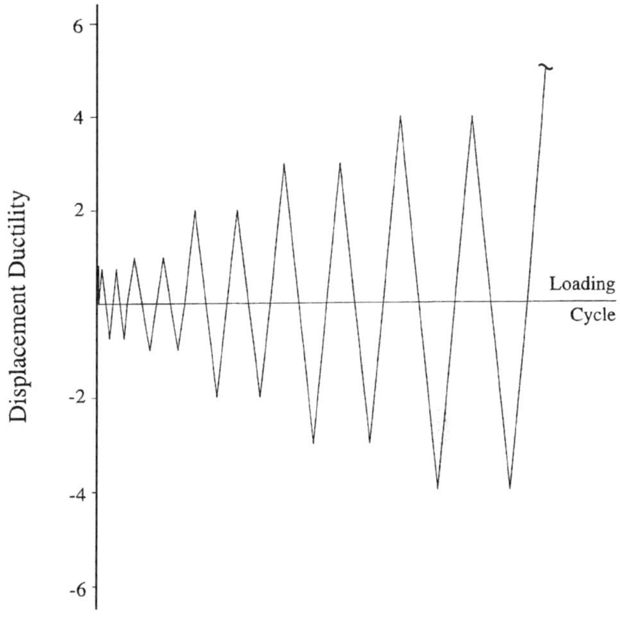

Fig. 7—Lateral displacement sequence.

distance and, hence, to an overestimation of the moment resisted by compression in the concrete. This is most serious for columns failing in compression, and with e/h ratios less than about 0.3, where e = eccentricity of axial load and h = overall thickness of the column cross-section.

Table 1 gives a comparison of calculated and experimental flexural strengths of five column specimens tested by Bing et al.[40] As indicated in this table, the ratios of the experimentally obtained flexural strength to that calculated according to the New Zealand Standard (NZS) 3101 procedures (the same as ACI 318-89 requirements) are less than 1, especially for columns subjected to higher axial load levels. Based on these tests, Bing et al. have suggested that an equivalent rectangular compressive stress block with an average stress, $\alpha_1 f_c'$, and a depth, $a = \beta_1 c$, be used in design of HSC column cross-sections, where:

$$\alpha_1 = 0.85, \text{ for } f_c' \leq 55 \text{ MPa}$$

and

$$\alpha_1 = 0.85 - 0.004(f_c' - 55) \geq 0.75, \text{ for } f_c' > 55 \text{ MPa}$$

Table 1 also gives the ratio of the experimentally obtained flexural strength for test columns to that calculated by the

Table 2—Comparison of calculated and experimental flexural strengths (adapted from Reference 42)

Specimen number	Axial load level P/P_o^*	f'_c, MPa	M_{EXP}/M_{ACI}	M_{EXP}/M_{MOD}
1	0.2	54	1.25	—
2	0.2	51	1.23	—
3	0.2	101	0.93	1.04
4	0.2	100	1.01	1.14
5	0.2	102	0.91	1.02
6	0.2	102	0.98	1.1
7	0.3	104	0.87	1.12

$^*P_o = 0.85f'_c(A_g - A_{st}) + A_{st}f_y$

modified procedure. As indicated in Table 1, the modified procedure gives a better estimation of test results. For the type of specimens tested by Bing et al., flexural strengths obtained from tests are usually 10 to 25% higher than the calculated values when NSC is used. This higher strength is attributed primarily to confinement provided by the beam stub in the critical region of the test column. See Fig. 6 for the type of test column used in that testing program.

Table 2 gives a comparison of experimentally obtained flexural strengths for some of the test columns reported in Ref. 42 to those calculated by ACI 318-89 requirements. As indicated in this table, the ACI 318-89 procedure results in reasonable calculation of flexural strengths for test columns with concrete compressive strengths equal to 54 and 51 MPa. The conservatism of the ACI 318-89 procedure in calculating flexural strength of these two test columns is similar to NSC columns. As stated earlier, NSC column tests usually give 10 to 25% higher flexural strength than that calculated by ACI 318-89 procedure for the type of test columns used. However, as concrete compressive strength or level of axial load increases, the ratio of experimentally obtained flexural strength to that calculated by ACI 318-89 procedures decreases and falls below 1, as indicated by Table 2. This is especially true for the test column with an axial load equivalent to 30% of the axial load strength of the column. Those authors[42] offer the following explanation for this observation.

Available test data indicate that typical stress-strain curves in compression for HSC are characterized by an ascending portion that is primarily linear, with maximum strength achieved at an axial strain between approximately 0.0024 and 0.003. Therefore, it may be more appropriate to use a triangular compression stress block having properties shown in Fig. 8 for calculating the flexural strength of HSC columns when f'_c exceeds approximately 70 MPa. In this approach, the maximum compressive stress is assumed to be $0.85f'_c$ at an axial compressive strain of 0.003. Considering the equilibrium of horizontal forces and moment equilibrium, it can be shown that the equivalent rectangular compression block shown in Fig. 8 has the following properties: intensity of compression stress equals $0.63f'_c$ rather than $0.85f'_c$, the value currently specified in ACI 318-89, and the depth of the rectangular compression block is equal to 0.67 times the depth of the neutral axis, corresponding approximately to current ACI 318-89 requirements for f'_c greater than 55 MPa.

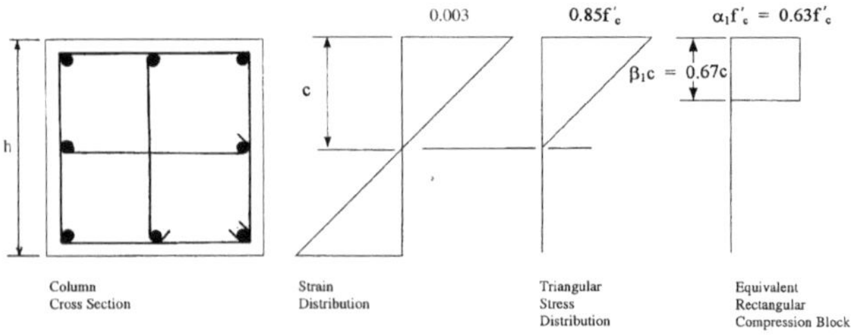

Fig. 8—Modified compression stress block.

Those authors[42] recommend that, until further research is conducted, the following equivalent rectangular compression block be adopted for calculating the nominal moment strength of concrete columns with f'_c exceeding 70 MPa and designed according to seismic provisions of ACI 318-89: When f'_c exceeds 70 MPa, the stress intensity of an equivalent rectangular compression block must be decreased linearly from 0.85 to 0.6, using the expression

$$\alpha_1 = 0.85 - 0.0073(f'_c - 69) \geq 0.6$$

for f'_c in MPa. Table 2 also gives the ratios of experimentally obtained flexural strength to the strength using the modified procedure described above for the five test columns having $f'_c \geq 100$ MPa.

A comprehensive investigation assessing the applicability of the rectangular compression block specified in ACI 318-89 for computing flexural strength of HSC columns is reported by Ibrahim and MacGregor.[49] The objective of the research project was to investigate the applicability of the rectangular stress block to HSC. The experimental phase of the investigation consisted of testing a total of 21 C-shaped specimens, 15 of which had rectangular cross-sections and six of which had triangular cross-sections having the extreme compression fiber at the tip of the triangle. The rectangular specimens included three plain specimens, while the triangular specimens included two plain specimens. The test specimens were loaded so that the entire cross section was subjected to compressive force, with the strain at one face remaining zero. The main variables were concrete strength, shape of cross section, and amount of transverse steel. The study was limited to relatively low longitudinal and transverse reinforcement ratios. The volumetric ratios of ties ranged from those required for non-seismic design to the minimum required for seismic design according to ACI 318-89.

Ibrahim et al.[49] compared the concrete component of the measured load and moment strengths of 94 tests of eccentrically loaded columns with the strengths computed using the ACI 318-89 for columns with concrete strengths ranging up to 130 MPa. For 55% of the tests, the concrete component of the strength was less than that calculated by ACI 318-89. There was a definite downward trend in the strength ratios as f'_c increased. Those authors concluded that the ACI 318-89 stress block needed revision for HSC.

Those authors[49] also reported that, for all specimens, the maximum concrete compressive strains before spalling were greater than 0.003, and concluded the following:

1. The rectangular stress block can be used to design HSC cross-sections with some modification to the parameters used to define the stress block.

2. The constant value of 0.85 for the compressive stress intensity factor as currently used by ACI 318-89 is unconservative for HSC and the following modified value should be used:

$$\alpha_1 = (0.85 - 0.00125f_c') \geq 0.725 \ (f_c' \text{ in MPa})$$

3. The distance from extreme compression face to centroid of the rectangular compression block (parameter $\beta_1 c/2$) as specified by ACI 318-89 leads to an overestimation of the lever arm. They proposed the following equation:

$$\beta_1 = (0.95 - 0.0025f_c') \geq 0.70 \quad (f_c' \text{ in MPa})$$

It has been reported[50,51] that ACI 318-89 provisions give a good estimation of flexural strengths of HSC beams. When a cross section is subjected to a bending moment only, the depth of the neutral axis at ultimate conditions is generally small and the shape of the compression block becomes less important. However, in the case of columns, the depth of the neutral axis is a significant portion of the member's overall depth, particularly if the level of axial load is relatively high, making the nominal moment capacity more sensitive to the assumed shape of the compression block.

The Canadian Code for Design of Concrete Structures[52] treats the flexural stress block for HSC in two ways. Design may be based on equations for the stress-strain curves of the concrete with peak stresses no greater than $0.9f_c'$. Alternatively, a modified rectangular stress block is defined by:

$$\alpha_1 = 0.85 - 0.0015f_c' \geq 0.67 \ (f_c' \text{ in MPa})$$
$$\beta_1 = 0.97 - 0.0025f_c' \geq 0.67 \ (f_c' \text{ in MPa})$$

These two equations were based in part on those proposed by Ibrahim et al.[49] with the further provision that they represent a stress-strain curve with peak stress not greater than $0.9f_c'$. Additionally, the Canadian code allows using 0.0035 as the maximum concrete strain.

The Canadian code (52) specifies that its strength equations and related detailing rules are applicable for concretes with f_c' ranging from 20 MPa to 80 MPa. Concrete strengths higher than 80 MPa are permitted if the designer can establish structural properties and detailing requirements for the concrete to be used. However, the Canadian code limits the range of f_c' in members resisting earthquake-induced forces to 20 MPa to 55 MPa for normal density concretes and 20 MPa to 30 MPa for structural low density concretes.

Test results on column specimens with compressive strength from about 50 to 55 MPa, and subjected to combined axial load and bending moment, have been reported by Sheikh.[53] Test columns were subjected to high axial loads ($0.6f_c'A_g$ to $0.7f_c'A_g$). Results indicate that, at this level of concrete compressive strength, ACI 318-89 procedures conservatively predict the flexural strength of the columns. Based on data reported by Sheikh and the two column tests (with f_c' = 54 and 51 MPa) given in Table 2, it could be concluded that the flexural strength of columns with a level of confinement prescribed by seismic provisions of ACI 318-89 and compressive strength below 55 MPa could be calculated conservatively by ACI 318-89 procedures.

3.2—Ductility of HSC columns under combined axial load and bending moment

Reinforced concrete columns in moment-resisting frames constructed in areas of high seismicity should be proportioned to have adequate curvature and displacement properties. This requirement has arisen, in part, as a result of observations[54-57] of field performance of columns after major earthquakes, which indicate that, despite following the strong-column weak-beam concept in design,[58-62] damage could occur at ends of the columns. Therefore, it becomes necessary for reinforced concrete columns to be proportioned in such a way that they are capable of inelastic response without appreciably losing load-carrying capacity.

One of the ways in which building codes, such as those in the U.S.,[63] ensure such ductility in columns is by specifying the amount of transverse reinforcement in critical regions of columns. However, these equations are empirical and based on strength criteria, though intended to provide ductility. Through experimental testing of NSC columns it has been shown that, although these equations are based on strength criteria, they also provide adequate ductility for reinforced NSC columns.[31,34] The extension of these equations to the case of HSC columns has been questioned.

Using the type of specimen shown in Fig. 6 and loading procedures depicted in Fig. 7, researchers have investigated effects of concrete compressive strength, type, spacing, amount and yield strength of transverse reinforcement, and level of axial loads on ductility of HSC columns. Following is a brief description of some parameters affecting the performance of HSC columns under combined and repeated loading.

3.2.1 *Effect of concrete compressive strength and axial load on ductility*—An increase in concrete compressive strength tends to result in lower ductility. Ductility also is affected adversely by an increase in the level of axial load applied to the column.

Lateral load versus lateral displacement diagrams are shown in Fig. 9 for two columns tested using the test setup shown in Fig. 6.[42] These results can be compared for effect of concrete compressive strength on ductility. The concrete compressive strength, the amount of transverse reinforcement in the critical regions of each column, and maximum interstory drift ratio for each column prior to failure are indicated in Fig. 9. Both columns were subjected to constant axial load level, equivalent to 20% of the axial load capacity of the columns. For both specimens, the spacing, amount, type, and yield strength of transverse reinforcement was the same. Both specimens had identical longitudinal steel arrangement. Both specimens used #4, Grade 60 (414 MPa yield strength) peripheral hoops at 64 mm spacing. Seismic provisions of ACI 318-89 require a larger amount of transverse reinforcement for Specimen 2 due to the higher concrete

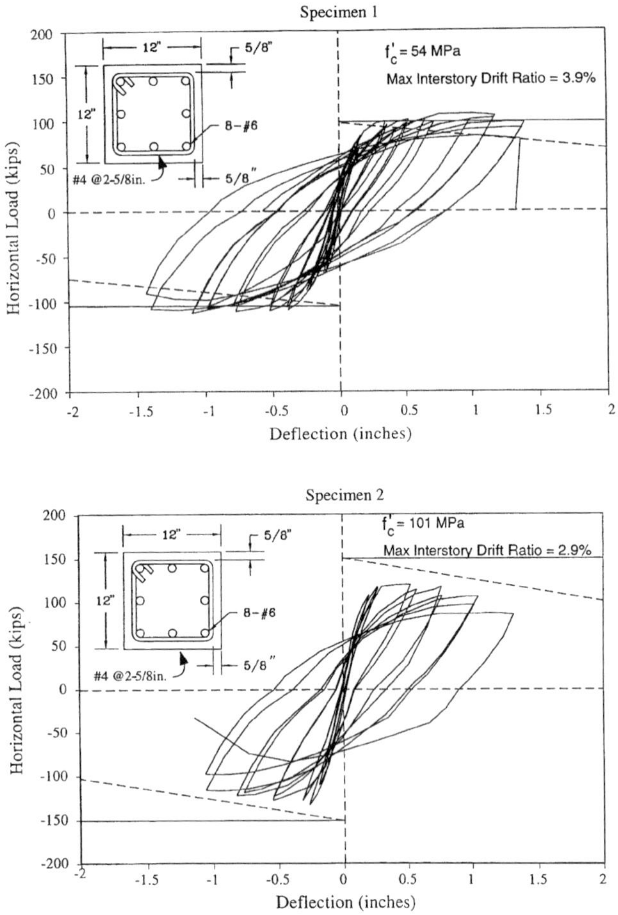

Note: Both specimens had identical transverse reinforcement arrangement.

Fig. 9—Effect of concrete compressive strength on ductility (adapted from Ref. 42).

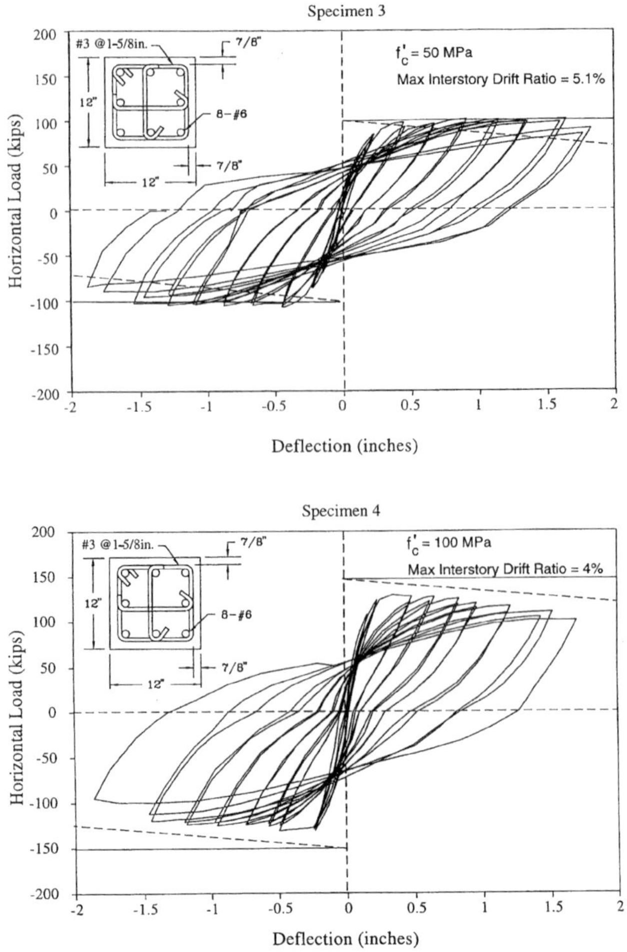

Note: Both specimens had identical transverse reinforcement arrangement.

Fig. 10—Effect of concrete compressive strength on ductility (adapted from Ref. 42).

compressive strength. As indicated in Fig. 9, increasing concrete compressive strength from 54 MPa to 101 MPa resulted in almost 25% reduction in the maximum interstory drift ratio of Specimen 2.

This reduced interstory drift ratio, however, should not be interpreted as evidence that HSC should not be constructed in areas of high seismicity. Assuming that a 4% interstory drift ratio represents a very good level of ductility, Azizinamini et al.[42] report that when axial load levels are below 20% of column axial load capacity (which is the case for most columns encountered in seismic design), adequate ductility exists for columns with transverse reinforcement levels that are even slightly below the seismic requirement of ACI 318-89. The type of test column used in their investigation was similar to that shown in Fig. 6. Figure 10 shows the cyclic lateral load versus lateral deflection relationships for two of the specimens tested. Specimens 3 and 4, whose response is shown in, had concrete compressive strengths of approximately 50 and 100 MPa, respectively, at the time of testing. Both specimens used #3, Grade 60 peripheral hoops and cross ties spaced at 38-mm. It can be observed from this figure that, although increasing concrete compressive strength resulted in a decrease in the maximum interstory drift ratio, Specimen 4, with 100 MPa concrete compressive strength, still exhibited a good level of ductility (4% interstory drift ratio). Both columns were subjected to constant axial load corresponding to 20% of their respective concentric axial load capacities.

Further evidence that HSC columns are able to behave in a ductile manner under relatively small axial load levels (below 20% of concentric axial load capacity) is provided by test results reported by Thomsen et al.[41] Those authors report results of tests on twelve relatively small columns (150-mm square cross section) with compressive strength of approximately 83 MPa. These specimens were subjected to constant axial load and repeated lateral loads. The level of axial load used by these investigators varied between 0% and 20% of the concentric axial load capacities of the columns. Those authors report that all columns were able to sustain 4% interstory drift ratio before failure, which was characterized by buckling of longitudinal bars.

Data are limited on ductility of HSC columns with axial load in the range of 20% to 30% of column axial load capacity.

In general, when the level of axial load is above 40% of column axial load capacity and concrete compressive strength is approximately 100 MPa, larger amounts of transverse reinforcement than specified in the seismic provisions of ACI 318-89 are needed. Test results indicate[36,37] that when the level of axial load is high, the use of transverse reinforcement having high yield strength could be necessary because of high confinement demands.

The behavior of HSC columns under combined bending moment and relatively high axial load was investigated by Bing et al.[40] Those authors report tests on five square HSC columns having an overall configuration similar to the specimen shown in Fig. 6. Each specimen had a 350 x 350-mm cross-section and was subjected to constant axial load and repeated lateral loads. Table 1 gives concrete compressive strength, level of applied constant axial load, and yield strength of transverse reinforcement for each test column. Figure 11 shows the lateral load vs. lateral displacement behavior for each column. The amounts of transverse reinforcement provided in the test regions of specimens 1, 2, 3, 4, and 5 (designated as unit 1, 2, 3, 4, and 5 in Fig. 11) were 133%, 103%, 131%, 108%, and 92%, respectively, of NZS 3101 requirements.[58] For Specimens 1 and 2, seismic provisions of ACI 318-89 would require 1.06 times as much transverse reinforcement as that specified by NZS 3101. For Specimens 3, 4, and 5, on the other hand, seismic provisions of ACI 318-89 would require 0.62 times as much transverse reinforcement as NZS 3101. This difference stems from the fact that NZS 3101 requirements include the effect of axial load level in calculating the required amount of transverse reinforcement for columns. As indicated in Table 1 and Fig. 11, the level of axial load on the test columns was relatively high (either $0.3f_c'A_g$ or $0.6f_c'A_g$). From these data, those authors concluded that the ductility of HSC columns designed on the basis of NZS 3101 is not adequate and that higher amounts of transverse reinforcement would be needed, especially when axial load levels are relatively high. Given the fact that at high axial load levels, ACI 318-89 seismic provisions require a lower amount of transverse reinforcement than that specified by NZS 3101 requirements, it could be concluded that for columns subjected to high axial load levels, the amount of transverse reinforcement specified by the seismic provisions of ACI 318-89 is not adequate.

Information is limited on columns with loads in excess of 0.6 times the axial load capacity and concrete compressive strength above 70 MPa. Muguruma et al.[39] report tests on twelve 200-mm square HSC columns, having geometry similar to that shown in Fig. 6, with concrete compressive strengths exceeding 120 MPa at the time of testing. Two of the variables investigated by these authors were concrete compressive strength (80 to 120 MPa) and level of axial load (25% to 63% of column's axial load capacities). One of the major conclusions drawn by those authors is that square HSC columns could be made to behave in a ductile manner, even at high axial load levels, by using high yield strength transverse reinforcement. However, two points deserve closer examination when interpreting their test results: (a) The amount of transverse reinforcement provided for test columns was as high as 230% of that required by seismic provisions of ACI 318-89; and (b) for a relatively small column cross-section (200 x 200 mm), Muguruma et al.[39] used an arrangement of 12 longitudinal bars with an extremely congested scheme of transverse reinforcement.

When concrete compressive strength is below 55 MPa, test data indicate that even at high axial load levels, ductility comparable to NSC columns could be achieved.[53] From a review of reported data, the following conclusions could be made with regard to the seismic behavior of HSC columns:

1. Columns with concrete compressive strength of approximately 55 MPa exhibited an acceptable level of ductility, even at high axial load levels.

2. Columns that had approximately 100-MPa concrete and axial loads below 20% of column axial load capacity, and that were designed based on seismic provisions of ACI 318-89, exhibited adequate ductility.

3. Data are limited to evaluate ductility of HSC columns with axial loads in the range of 20% to 30% of column axial load capacities.

4. Columns with concrete compressive strength of approximately 100 MPa and with axial loads above 30% of column axial load capacity require higher amounts of transverse reinforcement than that required by seismic provisions of ACI 318-89. Furthermore, in this range of axial load levels, higher yield strength transverse reinforcement might be necessary. Few data are available to provide design guidelines in this range of axial load levels.

3.2.2 *Effect of yield strength of transverse reinforcement*—High yield strength transverse reinforcement (yield strength exceeding 800 MPa) has been shown to be advantageous when the level of axial loads is high (above 40% of column axial load capacity). Figure 12 shows bending moment caused by lateral loads versus lateral displacement response of two test columns reported by Muguruma et al.,[39] with configuration similar to that shown in Fig. 6. All transverse reinforcement details, except for yield strength, were identical for both specimens. Both specimens were subjected to axial load levels of approximately 60% of the column axial load capacity. Compressive strength of concrete for both specimens was 85.7 MPa at the time of testing. Specimens AL-2 and AH-2 used 328 and 792 MPa yield-strength transverse reinforcement, respectively. Specimen AH-2, which used higher-yield-strength steel for transverse reinforcement, was able to sustain a 4% interstory drift ratio prior to failure. However, the amount of transverse reinforcement provided for specimen AH-2 was approximately 230% of that required by the seismic provisions of the ACI 318-89. Response of specimen AL-2, which had 328-MPa transverse reinforcement in an amount approximately equal to requirements of seismic provisions of ACI 318-89, was unsatisfactory, as indicated in Fig. 12.

When the level of axial load is relatively low (less than or equal to 20% of axial load capacity), the use of higher-yield-strength steel for transverse reinforcement may not result in any improvement in strength and ductility of HSC columns.[39,42] Table 3 shows results of two test columns that could be compared for effect of yield strength of transverse reinforcement on ductility of HSC columns. Both specimens

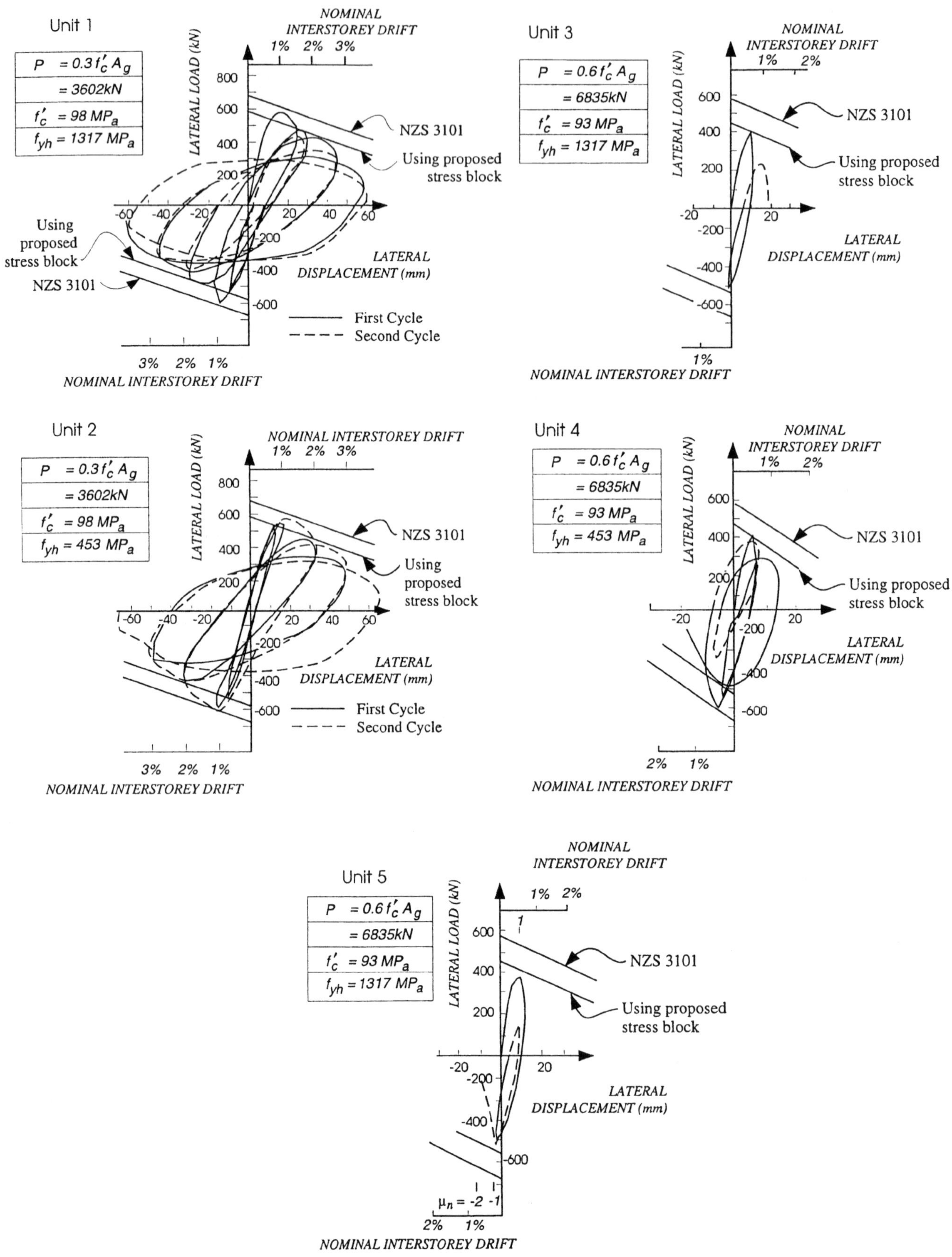

Fig. 11—Measured lateral load vs. lateral displacement hysteresis loops of columns (adapted from Ref. 40).

Table 3—Effect of yield strength of ties (adapted from Reference 42)

Specimen number	f'_c, MPa	f_y, MPa	Maximum drift index, percent
1	100	414	4
2	102	828	3.9

Table 4—Effect of yield strength of ties (adapted from Reference 42)

Specimen number	f'_c, MPa	f_y, MPa	Tie spacing, mm	Maximum drift index, percent
1	100	414	41	4
2	102	828	67	2.8

were subjected to constant axial load levels equivalent to 20% of the column capacity, and to repeated lateral loads. Both specimens failed at an interstory drift ratio of approximately 4%. This result indicates that increasing the yield strength of the transverse reinforcement has no influence on the ductility capacity of HSC columns at relatively low axial load levels. The yielding of transverse reinforcement for specimen #1 (using conventional-yield-strength transverse reinforcement) in Table 3 was observed just prior to achievement of 4% interstory drift ratio, coinciding with failure of the specimen. Therefore, it would be anticipated that increasing the yield strength of transverse reinforcement would have no significant effect.

When the level of axial load is low, an additional consideration should also be taken into account. One reason for using higher strength steel for transverse reinforcement in HSC columns is to allow larger spacing of ties. However, one should be very careful in using this approach, particularly when the level of axial load is low. This point could be explained by comparing the behavior of the two test columns given in Table 4. The spacing of transverse reinforcement for Specimens 1 and 2 in Table 4 was 41 and 67 mm, respectively. Because higher-yield-strength transverse steel (828 MPa) was used, Specimen 2 had larger spacing and, in addition, a larger transverse reinforcement ratio than that required by the seismic provisions of ACI 318-89 (105%). On the other hand, Specimen 1, which had 414 MPa transverse reinforcement, used smaller spacing and only 76% of the transverse reinforcement required by the seismic provisions of ACI 318-89. Table 4 indicates that Specimen 2 failed at an interstory drift ratio of 2.8%, while Specimen 1, which had a smaller amount of transverse reinforcement, failed at an interstory drift ratio of 4%. The use of high strength steel for transverse reinforcement could result in satisfying the transverse reinforcement requirements at a larger tie spacing. However, the larger tie spacing may lead to early buckling of longitudinal bars, as was the case for Specimen 2.

In summary, the reported data support the following conclusions regarding the use of high-yield-strength transverse reinforcement in seismic areas:

1. Use of high-yield-strength steel as transverse reinforcement is beneficial when axial loads are relatively high (above 40% of the axial load capacity of the column).

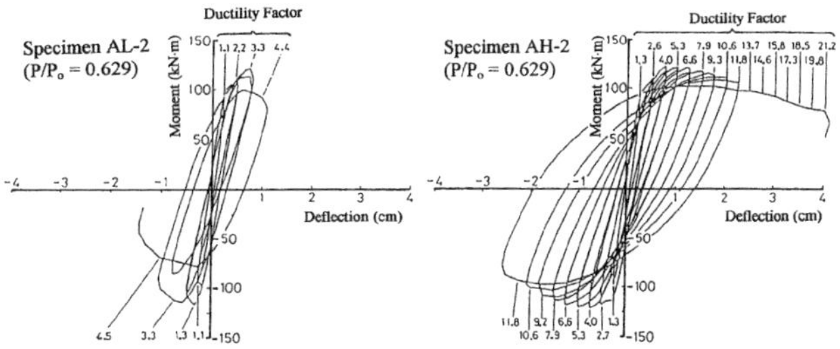

Fig. 12—Effect of yield strength of transverse reinforcement (adapted from Ref. 39).

Further, at these axial load levels, more transverse reinforcement than that required by the seismic provisions of ACI 318-89 is required to achieve adequate ductilities. The reported information is not sufficient to establish this required transverse reinforcement level.

2. When axial load is below 20% of the axial load capacity of the column, use of high-yield-strength steel for transverse reinforcement is not recommended. Although such practice could result in satisfying the code requirements, the larger permissible tie spacing may lead to premature buckling of the longitudinal steel and to lower ductility than would be achieved with lower strength steel.

CHAPTER 4—RECOMMENDED RESEARCH

Effects of the parameters related to concrete and transverse reinforcement and their influence on the behavior of HSC columns were discussed. The available information was divided into two general categories (columns under concentric loads and columns subjected to the combined effects of axial load and bending moment), and the effects of different parameters on the behavior of HSC columns were discussed separately for each category. Based on these discussions, additional areas of research are identified for increasing the understanding of behavior of HSC columns. These areas of research include: a) effect of sustained loads; b) effect of loading rate; c) effect of concrete mix proportions such as ratio of coarse to fine aggregate and silica fume; d) effect of different arrangements of transverse reinforcement on ductility under combined axial and repeated lateral loads; e) fire resistance of HSC columns; and f) additional physical tests on the effects of axial load, transverse reinforcement yield strength, longitudinal reinforcement arrangement, and transverse reinforcement layout on ductility of HSC columns under combined axial and repeated lateral loads.

CHAPTER 5—REFERENCES

1. Russell, H. G., "High-Strength Concrete in North America," *Proceedings* of Symposium on Utilization of High-Strength Concrete, Stavenger, Norway, June 15-18, 1987.

2. "Economical Composite System is Stiffened with Record 19,000 psi Mix," *Engineering News-Record*, Feb. 16, 1989.

3. "High-Strength Concrete for Pacific First Center," *Concrete International*, V. 11, No. 4, Apr. 1989.

4. Shah, S. P.; Zia, P.; and Johnston, D., "Economic Considerations for Using High-Strength Concrete in High Rise Buildings," a study prepared for Elborg Technology Co., Dec. 1983.

5. Kunze, W. E., "High-Strength Concrete in the United States," *Proceedings* of XIV Biennial Conference in Concrete, Concrete Institute of Australia, May 1989.

6. Luther, M. D., and Bauer, K. C., "Using High-Strength Concrete Simplifies Precast Column Design," *Concrete Construction*, June 1987.

7. Schmidt, W., and Hoffman, E. D., "9000 PSI Concrete—Why? Why Not?," *Civil Engineering*, ASCE, V. 45, No. 5, May 1985.

8. "Computer Cuts Column Costs," *Concrete Construction*, V. 31, No. 5, May 1986.

9. Moreno, J., and Zils, J., "Optimization of High-Rise Concrete Buildings," *Analysis and Design of High-Rise Buildings*, SP-97, J. Moreno, ed., American Concrete Institute, Farmington Hills, MI, 1985, pp. 25-92.

10. Smith, J. G., and Rad, F. N., "Economic Advantages of High-Strength Concrete in Columns," *Concrete International*, V. 11, No. 4, Apr. 1989, pp. 37-43.

11. Colaco, J. P., "75-Story Texas Commerce Plaza, Houston—The Use of High-Strength Concrete," *High-Strength Concrete*, SP-87, H. G. Russell, ed., American Concrete Institute, Farmington Hills, MI, 1985, pp. 1-9.

12. Muguruma, H.; Watanabe, F.; Iwashimizu, T.; and Mitsueda, R., "Ductility Improvement of High-Strength Concrete by Lateral Confinement," *Transactions of the Japan Concrete Institute,* V. 5, 1983, pp. 403-410.

13. Martinez, S.; Nilson, A. H.; and Slate, F. O., "Spirally Reinforced High-Strength Concrete Columns," *Research Report* No. 82-10, Department of Structural Engineering, Cornell University, Ithaca, Aug. 1982, 255 pp.

14. Martinez, S.; Nilson, A. H.; and Slate, F. O., "Spirally Reinforced High-Strength Concrete Columns," ACI JOURNAL, *Proceedings* V. 81, No. 5, Sept.-Oct. 1984, pp. 431-442.

15. Basset, R., "Behavior of Reinforced High Strength Lightweight Concrete Columns with Rectangular Ties," M.A.Sc. Thesis, University of Toronto, Toronto, ON, Canada, 1983, 108 pp.

16. Basset, R., and Uzumeri, S. M., "Effect of Confinement on the Behavior of High Strength Lightweight Concrete Columns," *Canadian Journal of Civil Engineering*, V. 13, No. 6, Dec. 1986, pp. 741-751.

17. Yong, Y. K.; Nour, M. G.; and Nawy, E. G., "Behavior of Laterally Confined High-Strength Concrete under Axial Loads," *Journal of Structural Engineering*, ASCE, V. 114, No. 2, 1988, pp. 332-351.

18. Abdel-Fattah, H., and Ahmed, S. H., "Behavior of Hoop Confined High-Strength Concrete under Axial and Shear Loads," *ACI Structural Journal*, V. 86, Nov.-Dec. 1989, pp. 652-659.

19. Bjerkeli, L.; Tomaszewica, A.; and Jensen, J. J., "Deformation Properties and Ductility of High-Strength Concrete," *High-Strength Concrete*, Second International Symposium, SP-121, W. T. Hester, ed., American Concrete Institute, Farmington Hills, MI, 1990, pp. 215-238.

20. Rangan, B. V.; Saunders, P.; and Seng, E. J., "Design of High-Strength Concrete Columns," *Evaluation and Rehabilitation of Concrete Structures and Innovations in Design*, SP-128, V. M. Malhotra, ed., American Concrete Institute, Farmington Hills, MI, 1991, pp. 851-862.

21. Nagashima, T.; Sugano, S.; Kimura, H.; Ichikawa, A., "Monotonic Axial Compression Test on Ultra-High-Strength Concrete Tied Columns," *Proceedings* of 10th World Conference on Earthquake Engineering, Madrid, Spain, 1992, pp. 2983-2988.

22. Itakura, Y., and Yagenji, A., "Compressive Test on High-Strength R/C Columns and Their Analysis Based on Energy Concept," *Proceedings* of 10th World Conference on Earthquake Engineering, Madrid, Spain, 1992, pp. 2599-2602.

23. Hatanaka, S., and Tanigawa, Y., "Lateral Pressure Requirements for Compressive Concrete," *Proceedings* of 10th World Conference on Earthquake Engineering, Madrid, Spain, 1992, pp. 2603-2608.

24. Galeota, D.; Giammatteo, M. M.; and Marino, R., "Strength and Ductility of Confined High-Strength Concrete," *Proceedings* of 10th World Conference on Earthquake Engineering, 1992, Madrid, Spain, pp. 2609-2613.

25. Cusson, D., and Paultre, P., "High-Strength Concrete Columns Confined by Rectangular Ties," *Journal of Structural Engineering*, ASCE, V. 120, No. 3, Mar. 1994, pp. 783-804.

26. Polat, M. B., Behavior of Normal and High-Strength Concrete under Axial Compression," M.A. Sc. Thesis, Department of Civil Engineering, University of Toronto, 1992, 175 pp.

27. Saatcioglu, M., and Razvi, S., "Behavior of Confined High-Strength Concrete Columns," *Proceedings* of CPCA/CSCE Structural Concrete Conference, Toronto, May 19-21, 1993, pp. 37-50.

28. Nishiyama, M.; Fukushima, I.; Watanabe, F.; and Muguruma, H., "Axial Loading Tests on High-Strength Concrete Prisms Confined by Ordinary and High-Strength Steel," *Symposium on Utilization of High-Strength Concrete*, Lillehammer, Norway, 1993, 1, pp. 322-329.

29. Collins, M. P.; Mitchell, D.; and MacGregor, J. G., "Structural Design Considerations for High-Strength Concrete," *Concrete International*, V. 15, No. 5, May 1993, pp. 27-34.

30. Razvi, S. R., and Saatcioglu, M., "Strength and Deformability of Confined High-Strength Concrete Columns," *ACI Structural Journal*, V. 91, No. 6, Nov.-Dec. 1994, pp. 678-687.

31. Sheikh, S. A., and Uzumeri, S. M., "Strength and Ductility of Tied Concrete Columns," *Journal of the Structural Division*, ASCE, V. 106, No. ST5, May 1980, pp. 1079-1102.

32. Sugano, S.; Nagashima, T.; Kimura, H.; Tamura, A.; and Ichikawa, A., "Experimental Studies on Seismic Behavior of Reinforced Concrete Members of High-Strength Concrete," *High-Strength Concrete*, Second International Symposium, SP-121, W. T. Hester, ed., American Concrete Institute, Farmington Hills, MI, 1990, pp. 61-87.

33. Saatcioglu, M., and Razvi, S., "Strength and Ductility of Confined Concrete," *Journal of Structural Engineering*, ASCE, V. 118, No. 6, June, 1992.

34. Sheikh, S. A., and Uzumeri, S. M., "Analytical Model for Concrete Confinement in Tied Columns," *Journal of the Structural Division*, ASCE, V. 108, ST12, Dec. 1982, pp. 2703-2722.

35. Watanabe, F.; Muguruma, H.; Matsutani, T.; and Sanda, D., "Utilization of High-Strength Concrete for Reinforced Concrete High-Rise Buildings in Seismic Areas," *Proceedings* of Symposium on Utilization of High-Strength Concrete, Stavanger, Norway, June 1987.

36. Muguruma, H., and Watanabe, F., "Ductility Improvement of High-Strength Concrete Columns with Lateral Confinement," *Proceedings* of Second International Symposium on Utilization of High-Strength Concrete, Berkeley, May 20-23, 1990.

37. Kabeyasawa, T.; Li, K. N.; and Huang, K., "Experimental Study on Strength and Deformability of Ultra High-Strength Concrete Columns," *Transactions of the Japan Concrete Institute*, V. 12, 1990, pp. 315-322.

38. Sakai, Y.; Hibi, J.; Otani, S.; and Aoyama, H., "Experimental Study of Flexural Behavior of Reinforced Concrete Columns Using High-Strength Concrete," *Transactions of the Japan Concrete Institute*, V. 12, 1990, pp. 323-330.

39. Muguruma, H.; Nishiyama, M.; Watanabe, F.; and Tanaka, H, "Ductile Behavior of High-Strength Concrete Columns Confined by High-Strength Transverse Reinforcement," *Evaluation and Rehabilitation of Concrete Structures and Innovations in Design*, SP-128, V. M. Malhotra, ed., American Concrete Institute, Farmington Hills, MI, V. II, 1991, p. 877-892.

40. Bing, Li; Park, R.; and Tanaka, H., "Effect of Confinement on the Behavior of High-Strength Concrete Columns under Seismic Loading," Pacific Conference on Earthquake Engineering, New Zealand, Nov. 1991.

41. Thomsen, J. H., and Wallace, J. W., "A Study of High-Strength Reinforced Concrete Columns Subjected to Lateral and Axial Loads," Open Paper Session, ACI Spring Convention, 1992.

42. Azizinamini, A.; Kuska, S.; Brungardt, P.; and Hatfield, E., "Seismic Behavior of Square High-Strength Concrete Columns," *ACI Structural Journal,* V. 91, No. 3, May-June 1994, pp. 336-345.

43. Lloyd, N. A., and Rangan, B. V., "High-Strength Concrete Columns under Eccentric Compression," *Research Report* No. 1/95, School of Civil Engineering, Curtin University of Technology, Perth, Western Australia, January 1995.

44. Mattock, A. H.; Kriz, L. B.; and Hognestad, E., "Rectangular Concrete Stress Distribution in Ultimate Strength Design," ACI JOURNAL, *Proceedings* V. 57, No. 8, Feb. 1961, pp. 875-928.

45. Hognestad, E.; Hanson, N. W.; and McHenry, D., "Concrete Stress Distribution in Ultimate Strength Design," ACI JOURNAL, *Proceedings* V. 52, No. 6, Dec. 1955, pp. 455-479.

46. Nedderman, H., "Flexural Stress Distribution in Very-High-Strength Concrete," MSc Thesis, University of Texas at Arlington, Dec. 1973, 182 pp.

47. Kaar, P. H.; Hanson, N. W.; and Capell, H. T., "Stress-Strain Characteristics of High-Strength Concrete," *Douglas McHenry International Symposium on Concrete and Concrete Structures*, SP-55, American Concrete Institute, Farmington Hills, MI, 1978, pp. 161-185.

48. Swartz, S. E.; Nikaeen, A.; Narayan Babu, H. D.; Periyakaruppan, N.; and Refai, T. M. E., "Structural Bending Properties of High-Strength Concrete," SP-87, H. G. Russell, ed., American Concrete Institute, Farmington Hills, MI, 1985, pp. 147-178.

49. Ibrahim, Hisham, and MacGregor, James G., "Flexural Behavior of High-Strength Concrete Columns," *Structural Engineering Report* No. 196, University of Alberta, Edmonton, AB, Canada, Mar., 1994.

50. Shin, S. W.; Ghosh, S. K.; and Moreno, J., "Flexural Ductility of Ultra-High-Strength Concrete Members," *ACI Structural Journal*, V. 86, No. 4, July-Aug. 1989, pp. 394-400.

51. Shin, S. W., "Flexural Behavior Including Ductility of Ultra-High-Strength Concrete Members," Ph.D. Dissertation, Department of Civil Engineering, University of Illinois at Chicago, 1986.

52. Canadian Standards Association, "A23.3-94 Design of Concrete Structures," Rexdale, ON, Canada, Dec. 1994, 199 pp.

53. Sheikh, S. A., "Deformability of High-Strength Concrete Columns," *Proceedings* of Third International Symposium on Utilization of High-Strength Concrete, Lillehammer, Norway, June 1993, pp. 346-353.

54. "Reducing Earthquake Hazards: Lessons Learned from Earthquakes," *Publication* No. 68-02, Earthquake Engineering Research Institute, El Cerrito, CA, Nov. 1986.

55. Rosenblueth, E., "The Mexico Earthquake: A First Hand Report," *Civil Engineering*, ASCE, 56(1), Jan. 1986.

56. Kreger, M. E., and Sozen, M. A., "Seismic Response of Imperial County Services Building in 1979," *Journal of Structural Engineering*, ASCE, V. 115, No. 12, Dec. 1989.

57. Ghosh, S. K., and Corley, W. G., "Behavior of Reinforced Concrete Framing Systems," *Proceedings* of International Conference on the Mexico Earthquake—1985, ASCE, Sept. 1986.

58. "Code of Practice for the Design of Concrete Structures, NZS 3101," Standards Association of New Zealand, Wellington, 1982 (Parts 1 and 2).

59. Uniform Building Code, International Conference of Building Officials, Whittier, CA, 1988.

60. "Tentative Provision for the Development of Seismic Regulations for Buildings," *Publication* No. ACT-3-06, Applied Technology Council, Palo Alto, Calif., June 1978.

61. "Recommended Lateral Force Requirements and Commentary," Seismology Committee, Structural Engineers Association of California, San Francisco, CA, 1980.

62. "CAN3-A23.3-M84 Design of Concrete Structures for Buildings," Canadian Standards Association, Rexdale, ON, Canada, 1984.

63. "Building Code Requirements for Reinforced Concrete (ACI 318-89)," American Concrete Institute, Farmington Hills, MI, 1989, 347 pp.

CHAPTER 6—NOTATION

A_g = gross area of column cross-section
A_{st} = cross-sectional area of longitudinal reinforcing steel
f_c' = concrete compressive strength
f_y = yield strength of non-prestressed steel
f_{yh} = yield strength of hoops
f_{yt} = yield strength of transverse reinforcing steel
P = axial load
P_o = pure axial load capacity of columns calculated according to nominal strength equations of ACI 318-89
α_1 = compression block stress intensity factor
β_1 = ratio of depth of compression stress block to depth of neutral axis
Δ = axial displacement
ε_{01} = axial strain corresponding to peak stress of plain concrete
ρ_s = ratio of volume of hoops to total volume of core concrete (out-to-out of hoops)

Report on Equivalent Rectangular Concrete Stress Block and Transverse Reinforcement for High-Strength Concrete Columns

Reported by Joint ACI-ASCE Committee 441

Sungjin Bae, Chair | Aly Said, Secretary

Ahmed Abd El Fattah, Perry Adebar, Shahria Alam, Bassem Andrawes, Oguzhan Bayrak, Muhammad A. Cheema, Rami Eid, Asad Esmaeily, Richard W. Furlong, Wael Mohammed Hassan, Riyadh A. Hindi, Mahmoud E. Kamara, Tony C. Liu, Mustafa A. Mahamid, S. Ali Mirza, Ronald L. O'Kane, Patrick Paultre, Hayder A. Rasheed, Murat Saatcioglu, Ayman E. Salama, Halil Sezen, Shamim A. Sheikh, Nadim I. Wehbe

Consulting Members

Alaa E. Elwi, Esko Hyttinen, Said Iravani, Chien-Hung Lin, Santiago Pujol, L. N. Ramamurthy

This report provides a research summary of equivalent rectangular concrete compressive stress blocks and transverse reinforcement design requirements for high-strength concrete (HSC) columns. Because ACI 318 code provisions for column design are mostly based on concrete strengths less than 10,000 psi (70 MPa), the use of equivalent rectangular concrete stress block factors given in the code has been questioned. As a result, many alternative expressions have been developed. This report provides a summary of various suggestions of equivalent rectangular concrete stress block and design guidelines for HSC columns.

The report also provides highlights of the research on the performance of HSC columns under various loading conditions, including monotonically increasing concentric or eccentric compression, and load reversals with increasing deformation and constant axial compression. The behavior of HSC columns subjected to combined axial load and bending moment is discussed. Various proposals for determining the amount and details of transverse reinforcement for seismic design are also reviewed.

Keywords: axial load; bending moment; columns; concrete stress block; ductility; flexural strength; high-strength concrete; longitudinal reinforcement; seismic design; transverse reinforcement.

CONTENTS

ACI 441.1R-18 was adopted and published July 2018.

CHAPTER 1—INTRODUCTION

High-strength concrete (HSC) has advantages over normal-strength concrete (NSC), especially for columns of high-rise buildings, as it reduces column sizes and increases the durability of concrete (Smith and Rad 1989). Moreover, HSC can be advantageous with regard to lateral stiffness and axial shortening (Colaco 1985). Another advantage cited by Colaco in the use of HSC columns is the cost reduction of formwork stemming from reduced cross-sectional dimensions. This economic advantage is achieved by using HSC in lower-story columns and reducing concrete strength over the height of the building while keeping the same column size over the building height.

Increased use of HSC has caused concerns over the applicability of the current building code requirements (ACI 318) for the design and detailing of HSC columns. Those concerns are mainly related to: 1) equivalent rectangular concrete stress distribution; and 2) transverse reinforcement requirements for seismic design. Chapter 22 of ACI 318-14 provides a concept of equivalent rectangular concrete stress distribution (ACI concrete stress block) for design of reinforced concrete columns. In the equivalent rectangular concrete stress block, an average stress of $0.85f_c'$ is used with a rectangle of depth $a = \beta_1 c$. The 1976 supplement to the 1971 code adopted a lower limit of β_1 equal to 0.65 based on research data from tests with concrete strengths exceeding 8000 psi (55 MPa). Several research studies reported that the use of current rectangular concrete stress block expressions of ACI 318 could produce overestimated flexural and axial strengths of HSC columns (Wahidi 1995; Ibrahim and MacGregor 1996; Lloyd and Rangan 1996). As a result, alternative concrete stress block expressions have been proposed (Ibrahim and MacGregor 1997; Bae and Bayrak 2003; Ozbakkaloglu and Saatcioglu 2004; Azizinamini et al. 1994).

Numerous research studies (Sakai and Sheikh 1989; Elwood et al. 2009a,b; Paultre and Légeron 2008) have been conducted in several countries to investigate the behavior of HSC columns, to evaluate similarities or differences between HSC and NSC columns, and to identify important parameters affecting performance of HSC columns designed for seismic, as well as nonseismic, areas. As a result, Chapter 18 of ACI 318-14 has updated the transverse reinforcement requirement to address the concern on the use of HSC columns for seismic design.

This document reports the results of recent studies on the equivalent rectangular concrete stress distribution, or concrete stress block, and transverse reinforcement requirements of HSC columns for seismic design.

CHAPTER 2—NOTATION AND DEFINITIONS

2.1—Notation

A_b = area of an individual longitudinal reinforcing bar, in.^{2} (mm^2)
A_c = area of core concrete measured out-to-out of transverse reinforcement, in.^{2} (mm^2)
A_{ch} = cross-sectional area of a member measured to the outside edges of transverse reinforcement, in.^{2} (mm^2)
A_g = gross area of concrete section, in.^{2} (mm^2). For a hollow section, A_g is the area of concrete only and does not include the area of void(s)
A_{sh} = total cross-sectional area of transverse reinforcement, including crossties, within spacing s (or s_h) and perpendicular to dimension b_c, in.^{2} (mm^2)
A_{st} = total area of nonprestressed longitudinal reinforcement including bars or steel shapes, and excluding prestressing reinforcement, in.^{2} (mm^2)
A_{te} = sum of the areas of tie legs used to provide lateral support against buckling for longitudinal reinforcement of column, in.^{2} (mm^2)
b = width of compression face of member, in. (mm)
b_c = cross-sectional dimension of member core measured to the outside edges of the transverse reinforcement composing area A_{sh}, in. (mm)
c = distance from extreme compression fiber to neutral axis, in. (mm)
d = distance from extreme compression fiber to centroid of longitudinal tension reinforcement, in. (mm)
d'' = depth of concrete core of column measured from center-to-center of peripheral rectangular hoop, circular hoop, or spiral in. (mm)
d_b = nominal diameter of bar, wire, or prestressing strand, in. (mm)
E_s = modulus of elasticity of reinforcement and structural steel, excluding prestressing reinforcement, psi (MPa)
f_c' = specified compressive strength of concrete, psi (MPa)
f_{cm} = measured compressive strength of concrete, psi (MPa)
f_s = transverse steel stress at or shortly after the attainment of maximum strength under concentric compression, psi (Eq. (4.5))
f_y = specified yield strength for nonprestressed reinforcement, psi (MPa)
f_{yh} (or f_{yt}) = specified yield strength of transverse reinforcement, psi (MPa)
h = overall height or depth of member, in. (mm)
h'' = dimension of concrete core of rectangular section, measured perpendicular to the direction of the hoop legs, measured to the outside of the peripheral hoop, in. (mm)
h_c = cross-sectional dimension of member core measured to outside edges of transverse reinforcement composing area A_{sh}, in. (mm), parallel to shear force in the member

h_x = maximum center-to-center spacing of longitudinal bars laterally supported by corners of crossties or hoop legs around the perimeter of the column, in. (mm)

k_f = concrete strength factor, which is calculated as $f_c'/25{,}000 + 0.6 \geq 1.0$, where f_c' is in psi

k_n = confinement effectiveness factor, which is calculated as $n_\ell/(n_\ell - 2)$, where n_ℓ is the number of longitudinal bars or bar bundles around the perimeter of a column core with rectilinear hoops that are laterally supported by the corner of hoops or by seismic hooks

M = maximum moment, in.-lb (N-mm) (Fig. 3.3)

M_n = nominal flexural strength of the section, in.-lb (N-mm)

N^* = design axial load at ultimate limit state; to be taken as positive for compression and negative for tension, lb (N)

P = maximum axial load; to be taken as positive for compression and negative for tension, lb (N) (Fig. 3.3)

P_n = nominal axial strength of member, lb (N)

$P_{n,max}$ = maximum nominal axial compressive strength of member, lb (N)

P_o = nominal axial strength at zero eccentricity, lb (N)

P_u = factored axial load; to be taken as positive for compression and negative for tension, lb (N)

s (or s_h) = center-to-center spacing of items, such as longitudinal reinforcement, transverse reinforcement, tendons, or anchors, in. (mm)

s_o = center-to-center spacing of transverse reinforcement within potential plastic hinge length, in. (mm)

α_1 = factor relating magnitude of uniform stress in equivalent rectangular concrete stress block to specified compressive strength of concrete

β_1 = factor relating depth of equivalent rectangular concrete stress block to depth of neutral axis

Δ = relative lateral deflection between the top and bottom of a column, in. (mm)

$\sum A_b$ = sum of areas of longitudinal reinforcing bars reliant on the tie, in.^{2} (mm^2)

ρ_s = ratio of the volume of spiral reinforcement to the total volume of core confined by spiral (measured out-to-out of spirals)

ρ_t = ratio of the area of nonprestressed longitudinal reinforcement to gross concrete area ($= A_{st}/A_g$)

ρ_{tc} = ratio of the area of transverse reinforcement, A_{sh}, to the area of core perpendicular to that transverse reinforcement

ϕ = strength reduction factor

2.2—Definitions

ACI provides a comprehensive list of definitions through an online resource, ACI Concrete Terminology. Definitions provided here complement that resource.

axial load ratio—ratio of axial load to the product of compressive strength of concrete and the gross area of concrete cross section.

confinement index—product of transverse reinforcement ratio, either by area or volume, and the yield strength of the transverse reinforcement divided by the compressive strength of concrete.

ductility—ability of a reinforced concrete section or member to maintain its strength under monotonic load or when subjected to repeated load reversals beyond the linear range of response.

killed steel—steel made by completely removing or tying up the oxygen in the liquid steel through the addition of elements such as aluminum or silicon before the ingot solidifies, with the objective of achieving maximum uniformity in the steel.

limiting drift—drift corresponding to a 20 percent reduction in lateral load resistance of a reinforced concrete member subjected to load reversals with increasing maximum displacements.

limiting drift ratio—ratio of limiting drift to column height.

story drift—horizontal deflection at the top of the story relative to the bottom of the story.

volumetric transverse reinforcement ratio—ratio of the volume of transverse reinforcement confining the concrete core of a potential plastic hinge region to the volume of concrete inside the confined core.

CHAPTER 3—EQUIVALENT RECTANGULAR CONCRETE STRESS BLOCK

Although the stress-strain characteristics of high-strength concrete (HSC) are different from those of normal-strength concrete (NSC), there is no well-defined compressive strength boundary between the two. Instead, there is a gradual change with increasing concrete strengths, as shown in Fig. 3 (ACI Innovation Task Group 4 2006). The figure shows that the ascending branch of the stress-strain relationship is steeper for higher-strength concretes, indicating higher elastic modulus. The stress-strain curve changes from approximately a second-order parabola for concretes within

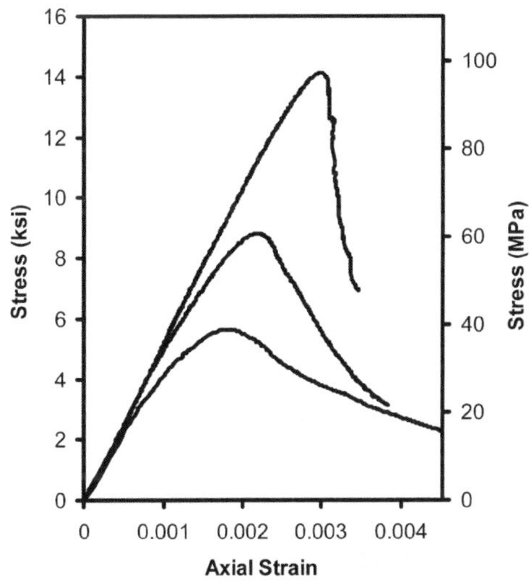

Fig. 3—Compressive stress-strain curves for plain normal-density concrete (ACI Innovation Task Group 4 2006).

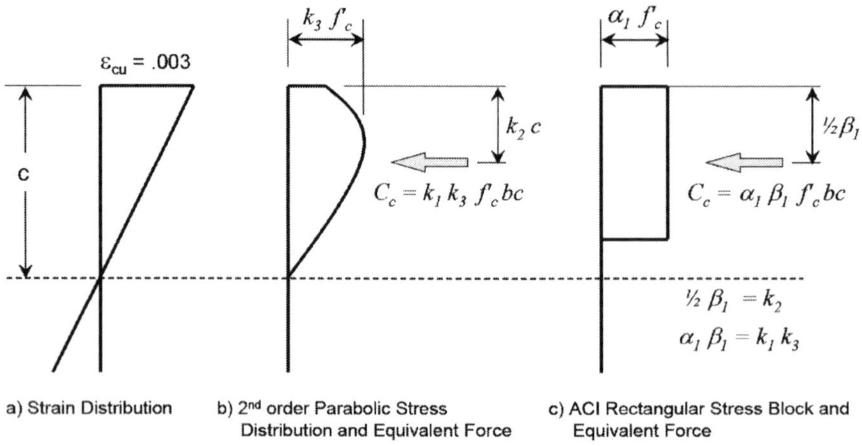

Fig. 3.1—Factors for equivalent rectangular concrete stress block.

Table 3.2—Rectangular concrete stress block expressions: codes and proposals

	f_c' in MPa	f_c' in psi
ACI 318-14	$\alpha_1 = 0.85$ $\beta_1 = 0.85 - 0.008(f_c' - 30)$ $0.85 \geq \beta_1 \geq 0.65$	$\alpha_1 = 0.85$ $\beta_1 = 0.85 - 0.05/1000 \times (f_c' - 4000)$ $0.85 \geq \beta_1 \geq 0.65$
CSA A23.3-04	$\alpha_1 = 0.85 - 0.0015f_c' \geq 0.67$ $\beta_1 = 0.97 - 0.0025f_c' \geq 0.67$	$\alpha_1 = 0.85 - 0.010/1000 \times f_c' \geq 0.67$ $\beta_1 = 0.97 - 0.017/1000 \times f_c' \geq 0.67$
NZS 3101:2006) (or Li et al. [1994])	$\alpha_1 = 0.85 - 0.004(f_c' - 55)$ $0.85 \geq \alpha_1 \geq 0.75$ $\beta_1 = 0.85 - 0.008(f_c' - 30)$ $0.85 \geq \beta_1 \geq 0.65$	$\alpha_1 = 0.85 - 0.028/1000 \times (f_c' - 8000)$ $0.85 \geq \alpha_1 \geq 0.75$ $\beta_1 = 0.85 - 0.05/1000 \times (f_c' - 4000)$ $0.85 \geq \beta_1 \geq 0.65$
Ibrahim and MacGregor (1997)	$\alpha_1 = 0.85 - 0.00125f_c' \geq 0.725$ $\beta_1 = 0.95 - 0.0025f_c' \geq 0.70$	$\alpha_1 = 0.85 - 0.0086/1000 \times f_c' \geq 0.725$ $\beta_1 = 0.95 - 0.0172/1000 \times f_c' \geq 0.70$
Bae and Bayrak (2003)	$\alpha_1 = 0.85 - 0.004(f_c' - 70)$ $0.85 \geq \alpha_1 \geq 0.67$ $\beta_1 = 0.85 - 0.004(f_c' - 30)$ $0.85 \geq \beta_1 \geq 0.67$	$\alpha_1 = 0.85 - 0.028/1000 \times (f_c' - 10{,}000)$ $0.85 \geq \alpha_1 \geq 0.67$ $\beta_1 = 0.85 - 0.028/1000 \times (f_c' - 4000)$ $0.85 \geq \beta_1 \geq 0.67$
Ozbakkaloglu and Saatcioglu (2004)	$\alpha_1 = 0.85 - 0.0014(f_c' - 30)$ $0.85 \geq \alpha_1 \geq 0.72$ $\beta_1 = 0.85 - 0.0020(f_c' - 30)$ $0.85 \geq \beta_1 \geq 0.67$	$\alpha_1 = 0.85 - 0.01/1000 \times (f_c' - 4000)$ $0.85 \geq \alpha_1 \geq 0.72$ $\beta_1 = 0.85 - 0.013/1000 \times (f_c' - 4000)$ $0.85 \geq \beta_1 \geq 0.67$
Azizinamini et al. (1994)	$\alpha_1 = 0.85 - 0.007(f_c' - 69)$ $0.85 \geq \alpha_1 \geq 0.60$ $\beta_1 = 0.85 - 0.008(f_c' - 30)$ $0.85 \geq \beta_1 \geq 0.65$	$\alpha_1 = 0.85 - 0.05/1000 \times (f_c' - 10{,}000)$ $0.85 \geq \alpha_1 \geq 0.60$ $\beta_1 = 0.85 - 0.05/1000 \times (f_c' - 4000)$ $0.85 \geq \beta_1 \geq 0.65$

the normal-strength range to almost a straight line as the strength approaches 14,000 psi (97 MPa). The strain at peak concrete stress (ε_o) increases with strength as well, varying approximately from 0.002 to 0.003 for 3000 to 14,000 psi (21 to 97 MPa) concrete, respectively. Failure becomes more brittle as the concrete strength increases and unloading beyond the peak stress becomes more rapid.

3.1—ACI 318-14 concrete stress block

Because it is not straightforward to calculate member strengths using a nonlinear stress-strain relationship, ACI 318-14 Section 22.2.2.4 provides an equivalent rectangular concrete stress block for ease in design calculations. This equivalent stress block is derived so that the area under the actual nonlinear stress distribution, or resultant force, and the centroid of this area, which is the point of application of resultant force, correspond closely to those of the equivalent stress block, as shown in Fig. 3.1. The stress block adopted by ACI 318 is of rectangular geometry. The stress block expression of ACI 318 stipulates that the α_1 factor is not sensitive to compressive strength and remains constant at 0.85, while the β_1 factor decreases from 0.85 for a compressive strength of 4000 psi (28 MPa) to 0.65 for a compressive strength of 8000 psi (55 MPa). The strain limit at the extreme compression fiber in the concrete at the onset of crushing is 0.003 as shown in Fig. 3.1(a).

3.2—Other concrete stress blocks

Test results of HSC columns indicated that the ACI 318 concrete stress block tends to overpredict flexural and axial strengths as concrete strength increases (Wahidi 1995; Ibrahim and MacGregor 1996; Lloyd and Rangan 1996). As a result, various expressions of concrete stress block have been developed (Table 3.2). Comparisons of the stress inten-

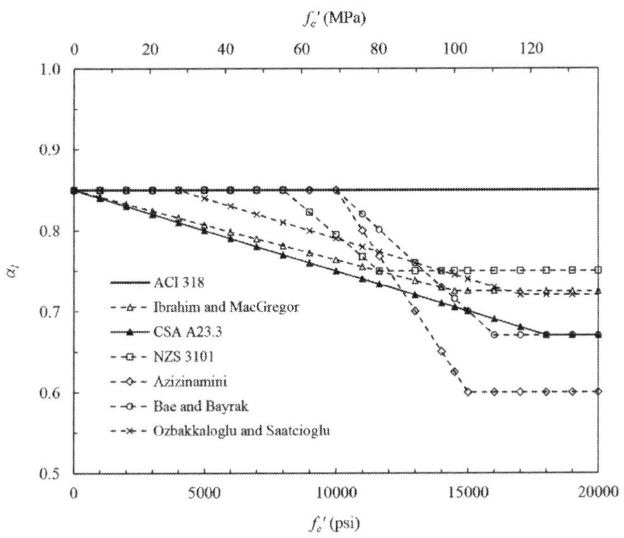

Fig. 3.2a—Comparison of proposed expressions for stress intensity factor α_1.

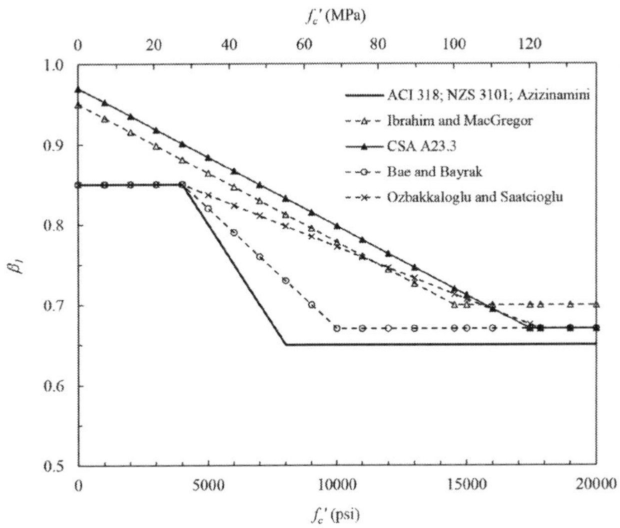

Fig. 3.2b—Comparison of proposed expressions for stress block depth factor β_1.

sity factor α_1 and the stress block depth factor β_1 in various stress blocks are shown in Fig. 3.2a and Fig. 3.2b. The stress intensity factor α_1 of ACI 318 does not vary with concrete strength. The α_1 factors of other codes and proposals in Table 3.2 reduce with increased concrete strength. The stress block depth factor β_1 of ACI 318, however, reduces with increased concrete strength; a similar trend can be observed for other codes and proposals. The stress block depth factor β_1 of ACI 318 is adopted in New Zealand Standard (NZS) code and used in a proposal by Azizinamini et al. (1994). Comparisons of α_1 and β_1 factors in Fig. 3.2a and Fig. 3.2b show that it is difficult to find consensus among different codes and proposals of various researchers.

A parametric study was conducted as part of this report to provide further insight into the differences among various proposals. Column interaction diagrams were calculated without the strength-reduction factors ϕ to compare ACI 318 stress block with the other code expressions and proposals of various researchers. The column cross section analyzed is shown in Fig. 3.2c, with the bending moment about the Y-Y axis. The column was analyzed for steel ratio of 1 percent and concrete compressive strengths of 4000, 8000, 12,000, and 16,000 psi (28, 55, 83, and 110 MPa). Results of the parametric study are given in Fig. 3.2d.

From Fig. 3.2d(a) and 3.2d(b), for concrete compressive strengths of 4000 and 8000 psi (28 and 55 MPa), the nominal column strength obtained by the ACI 318 stress block were noticeably different from those of the CSA A23.3-04 and Ibrahim and MacGregor (1997). The stress blocks of CSA A23.3-04 and Ibrahim and MacGregor (1997) resulted in progressively smaller estimates of nominal column strength as concrete compressive strength increased. Other stress blocks resulted in identical estimates of nominal strength to ACI 318 at a concrete compressive strength of 4000 psi (28 MPa), and produced slightly more conservative estimates

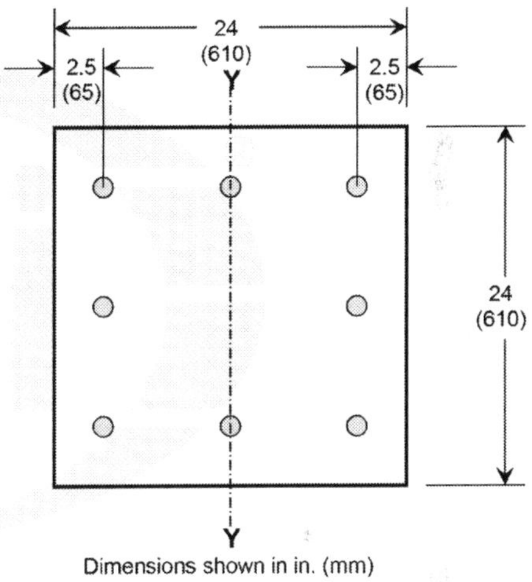

Fig. 3.2c—Column cross section used in parametric study.

than ACI 318 at a concrete compressive strength of 8000 psi (55 MPa).

For a concrete compressive strength of 12,000 psi (83 MPa) in Fig. 3.2d(c), all stress block expressions resulted in conservative estimates compared with the ACI 318 stress block. The stress block proposed by Bae and Bayrak (2003) produced less-conservative nominal strength estimates in comparison to other stress block expressions, except for ACI 318. For a concrete compressive strength of 12,000 psi (83 MPa), the stress blocks of NZS, Azizinamini et al. (1994), and Ibrahim and MacGregor (1997) produced identical or close to identically the same nominal strengths.

For a concrete compressive strength of 16,000 psi (110 MPa) in Fig. 3.2d(d), all stress block expressions yielded more-conservative results than those obtained by using the stress block of ACI 318. However, the model proposed by Azizinamini et al. (1994) resulted in severe underestimates of strength when compared to the other expressions.

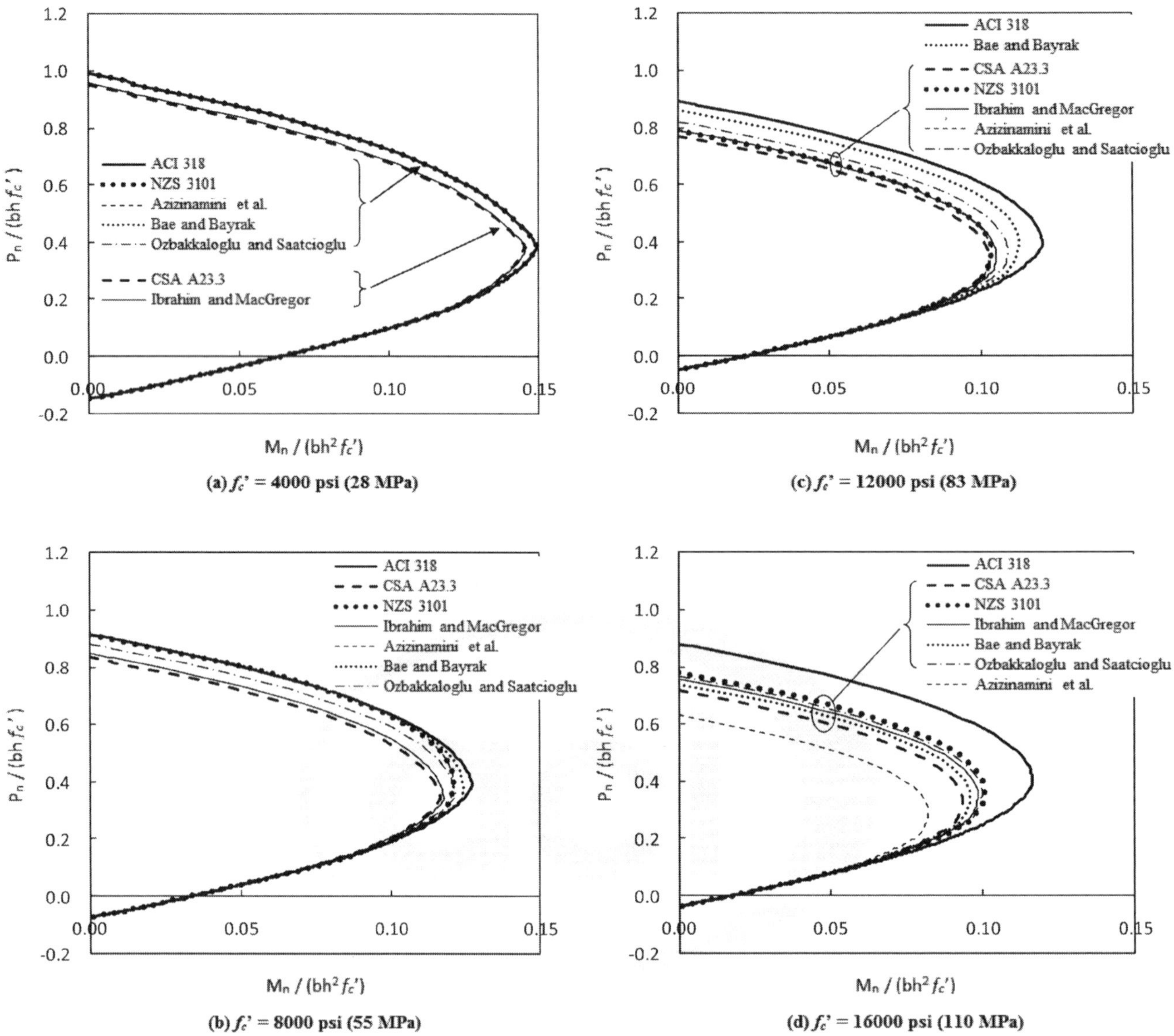

Fig. 3.2d—Column strength interaction diagrams comparing different stress blocks.

3.3—Performance of ACI concrete stress block

Bae and Bayrak (2013) evaluated the performance of the ACI 318 concrete stress block and other concrete stress blocks in Table 3.2. The performance of various stress block expressions for calculating flexural and axial strengths was evaluated using 224 tested column results.

Normalized *P-M* interaction curves were used to represent various column tests. Details of tested column database, their references, and the more detailed explanations of normalized interaction curves are found in Bae and Bayrak (2003, 2013).

The results for the comparison of ACI 318 concrete stress block expressions are shown in Fig. 3.3 with varying ranges of concrete strengths. ACI 318 limits the design axial strength in compression to $0.80\phi P_o$ for columns with tie reinforcement (Section 22.4 of ACI 318-14). This maximum allowable design compressive strength ($\phi P_{n,max}$) is also included in the figures. Column strengths are underestimated when data points are located outside the nominal interaction curve (M_n, P_n), whereas column strengths are overestimated when data points are placed within the nominal interaction curve.

Figure 3.3(a) indicates that ACI 318 concrete stress block is acceptable for columns with concrete strengths less than 8000 psi (55 MPa). For columns with concrete strengths greater than 8000 psi (55 MPa), Fig. 3.3(b) and (c) indicate that the ACI 318 concrete stress block can overestimate the column strengths gradually with the increase of concrete strength. As a result, safety margins of factored design strengths are reduced for columns with concrete strengths greater than 8000 psi (55 MPa). However, a careful examination of Fig. 3.3(b) suggests that the column strengths predicted by the ACI concrete stress block are acceptable up to the concrete strength of 12,000 psi (83 MPa), considering that axial loads are limited to the maximum allowable design compressive strength ($\phi P_{n,max}$). Therefore, the ACI

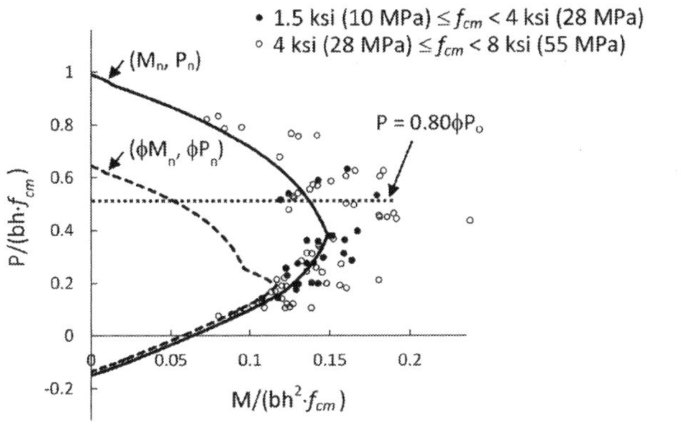

(a) Columns with f_{cm} < 8000 psi (55 MPa).

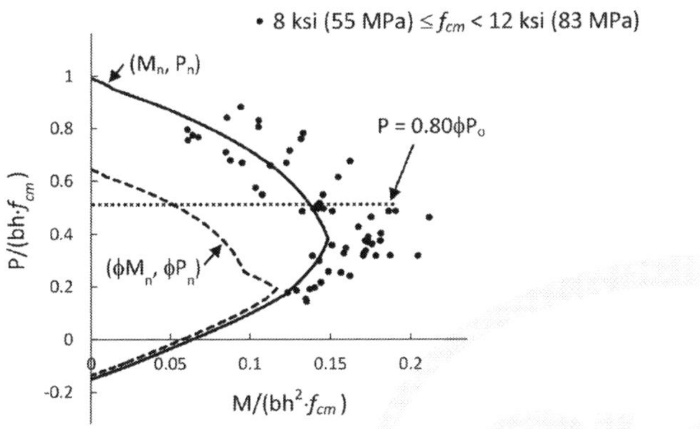

(b) Columns with 8000 psi (55 MPa) ≤ f_{cm} < 12,000 psi (83 MPa).

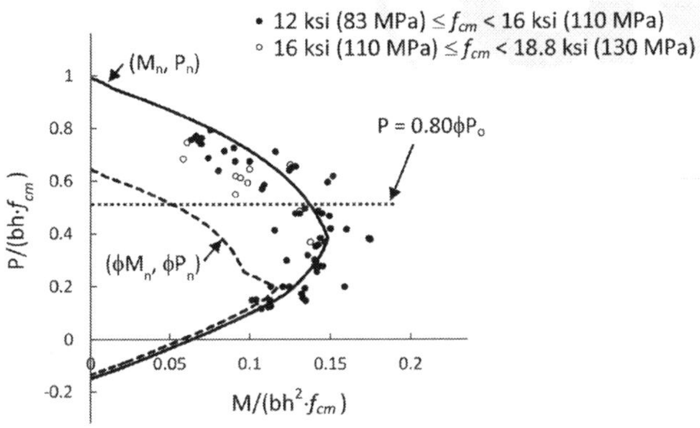

(c) Columns with 12,000 psi (83 MPa) ≤ f_{cm}.

Fig. 3.3—Performance of ACI 318-14 stress block expressions on predicted column strengths (Bae and Bayrak 2013).

318 stress block can be used for column design up to the concrete strength of 12,000 psi (83 MPa).

Bae and Bayrak (2013) stated that ACI 318 concrete stress block should either be avoided or used with caution for column design when the concrete strength exceeds 12,000 psi (83 MPa). The concrete stress blocks of CSA A23.3-04, New Zealand Standard (NSZ) 3101:2006, Ibrahim and MacGregor (1997), Bae and Bayrak (2003), or Ozbakkaloglu and Saatcioglu (2004) were recommended in such cases.

CHAPTER 4—TRANSVERSE REINFORCEMENT

The increasing brittleness of concrete with higher compressive strength is a major concern for seismic applications where toughness under repeated load reversals is of paramount importance. For this reason, proper confinement of concrete is essential for the safe use of high-strength concrete (HSC) in moderate to high seismic applications.

This chapter addresses concrete confinement for columns. Constitutive models for confined concrete, salient features of previous research, and design recommendations are provided in the following sections.

4.1—Constitutive models for confined concrete

Several researchers have indicated that constitutive models developed for normal-strength concrete (NSC) do not offer a good representation of the behavior of HSC, especially in the case of columns, where the characteristics of the constitutive model have the highest impact on the calculated response. Therefore, previously developed constitutive models have been modified to reflect the differences in behavior, and several additional analytical models have been developed specifically for HSC.

Ahmad and Shah (1982), Martinez et al. (1984), and Fafitis and Shah (1985) were among the first to develop models for high-strength confined concrete based on tests of spirally reinforced small cylinders. These models incorporate the effect of confinement through a lateral confining pressure that develops under hoop tension. The models were shown to produce good correlations with tests of spirally confined circular cylinders for concrete strengths of up to 12,000 psi (83 MPa).

Yong et al. (1988) developed a model based on small-scale square column tests with concrete strengths ranging between 12,000 and 13,600 psi (83 and 94 MPa). Their approach was similar to that originally proposed by Sargin et al. (1971) for NSC. Azizinamini et al. (1994) subsequently modified the model based on column tests under reversed cyclic loading.

Bjerkeli et al. (1990) proposed a generalized model for normalweight and lightweight aggregate confined concretes with compressive strengths of up to 13,000 and 10,000 psi (90 and 69 MPa), respectively. Their model is applicable to elements with circular, square, and rectangular section geometry.

Several confinement models were developed in Japan based on experimental results from the New RC project (Muguruma and Watanabe 1990; Muguruma et al. 1991; Nagashima et al. 1992).

Cusson and Paultre (1995) proposed a model based on tests of HSC columns. Their model uses the effectively confined core area concept that was originally proposed by Sheikh and Uzumeri (1982) and modified by Mander et al. (1988). This partially iterative-based model was later improved by Légeron and Paultre (2003), introducing a direct procedure to compute the strain in transverse reinforcement. Moreover, Paultre et al. (1996) presented an expression for calculating the delamination buckling of the concrete cover in HSC columns, while Cusson et al. (1996) showed that HSC columns loaded in axial compression are characterized by strain localization in the concrete. Cusson et al. proposed a model capable of predicting this post-peak load-displacement behavior.

Li et al. (1994) developed a constitutive model for confined concrete that covered a wide range of concrete compressive strengths, between 4000 and 19,000 psi (28 and 131 MPa). The model was comprehensive and elaborate, incorporating several parameters to reflect the effects of confinement.

Razvi and Saatcioglu (1999) developed a generalized confined model based on the equivalent uniform lateral pressure concept that they proposed earlier for a confined model of NSC (Saatcioglu and Razvi 1992). The model covers a wide range of concrete compressive strengths between 3000 and 19,000 psi (21 and 131 MPa), and incorporates the effects of different reinforcement geometry and arrangement while also incorporating the effect of high-strength transverse reinforcement.

Eid et al. (2007) introduced a fully analytical, yet design-oriented, confinement model based on the theories of elasticity and plasticity (Eid and Dancygier 2005, 2006). The model simulates the behavior of reinforced-concrete circular columns with compressive concrete strengths of 2600 to 15,000 psi (18 to105 MPa).

4.2—Previous research and general observations

The most challenging aspects of interpreting the results of the tests on beams and columns found in the literature are the differences in loading protocols, loading configurations, and scale and failure criteria used by different researchers. Thus, P-Δ effects, shear strengths, and drifts at failure are not directly comparable in some instances (Brachmann et al. 2004a,b). Despite these differences, there are some well-established common trends that have been observed about the behavior of beams and columns with HSC.

Larger section ductility and displacement ductility were obtained from HSC beams than NSC beams with similar amounts of longitudinal and transverse reinforcement, even though the ultimate compressive strain was smaller for HSC (Nilson 1985; Shin et al. 1990).

Transverse steel volumetric ratio and axial-load level have significant effects on the behavior of HSC columns. High-strength concrete columns can be used safely in seismic design for a wide range of axial loads when an adequate amount of transverse reinforcement is provided to confine the core concrete (Bayrak and Sheikh 1997, 1998a; Légeron and Paultre 2000; Paultre et al. 2001).

Column deformability decreases with an increase of compressive loads. When the axial load demand on the columns is below $0.2f_c'A_g$, drift ratios more than 3 percent can be obtained for HSC columns with transverse reinforcement conforming to the provisions in Chapter 22 of ACI 318-14 (Bjerkeli et al. 1990; Sheikh et al. 1994; Azizinamini et al. 1994; Sheikh and Khoury 1997; Xiao and Yun 1998; Xiao and Martirossyan 1998; Matamoros and Sozen 2003; Bae and Bayrak 2009).

A minimum of 5 percent drift capacity can be attained in HSC circular columns when the volumetric ratio of spiral reinforcement is at least equal to $0.17f_c'/f_{yt}$ and the columns are subjected to less than 43 percent of P_o as constant axial compression (Lipien and Saatcioglu 1997; Saatcioglu and Baingo 1999; Bae and Bayrak 2006).

The volume of transverse reinforcement required for proper confinement of HSC columns could be reduced with the use of high-strength transverse reinforcement, particularly for high compressive axial loads. However, the use of high-strength transverse reinforcement did not improve the behavior when axial loads were low (Razvi and Saatcioglu 1994).

The strength of confined concrete increased with the increase of the amount of transverse reinforcement. For a similar percent increase in strength, higher confinement pressure is required for HSC than for NSC. The distribution and spacing of the transverse reinforcement is another important parameter that affects the behavior. Although high-strength reinforcement can be used to decrease the volumetric transverse reinforcement ratio, the effectiveness of the transverse reinforcement decreases as spacing increases. High-strength transverse reinforcement may not develop its yield strength when the volumetric reinforcement ratio or the axial compression is low. In addition, increasing the spacing of the transverse reinforcement by using high-strength reinforcement increased the probability of failure due to buckling of the longitudinal reinforcement (Thomsen and Wallace 1994; Saatcioglu et al. 1998; Kato et al. 1998). The suggested upper limit on the strength of the transverse reinforcement was 100 ksi (690 MPa) by Kato et al. (1998) and 145 ksi (1000 MPa) by Saatcioglu and Baingo (1999).

Premature spalling of cover concrete under concentric loading was observed before reaching the crushing strength of unconfined concrete. This was attributed to the stability failure of the cover shell under high compressive stresses when a mesh of reinforcement consisting of longitudinal reinforcement and closely spaced transverse reinforcement separated the cover from the core (Saatcioglu and Razvi 1998; Razvi and Saatcioglu 1999).

A significant amount of experimental data for HSC and high-strength transverse reinforcement for columns has been generated under the sponsorship of the Ministry of Construction in Japan. Tests conducted in Japan focused on columns subjected to axial load ratios above 0.3 (Aoyama et al. 1990; Sakaguchi et al. 1990; Muguruma and Watanabe 1990; Sugano et al. 1990; Kimura et al. 1995; Hibi et al. 1991). These tests showed a strong correlation among axial load, amount of confinement, and the drift capacity of columns. A large amount of transverse reinforcement was required to obtain ductile behavior in columns subjected to axial loads greater than the balanced load. The researchers addressed this problem by incorporating high-strength steel as transverse reinforcement.

Sakaguchi et al. (1990) reported test results from eight HSC columns with compressive strengths of 11,200 and 13,600 psi (77 and 94 MPa) and a shear span-depth ratio of 1.1. A shear slip failure occurred soon after the formation of an inclined crack for HSC columns with very light amounts of transverse reinforcement. In specimens with intermediate and high amounts of transverse reinforcement, shear strength increased with the amount of reinforcement. A rela-

Table 4.3—Transverse reinforcement for columns of special moment frames (excerpt from Table 18.7.5.4 of ACI 318-14)

Transverse reinforcement	Conditions	Applicable expressions		
A_{sh}/sb_c for rectilinear hoop	$P_u \leq 0.3A_g f_c'$ and $f_c' \leq 10{,}000$ psi	Greater of Eq. (4.3a) and (4.3b)	$0.3\left(\frac{A_g}{A_{ch}}-1\right)\frac{f_c'}{f_{yt}}$	(4.3a)
	$P_u > 0.3A_g f_c'$ and $f_c' > 10{,}000$ psi	Greater of Eq. (4.3a), (4.3b), and (4.3c)	$0.09\frac{f_c'}{f_{yt}}$	(4.3b)
			$0.2k_f k_n \frac{P_u}{f_{yt}A_{ch}}$	(4.3c)
ρ_s for spiral or circular hoop	$P_u \leq 0.3A_g f_c'$ and $f_c' \leq 10{,}000$ psi	Greater of Eq. (4.3d) and (4.3e)	$0.45\left(\frac{A_g}{A_{ch}}-1\right)\frac{f_c'}{f_{yt}}$	(4.3d)
	$P_u > 0.3A_g f_c'$ and $f_c' > 10{,}000$ psi	Greater of Eq. (4.3d), (4.3e), and (4.3f)	$0.12\frac{f_c'}{f_{yt}}$	(4.3e)
			$0.35k_f \frac{P_u}{f_{yt}A_{ch}}$	(4.3f)

tively high amount of transverse reinforcement was needed to maintain ductile behavior after the formation of inclined cracks considering low shear span-depth ratio.

Muguruma and Watanabe (1990) tested eight specimens, varying the transverse reinforcement yield strength between 48 and 115 ksi (331 and 793 MPa) while maintaining a constant volumetric ratio of 1.6 percent. There was a strong correlation among the limiting drift ratio, axial load, and the yield strength of the transverse reinforcement. The limiting drift ratio decreased as the axial load ratio increased. Conversely, the limiting drift ratio increased with the increase of the yield strength of transverse reinforcement; it increased by a factor of 3 when the yield strength of the transverse reinforcement was increased from 48 to 115 ksi (331 and 793 MPa). They concluded it was possible to achieve a high ductility ratio in columns with HSC using high-strength transverse reinforcement.

A research program, motivated by the need to use high-strength materials in high-rise structures, was conducted in Tokyo. The research was comprised of eight column tests and ten beam tests (Sugano et al. 1990), and a second series of five column tests (Kimura et al. 1995). The tests showed excellent behavior for column specimens with an axial load ratio of 0.3, which achieved limiting drift ratios of 4 percent. The limiting drift ratio increased in proportion to the yield strength of the transverse reinforcement normalized by the concrete compressive strength. The beams that were tested had span-depth ratios of 1.5, concrete compressive strengths ranging from 5800 to 12,000 psi (40 to 83 MPa), longitudinal reinforcement ratios of 1.9 and 2.9 percent, transverse reinforcement with yield strengths of 44.3, 114.6, and 197 ksi (305, 790, and 1358 MPa), with confinement indexes ranging from 0.08 to 0.36. Beams with high confinement indexes above 0.15 had limiting drift ratios above 5 percent; the limiting drift ratio was not sensitive to the amount of transverse reinforcement or concrete compressive strength. The second series in the study concluded that the ductility of HSC columns was strongly affected by both the level of axial compression and yield strength of the transverse reinforcement. Sugano et al. (1990) and Kimura et al. (1995) reported that the yield strength of the transverse reinforcement normalized by the compressive strength of the concrete was an appropriate index to evaluate ductility.

4.3—Equations for determining transverse reinforcement required in columns

Table 18.7.5.4 of ACI 318-14 specifies the minimum amount of transverse reinforcement for columns of special moment frames, as presented in Table 4.3.

Expressions of Eq. (4.3a), (4.3b), (4.3d), and (4.3e) were developed to ensure that the strength of the confined core would be sufficient to compensate for the loss in axial strength that occurs when the concrete in the exterior shell of the column spalls off. Expressions of Eq. (4.3c) and (4.3f) were developed from a review of column tests. These equations imply that the confining stress provided by rectangular hoops is less effective than that provided by a similar volume of spiral reinforcement. A comparison between the volumetric ratios of rectangular hoop required to confine a similar volume of concrete in a circular column with spiral hoop, per ACI 318 expressions, indicates that spiral hoop is approximately 50 percent more effective than rectangular hoop. Section 20.2.2.4 of ACI 318-14 also permits a value of f_{yt} as high as 100,000 psi to be used in the prior expressions.

Like ACI 318, the NZS 3101:2006 recognizes the effect of axial load on column behavior. In potential plastic hinge regions for columns where circular transverse reinforcement is used, the design standard requires the volumetric ratio ρ_s must be equal to or greater than that given by the greater of Eq. (4.3g) and (4.3h).

$$\rho_s = \frac{(1-\rho_t m)}{2.4}\frac{A_g}{A_c}\frac{f_c'}{f_{yt}}\frac{N^*}{\varphi f_c' A_g} - 0.0084 \qquad (4.3g)$$

$$\rho_s = \frac{A_{st}}{155d''}\frac{f_y}{f_{yt}}\frac{1}{d_b} \tag{4.3h}$$

The total effective area of transverse reinforcement, A_{sh}, in each of the principal directions within spacing s shall be greater than that given by Eq. (4.3i) and (4.3j).

$$A_{sh} = \frac{(1-\rho_t m)s_h h''}{3.3}\frac{A_g}{A_c}\frac{f_c'}{f_{yt}}\frac{N^*}{\varphi f_c' A_g} - 0.0065 s_h h'' \tag{4.3i}$$

$$A_{te} = \frac{\Sigma A_b f_y}{135 f_{yt}}\frac{s_h}{d_b} \tag{4.3j}$$

where $A_g/A_c \leq 1.5$ and $\rho_t m \leq 0.4$; $m = f_y/0.85f_c'$; and $\sum A_b$ is the sum of areas of longitudinal reinforcing bars reliant on the tie, in.^{2} (mm^2).

For rectangular-shaped transverse reinforcement, the center-to-center spacing in potential plastic hinge regions should not exceed the smaller of one-fourth of the smaller dimension of the cross section, or six times the diameter of the smallest longitudinal bar. The spacing between adjacent hoop legs or crossties should not exceed the larger of one-third of the lateral dimension of the cross section in the direction of the spacing or 8 in. (200 mm).

The Canadian Standard CSA A23.3-04 provides the amount of transverse reinforcement in relation to the axial load level and the effectiveness of the transverse reinforcement in confining the concrete. For ductile reinforced-concrete columns, the transverse reinforcement amount is obtained by Eq. (4.3k) and (4.3l).

$$\rho_s = 0.4k_p\frac{f_c'}{f_{yh}} \text{ (circular columns)} \tag{4.3k}$$

$$A_{sh} = \max\left(0.2k_n k_p \frac{A_g}{A_{ch}}\frac{f_c'}{f_{yh}} sh_c, 0.09\frac{f_c'}{f_{yh}} sh_c\right) \tag{4.3l}$$

(rectangular columns)

where $k_p = P_f/P_o$, where P_f is the factored axial load and P_o is the nominal axial capacity under concentric compression.

For moderately ductile reinforced-concrete columns, the amount of transverse reinforcement is obtained by Eq. (4.3m) and (4.3n).

$$\rho_s = 0.3k_p\frac{f_c'}{f_{yh}} \text{ (circular columns)} \tag{4.3m}$$

$$A_{sh} = \max\left(0.15k_n k_p \frac{A_g}{A_{ch}}\frac{f_c'}{f_{yh}} sh_c, 0.09\frac{f_c'}{f_{yh}} sh_c\right) \tag{4.3n}$$

(rectangular columns)

Several other performance-based equations for the amount of transverse reinforcement were proposed by Sheikh and Khoury (1997), Bayrak and Sheikh (1998b), Saatcioglu and Razvi (2002), Li and Park (2004), Brachmann et al. (2004a,b), and Paultre and Légeron (2008).

4.4—Definition of limiting drift ratio on basis of expected drift demand

The seismic design provisions in Section 12.12 of ASCE/SEI 7-16 require that beams and columns of moment-resisting frames be proportioned for stiffness so that the story drift demand generated by the design earthquake forces is limited to 2.0 percent of story height for standard occupancy structures (Risk Category II). The design earthquake is defined in Section 11.4.5 of ASCE/SEI 7-16 as two-thirds of the risk-targeted maximum considered earthquake (MCE_R), which has a 2 percent probability of being exceeded in a period of 50 years (2475-year average return period). There is a possibility, albeit small, that the drift demands experienced during the life cycle of a standard occupancy structure may exceed the 2 percent limit established in ASCE/SEI 7.

Because of the implemented drift computation procedure of ASCE/SEI 7, drift demand can be underestimated. The most frequently used drift computation procedure in ASCE/SEI 7-16 (Section 12.8.6) involves an elastic analysis of the building structure using design earthquake forces. The design earthquake forces specified in Section 12.8.3 of ASCE/SEI 7-16 are obtained from an elastic design response spectrum that produces a seismic response coefficient C_s (Section 12.8.1), which is inversely proportional to the response modification factor R for most building designs.

Because proportioning the strength of the structure on the basis of reduced earthquake forces does not reduce the drift demands (Shimazaki and Sozen 1984; Shimazaki 1988; Lepage 1997; Browning 2001; Matamoros and Sozen 2003), the reduced displacement demands computed based on the forces specified in Section 12.8 of ASCE/SEI 7-16, with the inclusion of the response modification factor R, must be adjusted to obtain reasonable estimates of the displacement demands caused by the design earthquake. This is accomplished in Sections 12.8.6 and 12.9.2 of ASCE/SEI 7-16 using the deflection amplification factor C_d. Current values of R and C_d specified in Table 12.2-1 of ASCE/SEI 7-16 for special reinforced concrete moment-resisting frames are 8 and 5.5, respectively. However, there is a significant body of research based on nonlinear analyses of reinforced concrete frames and physical tests of small-scale specimens in earthquake simulators showing that these two factors are approximately equal for special reinforced concrete moment-resisting frames if the stiffnesses of the frames used in the linear analysis are calculated on the basis of cracked section properties (Shibata and Sozen 1976; Shimazaki and Sozen 1984; Lepage 1997; Browning 2001; Matamoros et al. 2004). Consequently, drift demands in special moment-resisting frames calculated using the R and C_d factors specified in Table 12.2-1 of ASCE/SEI 7-16 may underestimate the drift demand associated with the design earthquake by as much as 45 percent.

Also, as hinges form in columns, the nonlinear response tends to concentrate drift demands in the stories between plastic hinges in columns rather than distributing them evenly over the height of a building, as an elastic analysis would indicate. In special reinforced concrete moment frames, however, the strong column-weak beam provision guards against plastic hinges within columns from being close to one another—that is, plastic mechanisms over only a few stories, where large drifts are concentrated.

One of the criteria that must be considered in establishing a limiting drift for the purpose of determining the amount of confinement in columns is the performance objective outlined by design codes. The performance goal of the ASCE/SEI 7-16 seismic design procedures for Risk Category II structures is to provide no more than a 1 percent probability of structural collapse within 50-year period on the occurrence of MCE_R ground shaking. It is generally expected that the structures will provide life safety performance when subjected to Design Earthquake ground shaking. The amount of confinement is primarily determined by the need for providing life safety in the Design Earthquake while considering collapse prevention in the MCE_R. The MCE_R drift demands may be as high as 50 percent greater than the drift demand for the Design Earthquake.

The most common failure criterion adopted by researchers investigating the relationship between column performance and the amount of transverse reinforcement used to confine the concrete has been the point in the hysteresis curve corresponding to a 20 percent reduction in the maximum lateral load that was measured (Saatcioglu and Razvi 2002; Bayrak and Sheikh 1997; Légeron and Paultre 2000; Bae and Bayrak 2006, 2009). If the performance of a frame in the MCE_R is considered, the amount of confinement must be adequate to achieve collapse prevention at drift demands approximately 50 percent greater than the 2 percent story drift limit established in Section 12.12 of ASCE/SEI 7-16. Experimental results from columns tested to axial load failure at the University of California (Lynn 2001; Sezen 2002) show that specimens with significantly less transverse reinforcement than specified by the codes and proposals summarized in Table 4.3 were able to sustain drift demands prior to axial load failure exceeding 3.5 percent of the story height. Note, however, that all columns tested by Lynn (2001) and Sezen (2002) were made with NSC and that there were no references found addressing the axial load failure of columns with HSC.

4.5—Use of high-strength reinforcement for transverse reinforcement

Because the amount of transverse reinforcement required in columns is proportional to the compressive strength of concrete, congestion problems arise in potential plastic hinge regions of columns with HSC, particularly in the beam-column joints. Conversely, the amount of required transverse reinforcement is inversely proportional to the yield strength of the transverse reinforcement, which presents the possibility of decreasing the volume of transverse reinforcement, thereby relieving congestion. Section 20.2.2.4 of ACI 318-14 allows maximum yield strength of transverse reinforcement for columns of special moment frames as high as 100 ksi (689 MPa).

Several studies conducted at the University of Ottawa have investigated the use of high-strength reinforcement for the transverse reinforcement of HSC columns (Saatcioglu and Razvi 1998, 2002; Razvi and Saatcioglu 1999; Lipien and Saatcioglu 1997; Saatcioglu and Baingo 1999). The researchers tested a total of 66 circular and square columns, with concrete strengths ranging between 8700 psi and 18,000 psi (60 to 124 MPa), under either monotonically increasing concentric compression or a constant compression accompanied by incrementally increasing lateral deformation reversals. Three different grades of transverse reinforcement were used, with yield strengths of 60, 83, and 145 ksi (414, 572, and 1000 MPa). The researchers concluded that, given the right combination of parameters, transverse reinforcement with yield strengths up to 145 ksi (1000 MPa) can be effective in confining HSC columns, increasing the column lateral drift ratio up to a minimum of 5 percent in heavily loaded columns ($0.43P_o$) and 8 percent in lightly loaded columns ($0.22P_o$). They observed that the effectiveness of transverse reinforcement increased with confinement efficiency, the volumetric ratio of steel, and the level of axial compression.

Tests of columns under concentric compression indicated that square columns with 12,000 to 18,000 psi (83 to 124 MPa) concrete and confinement efficiency factor $k_{ve} \geq 0.5$ experienced yielding of transverse reinforcement with yield strength of 145 ksi (1000 MPa) when the volumetric ratio of reinforcement was approximately 2 percent. Circular columns with similar properties required a smaller volumetric ratio of 1.3 percent to trigger the yielding of 145 ksi (1000 MPa) reinforcement when spiral reinforcement of k_{ve} = 1.0 was used. The yielding of high-strength transverse reinforcement was recorded at or immediately after column strength, often just before the onset of significant strength degradation. Equation (4.5) was suggested by Razvi and Saatcioglu (1999) for the computation of transverse steel stress at or shortly after the attainment of maximum strength under concentric compression

$$f_s = E_s\left(0.0025 + 0.21\sqrt[3]{\frac{k_{ve}\,\rho_{tc}}{f'_{co}}}\right) \leq f_{yt} \tag{4.5}$$

where f_{co}' is the in-place strength of unconfined concrete in the column, in psi (often taken as $0.85f_c'$), and k_{ve} is the confinement efficiency factor, which is expressed as

$$0.15\sqrt{\left(\frac{b_c}{s}\right)\left(\frac{b_c}{s_l}\right)} \leq 1.0$$

in which s_l is the spacing of laterally-supported longitudinal reinforcement.

The level of axial load was found to be another parameter that affects the effectiveness of high-strength transverse reinforcement for columns subjected to lateral loading (Saatcioglu and Baingo 1999). Spirals with 145 ksi (1000 MPa) yield strength developed their tensile strength in columns with 18 ksi (124 MPa) concrete prior to significant strength decay when the level of axial load was $0.43P_o$. When the level of axial compression dropped to $0.22P_o$, the stress in spirals did not exceed 110 ksi (758 MPa). Steel with 90 ksi (621 MPa) yield strength was effective in all columns tested. Based on the results, Saatcioglu and Razvi (2002) recommended to limit the yield strength of transverse reinforcement to 90 ksi (621 MPa) in confinement design.

Otani et al. (1998) and Otani (1995) described the use of high-strength reinforcement in the seismic design guidelines for high-rise reinforced concrete buildings in Japan. According to Otani (1995), high yield strength is normally attained by heat treatment of hot-rolled, chemically controlled killed steel. The chemical composition of the reinforcing steel is carefully controlled to develop large elongations at fracture, especially when welding is used to splice closed hoops and stirrups. Shear reinforcement is provided in the form of rectangular hoops and stirrups with 135-degree hooks, circular or rectangular spirals, supplementary ties with 135- or 90-degree hooks, or welded closed hoops and stirrups. The yield strength is defined by the 0.2 percent permanent offset. The fracture strain is measured over a gauge length of eight times the nominal bar diameter, and must not be less than 0.05 at any region of the bar, including sections where bars have been connected through welding.

Otani et al. (1998) described the earthquake-resistant design guidelines for high-rise structures with high-strength materials, which have been developed as part of a Japanese national research project for "The Development of Advanced Reinforced Concrete Buildings using High-Strength Concrete and Reinforcement." According to Otani et al. (1998), these seismic design guidelines limit the yield strength of the longitudinal reinforcement to 102 ksi (703 MPa) and the concrete compressive strength to 8700 psi (60 MPa). The maximum yield strength of the transverse reinforcement allowed by the document is 189 ksi (1303 MPa).

The database used in the study by Brachmann et al. (2004a,b) had specimens with transverse reinforcement yield strengths ranging between 37 and 183 ksi (255 and 1262 MPa), and volumetric transverse reinforcement ratios ranging from 0.17 to 6.64 percent. Because specimens with transverse reinforcement with yield strengths of 180 ksi (1241 MPa) had significantly lower test/calculated ratios, they recommended establishing an upper limit of 120 ksi (827 MPa) on the yield strength of the transverse reinforcement. The ratio of measured to calculated limiting drift ratio according to the equation proposed by Brachmann et al. (2004b) versus the yield strength of the transverse reinforcement is shown in Fig. 4.5, where the yield strength of transverse reinforcement was limited to 120 ksi (827 MPa) in the calculation of the limiting drift ratio regardless of the actual yield strength. The dotted line in Fig. 4.5 is a linear regression line for the measured-to-calculated drift ratio. The calculated drift ratio is computed by limiting the yield strength of the transverse reinforcement to 120 ksi (827 MPa) when the actual yield strength of the transverse reinforcement exceeds the limit.

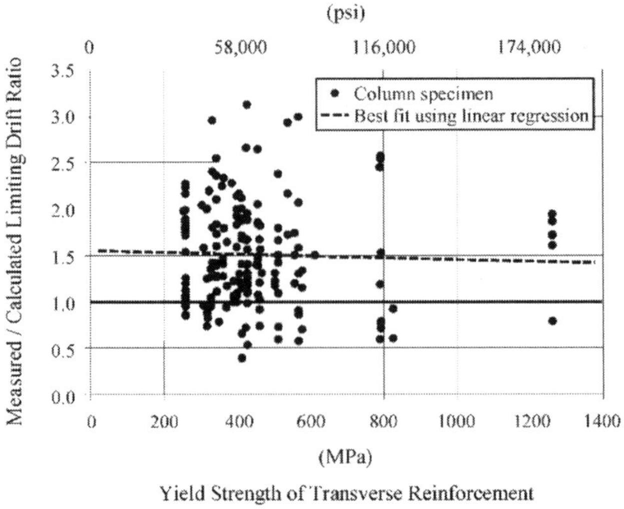

Fig. 4.5—Ratio of measured to calculated limiting drift ratio versus yield strength of the transverse reinforcement.

The suggestion by Brachmann et al. (2004a,b) to limit the yield strength of the transverse reinforcement to 120 ksi (827 MPa) is consistent with the NZS 3101:2006 design provision, which establishes an upper limit of 116 ksi (800 MPa) for the nominal yield strength of the transverse reinforcement.

4.6—Maximum hoop spacing requirements for columns

Section 18.7.5.3 of ACI 318-14 allows a maximum spacing of transverse reinforcement in regions of potential plastic hinging of one-fourth of the minimum column dimension, six times the diameter of the smallest longitudinal bar, and s_o as calculated by $4+\left(\frac{14-h_x}{3}\right)$. The value of s_o shall not exceed 6 in. (152 mm) and need not be taken less than 4 in. (102 mm).

The International Conference of Building Officials (ICBO) (now International Code Council [ICC]) *Evaluation Report ER-5536* (ICBO 2001) suggests that the maximum spacing of hoops within plastic hinge regions should be 5 in. (127 mm). The rationale for this provision stems from experimental research used as the basis for the aforementioned document (C^4 Committee 2000); satisfactory behavior was observed in specimens with a maximum hoop spacing of 6 in. (152 mm). Englekirk and Pourzanjani indicate in the C^4 Committee (2000) report, however, that test results by Azizinamini et al. (1994) contradict this observation. In specimens with an axial load ratio of 0.2 and concrete compressive strength of approximately 14,500 psi (100 MPa), Azizinamini et al. observed that the mode of failure changed from yielding of the transverse reinforcement to buckling of the longitudinal reinforcement as the hoop spacing was increased from 1.62 in. (41 mm) (which represents a hoop spacing of $d/6.6$, 2.2 longitudinal bar diameters, and 4.32 transverse bar diameters) to 2.62 in. (67 mm) (which represents a hoop spacing of $d/4.1$, 3.5 longitudinal bar diameters, and 7 transverse bar diameters), and the strength of the transverse reinforcement from 60 to 120 ksi (414 to 827 MPa). The database used by Brachmann et al. (2004a,b) had columns with hoop spacing ranging between 1 and 17 in. (25 to 432 mm). When specimens with concrete compressive strengths of 5000 psi (34 MPa) or above only were considered, however, most of the data had a maximum hoop spacing below 4 in. (102 mm). The data did not show any adverse effect due to increased spacing, and two specimens with hoop spacing of approximately 10 in. (254 mm) showed adequate performance. On this limited basis, there seems to be no conclusive experimental evidence justifying the reduction in maximum hoop spacing from 6 to 5 in. (152 to 127 mm), although the paucity of experimental data with maximum hoop spacing above 4 in. (102 mm) is a concern.

CHAPTER 5—SUMMARY AND RECOMMENDATIONS

Recent research in the area of rectangular stress block and transverse reinforcement of HSC columns was reviewed. The performance of ACI 318-14 provisions for rectangular stress block were evaluated by using a large database of tests conducted on reinforced concrete columns. This evaluation indicated that the ACI 318 stress block can be used for column designs up to a concrete strength of 12,000 psi (83 MPa). However, the use of ACI 318 stress block expressions was shown to result in gradually reducing safety margins for cases in which the compressive strength of concrete exceeds 12,000 psi (83 MPa), as shown in Fig. 3.3. The stress block expressions of CSA A23.3-04, NZS 3101:2006, Ibrahim and MacGregor (1997), Bae and Bayrak (2003), or Ozbakkaloglu and Saatcioglu (2004) are preferable for the compressive strength of concrete greater 12,000 psi (83 MPa).

Even though the failure of concrete columns become increasingly more brittle as the concrete strength increases, HSC columns can be used safely in seismic regions, provided that an adequate amount of transverse reinforcement is provided with proper details to confine the structural concrete core. Various proposals for determining the required amount and detailing of the transverse reinforcement were reviewed. Also reviewed was the use of high-yield-strength reinforcement for confinement, which may help relieve reinforcement congestion especially in potential plastic hinge regions of HSC columns. However, examination of various studies suggested that the yield strength of transverse reinforcement in design calculations for confining reinforcement should be limited to 90 ksi (621 MPa), as higher strength of reinforcement may not be mobilized prior to significant degradation of HSC columns.

CHAPTER 6—REFERENCES

ACI committee documents and documents published by other organizations are listed first by document number, full title, and year of publication followed by authored documents listed alphabetically.

American Concrete Institute (ACI)

ACI 318-14—Building Code Requirements for Structural Concrete and Commentary

American Society of Civil Engineers (ASCE)

ASCE/SEI 7-16—Minimum Design Loads and Associated Criteria for Buildings and Other Structures

CSA Group

CSA-A23.3-04—Design of Concrete Structures

Standards New Zealand

NZS 3101:2006—Concrete Structures Standard

Authored documents

ACI Innovation Task Group 4, 2006, "Materials and Quality Considerations for High-Strength Concrete in Moderate to High Seismic Applications (ITG-4.2R-06)," American Concrete Institute, Farmington Hills, MI, 26 pp.

Ahmad, S., and Shah, S., 1982, "Stress-Strain Curves of Concrete Confined by Spiral Reinforcement," *ACI Journal Proceedings*, V. 79, No. 6, June, pp. 484-490.

Aoyama, H.; Murota, T.; Hiraishi, H.; and Bessho, S., 1990, "Outline of the Japanese National Project on Advanced Reinforced Concrete Buildings with High-Strength and High-Quality Materials," *Proceedings of the Second International Symposium on High Strength Concrete*, SP-121, American Concrete Institute, Farmington Hills, MI, pp. 21-32.

Azizinamini, A.; Kuska, S. S. B.; Brungardt, P.; and Hatfield, E., 1994, "Seismic Behavior of Square High-Strength Concrete Columns," *ACI Structural Journal*, V. 91, No. 3, May-June, pp. 336-345.

Bae, S., and Bayrak, O., 2003, "Stress Block Parameters for High-Strength Concrete Members," *ACI Structural Journal*, V. 100, No. 5, Sept.-Oct., pp. 626-636.

Bae, S., and Bayrak, O., 2006, "Lateral Deformation Capacity of Concrete Bridge Piers," *Proceedings of the Fifth National Seismic Conference on Bridges and Highways*, San Francisco, CA, 11 pp.

Bae, S., and Bayrak, O., 2009, "Drift Capacity of Reinforced Concrete Columns," *ACI Structural Journal*, V. 106, No. 4, July-Aug., pp. 405-415.

Bae, S., and Bayrak, O., 2013, "Examination of Stress Block Parameters for High-Strength Concrete in the Context of ACI 318 Code," *Reinforced Concrete Columns with High Strength Concrete and Steel Reinforcement*, SP-293, American Concrete Institute, Farmington Hills, MI, pp. 59-77.

Bayrak, O., and Sheikh, S. A., 1997, "High Strength Concrete Columns under Simulated Earthquake Loading," *ACI Structural Journal*, V. 94, No. 6, Nov.-Dec., pp. 708-722.

Bayrak, O. and Sheikh, S. A., 1998a, "Earthquake Resistance of 100 MPa Concrete Columns," *Engineering Foundation Special Publication on High Strength Concrete*, American Society of Civil Engineers (ASCE), Reston, VA, June.

Bayrak, O., and Sheikh, S. A., 1998b, "Confinement Reinforcement Design Considerations for Ductile HSC Columns," *Journal of Structural Engineering*,

V. 124, No. 9, Sept., pp. 999-1010. doi: 10.1061/(ASCE)0733-9445(1998)124:9(999)

Bjerkeli, L.; Tomaszewicz, A.; and Jensen, A. A., 1990, "Deformation Properties and Ductility of High Strength Concrete," *Proceedings of the Second International Symposium on High-Strength Concrete*, SP-121, American Concrete Institute, Farmington Hills, MI, pp. 215-238.

Brachmann, I.; Browning, J.; and Matamoros, A., 2004a, "Drift-Dependent Confinement Requirements for R/C Columns under Cyclic Loading," *ACI Structural Journal*, V. 101, No. 5, Sept.-Oct., pp. 669-677.

Brachmann, I.; Browning, J.; and Matamoros, A., 2004b, "Relationship Between Drift and Confinement in Reinforced Concrete Columns Under Cyclic Loading," *Proceedings of the 13th World Conference in Earthquake Engineering*, Vancouver, BC, Canada, 15 pp.

Browning, J., 2001, "Proportioning of Earthquake-Resistant RC Building Structures," *Journal of Structural Engineering*, V. 127, No. 2, Feb., pp. 145-151. doi: 10.1061/(ASCE)0733-9445(2001)127:2(145)

C^4 Committee, 2000, "High Strength Concrete Research," Carpenters/Contractors Cooperation Committee, Inc., Los Angeles, CA.

Colaco, J. P., 1985, "75-Story Texas Commerce Plaza, Houston-The Use of High-Strength Concrete," *High-Strength Concrete*, SP-87, American Concrete Institute, Farmington Hills, MI, pp. 1-8.

Cusson, D.; Larrard, F. D.; Boulay, C.; and Paultre, P., 1996, "Strain Localization in Confined High Strength Concrete Columns," *Journal of Structural Engineering*, V. 122, No. 9, Sept., pp. 1055-1061. doi: 10.1061/(ASCE)0733-9445(1996)122:9(1055)

Cusson, D., and Paultre, P., 1995, "Stress-Strain Model for Confined High-Strength Concrete," *Journal of Structural Engineering*, V. 121, No. 3, Mar., pp. 468-477. doi: 10.1061/(ASCE)0733-9445(1995)121:3(468)

Eid, R., and Dancygier, A. N., 2005, "Partially Confined Circular Members Subjected to Axial Compression: Analysis of Concrete Confined by Steel Ties," *Structural Engineering and Mechanics*, V. 21, No. 6, June, pp. 737-765. doi: 10.12989/sem.2005.21.6.737

Eid, R., and Dancygier, A. N., 2006, "Confinement Effectiveness in Circular Concrete Columns," *Engineering Structures*, V. 28, No. 13, Nov., pp. 1885-1896. doi: 10.1016/j.engstruct.2006.03.015

Eid, R.; Dancygier, A. N.; and Paultre, P., 2007, "Elastoplastic Confinement Model for Circular Concrete Columns," *Journal of Structural Engineering*, V. 133, No. 12, Dec., pp. 1821-1831. doi: 10.1061/(ASCE)0733-9445(2007)133:12(1821)

Elwood, K. J.; Maffei, J.; Riederer, K. A.; and Telleen, K., 2009a, "Improving Column Confinement – Part 1: Assessment of Design Provisions," *Concrete International*, V. 31, No. 11, Nov., pp. 32-39.

Elwood, K. J.; Maffei, J.; Riederer, K. A.; and Telleen, K., 2009b, "Improving Column Confinement – Part 2: Proposed New Provisions for The ACI 318 Building Code," *Concrete International*, V. 31, No. 12, Dec., pp. 41-48.

Fafitis, A., and Shah, S. P., 1985, "Lateral Reinforcement for High Strength Concrete Columns," *High Strength Concrete*, SP-87, American Concrete Institute, Farmington Hills, MI, pp. 213-232.

Hibi, J.; Mihara, Y.; Otani, S.; and Aoyama, H., 1991, "Behavior of Reinforced Concrete Columns Using High Strength Concrete after Flexural Yielding," *Transactions of the Japan Concrete Institute*, V. 13, pp. 395-402.

Ibrahim, H. H. H., and MacGregor, J. G., 1996, "Flexural Behavior of Laterally Reinforced High-Strength Concrete Sections," *ACI Structural Journal*, V. 93, No. 6, Nov.-Dec., pp. 674-684.

Ibrahim, H. H. H., and MacGregor, J. G., 1997, "Modification of the ACI Rectangular Stress Block for High-Strength Concrete," *ACI Structural Journal*, V. 94, No. 1, Jan-Feb., pp. 40-48.

ICBO, 2001, "Seismic Design Utilizing High-Strength Concrete," ICBO ER-5536, ICBO Evaluation Service Inc., Whittier, CA.

Kato, D.; Watanabe, F.; Nishiyama, M.; and Sato, H., 1998, "Confined Concrete with High-Strength Materials," *High-Strength Concrete in Seismic Regions*, SP-176, American Concrete Institute, Farmington Hills, MI, pp. 85-104.

Kimura, H.; Sugano, S.; and Nagashima, T., 1995, "Study of Flexural Strength and Ductility of R/C Columns Using High Strength Concrete," *Takenaka Technical Research Report No. 51*, pp. 161-178.

Légeron, F., and Paultre, P., 2000, "Behavior of High-Strength Concrete Columns under Cyclic Flexure and Constant Axial Load," *ACI Structural Journal*, V. 97, No. 4, July-Aug., pp. 591-601.

Légeron, F., and Paultre, P., 2003, "Uniaxial Confinement Model for Normal- and High-strength Concrete Columns," *Journal of Structural Engineering*, V. 129, No. 2, Feb., pp. 241-252. doi: 10.1061/(ASCE)0733-9445(2003)129:2(241)

Lepage, A., 1997, "A Method for Drift-Control in Earthquake-Resistant Design of Reinforced Concrete Building Structures," PhD thesis, University of Illinois at Urbana-Champaign, Champaign, IL.

Li, B., and Park, R., 2004, "Confining Reinforcement for High-Strength Concrete Columns," *ACI Structural Journal*, V. 101, No. 3, May-June, pp. 314-324.

Li, B.; Park, R.; and Tanaka, H., 1994, "Strength; and Ductility of Reinforced Concrete Members and Frames Constructed Using High Strength Concrete," *Research Report 94-5*, Department of Civil Engineering, University of Canterbury, Christchurch, New Zealand, 373 pp.

Lipien, W., and Saatcioglu, M., 1997, "Tests of Square High-Strength Concrete Columns Under Reversed Cyclic Loading," *Research Report No. OCEERC 97-11*, Ottawa Carleton Earthquake Engineering Research Centre, University of Ottawa, Ottawa, ON, Canada, 184 pp.

Lloyd, N., and Rangan, B., 1996, "Studies on High-Strength Concrete Columns under Eccentric Compression," *ACI Structural Journal*, V. 93, No. 6, Nov.-Dec., pp. 631-638.

Lynn, A. C., 2001, "Seismic Evaluation of Existing Reinforced Concrete Building Columns," PhD dissertation, Department of Civil and Environmental Engineering, University of California, Berkeley, CA.

Mander, J. B.; Priestley, M. J. N.; and Park, R., 1988, "Theoretical Stress-Strain Model for Confined Concrete," *Journal of Structural Engineering*, V. 114, No. 8, Sept., pp. 1804-1826. doi: 10.1061/(ASCE)0733-9445(1988)114:8(1804)

Martinez, S.; Nilson, A.; and Slate, F., 1984, "Spirally Reinforced High-Strength Concrete Columns," *ACI Journal Proceedings*, V. 81, No. 5, Sept.-Oct., pp. 431-442.

Matamoros, A.; Garcia, L. E.; Browning, J.; and Lepage, A., 2004, "The Flat-Rate Design Method for Low- and Medium-Rise Reinforced Concrete Structures," *ACI Structural Journal*, V. 101, No. 4, July-Aug., pp. 435-446.

Matamoros, A., and Sozen, M., 2003, "Drift Limits of High-Strength Concrete Columns Subjected to Load Reversals," *Journal of Structural Engineering*, V. 129, No. 3, Mar., pp. 297-313. doi: 10.1061/(ASCE)0733-9445(2003)129:3(297)

Muguruma, H.; Nishiyama, M.; Watanabe, F.; and Tanaka, H., 1991, "Ductile Behavior of High Strength Concrete Columns Confined by High Strength Transverse Reinforcement," *Evaluation and Rehabilitation of Concrete Structures and Innovations in Design*, SP-128, V. M. Malhotra, ed., American Concrete Institute, Farmington Hills, MI, pp. 877-891.

Muguruma, H., and Watanabe, F., 1990, "Ductility Improvement of High-Strength Concrete Columns with Lateral Confinement," *Proceedings of Second International Symposium on High-Strength Concrete*, SP-121, W. T. Hester, ed., American Concrete Institute, Farmington Hills, MI, pp. 47-60.

Nagashima, T.; Sugano, S.; Kimura, H.; and Ichikawa, A., 1992, "Monotonic Axial Compression Test on Ultra-High-Strength Concrete Tied Columns," *Proceedings of the 10th World Conference on Earthquake Engineering*, Madrid, Spain, pp. 2983-2988.

Nilson, A., 1985, "Design Implications of Current Research on High-Strength Concrete," *High-Strength Concrete*, SP-87, American Concrete Institute, Farmington Hills, MI, pp. 85-118.

Otani, S., 1995, "Use of High-Strength Lateral Reinforcement in Japanese RC Construction," *Proceedings, Samsung Forum on Tall Buildings*, Seoul, South Korea.

Otani, S.; Teshigawara, M.; Murakami, M.; and Okada, T., 1998, "New RC Design Guidelines for High-Rise Reinforced Concrete Buildings using High-Strength Materials," *High-Strength Concrete in Seismic Regions*, SP-176, C. W. French and M. E. Kreger, eds., American Concrete Institute, Farmington Hills, MI, pp. 405-417.

Ozbakkaloglu, T., and Saatcioglu, M., 2004, "Rectangular Stress Block for High-Strength Concrete," *ACI Structural Journal*, V. 101, No. 4, July-Aug., pp. 475-483.

Paultre, P.; Khayat, K. H.; Langlois, Y.; Trudel, A.; and Cusson, D., 1996, "Structural Performance of Some Special Concretes," *Proceedings of the Fourth International Symposium on Utilization of High-Strength/High-Performance Concrete*, Presses de l'ENPC, Paris, pp. 787–796.

Paultre, P., and Légeron, F., 2008, "Confinement Reinforcement Design for Reinforced Concrete Columns," *Journal of Structural Engineering*, V. 134, No. 5, May, pp. 738-749. doi: 10.1061/(ASCE)0733-9445(2008)134:5(738)

Paultre, P.; Légeron, F.; and Mongeau, D., 2001, "Influence of Concrete Strength and Yield Strength of Ties on the Behavior of High-Strength Concrete Columns," *ACI Structural Journal*, V. 98, No. 4, July-Aug., pp. 490-501.

Razvi, S., and Saatcioglu, M., 1994, "Strength and Deformability of Confined High-Strength Concrete Columns," *ACI Structural Journal*, V. 91, No. 6, Nov.-Dec., pp. 678-687.

Razvi, S., and Saatcioglu, M., 1999, "Confinement Model for High-Strength Concrete," *Journal of Structural Engineering*, V. 125, No. 3, Mar., pp. 281-289. doi: 10.1061/(ASCE)0733-9445(1999)125:3(281)

Saatcioglu, M., and Baingo, D., 1999, "Circular High-Strength Concrete Columns Under Simulated Seismic Loading," *Journal of Structural Engineering*, V. 125, No. 3, Mar., pp. 272-280. doi: 10.1061/(ASCE)0733-9445(1999)125:3(272)

Saatcioglu, M.; Paultre, P.; and Ghosh, S. K., 1998, "Confinement of High-Strength Concrete," *High-Strength Concrete in Seismic Regions*, SP-176, American Concrete Institute, Farmington Hills, MI, pp. 105-136.

Saatcioglu, M., and Razvi, S., 1992, "Strength and Ductility of Confined Concrete," *Journal of Structural Engineering*, V. 118, No. 6, June, pp. 1590-1607. doi: 10.1061/(ASCE)0733-9445(1992)118:6(1590)

Saatcioglu, M., and Razvi, S., 1998, "High-strength Concrete Columns with Square Sections under Concentric Compression," *Journal of Structural Engineering*, V. 124, No. 12, Dec., pp. 1438-1447. doi: 10.1061/(ASCE)0733-9445(1998)124:12(1438)

Saatcioglu, M., and Razvi, S., 2002, "Displacement-Based Design of Reinforced Concrete Columns for Confinement," *ACI Structural Journal*, V. 99, No. 1, Jan.-Feb., pp. 3-11.

Sakaguchi, N.; Yamanobe, K.; Kitada, Y.; Kawachi, T.; and Koda, S., 1990, "Shear Strength of High-Strength Concrete Members," *Proceedings of the Second International Symposium on High-Strength Concrete*, SP-121, American Concrete Institute, Farmington Hill, MI, pp. 155-178.

Sakai, K., and Sheikh, S. A., 1989, "What Do We Know About Confinement in Reinforced Concrete Columns? (A Critical Review of Previous Work and Code Provisions)," *ACI Structural Journal*, V. 86, No. 2, Mar.-Apr., pp. 192-207.

Sargin, M.; Ghosh, S. K.; and Handa, V., 1971, "Effect of Lateral Reinforcement Upon the Strength and Deformation Properties of Concrete," *Magazine of Concrete Research*, V. 23, No. 75-76, Oct., pp. 99-110. doi: 10.1680/macr.1971.23.76.99

Sezen, H., 2002, "Seismic Response and Modeling of Reinforced Concrete Building Columns," PhD dissertation, Department of Civil and Environmental Engineering, University of California, Berkeley, Berkeley, CA.

Sheikh, S., and Khoury, S., 1997, "A Performance-Based Approach for the Design of Confining Steel in Tied Columns," *ACI Structural Journal*, V. 94, No. 4, July-Aug., pp. 421-431.

Sheikh, S.; Shah, D.; and Khoury, S., 1994, "Confinement of High-Strength Concrete Columns," *ACI Structural Journal*, V. 91, No. 1, Jan.-Feb., pp. 100-111.

Sheikh, S. A., and Uzumeri, S. M., 1982, "Analytical Model for Concrete Confinement in Tied Columns," *Journal of the Structural Division*, V. 108, No. 12, pp. 2703-2722.

Shibata, A., and Sozen, M., 1976, "Substitute-Structure Method for Seismic Design in Reinforced Concrete," *Journal of the Structures Division*, V. 102, pp. 1-18.

Shimazaki, K., 1988, "Strong Ground Motion Drift and Base Shear Strength Coefficient for R/C Structures," *Proceedings of the Ninth World Conference on Earthquake Engineering*, Tokyo and Kyoto, Japan, pp. 165-170.

Shimazaki, K., and Sozen, M., 1984, "Seismic Drift of Reinforced Concrete Structures," *Technical Research Report of Hazama-Gumi Ltd.*, pp. 145-166.

Shin, S. W.; Kamara, M.; and Ghosh, S. K., 1990, "Flexural Ductility, Strength Prediction, and Hysteretic Behavior of Ultra-High-Strength Concrete Members," *Proceedings of the Second International Symposium on High-Strength Concrete*, SP-121, American Concrete Institute, Farmington Hills, MI, pp. 239-264.

Smith, J. G., and Rad, F. N., 1989, "Economic Advantages of High-Strength Concrete in Columns," *Concrete International*, V. 11, No. 4, Apr., pp. 37-43.

Sugano, S.; Nagashima, T.; Kimura, H.; Tamura, A.; and Ichikawa, A., 1990, "Experimental Studies on Seismic Behavior of Reinforced Concrete Members of High Strength Concrete," *Proceedings of the Second International Symposium on High-Strength Concrete*, SP-121, American Concrete Institute, Farmington Hills, MI, pp. 61-87.

Thomsen, J., and Wallace, J., 1994, "Lateral Load Behavior of Reinforced Concrete Columns Constructed Using High-Strength Materials," *ACI Structural Journal*, V. 91, No. 5, Sept.-Oct., pp. 605-615.

Wahidi, S. A., 1995, "Strength and Behavior of Reinforced Concrete Columns Made from High Performance Materials," PhD dissertation, University of Texas at Austin, Austin, TX, 299 pp.

Xiao, Y., and Martirossyan, A., 1998, "Seismic Performance of High Strength Concrete Columns," *Journal of Structural Engineering*, V. 124, No. 3, Mar., pp. 241-251. doi: 10.1061/(ASCE)0733-9445(1998)124:3(241)

Xiao, Y., and Yun, H., 1998, "Full-Scale Experimental Studies on High-Strength Concrete Short Columns," *Report No. USC-SERP 98/05*, University of Southern California, Los Angeles, CA, 97 pp.

Yong, Y. K.; Nour, M. G.; and Nawy, E. G., 1988, "Behavior of Laterally Confined High-Strength Concrete under Axial Loads," *Journal of Structural Engineering*, V. 114, No. 2, Feb., pp. 332-351. doi: 10.1061/(ASCE)0733-9445(1988)114:2(332)

ACI 445R-99
(Reapproved 2015)

Recent Approaches to Shear Design of Structural Concrete

Reported by Joint ACI-ASCE Committee 445

J. A. Ramirez*
Chair

C. W. French
Secretary

P. E. Adebar*
J. F. Bonacci
M. P. Collins
D. Darwin
W. H. Dilger
A. B. Gogate

N. M. Hawkins
T. T. C. Hsu
G. J. Klein
T. Krauthammer
J. G. MacGregor

D. Mitchell*
R. G. Oesterle
M. A. Polak
K. S. Rajagopalan
K. H. Reineck*

D. M. Rogowsky*
G. M. Sabnis
D. H. Sanders
J. K. Wight
P. Zia

*Members of Subcommittee 445-1 who prepared this report.

Truss model approaches and related theories for the design of reinforced concrete members to resist shear are presented. Realistic models for the design of deep beams, corbels, and other nonstandard structural members are illustrated. The background theories and the complementary nature of a number of different approaches for the shear design of structural concrete are discussed. These relatively new procedures provide a unified, intelligible, and safe design framework for proportioning structural concrete under combined load effects.

Keywords: beams (supports); concrete; design; detailing; failure; models; shear strength; structural concrete; strut and tie.

CONTENTS

ACI 445R-99 became effective November 22, 1999.

CHAPTER 1—INTRODUCTION

1.1—Scope and objectives

Design procedures proposed for regulatory standards should be safe, correct in concept, simple to understand, and should not necessarily add to either design or construction costs. These procedures are most effective if they are based on relatively simple conceptual models rather than on complex empirical equations. This report introduces design engineers to some approaches for the shear design of one-way structural concrete members. Although the approaches explained in the subsequent chapters of this report are relatively new, some of them have reached a sufficiently mature state that they have been implemented in codes of practice. This report builds upon the landmark state-of-the-art report by the ASCE-ACI Committee 426 (1973), *The Shear Strength of Reinforced Concrete Members*, which reviewed the large body of experimental work on shear and gave the background to many of the current American Concrete Institute (ACI) shear design provisions. After reviewing the many different empirical equations for shear design, Committee 426 expressed in 1973 the hope that "the design regulations for shear strength can be integrated, simplified, and given a physical significance so that designers can approach unusual design problems in a rational manner."

The purpose of this report is to answer that challenge and review some of the new design approaches that have evolved since 1973 (CEB 1978, 1982; Walraven 1987; IABSE 1991a,b; Regan 1993). Truss model approaches and related theories are discussed and the common basis for these new approaches are highlighted. These new procedures provide a unified, rational, and safe design framework for structural concrete under combined actions, including the effects of axial load, bending, torsion, and prestressing.

Chapter 1 presents a brief historical background of the development of the shear design provisions and a summary of the current ACI design equations for beams. Chapter 2 discusses a sectional design procedure for structural-concrete one-way members using a compression field approach. Chapter 3 addresses several approaches incorporating the "concrete contribution." It includes brief reviews of European Code EC2, Part 1 and the Comité Euro-International du Béton–Fédération International de la Précontrainte (CEB-FIP) Model Code, both based on strut-and-tie models. The behavior of members without or with low amounts of shear reinforcement is discussed in Chapter 4. An explanation of the concept of shear friction is presented in Chapter 5. Chapter 6 presents a design procedure using strut-and-tie models (STM), which can be used to design regions having a complex flow of stresses and may also be used to design entire members. Chapter 7 contains a summary of the report and suggestions for future work.

1.2—Historical development of shear design provisions

Most codes of practice use sectional methods for design of conventional beams under bending and shear. ACI Building Code 318M-95 assumes that flexure and shear can be handled separately for the worst combination of flexure and shear at a given section. The interaction between flexure and shear is addressed indirectly by detailing rules for flexural reinforcement cutoff points. In addition, specific checks on the level of concrete stresses in the member are introduced to ensure sufficiently ductile behavior and control of diagonal crack widths at service load levels.

In the early 1900s, truss models were used as conceptual tools in the analysis and design of reinforced concrete beams. Ritter (1899) postulated that after a reinforced concrete beam cracks due to diagonal tension stresses, it can be idealized as a parallel chord truss with compression diagonals inclined at 457 with respect to the longitudinal axis of the beam. Mörsch (1920, 1922) later introduced the use of truss models for torsion. These truss models neglected the contribution of the concrete in tension. Withey (1907, 1908) introduced Ritter's truss model into the American literature and pointed out that this approach gave conservative results when compared with test evidence. Talbot (1909) confirmed this finding.

Historically, shear design in the United States has included a concrete contribution V_c to supplement the 45 degree sectional truss model to reflect test results in beams and slabs with little or no shear reinforcement and ensure economy in the practical design of such members. ACI Standard Specification No. 23 (1920) permitted an allowable shear stress of $0.025 f_c'$, but not more than 0.41 MPa, for beams without web reinforcement, and with longitudinal reinforcement that did not have mechanical anchorage. If the longitudinal reinforcement was anchored with 180 degree hooks or with plates rigidly connected to the bars, the allowable shear stress was increased to $0.03 f_c'$ or a maximum of 0.62 MPa (Fig. 1.1). Web reinforcement was designed by the equation

$$A_v F_v = V' s \sin \alpha / jd \tag{1-1}$$

where

A_v = area of shear reinforcement within distance s;
f_v = allowable tensile stress in the shear reinforcement;
jd = flexural lever arm;

V' = total shear minus $0.02f_c'bjd$ (or $0.03f_c'bjd$ with special anchorage);
b = width of the web;
s = spacing of shear steel measured perpendicular to its direction; and
α = angle of inclination of the web reinforcement with respect to the horizontal axis of the beam.

The limiting value for the allowable shear stresses at service loads was $0.06f_c'$ or a maximum of 1.24 MPa, or with anchorage of longitudinal steel $0.12f_c'$ or a maximum of 2.48 MPa. This shear stress was intended to prevent diagonal crushing failures of the web concrete before yielding of the stirrups. These specifications of the code calculated the nominal shear stress as $v = V/bjd$.

This procedure, which formed the basis for future ACI codes, lasted from 1921 to 1951 with each edition providing somewhat less-conservative design procedures. In 1951 the distinction between members with and without mechanical anchorage was omitted and replaced by the requirement that all plain bars must be hooked, and deformed bars must meet ASTM A 305. Therefore, the maximum allowable shear stress on the concrete for beams without web reinforcement (ACI 318-51) was $0.03f_c'$ and the maximum allowable shear stress for beams with web reinforcement was $0.12f_c'$.

ACI 318-51, based on allowable stresses, specified that web reinforcement must be provided for the excess shear if the shear stress at service loads exceeded $0.03f_c'$. Calculation of the area of shear reinforcement continued to be based on a 45 degree truss analogy in which the web reinforcement must be designed to carry the difference between the total shear and the shear assumed to be carried by the concrete.

The August 1955 shear failure of beams in the warehouse at Wilkins Air Force Depot in Shelby, Ohio, brought into question the traditional ACI shear design procedures. These shear failures, in conjunction with intensified research, clearly indicated that shear and diagonal tension was a complex problem involving many variables and resulted in a return to forgotten fundamentals.

Talbot (1909) pointed out the fallacies of such procedures as early as 1909 in talking about the failure of beams without web reinforcement. Based on 106 beam tests, he concluded that

> It will be found that the value of v [shear stress at failure] will vary with the amount of reinforcement, with the relative length of the beam, and with other factors which affect the stiffness of the beam.... In beams without web reinforcement, web resistance depends upon the quality and strength of the concrete.... The stiffer the beam the larger the vertical stresses which may be developed. Short, deep beams give higher results than long slender ones, and beams with high percentage of reinforcement [give higher results] than beams with a small amount of metal.

Unfortunately, Talbot's findings about the influence of the percentage of longitudinal reinforcement and the length-to-depth ratio were not reflected in the design equations until much later. The research triggered by the 1956 Wilkins warehouse failures brought these important concepts back to the forefront.

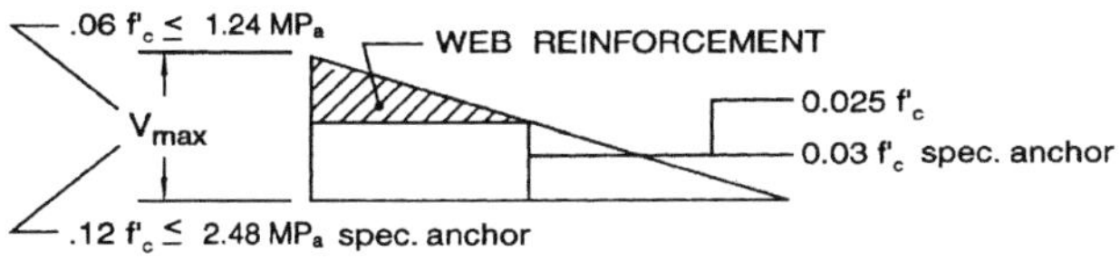

Fig. 1.1—American Specification for shear design (1920-1951) based on ACI Standard No. 23, 1920.

More recently, several design procedures were developed to economize on the design of the stirrup reinforcement. One approach has been to add a concrete contribution term to the shear reinforcement capacity obtained, assuming a 45 degree truss (for example, ACI 318-95). Another procedure has been the use of a truss with a variable angle of inclination of the diagonals. The inclination of the truss diagonals is allowed to differ from 45 degree within certain limits suggested on the basis of the theory of plasticity. This approach is often referred to as the "standard truss model with no concrete contribution" and is explained by the existence of aggregate interlock and dowel forces in the cracks, which allow a lower inclination of the compression diagonals and the further mobilization of the stirrup reinforcement. A combination of the variable-angle truss and a concrete contribution has also been proposed. This procedure has been referred to as the modified truss model approach (CEB 1978; Ramirez and Breen 1991). In this approach, in addition to a variable angle of inclination of the diagonals, the concrete contribution for nonprestressed concrete members diminishes with the level of shear stress. For prestressed concrete members, the concrete contribution is not considered to vary with the level of shear stress and is taken as a function of the level of prestress and the stress in the extreme tension fiber.

As mentioned previously, the truss model does not directly account for the components of the shear failure mechanism, such as aggregate interlock and friction, dowel action of the longitudinal steel, and shear carried across uncracked concrete. For prestressed beams, the larger the amount of prestressing, the lower the angle of inclination at first diagonal cracking. Therefore, depending on the level of compressive stress due to prestress, prestressed concrete beams typically have much lower angles of inclined cracks at failure than nonprestressed beams and require smaller amounts of stirrups.

Traditionally in North American practice, the additional area of longitudinal tension steel for shear has been provided by extending the bars a distance equal to d beyond the flexural cutoff point. Although adequate for a truss model with 45 degree diagonals, this detailing rule is not adequate for trusses with diagonals inclined at lower angles. The additional longitudinal tension force due to shear can be determined from equilibrium conditions of the truss model as $V \cot\theta$, with θ as the angle of inclination of the truss diagonals. Because the shear stresses are assumed uniformly distributed over the depth of the web, the tension acts at the section middepth.

The upper limit of shear strength is established by limiting the stress in the compression diagonals f_d to a fraction of the concrete cylinder strength. The concrete in the cracked web

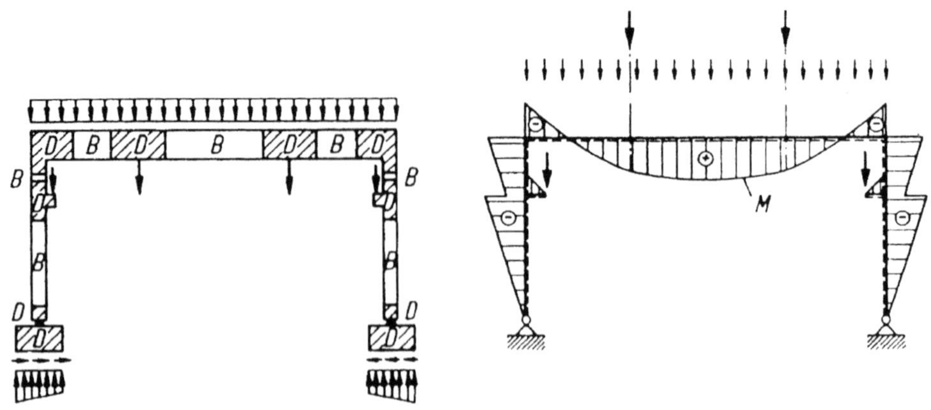

Fig. 1.2—Frame structure containing substantial part of B regions, its statical system, and bending moments (Schlaich et al. 1987).

of a beam is subjected to diagonal compressive stresses that are parallel or nearly parallel to the inclined cracks. The compressive strength of this concrete should be established to prevent web-crushing failures. The strength of this concrete is a function of 1) the presence or absence of cracks and the orientation of these cracks; 2) the tensile strain in the trans-verse direction; and 3) the longitudinal strain in the web. These limits are discussed in Chapters 2, 3, and 6.

The pioneering work from Ritter and Mörsch received new impetus in the period from the 1960s to the 1980s, and there-fore, in more recent design codes, modified truss models are used. Attention was focused on the truss model with diagonals having a variable angle of inclination as a viable model for shear and torsion in reinforced and prestressed concrete beams (Kupfer 1964; Caflisch et al. 1971; Lampert and Thurlimann 1971; Thurlimann et al. 1983). Further development of plasticity theories extended the applicability of the model to nonyielding domains (Nielsen and Braestrup 1975; Muller 1978; Marti 1980). Schlaich et al. (1987) extended the truss model for beams with uniformly inclined diagonals, all parts of the structure in the form of STM. This approach is particularly relevant in regions where the distribution of strains is significantly nonlinear along the depth. Schlaich et al. (1987) introduced the concept of D and B regions, where D stands for disconti-nuity or disturbed, and B stands for beam or Bernoulli. In D regions the distribution of strains is nonlinear, whereas the distribution is linear in B regions. A structural-concrete member can consist entirely of a D region; however, more often D and B regions will exist within the same member or structure [see Fig. 1.2, from Schlaich et al. (1987)]. In this case, D regions extend a distance equal to the member depth away from any discontinuity, such as a change in cross section or the presence of concentrated loads. For typical slender members, the portions of the structure or member between D regions are B regions. The strut-and-tie approach is discussed in detail in Chapter 6.

By analyzing a truss model consisting of linearly elastic members and neglecting the concrete tensile strength, Kupfer (1964) provided a solution for the inclination of the diagonal cracks. Collins and Mitchell (1980) abandoned the assumption of linear elasticity and developed the compression field theory (CFT) for members subjected to torsion and shear. Based on extensive experimental investigation, Vecchio and Collins (1982, 1986) presented the modified compression field theory (MCFT), which included a rationale for determining the tensile stresses in the diagonally cracked concrete. Although the CFT works well with medium to high percentages of transverse reinforcement, the MCFT provides a more realistic assessment for members having a wide range of amounts of transverse reinforcement, including the case of no web reinforcement. This approach is presented in Chapter 2. Parallel to these developments of the truss model with variable strut inclinations and the CFT, the 1980s also saw the further development of shear friction theory (Chapter 5). In addition, a general theory was developed for beams in shear using constitutive laws for friction and by determining the strains and deformations in the web. Because this approach considers the discrete formation of cracks, the crack spacing and crack width should be determined and equilibrium checked along the crack to evaluate the crack-slip mechanism of failure. This method is presented in Chapter 3. The topic of members without transverse reinforcement is dealt with in Chapter 4.

1.3—Overview of current ACI design procedures

The ACI 318M-95 sectional design approach for shear in one-way flexural members is based on a parallel truss model with 45-degree constant inclination diagonals supplemented by an experimentally obtained concrete contribution. The contribution from the shear reinforcement V_s for the case of vertical stirrups (as is most often used in North American practice), can be derived from basic equilibrium consider-ations on a 45-degree truss model with constant stirrup spacing s, and effective depth d. The truss resistance is supplemented with a concrete contribution V_c for both reinforced and prestressed concrete beams. Appendix A presents the more commonly used shear design equations for the concrete contribution in normalweight concrete beams, including effects of axial loading and the contribution from vertical stirrups V_s.

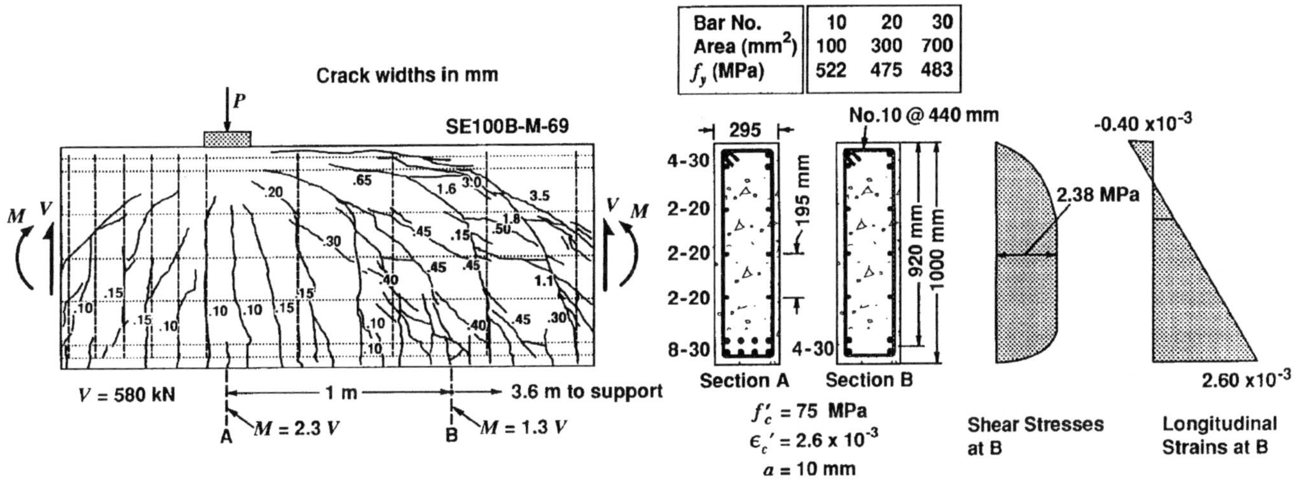

Fig. 2.1—Example of cracked web of beam failing in shear.

1.4—Summary

ACI 318 procedures have evolved into restricted, semiempirical approaches. The primary shortcomings of ACI 318M-95 are the many empirical equations and rules for special cases, and particularly the lack of a clear model that can be extrapolated to cases not directly covered. This situation would be improved if code approaches were based on clear and transparent physical models. Several of such models are discussed in subsequent chapters.

In Chapter 2, a sectional design approach using the MCFT is described. Chapter 3 discusses other truss models incorporating a concrete contribution and provides a brief review of some European code approaches.

The special case of members with no transverse shear reinforcement is addressed in Chapter 4. This chapter also presents an overview of the way the concrete contribution V_c is determined for beams. Chapter 5 presents a method of limit analysis in the form of a shear-friction mechanism. In Chapter 6, the generalized full member truss approach in the form of strut-and-tie systems for one-way flexural members is illustrated. Particular attention is given to the design approach in B and D regions, including the detailing of reinforcement ties, individual struts, and nodal zones.

The aim of this report is to describe these recent approaches to shear design and point out their common roots and complementary natures. This report does not endorse any given approach but provides a synthesis of these truss model-based approaches and related theories. The final goal of this report is to answer the challenge posed by Committee 426 over 20 years ago.

CHAPTER 2—COMPRESSION FIELD APPROACHES

2.1—Introduction

The cracked web of a reinforced concrete beam transmits shear in a relatively complex manner. As the load is increased, new cracks form while preexisting cracks spread and change inclination. Because the section resists moment as well as shear, the longitudinal strains and the crack inclinations vary over the depth of the beam (Fig. 2.1).

The early truss models of Ritter (1899) and Mörsch (1920, 1922) approximated this behavior by neglecting tensile stresses in the diagonally cracked concrete and assuming that the shear would be carried by diagonal compressive stresses in the concrete inclined at 45 degree to the longitudinal axis. The diagonal compressive concrete stresses push apart the top and bottom faces of the beam, while the tensile stresses in the stirrups pull them together. Equilibrium requires that these two effects be equal. According to the 45 degree truss model, the shear capacity is reached when the stirrups yield and will correspond to a shear stress of

$$v = \frac{A_v f_y}{b_w s} = \rho_v f_y \qquad (2\text{-}1)$$

For the beam shown in Fig. 2.1, this equation would predict a maximum shear stress of only 0.80 MPa. As the beam actually resisted a shear stress of about 2.38 MPa, it can be seen that the 45 degree truss equation can be very conservative.

One reason why the 45 degree truss equation is often very conservative is that the angle of inclination of the diagonal compressive stresses measured from the longitudinal axis θ is typically less than 45 degrees. The general form of Eq. (2-1) is

$$v = \rho_v f_y \cot\theta \qquad (2\text{-}2)$$

With this equation, the strength of the beam shown in Fig. 2.1 could be explained if θ was taken equal to 18.6 degrees. Most of the inclined cracks shown in Fig. 2.1 are not this flat.

Before the general truss equation can be used to determine the shear capacity of a given beam or to design the stirrups to resist a given shear, it is necessary to know the angle θ. Discussing this problem, Mörsch (1922) stated, "it is absolutely impossible to mathematically determine the slope of the secondary inclined cracks according to which one can design the stirrups." Just seven years after Mörsch made this statement, another German engineer, H. A. Wagner (1929), solved an analogous problem while dealing with the shear

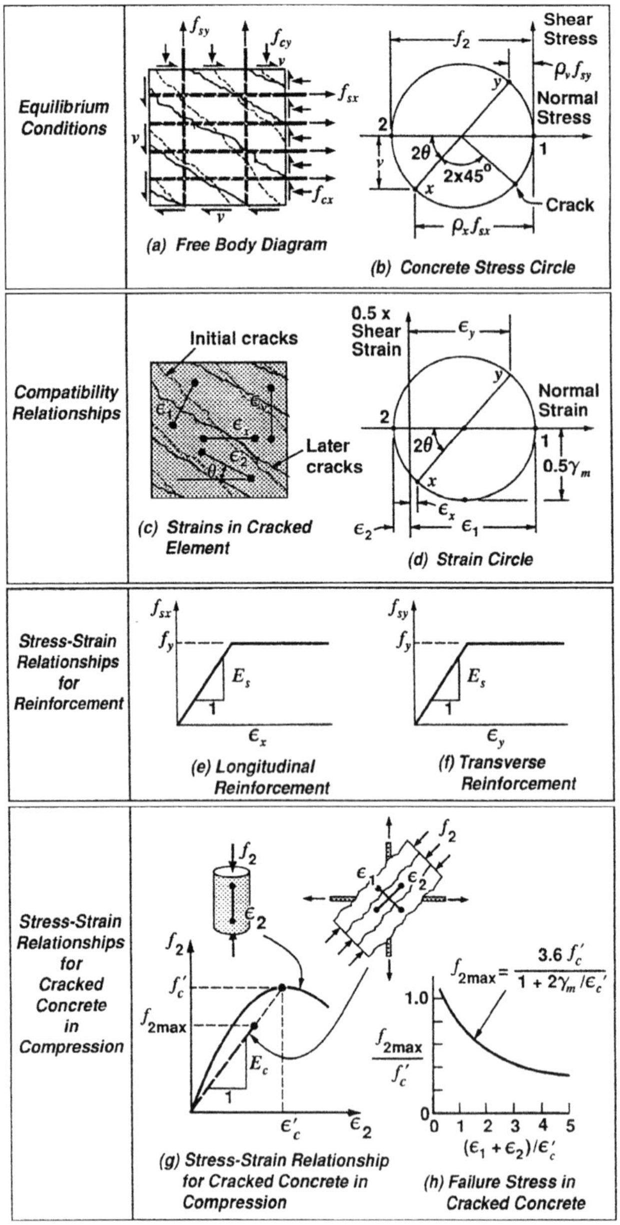

Fig. 2.2—Compression field theory (Mitchell and Collins 1974).

design of "stressed-skin" aircraft. Wagner assumed that after the thin metal skin buckled, it could continue to carry shear by a field of diagonal tension, provided that it was stiffened by transverse frames and longitudinal stringers. To determine the angle of inclination of the diagonal tension, Wagner considered the deformations of the system. He assumed that the angle of inclination of the diagonal tensile stresses in the buckled thin metal skin would coincide with the angle of inclination of the principal tensile strain as determined from the deformations of the skin, the transverse frames, and the longitudinal stringers. This approach became known as the tension field theory.

Shear design procedures for reinforced concrete that, like the tension field theory, determine the angle θ by considering the deformations of the transverse reinforcement, the longitudinal reinforcement, and the diagonally stressed concrete have become known as compression field approaches. With these methods, equilibrium conditions, compatibility conditions, and stress-strain relationships for both the reinforcement and the diagonally cracked concrete are used to predict the load-deformation response of a section subjected to shear.

Kupfer (1964) and Baumann (1972) presented approaches for determining the angle θ assuming that the cracked concrete and the reinforcement were linearly elastic. Methods for determining θ applicable over the full loading range and based on Wagner's procedure were developed by Collins and Mitchell (1974) for members in torsion and were applied to shear design by Collins (1978). This procedure was known as CFT.

2.2—Compression field theory

Figure 2.2 summarizes the basic relationships of the CFT. The shear stress v applied to the cracked reinforced concrete causes tensile stresses in the longitudinal reinforcement f_{sx} and the transverse reinforcement f_{sy} and a compressive stress in the cracked concrete f_2 inclined at angle θ to the longitudinal axis. The equilibrium relationships between these stresses can be derived from Fig. 2.2 (a) and (b) as

$$\rho_v f_{sy} = f_{cy} = v\tan\theta \tag{2-3}$$

$$\rho_x f_{sx} = f_{cx} = v\cot\theta \tag{2-4}$$

$$f_2 = v(\tan\theta + \cot\theta) \tag{2-5}$$

where ρ_x and ρ_v are the reinforcement ratios in the longitudinal and transverse directions.

If the longitudinal reinforcement elongates by a strain of ε_x, the transverse reinforcement elongates by ε_y, and the diagonally compressed concrete shortens by ε_2, then the direction of principal compressive strain can be found from Wagner's (1929) equation, which can be derived from Mohr's circle of strain (Fig. 2.2(d)) as

$$\tan^2\theta = \frac{\varepsilon_x + \varepsilon_2}{\varepsilon_y + \varepsilon_2} \tag{2-6}$$

Before this equation can be used to determine θ, however, stress-strain relationships for the reinforcement and the concrete are required. It is assumed that the reinforcement strains are related to the reinforcement stresses by the usual simple bilinear approximations shown in Fig. 2.2(e) and (f). Thus, after the transverse strain ε_y exceeds the yield strain of the stirrups, the stress in the stirrups is assumed to equal the yield stress f_y, and Eq. (2-3) becomes identical to Eq. (2-2).

Based on the results from a series of intensively instrumented beams, Collins (1978) suggested that the relationship between the principal compressive stress f_2 and the principal compressive strain ε_2 for diagonally cracked concrete would differ from the usual compressive stress-strain curve derived from a cylinder test (Fig 2.2(g)). He postulated that as the strain circle becomes larger, the compressive stress required to fail the concrete f_{2max} becomes smaller (Fig. 2.2(h)). The relationships proposed were

$$f_{2max} = \frac{3.6f_c'}{1 + 2\gamma_m/\varepsilon_c'} \tag{2-7}$$

where

γ_m = diameter of the strain circle (that is, $\varepsilon_1 + \varepsilon_2$); and

ε_c' = strain at which the concrete in a cylinder test reaches the peak stress f_c'.

For values of f_2 less than f_{2max}

$$\varepsilon_2 = \frac{f_2}{f_c'\varepsilon_c'} \tag{2-8}$$

It was suggested that the diagonally cracked concrete fails at a low compressive stress because this stress must be transmitted across relatively wide cracks. If the initial cracks shown in Fig. 2.2(a) formed at 45 degrees to the longitudinal reinforcement, and if θ is less than 45 degrees, which will be the case if ρ_v is less than ρ_x, then significant shear stresses should be transmitted across these initial cracks (Fig. 2.2(b)). The ability of the concrete to transmit shear across cracks depends on the width of the cracks, which, in turn, is related to the tensile straining of the concrete. The principal tensile strain ε_1 can be derived from Fig. 2.2(d) as

$$\varepsilon_1 = \varepsilon_x + (\varepsilon_x + \varepsilon_2)\cot^2\theta \tag{2-9}$$

For shear stresses less than that causing first yield of the reinforcement, a simple expression for the angle θ can be derived by rearranging the previous equations to give

$$\tan^4\theta = \left(1 + \frac{1}{n\rho_x}\right) / \left(1 + \frac{1}{n\rho_v}\right) \tag{2-10}$$

where

n = modular ratio E_s/E_c; and

E_c is taken as f_c'/ε_c'.

For the member shown in Fig. 2.1, ρ_x is 0.0303, ρ_v is 0.00154, and $n = 6.93$; therefore, Eq. (2-10) would give a θ value of 26.4 degrees. This would imply that the stirrups would yield at a shear stress of 1.62 MPa.

After the stirrups have yielded, the shear stress can still be increased if θ can be reduced. Reducing θ will increase the tensile stress in the longitudinal reinforcement and the compressive stress in the concrete. Failure will be predicted to occur either when the longitudinal steel yields or when the concrete fails. For the member shown in Fig. 2.1, failure of the concrete is predicted to occur when θ is lowered to 15.5 degrees, at which stage the shear stress is 2.89 MPa and ε_x is 1.73×10^{-3}. Note that these predicted values are for a section where the moment is zero. Moment will increase the longitudinal tensile strain ε_x, which will reduce the shear capacity. For example, if ε_x was increased to 2.5×10^{-3}, concrete failure would be predicted to occur when θ is 16.7 degrees and the shear stress is 2.68 MPa.

2.3—Stress-strain relationships for diagonally cracked concrete

Since the CFT was published, a large amount of experimental research aimed at determining the stress-strain characteristics of diagonally cracked concrete has been conducted. This work has typically involved subjecting reinforced concrete elements to uniform membrane stresses in special-purpose testing machines. Significant experimental studies have been conducted by Aoyagi and Yamada (1983), Vecchio and Collins (1986), Kollegger and Mehlhorn (1988), Schlaich et al. (1987), Kirschner and Collins (1986), Bhide and Collins (1989), Shirai and Noguchi (1989), Collins and Porasz (1989), Stevens et al. (1991), Belarbi and Hsu (1991), Marti and Meyboom (1992), Vecchio et al. (1994), Pang and Hsu (1995), and Zhang (1995). A summary of the results of many of these studies is given by Vecchio and Collins (1993).

These experimental studies provide strong evidence that the ability of diagonally cracked concrete to resist compression decreases as the amount of tensile straining increases (Fig. 2.3). Vecchio and Collins (1986) suggested that the maximum compressive stress f_{2max} that the concrete can resist reduces as the average principal tensile strain ε_1 increases in the following manner

$$f_{2max} = \frac{f_c'}{0.8 + 170\varepsilon_1} \le f_c' \tag{2-11}$$

The Norwegian concrete code (1989) recommended a similar relationship except the coefficient of 170 was reduced to 100. Belarbi and Hsu (1995) suggested

$$f_{2max} = \frac{0.9f_c'}{\sqrt{1 + 400\varepsilon_1}} \tag{2-12}$$

The various relationships for the reduction in compressive strength are compared with the experimental results from 73 element tests in Fig. 2.3. It can be seen that Eq. (2-11) lies near the middle of the data scatter band. For larger strains, Eq. (2-12) gives higher values to better fit some data at strains of up to 4%.

The compression field approach requires the calculation of the compressive strain in the concrete ε_2 associated with the compressive stress f_2 [Eq. (2-6)]. For this purpose, Vecchio and Collins (1986) suggested the following simple stress-strain relationship

$$f_2 = f_{2max}\left[2\left(\frac{\varepsilon_2}{\varepsilon_c'}\right) - \left(\frac{\varepsilon_2}{\varepsilon_c'}\right)^2\right] \tag{2-13}$$

where f_{2max} is given by Eq. (2-11).

Somewhat more complex expressions relating f_2 and ε_2 were suggested by Belarbi and Hsu (1995). They are

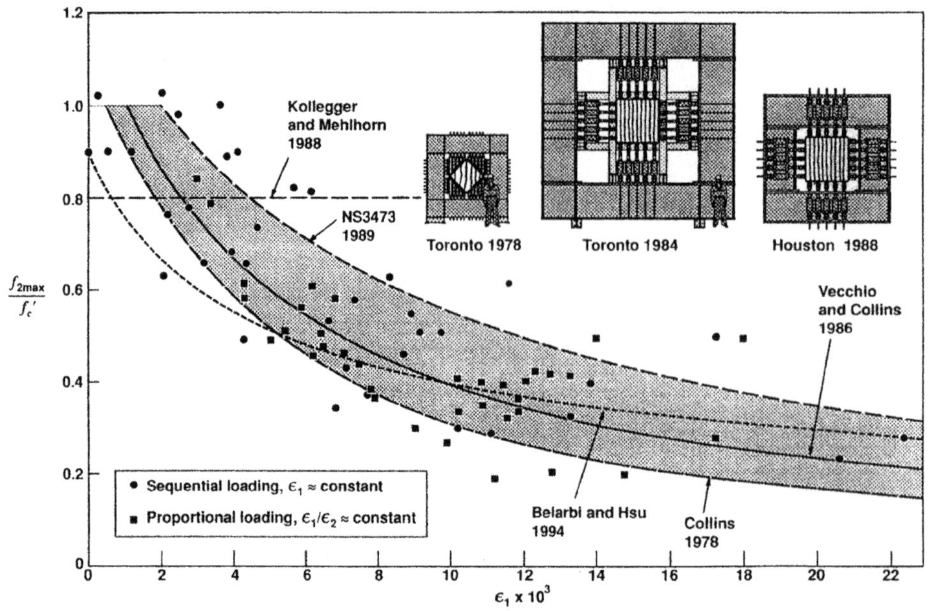

Fig. 2.3—*Maximum concrete compressive stress as function of principal tensile strain.*

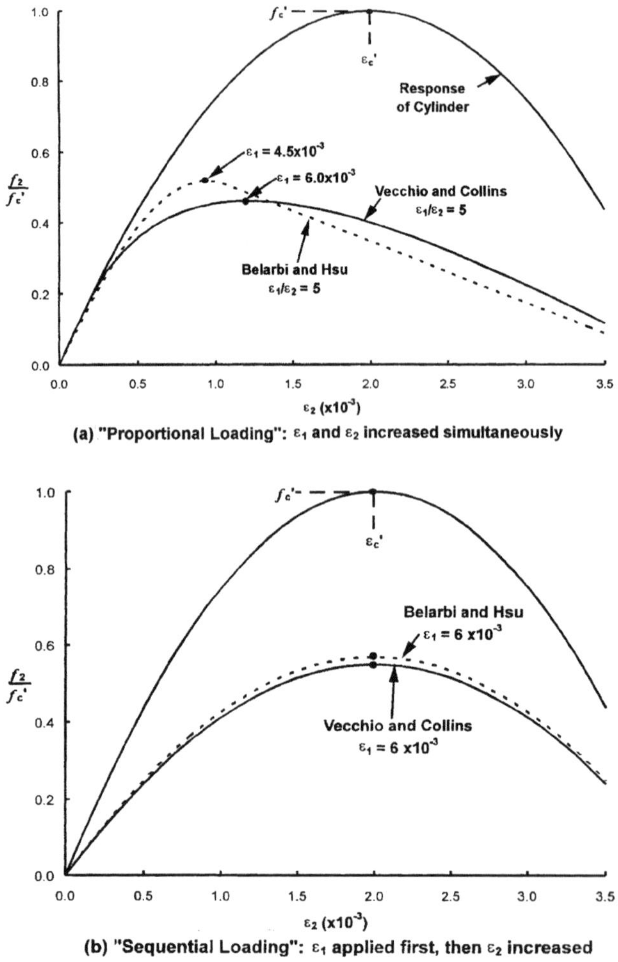

Fig. 2.4—Compressive stress-compressive strain relationships for diagonally cracked concrete: (a) proportion loading, ε_1 and ε_2 increased simultaneously; and (b) sequential loading ε_1 applied first then ε_2 increased.

$$f_2 = \zeta_{\sigma 0} f_c'\left[2\left(\frac{\varepsilon_2}{\zeta_{\varepsilon 0}\varepsilon_c'}\right) - \left(\frac{\varepsilon_2}{\zeta_{\varepsilon 0}\varepsilon_c'}\right)^2\right] \quad \text{if} \quad \frac{\varepsilon_2}{\zeta_{\varepsilon 0}\varepsilon_c'} \le 1 \qquad (2\text{-}14a)$$

$$f_2 = \zeta_{\sigma 0} f_c'\left[1 - \left(\frac{\varepsilon_2/\zeta_{\varepsilon 0}\varepsilon_c' - 1}{2/\zeta_{\varepsilon 0} - 1}\right)^2\right] \quad \text{if} \quad \frac{\varepsilon_2}{\zeta_{\varepsilon 0}\varepsilon_c'} > 1 \qquad (2\text{-}14b)$$

where, for "proportional loading"

$$\zeta_{\alpha 0} = \frac{0.9}{\sqrt{1 + 400\varepsilon_1}} \quad \text{and} \quad \zeta_{\varepsilon 0} = \frac{1}{\sqrt{1 + 500\varepsilon_1}} \qquad (2\text{-}14c)$$

and for "sequential loading"

$$\zeta_{\alpha 0} = \frac{0.9}{\sqrt{1 + 250\varepsilon_1}} \quad \text{and} \quad \zeta_{\varepsilon 0} = 1 \qquad (2\text{-}14d)$$

As the cracked web of a reinforced concrete beam is subjected to increasing shear forces, both the principal compressive strain ε_2 and the principal tensile strain ε_1 are increased. Before yield of the reinforcement, the ratio $\varepsilon_1/\varepsilon_2$ remains reasonably constant. Figure 2.4(a) shows that the compressive stress–compressive strain relationships predicted by Eq. (2-13) and (2-14) for the case where the ratio $\varepsilon_1/\varepsilon_2$ is held constant at a value of 5 are similar. Figure 2.4(b) compares the relationship for the less realistic situation of holding ε_1 constant while increasing ε_2. The predicted stress-strain relationships depend on the sequence of loading. Once again, the predictions of Eq. (2-13) and (2-14) are very similar.

For typical reinforced concrete beams, the percentage of longitudinal reinforcement ρ_x will greatly exceed the percentage of stirrup reinforcement ρ_v. In this situation there

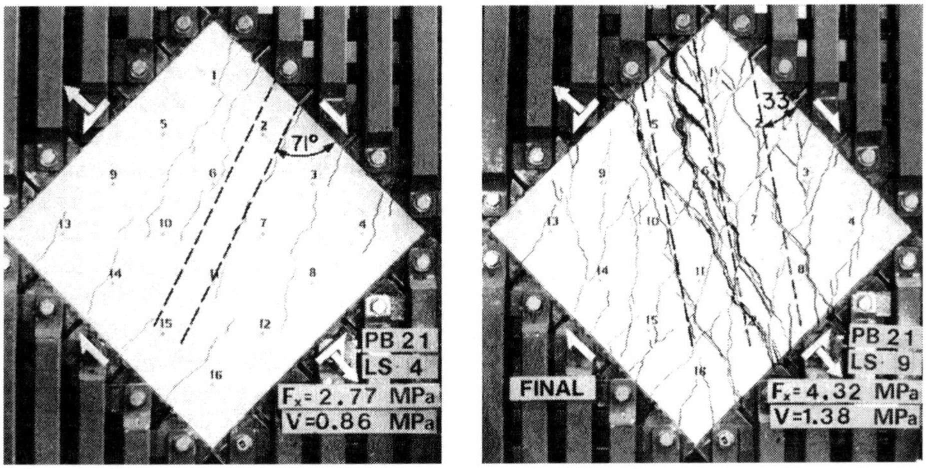

Fig. 2.5—Change of inclination of crack direction with increase in load.

will be a substantial reduction in the inclination θ of the principal compressive stresses after cracking. Figure 2.5 shows the observed crack patterns for a reinforced concrete element that contained reinforcement only in the direction of tension (x-direction) and was loaded in combined tension and shear. The first cracks formed at about 71 degrees to the x-axis. These initial cracks were quite narrow and remained reasonably constant in width throughout the test. As the load was increased, new cracks formed in directions closer to the reinforcement direction, and the width of these new cracks increased gradually. Failure was characterized by the rapid widening of the cracks that formed at about 33 degrees to the x-axis. For this extreme case, the direction of principal stress in the concrete differed by up to 20 degrees from the direction of principal strain (Bhide and Collins 1989). The predicted angle, based on Wagner's assumption that the principal stress direction coincides with the principal strain direction, lay about halfway between the observed strain direction and the observed stress direction. For elements with both longitudinal and transverse reinforcement, the directions of principal stress in the concrete typically deviated by less than 10 degrees from the directions of the principal strain (Vecchio and Collins 1986). Based on these results, it was concluded that determining the inclination of the principal stresses in the cracked concrete by Wagner's equation was a reasonable simplification.

The CFT assumes that after cracking there will be no tensile stresses in the concrete. Tests on reinforced concrete elements, such as that shown in Fig. 2.5, demonstrated that even after extensive cracking, tensile stresses still existed in the cracked concrete and that these stresses significantly increased the ability of the cracked concrete to resist shear stresses. It was found (Vecchio and Collins 1986; Belarbi and Hsu 1994) that after cracking the average principal tensile stress in the concrete decreases as the principal tensile strain increases. Collins and Mitchell (1991) suggest that a suitable relationship is

$$f_1 = \frac{0.33\sqrt{f_c'}}{1 + \sqrt{500\varepsilon_1}} \quad \text{(MPa) units} \qquad (2\text{-}15)$$

while Belarbi and Hsu (1994) suggest

$$f_1 = \frac{0.31\sqrt{f_c'}}{(12{,}500\varepsilon_1)^{0.4}} \quad \text{(MPa) units} \qquad (2\text{-}16)$$

Equation (2-16) predicts a faster decay for f_1 with increasing ε_1 than does Eq. (2-15). For example, for a 35 MPa concrete and an ε_1 value of 5×10^{-3}, Eq. (2-15) would predict an average tensile stress of 0.76 MPa, whereas Eq. (2-16) would predict 0.35 MPa.

2.4—Modified compression field theory

The MCFT (Vecchio and Collins 1986) is a further development of the CFT that accounts for the influence of the tensile stresses in the cracked concrete. It is recognized that the local stresses in both the concrete and the reinforcement vary from point to point in the cracked concrete, with high reinforcement stresses but low concrete tensile stresses occurring at crack locations. In establishing the angle θ from Wagner's equation, Eq. (2-6), the compatibility conditions relating the strains in the cracked concrete to the strains in the reinforcement are expressed in terms of average strains, where the strains are measured over base lengths that are greater than the crack spacing (Fig. 2.2(c) and (d)). In a similar manner, the equilibrium conditions, which relate the concrete stresses and the reinforcement stresses to the applied loads, are expressed in terms of average stresses; that is, stresses averaged over a length greater than the crack spacing. These relationships can be derived from Fig. 2.6(a) and (b) as

$$\rho_v f_{sy} = f_{cy} = v\tan\theta - f_1 \qquad (2\text{-}17)$$

$$\rho_x f_{sx} = f_{cx} = v\cot\theta - f_1 \qquad (2\text{-}18)$$

$$f_2 = v(\tan\theta + \cot\theta) - f_1 \qquad (2\text{-}19)$$

These equilibrium equations, the compatibility relationships from Fig. 2.2(d), the reinforcement stress-strain relationships

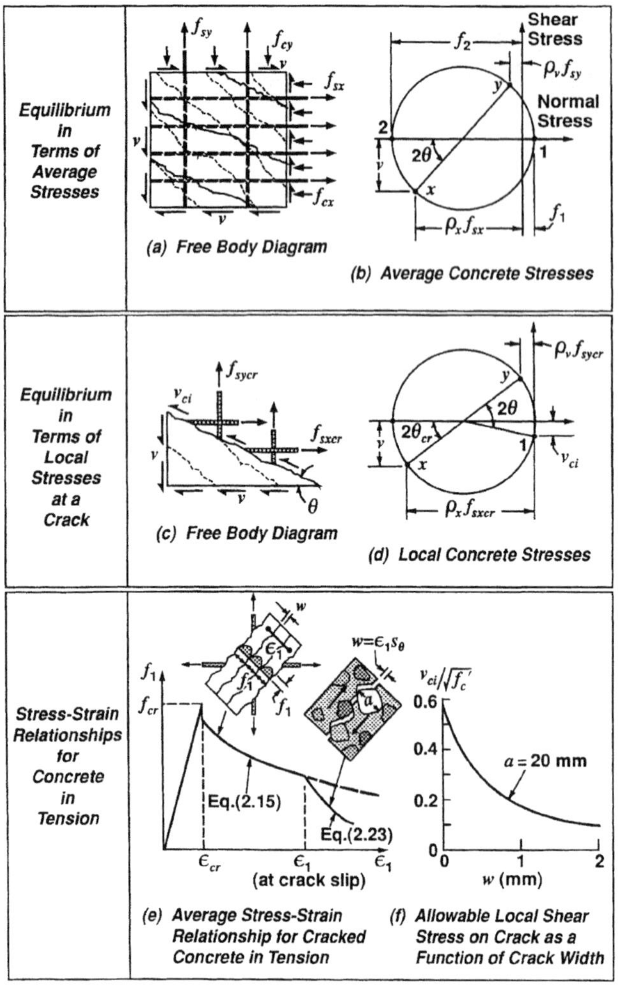

Fig. 2.6—Aspects of modified compression field theory.

from Fig. 2.2(e) and (f), and the stress-strain relationships for the cracked concrete in compression (Eq. (2-13)) and tension (Fig. 2.6(e)) enable the average stresses, the average strains, and the angle θ to be determined for any load level up to the failure.

Failure of the reinforced concrete element may be governed not by average stresses, but rather by local stresses that occur at a crack. In checking the conditions at a crack, the actual complex crack pattern is idealized as a series of parallel cracks, all occurring at angle θ to the longitudinal reinforcement and space a distance s_θ apart. From Fig. 2.6(c) and (d), the reinforcement stresses at a crack can be determined as

$$\rho_v f_{syer} = v\tan\theta - v_{ci}\tan\theta \qquad (2\text{-}20)$$

$$\rho_x f_{sxer} = v\cot\theta - v_{ci}\cot\theta \qquad (2\text{-}21)$$

It can be seen that the shear stress v_{ci} on the crack face reduces the stress in the transverse reinforcement but increases the stress in the longitudinal reinforcement. The maximum possible value of v_{ci} is taken (Bhide and Collins 1989) to be related to the crack width w and the maximum

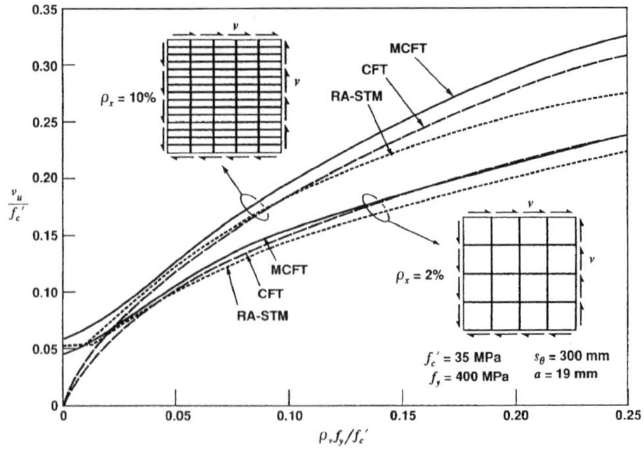

Fig. 2.7—Comparison of predicted shear strengths of two series of reinforced concrete elements.

aggregate size a by the relationship illustrated in Fig. 2.6(f) and given by

$$v_{ci} \le 0.18\sqrt{f_c'}/\left(0.3 + \frac{24w}{a+16}\right) \quad \text{(MPa, mm)} \qquad (2\text{-}22)$$

The crack width w is taken as the crack spacing times the principal tensile strain ε_1. At high loads, the average strain in the stirrups ε_y will typically exceed the yield strain of the reinforcement. In this situation, both f_{sy} in Eq. (2-17) and f_{sycr} in Eq. (2-20) will equal the yield stress in the stirrups. Equating the right-hand sides of these two equations and substituting for v_{ci} from Eq. (2-22) gives

$$f_1 \le 0.18\sqrt{f_c'}\tan\theta/\left(0.3 + \frac{24w}{a+16}\right) \qquad (2\text{-}23)$$

Limiting the average principal tensile stress in the concrete in this manner accounts for the possibility of failure of the aggregate interlock mechanisms, which are responsible for transmitting the interface shear stress v_{ci} across the crack surfaces.

Figure 2.7 illustrates the influence of the tensile stresses in the cracked concrete on the predicted shear capacity of two series of reinforced concrete elements. In this figure, RA-STM stands for rotating-angle softened-truss model. If tensile stresses in the cracked concrete are ignored, as is done in the CFT, elements with no stirrups ($\rho_v = 0$) are predicted to have no shear strength. When these tensile stresses are accounted for, as is done in the MCFT, even members with no stirrups are predicted to have significant postcracking shear strengths. Figure 2.7 shows that predicted shear strengths are a function not only of the amount of stirrup reinforcement, but also of the amount of longitudinal reinforcement. Increasing the amount of longitudinal reinforcement increases the shear capacity. Increasing the amount of longitudinal reinforcement also increases the difference between the CFT prediction and the MCFT prediction. For the elements with 2% of longitudinal reinforcement and with

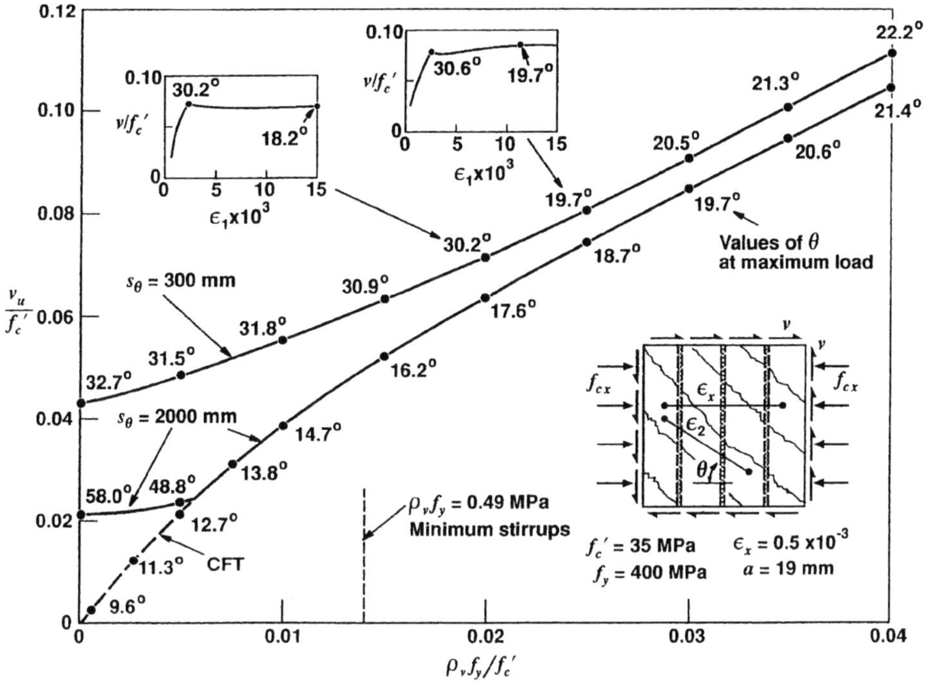

Fig. 2.8—Influence of crack spacing on predicted shear capacity.

$\rho_v f_y/f_c'$ greater than about 0.10, yielding of the longitudinal reinforcement at a crack (Eq. (2-21)) limits the maximum shear capacity. In this situation, the influence of the tensile stresses in the cracked concrete on the predicted shear strength is negligibly small. On the other hand, when the total longitudinal reinforcement is 10% of the web area, this longitudinal reinforcement remains well below yield stress, and the failure, for larger amounts of stirrups, is then governed by crushing of the concrete. The tensile stresses in the cracked concrete stiffen the element, reduce the concrete strains, and make it possible to resist larger shear stresses before failure. The predicted shear strength of elements that contain relatively small amounts of stirrups are influenced by the spacing of the diagonal cracks s_θ. If this spacing is increased, the crack width, *w,* associated with a given value of ε_1 increases, and the tension transmitted through the cracked concrete decreases (Eq. (2-23)). This aspect of behavior is illustrated in Fig. 2.8, which compares the shear capacities predicted by the MCFT for two series of elements. In one series, the crack spacing is taken as 300 mm, whereas in the other, it is taken as 2000 mm. It is assumed that the amount of longitudinal reinforcement and the axial loading of the elements is such that the longitudinal strain, εx, is held constant at 0.5×10^{-3}. It can be seen that the predicted shear capacity becomes more sensitive to crack spacing as the amount of stirrup reinforcement, ρ_v, is reduced. When ρ_v is zero, the element with a 2000 mm crack spacing is predicted to have only about half the shear capacity of the element with a 300 mm crack spacing.

2.5—Rotating-angle softened-truss model

A somewhat different procedure to account for tensile stresses in diagonally cracked concrete has been developed by Hsu and his coworkers at the University of Houston (Belarbi and Hsu 1991, 1994, 1995; Pang and Hsu 1995; Hsu 1993). This procedure is called the rotating-angle softened-truss model (RA-STM). Like the MCFT, this method assumes that the inclination of the principal stress direction θ in the cracked concrete coincides with the principal strain direction. For typical elements, this angle will decrease as the shear is increased, hence the name rotating angle. Pang and Hsu (1995) limit the applicability of the rotating-angle model to cases where the rotating angle does not deviate from the fixed angle by more than 12 degrees. Outside this range, they recommend the use of a fixed-angle model, which is discussed in Section 3.5.

The method formulates equilibrium equations in terms of average stresses (Fig. 2.6(b)) and compatibility equations in terms of average strains (Fig. 2.2(d)). The softened stress-strain relationship of Eq. (2-14) is used to relate the principal compressive stress in the concrete f_2 to the principal compressive strain ε_2, whereas Eq. (2-16) is used to relate the average tensile stress in the concrete f_1 to the principal tensile strain ε_1.

Instead of checking the stress conditions at a crack, as is done by the MCFT, the RA-STM adjusts the average stress-average strain relationships of the reinforcement to account for the possibility of local yielding at the crack. The relationships suggested are

$$f_s = E_s \varepsilon_s \text{ if } \varepsilon_s \le \varepsilon_n \tag{2-24a}$$

$$f_s = f_y\left[(0.91 - 2B) + (0.02 + 0.25B)\frac{E_s}{f_y}\varepsilon_s\right] \tag{2-24b}$$

$$\left[1 - \frac{2 - \alpha_2/45}{1000\rho}\right] \text{ if } \varepsilon_s > \varepsilon_n$$

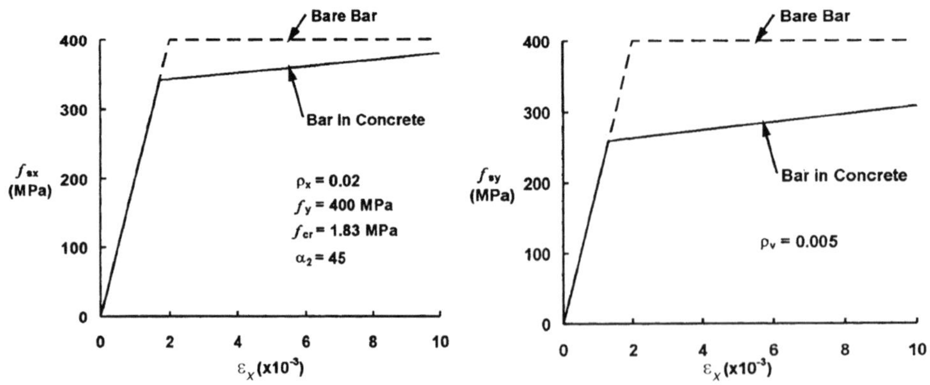

Fig. 2.9—Average-reinforcement-stress/average-reinforcement-strain relationships used in rotating-angle softened-truss model.

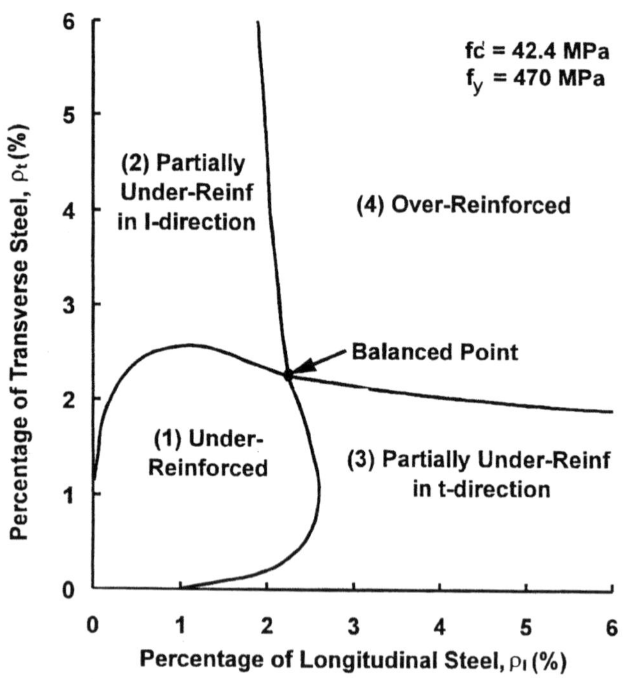

Fig. 2.10—Four failure modes predicted by rotating-angle softened-truss model for element loaded in pure shear (adapted from Pang and Hsu [1995]).

$$\varepsilon_n = \frac{f_y}{E_s}(0.93 - 2B)\left[1 - \frac{2 - \alpha_2/45}{1000\rho}\right] \tag{2-24c}$$

$$B = \left(\frac{f_{cr}}{fy}\right)^{1.5}/\rho \tag{2-24d}$$

where

ρ = reinforcement ratio; and

α_2 = angle between the initial crack direction and the longitudinal reinforcement.

The resulting relationships for a case where the longitudinal reinforcement ratio is 0.02 and the transverse reinforcement ratio is 0.005 are illustrated in Fig. 2.9.

Figure 2.7 shows the shear strength-predictions for the RA-STM for the two series of reinforced concrete elements. Although the MCFT and the RA-STM give similar predictions for low amounts of reinforcement, the predicted strengths using the RA-STM are somewhat lower than those given by the MCFT for elements with higher amounts of reinforcement.

Both the MCFT and the RA-STM are capable of predicting not only the failure load but also the mode of failure. Thus, as shown in Fig. 2.10, a reinforced concrete element loaded in pure shear can fail in four possible modes. Both the longitudinal and transverse reinforcement can yield at failure (Mode 1), only the longitudinal reinforcement yields at failure (Mode 2), only the transverse reinforcement yields at failure (Mode 3), or neither reinforcement yields at failure (Mode 4).

2.6—Design procedure based on modified compression field theory

The relationships of the MCFT can be used to predict the shear strength of a beam such as that shown in Fig. 2.11. Assuming that the shear stress in the web is equal to the shear force divided by the effective shear area $b_w d_v$, and that, at failure, the stirrups will yield, equilibrium equations (2-17) can be rearranged to give the following expression for the shear strength V_n of the section

$$V_n = V_c + V_s + V_p \tag{2-25a}$$

$$V_n = f_1 b_w d_v \cot\theta + \frac{A_v f_y}{s} d_v \cot\theta + V_p \tag{2-25b}$$

$$V_n = \beta\sqrt{f_c'}\, b_w d_v + \frac{A_v f_y}{s} d_v \cot\theta + V_p \tag{2-25c}$$

where

V_c = shear strength provided by tensile stresses in the cracked concrete;

V_s = shear strength provided by tensile stresses in the stirrups;

V_p = vertical component of the tension in inclined prestressing tendon;

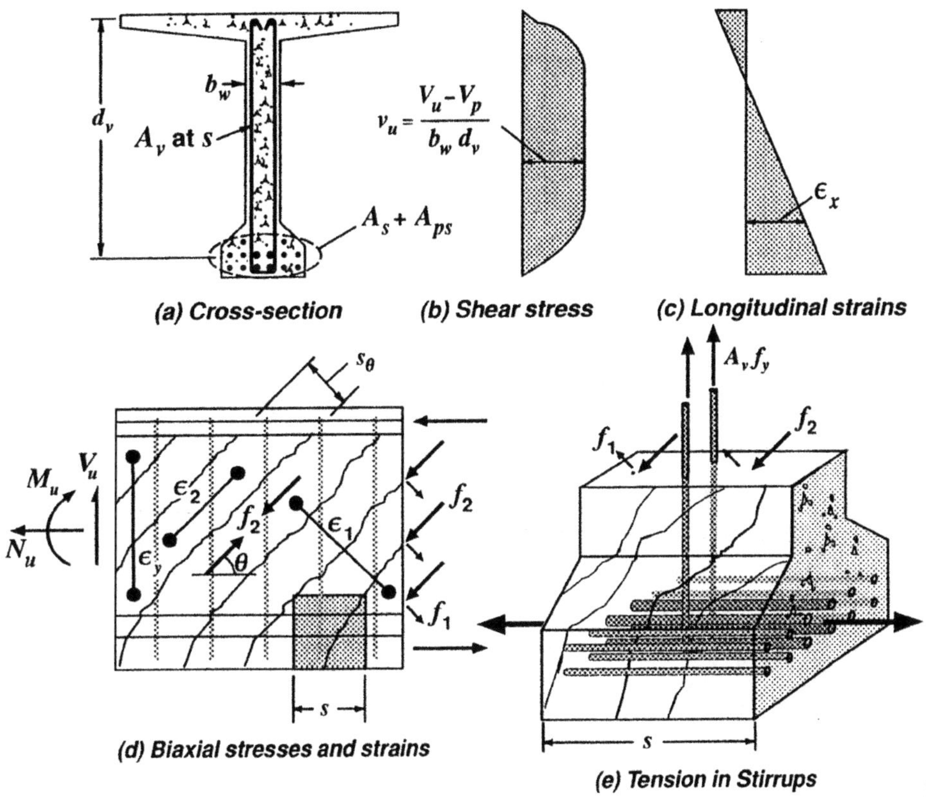

Fig. 2.11—Beam subjected to shear, moment, and axial load.

b_w = effective web width, taken as the minimum web width within the shear depth;

d_v; d_v = effective shear depth, taken as the flexural lever arm, but which need not be taken less than $0.9d$; and

β = concrete tensile stress factor indicating the ability of diagonally cracked concrete to resist shear.

The shear stress that the web of a beam can resist is a function of the longitudinal straining in the web. The larger this longitudinal straining becomes, the smaller the shear stress required to fail the web. In determining the shear capacity of the beam, it is conservative to use the highest longitudinal strain ε_x occurring within the web. For design calculations, ε_x can be approximated as the strain in the tension chord of the equivalent truss. Therefore

$$\varepsilon_x = \frac{(M_u/d_v) + 0.5N_u + 0.5V_u\cot\theta - A_{ps}f_{po}}{E_sA_s + E_pA_{ps}} \qquad (2\text{-}26)$$

but need not be taken greater than 0.002, where

f_{po} = stress in the tendon when the surrounding concrete is at zero stress, which may be taken as 1.1 times the effective stress in the prestressing f_{se} after all losses;

A_s = area of nonprestressed longitudinal reinforcement on the flexural tension side of the member;

A_{ps} = area of prestressed longitudinal reinforcement on the flexural tension side of the member;

M_u = moment at the section, taken as positive; and

N_u = axial load at the section, taken as positive if tensile and negative if compressive.

The determination of ε_x for a nonprestressed beam is illustrated in Fig. 2.12.

The longitudinal strain parameter, ε_x, accounts for the influence of moment, axial load, prestressing, and amount of longitudinal reinforcement on the shear strength of a section. If ε_x and the crack spacing s_θ are known, the shear capacity corresponding to a given quantity of stirrups can be calculated (Fig. 2.8). This is equivalent to finding the values of β and θ in Eq. (2-25).

Values of β and θ determined from the modified compression field theory (Vecchio and Collins 1986) and suitable for members with at least minimum web reinforcement are given in Fig. 2.13. In determining these values, it was assumed that the amount and spacing of the stirrups would limit the crack spacing to about 300 mm. The θ values given in Fig. 2.13 ensure that the tensile strain in the stirrups, ε_v, is at least equal to 0.002 and that the compressive stress, f_2, in the concrete does not exceed the crushing strength f_{2max}. Within the range of values of θ that satisfy these requirements, the values given in Fig. 2.13 result in close to the smallest amount of total shear reinforcement being required to resist a given shear.

2.6.1 *Minimum shear reinforcement*—A minimum amount of shear reinforcement is required to control diagonal cracking and provide some ductility. In ACI 318M-95, this amount is specified as

$$\frac{A_vf_y}{b_ws} > 0.33 \text{ MPa} \qquad (2\text{-}27)$$

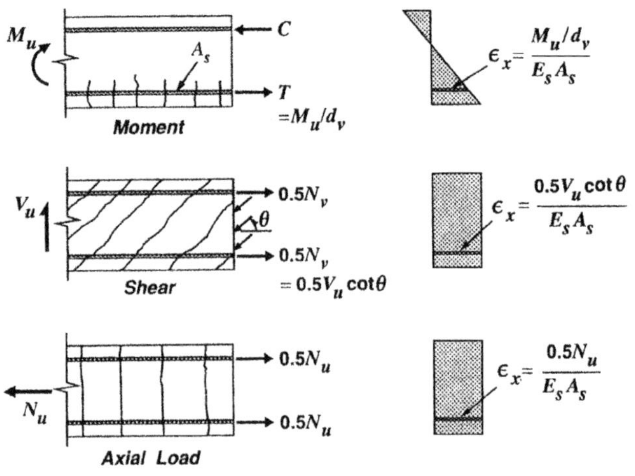

Fig. 2.12—Determination of strains, ε_x, for nonprestressed beam.

The AASHTO specifications (1994) relate the minimum reinforcement required to the concrete strength and require a larger quantity of stirrups for high-strength concrete. The AASHTO specifications require that

$$\frac{A_v f_y}{b_w s} > 0.083\sqrt{f'_c} \text{ MPa} \tag{2-28}$$

This type of relationship was found to give reasonable estimates compared with experiments conducted by Yoon et al. (1996). There is some concern, however, that the equation may not be conservative enough for large reinforced concrete members that contain low percentages of longitudinal reinforcement (ASCE-ACI 426 1973).

2.6.2 *Example: Determine stirrup spacing in reinforced concrete beam*—To illustrate the method, it will be used to determine the stirrup spacing at Section B for the member tested as shown in Fig. 2.1, which will result in a predicted shear strength of 580 kN with a capacity-reduction factor of 1.0

$$\frac{v_u}{f'_c} = \frac{V_u}{b_w d_v f'_c} = \frac{580{,}000}{295 \times 0.9 \times 920 \times 75} = 0.032$$

From Eq. (2-26)

$$\varepsilon_x = \frac{(M_u/d_v) + 0.5V_u \cot\theta}{E_s A_s}$$

$$\varepsilon_x = \frac{(580{,}000 \times 1300/828) + 0.5 \times 580{,}000 \cot\theta}{200{,}000\,(4 \times 700 + 3 \times 300)}$$

$$\varepsilon_x = 1.23 \times 10^{-3} + 0.329 \times 10^{-3} \cot\theta$$

Figure 2.13 shows that if v_u/f_c' is less than 0.05 and ε_x is between 1.5×10^{-3} and 2×10^{-3}, θ is about 42 degrees. With this value of θ, the calculated value of ε_x is 1.67×10^{-3} and β is approximately 0.155, according to Fig. 2.13. Equation (2-25) then becomes

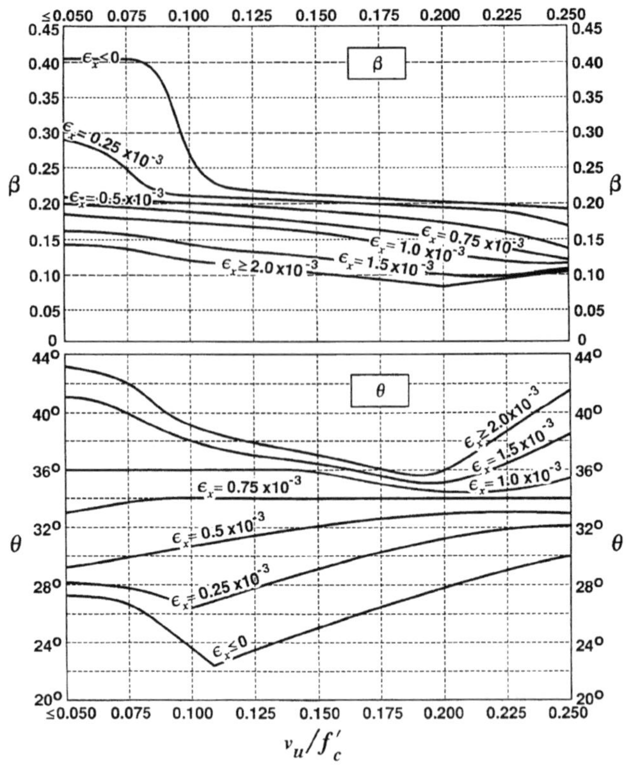

Fig. 2.13—Values of β and θ for members containing at least the minimum amount of stirrups.

$$580{,}000 = 0.155\sqrt{75} \times 295 \times 828 + \frac{200 \times 522}{s} \times 828 \cot 42 \text{ degrees}$$

which gives a required stirrup spacing of 381 mm. Note that because the tested beam contained stirrups at a spacing of 440 mm, the design method is a little conservative.

2.6.3 *Example: Determine stirrup spacing in a prestressed concrete beam*—If the member had also contained a straight post-tensioned tendon near the bottom face consisting of six 13 mm strands with an effective stress of 1080 MPa, the calculations for the required stirrup spacing would change in the following manner. Equation (2-26) becomes

$$\varepsilon_x \times \{[(580{,}000 \times 1300/828) + 0.5 \times 580{,}000 \cot\theta$$
$$- 6 \times 99 \times 1.1 \times 1080]/[200{,}000(4 \times 700 + 3 \times 300 + 6 \times 99)]\}$$

Figure 2.13 shows that if v_u/f_c' is less than 0.05 and ε_x is approximately 0.75×10^{-3}, θ is about 33 degrees. With this value of θ, the calculated value of ε_x is 0.76×10^{-3}. For this value of ε_x, the value of β is approximately 0.155, according to Fig. 2.13. Equation (2-25) then becomes

$$580{,}000 = 0.20\sqrt{75} \times 295 \times 828 + \frac{200 \times 522}{s} \times 828 \cot 33 \text{ degrees}$$

which gives a required stirrup spacing of 848 mm.

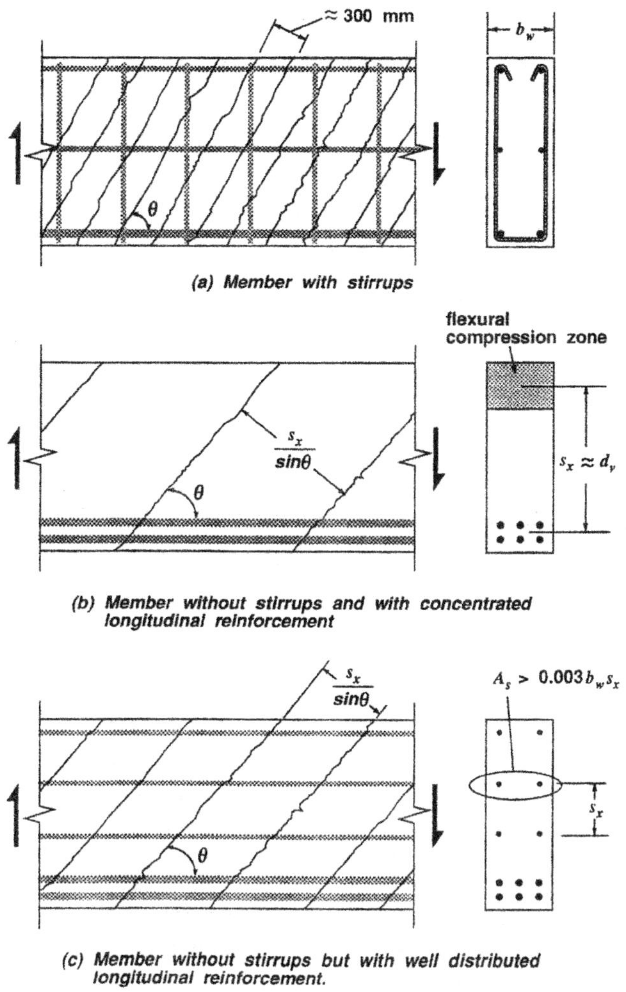

Fig. 2.14—Influence of reinforcement on spacing of diagonal cracks.

Stirrups spaced at 848 mm in the member shown in Fig. 2.1 would not provide adequate crack control, and the assumption that the crack spacing was about 300 mm would no longer be valid. The AASHTO specifications (1994) require that the stirrup spacing not exceed $0.8d_v$ or 600 mm.

To satisfy the minimum stirrup amount requirement of Eq. (2-28), the stirrup spacing for the beam shown in Fig. 2.1 should be

$$s < \frac{200 \times 522}{295 \times 0.083\sqrt{75}} = 492 \text{ mm}$$

2.6.4 *Design of member without stirrups*—For members without stirrups or with less than the minimum amount of stirrups, the diagonal cracks will typically be more widely spaced than 300 mm. For these members, the diagonal cracks will become more widely spaced as the inclination of the cracks θ is reduced (Fig. 2.5 and 2.14). The crack spacing when θ equals 90 degrees is called s_x, and this spacing is primarily a function of the maximum distance between the longitudinal reinforcing bars (Fig. 2.14). As shown in Fig. 2.5 the values of β and θ for members with less than the minimum amount of stirrups depend on the longitudinal strain parameter, ε_x, and the crack spacing parameter, s_x, where s_x need not be taken as greater than 2000 mm. As s_x increases, β decreases, and the predicted shear strength decreases.

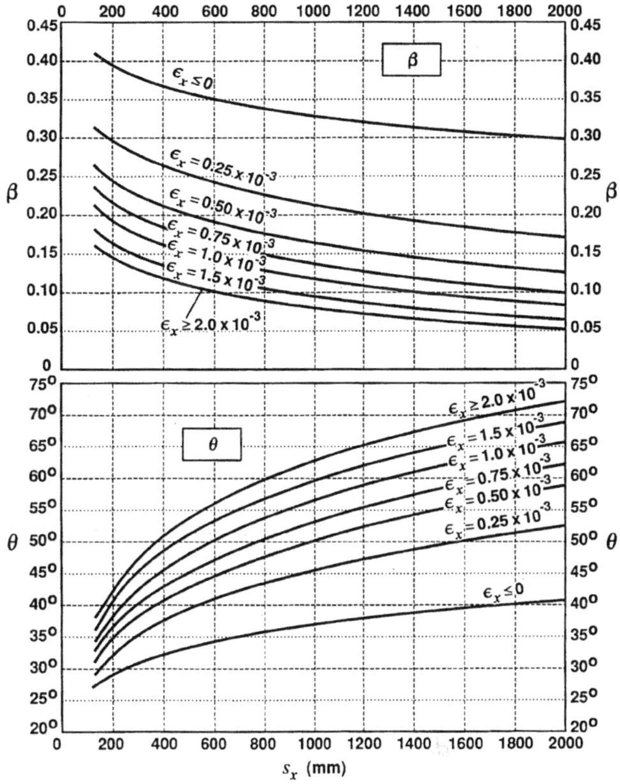

Fig. 2.15—Values of β and θ for members containing less than the minimum amount of stirrups.

The β and θ values given in Fig. 2.15 were calculated assuming that the maximum aggregate size *a* is 19 mm. The values can be used for other aggregate sizes by using an equivalent crack spacing parameter

$$s_{xe} = s_x \frac{35}{a + 16} \tag{2-29}$$

If the member shown in Fig. 2.1 did not contain any stirrups, the shear strength at Section B could be predicted from Fig. 2.15 in the following manner. As the longitudinal bars are spaced 195 mm apart and the maximum aggregate size is 10 mm, the crack spacing parameter is

$$s_{xe} = 195 \frac{35}{10 + 16} = 263 \text{ mm}$$

If ε_x is estimated to be 1.0×10^{-3}, then from Fig. 2.15, θ is about 41 degrees and β is about 0.18. Equation (2-25) then gives

$$V_n = 0.18\sqrt{75} \times 295 \times 828 = 381 \text{ kN}$$

whereas Eq. (2-26) gives

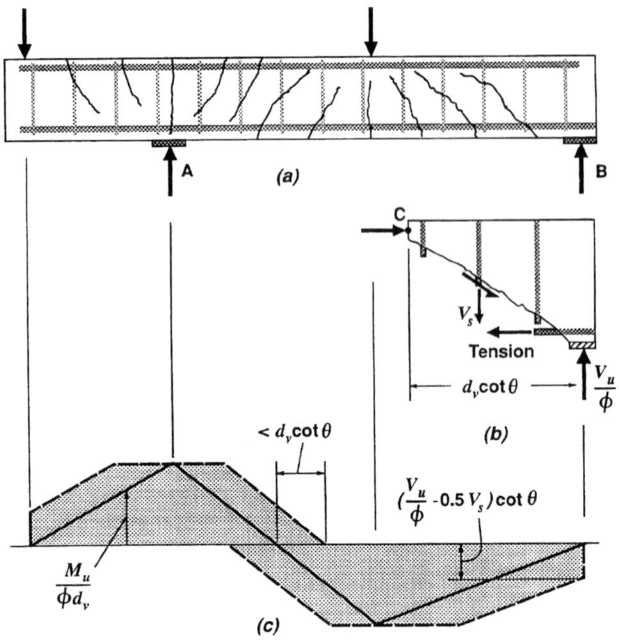

Fig. 2.16—Influence of shear on tension in longitudinal reinforcement.

$$\varepsilon_x = \frac{(381{,}000 \times 1300/828) + 0.5 \times 381{,}000 \cot 41 \text{ degree}}{200{,}000(4 \times 700 + 3 \times 300)}$$

$$= 1.10 \times 10^{-3}$$

Using the calculated value of ε_x of 1.10×10^{-3}, β would be about 0.17, and a more accurate estimate of the failure shear would be 360 kN.

If the member did not contain the six 20 mm diameter skin-reinforcement bars, the crack spacing parameter would increase to

$$s_{xe} = 828\frac{35}{10+16} = 1115 \text{ mm}$$

If ε_x is estimated to be 0.75×10^{-3}, then from Fig. 2.15, θ is about 55.5 degrees and β is about 0.13. With these values, V_n would be reduced to 275 kN and ε_x would be calculated as 0.71×10^{-3}. A second iteration would give the predicted shear strength as 283 kN.

2.6.5 *Additional design considerations*—In the design approach based on the MCFT, the stirrups required at a particular section can be determined from Eq. (2-25) as

$$V_s \geq \frac{V_u}{\phi} - V_c - V_p \qquad (2\text{-}30)$$

given the shear, moment, and axial load acting at the section. Although this calculation is performed for a particular section, shear failure caused by yielding of the stirrups involves yielding this reinforcement over a length of beam of about $d_v \cot\theta$. Therefore, a calculation for one section can be taken as representing a length of beam $d_v \cot\theta$ long with the calculated section being in the middle of this length. Near a support, the first section checked is the section $0.5d_v \cot\theta$ from the face of the support. In addition, near concentrated loads, sections closer than $0.5d_v \cot\theta$ to the load need not be checked. As a simplification, the term $0.5d_v \cot\theta$ may be approximated as d_v. The required amount of stirrups at other locations along the length of the beam can be determined by calculating sections at about every tenth point along the span, until it is evident that shear is no longer critical.

Shear causes tensile stresses in the longitudinal reinforcement as well as in the stirrups (Eq. (2-21)). If a member contains an insufficient amount of longitudinal reinforcement, its shear strength may be limited by yielding of this reinforcement. To avoid this type of failure, the longitudinal reinforcement on the flexural tension side of the member should satisfy the following requirement

$$A_s f_y + A_{ps} f_{ps} \geq \frac{M}{\phi d_v} + \frac{0.5N_u}{\phi} \qquad (2\text{-}31)$$

$$+ \left(\frac{V_u}{\phi} - 0.5V_s - V_p\right)\cot\theta$$

Figure 2.16 illustrates the influence of shear on the tensile force required in the longitudinal reinforcement. Whereas the moment is zero at the simple support, there still needs to be considerable tension in the longitudinal reinforcement near this support. The required tension T at a simple support can be determined from the free-body diagram in Fig. 2.16 as

$$T = \left(\frac{V_u}{\phi} - 0.5V_s - V_p\right)\cot\theta \qquad (2\text{-}32)$$

but

$$T \geq 0.5\left(\frac{V_u}{\phi} - V_p\right)\cot\theta \qquad (2\text{-}33)$$

The reinforcement provided at the support should be detailed in such a manner that this tension force can be safely resisted and that premature anchorage failures do not occur.

At the maximum moment locations shown in Fig. 2.16, the shear force changes sign, causing fanning of the diagonal compressive stresses as θ passes through 90 degrees. See also Fig. 2.1. Because of this, the maximum tension in these regions need not be taken as larger than that required for the maximum moment. The traditional North American procedure for accounting for the influence of shear on the longitudinal reinforcement involves extending the longitudinal reinforcement a distance d beyond the point at which it is no longer required to resist flexure, in addition to other detailing requirements. As can be seen from Fig. 2.16, the requirements of Eq. (2-31) can be satisfied in a conservative manner by simply extending the longitudinal reinforcement a

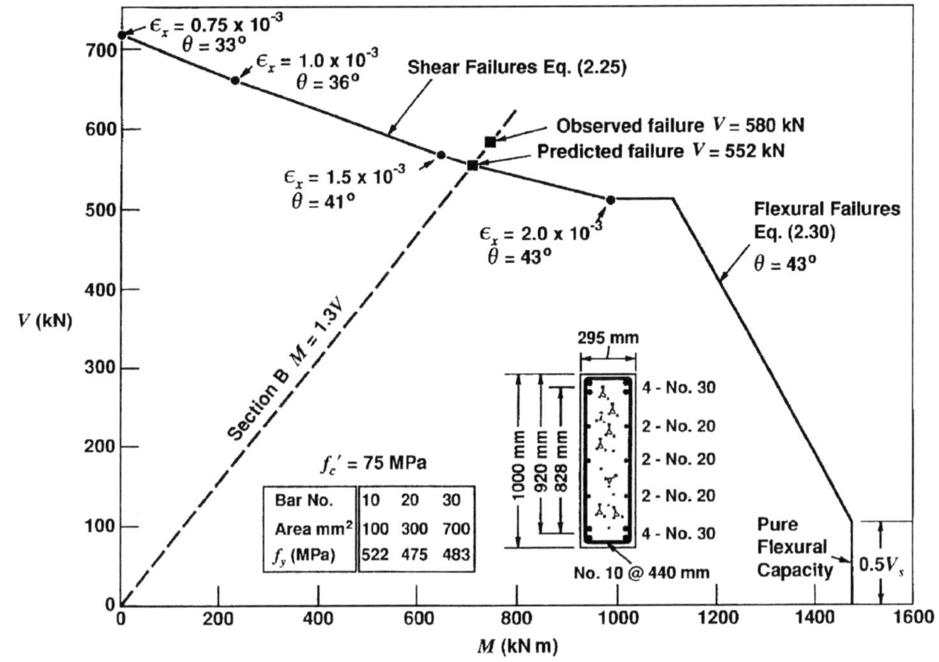

Fig. 2.17—Shear-moment interaction diagram for rectangular section.

distance d_v cot θ past the point at which it is no longer required for flexure alone.

For sections at least a distance d_v away from the maximum moment locations, the MCFT predicts that increasing the moment decreases the shear strength, while increasing the shear decreases the flexural strength. This point is illustrated in Fig. 2.17, which gives the shear-moment interaction diagram for Section B of the beam described in Fig. 2.1. The shear-failure line shown in this figure was determined by calculating V from Eq. (2-25) with the β and θ values corresponding to chosen values of ε_x and then using Eq. (2-26) to determine the corresponding values of M. The flexural-failure line, which corresponds to yielding of the longitudinal steel, was determined from Eq. (2-31) with ϕ taken as unity. A shear-moment interaction diagram, such as that shown in Fig. 2.17, can be used to determine the predicted failure load of a section when it is subjected to a particular loading ratio. Section B of the beam shown in Fig. 2.1 is loaded such that the moment-to-shear ratio equals 1.3 m (that is, M/Vd = 1.41). At this loading ratio, the section is predicted to fail in shear with yielding of the stirrups and slipping of the cracks when the shear reaches a value of 552 kN. For this beam, the experimentally determined failure shear was about 5% higher than this value.

CHAPTER 3—TRUSS APPROACHES WITH CONCRETE CONTRIBUTION

3.1—Introduction

The traditional truss model assumes that the compression struts are parallel to the direction of cracking and that no stresses are transferred across the cracks. This approach has been shown to yield conservative results when compared with test evidence. More recent theories consider one or both of the following two resisting mechanisms: 1) tensile stresses in concrete that exist transverse to the struts; or 2) shear stresses that are transferred across the inclined cracks by aggregate interlock or friction. Both mechanisms are interrelated and result in: 1) the angle of the principal compression stress in the web being less than the crack angle; and 2) a vertical component of the force along the crack that contributes to the shear strength of the member. The resisting mechanisms give rise to V_c, the concrete contribution. These theories typically assume that there is no transfer of tension across cracks.

In this chapter, several approaches incorporating the so-called concrete contribution are discussed, which begin with the assumptions for the angle and the spacing of the inclined cracks. Then, the principal tensile strain, ε_1, in the web and the widths of the inclined cracks are calculated, as in the MCFT, discussed in Chapter 2. The stress transfer across the cracks can then be determined, giving V_c. The state of stress in the web results in tensile stress in the web perpendicular to the principal compressive stresses.

This chapter also contains a brief review of the recent European codes EC2, Part 1 (1991) and CEB-FIP Model Code 1990 (1993), which are based on related approaches.

3.2—Overview of recent European codes

The strut-and-tie model approaches have influenced recent European codes including, to a lesser extent, EC2, Part 1 (1991) and, to a larger extent, CEB-FIP Model Code 1990 (1993). EC2, Part 1 is primarily based on the previous CEB-FIP Model Code 1978, although the design clauses contain several changes. The static or lower-bound approach of the theory of plasticity may be used in the design. Appropriate measures should be taken to ensure ductile behavior (EC2, Part 1, Section 2.5.3.6.3). Corbels, deep beams, and anchorage zones for post-tensioning forces (EC2, Part 1,

Section 2.5.3.7) may be analyzed, designed, and detailed in accordance with lower-bound plastic solutions. In this approach, the average design compressive stress may be taken as νf_{cd} with ν = 0.60 and $f_{cd} = f_{ck}/1.5$ the factored design strength (with $f_{ck} = 0.9f_c'$ for a concrete of about 28 MPa, the value of $0.60f_{cd}$ corresponds to about $0.54f_c'/1.5 = 0.36f_c'$). Them aximum strength to be used in axial compression is $0.85f_{cd}$, including an allowance for sustained load.

The shear design expressions use three different values for shear resistance: V_{Rd1}, V_{Rd2}, and V_{Rd3}. The shear resistance V_{Rd1} is for members without shear reinforcement and is based on an empirical formula. This formula incorporates the influence of concrete strength, reinforcing ratio, member depth, and axial forces.

The resistance V_{Rd2} is the upper limit of the shear strength intended to prevent web-crushing failures. The limiting value is a function of: 1) the inclination and spacing of the cracks; 2) the tensile strain in the transverse reinforcement; and 3) the longitudinal strain in the web. The limiting value of the shear strength is calculated using an effective diagonal stress in the struts $f_{dmax} = (\nu f_{cd})$. As a simplification, the effectiveness factor ν in EC2, Part 1 is given by the expression

$$\nu = 0.70 - f_{ck}/200 \qquad (3\text{-}1)$$

with f_{ck} (MPa) as the characteristic cylinder strength (approximately $0.9f_c'$). The stress in the inclined struts is calculated from equilibrium as

$$f_d = V_u/b_v d_v \cos\theta \sin\theta \qquad (3\text{-}2)$$

The resistance V_{Rd3}, provided by the shear reinforcement, may be determined from two alternative design methods—the standard method or the variable-angle truss method. The standard method is similar to the current U.S. design practice where a concrete contribution is added to that of the shear reinforcement

$$V_{Rd3} = V_{cd} + V_{wd} \qquad (3\text{-}3)$$

The concrete contribution V_{cd} is assumed equal to V_{Rd1}, the shear resistance of members without shear reinforcement. The resistance of the vertical shear reinforcement is given by

$$V_{wd} = A_v f_y 0.9d/s \qquad (3\text{-}4)$$

based on a flexural lever arm of $0.9d$ between the truss chords (d = effective depth).

The variable-angle truss method uses a truss with struts inclined at an angle θ. The strength provided by the shear reinforcement for beams with transverse stirrups follows from the vertical equilibrium

$$V_{Rd3} = V_s = (A_v/s)f_{yv}0.9d\cot\theta \qquad (3\text{-}5)$$

where

A_v = area of the transverse stirrups at spacing s;

f_{yv} = yield strength of the shear reinforcement;

$0.9d$ = flexural lever arm or effective truss depth; and

θ = angle of inclined struts.

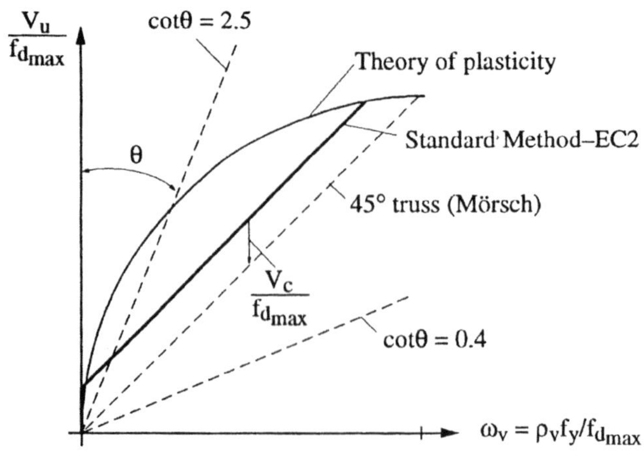

Fig. 3.1—Calculated shear strength as function of θ and shear reinforcement index ($\rho_v = A_v/b_w s$).

The theory of plasticity assumes that the capacity of the web is achieved by reaching simultaneously the yielding of the shear reinforcement and the limiting stress, f_d, in the inclined struts (web crushing), Eq. (3-2). This assumption yields a condition for the angle θ of the inclined struts and a function for the capacity depending on the amount of shear reinforcement. This function may be plotted in a dimensionless format, and it appears as a quarter circle (Fig. 3.1). This figure shows that, for given amounts of transverse reinforcement, much larger capacities are predicted than those based on the traditional Mörsch model, which assumes a strut inclination θ = 45 degrees. For very low amounts of transverse reinforcement, very flat angles θ are predicted so that mostly lower limits are given to avoid under-reinforced members. For comparison, Fig. 3.1 also shows the predictions using Eq. (3-3) according to the standard method of EC2. According to EC2, Section 4.3.2.4.4, the angle θ in Eq. (3-5) may be varied between:

- $-0.4 < \cot\theta < 2.5$ for beams with constant longitudinal reinforcement; and
- $-0.5 < \cot\theta < 2.0$ for beams with curtailed longitudinal reinforcement

There are different interpretations of these limits. For example, in Germany, the minimum inclination was increased to $\cot\theta = 1.75$ for reinforced concrete members. For members subjected also to axial forces, the following value is suggested

$$\cot\theta = (1.25 - 3\sigma_{cp}/f_{cd}) \qquad (3\text{-}6)$$

with σ_{cp} = axial stress in middle of web. In the presence of axial tension, larger strut angles are required, and the limits of EC2 are considered unsafe.

The provisions for member design have been based on strut-and-tie models instead of a separate sectional design for the different effects of bending, axial load, shear, and torsion. The shear design is based on the theory of plasticity using the variable-angle truss method. The strut angle may be assumed as low as about 18 degrees, corresponding to cot θ = 3. CEB-FIP Model Code 1990 (1993) limits the compressive stress in the struts in a form similar to that given by Eq. (3-1).

In EC2, Part 1 (1991), minimum reinforcement amounts are required in beams for the longitudinal reinforcement as well as for the shear reinforcement. The latter is made dependent on the concrete class and the steel class, and, for example, amounts to a minimum required percentage of shear reinforcement, min $\rho_v = 0.0011$ for steel with f_y = 500 MPa and medium concrete strengths between $f_c' = 23$ and 33 MPa. For higher concrete strengths, the value is increased to min $\rho_v = 0.0013$. For slabs, no minimum transverse reinforcement is required.

CEB-FIP Model Code 1990 requires a minimum longitudinal reinforcement in the tension zone of 0.15%. The required minimum amount for the shear reinforcement in webs of beams is expressed in terms of a mechanical reinforcing ratio rather than a geometrical value min ρ_v. The requirement is min $\omega v = \rho_v f_y / f_{ctm} = 0.20$. This correctly expresses that, for higher concrete strengths (that is, higher values of average tensile strengths f_{ctm}), more minimum transverse reinforcement is required. In terms of min ρ_v, this means about 0.090% for concrete having a compressive strength of about 22 MPa, 0.14% for concrete with 41 MPa compressive strength, and 0.20% for 67 MPa concrete.

3.3—Modified sectional-truss model approach

In the so-called "modified sectional-truss model" approach (CEB-FIP 1978; EC2 1991; Ramirez and Breen 1991) the nominal shear strength of nonprestressed or prestressed concrete beams with shear reinforcement is $V_n = V_c + V_s$, where V_c represents an additional concrete contribution as a function of the shear stress level, and V_s = strength provided by the shear reinforcement.

For nonprestressed concrete beams, the additional concrete contribution V_c has been suggested (Ramirez and Breen 1991) as

$$V_c = \frac{1}{2}(3v_{cr} - v)b_w d \tag{3-7}$$

where

v_{cr} = shear stress resulting in the first diagonal tension cracking in the concrete; and

v = shear stress level due to factored loads.

For prestressed concrete beams (Ramirez and Breen 1991), the additional concrete contribution takes the form of

$$V_c = K\left(\frac{\sqrt{f_c'}}{6}\right)b_w 0.9d \tag{3-8}$$

with f_c' in MPa, where K is the factor representing the beneficial effect of the prestress force on the concrete diagonal tensile strength and further capacity after cracking. The expression for the K factor can be derived from a Mohr circle analysis of an element at the neutral axis of a prestressed concrete beam before cracking and is

$$K = \left[1 + \frac{f_{pc}}{f_t}\right]^{0.5} \tag{3-9}$$

where

f_t = principal diagonal tension stress; and

f_{pc} = normal stress at the neutral axis.

This expression is the same one used in ACI 318M-95 as the basis for the web cracking criteria V_{cw}. The factor K is usually limited to 2.0, and is set equal to 1.0 in those sections of the member where the ultimate flexural stress in the extreme tension fiber exceeds the concrete flexural tensile strength. This limitation is similar to the provision in 318M-95 that limits the concrete contribution to the smaller of the two values, V_{cw} and V_{ci}.

The strength provided by the shear reinforcement, V_s, for beams with vertical stirrups represents the truss capacity in shear derived from the equilibrium condition by summing the vertical forces on an inclined crack free-body diagram. The resulting expression is given in Eq. (3-5). According to Ramirez and Breen (1991), the lower limit of angle of inclination θ for the truss diagonals is 30 degrees for nonprestressed concrete and 25 degrees for prestressed beams. The additional longitudinal tension force due to shear can be determined from equilibrium conditions of the truss model as V_u cot θ, where V_u = factored shear force at the section. Because the shear stresses are assumed uniformly distributed over the depth of the web, this force acts as the section middepth; thus, it may be resisted by equal additional tension forces acting at the top and bottom longitudinal truss chords, with each force being equal to $0.5V$ cot θ. In this approach, a limit of $2.5\sqrt{f_c'}$ (MPa) has also been proposed (Ramirez and Breen 1991) that represents the diagonal compressive strength as a function of the shear that can be carried along the diagonal crack surface. The diagonal compression stress can be calculated from equilibrium and geometry considerations in the diagonally cracked web, resulting in an expression similar to Eq. (3-2).

The term V_c is reflected in the design of the stirrup reinforcement, but it does not affect dimensioning of the longitudinal reinforcement or the check of diagonal compressive stresses. This is justified by the fact that experimental observations have shown that dimensioning of the shear reinforcement (stirrups) based entirely on the equilibrium conditions of the parallel chord truss model described in this section unduly penalizes the majority of members that are subjected to low levels of shear stress or that have no or low amounts of shear reinforcement. A similar effect on the longitudinal reinforcement, however, has not been fully verified. The fact that there are relatively few large-scale specimens, that most tests consist of beams with continuous longitudinal reinforcement, and that almost all of the test beams are simply supported under a point load would seem to justify a simple expression for the required amount of longitudinal reinforcement at this time. The check of diagonal compressive stresses is a

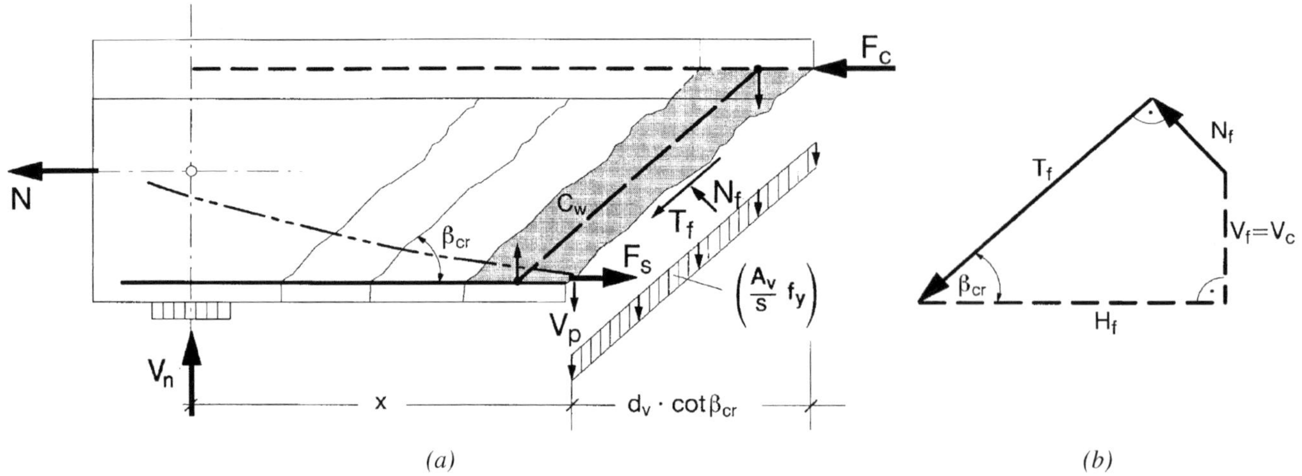

Fig. 3.2—Free-body diagram of B region at end support and forces at crack due to friction: (a) free-body diagram of end support region; and (b) forces due to friction (Reineck 1991a).

conservative check on the maximum shear force that can be carried by the reinforced concrete member without a web-crushing failure.

3.4—Truss models with crack friction

3.4.1 *Equilibrium of truss models with crack friction*—The truss model with crack friction starts with basic assumptions for the spacing and shape of cracks in a B region of a structural-concrete member subjected to shear. It is assumed that forces are transferred across the cracks by friction, which depends on the crack displacements (slips and crack widths); therefore, the strains in the member have to be calculated. This approach was developed for the shear design of webs by several researchers, including Gambarova (1979), Dei Poli et al. (1987, 1990), Kupfer et al. (1979, 1983), Kirmair (1987), Kupfer and Bulicek (1991), and Reineck (1990, 1991a).

The approach uses the free-body diagram in Fig. 3.2, which is obtained by separating the member along an inclined crack in the B region of a structural-concrete member with transverse reinforcement. Vertical equilibrium of this body gives the basic equation

$$V_n = V_s + V_c + V_p \tag{3-10}$$

where

V_n = total shear resisted (nominal shear strength);
V_s = shear force carried by the stirrups;
V_c = sum of the vertical components of the tangential friction forces at the crack T_f, and the normal force at the crack N_f (see Fig. 3.2(b)); and
V_p = vertical component of force in prestressing tendon.

The dowel force of the longitudinal reinforcement, which has a role in members without transverse reinforcement, is neglected (see Fig. 3.2). Furthermore, the chords are assumed to be parallel to the axis of the member so that there is no vertical component of an inclined compression chord of the truss. Leonhardt (1965, 1977), Mallee (1981), and Park and Paulay (1975) have proposed truss models with inclined chords, but these models are more complicated. For a nonprestressed concrete member of constant depth, Eq. (3-10) simplifies to

$$V_n = V_s + V_c \tag{3-11}$$

The shear force component carried by all the stirrups crossing the crack is

$$V_s = A_v f_y d_v \frac{\cot\beta_{cr}}{s} \tag{3-12}$$

where

β_{cr} = crack inclination;
d_v = inner level arm; and
s = stirrup spacing.

The force, V_f, is the vertical component of the combined friction forces, T_f and N_f, across the inclined crack in the web, as shown in Fig. 3.2(b). The tangential force T_f has two components: T_{fo}, the tangential force that may be resisted with N_f equal to zero; and $T_{f\sigma}$, the additional tangential force due to N_f. Equation (3-11) shows that the shear force component, V_f, due to friction is additive with the shear force carried by the stirrups. This is also the case for all design methods with a concrete contribution such as ACI 318M-95 or EC2 with either the standard method (Section 3.2) or the modified truss model approach (Section 3.3). So far, the concrete contribution has mainly been justified empirically, but Eq. (3-11) offers a clear physical explanation for the V_c term; that is, that it is the shear force component, V_f, transferred by friction across the cracks.

3.4.2 *Inclination and spacing of inclined cracks*—In the use of the truss model with crack friction, a necessary condition is that the crack inclination and the crack spacing should be assumed or determined by nonlinear analysis. The angle of the inclined cracks is normally assumed at 45 degrees for nonprestressed concrete members. Kupfer et al. (1983) has pointed out that this angle could be up to 5 degrees flatter, because of a reduction of Young's modulus caused by micro-

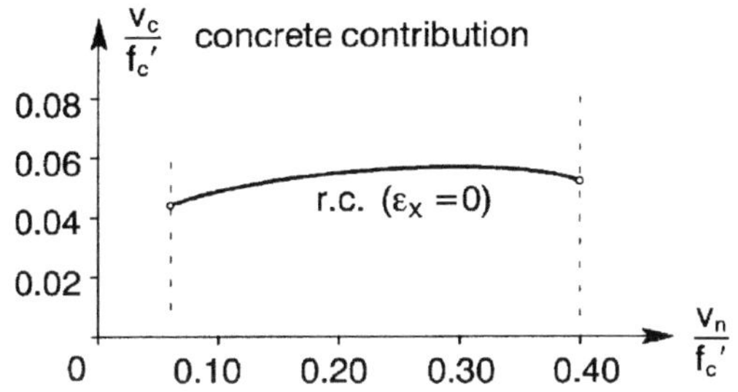

(a) reinforced concrete beams

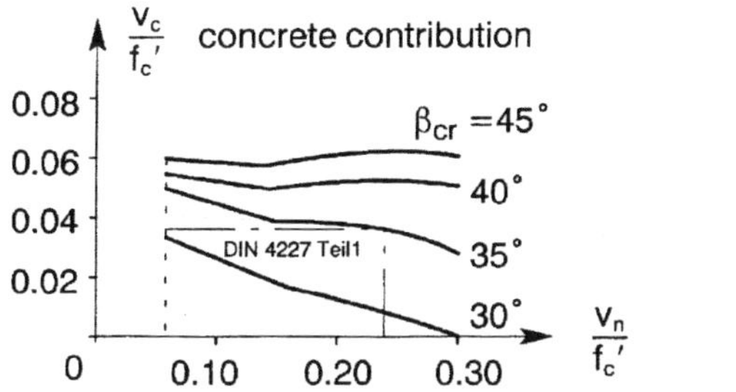

(b) prestressed concrete beams

Fig. 3.3—Shear force carried by crack friction versus ultimate shear force plotted in a dimensionless diagram.

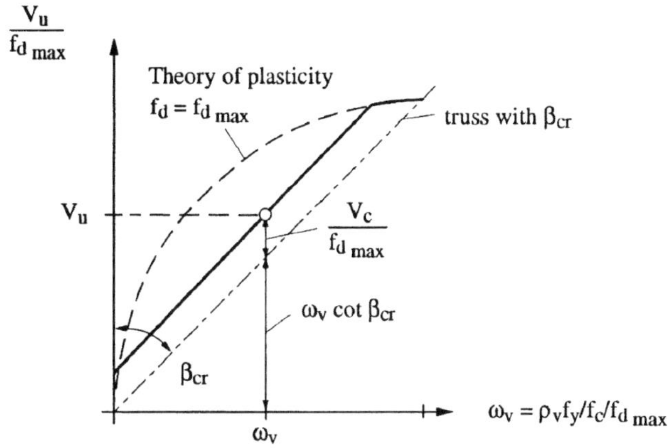

Fig. 3.4—Design diagram with simplified shear force component V_f *due to friction, proposed by Reineck (1990, 1991a).*

cracking. Flatter angles will appear for prestressed concrete members or for members with axial compression, and steeper angles will occur for members with axial tension. For such members, the angle of the principal compressive stress at the neutral axis of the uncracked state is commonly assumed as the crack angle (Loov and Patniak 1994).

The spacing of the inclined cracks is primarily determined by the amount and spacing of reinforcement, and relevant formulas have been proposed by Gambarova (1979), Kupfer and Moosecker (1979), Kirmair (1987), and Dei Poli et al. (1990).

3.4.3 *Constitutive laws for crack friction*—The truss model with crack friction requires constitutive laws for the transfer of forces across cracks by friction or interface shear. Before 1973, this shear transfer mechanism was known and clearly defined by the works of Paulay and Fenwick, Taylor, and others (see Section 4.3.3), but only a few tests and no theories were available for formulating reliable constitutive laws. This has changed considerably in the last 20 years due to the work of Hamadi (1976), Walraven (1980), Walraven and Reinhardt (1981), Gambarova (1981), Daschner and Kupfer (1982), Hsu et al. (1987), Nissen (1987), and Tassios and Vintzeleou (1987). An extensive state-of-the-art report on interface shear was recently presented by Gambarova and Prisco (1991).

The constitutive law proposed by Walraven (1980) has often been used by others because it describes not only the shear stress-slip relation for different crack widths but also the associated normal stresses. It was based on a physical model for the contact areas between crack surfaces, and the proposed laws were corroborated with tests on concrete with normal as well as lightweight aggregates (see Section 4.3.3).

3.4.4 *Determining shear resistance* $V_f = V_c$ *due to crack friction*—The shear force component $V_f = V_c$ in Eq. (3-10) and (3-11) transferred by friction across the cracks depends on the available slip and on the crack width, requiring that the strains in the chords and in the web be determined. In addition, the displacements and the strains should be compatible with the forces in the model according to the constitutive laws for the shear force components. Often the capacity of the crack friction mechanism is reached before crushing of the concrete struts between the cracks. Figure 3.3 gives the results of different calculations of the shear force component $V_f = V_c$ in terms of stress from Dei Poli et al. (1987) and Kupfer and Bulicek (1991), but similar results have been obtained by Leonhardt (1965) and Reineck (1990, 1991a). The shear stress, v_c, in Fig. 3.3 depends on the magnitude of the shear, the strains of the struts and stirrups on the longitudinal strain ε_x in the middle of the web, and on the crack spacing. The results in Fig. 3.3 support the notion that, for a wide range of applications and for code purposes, a constant value of the concrete contribution may be assumed. The practical result for the shear design is the dimension-free design diagram shown in Fig. 3.4, which is well known and used in many codes (see also Fig. 3.1). Crack friction governs the design for low and medium shear. For very high shear, the strength of the compression struts governs, which is characterized by the quarter circle in Fig. 3.4, as explained in Fig. 3.1. The crack friction approach considers the influence of axial forces (tension and compression) as well as prestress, as shown in Fig. 3.5. For a member with axial tension (Fig. 3.5(a)), steeper inclined cracks occur in the web and the term V_c is reduced so that more transverse reinforcement is required in comparison with a member without axial forces (Fig. 3.4). The opposite is the case for prestressed members or members with axial compression (Fig. 3.5(b)); the cracks are flatter, so that less transverse reinforcement is required than for a member without axial force (Fig. 3.4), although the term V_c is reduced and may even disappear.

3.4.5 *Stresses and strength of concrete between cracks*—The main function of the concrete between the cracks is to act as the struts of a truss formed together with the stirrups as described by Mörsch (1920, 1922) (see Fig. 3.6(a)). The additional friction forces acting on the crack surfaces result

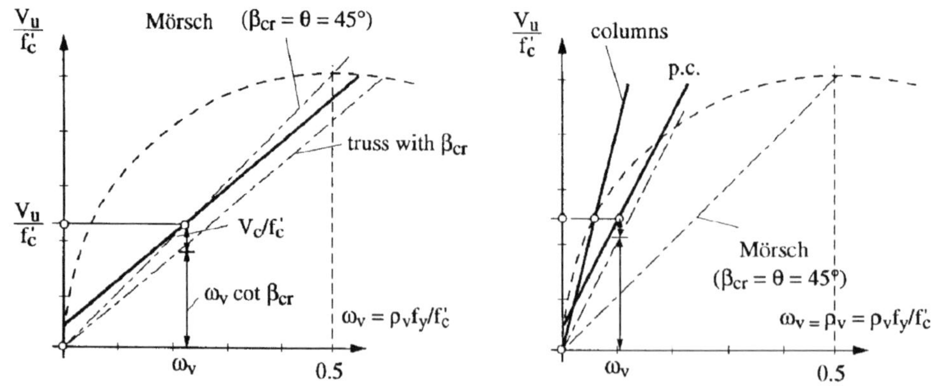

(a) reinforced concrete beams *(b) prestressed concrete beams*

Fig. 3.5—Simplified design diagram for members subjected to shear and axial forces as proposed by Reineck (1991).

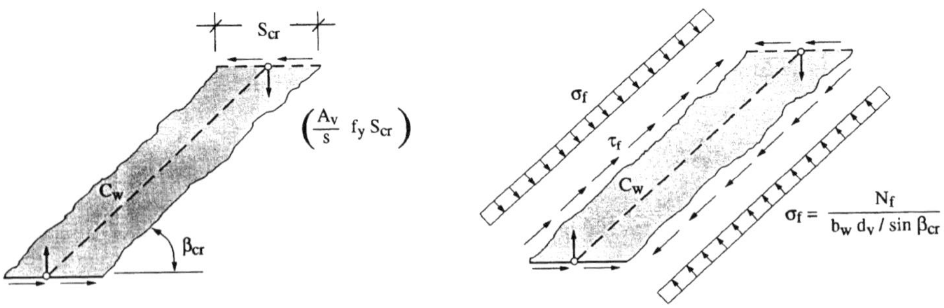

(a) stirrup forces acting on strut *(b) friction forces acting on strut*

Fig. 3.6—Forces and stress fields in discrete concrete struts between cracks (Kupfer et al. 1983; Reineck 1991a).

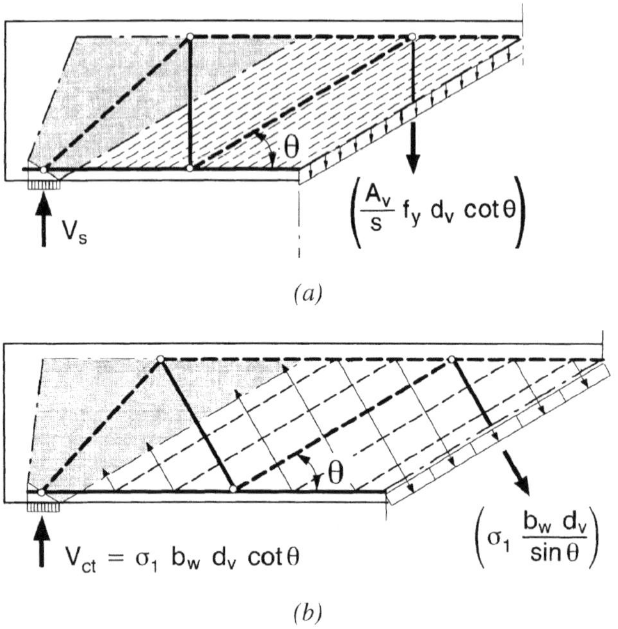

Fig. 3.7—Smeared truss-models corresponding to principal struts between cracks as proposed by Reineck (1991a): (a) truss with uniaxial compression field; and (b) truss with biaxial tension-compression field.

in a biaxial state of stress with a principal compression field at a flatter inclination than the crack angle β_{cr}.

The minor principal stress is tensile for small shear forces so the state of stress may be visualized by the two trusses shown in Fig. 3.7. The usual truss model with uniaxial compression inclined at the angle θ in Fig. 3.7(a) is superimposed on a truss with concrete tension ties perpendicular to the struts (Fig. 3.7(b)). Thus, there are two load paths for the shear transfer, as defined by Schlaich et al. (1987) and as earlier shown by Reineck (1982), and with different explanations by Lipski (1971, 1972) and Vecchio and Collins (1986) in their MCFT. The model in Fig. 3.7(b) is the same as that proposed by Reineck (1989; 1991a,b) for members without transverse reinforcement, so that the transition from members with to members without transverse reinforcement is consistently covered.

For high shear forces, the minor principal stress is compressive. These compressive stresses are so small, however, that they are usually neglected, and only the truss of Fig. 3.7(a) remains with a uniaxial compression field. This is the model used in the theory of plasticity.

The concrete between the cracks is uncracked and forms the strut. Apart from the compressive stresses due to the truss action, however, there is also transverse tension in the struts due to the friction stresses and forces induced by the bonded stirrups. This reduces the strength below that allowed in

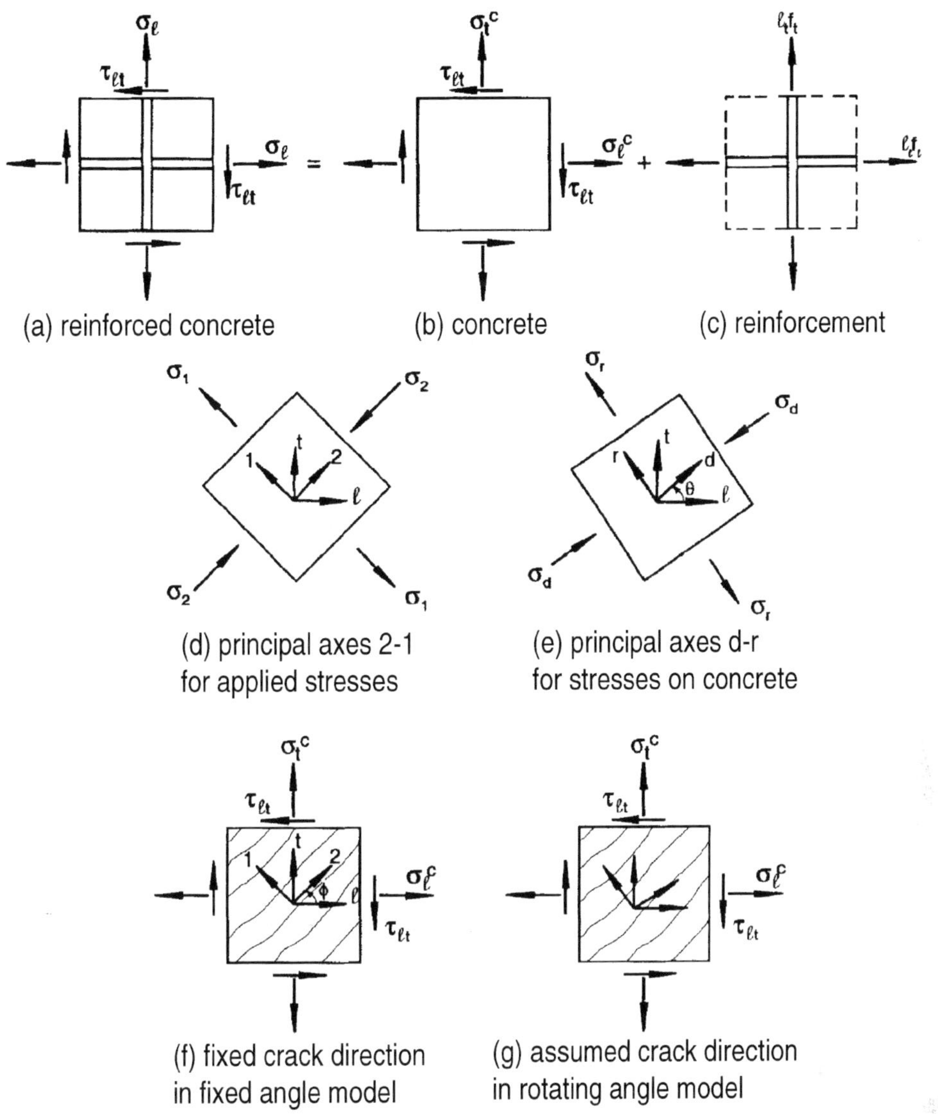

Fig. 3.8—Reinforced concrete membrane elements subjected to in-plane stresses.

compression chords. Further reasons for such a reduction are also the smaller effective width of the strut (rough crack surfaces) and the disturbances by the crossing stirrups. Schlaich and Schafer (1983), Eibl and Neuroth (1988), Kollegger/Mehlhorn (1990), and Schafer et al. (1990) reached the conclusions that the effective concrete strength of the struts could be assumed as

$$f_{dmax} = 0.80 f_{ck} = 0.72 f_c' \quad (3\text{-}13)$$

This is a relatively high value compared to those discussed in Chapters 2 and 6 and the so-called "effective strengths" of, for example, $0.60 f_{ck} = 0.54 f_c'$ or $0.50 f_{ck} = 0.45 f_c'$ used in the theory of plasticity. The results of the higher value for compressive strength is that, for high ratios of transverse reinforcement, excessively high shear capacities may be predicted.

3.5—Fixed-angle softened-truss models

Hsu and his colleagues have proposed two different softened-truss models to predict the response of membrane elements subjected to shear and normal stress (Pang and Hsu 1992, 1996; Zhang 1995). The rotating angle softened-truss model considers the reorientation of the crack direction that occurs as the loads are increased from initial cracking up to failure (see Fig. 3.8(g)). This first model has been discussed in Chapter 2. The second model, known as the fixed-angle softened-truss model, assumes that the concrete struts remain parallel to the initial cracks. This initial crack direction depends on the principal concrete stress directions just before cracking (see Fig. 3.8(f)). With the terms defined as in Fig. 3.8, the equilibrium and compatibility equations for this fixed-angle softened-truss model are as follows

- Equilibrium equations

$$\sigma_l = \sigma_2^c \cos^2\phi + \sigma_1^c \sin^2\phi + \tau_{21}^c 2\sin\phi\cos\phi + \rho_l f_l \quad (3\text{-}14)$$

$$\sigma_t = \sigma_2^c \sin^2\phi + \sigma_1^c \cos^2\phi - \tau_{21}^c 2\sin\phi\cos\phi + \rho_t f_t \quad (3\text{-}15)$$

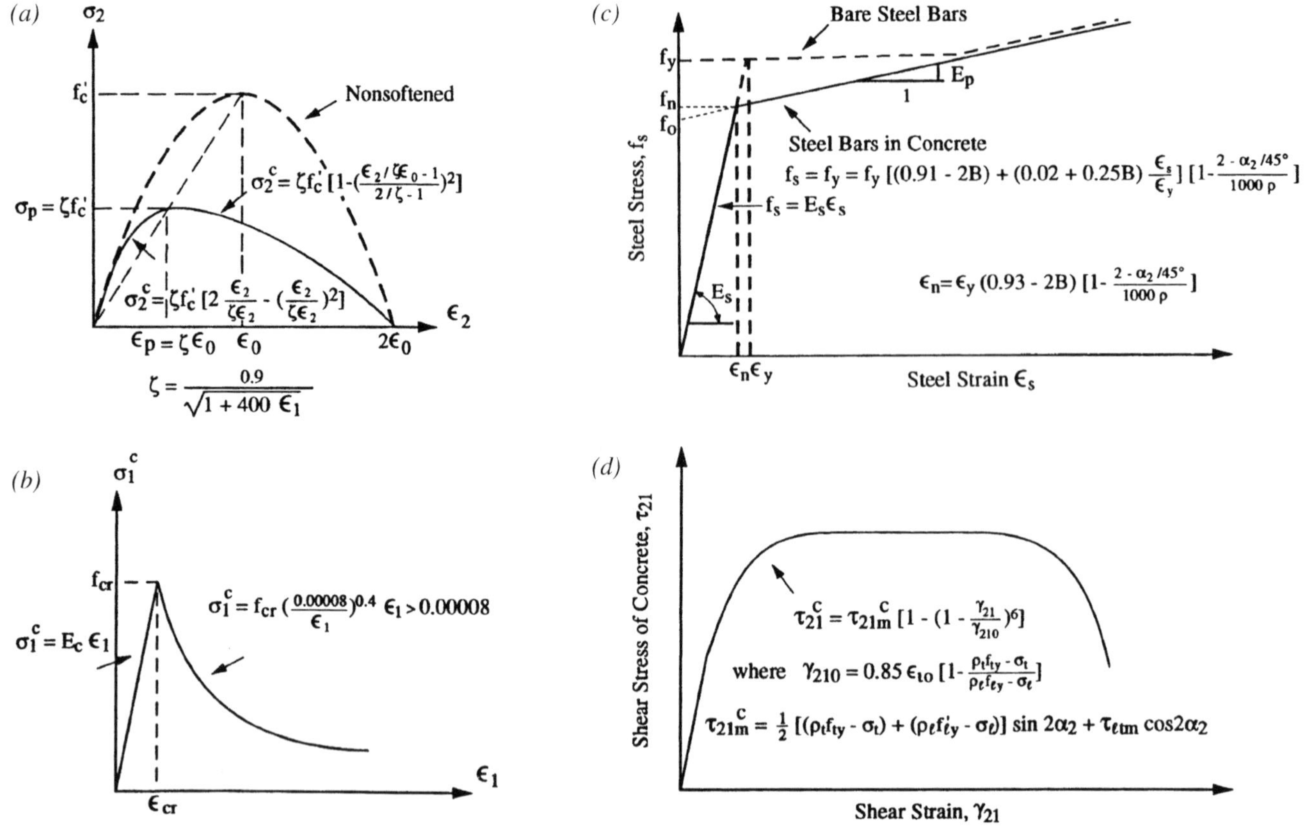

Fig. 3.9—Constitutive laws of concrete and steel: (a) softened stress-strain curve of concrete in compression; (b) average stress-strain curve of concrete in tension; (c) average stress-strain curve of steel bars in concrete; (d) average stress-strain curve of concrete in shear.

$$\tau_{lt} = (-\sigma_2^c + \sigma_1^c)\sin\phi\cos\phi + \tau_{21}^c(\cos^2\phi - \sin^2\phi) \quad (3\text{-}16)$$

where

σ_2^c, σ_1^c = average normal stresses of concrete in Directions 2 and 1, respectively;

τ_{21}^c = average shear stress of concrete in 2-1 coordinate; and

ϕ = angle of inclined cracks.

[In the following equations, the symbol ϕ is kept, following previous works by Pang and Hsu (1992, 1996) and Zhang (1995); ϕ actually is the same as β_{cr} in this chapter.]

- Compatibility equations

$$\varepsilon_l = \varepsilon_2\cos^2\phi + \varepsilon_1\sin^2\phi + \gamma_{21}\sin\phi\cos\phi \quad (3\text{-}17a)$$

$$\varepsilon_t = \varepsilon_2\sin^2\phi + \varepsilon_1\cos^2\phi - \gamma_{21}\sin\phi\cos\phi \quad (3\text{-}17b)$$

$$\gamma_{lt} = 2(-\varepsilon_2 + \varepsilon_1)\sin\phi\cos\phi + \gamma_{21}(\cos^2\phi - \sin^2\phi) \quad (3\text{-}18)$$

where

$\varepsilon_1, \varepsilon_2$ = average normal strain of element in Directions 2 and 1, respectively; and

γ_{21} = average shear strain of element in 2-1 coordinate.

Because Eq. (3-14) to (3-18) are based on the 2-1 coordinate, they include the terms for shear stress, τ_{21}^c, and shear strain, γ_{21}, of cracked concrete. After cracking, the principal compressive stress direction in the concrete struts will typically not coincide with the crack direction; that is, $\theta \neq \phi$. This corresponds to the concept of friction across the cracks as discussed in Section 3.4. In applying these equations, the stresses are determined from the strains using the constitutive laws shown in Fig. 3.9. An efficient algorithm for the solution of the equations in the fixed-angle softened-truss model has been developed (Zhang 1995).

The shear yield strength of a membrane element has been derived from the fixed-angle softened truss model (Pang and Hsu 1996) to be

$$\tau_{lt} = \frac{(\tau_{21}^c)^2}{2\sqrt{\rho_l f_{ly} y \rho_t f_{ty}}} + \sqrt{\rho_l f_{ly} \rho_t f_{ty}} \quad (3\text{-}19)$$

The right-hand side of Eq. (3-19) consists of two terms: the first term is the contribution of concrete, V_c, attributable to τ_{21}^c the shear stress of cracked concrete in the 2-1 coordinate; and the second term is the contribution of steel V_s.

3.6—Summary

Shear design codes require a simple means of computing a realistic V_c term. This so-called concrete contribution is important in the design of beams where the factored shear force is near the value of the shear force required to produce diagonal tension cracking. This term is necessary for the economic design of beams and slabs with little or no shear reinforcement.

In this chapter, several truss model approaches incorporating a concrete contribution term have been discussed. The implementation of this philosophy in recent European codes has also been illustrated. In this chapter, a physical explanation is given of the concrete contribution term, V_c, in the form of the vertical component of the forces transferred across the inclined cracks. This crack leads to a sliding failure through the compression strut. In other approaches, this phenomenon is accounted for by permissible stresses for the struts, that is, reduction factors ν as explained, for example, in Section 3.2.

The approaches discussed in this chapter as well as the MCFT in Chapter 2 describe the structural behavior of beams under increasing load. This behavior cannot be covered by the truss model alone. Furthermore, the truss model with crack friction discussed in Section 3.4 can be extended to regions of geometrical discontinuities (D regions). In those regions, the formation of a single crack indicating a failure surface is very likely. Common examples are beams with openings, dapped end-supports, frame corners, and corbels (Marti 1991). In all of these cases, a crack will start from the corner, and a premature failure may occur before the steel yields.

The fixed-angle softened-truss model proposed by Hsu and his colleagues assumes that the initial crack direction remains unchanged. The orientation of these initial cracks is dictated by the orientation of principal stresses just prior to cracking. After cracking, shear stresses due to friction develop in these cracks so that the principal compressive stress direction is not the same as the crack direction. Such was the case also in the truss model with crack friction approach. The fixed-angle method also results in a concrete contribution term.

CHAPTER 4—MEMBERS WITHOUT TRANSVERSE REINFORCEMENT

4.1—Introduction

Structural-concrete members with sufficient reinforcement in both directions (longitudinal and transverse) can be designed using the simple strut-and-tie models described in Chapter 6 or the theories described in Chapters 2 and 3. Many structural-concrete members are constructed without transverse reinforcement (that is, no stirrups or bent-up bars), such as slabs, footings, joists, and lightly stressed members. The application of a simple strut-and-tie model, as shown in Fig. 4.1, may result in an unsafe solution. Assuming that the shear in this slender member without transverse reinforcement is carried by a flat compression strut predicts that failure will be due to yielding of the longitudinal reinforcement, whereas in reality, a brittle shear failure occurred due a diagonal crack.

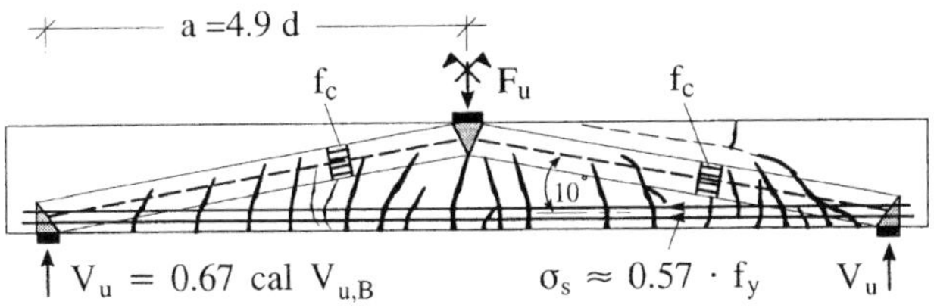

Fig. 4.1—Crack pattern at shear failure of beam with unsafe strut-and-tie model (Kupfer and Gerstle 1973).

The 1973 ASCE-ACI Committee 426 report gave a detailed explanation of the behavior of beams without transverse reinforcement, including the different shear-transfer mechanisms and failure modes. The main parameters influencing shear failure were discussed, and numerous empirical formulas were given. These equations continue to be the basis of shear design rules in many building codes around the world.

The main goal of this chapter is to review design models and analytical methods for structural-concrete members without transverse reinforcement in light of new developments that have occurred since the early 1970s. Only the two-dimensional problem of shear in one-way members is considered. Punching shear, which is a complex three-dimensional problem, is not discussed here; however, it should be noted that an important reason for the continued research on structural-concrete beams without transverse reinforcement is the development of more rational methods for punching shear. Information on punching shear can be found in the CEB state-of-the-art report prepared by Regan and Braestrup (1985).

4.2—Empirical methods

The simplest approach, and the first to be proposed (Mörsch 1909), is to relate the average shear stress at failure to the concrete tensile strength. This empirical approach is presented first because it forms the basis of ACI 318M-95 and several other codes of practice. Experimental results have shown that the average principal tensile stress to cause secondary diagonal cracking (that is, flexure-shear cracking) is usually much less than concrete tensile strength. One reason is the stress concentration that occurs at the tip of initial cracks. Another factor is the reduction in cracking stress due to coexisting transverse compression (Kupfer and Gerstle 1973). Woo and White (1991) have suggested recently that the reason for the low average stress at flexure-shear cracking is a nonuniform shear stress distribution at the outermost flexural crack as a result of a concentration of bond stresses and a reduction of the internal lever arm due to arch action in the flexurally cracked zone.

A simple lower-bound average shear stress at diagonal cracking is given by the following equation (in MPa units)

$$\frac{V_c}{bd} = v_c = \frac{\sqrt{f_c'}}{6} \tag{4-1}$$

This well-known equation is a reasonable lower bound for smaller slender beams that are not subjected to axial load and have at least 1% longitudinal reinforcement (Fig. 4.2).

The 1962 ASCE-ACI Committee 326 report presented a more complex empirical equation for calculating the shear

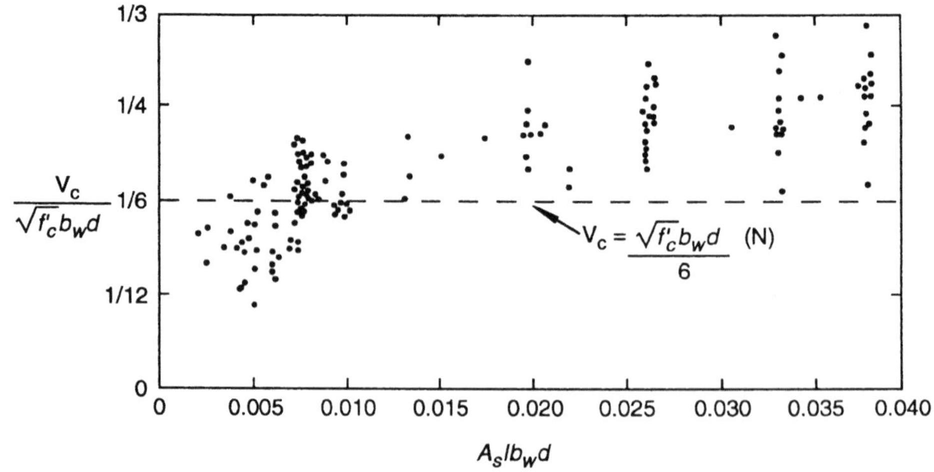

Fig. 4.2—Influence of longitudinal reinforcement amount on shear strength of members without transverse reinforcement (MacGregor 1992).

capacity of beams without web reinforcement (Eq. (11-5) of ACI 318M-95). For a number of reasons, this equation is now considered inappropriate (MacGregor 1992), and in 1977, ASCE-ACI Committee 426 suggested this equation no longer be used. Zsutty (1971) presented the following empirical equation

$$v_c = 59\left(f'_c\, \rho\frac{d}{a}\right)^{1/3} \quad (4\text{-}2)$$

where

f'_c = psi (1 MPa = 145.03 psi).

This equation is a significant improvement over Eq. (11-5) in ACI 318M-95.

Many other empirical equations have since been proposed. These equations typically contain the following parameters: the concrete tensile strength, usually expressed as a function of f'_c; the longitudinal reinforcement ratio $\rho = A_s/b_wd$; the shear span-to-depth ratio a/d or M/Vd; the axial force or amount of prestress; and the depth of the member, to account for size effect. Bažant and Kim (1984) also included the maximum size of the aggregates in his formula that is based on fracture mechanics. The empirical formula by Okamura and Higai (1980) and Niwa et al. (1986) considers all the main parameters

$$v_c = 0.20\frac{\rho^{1/3}}{d^{1/4}}(f'_c)\left(0.75 + \frac{1.40}{a/d}\right) \quad (4\text{-}3)$$

where ρ = a percentage; d in meters; and f'_c in MPa. This equation may be considered one of the more reliable empirical formulas as recent test results on large beams were considered for the size effect.

With respect to the various empirical formulas, considerable differences exist as a result of the following factors: the uncertainty in assessing the influence of complex parameters in a simple formula; the scatter of the selected test results due to inappropriate tests being considered (for example, bending failures or anchorage failures); the poor representation of some parameters in tests (for example, very few specimens with a low reinforcement amount or high concrete strength); and finally, the concrete tensile strength often not being evaluated from control specimens. These issues limit the validity of empirical formulas and increase the necessity for rational models and theoretically justified relationships.

4.3—Mechanisms of shear transfer

4.3.1 *Overview*—Before shear failure, the state of stress in the web of a cracked reinforced concrete member (that is, the zone between the flexural tension and flexural compression zones) differs considerably from what is predicted by the theory of linear elasticity. Therefore, the question of how a cracked concrete member transmits shear (combined with axial load and bending moment) should be considered. The 1973 ASCE-ACI Committee 426 report identified the following four mechanisms of shear transfer: 1) shear stresses in uncracked concrete, that is, the flexural compression zone; 2) interface shear transfer, often called aggregate interlock or crack friction; 3) dowel action of the longitudinal reinforcing bars; and 4) arch action. Since that report was issued, a new mechanism has been identified, namely 5) residual tensile stresses transmitted directly across cracks.

The question of what mechanisms of shear transfer will contribute most to the resistance of a particular beam is difficult to answer. A cracked concrete beam is a highly indeterminate system influenced by many parameters. Different researchers assign a different relative importance to the basic mechanisms of shear transfer, resulting in different models for members without transverse reinforcement. In the following, the different mechanisms of shear transfer are briefly reviewed, before discussion of the different models in Section 4.4.

4.3.2 *Uncracked concrete and flexural compression zone*—In uncracked B regions of a member, the shear force is transferred by inclined principal tensile and compressive stresses, as visualized by the principal stress-trajectories. In cracked B regions, this state of stress is still valid in the

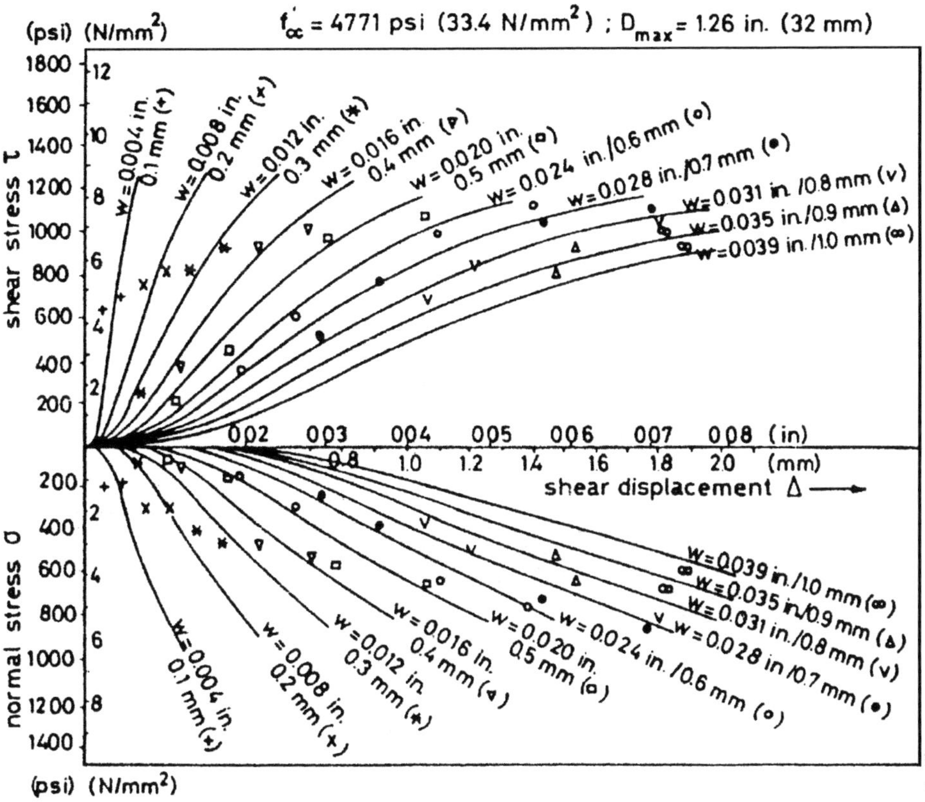

Fig. 4.3—Comparison of Walraven's experimental results and predictions for crack interface shear transfer (Walraven 1981).

uncracked compression zone. The integration of the shear stresses over the depth of the compression zone gives a shear force component, which is sometimes thought to be the explanation for the concrete contribution. Note that the shear force component referred to above is different from the vertical component of an inclined compression strut, which is discussed in Section 4.4.4.

In a slender member without axial compression, the shear force in the compression zone does not contribute significantly to the shear capacity because the depth of the compression zone is relatively small (Taylor 1974; Reineck 1991c). On the other hand, at locations of maximum movement, much of the shear is resisted in the compression zone, particularly after significant yielding of the longitudinal reinforcement.

4.3.3 *Interface shear transfer*—The interface shear transfer mechanism was clearly described in the 1973 ASCE-ACI Committee 426 report, based on work by Fenwick and Paulay (1968), Mattock and Hawkins (1972), and Taylor (1974). The physical explanation for normal-density concrete was aggregate interlock; that is, aggregates protruding from the crack surface provide resistance against slip. Because the cracks go through the aggregate in lightweight and high-strength concrete and still have the ability to transfer shear, however, the term "friction" or "interface shear" is more appropriate. These latter terms also indicate that this mechanism depends on the surface conditions and is not merely a material characteristic.

Significant progress (Gambarova 1981; Walraven 1981; Millard and Johnson 1984; Nissen 1987) has been made in the last two decades toward understanding this mechanism, which involves the relationships between four parameters: crack interface shear stress, normal stress, crack width, and crack slip. Walraven (1981) developed a model that considered the probability that aggregate particles (idealized as spheres) will project from the crack interface. Numerous test results are shown in Fig. 4.3. In Walraven's model, the relationship between stresses and displacements are a function of the concrete compressive strength; however, these relationships were developed for a range of normal (compressive) stresses beyond the range that is relevant for shear transfer in beams without stirrups. Other relationships have been proposed based on Walraven's experimental data (Kupfer et al. 1983; Vecchio and Collins 1986), which assume that the shear that can be transferred is a function of $\sqrt{f'_c}$. Although large differences may still occur between the constitutive laws of different researchers, it may be said that this mechanism is now well known and is widely accepted as an important shear-transfer mechanism.

The important role of interface shear transfer in the redistribution of diagonal compression fields in beams with stirrups is well known (Collins 1978; Kupfer et al. 1983; Dei Poli et al. 1990). In the compression field approaches for members with stirrups (Chapter 2), this effect is implicit within the diagonal crushing strength of concrete. In a number of the models discussed later in this chapter, the ability of diagonal cracks to transfer shear explicitly controls the capacity of members without stirrups.

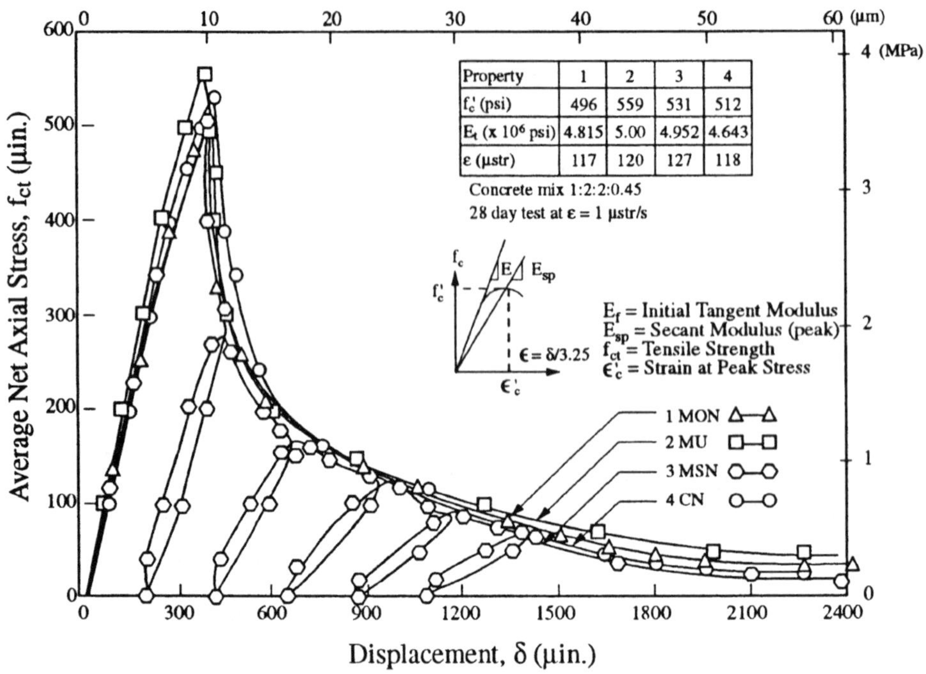

Fig. 4.4—Response of plain concrete loaded in uniaxial tension (Gopalaratnam and Shah 1985).

4.3.4 *Dowel action of longitudinal reinforcement*—Recent work by Vintzeleou and Tassios (1986) and Chana (1987) has reaffirmed the well-known work by Baumann and Rüsch (1970) on the resistance of dowels near a surface. Normally, dowel action is not very significant in members without transverse reinforcement, because the maximum shear in a dowel is limited by the tensile strength of the concrete cover supporting the dowel. Dowel action may be significant in members with large amounts of longitudinal reinforcement, particularly when the longitudinal reinforcement is distributed in more than one layer.

4.3.5 *Residual tensile stresses across cracks*—The basic explanation of residual tensile stresses is that when concrete first cracks, a clean break does not occur. Small pieces of concrete bridge the crack and continue to transmit tensile force up to crack widths in the range of 0.05 to 0.15 mm. The fact that there is a significant descending (softening) branch after the peak tensile stress is reached has been known for some time (Evans and Marathe 1968); however, methods for the reliable measurement of the softening branch have only recently been developed (Gopalaratnam and Shah 1985; Reinhardt et al. 1986). Figure 4.4 shows a typical response of concrete loaded in tension. The deformations are localized in a very small region (the fracture zone); therefore, the response should be expressed in terms of a stress-crack opening relationship and not strain (Evans and Marathe 1968; Gopalaratnam and Shah 1985; Reinhardt et al. 1986; ACI 446.1R).

The application of fracture mechanics (ACI 446.1R) to shear design is based on the premise that residual tensile stress is the primary shear-transfer mechanism. Other methods consider the contribution from residual tensile stresses such as Reineck's tooth model (1991c), which indicates that residual tensile stresses provide a significant portion of the shear resistance of very shallow members (that is, depths less than about 100 mm) where the width of flexural and diagonal cracks are small (Reineck 1991b).

4.4—Models for members without transverse reinforcement

4.4.1 *Introduction*—The different models that can be used to design members without transverse reinforcement may have to be classified into the following groups: 1) mechanical or physical models for structural behavior and failure; 2) fracture mechanics approaches; and 3) nonlinear finite element analysis. In recent decades, the importance of developing relatively simple physical models to explain empirical trends has become increasingly realized.

The shear resistance of a concrete member can be divided into two separate modes: beam action, where the tension force in the longitudinal reinforcement acting on a constant lever arm changes to balance the external moment; and arch action, where the internal lever arm (location of longitudinal compression stress resultant in concrete) changes to balance the moment. The method of choice for cases dominated by arch action is the strut-and-tie model, which is described in detail in Chapter 6. Some of the special considerations required when no transverse reinforcement is provided are discussed in this chapter. Regarding beam action, the physical/mechanical models can be classified further into tooth models, which begin with an assumed arrangement of cracks, and truss models with (smeared) concrete tension fields or concrete ties.

In the following, the empirical methods are presented first because they are the basis of the ACI 318M-95 approach. Fracture mechanics approaches for shear design, which are only briefly discussed here, are presented first. The nonlinear finite-element method is not discussed in this report. Information on this approach can be found elsewhere (Isenberg 1993). The remaining physical/mechanical models are presented in order of historical development: tooth models, simple strut-and-tie models, and truss models with concrete ties.

4.4.2 *Fracture mechanics approaches*—Fracture mechanics approaches account for the fact that there is a peak tensile stress near the tip of a crack and a reduced tensile stress (softening) in the cracked zone. For the case of a beam that fails in shear due to the propagation of a single critical diagonal crack, fracture mechanics can be considered a more rational approach than empirical methods. This approach offers a possible explanation for the size effect in shear.

A number of different fracture mechanics models have been proposed over recent years. Among them, two well known ones are the fictitious crack model, by Hillerborg et al. (1976), and the crack band model, by Bažant and Oh (1983). A survey of fracture mechanics applications was recently given by Reinhardt (1986). Further information on fracture mechanics can be found in ACI 446.1R.

Fracture mechanics approaches are usually numerically demanding because of the complexity of the tensile stress-crack displacement relationships. As a result, empirical formulas are sometimes developed in terms of fracture mechanics parameters, like that by Bažant and Kim (1984) or Remmel (1994). These equations give little explanation of structural behavior, so the end result is often very similar to empirical formulas, as was pointed out by Walraven (1987).

4.4.3 *Simple strut-and-tie models*—The application of strut-and-tie models, which has its theoretical basis from the lower-bound theorem of plasticity (Drucker 1961, Nielsen et al. 1978, Marti 1985), requires a minimum amount of distributed reinforcement in all directions, including transverse reinforcement, to ensure sufficient ductility for redistribution of internal stresses after cracking. In the elastic stress distribution of deep members, significant shear is transmitted directly to the support by diagonal compression. This means that less redistribution is required after cracking, and it would seem reasonable to apply strut-and-tie models (cautiously) to deep members without transverse reinforcement. When members are very deep, all of the shear will be transferred directly to the support by compression stresses; however, premature failure of a compression strut without minimum distributed reinforcement may result from transverse splitting due to spreading of compression stress (Schlaich et al. 1987). Walraven and Lehwalter (1994) carried out tests to investigate the capacity increase beyond that due to spreading of the compression stress; in a bottle-like shape, they proposed a failure criterion for the compression strut that included a size effect.

The simple strut-and-tie approach has also been suggested for more slender members without transverse reinforcement; however, this may result in an unsafe solution, as discussed in Section 4.1 and shown in Fig. 4.1. Solutions have been suggested using a reduced compressive strength of the strut. One method (Collins and Mitchell 1986) for accomplishing this involves considering strain compatibility between the

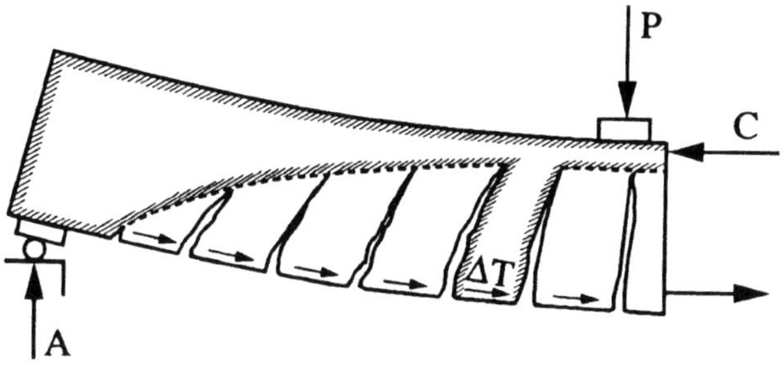

Fig. 4.5—Kani's tooth model (Kani 1964).

concrete strut and tension tie. As the strut becomes flatter, the transverse tensile strain in the strut increases, which reduces the diagonal crushing strength of concrete (Vecchio and Collins 1986). An alternate approach is that of Braestrup (1990), where the maximum diagonal compressive stress is not related to the strut inclination, but the shear capacity reduces with increasing shear span because of the geometry of the nodal zone, which depends on the support dimension, the cover-to-longitudinal reinforcement, and an assumed hydrostatic stress in the nodal zones.

4.4.4 *Tooth model for slender members*—An early attempt to develop a rational model to explain flexure-shear cracking was Kani's tooth model (1964), in which the secondary diagonal cracks were believed to result from bending of concrete "teeth." The concrete between two adjacent flexural cracks was considered to be analogous to a tooth in a comb (Fig. 4.5). The concrete teeth were assumed to be free cantilevers fixed in the compression zone of the beams and loaded by the horizontal shear from bonded reinforcement.

Kani's model was evaluated by Fenwick and Paulay (1968) as well as Taylor (1974), who pointed out that the teeth are restricted from bending freely by the friction of the crack faces and the dowel action of the longitudinal reinforcement. Tooth models have been developed further to include these mechanisms (MacGregor and Walters 1967, Hamadi and Regan 1980, Reineck 1991c). With his tests, Chana (1987) confirmed the basic mechanisms of the tooth model by extensive measurements of the deformations before failure.

The main features of the tooth model are that discrete cracks are assumed and assumptions have to be made for the inclination and spacing of the cracks. Hamadi and Regan (1980) assume that the cracks are vertical and that their spacing is equal to half the effective depth ($s = d/2$) for a particular beam, whereas Reineck (1991c) assumes the cracks are inclined at 60 degrees with a spacing of 70% of the crack height determined from a flexural analysis; that is, $s = 0.7(d - c)$, where c = depth of compression. For a given crack spacing, the shear displacement and crack width can be calculated if the strains are known. This requires kinematic considerations, but as a result, gives the actual contributions of the different shear transfer mechanisms in the highly indeterminate system.

Based on extensive experimental work in interface shear, Hamadi and Regan (1980) developed a tooth model with simplified assumptions for the interface shear and the geometry of the tooth. Reineck (1991c) further developed the tooth model, considering all the shear-transfer mechanisms and

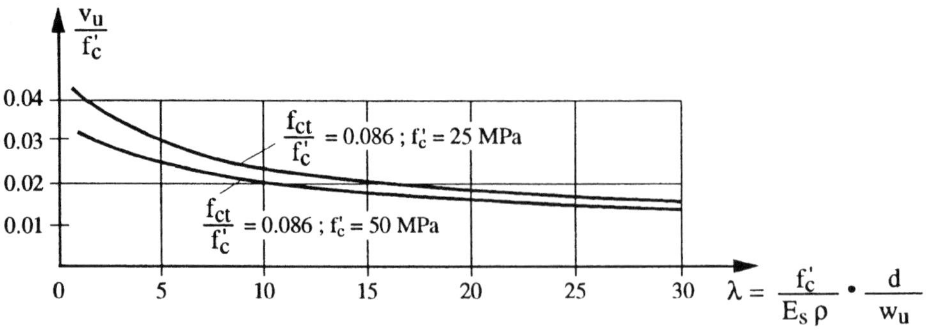

Fig. 4.6—Dimensionless diagram for ultimate shear force derived from mechanical model by Reineck (1991c).

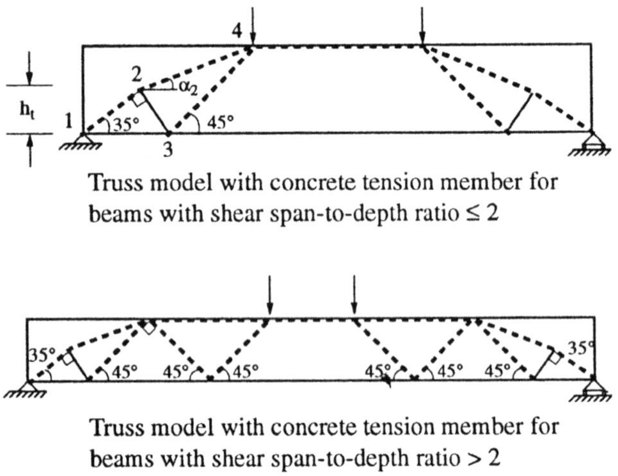

Fig. 4.7—Refined strut-and-tie models proposed by Al-Nahlawi and Wight (1992).

carried out a full nonlinear calculation including compatibility. The distribution of the shear force on the different shear mechanisms was similar to those found earlier by Taylor (1974); interface shear was the dominant shear mechanism, and dowel action was important for heavily reinforced members.

Based on his mechanical model, Reineck (1991c) derived an explicit formula for the ultimate shear force, which for members without axial force or prestress is

$$V_u = \frac{0.4 b_w d f_{ct} + V_{du}}{(1 + 0.054\lambda)} \quad (4\text{-}4)$$

where

f_{ct} = axial tensile strength = $0.246 f_c'^{2/3}$ (MPa); and

V_{du} = dowel force [$V_{du}/b_w d f_c' = 1.33\rho^{8/9}/(f_c')^{2/3}$, with f_c' in MPa units and d in m].

The parameter l is a dimension-free value for the crack width, which determines the friction capacity, and is defined by

$$\lambda = f_c' d / E_s \rho w_u \quad (4\text{-}5)$$

where

w_u = 0.09 mm = a limiting crack width;

d = effective depth (mm);

E_s = 2,000,000 (MPa); and

ρ = $A_s/b_w d$.

This parameter comprises the reinforcing ratio, ρ, and the depth, *d*, of the member; therefore, the formula can be presented in a simple diagram for the dimensionless shear force $v_u/f_c' = Vu/b_w d f_c'$ (Fig. 4.6), in which the concrete tensile strength is the only parameter. Similar diagrams could be developed using the more common parameters ρ and *d* (Reineck 1991c). Comparisons showed that this theoretically derived formula matches the tests as well as many empirical formulas.

4.4.5 *Truss models with concrete ties*—Although it may be possible to extend the simple strut-and-tie model approach to more slender members, clearly a different approach is needed to capture the shear failure of very slender members without transverse reinforcement where the concrete tensile stresses have the major role. Marti (1980) extended the plasticity approach by using a Coulomb-Mohr yield criterion for concrete that includes tensile stresses. Schlaich et al. (1987) suggested a refined strut-and-tie approach that includes concrete tension ties. Al-Nahlawi and Wight (1992) proposed a truss model with concrete compression struts inclined at either 45 or 35 degrees and concrete tension ties perpendicular to the struts (Fig. 4.7). An empirical rule was used for limiting the tension tie force based on the geometry of the truss. Muttoni (1990) proposed a truss model for less slender members (Fig. 4.8) in which, rather than going directly from the load point to the support, the inclined compression is bent around the initial flexural compression zone.

Reineck (1991c, 1982) has shown that such truss models with concrete ties fully comply with the tooth model discussed above. In the tooth model, the state of the stress in the B region is defined by the stresses at the crack, but from these stresses, the principal stresses in the tooth between the cracks may also be determined. The dominant action due to the friction along the crack surfaces then results in a biaxial tension-compression field, shown in Fig. 4.9(a) for constant friction stresses (Reineck 1982). The inclination of the principal compressive stress is equal to half the inclination of the cracks, which Reineck assumes to be at 60 degrees. The dowel action induces a concentrated tension force in the lower part of the tooth and results in bending of the tooth, which is resisted by friction stresses. Therefore, the strut-and-tie model of Fig. 4.9(b)

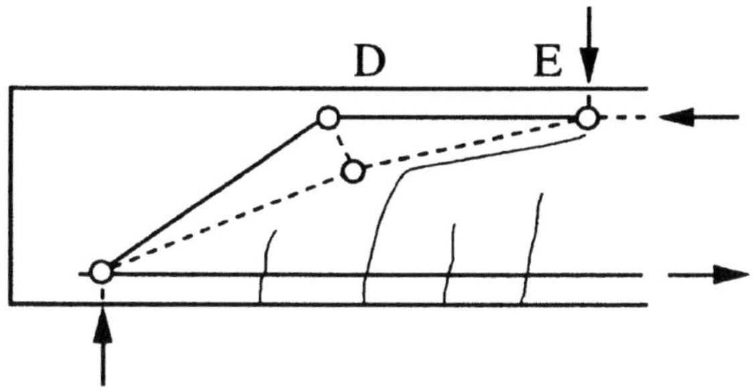

Fig. 4.8—Refined strut-and-tie model proposed by Muttoni (1990).

for the resulting state of stress shows, in the upper part of the tooth, an inclined stress field similar to Fig. 4.9(a).

4.4.6 *Modified compression field theory*—The MCFT (Vecchio and Collins 1986) can be implemented in different ways with varying levels of complexity, from a full, nonlinear finite-element analysis, to a multilayer sectional analysis that accounts for the variation of crack width over the section (Vecchio and Collins 1988), to the simplest case where only the crack width at the level of the flexural tension reinforcement is considered (Collins and Mitchell 1991). The latter case, which is most suitable for design, is analogous to a variable-angle truss model with diagonal concrete tension ties.

The MCFT is usually formulated in terms of average stresses; however, the method includes a check to ensure that the loads resisted by the average stresses can be transmitted across cracks. For members without transverse reinforcement, the stresses locally at a crack always control the capacity of the member, and the average stress calculation is used only in estimating the inclination of the critical diagonal crack. In calculating the stresses locally at a crack, it is assumed that the crack plane can resist only shear stress (no normal stress). As the horizontal plane should also be free of normal stress, the principal directions of the stresses locally at a crack should bisect the angles between the crack plane and the horizontal plane. In other words, the inclination of the principal compression stress (which is perpendicular to the direction of the principal tension stress) is half the inclination of the diagonal cracks, as summarized by the Mohr's circle in Fig. 4.10. One important consequence is that the shear stress on the crack plane is equal to the shear stress on the vertical and horizontal planes. Thus, the shear strength of a member is equal to the shear stress that can be resisted by the critical crack. Based on Walraven's experimental results (1981) for the maximum shear stress that can be resisted by a crack without the beneficial effects of compression, the shear resistance of a diagonally cracked member is given by

$$v_c \leq \frac{0.18\sqrt{f_c'}}{0.3 + \dfrac{24w}{(a+16)}} \quad (4\text{-}6)$$

where

w = crack width;

a = maximum aggregate size in mm; and

f_c' is in MPa (Vecchio and Collins 1986).

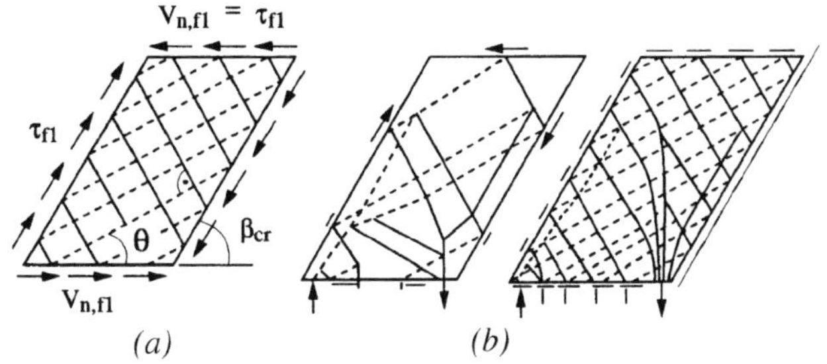

Fig. 4.9—Refined strut-and-tie model for crack friction and dowel action (Reineck 1991c): (a) constant friction stresses; and (b) simple and refined strut-and-tie model for dowel action combined with friction.

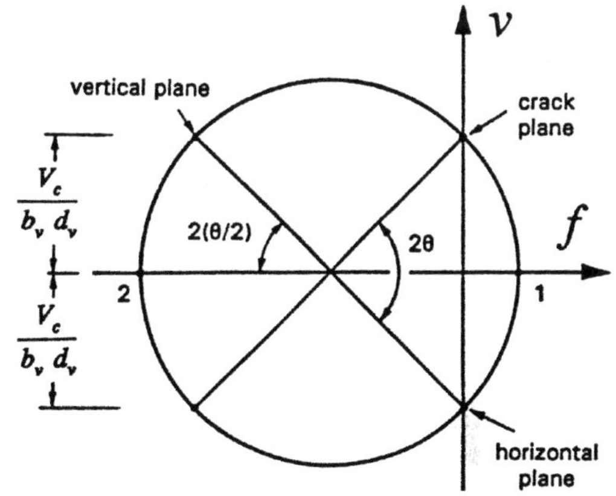

Fig. 4.10—Mohr's circle for biaxial stresses locally at crack in member without transverse reinforcement (Adebar and Collins 1996).

As in the tooth models, the width of a crack depends on the assumed crack spacing; however, unlike in the tooth models, the crack spacing and the inclination of the diagonal cracks are not assumed a priori in the MCFT. The crack spacing depends on the crack inclination. Cracks that are inclined at 90 degrees to the member axis are assumed to have a spacing $s_x = d_v$ (for members with one layer of reinforcement), whereas cracks at other inclinations β are assumed to have a spacing $s_\theta = s_x/\sin\beta$. This simple expression reflects the fact that longitudinal reinforcement has less ability to control cracks as they become flatter (smaller β).

In the MCFT, the inclination of diagonal cracks is calculated using a strain compatibility procedure involving average stresses and average strains. The diagonal cracks are assumed to be perpendicular to the principal tensile average stress direction. In that way, the influence of the loading (for example, the axial load and bending moment) on the inclination of the diagonal cracks (initial diagonal crack as well as the more important secondary diagonal cracks) are properly accounted for. For members without transverse reinforcement, the inclination of the diagonal cracks is given by

$$\theta = \tan^{-1}\left[\frac{\dfrac{0.33\sqrt{f_c'}}{1+\sqrt{500\varepsilon_1}}}{v_c}\right] \text{ (MPa)} \quad (4\text{-}7)$$

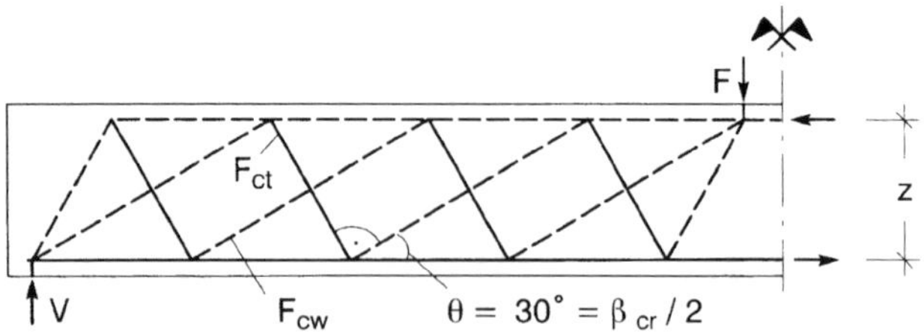

Fig. 4.11—Refined strut-and-tie model for diagonally cracked member without stirrups (Reineck 1991c).

where the numerator within the brackets is an empirical expression (Collins and Mitchell 1991) for the principal tensile average stress as a function of the principal tensile average strain ε_1.

A description of how the MCFT can be used to design members with or without transverse reinforcement is given in Chapter 2, and further information can be found in Collins and Mitchell (1991). Some additional discussion on the application of the modified compression field theory specifically to members without transverse reinforcement is given in Adebar and Collins (1996).

4.4.7 *Toward a consistent method*—Although the refined tooth models (Section 4.4.4) and the MCFT (Section 4.4.6) approach the problem from different directions, the end result of these two methods is very similar for members without transverse reinforcement. Although the two methods use different approaches to estimate the crack inclination and crack width, both methods consider that the ability of diagonal cracks to transfer interface shear stress is most important in determining the shear strength of members without transverse reinforcement.

The biaxial tension-compression field in the concrete (teeth) between the cracks for a constant shear stress along the cracks may be visualized by a truss model. This representation of the concrete stresses due to interface shear stresses was first developed by Reineck (1982) based on his tooth model (Reineck 1991c). As discussed earlier, the principal stresses locally at a crack (that is, the compression struts and tension ties of the truss model) bisect the angles between the diagonal crack and the horizontal plane. Reineck assumes that the inclination of the critical diagonal cracks will be 60 degrees; therefore, the inclination of his compression struts are at 30 degrees (Fig. 4.11). In the MCFT approach, the inclination of the critical diagonal cracks depends on the loading and the ability of the longitudinal reinforcement to control the secondary diagonal cracks. A generalization of Reineck's model involves a truss with compression struts inclined at half the inclination of the diagonal cracks, which may have any inclination (Adebar and Collins 1996). Thus, the modeling for members without transverse reinforcement is fully compatible with the strut-and-tie models explained in Chapter 6.

4.5—Important parameters influencing shear capacity

4.5.1 *Depth of member or size effect*—It was shown by Kani (1967) as early as 1967 that there is a very significant size effect on the shear strength of members without transverse reinforcement. Shioya et al. (1989) reaffirmed this fact and extended the available data to beam depths of 3000 mm. As shown in Fig. 4.12, the average shear stress to cause failure of the largest beam was about one-third the average shear stress to cause failure of the smallest beam.

There is general agreement that the main reason for this size effect is the larger width of diagonal cracks in larger beams; however, there is disagreement on how best to model this phenomenon. Some (Bažant and Kim 1984) believe that the most important consequence of wider cracks is reduced-residual tensile stresses and, as a result, propose using a size-reduction factor

$$\lambda = \frac{1}{\sqrt{1 + \frac{d}{d_o}}} \tag{4-8}$$

based on nonlinear fracture mechanics where d_o is an empirical parameter. Others believe that the most important consequence of wider cracks is a reduced ability to transmit crack interface shear stresses. As the crack spacing used to determine the limiting crack interface shear stress is a function of the specimen depth, no special factor is required to account for the size effect. Perhaps the strongest argument for this latter approach is that it leads to a consistent treatment of members with different arrangements of longitudinal reinforcement. Tests (Collins et al. 1993) have demonstrated that the size effect disappears when beams without stirrups contain well distributed longitudinal reinforcement. Figure 4.12 compares the results from the tests by Shioya et al. (1989) with the design method described in Chapter 2, whereas Fig. 4.13 compares the predictions of different formulas for the influence of size effect. The proposals by Okamura and Reineck consider the tests in Shioya et al. (1989) and, therefore, represent reasonable lower bounds, whereas the predictions using Zsutty's approach (1971) and ACI 318M-95 do not account for the size effect.

4.5.2 *Shear span-to-depth ratio (a/d) and support conditions*—It has long been recognized that, as members become deep (the shear span-to-depth ratio decreases below about 2.5), the average shear stress at failure becomes progressively larger than in slender beams. The reason for this is that, as members become deeper, it is easier for the shear to be transmitted directly to the support by a compression strut. The support conditions also strongly influence whether a direct compression strut can form. A compression strut is more likely to form if a member is loaded on the top face and supported on the bottom face (Adebar 1994).

Members in which a direct compression strut is likely to form should be designed using the strut-and-tie model approach described in Chapter 6, rather than a sectional design procedure. Collins and Mitchell (1991) demonstrated how the two methods can be combined to predict the strength of beams without stirrups over a wide range of shear span-to-depth ratios.

The simple strut-and-tie model assumes that all of the shear will be transmitted by compressive stresses, which

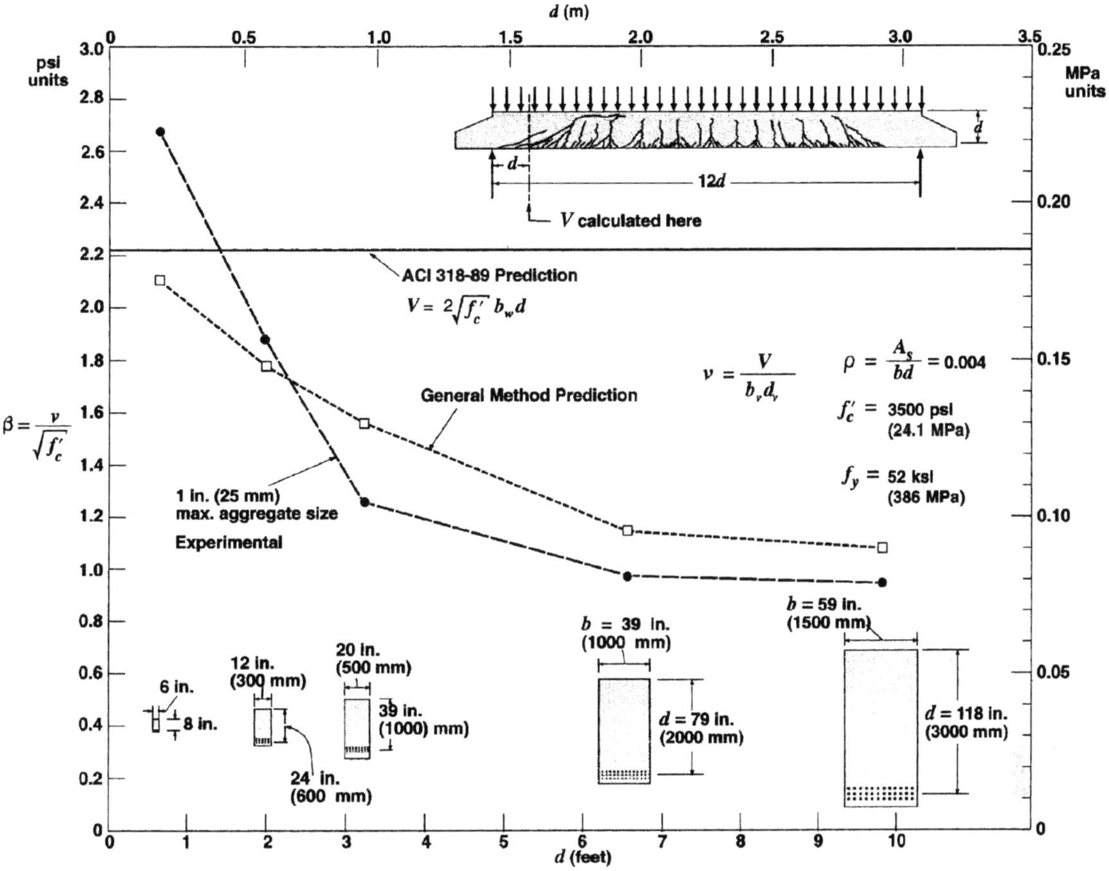

Fig. 4.12—Comparison of large-scale beam tested by Shioya et al. (1989) with predictions from ACI Code and MCFT (Collins and Mitchell 1991).

suggests that there will not be a size effect for such members. Experimental results (Walraven and Lehwalter 1990, 1994), however, indicate that deep members without transverse reinforcement do experience a very significant size effect. Thus, refined strut-and-tie models that consider concrete tensile stresses should be used to design deep members, unless minimum distributed longitudinal and transverse reinforcement is provided.

4.5.3 *Longitudinal reinforcement*—As can be seen in Fig. 4.14, members with low amounts of longitudinal reinforcement may fail at very low shear stresses. The 1973 ASCE-ACI Committee 426 report suggested that the following equation, incorporating the percentage of longitudinal tension reinforcement, be used to estimate the average shear stress at diagonal cracking

$$v_c = (0.8 + 100\rho)\frac{\sqrt{f'_c}}{12} \leq 0.179\sqrt{f'_c} \quad \text{(MPa)} \qquad (4\text{-}9)$$

The report further suggested that the reduction in shear capacity can be explained by an increased crack width, resulting in lower interface shear transfer, correspondingly longer flexural cracks that reduce the size of the compression zone, and reduced dowel action.

In the design method presented in Chapter 2, the influence of the area of longitudinal reinforcement A_s is reflected in

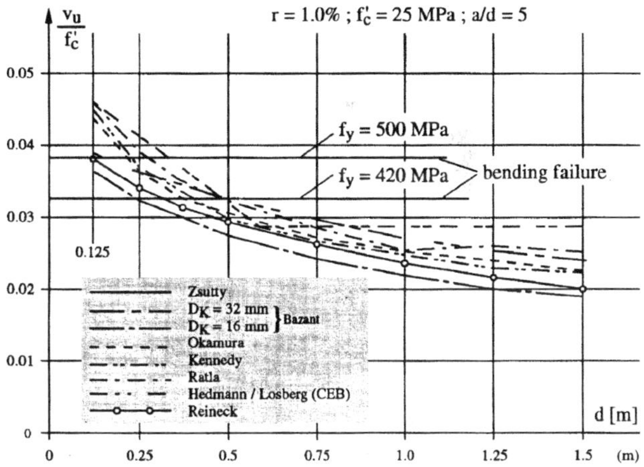

Fig. 4.13—Decrease of ultimate shear force (dimensionless value) with increasing depth of member size (size effect) according to different proposals (Reineck 1991c).

Eq. (2-9), which predicts larger tensile strains (and larger crack widths) in members with less longitudinal reinforcement and, as a result, lower shear strengths. The design method also requires that additional longitudinal reinforcement be provided to resist the large tensile force due to shear in diagonally cracked members without stirrups, as expressed by Eq. (2-12).

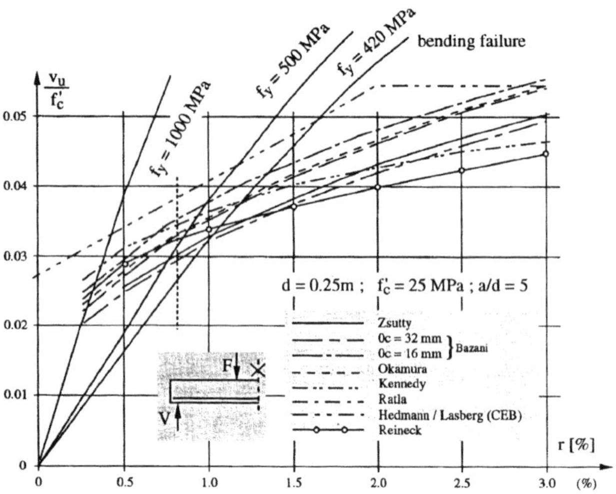

Fig. 4.14—Increase in ultimate shear capacity (dimensionless value) with increasing reinforcing ratio according to different proposals (Reineck 1991c).

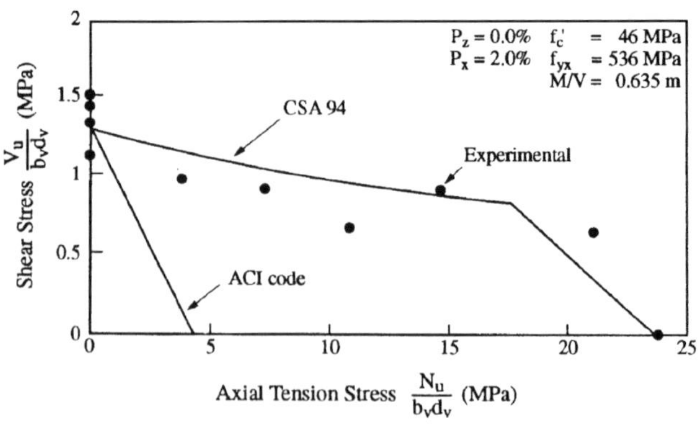

Fig. 4.15—Influence of axial load on shear strength of members without stirrups.

Figure 4.14 compares the predicted influence of the quantity of longitudinal reinforcement from a number of empirical formulas. The flexural capacity is also shown for different strengths of longitudinal reinforcement. From this figure follows the well-known fact that flexural failures rather than shear failures will govern the capacity of moderately long beams ($a/d \approx 5.0$) with low amounts of longitudinal reinforcement ($r < 1.0\%$) if the reinforcement is constant throughout the span.

4.5.4 *Axial force*—It is well known that axial tension decreases the shear strength of members without transverse reinforcement, and axial compression (due to an applied load or prestressing) in-creases the shear resistance. What is perhaps not very well understood is how much the shear capacity is influenced by axial load and what the influence is on the ductility of the member.

Members without shear reinforcement subjected to large axial compression and shear may fail in a very brittle manner at the instance of first diagonal cracking. As a result, a conservative approach should be used for such members. Recent tests by Gupta and Collins (1993) have demonstrated that the ACI 318 approach may be unconservative for members subjected to axial compression and shear.

On the other hand, for members without shear reinforcement but containing appropriate longitudinal reinforcement, the ACI 318M-95 procedure for members subjected to shear and tension can be very conservative. The code procedure, although intended for members subjected to significant axial tension, is based upon test results from members that did not contain appropriate longitudinal reinforcement to resist axial tension (Bhide and Collins 1993). Figure 4.15 compares the results from a series of beams without stirrups subjected to shear and tension (Adebar and Collins 1996) with predictions from ACI 318M-95 and the design method based on the modified compression field theory (Chapter 2). In Fig. 4.15, d_v represents the effective shear depth, b_v is the web width, and P_x and P_z are the reinforcement ratio in the orthogonal directions. A similar result is achieved with Reineck's tooth model (1982).

A typical crack pattern for a member subjected to large axial tension and shear is shown in Fig. 4.16. The initial cracks are very steep (close to 90 degrees) and extend over the full depth of the member. Thus, longitudinal reinforcement is required at the top as well as the bottom of the member, and the initial cracks, which are initially perpendicular to this reinforcement, are well controlled. As the loading is increased, new, flatter diagonal cracks form. Failure occurs only after the diagonal cracks become too flat to be controlled by the longitudinal reinforcement; therefore, members subjected to tension and shear are comparatively ductile.

4.6—Conclusions

More than 20 years after the 1973 ASCE-ACI Committee 426 state-of-the-art report expressed the hope that more rational methods could be developed for shear design, it can now be concluded that methods are available to explain the structural behavior and failure load of members without transverse reinforcement. Despite differences in the approaches, the main conclusion from most models is that concrete tensile stresses should be considered explicitly. Therefore, the empirical approaches in current codes can now be replaced by relationships based on truss models with concrete ties. The failure is not explained simply by the tensile stresses attaining the tensile strength, rather; it involves the breakdown of the shear transfer mechanisms at the crack. For smaller-sized members, the residual tensile stresses are very important, whereas for larger members (which have wide cracks prior to failure), the interface friction has the dominant role.

A number of parameters influencing the shear capacity of members without transverse reinforcement have been well explored during the last 20 years. It is certainly well known that there is a size effect on the shear strength of the member, and this fact should be addressed in ACI 318M-95. The strength of members with low reinforcing ratios was rarely investigated in the past and is often overestimated in present codes. The influence of both the size effect and the longitudinal reinforcement amount can be explained by the effect on the width of inclined cracks. The influence of the shear

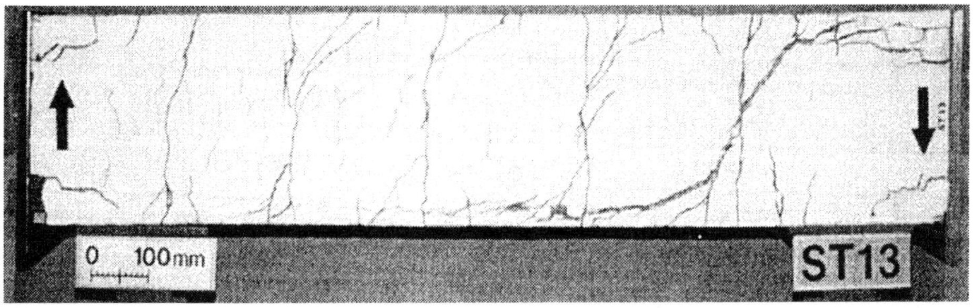

Fig. 4.16—Typical crack pattern without transverse reinforcement subjected to axial tension and shear (Adebar and Collins 1996).

span can be accounted for very simply by using a sectional approach for slender members and a strut-and-tie approach for short spans where a direct compression strut can form between the load and the support. Some caution is needed when applying strut-and-tie models to members without distributed transverse reinforcement.

Recent information on the behavior of members subjected to shear and axial tension illustrates a number of important issues to be considered in the choice of a consistent shear design model. First, as the initial diagonal cracks in a member subjected to tension and shear extend over the full depth of the member (Fig. 4.16), there is no compression zone to resist shear. Second, there is no fracture zone where a concentration of residual tensile stresses exists. Finally, the need to consider a changing critical diagonal crack inclination is apparent.

CHAPTER 5—SHEAR FRICTION

5.1—Introduction

Shear friction is treated as a separate chapter in this report because in ACI 318 it is treated separately from the concept of diagonal tension used for beam design. It is presented here to show a complete picture of models for shear design. The shear-friction concept permits one to treat problems where failure occurs by sliding along a plane of discontinuity. Construction joints and discrete cracks are common situations where sliding failures can be assessed readily by shear-friction theory. These discontinuities represent planes of weakness that may behave quite differently than the parent concrete. The shear-friction approach can account for the nature of the discontinuity (that is, roughness and interface with materials such as structural steel shapes). For this class of problem, shear friction is the preferred method of solution.

It should be emphasized that shear friction, as used in North America, is not another approach to shear design of slender beams. It may be regarded as a simplified version of the more refined and more recent methods described in Section 3.4.

5.2—Shear-friction hypothesis

The shear-friction approach, as it is known today, was first introduced by Birkeland and Birkeland (1966). It was originally developed to deal with force transfer across joints in precast-concrete construction. Figure 5.1 illustrates the shear-friction hypothesis. Assuming sliding along the failure plane *m-m*, and simple Coulomb friction, the shear force, V, required to produce sliding is equal to μP, where μ is the coefficient of friction between the two elements and P is the clamping force perpendicular to the sliding plane. The roughness of crack *m-m* will create a separation δ between the two halves (see Fig. 5.1(b)). If reinforcement is placed across the interface, the separation will develop tension T in the reinforcement. The tension provides an external clamping force on the concrete resulting in compression across the interface of equal magnitude. The roughness may be visualized as a series of frictionless fine sawtooth ramps having a slope of tan ϕ. Comparing Fig. 5.1(a) and (b), it is seen that $T \tan \phi$ is equivalent to the frictional force μP. Tests revealed that the separation δ is usually sufficient to yield the reinforcement crossing the crack. Thus

$$V_n = \mu A_s f_y \qquad (5\text{-}1)$$

where μ = 1.7 for monolithic concrete, 1.4 for artificially roughened construction joints, and 0.8 to 1.0 for ordinary construction joints and for concrete to steel interfaces.

To prevent crushing of the concrete in the crack (caused by the clamping force from the steel), before yielding of the steel, the steel ratio across the crack was limited to 0.015 and the shearing stress was limited to not more than 5.5 MPa for concrete with f_c' = 27.6 MPa.

5.3—Empirical developments

The simple physical model put forward by Birkeland and Birkeland (1966) has proven remarkably useful and robust. It has formed a logical framework for further study and refinement. Hofbeck et al. (1969) conducted an extensive testing program and found that shear-friction theory provided conservative estimates for the interface shear transfer strength.

Mattock and Hawkins (1972) showed that the effects of direct stresses perpendicular to the crack agreed with predictions based on the models in Fig. 5.1. Mattock (1974) subsequently showed that the effects of reinforcement at an angle to the shear plane could also be predicted based on the models of Fig. 5.1.

Paulay et al. (1974) investigated various treatments for construction joints. With well-designed and well-constructed joints, one can produce an interface shear strength that is equal to or larger than the corresponding load capacity of the

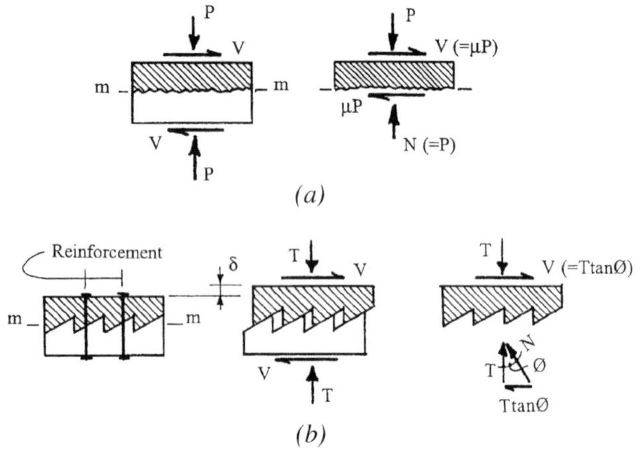

Fig. 5.1—Shear friction hypothesis (Birkeland and Birkeland 1966).

remainder of the member. For members cast in lifts, failure was found to occur in the zone immediately below the construction, that is, in the top-cast concrete in the lower lift. As such, one would expect to have weaker concrete in this region.

Mattock et al. (1976) investigated shear friction in lightweight concrete and found the friction values to be lower than in corresponding normalweight concrete. Mattock (1976) presented empirical equations that produce better predictions for monolithic uncracked concrete and low amounts of transverse reinforcement. For these situations, Eq. (5-1) is quite conservative. In effect, he introduced a constant cohesion term to supplement the friction term for Eq. (5-1). Mattock's equations have been found to be valid but somewhat conservative for concrete strengths up to 68 MPa (Walraven et al. 1987).

Walraven (1981), Walraven and Reinhardt (1981), and Walraven et al. (1987) have extended the range of variables studied and developed empirically based constitutive relationships for cracks that relate normal and shear displacements and stresses. Load history, such as previous sustained or repeated cyclic loading, was found to have little effect on the ultimate shear strength. It is believed that as ultimate load is approached, greater and virgin contact areas within the crack are mobilized. These virgin areas are not affected by the previous load history (Walraven et al. 1987). Loebert (1970), Tassios and Vintzeleou (1986), and others have also contributed to the empirical database. It is clear that for low normal stresses (less than about 1.5 MPa) the coefficient of friction is higher than previously thought. Thus, the apparent cohesion intercept introduced by Mattock is primarily due to an increase in μ rather than to the tension strength of uncracked monolithic concrete. Barton (1973), in effect, developed an empirical expression for μ as a function of clamping stress on the joint. As the clamping stress increases, the asperities that are in contact within the crack crush, producing a smoother crack and hence a lower μ.

Loov and Patnaik (1994) investigated the horizontal shear strength of composite concrete beams with a rough interface. Based on a review of the literature, and a series of tests, they proposed a single empirical equation that accounted for the concrete strength and the clamping stress.

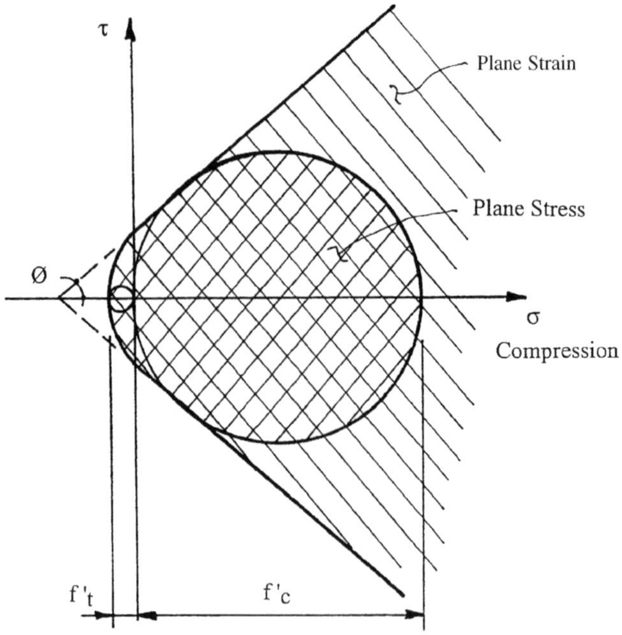

Fig. 5.2—Modified Mohr-Coulomb yield criterion for concrete with nonzero tension cutoff.

5.4—Analytical developments

In parallel with experimental developments, analytical developments by Nielsen (1971, 1984) and Nielsen et al. (1978) provide a theoretical basis for integrating the shear-friction approach in the theory of plasticity. They show that shear friction is an upper-bound solution within the general theory of plasticity. Shear friction is similar to limit equilibrium methods used in soil and rock mechanics. The failure planes used to determine active and passive earth pressures behind a retaining wall are examples of shear-friction planes. Using a modified Mohr-Coulomb yield criterion as shown in Fig. 5.2, one can readily predict shear-friction failures. It is worth noting that Nielsen had to introduce v, an efficiency factor into the plasticity analysis. The Mohr-Coulomb yield criterion is based on the tensile and compressive strengths of vf_t' and vf_c' respectively. Figure 5.3 illustrates typical results from plasticity theory. Even though the typical shear-friction test specimens are more representative of plane-stress conditions, one usually finds that the experimental results are bounded by predictions based on plane-strain and plane-stress conditions. Plasticity theory is particularly useful when the likely failure plane cannot otherwise be identified readily.

5.5—Code developments

Shear friction was adopted swiftly by ACI 318 and first appeared in the 1971 code. The values of μ specified in ACI 318M-95 have been reduced from those suggested by Birkeland and Birkeland (1966) to provide an appropriate margin of safety against shear failure. ACI 318M-95 recognizes the reduced friction values that exist with lightweight concrete through the introduction of the factor λ for lightweight concrete. Inclined reinforcement and the effects of external forces are accounted for. The maximum shear stress, v_n, is limited to the lesser of $0.2f_c'$ or 5.5 MPa to ensure that the section is not

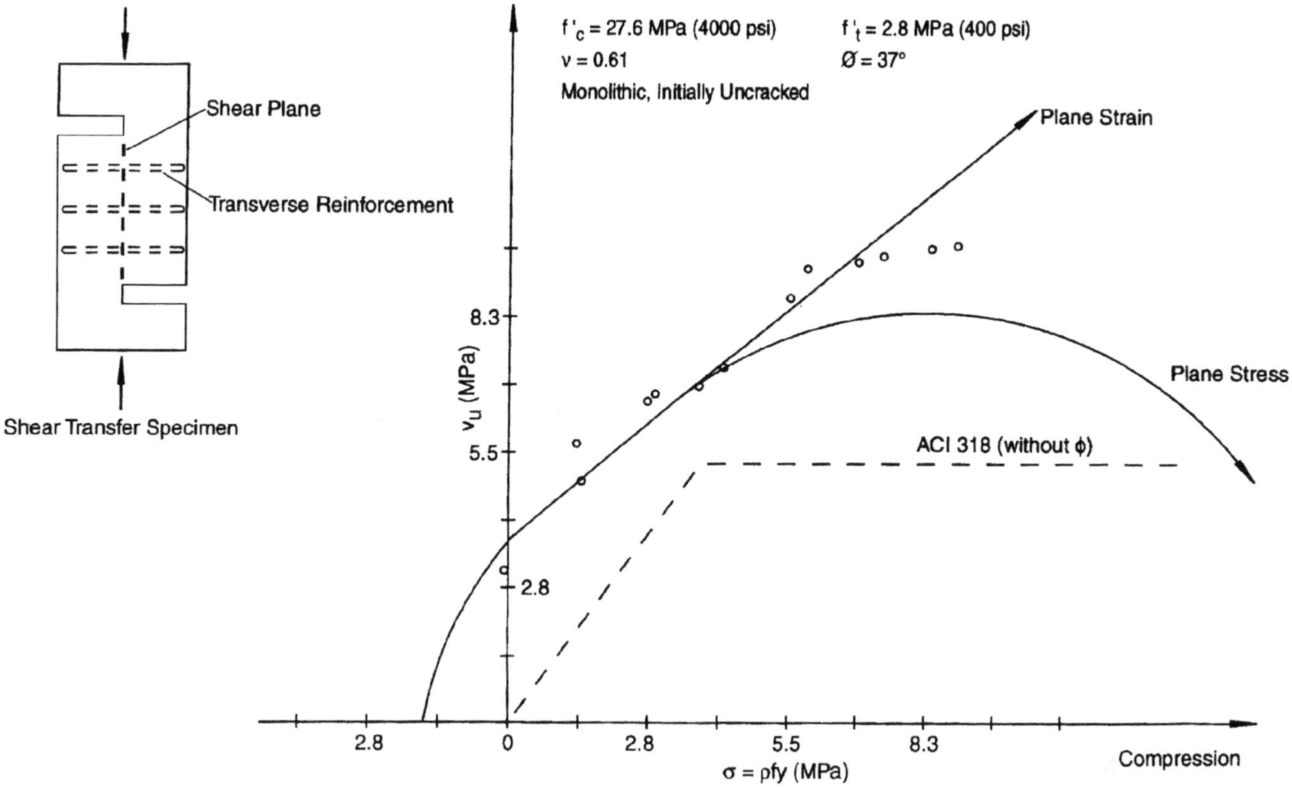

Fig. 5.3—Shear friction strength predictions (date from Hofbeck et al. 1969).

over-reinforced. It was recognized that the simple shear-friction provision of ACI 318M-95 was somewhat conservative, but relief was provided in Clause 11.7.3, which explicitly permits one to use "any other shear transfer design methods that result in prediction of strength in substantial agreement with results of comprehensive tests." As evident from Fig. 5.3, it would be particularly advantageous to do so when one has uncracked concrete and low amounts of reinforcement. In the case of composite concrete beams, Clause 17.5 may be used to advantage. In other situations, one should resort to the various approaches in the literature.

The Canadian code ("Design" 1994) has recently introduced two modified shear-friction formulations. Both approximate a Mohr-Coulomb yield envelope. One formulation, based on Loov and Patnaik (1994), produces a curve envelope that recognizes the benefits of increased concrete strength. The other formulation is similar to that proposed by Mattock (1976). It uses an initial cohesion term and a constant friction angle. Although it does not recognize the benefits of increased concrete strength, it can accommodate a wider range of surface roughness and cohesion. Adjusting the equations of CSA Standard A23.3-94 ("Design" 1994) to suit ACI 318M-95 load and resistance factors gives approximately

$$V_n = cA_c + A_{vf} f_y \mu \qquad (5\text{-}2)$$

where c, denoting cohesion, is equal to 0.3 MPa for concrete cast against clean hardened concrete, 0.55 MPa for concrete cast against clean hardened concrete intentionally roughened to a full amplitude of 6.25 mm, and 1.10 MPa for concrete cast monolithically. The quantity μ is the coefficient of friction per ACI 318M-95. The influence of inclined reinforcement is taken into account in the same manner as in ACI 318M-95.

CHAPTER 6—DESIGN WITH STRUT-AND-TIE MODELS

6.1—Introduction

As described in Chapter 1, truss models for the design of members have been used since the turn of the century (Ritter 1899; Mörsch 1920, 1922). One of the main advantages of using truss members to represent key resisting elements of a member is that the flow of forces can be easily visualized by the designer. The flow of compressive stresses is idealized with compression members called struts, and tension is taken by tension ties. Figure 6.1 shows how truss models using struts and ties can be used to idealize the flow of forces in members with a variety of span-to-depth ratios. This figure also shows the reinforcement required for a slender beam, a beam with a shear span-to-depth ratio *a/d* of 2.5, and a deep beam. Another advantage of using a truss model to idealize the flow of forces is that the influences of both shear and moment are accounted for simultaneously and directly in the design.

In the case of the slender beam, the concentrated load and reactions cause a disturbance in the internal flow of the stresses near the ends and the center of the beam (Fig. 6.1(a)). In current design procedures, these disturbed regions, or D regions, are treated separately in the design process. The regions of the slender beam where the flow of compressive stresses is uniform are referred to as beam regions, or B regions. The slender beam is modeled by a parallel chord

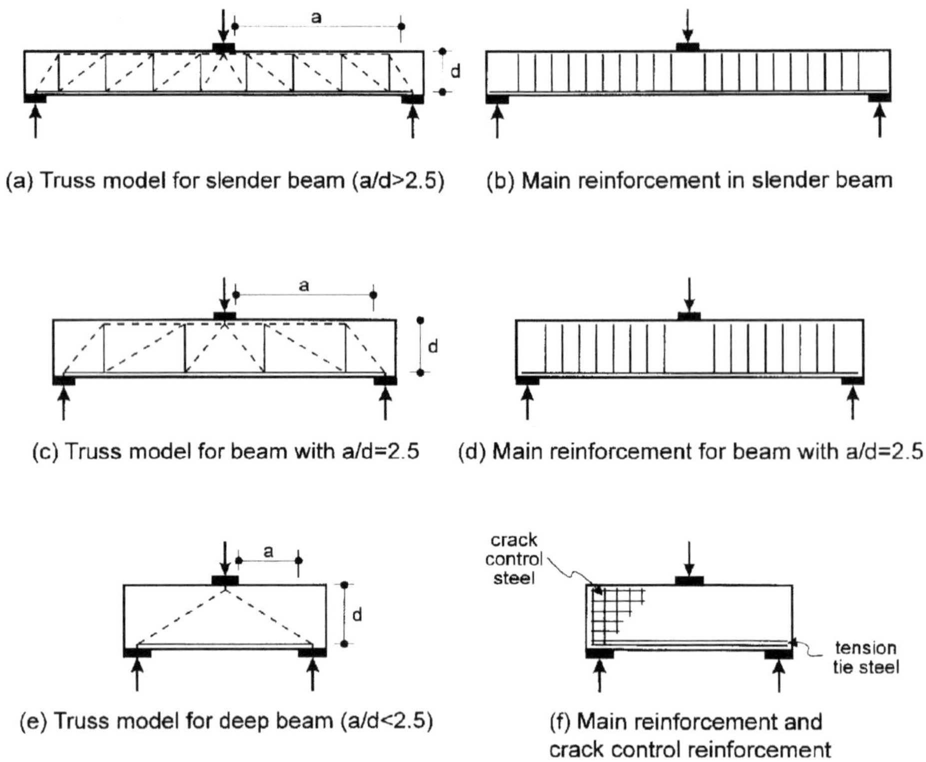

Fig. 6.1—Truss models showing transition from slender beams to deep beams.

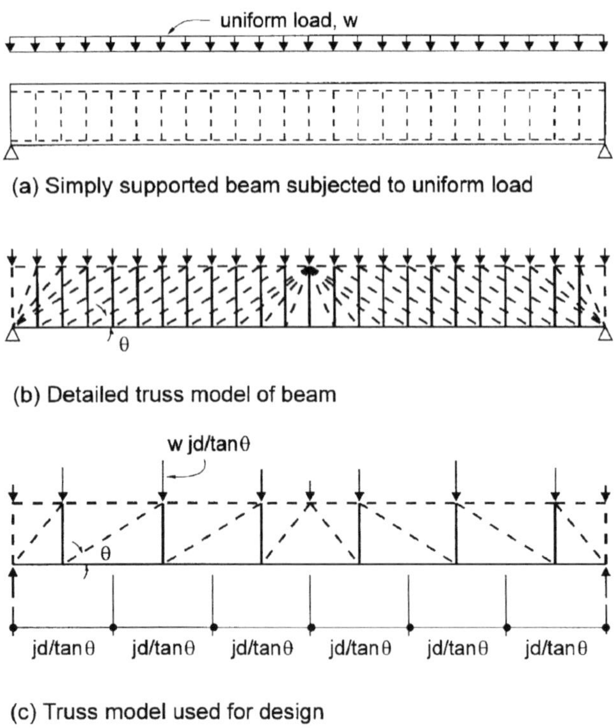

Fig. 6.2—Truss models for full member design.

truss with the flow of compressive stresses in the B regions idealized by a series of parallel compressive struts. The design can be carried out for the entire member using a full-member truss model. An example of this approach is given by Ramirez (1994) for the case of a prestressed concrete member. Alternatively, the design of the B regions can be carried out using a sectional design model, as described in Chapters 2 and 3.

Figure 6.1 shows the transition from a slender member to a deep beam. For low *a*/*d* ratios, the entire member becomes a disturbed region, where the normal assumptions of plane strain and uniform shear stress distribution are inappropriate. The strut-and-tie model is particularly useful in the design of D regions, which are characterized by a complex flow of internal stresses. The design of D regions using strut-and-tie models is discussed in Section 6.3.

6.2—Design of B regions

Figure 6.2 illustrates the manner in which a truss model is developed for the design of a member, according to the procedure recommended by Marti (1985). The design steps are as follows:

1. A reasonable angle of principal compression θ is chosen. Typical angles vary from about 18 to about 65 degrees.

2. The truss model used for design, having a depth equal to the flexural lever arm jd is sketched. Each vertical member represents a group of stirrups within a length $jd\cot\theta$. Each diagonal member of the simplified truss represents a zone of diagonal compression. Once the diagonal force D is found, then, using b_w as the effective web width resisting shear, the diagonal compressive stress in the concrete can be found from

$$f_2 = \frac{D}{b_w jd\cos\theta} = \frac{V}{b_w jd}\cdot\frac{1}{\sin\theta\sin\theta} \quad (6\text{-}1)$$

3. Solve for the forces in the truss members and design the transverse and longitudinal reinforcement. Check the compressive stress in the concrete diagonals. Check that the reinforcement is sufficiently anchored at all critical points so that it can develop the necessary force in the steel.

Typically, the concrete contribution to the shear resistance, V_c, is conservatively ignored, $V_c = 0$, and the shear reinforcement is designed to resist all of the shear. There is no significant economic penalty arising from this conservatism for many situations where stirrups are typically provided. For members such as slabs and some joists that are constructed without transverse reinforcement, a truss model accounting for tensile stresses in the concrete can be used for design (see Chapters 3 and 4). Alternatively, a simplified approach with an empirical concrete contribution, such as in ACI 318M-95, could be used.

6.3—Design of D regions

6.3.1 *Definition of D region*—Strut-and-tie models are most appropriately used for the design of disturbed regions, or D regions, as shown in Fig. 6.3 (Schlaich et al. 1987). These D regions are characterized by a complex flow of internal stresses and include regions adjacent to discontinuities caused by abrupt changes of cross section or the presence of concentrated loads or reactions. In the design of these disturbed regions, it is inappropriate to assume that plane sections remain plane or that the shear stress is uniform over the depth of the member.

Figure 6.4 compares the experimentally determined shear strengths of a series of beams tested by Kani (1979) with the strength predictions using a sectional design model and a strut-and-tie model (Collins and Mitchell 1991). In these tests, the shear span-to-depth ratio *a*/*d* was varied from 1 to 7 and no web reinforcement was provided. At *a*/*d* values less than 2.5, the resistance is governed by strut-and-tie action, with the resistance dropping off rapidly as *a*/*d* increases. The failures in this region were governed by crushing the compressive struts that were sensitive to the size of the bearing plates. For *a*/*d* values greater than 2.5, the strength is governed by the conditions away from the supports and concentrated loads. In these regions, failure is governed by shear, owing to the large amount of longitudinal reinforcement in the beams. It is clear from this figure that for *a*/*d* values less than about 2.5, a strut-and-tie model more accurately predicts the strength, whereas for *a*/*d* values greater than about 2.5, a sectional design model that includes V_c caused by tensile stresses in the concrete is more appropriate.

Figure 6.5 gives examples of disturbed regions in which the flow of the compressive stresses is indicated by dashed lines and the tension ties are indicated by solid lines. Fig. 6.5(a) illustrates how the presence of the support reaction interrupts the uniform field of compressive stresses in the concrete, causing a disturbed region where the compressive stresses fan into the support. The corbel shown in Fig. 6.5(b) is a disturbed region characterized by compressive struts anchored by a tension tie through the column. The entire deep beam shown in Fig. 6.5(c) is a D region in which the concentrated load is transmitted directly to the supports by concentrated unidirectional compressive stresses or struts. The resisting mechanism resembles a tied arch with the reinforcement acting as a tension tie between the supports. Figure 6.5(d) illustrates the high compressive stresses from the concentrated loads fanning into a uniform field of compressive stresses in a wall.

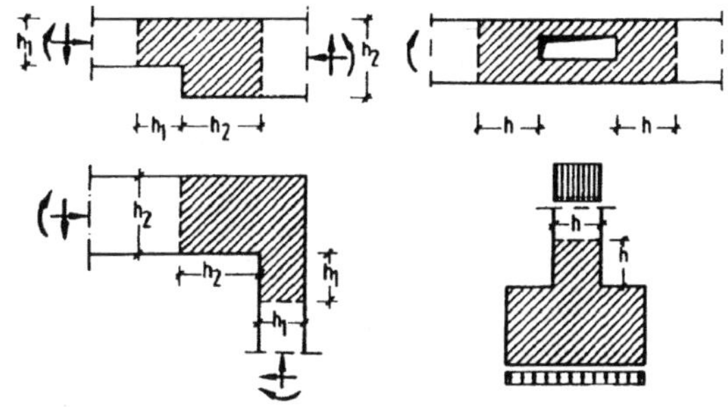

(a) D-regions caused by geometric discontinuities

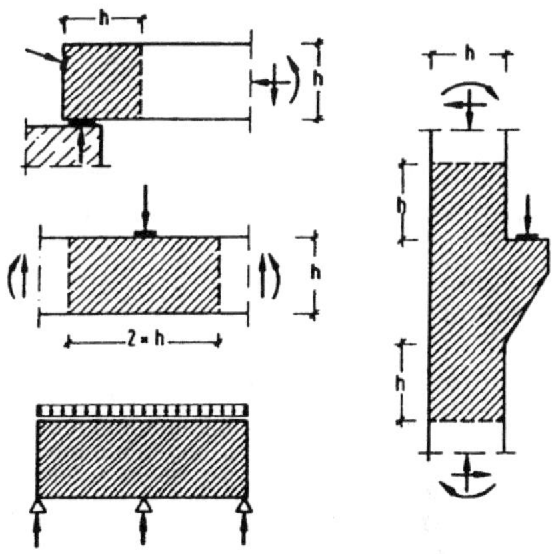

(b) D-regions caused by concentrated loads and/or geometric discontinuities

Fig. 6.3—D regions shown as shaded areas, adapted from Schlaich et al. (1987).

6.3.2 *Choosing strut-and-tie model*—The first step in design is to visualize the flow of the forces and identify B regions with parallel compressive fields and D regions with compressive struts representing the flow of concentrated compressive stresses in the concrete and tension ties representing the reinforcing steel. The visualization of the flow of compressive stresses in the concrete allows the development of a truss idealization consisting of compressive struts and the tension ties necessary for equilibrium. It takes experience to determine the most efficient strut-and-tie models for different situations. For any given situation, many strut-and-tie models are possible and no unique solution exists. Schlaich and Shäfer (1984) and Schlaich et al. (1987) have suggested that a strut-and-tie model be chosen after carrying out an elastic analysis. They suggested choosing the geometry of the truss model so that the angles of the compression diagonals are within ±15 degrees of the angle of the resultant of the compressive stresses obtained from an elastic analysis. Although this approach can give some guidance in choosing the geometry of the strut-and-tie model, it should be recognized that considerable redistribution of stresses takes place after cracking if the member is ductile. Figure 6.6(a) illustrates

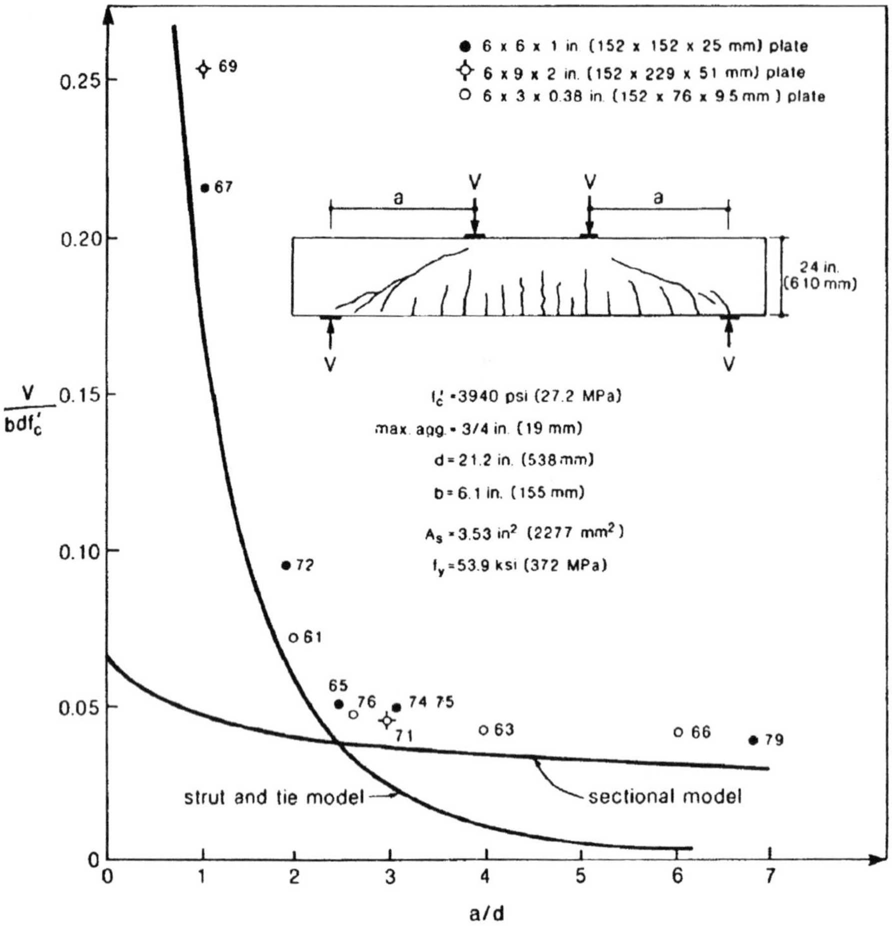

Fig. 6.4—Applicability of sectional design model and strut-and-tie model for series of beams tested by Kani (1979), adapted from Collins and Mitchell (1991).

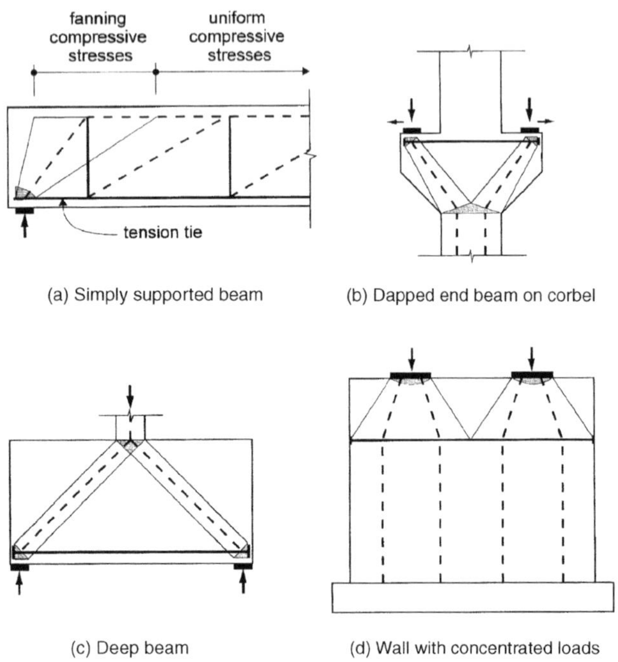

Fig. 6.5—Example of D regions modeled with compressive struts and tension ties.

the compressive stress fields for a deep beam. The compressive struts are actually bulging between the loading point and the supports, causing transverse tension idealized by the tension ties shown in Fig. 6.6(b). A simpler approach is to assume that the compressive struts can be idealized as straight line truss members following the centerline of the compressive struts, provided that additional reinforcement is provided to control cracking. To control cracking, most codes require that a minimum amount of reinforcement be provided in both the transverse and the longitudinal direction. For example, CSA Standard CAN3-A23.3-M84 ("Design" 1984) requires that a minimum reinforcement ratio equal to 0.002 be provided in both the transverse and the longitudinal direction. This small amount of uniformly distributed reinforcement is usually neglected in the strut-and-tie model and the truss idealization (Fig. 6.6(c)).

A simple truss idealization of the flow of the forces can be developed in which the zones of high unidirectional compressive stresses are modeled as compressive struts, and tension ties are used to model the principal tension reinforcement. Figure 6.7(a) illustrates the flow of the forces in a deep beam subjected to two point loads. The loads are assumed to be transferred to the supports by compressive struts, requiring a tension tie between the supports. The internal resisting mech-

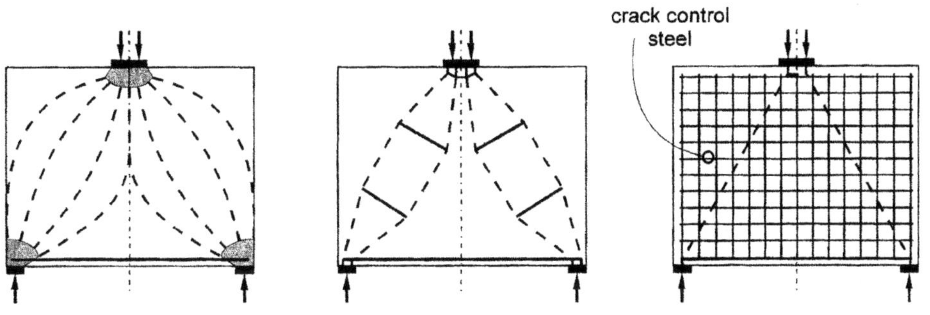

(a) Flow of compressive stresses (b) Required tension ties (c) Assumption of straight compressive struts

Fig. 6.6—Crack control reinforcement required with assumed straight-line compressive struts, adapted from Schlaich et al. (1987).

anism is shown as a truss model in Fig. 6.7. The strut-and-tie design method described below is based primarily on the design approach developed for the CSA Standard CAN3-A23.3-M84 ("Design" 1984; Collins and Mitchell 1985, 1984). The behavior of a variety of heavily instrumented D regions has been presented by Leonhardt and Walther (1966), Mattock et al. (1976), Rogowsky et al. (1986), Cook and Mitchell (1988), Adebar et al. (1990), and Jirsa et al. (1991).

6.3.3 *Checking compressive stresses in struts*—In the design using strut-and-tie models, it is necessary to check that crushing of the compressive struts does not occur. Marti (1985) pointed out that the cross-sectional area of the compressive struts is highly dependent on the details at their ends. Figure 6.8 gives examples of details that affect the geometry of the struts. Marti suggested that the compressive stress in the struts be limited to $0.6f_c'$ whereas Ramirez and Breen (1991) suggest a compressive stress limit of $2.5\sqrt{f_c'}$ in MPa units. Schlaich et al. (1987) and MacGregor (1997) have proposed compressive stress limits for the struts that depend on the stress conditions and angle of cracking surrounding the strut. Based on tests results, Bergmeister et al. (1991) proposed an equation for the stress limit of concrete struts as

$$f_c = \left(0.5 + \frac{1.25}{\sqrt{f_c'}}\right)f_c' \qquad (6\text{-}2)$$

where f_c' is in MPa. This equation is proposed for concrete having a compressive strength ranging from 20 to 80 MPa.

Alshegeir and Ramirez (1990) proposed the stress limits listed in Table 6.1 to account for the level of concrete confinement by reinforcement, the disturbance resulting from cracking, and the existence of prestressing forces.

CSA Standard CAN3-A23.3-M84 ("Design" 1984) uses a compressive stress limit based on the work of Vecchio and Collins (1986), which considers the strain compatibility of the struts and the strain softening of diagonally cracked concrete. The compressive strength f_{2max} of the diagonally cracked concrete is given as

$$f_{2max} = \frac{f_c'}{0.8 + 170\varepsilon_1} \le 0.85f_c' \qquad (6\text{-}3)$$

where

ε_1 = the principal tensile strain; and

f_c' = the cylinder compressive strength.

Compatibility of strains is used to determine the principal tensile strain from

$$\varepsilon_1 = \varepsilon_s + (\varepsilon_s + 0.002)\cos^2\alpha_s \qquad (6\text{-}4)$$

where

ε_s = the required strain in the tension tie (usually taken as ε_y), the value 0.002 is the assumed strain in the compressive strut at crushing; and

α_s = the angle between the strut and the tie that crosses the strut.

It is necessary to determine ε_1 from Eq. (6-4) where α_s is the angle between the strut and the main tension tie reinforcement shown in Fig. 6.7(b). After determining f_{2max} from Eq. (6-3), the required capacity of the compressive strut N_u can be expressed as

$$\phi A_{cs} f_{2max} \ge N_u \qquad (6\text{-}5)$$

where

A_{cs} = the effective cross-sectional area of the strut; and

ϕ = the capacity reduction factor for axial compression, equal to 0.70.

In some circumstances, it may be advantageous to provide reinforcing bars of area A_{ss} that are parallel to the strut and are restrained against buckling by the provision of ties. In this case, the design of the strut can be checked by

$$\phi(A_{cs} f_{2max} + A_{ss} f_y) \ge N_u \qquad (6\text{-}6)$$

where f_y is the yield strength of the reinforcing bars.

6.3.4 *Design of nodal zones*—Figure 6.7 shows the strut-and-tie model and the idealized truss model for a deep beam. The locations of the intersection of the truss members, or nodes of the truss, represent regions of multidirectionally stressed concrete or nodal zones of the strut-and-tie model. The compressive strength of the nodal zone concrete depends on many factors, including the tensile straining

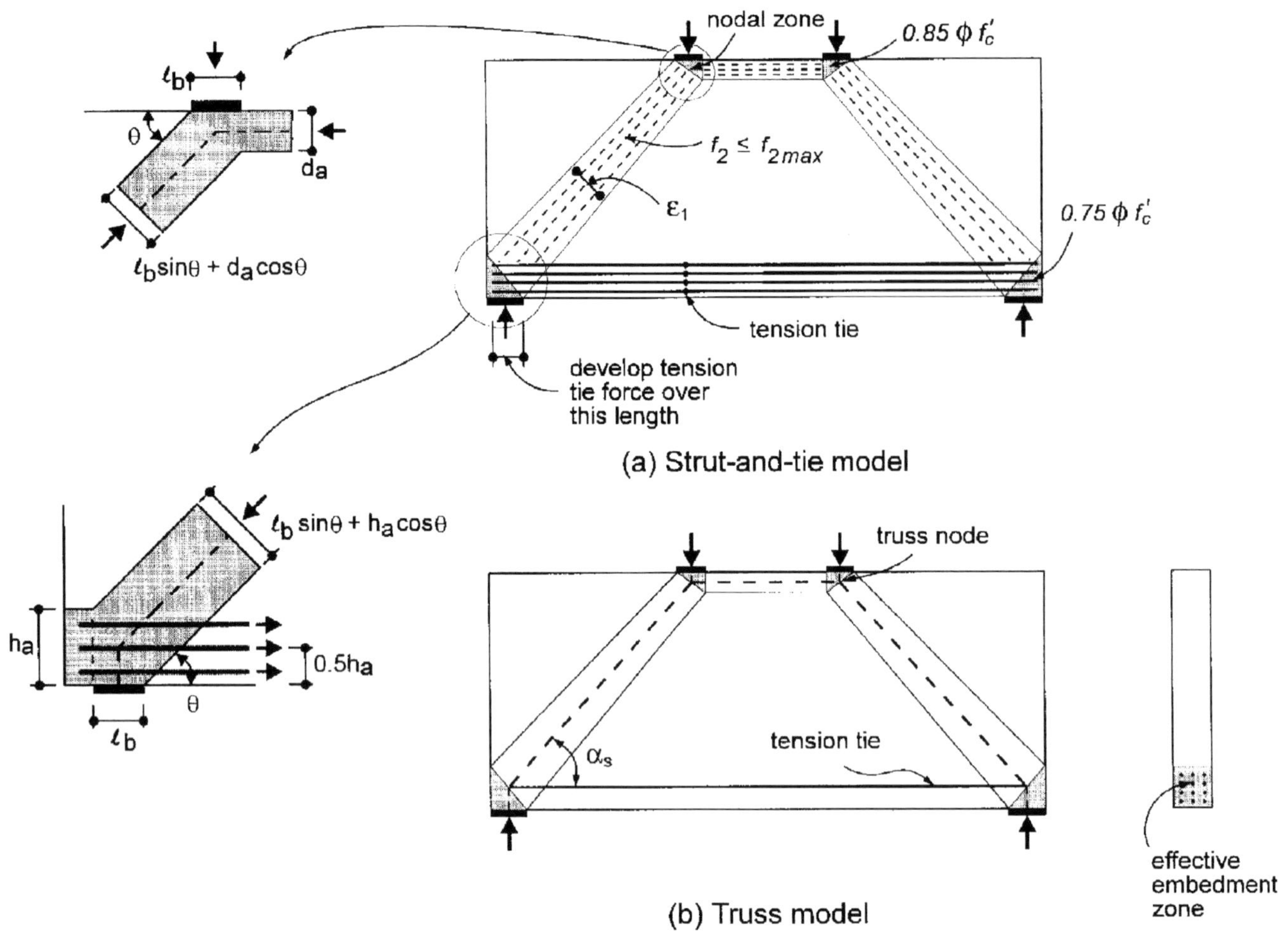

Fig. 6.7—Strut-and-tie model and idealized truss model for deep beam.

Table 6.1—Effective stress levels in concrete struts

Effective stress level (1)	Concrete struts (2)	Proposed by (3)
$0.8f_c'$	Undisturbed and uniaxial state of compressive stress that may exist for prismatic struts	Schlaich et al. (1987)
$0.68f_c'$	Tensile strains and/or reinforcement perpendicular to the axis of the strut may cause cracking parallel to the strut with normal crack width	Schlaich et al. (1987)
$0.51f_c'$	Tensile strains causing skew cracks and/or reinforcement at skew angles	Schlaich et al. (1987)
$0.34f_c'$	For skew cracks with extraordinary crack width. Skew cracks would be expected if modeling of the struts departed significantly from the theory of elasticity's flow of internal forces	Schlaich et al. (1987)
$0.85f_c'$	Moderately confined diagonal struts going directly from point load to support with shear span-to-depth ratio less than 2.0	Alshegeir and Ramirez (1990)
$0.75f_c'$	Struts forming arch mechanism	Alshegeir and Ramirez (1990)
$0.50f_c'$	Arch members in prestressed beams and fan compression members	Alshegeir and Ramirez (1990)
$0.95f_c'$	Undisturbed and highly stressed compression struts	Alshegeir and Ramirez (1990)
v_2f_c'	Uncracked uniaxially stressed struts or fields	MacGregor (1997)
$v_2(0.80)f_c'$	Struts cracked longitudinally in bulging compression fields with transverse reinforcement	MacGregor (1997)
$v_2(0.65)f_c'$	Struts cracked longitudinally in bulging compression fields without transverse reinforcement	MacGregor (1997)
$v_2(0.60)f_c'$	Struts in cracked zone with transverse tensions from transverse reinforcement	MacGregor (1997)
$v_2(0.30)f_c'$	Severely cracked webs of slender beams with θ = 30 degrees	MacGregor (1997)
$v_2(0.55)f_c'$	Severely cracked webs of slender beams with θ = 45 degrees	MacGregor (1997)

Note: $v_2 = 0.5 + 1.25/\sqrt{f_c'}$ in MPa after Bergmeister et al. (1991).

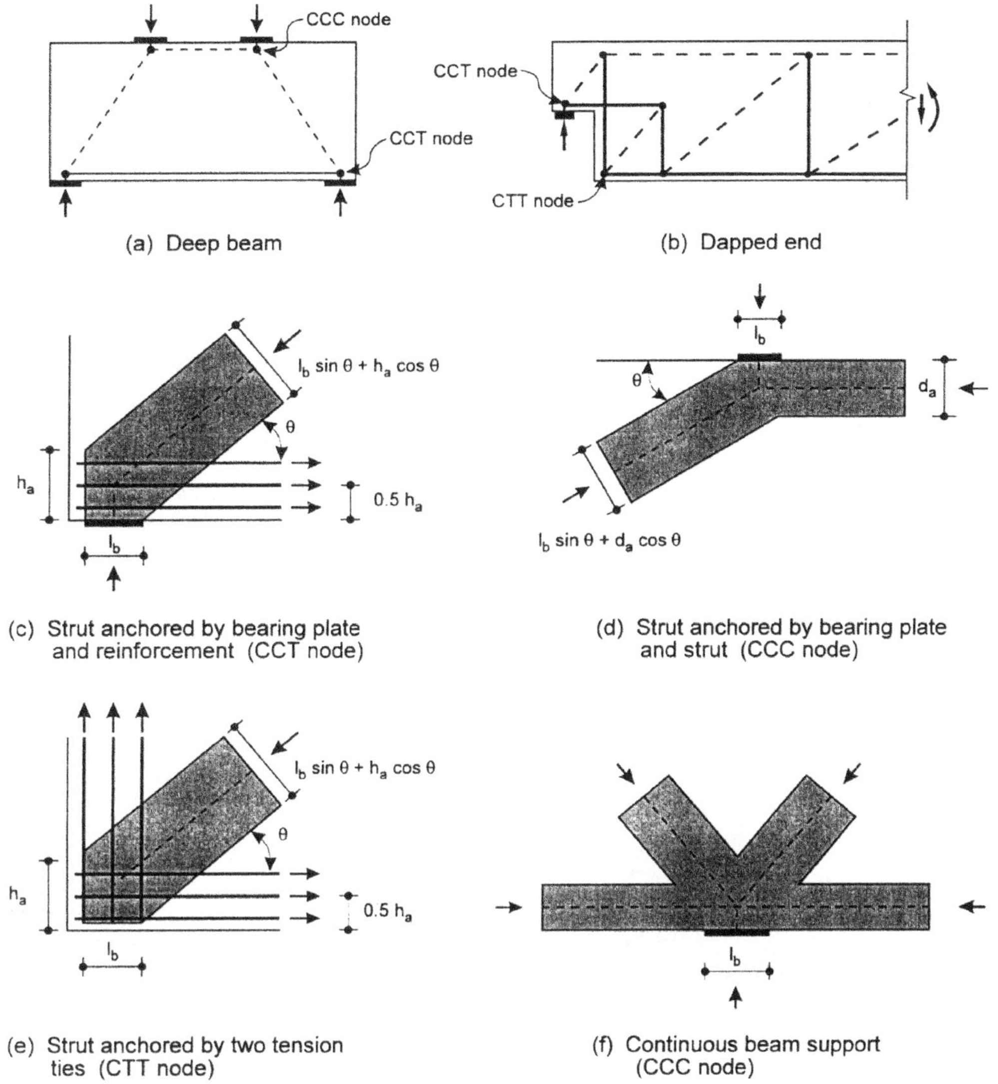

Fig. 6.8—Type of nodal zones (Jirsa et al. 1991) and dimensions of struts ("Design" 1984).

from intersecting tension ties, confinement provided by compressive reactions, and confinement provided by transverse reinforcement. To distinguish between the different straining and confinement conditions for nodal zones it is helpful to identify these zones as follows:

- CCT—nodal zone bounded by compression struts and one tension tie;
- CCC—nodal zone bounded by compression struts only; and
- CTT—nodal zone bounded by a compression strut and tension ties in two or more directions.

Figure 6.8 illustrates the different types of nodal zones for a deep beam and a beam with a dapped end together with examples of reinforcement details and dimensions required for determining the geometry of struts and nodal zones (Jirsa et al. 1991; "Design" 1984).

The Canadian standard ("Design" 1984) limits the concrete stress in the nodal zones to the following values:

- $0.85\phi f_c'$ in node regions bounded by compressive struts and bearing areas (CCC nodes);
- $0.75\phi f_c'$ in node regions anchoring a tension tie in only one direction (CCT nodes); and
- $0.60\phi f_c'$ in node regions anchoring tension ties in more than one direction (CTT nodes), where ϕ is the capacity-reduction factor for bearing (taken as 0.70 in ACI 318M-95).

In addition, the beneficial effects of confinement provided by transverse reinforcement can be accounted for by increasing these stress limits.

Marti (1985) described a procedure for evaluating stresses of nodal zones bounded by three or more intersecting struts. In this approach, the tie forces can be converted to compressive forces acting on the back face of the nodal zone. Also, the state of stress of nodal zones is assumed to be hydrostatic. If the strut widths are chosen so that the stresses in all struts are equal to $-\sigma$, then the stresses in the nodal zone are $\sigma_x = \sigma_y = \sigma_1 = \sigma_2 = -\sigma$, $\tau_{xy} = 0$, and $\sigma_z = \sigma_3 = 0$ and the sides of the nodal zone are perpendicular to the struts. He proposed that nodal zones and compressive struts could be stressed up to a limit of $0.6f_c'$.

Table 6.2—Effective stress levels in nodal zones

Effective stress level (1)	Nodes (2)	Proposed by (3)
$0.85f'_c$	CCC-nodes	Schlaich et al. (1987)
$0.68f'_c$	Nodes where reinforcement is anchored in or crossing the node	Schlaich et al. (1987)
$\nu_2 f'_c$	Nodes bounded by compressive struts and bearing areas	MacGregor (1997)
$\nu_2(0.85)f'_c$	Nodes anchoring one tension tie	MacGregor (1997)
$\nu_2(0.75)f'_c$	Nodes anchoring tension ties in more than one direction	MacGregor (1997)
$0.8f'_c$ for $f'_c \leq 30$ MPa	Unconfined nodes without bearing plates	Bergmeister et al. (1991)
$(0.9 - 0.25f'_c/70)f'_c$ MPa for $30 < f'_c \leq 70$ MPa	Unconfined nodes without bearing plates	Bergmeister et al. (1991)
$0.65f'_c$ for $f'_c \geq 70$ MPa	Unconfined nodes without bearing plates	Bergmeister et al. (1991)
$\nu f'_c(A/A_b)^{0.5} + \alpha(A_{core}/A_b)f_{lat}(1 - s/d)^2 \leq 2.5f'_c$	Confined nodes	Bergmeister et al. (1991)
$\nu f'_c(A/A_b)^{0.5}$	Unconfined nodes with bearing plates	Bergmeister et al. (1991)
$2.5f'_c$	Triaxially confined nodes	Bergmeister et al. (1991)

Note: A, A_b, and A_{core} are, respectively, area of confined concrete, bearing plate, and the confined strut; f_{lat} = lateral pressure $(2f_yA_s/(ds))$; A_s = area of one leg of confining reinforcement; μ = efficiency factor; s = pitch or spacing of confining reinforcement; d = diameter of core; α = parameter (4.0 for spiral confinement, 2.0 for square closed hoop confinement anchored with longitudinal reinforcement, and 1.0 for square closed hoop confinement without longitudinal reinforcement anchorage); and $\nu_2 = 0.5 + 1.25/\sqrt{f'_c}$ in MPa after Bergmeister et al. (1991).

Schlaich et al. (1985) and MacGregor (1997) also proposed values of effective stress levels of nodal zones. Based on the test results of 10 isolated CCT nodes and nine CTT nodes, Jirsa et al. (1991) recommended an effective concrete strut stress level of $0.8f'_c$.

Bergmeister et al. (1991) proposed several equations of effective concrete strength for various kinds of nodes, including reinforcement, unconfined nodes with bearing plates, and triaxially confined nodes. The proposed equations are listed in Table 6.2. An upper limit of $2.5f'_c$ was recommended for situations with a large amount of confinement.

Adebar and Zhou (1993) suggested stress limits for the concrete compressive struts confined by plain concrete based on the results of analytical and experimental studies. The limiting stress for deep members without reinforcement distributed over the depth (for example, pile caps) is limited to

$$f_c = \phi 0.6f'_c(1 + 2\alpha\beta) \quad (6\text{-}7)$$

where

$$\alpha = 0.33(\sqrt{A_2/A_1} - 1) \leq 1.0 \quad (6\text{-}8)$$

$$\beta = 0.33\left(\left(\frac{l_s}{b_s}\right) - 1\right) \leq 1.0 \quad (6\text{-}9)$$

The ratio l_s/b_s, which represents the slenderness ratio (length/width) of the compression strut, should not be taken less than 1. The parameter α accounts for the degree of confinement, whereas the parameter β accounts for the geometry of the compressive stress field and ϕ is a capacity-reduction factor for axial compression. In the expression for α, A_1, and A_2 resent the loaded area and the supporting surface area, respectively. The lower bearing-stress limit of $\phi 0.6f'_c$ was suggested if there is no confinement, regardless of the length of the compressive strut. For relatively short compressive struts with a large degree of confinement, the upper limit suggested was $\phi 1.8f'_c$.

6.3.5 *Design of tension ties*—Once the required force N_u in the tension tie has been determined from the statics of the truss model, the area of reinforcement can be determined from

$$\phi(A_{st}f_y + A_{ps}(f_{ps} - f_{se})) \geq N_u \quad (6\text{-}10)$$

where

A_{st} = area of reinforcing bars;
A_{ps} = area of prestressed reinforcement;
f_y = yield strength of reinforcing bars;
f_{ps} = stress in prestressed reinforcement at ultimate;
ϕ = capacity-reduction factor for axial tension (0.9); and
f_{se} = stress in prestressing steel after all losses.

Note that Eq. (6-10) assumes that a portion of the prestress has been accounted for using an external compression on the member of $A_{ps}f_{se}$. If the prestressing is treated as a tension tie having an initial tensile force of $A_{ps}f_{se}$, then the full force $A_{ps}f_{ps}$ can be used in Eq. (6-10).

6.3.6 *Anchorage of tension ties*—An important aspect of the design of D regions is the anchorage of tension ties to develop the required stress in the reinforcement. The tension tie reinforcement shown in Fig. 6.7 should develop the required stress in the steel at the inner face of the bearing to avoid failure by loss of anchorage. Anchorage may be achieved by an appropriate embedment length over the bearing areas or by mechanical anchorage. Another important aspect of the design of the anchorage regions is to spread the anchorage of the tie reinforcement over a large enough area to avoid nodal zone crushing. Therefore the distance h_a, shown in Fig. 6.7, should be such that the tension tie force divided by the effective embedment area does not exceed the appropriate nodal zone stress limit.

6.3.7 *Design procedure*—The steps in the design of a D region, such as the deep beam shown in Fig. 6.7, are as follows:

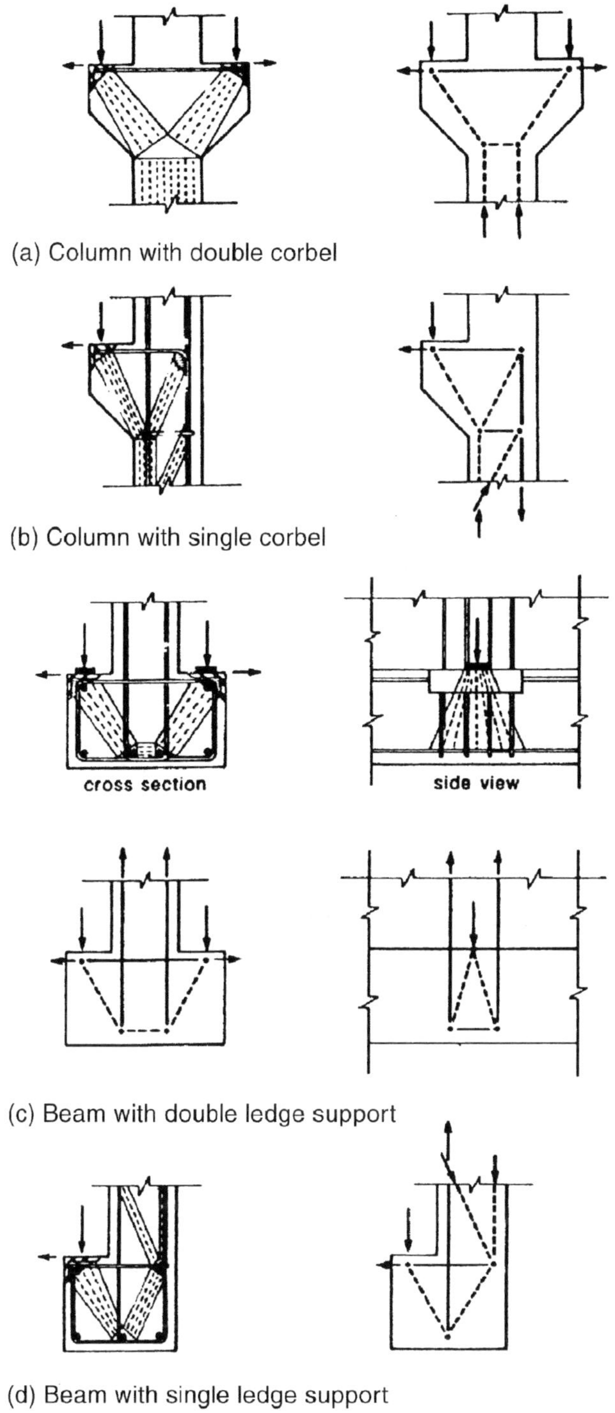

Fig. 6.9—Strut-and-tie model idealizations for brackets, ledges, and corbels (Cook and Mitchell 1988).

1. Sketch the flow of the forces in the D region and locate nodal zones that are the regions bounded by struts, tension ties, or bearing areas;

2. Determine the dimensions of the loading and reaction areas such that the nodal zone stresses are less than the permissible limits;

3. Determine the geometry of the truss model. The compressive struts are represented by truss members located

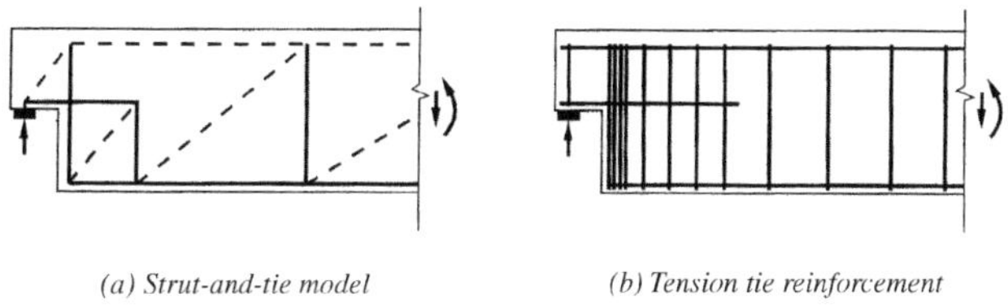

Fig. 6.10—Beam with dapped end, adapted from Schlaich et al. (1987): (a) strut-and-tie model; and (b) tension tie reinforcement.

at their centerlines, and the tension ties are represented by truss members at the centroid of the reinforcement. The nodes of the truss are located at the intersections of struts, tension ties, and applied loads or reactions. In determining the geometry, it is important to account for the actual dimensions of the nodal zones. The dimensions of the nodal zones are dependent on the sizes of the bearing plates, where used, and the anchorage details of the tension ties. (Note: Because the force in the tension tie is not known at this stage, it will be necessary to make an estimate of the required anchorage area so that the size of the nodal zone and hence the geometry of the truss model can be determined.);

4. Determine the forces in the truss members. If the truss is statically indeterminate (for example, a continuous deep beam) it will be necessary to estimate the relative stiffnesses of the truss members to solve for the forces in the struts and ties;

5. Determine the required areas of the tension tie reinforcement (see Eq. (6-10));

6. Ensure that the anchorages of the tension ties are sufficient and that the tension ties are spread over a sufficient anchorage area to satisfy the nodal zone stress limits; and

7. Calculate the compressive stresses in the struts at critical locations using the effective strut areas. Check that the calculated compressive stresses in the struts are less than f_{2max} (see Eq. (6-3) and (6-4) or Table 6.1).

The design of D regions using the above steps may involve an iterative procedure.

6.3.8 *Examples of strut-and tie models*—Figures 6.9 to 6.11 show strut-and-tie models suggested by Cook and Mitchell (1988) and Schlaich et al. (1987) with the corresponding truss idealizations for different examples of D regions including brackets, corbels, ledges, a dapped end, and deep beams. The truss models for the continuous deep beams shown in Fig. 6.11(b) and (c) are statically indeterminate; therefore, it is necessary to assume the relative stiffness of the truss members to find the forces in the truss member. Guidance on more sophisticated truss models for deep beams is given by Rogowsky and MacGregor (1986).

Strut-and-tie models are well suited to designing the reinforcement required in anchorage regions of post-tensioned tendons (Schlaich et al. 1987; Collins and Mitchell 1991; Burdet et al. 1991). Figure 6.12 illustrates strut-and-tie models used to model the flow of forces for the case of the anchorage of two post-tensioned tendons (Burdet et al. 1991). For the case where the tendon anchorages are within the quarter points of the section, the flow of forces in the anchorage zone will be similar to that induced by a single

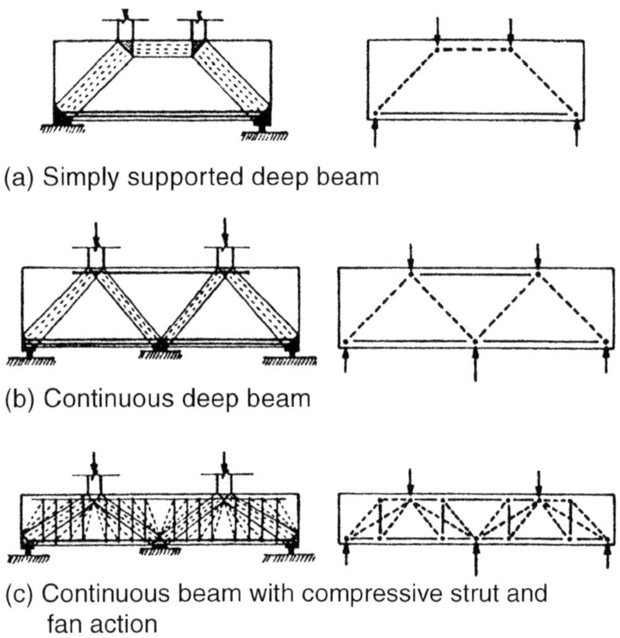

Fig. 6.11—Strut-and-tie models and truss idealizations for deep beams (Cook and Mitchell 1988).

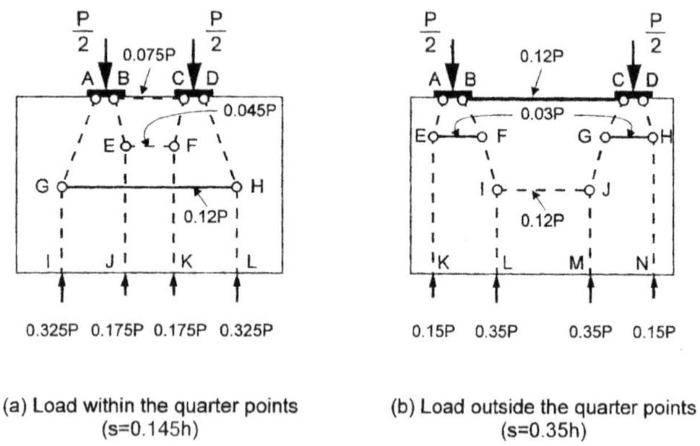

Fig. 6.12—Strut-and-tie models for design of anchorage zones with two post-tensioned tendons (Burdet et al. 1991): (a) load within quarter points (s = 0.145h) and (b) load outside quarter points (s = 0.35h).

anchorage device (Fig. 6.12(a)). For the case with the tendons anchored outside of the kern, there is a different flow of forces with a tension tie required near the end face of the member (Fig. 6.12(b)).

Figure 6.13 illustrates the strut-and-tie model and the three-dimensional truss idealization suggested by Adebar et al. (1990) for a column supported by a pile cap. Schlaich et al. (1987) and Muttoni et al. (1997) have presented examples of the strut-and-tie modeling of deep beams and walls with openings, post-tensioned buttress anchorage regions, bridge piers, and bridge diaphragms. Detailed design examples of D regions using the design approach described in this report are given by Collins and Mitchell (1991) and MacGregor (1992).

More research is required to be able to predict service load behavior of D regions.

CHAPTER 7—SUMMARY

7.1—Introduction

This report focused on the presentation of new shear design approaches. The work presented is limited to monotonic loads on one-way flexural members covering both B and D regions. The interaction of shear and torsional effects is not covered in this report. Although the methods presented can be applied to shear and torsion, in several instances this fact remains to be shown and verified with sufficient tests.

The approaches reviewed cover the entire range of structural concrete from reinforced to fully prestressed concrete members. The so-called concrete contribution to the shear strength has been discussed. Thus, this report addresses members with and without shear reinforcement.

This report does not attempt to endorse any of the individual approaches presented. Instead, the goal is to illustrate the common base of these new approaches and how they supplement each other. By doing so, the report attempts to answer the challenge put forth by ASCE-ACI Committee 426 in 1973.

7.2—Truss models

Since the early 1900s, engineers have used truss models to follow the flow of internal forces in structural-concrete members and provide structural systems made out of concrete and reinforcement that ensure equilibrium. The original 45 degree truss model of Ritter (1899) and Mörsch (1920, 1922) has been adopted either explicitly or implicitly by most major codes as the basis for their shear and torsion design specifications. Throughout this report, the reader has been exposed to this concept in forms ranging from sectional truss models and compression field approaches for beams to generalized strut-and-tie models. Their application extended to the full range of structural concrete in members with and without shear reinforcement.

As explained in Chapter 1, beginning with the earlier concept of a sectional design approach based on a constant con-crete contribution supplementing a constant 45 degree angle truss, the truss model approach was first introduced by M. O. Withey into the American literature in 1907. At this time Withey noted that the 45 degree truss model alone yielded conservative results when compared with test evidence. This observation was confirmed by Talbot in 1909. As a result, since its origins, shear design in the United States has included a concrete contribution, V_c, to supplement the 45 degree truss model. More modern versions of design specifications, such as the 1978 Comité Euro-International du Béton-Fédération International de la Précontrainte (CEB-FIP) Model Code for Concrete Structures, have extended this concept by recognizing the capability for redistribution of internal forces of reinforced and prestressed concrete beams containing stirrup reinforcement. This approach has been set up as a variable-angle truss with some established semiempirical limits for the angle of inclination supplemented with a concrete contribution. The concrete contribution included the presence of prestressing as well as the possibility of a diminished contribution after significant redistribution of internal forces. This reduction was deemed to be the result of crack

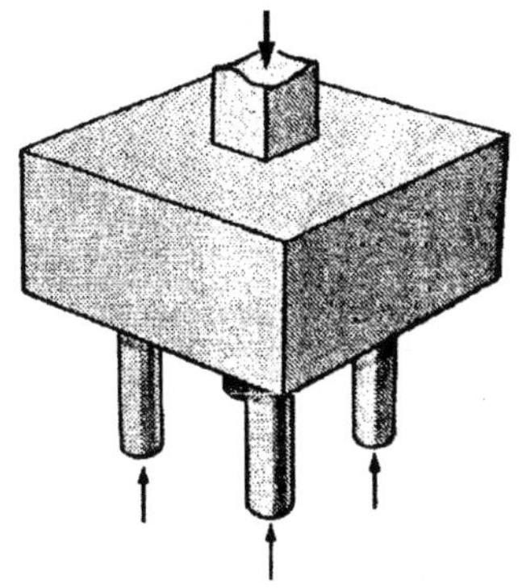

Pile cap supported on four piles

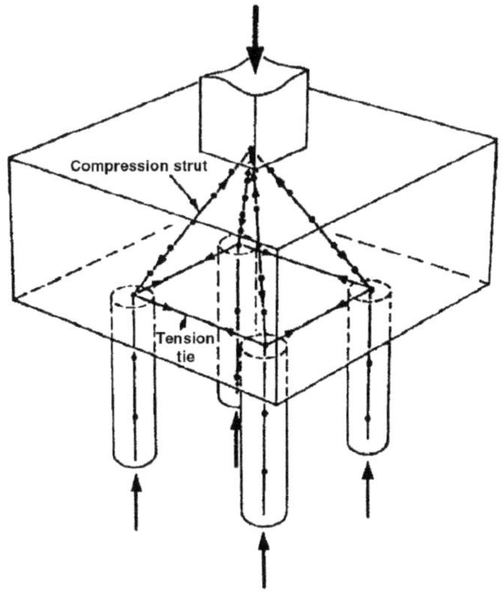

Simple three-dimensional truss model for a four-pile cap

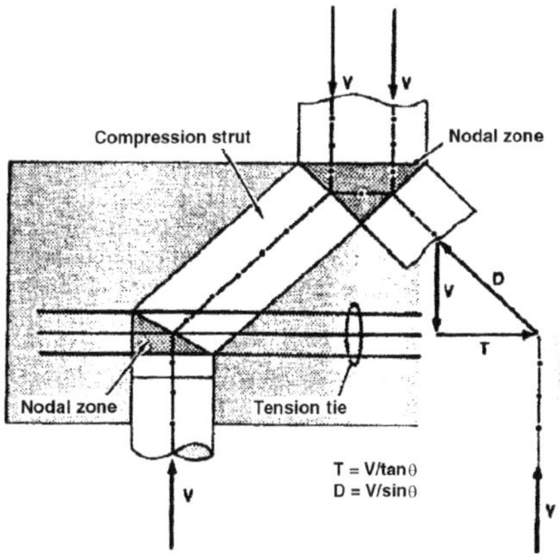

Components of the idealized load-resisting strut-and-tie model

Fig. 6.13—Strut-and-tie models and three-dimensional truss model for column supported on four-pile cap (adapted from Adebar et al. [1993]).

inclinations deviating significantly from 45 degrees and the resulting wider crack widths from this redistribution.

Kupfer (1964) provided a solution for the angle of inclination of the diagonal cracks from the analysis of a truss model consisting of linearly elastic members and neglecting the tensile strength of the concrete. Baumann (1972) derived a similar solution for wall elements. Mitchell and Collins (1974) abandoned the notion of cracked linear behavior and developed the CFT for torsion in 1974 to allow for evaluation of the angle of inclination of the diagonally cracked concrete through the entire range of loading (see Chapter 2). This approach, extended by Collins (1978) to the case of shear, neglected the presence of tensile stresses in the concrete between the cracks. The compression field approach views the cracked concrete as a material where the principal stress and strain axes coincide. Under a given loading condition, the direction of the uniaxial compression field will rotate as needed to satisfy equilibrium. Marti (1992) has shown that for a wall element with orthogonal reinforcement in the x and y directions, and assuming linear elastic behavior for both steel and concrete, the compression field approach leads to Baumann's and originally Kupfer's solutions for the angle of inclination of the diagonally cracked concrete. These approaches essentially neglected the so-called concrete contribution term.

The truss-model-related theories discussed in the preceding paragraph are based on the assumption that the compression struts are parallel to the principal compressive stress orientation and that no stresses are transferred across the cracks. Vecchio and Collins (1986) (see Chapter 2) introduced the MCFT based on the assumption that tensile stresses in the concrete between the cracks contribute significantly to the shear resistance. The basic assumption was that these tensile stresses transverse to the axis of the strut exist at points between the inclined cracks, but are zero at the cracks. In their derivation, the tensile strength of the concrete strut diminishes as the principal strain ε_1 increases. The explanation is that the tensile strength of the concrete may be limited by the ability of the crack surfaces to transmit shear stresses that will depend on the crack width and the maximum size of the aggregate.

Refined methods have attempted to supplement the truss model strength with a concrete contribution term. Such methods are discussed in Chapter 3. An example of such approaches has been called the truss analogy with crack friction. In this approach, shear stresses are transferred across the cracks by friction. It can be regarded as a development based on the shear-friction theory discussed briefly in Chapter 5. As a failure mechanism approach, it starts with basic assumptions for the spacing and the shape of the cracks in the B region of a structural-concrete member subjected to shear forces. In this sense, it is a discrete approach with respect to cracking. Two results from the assumption that shear stresses are transferred across the cracks are that the angle of the principal compression stress in the web does not coincide with the angle of the cracks, θ, and there is a vertical component of the frictional forces, V_f, along the crack face contributing to the shear strength. Several researchers postulate that V_f could represent the concrete contribution appearing in ACI 318M-95.

Truss models approaches have been generalized to all parts of the structure in the form of strut-and-tie models. Structural-concrete members with an adequate distribution of minimum transverse and longitudinal reinforcement can be designed using the simple strut-and-tie models described in Chapter 6. Three key advantages of this approach are that 1) the flow of internal forces is clearly visualized by the designer; 2) the effects of both shear and moment are accounted for simultaneously and directly in the design; and 3) the design of B-and D-type regions and the transitions between them can be conducted in a unified and consistent manner. Thus, the design can be carried out for the entire member using a full-member truss model as a generalized strut-and-tie model. Alternatively, the design of B regions can be carried out using a sectional model, such as the ones described in Chapters 1, 2, and 3.

The use of strut-and-tie models is especially advantageous in the design of D regions characterized by a complex flow of internal forces resulting in significant nonlinear distribution of strains. Although this report is directed at the design of one-way flexural members under monotonic loads, the design of D regions appearing in this type of member is discussed thoroughly in Chapter 6. The same basic principles presented in this report could be extended to other types of D regions such as beam-column joints. The effects of dynamic load and torsion in this and other relevant cases, however, are not within the scope of this report.

7.3—Members without transverse reinforcement

Many structural elements such as slabs, footings, and joists are built without transverse reinforcement (that is, stirrups or bent bars). The design of these members cannot be covered by a truss approach that ignores the tensile strength of the concrete. In this report, an extensive review of design models and analytical methods for structural members with no transverse reinforcement has been given (see Chapter 4). Although the case of two-way shear is not covered, much of the work presented and presently under way in members without transverse reinforcement is geared toward the development of more rational methods for punching shear.

Since the 1973 ASCE-ACI Committee 426 report was issued, an additional mechanism of shear transfer in the form of residual tensile stresses transmitted directly across cracks has been identified. The 1973 ACI 426 report identified four mechanisms of shear transfer: 1) shear stresses in the uncracked concrete; 2) aggregate interlock or shear friction; 3) dowel action of the longitudinal reinforcement; and 4) arch action. The question of which mechanisms will contribute most to the shear resistance of diagonally cracked concrete beam does not have a unique answer. As discussed in Chapter 4, the importance of the contribution of the individual mechanisms will depend on the specific member and loading conditions. For example, the amount of the shear stress carried across the uncracked concrete that can be significant in columns under compressive axial load, bending and shear, is relatively small in slender members without axial compression because the depth of the compression zone is smaller. Dowel action, which is not very significant when the dowels are near the surface, cannot be neglected when the member has high reinforcement ratios. It may be very significant, even in members with lower reinforcement ratios, when the longitudinal reinforcement is distributed in layers. Researchers have recognized this complexity and the importance of developing physical models to explain this behavior. The existing mechanical or physical models are presented in Chapter 4.

Despite differences in the approaches, the main conclusion from the models discussed in Chapter 4 is that concrete tensile stresses should be considered explicitly. Failure is not explained simply by the tensile stress reaching the tensile strength of the concrete, however, but involves the mechanism of shear transfer at the cracks. In other words, for smaller-sized members, the residual stresses are very important, whereas for larger members, with several cracks before failure, the friction across the cracks has a more important role. Another important conclusion is that relationships based on smeared or discrete truss models considering the tensile strength of the concrete are now available to replace the largely empirical approaches in current codes of practice. This is particularly true for ACI 318M-95, where the observation that the shear stress at failure decreases as the depth of the member increases is not properly recognized.

7.4—Additional work

During the past 20 years, considerable improvements have been made in understanding the behavior and improving the design of structural-concrete members subject to shear to the extent that there is now a solid foundation for advancing, unifying, and simplifying our design specifications. Such solid foundation is provided by truss model approaches and related theories. The challenge presented by the 1973 ASCE-ACI Committee 426 report has been answered. Realistic models exist now for deep beams, corbels, shear walls, and other unusual design problems. Much work remains to be done, however. The effects of dynamic loading on the truss model approach still need to be accounted for. Rational models for the complex problem of punching shear are still needed in codes of practice. The enhanced shear strength of prestressed concrete members calls for additional study. Additional experimental verification is needed for the nodal zones of strut-and-tie models. Finally, a survey on the state-of-the-art for combined shear and torsion remains to be conducted. The next challenge is to integrate, simplify, and give a physical significance to the design regulations for torsional strength.

APPENDIX A—ACI 318M-95 SHEAR DESIGN APPROACH FOR BEAMS

$$V_n = V_c + V_s$$

$$V_s = A_v f_y d/s \le (2/3)\sqrt{f_c'}\, b_w d$$

$$V_c = \text{concrete contribution}$$

For nonprestressed beams

$$(\sqrt{f_c'}/6) b_w d$$

For prestressed beams

$$[(\sqrt{f_c'}/20) + 5(V_u d/M_u)] b_w d$$

$$\text{but } (1/6)\sqrt{f_c'}\, b_w d \le V_c \le 0.4\sqrt{f_c'}\, b_w d$$

or the lesser of V_{ci} or V_{cw}

$$V_{ci} = (\sqrt{f_c'}/20) b_w d + V_d + (V_i M_{cr}/M_{max})$$

$$\text{but } V_{ci} \ge (1/7)\sqrt{f_c'}\, b_w d$$

$$V_{cw} = 0.3(\sqrt{f_c'} + f_{pc}) b_w d + V_p$$

For axial compression and shear

$$V_c = [1 + (N_u/14A_g)](\sqrt{f_c'}/6) b_w d$$

For axial tension and shear

$$V_c = [1 + (0.3N_u/A_g)](\sqrt{f_c'}/6) b_w d$$

Detailing rules are as follows:

- Reinforcement shall extend beyond the point at which it is no longer required to resist flexure for a distance equal to the effective depth of the member or $12d_b$, whichever is greater.
- Flexural reinforcement shall not be terminated in a tension zone unless shear at cutoff ≤ 2/3 shear permitted; stirrup area A_v in excess of that required for shear and torsion, is provided [$A_v \ge 0.4 b_w s/f_y$; $s \le d/8\beta_b$]; or for No. 36 bars or smaller, shear at the cutoff ≤3/4 of the permitted, and continuing reinforcement provides double the area required for flexure at the cutoff.
- At simple supports and points of inflection, the diameter of the positive moment tension reinforcement shall be limited so that $l_d \le (M_n/V_u) + l_a$.

A-1—Notation

A_g = gross area of section (mm^2)

A_v = area of shear reinforcement within a distance s (mm^2)

b_w = web width or diameter of circular section (mm)

d = distance from extreme compression fiber to centroid of longitudinal tension reinforcement, which need not be less than $0.80h$ for prestressed members (mm) (for circular sections, d need not be less than the distance from extreme compression fiber to centroid of tension reinforcement in opposite half of member)

f_c' = specified compressive strength of concrete (MPa)

$\sqrt{f_c'}$ = square root of specified compressive strength of concrete (MPa)

f_{pc} = compressive stress in concrete (after allowance for all prestress losses) at centroid of cross section resisting externally applied loads or at junction of web and flange when the centroid lies within the flange (MPa) (in a composite member, f_{pc} is resultant compressive stress at centroid of composite section, or at junction of web and flange when the centroid lies within the flange, due to both prestress and moments resisted by precast member acting alone)

f_y = specified yield strength of nonprestressed reinforcement (MPa)

h = overall thickness of member (mm)

l_a = additional embedment length at support or at point of inflection (mm)

l_d = development length (mm)

M_{cr} = moment causing flexural cracking at section due to externally applied loads

M_{max} = maximum factored moment at section due to externally applied loads

M_u = factored moment at section

N_u = factored axial load normal to cross section occurring simultaneously with V_u (to be taken as positive for compression, negative for tension, and to include effects of tension due to creep and shrinkage)

s = spacing of shear or torsion reinforcement in direction parallel to longitudinal reinforcement (mm)

V_c = nominal shear strength provided by concrete

V_{ci} = nominal shear strength provided by concrete when diagonal cracking results from combined shear and moment

V_{cw} = nominal shear strength provided by concrete when diagonal cracking results from excessive principal tensile stress in web

V_d = shear force at section due to unfactored dead load

V_i = factored shear force at section due to externally applied loads occurring simultaneously with M_{max}
V_s = nominal shear strength provided by shear reinforcement
V_p = vertical component of the effective prestress force at the section
V_u = factored shear force at section
β = ratio of area of tension reinforcement cut off to total area of tension reinforcement at the section

A-2—Acknowledgments

This report is dedicated to the memory of Professor Peter Gergely, who chaired Committee 445 when this work was started. This report was written by members of a subcommittee of ASCE-ACI Committee 445 on Shear and Torsion. It was reviewed and balloted by the voting members of Committee 445. The subcommittee (445-1) acknowledges the guidance of Peter Marti, who as a former chairman of Committee 445 was instrumental in supporting the work on this report.

APPENDIX B—REFERENCES

B-1—Referenced standards and reports

The standards and reports listed below were the latest editions at the time this document was prepared. Because these documents are revised frequently, the reader is advised to contact the proper sponsoring group if it is desired to refer to the latest version.

American Concrete Institute

318M Building Code Requirements for Structural Concrete and Commentary

446.1R Fracture Mechanics of Concrete: Concepts, Models, and Determination of Material Properties

These publications may be obtained from this organization:

American Concrete Institute
38800 Country Club Dr.
Farmington Hills, MI 48331

B-2—Cited references

ACI Committee 318, 1951, "Building Code Requirements for Reinforced Concrete (ACI 318-51)," ACI JOURNAL, *Proceedings* V. 48, Apr. 1951, pp. 589-652.

ACI Committee 318, 1971, "Building Code Requirements for Reinforced Concrete (ACI 318-71)," American Concrete Institute, Farmington Hills, Mich., 78 pp.

Adebar, P., 1994, "Testing Structural Concrete Beam Elements," *RILEM: Materials and Structures*, V. 27, No. 172, pp. 445-451.

Adebar, P., and Collins, M. P., 1996, "Shear Strength of Members without Transverse Reinforcement," *Canadian Journal of Civil Engineering*, V. 23, No. 1, pp. 30-41.

Adebar, P.; Kuchma, D.; and Collins, M. P., 1990, "Strut-and-Tie Models for the Design of Pile Caps," *ACI Structural Journal*, V. 87, No. 1, Jan.-Feb., pp. 81-92.

Adebar, P., and Zhou, Z., 1993, "Bearing Strength of Compressive Struts Confined by Plain Concrete," *ACI Structural Journal*, V. 90, No. 5, Sept.-Oct., pp. 534-541.

Al-Nahlawi, K. A., and Wight, J. K., 1992, "Beam Analysis using Concrete Tensile Strength in Truss Models," *ACI Structural Journal*, V. 89, No. 3, May-June, pp. 284-289.

Alshegeir, A., and Ramirez, J. A., 1990, "Analysis of Disturbed Regions with Strut-and-Tie Models," *Structural Engineering Report* No. CE-STR-90-1, Purdue University, West Lafayette, Ind.

Aoyagi, Y., and Yamada, K., 1983, "Strength and Deformation Characteristics of Reinforced Concrete Shell Elements Subjected to In-Plane Forces," *Proceedings,* Japanese Society of Civil Engineers, Tokyo, 331, pp. 167-190.

ASCE-ACI Committee 326, 1962a, "Shear and Diagonal Tension," ACI JOURNAL, *Proceedings* V. 59, No. 1, Jan., pp. 1-30.

ASCE-ACI Committee 326, 1962b, "Shear and Diagonal Tension," ACI JOURNAL, *Proceedings* V. 59, No. 1, Jan., pp. 277-344.

ASCE-ACI Committee 326, 1962c, "Shear and Diagonal Tension," ACI JOURNAL, *Proceedings* V. 59, No. 1, Jan., pp. 352-396.

ASCE-ACI Committee 426, 1973, "The Shear Strength of Reinforced Concrete Members," *Journal of Structural Division*, ASCE, V. 99, No. 6, pp. 1091-1187.

ASCE-ACI Committee 426, 1978, "Suggested Revisions to Shear Provisions for Building Codes," American Concrete Institute, Farmington Hills, Mich.

Barton, N., 1973, "Review of New Shear-Strength Criterion for Rock Joints," *Engrg. Geol.*, V. 7, pp. 287–332.

Baumann, T., 1972, "Zur Frage der Netzbewehrung von Flachen trag-werken (On the Problem of Net Reinforcement of Surface Structures)," *Bauingenieur*, V. 47, No. 10, pp. 367-377.

Baumann, T., and Rüsch, H., 1970, "Versuche zum Studium der Ver-dübelungswirkung der Biegezugbewehrung eines Stahlbetonbalkens," *DAfStb H.210*, Wilhelm Ernst und Sohn, Berlin, pp. 43-83.

Bažant, Z. P., and Kim, J.-K., 1984, "Size Effect in Shear Failure of Longitudinally Reinforced Beams," ACI JOURNAL, *Proceedings* V. 81, No. 5, Sept.-Oct., pp. 456-468.

Bažant, Z. P., and Oh, B. H., 1983, "Crack Band Theory for Fracture of Concrete," *RILEM*, V. 16, No. 93, pp. 155-177.

Belarbi, A., and Hsu, T. T. C., 1991, "Constitutive Laws of Reinforced Concrete in Biaxial Tension-Compression," *Research Report UHCEE 91-2*, University of Houston, Tex.

Belarbi, A., and Hsu, T. T. C., 1994, "Constitutive Laws of Concrete in Tension and Reinforcing Bars Stiffened by Concrete." *ACI Structural Journal*, V. 91, No. 4, July-Aug., pp. 465-474.

Belarbi, A., and Hsu, T. T. C., 1995, "Constitutive Laws of Softened Concrete in Biaxial Tension-Compression," *ACI Structural Journal*, V. 92, No. 5, Sept.-Oct., pp. 562-573.

Bergmeister, K.; Breen, J. E.; and Jirsa, J. O., 1991, "Dimensioning of the Nodes and Development of Reinforcement," *Report IABSE Colloquium on Structural Concrete*, IABSE, Zurich, pp. 551-564.

Bhide, S. B., and Collins, M. P., 1989, "Influence of Axial Tension on the Shear Capacity of Reinforced Concrete Members," *ACI Structural Journal*, V. 86, No. 5, Sept.-Oct., pp. 570-581.

Birkeland, P. W., and Birkeland, H. W., 1966, "Connections in Precast Concrete Construction." ACI JOURNAL, *Proceedings* V. 63, No. 3, Mar., pp. 345-368.

Braestrup, M. W., 1990, "Shear Strength Prediction—Plastic Method," *Reinforced Concrete Deep*, F. K. Kong, ed., Blackie and Son, London/Van Nostrand Reinhold, New York, pp. 182-203.

Bridge Design Specifications and Commentary, 1994, 1st Edition, American Association of State Highway and Transportation Officials, Washington, D.C.

Brock, G. et al., 1964, discussions of "The Riddle of Shear Failure and its Solution" by G. N. J. Kani, ACI JOURNAL, *Proceedings* V. 61, No. 12, Dec., pp. 1587-1636.

Burdet, O. L.; Sanders, D. H.; Roberts, C. L.; Breen, J. E.; and Fenves, G. L., 1991, "Models and Tests of Anchorage Zones of Post-Tensioning Tendons," *IABSE Colloquium on Structural Concrete*, Zurich, pp. 545–550.

Caflisch, R.; Krauss, R.; and Thurlimann, B., 1971, "Biege-und Schub-versuche and teilweise Vorgespannten Betonbalken," *Serie C., Bericht 6504-3*, Institut fur Baustatik, ETH, Zurich.

"CEB-FIP Model Code for Concrete Structures," 1978, *International System of Unified Standard Codes of Practice for Structures*, Comité Euro-International du Béton, Paris, V. II.

Chana, P. S., 1987, "Investigation of the Mechanism of Shear Failure of Reinforced Concrete Beams," *Magazine of Concrete Research*, V. 39, No. 12, pp. 196-204.

Collins, M. P., 1978, "Toward a Rational Theory for RC Members in Shear," *Journal of Structural Division*, ASCE, V. 104, No. 4, pp. 649-666.

Collins, M. P., and Mitchell, D., 1980, "Shear and Torsion Design of Prestressed and Nonprestressed Concrete Beams," *PCI Journal*, V. 25, No. 5, pp. 32-100.

Collins, M. P., and Mitchell, D., 1981a, Discussion of "Shear and Torsion Design of Non-prestressed Concrete Beams," *PCI Journal*, V. 26, No. 6, pp. 96-118.

Collins, M. P., and Mitchell, D., 1985, "Chapter 4—Shear and Torsion," *CPCA Concrete Design Handbook*, Canadian Portland Cement Association, Ottawa, Canada, pp. 4-1 to 4-51.

Collins, M. P., and Mitchell, D., 1986, "A Rational Approach to Shear Design—The 1984 Canadian Code Provisions," ACI JOURNAL, *Proceedings* V. 83, No. 6, June, pp. 925-933.

Collins, M. P., and Mitchell, D., 1991, *Prestressed Concrete Structures*, Prentice-Hall, Englewood Cliffs, N.J.

Collins, M. P.; Mitchell, D.; Adebar, P. E.; and Vecchio, F. J., 1996, "A General Shear Design Method," *ACI Structural Journal*, V. 93, No. 1, Jan.-Feb., pp. 36-45.

Collins, M. P.; Mitchell, D.; and MacGregor, J. G., 1993, "Structural Design Considerations for High-Strength Concrete," *Concrete International*, V. 15, No. 5, May, pp. 27-34.

Collins, M. P., and Porasz, A., 1989, "Shear Design for High-Strength Concrete," *Proceedings, Workshop on Design Aspects of High-Strength Concrete, Comité Euro-International du Béton Bulletin d'Information*, CEB, Paris, 193, pp. 77-83.

Comité Euro-International du Beton (CEB), 1978, "Shear and Torsion, June: Explanatory and Viewpoint Papers on Model Code Chapters 11 and 12, prepared by CEB Committee V," CEB, Paris, *CEB* Bulletin 12.

Comité Euro-International du Beton (CEB), 1982, "Shear, Torsion, and Punching: Progress Report by CEB Committee IV," *Member Design*, CEB, Paris, CEB Bulletin 146.

Comité Euro-International du Beton (CEB)-Fédération International de la Precontrainte (FIP), 1993, *Model Code 1990, 1993, (MC90)*, Thomas Telford, London.

Cook, W. D., and Mitchell, D., 1988, "Studies of Disturbed Regions near Discontinuities in Reinforced Concrete Members," *ACI Structural Journal*, V. 85, No. 2, Mar.-Apr., pp. 206-216.

Daschner F., and Kupfer, H., 1982, "Versuche zur Schubkraftubertra-gung an Rissen von Nornal- und Leichtbeton," *Bauingenieur*, 59, pp. 57-60.

Dei Poli, S.; Gambarova, P. G.; and Karakoc, C., 1987, "Aggregate Interlock Role in RC Thin-Webbed Beams in Shear," *Journal of Structural Engineering*, ASCE, V. 113, No. 1, pp. 1-19.

Dei Poli, S.; Prisco, M. D.; and Gambarova, P. G., 1990, "Stress Field in Web of RC Thin-Webbed Beams Failing in Shear," *Journal of Structural Engineering*, ASCE, V. 116, No. 9, pp. 2496-2515.

"Design of Concrete Structures for Buildings," 1984, *CAN3-A23.3-M84*, Canadian Standards Association, Rexdale, ON, Canada.

"Design of Concrete Structures," 1994, Canadian Standards Association, Rexdale, Ontario, Canada.

Drucker, D. C., 1991, "On Structural Concrete and the Theorems of Limit Analysis: IABSE Report 2, Zürich, EC2, 1991, Eurocode No. 2," *Design of Concrete Structures, Part 1: General Rules and Rules for Buildings*, Thomas Telford, London.

Eibl, J., and Neuroth, 1988, "Untersuchungen zur Druckfestigkeit von bewehrtem Beton bei gleichzeitig wirkendem Querzug," Institut fur Massivbau und Baustofftechnologie, University of Karlsruhe, Germany.

Eurocode No. 2, 1991, "Design of Concrete Structures, Part 1: General Rules and Rules for Buildings," Thomas Telford, London.

Evans, R. H., and Marathe, M. S., 1968, "Microcracking and Stress-Strain Curves for Concrete in Tension," *Materials and Structures, Research and Testing*, RILEM, Paris, V. 1, No. 1, pp. 61-64.

Fenwick, R. C., and Paulay, T., 1968, "Mechanisms of Shear Resistance of Concrete Beams," *Journal of Structural Division*, ASCE, V. 94, No. 10, pp. 2325-2350.

Gambarova, P. G., 1979, "Aggregate Interlock Role in RC Thin-Webbed Beams in Shear," *Journal of Structural Division*, ASCE, V. 113, No. 1, pp. 1-19.

Gambarova, P. G., 1981, "On Aggregate Interlock Mechanism in Reinforced Concrete Plate with Extensive Cracking," *IABSE Colloquium*, Zurich, pp. 105-134.

Gambarova, P. G., and di Prisco, M., 1991, "Interface Behaviour," *Behavior and Analysis of Reinforced Concrete Structures under Alternate Actions including Inelastic Response*, V. 1, General Models, CEB Bulletin 210, Lausanne, Switzerland.

Gopalaratnam, V. S., and Shah, S. P., 1985, "Softening Response of Plain Concrete in Direct Tension," ACI JOURNAL, *Proceedings* V. 82, No. 2, Mar.-Apr., pp. 310-323.

Gupta, P., and Collins, M. P., 1993, "Behaviour of Reinforced Concrete Members Subjected to Shear and Compression," *Report*, Department of Civil Engineering, University of Toronto, Ontario, Canada.

Hamadi, Y. D., 1976, "Force Transfer across Cracks in Concrete Structures," PhD thesis, Polytechnic of Central London.

Hamadi, Y. D., and Regan, P. E., 1980, "Behaviour in Shear of Beams with Flexural Cracks," *Magazine of Concrete Research*, V. 32, No. 1, pp. 67-77.

Hillerborg, A.; Modéer M.; and Petersson, P. E.; 1976, "Analysis of Crack Formation and Crack Growth in Concrete by Means of Fracture Mechanics and Finite Elements," *Cement and Concrete Research,* V. 6, pp. 773-782.

Hofbeck, J. A.; Ibrahim, I. A.; and Mattock, A. H., 1969, "Shear Transfer in Reinforced Concrete," ACI JOURNAL, *Proceedings* V. 66, No. 2, Feb., pp. 119-128.

Hsu, T. T. C., 1993, *Unified Theory of Reinforced Concrete*, CRC Press, Boca Raton, Fla., 336 pp.

Hsu, T. T. C.; Mau, S. T.; and Chen, B, 1987, "A Theory of Shear Transfer Strength of Reinforced Concrete," *ACI Structural Journal*, V. 84, No. 2, Mar.-Apr., pp. 149-160.

IABSE, 1991a, "IABSE—Colloquium Stuttgart 1991: Structural Concrete IABSE Report," IABSE, Zurich, 62, pp. 1-872.

IABSE, 1991b, "IABSE—Colloquium Stuttgart 1991: Structural Concrete—Summarizing Statement," *Structural Engineering Int.*, V. 1, No. 3, pp. 52-54.

Isenberg, J., 1993, "Finite Element Analysis of Reinforced Concrete Structures II," *Proceedings, International Workshop*, New York.

Jirsa, J. O.; Breen, J. E.; Bergmeister, K.; Barton, D.; Anderson, R.; and Bouadi, H., 1991, "Experimental Studies of Nodes in Strut-and-Tie Models," *IABSE Colloquium, Structural Concrete*, Zurich, pp. 525-532.

Kani, G. N. J., 1964, "The Riddle of Shear Failure and its Solution," ACI JOURNAL, *Proceedings* V. 61, No. 4, Apr., pp. 441-467.

Kani, G. N. J., 1967, "How Safe Are Our Large Reinforced Concrete Beams?" ACI JOURNAL, *Proceedings* V. 64, No. 3, Mar., pp. 128-141.

Kani, M. W.; Huggins, M. W.; and Wiltkopp, P. F., 1979, *Kani on Shear in Reinforced Concrete*, Department of Civil Engineering, University of Toronto, Ontario, Canada.

Kirmair, M., 1987, "Das Schubtragverhalten Schlanker Stahlbeton-balken—Theoretische und Experimentele Untersuchungen fur Leicht-und Normalbeton," Dissertation, University of Munich, Germany.

Kirschner, U., and Collins, M. P., 1986, "Investigating the Behaviour of Reinforced Concrete Shell Elements," *Publication* No. 86-09, Department of Civil Engineering, University of Toronto, Canada.

Kollegger, J., and Mehlhorn, G., 1988, "Biaxial Tension-Compression Tests on Reinforced Concrete Panels," *Forschungsberichte aus dem Fachgebeit Massivbau,* No. 6, Gesamthochschule Kassel, Kassel, West Germany.

Kollegger, J., and Mehlhorn, G., 1990, "Experimentelle Untersuchungen zur Bestimmung der Druckfestigkeit des gerissenen Stahlbetons bei einer Querzugbeanspruchung," *DAtStb H.413*, Beuth, Berlin.

Kupfer, H., 1964, "Erweiterung der Mörsch schen Fachwerkanalogie mit Hilfe des Prinzips vom Minimum der Formanderungsarbeit (Generalization of Mörsch's Truss Analogy Using the Principle of Minimum Strain Energy)," *Comite Euro-International du Beton, Bulletin d'Information,* No. 40, CEB, Paris, pp. 44-57.

Kupfer, H., and Bulicek, H., 1991, "Comparison of Fixed and Rotating Crack Models in Shear Design of Slender Concrete Beams," *Progress in Structural Engineering*, D. E. Grierson, et al., eds., Kluwer, Dordrecht, The Netherlands, pp. 129-138.

Kupfer, H. B., and Gerstle, K., 1973, "Behaviour of Concrete under Biaxial Stresses," *Journal of the Engineering Mechanics Division*, ASCE, V. 99, No. 4, pp. 853-866.

Kupfer, H., Mang, R., and Karavesyrouglou, M., 1983, "Failure of the Shear-Zone of R.C. and P.C. Girders—An Analysis with Consideration of Interlocking of Cracks," *Bauingenieur*, 58, pp. 143-149. (in German)

Kupfer, H., and Moosecker, W., 1979, "Beanspruchung und Verformung der Schubzone des schlanken profilierten Stahlbeton balkens," *For-schungsbeitrage fur die Baubraxis*, Wilhelm Ernst und Sohn, Berlin, pp. 225-236.

Lampert, P., and Thurlimann, B., 1971, "Ultimate Strength and Design of Reinforced Concrete Beams in Torsion and Bending," *IABSE*, No. 31-I, pp. 107-131.

Leonhardt, F., 1965, "Reducing the Shear Reinforcement in Reinforced Concrete Beams and Slabs," *Magazine of Concrete Research*, V. 17, No. 53, pp. 187-198.

Leonhardt, F., 1977, "Schub bei Stahlbeton und Spannbeton: Grundlagen der neueren Schubbemessung," BuStb 72 H.11, 270–277 und H.12, pp. 295-302.

Leonhardt, F., and Walther, R., 1966, "Wardetiger Trager," *Deutscher Ausschuss fur Stahlbeton, Bulletin* No. 178, Wilhelm Ernst und Sohn, Berlin.

Lipski, A., 1971/1972, "Poutres a ame mince en beton arme ou precon-traint," *Ann. d. Travaux Publ. d. Belgique* No. 1-2.

Loeber, P. J., 1970, "Shear Transfer by Aggregate Interlock," MS thesis, University of Canterbury, Christchurch, New Zealand.

Loov, R. E., and Patniak, A. K., 1994, "Horizontal Shear Strength of Composite Concrete Beams with a Rough Interface," *PCI J.*, V. 39, No. 1, pp. 48-109.

MacGregor, J. G., 1992, *Reinforced Concrete Mechanics and Design*, 2nd Edition, Prentice-Hall, Englewood Cliffs, N.J., 848 pp.

MacGregor, J. G., and Walters, J. R. V., 1967, "Analysis of Inclined Cracking Shear in Slender RC Beams," ACI JOURNAL, *Proceedings* V. 64, pp. 644-653.

Mallee, R., 1981, "Zum Schubtragverhalten stabformiger Stahlbeton-elemente," *DAfStbH.323*, Berlin.

Marti, P., 1980, "Zur Plastischen Berechnung von Stahlbeton (On Plastic Analysis of Reinforced Concrete)," *Report* No. 104, Institute of Structural Engineering, ETH, Zurich.

Marti, P., 1985, "Basic Tools of Reinforced Concrete Beam Design," ACI JOURNAL, *Proceedings* V. 82, No. 1, Jan.-Feb., pp. 46-56.

P. Marti, 1985, discussion of "Basic Tools of Reinforced Concrete Design," ACI JOURNAL, *Proceedings* V. 82, No. 6, Nov.-Dec., pp. 933-935.

Marti, P., 1991, "Dimensioning and Detailing," *IABSE Colloquium on Structural Concrete, IABSE Report*, Zurich, 62, pp. 411-443.

Marti, P., and Meyboom, J., 1992, "Response of Prestressed Elements to In-Plane Shear," *ACI Structural Journal*, V. 89, No. 5, Sept.-Oct., pp. 503-514.

Mattock, A. H., 1974, "Shear Tansfer in Concrete Having Reinforcement at an Angle to the Shear Plane," *Shear in Reinforced Concrete*, V. 1, SP-42, American Concrete Institute, Farmington Hills, Mich., pp. 17-42.

Mattock, A. H., 1976, "Design Proposal for Reinforced Concrete Corbels," *PCI Journal*, V. 21, No. 3, pp. 18-42.

Mattock, A. H.; Chen, K. C.; and Soongswang, K., 1976, "The Behavior of Reinforced Concrete Corbels," *PCI Journal*, V. 21, No. 2, pp. 52-77.

Mattock, A. H., and Hawkins, N. M., 1972, "Research on Shear Transfer in Reinforced Concrete.," *PCI Journal*, V. 17, No. 2, pp. 55-75.

Mattock, A. H.; Li, W. K.; and Wang, T. C., 1976, "Shear Transfer in Lightweight Reinforced Concrete," *PCI Journal*, V. 21, No. 1, pp. 20-39.

Millard, S. G., and Johnson, R. P., 1984, "Shear Transfer in Cracked Reinforced Concrete," *Magazine of Concrete Research*, V. 37, No. 130, pp. 3-15.

Millard, S. G., and Johnson, R. P., 1986, Discussion of "Shear Transfer in Cracked Reinforced Concrete," *Magazine of Concrete Research*, V. 38, No. 134, pp. 47-51.

Mitchell, D., and Collins, M. P., 1974, "Diagonal Compression Field Theory—A Rational Model for Structural Concrete in Pure Torsion," ACI JOURNAL, *Proceedings* V. 71, pp. 396-408.

Mörsch, E., 1909, *Concrete-Steel Construction*, McGraw-Hill, New York. (English translation by E. P. Goodrich).

Mörsch, E., 1920, "Der Eisenbetonbau-Seine Theorie und Anwendung (Reinforced Concrete Construction—Theory and Application)," 5th Edition, Wittwer, Stuttgart, V. 1, Part 1.

Mörsch, E., 1922, "Der Eisenbetonbau-Seine Theorie und Anwendung," 5th Edition, Wittwer, Stuttgart, V. 1, Part 2.

Muller, P., 1978, "Plastische Berechnung von Stahlbetonscheiben und-balken (Plasticity Analysis of Reinforced Concrete Wall and Beams), *Report* No. 83, Institute of Structural Engineering, ETH, Zurich.

Muttoni, A., 1990, "Applicability of the Theory of Plasticity for Dimensioning Reinforcing Concrete," PhD thesis, ETH Zurich Birkhäuser, Basel. (in German)

Nielsen, M. P., 1971, "Om jernbetonskivers styrke," *Polyteknisl Forlag*, Copenhagen (On the Strength of Reinforced Concrete Discs), *Acta Polytech. Scand.*, Ci-70, Copenhagen.

Nielsen, M. P., 1984, *Limit Analysis and Concrete Plasticity*, Prentice-Hall, Englewood Cliffs, N.J.

Nielsen, M. P., and Braestrup, N. W., 1975, "Plastic Shear Strength of Reinforced Concrete Beams," *Technical Report* 3, Bygningsstatiske Meddelelser, V. 46.

Nielsen, M. P.; Braestrup, M. W.; Jensen, B. C.; and Bach, F., 1978, "Concrete Plasticity, Beam Shear–Shear in Joints–Punching Shear," *Special Publication*, Danish Society for Structural Science and Engineering, Lyngby, Denmark.

Nissen, I., 1987, "Riβverzahnung des Betons-gegenseitige Rißuferverschie-bungen und ubertragbare Krafte," Dissertation, University of Munich.

Niwa, J., Yamada, K., Yokozawa, K., and Okamura, M., 1986, "Reevaluation of the Equation for Shear Strength of RC-Beams without Web Reinforcement," Translated from *Proceedings, Japan Society of Civil Engineering*, V. 5, No. 372, pp. 1986-1988.

Norges Byggstandardiseringsrad (NBR), 1989, "Prosjektering av betongkonstruksjoner Beregnings-og konstruksjonsregler (Concrete Structures Design Rules)," *NS 3743*, Norges Standardiseringsforbund, Oslo.

Okamura, H., and Higai, T., 1980, "Proposed Design Equation for Shear Strength of R.C. Beams without Web Reinforcement," *Proceedings, Japan Society of Civil Engineering*, 300, pp. 131-141.

Pang, X.-B. D., and Hsu, T. T. C., 1992, "Constitutive Laws of Reinforced Concrete in Shear," *Res. Rep. UHCEE 92-1*, Department of Civil and Environmental Engineering, University of Houston, Tex.

Pang, X.-B. D., and Hsu, T. T. C., 1995, "Behavior of Reinforced Concrete Membrane Elements in Shear," *ACI Structural Journal*, V. 92, No. 6, Nov.-Dec., pp. 665-679.

Pang, X.-B. D., and Hsu, T. T. C., 1996, "Fixed-Angle Softened-Truss Model for Reinforced Concrete," *ACI Structural Journal*, V. 93, No. 2, Mar.-Apr., pp. 197-207.

Park, R., and Paulay, T., 1975, *Reinforced Concrete Structures*, Wiley, New York.

Paulay, T.; Park, R.; and Phillips, M. H., 1974, "Horizontal Construction Joints in Cast-in-Place Reinforced Concrete," *Shear in Reinforced Concrete*, SP-42, American Concrete Institute, Farmington Hills, Mich., pp. 599-616.

Ramirez, J. A., 1994, "Strut-and-Tie Design of Pretensioned Members," *ACI Structural Journal*, V. 91, No. 5, Sept.-Oct., pp. 572-578.

Ramirez, J. A., and Breen, J. E., 1991, "Evaluation of a Modified Truss-Model Approach for Beams in Shear," *ACI Structural Journal,* V. 88, No. 5, Sept.-Oct., pp. 562-571.

Regan, P., 1993, "Research on Shear: A Benefit to Humanity or a Waste of Time?" *Structural Engineering*, 71(19/5), pp. 337-347.

Regan, P. E., and Braestrup, M. W., 1985, "Punching Shear in Reinforced Concrete—A State of the Art Report," *CEB* Bulletin 168, Lausanne, Switzerland.

Reineck, K.-H., 1982, "Models for the Design of Reinforced and Prestressed Concrete Members," *CEB Bulletin* 146, Paris, pp. 43-96.

Reineck, K.-H., 1989, "Theoretical Considerations and Experimental evidence on Web Compression Failures of High-Strength Concrete Beams," *CEB Bulletin* 193, Lausanne, Switzerland, pp. 61-73.

Reineck, K.-H., 1990, "Mechanical Model for the Behaviour of Reinforced Concrete Members in Shear," PhD thesis, University of Stuttgart.

Reineck, K.-H., 1991a, "Modelling of Members with Transverse Reinforcement," *IABSE Colloquium on Structural Concrete, IABSE Report,* IABSE, Zurich, 62, pp. 481-488.

Reineck, K.-H., 1991b, "Model for Structural Concrete Members without Transverse Reinforcement," *IABSE Colloquium Structural Concrete, IABSE Report*, IABSE, Zurich, 62, pp. 643-648.

Reineck, K.-H., 1991c, "Ultimate Shear Force of Structural Concrete Members without Transverse Reinforcement Derived from a Mechanical Model," *ACI Structural Journal*, V. 88, No. 5, Sept.-Oct., pp. 592-602.

Reinhardt, H. W., 1986, "The Role of Fracture Mechanics in Rational Rules for Concrete Design," *IABSE Survey S-34/86 in IABSE Per. 1/1986*, IABSE, Zurich.

Reinhardt, H. W.; Cornelissen, H. A. W.; and Hordijk, D. A., 1986, "Tensile Tests and Failure Analysis of Concrete," *Journal of Structural Enginering*, ASCE, V. 112, No. 11, pp. 2462-2477.

Remmel, G., 1994, '"Zum Zug- und Schubtragverhalten von Bauteilen aus hochfestem Beton," *DAfStb, H.444*, Beuth, Berlin.

Ritter, W., 1899, "Die bauweise hennebique," *Schweizerische Bauzei-tung*, V. 33, No. 7, pp. 59-61.

Rogowsky, D. M., and MacGregor, J. G., 1986, "Design of Reinforced Concrete Deep Beams," *Concrete International: Design and Construction*, V. 8, No. 8, Aug., pp. 49-58.

Rogowsky, D. M.; MacGregor, J. G.; and Ong, S. Y., 1986, "Tests of Reinforced Concrete Deep Beams," ACI JOURNAL, *Proceedings* V. 83, No. 4, July-Aug., pp. 614-623.

Schäfer, K.; Schelling, G.; and Kuchler, T., 1990, "Druck- und Querzug in bewehrten Betonelementen," *DAfStb H.408*, Beuth, Berlin.

Schlaich, J., and Schäfer, K., 1983, "Zur Druck-Querzug-Festigkeit des Stahlbetons," *BuStb 78*, H.3, pp. 73-78.

Schlaich, J., and Schäfer, K., 1984, "Konstrulerenim Stahlbeton-bau," *Beton-Kalender 1984*, Wilhelm Ernst und Sohn, Berlin, pp. 787-1004.

Schlaich, J.; Schäfer, I.; and Jennewein, M., 1987, "Towards a Consistent Design of Structural Concrete," *Journal of the Prestressed Concrete Institute*, V. 32, No. 3, pp. 74-150.

Shioya, T.; Iguro, M.; Nojiri, Y.; Akiayma, H.; and Okada, T., 1989, "Shear Strength of Large Reinforced Concrete Beams, Fracture Mechanics: Application to Concrete," *Fracture Mechanics: Application to Concrete*, SP-118, V. C. Li and A. P. Bažant, eds., American Concrete Institute, Farmington Hills, Mich., pp. 259-279.

Shirai, S., and Noguchi, H., 1989, "Compressive Deterioration of Cracked Concrete," *Proceedings, ASCE Structural Congress 1989: Design, Analysis, and Testing*, ASCE, New York, pp. 1-10.

Stevens, N. J.; Uzumeri, S. M.; and Collins, M. P., 1991, "Reinforced Concrete Subjected to Reversed Cyclic Shear—Experiments and Constitutive Model," *ACI Structural Journal*, V. 88, No. 2, Mar.-Apr., pp. 135-146.

"Standard Building Regulations for the Use of Reinforced Concrete," 1920, *American Standard Specification* No. 23, American Concrete Institute, Farmington Hills, Mich., pp. 283-322.

Talbot, A. N., 1909, "Tests of Reinforced Concrete Beams: Resistance to Web Stresses Series of 1907 and 1908," *Bulletin* 29, University of Illinois Engineering Experiment Station, Urbana, Ill.

Tassios, T. P., and Vintzeleou, E. N., 1986, "Concrete-to-Concrete Friction," *Journal of Structural Engineering*, ASCE, V. 113, No. 4, pp. 832-849.

Tassios, T. P., and Vintzeleou, E. N., 1987, "Concrete-to-Concrete Friction," *Journal of Structural Engineering*, ASCE, V. 113, No. 4, pp. 832-849.

Taylor, H. P. J., 1974, "The Fundamental Behaviour of Reinforced Concrete Beams in Bending and Shear," *Shear in Reinforced Concrete*, SP-42, American Concrete Institute, Farmington Hills, Mich., pp. 43-77.

Thurlimann, B.; Marti, P.; Pralong, J.; Ritz, P.; and Zimmerli, B., 1983, "Vorlesung Zum Bortbildungs kurs fur Bauingenieure," Institute fur Bautecknik und Konstruktion, ETH, Zurich.

Vecchio, F. J., and Collins, M. P., 1982, "The Response of Reinforced Concrete to In-Plane Shear and Normal Stresses," *Publication* No. 82-03, Department of Civil Engineering, University of Toronto, Canada.

Vecchio, F. J., and Collins, M. P., 1986, "The Modified Compression Field Theory for Reinforced Concrete Elements Subjected to Shear," ACI JOURNAL, *Proceedings* V. 83, No. 2, Mar.-Apr., pp. 219-231.

Vecchio, F. J., and Collins, M. P., 1988, "Predicting the Response of Reinforced Concrete Beams Subjected to Shear Using the Modified Compression Field Theory," *ACI Structural Journal*, V. 85, No. 4, July-Aug., pp. 258-268.

Vecchio, F. J., and Collins, M. P., 1993, "Compression Response of Cracked Reinforced Concrete," *Journal of Structural Engineering*, ASCE, V. 119, No. 12, pp. 3590-3610.

Vecchio, F. J.; Collins, M. P.; and Aspiotis, J., 1994, "High-Strength Concrete Elements Subjected to Shear," *ACI Structural Journal*, V. 91, No. 4, July-Aug., pp. 423-433.

Vintzeleou, E. N., and Tassios, T. P., 1986, "Mathematical Models for Dowel Action under Monotonic Conditions," *Magazine of Concrete Research*, 38, pp. 13-22.

Wagner, H., 1929, "Ebene Blechwandträger mit sehr dünnem Stegblech (Metal Beams with Very Thin Webs)," *Zeitschrift für Flugtechnik und Motorloftschiffahr*, 20 (8-12), Berlin.

Walraven, J. C., 1980, "Aggregate Interlock: A Theoretical and Experimental Analysis," PhD thesis, Delft University, The Netherlands.

Walraven, J. C., 1981, "Fundamental Analysis of Aggregate Interlock," *Journal of Structural Division*, ASCE, V. 108, pp. 2245-2270.

Walraven, J. C., 1987, "Shear in Prestressed Concrete Members," *A State-of-the-Art Report; CEB Bulletin* 180, Paris.

Walraven, J. C.; Frenay, J.; and Pruijssers, A., 1987, "Influence of Concrete Strength and Load History on the Shear Friction Capacity of Concrete Members," *PCI Journal*, V. 21, No. 1, pp. 66-84.

Walraven, J., and Lehwalter, N., 1990, "Einfluß des Maßstabs in schub-beanspruchten Bauteilen ohne Schubbewehrung," *BuStb 85*, H.9, pp. 228–232.

Walraven, J., and Lehwalter, N., 1994, "Size Effects in Short Beams Loaded in Shear," *ACI Structural Journal*, V. 91, No. 5, Sept.-Oct., pp. 585-593.

Walraven, J. C., and Reinhardt, H. W., 1981, "Concrete Mechanics Port A—Theory and Experiment on the Mechanical Behaviour of Cracks in Plain and Reinforced Concrete Subjected to Shear Loading," University of Delft, The Netherlands, 26(1A).

Withey, M. O., 1907, "Tests of Plain and Reinforced Concrete Series of 1906," *Bulletin of the University of Wisconsin, Engineering Series*, V. 4, No. 1, pp. 1-66.

Withey, M. O., 1908, "Tests of Plain and Reinforced Concrete Series of 1907," *Bulletin of the University of Wisconsin, Engineering Series*, V. 4, No. 2, pp. 1-66.

Woo, K., and White, R. N., 1991 "Initiation of Shear Cracking in Reinforced Concrete Beams with No Web Reinforcement," *ACI Structural Journal*, V. 88, No. 3, May-June, pp. 301-308.

Yoon, Y. S.; Cook, W. D.; and Mitchell, D., 1996, "Minimum Shear Reinforcement in Normal, Medium and High-Strength Concrete Beams," *ACI Structural Journal*, V. 93, No. 5, Sept.-Oct., pp. 576-584.

Zhang, L. X., 1995, "Constitutive Laws of Reinforced Membrane Elements with High-Strength Concrete," PhD dissertation, Department of Civil and Environmental Engineering, University of Houston, Tex.

Zsutty, T. C., 1971, "Shear Strength Prediction for Separate Categories of Simple Beams Tests," ACI JOURNAL, *Proceedings* V. 68, No. 2, Feb., pp. 138-143.

ACI 445.1R-12

Report on Torsion in Structural Concrete

Reported by Joint ACI-ASCE Committee 445

Daniel A. Kuchma, Chair

Robert W. Barnes Jr., Secretary

Perry Adebar
Neal S. Anderson
Robert B. Anderson
Mark A. Ascheim
Oguzhan Bayrak
Zdenek P. Bažant
Abdeldjelil Belarbi[*†]
Evan C. Bentz
John F. Bonacci

Hakim Bouadi
Michael D. Brown
Michael P. Collins
David Darwin
Walter H. Dilger[*]
Marc O. Eberhard
Catherine E. French
Robert J. Frosch
Gary G. Greene[*]

Neil M. Hawkins
Thomas T. C. Hsu[*]
Gary J. Klein
Zhongguo John Ma
Adolfo B. Matamoros
Denis Mitchell
Yi-Lung Mo[*]
Lawrence C. Novak
Carlos E. Ospina

Stavroula J. Pantazopoulou
Maria A. Polak
Julio A. Ramirez
Karl-Heinz Reineck
David H. Sanders[*]
Raj Valluvan
James K. Wight

[*]Subcommittee members who produced this report.

[†]Subcomittee Chair.

The committee would like to thank the following individuals for their contribution to this report: Mohammad Ali, Neal S. Anderson, Shri Bhide, Michael D. Collins, Maria Cristina Vidigal de Lima, Leonard Elfgren, Christos Karayannis, Liang-Jenq Leu, Mohammad Mansour, Basile Rabbat, Khaldoun Rahal, and Paul Zia.

A clear understanding of the effects of torsion on concrete members is essential to the safe, economical design of reinforced and prestressed concrete members. This report begins with a brief and systematic summary of the 180-year history of torsion of structural concrete members, new and updated theories and their applications, and a historical overview outlining the development of research on torsion of structural concrete members. Historical theories and truss models include classical theories of Navier, Saint-Venant, and Bredt; the three-dimensional (3-D) space truss of Rausch; the equilibrium (plasticity) truss model of Nielson as well as Lampert and Thürlimann; the compression field theory (CFT) by Collins and Mitchell; and the softened truss model (STM) by Hsu and Mo.

This report emphasizes that it is essential to the analysis of torsion in reinforced concrete that members should: 1) satisfy the equilibrium condition (Mohr's stress circle); 2) obey the compatibility condition (Mohr's strain circle); and 3) establish the constitutive relationships of materials such as the "softened" stress-strain relationship of concrete and "smeared" stress-strain relationship of steel bars.

The behavior of members subjected to torsion combined with bending moment, axial load, and shear is discussed. This report deals with design issues, including compatibility torsion, spandrel beams, torsional limit design, open sections, and size effects. The final two chapters are devoted to the detailing requirements of transverse and longitudinal reinforcement in torsional members with detailed, step-by-step design examples for two beams under torsion using ACI (ACI 318-11), European (EC2-04), and Canadian Standards Association (CSA-A23.3-04) standards. Two design examples are given to illustrate the steps involved in torsion design. Design Example 1 is a rectangular reinforced concrete beam under pure torsion, and Design Example 2 is a prestressed concrete girder under combined torsion, shear, and flexure.

Keywords: combined action (loading); compatibility torsion; compression field theory; equilibrium torsion; interaction diagrams; prestressed concrete; reinforced concrete; shear flow zone; skew bending; softened truss model; spandrel beams; struts; torsion detailing; torsion redistribution; warping.

CONTENTS

ACI 445.1R-12 was adopted and published April 2013.

CHAPTER 1—INTRODUCTION AND SCOPE

1.1—Introduction

Accounting for the effects of torsion is essential to the safe design of structural concrete members, requiring a full knowledge of the effects of torsion and a sound understanding of the analytical models that can easily be used for design. For over three decades, considerable research has been conducted on the behavior of reinforced concrete members under pure torsion and torsion combined with other loadings. Likewise, analytical models have been developed based on the truss model concept. Several of these models were developed to predict the full load history of a member, whereas others are simplified and used only to calculate torsional strength. Many models developed since the 1980s account for softening of diagonally cracked concrete.

This report reviews and summarizes the evolution of torsion design provisions in ACI 318, followed with a summary of the present state of knowledge on torsion for design and analysis of structural concrete beam-type members. Despite a vast amount of research in torsion, provisions of torsion design did not appear in ACI 318 until 1971 (ACI 318-71), although ACI 318-63 included a simple clause regarding detailing for torsion. Code provisions in 1971 were based on Portland Cement Association (PCA) tests (Hsu 1968b).

These provisions were applicable only to rectangular nonprestressed concrete members. In 1995, ACI 318-95 adopted an approach based on a thin-tube, space truss model previously used in the Canadian Standards Association (CSA-A23.3-77) code and the Comité Euro-International du Béton (CEB)-FIP code (1978). This model permitted treatment of sections with arbitrary shape and prestressed concrete (Ghoneim and MacGregor 1993; MacGregor and Ghoneim 1995). The ACI 318-02 code extended the application of the (ACI 318) 1995 torsion provisions to include prestressed hollow sections. ACI 318 allows the use of alternative design methods for torsional members with a cross section aspect ratio of 3 or greater, like the procedures of pre-1995 editions of ACI 318 or the Prestressed Concrete Institute (PCI) method (Zia and Hsu 1978).

This report reviews and summarizes the present state of knowledge on torsion and reviews their use as a framework for design and analysis of structural concrete beam-type members. Chapter 3 presents a historical background outlining the development of research on torsion of structural concrete members. The general behavior of reinforced and prestressed concrete members under pure torsion is discussed in Chapter 4. In Chapter 5, the compression field theory (CFT) and softened truss model (STM) are presented in detail. Chapter 5 also includes a description of two graphical methods (Rahal 2000a,b; Leu and Lee 2000). The behavior of members subjected to torsion combined with shear, flexure, and axial load is discussed in Chapter 6. Chapter 7 introduces additional design issues related to torsion, such as precast spandrel beams, torsion limit design, size effect, open sections, and torsional moment distribution. Detailing of torsional members is described in Chapter 8. Chapter 9 covers detailed design examples of several beams subjected to torsion using ACI 318, EC2-04, and CSA-A23.3-04 design equations, and additional graphical design methods reported by researchers.

1.2—Scope

Theories presented in this report were developed and verified for building members of typical size. For application to large-scale members, size effects should be considered. They could present a serious safety issue when using the shear strength equations provided in the design standard, which cannot take into account the shear strength reduction in large-scale members caused by loss of aggregate interlock behavior. Experimental information on large-scale torsional members is lacking.

CHAPTER 2—NOTATION AND DEFINITIONS

The material presented is a summary of research carried out worldwide and spanning more than four decades, making unification of the symbols and notations used by the various researchers and design codes a challenge. In some cases, mostly for graphs and figures, the notation is kept as originally published.

2.1—Notation

a = moment arm for bending, mm (in.)
a_c = geometric property index
a_o = depth of equivalent rectangular stress block in concrete strut of torsional member, mm (in.)
A = area of yield surface, mm^2 (in.^{2})
A_{cp} = area enclosed by outside perimeter of concrete cross section, mm^2 (in.^{2})
A_ℓ = total area of longitudinal reinforcement to resist torsion, mm^2 (in.^{2})
A_o = gross area enclosed by shear flow path, mm^2 (in.^{2}) (noted as A_{tb} in Eq. (7.2.6))
A_{oh} = area enclosed by centerline of outermost closed transverse torsional reinforcement, mm^2 (in.^{2})
A_{ps} = area of prestressing reinforcement in flexural tension zone, mm^2 (in.^{2})
A_s = area of nonprestressed longitudinal tension reinforcement, mm^2 (in.^{2})
A_s' = area of longitudinal compression reinforcement, mm^2 (in.^{2})
A_t = area of one leg of a closed stirrup resisting torsion within spacing s, mm^2 (in.^{2}) (noted as A_{tb} in Eq. (7.2.6))
b = width of compression face of member, mm (in.)
b_c = width of stirrups, mm (in.)
B = integral of T_w
C = cross-sectional constant to define torsional properties of a beam
d_v = distance between top and bottom longitudinal reinforcement, mm (in.)
D = cross-sectional depth used in fracture mechanics calculations, mm (in.)
D_0 = size effect constant for computing σ_N for plain concrete section
D_1 = normalized constant to represent characteristic structural dimensions used in fracture mechanics calculations
D_b = size effect constant for computing σ_N for reinforced concrete section
D_c = total energy dissipated on discontinuous concrete yield surface
D_s = total energy dissipated by reinforcement
e = moment arm for torsion, mm (in.)
E_c = modulus of elasticity of concrete, MPa (psi)
E_{ps} = modulus of elasticity of prestressed reinforcement in flexural tension zone, MPa (psi)
E_{ps}' = tangential modulus of Ramberg-Osgood curve at zero load MPa (psi)
E_s = modulus of elasticity of reinforcement and structural steel, MPa (psi)
EJ_w = rigidity of beam under warping torque, N·m^2 (lb-in.^{2})
f_c' = characteristic concrete cylinder compressive strength, MPa (psi)
f_c^* = concrete effective (plastic) compressive stress, MPa (psi)
f_{ck} = characteristic compressive strength of concrete, MPa (psi); $f_{ck} = f_{cm} - 8$ MPa ($f_{ck} = f_{cm} - 1200$ psi)
f_{cm} = mean compressive strength of concrete, MPa (psi)
f_d = diagonal concrete stress, MPa (psi)
f_{ds} = diagonal concrete stress corresponding to strain ε_{ds}, MPa (psi)
f_ℓ = reinforcement stress in ℓ direction, MPa (psi)
$f_{\ell p}$ = prestressing reinforcement stress in the ℓ direction, MPa (psi)
$f_{\ell y}$ = specified yield strength of longitudinal reinforcement, MPa (psi)
f_p = stress in prestressing reinforcement; f_p becomes $f_{\ell p}$ or f_{tp} when applied to longitudinal and transverse reinforcement, respectively, MPa (psi)
$f_{p0.1}$ = characteristic yield strength of prestressing reinforcing strands, MPa (psi); $f_{p0.1} = 0.9f_u$
f_{pc} = compressive stress in concrete due to prestress, MPa (psi)
f_{pk} = characteristic tensile strength of prestressing reinforcing strands, MPa (psi); $f_{pk} = f_{pu}$
f_{po} = effective prestress after losses in prestressing reinforcement, MPa (psi)
f_{pu} = specified tensile strength of prestressing reinforcement, MPa (psi)
$f_{p,ud}$ = design ultimate strength of prestressing reinforcing strands, MPa (psi); $f_{p,ud} = f_{pk}/\gamma_s$ ($\gamma_s = 1.15$)
f_r = modulus of rupture of concrete, MPa (psi)
f_t = reinforcement stress in t direction, MPa (psi)
f_t' = uniaxial tensile strength of concrete, MPa (psi)
f_t^* = concrete effective (plastic) tensile stress, MPa (psi)
f_{tp} = prestressing reinforcement stress in t direction, MPa (psi)

f_{ty} = specified yield strength of transverse reinforcement, MPa (psi)
f_y = specified yield strength of reinforcement, MPa (psi)
f_{yd} = design yield strength reinforcing steel, MPa (psi); $f_{yd} = f_y/\gamma_s$ ($\gamma_s = 1.15$)
$f_{y\ell}$ = yield strength of the torsional longitudinal reinforcement, MPa (psi)
f_{yv} = torsional hoop yield strength reinforcement, MPa (psi)
G = shear modulus, MPa (psi)
h = overall thickness or height of a member, mm (in.)
H_o = horizontal force in radial direction, N (lb) (Chapter 7)
I_p = polar moment of inertia, mm^4 (in.^{4})
k_1 = ratio of average stress to peak stress
K = value from Mohr-Coulomb yield criterion
K_f = flexural stiffness of floor beams, N·m^2 (lb-in.^{2})
K_{ts} = torsional stiffness of spandrel beam, N·m/rad (in.-lb/rad)
ℓ = span length of beam, mm (in.)
ℓ_f = length of flexural beam, mm (in.)
ℓ_q = width of shear flow q along top wall (Fig. 4.2(a) and (b)), mm (in.)
m = ratio of effective (plastic) compressive stress to effective (plastic) tensile stress of concrete
M = applied flexural moment at section, N·m (in.-lb)
M_o = pure flexural strength of section, N·m (in.-lb)
n = integer value
n_R = number of redundants
n_V = coefficient describing an under-reinforced, partially under-reinforced, or completely over-reinforced section
N = applied axial load at section, N (lb)
N_o = pure axial strength of section, N (lb)
p_h = perimeter of centerline of outermost closed transverse torsional reinforcement, mm (in.)
p_o = perimeter of outer concrete cross section, mm (in.) (sometimes noted as p_{cp})
P = applied concentrated load, N (lb)
q = shear flow, N/m (lb/in.)
r = ratio of top-to-bottom yield forces of the longitudinal reinforcement
r = size effect constant for computing σ_N
R = shape parameter used in Ramberg-Osgood
s = center-to-center spacing of longitudinal and transverse reinforcements, mm (in.)
s_1 = center-to-center spacing of longitudinal reinforcement, mm (in.)
s_t = center-to-center spacing of transverse reinforcement, mm (in.)
t = wall thickness of hollow section, mm (in.)
t_d = thickness of shear flow zone, mm (in.)
T = applied torsional moment at section, N·m (in.-lb)
T_c = nominal torsional strength provided by concrete, N·m (in.-lb)
T_{cr} = torsional cracking resistance of cross section, N·m (in.-lb)
T_f = applied torsional moment, N·m (in.-lb) (Chapter 9)
T_{max} = maximum torsional moment, N·m (in.-lb) (Chapter 7)
T_n = nominal torsional moment strength, N·m (in.-lb)
T_o = pure torsional strength of section, N·m (in.-lb)
T_s = nominal torsional strength provided by reinforcement, N·m (in.-lb)
T_u = factored torsional moment at section, N·m (in.-lb)
T_w = warping torsional moment, N·m (in.-lb)
T_{xu} = factored balanced torsional strength, N·m (in.-lb)
T_{xub} = balanced torsional strength, N·m (in.-lb)
$\overline{T}_{xub}$ = nondimensional balanced torsional strength, N·m (in.-lb)
v = shearing stress due to shear, MPa (psi)
v^* = plastic flow rate (Chapter 7)
v_u = ultimate shear stress, MPa (psi)
V = applied shear force at section, N (lb)
V_c = nominal shear strength provided by concrete, N (lb)
V_o = pure shear strength of section, N (lb)
V_u = factored shear force at section, N (lb)
w = ultimate distributed load on helical stair, N/m (lb/ft) (Chapter 7)
W = external work, N/m (lb/ft)
x = shorter overall dimension of rectangular part of cross section, mm (in.)
x_1 = distance section centroid and an infinitesimally small area of yield surface, mm (in.)
y = longer overall dimension of rectangular part of cross section, mm (in.)
z = distance along axis of beam, mm (in.)
α, β = Saint-Venant's coefficients for homogeneous torsional section
α^*, β^* = rotational angles in beam subjected to torsion (Chapter 7)
α_1 = stress block factor given as ratio of f_d to f_c' (Chapter 5)
β = factor relating effect of longitudinal strain on shear strength of concrete (American Association of State Highway and Transportation Officials (AASHTO) LRFD (general message)
β_1 = factor relating depth of equivalent rectangular compressive stress block to neutral axis depth; also, block factor given as ratio of a_o to t_d (Fig. 4.5)
γ_1 = angle along helical stair (in plan) at which maximum torsional moment is assumed to occur
γ_2 = angle along helical stair (in plan) at which vertical moment is assumed to be zero
$\gamma_{\ell t}$ = shear strain
ε_d = strain in d direction
ε_{dec} = strain in prestressing reinforcement at decompression of concrete
ε_{ds} = maximum strain at concrete strut surface (Fig. 4.3)
ε_h = strain in hoop direction ε_ℓ
$\varepsilon_{\ell y}$ = yield strain in ℓ direction
ε_o = strain at peak compressive stress f_c' in concrete
ε_p = peak strain in concrete
ε_r = strain in r direction
ε_s = strain in nonprestressed reinforcement; ε_s becomes ε_ℓ or ε_t when applied to longitudinal or transverse reinforcement, respectively
ε_t = strain in t direction
ε_{ty} = yield strain in t direction

ε_{uk} = characteristic total elongation of reinforcing steel at ultimate load
ε_x = longitudinal strain at midheight of concrete section
ζ = softening coefficient of concrete strut
η_ℓ = normalized reinforcement ratio of longitudinal reinforcement
η_{lb} = balanced normalized reinforcement ratio of longitudinal reinforcement
η_t = normalized reinforcement ratio of transverse steel reinforcement
η_{tb} = balanced normalized reinforcement ratio of transverse steel reinforcement
θ = angle between axis of strut, compression diagonal, or compression field and tension chord of the member; also, the angle between ℓ-t direction/axis and d-r direction/axis, radians
ξ = coefficient equal to 1 for rectangular sections and to $\pi/4$ for circular cross sections; ξ can be taken as unity for all shapes of cross sections with only negligible loss of accuracy for A_o and p_o
ρ_ℓ = reinforcement ratio in ℓ direction
$\rho_{\ell p}$ = prestressing reinforcement ratio in ℓ direction
ρ_t = reinforcement ratio in t direction
ρ_{tp} = prestressing reinforcement ratio in t direction
σ = compressive stress acting in combination with torsional moment, psi (MPa)
σ_0 = nominal torsional strength according to the current code specifications based on plastic limit analysis, MPa (psi)
σ_d = principal stress in d direction for concrete struts, MPa (psi)
σ_ℓ = normal stress in longitudinal direction for reinforced concrete, MPa (psi)
σ_{max} = maximum principal tensile stress, MPa (psi)
σ_N = nominal strength of structure, MPa (psi)
σ_r = principal stress in r direction for the concrete struts, MPa (psi)
σ_t = normal stress in the transverse direction for reinforced concrete, MPa (psi)
σ_∞ = strength of plain beams according to elastic analysis with maximum stress limited by material strength, MPa (psi)
τ = shearing stress due to torsion and shear, MPa (psi)
τ_{max} = maximum shear stress, MPa (psi)
$\tau_{\ell t}$ = applied shear stress in ℓ-t coordinate for reinforced concrete, MPa (psi)
ν = uniform plastic effectiveness factor (Chapter 7)
ν_c = plastic effectiveness factor for compression (Chapter 7)
ν_t = plastic effectiveness factor for tension (Chapter 7)
φ = friction angle
ϕ = strength reduction factor
ϕ_c = strength reduction factor for concrete (0.65 for cast-in-place, 0.70 for precast concrete)
ϕ_p = strength reduction factor for prestressing tendons (0.90)
ϕ_s = strength reduction factor for nonprestressed reinforcing bars (0.85)
Φ = angle of twist in torsional beam, radians/m (radians/in.)
Φ' = second derivative of rotation with respect to beam's axis z
Φ'' = third derivative of rotation with respect to beam's axis z
Ψ = bending curvature of concrete strut
ω_ℓ = reinforcement index in ℓ direction
ω_s = functional indicator of an index of reinforcement
$\omega_{s\ell}$ = reinforcement ratio index
ω_t = reinforcement index in t direction

2.2—Definitions

ACI provides a comprehensive list of definitions through an online resource, "ACI Concrete Terminology," http://terminology.concrete.org.

CHAPTER 3—HISTORICAL OVERVIEW OF TORSION THEORIES AND THEORETICAL MODELS

3.1—Navier's theory

A theory for torsion of elastic homogeneous members was first developed by C. L. Navier (1826) for circular cross sections. His theory, which was based on equilibrium conditions, compatibility conditions, and a linear stress-strain relationship like Hooke's Law, has guided the development of various theories about the behavior of reinforced concrete members subjected to torsion after cracking.

3.2—Thin-tube theory

Navier's torsion theory for members of circular sections was followed by Saint-Venant's (1856) solution for rectangular sections. Saint-Venant's torsional constants considered warping of rectangular cross sections. According to Saint-Venant's circulatory shear flow theory, the most efficient cross section to resist torsion is a thin tube. Bredt (1896) was able to derive simple equations for thin tubes. His thin-tube theory states that the shear stress multiplied by wall thickness has a constant value around the perimeter and that this shear flow is found by dividing the torsion by twice the area enclosed by the shear flow path. Bredt's theory has served as the basis for modern theories of cracked reinforced concrete members subjected to Saint-Venant torsion.

3.2.1 *Two- and three-dimensional plane truss models*—The first theoretical models for shear in cracked reinforced concrete members date back to the turn of the century when Ritter (1899) and Mörsch (1902) formulated the two-dimensional (2-D) plane truss model concept, where reinforced concrete members were modeled as an assembly of two types of linear elements—concrete struts and reinforcement ties. The axis of concrete struts in the model was assumed to be inclined at 45 degrees to longitudinal members, and shear strength was assumed to be controlled by the yielding of transverse reinforcement ties. By extending the 2-D plane truss model, Rausch (1929) developed a three-dimensional (3-D) space truss model for torsion that consisted of longitudinal and hoop reinforcement-resisting tension and concrete struts-resisting compression. He also assumed that the shear flow path would follow the centerline of the hoop reinforcement.

3.2.2 *Skew-bending and space truss theories*—In later years, research on torsion followed two theoretical tracks—skew-bending and space truss. Lessig (1959) first proposed a skew-bending theory for reinforced concrete members with two modes of failure, Mode 1 and Mode 2, as explained in 3.3.5. The skew-bending theory used only equilibrium equations and assumed that all reinforcement yielded before failure. Lessig's research was followed by the skew-bending theory of Walsh et al. (1966) and Collins et al. (1968a,b), who proposed a third failure mode, Mode 3, and used all three modes to derive nondimensional torsion-bending moment interaction equations (Walsh et al. 1967), as described in 3.3.5. Based on three modes of failure, a nondimensional interaction surface of torsion, shear, and flexure was derived by Elfgren (1972a,b) and Elfgren et al. (1974a,b). Rausch's space truss theory for torsion was generalized by Lampert and Thürlimann (1969, 1971), who showed how the angle of inclination of the compression diagonals at failure could be determined from equilibrium if both the hoops and longitudinal reinforcement were assumed to yield. Lampert and Collins (1972) showed that predictions of skew-bending and space truss theories were in close agreement.

3.2.3 *Compression field theory (CFT)*—The truss model with linear elements developed by Rausch was replaced in the 1960s by a new type of truss model with membrane elements that were subjected to in-plane normal and shearing stresses. In determining the torsional strength of members where some reinforcement does not yield, consider compatibility conditions. Such conditions were introduced by Baumann (1972) for shear and by Collins (1973) for torsion. Mitchell and Collins (1974) incorporated compatibility conditions in their CFT, which also relied on equilibrium equations and nonlinear material models for concrete and reinforcement. Unlike previous models, the CFT calculates cracked member torsional behavior up to the peak torque. A compatibility condition derived by minimizing the strain energy in the system is used to calculate the angle of inclination in the truss model struts.

3.2.4 *Softened truss model (STM)*—In 1985, Hsu and Mo (1985a,b,c) proposed the STM by softening the concrete stress-strain curve. All of the aforementioned models satisfy Navier's theory. Earlier models overestimated test strengths (Hsu 1968c), whereas the CFT, which uses spalling of concrete cover, and STM, which uses softening of concrete, have been shown to predict test results accurately (McMullen and El-Degwy 1985).

3.3—Historical development of theories for reinforced concrete members subjected to torsion

3.3.1 *General*—Section 3.1 summarizes the historical models developed to describe reinforced concrete members subjected to torsion, covering almost two centuries of research from 1826 to the early twenty-first century (2007). Classical theories include Navier (1826), Saint-Venant (1856), Bredt (1896), and Bach (1911). This review shows that the three principles of mechanics of materials (equilibrium, compatibility conditions, and materials stress-strain relationships) have been the basis for research in torsion of reinforced concrete members. (Equation notation in this section are provided in Chapter 2.)

Fig. 3.3.2—ACI Committee 438—Torsion, Mexico City, October 1976: Tom Hsu, Lennart Elfgren, Phil Ferguson, Art McMullen, Emory Kemp, Gordon Fisher, Paul Zia, and Michael Collins.

3.3.2 *Twentieth century*—In the first 60 years of the twentieth century, progress in reinforced concrete theories was made primarily on flexure. Early flexural theories for reinforced concrete assumed plane sections remained plane and stress-strain relationships of concrete and reinforcement were linear. Equilibrium conditions for longitudinal stresses were used to determine the location of the neutral axis and stresses in concrete and reinforcement caused by the moment. The contribution of concrete tensile stresses was disregarded if concrete cracking was expected. Later flexural theories accounted for the nonlinear stress-strain response of the concrete and steel reinforcement so that the complete moment-curvature relationship for a section could be predicted. In terms of shear and torsion research, a significant achievement was made with the development of truss models (Ritter 1899; Mörsch 1902; Rausch 1929). Research in torsion of reinforced concrete underwent significant advances during the last 40 years of the twentieth century. Two theories were developed—skew-bending and truss models with membrane elements. Skew-bending includes the theories of Lessig (1959), Yudin (1962), Collins et al. (1968a), Hsu (1968a), and Elfgren (1972a,b). Truss models include the theories of Nielsen (1967), Lampert and Thürlimann (1968, 1969), Collins (1973), Mitchell and Collins (1974), Elfgren et al. (1974a,b), Collins and Mitchell (1980), and Hsu and Mo (1985a,b,c). Several researchers involved in the development of theories for torsion were members of ACI Committee 438 for Torsion, which is now the Joint ACI-ASCE Committee 445 for Shear and Torsion (Fig. 3.3.2).

Development of these modern truss models was based on the same three principles of mechanics, which, in terms of torsion and shear, include the softened stress-strain relationship of concrete.

3.3.3 *Classical torsion theory for homogeneous members*—Navier (1826) derived a theory for torsion of homogeneous elastic members with circular cross sections. His theory is based on the three principles of mechanics of materials:

equilibrium, compatibility conditions, and Hooke's Law. Navier's work also includes the linear theory for flexure. His book is recognized as the first on the mechanics of materials The three principles of the mechanics of materials have become well known as Navier's theory. His theory defines the torsional rigidity for circular sections as GI_p. By extending the formulas for the polar moment of inertia of circular sections to square sections, Navier found that the calculated strength for specimens tested by Duleau (1820) overestimated measured values by approximately 20 percent. He acknowledged that "the formulae for square members do not depict as accurately the behavior as those for circular members." This inconsistency was explained three decades later by Saint-Venant (1856), who recognized that Navier's polar moment of inertia could not reflect the warping deformation of rectangular cross sections. To obtain the correct solution, Saint-Venant developed the semi-inverse method to solve all 15 differential and algebraic equations in the theory of elasticity developed by Cauchy (1828). By satisfying equilibrium, compatibility, and Hooke's Law at each differential element of a member, Saint-Venant developed a solution that considered the warping displacements of rectangular cross sections. In Saint-Venant's rigorous derivation, torsional rigidity is defined as GC. The torsional constant C is taken as $\beta x^3 y$, where the coefficient β is a function of the ratio y/x and varies between 0.141 (y/x = 1) and 0.333 ($y/x = \infty$). The maximum shear stress τ_{max} occurs on the outside face of the rectangular section at midpoint of each long side and is equal to $T/\alpha x^2 y$, where α is a coefficient that varies between 0.208 (y/x = 1) and 0.333 ($y/x = \infty$). The relationship between α and β is $\alpha = \beta/k$, where

$$k = 1 - \frac{8}{\pi^2} \sum_{n=1,3,5,\ldots}^{\infty} \frac{1}{n^2 \cosh \frac{n\pi y}{2x}}$$

According to Saint-Venant's circulatory shear flow pattern, the maximum shear stresses occur at the outer periphery of a cross section, and the most efficient cross section to resist torsion is a thin tube, as shown in Fig. 3.3.3.

Bredt (1896) was able to derive a simple equilibrium equation for thin tubes by assuming the entire tube to be uniformly and fully stressed

$$q = \frac{T}{2A_o} \qquad (3.3.3)$$

The area A_o is formed by sweeping the lever arm (symbol a in Fig. 3.3.3) around the axis of twist, a term later called the "lever arm area" (Hsu 1988). The torsional constant C for thin tubes with constant thickness was also simplified to $C = 4A_o^2 t/p_o$. Bach's formula (1911) is a simplification of Saint-Venant's theory for thin-walled open sections, such as T, L, and I sections. Because the coefficient β for each rectangular component of such sections can be approximated as 1/3, the torsional constant for the entire section can be taken as the sum (Σ) of the components ($C = 1/3\Sigma x^3 y$). The theories of

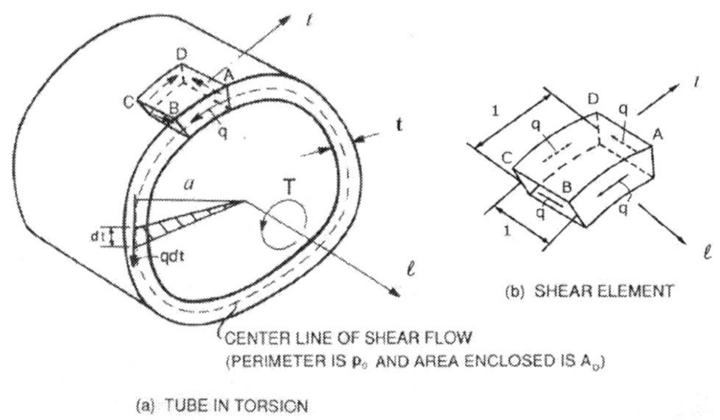

Fig. 3.3.3—Torsion of thin tube and lever arm area A_o *(Hsu 1993 after Bredt 1896).*

Navier, Saint-Venant, Bredt, and Bach are applicable to reinforced concrete beams before cracking. They also laid the foundation for developing theories to calculate the behavior of cracked reinforced concrete members subjected to torsion.

3.3.4 *Theories for reinforced concrete under flexure, shear, and torsion*—Reinforced concrete was first developed in 1867 when Joseph Monier obtained a patent for reinforcing his concrete flowerpots with wrought iron wires. The concept of using steel reinforcement to overcome the weakness of concrete in tension was quickly adapted to buildings and bridges, making the use of reinforced concrete for construction a widely accepted application in the last quarter of the nineteenth century. Such growth in applications gave rise to the demand for theories to analyze and design reinforced concrete structures.

3.3.4.1 *Flexure theory*—As reported by Delhumeau (1999), the first flexure theory to emerge was the linear flexure theory developed by Hennebique's firm near the end of the nineteenth century. This theory served as the basis for the allowable stress design method addressed in the first ACI code [National Association of Cement Users (NACU) 1910]. Theories for nonlinear flexure occupied the attention of researchers until 1963 when strength design was incorporated in ACI 318 (ACI 318-63). Both linear and nonlinear theories for flexure satisfy the three principles of the mechanics of materials: equilibrium of parallel coplanar forces, Bernoulli's linear strain compatibility, and the constitutive laws of materials. A linear stress-strain curve of reinforcement and concrete is used for the linear flexure theory and a nonlinear stress-strain curve of concrete is used for the nonlinear flexure theory.

3.3.4.2 *Torsion and shear theory*—Following the adoption of flexural strength design in ACI 318-63, the attention of researchers turned to more complex problems of torsion and shear in beams. Computer analysis of structures was becoming available, making it feasible for design engineers to compute the magnitude of torsions in their buildings. In addition, concrete box-girder bridges, which were becoming a competitive bridge type, needed to be designed for torsion. Shear is essentially a two-dimensional problem, requiring an understanding of the interaction of two principal stresses and strains in a membrane element. Torsion is complicated because it is a three-dimensional problem involving

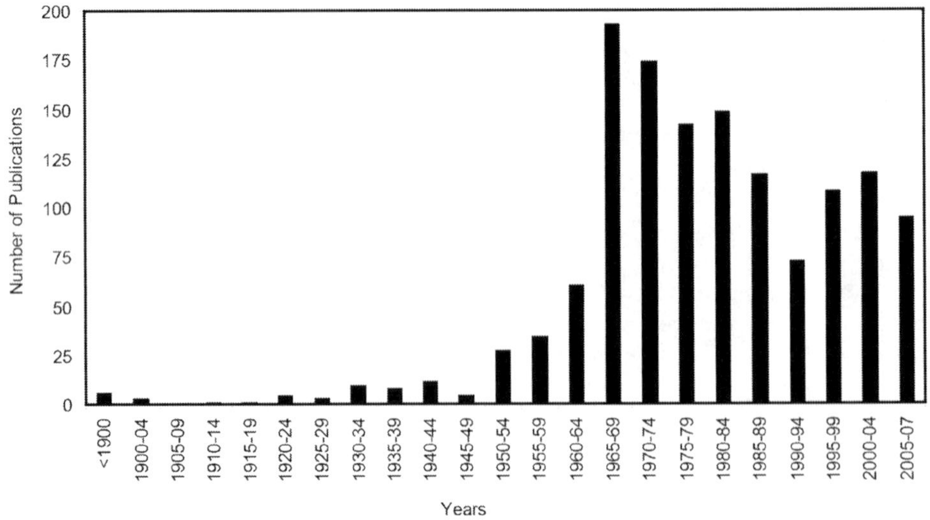

Fig. 3.3.4.2—Number of published papers on torsion from 1900 to 2007 (plot prepared by A. Belarbi).

the shear problem of membrane elements in a tube and the warping of tube walls that cause flexure in concrete struts. Figure 3.3.4.2 shows that the number of globally published papers on torsion began to surge around 1960 and peaked around 1970. The principles of equilibrium, compatibility conditions, and materials stress-strain relationships that were needed to solve torsion problems in reinforced concrete members were primarily developed between 1960 and 1985 (Lampert and Thürlimann 1971; Lampert and Collins 1972; Elfgren et al. 1974a,b; Collins and Mitchell 1980; Hsu and Mo 1983). Theories and tests produced before 1980 are summarized in detail by Hsu (1984).

By 1985, researchers solved the basic problems of reinforced concrete design by applying Navier's theory. Further research was necessary to refine the constitutive laws of materials for torsion and shear. The experimental work needed to generate new advancements is tedious and requires highly sophisticated testing equipment. Only two universities in North America—the University of Toronto and University of Houston—are capable of studying the behavior of reinforced concrete shell and panel elements subjected to in-plane shear and normal stresses. The study of softened concrete in shear elements, which has been the subject of extensive research worldwide in the last three decades, continues to be a major research topic. The need for larger and more complex specimens has increased ongoing work at both universities. The high cost of experimental research needed for new developments in torsion imposes a limiting constraint on new research. For example, studying the behavior of full-scale girders with open sections that involve Saint-Venant and warping torsion is expensive. The future of torsion research is largely tied to available equipment or to the combined efforts of many institutions, or both.

3.3.5 *Space truss model using struts and ties*—The first theory for shear design of reinforced concrete was developed at the turn of the twentieth century when Ritter (1899) and Mörsch (1902) formulated the concept of plane trusses with struts and ties. They modeled a reinforced concrete member as a truss with two types of linear elements: struts made out of concrete and ties made out of steel reinforcement. The Ritter and Mörsch model represents the struts and ties as lines without cross-sectional dimensions, where forces satisfy equilibrium at points of intersection—a model with the advantage of conceptual clarity. Extending the 2-D plane truss model to a 3-D space truss model, Rausch (1929) developed a theory for torsion of reinforced concrete. Rausch's space truss model, as shown in Fig. 3.3.5a, is made up of 45-degree diagonal concrete struts, longitudinal reinforcing bars, and hoop reinforcing bars connected at the joints by hinges. Torsional moment is carried by the concrete struts in axial compression (dotted lines), and by the straight reinforcing bars in axial tension (solid lines) in the longitudinal (horizontal) and lateral (hoop) directions. Equilibrium of the joints in the longitudinal, lateral, and radial directions requires that the forces in the longitudinal bars (X), in the hoop bars (Q), and in the inclined struts (D) should be evenly distributed among all cells and joints. To satisfy equilibrium, the relationship between these forces should be $X = Q = D/\sqrt{2}$. As shown in Fig. 3.3.5a, the series of hoop forces Q at the joints constitute a shear flow $q = Q/s$. Using Bredt's lever arm area concept, T can be related to q (or Q/s) by $2A_o$, as expressed by Bredt's equation (Eq. (3.3.3)). The term A_o refers to the area enclosed by a series of straight lines connecting joints of the cross section.

Assuming that ultimate torque is reached when the forces in the transverse reinforcement reach the yield stress, then $q = Q/s = A_t f_{ty}/s_t$ and Eq. (3.3.3) becomes

$$T_n = 2A_{oh}\frac{A_t f_{ty}}{s_t} \tag{3.3.5}$$

Although the space truss model has the advantage of conceptual clarity in terms of simple assemblage of compression struts and tension ties, Rausch's equation (Eq.

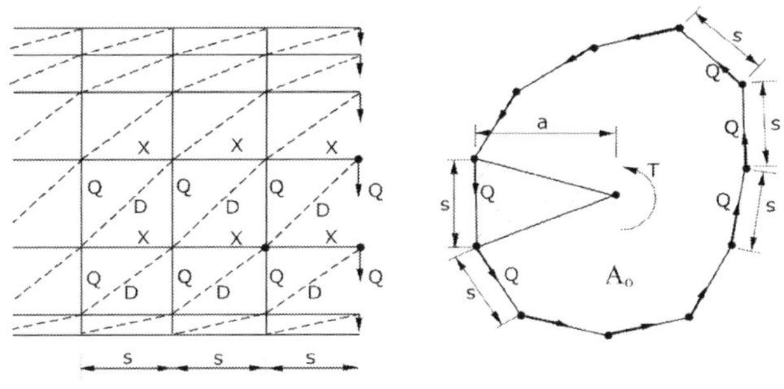

(a) Member under torsion (b) Section
struts = dashed lines, ties = solid lines

Fig. 3.3.5a—Rausch's space truss model with struts and ties (Hsu 1984 after Rausch 1929).

(3.3.5)) was found to greatly overestimate the experimentally measured torsional strengths of reinforced concrete members. As a result, empirical formulas proposed to correct limitations of Rausch's equation were incorporated in the design codes of several countries until the 1990s. Figure 3.3.5b compares T_n with $A_{oh}A_t f_{ty}/s_t$ for a series of PCA tests (Hsu 1968b,c). The straight solid line represents calculated strengths using Eq. (3.3.5) (that is, $\alpha_t = 2.0$) and the dashed line represents calculated strengths using the torsion provisions of ACI 318-71. The lever arm area A_o is defined by the centerline of closed stirrups and denoted as A_{oh}. In view of the unconservative nature of the values calculated with Eq. (3.3.5) (Fig. 3.3.5b), in particular when T_n is large, an empirical formula was proposed for ACI 318-71. This formula is simply a modification of Eq. (3.3.5), which reduced the slope of 2, shown in Fig. 3.3.5b, down to 1.2, and added a "concrete contribution T_c," which represents the vertical intercept of the experimental curve. The background of "concrete contribution T_c," which is included in early ACI Codes (1971 through 1995) and later dropped, is provided in Hsu (1997). From the experimental tests shown in Fig. 3.3.5b, Lampert and Thürlimann deduced that A_o should be less than A_{oh}. They recommended that A_o be taken as the area enclosed by a line joining the centers of the longitudinal bars in the corners of the hoops. With this change in the definition of A_o, the agreement between predictions and experimental results was considerably improved. This new definition of A_o was subsequently incorporated into the CEB-FIP code (1978). A different approach, to reconcile Rausch's space truss model with test results, was achieved in 1983 with development of the STM (Hsu and Mo 1983) for torsion, using a softened stress-strain relationship of concrete developed by Vecchio and Collins (1981). A more recent STM is provided in 5.6.

3.3.6 *Skew-bending theory for torsion*

3.3.6.1 *Beams without web reinforcement*—When beams without transverse reinforcement are subjected to torsion, they fail quickly and abruptly at cracking. Their torsional behavior is similar to that of plain concrete beams and, therefore, can be predicted by Saint-Venant's theory. The experimental torsional strength of a plain concrete beam, however, was found to be greater than the theoretical strength calculated with Saint-Venant's theory ($T_n = \alpha x^2 y f_t'$), assuming the principal tensile stress reaches a value of f_t'. To find a reason for this discrepancy, Hsu (1968a) used high-speed photography to record failure process and observed that plain concrete members failed abruptly in a skew-bending mode. Based on the skew-bending failure, the torsional strength was derived as $T_n = (1/3)x^2 y(0.85 f_r)$. As a result, the constant 1/3 is an approximation of Saint-Venant's coefficient α when the x/y ratio becomes large. This coefficient was used as the calibration for the torsional strength of beams without reinforcement, and for the "concrete contribution" of the torsional strength of beams with web reinforcement (Hsu 1968b,c). Formulas based on the parameter $(1/3)x^2 y f_t'$ were incorporated into the ACI Code from 1971 to 1995.

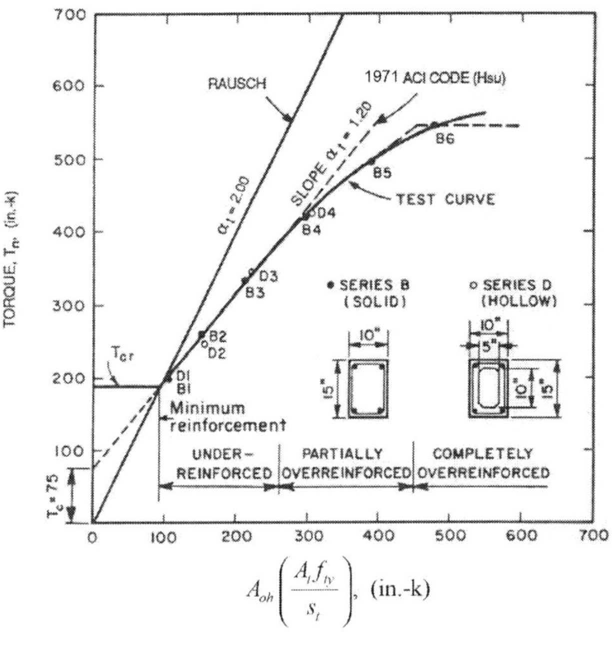

Fig. 3.3.5b—Experimental T_n *versus* $A_{oh}(A_t f_{ty}/s_t)$ *curve compared with Rausch's formula and 1971 ACI code (Hsu 1984).*

3.3.6.2 *Beams with web reinforcement*—Skew-bending theory was also applied to concrete beams reinforced with both longitudinal and transverse reinforcement subjected to torsion, shear, and flexure. Such beams could fail in three modes (Fig. 3.3.6.2). The first mode has a compression zone along the top face and failure is caused by yielding of the bottom longitudinal reinforcement and transverse reinforcement along the three remaining faces. The second mode has a compression zone along a side face and failure is caused by the yielding of the longitudinal reinforcement on the opposite side face, and in the transverse reinforcement along the three faces in tension. The third mode has a compression zone along the bottom face where failure is caused by yielding along the top longitudinal reinforcement and transverse reinforcement along the three remaining faces.

Lessig (1959) was the first to propose a skew-bending theory in connection with Mode 1 and Mode 2 failures. Equilibrium equations were established by taking moment equilibrium about an axis AB (Fig. 3.3.6.2) parallel to the

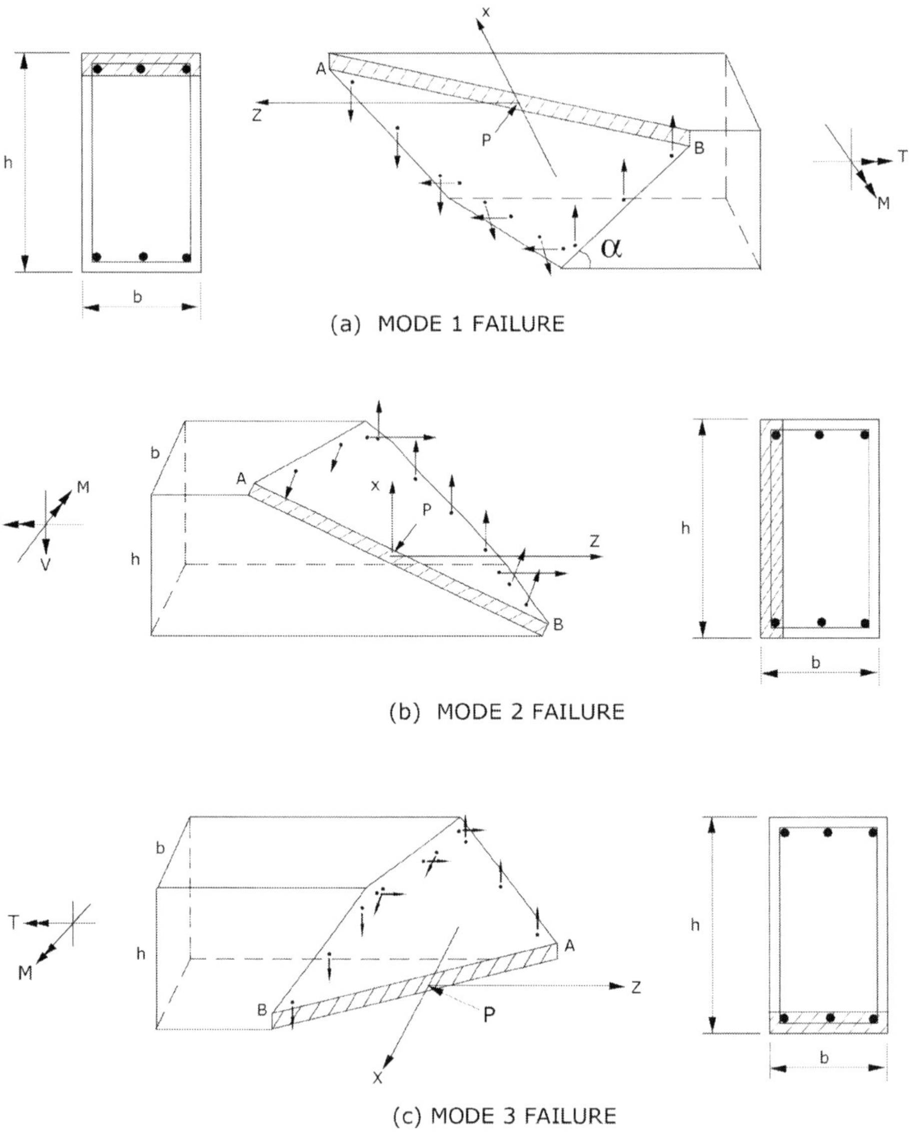

Fig. 3.3.6.2—Three failure modes in skew-bending theories (Lessig 1959; Yudin 1962; Collins et al. 1968a; Elfgren et al. 1974a,b).

skew compression zone and force equilibrium about the axis perpendicular to the center of the compression zone. The crack angles defining the skew failure surfaces were determined by minimizing the torsional resistance. The resulting equations were complicated.

Yudin (1962) took a different approach by deriving three equilibrium equations: moment equilibrium about the longitudinal and transverse axes through the compression zone center, and force equilibrium about the axis perpendicular to the compression zone center. Assuming all the crack angles to be 45 degrees, he showed that his equation for ultimate torque was the same as Rausch's. Lessig and Yudin's research was followed by the skew-bending theory of Walsh et al. (1966, 1967) and Collins et al. (1968a,b), who also identified Mode 3 failure. The three modes of failure—1, 2, and 3—are governed by the dominant effect of positive flexure moment, torsion shear, and negative flexure moment, respectively. For combined torsion and flexure, Lampert and Collins (1972) showed that both the skew-bending theory and the space truss theory predict the same parabolic shape for the interaction curve. Elfgren et al. (1974a,b) then derived a nondimensional interaction surface for combined torsion, shear, and flexure based on the three modes of skew-bending failure. If the inclination angle of cracks in each wall is adjusted so that both longitudinal and transverse reinforcement can yield, the interaction surface becomes identical to that derived from the equilibrium (plasticity) truss model (Elfgren 1972a,b), as explained in 5.5.

3.3.7 *Truss models with membrane elements*

3.3.7.1 *Equilibrium (plasticity) truss model for torsion*—Rausch's space truss model with struts and ties (1929) was replaced by a new and more realistic type of truss model in the 1960s. As shown in Fig. 3.3.7.1a, the new model is made up of membrane elements treated as trusses made up

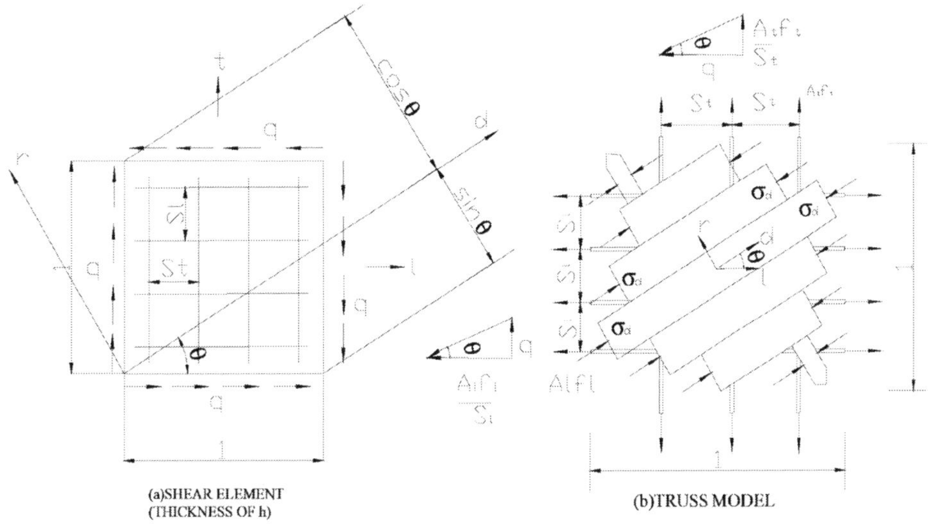

Fig. 3.3.7.1a—Membrane element in shear (Hsu 1993).

of struts and ties after cracking. Three equilibrium equations for the elements were first derived by Nielsen (1967) and Lampert and Thürlimann (1968).

$$\sigma_\ell = \sigma_d \cos^2\theta + \sigma_r \sin^2\theta + \rho_\ell f_\ell \quad (3.3.7.1a)$$

$$\sigma_t = \sigma_d \sin^2\theta + \sigma_r \cos^2\theta + \rho_t f_t \quad (3.3.7.1b)$$

$$\tau_{\ell t} = (-\sigma_d + \sigma_r)\sin\theta\cos\theta \quad (3.3.7.1c)$$

Because Eq. (3.3.7.1a) through (3.3.7.1c) describe the transformation of stresses, they form a system of stresses that should satisfy Mohr's stress circle, a type of equilibrium condition called Mohr's circular stress equilibrium. The longitudinal and transverse reinforcement stresses are assumed to yield at failure ($f_\ell = f_{\ell y}$, $f_t = f_{ty}$), and the concrete tensile stress is neglected ($\sigma_r = 0$). Equations (3.3.7.1a) through (3.3.7.1c) were simplified to yield the shear stress $\tau_{\ell t}$ in the case of pure shear without normal stresses ($\sigma_\ell = \sigma_t = 0$)

$$\tau_{\ell t} = \sqrt{\rho_\ell f_{\ell y} \rho_t f_{ty}} \quad (3.3.7.1d)$$

A reinforced concrete tube subjected to torsion can be visualized as an assembly of shear elements similar to the one shown in Fig. 3.3.3.

The torsional moment at yielding, T_n, can then be derived by substituting the shear stress $\tau_{\ell t}$ ($\tau_{\ell t} = q/t$) into Eq. (3.3.3) and substituting $\rho_\ell = A_\ell/s_\ell t$ and $\rho_t = A_t/s_t t$

$$T_u = 2A_o\sqrt{\frac{A_\ell f_{\ell y}}{s_\ell}\frac{A_t f_{ty}}{s_t}} \quad (3.3.7.1e)$$

Equation (3.3.7.1e) was first derived by Lampert and Thürlimann (1968, 1971). In the case of equal yield force in both longitudinal and transverse reinforcement ($A_\ell f_{\ell y}/s_\ell = A_t f_{ty}/s_t$), Lampert and Thürlimann's Eq. (3.3.7.1e) looks identical to Rausch's Eq. (3.3.5), except the term A_o is defined differently. The former is based on the centerline of longitudinal reinforcement and the latter is based on the centerline of hoop reinforcement. Equation (3.3.7.1e) was first used in the European CEB-FIP code in 1978 and ACI 318-95. By combining Eq. (3.3.7.1e) with equations for shear and flexure, Elfgren (1972a,b) and Elfgren et al. (1974a) established a nondimensional interaction surface as shown in Fig. 3.3.7.1b, based on the equilibrium (plasticity) truss model as explained in 6.5. Elfgren also showed the interaction surface derived from the truss model is identical to the skew-bending theory. Actual testing has shown, however, that Elfgren's interaction surface overestimates the experimental results in the region adjacent to torsion and shear because softening concrete was not considered.

3.3.7.2 *Compatibility truss model for torsion*

3.3.7.2.1 *General*—According to Navier, the strain compatibility (or geometric) condition should be considered if the angle of twist is to be related to the torsional moment. Assuming the membrane element to behave as a truss as shown in Fig. 3.3.7.1a(b), three strain compatibility equations have been derived by Baumann (1972) and Mitchell and Collins (1974).

$$\varepsilon_\ell = \varepsilon_d \cos^2\theta + \varepsilon_r \sin^2\theta \quad (3.7.2.1a)$$

$$\varepsilon_t = \varepsilon_d \sin^2\theta + \varepsilon_r \cos^2\theta \quad (3.7.2.1b)$$

$$\frac{\gamma_{\ell t}}{2} = (-\varepsilon_d + \varepsilon_r)\sin\theta\cos\theta \quad (3.7.2.1c)$$

Because Eq. (3.7.2.1a) through (3.7.2.1c) describe the transformation and distribution of strains, they should satisfy Mohr's strain circle. This type of compatibility condition is therefore called Mohr's circular strain compatibility.

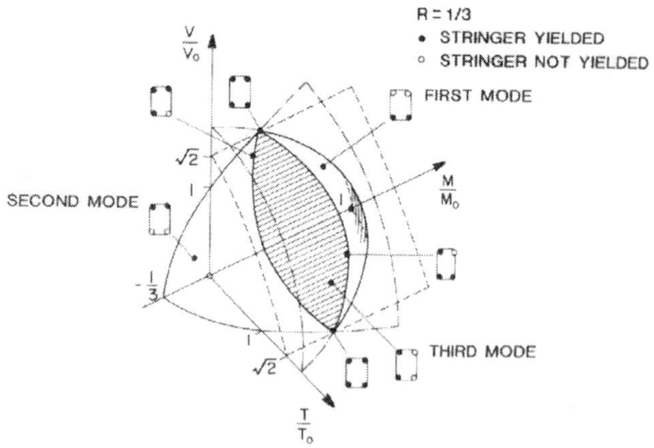

Fig. 3.3.7.1b—Nondimensional interaction surface for torsion, shear, and flexure (Hsu 1984). (The numbering of the failure modes is different than in Fig. 3.3.6.2.)

In addition, there are two compatibility equations for a reinforced concrete tube. One relates the angle of twist, Φ, to the shear strain γ_{lt} of the membrane element

$$\Phi = \frac{p_o}{2A_o}\gamma_{lt} \tag{3.7.2.1d}$$

The other describes the flexure of diagonal concrete struts. The curvature of concrete struts Ψ at an angle θ is related to the angle of twist Φ by

$$\Psi = \Phi \sin 2\theta \tag{3.7.2.1e}$$

3.3.7.2.2 *Compression field theory (CFT)*—In a major development of the truss model for torsion, Mitchell and Collins (1974) established the CFT to calculate the complete torque-twist response of reinforced concrete members, irrespective of sectional shape or amount of reinforcement. The CFT combines four equilibrium equations [Eq. (3.3.3), (3.3.7.1a), (3.3.7.1b), and (3.3.7.1c)], five compatibility equations [Eq. (3.7.2.1a) through (3.7.2.1e)], and constitutive equations of concrete and reinforcement. Concrete diagonals thickness t_d and the area enclosed by the shear flow, A_o, are determined as described in 5.5. The assumption that the diagonal compressive strains are reduced linearly with depth below the surface was experimentally verified (Collins 1973). Because the concrete cover is assumed to have spalled before the section reaches the maximum strength, this theory is also known as the "spalling model." Details of the CFT are provided in 5.5.

3.3.7.2.3 *Softened truss model*—A reinforced concrete membrane element subjected to shear is a two-dimensional problem; the shear stress can be resolved into a principal tension stress and a principal compressive stress in the 45-degree direction. Robinson and Demorieux (1972) found that when a membrane element is subjected to a biaxial stress condition, the principal compressive stress is reduced (or "softened") by the principal tensile stress in the perpendicular direction. Performing biaxial tests of shear elements, Collins and Chockalingam (1979) and Vecchio and Collins (1981, 1982, 1986) showed that the concrete softening coefficient in the principal compressive direction is related to the principal tensile strain, not the principal tensile stress. Softening was further quantified experimentally by researchers at the University of Houston (Belarbi and Hsu 1994, 1995; Pang and Hsu 1995; Hsu et al. 1995a,b; Zhang and Hsu 1998), who confirmed the importance of principal tensile strain and identified concrete strength as an additional essential variable. An STM (Hsu and Mo 1985a,b,c; Hsu 1993) was developed based on a softened concrete stress-strain curve. Details of this theory are provided in 4.6. In the STM (Hsu and Mo 1985a,b,c), the softened concrete stress-strain curve is used in the flexural compression zone of the concrete struts to determine the neutral axis position. The distance from the neutral axis to the extreme compression fiber is known as thickness of shear flow zone, t_d; the lower the softening coefficient, the greater the thickness t_d. Because the softening coefficient varies from approximately 1/4 to 1/2, the thickness t_d, obtained from the softened stress-strain curve, is expected to be approximately two to four times the thickness from the non-softened stress-strain curve. Based on Vecchio and Collins's softened stress-strain curve (1981), Hsu and Mo (1985a,b,c) studied the torsional behavior of cracked reinforced concrete members. They found that Rausch's Eq. (3.3.5) is valid if the thickness t_d and the lever arm area A_o are calculated by a truss model that incorporates a softened stress-strain curve. Hsu and Mo's STM satisfies the Mohr's stress circle, the Mohr's strain circle, and the softened stress-strain curve of concrete. It was further shown to provide accurate estimates of experimental results available at the time (Hsu and Mo 1985a,b,c), including the PCA tests in Fig. 3.3.5b. When accounting for the modifications proposed by Hsu and Mo, the equation by Rausch becomes a curve that starts at the origin, closely tracing the PCA test points. This occurs because A_o decreases with increasing T_n and an increasing percentage of reinforcement. The dependency of t_d and A_o on T_n is expressed by simple equations (Hsu 1990, 1993)

$$t_d = \frac{4T_n}{A_{cp}f_c'} \tag{3.3.7.2.3a}$$

$$A_o = A_{cp} - \frac{t_d p_o}{2} \approx A_{cp} - \frac{t_d p_{cp}}{2} = A_{cp} - \frac{2T_n p_{cp}}{A_{cp}f_c'} \tag{3.3.7.2.3b}$$

As indicated by Eq. (3.3.7.2.3b), the lever arm area A_o is decreased by increases in the thickness t_d. Because the torsional strength of a reinforced concrete member is directly proportional to A_o, the strong effect of the softening coefficient on the torsional strength is evident, explaining why Rausch's equation for T_n (Eq. (3.3.5)) overestimated the test results, as shown in Fig. 3.3.5b. Because the shear flow zone thickness t_d occupies such a crucial position in the STM, torsion research is now focused on a better understanding of the shear flow zone (Alkhrdaji and Belarbi 2003).

3.3.8 *Derivation of ACI 318: torsion design equations*—As discussed previously, ACI equations for designing members under torsion are based on the space truss model theory. There are two basic design equations: one for torsional hoop reinforcement, and another for torsional longitudinal reinforcement.

From the truss model (Fig. 3.3.7.1a(b)), equilibrium of the top face of the element relates the shear flow q to the force in the torsional hoop reinforcement per unit length ($A_t f_t/s$) through the equation

$$q = \frac{A_t f_t}{s} \cot\theta \qquad (3.3.8a)$$

Substituting Eq. (3.3.8a) into Bredt's Eq. (3.3.3) and assuming yielding of the hoop reinforcement renders the ACI 318 requirement for transverse reinforcement for torsion.

$$\frac{A_t}{s} = \frac{T_u}{\phi 2 A_o f_{yv} \cot\theta} \qquad (3.3.8b)$$

Similarly, from the equilibrium of the left face of the element, shown in Fig. 3.3.7.1a(b), the shear flow q can be related to the force in the torsional longitudinal reinforcement per unit length ($A_\ell f_\ell/s_\ell$) through the equation

$$q = \frac{A_\ell f_\ell}{s_\ell} \tan\theta \qquad (3.3.8c)$$

The term A_l is defined in ACI 318 as the total area of torsional longitudinal reinforcement in the cross section, assuming that $p_o = p_h$, where p_h is the perimeter of the centerline of the outermost hoop bars. Equation (3.3.8c) then becomes

$$q = \frac{A_\ell f_\ell}{p_h} \tan\theta \qquad (3.3.8d)$$

An expression for the amount of torsional longitudinal reinforcement in ACI 318 is obtained by equating the shear flow q of Eq. (3.3.8d) and (3.3.8a)

$$A_\ell = \left(\frac{A_t}{s}\right) p_h \frac{f_{yv}}{f_{y\ell}} \cot^2\theta \qquad (3.3.8e)$$

3.4—Concluding remarks

This historical review identifies research studies that have provided the foundation for the current understanding of torsion in reinforced concrete members. It took almost the entire twentieth century for researchers to successfully apply Navier's theory to cracked reinforced concrete. The application of Navier's theory to the flexure theory of reinforced concrete was relatively easy because of the one-dimensional condition of equilibrium and compatibility, combined with the non-softened constitutive laws of materials. Shear and torsion proved to be much more difficult because they were, respectively, two-dimensional and three-dimensional problems. From a historical perspective, the development of the current torsional theories spanned over 40 years. Equilibrium equations were developed in the 1960s, compatibility conditions were added in the 1970s, and the formulation and expressions of softening coefficients were introduced in the 1980s and 90s.

Since 1960, the bulk of work has focused on beams subjected to monotonic torsional loading. Early tests on members subjected to reverse cyclic loading were conducted by Collins and Chockalingam (1979) on reinforced concrete beams subjected to pure torsion and by Stevens et al. (1991) on membrane elements subjected to reversed cyclic shear up to reinforcement yielding. Significant recent advances have applied rational models such as the STM and the modified CFT to shear elements subjected to cyclic loading. Vecchio (1999) and Palermo and Vecchio (2003, 2004) applied the MCFT for cyclic loading up to yielding of the reinforcement. Mansour et al. (2001), Zhu et al. (2001), Zhu and Hsu (2002), and Mansour and Hsu (2005a,b) studied cyclic torsion and shear, including the unloading and reloading branches of the softened stress-strain curves (hysteretic loop) and the effect of Poisson's ratio in a membrane element. The latter was crucial in determining the descending branch of the softened stress-strain curve.

Belarbi and Greene tested several full-scale box girders under cyclic torsion and combined torsion and shear (Belarbi and Greene 2004; Greene and Belarbi 2008). The cyclic shear tests of membrane elements (Mansour and Hsu 2005a,b) permitted the establishment of cyclic stress-strain curves of concrete and reinforcement. These cyclic constitutive laws laid the foundation for the Cyclic Softened Membrane Model (Mansour and Hsu 2005a,b), which could be used for the design of seismic-resistant reinforced concrete structures (Hsu et al. 2006).

CHAPTER 4—BEHAVIOR OF MEMBERS SUBJECTED TO PURE TORSION

4.1—General

The torsional strength of concrete beams depends on many factors, such as the amount and distribution of transverse and longitudinal reinforcement, concrete strength, cross section shape, and, for rectangular cross sections, aspect ratio (depth-to-width ratio). Concrete strength and amount of reinforcement can change the failure characteristics. The failure mechanism varies from ductile to sudden and brittle, depending on the longitudinal and transverse reinforcement volumetric ratios. Chapter 4 details the behavior of plain concrete, reinforced concrete, and prestressed concrete under pure torsion. The behavior of high-strength concrete members is also discussed. The effects of other stress resultants, such as shear and flexure, are discussed in Chapter 6.

4.2—Plain concrete

Saint-Venant's elastic theory accurately describes the torsional behavior of plain concrete members at low torque before cracking. This theory assumes that the cross section

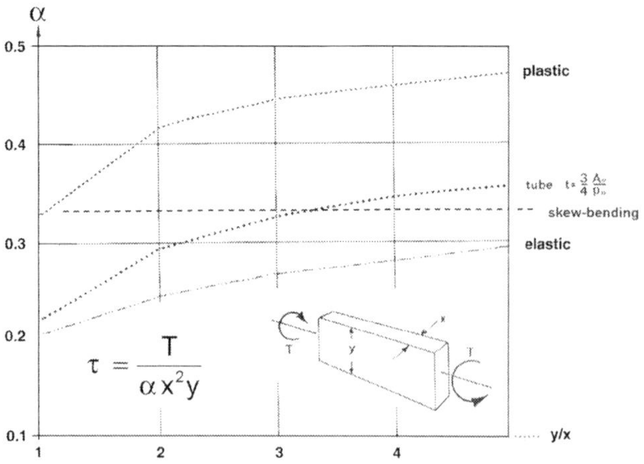

Fig. 4.2—Torsion stress factors for solid rectangular section (Collins 1973).

shape remains unchanged after twisting, the angle of twist per unit length is constant along the element length, warping deformation perpendicular to the cross section is identical throughout the member length, and failure occurs when the maximum principal stress equals the direct concrete tensile strength. Member shear stresses resist applied torsional moment. For solid circular shapes, the shear stresses are zero at the member center, increasing linearly to a maximum at the outside surface. In solid rectangular shapes, the shear stresses increase from zero at the center to a maximum at the center of the longer faces, but they decrease along the edge and are zero again at the corners. When considering a small element in the member subjected to pure torsion, only shear stresses develop and, therefore, the principal tension and compression stresses equal the shear stress. Applying Saint-Venant's theory to concrete, the member would fail in a brittle manner once the maximum shear stress equals the concrete tensile cracking strength.

After cracking, the elastic theory does not accurately describe the concrete behavior because various nonlinearities exist, such as the stress-strain behavior in compression, post-peak softening in tension, and tensile cracking. When the elastic theory is used, it considerably underestimates the torque at failure of a plain concrete member because concrete exhibits a complex structural response (Hsu 1984; Karayannis 2000). Other theories, such as the plastic theory and skew-bending theory, were developed to estimate the failure strength of plain concrete members. The plastic theory is not theoretically satisfactory because it assumes that concrete develops plasticity. As in the elastic theory, failure is assumed to occur when the maximum principal tensile stress of concrete reaches the tensile strength. Concrete, however, does not develop significant plastic behavior in tension and the torsional failure of plain concrete is brittle (Hsu 1984). Unlike the elastic theory, the plastic theory tends to overestimate the failure strength. The skew-bending theory assumes that plain concrete members subjected to torsion fail by bending about an axis parallel to the wider cross-sectional face and inclined at an angle of 45 degrees with respect to the longitudinal axis. Failure occurs when the tensile stress on the inclined face due to flexure reaches a reduced modulus of rupture of concrete (Hsu 1968b). The formula derived is $T_u = (1/3)x^2y(0.85f_r)$. The torsional stress factor is a constant 1/3, which lies between those of the elastic theory and the plastic theory, as shown in Fig. 4.2. The skew-bending theory accurately describes the failure of concrete elements with rectangular cross sections, but it should be considered as an assembly of component rectangles when applied to flanged sections (Hsu 1968b, 1984). For thin-walled tubes, the relationship between torsional shear stress τ (= q/t_d) and the torsional moment T can be found from Bredt's Eq. (4.2). Collins (1973) suggested that if a solid section was replaced by a tube with the same exterior dimensions but with a wall thickness of $0.75A_o/p_o$, the cracking torques could still be determined from Eq. (4.2). Figure 4.2 compares the torsional shear stress τ for solid rectangular sections predicted by elastic theory, plastic theory, skew-bending, and thin-walled tube model. By further assuming that the wall thickness $t_d = 0.75A_{cp}/p_{cp}$ and that $A_o = (2/3)A_{cp}$, the torsional stress τ can be related to T by the parameter (A_{cp}^2/p_{cp})

$$\tau = \frac{T}{2A_o t_d} = \frac{T}{(A_{cp}^2 / p_{cp})} \tag{4.2}$$

Equation (4.2) was embraced by CSA-A23.3-77 and later adopted by ACI 318-95. For example, the ACI code uses it to address compatibility torsion of spandrel beams. When the nominal torsional stress τ_n needed to form a torsional plastic hinge was found to be $0.33\sqrt{f_c'}$ MPa ($4\sqrt{f_c'}$ psi) (Hsu and Burton 1974; Hsu and Hwang 1977), the design torsional moment T_n of the spandrel beam was $0.33\sqrt{f_c'}\,(A_{cp}^2/p_{cp})$ MPa (equivalent to $4\sqrt{f_c'}\,(A_{cp}^2/p_{cp})$ psi).

In addition to the aforementioned theories, there are approaches to the torsional behavior of concrete elements that are based on truss models. Because these models are developed on modeling the flow of forces in a cracked reinforced concrete element, as stated by Karayannis (1995), these approaches are more suitable for the description of the post-elastic behavior of reinforced concrete elements, with rectangular cross sections reinforced in the longitudinal and transverse directions.

Karayannis and Chalioris (2000) confirmed that a plain concrete member subjected to pure torsion fails in tension due to shear. The behavior of concrete elements in torsion is therefore primarily governed by the material tensile response, particularly its tensile cracking characteristics. Tension cracking makes the plain concrete behavior in torsion nonlinear. Before the peak torque is reached, some microcracking and softening regions are developed.

The cracking is a localized phenomenon, and its boundary is the strain softening region called the fracture process zone. Bažant and Oh (1983) regarded this tension-softening region as being caused not only by microcracking but also by bond rupture; therefore, the fracture process zone has to be assumed wider than the region of visible microcracks. The zone width can be considered as a material property that, except for the influence of the stress distribution, depends on the nature and size of the aggregate. The role of the size effect on the fracture process zone width should also be considered. Between the microcracks of this zone, there are less damaged or even elastic parts; this can be considered a result of the strong tendency for eventual localization of the damage and the simultaneous recovery of the undamaged or less-damaged parts between the final cracks.

4.3—Reinforced concrete

Before the concrete cracks, a member under pure torsion is assumed to behave elastically with a nearly linear torque-twist curve. After cracking, the resistance mechanism changes and stirrups and longitudinal reinforcement carry most of the tensile load whereas the concrete carries the compression. The longitudinal reinforcement tensile strains cause the member to lengthen. Increased torque causes a nearly linear increase in the beam length. A longitudinal restraint acts as a compressive prestress force and increases the torsional strength.

Members reinforced for torsion typically have closed stirrups and at least one longitudinal bar in each stirrup corner, as originally required in ACI 318-63. Members with only longitudinal reinforcement behave similarly to a plain concrete member in regards to torsional rigidity and torsional strength, and they fail after cracking in a brittle manner. An uncracked beam loaded in pure torsion has only shear stresses on a plane perpendicular to the longitudinal axis. The associated principal tensile stress causes inclined cracks on the beam surface in the principal compressive stress direction. These cracks spiral around the beam. The tensile cracks are due to strain in the principal tensile stress direction, and they affect the concrete acting in compression. The tensile strains have the effect of weakening the concrete's ability to withstand compression in the perpendicular direction. This is known as "softening" of the concrete compressive strength. Tests performed by Vecchio and Collins (1982, 1986) identified the principal tensile strains as the primary variable influencing tension softening and quantified them. Continued work by Belarbi and Hsu (1994, 1995), Pang and Hsu (1995), Hsu and Zhang (1997), Hsu and Zhu (2002), and Hsu and Mo (2010) has improved the understanding of compression softening. Other studies, such as those of Vecchio and Collins (1993) and Koutchoukali and Belarbi (2001), have also been reported based on further calibration of their original models. After cracking, the longitudinal and transverse reinforcement are in tension, and the concrete struts between the diagonal cracks are in compression, as shown in Fig. 4.3a. The faces of the beam are warped as the beam is twisted, causing bending stresses, in addition to compressive stresses, in the concrete struts (Mitchell and Collins 1974). The concrete between inclined cracks is capable of acting in tension and will increase the overall beam torsional stiffness. This effect, known as tension stiffening, is most predominant immediately after the first cracks appear and decreases with increased torsion.

The volumetric ratios ρ_l and ρ_t have a significant effect on the behavior of reinforced concrete beams subject to pure torsion. The ratios affect the angle of the inclined cracks on

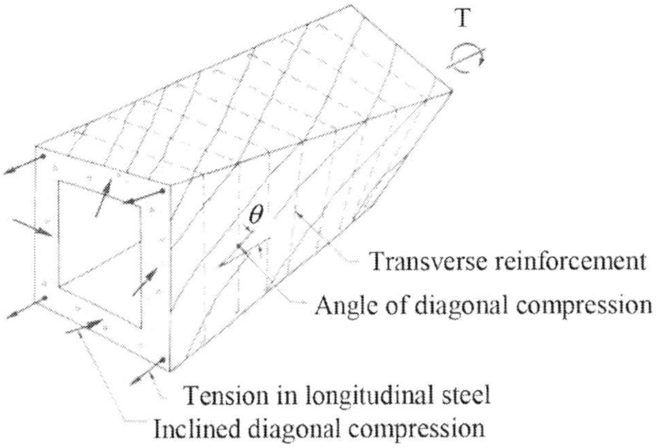

Fig. 4.3a—Diagonal cracks in members under pure torsion (Mitchell and Collins 1974).

the surface, crack width, beam strength and stiffness, and failure mode. The reinforcement ratios have a strong influence on the inclined crack angles and the size of the cracks. The crack width is related to the principal tensile strain. When the volumetric ratios ρ_ℓ and ρ_t are equal, θ is approximately 45 degrees and the cracks have the smallest width because the strains in the longitudinal and transverse directions are also equal. For beams with ρ_t larger than ρ_ℓ, the angle is greater than 45 degrees relative to the horizontal and the strain larger in the longitudinal direction, which causes wider cracks. Similarly, in beams with ρ_t less than ρ_ℓ, the angle is less than 45 degrees relative to the horizontal, the transverse strain is larger, and the cracks are also wider than when θ is 45 degrees (Mitchell and Collins 1974; Hsu 1993).

The volumetric ratios of longitudinal and transverse reinforcement to concrete affect the failure mechanism. Large ρ_ℓ and ρ_t result in reinforcement stresses less than yield when the concrete reaches its ultimate compressive strain at the surface. In this condition, the concrete is crushed and the beam experiences sudden and brittle failure. Beams of this type are called over-reinforced. Figure 4.3b shows an example of beam failure boundaries (Leu and Lee 2000). The condition known as balanced failure point is met when concrete reaches its ultimate strain at the point where both transverse and longitudinal reinforcement begin to yield simultaneously. Beams with ρ_ℓ and ρ_t less than the reinforcement ratios at the balanced condition are called under-reinforced (Region I). Under-reinforced beams are capable of continued twist as the reinforcement yields, producing a ductile failure. Over-reinforced beams are in Region IV where the failure is brittle. For lower ratios of either ρ_ℓ or ρ_t, the longitudinal or transverse reinforcement yields before the concrete is crushed. Beams of this type are called partially under-reinforced (Regions II and III).

The volumetric ratios of longitudinal and transverse reinforcement have a strong influence on the beam's postcracking torsional stiffness and strength as shown in Fig. 4.3c. For under-reinforced beams with equal volumetric ratios ρ_l and ρ_t, an increase in the total reinforcement increases the torsional rigidity, which is the slope of the torque-twist curve

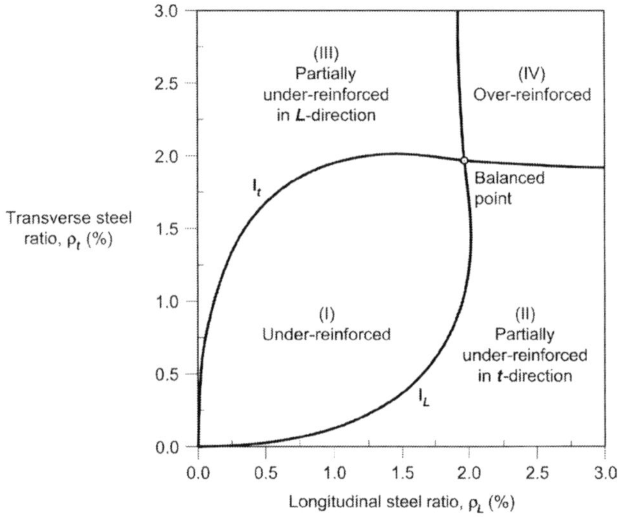

Fig. 4.3b—Failure boundaries for f_c' = 4.0 ksi (27.5 MPa) and f_y = 45.2 ksi (311.9) MPa (Leu and Lee 2000).

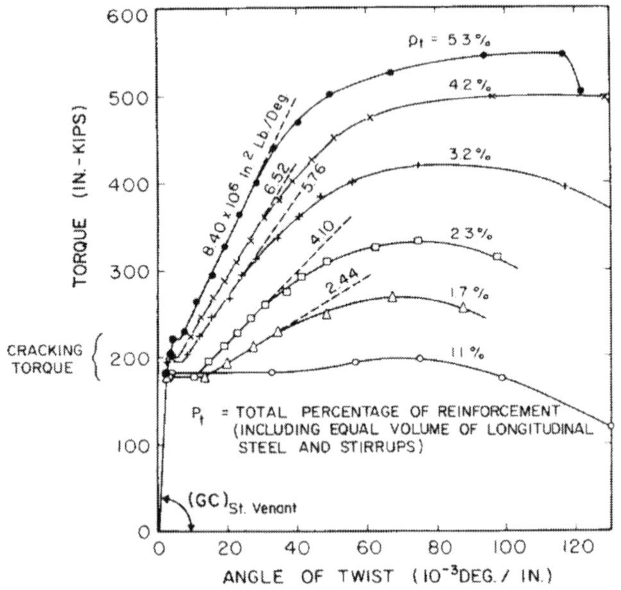

Fig. 4.3c—Torque-twist curves of beams with various percentages of reinforcement (Hsu 1968a). (Note: 1 in.-kip = 0.113 kN·m; 1 in. = 0.0254 m.)

after cracking. It also increases the torsional yield strength, which is the amount of torque the beam can sustain at the onset of yielding. As the total amount of reinforcement increases, however, the post-yield twist decreases. Therefore, an increase in the total amount of reinforcement results in a stiffer and stronger beam, but at the cost of decreased ductility. The minimum amount of reinforcing steel is the amount causing the reinforcement to yield at the same torque that causes cracking. Where this occurs, the total reinforcement ratio, including both longitudinal and transverse reinforcement, is approximately 0.01 for normal-strength concrete (Hsu 1984). A beam with less reinforcement than the minimum fails in a brittle manner after cracking. The behavior of solid members compared with hollow ones

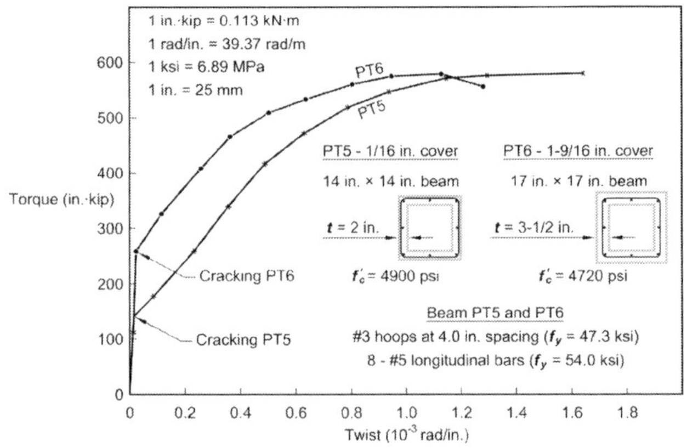

Fig. 4.3d—Effects of concrete cover on the torque-twist behavior (Mitchell and Collins 1974). (Note: 1 in.-kip = 0.113 kN·m.)

shows that the inner core of solid members is not effective after cracking. The cracking strength of hollow members is less than that of solid members because, before cracking, the shearing stresses are acting over the entire section including the inner core. After cracking, however, the outer perimeter of the cross section resists the larger part of the torque with minimum contribution from the inside core; torsional strength and general behavior are similar in solid and hollow sections after cracking as long as wall thickness is large enough to accommodate the shear flow zone (Hsu 1968a). Leonhardt and Schelling (1974) confirmed these findings when they performed an extensive test series on beams in pure torsion where they systematically varied the cross section type, including rectangular with different aspect ratios, circular, and hollow box, T-shaped, and inverted L-shaped beams. They made two observations: 1) the importance of detailing by properly anchoring the stirrups and using small stirrup spacings; and 2) the importance of softening concrete in the inclined struts.

High concrete shearing stresses outside of the stirrups create potential for concrete spalling under high torques. To satisfy equilibrium, the concrete compressive stress that resists torsion after cracking changes directions at the outside corners of the beam. The resultant tensile stress in the concrete causes the concrete cover to spall off when the tensile stress is high enough. Typically the spall occurs at the stirrups. The resulting section is smaller and has a reduced torsional strength (Mitchell and Collins 1974). Mitchell and Collins (1974) tested two beams, PT5 and PT6, to investigate the effects of spalling. The beams had similar reinforcing cages and concrete strength, but PT5 had a cover of 1.5 mm (1/16 in.) and PT6 had a cover of 40 mm (1-9/16 in.). The results of the tests showed that, although the concrete outside the hoop reinforcement had a significant effect on the cracking torque, it had little effect on the peak torque after significant spalling occurred (Fig. 4.3d).

In tests performed by Rahal and Collins (1995b), specimens with small covers did not experience spalling until after the torsional strength had been reached. Beams with larger covers spalled before reaching the torsional strength.

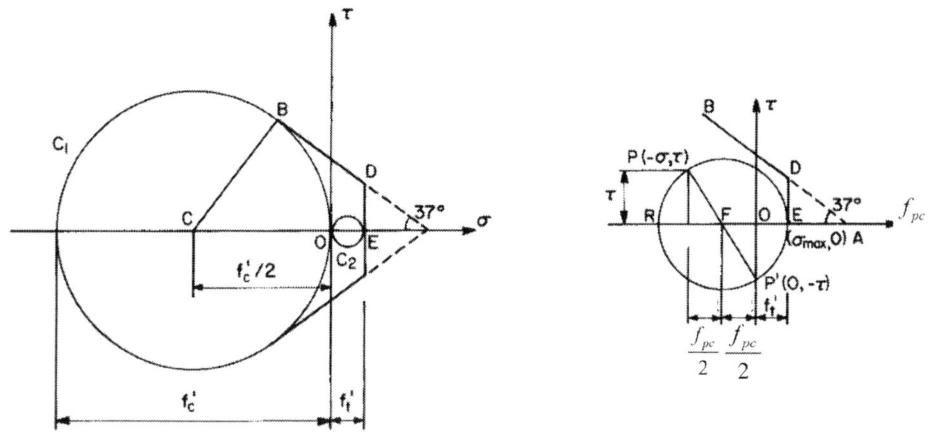

(a) Cowan's failure envelope (b) Primary tension failure

Fig. 4.4a—Cowan's failure criteria (Hsu 1984).

In these tests, the beams had stirrup dimensions of 245 x 545 mm (9.65 x 21.5 in.), and cover thicknesses of 22.5 and 42.5 mm (0.89 and 1.637 in.) for the thin and thick concrete covers, respectively. The concrete cover thickness is the critical parameter for spalling. Experiments have shown that spalling occurs if the cover thickness is greater than 30 percent of the ratio of area to the perimeter of the cross section. For this calculation, the sectional area and perimeter do not consider the area removed by voids in hollow sections (Rahal and Collins 1996).

4.4—Prestressed concrete

Adding longitudinal prestressing to a concrete member subjected to torsion will increase its cracking torque and, to a lesser extent, its failure torque. The longitudinal prestressing force creates a compression-shear biaxial state of stress that delays cracking of concrete. This effect is illustrated by Cowan's failure criteria, shown in Fig. 4.4a. The Mohr-Coulomb failure criterion has been simplified into two straight lines in Fig. 4.4a(a). If the Mohr circle of stresses touches the straight line BD described by $t = c - \tan\varphi$, where φ is the friction angle (a typical value for concrete is 37 degrees), failure is by sliding. If the Mohr circle touches the tension cutoff line DE described by $\sigma = f_t'$, failure is by separation. In Fig. 4.4a(b), $\sigma_{max} = OE = FE - FO = FP - FO$. The variable σ_{max} can then be expressed as

$$\sigma_{max} = \sqrt{\left(\frac{f_{pc}}{2}\right)^2 + \tau^2} - \frac{f_{pc}}{2} \tag{4.4a}$$

$$\tau = f_t'\sqrt{1 + \frac{f_{pc}}{f_t'}} \tag{4.4b}$$

Cowan's failure envelope using Mohr's circle is a simple and concise model for explaining the compression-shear state of stress. Equation (4.4b) for shear stress is derived from the geometry of Cowan's failure envelope, and it is a function of only the prestress and the concrete tensile strength. In effect, the longitudinal prestress increases the cracking strength, as shown in Fig. 4.4b. Expressions for prestressed member torsional strength are similar to those of reinforced concrete members with a factor that increases the strength due to the prestress. Pure torsion tests have shown that the prestressed beams appear to require approximately the same minimum torsional reinforcement as the nonprestressed beams (Hsu 1984). Similar to beams without prestressing, prestressed beams without web reinforcement fail shortly after cracking.

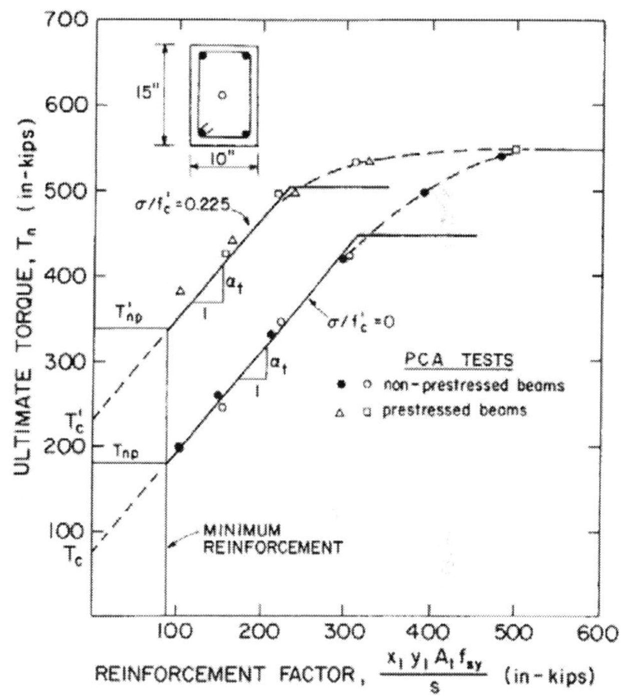

Fig. 4.4b—Ultimate torque-versus-reinforcement factor for prestressed and reinforced beams under pure torsion (Hsu 1984). (Note: 1 in.-kip = 0.113 kN·m.)

Testing has shown that as the amount of longitudinal reinforcement and prestress increase, the torque at cracking,

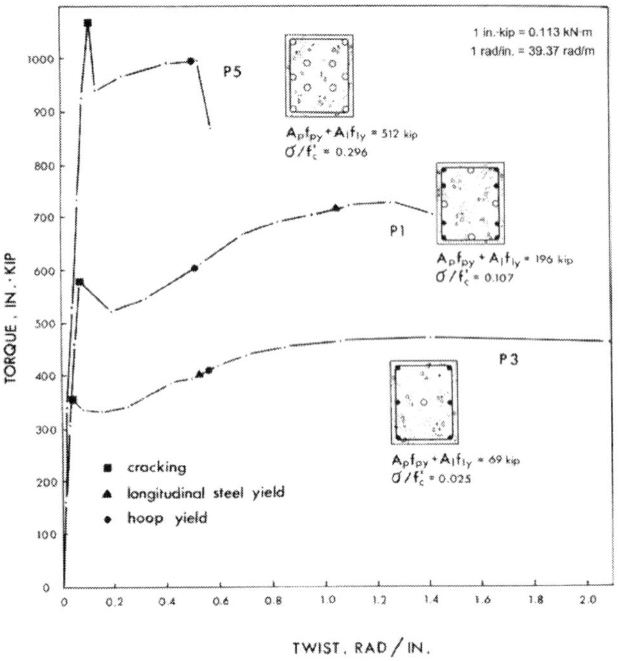

Fig. 4.4c—Effect of longitudinal reinforcement (Mitchell and Collins 1978). (Note: 1 in.-kip = 0.113 kN·m; 1 in. = 0.0254 m.)

hoop yield, and ultimate increase, whereas the twist at failure decreases. This effect is illustrated by the experimental work shown in Fig. 4.4c (Mitchell and Collins 1978). Prestressing does not influence ultimate torsional strength if the longitudinal reinforcement does not yield before failure, as shown in Specimen P5 of Fig. 4.4c. Prestressed beams under sustained, long-term torque exhibit increased rotation due to concrete creep. Beams tested to failure after being subjected to sustained torque showed no reduction in strength or stiffness. After a sustained torque, the ratio of beam rotation to the instantaneous rotation increases with larger values of torque and decreases with increased amounts of transverse reinforcement (Allos and Rashid 1989).

4.5—High-strength concrete

High-strength concrete (HSC) is more brittle than normal-strength concrete (NSC). In some ways, NSC can be treated as a perfectly plastic material, whereas HSC very much resembles a perfect elastic and brittle material (Rasmussen and Baker 1995; Iravani 1996; ACI Committee 363 1992). The ascending portion of the stress-strain curve is steep and approximately linear, whereas the descending curve almost vanishes and shows no strain softening. High-strength concrete has a peak load strain that is higher than the typical 0.0020 of NSC and an ultimate load strain less than the typical 0.0035 (Wafa et al. 1995).

In the case of pure torsion, the use of HSC allows a higher service stress in the reinforcing bars and produces a higher stiffness in the structure, enhancing both the strength and stiffness of a member. For a given amount of reinforcement, the initial crack width is larger for HSC because larger reinforcement strains are induced at cracking, but the number of cracks is fewer in HSC. Wider cracks in HSC could also be the result of reduced aggregate interlock. Shortly after the longitudinal reinforcement yield, the width of the largest crack increases rapidly. After this point, all further deformation occurs at this crack. This behavior is exaggerated in high-strength beams due to the smooth-faced cracks that are less effective in transmitting shear (Koutchoukali and Belarbi 2001). Rasmussen and Baker (1995) observed diagonal cracks in both NSC and HSC beams on the larger faces of the cross section because the shear stresses are largest on these faces. This is similar to the assumptions of the elastic theory. After first cracking, the number of cracks parallel to the initial cracks increases as the applied torque increases. In this way, characteristic spiral cracks develop around the beam. Cracking of an HSC beam is more brittle than that of an NSC beam because in HSC, not only the matrix but also the aggregate splits (the paste in HSC is strengthened relative to the aggregate), producing a more brittle response. That is, cracks in NSC matrix propagate around the aggregate, whereas cracks in HSC also pass through the aggregate. Before cracking, surface compressive principal strains for HSC beams are smaller compared with those of NSC beams because of the higher elastic modulus for HSC. Whereas HSC delays beam cracking, after cracking the shape and angle of inclination of cracks are the same for all concrete strengths (Rasmussen and Baker 1995). An HSC beam has a higher torsional stiffness than an NSC beam for a given torque and a given cross section because of the higher modulus of elasticity for HSC. Figure 4.5 shows the torque-twist diagrams for an NSC (36.3 MPa [5.1 ksi]) beam and an HSC (109.8 MPa [15.9 ksi]) beam under pure torsion, reported by Rasmussen and Baker (1995). They showed that the torque-twist curves are linear in the uncracked range and the HSC beam has greater torsional stiffness. After cracking, the torsional stiffness is reduced significantly in both beams. The post-cracking stiffness of an HSC beam remains almost linear up to failure, whereas the NSC beam curve is nonlinear.

The surface compression strain at failure for an HSC beam is slightly less than that for an NSC beam, but because of the higher modulus of rupture and consequent higher torsional stiffness for HSC, the twist at peak torque for both NSC and HSC beams is approximately the same (Rasmussen and Baker 1995). The increase in both longitudinal and transverse reinforcement stresses immediately after cracking is larger for an HSC beam than an NSC beam, resulting in brittle cracking of HSC. For the same torque, the reinforcement stresses in an HSC beam are less than in an NSC beam. Therefore, for a given torque, the required amount of beam reinforcement is less using HSC. A general comparison of the ratio of transverse reinforcement stresses and longitudinal reinforcement stresses at failure shows that this ratio increases when concrete strength increases. For HSC, the bond between concrete and reinforcement is better due to denser concrete. Therefore, the vertical component of diagonal concrete compression is easily transferred to the transverse reinforcement so that it attracts a relatively larger amount of torque using HSC. The lower bond between the stirrup and concrete using NSC results in a portion of the vertical component of the diagonal concrete compression being carried by longitudinal reinforcement bending and shear (Rasmussen and Baker 1995). Beams elongate because of increased cracking state under increasing torque. The longitudinal strains at failure increase as concrete strength increases. For the beams tested by Rasmussen and Baker (1995), the longitudinal elongation at failure for the HSC beam (B110.1) was approximately 35 percent larger than for the NSC beam (B30.3). For a given torque and given cross section, however, the longitudinal elongation of an HSC beam was approximately 50 percent lower than that of an NSC beam. Accounting for the spalling effect is essential for a correct evaluation of the ultimate torque and failure mechanism of over-reinforced HSC beams. In the case of some over-reinforced NSC concrete beams, the phenomenon of spalling does not take place, and the torsional strength is evaluated considering the entire cross section. In the case of over-reinforced HSC beams, the cover is reasonably neglected. In fact, due to the higher load-carrying strength of HSC with respect to NSC, spalling of the concrete cover from the stirrups appears to occur more frequently (Cerioni et al. 1998). This is because, even though HSC is stronger, the tensile strength does not increase in the same proportion as the compressive strength. Based on the results of 14 HSC prestressed beams tested under pure torsion, Wafa et al. (1995) concluded that increasing the concrete strength and the prestressing force increases the cracking strength, ultimate torsional strength, and torsional stiffness. Wafa et al. reported that among the several models used—the space truss model with softening of concrete (Hsu and Mo 1985a), space truss model with spalling of concrete cover (Mitchell and Collins 1974), and skew-bending theory (Hsu 1984)—the space truss theory with softening of concrete gave the best estimate of the test beams' torsional strength. Wafa et al. (1995) observed that the tensile strength increase is not in direct proportion to the compressive strength. The ratios of the modulus of rupture to the compressive strength were 9 percent and 11.4 percent, respectively, for the HSC and NSC.

The post-cracking torsional stiffness, as represented by the slope of the torque-twist curve, is influenced by the concrete strength, the amount of torsional reinforcement, and the aspect ratio. The higher the concrete compressive strength, the higher the stiffness is, irrespective of the aspect ratio and the prestressing force. Torsional strength increases with an increase in concrete strength and prestressing level and with a decrease in aspect ratio. All tested beams exhibited compression zones located on the longer sides of the beams. Tension cracks initially formed on one longer face and progressed to form spirals. Crushing on the other longer face characterized the failure pattern.

4.6—Concluding remarks

Saint-Venant's elastic theory accurately describes plain concrete members' torsional behavior at low torque and before cracking. Applying Saint-Venant's theory to concrete,

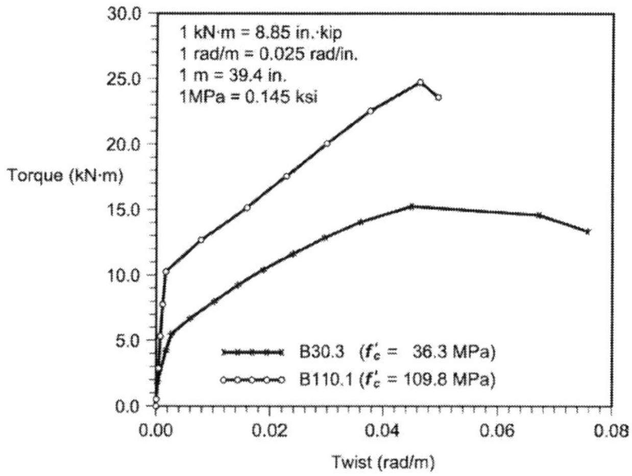

Fig. 4.5—Torque-twist relation for NSC and HSC beams (Rasmussen and Baker 1995). (Note: 1 m = 39.4 in., 1 kN·m= 8.85 in.-kip; 1 MPa = 0.145 ksi.)

the plain concrete member fails in a brittle manner once the maximum shear stress equals the concrete tensile cracking strength. For reinforced concrete members, because the elastic theory does not accurately describe the member behavior after cracking, other theories such as truss model and skew-bending were developed to estimate the failure strength.

In members with mild reinforcement, the longitudinal and transverse reinforcement are in tension after cracking and the concrete struts between diagonal cracks are in compression. The beam faces are warped as the beam is twisted, causing bending and compressive stresses in the concrete strut. The total amount and the relative amount of longitudinal and transverse steel reinforcement have a significant effect on the behavior of reinforced beams subject to pure torsion. The behavior of solid members compared with hollow ones shows the inner core of solid members is ineffective after cracking, making the cracking strength of hollow members less than solid ones. Torsional strength and general behavior, however, are similar in solid and hollow sections. The most significant effect of adding longitudinal prestressing to a concrete member subjected to torsion is the increase in its cracking strength. Testing has shown that as the amount of longitudinal reinforcement and the prestress force increase, the torque at cracking, hoop yield, and ultimate strength all increase whereas the twist at failure decreases. The use of HSC in torsional members allows a higher service stress in the reinforcing bars and produces a higher sectional stiffness in the structure. Cracking of an HSC beam is more brittle than that of an NSC beam; in HSC, both the matrix and aggregate split, producing a more brittle response. An increase in twist immediately after cracking is more significant for HSC beams because of the higher energy released, with more brittle crushing as concrete strength increases. The post-cracking stiffness of an HSC beam remains almost linear up to failure, whereas the curve for an NSC beam is nonlinear.

CHAPTER 5—ANALYTICAL MODELS FOR PURE TORSION

5.1—General

In 1929, using space truss concepts, Rausch developed an analytical model to predict the torsional strength of reinforced concrete members. Unfortunately, Rausch's equation was unconservative by more than 30 percent for under-reinforced beams (Hsu 1968a,b). Although Rausch's model assumed the lever arm area A_o to be the area within the hoop reinforcement centerline, A_{oh}, the actual A_o would be much smaller in members with a high percentage of reinforcement when concrete softening is considered. The correct determination of the lever arm area A_o and the centerline of shear flow depend on the means used to determine the thickness of the shear flow zone, t_d.

Since the late 1960s, the truss model theory has undergone four major developments. First, Lampert and Thürlimann (1968, 1969) introduced the variable angle truss model and discovered the bending phenomenon in diagonal concrete struts. Second, Collins (1973) derived compatibility equations to determine the angle of the diagonal concrete struts. It should be noted that Mitchell and Collins (1974) developed a space truss model with concrete cover spalling to determine the shear flow zone thickness. Third, the softening phenomenon in the concrete struts, discovered by Robinson and Demorieux (1972), was quantified by Vecchio and Collins (1981) using a softened coefficient. Fourth, combining the equilibrium, compatibility, and softened stress-strain relationships, Hsu and Mo (1985a) developed a softened truss model (STM) theory to determine the shear and torsional behavior of reinforced concrete members throughout the post-crack loading history up to the peak strength. Using the STM theory, shear flow zone thickness t_d can be determined accurately to calculate the torsional strength of reinforced concrete members. The test data for solid and hollow beams suggests that, once cracking has occurred, concrete in the center member has little effect on torsional strength of the cross section and is ignored (Hsu 1968b; Lampert and Thürlimann 1968; Leonhardt and Schelling 1974). Beams are considered equivalent tubular members. This observation is the basis of the torsion design procedures introduced in ACI 318-95 (MacGregor and Ghoneim 1995).

5.2—Equilibrium conditions

5.2.1 *Shear elements in shear flow zone*—A prismatic reinforced concrete member subjected to an external torque T is shown in Fig. 5.2.1(a). This external torque is resisted by an internal torque formed by the circulatory shear flow q along the cross section periphery. This shear flow q occupies a zone, called the shear flow zone, with a thickness denoted by t_d. The thickness is a variable determined from equilibrium and compatibility conditions; it is different than the given wall thickness t of a hollow member.

An element A in the shear flow zone (Fig. 5.2.1(a)) is subjected to a shear stress $\tau_{\ell t} = q/t_d$, as shown in Fig. 5.2.1(b). The in-plane equilibrium of this element should satisfy Eq. (5.2.1a) through (5.2.1c) as originally derived by Nielsen (1967), Lampert and Thürlimann (1968, 1969), and Mitchell

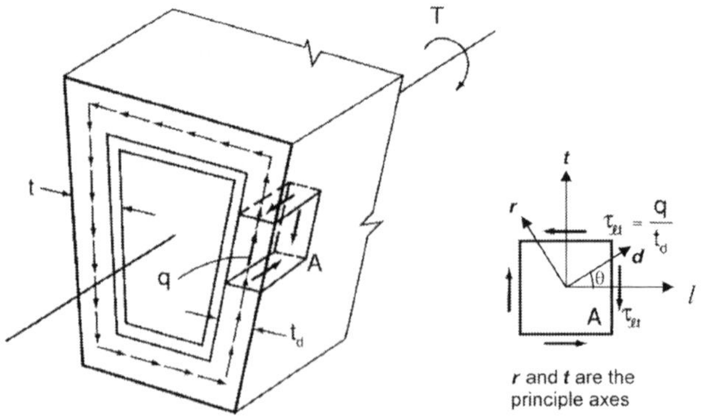

(a) Element in shear flow zone (b) Shear stress on element A

Fig. 5.2.1—Hollow box subjected to torsion (Hsu 1993).

and Collins (1974), but given herein in the most general form (Hsu 1993)

$$\sigma_\ell = \sigma_d \cos^2\theta + \sigma_r \sin^2\theta + \rho_\ell f_\ell + \rho_{\ell p} f_{\ell p} \quad (5.2.1a)$$

$$\sigma_t = \sigma_d \sin^2\theta + \sigma_r \cos^2\theta + \rho_t f_t + \rho_{tp} f_{tp} \quad (5.2.1b)$$

$$\tau_{\ell t} = (-\sigma_d + \sigma_r)\sin\theta\cos\theta \quad (5.2.1c)$$

In Eq. (5.2.1a) through (5.2.1c) the reinforcement ratios ρ_ℓ, ρ_t, $\rho_{\ell p}$, and ρ_{tp}, should be taken with respect to the shear flow zone thickness t_d as described in 5.3. These three equations are expressed in the form of stress transformation to show that they satisfy Mohr's stress circle. Because Eq. (5.2.1a) and (5.2.1b) involve the reinforcement ratios ρ_ℓ, ρ_t, $\rho_{\ell p}$, and ρ_{tp}, they are coupled with the compatibility equations through the variable t_d. The thickness t_d is a geometric variable that is determined by equilibrium conditions as well as compatibility conditions. The determination of t_d is similar to that of the flexural neutral axis, which requires plane section compatibility conditions.

5.2.2 *Bredt's equilibrium equation*—Equations (5.2.1a) through (5.2.1c) were derived from the equilibrium of a member element in the shear flow zone. To maintain equilibrium of the whole cross section, a fourth equation derived by Bredt (1896) should be satisfied. For a shear flow zone thickness t_d, the shear stress $\tau_{\ell t}$ is

$$\tau_{\ell t} = \frac{T}{2A_o t_d} \quad (5.2.2)$$

This also introduces an additional variable, the torque T. The shear flow zone thickness t_d is incorporated in Eq. (5.2.2) explicitly, as well as implicitly through A_o, which is a function of t_d.

5.3—Compatibility conditions

5.3.1 *Shear elements*—As shown in Fig. 5.2.1(a), the element A in the shear flow zone is subjected to a shear stress. The in-plane deformation of this element should satisfy three compatibility equations (Eq. (5.3.1a) through (5.3.1c)) as originally derived by Baumann (1972) and Mitchell and Collins (1974). The equations are in the form of strain transformation to show they satisfy Mohr's strain circle (Hsu 1993)

$$\varepsilon_\ell = \varepsilon_d \cos^2\theta + \varepsilon_r \sin^2\theta \qquad (5.3.1a)$$

$$\varepsilon_t = \varepsilon_d \sin^2\theta + \varepsilon_r \cos^2\theta \qquad (5.3.1b)$$

$$\frac{\gamma_{\ell t}}{2} = (-\varepsilon_d + \varepsilon_r)\sin\theta\cos\theta \qquad (5.3.1c)$$

Element A is subjected to in-plane shear as well as out-of-plane deformation as demonstrated (5.3.2).

5.3.2 *Shear strain due to twisting*—When a tube is subjected to torsion, the relationship between the shear strain $\gamma_{\ell t}$ in the tube wall and the member angle of twist, Φ, can be derived from the compatibility condition of warping deformation (Bredt 1896).

$$\Phi = \frac{p_o}{2A_o}\gamma_{\ell t} \qquad (5.3.2)$$

It is clear from Eq. (5.3.2) that the angle of twist, Φ, will produce a shear strain $\gamma_{\ell t}$ in the shear flow zone elements. This shear strain $\gamma_{\ell t}$ will induce the reinforcement strains ε_ℓ and ε_t in the *d-r* direction. The relationships between the strains in the *ℓ-t* direction (ε_ℓ, ε_t, and $\gamma_{\ell t}$) and the strains in the *d-r* direction (ε_d and ε_r) are described by the three transformation equations, Eq. (5.3.1a) through (5.3.1c).

5.3.3 *Bending of diagonal concrete struts*—In a torsional member, the angle of twist, Φ, also produces warping in member walls that in turn causes bending in the concrete struts. The concrete struts are therefore subjected to compression due to circulatory shear and bending due to warping of walls. The relationship between the angle of twist, Φ, and the bending curvature of concrete struts, Ψ, is given in Eq. (5.3.3).

Figure 5.3.2(a) shows a box member with four walls of thickness t subjected to torsional moment T. Each wall contains a shear flow zone with a thickness of t_d. The perimeter of the centerline of shear flow q has a width of ℓ_q along the top wall. The length of the member in the longitudinal direction is taken to be $\ell_q\cos\theta$, so that the diagonal line in the center plane of the shear flow in the top wall OABC has an angle of inclination θ with respect to the longitudinal axis. When this member is subjected to an angle of twist, Φ, the center plane OABC becomes a hyperbolic paraboloid surface OADC, as shown in Fig. 5.3.2(b). The plane edge CB rotates to the position CD through an angle $\Phi\ell_q\cos\theta$.

The curve OD has a curvature of Ψ that is related to Φ and θ by

$$\Psi = \Phi\sin 2\theta \qquad (5.3.3)$$

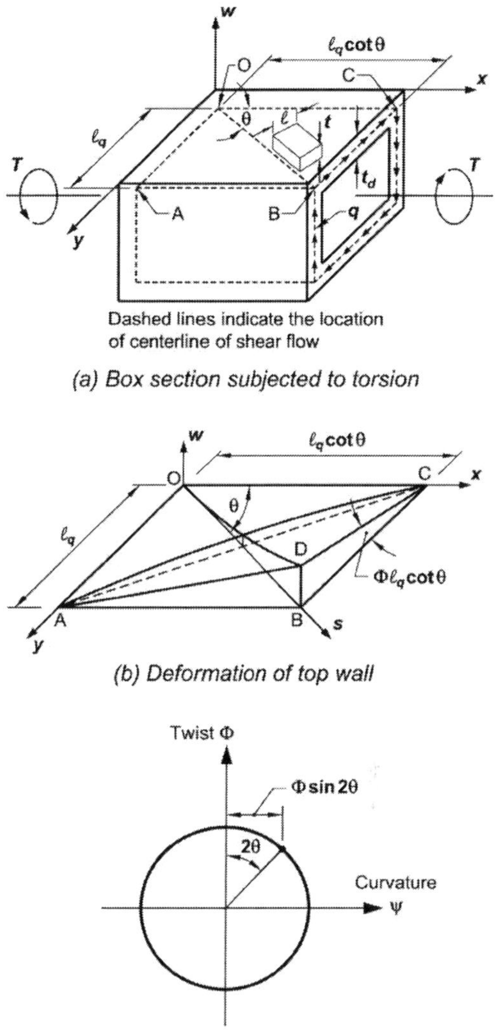

(a) Box section subjected to torsion

(b) Deformation of top wall

(c) Relating curvature to twist

Fig. 5.3.2—Bending of a concrete strut in the wall of a box section subjected to torsion (Mitchell and Collins 1974; Hsu and Mo 1985a).

Equation (5.3.3) was originally derived by Mitchell and Collins (1974) using Mohr's circle, as shown in Fig. 5.3.2(c).

The illustration in Fig. 5.3.2 and the derivation of bending curvature are based on Hsu (1993). Although the imposed curvature is illustrated by a rectangular box section, this equation is applicable for any arbitrary, bulky section with multiple walls.

5.3.4 *Strain distribution in concrete struts*—The curvature Ψ derived in Eq. (5.3.3) produces a nonuniform strain distribution in the concrete struts. Figure 5.3.4(a) shows a concrete strut unit width in a hollow section with a wall thickness of t. The tension area in the cross section inner portion is disregarded. The area in the outer portion, which is in compression, is considered effective in resisting the shear flow. The compression zone depth from the neutral axis to the extreme compression fiber is defined as the shear flow zone thickness t_d. Within this thickness t_d, the strain distribution is assumed to be linear, as shown in Fig. 5.3.4(b), based on Bernoulli's

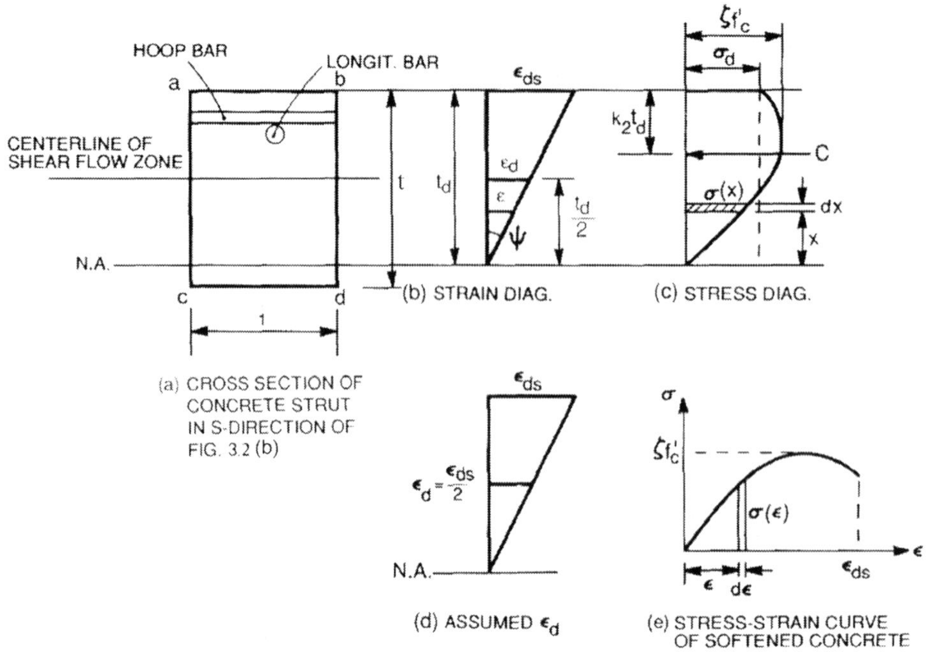

Fig. 5.3.4—Strains and stresses in concrete struts (Hsu and Mo 1985a).

plane section hypothesis used in bending theory (Mitchell and Collins 1974). The thickness t_d can therefore be related to the curvature Ψ and the maximum surface strain ε_{ds} by the simple relationship

$$t_d = \frac{\varepsilon_{ds}}{\Psi} \quad (5.3.4a)$$

The average strain ε_d can be related to the maximum strain ε_{ds}, as shown in Fig. 5.3.4(d), by

$$\varepsilon_d = \frac{\varepsilon_{ds}}{2} \quad (5.3.4b)$$

Equations (5.3.2), (5.3.3), (5.3.4a), and (5.3.4b) are the four additional compatibility equations for torsion. They introduce four additional variables, Φ, Ψ, t_d, and ε_{ds}.

5.4—Stress strain relationships

5.4.1 *Concrete struts*—The stress-strain relationship of concrete in compression, as shown in Fig. 5.3.4(e), can be expressed analytically by two branches of parabolic curves (Hsu and Mo 1985a)

Ascending branch:

$$\sigma_d = \zeta f_c'\left[2\left(\frac{\varepsilon_d}{\zeta\varepsilon_o}\right)-\left(\frac{\varepsilon_d}{\zeta\varepsilon_o}\right)^2\right] \quad \frac{\varepsilon_d}{\zeta\varepsilon_o}<1 \quad (5.4.1a(a))$$

Descending branch:

$$\sigma_d = \zeta f_c'\left[1-\left(\frac{\varepsilon_d/\zeta\varepsilon_o-1}{2/\zeta-1}\right)^2\right] \quad \frac{\varepsilon_d}{\zeta\varepsilon_o}>1 \quad (5.4.1a(b))$$

where ζ is a softening coefficient. Section 5.6.1 discusses the softening coefficient as presented in Eq. (5.6.2d) with values ranging between 0.2 and 0.6.

The stress-strain relationship of concrete in tension is irrelevant, if the concrete tensile stress σ_r is assumed to be zero in the equilibrium Eq. (5.2.1a) through (5.2.1c).

5.4.2 *Mild reinforcement*—When the tensile strength of the concrete is neglected, the stress-strain relationship of mild reinforcement is taken as the elastic perfectly plastic relationship, expressed as

$$f_\ell = E_s\varepsilon_\ell \quad \varepsilon_\ell < \varepsilon_{\ell y} \quad (5.4.2a(a))$$

$$f_\ell = f_{\ell y} \quad \varepsilon_\ell \geq \varepsilon_{\ell y} \quad (5.4.2a(b))$$

$$f_t = E_s\varepsilon_t \quad \varepsilon_t < \varepsilon_{ty} \quad (5.4.2b(a))$$

$$f_t = f_{ty} \quad \varepsilon_t \geq \varepsilon_{ty} \quad (5.4.2b(b))$$

5.4.3 *Prestressing reinforcement*—In the case of prestressed concrete, the nonlinear stress-strain relationship of prestressed strands is given (Mattock 1979)

$$f_p = E_{ps}(\varepsilon_{dec} + \varepsilon_s) \quad f_p \leq 0.7f_{pu} \quad (5.4.3a(a))$$

$$f_p = \frac{E'_{ps}(\varepsilon_{dec} + \varepsilon_s)}{\left[1 + \left\{\left(E'_{ps}(\varepsilon_{dec} + \varepsilon_s)\right)/f_{pu}\right\}^R\right]^{1/R}} \quad f_p > 0.7f_{pu} \tag{5.4.3a(b)}$$

In structures where prestress is applied in both directions, such as nuclear containment vessels, Eq. (5.4.3a(a)) and (5.4.3a(b)) are applicable to prestressing reinforcement in both the longitudinal and transverse directions. When applied to longitudinal reinforcement, f_p is equal to f_{lp}, and ε_s is equal to ε_l. When applied to transverse reinforcement, f_p is equal to f_{tp}, and ε_s is equal to ε_t.

The equilibrium equations, compatibility conditions, and stress-strain relationships described are used by two main theories for torsional analysis, namely, the compression field theory (CFT) (Mitchell and Collins 1974) and the STM (Hsu and Mo 1985a). The primary difference between these two theories lies in the stress-strain relationship of concrete in compression. The STM uses a softened stress-strain curve as shown in Eq. (5.4.1a(a)) and (5.4.1a(b)), where ζ is a softening coefficient that varies from approximately 0.2 to 0.6. In CFT, the stress-strain curve is not softened, and the softening coefficients ζ in Eq. (5.4.1a(a)) and (5.4.1a(b)) are taken as unity. To compensate for the unconservative nature of this non-softening assumption, CFT also assumes spalling concrete cover before reaching the peak torque. As a result, CFT is also called the "spalling theory." The CFT and the STM are described in detail in 5.5 and 5.6, respectively, by combining the stress-strain relationships of concrete and reinforcement given in Eq. (5.4.1a(a)) and (5.4.1a(b)), and through (5.4.3a(b)) with the equilibrium Eq. (5.2.1a) through (5.2.1c) and (5.2.2) and compatibility Eq. (5.3.1a) through (5.3.4b). The solution algorithms of the two theories are also different.

5.5—Compression field theory

Collins and Mitchell (1980) and Mitchell and Collins (1974, 1978) proposed the CFT for structural concrete in torsion. This section provides a basic summary of the theory. Before the truss analogy equilibrium equations can be used to design a member for torsion, the diagonal struts' inclination should be determined from the compatibility equations. Wagner (1929) dealt with an analogous problem in studying the post-buckling shear resistance of thin-webbed metal girders. He assumed that after buckling, the thin webs would not resist compression and the shear would be carried by a field of diagonal tension. To determine the angle of inclination of the diagonal tension, Wagner considered the systems' deformations. He assumed that the angle of inclination of the diagonal tensile stress would coincide with the angle of inclination of the principal tensile strain. This approach became known as Tension Field Theory. Applying Wagner's approach to reinforced concrete, the concrete is assumed to be unable to carry tension, and the shear is assumed to be carried by a diagonal compression field. The inclination θ of the diagonal compression represents the principal strain

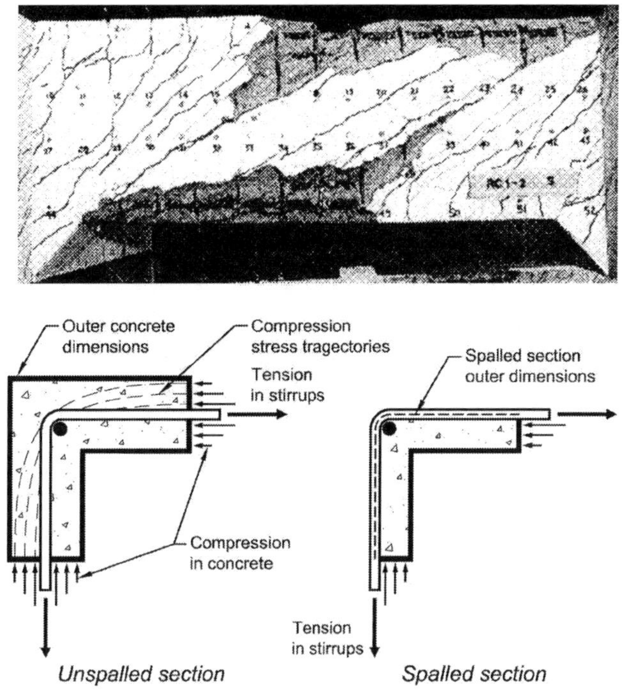

Fig. 5.5a—Spalling of the concrete cover due to torsion (Mitchell and Collins 1974).

direction; therefore, it can be derived from compatibility Eq. (5.3.1a) and (5.3.1b), by combining them to form

$$\tan^2\theta = \frac{\varepsilon_\ell + \varepsilon_d}{\varepsilon_t + \varepsilon_d} \tag{5.5a}$$

Equation (5.5a) shows the angle θ can be expressed as a function of the strain in the concrete diagonals (ε_d), the strain in the longitudinal reinforcement (ε_ℓ), and the strain in the transverse reinforcement (ε_t). The full behavioral response of reinforced concrete members subjected to torsion can be predicted by using compatibility (Eq. (5.5)), equilibrium (Eq. (5.2.1a) through (5.2.1c)), the stress-strain relationships of concrete [Eq. (5.4.1a(a)) and (5.4.1a(b))], and those for mild reinforcement [Eq. (5.4.2a(a)) through (5.4.2b(b))]. Besides these fundamental equations, additional torsional aspects were considered in the CFT. In resisting the torsion, not all of the concrete is effective in providing diagonal compressive stresses. Estimating the equilibrium of a corner element for a beam in torsion (Fig. 5.5a) reveals that the compression in the concrete tends to push off the corner whereas the tension in the hoops holds it in place. Because concrete is weak in tension, the concrete outside of the hoops spalls off at higher torsions. Because of concrete cover spalling, the effective outer surface of the concrete is assumed to coincide with the hoop centerline.

Examining the deformed shape of the twisted beam in Fig. 5.5b reveals that the diagonal concrete stresses vary in magnitude over the thickness of the effective concrete tube, from zero at the inside to a value f_{ds} corresponding to the strain ε_{ds} at the effective outer surface. As in the case of

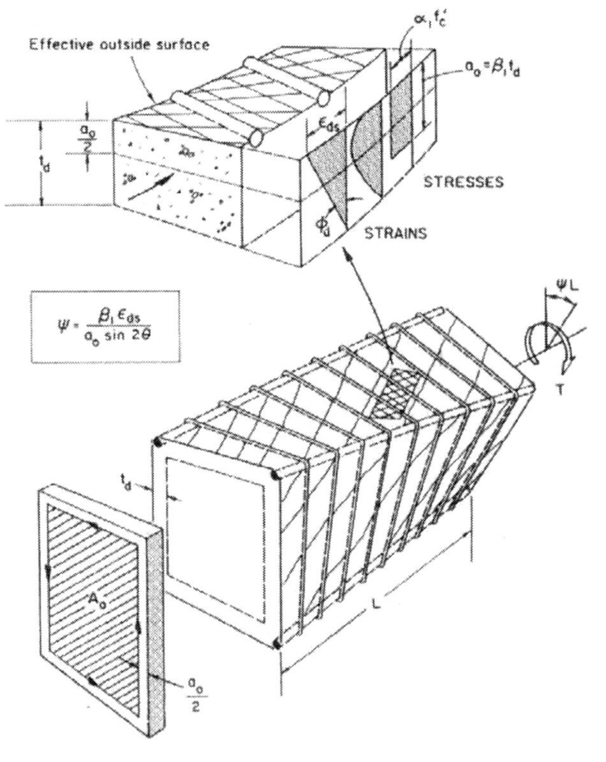

Fig. 5.5b—Effective wall thickness of a twisted beam (Collins and Mitchell 1980).

flexure, this actual stress distribution can be replaced by a uniform stress of $\alpha_1 f_c' = f_d$ acting over a depth of $\beta_1 t_d = a_o$, where the stress block factors α_1 and β_1 depend on the shape of the concrete stress-strain curve and the value of surface compression strain ε_{ds}. The centerline dimensions of the resulting tube of uniformly stressed concrete of thickness a_o are assumed to define the shear flow path q. This path lies $a_o/2$ inside the centerline of the hoop, as shown in Fig. 5.5b.

Given the shear flow path, the terms A_o (the area enclosed by the shear flow) and p_o (the perimeter of the shear flow path) can be determined using Eq. (5.5b) and (5.5c), which assume a thin tube

$$A_o = A_{oh} - \frac{a_o}{2} p_h \tag{5.5b}$$

$$p_o = p_h - 4a_o \tag{5.5c}$$

Based on the equilibrium of an element in the shear flow zone, shown in Fig. 5.2.1(b), the shear flow, q, in a box section can be expressed in terms of the longitudinal reinforcement force, ΔN, and the transverse reinforcement force, $A_t f_t$. The longitudinal reinforcement force ΔN is assumed to be distributed uniformly along the shear flow path p_o. Equilibrium Eq. (5.2.1a) through (5.2.1b) are simplified in three respects: 1) for pure torsion, $\sigma_\ell = \sigma_t = 0$; 2) concrete tensile stress is neglected, $\sigma_r = 0$; and 3) when no prestressing reinforcement is used, $\rho_{\ell p} = \rho_{tp} = 0$. In the CFT, several approximate algorithms have been developed to plot the torque-twist curve using the equilibrium and compatibility relationships given in 5.2 and 5.3. The following solution algorithm is the approach used by Rahal (2000a,b). From the equilibrium of an element in the shear flow zone, shown in Fig. 5.2.1(b), two equations can be derived

$$q = \sqrt{\frac{\Delta N}{p_o}\frac{A_t f_t}{s}} \tag{5.5d}$$

and

$$\tan\theta = \sqrt{\frac{A_t f_t}{s}\frac{p_o}{\Delta N}} \tag{5.5e}$$

Substituting Eq. (5.5d) into Bredt's Eq. (5.2.2), the torsional moment T can be related to the longitudinal and transverse reinforcement forces, ΔN and $A_t f_t$, as

$$T = 2A_o\sqrt{\frac{\Delta N}{p_o}\frac{A_t f_t}{s}} \tag{5.5f}$$

Like flexure, the compression zone depth is a function of reinforcement tensile forces, and is derived from equilibrium conditions. In a shear element, however, the reinforcement includes both longitudinal and transverse reinforcement forces, ΔN and $A_t f_t$. Eliminating θ from Eq. (5.2.1a) through (5.2.1c) results in

$$a_o = \frac{\Delta N}{\alpha_1 f_c' p_o} + \frac{A_t f_t}{\alpha_1 f_c' s} \tag{5.5g}$$

Once the compression block depth a_o is known, the terms A_o and p_o in Eq. (5.5b) and (5.5c) can be calculated.

To determine the longitudinal and transverse beam strains, ε_l and ε_t, which correspond to the chosen value of ε_{ds}, the basic compatibility Eq. (5.3.1a) through (5.3.4a) were manipulated to develop two expressions

$$\varepsilon_t = \left[\frac{\alpha_1\beta_1 f_c' A_{oh} s}{2 p_h A_t f_t} - 1\right]\varepsilon_{ds} \tag{5.5h}$$

$$\varepsilon_\ell = \left[\frac{\alpha_1\beta_1 f_c' A_{oh} p_o}{2 p_h \Delta N} - 1\right]\varepsilon_{ds} \tag{5.5i}$$

A trial-and-error process can be used to solve the eight variables (θ, a_o, ε_t, ε_ℓ, A_o, p_o, ΔN, and $A_t f_t$) using Eq. (5.5a) through (5.5c), (5.5e), and (5.5h) through (5.5i). First, select a value of ε_{ds} and assume a pair of forces, $A_t f_t$ and ΔN, from which the four variables θ, a_o, ε_t, and ε_ℓ can be calculated from Eq. (5.5e), (5.5g), (5.5h), and (5.5i), respectively. Given the value of a_o, the variables A_o and p_o can be calculated from Eq. (5.5b) and (5.5c). Based on the values of ε_t, ε_ℓ, and ε_d (related to ε_{ds} by geometry), another θ can be calculated from Eq. (5.5a). If the θ from Eq. (5.5a) is not equal to the θ from Eq. (5.5e), assume another pair of $A_t f_t$ and ΔN, and repeat the cycle until Eq. (5.5a) and Eq. (5.5e) produce the same θ.

Once these eight variables are solved, the torque T can be calculated from Eq. (5.5f) and the angle of twist, Φ, from Eq. (5.3.3). In Eq. (5.3.3), the curvature of the concrete struts, Ψ, can be obtained from Eq. (5.3.4a), where the neutral axis depth t_d is equal to a_o/β_1, and ε_{ds} is the chosen value. The torque-twist response (T-versus-Φ curve) is then determined by repeating these calculations for a series of ε_{ds} values.

5.6—Softened truss model

5.6.1 *General*—The softened truss model is built on satisfying the equilibrium Eq. (5.2.1a) through (5.2.2), the compatibility Eq. (5.3.1a) through (5.3.4b), the stress-strain materials relationships in Eq. (5.4.1a(a)) and (5.4.1a(b)) for concrete, and Eq. (5.4.2a(a)) through (5.4.2b(b)) for reinforcement. The properties of concrete stress block and shear flow zone are derived from the softened stress-strain relationships of concrete.

5.6.2 *Concrete stress block*—As shown in Fig. 5.3.2, the diagonal concrete struts in the shear flow zone are subjected to axial stress and bending. The compressive strains in the concrete struts are assumed to vary linearly from ε_{ds} at the surface to zero at the neutral axis, as shown in Fig. 5.3.4(b). The compression zone depth t_d is defined as the shear flow zone. The distribution of compressive stresses within the thickness t_d is shown by the solid curve in Fig. 5.3.4(c), based on the softened stress-strain relationship shown in Fig. 5.3.4(e). The peak stress is $\zeta f_c'$ and the average compressive stress is σ_d. The average stress σ_d of the concrete stress block in Fig. 5.3.4(c) can be expressed as follows

$$\sigma_d = \zeta f_c' \tag{5.6.2a}$$

where the coefficient k_1 is the ratio of the average stress to the peak stress. By integrating the stress-strain curve in Eq. (5.4.1a(a)) and (5.4.1a(b)), the coefficient k_1 can be expressed

$$k_1 = \frac{\varepsilon_{ds}}{\zeta\varepsilon_o}\left(1 - \frac{1}{3}\frac{\varepsilon_{ds}}{\zeta\varepsilon_o}\right) \quad \frac{\varepsilon_{ds}}{\varepsilon_p} \le 1 \tag{5.6.2b(a)}$$

$$k_1 = \left[1 - \frac{\zeta^2}{(2-\zeta)^2}\right]\left(1 - \frac{1}{3}\frac{\varepsilon_{ds}}{\zeta\varepsilon_o}\right) + \frac{\zeta^2}{(2-\zeta)^2}\frac{\varepsilon_{ds}}{\zeta\varepsilon_o}\left(1 - \frac{1}{3}\frac{\varepsilon_{ds}}{\zeta\varepsilon_o}\right) \quad \frac{\varepsilon_{ds}}{\varepsilon_p} > 1 \tag{5.6.2b(b)}$$

The coefficient k_1 is tabulated as a function of ε_{ds} and ζ in the work by Hsu (1993), and equals approximately 0.8. The softening coefficient was first proposed by Zhang and Hsu (1998) as a function of tensile strain ε_r and concrete strength f_c'

$$\zeta = \frac{5.8}{\sqrt{f_c'(\text{MPa})}}\frac{1}{\sqrt{1+400\varepsilon_r}} \text{ and } \frac{5.8}{\sqrt{f_c'(\text{MPa})}} \le 0.9 \tag{5.6.2c}$$

This equation is valid for HSC up to 100 MPa (14.5 ksi). As proposed by Belarbi and Hsu (1995), for NSC up to 42 MPa (6.0 ksi), ζ becomes

$$\zeta = \frac{0.9}{\sqrt{1+400\varepsilon_r}} \tag{5.6.2d}$$

The softening coefficient ζ usually varies in the range of 0.2 (0.045) to 0.6 (0.003).

5.6.3 *Centerline of shear flow zone and formulas for* A_o and p_o—As shown in Fig. 5.3.4(c), the resultant force of the compression stress block C is located at a distance $k_2 t_d$ from the surface, where the coefficient k_2 defines the location of the resultant force C. By integrating the concrete stress-strain curve given in Eq. (5.4.1a(a)) and (5.4.1a(b)), the coefficient k_2 is found to vary in the range of 0.40 to 0.45, depending on concrete strength. To simplify, the centerline of the shear flow is assumed to coincide with the centerline of the shear flow zone, located at a distance $0.5t_d$ from the extreme compression fiber. This assumption also results in a compatible agreement between theory and tests.

Based on this simple centerline assumption of the shear flow, formulas for calculating the lever arm area A_o and the perimeter p_o for a thick tube, whose thickness is usually quite large with respect to the overall dimension of the cross section and different from the thin tube assumption in CFT when the softening of concrete is considered, are

$$A_o = A_c - \frac{1}{2}p_c t_d + t_d^2 \tag{5.6.3a}$$

$$p_o = p_c - 4t_d \tag{5.6.3b}$$

When the softening coefficient ζ of concrete is considered, the shear flow zone thickness t_d is usually large with respect to the overall cross section dimensions. The formula for calculating A_o, therefore, is expressed in Eq. (5.6.3a) and should include the third term t_d^2. The thin tube formula used in the CFT (Eq. (5.5b)) is not used because it neglects the a_o^2 term.

5.6.4 *Solution algorithm*—The 18 governing equations for a torsional member are: four equilibrium, (Eq. (5.2.1a) through (5.2.2); seven compatibility, (Eq. (5.3.1a) through (5.3.4b)); seven constitutive (Eq. (5.4.2a(a)) through (5.4.3a(b)) and (5.6.2a) through (5.6.2d)).

For a member subjected to pure torsion, the normal stresses σ_ℓ and σ_t acting on an element in the shear flow zone are equal to zero ($\sigma_\ell = \sigma_t = 0$). When the concrete tensile strength is neglected, $\sigma_r = 0$. Disregarding these three variables (σ_ℓ, σ_t, and σ_r) in the analysis, the 18 equations contain 19 unknown variables as follows: seven stress or force variables, including $\tau_{\ell t}$, σ_d, f_ℓ, f_t, $f_{\ell p}$, f_{tp}, and T; 10 strain or geometry variables, including ε_ℓ, ε_t, $\gamma_{\ell t}$, ε_d, ε_r, θ, Φ, Ψ, t_d, and ε_{ds}; and two material coefficients, including ζ and k_1. If one of the unknown variables is selected—for example, ε_d—then the remaining 18 unknown variables can be solved using the 18 equations.

Next, select the concrete strain ε_d as the independent variable because it varies monotonically from zero to maximum. The solutions for a series of ε_d values make it possible to trace the loading history. A solution algorithm is given by Hsu (1993) and is briefly described. This method requires the derivations of Eq. (5.6.5) through (5.6.8b).

5.6.5 *Thickness* t_d *as a function of strain*—The thickness of the shear flow zone t_d can be expressed in terms of strain using the seven compatibility equations: Eq. (5.3.1a) through (5.3.4b). The substitutions and manipulations result in

$$t_d = \frac{A_o}{p_o}\left[\frac{(-\varepsilon_d)(\varepsilon_r - \varepsilon_d)}{(\varepsilon_\ell - \varepsilon_d)(\varepsilon_t - \varepsilon_d)}\right] \tag{5.6.5}$$

The variable t_d is expressed in terms of strains in all d, r, ℓ, and t directions (ε_d, ε_r, ε_ℓ, ε_t). The compressive strain is defined as negative and tensile as positive. The variable t_d is also involved in equilibrium equations through the terms A_o, p_o, ρ_ℓ, and ρ_t. Therefore, the variable t_d should first be assumed and then checked by Eq. (5.6.5).

5.6.6 *Longitudinal strain* ε_l *as a function of longitudinal stresses* f_ℓ, $f_{\ell p}$—The strain ε_ℓ can be related to the reinforcement stresses f_ℓ and $f_{\ell p}$ by eliminating the angle α from the equilibrium Eq. (5.2.1a) using compatibility equations (Eq. (5.3.1c) to (5.3.4b)). These manipulations result in

$$\varepsilon_\ell = \varepsilon_d + \frac{A_o(-\varepsilon_d)(-\sigma_d)}{(A_\ell f_\ell + A_{\ell p} f_{\ell p})} \tag{5.6.6}$$

The three unknown variables ε_ℓ, f_ℓ, and $f_{\ell p}$ can be solved using Eq. (5.6.6) and the stress-strain relationships, Eq. (5.4.2a), (5.4.2b), (5.4.3a), and (5.4.3b), for longitudinal reinforcement.

5.6.7 *Transverse strain* ε_t *as a function of transverse stresses* f_t, f_{tp}—Similarly, the strain ε_t can be related to the reinforcement stresses f_t and f_{tp} by eliminating the angle θ from equilibrium Eq. (5.2.1b) using the compatibility equations (Eq. (5.3.1b) through (5.3.4b)).

$$\varepsilon_t = \varepsilon_d + \frac{A_o s(-\varepsilon_d)(-\sigma_d)}{p_o(A_t f_t + A_{tp} f_{tp})} \tag{5.6.7}$$

The three unknown variables ε_t, f_t, and f_{tp} can be solved using Eq. (5.6.7) and the two stress-strain relationships, Eq. (5.4.2c) and (5.4.2d) and (5.4.3a) and (5.4.3b), for transverse reinforcement.

5.6.8 *Useful compatibility equations*—By combining compatibility Eq. (5.3.1a) and (5.3.1b), the two variables ε_r and θ can be expressed directly in terms of strains ε_ℓ, ε_t, and ε_d as

$$\varepsilon_r = \varepsilon_\ell + \varepsilon_t - \varepsilon_d \tag{5.6.8a}$$

$$\tan^2\theta = \frac{\varepsilon_\ell + \varepsilon_d}{\varepsilon_t + \varepsilon_d} \tag{5.6.8b}$$

Equations (5.6.8a) and (5.6.8b) will facilitate solution procedures (5.6.9).

5.6.9 *Solution procedure and flow chart*—The solution procedure is illustrated by a flow chart shown in Fig. 5.6.9. This solution procedure uses 12 equations, Eq. (5.4.2a) through (5.4.3b) and (5.6.2a) through (5.6.8b), to solve 12 unknown variables (ζ, k_1, σ_d, f_ℓ, f_t, $f_{\ell p}$, f_{tp}, ε_ℓ, ε_t, ε_r, ε_{ds}, and t_d) for each selected value of ε_d. From these stresses and strains, the remaining six variables (θ, $\tau_{\ell t}$, T, $\gamma_{\ell t}$, Φ, Ψ) can be calculated using the appropriate equilibrium or compatibility equations. Selecting a series of ε_d values produces the entire loading history of the 18 variables. The efficiency of this solution procedure arises from the fact that the angle θ is eliminated in the expressions of Eq. (5.6.5) through (5.6.7) and is, therefore, not involved in the iteration process of t_d. In addition, the derivation of Eq. (5.6.6) and (5.6.7) allows the iteration to include nonlinear stress-strain relationships (Eq. (5.4.3a) and (5.4.3b)) of prestressing strands, therefore making this solution algorithm applicable to prestressed concrete beams. A hand-calculation example illustrating this solution procedure is given by Hsu (1993). The STM theory was used by Hsu and Mo (1983, 1985a) to calculate the strength and behavior of 108 torsional beams. For the 61 under-reinforced beams with stirrup spacings within ACI code limits, the experiment-to-calculated torsional strengths had a mean value of 1.014 and a standard deviation of 0.051. The post-cracking deformations (angle of twist and strains in reinforcement and concrete) were also correctly evaluated. McMullen and El-Degwy (1985) compared the STM (softened theory) and the CFT (spalling theory) using their extensive torsion tests. They concluded that "the space truss model with softening of concrete gives a better prediction of maximum torque for the beams tested in this investigation (average $T_{u,exp}/T_{u,th}$ = 1.03 and standard deviation = 0.091)." The space truss theory with spalling of concrete cover gave an average $T_{u,exp}/T_{u,th}$ of 1.10 and a standard deviation of 0.129. McMullen and El-Degwy also observed that concrete cover spalled only after maximum torque was reached.

5.7—Graphical methods

5.7.1 *Rahal's method*—Rahal developed a simplified method for design and analysis of reinforced concrete panels subjected to pure shear (Rahal 2000a) and extended the method to reinforced concrete beams subjected to torsional moments (Rahal 2000b). The original method for calculating shear relates the normalized ultimate shear strength v_u/f_c' of reinforced concrete panels to the reinforcement indexes ω_t and ω_L in the transverse and longitudinal directions, respectively. The reinforcement indexes are given by

$$\omega_t = \frac{\rho_t f_{ty}}{f_c'}; \quad \omega_L = \frac{\rho_\ell f_{\ell y}}{f_c'} \tag{5.7.1a}$$

Figure 5.7.1 gives the relationship between the indices and the normalized shear strength obtained using the results of the modified CFT (Collins and Mitchell 1991). Each curve in the figure represents the relationship between v_u/f_c' and ω_L at a given ω_t.

At relatively low ω_t values, the strains in the transverse reinforcement exceed the yield strains before the ultimate strength is reached. Beyond a specific level of reinforcement, the concrete is crushed before the reinforcement yields. Figure 5.7.1 shows a curve passing through those points beyond which concrete is crushed before the transverse reinforcement yields (over-reinforced case). The figure also shows a similar curve for the over-reinforced case in the longitudinal direction. The two "yield" curves divide Fig. 5.7.1 into four regions. The relative position of a point of coordinates (ω_t; ω_L), with respect to these curves or regions, indicates the expected failure mode of an element with these reinforcement ratios. Four failure modes are possible:

1) Partially over-reinforced; only longitudinal reinforcement yields (Zone III)

2) Only transverse reinforcement yields (Zone II)

3) Completely over-reinforced; concrete crushing before reinforcement yielding (Zone IV) or

4) Completely under-reinforced; longitudinal and transverse reinforcement yield (Zone I).

- Because v_u/f_c', ω_t, and ω_L are dimensionless, they provide values that can be applied to any system of units. The results of this method have been compared with experimental results from 46 reinforced concrete panels and were found to give accurate calculations of the ultimate shear strength and failure mode. The average ratio of experimental shear strength to predicted strength was 1.01 with a coefficienct of variation (COV) of 0.125 (Rahal 2000a). This method is applied to beams subjected to torsion by idealizing the section as a hollow tube and adopting simplified assumptions regarding the hollow tube thickness and shear flow zone size. Tube walls are treated as the shear panels to which Fig. 5.7.1 applies. Based on assumptions of the method, the reinforcing indexes are given by

$$\omega_t = \frac{A_t f_{ty} p_o}{0.42 s A_{cp} f_c'} \tag{5.7.1b}$$

$$\omega_L = \frac{A_t f_{ty} + A_{ps} f_{py}}{0.375 s A_{cp} f_c'} \tag{5.7.1c}$$

The ultimate torsional moment is also related to ultimate shear strength in walls

$$T_u = 0.67 \frac{A_{cp}^2}{p_{cp}} v_u \tag{5.7.1d}$$

In contrast to the STM and CFT, Rahal's method is used to calculate approximate strength only and not the entire torque-twist relationship. Equations (5.7.1b), (5.7.1c), and (5.7.1d) depend on the gross area and outer cross section

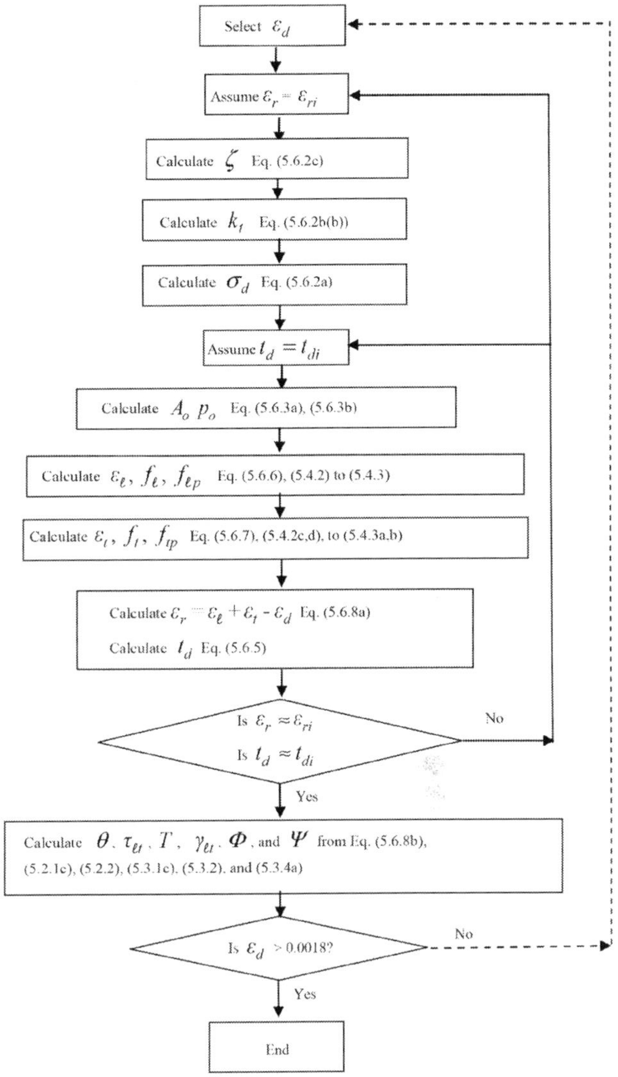

Fig. 5.6.9—Flow chart for torsion analysis using STM.

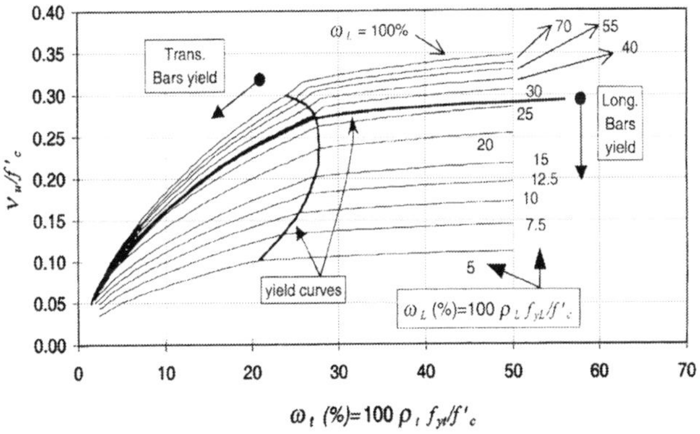

Fig. 5.7.1—Normalized shear strength curve for reinforced concrete members (Rahal 2000a).

perimeter, and therefore can be applied to closed sections of various shapes.

5.7.2 *Leu and Lee's Method*—Leu and Lee (2000) proposed a graphical solution to the STM. Their strength

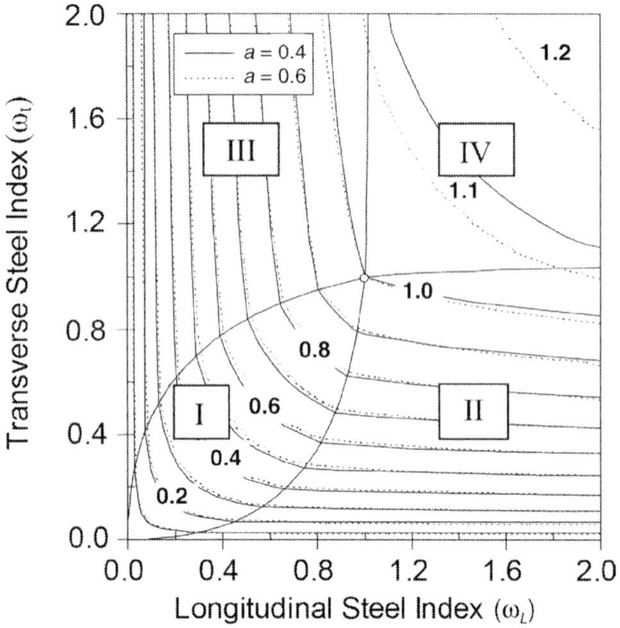

Fig. 5.7.2—Strength contour (ω_s) diagram (Leu and Lee 2000).

contour diagram, shown in Fig. 5.7.2, is constructed to give the torsional strength of a rectangular member when their longitudinal and transverse reinforcement indexes, ω_L and ω_t, are calculated. Because this graph is built on the failure mode diagram of Hsu (1993), it can also predict the four failure modes:

1) Under-reinforced
2) Partially under-reinforced in the transverse direction
3) Partially under-reinforced in the longitudinal direction
4) Over-reinforced. The diagram was constructed based on STM and, therefore, its accuracy is as good as STM prediction.

In Fig. 5.7.2, $a = 0.4$ and $a = 0.6$ correspond to reinforcement grades of 40 and 60 ksi (280 and 420 MPa), respectively. The diagrams are almost identical for different reinforcement grades, except in the over-reinforced region of failure mode of Region IV. Fortunately, an over-reinforced design is usually not permitted.

1. Calculate the normalized reinforcement ratios η_ℓ and η_t

$$\eta_\ell = \frac{f_y}{f_c'}\left(\frac{A_\ell}{A_{cp}}\right) \tag{5.7.2a}$$

$$\eta_t = \frac{f_y}{f_c'}\left(\frac{A_t p_{cp}}{A_{cp} s}\right) \tag{5.7.2b}$$

where $f_y = (f_{\ell y} + f_{ty})/2$ is the average yield strength of longitudinal and transverse reinforcement bars, $f_{\ell y}$ and f_{ty}.

2. Calculate the balanced normalized reinforcement ratios when both longitudinal and transverse reinforcements yield simultaneously with crushing of diagonal struts $\eta_{\ell b}$ and η_{tb}

$$\eta_{\ell b} = \frac{76}{200 + f_y} \text{ } (f_y \text{ in MPa) or } \eta_{\ell b} = \frac{11{,}000}{30{,}000 + f_y} \text{ } (f_y \text{ in psi)} \tag{5.7.2c}$$

$$\eta_{tb} = \frac{76}{100 + f_y} \text{ } (f_y \text{ in MPa) or } \eta_{tb} = \frac{11{,}000}{15{,}000 + f_y} \text{ } (f_y \text{ in psi)} \tag{5.7.2d}$$

3. Calculate the reinforcement indexes ω_L and ω_t

$$\omega_L = \frac{\eta_\ell}{\eta_{\ell b}} \tag{5.7.2e}$$

$$\omega_t = \frac{\eta_t}{\eta_{tb}} \tag{5.7.2f}$$

4. Calculate the nondimensional balanced torsional strength that corresponds to the case when both longitudinal and transverse reinforcements yield simultaneously with crushing of diagonal struts, $\overline{T}_{xub}$

$$\overline{T}_{xub} = \frac{140}{300 + f_y} \text{ } (f_y \text{ in MPa) or } \overline{T}_{xub} = \frac{20{,}000}{42{,}000 + f_y} \text{ } (f_y \text{ in psi)} \tag{5.7.2g}$$

5. Calculate the balanced torsional strength T_{xub}

$$T_{xub} = \frac{\overline{T}_{xub} f_c' A_{cp}^2}{p_{cp}} \tag{5.7.2h}$$

6. Use ω_L and ω_t to determine the value of ω_s from Fig. 5.7.2. The coordinate (ω_L, ω_t) in the four failure mode regions also determines the member failure mode; and

7. Determine the torsional strength.

$$T_{xu} = \omega_s T_{xub} \tag{5.7.2i}$$

The graphical solution using Fig. 5.7.2 is not exact. Its accuracy is affected by the aspect ratio of the cross section and the difference in the yield stresses, $f_{\ell y}$ and f_{ty}, when calculating $f_y = (f_{\ell y} + f_{ty})/2$. Thirty-eight specimens from Hsu (1968a) were used to evaluate the accuracy of the design charts. The mean and maximum errors of torsional strength were only 1.5 percent and 4 percent, respectively, when compared with the strengths of those obtained using the STM.

CHAPTER 6—MEMBERS SUBJECTED TO TORSION COMBINED WITH OTHER ACTIONS

6.1—General

Torsion rarely occurs in concrete structures unaccompanied by other stress resultants. In the general case of loading, a beam cross section can be subjected to a maximum of six stress resultants, in the orthogonal system of coordinates, as shown in Fig. 6.1. Three forces and three moments are possible: one torsional moment, two shearing forces along the major and minor axes, two bending moments along the major and minor axes, and an axial force (tension or compression).

Early torsion research concerned with pure torsion was later extended to torsion combined with flexure (Hsu 1968d). This combination has attracted more attention than all other combinations of stress-resultants combined. The torsion and shear combination attracted less attention regardless of its practical importance in design. Torsion and compression is the least studied regardless of its occurrence in design, like in the case of prestressed members or columns subjected to earthquake loading. Early experimental work, which focused on the ultimate strength and stiffness of concrete beams, mostly recommended interaction curves between torsion and flexure and, in some cases, with shear. Although early theoretical work focused on the ultimate strength and shape of interaction curves, most research was applicable only to under-reinforced members in which all the reinforcement yields at ultimate state. The first model to consider strain compatibility and that was capable of obtaining the full response of concrete members subjected to pure torsion was the CFT by Mitchell and Collins (1974). This model was further developed by Onsongo and Collins (Onsongo 1978) to calculate the full response under combined torsion, flexure, and axial load. These two models apply to both under-reinforced and over-reinforced sections, a feature unavailable in previous models concerned only with ultimate strength and not strain compatibility. Various models are capable of calculating the full response of sections subjected to some combination of the six stress resultants (Fig 6.1). One such model was developed in the 1970s by Rabbat and Collins (1977, 1978) in parallel with the development of CFT for torsion. Another model developed by Rahal and Collins (1995a), considers information on torsion and shear published in the 1980s and 1990s, like tension stiffening and concrete softening. Both models provide a unique perspective on the interaction between torsion and various stress resultants. Ewida and McMullen (1981) also developed a model to calculate the post-cracking response of sections subjected to combined torsion, shear, and flexure. Their model explained other phenomena like shear lag and effects of reinforcement on the interaction curve. Chapter 6 summarizes work on the interaction between torsion and other stress resultants. Although members reinforced only in the longitudinal direction are uncommon in practice, their behavior is discussed because in some cases they behave similarly to members with transverse and longitudinal reinforcement.

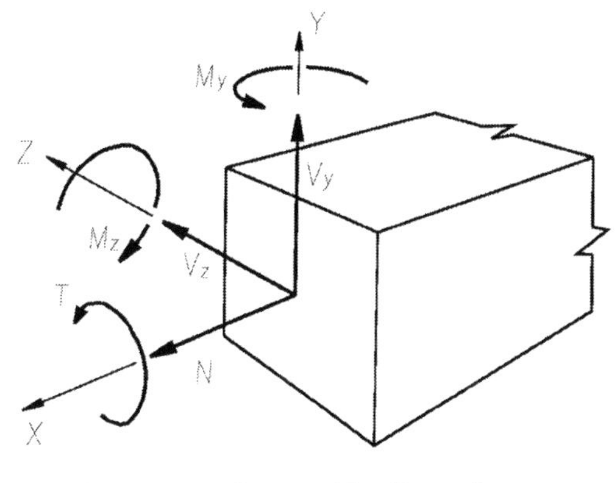

Fig. 6.1—General case of loading a beam cross section.

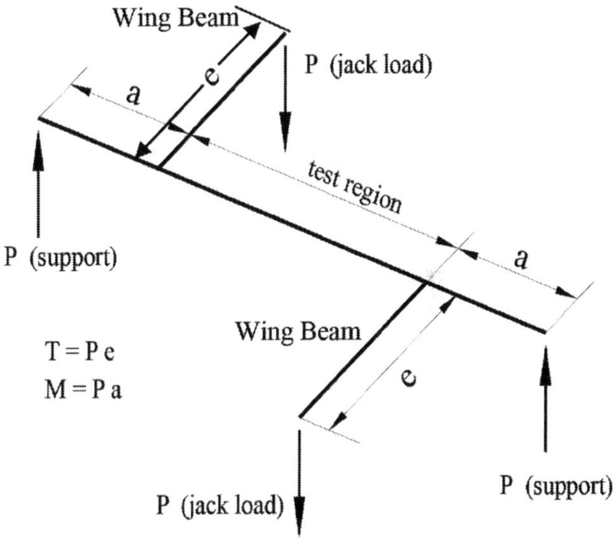

Fig. 6.2.1—Test setup for combined torsion and flexure.

6.2—Torsion and flexure

6.2.1 *General*—The combination of torsional (T) and flexural (M) moments can be critical in design. Aside from the case of pure torsion, this combination has prompted most of the torsion research combined with other stress-resultants for two reasons: first, researchers understood that such studies would be prudent. Second, the beams subjected to combined T and M are easier than other combinations in testing. In the absence of shearing forces, a four-point loading arrangement is easily modified to apply constant torsional moment along with the bending moment in the central portion of a beam specimen. Figure 6.2.1 shows that the central loads in a four-point load test setup can be applied at an eccentricity e, subjecting the test region of the specimen to a uniform combination of torsion and flexure. Torsional moment to bending moment ratio is controlled by distances e and a, as shown in Fig. 6.2.1. The most significant factors affecting the behavior of members subjected to combined torsion and flexure are the presence of transverse reinforcement, the torque-to-bending ratio (T/M), and the amount and distribution of the longitudinal reinforcement. Other factors include the cross section shape and concrete strength.

The presence of transverse reinforcement causes a considerable change in the behavior of members subjected to torsion. Sections 6.2.1 and 6.2.2 discuss the T-M interaction for members with and without this reinforcement.

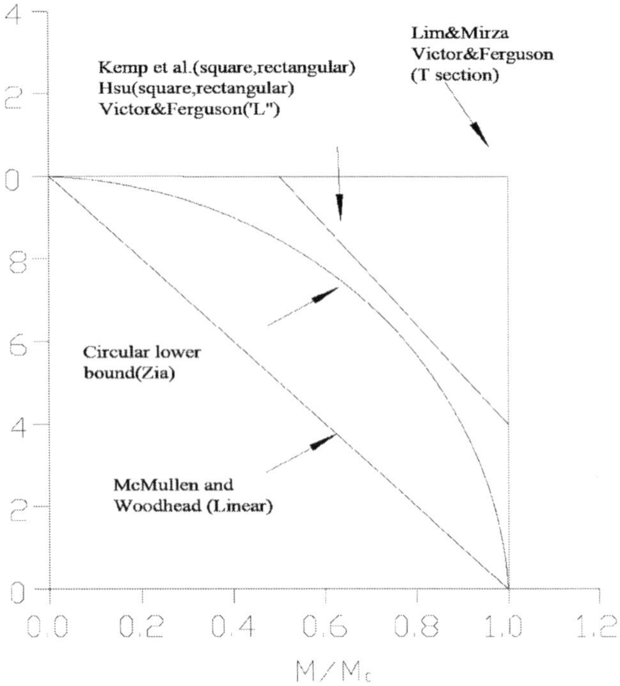

Fig. 6.2.2—Normalized T-M *interaction curves for members without transverse reinforcement.*

6.2.2 *Members without transverse reinforcement*—The *T-M* interaction in members containing only longitudinal reinforcement has limited practical importance, attracting a significant amount of experimental research, mainly due to its importance in understanding the behavior of members with transverse reinforcement.

Behavior and ultimate strength of these members is significantly influenced by the shape of the beam cross section, the amount and distribution of the longitudinal reinforcement, and the *T-M* ratio (Zia 1970). In prestressed members, the amount and eccentricity of the prestressing also affect the behavior because it causes a strain variation similar to that due to a compressive force or a flexural moment (Hsu 1968e).

Research on the *T-M* interaction in members containing only longitudinal reinforcement is mainly experimental. Figure 6.2.2 shows examples of the interaction curves recommended by various researchers. For square and rectangular sections, Kemp et al. (1961) and Hsu (1968a) independently suggested the trilinear interaction. Victor and Ferguson (1968) suggested a similar trilinear interaction for L-sections, and a square interaction for T-beams. Lim and Mirza (1968) also proposed a square interaction curve for T-beams. Based on evaluation of numerous experimental results, Zia (1970) observed that significant reduction in the torsion strength occurs at M/M_o between 0.5 and 1.0, and a circular interaction curve serves as a lower bound for the larger portion of the experimental results. The scatter in the results, however, is relatively large. Some test results have suggested a linear interaction (McMullen and Woodhead 1973).

The interaction between T and M in prestressed members depends largely on the eccentricity of the prestressing force.

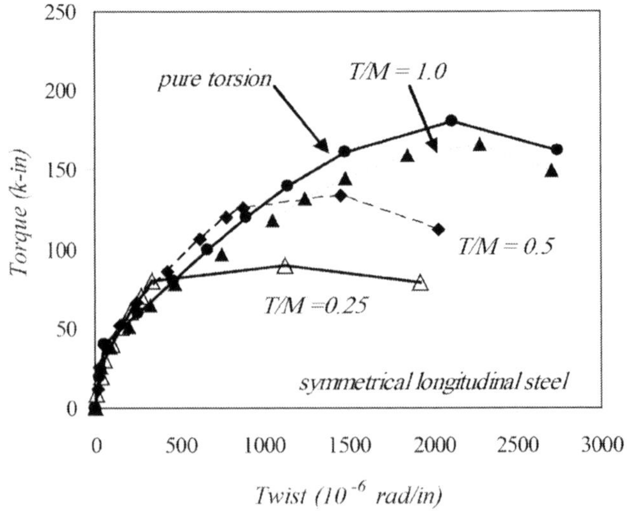

Fig. 6.2.3a—Torque-twist relationships at various T/M *ratios for sections with symmetrical longitudinal reinforcement (McMullen and Warwaruk 1967). (Note: 1 in.-kip = 0.113 kN·m; 1 in. = 0.0254 m.)*

The interaction for concentrically prestressed members is similar to that of nonprestressed members. For eccentrically prestressed beams, the induced flexural moment increases torsional strength when the eccentric prestressing force counteracts tension caused by the bending moment. ACI 318 gives no direct information on the *T-M* interaction in members without transverse reinforcement. The minimum amounts of longitudinal reinforcement for torsion and flexure and transverse reinforcement for torsion required for the combination are usually calculated independently for each stress resultant, and the sole interaction is from superposition of the amount of longitudinal reinforcement required for each stress-resultant.

6.2.3 *Members with transverse reinforcement*—Members reinforced in the longitudinal and transverse directions exhibit significant post-cracking strength and ductility compared with members reinforced in the longitudinal direction only. Behavior of these members also depends significantly on the distribution and amount of longitudinal reinforcement. Cases of symmetrically and unsymmetrically reinforced beams are discussed later. Figure 6.2.3a shows the torque-twist relationship of four 152 x 305 mm (6 x 12 in.) under-reinforced solid reinforced concrete specimens with symmetrical longitudinal reinforcement tested by McMullen and Warwaruk (1967) at variable *T/M* ratios ranging from pure torsion to *T/M* = 0.25. As shown in Fig. 6.2.3a, the addition of flexural moment significantly reduced the torsional strength, ductility, and post-cracking stiffness.

Figure 6.2.3b shows the experimentally observed normalized *T-M* interaction from two series of symmetrically reinforced specimens. Five under-reinforced beams, each 152 x 305 mm (6 in. x 12 in.), were tested by McMullen and Warwaruk (1967). The torque-twist results of four of those beams are shown in Fig. 6.2.3b. In addition, five symmetrically prestressed beams, each 305 x 432 mm (12 x 17 in.), were tested by Mardukhi and Collins (Mardukhi 1974). This

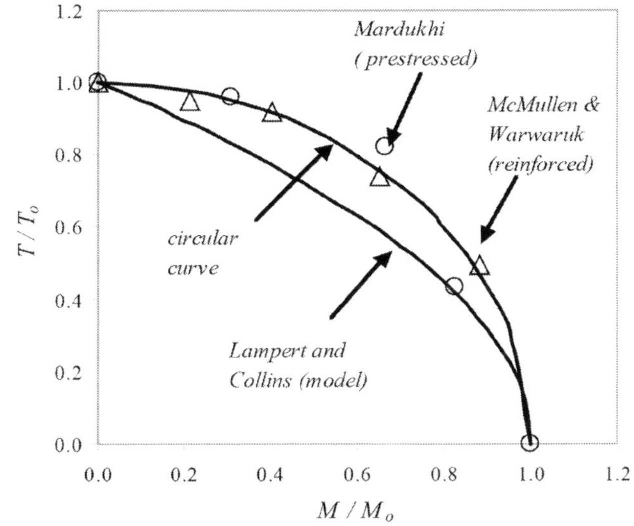

Fig. 6.2.3b—Normalized T-M *interaction curves in symmetrically longitudinally reinforced members with transverse reinforcement.*

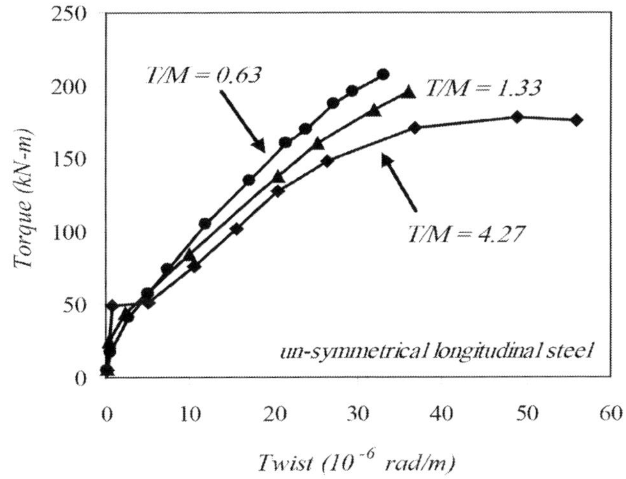

Fig. 6.2.3c—Torque-twist relationships at various T/M *ratios for sections with unsymmetrical longitudinal reinforcement (Onsongo 1978). (Note: 1 kN·m= 8.85 in-kip; 1 m = 39.4 in.)*

study showed that a flexural moment equal to 60 percent of the ultimate flexural strength caused only about a 10 percent reduction in the torsional strength. Similarly, a torsional moment equal to 40 percent of the ultimate pure torsion strength caused an approximately 20 percent reduction in the flexural strength.

The behavior of unsymmetrically reinforced beams differs significantly from the previous case. Figure 6.2.3c shows the torque-twist relationship of three under-reinforced hollow reinforced concrete specimens tested by Onsongo and Collins (Onsongo 1978) at variable T/M ratios from 0.63 to 4.27. Figure 6.2.3d shows the observed interaction from four series of unsymmetrically reinforced beams, one series tested by McMullen and Warwaruk (1967), two by Onsongo and Collins (Onsongo 1978), and one by the Zurich group (Lampert and Thürlimann 1968, 1969). The ratio of compression to tension longitudinal reinforcement yield force $r = (A_s'f_y/A_sf_y)$ ranged from 0.1 to 0.27.

Figures 6.2.3c and 6.2.3d show that the addition of a small flexural moment can significantly increase the torsional strength and post-cracking stiffness of unsymmetrically reinforced beams. In pure torsion, the additional bottom longitudinal reinforcement available in unsymmetrically reinforced sections does not increase the ultimate strength because the weaker top reinforcement is critical. The addition of a flexural moment introduces compression in the weaker top reinforcement and increases its resistance to torsional shear stresses. Based on the same experimental study (Onsongo 1978), it appears that a small axial compressive force substantially increases post-cracking stiffness of a torsionally cracked member with strength also enhanced. This is because the compressive axial load counteracts the widening of spiral cracks. This point may be particularly relevant in the case of structures under seismically-induced torsion.

The peak in the interaction curve depends on the ratio r and if the beam is under-reinforced or over-reinforced in the transverse direction. An increase of up to 30 percent in torsional strength was observed with the addition of flexural moment equal to 40 percent of the pure flexural strength in the under-reinforced tests by McMullen and Warwaruk (1967). Onsongo (1978), however, observed a 25 percent and 6 percent increase in under-reinforced and over-reinforced beams, respectively. The interaction curve seems to be less affected by the ratio r in over-reinforced members, where Onsongo reported only a 6 percent increase in torsional strength in spite of the significantly low value of r (approximately 0.1). Zararis and Penelis (1986), who tested unsymmetrically reinforced T-beams under combined flexure and torsion, observed a trend similar to that shown in Fig. 6.2.3d. Due to weakness in the smaller top reinforcement, the presence of a flexural moment increased torsional strength by approximately 18 percent. Presence of a flexural moment reduces the torsional ductility of a member with symmetrical or unsymmetrical longitudinal reinforcement, as shown in Fig. 6.2.3a and 6.2.3c.

The torque-twist diagrams show that the ultimate torsional strength was reached at a smaller twist. Similarly, the presence of relatively small flexural moment, $T/M = 4.27$ in Fig. 6.2.3c, also affected the curvature of unsymmetrically reinforced members subjected to torsion. Figure 6.2.3e shows the moment-curvature diagrams of two unsymmetrically reinforced hollow specimens tested by Onsongo and Collins (Onsongo 1978). The flexural moment was predominant in one specimen ($T/M = 0.63$), whereas torsion was predominant in the other ($T/M = 4.27$). Torsional shearing stresses were equilibrated by tensile stresses in the longitudinal direction. Upon cracking, the weaker top reinforcement elongates more than the stronger bottom reinforcement, causing greater reverse curvature than that caused by less-predominant flexural moment in the torsion predominant beam ($T/M = 4.27$). Therefore, the section is subjected to an

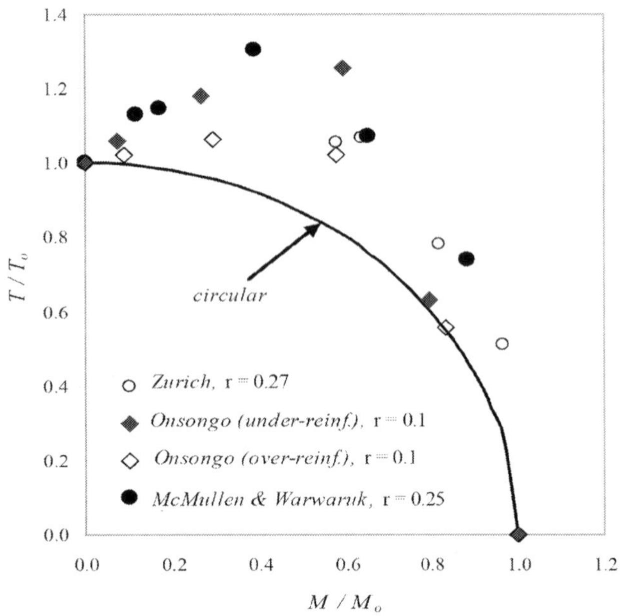

Fig. 6.2.3d—Normalized T-M *interaction curves in unsymmetrically longitudinally reinforced members with transverse reinforcement.*

overall curvature opposite in direction to the relatively small flexural moment.

The *T/M* ratio affects the diagonal compression angle and the beam crack pattern. The presence of flexural moment introduces tensile and compressive strains in the bottom and top faces of the section, respectively. Compression in the top face delays cracking, in some cases until ultimate strength is reached (Johnston and Zia 1975). In the case of pure torsion, the crack angle on the bottom face is diagonal, and in the case of pure flexure, the crack angle becomes normal to the beam longitudinal axis. The angle on the sides is steep near the bottom of the section due to tensile strains and becomes flatter near the top from the compressive strains caused by flexural moment. Numerous theories (McMullen and Warwaruk 1967; Walsh et al. 1967; Lampert and Collins 1972; Elfgren et al. 1974a) recognize the phenomenon shown in Fig. 6.2.3d. Interaction curves recommended by Lampert and Collins (1972) have been simplified. They suggested that in under-reinforced sections, an increase in torsional strength is related to the ratio of the yield force of the compression reinforcement to tension reinforcement, $r = (A_s'f_y/A_sf_y)$. The flexural moment creates a tensile force M/d_v in the bottom stringer and an equal compressive force in the top stringer. The torsion induces a total tensile force of $Tp_o/2A_o\cot\theta$ in the longitudinal reinforcement, where d_v is the distance between the top and bottom longitudinal reinforcement, p_o and A_o are the shear flow perimeter and area, respectively, and θ is the diagonal crack angle. The longitudinal reinforcement forces are additive, as shown in Fig. 6.2.3f.

Two simple equations (Eq. (6.2.3a) and (6.2.3b)) are derived based on this theory. Equation (6.2.3a) is for the case where the bottom longitudinal reinforcement yields along with the stirrups

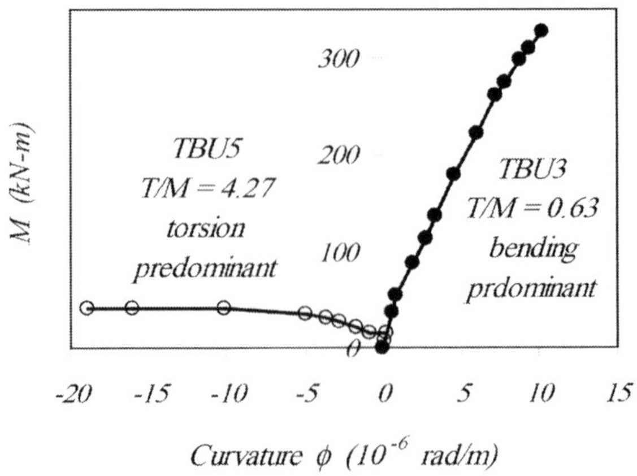

Fig. 6.2.3e—Observed flexural moment-curvature relationships at various T/M *ratios (Onsongo 1978). (Note: 1 kN·m= 8.85 in.-kip; 1 m = 39.4 in.)*

$$r\left(\frac{T}{T_o}\right)^2 + \frac{M}{M_o} = 1 \tag{6.2.3a}$$

Equation (6.2.3b) applies when the weaker top longitudinal reinforcement yields along with the stirrups

$$\left(\frac{T}{T_o}\right)^2 - \frac{1}{r}\frac{M}{M_o} = 1 \tag{6.2.3b}$$

Figure 6.2.3g shows the interaction curves for members with *r* values of 0.3, 0.5, and 1. The increase in torsional strength calculated using both equations is larger than the experimentally-observed increase shown in Fig. 6.2.3d. Both equations give acceptable results when pure torsion strength T_o is taken as the conservatively calculated value, as demonstrated by Lampert and Collins (1972).

Since 1971, ACI code procedures for the design of members subjected to flexure and torsion are based on the superposition of reinforcement required to resist torsion (longitudinal and transverse) and flexural moment (longitudinal). Experience has proved that these procedures are conservative and suitable for design. Calculating cross section strength using the code equation is not a simple task, especially in common cases where shear is present and the longitudinal reinforcement is not symmetrically placed in the top and bottom sides of the cross section.

Research by Greene and Belarbi (Greene and Belarbi 2006a; Greene 2006) expands the STM described in Chapter 5 to members under torsion combined with shear and flexure. The combined-action STM can be used to predict the load-deformation response for hollow or solid reinforced concrete members and also to generate torque-flexure interaction curves for members under torsion combined with shear and flexure. The model considers a softened stress-strain relationship for concrete and the tension-stiffening effect of concrete surrounding the reinforcement to predict service-level deformation (Greene and Belarbi 2006b). Figure 6.2.3h shows the interaction curves created using the combined-action STM

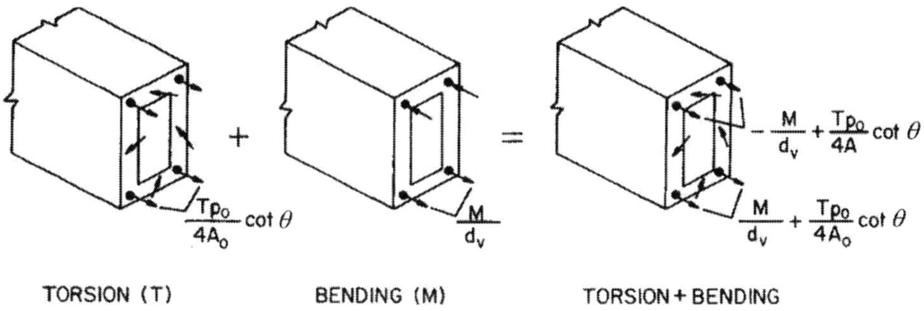

Fig. 6.2.3f—Superposition of forces due to torsion and bending (Hsu 1993).

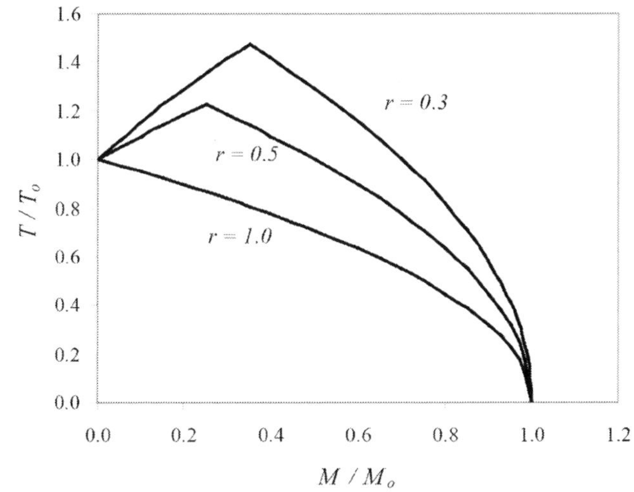

Fig. 6.2.3g—Normalized T-M *interaction curves for members with unsymmetrical longitudinal reinforcement (Lampert and Collins 1972).*

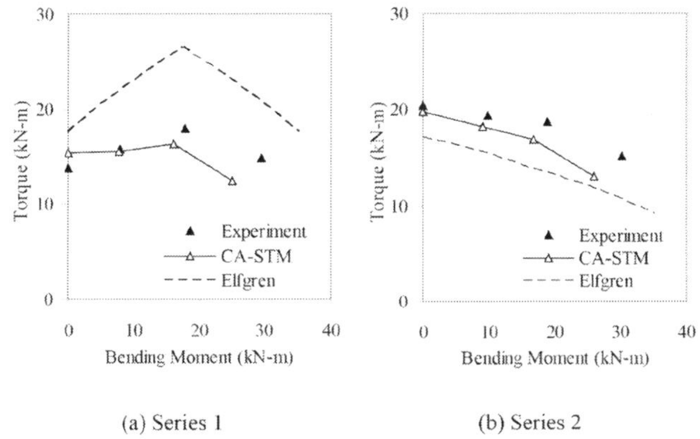

Fig. 6.2.3h—Interaction curves predicted by combined-action STM and Elfgren Model for McMullen and Warwaruk (1967). Series 1 and 2 (Greene 2006). (Note: 1 kN·m= 8.85 in.-kip.)

for two series of beams under combined torsion and flexure tested by McMullen and Warwaruk (1967). Members of Series 1 in Fig. 6.2.3h(a) were symmetrically reinforced and members of Series 2 in Fig. 6.2.3h(b) had more reinforcement on the flexural tension side. For comparison, the model by Elfgren et al. (1974a) was also used to predict the interaction curve. ACI code equations were used to calculate T_o and M_o for use in Elfgren's model. Figure 6.2.3h(a) shows that for this test series, ACI 318-11 is unconservative in its calculation of T_o, which affects the calculation of the interaction curves.

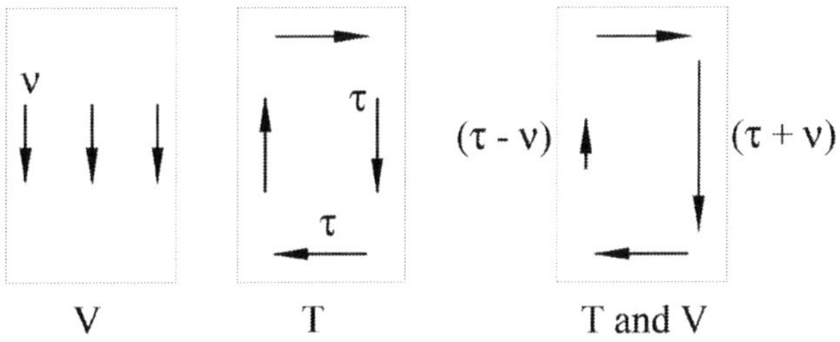

Fig. 6.3.1—Shearing stress due to shear, torsion, and combined shear and torsion.

6.3—Torsion and shear

6.3.1 *General*—Torsion and shear cause shear stresses across a section, as shown in Fig. 6.3.1. The shear stresses τ due to the torsional moment T circulate around the section, whereas the vertical shear force V induces shear stress v. The side of the section where the stresses are additive ($\tau + v$) is critical in design due to the great intensity of the shearing stresses. Intense stresses increase tensile strains in transverse and longitudinal steel reinforcement and compressive and tensile strains in the concrete.

In spite of its practicality, work on *T-V* interaction is relatively limited mainly because shear cannot be induced without causing a flexural moment along the beam's test region. Therefore, the torsion-shear condition is achieved only at the point of inflection. Otherwise, an experimentally observed *T-V* interaction curve is only a projection of a part of the *T-V-M* curve on the *T-V* coordinate plane. Additionally, the pure shear strength of a beam cannot be determined experimentally. Typically, this maximum should be defined from a calculated value.

6.3.2 *Members without transverse reinforcement*—Nylander (1945) reported that tests on members containing only longitudinal reinforcement showed considerable scatter, with most tests falling between a linear and circular interaction curve. Other tests by Birkland (1965), Hamilton (1966), and Ersoy and Ferguson (1968) have shown that a circular interaction is more accurate. Nylander's model (1945) and

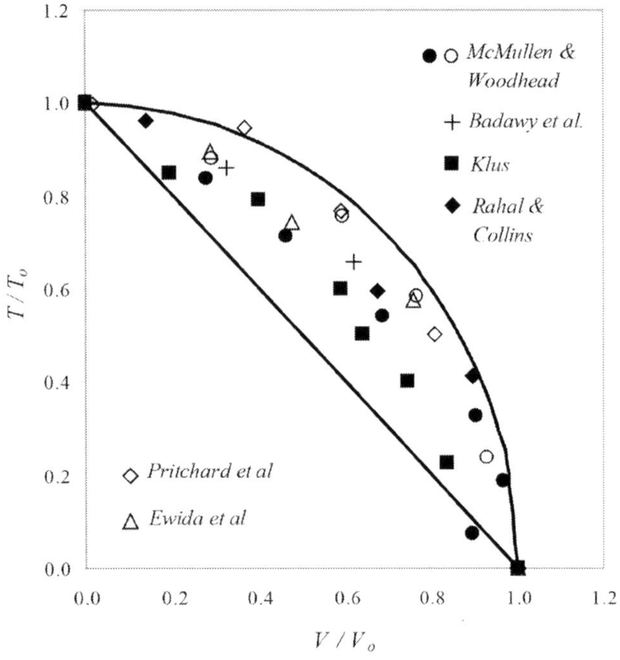

Fig. 6.3.3a—Normalized T-V *linear and circular interaction curves for members with transverse reinforcement.*

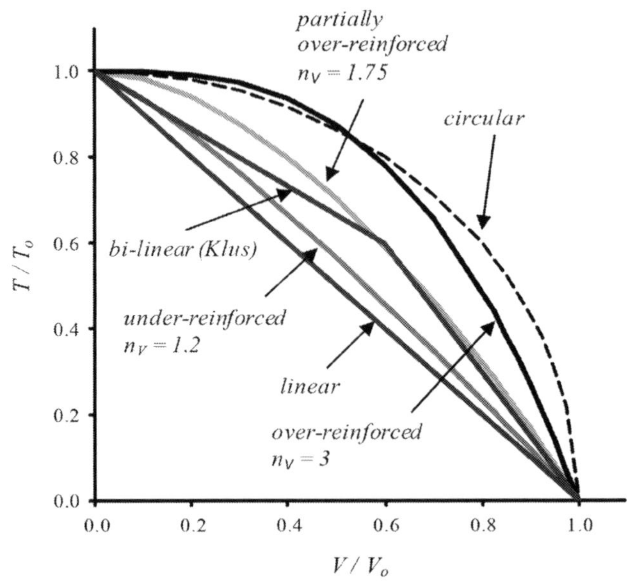

Fig. 6.3.3b—Various normalized T-V *interaction curves in the literature.*

those proposed by other researchers conservatively recommend a linear interaction.

Mirza and McCutcheon (1968) tested quarter-scale specimens and found that the longitudinal reinforcement has a significant effect on the interaction curve. The result's scatter was considerable and a lower-bound linear interaction was recommended. McMullen and Woodhead (1973) tested eccentrically prestressed beams under various combinations of torsion, shear, and flexure and found that Eq. (6.3.2) adequately fit the results

$$\frac{V}{V_o}+\left(\frac{T}{T_o}\right)^2=1 \tag{6.3.2}$$

When torsion design provisions were introduced in ACI 318-71, a circular interaction between V_c and T_c was assumed. This approach slightly increased the design time in members with transverse reinforcement. Abandoned in ACI 318-95, concrete contribution to torsion, T_c, was disregarded to simplify design. Considerable scatter produced by test specimens without transverse reinforcement confirms the sound principle that members subjected to such combined actions should contain a minimum amount of transverse and longitudinal reinforcement.

6.3.3 *Members with transverse reinforcement*—Development of models for torsion and shear has made it possible to calculate shear strength with accuracy similar to the flexure theory's. Early work on *T-V* interaction was largely experimental. Figure 6.3.3a shows some of the experimental results and Fig. 6.3.3b presents some proposed interaction curves. Experimental results were scattered, with most strength values falling between linear and circular curves (Fig. 6.3.3a). Because the pure shear strength V_o is significantly affected by the accompanying bending moment, it cannot be accurately represented and a calculated value is used. Different researchers use varying definitions of V_o, making it difficult to compare the suggested curves in Fig. 6.3.3b. Figure 6.3.3a shows that the presence of torsional moment reduces shear strength, especially if the torque is more than 25 percent of pure torsional strength. The amount of transverse reinforcement is considered the main factor affecting the shape of the curve.

Based on an experimental testing program, Klus (1968) suggested a bilinear interaction curve as shown in Fig. 6.3.3b. The theoretical model and the experimental results of Ewida and McMullen (1981) showed the amount of reinforcement has a considerable effect on the interaction curve. They suggested a simplified equation

$$\frac{T}{T_o}+\left(\frac{V}{V_o}\right)^{n_V}=1 \tag{6.3.3a}$$

where

n_V = 1.2 for under-reinforced sections in which all reinforcement (both longitudinal and transverse) yields when the section reaches ultimate strength

= 1.75 for partially under-reinforced sections in which only stirrups yield or only the longitudinal reinforcement yields when the section reaches ultimate strength

= 3.0 for completely over-reinforced sections in which concrete crushes before yielding in any of the reinforcement.

All three curves are shown in Fig. 6.3.3b. For under-reinforced sections (n_V = 1.2), the curve is almost linear, whereas for completely over-reinforced sections, the interaction is closer to a circular curve. Klus's (1968) bilinear interaction curve lies between the linear and circular curves.

Elfgren et al. (1974b) developed a theoretical model for the ultimate strength of members subjected to combined torsion, shear, and flexure as discussed in 6.5. The *T-V* interaction curve is circular when Failure Modes 1 and 2 govern, but it is close to a straight line when Failure Mode 3 governs. These three failure modes were introduced in 3.3.5. Ewida and McMullen (1981) observed that the addition of a relatively small shear force (20 percent of the shear strength calculated using ACI equations) caused a slight torsional strength increase of 3 percent, a phenomenon they called "shear lag." Rahal and Collins (1995a) confirmed the significant effect the amount of reinforcement has on the interaction curve. They also showed that in completely over-reinforced sections, the addition of shear causes redistribution in the shear stresses and a subsequent slight increase in the torsional strength. ACI 318 adopts a superposition approach in which the reinforcement required to resist *T* is added to that required to resist *V*. To control the diagonal cracks and ensure that the stirrups yield before crushing occurs in the concrete, ACI 318 sets an upper limit on the shear stresses due to torsion and shear.

Figure 6.3.3c (Rahal and Collins 2003b) shows a comparison between the calculations of ACI 318-02 and the 1998 AASHTO LRFD Bridge Design Specifications with test results from a series of large-scale beams subjected to a combination of *T*, *V*, and a relatively small *M* (Rahal and Collins 1995b). The test setup for this study ensured failure near the point of inflection to accurately represent the interaction of torsion and shear. Loads were introduced to the test region through wing beams instead of by typical concentrated loads applied at the beam top. This loading minimized disturbance in the test region, representing the actual situation better than the typical setup.

ACI design provisions give conservative results if the recommended value of $\theta = 45$ degrees is used for the angle of inclination of the diagonal struts. Using the lowest allowable value of $\theta = 30$ degrees gives less consistent results, overestimating failure loads at high torsion-to-shear ratios. Nonetheless, the AASHTO (1998) provisions resulted in a calculated value for θ of approximately 36 degrees for the series of beams studied. This value compared well with the inclinations of cracks observed in the tests, producing better correlation with failure loads. Calculations from ACI and AASHTO resulted in a nearly linear interaction curve, with a cut-off horizontal plateau defined by the limit set on torsional strength to avoid concrete crushing before the reinforcement yields. Variation in the intensity and direction of shear stresses (Fig. 6.3.1) affects the concrete crack patterns and reinforcement strains. Figure 6.3.3d shows the angle of diagonal compressive stresses and crack orientation at variable *T*/*V* ratios calculated by Rabbat and Collins (1977, 1978) using a Variable Angle Space Truss Model. In pure shear, the side cracks are parallel, where in pure torsion, they are nearly perpendicular. Crack orientation on the side where shear stresses are subtractive ($\tau - v$) depends on the relative intensity of stresses τ and v. This trend was confirmed in tests by Rahal and Collins (1995b), Greene (2006), and Greene and Belarbi (2008).

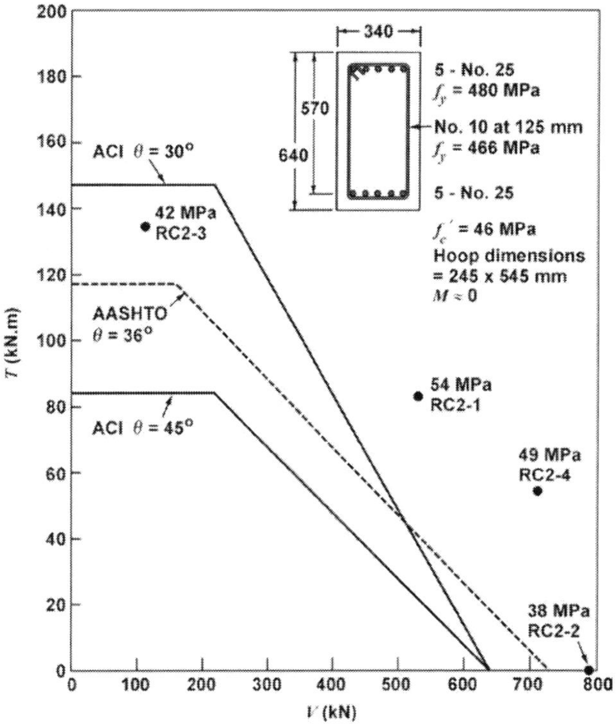

Fig. 6.3.3c—Comparison between observed and calculated T-V *interaction curves (Rahal and Collins 2003b). (Note: 1 kN·m= 8.85 in.-kip; 1 MPa = 0.145 ksi.)*

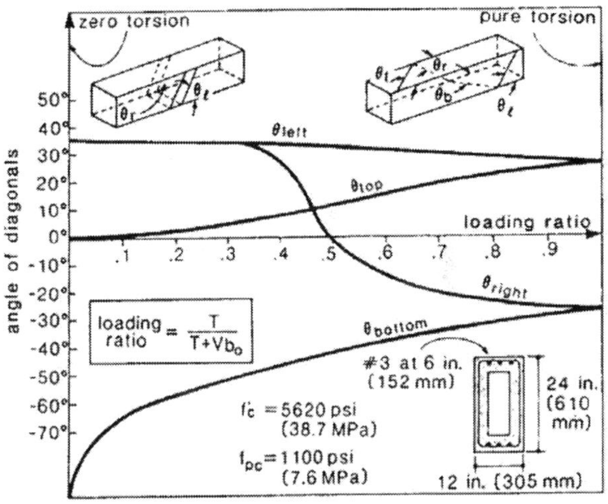

Fig. 6.3.3d—Variation of angle of diagonal compressive stresses depending on T/V *(Rabbat and Collins 1978).*

Stresses in the transverse and longitudinal reinforcement are also significantly larger on the side where the shear stresses are additive ($\tau + v$) rather than subtractive ($\tau - v$). The difference in longitudinal stress magnitude on vertical sides of the section shown in Fig. 6.3.1 causes a lateral curvature, as shown in Fig. 6.3.3e, tested by Rahal and Collins (1995b). They observed lateral curvatures in two of the specimens tested at a *T*/*V* of 76 mm (3.0 in.) (Specimen RC2-4) and 156 mm (6.14 in.) (Specimen RC2-1). Calculations of the model (Rahal and Collins 1995a) plotted with

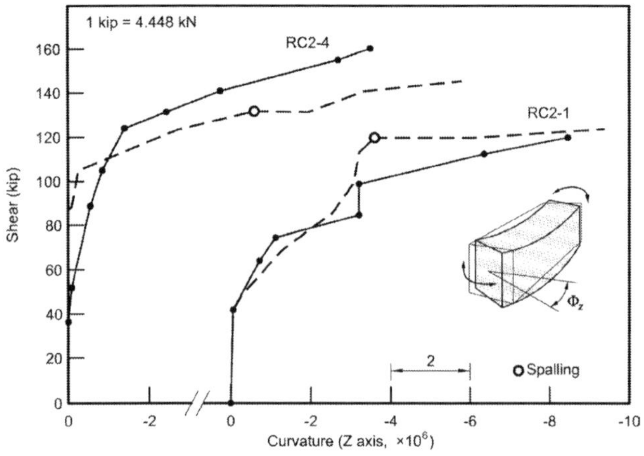

Fig. 6.3.3e—Observed and calculated lateral curvature in sections subjected to combined torsion and shear (Rahal and Collins 1995a).

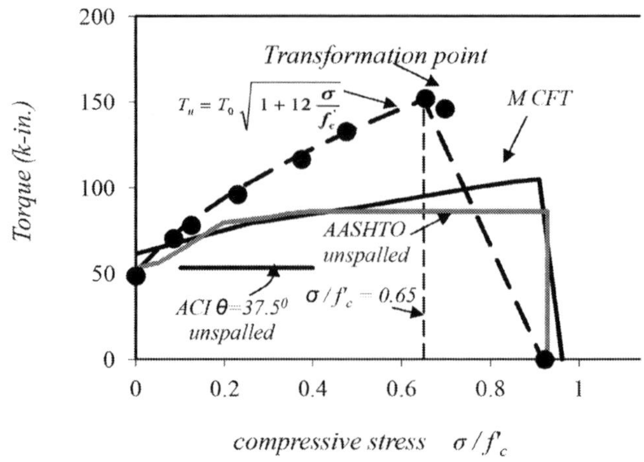

Fig. 6.4a—Torsion-axial compression interaction in reinforced concrete (tests from Bishara and Peir [1968]). (Note: 1 in.-kip = 0.113 kN·m.)

dashed lines show that available theoretical models capture this phenomenon.

6.4—Torsion and axial load

Elements subjected to combined torsion and compressive forces, N, include bridge piers under gravity and horizontal loads, and prestressed concrete beams subjected to torsion. Cracked spandrel beams subjected to torsion tend to elongate. Columns and walls restraining the elongation induce beam compression acting in combination with the compatibility torsion and other stress-resultants. Though the combination of torsional moment (T) and axial force (N) can be critical in design for these cases, it has not been adequately studied, particularly in experimental research on nonprestressed elements and combined torsion and axial tension. Lack of adequate studies is due to the relative rarity of this combination compared with others discussed. Torsion tests on concrete beams subjected to various prestressing levels simulate conditions similar to those of reinforced concrete beams subjected to combined torsion and axial compression. Experimental results on members without stirrups by Humphreys (1957) and Zia (1961) showed trends similar to those observed by Bishara and Peir (1968), which are given in Fig. 6.4a. A drop in torsional strength beyond the "transformation point" was clear only in Zia's tests on prestressed concrete beams subjected to torsion. Humphreys reported a 170 percent increase in torsional strength and Zia a 190 percent increase in torsional strength due to prestressing. Based on their experimental findings, Bishara and Peir (1968) recommended Eq. (6.4) to calculate an increase in torsional strength in the presence of an axial compressive load

$$T = T_o\sqrt{1+12\frac{\sigma}{f_c'}} \tag{6.4a}$$

The descending part is taken as a line joining the transformation point (defined as the point at $\sigma = 0.65f_c'$ and $T_u = 3T_o$) and pure compression strength. Results from Eq. (6.4) and the descending branch plotted in Fig. 6.4a demonstrate suitability to capturing the behavior of Bishara's tests. The results provided (Bishara and Peir 1968), however, were less conservative than experimental results from prestressed beams without stirrups tested by Humphreys (1957) and Zia (1961).

The square root factor in Eq. (6.4a) was labeled the prestressed factor by Hsu (1968b). He showed that this prestressed factor can be derived theoretically from an element subjected to shear and compression, assuming the principal tensile stress reached the tensile strength of concrete. The prestressed factor can also be derived from the skew-bending theory of a beam in pure torsion. To ensure best fit with the test results, the constant 12 in the prestressed factor was reduced to 10, resulting in Eq. (6.4b).

$$\sqrt{1+10(\sigma/f_c')} \tag{6.4b}$$

The torsional strength based on a 45-degree angle of the compression diagonals and spalled dimensions of the section described previously is 2.1 kN·m (18.6 in.-kip)—38 percent of experimental strength. This strength is increased to 4.62 kN·m (40.9 in.-kip)—84 percent of experimental strength—if unspalled dimensions are used. For prestressed members, ACI 318 allows a 37.5 degrees value for the angle of the compression diagonals, leading to a strength of 6.05 kN·m (53.3 in.-kip). This value, which is plotted in Fig. 6.4a, remains considerably smaller than the experimentally observed maximum torque of 17.2 kN·m (151.9 in.-kip) acting in combination with a compressive stress of approximately $0.65f_c'$. The general procedure for the torsion and shear design method in AASHTO (1998) specifications accounts for the effect of compression and prestressing on torsional strength.

Figure 6.4a shows calculations for the AASHTO LRFD method (general method) based on unspalled dimensions, which are more likely for small concrete cover (19 mm [0.75 in.]) used in Bishara and Peir's (1968) tests. Although this method underestimates an increase in torsional strength at

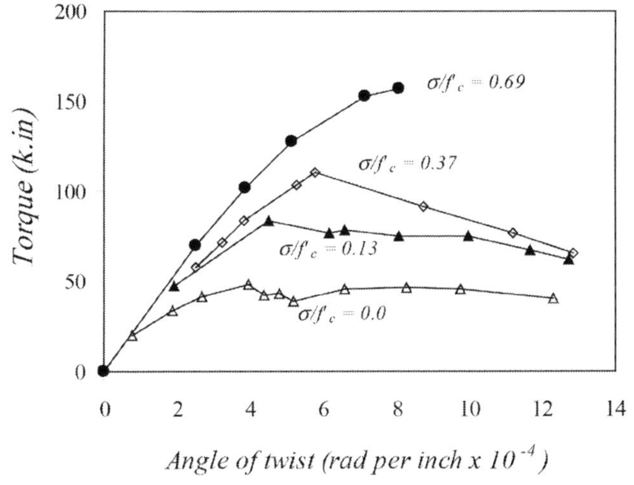

Fig. 6.4b—Torque-twist relationships of reinforced concrete columns under different compressive stresses (Bishara and Peir 1968). (Note: 1 in.-kip = 0.113 kN·m; 1 in. = 0.0254 m.)

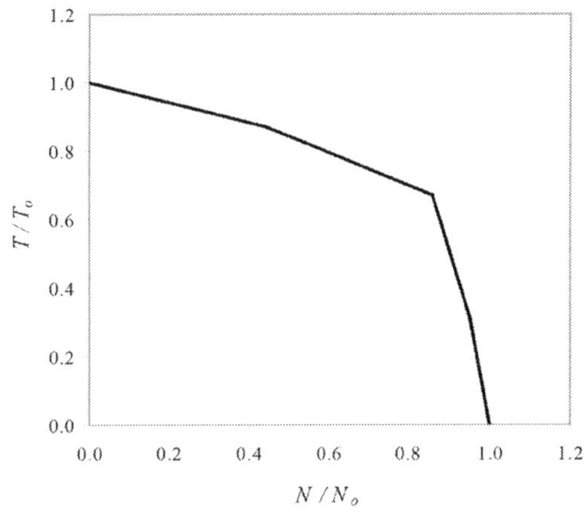

Fig. 6.4c—Torque-axial tension calculated using diagonal compression field theory (Rahal and Collins 1995a).

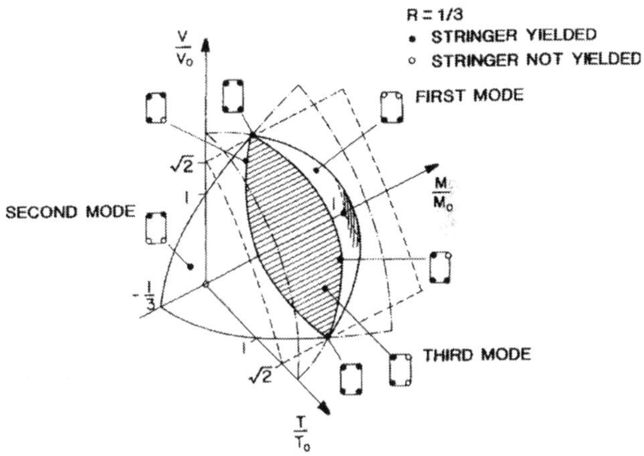

Fig. 6.5a—Interaction surface for torsion, shear, and flexure (Hsu 1993).

higher compressive loads, it produces slightly better predictions than ACI 318. Figure 6.4a also shows the results of a detailed analysis based on the modified compression field theory (CFT) (Rahal and Collins 1995a, 2003a). Similar to the AASHTO LRFD method (general method), this model underestimates the torsional strength increase at high compressive loads while providing generally conservative results. Figure 6.4b shows the torque-twist relationships for some tests by Bishara and Peir (1968). Compressive stress increased torsional strength and torque at peak strength, but decreased torsional ductility in members with compressive stresses greater than approximately $0.4f_c'$.

Combined torsion and axial tension in beams and columns is uncommon. The rarity of this combination and difficulties testing reinforced concrete sections under this forced combination account for little experimental data on the subject. Figure 6.4c shows the torsion-axial tension interaction calculated using diagonal CFT (Rahal and Collins 1995a, 2003a) for an under-reinforced section. Axial tension reduced torsional strength most significantly after tensile forces reach 85 percent of pure tensile strength.

6.5—Torsion, shear, and flexure

In most cases of practical importance, torsion acts in combination with shear and flexure. Research on pure torsion, torsion combined with flexure, and torsion combined with shear in members with or without reinforcement is aimed at understanding the general behavior of the three stress-resultants T, M, and V. Similar to work on the combinations discussed above, early research (Hsu 1968a; Johnston 1971; McGee and Zia 1973; Elfgren et al. 1974a) focused on studying ultimate strength and sought to produce three-dimensional interaction surfaces. Further work created the ability to obtain the full response of reinforced and prestressed concrete beams subjected to various stress combinations (Rabbat and Collins 1978; Rahal and Collins 1995a, 2003a).

Elfgren et al.'s (1974a) theory assumes that a rectangular box section subjected to torsion, shear, and flexure could fail in one of three modes. In first mode failure, the bottom longitudinal reinforcement and transverse reinforcement yield on the side where shear and torsional stresses are additive. In second mode failure, the top longitudinal reinforcement and transverse reinforcement yield on the additive side. Third mode failure occurs when longitudinal and transverse reinforcement on the additive side yield. Nondimensionalized interaction relationships for M, V, and T for the three failure modes are given by Eq. (6.5a) through (6.5c). Figure 6.5a illustrates the three interaction relationships for the case of $r = 1/3$, where $r = (A_s'f_y/A_sf_y)$. A systematic derivation of Elfgren's nondimensional interaction surface is found in Hsu's (1993) book.

$$\text{Mode 1:} \quad \frac{M}{M_o} + r\left(\frac{V}{V_o}\right)^2 + r\left(\frac{T}{T_o}\right)^2 = 1 \qquad (6.5a)$$

$$\text{Mode 2:} \quad -\frac{1}{r}\left(\frac{M}{M_o}\right) + \left(\frac{V}{V_o}\right)^2 + \left(\frac{T}{T_o}\right)^2 = 1 \qquad (6.5b)$$

$$\text{Mode 3:} \quad \left(\frac{V}{V_o}\right)^2 + \left(\frac{T}{T_o}\right)^2 + 2\left(\frac{VT}{V_oT_o}\right)\sqrt{\frac{2d_v}{p_o}} = \frac{1+r}{2r} \qquad (6.5c)$$

Researchers in Zürich carried out work in this area. Their results are reported in Lüchinger (1977), Müller (1976, 1978), and Thürlimann (1978). Based on their work, Elfgren (1979) provided a kinematics model that gives the same equations, Eq. (6.5a) through (6.5c), derived earlier with equilibrium methods (Elfgren 1972a,b; Elfgren et al. 1974a,b). The kinematics model is illustrated in Fig. 6.5b and Fig. 6.5c. Beam deformations under combined torsion, shear, and flexure were studied by Petersson (1972), Karlsson (1973), and Gabrielsson (1999). Both theoretical and experimental work has been carried out by Teutsch (1980) and Kordina and Teutsch (1983, 1985).

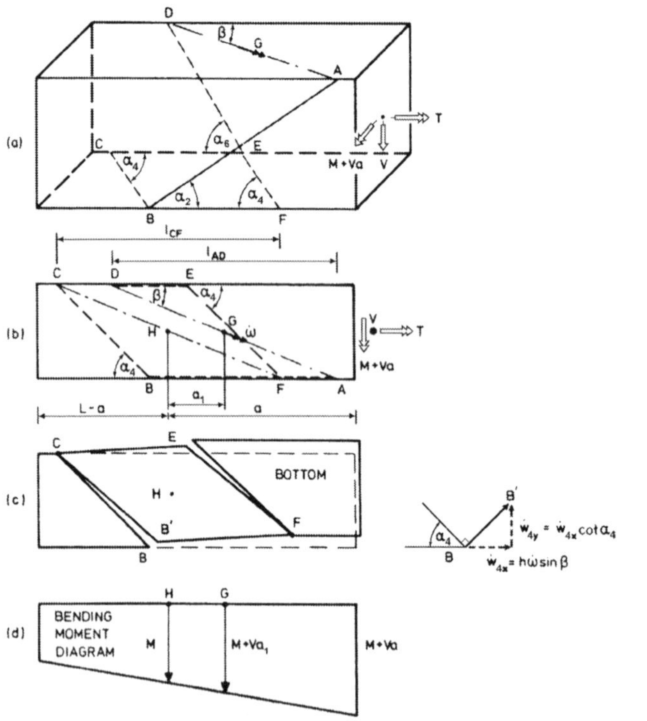

Fig. 6.5b—Kinematics failure model: (a) General view; (b) Model seen from above; (c) Deformations in bottom; (d) Bending moment diagram (Elfgren 1979).

The beam shown in Fig. 6.5c was loaded at midspan with an eccentric point load acting downward. The beam was rectangular, with dimensions 100 x 200 x 3300 mm (3.94 x 7.87 x 130 in.) (width x height x length), and the stirrup strength was 0.236 MN/m (13.5 kip/in.). The relationship among torsional moment T, vertical shear force V, and flexural moment M in the failure section was M:T:V = 0.1:0.5:0.2 (Elfgren 1979). The numerals shown along the cracks in Fig. 6.5c refer to the applied load when this part of the crack became visible (units in multiples of 10 kN [2.25 kip]). In the left beam, two failure cracks, ABC and FED, are indicated as well as a rotation hinge AD (compare with Fig. 6.5b).

The combined-action STM (Greene and Belarbi 2006a, 2009a,b) can also generate torsional-flexural moment interaction curves for members under torsion combined with shear and flexure. Figure 6.5d shows the interaction curves for two series of beams under combined torsion, shear, and flexure tested by McMullen and Warwaruk (1967). Members of Series 5 and 6 were unsymmetrically reinforced with more longitudinal reinforcement in the bottom side. For comparison, the model by Elfgren et al. (1974a) was also used to predict the interaction curve. ACI 318 code equations were used to calculate T_o, M_o, and V_o for use in Elfgren's model. Figure 6.5d(a) shows that for this test series, ACI 318-11 is unconservative in its calculation of T_o, which affects the calculation of the interaction curves, similar to the observation related to Fig. 6.2.3h(a).

Leonhardt et al. (1968) performed large-scale tests on two prestressed concrete box beams representing hollow box bridges loaded in torsion, shear, and flexure. Figure 6.5e shows crack patterns after failure. The beams had a depth of 0.78 m (2.6 ft) and a span of 6.00 m (19.7 ft), and web, top, and bottom flange widths of only 80 mm (3.1 in.). The beams were eccentrically loaded at a midspan diaphragm, and the bottom slab reinforcement was varied between an orthogonal mesh and mesh inclined at 45 degrees. In both beams, the web near the eccentrically applied load failed in

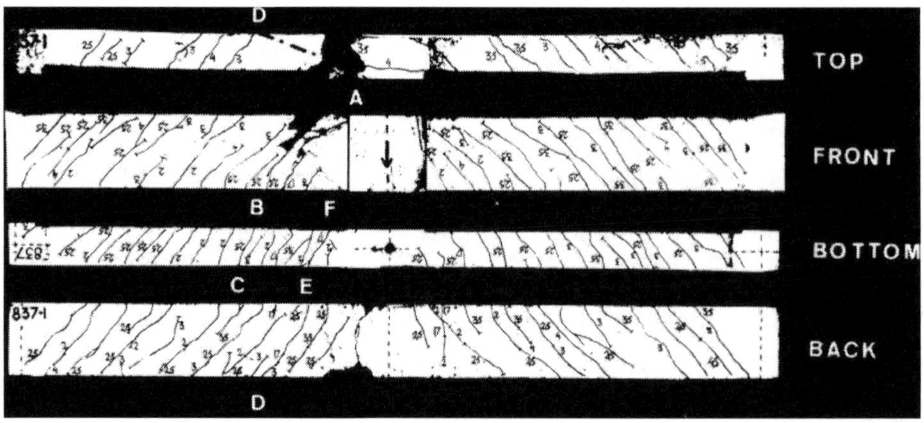

Fig. 6.5c—Crack pattern and failure mechanism for a beam loaded in combined torsion, shear, and flexure (Elfgren 1979).

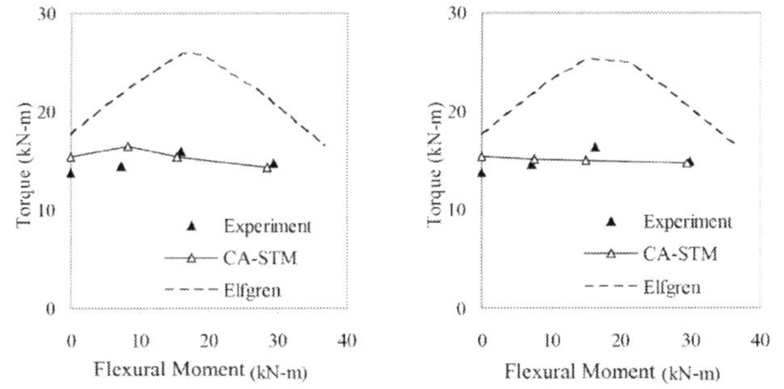

(a) Series 5 (M/V = 1.73 m) and Specimen 1-1 (b) Series 6 (M/V = 0.86 m) and Specimen 1-1

Fig. 6.5d—Interaction curves predicted by combined-action TM and the Elfgren model for McMullen and Warwaruk (1967) Series 5 and 6 (torsion, shear, and flexure) and Specimen 1-1 (pure torsion) (Greene 2006). (Note: 1 kN·m= 8.85 in.-kip; 1 m = 39.4 in.)

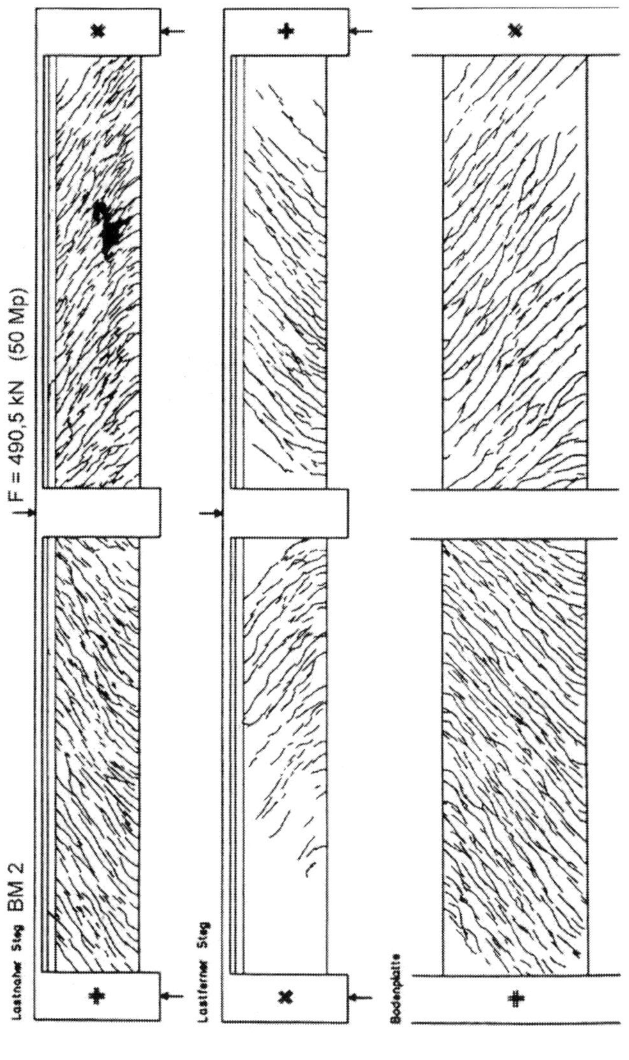

a) web near load b) web c) bottom slab

Fig. 6.5e—Crack pattern and failure of prestressed hollow box-beam BM 2 with high shear and torsion loaded eccentrically (Leonhardt et al. 1968).

inclined compression (Fig. 6.5e(a)), confirming the importance of determining the strength of struts considering the web strain and cracking conditions.

CHAPTER 7—ADDITIONAL DESIGN ISSUES RELATED TO TORSION

7.1—General

Chapter 7 addresses additional issues encountered in the design of structural concrete members subjected to torsion. When considering torsion, it is necessary to distinguish between when the torsional moment cannot be reduced by redistribution of internal forces (equilibrium torsion) and when the torsional moment can be reduced by redistribution of internal forces (compatibility torsion) after cracking. To allow torsional moment redistribution, a limit analysis method similar to moment redistribution is addressed. This method requires the formation of torsional plastic hinges at critical sections. Also addressed are precast spandrel beams, torsion limit design, treatment of open concrete sections subjected to torsion, and the size effect on torsional members.

7.2—Compatibility torsion and torsional moment redistribution

7.2.1 *Basic concepts of torsional moment redistribution*

7.2.1.1 *General*—Torsion and flexural moment redistribution are similar in their dependency on adequate ductility in plastic hinge regions. No fundamental difference, therefore, exists between flexural analysis for the case of pure flexure and of torsion. Three standard cases of torsion redistribution have been reported:

1) A spandrel beam loaded by a floor beam
2) An L-beam
3) A spandrel beam in a frame subjected to lateral load.

7.2.1.2 *Spandrel beam loaded by floor beam (Collins and Lampert 1973; Hsu and Burton 1974; Hsu and Hwang 1977; Abul Mansur and Rangan 1978)*—Figure 7.2.1.2 shows a portion of a three-dimensional structural frame, including columns, spandrel, and floor beams. A load P, applied to the floor beam, produces end rotation, which in turn produces a torsional moment in the spandrel beam. The interaction of floor beam and spandrel beam can be studied using the shaded portion in the shape of a T-specimen. This T-specimen is cut at the flexural inflection points, which can be simulated by hinges. When the T-specimen is loaded, a concentrated torsional moment will occur at the spandrel beam midspan. Whereas cracking reduces flexural stiffness by a factor of approximately 2, it reduces torsional stiffness by a factor of approximately 10. Therefore, a significant redistribution of internal actions begins to occur at cracking. When the spandrel beam develops torsional hinges at its ends, additional load will result in higher flexural moment in the floor beam. This response is equivalent to moment redistribution from the spandrel to floor beam. Designing the spandrel beam based on this redistribution could be economical.

7.2.1.3 *L-beam (Bishara and Londot 1979)*—Figure 7.2.1.3 shows a floor system with an L-shaped spandrel beam. Bishara and Londot (1979) studied the response of L-beams under combined torsion and flexure. Due to torsional stiff-

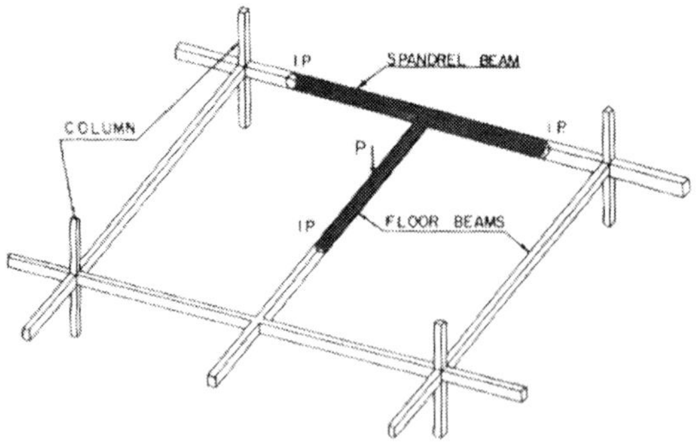

Fig. 7.2.1.2—Spandrel beam under concentrated torsion (Hsu and Burton 1974).

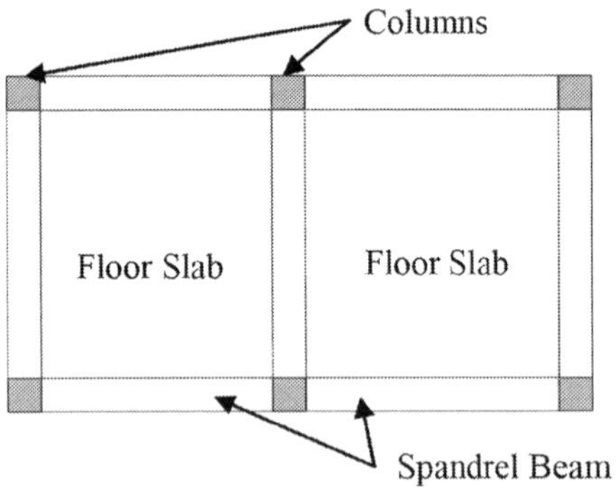

Fig. 7.2.1.3—Floor system with spandrel beams.

ness of the spandrel beam, the slab cannot be assumed to be hinged at the spandrel beam, and vertical loading on the slab causes distributed compatibility torsion in the spandrel beam. When torsion redistribution occurs, a plastic hinge forms at a yield line that defines the slab-beam connection. This could potentially result in a torsional moment that affects the spandrel beam and that could be lower than that computed by elastic analysis. These tests also showed that before yielding, torsional cracking caused a reduction in both torsional and flexural stiffnesses, leading to redistribution of internal forces. Reductions in flexural stiffness range from 40 to 60 percent to as much as 80 to 90 percent for torsional stiffness in beams with transverse reinforcement.

7.2.1.4 *Spandrel beam in a frame subjected to lateral load (Pantazopoulou and Moehle 1990)*—When a typical building frame is subjected to lateral sway, beam flexural action results in contraflexure along the span length, as shown in Fig. 7.2.1.4a. Therefore, in each bay, there is a region where the slab is in tension and another where the slab is in compression. The beam is likely to respond asymmetrically. The surface of the floor beam or main beam elongates as the beam is subjected to negative moment. The same elongation develops in the slab at the slab-beam interface. Elongation distorts the slab in shear to satisfy deformation compatibility and the slab reinforcement develops tensile stresses that are transferred to the spandrel (transverse) beams. These actions are illustrated by an equivalent structure shown in Fig. 7.2.1.4b. Slab action is modeled as truss elements positioned eccentric to the centroidal axes of the beam elements and connected to the beams through short rigid segments. For any curvature distribution along the longitudinal or main beams as identified in the figure, this configuration creates a corresponding elongation at mid-depth of the slab. Truss forces are transferred to the transverse beam by modeled truss elements in the slab, creating a torsional lever arm in the transverse (spandrel) beam.

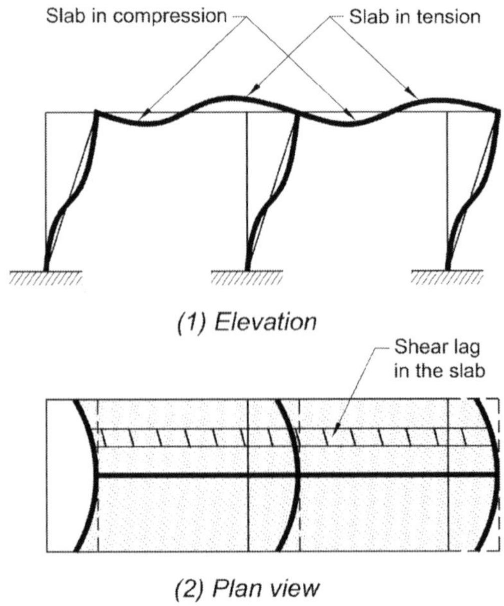

Fig. 7.2.1.4a—Spandrel beam under lateral sway (Pantazopoulou and Moehle 1990).

The result of these stresses on the spandrel beam (the transverse beam in Fig. 7.2.1.4b), is twofold: 1) flexural moments in the slab plane develop in the spandrel beam; and 2) torsional moments develop in the spandrel beam as a result of the eccentric action of the slab relative to the shear center of the spandrel beam cross section. Torsion and flexure in the spandrel beam are maximum at the column, and decrease with distance from the joint. This combination of torsion and flexure in the spandrel beams of exterior connections is commonly responsible for the diagonal cracking often observed on the exterior and interior faces of such beams. These actions can also be tested as shown in Fig. 7.2.1.4c. Two separate actions can cause moment redistribution in this case:

1. Plasticity could occur in the slab under axial load, a condition often idealized as plastic deformation in the truss members modeling the slab of Fig. 7.2.1.4b. Alternatively, the slab could reach its maximum load strength and redistribution could occur before reinforcement in the spandrel beam yields (Collins and Lampert 1973; Pantazopoulou and Moehle 1990). This redistribution could result in less torsion on the spandrel beam than that computed from elastic analysis; and

2. A torsional plastic hinge could form at a location across the spandrel beam axis, resulting in redistribution of torsion along the spandrel beam axis. For example, formation of the plastic hinge at the connection between the spandrel beam and

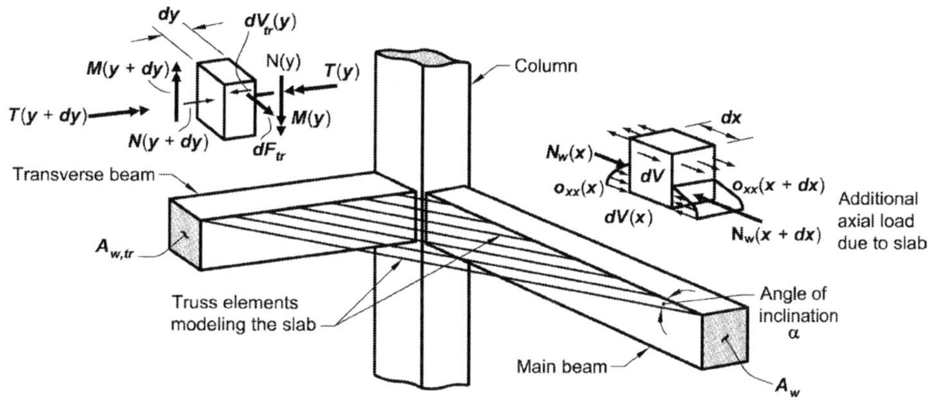

Fig. 7.2.1.4b—Idealization of spandrel beam under lateral sway (Pantazopoulou and Moehle 1990).

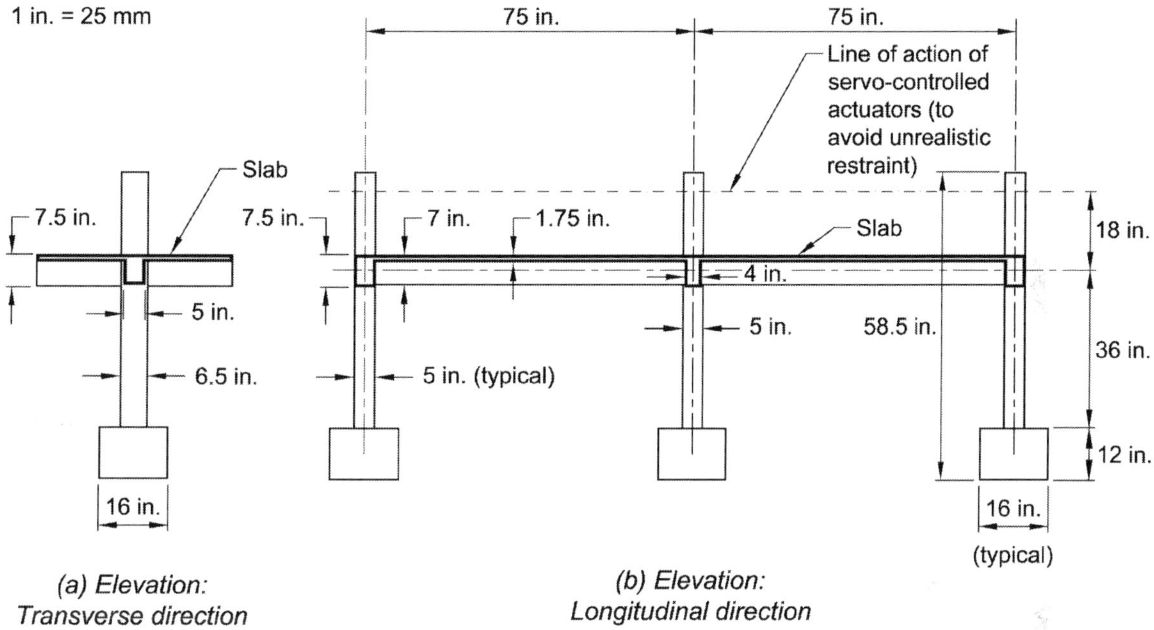

Fig. 7.2.1.4c—Testing of spandrel beams under lateral sway (Pantazopoulou and Moehle 1990).

the column (location of maximum torsion in the elastic analysis) would reduce the torsion affecting the spandrel beam.

7.2.2 *Design benefits from torsion redistribution*—Plastic hinges and moment redistribution permit full strength use of the strength of the flexural member's cross section. With torsion, elastic analysis typically results in high torsional moments, which in turn require both stirrups and longitudinal reinforcement beyond minimum requirements. Torsion redistribution can reduce design torsional moments, allowing for more economical designs.

7.2.3 *Disregarding structural torsion*—Although one common practice is to ignore spandrel beams' torsional stiffness and similar members under gravity loads, effectively assuming full torsional redistribution, it is also common to disregard spandrel beams' torsional moments resulting from wind or earthquake actions. Sections 7.2.4 through 7.2.6 address codes, experimental evidence, and discussions of their limitations.

7.2.4 *Existing code status*—In ACI 318, a flexural moment redistribution formula has been maintained since 1971. In 1977, ACI 318 introduced the torsional limit design, allowing the torsional plastic hinge to have a torsional shear stress of $0.33\sqrt{f_c'}$ MPa ($4\sqrt{f_c'}$ psi). Alternatively, Section 8.6.1 of ACI 318-77 specifies that any set of reasonable assumptions can be used for computing relative torsional stiffnesses of structural elements. Commentary indicates that member stiffnesses should reflect the degree of cracking and inelastic action occurring along the member length before yielding. The stress limit of $0.33\sqrt{f_c'}$ MPa ($4\sqrt{f_c'}$ psi), however, has been extended to prestressed and nonprestressed members subjected to an axial tensile or compressive force in the subsequent editions, including ACI 318-11. In commentary Section 8.6.1 of ACI 318-77, in the case of compatibility torsion (torsion not required to satisfy equilibrium), torsional stiffness can be disregarded. Section 8.6.1 commentary effectively permits up to 100 percent torsional moment redistribution from the elastic analysis without relating it to specific ductility requirements to satisfy a high redistribution value. The gap between flexural and torsional redistribution provisions is due primarily to limited research in literature for torsional moment redistribution. There are three areas where the gap is manifested:

1. Testing relating torsional redistribution to member ductility is limited
2. Factors affecting torsional ductility and, therefore, the permissible torsional redistribution, are not understood
3. The concept of permissible torsional redistribution not affecting service load deflections has not been experimentally verified.

In the following sections, literature available in these three areas is reviewed and specific recommendations for needed research are given.

7.2.5 *Torsion redistribution experiments*

7.2.5.1 *Early tests*—Collins and Lampert (1973) tested six spandrel beam specimens with a concentrated load applied at midspan of the floor beam. They found that specimens designed assuming the members had zero torsional stiffness behaved as satisfactorily as specimens designed assuming uncracked stiffness values. They also showed that the ratio of torsional-to-flexural stiffness will drop at cracking, causing redistribution of the torsion and flexural moments. They concluded that for compatibility torsion, design should be for a twist and not a torque, and the main function of torsional reinforcement is to distribute the cracks caused by twist. The compatibility torsion magnitude is overestimated if gross stiffness is used. If zero torsional stiffness is assumed, design procedures become simple, and only minimum torsional reinforcement is needed to ensure ductility and limit crack width.

Hsu and Burton (1974) focused on addressing three questions for spandrel beams subjected to torsion:

1. Would torsional strength be reached if a spandrel beam were designed by the ACI philosophy of neglecting torsional stiffness and assuming 100 percent redistribution?
2. Can other values of torsion redistribution be used effectively?
3. How serious is service load cracking?

To address these questions, Hsu and Burton (1974) tested two series (A and B) of spandrel beams:

1. Series A: Concentrated load at midspan of the floor beam.
2. Series B: Uniform loads on the floor beam simulated by four concentrated loads. Based on elastic stiffness analysis, the torsional moment distribution to spandrel beam at its joint with the floor beam is calculated according to Eq. (7.2.5.1a)

$$M = \frac{\frac{3}{16}P\ell_f}{1+\frac{3}{4}\frac{K_f}{K_u}} \qquad (7.2.5.1a)$$

Assuming that spandrel beam is torsionally fixed at both ends, the spandrel beam torsional moment magnitude is

$$T = \frac{M}{2} \qquad (7.2.5.1b)$$

In each series, the spandrel beam torsional moment is calculated assuming varying ratios between the spandrel beam torsional stiffness and the floor beam flexural stiffness. These ratios varied from zero to 100 percent torsional redistribution. The latter condition is where torsional moment does not affect the spandrel beam, but is redistributed to positive flexure in the floor beam. Figure 7.2.5.1a shows the results for Series A tests and Fig. 7.2.5.1b for Series B.

Three pairs of companion specimens from this testing program are examined: A1 and B1, A3 and B2, and A5 and B3. Specimens A1 (Fig. 7.2.5.1a) and B1 (Fig. 7.2.5.1b) were designed assuming uncracked stiffness of both floor and spandrel beams. Specimens A3 and B2 were designed according to the limit design theory, assuming the joint moment to be 48 percent of that calculated from the elastic moment distribution using the uncracked stiffnesses of the floor and spandrel beams (and therefore 52 percent torsional moment redistribution). Specimens A5 and B3 were designed by assuming 100 percent redistribution, meaning the spandrel beam in these two specimens was not designed to sustain any torsion.

7.2.5.2 *Strength and adequate values of torsion redistribution*—When comparing torsional ductility and redistribution among the three specimen sets, review the experimental load-torque relationships. These curves are presented in Sets A (Fig. 7.2.5.1a) and B (Fig. 7.2.5.1b). Curves for Specimens A1 and B1 (no torsional redistribution) can be divided into three stages:

1. The first stage represents elastic behavior before cracking.
2. The second stage is where crack development caused the torsional stiffness to decrease and the torsional moment remained essentially constant whereas the load increased. When the load reached a magnitude at which the floor beam bottom longitudinal reinforcement yielded, as indicated by in the figures, the third stage began.
3. In the third stage, a torsional plastic hinge developed under the load and accelerated the twist deformation of the spandrel beam. A new load-carrying mechanism evolved in which the load was primarily sustained by the spandrel beam torsional strength. The torsional moment again increased, primarily resulting in a slight increase in the load-carrying strength of the assembly. Spandrel beam stirrups and longitudinal bars then yielded before reaching assembly failure.

Specimens A3 and B2 (52 percent redistribution) exhibited behavior similar to that of Specimens A1 and B1, despite the assumed higher value of torsion redistribution. In both Specimens A3 and B2, the load-torque curve had the same three stages exhibited in Specimens A1 and B1. Both specimens also exhibited the anticipated level of torsion redistribution. Although A3 and B2 failed at almost the same ultimate failure loads as Specimens A1 and B1, the failure load corresponded to about 50 percent less torsion at the joint between the floor beam and the spandrel beam. These results indicate the torsional redistribution assumptions for

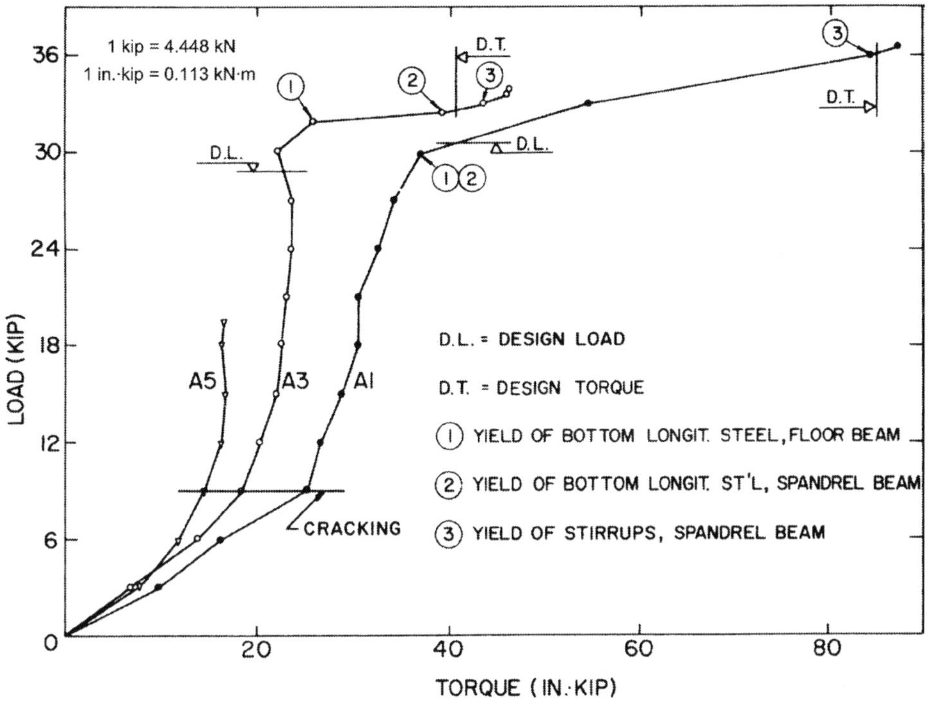

Fig. 7.2.5.1a—Series A (Hsu and Burton 1974). (Note: 1 in.-kip = 0.113 kNm; 1 kip = 4.45 kN.)

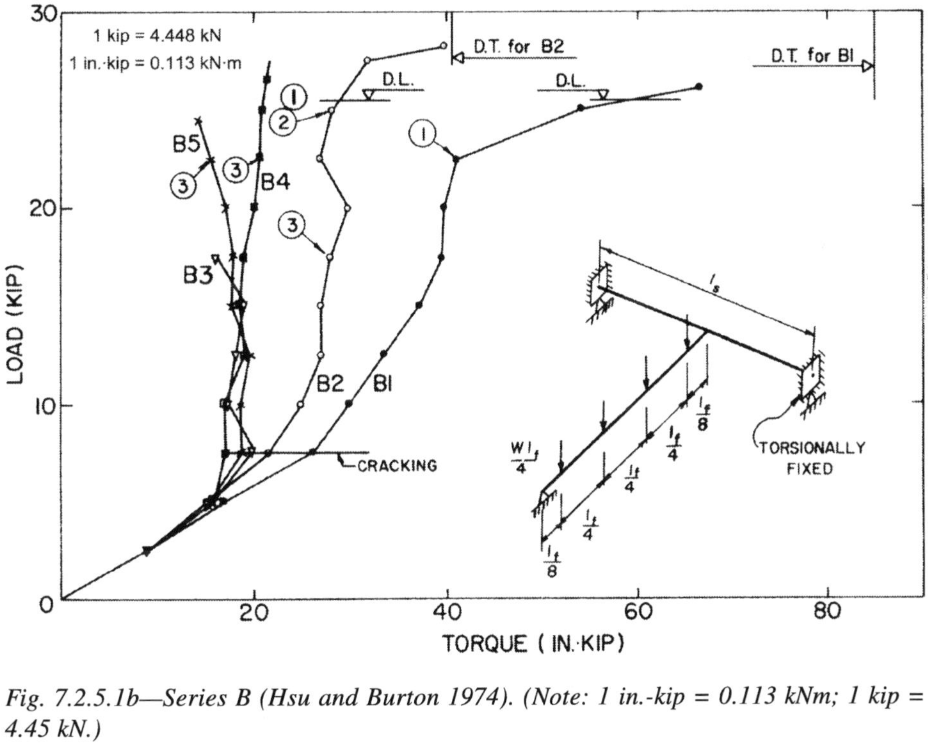

Fig. 7.2.5.1b—Series B (Hsu and Burton 1974). (Note: 1 in.-kip = 0.113 kNm; 1 kip = 4.45 kN.)

these two specimens closely matched the design assumption of 52 percent torsional redistribution. Specimens A5 and B3 were designed for 100 percent or complete torsion redistribution, as implied in ACI 318 commentary. As shown in Fig. 7.2.5.1a and 7.2.5.1b, torsion resisted at the joint stayed constant after cracking occurred. After cracking, the load increased significantly, almost doubling, at the same torsion value until failure. Failure occurred at a load far below design strength, indicating that the design assumption for torsion redistribution was invalid. Figures 7.2.5.2a and 7.2.5.2b show the torque-versus-measured twist deformation for Series A and B, respectively. Both figures support conclu-

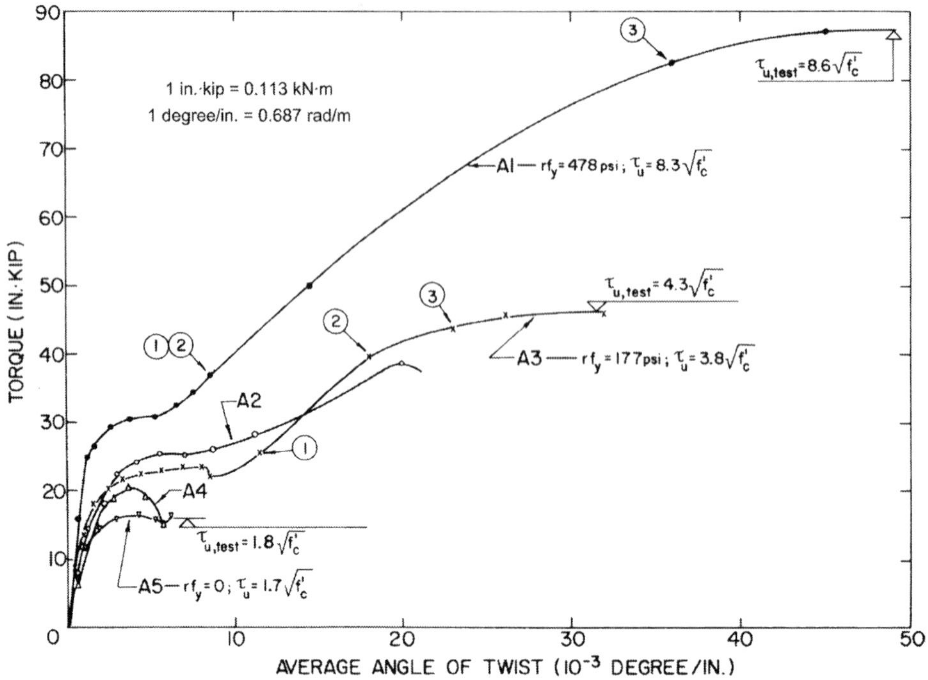

Fig. 7.2.5.2a—Series A torque-twist relationship (Hsu and Burton 1974). (Note: 1 in.-kip = 0.113 kN·m; 1 in. = 0.0254 m.)

Table 7.2.5.3—Maximum crack width at service load

Specimen	Design torsion redistribution, %	Maximum crack width, in. (mm)
A1	0	0.004 (0.10)
A3	52	0.005 (0.13)
A5	100	—
B1	0	0.005 (0.13)
B2	52	0.005 (0.13)
B3	100	—

sions similar to those previously mentioned. Specimens A1 and B1 exhibited high ductility, as indicated by the torque-twist relationship. Specimens A3 and B2 exhibited a twist angle 50 percent greater at failure than that of Specimens A1 and B1, which roughly corresponds to the designed torsion redistribution. Specimens A5 and B3, designed for 100 percent torsion redistribution, exhibited far lower twisting ductility. In fact, both specimens failed at 0.3×10^{-3} degrees/mm (8×10^{-3} degrees/in.), far less than the angles exhibited by Specimens A1 and B1.

7.2.5.3 *Cracking service load*—Hsu and Burton (1974) defined service load as one-half the ultimate design load or actual ultimate test load, whichever was larger. Table 7.2.5.3 shows the maximum crack widths measured at the service load for specimen Series A and B. No crack width readings were recorded while testing Specimens A5 and B3 because these two specimens failed prematurely.

A comparison of Specimens A1 and A3 shows that maximum crack widths increased only slightly for the 52 percent torsional redistribution assumption. This slight increase may be within the measurement tolerance. Specimens B1 and B2 had the same crack widths. These tests indicate that significant torsion redistribution can be performed without affecting serviceability. Nevertheless, the results were not sufficient to support firm conclusions.

Following Hsu and Burton's (1974) approach, Hsu and Hwang (1977) tested T-shaped specimens with short spandrel beams to study the moment distributions between floor beam and spandrel in the case of high shear-to-torsion ratio. They found that the torsional limit design method can be applied to spandrel beam end regions, adjacent to the columns, where high shear stresses are combined with torsional moments. As a result, the torsional limit design was added to ACI 318-77 code. Torsional hinge moment was calculated based on a torsional shear stress of $0.33\sqrt{f'_c}$ in MPa ($4\sqrt{f'_c}$ in psi). Conclusions reached by Hsu and Burton (1974) were validated by Abul Mansur and Rangan (1978), who conducted experiments on seven beams with configurations shown in Fig. 7.2.5.3. In a related set of experiments, Abul Mansur and Rangan (1978) studied the effect of applying a concentrated load directly on the joint. This work showed that applying a concentrated load in this fashion did not affect torsion redistribution results, which were generally in agreement with Hsu and Burton's work (1974).

7.2.6 *Factors affecting torsion redistribution*—Section or member ductility is a primary factor affecting permissible redistribution. With flexure, ACI 318-95 relates ductility to the amount of longitudinal reinforcement present in the section. A similar torsional ductility reinforcement ratio would be appropriate, accounting for transverse and longitudinal reinforcement areas coupled with torsion support

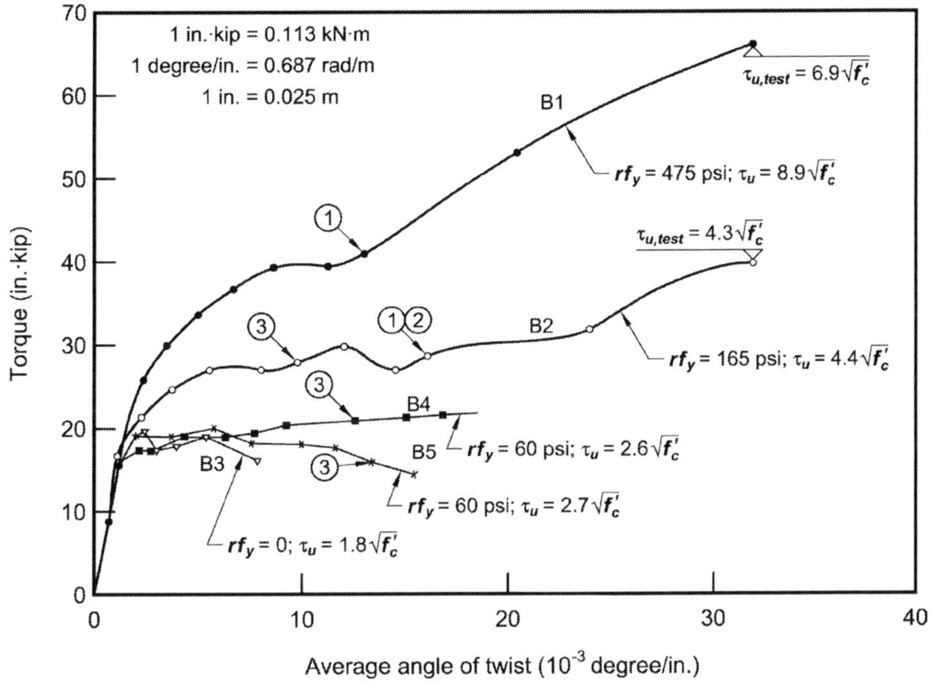

Fig. 7.2.5.2b—Series B torque-twist relationship (Hsu and Burton 1974). (Note: 1 in.-kip = 0.113 kN·m; 1 in. = 0.0254 m.)

conditions. Before 1995, ACI 318 adopted the skew-bending theory for idealizing member behavior under torsion. The approach was complex and produced several inconsistencies, which led to the introduction of space truss analogy to ACI 318-95. Before this change in philosophy, when the skew-bending theory was prevalent, Chakraborty (1977) proposed a balanced torsional reinforcement concept. He used a space-truss analogy for developing the balanced reinforcement concept, referring to early work on the space-truss analogy by Mitchell and Collins (1974). Chakraborty's work was the subject of commentary in several professional journal discussions, most notably by Collins and Mitchell (1978). Discussions focused primarily on the difficulty of using Chakraborty's approach and an apparent difficulty in deriving similar conclusions from the specimens tested. These difficulties stemmed primarily from the fact that Chakraborty's (1977) approach considered both the truss-analogy from the modified compression field theory (CFT) of Mitchell and Collins, and compatibility between the concrete and reinforcement. This approach resulted in a complex equation (Eq. (7.2.6)) for the balanced stirrup reinforcement area for one leg. The equation proposed by Chakraborty is

$$\frac{A_{tb}p_h}{A_o s} = \frac{\dfrac{0.85 f'_c \beta_1}{2}}{f_{ty}} \cdot \frac{0.003E_s}{f_{ty} + \dfrac{0.003E_s\left(1-\dfrac{\beta_1}{2}\right)p_o}{p_h}} \qquad (7.2.6)$$

The above approach of combining both compatibility and the space-truss analogy to compute torsional strength, or

Series	SA	SB
Parameter	Low shear	High shear
Specimen and loading	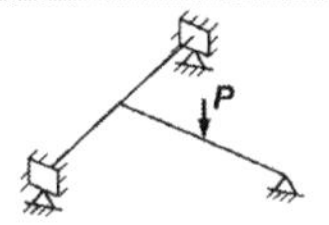	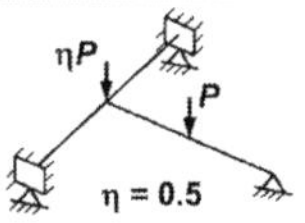
Dimensions (1 mm = 0.039 in.)	ℓ_f = 3000 mm ℓ_s = 3000 mm X-section = 300 mm × 180 mm	ℓ_f = 3000 mm ℓ_s = 1500 mm X-section = 300 mm × 180 mm

Fig. 7.2.5.3—Test configuration (Abul Mansur and Rangan 1978). (Note: 1 mm = 0.0394 in.)

balanced reinforcement, was not adopted by ACI 318 when the committee changed to the space-truss analogy in 1995. Numerous articles have reviewed ductility and serviceability, and their relationship to the angle θ. The crack width model of Thürlimann et al. (1983) is shown in Fig. 7.2.6a. This model minimizes strains in the entire concrete medium. Thürlimann also suggested that within these limits of strut inclination, the maximum diagonal compressive stress is independent of the angle of inclination. Further, the minimum crack strain corresponds to an inclination angle of 45 degrees that in turn corresponds to the simultaneous yielding of both the longitudinal (main) and transverse (stirrups) reinforcement (Fig. 7.2.6b).

Their analysis was based on shallow beams to which Bernoulli's hypothesis of plane strain is applicable. In such beams, the elastic stress distribution implies that maximum diagonal compression stress is inclined at 45 degrees to the beam axis. Another interpretation of angle limits suggested by Thürlimann et al. (1983) is that as the truss-model stress field deviates from the elastic stress field that the crack strains,

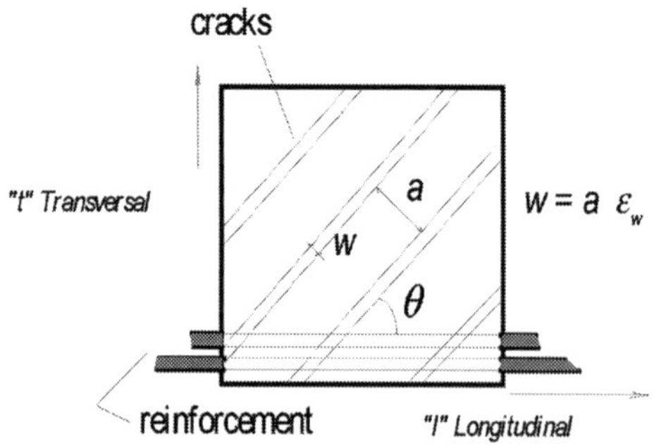

$\varepsilon_w = \varepsilon_t(1+\tan^2\theta)$ Yielding in the Transverse Reinforcement

$\varepsilon_w = \varepsilon_l(1+\cot^2\theta)$ Yielding in the Transverse Reinforcement

Fig. 7.2.6a—Crack width model (Thürlimann et al. 1983).

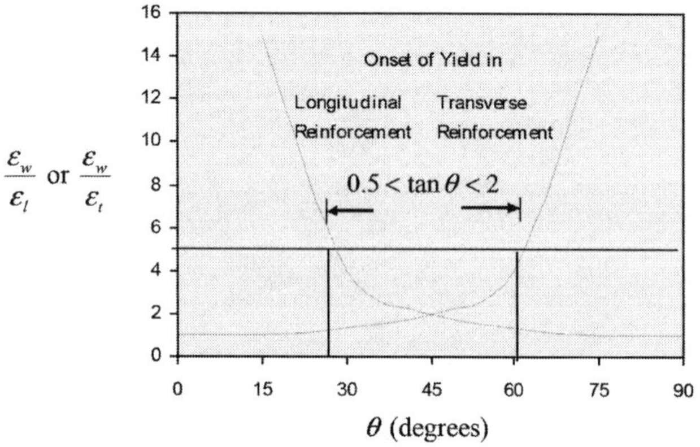

Fig. 7.2.6b—Cracking strain and inclination angle θ (Thürlimann et al. 1983).

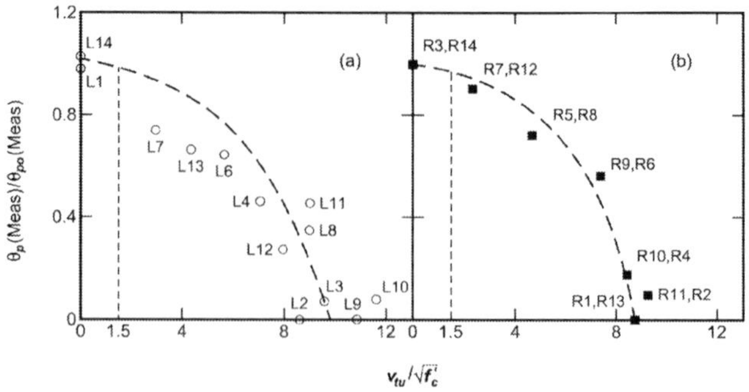

θ_p *(Meas)* is the measured flexural curvature for combined torsion and flexure

θ_{po} *(Meas)* is the measured flexural curvature when no torsion is present

v_{tu} is the ultimate torsional stress

Fig. 7.2.6c—Effect of torsion-flexure interaction on ductility (Bishara and Londot 1979).

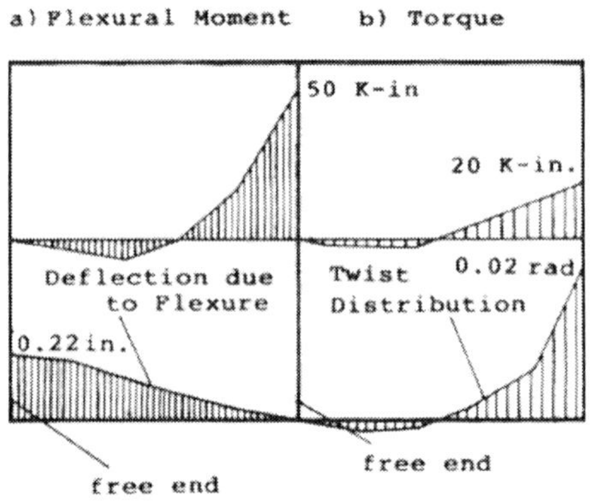

Fig. 7.2.7—Torsion in spandrel beam due to lateral load (Pantazopoulou and Moehle 1990). (Note: 1 in.-kip = 0.113 kN·m; 1 in. = 0.0254 m.)

thereby increasing crack widths. Ali and White (1997) showed that for disturbed (D) regions in general, where the Bernoulli hypothesis is not applicable, large deviations from the elastic stress distribution result in loss of ductility and serviceability due to increased concrete strains. This deviation corresponds to large deviations from the 45-degree angle between diagonal compression and longitudinal reinforcement for shallow beams. The interaction between torsion and flexure has been shown to affect flexural ductility of members subjected to both torsion and flexure. An extensive testing program of L-beams conducted by Bishara and Londot (1979) showed that increased torsional loads resulted in reduced flexural rotations. This reduced flexural rotation is shown in Fig. 7.2.6c, where the flexural rotation is plotted with respect to a normalized torsional stress.

7.2.7 *Needed research*—As in the case of flexure, a preferred design procedure would specify the relationship between permissible moment redistribution and measure of ductility. To allow for complete yielding of reinforcement, ACI 318 limits the angle θ in the range of 30 to 60 degrees. The angle θ is generally taken as 45 degrees for reinforced concrete beams and 37.5 degrees for prestressed beams. A comprehensive study of prestressed and nonprestressed beams, however, is needed to understand the relationship between permissible moment redistribution and θ. Understanding the direct relationship between θ and ductility is also essential. It is important to include the interaction among torsion, shear, and flexure and its effect on ductility. The beam configuration in Fig. 7.2.1.2 (described in more detail in Hsu and Burton (1974)) is ideally suited for such a study. Curves similar to those in Fig. 7.2.6c, once developed, could be readily used in code provisions. Simple provisions for calculating the slab-action torsion developed in a spandrel beam under lateral loads (Pantazopoulou and Moehle 1990) would be equally beneficial.

Tests by Pantazopoulou and Moehle (1990) showed that, due to slab action, torsional moments can develop in the spandrel beam. Figure 7.2.7 shows flexural and torsional moment distribution and the corresponding deflection and twist along the longitudinal axis of a spandrel beam due to slab action under lateral loads. The left side shows measured flexure with moment at the top and deflection at the bottom, and the right side shows torque with torque and twist at the bottom. The supported end is to the right of each graph whereas the free end is to the left.

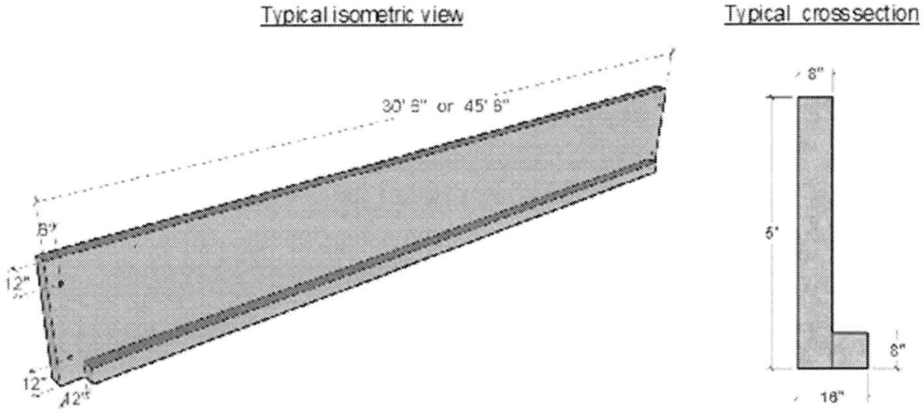

Fig. 7.3.1a—Typical L-shaped spandrel beam (Lucier et al. 2007). (Note: 1 in. = 0.0254 m.)

7.3—Precast spandrel beams

7.3.1 *General*—To this point, the discussion has focused on monolithic, cast-in-place concrete structures in which torsional moment redistribution could occur when members subjected to torsion are designed with adequate ductility. The situation is different in precast structures where precast members, either reinforced or more commonly prestressed, are usually simply supported and, without continuity, there is no torsional moment redistribution. These members are designed for the full torsional moment to maintain equilibrium. Precast L-shaped spandrel beams, which are common to their type, are characterized by a plate- or wall-like web with a continuous ledge running along the bottom of one web side. This ledge provides deck beam support, as shown in Fig. 7.3.1a. A variation on the L-shaped spandrel beam is the corbel spandrel beam on which the continuous ledge is replaced by a series of discrete corbels that support deck beams. Another variation is the pocketed spandrel beam in which the discrete corbels are replaced by a series of pockets (recesses) cast in the spandrel that support deck beams. All these spandrel beam types are subjected to significant torsion caused by the series of large, concentrated, eccentric loads along the span.

Typically, L-shaped spandrel beams are 1.5 to 2.1 m (5 to 7 ft) deep with spans ranging from 9.1 to 15.2 m (30 to 50 ft) in length. They are usually simply supported at the columns. Beam ends are connected to columns to prevent torsional rotation. Deck beams are often connected to the spandrel web to provide lateral restraint along the span. These end and intermediate connections greatly affect the L-shaped spandrel beam torsional behavior. Eccentrically applied loads on unsymmetrical L-shaped spandrel beams cause vertical displacement as well as significant lateral displacement and rotation. The torsional and shear effects are largest near the spandrel end. This complex structural behavior, coupled with heavy loadings, often results in heavy reinforcement in the end regions. Similar significant torsion effects may occur in inverted T-beams with severely unbalanced loads, as shown in Fig. 7.3.1b.

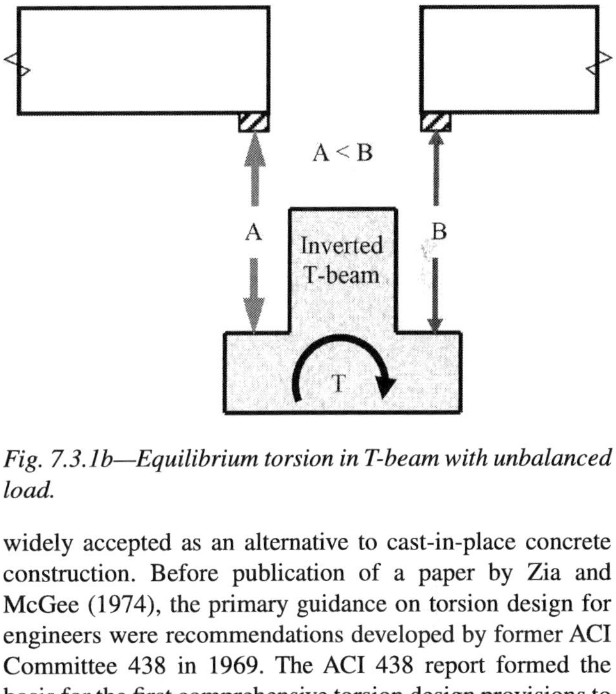

Fig. 7.3.1b—Equilibrium torsion in T-beam with unbalanced load.

7.3.2 *Torsion design*—Design for torsion in precast concrete members received much attention in the late 1950s and early 1960s when precast prestressed concrete became widely accepted as an alternative to cast-in-place concrete construction. Before publication of a paper by Zia and McGee (1974), the primary guidance on torsion design for engineers were recommendations developed by former ACI Committee 438 in 1969. The ACI 438 report formed the basis for the first comprehensive torsion design provisions to be included in the ACI code (ACI 318-71). The Zia-McGee paper extended the ACI code torsion provisions to prestressed concrete. In 1978, Zia and Hsu updated the Zia and McGee (1974) paper to conform to ACI 318-77 code format and also introduced refinements. Papers by Zia and Hsu (1978, 2004) were widely used by the precast prestressed concrete industry for nearly 30 years for its reliability and ease of use. The ACI 318 code and the Zia-Hsu approach present a unified method for the torsion and shear design of prestressed and nonprestressed concrete flexural members following similar design procedures. The ACI method was developed with a primary focus on compact closed sections generally found in cast-in-place reinforced concrete structures and was based on a thin-walled tube, space-truss analogy. The Zia-Hsu method was developed with a primary focus on noncompact flanged sections more common in precast structural systems and the skewed-bending concept. In cast-in-place structures, the torsional effect is often minimized due to stress redistribution in an indeterminate system, whereas in precast structures, the torsional effect is more severe and connection details can significantly affect the torsional behavior of structural members (Klein 1986; Raths 1984). The ACI torsion design procedure produces design results comparable to those offered by the Zia-Hsu method (Prestressed Concrete Industry (PCI) 1999), which was updated based on the Zia-McGee method (PCI 1978) developed from the tests of small laboratory specimens. The aspect ratio of their component rectangles rarely exceeded 1:3, in contrast to large precast members such as L-shaped spandrel beams. Large spandrel beams may develop plate-bending action in addition to torsion, and therefore behave differently from the small laboratory specimens. The Zia-Hsu method has been calibrated by two tests (Klein 1986) of full-size members with support conditions similar to those used in practice. Torsion behavior of slender spandrel beams in these tests was much different from compact beams. Spandrel beams showed no signs of cover spalling or stirrup debonding, for which closed stirrups are required. End region behavior of slender precast spandrel beams is dictated by connection details. Upper horizontal reaction prevents torsional rotation and causes out-of-plane web bending. Vertical shear force from the reaction is carried across the same diagonal crack. Out-of-plane bending caused by horizontal forces is essentially the same as bending produced by a concentrated load or reaction in the corner of a structural slab. Tests conducted by Logan (2007) and by researchers at North Carolina State University (Lucier et al. 2007; Hassan et al. 2007) confirmed that out-of-plane bending dominates torsional response in slender spandrel beams. Lucier et al. (2007) evaluated the need for closed ties in slender spandrel beams using full-scale tests. Twelve precast spandrel beams, each nearly 14 m (15.3 yd) long, were tested to failure in the laboratory. End regions of the beams were designed for plate-bending and shear, with and without closed ties. Several beams failed in flexure or ledge punching. The ledge punching failures were primarily due to interaction with global tension and shear forces in the ledge. Beams with reinforcement to prevent such flexure and ledge failures failed due to combined torsion and shear. All torsion/shear failures occurred just inside the bearing reaction along a failure plane inclined at an angle of approximately 45 degrees. Although the beams with closed ties performed somewhat better, all of the beams sustained test loads well in excess of their calculated nominal strengths. These test results suggest that slender spandrel beams can be designed based on the following four principles:

1. Torsion acting on the 45-degree failure plane can be divided into plate-bending and twist components, each equal to $T_u/\sqrt{2}$.

2. Plate-bending component of torsion requires equal amounts of longitudinal and transverse reinforcement evenly distributed over the height of the beam.

3. The twist component of torsion is resisted by out-of-plane shear stresses, which are greatest near the section's top and bottom, as predicted by Saint-Venant's equations and the "soap bubble" analogy. Closed ties are required only where the concrete section cannot safely resist the twist component.

4. Reinforcement requirements on the inside face and outside face may be considered separately; shear increases the required reinforcement on the inside face and decreases it on the outside face.

7.4—Torsion limit design

7.4.1 *Basic concepts of limit design*—There are two fundamental limit design theories (Gvozdev 1938, 1960; Melan 1938; Horne 1949; Lubliner 1990):

1. Lower-bound approach
2. Upper-bound approach

The lower-bound approach requires a stress field that satisfies equilibrium everywhere and that does not violate appropriate yield conditions or material failure criterion at any location.

The classic example of lower-bound approach is the truss model in which a stress field is assumed and the ultimate strength is calculated accordingly. For members made of plastic materials (those that satisfy the basic theory of plasticity assumptions), the resulting ultimate strength is easily proved to provide a conservative, lower-bound solution to the true ultimate strength of the structural member under consideration.

In the upper-bound approach, a strain field or failure mechanism is developed that does not violate strain compatibility conditions or appropriate yield conditions and that does not exceed material failure criterion at any location. A classic example of the upper-bound approach is the yield line theory of slabs in which a failure mechanism is assumed and the ultimate strength is calculated accordingly. For members made of materials satisfying basic theory of plasticity assumptions, the resulting ultimate strength can easily be proved to provide an upper-bound, or unconservative, solution to the true ultimate strength of the structural member under consideration. Exact solutions are obtained if the results of the upper-bound and lower-bound solutions are identical.

The advantage of applying either the lower- or upper-bound approach lies in their simplicity. Relaxation of either equilibrium or strain compatibility conditions in solving the governing equations significantly simplifies the solution procedure. There are, however, disadvantages:

1. The potential exists for large over- or under-estimation of true ultimate strength; and

2. No information is provided for important quantities, such as deflections and rotations at service or failure load limit states; these are quantities that require classical deflection analysis considering all conditions of equilibrium, compatibility, and constitutive relations governing the material behavior.

Research in limit design faces the following challenges:

1. The appropriate stress or strain field needs to be derived and validated from ultimate strength calculations that are neither too conservative nor too unconservative for the cases of lower and upper bounds, respectively.

2. Parameters should be developed for the stress or strain field such that optimization techniques can be used to achieve realistic lower and upper ultimate strength estimations.

3. Both the lower- and upper-bound approaches should be recognized as valid for plastic materials; any approach is valid if it satisfies the basic assumptions of the theory of plasticity of either perfectly rigid plastic or linear-elastic, perfectly-plastic materials. In both cases, the material should ideally exhibit unlimited strains under the same yield stress. Reinforcement—in particular, mild reinforcement—exhibits behavior closely satisfying these requirements. Certain assumptions and limitations, however, should be applied to achieve proper application of the plasticity theory to structural concrete members.

The first two challenges are explored in 7.4.2 and 7.4.3 because they pertain to reinforced concrete members subjected to torsion. The third challenge is far-reaching, as it relates to the general applicability of the theory of plasticity to plain concrete and reinforced concrete members. A thorough discussion of this challenge is provided in Chen (1982).

7.4.2 *Lower-bound approach*

7.4.2.1 *General*—The lower-bound method requires a stress field that satisfies equilibrium everywhere and does not violate the appropriate yield conditions at any point. From a design perspective, this approach means that if every region of a structure is proportioned and detailed for a stress-resultant field that satisfies all equilibrium conditions, then the structure should safely carry the design ultimate load. A well-known example of this approach is the Hillerborg's (1960) strip method for reinforced concrete slab design. Rangan et al. (1978) proposed a closed-form, lower-bound method for the torsion design of reinforced concrete structures with the following steps:

1. Set up all equilibrium equations in a statically indeterminate structure in terms of the redundancy. Let the number of redundancies be represented as n.

2. Assign the maximum torsional moment, which is normally near a torsionally-restrained support, equal to the limiting value T_o defined as

$$T_o = 0.33\sqrt{f_c'}(x^2y/3) \text{ (N-mm)}$$
$$T_o = 4\sqrt{f_c'}(x^2y/3) \text{ (in.-lb)} \quad (7.4.2.1)$$

Here, x and y are the shorter and longer overall rectangular cross section dimensions, respectively, and f_c' is the characteristic concrete cylinder compressive strength. As an alternative, the value of T_o can be set by the designer. This condition will result in an equation to solve the redundancies.

3. To obtain the other (n_R – 1) equations, introduce (n_R – 1) known quantities in the stress-resultant field. For this purpose, the position of inflection points and the maximum values of any stress resultants other than the torsional moment can be assumed. All or some of these can then be used to obtain the required (n_R – 1) equations.

4. The above n_R equations can be solved simultaneously, and the resultant stress field is completely formulated.

7.4.2.2 *Helical stair example*—Rangan et al. (1978) applied the four steps (7.4.2.1) to the design of a helical stair with fixed support, subject to uniformly distributed loading. Elastic theory indicated that a helical stair is indeterminate to the sixth degrees. When the loading is symmetrical, however, the resultant-stress field is fully described in terms of only two redundancies: M_o and H_o. The term M_o is the flexural moment acting in a tangential plane, and H_o is the horizontal force in the radial direction, both acting at midspan, as shown in Fig. 7.4.2.2.

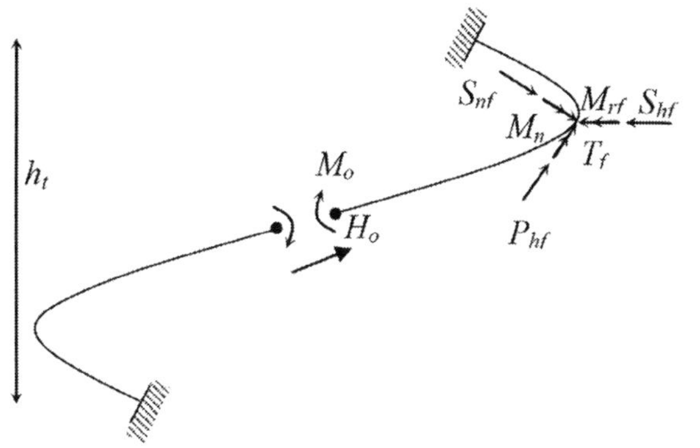

Fig. 7.4.2.2—Elevation of helical stair (Rangan et al. 1978).

For this problem, six equilibrium equations were first formulated in terms of M_o and H_o, along with reactions at the two supports. In their paper, Rangan et al. (1978) assumed the maximum torsional moment to occur at an angle $\Phi = \gamma_1$ on the helical stair. The maximum T_o is therefore expressed as

$$(T_f)_{\Phi=\gamma_1} = T_o = 0.33\sqrt{f_c'}\frac{x^2y}{3} \text{ (N-mm)}$$
$$(T_f)_{\Phi=\gamma_1} = T_o = 4\sqrt{f_c'}\frac{x^2y}{3} \text{ (in.-lb)} \quad (7.4.2.2a)$$

Development of another equilibrium equation required solving for the redundancies. The moment M_{rf} was assumed to be zero at $\Phi = \gamma_2$ (inflection point). The resulting equations were in the form

$$w = f(\gamma_1, \gamma_2) \quad (7.4.2.2b)$$

For any selection of angle γ_1 or γ_2, the resulting ultimate load provided a lower-bound solution to the true ultimate strength of the stair. The most realistic estimate of ultimate load is obtained when the ultimate load in Eq. (7.4.2.2b) is maximized, or when

$$\frac{\partial w}{\partial \gamma_1} = 0 \text{ and } \frac{\partial w}{\partial \gamma_2} = 0 \quad (7.4.2.2c)$$

Substituting Eq. (7.4.2.1) and (7.4.2.2a) through (7.4.2.2c) into the equilibrium equations for the helical stairs and proceeding with additional mathematical derivations (Rangan et al. 1978), an equation for the ultimate helical stair strength was provided as a function of the stair dimensions and the characteristic concrete cylinder compressive strength f_c'.

Fig. 7.4.2.3—*Curved guideway structure.*

7.4.2.3 *Curved girder example*—Collapse load cannot be determined using a closed-form solution for all torsional cases, except for possibly the simplest cases, such as the helical stair case solved by Rangan. A trial-and-error procedure, therefore, is usually employed to solve for more complex members subjected to torsion, such as the case of curved girders. Geometrical techniques have been presented for the trial-and-error procedure (Boulton and Boonsukha 1959; Jordaan et al. 1974), but most published studies use computer analyses (Yoo and Heins 1972) to obtain the collapse load. Representative solutions from the literature provide various iterative procedures. For example, Yoo and Heins (1972) determined the sequence of hinge formation leading to collapse using an elastoplastic analysis and finite-difference techniques. Badawy and Jordaan (1977) used a methodical technique to set up a system of nonlinear, simultaneous equations for each failure mode. The equations satisfied both equilibrium and yield conditions, and provide a lower-bound estimate of ultimate strength. Badawy and Jordaan (1977) indicated that:

> "In some cases, the number of equations that result from a lower bound analysis is less than the number of unknowns, and the required solution should correspond to the maximum load. The problem is generally reduced to one requiring an optimization type procedure, in which the load is maximized under the conditions or constraints of equilibrium and yield criterion."

A typical optimization technique normally employed to satisfy those constraints simultaneously, and therefore solve for the ultimate carrying strength, is the Lagrange multiplier method as described in Badawy and Jordaan (1977). The curved girder (Fig. 7.4.2.3) and associated torsion solution problem represent a condition examined by several researchers because of its frequent occurrence in design. Separate studies by Boulton and Boonsukha (1959), Jordaan et al. (1974), Yoo and Heins (1972), and Badawy and Jordaan (1977) all shared a similar general lower-bound framework as applied to a curved girder subjected to torsion.

7.4.3 *Upper-bound approach*—The use of upper-bound approaches in the design of reinforced concrete structures is common, particularly in slab design using the yield line method. Studies of the general upper-bound approach applicability to reinforced concrete structures are numerous. For example, Wang (1995) and Wang et al. (1993) have discussed its applicability to reinforced concrete deep beams. Wang and Hsu (1997) extended that work to reinforced concrete beams subjected to pure torsion. Both deep beams and torsional members can be viewed as special cases of shear failure. In the case of pure torsion, shear stresses develop in the member to resist the applied torque. The application of plasticity theory may then successfully lead to analytical solutions of ultimate torsional strength of prismatic elements made of homogeneous and isotropic materials. One analytical solution method is the "sand-heap analogy" developed by Sadowsky (1949), which is based on the theory of plasticity. This theory has proven effective for estimating the ultimate torsional strength of metal elements. The same approach for structures made of reinforced concrete, however, is complex and not as straightforward.

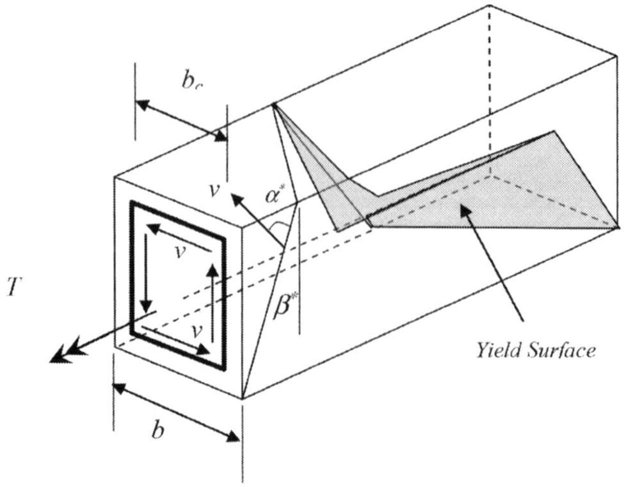

Fig. 7.4.3—*Square prismatic reinforced concrete torsional member (Wang and Hsu 1997).*

Wang and Hsu (1997) used the modified Coulomb-Mohr failure criterion as the constitutive law of concrete and assumed that the plastic flow of concrete after yielding is associated with this failure surface. They employed a generalized energy dissipation rate formulation along a yield line in accordance with the failure criterion, as well as the upper-bound approach, to calculate the ultimate strength of a beam subjected to pure torsion. They compared their approach to the results of experiments and found substantial agreement with them. Expanding on this work, Wang and Hsu (1997) considered a square prismatic reinforced concrete beam subjected to pure torsion, as shown in Fig. 7.4.3. By introducing a permissible failure mechanism, as illustrated in Fig. 7.4.3, the work equation along the yield surface is formulated as

$$v^* = \Phi\, x_1 \sec(\alpha^* - \beta^*) \qquad (7.4.3a)$$

Refer to Fig 7.4.3 for identification of $\alpha^* - \beta^*$.

Therefore, on an infinitesimally small area dA of the yield surface, Eq. (7.4.3b) applies

$$dA = 2x_1 \sec\beta^* dx_1 \quad (7.4.3b)$$

By counting all reinforcement bars on the relevant section, the total energy dissipated by the reinforcement is represented by

$$D_s = 2b_c^2\Phi\left[\rho_l f_{ly}\tan(\alpha^* - \beta^*) + \rho_t f_{ty}\tan\beta^*\right] \quad (7.4.3c)$$

The energy dissipated on the discontinuous concrete yield surface can also be found by integration, resulting in the equation

$$D_c = \frac{\Phi}{3}b^3\sec(\alpha^* - \beta^*)f_t^*(\sin\alpha^* + K\cot\alpha^*\cos\alpha^*) \quad (7.4.3d)$$

K is obtained from the Mohr-Coulomb yield criterion and takes the form

$$K = \frac{1}{4}\left[m + 2(1 - \sqrt{m+1})\right] \quad (7.4.3e)$$

where

$$m = \frac{f_c^*}{f_t^*} \quad (7.4.3f)$$

The quantities f_c^* and f_t^* are related to the concrete compressive and tensile strength, f_c' and f_t', using plastic effectiveness factors ν_c and ν_t

$$m = \frac{f_c^*}{f_t^*} = \frac{\nu_c f_c'}{\nu_t f_t'} \quad (7.4.3g)$$

Although a discussion on plastic effectiveness factors is generic to concrete structures and outside the scope of this report, it is noteworthy that plastic effectiveness factors are direct functions of concrete ductility. The higher the strains withstood by the concrete medium, the higher the plastic effectiveness factors. Many formulas give the plastic effectiveness factors as inversely proportional to concrete strengths because ductility decreases as concrete strength increases.

Another factor affecting plastic effectiveness is the degrees of concrete confinement. Higher confinement levels, due either to applied triaxial pressure or to the presence of confining reinforcement, increases concrete ductility and increases the plastic effectiveness factors. For a complete discussion of the plastic effectiveness factors, refer to Chen (1982). Equations (7.4.3c) through (7.4.3g) are related to applied torsion by the equation of externally applied work to the dissipated work, where the external work is given as

$$W = T\Phi \quad (7.4.3h)$$

The lowest upper-bound solution can then be found by

$$\partial T = 0 \quad (7.4.3i)$$

The solution of Eq. (7.4.3i) results in an expression of the form

$$T = f(m, f_t^*, \omega_s, a_c) \quad (7.4.3j)$$

Where the first two functional terms represent concrete properties, ω_s is a reinforcement index, and the last term represents geometric properties. Wang and Hsu (1997) obtained a closed form solution for Eq. (7.4.3j), which they compared with experimental results. Thirty-nine test results from previous empirical studies conducted at the University of Houston were analyzed and compared with the proposed theoretical solution. Researchers found that the best fit was obtained when a uniform plastic effectiveness factor for concrete under torsion was used, as given by

$$\nu = 0.75 + 0.125\omega_{st} \quad (7.4.3k)$$

This equation indicates that an increase in reinforcement ratio provides better confinement to the concrete medium, allowing higher ductility and, therefore, higher plastic effectiveness. The measured-versus-predicted ultimate torsional strength of the 39 tested beams averaged 0.98 with a standard deviation of 0.09, which is a good indicator of the applicability of the upper-limit approach derived by Wang and Hsu (1997).

7.4.4 *Approach comparison*—Marti and Kong (1987) provided a similar yield line approach to predict the overall load-deformation response of orthogonally reinforced concrete slabs subjected to pure torsional moments in the reinforcement direction. In a companion paper, Marti et al. (1987) presented the results of torsion tests for nine orthogonally reinforced concrete slabs, comparing the results to various theoretical predictions, including the yield line approach provided by Marti and Kong (1987) . They found the measured ultimate resistances to be 5 to 46 percent below yield line theory predictions. They also found that a close prediction was achieved by a lower-bound limit analysis approach with an assumed effective plastic concrete strength equal to 45 percent of the measured uniaxial compressive strength. No similar research comparing lower- and upper-bound approaches to each other or to existing codes for beams subjected to torsion, either pure or combined with flexure, has been published. Such a study would have shed more light on the validity of upper- and lower-bound assumptions, as well as the degrees of conservatism or unconservatism in the lower- and upper-bound approaches, respectively.

7.5—Treatment of open sections

7.5.1 *Types of torsional resistance*—When subjected to torsion, a member develops two types of torsional resistance—Saint-Venant and warping. For solid and hollow members with a relatively bulky cross section, Saint-Venant (1856) torsion normally predominates with warping torsion generally neglected. In contrast, for members with a thin-

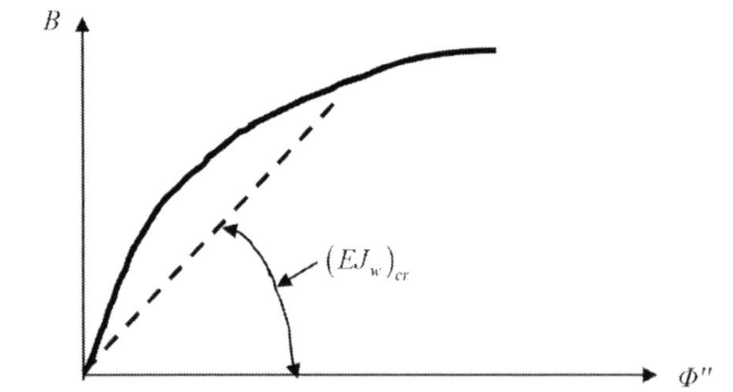

Fig. 7.5.2—Typical B-Φ'' curve (Hwang and Hsu 1983).

walled open section, warping torsional resistance predominates and the Saint-Venant torsional resistance does not occur. Past research has usually ignored secondary warping phenomenon (Vlasov 1961; Zbirohowski-Koscia 1968), mainly due to the mathematical difficulties encountered in formulating this type of torsional resistance and the very narrow thickness of metal structures. The wall thickness of reinforced concrete open sections, however, is often substantial compared with overall cross-sectional dimensions. Therefore, this secondary warping may play a less significant role in reinforced concrete structures.

The torsional behavior of a homogeneous elastic member with a thin-walled open section is predicted by Vlasov's theory. The theory was later generalized by Hwang (1978) to predict the torsional behavior of inelastic reinforced concrete members with open sections in the post-cracking stage. The mathematical formulation proposed by Hwang (1978) and Hwang and Hsu (1983) is given Eq. (7.5.2a).

7.5.2 *Consideration of warping torsion-formulation*—Applying the concept of sectorial area, Vlasov (1961) derived the basic equation for warping torsion, which relates the warping torque T_w to the third derivative of the rotation with respect to z, Φ''

$$T_w = -EJ_w\Phi'' \quad (7.5.2a)$$

The quantity EJ_w is used to characterize rigidity of the beam subjected to warping torsion. The term J_w is called the sectional moment of inertia and is calculated by applying procedures summarized by the original work of Vlasov.

Integrating Eq. (7.5.2a) results in the equation

$$B = -EJ_w\Phi'' \quad (7.5.2b)$$

where B is the integral of $T_{w.}$

A typical B-Φ'' curve is shown as a solid line in Fig. 7.5.2. This B-Φ'' curve can be constructed by a method proposed by Hwang (1978) and Hwang and Hsu (1983) based on a bimaterial model. A point on the curve signifies a certain load level. At the specific load level, the slope of a straight line connecting this point with the origin is defined as the post-cracking warping rigidity. The presence of an end diaphragm provides a warping restraint, which is considered in their model by incorporating an applied bidirectional moment.

Hwang and Hsu (1983) used the Fourier Series method to solve the warping torsion problem. Under torsional stresses, the angle of rotation was assumed to be a continuous function between the supports, expressed as

$$\Phi(z) = B_o + \sum_{n=1}^{\infty} B_n \sin\frac{n\pi}{L}z + \sum_{n=1}^{\infty} B_n \cos\frac{n\pi}{L}z \quad (7.5.2c)$$

where B_o, B_n, and B_n' are constant in the Fourier series expansion of a beam's curvature under warping.

Both the Saint-Venant torsion (1856) and warping torsion expressed the basic governing equation for mixed torsion as

$$T(z) = GC_{(z)}\Phi'_{(z)} - EJ_w\Phi'''_{(z)} \quad (7.5.2d)$$

The solution for Eq. (7.5.2d) was calculated using matrix notation, which is discussed in Hwang (1978). Before cracking, the GC and EJ_w terms are constants for any given section and the solution is easily identified. After cracking, however, both GC and EJ_w vary as functions of the load level, making it essential to employ a trial-and-error procedure to obtain the solution for mixed torsion (Hwang 1978). Krpan and Collins (1981a) also developed an analytical procedure using Vlasov's theory and incorporating CFT to predict the elastic pre- and post-cracking response of thin-walled reinforced concrete beams. The post-cracking response was categorized as either a cracked elastic response or an inelastic warping torsion response. The authors recommended using the post-cracking section properties rather than the uncracked elastic when analyzing the overall structure values to achieve a more accurate assessment of the actual force distributions in the structure.

7.5.3 *Experimental verification*—To verify the proposed solution, Hwang and Hsu (1983) designed two reinforced concrete beam specimens with open sections and tested their torsional behavior. The results were used to evaluate the Fourier Series method proposed for mixed torsion analysis. A comparison of experimental and predicted data revealed that the Saint-Venant torsional resistance had to be modified by certain empirical factors, as addressed in Hwang (1978). Incorporating these modification factors into the Saint-Venant torsional portion, they found that a mixed torsion analysis using the Fourier Series method predicted with reasonable accuracy both the pre- and post-cracking stages of the T-Φ' curves of the two tested specimens.

Figure 7.5.3 shows test results and analytical predictions for a thin-walled reinforced concrete specimen tested by Krpan and Collins (1981b). That work indicated that to accurately predict the strain in transverse reinforcement, it was necessary to account for the interaction between Saint-Venant's torsion and warping torsion. Test results obtained by the authors, shown as numbered points in Fig. 7.5.3, were compared with analytical predictions made according to the method proposed by Krpan and Collins (1981a). Figure 7.5.3 shows these results and confirms the ability of the developed analytical method to predict measured torque and twist. This method also accurately predicted the elastic pre- and post-cracking response of thin-walled reinforced concrete beams. The post-cracking response for the cracked elastic response, or inelastic warping torsion response, was also well predicted.

7.6—Size effect on the strength of concrete beams in torsion

Concrete size effect (Bažant 1984; Bažant and Planas 1998; RILEM Committee QFS 2004) is explained by energy release due to stress redistribution caused by a large crack or cracking zone formed before the point of maximum load. This size effect is transitional between the small-size asymptotic case of no size effect corresponding to plastic limit analysis, and the large-size asymptotic size effect corresponding to linear elastic fracture mechanics (LEFM), in which the fracture process is assumed to take place at one point only at the tip of a sharp crack. In small concrete structures, typical of normal laboratory tests, the size effect is normally almost undetectable, whereas in extrapolations to large structures in the field, the size effect may be strong. Size effect is expected for all the brittle failures of concrete structures, which include torsion. For reinforced concrete beams with little or no shear reinforcement, size effect for shear has been observed (Bažant and Yu 2005a,b; Bentz 2005). Torsional failure is a similar phenomenon, making size effect expected in that case as well. There is significantly less experimental data on size effect in torsion than for that of shear failure, although some experimental data do exist (Humphreys 1957; Hsu 1968a; McMullen and Daniel 1975). These were collected and evaluated by Bažant and Sener (1987) and reproduced in Fig. 7.6a (top and middle) as plots of log σ_N/σ_1 versus log D/D_1. Term D is the characteristic structural dimension, chosen here as the cross section size, whereas σ_1 and D_1 are normalizing constants that are different for each plot. The term σ_N is the nominal strength of the structure; for beams it is geometrically similar in three dimensions. It is defined as $\sigma_N = T_{max}/D^3$, where T_{max} is the maximum torsional moment, and D is the cross section depth defined in Bažant et al. (1988). The size range of these data, which does not exceed 1:2.7, is too limited comparing to the scatter width. If the plastic limit analysis approach to torsion was valid, the size effect would have to be absent and the trends in Fig. 7.6a would be horizontal. Clearly this is not the case. Although these data are too scattered to confirm any particular formula, they nevertheless indicate the downward trend of size effect. Bažant et al. (1988) obtained further test data, with strict geometrical scaling in three dimensions and a greater size range (1:4). They used square beams made of microconcrete with a maximum aggregate size of 4.8 mm (0.18 in.) and cross-sectional side dimensions of 25.4, 50.8, and 101.6 mm (1, 2, and 4 in.) (Fig. 7.6b). They used standard reduced-scale deformed reinforcement taken from PCA tests (Hsu 1968a) and tested both plain concrete beams and those with longitudinal reinforcement without stirrups (except at beam ends). Results are shown in the bottom of Fig. 7.6a. There was much less scatter observed in this study than the data in the top and middle of Fig. 7.6a. Decreasing strength with the increase of size is clearly confirmed due to size effect. The straight lines of slope (–1/2) in these plots represent the size effect for similar fractures according to LEFM, which is the strongest possible size effect. The data trend closely approaches this slope, which means that the size effect is strong for plain concrete beams and beams with longitudinal reinforcement without stirrups.

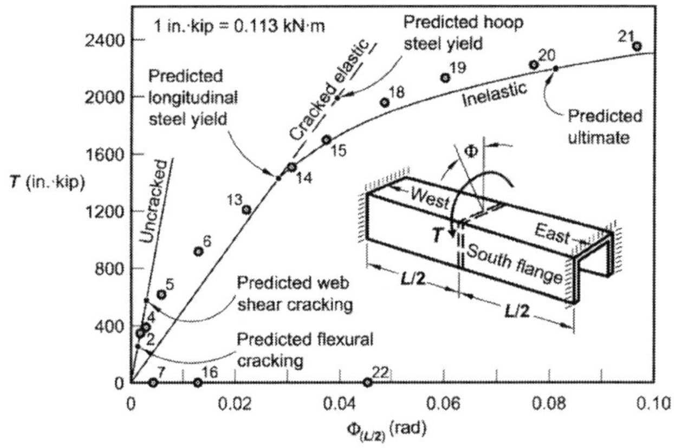

Fig. 7.5.3—Comparison of predicted and observed torque-twist relationship (Krpan and Collins 1981b).

Based on the works of Bažant and Planas (1998) and Bažant (2002), theoretically-justified formulas were developed to account for the size effect in beams subjected to torsion. The equation for reinforced concrete beams, which fail after significant fracture growth, is given in simplified form as

$$\sigma_N = \sigma_0(1 + D/D_0)^{-1/2} \quad (7.6a)$$

and for plain concrete beams, which fail at fracture initiation, is given in simplified form as

$$\sigma_N = \sigma_\infty(1 + rD_b/D)^{1/r} \quad (7.6b)$$

where σ_0 is the nominal torsional strength according to the current code specifications based on plastic limit analysis; σ_∞ is the plain beams strength according to elastic analysis with the maximum stress limited by material strength; and D_0, D_b, and r are constants. Simple prediction of these three constants and their dependence on the longitudinal reinforcement ratio, stirrup ratio, stirrup spacing, longitudinal and transverse prestress (if any), and some other geometrical ratios is lacking at present. Experiments should be conducted to verify and calibrate the theory to predict these constants. For very large beams, the Weibull statistical size effect should be supplanted to Eq. (7.6b), as shown in general in Bažant (2002).

CHAPTER 8—DETAILING FOR TORSIONAL MEMBERS

8.1—General

Torsional moment in a reinforced concrete member is resisted by a circulatory shear flow in a tube along the cross section periphery. The tube can be idealized as a space truss made up of reinforcement ties and concrete struts, as shown in Fig. 3.3.5a. The shear flow induces tensile forces in both the hoop reinforcement and longitudinal reinforcement.

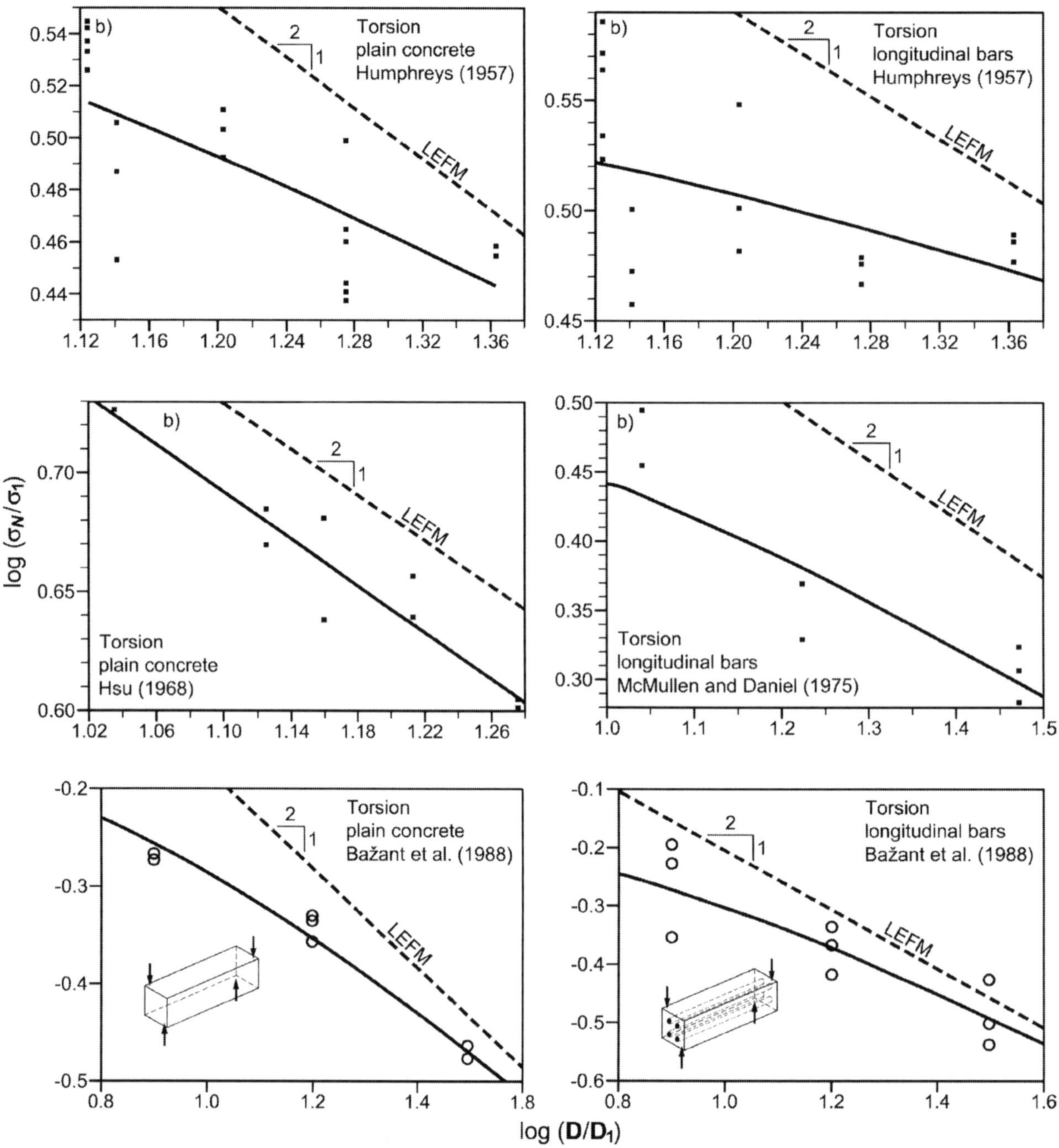

Fig. 7.6a—Nominal torsional failure strengths of beams with rectangular cross section and various sizes. Left: plain concrete beams. Right: longitudinally reinforced concrete beams without stirrups (Bažant and Sener 1987). Top: test data of Humphreys (1957). Middle: data of Hsu (1968a) and of McMullen and Daniel (1975). Bottom: Data on micro-concrete beams tested by Bažant et al. (1988).

Good reinforcement detailing is required to ensure that the hoop and longitudinal reinforcement can develop their yield strength to resist circulatory shear flow.

Good detailing demands consideration of the interaction between the member longitudinal and transverse reinforcement. Although each member type brings about different detailing conditions, the designer should be mindful of this overall force interaction in the member. Transverse reinforcement, oriented either horizontally or vertically, should contain a longitudinal bar at the corners. Enclosure of the longitudinal reinforcement by the transverse reinforcement provides the necessary equilibrium at the joint in the three principal directions, where the three-dimensional force flow is equilibrated.

8.2—Transverse reinforcement

8.2.1 *General*—Once proportioned for torsion and shear, the transverse reinforcement is laid out at a specific longitudinal spacing along the member span. The objective of transverse

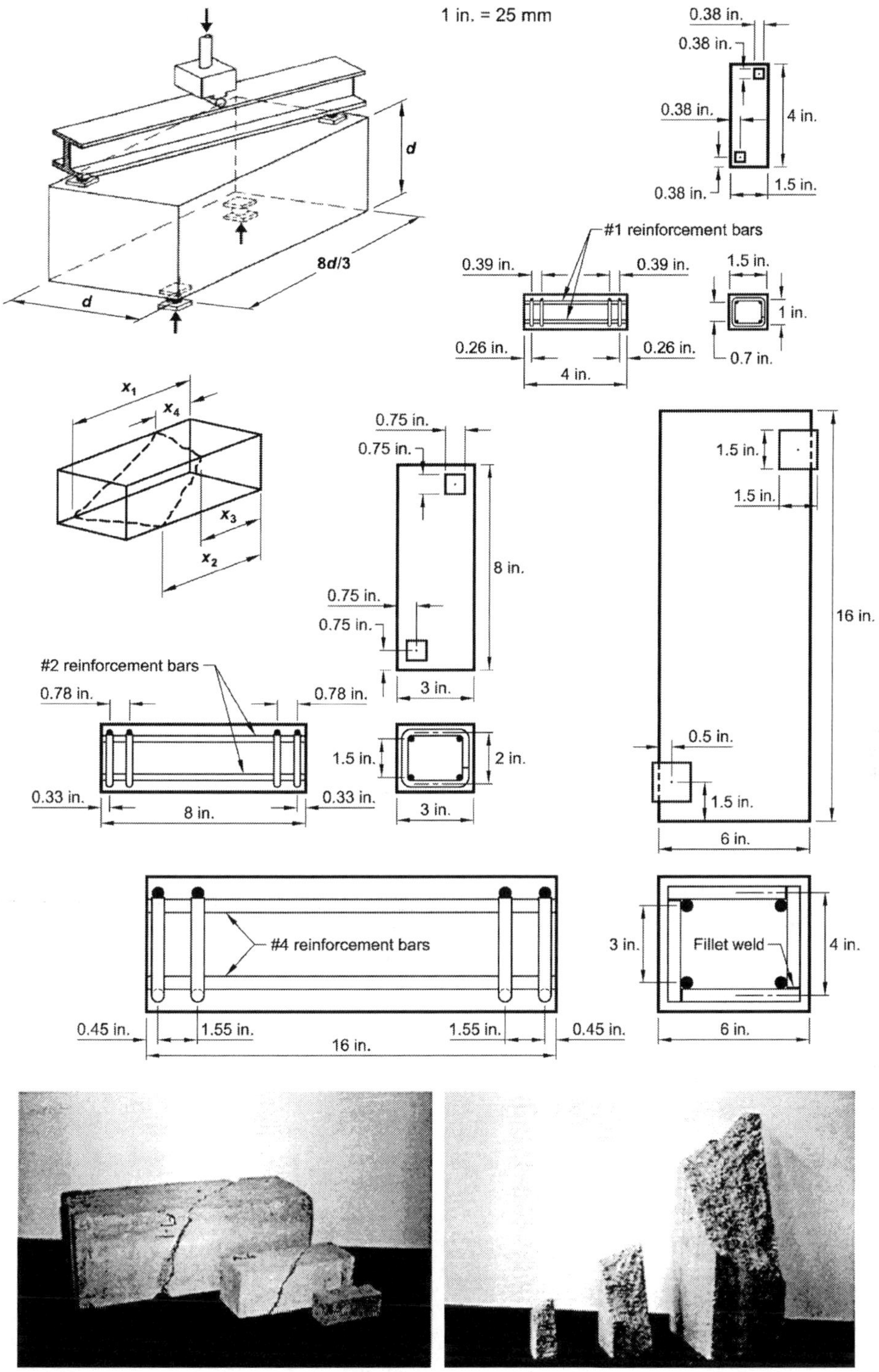

Fig. 7.6b—Plain and longitudinally reinforced microconcrete beams tested by Bažant et al. (1988) and the typical mode of torsional failure.

reinforcement for torsion and shear is to provide the reinforcement around the perimeter to enclose the member core. Typically, this reinforcement has a smaller diameter than the longitudinal reinforcement due to spacing, placement, bending, and proportioning needs. The transverse reinforcement should enclose the perimeter as closely as possible while maintaining clear cover requirements. A closed stirrup is imperative for torsional detailing. In the simplest case—a basic rectangular member cross section—stirrups are provided in a closed rectangular shape to encase the rectangular member core. The hooks of the closed stirrup are developed into the core with 135-degree bends. These bends ensure the hooks are well-anchored to the member core and prevent hook pullout under high torsional loads. Figure 8.2.1a provides an example of a simple rectangular closed stirrup.

Other common examples of cast-in-place member types subject to torsional loads are shown in Fig. 8.2.1b. The key to providing transverse reinforcement in a member subject to torsion is to start with the largest rectangular cross section and provide a rectangular closed stirrup in that section. Alternately, multiple-leg configurations can also be used for this purpose with single or multi-leg pieces or bar layouts to reinforce the cross section. Any protrusions, apertures, ledges, corbels, or other geometric outcroppings are provided with supplemental ties or stirrups developed back into the rectangular core of the individual member.

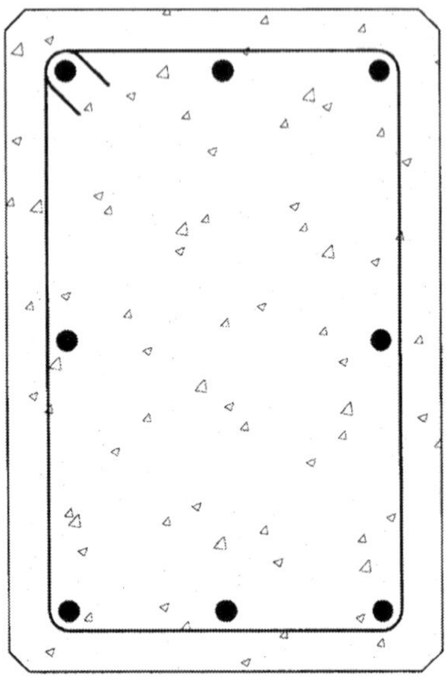

Fig. 8.2.1a—A typical closed stirrup used in a simple rectangular cross section.

As shown in Fig. 8.2.1c, good detailing usually dictates that additional ties or stirrups reinforce any protrusions. These ties also have a semi-closed detail with 135-degree or greater hooks developed into the core, which is the region enclosed by the closed stirrup shape. In addition to the closed stirrups and longitudinal bars shown in Fig. 8.2.1b and 8.2.1c, local reinforcement in the disturbed regions or D-regions should accommodate specific load concentrations. To be effective in any size member subjected to torsion, spacing between the closed ties should not exceed about one-half of the smallest dimension of the member, except for slender precast spandrel beams, such as those used in parking structures. In these members, torsional forces cause out-of-plane bending in the web. As described in 7.3, limited testing of such members has not shown signs of spalling or stirrup debonding for which closed stirrups are required. In load tests, slender precast spandrel beams have performed exceptionally well without closed ties. The current state of practice on spandrel beam behavior is contained in a recent study at North Carolina State University (Lucier et al. 2010).

8.2.2 *Hooks and development considerations*—Stirrups or ties are best terminated with 135- or 180-degree bends. Hooks should be developed into the main core of the member, where greater confinement is present. This detail is important in isolated members, where hook confinement is only provided by the member core, and no other external geometric conditions provide confinement. Practical considerations might dictate the use of simpler stirrup geometry, usually employing 90-degree hooks. When 90-degree hooks are used, confinement should be provided at locations where a slab frames into the beam side or elsewhere as needed.

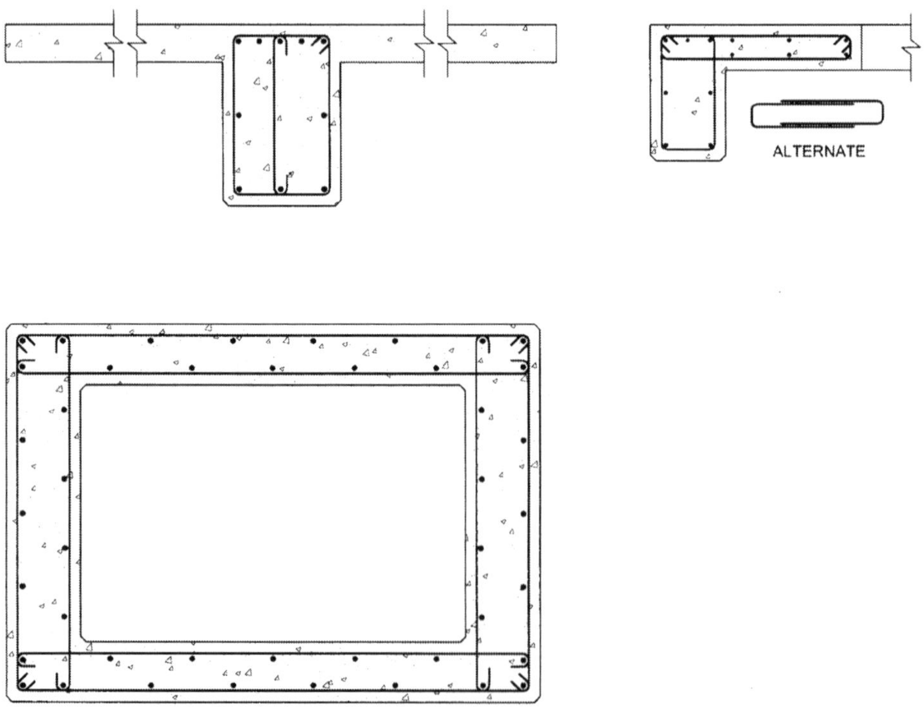

Fig. 8.2.1b—Examples of transverse torsional detailing in cast-in-place concrete members.

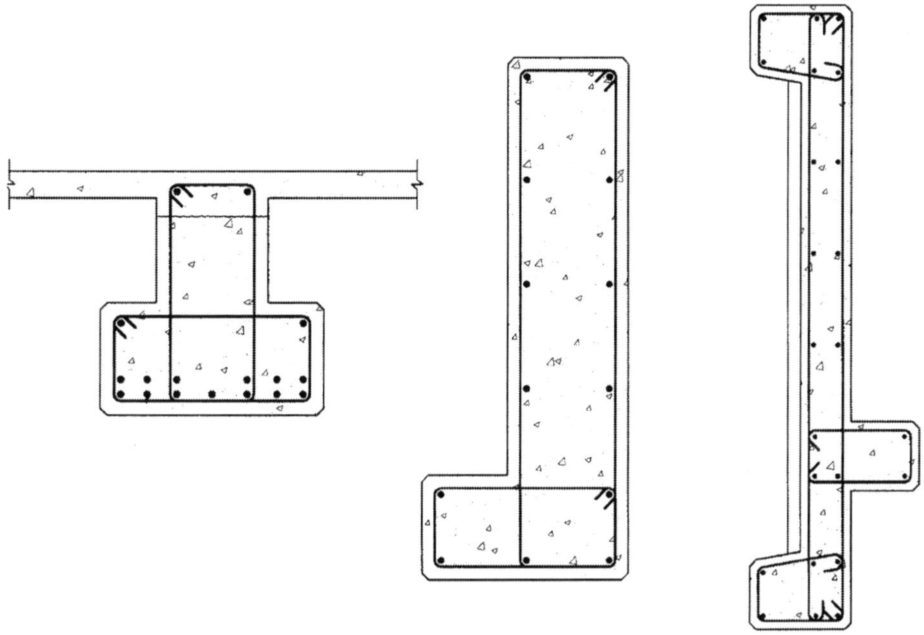

Fig. 8.2.1c—Examples of transverse torsional detailing in precast concrete members.

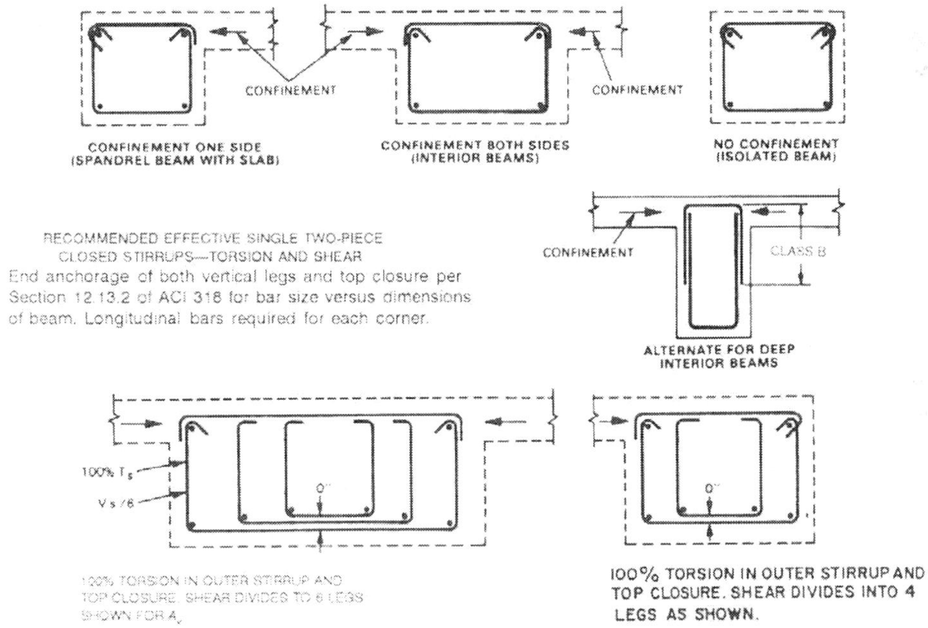

Fig. 8.2.2a—Recommended two-piece closed single and multiple U-stirrups for members subjected to torsion (ACI Committee 315 2004).

In Fig. 8.2.2a, examples from the *ACI Detailing Manual* (ACI Committee 315 2004) suggest using 90-degrees hooks under various side confinement conditions. Examples of poor detailing are reproduced in Fig. 8.2.2b. These recommendations are adopted from the research of Mitchell and Collins (1976).

8.3—Longitudinal reinforcement

Longitudinal reinforcement is also proportioned according to torsional requirements and provided around the member cross section perimeter. In beam regions or B-regions, special details need not be provided aside from equal spacing or proportioning of the reinforcement around the perimeter. Splices are proportioned in accordance with ACI requirements. At the end of a cast-in-place member, the perimeter longitudinal reinforcement may have to be developed into a column or other type of rigid vertical member providing torsional restraint.

CSA-A23.3-04 also includes the requirement that "A longitudinal reinforcing bar or bonded prestressing tendon

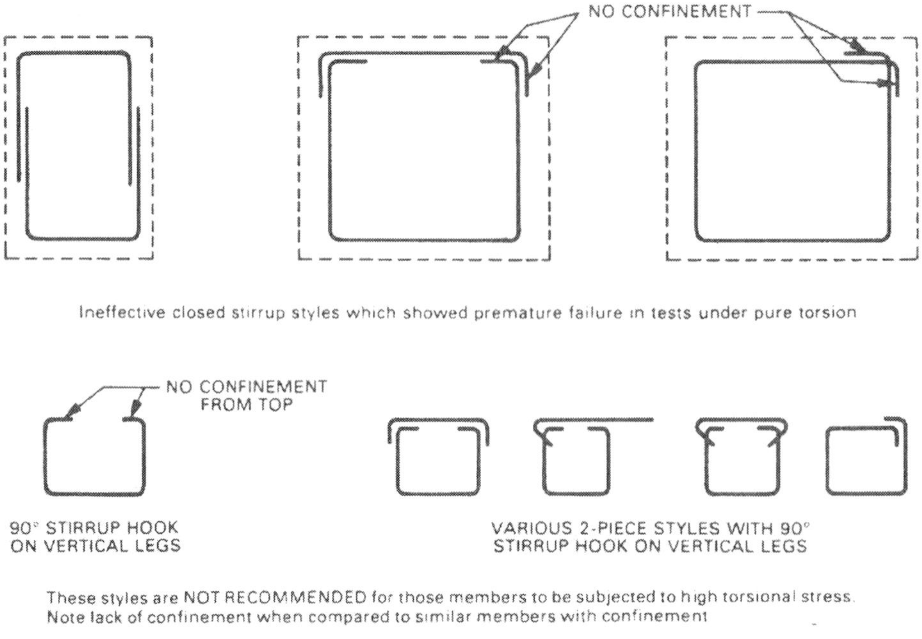

Fig. 8.2.2b—Ineffective closed stirrup types for members subjected to torsion (ACI Committee 315 2004).

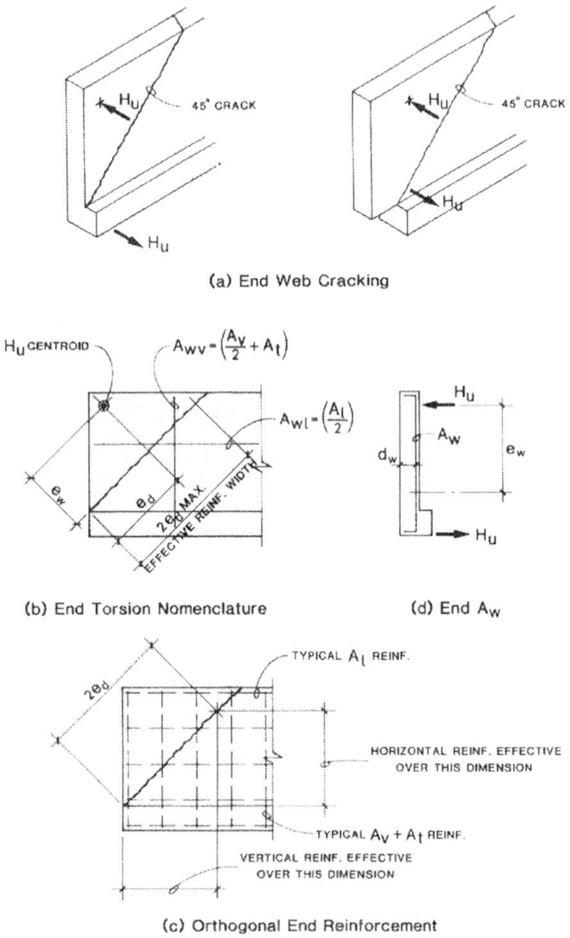

Fig. 8.4—Support detailing requirements in a precast spandrel member, dependent on the support connection locations (Raths 1984).

shall be placed in each corner of closed transverse steel reinforcement required for torsion. The nominal diameter of the bar or tendon shall not be less than $s/16$." The corner bars help to support the outward thrusts in the zones between the hoops (Mitchell and Collins 1976). Precast concrete members usually have semi-rigid connections consisting of field-welded angles and plates. Although this is not a full torsionally restrained connection, the end region should be examined along with subsequent development of the longitudinal, perimeter reinforcement at the member ends. One common way of developing reinforcement is to use conventional hooks at the member end. In some cases, the hooks might not fit in typically thin precast concrete members. Another common detail is the use of U-bars placed horizontally at the member end and lapped with the longitudinal reinforcement. The U-bars provide additional end confinement to a given precast member. Likewise, such reinforcement can provide sufficient confinement around the connection plate studs or tail bar reinforcement.

8.4—Detailing at supports

Precast concrete members require other special detailing considerations due to their horizontal support conditions. Torsional forces in precast members are often equilibrated by out-of-plane, horizontal, or sometimes vertical, reactions at discrete locations along the member depth. The member end conditions and subsequent details are highly dependent on the support configuration.

Figure 8.4 illustrates a common precast spandrel beam and the horizontal force couple that is typically developed at the end. Additional reinforcement is thereby required at the member end near the top to accommodate a potential 45-degree crack that typically develops at the end location due to the couple resisting torsion, as shown in Fig. 8.4(a) and 8.4(b). A possible reinforcement scheme to address this condition is shown in Fig. 8.4(c) and 8.4(d). Similar conditions often exist in other precast members due to their specific support and horizontal restraint conditions. Although these are D-region locations, they are complicated by the three-dimensional or out-of-plane nature of the problem.

CHAPTER 9—DESIGN EXAMPLES

9.1—Torsion design philosophy

Design philosophy for torsion in the ACI 318-11 building code is based on a thin-walled tube, space truss analogy in which compression diagonals wrap around the tube and the tensile contribution of concrete is neglected. Both solid and hollow members are considered tubes in accordance with Saint-Venant's circulatory shear flow pattern both before and after cracking. The outer part of the cross section centered along the stirrups is assumed to provide torsional resistance. The contribution of core concrete cross section is neglected. Once a reinforced concrete beam has cracked in torsion, the torsional resistance is provided primarily by closed stirrups and longitudinal bars located near the member's surface and diagonal compression struts. The inclined angle of the diagonal compression struts is permitted to be taken as 45 degrees for nonprestressed and lightly prestressed members, and 37.5 degrees for most prestressed members. Accordingly, ACI 318-11 makes the specific assumptions in torsion design that:

a) Concrete tensile strength in torsion is neglected

b) Torsion has no effect on the shear strength of concrete

c) Torsion stress determination is based on the closed thin-walled tube with uniform stress distribution and specific thickness, called shear flow

d) The torsional, flexural, and shear strength are accounted for by adding longitudinal reinforcement calculated for torsion and flexure

e) The longitudinal reinforcement are calculated for torsion and shear

The design of torsional resistance in Section 6.3 of EC2-04 is also based on a truss model using the thin-walled closed section theory with an effective wall thickness. The angle between the concrete compression strut and the member's longitudinal axis, θ, may be taken between 22 and 45 degrees. Both the solid and hollow cross section can be modeled by an equivalent hollow section neglecting the core concrete contribution to calculate the torsional resistance, which is limited by the strength of the concrete struts. The longitudinal and transverse reinforcement contributions to torsional resistance are accounted for after the thin-wall cracks. Effects of combined torsion and shear may be superimposed assuming the same value for the strut inclination angle. The required longitudinal and transverse reinforcement for torsion should be added to the existing longitudinal reinforcement for bending and transverse reinforcement for shear, respectively.

The Canadian code (CSA-A23.3-04) provides a General Design Method for torsion derived from the modified compression field theory (CFT) and represents solid cross sections by an equivalent thin-walled tube with a wall thickness determined by cross section dimensions. The space truss analogy provides the basic concept for torsion design assuming that the tension contribution in concrete is neglected and the diagonal compression struts spiraling around the member with variable inclined angle that depend on the loading condition and reinforcement ratio. The bending moment and longitudinal forces due to torsion and shear are considered resistant to four chords, one in each corner of the space truss and the shears by the shear flows in the walls. Dimensions of a cross section are limited to prevent crushing of the diagonal compression struts. In addition to this AASHTO LRFD method (general method), CSA-23.3-04 provides a simplified method for a restricted group of structural members, which states that the inclined angle of diagonal concrete compression strut is fixed at 35 degrees.

9.2—Torsion design procedures

9.2.1 *Torsion design in ACI 318-11*—According to Saint-Venant's circulatory shear flow pattern, the most efficient cross section to resist torsion is tube-shaped. Therefore, torsion of a reinforced concrete member is a three-dimensional (3-D) problem because it involves the shear in a reinforced concrete two-dimensional (2-D) wall element of a hollow tube and the out-of-wall bending of the concrete struts. In ACI 318-11, two simplifications are made. First, the concrete strut bending is neglected and the amount of hoop steel required in the tube determined from Bredt's (1896) equilibrium equation of a cross section

$$q_y = T_u/2A_o \tag{9.2.1a}$$

where the symbol q_y is the shear flow at yield (N/mm [lb/in.]); T_u is the torsional moment (N·mm [in.-lb]); and A_o [mm^2 (in.^{2})] is the lever arm area enclosed by the centerline of the shear flow.

Second, the hoop and longitudinal steel are assumed to yield at ultimate strength. To design steel reinforcement in a 2-D shear element, it is possible to use only three equilibrium equations (Hsu 1993). Combining the three equations creates a simple equation for yield shear flow

$$q_y = \sqrt{(A_t f_y / s_t)(A_t f_y / s_t)} \tag{9.2.1b}$$

where f_y is yield stress of hoop steel and longitudinal steel (MPa [psi]); A_t, A_l are area of hoop steel and longitudinal steel (mm^2 [in.^{2}]), respectively; and s_t, s_l are spacing of hoop steel and longitudinal steel (mm [in.]), respectively. Substituting the shear flow q_y into Bredt's (1896) equation gives

$$T_u = 2A_0\sqrt{(A_t f_y / s_t)(A_l f_y / s_l)} \tag{9.2.1c}$$

which is the essence of the ACI code provision.

The lever arm area A_o (mm^2 [in.^{2}]) is formed by sweeping the lever arm of shear flow one full circle around the axis of twist. The centerline of shear flow was taken by Rausch (1929) to be the centerline of the hoop steel bar, and the corresponding lever arm area is denoted as A_{oh} (mm^2 [in.^{2}]).

However, this definition of area A_{oh} was found to overestimate the torsional strength by as much as 30 percent. Therefore, the ACI code provides a simple, approximate formula for calculating the lever arm area as

$$A_o = 0.85A_{oh} \quad (9.2.1d)$$

To provide a more accurate formula for the ultimate torsional strength, consider the softening of concrete struts in the reinforced concrete 2-D wall elements of a tube. Under a biaxial tension-compression stress condition, the compressive stress-strain curve of the 2-D elements should be multiplied by a softening coefficient. This softening coefficient is a function of the principal tensile strain (Zhang and Hsu 1998) and varies from approximately 0.25 to 0.50. Applying this softened stress-strain curve of concrete to the study of reinforced concrete tubes under torsion (Hsu 1990, 1993), the thickness t_d (mm [in.]) of the shear flow zone and lever arm area can be determined as

$$\begin{aligned} t_d &= 4T_u / A_{cp} f_c' \\ A_o &= A_{cp} - (2T_u p_{cp} / A_{cp} f_c') \end{aligned} \quad (9.2.1e)$$

where A_{cp} is the area enclosed by the outer boundary of cross section (mm² [in.²]); and p_{cp} is the periphery of the outer boundary (mm [in.]). These formulas are given in the ACI code commentary, and the background was given in a paper by Hsu (1997).

9.2.2 *Torsion design in EC2-04*—Section 6.3 of EC2-04 requires a full design procedure for a reinforced concrete member under torsion covering both ultimate and serviceability limit states in cases where the static equilibrium of the structure depends on torsional resistance of the elements. In conventional statically indeterminate reinforced concrete structures, torsion arises from consideration of compatibility and it is normally unnecessary to consider torsion at the ultimate limit state. However, even if torsion arises from consideration of compatibility only, it may lead to excessive cracking in the serviceability limit state. Therefore, a minimum reinforcement of stirrups and longitudinal bars should be provided to prevent excessive cracking, as indicated in EC2-04 for cracking control (Section 7.3) and detailing beams (Section 9.2).

In normal slab-and-beam or framed structures, specific calculations for torsion are usually unnecessary when torsional cracking is being adequately controlled by shear and minimum flexural reinforcement. Where torsion is essential for equilibrium of the structure, EC2-04 should be consulted. One example of this is when structure arrangement is such that loads are imposed mainly on one face of a beam without corresponding rotational restraints provided.

The design of torsional resistance moment is based on a truss model using a thin-walled closed section theory with inclined angle θ between the concrete compression strut and the beam axis. The angle θ should be limited and recommended limits are: $1 \leq \cot\theta \leq 2.5$ (45 degrees $\geq \theta \geq$ 22 degrees). With a solid section, the section can be modeled by an equivalent hollow section from which the torsional resistance is calculated. Complex shapes, such as T-sections, can be divided into a series of subsections modeled as an equivalent thin-walled section, and the total torsional resistance taken as the sum of the capacities of each individual element. The effects of combined torsion and shear for both hollow and solid members can be superimposed assuming the same value for the strut inclination angle θ.

A common value for angle θ is 45 degrees. Eurocode 8 (EN 1998-1:2004) determines that: "In the critical regions of primary seismic beams, the strut inclination θ in the truss model shall be 45 degrees" (Paragraph 5.5.3.1.2(2) of EC2-08). However, in members not designated to resist seismic actions, a reduced value of angle θ could be considered to decrease the required transverse reinforcement and required longitudinal reinforcement. This way, fewer stirrups and more longitudinal bars could be provided. Required torsional reinforcement is added to the required stirrups and bars calculated from the shear and flexural design, respectively. Strength of materials used in EC2-04 is based on characteristic values and depend on whether the value is used for strength or stiffness. The characteristic value used to calculate strength corresponds to the 95 percent fractile of strength from material tests. The characteristic value for stiffness corresponds to mean strength from material tests. Design values are based on multiplying the characteristic value for resistance by the safety factors α and β.

Design procedure in accordance with EC2-04:

- Step 1: Calculation of the equivalent thin-walled section characteristics such as t_{ef}, A_k, and u_k (also refer to Fig. 6.11 of EC2-04 for notation)

$t_{ef} = A/u \geq 2c$ and, in the case of a hollow section, $t_{ef} < t_{real}$

where

t_{ef} = effective wall thickness of the equivalent thin-walled section [mm (in.)]

A = total area of the cross section within the outer circumference, including inner hollow areas (for example, $A = bh$ in a rectangular cross section with width and height equal to b and h, respectively) [mm² (in.²)]

u = outer circumference of the cross section (for example, $u = 2(b + h)$ in a rectangular cross section)

c = distance between edge and center of the longitudinal reinforcement (centroid cover) [mm (in.)]

t_{real} = real thickness of a hollow section [mm (in.)]

A_k = area enclosed by the centerlines of connecting walls, including inner hollow areas (for example, $A_k = (b - t_{ef})(h - t_{ef})$ in a rectangular cross section) [mm² (in.²)]

u_k = perimeter of the area A_k (for example, $u_k = 2(b + h - 2t_{ef})$ in a rectangular cross section)

- Step 2: Assume the value of angle of compression struts, θ, based on the expression: $1 \leq \cot\theta \leq 2.5$ (45 degrees $\geq \theta \geq$ 22 degrees). For combined shear and torsion, the same value of θ should be assumed and the common value is 45 degrees.

- Step 3: Check the maximum resistance of the member subjected to torsion and shear. This is limited by the strength of the concrete struts. If the following relationship is not satisfied, the member cross section dimensions, the concrete compressive strength, or both, should be increased

$$\frac{T_{Ed}}{T_{Rd.max}} + \frac{V_{Ed}}{V_{Rd.max}} \leq 1 \quad (9.2.2a)$$

where

T_{Ed} = design torsional moment [N·m (in.-lb)]

$T_{Rd.max}$ = design torsional resistance moment according to the following relationship [N·m (in.-lb)]

$$T_{Rd,max} = 2v\alpha_{cw}f_{cd}A_k t_{ef}\sin\theta\cos\theta \quad (9.2.2b)$$

V_{Ed} = design shear force [N (lb)]

$V_{Rd.max}$ = maximum design shear resistance according to the following relationship [N·m (in.-lb)]

$$V_{Rd,max} = \frac{\alpha_c b_w z v f_{cd}}{(\cot\theta + \tan\theta)} \quad (9.2.2c)$$

b_w = width of the web of the cross section [mm (in.)]

z = inner lever arm, for a member with constant depth, corresponding to the bending moment in the element under consideration. In the shear analysis of reinforced concrete without axial force, the approximate value $z = 0.9d$ may normally be used (d is the effective depth of the cross section) [mm (in.)]

v = strength reduction factor for concrete cracked in shear, recommended values (values for use in a country may be found in its National Annex):

$$\begin{aligned} v &= 0.6(1 - f_{ck}/250)\ [f_{ck} \text{ in MPa}] \\ v &= 0.6(1 - f_{ck}/36.26)\ [f_y \text{ in ksi}] \end{aligned} \quad (9.2.2d)$$

α_{cw} = coefficient taking into account the state of compressive stress

$$\alpha_{cw} = \begin{Bmatrix} 1 & \text{non-prestressed} \\ 1+\sigma_{cp}/f_{cd} & 0 < \sigma_{cp} \leq 0.25f_{cd} \\ 1.25 & 0.25f_{cd} < \sigma_{cp} \leq 0.5f_{cd} \\ 2.5(1-\sigma_{cp}/f_{cd}) & 0.5f_{cd} < \sigma_{cp} \leq f_{cd} \end{Bmatrix} \quad (9.2.2e)$$

σ_{cp} = mean compressive concrete stress due to design axial force (measured positive) [MPa (psi)]

f_{ck} = characteristic compressive concrete strength [MPa (psi)]

f_{cd} = design compressive concrete strength (= f_{ck}/γ_c, where γ_c is the partial factor for concrete equal to 1.5 for ultimate limit state and persistent and transient design situations) [MPa (psi)]

- Step 4: Calculation of the required cross-sectional area of the longitudinal reinforcement for torsion, ΣA_{st}:

$$\sum A_{st} = \frac{T_{Ed}u_k \cot\theta}{2A_k f_{yld}} \quad (9.2.2f)$$

where f_{yld} is the design yield stress of the longitudinal reinforcement [MPa (psi)].

Notes: The longitudinal reinforcement for torsion should be added to the required longitudinal reinforcement for flexure. The longitudinal reinforcement should generally be distributed over the length of side, z_i, (z_i is the side length of wall i defined by the distance between intersection points with the adjacent walls [refer to Fig. 6.11 of EC2-04]), but for smaller sections it may be concentrated at the ends of this length. According to EC2-04 provisions (Section 9.2.3(4)), longitudinal bars for torsion should be arranged such that there is at least one bar at each corner, with the others being distributed uniformly around the inner periphery of the links, with a spacing not greater than 350 mm (14 in.).

- Step 5: Calculation of the required cross-sectional area of the transversal reinforcement for torsion

$$\frac{A_{sw}}{s} = \frac{T_{Ed}}{2A_k f_{ywd} \cot\theta} \quad (9.2.2g)$$

where

A_{sw} = cross-sectional area of the transversal reinforcement (stirrups) [mm^2 (in.^{2})]

s = spacing of the stirrups [mm (in.)]

f_{ywd} = design yield stress of transversal reinforcement [MPa (psi)]

Notes: The transversal reinforcement for torsion should be added to the existing transverse reinforcement for shear. The torsion links (stirrups) should be closed and anchored by means of laps or hooked ends and form an angle of 90 degrees with the axis of the structural element. Refer to Fig. 9.6 of EC2-04 for recommended shapes. According to provisions of EC2-04 (Section 9.2.3(3)), longitudinal spacing of the torsion stirrups should not exceed $u/8$, or the requirements about the maximum longitudinal spacing between shear assemblies (Section 9.2.2(6) of EC2-04) or the lesser dimension of the beam cross section.

- Step 6: Check the value of the angle of compression struts, θ, based on the calculated and provided longitudinal and transversal reinforcement from Steps 4 and 5

$$\theta_{calc} = \tan^{-1}\sqrt{\frac{A_{sw}u_k f_{ywd}}{s\sum A_{sl} f_{yld}}} \quad (9.2.2h)$$

Note: In case of a significant difference between the calculated and the initially assumed angle of compression struts, reassume the angle θ (Step 2) and recalculate Step 3 through 6.

9.2.3 *Torsion design in CSA-A23.3-04*—The Canadian code stipulates that the effect of torsion should be considered in design only if the torsion due to factored loads, T_f [N·m (in.-lb)], exceeds $0.25T_{cr}$. The cracking torque T_{cr} [N·m (in.-lb)] is assumed to be reached when the principal tensile stress f_l [MPa (psi)] (equivalent to the shear stress v in pure torsion) equals the factored tensile strength of the concrete, f_{cr} [MPa (psi)]. For the calculation of T_{cr}, the following assumptions are made:

- Solid cross sections are represented by an equivalent thin-walled tube with a wall thickness [mm (in.)]

$$t_c = 0.75A_c/p_c \quad (9.2.3a)$$

- Bredt's classical equation for tubular section applies [MPa (psi)]

$$v = T_f/(2A_o t_c) \quad (9.2.3b)$$

- Area enclosed by shear flow path [mm² (in.²)]

$$A_o = 2/3A_c \quad (9.2.3c)$$

- Factored design tensile strength for normal concrete [MPa (psi)]

$$f_{cr} = 0.38\phi_c\sqrt{f_c'}\ [f_c' \text{ in MPa}]$$
$$f_{cr} = 0.38\phi_c 12\sqrt{f_c'}\ [f_c' \text{ in psi}] \quad (9.2.3d)$$

For non-prestressed concrete members, this results in the following expression

$$T_{cr} = (A_c / p_c)0.38\phi_c\sqrt{f_c'}\ [f_c' \text{ in MPa}]$$
$$T_{cr} = (A_c / p_c)0.38\phi_c 12\sqrt{f_c'}\ [f_c' \text{ in psi}] \quad (9.2.3e)$$

The symbols not defined in the above equations are as follows:

A_c = area enclosed by outside perimeter p_c of concrete section [mm² (in.²)]
f_c' = specified compressive strength of concrete
ϕ_c = resistance factor for concrete (= 0.65)

If torsion is not negligible ($T_f > T_{cr}$), torsion reinforcement should be provided. The General Design Method (CSA-A23.3-04) for torsion was derived from the MCFT, which represents a holistic approach for both shear and torsion design. For torsion, the basic concept is the space truss analogy, originally envisioned by Rausch (1929), assuming a 45-degree angle for the compression struts. This AASHTO LRFD (general method), originally developed for shear, requires a longitudinal strain indicator ε_x and the level of normalized shear stress v_u/f_c' to estimate θ and β. In the case of a member subjected to pure torsion, it is not necessary to consider β. For torsion, the General Design Method (CSA-A23.3-04) assumes that:

a) Concrete in the cracked member carries no tension

b) The angle θ of the diagonal compression struts spiraling around the member is variable and depends on longitudinal strain at mid-depth of section, ε_x

The longitudinal strain ε_x is affected by the bending moment: shear, torsion, and if present, by axial load and prestressing in the member. In the presence of bending moment, shear, and torsion, the strain at mid-depth of the section, ε_x, is computed from the expression

$$\varepsilon_x = (M_f / d + \sqrt{V_f^2 + [0.9p_h T_f / 2A_o]^2}) / (2E_s A_s) \quad (9.2.3f)$$

where

M_f = moment due to factored loads [N·m (in.-lb)]
V_f = shear force due to factored loads [N (lb)]
p_h = perimeter of the centerline of the closed transverse reinforcement [mm (in.)]
E_s = modulus of elasticity of the reinforcement [MPa (psi)]
A_s = area of flexural reinforcement on the flexural tension side of the member [mm² (in.²)]
θ = 29 + 7000ε_x (degrees)

For a given angle θ, the transverse reinforcement to resist the factored torque T_f is derived from equilibrium and given by

$$A_t = \frac{T_f s}{2A_o\phi_s f_y \cot\theta} \quad (9.2.3g)$$

where

s = spacing of transverse reinforcement measured parallel to the axis of the member [mm (in.)]
ϕ_s = resistance factor for non-prestressed reinforcing bars
f_y = specified yield strength of transverse reinforcement [MPa (psi)]

Additional longitudinal reinforcement is required to resist the longitudinal forces generated by torsion. As usual, the transverse reinforcement due to torsion should be added to the shear reinforcement.

Dimensions of the cross section of the member have to be such that crushing of the diagonal compression struts is prevented. This is achieved if the combined stress due to shear and torsion does not exceed 25 percent of the factored compressive strength of the concrete. This is expressed by Eq. (9.2.3h)

$$\frac{V_f}{b_w d_v} + \frac{T_f p_h}{1.7A_{oh}^2} \le 0.25\phi_c f_c' \quad (9.2.3h)$$

9.2.4 *A comparison of torsion design procedures for ACI, EC2, and CSA*—The design philosophy and procedures for pure torsion and combined loads including bending, shear, and torsion are discussed previously according to ACI 318-11, EC2-04, and CSA-A23.3-04. All the design procedure and equation citations from these codes are summarized and compared in Table 9.2.4.

9.3—Introduction to design examples

Two examples were selected to illustrate the steps involved in torsion design: 1) a solid reinforced concrete rectangular beam under pure torsion; and 2) A prestressed box girder under combined loading including torsion. Although this

Table 9.2.4—Comparison of torsion design procedures for ACI, EC2, and CSA codes

Pure torsion design procedure		ACI 318-11	EC2-04	CSA-A23.3-04
1	Determine the factored torsional moment and if torsion effects can be disregarded	Section 11.5.1, $T_u < \phi\lambda\left(\frac{\sqrt{f'_c}}{12}\right)\left(\frac{A_{cp}^2}{p_{cp}}\right)$ [f'_c in MPa] $T_u < \phi\lambda\sqrt{f'_c}\left(\frac{A_{cp}^2}{p_{cp}}\right)$ [f'_c in psi]		Section 11.2.9.1, Eq. 11-2 $T_{cr} = \left(\frac{A_c^2}{p_c}\right)0.38\lambda\phi_c\sqrt{f'_c}$ [f'_c in MPa] $T_{cr} = \left(\frac{A_c^2}{p_c}\right)0.38\lambda\phi_c 12\sqrt{f'_c}$ [f'_c in psi]
2	Calculate properties of the equivalent thin-walled section		Section 6.3.2(1) Eq. 6-26 and 6-27 $t_{ef} = \max\begin{Bmatrix} A/p_c \\ 2c_t \end{Bmatrix}$ $\tau_t t_{ef} = \frac{T_{Ed}}{2A_k}$, $V_{Ed} = \tau_t t_{ef} z$	
3	Determine if dimensions of the cross section are adequate	Section 11.5.3.1, Eq. (11-18) and Section 11.2.1.1, Eq. (11-3) $\frac{T_u p_h}{1.7A_{oh}^2} \le \phi\left(\frac{V_c}{b_w d} + \frac{2\sqrt{f'_c}}{3}\right)$ [f'_c in MPa] $\frac{V_c}{b_w d} = \frac{2\lambda\sqrt{f'_c}}{12}$ [f'_c in MPa] $\frac{T_u p_h}{1.7A_{oh}^2} \le \phi\left(\frac{V_c}{b_w d} + 8\sqrt{f'_c}\right)$ [f'_c in psi] $\frac{V_c}{b_w d} = 2\lambda\sqrt{f'_c}$ [f'_c in psi]	Section 6.3.2(4) Eq. 6-29 and 6-30 $\frac{T_{Ed}}{T_{Rd,max}} + \frac{V_{Ed}}{V_{Rd,max}} \le 1 \xrightarrow{\text{pure torsion}} T_{Ed} <= T_{Rd,max}$ $T_{Rd,max} = 2\nu\alpha_c f_{cd} A_k t_{ef} \sin\theta\cos\theta$	Section 11.3.10.4, Eq.11-19 $\sqrt{\left(\frac{V_f - V_p}{b_w d_v}\right)^2 + \left(\frac{T_f p_h}{1.7A_{oh}^2}\right)^2} \le 0.25\phi_c f'_c$
4	Calculate the amount of stirrups required for pure torsion Select and check the stirrups' details for the cross section	Section 11.5.3.5, Eq. (11-20) and Section 11.5.3.6, Eq. (11-21) transformed into $\frac{A_t}{s} \ge \frac{T_u}{\phi 2 f_{yt} A_o \cot\theta}$ and Section 11.5.5.2, Eq. (11-23) can be expressed by $\left(\frac{A_t}{s}\right)_{min} = \frac{0.75}{2}\frac{\sqrt{f'_c}}{12}\frac{b_w}{f_{yv}}$ [f'_c in MPa] $\left(\frac{A_t}{s}\right)_{min} = \frac{0.75}{2}\sqrt{f'_c}\frac{b_w}{f_{yv}}$ [f'_c in psi]	Section 6.3.2 (2) and 9.2.3 Eq. 9.5N $\rho_l = \left(0.08\frac{\sqrt{f'_c}}{12}\right)/f_{yk}$ [f'_c in MPa] $\rho_l = \left(0.08\sqrt{f'_c}\right)/f_{yk}$ [f'_c in] and Eq. 9.6N $s_{max} = \min\begin{Bmatrix} u/8 \\ 0.75d(1+\cot\alpha) \\ \min(b_w, h) \end{Bmatrix}$	Section 11.3.10.3, 11.3.8.1, and 11.3.6.4 Eq.11-1, 11-12, 11-13, and 11-17 $\left(\frac{A_v}{s}\right)_{min} = 0.06\sqrt{f'_c}\frac{b_w}{f_{yv}}$ [f'_c in MPa] $\left(\frac{A_v}{s}\right)_{min} = 0.72\sqrt{f'_c}\frac{b_w}{f_{yv}}$ [f'_c in psi] $\theta = 29 + 7000\varepsilon_x$ $\varepsilon_x = \frac{\frac{M_f}{d_v} + \sqrt{(V_f - V_p)^2 + \left(\frac{0.9p_h T_f}{2A_o}\right)^2} + 0.5N_f - A_p f_{po}}{2[E_s A_s + E_p A_p]}$ $\frac{A_t}{s} \ge \frac{T_f}{2A_o\phi_s f_y \cot\theta}$ $s_{max} = \min\begin{Bmatrix} 0.7d_v \\ 600\text{ mm }(23.62\text{ in.}) \end{Bmatrix}$

5	Calculate the longitudinal bars required for pure torsion Select and check the longitudinal bar details for the cross section	Section 11.5.3.7, $A_l = \frac{A_t}{s} p_h \frac{f_{yt}}{f_{yl}} \cot^2 \theta$ Eq. (11-22): $A_{\ell,min} = \frac{5\sqrt{f'_c}A_{cp}}{12 f_{yt}} - \left(\frac{A_t}{s}\right) p_h \frac{f_{yt}}{f_{yt}}$ [f'_c in MPa] and Section 11.5.5.3, Eq. (11-24) $A_{\ell,min} = \frac{5\sqrt{f'_c}A_{cp}}{f_{yl}} - \left(\frac{A_t}{s}\right) p_h \frac{f_{yt}}{f_{yl}}$ [f'_c in psi]	Section 6.3.2(3) and 9.2.1 Eq. 6-28 $\Sigma A_l \geq \frac{T_u u_k \cot\theta}{2A_k f_{yd}}$ and Eq. 9.1N $A_{\ell,min} = 0.26 \frac{f_{ctm}}{f_{yk}} b_t d$	Section 11.3.10.6 and 11.3.9.2, Eq. 11-14 and 1-21 can be transformed into $F_n = \frac{M_f}{d_v} + \cot\theta \sqrt{(V_f - 0.5V_s - V_p)^2 + \left(\frac{0.45 p_h T_f}{2A_o}\right)^2}$ $= \cot\theta \frac{0.45 p_h T_f}{2A_o}$
Design procedure for prestressed concrete members under combined loadings (Bending, shear, and torsion)		**ACI 318-11**	**EC2-04**	**CSA-A23.3-04**
1	Determine the material and cross-sectional properties		Concrete compressive strength: Section 3.1.2 and 3.1.6 Concrete tensile strength: Section 3.1.2, 3.1.6 and 3.1.7 Reinforcing steel properties: Section 2.4.2.4 and 3.2.7 Pre-stressed reinforcing strands: Section 2.4.2.4, 3.3.6 and 3.3.3 Concrete covers: Section 4.4.1.2	
2	Determine the factored loads and calculate factored shear, torque, and bending moment	Section 8.3 and 9.2.1	Section 2.4.3 and 5.1.3	Annex C
3	Determine if torsion effects can be disregarded	Section 11.5.1 $T_u < \phi\lambda \left(\frac{\sqrt{f'_c}}{12}\right)\left(\frac{A_{cp}^2}{p_{cp}}\right)$ [f'_c in MPa] $T_u < \phi\lambda\sqrt{f'_c}\left(\frac{A_{cp}^2}{p_{cp}}\right)$ [f'_c in psi]		Section 11.2.9.1, Eq. 11-2 $T_{cr} = \frac{(1.5A_e)^2}{p_c} 0.38\lambda\phi_c\sqrt{f'_c}\sqrt{1+\frac{\phi_p f_{cp}}{0.38\lambda\phi_c\sqrt{f'_c}}}$ [f'_c in MPa] $T_{cr} = \frac{(1.5A_e)^2}{p_c} 0.38\lambda\phi_c 12\sqrt{f'_c}\sqrt{1+\frac{\phi_p f_{cp}}{0.38\lambda\phi_c 12\sqrt{f'_c}}}$ [f'_c in psi]

4	Determine if dimensions of the cross section are adequate	Section 11.2.1.1, Eq. (11-3), Section 11.3.2, Eq. (11-9), and Section 11.5.3.1, Eq. (11-18) $\frac{V_c}{b_w d} = \frac{2\lambda\sqrt{f'_c}}{12}$ [f'_c in MPa] $\frac{V_c}{b_w d} = 2\lambda\sqrt{f'_c}$ [f'_c in psi] $V_c = \left(0.6\lambda\frac{\sqrt{f'_c}}{12} + 700(0.00689)\frac{V_u d}{M_u}\right)b_w d$ [f'_c in MPa] $V_c = \left(0.6\lambda\sqrt{f'_c} + 700\frac{V_u d}{M_u}\right)b_w d$ [f'_c in psi] $\frac{T_u p_h}{1.7A_{oh}^2} \le \phi\left(\frac{V_c}{b_w d} + \frac{2\sqrt{f'_c}}{3}\right)$ [f'_c in MPa] $\frac{T_u p_h}{1.7A_{oh}^2} \le \phi\left(\frac{V_c}{b_w d} + 8\sqrt{f'_c}\right)$ [f'_c in psi]	Section 6.3.2(4) Eq. 6-14, 6-29, and 6-30 $V_{Rd,max} = \frac{\alpha_c b_w z v f_{cd}}{(\cot\theta + \tan\theta)}$ $\frac{T_{Ed}}{T_{Rd,max}} + \frac{V_{Ed}}{V_{Rd,max}} \le 1$ $T_{Rd,max} = 2v\alpha_c f_{cd} A_k t_{ef} \sin\theta\cos\theta$	Section 11.3.10.4, Eq. 11-18 $\frac{V_f - V_p}{b_w d_v} + \frac{T_f}{1.7A_{oh}t} \le 0.25\phi_c f'_c$
5	Shear design to determine transverse reinforcement for shear force	Shear resistance of concrete: Section 11.2.1.1, Eq. (11-3) and Section 11.3.2, Eq. (11-9) $\frac{V_c}{b_w d} = \frac{2\lambda\sqrt{f'_c}}{12}$ [f'_c in MPa], $\frac{V_c}{b_w d} = 2\lambda\sqrt{f'_c}$ [f'_c in psi] $V_c = \left(0.6\lambda\frac{\sqrt{f'_c}}{12} + 700(0.00689)\frac{V_u d}{M_u}\right)b_w d$ [f'_c in MPa] $V_c = \left(0.6\lambda\sqrt{f'_c} + 700\frac{V_u d}{M_u}\right)b_w d$ [f'_c in psi] Required shear reinforcement: Section 11.4.7 $\frac{A_v}{s} = \frac{V_u - \phi V_c}{\phi d f_{yv}}$ Minimum shear reinforcement: Section 11.4.6.3, Eq. 11-13 $A_{v,min} = 0.75\frac{\sqrt{f'_c}}{12}\frac{b_w}{f_{yt}}$ [f'_c in MPa] $A_{v,min} = 0.75\sqrt{f'_c}\frac{b_w}{f_{yt}}$ [f'_c in psi]	Design shear resistance of a member without shear reinforcement: Section 6.2.2(1). If not adequate, Required shear reinforcement: Section 6.2.3(4); Eq. 6-13 and 6-14 $V_{Rd,s} = \frac{A_{sw}}{s} z f_{ywd} \cot\theta \ge V_{Ed} \Rightarrow \frac{A_{sw}}{s} \ge \frac{V_{Ed}}{z f_{ywd} \cot\theta}$ $V_{Rd,max} = \frac{\alpha_c b_w z v f_{cd}}{(\cot\theta + \tan\theta)}$ Minimum shear reinforcement: Section 9.2.2(5); Maximum effective cross-sectional area of the shear reinforcement: Section 6.2.3(3)	The angle of diagonal compression strut: Section 11.3.6.4, Eq. 11-12 and 11-13 $\varepsilon_x = \frac{\frac{M_f}{d_v} + \sqrt{(V_f - V_p)^2 + \left(\frac{0.9p_h T_f}{2A_o}\right)^2} + 0.5N_f - A_p f_{po}}{2[E_s A_s + E_p A_p]}$ $\frac{A_t}{s} \ge \frac{T_f}{2A_o \phi_s f_y \cot\theta}$ Shear resistance of concrete: Section 11.3.4, Eq. 11-6 and 11-11 $V_c = \phi_c \lambda\beta\sqrt{f'_c} b_w d_v$ [f'_c in MPa] $V_c = \phi_c \lambda\beta 12\sqrt{f'_c} b_w d_v$ [f'_c in psi] $\beta = \frac{0.40}{1 + 1500\varepsilon_x} \times \frac{1300}{1000 + s_{ze}}$ Required shear reinforcement: Section 11.3.5.1, Eq. 11-7 can be transformed into $\frac{A_v}{s} = \frac{V_s}{\phi_s f_y d_v \cot\theta}$

6	Torsion design to determine transverse reinforcement for torsion	Assume the angle of diagonal compression strut. Required transverse reinforcement for torsion: Section 11.5.3.6 and 11.5.6, Eq. (11-21) transformed into $\frac{A_t}{s} = \frac{T_u}{\phi 2 A_o f_{yv} \cot\theta}$	Calculate the terms of the equivalent thin-walled section for torsion design: Section 6.3.2(1) Required transverse reinforcement for torsion: Section 6.3.2(2) $\frac{A_{sw}}{s} \geq \frac{T_{Ed}}{2 A_k f_{ywd} \cot\theta}$	Required transverse reinforcement for torsion: Section 11.3.10.3, Eq. 11-17 can be transformed into $\frac{A_t}{s} \geq \frac{T_f}{2 A_o \phi_s f_y \cot\theta}$
7	Determine the total transverse reinforcement for shear and torsion Select and check transverse reinforcement details for cross section	Add the shear and torsion transverse reinforcement in one web: Section 11.5.3.8 Minimum transverse reinforcement: Section 11.5.5, Eq. (11-23) $\left(\frac{A_t}{s} + \frac{1}{2}\frac{A_v}{s}\right)_{min} = 0.375\frac{\sqrt{f'_c}}{12}\frac{b_w}{f_{yt}}$ [f'_c in MPa] $\left(\frac{A_t}{s} + \frac{1}{2}\frac{A_v}{s}\right)_{min} = 0.375\sqrt{f'_c}\frac{b_w}{f_{yt}}$ [f'_c in psi] Maximum longitudinal spacing of transverse reinforcement for shear: ACI 318-11, Section 11.4.5 Maximum longitudinal spacing of transverse reinforcement for torsion: ACI 318-11, Section 11.5.6	Add the shear and torsion transverse reinforcement in one web: Section 6.3.2(2) Maximum longitudinal spacing of the stirrups for shear: Eq. 9.6N in EC2-04 $s_{max} = 0.75d(1+\cot\alpha) \xrightarrow{\alpha=90°} s_{max} = 0.75d$ Maximum longitudinal spacing of the stirrups for torsion: Section 9.2.3(3) $s_{max} = \min\left\{\begin{matrix} u/8 \\ 0.75d(1+\cot\theta) \\ \min(b,h) \end{matrix}\right\}$	Minimum transverse reinforcement: Section 11.2.8.2, Eq. 11-1 $\frac{A_{v,\min}}{s} = 0.06\sqrt{f'_c}\frac{b_w}{f_y}$ [f'_c in MPa] $\frac{A_{v,min}}{s} = 0.72\sqrt{f'_c}\frac{b_w}{f_y}$ [f'_c in psi] Maximum longitudinal spacing of transverse reinforcement: Section 11.3.8.1 $s_{max} = \min\left\{\begin{matrix} 0.7d_v \\ 600 \text{ mm } (23.62 \text{ in.}) \end{matrix}\right\}$ Add the shear and torsion transverse reinforcement in one web
8	Determine longitudinal reinforcement for torsion Select and check longitudinal reinforcement details for cross section	Required torsional longitudinal reinforcement: Section 11.5.3.7, Eq. (11-22) $A_l = \frac{A_t}{s} p_h \left(\frac{f_{yv}}{f_{yl}}\right) \cot^2\theta$ Minimum longitudinal reinforcement for torsion: Section 11.5.5.3, Eq. (11-24) $A_{\ell,min} = \frac{5\sqrt{f'_c}A_g}{12 f_{yl}} - \left(\frac{A_t}{s}\right) p_h \left(\frac{f_{yv}}{f_{yl}}\right)$ [f'_c in MPa] $A_{\ell,min} = \frac{5\sqrt{f'_c}A_g}{f_{yl}} - \left(\frac{A_t}{s}\right) p_h \left(\frac{f_{yv}}{f_{yl}}\right)$ [f'_c in psi] Spacing of longitudinal reinforcement for torsion: Section 11.5.6.2 and 11.5.6.3	Required torsional longitudinal reinforcement: Section 6.3.2(3), Eq. 6-28 $\Sigma A_{sl} \geq \frac{T_{Ed} u_k \cot\theta}{2 A_k f_{ywd}}$ Spacing of longitudinal reinforcement for torsion: Section 9.2.3 (4)	Required torsional longitudinal reinforcement: Section 11.3.10.6 11.3.9.2, Eq. 11-14 and 11-21 can be transformed into $F_{lt} = \frac{M_f}{d_v} + \cot\theta\sqrt{(V_f - 0.5V_s - V_p)^2 + \left(\frac{0.45 p_h T_f}{2A_o}\right)^2}$ $A_p = \frac{F_{lt}}{f_p f_{pr}}$

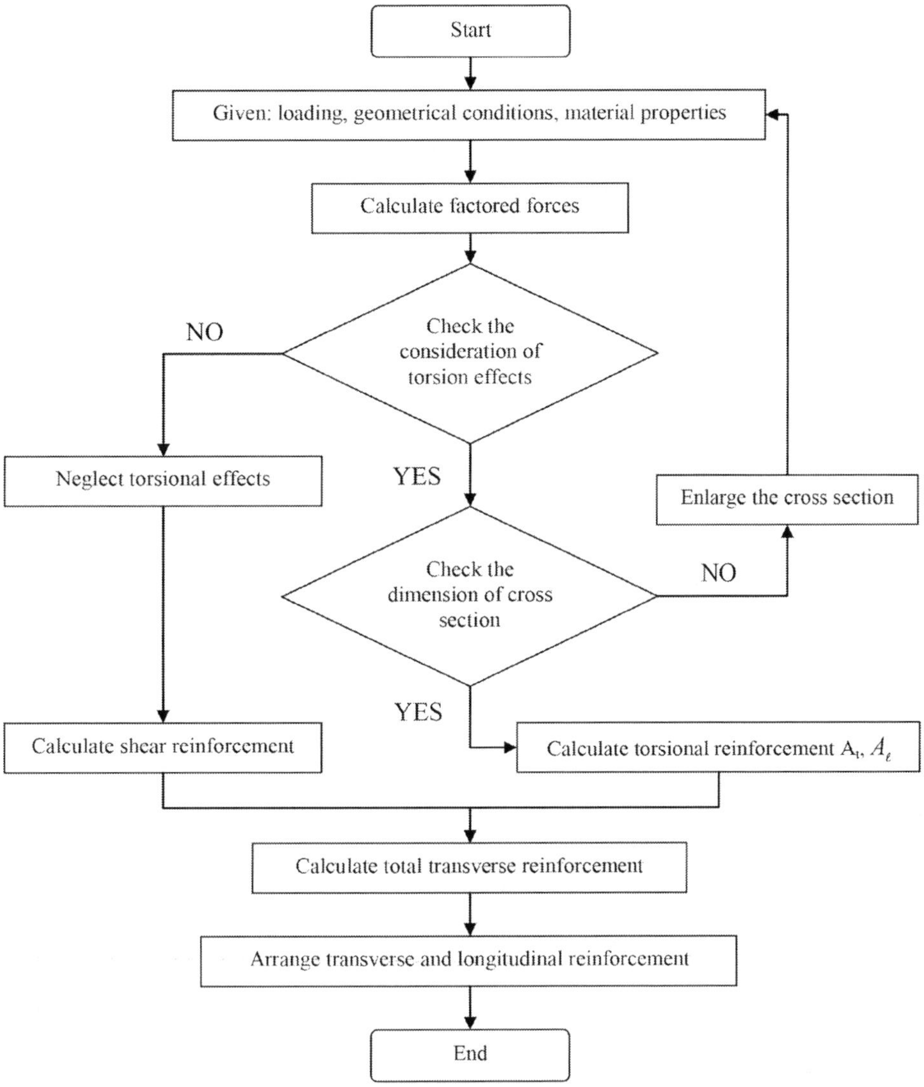

Fig. 9.3—Design process flow chart for combined shear and torsion effects.

report focuses on recent torsion developments and theories, the design examples are solved by three major building codes: ACI 318-11, EC2-04, and CSA-A23.3-04. Example 1 was also solved by the two graphical methods proposed by Rahal (2000b) and Leu and Lee (2000). These graphical methods are not suitable to treat prestressed concrete sections under combined loads and, therefore, were not used for Example 2. Following each design example, a comparison table summarizes results of the various codes and methods. Design equations and expressions may vary slightly from one code to another, but the general design procedure is the same. In the case of torsion combined with shear, design procedures for all three codes follow the flow chart given in Fig. 9.3. Hsu (1997) provides background on ACI 318-95 that is still applicable to the updated ACI 318-11. His paper includes a detailed design example of a prestressed hollow box girder subjected to torsion, shear, and flexure that constitutes Example 2.

9.4—Design Example 1: solid rectangular reinforced concrete beam under pure torsion

9.4.1 *Design problem statement*—As shown in Fig. 9.4.1, the cross-sectional dimensions of beam are b_w = 300 mm (12 in.) and h = 500 mm (20 in.). The characteristic concrete cylinder compressive strength is f_c' = 20 MPa (2900 psi) [Class C20/25], and the characteristic steel yield strength is f_y = 420 MPa (60,000 psi). The applied torsional moment is T_u = 30 kN·m (266 in.-kip). The mean cylinder strength of the concrete: $f_{cm} = f_c' + 8$ MPa $\Rightarrow f_{cm}$ = 28 MPa (4000 psi). Assume 40 mm (1.5 in.) from exterior face to stirrup centerline typical.

9.4.2 *Solution according to ACI 318-11*

1. Determine if torsion effects may be disregarded (ACI 318-11, Section 11.5.1).

The torsion effects can be disregarded if the following expression is valid

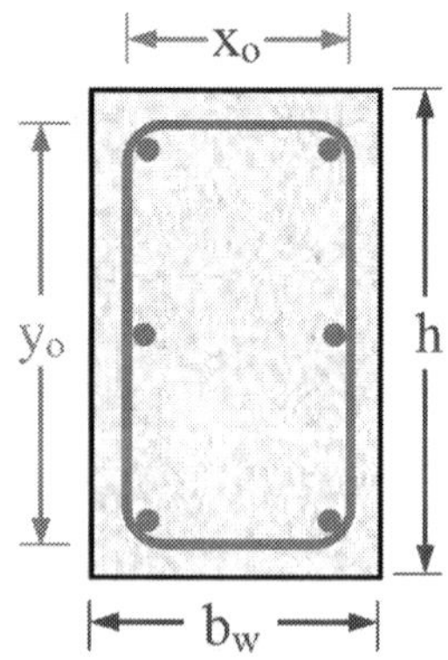

Fig. 9.4.1—Rectangular reinforced concrete cross section subjected to pure torsion.

$$T_u < \phi\lambda\left(\frac{\sqrt{f_c'}}{12}\right)\left(\frac{A_{cp}^2}{p_{cp}}\right)$$

where

ϕ = strength reduction factor for torsion and shear = 0.75

λ = modification factor of lightweight concrete (normalweight concrete, $\lambda = 1.0$)

p_{cp} = perimeter of outer concrete cross section, mm (in.) $= 2 \times (b_w + h) = 1600$ mm (64.0 in.)

A_{cp} = total area enclosed by the outside perimeter of concrete cross section, mm (in.) $= b_w \times h = 150{,}000$ mm^2 (240 in.^{2})

Therefore,

$$\phi\lambda\left(\frac{\sqrt{f_c'}}{12}\right)\left(\frac{A_{cp}^2}{p_{cp}}\right) = 0.75(1.0)\left(\frac{\sqrt{20}}{12}\right)\left(\frac{150{,}000^2}{1600}\right)\text{ N·mm}$$

= 3.93 kN·m (36.3 in.-kip)

T_u = 30.0 kN·m (266 in.-kip).

Torsion effects must be considered.

2. Determine if dimensions of the cross section are adequate (ACI 318-11, Section 11.5.3.1).

Dimensions of the cross section are adequate if

$$\sqrt{\left(\frac{V_u}{b_w d}\right)^2 + \left(\frac{T_u p_h}{1.7A_{oh}^2}\right)^2} \le \phi\left(\frac{V_c}{b_w d} + 8\frac{\sqrt{f_c'}}{12}\right)$$

In case of pure torsion, the expression becomes

$$\frac{T_u p_h}{1.7A_{oh}^2} \le \phi\frac{5\sqrt{f_c'}}{6}, \text{ taking } \frac{V_c}{b_w d} = \frac{2\lambda\sqrt{f_c'}}{12}$$

where

p_h = perimeter of centerline of outermost closed transverse torsional reinforcement, mm (in.)

p_h = $2(x_o + y_o) = 1280$ mm (50.4 in.)

x_o, y_o = horizontal and vertical dimension of the centerline of outermost closed transverse torsional reinforcement, respectively

x_o = $(300 - 2 \times 40) = 220$ mm (9 in.)

y_o = $(500 - 2 \times 40) = 420$ mm (17 in.)

A_{oh} = area enclosed by centerline of outermost closed transverse torsional reinforcement, mm (in.)

A_{oh} = $x_o y_o = 92{,}400$ mm^2 (153 in.^{2})

Therefore

$$\frac{T_u p_h}{1.7A_{oh}^2} = \frac{30 \times 10^6 \times 1280}{1.7 \times 92{,}400^2} = 2.65 \text{ MPa (347.6 psi)}$$

whereas

$$\phi\frac{5\sqrt{f_c'}}{6} = 0.75\frac{5\sqrt{20}}{6} = 2.80 \text{ MPa (403 psi)}$$

Because the expression in Eq. (11-18) (ACI 318-11) is valid, the cross section dimensions are adequate.

3. Calculate the amount of stirrups required for pure torsion (ACI 318-11, Section 11.5.3.3).

To calculate transverse reinforcement for torsion, Eq. (11-20) and (11-21) in ACI 318 can be transformed into

$$\frac{A_t}{s} \ge \frac{T_u}{\phi 2 f_{yt} A_o \cot\theta}$$

where

A_t = area of one leg of a closed stirrup resisting torsion

s = spacing of the stirrups

f_{yt} = design strength of torsion transverse reinforcement = 420 MPa (60,000 psi)

A_o = gross area enclosed by shear flow path $= 0.85A_{oh}$

θ = angle of compression diagonals in truss analogy for torsion = 45 degrees

Therefore,

$$\frac{A_t}{s} \ge \frac{T_u}{\phi 2 f_{yt} A_o \cot\theta} = \frac{30 \times 10^6}{0.75 \times 2 \times 420 \times 0.85 \times 92{,}400 \times 1}$$

= 0.61 mm^2/mm (0.0227 in.^{2}/in.)

The minimum transverse reinforcement is computed by Eq. (11-23) in ACI 318.

In the case of pure torsion, this minimum is expressed by

$$\left(\frac{A_t}{s}\right)_{min} = \frac{0.75}{2}\frac{\sqrt{f_c'}}{12}\frac{b_w}{f_{yv}} = \frac{0.75 \times \sqrt{20} \times 300}{24 \times 420}$$

= 0.1 mm^2/mm (0.0040 in.^{2}/in.)

but shall not be less than

$$\left(\frac{A_t}{s}\right)_{min} = \frac{0.35}{2}\frac{b_w}{f_{yv}} = \frac{0.175\times 300}{420}$$

$= 0.125$ mm²/mm (0.005 in.²/in.)

The maximum spacing of the stirrups is given by (ACI 318-11, Section 11.5.6.1)

$$s_{max} = \min\begin{Bmatrix} p_h/8 \\ 300\text{ mm}\end{Bmatrix} = 160\text{ mm (6.50 in.)}$$

For stirrups ∅8 (d_s = 8.0 mm [0.315 in.]), A_t = 50.3 mm² (0.078 in.²) and $s \leq 81$ mm (3.20 in.). Select: s = 80 mm (3.15 in.).

For stirrups No. 3 (d_s = 9.5 mm (0.375 in.)), A_t = 71.0 mm² (0.11 in.²) and $s \leq 115$ mm (4.55 in.). Select: s = 110 mm (4.50 in.).

4. Calculate the longitudinal bars required for pure torsion (ACI 318-11, Section 11.5.3.7).

Total longitudinal reinforcement for torsion (A_ℓ) (Eq. (11-22) in ACI 318) is calculated by

$$A_\ell = \frac{A_t}{s}p_h\frac{f_{yt}}{f_{yt}}\cot^2\theta = 0.610\times 1280\times 1\times 1$$

$= 780.8$ mm² (1.18 in.²)

Minimum longitudinal reinforcement (Eq. (11-24) in ACI 318) is calculated by

$$A_{\ell,min} = \frac{5\sqrt{f_c'}A_{cp}}{12 f_{yt}} - \left(\frac{A_t}{s}\right)p_h\frac{f_{yt}}{f_{yt}} = \frac{5\sqrt{20}\times 150{,}000}{12\cdot 420} - 0.610\times 1280\times 1$$

< 0 mm²

Therefore, A_ℓ = 780.8 mm² (1.18 in.²)

At least one longitudinal bar is required in each stirrup's corner and bar spacing distributed around the perimeter of the closed stirrups is 305 mm (12 in.). Therefore, the number of longitudinal bars is at least six and each bar requires780.8/6 = 130 mm² (0.20 in.²).

For six longitudinal bars ∅14 (d_s = 14.0 mm [0.551 in.]), $A_\ell = 6 \times 153.9 = 923$ mm² (1.43 in.²).

For six longitudinal bars No. 5 (d_s = 15.9 mm [0.625 in.]), $A_\ell = 6 \times 200 = 1200$ mm² (1.86 in.²).

9.4.3 *Solution according to EC2-04 code*

1. Calculate the terms of the equivalent thin-walled section (EC2-04, Section 6.3.2(1)).

Effective wall thickness is calculated by

$$t_{ef} = \max\begin{Bmatrix} A/p_c \\ 2c_\ell \end{Bmatrix} = \max\begin{Bmatrix} 150{,}000/1600 = 94\text{ mm} \\ 2\times 50 = 100\text{ mm}\end{Bmatrix}$$

$\Rightarrow t_{ef}$ = 100 mm (4.0 in.)

where

A = total area of the outside perimeter of the concrete cross section
$= b_w h = 300 \times 500 = 150{,}000$ mm² (240 in.²)
p_c = outside perimeter of the concrete cross section
$= 2(b_w + h) = 2(300 + 500) = 1600$ mm (64.0 in.)
c_ℓ = distance between edge of member and center of the longitudinal reinforcement
= 50 mm (2.0 in.)

The area enclosed by the centerlines of the connecting thin-walls is calculated by

$$A_k = (b_w - t_{ef})(h - t_{ef}) = 200 \times 400 = 80{,}000\text{ mm}^2\text{ (124.0 in.}^2)$$

The perimeter of the area enclosed by the centerlines of the connecting thin-walls is determined by

$$u_k = 2\left[(b_w - t_{ef}) + (h - t_{ef})\right] = 2(200 + 400)$$

$= 1200$ mm (48.0 in.)

2. Determine if dimensions of the cross section are adequate (evaluate the strength of concrete struts) (EC2-04, Section 6.3.2(4)).

Dimensions of the cross section are adequate if (Eq. (6-29) in EC2-04)

$$\frac{T_{Ed}}{T_{Rd,max}} + \frac{V_{Ed}}{V_{Rd,max}} \leq 1 \xrightarrow{\text{pure torsion}} T_{Ed} <= T_{Rd,max}$$

where

T_{Ed} = design torsional moment = T_u
V_{Ed} = design shear force = V_u
$T_{Rd,max}$ = design torsional resistance moment
$V_{Rd,max}$ = design shear resistance force

The design torsional resistance moment is given as (Eq. (6-30) in EC2-04)

$$T_{Rd,max} = 2\nu\alpha_c f_{cd} A_k t_{ef}\sin\theta\cos\theta$$

where

ν = strength reduction factor for concrete cracked in shear
= $0.6(1 - f_c'/250) = 0.6(1 - 20/250) = 0.552$ (f_c' in MPa)
α_c = 1 (for non-prestressed elements)
θ = angle of compression diagonals in truss analogy for torsion

$1 \leq \cot\theta \leq 2.5$; assume θ = 35 degrees

f_{cd} = design compressive strength of concrete
= $f_{ck}/\gamma_c = 20/1.5 = 13.3$ MPa (1933 psi)

Therefore,

$$T_{Rd,max} = 2\nu\alpha_c f_{cd} A_k t_{ef}\sin\theta\cos\theta$$
$$= 2\times 0.552\times 1\times 13.3\times 80{,}000\times 100\times \sin 35°\cos 35°(10^{-6})$$
$$= 55.2\text{ kN·m (513.4 in.-kip)}$$
$$\Rightarrow T_{Rd,max} = 55.2\text{ kN·m (513.4 in.-kip)} > T_u = 30\text{ kN·m (266 in.-kip)}$$

Because the above expression is valid, dimensions of the cross section are adequate.

3. Calculate the amount of stirrups required for pure torsion (EC2-04, Section 6.3.2 (2)).

The amount of stirrups required is calculated using the equation

$$\frac{A_t}{s} \geq \frac{T_u}{2A_k f_{ywd} \cot\theta}$$

where

A_t = area of one leg of a closed stirrup resisting torsion
s = spacing of the stirrups
f_{ywd} = design strength of transverse reinforcement
= f_{ty}/γ_s = 420/1.15 = 365.2 MPa (52,174 psi)

Therefore,

$$\frac{T_u}{2A_k f_{ywd} \cot\theta} = \frac{30\times10^6}{2\times80{,}000\times365.2\cot35°}$$
$$= 0.359 \text{ mm}^2\text{/mm } (0.014 \text{ in.}^2\text{/in.})$$

$$\Rightarrow \frac{A_t}{s} \geq 0.359 \text{ mm}^2\text{/mm } (0.014 \text{ in.}^2\text{/in.})$$

The ratio of the required transverse reinforcement is given by

$$\rho_t = \frac{A_t}{sb_w \sin\alpha} \xrightarrow{\alpha = 90} \rho_t = \frac{A_t}{sb_w} = \frac{0.359}{300} = 0.0012$$

where α is angle between the stirrups and longitudinal axis.

The ratio of the minimum stirrups is (Eq. (9.5N) in EC2-04)

$$\rho_t = \left(0.08\sqrt{f_c'}\right)/f_{yk} = 0.08\sqrt{20}/420 = 0.00085$$

The maximum longitudinal spacing of the stirrups is given as (Eq. (9.6N) and Section 9.2.3(3) in EC2-04).

$$s_{max} = \min\begin{Bmatrix} u/8 \\ 0.75d(1+\cot\alpha) \\ \min(b_w, h) \end{Bmatrix} \Rightarrow$$

$$s_{max} = \min\begin{Bmatrix} 1{,}600/8 = 200 \\ 0.75\cdot450 = 338 \\ 300 \end{Bmatrix} = 200 \text{ mm } (8.0 \text{ in.})$$

For stirrups Ø8 (d_s = 8.0 mm [0.315 in.]), A_t = 50.3 mm² (0.078 in.²) and $s \leq$ 137 mm (5.40 in.). Select: s = 125 mm (5.0 in.).

For stirrups No. 3 (d_s = 9.5 mm [0.375 in.]), A_t = 71 mm² (0.110 in.²) and $s \leq$ 194 mm (7.66 in.). Select: s = 180 mm (7.1 in.).

4. Calculate the longitudinal bars required for pure torsion (EC2-04, Section 6.3.2(3)).

The total longitudinal reinforcement needed for torsion (ΣA_ℓ) is given as (Eq. (6-28) in EC2-04)

$$\Sigma A_\ell \geq \frac{T_u u_k \cot\theta}{2A_k f_{yd}} = \frac{30\times10^6\times1200\times\cot35°}{2\times80{,}000\times365.2} \Rightarrow \Sigma A_\ell$$
$$\geq 880 \text{ mm}^2 \text{ } (1.37 \text{ in.}^2)$$

Minimum longitudinal reinforcement is determined by (Eq. (9.1N) in EC2-04)

$$A_{\ell,min} = 0.26\frac{f_{ctm}}{f_{yk}}b_t d \geq 0.0013 b_t d$$

where

f_{ctm} = mean tension strength of concrete
= $0.30 f_{ck}^{2/3} = 0.3(20^{2/3})$ = 2.21 MPa (319 psi)
b_w = mean width of tension zone = 300 mm (12 in.)

Therefore,

$$A_\ell = \max\begin{Bmatrix} 0.26(2.21/420)300\times450 = 184.7 \text{ mm}^2 \\ 0.0013\times300\times450 = 175 \text{ mm}^2 \end{Bmatrix}$$
$$= 185 \text{ mm}^2 \text{ } (0.287 \text{ in.}^2)$$

Longitudinal bars are arranged with at least one bar at each corner of the stirrups and the others distributed uniformly around the inner periphery of the torsion links (closed stirrups) with a maximum spacing of 350 mm (13.8 in.). Therefore, the number of longitudinal bars is at least six and each bar requires 880/6 = 147 mm² (0.23 in.²).

For six longitudinal bars Ø14 (d_s = 14 mm [0.55 in.]), A_l = 6 × 154 = 924 mm² (1.43 in.²).

For six longitudinal bars No. 5 (d_s = 15.8 mm [0.625 in.]), A_l = 6 × 198 = 762 mm² (1.88 in.²).

9.4.4 *Solution according to CSA-A23.3-04 code*

1. Determine if torsion effects may be disregarded (CSA-A23.3-04, Section 11.2.9.1).

If the magnitude of the torsion, T_f, satisfies the following expressions (Eq. (11-2) in CSA-A23.3-04), torsional effects need not be considered

$$T_f < 0.25 T_{cr}$$

$$T_{cr} = \left(\frac{A_c^2}{p_c}\right)0.38\lambda\phi_c\sqrt{f_c'}\sqrt{1+\frac{f_p f_{cp}}{0.38\lambda\phi_c\sqrt{f_c'}}} = \left(\frac{A_c^2}{p_c}\right)0.38\lambda\phi_c\sqrt{f_c'}$$

$$= \left(\frac{150{,}000^2}{1600}\right)0.38\times1\times0.65\sqrt{20} \text{ N·mm}$$
$$= 15.5 \text{ kN·m } (144 \text{ in.-kip})$$

where

A_c = A_{cp} = 150,000 mm² (240 in.²)

p_c = p_{cp} = outside perimeter of concrete cross section = 1600 mm (64.0 in.)
λ = factor to account for low-density concrete
= $1.0\phi_c$ = resistance factor for concrete = 0.65
f_c' = specified compressive strength of concrete, which is the same as the characteristic concrete cylinder compressive strength in previous definition = 20 MPa (2.9 ksi)

Therefore,

$$T_f (= 30.0 \text{ kN·m}) > 0.25T_{cr} (= 0.25 \times 15.5 = 3.9 \text{ kN·m})$$

Torsion effect, therefore, must be considered.

2. Determine if dimensions of the cross section are adequate (CSA-A23.3-04, Section 11.3.10.4).

Dimensions of the cross section are adequate if the equation below (Eq. (11-19) in CSA-A23.3-04) is satisfied

$$\sqrt{\left(\frac{V_f - V_p}{b_w d_v}\right)^2 + \left(\frac{T_f p_h}{1.7A_{oh}^2}\right)^2} \leq 0.25\phi_c f_c'$$

In case of pure torsion, the expression is simplified to

$$\frac{T_f p_h}{1.7A_{oh}^2} \leq 0.25\phi_c f_c'$$

With a concrete cover of 40 mm (1-1/2 in.) and 10M (No. 3) stirrups (diameter 12 mm [1/2 in.]), the following calculations/values apply

$$A_{oh} = (300 - 2 \times 46)(500 - 2 \times 46) = 84{,}864 \text{ mm}^2 \text{ (141.0 in.}^2\text{)}$$

$$p_h = 2[(300 - 2 \times 46) + (500 - 2 \times 46)] = 1232 \text{ mm (50.0 in.}^2\text{)}$$

Therefore,

$$\frac{T_u p_h}{1.7A_{oh}^2} = \frac{30 \times 10^6 \times 1232}{1.7 \times 84{,}860^2} = 3.02 \text{ MPa (394 psi)}$$

$$0.25 f_c f_c' = 0.25 \times 0.65 \times 20 = 3.25 \text{ MPa (471 psi)}$$

Because 3.02 MPa < 3.25 MPa, dimensions of the cross section are adequate.

3. Calculate the stirrups required for pure torsion (CSA-A23.3-04, Section 11.3.10.3);

The equation of nominal torsional strength T_n is same in all codes except for differences in strength reduction factors. For design, the equation for T_n is rearranged to express the required area of transverse reinforcement per unit length (Eq. (11-7) in CSA-A23.3-04)

$$\frac{A_t}{s} \geq \frac{T_f}{2A_o \phi_s f_y \cot\theta}$$

where
A_t = area of one leg of a closed stirrup resisting torsion
s = spacing of the stirrups
A_o = gross area enclosed by shear flow path = $0.85A_{oh}$
θ = angle of inclination of compression stresses to the longitudinal member axis

4. Calculate angle of inclination of compression strut (CSA-A23.3-04, Section 11.3.6.4).

The angle of inclination of the diagonal compression strut is given by the expression (Eq. (11-12) in CSA-A23.3-04)

$$\theta = 29 + 7000\varepsilon_x$$

The longitudinal strain indicator ε_x is defined by (Eq. (11-13) in CSA-A23.3-04)

$$\varepsilon_x = \frac{\frac{M_f}{d_v} + \sqrt{(V_f - V_p)^2 + \left(\frac{0.9p_h T_f}{2A_o}\right)^2} + 0.5N_f - A_p f_{po}}{2[E_s A_s + E_p A_p]}$$

For non-prestressed sections subjected to pure torsion, the expression for ε_x is simplified to

$$\varepsilon_x = \frac{\frac{0.9p_h T_f}{2A_o}}{2E_s A_s} = \frac{0.9p_h T_f}{4E_s A_s A_o}$$

with $A_o = 0.85A_{oh} = 0.85 \times 84{,}860 = 72{,}130$ mm^2 (120 in.^{2}) and the longitudinal reinforcement $A_s = 413$ mm^2 (0.64 in.^{2}) (established in Section 6 of this example), the following is obtained

$$\varepsilon_x = \frac{0.9 \times 1232 \times 30 \times 10^6}{4 \times 200{,}000 \times 413 \times 72{,}130} = 0.00140$$

Therefore

$$\theta = 29 + 7000 \times 0.00140 = 38.8 \text{ degrees}$$

5. Calculate stirrups required for pure torsion.

The required stirrup area per unit length is calculated using the equation (Eq. (11-17) in CSA-A23.3-04)

$$\frac{A_t}{s} \geq \frac{T_f}{2A_o \phi_s f_y \cot\theta} = \frac{30 \times 10^6}{2 \times 72{,}130 \times 0.85 \times 420 \times \cot 38.8°}$$
$$= 0.468 \text{ mm}^2\text{/mm (0.0175 in.}^2\text{/in.)}$$

The minimum transverse reinforcement is (Eq. (11-1) in CSA-A23.3-04)

$$\left(\frac{A_v}{s}\right)_{min} = 0.06\sqrt{f_c'}\frac{b_w}{f_{yv}} = 0.06 \times \sqrt{20} \times \frac{300}{420}$$
$$= 0.192 \text{ mm}^2\text{/mm (0.00775 in.}^2\text{/in.)}$$

The maximum spacing of the stirrups is (CSA-A23.3-04, Section 11.3.8.1)

$$s_{max} = \min\begin{Bmatrix} 0.7d_v \\ 600 \text{ mm} \end{Bmatrix} = 277 \text{ mm (11.2 in.)}$$

where

d_v = effective shear depth = max(0.9d, 0.72h) = max(396 mm, 360 mm) [(16.0 in., 14.4 in.)]

For stirrups 10M (d_s = 11.3 mm [0.444 in.]), A_t = 100 mm^2 (0.155 in.^{2}) → $s \leq$ 210 mm (8.32 in.). Select s = 200 mm (8.0 in.).

For stirrups No. 3 (d_s = 9.5 mm [0.375 in.]), 71.0 mm^2 (A_t = 0.110 in.^{2}) → $s \leq$ 150 mm (5.89 in.). Select: s = 150 mm (5. 9 in.).

6. Calculate the amount of longitudinal bars required for pure torsion (CSA-A23.3-04, Section 11.3.9.2 and 11.3.10.6).

Total longitudinal reinforcement is calculated using the following equation (Eq. (11-21) and (11-14) in CSA-A23.3-04)

$$F_{\ell t} = \frac{M_f}{d_v} + \cot\theta\sqrt{(V_f - 0.5V_s - V_p)^2 + \left(\frac{0.45p_hT_f}{2A_o}\right)^2} = \cot\theta\frac{0.45p_hT_f}{2A_o}$$

where

$F_{\ell t}$ = required tension force in longitudinal reinforcement = $\phi A_\ell f_y$

Therefore

$$A_\ell \geq \frac{0.45p_hT_f}{2\phi_s A_o f_y}\cot\theta = \frac{0.45\times1232\times30\times10^6}{2\times0.85\times72{,}130\times420}\cot 38.8°$$

$$= 402 \text{ mm}^2 \text{ (0.61 in.}^2\text{)}$$

One longitudinal reinforcing bar should be placed in each corner of the closed transverse reinforcement required for torsion. The nominal diameter of the corner bars should be no less than s/16. If 10M stirrups are used, the bar nominal diameter can be no less than 200/16 = 12.5 mm (8/16 = 0.5 in.), and for No. 3: 150/16 = 9.38 mm (6/16 = 0.375 in.).

Select: four longitudinal bars 15M (d_s = 16.0 mm [0.628 in.]) at the corners and two 10M bars at mid-depth: A_ℓ = 4 × 200 + 2 × 100 = 1000 mm^2 (1.55 in.^{2}).

Or, six longitudinal bars No. 3 (d_s = 9.5 mm [0.375 in.]), A_ℓ = 6 × 71 = 426 mm^2 (0.66 in.^{2})

Or, six longitudinal bars No. 4 (d_s = 12.7 mm [0.50 in.]), A_ℓ = 6 × 127 = 762 mm^2 (1.18 in.^{2}).

9.4.5 *Rahal's graphical method*—Rahal's graphical method uses the ACI general requirements, such as an upper limit on spacing of transverse and longitudinal reinforcement, an upper limit on yield strength of reinforcing rebars, and a minimum of four corner longitudinal bars. Based on ACI requirements, it also disregards the torque effect if it is smaller than 25 percent of the cracking torque. The cross section outer perimeter and area enclosed within this perimeter were calculated in the ACI design example 9.4.1 as: A_c = 150,000 mm^2 (240 in.^{2}) and p_c = 1600 mm (64.0 in.).

1. Determine if torsion effects may be disregarded.

The torsion effects can be disregarded if this expression is valid: $T_u \leq 0.25\phi T_{cr}$

$$T_{cr} = 0.328\sqrt{f_c'}\frac{A_c^2}{p_c} = 0.328\sqrt{20}\frac{150{,}000^2}{1600} \text{ N•mm}$$

$$= 20.\ 6 \text{ kN·m (193.87 in.-kip)}$$

The torque is disregarded if T_u = 30 kN·m (266 in.-kip) is smaller than $0.25\phi T_{cr}$ = 0.25 × 0.75 × 20.6 = 3.9 kN·m (35.8 in.-kip). Torsion effects, therefore, must be considered.

2. Calculate normalized shear stress and determine if size of cross section is adequate.

Shear stress in the walls of the cross section is calculated by

$$\frac{v}{f_c'} = \frac{\frac{T_u}{\phi}p_c}{0.67A_c^2 f_c'} = \frac{\frac{30\times10^6}{0.75}\times1600}{0.67(150{,}000)^2 20} = 0.212$$

The normalized shear stress fits well within Region I in Fig. 5.7.1. Therefore, the section can be designed as under-reinforced, and the section dimensions are adequate.

3. Calculate the required reinforcement.

The most straightforward design of an under-reinforced section uses equal amounts of longitudinal and transverse reinforcement indexes ($\omega_l = \omega_t = v/f_c'$). Therefore, ω_l = 0.212 and ω_t = 0.212. From Eq. (5.7.1b) and (5.7.1c), the longitudinal and transverse reinforcement are

$$\frac{A_t}{s} = \frac{0.42A_cf_c'}{f_{ty}p_c}\omega_t = \frac{(0.42)(150{,}000)(20)}{(420)(1600)}0.212$$

$$= 0.40 \text{ mm}^2\text{/mm (0.016 in.}^2\text{/in.)}$$

$$A_\ell = \frac{0.375A_cf_c'}{f_{\ell y}}\omega_l = \frac{(0.375)(150{,}000)(20)}{420}0.212$$

$$= 568 \text{ mm}^2 \text{ (0.92 in.}^2\text{)}$$

The stirrups' maximum spacing is taken as 160.0 mm (6.30 in.), which is the smaller of p_h/8 = 2(220 + 420)/8 = 160.0 mm (6.50 in.) and 300.0 mm (12.0 in.), as per ACI code.

For stirrups ∅8 (d_s = 8.0 mm [0.315 in.]), A_t = 50.3 mm^2 (0.08 in.^{2}), the maximum spacing is calculated as $s \leq$ 124.5 mm (4.9 in.), and it can be used: s = 120 mm (4.50 in.).

For stirrups No. 3 (d_s = 9.5 mm [0.375 in.]), A_t = 71.0 mm^2 (0.11 in.^{2}), the maximum spacing is calculated as $s \leq$ 175.7 mm (6.9 in.), and it can be used: s = 160 mm (6.50 in.).

To provide four longitudinal corner bars and limit the spacing to 300 mm (12 in.), six bars are needed (ACI design in 8.2.2). The minimum bar area is of 578/6 = 96 mm^2 (0.15 in.^{2}).

For six longitudinal bars ∅12 (d_s = 12.0 mm [0.472 in.]), A_ℓ = 6 × 113 = 678 mm^2 (1.05 in.^{2}).

For six longitudinal bars No. 4 (d_s = 12.7 mm [0.50 in.]), A_ℓ = 6 × 129 = 774 mm^2 (1.20 in.^{2}).

9.4.6 *Leu and Lee's graphical method*

1. Calculate the required torsional strength.

Table 9.4.6—Summary of design solution of Example 1 using all five solution methods

Code		Transverse reinforcement	Longitudinal reinforcement
ACI 318	required:	0.61 mm²/mm (0.0227 in.²/in.)	780.8 mm² (1.18 in.²)
	minimum:	*0.125 mm²/mm (0.0042 in.²/in.)*	*0 mm² (0 in.²)*
	provided:	Ø8/80 mm (No. 3 at 4.50 in.) 0.625 mm²/mm (0.0244 in.²/in.)	6Ø14 (6 No. 5) 923 mm² (1.86 in.²)
EC2-04	required:	0.359 mm²/mm (0.014 in.²/in.)	880 mm² (1.37 in.²)
	minimum:	*0.261 mm²/mm (0.0103 in.²/in.)*	*185 mm² (0.287 in.²)*
	provided:	Ø8/125 mm (No. 3 at 7.10 in.) 0.402 mm²/mm (0.0155 in.²/in.)	6Ø14 (6 No. 5) 924 mm² (1.88 in.²)
CSA A23.3-04	required:	0.468 mm²/mm (0.0175 in.²/in.)	402 mm² (0.61 in.²)
	minimum:	*0.192 mm²/mm (0.00775 in.²/in.)*	—
	provided:	10M/200.0 mm (No. 3 at 5.90 in.) 0.500 mm²/mm (0.0186 in.²/in.)	Four 15M + two 10M (6 No. 3 or 6 No. 4) 1000 mm² (0.66 or 1.18 in.²)
Rahal (2000b)	required:	0.40 mm²/mm (0.0160 in.²/in.)	568.1 mm² (0.92 in.²)
	minimum:	—	—
	provided:	Ø8/120.0 mm (No. 3 at 6.50 in.) 0.417 mm²/mm (0.0183 in.²/in.)	6Ø12 (6 No.4) 678 mm² (1.18 in.²)
Leu and Lee (2000)	required:	0.470 mm²/mm (0.019 in.²/in.)	615 mm² (1.0 in.²)
	minimum:	—	—
	provided:	Ø8/100.0 mm (No. 3 at 5.50 in.) 0.503 mm²/mm (0.0200 in.²/in.)	6Ø12 (6 No. 4) 678 mm² (1.18 in.²)

Nondimensional balanced torsional strength is calculated using Eq. (5.7.2g)

$$\bar{T}_{xub} = \frac{140}{300+f_y}\ (f_y \text{ in MPa}) = \frac{140}{300+420} = 0.196$$

Balanced torsional strength T_{xub} is determined by Eq. (5.7.2h)

$$T_{xub} = \frac{\bar{T}_{xub} f'_c A_c^2}{p_c} = \frac{0.196 \times 20 \times 150{,}000^2}{1600}\ \text{N·mm}$$

$$= 55.1 \text{ kN·m (512 in.-kip)}$$

where

A_c = total area of the outside perimeter of the concrete cross section
= $b_w h$ = 300 × 500 = 150,000 mm² (240 in.²)

p_c = outside perimeter of the concrete cross section
= $2(b_w + h)$ = 2(300 + 500) = 1600 mm (64.0 in.)

The torsional strength indicated by Eq. (5.7.2h) is greater than required

$$\omega_s T_{xub} \geq \frac{T_u}{f} \rightarrow \omega_s \times 55.1 \geq \frac{30}{0.75} \rightarrow f_s \geq 0.73\omega$$

where

T_u = applied torsional moment = 30.0 kN·m (266 in.-kip)

ϕ = strength reduction factor for shear and torsion; which is assumed to be the same as that in ACI 318
= $0.75\omega_s$ = strength contour value (Fig. 5.7.2)

2. Calculate the longitudinal bars and stirrups required for pure torsion.

Required reinforcement indexes (Fig. 5.7.2): ω_l and ω_t

For convenience, assume $\omega_\ell = \omega_t$.

Referring to (Fig. 5.7.2), $\omega_s \geq 0.73 \rightarrow \omega_\ell = \omega_t \geq 0.70$.

Balanced normalized reinforcement ratios $\eta_{\ell b}$ and η_{tb} are calculated by Eq. (5.7.2c) and (5.7.2d)

$$\eta_{\ell b} = \frac{76}{200+f_y}\ (f_y \text{ in MPa}) = \frac{76}{200+420} = 0.125$$

$$\eta_{tb} = \frac{76}{100+f_y}\ (f_y \text{ in MPa}) = \frac{76}{100+420} = 0.135$$

Required normalized reinforcement ratios η_ℓ and η_t are determined using Eq. (5.7.2e) and (5.7.2f)

$$\eta_\ell = \omega_t \eta_{\ell b} \geq 0.70 \times 0.125 \rightarrow \eta_\ell \geq 0.0861$$

$$\eta_t = \omega_t \eta_{tb} \geq 0.70 \times 0.135 \rightarrow \eta_t \geq 0.105$$

Required longitudinal bars are calculated with Eq. (5.7.2a)

$$\eta_\ell \geq 0.0868 \rightarrow \frac{f_y}{f'_c}\frac{A_\ell}{A_{cp}} \geq 0.0868 \rightarrow \frac{420}{20}\frac{A_\ell}{150{,}000} \geq 0.0861$$

$$\rightarrow A_\ell \geq 615 \text{ mm}^2 \text{ (1.0 in.}^2\text{)}$$

At least one longitudinal bar is placed at each corner of the stirrups with the others distributed uniformly around the inner periphery of the closed stirrups with a maximum spacing of 300 mm (12 in.), as indicated in ACI 318. Therefore, the number of longitudinal bars is at least six and each bar requires 615/6 = 102.5 mm² (0.17 in.²).

For six longitudinal bars Ø12 (d_s = 12.0 mm [0.472 in.]), $A_\ell = 6 \times 113 = 678$ mm² (1.05 in.²).

For six longitudinal bars No. 4 (d_s = 12.7 mm [0.500 in.]), $A_\ell = 6 \times 127 = 762$ mm² (1.18 in.²).

Required stirrups are calculated using Eq. (5.7.2b)

$$\eta_t \geq 0.1036 \rightarrow \frac{f_y}{f_c'}\frac{A_t p_{cp}}{A_{cp} s} \geq 0.1036$$

$$\rightarrow \frac{420}{20}\frac{1600 A_t}{150{,}000 s} \geq 0.1036$$

$$\rightarrow \frac{A_t}{s} \geq 0.470 \text{ mm}^2/\text{mm } (0.019 \text{ in.}^2/\text{in.})$$

Using ACI 318, the maximum stirrups' spacing is calculated to be 160 mm (6.3 in.).

For stirrups Ø8 (d_s = 8.0 mm [0.315 in.]) → A_t = 50.3 mm² (0.08 in.²) and $s \leq 107$ mm (4.2 in.) and it can be used: s = 100 mm (4.0 in.).

For stirrups No. 3 (d_s = 9.5 mm [0.375 in.]) → A_t = 71 mm² (0.11 in.²) and $s \leq 151$ mm (5.9 in.) and it can be used: s = 150 mm (5.5 in.).

9.5—*Design Example 2: Prestressed concrete box girder under combined torsion, shear, and flexure*

9.5.1 *Design problem statement*

9.5.1.1 *Design problem description*—Design the shear and torsional reinforcement of a box girder. A 3658 mm (12 ft) wide and 1270 mm (4 ft 2 in.) deep box girder with overhanging flanges (Fig. 9.5.1.1(a)) was designed as an alternative to the double-tee girder in Dade County, FL (Hsu 1997). The standard prestressed box girder is simply supported, 24.00 m (79.00 ft) long, and prestressed with 64 strands at 1860 MPa (270 ksi), 13.0 mm (1/2 in.), seven-wire strands as shown in Fig. 9.5.1.1(b). Total prestress force is 6076 kN (1366 kips) after prestress loss. The design of flexural reinforcement is omitted for simplicity. The concrete cover is 40 mm (1.5 in.), and material strengths are normalweight concrete: f_c' = 48.0 MPa (7000 psi) and f_y = 420 MPa (60,000 psi).

9.5.1.2 *Sectional properties*

L = 24.00 m (79.00 ft)
h = 1270 mm (50.00 in.)
d = 1016 mm (40.00 in.) at 0.3L from support
t = 251 mm (9.88 in.) (average of stem width)
b_w = 502 mm (20 in.)
A = 1.523 × 10⁶ mm² (2361.4 in.²)
I = 319.8 × 10⁹ mm² (768,336 in.⁴)
y_t = 516 mm (20.34 in.)
y_b = 753 mm (29.66 in.)
λ = modification factor of lightweight concrete (λ = 1.0)

9.5.1.3 *Loading criteria*—The standard girders are designed to carry a train of cars, each 22.86 m (75 ft 0 in.) long. Each car has two trucks with a center-to-center distance of 16.46 m (54 ft 0 in.). Each truck consists of two axles 1981 mm (6 ft 6 in.) apart. The crush live load of each car is 513.8 kN (115.5 kip). The maximum web reinforcement amount was obtained at section 0.3L from the support under a derailment load, which consists of two truckloads located symmetrically at a distance 3200 mm (10 ft 6 in.) from midspan. Each axle load is taken as 24 percent of the crush live load (513.8 kN/4) with 100 percent impact and a maximum side shift of 914 mm (36.0 in.). The self-weight of the girder is 34.4 kN/m (2.36 kip/ft). The girder is also subjected to a superimposed dead load caused by the track rails' weight, rail plinth pads, power rail, guard rail, cableway, acoustic barrier, and other permanent loads. At derailment, this superimposed dead load is assumed to produce a uniform vertical load of 12.8 kN/m (0.88 kip/ft) and a uniformly distributed torque of 3.16 kN-m/m (0.71 ft-kip/ft). This torque is neglected in the calculation because the magnitude of the distributed torque is small, and the torque is acting in a direction opposite to the derailment torque.

9.5.2 *Solution according to ACI 318-11*

1. Determine the factored forces for (ACI 318-11, Section 9.2.1).

Factored dead and live loads—

The load factor for live loads is taken as 1.6.

The derailment load per axle is calculated by

$$P_{u,L} = 16 \times \frac{513.8}{4} \times 2 = 411 \text{ kN/axle (92.4 kip/axle)}$$

The derailment torque per axle is calculated by

$$T_{u,L} = 16 \times \frac{513.8}{4} \times 2 \times 0.914$$

$$= 375.7 \text{ kN·m/axle (277 ft-kip/axle)}$$

The load factor for dead loads is taken as 1.2.

The girder weight is calculated by

$$w_{u,g} = 1.2(34.4) = 41.3 \text{ kN/m (2.83 kip/ft)}$$

The superimposed dead weight is calculated by

$$w_{u,s} = 1.2(12.8) = 15.4 \text{ kN/m (1.05 kip/ft)}$$

Factored shear, torque, and bending moment—

The V_u, T_u, and M_u at 0.3L from the support are

V_u = $(w_{u,g} + w_{u,s})(0.2L) + 2P_{u,L}$ = (41.3 + 15.3) × 0.2 × 24.00 + 2 × 411 = 1094 kN (246 kips)
T_u = $2T_{u,L}$ = 2 × 375.7 = 752 kN·m (554 ft-kip)
M_u = $0.5(w_{u,g} + w_{u,s})(L - 0.3L)(0.3L) + 2P_{u,L}(0.3L)$
= 0.5(41.3 + 15.4)(24.00 – 7.2) × 7.2 + 2 × 411 × 7.2
= 9347 kN·m (6922 ft-kip)

2. Determine if torsion effects may be disregarded.

Check outstanding flanges—

As indicated by Fig. 9.5.1.1(a), the parameter A_{cp}^2/p_{cp} is determined by (disregard overhanging flanges)

A_{cp} = 1854 × 203 + 0.5(1854 + 1791) × 1067 = 2.32 × 10⁶ mm² (3597 in.²)
p_{cp} = 1854 + 1791 + 2 × 1270 = 6185 mm (243.5 in.)
A_{cp}^2/p_{cp} = (2.32 × 10⁶)²/6185 = 8.70 × 10⁸ mm³ (53,090 in.³)

Check threshold torque (ACI 318-11, Section 11.5.1)—

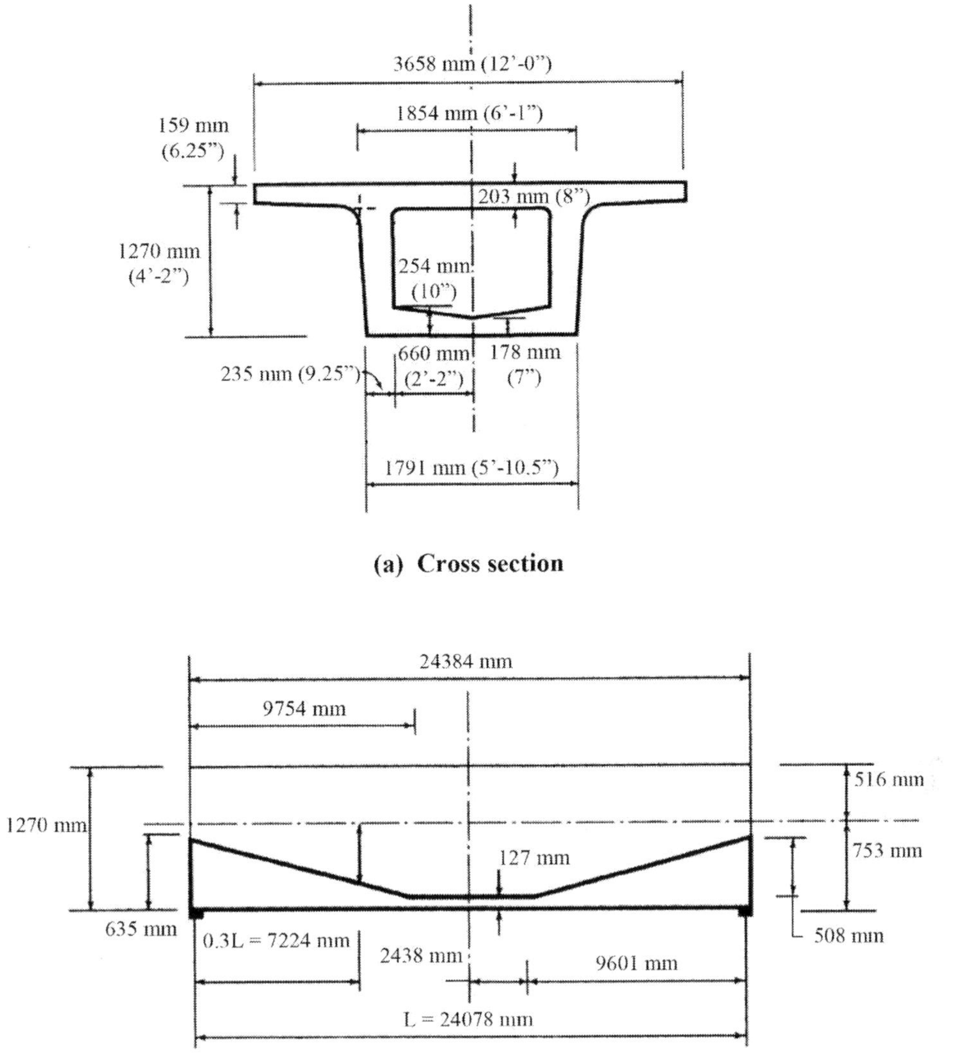

(a) Cross section

(b) Elevation and prestress profile

Fig. 9.5.1.1—Cross section and elevation of box girder.

A_g = $2.32 \times 10^6 - (1270 - 419)(1320) = 1.20 \times 10^6$ mm² (1855 in.²)($fT_{cr}/4$)

=

$$\phi(0.083)\lambda\sqrt{f_c'}\frac{A_g^2}{p_{cp}}\left(1+\frac{f_{pc}}{0.33\sqrt{f_c'}}\right)=\phi(0.083)\lambda\sqrt{f_c'}\frac{A_g^2}{p_{cp}}\left(1+\frac{6076/A}{0.33\sqrt{f_c'}}\right)$$

$$=0.75(0.083)(1.0)\sqrt{48.0}\,\frac{\left(1.20\times10^6\right)^2}{6185}\left(1+\frac{6076/1523}{0.33\sqrt{48.0}}\right)10^{-6}$$

= 276 kN·m (202 ft-kip) < 752 kN·m (554 ft-kip)

Factored torsional moment should be considered in design.

3. Determine if dimensions of the cross section are adequate (ACI 318-11, Section 11.5.3.1).

Check cross section—

Assume a clear concrete cover of 40 mm (1.5 in.) and 13 mm No. 4 bars for web reinforcement

A_{oh} = 0.5[(1854 – 93) + (1791 – 93)](1270 – 93)
= 2.04×10^6 mm² (3177 in.²)

p_h = (1854 – 93) + (1791 – 93) + 2(1270 – 93)
= 5813 mm (229.5 in.)

$$e = (753-127)-\frac{9754-7224}{9754}508$$

= 496 mm (19.47 in.) at 0.3L from support

d = $y_t + e = 516 + 496 = 1012$ mm (39.81 in.) at 0.3L from support

d = $0.8h = 0.8 \times 1270 = 1016$ mm (40.00 in.) governs

b_w = $2t = 2 \times 251 = 502$ mm (20.0 in.)

b_wd = $502 \times 1016 = 510 \times 10^3$ mm² (790 in.²)

The interaction equation for hollow box sections is (Eq. (11-18) in ACI 318-11)

$$\left(\frac{V_u}{b_w d}\right)+\left(\frac{T_u}{1.7A_{oh}t}\right)\leq\phi\left(\frac{V_c}{b_w d}+8\times0.083\sqrt{f_c'}\right)\text{ when }t<\frac{A_{oh}}{p_h}$$

$$\frac{A_{oh}}{p_h}=\frac{2.04\times10^6}{5829}=351\text{ mm (13.85 in.)}>t=251\text{ in.}$$

$$\left(\frac{V_u}{b_w d}\right)+\left(\frac{T_u}{1.7A_{oh}t}\right)=\frac{1094\times1000}{510\times10^3}+\frac{751.4\times10^6}{1.7(2.04\times10^6)251}$$

$$=2.15+0.86=3.01\text{ MPa (437 psi)}$$

$$V_c=\left(0.6\times0.083\lambda\sqrt{f_c'}+700\times0.0689\frac{V_u d}{M_u}\right)b_w d,$$

where $\frac{V_u d}{M_u}\leq1$ (Eq. (11-9) in ACI 318-11)

$$\frac{V_u d}{M_u}=\frac{1094\times1016}{9347\times1000}=0.119<1\quad\text{OK}$$

$$V_c=(0.6\times0.083(1.0)\sqrt{48.0}+700\times0.00689\times0.119)\left(\frac{510\times10^3}{1000}\right)=469\text{ kN (105.4 kip)}$$

$$V_{c,min}=2\times0.083\lambda\sqrt{f_c'}\,b_w d$$

$$=2\times0.083(1.0)\left(\sqrt{48.0}\frac{510\times10^3}{1000}\right)$$

(Eq. (11-3) in ACI 318-11)

= 587 kN (132.13 kip) governs

$$\phi\left(\frac{V_c}{b_w d}+8\times0.083\sqrt{f_c'}\right)=0.75\left(\frac{588\times1000}{510\times10^3}+8\times0.083\sqrt{48.0}\right)$$

= 0.75(1.15 + 4.61) = 4.31 MPa (627 psi) > 3.01 MPa (437 psi) OK

4. Calculate number of transverse bars required

Design of shear reinforcement (ACI 318-11, Section 11.4.7)—

$V_c = V_{c,min}$ = 587 kN (132.2 kip)

$$\frac{A_v}{s}=\frac{V_u-\phi V_c}{\phi d f_{yv}}=\frac{(1094-0.75\times587)1000}{0.75\times1016\times420}$$

= 2.043 mm²/mm (0.0818 in.²/in.)

$$>\frac{A_{v,min}}{s}=0.75\frac{\sqrt{f_c'}}{12}\frac{b_w}{f_{yt}}=0.51\text{ mm}^2\text{/mm (0.021 in.}^2\text{/in.)}$$

s_{max} = 305.0 mm (12.00 in.) for torsion governs

Design of torsional hoop reinforcement (ACI 318-11, Section 11.5.3.6)—

$$A_o=A_{cp}-\frac{2T_u p_{cp}}{\phi f_c' A_{cp}}=2.32\times10^6-\frac{2\times752\times6185(1000)^2}{0.75\times48\times2.32\times10^6}$$

= 2.21×10^6 mm² (3582.7 in.²)

Assume θ = 37.5 degrees, as recommended by the code provision for prestressed members:

$$\frac{A_{ts}}{s}=\frac{T_u}{\phi2A_o f_{yv}\cot\theta}=\frac{752(1000)^2}{0.75\times2(2.21\times10^6)\times420\times1.303}$$

= 0.414 mm²/mm (0.0166 in.²/in.)

s_{max} = $p_h/8$ = 5813/8 = 727 mm (28.7 in.) > 305 mm (12 in.) (ACI 318-11, Section 11.5.6);

These calculations indicate that 305 mm (12 in.) spacing governs.

Transverse reinforcement for vertical walls (ACI 318-11, Section 11.5.3.8)—

Transverse reinforcement in the vertical walls is contributed by both torsion and shear:

$$\frac{A_t}{s}+\frac{1}{2}\frac{A_v}{s}=0.414+0.5\times2.043$$

= 1.44 mm²/mm (0.0575 in.²/in.)

$$\left(\frac{A_t}{s}+\frac{1}{2}\frac{A_v}{s}\right)_{min}=0.375\frac{\sqrt{f_c'}}{12}\frac{b_w}{f_{yt}}$$

= 0.259 mm²/mm (0.01 in.²/in.)
< 1.44 mm²/mm (0.0575 in.²/in.) OK

(Eq. (11-23) in ACI 318-11)

Select two layers of ∅18 mm bars (No. 6 bars) in each vertical wall at 305 mm (12 in.) spacing

$$\frac{2(254)}{(305)}=1.666\text{ mm}^2\text{/mm (0.0656 in.}^2\text{/in.)}$$

> 1.440 mm²/mm (0.0575 in.²/in.) OK

Transverse reinforcement for horizontal walls—

Transverse reinforcement in horizontal walls is contributed by torsion only:

$$\frac{A_t}{s}=0.414\text{ mm}^2\text{/mm (0.0166 in.}^2\text{/in.)}$$

Select two layers of ∅10 mm bars (No. 3 bars) in each horizontal wall at 305 mm (12 in.) spacing:

$$\frac{2\times 78.5}{305} = 0.515 \text{ mm}^2/\text{mm } (0.0203 \text{ in.}^2/\text{in.})$$

$$> 0.259 \text{ mm}^2/\text{mm } (0.01 \text{ in.}^2/\text{in.})$$

The transverse reinforcement in the top wall should be added to the flexural reinforcement required in the top flange acting as a transverse continuous slab.

5. Calculate number of longitudinal bars required.

Design of torsional longitudinal reinforcement (Eq. (11-22) in ACI 318-11)—

$$A_\ell = \frac{A_t}{s} p_h \left(\frac{f_{yv}}{f_{y\ell}}\right)\cot^2\theta = 0.414 \times 5813 \times 1 \times 1.303^2$$

$$= 4085.9 \text{ mm}^2 \ (6.46 \text{ in.}^2)$$

Check minimum limitation for A_t/s and $A_{l,min}$ (Eq. (11-23) in ACI 318-11)

$$\frac{A_t}{s} = 0.375\sqrt{f_c'}\frac{b_w}{f_{yt}} = 0.259 \text{ mm}^2/\text{mm } (0.01 \text{ in.}^2/\text{in.})$$

$$< 0.414 \text{ mm}^2/\text{mm } (0.0166 \text{ in.}^2/\text{in.})$$

$$A_{\ell,\min} = \frac{5\sqrt{f_c'}A_g}{f_{yl}} - \left(\frac{A_t}{s}\right)p_h\left(\frac{f_{yv}}{f_{yl}}\right)$$

(Eq. (11-24) in ACI 318-11)

$$= \frac{5\times 0.083\sqrt{48.0}\left(1.2\times 10^6\right)}{420} - 0.414 \times 5813 \times 1$$

$$= 8215 - 2407 = 5808 \text{ mm}^2 \ (9.17 \text{ in.}^2) \text{ governs}$$

Select 36 ∅16 bars (No. 5 bars) longitudinal bars

$A_\ell = 36 \times 201 = 7236$ mm² (11.22 in.²) > 5808 mm² (9.17 in.²)

Arrangement of reinforcing bars—

The arrangement of the reinforcing bars for torsion and shear is summarized in Table 9.4.6. This reinforcement arrangement could be conservatively used throughout the girder length.

9.5.3 *EC2-04 code*

9.5.3.1 *Material properties*

9.5.3.1.1 *Concrete*

Concrete compressive cylinder strength (EC2-04, Section 3.1.2 and 3.1.6)—

- Specified compressive strength (The characteristic concrete cylinder compressive strength)

$$f_c' = 48.0 \text{ MPa } (7000 \text{ psi})$$

- Required average compressive strength (psi)

$$f_{cr}' = f_c' + 1400 \text{ (for } f_c' > 5000 \text{ psi)}$$

$$f_{cr}' = 7000 + 1400 = 8400 \text{ psi } (57.9 \text{ MPa})$$

- Mean compressive strength (based on f_{cr}')

$$f_{cm} = 57.9 \text{ MPa } (8400 \text{ psi})$$

- Characteristic compressive strength (MPa) (EC2-04, Section 3.1.2(5))

$$f_c' = f_{cm} - 8 \Rightarrow f_{ck} = 49.9 \text{ MPa } (7240 \text{ psi})$$

- Strength class for concrete according to Section 3.1.2 (Table 3.1) in EC2-04 corresponding to mean compressive strength

Strength Class C50, therefore: $f_c' = 50$ MPa (7250 psi)

- Design compressive strength (EC2-04 , Section 3.1.6(1))

$$f_{cd} = \alpha_{cc}\frac{f_c'}{\gamma_c} = 1\left(\frac{50}{1.5}\right) = 33.3 \text{ MPa } (4833 \text{ psi})$$

- Mean compressive strength according to EC2-04 (Section 3.1.6(5)) and the strength class of concrete

$$f_c' = f_{cm} + 8 \Rightarrow f_{cm} = 58.0 \text{ MPa } (8410 \text{ psi})$$

Concrete tensile strength (EC2-04, Section 3.1.6 (2)—

- Mean tensile strength

$$f_t' = 0.3 f_c'^{2/3} = 4.07 \text{ MPa } (590 \text{ psi})$$

- Design tensile strength

$$f_{ctd} = \alpha_{ct}\frac{f_{ctk,0.05}}{\gamma_c} = 1\times\frac{0.7f_t'}{1.5} = 1.89 \text{ MPa } (275 \text{ psi})$$

- Modulus of elasticity

$$E_m = 22\left(\frac{f_{cm}}{10}\right)^{0.3} = 37.28 \text{ GPa } (5405 \text{ ksi}) \text{ (EC2-04, Section}$$

3.1.3, Table 3.1)

Parabola-rectangle diagram for concrete under compression, as shown in Fig. 9.5.3.1.1 *(EC2-04, Section 3.1.7)*

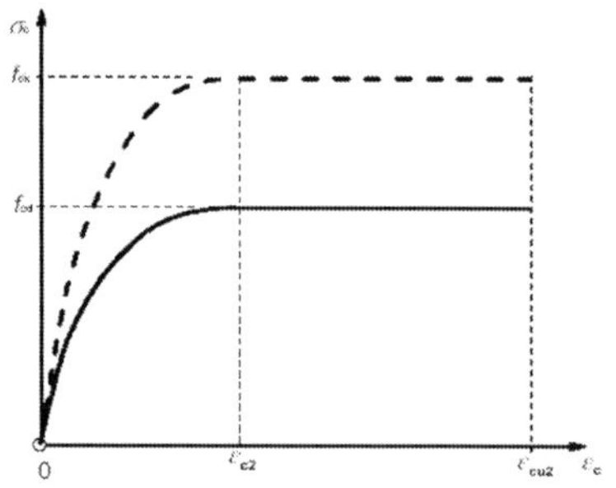

Fig. 9.5.3.1.1—Parabola-rectangle diagram for concrete under compression.

$$\sigma_c = f_{cd}\left[1-\left(1-\frac{\varepsilon_c}{\varepsilon_{c2}}\right)^n\right] \text{ for } \varepsilon_c < \varepsilon_{c2}$$

$\sigma_c = f_{cd}$ for $\varepsilon_c \geq \varepsilon_{c2}$

where

$$n = 1.4 + 23.4\left(\frac{90-f_c'}{100}\right)^4 = 2.00$$

$$\varepsilon_{c2} = (2 + 0.085(f_c' - 50)^{0.53})/100 = 0.0020$$

$$\varepsilon_{cu2} = \left(2.6 + 35\left(\frac{90-f_c'}{100}\right)^4\right)/100 = 0.0035$$

9.5.3.1.2 *Reinforcing steel*
Type: Ribbed reinforcing bars (high bond)—
Diameter for longitudinal bars = 12 mm [Ø12] (0.472 in.)
Diameter for closed stirrups = 10 mm [Ø10] (0.394 in.)
Characteristic yield strength f_y = 420 MPa (60 ksi)
Design yield strength (EC2-04, Section 2.4.2.4)—

$$f_{yd} = \frac{f_y}{\gamma_s} = \frac{420}{1.15} = 365 \text{ MPa (52 ksi)}$$

Modulus of elasticity (EC2-04, Section 3.2.7(4))

$$E_s = 200 \text{ GPa (29,000 ksi)}$$

Strain at design yield strength

$$\varepsilon_y = \frac{f_{yd}}{E_s} = 0.0018$$

Ultimate design strain (EC2-04, Section 3.2.7(2))

$$\varepsilon_{ud} = 0.9\varepsilon_{uk} = 0.0200$$

9.5.3.1.3 *Prestressing reinforcing strands*
Type: Low relaxation, 270K, seven-wire strands—
Diameter = 13.0 mm (0.50 in.)
Area A_s = 99 mm² (0.153 in.²)
Characteristic tensile strength $f_{pk} = f_{pu}$ = 1860 MPa (270 ksi)
Design ultimate strength (EC2-04, Section 2.4.2.4)

$$f_{p,ud} = \frac{f_{pk}}{\gamma_s} = \frac{1860}{1.15} = 1617 \text{ MPa (234.78 ksi)}$$

Characteristic yield strength (EC2-04, Section 3.3.3(1))

$$f_{p0.1k} = 0.9f_{pu} = 1674 \text{ MPa (243 ksi)}$$

Design yield strength (EC2-04, Section 3.3.6(6))

$$f_{pd} = \frac{f_{p0.1k}}{\gamma_s} = \frac{1674}{1.15} = 1456 \text{ MPa (211 ksi)}$$

Modulus of elasticity (EC2-04, Section 3.3.6(3))

$$E_p = 196 \text{ GPa (28,420 ksi)}$$

Strain at design yield strength

$$\varepsilon_{p,yd} = \frac{f_{pd}}{E_p} = 0.0074$$

Ultimate design strain (EC2-04, Section 3.3.6(7))

$$\varepsilon_{p,ud} = 0.0200$$

9.5.3.2 *Sectional areas and concrete cover*
9.5.3.2.1 *Gross concrete area*
- Area of the gross cross section including overhanging flanges

= 2.65 × 10⁶ mm² (4107 in.²)

- Area of the gross cross section (disregarding overhanging flanges)

= 2.32 × 10⁶ mm² (3596 in.²)

- Area of the concrete cross section (disregarding overhanging flanges)

A_c = 1.20 × 10⁶ mm² (1860 in.²)

9.5.3.2.2 *Concrete covers (EC2-04, Section 4.4.1.2);*
- Minimum cover with regard to bond

$c_{min,b}$ = diameter of bar = 12 mm (0.47 in.) (reinforcing steel)
= 13 mm (0.51 in.) (prestressing steel)

- Exposure class related to environmental conditions

Exposure Class XD3 (Cyclic wet and dry. Parts of bridges exposed to spray containing chlorides.)

- Structural class (XD3) = 4 + 2 (service life of 100 years) – 1 (concrete class ≥ C45) = 5

- Minimum cover with regard to durability

$$c_{min,dur} = 50 \text{ mm (1.97 in.) (reinforcing steel)}$$
$$= 60 \text{ mm (2.36 in.) (prestressing steel)}$$

- Minimum cover

$$c_{min} = \max(c_{min,b};\ c_{min,dur};\ 10 \text{ mm [0.39 in.]})$$
$$= 50 \text{ mm (1.97 in.) for reinforcing steel}$$
$$= 60 \text{ mm (2.36 in.) for prestressing steel}$$

- Nominal cover

$$c_{nom} = c_{min} + 10 \text{ mm (0.39 in.)}$$
$$= 60 \text{ mm (2.36 in.) for reinforcing steel}$$
$$= 70 \text{ mm (2.76 in.) for prestressing steel}$$

- Distance from center of longitudinal reinforcing bars to extreme concrete fiber (cover from bars centroid)

$$= 66 \text{ mm (2.56 in.)} > c_{min} + 12 \text{ mm/2 (0.47 in./2)}$$

- Distance from center of prestressed tendons to extreme bottom fiber (cover from tendons centroid)

$$= 102 \text{ mm (3.94 in.)} > c_{min} + 13 \text{ mm/2 (0.51 in./2)}$$

9.5.3.3 *Factored shear, torque, bending moment, and prestress*

Factored shear force

$$V_u = 1.4 \times (34.4 + 12.8) \times 0.2 \times 24.0 + 1.5 \times (513.8/2) \times 2 = 1089 \text{ kN (245 kip)}$$

Factored torsional moment

$$T_u = 1.5 \times (513.8/2)2 \times 0.914 = 705 \text{ kN·m (520 ft-kip)}$$

Factored bending moment

$$M_u = 0.5 \times 1.4 \times (34.4 + 12.8) \times 0.7 \times 0.3\times (24.00)^2 + 1.5(513.8/2)(2 \times 7.2) = 9561 \text{ kN·m (7074 ft-kip)}$$

Prestress force at time $t = \infty$

$$P_t = 6076 \text{ kN (1366 kip)}$$

Total prestress losses

$$20 \text{ percent or } \omega = P_t/P_o = 0.80$$

Therefore, prestress force at time $t = 0$

$$P_o = P_t/\omega = 7595 \text{ kN (1708 kip)}$$

9.5.3.4 *Prestressed tendons requirements*

Minimum longitudinal reinforcement (EC2-04, Section 9.2.1.1)

$$A_\ell = 0.26\frac{f_t'}{f_{yk}}b_t d \geq 0.0013 b_t d$$

where b_t = 470 mm (18.5 in.) mean width of tension zone, and d = 1172 mm (46.1 in.) distance from extreme top fiber to the centroid of the reinforcement.

Therefore,

$$A_{\ell,min} = 716 \text{ mm}^2 \text{ (1.1 in.}^2\text{)}$$

The area of the prestressed tendons is

$$A_P = 64 \times 99 = 6336 \text{ mm}^2 \text{ (9.82 in.}^2\text{)} \geq A_{s\ell,min} \quad \text{OK}$$

9.5.3.5 *Shear design*

9.5.3.5.1 *Design shear resistance of a member without shear reinforcement (EC2-04, Section 6.2.2(1))*

$$V_{Rd,c} = \left[C_{Rd,c}k(100\rho_\ell f_{ck})^{1/3} + k_1\sigma_{cp}\right]b_w d \geq (v_{min} + k_1\sigma_{cp})b_w d$$

where

d = 517 + 494 = 1011 mm (39.8 in.) at 0.3L = 7.2 m (23.7 ft) (9.5.1.1(b))

b_w = 470 mm (18.50 in.)

$C_{Rd,c}$ = $0.18/\gamma_c = 0.18/1.5 = 0.12$

k = $1+\sqrt{200/d} \leq 2 \Rightarrow k = 1.44$

$$\rho_\ell = \frac{A_p}{b_w d} \leq 0.02 \Rightarrow \rho_\ell = 0.0133$$

$$\sigma_{cp} = \frac{N_{Ed}}{A_c} = \frac{P_t}{A_c} = 5.1 \text{ MPa (734.4 psi) (compressive)}$$

k_1 = 0.15

v_{min} = $0.035k^{3/2}f_{ck}^{1/2} \Rightarrow v_{min} = 0.428$

Therefore

$$V_{Rd,c} = 698 \text{ kN (156.2 kip)}$$

Because

$$V_{Rd,c} < V_{Ed} = 1089 \text{ kN (245 kip),}$$

shear reinforcement must be provided

9.5.3.5.2 *Minimum shear reinforcement (EC2-04, Section 9.2.2(5))*

The ratio of the minimum stirrups is

$$\rho_{w,min} = \left(0.08\sqrt{f_{ck}}\right)/f_{yk} = 0.08\sqrt{49.9}/(1.2\times 420) = 0.00113$$

Therefore

$$\left(\frac{A_{sw}}{sb_w}\right)_{min} \geq 0.00113 \rightarrow \left(\frac{A_{sw}}{s}\right)_{min} \geq 0.5311 \text{ mm}^2/\text{mm } (0.0209 \text{ in.}^2/\text{in.})$$

9.5.3.5.3 *Maximum effective cross-sectional area of the shear reinforcement (EC2-04, Section 6.2.3(3))*

$$\frac{A_{sw.max} f_{yd}}{sb_w} \leq \frac{1}{2}\alpha_c v f_{cd} \rightarrow \left(\frac{A_{sw}}{s}\right)_{max} \leq \frac{\alpha_c v b_w f_{cd}}{2 f_{yd}}$$

where
$v = 0.6(1 - f_{ck}/250) = 0.48$
$\alpha_c = 1 + \sigma_{cp}/f_{cd} = 1.154$ (because $\sigma_{cp} \leq 0.25 f_{cd}$)
Therefore

$$\left(\frac{A_{sw}}{s}\right)_{max} \leq 11.86 \text{ mm}^2/\text{mm } (0.47 \text{ in.}^2/\text{in.})$$

9.5.3.5.4 *Required shear reinforcement (EC2-04, Section 6.2.3(4))*

$$V_{Rd} = \min\begin{Bmatrix} V_{Rd,s} \\ V_{Rd,max} \end{Bmatrix} \geq V_{Ed}$$

$$V_{Rd,max} = \frac{\alpha_c b_w z v f_{cd}}{(\cot\theta + \tan\theta)} \qquad \text{(Eq. (6.14) in EC2-04)}$$

With the above data and

$$z = 0.9d \Rightarrow z = 910 \text{ mm (35.8 in.)}$$

$$f_{cd} = 33.3 \text{ MPa (4830 psi)}$$

$$1 \leq \cot\theta \leq 2.5 \Rightarrow 22 \text{ degrees} \leq \theta \leq 45 \text{ degrees}$$
(mean value: $\theta = 35$ degrees)

$$V_{Rd,max} = \frac{1.154 \times 470 \times 910 \times 0.48 \times 33.3}{(\cot\theta + \tan\theta)}$$

For
$\theta = 45$ degrees

$$V_{Rd,max} = 3945 \text{ kN (886 kips)} > V_{Ed},$$

$\theta = 35$ degrees

$$V_{Rd,max} = 3707 \text{ kN (832 kips)} > V_{Ed}, \text{ and}$$

$\theta = 22$ degrees

$$V_{Rd,max} = 2740 \text{ kN (615 kips)} > V_{Ed}.$$

Thus for (Eq. (6.13) in EC2-04)

$$V_{Rd} = V_{Rd,s} \geq V_{Ed}$$

$$V_{Rd,s} = \frac{A_{sw}}{s} z f_{yd} \cot\theta \geq V_{Ed} \Rightarrow \frac{A_{sw}}{s} \geq \frac{V_{Ed}}{z f_{yd} \cot\theta} \Rightarrow \frac{A_{sw}}{s} \geq \frac{1089 \times 10^3}{910 \times 365 \times \cot\theta}$$

$\theta = 45$ degrees

$$\frac{A_{sw}}{s} \geq 3.28 \text{ mm}^2/\text{mm } (0.131 \text{ in.}^2/\text{in.}),$$

$\theta = 35$ degrees

$$\frac{A_{sw}}{s} \geq 2.30 \text{ mm}^2/\text{mm } (0.092 \text{ in.}^2/\text{in.}), \text{ and}$$

$\theta = 22$ degrees

$$\frac{A_{sw}}{s} \geq 1.32 \text{ mm}^2/\text{mm } (0.053 \text{ in.}^2/\text{in.}).$$

Check

$$\left(\frac{A_{sw}}{s}\right)_{min} < \frac{A_{sw}}{s} < \left(\frac{A_{sw}}{s}\right)_{max} \qquad \text{OK}$$

9.5.3.6 *Torsion design*

9.5.3.6.1 *Terms of the equivalent thin-walled section (EC2-04, Section 6.3.2(1));*

$$t_{ef} = \max\begin{Bmatrix} A/u \\ 2c_\ell \end{Bmatrix} \leq t_{real} = \max\begin{Bmatrix} 375 \text{ mm} \\ 134 \text{ mm} \end{Bmatrix} \leq 178 \text{ mm}$$

$$\Rightarrow t_{ef} = 178 \text{ mm (7 in.)}$$

where
A = 2.32×10^6 mm² (3597 in.²) total area of the cross section within the outer circumference, including inner hollow areas (disregarding overhanging flanges)
u = 6185 mm (243.5 in.) outer circumference of the cross section (disregarding overhanging flanges)
c_ℓ = 66 mm (2.56 in.) distance from edge to center of the longitudinal reinforcement
t_{real} = 178 mm (7 in.) minimum wall thickness of the real concrete thin-walled section.

The continuous area and perimeter enclosed by centerlines of the connecting thin-walls, as shown in Fig. 9.5.3.6.1, are: $A_k = 1.8 \times 10^6$ mm² (2794 in.²) and $u_k = 5486$ mm (216 in.).

9.5.3.6.2 *Determine if cross-sectional dimensions (for example, strength of concrete struts under torsion) are adequate (EC2-04, Section 6.3.2(4))*

Dimensions of the cross section are adequate if (Eq. (6.29) in EC2-04)

$$\frac{T_{Ed}}{T_{Rd,max}} + \frac{V_{Ed}}{V_{Rd,max}} \leq 1$$

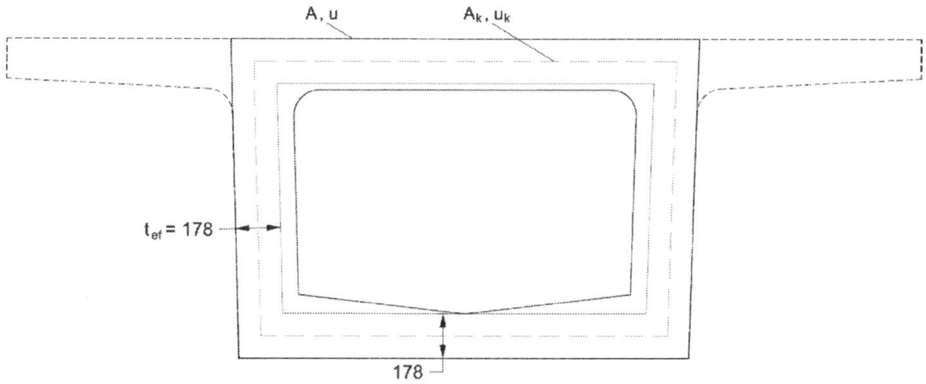

Fig. 9.5.3.6.1—Assumed effective thin-wall section.

The design torsional resistance moment is calculated as (Eq. (6.30) in EC2-04)

$$T_{Rd,max} = 2v\alpha_c f_{cd} A_k t_{ef} \sin\theta\cos\theta$$

where

$$v = 0.6(1 - f_{ck}/250) = 0.48 \text{ and } \alpha_c = 1 + \sigma_{cp}/f_{cd} = 1.154$$

and for

θ = 45 degrees

$$T_{Rd,max} = 5915 \text{ kN·m (4356 ft-kip)},$$

θ = 35 degrees

$$T_{Rd,max} = 5559 \text{ kN·m (4093 ft-kip)}$$

θ = 22 degrees

$$T_{Rd,max} = 4109 \text{ kN·m (3026 ft-kip)}$$

Design shear resistance force is determined by (Eq. (6.14) in EC2-04)

$$V_{Rd,max} = \frac{\alpha_c b_w z v f_{cd}}{(\cot\theta + \tan\theta)}$$

Thus for

θ = 45 degrees

$$\frac{T_{Ed}}{T_{Rd,max}} + \frac{V_{Ed}}{V_{Rd,max}} = \frac{705}{5910} + \frac{1089}{3945} = 0.39 \leq 1 \quad \text{OK}$$

θ = 35 degrees

$$\frac{T_{Ed}}{T_{Rd,max}} + \frac{V_{Ed}}{V_{Rd,max}} = \frac{705}{5554} + \frac{1089}{3707} = 0.42 \leq 1 \quad \text{OK}$$

θ = 22 degrees

$$\frac{T_{Ed}}{T_{Rd,max}} + \frac{V_{Ed}}{V_{Rd,max}} = \frac{705}{4085} + \frac{1089}{2740} = 0.57 \leq 1 \quad \text{OK}$$

Dimensions of the cross section, therefore, are adequate for every case (every value of θ).

9.5.3.6.3 *Calculate the required stirrup area for torsion. (EC2-04, Section 6.3.2(2))*

The required stirrup area per unit length is calculated by

$$\frac{A_{sw}}{s} \geq \frac{T_{Ed}}{2A_k f_{yd}\cot\theta} \Rightarrow \frac{A_{sw}}{s} \geq \frac{705\times10^6}{2\times1.8\times10^6\times365\times\cot\theta}$$

For

θ = 45 degrees

$$\frac{A_{sw}}{s} \geq 0.54 \text{ mm}^2\text{/mm (0.021 in.}^2\text{/in.)},$$

θ = 35 degrees

$$\frac{A_{sw}}{s} \geq 0.38 \text{ mm}^2\text{/mm (0.015 in.}^2\text{/in.), and}$$

θ = 22 degrees

$$\frac{A_{sw}}{s} \geq 0.22 \text{ mm}^2\text{/mm (0.009 in.}^2\text{/in.)}$$

where A_{sw} and s are the area of one leg of a closed stirrup resisting torsion and the spacing of the stirrups, respectively.

9.5.3.6.4 *Add stirrup areas for torsion and shear, and select the stirrups (EC2-04, Section 6.3.2(2))*

$$\frac{A_{sw,S+T}}{s} = \frac{1}{2}\frac{A_{sw,S}}{s} + \frac{A_{sw,T}}{s} \quad \text{(using two single-legged stirrups)}$$

For

θ = 45 degrees

$$\frac{A_{sw,S+T}}{s} \geq 2.18 \text{ mm}^2/\text{mm } (0.087 \text{ in.}^2/\text{in.}),$$

θ = 35 degrees

$$\frac{A_{sw,S+T}}{s} \geq 1.53 \text{ mm}^2/\text{mm } (0.061 \text{ in.}^2/\text{in.}), \text{ and}$$

θ = 22 degrees

$$\frac{A_{sw,S+T}}{s} \geq 0.88 \text{ mm}^2/\text{mm } (0.035 \text{ in.}^2/\text{in.}).$$

Maximum longitudinal spacing of the stirrups for shear (Eq. (9.6N) in EC2-04)

$$s_{max} = 0.75d(1+\cot\alpha) \xrightarrow{\alpha=90^\circ} s_{max} = 0.75d$$
$$= 758 \text{ mm } (29.8 \text{ in.})$$

where α is the inclination of the stirrups.

Maximum longitudinal spacing of the stirrups for torsion (EC2-04, Section 9.2.3(3))

$$s_{max} = \min\begin{Bmatrix} u/8 \\ 0.75d(1+\cot\theta) \\ \min(b,h) \end{Bmatrix} = \min\begin{Bmatrix} 773 \text{ mm} \\ 758 \text{ mm} \\ 235 \text{ mm} \end{Bmatrix}$$
$$= 235 \text{ mm } (9.25 \text{ in.})$$

For two single-legged ∅10 stirrups, spacing is calculated from shear and torsion

For

θ = 45 degrees

$$s \leq 72 \text{ mm } (2.8 \text{ in.}) < s_{max} \quad \text{OK}$$

θ = 35 degrees

$$s \leq 102 \text{ mm } (4.0 \text{ in.}) < s_{max} \quad \text{OK}$$

θ = 22 degrees

$$s \leq 178 \text{ mm } (7.0 \text{ in.}) < s_{max} \quad \text{OK}$$

Total selection (θ = 35 degrees): two single-legged stirrups of ∅10/100 mm (diameter 0.39 in. at 4.0 in.).

9.5.3.6.5 *Longitudinal reinforcement required for torsion (EC2-04, Section 6.3.2(3))*

Total longitudinal reinforcement for torsion (Eq. (6-28) in EC2-04)

$$\Sigma A_{sl} \geq \frac{T_{Ed} u_k \cot\theta}{2A_k f_{yd}} \Rightarrow \Sigma A_{sl} \geq \frac{705\times10^6 \times 5486 \times \cot\theta}{2\times1.8\times10^6\times365}$$

For

θ = 45 degrees

$$\Sigma A_{sl} \geq 2943 \text{ mm } (4.62 \text{ in.}^2),$$

θ = 35 degrees

$$\Sigma A_{sl} \geq 4204 \text{ mm}^2 \, (6.60 \text{ in.}^2), \text{ and}$$

θ = 22 degrees

$$\Sigma A_{sl} \geq 7285 \text{ mm}^2 \, (11.50 \text{ in.}^2).$$

Longitudinal bars shall be arranged so that at least one bar is placed at each corner of the stirrups and the others are distributed uniformly around the inner periphery of the torsion links (closed stirrups) with a maximum spacing of 350 mm (13.8 in.) (EC2-04, Section 9.2.3(4)). Therefore, the number of the longitudinal bars is at least 36.

For

θ = 45 degrees

$$38∅10 \rightarrow \Sigma A_{sl} = 2985 \text{ mm } (4.63 \text{ in.}^2) > 2943 \text{ mm } (4.63 \text{ in.}^2),$$

θ = 35 degrees

$$38∅12 \rightarrow \Sigma A_{sl} = 4298 \text{ mm}^2 \, (6.66 \text{ in.}^2) > 4204 \text{ mm}^2 \, (6.61 \text{ in.}^2), \text{ and}$$

θ = 22 degrees

$$38∅16 \rightarrow \Sigma A_{sl} = 7640 \text{ mm}^2 \, (11.84 \text{ in.}^2) > 7285 \text{ mm}^2 \, (11.58 \text{ in.}^2).$$

Final selection of longitudinal bars (θ = 35 degrees)

$$∅12 \, (0.47 \text{ in.}) \rightarrow A_{sl} = 113 \text{ mm}^2 \, (0.175 \text{ in.}^2).$$

Due to uniform distribution and symmetry

$$40∅12 \rightarrow \Sigma A_{sl} = 4524 \text{ mm}^2 \, (7.0 \text{ in.}^2).$$

The torsional longitudinal reinforcement is in addition to the prestressing tendons. Low values of angle θ (θ = 22 degrees) lead to a design with lower area requirements of transverse reinforcement (∅10/175 mm) and higher area requirements of longitudinal reinforcement (40∅16). High values of angle θ (θ = 45 degrees) lead to higher area requirements of stirrups (∅10/70 mm) and lower area requirements of longitudinal bars (40∅10).

The selection of θ = 45 degrees maximizes the concrete strength components, such as $V_{Rd,max}$ = 3945 kN and $T_{Rd,max}$ = 5910 kN·m. This value of angle θ could be used when checking the adequacy of the cross section dimensions.

The following relationship between the strength of concrete struts under torsion is satisfied

$$\frac{T_{Ed}}{T_{Rd,max}} + \frac{V_{Ed}}{V_{Rd,max}} \leq 1$$

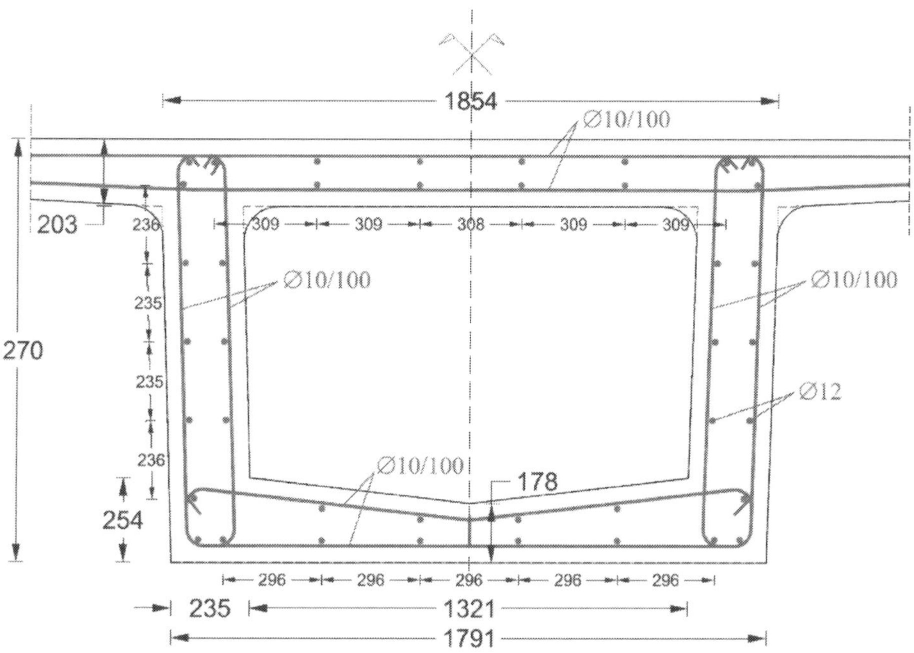

Fig. 9.5.3.6.6—Reinforcement configuration (EC2-04).

9.5.3.6.6 *Arrangement of reinforcing bars*—Total reinforcement of the member with a hollow cross section under combined loading of prestressing, torsion, shear, flexure, and axial force is depicted in detail in Fig. 9.5.3.6.6.

9.5.4 *Design solution using CSA-A23.3-04 code*—Load factors of the CSA-A23.3-04 code are used here to establish the forces due to factored loads. These factors are 1.25 for dead loads and 1.50 for live loads. The cross-sectional area of the prestressing strands (f_{pu} = 1860 MPa [270 ksi]) is 6336 mm^2 (9.82 in.^{2}), the effective prestressing force is 6076 kN (1366 kip), and the average prestress is 3.99 MPa (589 psi). The concrete strength is 48.0 MPa (7000 psi), and the yield strength of the non-prestressed reinforcement is 420 MPa (60,000 psi).

1. Determine the factored forces (CSA-A23.3-04, Annex C).

Factored dead and live loads—

The derailment load per axle is calculated by

$$P_{u,L} = 1.5\left(\frac{513.8}{4}\right)2 = 385 \text{ kN/axle (86.6 ft/axle)}$$

The derailment torque per axle is determined by

$$T_{u,L} = 1.5\left(\frac{513.8}{4}\right)(2 \times 0.914) = 352 \text{ kN·m (259.8 ft-kip)}$$

The girder weight (with a load factor of 1.25) is determined by

$$w_{u,g} = 1.25 \times 34.3 = 43.0 \text{ kN/mm (2.95 kip/ft)}$$

The superimposed dead weight (with a load factor of 1.25) is calculated by

$$w_{u,s} = 1.25 \times 12.8 = 16.0 \text{ kN/mm (1.10 kip/ft)}$$

Factored shear, torque, and bending moment—

At distance 0.3L from support, the following values are obtained

$$V_u = (43.0 + 16.0)(0.2 \times 24.00) + 2 \times 385 = 1053 \text{ kN (237 kip)}$$

$$T_u = 2 \times 352.3 = 705 \text{ kN·m (520 ft-kip)}$$

$$M_u = 0.5(43.0 + 16.0)(24.00 - 7.2)7.2 + 2 \times 385 \times 7.2$$
$$= 9112 \text{ kN·m (6759 ft-kip)}$$

2. Determine if torsion effects can be disregarded (CSA-A23.3-04, Section 11.2.9.1).

Threshold torque—

For a hollow section with a wall thickness of less than $0.75A_c/p_c$ (= 281 mm > 178 mm), torsion must be considered if the torque due to factored loads, T_f, exceeds $0.25T_{cr}$

$$T_{cr} = \frac{(1.5A_g)^2}{p_c}0.38\lambda\phi_c\sqrt{f_c'}\sqrt{1+\frac{\phi_p f_{cp}}{0.38\lambda\phi_c\sqrt{f_c'}}}$$

where

A_c = A_{cp} = 2.32 × 10^6 mm^2 (3597 in.^{2})

p_c = p_{cp} = outside perimeter of concrete cross section = 6185 mm (243.5 in.)

A_g = gross area of section (without flanges) = 1.20 × 10^6 mm^2 (1855 in.^{2})

λ = factor to account for low-density concrete = 1.0

ϕ_c = resistance factor for concrete = 0.70 for precast concrete

ϕ_p = resistance factor for prestressing reinforcement = 0.90

f_c' = specified compressive strength of concrete = 48.0 MPa (7.0 ksi)

f_{cp} = compression stress in concrete due to effective prestress = 3.99 MPa (0.6 ksi)

With this information, the following calculation can be made

$$T_{cr} = \frac{(1.5\times1.20\times10^6)^2}{6185} 0.38\times0.70\sqrt{48.0}\sqrt{1+\frac{0.9\times3.99}{0.38\times0.70\sqrt{48.0}}}$$

$$= 1658 \text{ kN·m (1216 ft-kip)}$$

Because $0.25T_{cr}$ = 415 kN·m (304 ft-kip) < 705 kN·m (520 ft-kip), torsion must be considered in the design.

3. Determine if dimensions of the cross section are adequate.

Check cross-sectional dimensions (CSA-A23.3-04, Section 11.3.10.4)—

For box sections with a wall thickness of less than A_{oh}/p_h, the cross-sectional dimensions must satisfy the following criterion (Eq. (11-19) in CSA-A23.3-04)

$$\frac{V_f - V_p}{b_w d_v} + \frac{T_f}{1.7A_{oh}t} \le 0.25\phi_c f_c'$$

where

V_f = V_u = shear force due to factored loads = 1054 kN (237.2 kip)

V_p = shear force due to prestressing factored by ϕ_p
= 0.9(6076)508/9754 = 284 kN (64.0 kip)

A_{oh} = [(1854 + 1791)/2 – 2 × 48)](1270 – 2 × 48) = 2.03 × 10^6 mm^2 (3142 $in.^2$)

p_h = (1854 – 96) + (1791 – 96) + 2(1270 – 96) = 5801 mm (228.4 in.)

A_{oh}/p_h = 350 mm (13.8 in) > 178 mm (7.0 in.) for bottom flange
> 235 mm (9.25 in.) for web
> 203 mm (8 in.) for top flange

d_v = larger of $0.9d$ = 0.9 × 1016 = 914 mm and $0.72h$ = 0.72 × 1270 = 914 mm (36 in.)

t = minimum wall thickness = 178 mm (7.0 in.)

T_u = torque due to factored loads = 705 kN·m (520 ft-kip)

Therefore,

$$\frac{(1054-284)\times10^3}{235\times2\times914} + \frac{705\times10^6}{1.7\times2.03\times10^6\times235}$$

= 2.66 MPa (386 psi) < 0.25 × 0.70 × 48.0 = 8.40 MPa (1223 psi) OK

Because the web governs design, the above equation was the web thickness t = 235 mm, not the bottom flange thickness.

4. Calculate θ and β (CSA-A23.3-04, Section 11.3.6.4).

Angle of diagonal compression strut and shear resistance of concrete—

The angle of inclination of the diagonal compression strut is given by the expression (Eq. (11-12) in CSA-A23.3-04)

$$\theta = 29 + 7000\varepsilon_x$$

In the absence of an axial load normal to the cross section, the strain at mid-depth of the section is defined by (Eq. (11-13) in CSA-A23.3-04)

$$\varepsilon_x = \frac{\frac{M_f}{d_v} + \sqrt{(V_f - V_p)^2 + \left(\frac{0.9p_h T_f}{2A_o}\right)^2} - A_p f_{po}}{2[E_s A_s + E_p A_p]}$$

The terms not defined above are

M_f = M_u = moment due to factored loads = 9166 kN·m (6761 ft-kip)

A_o = $0.85A_{oh}$ = 0.85 × 2.04 × 10^6 mm^2 = 1.74 × 10^6 mm^2 (2697 $in.^2$)

A_p = area of prestressing reinforcement = 6336 mm^2 (9.8 $in.^2$)

f_{po} = stress in prestressing tendons (may be taken as $0.7f_{pu}$ = 1302 MPa [189.0 ksi])

E_s = 200,000 MPa (29,000 ksi)

A_s = area of non-prestressed reinforcement in tension zone (assume fourteen 15M bars with 2800 mm^2 (4.34 $in.^2$)

E_p = 190,000 MPa (28,000 ksi)

With this information, the following calculation can be made

$$\varepsilon_x = \frac{\frac{9112\times10^6}{914} + \sqrt{[(1054-284)\times10^3]^2 + \left(\frac{0.9\times5801\times705\times10^6}{2\times1.74\times10^6}\right)^2} - 1302\times6336}{2(200{,}000\times2800 + 190{,}000\times6336)} = 0.00086$$

Therefore

$$\theta = 29 + 7000 \times 0.00086 = 35.0°$$

The shear force resisted by the concrete is (Eq. (11-6) in CSA-A23.3-04)

$$V_c = \phi_c \lambda \beta \sqrt{f_c'} b_w d_v$$

where (Eq. (11-11) in CSA-A23.3-04)

$$\beta = \frac{0.40}{1+1500\varepsilon_x} \times \frac{1300}{1000+s_{ze}}$$

Because minimum transverse reinforcement is provided, s_{ze} = 300 mm (12 in.)

Therefore

$$\beta = \frac{0.40}{1+1500(0.00086)} \times \frac{1300}{1000+300} = 0.175$$

and

$$V_c = 0.70 \times 1.0 \times 0.175(\sqrt{48.0})235 \times 2 \times 914 \text{ N} = 365 \text{ kN (80.7 kip)}$$

5. Calculate transverse reinforcement.

Design of transverse reinforcement for shear (CSA-A23.3-04, Section 11.3.5.1)—

Shear with $V_s = V_f - V_c - V_p = 1054 - 365 - 284 = 405$ kN (92.6 kip)

$$\frac{A_v}{s} = \frac{V_s}{\phi_s f_y d_v \cot\theta} = \frac{405 \times 10^3}{0.85 \times 420 \times 914 \times \cot 35.0°}$$

$$= 0.87 \text{ mm}^2\text{/mm } (0.0356 \text{ in.}^2\text{/in.})$$

Design of transverse reinforcement for torsion (CSA-A23.3-04, Section 11.3.10.3)—

The required transverse reinforcement for torsion is given by (Eq. 11-17 in CSA-A23.3-04)

$$\frac{A_t}{s} = \frac{T_u}{2A_o \phi_s f_y \cot\theta} = \frac{705 \times 10^6}{2 \times 1.74 \times 10^6 \times 0.85 \times 420 \times \cot 35.0°}$$

$$= 0.40 \text{ mm}^2\text{/mm } (0.0160 \text{ in.}^2\text{/in.})$$

This amount of transverse reinforcement must be provided in the top and bottom slabs.

Total transverse reinforcement in one web is calculated as

$$\frac{A_t}{s} + \frac{A_v}{2s} = 0.40 + 0.87/2 = 0.835 \text{ mm}^2\text{/mm } (0.0338 \text{ in.}^2\text{/in.})$$

The selection of two 10M bars ($A_{bar} = 100$ mm^2) per web yields the following bar spacing

$s = 2(100)/0.835 = 240$ mm; select $s = 225$ mm (9.17 in.)

Minimum transverse reinforcement (CSA-A23.3-04, Section 11.2.8.2)—

The minimum transverse reinforcement is established with the maximum web thickness at the top of the web, b_w = 534 mm (21.0 in.)

$$\frac{\min A_v}{s} = 0.06\sqrt{f'_c}\frac{b_w}{f_y} = 0.06\sqrt{48.0}\left(\frac{534}{420}\right)$$

$$= 0.528 \text{ mm}^2\text{/mm } (0.0212 \text{ in.}^2\text{/in.})$$
$$< 0.835 \text{ mm}^2\text{/mm } (0.0338 \text{ in.}^2\text{/in.})$$

For $T_f > 0.25T_{cr}$, max $s = 0.35d_v = 0.35(914) = 320$ mm (12.6 in.) > 225 mm. (9.17 in.) OK

6. Calculate the longitudinal reinforcement.

Design of longitudinal reinforcement for torsion (CSA-A23.3-04, Section 11.3.10.6 and 11.3.9.2)—

Longitudinal reinforcement on the tension side of the girder shall have dimensions such that its factored resistance shall not be less than that given by the following expression. In the absence of an axial load normal to the cross section, the required longitudinal force due to torsion, shear, and flexure is

$$F_{lt} = \frac{M_f}{d_v} + \cot\theta\sqrt{(V_f - 0.5V_s - V_p)^2 + \left(\frac{0.45p_h T_f}{2A_o}\right)^2}$$

$$= \frac{9112 \times 10^6}{914} + \cot 35.0°\sqrt{\left[\left(1054 - \frac{405}{2} - 284\right)(10^3)\right]^2 + \left(\frac{0.45 \times 5801 \times 705 \times 10^6}{2 \times 1.74 \times 10^6}\right)^2}$$

$$= 9969 + 1108 = 11{,}077 \text{ kN (2503 kip)}$$

Factored tension resistance provided by the prestressing reinforcement with $f_{pr} = 0.96f_{pu} = 1786$ MPa (259 ksi) (CSA-A23.3-04, Section 18.6.2)

$$F_{ltr} = \phi_p A_p f_{pr} = 0.9 \times 6336 \times 1786 \times 10^{-3}$$
$$= 10{,}184 \text{ kN (2290 kips)}$$

The non-prestressed reinforcement required in tension zone is calculated by

$$A_s = \frac{(11{,}077 - 10{,}184)(10^3)}{0.85 \times 420} = 2501 \text{ mm}^2 \text{ (4.26 in.}^2\text{)}$$

These calculations indicate that fourteen 15M bars are required, providing $A_s = 2800$ mm^2. This reinforcement is to be distributed in the tension zone.

Because the compression due to moment is much greater than the tension due to torsion and shear, there is no need to check the tension reinforcement required in the compression zone.

9.5.5 *Comparison of design solutions according to ACI 318, CSA-A23.3-04, and EC2-04*—The required longitudinal and transverse reinforcement calculated using ACI 318, CSA-A23.3-04, and EC2-04 is shown in Table 9.5.5. The longitudinal reinforcement ("L" Bars) given is the total amount required for the entire cross section. The required transverse reinforcement is given for the top wall ("T-1" Bars), side walls ("T-2" Bars), and bottom walls ("T-3" Bars). The transverse reinforcement in the top wall should be added to the flexural reinforcement required in the top wall acting as a transverse continuous slab.

Arrangement of reinforcing bars—

Arrangement of the reinforcing bars for torsion and shear is shown in Fig. 9.5.5 and summarized in Table 9.5.5.

The general location of longitudinal and transverse reinforcement in the section, which is shown in Fig. 9.5.5, is based on all three design codes.

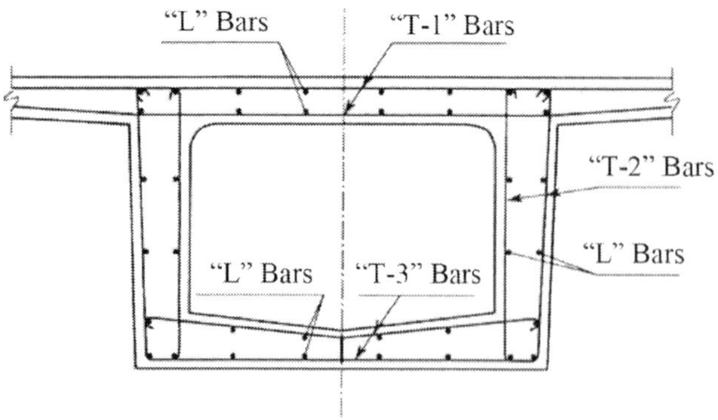

Fig. 9.5.5—Reinforcement locations for box girder under combined torsion, shear, and flexure.

Table 9.5.5—Summary of design solution of Example 2 using design codes ACI 318-11, EC2-04, and CSA-A23.3-04

Reinforcing bars	Design code		
	ACI 318-11	EC2-04 (45, 35, and 22 degrees)	CSA-A23.3-04
"L" bars	5808 mm^2 (9.2 in.^{2})	4204 mm^2 (6.61 in.^{2})	2503 mm^2 (4.26 in.^{2})
"T-1" and "T-3" bars (torsion)	0.414 mm^2/mm (0.0166 in.^{2}/in.)	2.18, 1.53, 0.88 mm^2/mm (0.087, 0.061, 0.035 in.^{2}/in.)	0.40 mm^2/mm (0.0160 in.^{2}/in.)
"T-2" bars (torsion + shear)	1.44 mm^2/mm (0.0575 in.^{2}/in.)	2.18, 1.53, 0.88 mm^2/mm (0.087, 0.061, 0.035 in.^{2}/in.)	0.835 mm^2/mm (0.0338 in.^{2}/in.)

CHAPTER 10—REFERENCES

Committee documents are listed first by document number and year of publication followed by authored documents listed alphabetically.

American Concrete Institute

318-63—Building Code Requirements for Reinforced Concrete

318-71—Building Code Requirements for Reinforced Concrete

318-77—Building Code Requirements for Reinforced Concrete

318-95—Building Code Requirements for Structural Concrete

318-02—Building Code Requirements for Structural Concrete

318-08—Building Code Requirements for Structural Concrete (ACI 318-08) and Commentary

318-11—Building Code Requirements for Structural Concrete (ACI 318-11) and Commentary

European Committee for Standardization

Eurocode 2-04—Design of Concrete Structures—Part 1-1

Eurocode 8-04—Design of Structures for Earthquake Resistance—Part 1

Canadian Standards Association

CSA-23.3-77—Design of Concrete Structures

CSA-23.3-04—Design of Concrete Structures

Abul Mansur, M., and Rangan, B., 1978, "Torsion in Spandrel Beams," *Journal of the Structural Division*, V. 104, pp. 1061-1075.

ACI Committee 315, 2004, *ACI Detailing Manual*, SP-66, American Concrete Institute, Farmington Hills, MI, 166 pp.

ACI Committee 363, 1992, "Report on High-Strength Concrete (ACI 363R-92)," American Concrete Institute, Farmington Hills, MI, 55 pp.

ACI Committee 438, 1969, "Tentative Recommendations for the Design of Reinforced Concrete Members to Resist Torsion," *ACI Journal*, V. 66, No. 1, Jan., pp. 1-8.

Ali, M. A., and White, R. N., 1997, "On Extending ACI 318 to High-Strength Concrete," *Proceedings of the First International Conference on High Strength Concrete*, Kona, HI, July 13-18, pp. 554-567.

Alkhrdaji, T., and Belarbi, A., 2003, "Shear Flow Zone in Rectangular RC Members Subjected to Pure Torsion," *Proceedings of the International Conference on the Performance of Construction Materials in the New Millennium*, Cairo, Egypt, Feb. 17-20, pp. 555-564.

Allos, A. E., and Rashid, A. H., 1989, "Prestressed Concrete Rectangular Beams Subjected to Sustained Torque," *ACI Structural Journal*, V. 86, No. 4, July-Aug., pp. 469-472.

American Association of State Highway and Transportation Officials, 1998, *AASHTO LRFD Bridge Design Specifications and Commentary*, SI units, second edition, Washington DC, 1091 pp.

Bach, B., 1911, *Elastizität und Festigkeit*, sixth edition. (in German)

Badawy, H., and Jordaan, I., 1977, "Plastic Analysis of Curved Beams under Point Loads," *Journal of the Structural Division*, V. 103, pp. 1429-1455.

Baumann, T., 1972, "Zur Frage der Netzbewehrung von Flachentragwerken (On the Problem of Net Reinforcement of Surface Structures)," *Der Bauingenieur*, V. 47, No. 10, pp. 367-377.

Bažant, Z. P., 1984, "Size Effect in Blunt Fracture: Concrete, Rock, Metal," *Journal of Engineering Mechanics*, V. 110, No. 4, pp. 518-535.

Bažant, Z. P., 2002, *Scaling of Structural Strength*, Hermes Penton Science, Ltd. (Kogan Page Science), London, UK, 336 pp.

Bažant, Z. P., and Oh, B. H., 1983, "Crack Band Theory for Fracture of Concrete," *Matériauxet Constructions*, V. 16, May-June, pp. 155-177.

Bažant, Z. P., and Planas, J., 1998, *Fracture and Size Effect in Concrete and Other Quasibrittle Materials,* CRC Press, Boca Raton, FL and London, UK, 640 pp.

Bažant, Z. P., and Sener, S., 1987, "Size Effect in Torsional Failure of Concrete Beams," *Journal of Structural Engineering*, V. 113, No. 10, pp. 2125-2136.

Bažant, Z. P.; Sener, S.; and Prat, P. C., 1988, "Size Effect Tests of Torsional Failure of Plain and Reinforced Concrete Beams," *Materials and Structures*, V. 21, pp. 425-430.

Bažant, Z. P., and Yu, Q., 2005a, "Designing Against Size Effect on Shear Strength of Reinforced Concrete Beams Without Stirrups: I. Formulations," *Journal of Structural Engineering*, V. 131, No. 12, pp. 1877-1885.

Bažant, Z. P., and Yu, Q., 2005b, "Designing Against Size Effect on Shear Strength of Reinforced Concrete Beams Without Stirrups: II. Verification and Calibration," *Journal of Structural Engineering*, V. 131, No. 12, pp. 1886-1897.

Belarbi, A., and Greene, G. G., 2004, "Reinforced Concrete Box Girders Under Cyclic Torsion," *Proceedings of the 13th World Conference on Earthquake Engineering*, Vancouver, BC, Canada, Aug. 1-6, 2004, 15 pp.

Belarbi, A., and Hsu, T. T. C., 1994, "Constitutive Laws of Concrete in Tension and Reinforcing Bars Stiffened by Concrete," *ACI Structural Journal*, V. 91, No. 4, July-Aug., pp. 465-474.

Belarbi, A., and Hsu, T. T. C., 1995, "Constitutive Laws of Softened Concrete in Biaxial Tension-Compression," *ACI Structural Journal*, V. 92, No. 5, Sept.-Oct., pp. 562-573.

Bentz, E. C., 2005, "Empirical Modeling of Reinforced Concrete Shear Strength Size Effect for Members without Stirrups," *ACI Structural Journal*, V. 102, No. 2, Mar.-Apr., pp. 232-241.

Birkland, C. J., 1965, "Strength of Rectangular Beams with Longitudinal Reinforcement in Combined Torsion, Bending, and Shear," Master's thesis, Department of Civil Engineering, University of Washington at Seattle, Seattle, WA.

Bishara, A., and Londot, L., 1979, "Flexural Rotational Capacity of Spandrel Beams," *Journal of the Structural Division*, V. 105, No. 1, pp. 147-161.

Bishara, A., and Peir, J. C., 1968, "Reinforced Concrete Rectangular Columns in Torsion," *Journal of the Structural Division*, V. 94, Dec, pp. 2913-2934.

Boulton, N., and Boonsukha, B., 1959, "Plastic Collapse Loads for Circular Arc Girders," *Proceedings of the Institution of Civil Engineers*, June, London, UK.

Bredt, R., 1896, *Kritische Bemerkungen zur Drehungselastizität, Zeitschrift des Vereines Deutscher Ingenieure*, Band 40, No. 28, July 11, pp. 785-790; No. 29, July 18, pp. 813-817. (in German)

Cauchy, A., 1828, "Sur les Equations qui Experimentent les Conditions d'Equilibre ou les Lois de Mouvement Interieur d'un Corps Solide," *Exercices de Mathematique*, Paris, France. (in French)

CEB-FIP, 1978, *Model Code for Concrete Structures*, CEB-FIP International Recommendation, third edition, Comite Euro-International du Beton (CEB), 348 pp.

Cerioni, R.; Ferreti, D.; and Iori, I., 1998, "Torsional Behavior of Prestressed High-Strength Concrete Beams," *Computational Modeling of Concrete Structures*, pp. 797-806.

Chakraborty, M., 1977, "Torsional-Balanced Steel in Concrete Beams," *Journal of the Structural Division*, V. 103, pp. 2181-2191.

Chen, W., 1982, *Plasticity in Reinforced Concrete*, McGraw-Hill Book, New York, 474 pp.

Collins, M. P., 1973, "Torque-Twist Characteristics of Reinforced Concrete Beams," *Inelasticity and Non-Linearity in Structural Concrete*, SM Study No. 8, University of Waterloo Press, Waterloo, ON, Canada, pp. 211-231.

Collins, M. P., and Chockalingam, S., 1979, "Reinforced Concrete Beams Under Reversed Cyclic Torsional Loading," Structural Concrete Under Seismic Actions, *CEB Bulletin D' Information*, No. 132, pp. 181-187.

Collins, M. P., and Lampert, P., 1973, "Redistribution of Moments at Cracking—The Key to Simpler Torsion Design?" *Analysis of Structural Systems for Torsion*, SP-35, American Concrete Institute, Farmington Hills, MI, pp. 343-383.

Collins, M. P., and Mitchell, D., 1978, "Torsional-Balanced Steel in Concrete Beams – Discussion by Michael P. Collins and Denis Mitchell," *Journal of the Structural Division*, V. 114, pp. 1684-1687.

Collins, M. P., and Mitchell, D., 1980, "Shear and Torsion Design of Prestressed and Non-Prestressed Concrete Beams," *PCI Journal*, V. 25, No. 5, pp. 32-100.

Collins, M. P., and Mitchell, D., 1991, *Prestressed Concrete Structures*, Prentice Hall, Englewood Cliffs, NJ, 766 pp.

Collins, M. P.; Walsh, P. F.; Archer, F. E.; and Hall, A. S., 1968a, "Ultimate Strength of Reinforced Concrete Beams Subjected to Combined Torsion and Bending," *Torsion of Structural Concrete*, SP-18, American Concrete Institute, Farmington Hills, MI, pp. 279-402.

Collins, M. P.; Walsh, P. F.; Archer, F. E.; and Hall, A. S., 1968b, "Reinforced Concrete in Torsion," *UNICIV Report* No. R-31, University of New South Wales, NSW, Australia, 338 pp.

Delhumeau, G., 1999, *L'Invention du Beton arme: Hennebique, 1890-1914*, Horma, Institut Francais d'Architecture, Paris, France, 344 pp.

Duleau, A., 1820, *Essai Theorique et Experimental sur la Resistance du Fer Forge*, Paris, France (in French).

Elfgren, L., 1972a, "Reinforced Concrete Beam Loaded in Torsion, Bending and Shear," *Publication 71:3*, Division of Concrete Structures, Chalmers University of Technology, Göteborg, Sweden, 249 pp.

Elfgren, L., 1972b, "Reinforced Concrete Beams Loaded in Combined Torsion, Bending and Shear. A Study of the Ultimate Load-Carrying Capacity," PhD thesis, *Publication 71:1*, Division of Concrete Structures, Chalmers University of Technology, second edition, Göteborg, Sweden, 230 pp.

Elfgren, L., 1979, "Torsion-Bending-Shear in Concrete Beams: A Kinematic Model," *Colloquium Copenhagen 1979: Plasticity in Reinforced Concrete, IABSE Reports, Zürich*, V. 29, pp. 111-118.

Elfgren, L.; Karlsson, I.; and Losberg, A., 1974a, "Torsion-Bending-Shear Interaction for Concrete Beams," *Journal of the Structural Division*, V. 100, No. 8, pp. 1657-1676.

Elfgren, L.; Karlsson, I.; and Losberg, A., 1974b, "Nodal Forces in the Analysis of the Ultimate Torsional Moment for Rectangular Beams," *Magazine of Concrete Research*, V. 26, No. 86, pp. 21-28; discussion in V. 27, No. 90, pp. 42-45.

Ersoy, U., and Ferguson, P. M., 1968, "Concrete Beams Subjected to Combined Torsion and Shear—Experimental Trends," *Torsion of Structural Concrete*, SP-18, American Concrete Institute, Farmington Hills, MI, pp. 441-460.

Ewida, A. A., and McMullen, A. E., 1981, "Torsion-Shear Interaction in Reinforced Concrete Members," *Magazine of Concrete Research*, V. 23, No. 115, pp. 113-122.

Gabrielsson, H., 1999, "Ductility in High Performance Concrete Structures: An Experimental Investigation and a Theoretical Study of Prestressed Hollow Core Slabs and Prestressed Cylindrical Pole Elements," PhD thesis 1999:15, Division of Structural Engineering, Luleå University of Technology, Luleå, 283 pp.

Ghoneim, M. G., and MacGregor, J. G., 1993, "Evaluation of Design Procedure for Torsion in Reinforced and Prestressed Concrete," *Report* No. 184, Department of Civil Engineering, University of Alberta, Edmonton, AB, Canada, 301 pp.

Greene, G., and Belarbi, A., 2009a, "Model for Reinforced Concrete Members under Torsion, Bending, and Shear. I: Theory," *Journal of Engineering Mechanics*, V. 135, No. 9, pp. 961-969.

Greene, G., and Belarbi, A., 2009b, "Model for Reinforced Concrete Members under Torsion, Bending, and Shear. II: Model Application and Validation," *Journal of Engineering Mechanics*, V. 135, No. 9, pp. 970-977.

Greene, G. G., 2006, "Behavior of Reinforced Concrete Girders under Cyclic Torsion and Torsion Combined with Shear: Experimental Investigation and Analytical Models," PhD thesis, Department of Civil, Architectural, and Environmental Engineering, University of Missouri at Rolla, Rolla, MO, 227 pp.

Greene, G. G., and Belarbi, A., 2006a, "Softened Truss Model for RC Torsional Members Under Combined Action," SEI Structures Congress, St. Louis, MO, 8 pp.

Greene, G. G., and Belarbi, A., 2006b, "Tension-Stiffened Softened Truss Model for RC Members under Torsion," *Proceedings of Second FIB Congress*, Naples, Italy, June 5-8, 12 pp.

Greene, G. G., and Belarbi, A., 2008, "Performance of Box Bridge Girders under Cyclic Torsion and Shear," *Proceedings of PCI National Bridge Conference*, Orlando, FL, Paper No. 73, Oct. 4-7, 25 pp.

Gvozdev, A. A., 1938, "Determination of the Value of Failure Load for Statically Indeterminate Systems Subject to Plastic Deformations (in Russian)," *Conference on Plastic Deformations 1936*, B. G. Galerkin, ed., Akademia Nauk SSSR, Moscow and Leningrad, pp. 19-38.

Gvozdev, A. A., 1960, "The Determination of the Value of the Collapse Load for Statically Indeterminate Systems Subject to Plastic Deformations," *International Journal of Mechanical Sciences*, V. 1, No. 1, pp. 322-335.

Hamilton, M. E., 1966, "Strength of T- and L- Section Beams with Longitudinal Reinforcement in Combined Torsion, Bending, and Shear," Master's thesis, Department of Civil Engineering, University of Washington at Seattle, Seattle, WA.

Hassan, T.; Lucier, G.; Rizkalla, S.; and Zia, P., 2007, "Modeling of L-Shaped Precast Prestressed Concrete Spandrels," *PCI Journal*, V. 52, No. 2, pp. 78-92.

Hillerborg, A., 1960, "A Plastic Theory for the Design of Reinforced Concrete Slabs," *Sixth ABSE Congress International Association for Bridges and Structural Engineering*, Zurich, Switzerland.

Horne, M. R., 1949, "Fundamental Propositions in the Plastic Theory of Structures," *Journal of the Institute of Civil Engineers (London)*, V. 34, pp. 174-177.

Hsu, T. T. C., 1968a, "Torsion of Structural Concrete—Plain Concrete Rectangular Sections," *Torsion of Structural Concrete*, SP-18, American Concrete Institute, Farmington Hills, MI, pp. 203-238.

Hsu, T. T. C., 1968b, "Torsion of Structural Concrete—Behavior of Reinforced Concrete Rectangular Members," *Torsion of Structural Concrete*, SP-18, American Concrete Institute, Farmington Hills, MI, Jan., pp. 261-306.

Hsu, T. T. C., 1968c, "Ultimate Torque of Reinforced Rectangular Beams," *Journal of the Structural Division*, V. 94, pp. 485-510.

Hsu, T. T. C., 1968d, "Torsion of Structural Concrete—Interaction Surface for Combined Torsion, Shear and Bending in Beams without Stirrups," *ACI Journal*, V. 65, No. 1, Jan., pp. 51-60.

Hsu, T. T. C., 1968e, "Torsion of Structural Concrete—Uniformly Prestressed Rectangular Members without Web Reinforcement," *PCI Journal*, V. 13, No. 2, pp. 34-44.

Hsu, T. T. C., 1984, *Torsion of Reinforced Concrete*, Van Nostrand Reinhold, New York, pp. 516-523.

Hsu, T. T. C., 1988, "Softening Truss Model Theory for Shear and Torsion," *ACI Structural Journal*, V. 85, No. 6, Nov.-Dec., pp. 624-635.

Hsu, T. T. C., 1990, "Shear Flow Zone in Torsion of Reinforced Concrete," *Journal of Structural Engineering*, V. 116, No. 11, pp. 3206-3226.

Hsu, T. T. C., 1993, *Unified Theory of Reinforced Concrete*, CRC Press, Boca Raton, FL, 313 pp.

Hsu, T. T. C., 1997, "ACI Shear and Torsion Provisions for Prestressed Hollow Girders," *ACI Structural Journal*, V. 94, No. 6, Nov.-Dec., pp. 787-799.

Hsu, T. T. C.; Belarbi, A.; and Pang, X. B., 1995a, "A Universal Panel Tester," *Journal of Testing and Evaluation*, V. 23, No. 1, pp. 41-49.

Hsu, T. T. C., and Burton, K., 1974, "Design of Reinforced Concrete Spandrel Beams," *Journal of the Structural Division*, V. 100, No. 1, pp. 209-229.

Hsu, T. T. C., and Hwang, C. S., 1977, "Torsional Limit Design of Spandrel Beams," *ACI Journal*, V. 74, No. 2, Feb., pp. 71-79.

Hsu, T. T. C.; Mansour, M. Y.; Mo, Y. L.; and Zhong, J., 2006, "Cyclic Softened Membrane Model for Nonlinear Finite Element Analysis of Concrete Structures," *Finite Element Analysis of Reinforced Concrete Structures*, SP-237, American Concrete Institute, Farmington Hills, MI, pp. 71-98.

Hsu, T. T. C., and Mo, Y. L., 1983, "Softening of Concrete in Torsional Members," *Research Report No. ST-TH-001-83*, Department of Civil Engineering, University of Houston, Houston, TX, Mar., 107 pp.

Hsu, T. T. C., and Mo, Y. L., 1985a, "Softening of Concrete in Torsional Members—Theory and Tests," *ACI Journal*, V. 82, No. 3, May-June, pp. 290-303.

Hsu, T. T. C., and Mo, Y. L., 1985b, "Softening of Concrete in Torsional Members—Prestressed Concrete," *ACI Journal*, V. 82, No. 5, Sept.-Oct., pp. 603-615.

Hsu, T. T. C., and Mo, Y. L., 1985c, "Softening of Concrete in Torsional Members—Design Recommendations," *ACI Journal*, V. 82, No. 4, July-Aug., pp. 443-452.

Hsu, T. T. C., and Mo, Y. L., 2010, *Unified Theory of Concrete Structures*, John Wiley and Sons, Ltd, London, UK, pp. 60-69.

Hsu, T. T. C., and Zhang, L. X., 1997, "Nonlinear Analysis of Membrane Elements by Fixed- Angle Softened-Truss Model," *ACI Structural Journal*, V. 94, No. 5, Sept.-Oct., pp. 483-492.

Hsu, T. T. C.; Zhang, L. X.; and Gomez, T., 1995b, "A Servo-Control System for Universal Panel Tester," *Journal of Testing and Evaluation*, V. 23, No. 6, pp. 424-430.

Hsu, T. T. C., and Zhu, R. R. H., 2002, "Softened Membrane Model for Reinforced Concrete Elements in Shear," *ACI Structural Journal*, V. 99, No. 4, July-Aug., pp. 460-469.

Humphreys, R., 1957, "Torsional Properties of Prestressed Concrete," *The Structural Engineer*, V. 35, No. 6, pp. 213-224.

Hwang, C. S., 1978, *Warping Torsion of Reinforced Concrete Beams with Thin Walled Open Sections*, PhD dissertation, Department of Civil Engineering, University of Miami, Coral Gables, FL, 286 pp.

Hwang, C. S., and Hsu, T. T. C., 1983, "Mixed Torsion Analysis of Reinforced Concrete Channel Beams-A Fourier Series Approach," *ACI Journal*, V. 80, No. 5, Sept.-Oct., pp. 377-386.

Iravani, S., 1996, "Mechanical Properties of High-Performance Concrete," *ACI Materials Journal*, V. 93, No. 5, Sept.-Oct., pp. 416-426.

Johnston, D. W., 1971, "Hollow Prestressed Concrete Beams Under Combined Torsion, Bending and Shear," PhD thesis, Department of Civil Engineering, North Carolina State University, Raleigh, NC, 177 pp.

Johnston, D. W., and Zia, P., 1975, "Prestressed Box Beams Under Combined Loading," *Journal of the Structural Division*, V. 101, No. 7, pp. 1313-1331.

Jordaan, I.; Khalifa, M.; and McMullen, A. E., 1974, "Collapse of Curved Reinforced Concrete Beams," *Journal of the Structural Division*, V. 100, No. 11, pp. 2255-2269.

Karayannis, C. G., 1995, "Torsional Analysis of Flanged Concrete Elements with Tension Softening," *Computers & Structures*, V. 54, No. 1, pp. 97-110.

Karayannis, C. G., 2000, "Smeared Crack Analysis for Plain Concrete in Torsion," *Journal of Structural Engineering*, V. 126, No. 6, pp. 638-645.

Karayannis, C. G., and Chalioris, C. E., 2000, "Experimental Validation of Smeared Analysis for Plain Concrete in Torsion," *Journal of Structural Engineering*, V. 126, No. 6, pp. 646-653.

Karlsson, I., 1973, "Stiffness Properties of Reinforced Concrete Beams Loaded in Combined Torsion, Bending and Shear," PhD thesis, *Publication 73:1*, Division of Concrete Structures, Chalmers University of Technology, Göteborg, Sweden, 203 pp.

Kemp, E. L.; Sozen, M. A.; and Siess, C. P., 1961, "Torsion of Reinforced Concrete," *Structural Research Series No. 226*, University of Illinois, Urbana, IL.

Klein, G. J., 1986, "Design of Spandrel Beams," *PCI Specially Funded R & D Program Research Project No. 5 (PCISFRAD #5)*, Prestressed Concrete Institute, Chicago, IL; summary paper in *PCI Journal*, V. 31, No. 5, Sept.-Oct., pp. 76-124.

Klus, J., 1968, "Ultimate Strength of Reinforced Concrete Beams in Combined Torsion and Shear," *ACI Journal*, V. 65, No. 3, Mar., pp. 210-216.

Kordina, K., and Teutsch, M., 1983, "Bemessung von Stahlbeton- und Spannbetonbalken unter kombinierter Beanspruchung infolge von Biegung, Querkraft und Torsion," *Beton- und Stahlbetonbau*, V. 78, Heft 1, pp. 1-6. (in German)

Kordina, K., and Teutsch, M., 1985, "Versuche an Konstruktionsleichtbetonbauteilen unter kombinierter Beanspruchung aus Torsion, Biegung und Querkraft," *Deutscher Ausschuß für Stahlbeton (DafStb)*, H. 362, W. Ernst u. Sohn, Berlin, 83 pp.

Koutchoukali, N. E., and Belarbi, A., 2001, "Torsion of High-Strength Reinforced Concrete Beams and Minimum Reinforcement Requirement," *ACI Structural Journal*, V. 98, No. 4, July-Aug., pp. 462-469.

Krpan, P., and Collins, M. P., 1981a, "Predicting Torsional Response of Thin-Walled Open RC Members," *Journal of the Structural Division*, V. 107, pp. 1107-1127.

Krpan, P., and Collins, M. P., 1981b, "Testing Thin-Walled Open RC Structure in Torsion," *Journal of the Structural Division*, V. 107, pp. 1129-1140.

Lampert, P., and Collins, M. P., 1972, "Torsion, Bending, and Confusion—An Attempt to Establish the Facts," *ACI Journal*, V. 69, No. 8, Aug., pp. 500-504.

Lampert, P., and Thürlimann, B., 1968, "Torsionsversuch an Stahlbetonbalken (Torsion Tests of Reinforced Concrete Beams)," *Bericht Nr. 6506-2*, 101 pp. (in German)

Lampert, P., and Thürlimann, B., 1969, "Torsion-Biege-Versuche an Stahlbetonbalken (Torsion-Bending Tests on Reinforced Concrete Beams)," *Bericht Nr. 6506-3*, Institute of Baustatik, ETH, Zurich, Switzerland (in German).

Lampert, P., and Thürlimann, B., 1971, "Ultimate Strength and Design of Reinforced Concrete Beams In Torsion and Bending," *Publications*, International Association of Bridge and Structural Engineering (Zurich), V. 31-I, pp. 107-131.

Leonhardt, F., and Schelling, G., 1974, "Torsionsversuche an Stahlbetonbalken," *Deutscher Ausschuß für Stahlbeton (DAfStb)*, Heft 239, W. Ernst u. Sohn, Berlin, 122 pp.

Leonhardt, F.; Walther, R.; and Vogler, O., 1968, "Torsions- und Schubversuche an vorgespannten Hohlkastenträgern," *Deutscher Ausschuß für Stahlbeton (DAfStb)*, Heft 202, Wilhelm Ernst & Sohn, Berlin, pp. 1-78.

Lessig, N. N., 1959, "Determination of Load-Carrying Capacity of Rectangular Reinforced Concrete Elements Subjected to Flexure and Torsion," Trudy No. 5, Institut Betona i Zhelezobetona (Concrete and Reinforced Concrete

Institute), Moscow, pp. 5-28 (in Russian). Translated by Portland Cement Association, Foreign Literature Study No. 371 (Available from S. L. A. Translation Center, John Crerar Library Translation Center, Chicago, IL).

Leu, L. J., and Lee, Y. S., 2000, "Torsion Design Charts for Reinforced Concrete Rectangular Members," *Journal of Structural Engineering*, V. 126, No. 2, pp. 210-218.

Lim, S. N., and Mirza, M. S., 1968, discussion of "Reinforced Concrete T-Beams without Stirrups under Combined Moment and Torsion," by D. J. Victor and P. M. Fergusson, *ACI Journal*, V. 65, No. 7, July, pp. 560-563.

Logan, D. R., 2007, "L-Spandrels: Can Torsional Distress Be Induced by Eccentric Vertical Loading?" *PCI Journal*, V. 52, No. 2, Mar.-Apr., pp. 46-61.

Lubliner, J., 1990, *Plasticity Theory*, Macmillan Publishing Co., New York, 495 pp.

Lüchinger, P., 1977, "Bruchwiderstand von Kastenträgern aus Stahlbeton unter Torsion, Biegung und Querkraft (Ultimate Strength of Box-Girder Beams in Reinforced Concrete loaded in Torsion, Bending and Shear)," Institut für Baustatik und Konstruktion, ETH Zürich, Bericht Nr 69, Birkhäuser Verlag, Basel und Stuttgart, 107 pp.

Lucier, G.; Rizkalla, S.; Zia, P.; and Klein, G., 2007, "Precast Concrete, L-Shaped Spandrels Revisited: Full-Scale Tests," *PCI Journal*, V. 52, No. 2, pp. 62-76.

Lucier, G.; Walter, C.; Rizkalla, S.; and Zia, P., 2010, "Development of a Rational Design Methodology for Precast Slender Spandrel Beams," *Research Report No. IS-09-10*, Construction Facilities Laboratory, Department of Civil, Construction, and Environmental Engineering, North Carolina State University, Raleigh, NC.

MacGregor, J. G., and Ghoneim, M. G., 1995, "Design for Torsion," *ACI Structural Journal*, V. 92, No. 2, Mar.-Apr., pp. 211-218.

Mansour, M., and Hsu, T. T. C., 2005a, "Behavior of Reinforced Concrete Elements under Cyclic Shear: Part I—Experiments," *Journal of Structural Engineering*, V. 131, No. 1, pp. 44-53.

Mansour, M., and Hsu, T. T. C., 2005b, "Behavior of Reinforced Concrete Elements under Cyclic Shear: Part II—Theoretical Model," *Journal of Structural Engineering*, V. 131, No. 1, pp. 54-65.

Mansour, M.; Lee, J. Y.; and Hsu, T. T. C., 2001, "Cyclic Stress-Strain Curves of Concrete and Steel Bars in Membrane Elements," *Journal of Structural Engineering*, V. 127, No. 12, pp. 1402-1411.

Mardukhi, J., 1974, *The Behaviour of Uniformly Prestressed Concrete Box Beams in Combined Torsion and Bending*, Master's thesis, University of Toronto, Toronto, ON, Canada, 73 pp.

Marti, P., and Kong, K., 1987, "Response of Reinforced Concrete Slab Elements to Torsion," *Journal of Structural Engineering*, V. 113, No. 5, pp. 976-993.

Marti, P.; Leesti, P.; and Khalifa, W., 1987, "Torsion Tests on Reinforced Concrete Slab Elements," *Journal of Structural Engineering*, V. 113, No. 5, pp. 994-1010.

Mattock, A. H., 1979, "Flexural Strength of Prestressed Concrete Sections by Programmable Calculator," *PCI Journal*, V. 24, No. 1, pp. 32-54.

McGee, W. D., and Zia, P., 1973, "Prestressed Concrete Members Under Torsion, Shear, and Bending," *Research Report*, Department of Civil Engineering, North Carolina State University, Raleigh, NC, 212 pp.

McMullen, A. E., and Daniel, H. R., 1975, "Torsional Strength of Longitudinally Reinforced Concrete Beams Containing an Openings," *ACI Journal*, V. 72, No. 8, Aug., pp. 415-420.

McMullen, A. E., and El-Degwy, W. M., 1985, "Prestressed Concrete Tests Compared with Torsion Theories," *PCI Journal*, V. 30, No. 5, pp. 96-127. (discussion by Hsu and Mo, V. 31, No. 5, Oct., pp. 160-164).

McMullen, A. E., and Warwaruk, J., 1967, "The Torsional Strength of Rectangular Reinforced Concrete Beams Subjected to Combined Loading," *Report* No. 2, Department of Civil Engineering, University of Alberta, Edmonton, AB, Canada, 162 pp.

McMullen, A. E., and Woodhead, H. R., 1973, "Experimental Study of Prestressed Concrete Under Combined Torsion, Bending, and Shear," *PCI Journal*, V. 18, No. 5, pp. 85-100.

Melan, E., 1938, Der Spannungszustand eines Mieseschen-Henckyschen Kontinuums bei veränderlicher Belastung. Sitzungsberichte, Akademie der Wissenschaften in Wien, Abt. II/a, V. 147, pp. 73-87.

Mirza, M. S., and McCutcheon, J. O., 1968, "discussion of "Torsion of Structural Concrete—Interaction Surface for Combined Torsion, Shear and Bending in Beams without Stirrups,"," *ACI Journal*, V. 65, No. 7, July, pp. 567-570.

Mitchell, D., and Collins, M. P., 1974, "Diagonal Compression Field Theory—A Rational Model for Structural Concrete in Pure Torsion," *ACI Journal*, V. 71, No. 8, Aug., pp. 396-408.

Mitchell, D., and Collins, M. P., 1976, "Detailing for Torsion," *ACI Journal*, V. 73, No. 9, Sept., pp. 506-511.

Mitchell, D., and Collins, M. P., 1978, "Influence of Prestressing on Torsional Response of Concrete Beams," *PCI Journal*, V. 23, No. 3, pp. 54-73.

Mörsch, E., 1902, *Der Eisenbetonbau, seine Anwendung und Theorie*, Wayss and Freytag, A. G., Im Selbstverlag der Firma, Neustadt, a. d.; Haardt, 118 pp. (English translation by E. P. Goodrich, McGraw-Hill Book Co., New York, 368 pp.).

Müller, P., 1976, "Failure Mechanisms for Reinforced Concrete Beams in Torsion and Bending," *Publications*, International Association for Bridge and Structural Engineering (IABSE), V. 36-II, Zürich, pp. 147-163.

Müller, P., 1978, "Platische Berechnung von Stahlbetonscheiben und Balken (Plastic Analysis of Walls and Beams of Reinforced Concrete)," Institut für Baustatik und Konstruktion, ETH Zürich, Bericht Nr 83, Birkhauser Verlag, Basel und Stuttgart, 160 pp.

National Association of Cement Users, 1910, "Standard Building Regulations for the Use of Reinforced Concrete," *NACU Standard* No. 4, American Concrete Institute, Farmington Hills, MI, pp. 349-361.

Navier, C. L., 1826, "Résume des Leçons Données a l'Ecole des Ponts et Chaussées sur l'Application de la Mécanique a l'Etablissement des Constructions et des Machines," Première partie, "Continuant les Leçons sur la Résistance des Matériaux et sur l'Etablissement des Constructions en Terre, en Maçonnerie et en Charpente," Firmin Didet, Paris (in French).

Nielsen, M. P., 1967, "Om Forskydningsarmering i Jernbetonbjaelker (On Shear Reinforcement in Reinforced Concrete Beams)," *Bygningsstatiske Meddelelser*, V. 38, No. 2, pp. 33-58.

Nylander, H., 1945, "Torsion and Torsional Restraint of Concrete Structures (Vrindning och Vridningsinspanning vid Betong-Konstruktioner)," *Bulletin* No. 3, Statens Kommittee for Byggnadsforskning, Stockholm, Sweden.

Onsongo, W. M., 1978, "The Diagonal Compression Field Theory for Reinforced Concrete Beams Subjected to Combined Torsion, Flexure, and Axial Load," PhD thesis, Department of Civil Engineering, University of Toronto, Toronto, ON, Canada, 246 pp.

Palermo, D., and Vecchio, F. J., 2003, "Compression Field Modeling of Reinforced Concrete Subjected to Reversed Loading: Formulation," *ACI Structural Journal*, V. 100, No. 5, Sept.-Oct., pp. 616-625.

Palermo, D., and Vecchio, F. J., 2004, "Compression Field Modeling of Reinforced Concrete Subjected to Reversed Loading: Verification," *ACI Structural Journal*, V. 101, No. 2, Mar.-Apr., pp. 155-164.

Pang, X. B., and Hsu, T. T. C., 1995, "Behavior of Reinforced Concrete Membrane Elements in Shear," *ACI Structural Journal*, V. 92, No. 6, Nov.-Dec., pp. 665-679.

Pantazopoulou, S., and Moehle, S., 1990, "Truss Model for 3-D Behavior of R.C. Exterior Connections," *Journal of Structural Engineering*, V. 116, No. 2, pp. 298-315.

PCI, 1978, *PCI Design Handbook*, second edition, Prestressed Concrete Institute, Chicago, IL.

PCI, 1999, *PCI Design Handbook*, fifth edition, Prestressed Concrete Institute, Chicago, IL, 540 pp.

Petersson, T., 1972, "Shear and Torsion in Reinforced Concrete Beams. A Study of Critical Load and Torsional Stiffness," *Document D10.1972*, National Swedish Building Research, Stockholm, Sweden, 276 pp.

Rabbat, B. G., and Collins, M. P., 1977, "The Computer Aided Design of Structural Concrete Sections Subjected to Combined Loading," *Computers & Structures*, V. 7, No. 2, pp. 229-236.

Rabbat, B. G., and Collins, M. P., 1978, "Variable Angle Space Truss Model for Structural Concrete Members Subjected to Complex Loading," *Douglas McHenry Symposium on Concrete and Concrete Structures*, SP-55, American Concrete Institute, Farmington Hills, MI, pp. 547-587.

Rahal, K. N., 2000a, "Shear Strength of Reinforced Concrete, Part I: Membrane Elements Subjected to Pure Shear," *ACI Structural Journal*, V. 97, No. 1, Jan.-Feb., pp. 86-93 (closure to discussion, V. 97, No. 6, Nov.-Dec., pp. 910-913.)

Rahal, K. N., 2000b, "Torsional Strength of Reinforced Concrete Beams," *Canadian Journal of Civil Engineering*, V. 27, No. 3, pp. 445-453.

Rahal, K. N., and Collins, M. P., 1995a, "Analysis of Sections Subjected to Combined Shear and Torsion—A Theoretical Investigation," *ACI Structural Journal*, V. 92, No. 4, July-Aug., pp. 459-469.

Rahal, K. N., and Collins, M. P., 1995b, "Effect of the Thickness of Concrete Cover on the Shear-Torsion Interaction – An Experimental Investigation," *ACI Structural Journal*, V. 92, No. 3, May-June, pp. 334-342.

Rahal, K. N., and Collins, M. P., 1996, "Simple Model for Predicting Torsional Strength of Reinforced and Prestressed Concrete Sections," *ACI Structural Journal*, V. 93, No. 6, Nov.-Dec., pp. 658-666.

Rahal, K. N., and Collins, M. P., 2003a, "Combined Torsion and Bending in Reinforced and Prestressed Concrete Beams," *ACI Structural Journal*, V. 100, No. 2, Mar.-Apr., pp. 157-165.

Rahal, K. N., and Collins, M. P., 2003b, "An Experimental Evaluation of the ACI and AASHTO-LRFD Design Provisions for Combined Shear and Torsion," *ACI Structural Journal*, V. 100, No. 3, May-June, pp. 277-282.

Rangan, B. V.; Natarajan, P. R.; and Parameswaran, V. S., 1978, "Lower Bound Limit Design of Concrete Structures," *Journal of the Structural Division*, V. 104, No. 12, pp. 1901-1906.

Rasmussen, L. J., and Baker, G., 1995, "Torsion in Reinforced Normal and High-Strength Concrete Beams—Part 1: Experimental Test Series," *ACI Structural Journal*, V. 92, No. 1, Jan.-Feb., pp. 55-63.

Raths, C. H., 1984, "Spandrel Beam Behavior and Design," *PCI Journal*, V. 29, No. 2, pp. 62-131.

Rausch, E., 1929, *Design of Reinforced Concrete in Torsion and Shear*, (Berechnung des Eisenbetons gegen Verdrehung [Torsion] und Abscheren), Technische Hochschule, Berlin, Germany, 53 pp. (in German); second edition in 1938; third edition in 1953, "Drillung (Torsion) Schub und Scheren in Stahlbelonbau, Deutcher Ingenieur-Verlag GmbH," Dusseldorf, Germany, 168 pp.

RILEM Committee QFS, 2004, "Quasibrittle Fracture Scaling and Size Effect," *Materials and Structures*, V. 37, No. 272, pp. 547-586.

Ritter, W., 1899, "Die Bauweise Hennebique," *Schweizerische Bauzeitung*, V. 33, No. 7, pp. 59-61.

Robinson, J. R., and Demorieux, J. M., 1972, "Essais de Traction-Compression sur Modèles d'Ame de Poutre en Béton Arme," *IRABA Report*, Institut de Recherches Appliquées du Béton de L'Ame, Part 1, June 1968, 44 pp; "Resistance Ultime du Béton de L'âme de Poutres en Double Te en Béton Arme," Part 2, May, 53 pp.

Sadowsky, M., 1949, "An Extension of the Sand Heap Analogy in Plastic Torsion Application to Cross-Section Having One or More Holes," *Journal of Applied Mechanics*, V. 62.

Saint-Venant, B., 1856, "Mémoire sur la torsion des prismes (lu à l'Académie le 13 juin 1853). *Mémoires Des Savants Etrangers*, Mémoires présentes par divers savants a l'Académie des Sciences, de l'institut Imperial de France et imprime par son ordre, V. 14, Imprimerie Impériale, Paris, pp.233-560.

Stevens, N. J.; Uzumeri, S. M.; and Collins, M. P., 1991, "Reinforced Concrete Subjected to Reversed Cyclic Shear—Experiments and Constitutive Model," *ACI Structural Journal*, V. 88, No. 2, Mar.-Apr., pp. 135-146.

Teutsch, M., 1980, "Trag- und Verformungsverhalten von Stahlbeton- und Spannbetonbalken mit rechteckigem Querschnitt unter kombinierter Beanspruchung aus Biegung, Querkraft und Torsion," dissertation, TU Carolo Wilhelmina Braunschweig, Fakultät für Bauwesen, 181 pp.

Thürlimann, B., 1978, "Plastic Analysis of Reinforced Concrete Beams," *Introductory Report*, IABSE Colloquium, Reports of the Working Commissions, International Association for Bridges and Structural Engineering (IABSE), Zürich, Switzerland, V. 28, Oct., pp. 71-90.

Thürlimann, B.; Marti, P.; Pralong, J.; Ritz, P.; and Zimmerli, B., 1983, *Anwendung der Plastizitatstheorie auf Stahlbeton (Plasticity in Concrete Structures)*, Vorlesung, ETH-Zurich.

Vecchio, F. J., 1999, "Towards Cyclic Load Modeling of Reinforced Concrete," *ACI Structural Journal*, V. 96, No. 2, Mar.-Apr., pp. 193-201.

Vecchio, F. J., and Collins, M. P., 1981, "Stress-Strain Characteristic of Reinforced Concrete in Pure Shear," *IABSE Colloquium, Advanced Mechanics of Reinforced Concrete*, Delft, Final Report, International Association of Bridge and Structural Engineering, Zurich, Switzerland, pp. 221-225.

Vecchio, F. J., and Collins, M. P., 1982, "The Response of Reinforced Concrete to In-Plane Shear and Normal Stresses," *Publication* No. 82-03, Department of Civil Engineering, University of Toronto, Toronto, ON, Canada, pp. 332.

Vecchio, F. J., and Collins, M. P., 1986, "The Modified Compression Field Theory for Reinforced Concrete Elements Subjected to Shear," *ACI Journal*, V. 83, No. 2, Mar.-Apr., pp. 219-231.

Vecchio, F. J., and Collins, M. P., 1993, "Compression Response of Cracked Reinforced Concrete," *Journal of Structural Engineering*, V. 119, No. 12, pp. 3590-3610.

Victor, D. J., and Ferguson, P. M., 1968, "Reinforced Concrete T-Beams without Stirrups under Combined Moment and Torsion," *ACI Journal*, V. 65, No. 1, Jan., pp. 29-36.

Vlasov, V. Z., 1961, *Thin Walled Elastic Beams*; second revised and augmented edition (1985) (in Russian); English translation, National Technical Information Service, Springfield, VA, 493 pp.

Wafa, F. F.; Shihata, S. A.; Akhtaruzzaman, A. A.; and Ashour, S. A., 1995, "Prestressed High-Strength Concrete Beams under Torsion," *Journal of Structural Engineering*, V. 121, No. 9, pp. 1280-1286.

Wagner, H., 1929, "Ebene Blechwandträger mit sehr dünnem Stegblech," *Zeitschrift für Flugtechnik und Motorluftschiffahrt*, V. 20, No. 8 to 12, Berlin, Germany.

Walsh, P. F.; Collins, M. P.; and Archer, F. E., 1967, "The Flexure-Torsion and Shear-Torsion Interaction Behaviour of Reinforced Concrete Beams," *Civil Engineering Transactions*, The Institution of Engineers, Australia, V. CE9, No. 2, pp. 313-320.

Walsh, P. F.; Collins, M. P.; Archer, F. E.; and Hall, A. S., 1966, "The Ultimate Strength Design of Rectangular Reinforced Concrete Beams Subjected to Combined Torsion, Bending and Shear," *Civil Engineering Transactions*, The Institution of Engineers, Australia, V. CE8, No. 2, pp. 143-157.

Wang, W., 1995, "Shear Strength and Fatigue Crack Propagation in Concrete by Energy Methods," PhD dissertation, Department of Civil and Environmental Engineering, New Jersey Institute of Technology, Newark, NJ.

Wang, W., and Hsu, T. T. C., 1997, "Limit Analysis of Reinforced Concrete Beams Subjected to Pure Torsion," *Journal of Structural Engineering*, V. 123, No. 1, pp. 86-94.

Wang, W.; Jiang, D.; and Hsu, T. T. C., 1993, "Shear Strength of Reinforced Concrete Deep Beams," *Journal of Structural Engineering*, V. 119, No. 8, pp. 2294-2312.

Yoo, C., and Heins, C., 1972, "Plastic Collapse of Horizontally Curved Girders," *Journal of the Structural Division*, V. 98, No. 4, pp. 899-913.

Yudin, V. K., 1962, "Determination of the Load-Carrying Capacity of Rectangular Reinforced Concrete Elements Subjected to Combined Torsion and Bending," *Beton i Zhelezobeton (Concrete and Reinforced Concrete)*, Moscow, No. 6, pp. 265-269.

Zararis, P. D., and Penelis, G. G., 1986, "Reinforced Concrete T-Beams in Torsion and Bending," *ACI Journal*, V. 83, No. 1, Jan.-Feb., pp. 145-155.

Zbirohowski-Koscia, K. F., 1968, "Stress Analysis of Cracked Reinforced Concrete Thin Walled Beams and Shells," *Magazine of Concrete Research*, V. 20, No. 65, pp. 213-220.

Zhang, L. X., and Hsu, T. T. C., 1998, "Behavior and Analysis of 100 MPa Concrete Membrane Elements," *Journal of Structural Engineering*, V. 124, No. 1, pp. 24-34.

Zhu, R. R. H., and Hsu, T. T. C., 2002, "Poisson Effect of Reinforced Concrete Membrane Elements," *ACI Structural Journal*, V. 99, No. 5, Sept.-Oct., pp. 631-640.

Zhu, R. R. H.; Hsu, T. T. C.; and Lee, J. Y., 2001, "Rational Shear Modulus for Smeared-Crack Analysis of Reinforced Concrete," *ACI Structural Journal*, V. 98, No. 4, July-Aug., pp. 443-451.

Zia, P., 1961, "Torsional Strength of Prestressed Concrete Members," *ACI Journal*, V. 57, No. 10, pp. 1337-1359.

Zia, P., 1970, "What Do We Know About Torsion in Concrete Members?" *Journal of the Structural Division*, V. 96, No. 6, pp. 1185-1199.

Zia, P., and Hsu, T. T. C., 1978, 2004, "Design for Torsion and Shear in Prestressed Concrete," Preprint No. 3424, a paper presented at ASCE Convention, Chicago, IL, 17 pp.; Updated version: *PCI Journal*, 2004, Precast/Prestressed Concrete Institute, V. 49, No. 3, pp. 34-42.

Zia, P., and McGee, W. D., 1974, "Torsion Design of Prestressed Concrete," *PCI Journal*, V. 19, No. 2, pp. 46-65.

ACI 446.1R-91

(Reapproved 1999)

An abstract:

Fracture Mechanics of Concrete: Concepts, Models and Determination of Material Properties

Reported by ACI Committee 446[§]

Zdenek P. Bažant[*†]
Chair

Vellore S. Gopalaratnam[*‡]
Secretary

Oral Buyukozturk[*]	Hideyuki Horii[*]	Sidney Mindess	Robert L. Sierakowski
Luigi Cedolin[*]	Jeremy Isenberg	Antoine E. Naaman[*]	Wimal Suaris[*]
David Darwin[‡]	Victor C. Li[*]	C. Dean Norman	Stuart E. Swartz[*†]
Manuel Elices[*‡]	Feng-Bao Lin[*]	Phillip A. Pfeiffer	Tatsuya Tsubaki
Shu-Jin Fang	Steven L. McCabe	Gilles Pijaudier-Cabot[‡]	C. Vipulanandan[*]
Walter Gerstles	Sheng-Taur Mau[‡]	Victor Saouma[*‡]	Methi Wecharatana[*]
Neil M. Hawkins	Jacky Marzas[‡]	Surendra P. Shah[*‡]	

The committee wishes to recognize the contributions of the following non-voting members:

Farhad Ansari[*]	Arne Hillerborg[*]	Pere C. Prat[*]	Hans W. Reinhardt[*]
Ravindra Gettu[‡]	B. L. Karihaloo[*]		

[*]Members of Subcommittee I (chaired by Bažant) that prepared this report.
[†]Principal authors.
[‡]Contributing authors.

[§]This report was approved by a vote of the full ACI Committee 446 in December 1989. It does not reflect research after that date.

The full report of ACI 446.1R-91 is available as a separate publication.

SYNOPSIS

In the first of its series of four state-of-the-art reports under preparation, the Committee describes the basic concepts of fracture mechanics of concrete, the existing theoretical models, and the methods for determining the material fracture parameters. Chapter 1 offers five reasons for introducing fracture mechanics into certain aspects of design of concrete structures, including some provisions: 1) a theoretical energy argument; 2) the need to achieve objectivity of finite element solutions, that is, eliminate spurious mesh sensitivity; 3) the progressive (propagating) nature of failure, implied whenever the load-deflection diagram lacks a yield plateau; 4) the need to rationally predict ductility and energy absorption capability; and most importantly, 5) the effect of structure size on the nominal strength (that is, nominal stress at maximum or ultimate load) as well as on ductility and energy absorption capability. The size effect is due to stored energy release into the fracture front, and is not governed by the Weibull-type statistical theory. Experimental evidence on the existence of the size effect, hitherto ignored in design practice and code provisions, is documented.

Chapter 2 gives a brief review of the necessary basic results of linear elastic fracture mechanics (LEFM). In concrete, departures from this classical theory are caused by the existence of distributed cracking (or damage) in a progressively softening fracture process zone that surrounds the tip of a continuous crack. In Chapter 3, nonlinear fracture models characterizing the softening stress-displacement or stress-strain relations (such as those of Hillerborg's fictitious crack model, crack band model, nonlocal strain-softening models, etc.) are described, and random particle simulation of aggregate microstructure is discussed. The principles of implementation of these models in finite element programs are also outlined. Chapter 4 presents simpler nonlinear fracture models that represent adaptations of linear elastic fracture mechanics, such as Jenq and Shah's model and the *R*-curve, along with the determination of geometry-dependent *R*-curves from the size effect law proposed by Bažant. This law, describing the approximate dependence of the nominal stress at maximum load on structure size, is discussed in Chapter 5, and structural response is characterized by the brittleness number.

Chapter 6 present in considerable detail the current methods for experimental and analytical determination of material fracture parameters, including the quasi-LEFM methods, RILEM (work-of-fracture) method, the Jenq-Shah and Karihaloo-Nallathambi methods, and the size-effect method. Experimental determination of the characteristic length for nonlocal continuum models and the strain-softening properties are then examined, and material parameters for Modes II and III, shear fractures and mixed-mode fracture, are also discussed. Chapter 7 then proceeds to describe various influencing factors, such as the loading rate, humidity and temperature, as well as the effect of cyclic loading. Chapter 8 is devoted to the effect of reinforcing bars and their bond slip on fracture propagation, and to fracture of fiber-reinforced concrete. Chapter 9 deals with more theoretical problems of modeling systems of interacting cracks. Attention is focused on systems of parallel growing cracks. Their stability decides the spacing and width of the cracks from a mechanics viewpoint.

It is concluded that, after a decade of rapid progress in research, the time appears ripe for introducing fracture mechanics into design practice. This should not only bring about more uniform safety margins, thus improving safety and economy of design, but also pave the way for safer and more efficient use of high-performance concretes and permit design extrapolations beyond the range of previous experiments and design.

Keywords: brittleness; concrete structures; cracking spacing and width; cracking; damage mechanics; design codes; ductility; failure; fiber-reinforced concrete; nonlocal continuum models; reinforced concrete; size effect; strain softening; structural design; testing methods; ultimate loads.

Finite Element Analysis of Fracture in Concrete Structures

Reported by ACI Committee 446

Vellore Gopalaratnam[1]
Chairman

Walter Gerstle[1,2]
Secretary and Subcommittee Co-Chairman

David Darwin[1,2]
Subcommittee Co-Chairman

Farhad Ansari
Zdenek P. Bažant[1,3]
Oral Buyukozturk[1,3]
Ignacio Carol[1]
Rolf Eligehausen
Shu-Jin Fang[1,3]
Ravindra Gettu
Toshiaki Hasegawa
Neil M. Hawkins
Anthony R. Ingraffea[1,3]
Jeremy Isenberg
Yeou-Sheng Jenq
Mohammad T. Kazemi
Neven Krstulovic
Victor C. Li
Jacky Mazars
Steven L. McCabe[1,3]
Christian Meyer
Hirozo Mihashi[1]
Richard A. Miller
Sidney Mindess
C. Dean Norman
Philip C. Perdikaris
Gilles Pijaudier-Cabot
Victor E. Saouma[1,3]
Wimal Suaris
Stuart E. Swartz[1,3]
Tianxi Tang
Tatsuya Tsubaki
Cumaraswamy Vipulanandan
Methi Wecharatana
Yunping Xi
Former member: Sheng-Taur Mao[1,3]

[1]Members of the Subcommittee that prepared this report
[2]Principal authors
[3]Contributing authors

Fracture is an important mode of deformation and damage in both plain and reinforced concrete structures. To accurately predict fracture behavior, it is often necessary to use finite element analysis. This report describes the state-of-the-art of finite element analysis of fracture in concrete. The two dominant techniques used in finite element modeling of fracture—the discrete and the smeared approaches—are described. Examples of finite element analysis of cracking and fracture of plain and reinforced concrete structures are summarized. While almost all concrete structures crack, some structures are fracture sensitive, while others are not. Therefore, in some instances it is necessary to use a consistent and accurate fracture model in the finite element analysis of a structure. For the most general and predictive finite element analyses, it is desirable to allow cracking to be represented using both the discrete and the smeared approaches.

Keywords: concrete; crack; cracking; damage; discrete cracking; finite element analysis; fracture; fracture mechanics; reinforced concrete; size effect; smeared cracking; structures.

CONTENTS

ACI 446.3R-97 became effective October 16, 1997.

CHAPTER 1—INTRODUCTION

In this report, the state-of-the-art in finite element modeling of concrete is viewed from a fracture mechanics perspective. Although finite element methods for modeling fracture are undergoing considerable change, the reader is presented with a snapshot of current thinking and selected literature on the topic.

1.1—Background

As early as the turn of the 19th century, engineers realized that certain aspects of concrete behavior could not be described or predicted based upon classical strength of materials techniques. As the discipline of fracture mechanics has developed over the course of this century (and indeed, is still developing), it has become clear that a correct analysis of many concrete structures must include the ideas of fracture mechanics.

The need to apply fracture mechanics results from the fact that classical mechanics of materials techniques are inadequate to handle cases in which severe discontinuities, such as cracks, exist in a material. For example, in a tension field, the stress at the tip of a crack tends to infinity if the material is assumed to be elastic. Since no material can sustain infinite stress, a region of inelastic behavior must therefore surround the crack tip. Classical techniques cannot, however, handle such complex phenomena. The discipline of fracture mechanics was developed to provide techniques for predicting crack propagation behavior.

Westergaard (1934) appears to have been the first to apply the concepts of fracture mechanics to concrete beams. With the advent of computers in the 1940s, and the subsequent rapid development of the finite element method (FEM) in the 1950s, it did not take long before engineers attempted to analyze concrete structures using the FEM (Clough 1962; Ngo and Scordelis 1967; Nilson 1968; Rashid 1968; Cervenka and Gerstle 1971, 1972). However, even with the power of the FEM, engineers faced certain problems in trying to model concrete structures. It became apparent that concrete structures usually do not behave in a way consistent with the assumptions of classical continuum mechanics (Bažant 1976).

Fortunately, the FEM is sufficiently general that it can model continuum mechanical phenomena as well as discrete phenomena (such as cracks and interfaces). Engineers performing finite element analysis of reinforced concrete structures over the past thirty years have gradually begun to recognize the importance of discrete mechanical behavior of concrete. Fracture mechanics may be defined as that set of ideas or concepts that describe the transition from continuous to discrete behavior as separation of a material occurs. The two main approaches used in FEM analysis to represent cracking in concrete structures have been to 1) model cracks discretely (discrete crack approach); and 2) model cracks in a smeared fashion by applying an equivalent theory of continuum mechanics (smeared crack approach). A third approach involves modeling the heterogeneous constituents of concrete at the size scale of the aggregate (discrete particle approach) (Bažant et al. 1990).

Kaplan (1961) seems to have been the first to have performed physical experiments regarding the fracture mechanics of concrete structures. He applied the Griffith (1920) fracture theory (modified in the middle of this century to become the theory of linear elastic fracture mechanics, or LEFM) to evaluate experiments on concrete beams with crack-simulating notches. Kaplan concluded, with some reservations, that the Griffith concept (of a critical potential energy release rate or critical stress intensity factor being a condition for crack propagation) is applicable to concrete. His reservations seem to have been justified, since more recently it has been demonstrated that LEFM is not applicable to typical concrete structures. In 1976, Hillerborg, Modeer and Petersson studied the fracture process zone (FPZ) in front of a crack in a concrete structure, and found that it is long and narrow. This led to the development of the fictitious crack model (FCM) (Hillerborg et al. 1976), which is one of the simplest nonlinear discrete fracture mechanics models applicable to concrete structures.

Finite element analysis was first applied to the cracking of concrete structures by Clough (1962) and Scordelis and his coworkers Nilson and Ngo (Nilson 1967, 1968; Ngo and Scordelis 1967). Ngo and Scordelis (1967) modeled discrete cracks, as shown in Fig. 1.1, but did not address the problem of crack propagation. Nilson (1967) modeled progressive discrete cracking, not by using fracture mechanics techniques, but rather by using a strength-based criterion. The stress singularity that occurs at the crack tip was not modeled. Thus, since the maximum calculated stress near the tip of a crack depends upon element size, the results were mesh-dependent (nonobjective). Since then, much of the research and development in discrete numerical modeling of fracture of concrete structures has been carried out by Ingraffea and his coworkers (Ingraffea and Manu 1980; Saouma 1981; Ingraffea 1977, 1983, 1990; Gerstle 1982, 1986; Wawrzynek and Ingraffea 1987, 1989; Swenson and Ingraffea 1988; Martha et al. 1991) and by Hillerborg and coworkers (Hillerborg et al. 1976; Petersson 1981; Gustafsson 1985).

Another important approach to modeling of fracture in concrete structures is called the smeared crack model (Rashid 1968). In the smeared crack model, cracks are modeled by changing the constitutive (stress-strain) relations of the solid continuum in the vicinity of the crack. This approach has been used by many investigators (Cervenka and Gerstle 1972, Darwin and Pecknold 1976, Bažant 1976, Meyer and Bathe 1982, Chen 1982, Balakrishnan and Murray 1988). Bažant (1976) seems to have been the first to realize that, because of its strain-softening nature, concrete cannot be modeled as a pure continuum. Zones of damage tend to localize to a size scale that is of the order of the size of the aggregate. Therefore, for concrete to be modeled as a continuum, account must be taken of the size of the heterogeneous structure of the material. This implies that the maximum size of finite elements used to model strain softening behavior should be linked to the aggregate size. If the scale of the structure is small, this presents no particular problem. However, if the scale of the structure is large compared to the size of its internal structure (aggregate size), stress intensity factors (fundamental parameters in LEFM) may provide a more efficient method for modeling crack propagation than the smeared crack approach (Griffith 1920; Bažant 1976). Most structures of interest are of a size between these two extremes, and controversy currently exists as to which of these approaches (discrete fracture mechanics or smeared cracking continuum mechanics) is more effective. This report describes both the discrete and smeared cracking methods. These two approaches, however, are not mutually exclusive, as shown, for example, by Elices and Planas (1989).

When first used to model concrete structures, it was expected that the FEM could be used to solve many problems for which classical solutions were not available. However, even this powerful numerical tool has proven to be difficult to apply when the strength of a structure or structural element is controlled by cracking. When some of the early finite element analyses are studied critically in light of recent developments, they are found to be nonobjective or incorrect in terms of the current understanding of fracture mechanics, although many produced a close match with experimental results. It is now clear that any lack of success in these models was not due to a weakness in the FEM, but rather due to incorrect approaches used to model cracks. In many cases, success can be achieved only if the principles of fracture mechanics are brought to bear on the problem of cracking in plain and reinforced concrete. These techniques have not only proven to be powerful, but have begun to provide explanations for material behavior and predictions of structural response that have previously been poorly or incorrectly understood.

While some preliminary research has been performed in the finite element modeling of cracking in three-dimensional structures (Gerstle et al. 1987; Wawrzynek and Ingraffea 1987; Ingraffea 1990; Martha et al. 1991), the state-of-the-art in the fracture analysis of concrete structures seems currently to be generally limited to two-dimensional models of structures.

1.2—Scope of report

Several previous state-of-the-art reports and symposium proceedings discuss finite element modeling of concrete structures (ASCE Task Committee 1982; Elfgren 1989, *Computer-Aided* 1984, 1990; *Fracture Mechanics* 1989; Firrao 1990; van Mier et al. 1991; *Concrete Design* 1992, *Fracture* 1992; *Finite Element* 1993; *Computational Modeling* 1994; *Fracture and Damage* 1994; *Fracture* 1995). This report provides an overview of the topic, with emphasis on the application of fracture mechanics techniques. The two most commonly applied approaches to the FEM analysis of fracture in concrete structures are emphasized. The first approach, described in Chapter 2, is the discrete crack model. The second approach, described in Chapter 3, is the smeared crack model. Chapter 4 presents a review of the literature of applications of the finite element technique to problems involving cracking of concrete. Finally, some general conclusions and recommendations for future research are given in Chapter 5.

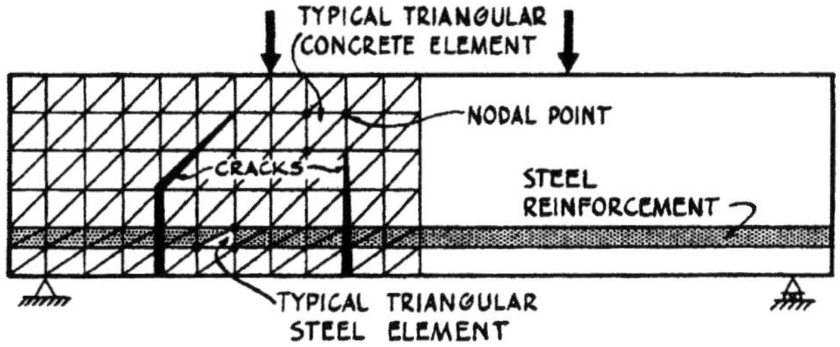

Fig. 1.1—The first finite element model of a cracked reinforced concrete beam (Ngo and Scordelis 1967).

No attempt is made to summarize all of the literature in the area of FEM modeling of fracture in concrete. There are several thousand references dealing simultaneously with the FEM, fracture, and concrete. An effort is made to crystallize the confusing array of approaches. The most important approaches are described in detail sufficient to enable the reader to develop an overview of the field. References to the literature are provided so that the reader can obtain further details, as desired. The reader is referred to ACI 446.1R for an introduction to the basic concepts of fracture mechanics, with special emphasis on the application of the field to concrete.

CHAPTER 2—DISCRETE CRACK MODELS

A discrete crack model treats a crack as a geometrical entity. In the FEM, unless the crack path is known in advance, discrete cracks are usually modeled by altering the mesh to accommodate propagating cracks. In the past, this remeshing process has been a tedious and difficult job, relegated to the analyst. However, newer software techniques now enable the remeshing process, at least in two-dimensional problems, to be accomplished automatically by the computer. A zone of inelastic material behavior, called the fracture process zone (FPZ), exists at the tip of a discrete crack, in which the two sides of the crack may apply tractions to each other. These tractions are generally thought of as nonlinear functions of the relative displacements between the sides of the crack.

2.1—Historical background

Finite element modeling of discrete cracks in concrete beams was first attempted by Ngo and Scordelis (1967) by introducing cracks into the finite element mesh by separating elements along the crack trajectory, as shown in Fig. 1.1. They did not, however, attempt to model crack propagation. Had they done so, they would have found many problems, starting with the fact that the stresses at the tips of the cracks increase without bound as the element size is reduced, and no convergence (of crack tip stresses) to a solution would have been obtained. Also, in light of the findings of Hillerborg et al. (1976) that a crack in concrete has a gradually softening region of significant length at its tip, it was inaccurate to model cracks with traction-free surfaces. It is notable that Ngo and Scordelis also grappled with the theoretically difficult issue of connecting the reinforcing elements with the concrete elements via "bond-link" elements.

Nilson (1967, 1968) was the first to consider a FEM model to represent *propagation* of discrete cracks in concrete structures. Quoting from his thesis (Nilson 1967):

> The present analysis includes consideration of progressive cracking. The uncracked member is loaded incrementally until previously defined cracking criteria are exceeded at one or more locations in the member. Execution terminates, and the computer output is subjected to visual inspection. If the average value of the principal tensile stress in two adjacent elements exceeds the ultimate tensile strength of the concrete, then a crack is defined between those two elements along their common edge. This is done by establishing two disconnected nodal points at their common corner or corners where there formerly was only one. When the principal tension acts at an angle to the boundaries of the element, then the crack is defined along the side most nearly normal to the principal tension direction.
>
> The newly defined member, with cracks (and perhaps partial bond failure), is then re-loaded from zero in a second loading stage, also incrementally applied to account for the nonlinearities involved. Once again the execution is terminated if cracking criteria are exceeded. The incremental extension of the crack is recorded, and the member loaded incrementally in the third stage, and so on. In this way, crack propagation may be studied and the extent of cracking at any stage of loading is obtained.

The problems associated with this approach to discrete crack propagation analysis are three-fold: (1) cracks in concrete structures of typical size scale develop gradually (Hillerborg et al. 1976), rather than abruptly; (2) the procedure forces the cracks to coincide with the predefined element boundaries; and (3) the energy dissipated upon crack propagation is unlikely to match that in the actual structure, resulting in a spurious solution.

In the 1970s, great strides were made in modeling of LEFM using the FEM. Chan, Tuba, and Wilson (1970) pointed out that a large number of triangular constant stress finite elements are required to obtain accurate stress intensity factor solutions using a displacement correlation technique (about 2000 degrees of freedom are required to obtain 5 percent accuracy in the stress intensity factor solution). At this time, singular finite elements had not yet been developed (singular elements exactly model the stress state at the tip of a crack). In their paper, Chan et al. pointed out that there were then three ways to obtain stress intensity factors from a finite element solution: (1) displacement correlation; (2) stress correlation; and (3) energy release rate methods (line integral or potential energy derivative approaches).

Wilson (1969) appears to have been the first to have developed a singular crack tip element. Shortly thereafter, Tracey (1971) developed a triangular singular crack tip finite element that required far fewer degrees of freedom than analysis with regular elements to obtain accurate stress intensity factor. At about the same time, Tong, Pian, and Lasry (1973) developed and experimented with hybrid singular crack tip elements (including stress-intensity factors, as well as displacement components, as degrees of freedom).

Jordan (1970) noticed that shifting the midside nodes along adjacent sides of an eight-noded quadrilateral toward the corner node by one-quarter of the element's side length caused the Jacobian of transformation to become zero at the corner node of the element. This led to the discovery by Henshell and Shaw (1975) and Barsoum (1976) that the shift allowed the singular stress field to be modeled exactly for an elastic material. Thus, standard quadratic element with midside nodes shifted to the quarter-points can be used as a $r^{-1/2}$ singularity element for modeling stresses at the tip of a crack in a linear elastic medium.

The virtual crack extension method for calculating Mode I stress intensity factors was developed independently by Hellen (1975) and Parks (1974). In this method, G, the rate of change of potential energy per unit crack extension, is calculated by a finite difference approach. This approach does not require the use of singular elements to obtain Mode I (opening mode) stress-intensity factors. Recently, it has been found that by decomposing the displacement field into symmetric and antisymmetric components with respect to the crack tip, the method may also be extended to calculate Mode II (sliding) and Mode III (tearing) energy release rates and stress-intensity factors (Sha and Yang 1990, Shumin and Xing 1990, Rahulkumar 1992).

Having developed the capability to compute stress intensity factors using the FEM, the next big step was to model linear elastic crack *propagation* using fracture mechanics principles. This was started for concrete by Ingraffea (1977), and continued by Ingraffea and Manu (1980), Saouma (1981), Gerstle (1982, 1986), Wawrzynek and Ingraffea (1986), and Swenson and Ingraffea (1988). These attempts were primarily aimed at facilitating the process of discrete crack propagation through automatic crack trajectory computations and semi-automatic remeshing to allow discrete crack propagation to be modeled. Currently, the main technical difficulties involved in modeling of discrete LEFM crack propagation are in the 3D regime. In 2D applications, automatic propagation and remeshing algorithms have been reasonably successful and are improving. In three-dimensional modeling, automatic remeshing algorithms are on the verge of being sufficiently developed to model general crack propagation, and computers are just becoming powerful enough to accurately solve problems with complex geometries caused by the propagation of a number of discrete cracks.

Another development in discrete crack modeling of concrete structures has been the realization that LEFM does not apply to structural members of normal size, because the FPZ in concrete is relatively large compared to size of the member. This has led to the development of finite element modeling of nonlinear discrete fracture—usually as the implementation of the fictitious crack model (FCM) (Hillerborg et al. 1976), in which the crack is considered to be a strain softening zone modeled by cohesive nodal forces or by interface elements [first developed by Goodman, Taylor, and Brekke (1968)].

Finally, there appear to be situations in which even the FCM seems inadequate to model realistic concrete behavior in the FPZ. In this case, a smeared crack model of some kind, as described in Chapter 3, becomes necessary.

2.2—Linear elastic fracture mechanics (LEFM)

Linear elastic fracture mechanics (LEFM) is an important approach to the fracture modeling of concrete structures, even though it is only applicable to very large (say several meters in length) cracks. For cracks that are smaller than this, LEFM over-predicts the load at which the crack will propagate. To determine whether LEFM may be used or whether nonlinear fracture mechanics is necessary for a particular problem, one must determine the size of the steady state fracture process zone (FPZ) compared to the least dimension associated with the crack tip (ACI 446.1R). The FPZ size and the crack tip least dimension are discussed next.

The FPZ may be defined as the area surrounding a crack tip within which inelastic material behavior occurs. The FPZ size grows as load is applied to a crack, until it has developed to the point that the (traction-free) crack begins to propagate. If the size of the FPZ is small compared to other dimensions in the structure, then the assumptions of LEFM lead to the conclusion that the FPZ will exhibit nonchanging characteristics as the crack propagates. This is called the steady state FPZ. The size of the steady state FPZ depends only upon the material properties. In concrete, as opposed to metals, the FPZ can often be thought of as an interface separation phenomenon, with little accompanying volumetric damage. The characteristics of the steady state FPZ depend upon the aggregate size, shape and strength, and upon microstructural details of the particular concrete under consideration. The FPZ was first studied in detail by Hillerborg, Modeer, and Petersson (1976). The size of the FPZ depends on the model used in the study. For example, in the analysis carried out by Ingraffea and Gerstle (1985) for normal strength concrete, the steady state FPZ ranged from 6 in. (150 mm) to 3 ft (1 m) in length.

The least dimension (L.D.) associated with a crack tip is best defined with the aid of Fig. 2.1 (Gerstle and Abdalla 1990). The least dimension is used to calculate an approximate radius surrounding the crack tip within which the singular stress field can be guaranteed to dominate the solution. The least dimension can be defined as the distance from the crack tip to the nearest discontinuity that might cause a local disturbance in the stress field. Fig. 2.1(a) shows the case where the crack tip L.D. is controlled by the proximity to the crack tip of a free surface. Fig. 2.1(b) shows the case where the least dimension is the crack length itself. Fig. 2.1(c) shows the case where the least dimension is controlled by the crack tip passing close by a reinforcing bar. [Of course, if the reinforcement is considered as a smeared (rather than discrete) constituent of the reinforced concrete composite, then it need not be modeled discretely, and the constitutive relations and the FPZ must correspondingly include the effect of the smeared reinforcing bars.] Fig. 2.1(d) shows the case where the least dimension is controlled by the size of the ligament (the remaining uncracked dimension of the member). In Fig. 2.1(e), the least dimension is governed by a kink in the crack. Finally, Fig. 2.1(f) shows an example of the least dimension being controlled by the radius of curvature of the crack.

As explained in Chapter 2 of ACI 446.1R, one of the fundamental assumptions of LEFM is that the size of the FPZ is negligible (say, no more than one percent of the least dimension associated with the crack tip). It is this assumption that allows for a theoretical stress distribution near the crack tip in linear elastic materials, in which the stress varies with $r^{-1/2}$, in which r is the distance from the crack tip. Stress-intensity factors K_I, K_{II}, and K_{III} are defined as the magnitudes of the singular stress fields for Mode I, Mode II, and Mode III cracks, respectively. If the FPZ is not small compared to the least dimension, then singular stress fields may not be assumed to exist, and consequently, K_I, K_{II}, and K_{III} are not defined for such a crack tip. In such a case, the FPZ must be modeled explicitly and a nonlinear fracture model is required.

As mentioned earlier, fracture process zones in concrete can be on the order of 1 ft (0.3 m) or more in length. For the great majority of concrete structures, least dimensions are less than several feet. Therefore, fracture in these types of structures must be modeled using nonlinear fracture mechanics. Only in very large concrete structures, for example, dams, is it possible to apply LEFM appropriately. For dams with large aggregate, possibly on the size scale of meters, LEFM may not be applicable because of the correspondingly larger size of the FPZ.

Even though it is recognized that LEFM is not applicable to typical concrete structures, it is appropriate to review the details of the finite element analysis of LEFM. Then, in Section 2.3, the finite element analysis of nonlinear discrete fracture mechanics will be presented.

2.2.1 *Fracture criteria:* K, G, *mixed-mode models*—Stress-intensity factors K_I, K_{II}, and K_{III} or energy release rates G_I, G_{II}, and G_{III} may be used in LEFM to predict crack equilibrium conditions and propagation trajectories. There are several theories that can be used to predict the direction of crack propagation. These include, for quasistatic problems, the maximum circumferential tensile stress theory (Erdogan and Sih 1963), the maximum energy release rate theory (Hussain et al. 1974), and the minimum strain energy density theory (Sih 1974). These theories all give practically the same crack trajectories and loads at which crack extension takes place, and therefore the theory of choice depends primarily upon convenience of implementation. Each of these theories may also be applied to dynamic fracture propagation problems (Swenson 1986). As in metals, cyclic fatigue crack propagation in concrete may be modeled with the Paris Model (Barsom and Rolfe 1987) in conjunction with the mixed-mode crack propagation theories just mentioned. However, it is rare that an unreinforced concrete structure is both (1) large enough to merit LEFM treatment and (2) subject to fatigue loading.

In most of the literature on discrete crack propagation in concrete structures, it has been considered necessary to model the stress singularity at a crack tip using singular elements. However, accurate results can also be obtained without modeling the stress singularity, but rather by calculating the energy release rates directly (Sha and Yang 1990; Rahulkumar 1992). However, for a comprehensive treatment, we discuss modeling of stress singularities next.

2.2.2 *FEM modeling of singularities and stress intensity factors*—Special-purpose singular finite elements have been created with stress-intensity factors included explicitly as

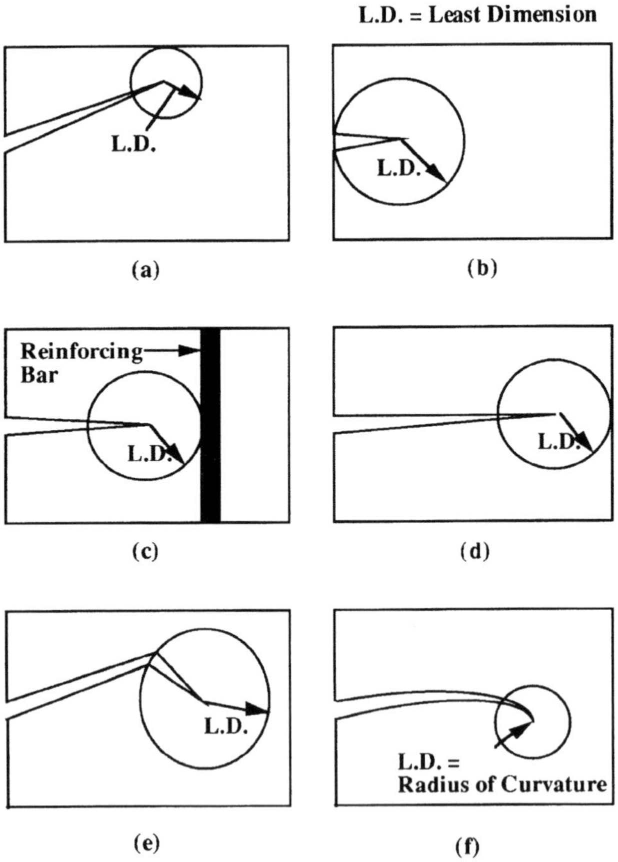

Fig. 2.1—Examples illustrating the concept of "least dimension (L.D.)" associated with a crack tip (Gerstle and Abdalla 1990).

degrees-of-freedom (Byskov 1970; Tong and Pian 1973; Atluri et al. 1975; Mau and Yang 1977). However, these are special-purpose hybrid elements that are not usually included in standard displacement-based finite element codes, and will not be discussed in further detail here. The most successful displacement-based elements are the Tracey element (Tracey 1971) and the quarterpoint quadratic triangular isoparametric element (Henshell and Shaw 1975; Barsoum 1976; Saouma 1981; Saouma and Schwemmer 1984). Most general purpose finite element codes unfortunately do not include the Tracey element, but they do include six noded triangular elements, which can then be used as singular quarterpoint crack tip elements.

After a finite element analysis has been completed, stress-intensity factors can be extracted by several approaches. The most accurate methods are the energy approaches: the J-integral, virtual crack extension, or stiffness derivative methods. However, these approaches are not as easy to apply for the case of mixed-mode crack propagation, and have been applied only rarely to three-dimensional problems (Shivakumar et al. 1988). Simpler to apply (for mixed-mode fracture mechanics) are the displacement correlation techniques. Because these techniques sample local displacements at various points, and correlate these with the theoretical displacement field associated with a crack tip, they are generally not as accurate as the energy approaches, which use integrated information. The displacement correlation techniques are usually used only when singular elements are employed, while the energy approaches are used for determining energy release rates for cracks that may or may not be discretized with the help of singular elements.

The displacement and stress correlation techniques assume that the finite element solution near the crack tip is of the same form as the singular near-field solution predicted by LEFM (Broek 1986). By matching the (known) finite element solution with the (known, except for K_I, K_{II}, and K_{III}) theoretical near-field LEFM solution, it is possible to calculate the stress-intensity factors. Since only three equations are needed to obtain the three stress-intensity factors, while many points that can be matched, there are many possible schemes for correlation. These include matching nodal responses only on the crack surfaces and least-squares fitting of all of the nodal responses associated with the singular elements.

The displacement correlation approach is more accurate than the stress correlation approach because displacements converge more rapidly than stresses using the FEM. Therefore only the displacement correlation approach is discussed in detail here (Shih et al. 1976).

Consider a linear elastic isotropic material with Young's modulus E and Poisson's ratio ν. For the case of plane strain, the near-field displacements (u,v), in terms of polar coordinates r and θ, shown in Fig. 2.2, are given by:

$$u = \frac{2K_I(1+\nu)}{E}\sqrt{\frac{r}{2\pi}}\cos\frac{\theta}{2}\left[1-2\nu+\sin^2\frac{\theta}{2}\right] + \tag{2.1}$$

$$\frac{2K_{II}(1+\nu)}{E}\sqrt{\frac{r}{2\pi}}\sin\frac{\theta}{2}\left[2-2\nu+\cos^2\frac{\theta}{2}\right]$$

$$v = \frac{2K_I(1+\nu)}{E}\sqrt{\frac{r}{2\pi}}\sin\frac{\theta}{2}\left[2-2\nu-\cos^2\frac{\theta}{2}\right] + \tag{2.2}$$

$$\frac{2K_{II}(1+\nu)}{E}\sqrt{\frac{r}{2\pi}}\cos\frac{\theta}{2}\left[-1+2\nu+\sin^2\frac{\theta}{2}\right]$$

in which u and v are parallel and perpendicular to the crack face, respectively.

Now consider a crack tip node surrounded by quarter point triangular elements shown in Fig. 2.2. Interpolating the radial coordinate, r, along the side AC, by using quadratic shape functions associated with nodes A, B, and C, and solving for the natural triangular area coordinate ξ_1 in terms of r, we obtain:

$$\xi_1 = 1 - \sqrt{\frac{r}{L}} \tag{2.3}$$

where L is the length of the side AC. Now interpolating the displacements along the side AC by using the computed displacement components at nodes A, B, and C, and using Eq. (2.3), we obtain the displacements along the crack surface AC in terms of r. These are given by:

$$u = u_A + (-3u_A + 4u_B - u_C)\sqrt{\frac{r}{L}} + (2u_A - 4u_B + 2u_C)\frac{r}{L} \quad (2.4)$$

$$v = v_A + (-3v_A + 4v_B - v_C)\sqrt{\frac{r}{L}} + (2v_A - 4v_B + 2v_C)\frac{r}{L} \quad (2.5)$$

Similarly the displacements alongside AE can be written as:

$$u = u_A + (-3u_A + 4u_D - u_E)\sqrt{\frac{r}{L}} + (2u_A - 4u_D + 2u_E)\frac{r}{L} \quad (2.6)$$

$$v = v_A + (-3v_A + 4v_D - v_E)\sqrt{\frac{r}{L}} + (2v_A - 4v_D + 2v_E)\frac{r}{L} \quad (2.7)$$

Subtracting Eq. 2.6 from 2.4 and subtracting 2.7 from 2.5, the crack opening displacement (COD) and crack sliding displacement (CSD) are computed as:

$$COD = (4v_B - v_C - 4v_D + v_E)\sqrt{\frac{r}{L}} + (-(4v_B + 2v_C) + 4v_D - 2v_E)\frac{r}{L} \quad (2.8)$$

$$CSD = (4u_B - u_C - 4u_D + u_E)\sqrt{\frac{r}{L}} + (-4u_B + 2u_C + 4u_D - 2u_E)\frac{r}{L} \quad (2.9)$$

Analytical solutions for COD and CSD can be obtained by evaluating the displacement components u and v given by Eq. (2.1) and (2.2) for $\theta = +\pi$ and $\theta = -\pi$ and subtracting the values at $\theta = -\pi$ from the values at $\theta = +\pi$. Equating the like terms in the finite element and the analytical COD and CSD profiles, the stress intensity factors are given by:

$$K_I = \sqrt{\frac{2\pi}{L}}\frac{E}{2(1+\nu)(3-4\nu)}[4(v_B - v_D) + v_E - v_C] \quad (2.10)$$

$$K_{II} = \sqrt{\frac{2\pi}{L}}\frac{E}{2(1+\nu)(3-4\nu)}[4(u_B - u_D) + u_E - u_C] \quad (2.11)$$

Thus by meshing the crack tip region with quarter-point quadratic triangular elements and solving for the displacements, the stress intensity factors can be computed by using Eq. (2.10) and (2.11). This technique does not require any special subroutines to develop the stiffness matrix for the singular elements. A single subroutine can be written to calculate the length L of the sides AC and AE, retrieve the displacement components at the nodes A, B, C, D, and E and thereby compute the stress-intensity factors using Eqs. 2.10 and 2.11.

Ingraffea and Manu (1980) have developed similar equations for the computation of stress-intensity factors in three dimensions with quarterpoint quadratic elements. In three dimensions, the crack tip is replaced by the crack front, the crack edge by the crack face.

Energy approaches for extracting stress-intensity factors make use of the fact that $K_I = [EG_I]^{1/2}$, $K_{II} = [EG_{II}]^{1/2}$, $K_{III} = [EG_{III}/(1+\nu)]^{1/2}$ for plane stress and $K_I = [EG_I/1-\nu^2)]^{1/2}$, $K_{II} = [EG_{II}/(1-\nu^2)]^{1/2}$, $K_{III} = [EG_{III}/(1+\nu)]^{1/2}$ for plane strain. Here, G_I, G_{II}, and G_{III} are the potential energy release rates created by collinear crack extension due to Mode I, Mode II, and Mode III deformations, respectively. In the simplest approach, the total energy release rate, $G = G_I + G_{II} + G_{III}$ can be calculated by performing an analysis, calculating the total potential energy, π_A, collinearly extending the crack by a small amount ∂a, reperforming the analysis to obtain π_B, and then using a finite difference to approximate G as $G = (\pi_A - \pi_B)/\partial a$. If G_I, G_{II}, and G_{III} are required separately, they can be calculated by decomposing the crack tip displacement and the stress fields into Mode I, Mode II, and Mode III components (Rahulkumar 1992).

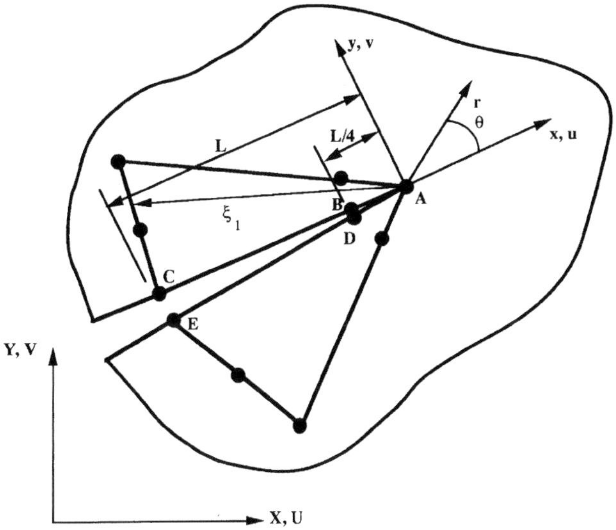

Fig. 2.2—Nomenclature for 2D quarter point singular isoparametric elements.

The stiffness derivative method for determination of the stress-intensity factor for Mode I (2D and 3D) crack problems was introduced by Parks (1974). The method is equivalent to the J-integral approach (described later).

With reference to Fig. 2.3, any set of finite elements that forms a closed path around the crack tip may be chosen. The simplest set to choose is the set of elements around the crack tip.

The stiffness derivative method involves determination of the stress-intensity factor from a calculation of the potential energy decrease per unit crack advance, G. For plane strain and unit thickness, the relation between K_I and G is

$$G = -\left[\frac{\partial\pi}{\partial a}\right]_{load} = \frac{(1-\nu^2)}{E}K_I^2 \quad (2.12)$$

in which P is the potential energy, a is the crack length, E is Young's modulus, and ν is Poisson's ratio.

Parks (1974) shows that the potential energy, π, in the problem is given by:

$$\pi = \frac{1}{2}\{u\}^T[K]\{u\} - \{u\}^T\{f\} \quad (2.13)$$

in which $[K]$ is the global stiffness matrix, and $\{f\}$ is the vector of prescribed nodal loads. Equation (2.13) is differen-

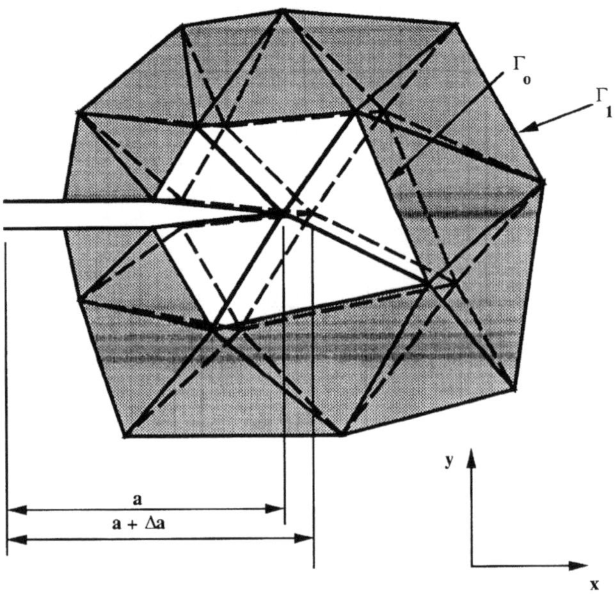

Fig. 2.3—Stiffness derivature approach for advancing nodes (Parks 1978).

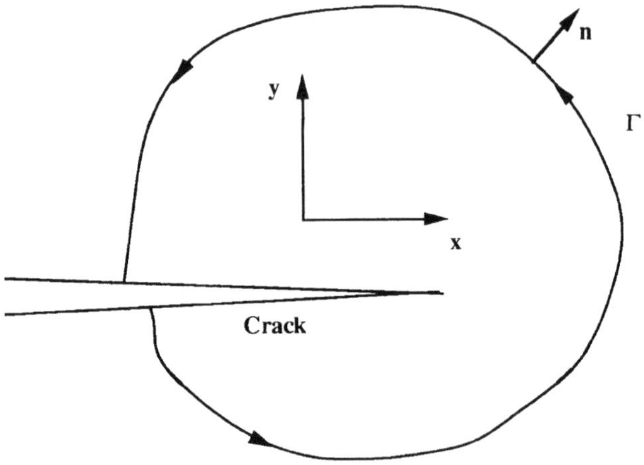

Fig. 2.4—J-Integral nomenclature (Rice 1968).

tiated with respect to crack length, a, to obtain the energy release rate as

$$\left[\frac{-\partial\pi}{\partial a}\right]_{load} = -\frac{1}{2}\{u\}^T\frac{\partial[K]}{\partial a}\{u\} - \{u\}^T\frac{\partial\{f\}}{\partial a} = K_I^2\frac{(1-\nu^2)}{E} \quad (2.14)$$

The matrix $\partial[K]/\partial a$ represents the change in the structure stiffness matrix per unit of crack length advance. The term $\partial\{f\}/\partial a$ is nil if the crack tip area is unloaded. The key to understanding the stiffness derivative method is to imagine representing an increment of crack advance with the mesh shown in Fig. 2.3 by rigidly translating all nodes on and within a contour Γ_o (see Fig. 2.3) about the crack tip by an infinitesimal amount Δa in the x-direction. All nodes on and outside of contour Γ_1 remain in their initial position. Thus the global stiffness matrix $[K]$, which depends on only individual element geometries, displacement functions, and elastic material properties, remains unchanged in the regions interior to Γ_o and exterior to Γ_1, and the only contributions to the first term of Eq. (2.14) come from the band of elements between the contours Γ_o and Γ_1. The structure stiffness matrix $[K]$ is the sum over all elements of the element stiffness matrices $[K_i]$. Therefore,

$$\frac{1}{2}\{u\}^T\frac{\partial[K]}{\partial a}\{u\} = \frac{1}{2}\{u\}^T\sum_{i=1}^{N_c}\frac{\partial[K_i^c]}{\partial a}\{u\} \quad (2.15)$$

in which $[K_i^c]$ is the element stiffness matrix of an element between the contours Γ_o and Γ_1, and N_c is the number of such elements. The derivatives of the element stiffness matrices can be calculated numerically by taking a finite difference:

$$\frac{\partial[K_i^c]}{\partial a} = \frac{\Delta[K_i^c]}{\Delta a} = \frac{1}{\Delta a}[[K_i^c]_{a+\Delta a} - [K_i^c]_a] \quad (2.16)$$

The method may be extended to mixed-mode cracks.

The J-Integral method (Rice 1968) for determining the energy release rate of a Mode I crack is useful for determining energy release rates, not only for LEFM crack propagation, but also for nonlinear fracture problems. For a two-dimensional problem, a path Γ is traversed in a counterclockwise sense between the two crack surfaces, as shown in Fig. 2.4. The J-integral is defined as:

$$J \equiv \int_\Gamma\left(w dx_2 - p_i\frac{\partial u_i}{\partial x_1}ds\right) \quad (2.17)$$

where summation over the range of repeated indices is understood.

Here, $w \equiv \int_0^\varepsilon \sigma_{ij} d\varepsilon_{ij}$, i,j = 1, 2, 3 is the strain energy density, s is the arc length, and p_i is the traction exerted on the body bounded by Γ and the crack surface. The J-integral is equal to the energy release rate G of the crack (Rice 1968).

The J-integral method can be relatively easily applied to a crack problem whose stress and displacement solution is known, and is not limited to linear materials. However, elasticity or pseudoelasticity along the contour, Γ, is a requirement (Rice 1968).

Alternate energy approaches for extraction of stress-intensity factors from three-dimensional problems have been developed (Shivakumar et al. 1988). Bittencourt et al. (1992) provide a single reference that compares the displacement correlation, the J-integral, and the modified crack closure integral techniques for obtaining stress-intensity factors.

When using triangular quarter-point elements to model the singularity at a crack tip, meshing guidelines have been suggested by a number of researchers (Ingraffea 1983; Saouma and Schwemmer 1984; Gerstle and Abdalla 1990). When using the displacement correlation technique to extract stress-intensity factors, the guidelines are summarized as follows:

1. Use a 2 x 2 (reduced) integration scheme (Saouma and Schwemmer 1984).

2. To achieve 5 percent maximum expected error in any stress component due to any mixed-mode problem, use at least eight approximately equiangular singular elements adjacent to the crack tip node. For 1 percent error, 16 singular elements should be used (Gerstle and Abdalla 1990).

3. There is an optimal size for the crack tip elements. If they are too small, they do not encompass the near-field region of the solution, and surrounding regular elements will be "wasted" modeling the near field. If they are too big, they do not model the far-field solution accurately. The singular elements should be related to the size of the region within which near-field solution is valid. For 5 percent accuracy in stress-intensity factors, the singular elements should be about 1/5 of the size of the least dimension associated with the crack tip. For one percent accuracy, the singular elements should be about 1/20 of the size of the least dimension associated with the crack tip (Gerstle and Abdalla 1990).

4. Regular quadratic elements should be limited in size, s, by their clear distance, b, from a crack tip. The ratio of s/b should not exceed unity to achieve 30 percent error, and should not exceed 0.2 to achieve 1 percent error in the near field solution (Gerstle and Abdalla 1990).

The meshing criteria given above show that a large number of elements are required at a crack tip to obtain accurate near-field stresses. Experience shows that 300 degrees of freedom are required per crack tip to reliably obtain 5 percent accuracy in the near field stresses (Gerstle and Abdalla 1990). This becomes prohibitive from a computational standpoint for problems with more than one crack tip.

Fortunately, it is not necessary to accurately model near-field stresses to calculate accurate stress intensity factors. In fact, using no singular elements, energy methods can be used, as described above, to obtain accurate stress intensity factors with far fewer than 300 degrees of freedom per crack tip.

2.3—Fictitious crack model (FCM)

Since 1961, there has been a growing realization that LEFM is not applicable to concrete structures of normal size and material properties (Kaplan 1961; Kesler et al. 1972; Bažant 1976). The FPZ ranges from a few hundred millimeters to meters in length, depending upon how the FPZ is defined and upon the properties of the particular concrete being considered (Hillerborg et al. 1976; Ingraffea and Gerstle 1985; Jenq and Shah 1985). The width of the FPZ is small compared to its length (Petersson 1981). LEFM, although not applicable to small structures, may still be applicable to large structures such as dams (Elfgren 1989). However, even for very large structures, when mixed-mode cracking is present the FPZ may extend over many meters; this is due to shear and compressive normal forces (tractions) caused by friction, interference, and dilatation (expansion) between the sides of the crack, far behind the tip of the FPZ. To clarify this notion, Gerstle and Xie (1992) have used an "interface process zone (IPZ)" to model the FPZ.

The fictitious crack model (FCM) has become popular for modeling fracture in concrete (Hillerborg et al. 1976; Petersson 1981; Ingraffea and Saouma 1984; Ingraffea and Gerstle 1985; Gustafsson 1985; Gerstle and Xie 1992; Feenstra et al. 1991a,b; Bocca et al. 1991; Yamaguchi and Chen 1991; Klisinski et al. 1991; Planas and Elices 1992, 1993a,b). Figure 2.5 shows the terminology and concepts associated with the FCM. This model assumes that the FPZ is long and infinitesimally narrow. The FPZ is characterized by a "normal stress versus crack opening displacement curve," which is considered a material property, as shown in Fig. 2.5.

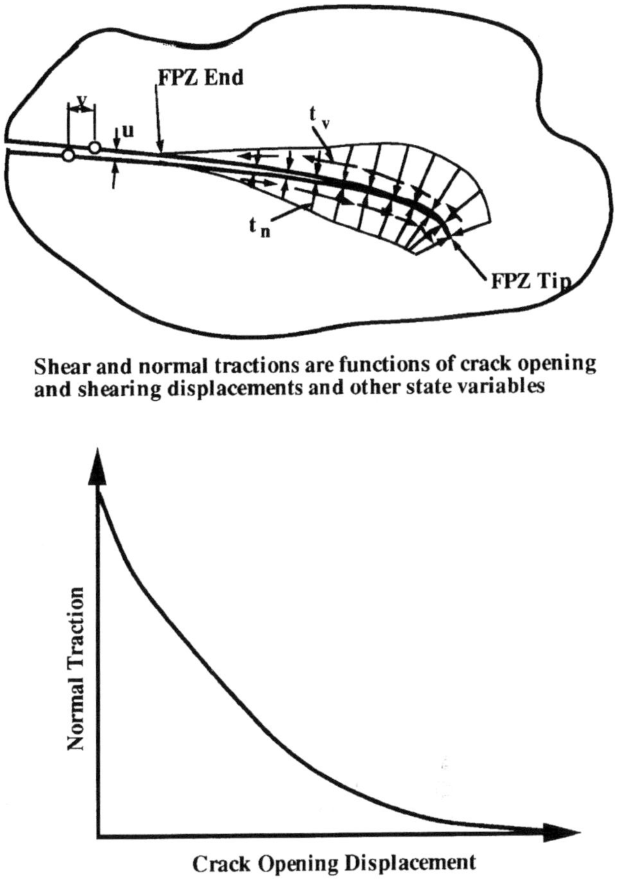

Fig. 2.5—Terminology and concepts associated with the fictitious crack model (FCM) (Hillerborg et al. 1976).

The FCM assumes that the FPZ is collapsed into a line in 2D or a surface in 3D. A natural way to incorporate the model into the finite element analysis is by employing interface elements. The first interface element was formulated by Goodman et al. (1968) and was used in the modeling of rock joints. Since then, many types of interface and thin layer elements have been developed and are widely used in geotechnical engineering (Heuze and Barbour 1982; Desai et al. 1984). Zero-thickness elements are the most widely used type of interface, with normal and shear stresses and relative displacements across the interface as constitutive variables. Unrealistic jumps in the results of adjacent integration points of contiguous interfaces have been reported by some authors depending on initial stiffness and load conditions, although most of these problems seem to disappear with the appropriate selection of integration points and integration rule (Gens et al. 1988; Hohberg 1990; Rots and Schellenkens 1990; Schellenkens and De Borst 1993). Other investigators have implemented a semi-discrete FCM by including strain discontinuities (Ortiz et al. 1987; Fish and Belytschko 1988; Belytschko et al. 1988; Dahlblom and Ottosen 1990; Klisinski et al. 1991) or displacement discontinuities (Dvorkin et al. 1990; Lotfi 1992; Lotfi and Shing 1994, 1995) within continuum elements.

The FCM has been incorporated into finite element codes with the use of interface elements. Ingraffea and coworkers (Ingraffea et al. 1984; Ingraffea and Saouma 1984; Ingraffea and Gerstle 1985; Bittencourt et al. 1992) extended the FCM to simulate mixed-mode crack propagation analysis employing six-noded interface elements. Swenson and Ingraffea (1988) used six-noded interface elements to model mixed-mode dynamic crack propagation. Bocca et al. (1991) have published similar work. Gerstle and Xie (1992) used a simple four-noded linear displacement interface element that was modified to allow an arbitrary distribution of tractions along its length.

Other references that implement the FCM include Rots (1988), Stankowski (1990), Stankowski et al. (1992), Hohberg (1992a), Lotfi (1992), Vonk (1992), Lotfi and Shing (1994), Garcia-Alvarez et al. (1994), Lopez and Carol (1995), and Bažant and Li (1995).

In the FCM, the stiffness of the interface element is a nonlinear function of the crack opening displacement, so that a nonlinear solution procedure is required. As with any other kind of nonlinear constitutive relation, FEM calculations with interface elements behaving in accordance the FCM require a nonlinear solution strategy. The various existing techniques such as classic Newton iteration, dynamic relaxation, and arc-length procedures have been used, with satisfactory results reported in the literature (Swenson and Ingraffea 1988; Gerstle and Xie 1992; Papadrakis 1981; Underwood 1983; Bathe 1982).

When using interface elements to model the FPZ, the elements must be very stiff prior to crack initiation to represent an uncracked material (i.e., to keep the two sides of the potential crack together). However, care must be taken not to use a stiffness so high as to cause nonconvergent numerical behavior in the finite element solution. Brown et al. (1993) successfully used interface elements with an axial stiffness equal to 50 times the stiffness of the adjacent concrete elements without numerical difficulties. Gerstle and Xie (1992) suggested using a precrack stiffness equal to the secant stiffness to a point on the descending normal traction-*COD* curve (Fig. 2.5) equal to a *COD* of 1/20 to 1/30 of the *COD* at which the normal traction drops to zero.

Some investigators have dispensed with interface elements and have instead simulated the FPZ using an influence function approach (Li and Liang 1986; Planas and Elices 1991) in which cohesive forces are applied to the crack faces (Gopalaratnam and Ye 1991). Weighted multipliers are used in the superposition of FEM solutions to satisfy overall equilibrium, compatibility and stress-crack width relations within the FPZ. This approach results in the solution of a set of nonlinear algebraic equations to determine the multipliers. In some cases, it may be appropriate to linearize the relationship between the *COD* and the tractions on the FPZ. Then linear equations can be solved to obtain the solution efficiently (Gopalaratnam and Ye 1991; Li and Bažant 1994).

Extension of the FCM with interface elements to mixed mode cracking requires a constitutive relation for the interface, in which normal and shear stresses and relative displacements are fully coupled. Crack opening and closing conditions are expressed with a biaxial failure surface in the normal-shear stress space. Crack surface displacements, thus, have two components: opening and sliding. Several models of this kind have been proposed recently, all based on the framework of non-associated work hardening plasticity, to obtain a formulation that is fully consistent and contains fracture energy parameters (Stankowski 1990; Stankowski at al. 1993; Lotfi 1992; Lotfi and Shing 1994; Hohberg 1992a,b; Vonk 1992; Garcia-Alvarez et al. 1994). After the crack is completely open, the models prevent interpenetration and provide Coulomb-type friction between crack surfaces. None of the models, however, provides secant unloading in pure tension, as is usual in the classic FCM, because this would complicate the model considerably, as discussed by Carol and Willam (1994).

An approach to nonlinear mixed-mode discrete crack propagation analysis was proposed by Ingraffea and Gerstle (1985). In this approach, the FPZ is modeled by interface elements, with the singular elements used in LEFM placed around the fictitious crack tip to predict the direction of the crack propagation. However, singular elements are not necessary for this purpose if energy release rates are calculated directly (Rice 1968; Parks 1974; Sha and Yang 1990; Rahulkumar 1992).

Another potential approach to determining the direction of a crack is based on the very reasonable assumption that the crack will propagate when the maximum tensile principal stress at the crack tip reaches the strength of the material (Petersson 1981; Gustafsson 1985; Hillerborg and Rots 1989, Bocca et al. 1991; Gerstle and Xie 1992). The direction of crack propagation is assumed to be perpendicular to the maximum tensile principal stress. The problem with this approach is that when the FPZ becomes small compared to the crack tip element size, objectivity of the results is lost. Therefore it makes more sense to use an energy-based approach to determine crack propagation. A basis for such an approach has been developed by Li and Liang (1992). Promising results have been obtained by using energy release rate approaches to determine both the direction and the load level at which a fictitious crack will propagate (Xie et al. 1995).

2.4—Automatic remeshing algorithms

In 1981, work was completed on a two-dimensional fracture propagation code that used simple interactive computer graphics to interactively model crack propagation (Saouma 1981). Many of the tasks that Ingraffea (1977) had performed by editing files manually were now performed interactively. These tasks included semi-automatic remeshing to allow the crack to advance, limited post-processing to view stresses, deformed mesh, and stress-intensity factors, and predictions based upon the mixed-mode crack propagation theories of the crack trajectory.

Subsequently, Wawrzynek and Ingraffea (1987) developed a second generation two-dimensional interactive graphical finite element fracture simulation code based upon a winged-edge topological data structure. This program demonstrated the value of using a topological data structure in fracture simulation codes. More recently, Gerstle and Xie (1992), in collaboration with others, developed an interactive graphical finite element code that is capable of representing and automatically propagating cracks in two dimensional problems.

Procedures have also been developed to handle numerical discretization and arbitrary fracture simulation in three dimensions (Martha 1989; Sousa et al. 1989; Ingraffea 1990; Martha et al. 1991).

A number of algorithms have been introduced for automatic meshing of solid models (Shepard 1984; Cavendish et al. 1985; Schroeder and Shepard 1988; Perucchio et al. 1989). These algorithms can be categorized into three broad families: element extraction, domain triangulation, and recursive spatial decomposition. Although substantial differences exist between these families, the algorithms involve the development of a geometric representation of the structure, which provides the basis for the construction of the finite element mesh (Sapadis and Perucchio 1989). Automatic modeling of discrete crack propagation in three dimensions remains a challenge.

It is worth noting that remeshing may not be required 1) if the crack path is known in advance due to symmetry or due to previous experimental or analytical experience with the same geometry; or 2) if interface elements are placed along all possible crack paths.

CHAPTER 3—SMEARED CRACK MODELS

Early in the application of finite element analysis to concrete structures (Rashid 1968), it became clear that it is often much more convenient to represent cracks by changing the constitutive properties of the finite elements than to change the topography of the finite element grid. The earliest procedure involved dropping the material stiffness to zero in the direction of the principal tensile stress, once the stress was calculated as exceeding the tensile capacity of the concrete. Simultaneously, the stresses in the concrete were released and reapplied to the structure as residual loads. Models of this type exhibit a system of distributed or "smeared" cracks. Ideally, smeared crack models should be capable of representing the propagation of a single crack, as well as a system of distributed cracks, with reasonable accuracy.

Over the years, a number of numerical and practical problems have surfaced with the application of smeared crack models. Principal among these involve the phenomenon of "strain localization." When microcracks form, they often tend to grow nonuniformly into a narrow band (called a "crack"). Under these conditions, deformation is concentrated in a narrow band, while the rest of the structure experiences much smaller strains. Because the band of localized strain may be so narrow that conventional continuum mechanics no longer applies, various "localization limiters" have been developed. These localization limiters are designed to deal with problems associated with crack localization and spurious mesh sensitivity that are inherent to softening models in general and smeared cracking in particular. Critical reviews of the practical aspects of smeared crack models are presented by ASCE Task Committee (1982) and Darwin (1993).

3.1—Reasons for using smeared crack models

The smeared crack approach, introduced by Rashid (1968), has become the most widely used approach in practice. Three reasons may be given for adopting this approach:

1. The procedure is computationally convenient.

2. Distributed damage in general and densely distributed parallel cracks in particular are often observed in structures (measurements of the locations of sound emission sources provide evidence of a zone of distributed damage in front of a fracture).

3. At many size scales, a crack in concrete is not straight but highly tortuous, and such a crack may be adequately represented by a smeared crack band.

There are, however, serious problems with the classical smeared crack models. They are in principle nonobjective, since they can exhibit spurious mesh sensitivity (Bažant 1976), i.e., the results may depend significantly on the choice of the mesh size (element size) by the analyst. For example, in a tensioned rectangular plain concrete panel with a rectangular finite element mesh, the cracking localizes into a one element wide band. The crack band becomes narrower and increases in length as the mesh is refined, as shown in Fig. 3.1 (Bažant and Cedolin 1979, 1980; Bažant and Oh 1983; Rots et al. 1984; Darwin 1985). Consequently, if not accounted for in the crack model, the load needed to extend the crack band into the next element is less for a finer mesh (for a crack model controlled by tensile strength alone, it decreases roughly by a factor of $\sqrt{2}$ if the element size is halved, and tends to zero as the element size tends to zero). Thus, the maximum load decreases as the mesh is refined (Bažant and Cedolin 1980). Furthermore, the apparent energy that is consumed (and dissipated) during structural failure depends on the mesh size, and tends to zero as the mesh size tends to zero. Such behavior, which is encountered not only for cracking with a sudden stress drop but also for gradual crack formation with a finite slope of the post-peak tensile strain-softening stress-strain diagram that is independent of element size (Bažant and Oh 1983), is nonobjective. These problems make the classical smeared crack approach unacceptable, although in some structures such lack of objectivity might be mild or even negligible [this latter behavior occurs especially when the failure is controlled by yielding of reinforcement rather than cracking of concrete (Dodds et al. 1984).

To avoid the unobjectivity or spurious mesh sensitivity, a mathematical device called a "localization limiter" must be introduced.

3.2—Types of localization limiters

Several types of localization limiters have been proposed:

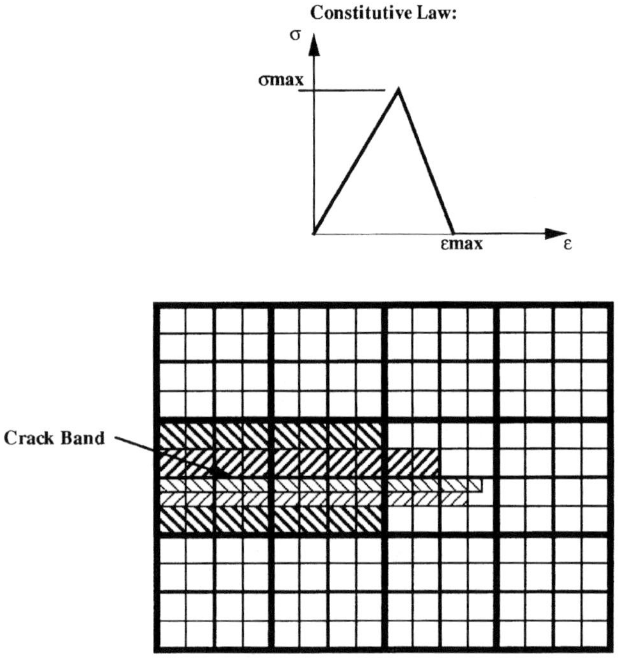

Fig. 3.1—If the constitutive model is independent of element size, the crack band becomes longer and narrower as the mesh is refined.

3.2.1 *Crack band model*—The simplest localization limiter is a relationship between the element size and the constitutive model so that the total energy dissipated will match that of the material being modeled. This can be done by adjusting the downward slope of the stress-strain curve, or, equivalently, the value of ε_{max}, shown in Fig. 3.1, as the element size is altered. ε_{max} is increased as the element size is decreased. This procedure, known as the crack band model, has a limitation for coarse meshes (large elements) ε_{max} cannot be conveniently reduced below the value of strain corresponding to the peak stress σ_{max}.

More advanced constitutive models based on the theories of plasticity and damage may not exhibit the simple stress-strain relationship shown in Fig. 3.1. Crack band modes can still be applied as long as a local fracture energy is included among the model's parameters (Pramono and Willam 1987, Carol et al. 1993).

Practically, the most important feature of the crack band model is that it can represent the effect of the structure size on: 1) the maximum capacity of the structure (Bažant 1984); and 2) the slope of the post-peak load-deflection diagram. However, from the physical viewpoint, the width of the cracking zone at the front of a continuous fracture (i.e., the FPZ) is represented by a single, element-wide band and cannot be subdivided further; consequently, possible variations in the process zone size, which cause variation of effective fracture energy (i.e., R-curve behavior) cannot be captured and the stress and strain states throughout the FPZ cannot be resolved. The procedure, however, has been widely and successfully applied.

3.2.2 *Nonlocal continuum*

a) *Phenomenological approach*

A general localization limiter is provided by the nonlocal continuum concept and spatial averaging (Bažant 1986).

A nonlocal continuum is a continuum in which some field variables are subjected to spatial averaging over a finite neighborhood of a point. For example, average (nonlocal) strain is defined as

$$\bar{\varepsilon}(x) = \frac{1}{V_r(x)}\int_V \alpha(x-s)\varepsilon(s)dV(s) = \int_V \alpha'(x,s)\varepsilon(s)dV(s) \quad (3.1)$$

in which $V_r(x) = \int_V \alpha(x-s)dV(s)$ and $\alpha'(x,s) = \alpha(x-s)/(V_r x)$; $\varepsilon(x)$ is the strain at the point in space defined by coordinate vector x; V is the volume of the structure; V_r is the representative volume of the material, as shown in Fig. 3.2, understood to be the smallest volume for which the heterogeneous material can be treated as a continuum (the size of V_r is determined by a characteristic length, L, which is a material property; α is a weighting function, which decays with distance from point x and is zero or nearly zero at points sufficiently remote from x; and the superimposed bar denotes the averaging operator. The dummy variable s represents the spatial coordinate vector in the integral.

As the simplest form of the weighting function, one may consider $\alpha = 1$ within a certain representative volume V_0 centered at point x and $\alpha = 0$ outside this volume. Convergence of numerical solutions, however, is better if α is a smooth bell-shaped function. An effective choice is

$$\alpha = \left[1 - \left(\frac{r}{\rho_o L}\right)^2\right]^2 \text{ if } |r| < \rho_o L;\ \alpha = 0 \text{ if } |r| > \rho_o L \quad (3.2)$$

in which $r = |x - s|$ = distance from point x, L = characteristic length (material property), and ρ_o = coefficient chosen in such a manner that the volume under function α given by Eq. 3.2 is equal to the volume under the function $\alpha = 1$ for $r < L/2$ and $\alpha = 0$ for $r > L/2$ (which represents a line segment in 1D, a circle in 2D, and a sphere in 3D). Alternatively, the normal distribution function has been used in place of Eq. 3.2 and found to work well enough, although its values are nowhere exactly zero (Bažant 1986).

For points whose distance from all the boundaries is larger than $\rho_o L$, $V_r(x)$ is constant; otherwise the averaging volume protrudes outside the body, and $V_r(x)$ must be calculated for each point to account for the locally unique averaging domain (Fig. 3.2b).

In finite element computations, the spatial averaging integrals are evaluated by finite sums over all integration points of all finite elements of the structure. For this purpose, the matrix of the values of α' for all integration points is computed and stored in advance of the finite element analysis.

This approach makes it possible to refine the mesh as required by structural considerations. Since the representative volume over which structural averaging takes place is treated as a material property, convergence to an exact continuum solution becomes meaningful and the stress and strain distributions throughout the FPZ can be resolved.

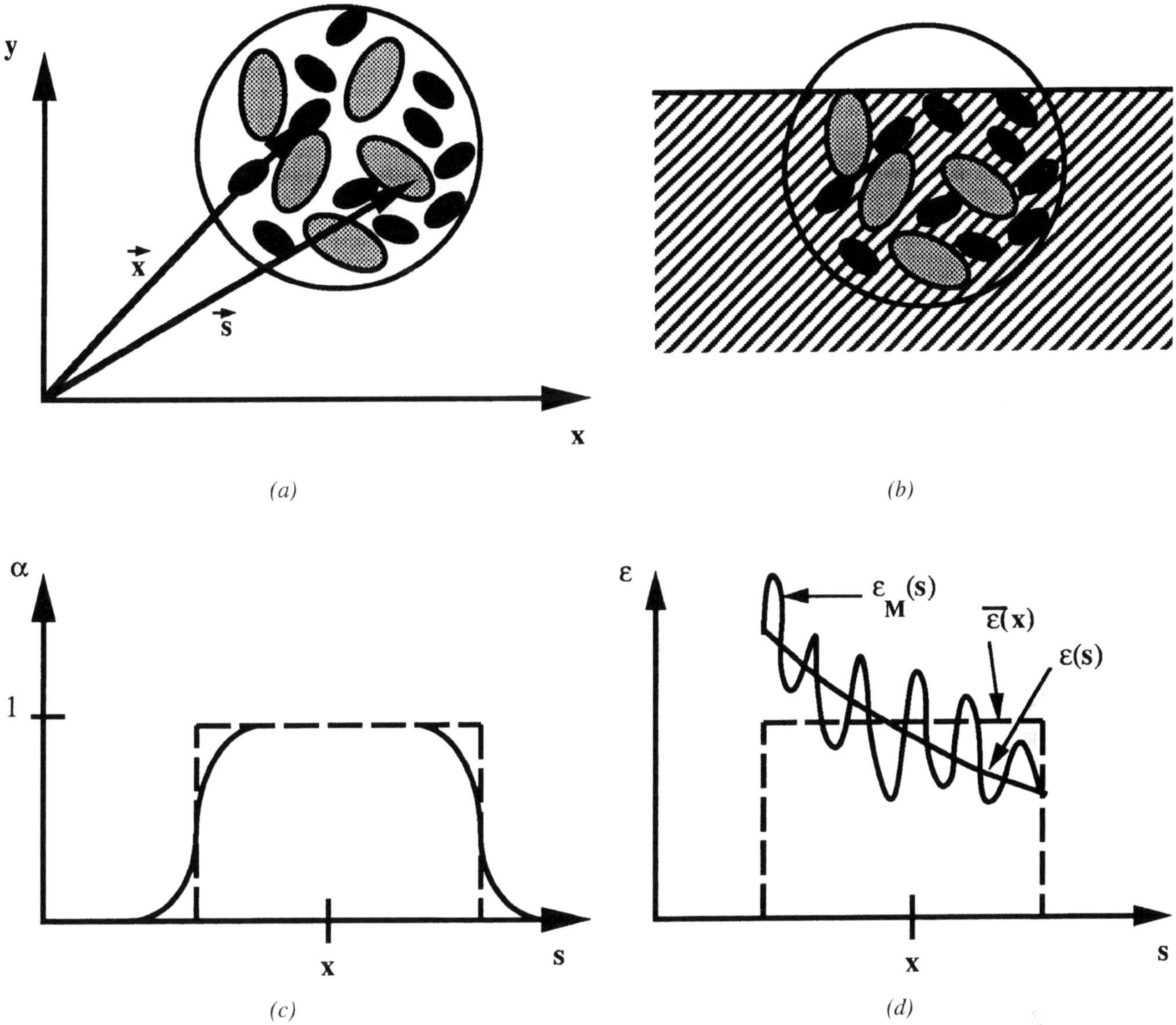

Fig. 3.2—Heterogeneity of concrete at the size scale of the aggregate.

The nonlocal continuum model for strain-softening of Bažant et al. (1984) involves the nonlocal (averaged) strain $\bar{\varepsilon}$ as the basic kinematic variable. This corresponds to a system of imbricated (i.e., overlapping in a regular manner, like roof tiles) finite elements, overlaid by a regular finite element system. Although this imbricate model limits localization of strain softening and guarantees mesh insensitivity, the programming is complicated, due to the nonstandard form of the differential equations of equilibrium and boundary conditions, i.e., energy considerations involve the nonlocal strain $\bar{\varepsilon}$.

These problems led to the idea of a partially nonlocal continuum in which stress is based on nonlocal strain, but local strains are retained. Such a nonlocal model, called "the nonlocal continuum with local strain" (Bažant et al. 1987; Bažant and Lin 1988; Bažant and Pijaudier-Cabot 1988, 1989) is easier to apply in finite element programming. In this formulation, the usual constitutive relation for strain softening is simply modified so that all of the state variables that characterize strain softening are calculated from nonlocal rather than local strains. Then, all that is necessary to change in a local finite element program is to provide a subroutine that delivers (at each integration point of each element, and in each iteration of each loading step) the value of $\bar{\varepsilon}$ for use in the constitutive model.

Practically speaking, the most important feature of a nonlocal finite element model is that it can correctly represent the effect of structure size on the ultimate capacity, as well as on the post-peak slope of the load-deflection diagram. Nonlocal models can also offer an advantage in the overall speed of solution (Bažant and Lin 1988). Although the numerical effort is higher for each iteration when using a nonlocal model, the formulation lends a stabilizing effect to the solution, allowing convergence in fewer iterations.

b) *Micromechanical approach*

Another nonlocal model for solids with interacting microcracks, in which nonlocality is introduced on the basis of microcrack interactions, was developed by Bažant and Jirasek (Bažant 1994; Jirasek and Bažant 1994; Bažant and Jirasek 1994) and applied to the analysis of size effect and localization of cracking damage (Jirasek and Bažant 1994). The model represents a system of interacting cracks using an integral equation that, unlike the phenomenological nonlocal model, involves a spatial integral that represents microcrack interaction based on fracture mechanics concepts. At long range, the integral weighting function decays with the square or cube of the distance in two or three dimensions, respectively. This model, combined with a microplane model, has provided consistent results in the finite element analysis of fracture and structural test specimens (Ozbolt and Bažant 1995).

3.2.3 *Gradient models*—Another general way to introduce a localization limiter is to use a constitutive relation in which the stress is a function of not only the strain but also the first or second spatial derivatives (or gradients) of strain. This idea appeared originally in the theory of elasticity. A special form of this idea, called Cosserat continuum and characterized by the presence of couple stresses, was introduced by Cosserat and Cosserat (1909) as a continuum approximation of the behavior of crystal lattices on a small scale. A generalization of Cosserat continuum, involving rotations of material points, is the micropolar continuum of Eringen (1965, 1966).

Bažant et al. (1984) and Schreyer (1990) have pointed out that spatial gradients of strains or strain-related variables can serve as a localization limiters. It has been shown (Bažant et al. 1984) that expansion of the averaging integral (Eq. 3.1) into a Taylor series generally yields a constitutive relation in which the stress depends on the second spatial derivatives of strain, if the averaging domain is symmetric and does not protrude outside the body, and on the first spatial derivatives (gradients) of strain, if this domain is unsymmetric or protrudes outside the domain, as happens for points near the boundary (Fig. 3.2b). Since the dimension of a gradient is length^{-1} times that of the differentiated variable, introduction of a gradient into the constitutive equation inevitably requires a characteristic length, L, as a material property. Thus, the use of spatial gradients can be regarded as an approximation to nonlocal continuum, or as a special case.

In material research of concrete, the idea of a spatial gradient appeared in the work of L'Hermite et al. (1952), who found that, to describe differences between observations on small and large specimens, the formation of shrinkage cracks needs to be assumed to depend not only on the shrinkage stress but also on its spatial gradient [see also L'Hermite et al. (1952) and a discussion by Bažant and Lin (1988)]. This was probably the first appearance of the nonlocal concept in fracture. The idea that the stress gradient (or equivalently the strain gradient) influences the material response has also been demonstrated by Sturman, Shah, and Winter (1965) and by Karsan and Jirsa (1969) for combined flexure and axial loading.

The nonlocal averaging integral is meaningful only if the finite elements are not larger than about one third of the representative volume, V_r (Droz and Bažant 1989). The gradient approach, on the other hand, offers the possibility of using finite elements with volumes as large as the representative volume, V_r. Thus, the gradient approach offers the possibility of using a smaller number of finite elements in the analysis. It appears, however, that the programming may be more complicated and less versatile than for the spatial averaging integrals. The problem is that interelement continuity must be enforced not only for the displacements but also for the strains. This requires the use of higher-order elements, or alternatively, if first-order elements are preferred, the use of an independent strain field with separate first-order finite elements.

Nonlocal constitutive models for concrete are reviewed in ACI 446.1R.

CHAPTER 4—LITERATURE REVIEW OF FEM FRACTURE MECHANICS ANALYSES

4.1—General

To help provide an overview of the state-of-the-art in finite element modeling of plain and reinforced concrete structures, a number of representative analyses are summarized in this chapter. Emphasis is placed on easily available, published analyses that attempt to address the fracture behavior of concrete structures. Wherever possible, problems solved using the discrete cracking approach are compared to solutions using the smeared crack approach. Symposium proceedings (Mehlhorn et al. 1978, *Computer-Aided* 1984, 1990, Firrao 1990, *Fracture Mechanics* 1989, van Mier et al. 1991, *Concrete Design* 1992, *Size Effect* 1994, *Computational Modeling* 1994, *Fracture and Damage* 1994, *Fracture Mechanics* 1995) provide the readers with the changing flavor of the state-of-the-art over the years. Many more published FEM fracture mechanics analyses exist than are presented in this chapter; however, it is fair to say that those referenced here are representative of the state-of-the-art.

The term "size effect," used throughout this chapter, is a term that has taken on a special meaning for quasibrittle materials, such as concrete and rock. It describes the *decrease in average stress at failure with increasing member size* that is directly attributable to the well-established fact that fracture is governed by a fracture parameter(s) that depends on the dimensions of the crack (which is tied to structure size) as well as some measure of stress. For concrete, it is not clear if the fracture parameter that governs failure is a material property or if it also depends on the structure size. However, this last point does not alter the general meaning of the term "size effect." In contrast, in the field of fracture of metals, the term "size effect" is understood to apply only to the dependence of fracture parameter (not the average stress) on the structural dimensions.

4.2—Plain concrete

Unreinforced concrete structures are the most fracture sensitive. While usually reinforced with steel in design practice to provide adequate tensile strength, many structures (or parts of structures), for one reason or another, are unreinforced. It makes sense to study the analysis of unreinforced concrete structures, because these provide the most severe tests of fracture behavior, and because the results of these analyses can add insight into the more complex behavior evidenced in reinforced concrete structures. In what follows, short descriptions (listed in chronological order) of finite element analyses of unreinforced concrete structures are presented.

4.2.1 *Tensile failure*—The plain concrete uniaxial tension specimen is probably more sensitive to fracture than any other type. For this reason, it provides a good test of the fracture-sensitivity of a finite element model for concrete. Although a number of such analyses have been reported, the following references are representative of first a discrete, then a local smeared, then a nonlocal smeared, and finally a micromechanics (random particle) approach.

Gustaffson (1985) used the fictitious crack model (discrete crack) to study the uniaxial tension specimen tested by Petersson (1981). Except for the fictitious crack, he assumed the continuum was linear elastic. He included the influence of initial stresses in his model. Good agreement with test results was obtained. Gustaffson (1985) also studied a number of other specimen geometries using the finite element method. Only Mode I cracks, with trajectories known in advance, were modeled. Interestingly, for larger specimens this analysis approach would still be expected to yield fracture-correct results (as long as a sufficient number of degrees of freedom are used to accurately represent the distribution of tractions along the FPZ). This type of model is extremely powerful for the class of problems in which the crack trajectory is known in advance, and the failure mechanism involves a single discrete crack.

Bažant and Prat (1988a,b) developed and used a (smeared crack) microplane model for brittle-plastic materials. The results were compared with Petersson's experimental uniaxial tension results (Petersson 1981), along with eight other series. The term microplane model is used to describe a class of constitutive models in which the material behavior is specified independently for planes of many orientations (called microplanes); the macroscopic stresses and strains are obtained through the use of a suitable superposition technique, such as the principle of virtual work. In the study by Bažant and Prat, the agreement between the numerical and experimental results was good. However, the analysis was conducted only at the constitutive level that, by its nature, is size-independent. The parameters obtained for these tests were applicable only to the particular specimen size analyzed.

A nonlocal microplane model (see ACI 446.1R) was used to analyze a rectangular concrete tension specimen in plane strain. The specimen was loaded by a prescribed uniform displacement and analyzed using two different meshes (Bažant and Ozbolt 1990, Droz and Bažant 1989). The model can represent tensile cracking in multiple directions, as well as nonlinear triaxial behavior in compression and shear, including post-peak response. The calculated load-displacement diagrams were obtained by local and nonlocal analysis for two meshes. While the local results differed substantially (spurious mesh sensitivity), the nonlocal results were nearly mesh independent. While producing excellent results, the nonlocal microplane model is very costly in terms of computer disk space and computer time.

Schlangen and van Mier (1992) employed a micromechanics approach to model tensile fracture using a triangular lattice of brittle beam elements with statistically varying properties to represent the heterogeneity of the material. Fracturing of the material takes place by removing, in each load step, the beam element with the highest stress (relative to its strength). The input parameters are, however, not directly observable material properties. Thousands of degrees of freedom are required for the analysis, and larger specimens would have required an unreasonable number of degrees of freedom.

Another lattice model, developed by Bažant et al. (1990) and refined by Jirasek and Bažant (1995a), involves lattices with only central force interactions between adjacent particles (avoiding the use of beams that are subject to bending and do not realistically reflect deformation modes on the microstructural level). The representation was further extended by Jirasek and Bažant (1995b) with a very efficient numerical algorithm of the central difference type. With this algorithm, it was possible to solve the nonlinear response of a lattice model with over 120,000 degrees of freedom by explicit dynamic integration on a desk-top workstation. The studies demonstrate that the transitional size effect can be reproduced by a lattice model.

Other examples of FEM analysis of tensile specimens are presented by Roelfstra et al. (1985), Rots (1988), Stankowski (1990), Stankowski et al. (1992), Vonk (1992), and Lopez and Carol (1995).

4.2.2 *Compressive failure*—Although plain concrete compression specimens do not exhibit as significant a fracture mechanics size effect as notched tension specimens, a good model for concrete should be capable of predicting the behavior of such a specimen. A significant fracture mechanics size effect is found in the descending branch of the compression stress-strain curve (van Mier and Vonk 1991). Some fracture mechanics-based models have been used to attempt to predict the compressive strength of compression specimens. More test results exist for this type of specimen than for any other. All but one of the following examples used a smeared crack approach to analyze this specimen type.

Bažant and Prat (1988b) compared their local smeared crack microplane model for brittle-plastic materials with experimental uniaxial compression test results. Excellent agreement was found (although, once again, the five microplane model parameters were selected to provide the best fit to the experimental data). As with the tensile experiments, calculations with this model were purely constitutive, and therefore size independent and fracture insensitive. To obtain some size effect in the softening branches with the model as described (without going into a non-local implementation), a new set of parameters would be needed for each specimen size.

Gajer and Dux (1990) used a "simplified nonorthogonal crack model" to analyze a plain concrete prism in compression. The model was based on the crack-band concept, featuring simplified representations of both nonorthogonal cracking and the influence of cracking on local material stiffnesses. At an integration point, only one primary crack of fixed direction and one rotating crack are allowed to develop. The model makes use of the crack band concept to assure constant energy release for elements of different sizes during the fracture process. The model would be expected to give correct fracture mechanics results as long as the element size is larger than the width of the FPZ and smaller than the length of the FPZ.

Tasdemir, Maji, and Shah (1990) modeled mixed-mode crack initiation and propagation in the cement paste matrix, starting from the interface of a single rigid rectangular inclusion in a concrete specimen under uniaxial compression, using both a discrete LEFM approach and a FCM approach, with singularities modeled at the crack tips in both approaches. The FEM results were compared with experimental results and satisfactory local cracking results were obtained. The approach is applicable for the study of micromechanical behavior in the neighborhood of a single aggregate particle, but inappropriate as a method for determining the gross response of the compression specimen.

The original microplane model of Bažant and Prat (1988a,b) was reformulated as a continuum damage model (Carol et al. 1991) and modified to make it computationally more efficient (Carol et al. 1992). In both cases, the original results of uniaxial compression specimens were reproduced with similar accuracy. The reformulation of the model in terms of a damage tensor, completely independent of the material rheology, allowed the authors to then replace elasticity with linear aging viscoelasticity and obtain a realistic approach to Rush curves (Carol and Bažant 1991; Carol et al. 1992). As discussed earlier, these results were purely constitutive and therefore size-independent and fracture-insensitive.

Bažant and Ozbolt (1992) used the nonlocal microplane model to investigate the effects of boundary conditions and size on the strength of a plane strain compression specimen. Meshes with 180 four-noded finite elements, enough to capture strain localizations, were employed. Many interesting and plausible damage localization patterns were displayed. Although their model is capable of predicting a fracture mechanics size effect, they found no size effect for this specimen type. The model would require great computational effort for realistic engineering structures.

Ozbolt and Bažant (1992) employed a nonlocal microplane model to simulate a uniaxial compression specimen subject to cyclic load. The rate effect was introduced by combining the damage model on each microplane with the Maxwell rheologic model. Nine four-noded isoparametric plane stress elements were employed. The model was shown to produce qualitatively correct results for cyclic compression, including hysteresis loops. This model, probably the most physically complete mechanical model developed for plain concrete to this time, requires a large computational effort for the analysis of even simple engineering structures.

Additional microstructural analyses of uniaxial compression specimens are reported by Stankowski (1990), Stankowski et al. (1992), and Vonk (1992, 1993).

4.2.3 *Fracture specimens*—The unreinforced three-point bend specimen is an important bench mark problem for several reasons. First, LEFM solutions for some span-to-depth ratios are known to better than 0.5 percent accuracy (Tada et al. 1973). Second, it is a standard fracture toughness testing geometry, which means that experimental results are available (Kaplan 1961, Catalano and Ingraffea 1982, Jenq and Shah 1985). Third, it is a symmetrical Mode I problem, and thus, the crack trajectory is known in advance. Symmetry can be used to reduce the cost of analysis. Many finite element analyses of this specimen are reported in the literature. It should be noted that most of the reported analyses were of relatively small specimens, and therefore LEFM would not be expected to apply.

The first analysis using the fictitious crack model of a three-point bend specimen was reported by Hillerborg, Modeer, and Petersson (1976). Finite element results regarding the effect of beam depth on flexural strength were presented. These theoretical results indicated that the flexural strength decreases with increased depth of the beam and that nonuniform shrinkage strains have a greater effect on the strength of deep beams than on the strength of shallow beams. A comparison between the theoretical predictions and a large number of experimental results showed good agreement. Similar analyses were presented by Modeer (1979), Petersson and Gustafsson (1980) for a discrete crack model and by Leibengood, Darwin, and Dodds (1986) for a smeared crack model.

Gerstle (1982) performed a series of discrete cracking analyses, employing a nonlinear FCM analysis, together with an LEFM fictitious crack tip assumption $K_{Ic(tip)}$. Nonlinear interface elements were used to model the fictitious crack. A secant stiffness iterative scheme was used to solve the nonlinear equations. Satisfactory agreement with experimental results (Catalano and Ingraffea 1982) was found using a pure FCM. When it was assumed that $K_{Ic(tip)}$ was non-zero, however, the strength of the beam was significantly over-predicted.

Gustafsson (1985) analyzed a three-point bend specimen using the FCM. He used a superposition of nonlinearly varying cohesive nodal forces to represent the fictitious crack, and a substructuring technique to reduce the number of unknowns in the iterative solution. He found the results to be consistent with theoretical expectations regarding the state of initial stress and specimen size.

De Borst (1986) analyzed a pre-notched, three-point bend specimen. The example was used mainly to demonstrate the efficiency of quasi and secant-Newton methods in the solution of nonlinear finite element equations. A local, smeared crack analysis was performed. He discussed several interesting points regarding the stability of the solution, and developed arc length control algorithms to increase solution stability for problems in which the post-peak response is important.

A three-point bending test according to a RILEM recommendation was numerically simulated using a discrete FCM with a bilinear stress versus crack opening displacement curve by Carpinteri et al. (1987) and Carpinteri (1989). The fictitious crack was represented by closing forces computed through the solution of a set of nonlinear equations. Many load-deflection curves for beams of various size were calculated and reported.

Yamaguchi and Chen (1990) used a smeared crack, crack-band-like model to simulate a notched three-point bend specimen. Eight-node isoparametric quadrilateral elements with 2 x 3 Gauss quadrature (3 in direction of crack propagation) were used in this analysis, and the notch was modeled as an element-wide gap. An extremely coarse mesh was used, with only two elements to model the 100 mm (3.9 in.) deep ligament. Good agreement was found with previous analysis results by others.

Gopalaratnam and Ye (1991) performed a fictitious crack analysis of a three-point bend specimen to determine the characteristics of the nonlinear FPZ in concrete. They used a linearized solution process to solve the nonlinear equations obtained in the FCM. The fictitious crack was modeled through appropriate application of cohesive point loads along the fictitious crack. Using a very refined mesh, they were able to obtain results that demonstrate the fracture mechanics size effect. They report extensively upon the process zone characteristics calculated.

Liang and Li (1991) used a boundary element program based on a discrete FCM to study the size effect in a plain concrete beam with a bending crack. The size effect was observed and verified. Justification for R-curve analysis (see ACI 446.1R) for such specimens was given. A double cantilever beam specimen is also analyzed using the same techniques.

Bolander and Hikosaka (1992) used a nonlocal smeared crack finite element model to simulate fracture of a three-point bend specimen. As shown by their finite element mesh, a large number of finite elements is required near the crack tip to adequately represent the development of the FPZ. They were able to obtain good agreement with experimental results using this approach, but it should be noted that, in their analysis, the FPZ was of significant size compared to the size of the beam specimen. Had the specimen been larger, the approach might not have been feasible because of the large number of elements required.

A nonlocal smeared crack analysis of a beam was reported by Bažant and Lin (1988). Geometrically similar notched beams of three different sizes of ratio 1:2:4 were analyzed using the meshes of four-node quadrilaterals. The concrete was described by the nonlocal smeared crack model, which is the same as the classical smeared crack model, except that the cracking strain was calculated from the maximum principal value of the spatially averaged nonlocal strain, $\bar{\varepsilon}$, rather than local strain, ε. The stress was assumed to decrease, either linearly or exponentially, with the maximum nonlocal principal strain. The characteristic length was taken in the first case as 2.3 in. (58 mm) and in the second case as 3.2 in. (81 mm), while the maximum aggregate size is $d_a = 0.5$ in. (13 mm). The maximum values of the nominal stress σ_N (load divided by beam depth and thickness) obtained by nonlocal finite element analysis with step-by-step loading are plotted as a function of the beam depth relative to the size of the aggregate, d/d_a, for both linear and exponential softening laws. Their work demonstrates that a nonlocal finite element model can represent the size effect quite well.

Bažant and Lin (1988) also compared results obtained with slanted square meshes to results obtained with aligned meshes. When the element size was less than about 1/3 of the characteristic length L, they observed no mesh bias with regard to crack direction. In contrast, with the crack band model, it was impossible to analytically produce fractures in specimens with slanted meshes that run vertically, as they should.

Additional analyses of three-point bend specimens are reported by Rots (1988), Carol et al. (1993), Garcia-Alvarez et al. (1994), and Lotfi and Shing (1994).

4.2.4 *Shear failure in plain concrete beams*—The finite element analysis of the single-notched beam subject to four-point loading is a benchmark problem that has been investigated by many researchers. It provides a good test of the ability of a finite element code to model mixed-mode fracture. Tests have been performed by Arrea and Ingraffea (1982). No accurate closed-form LEFM solution to this problem has been reported. The following are descriptions of several of these analysis attempts.

Ingraffea and Gerstle (1985) performed several mixed-mode FCM analyses of the beam tested by Arrea and Ingraffea (1982) using interface elements. They used a discrete aggregate interlock model to simulate shear stress transfer across the fictitious crack. They also used quadratic elements, requiring a total of approximately 600 degrees of freedom. The predicted crack trajectory and load-displacement curve were in close agreement with the experimental results.

Ingraffea and Panthaki (1986) performed a FCM mixed-mode fracture analysis of the four point shear beam tested by Bažant and Pfeiffer (1985). They concluded that, although shear fracture (fracture forming by shear slip, rather than by tensile opening) can occur under certain conditions, tensile and not shear fracture occurred in the specimens.

A smeared crack analysis of the four point shear beam tested by Arrea and Ingraffea (1982) was performed by de Borst (1986). A rotating crack band model was used, with adjustment of the constitutive model with element size to maintain a constant fracture energy. Loading was controlled by the arc-length method to prevent solution instabilities. The analysis required approximately 600 degrees of freedom, using eight-noded elements. The finite element results agree closely with the experimental results.

A smeared crack band model that covers tensile softening in Mode I and shear softening in Mode II fracture was used by Rots and de Borst (1987) to analyze the specimen tested by Arrea and Ingraffea (1982). Interesting snap-back behavior was modeled using an arc-length control procedure.

Carpinteri (1989) used a FCM to simulate a double-edge notched four-point loaded shear specimen. Snapback instability was modeled for some cases, using a crack length control approach. A number of other problems were also analyzed.\\

A discrete cracking FCM was used by Gerstle and Xie (1992) to model the single-edge notched four-point loaded specimen of Arrea and Ingraffea (1982). A dynamic relaxation solution scheme was employed, with stress-based criteria for crack propagation and an automated remeshing algorithm. The analysis is significant in that it uses a very coarse mesh (200 degrees of freedom), yet seems to capture the significant behavior.

Malvar (1992) modeled the Arrea and Ingraffea (1982) beam using a local smeared crack approach with and without considering the transfer of shear stresses across the crack. A better prediction of the experimental results was obtained when shear stress transfer was modeled. However, up to the peak load, the results were not sensitive to presence or absence of shear stress transfer across the crack. Finally, it was shown that inadmissible results are obtained if both tensile and shear stresses are assumed to completely vanish upon cracking.

Schlangen and van Mier (1992) accurately modeled shear fracture with the triangular lattice model of brittle beam elements described in Section 4.2.1, which represent the heterogeneity of the material.

In conclusion, the four-point loaded shear beam has been analyzed by numerous investigators. The discrete cracking, the nonlocal smeared cracking, and the triangular lattice model all performed well in simulating the fracture behavior of the beam. All of the methods have particular advantages and limitations, and it is therefore difficult to decide which of the approaches is preferable although the discrete cracking approach appears to be more efficient for beams that approach the size at which LEFM becomes applicable.

4.2.5 *Dams*—Dams belong to an important class of concrete structures that are fracture sensitive, and may in some cases even be analyzed using LEFM. A collection of nine papers presented at the International Conference held in Vienna in 1988 concerning fracture of dams is given in a special issue of *Engrg. Frac. Mech.*, V. 35, 1/2/3, 1990. Other valuable sources include *Dam Fracture* (1991, 1994) and six papers on fracture of dams in *Fracture* (1987).

Several finite element studies of dams (considering cracking, but not using fracture mechanics concepts) were conducted in the 1960's, notably by Clough (1962) and Zienkiewicz and Cheung (1964).

Apparently, the first true FEM fracture mechanics analysis of a dam is presented by Ingraffea and Chappel (1981). The Fontana Dam had developed a crack, and finite element analyses using LEFM were used to model the discrete crack. A rational explanation for the crack was thus obtained.

A discrete crack FEM approach was employed by Skrikerud (1986) with a simple remeshing scheme to analyze a dam under dynamic excitation. However, despite his use of a discrete crack approach, no fracture mechanics based criteria were used to predict crack extension, and tensile stresses at the crack tip were simply compared with the tensile strength. This approach would produce nonobjective results, because the stress magnitudes produced by the crack tip elements would be strongly dependent upon their size.

A comprehensive investigation into the behavior of cracked concrete gravity dams using fracture mechanics concepts is presented in Saouma, Ayari and Boggs (1987). The work includes the results of discrete LEFM fracture mechanics analyses. The effect of various forms of loading, concrete age and anisotropy on the stress intensity factors, direction of crack profiles, crack lengths and stress redistribution is considered.

A plane strain LEFM discrete cracking finite element analysis of the Kolnbrein dam, a doubly curved arch dam in Austria, is presented by Wawrzynek and Ingraffea (1987) and Linsbauer et al. (1989a, 1989b). Cracks are loaded by water pressure, as well as by remotely applied stress resultants from a three-dimensional arch analysis. The cracks are approximately 14 m long, and therefore the assumption of LEFM seems justified. Actual fractures in the dam seem to have been accurately modeled using LEFM.

Cervera et al. (1990) describe a smeared crack approach to the analysis of progressive cracking in large dams. They account for crack pressurization by the water by including an effective stress in the model. Thermal and water pressure effects are also accounted for. 2D and 3D models are presented.

Ayari and Saouma (1990) developed discrete LEFM cracking models for the efficient simulation of transient dynamic discrete crack closure and crack propagation in dams. After presenting simple validation problems, these models are integrated into an interactive graphical program that was used to analyze the Koyna Dam.

Ingraffea (1990) describes the analyses of the Fontana Dam, a generic gravity dam, and the Kolnbrein Dam. The studies employ mixed-mode LEFM implemented within a discrete crack model, automatic rezoning, finite element method. The study of the Fontana dam elucidated the mechanisms for crack initiation, accurately reproduced the observed trajectory, and evaluated the effectiveness of interim repair measures. The study of the generic gravity dam has as its objective the evaluation of usefulness of LEFM for design and quality control during construction. An envelope of safe lengths, heights, and orientations of cracks potentially growing from cold lift joints on the upstream face is derived. It is also shown that, by neglecting the toughness of the foundation contact, classical design methods predict a conservative factor-of-safety against sliding, and that, when toughness is set to zero, LEFM predictions are in good agreement with the classical method.

Carpinteri et al. (1992) performed testing and analysis of a gravity dam. Two scaled-down (1:40) models of a gravity dam were subjected to equivalent hydraulic and self-weight loading. From an initial notch, a crack propagated during the loading process towards the foundation. The numerical simulation of the experiments used a FCM. The structural behavior of the models and the crack trajectories are reproduced satisfactorily.

4.2.6 *Bond-slip*—Modeling of bond-slip is one of the most difficult and controversial aspects of finite element analysis of reinforced concrete structures. Much of the controversy lies in the fact that many finite element models of reinforced concrete have provided excellent representations of experimental behavior while allowing no bond-slip to occur (i.e., perfect bond) (Darwin 1993). While it is known that bond-slip behavior arises at least partially from fracture of the concrete, other types of nonlinearity such as crushing in front of ribs, chemical adhesion, and friction between the concrete and the steel play a role. The detailed modeling of bond-slip associated with even one reinforcing bar is an extremely complex problem.

The problem of modeling the mechanical interaction between a ribbed steel reinforcing bar and the surrounding concrete has been addressed in a number of different ways. When the steel is modeled as smeared, the reinforced concrete is considered as a homogeneous composite material with constitutive relations that include (implicitly or explicitly) the bond-slip behavior. These models often represent perfect bond. At a more detailed level, individual reinforcing bars may be represented by truss elements. In this case, either the truss elements are connected directly to the concrete elements, where bond-slip is zero, or nonlinear link elements are used to connect the steel truss elements to the concrete to represent the effect of bond-slip. This is a cumbersome modeling approach, theoretically unsatisfactory from a fracture mechanics viewpoint, and nearly always ignores the splitting effects of bar movement. At a much more detailed scale, both the steel and the concrete can be modeled as continua, with interface elements placed between the concrete and the steel. Recognizing bond-slip as a fracture mechanics problem, the cracking of the concrete can be modeled using either discrete or smeared fracture mechanics. In this approach, the concrete can be considered as an unreinforced continuum surrounding the reinforcing bar. Several analyses of this type are described next.

In pioneering work, a singly reinforced, axisymmetric, "tension-pull" specimen was analyzed using the FEM by Bresler and Bertero (1968). To simulate cracking and other nonlinear effects, a soft "homogenized boundary layer" of prespecified width was assumed surrounding the steel. As expected, the existence of the boundary layer significantly decreased the overall stiffness of the specimen. However, this approach did not use fracture mechanics principles. The approach was of limited predictive utility, as both the width of the boundary layer and its constitutive properties were not determined from any fundamental theoretical or experimental information.

The first fracture mechanics approach to the analysis of bond slip was taken by Gerstle (1982) and Ingraffea et al. (1984). An axisymmetric discrete FCM analysis of a single concentrically reinforced axisymmetric tension-pull specimen was presented. No slip was assumed between the concrete and the steel; instead, "bond-slip" occurred as the result of the development of radial cracks propagating outward from each of the ribs on the reinforcing bar. The effect of longitudinal splitting cracks was neglected due to the axisymmetric modeling assumptions.

Vos (1983) reported on several smeared crack FEM bond-slip calculations. To account for the complex cracking/crushing situation surrounding a ribbed steel bar, the bar was assumed to be surrounded by a softening layer of concrete, approximately one bar diameter in width. The elements within this softening layer were assumed to be equal in length to the rib spacing on the bar. This approach suffers from the fact that the input parameters used to describe the softening layer are not based on fundamental or experimentally obtainable mechanical properties (such as tensile strength and fracture toughness). Additionally, mesh sensitivity would be expected because of the (local) softening model used to describe the concrete continuum. However, very close agreement between analysis and experiment was obtained.

Kaiser and Mehlhorn (1987) studied and compared the bond-link and the interface element approaches for modeling bond-slip between concrete and a single steel reinforcing bar. They analyzed a reinforced tensile specimen and compared the numerical solution with tests and the numerical solutions of others. They made the questionable assumption that bond-slip can be considered to be a property of the interface, rather than a property that results from damage to the structure.

Kay et al. (1992) developed a smeared crack model to represent accumulated damage in plain concrete and bond between concrete and steel reinforcing bars. The model used a continuum approach to describe microcracks and crack coalescence in plain concrete for the prediction of concrete tensile fracture. Concrete-reinforcing bar interaction was achieved through the use of one-dimensional beam elements that interact with three-dimensional continuum elements through a one-dimensional contact algorithm. Stick/slip interactions between the concrete and the reinforcing bars were expressed in terms of the interface stress or the internal damage variables of the concrete damage model.

Ozbolt and Eligehausen (1992) analyzed the pullout of a deformed steel bar embedded in a concrete cylinder under monotonic and cyclic loading. The analysis was performed using axisymmetric finite elements and a 3D microplane (smeared crack) model for concrete. Instead of the classical interface element approach, a more general approach with spatial discretization modeling the ribs of a deformed steel bar was employed. The pull-out failure mechanism was analyzed. Comparison between numerical results and test results indicated good agreement. Their approach was able to correctly predict the monotonic as well as cyclic behavior, including friction and degradation of pull-out resistance due to the previous damage.

Brown et al. (1993) and Darwin et al. (1994) modeled the bond-slip/strength behavior of individual bars in beam-end specimens. The 3D analyses represented the individual ribs on the bar, and used the FCM to represent concrete fracture and a Mohr-Coulomb model to represent steel-concrete interaction at the face of the ribs. The analyses, aimed at better understanding bond behavior, gave an excellent match with empirical relationships between bond strength and the effects of embedded length and concrete cover.

More work is required to develop credible methods for the numerical modeling of bond-slip.

4.2.7 Other types of plain concrete structures—Several other types of plain concrete structures have been extensively analyzed using fracture mechanics and FEM approaches. Pullout of anchor bolts is an important fracture-controlled problem. Analyses of this problem has been reported by Hellier et al. (1987), Bittencourt, Ingraffea and Llorca (1992), Eligehausen and Ozbolt (1990), Cervenka et al. (1991), Eligehausen and Ozbolt (1992). Elfgren and Swartz (Elfgren 1992) have reported results of a round-robin analysis of anchor bolts organized by RILEM TC 90-FMA. Fifteen of the 16 analyses were based on fracture mechanics concepts. Variations in the analysis assumptions included: LEFM, FCM, and smeared crack models. The responses included some 82 solutions. Punching shear in slabs and anchorage of tendons in prestressed members can also often be considered as the fracture of a plain concrete structure.

Other types of plain concrete structures that have been analyzed include thick walled rings (Pukl et al. 1992) and many analyses of the double cantilever beam (DCB) fracture toughness testing specimen (Gustafsson 1985; Sluys and de Borst 1992; Liang and Li 1991; Liaw et al. 1990a,b; Dahlblom and Ottosen 1990; Du et al. 1990; Bertholet and Robert 1990).

4.3—Reinforced concrete

The fracture mechanics analysis of reinforced concrete structures is often much more difficult than the analysis of plain concrete structures. The reinforcing steel may be modeled discretely (usually using truss elements) or in a smeared fashion, with the steel included in the constitutive model used to calculate the element stiffness matrices. In both cases, it is difficult to correctly model the effect of bond-slip between the concrete and the steel, although modeling bond-slip does not appear to be critical to the solution of many problems (Darwin 1993).

Often, but not always, the addition of steel reinforcement to a structure makes it fracture insensitive. Because of this fact, early finite element analyses of reinforced concrete structures were often successful, even without realistically modeling fracture of the concrete. Sometimes, however, structures will fail by fracturing in unanticipated ways. This is the reason for developing finite element methods that are capable of predicting fracturing modes of behavior. A summary of recent techniques used to model reinforced concrete structures using finite element analysis is presented by Darwin (1993).

In the remainder of this section, the literature is surveyed to illustrate various methods of analysis, including fracture mechanics effects, of reinforced concrete structures. In the course of the chapter, the term "tension stiffening" will be used to describe some models. The term refers to a numerical device used to limit the rate at which the stress across a smeared crack drops to zero once a crack forms. Tension stiffening represents the physical behavior of cracks crossed by reinforcing steel and serves to stabilize the numerical solution. Models that include tension stiffening usually do not take into account the fact that the method inherently represents fracture energy. Therefore, application of tension stiffening does not provide an accurate fracture mechanics representation and serves principally to slow the rate at which residual stresses are reimposed on the structure being modeled.

4.3.1 *Reinforced concrete membranes*—Shear walls in buildings are usually considered as membrane elements because the major loads are in the plane of the wall. Cervenka and Gerstle (1972) reported on an early smeared crack analysis of reinforced concrete panels that compared very favorably with their experimental results under monotonic loading but not under cyclic loading. Load-displacement diagrams, crack patterns and failure mechanisms of shear wall specimens were examined. The excellent agreement with the monotonic results probably resulted from the fact that these panels had a reasonable amount of reinforcing steel, and therefore were insensitive to fracture mechanics and did not exhibit the fracture mechanics size effect. The cracking of concrete in these panels was of a distributed, rather than a localized, nature. The lack of agreement under cyclic load was due to other (noncrack) aspects of their modeling scheme.

Another (local) smeared crack analysis of several Cervenka and Gerstle (1972) shear panels was conducted by Darwin and Pecknold (1976), using both smeared crack and smeared modeling of steel. The observed behavior of the panels was replicated very well by the analysis for both monotonic and cyclic loading. Other (local) smeared crack analyses of the same shear panel are presented by Schnobrich (1977) and Bergan and Holand (1979), giving excellent agreement between experimental and analytical results.

Bažant and Cedolin (1980) performed a smeared crack analysis of a center cracked reinforced concrete membrane in uniaxial tension. They did not, however, compare the analytical results to tests. They used the crack band model to investigate the effect of mesh size on the results. They observed that, if bond-slip is not correctly modeled, lack of objectivity results. In a subsequent paper, Bažant and Cedolin (1983) investigated the same problem, this time with respect to meshes at an angle to the crack propagation direction. They concluded that the blunt crack band approach is valid only if fine meshes are used, although correct convergence appears to occur as the mesh size tends to zero.

Gerstle (1982) and Ingraffea et al. (1984) employed a discrete FCM model to simulate a cracked membrane in uniaxial tension. Bond-slip was modeled using special "tension-softening" elements which bridge the primary crack. The effect of varying assumptions regarding bond-slip were studied.

Bédard and Kotsovos (1985) used a (local) smeared crack approach to model four reinforced concrete panels subjected to loading in pure shear, shear and uniaxial compression, and shear and biaxial compression. The panels were reinforced with different steel ratios in each direction, varying from 0.00713 to 0.01785. Good correlation was obtained with experimental results. In this case, fracture was not a controlling mechanism, and therefore correct modeling of fracture was unnecessary to obtain reasonable results.

Chang, Taniguchi and Chen (1987) applied a (local) smeared crack model to the analysis of reinforced concrete panels tested and analyzed by Vecchio (1986). The "fracture criteria" in the sophisticated constitutive model were based upon limiting stresses and strains, and therefore are not true fracture criteria. However, very good agreement between experiment and numerical results was obtained. The panels were heavily reinforced, and therefore fracture criteria, again, probably did not control the behavior of the panels.

Channakeshava and Iyengar (1988) presented a (local) smeared crack constitutive model used to model shear panels tested by Vecchio (1986). An elasto-plastic cracking model for reinforced concrete was presented. The model included concrete cracking in tension, plasticity in compression, aggregate interlock, tension softening, elasto-plastic behavior of the steel, bond-slip, and tension stiffening. A procedure for incorporating bond-slip in smeared steel elements was described. Good agreement with experimental results was obtained.

Barzegar (1989) examined the effect of skew, anisotropic reinforcement on the post-cracking response and ultimate capacity of reinforced concrete membrane elements under monotonically increasing proportional loading. Appropriate constitutive models, including post-cracking behavior, were discussed. Several shear panels were analyzed and compared with experimental results.

Crisfield and Wills (1989) used smeared crack models to analyze the concrete panels tested by Vecchio and Collins (1982). Different models involved both fixed and rotating cracks, with and without allowance for the tensile strength of concrete. A simple plasticity model was also applied. Because the numerical solutions were obtained using the arc-length analysis procedure, the complete (including post-maximum) load-deflection responses were traced, and consequently it was possible to clearly identify not only the computed collapse load but also the collapse modes.

Gupta and Maestrini (1989a,b) used a (local) smeared approach to model the post-cracking behavior of reinforced concrete panels tested by Vecchio (1986). Good agreement was obtained.

Massicote, Elwi and MacGregor (1990) presented a practical two-dimensional hypoelastic model for reinforced concrete, with emphasis on a new approach for the description of tension stiffening, using a smeared crack band model. The model predictions were compared with test results of reinforced concrete members in uniaxial tension, reinforced concrete panels in shear, and reinforced concrete plates, simply supported along four sides and loaded axially and transversely.

Wu, Yoshikawa and Tanabe (1991) presented a complex smeared crack and damage mechanics approach used to model the tension stiffening effect for cracked reinforced concrete. Several membranes in various states of shear and tension were analyzed, and the results were compared with the experimental results. Nonlocal effects appear to have been ignored.

4.3.2 *Beams and frames*—A number of finite element analyses of beams and frames include correct fracture mechanics effects. In general, early models did not account for the fracture effects correctly, but since about 1980, most of the work has been quite sophisticated in this regard. Neither the smeared nor the discrete approach to cracking is clearly dominant. Some of the studies described below have provided useful insights into the structural behavior of beams. However, it is clear that the fracture analysis of beams and frames is not yet at a state where it can be used routinely in the design office.

Ngo and Scordelis (1967) reported the first discrete cracking FEM analysis of a reinforced concrete beam. Cracks were modeled but not propagated; their positions were assumed a priori. The analysis was linear elastic, and stresses at the crack tips were not accurately modeled. This pioneering paper did not make use of fracture mechanics principles.

Nilson (1967, 1968) discussed the use of the finite element method to represent reinforced concrete beams, including bond-slip and discrete cracking. He suggested an approach to modeling discrete cracks by remeshing to represent the development of cracks. His criteria for cracking were stress-based, rather than fracture mechanics-based, and yet, for the example that he studied (eccentrically reinforced tension-pull specimen), reasonable results were obtained. (The behavior of the tension-pull specimen that he studied was not primarily controlled by fracture of the concrete, but rather by the stiffness of the reinforcement).

Valliapan and Doolan (1972) discussed a smeared crack approach to the finite element analysis of reinforced concrete. Their model assumed elastoplastic response of the concrete with a tension cut-off, discrete modeling of the steel using truss elements, and no bond-slip. They analyzed several reinforced concrete beams with reasonable success, even without using correct fracture mechanics principles.

Colville and Abbasi (1974) employed an approach very similar to that of Valliapan and Doolan (1972), with the exception that the steel was considered to be smeared. For the examples presented in the paper, reasonable results were obtained. Again, their examples appear to have been fracture insensitive.

Nam and Salmon (1974) reported upon the smeared crack analysis of reinforced concrete beams. Their approach was similar to that of Colville and Abassi (1974). One of their most significant conclusions was that an incremental tangent stiffness rather than an iterative initial stiffness method of solution was necessary to solve such problems. While cracks were predicted, fracture mechanics principles were not included in their model.

Salah El-Din and El-Adawy Nassef (1975) were among the first to combine fracture mechanics and the finite element method to analyze reinforced concrete beams. They assumed (incorrectly) that the conditions of LEFM apply to relatively small concrete beams. Discrete cracks were represented by setting the stiffnesses of the concrete elements equal to zero. They modeled the reinforcement discretely, and connected the steel elements directly to the concrete elements, modeling zero bond-slip. They used the compliance derivative method to calculate the energy release rate of a single vertical crack. Their derivation predicted the fracture mechanics size effect. By comparing their numerical results to the results of an experimental investigation, they found that the fracture mechanics-based approach was more accurate than the stress-based approach for the prediction of cracking in reinforced concrete beams.

Although they recognized that their smeared crack approach to the modeling of shear-critical reinforced concrete beams was not theoretically correct from a fracture mechanics viewpoint, Cedolin and Dei Poli (1977) concluded that their local nonlinear constitutive model for concrete was adequate to predict most of the significant responses. Studying a reinforced concrete beam 22 in. (560 mm) deep with approximately 1.8 percent flexural steel, they compared the numerical and experimental results. Their finite element model, which incorporated a nonlinear representation for concrete under biaxial stress, was able to predict the load-deflection curve, crack pattern, and failure load of the reinforced concrete beams. The model, however, was not able to represent dowel action and crack propagation at failure.

Bergan and Holand (1979), in a comprehensive discussion of finite element analysis of reinforced concrete structures, presented an analysis using local smeared crack techniques similar to those just described by Cedolin and Dei Poli (1977). The computed load-deflection curve agreed with the experimental results, except that the ultimate load was overpredicted, and the actual beam failed by a rapid diagonal tension failure mechanism that could not be captured by their smeared crack model.

A smeared crack approach was also employed by Bédard and Kotsovos (1985), who analyzed a reinforced concrete deep beam with web openings and web reinforcement, using plane stress analysis. Comparison with experimental results was good. They stated that the smallest finite element size should be greater than two to three times the maximum aggregate size. However, there is no evidence that such a rule would produce universally successful analyses.

Bažant, Pan, and Pijaudier-Cabot (1987) and Bažant, Pijaudier-Cabot, and Pan (1987) analyzed the softening post-peak load-deflection relation for reinforced concrete beams and frames using layered finite elements. Concrete was modeled as a strain softening material in both tension and in compression; the steel reinforcement was modeled as elastic-plastic. Standard bending theory assumptions were used and bond-slip of reinforcement was neglected. The model could approximate existing test results for beams and frames. At the same time, constitutive laws with strain softening, including those of continuum damage mechanics, were shown, in general, to lead to spurious sensitivity of results with respect to the chosen finite element size, similar to that documented for other strain-softening problems. In analogy to the finite element crack-band model (Section 3.2.1), they recommend the use of a minimum admissible element size, specified as a cross sectional property.

One of the major problems in the analysis of reinforced concrete structures is the representation of reinforcing bars. Typically, either the steel is represented in a smeared manner, in which case the concrete and steel are combined and treated as a composite material, or the steel is represented discretely, in which case nodes must be collocated with the steel bars, which severely constrains the finite element mesh. Allwood and Bajarwan (1989) presented an interesting approach for modeling reinforcing steel that allows the steel to be represented independently of the concrete (using separate nodes and elements). An iterative solution scheme is used to bring the steel elements into equilibrium with the concrete elements.

Balakrishnan and Murray (1988) described a practical stress-strain relationship for concrete for use in a smeared crack model, which was effective in predicting the behavior of reinforced concrete beams and panels. The model includes homogenized material properties that incorporate the estimated effects of strain localization. Methods of estimating these homogenized properties are presented in their paper.

Channakeshava and Iyengar (1988) used a smeared crack model for a singly reinforced concrete beam without shear reinforcement. The shear critical beam failed by diagonal cracking. In the analysis, although the exact mode of failure could not be assessed, extensive diagonal cracking with large crack strains was observed prior to failure. Their model took into account the width of the element/Gauss point over which a crack is assumed to be smeared, and thus attempted to use a form of the crack band model to objectively represent the fracture energy associated with discrete cracking. They concluded that an accurate response was obtained using the technique.

Gustafsson and Hillerborg (1988) used the finite element method to model size effects due to discrete cohesive diagonal tension cracks in singly reinforced beams without shear reinforcement. They found a significant fracture mechanics size effect. They carried out studies of the sensitivity of shear strength to reinforcement ratio and depth-to-span ratio. Their work, aside from being correct from a fracture mechanics viewpoint, serves as an example of how the finite element method can be used to great advantage in developing design code equations. More recently, Ma, Niwa, and McCabe (1992) performed similar fictitious crack analyses. Their results show conclusively that a size effect exists for shear critical beams. They therefore recommend that the ACI 318 code be modified to reflect this size effect.

Cervenka et al. (1991) used a smeared crack approach to model shear failure in simply supported, reinforced and unreinforced concrete beams without shear reinforcement, using a crack band model. They performed two plane stress analyses of a fracture-sensitive unreinforced beam. One analysis used a fine mesh, and the other used a coarse mesh. They found that the same peak load was predicted by both meshes, but that the post-peak behavior of the beam was sensitive to mesh density.

Feenstra et al. (1991b) used the fictitious crack approach, including a representation for aggregate interlock, to model a moderately deep beam with several predefined discrete crack locations. Eight-noded continuum elements, six-noded interface elements and three-noded truss elements were used to model steel, with interface elements used to model bond-slip. A Newton-Raphson method with indirect displacement control was used. The computed response of the beam matched the experimental results reasonably well. This, of course, was not a completely predictive analysis because they inserted fictitious cracks at predetermined locations.

Gajer and Dux (1990, 1991) analyzed two reinforced concrete beams using a crack band model, featuring simplified representations of both nonorthogonal cracking and the influence of cracking on local material stiffnesses. At an integration point, one primary crack of fixed direction and one rotating crack are allowed to develop. The beams analyzed by de Borst and Nauta (1985) and Bédard and Kotsovos (1985) were analyzed, with satisfactory results.

Ozbolt and Eligehausen (1991) studied the shear resistance of four similar reinforced concrete beams of different sizes without shear reinforcement using a nonlocal microplane model and plane stress finite elements to study the size effect. The calculated results demonstrated a decrease in average strength with an increase in member size. It was demonstrated that the results were insensitive to mesh size and load path, in contrast to the results of local continuum analyses. They stated that element size must not be larger than one-half of the characteristic length of the nonlocal continuum. Otherwise, the analysis is equivalent to the classical local continuum analysis. Thus, large numbers of elements were required in these analyses, especially for the larger beams.

Pagnoni, Slater, Ameur-Moussa, and Buyukozturk (1992) formulated a finite element procedure for the nonlinear analysis of general three-dimensional reinforced concrete structures. They employed a three-dimensional version of the crack band model to objectively represent fracture. The model was demonstrated using detailed analyses of a singly reinforced deep beam and a prestressed concrete reactor vessel.

4.3.3 *Containment vessels*—Concrete containment vessels present challenging analysis problems. However, because of the severe consequences of a failure, detailed analysis is required to assure a safe design. These vessels are usually heavily reinforced, and therefore, fracture mechanics issues are of secondary importance.

Rashid (1968), in the first application of the smeared crack approach, analyzed a reinforced concrete pressure vessel. The steel elements were assumed to be elastic-perfectly-plastic. Concrete and steel were modeled using separate finite elements and an axisymmetric condition was assumed. Limited experimental comparisons were made. Because the concrete cracking was expected to be distributed, rather than localized, due to the presence of heavy reinforcement, no fracture mechanics issues arose from this approach. In essence, the composite reinforced concrete material was fracture insensitive.

Meyer and Bathe (1982) used smeared crack and smeared steel models to address some of the questions facing the engineer charged with assessing the safety of concrete structures for which conventional linear analysis is deemed inadequate. Guidelines were given to aid the engineer in selecting the material model most appropriate for this purpose. Questions related to finite element analysis were discussed, including numerical solution techniques, their accuracy, efficiency and applicability. An example of a prestressed concrete reactor vessel illustrated the application of nonlinear finite element analysis to concrete structures in engineering practice.

Clauss (1987) presented a summary of the analyses of a one-sixth scale model of a reinforced concrete containment vessel. The analyses were conducted by ten organizations in the United States and Europe. Each organization was supplied with a standard information package, which included construction drawings and actual material properties for most of the materials used in the model. Each organization worked independently using its own analytical methods. The report includes descriptions of the various analytical approaches and pretest predictions. Significant milestones that occur with increasing pressure, such as cracking and crushing of the concrete, yielding of the steel components, and the failure pressure and failure mechanism were described.

4.3.4 *Plates and shells*—Plates and shells are common components of reinforced concrete structures. Cracking can significantly influence both the stiffness and strength of these members. To model such a structure as a three-dimensional solid is computationally expensive. It is therefore necessary to use special plate and shell finite elements. Various approaches have been taken to model the effects of reinforcement and cracking, as the following references demonstrate. In general, a close match has been obtained between analytical and experimental results, without special regard for fracture concepts. The lack of importance of fracture mechanics is likely due to the dominant role played by reinforcement in the structures analyzed.

Hand, Pecknold, and Schnobrich (1973) used layered elements, in which material properties changed from layer to layer, to model several reinforced concrete plates under uniform stress, a corner-propped reinforced concrete plate under a concentrated central load, and a reinforced concrete shell under uniform load. The numerical results closely match the experimental results for all problems. Cracking was governed by the principal tensile stress.

Lin and Scordelis (1975) took a layered shell element approach to model both the reinforcement and the concrete. The concrete was assumed to behave plastically in compression, and as a gradually softening material in tension. Three examples, including a circular slab, a square slab, and a hyperbolic paraboloid shell, were analyzed, with the predicted results closely matching the experimental data.

Schnobrich (1977) used a (local) smeared crack and smeared steel model to solve several plate and shell problems. The agreement between analysis and experiment was excellent. Specimens modeled include a plate subject to twist and uniaxial moment, a corner-supported two-way slab, a three-hole conduit, a shear panel, and a hyperbolic paraboloid shell roof subjected to vertical loading.

Bashur and Darwin (1978) used a (local) smeared crack model to represent cracking as a continuous process through the depth of plate elements. Reinforcing steel was represented as a smeared material at the appropriate position(s) through the depth. Element stiffness was determined using numerical integration, removing the need to use layered elements. A close match was obtained with experimental results for a series of one-way and two-way reinforced concrete slabs.

Bergan and Holand (1979) analyzed reinforced concrete plates, using a smeared crack and smeared reinforcing model. Initially, zero tensile strength was assumed, which produced deflections that were too large. A gradually softening tensile behavior, however, produced a good comparison with experimental data. Neither of the plates analyzed was fracture sensitive.

Barzegar (1988) introduced the use of a layering technique in general nonlinear FEM analysis of reinforced concrete flat shell elements. A stack of simple four-node plane stress and Mindlin plate elements were used to simulate membrane and bending behavior. The method, together with nonlinear constitutive models for concrete and steel, was described and implemented. Bond degradation between concrete and steel was simulated. Several examples were presented.

Hu and Schnobrich (1991) proposed plane stress constitutive models for the nonlinear FEM analysis of reinforced concrete structures under monotonic loading. An elastic strain hardening plastic stress-strain relationship with a nonassociated flow rule was used to model concrete in the compression dominated region and an elastic brittle fracture behavior was assumed for concrete in the tension dominated area. After cracking took place, the smeared crack approach, together with the rotating crack concept, was employed. Together with layered elements, these material models were used to model the flexural behavior of reinforced concrete plates and shells. Good agreement with test results was obtained. Fracture mechanics issues were not considered.

Lewinski and Wojewodzki (1991) introduced a nonlinear (local, smeared crack) procedure to solve the elasticity problem for a doubly reinforced concrete slab was found. Ten different cracking patterns were assumed in this formulation. Use of each pattern implied the division of the slab thickness into several layers (maximum three layers of concrete). Numerical integration was used through the total thickness. Results obtained with this approach compared favorably with experimental results for deep beams and plates.

4.3.5 *Other reinforced concrete structures*—Gerstle (1987) used a plane strain (local) smeared crack finite element analysis technique for the analysis of reinforced concrete culverts. Since the steel was smeared, the reinforced concrete was treated as a composite material. No post-cracking strength was included in the model. The FEM analyses matched the experimental results reasonably well. However, the analyses tended to underestimate the deflections and overestimate the strength of the ring sections.

Swenson and Ingraffea (1991) used discrete cracking models to analyze the Schoharie Creek Bridge. The bridge collapsed in 1987 due to a flood that scoured the soil under a pier, which led to unstable cracking of the minimally reinforced pier section and precipitated the actual collapse. The analyses included linear and nonlinear fracture mechanics evaluations of the initiation, stability, and trajectory of the crack that precipitated failure. A principal motivation of the effort was to apply the results of recent concrete fracture mechanics research to a problem of practical significance. The authors concluded that cracking of the bridge foundation could be rationally explained through the use of fracture mechanics principles.

4.4—Closure

Over the past 30 years, research into finite element modeling of plain and reinforced concrete structures has been intense. The early models that did not take fracture mechanics into account proved quite satisfactory to solve the problems to which they were applied. While the possibility exists that the problems that have been presented in the literature were only the most successful (and fracture sensitive structures for which correspondence between analysis and test results was poor were simply not reported), it is clear that the published problems that were solved without the proper application of fracture mechanics were largely fracture insensitive. Further, nearly all of the analyses fail to include a convergence study to determine the effect of mesh refinement upon the solution.

Discrete and smeared crack FEM analyses have successfully solved problems that are known to be fracture sensitive. However, most of the problems solved to date have been almost trivial in their geometric simplicity. The challenge now is to apply correct fracture mechanics principles to the efficient analysis of nontrivial structures, such as reinforced concrete solids, frames, plates, and shells.

It does not appear that the discrete approach is "better" than the smeared approach or vice versa. On the contrary, it appears that it would be wise to move toward a marriage of these two approaches to better model the entire regime of concrete behavior from multiaxial compression of small specimens to LEFM in large structures. This is surely one of the major challenges of the coming years.

CHAPTER 5—CONCLUSIONS

5.1—General summary

The purpose of this report has been to describe the state-of-the-art of finite element analysis of fracture in reinforced concrete structures. Chapter 1 provides general background and historical perspective, and highlights the two most popular methods used in finite element analysis to represent fracture: the discrete crack approach and the smeared crack approach. Chapters 2 and 3 present the two approaches in greater detail. In each case, the goal has been to provide a broad view of the field and to provide references to the literature so that the interested reader may go directly to the primary sources for additional information.

Chapter 4 provides examples (with special attention to fracture mechanics) of finite element analyses of concrete structures that are available in the archival literature. Attention is focused on twelve types of structures, consisting of both plain and reinforced concrete. The review includes almost 100 papers that describe finite element analyses of these structural types. While most of the early analyses of concrete were inadequate from a theoretical fracture mechanics viewpoint, the results proved more than adequate, probably because of the nature of the problems selected. These early models were principally research tools, not applicable to design. The more recent efforts, although largely theoretically correct from a fracture mechanics viewpoint, generally describe analysis methods that remain too cumbersome to apply in routine design environments. For the most part, only the most readily available archival literature has been referenced.

From Chapter 4, it is clear that concrete structures can be categorized into two classes: those with behavior that is not governed by fracture mechanics and those with behavior that is. It is clear that special fracture mechanics considerations are not necessary for a large class of reinforced concrete structures. However, it is also clear that some reinforced concrete structures are fracture sensitive. To predict fracture behavior, a finite element code must include the capability to model fracture. As far as the engineer is concerned, if the structure being analyzed is known to be fracture insensitive, then the analysis may proceed without resorting to fracture mechanics. On the other hand, for structures that are known to be fracture sensitive, the engineer is well-advised to use finite element codes that correctly model fracture. There are, of course, times when an engineer is uncertain whether or not a structure is fracture sensitive. In these cases, the use of a finite element code capable of predicting fracture is advisable.

Fracture sensitivity is similar to buckling in that both behaviors may or may not exist, but each can only be detected through the use of analytical tools capable of predicting such effects. It seems likely that, in the next ten years, reliable codes with the capability to predict fracture of concrete structures will be developed, just as reliable codes with geometrically nonlinear capabilities have developed over the past ten years.

5.2—Future work

Very few successful commercial finite element codes can handle the nonlinear analysis of concrete structures. Those commercial codes that can handle the analysis and design of reinforced concrete structures typically perform linear analyses, or at most nonlinear geometric analyses, and then use the results of these analyses, together with design provisions (ACI 318), to determine the adequacy of the members. In other words, current design practice does not account for nonlinear material behavior, much less fracture behavior, during the analysis phase.

It therefore seems that the research community is far ahead of (and out of step with) the practicing community in the analysis of concrete structures. Since practicing engineers do not have a routine need to predict fracture behavior, they are not willing to pay for such an analysis capability. When they do have a need to predict or evaluate the fracture behavior of a structure, it usually involves a special application requiring so much of their time and energy that they often go to researchers to provide the solution.

The research community should continue to develop generally accepted methods for numerically modeling fracture behavior. But it is clear that, at least for now, these analysis methods will not be used directly by practicing design engineers, but rather by other researchers who may use the codes to study aspects of concrete structural behavior. These codes may be used, for instance, to develop new and better design expressions for the ACI Building Code (ACI 318).

A fruitful area of research involves combining the smeared crack and the discrete crack approaches. Both approaches have limitations, and by combining the two approaches, a high-fidelity numerical representation of concrete behavior may be achieved.

The fracture analysis of three-dimensional structures is still in its infancy. Both the smeared and discrete crack approaches require further development to be useful for the analysis of three-dimensional solids, not to mention shells, plates, and beams. The numerical approaches to the representation of the interaction between reinforcement and concrete (bond-slip) are currently unsatisfactory and primitive. Also, the effects of concrete creep, shrinkage, and cyclic behavior are usually ignored in numerical models of concrete behavior, even though they may have large effects upon structural response. Although much progress has been made, the goal of developing accurate and general numerical models for plain and reinforced concrete structures remains.

CHAPTER 6—REFERENCES

6.1—Recommended references

American Concrete Institute

318 Building Code Requirements for Structural Concrete

446.1R Fracture Mechanics of Concrete: Concepts, Models and Determination of Material Properties

6.2—Cited references

Allwood, R. J., and Bajarwan, A. A., 1989, "A New Method for Modeling Reinforcement and Bond in Finite Element Analyses in Reinforced Concrete," *International J. Num. Meth. in Engrg.*, V. 28, pp. 833-844.

Arrea, M., and Ingraffea, A. R., 1982, "Mixed-Mode Crack Propagation in Mortar and Concrete," *Report* No. 81-13, Department of Structural Engineering, Cornell University, Ithaca, N.Y.

ASCE Task Committee on Finite Element Analysis of Reinforced Concrete Structures, 1982, *Finite Element Analysis of Reinforced Concrete*, A. H. Nilson, Chmn., American Society of Civil Engineers, New York, 545 pp.

Atluri, S. N.; Kobayashi, A. S.; and Nakagaki, M., 1975, "An Assumed Displacement Hybrid Element Model for Linear Fracture Mechanics," *International Journal of Fracture*, V. 11, No. 2, Apr., pp. 257-271.

Ayari, M. L., and Saouma, V. E., 1990, "A Fracture Mechanics Based Seismic Analysis of Concrete Gravity Dams Using Discrete Cracks," *Engrg. Fracture Mechanics*, V. 35, No. 1/2/3, pp. 587-598.

Balakrishnan, S., and Murray, D. W., 1988, "Concrete Constitutive Model for NLFE Analysis of Structures," *Journal of Structural Engineering*, ASCE, V. 114, No. 7, July, pp. 1449-1466.

Balakrishnan S.; Elwi, A. E.; and Murray, D. W., 1988, "Effect of Modeling on NLFE Analysis of Concrete Structures," *Journal of Structural Engineering*, ASCE, V. 114, No. 7, July, pp. 1467-1487.

Barsom, J. M., and Rolfe, S. T., 1987, *Fracture and Fatigue Control in Structures*, 2nd Edition, Prentice-Hall, New York.

Barsoum, R. S., 1976, "On the Use of Isoparametric Finite Elements in Linear Fracture Mechanics," *Int. J. Num. Meth. in Engrg.*, V. 10, pp. 25-37.

Barzegar, F., 1988, "Layering of RC Membrane and Plate Elements in Nonlinear Analysis," *Journal of Structural Engineering*, ASCE, V. 114, No. 11, Nov., pp. 2474-2492.

Barzegar F., 1989, "Analysis of RC Membrane Elements with Anisotropic Reinforcement," *Journal of Structural Engineering*, ASCE, V. 115, No. 3, Mar., pp. 647-665.

Bashur, F. K., and Darwin, D., 1978, "Nonlinear Model for Reinforced Concrete Slabs," *Journal of the Structural Division*, ASCE, V. 104, No. ST1, Jan., pp. 157-170.

Bathe, K.-J., 1982, *Finite Element Procedures in Engineering Analysis*, Prentice-Hall, Englewood Cliffs, N.J., 736 pp.

Bažant, Z. P., 1976, "Instability, Ductility and Size Effect in Strain-Softening Concrete," *Journal of Engineering Mechanics Division*, ASCE, V. 102, No. EM2, Apr., pp. 331-344.

Bažant, Z. P., 1984, "Size Effect in Blunt Fracture: Concrete, Rock, Metal," *Journal of Engineering Mechanics*, ASCE, V. 110, No. 4, Apr., pp. 518-535.

Bažant, Z. P., 1986, "Mechanics of Distributed Cracking," *ASME Appl. Mech. Rev.*, V. 39, No. 5, May, pp. 675-705.

Bažant, Z. P., 1994, "Nonlocal Damage Theory Based on Micromechanics of Crack Interactions," *Journal of Engineering Mechanics*, ASCE, V. 120, No. 3, Mar., pp. 593-617. Addendum and Errata, V. 120, No. 6, June, pp. 1401-1402.

Bažant, Z. P., and Cedolin, L., 1979, "Blunt Crack Band Propagation in Finite Element Analysis," *Journal of Engineering Mechanics*, ASCE, V. 105, No. 2, Feb., pp. 297-315.

Bažant, Z. P., and Cedolin, L., 1980, "Fracture Mechanics of Reinforced Concrete," *Journal of Engineering Mechanics,* ASCE, V. 106, No. 6, Dec., pp. 1287-1306.

Bažant, Z. P., and Cedolin, L., 1983, "Finite Element Modeling of Crack Band Propagation," *Journal of Structural Engineering*, ASCE, V. 109, No. 2, Feb., pp. 69-92.

Bažant, Z. P., and Oh, B. H., 1983, "Crack Band Theory for Fracture of Concrete," *Mater. Struct.,* RILEM, Paris, France, V. 16, Feb., pp. 155-177.

Bažant, Z. P.; Belytschko T. B.; and Chang, T.-P., 1984, "Continuum Theory for Strain-Softening," *Journal of Engineering Mechanics*, ASCE, V. 110, No. 12, pp. 1666-1692.

Bažant, Z. P., and Pfeiffer, P. A., 1985, "Tests of Shear Fracture and Strain Softening in Concrete," *Proc.*, 2nd Symposium on the Interaction of Non-Nuclear Munitions on Structures, Panama City Beach, Florida, Apr., 15-19.

Bažant, Z. P.; Pan, J.; and Pijaudier-Cabot, G., 1987, "Softening in Reinforced Concrete Beams and Frames," *Journal of Structural Engineering*, ASCE, V. 113, No. 12, Dec., pp. 2333-2347.

Bažant, Z. P.; Pijaudier-Cabot, G.; and Pan, J., 1987, "Ductility, Snapback, Size Effect, and Redistribution in Softening Beams or Frames," *Journal of Structural Engineering*, ASCE, V. 113, No. 12, Dec., pp. 2348-2364.

Bažant, Z. P., and Lin, F. B., 1988, "Nonlocal Smeared Crack Model for Concrete Fracture," *Journal of Engineering Mechanics*, ASCE, V. 114, No. 11, Nov., pp. 2493-2510.

Bažant, Z. P., and Pijaudier-Cabot, G., 1988, "Nonlocal Continuum Damage, Localization Instability and Convergence," *Journal of Applied Mechanics*, ASME, V. 55, No. 2, pp. 287-293.

Bažant, Z. P., and Prat, P. C., 1988a, "Microplane Model for Brittle-Plastic Material: I. Theory," *Journal of Engineering Mechanics*, ASCE, V. 114, No. 10, Oct., pp. 1672-1688.

Bažant, Z. P., and Prat, P. C., 1988b, "Microplane Model for Brittle-Plastic Material: II. Verification," *Journal of Engineering Mechanics*, V. 114, No. 10, Oct., pp. 1689-1702.

Bažant Z. P., and Pijaudier-Cabot, G., 1989, "Measurement of Characteristic Length of Nonlocal Continuum," *Journal of Engineering Mechanics*, ASCE, V. 115, No. 4, Apr., pp. 755-767.

Bažant, Z. P., and Ozbolt, J., 1990, "Nonlocal Microplane Model for Fracture, Damage, and Size Effect in Structures," *Journal of Engineering Mechanics*, ASCE, V. 116, No. 11, Nov., pp. 2485-2505.

Bažant, Z. P.; Tabbara, M. R.; Kazemi, M. T.; and Pijaudier-Cabot, G., 1990, "Random Particle Model for Fracture of Aggregate or Fiber Composites," *Journal of Engineering Mechanics*, ASCE, V. 116, No. 8, Aug., pp. 1686-1705.

Bažant, Z. P., and Ozbolt, J., 1992, "Compression Failure of Quasibrittle Material: Nonlocal Microplane Model," *Journal of Engineering Mechanics*, ASCE, V. 118, No. 3, Mar., pp. 540-556.

Bažant, Z. P., and Jirasek, M., 1994, "Damage Nonlocality due to Microcrack Interactions: Statistical Determination of Crack Influence Function," *Fracture and Damage in Quasibrittle Structures: Experiment, Modeling, and Computer Analysis*, Z. P. Bažant, Z. Bittnar, M. Jirasek, and J. Mazars, eds., E&FN Spon, London, pp. 3-17.

Bažant, Z. P., and Li, Y.-N., 1995, "Stability of Cohesive Crack Model," *Journal of Applied Mechanics*, ASME, V. 62, No. 4, pp.

Bédard, C., and Kotsovos, M. D., 1985, "Application of NLFEA to Concrete Structures," *Journal of Structural Engineering*, ASCE, V. 111, No. 12, Dec., pp. 2691-2707.

Belytschko, T.; Fish, J.; and Engelmann, B. E., 1988, "A Finite Element with Embedded Localization Zones," *J. Comp. Meth. Appl. Mech. Engrg.*, V. 70, pp. 59-89.

Bergan, P., and Holand, I., 1979, "Nonlinear Finite Element Analysis of Concrete Structures," *J. Comp. Meth Appl. Mech Engrg.*, V. 17/18, pp. 443-467.

Bertholet, J. M., and Robert, J. L., 1990, "Damage Evaluation of Concrete Test Specimens Related to Failure Analysis," *Journal of Engineering Mechanics*, ASCE, V. 116, No. 3, Mar., pp. 587-605.

Bittencourt, T. N.; Barry, A.; and Ingraffea, A. R., 1992, "Comparison of Mixed-Mode Stress Intensity Factors Obtained Through Displacement Correlation, J-Integral Formulation, and Modified Crack-Closure Integral," *Fracture Mechanics: Twenty Second Symposium*, ASTM STP 1131, American Society for Testing and Materials, Philadelphia, V. II, pp. 69-82.

Bittencourt, T.; Ingraffea, A. R.; and Llorca, J., 1992, "Simulation of Arbitrary, Cohesive Crack Propagation," *Fracture Mechanics of Concrete Structures*, edited by Z. Bažant, Elsevier, London, pp. 339-350.

Bocca, P.; Carpinteri, A.; and Valente, S., 1991, "Mixed Mode Fracture of Concrete," *Int. J. Solids Structures*, V. 27, No. 9, pp. 1139-1153.

Bolander, J., and Hikosaka, H., 1992, "Applying Nonlocal Smeared Crack Concepts to Modeling Concrete Fracture," *Fracture Mechanics of Concrete Structures*, Z. P. Bažant, ed., Elsevier, London, pp. 198-203.

Bresler, B., and Bertero, V., 1968, "Behavior of Reinforced Concrete under Repeated Load," *Journal of the Structural Division*, ASCE V. 94, No. ST6, June, pp. 1567-1589.

Broek, D., 1986, *Elementary Engrg. Frac. Mech.*, 4th Ed., Klewer Academic Publishers, Dordrecht.

Brown, C. J.; Darwin, D.; and McCabe, S. L., 1993, "Finite Element Fracture Analysis of Steel-Concrete Bond," *SM Report* No. 36, University of Kansas Center for Research, Lawrence, Kansas.

Byskov, E., 1970, "The Calculation of Stress Intensity Factors Using the Finite Element Method with Cracked Elements," *Int. J. Fracture Mechanics*, V. 6, No. 2, June, pp. 129-267.

Carol, I., and Bažant, Z., 1991, "Damage-Rheology Uncoupling for Microplane Damage Tensor, with Application to Concrete with Creep," *Constitutive Laws for Engineering Materials*, C. Desai, E. Krempl, G. Frantziskonis, and H. Saadatmanesh, eds., ASME Press, Tucson, pp. 391-394.

Carol, I.; Bažant, Z. P.; and Prat, P. C., 1991, "Geometric Damage Tensor Based on Microplane Model," *Journal of Engineering Mechanics*, ASCE, V. 117, No. 10, Oct., pp. 2429-2448.

Carol, I.; Bažant, Z.; and Prat, P., 1992, "Microplane-Type Constitutive Models for Distributed Damage and Localized Cracking in Concrete Structures," *Fracture Mechanics of Concrete Structures*, edited by Z. P. Bažant, Elsevier, London, pp. 299-304.

Carol, I.; Prat, P.; and Bažant, Z., 1992, "New Explicit Microplane Model for Concrete: Theoretical Aspects and Numerical Implementation," *Int. J. Solids Structures*, V. 29, No. 9, pp. 1173-1191.

Carol, I.; Prat, P.; and Gettu, R., 1993, "Numerical Analysis of Mixed-Mode Fracture of Quasibrittle Materials Using a Multicrack Constitutive Model," *Mixed-Mode Fatigue and Fracture*, H. Rossmanith and K. Miller, eds., ESIS Publ. 14, Mechanical Engineering Publications Ltd., London, pp. 319-332.

Carol, I., and Willam, K., 1994, "Microcrack Opening/Closure Effects in Elastic-Degrading Models," *Fracture and Damage in Quasibrittle Structures: Experiment, Modeling, and Computer Analysis*, Z. P. Bažant, Z. Bittnar, M. Jirasek and J. Mazars, eds., E&FN Spon, London, pp. 41-52.

Carpinteri, A.; Colombo, G.; Ferrara, G.; and Giuseppetti, 1987, "Numerical Simulation of Concrete Fracture Through a Bilinear Softening Stress-Crack Opening Displacement Law," *Fracture of Concrete and Rock*, Proc. SEM-RILEM Int. Conf., pp. 178-191.

Carpinteri, A., 1989, "Size-Scale Effects on the Brittleness of Concrete Structures: Dimensional Analysis and Snap-Back Instability," *Fracture Mechanics: Application to Concrete*, SP-118, V. C. Li and Z. P. Bažant, eds., American Concrete Institute, Detroit, pp. 197-235.

Carpinteri, A.; Valente, S.; Ferrara, G.; and Imperato, L., 1992, "Experimental and Numerical Fracture Modelling of a Gravity Dam," *Fracture Mechanics of Concrete Structures*, Z. P. Bažant, ed., Elsevier, London, pp. 351-360.

Catalano, D., and Ingraffea, A. R., 1982, "Concrete Fracture: A Linear Elastic Fracture Mechanics Approach," MS thesis, *Department of Structural Engineering Report* 82-1, School of Civil and Environmental Engineering, Cornell University, Ithaca, N.Y.

Cavendish, J. C.; Field, D. A.; and Frey, W. H., 1985, "An Approach to Automatic Three-dimensional Finite Element Mesh Generation," *Int. J. Num. Meth. in Engrg.*, V. 21, pp. 329-347.

Cedolin, L., and Dei Poli, S., 1977, "Finite Element Studies of Shear-Critical R/C Beams," *Journal of Engineering Mechanics Division*, ASCE, V. 103, No. EM3, June, pp. 395-410.

Cervenka, V., and Gerstle, K. H., 1971, "Inelastic Analysis of Reinforced Concrete Panels, Part I: Theory," *IABSE Publications*, V. 31-II, pp. 31-45.

Cervenka, V., and Gerstle, K. H., 1972, "Inelastic Analysis of Reinforced Concrete Panels, Part II: Experimental Verification and Application," *IABSE Publications*, V. 32-II, pp. 25-39.

Cervenka, V.; Eligehausen, R.; and Pukl, R., 1991, "Computer Models of Concrete Structures," *IABSE Colloquium*, Stuttgart, pp. 311-320.

Cervera, M.; Oliver, J.; Herrero, E.; and Onate, E., 1990, "A Computational Model For Progressive Cracking in Large Dams Due to the Swelling of Concrete," *Engrg. Frac. Mech.*, V. 35, No. 1/2/3, pp. 573-585.

Chan, S. K.; Tuba, I. S.; and Wilson, W. K., 1970, "On the Finite Element Method in Linear Fracture Mechanics," *Engrg. Frac. Mech.*, V. 2, pp. 1-17.

Chang, T. Y.; Taniguchi, H.; and Chen, W. F., 1987, "Nonlinear Finite Element Analysis of Reinforced Concrete Panels," *Journal of Structural Engineering*, ASCE, V. 113, No. 1, Jan., pp. 122-140.

Channakeshava, C., and Iyengar, K. T. S. R., 1988, "Elasto-Plastic Cracking Analysis of Reinforced Concrete," *Journal of Structural Engineering*, ASCE, V. 114, No. 11, Nov., pp. 2421-2438.

Chen, W. F., 1982, *Plasticity in Reinforced Concrete*, McGraw-Hill, New York, 474 pp.

Clauss, D. B., 1987, "Round-Robin Pretest Analyses of a 1:6-Scale Reinforced Concrete Containment Model Subject to Static Internal Pressurization," *Report* SAND87-0891, May, Sandia Natl. Labs., Albuquerque, N.M.

Clough, R. W., 1962, "The Stress Distribution of Norfolk Dam," Series 100, Issue 19, Institute of Engineering Research, University of California, Berkeley, Aug.

Colville, J., and Abbasi, J., 1974, "Plane Stress Reinforced Concrete Finite Elements," *J. Struct. Div.*, ASCE, V. 100, No. ST5, May, pp. 1067-1083.

Computational Modelling of Concrete Structures, 1994, H. Mang, N. Bicanic, and R. deBorst, eds., Pineridge Press, London.

Computer-Aided Analysis and Design of Concrete Structures, 1984, F. Damjanic et al., eds., Pineridge Press, London.

Computer-Aided Analysis and Design of Concrete Structures, 1990, N. Bicanic and H. Mang, eds., Pineridge Press, London.

Concrete Design Based on Fracture Mechanics, 1992, SP-134, W. H. Gerstle and Z. P Bažant, eds., American Concrete Institute, Detroit, 176 pp.

Cosserat, E., and Cosserat, F., 1909, *Theorie des Corps Deformables*, Hermann, Paris.

Crisfield, M. A., and Wills, J., 1989, "Analysis of R/C Panels Using Different Concrete Models," *Journal of Engineering Mechanics*, ASCE, V. 115, No. 3, Mar., pp. 578-597.

Dahlblom, O., and Ottosen, N., 1990, "Smeared Crack Analysis Using a Generalized Fictitious Crack Model," *Journal of Engineering Mechanics*, ASCE, V. 116, No. 1, Jan., pp. 55-76.

Dam Fracture, 1991, V. Saouma, R. Dunger and D. Morris, eds., Electrical Power Research Inst., Palo Alto.

Dam Fracture and Fatigue, 1994, E. Bourdarot, J. Mazars, and V. Saouma, eds., Balkema, Rotterdam.

Darwin, D., 1985, "Concrete Crack Propagation-Study of Model Parameters," *Finite Element Analysis of Reinforced Concrete Structures*, C. Meyer and H. Okamura, eds., ASCE, New York, pp. 184-203.

Darwin, D., 1993, "Reinforced Concrete," Chapter 4, *Finite Element Analysis of Reinforced Concrete Structures II*, edited by J. Isenberg, American Society of Civil Engineers, pp. 203-232.

Darwin, D., and Pecknold, D. A. W., 1976, "Analysis of RC Shear Panels Under Cyclic Loading," *Journal of Structural Division*, ASCE, V. 102, No. ST2, Feb., pp. 355-369.

Darwin, D.; McCabe, S. L.; Brown, C. J.; and Tholen, M. L., 1994, "Fracture Analysis of Steel-Concrete Bond," *Fracture and Damage in Quasibrittle Structures: Experiment, Modeling, and Computer Analysis*, Z. P. Bažant, Z. Bittnar, M. Jirasek and J. Mazars, eds., E&FN Spon, London, pp. 549-556.

De Borst, R., and Nauta, P., 1985, "Nonorthogonal Cracks in a Smeared Finite Element Model," *Engrg. Comp.*, V. 2, No. 1, pp. 35-46.

De Borst, R., 1986, "Nonlinear Analysis of Frictional Materials," PhD Thesis, Technical University, Delft, Netherlands.

Desai, C. S.; Zaman, M. M.; Lightner, J. G.; and Siriwardane, H. J., 1984, "Thin-Layer Element for Interfaces and Joints," *Int. J. Numer. Anal. Meth. Geotech.*, V. 8., pp. 19-43.

Dodds, R. H.; Darwin, D.; and Leibengood, L. D., 1984, "Stress Controlled Smeared Cracking in R/C Beams," *Journal of Structural Engineering*, ASCE, V. 110, No. 9, Sept., pp. 1959-1976.

Droz, P., and Bažant, Z. P., 1989, "Nonlocal Analysis of Stable States and Stable Paths of Propagation of Damage Shear Bands," *Cracking and Damage*, J. Mazars and Z. P. Bažant, eds., Elsevier, London, pp. 415-425.

Du, J.; Kobaysh, A. S.; and Hawkins, N. M., 1990, "Direct FEM Analysis of Concrete Fracture Specimens," *Journal of Engineering Mechanics*, ASCE, V. 116, No. 3, Mar., pp. 605-619.

Dvorkin, E.; Cuitino, A.; and Gioia, G., 1990, "Finite Elements with Displacement Interpolated Embedded Localization," *Int. J. Num. Meth. in Engrg.*, V. 30, pp. 541-564.

Elfgren, L., 1989, *Fracture Mechanics of Concrete Structures*, RILEM Report, Chapman and Hall, London, 256 pp.

Elices, M., and Planas, J., 1989, "Material Models," Chapter 3 in *Fracture Mechanics of Concrete Structures*, L. Elfgren, ed., Chapman and Hall, London, pp. 16-66.

Elfgren, L., 1992, "Round Robin Analysis and Tests of Anchor Bolts," *Fracture Mechanics of Concrete Structures*, Z. P. Bažant, ed., Elsevier, London, pp. 865-869.

Eligehausen, R., and Ozbolt, J., 1990, "Size Effect in Anchorage Behavior," *Proc.*, 8th Biennial European Conf. on Fracture, ECF8, Oct. 1-5, Torino, pp. 2671-2677, EMAS, U.K.

Eligehausen, R., and Ozbolt, J., 1992, "Influence of Crack Width on the Concrete Cone Failure Load," *Fracture Mechanics of Concrete Structures*, Z. P. Bažant, ed., Elsevier, London, pp. 876-881.

Erdogan, F., and Sih, G. C., 1963, "On the Crack Extension in Plates Under Plane Loading and Transverse Shear," *Journal of Basic Engineering*, ASME, V. 85, pp. 519-527.

Eringen, A. C., 1965, "Theory of Micropolar Continuum," *Proceedings*, 9th Midwestern Mechanics Conf., Univ. of Wisconsin, Madison, pp. 23-40.

Eringen, A. C., 1966, "A Unified Theory of Thermomechanical Materials," *Int. J. Engrg. Sci*, V. 4, pp. 179-202.

Feenstra, P. H.; de Borst, R.; and Rots, J. G., 1991a, "Numerical Study on Crack Dilatancy. I: Models and Stability Analysis," *Journal of Engineering Mechanics*, ASCE, V. 117, Apr., No. 4, pp. 733-753.

Feenstra, P. H.; de Borst, R.; and Rots, J. G., 1991b, "Numerical Study on Crack Dilatancy. II: Applications," *Journal of Engineering Mechanics*, ASCE, V. 117, No. 4, Apr., pp. 754-769.

Finite Element Analysis of Reinforced Concrete II, 1993, J. Isenberg, ed., American Society of Civil Engineers, New York, 544 pp.

Firrao, D., Oct., 1990, "Fracture Behaviour and Design of Materials and Structures," *Proceedings*, 8th Biennial European Conference on Fracture, ECF8, Torino, Italy.

Fish, J., and Belytschko, T., 1988, "Elements with Embedded Localization Zones," *J. Comp. and Structures*, V. 30, pp. 247-256.

Fracture and Damage in Quasibrittle Structures: Experiment, Modeling, and Computer Analysis, 1994, Z. P. Bažant, Z. Bittnar, M. Jirasek, and J. Mazars, eds., E&FN Spon, London, 646 pp.

Fracture Mechanics: Application to Concrete, 1989, SP-118, V. C. Li and Z. P Bažant, eds., American Concrete Institute, Detroit.

Fracture Mechanics of Concrete Structures, 1992, Proceeding of the 1st International Conference on Fracture Mechanics of Concrete Structures (FraMCoS-1), Z. P. Bažant, ed., Elsevier, London.

Fracture Mechanics of Concrete Structures, 1995, Proc. 2nd Int. Conf. Fract. Mech. of Conc. and Conc. Struct. (FraMCoS-2), F. H. Wittmann, ed., Aedificatio Publications, Freiburg, Germany.

Fracture of Concrete and Rock, 1987, S. P. Shah and S. E. Swartz, eds., Proceedings of the SEM-RILEM International Conference, June 17-19, 1987, Houston.

Gajer, G., and Dux, P. F., 1990, "Crack Band Based Model for FEM Analysis of Concrete Structures," *Journal of Structural Engineering*, ASCE, V. 116, No. 6, June, pp. 1696-1714.

Gajer, G., and Dux, P. F., 1991, "Simplified Nonorthogonal Crack Model for Concrete," *Journal of Structural Engineering*, ASCE, V. 117, No. 1, Jan., pp. 149-164.

Garcia-Alvarez, V. O.; Carol., I.; and Gettu, R., 1994, "Numerical Simulation of Fracture in Concrete Using Joint Element," *Anales de Mecanica de la Fractura*, V. 11, pp. 75-80.

Gens, A.; Carol, I.; and Alonso, E., 1988, "An Interface Element Formulation for the Analysis of Soil-Reinforcement Interaction," *Computers and Geotechnics*, V. 7, pp. 133-151.

Gerstle, K. H., 1987, "Strength Design of Reinforced-Concrete Hydraulic Structures," *Technical Report* SL-80-4, U.S. Army Corps of Engineers, July.

Gerstle, W. H., 1982, "The Fracture Mechanics of Bond in Reinforced Concrete," MS Thesis, Department of Structural Engineering, Cornell University, Ithaca, N.Y.

Gerstle, W. H., 1986, "Finite and Boundary Element Modelling of Crack Propagation in Two- and Three-Dimensions Using Interactive Computer Graphics," Department of Structural Engineering, Cornell University, Ithaca, N.Y.

Gerstle, W. H.; Martha, L. F.; and Ingraffea, A. R., 1987, "Finite and Boundary Element Modelling of Crack Propagation in Two- and Three-Dimensions," *Engrg. with Comp.*, V. 2, pp. 167-183.

Gerstle, W. H., and Abdalla, J. E., Jr., 1990, "Finite Element Meshing Criteria for Crack Problems," *Fracture Mechanics: Twenty-First Symposium*, ASTM STP 1074, American Society for Testing and Materials, Philadelphia, pp. 509-521.

Gerstle, W. H., and Xie, M., 1992, "FEM Modeling of Fictitious Crack Propagation in Concrete," *Journal of Engineering Mechanics*, ASCE, V. 118, Feb., No. 2, pp. 416-434.

Goodman, R. E.; Taylor, R. L.; and Brekke, T. L., 1968, "A Model for the Mechanics of Jointed Rock," *J. Soil Mech. and Found. Div.*, ASCE, V. 94, No. 3, pp. 637-659.

Gopalaratnam, V. S., and Ye, B. S., 1991, "Numerical Characterization of Nonlinear Fracture Process in Concrete," *Engrg. Fract. Mech.*, V. 40, No. 6, pp. 991-1006.

Griffith, A. A., 1920, "The Phenomenon of Rupture and Flow in Solids," *Philosophical Transactions*, Royal Society of London, Series A, V. 221, pp. 163-198.

Gupta, A. K., and Maestrini, S. R., 1989a, "Post-Cracking Behavior of Membrane Reinforced Concrete Elements Including Tension Stiffening," *Journal of Structural Engineering*, ASCE, V. 115, No. 4, Apr., pp. 957-976.

Gupta, A. K., and Maestrini, S. R., 1989b, "Unified Approach to Modeling Post-Cracking Membrane Behavior of Reinforced Concrete," *Journal of Structural Engineering*, ASCE, V. 115, No. 4, Apr., pp. 977-993.

Gustafsson, P. J., and Hillerborg, A., 1984, "Improvements in Concrete Design Achieved through the Application of Fracture Mechanics," *Applica-*

tion of Fracture Mechanics to Cementitious Materials, NATO Advanced Research Workshop, Northwestern University, Evanston, Ill.

Gustafsson, P. J., 1985, *Fracture Mechanics Studies of Non-Yielding Materials Like Concrete*, TVBM-1007, Division of Building Materials, University of Lund, Sweden.

Gustafsson, P. J., and Hillerborg, A., 1988, "Sensitivity in Shear Strength of Longitudinally Reinforced Concrete Beams to Fracture Energy of Concrete," *ACI Structural Journal*, V. 85, No. 3, May-June, pp. 286-294.

Hand, F. R.; Pecknold, D. A.; and Schnobrich, W. C., 1973, "Nonlinear Layered Analysis of RC Plates and Shells," *Journal of Structural Division*, ASCE, V. 99, No. ST7, July, pp. 1491-1505.

Hellen, T. K., 1975. "On the Method of Virtual Crack Extensions," *Int. J. Num. Meth. in Engrg.*, V. 9, pp. 187-207.

Hellier, A. K.; Sansalone, M.; Ingraffea, A. R.; Carino, N. J.; and Stone, W. C., 1987, "Finite Element Analysis of the Pullout Test Using a Nonlinear Discrete Cracking Approach," *Cement, Concrete and Aggregates*, V. 9, No. 1, pp. 20-29.

Henshell, R. D., and Shaw, K. G., 1975, "Crack Tip Finite Elements are Unnecessary," *Int. J. Num. Meth. in Engrg.*, V. 9, pp. 495-507.

Heuze, F. E., and Barbour, T. G., 1982, "New Models for Rock Joints and Interfaces," *Journal of the Geotechnical Engineering Division*, ASCE, V. 108, No. GT5, pp. 757-776.

Hillerborg, A.; Modeer, M.; and Petersson, P. E., 1976, "Analysis of Crack Formation and Crack Growth in Concrete by Means of Fracture Mechanics and Finite Elements," *Cem. and Conc. Res.*, V. 6, 773-782.

Hillerborg, A., and Rots, J., 1989, "Crack Opening and Numerical Modelling," *Fracture Mechanics of Concrete Structures*, L. Elfgren, ed., Chapman and Hall, London, pp. 128-146.

Hohberg, J., 1990, "A Note on Spurious Oscillations in FEM Joint Elements," *Earthquake Engrg. and Struct. Dyn.*, V. 19, pp. 773-779.

Hohberg, J., 1992a, "A Joint Element for the Nonlinear Dynamic Analysis of Arch Dams," *Technical Report* 186, Institute of Structural Engineering, ETH Zurich.

Hohberg, J., 1992b, "Multimechanism Plasticity with Coupled Damage in Tension and Shear," *Computational Plasticity* (COMPLAS III), D. Owen, E. Onate, and E. Hinton, eds., V. 2, Pineridge Press, London, pp. 1503-1514.

Hu, H. T., and Schnobrich, W. C., 1991, "Nonlinear Finite Element Analysis of Reinforced Concrete Plates and Shells under Monotonic Loading," *Computers and Structures*, V. 38, No. 5/6, pp. 637-651.

Hussain, M. A.; Pu, S. L.; and Underwood, J. H., 1974, "Strain Energy Release Rate for a Crack Under Combined Mode I and Mode II," *Fracture Analysis*, ASTM, STP 560, American Society for Testing and Materials, pp. 2-28.

Ingraffea, A. R., 1977, "Discrete Fracture Propagation in Rock: Laboratory Tests and Finite Element Analysis," PhD Dissertation, University of Colorado, Boulder.

Ingraffea, A. R., and Manu, C., 1980, "Stress-Intensity Factor Computation in Three Dimensions with Quarter-Point Elements," *Int. J. Num. Meth. in Engrg.*, V. 15, pp. 1427-1445.

Ingraffea, A. R., and Chappell, J. F., 1981, "A Fracture Mechanics Investigation of the Cracking of Fontana Dam," *Technical Report* No. 81-7, Cornell University, Ithaca, NY.

Ingraffea, A. R., 1983, "Numerical Modelling of Crack Propagation," *Rock Fracture Mechanics*, Springer-Verlag, Wien-New York, pp. 151-208.

Ingraffea, A. R.; Gerstle, W. H.; Gergely, P.; and Saouma, V., 1984, "Fracture Mechanics of Bond in Reinforced Concrete," *Journal of Structural Engineering*, ASCE, V. 110, No. 4, Apr., pp. 871-890.

Ingraffea, A. R., and Saouma, V., 1984, "Numerical Modeling of Discrete Crack Propagation in Reinforced and Plain Concrete," *Fracture Mechanics of Concrete: Structural Application and Numerical Calculation*, G. Sih and A. DiTommaso, eds., Martinus Nijhoff, Hingham, Mass., pp. 171-225.

Ingraffea, A. R., and Gerstle, W. H., 1985, "Non-Linear Fracture Models for Discrete Crack Propagation," *Application of Fracture Mechanics to Cementitious Composites*, NATO-ARW, S. P. Shah, ed., Sept. 4-7, 1984, Martinus-Nijhoff, pp. 247-285.

Ingraffea, A. R., and Panthaki, M., 1986, "Analysis of Shear Fracture Tests of Concrete Beams," *Finite Element Analysis of Reinforced Concrete Structures*, C. Meyer and H. Okamura, ed., American Society of Civil Engineers, New York, pp. 151-173.

Ingraffea, A. R., 1990, "Case Studies of Simulation of Fracture in Concrete Dams," *Engrg. Fracture Mechanics*, V. 35, No. 1/2/3, pp. 553-564.

Jenq Y. S., and Shah, S. P., 1985, "Nonlinear Fracture Parameters for Cement Based Composites: Theory and Experiments," *Application of Fracture Mechanics to Cementitious Composites*, Proc. NATO Adv. Res. Workshop, V. 15, pp. 319-360.

Jirasek, M., and Bažant, Z. P., 1994, "Localizaton Analysis of Nonlocal Model Based on Crack Interactions," *Journal of Engineering Mechanics*, ASCE, V. 120, No. 3, Mar., pp. 1521-1542.

Jirasek, M., and Bažant, Z. P., 1995a, "Macroscopic Fracture Characteristics of Random Particle Systems," *Int. J. Fracture*, V. 69, pp. 201-228.

Jirasek, M., and Bažant, Z. P., 1995b, "Particle Model for Quasibrittle Fracture," *Journal of Engineering Mechanics*, ASCE, V. 121, No. 9, Sept., pp. 1016-1025.

Jordan, W. B., 1970, "The Plane Isoparametric Structural Element," General Electric Co., Knolls Atomic Power Lab, *Report* No. KAPL-M-7112.

Kaplan, M. F., 1961, "Crack Propagation and the Fracture of Concrete," ACI JOURNAL, *Proceedings* V. 58, No. 5, May, pp. 591-610.

Karsan, I. D., and Jirsa, J. O., 1969, "Behaviour of Concrete Under Compressive Loadings," *Journal of Structural Engineering*, ASCE, V. 95, No. ST12, Dec., pp. 2543-2563.

Kay, G. J.; Govindjee, S.; Laursen, T.; and Maker, B., 1992, "Development of Concrete Tensile Fracture and Concrete/Rebar Bond Strength Models," *Fracture Mechanics of Concrete Structures*, Z. P. Bažant, ed., Elsevier, London, pp. 849-852.

Kesler, C.; Naus, D.; and Lott, J., 1972, "Fracture Mechanics—Its Applicability to Concrete," *Proc.*, 1971 Int. Conf. on Mechanical Behavior of Materials, V. IV, pp. 113-124.

Keuser, M., and Mehlhorn, G., 1987, "Finite Element Models for Bond Problems," *Journal of Structural Engineering*, ASCE, V. 113, No. 10, Oct., pp. 2160-2173.

Klisinski, M.; Runesson, K.; and Sture, S., 1991, "Finite Element with Inner Softening Band," *Journal of Engineering Mechanics*, ASCE, V. 117, No. 3, Mar., pp. 575-587.

Leibengood, L. D.; Darwin, D.; and Dodds, R. H., 1986, "Parameters Affecting FE Analysis of Concrete Structures," *Journal of Structural Engineering*, ASCE, V. 112, No. 2, Feb., pp. 326-341.

Lewinski, P. M., and Wojewodzki, W., 1991, "Integrated Finite Element Model for Reinforced Concrete Slabs," *Journal of Structural Engineering*, ASCE, V. 117, No. 4, Apr., pp. 1017-1038.

L'Hermite, R. et al., 1952, "Déformation du Béton," *Batir*, V. 27, pp. 24-28.

Li, V. C., and Liang, E., 1986, "Fracture Processes in Concrete and Fiber Reinforced Cementitious Composites," *Journal of Engineering Mechanics*, ASCE, V. 112, No. 6, June, pp. 566-586.

Li, Y. N., and Liang, R. Y., 1992, "Stability Theory of Cohesive Crack Model," *Journal of Engineering Mechanics*, ASCE, V. 118, No. 3, Mar., pp. 587-603.

Li, Y.-N., and Bažant, Z. P., 1994, "Eigenvalue Analysis of Size Effect for Cohesive Crack Model," *Int. J. Fracture*, V. 66, No. 4, pp. 213-226.

Liang, R. Y., and Li, Y.-N., 1991, "Study of Size Effect in Concrete Using Fictitious Crack Model," *Journal of Engineering Mechanics*, ASCE, V. 117, No. 7, July, pp. 1631-1651.

Liaw, B. M.; Jeang, F. L.; Hawkins, N. M.; and Kobayashi, A. S., 1990a, "Fracture-Process Zone for Mixed-Mode Loading of Concrete," *Journal of Engineering Mechanics*, ASCE, V. 116, No. 7, July, pp. 1560-1579.

Liaw, B. M.; Jeang, F. L.; Du, J. J.; Hawkins, N. M.; and Kobayashi, A. S., 1990b, "Improved Nonlinear Model for Concrete Fracture," *Journal of Engineering Mechanics*, ASCE, V. 116, No. 2, Feb., pp. 429-445.

Lin, C. S., and Scordelis, A. C., 1975, "Nonlinear Analysis of RC Shells of General Form," *Journal of Structural Division*, ASCE, V. 101, No. ST3, Mar., pp. 523-538.

Linsbauer, H. N.; Ingraffea, A. R.; Rossmanith, H. P.; and Wawrzynek, P. A., 1989a, "Simulation of Cracking in Large Arch Dam: Part I," *Journal of the Structural Division*, ASCE, V. 115, No. 7, July, pp. 1599-1615.

Linsbauer, H. N.; Ingraffea, A. R.; Rossmanith, H. P.; and Wawrzynek, P. A., 1989b, "Simulation of Cracking in Large Arch Dam: Part II," *J. Struct. Div.*, ASCE, V. 115, No. 7, July, pp. 1616-1630.

Lopez, C. M., and Carol, I., 1995, "Fracture Analysis of Concrete Microstructure Using Interface Elements," *Anales de Mecanica de la Fractura*, V. 12, pp. 75-80.

Lotfi, H. R., 1992, "Finite Element Analysis of Fracture of Concrete and Masonry Structures," PhD thesis, Dept. of CEAE, University of Colorado, Boulder.

Lotfi, H. R., and Shing, P. B., 1994, "Interface Model Applied to Fracture of Masonry Structures," *Journal of Engineering Mechanics*, ASCE, V. 120, No. 1, Jan., pp. 63-80.

Lotfi, H. R., and Shing, P. B., 1995, "Embedded Representation of Fracture in Concrete with Mixed Finite Elements," *Int. J. Num. Meth. in Engrg*, in press.

Ma, T.; Niwa, J.; and McCabe, S. L., 1992, "Evaluation of the Shear Behavior of Reinforced Concrete Beams—A Comparison of Code and Fracture Mechanics Predictions," *Fracture Mechanics of Concrete Structures*, Z. P. Bažant, ed., Elsevier, London, pp. 788-793.

Malvar, L. J., 1992, "Mixed Moded Fracture in Concrete," *Fracture Mechanics of Concrete Structures*, Z. P. Bažant, ed., Elsevier, London, pp. 677-688.

Martha, L., 1989, "Topological and Geometrical Modeling Approach to Numerical Discretization and Arbitrary Fracture Simulation in Three-Dimensions," PhD Dissertation, School of Civil Engineering, Cornell University, Ithaca, N.Y.

Martha, L. F.; Llorca, J.; Ingraffea, A. R.; and Elices, M., 1991, "Numerical Simulation of Crack Initiation and Propagation in an Arch Dam," *Dam Engineering*, V. 2, No. 3, pp. 193-214.

Massicote, B.; Elwi, A. E.; and MacGregor, J. G., 1990, "Tension-Stiffening Model for Planar Reinforced Concrete Members," *Journal of Structural Engineering*, ASCE, V. 116, No. 11, Nov., pp. 3039-3058.

Mau, S. T., and Yang, M. S., 1977, "Some Applications of a Hybrid Finite Element Method to Crack Problems," *Fracture Mechanics and Technology*, edited by George Sih, Noordhoff, Netherlands, V. 11, pp. 1467-1482.

Mehlhorn G.; Herrmann, R.; and Zerna, W., eds., 1978, "Nonlinear Behaviour of Reinforced Concrete Spatial Structures," *Proc.*, Int. Assoc. of Shell Structures, July 3-7, Darmstadt, Werner-Verlag, Dusseldorf, Germany.

Meyer, C.; and Bathe, K.-J., 1982, "Nonlinear Analysis of R/C Structures in Practice," *Journal of the Structural Division*, ASCE, V. 108, No. ST7, July, pp. 1605-1622.

Modeer, M., 1979, "A Fracture Mechanics Approach to Failure Analyses of Concrete Materials," *Report* TVBM-1001, thesis, Div. of Build. Mat., Univ. of Lund, Sweden.

Nam, C.-H., and Salmon, C. G., 1974, "Finite Element Analysis of Concrete Beams," *Journal of the Structural Division*, ASCE, V. 100, No. ST12, Dec., pp. 2419-2432.

Ngo, D., and Scordelis, A. C., 1967, "Finite Element Analysis of Reinforced Concrete Beams," ACI JOURNAL, *Proceedings* V. 64, No. 3, Mar., pp. 152-163.

Nilson, A. H., 1967, "Finite Element Analysis of Reinforced Concrete," PhD Thesis, Dept. of Civil Engineering, University of California, Berkeley, Calif.

Nilson, A. H., 1968, "Nonlinear Analysis of Reinforced Concrete by Finite Element Method," ACI JOURNAL, *Proceedings* V. 65, No. 9, Sept., pp. 757-766.

Ortiz, M.; Leroy, Y.; and Needleman, A., 1987, "A Finite Element Method for Localized Failure Analyis," *J. Comp. Meth in Appl Mech. and Engrg.*, V. 61, pp. 189-214.

Ozbolt, J., and Eligehausen, R., 1991, "Analysis of Reinforced Concrete Beams without Shear Reinforcement Using Non-Local Microplane Model," *Fracture Processes in Concrete, Rock and Ceramics*, RILEM, pp. 919-930.

Ozbolt, J., and Eligehausen, R., 1992, "Simulation of Cyclic Bond-Slip Behavior," *Fracture Mechanics of Concrete Structures*, Z. P. Bažant, ed., Elsevier, London, pp. 912-917.

Ozbolt, J., and Bažant, Z. P., 1992, "Microplane Model for Cyclic Triaxial Behavior of Concrete," *Journal of Engineering Mechanics*, ASCE, V. 118, No. 7, July, pp. 1365-1386.

Ozbolt, J., and Bažant, Z. P., 1996, "Numerical Smeared Fracture Analysis: Nonlocal Microcrack Interaction Approach," *Int. J. Num. Meth. Engrg.*, V. 39, pp. 635-661.

Pagnoni, T.; Slater, J.; Ameur-Moussa, R.; and Buyukozturk, O., 1992, "A Nonlinear Three-Dimensional Analysis of Reinforced Concrete Based on a Bounding Surface Model," *Computers and Structures*, V. 43, No. 1, pp. 1-12.

Papadrakis, M., 1981, "A Method for the Automatic Evaluation of the Dynamic Relaxation Parameters," *Comp. Meth. Appl. Mech. Engrg.*, V. 25, No. 1, pp. 35-48.

Parks, D. M., 1974, "A Stiffness Derivative Finite Element Technique for Determination of Crack Tip Stress Intensity Factors," *Int. J. of Fract.*, V. 10, No. 4, pp. 487-502.

Peruchio, R.; Saxena, M.; and Kela, A., 1989, "Automatic Mesh Generation from Solid Models Based on Recursive Spatial Decompositions," *Int. J. Num. Methods in Engrg.*, V. 28, pp. 2469-2501.

Petersson, P.-E., and Gustafsson, P. J., 1980, "A Model for Calculation of Crack Growth in Concrete-Like Materials," *Numerical Methods in Fracture Mechanics*, edited by Owen and Luxmoore, Pineridge Press, Swansea, U.K., pp. 707-719.

Petersson, P. E., 1981, "Crack Growth and Development of Fracture Zones in Plain Concrete and Similar Materials," *Report* TVBM-1006, Division of Building Materials, Lund Institute of Technology, Sweden.

Planas, J., and Elices, M., 1991, "Nonlinear Fracture of Cohesive Materials," *Int. J. Fracture*, V. 51, No. 2, pp. 139-158.

Planas, J., and Elices, M., 1992, "Asymptotic Analysis of a Cohesive Crack: 1. Theoretical Background," *Int. J. Fracture*, V. 55, No. 2, pp. 153-178.

Planas, J., and Elices, M., 1993a, "The Equivalent Elastic Crack: 1. Load-Y Equivalences," *Int. J. Fracture*, V. 61, No. 2, pp. 159-172.

Planas, J., and Elices, M., 1993b, "Asymptotic Analysis of a Cohesive Crack: 2. Influence of the Softening Curve," *Int. J. Fracture*, V. 64, No. 3, pp. 221-237.

Pramono, E., and Willam, K, 1987, "Fracture Energy-Based Plasticity Formulation of Plain Concrete," *Journal of Engineering Mechanics*, ASCE, V. 115, No. 6, June, pp. 1183-1204.

Pukl, R.; Schlottke, B.; Ozbolt, J.; and Eligehausen, R., 1992, "Computer Simulation: Splitting Tests of Concrete Thick-Walled Rings," *Fracture Mechanics of Concrete Structures,* edited by Z. P. Bažant, Elsevier, London.

Rahulkumar, P., 1992, "FEM Modeling of Crack Propagation in Concrete Using an Energy Approach," MS thesis, Dept. of Civil Engineering, Univ. of New Mexico, Albuquerque.

Rashid, Y. R., 1968, "Ultimate Strength Analysis of Prestressed Concrete Pressure Vessels," *Nuclear Engineering and Design*, V. 7, pp. 334-344.

Rice, J. R., 1968, "A Path Independent Integral and the Approximate Analysis of Strain Concentrations by Notches and Cracks," *Journal of Applied Mechanics*, V. 35, pp. 379-386.

Roelfstra, P. E.; Sadouki, H.; and Wittmann, F. H., 1985, "Le Beton Numerique," *Mater. Struct.*, RILEM, V. 18, pp. 309-317.

Rots, J., 1988, "Computational Modelling of Concrete Fracture," PhD Thesis, Delft University of Technology, the Netherlands.

Rots, J. G.; Kusters, G. M. A.; and Blaawendraad, J., 1984, "The Need for Fracture Mechanics Options in Finite Element Models for Concrete Structures," *Computer-Aided Analysis and Design of Concrete Structures*, Part I, pp. 19-32.

Rots, J. G., and de Borst, R., 1987, "Analysis of Mixed-Mode Fracture in Concrete," *Journal of Engineering Mechanics*, V. 113, No. 11, Nov., pp. 1739-1758.

Rots, J., and Schellekens, J., 1990, "Interface Elements in Concrete Mechanics," N. Bicanic and H. Mang, eds., *Computer Aided Analysis and Design of Concrete Structures*, Pineridge Press, London, V. 2, pp. 909-918.

Salah El-Din, A. S., and El-Adawy Nassef, M. M., 1975, "A Modified Approach for Estimating the Cracking Moment of Reinforced Concrete Beams," ACI JOURNAL, *Proceedings* V. 72, No. 7, July, pp. 356-360.

Saouma, V. E., 1981, "Interactive Finite Element Analysis of Reinforced Concrete: A Fracture Mechanics Approach," PhD thesis, Dept. of Struct. Engrg., Cornell University, Ithaca, N.Y.

Saouma, V. E., and Schwemmer. D., 1984, "Numerical Evaluation of the Quarter-Point Crack Tip Element," *Int. J. Num. Meth. in Engrg.*, V. 20, pp. 1629-1641.

Saouma, V. E.; Ayari, M. L.; and Boggs, H., 1987, "Fracture Mechanics of Concrete Gravity Dams," *Proceedings*, SEM-RILEM Int. Conf. of Fracture and Concrete Rock, Houston, June 17-19, pp. 496-519.

Sapadis, N., and Perucchio, R., 1989, "Advanced Techniques for Automatic Finite Element Meshing from Solid Models," *Computer-Aided Design*, V. 21, pp. 248-253.

Schellekens, J., and de Borst, R., 1993, "On the Numerical Integration of Interface Elements," *Int. J. Num. Meth. Engrg.*, V. 36, pp. 43-66.

Schlangen, E., and van Mier, J. G. M., 1992, "Shear Fracture in Cementitious Composites Part II: Numerical Simulations," *Fracture Mechanics of Concrete Structures*, Z. P. Bažant, eds., Elsevier, London, pp. 671-676.

Schnobrich, W. C., 1977, "Behavior of Reinforced Concrete Structures Predicted by the Finite Element Method," *Computers and Structures*, V. 7, pp. 265-376.

Schreyer, H. L., 1990, "Analytical Solutions for Nonlinear Strain-Gradient Softening and Localization," *Journal of Applied Mechanics*, V. 57, pp. 522-528.

Schroeder, W. J., and Shepard, M. S., 1988, "Geometry-based Fully Automatic Mesh Generation and the Delauney Triangulation," *Int. J. Num. Meth. in Engrg.*, V. 26, pp. 2503-2515.

Sha, G. T., and Yang, C.-T., 1990, "Use of Analytically Decoupled Near-Tip Displacement Solutions for Calculating the Decoupled Weight Functions of Mixed Fracture Modes," *Engrg. Fract. Mech.*, V. 37, No. 3, pp. 591-610.

Shepard, M. S., Oct., 1984, "Automated Analysis Model Generation," *Computer Aided Design in Civil Engineering*, American Society of Civil Engineers, New York, pp. 92-99.

Shih, C. F.; deLorenzi, H. G.; and German, M. D., 1976, "Crack Extension Modelling with Singular Quadratic Isoparametric Elements," *Int. J. Fract.*, V. 12, pp. 647-651.

Shivakumar, K. N.; Tan, P. W.; and Newman, J. C., 1988, "A Virtual Crack-Closure Technique for Calculating Stress Intensity Factors for Cracked Three-Dimensional Bodies," *Int. J. Fracture*, V. 36, pp. R43-R50.

Shumin, C., and Xing, Z., 1990, "Generalized Stiffness Derivative Method for Mixed Mode Crack Problems," *Mechanics Res. Comm.*, V. 17, No. 6, pp. 437-444.

Sih, G. C., 1974, "Strain Energy Density Factor Applied to Mixed-Mode Crack Problems," *Int. J. of Fracture*, V. 10, pp. 305-321.

Size Effect in Concrete Structures, 1994, H. Mihashi, H. Okamma, and Z. P. Bažant, eds., E & F N Spon, London.

Skrikerud, P., 1986, "Discrete Crack Modelling for Dynamically Loaded, Unreinforced Concrete Structures," *Earthquake Engrg. and Struct. Dyn.*, V. 14, pp. 297-315.

Sluys, L. J., and de Borst, R., 1992, "Analysis of Impact Fracture in a Double-Notched Specimen Including Rate Effects," *Fracture Mechanics of Concrete Structures*, edited by Z. P. Bažant, Elsevier, London, pp. 610-615.

Sousa, J. L.; Martha, L. F.; Wawrzynek, P. A.; and Ingraffea, A. R., 1989, *Simulation of Non-Planar Crack Propagation in Three-Dimensional Structures in Concrete and Rock Fracture of Concrete and Rock Recent Developments*, S.P. Shah, S. E. Swartz, and B. I. G. Barr, eds., Elsevier Science Publishing, New York, pp. 254-264.

Stankowski, T., 1990, "Numerical Simulation of Progressive Failure in Particle Composites," PhD Thesis, Dept. of CEAE, Univ. of Colorado, Boulder.

Stankowski, T.; Etse, G.; Runesson, K.; Sture, S.; and Willam, K., 1992, "Composite Analysis with Discrete and Smeared Crack Concepts," *Fracture Mechanics of Concrete Structures*, edited by Z. P. Bažant, Elsevier, London, pp. 269-274.

Stankowski, T.; Runesson, K.; and Sture, S., 1993, "Fracture and Slip of Interfaces in Cementitious Composites. I: Characteristics, II: Implementation," *Journal of Engineering Mechanics* ASCE, V. 119, No. 2, Feb., pp. 292-327.

Struman, G. M.; Shah, S. P.; and Winter, G., 1965, "Effect of Flexural Strain Gradients on Microcracking and Stress-Strain Behavior of Concrete," ACI JOURNAL, *Proceedings* V. 62, No. 7, July, pp. 805-822.

Swenson, D. V., 1986, "Modeling Mixed-Mode Dynamic Crack Propagation Using Finite Elements," PhD Dissertation, School of Civil and Environmental Engineering, Cornell University, Ithaca.

Swenson, D. V., and Ingraffea, A. R., 1991, "The Collapse of the Schoharie Creek Bridge: A Case Study in Concrete Fracture Mechanics," *Int. J. of Fracture*, V. 51, pp. 73-92.

Swenson, D. V., and Ingraffea, A. R., 1988, "Modeling Mixed-Mode Dynamic Crack Propagation Using Finite Elements: Theory and Applications," *Comput. Mech.*, V. 3, pp. 381-397.

Tada, H.; Paris, P. C.; and Irwin, G. R., 1973, *The Stress Analysis of Cracks Handbook*, Del Research Corp., Hellertown.

Tasdemir, M. A.; Maji, A. K.; and Shah, S. P., 1990, "Crack Propagation in Concrete under Compression," *Journal of Engineering Mechanics*, ASCE, V. 116, No. 5, May, pp. 1058-1077.

Tong, P., and Pian, T. H. H., 1973, "On the Convergence of the Finite Element Method for Problems with Singularity," *Int. J. Solids Structures*, V. 9, pp. 313-321.

Tong, P.; Pian, T. H. H.; and Lasry, S. J., 1973, "A Hybrid Element Approach to Crack Problems in Plane Elasticity," *Int. J. Num. Meth. in Engrg.*, V. 7, pp. 293-308.

Tracey, D. M., 1971, "Finite Elements for the Determination of Crack Tip Elastic Stress Intensity Factors," *Engrg. Fract. Mech.*, V. 3, pp. 255-265.

Underwood, P., 1983, "Dynamic Relaxation," *Computational Methods in Mechanics for Transient Analysis*, edited by T. Belytschko and T. J. R. Hughes, North-Holland Publishing, Amsterdam, pp. 245-265.

Valliappan, S., and Doolan, T., 1972, "Nonlinear Stress Analysis of Reinforced Concrete," *Journal of Structural Division*, ASCE, V. 98, No. ST4, Apr., pp. 885-898.

Van Mier, J. G. M.; Rots, J. G.; and Bakker, A., 1991, "Fracture Processes in Concrete, Rock and Ceramics," *Proceedings,* Intl. RILEM/ESIS Conf., June, Noordwijk, The Netherlands.

Van Mier, J. G. M., and Vonk, R., 1991, "Fracture of Concrete and Multiaxial Stress-Recent Developments," *Mater. Struct.*, RILEM, V. 24, pp. 61-65.

Vecchio, F. J., 1986, "Nonlinear Finite Element Analysis of Reinforced Concrete Membranes," *ACI Structural Journal*, V. 86, No. 1, Jan.-Feb., pp. 26-35.

Vecchio, F. J., and Collins, M. P., 1982, "The Response of Reinforced Concrete to Inplane Shear and Normal Stress," *Publication* 82-03, Dept. of Civil Engrg., Univ. of Toronto.

Vonk, R., 1992, "Softening of Concrete Loaded in Compression," PhD thesis, Technische Univeriteit Eindhoven, Eindhoven, the Netherlands.

Vonk, R., 1993, "A Micromechanical Investigation of Softening of Concrete Loaded in Compression," *HERON*, V. 38.

Vos, E., 1983, "Influence of Loading Rate and Radial Pressure on Bond in Reinforced Concrete," PhD Thesis, Delft University Press, The Netherlands.

Wawrzynek, P. A., and Ingraffea, A. R., 1986, "Local Remeshing around a Propagating Crack Tip Using Interactive Computer Graphics," *Finite Element Method, Modeling, and New Applications*, E. M. Patten et al., eds., ASME CED-1, PVP-101, pp. 33-38.

Wawrzynek, P. A., and Ingraffea, A. R., 1987, "Interactive Finite Element Analysis of Fracture Processes: An Integrated Approach," *Theo. and Appl. Fract. Mech.*, V. 8, pp. 137-150.

Wawrzynek, P. A., and Ingraffea, A. R., 1989, "An Interactive Approach to Local Remeshing Around a Propagating Crack," *Finite Elements in Analysis and Design*, V. 5, pp. 87-96.

Westergaard, H. M., 1934, "Stresses at a Crack, Size of the Crack, and the Bending of Reinforced Concrete," ACI JOURNAL, *Proceedings* V. 30, pp. 93-102.

Wilson, W. K., 1969, "On Combined Mode Fracture Mechanics," PhD Dissertation, University of Pittsburgh.

Wu, Z.; Yoshikawa, H.; and Tanabe, T.-A., 1991, "Tension Stiffness Model for Cracked Reinforced Concrete," *Journal of Structural Engineering*, ASCE, V. 117, No. 3, Mar., pp. 715-732.

Xie, M.; Gerstle, W.; and Rahulkumar, P., 1995, "Energy-Based Automatic Mixed-Mode Crack Propagation Modeling," *Journal of Engineering Mechanics*, ASCE, V. 121, No. 8, Aug., pp. 914-923.

Yamaguchi, E., and Chen, W.-F., 1990, "Cracking Model for Finite Element Analysis of Concrete Materials," *Journal of Engineering Mechanics*, ASCE, V. 116, No. 6, June, pp. 1242-1261.

Yamaguchi, E., and Chen, W.-F., 1991, "Microcrack Propagation Study of Concrete under Compression," *Journal of Engineering Mechanics*, ASCE, V. 117, No. 3, Mar., pp. 653-673.

Zienkiewicz, O. C., and Cheung, Y. K., 1964, "Buttress Dams on Complex Rock Foundations," *Water Power*, V. 16, p. 193.

APPENDIX—GLOSSARY

bond-slip—the force-displacement relationship describing the mechanical interaction between a reinforcing bar and the surrounding concrete. The bond-slip relationship is a structural rather than materials phenomenon.

continuum mechanics—a view of mechanical behavior that assumes that all responses can be represented using continuous functions.

continuum damage mechanics (CDM)—a form of continuum mechanics that assumes that all damage is in the form of continuously distributed damage (microcracks).

cosserat continuum—a particular view of continuum mechanics that includes the assumption of couple stresses as a continuum approximation of the behavior of crystal lattices on a small scale (Cosserat and Cosserat 1909).

crack—A conceptual model of damage within a solid body. A crack is usually understood as a surface within the body, across which tensile tractions cannot be transmitted.

crack band model—a method for objectively modeling cracks using the finite element method, in which the constitutive models for continuum finite elements are adjusted to correctly account for the formation of a discrete crack within the element (Bažant 1984).

critical stress intensity factor (K_{Ic})—In LEFM, for a given material, the mode I stress-intensity factor at which a crack will propagate (Broek 1986). The critical stress-intensity factor is a material property.

damage—breaking of bonds within a material. Damage can be manifested as continuum damage (distributed microcracks), or as a single macrocrack. Other forms of damage are also conceivable.

deformation—relative movement of particles within a body, due to a variety of causes.

discrete cracking—damage of a deformable body which results in one or more distinct cracks. Discrete cracking models usually assume no distributed damage in the form of microcracks.

dynamic relaxation—a numerical solution method, particularly suited to nonlinear statical problems, in which a static problem is viewed as a damped dynamic problem which converges to a steady-state static solution. The damping coefficients are adaptively adjusted to achieve rapid convergence (Papadrakis 1981).

energy release rate (G)—the rate at which potential energy is released with respect to change in area of a discrete crack (Broek 1986). The rate of energy release may also be considered with respect to other variables.

fictitious crack model (FCM)—a view of fracture mechanics that assumes that a discrete crack forms through gradual breaking of bonds across its advancing path (Hillerborg et al. 1976).

finite element method (FEM)—a numerical technique for solving boundary value problems (such as structural analyses) by representing a continuum (a structure or structural member) by a group of discrete regions (finite elements) for which the behavior is described by an assumed set of functions (which, in structural analysis, relate displacements and forces).

fracture—a fracture is a synonym with "crack." Each word may be used as a noun or a verb.

fracture mechanics—the study of the formation of cracks (Broek 1986). Recently, fracture mechanics also includes concepts of continuum damage mechanics.

fracture-sensitive—a structure is fracture-sensitive if its global mechanical response is sensitive to the presence of cracks.

fracture toughness (G_{Ic})—in LEFM, for a given material, the Mode I energy release rate, above which a crack will propagate (Broek 1986). Fracture toughness is a material property. Sometimes confused with critical stress-intensity factor.

fracture process zone (FPZ)—The zone at the tip of a discrete crack within which damage (or some contexts, any nonlinear material behavior) occurs.

hybrid finite element—a finite element that includes both stresses and displacements or strains as primary responses (Tong et al. 1973).

interface—an idealized surface representing the mechanical interaction between two adjacent bodies.

interface element—a finite element designed to model an interface (Goodman et al. 1968).

linear elastic fracture mechanics (LEFM)—A classical branch of fracture mechanics, in which material behavior is assumed to be linear elastic, damage is considered to be only in the form of discrete cracks, and the fracture process zone is considered to be negligibly small (Broek 1986).

J-integral—an integral method for determining energy release rates of discrete cracks (Rice 1968).

least dimension—a parameter characterizing the geometrical size associated with a crack tip (Gerstle and Abdalla 1990).

localization limiter—a mathematical device used to avoid unobjectivity or spurious mesh sensitivity when modeling continuum damage mechanics (Bažant and Pijaudier-Cabot 1988).

mode—type of loading in which stresses at a crack tip are (1) normal to the crack surface, giving rise to opening, or mode I, crack deformation perpendicular to the plane of the crack; (2) shear stresses in the crack plane, giving rise to sliding, or mode II, crack deformation in the plane of the crack, perpendicular to the edge of the crack; or (3) shear stresses out of the crack plane, giving rise to tearing, or mode III, crack deformation in the plane of the crack, parallel to the edge of the crack (Broek 1986).

mixed-mode—type of loading, often resulting in a curved crack, due to the simultaneous application of two or more loading modes. In most cases, mixed-mode cracks rapidly follow a direction that places them under mode I loading (Broek 1986).

nonlocal continuum—a form of continuum damage mechanics in which localization limiters are employed to preserve solution objectivity with respect to mesh refinement (Bažant 1986).

propagation—a term used to describe the evolution or growth of a discrete crack.

size effect—a phenomenon associated with cracks in which the apparent nominal strength of a structure decreases as a function of structure size (Bažant and Oh 1983).

smeared cracking—a synonym for "continuum damage mechanics." Some smeared cracking models have been found to be nonobjective with respect to mesh refinement.

singularity—a mathematical concept in which a field tends asymptotically to infinity at a point.

singular crack tip element—a finite element with a strain singularity included in its shape function, for use in finite element modeling of LEFM problems (Barsoum 1976).

stiffness-derivative method—a finite-difference method for obtaining the energy release rate of a discrete crack using the finite element method (Parks 1974).

stress intensity factor (K_I, K_{II}, K_{III})—factors that characterize the stress field a the tip of a crack using the assumptions of LEFM (Broek 1986).

strain gradient—spatial rate of change of strain, used in a method of limiting localization in continuum damage mechanics (Schreyer 1990).

strain localization—a phenomenon associated with continuum damage mechanics, in which damage tends to concentrate along surfaces, resulting in the formation of cracks (Bažant 1976).

strain-softening—a feature of some constitutive models, in which stress decreases with increasing strain. Constitutive models that include strain-softening must include the concept of nonlocal continuum to obtain objective solutions (Bažant 1976).

traction—force per unit area acting upon a specified surface. Traction is a vector quantity.

virtual crack extension (VCE)—a finite-difference method for obtaining the energy release rate of a discrete crack using the finite element method (Hellen 1975).

This report was submitted to letter ballot by the committee and was approved in accordance with ACI balloting procedures.

Design Guide for Twisting Moments in Slabs

Reported by Joint ACI-ASCE Committee 447

Ganesh Thiagarajan, Chair

Jian Zhao, Secretary

Riadh S. Al-Mahaidi
Gangolu Appa Rao
Ashraf S. Ayoub
Zdeněk P. Bažant
Allan P. Bommer
Mi-Geum Chorzepa
Carlos Arturo Coronado
Gianluca Cusatis
Mukti L. Das
James B. Deaton
Jason L. Draper
Serhan Guner
Trevor D. Hrynyk
John F. Jakovich
Song F. Jan
Ioannis Koutromanos
Laura N. Lowes
Yong Lu
Yi-Lung Mo
Abbas Mokhtar Zadeh
Wassim I. Naguib
Dan Palermo
Guillermo Alberto Riveros
Mohammad Sharafbayani
Hazim Sharhan
Sri Sritharan

Consulting Members

Ahmet Emin Aktan
Sarah L. Billington
Johan Blaauwendraad
Oral Buyukozturk
Ignacio Carol
Luigi Cedolin
Wai F. Chen
Christopher H. Conley
Robert A. Dameron
Filip C. Filippou
Kurt H. Gerstle
Walter H. Gerstle
Robert Iding
Anthony R. Ingraffea
Feng-Bao Lin
Christian Meyer
Hiroshi Noguchi
Gilles Pijaudier-Cabot
Syed Mizanur Rahman
Victor E. Saouma
Frank J. Vecchio
Kaspar J. Willam

The committee would like to thank B. Siljenberg for his contribution to this guide.

This guide assists practitioners in understanding: 1) twisting moments in two-way slabs, when twisting moments are an essential consideration; 2) methods that can be used to account for twisting moments in design; and 3) the options available for each method of the various system geometries. Descriptions of twisting moments are provided theoretically and visually in the guide, and six methods of accounting for twisting moments in design are discussed. Applicability of the various methods is evaluated through a comparison of designs resulting from each method for a variety of two-way slab types and geometries. The theories described in the guide also apply to the design of two-way wall and two-way dome systems.

Keywords: finite element analysis; shell design; slab design; torsion; twist; twisting moments; wall design.

CONTENTS

ACI 447R-18 was adopted and published April 2018.

CHAPTER 1—INTRODUCTION AND SCOPE

1.1—Introduction

Section 8.2.1 of ACI 318-14 allows slabs to be designed by any procedure that satisfies equilibrium and geometric compatibility, and requires that, at each section, the design strength exceeds the required strength and serviceability requirements are fulfilled.

Traditional strip design methods for slabs are based on approximate analysis and provide neither a complete equilibrium load path or satisfy geometric compatibility. Nonetheless, these methods have been used successfully for many years to design slabs with supports arranged in a rectangular grid.

From 1995 to 2015, design engineers transitioned from predominantly using traditional slab analysis methods to using finite element analysis (FEA). More recently, engineers use FEA to assist in the structural design of two-way concrete members. Twisting moments in two-way slabs can require additional reinforcement from those proportioned for bending moments, yet they are often misunderstood and sometimes ignored, neglected, or both, by practitioners in design. This is most likely due to their lack of being discussed comprehensively in design codes and frequent exclusion from college concrete design course curricula.

Although FEA solutions provide a full equilibrium load path and satisfy geometric compatibility, they determine load paths that require twisting moments for equilibrium (Shin et al. 2009). Many designers using FEA have ignored these twisting moments—a possible unconservative assumption where twisting moments are high (Park and Gamble 2000). To provide designers with guidance related to this issue, methods for explicitly incorporating twisting moments determined from FEA in the design of slabs are discussed in this guide.

The purpose of this design guide is to provide advice to design engineers who analyze slab systems with finite element methods and who need to ensure their designs are satisfactory for the twisting moments predicted by the analysis. This guide provides background information regarding twisting moments and describes multiple approaches for consideration of twisting moments in design. It also provides advice for designers of walls and shells with twisting moment conditions similar to those in slabs.

1.2—Scope

This design guide applies to slabs of both uniform and nonuniform thicknesses, including drop caps and drop panels, except where noted in the text. This guide does not apply to waffle slabs, or the beams of beam-and-slab floor systems. Chapters 3 through 6 address slabs and walls in which the response is determined purely by bending. Chapter 7 addresses shells for which the response is determined by bending and membrane action. Chapter 6 and the theory sections of this guide are applicable to walls. Chapter 7 and the theory sections of this guide are applicable to shells, with the caveat that equations presented in Chapter 3 are not valid for curved shells.

CHAPTER 2—NOTATION AND DEFINITIONS

2.1—Notation

$c_{i,j}$ = fraction for consideration of sections partially crossing element to apply to forces in local node j in element i
D = flexural rigidity of plate, in.-lb (N·mm)
E = Young's modulus, psi (MPa)
F = force vector, lb (N)
$f_{i,j}$ = nodal force vector for local node j in element i
h = thickness of slab or plate, in. (mm)
L = width of design section, in. (mm)
M = bending moment, or moment vector, in.-lb (N·mm)
M_d = design bending moment, in.-lb (N·mm)
M_i = bending moment from isotropic analysis, in.-lb (N·mm)
M_{tf} = bending moment from twist-free analysis, in.-lb (N·mm)
M_u = design moment for slab cross section, in.-lb (N·mm)
$m_{i,j}$ = nodal moment vector for local node j in element i, in.-lb/in. (N·mm/mm)
m_r = bending moment causing stresses parallel to r-axis, per unit length of slab or plate, in.-lb/in. (N·mm/mm)
m_{rs} = twisting moment relative to r-s-axes per unit length of slab or plate, in.-lb/in. (N·mm/mm)
m_s = bending moment causing stresses parallel to s-axis, per unit length of slab or plate, in.-lb/in. (N·mm/mm)

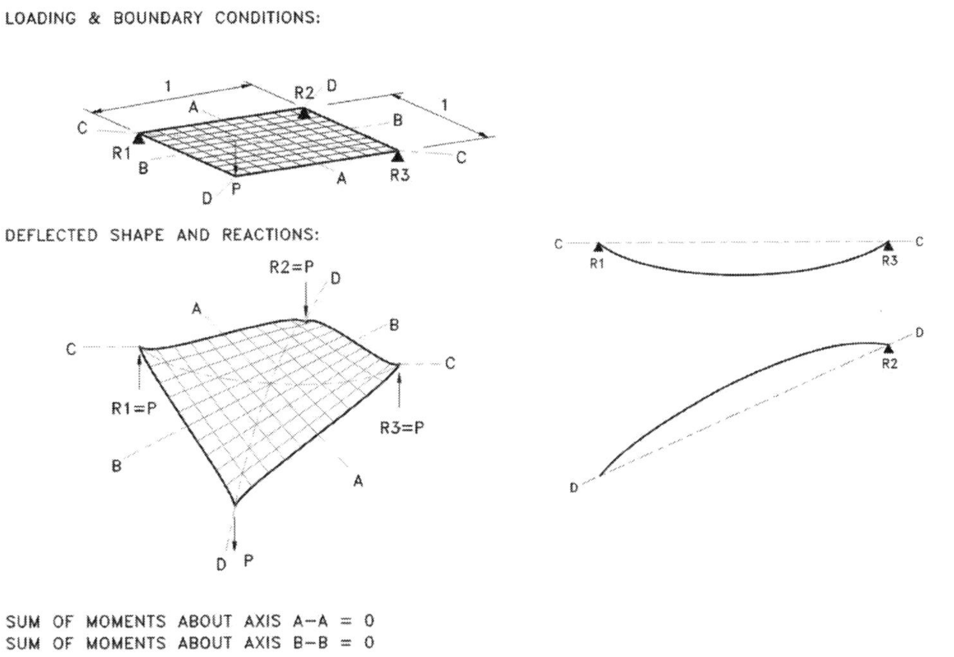

Fig. 3.1—Twisting moment example free body diagram.

m_{ux} = design moment causing stresses parallel to x-axis, per unit length of slab or plate, in.-lb/in. (N·mm/mm)

m_{ux}^+ = positive design moment causing stresses parallel to x-axis, per unit length of slab or plate, in.-lb/in. (N·mm/mm)

m_{ux}^- = negative design moment causing stresses parallel to x-axis, per unit length of slab or plate, in.-lb/in. (N·mm/mm)

m_{uy} = design moment causing stresses parallel to y-axis, per unit length of slab or plate, in.-lb/in. (N·mm/mm)

m_{uy}^+ = positive design moment causing stresses parallel to y-axis, per unit length of slab or plate, in.-lb/in. (N·mm/mm)

m_{uy}^- = negative design moment causing stresses parallel to y-axis, per unit length of slab or plate, in.-lb/in. (N·mm/mm)

m_x = bending moment causing stresses parallel to x-axis, per unit length of slab or plate, in.-lb/in. (N·mm/mm)

m_{xy} = twisting moment relative to x-y-axes per unit length of slab or plate, in.-lb/in. (N·mm/mm)

m_y = bending moment causing stresses parallel to y-axis, per unit length of slab or plate, in.-lb/in. (N·mm/mm)

n_x = membrane tension in x-axis direction per unit length, lb/in. (N/mm)

n_{xy} = membrane in plane-shear in x-y-axes direction per unit length, lb/in. (N/mm)

n_y = membrane tension in y-axis direction per unit length, lb/in. (N/mm)

q = transverse load per unit area, lb/in.^{2} (N/mm^2)

T = torsional moment, in.-lb (N·mm)

V = shear force, lb (N)

V_d = design shear force, lb (N)

v_x = transverse shear on x-face per unit length, lb/in. (N/mm)

v_y = transverse shear on y-face per unit length, lb/in. (N/mm)

w = transverse deflection, in. (mm)

$x_{i,j}$ = distance vector from section centroid to local node j in element i, in. (mm)

ν = Poisson's ratio

2.2—Defintions

ACI provides a comprehensive list of definitions through an online resource, ACI Concrete Terminology. Definitions provided herein complement that source.

anticlastic bending—curvature caused by the Poisson effect and curvature about a perpendicular axis.

strip design method—a method of designing slabs by dividing them into two sets of approximately perpendicular strips, with each strip analyzed and designed independently from each other.

CHAPTER 3—BACKGROUND

3.1—Qualitative introduction to twisting moments in slabs

Twist exists in most every slab, except those theoretical-case-only slabs whose moments at any point are identical about any axis. Figure 3.1 illustrates an extreme case of twist; a square slab with supports at three corners and a load at the fourth corner.

From equilibrium, it can be shown that the bending moment about the A-A and B-B axes in the figure is zero, although this slab is clearly supporting a load and needs to be reinforced. Looking at the C-C axis, whose bending moment is nonzero, gives us insight to the load-carrying mechanism of the slab. What appears as twist about A-A and B-B is bending about C-C and D-D, as shown in the deflected shapes along C-C and D-D.

3.2—Behavior of linear-elastic isotropic slabs

Like other components in concrete structures, the design forces for slabs are typically determined using some form of a linear-elastic analysis. For slabs that are not thick (span-to-thickness ratio of more than 10), linear-elastic slab behavior can be predicted by well-known plate theory.

The deflection of a thin, linear-elastic, isotropic plate subjected to loads perpendicular to its plane can be expressed by a fourth-order partial differential equation (Timoshenko and Woinowski-Krieger 1959)

$$\frac{\partial^4 w}{\partial x^4} + 2\frac{\partial^4 w}{\partial x^2 \partial y^2} + \frac{\partial^4 w}{\partial y^4} = \frac{q}{D} \quad (3.2a)$$

where $w = w(x,y)$ is the transverse (out-of-plane) displacement field in the direction of loading; $q = q(x,y)$ per unit area; and D is the flexural rigidity of the plate, which is expressed as

$$D = \frac{Eh^3}{12(1-\nu^2)} \quad (3.2b)$$

in which E is the modulus of elasticity, h is the thickness of the plate, and ν is Poisson's ratio.

The deformations of the plate can be interpreted as two curvatures $\frac{\partial^2 w}{\partial x^2}$ and $\frac{\partial^2 w}{\partial y^2}$, and one twist $\frac{\partial^2 w}{\partial x \partial y}$. Corresponding to these three deformations, there are two bending moments and one twisting moment as shown in Eq. (3.2c).

$$\begin{Bmatrix} m_x \\ m_y \\ m_{xy} \end{Bmatrix} = -\begin{bmatrix} D & \nu D & 0 \\ \nu D & D & 0 \\ 0 & 0 & (1-\nu)D/2 \end{bmatrix} \begin{Bmatrix} \frac{\partial^2 w}{\partial x^2} \\ \frac{\partial^2 w}{\partial y^2} \\ 2\frac{\partial^2 w}{\partial x \partial y} \end{Bmatrix} \quad (3.2c)$$

From Eq. (3.2c), note that unless $\frac{\partial^2 w}{\partial x \partial y}$ is zero, there will be twisting moments in the plate. In general, $\frac{\partial^2 w}{\partial x \partial y}$ is zero in only a small subset of locations in a slab, so twisting moments will exist almost everywhere in a linear-elastic isotropic slab.

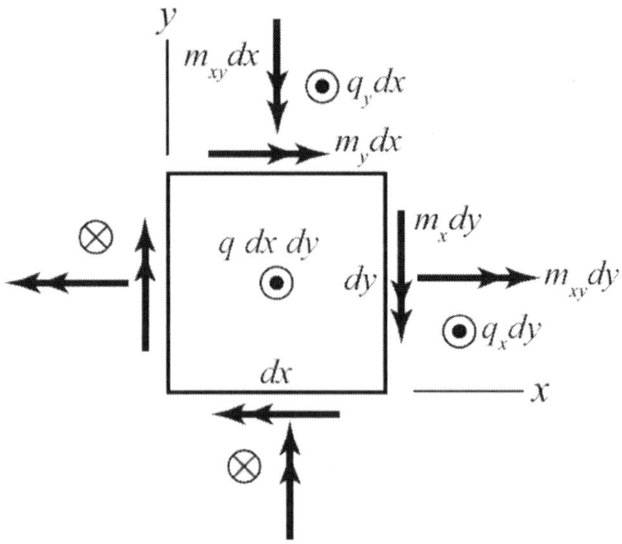

Fig. 3.3—An infinitesimal plate element shown with resulting shear forces and moments due to transverse loading.

3.3—Equilibrium in slabs

By considering the rotational equilibrium of the slab element shown in Fig. 3.3, Eq. (3.3), describing the interrelationship of m_x, m_y, and m_{xy}, can be derived.

$$\frac{\partial^2 m_x}{\partial x^2} + 2\frac{\partial^2 m_{xy}}{\partial x \partial y} + \frac{\partial^2 m_y}{\partial y^2} = -q \quad (3.3)$$

Equation (3.3), which holds for all slabs regardless of material behavior, reinforcement, or cracking, shows that any change in m_{xy} must be offset by some change in m_x, m_y, or both. Hence, m_{xy} can only be reduced to zero by changing the load path of the slab.

3.4—Principal axes

The values of m_x, m_y, and m_{xy} are related via Mohr's circle (Timoshenko and Woinowski-Krieger 1959); the moments m_r, m_s, and m_{rs} for any set of perpendicular *r-s* axes can be determined from m_x, m_y, and m_{xy}, as shown in Eq. (3.4a) through (3.4c).

$$m_r = \frac{1}{2}(m_x + m_y) + \frac{1}{2}(m_x - m_y)\cos 2\theta + m_{xy}\sin 2\theta \quad (3.4a)$$

$$m_s = \frac{1}{2}(m_x + m_y) - \frac{1}{2}(m_x - m_y)\cos 2\theta - m_{xy}\sin 2\theta \quad (3.4b)$$

$$m_{rs} = -\frac{1}{2}(m_x - m_y)\sin 2\theta + m_{xy}\cos 2\theta \quad (3.4c)$$

For every point in a plate, there is one set of principal axes, where m_{rs} is zero and about which the slab is in pure bending. The orientation of these axes changes from point to point in the slab, and the principal axes are typically more valuable for general understanding than for determining design forces. Figure 3.4 illustrates the construction of Mohr's circle for Eq. (3.4a) through (3.4c).

3.5—Orthogonal reinforcement and equilibrium for twisting moments

How can reinforcement be used to resist twisting moments? This question is most conveniently addressed by considering the reinforcement required to resist the moments about the principal axes. Figure 3.5 illustrates the conceptual approach.

Figures 3.5(a) and 3.5(b) illustrate the slab differential element to be reinforced, and its principal moments *m*1 and

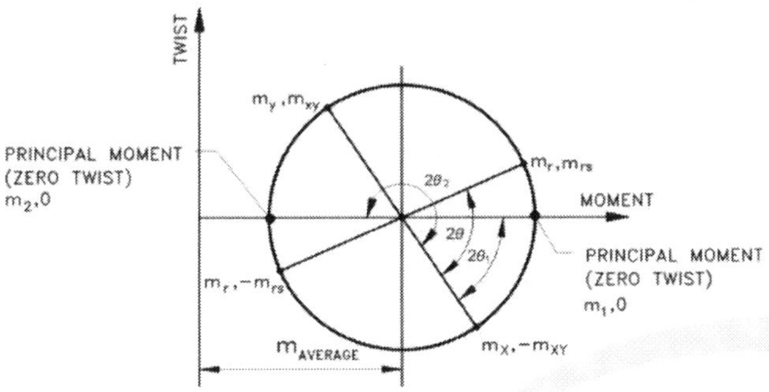

Fig. 3.4—Mohr's circle for situation where m_x *and* m_y *are both positive and* $m_x > m_y$.

m_2. The values of m_1 and m_2 cause tension in the bottom of the slab. Figure 3.5(c) shows bottom reinforcement forces aligned with the principal axes, selected using standard bending design approaches, and the *x*- and *y*-axis force vectors the reinforcement provides. Figure 3.5(d) shows the reinforcement forces along the 1- and 2-axes replaced by reinforcement forces along the *x*- and *y*-axes that, when combined, provide vector forces equivalent to the 1- and 2-axes reinforcement forces.

3.6—Effects of slab geometry on twisting moments

The slab in Fig. 3.6a illustrates how different slab regions and geometries affect the importance of the consideration of twisting moments when designing a slab. The figure highlights three slab regions that are discussed as follows. The slab is assumed to be designed with reinforcing bars parallel to the *x*- and *y*-axes.

Region 1 of the slab is regular with all columns arranged in a rectangular grid. Regions similar to this are not twist-sensitive. Ignoring twisting moments in these regions can be proved safe through yield line theory. Although yield line theory is an upper-bound approach, for a slab with regular supports such as those in Region 1, the number of yielding patterns to consider is small. If top reinforcement is clustered at the columns, only the folded-plate pattern in each

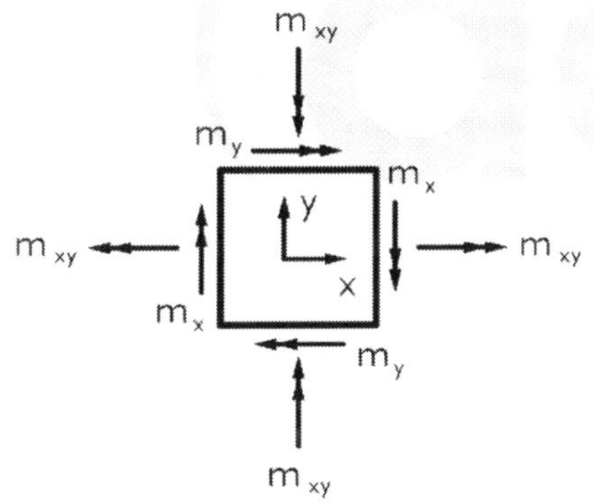

(a) Moments and twists about *x*-*y* axes

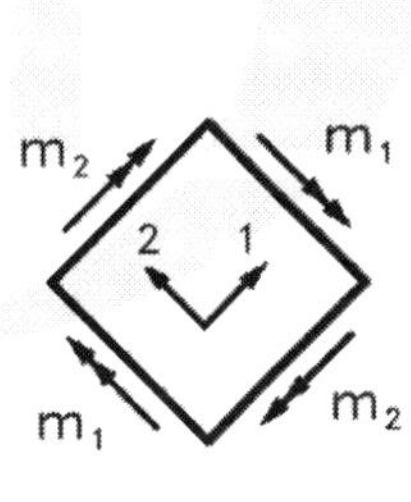

(b) Principle axes moments

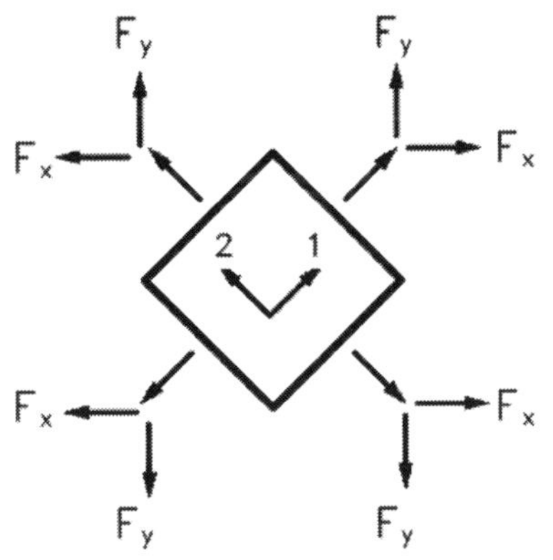

(c) Principle axes bottom reinforcement

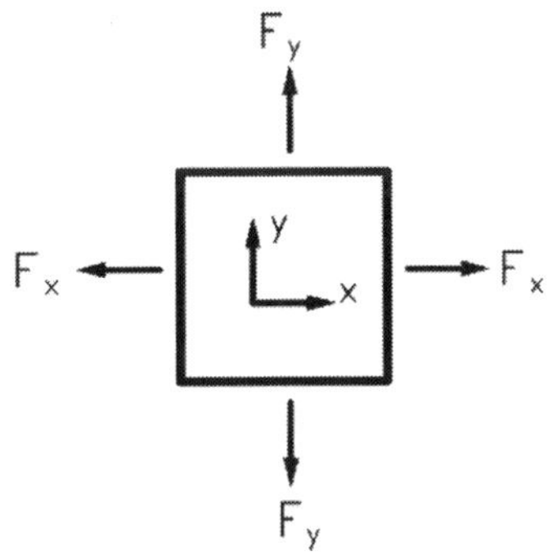

(d) The *x*-*y* axes bottom reinforcement

Fig. 3.5—Orthogonal reinforcement from twisting moments.

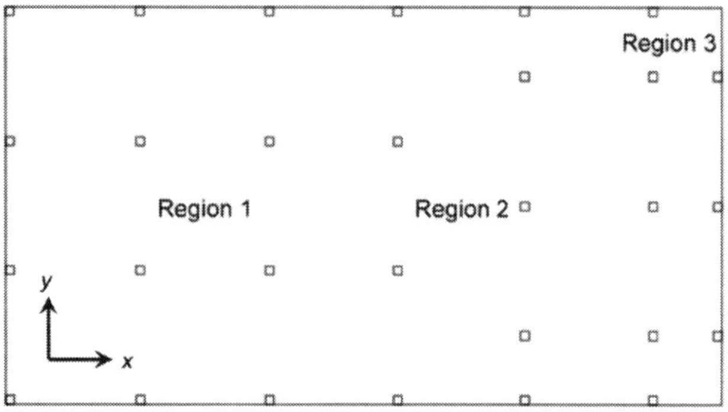

Fig. 3.6a—Representative regions of a slab floor plan. Region 1 is twist-sensitive, Region 2 moderately twist-sensitive, and Region 3 highly twist-sensitive.

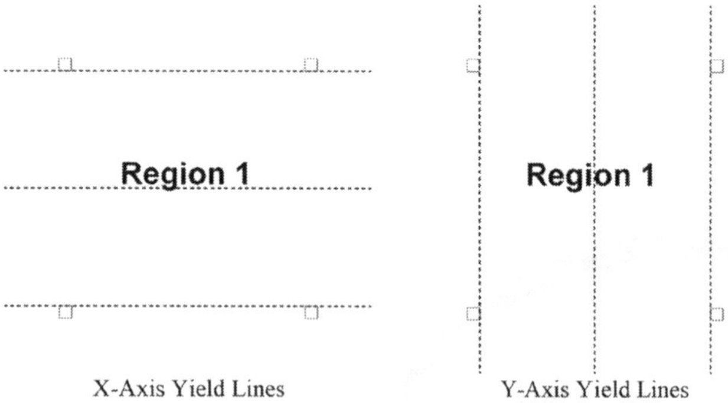

Fig. 3.6b—Folded plate yield lines.

direction shown in Fig 3.6b need be considered (Kennedy and Goodchild 2003). The folded plate yield line patterns show that any reinforcement design that supports the total span static moment ($wL^2/8$ for a uniform load) in each direction is safe.

Region 2 of the slab is where there is a major shift in the column layout. Regions similar to this are moderately twist-sensitive. The slab tends to span along a 45-degree line between the closest columns. This diagonal spanning reduces the bending moments along the *x*- and *y*-axes while increasing twisting moments. Ignoring twisting moments in design of similar regions could lead to a slab load capacity that is approximately 20 percent less than calculated. This percentage is calculated by comparing integrated slab strip bending moments using isotropic plates and using orthotropic plates with zero twist stiffness. This calculation assumes that the slab has no reliable twist capacity if twist has not been considered in the design of reinforcement, and is performed by dividing the bending moment determined by a twist-free analysis, M_{tf}, by the bending moment determined by an isotropic analysis, M_i. For Region 2, $M_{tf}/M_i = 1.2$.

Region 3 of the slab is a cantilevered corner. Regions similar to this are highly twist-sensitive. The slab tends to span between, and cantilever off, the line between the two columns. Although this diagonal cantilevering reduces the bending moments along the *x*- and *y*-axes, it adds large twisting moments. Ignoring twisting moments in design of similar regions could lead to a slab load capacity that is approximately 40 percent less than calculated. This percentage is calculated by comparing integrated slab strip bending moments using isotropic plates and using orthotropic plates with zero-twist stiffness. This calculation assumes that the slab has no reliable twist capacity if twist has not been considered in the design of reinforcement ($M_{tf}/M_i = 1.4$). The strip design methods referred to in this document are those commonly used by practicing engineers, not the more sophisticated strip approaches discussed in academic literature.

3.7—Traditional slab design methods

Traditionally, slab design is performed using design forces determined by simplified analysis methods that idealize the slab as a set of intersecting strips (effectively supports and wide beams) in two perpendicular directions. The required reinforcement is calculated using the strip bending moments and standard beam bending design approaches.

These methods have two primary flaws. The first is that equilibrium is not fully satisfied because there is no consistency between the load paths of the two sets of strips. The second flaw is that deformation compatibility between both parallel and perpendicular strips is ignored.

Despite these flaws, slabs designed using these methods have generally performed well. For slabs with supports arranged in a rectangular grid it is often shown, using yield-line theory or lower-bound methods, that the traditional strip methods provide adequate capacity when reinforcement is distributed appropriately. (Burgoyne 2004; Kennedy and Goodchild 2003). For those slabs with supports not arranged in a rectangular grid, engineering judgment is necessary to determine if the traditional methods will produce a safe design.

3.8—Finite element analysis (FEA)-based slab design resultants

To complete slab design, an engineer should determine the quantity of reinforcement required for each design cross section. Although design cross section locations and lengths are often guided by code rules, in general, sections are needed at peak stress locations and their lengths based on the extent of the slab that can be assumed to act as a unit in resisting internal forces. The width of each section should be chosen so that the moment distribution along the section is reasonably uniform, does not change sign, and can be resisted by uniformly distributed reinforcements.

When using FEA to support slab design, the engineer should convert the slab analysis element results to resultant forces and moments acting on these sections. At this stage, the engineer should first transform the results of FEA into a coordinate system orthogonal to the section, and then integrate/sum all forces and moments acting on the section to determine the design moment. Consideration of twisting moments is the most difficult aspect of this conversion. ACI 318 does not explicitly address twisting moments, nor does the commentary provide guidance on their consideration. However, Section 6.1.1 of ACI 318.2-14, which addresses shell reinforcement, states that reinforcement must be designed to resist tension from bending and twisting moments.

Design can also be performed considering the FEA results at each point in the slab, independently and foregoing the integration along a section. A zero-length section should still be considered to provide the correct axes for the design resultants and reinforcement. Design code requirements, such as those in ACI 318, can be difficult to apply to these point results, as the codes are written for section-based results. Another potential difficulty with using FEA results at a point for design is that the results at certain locations, such as under a point load, will approach infinity as the mesh is refined. In general, the approaches described in this guide apply to either point-based or section-based resultants.

CHAPTER 4—AVAILABLE DESIGN METHODS

This chapter discusses various options to consider for twisting moments. Chapters 5 through 7 evaluate some of these options in sample structures. Design methods discussed in this chapter are typically used with the results from linear-elastic analyses.

4.1—Finite element analysis (FEA)-based design ignoring twist

A simple, commonly-used approach to determine the design moment for a slab cross section is to integrate the FEA-predicted moment about the axis of the section along the length of the section.

$$M_u = \int m_r ds \tag{4.1}$$

In Eq. (4.1), m_r is the moment per unit length along the section. This approach ignores twisting moments in the equilibrium load path and may significantly underestimate the design demand.

The effect of twisting moments can be safely ignored in this manner when sections are approximately aligned orthogonally to the principal bending directions at all locations, as the twist on the section vanishes in this case. An instructive rule of thumb is that twist can be ignored if it is smaller than 10 percent of the primary bending moment (Deaton 2005). If this criterion is not satisfied, neglecting the effect of twist could lead to unconservative results.

4.2—Design using the Wood and Armer method

Wood (1968) and Armer (1968) (referred to in this document as Wood and Armer) proposed one of the most popular methods for explicitly incorporating twisting moments in slab design. The method seeks to prevent yielding in all directions and was developed by considering the normal moment and Johansen's yield criteria (Park and Gamble 2000). At any point in the slab, for any arbitrary direction, the design moment determined from FEA results must not exceed the ultimate normal resisting moment in that direction. The ultimate normal resisting moments calculated from the reinforcement in the x- and y-directions are typically compared to adjusted design moments m_{ux} and m_{uy}. Design moments based on the Wood and Armer method are computed

$$m_{ux} = m_x \pm |m_{xy}| \tag{4.2a}$$

$$m_{uy} = m_y \pm |m_{xy}| \tag{4.2b}$$

where all plus signs are used to compute required bottom reinforcement, all minus signs are used to compute required top reinforcement, and m_{ux} and m_{uy} will be negative when the top reinforcement is in tension. This assumes a sign convention where positive moment causes tension on the bottom surface of the slab and negative moment causes tension on the top surface of the slab. For non-slab systems such as walls and domes, top and bottom should be defined such that they are consistent throughout and consistent with loading assumptions.

Where m_{ux} or m_{uy} is found to have the opposite sign from what is expected (negative for bottom reinforcement or positive for top reinforcement), the design moment can be conservatively set to zero. A less conservative calculation option for this case is shown in Table 4.2.

The Wood and Armer approach requires that a fine regular mesh be used to produce accurate slab moment and twist predictions. Also, because moments, twists, or both, can be theoretically infinite at concentrated loads or reactions (such as slab column connections), integration of the moments or twists at these locations is difficult to perform accurately. Previous studies have indicated that the Wood and Armer method could produce unconservative results for slabs with high reinforcement ratios (approximately more than 0.75 percent) at regions of significant twisting moments, especially near restrained slab corners (May and Lodi 2005). Note also that this method cannot be directly applied to slabs with beams or drop panels. The Wood and Armer method has been modified and implemented in more conservative forms in both CSA-A23.3-04 and EN 1992-1-1:2004.

4.3—Design using the sandwich model

Designs based on the normal moment yield criterion, such as the Wood and Armer method, do not account for transverse shear or membrane forces within the slab. The sandwich model approach allows for the computation of reinforcement to resist not only bending and twisting moments, but also the effects of shear and membrane stresses. In this approach, membrane effects (n_x, n_y, and n_{xy}) and bending and twisting moments (m_x, m_y, and m_{xy}) are resisted by the sandwich exterior layers, whereas the shear effects (v_x and v_y) are resisted via the sandwich core. Figure 4.3 (Marti 1990) shows this concept.

To design by using the sandwich model, a slab section is divided into three layers; the depth of each layer can be determined such that the middle planes of the outer layers coincide with the center of the top and bottom reinforcement layers. Then, bending moments are decomposed into a couple of tensile and compressive normal forces, and twisting moments are decomposed into a couple of in-plane shear forces acting at top and bottom layers of the slab. The sandwiched inner layer is used to resist transverse shear. These normal and shear forces due to moments are combined with membrane (in-plane) forces. Finally, slab reinforce-

Table 4.2—Less conservative design moments (Wood 1968; Armer 1968)

	$m_x > \lvert m_{xy}\rvert$	$m_x \leq \lvert m_{xy}\rvert$	$m_x < -\lvert m_{xy}\rvert$
$m_y > \lvert m_{xy}\rvert$	$m_{ux}^{bot} = m_x + \lvert m_{xy}\rvert$ $m_{ux}^{top} = 0$ $m_{uy}^{bot} = m_y + \lvert m_{xy}\rvert$ $m_{uy}^{top} = 0$	$m_{ux}^{bot} = m_x + \lvert m_{xy}\rvert$ $m_{ux}^{top} = m_x - \left\lvert \frac{m_{xy}^2}{m_y} \right\rvert$ $m_{uy}^{bot} = m_y + \lvert m_{xy}\rvert$ $m_{uy}^{top} = 0$	$m_{ux}^{bot} = 0$ $m_{ux}^{top} = m_x - \left\lvert \frac{m_{xy}^2}{m_y} \right\rvert$ $m_{uy}^{bot} = m_y + \left\lvert \frac{m_{xy}^2}{m_x} \right\rvert$ $m_{uy}^{top} = 0$
$m_y \leq \lvert m_{xy}\rvert$	$m_{ux}^{bot} = m_x + \lvert m_{xy}\rvert$ $m_{ux}^{top} = 0$ $m_{uy}^{bot} = m_y + \lvert m_{xy}\rvert$ $m_{uy}^{top} = m_y - \left\lvert \frac{m_{xy}^2}{m_x} \right\rvert$	$m_{ux}^{bot} = m_x + \lvert m_{xy}\rvert$ $m_{ux}^{top} = m_x - \lvert m_{xy}\rvert$ $m_{uy}^{bot} = m_y + \lvert m_{xy}\rvert$ $m_{uy}^{top} = m_y - \lvert m_{xy}\rvert$	$m_{ux}^{bot} = 0$ $m_{ux}^{top} = m_x - \lvert m_{xy}\rvert$ $m_{uy}^{bot} = m_y + \left\lvert \frac{m_{xy}^2}{m_x} \right\rvert$ $m_{uy}^{top} = m_y - \lvert m_{xy}\rvert$
$m_y < -\lvert m_{xy}\rvert$	$m_{ux}^{bot} = m_x + \left\lvert \frac{m_{xy}^2}{m_y} \right\rvert$ $m_{ux}^{top} = 0$ $m_{uy}^{bot} = 0$ $m_{uy}^{top} = m_y - \left\lvert \frac{m_{xy}^2}{m_x} \right\rvert$	$m_{ux}^{bot} = m_x + \left\lvert \frac{m_{xy}^2}{m_y} \right\rvert$ $m_{ux}^{top} = m_x - \lvert m_{xy}\rvert$ $m_{uy}^{bot} = 0$ $m_{uy}^{top} = m_y - \lvert m_{xy}\rvert$	$m_{ux}^{bot} = 0$ $m_{ux}^{top} = m_x - \lvert m_{xy}\rvert$ $m_{uy}^{bot} = 0$ $m_{uy}^{top} = m_y - \lvert m_{xy}\rvert$

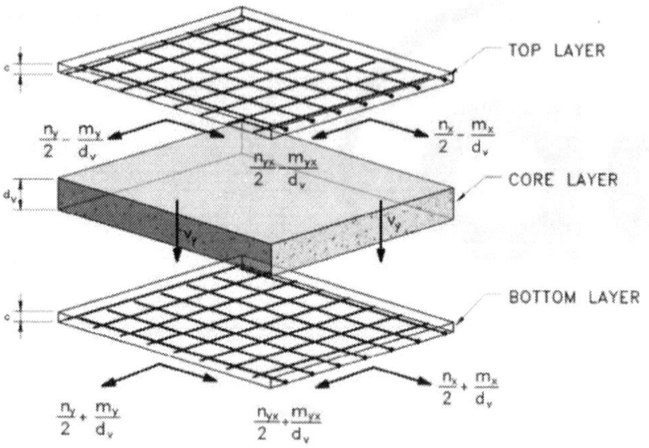

Fig. 4.3—Sandwich model.

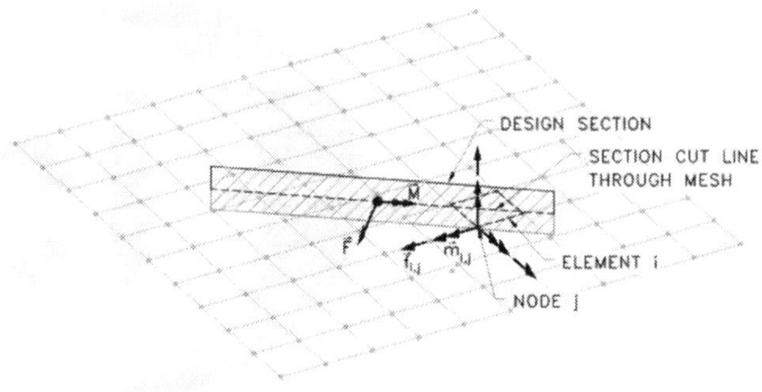

Fig. 4.4.2—Design section forces derived from element nodal forces.

ment is proportioned to satisfy equilibrium conditions at each top and bottom layer in the two orthogonal directions. Further descriptions of this method have been presented by *fib* Bulletin 45 (Comité Euro-International du Béton 2008), Brøndum-Nielsen (1985), Lourenço and Figueiras (1995), and the CEB-FIP Model Code (Comité Euro-International du Béton 1993).

The sandwich model is not typically used for slab design for commercial structures or generally available in slab design software. The assumed lever arm between the outer layers is very conservative for thin slabs, as the ratio of true ultimate lever arm to assumed lever arm is significantly greater than 1.0 (although this conservatism is reduced as the slab thickness is increased).

4.4—Design using element nodal forces

4.4.1 *Nodal forces approach*—The methods discussed in 4.1, 4.2, and 4.3 use slab stresses as the analysis quantities from which slab design forces are calculated. Another approach for slab design using FEA results is to compute results for the design cross section from element nodal forces and moments. Using this approach, the section design forces are guaranteed to be in equilibrium with the externally applied nodal loads (Deaton 2005). Slab design methods based on element nodal forces have been implemented in various FEA software. These methods are attractive because results are relatively accurate (even for very coarse or irregular meshes), as they can be used for slabs containing beams or drop panels and are easily extended to design post-tensioned floors.

4.4.2 *Calculation of nodal forces*—The nodal forces approach transforms all element nodal forces and moments acting at nodes on the design cross section into resultant forces and moments acting at the centroid of the section, as shown in Fig. 4.4.2. Because transformation of nodal forces to the centroid considers the eccentricity of the nodes from the centroid, equilibrium of all acting forces and moments is provided.

The equilibrium equations in vector form are:

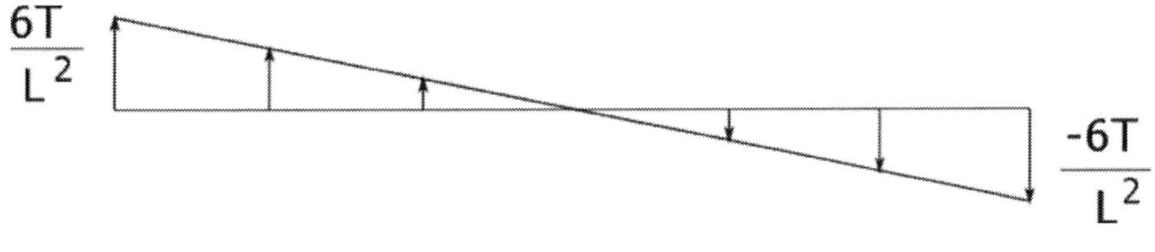

Fig. 4.4.4—Assumed shear stress distribution causing torsion.

$$\vec{F} = \sum_i \sum_j c_{i,j} \cdot \vec{f}_{i,j} \tag{4.4.2a}$$

$$\vec{M} = \sum_i \sum_j c_{i,j} \cdot (\vec{m}_{i,j} + \vec{x}_{i,j} \times \vec{f}_{i,j}) \tag{4.4.2b}$$

where i is the iterator over the elements crossed; j is the iterator over the local nodes in element i (on one side of the section); $\vec{F}$ and $\vec{M}$ are the force and moment vectors at the section centroid, respectively; $\vec{f}_{i,j}$ and $\vec{m}_{i,j}$ are nodal force and moment vectors, respectively; $\vec{x}_{i,j}$ is a vector distance from the section centroid to node i,j; and $c_{i,j}$ is a multiplier that equals 1.0 unless the section does not fully cross element i, in which case it varies between 0.0 and 1.0, typically based on the fraction of the element crossed.

The forces and moments to be used in the slab design are found by transforming the $\vec{F}$ and $\vec{M}$ vectors into the coordinate system of the design cross section. While six resultants are determined, typically only transverse shear V, bending moment M, and torsion T are of interest. There is no resultant quantity that directly corresponds to the twisting moments on the section. Methods available for consideration of the calculated torsion are as follows.

4.4.3 *Torsion as bending*—One approach to incorporating T is to directly combine M with $\pm T$, as shown in Eq. (4.4.3). Effectively, this method uses T as a proxy for integrated twist and then applies the Wood and Armer approach.

$$M_d = M \pm T \tag{4.4.3}$$

This approach can lead to unconservative design moments in two conditions:

(1) Condition 1—Twisting moments along the cross section change sign and partially cancel in their integration

(2) Condition 2—Torsion due to shear forces is of the opposite sign as torsion due to twist

Condition 1 can be detected visually and avoided by breaking the design section into two design sections; however, this can be difficult when considering numerous loading conditions, as the location of the sign change could vary. Condition 2 rarely occurs without Condition 1 also occurring. When neither Condition 1 or 2 occur, the design moments derived from considering torsion as bending can be conservative. For a simple torsional cantilever slab, the design moments are twice those derived from the Wood and Armer approach (Den Hartog 2014).

4.4.4 *Torsion as shear*—The total torsion T acting on the design cross section can be decomposed into an assumed linearly varying transverse shear per unit length, as shown in Fig. 4.4.4. This torsional shear stress has a maximum value equal to $6T/L^2$, where L is the width of the section—that is, the width of section cut in Fig. 4.4.2. This is analogous to the bending stress in a rectangular beam being calculated as $6M/bh^2$.

By multiplying this maximum shear stress assumed due to torsion by the width of the section, an equivalent, but very conservative, resultant design shear force due to torsion, which is equal to $6T/L$, can be considered during the shear design of the section, as is shown in Eq. (4.4.4).

$$V_d = V \pm \frac{6T}{L^2} L = V \pm \frac{6T}{L} \tag{4.4.4}$$

Considering torsion as shear is generally not applicable when significant twisting moments exist. However, applying the torsion-as-shear design approach to the results of a twist-free analysis (3.5) has the potential for providing improved vertical shear designs while simultaneously fully considering twisting moments.

4.4.5 *Nodal forces with twisting moment integrations*—While the nodal forces approach does not provide twisting moments, it can be supplemented with element twisting moment integrations to provide a full set of design forces suitable for the Wood and Armer method or other design methods. The benefit of using such a hybrid approach is to gain accuracy of the nodal forces approach for all quantities to which it applies.

4.5—Design using twist-free analysis

Slab analysis and design software capable of analyzing orthotropic slabs (with different properties in two orthogonal directions) typically allows engineers to separately modify the stiffness corresponding to each of the bending moments about the two orthogonal axes and the twisting moment (Shin et al. 2009). Setting the twist stiffness to zero leads to an analytical solution and load path in which all the twisting moments are zero. The behavior and load path of a twist-free slab is somewhat similar to that of a grillage of beams. The design philosophy behind twist-free analysis is similar to that for the compatibility torsion approach that has long been accepted in ACI 318.

Bending and twisting moments are interrelated through a Mohr's circle equilibrium constraint. Because of this constraint, the twist stiffness can only be set to zero about a single set of perpendicular axes at each location. The most practical axes for this purpose are the axes of the design cross sections, which are perpendicular to the reinforcement. In a typical slab that is reinforced in only two perpendicular direc-

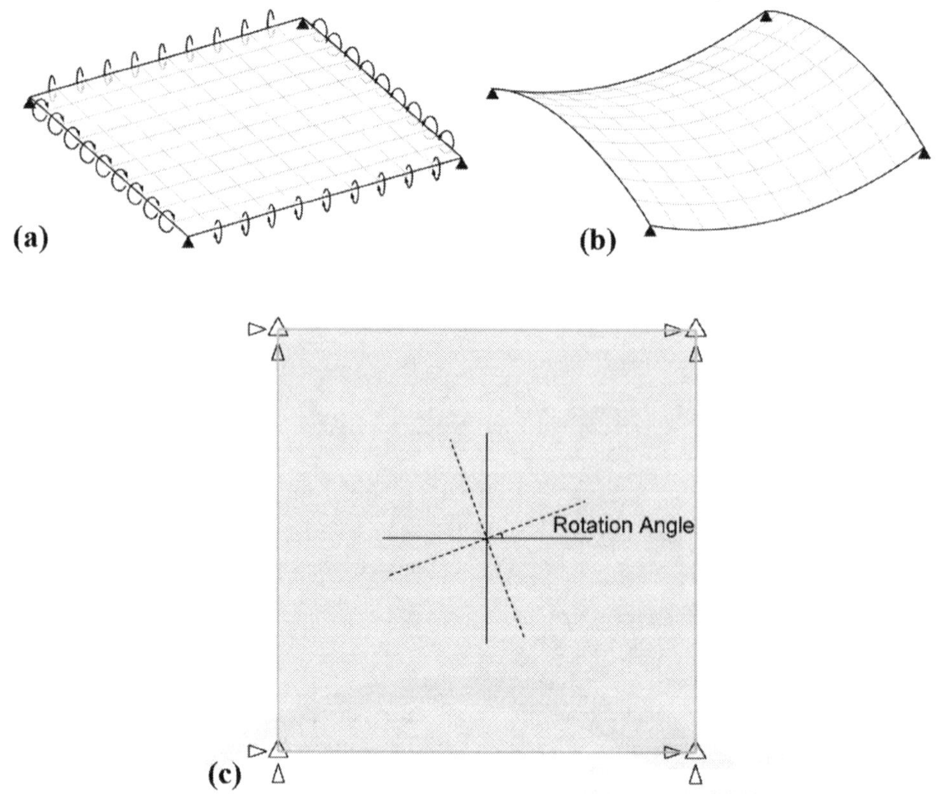

Fig. 5.1a—(a) Uniform moment field; (b) deflected shape; and (c) design sections to rotate in evaluation of twist-sensitivity.

tions, the entire slab can be given the same modified stiffness. For slabs where the reinforcement directions vary from region to region, multiple stiffness modifications are required.

Twist-free slab analyses result in larger deflections than those predicted by analyses with nonzero twist stiffness. This is to be expected, as the load path with nonzero twist stiffness is more efficient in terms of elastic energy than one with zero twist stiffness. Comparing slab deflections computed by analyses with and without twist stiffness is a quick means of gauging how much the forces will be redistributed to mobilize the twist-free load path. If the deflections from the twist-free analysis are significantly greater than those from the analysis with twist stiffness—for example, by a factor of 2—the potential for twist-related cracking should be investigated. Twist-free analysis can be used in combination with nodal force or moment resultant methods.

CHAPTER 5—COMPARISON OF DESIGN METHODS

This chapter compares the available twisting moment design approaches in various scenarios. Although numerical comparisons are provided, the intent is to qualitatively illustrate how well or poorly each design method considers support configurations and design section orientations. The slabs investigated in this chapter are thin (8 in. [200 mm]), with small bars (0.5 in. [12 mm]) and moderate cover (1 in. [25 mm]); other slab parameters have little influence over the comparative results.

5.1—Sensitivity to angle of principal axes

This section investigates the sensitivity of the considered design methods to the angle between the design section and the principal zero-twist axes for an isotropic linear-elastic slab. Figure 5.1a shows a slab with uniform moments (and zero twist) about the x- and y-axes (one moment causing deflection upward and the other causing deflection downward), causing uniform twist and zero bending moments along the 45-degree axis.

The design sections shown in Fig. 5.1a(c) are rotated about their crossing point and the design quantities and results investigated. At angle zero, there is pure bending and no twist. As shown in Fig. 5.1b, all methods, except for the sandwich method, lead to the same results. The sandwich method requires additional reinforcement due to its conservative lever arm assumption. At 45 degrees, there is pure twist and no bending. Between zero and 45 degrees, there is a combination of twist and bending.

Figure 5.1b shows the relative total quantities of reinforcement required as the design sections are rotated from zero to 45 degrees. The y-axis is normalized by the required reinforcement for the no-twist zero angle. Although this plot shows reinforcement quantities for a particular design scenario, general behaviors can be observed.

The Wood and Armer method and the torsion-as-bending method have similar curves that approach a ratio of 2 as the slab approaches pure twist. As explained in 3.2, the Wood and Armer approach has a strong theoretical basis that ensures a safe design.

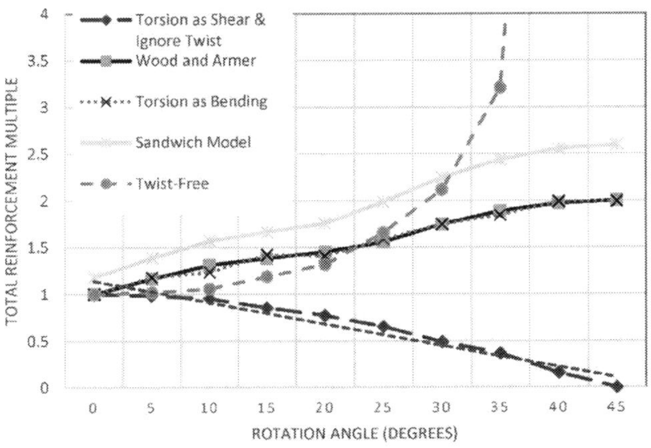

Fig. 5.1b—Relative calculated total reinforcement versus section angle.

It is expected that the torsion-as-bending approach will give similar results to the Wood and Armer approach, provided that the total torsion is approximately the same as the total twist. Figure 5.1b shows that the torsion-as-bending gives the same design as Wood and Armer for pure twist and pure bending, and similar designs for cases between those two extremes. The differences between Wood and Armer and torsion-as-bending in the plot are caused by the of the $c_{i,j}$ factor of Eq. (4.4.2b) not exactly matching the m_{xy} stress integration results.

Ignoring torsion and treating torsion-as-shear have identical curves that approach zero as the slab approaches pure twist. In this case, the torsional moments were not large enough for the torsion-as-shear method to require transverse reinforcement. Therefore, it is expected that its design will be the same as ignoring torsion. Both methods are unsafe for the pure twist condition.

Although the sandwich model follows a similar pattern to Wood and Armer, it requires more reinforcement due to its conservative lever-arm assumption. The test slab is thin (8 in. [200 mm]), which accentuates this conservatism. As explained in 4.3, the sandwich method has a strong theoretical basis that ensures a safe design.

The twist-free method provides a clear, safe equilibrium load path and is efficient for angles less than approximately 25 degrees. However, at angles greater than 25 degrees, the reinforcement layout creates an inefficient load path, where it becomes questionable if this load path can be attained without some failure or serviceability distress. At 45 degrees, the requirements theoretically approach infinity and no equilibrium is found.

Figure 5.1c shows the predicted relative maximum deflections as sections (and the slab behavior in the twist-free case) are rotated from zero to 45 degrees. The y-axis is normalized by the isotropic deflection value. For all methods, except twist-free analysis, the predicted deflections are constant, as the design method does not affect the isotropic analysis. For twist-free analysis, the slab orthotropic properties are oriented parallel to the design sections, and the slab becomes increasingly more flexible as the axes approach the pure-twist axes of 45 degrees. Above approximately 25 degrees,

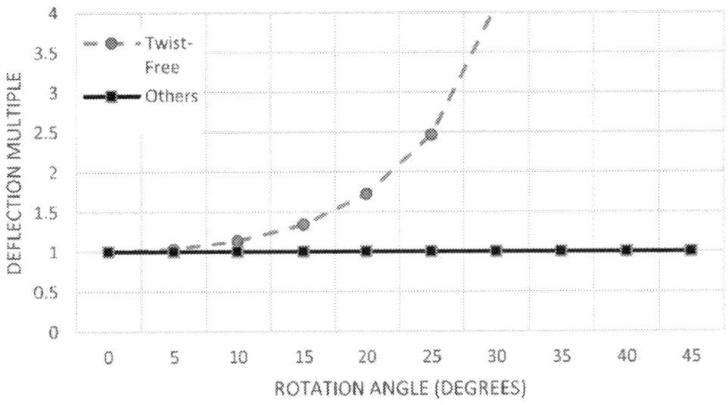

Fig. 5.1c—Relative maximum deflection versus section angle.

the deflections become very large, and significant cracking would be expected for the slab to deform from the uncracked isotropic load path to the cracked orthotropic load path. This plot shows that the ratio of twist-free deflection to isotropic deflection can be used as a proxy to determine qualitatively the amount of redistribution and cracking necessary to reach a twist-free load path.

5.2—Typical design conditions

This section investigates regions of slabs with plan geometries that occur frequently, comparing how the considered design methods differ for each of the geometries. Isotropic and orthotropic (twist-free) linear-elastic behaviors are investigated. While the slab configurations discussed do not cover every scenario, they provide a range of conditions that expose the strengths and weaknesses of each design method. Figure 5.2 locates the typical conditions on a slab plan and lists the section numbers where the conditions are discussed.

Note that while many design codes, including ACI 318, would suggest using column strips and middle strips for slab design, the design sections investigated in this section have been chosen for illustrative purposes and might not follow any particular code requirements.

The slab regions analyzed for in 5.2 are simplified and are intended to represent typical behavior patterns. Only gravity loading effects are evaluated. Lines of symmetry have been used where the slab continues into an adjacent region.

5.2.1 *Notes on figures and tables used in 5.2*—Sections 5.2.2 through 5.2.7 each contain one figure with four images. The legend for each set of four images is the same and presented in Fig. 5.2.1.

The tables presented in 5.2.2 through 5.2.7 contain the same rows and columns presented for each of the design conditions. Each row reports for a design method, and the meaning of the columns is:

a) M represents the bending moment at the cross section. This will be the same for all methods except for twist-free. The value is normalized by the isotropic result.

b) T represents the torsion (due to both twist and eccentric shear) at the cross section. This will be the same for all methods except for twist-free. The value is normalized by the isotropic M value.

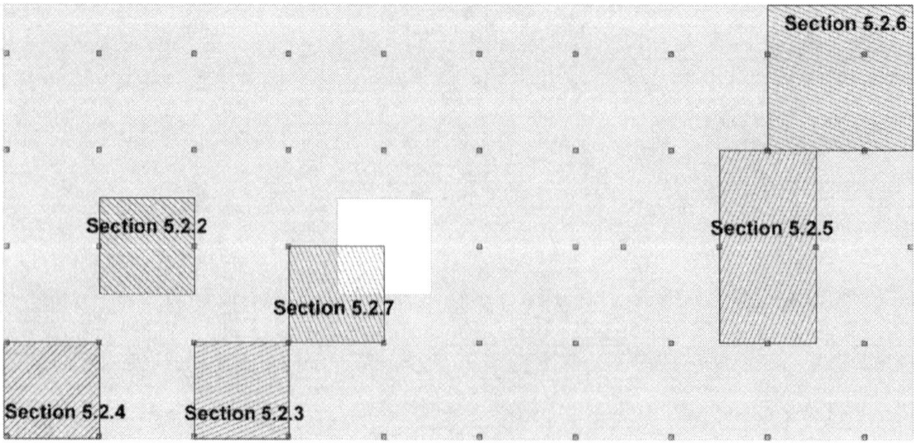

Fig. 5.2—Key map for typical design conditions discussions.

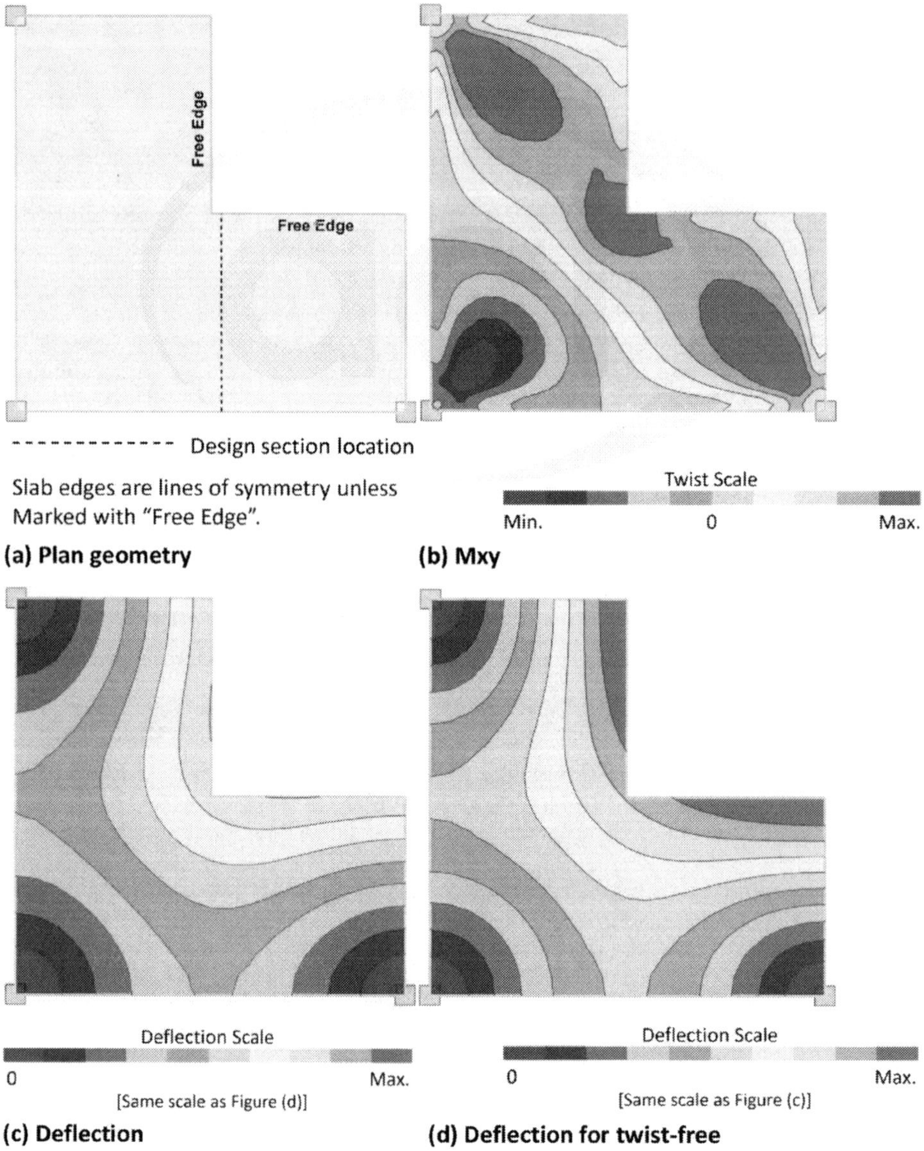

Fig. 5.2.1—Legends for figures in 5.2.2 to 5.2.7.

c) |*Twist*| represents the integral of the absolute value of twist ($\int|m_{xy}|$) along the cross section. This will be the same for all methods except for twist-free. The value is normalized by the isotropic *M* value.

d) *Defl* represents maximum deflection in the slab region. This will be the same for all methods except for twist-free. The value is normalized by the isotropic deflection value.

e) *Md* represents design moment. The value is normalized by the Wood and Armer design moment. Design moments are not provided for the sandwich model, as it reduces the design demand to a design force instead of a design moment.

f) *Ab* represents the bottom reinforcement area. The value is normalized by the maximum of the Wood-Armer top and bottom reinforcement areas.

g) *At* represents the top reinforcement area. The value is normalized by the maximum of the Wood-Armer top and bottom reinforcement areas.

h) *Av* represents the requirement for shear reinforcement. The only method that can require shear reinforcement is torsion-as-shear.

i) *Appl* summarizes the applicability of the method to the configuration. The values used are:

i. *Yes* — The method is applicable, and within 5 percent of the most efficient applicable method.

ii. *Cons* — The method is applicable, and within 20 percent of the most efficient applicable method.

iii. *Ex. Cons* — The method is applicable, although it is 20 percent or more conservative than the most efficient applicable method.

iv. *Serv?* — The method is applicable for strength, but there are serviceability concerns (this value only applies to the twist-free analysis approach).

v. *?* — The reinforcement calculated is not less than one of the applicable methods, but the assumed mechanisms of the method do not match the slab behavior, so the applicability of the method to similar configurations is questionable.

vi. *No* — The method is not applicable.

The discussion in 5.2.2 through 5.2.7 references the correct quantity of reinforcement. In the context of those discussions, the correct quantity of reinforcement is the minimum quantity of reinforcement required by a method that is known to be appropriate for the conditions being discussed.

5.2.2 *Regular slab – interior panel at column*—Figure 5.2.2 shows an interior panel of a regular slab and the location of a critical design section at the support in a region with large bending moments and significant twist.

Table 5.2.2 notes the normalized key design quantities and results for the design section shown in Fig. 5.2.2.

While there is a significant twist peak near the support, overall the integrated twist for this section is only 9 percent of the integrated moment. Due to symmetry of the slab configuration, torsion on the section is zero. Note the following regarding design methods:

a) **Ignore twist:** Ignoring twist results in the correct quantity of reinforcement, although it is questionable, as it does not guarantee a mechanism exists to resist the twist.

b) **Wood and Armer:** This approach is applicable, although it results in a 9 percent excess reinforcement penalty.

c) **Torsion as bending:** Condition 1 discussed in 4.4.3 applies here; the twisting moments change sign along the section, so torsion-as-bending is not expected to apply. Treating torsion as bending results in the correct quantity of reinforcement. Due to symmetry, however, the total torsion is zero; therefore, torsion is a poor proxy for integrated twist and torsion-as-bending questionable.

d) **Torsion-as-shear:** Treating torsion as shear results in the correct quantity of reinforcement. However, it does not guarantee a mechanism for resisting the twist that exists; therefore, torsion-as-shear is questionable.

e) **Sandwich model:** Although this approach is applicable, it results in a 32 percent excess reinforcement penalty. This excess would be smaller in thicker slabs.

f) **Twist-free analysis:** This approach results in deflection predictions that are 4 percent larger than a regular (isotropic) analysis. This increase is very small, so there are no concerns of significant cracking required to achieve the twist-free load path. Twist-free analysis is the most efficient applicable method for this case.

5.2.3 *Regular slab – edge panel at column*—Figure 5.2.3 shows an edge panel of a regular slab and the location of a design section at the support in a region with large bending moments and significant twist.

Table 5.2.3 notes the normalized key design quantities and results for the design section shown in Fig. 5.2.3.

While there is a significant twist peak near the support, overall the integrated twist for this section is only 14 percent of the integrated moment. Due to support at one end of the design section, torsion on this section is large—149 percent of the integrated moment. This torsion is almost entirely caused by eccentric shear and not twisting. Note the following regarding design methods:

a) **Ignore twist:** Ignoring twist results in the correct quantity of reinforcement, although it is questionable as it does not guarantee a mechanism exists to resist the twist.

b) **Wood and Armer:** This approach is applicable, although it results in a 15 percent excess reinforcement penalty.

c) **Torsion-as-bending:** Neither Condition 1 or 2 discussed in 4.4.3 applies here, so torsion-as-bending is expected to apply. Treating torsion-as-bending results in a very conservative quantity of reinforcement, as torsion in this case is a very conservative proxy for integrated twist.

d) **Torsion-as-shear:** Because the torsion is primarily caused by eccentric shear, torsion-as-shear largely matches the actual slab behavior and provides a much more appropriate shear design than the other approaches. However, although this approach results in the correct quantity of reinforcement, it does not provide a mechanism for resisting the 10 percent of torsion caused by twist; therefore, torsion-as-shear is considered questionable.

e) **Sandwich model:** This approach is applicable but results in a 40 percent excess reinforcement penalty. This excess would be smaller in thicker slabs.

f) **Twist-free analysis:** This approach results in deflection predictions that are 6 percent larger than a regular (isotropic) analysis. The increase is small, so there are no concerns of significant cracking required to achieve the twist-free load

Table 5.2.2—Analysis and design quantities for regular slab interior panel

Design approach	M	T	|twist|	Defl.	Md	Ab	At	Av	Appl.
Ignore twist	1.00	0.00	0.09	1.00	0.92	0.00	0.92	No	?
Wood and Armer	1.00	0.00	0.09	1.00	1.00	0.00	1.00	No	Cons.
Torsion as bending	1.00	0.00	0.09	1.00	0.92	0.00	0.92	No	?
Torsion as shear	1.00	0.00	0.09	1.00	0.92	0.00	0.92	No	?
Sandwich model	1.00	0.00	0.09	1.00	N/A	0.00	1.21	No	Ex. Cons
Twist-free analysis	1.00	0.00	0.00	1.04	0.92	0.00	0.92	No	Yes

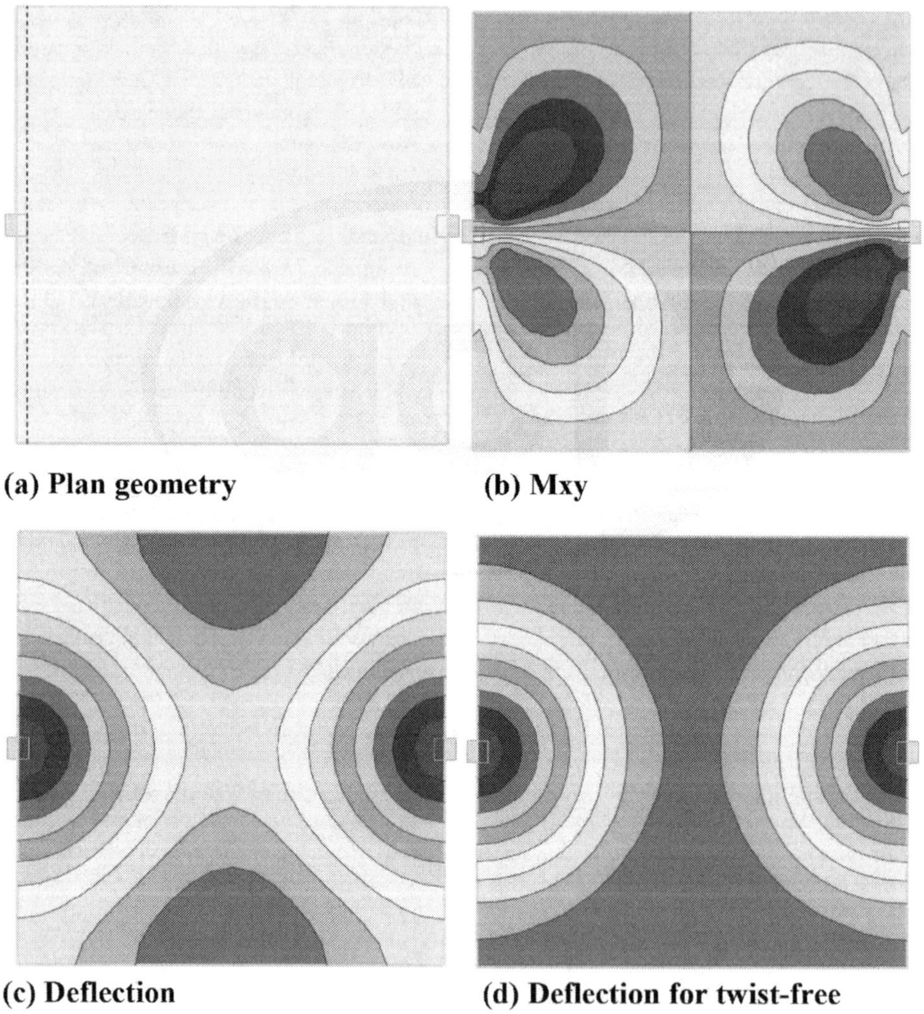

Fig. 5.2.2—Regular slab, interior panel.

path. Twist-free analysis is the most efficient applicable method for this case.

5.2.4 *Regular slab – corner panel at column*—Figure 5.2.4 shows a corner panel of a regular slab and the location of a critical design section at the support in a region with small bending moments and significant twist.

Table 5.2.4 notes the normalized key design quantities and results for the design section shown in Fig. 5.2.4.

There is a significant twist peak near the support and the bending moment is very small; overall, the integrated twist for this section is 157 percent of the integrated (small) moment. Due to the support at one end of the design section,

Table 5.2.3—Analysis and design quantities for regular slab, edge panel

Design approach	M	T	|twist|	Defl.	Md	Ab	At	Av	Appl
Ignore twist	1.00	1.49	0.14	1.00	0.87	0.00	0.87	No	?
Wood and Armer	1.00	1.49	0.14	1.00	1.00	0.00	1.00	No	Cons
Torsion as bending	1.00	1.49	0.14	1.00	2.18	0.42	2.23	No	Ex. Cons
Torsion as shear	1.00	1.49	0.14	1.00	0.87	0.00	0.87	No	?
Sandwich model	1.00	1.49	0.14	1.00	N/a	0.00	1.22	No	Ex. Cons
Twist-free analysis	1.00	1.43	0.00	1.06	0.87	0.00	0.87	No	Yes

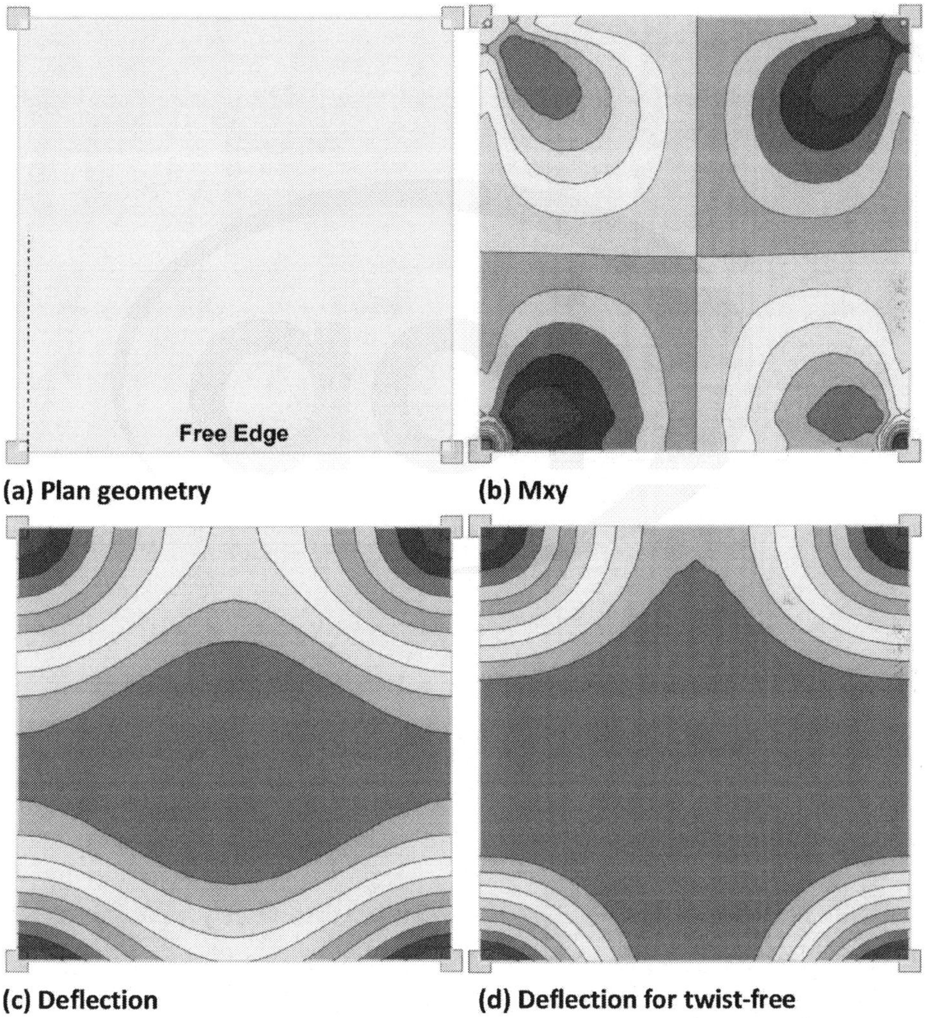

Fig. 5.2.3—Regular slab, edge panel.

the torsion on the section is very large at 525 percent of the integrated moment. Although torsion is mostly caused by eccentric shear, the twisting component is significant. Due to the corner column being much less stiff than the slab, the overall reinforcement demands at the corner column are very small; minimum reinforcement provisions can govern in this region. Note the following regarding design methods:

a) **Ignore twist:** Ignoring twist results in the correct quantity of reinforcement, although it is questionable as it does not guarantee a mechanism exists to resist the twist.

Table 5.2.4—Analysis and design quantities for regular slab, corner panel

Design approach	M	T	|twist|	Defl.	Md	Ab	At	Av	Appl.
Ignore twist	1.00	5.25	1.57	1.00	0.39	0.00	0.39	No	?
Wood and Armer	1.00	5.25	1.57	1.00	1.0	0.22	1.00	No	Ex. Cons
Torsion as bending	1.00	5.25	1.57	1.00	2.44	1.67	2.47	No	Ex. Cons
Torsion as shear	1.00	5.25	1.57	1.00	0.39	0.00	0.39	No	?
Sandwich model	1.00	5.25	1.57	1.00	N/a	0.34	1.29	No	Ex. Cons
Twist-free analysis	0.99	5.10	0.00	1.10	0.38	0.00	0.38	No	Yes

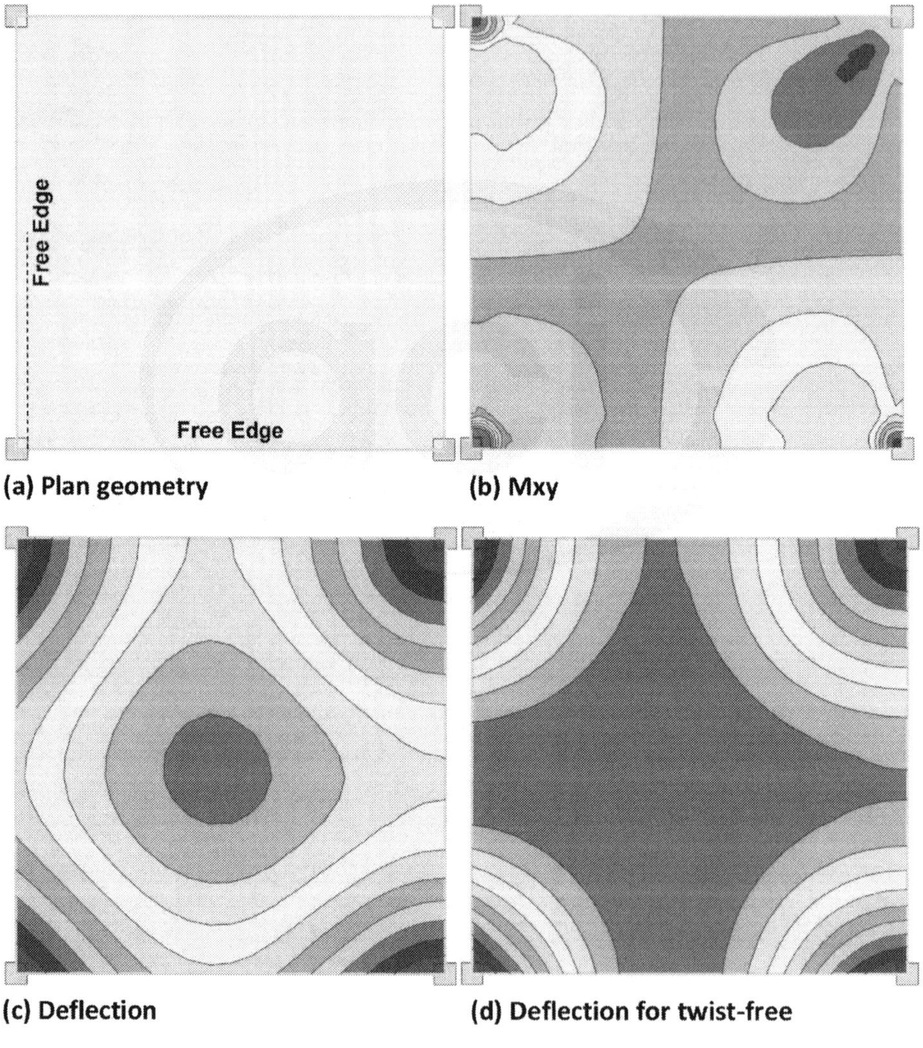

Fig. 5.2.4—Regular slab, corner panel.

b) **Wood and Armer:** This approach is applicable, although it results in a 220 percent excess reinforcement penalty.

c) **Torsion-as-bending:** Neither Condition 1 or 2 discussed in 4.4.3 applies here, so torsion-as-bending is expected to apply. Treating torsion as bending results in a very conservative quantity of reinforcement, as torsion in this case is a very conservative proxy for integrated twist.

d) **Torsion-as-shear:** Because the torsion is primarily caused by eccentric shear, the torsion-as-shear approach largely matches the actual slab behavior and provides a much more appropriate shear design than the other approaches.

Table 5.2.5—Analysis and design quantities for slab with misaligned supports, interior panel

Design approach	M	T	|twist|	Defl.	Md	Ab	At	Av	Appl.
Ignore twist	1.00	0.00	0.11	1.00	0.90	0.00	0.89	No	No
Wood and Armer	1.00	0.00	0.11	1.00	1.00	0.00	1.00	No	Yes
Torsion as bending	1.00	0.00	0.11	1.00	0.90	0.00	0.89	No	No
Torsion as shear	1.00	0.00	0.11	1.00	0.90	0.00	0.89	No	No
Sandwich model	1.00	0.00	0.11	1.00	N/a	0.00	1.29	No	Ex. Cons
Twist-free analysis	1.21	0.00	0.00	1.18	1.09	0.00	1.09	No	Cons

However, it does not guarantee a mechanism for resisting twist, which causes 30 percent of the torsion; therefore, torsion-as-shear is considered questionable.

e) **Sandwich model:** This approach is applicable but results in a 340 percent excess reinforcement penalty.

f) **Twist-free analysis:** This approach results in deflection predictions that are 10 percent larger than a regular (isotropic) analysis. This increase is small, so there are no concerns of significant cracking to achieve the twist-free load path. Twist-free analysis is the most efficient applicable method for this case.

5.2.5 *Slab with unaligned supports – interior panel at column*—Figure 5.2.5 shows an interior panel of an irregular slab with supports staggered by half the span length and the location of a critical design section at the support in a region with large bending moments and significant twist.

Table 5.2.5 notes the normalized key design quantities and results for the design section shown in Fig. 5.2.5.

While there is a significant twist peak near the support, overall the integrated twist for this section is only 11 percent of the integrated moment. Due to the symmetry of the slab configuration, torsion on the section is zero. Note the following regarding design methods:

a) **Ignore twist:** Ignoring twist neither provides the correct quantity of reinforcement or a mechanism to resist the twisting moments, so it is not applicable for this configuration.

b) **Wood and Armer:** The Wood and Armer method is the most efficient applicable method for this case.

c) **Torsion-as-bending:** Condition 1 discussed in 4.4.3 applies here; the twisting moments change sign along the section, so torsion-as-bending is not expected to apply. Treating torsion as bending does not provide the correct amount of reinforcement, as torsion in this case is a poor proxy for integrated twist. Torsion-as-bending is not applicable for this configuration.

d) **Torsion-as-shear:** Treating torsion as shear neither provides the correct quantity of reinforcement nor a mechanism to resist the twisting moments, so it is not applicable for this configuration.

e) **Sandwich model:** This approach is applicable but results in a 21 percent excess reinforcement penalty. This excess would be smaller in thicker slabs.

f) **Twist-free analysis:** The twist-free analysis approach for this slab results in deflection predictions that are 18 percent larger than a regular (isotropic) analysis. This increase is small, so there are no concerns of significant cracking to achieve the twist-free load path. However, twist-free analysis results in a 9 percent reinforcement penalty.

5.2.6 *Two-way cantilever slab*—Figure 5.2.6 shows a two-way cantilever and the location of a critical design section at the support in a region with very large bending moments and significant twist.

Table 5.2.6 notes the normalized key design quantities and results for the design section shown in Fig. 5.2.6.

While there is a significant twist peak near the support, overall, the integrated twist for this section is only 4 percent of the integrated moment. Due to the length of the cantilever (creating a large moment), the torsion on the section is also small: 3 percent of the integrated moment. Note the following regarding design methods:

a) **Ignore twist:** Ignoring twist results in the correct quantity of reinforcement, although it is questionable as it does not guarantee a mechanism exists to resist the twist.

b) **Wood and Armer:** This approach is also applicable, although it results in a 4 percent excess reinforcement penalty.

c) **Torsion-as-bending:** Condition 1 discussed in 4.4.3 applies here; the twisting moments change sign along the section, so torsion-as-bending is not expected to apply. Treating torsion as bending results in a safe quantity of reinforcement, but the total torsion is a little less than integrated twist. Therefore, torsion-as-bending is questionable.

d) **Torsion-as-shear:** Treating torsion as shear results in the correct quantity of reinforcement, but eccentric shear only causes 25 percent of total torsion. Therefore, torsion-as-shear is questionable.

e) **Sandwich model:** This approach is applicable but results in a 24 percent excess reinforcement penalty. This excess would be smaller in thicker slabs.

f) **Twist-free analysis:** Unexpectedly, the twist-free analysis for this slab results in a peak deflection prediction that is 15 percent less than a regular (isotropic) analysis. One asks the question of how can this be; how can removing a slab stiffness reduce the deflection? The answer is that the average deflection for the twist-free analysis is 12 percent

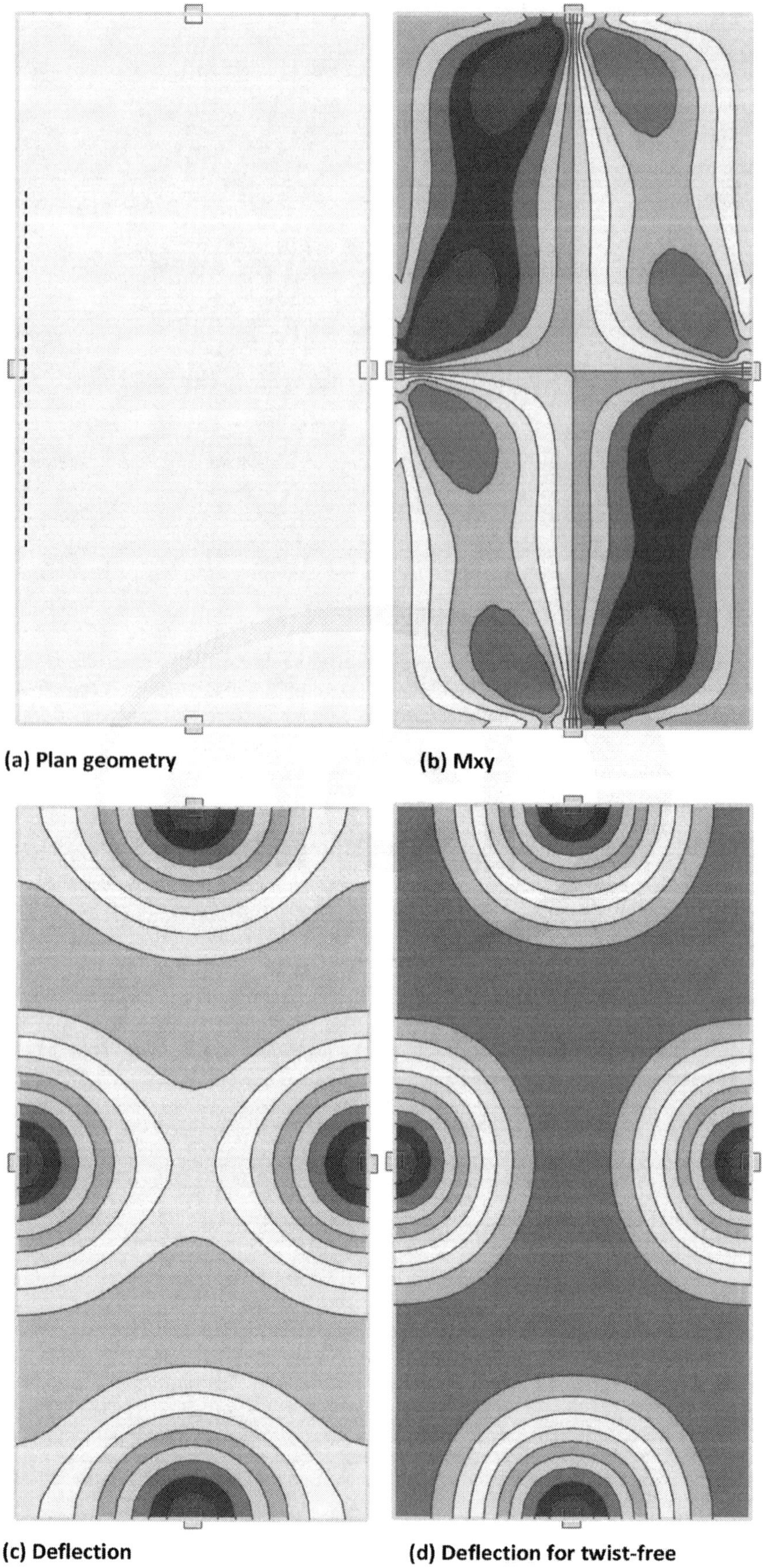

Fig. 5.2.5—Slab with misaligned supports, interior panel.

Table 5.2.6—Analysis and design quantities for two-way cantilever

Design approach	M	T	|twist|	Defl.	Md	Ab	At	Av	Appl.
Ignore twist	1.00	0.03	0.04	1.00	0.96	0.00	0.96	No	?
Wood and Armer	1.00	0.03	0.04	1.00	1.00	0.00	1.00	No	Yes
Torsion as bending	1.00	0.03	0.04	1.00	0.98	0.00	0.98	No	?
Torsion as shear	1.00	0.03	0.04	1.00	0.96	0.00	0.96	No	?
Sandwich model	1.00	0.03	0.04	1.00	N/a	0.00	1.19	No	Ex. Cons
Twist-free analysis	1.00	0.03	0.00	0.85	0.96	0.00	0.96	No	Yes

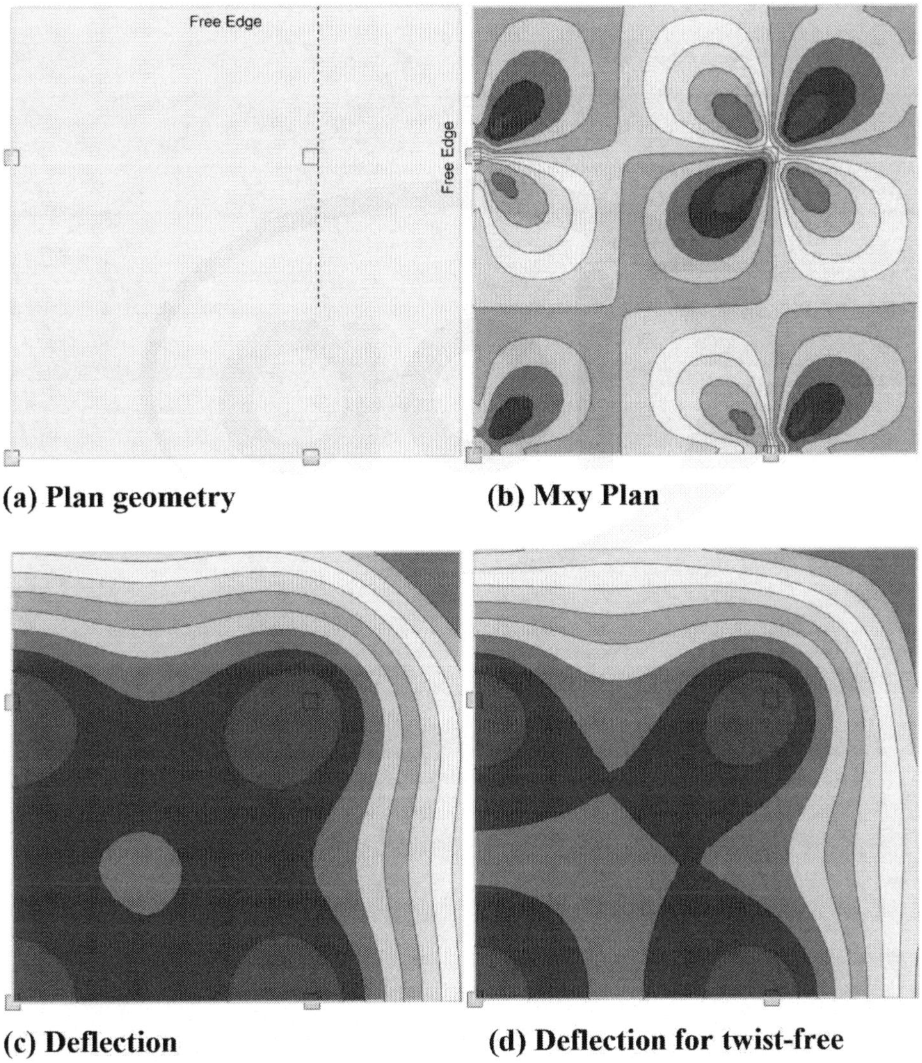

Fig. 5.2.6—Two-way cantilever.

more than for the regular (isotropic) analysis; however, the load path is different and there is less of a peak deflection at the cantilever tip. Overall, there are no concerns of significant cracking to achieve the twist-free load path. Twist-free analysis is the most efficient applicable method for this case.

5.2.7 *Reentrant corner*—Figure 5.2.7 shows a reentrant unsupported slab corner with the design section at the corner crossing a region of moderate bending moments and significant twists.

Table 5.2.7—Analysis and design quantities for reentrant corner

Design approach	M	T	|twist|	Defl.	Md	Ab	At	Av	Appl.
Ignore twist	1.00	0.47	0.30	1.00	0.77	0.76	0.00	No	No
Wood and Armer	1.00	0.47	0.30	1.00	1.00	1.00	0.00	No	Yes
Torsion as bending	1.00	0.47	0.30	1.00	1.13	1.13	0.00	No	Cons.
Torsion as shear	1.00	0.47	0.30	1.00	0.77	0.76	0.00	No	No
Sandwich model	1.00	0.47	0.30	1.00	N/a	1.23	0.00	No	Ex. Cons
Twist-free analysis	0.87	0.09	0.00	1.38	0.66	0.66	0.00	No	Serv?

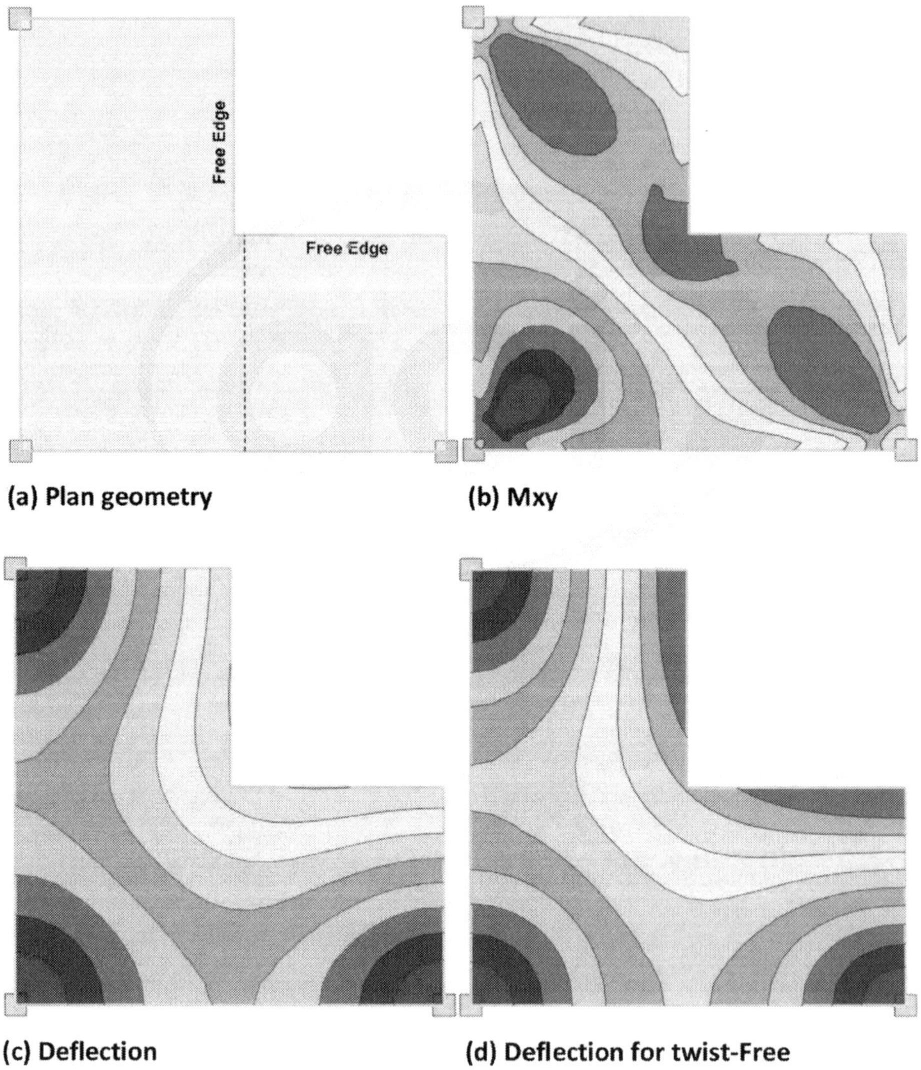

Fig. 5.2.7—Reentrant corner.

Table 5.2.7 notes the normalized key design quantities and results for the design section shown in Fig. 5.2.7.

There is a significant twist peak near the notch and the bending moment is moderate; overall, the integrated twist for this section is 30 percent of the integrated moment. Torsion, 47 percent of the integrated moment, is caused primarily by twist, but also has a significant component due to eccentric shear. Note the following regarding design methods:

Table 5.2.8—Summary of method applicability

Design approach	Regular slab interior	Regular slab edge	Regular slab corner	Unaligned supports	Two-way cantilever	Reentrant corner
Ignore twist	?	?	?	No	?	No
Wood and Armer	Cons.	Cons	Ex. Cons	Yes	Yes	Yes
Torsion as bending	?	Ex. Cons	Ex. Cons	No	?	Cons.
Torsion as shear	?	?	?	No	?	No
Sandwich model	Ex. Cons	Ex. Cons	Ex. Cons	Ex. Cons	Ex. Cons	Ex. Cons
Twist-free analysis	Yes	Yes	Yes	Cons	Yes	Serv?

a) **Ignore twist:** Ignoring twist neither provides the correct quantity of reinforcement or a mechanism to resist the twisting moments, so it is considered not applicable for this configuration.

b) **Wood and Armer:** The Wood and Armer method is the most efficient applicable method for this case.

c) **Torsion-as-bending:** Neither Condition 1 or 2 discussed in 4.4.3 applies here, so torsion-as-bending is expected to apply. Treating torsion as bending is applicable for this case, although it requires 13 percent excess reinforcement. Torsion is a conservative proxy for twist.

d) **Torsion-as-shear:** Treating torsion as shear neither provides the correct quantity of reinforcement nor a mechanism to resist the twisting moments, so it is considered not applicable for this configuration. While eccentric shear causes significant torsion, twisting moments are still large.

e) **Sandwich model:** This approach is applicable but results in a 23 percent excess reinforcement penalty. This excess would be smaller in thicker slabs.

f) **Twist-free analysis:** The twist-free analysis approach for this slab requires the least reinforcement but results in deflection predictions that are 38 percent larger than a regular (isotropic) analysis. This increase is somewhat large, so there are some concerns of noticeable cracking to achieve the twist-free load path. That makes twist-free analysis questionable for this configuration. Adding diagonal reinforcement in the twist-critical reentrant corner might be adequate to ease the serviceability concerns while still requiring less reinforcement than the other methods. This, however, requires engineering judgment beyond the scope of this guide.

5.2.8 *Summary of typical design conditions*—Table 5.2.8 displays a summary of the applicability of design methods for each condition discussed in 5.2. Note that no method is optimal for all cases, but that some conclusions can be drawn.

Although the Wood and Armer method is always applicable, it can be more conservative than required.

The sandwich model is always applicable but is very conservative for thin slabs such as the one investigated. For significantly thicker slabs, the sandwich model will have results similar to the Wood and Armer method. The strengths of the sandwich model are generally not relevant for typical slab design.

Twist-free analysis, where applicable, is usually among the most efficient methods. Twist-free analysis is applicable to most configurations; if the twist-free deflection is less than 25 percent greater than the isotropic deflection, twist-free analysis is generally applicable.

The other methods, while appropriate in some circumstances, are either inappropriate or excessively conservative in too many cases to be used indiscriminately.

While not investigated in detail in this chapter, applying the torsion-as-shear design approach to a twist-free analysis is applicable whenever the twist-free analysis approach is applicable. This combination approach has the potential for improving vertical shear design while retaining the advantages of the twist-free analysis approach.

CHAPTER 6—TWO-WAY WALLS

6.1—General considerations

The discussion of twisting moments to this point has focused on two-way slabs supported by discrete columns. Twisting moments in two-way walls can also impact the out-of-plane design moments. Generally, walls are fundamentally different than slabs because: 1) out-of-plane flexure in the wall results from horizontal loads rather than vertical loads; and 2) boundary conditions are typically continuous rather than discrete. The former difference is largely irrelevant to this discussion; however, the later difference can have an impact on the importance of twisting moment on the magnitude of design moments.

Multiple combinations of boundary and loading conditions acting on walls can result in twisting moments. Many infrastructure projects involve wall elements with continuous boundary conditions along two adjacent edges and out-of-plane loading along one face, resulting in a behavior similar to that shown in Fig. 3.1. This geometry is common in dam structures involving piers, as shown in Fig. 6.1a. Tanks often exhibit walls that are continuously supported on three sides, as shown in Fig. 6.1b. This geometry can also create the conditions that result in high twisting moments occurring at locations of high bending moment.

6.2—Impact of twisting moment on walls exhibiting two adjacent fixed edges

The pier shown in Fig. 6.1a demonstrates a geometry and boundary conditions common in water control structures such as dam spillways. In this section, the impacts of twisting moment will be explored for wall-like spillway piers. Three different aspect ratios, as shown in Fig. 6.2a, are considered to quantify the impact of geometry. Walls with a

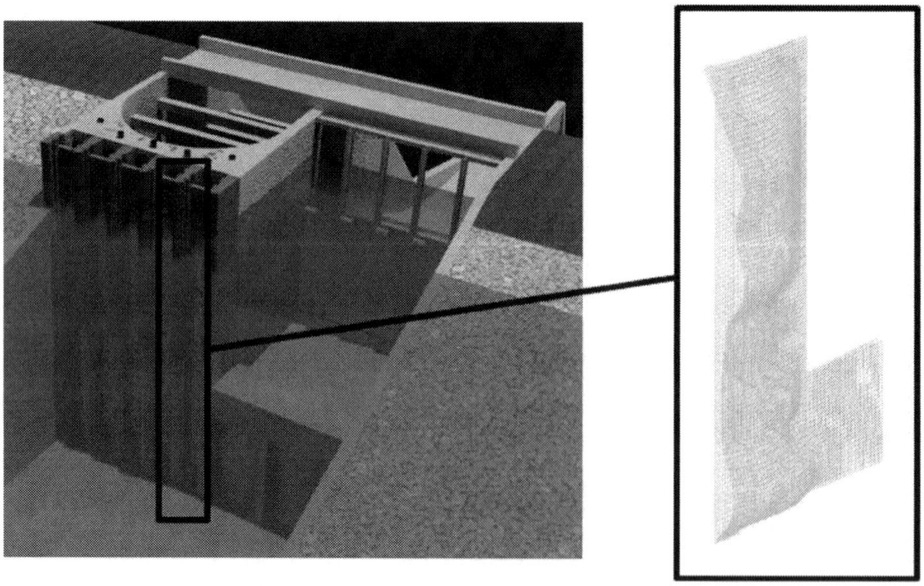

Fig. 6.1a—Dam pier and resulting deflected shape of the twist-sensitive element.

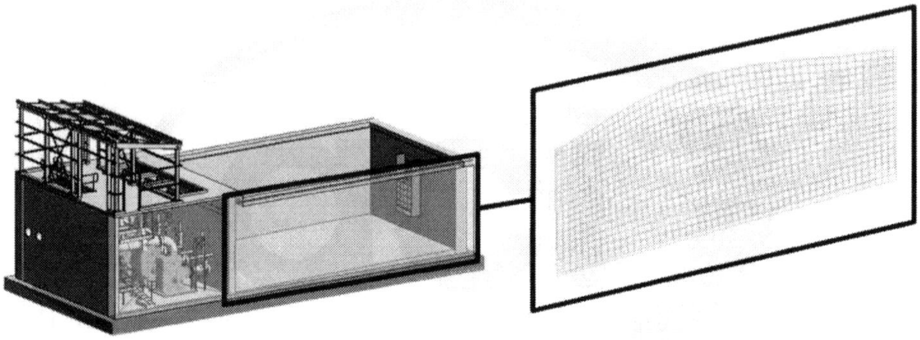

Fig. 6.1b—Rectangular tank and resulting deflected shape of the twist-sensitive element.

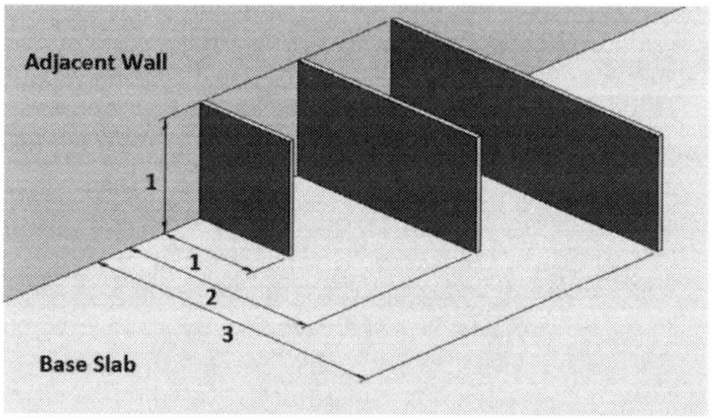

Fig. 6.2a—Twist-sensitive wall examples.

ratio close to 1:1 generally exhibit two-way behavior. As the aspect ratio increases, the wall starts behaving more like a one-way element near the free edge.

The walls in this analysis are subject to a uniform lateral load. To provide simpler behavior that can lead to better understanding, Poisson's ratio has been set to 0 to eliminate anticlastic bending. Results are produced similar to the results of the examples in Chapter 5. However, only the following design approaches are considered: 1) ignore twist; 2) Wood and Armer; and 3) twist-free. For Chapter 5, the assumed design strip is the width of a slab cross section and for Chapter 6, the point values are used to produce the design moment. Results of these analyses for the three different aspect ratios are shown in Fig. 6.2b. Location 1 exhibits the largest twist and Location 2 exhibits the largest bending moment. Deflection and required steel areas are determined using the forces obtained in these analyses. The contours shown in Fig. 6.2b are similar to the assumptions outlined in Fig. 5.2.1, with the exception that 0 is centered on the scale allowing for negative values in this chapter.

The forces at Locations 1 and 2 changed significantly between the 1:1 and 2:1 aspect ratios for the twist analysis. However, they change little between the 2:1 and 3:1 aspect ratios.

The deflections and required area of steel for the design moments are presented in Table 6.2 for the three walls and three design approaches. Whereas the largest twist occurs at Location 1 for each aspect ratio, the controlling design moment is still largely controlled by moment at the fixed

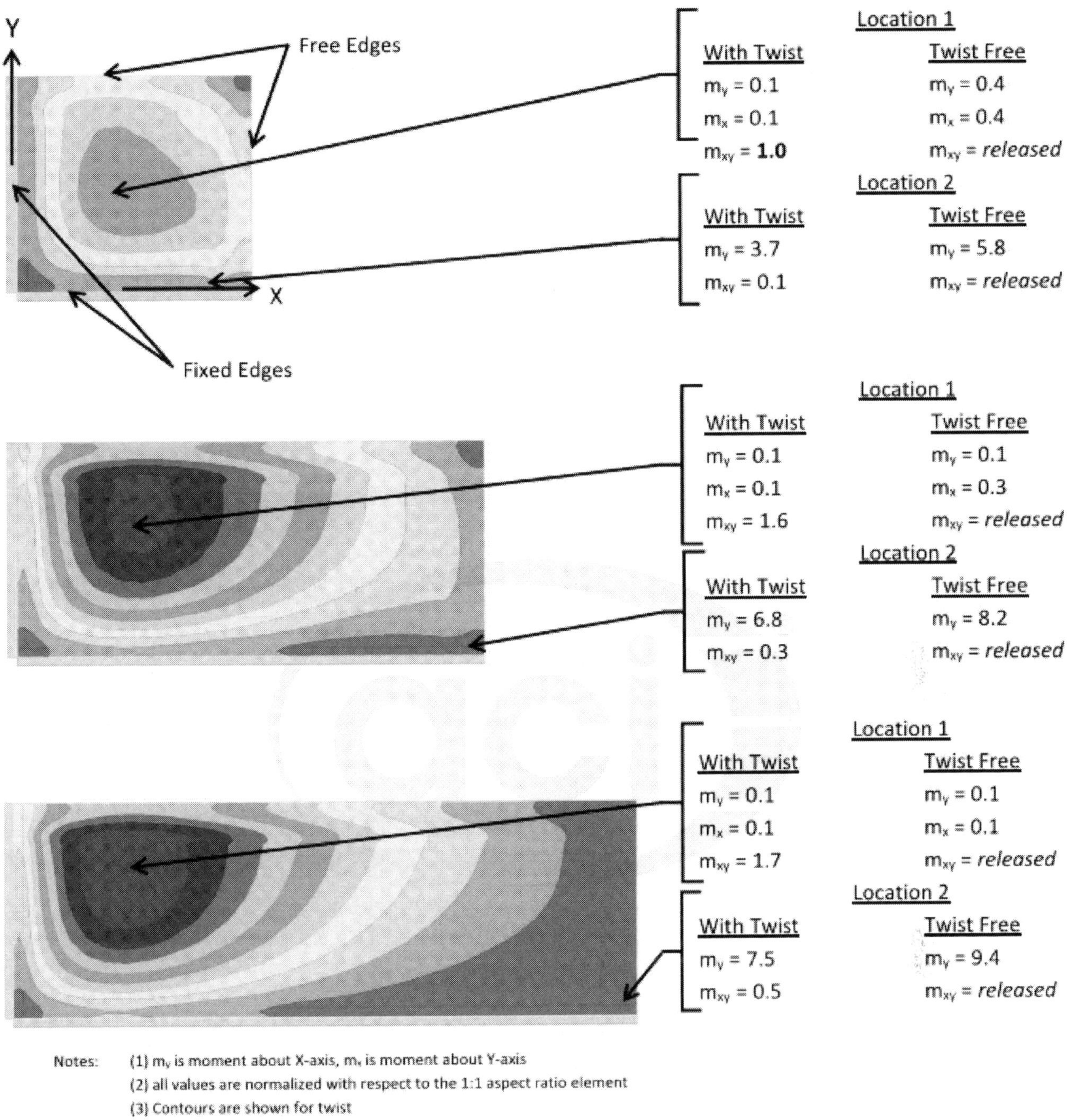

Fig. 6.2b—Twist-sensitive wall study showing twisting moment trends.

boundary condition at Location 2. Therefore, the values that populate Table 6.2 are at Location 2.

The following is a summary of each of the design approaches considering all three aspect ratios:

(1) **Ignore twist**: For the examples of this study, ignoring twist was only unconservative by approximately 5 percent. However, other wall geometries and boundary conditions can create locations of higher twist and moment acting at the same location. Therefore, accounting for twist in the element stiffness formulation and then ignoring the results is not recommended.

(2) **Wood and Armer**: This approach was applicable for all aspect ratios. However, the resulting design for the ignore twist method starts to approach the Wood and Armer method for larger aspect ratios. Twisting stiffness enables a more efficient load distribution because the twist is largely concentrated at locations of lower moment for uniformly reinforced walls. Therefore, this approach is generally recommended as a safe and efficient design method.

(3) **Twist-free analysis**: This approach is excessively conservative until you approach the 3:1 ratio, where values of twist are small and the moment increases less where

Table 6.2—Analysis and design quantities for various aspect ratios

1:1 ratio design approach	**M**	**Twist**	**Defl.**	**Md**	**As**	**Applicability**
Ignore twist	1.0	0.08	1.00	0.92	0.92	No
Wood and Armer	1.0	0.08	1.00	1.00	1.00	Yes
Twist-free analysis	1.6	0.00	2.26	1.44	1.43	Ex. Cons
2:1 ratio design approach	**M**	**Twist**	**Defl.**	**Md**	**As**	**Applicability**
Ignore twist	1.0	0.02	1.00	0.98	0.98	No
Wood and Armer	1.0	0.02	1.00	1.00	1.00	Yes
Twist-free analysis	1.4	0.00	1.52	1.36	1.41	Ex. Cons
3:1 ratio design approach	**M**	**Twist**	**Defl.**	**Md**	**As**	**Applicability**
Ignore twist	1.0	0.005	1.00	1.00	0.995	Yes
Wood and Armer	1.0	0.005	1.00	1.00	1.000	Yes
Twist-free analysis	1.1	0.000	1.12	1.09	1.087	Cons

twisting stiffness does not contribute to redistribution of the force. Although it could be used for large aspect ratios, it is recommended to avoid this method. The large deflection multiple for the 1:1 ratio wall and, to a lesser extent, the 3:1 wall, is indicative of a large alteration of the load path. This alteration in load path can lead to service cracking.

For the 1:1 aspect ratio, behavior observed is close to a pure twist (Fig. 3.1) in which the flow of force is most efficient at a 45-degree angle to the edges and assumed reinforcement direction. By eliminating twisting stiffness, the flow of force is confined to the far more inefficient load path, which is orthogonal to the edges and, basically, creating two cantilevers. For larger aspect ratios, the most efficient flow of force is naturally more closely aligned to the edges and assumed reinforcement direction. In these situations, the ignore twist analysis starts to produce designs similar to the Wood and Armer design method.

In conclusion, the Wood and Armer design approach is generally recommended for all wall analysis with two adjacent fixed boundary conditions and orthogonally placed reinforcement. Twist can be ignored for larger aspect ratios, but this is generally not recommended.

CHAPTER 7—SHELL STRUCTURES

7.1—General considerations

While shells, which are curved thin structures, behave very differently than flat slabs, they do exhibit twisting moments similar to flat slabs. Shell behavior is typically dominated by membrane actions, although flexural behavior (including twisting moments) can be significant at boundary conditions, near loading points, and at openings and other discontinuities. These twisting moments should be considered in design, as required in Section 6.1.1 of ACI 318.2-14. This chapter investigates twisting moment considerations in hemispherical domes, which are commonly shell-shaped. The conclusions may or may not be relevant for other shell shapes, in which twisting may be more or less important in the load paths.

7.2—Typical bulk material storage hemisphere

Hemispherical shells are commonly used as an efficient structural form for bulk granular material storage. This chapter considers the effects of twisting moments in the design of a typical bulk storage hemisphere. The hemisphere studied has a thickness to radius ratio of 0.01.

7.2.1 *Finite element modeling*—The finite element model of the hemisphere that was studied uses quadrilateral shell elements (Fig. 7.2.1a). Isotropic elements are used to evaluate the ignore twist, and the Wood and Armer design approaches. Orthotropic elements are used to evaluate the twist-free design approach. The analyses performed are all linear-elastic.

One significant difference between shells and slabs is the varied orientation of the shell local axes from location-to-location; therefore, it is not parallel to any global axes. To account for this, a clear axes convention needs to be established. For the hemispheric shell case investigated, the local axes are defined in Fig. 7.2.1b. Except in the apex region, reinforcement is placed parallel to the local axes shown in the figure.

The base of the hemisphere is constrained against vertical translation (local x-axis direction) and against translation tangential to the radius from the center vertical axis of the hemisphere (local y-axis direction). The base is allowed to translate radially (local z-axis direction). Figure 7.2.1c illustrates constraints at the base of the dome near the opening. The figure shows the undeformed and deformed shape of the model. Arrows at the base of the hemisphere depict the constraint against translation used in the model. The cylindrical coordinate system is provided to clarify the boundary conditions. The model is constrained against translation in

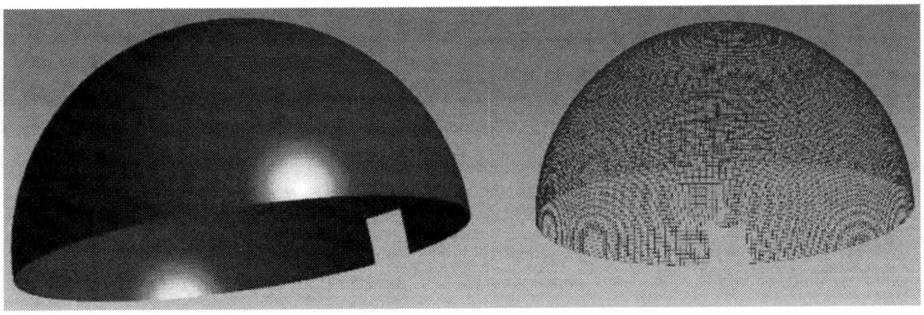

Fig. 7.2.1a—Finite element model (geometry and mesh).

the θ- and Z-directions, as shown in the figure. The figure illustrates how the model is only allowed to translate radially. While the region of the opening is shown, the constraints are typical all around the base. The ring foundation continues under the opening to provide the constraint in the θ direction.

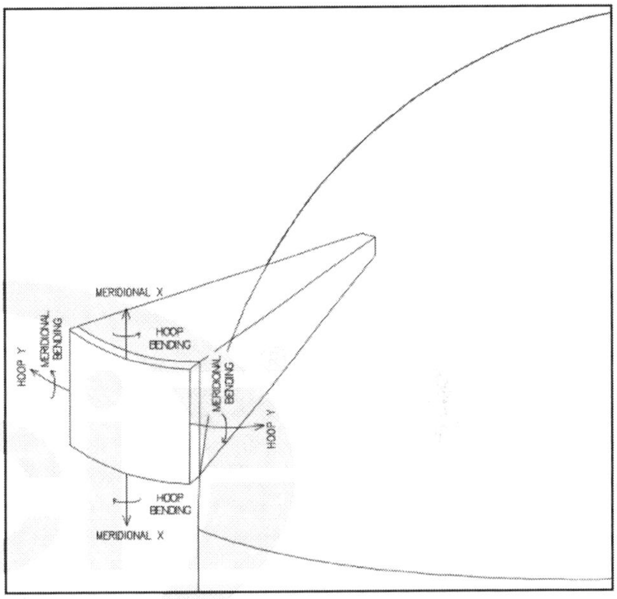

Fig. 7.2.1b—Local shell axes.

7.3—Typical loading conditions

Three common loadings are considered for the hemisphere: 1) self-weight; 2) pseudo-hydrostatic; and 3) a point load at the apex.

7.3.1 *Self-weight loading*—The self-weight of the hemisphere due to gravity is a significant load for this type of structure. The effects of this loading are considered in 7.4.1 and 7.4.2.

7.3.2 *Pseudo-hydrostatic loading*—Bulk granular materials stored in the hemisphere apply a load to the shell similar to a hydrostatic load. This loading is different from a hydrostatic pressure in that it acts horizontally and not perpendicularly to the shell. Figure 7.3.2 illustrates the load considered. The effects of this loading are considered in 7.4.1 and 7.4.2.

7.3.3 *Apex point loading*—Point loading at the apex of the hemisphere is typical for bulk material storage facilities. This load comes from the equipment to fill the hemisphere. The effects of this loading are considered in 7.4.3.

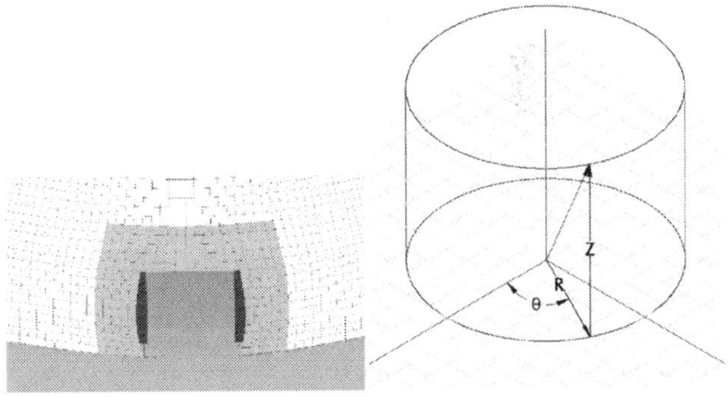

Fig. 7.2.1c—Hemisphere base boundary conditions and cylindrical coordinate system.

7.4—Typical design regions

Three regions of the hemisphere are considered: 1) the apex; 2) an opening at the base; and 3) a typical region at the base. Figure 7.4 identifies the three regions.

For each region the ignore-twist, Wood and Armer, and twist-free analysis approaches are evaluated. All results are normalized by the Wood and Armer results; bending and twisting analysis results are normalized together so the relative magnitude of twist and bending is apparent. The reported deflections are the maximum deflection in any direction. Note that scales are intentionally omitted from the plots. The importance of the figure is to show qualitative behaviors. Refer to Fig. 5.2.1 for scale information.

7.4.1 *Typical base region*—At the typical base section away from the opening, the results are observed to be similar to those of a pure hemisphere with no discontinuities. As the loadings are all radially symmetric, no significant twist is observed. Figure 7.4.1 shows this region subjected to the self-weight and hydrostatic loadings; the contours shown in Fig. 7.4.1(b) are all nearly zero and are primarily a result of numerical rounding in the FEA analysis. The section cut of the area analyzed is shown in the figure. The length of the analyzed strip is approximately 20 percent of the radius of curvature of the shell. The analyzed strip location was chosen at the location of highest Mx moment. This is different than previous chapters because the twisting moment is very small.

The calculation of the values in Table 7.4.1 is calculated the same way as in previous chapters. Note the following regarding design methods:

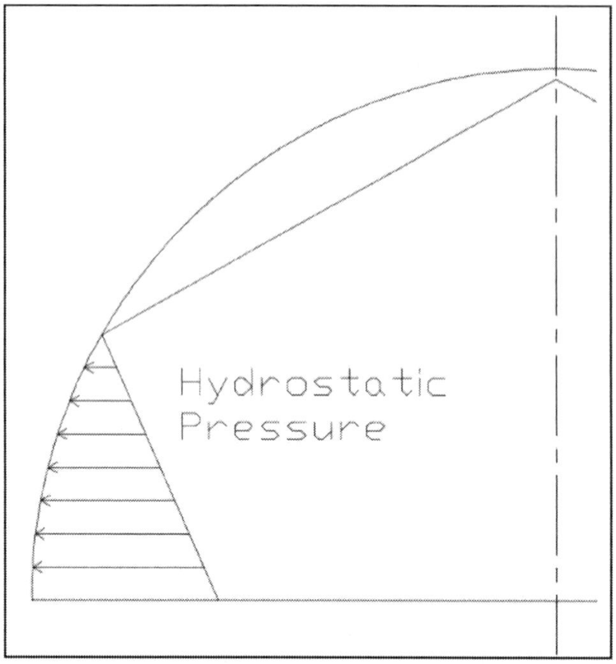

Fig. 7.3.2—Pseudo-hydrostatic loading.

a) **Ignore twist:** Ignoring twist results in the correct quantity of reinforcement, although it is questionable as it does not guarantee a mechanism exists to resist the twist.

b) **Wood and Armer:** This approach is applicable, although it results in a 4 percent excess reinforcement penalty.

c) **Twist-free analysis:** This approach results in deflections that are similar to a regular (isotropic) analysis. There are no concerns of significant cracking required to achieve the twist-free load path. Twist-free analysis is the most efficient applicable method for this case.

7.4.2 *Opening base region subject to self-weight loading*—Figure 7.4.2 shows this region subjected to the self-weight loading. The section cut of the area analyzed is shown in the figure. The length of the analyzed strip is approximately 5 percent of the radius of curvature of the shell. The analyzed strip location was chosen at the location of highest twisting moment.

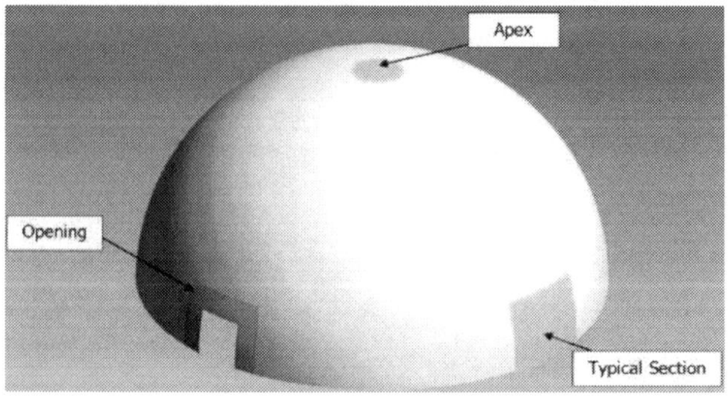

Fig. 7.4—Design regions.

Table 7.4.1—Comparisons for typical base section

Design approach	M	|twist|	Md	Deflection	Applicable
Ignore twist	1.00	0.03	0.97	1.00	?
Wood and Armer	1.00	0.03	1.00	1.00	Cons
Twist-free analysis	0.99	0.00	0.96	1.00	Yes

The area around the opening of the curved shell shows a large amount of twisting moment. This is illustrated in Table 7.4.2.

a) **Ignore twist:** This approach is not applicable in this area, as the twisting moment is significant and the isotropic load path requires twisting moments. The ignore twist design moment is only 44 percent of the most efficient applicable design method and, hence, could lead to unsafe designs.

b) **Wood and Armer:** This method is excessively conservative because it results in a 47 percent excess reinforcement penalty.

c) **Twist-free analysis:** This approach results in deflection predictions that are 15 percent larger than a regular (isotropic) analysis. This increase is relatively small, so there are no concerns of significant cracking required to achieve the twist-free load path. Twist-free analysis is the most efficient applicable method for this case.

7.4.3 *Opening base region subject to pseudo-hydrostatic loading*—Figure 7.4.3 shows this region subjected to pseudo-hydrostatic loading. The section cut of the area analyzed is shown in the figure. The length of the analyzed strip is approximately 5 percent of the radius of curvature of the shell. The analyzed strip location was chosen at the location of highest twisting moment.

The area around the opening of the curved shell shows a noticeable amount of twisting moment. This is illustrated in Table 7.4.3.

a) **Ignore twist:** This approach is not applicable in these areas, as the twisting moment is significant and the isotropic load path requires twisting moments. The ignore twist design moment is only 58 percent of the most efficient applicable design moment and, therefore, could lead to unsafe designs.

b) **Wood and Armer:** This method is excessively conservative because it results in a 75 percent excess reinforcement penalty.

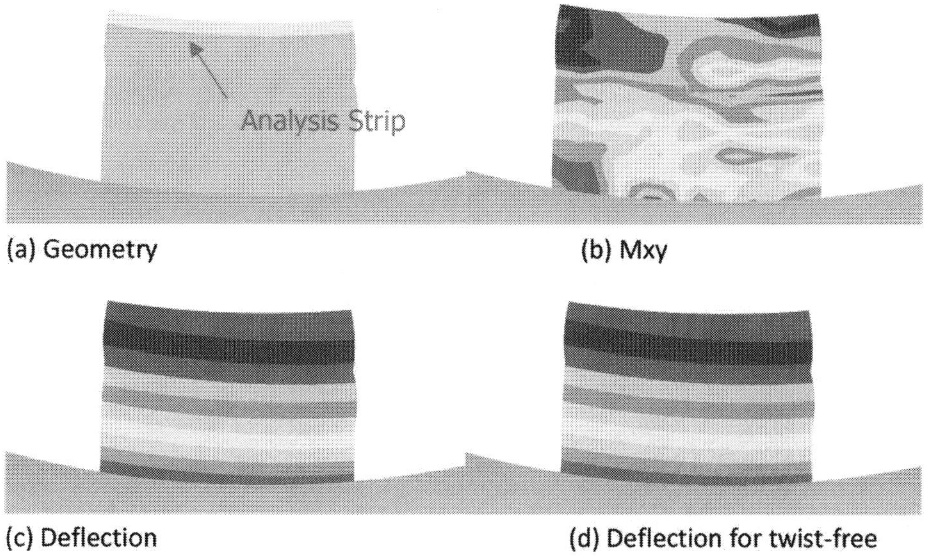

Fig. 7.4.1—Typical section of hemisphere at base.

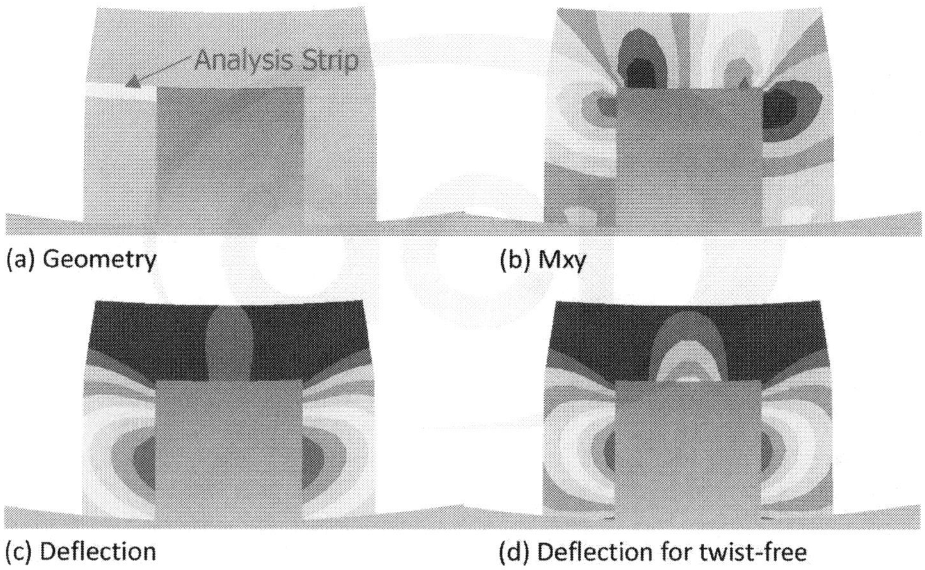

Fig. 7.4.2—Base at opening subject to self-weight.

Table 7.4.2—Comparisons for base at opening subject to self-weight

Design approach	M	|twist|	Md	Deflection	Applicable
Ignore twist	1.00	2.33	0.30	1.000	No
Wood and Armer	1.00	2.33	1.000	1.000	Ex. cons
Twist-free analysis	2.26	0.000	0.68	1.15	Yes

c) **Twist-free analysis:** This approach results in deflection predictions that are 11 percent larger than a regular (isotropic) analysis. This increase is relatively small, so there are no concerns of significant cracking required to achieve the twist-free load path. Twist-free analysis is the most efficient applicable method for this case.

Table 7.4.3—Comparisons for base at opening subject to pseudo-hydrostatic loading

Design approach	M	|twist|	Md	Deflection	Applicable
Ignore twist	1.00	2.02	0.33	1.00	No
Wood and Armer	1.00	2.02	1.000	1.00	Ex. cons
Twist-free analysis	1.72	0.000	0.57	1.11	Yes

7.4.4 *Apex region*—Due to the radial symmetry of the loadings and the structure in the region of the apex (that is, the base opening is far away), the behavior in this region is largely radially symmetric. Therefore, there are no significant twisting moments about the meridional and hoop axes.

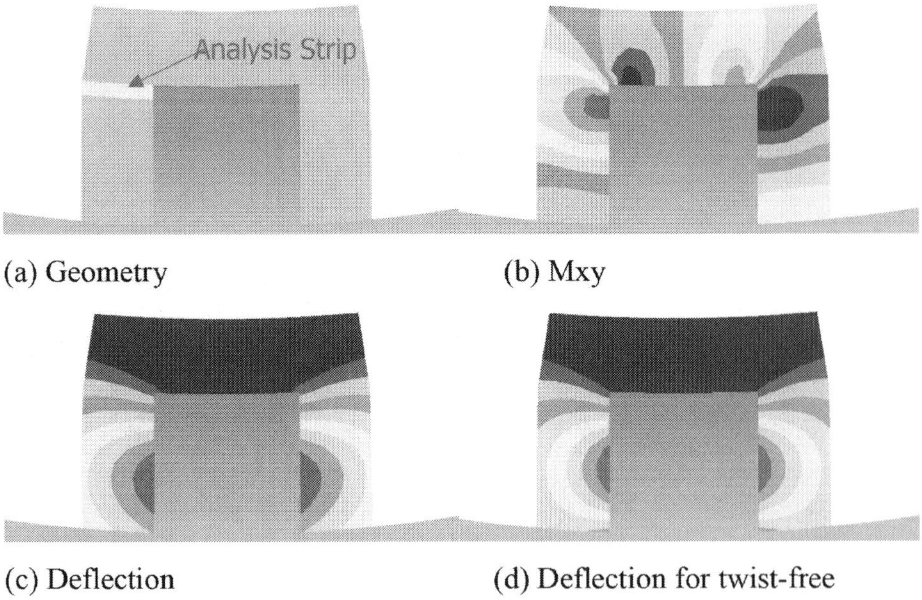

Fig. 7.4.3—Base at opening subject to pseudo-hydrostatic loading.

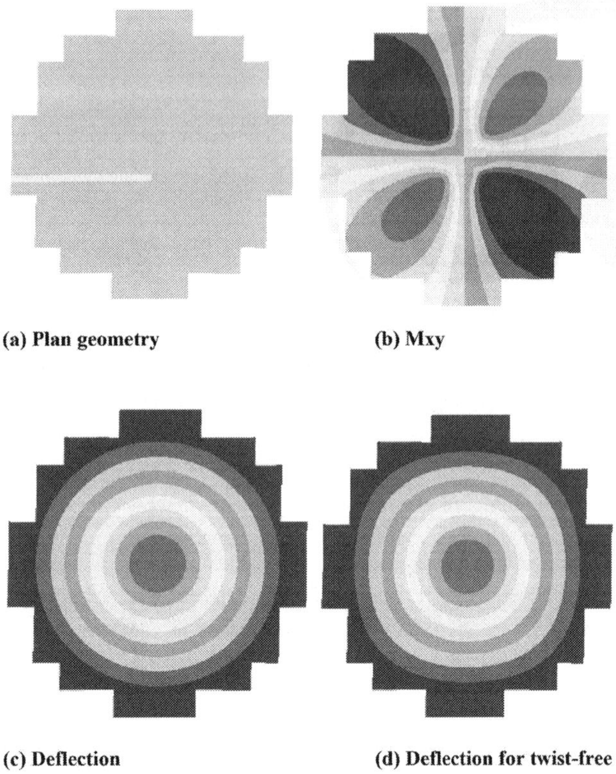

Fig. 7.4.4—Apex of hemisphere under point loading.

The reinforcement at the apex, however, cannot be placed in a radially symmetric pattern due to constructability constraints. Reinforcement in this region is placed along orthogonal axes that are approximately in a horizontal plane. When the moments about the radial axes are transformed to the reinforcement axes, significant twisting moments arise.

Figure 7.4.4 shows the apex region. Only the apex point loading is considered, as that is the most significant loading in this region. Note that moments, normal and twisting, are theoretically infinite under a point load.

The section cut of the area analyzed is shown in the figure. The length of the analyzed strip is approximately 10 percent of the radius of curvature of the shell. The analyzed strip is slightly off center of the apex. This is so that the twisting moment is not exactly zero along the length of the analysis strip.

Table 7.4.4—Analysis and design comparisons for apex loading

Design approach	M	|twist|	Md	Deflection	Applicable
Ignore twist	1.000	0..27	0..79	1.000	?
Wood and Armer	1.000	0.27	1.00	1.000	Cons
Twist-free analysis	1.07	0.00	0.84	1.03	Yes

Table 7.4.4 notes the normalized key design quantities and results for the apex region shown in Fig. 7.4.4.

a) **Ignore twist:** This approach is not applicable in these areas, as the twisting moment is significant and the isotropic load path requires twisting moments. The ignore twist design moment is only 94 percent of the most efficient applicable design moment and, therefore, could lead to unsafe designs.

b) **Wood and Armer:** This approach is applicable, although it results in a 19 percent excess reinforcement penalty.

c) **Twist-free analysis:** This approach results in deflection predictions that are 3 percent larger than a regular (isotropic) analysis. This increase is very small, so there are no concerns of significant cracking required to achieve the twist-free load path. Twist-free analysis is the most efficient applicable method for this case.

Table 7.4.5—Summary of method applicability

Design approach	Typical	Opening with in-plane loading	Opening with out-of-plane loading	Point load at apex
Ignore twist	?	No	No	?
Wood and Armer	Cons	Ex. cons	Ex. cons	Cons
Twist-free analysis	Yes	Yes	Yes	Yes

7.4.5 *Summary*—A hemispherical shell is modeled with several different loading scenarios. The response of the shell under a point load at the apex, near an opening with a pseudo-hydrostatic force, and in a typical section with pseudo-hydrostatic forces is investigated. These loads are additional to the load caused by the self-weight of the structure. Generally, the twisting moments in this shell are more influenced by geometric discontinuities than by loading discontinuities. The following is a written summary of the behavior of the shell for twisting moments, which is also summarized in Table 7.4.5.

Ignore twist: The analyses found that the ignore twist method was not a valid methodology for all the investigated loading scenarios. The stresses in the shell away from openings do not exhibit large twisting moments because the membrane action dominates the behavior of the shell in these areas. However, it does not guarantee a mechanism exists to resist the twisting moments.

Wood and Armer: The Wood and Armer method is an appropriate analysis method for shells in all locations. However, it also results in excess reinforcement in areas of geometric or loading discontinuities.

Twist-free analysis: The twist-free analysis is an appropriate analysis method. The deflections are very similar to the deflections produced when considering twist. This small difference in deflections create no concern of significant cracking required to achieve the twist-free load path. Differences between the shells and slabs are due to the membrane capabilities of the shells. The twist-free analysis results in different membrane forces in the shell that need to be considered in design.

CHAPTER 8—REFERENCES

Committee documents are listed first by document number and year of publication followed by authored documents listed alphabetically.

American Concrete Institute

ACI 318-14—Building Code Requirements for Structural Concrete and Commentary

ACI 318.2-14—Building Code Requirements for Concrete Thin Shells and Commentary

CSA Group

CSA-A23.3-04—Design of Concrete Structures

European Committee for Standardization

EN 1992-1-1:2004—Eurocode 2: Design of Concrete Structures, General Rules and Rules for Buildings

Authored documents

Armer, G. S. T., 1968, "The Reinforcement of Slabs in Accordance with a Pre-Determined Field of Moments," by R. H. Wood, discussion of reference 12.15, *Concrete (London)*, V. 2, No. 8, Aug., pp. 319-320.

Brøndum-Nielsen, T., 1985, "Optimization of Reinforcement in Shells, Folded Plates, Walls, and Slabs," *ACI Journal Proceedings*, V. 82, No. 3, May-June, pp. 304-309.

Burgoyne, C., 2004, "Are Structures Being Repaired Unnecessarily?" *The Structural Engineer*, V. 82, No. 1, Jan., pp. 22-26.

Comité Euro-International du Béton *fib*, 1993, "CEB-FIP Model Code 1990," Lausanne, Switzerland, 460 pp.

Comité Euro-International du Béton *fib*, 2008, "Practitioners' Guide to Finite Element Modeling of Reinforced Concrete Structures," *Bulletin* 45, Lausanne, Switzerland, 344 pp.

Deaton, J. B., 2005, "A Finite Element Approach to Reinforced Concrete Slab Design," MS thesis, Georgia Institute of Technology, Atlanta, GA, 170 pp.

Den Hartog, J. P., 2014, *Advanced Strength of Materials*, Courier Corporation, Dover Publications Inc., New York, 379 pp.

Kennedy, G., and Goodchild, C., 2003, *Practical Yield Line Design*, British Cement Association, 171 pp.

Lourenço, P. B., and Figueiras, J. A., 1995, "Solution for the Design of Reinforced Concrete Plates and Shells," *Journal of Structural Engineering*, V. 121, No. 5, May, pp. 815-823. doi: 10.1061/(ASCE)0733-9445(1995)121:5(815)

Marti, P., 1990, "Design of Concrete Slabs for Transverse Shear," *ACI Structural Journal*, V. 87, No. 2, Mar.-Apr., pp. 180-190.

May, I. M., and Lodi, S. H., 2005, "Deficiencies of the Normal Moment Yield Criterion for RC Slabs," *Proceedings of the Institution of Civil Engineers—Structures and Buildings*, V. 158, No. 6, Dec., pp. 371-380.

Park, R., and Gamble, W. L., 2000, *Reinforced Concrete Slabs*, second edition, John Wiley and Sons, New York, 716 pp.

Shin, M.; Bommer, A.; Deaton, J. B.; and Alemdar, B. N., 2009, "Twisting Moments in Two-Way Slabs," *Concrete International*, V. 31, No. 7, July, pp. 35-40.

Timoshenko, S., and Woinowski-Krieger, S., 1959, *Theory of Plates and Shells*, second edition, McGraw-Hill Book Co.

Wood, R. H., 1968, "The Reinforcement of Slabs in Accordance with a Pre-Determined Field of Moments," *Concrete (London)*, V. 2, No. 2, pp. 69-76.

Report on the Modeling Techniques Used in Finite Element Simulations of Concrete Structures Strengthened Using Fiber-Reinforced Polymer (FRP) Materials

Reported by Joint ACI-ASCE Committee 447

Ganesh Thiagarajan, Chair

Jian Zhao, Secretar

Riadh S. Al-Mahaidi*
Gangolu Appa Rao
Ashraf S. Ayoub
Zdenek P. Bažant
Allan P. Bommer
Mi-Geum Chorzepa
Christopher H. Conley
Carlos Arturo Coronado
Gianluca Cusatis
Mukti L. Das
James B. Deaton
Jason L. Draper
Serhan Guner
Trevor D. Hrynyk
John F. Jakovich
Song F. Jan
Ioannis Koutromanos
Feng-Bao Lin
Laura N. Lowes
Yong Lu
Christian Meyer
Yi-Lung Mo
Abbas Mokhtar Zadeh
Wassim I. Naguib
Dan Palermo
Guillermo Alberto Riveros
Mohammad Sharafbayani
Hazim Sharhan
Sri Sritharan

Consulting Members

Ahmet Emin Aktan
Sarah L. Billington
Johan Blaauwendraad
Oral Buyukozturk
Ignacio Carol
Luigi Cedolin
Wai F. Chen
Robert A. Dameron
Filip C. Filippou
Kurt H. Gerstle
Walter H. Gerstle
Robert Iding
Anthony R. Ingraffea
Hiroshi Noguchi
Gilles Pijaudier-Cabot
Syed Mizanur Rahman
Victor E. Saouma
Frank J. Vecchio
Kaspar J. Willam

*Primary author.

The committee acknowledges R. Kalfat for contributing to this report.

The strengthening of reinforced concrete (RC) members using fiber-reinforced polymers (FRPs) as externally bonded reinforcement has been widely used to enhance the flexural, shear, and axial capacity, or any combination thereof, of structural elements. Although experimental testing has been used predominantly as the sole method of investigation, numerical techniques such as the finite element (FE) method have also been gradually developed to provide predictive models for structural characterization. Well-calibrated FE models have the potential to expand the range of experimental data, provide information on important parameters difficult to measure using experimental instrumentation, and aid in the design of systems requiring complex FRP strengthening where testing may not be possible. This report provides a state-of-the-art review in the area of modeling of FRP-strengthened RC members and provides general guidelines on the best modeling practices that capture the complex phenomenon of concrete cracking and crushing, concrete shear retention, concrete fracture energy, steel-to-concrete bond behavior, FRP-to-concrete interface, FRP debonding failure modes, and FE mesh dependency.

Keywords: bond; fiber-reinforced polymer; finite element modeling; fracture energy; interface; reinforced concrete; shear retention.

ACI 447.1R-18 was adopted and published August 2018.

CONTENTS

CHAPTER 1—INTRODUCTION AND SCOPE

1.1—Introduction

Alongside the escalating demand to increase the strength of existing structures, new strengthening technologies have evolved, such as fiber-reinforced polymer (FRP) materials, which can be used as externally bonded reinforcement. Fiber-reinforced polymer is used not only to increase strength, but also to increase stiffness and provide confinement in existing structures. The technology has found significant success in applications to reinforced concrete (RC) and post-tensioned (PT) structures due to the FRP's strength-to-weight ratio, stiffness-to-weight ratio, excellent durability performance, resistance to corrosion, cost-effectiveness, ability to conform to various shapes, and ease of application.

Fibers are most commonly manufactured using carbon, glass, aramid, and basalt, and are produced in the form of loosely woven mats, pultruded laminates, or bars that are applied to structural elements using high-strength epoxy resins (ACI 440.2R). Research has demonstrated that the use of externally-bonded FRP composites can improve the flexural, shear, torsional, and axial performance of concrete members. In spite of their potential benefits, complete fiber use is often not realized due to the occurrence of premature debonding, which can take one of several forms: concrete cover separation failure; plate-end interfacial debonding; intermediate flexural; or flexural-shear, crack-induced interfacial debonding that is otherwise known as IC debonding (Hollaway and Teng 2008), and shear-induced debonding (also referred to as critical diagonal crack (CDC) debonding [Wang and Zhang 2008]). However, debonding failures involve complex mechanisms and remains a subject of research.

Extensive numerical studies using the finite element (FE) method have been conducted to simulate the various modes of FRP debonding (Kotynia et al. 2008). Finite element simulations have the potential to provide a predictive model for structural failure, expand the range of experimental data, and provide information on key phenomena in the absence of experimental data (Zhang and Teng 2014). However, the simulation of FRP-strengthened RC members is numerically demanding due to the complex nature of concrete, as well as the bond between the externally-bonded FRP and concrete and the relative size of the bond critical zone to the overall member size. As a result, a variety of material models and modeling techniques have been introduced by researchers to quantify the concrete material behavior, the bond properties between the concrete and steel reinforcement, and the bond properties between the FRP and concrete resulting in numerical predictions to various degrees of correlation with experimental data.

1.2—Scope

This report summarizes the latest research for the FE modeling techniques of FRP-strengthened RC members, and attempts to provide general guidelines and recommendations on the best modeling practices that capture the complex phenomenon of concrete cracking and crushing, concrete shear retention, concrete fracture energy, steel-to-concrete bond behavior, FRP-to-concrete interface, FRP debonding failure modes, and issues related to FE mesh dependency.

CHAPTER 2—NOTATION AND DEFINITIONS

2.1—Notation

A_g = aggregate size
c = cohesion
f_c' = compressive strength of concrete
f_{ct} = tensile strength of concrete
G_c = shear modulus of concrete
G_F^I = Mode I concrete fracture energy
G_F^{II} = Mode II concrete fracture energy
G_F^{III} = Mode III concrete fracture energy
K_{tt} = tangential stiffness of bond slip curve

L = length
L_{ch} = crack band width used to calculate shear retention factor
r_g = shear retention factor
s_i = slip at location i, in bond slip curve of FRP to concrete interface
s_F = shear factor coefficient relating the normal and shear stiffness of a crack
t_f = thickness of FRP plate
ν = Poisson's ratio
ε_F = normal crack opening strain
$\varepsilon_{f,i}$ = fiber-reinforced polymer strains measured in the direction of the fibers at location i
$\varepsilon_{f,i+1}$ = fiber-reinforced polymer strains measured in the direction of the fibers at location i+1
σ = normal stress within FRP to concrete interface
ϕ = friction coefficient
τ = shear stress at location i, in bond slip curve of FRP to concrete interface

2.2—Definitions

Please refer to the latest version of "ACI Concrete Terminology" for a comprehensive list of definitions.

CHAPTER 3—FINITE ELEMENT MODELING APPROACHES FOR FRP-STRENGTHENED RC MEMBERS

3.1—Modeling concrete compression

This chapter outlines some of the challenges facing researchers to numerically quantify the behavior of various materials such concrete, FRP, adhesive, and their interaction using the FE method.

Concrete is a quasi-brittle heterogeneous material that can incur tensile cracking and compressive crushing. The compressive response of concrete is highly nonlinear and can be described numerically using several approaches: 1) representation of the stress-strain behavior by curve-fitting methods (Chen 2007); 2) linear and nonlinear elasticity theories (Ortiz 1985); 3) elastoplastic models (Han and Chen 1985; Grassl et al. 2002); and 4) the endochronic theory of plasticity (Bažant 1978). Of the available approaches used to represent the stress-strain behavior of concrete under multiaxial stress states, damage plasticity models have become the most popular; however, success has also been achieved using other approaches (Wong and Vecchio 2003). Damage plasticity models apply the plasticity theory in the compression zone and fracture mechanics to represent the damage behavior due to cracking. The plasticity theory captures the accumulation of irreversible strains resulting from loading beyond the yield limit. The yield limit in concrete is defined as the compressive strain level beyond which behavior ceases to be linear elastic and permanent deformation occurs. Plasticity models are also useful for the modeling of concrete subjected to triaxial stress states because the yield surface at a certain stage of hardening can be correlated with the strength envelope of concrete. Furthermore, the total concrete strain is usually split into elastic and plastic components, which has been found to realistically represent the observed deformations in confined compression so that unloading can be described well (Grassl et al. 2013). When defining the parameters of concrete compressive strength within a numerical model, standard test procedures, such as crushing of concrete cylinders or cubes, should always be used to determine the average concrete compressive strength that is representative of the concrete used. The Poisson's ratio of concrete has been found to range between 0.2 (AS 3600-2009; BS EN 1992-1-1:2005) to 0.15 (CEB-FIB Model Code 2010/2012) for normal-strength concretes. Further, ACI 363R reports that Poisson's ratio for high-strength concrete within the elastic range is comparable to the expected values for normal-strength concrete.

3.2—Modeling of concrete cracking

The tensile behavior of concrete is considered as approximately linear elastic until fracture is reached, resulting in a sudden loss of strength. Concrete cracking is a highly localized phenomenon that can be modeled using either discrete crack or smeared crack models.

3.2.1 *Discrete crack models*—Discrete crack models rely on simulating individual cracks as geometrical identities within a model by introducing discontinuities within an FE mesh at element boundaries. As a result, crack location and orientation are dependent on the geometry and topology of the mesh that inevitably introduces mesh bias. To capture crack propagation, a continuous change in nodal connectivity is required, which is inconsistent with the nature of the FE displacement method (Rots and Blaauwendraad 1989). This drawback can be overcome by introducing automatic remeshing algorithms; however, these approaches are demanding on computational resources (Rabczuk et al. 2008) and plagued by numerical difficulties associated with topology changes due to remeshing (Chen et al. 2012). Higher success has been achieved by incorporating interface elements within the original mesh along predefined locations of potential cracking. Using this approach, the discrete cracks are mobilized by assigning interface elements with crack initiation criteria such as a stress violation condition, which is where the interface properties are changed when a maximum stress limit is reached using a constitutive model. Modeling approaches include predefining the potential crack locations at all element boundaries (as opposed to a limited number of probable crack locations) and permitting the fracture to propagate anywhere between the element boundaries (De Borst 1997).

3.2.2 *Smeared crack models*—In the smeared crack models, the cracked material is treated as a continuum and the deterioration caused by the crack is spread across the element by changing the element stiffness. In general, the smeared crack approach has grown more popular and demonstrated greater advantages than the discrete crack method. However, the smeared crack strategy tends to spread crack formation over a band of elements and fails to predict localized fracture (Kalfat 2014). Smeared crack models can be divided into two main categories—fixed and rotating. Fixed crack models use a constant crack orientation during the entire computational process. In rotating crack models, however, the crack direction changes during loading in accordance with the principal stress directions. Important limitations of the fixed-smeared crack model have been discovered relating to excessive stress rotation where rotation of the axes of principal stress ceases to coincide with the axes of principal strain and stress buildup after cracking when relatively high shear retention factors are employed, resulting in models that are too stiff (Rots and Blaauwendraad 1989). Shear retention refers to the ability of a concrete crack to transfer shear stresses across its width due to the presence of aggregate interlock. Researchers have found that acceptable fixed crack results can only be achieved when a near-zero shear retention factor is used, as this has been found to reduce the excessive stress rotation. In contrast, the rotating smeared crack concept provides acceptable stress/strain rotations and a better prediction of model stiffness because of the shear retention function that provides coaxiality between the principal stresses and strains. One disadvantage of the general rotating crack concept is its inability to retain memory of damage orientation, which can lead to errors when inactive cracks are erased from the model on activation of new cracks. This is particularly problematic when the model is subjected to loading and unloading, or when the loading in multiple directions is not proportional to each other (Rots and Blaauwendraad 1989). However, in some of the newer models such as Moharrami and Koutromanos (2016), this has been resolved. This also has been overcome to some extent by the introduction of various plastic damage models developed for the purpose of retaining plastic damage development (magnitude and orientation) in rotating crack procedures, such as Vecchio and Bucci (1999).

3.2.3 *Discrete crack versus smeared crack approach*—Using appropriate modeling techniques, researchers have successfully replicated the various failure modes of FRP-strengthened RC beams and the modes of FRP debonding obtained from experimental tests. Both intermediate crack-induced debonding (shearing of concrete along the bond surface from the tip of a diagonal crack) and concrete cover separation failure (shearing of concrete along the tension reinforcement level) have been successfully simulated using both smeared and discrete crack methods with good correlations with the experimental data; it was demonstrated that both crack modeling approaches could yield similar results (Pham et al. 2006). This observation was also confirmed by Chen et al. (2011) who demonstrated that when the crack band model is adopted, the discrete crack and smeared crack models yield approximately the same results, provided that the crack opening displacement in the discrete crack model is taken as the cracking strain accumulated over the width of the crack band in a smeared crack model. This will overcome strain localization and mesh objectivity problems present in early smeared crack models (Bažant and Planas 1998).

For accurate modeling of RC members retrofitted with FRP composites, including debonding failure modes, modeling of localized shear and flexural cracking is imperative. To accurately model discrete cracks using the smeared crack methodology, Chen et al. (2011) summarized three criteria that have to be fulfilled: 1) the constitutive model for cracked concrete has to be accurate and realistically capture the post-cracking behavior of concrete; 2) an accurate bond-slip model representing the bond behavior between concrete and FRP has to be defined; and 3) an accurate bond-slip model between the internal steel reinforcement and the concrete should be defined. Without all the above criteria being fulfilled, the FE model cannot be relied on to provide a realistic prediction of results.

3.3—Defining concrete fracture energy

To determine the load at which FRPs debond from the concrete substrate using fracture mechanics principles, the single most important parameter is the concrete fracture energy G_f. Fracture energy is defined as the energy required to open a unit area of crack surface. The fracture energy can be calculated using the area under the descending branch of the bond-slip curve. Fracture mechanics have demonstrated three modes in which a crack can propagate: Mode-I fracture (G^I_F) is classified as an opening mode where the tensile stresses are normal to the plane of the crack. Mode-II (G^{II}_F) is a sliding mode where crack propagation is propelled by shear stresses acting parallel to the plane of the crack and normal to the crack front. Mode-III fracture (G^{III}_F) is classified as a tearing mode with shear stresses acting parallel to the crack plane and parallel to the crack front. Research into FRP debonding has proven that, despite the FRP-to-concrete interfacial bond stresses being predominantly through shear, the initiation of debonding is still regarded as a Mode-I fracture of concrete (Achintha and Burgoyne 2013). Furthermore, researchers have investigated the use of both Mode-I and Mode-II fracture energy values in various numerical simulations and concluded that Mode-I fracture energy gives strength predictions that result in closer correlations with the test results in the majority of cases (Aram et al. 2008; Chen et al. 2011; Zhang and Teng 2014).

Currently available Mode-I fracture energy correlations are empirical formulations derived from experimental procedures (Bažant and Planas 1998). Consequently, most fracture energy models are based on maximum aggregate size, concrete strength, and water-cement ratio (w/c) (Van Mier 1997; Trunk and Wittmann 1998; Neubauer and Rostasy 1999; Bažant and Becq-Giraudon 2002; Ulaga et al. 2003; Elsayed et al. 2007; Freddi and Savoia 2008). Such models may be used in numerical simulations in the absence of experimental data. However, a preferred approach is to use standard test procedures such as a notched beam test (JCI-S-001 2003) to obtain a more accurate estimation of concrete fracture energy. Further, Hoover and Bažant (2014) investigated the effect of varying member sizes on the fracture process zone and found it to be mode-dependent, and that the impact can be eliminated only when the test is conducted at a variety of member sizes. This should be considered when developing experimental procedures to estimate concrete fracture parameters.

3.4—Influence of shear retention factor

Shear retention in concrete refers to the ability of a concrete crack to transfer shear stresses across its width. Shear transfer is dependent on the crack width and aggregate interlock, which governs the level of friction between two surfaces of the crack. The concept of a shear retention factor has been introduced in many damage plasticity models to account for the shear retention of concrete after concrete cracking. This is a coarse method due to the shear retention factor usually being assigned as a constant value. Shear retention should, theoretically, decrease for large crack strains, as less shear stress is capable of being transmitted across the crack. In general, the use of a constant shear retention factor has tended to predict a stiffer post-peak response and too little structural softening (Rots and de Borst 1987). Furthermore, researchers have found that those models that used a shear retention factor of zero or a constant factor were unable to predict the measured failure load accurately (Scotta et al. 2001). Scotta et al. (2001) further stipulated that only the use of a variable shear retention factor, where the crack shear resistance decreases with the crack opening, provides satisfactory correlations between numerical and experimental results. Several models are currently available to predict the shear retention factor or shear stiffness of cracked concrete based on the level of strain normal to the crack (Al-Mahaidi 1978; Rots et al. 1984; Kolmar 1986; Kang 1996; HKS 2001); many of these models have been incorporated into the latest FE platforms such as ATENA (Kolmar 1986) and HKS (2001).

3.5—Effect of bond between steel reinforcement and concrete

The bond between the steel reinforcement and concrete needs to be carefully considered in any FE model, as it directly influences the force transferred from the steel bar to the concrete or vice versa. Although perfect compatibility between the steel and the concrete has been assumed by some researchers, the connection between the steel and the concrete is more accurately represented in terms of a peak bond stress and corresponding slip.

In the case of longitudinal reinforcement, researchers have found that the assumption of perfect bond between the steel reinforcement and the concrete results in a more distributed diagonal crack pattern than when a bond relationship is assumed (Chen et al. 2012). The effects of distributed cracking are seen as lesser crack widths and a delay in the debonding failure load in the case of FRP-strengthened beams. The assumption of perfect bond and the subsequent reduction of crack widths can also result in higher shear retention and an artificial increase in the shear capacity of the beam.

Two different methods have been used to include the bond-slip effect in the FE analysis of RC structures. The first method is to use a contact element with a predefined bond-slip law between the discrete reinforcement elements and the concrete. The second method is to apply to a bond-slip behavior using one-dimensional bond link elements or spring elements at the nodes connecting the reinforcement to the concrete (Youai 2000).

Several steel reinforcement bond slip models have been proposed by researchers and many are employed in FE software programs such as ABAQUS (HKS 2001), ATENA (Červenka et al. 2002), and DIANA (de Witte and Kikstra 2003), which include reinforcement bond-slip models proposed by CEB-FIB Model Code (2010/2012) and Bigaj (1999). The laws are generated based on concrete compressive strength, reinforcement diameter, and reinforcement type (bond quality). Other important parameters are confinement conditions and concrete quality. User-defined reinforcement bond slip models can also be specified. These could be more suitable under certain loading conditions, such as Eligehausen et al. (1983), which was developed primarily for cyclic loading, although it works well for monotonic loading as well.

In contrast to longitudinal steel reinforcement, researchers have found that the predicted load displacement response for RC beams was less sensitive to the assumed bond condition between the steel stirrups and concrete. Although many existing studies conducted in the area of flexural strengthening of RC beams adopted an accurate bond-slip model between the internal steel stirrup reinforcement and the concrete; the practice was less common for shear strengthened beams. Only a single study by Chen et al. (2012) was found to investigate alternative bond-slip models between internal steel stirrup reinforcement and surrounding concrete. Both weakly bonded, strongly bonded, and perfectly bonded scenarios were investigated to represent the bond behavior between the internal steel stirrups and the concrete. When examining the effects of the assumed bond condition with respect to the peak load predictions, the weakly bonded stirrups provided an underestimation of the failure load while a strongly bonded scenario provided good predictions that matched the experimental data in terms of peak load and ductility. The assumption of perfect bond resulted in a similarly good prediction of peak load, but produced greater post-peak ductility behavior, which was absent from the experimental results. Furthermore, it was found that the bond condition affected the crack pattern, and the process of crack propagation in a complex manner and further investigation was required to achieve a better understanding.

3.6—Modeling of FRP-to-concrete interface

Debonding of FRP-to-concrete joints is generally governed by bond characteristics of the FRP-to-concrete interface. Three general approaches used by researchers to numerically represent the interfacial bond properties between FRP and concrete are summarized by Kalfat (2014): 1) perfect bond at the interface; 2) the use of one-dimensional nonlinear spring elements between the adjacent concrete and adhesive layers (Luo at al. 2012); and 3) a layer of interface elements between the FRP and the concrete (Wu et al. 2009).

When a perfect bond is assumed between the FRP and concrete materials, damage degradation of the bond line and the occurrence of FRP debonding occurs through the layer of concrete elements in direct contact with the FRP material (Cui et al. 2009; Hajsadeghi et al. 2011). As a result, this approach is highly sensitive to the accuracy of the concrete constitutive model and size of the concrete elements directly in contact with the FRP material. The FRP material should be sufficiently refined to capture stress concentrations at the interface. Although researchers have achieved acceptable results using this technique, the results obtained using this approach are found to be highly mesh-dependent and non-objective. An alternative approach for modeling the FRP-to-concrete interface is by assigning interfacial bond-slip parameters through a series of nonlinear, one-dimensional spring elements between the FRP and concrete layers (Luo et al. 2012). The spring elements are capable of introducing strength and stiffness of the interface derived from a damage-type constitutive law. When a well-defined and calibrated bond-slip law is assigned to the spring elements, the prepeak and post-peak properties of the FRP-to-concrete interface can be well represented (Hii 2006).

Lastly, the final approach used to represent the FRP-to-concrete interface and the occurrence FRP debonding is to assign interface elements between the FRP and concrete (Fig. 3.6a and Fig. 3.6b). The interface elements are typically assigned with a constitutive bond-slip model representing the strength and stiffness of the interface, which can be derived from the experimental data or available theoretical models based on linear or nonlinear fracture mechanics (Ferracuti et al. 2007). A typical FRP-to-concrete bond-slip relationship consists of two stages: 1) an initially elastic stage where the interfacial stresses increase with the slip until the strength of the interface is reached; and 2) a softening stage in which interfacial stress decreases with the slip resulting in debonding.

To obtain the bond-slip information from experimental tests, axial strains should be measured along FRP laminate length in the direction of the fibers at different loading levels and used to calculate bond-slip relation. Considering an elastic behavior for the FRP, the average shear stress values between two subsequent strain gauges, or any two points along the laminate length, can be written as a function of the difference of measured strains ε_i and ε_{i+1}. This equates to dividing the force difference between two points by the total area of the FRP laminate, as represented in Eq. (3.6a)

$$\tau = \frac{E_f(\varepsilon_{f+1} - \varepsilon_{f,i})t_f}{\Delta L} \quad (3.6a)$$

where E_f and t_f are FRP elastic modulus and thickness; $\varepsilon_{f,i+1}$ and $\varepsilon_{f,i}$ are axial FRP strains measured in the direction of the fibers at two discrete locations (i, i+1); and ΔL is the distance between strain gauges. Note that the width of the FRP exists on both the numerator and denominator, thus canceling each other in Eq. (3.6a).

The average slip should be calculated as the incremental accumulation of strains along the FRP length. This is expressed in Eq. (3.6b)

$$s_i = \frac{\varepsilon_{f,i+1} + \varepsilon_{f,i}}{2}\Delta L + s_{i-1} \quad (3.6b)$$

The bond-slip properties of the FRP-to-concrete interface can be used to determine important parameters such as maximum shear strength of the interface (τ_{max}), tangential stiffness (K_{tt}), cohesion (c) and friction coefficient (ϕ) for the shear-displacement function (Fig. 3.6c(a)). Such parameters are typically required for numerical definition of the interface properties within the FE software (Červenka et al. 2002). However, upon execution of the FE model using the interface input parameters, the bond slip data should be verified by back-calculation from the FE results using the axial strains on the FRP laminate measured at discrete monitoring point locations that correspond to locations used in the experiments. This procedure has been presented in Kalfat and Al-Mahaidi (2015) as depicted in Fig. 3.6c(b), where an experimental bond-slip curve was calculated from the experimental data and used to define the properties of the interface as "Exp. Fitted. Bond-slip." The back-calculated bond-slip curve, or "FEM," also depicted in Fig. 3.6c(b), shows close correlations with the experimental bond-slip curve, which provides some verification of the parameters used to define the FRP-to-concrete interface in the FE model.

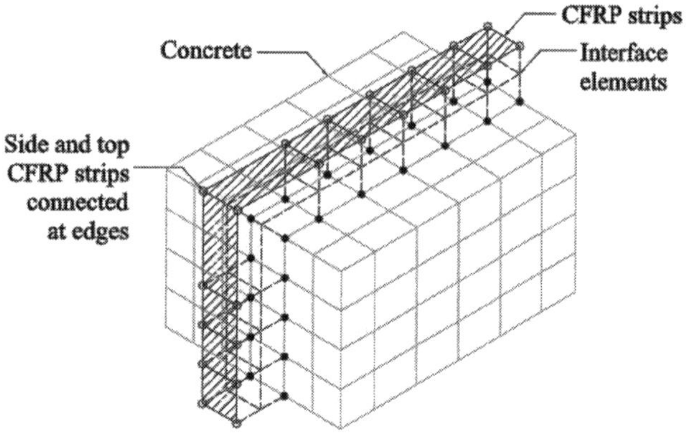

Fig. 3.6a—Connectivity between FRP and concrete through interface elements (Hii 2006).

3.7—Influence of mesh size and element type

Researchers found that the accuracy of results obtained from FE simulations of concrete structures are strongly mesh-dependent and that a finer mesh predicts the crack

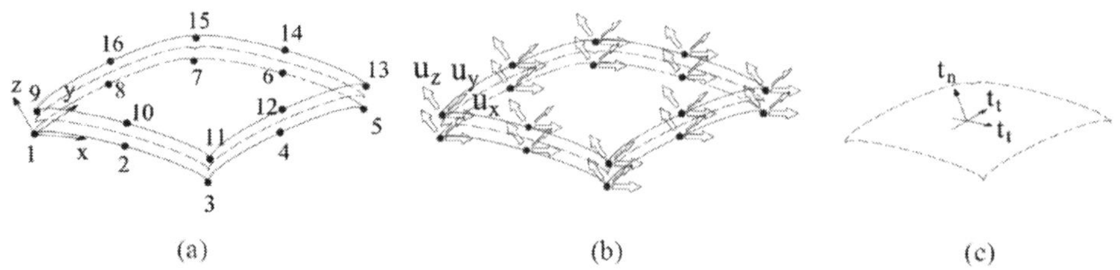

Fig. 3.6b—(a) Typical 8+8 nodes structural interface element: (a) topology, (b) displacements, and, (c) tractions.

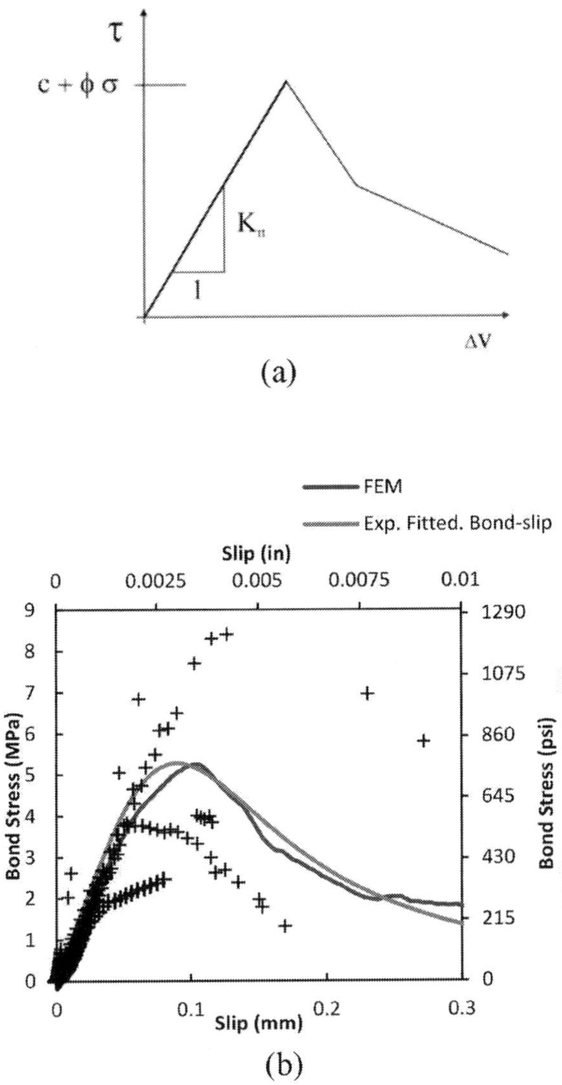

Fig. 3.6c—Typical interface model behavior in shear with cohesion softening law; (a) numerical definition (Červenka and Červenka 2007); (b) shear-slip curve for interface derived from experimental data, where τ is interfacial shear stress, σ is normal stress, ϕ is friction angle, and K_{tt} *is tangential stiffness (Kalfat and Al-Mahaidi 2015).*

patterns more realistically (Gribniak et al. 2010). Coarser meshes have been found to result in stiffer model behaviors and higher predicted failure loads. Particular attention should be paid to the size of the concrete elements to which the FRP material is bonded and other zones of high stress to ensure that the element sizes are sufficiently small to capture the stress concentrations that initiate debonding failure modes. Parametric studies on mesh dependency of FRP-to-concrete bonded joints conducted by Kalfat (2014) highlighted that a concrete element size of 0.2 in. (5 mm) directly below the FRP laminate was necessary to accurately capture the bond-slip behavior realistically. Finer meshes, however, are computationally expensive. Insufficient mesh discretization is a major source of error that should be avoided by sufficient mesh refinement. Finite element methods should be objec-

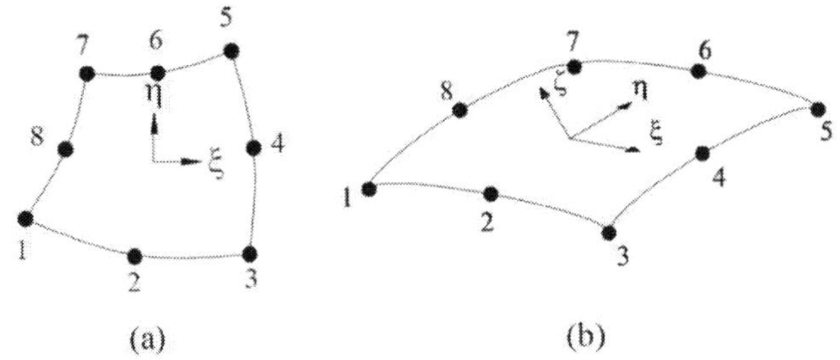

Fig. 3.8a—(a) Typical eight-node quadrilateral isoparametric plane stress element; and (b) typical eight-node quadrilateral isoparametric curved shell element (Hu et al. 2004).

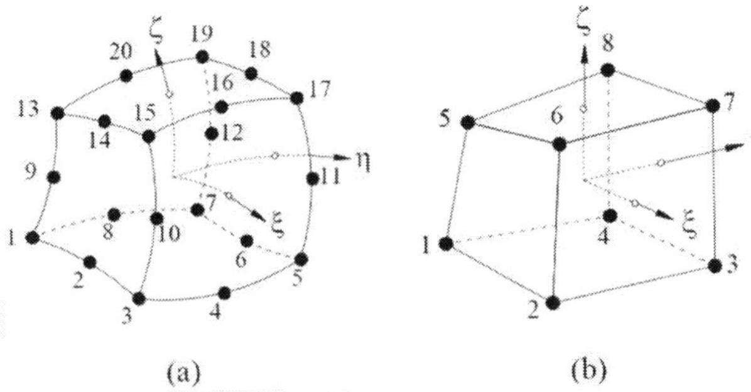

Fig. 3.8b—(a) Typical 20-node isoparametric solid brick element; and (b) typical eight-node isoparametric solid brick element (Kalfat and Al-Mahaidi 2015).

tive and the results obtained should not be dependent on the choice of mesh topology and element size. As a result, the objectivity of any numerical solution should be confirmed by parametric studies on mesh size to ensure that the discretization error is low and that the solution is not mesh-dependent.

3.8—Modeling FRP sheets and fabrics

Fiber-reinforced polymer materials generally consist of unidirectional fibers embedded within a resin matrix. These fibers may be pre-impregnated with resin to form pultruded FRP laminates or applied to concrete structures using a wet layup procedure where dry fiber sheets are saturated with resin and applied to the surface of a concrete member as externally bonded reinforcement. Fiber-reinforced polymer systems are also available in the form of FRP bars, which can be near-surface mounted and placed into cuts or grooves in the concrete cover zone, later to be filled with adhesive.

Fiber-reinforced polymer materials have a very high tensile strength but are relatively weak in compression and possess negligible tensile strength and stiffness transverse to the principle fiber orientation. In three-dimensional FE models, researchers have defined the FRP material using several approaches. including: 1) an orthotropic material in a shell element with or without rotational degrees of freedom, as depicted in Fig. 3.8a (Hu et al. 2004); 2) isoparametric brick elements defined with an isotropic three-dimensional material (Fig. 3.8b) representing the adhesive and unidirectional smeared reinforcement for the fibers (Kalfat and Al-Mahaidi 2015); and 3) one-dimensional beam element in the case of two-dimensional FE models (Yang et al. 2003). The use of plane stress elements to represent the FRP material is advantageous because the forces carried by the FRP are mainly in-plane; out-of-plane forces are considered negligible. Shell elements, therefore, are well-suited to represent the FRP material and have been successfully used to model FRP laminates in addition to fabrics (Al-Mahaidi et al. 2001; Hii and Al-Mahaidi 2006). The use of a three-dimensional isotropic material containing embedded smeared reinforcement is more suitable, and found to be more successful, when modeling bidirectional FRP fabrics with multiple fiber orientations, embedded within a saturant matrix. Acceptable results have been obtained using all the above modeling techniques.

3.9—Calibration of material parameters

Material models within FE software consist of various input parameters that require definition; the choice of parameters can significantly influence the results produced by the analysis and should be informed directly from experimental tests for the highest accuracy. In selecting material parameters, the average values from experimental tests should be used as opposed to the characteristic values. For defining the properties of concrete, cylinders or cubes should be tested at the time of testing to determine the average concrete compressive strength. The concrete tensile strength can be found using the concrete compressive strength and established formulations derived from existing codes (ACI 301-16; EN 1992-1-1:2005; AS 3600-2009); however, the use of test procedures such as JCI-S-001 are expected to provide higher accuracy of results. The tensile strength and modulus of elasticity of other materials such as steel reinforcement, FRP, and epoxy should also be established using standard testing procedures, with the results used to inform relevant parameters in the FE model.

CHAPTER 4—EXISTING STUDIES ON THE NUMERICAL MODELING OF DEBONDING FAILURE IN FRP-PLATED RC MEMBERS

4.1—Finite element (FE) simulations of debonding failure models for beams strengthened in flexure

This chapter reviews existing studies conducted to simulate FRP debonding failure modes and their respective modeling techniques, assumptions, and results. Prior to 2005, many researchers did not implement bond-slip interactions between the FRP and concrete in their finite element (FE) models, as modeling techniques and software were still in developmental stages. Furthermore, few studies since then have included the important elements of an accurate concrete model and well-defined bond-slip models between the internal steel reinforcement, FRP, and concrete. However, when these elements are considered properly, numerical predictions can be achieved using FE simulations (Table 4.1). Modeling conducted by Pham et al. (2006) using DIANA (de Witte and Kikstra 2003) incorporated a well-calibrated damage plasticity model for concrete using four-node quadrilateral isoparametric plane stress elements. A fracture-energy-based rotating smeared crack model was used for the concrete elements below the tension reinforcement, where the shear strength of cracked concrete as a result of aggregate interlock was ignored, resulting in zero shear retention. However, the concrete above the level of the tensile steel reinforcement was modeled using a fixed-crack model and adopted a shear retention factor of 0.05. The discrete crack model made use of an interface crack model in which concrete degradation in both tension and shear was modeled. The reinforcement elements were connected to concrete elements through structural interface elements, and a cubic function proposed by Dorr (1980) was used to describe the bond-slip relation between the reinforcement and concrete. A bond-slip relation was used to describe the bond between the FRP and concrete, and was derived experimentally from shear-lap tests by installing a series of strain gauges at 0.71 in. (18 mm) spacing along the length of the FRP laminate and using Eq. (3.6a) and (3.6b) to derive bond slip data. The simulation results are similar between the discrete and smeared crack models, where both debonding failure modes of intermediate crack-induced debonding and end cover separation failure were reproduced at similar load levels to those achieved in the experiments. Figures 4.1a and 4.1b demonstrate the capabilities of discrete and smeared crack methods to predict the various modes of FRP debonding. Although the study made use of parametric studies on the shear retention factor to determine the appropriate value, the method provided poor predictions; a variable shear retention factor based on the level of crack opening would have improved the model greatly.

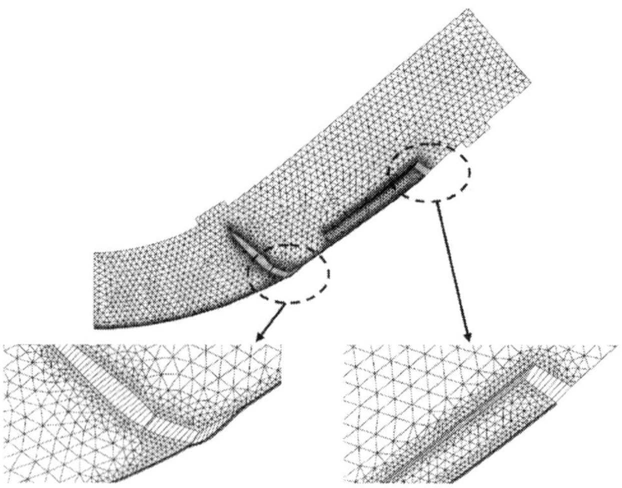

Fig. 4.1a—Intermediate crack-induced debonding and concrete cover separation failure modeled using the discrete crack approach (Pham 2005).

Other studies such as Camata et al. (2007) combined the discrete crack method with a smeared crack model in areas outside of discrete crack and achieved similarly good predictions that correlated well with the experimental results envelope, as shown in Fig. 4.1c(b), where B_r^n and B_s^n represent numerical predictions for the control and strengthened beams, respectively. Predictions of intermediate crack-induced debonding were also successfully modeled, as

Table 4.1—Summary of finite element (FE) studies conducted into the flexural strengthening of RC

Researchers	Strengthening method	Platform	Crack method	Rotating or fixed?	Shear retention	Fracture energy	Plasticity model	Reinforcement bond slip	FRP bond slip	Peak load error, %	Predicted failure mode	Stiffness	Post-peak prediction
Aram et al. (2008)	Flexural	ATENA (Červenka et al. 2016)	Smeared crack band	Fixed	Variable according to Červenka et al. (2016)	Program calculated	Concrete model proposed by Menetrey and Willam (1995)	Unknown	Bond slip according to Ulaga et al. (2003)	< 15%		+3	NA
Kishi et al. (2005)	Flexural	DIANA (de Witte and Kikstra 2003)	Discrete crack combined with smeared crack	None	0.01	CEB-FIB Model Code 2010/2012	Concrete model proposed by Thorenfeldt et al. (1987)	CEB-FIB Model Code 2010/2012	Bond-slip parameters determined by parametric study to match experiments.	< 5%	CDC opening and FRP sheet debonding	0	Good
Chen et al. (2011)	Flexural	ABAQUS (HKS 2001)	Smeared crack band	Rotating	Variable according to: (Rots 1988)	CEB-FIB Model Code 2010/2012	Concrete model proposed by Lubliner et al. (1989); Lee and Fenves (1998)	CEB-FIB Model Code 2010/2012	Bond-slip model proposed by Lu et al. (2005)	< 2.9%	IC debonding	0	Good
Lu et al. (2007)	Flexural	MSC. MARC® (MSC Software Corp. 2005)	Smeared crack band	None	Variable for beams without steel stirrups according to Walraven and Reinhardt (1981). However, constant for beams without steel stirrups (β = 0.4)	CEB-FIB Model Code 2010/2012	Concrete model proposed by Hognestad (1951)	Perfect bond	Bond slip derived from calibrated mesoscale FE model	< 9.1% ave. < 27% max	IC debonding	+1	NA
Camata et al. (2007)	Flexural	Merlin (Camata et al. 2007)	Discrete crack combined with smeared crack. The cohesive crack model adopted here was presented by Červenka (1994)	Rotating	The cohesive crack model adopted here was presented by: (Červenka et al. 1998)	CEB-FIB Model Code 2010/2012	Concrete model proposed by Menetrey and Willam (1995)	Not mentioned	Interface elements, unclear definition of bond-slip properties	< 12%	IC debonding	+2	Poor
Hu et al. (2010)	Flexural	ABAQUS (HKS 2001)	Smeared crack band	Fixed	1.0	CEB-FIB Model Code 2010/2012	Concrete model proposed by Saenz (1964)	Perfect bond	Perfect	< 4%		–2	NA
Hashemi and Al-Mahaidi (2012)	Flexural	ATENA (Červenka et al. 2016)	Smeared crack band	Rotating	Variable according to: Červenka et al. 2016	None	Concrete model proposed by Menetrey and Willam (1995)	None	None	< 3.5%	IC debonding and end debonding	+2	Good
Pham (2005)	Flexural	DIANA (de Witte and Kikstra 2003)	Discrete crack combined with smeared crack	Rotating below tension steel, fixed crack above tension steel	0.05	Trunk and Wittmann (1998)	Concrete model proposed by: Thorenfeldt et al. (1987)	Interface elements and a cubic function proposed by Dorr (1980)	Bond-slip relation was used to describe the bond between the FRP and was derived experimentally from shear-lap tests.	< 10%	IC debonding and end debonding	+3	Poor
Pham (2005)	Flexural	DIANA (de Witte and Kikstra 2003)	Smeared crack band	Rotating below tension steel, fixed crack above tension steel	0.05	Trunk and Wittmann (1998)	Concrete model proposed by Thorenfeldt et al. (1987)	Interface elements and a cubic function proposed by Dorr (1980)	Bond-slip relation was used to describe the bond between the FRP and was derived experimentally from shear-lap tests.	< 9%	IC debonding and end debonding	0	Good
Zhang and Teng (2014)	Flexural	MSC. MARC® (MSC Software Corp. 2005)	Smeared crack band	Fixed	Variable according to Okamura and Kohichi (1991)	CEB-FIB Model Code 2010/2012	Elwi and Murray (1979)	CEB-FIB Model Code 2010\2012	Bond-slip model proposed by Lu et al. (2005)	6% ave.	IC debonding and end debonding	0	Satisfac-tory

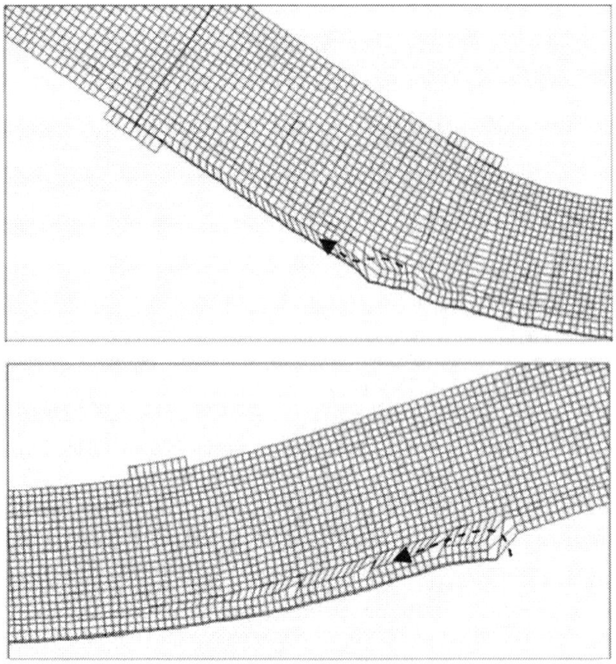

Fig. 4.1b—Intermediate crack-induced debonding and concrete cover separation failure modeled using the smeared crack approach (Pham 2005).

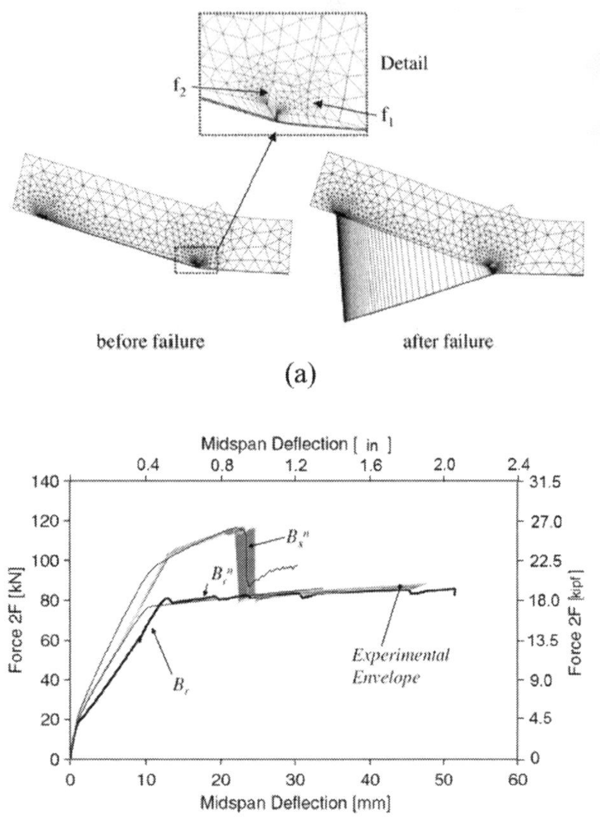

Fig. 4.1c—Comparison of experimental and numerical results for Dong et al. (2002) tests. (a) Failure mode for beam B3; (b) Load deflection results for the beams (Camata et al. 2007).

shown in Fig. 4.1c(a), where debonding was observed to initiate at flexural cracks f_1 and f_2 and progress toward the plate end. However, most of the FE analysis (FEA) models summarized in Table 4.1 were based on the smeared crack band concept and able to achieve relatively low errors in peak load predictions, predicted the critical debonding failure mode, provided good predictions of beam stiffness, and captured post-peak responses fairly well.

Although the use of a variable shear retention factor has a better theoretical basis, where a discrete crack model was used in conjunction with a smeared crack model (Aram et al. 2008; Chen et al. 2011; Zhang and Teng 2014), a low shear retention factor (0.01 to 0.05) was found to provide good results (Kishi et al. 2005; Pham 2005), even when the tangential stiffness of the discrete crack interface was assumed to be zero to ignore the effects of aggregate interlock (Kishi et al. 2005). Despite the wide range of concrete fracture energy models available in the literature, the summary provided in Table 4.1 indicates the model provided in CEB-FIB Model Code 2010/2012 has been most commonly implemented by the latest FEA simulations conducted by researchers.

While many researchers implemented bond-slip properties between the FRP and concrete that were derived from experimental measurements (Kishi et al. 2005), many researchers adopted the theoretical bond-slip model proposed by Lu et al. (2005), which provided a good representation of FRP-to-concrete bond behavior and debonding failure mode predictions.

4.2—Modeling of shear-strengthened RC members using FRP

Extensive research has been conducted on the application of fiber-reinforced polymer (FRP) composites in the strengthening of reinforced concrete (RC) structures in shear and a summary of numerical studies has been provided in Table 4.2. Chen et al. (2012) constructed a FE model using ABAQUS (HKS 2001) to describe the behavior of an RC beam strengthened in shear using FRP. The concrete was modeled using plane stress elements incorporating the crack band model to represent concrete cracking and Saenz (1964) stress-strain relationship to represent the compressive behavior of concrete. The concrete model was also capable of including tension softening of concrete and the concrete fracture energy that was defined using a relationship proposed by Rots (1988). The internal steel reinforcement was modeled using truss elements with interface elements defined between the steel and concrete assigned with bond-slip model in accordance with CEB-FIP Model Code (1990). The bond between FRP and the concrete was also modeled using interface elements and the properties of the interface determined from a simplified bond-slip model defined by Lu et al. (2005). A mesh sensitivity study proved that the model was not mesh-dependent, and changes to element size had little influence on results. The model developed was verified against test results from the literature with a satisfactory agreement between failure loads and failure modes when bond-slip models between the internal steel reinforcement and FRP were implemented. When the bond between

Table 4.2—Summary of FE studies conducted into FRP-to-concrete joints and the shear/torsional strengthening of RC members

Researchers	Strengthening method	Platform	Crack method	Rotating or fixed?	Shear retention	Fracture energy	Plasticity model	Reinforcement bond slip	FRP bond slip	Peak load error, %	Predicted failure mode	Stiffness	Post-peak prediction
Chen et al. (2012)	Shear	ABAQUS	Smeared crack band	Rotating	Variable following Rots (1988)	CEB-FIB Model Code 2010/2012	Concrete model proposed by: Lubliner et al. (1989); Lee and Fenves (1998)	CEB-FIB Model Code 2010/2012	Bond-slip model proposed by Lu et al. (2005)	< 10.9%	Debonding in agreement with tests	0	Good
Ibrahim and Sh Mahmood (2009)	Shear	ANSYS	Smeared crack band	Rotating	Variable: Shear transfer coefficient (βt)	None	Plasticity model based on Desayi and Krishnan (1964)	Perfect bond	Perfect	< 8%	Debonding not modelled	+2	Good
Godat et al. (2007)	Shear	ADINA (2003)	Smeared crack band	Fixed	0.5	Fracture energy model proposed by Lu et al. (2005)	Plasticity Adina	Perfect bond	Bond Slip model proposed by (Lu et al. 2005)	< 7%	Debonding	0	Good
You et al. (2011)	Shear	DIANA	Smeared crack band	Hybrid rotating fixed crack	DSFM (Vecchio 2000)	fracture energy of concrete using Sato and Vecchio (2003)	Concrete model proposed by Thorenfeldt et al. (1987)	Perfect	Bilinear bond stress-slip model according to: (Sato and Vecchio 2003)	< 10%	Debonding	+3	NA
Lee (2003)	Shear	DIANA	Smeared crack band	Fixed	0.05	CEB-FIB Model Code 2010/2012	Behavior of the concrete in compression is described by the function of Thorenfeldt et al. (1987)	Perfect bond	Bond-slip relation was derived numerically from 2D bond specimen models	< 5%	Debonding	0	Satisfactory
Godat et al. (2012b)	Shear	ADINA (2003)	Smeared crack band	Fixed	0.5	Fracture energy model proposed by Lu et al. (2005)	Plasticity Adina	Perfect bond	Bond Slip model proposed by (Lu et al. 2005)	< 7%	Debonding	0	Good
Godat et al. (2012b)	Shear	ADINA (2003)	Smeared crack band	Fixed	Linearly decreasing from 1 to 0.5 after cracking, then constant at 0.5	Lu et al. (2005)	Hypoelastic (incremental) model	Perfect connection	Bond Slip model proposed by (Lu et al. 2005)	< 5%	Debonding	0	Good
Hii (2006)	Torsional	DIANA	Smeared crack band	Rotating	Parametric studies 0.03 - 0.36	Trunk and Wittmann (1998)	Concrete model proposed by Thorenfeldt et al. (1987)	Perfect bond	Bond slip based on shear lap tests conducted by: (Pham 2005)	< 7.1%	Rupture of CFRP	0	NA
Kalfat and Al-Mahaidi (2015)	FRP-to-concrete joint	ATENA	Smeared crack band	Fixed	Variable according to Červenka et al. (2016)	Trunk and Wittmann (1998)	Concrete model proposed by Menetrey and Willam (1995)	Perfect connection	Derived from experiments	< 8.3% ave. < 16.2% max	Debonding	+1	NA
(Chen and Tao 2011)	FRP-to-concrete joint	ABAQUS	Smeared crack band	None	Variable according to: Rots (1988)	(CEB-FIB Model Code 2010/2012)	Concrete model proposed by Lubliner et al. (1989); Lee and Fenves (1998)	Perfect connection	Perfect connection	< 3.6%	Failure by damage in concrete	0	Good
(Teng et al. 2013)	FRP-to-concrete joint	MSC.MARC® (MSC Software Corp. 2005)	Orthogonal fixed smeared crack model	Fixed	Model by Okamura and Kohichi (1991) was adopted	CEB-FIB Model Code 2010/2012	Concrete model proposed by Buyukozturk (1977)	None	Derived from strain gauge results at discrete locations	< 4% ave. < 20% max	Debonding	+1	Good

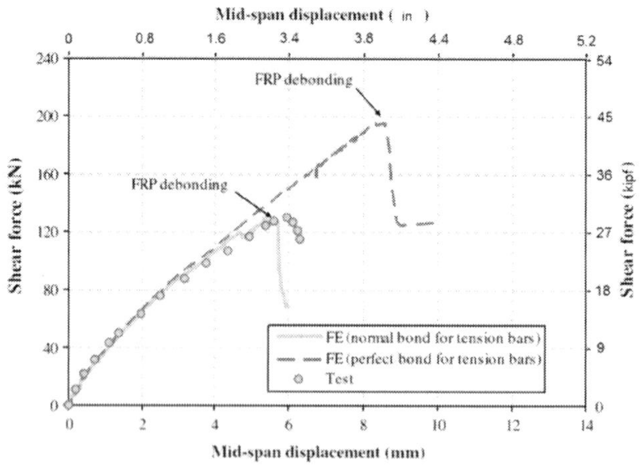

(a)

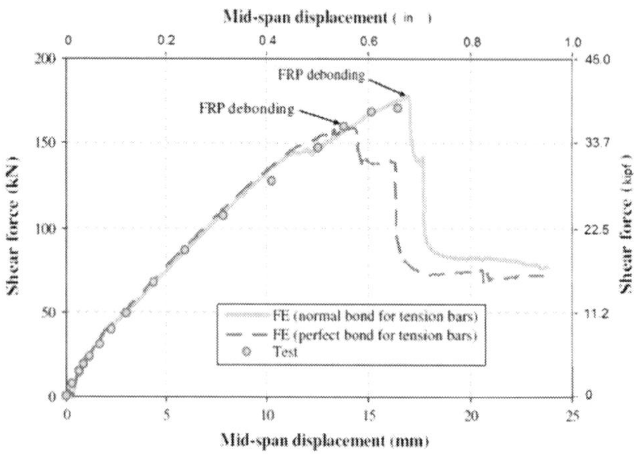

(b)

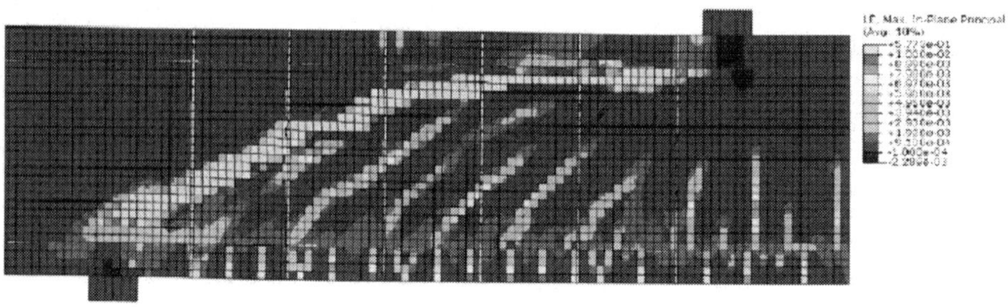

(c)

Fig. 4.2a—Effect of bond between steel tension bars and concrete: (a) specimen SO3-2; (b) specimen BS5; (c) crack pattern of beam with perfectly bonded steel tension bars (Chen et al. 2012).

the steel reinforcement and concrete was also studied, it was discovered that the assumption of a perfect bond could increase the shear capacity of the beam due to a more distributed crack pattern, or it could decrease the shear capacity of the beam if the angle of the critical diagonal shear crack was increased. Typically, the assumption of perfect steel reinforcement bond resulted in an increase or decrease in the predicted failure load when compared to the experimental test results. However, when a bond model was defined, the predictions matched the experimental results very well. This is shown in Fig. 4.2a, where specimen SO3-2 was predicted to fail due to FRP debonding at a much higher load than the test value when the steel tension bars were assumed to be perfectly bonded to the concrete. In contrast, the predicted failure load was underestimated for Specimen BS5 when a perfect bond was assumed between the steel tension bars and the concrete. However, when an appropriate bond-slip model between the steel reinforcement and concrete was

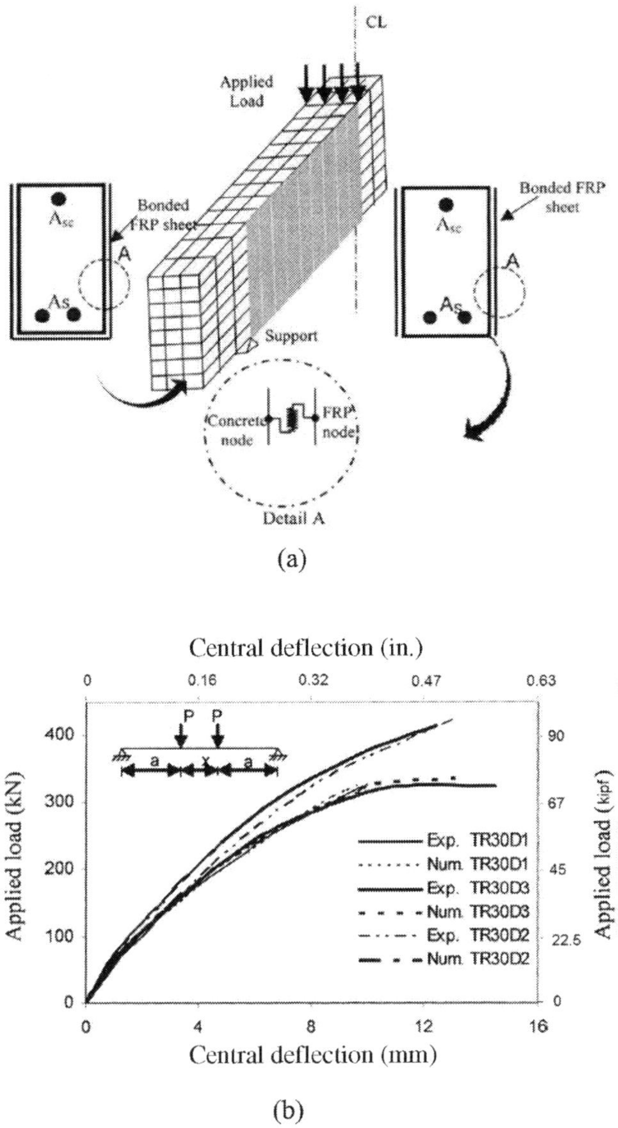

Fig. 4.2b—(a) Finite-element model proposed by (Godat et al. 2007); and (b) applied load–central deflection relationships for Pellegrino and Modena (2002) beams (Godat et al. 2007).

defined, the numerical predictions matched very closely with the experimental results.

Other researchers such Godat et al. (2007) investigated the numerical modeling of FRP shear-strengthened RC beams using the FE platform ADINA (2003). Both U-wrapped and side-bonded strengthening configurations were investigated, and the complex behavior of concrete was implemented using a fracture-plastic model where concrete cracking was simulated using a smeared crack band model with fixed crack orientations and a shear retention factor of 0.5. A perfect bond was assumed between the internal steel reinforcement and concrete that ignored any bond-slip mechanisms. Although inclusion of a steel reinforcement bond-slip relationship may have improved the accuracy of the model, based on studies conducted by Chen et al. (2012), the assumption of perfect bond between the steel stirrups and

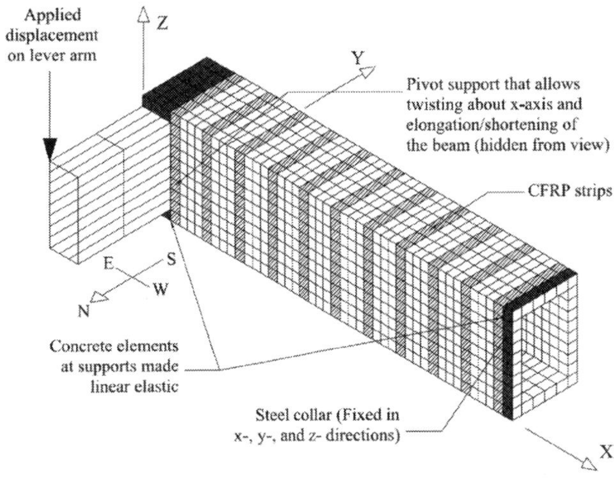

Fig. 4.3—Boundary conditions of finite element (FE) mesh of FH050D1/FH050D2 (Hii 2006).

the concrete may not significantly alter the peak load predictions. The concrete fracture energy and bond-slip model between the FRP and concrete were defined in accordance with Lu et al. (2005). The bond slip between the concrete and FRP laminates was implemented in the model using nonlinear link elements. Both the FE model proposed by Godat et al. (2007) and results of the model when compared with experimental data from Pellegrino and Modena (2002) are presented in Fig. 4.2b. Although the models were developed using the same boundary conditions and load patterns, only half of the beam was modeled due to symmetry. The peak failure loads were predicted within an error of less than 7 percent and provided a good estimation of model stiffness.

4.3—Reinforced concrete members strengthened in torsion using FRP

Limited studies have been conducted into the FRP retrofitting of RC beams in torsion. Hii and Al-Mahaidi (2006) and Al-Mahaidi and Hii (2007) investigated the simulation of both medium-scale, thin-walled, box sections and solid RC beams strengthened using FRP. The models were implemented using DIANA (de Witte and Kikstra 2003) as depicted in Fig. 4.3, where FH050D1/FH050D2 denotes the use of full hoop strips using a hollow section and 0.5*D* spacing between FRP strips. The constitutive model for the concrete was represented using a smeared crack band approach that included hardening and softening of concrete in compression modeled using the Thorenfeldt et al. (1987) function. The shear stiffness of cracked concrete due to aggregate interlock was defined using a shear retention factor ranging from 0.03 to 0.06. The results indicated that a linear relationship existed between the predictions of torsional capacities and shear retention factor. Better results were obtained when increasing the shear retention factor for beams with increasing FRP strengthening. This was attributed to the smaller crack widths developed in the concrete at a given level of loading and a corresponding increase of aggregate interlock for the beams with a high amount to strengthening.

The model was found to be less sensitive to the bond between the internal steel ties and the concrete; as a result, both a perfect bond and a bond-slip model was implemented in the analysis. Furthermore, a bond-slip relationship was implemented between the FRP and the concrete based on shear lap tests conducted by Pham (2005). It was found that the inclusion of interface elements with bond-slip properties between the FRP and concrete resulted in softer torque-twist responses compared to when a perfect bond was assumed. Using a bond slip model between the FRP and the concrete, it was found that local debonding between the FRP and the concrete could be replicated in the vicinity of torsional cracks. Overall, the modeling techniques implemented in the study resulted in models capable of predicting the responses of the experimental beams with very good correlations with test results.

4.4—Modeling of FRP-to-concrete joints

Several numerical studies have been conducted to simulate the behavior of FRP-to-concrete joints such as those depicted in Fig. 4.1a (Chen and Tao 2011; Teng et al. 2013; Kalfat 2014) and a summary of the models is provided in Table 4.2. A comprehensive study into the FE modeling of FRP-to-concrete joints was conducted by Kalfat and Al-Mahaidi (2015) and was implemented using ATENA 3D. In this study, a smeared crack model based on refined crack band theory was specified. The rotating crack model was based on a nonlinear, plasticity-fracture material model using fracture energy and a crack opening law. The model can simulate the nonlinear behavior of concrete in compression, including hardening and softening based on an equivalent uniaxial stress-strain law.

A variable shear retention factor was used that expresses the shear stiffness as a function of the normal crack opening and scaled using the shear factor coefficient s_F, which relates the normal and shear stiffness of a crack most favorably. A shear factor coefficient of 20 was specified based on experimental work by Walraven and Reinhardt (1981). The equation used to determine the shear retention factor r_g is depicted in Eq. (4.4), where G_c is the initial shear modulus of concrete, L_{ch} is the crack band width, ε_F is the normal crack opening strain, and $f()_F$ refers to the Hordijk's softening formula (Hordijk 1991).

$$r_g = \frac{1}{1 + \dfrac{G_c}{s_F f(\varepsilon_F L_{ch})_F}\varepsilon_F} \tag{4.4}$$

The input parameters required by the concrete material model were Young's Modulus (E_s), compressive strength (f_c'), tensile strength (f_{ct}), Poisson's ratio (v), specific fracture energy (G^I_F), shear factor coefficient (s_F) and maximum aggregate size (Ag). The first four parameters can be determined from the experimental measurements. The fracture energy, however, based on Trunk and Wittmann (1998), was found to match the experimental results most favorably. Hence, determination of the variable shear retention factor in ATENA (Červenka et al. 2016) first uses the normal crack opening width to estimate the normal crack stiffness, which is subsequently translated into a shear crack stiffness using the shear factor coefficient.

Bidirectional patch anchorages were used to anchor FRP laminates and defined using a matrix of carbon fibers orientated in the ±45-degree directions. The dry fibers were applied using a wet layup technique, which resulted in fiber embedment within the saturant matrix; the FRP sheet was modeled using an orthotropic linear elastic base material with von-Mises plasticity hardening (ATENA) (Červenka et al. 2016). This was defined using saturant properties as the base material with smeared reinforcement in perpendicular orientations representing the embedded fibers. A thickness of 0.0034 in. (0.86 mm) was used for a single layer of sheet with a fiber fraction of 19.7 percent in each direction representing the orthogonal fibers. The FRP laminate was modeled as an orthotropic linear elastic material. Interface elements were used to model the contact between two surfaces (concrete and adhesive). The constitutive bond-slip relation was defined in terms of tractions on interface planes and relative sliding and opening displacements. The FE results were later used to recalculate the numerical bond-slip curve for comparison with the experimental data with satisfactory agreement. The results achieved using the above modeling technique provided very good correlations with experimental results and could correctly predict the debonding failure plane, as depicted in Fig. 4.4a.

Chen and Tao (2011) investigated the bond behavior between FRP and concrete by proposing a robust two-dimensional FE model of an FRP-to-concrete joint (pull-off test) implemented in ABAQUS (HKS 2001). A concrete damage plasticity model was proposed based on the plastic degradation theory, with the plasticity component of the concrete modeled according to Lubliner et al. (1989), Lee and Fenves (1998), and the damage implemented using a smeared crack band model with fixed crack orientation. The authors used a variable shear retention factor in accordance with Rots (1988) and a fracture energy determined in accordance with CEB-FIB odel Code 2010/2012. A perfect connection was assumed between the FRP and concrete, and the debonding failure was simulated through damage of the concrete elements attached to the FRP composite. A summary of the results obtained from the model with a comparison with the experimental data is presented in Fig. 4.4b, where Series II-5 contained the following properties: concrete compressive strength of 3321 psi (22.9 MPa), width of the concrete prism of 5.9 in. (150 mm), FRP width of 0.98 in. (25 mm), FRP bond length of 7.48 in. (190 mm), and height of the concrete free edge of 4.72 in. (120 mm).

Although the model proves that it is possible to obtain good results when damage (debonding) occurs within the concrete cover zone by assuming a perfect bond between the FRP and concrete, the approach is dependent on the accuracy and calibration of the concrete damage model and size of the concrete elements directly below the FRP, which should be small enough to capture local damage at the interface. Based on the experience of prior researchers, greater success has

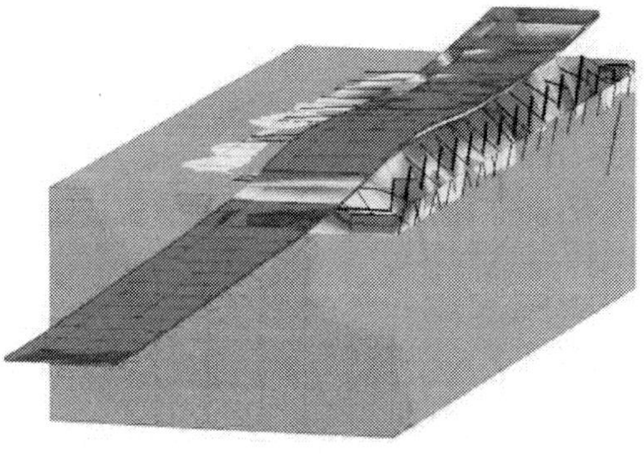

(a)

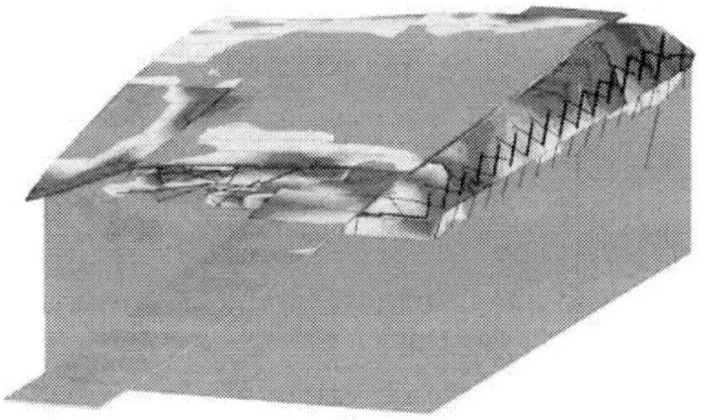

(b)

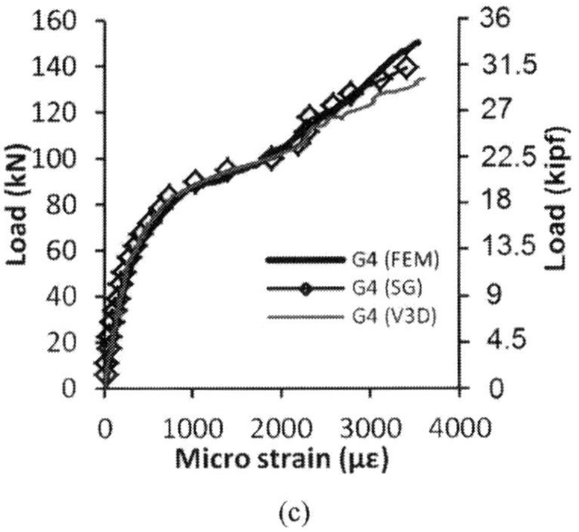

(c)

Fig. 4.4a—Failure model of control specimen (FEM model) depicting exaggerated deformations: (a) concrete cover separation failure; (b) patch anchor debond; and (c) experimental and predicted load versus strain curve (Kalfat and Al-Mahaidi 2015).

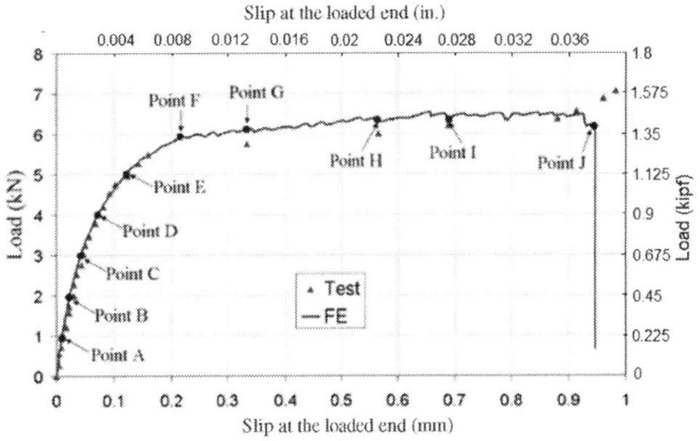

(a)

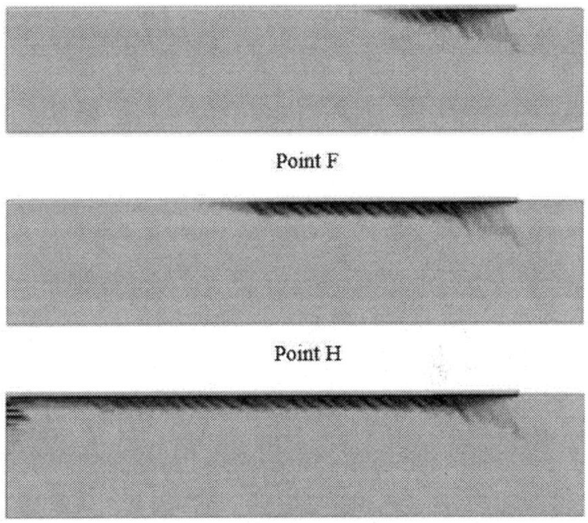

Point F

Point H

Point J

(b)

Fig. 4.4b—(a) Comparison of FE prediction with test for specimen II-5 in Yao et al (2005); and (b) predicted crack pattern at different stages: specimen II-5 in Yao et al. (2005); (Chen and Tao 2011).

been achieved when implementing interface elements with an appropriate bond-slip relationship due to a greater flexibility to control the interface properties. Examples would be bond strength and stiffness.

4.5—Axially strengthened RC columns using FRP

Fiber-reinforced polymer is also gaining use for the confinement of RC columns to improve their structural performance. Use of the FE method to investigate the effectiveness of confined circular and rectangular sections has been assessed by several studies. Most studies have implemented plasticity models to simulate the behavior of concrete when subject to compression. Doran et al. (2009) conducted studies into FEA simulations square/rectangular concrete columns wrapped with FRP using the LUSAS (Finite Element Analysis Ltd. 2006) model. Columns were constructed using eight-node isoparametric hexahedral solid elements with a mesh size of 1.96 in. (50 mm). A Drucker-Prager (D-P)-type plasticity model (Jiang and Wu 2012) was implemented for the concrete with the material parameters of the model defined using the mean concrete compressive strength. The FRP jacket was modeled using a linear elastic shell element and a perfect bond assumed between the FRP and concrete. This assumption is valid for confined elements, as failure of the strengthened element is usually by FRP rupture as opposed to FRP debonding. Their model was compared to 33 specimens from the literature (Harajli et al. 2006; Pan et al. 2007; Turgay 2007; Masia et al. 2004) and the model predictions demonstrated good correlations with the experimental results. The model was also able to capture the confinement pressure for square/rectangular columns and provide predictions that matched closely with those obtained by classical mechanics. Koksal et al. (2009) also investigated FE modeling of FRP-confined columns in LUSAS (Finite Element Analysis Ltd. 2006) using a similar model to that adopted by Doran et al. (2009). However, Koksal et al. (2009) verifies the FEA model results with 42 small- and large-scale column specimens tested by eight different researchers. The modeling approach and FE predictions also correlated very well with the experimental results. From the available research on axially strengthened members, the accuracy of a concrete plasticity model is one of the most critical component leading to successful predictions. The inclusion of bond-slip models between the FRP and concrete, or between the longitudinal reinforcement and concrete, are not required when modeling such elements; the assumption of a perfect bond may be generally acceptable.

CHAPTER 5—CONCLUSIONS AND RECOMMENDED MODELING APPROACHES FOR FIBER-REINFORCED, POLYMER-STRENGTHENED REINFORCED CONCRETE MEMBERS

This report summarizes the significant concepts to consider when developing a finite element (FE) model to simulate the behavior of reinforced concrete (RC) members strengthened with fiber-reinforced polymer (FRP) and subjected to various loading actions. General guidance is provided for the most appropriate definition of concrete material models, choice of elements, mesh size, bond between internal steel reinforcement, and concrete and bond between FRP and concrete. Examples given from the literature illustrate results that can be achieved when the recommended modeling guidelines as follows are implemented.

(a) The concrete material model should consist of a suitable damage plasticity model, capable of simulating both the nonlinear response of concrete crushing and concrete fracture. However, nonlinear elasticity models for concrete in compression can also be used.

(b) Concrete cracking can be simulated using the discrete crack or smeared crack approaches. If the smeared crack approach is used, the crack band model should be used to avoid strain localization and mesh dependency. When used correctly, both the discrete and smeared crack modeling approaches are capable of yielding good predictions of results.

(c) A variable shear retention factor should be used to most accurately represent the shear behavior of cracked concrete. Variable or high shear retention factors should be used in conjunction with a rotating crack model, where the crack direction is capable of changing with loading history. If a fixed crack model is adopted, then a near zero shear retention factor is recommended.

(d) Mode-I fracture energy should be used to represent the phenomenon of FRP debonding. Standard testing procedures are recommended for determining the concrete fracture energy most accurately; alternatively, fracture energy models such as CEB-FIB Model Code 2010/2012) have been used with success.

(e) A bond-slip model between the internal steel reinforcement and concrete should be implemented in the FE model. The assumption of perfect bond has been shown to be invalid and results in highly distributed cracking and in overestimation of beam stiffness and failure load.

(f) The FRP-to-concrete interface should be defined using an appropriate bond-slip law derived from the experimental data. The interface behavior may be implemented in the FE model using either interface elements or one-dimensional spring elements. A perfect bond between the FRP and concrete is only recommended where FRP debonding is not the controlling failure mode.

(g) If debonding is simulated in the FE analysis by fracturing of the concrete, the element sizes directly in contact with the FRP and within the debonding failure plane should be sufficiently refined to adequately capture the local stress concentrations. Parametric studies should be conducted to verify mesh independency of the proposed solution.

(h) All parameters used to inform the various material models used in the FE analysis should have physical basis. Concrete tensile strength, compressive strength, and fracture energy; and properties of steel reinforcement, FRPs, and adhesives should be determined by experimental testing. The bond-slip properties between the FRP and concrete should also have physical basis and be determined from the experimental tests by appropriate monitoring of strains along the length of the FRP-to-concrete bond line.

CHAPTER 6—REFERENCES

ACI committee documents and documents published by other organizations are listed first by document number, full title, and year of publication followed by authored documents listed alphabetically.

American Concrete Institute (ACI)

ACI 301-16—Specifications for Structural Concrete

ACI 363R-10—Report on High-Strength Concrete

ACI 440.2R-17—Guide for the Design and Construction of Externally Bonded FRP Systems for Strengthening Concrete Structures

European Committee for Standardization

EN 1992-1-1 (2005)—Eurocode 2: Design of Concrete Structures. Part 1-1 General Rules and Rules for Buildings

Japan Concrete Institute Standard

JCI-S-001 (2003)—Method of Test for Fracture Energy of Concrete by Use of Notched Beam

Standards Australia

AS 3600-2009—Concrete Structures

Authored documents

Achintha, M., and Burgoyne, C., 2013, "Fracture Energy of the Concrete-FRP Interface in Strengthened Beams," *Engineering Fracture Mechanics*, V. 110, pp. 38-51. doi: 10.1016/j.engfracmech.2013.07.016

ADINA, 2003, "Automatic Dynamic Incremental Nonlinear Analysis," finite element software, Version 8.1, ADINA, Watertown, MA.

Al-Mahaidi, R., 1978, "Nonlinear Finite Element Analysis of Reinforced Concrete Deep Members," PhD thesis, Dept. of Structural Engineering, Cornell University, Ithaca, NY, pp. 1-374.

Al-Mahaidi, R., and Hii, A., 2007, "Bond Behaviour of CFRP Reinforcement for Torsional Strengthening of Solid and Box-Section RC Beams," *Composites. Part B, Engineering*, V. 38, No. 5-6, pp. 720-731. doi: 10.1016/j.compositesb.2006.06.018

Al-Mahaidi, R.; Lee, K.; and Taplin, G., 2001, "Behavior and Analysis of RC T-Beams Partially Damaged in Shear and Repaired with CFRP Laminates," *Structures: A Structural Engineering Odyssey*, pp. 1-8.

Aram, M. R.; Czaderski, C.; and Motavalli, M., 2008, "Debonding Failure Modes of Flexural FRP-Strengthened RC Beams," *Composites. Part B, Engineering*, V. 39, No. 5, pp. 826-841. doi: 10.1016/j.compositesb.2007.10.006

Bažant, Z. P., 1978, "Endochronic Inelasticity and Incremental Plasticity," *International Journal of Solids and Structures*, V. 14, No. 9, pp. 691-714. doi: 10.1016/0020-7683(78)90029-X

Bažant, Z. P., and Becq-Giraudon, E., 2002, "Statistical Prediction of Fracture Parameters of Concrete and Implications for Choice of Testing Standard," *Cement and Concrete Research*, V. 32, No. 4, pp. 529-556. doi: 10.1016/S0008-8846(01)00723-2

Bažant, Z. P., and Planas, J., 1998, *Fracture and Size Effect in Concrete and Other Quasibrittle Materials*, CRC Press, Boca Raton, FL.

Bigaj, A., 1999, "Structural Dependence of Rotation Capacity of Plastic Hinges in RC Beams And Slabs," PhD thesis, Department of Civil Engineering, Delft University of Technology, the Netherlands.

Buyukozturk, O., 1977, "Nonlinear Analysis of Reinforced Concrete Structures," *Computers & Structures*, V. 7, No. 1, pp. 149-156. doi: 10.1016/0045-7949(77)90069-4

Camata, G.; Spacone, E.; and Zarnic, R., 2007, "Experimental and Nonlinear Finite Element Studies Of RC Beams Strengthened with FRP Plates," *Composites. Part B, Engineering*, V. 38, No. 2, pp. 277-288. doi: 10.1016/j.compositesb.2005.12.003

CEB-FIB Model Code, 2010/2012, Comite Euro-International du Beton, "CEB-FIB Model Code 2010," Lausanne, Switzerland, Ernst & Sohn, Berlin, Germany.

CEB-FIP Model Code, 1990, Comite Euro-International du Beton, "CEB-FIP Model Code 1990," Lausanne, Switzerland, Thomas Telford Publishing, London, UK, 436 pp.

Červenka, J., 1994, "Discrete Crack Modeling in Concrete Structures," PhD thesis, University of Colorado, Boulder, CO.

Červenka, V., and Červenka, J., 2007, "ATENA Program Documentation: Part 2-1," *User's Manual for ATENA 2D*, Červenka Consulting, Prague.

Červenka, V.; Červenka, J.; and Pukl, R., 2002, "ATENA—A Tool for Engineering Analysis of Fracture in Concrete," *Sadhana*, V. 27, No. 4, pp. 485-492. doi: 10.1007/BF02706996

Červenka, V.; Jendele, L.; and Červenka, J., 2016, "ATENA Program Documentation Part 1—Theory," Červenka Consulting sro, Prague, Czechia.

Červenka, J.; Chandra Kishen, J. M.; and Saouma, V. E., 1998, "Mixed Mode Fracture of Cementitious Bimaterial Interfaces: Part II: Numerical Simulation," *Engineering Fracture Mechanics*, V. 60, No. 1, pp. 95-107. doi: 10.1016/S0013-7944(97)00094-5

Chen, W. F., 2007, *Plasticity in Reinforced Concrete*, J. Ross Publishing Classics, J. Ross Publishing, 474 pp.

Chen, G.; Teng, J.; and Chen, J., 2011, "Finite-Element Modeling of Intermediate Crack Debonding in FRP-Plated RC Beams," *Journal of Composites for Construction*, V. 15, No. 3, pp. 339-353. doi: 10.1061/(ASCE)CC.1943-5614.0000157

Chen, G. M.; Chen, J. F.; and Teng, J. G., 2012, "On the Finite Element Modeling of RC Beams Shear-Strengthened with FRP," *Construction & Building Materials*, V. 32, pp. 13-26. doi: 10.1016/j.conbuildmat.2010.11.101

Chen, J. F., and Tao, Y., 2011, "Finite Element Modelling of FRP-to-Concrete Bond Behaviour Using the Concrete Damage Plasticity Theory Combined with a Plastic Degradation Model," *Advances in FRP Composites in Civil Engineering*, L. Ye, P. Feng, and Q. Yue, eds., Springer Berlin, Heidelberg, Germany, pp. 45-50.

Cui, S.; Wang, J.; Shi, L.; and Liu, Y., 2009, "Test and FEM Analysis of Debonding Failure of RC Beam Strengthened with CFRP," *Journal of Reinforced Plastics and Composites*, V. 28, No. 17, pp. 2151-2160. doi: 10.1177/0731684408091932

de Witte, F. C., and Kikstra, W. P., 2003, *DIANA User's Manual*, TNO DIANA BV, Delft University, the Netherlands.

Desayi, P., and Krishnan, S., 1964, "Equation for the Stress-Strain Curve of Concrete," *ACI Journal Proceedings*, V. 61, No. 3, Mar., pp. 345-350.

De Borst, R., 1997, "Some Recent Developments in Computational Modelling of Concrete Fracture," *International Journal of Fracture*, V. 86, No. 1/2, pp. 5-36. doi: 10.1023/A:1007360521465

Dong, Y.; Zhao, M.; and Ansari, F., 2002, "Failure Characteristics of Reinforced Concrete Beams Repaired with CFRP Composites," Proceeding of the Third International Conference on Composites in Infrastructure, San Francisco, CA, pp. 26-40.

Doran, B.; Koksal, H. O.; and Turgay, T., 2009, "Nonlinear finite Element Modeling of Rectangular/Square Concrete Columns Confined with FRP," *Materials & Design*, V. 30, No. 8, pp. 3066-3075. doi: 10.1016/j.matdes.2008.12.007

Dorr, K., 1980, "Ein Beitrag zur Berechnung von Stahlbetonscheiben unter Besonderer Berücksichtigung des Verbundverhaltens," PhD thesis, University of Darmstadt, Germany. (in German)

Eligehausen, R.; Popov, E. P.; and Bertero, V. V., 1983, "Local Bond Stress-Slip Relationships of Deformed Bars under Generalized Excitations," Report No. UCB/EERC-

83/23, Earthquake Engineering Research Center, University of California, Berkeley, Berkeley, CA, 180 pp.

Elsayed, W.; Ebead, U. A.; and Neale, K. W., 2007, "Interfacial Behavior and Debonding Failures in FRP-Strengthened Concrete Slabs," *Journal of Composites for Construction*, V. 11, No. 6, pp. 619-628. doi: 10.1061/(ASCE)1090-0268(2007)11:6(619)

Elwi, A. A., and Murray, D. W., 1979, "A 3D Hypoelastic Concrete Constitutive Relationship," *Journal of the Engineering Mechanics Division*, V. 105, No. 4, Oct., pp. 623-641.

Ferracuti, B.; Savoia, M.; and Mazzotti, C., 2007, "Interface Law for FRP-concrete Delamination," *Composite Structures*, V. 80, No. 4, pp. 523-531. doi: 10.1016/j.compstruct.2006.07.001

Finite Element Analysis Ltd, 2006, "LUSAS Finite Element System," Version 13.6, Surrey, UK.

Freddi, F., and Savoia, M., 2008, "Analysis of FRP–concrete Debonding Via Boundary Integral Equations," *Engineering Fracture Mechanics*, V. 75, No. 6, pp. 1666-1683. doi: 10.1016/j.engfracmech.2007.05.016

Godat, A.; Labossière, P.; and Neale, K. W., 2012a, "Numerical Investigation of the Parameters Influencing the Behavior of FRP Shear-strengthened Beams," *Construction & Building Materials*, V. 32, No. 7, pp. 90-98. doi: 10.1016/j.conbuildmat.2010.11.110

Godat, A.; Labossière, P.; Neale, K. W.; and Chaallal, O., 2012b, "Behavior of RC Members Strengthened in Shear with EB FRP: Assessment of Models and FE Simulation Approaches," *Computers & Structures*, V. 92–93, No. 2, Feb., pp. 269-282. doi: 10.1016/j.compstruc.2011.10.018

Godat, A.; Neale, K.; and Labossière, P., 2007, "Numerical Modeling of FRP Shear-Strengthened Reinforced Concrete Beams," *Journal of Composites for Construction*, V. 11, No. 6, pp. 640-649. doi: 10.1061/(ASCE)1090-0268(2007)11:6(640)

Grassl, P.; Lundgren, K.; and Gylltoft, K., 2002, "Concrete in Compression: A Plasticity Theory with a Novel Hardening Law," *International Journal of Solids and Structures*, V. 39, No. 20, pp. 5205-5223. doi: 10.1016/S0020-7683(02)00408-0

Grassl, P.; Xenos, D.; Nyström, U.; Rempling, R.; and Gylltoft, K., 2013, "CDPM2: A Damage-Plasticity Approach to Modeling the Failure of Concrete," *International Journal of Solids and Structures*, V. 50, No. 24, pp. 3805-3816. doi: 10.1016/j.ijsolstr.2013.07.008

Gribniak, V.; Kaklauskas, G.; Idnurm, S.; and Bacinskas, D., 2010, "Finite Element Mesh Size Effect on Deformation Predictions of Reinforced Concrete Bridge Girder," *The Baltic Journal of Road and Bridge Engineering*, V. 5, No. 1, pp. 19-27. doi: 10.3846/bjrbe.2010.03

Han, D. J., and Chen, W. F., 1985, "A Nonuniform Hardening Plasticity Model for Concrete Materials," *Mechanics of Materials*, V. 4, No. 3-4, Dec., pp. 283-302. doi: 10.1016/0167-6636(85)90025-0

Harajli, M. H.; Hantouche, E.; and Soudki, K., 2006, "Stress-Strain Model for Fiber-reinforced Polymer Jacketed Concrete Columns," *ACI Structural Journal*, V. 103, No. 5, Sept.-Oct., pp. 672-682.

Hashemi, S., and Al-Mahaidi, R., 2012, "Experimental and Finite Element Analysis of Flexural Behavior of FRP-strengthened RC Beams using Cement-Based Adhesives," *Construction & Building Materials*, V. 26, No. 1, pp. 268-273. doi: 10.1016/j.conbuildmat.2011.06.021

Hii, A. K. Y., 2006, "Torsional Strengthening of Reinforced Concrete Members using Carbon Fibre Reinforced Polymer Composites," doctoral thesis, Department of Civil Engineering, Monash University, Melbourne, Australia.

Hii, A. K. Y., and Al-Mahaidi, R., 2006, "An Experimental and Numerical Investigation on Torsional Strengthening of Solid and Box-section RC Beams using CFRP Laminates," *Composite Structures*, V. 75, No. 1-4, pp. 213-221. doi: 10.1016/j.compstruct.2006.04.050

HKS, 2001, *ABAQUS/Standard User's Manual*, HKS Corporation, Atlanta, GA.

Hajsadeghi, M.; Alaee, F. J.; and Shahmohammadi, A., 2011, "Investigation on Behaviour of Square/Rectangular Reinforced Concrete Columns Retrofitted with FRP Jacket," *Journal of Civil Engineering and Management*, V. 17, No. 3, pp. 400-408. doi: 10.3846/13923730.2011.594155

Hognestad, E., 1951, "Study of Combined Bending and Axial Load in Reinforced Concrete Members," *Bulletin Series No. 399*, Engineering Experiment Station, University of Illinois at Urbana-Champaign, Champaign, IL.

Hollaway, L. C., and Teng, J. G., 2008, *Strengthening and Rehabilitation of Civil Infrastructures using Fibre-reinforced Polymer (FRP) Composites*, Woodhead Publishing, Cambridge, UK.

Hoover, C. G., and Bažant, Z. P., 2014, "Cohesive Crack, Size Effect, Crack Band and Work-of-Fracture Models Compared to Comprehensive Concrete Fracture Tests," *International Journal of Fracture*, V. 187, No. 1, Jan., pp. 133-143. doi: 10.1007/s10704-013-9926-0

Hordijk, D. A., 1991, "Local Approach to Fatigue of Concrete," doctor thesis, Delft University of Technology, the Netherlands, Oct.

Hu, H.-T.; Lin, F. M.; and Jan, Y. Y., 2004, "Nonlinear Finite Element Analysis of Reinforced Concrete Beams Strengthened by Fiber-reinforced Plastics," *Composite Structures*, V. 63, No. 3-4, pp. 271-281. doi: 10.1016/S0263-8223(03)00174-0

Hu, H. T.; Lin, F. M.; Liu, H. T.; Huang, Y. F.; and Pan, T. C., 2010, "Constitutive Modeling of Reinforced Concrete and Prestressed Concrete Structures Strengthened by Fiber-Reinforced Plastics," *Composite Structures*, V. 92, No. 7, pp. 1640-1650. doi: 10.1016/j.compstruct.2009.11.030

Ibrahim, A. M., and Sh Mahmood, M., 2009, "Finite Element Modeling of Reinforced Concrete Beams Strengthened with FRP Laminates," *European Journal of Scientific Research*, V. 30, No. 4, pp. 526-541.

Jiang, J. F., and Wu, Y. F., 2012, "Identification of Material Parameters for Drucker–Prager Plasticity Model for FRP Confined Circular Concrete Columns," *International Journal of Solids and Structures*, V. 49, No. 3-4, pp. 445-456. doi: 10.1016/j.ijsolstr.2011.10.002

Kalfat, R., 2014, "Anchorage Systems in Concrete Structures Strengthened With Carbon Fiber Reinforced Polymer Composites," PhD thesis, Swinburne University of Technology, Department of Engineering and Industrial Sciences, Melbourne, Australia.

Kalfat, R., and Al-Mahaidi, R., 2014, "Numerical and Experimental Validation of FRP Patch Anchors Used to Improve the Performance of FRP Laminates Bonded to Concrete," *Journal of Composites for Construction*, V. 18, No. 3, p. A4013008 doi: 10.1061/(ASCE)CC.1943-5614.0000437

Kalfat, R., and Al-Mahaidi, R., 2015, "Finite Element Investigation into the Size Effect of Bidirectional Fibre Patch Anchors Used to Enhance the Performance of FRP-to-Concrete Joints," *Composite Structures*, V. 121, pp. 27-36. doi: 10.1016/j.compstruct.2014.09.016

Kang, Q., 1996, *Finite Element Analysis for Reinforced Concrete*, China Water Power Press, Beijing, pp. 120-126.

Kishi, N.; Zhang, G.; and Mikami, H., 2005, "Numerical Cracking and Debonding Analysis of RC Beams Reinforced with FRP Sheet," *Journal of Composites for Construction*, V. 9, No. 6, pp. 507-514. doi: 10.1061/(ASCE)1090-0268(2005)9:6(507)

Koksal, H. O.; Doran, B.; and Turgay, T., 2009, "A Practical Approach for Modeling FRP Wrapped Concrete Columns," *Construction & Building Materials*, V. 23, No. 3, pp. 1429-1437. doi: 10.1016/j.conbuildmat.2008.07.008

Kolmar, W., 1986, "Beschreibung der Kraftuebertragung über Risse in nichtlinearen Finite-Element-Berechnungen von Stahlbetontragwerken," dissertation, T. H. Darmstadt, ed., 94 pp.

Kotynia, R.; Abdel Baky, H.; Neale, K.; and Ebead, U., 2008, "Flexural Strengthening of RC Beams with Externally Bonded CFRP Systems: Test Results and 3D Nonlinear FE Analysis," *Journal of Composites for Construction*, V. 12, No. 2, pp. 190-201. doi: 10.1061/(ASCE)1090-0268(2008)12:2(190)

Lee, J., and Fenves, G., 1998, "Plastic-Damage Model for Cyclic Loading of Concrete Structures," *Journal of Engineering Mechanics*, V. 124, No. 8, pp. 892-900. doi: 10.1061/(ASCE)0733-9399(1998)124:8(892)

Lee, T. K., 2003, "Shear Strengthening of Reinforced Concrete T-Beams Strengthened using Reinforced Carbon Fibre Polymer (CFRP) Laminates," doctoral thesis, Department of Civil Engineering, Monash University, Melbourne, Australia, pp. 161-162.

Lu, X.; Teng, J.; Ye, L.; and Jiang, J., 2007, "Intermediate Crack Debonding in FRP-Strengthened RC Beams: FE Analysis and Strength Model," *Journal of Composites for Construction*, V. 11, No. 2, pp. 161-174. doi: 10.1061/(ASCE)1090-0268(2007)11:2(161)

Lu, X. Z.; Teng, J. G.; Ye, L. P.; and Jiang, J. J., 2005, "Bond-slip Models for FRP Sheets/Plates Bonded to Concrete," *Engineering Structures*, V. 27, No. 6, pp. 920-937. doi: 10.1016/j.engstruct.2005.01.014

Lubliner, J.; Oliver, J.; Oller, S.; and Oñate, E., 1989, "A Plastic-damage Model for Concrete," *International Journal of Solids and Structures*, V. 25, No. 3, pp. 299-326. doi: 10.1016/0020-7683(89)90050-4

Luo, Y.; Li, A.; and Kang, Z., 2012, "Parametric Study of Bonded Steel–concrete Composite Beams by using Finite Element Analysis," *Engineering Structures*, V. 34, pp. 40-51. doi: 10.1016/j.engstruct.2011.08.036

Masia, M. J.; Gale, T. N.; and Shrive, N. G., 2004, "Size Effects in Axially Loaded Square-section Concrete Prisms Strengthened using Carbon Fibre Reinforced Polymer Wrapping," *Canadian Journal of Civil Engineering*, V. 31, No. 1, pp. 1-13. doi: 10.1139/l03-064

Menetrey, P., and Willam, K. J., 1995, "Triaxial Failure Criterion for Concrete and its Generalization," *ACI Structural Journal*, V. 92, No. 3, pp. 311-318.

Moharrami, M., and Koutromanos, I., 2016, "Triaxial Constitutive Model for Concrete under Cyclic Loading," *Journal of Structural Engineering*, V. 142, No. 7, p. 4016039 doi: 10.1061/(ASCE)ST.1943-541X.0001491

MSC Software Corp., 2005, *MSC.MARC® (2005) User's Manual*, MSC Software Corp., Santa Ana, CA.

Neubauer, U., and Rostasy, F. S., 1999, "Bond Failure of Concrete Fiber Reinforced Plates at Inclined Cracks-Experiments and Fracture Mechanics Model," *Proceedings of the Fourth International Symposium on Fiber-Reinforced Polymer Reinforcement for Reinforced Concrete Structures (FRPRCS-4)*, SP-188, U. Neubauer, and F. S. Rostasy, eds., American Concrete Institute, Farmington Hills, MI, pp. 369-381.

Okamura, H., and Kohichi, M., 1991, "Nonlinear Analysis and Constitutive Models of Reinforced Concrete," Gihodo, Tokyo, 10 pp.

Ortiz, M., 1985, "A Constitutive Theory for the Inelastic Behavior of Concrete," *Mechanics of Materials*, V. 4, No. 1, pp. 67-93. doi: 10.1016/0167-6636(85)90007-9

Pan, J. L.; Xu, T.; and Hu, Z. J., 2007, "Experimental Investigation of Load-Carrying Capacity of the Slender Reinforced Concrete Columns Wrapped with FRP," *Construction & Building Materials*, V. 21, No. 11, Nov., pp. 1991-1996. doi: 10.1016/j.conbuildmat.2006.05.050

Pellegrino, C., and Modena, C., 2002, "Fiber-Reinforced Polymer Shear Strengthening of Reinforced Concrete Beams with Transverse Steel Reinforcement," *Journal of Composites for Construction*, V. 6, No. 2, May, pp. 104-111. doi: 10.1061/(ASCE)1090-0268(2002)6:2(104)

Pham, H. B., 2005, "Debonding Failure In Concrete Members Retrofitted With Carbon Fibre Reinforced Polymer Composites," doctoral thesis, Department of Civil Engineering, Monash University, Melbourne, Australia.

Pham, H. B.; Al-Mahaidi, R.; and Saouma, V., 2006, "Modelling of CFRP–concrete Bond using Smeared and Discrete Cracks," *Composite Structures*, V. 75, No. 1-4, pp. 145-150. doi: 10.1016/j.compstruct.2006.04.039

Rabczuk, T.; Zi, G.; Bordas, S.; and Nguyen-Xuan, H., 2008, "A Geometrically Non-linear Three-Dimensional Cohesive Crack Method for Reinforced Concrete Structures," *Engineering Fracture Mechanics*, V. 75, No. 16, pp. 4740-4758. doi: 10.1016/j.engfracmech.2008.06.019

Rots, J., 1988, "Computational Modeling of Concrete Fracture," PhD thesis, Delft University of Technology, the Netherlands.

Rots, J., and Blaauwendraad, J., 1989, "Crack Models for Concrete, Discrete or Smeared? Fixed, Multi-directional or Rotating?" *Heron*, V. 34, No. 1, pp. 1-59.

Rots, J., and de Borst, R., 1987, "Analysis of Mixed-Mode Fracture in Concrete," *Journal of Engineering Mechanics*, V. 113, No. 11, pp. 1739-1758. doi: 10.1061/(ASCE)0733-9399(1987)113:11(1739)

Rots, J.; Kusters, G. M. A.; and Nauta, P., 1984, "Variable Reduction Factor for the Shear Resistance of Cracked Concrete," *TNO-report BI-84-33/68.8.2001*, Rijswijk, the Netherlands.

Saenz, L., 1964, discussion of "Equation for the Stress-strain Curve of Concrete by Desayi and Krishnan," *ACI Journal*, V. 61, No. 9, pp. 1229-1235.

Sato, Y., and Vecchio, F., 2003, "Tension Stiffening and Crack Formation in Reinforced Concrete Members with Fiber-Reinforced Polymer Sheets," *Journal of Structural Engineering*, V. 129, No. 6, pp. 717-724. doi: 10.1061/(ASCE)0733-9445(2003)129:6(717)

Scotta, R.; Vitaliani, R.; Saetta, A.; Oñate, E.; and Hanganu, A., 2001, "A Scalar Damage Model with a Shear Retention Factor for the Analysis of Reinforced Concrete Structures: Theory and Validation," *Computers & Structures*, V. 79, No. 7, pp. 737-755. doi: 10.1016/S0045-7949(00)00178-4

Teng, J. G.; Zhang, S. S.; Dai, J. G.; and Chen, J. F., 2013, "Three-Dimensional Meso-scale Finite Element Modeling of Bonded Joints Between a Near-surface Mounted FRP Strip and Concrete," *Computers & Structures*, V. 117, pp. 105-117. doi: 10.1016/j.compstruc.2012.12.002

Thorenfeldt, E.; Tomaszewicz, A.; and Jensen, J. J., 1987, "Mechanical Properties of High-Strength Concrete and Applications in Design," Proceedings of the Symposium on Utilization of High-Strength Concrete, Stavanger, Norway, Tapir Publishers, Trondheim, Norway.

Trunk, B., and Wittmann, F. H., 1998, "Experimental Investigation into the Size Dependence of Fracture Mechanics Parameters," Third International Conference of Fracture Mechanics of Concrete Structures, Aedificatio Publishers, Freidburg, Germany, pp. 1937-1948.

Turgay, T., 2007, "CFRP Uygulanmıs Betonarme Elemanların Performansı (The Performance of FRP Strengthened Structural Members)," PhD thesis, submitted to Yıldız Technical University.

Ulaga, T.; Vogal, T.; and Meier, U., 2003, "The Bilinear Stress-slip Bond Model: Theoretical Background and Significance," Sixth International Conference on Fibre-reinforced Polymer Reinforcement for Concrete Structures, FRPRCS-6, Singapore, July, pp. 153-162.

Van Mier, J. G. M., 1997, *Fracture Processes of Concrete. Assessment of Material Parameters for Fracture Models*, CRC Press, Boca Raton, FL, 446 pp.

Vecchio, F., and Bucci, F., 1999, "Analysis of Repaired Reinforced Concrete Structures," *Journal of Structural Engineering*, V. 125, No. 6, pp. 644-652. doi: 10.1061/(ASCE)0733-9445(1999)125:6(644)

Vecchio, F. J., 2000, "Disturbed Stress Field Model for Reinforced Concrete: Formulation," *Journal of Structural Engineering*, V. 126, No. 9, pp. 1070-1077. doi: 10.1061/(ASCE)0733-9445(2000)126:9(1070)

Walraven, J., and Reinhardt, H., 1981, "Theory and Experiments on the Mechanical Behavior of Cracks in Plain and Reinforced Concrete Subject to Shear Load," *Heron*, V. 26, No. 1A, 68 pp.

Wang, J., and Zhang, C., 2008, "Nonlinear Fracture Mechanics of Flexural-Shear Crack Induced Debonding of FRP Strengthened Concrete Beams," *International Journal of Solids and Structures*, V. 45, No. 10, pp. 2916-2936. doi: 10.1016/j.ijsolstr.2008.01.004

Wong, R. S., and Vecchio, F. J., 2003, "Towards Modeling of Reinforced Concrete Members with Externally Bonded Fiber-reinforced Polymer Composites," *ACI Structural Journal*, V. 100, No. 1, Jan.-Feb., pp. 47-55.

Wu, Y. F.; Wang, Z.; Liu, K.; He, W.; and Wang, Z., 2009, "Numerical Analyses of Hybrid-Bonded FRP Strengthened Concrete Beams," *Computer-Aided Civil and Infrastructure Engineering*, V. 24, No. 5, pp. 371-384. doi: 10.1111/j.1467-8667.2009.00596.x

Yang, Z. J.; Chen, J. F.; and Proverbs, D., 2003, "Finite Element Modeling of Concrete Cover Separation Failure in FRP Plated RC Beams," *Construction & Building Materials*, V. 17, No. 1, pp. 3-13. doi: 10.1016/S0950-0618(02)00090-9

Yao, J.; Teng, J. G.; and Chen, J. F., 2005, "Experimental Study on FRP-to-Concrete Bonded Joints," *Composites. Part B, Engineering*, V. 36, No. 2, pp. 99-113. doi: 10.1016/j.compositesb.2004.06.001

You, Y.-M.; Ayoub, A.; and Belarbi, A., 2011, "Three-Dimensional Nonlinear Finite-Element Analysis of Prestressed Concrete Beams Strengthened in Shear with FRP Composites," *Journal of Composites for Construction*, V. 15, No. 6, pp. 896-907. doi: 10.1061/(ASCE)CC.1943-5614.0000226

Youai, G., 2000, "Bond Stress and Slip Modeling in Nonlinear Finite Element Analysis of Reinforced Concrete Structures," MSc thesis, Department of Civil Engineering, The University of Toronto, Toronto, ON, Canada.

Zhang, S. S., and Teng, J. G., 2014, "Finite Element Analysis of End Cover Separation in RC Beams Strengthened in Flexure with FRP," *Engineering Structures*, V. 75, pp. 550-560. doi: 10.1016/j.engstruct.2014.06.031